T0255742

New Frontiers

in SCIENCES, ENGINEERING

and the ARTS

Vol. II

The Chemistry of Initiation of Non-Ringed
Monomers/Compounds

Sunny N.E. Omorodion

New Frontiers
in SCIENCES,
ENGINEERING
and ARTS
the

Vol. II

The Chemistry of Initiation of Non Ringed Monomers/Compounds

Sunny N.E. Omorodion

Professor of Chemical Engineering
University of Benin.

authorHOUSE®

AuthorHouse™
1663 Liberty Drive
Bloomington, IN 47403
www.authorhouse.com
Phone: 1 (800) 839-8640

© 2017 Sunny N.E. Omorodion. All rights reserved.

No part of this book may be reproduced, stored in a retrieval system, or transmitted
by any means without the written permission of the author.

Published by AuthorHouse 11/13/2017

ISBN: 978-1-5462-0203-5 (sc)
ISBN: 978-1-5462-0202-8 (e)

Print information available on the last page.

Any people depicted in stock imagery provided by Thinkstock are models,
and such images are being used for illustrative purposes only.
Certain stock imagery © Thinkstock.

This book is printed on acid-free paper.

Because of the dynamic nature of the Internet, any web addresses or links contained in this book may have changed
since publication and may no longer be valid. The views expressed in this work are solely those of the author and do
not necessarily reflect the views of the publisher, and the publisher hereby disclaims any responsibility for them.

Contents

SECTION B
INTER-MOLECULAR ADDITION MONOMERS NON-OLEFINS

APPENDICES

Preface

Without Volumes (II), (III) and above, some of the new concepts in Chemistry which were introduced in Volume (I) when new classifications for Polymerization systems were provided, would not have been possible. In Volume (I) one began to introduce the new classifi-cations for Radicals. Therein, one began to show that what universally up to the present moment are called "ELECTRONS" are indeed "RADICALS". Therefore, to radically change from the use of the word "radical" in place of the word "electron" is not an easy task as was seen in Volume (I) and the same will apply to Volume (II), but not in Volume (III) and thereafter. The same too applies to the use of the word "IONS" instead of "CHARGES", when indeed there are four kinds of charges in which an ion is only one of them.

Volumes (II) and (III) are in four sections, three sections in Volume (II) and the last section in Volume (III). Both volumes deal with the chemistry of initiation of compounds. Volume (II) titled "The Chemistry of Initiation of Non-ringed monomers/compounds", contains nine chapters. All the chapters deal with entirely new concepts. How chemical reactions take place are for the time being explained from a different point of view. Most of the reactions which could not adequately be explained for so many years are now clearly explained, based on the new definitions provided for an ion, charged species, radicals, atoms, compounds, monomers all of which have characters (male or female or even both). For the first time, the concepts of initiations, activations, molecular rearrangement phenomena, resonance stabilization phenomena, transfer of transfer species, transfer species, "electron"-pulling and pushing groups, "electron"-donating and withdrawing groups, etc. are being clearly defined and introduced to explain mechanisms of chemical and polymeric reactions.

Section A which contains four Chapters begins with olefins of different types. It is in this section one of the most important laws for chemical and polymeric systems was introduced. This law which is analogous to the laws of conservations of mass, momentum and energy, is the law of conservation of transfer of transfer species. It was through application of this law that most other developments followed, after definitions already provided for Addition monomers, Step monomer, and monomers in general. With its application, different types of transfer species that exist in substituent groups, monomer, growing polymer chains, compounds, etc. have begun to be clearly identified under Equilibrium, Combination and Decomposition mechanisms.

Section B containing three Chapters, deals largely with non-olefin monomers. These include aldehydes, ketones and related monomers, acetylenes, nitriles, aldimines, ketimines, cyanates, diazoalkanes and related monomers, sulfur dioxide, phosphines, carbon monoxide, nitroso compounds, quinones, etc. Not only were new transfer species identified for these systems, so also were new concepts developed, in trying to provide the characters, capacities, and chemical (and physical) properties of these compounds.

Section C containing two Chapters, deals with unfamiliar monomers or specialty monomers, but familiar compounds. These include the cumulenes, ketenes, isocyanates, imines, dimines and diketones. Without consideration of these compounds, the order of nucleophilic and electrophilic capacities of the compounds, or the order of "electron"-pushing or pulling capaci-ties of groups would have been impossible to know. So also are the developments on ringed compounds or monomers in Volume (III). For the first time, new different types of groups and their capacities have been identified. New types of molecular rearrangement and resonance stabilization phenomena were identified. New concepts on ceiling temperatures, propagation/ depropagation phenomena were introduced.

Why the monomers favor the routes which they have been observed to favor over the years, but could not clearly be explained, have been provided, based on application of natural laws. At least fifteen different types of transfer species have so far been identified. At the end of every chapter, rules were proposed. Some of the rules are specific to families of compounds or monomers, while others are specific to phenomena and others of very general characters. None of the rules has exception, since the rules are based on NATURE (Natural laws).

The problems sections in all these chapters are not just ordinary problems, but are continuations on the new developments, for better understanding. Before understanding this Volume, Volume (I) must be read – particularly sections B and C whether you are a polymer chemist or not. These Volumes are meant for applications by Chemists, Biochemists, Natural Scientists, Medical Scientists, Pharmaceutical Scientists as well as Engineers or any other discipline that deals with Nature, noting that no discipline where laws of Nature are not applied directly or indirectly exists.

While all the works are original to the author, one will not forget to thank many publishers whose works have helped to open these New Frontiers. With all due respect to institutions and persons mentioned in the first volume, this volume still remains dedicated to humanity.

Univ. of. Benin.   Sunny. N.E. Omorodion
                                                                                (ETG)

"Humanity is homo-sapience, the chosen caretaters of Animals and Plants in our world, the solutions to all our problems, since Life is Problem solving wherein To err is human"

SECTION A

Inter – Molecular
Addition Monomers (Olefins)

Chapter 1

RULES OF ACTIVATION IN ADDITION MONOMERS

1.0 Introduction

More than ten rules covering the direction of attack on an activated monomer based on the types of substituent groups carried by the monomer have been proposed herein. Markov-nikov's rule of addition of alkenes is not in order in so many ways and therefore not one of the rules. The rules cover the whole spectrum of olefinic monomers, the two first members of the extremes of the spectrum being ethylene (alkenes) and tetra-fluoro ethylene (perfluroroalkenes). This will mark the beginning of proposition of new laws or rules of Chemistry in Addition polymerization systems. These rules which will be used to show how different types of polymeric products are obtained from different monomers or same monomer in different kinetic routes favored, mark the beginning of a New Science and will cover:-

    (a) Activation of monomers and unsaturated dead polymers.
    (b) Initiation of monomers and unsaturated dead polymers.
    (c) Propagation of growing polymer chains.
    (d) Transfer of transfer species.
    (e) Termination of growing polymer chains.
    (f) Branching of growing polymer chains.
    (g) Driving forces for opening of ringed monomers.
    (h) Etc.

Some of these rules have already been applied when the definitions of an Addition monomer, a free-radical, a non-free-radical and Z/N initiators and more were provided.[1,2]

There is urgent need for these rules to be proposed, since.

    (i)   Different types of monomers favor different sub-steps and steps during polymerization.
    (ii)   Based on the new classifications of homopolymerization kinetics and Addition homopolymers,[1-3] the new order has to be clearly spelt out.
    (iii) It is not possible to transfer a transfer species <u>from a monomer or growing chain </u>during polymerization, except during initiation or termination. This has been thought to be the case for many years.[4]
    (iv) It is not possible to abstract an anion from a cationically growing polymer chain if it exists or a cation from an anionically growing polymer chain. Hydrogen cannot even be abstracted

as an anion from carbon-chain polymers either in the presence of Z/N catalyst components[5] or any other component.

(v)   It is not possible to abstract any species directly connected to the main chain backbone of a dead polymer. This has also been thought to be the case for years.[6] Even Z/N catalyst components cannot abstract hydrogen as an anion or cation from side chain of a dead polymer. This is not possible under any operating conditions. It is only possible for a living chain radically under very harsh operating conditions as is the case with ethylene nucleo-free-radical polymeriza-tion where H is released as a hydride.

(vi)   It will make modeling of polymerization reactors easier to handle.

(vii)   It will make it possible for us to know how new polymers can be made.

(viii)   Etc.

The major steps which are involved during polymerization of Addition monomers which consist of ring-opening monomers, traditional Addition monomers (mono-olefins, di-olefins etc.) and pseudo-addition monomers, have been identified.[3] The re-classification of some of the steps and their sub-steps will be put into rules of Chemistry in Addition polymerization systems. In this Volume, only rules guiding activation and initiation of monomers will be emphasized on. When a monomer is activated, there is need to know where the free and non-free opposite charges or free and non-free-radicals will be placed, based on the substituent groups carried by the carbon or other centers. Only mono-olefinic monomers will be considered in this chapter, since those of di-olefinic monomers and some similar monomers are extension of the rules.

As has been said in the first Volume, the true functions of the substituent groups can be better understood when "electron"-pushing and "electron"-pulling terminologies are used in place of "electron"-donating and "electron"-withdrawing terminologies respectively as done in Present-day Science. On the other hand, they both have different meanings. The capacities of "electron"-pulling groups have been noted to be very high compared to the capacities of "electron"-pushing groups.[7] Hence, there are by far less electrophiles, than nucleophiles. Indeed, commercially, there are by far more nucleophile-type of monomer-processes on a large scale than electrophile-type of monomer-processes, since it has been far easier to obtain initiators for the latter than for the former. The reasons why this is so have been partly largely due to the high "electron"-pulling capacities of the substituent group carried by these monomers and the paired unbonded or unshared "electrons" on them (electrophilic type of monomers).

Before moving forward however, one will notice that in Volume (I), the concept of States of existences were begun to be introduced. So also was the concept of Mechanisms of reactions or for any system. Since their great importance will now begin to emerge, shown below in Figures 1.1 and 1.2 are the new classifications of States of existence and Mechanism of Systems.

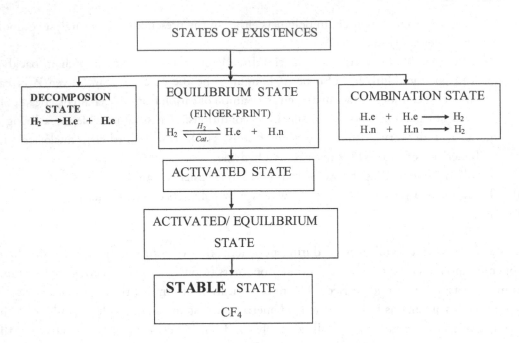

Figure 1.1 Types of States of Existenes

The three states of existences are <u>Equilibrium, Decomposition and Combination</u>. If anything is not in Equilibrium or cannot be in Equilibrium state, it can be in <u>Activated</u> or Activated/ Equilibrium states if **it has activation center(s)**. If not Activated or in Equilibrium, it can be <u>Stable</u>. For anything like chemical compounds or molecules, these states are fixed and different for each existence and for each compound. Consider a system containing only two compounds, A and B. Before they can react with each other to give **a productive or non-productive** stage or stages, one of them or both of them must first exist in Equilibrium or Decomposition state, otherwise there will be no reaction. If all remain in Stable States such as the components of air, they can either **dissolve or miscibilize** or remain inactive with themselves as has been shown.

Some compounds can readily exist in Equilibrium state (The Finger Print), while others cannot. To put some in Equilibrium state may require for example special **operating conditions** such as high temperatures (For example with propylene) or a special neighbor such as **a Passive catalyst** (For example hydrogen catalyst for hydrogen), noting that there are Active and Passive catalysts and ***Enzymes are Active in character***. In polymerization reactions, parts of the so-called catalysts, which indeed are Initiators or generators of Initiators, are part from the beginning of the polymeric product formed. There are others that remain in Equilibrium state all the time (in the absence of any Passive catalysts) to make us smell and perceive them as will become clear.

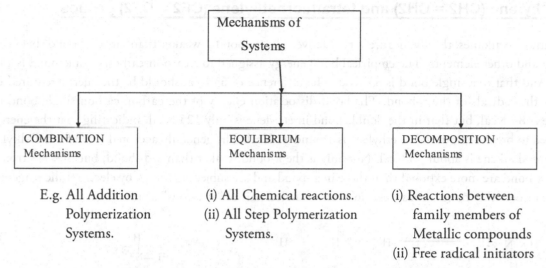

Figure 1.2 Classifications of Mechanisms of Systems

Note that the examples shown below the figure above, use only chemical and polymeric reactions. All the three mechanisms lead to positive or negative results, depending on the **_operating conditions_** and the type of products desired and much more. Decomposition should not be seen as something negative. Useful products can be obtained under such mechanisms. Combination mechanism like all the other mechanisms, take place in living and non-living systems. ***All things that exist in life or our world have their goods and bads, e.g., water, fire, air, wind, food, good, bad, thief, criminal, and so on.***

Based on universal data, unlike what is presently known, metals can also and indeed be ***reclassified*** as-

(a) *Ionic Non-Transition metals (Groups 1A and IIA),*
(b) *Ionic Transition metals (Group IIIA),*
(c) *Non-ionic Transition metals (Groups IVA, VA, VIA and VIIA),*
(d) *Non-ionic Transition-transition metals (Groups VIIIA), and*
(e) *Non-ionic Non-Transition metals (Groups IB, IIB, IIIB and some of Groups IVB, VB, VIB, and VIIB).*

These make up the group of metallic compounds alluded to in Figure 2.2 under Decomposition mechanisms.

All systems whether mechanical, electrical, chemical or otherwise, operate using one or more of the mechanisms above in a stage or many stages. In a single-stage system, only one mechanism operates. In multi-stage systems, one or more mechanisms can be involved. In polymerization reactions, one, two or three mechanisms are involved. Combination mechanism takes place in stages, while the other two take place in a single stage or in many stages. There are also *different types* of Equilibrium, Decomposition and Combination mechanisms. So far, only the mechanisms for Decomposition and Combination have partly been shown. So far one has not yet bothered to define the symbols used in representation of States of Existences. However, notice the symbols used in Figure 1.1 when H_2 was used as an example. To keep H_2 in Equilibrium State of existence, a passive catalyst must be used or this be done via some Physical means. Oxygen cannot be kept in Equilibrium State of existence, but can only be activated. The same applies to N_2. The symbols in Figure 1.1 above are completely different from the real symbol for Reversible reactions and from the ones used universally.

1.1 Ethylene (CH2 = CH2) and tetrafluoroethylene (CF2 = CF2) groups

In contrast to alkanes, the alkenes are very reactive. The π - bond is weaker than most σ - bonds between carbon and other elements. The empirical bond energy assigned to a carbon-carbon double bond is 146 Kcal. And that to a single bond is 83 Kcal. The difference of 63 Kcal should be the energy required to unpair the radicals of the π-bond. The bond dissociation energy of the carbon-carbon single bond in ethane is 84 Kcal, but that in the double bond in ethylene is only 125 Kcal, indicating that the energy required to break the π-bond in ethylene is around 41 Kcal. The generally accepted value for a dialkyl – substituted alkene is about 58 Kcal. Not only is the π-bond weaker than a σ -bond, but the "electrons" of the π-bond are more exposed than those in a -bond and are subject to attack by electrophilic reagents. In this respect the -bond almost resembles an unshared pair of "electrons" as in

| (Ammonia) | (Nucleophile) | (amine anion) | 1.1 |

Thus, alkenes are nucleophiles. They become stronger nucleophiles when there are more alkyl groups in place of hydrogen atoms located more on one carbon atom such as in $CH_2 = CHR_1$ or $CH_2 = CR_1R_2$ because alkyl groups are more electron-pushing than hydrogen. Therefore, the first step of an ionic or charged or radical reaction is the combination of the "π-electrons" with an attacking electrophilic reagent.[8] This applies only to the nucleophiles.

If in place of the alkyl groups or hydrogen which are "electron"-pushing, one has "electron"-pulling groups such as in tetrafluoro ethylene ($CF_2 = CF_2$), the alkene loose its nucleophilicity and becomes more of an electrophile. Free-radically, these types of halogenated monomers remain as nucleophiles or without any character. This can be illustrated by writing the different possible "electronic electron-dot" configurations of $CH_2 = CH_2$ and $CF_2 = CF_2$ their first members as follows:-

For ethylene

| (I)a | (II)a | (III)a | (IV)a (Existence not favored) | 1.2 |

(Free-radical activation) (Charged activation)

OR

| (I)b | (II)b | (III)b | (IV)b | 1.3 |

7

For perfluoro ethylene

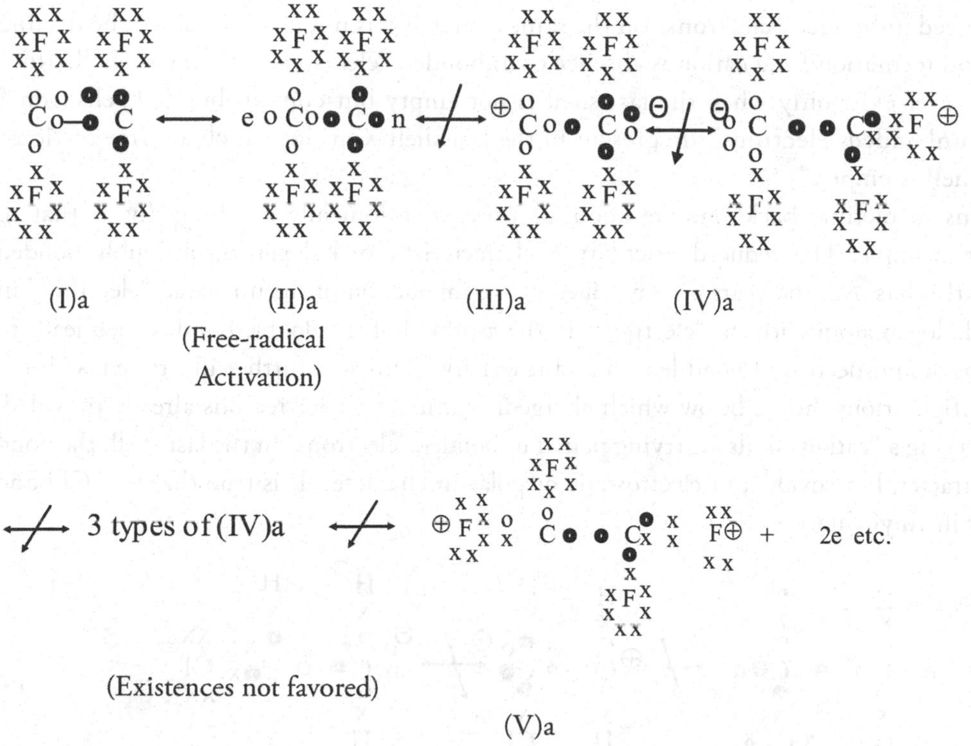

$$
\text{(I)a} \longleftrightarrow \text{(II)a} \xslash \text{(III)a} \xslash \text{(IV)a}
$$

(I)a   (II)a   (III)a   (IV)a

(Free-radical
Activation)

$$
\xslash \text{3 types of (IV)a} \xslash \text{(V)a} + 2e \text{ etc.} \qquad 1.4
$$

(Existences not favored)

(V)a

OR

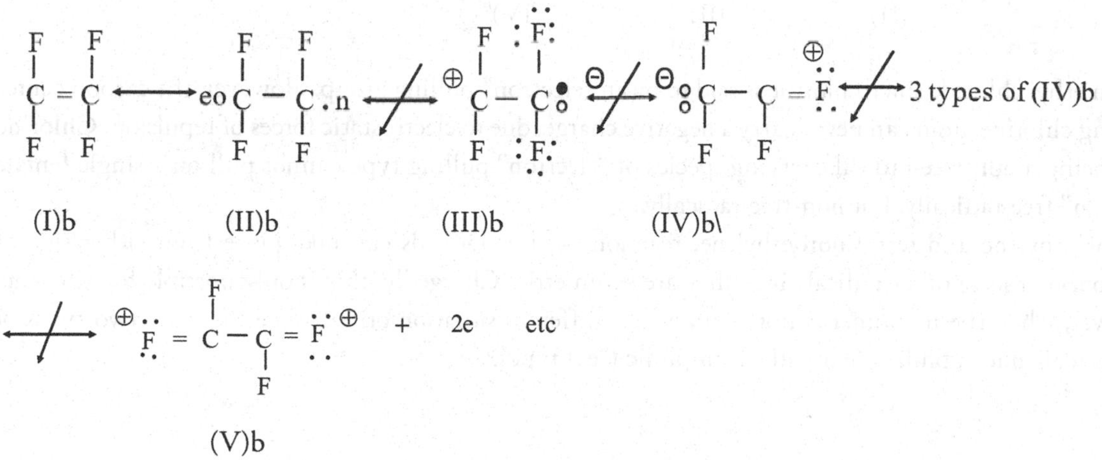

$$
\text{(I)b} \longleftrightarrow \text{(II)b} \xslash \text{(III)b} \xslash \text{(IV)b} \xslash \text{3 types of (IV)b}
$$

(I)b   (II)b   (III)b   (IV)b\

$$
\xslash \;\; \overset{\oplus}{\ddot F} = C - \underset{\underset{F}{|}}{C} = \overset{\oplus}{\ddot F} \;\; + \;\; 2e \quad \text{etc}
$$

(V)b   1.5

The "e" and "n" denote electrophilic and nucleophilic types of free-radicals. As shown by the configuration above, reaction of a π- bond may take place in several ways. The (II)a for both ethylene and perfuoroethylene indicate that they both have two free radicals located at both ends of the carbon atoms of the monomers, that for perfluoroethylene being far easier to obtain than for ethylene. This implies that they can both be attacked by free-radicals when the conditions exist. The (III)a for only ethylene indicates that it can undergo "anionic" and "cationic" polymer-ization under suitable conditions. (IV)a for ethylene clearly indicates the nucleophilic nature of this monomer though its existence is impossible. The (III)a for perfluoroethylene indicates that its existence is not possible since the "anion" on the carbon center cannot be adjacently located to a center (F) carrying paired

8

unbonded "electrons", due to electrostatic forces of repulsion. (IV)a and (V)a for perfluoroethylene indicate that existence of a non-free "cationic" center, that is, one in which a "cation" is carried in the presence of paired unbonded "electrons" on the same center (F), is not possible, since the driving force for ionic bond formation for a cation is absence of unbonded "electrons" in the last shell. Non-free "cations" however exist only when the last shell is not empty but contain bonded "electrons" and when paired unbonded "electrons" are present in the last shell. Cations which are free are ions, because the last shell is empty.

Halogen atoms in olefinic backbone are about as unreactive ionically or chargedly as that in chlorobenzene for example. The reduced reactivity is characteristic of halogen on a doubly bonded carbon atom, and this has over the years been ascribed to the interaction of the unshared "electrons" in a p-orbital of the halogen atom with the "electrons" in the π-orbital of the double bond, which leads to greater bond strength and decreased bond length and reactivity.[9] This was further illustrated with the (III) and (IV) configurations shown below which chargedly cannot exist for reasons already provided. When a center carrying a "cation" is also carrying paired unbonded "electrons" in the last shell, the bond is not ionic in character, but covalent or electrostatic or polar in character. It is true that C – Cl bond distance is shorter in vinyl chloride

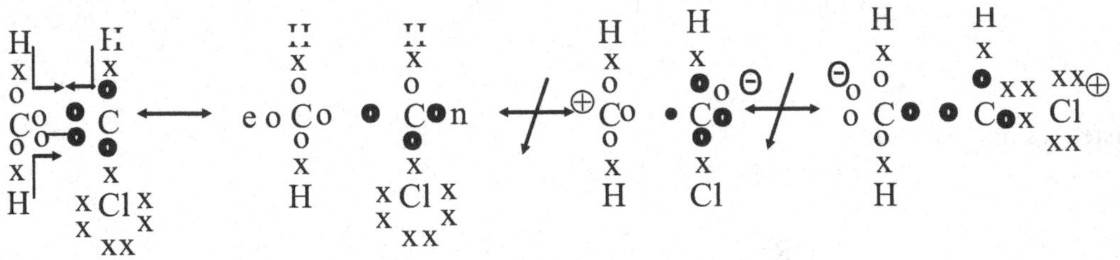

(I)　　　　(II)　　　(III)　　　(IV)[9]　　　　　　　1.6

than in ethyl chloride since chlorine is said to be an "electron"-pulling group. However, the carbon center carrying chlorine atom can never carry a negative charge, due to electrostatic forces of repulsion. Chlorine atom being a non-free-radical carrying species of "electron"-pulling type cannot pull on a single female "electron" free-radically, but non-free-radically.

For ethylene and tetrafluoroethylene, reactions of free-radicals can take place from either side of the monomer to another radical since they are symmetric. Chargedly, this is only possible for ethylene. However, when the monomer is not symmetric a different situation arises, since there are two types of free-radicals-nucleophilic ("n") and electrophilic ("e") types.[2]

1.2 CH2 = CHF (CH2 = CHX) and CF2 = CFH groups

For alkenes, if a free radical of the electrophilic type ("E") and vinyl chloride are combined together, the following reactions will occur:-

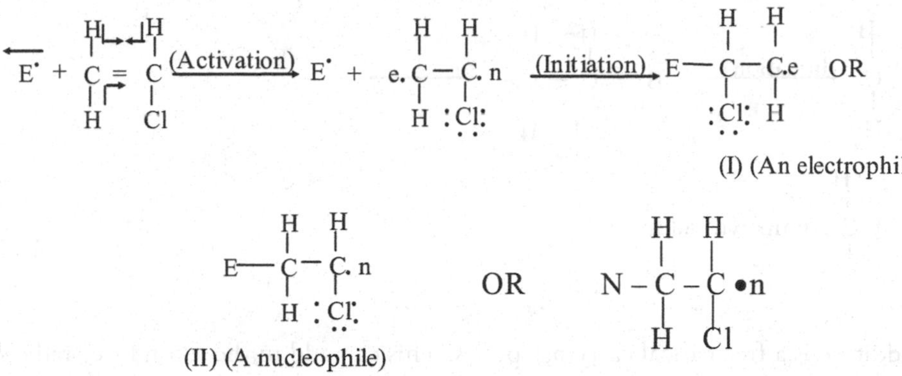

$$E^{\bullet} + \underset{\underset{H}{|}}{\overset{\overset{H}{|}}{C}} = \underset{\underset{Cl}{|}}{\overset{\overset{H}{|}}{C}} \xrightarrow{\text{(Activation)}} E^{\bullet} + e.\underset{\underset{H}{|}}{\overset{\overset{H}{|}}{C}} - \underset{\underset{:\overset{..}{C}l:}{|}}{\overset{\overset{H}{|}}{C}}.n \xrightarrow{\text{(Initiation)}} E - \underset{\underset{:\overset{..}{C}l:}{|}}{\overset{\overset{H}{|}}{C}} - \underset{\underset{H}{|}}{\overset{\overset{H}{|}}{C}}.e \quad \text{OR}$$

(I) (An electrophile)

$$E - \underset{\underset{H}{|}}{\overset{\overset{H}{|}}{C}} - \underset{\underset{:\overset{..}{C}l:}{|}}{\overset{\overset{H}{|}}{C}}.n \qquad \text{OR} \qquad N - \underset{\underset{H}{|}}{\overset{\overset{H}{|}}{C}} - \underset{\underset{Cl}{|}}{\overset{\overset{H}{|}}{C}} \bullet n$$

(II) (A nucleophile)

[Impossible Reaction]   (III) (A Nucleophile) [A Possible Reaction]   1.7

Since the reaction between a nucleophile and an electrophile is favored, (I) will be formed instead of (II), though two electro-free-radicals or two nucleo-free radicals can combine together to form a stable compound or molecule only in the absence of the opposite type and under certain conditions.

On the other hand, with (II) above the equation cannot be balanced. (III) which is balanced is the route not Natural to the monomer, the monomer being a Nucleophile. For fluoro alkenes, the corresponding reaction to Equation 1.7 is as follows-

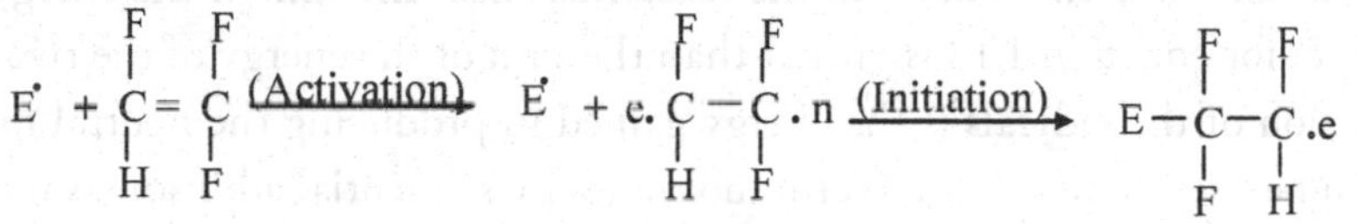

$$E^{\bullet} + \underset{\underset{H}{|}}{\overset{\overset{F}{|}}{C}} = \underset{\underset{F}{|}}{\overset{\overset{F}{|}}{C}} \xrightarrow{\text{(Activation)}} E^{\bullet} + e.\underset{\underset{H}{|}}{\overset{\overset{F}{|}}{C}} - \underset{\underset{F}{|}}{\overset{\overset{F}{|}}{C}}.n \xrightarrow{\text{(Initiation)}} E - \underset{\underset{F}{|}}{\overset{\overset{F}{|}}{C}} - \underset{\underset{H}{|}}{\overset{\overset{F}{|}}{C}}.e$$

1.8

Now consider the reactions with a nucleo-free-radical N For the alkenes, the following is obtained:-

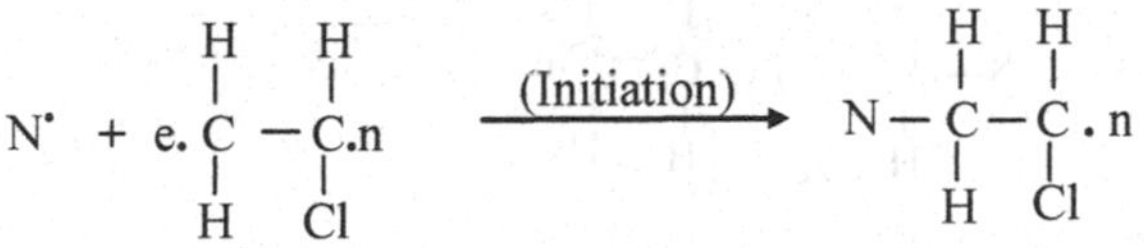

$$N^{\bullet} + e.\underset{\underset{H}{|}}{\overset{\overset{H}{|}}{C}} - \underset{\underset{Cl}{|}}{\overset{\overset{H}{|}}{C}}.n \xrightarrow{\text{(Initiation)}} N - \underset{\underset{H}{|}}{\overset{\overset{H}{|}}{C}} - \underset{\underset{Cl}{|}}{\overset{\overset{H}{|}}{C}}.n$$

1.9

The corresponding equation to Equation 1.9 for the perfluoro alkene is as follows:-

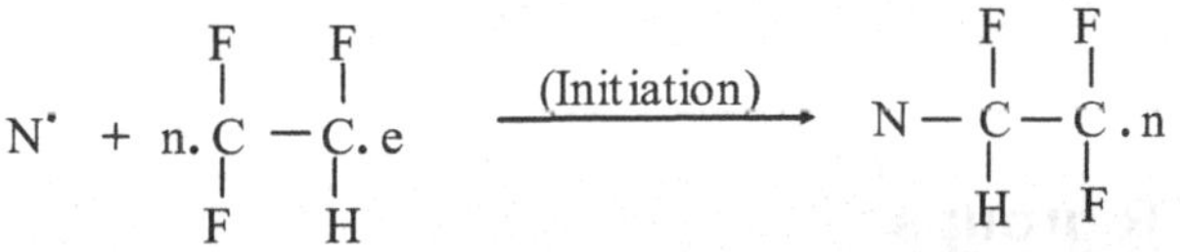

$$N^{\bullet} + n.\underset{\underset{F}{|}}{\overset{\overset{F}{|}}{C}} - \underset{\underset{H}{|}}{\overset{\overset{F}{|}}{C}}.e \xrightarrow{\text{(Initiation)}} N - \underset{\underset{H}{|}}{\overset{\overset{F}{|}}{C}} - \underset{\underset{F}{|}}{\overset{\overset{F}{|}}{C}}.n$$

1.10

Thus, just as in charged systems, monomers in free-radical systems add head-to-tail. The only situation where head-to-head or tail-to-tail addition takes place is during termination by combination under certain conditions and this is only possible radically since radicals do not repel or attract with themselves under any conditions. Radicals carry identities. Two or more rules can be identified at this point. These concern the symmetric and non-symmetric nature of the monomers. Free radicals were thought to add from any side of a monomer whether symmetric or unsymmetric and this is not true. The two rules will be identified at the end.

In free radical polymerization, breaking and creation of bonds are done homolytically.

$$Energy + A./.B \longrightarrow A.n + B.e \text{ (homolytic breaking)} \qquad 1.11$$

$$B.e + \underset{\substack{| \\ H}}{\overset{\substack{Cl \\ |}}{C}} \overset{\pi}{\underset{\sigma}{=}} \underset{\substack{| \\ H}}{\overset{\substack{H \\ |}}{C} \xrightarrow{\text{(Initiation)}}} \quad B - \underset{\substack{| \\ Cl}}{\overset{\substack{H \\ |}}{C}} - \underset{\substack{| \\ H}}{\overset{\substack{H \\ |}}{C}}.e \longrightarrow$$

$$\longrightarrow \quad B \overset{\sigma}{-} \underset{\substack{| \\ Cl}}{\overset{\substack{H \\ |}}{C}} \overset{\sigma}{-} \underset{\substack{| \\ H}}{\overset{\substack{H \\ |}}{C}}.e \text{ (homolytic addition)} \qquad 1.12$$

The product after the addition is a free radical carrying species. This can add to the second free radical to form a compound or molecule, since AB is not an ideal free-radical catalyst.

$$B \overset{\sigma}{-} \underset{\substack{| \\ Cl}}{\overset{\substack{H \\ |}}{C}} \overset{\sigma}{-} \underset{\substack{| \\ H}}{\overset{\substack{H \\ |}}{C}}.e + n.A \longrightarrow A \overset{\sigma}{-} \underset{\substack{| \\ H}}{\overset{\substack{H \\ |}}{C}} \overset{\sigma}{-} \underset{\substack{| \\ H}}{\overset{\substack{Cl \\ |}}{C}} \overset{\sigma}{-} B \qquad 1.13$$

In the reactions above (equations 1.12 and 1.13), the driving force of the reaction is the energy evolved because the sum of the energies of the two new σ- bonds in the product for equation 1.12 is greater than the sum of the energy of the π-bond and the σ bond in the reactants. Also, the sum of the energy of the three new bonds in the product for equation 1.13 is greater than the sum of the energy of the two bonds. The energy lost in the formation of the radicals is the energy gained in producing the neutral molecule in Eqn. 1.13. This energy can be used to react with olefin molecules by sequential addition as follows in the presence of an ideal initiator, N.n.

$$N - \underset{\substack{| \\ H}}{\overset{\substack{H \\ |}}{C}} - \underset{\substack{| \\ Cl}}{\overset{\substack{H \\ |}}{C}}.n + (m)e.\underset{\substack{| \\ H}}{\overset{\substack{H \\ |}}{C}} \underset{\substack{| \\ H}}{\overset{\substack{Cl \\ |}}{C}}.n \xrightarrow{\text{(Propagation)}} N - (\underset{\substack{| \\ H}}{\overset{\substack{H \\ |}}{C}} - \underset{\substack{| \\ H}}{\overset{\substack{Cl \\ |}}{C}})_m \underset{\substack{| \\ H}}{\overset{\substack{H \\ |}}{C}} - \underset{\substack{| \\ H}}{\overset{\substack{Cl \\ |}}{C}}.n \qquad 1.14$$

Equation 1.14 would predominate during propagation when A.n is absent in the system, whereas equation 1.13 would predominate when A.n is present in the system. Hence, these reactions are all exothermic in nature.

1.3 CH2 =CHRF and CF2 = CFR groups

With charged systems, for ethylene and not tetrafluoro, attack can also take place from any side, since the carbon centers are symmetric, noting that tetrafluoro cannot be activated chargedly. Like free-radicals, however, when the hydrogen or fluorine is changed for another group, electrophilic attack on the π-bond can only take place according to some rules. Without reference to any rule, replacing the chlorine atom in vinyl chloride with a "free electron-pulling" group such as CN, ionic/charged reactions would occur-

$$A:/B \longrightarrow A^{\ominus} + B^{\oplus} \quad (\text{heterolytic breaking})$$
$$(\text{anion}) \quad (\text{free-cation}) \tag{1.15}$$

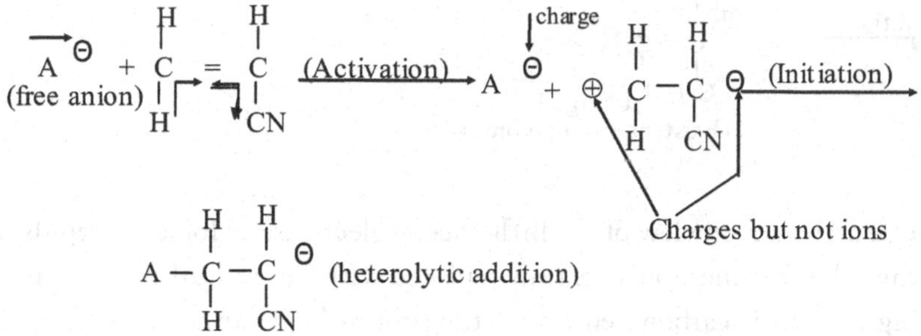

$$\xrightarrow{\quad} A^{\ominus} + \underset{\substack{\text{H} \\ \text{C} \\ \text{H}}}{\overset{\substack{\text{H}}}{|}} = \underset{\substack{\text{C} \\ \text{CN}}}{\overset{\substack{\text{H}}}{|}} \xrightarrow{\text{(Activation)}} A \overset{\ominus}{} + \overset{\oplus}{\underset{\substack{\text{H}}}{\text{C}}} - \underset{\substack{\text{C}}}{\text{C}} \overset{\ominus}{} \xrightarrow{\text{(Initiation)}}$$

(free anion) ... (charge) ... Charges but not ions

$$A - \underset{\substack{\text{H} \\ \text{CN}}}{\overset{\substack{\text{H} \\ |}}{\text{C}}} - \underset{\substack{\text{H} \\ \text{CN}}}{\overset{\substack{\text{H} \\ |}}{\text{C}}} \overset{\ominus}{} \quad (\text{heterolytic addition}) \tag{1.16}$$

$$\overset{\quad}{\underset{\longleftarrow}{B^{\oplus}}} + \underset{\substack{\text{H} \\ \text{CF}_3}}{\overset{\substack{\text{H}}}{|}} = \underset{\substack{\text{CF}_3}}{\overset{\substack{\text{H}}}{|}} \xrightarrow{\text{(Activation)}} B^{\oplus} + \overset{\oplus}{\underset{\substack{\text{H}}}{\text{C}}} - \underset{\substack{\text{CF}_3}}{\text{C}} \overset{\ominus}{} \xrightarrow{\text{(Initiation)}} BF + \overset{\oplus}{\underset{\substack{\text{F}}}{\text{C}}} - \underset{\substack{\text{F}}}{\text{C}} = \underset{\substack{\text{H}}}{\text{C}}$$

$$\text{OR} \quad \underset{\substack{\text{F} \\ \text{F}}}{\overset{\substack{\text{F} \\ |}}{\text{C}}} = \underset{\substack{\text{H}}}{\text{C}} - \underset{\substack{\text{H}}}{\text{C}} \overset{\oplus}{} \quad (\text{heterolytic addition}) \tag{1.17}$$

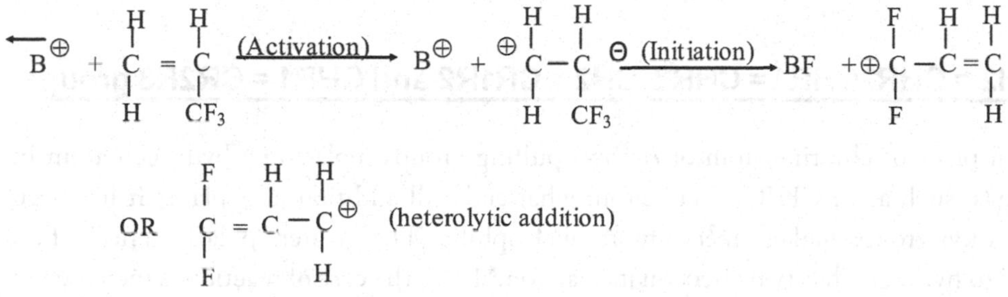

$A^{\ominus}$ above is assumed to be a free-anion, instead of a non-free anion $A^{\ominus}$ such as $CH_3 O^{\ominus}$. With a non-free anion, the equation will not be chargedly balanced after the initiation step, since the negatively charged active carbon center on the activated monomer is of the free type. When the $\ominus$ in A is a free one such as $^{\ominus}CH_3$, then Equation 1.16 is favored. The second monomer above does not favor the "cationic" route in view of presence of transfer species on one of the active carbon centers (transfer from monomer step). When the only hydrogen atom on the active carbon center is replaced by a resonance stabilization group, the initiation step will be favored as shown below *only if resonance stabilization can take place chargedly. Meanwhile, one will assume it can take place chargedly until a point is reached when it will be shown very clearly that it cannot take place.*

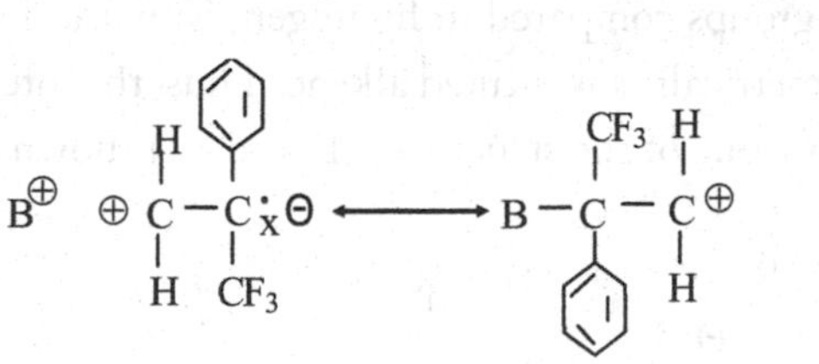

$$B^{\oplus} + \overset{\oplus}{\underset{\substack{\text{H} \\ \text{H}}}{\text{C}}} - \underset{\substack{\text{CF}_3}}{\text{C}} \overset{:\ominus}{\underset{x}{}} \longleftrightarrow B - \underset{\substack{\text{CF}_3}}{\overset{\substack{\text{CF}_3}}{\text{C}}} - \underset{\substack{\text{H}}}{\text{C}} \overset{\oplus}{} \tag{1.18}$$

$$(CH_2 - CR\ R_F)$$

The transfer species of the first kind cannot be abstracted here, in view of the presence of resonance stabilization group on the carbon center carrying the substituent group, CF_3. While CF_3 group is a substituent group, the other groups without transfer species are not. As we move along, we will find that this can only be done radically and not chargedly. Just like the case of Equation 1.17, BF must be formed chargedly, in Equation 1.18.

It is important to note in reactions above, that ionic or charge breaking and formation of bonds involve heterolytic action. The corresponding equations to Equations 1.16 and 1.17, for perfluoroalkenes are as follows:

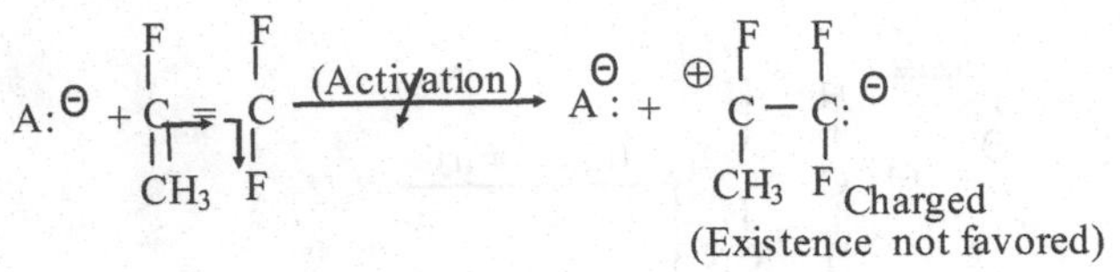

$$
A:^{\ominus} + \underset{\underset{CH_3}{|}}{\overset{\overset{F}{|}}{C}} = \underset{\underset{F}{|}}{\overset{\overset{F}{|}}{C}} \xrightarrow{\text{(Activation)}} A:^{\ominus} + {}^{\oplus}\underset{\underset{CH_3}{|}}{\overset{\overset{F}{|}}{C}} - \underset{\underset{F}{|}}{\overset{\overset{F}{|}}{C}}:^{\ominus}
$$

Charged
(Existence not favored)

1.19

Chargedly, the monomer cannot be activated, in view of the influence of electrostatic forces of repulsion. Activation will be favored only when the fluorine atoms are fully replaced with free radical-pulling groups such as CN or $COOCH_3$, noting that their carbon centers on the groups here cannot carry negative charges unlike CH_3.

1.4 Stability in CH2 = CHR, CHR1 = CHR2, CH2 = CR1R2 and CHR1 = CR2R3 groups.

Assuming that, in place of chlorine atom or *electron*-pulling groups replacing a hydrogen atom in ethylene, one has alkyls, such as in $CHR = CH_2$. From what end will addition take place? It has been said that presence of alkyl groups makes alkenes more nucleophilic. The greater "polarizability" of an alkyl group compared to hydrogen has two effect on its reaction.[8] First, the ease of reaction is increased as hydrogens attached to the doubly bonded carbon atoms are replaced by alkyl groups because the positive charge resulting from the combination of the attacking electrophile with the "*π-electrons*" is relieved by polarization potentials of the alkenes.[8] Ionization potential is the energy necessary to remove an "electron" from an atom or molecule. The energy necessary to remove an "*electron*" within any group of isomers decreases as the number of alkyl groups attached to especially one of the doubly bonded carbon atoms increases, thus reflecting the "*electron*"-pushing power of the alkyl groups. On the other hand, one should expect the reverse with perfluoro alkenes in which replacement of one or two of the fluorine atoms with fluoroalkyl groups or chloride atoms, the monomer becomes more electrophilic, since chloride is more "*electron*"-pulling than fluorine. Since these halogen atoms are "*electron*"-pulling, one should expect lower energy to remove "an electron" from $CF_2 = CCl_2$ than from $CF_2 = CF_2$.

The second effect of greater polarizability of alkyl groups compared to hydrogen, is that it determines the point of attack of a polarized molecule on an unsymmetrically substituted alkene. Thus, the intermediate "carbonium ion" in the following reaction will favor only one of the structures (I) and (II) shown below.

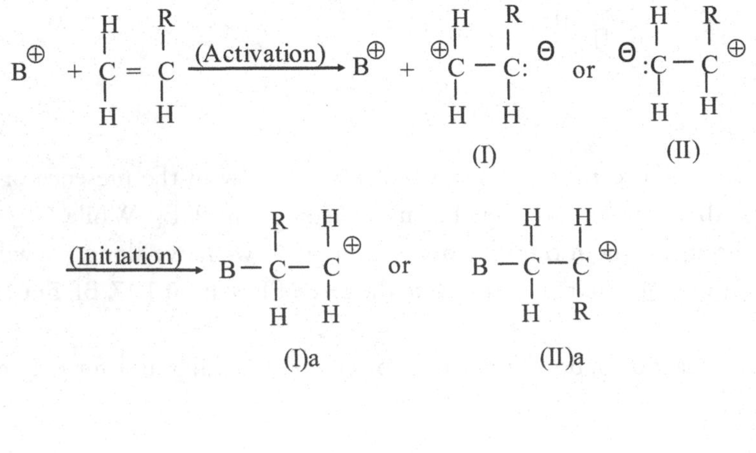

$$
B^{\oplus} + \underset{\underset{H}{|}}{\overset{\overset{H}{|}}{C}} = \underset{\underset{H}{|}}{\overset{\overset{R}{|}}{C}} \xrightarrow{\text{(Activation)}} B^{\oplus} + {}^{\oplus}\underset{\underset{H}{|}}{\overset{\overset{H}{|}}{C}} - \underset{\underset{H}{|}}{\overset{\overset{R}{|}}{C}}: {}^{\ominus} \quad \text{or} \quad {}^{\ominus}:\underset{\underset{H}{|}}{\overset{\overset{H}{|}}{C}} - \underset{\underset{H}{|}}{\overset{\overset{R}{|}}{C}}{}^{\oplus}
$$

(I)   (II)

$$
\xrightarrow{\text{(Initiation)}} B - \underset{\underset{H}{|}}{\overset{\overset{R}{|}}{C}} - \underset{\underset{H}{|}}{\overset{\overset{H}{|}}{C}}{}^{\oplus} \quad \text{or} \quad B - \underset{\underset{H}{|}}{\overset{\overset{H}{|}}{C}} - \underset{\underset{R}{|}}{\overset{\overset{H}{|}}{C}}{}^{\oplus}
$$

(I)a   (II)a

1.20

Sunny N.E. Omorodion

Since the R group is more "*electron*"-pushing than hydrogen, the "π-*electrons*" will be pushed away from the carbon atom carrying the R group. Hence (II)a should be the intermediate most readily formed. In other words, a secondary alkyl "cation" will be more stable than a primary alkyl "cation". A tertiary alkyl "cation" should be more stable than a secondary alkyl "cation", whether molecular rearrangement is favored or not. On the other hand, stability of an alkyl "cation" will depend on where the R group(s) is located. For the case where they are located on one carbon center, the following is obtained:-

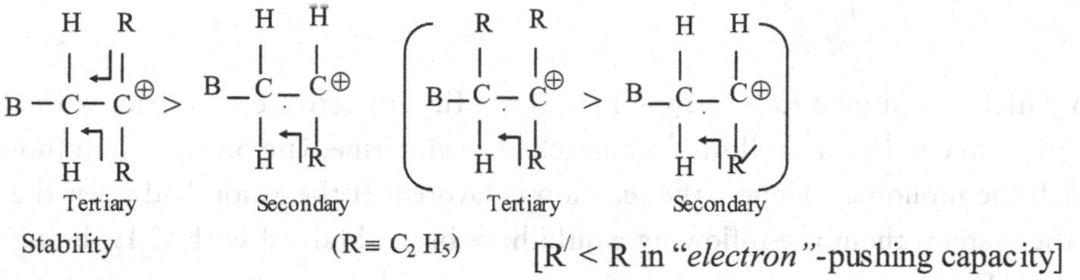

$$
\begin{array}{ccc}
\underset{\substack{\text{Tertiary}}}{\overset{\substack{H \quad R}}{B-C-C^{\oplus}}} & > & \underset{\substack{\text{Secondary}}}{\overset{\substack{H \quad H}}{B-C-C^{\oplus}}} \\
\end{array}
\left(
\begin{array}{ccc}
\underset{\substack{\text{Tertiary}}}{\overset{\substack{R \quad R}}{B-C-C^{\oplus}}} & > & \underset{\substack{\text{Secondary}}}{\overset{\substack{H \quad H}}{B-C-C^{\oplus}}}
\end{array}
\right) \qquad 1.21
$$

Stability $\qquad\qquad (R \equiv C_2H_5) \qquad [R' < R$ in "*electron*"-pushing capacity]

The equation above is also valid electro-free-radically. Removal of an "*electron*" from a carbon center leads to carbonion or cation. The more readily this "*electron*" is removed, the greater the stability of the "cation". The ionization potentials for any group of isomeric free radicals decrease with increasing number of alkyl substituent groups on a carbon center. As shown in Equation 1.21, even when weak initiators are involved, the tertiary alkyl "cation" is more stable than the secondary alkyl "cation". On the other hand, the stability indicated in Equation 1.21 is more favored – when coordination catalysts are employed. Note the direction of the resultant force or arrow, bearing in mind that R is more "*electron*" pushing than R' which in turn is more "*electron*"-pushing than H. R groups are usually alkyls, alkenyl, phenyls, alkoxy, etc. Therefore, the reaction in favor of "cationic" polymerization will be stronger for $CH_2 = CR_1R_2$ than $CH_2 = CHR_1$.

Of great interest are some reactions *in organic chemistry* which involve addition of substituted ethylenic monomers to sulfuric acid, halogen acid, hydrogen peroxide, aldehydes, bisulfite, hydrogen sulfide in the presence of peroxides or light. Most of the reactions were found to obey Markovnikov's rule of addition while some in the presence of peroxides and oxygen were said to obey Anti-Markovnikov's free-radical rule of addition.[8] Now in order to see the pattern more clearly, one will begin by considering the reactions between $CHR_1 = CR_2R_3$, a substituted alkene (where Rs are alkyl groups) and sulphuric acid. When the monomer is activated, the following is obtained.

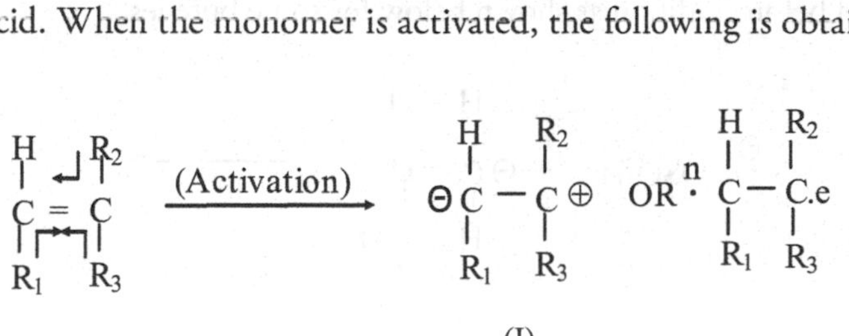

$$
\underset{R_1 \quad R_3}{\overset{H \quad R_2}{C=C}} \xrightarrow{\text{(Activation)}} \underset{R_1 \quad R_3}{\overset{H \quad R_2}{\ominus C - C^{\oplus}}} \text{ OR } \overset{n}{\cdot} \underset{R_1 \quad R_3}{\overset{H \quad R_2}{C - C.e}} \qquad 1.22
$$

(I)

where the "*electron*"-pushing capacities of R_2 and R_3 put together is greater than the "*electron*"-pushing capacities of R_1 and H. The monomer being a nucleophile, will favor the following reactions with sulphuric acid:-

14

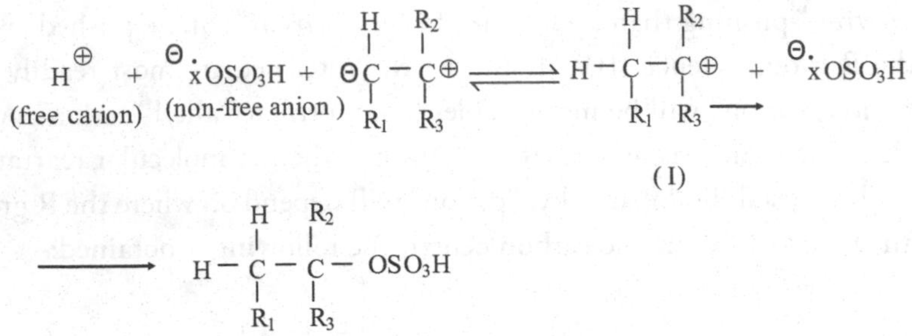

$$H^{\oplus} + {}^{\ominus}\!:\!xOSO_3H + {}^{\ominus}\overset{\overset{\displaystyle H}{|}}{C}-\overset{\overset{\displaystyle R_2}{|}}{\underset{\underset{\displaystyle R_3}{|}}{C}}{}^{\oplus} \rightleftharpoons H-\overset{\overset{\displaystyle H}{|}}{\underset{\underset{\displaystyle R_1}{|}}{C}}-\overset{\overset{\displaystyle R_2}{|}}{\underset{\underset{\displaystyle R_3}{|}}{C}}{}^{\oplus} + {}^{\ominus}\!:\!xOSO_3H \longrightarrow$$

(free cation) (non-free anion)

(I)

$$\longrightarrow \quad H-\overset{\overset{\displaystyle H}{|}}{\underset{\underset{\displaystyle R_1}{|}}{C}}-\overset{\overset{\displaystyle R_2}{|}}{\underset{\underset{\displaystyle R_3}{|}}{C}}-OSO_3H$$

1.23

The cation which is assumed to be strong here is the first to activate and react with the monomer under Equilibrium conditions, followed by attack of the intermediate on the anion (non-free) which cannot attack the monomer. Hence, the reaction is favored. If the anion had been the only species present in the system, them the following would have been obtained with C_2H_5 being used for R_3. (Note that $R_3 \geq R_2$).

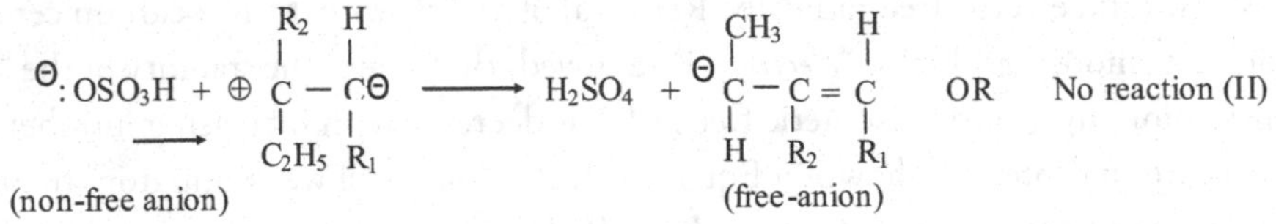

$$^{\ominus}\!:OSO_3H + {}^{\oplus}\overset{\overset{\displaystyle R_2}{|}}{\underset{\underset{\displaystyle C_2H_5}{|}}{C}}-\overset{\overset{\displaystyle H}{|}}{\underset{\underset{\displaystyle R_1}{|}}{C}}{}^{\ominus} \longrightarrow H_2SO_4 + {}^{\ominus}\overset{\overset{\displaystyle CH_3}{|}}{\underset{\underset{\displaystyle H}{|}}{C}}-\overset{\overset{\displaystyle H}{|}}{\underset{\underset{\displaystyle R_2}{|}}{C}}=\overset{\overset{\displaystyle H}{|}}{\underset{\underset{\displaystyle R_1}{|}}{C}} \quad OR \quad \text{No reaction (II)}$$

(non-free anion)   (free-anion)

(I)–Not chargedly balanced.

1.24

While $CH_3O^{\ominus}$ or $HO_3SO^{\ominus}$ a non-free anion will combine with (I) of Equation 1.23 to produce a stable molecule, it cannot combine with the activated monomer since the equation will not be chargedly balanced. $HO_3SO^{\ominus}$ or $CH_3O^{\ominus}$ being more nucleophilic than the monomer and since the non-free anion cannot readily be isolated from the system, one can observe why free cationic initiators cannot be used with these types of nucleophiles for polymerization. To use it for poly-merization is impossible, whether $CH_3O^{\ominus}$ is isolated from the system or be paired covalently or electrostatically. Only $H_3C^{\ominus}$ type can be used.

It is possible that the activated monomer represented by (I) of Equation 1.22 could favor some molecular rearrangement if "weak initiators" are involved before addition as shown below for some butenes.

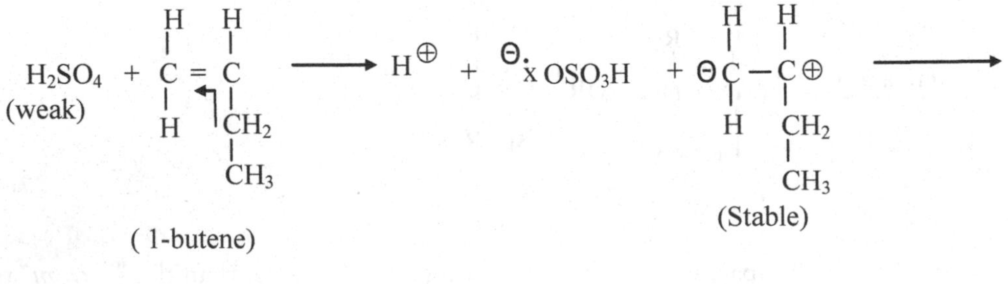

$$H_2SO_4 + \overset{\overset{\displaystyle H}{|}}{\underset{\underset{\underset{\underset{\displaystyle CH_3}{|}}{\displaystyle CH_2}}{|}}{\underset{\displaystyle H}{|}}}{C}=\overset{\overset{\displaystyle H}{|}}{\underset{\underset{\displaystyle CH_2}{|}}{C}} \longrightarrow H^{\oplus} + {}^{\ominus}\!:\!xOSO_3H + {}^{\ominus}\overset{\overset{\displaystyle H}{|}}{\underset{\underset{\displaystyle H}{|}}{C}}-\overset{\overset{\displaystyle H}{|}}{\underset{\underset{\underset{\displaystyle CH_3}{|}}{\displaystyle CH_2}}{|}}{C}{}^{\oplus} \longrightarrow$$

(weak)

(1-butene)   (Stable)

$$H-\underset{\underset{H}{|}}{\overset{\overset{H}{|}}{C}}-\overset{\overset{H}{|}}{\underset{\underset{CH_3}{|}}{\underset{CH_2}{|}}}C\oplus \;+\; \overset{\ominus}{\underset{x}{:}}OSO_3H \longrightarrow H-\underset{\underset{H}{|}}{\overset{\overset{H}{|}}{C}}-\underset{\underset{\underset{CH_3}{|}}{CH_2}}{\overset{\overset{H}{|}}{C}}-OSO_3H$$

$$\text{(s – Butyl hydrogen sulfate – } CH_3 \overset{\overset{OSO_3H}{|}}{CH}C_2H_5)$$

1.25

$$H_2SO_4 + \underset{\text{(weak)}}{\underset{\underset{CH_3}{|}}{\overset{\overset{H}{|}}{C}}} = \underset{\underset{CH_3}{|}}{\overset{\overset{H}{|}}{C}} \longrightarrow H^\oplus + {}^\ominus :OSO_3H + \oplus \underset{\underset{CH_3}{|}}{C}-\overset{\ominus}{C}-CH_3 \xrightarrow{\underset{\text{rearrangement}}{\text{(molecular}}}$$

$$\text{(cis-2-butene))} \qquad\qquad \text{(less stable)}$$

$$H^\oplus + {}^\ominus OSO_3H + \ominus \underset{\underset{H}{|}}{\overset{\overset{H}{|}}{C}}-\underset{\oplus}{\overset{\overset{H}{|}}{C}}-\underset{\underset{H}{|}}{\overset{\overset{H}{|}}{C}}-CH_3 \longrightarrow H-\underset{\underset{H}{|}}{\overset{\overset{H}{|}}{C}}-\underset{\underset{\underset{CH_2}{|}}{CH_2}}{\overset{\overset{H}{|}}{C}}\oplus + {}^\ominus :OSO_3H$$

$$\text{(more stable)}$$

$$H-\underset{\underset{H}{|}}{\overset{\overset{H}{|}}{C}}-\underset{\underset{\underset{CH_3}{|}}{CH_2}}{\overset{\overset{H}{|}}{C}}-OSO_3H$$

$$\text{(s – Butyl hydrogen sulfate – } CH_3 \overset{\overset{OSO_3H}{|}}{CH}C_2H_5)$$
$$\text{(molecular rearrangement favored)}$$

1.26

$$H_2SO_4 + \underset{\text{(weak)}}{\underset{\underset{H}{|}}{\overset{\overset{CH_3}{|}}{C}}} = \underset{\underset{CH_3}{|}}{\overset{\overset{H}{|}}{C}} \longrightarrow H^\oplus + {}^\ominus :OSO_3H + \oplus \underset{\underset{H}{|}}{C}-\overset{\overset{CH_3}{|}}{\underset{\underset{CH_3}{|}}{\overset{\ominus}{C}}}-H \xrightarrow{\underset{\text{Rearrangement}}{\text{(molecular}}}$$

$$\text{(trans-2-butene)} \qquad\qquad \text{(less stable)}$$

$$H^\oplus + {}^\ominus :OSO_3H + \ominus \underset{\underset{H}{|}}{\overset{\overset{H}{|}}{C}}-\underset{\oplus}{\overset{\overset{H}{|}}{C}}-\underset{\underset{H}{|}}{\overset{\overset{H}{|}}{C}}-CH_3 \longrightarrow H-\underset{\underset{H}{|}}{\overset{\overset{H}{|}}{C}}-\underset{\underset{\underset{CH_3}{|}}{CH_2}}{\overset{\overset{H}{|}}{C}}\oplus + {}^\ominus :OSO_3H$$

$$\text{(more stable)}$$

$$\longrightarrow H-\underset{\underset{H}{|}}{\overset{\overset{H}{|}}{C}}-\underset{\underset{\underset{CH_3}{|}}{CH_2}}{\overset{\overset{H}{|}}{C}}-OSO_3H$$

$$\text{(s – Butyl hydrogen sulfate)}$$

1.27

$$\text{(Molecular rearrangement favored)}$$

16

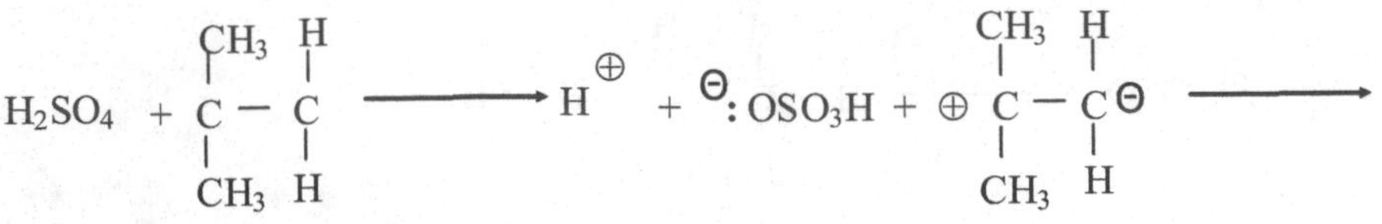

$$H_2SO_4 + \underset{\underset{CH_3}{|}}{\overset{\overset{CH_3}{|}}{C}} - \underset{\underset{H}{|}}{\overset{\overset{H}{|}}{C}} \xrightarrow{\hspace{1cm}} H^{\oplus} + {}^{\ominus}:OSO_3H + {}^{\oplus}\underset{\underset{CH_3}{|}}{\overset{\overset{CH_3}{|}}{C}} - \underset{\underset{H}{|}}{\overset{\overset{H}{|}}{C}}{}^{\ominus} \xrightarrow{\hspace{1cm}}$$

(methyl propene
or isobutylene)

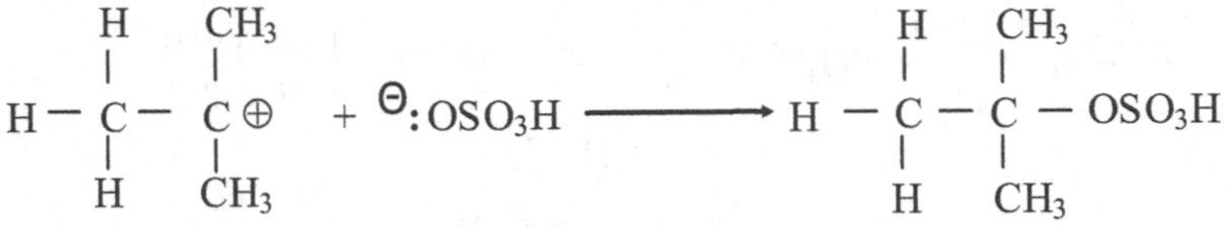

$$H - \underset{\underset{H}{|}}{\overset{\overset{H}{|}}{C}} - \underset{\underset{CH_3}{|}}{\overset{\overset{CH_3}{|}}{C}}{}^{\oplus} + {}^{\ominus}:OSO_3H \xrightarrow{\hspace{1cm}} H - \underset{\underset{H}{|}}{\overset{\overset{H}{|}}{C}} - \underset{\underset{CH_3}{|}}{\overset{\overset{CH_3}{|}}{C}} - OSO_3H \qquad 1.28$$

(t-Butyl hydrogen sulfate)

Indeed these reactions generally take place radically and not chargedly as shown above. When an initiator is paired, they largely take place chargedly, the charges being all radicals in character. How they do take place radically will be explained later in the Series after laying the founda-tions.

It may seem that molecular rearrangement is favored only when there is a cationic transfer species (such as) or an electro-free-radical transfer species on a substituent close to a free-negatively charged/ nucleo-free-radical center of the activated Nucleophilic monomer. No, nucleo-non-free-radical or anionic (i.e. non-free- negatively charged) transfer species can be transferred when they exist. However, one can see how the reported products are obtained. The charged centers of the activated monomers are fixed. So also are the radical centers. It has been reported in the past that when Ziegler-Natta (Z/N) catalysts are employed for 1,2-disubstituted ethylenes, there is no polymerization due to steric hindrance. At the same time, polymers have been said to be obtained from some 1,2-disubstituted ethylenes, but through so-called isomerization of the monomer to a 1-substituted ethylene followed by polymerization, e.g. 2-butene yields poly (1-butene).[10] From the reactions above and depending on the characters and ratios of the components of Z/N catalysts used, one can see why the reported observations are possible. Weak and strong centers and number of vacant orbitals involved are important variables. To obtain polymers of 2-butene, it has however been shown that special but different Z/N initiators are required for both cis- and trans-configuration of this monomer.[2] The driving forces that favor molecular rearrangement of monomers (not isomerization) when activated cannot yet be fully disclosed, since they are numerous. Though a different more stable isomer is favored, the phenomenon involved is a molecular rearrangement phenomenon as will be shown in the Series and Volumes and not Markovnikov's rule. Indeed, there are also other types of rearrangements which do not involve movement of atomic or molecular species, but only electro-radicals (i.e. free or non-free).

1.5. Non-existence of Anti-Markovnikov's rule for propylene (α-olefins)

Now consider the reactions of these alkenes with halogen acids. Olefins are said to add to halogen acids to give alkyl halides as follows:

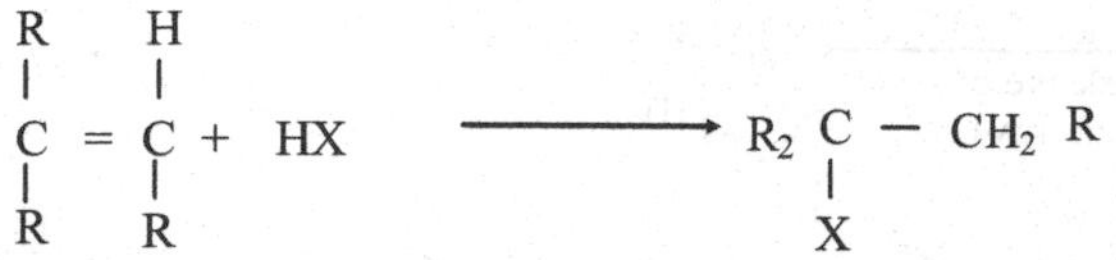

$$R_2C = CHR + HX \longrightarrow R_2C - CH_2R \quad\text{(with } X \text{ on central carbon)} \qquad 1.29$$

(Equation 1.29 shown structurally)

Using propylene as an example, i-propyl bromide is obtained as follows:-

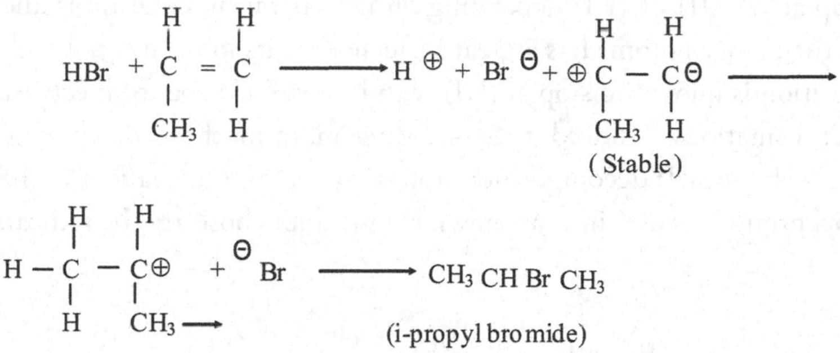

$$HBr + CH_3CH = CH_2 \longrightarrow H^\oplus + Br^\ominus + {}^\oplus CH(CH_3) - CH_2^\ominus \text{ (Stable)} \longrightarrow$$

$$H_3C - CH^\oplus(CH_3) + {}^\ominus Br \longrightarrow CH_3\,CHBr\,CH_3 \qquad 1.30$$

(i-propyl bromide)

These reactions largely take place radically and via Equilibrium mechanisms.

However, in the presence of peroxides or oxygen, only hydrogen bromide amongst all the halides favored producing n-propyl bromide instead of i-propyl bromide.[8] These types of abnormal additions gave rise to Anti-Markovnikov's rule, for which a different mechanism from the ionic case above was proposed, that is free-radical mechanism. If the activated monomer had been what is shown below,

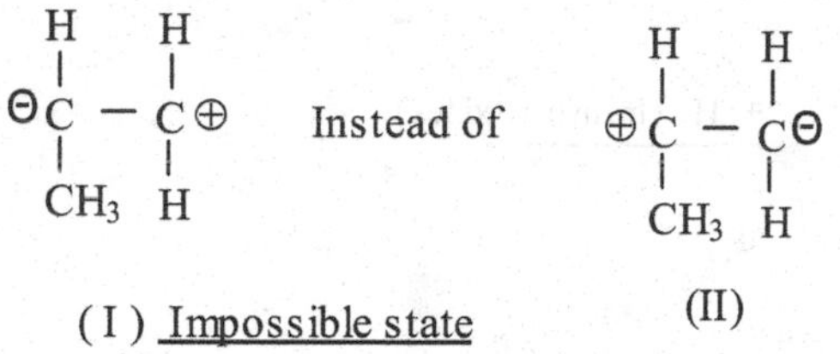

$$^\ominus C(CH_3)H - C^\oplus H_2 \quad\text{Instead of}\quad {}^\oplus C(CH_3)H - C^\ominus H_2 \qquad 1.31$$

(I) <u>Impossible state</u>   (II)

then the said product would have been obtained as follows with or without another type of impossible molecular rearrangement.

$$H^\oplus + Br^\ominus + {}^\ominus C(CH_3)H - C^\oplus H_2 \longrightarrow H - CH_2 - CH_2 - CH_2 - Br \qquad 1.32$$

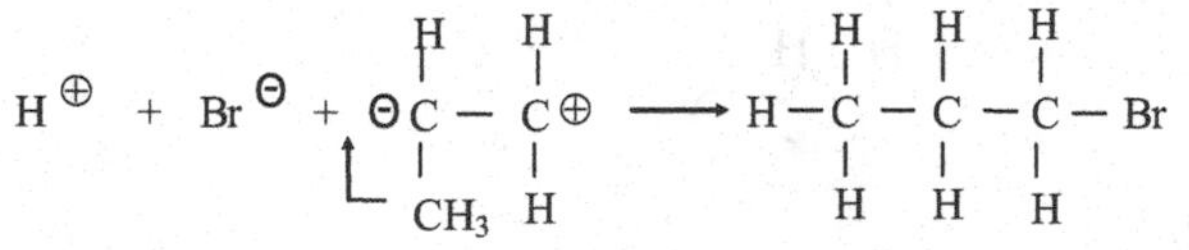

(N-propyl bromide –
$CH_3CH_2CH_2Br$) [Impossible reaction]

The molecular rearrangement indicated above is impossible. The addition represented by Equation 1.32 was said to be anti-Markovnikov. However, there is never a time the activated monomer of propylene can be (I) since the carbon center carrying CH_3 above cannot carry a negative charge. With the presence of peroxides or oxygen, nucleo-free-radicals ($H^{\cdot n}$) may be said to be produced.

Based on the newly proposed definition of a radical,[1] the general steps involved in the de-composition of halogen acids are as follows using present-day methods which are still (ART)s, but not a SCIENCE :-

$$2HX \xrightarrow[\text{environment})]{\text{(ionic}} 2H^{\oplus} + 2^{\ominus}:X \xrightarrow[\text{energy)}]{\text{(release of}} 2H^{\cdot e} + 2\,\ddot{X}\cdot nn$$
$$\qquad\qquad\qquad (I) \qquad\qquad\qquad\qquad\qquad\qquad (II)$$

$$\xrightarrow{\qquad} H_2 + X_2$$
$$\qquad (III) \qquad\qquad\qquad\qquad\qquad\qquad\qquad\qquad\qquad\qquad 1.33$$

The reactions above could stop at (I), (II) or (III) depending on the operating conditions and type of halide, noting that the size of the halogen atoms has a great influence on its ionic and polar characters. In fact, with HI, the decomposition is known to stop at (III) with impossible types of mechanisms past proposed to favor such product formations.[11] Indeed, there are three main mechanisms in which one of them is DECOMPOSITION mechanism. If decomposition stops at (II), for all the halides in the absence of peroxides, then same i-propyl bromide is obtained as shown below, since those are the radicals, largely present in the system.

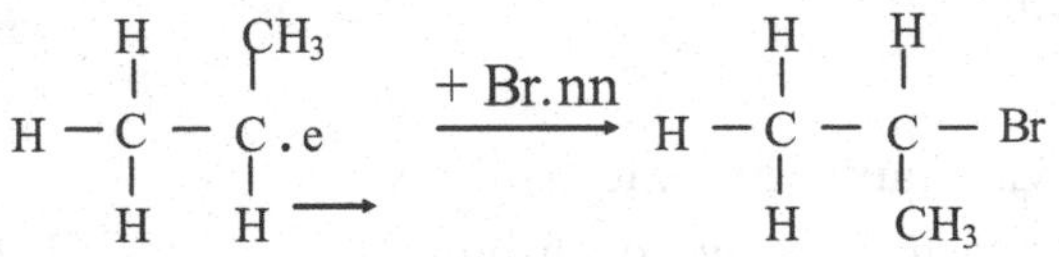

$$
\begin{array}{c}
\quad H \quad\ CH_3 \\
\quad | \quad\ \ | \\
H - C - C\cdot e \qquad \xrightarrow{+\ Br.nn} \\
\quad | \quad\ \ | \\
\quad H \quad\ H \longrightarrow
\end{array}
\qquad
\begin{array}{c}
\quad H \quad\ H \\
\quad | \quad\ | \\
H - C - C - Br \\
\quad | \quad\ | \\
\quad H \quad\ CH_3
\end{array}
\qquad 1.34
$$

Free-radically and ionically/chargedly in the absence of peroxides, i-propyl bromide is obtained. With the presence of peroxides which may favor the existence of a hydrogen center carrying a nucleo-free-radical, the following steps are involved in producing the n-propyl bromide.

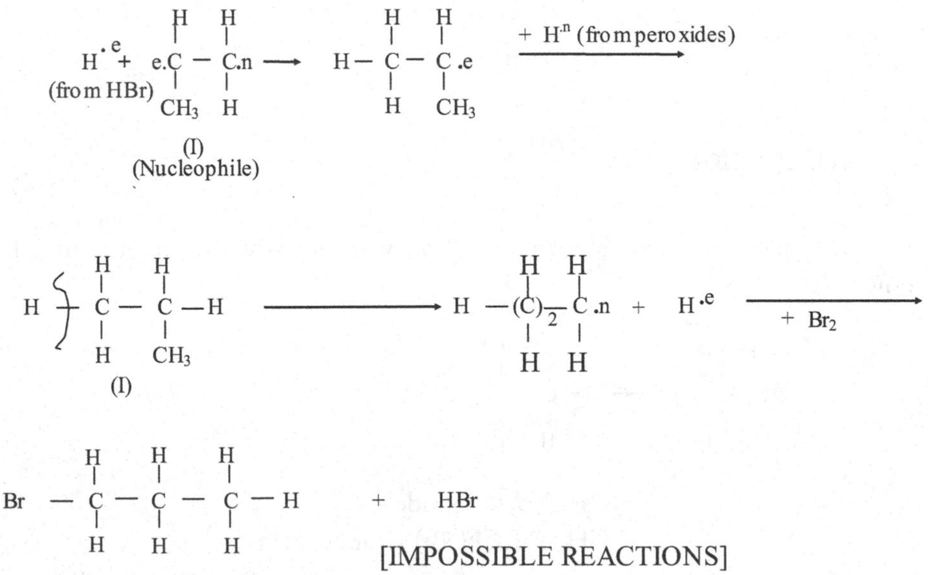

$$
\begin{array}{c}
\qquad\qquad H \ H \\
\qquad\qquad | \ \ | \\
H^{\cdot e} + e.C - C.n \longrightarrow \\
\text{(from HBr)} \ \ | \ \ | \\
\qquad\qquad CH_3 \ H \\
\qquad\quad (I) \\
\quad \text{(Nucleophile)}
\end{array}
\begin{array}{c}
\ H \ \ H \\
\ | \ \ | \\
H - C - C\ .e \qquad \xrightarrow{+\ H^n\ (\text{from peroxides})} \\
\ | \ \ | \\
\ H \ \ CH_3
\end{array}
$$

$$
H
\left\{
\begin{array}{c}
H \ \ H \\
| \ \ | \\
C - C - H \\
| \ \ | \\
H \ \ CH_3 \\
(I)
\end{array}
\right.
\xrightarrow{\qquad\qquad}
\begin{array}{c}
H \ H \\
| \ | \\
H - (C)_2 - C\ .n \ + \ H^{\cdot e} \quad \xrightarrow{+\ Br_2} \\
| \ | \\
H \ H
\end{array}
$$

$$\qquad\qquad\qquad\qquad\qquad\qquad\qquad\qquad\qquad\qquad\qquad\qquad\qquad 1.35$$

$$
\begin{array}{c}
\ H \ \ H \ \ H \\
\ | \ \ | \ \ | \\
Br - C - C - C - H \qquad + \qquad HBr \\
\ | \ \ | \ \ | \\
\ H \ \ H \ \ H \\
\text{n - propyl bromide}
\end{array}
\qquad \text{[IMPOSSIBLE REACTIONS]}
$$

As shown above the free-radicals activated the monomer i.e. attack by the (opposite sex)- electro-free-radicals, the monomer being a nucleophile; then followed by addition to nucleo-free-radicals to form (I) which further combines with bromine to form the n-propyl bromide. The steps involved are worthy of note, where the driving force is to balance the equation. This is different from the reaction between propylene and chlorine in the vapor phase to produce allyl chloride,[9] for which the mechanism also proposed is as follows:-

$$Cl^{.en} + Cl\,.nn + H^{.e} + n.\underset{\underset{H}{|}}{\overset{\overset{H}{|}}{C}} - \underset{\underset{H}{|}}{\overset{\overset{H}{|}}{C}} = \overset{\overset{H}{|}}{C} \xrightarrow{(500-600^{o}C)} HCl +$$

$$Cl - \underset{\underset{H}{|}}{\overset{\overset{H}{|}}{C}} - \underset{\underset{H}{|}}{\overset{\overset{H}{|}}{C}} = \overset{\overset{H}{|}}{C} \qquad 1.36$$

(allyl chloride)

Even at such high temperatures, chlorine molecules cannot dissociate to produce cations and anions. The chlorine nucleo-non-free- or electro-non-free-radical produced can activate and combine with the monomer. Hence, the simultaneous occurrence of the reaction above to give 85 – 90% of the allyl chloride. Nevertheless, all along, we have been using the ART as currently used today to move forward after knowing what a radical is. The real Science has not yet emerged. ***The full mechanisms of the reactions of Equation 1.35 will be provided in Chapter 6.***

Thus, it can be observed that with these monomers, ***Markovnikov's rule of addition which states that the more negative element or group adds to the less hydrogenated carbon atom or the more positive element or group adds to the less-alkyl substituted end of the polarized double bond*** is meaningless and not true in the sense that the first statement, that is the more negative element adding to the less hydrogenated carbon atom is not possible. It is the "cation" that diffuses to the nucleophile in all cases, before the product diffuses to the anion. As can be observed so far, it implies that even the Anti-Markovnikov's rule of addition (free-radical or ionic or charged) does not exist, not even for a monomer such as propylene (the name of which will be changed downstream when the real propylene is identified).

1.6 Stability of CF2 = CFRF, CF2 = CRF1RF2, CFRF1 = CFRF2, CFRF1 = CRF2RF3, (CF2 = CR1R2).

Now coming back to the corresponding equation for tetrafluoroethylene of Equation 1.20 the following equations can be written:-

$$A:^{\ominus} + \underset{\underset{F}{|}}{\overset{\overset{F}{|}}{C}} = \underset{\underset{F}{|}}{\overset{\overset{CF_3}{|}}{C}} \xrightarrow{(Activation)} A:^{\ominus} + \overset{\oplus}{\underset{\underset{F}{|}}{\overset{\overset{F}{|}}{C}}} - \underset{\underset{:\overset{..}{F}:}{|}}{\overset{\overset{CF_3}{|}}{C}}^{\ominus} \qquad 1.37a$$

(Not possible chargedly)

Chargedly, activation is impossible. Free-radically, the following is obtained

$$N^{\bullet} + \underset{\underset{F}{|}}{\overset{\overset{F}{|}}{C}} = \underset{\underset{F}{|}}{\overset{\overset{CF_3}{|}}{C}} \xrightarrow{Activation} N^{\bullet} + n.\underset{\underset{F}{|}}{\overset{\overset{F}{|}}{C}} - \underset{\underset{F}{|}}{\overset{\overset{CF_3}{|}}{C}}.e \longrightarrow N - \underset{\underset{F}{|}}{\overset{\overset{CF_3}{|}}{C}} - \underset{\underset{F}{|}}{\overset{\overset{F}{|}}{C}}.n \qquad 1.37b$$

Electro-free-radically and nucleo-free-radically, these monomers can readily be polymerized, noting that while $CF_2 = CF_2$ has no character more so than $CH_2 = CH_2$ (yet both are nucleophiles since they have only one activation center- C = C), $CF_2 = CFCF_3$ is almost the same free-radically. While *CF_3 free-radically is an "electron"-pushing group, COOCH$_3$ is a free-radical "electron"-pulling group*. Why this is so will be seen and further explained as we move along in the Series and Volumes. Chargedly, they are *"electron"*-pulling groups of different capacities. CH_3, CH_2F, CHF_2 and CF_3, are all free-radical *"electron"*-pushing groups, their capacities decreasing in the order in which they have been listed with CH_3 being the highest. Therefore, provision of the corresponding equation to Equation 1.21, can only be done free-radically and separately for CF_3 and $COOCH_3$ or $CONH_2$ or CN or $COCH_3$ etc. types of groups, when they replace F in $CF_2 = CF_2$. The stability of their primary, secondary and tertiary based fluoroalkanes are shown below, via the route natural to them. Nevertheless as already said, it must be noted that (I), (II) and (III) of the equations below can be activated chargedly, since the carbon center carrying fluorine atom(s) cannot carry the negative charge. However, only (I) cannot favor electro-free-radical route as wrongly represented in the first equation below, because of presence of transfer species. They have been used only to distinguish between primary, secondary and tertiary.

$$R_F \equiv CF_3, CHF_3, CH_2F, R \equiv CH_3; \quad (CF_2 = C(R)_2, CF_2 = CFR, CFR = CR_1R_2)$$

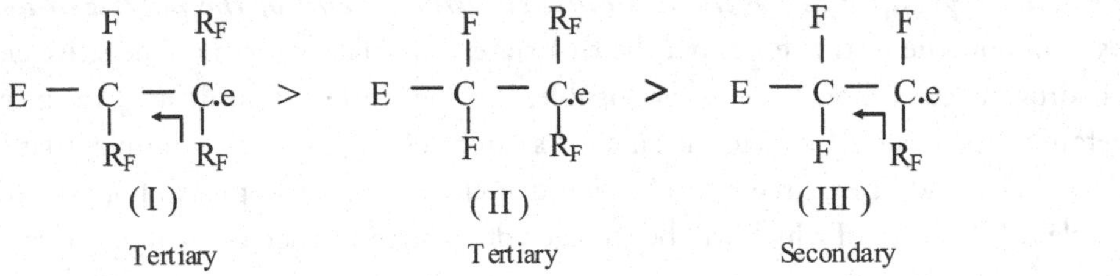

$$1.38a$$

$$R_F \equiv COOCH_3, CONH_2, COCH_3 \text{ etc. } (CF_2 = C(R_F)_2), CF_2 = CFR_F, CFR_F = C(R_F)_2$$

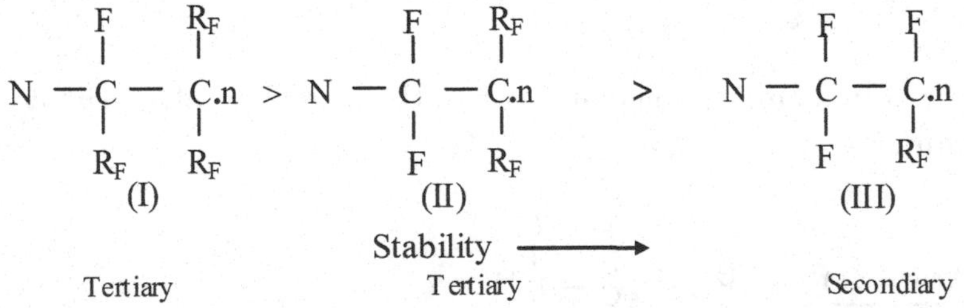

$$1.38b$$

Unlike the first cases above which are all Nucleophiles, the second cases are all Electrophiles.

Consider, for example, the addition of 1,1-difluoro-2,2-dichloroethylene or 1,1-difluoro-2,2 ditrifluoromethyl ethylene to ethanol in the presence of sodium ethoxide.[12] –

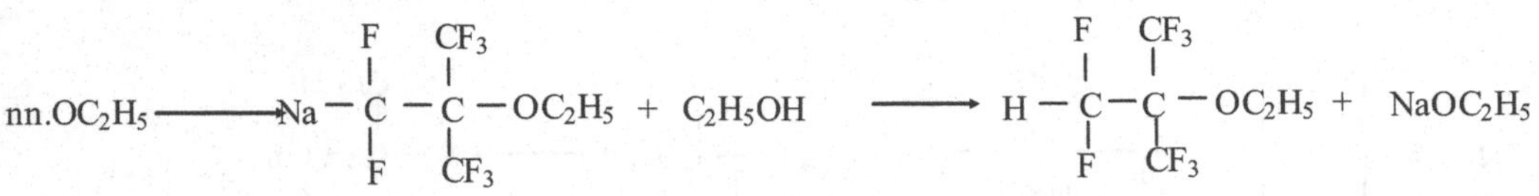

$$1.39$$

Latter in the Series as the development continues, one will explain the complete mechanisms of these reactions. Note the carbon centers carrying the electro-free and nucleo-free-radicals above. Chargedly, the reactions above cannot take place, because while CF_3 is radical-pushing radically, chargedly, it is charged-pulling, unlike COOR, $CONH_2$ $COCH_3$ types of groups, which are both pulling groups chargedly or radically.

1.7 CFR1= CFRF1, CHR1 = CHRF1, CFR1 = CFR2, CHRF1 = CFRF2, (CHR1 = CHR2, CH2 = CR1RF, CH2 = CR,R2)

Other cases of interest include $CFR_1 = CFR_1$ and $CHR_1 = CHRF_1$, $CFR_1 = CFR_2$ and $CHRF_1 = CHRF_2$, where RF_1 and RF_2 are "*electron*"-pulling groups and $RF_2 > RF_1$ in that capacity, R_1 and R_2 are "*electron*"-pushing groups and $R_2 > R_1$ in that capacity. All existing cases above are nucleophiles free-radically with $RF \equiv CF_3$ type of groups or electrophiles with $RF \equiv COOCH_3$ types. For the last two cases, when $R_1 = R_2$ or $RF_1 = RF_2$, addition can occur from any side of the monomer in the absence of molecular rearrangement. When they are both different, in what direction will addition occur free radically and chargedly? For $CHR_1 = CHR_{F1}$ case, since R_1 and RF_1 are different, the monomers cannot be symmetric.

Now consider the addition of β-nitrostyrene to an anion.[13] The following reactions will take place:-

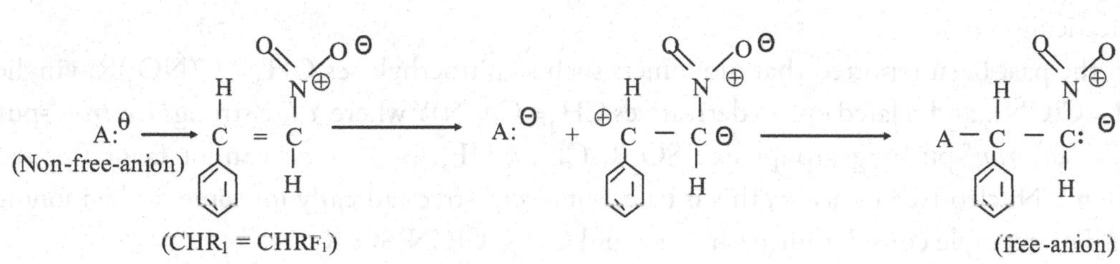

(CHR₁ = CHRF₁)   (free-anion)

NOT FAVORED CHARGEDLY (NOT CHARGEDLY BALANCED)   1.40

When the phenyl group is replaced with for example CH_3, the so-called free-anion which indeed is negatively charged route, will not initiate the monomer, because while the CH_3 group has transfer species, the phenyl group a resonance stabilization group has no transfer species under free-radical conditions as will be seen downstream in this Volume.

In the monomer here, $R_1 \equiv$ phenyl group is "*electron*"-pushing and of little greater capacity than hydrogen while $RF_1 \equiv NO_2$, chargedly is a strong "*electron*"-pulling group.[7] Hence, when activated chargedly (not ionically), the opposite charges are fixed on the two carbon atoms as shown above. The charges here like others are covalent charges. Cationically or when positively charged centers are involved, it is the N = O center that is activated, the monomer being an Electrophile. Hence anionically or negatively, only the C = C center can be activated. Free-radically, the following reactions take place:-

22

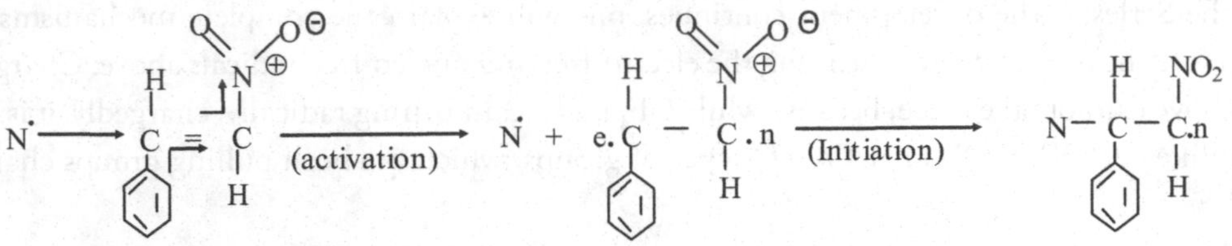

(I) ($CHR_1 = CHR_2$; $R_2 \gg R_1$)

1.41

($CHR_2 = CHR_1$)

FAVORED

1.42

Radically, initiation is favored both ways via the C=C cenetr. Indeed, it is the C=C center that is activated electro-free-radically, since the $^{\oplus}N = O$ center is more nucleophilic than the C=C center, and radically the monomer is a Nucleophile. However, the radical carried by the centers are fixed, since the monomer is not symmetric.

It has in the past been reported that monomers such as nitroethylenes $CH_2 = C(NO_2)R$, vinylidene cyanide $CH_2 = C(CN)_2$ and related cyano derivatives $CH_2 = C(CN)Y$ where Y are strong *"electron"*-pulling or very weak *"electron"*-pushing groups e.g. SO_2R, CF_3, CHF_2, NO_2, etc., cannot favor free-radical polymerization.[14] Nucleo-free-radically, this is true, but electro-free-radically for some of the monomers, it is not true. For example consider nitroethylenes and $CH_2 = C(CN)SO_2R$.

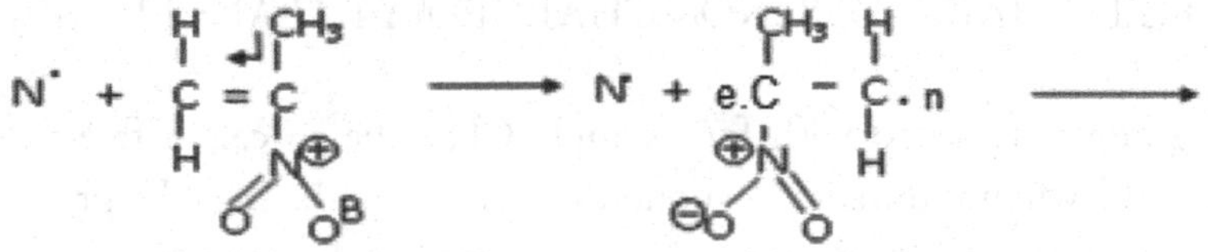

($CH_2 = CR_1 R_{F2}$)

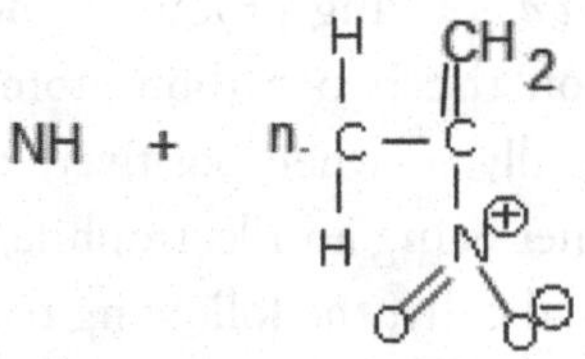

No polymerization

1.43

Radically the monomer is a Nucleophile, while chargedly it is an Electrophile. Nucleo-free-radically the route not natural to the center in the monomer, initiation is not favored.

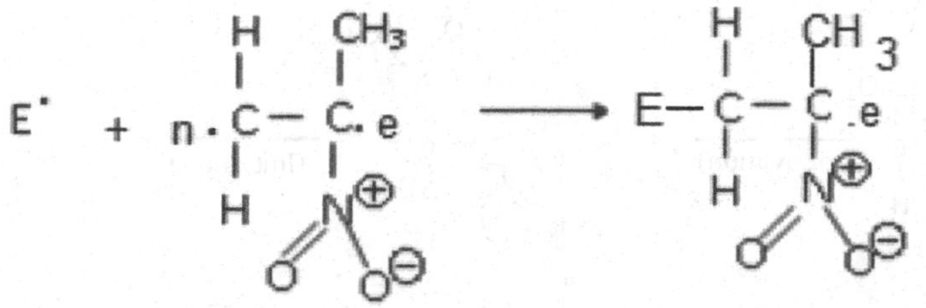

[FAVORED] 1.44

The electro-free-radical route can only be favored above, if the C=C center is less nucleophilic than the $^\oplus$N=O center radically, that which is the case. Hence the last equation above is favored. If it is just the N = O center without the polar bond, then it will be less nucleophilic than the C=C center and that will be the center that is first activated. Chargedly, the activation is different, the monomer being an Electrophile. Hence it is known to favor only negatively charged (not anionic) polymerization. For the second case, the followings are to be expected.

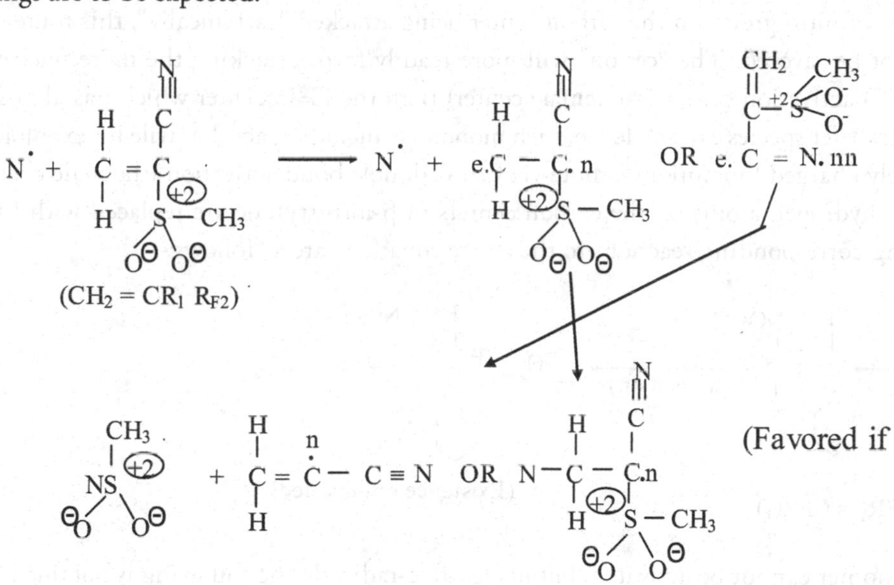

(Favored if activated) 1.45

It is said to be favored if activated, because $S^\oplus$- O^θ center is less nucleophilic than C = C center and for this case, the $S^\oplus$-O^θ cannot be activated due to boundary laws. These are unlike monomers such as CH_2 = $C(CN)_2$, CH_2 = $C(CN)COOCH_3$ etc. which do not carry substituted groups that have polar bonds, These two should favor the nucleo-free-radical route. But they do not for the same reason as above, being that they cannot be activated radically as will be shown downstream. *It is one's belief as will be confirmed downstream that while, NO_2, SO_2R with polar bonds are weak "electron"-pushing groups free-radically, chargedly they are strong "electron"-pulling groups, just like CF_3.* If indeed, it is the $S^\oplus$- O center that is first activated radically, the CH_3 group on the S center is the provider of transfer species. No initiation will be favored nucleo-free-radically. Now considering the "cationic" polymerization of (I) of Equation 1.41, the followings are obtained:-

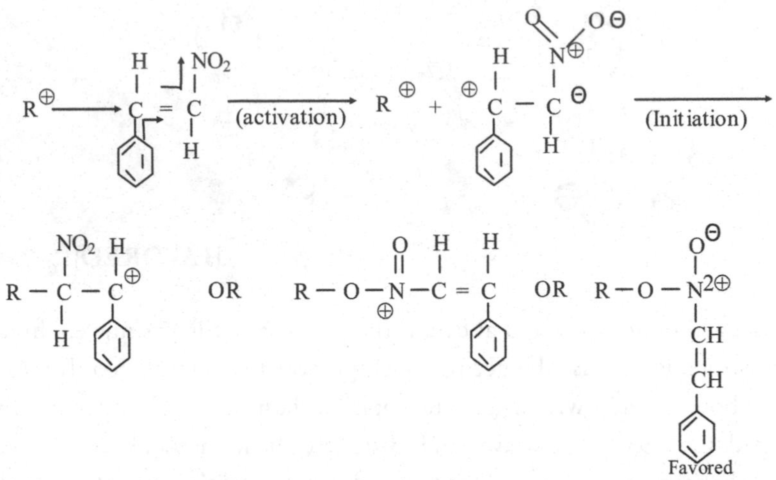

$$\text{(Scheme 1.46)}$$

1.46

Due to the presence of nitro group on the carbon center being attacked "cationically", this route via the C=C center will not be favored. The "cation" will more readily favor attacking the more nucleophilic center which is N=O activation center (the female center) than the C=C center which has almost dual character since no transfer species exists. Hence, such monomers including acrylonitrile for example favor largely the negatively charged (not anionic) route via C = C double bond activation (the male center).

Now when the hydrogen atoms on the carbon centers in β-nitrostyrene are replaced with fluorine atoms, the following corresponding reactions to the above equations are as follows:-

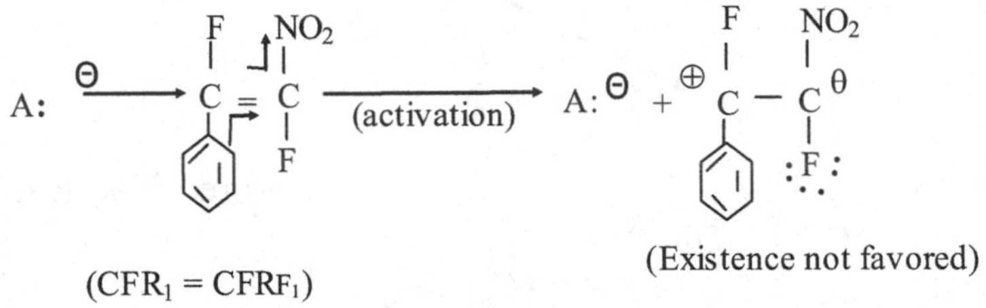

$$\text{(Scheme 1.47)}$$

$(CFR_1 = CFRF_1)$ (Existence not favored)   1.47

Chargedly, the monomer cannot be activated, but nucleo-free-radically the following is obtained.

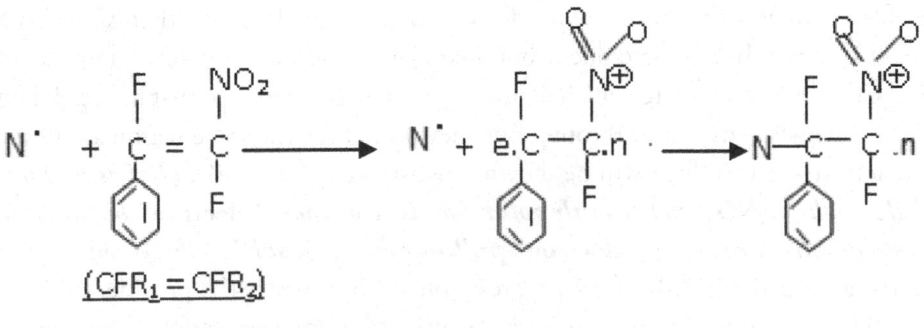

$$\text{(Scheme 1.48)}$$

$(CFR_1 = CFR_2)$

1.48

When the F on the carbon center where NO_2 is located, is replaced with CF_3, the fluoroalkene monomer will then favor charged activation.

1.8 CHRF1 = CR1R2, CHR1 = CR2R3, CHRF = CHF, CHF = CFR

For an alkene case, when the hydrogen atom on the carbon center carrying the phenyl group is now replaced, the followings are obtained:-

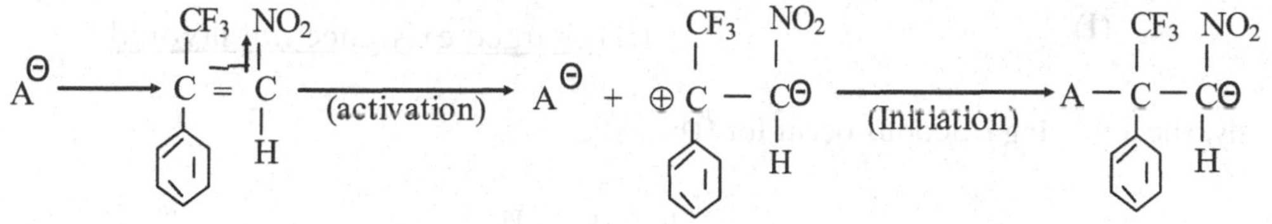

$$1.49$$

(I)

$(CR_{F2}R_1 = CHR_{F1})$ $R_{F2} \equiv CF_3$, $R_1 \equiv$ phenyl group, $R_{F1} \equiv NO_2$, $R_{F1} \gg R_{F2})$

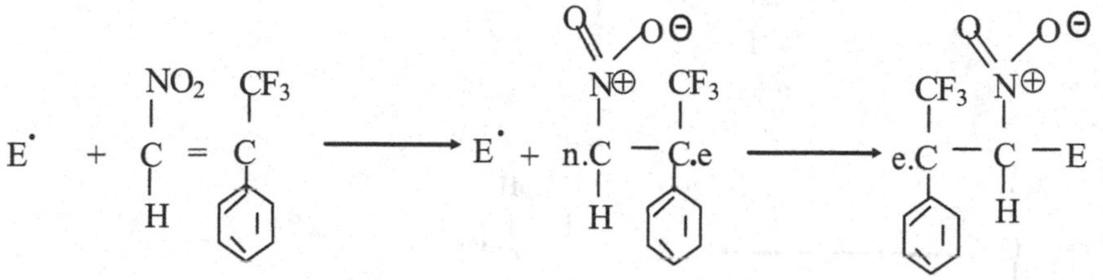

$$1.50$$

$(CHR_1 = CR_2R_3)$ $R_1 \equiv NO_2$, $R_2 \equiv CF_3$, $R_3 \equiv$ phenyl group, $R_2 > R_3 > R_1)$

Nucleo-free-radically, the route is favored, and same electro-free-radically as shown above. NO_2 group chargedly is a strong "electron"-pulling group; hence activation represented by (I) of Equation 1.49 is the favored activation. Free radically, all the groups on the monomer are "electron"-pushing groups. That is why it behaved as shown in Equation 1.43 as a nucleophile which indeed it is, and the same is the case above. Chargedly, the monomer is a strong Electrophile, while radically it is a Nucleophile, favoring both routes. Though there is no transfer species to abstract "cationically", in view of the strong presence of NO_2 group, initiation step is not readily favored via $C = C$ bond. Thus, several rules of Addition for activation of monomers can so far be identified, covering alkenes and perfluoroalkenes of the types, $CH_2 = CRH$, $CH_2 = CR_1R_2$, $CHR_1 = CHR_2$, $CF_2 = CRFF$, $CF_2 = CRF$, $CF_2 = C(R_F)_2$, $CFRF = CFRF$, $CHR_1 = CR_2R_3$, $CR_1RF_2 = CHRF_1$, $CHR_1 = CHRF_1$ and $CFR_1 = CFRF_1$ etc.

Now considering the following two monomers shown below.

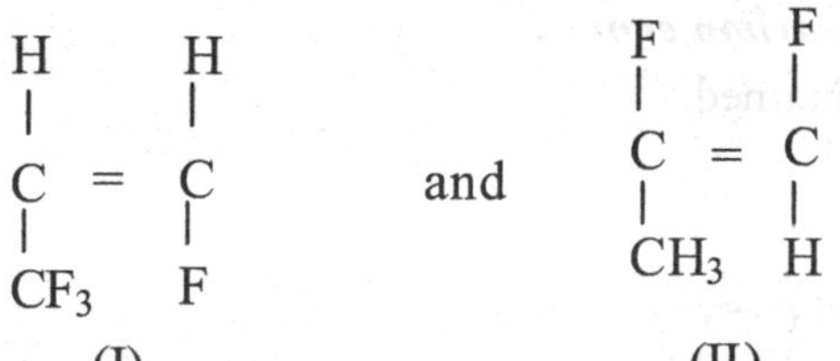

$$1.51$$

(I)   (II)

When activated chargedly, the monomers assume the following configurations:-

26

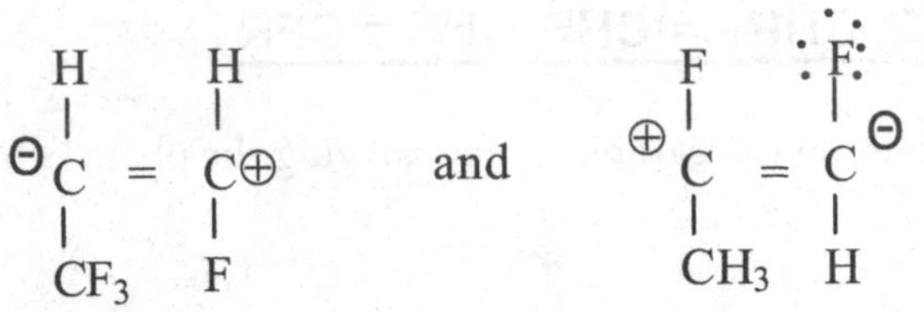

$$
\begin{array}{ccc}
& H & H \\
& | & | \\
\ominus C & = & C \oplus \\
& | & | \\
& CF_3 & F
\end{array}
\quad \text{and} \quad
\begin{array}{ccc}
& F & \ddot{F}\!: \\
& | & | \\
\oplus C & = & C \ominus \\
& | & | \\
& CH_3 & H
\end{array}
\qquad\qquad 1.52
$$

(I)   (II) <u>charged existence not favored</u>

Chargedly, the following reactions occur for (I):-

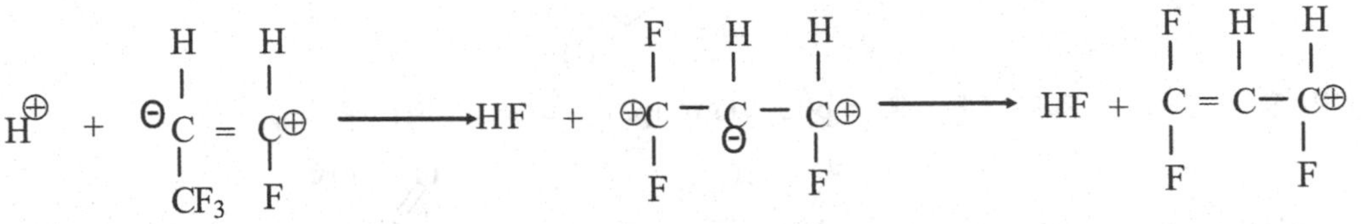

$$
H^{\oplus} + \;
\begin{array}{ccc}
H & H \\
| & | \\
\ominus C = & C\oplus \\
| & | \\
CF_3 & F
\end{array}
\longrightarrow HF + \;
\begin{array}{ccc}
F & H & H \\
| & | & | \\
\oplus C - & C - & C\oplus \\
| & \ominus & | \\
F & & F
\end{array}
\longrightarrow HF + \;
\begin{array}{ccc}
F & H & H \\
| & | & | \\
C = & C - & C\oplus \\
| & & | \\
F & & F
\end{array}
\qquad 1.53
$$

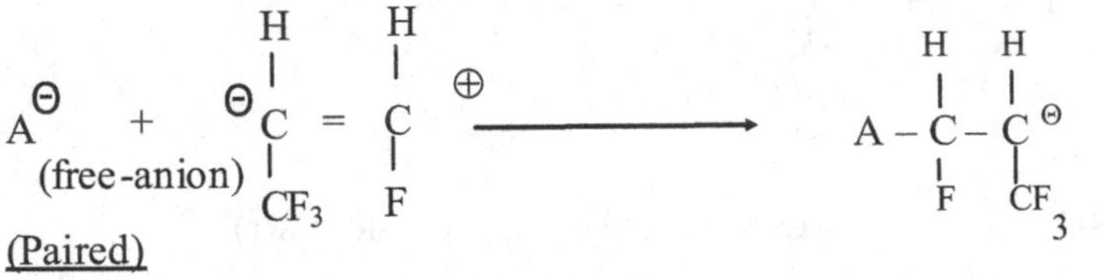

$$
A^{\ominus} + \;
\begin{array}{ccc}
H & H \\
| & | \\
\ominus C = & C & \oplus \\
| & | \\
CF_3 & F
\end{array}
\longrightarrow \;
\begin{array}{ccc}
H & H \\
| & | \\
A - C - & C^{\ominus} \\
| & | \\
F & CF_3
\end{array}
$$

(free-anion)

(Paired)   1.54

When attacked by an electrophilic or nucleophilic type of free-radical, the following reactions occur:-

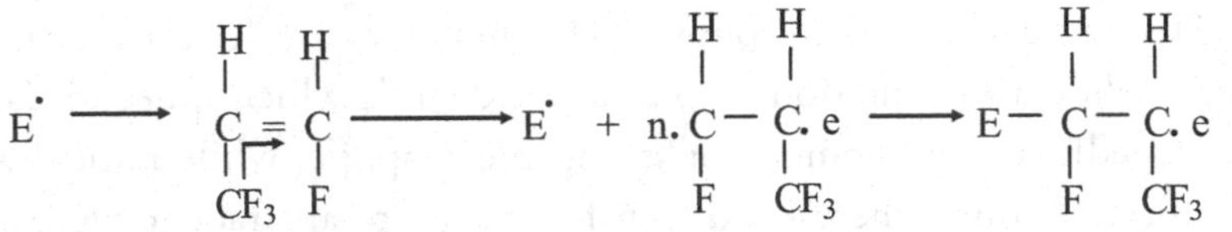

$$
E^{\cdot} \longrightarrow \;
\begin{array}{ccc}
H & H \\
| & | \\
C \equiv & C \\
| & | \\
CF_3 & F
\end{array}
\longrightarrow E^{\cdot} + n.
\begin{array}{cc}
H & H \\
| & | \\
C - & C.e \\
| & | \\
F & CF_3
\end{array}
\longrightarrow E -
\begin{array}{cc}
H & H \\
| & | \\
C - & C.e \\
| & | \\
F & CF_3
\end{array}
$$

(I)   1.55

$$
N^{\cdot} \longrightarrow \;
\begin{array}{ccc}
H & H \\
| & | \\
C = & C \\
| & | \\
CF_3 & F
\end{array}
\longrightarrow \;
\begin{array}{cc}
H & H \\
| & | \\
N - C - & C.n \\
| & | \\
CF_3 & F
\end{array}
$$

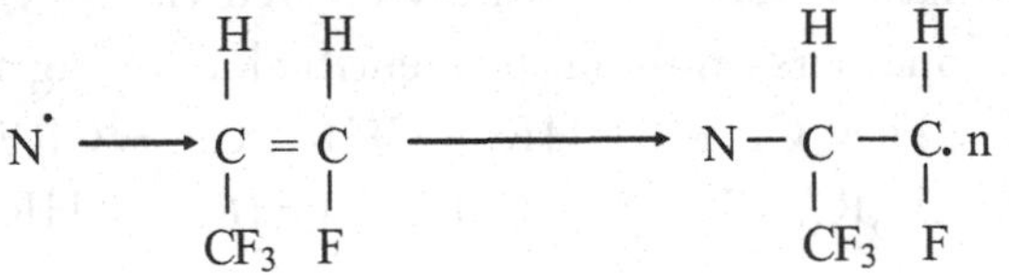

1.56

Chargedly and radically, the activations of the monomer (I) are different, in view of the character of CF₃ group and presence of F as one of the groups on a carbon center.
For the second monomer (II) radically, the followings are obtained.

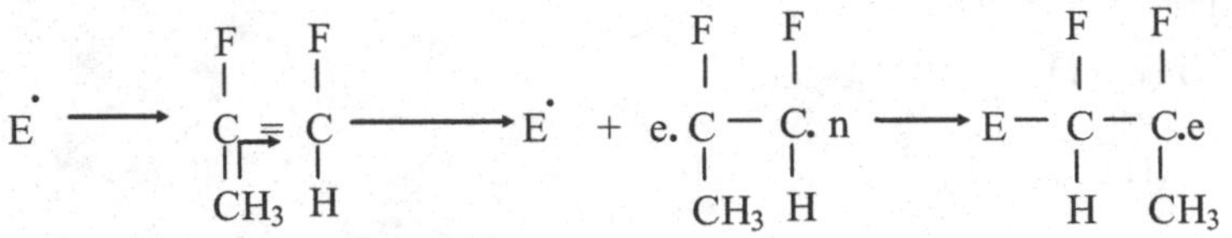

$$
E^{\cdot} \longrightarrow \;
\begin{array}{ccc}
F & F \\
| & | \\
C \equiv & C \\
| & | \\
CH_3 & H
\end{array}
\longrightarrow E^{\cdot} + e.
\begin{array}{cc}
F & F \\
| & | \\
C - & C.n \\
| & | \\
CH_3 & H
\end{array}
\longrightarrow E -
\begin{array}{cc}
F & F \\
| & | \\
C - & C.e \\
| & | \\
H & CH_3
\end{array}
$$

(II)   1.57

Radically, these monomers are nucleophiles, with (II) of Equation 1.51 being more nucleophilic than (I). Hence, as shown below,

$$
N^{\cdot} \longrightarrow \underset{\substack{| \\ H}}{\overset{\substack{F \\ |}}{C}} = \underset{\substack{| \\ CH_3}}{\overset{\substack{F \\ |}}{C}} \longrightarrow N^{\cdot} + n.\underset{\substack{| \\ H}}{\overset{\substack{F \\ |}}{C}} - \underset{\substack{| \\ CH_3}}{\overset{\substack{F \\ |}}{C}}e \longrightarrow NH + n.\underset{\substack{| \\ H}}{\overset{\substack{F \\ |}}{C}} - \underset{\substack{| \\ H}}{\overset{\substack{F \\ |}}{C}} = \underset{\substack{| \\ H}}{\overset{\substack{H \\ |}}{C}} \qquad 1.58
$$

The nucleo-free-radical route is not favored by it.

The character of (I) of Equation 1.52 will remain almost the same chargedly when CF_3 is replaced with $COOCH_3$ type of groups, but not the same free-radically. It is important to note how the character of the groups carried by a monomer differ free-radically and chargedly for some specific monomers and how these influence the charges or free-radical types carried by the active centers of the monomers when activated. Groups such as halogens (F, Cl, Br, I), haloge-nated alkyls (CF_3, CF_2H, CFH_2), nitro groups (NO_2), SO_2R groups, $OCOCH_3$ (acetate groups) etc. are very unique groups and they are uniquely different from each other. Because of some of them, ionic existence is not favored for some monomers. Also in their presence, some of their monomers are nucleophiles free-radically, while chargedly some are electrophilic. Groups such as NH_2, OR etc. which are also unique, do not however give different characters to their monomers radically or chargedly.

Thus, from all the above considerations, there is need to establish absolute measures of the "electron-pulling and pushing" capabilities of these substituent/substituted groups with respect to $CH_2 = CH_2$ and $CF_2 = CF_2$ as has been done for aromatic benzene, through measure-ment of electric dipole moments and their directions.[7] Nevertheless Table1.1 below shows some relative measures of the" electron-pulling and electron-pushing" capacities of some groups on $CH_2 = CHR$, $CH_2 = CHRF$, $R_1CH = CHR_2$ and $CHR = CHR_F$ with respect to bromine in acetic acid.[15,16] CN, COOH, NO_2, NO, SO_2R, Cl, Br etc., are known to be strong charged "electron-pulling" groups with the first four being of the free type and the halogens non-free type; while C_6H_5, CH_3 etc. are "free electron-pushing" groups. In the presence of these "electron-pulling" groups, the nucleophilicity of the alkene is reduced and one should therefore expect the monomers to favor "cationic" attack in decreasing order of presence of more *electron*-pulling groups. Nevertheless, as has been observed, each monomer is uniquely different. Since these measures are important in explaining how homopolymers and copolymers of different kinds are favored, some relative measures will be provided before considering polymerization phenomena.

Table1.1: Relative "electron-pulling and –pushing" capacities of some groups towards bromine in acetic acid at 24ºC

Compound		Relative Reactivity
CHR = CH$_2$ CH$_2$ = CHR$_F$		Very fast
C$_6$H$_5$ C$_6$H$_5$		1.6
CH$_2$Cl		1.0
CH$_2$Br		0.0011
Br		

$CHR_1 = CHR_2$	$CHR = CHR_F$	
C_6H_5, C_6H_5	C_6H_5, C_6H_5	18
	C_6H_5, COC_6H_5	0.33
	C_6H_5, Br	0.11
	Cis-C_6H_5, COOH	0.063
	Trans-C_6H_5, COOH	0.017

Nevertheless, it is important to note that the reactions above with bromine from bromine molecules are only radical in character. Where charged reactions are involved, it is the acetic acid that provides the positively charged species. It can also do it radically.

1.9 Proposition of Rules of Chemistry

To consider all the possible types of olefinic monomers both existing and non-existing, in order to determine the type of charges or free-radicals carried by the active centers when activated, will virtually be impossible in the absence of the NEW SCIENCE. Nevertheless unlike the modified or unmodified Markovnikov's rule, most of olefinic monomers were considered in the rules proposed herein. In proposing the rules, one has tried as much as possible to put many groups together since hundreds of rules will be proposed downstream in the Series and Volumes.

Before the proposition of rules for activation of alkenes, there was need to go back to the first volume where so many of the concepts introduced have started to be first put into rules and this will continue to be done at appropriate times downstream up to the sixth Volume. One cannot start proposing rules without bringing to realization the existence of free and non-free-radicals, ions and other charges, since they are involved with the active centers and they form the foundation under which the present concepts rely on and indeed all things depend on. For the activation of Addition monomers, only alkenes have been considered here. These can be extended to di-olefins. More rules with respect to activation of Addition monomers will continue to emerge when each member of Addition monomers is being considered in this Volume.

Rule 1: This rule of Chemistry for **Atoms,** states that, of all atoms, the only atom that is the smallest, indivisible, cannot be created or destroyed is the HYDROGEN atom for without it, nothing exists in our universe and without the "ZERO" atom there is no HYDROGEN atom.
(Laws of Creations for Hydrogen atom)

Rule 2: This rule of Chemistry for **All atoms,** states that, every atom and those that have elements like everything in life, have **their domains** marked by the distance between the nucleus and boundary and **a boundary** whose limitations are marked by the inert elements of the Period to which they belong to in the Periodic Table, the greatest TABLE in humanity.
(Laws of Creations for Domains and Boundaries)

Rule 3: This rule of Natural Sciences for **All things in Life,** states that, **MATHEMATICS, PHYSICS and CHEMISTRY the trios called THE NATURAL SCIENCES** were already in place before all things in life came to manifest.
(Laws of Creations for Natural Sciences)

Rule 4: This rule of Natural Sciences for **All things in Life**, states that, of all the trios-Mathematics, Physics and Chemistry, the greatest of them all is CHEMISTRY, for without it no discipline can be manifested in Humanity.
(Laws of Creations for CHEMISTRY-MOTHER OF ALL DISCIPLINES)

Rule 5: This rule of Chemistry for **All atoms, elements, molecules and their species,** states that, ELECTRONS do not exist OUTSIDE the nucleus, but INSIDE in part as one of the four sub-atomic particles of the Matter side with POSITRONS as its mirror counterpart in the Anti-matter side of THE NUCLEUS; and what exist outside THE NUCLEUS are called RADICALS.
(Laws of Creations for Electrons and Positrons)

Rule 6: This rule of Chemistry for **All building blocks of FAMILY TREES of Compounds and all things in Life**, states that, **the first member of all of them** is always uniquely different from all the other members in the same family, for which one can never find two FAMILY TREES with the same characters.
(Laws of Creations for Family Trees)

Rule 7: This rule of Chemistry for **All things** - Living or so-called Non-living based on the manners by which Chemical Reactions take place, states that, *not all things in Life including humans, are created equal, for which the only things created equal for all to apply are the Laws of Nature where no exceptions exist.*
(Laws of Creations-The Zeroth Law of Nature)

Rule 8: This rule of Natural Sciences for **All things in Life**, states that, they all have their **REAL and IMAGINARY DOMAINS**, because *MATHETHEMATICS is the NATURAL LANGUAGE OF COMMUNICATION in the Real and Imaginary domains and PHYSICS is the study of the FORCES OF NATURE, that is, the Senses – SIGHT, SOUND, SMELL, TASTE AND FEELINGS in the Real and Imaginary domains and CHEMISTRY is study of THE LAWS OF NATURE in the real and Imaginary domains.*
(Laws of Creations Mathematics, Physics and Chemistry)

Rule 9: This law of Chemistry for **All things**, states that, in general there are three major STATES OF EXISTENCE- **Equilibrium, Decomposition and Combination States of Existence,** for which when not in Equilibrium State, it is in **Stable State** and if not in Stable State, it is in **Activated State of Existence** and if not in Activated State, then it is in **Activated/ Equilibrium State of Existence.**
(Laws of Creations for States of Existence)

Rule 10: This rule of Chemistry **for All things,** states that, in general there are three major MECHANISMS –**Equilibrium, Decomposition and Combination mechanisms,** all of which are uniquely different and wherein they can take place in one or in many stages either in Series or Parallel or both.
(Laws of Creations Mechanisms of Systems)

Rule 11: This rule of Chemistry for **All Living things in Life,** states that, though H is the first member of all atoms in the Periodic Table for without which other atoms do not exist, based on Oxygen being

one of the largest components of AIR implies that it is one of the sources of Life and Living, followed by Nitrogen.
(Laws of Creations for Nitrogen and Oxygen)

Rule 12: This rule of Nature for **All systems,** states that, there are in nature two types of radicals, **the free and non-free-radicals,** for which free-radicals when present are alone with no paired unbonded radicals with opposite spin in the last shell and the non-free-radicals when present have paired unbonded radicals in the last shell of any Atom or Central atom.
(The first law of Nature from Radicals- To be free and not to be free)

Rule 13: This rule of Nature for **All systems,** states that, there are in nature two types of charges, **the free and non-free-charges**, for which free charges when present have no paired unbonded radicals in the last shell and non-free charges when present have paired unbonded radicals in the last shell of any Atom or Central atom.
(The first law of Nature from charges- To be free and not to be free)

Rule 14: This rule of Nature for **All systems,** states that, in nature for all free and non-free-radicals, there are **male and female counterparts** for which for free-radicals, the males are called **electro-free-radicals** and the females are called the **nucleo-free-radicals**, while for non-free-radicals, the males are called **electro-non-free-radicals** and the females are called the **nucleo-non-free-radicals.**
(Second Law of Nature from Radicals-The law of Duality)

Rule 15: This rule of Nature for **All systems,** states that, in nature for all free and non-free charges, there are **male and female counterparts** for which for free-charges the males are called **cations** when the last shell is empty of radicals and the females are called **negatively charged** when the last shell is empty of paired unbonded radicals, while for non-free-charges, the males are called **positively charged** when the last shell contains paired bonded or *unbonded radicals (Polarity)* and the females are called **Anions** when the last shell is full with *at least one paired unbonded radicals that makes it Polar in character,* with or without bonded radicals.
(Second Law of Nature from Charges- The law of Duality)

Rule 16: This rule of Chemistry for **All Charges**, states that, there are two types of Charges **the Real and Imaginary charges**; for which the *Ionic and Covalent* charges are REAL while the *Electrostatic and most or all Polar* charges are COMPLEX (i.e., Real and imaginary).
(Laws of Physics-Real and Imaginary worlds- the Complex world)

Rule 17: This rule of Chemistry for **Radicals and Charges**, states that, while RADICALS and IONIC charges can be *fully isolated and selectively paired*, COVALENT, ELECTROSTATIC and POLAR charges *cannot be isolated, but only paired*.
(Laws of Physics/Free- and Paired media)

Rule 18: This rule of Chemistry for **All Bonds,** states that, there are two types of bonds- **the Real and Imaginary**, for which *Dative, Covalent and Ionic bonds* are REAL while *Electrosta- tic (containing some so-called hydrogen bonds), and some Polar bonds* are COMPLEX (i.e.,
Real and Imaginary).
(Laws of Physics-Real and Imaginary worlds- the Complex world)

Rule 19: This rule of Chemistry for **All Compounds,** states that, there are three main types of Compounds- **Polar/Ionic, Polar/Non-Ionic, and Non-polar/Non-ionic,** for which the word polar refers to presence of elements carrying paired unbonded radicals in its last shell in the compound and for which for a compound to be ionic, it must first be polar.
(Laws of Creation for Classification of Compounds)

Rule 20: This rule of Chemistry for **All Compounds,** states that, every compound whether it carries one atom or more has what is called a **CENTRAL ATOM,** which is the carrier of all other atoms or species and from which all states of existences depend on directly or indirectly, and in general always centrally located when more than two atoms are present.
(Laws of Creations for Central atoms in Compounds)

Rule 21: This rule of Chemistry for **All systems,** states that, while radicals and charges add in a **Head–to –Tail** manner (that is male to female radicals or charges addition –Unlike Poles attract- Physics)), *only Radicals can add in Head– to –Head or Tail– to -Tail manner* (that is male to male or Female to female radicals addition-Like Poles attract- Metaphysics) only in the last step of a Productive Stage in Equilibrium or Decomposition mechanisms (only for males and females on an electropositive or electronegative almost filled carrier respectively when two of them exist).
(Laws of Physics, Metaphysics and Mathematics in Nature)

Rule 22: This rule of Chemistry for **All systems,** states that, when *Combination mechanisms* take place during Addition polymerization, monomers add only by **HEAD– to –TAIL** additions, for which the chemical equations must be radically or chargedly balanced; for example a nucleo-free-radical growing polymer chain must remain as such, otherwise the chain can never grow or a positively charged growing polymer chain must remain as such (and not cationically charged), otherwise the chain can never grow.
(Laws of Physics)

Rule 23: This rule of Chemistry for **All systems,** states that, when Decomposition mechanisms take place during Chemical reactions, they take place *in stages just like in Combination mechanism and only radically* unlike in Combination mechanism, for which in any stage the chemical equations must be *radically balanced.*
(Laws of Mathematics)

Rule 24: This rule of Chemistry for **All systems,** states that, when Decomposition mechanisms take place during Chemical reactions, a molecule can be produced either between two electro-radicals only if the *Central atoms* carrying the electro-radicals are electro-positive and full in character, or between two nucleo-radicals only if the *Central atoms* carrying the nucleo-radicals are electro negative and full in character; for which for the latter, nucleo-radicals can be produced while for the former electro-radicals can be produced exclusively.
(Laws of Metaphysics and Mathematics)

Rule 25: This rule of Chemistry for **All systems,** states that, when *decomposition takes place between two compounds, it begins by one of them existing in Decomposition State of Existence,* and this continues under **Decomposition mechanisms,** for which no compound can exist in Equilibrium State of Existence unless *reversely* where necessary taking place only inside or the last step of a stage in Decomposition mechanism when a new compound is being formed.
(Laws of Mathematics)

Rule 26: This rule of Chemistry for **All systems,** states that, when compounds are **Polar/Ionic in character**, they are the only ones that can exist in ionic Equilibrium State of Existence; for which no productive *chemical reactions* can take place ionically noting that all productive and non-Productive *chemical reactions* take place radically, while for *polymeric reactions* they take place radically and chargedly.
(Laws of Physics)

Rule 27: This rule of Chemistry for **All systems,** states that, when compounds are **Polar/Ionic, Polar/ Non -ionic and Non-polar/Non-ionic**, they can exist in Decomposition or Equilibrium States of Existence only radically during chemical and polymeric reactions.
(Laws of States of Existences)

Rule 28: This rule of Chemistry for **Addition monomers,** states that, in general, there are three kinds of **Activation centers,** i) *π-bond/some polar bonds (visible and invisible)*, ii) *presence of vacant orbitals and paired unbonded radicals in the last shell of an atom or a central atom, and iii) functional centers carried by some ringed compounds identified by the presence of paired unbonded radicals in the last shell of a hetero-atom in the ring.*
(Laws of Creations for Activation Centers)

Rule 29: This rule of Chemistry for **Addition monomers,** states that, when two Activation centers of *different capacities* and from two *different families* in which one is carrying a heroatom exist on a compound or monomer specially adjacently (i.e.,conjugatedly) or cummulatively placed, then two types of **Activation centers are created,** *X and Y activation centers, where X is Nucleophilic (Female), Y is Electrophilic (Male); for which when both are present on a monomer or Compound, the monomer or compound is said to be an ELECTROPHILE or MALE, and when only X and more are present not favoring most or all the conditions above, then the monomer or compound is said to be a NUCLEOPHILE or FEMALE.*
(Laws of Creations for Addition monomers)

Rule 30: This rule of Chemistry for **Step monomers,** states that, in general most Step monomers are **Polar/Ionic** in character, each carrying at least two of what are called **Functional groups** which identify with the families to which they belong; and yet their reactions to produce polymers take place only radically and never ionically or chargedly.
(Laws of Creations for Step monomers)

Rule 31: This rule of Chemistry for **Polymerization systems,** states that, while for Ideal Addition monomers *no small by-molecular products* are produced during polymerization, for Ideal Sep monomers, *small molecular by-products* are produced for every Stage during polymerization; the word "Ideal" indicating that there are Pseudo Addition and Step monomers which are Non-ideal.
(Laws of Creations for Addition and Step monomers)

Rule 32: This rule of Chemistry for **All systems,** states that, *when Chemical reactions take place,* it is always the electro-radical or positively-charged species (the male species) that diffuses all the time, the female species diffusing only in the absence of male species in the system or when paired to a male species. [Part of Diffusion Controlled mechanism]
(Laws of Creations for Chemical reactions)

Rule 33: This rule of Chemistry for **Initiation for Addition monomers during polymeriza-tion,** states that, **Initiators** are living chemical species which when produced from a single symmetric chemical compound by Decomposition mechanism gives two same type of radicals and when produced from a single or more non-symmetric chemical compounds radically by Decomposition and Equilibrium mechanisms gives paired covalently, ionically or electro-statically charged and radical initiators carrying **one or two active centers.**
(Laws of Creations for Initiators-Birth and Re-birth)

Rule 34: This rule of Chemistry for **Initiation for Addition monomers during polymeriza-tion,** states that, *because of the uniqueness of Activation centers, the Nucleophilic (Female) and Electrophilic (Male) character of a monomer is determined by the route favored during Initiation*; for which we have monomers that favor only the route opposite to their characters radically or chargedly (The Natural route), we have monomers that favor all routes radically or chargedly (Natural and Unnatural routes), those that favor only the route opposite to their characters radically (Natural route), and those that favor only the route same to their characters radically or chargedly (Unnatural route).
(Laws of Creations for Initiations)

Rule 35: This rule of Chemistry for **Propagation for Addition monomers,** states that, during addition of monomers to growing chains during polymerization, *Propagation can take place either via Combination mechanism that which is very fast or via Equilibrium mechanism, that which is very slow;* to give a Living or Dead polymer; for which when Dead polymers are produced, there could be re-initiation or no-re-initiation depending on the type of initiator used and presence of monomers in the system.
(Laws of Creations for Propagation-Growth)

Rule 36: This rule of Chemistry for **Initiation/Propagation for Step monomers,** states that, when polymers are produced, there is *INITIATION step with one or more stages that which is Dimerization and Imaginary since no Initiators are involved, followed largely by PROPAGA-TION step with many stages,* all of which take place only via EQUILIBRIUM mechanism, that which is slow; to give mostly Living polymers.
(Laws of Creations for Propagation-Growth)

Rule 37: This rule of Chemistry for **Termination for Addition and Step monomers,** states that, *when monomers have added to a chain to the point where the optimum chain length dictated by the Glass Transition temperature of the polymer at the operating conditions has been reached, no monomer can be added, for which the Living chain must be killed for it to become an environmentally friendly product.*
(Laws of Creations for Termination-Transition)

Rule 38: This rule of Chemistry for **Termination for Addition and Step monomers,** states that, *while Step and Addition-produced polymers via Equilibrium mechanisms can be killed by replacement of a terminal species, Living Addition-produced polymers can be killed either by Combination mechanism only radically or by Starvation radically and chargedly if transfer species exist or by introduction of a terminating agent radically or chargedly.*
(Laws of Creations for Termination- Transition)

Rule 39: This rule of Chemistry for **All things in Life,** states that, based on the fact that laws of Nature are equal for all to apply with no exception and the first law of Nature, NOT ALL...... IS OR ARE........., wherein any same thing put in the......., the rule remains the same.
(The third law of Nature)

Rule 40: This rule of Chemistry **for All things in Life** - Living or so-called Non-living based on the manners by which Chemical Reactions take place, states that, it is the **OPERATING CONDITIONS** that determines the ACTIONS and REACTIONS of all things in Life.
(The fourth law of Nature)

Rule 41: This rule of Chemistry for **Symmetric and Non-symmetric Addition monomers Olefinic in character,** states that, when the monomers are activated *where possible* radically or chargedly, for symmetric monomers the charges or radicals carried by the active centers are not fixed, while for non-symmetric monomers, the charges or radicals are fixed; the charges or radicals carried determined by the characters and capacities of the groups carried by the active Carbon centers.
(Laws of Physics)

Rule 42: This rule of Chemistry for **Radical-pushing and radical-pulling groups**, states that, -R, -CH$_2$X, -CHX$_2$, and more (Free), -OR, -NR$_2$, -ONO$_2$, -OSO$_2$R, and more (Non-free) groups are *radical-pushing groups of greater capacity than H* and, -CX$_3$, -NO$_2$, -SO$_2$R, groups (Free) are also *radical-pushing groups but of lesser capacity than H;* while -COOR, -COR, -CONR$_2$, -CN, (Free) groups, -NO, -OCOR, (Non-free) groups are *radical-pulling groups of greater capacity than F,* where R is an alkyl group and X is a halogen atom (F, Cl, Br), noting that I is not in the group of X, because I$_2$ is not a gaseous halogen molecule, but a liquid metal.
(Laws of Creations for Molecular Species on Monomers)

Rule 43: This rule of Chemistry for **Charged-pushing and charged-pulling groups**, states that, while -R, (Free), -OR, -NR$_2$ (Non-Free) groups are *charged-pushing groups of greater capacity than H (i.e., Cation)*; -CX$_3$, -CX$_2$H, -CXH$_2$, -NO$_2$, -SO$_2$R, -SO$_3$R, -COOR, -COR, -CONR$_2$, -CN, (Free groups), and –OCOR, - NO, -ONO$_2$ (Non-free-groups, i.e., Anions) groups are *charged-pulling groups of greater capacity than F (i.e., Anion).*
(Laws of Creations for Molecular Species on Monomers)

Rule 44: This rule of Chemistry for **Addition monomers** which are **symmetric on both active centers and are the first members of their families** (ethene, tetra halogenated ethene, cyclopropane, cyclobutane, acetylene, halogenated acetylene), states that, since these monomers favor both natural and unnatural routes whether radically or chargedly or both under different operating conditions, clearly indicates that they are **Weak Nucleophiles (Females),** with the following order of capacity -

$$H-C \equiv C-H \quad > \quad Cl-C \equiv C-Cl \quad > \quad \begin{matrix} H & H \\ | & | \\ C & = & C \\ | & | \\ H & H \end{matrix} \quad > \quad \begin{matrix} Cl & Cl \\ | & | \\ C & = & C \\ | & | \\ Cl & Cl \end{matrix} \quad >$$

ORDER OF NUCLEOPHILIC CHARACTER
(Laws of Creations for First members of Families of Addition monomers)

Rule 45: This rule of Chemistry for **Addition monomers** which are non-symmetric olefins of the types H$_2$C = CHR$_F$ where R$_F$ are non-free-radical-pulling groups such as halogens types, *(but not -OCOR types)* and R is an alkyl group, states that, since these monomers favor **both Natural and Unnatural routes free-radically,** clearly indicates that they are **Weak nucleophiles** with the following capacities-

$$\underset{\substack{| \quad | \\ H \quad H}}{\overset{\substack{CH_3 \quad CH_3 \\ | \quad |}}{C = C}} \gg \underset{\substack{| \quad | \\ H \quad H}}{\overset{\substack{H \quad H \\ | \quad |}}{C = C}} > \underset{\substack{| \quad | \\ H \quad Cl}}{\overset{\substack{H \quad H \\ | \quad |}}{C = C}} > \underset{\substack{| \quad | \\ H \quad Cl}}{\overset{\substack{H \quad Cl \\ | \quad |}}{C = C}} > \underset{\substack{| \quad | \\ Cl \quad Cl}}{\overset{\substack{H \quad Cl \\ | \quad |}}{C = C}} > \underset{\substack{| \quad | \\ Cl \quad Cl}}{\overset{\substack{Cl \quad Cl \\ | \quad |}}{C = C}}$$

ORDER OF NUCLEOPHILIC CHARACTER
(Laws of Creations for Nucleophilic Olefins)

Rule 46: This rule of Chemistry for **Addition monomers** which **are symmetric on both centers** (H_3CC ≡ CCH_3, $H_3CCH = CHCH_3$, $H_2C = C = CH_2$, Dimethyl cyclopropane, cyclo-propene, etc.), states that, whether they rearrange or not, since monomers when activated favor only the natural route chargedly and free-radically, clearly indicates that they are **Strong Nucleophiles (Females)** with the following order of capacity.

$$H_3CC \equiv CCH_3 > \underset{\substack{| \quad | \\ H_2C - CH_2}}{HC = CH} > H_2C = C = CH_2 > \underset{\substack{\diagdown \quad \diagup \\ CH_2}}{HC = CH} > \underset{\substack{\diagdown \quad \diagup \\ CH_2}}{HC - CH} > \underset{\substack{| \quad | \\ H \quad H}}{\overset{\substack{CH_3 \quad CH_3 \\ | \quad |}}{C = C}}$$

ORDER OF NUCLEOPHILIC CHARACTER
(Laws of Creations for Nucleophilic Hydrocarbons)

Rule 47: This rule of Chemistry for **Activation of *non-symmetric alkenes*** *of the types* $CH_2 = CHR$, $CH_2 = CR_1R_2$, $CHR_1=CHR_2$, and $CHR_1=CR_2R_3$ **(Nucleophiles or Females, chargedly and free-radically)** where the Rs are all free-radical/free-charged-pushing groups and non-free-radical/charged-pushing groups (e.g., OR, NR_2) and $R_3 > R_2 > R_1$ in that capacity, states that, an active electro-free-radical or positively charged initiator or center **(Males)** adds to the carbon center carrying less radical-pushing group; and the "carbonium" or electro-free-radical carbon center that results increases in stability in the following order in the absence of rearrangement:-

$$\underset{\substack{| \\ R_1 \quad R_3}}{\overset{\substack{H \quad R_2 \\ | \quad |}}{X - C - C \oplus \ominus}} > \underset{\substack{| \quad | \\ H \quad R_3}}{\overset{\substack{H \quad R_2 \\ | \quad |}}{X - C - C \oplus \ominus}} > \underset{\substack{| \quad | \\ R_1 \quad H}}{\overset{\substack{H \quad R_2 \\ | \quad |}}{X - C - C \oplus \ominus}} > \underset{\substack{| \quad | \\ H \quad H}}{\overset{\substack{H \quad R_2 \\ | \quad |}}{X - C - C \oplus \ominus}}$$

$$\text{Tertiary} \qquad\qquad \text{Tertiary} \qquad\qquad \text{Secondary} \qquad\qquad \text{Secondary}$$

where X is a positively charged center or electro-free-radical such as $H_3C^{\oplus}$ or H.e.
(Laws of Creations for Hydrocarbon groups)

Rule 48: This rule of Chemistry for **Activation of fluoroalkenes** of the types $CF_2 =CFR_{F1}$, $CF_2 = CR_{F1}R_{F2}$, $CFR = CFR_{F1}$, $CFR = CR_{F2}R_{F3}$ where the R_Fs, are free-radical-pushing groups such as CF_2H, CFH_2, $CH_3(R)$ groups etc. **(Nucleophiles)** and $R_{F3} > R_{F2} > R_{F1}$ in that capacity, states that, an active electro-free-radical initiator or center adds to the carbon center carrying less of the groups; and the electro-free-radical center generated increases in stability in the following order in the absence of rearrangement:-

$$\underset{\text{Tertiary}}{E-\overset{\overset{\displaystyle F}{|}}{\underset{\underset{\displaystyle R}{|}}{C}}-\overset{\overset{\displaystyle R_{F2}}{|}}{\underset{\underset{\displaystyle R_{F3}}{|}}{C}}.e} \quad > \quad \underset{\text{Tertiary}}{E-\overset{\overset{\displaystyle F}{|}}{\underset{\underset{\displaystyle F}{|}}{C}}-\overset{\overset{\displaystyle R_{F2}}{|}}{\underset{\underset{\displaystyle R_{F3}}{|}}{C}}.e} \quad > \quad \underset{\text{Secondary}}{E-\overset{\overset{\displaystyle F}{|}}{\underset{\underset{\displaystyle R}{|}}{C}}-\overset{\overset{\displaystyle R_{F2}}{|}}{\underset{\underset{\displaystyle F}{|}}{C}}.e} \quad > \quad \underset{\text{Secondary}}{E-\overset{\overset{\displaystyle F}{|}}{\underset{\underset{\displaystyle F}{|}}{C}}-\overset{\overset{\displaystyle R_{F2}}{|}}{\underset{\underset{\displaystyle F}{|}}{C}}.e}$$

where E is an electro-free-radical initiator such as H.e.
(Laws of Creations for Fluorinated Alkenes)

Rule 49: This rule of Chemistry for **Activation of fluoroalkenes** of the types $CF_2 = CFR_{F1}$, $CF_2 = CR_{F1}R_{F2}$, $CFR_{F1} = CFR_{F2}$, $CFR_{F1} = CR_{F2}R_{F3}$ where the R_Fs, are free-charged-pulling groups such as CF_3, CF_2H, CFH_2 groups, etc. **(Nucleophiles)** and $R_{F3} > R_{F2} > R_{F1}$ in that capacity, states that, chargedly, $CF_2 = CFR_{F1}$, $CFR_{F1} = CFR_{F2}$ cannot be activated due to electrostatic forces of repulsion, while for the other two cases, an active positively charged initiator or center cannot initiate them due to presence of transfer species (F on the groups), but a negatively charged initiator may initiate some depending on the magnitude of electrodynamic forces of repulsion.
(Laws of Creations for Fluorinated Alkenes).

Rule 50: This rule of Chemistry for **Activation of fluoroalkenes** of the types $CF_2 = CFR_{F1}$, $CF_2 = CR_{F1}R_{F2}$, $CFR_{F1} = CFR_{F1}$, $CFR_{F1} = CR_{F2}R_{F3}$ where the R_Fs, are free-charged-pulling groups such as NO_2 group, SO_2R groups, $COOCH_3$, $CONH_2$, $COCH_3$ groups **(Electrophiles)** and $RF_3 > RF_2 > R_{F1}$ in that capacity, states that, chargedly only $CFR_{F1} = CFR_{F1}$ cannot be activated due to electrostatic forces of repulsion, while for the other three cases, an active negatively charged initiator or center adds to the carbon center carrying less of the groups or carrying the less charged-pulling groups in the absence of electrodynamic forces of repulsion; and the "carbanion" generated increases in stability in the following order in the absence of Rearrangement.

$$\underset{\text{Tertiary}}{X-\overset{\overset{\displaystyle F}{|}}{\underset{\underset{\displaystyle R_{F1}}{|}}{C}}-\overset{\overset{\displaystyle R_{F2}}{|}}{\underset{\underset{\displaystyle R_{F3}}{|}}{C}}\ominus} \quad > \quad \underset{\text{Tertiary}}{X-\overset{\overset{\displaystyle F}{|}}{\underset{\underset{\displaystyle F}{|}}{C}}-\overset{\overset{\displaystyle R_{F1}}{|}}{\underset{\underset{\displaystyle R_{F2}}{|}}{C}}\ominus} \quad > \quad \underset{\text{Secondary}}{X-\overset{\overset{\displaystyle F}{|}}{\underset{\underset{\displaystyle F}{|}}{C}}-\overset{\overset{\displaystyle F}{|}}{\underset{\underset{\displaystyle R_{F1}}{|}}{C}}\ominus}$$

where X is a negatively charged end of the paired initiator.
(Laws of Creations for Fluorinated Alkenes)

Rule 51: This rule of Chemistry for **Activation of olefins** of the type $H_2C = CRNO_2$ and $H_2C = CRSO_2R$, where R is an alkyl group and NO_2 and SO_2R groups are free-charged pulling groups, states that, since these monomers are Electrophiles, they favor their natural route-negatively-charged-paired initiators; while radically, the situation is different, because the NO_2 and SO_2R groups are weak free-radical-pushing groups of lower capacity than H.
(Laws of Creations for Special Alkenes)

Rule 52: This rule of Chemistry for **Activation of fluoroalkenes** of the types $CF_2 = CFRF_1$, $CF_2 = CRF_1RF_2$, $CFR_{F1} = CFR_{F2}$ and $CFRF_1 = CFR_2RF_3$ where the R_Fs are free-radical/charged-pulling groups such as $COOCH_3$, $CONH_2$, $COCH_3$ etc. **(Electrophiles)** and $RF_3 > RF_2 > RF_1$, in that capacity, states that, radically, an active nucleo-free-radical adds to the carbon center carrying less of the groups or carrying the less radical-pulling group; and the nucleo-free-radical center generated increases in stability in the following order-

$$
\underset{\text{Tertiary}}{\underset{\overset{\displaystyle R_{F1}}{|}}{N-C}\overset{\overset{\displaystyle R_{F3}}{|}}{\underset{\overset{\displaystyle R_{F2}}{|}}{C}}.n}
\quad > \quad
\underset{\text{Tertiary}}{\underset{\overset{\displaystyle F}{|}}{N-C}\overset{\overset{\displaystyle R_{F2}}{|}}{\underset{\overset{\displaystyle R_{F1}}{|}}{C}}.n}
\quad > \quad
\underset{\text{Secondary}}{\underset{\overset{\displaystyle R_{F1}}{|}}{N-C}\overset{\overset{\displaystyle R_{F2}}{|}}{\underset{\overset{\displaystyle F}{|}}{C}}.n}
\quad > \quad
\underset{\text{Secondary}}{\underset{\overset{\displaystyle F}{|}}{N-C}\overset{\overset{\displaystyle R_{F1}}{|}}{\underset{\overset{\displaystyle F}{|}}{C}}.n}
$$

where N is a nucleo-free-radical initiator.
(Laws of Creations for Fluorinated Alkenes)

Rule 53: This rule of Chemistry for **Activation of perfluoroalkenes** of the types $CF_2=CFR_1$, $CF_2=CR_1R_2$, $CFR_1=CFR_2$ and $CFR_1=CR_2R_3$ where the Rs are free-radical-pushing groups such as alkyl groups, and non-free-radical-pushing group such as -OR groups etc. **(Nucleophiles)** and $R_3 > R_2 > R_1$ in that capacity, states that, chargedly none of the monomers can be activated due to electrostatic forces of repulsion, while free-radically, an electro-free-radical initiator or center adds to the carbon center carrying less of the groups or carrying the less radical-pushing group; and the electro-free-radical center generated increases in stability in the following order:-

$$
\underset{\text{Tertiary}}{\underset{\overset{\displaystyle R_1}{|}}{E-C}\overset{\overset{\displaystyle R_3}{|}}{\underset{\overset{\displaystyle R_2}{|}}{C}}.e}
\quad > \quad
\underset{\text{Tertiary}}{\underset{\overset{\displaystyle F}{|}}{E-C}\overset{\overset{\displaystyle R_2}{|}}{\underset{\overset{\displaystyle R_1}{|}}{C}}.e}
\quad > \quad
\underset{\text{Secondary}}{\underset{\overset{\displaystyle R_1}{|}}{E-C}\overset{\overset{\displaystyle R_2}{|}}{\underset{\overset{\displaystyle F}{|}}{C}}.e}
\quad > \quad
\underset{\text{Secondary}}{\underset{\overset{\displaystyle F}{|}}{E-C}\overset{\overset{\displaystyle R_2}{|}}{\underset{\overset{\displaystyle F}{|}}{C}}.e}
$$

where E is an electro-free-radical initiator.
(Laws of Creations for Fluorinated Alkenes)

Rule 54: This rule of Chemistry for **Activation of alkenes** of the types $CH_2=CHR_1$, $CH_2=CR_1R_2$, $CHR_1=CHR_2$ and $CHR_2=CR_1R_3$ where the R_1 is a radical-pushing group of equal or lesser capacity than H, also that with no activation center, and the remaining Rs are also radical-pushing groups of greater capacity than H and $R_3 > R_2 > R_1$ in that capacity **(Nucleophiles)**, states that, an active electro-free-radical initiator or center adds to the carbon center carrying the less radical-pushing group and the electro-free-radical carbon center that results increases in stability in the following order-

$$
\underset{\text{Tertiary}}{\underset{\overset{\displaystyle R_2}{|}}{E-C}\overset{\overset{\displaystyle R_1}{|}}{\underset{\overset{\displaystyle R_3}{|}}{C}}.e}
\quad > \quad
\underset{\text{Tertiary}}{\underset{\overset{\displaystyle H}{|}}{E-C}\overset{\overset{\displaystyle R_1}{|}}{\underset{\overset{\displaystyle R_2}{|}}{C}}.e}
\quad > \quad
\underset{\text{Secondary}}{\underset{\overset{\displaystyle R_1}{|}}{E-C}\overset{\overset{\displaystyle H}{|}}{\underset{\overset{\displaystyle R_2}{|}}{C}}.e}
\quad > \quad
\underset{\text{Primary}}{\underset{\overset{\displaystyle H}{|}}{E-C}\overset{\overset{\displaystyle H}{|}}{\underset{\overset{\displaystyle H}{|}}{C}}.e}
$$

where E is an electro-free-radical initiator.
(Laws of Creations for Alkenes)

Rule 55: This rule of Chemistry for **Activation of alkenes** of the types $CH_2 = CHR_1$, $CHR_1 = CHR_2$, $CH_2 = CR_1R_2$, $CHR_1 = CR_2R_3$ where the Rs are free-radical-pushing groups of types CF_3, CF_2H, CFH_2, CH_2OCOCH_3, CH_2OR, etc. and $R_3 > R_2 > R_1$ in that capacity **(Nucleophiles)**, states that, an active electro-free-radical initiator or center adds to the carbon center carrying less of the groups or carrying the less radical-pushing group; and the electro-free-radical carbon center that results increases in stability in the following order:-

Tertiary   Tertiary   Secondary   Secondary

where E is an electro-free-radical initiator.
(Laws of Creations for Alkenes)

Rule 56: This rule of Chemistry for **Activation of alkenes** of the types $CH_2 = CHR_{F1}$, $CH_2 = CR_{F1}R_1$, $CH_2 = CR_{F1}R_{F2}$, $CHR_1 = CHR_{F1}$, $CHR_{F1} = CHR_{F2}$, $CHR_1 = CR_{F1}R_{F2}$, $CHR_{F1} = CR_1R_{F2}$, $CHR_{F1} = CR_{F2}R_{F3}$ where the R_1 is a free-radical/charged-pushing group such as alkyl groups and R_Fs are free-radical/charged-pulling groups such as CN, $COOCH_3$, $CONH_2$, etc. and $R_{F3} > R_{F2} > R_{F1}$ in that capacity **(Electrophiles)**, states that, an active nucleo-free-radical or a negatively-charged initiator or center adds to the carbon center carrying less of the groups or carrying the groups with less capacity and the nucleo-free-radical or "carbanion" center that results increases in stability in the following order:-

Tertiary   Tertiary   Secondary   Primary

where X is a nucleo-free-radical or negatively charged center of the paired type and the double stars is to indicate that either these and more cannot be initiated nucleo-free-radically, but only chargedly when specially trans-placed if R_1 is a group such as CH_3, CH_2F or when R_1 is e.g., CF_3.
(Laws of Creations for Olefins)

Rule 57: This rule of Chemistry for **Activation of fluoroalkenes** of the types $CF_2 = CR_{F1}R_1$, $CFR_1 = CFR_{F1}$, $CFR_{F1} = CR_1R_2$ and $CFR_1 = CR_{F1}R_2$ where R_{F1} is a free-radical-pulling group (such as $COOCH_3$) and the Rs are free-radical-pushing groups, in which the last three favor no free-radical route via the C = C center in question **(Electrophiles),** states that, only for the first will an active nucleo-free-radical generating initiator or center add to the carbon center carrying none of the groups.
(Laws of Creations for Fluorinated Alkene)

Rule 58: This rule of Chemistry for **Activation of fluoroalkenes** of the types $CF_2 = CR_{F1}R_1$, $CFR_1 = CFR_{F1}$, $CFR_{F1} = CR_1R_2$ and $CFR_1 = CR_{F1}R_2$, where R_{F1} is a free-charged-pulling group of the type COOR and the Rs are free-charged-pushing groups, in which chargedly the second and the third cannot be activated, states that, even the fourth that can be activated does not favor any charged route via the C

= C center in question when R_{F1} and R_1 are cis-placed, while the first will favor one negatively charged route in the absence of electrodynamic forces of repulsion.
(Laws of Creations for Fluorinated Alkenes)

1.10 Conclusions

More rules than expected have been proposed, for the purpose of making sure that the new foundations being laid were followed with the use of the "eye of the needle". On the whole, ***fifty-eight rules*** have been proposed, for which only four laws of NATURE have appeared. Why alkene monomers favor the routes which literature data have shown them to favor, but could not previously be explained, have begun to be explained.

Though the concept of transfer of transfer species has not yet been fully introduced, one has observed why in view of their presence, some monomers cannot be initiated (transfer from monomer step). In trying to determine the activation of these monomers, the past had to be revisited especially with respect to countless organic reactions and their proposed mechanisms. Without the past, all these however would not have been possible. Nevertheless, the visit to the past was necessitated by the fact that most of the mechanisms proposed in the past are not only questionable, but limited in applications and indeed meaningless. So also are countless numbers of rules such as Markovnikov's rule meaningless. This is to be expected since when rules and mechanisms were proposed, nothing was known about RADICALS, CHARGES, ACTIVATION CENTERS, NUCLEOPHILICITY, ELECTROPHILICITY, ACTIVE CENTERS, INITIATORS, and too much too countless to list. All along, up to the present moment, Science has been just an ART.

In proposing the rules for activation of the monomers, the question as to whether the monomer will favor the initiation step or not was not taken into serious consideration, since this will also depend on the polymerization techniques involved and type of polymers produced. For example, in Emulsion polymerization systems, in one or two of the stages, it is the monomers that diffuse to the mini-reactors containing a single growing polymer chain as opposed to the phenomena of diffusion controlled mechanism, for very clear and specific reasons. Downstream as we move along, some of the laws will be clearly displayed figuratively. One has only just begun a long endless journey, because for us to move forward, the wheel has to be turned backwards to correct so many things of the past. This can only be done through Chemistry, the greatest of all disciplines, wherein the LAWS OF NATURE are embedded in, and wherein all things in our planet in its universe and all other universes were created.

References

1. S.N.E. Omorodion, "Introduction to New Classification Methods in Polymeric Systems and New concepts in Chemistry" Vol. 1, Chaps 6.7. In Press.

2. S.N.E. Omorodion, "Introduction to New Classification Methods in Polymeric Systems and New concepts in Chemistry" Vol. 1, Chaps 8 – 10. In Press.

3. S.N.E. Omorodion, "Introduction to New Classification Methods in Polymeric Systems and New concepts in Chemistry" Vol. 1, Chaps 3,5. In Press.

4. G. Odian, "Principles of Polymerization", McGraw-Hill Book Company, 1970, pg.209.

5. R.B. Seymour, "Introduction to Polymer Chemistry", International Student Edition, McGraw-Hill Book Company, 1971, pg. 167.

6. R.B. Seymour, "Introduction to Polymer Chemistry", International Student Edition, McGraw-Hill Book Company, 1971, pg.146.

7. C.R. Noller, "Textbook of Organic Chemistry", 3rd Edition, W.B. Saunders Company, 1966, pg. 380.

8. C.R. Noller, "Textbook of organic Chemistry", 3rd Edition, W.B. Saunders Company, 1966, pg. 82-85.

9. C.R. Noller, "Textbook of Organic Chemistry," 3rd Edition, W.B. Saunders Company, 1966, pg. 592-593.

10. G. Odian, "Principles of Polymerization", McGraw-Hill Book Company, 1970, pg. 579.

11. C.R. Noller, "Textbook of Organic Chemistry," 3rd Edition, W.B. Saunders Company, 1966, pg. 37.

12. J. Hines, "Physical Organic Chemistry", 2nd Edition, McGraw-Hill Book Company, Inc. New York (1962), pg. 230.

13. F.G. Bordwell and K. Rohde, J. Am. Chem. Soc., 70, 1191 (1948).

14. F. Rodriguez, "Principles of Polymer Systems", McGraw-Hill Book Company, 1970, pg. 72.

15. P.B.D. del la Mare, Quart. Revs. (London), 3, 126 (1949).

16. J. Hines, "Physical Organic Chemistry", 2nd Edition, McGraw-Hill Book Company, Inc. New York (1962), pg. 221.

Problems

1.1. Distinguish between "Markovnikov's rule of Addition and Anti-Markovnikovs phenomenon" for monomers such as

(a)
$$
\begin{array}{cc}
H & H \\
| & | \\
C & = C \\
| & | \\
H & CH_3
\end{array}
$$

(b)
$$
\begin{array}{cc}
H & H \\
| & | \\
C & = C \\
| & | \\
H & CH_2Cl
\end{array}
$$

(c)
$$
\begin{array}{cc}
H & H \\
| & | \\
C & = C \\
| & | \\
H & N^{\oplus} \\
 & O \quad O^{\ominus}
\end{array}
$$

1.2. (a) According to "Anti-Markovnikov's rule" when propylene is so-called activated "free-radically and ionically", the followings are to be expected:-

$$
\begin{array}{cc}
H & H \\
| & | \\
e.\,C & - C.n \\
| & | \\
H & CH_3
\end{array}
\quad \text{and} \quad
\begin{array}{cc}
H & H \\
| & | \\
{}^{\oplus}C & - C^{\ominus} \\
| & | \\
H & CH_3
\end{array}
$$

Explain how these came about.

(b) Is it possible to have a transfer species transferred from a substituent group to an active center to which it is connected to or to another active center adjacently located during molecular rearrangement phenomenon? Explain.

(c) Why are the products obtained when cis- or trans- 2-butene and 1-butene add to sulphuric acid the same? Explain the mechanism and identify the product.

1.3. (a) Based on the existence of free and non-free charges and radicals, explain how hydrogen and iodine can be obtained from the decomposition of hydrogen iodide.

(b) Why is the measured "activation energy" of the process – (39 kcal), less than the dissociation energy of the hydrogen iodide – (71 kcal)? What is indeed called Activation energy?

(c) Compare the newly proposed mechanism in (a) with previously known mechanisms (e.g. four-centered type).

(d) Compare the decomposition above with those of the other halogen acids-hydrogen fluoride, hydrogen bromide and hydrogen chloride.

1.4 Display figuratively the laws pertaining to New Classifications for Radicals, Charges, Bonds, Mechanisms of Systems, States of Existence and for Compounds.

1.5. (a) Distinguish between the following centers and show whether they exist or not:

$$:\overset{..}{\underset{..}{Cl}}\,{}^{\oplus} \quad ; \quad H-O-\overset{\overset{H}{|}}{\underset{\underset{H}{|}}{C}}-\overset{\overset{H}{|}}{\underset{\underset{H}{|}}{C}}{}^{\oplus} \quad ; \quad H^{\oplus} \quad ; \quad Na^{\oplus}$$

      (i)   (ii)   (iii)   (iv)

(b) Is it possible to have the following centers?

$$\overset{\displaystyle H_2C{-}{-}{-}H_2C}{\underset{\displaystyle\underset{\displaystyle H}{|}}{\underset{\displaystyle :\overset{}{O}\,\oplus}{\diagdown\diagup}}} \qquad ; \qquad \overset{\displaystyle H_2C{-}{-}{-}CH_2}{\underset{\displaystyle\underset{\displaystyle H}{|}}{\underset{\displaystyle \oplus N{-}R}{\diagdown\diagup}}}$$

Explain.

(c) Distinguish between the following centers and show whether they exist or not:-

$$:\overset{..}{\underset{..}{Cl}}x^{\ominus} \quad ; \quad CH_3-O-\overset{\overset{H}{|}}{\underset{\underset{H}{|}}{C}}-\overset{\overset{H}{|}}{\underset{\underset{H}{|}}{C}}{}^{\ominus} \quad ; \quad Hx^{\ominus}$$

      (i)   (ii)   (iii)

(d) How does (i) in (a) combine with (i) in (c) or (iii) in (a) combine with (iii) in (c) or (ii) in (a) combine with (ii) in (c) to produce bonds where possible?

1.6 (a) Distinguish between the following growing polymer chains of ethylene and tetrafluoroethylene showing which one exist.

(i)

$$N-\underset{\underset{H}{|}}{\overset{\overset{H}{|}}{C}}-\underset{\underset{H}{|}}{\overset{\overset{H}{|}}{C}}-\underset{\underset{H}{|}}{\overset{\overset{H}{|}}{C}}-\underset{\underset{H}{|}}{\overset{\overset{H}{|}}{C}}\left(\underset{\underset{H}{|}}{\overset{\overset{H}{|}}{C}}-\underset{\underset{H}{|}}{\overset{\overset{H}{|}}{C}}\right)_n\underset{\underset{H}{|}}{\overset{\overset{H}{|}}{C}}-\underset{\underset{H}{|}}{\overset{\overset{H}{|}}{C}}.n$$

$$E-\underset{\underset{H}{|}}{\overset{\overset{H}{|}}{C}}-\underset{\underset{H}{|}}{\overset{\overset{H}{|}}{C}}-\underset{\underset{H}{|}}{\overset{\overset{H}{|}}{C}}-\underset{\underset{H}{|}}{\overset{\overset{H}{|}}{C}}\left(\underset{\underset{H}{|}}{\overset{\overset{H}{|}}{C}}-\underset{\underset{H}{|}}{\overset{\overset{H}{|}}{C}}\right)_n\underset{\underset{H}{|}}{\overset{\overset{H}{|}}{C}}-\underset{\underset{H}{|}}{\overset{\overset{H}{|}}{C}}.e$$

$$H-\underset{\underset{H}{|}}{\overset{\overset{H}{|}}{C}}-\underset{\underset{H}{|}}{\overset{\overset{H}{|}}{C}}-\underset{\underset{H}{|}}{\overset{\overset{H}{|}}{C}}-\underset{\underset{H}{|}}{\overset{\overset{H}{|}}{C}}\left(\underset{\underset{H}{|}}{\overset{\overset{H}{|}}{C}}-\underset{\underset{H}{|}}{\overset{\overset{H}{|}}{C}}\right)_n\underset{\underset{H}{|}}{\overset{\overset{H}{|}}{C}}-\underset{\underset{H}{|}}{\overset{\overset{H}{|}}{C}}\oplus$$

$$CH_3O-\underset{\underset{H}{|}}{\overset{\overset{H}{|}}{C}}-\underset{\underset{H}{|}}{\overset{\overset{H}{|}}{C}}-\underset{\underset{H}{|}}{\overset{\overset{H}{|}}{C}}-\underset{\underset{H}{|}}{\overset{\overset{H}{|}}{C}}\left(\underset{\underset{H}{|}}{\overset{\overset{H}{|}}{C}}-\underset{\underset{H}{|}}{\overset{\overset{H}{|}}{C}}\right)_n\underset{\underset{H}{|}}{\overset{\overset{H}{|}}{C}}-\underset{|}{\overset{\overset{H}{|}}{C}}\overset{..}{x}{}^{\ominus}$$

$$C_4H_9-\underset{\underset{H}{|}}{\overset{\overset{H}{|}}{C}}-\underset{\underset{H}{|}}{\overset{\overset{H}{|}}{C}}-\underset{\underset{H}{|}}{\overset{\overset{H}{|}}{C}}-\underset{\underset{H}{|}}{\overset{\overset{H}{|}}{C}}\left(\underset{\underset{H}{|}}{\overset{\overset{H}{|}}{C}}-\underset{\underset{H}{|}}{\overset{\overset{H}{|}}{C}}\right)_n\underset{\underset{H}{|}}{\overset{\overset{H}{|}}{C}}-\underset{|}{\overset{\overset{H}{|}}{C}}\overset{..}{x}{}^{\ominus}\ldots\ldots{}^{\oplus}Li$$

(ii)

$$N-\underset{\underset{F}{|}}{\overset{\overset{F}{|}}{C}}-\underset{\underset{F}{|}}{\overset{\overset{F}{|}}{C}}-\underset{\underset{F}{|}}{\overset{\overset{F}{|}}{C}}-\underset{\underset{F}{|}}{\overset{\overset{F}{|}}{C}}\left(\underset{\underset{F}{|}}{\overset{\overset{F}{|}}{C}}-\underset{\underset{F}{|}}{\overset{\overset{F}{|}}{C}}\right)_n\underset{\underset{F}{|}}{\overset{\overset{F}{|}}{C}}-\underset{\underset{F}{|}}{\overset{\overset{F}{|}}{C}}.n$$

$$E-\underset{\underset{F}{|}}{\overset{\overset{F}{|}}{C}}-\underset{\underset{F}{|}}{\overset{\overset{F}{|}}{C}}-\underset{\underset{F}{|}}{\overset{\overset{F}{|}}{C}}-\underset{\underset{F}{|}}{\overset{\overset{F}{|}}{C}}\left(\underset{\underset{F}{|}}{\overset{\overset{F}{|}}{C}}-\underset{\underset{F}{|}}{\overset{\overset{F}{|}}{C}}\right)_n\underset{\underset{F}{|}}{\overset{\overset{F}{|}}{C}}-\underset{\underset{F}{|}}{\overset{\overset{F}{|}}{C}}.e$$

(b) Why is it not possible to have the charged version for tetrafluoroethylene?

1.7. (a) What is the meaning of the First Law of Nature in terms of applications to all disciplines including Religion?

(b) Consider the following monomers

$$
\begin{array}{cccc}
\underset{\underset{CH_3}{|}}{\overset{\overset{H}{|}}{1.\ C}} = \underset{\underset{CH_3}{|}}{\overset{\overset{CH_3}{|}}{C}} &
\underset{\underset{\underset{CH_3}{|}}{\overset{}{CH_2}}}{\overset{\overset{H}{|}}{2.\ C}} = \underset{\underset{H}{|}}{\overset{\overset{CH_3}{|}}{C}} &
\underset{\underset{CH_3}{|}}{\overset{\overset{H}{|}}{3.\ C}} = \underset{\underset{CH_3}{|}}{\overset{\overset{H}{|}}{C}} &
\underset{\underset{H}{|}}{\overset{\overset{H}{|}}{4.\ C}} = \underset{\underset{\underset{CH_3}{|}}{\overset{}{CH_2}}}{\overset{\overset{H}{|}}{C}}
\end{array}
$$

(i) Show their activations free-radically.

(ii) If a very weak free-radical initiator (type?) is involved wherein molecular rearrangement where possible takes place, establish the order of their stabilities.

(iii) If a very strong free-radical initiator that which does not give time to rearrange is involved, establish the order of their stabilities.

(iv) When a weak initiator is involved, what happens latter during propagation for these monomers via their natural route?

1.8 (a) For the same monomers in Q 1.7 (b) above, answer (i) to (iv) for charged activation.

(b) What are the characters of these monomers free-radically and chargedly?

(c) When the monomers above are reacted with sulfuric acid, what types of products will be obtained?

1.9. Consider the following monomers:-

$$
\begin{array}{cccc}
\underset{\underset{\underset{CH_3}{|}}{\overset{}{CH_2}}}{\underset{\underset{O}{|}}{\overset{\overset{H}{|}}{(i)\ C}}} = \underset{\underset{H}{|}}{\overset{\overset{H}{|}}{C}} &
\underset{\underset{\underset{CH_3}{|}}{\overset{}{CH_2}}}{\underset{\underset{O}{|}}{\overset{\overset{H}{|}}{(ii)\ C}}} = \underset{\underset{CH_3}{|}}{\overset{\overset{H}{|}}{C}} &
\underset{\underset{H}{|}}{\overset{\overset{H}{|}}{(iii)\ C}} = \underset{\underset{\underset{CH_3}{|}}{\overset{}{O}}}{\overset{\overset{\overset{CH_3}{|}}{O}}{C}} &
\underset{\underset{CH_3}{|}}{\overset{\overset{H}{|}}{(iv)\ C}} = \underset{\underset{H}{|}}{\overset{\overset{\overset{CH_3}{|}}{O}}{C}}
\end{array}
$$

(a) Show their activations free-radically and chargedly.

(b) First establish what free-radical and charged routes will be favored by the monomers.

(c) If a very weak initiator is involved free-radically and chargedly, establish the order of their stabilities.

(d) If a very strong initiator is involved free-radically and chargedly, establish the order of their stabilities.

1.10. Consider the following monomers:-

$$(i) \quad \begin{array}{cc} F & CFH_2 \\ | & | \\ C & = & C \\ | & | \\ F & CF_3 \end{array} \qquad (ii) \quad \begin{array}{cc} F & CH_2F \\ | & | \\ C & = & C \\ | & | \\ F & CH_2 \\ & | \\ & O \\ & | \\ & C=O \\ & | \\ & CH_3 \end{array} \qquad (iii) \quad \begin{array}{cc} & O & O^{\ominus} \\ & \diagdown & \diagup \\ F & N^{\oplus} \\ | & | \\ C & = & C \\ | & | \\ F & F \end{array}$$

(a) Show their activations free-radically and chargedly,

(b) Identify the free-radical and charged characters of the substituted groups.

(c) First establish what free-radical and charged routes that will be favored by the monomers.

(d) If a very strong initiator is involved, determine the order of stability

1.11. Consider the following pairs of monomers:

$$(i) \quad \begin{array}{cc} F & F \\ | & | \\ C & = & C \\ | & | \\ F & C=O \\ & | \\ & O \\ & | \\ & CH_3 \end{array} \quad and \quad \begin{array}{cc} H & H \\ | & | \\ C & = & C \\ | & | \\ H & C=O \\ & | \\ & O \\ & | \\ & CH_3 \end{array} \quad ; \quad (ii) \quad \begin{array}{cc} F & F \\ | & | \\ C & = & C \\ | & | \\ F & Cl \end{array} \quad and \quad \begin{array}{cc} H & H \\ | & | \\ C & = & C \\ | & | \\ H & Cl \end{array}$$

(a) Identify the family to which the monomers belong.

(b) Show their activations free-radically and chargedly.

(c) Identify the characters and the routes favored by the monomers.

(d) What are the substituted groups carried by the monomers? Which of the substituted groups are substituent groups? What are the characters of the groups?

1.12. Consider the following group of monomers:-

$$(i) \quad \begin{array}{cc} F & F \\ | & | \\ C & = & C \\ | & | \\ F & CH_3 \end{array} \quad (ii) \quad \begin{array}{cc} F & CH_3 \\ | & | \\ C & = & C \\ | & | \\ F & CH_3 \end{array} \quad , (iii) \quad \begin{array}{cc} F & H \\ | & | \\ C & = & C \\ | & | \\ CH_3 & CH_3 \end{array} \quad (iv) \quad \begin{array}{cc} F & CH_3 \\ | & | \\ C & = & C \\ | & | \\ CH_3 & CH_3 \end{array}$$

(a) Show the activations of monomers free-radically and chargedly.

(b) Identify the family to which the monomers belong.

(c) Identify the characters of the monomers and the routes favored by them.

(d) Determine the order of stability of the monomers.

1.13. Distinguish between the following monomer pairs free-radically and chargedly.

(i)

$$
\begin{array}{cc}
\text{H} & \text{CH}_3 \\
| & | \\
\text{C} & = \text{C} \\
| & | \\
\text{H} & \text{N}^{\oplus} \\
& \diagup \diagdown \\
& \text{O} \quad \text{O}^{\ominus}
\end{array}
\qquad \text{and} \qquad
\begin{array}{cc}
\text{H} & \text{H} \\
| & | \\
\text{C} & = \text{C} \\
| & | \\
\text{H} & \text{N}^{\oplus} \\
& \diagup \diagdown \\
& \text{O} \quad \text{O}^{\ominus}
\end{array}
$$

(ii)

$$
\begin{array}{cc}
\text{F} & \text{F} \\
| & | \\
\text{C} & = \text{C} \\
| & | \\
\text{F} & \text{F}
\end{array}
\qquad \text{and} \qquad
\begin{array}{cc}
\text{H} & \text{H} \\
| & | \\
\text{C} & = \text{C} \\
| & | \\
\text{H} & \text{H}
\end{array}
$$

1.4. Distinguish between the following monomer pairs free-radically and chargedly.

(i)

$$
\begin{array}{cc}
\text{H} & \text{H} \\
| & | \\
\text{C} & = \text{C} \\
| & | \\
\text{H} & \text{CH}_2 \\
& | \\
& \text{O} \\
& | \\
& \text{C}=\text{O} \\
& | \\
& \text{CH}_3
\end{array}
\qquad \text{and} \qquad
\begin{array}{cc}
\text{H} & \text{H} \\
| & | \\
\text{C} & = \text{C} \\
| & | \\
\text{H} & \text{CH}_3
\end{array}
$$

(ii)

$$
\begin{array}{cc}
\text{H} & \text{CFH}_2 \\
| & | \\
\text{C} & = \text{C} \\
| & | \\
\text{H} & \text{CF}_3
\end{array}
\qquad \text{and} \qquad
\begin{array}{cc}
\text{H} & \text{CH}_3 \\
| & | \\
\text{C} & = \text{C} \\
| & | \\
\text{H} & \text{CH}_3
\end{array}
$$

1.15 Determine the order of stability of the following monomers, free-radically and chargedly,

(i)

$$
\begin{array}{cc}
& \text{N} \\
& ||| \\
\text{H} & \text{C} \\
| & | \\
\text{C} & = \text{C} \\
| & | \\
\text{C}=\text{O} & \text{C} \\
| & ||| \\
\text{O} & \text{N} \\
| & \\
\text{CH}_3 &
\end{array}
$$

(ii)

$$
\begin{array}{cc}
& \text{N} \\
& ||| \\
\text{H} & \text{C} \\
| & | \\
\text{C} & = \text{C} \\
| & | \\
\text{CH}_3 & \text{C} \\
& ||| \\
& \text{N}
\end{array}
$$

(iii)

$$
\begin{array}{cc}
& \text{N} \\
& ||| \\
\text{H} & \text{C} \\
| & | \\
\text{C} & = \text{C} \\
| & | \\
\text{H} & \text{C} \\
& ||| \\
& \text{N}
\end{array}
$$

(iv)

$$
\begin{array}{cc}
& \text{N} \\
& ||| \\
\text{H} & \text{C} \\
| & | \\
\text{C} & = \text{C} \\
| & | \\
\text{C}=\text{O} & \text{H} \\
| & \\
\text{CH}_3 &
\end{array}
$$

$$
\begin{array}{ccc}
& & \text{N} \\
& & \text{|||} \\
\text{H} & & \text{C} \\
\text{|} & & \text{|} \\
(\text{v}) \quad \text{C} & = & \text{C} \\
\text{|} & & \text{|} \\
\text{CH}_3 & & \text{H}
\end{array}
\qquad
\begin{array}{ccc}
& & \text{O} \\
& & \text{|} \\
\text{H} & & \text{C} =\text{O} \\
\text{|} & & \text{|} \\
(\text{vi}) \quad \text{C} & \approx & \text{C} \\
\text{|} & & \text{|} \\
\text{H} & & \text{H}
\end{array}
\qquad
\begin{array}{ccc}
\text{H} & & \text{H} \\
\text{|} & & \text{|} \\
(\text{vii}) \quad \text{C} & = & \text{C} \\
\text{|} & & \text{|} \\
\text{H} & & \text{C} \\
& & \text{|||} \\
& & \text{N}
\end{array}
$$

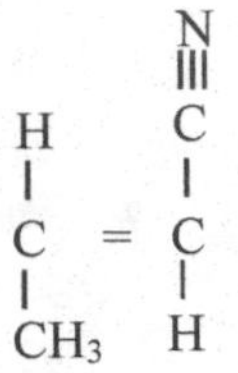

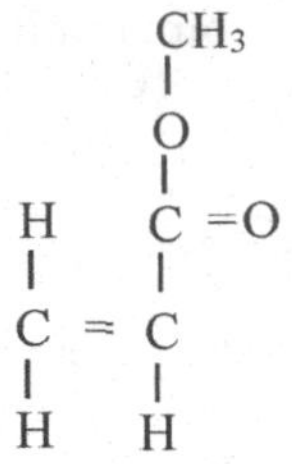

(vi) top: CH₃—O

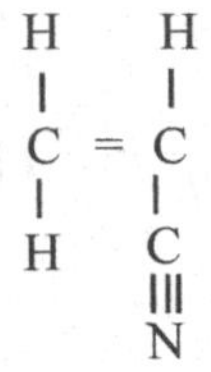

noting that $C \equiv N$ is more radical pulling than $COOCH_3$ group chargedly and free-radically.

1.16 What are the significances of the Second and Third Laws of Nature with respect to Life in humanity? Discuss.

1.17 Consider the following two groups of monomers:-

(i) 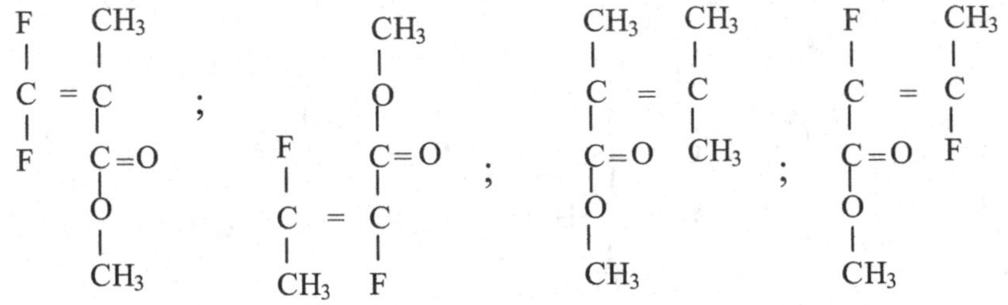 [structures of monomers]

Without any molecular rearrangement,

(a) Identify the routes favored free-radically and chargedly by the monomers.
(b) Whether a route is favored or not, identify the characters of the monomers in the two groups.
(c) What distinguishes between the two groups in terms of transfer species?

1.18. Identify the character of the transfer species carried by the substituent groups on the carbon centers of the following monomers when activated free-radically and chargedly.

(i) 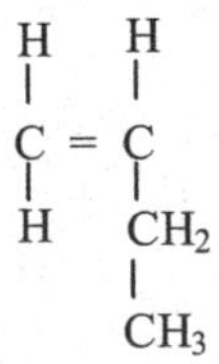 [structure]   (ii) 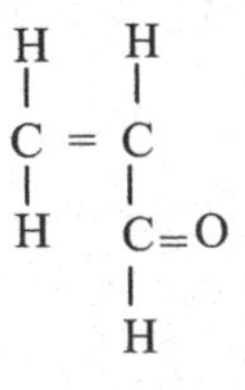 [structure]   (iii) 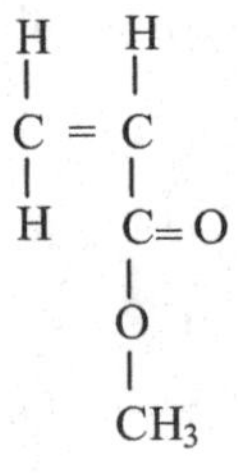 [structure]

(iv) 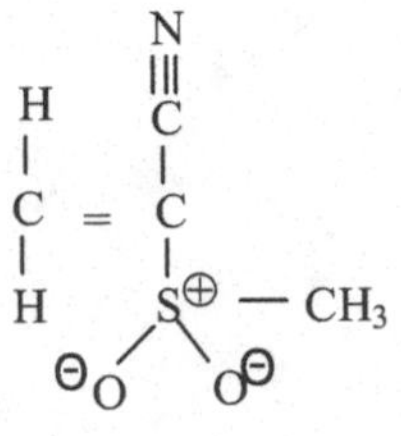 [structure]   (v) 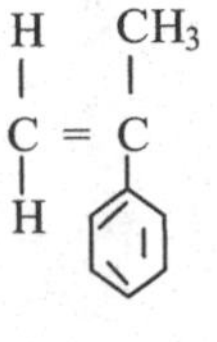 [structure]   (vi) 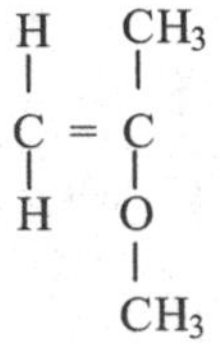 [structure]

$$\text{(vii)} \quad \begin{matrix} F & CH_3 \\ | & | \\ C & = & C \\ | & | \\ CH_3 & F \end{matrix} \qquad \text{(viii)} \quad \begin{matrix} H & CH_3 \\ | & | \\ C & = & C \\ | & | \\ \overset{\oplus}{N} & H \\ \diagup \diagdown \\ O & O^{\ominus} \end{matrix} \qquad \text{(ix)} \quad \begin{matrix} F & F \\ | & | \\ C & = & C \\ | & | \\ C=O & C=O \\ | & | \\ O & O \\ | & | \\ CH_3 & CH_3 \end{matrix}$$

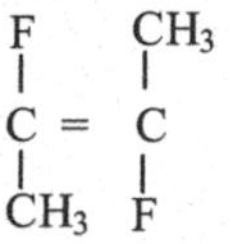

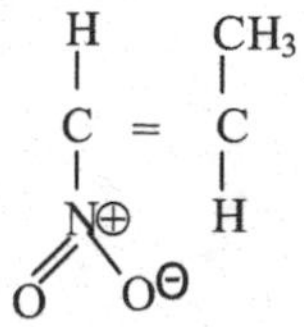

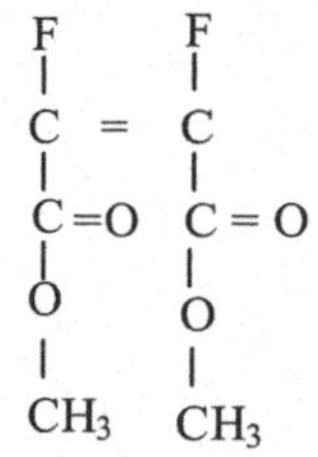

$$\text{(x)} \quad \begin{matrix} H & H \\ | & | \\ C & = & C \\ | & | \\ H & CH_2 \end{matrix}$$

(with phenyl ring and CH_3)

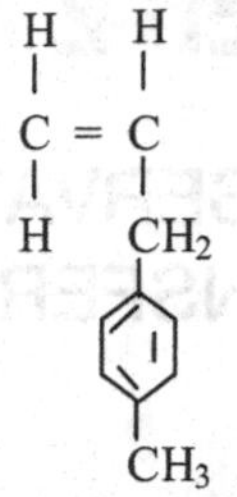

When there is no transfer species, say so. Consider only the C=C activation centers.

1.19. The addition of ethanol in the presence of sodium ethoxide occurs readily with

    (a) Tetrafluoroethylene
    (b) Trifluorochloroethylene
    (c) 1,1-difluoro-2,2-dichloroethylene and
    (d) 1,1-difluoro-2-chloroethylene

Show the mechanisms of the reactions radically and chargedly where possible.

1.20. (a) Write the structural configurations of the alkene monomers with the following groups:-

        (i) SO_2
        (ii) NO_2
        (iii) SO_2R
        (iv) SO_3H

    (b) Identify the characters of the groups and monomers radically and chargedly.
    (c) Identify which of the groups can be used in a monomer alone.
    (d) Show if SO_3 and SO_3H_2 can be used as groups? Explain.

CHAPTER 2

THE LAW OF CONSERVATION OF
TRANSFER OF TRANSFER SPECIES

2.0 Introduction

Since the law of conservation of transfer of transfer species was alluded to in the first Volume there is need to clearly identify it and state the law. The law will find very useful applications in this Volume when dealing with different types of Addition monomers. Though presently, mostly olefins are being considered, the law applies to all Addition monomers. It is from the law that amongst others one will be able to identify the different characteristic properties of Addition monomers and their polymeric products.

Transfer species one is concerned with are those internally generated within the system, that is, from the monomers and polymeric chains; and not those externally generated from foreign agents such as solvents, chain transfer agents, terminating agents and so on. The transfer species are those located on the substituent groups on active centers and those directly connected to the active centers. There are different kinds of transfer species in Addition monomers depending on their States of Existence (Stable, Activated or Activated/Equilibrium States). For alkenes, there are many kinds based on their States of Existence. Of greatest interest are those based on Activated States of Existence of which there are three kinds, of different types for alkenes, dialkenes and cumulenes and other types for some specific non-ringed monomers such as Carbon monoxide (CO) and ringed monomers.

Though largely initiation of monomers step is the main thrust in this volume, in view of the importance of the law of conservation of transfer of transfer species, some other substeps in some of the three major steps will be considered in passing in order to illustrate some basic principles and correct some ideas of the past with respect to transfer of transfer species. For example, ***there is no transfer of transfer species, from a monomer during propagation in polymerization;*** for if there was any transfer whatsoever, the monomer cannot go beyond the initiation step. This has partly been illustrated so far in the Series.

How and when molecular rearrangement is favored for some monomers will also be explained using the law of conservation of transfer of transfer species for different activation centers.

2.1 Transfer of Transfer Species

Reaction substeps involving transfer of transfer species are very important in two respects. First during polymerization, there could be transfer of transfer species from a dead polymer to another acceptor species which could be an active center such as an initiator, or a growing polymer chain or from a growing polymer chain to a dead polymer or transfer of transfer species from a growing polymer chain to itself all leading most exclusively to branched polymers. Secondly, during polymerization, there could be transfer of transfer species from a growing polymer chain to a monomer or from a growing polymer chain to an initiator's counter ion or another growing polymer chain, to produce dead polymers of different types. The possibilities could be more. By growing or living is meant a polymer having either free radical, or non-free radical or free or non-free negative charge or free positive charge at its end or along the backbone. This is the attacking nucleophile or electrophile which has grown with time to form a long chain through addition of monomers via Diffusion controlled mechanisms. By dead or neutral, is meant a polymer or molecule (e.g., unactivated monomer) without charges or radicals.

Now that transfer reactions have been partly defined, the next questions should be what can be transferred? In addition, from what positions can a transfer species be transferred?

2.1.1 Transfer from growing Chains/Transfer species of Second Kind

Hydrogen is the smallest transfer species that can be transferred as a cation or nucleo-free-radical (hydride) or electro-free-radical (the atom) in polymerization systems. It cannot be transferred as an anion as a hydride, since $H^{\ominus}$ does not exist. Hydrogen transfer is favored more in free-radical systems than charged systems, since homolysis requires less energy than needed in heterolysis. Where more than one transfer species exists in a particular location, it is the least charged or electro-radical species that is transferred.

Hydrogen is so small that it can be transferred readily even from an unsaturated monomer when properly located e.g. when the monomer is carrying substituent groups that have free hydrogen atoms bonded to a carbon atom with no double bond attached to it or from a growing polymer chain. When the former happens in the route not natural to the monomer, usually there is no polymerization. For the latter, consider the following reactions:-

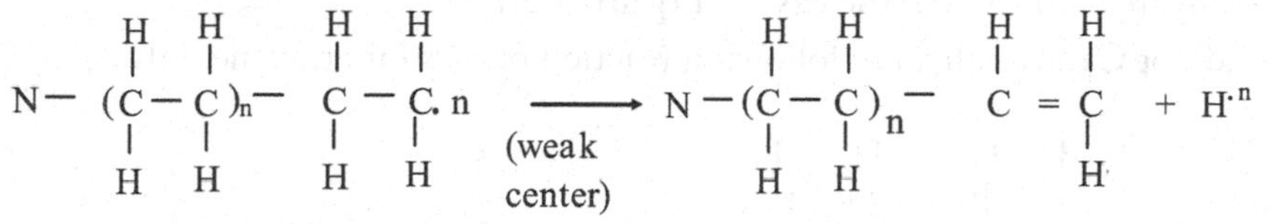

$$
\begin{array}{c}
\text{H H} \quad\quad \text{H H} \\
| \ | \quad\quad | \ | \\
N - (C - C)_n - C - C \cdot n \\
| \ | \quad\quad | \ | \\
\text{H H} \quad\quad \text{H H}
\end{array}
\;\underset{\substack{\text{(weak}\\\text{center)}}}{\longrightarrow}\;
\begin{array}{c}
\text{H H} \quad\quad \text{H H} \\
| \ | \quad\quad | \ | \\
N - (C - C)_n^- \ \ C = C \ \ + \ H^{\cdot n} \\
| \ | \quad\quad\quad | \\
\text{H H} \quad\quad\quad \text{H}
\end{array}
\qquad 2.1
$$

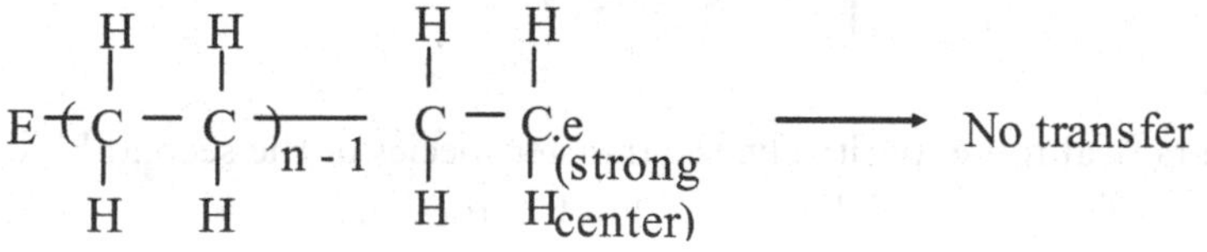

$$
\begin{array}{c}
\text{H H} \quad\quad \text{H H} \\
| \ | \quad\quad | \ | \\
E + C - C +_{n-1} \ \ C - C \cdot e \\
| \ | \quad\quad | \ | \\
\text{H H} \quad\quad \text{H H}
\end{array}
\substack{\text{(strong}\\\text{center)}}
\;\longrightarrow\; \text{No transfer}
\qquad 2.2
$$

In the reactions above involving growing or living polymer chains carrying terminal radicals, the only possible transfer species ($H^{\cdot n}$) has been rejected from one of the two groups on the terminal β-carbon atom to form a dead terminal double bond polymer, that is, a hydrogen nucleo-free-radical only for the nucleo-free-radical growing polymer chain. This is made possible based on the type of driving forces present in the system some of which include:

(i)   Long growing polymer chains of polyethylene.
(ii) Most importantly high operating conditions-High temperatures and pressures.

(iii) The decreasing strength of the active center for every addition of monomer nucleo-free-radically, the monomer being a nucleophile. For the electro-free-radical growing polymer chain of ethylene above, whose strength is increasing, H.e cannot be rejected no matter what the operating condition is, because the β-carbon center from where it is rejected can never carry a nucleo-free-radical when activated as a dead terminal double bonded polymer formed after rejection. In the first case H.n can be rejected, because the β-carbon center can carry what it should carry when the dead terminal double bonded polymer obtained after rejection is activated. The transfer species involved above is called **transfer species of the second kind** since it does not disturb the monomer from undergoing polymerization electro-free-radically. This type of transfer species is not the type of transfer species involved in the law of conservation of transfer of transfer species, since this type of transfer species is only involved in the route which is not natural to the monomer for all Nucleophiles. Transfer species of the second kind is common with olefinic monomers which favor polymerization via both routes and only via free-radical polymerizations.

Now consider perfluoroethylene. The corresponding equation to Equation 2.1 above is as follows:-

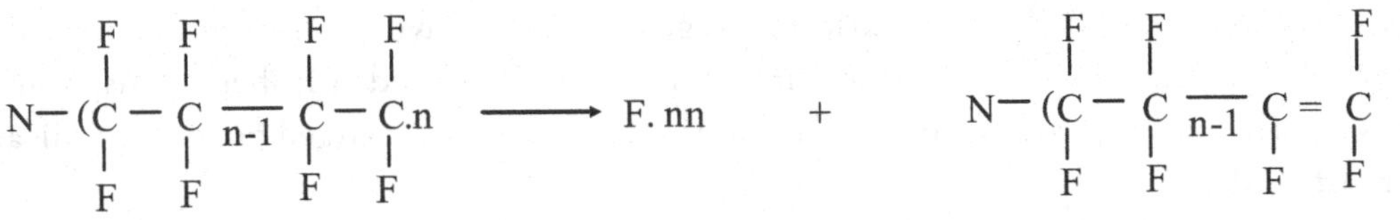

$$N - (\underset{\underset{F}{|}}{\overset{\overset{F}{|}}{C}} - \underset{\underset{F}{|}}{\overset{\overset{F}{|}}{C}})_{n-1} \underset{\underset{F}{|}}{\overset{\overset{F}{|}}{C}} - \underset{\underset{F}{|}}{\overset{\overset{F}{|}}{C}} . n \longrightarrow F . nn \quad + \quad N - (\underset{\underset{F}{|}}{\overset{\overset{F}{|}}{C}} - \underset{\underset{F}{|}}{\overset{\overset{F}{|}}{C}})_{n-1} \underset{\underset{F}{|}}{\overset{\overset{F}{|}}{C}} = \underset{F}{\overset{\overset{F}{|}}{C}} \qquad 2.3$$

NOT FAVORED Not radically balanced

Because the fluorine atom unlike hydrogen atom is carrying a nucleo-non-free-radical and the growing chain is nucleo-free-radical in character, F cannot be released as transfer species of the second kind, since the equation will not be radically balanced. However, note that when F is released, the β-carbon center can carry an electro-free-radical like the case of Equation 2.1.

Positively and not Cationically, the following reaction occurs for ethylene only.

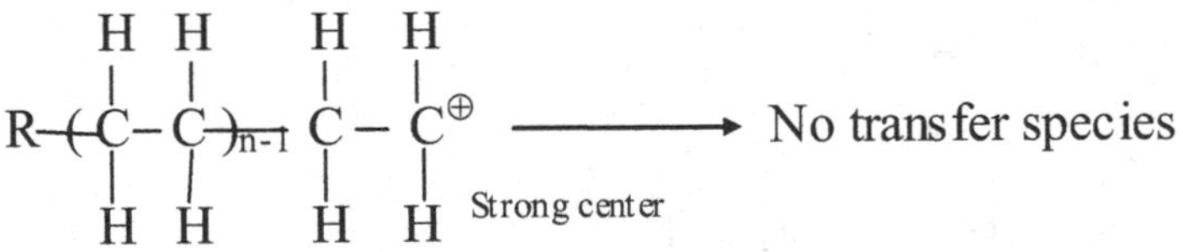

$$R - (\underset{\underset{H}{|}}{\overset{\overset{H}{|}}{C}} - \underset{\underset{H}{|}}{\overset{\overset{H}{|}}{C}})_{n-1} \underset{\underset{H}{|}}{\overset{\overset{H}{|}}{C}} - \underset{H}{\overset{\overset{H}{|}}{C}}^{\oplus} \longrightarrow \text{No transfer species} \qquad 2.4$$

Strong center

Like the electro-free-radical route, positively, the transfer species of the second kind cannot be rejected. In fact, there is no transfer species of the second kind in these natural routes for the monomers. They only exist in the route not natural to them for Nucleophiles. Why this is so will become obvious as we progress in the Series and Volumes.

Chargedly negatively, the following reaction occurs for the alkene:-

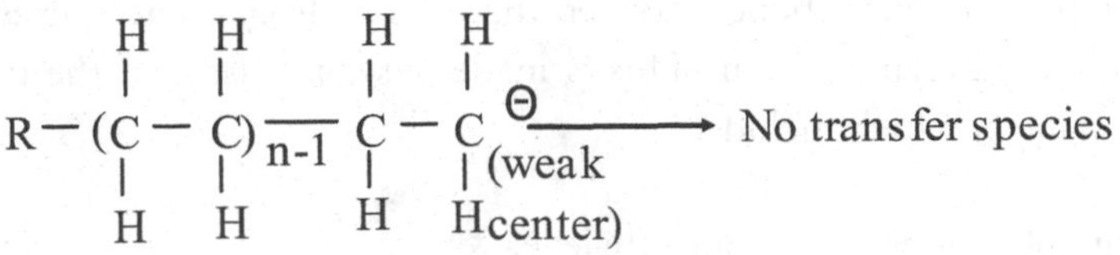

$$R - (\underset{\underset{H}{|}}{\overset{\overset{H}{|}}{C}} - \underset{\underset{H}{|}}{\overset{\overset{H}{|}}{C}})_{n-1} \underset{\underset{H}{|}}{\overset{\overset{H}{|}}{C}} - \underset{\underset{H}{|}}{\overset{\overset{H}{|}}{C}}^{\ominus} \longrightarrow \text{No transfer species} \qquad 2.5$$

(weak center)

53

Unlike the nucleo-free-radical case, these is no reaction here, since there are no anions or negatively charged component on the terminal β-carbon atom. Anionic hydrogen, $H^{\ominus}$ which like for Li, Na, K, and its members with no paired unbonded radicals on its center does not exist and cannot therefore be transferred from a growing polymer chain under any polymerization conditions for carbon-chain polymers. Electronegativity of hydrogen is small enough for it to gain a radical from more electropositive ionic metals. But this transfer is rarely possible. If the transfer was possible, then H will not be the first member of the Periodic Table, based on how Nature operates. H like other members below her in the Group IA of the Periodic Table can never carry a negative charge. However, as the first member, a mark of great distinction, she is the only one in the Group allowed to carry a nucleo-free-radical. Other members cannot; hence they (Li, Na, K, etc.) are more electropositive than H. In fact H like the others is a METAL In the Liquid and Solid States, H is obviously a metal. In the gaseous State, under Standard Temperature and Pressure (STP), being the first member of the Periodic Table, H is the only GASEOUS METAL OF ALL ATOMS known to exist.

For a polymerizable perfluoroalkene monomer, the following is also obtained:-

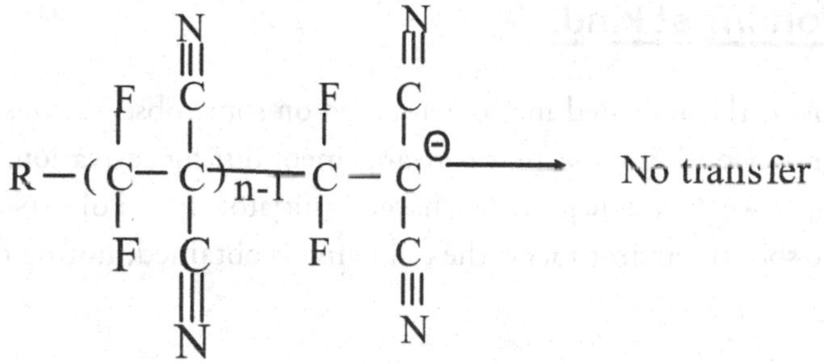

$$\text{2.6a}$$

where R is free-negatively charged initiator which cannot be isolated; in order words it is paired. Note that the negative charges carried by the growing polymer chains are free and not of the non-free-types. How can the chain release a non-free-type of charge from a free-type of growing chain into emptiness without a center to receive it and form a stable molecule as shown below, noting that free-negative charges cannot be isolated? Only the non-free-negative charges called Anions can be isolated. If F had been H, it will never be released via Combination mechanisms.

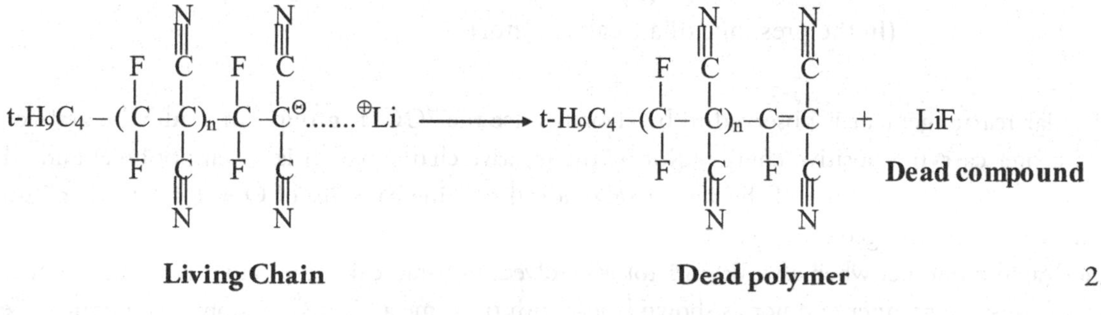

Living Chain   **Dead polymer**   2.6b

Thus, one can observe the conditions under which transfer species of the Second kind of the second type are released nucleo-free-radically and chargedly in the route natural to the monomer only for fluorinated alkenes which are Electrophilic in character. *In the Equation above, notice that dead products are released on the right-hand-side (RHS) of the equation in order to balance the equation unlike in Equation 2.6a.* Radically, with Equation 2.6a, F•nn will be released when a stable dead polymer is produced without a receiving end via Decomposition mechanism. If the monomer was non-fluorinated, it cannot be released. These non-fluorinated ones are less Electrophilic than the fluorinated case. Like H which is the smallest cationic and free-radical transfer species, F is the smallest anionic and non-free-radical transfer species. Though the monomer is an electrophile being polymerized via the route natural to it, F^Θ can be released as transfer species of the second kind *only when there is a receiver, but radically a receiver is not needed.* Note that the route above is natural to the monomer (a Male) on its C = C center. That was why there was no transfer species in Equation 2.6a chargedly. It is only via Decomposition mechanism which takes place only radically, that initiators can be obtained..

The transfer species (F^Θ of F•nn) is said to be of second kind, because the transfer species is coming from the β-carbon center in the absence of any on the α-carbon center. It is of the *second type,* because it is coming from the route natural to the monomer. The one coming from the monomer in the route not natural to the monomer is of the *first type.*

2.1.2 Transfer from charged polymer chains/from Monomer/Transfer species of First kind and second/first kind.

There is need to look at the activated monomer, based on some observations and issues which were raised in the last chapter concerning molecular rearrangement during activation. When methyl acrylate is activated chargedly by a weak free-negatively-charged initiator, in a Polar/Non-ionic or Non-polar/Non-ionic or Polar/Ionic solvent environment, the following is obtained, noting that the best solvent for it is Polar/Non-ionic:-

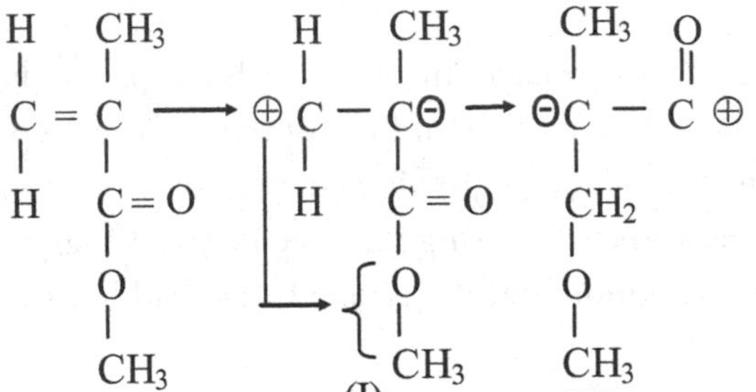

(In the presence of a weak initiator)

2.7

Molecular rearrangement is observed to be favored since the $^\Theta OCH_3$ group can readily be grabbed by the C center carrying positive charge (Not by the positive charge which is an empty hole) and (II) is more stable than (I) (because of the very strong radical-pushing capacity of O =, RN= types of group) and the covalently charged characters of the centers are not changed. Therefore (II) is the favored state of the activated monomer when a weak initiator is involved. Free-radically, molecular rearrangement takes place in the same manner and not as shown below with the same activated monomer (Ignorance) being obtained in the presence of a weak initiator. What is shown below is not favored, since H.n cannot be **abstracted or transferred** to create an electro-free-radical center. H.n can only be **released** to create an electro-free-radical center. Notice that H cannot also be **transferred** as H^Θ, hence the transfer of $^\Theta OCH_3$ (non-free anion),

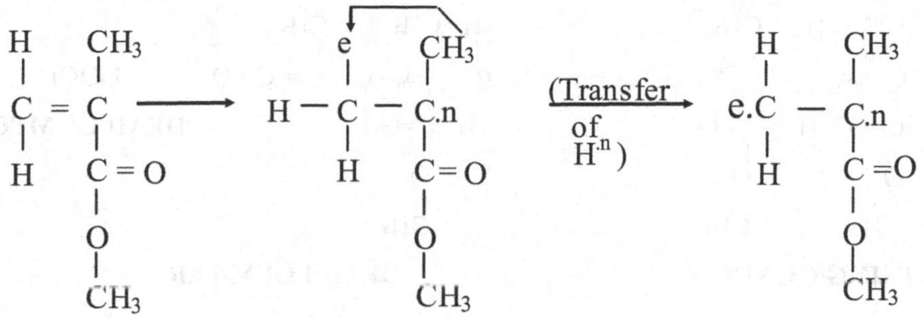

$$
\begin{array}{ccccc}
\text{H} & \text{CH}_3 & & \text{e} & \text{CH}_3 \\
| & | & & | & | \\
\text{C} = \text{C} & \longrightarrow & \text{H} - \text{C} - \text{C.n} & \xrightarrow[\text{of } H^n]{\text{(Transfer}} & \text{e.C} - \text{C.n} \\
| & | & & | \quad | & | \quad | \\
\text{H} & \text{C}=\text{O} & & \text{H} \quad \text{C}=\text{O} & \text{H} \quad \text{C}=\text{O} \\
& | & & | & | \\
& \text{O} & & \text{O} & \text{O} \\
& | & & | & | \\
& \text{CH}_3 & & \text{CH}_3 & \text{CH}_3 \qquad 2.8
\end{array}
$$

(Impossible transfer- Not Favored)

being an Electrophile. If it was a Nucleophile, then H will be transferred as an atom or electro-free-radical carrying species. If $^{\ominus}$OCH$_3$ or nn•OCH$_3$ can be **transferred** for molecular rearrange-ment, it should therefore be **abstracted** in the route not natural to it and should therefore be **released** to kill the growing chain from within in the route natural to it. In the route natural to it, its release can only be done like the case of Equation 2.6b when charged-paired initiators are involved to form stable compounds including a dead terminal double bond polymer of almost the same or lesser or greater character as the monomer. Imagine what the situation will be if F was put in place of H in the monomer above. The situation will still remain the same with OCH$_3$ group still remaining being the transfer species. The monomer in Equation 2.6b cannot undergo the molecular rearrangement in Equation 2.7 above, because the C≡N group unlike COOCH$_3$ group has no transfer species.

Therefore, considering the growing polymer chain of methyl methacrylate, the followings are obtained when molecular rearrangement is favored, that is, the initiator and the growing chain's active center is weak and the same for a long time:-

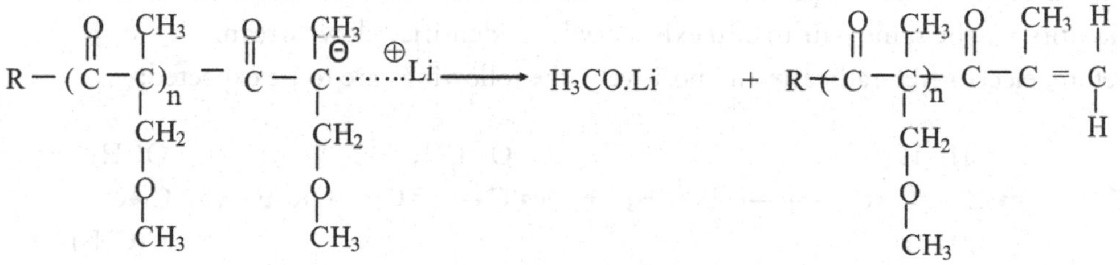

$$
\text{R} - (\overset{\overset{\text{O}}{\|}}{\underset{\underset{\text{CH}_2}{|}}{\text{C}}} - \overset{\overset{\text{CH}_3}{|}}{\underset{\underset{\text{O}}{|}}{\text{C}}})_n - \overset{\overset{\text{O}}{\|}}{\text{C}} - \overset{\overset{\text{CH}_3}{|}}{\underset{\underset{\text{CH}_2}{|}}{\underset{\text{CH}_3}{\text{C}}}}\overset{\ominus}{}\cdots\cdots\overset{\oplus}{\text{Li}} \longrightarrow \text{H}_3\text{CO.Li} \quad + \quad \text{R} +\text{C} - \text{C}\text{)}_n \ \text{C} - \text{C} = \text{C}
$$

$$ \qquad\qquad\qquad\qquad\qquad\qquad\qquad\qquad\qquad\qquad\qquad\qquad\qquad\qquad\qquad\qquad\qquad\qquad\qquad 2.9 $$

Negatively-charged-paired polymerization (Natural route)

The route being natural to it would indicate that rearrangement cannot continue all along during propagation, because the strength of the active center will be increasing or remain the same for every monomer added. However as shown above, virtually almost the same type Giant methyl methacrylate dead polymer is produced. Note should be taken of the type of dead terminal double bond polymer produced. Also, note that if anionic-ion-paired initiator (e.g. RO$^{\ominus}$.......$^{\oplus}$Na) had been used, the Initiation Step will not be favored, since the equation will not be chargedly balanced. With the case above, the Initiation Step is favored and so also is Propagation Step since the equations are chargedly balanced. With the monomers above, no re-initiation can take place. With the initiators above, the cationic centers cannot be used due to charge balancing.

Now when molecular rearrangement is not allowed to take place because of the use of a strong initiator, the followings are obtained.

$$R-(\overset{\overset{\displaystyle H}{|}}{\underset{\underset{\displaystyle H}{|}}{C}}-\overset{\overset{\displaystyle CH_3}{|}}{\underset{\underset{\displaystyle \underset{\displaystyle CH_3}{|} O}{|} C=O}{C}})_n-\overset{\overset{\displaystyle H}{|}}{\underset{\underset{\displaystyle H}{|}}{C}}-\overset{\overset{\displaystyle CH_3}{|}}{\underset{\underset{\displaystyle \underset{\displaystyle CH_3}{|} O}{|} C=O}{C}}{}^{\ominus}........{}^{\oplus}Li \longrightarrow R-(\overset{\overset{\displaystyle H}{|}}{\underset{\underset{\displaystyle H}{|}}{C}}-\overset{\overset{\displaystyle CH_3}{|}}{\underset{\underset{\displaystyle \underset{\displaystyle CH_3}{|} O}{|} C=O}{C}})_n-\overset{\overset{\displaystyle H}{|}}{\underset{\underset{\displaystyle H}{|}}{C}}-\overset{CH_3}{\underset{H}{C}}=C=O \quad + \quad LiOCH_3$$

$$\textbf{DEAD COMPOUND}$$

2.10

LIVING CHAIN   **DEAD POLYMER**

A ketenic type of dead terminal double bond polymer is produced. This was the monomer obtained after molecular rearrangement. While free-chargedly this transfer species can be released being paired, free-radically it can only be released, if there is center to receive it to form a stable compound. In today's Science, paired radical initiators have not yet been identified, because the types of components involved in forming them are unique. Unknown, just like for electro-free-radical initiators not known to exist, they exist and are common. In fact, the so-called "anionic-ion-paired initiator" used above which cannot be used for alkenes except ethylene, can be paired radically in non-ionic environment, because the electro-potential difference between the Li and the C center is high, unlike between two like metals (Al or Zn). For if it (Li/C) was zero, no pairing can take place radically. A re-initiation of a monomer when present is not possible here, since the former ion-paired initiator cannot be obtained. $R^{\ominus}$ is a free centered negative charge, while $CH_3O^{\ominus}$ is a non-free centered negative charge (the anion). Instead of a re-initiation, $LiOCH_3$ a stable molecule is produced for Nature abhors a vacuum. Since this transfer species can be released when charged-paired initiators are used, then it can also always be abstracted in the route not natural to it, that is, positively. The same should apply when radical paired types of initiators are identified and used. From the considerations above, it can be seen that free-negatively charged initiator such as $^{\ominus}CH_3$ or $^{\ominus}t\text{-}C_4H_9$ cannot exist in isolation. $^{\ominus}OCH_3$ or $^{\ominus}Cl$ can exist in isolation. Yet, they can be paired with $Na^{\oplus}$ which can also exist in isolation. Since ion-paired initiators such as $H_3CO^{\ominus}......^{\oplus}Na$ exist, one should expect radical paired initiators for the same pair to also exist as will be identified downstream.

Therefore, electro-free-radically and positively, the followings are to be expected:-

$$E^{\bullet} + n\bullet\overset{\overset{\displaystyle CH_3}{|}}{\underset{\underset{\underset{\displaystyle CH_3}{|}}{\underset{\displaystyle O}{|}}}{\underset{\displaystyle C=O}{C}}}-\overset{\overset{\displaystyle H}{|}}{\underset{\underset{\displaystyle H}{|}}{C}}\bullet e \longrightarrow EOCH_3 + e\bullet\overset{\overset{\displaystyle O}{\|}}{C}-\overset{CH_3}{\underset{}{C}}=CH_2 \quad OR \quad E-O-\overset{\overset{\displaystyle OCH_3}{|}}{\underset{\underset{\displaystyle CH_2}{|}}{\underset{\displaystyle C(CH_3)}{C}}}\bullet e$$

$$\textbf{(Favored)} \qquad \textbf{(Favored)} \qquad 2.11$$

$$R^{\oplus} + {}^{\ominus}\overset{\overset{\displaystyle CH_3}{|}}{\underset{\underset{\underset{\displaystyle CH_3}{|}}{\underset{\displaystyle O}{|}}}{\underset{\displaystyle CH_2}{C}}}-\overset{\overset{\displaystyle O}{\|}}{C}{}^{\oplus} \longrightarrow ROCH_3 + {}^{\oplus}\overset{\overset{\displaystyle H}{|}}{\underset{\underset{\displaystyle H}{|}}{C}}-\overset{CH_3}{\underset{}{C}}=C=O \quad OR \quad \overset{\overset{\displaystyle H}{|}}{\underset{\underset{\displaystyle H}{|}}{C}}=\overset{CH_3}{\underset{}{C}}-\overset{\overset{\displaystyle O}{\|}}{C}{}^{\oplus}$$

$$\underline{\textbf{Favored}} \qquad 2.12a$$

$$R^{\oplus} + {}^{\ominus}\overset{\overset{\displaystyle H_3C}{|}}{\underset{\underset{\underset{\displaystyle CH_3}{|}}{\underset{\displaystyle O}{|}}}{\underset{\displaystyle O=C}{C}}}-\overset{\overset{\displaystyle H}{|}}{\underset{H}{C}}{}^{\oplus} \quad OR \quad {}^{\ominus}\overset{\overset{\displaystyle H_3C}{|}}{\underset{\underset{\underset{\displaystyle CH_3}{|}}{\underset{\displaystyle O}{|}}}{\underset{\displaystyle CH_2}{C}}}-\overset{\overset{\displaystyle O}{\|}}{C}{}^{\oplus} \longrightarrow R-O-\overset{\overset{\displaystyle OCH_3}{|}}{\underset{\underset{\displaystyle CH_2}{\|}}{\underset{\displaystyle C(CH_3)}{C}}}{}^{\oplus} \quad OR \quad R-O-\overset{\overset{\displaystyle OCH_3}{|}}{\underset{\underset{\underset{\displaystyle OCH_3}{|}}{\underset{\displaystyle CH_2}{|}}}{\underset{\displaystyle C(CH_3)}{C}}}{}^{\oplus}$$

2.12b

57

Thus, electro-free-radically and positively, the monomer cannot be polymerized via the C =C center (Y) whether the initiator is weak or strong, since the monomer is an electrophile. It is the X center-C = O that should be the active centers to be attacked electro-free-radically or "cation-ically". After molecular rearrangement where favored, transfer species are abstracted. Whether there is molecular rearrangement or not, the same transfer species is involved during polymer-ization. Nucleo-free-radically and with negatively charged-paired initiators, it can be observed that the routes are favored and that the C = O center cannot be used. For **methyl acrylate,** initiation electro-free-radically and positively are not favored. For the same reasons as for methyl methacrylate and methyl acrylate, acrylamide and **methyl acrylamide** will not favor electro-free-radical and positively charged routes. However, it can be observed that in the presence of adequate driving forces, different dead terminal double bond polymers can be produced nucleo-free-radically and anionically, the routes being natural to the monomer. In addition, it is the transfer species which did not allow for the polymerization of these monomers electro-free-radically and positively ($^{\ominus}OCH_3$), that are abstracted from the growing polymer chains nucleo-free-radically when paired and negatively charged when paired respectively to kill the chains. It is the same transfer species involved in molecular rearrangements of the activated monomers when the operating conditions exist.

Now, consider a more complex monomer with bulkier and fully substituted type of substi-tuent groups shown below for its growing polymer chain, hoping that steric forces do not prevent its polymerization. Using electro-free-radicals, the following will occur, for its growing polymer chain.

$$
E-(\underset{\underset{\underset{CH_3}{|}}{\overset{|}{C(CH_3)_2}}}{\overset{H}{\underset{|}{C}}} - \overset{H}{\underset{|}{C}})_n \quad \underset{\underset{\underset{CH_3}{|}}{\overset{|}{C(CH_3)_2}}}{\overset{H}{\underset{|}{C}}} - \overset{H}{\underset{}{C}}.e \longrightarrow E-(\underset{\underset{\underset{CH_3}{|}}{\overset{|}{C(CH_3)_2}}}{\overset{H}{\underset{|}{C}}} - \overset{H}{\underset{}{C}})_n -\overset{H}{\underset{|}{C}} - \overset{H}{C} = \overset{CH_3}{\underset{\underset{CH_3}{|}}{C}} \quad + H_3C^{.e}
$$

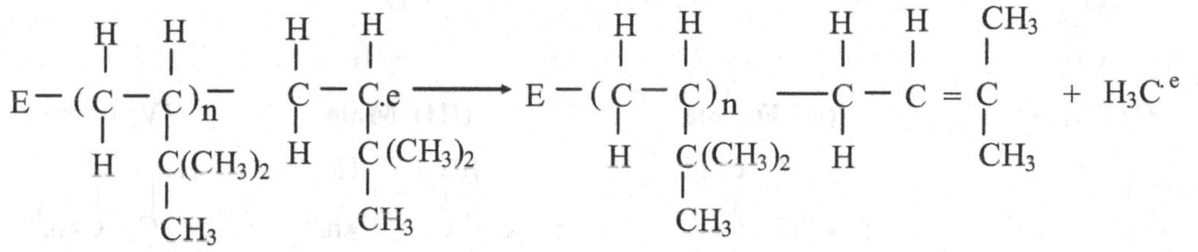

$$2.13$$

It is important to note that chargedly and radically there is no molecular rearrangement here, since as will be shortly explained, it is impossible. This monomer cannot undergo nucleo-free-radical polymerization since the same transfer species involved above - e.CH$_3$ will prevent the initiation of the monomer. The same also applies when attacked with negative free-charges. It can only undergo electro-free-radical and positively charged polymerization in the absence of steric limitations. Here is another case where the law of conservation of transfer species apply, but with no molecular rearrangement allowed to take place.

For propylene which presently is known to favor only the use of coordination initiators which are charged and paired, their growing polymer chains undergo the following reaction shown below:-

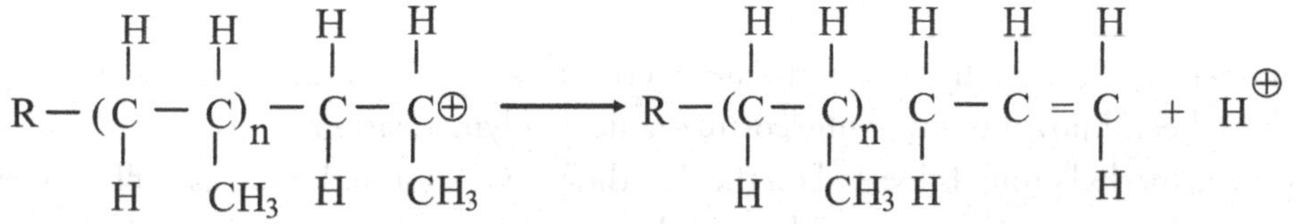

$$
R-(\underset{\underset{H}{|}}{\overset{H}{\underset{|}{C}}} - \overset{H}{\underset{\underset{CH_3}{|}}{C}})_n - \overset{H}{\underset{\underset{H}{|}}{C}} - \overset{H}{\underset{\underset{CH_3}{|}}{C}} \oplus \longrightarrow R-(\underset{\underset{H}{|}}{\overset{H}{\underset{|}{C}}} - \overset{H}{\underset{\underset{CH_3}{|}}{C}})_n \overset{H}{\underset{\underset{H}{|}}{C}} - \overset{H}{C} = \overset{H}{\underset{\underset{H}{|}}{C}} \quad + H^{\oplus}
$$

$$2.14$$

__(Natural route)__ (Note that the initiator is paired)

The hydrogen cation removed is the same hydrogen abstracted during neagatively charged initiation of the monomer. It is the same type of transfer species nucleo-free-radically that prevented its initiation. Electro-free-radically, the following is obtained:-

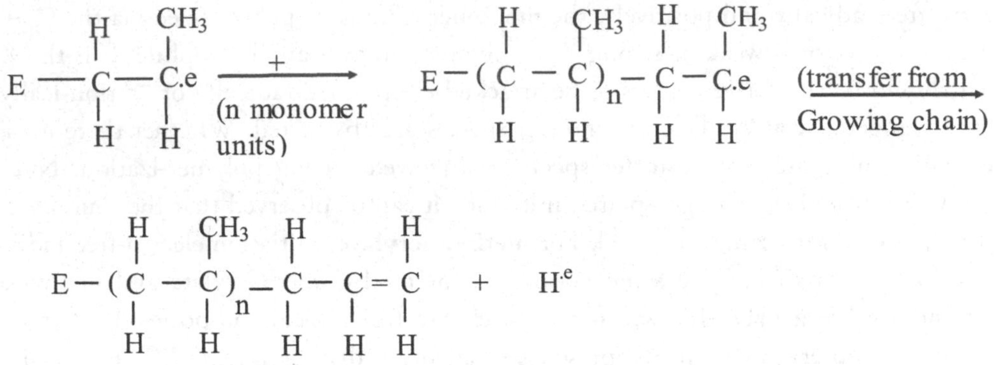

$$E - \underset{\underset{H}{|}}{\overset{\overset{H}{|}}{C}} - \underset{\underset{H}{|}}{\overset{\overset{CH_3}{|}}{C}}.e \quad \xrightarrow{\;+\;} \quad E - (\underset{\underset{H}{|}}{\overset{\overset{H}{|}}{C}} - \underset{\underset{H}{|}}{\overset{\overset{CH_3}{|}}{C}})_n - \underset{\underset{H}{|}}{\overset{\overset{H}{|}}{C}} - \underset{\underset{H}{|}}{\overset{\overset{CH_3}{|}}{C}}.e \qquad \underset{\text{Growing chain}}{(\text{transfer from}}$$

(n monomer units)

$$E - (\underset{\underset{H}{|}}{\overset{\overset{H}{|}}{C}} - \underset{\underset{H}{|}}{\overset{\overset{CH_3}{|}}{C}})_n - \underset{\underset{H}{|}}{\overset{\overset{H}{|}}{C}} - \underset{\underset{H}{|}}{\overset{\overset{H}{|}}{C}} = \underset{\underset{H}{|}}{\overset{\overset{H}{|}}{C}} \quad + \quad H^e \qquad\qquad 2.15$$

(Natural route)

So far, notice that the α-carbon center for Electrophiles is different from the α-carbon center in Nucleophiles as shown below. While for Electrophiles, the α-carbon center is fixed when R groups are placed on the β-carbon center, for Nucleophiles the α-carbon center is not fixed.

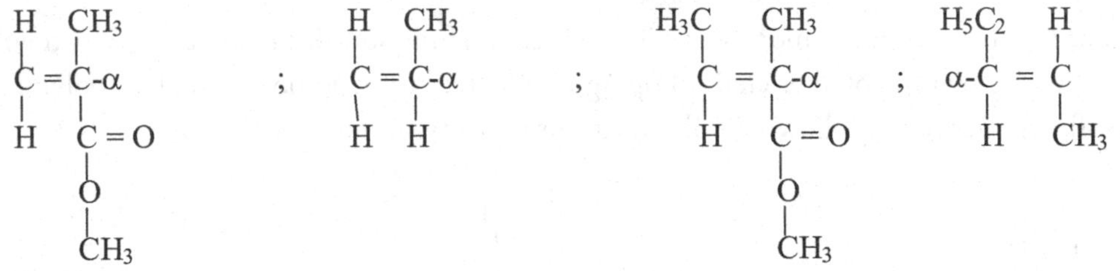

$$\underset{\underset{\underset{\underset{CH_3}{|}}{O}}{\underset{|}{C=O}}}{\overset{\overset{CH_3}{|}}{\underset{|}{\overset{H}{C}}=\overset{|}{C}\text{-}\alpha}}}\quad ; \quad \overset{\overset{H\;\;CH_3}{|\;\;\;|}}{\underset{\underset{H\;\;H}{|\;\;\;|}}{C=C\text{-}\alpha}} \quad ; \quad \underset{\underset{\underset{CH_3}{|}}{O}}{\overset{\overset{H_3C\;\;CH_3}{|\;\;\;\;\;|}}{\underset{\underset{H}{|}\;\;\underset{C=O}{|}}{C=C\text{-}\alpha}}} \quad ; \quad \overset{\overset{H_5C_2\;\;H}{|\;\;\;\;\;|}}{\underset{\underset{H\;\;\;CH_3}{|\;\;\;\;\;|}}{\alpha\text{-}C=C}} \qquad 2.16a$$

(I) Male ; **(II) Female** ; **(III) Male** ; **(IV) Female**

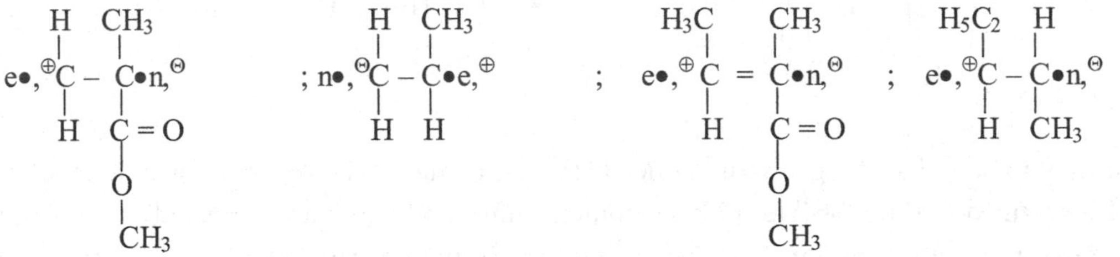

$$e\bullet,{}^{\oplus}\underset{\underset{\underset{\underset{CH_3}{|}}{O}}{\underset{|}{C=O}}}{\overset{\overset{H\;\;\;CH_3}{|\;\;\;\;\;|}}{C}} - C\bullet n,{}^{\ominus} \quad ; \; n\bullet,{}^{\ominus}\overset{\overset{H\;\;CH_3}{|\;\;\;|}}{\underset{\underset{H\;\;H}{|\;\;\;|}}{C}} - C\bullet e,{}^{\oplus} \quad ; \quad e\bullet,{}^{\oplus}\underset{\underset{\underset{\underset{CH_3}{|}}{O}}{\underset{|}{C=O}}}{\overset{\overset{H_3C\;\;\;CH_3}{|\;\;\;\;\;|}}{C}} = C\bullet n,{}^{\ominus} \quad ; \; e\bullet,{}^{\oplus}\overset{\overset{H_5C_2\;\;H}{|\;\;\;\;\;|}}{\underset{\underset{H\;\;\;CH_3}{|\;\;\;\;\;|}}{C}} - C\bullet n,{}^{\ominus} \qquad 2.16b$$

THEIR ACTIVATED STATES OF EXISTENCES

(II) and (IV) will give the same monomer when molecular rearrangement is favored. All of them will favor their natural routes using free-paired charged initiators or free-radical initiators, since H cannot be abstracted or transferred as H•n which exists or $H^{\ominus}$ which does not exist. From all the considerations so far, the followings are worthy of note-

(i) Monomers such as methyl methacrylate, methacrylamide, methacrylonitrile, α-methyl styrene etc. have been known to favor nucleo-free-radical polymerization.

(ii) When the methyl group is located on the β-carbon center, for males the use of free-media nucleo-free-radical initiator is not possible, but the use of negatively free-charged-paired initiator is favored only under specific placements of the groups, while for females, the use of electro-free-radical initiators or positively charged paired initiators still remain favored.

All the transfers species from substituent groups identified so far are of the same type and ***these are known as transfer species of the first kind of the first type,*** since they are located on the group carried by the α-carbon centers and most importantly, they are coming from alkylane groups (C_nH_{2n+1}), and groups such as COOR, $CONH_2$ and more.

Now consider what happens when vinyl acetate growing polymer chains from the two types of free-radicals are involved in transfer of transfer species reactions. With a nucleo-free-radical and an electro-free-radical initiator, the following reactions are obtained:-

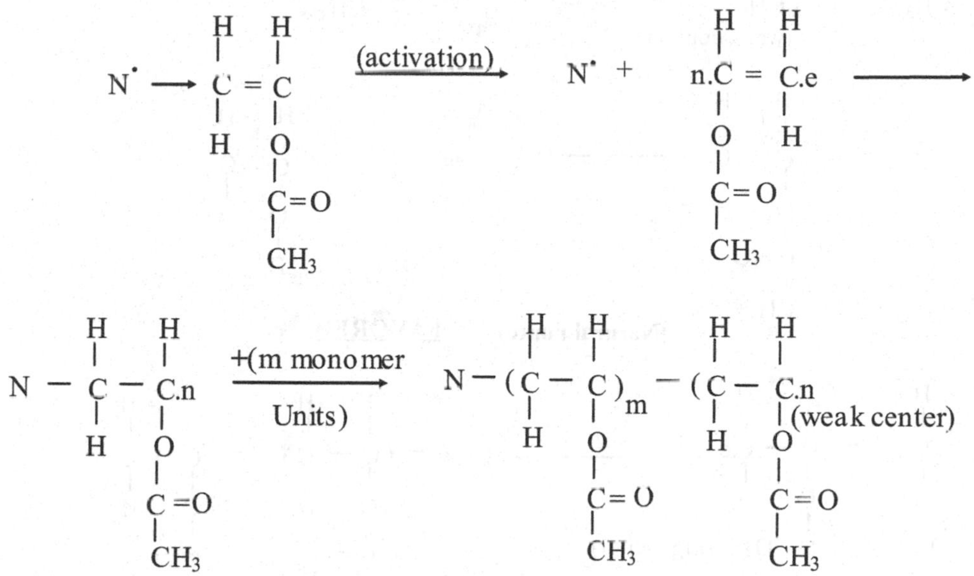

$$
N^\bullet \longrightarrow \underset{\underset{\underset{\underset{CH_3}{|}}{\underset{C=O}{|}}}{\overset{\overset{H}{|}}{\underset{\underset{|}{H}}{C}}} = \underset{\underset{O}{|}}{\overset{\overset{H}{|}}{C}} \quad \xrightarrow{\text{(activation)}} \quad N^\bullet \; + \; n.\underset{\underset{\underset{\underset{CH_3}{|}}{\underset{C=O}{|}}}{\underset{O}{|}}}{\overset{\overset{H}{|}}{C}} = \underset{\underset{H}{|}}{\overset{\overset{H}{|}}{C}}.e \longrightarrow
$$

$$
N - \underset{\underset{\underset{\underset{CH_3}{|}}{\underset{C=O}{|}}}{\underset{H}{|}}}{\overset{\overset{H}{|}}{C}} - \underset{\underset{O}{|}}{\overset{\overset{H}{|}}{C}}.n \quad \xrightarrow[\text{Units)}]{+(m\ \text{monomer}} \quad N - (\underset{\underset{\underset{\underset{CH_3}{|}}{\underset{C=O}{|}}}{\underset{H}{|}}}{\overset{\overset{H}{|}}{C}} - \underset{\underset{O}{|}}{\overset{\overset{H}{|}}{C}})_m - (\underset{\underset{\underset{\underset{CH_3}{|}}{\underset{C=O}{|}}}{\underset{H}{|}}}{\overset{\overset{H}{|}}{C}} - \underset{\underset{O}{|}}{\overset{\overset{H}{|}}{C}}.n \quad \text{(weak center)} \qquad 2.17
$$

(Unnatural route)

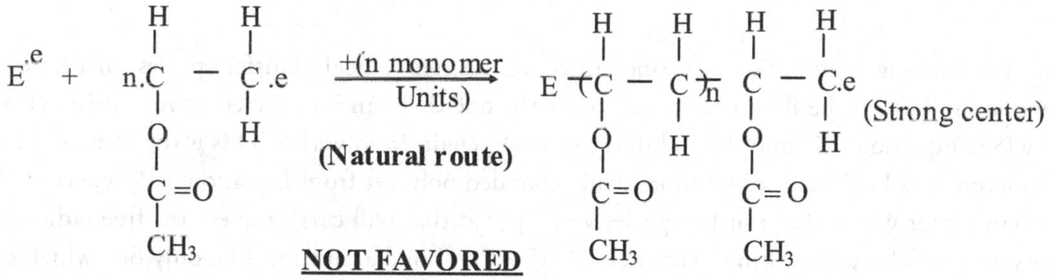

$$
E^{\bullet e} \; + \; n.\underset{\underset{\underset{\underset{CH_3}{|}}{\underset{C=O}{|}}}{\underset{O}{|}}}{\overset{\overset{H}{|}}{C}} - \underset{\underset{H}{|}}{\overset{\overset{H}{|}}{C}}.e \quad \xrightarrow[\text{Units)}]{+(n\ \text{monomer}} \quad E \overset{}{(}\underset{\underset{\underset{\underset{CH_3}{|}}{\underset{C=O}{|}}}{\underset{O}{|}}}{\overset{\overset{H}{|}}{C}} - \underset{\underset{H}{|}}{\overset{\overset{H}{|}}{C}})_n \; \underset{\underset{\underset{\underset{CH_3}{|}}{\underset{C=O}{|}}}{\underset{O}{|}}}{\overset{\overset{H}{|}}{C}} - \underset{\underset{H}{|}}{\overset{\overset{H}{|}}{C}}.e \quad \text{(Strong center)}
$$

(Natural route)

$$\underline{\textbf{NOT FAVORED}} \qquad\qquad\qquad\qquad 2.18$$

Since free-radically, the monomer is a nucleophile, its natural route is electro-free-radical route which as it seems looks favored as shown above. It was shown above because the route is supposed to be natural to it. But as will shortly be shown the monomer is very unique and ***the route is not indeed favored.*** The monomer favoring the nucleo-free-radical and not the electro-free-radical routes is a clear indication that the monomer is a "very weak" nucleophile.

The monomer can only be readily polymerized using nucleo-free-radical initiators since no transfer species exists on the carbon center being attacked and this is in line with current practice in manufacturing industries. 2-azo-bis-isobutyronitrile (AIBN) is the commonly used initiator for this monomer and the acrylates. When the growing polymer chain is subjected to transfer of transfer species, the following reactions occur:-

$$
\begin{array}{c}
\text{H} \quad \text{H} \qquad\qquad \text{H} \quad \text{H} \\
| \quad\ | \qquad\qquad\ \ | \quad\ | \\
\text{N} - (\text{C} - \text{C})_m - \text{C} - \text{C.n} \\
| \quad\ | \qquad\qquad\ \ | \quad\ | \\
\text{H} \quad \text{O} \qquad\qquad \text{H} \quad \text{O} \\
\qquad\quad | \qquad\qquad\qquad\quad | \\
\qquad \text{C}=\text{O} \qquad\qquad\quad \text{C}=\text{O} \\
\qquad\quad | \qquad\qquad\qquad\quad | \\
\qquad \text{CH}_3 \qquad\qquad\quad\ \text{CH}_3 \\
\qquad\qquad\qquad\quad \text{(weak center)}
\end{array}
\longrightarrow
\begin{array}{c}
\text{H} \quad \text{H} \qquad\qquad \text{H} \quad \text{H} \\
| \quad\ | \qquad\qquad\ \ | \quad\ | \\
\text{N} - (\text{C} - \text{C})_m - \text{C} = \text{C} \\
| \quad\ | \qquad\qquad\ \ | \\
\text{H} \quad \text{O} \qquad\qquad \text{O} \\
\qquad\quad | \qquad\qquad\quad | \\
\qquad \text{C}=\text{O} \qquad\quad \text{C}=\text{O} \\
\qquad\quad | \qquad\qquad\quad | \\
\qquad \text{CH}_3 \qquad\qquad \text{CH}_3
\end{array}
+\ \text{H}^{\cdot\text{n}}
\qquad 2.19
$$

$$
\text{E .e} \quad + \quad
\begin{array}{c}
\text{H} \quad \text{H} \\
| \quad\ | \\
\text{C} = \text{C} \\
| \quad\ | \\
\text{O} \quad \text{H} \\
| \\
\text{C}=\text{O} \\
| \\
\text{CH}_3
\end{array}
\longrightarrow
\begin{array}{c}
\text{O} \\
\| \\
\text{H}_3\text{C} - \text{C}\bullet\text{e}
\end{array}
\quad + \quad
\begin{array}{c}
\text{H} \quad \text{H} \\
| \quad\ | \\
\text{C} = \text{C} \\
| \quad\ | \\
\text{O} \quad \text{H} \\
| \\
\text{E}
\end{array}
\qquad 2.20a
$$

(Natural route)   <u>**FAVORED**</u>

$$
\begin{array}{c}
\text{H} \quad \text{H} \qquad\qquad \text{H} \quad \text{H} \\
| \quad\ | \qquad\qquad\ \ | \quad\ | \\
\text{E} - (\text{C} - \text{C})_m - \text{C} - \text{C.e} \\
| \quad\ | \qquad\qquad\ \ | \quad\ | \\
\text{O} \quad \text{H} \qquad\qquad \text{O} \quad \text{H} \\
| \qquad\qquad\qquad\quad | \\
\text{C}=\text{O} \qquad\qquad \text{C}=\text{O (Strong center)} \\
| \qquad\qquad\qquad\quad | \\
\text{CH}_3 \qquad\qquad\quad \text{CH}_3
\end{array}
\longrightarrow
\begin{array}{c}
\text{H} \quad \text{H} \qquad\qquad \text{H} \\
| \quad\ | \qquad\qquad\ \ | \\
\text{E} - (\text{C} - \text{C})_m - \text{C} = \text{C} \\
| \quad\ | \qquad\qquad\ \ | \quad\ | \\
\text{O} \quad \text{H} \qquad\qquad \text{O} \quad \text{H} \\
| \qquad\qquad\qquad\quad | \\
\text{C}=\text{O} \qquad\qquad \text{C}=\text{O} \\
| \qquad\qquad\qquad\quad | \\
\text{CH}_3 \qquad\qquad\quad \text{CH}_3
\end{array}
+\ \text{H.e}
$$

(Natural route)   <u>***NOT POSSIBLE***</u>   2.20b

For the two growing polymer chains, since there are no free-radical transfer species on the terminal α-carbon atom, but on the β-carbon atom, this is the one that can be rejected or released nucleo-free-radically (See Equation 2.1) in order to kill the growing chain from within. This is the route not natural to the monomer. When the dead terminal double bonded polymer from Equation 2.19 is activated, it is that carbon center where the transfer species was rejected that will carry the electro-free-radical. Only transfer species of the second kind exists here via C = C π-bond activation (the only one which can be activated), because it is less nucleophile than the C = O center. The same will also apply to vinyl chloride. For ethene, the release is at very high operating condition because of the symmetric character of ethene compared to vinyl chloride and vinyl acetate, noting that while H is weakly radical-pushing, $OCOCH_3$ and Cl groups are strongly radical-pulling. Nevertheless, notice the types of dead terminal double bond polymers produced in the first equation above. It bears the same character as the original monomer. For the second equation, notice that the monomer itself became the transfer species, just like the OH group in acetic acid H_3CCOOH. This is only possible when the monomer is in Stable state of existence. However, electro-free-radically, the monomer cannot be polymerized as against what is shown by the last equation above. This is unique, because, for the first time as shown by the second equation we are encountering a different kind of transfer species which is not of the first kind but of the ***second first kind with respect to the C = O activation center*** which can never be activated radically, being more nucleophilic than the C=C center. Chargedly as already said, the monomer unlike ethene cannot be polymerized through the C = C π-bond, since it cannot be activated due to electrostatic forces of repulsion between the negative charge on the C center and the adjacently located paired unbonded radicals on the O center. Polymerization through the C = O π-bond is not favored because first and foremost as will shortly be shown and already said, **the C = C center is less nucleophilic than the C = O center**, and **any initiator will first activate the weaker center.** However, since the C = C center cannot be activated chargedly, then the C = O center should be activated chargedly. With the use of negatively charged initiators, due to presence of transfer species of the first kind of the first type, no initiation is favored via the C = O center. *With positively charged initiator, no initiation is also favored via the C = O center,* R *because of the same transfer species above* $(-O - CH=CH_2)$ *transfer species of second first kind.*

$$\ddot{R_x}^{\ominus} \quad + \quad \begin{array}{c} CH_2 \\ \| \\ CH \\ | \\ O \\ | \\ \oplus C - O^{\ominus} \\ | \\ CH_3 \end{array} \quad \longrightarrow \quad {}^{\ominus}O - \underset{\underset{CH_3}{|}}{C} = O \quad + \quad RCH=CH_2$$

$$2.21$$

(Transfer from monomer to initiator)-1st kind of 1st type

Note however that, it is not the center activated above that is activated with $R^{\ominus}$. It is the C=C center. that is actually activated being less nucleophilic than the C=O center. It has been used to illustrate some fundamental principles. Notice the transfer species abstracted above and the group supplying it. The group supplying it ($OCH=CH_2$) is more radical/chargedly-pushing than CH_3 group. Indeed, this is taking from the rich and not from the poor, for it was not CH_3 group supplying H•e, but $OCH=CH_2$ supplying e•$CH=CH_2$ whose capacity is greater than that of H•e. Notice that the initiator involved above is of the non-free type ($H_3CO^{\ominus}$.......$^{\oplus}$Na or just $H_3CO^{\ominus}$) and not the free type ($H_9C_4^{\ominus}$.......$^{\oplus}$Li). If the initiator had been the latter, then the CH_3 group will now provide $H^{\oplus}$ since the equation will now be chargedly balanced, i.e., taking from the poor and not from the rich, to balance the equation via Combination mechanism. Via Equilibrium mechanism, it is taking from the rich and not the poor, since same transfer species (e•$CH=CH_2$) will be involved. (See Equation 2.22 below). At the end however, no initiation is favored. The transfer species can either be removed via Combination or Equilibrium mechanism as along it exists. But for the case above it is via Combination mechanism when the two different types of initiators above are used. The situation where the Poor get poorer in the presence of the Rich does not exist. NATURE abhors a vacuum and will also not allow taking from the Poor to give to the Rich under Equilibrium conditions. Instead the order of flow is Richest to the Richer and to the Rich and to the Poor and to the Poorer and to the Poorest, just like temperature gradient, concentration gradient and whatever gradient. No matter the operating conditions, the reverse will never take place. It is only in the Physical world that it seems to take place. That is, Chemistry is indeed the study of the laws of Nature.

$$H_9C_4^{\ominus}.......^{\oplus}Li \quad + \quad \begin{array}{c} OCH=CH_2 \\ | \\ \oplus C - O^{\ominus} \\ | \\ CH_3 \end{array} \quad \rightleftharpoons \quad H_9C_4-CH=CH_2 \quad + \quad H_3C - \overset{\overset{O}{\|}}{C} - O^{\ominus}.......^{\oplus}Li$$

(FAVORED)

$$2.22$$

$$H_9C_4^{\ominus}.......^{\oplus}Li \quad + \quad \begin{array}{c} OCH=CH_2 \\ | \\ \oplus C - O^{\ominus} \\ | \\ CF_3 \end{array} \quad \longrightarrow \quad \text{Same as above.}$$

$$2.23$$

Based on Equations 2.21 and 2.22, initiation is not favored. Imagine if instead of the negatively charged initiator above, it was a nucleo-free-radical; both equations will remain the same. In the last equation, chargedly, F cannot be abstracted as an anion by a negatively active charged center. Using the "eye of the needle", one can see the significance of Law #36 in Chapter 1, the third law- a philosophical law, because *to be a Philosopher, you must be a Natural Scientist and nothing else.* Indeed, one can see that not all Poor beings are Poor beings and not all Rich beings are Rich beings. Not all Electrophiles are Electrophiles. So also, not all THIEVES are THIEVES. Not all Criminals are Criminals. This is why for example "Thou shall not steal" is not a Natural Law as it is already inherently embedded in the first law of Nature, just like the TEN Commandments. These, one can see by the ways chemical reactions take place.

Positively with the C = O center, the following is obtained.

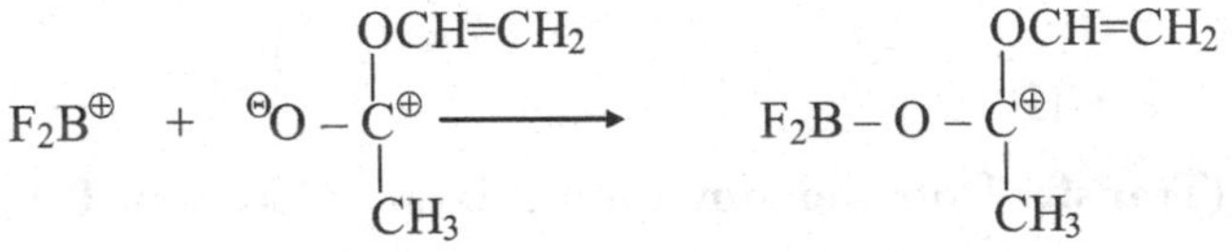

$$F_2B^{\oplus} \quad + \quad {}^{\ominus}O - \overset{\overset{\displaystyle OCH=CH_2}{|}}{\underset{\underset{\displaystyle CH_3}{|}}{C^{\oplus}}} \longrightarrow \quad F_2B - O - \overset{\overset{\displaystyle OCH=CH_2}{|}}{\underset{\underset{\displaystyle CH_3}{|}}{C^{\oplus}}}$$

$$\text{2.24}$$

In the absence of transfer species on the O center, the free-positively charged-paired center is favored when the C=O center is activated.

Note that whether it is $H^{\oplus}$ or $H_3C^{\oplus}$, the positive charge is free except that the latter cannot be isolated. The former is a cation but not the latter. $\underset{3\bullet en}{Fe^{3\oplus}_{\overset{2\bullet\bullet}{}}}$ or $Cl_2\underset{\bullet en}{Fe^{\oplus}_{\overset{2\bullet\bullet}{}}}$ is non-free positively charged center which cannot be isolated. It is non-free, because of the presence of paired unbonded radicals in its boundary which does not exist with $Li^{\oplus}$ or with $H_3C^{\oplus}$. While with $Li^{\oplus}$ the last shell is empty, with $H_3C^{\oplus}$, the last shell is not empty. One can see three different kinds of positive charges here. In the last equation above, the positive charge is not a cation. With a cation, the followings are obtained.

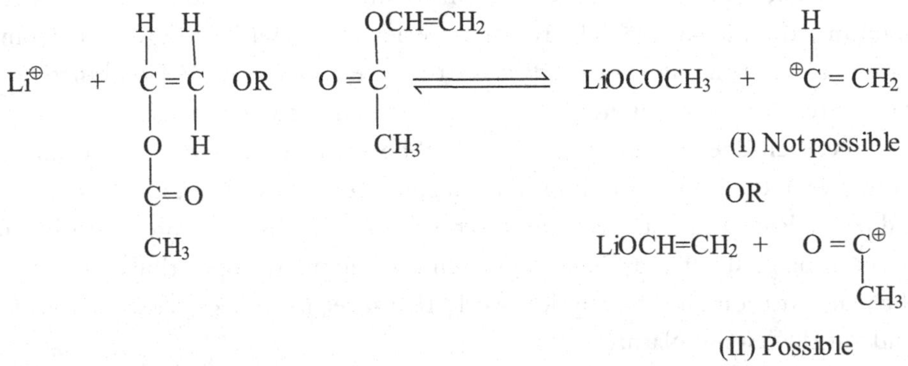

$$Li^{\oplus} \quad + \quad \begin{matrix} H & H \\ | & | \\ C & = & C \\ | & | \\ O & H \\ | \\ C=O \\ | \\ CH_3 \end{matrix} \quad OR \quad \begin{matrix} OCH=CH_2 \\ | \\ O = C \\ | \\ CH_3 \end{matrix} \rightleftharpoons \quad LiOCOCH_3 \quad + \quad {}^{\oplus}\overset{\overset{\displaystyle H}{|}}{C}=CH_2$$

(I) Not possible

OR

$$LiOCH=CH_2 \quad + \quad O=\underset{\underset{\displaystyle CH_3}{|}}{C^{\oplus}}$$

(II) Possible

(Transfer Species of Second first kind of the first type)   2.25

While transfer species of **the first kind of the first type** is involved in the reactions of Equations 2.21 and 2.22, when the monomers are in Activated State of Existence taking place under Com-bination mechanisms negatively, in the reactions above positively, a different kind of transfer species is involved when the monomer is in Stable State of Existence taking place under Equilibrium mechanism as shown above. The same applies when in Activated state of existence under Combination mechanism. Between (I) and (II) above, we shall shortly know which the favored one is. Nevertheless the transfer species involved in (II) under Equilibrium conditions is of **the second first kind of the first type, the same as that shown**

Sunny N.E. Omorodion

in Equation 2.20a. The one shown in (I) does not exist. It has already been shown that vinyl acetate cannot be polymerized chargedly via C = C centers, since it cannot be activated. Nevertheless based on Equation 2.24, since positively free charged-paired polymerization is favored for the C = O center, the following is obtained for its growing polymer chain.

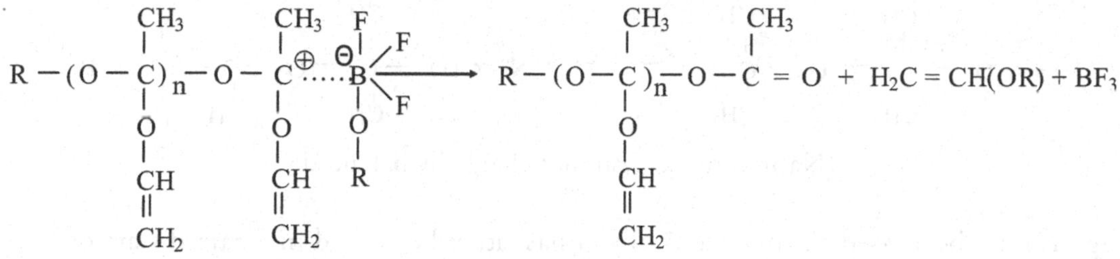

$$R-(O-C)_n-O-C^{\oplus}\cdots B^{\ominus} \longrightarrow R-(O-C)_n-O-C=O \ + \ H_2C=CH(OR) \ + \ BF_3 \qquad 2.26$$

The transfer species removed to kill the chain from within is the same involved in Equation 2.21, that is, transfer species of the first kind of the first type wherein the negatively charged initiator involved is of the non-free type. In order words, it is (II) of Equation 2.25 that is favored. How-ever, based on what is shown below, (I) of Equation 2.25 cannot be favored either via Equili-brium or Combination mechanism. From the first equation below, the (I) obtained therein can do something else. Depending on the operating conditions, allene can be produced and this can rearrange to give methyl acetylene as will be shown downstream.

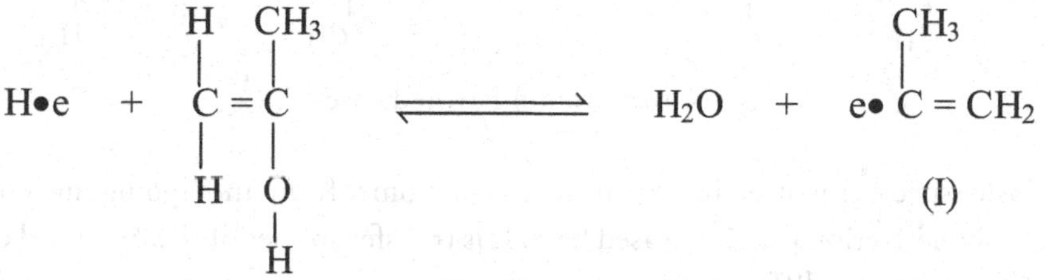

$$H\bullet e \ + \ \begin{matrix} H \ CH_3 \\ C=C \\ H \ O \\ H \end{matrix} \ \rightleftharpoons \ H_2O \ + \ e\bullet \begin{matrix} CH_3 \\ C=CH_2 \end{matrix} \qquad (I) \qquad 2.27$$

(Female)   **Transfer species of Second first kind of different type**

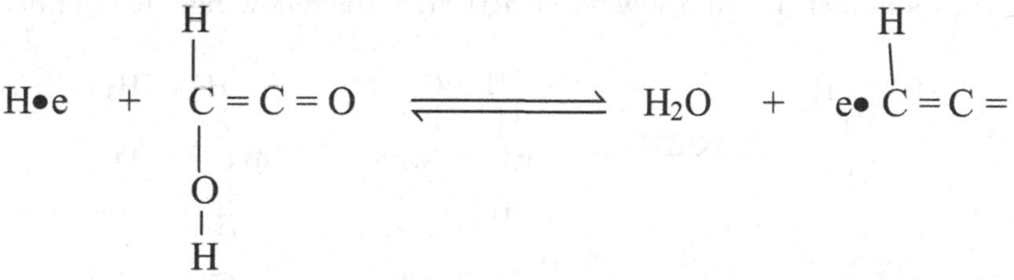

$$H\bullet e \ + \ \begin{matrix} H \\ C=C=O \\ O \\ H \end{matrix} \ \rightleftharpoons \ H_2O \ + \ e\bullet \begin{matrix} H \\ C=C= \end{matrix} \qquad 2.28$$

(Male)   **Transfer species of Second first kind of another type**

One can observe how Complex Nature can be, that which has been built in such a way that no law is contravened with an exception. The great distinction between ionic and non-ionic metals has partly been shown above. When a component is abstracted, it is never the one that is held when the compound is made to exist in Equilibrium state of existence, but something else. For the cases above H is held when in Equilibrium state of existence, The female shown above in the last but one equation can molecularly rearrange to give **acetone** under a particular operating condition and rearrange back to its former self under another operating condition. In fact, the two monomers shown above may be placed in Equilibrium state of existence both being strongly Polar/Ionic, such that no abstraction can take place with no

64

productive reaction. To the non-Scientist, these are mysteries, and to the Present-day Scientist, the same also applies. But as shown in the NEW FRONTIERS, they are no mysteries, but Nature.

For acetone, it has been said and shown that it favors chargedly the "cationic" route. Radically, it can be activated. For its growing polymer chain cationically, the following is obtained:-

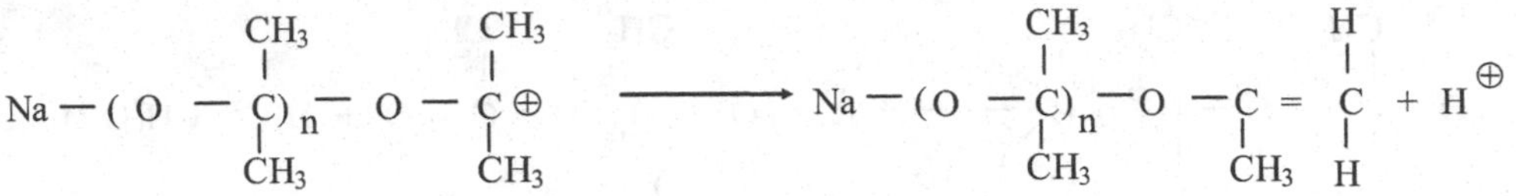

$$Na-(O-\underset{\underset{CH_3}{|}}{\overset{\overset{CH_3}{|}}{C}})_n-O-\underset{\underset{CH_3}{|}}{\overset{\overset{CH_3}{|}}{C}}\oplus \longrightarrow Na-(O-\underset{\underset{CH_3}{|}}{\overset{\overset{CH_3}{|}}{C}})_n-O-\underset{\underset{CH_3}{|}}{\overset{\overset{CH_3}{|}}{C}}=\underset{\underset{H}{|}}{\overset{\overset{H}{|}}{C}}+H^{\oplus} \qquad 2.29$$

(Natural route) ; But not chargedly balanced

Although Na has been used above cationically, it has either been used in a paired state or in a state where its anion has been isolated or just as it exist on its own, that is, as an electro-free-radical. As a cation, the reaction above is not possible. Since there exists transfer species on the terminal α-carbon atom, formation of dead terminal double bond polymers is highly favored in the presence of the driving forces and when positively charged-paired initiator is used as opposed to the Na used above. This is the same transfer species that prevented the monomer from anionic initiation. Electro-free-radically, the following is obtained.

$$Na(-O-\underset{\underset{CH_3}{|}}{\overset{\overset{CH_3}{|}}{C}})_n-O-\underset{\underset{CH_3}{|}}{\overset{\overset{CH_3}{|}}{C}}.e \longrightarrow Na(-O-\underset{\underset{CH_3}{|}}{\overset{\overset{CH_3}{|}}{C}})_n-O-\underset{}{\overset{\overset{CH_3}{|}}{C}}=\underset{\underset{H}{|}}{\overset{\overset{CH_3\ H}{|\ |}}{C}}+H^{.e}$$

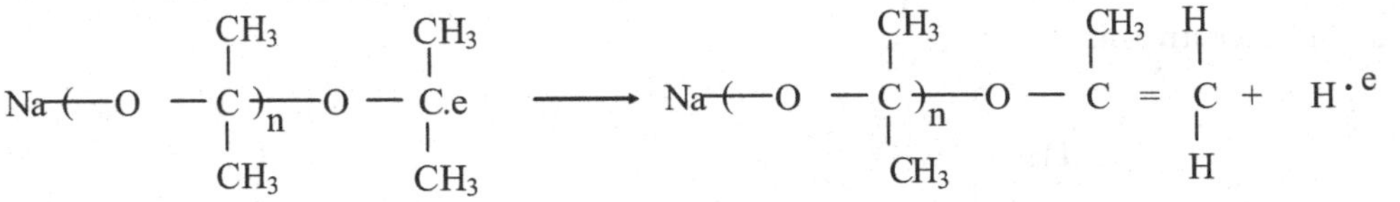

$$2.30$$

(Natural route) Favored case.

It is the same transfer species involved in preventing the monomer from undergoing nucleo-free or nucleo-non-free-radical polymerization that is released here. It is transfer species of the first kind of the first type. It can be used differently via a different rearrangement mechanism to go back to its former self. This will be shown in this Volume downstream.

Considering the monomer shown below, when activated, the followings are obtained.

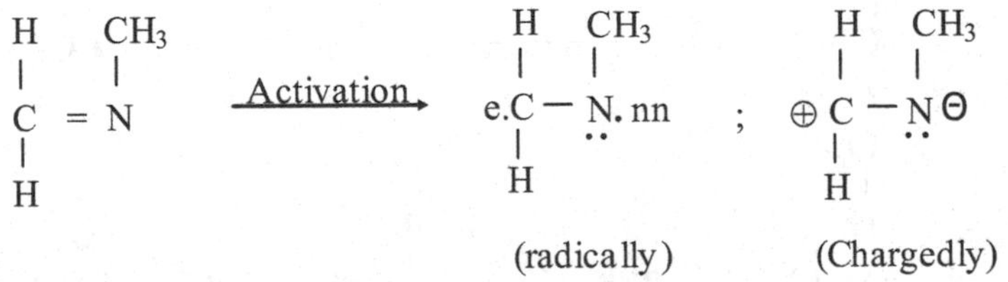

$$\underset{\underset{H}{|}}{\overset{\overset{H}{|}}{C}}=\overset{\overset{CH_3}{|}}{N} \quad \xrightarrow{\text{Activation}} \quad e.\underset{\underset{H}{|}}{\overset{\overset{H}{|}}{C}}-\underset{\overset{..}{}}{N}.nn \quad ; \quad \oplus\underset{\underset{H}{|}}{\overset{\overset{H}{|}}{C}}-\underset{\overset{..}{}}{N}\ominus \qquad 2.31$$

(radically)   (Chargedly)

Molecular rearrangement cannot take place here. Radically the followings are to be expected.

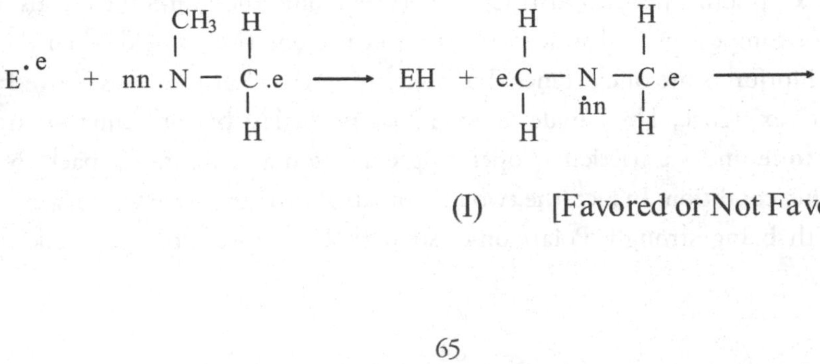

$$E^{.e} + nn.\underset{\underset{H}{|}}{\overset{\overset{CH_3\ H}{|\ |}}{N}-C}.e \longrightarrow EH + e.\underset{\underset{H}{|}}{\overset{\overset{H}{|}}{C}}-\underset{\hat{n}n}{N}-\underset{\underset{H}{|}}{\overset{\overset{H}{|}}{C}}.e \longrightarrow$$

(I)   [Favored or Not Favored]

$$\text{OR} \quad \text{E—N—C.e} \quad \begin{matrix} CH_3 & H \\ | & | \\ & \\ & H \end{matrix}$$

2.32

(II)   [Favored]

For the fact that H cannot be removed as a hydride, (I) is not favored, unless E is an ionic metal or some non-ionic metals, which are more electropositive than H.

$$N \cdot^n + \begin{matrix} H & CH_3 \\ | & | \\ e.C & — N.nn \\ | \\ H \end{matrix} \longrightarrow \begin{matrix} H & CH_3 \\ | & | \\ N — C — N.nn \\ | \\ H \end{matrix} \longrightarrow \text{No reaction}$$

(Free)   (Non-free)

(Not Radically balanced)

2.33

$$\ddot{N}\cdot nn + \begin{matrix} H & CH_3 \\ | & | \\ e.C — N.nn \\ | \\ H \end{matrix} \longrightarrow \begin{matrix} H & CH_3 \\ | & | \\ \ddot{N} — C — N. nn \\ | \\ H \end{matrix} \xrightarrow[\text{Units)}]{+(n\ monomer}$$

$$\ddot{N}\!\!-\!\!\left(\!\begin{matrix} H & CH_3 \\ | & | \\ C — N \\ | \\ H \end{matrix}\!\right)_{\!n}\!\!\begin{matrix} H & CH_3 \\ | & | \\ C — N. nn \\ | \\ H \end{matrix} \xrightarrow[\text{Growing chain}]{\text{Transfer from}} \ddot{N}\!\!-\!\!\left(\!\begin{matrix} H & CH_3 \\ | & | \\ C — N \\ | \\ H \end{matrix}\!\right)_{\!n}\!\!\!\begin{matrix} CH_3 \\ | \\ — C = N \\ | \\ H \end{matrix} + \quad H\bullet n$$

2.34

(Favored)

$$\text{(benzene ring)}\begin{matrix} O \\ \| \\ C — O \bullet nn \end{matrix} \longrightarrow CO_2 + \text{(benzene ring)} \bullet n$$

2.35a

$$\begin{matrix} H \\ | \\ H — C — O \bullet nn \\ | \\ H \end{matrix} \longrightarrow H\bullet n + H_2C = O$$

2.35b

In the same manner in which nucleo-free-radicals are obtained from nucleo-non-free-radicals by decomposition, so also is the $H\cdot^n$ released from a nucleo-non-free-radical growing polymer chain only as long as a dead polymer or component is produced. Hence the equations above are balanced. The nucleo-non-free-radical route is not natural to the monomer. Hence transfer species species of the second kind of the first type is rejected to kill the chain from within. In addition, since CH_3 can be used electro-free-radically (See (I) of Equation 2.32) for some $E^{\cdot e}$ initiators, hence transfer species of the second kind of the second type can be involved when initiation is favored in the route natural to the monomer. This however is not one of the essences of the law of Conservation of transfer species, since there is no transfer species of the first kind of the first type for the case above (A Formaldimine).

Chargedly also, the monomer is either a nucleophile or an electrophile, since CH_3 group cannot provide transfer species. When the CH_3 group is replaced with H, the followings are obtained only radically. Chargedly, electrostatic forces of repulsion will not allow it if the monomer is kept in Equilibrium state of existence.

$$R^{\cdot n} \;+\; \underset{\underset{H}{|}}{\overset{\overset{H\;\;H}{|\;\;\;|}}{C}}=N \;\;\rightleftharpoons\;\; H_2C=N^{\cdot nn} \;+\; H^{\cdot e} \;+\; R^{\cdot n}$$

(Favored)

$$\longrightarrow \;\; RH \;+\; \underset{\underset{H}{|}}{\overset{\overset{H}{|}}{C}}=N^{\cdot nn} \qquad\qquad 2.36a$$

(I) (Transfer from monomer step without activation)

The transfer species above is not of any kind or any type, since the H is the species held in Equilibrium state of existence. It is the one going to add or abstract a transfer species. In order to remove the problem of instability above, in the absence of a suppressing agent (called a passive catalyst), the H must be replaced with R.

$$R^{\oplus} \;+\; {}^{\ominus}\underset{\underset{H}{|}}{\overset{\overset{R\;\;H}{|\;\;\;|}}{N}}-C^{\oplus} \;\longrightarrow\; R-\underset{\underset{H}{|}}{\overset{\overset{R\;\;H}{|\;\;\;|}}{N}}-C^{\oplus} \;\xrightarrow{\;\;+\;\;n\,(I)\;\;}$$

(I)

$$R\left[\underset{\underset{H}{|}}{\overset{\overset{R}{|}}{N}}-\underset{\underset{H}{|}}{\overset{\overset{H}{|}}{C}}\right]_{n-1}\underset{\underset{H}{|}}{\overset{\overset{R}{|}}{N}}-\underset{\underset{H}{|}}{\overset{\overset{H}{|}}{C}}{}^{\oplus} \;\longrightarrow\; R\left[\underset{\underset{H}{|}}{\overset{\overset{R}{|}}{N}}-\underset{\underset{H}{|}}{\overset{\overset{H}{|}}{C}}\right]_{n-1}\overset{\overset{R}{|}}{N}=\underset{\underset{H}{|}}{\overset{\overset{H}{|}}{C}} \;+\; R^{\oplus} \qquad 2.36b$$

The transfer species released above is of the *second kind of the second type*, noting that the initiator above is paired, since it cannot be isolatedly placed. Unlike aziridines as will be shown, the growing polymer chain above will not favor branch formations along the chain, since the R group unlike hydrogen atom on nitrogen is not chargedly or radically loosely bonded to it. It is important to note that, positively charged polymerization is not favored until the H on N center has been replaced. Though the route is natural to the monomer, unlike the case of vinyl acetate, transfer species of the second kind is rejected in the route natural to the monomer. This is like the case of Equation 2.6b for an Electrophile (Fluorinated alkene). Anionically, the monomer above with the H replaced can also be polymerized. With negative charges, it cannot be polymerized. The H in Equation 2.36a is not a transfer species, because the monomer was not in a Stable state, *but in Equilibrium State of Existence*. Aldimines and ketimines are stronger nucleophiles than the monomer above- a formaldimine. Free-radically, no polymerization is favored until the H atom chargedly or radically loosely bonded to the nitrogen center, has been replaced. If one H on the C center is replaced by CH_3 and larger groups (Aldimines), the monomer becomes more nucleophilic and never favor nucleo-non-free-radical and negatively charged routes.

Radically, transfer species of the second first kind of the first or different type (i.e. the case where the monomer is stable) as already shown in Equations 2.27 and 2.28 exist. Consider the radical polymerization of the fluorinated aldehyde shown below.

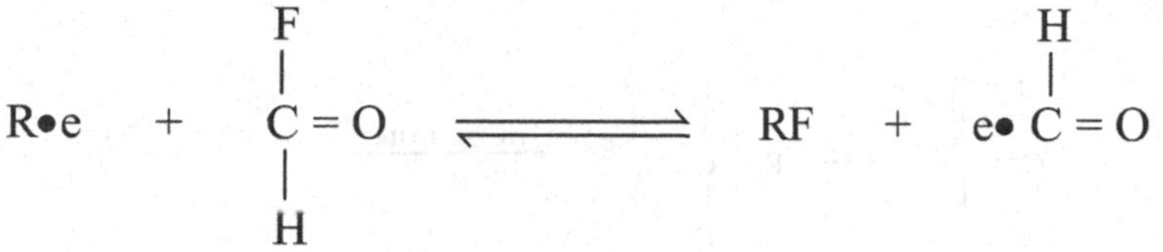

$$
R{\bullet}e \quad + \quad
\underset{\underset{H}{|}}{\overset{\overset{F}{|}}{C}} = O
\quad \rightleftharpoons \quad
RF \quad + \quad
e{\bullet}\;\underset{}{\overset{\overset{H}{|}}{C}} = O
$$

2.37a

$$
R{\bullet}e \quad + \quad n\,[\,nn{\bullet}O - \underset{\underset{H}{|}}{\overset{\overset{F}{|}}{C}}\,{\bullet}e\,]
\quad \longrightarrow \quad
R - (O - \underset{\underset{H}{|}}{\overset{\overset{F}{|}}{C}}\,)_{n-1} - O - \underset{\underset{H}{|}}{\overset{\overset{F}{|}}{C}}{\bullet}e
$$

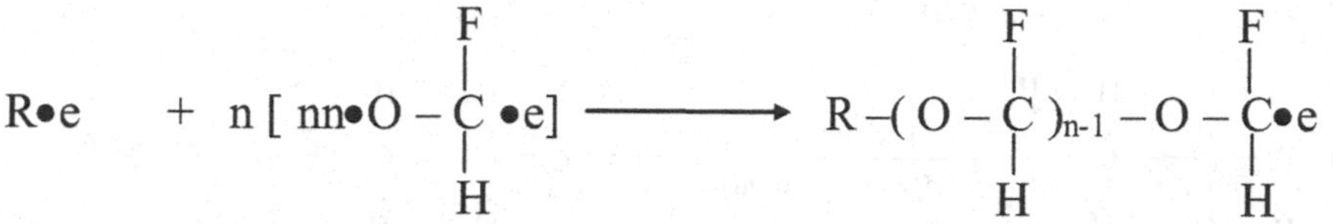

2.37b

Note that in the last two equations above, the monomer was not in Equilibrium State of Existence, for if it was, the situation would have been completely different. The transfer species for the first equation above is of **the second first kind of second type**. For the case above where F as opposed to Cl and Br is carried, it can rarely exist in Equilibrium state of existence. Different operating conditions would be required however for the two reactions above..

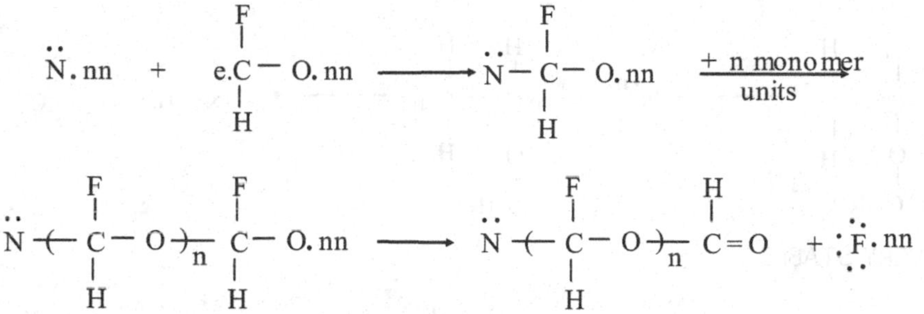

$$
\overset{..}{N}.nn \quad + \quad
e.\underset{\underset{H}{|}}{\overset{\overset{F}{|}}{C}} - O.nn
\quad \longrightarrow \quad
\overset{..}{N} - \underset{\underset{H}{|}}{\overset{\overset{F}{|}}{C}} - O.nn
\quad \xrightarrow{\;+\,n\ \text{monomer units}\;}
$$

$$
\overset{..}{N} \!\!\left(\!\! \underset{\underset{H}{|}}{\overset{\overset{F}{|}}{C}} - O \!\!\right)_{\!n} \underset{\underset{H}{|}}{\overset{\overset{F}{|}}{C}} - O.nn
\quad \longrightarrow \quad
\overset{..}{N} \!\!\left(\!\! \underset{\underset{H}{|}}{\overset{\overset{F}{|}}{C}} - O \!\!\right)_{\!n} \underset{}{\overset{\overset{H}{|}}{C}} = O
\quad + \quad :\!\overset{..}{\underset{..}{F}}.nn
$$

2.38

The electro-free-radical route is the natural route. In Equation 2.37a, F was only removed under Stable State of Existence via Equilibrium mechanism. If activated as shown in the next equation, for its growing polymer chain electro-free-radically, there is no transfer species to reject. Hence nucleo-non-free-radical polymerization is favored with transfer species of the **second kind** released from its growing polymer chain. However, both route being favored clearly indicates that the monomer is a weak Nucleophile. It will remain so even when the H is changed to CF_3. But when H is changed to CH_3, the monomer becomes a complete Nucleophile as already shown.

While transfer species of the first kind of the first type are located on a substituent group carried by an ideal center such as the α-carbon center in the sense that the active center carrying it, is the center of attack in the route not natural to the monomer, transfer species of the second first kind of first and second types are located directly on the center not supposed to be attacked by the initiator. Hence in general, when transfer species of the second first kind exist, the monomers are not in Activated State of Existence, but are in Stable State of Existence.

Now consider the polymerization of alkyl vinyl ether like the type shown in Equation 2.27.

$$R^{\ominus} + \underset{\substack{| \\ O \\ | \\ C_2H_5}}{\overset{\substack{H \quad H \\ | \quad | }}{C = C}} \longrightarrow R^{\ominus} + \oplus\underset{\substack{| \\ O \\ | \\ C_2H_5}}{\overset{\substack{H \quad H \\ | \quad | }}{C - C}}\ominus \longrightarrow RC_2H_5 + x\ddot{O}\overset{\ominus}{-}\underset{\substack{| \\ H}}{\overset{\substack{H \quad H \\ | \quad | }}{C = C}}$$

(Non-free anion)

2.39

$$R^{\oplus} + \overset{\ominus}{\underset{\substack{| \\ O \\ | \\ C_2H_5}}{\overset{\substack{H \quad H \\ | \quad | }}{C - C}}}{}^{\oplus} \longrightarrow R-\underset{\substack{| \\ O \\ | \\ C_2H_5}}{\overset{\substack{H \quad H \\ | \quad | }}{C - C}}{}^{\oplus} \xrightarrow{\substack{+(n\ monomer \\ units)}}$$

$$R\left(\underset{\substack{| \\ H}}{\overset{\substack{H \quad H \\ | \quad | }}{C - C}}\right)_n \underset{\substack{| \\ O \\ | \\ C_2H_5}}{\overset{\substack{H \quad H \\ | \quad | }}{C - C}}{}^{\oplus} \xrightarrow[\substack{\text{Growing chain}}]{\text{(Transfer from}} R\left(\underset{\substack{| \\ H}}{\overset{\substack{H \quad H \\ | \quad | }}{C - C}}\right)_n \underset{\substack{| \\ H}}{\overset{\substack{H \quad H \\ | \quad | }}{C - C}}=O \ +^{\oplus}C_2H_5$$

2.40

Full polymerization is favored only when charged-paired initiators are involved as is already becoming obvious, because $R^{\oplus}$ cannot be isolatedly placed. Nevertheless, the law of conser-vation of transfer of transfer species can be seen to also apply here with the transfer species being of the first kind of the first type. Free-radically, the followings are obtained for the monomer.

$$N^{\bullet n} + \underset{\substack{| \\ O \\ | \\ C_2H_5}}{\overset{\substack{H \quad H \\ | \quad | }}{C = C}} \longrightarrow N^{\bullet n} + e.\underset{\substack{| \\ O \\ | \\ C_2H_5}}{\overset{\substack{H \quad H \\ | \quad | }}{C - C}}.n \longrightarrow NC_2H_5 + \underset{\substack{| \\ H}}{\overset{\substack{H \quad H \\ | \quad | }}{O = C - C}}.n$$

VERY STABLE

2.41

$$E.e + n.\underset{\substack{| \\ O \\ | \\ C_2H_5}}{\overset{\substack{H \quad H \\ | \quad | }}{C - C}}.e \longrightarrow E-\underset{\substack{| \\ O \\ | \\ C_2H_5}}{\overset{\substack{H \quad H \\ | \quad | }}{C - C}}.e \xrightarrow{\substack{+(n\ monomer \\ Units)}}$$

$$E\left(\underset{\substack{| \\ O \\ | \\ C_2H_5}}{\overset{\substack{H \quad H \\ | \quad | }}{C - C}}\right)_n \underset{\substack{| \\ O \\ | \\ C_2H_5}}{\overset{\substack{H \quad H \\ | \quad | }}{C - C}}.e \xrightarrow[\substack{\text{Growing chain}}]{\text{Transfer from}} E\left(\underset{\substack{| \\ O \\ | \\ C_2H_5}}{\overset{\substack{H \quad H \\ | \quad | }}{C - C}}\right)_n \underset{\substack{| \\ H}}{\overset{\substack{H \quad H \\ | \quad | }}{C - C}}=O + e.C_2H_5$$

2.42

Electro-free-radically, full polymerization is favored. The same transfer species is involved chargedly and free-radically, but with a difference. Radically, it is free, but chargedly it is paired. However, as is already clear, while here paired media initiator may not be required since the same type of radical is released, in many cases there must be a receiving center for the free-positively charged species released for the system to exist (use of charged-paired initiators).

With mono-olefin carrying $N(CH_3)_2$ substituent group, the followings are obtained chargedly, if activation is favored, noting that this type of monomer which exists is not commonly KNOWN in present-day Science. Since the existence of $H_2C = CH(NH_2)$ which is very unstable is UNKNOWN that which quickly molecularly rearranges to give $CH_3CH = NH$, then how can its stable forms be known *in a world of illusions, a world that is color blind, a world that cannot see the real and imaginary, a world of IGNORANCE, a world where some "Great Philosophers" will say that Boundaries are meaningless when the thing that is meaningless is the statement itself, a world........?* How can boundaries be meaningless when everything in life has a domain and boundary? (See Rule #2). Every country has *her* own domain and boundary, otherwise *she* will not exist as a country.

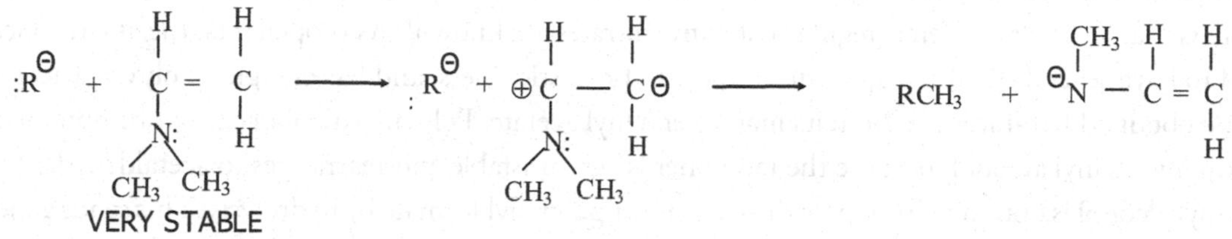

$$\text{2.43}$$

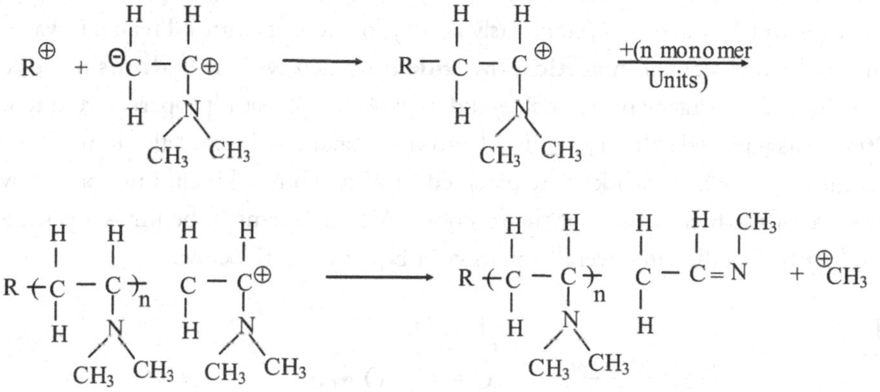

$$\text{2.44}$$

Full polymerization is favored with only free-charged-paired initiators. Thus, this is the only condition in which CH_3 can be abstracted from $-N(CH_3)_2$ type of group as a charged species, not as a cation. On the other hand, it is no surprise why $N(CH_3)_3$ for example are used as solvating agents to shield free cationic metallic centers that have vacant orbitals through dative bonds.

The transfer species that exist "anionically" are also limited, but far more than the "cationic" ones. Some important ones include $F^{\ominus}$, $CH_3O^{\ominus}$, $H_2N^{\ominus}$, $RHN^{\ominus}$, $^{\ominus}NR_2$, $^{\ominus}O-NO_2$. The last one is not popular but seems to be known. It can be found with monomers as shown below. It is in view of the far more abundance of radical species, that most polymerization reactions are of radical and charged characters, while simple chemical reactions are of radical characters, but never ionic in character.

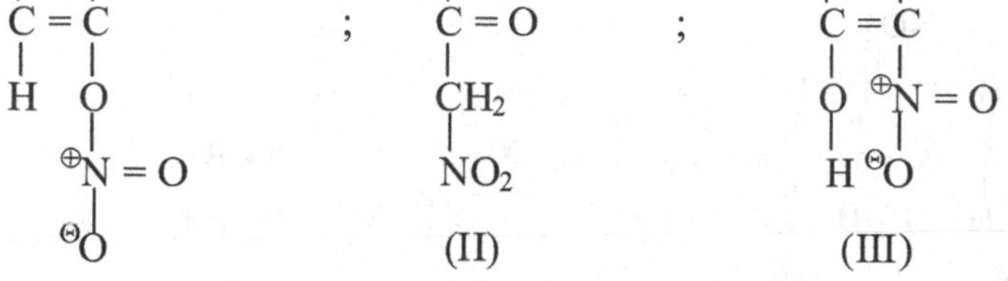

$$\text{2.45}$$

(I) Vinyl nitrate (Unstable)   (Stable)   (Unstable)

(I) above which is the real structure of vinyl nitrate cannot be activated chargedly if ONO_2 group is a non-free-charged-pulling group, *which it is.* **(I) is a nucleophile wherein the capacity of N = O is stronger than that of C = C chargedly.** C = C center cannot be activated chargedly, because, it is believed that like OH, and NH_2 groups, the ONO_2 group is also a non-free-radical-pushing group of greater capacity than H radically. Chargedly, only ONO_2 is non-free-charged-pulling amongst the three groups. Hence the monomer molecularly rearranges only radically to give (II) above, because the capacity of NO_2 is less than that of H. Radically, it can rearrange, because while N=O is less nucleophilic than C=C center, $^{\oplus}N = O$ is more nucleophilic than C = C. It is the C = C center that is first activated radically. Hence, it cannot be polymerized nucleo-free-radically. (III) which is an electrophile is an isomer which cannot be obtained by any form of rearrangement as will be seen downstream. Because of the OH group, it is unstable. Nevertheless, interesting to note is that atactic and stereo-regular polyvinyl nitrates are known[1]. As propellants, the atactic ones were found to have several disadvantages with respect to flow, stickiness, and less energies. Polyvinyl acetate can only be obtained free-radically from its monomer, vinyl acetate. Polyvinyl alcohol cannot be obtained from its monomer, vinyl alcohol, because the monomer is very unstable and rearranges to acetaldehyde. Instead, polyvinyl alcohol is obtained from polyvinyl acetate or polyvinyl formate by hydrolysis. This sends a message, that which is obvious, that is, (I) above is unstable and more so than vinyl alcohol, despite the size of NO_2 compared to H. It rearranges to (II) above instantaneously at very low temperatures. Hence, it was found that, when stereo-regular polyvinyl alcohol in **acetic anhydride** is reacted with **anhydrous** nitric acid at temperatures of -15^0C to -10^0C, a substance of surprising and unexpected physical properties resulted. The substance, polyvinyl nitrate was reported to be crystalline, dry, dimensionally stable and also found to possess powerful propellant energies for which the works were patented.[1] Unknown to all including the discoverers and Patentees, is the Science of the mechanisms of the reactions. All that seems to be known is the ART. Many Stages are involved with the following overall equation in Equation 2.47 below.

$$
\begin{array}{ccc}
\underset{\overset{|}{\underset{H}{C}}}{\overset{\overset{H}{|}}{C}} = \underset{\overset{|}{\underset{O}{C}}}{\overset{\overset{H}{|}}{C}} & \xrightleftharpoons[\text{Existence}]{\text{Equilibrium State of}} & \underset{\overset{|}{\underset{H}{C}}}{\overset{\overset{H}{|}}{C}} = \underset{}{\overset{\overset{H}{|}}{C}} - O \bullet nn \quad + \quad e \bullet NO_2
\end{array}
\qquad 2.46
$$

Notice that the equation above is the Equilibrium State of Existence of vinyl nitrate. Notice that what is held under this state of Existence is different from what is abstracted (NO_3) when in Stable State of Existence. The same applies to all including vinyl alcohol and vinyl acetate.

$$
E - \underset{\overset{|}{H}}{\overset{\overset{H}{|}}{C}} - \underset{\overset{|}{O}}{\overset{\overset{H}{|}}{C}} - \underset{\overset{|}{H}}{\overset{\overset{H}{|}}{C}} - \underset{\overset{|}{H}}{\overset{\overset{O}{|}}{C}} - (\underset{\overset{|}{H}}{\overset{\overset{H}{|}}{C}} - \underset{\overset{|}{O}}{\overset{\overset{H}{|}}{C}})_{n-2} \dots X \quad + \quad n\,H - O - NO_2 \longrightarrow
$$

$$
E - \underset{\overset{|}{H}}{\overset{\overset{H}{|}}{C}} - \underset{\overset{|}{O}}{\overset{\overset{H}{|}}{C}} - \underset{\overset{|}{H}}{\overset{\overset{H}{|}}{C}} - \underset{\overset{|}{H}}{\overset{\overset{O}{|}}{C}} - (\underset{\overset{|}{H}}{\overset{\overset{H}{|}}{C}} - \underset{\overset{|}{O}}{\overset{\overset{H}{|}}{C}})_{n-2} \dots X \quad + \quad \textbf{n H}_2\textbf{O}
\qquad 2.47
$$

(Water)

(Syndiotactic polyvinyl nitrate)

71

Just from one molecular chain, the OH groups in the polymer's Stable State of existence at such low temperatures are removed one at a time in many stages via Equilibrium mechanism until completed. The alcohol was the component that existed in Equilibrium State of Existence to initiate the removal of OH groups. Invariably, for the job to be well done, large moles of HNO_3 must be used, otherwise it will either be incomplete if less or completed with excess HNO_3. If the excess is more than the moles of water formed, then there is no problem of hydrolysis taking place. But if less, then the water must be removed. Nature is so complex to comprehend.

The presence of water above has been highlighted, because if too much of it as above is present just like what takes place during Condensation polymerization when the operating conditions are not adequate, then it will start hydrolyzing the poly vinyl nitrate formed to make it loose its unique characters. The **unique energy character** is from N center carrying polar bonds (ORIGIN OF THE BIG BANG THEORY). Hence, the acetic anhydride is present to remove the water as follows at such low temperatures.

Stage N +1:

$$H_2O \; \rightleftharpoons \; H{\bullet}e \;\; + \;\; nn{\bullet}OH$$

$$H{\bullet}e \;\; + \;\; H_3C - CO - O - CO - CH_3 \; \rightleftharpoons \; H - O - CO - CH_3 \;\; + \;\; e{\bullet}CO - CH_3$$

$$H_3C - \overset{\bullet e}{C} = O \;\; + \;\; nn{\bullet}OH \; \longrightarrow \; H - O - CO - CH_3 \qquad 2.48$$

Overall Equation: $n\, H_2O \;\; + \;\; n\,(H_3CCOOOCCH_3) \; \longrightarrow \; 2n\, H_3CCOOH \qquad 2.49$

Nitryl chloride or nitronium tetrafluoroborate may be used in place of the anhydrous HNO_3 where no water as above is formed. Based on the considerations above, vinyl nitrate a monomer not commonly mentioned in textbooks, indeed exists. One can imagine what will emerge if we look at other members where their existences are established, but yet unknown just as the case of vinyl nitrate.

2.1.3. Transfer from dead polymers/Transfer species of a third kind

One of the most important sub-steps under Addition transfer step is transfer from dead polymers to either a growing chain or an initiator to form a branching site. Transfer of transfer species from a dead polymer can only take place with specific polymers and with specific initiators.

Linear dead unsaturated polyethylene for example can be considered to be a very large or gigantic member of the alkene family. It is very well known, that alkenes are very inert to most reagents chemically. *"Usually, high chemical reactivity is associated with ions, free-radicals, or with compounds that contain an unshared or weakly held pair of "electrons", an incomplete valence shell, or a partially polarized covalent bond".[2]* None of these features is present in alkenes or linear polyethylenes, which are neutral giant molecules. They are not ionic; there are no "electrons", but radicals; all radicals are bonded covalently; all valence shells are filled. The carbon-carbon bonds in alkanes are non-ionic, non-polar and the carbon-hydrogen bonds are practically so. Hence, it is not surprising that alkenes are relatively unreactive. At room temperature they are not affected by concentrated acids, or alkalis, strong oxidizing or reducing agents or most other reagents and they frequently resist reaction under more drastic conditions. It is this inertness that gave rise to the name paraffins-(L. parum little, affinitas affinity).[2]

At high temperatures (500-600°C), homolytic bond breaking takes place with subsequent recombinations of the free-radical- a process known as cracking or pyrolysis. It is an endother-mic process involving lots of energy, the type of energy that exists only in the nucleo-free-radical polymerization of ethylene (harsh operating conditions), for which branch formations can be favored. Under most operating conditions of polymerization systems, it is impossible to transfer a transfer species from a polymer backbone such as in dead unsaturated polyethylene or poly (vinyl chloride). The same applies to polypropylene as shown below.

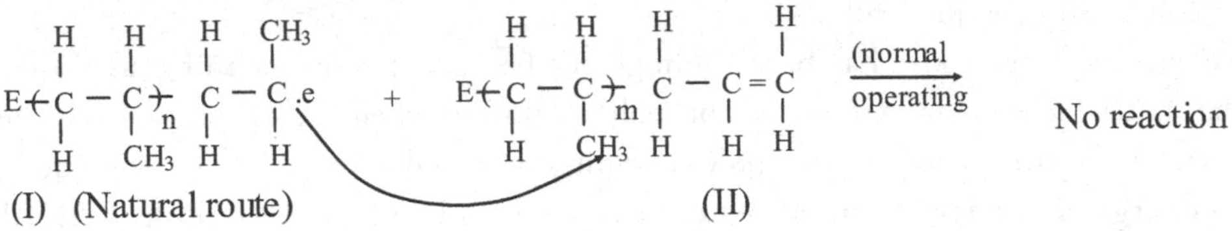

$$(I) \text{ (Natural route)} \qquad (II) \qquad \xrightarrow[\text{operating}]{\text{(normal}} \qquad \text{No reaction} \qquad 2.50a$$

No branching site formation can be favored here (except at the terminal end) as shown below, since the electro-free-radical active center cannot provide the energy to dissociate the C-H bond in CH_3 homolytically in the absence of an activation center **adjacently** located nearby.

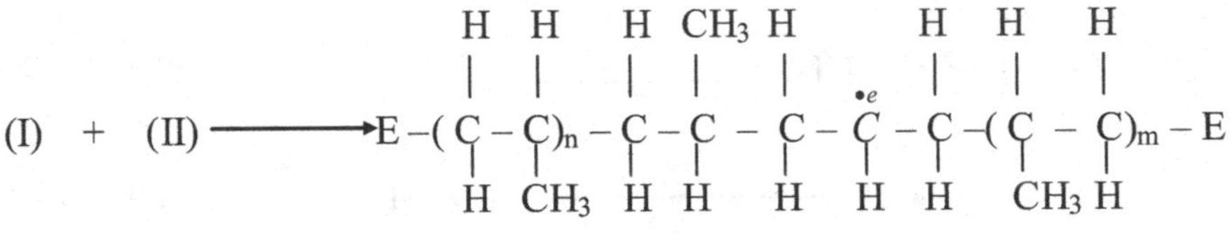

$$(I) + (II) \longrightarrow$$

NATURAL ROUTE

$$2.50b$$

Notice that, first and foremost the growing polymer chain has lost its identity with no activation center (X). This can only be introduced terminally when excess monomers are present. Secondly, free-radically, the initiator can be of the isolated type or paired. When paired, the situation is more difficult due to steric limitations. Chargedly, positively charged-paired initiators (what mistakenly in present-day Science is called cationic ion-paired initiator, the structure of which is unknown to them) can be used only for homopolymerization. Thirdly the branching site formed above can be used either horizontally or vertically depending on certain operating conditions. If the chain has reached its optimum length at its operating conditions in its horizontal growth, then it can only grow vertically; otherwise it will grow horizontally up to a limit- the boundary set by its Glass Transition temperature. Fourthly, the dead terminal double polymer, (II), has the same character as the originating monomer when transfer species of the first kind of the first type is involved. The same will apply when the living branched chain above is killed either by starvation or use of H_2 which always is in Stable State of Existence in the absence of a passive catalyst,

The H on CH_3 which is a transfer species of the first kind of the first type does not usually favor being used for branching site formation not even when the CH_3 group is connected directly to another activation center as shown below for methyl vinyl ketone.

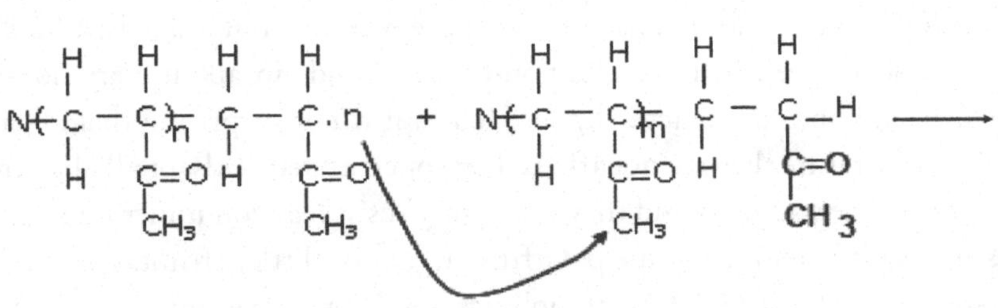

(Natural route) $\qquad\qquad$ (I)

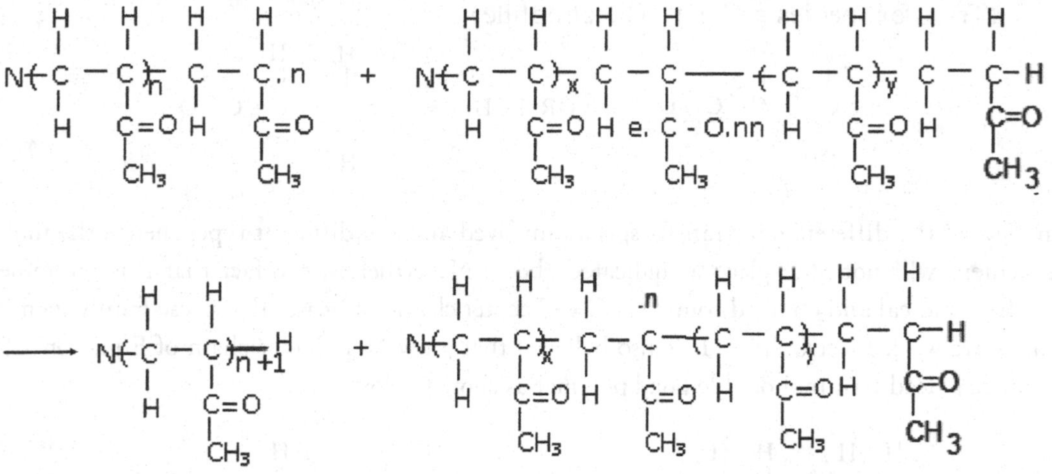

(II)   2.51

As shown by (II), a branching site is generated when the C = O activation center is activated, leading to abstraction of transfer species of the first kind of the first type with respect to C = O activation center. But since polymers are involved instead of monomers, the transfer species which allows for a branching site to be formed to continue polymerization is herein called transfer species of **the <u>third kind of the first type</u>**. The CH_3 group on the side chain has no part to play in the branching site formation in view of its lower capacity. Electro-free-radically, the followings are to be expected.

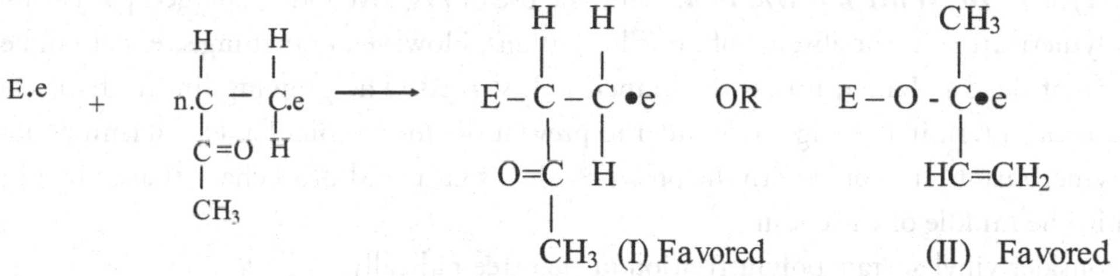

2.52

As a matter of fact, both (I) and (II) will be favored electro-free-radically along the chain with more of (II) than (I) for this monomer. Just like H cannot be a transfer species as H.n, so also CH_3 group cannot be abstracted as $n.CH_3$. Imagine if CH_3 was to be transferred as $n.CH_3$, then the followings would have been obtained when molecular rearrangement is allowed to take place, noting that H.n is greater than $n.CH_3$ in radical-pushing capacity as nucleo-free-radicals, as will be shown downstream..

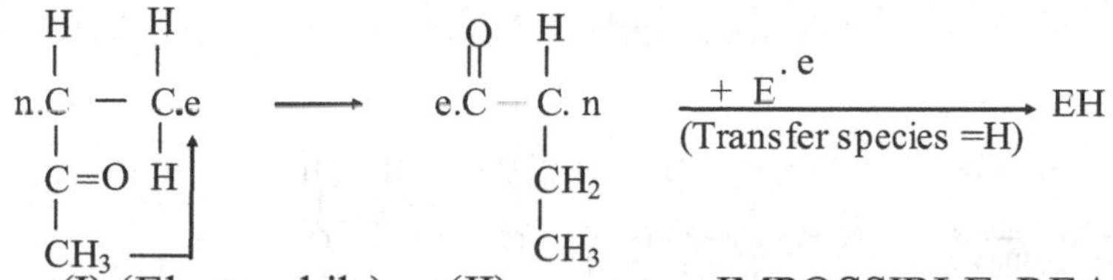

IMPOSSIBLE REARRANGEMENT

(Transfer species = CH_3)   (Electrophile)

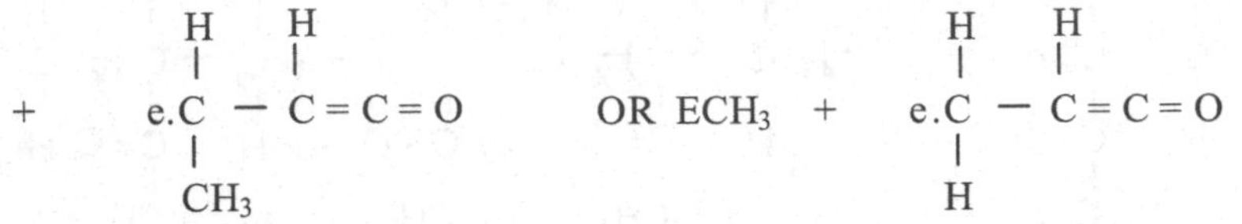

$$+ \quad \overset{\overset{\displaystyle H}{|}}{e.C} - \overset{\overset{\displaystyle H}{|}}{C} = C = O \qquad OR \ ECH_3 \ + \quad \overset{\overset{\displaystyle H}{|}}{e.C} - \overset{\overset{\displaystyle H}{|}}{C} = C = O \qquad\qquad 2.53$$
(with CH_3 below first C, and H below second $e.C$)

It is in view of the difference in transfer species involved and the different types, hence the molecular rearrangement will not take place as indicated above. Nevertheless, the fact that this monomer can favor all free- radical and charged routes via C = C center clearly indicates that these family members of monomers are weak Electrophiles. It is also believed that growing polymer chain of Equation 2.51 can also produce a dead terminal double bond polymer as shown below.

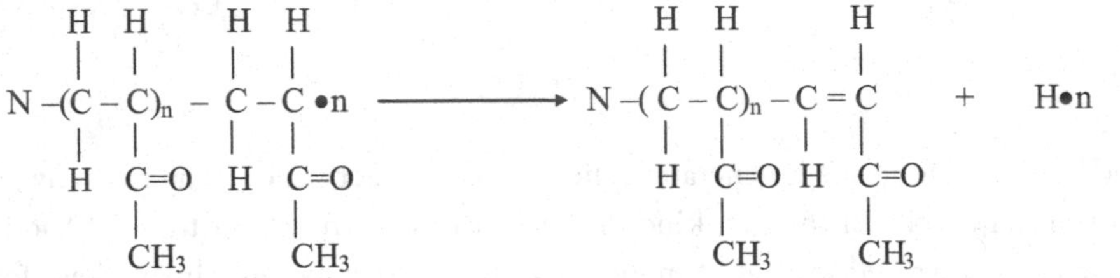

$$N-(C-C)_n - C - C \bullet n \longrightarrow N-(C-C)_n - C = C \quad + \quad H\bullet n \qquad 2.54a$$

The transfer species involved is of the **second kind of the second type** in the route natural to the monomer. The dead terminal double bond polymer above, has the same character as the monomer, except that it can no longer favor its natural route, ***because of the presence of transfer species of the first kind of the first type in the route natural to it.*** Only the use of negatively free-charged-paired initiator will favor its polymerization in the absence of steric limitations. However, branching sites cannot be obtained for this type of dead polymer, unless the branch is L-shaped. The growing linear chain can only be terminated using a terminating agent, in order to prevent the formation of a dead terminal double bond during polymerization. It is only with the presence of this saturated dead chain that branching site can be formed in the middle of the chain

Now consider vinyl acetate polymerization nucleo-free-radically.

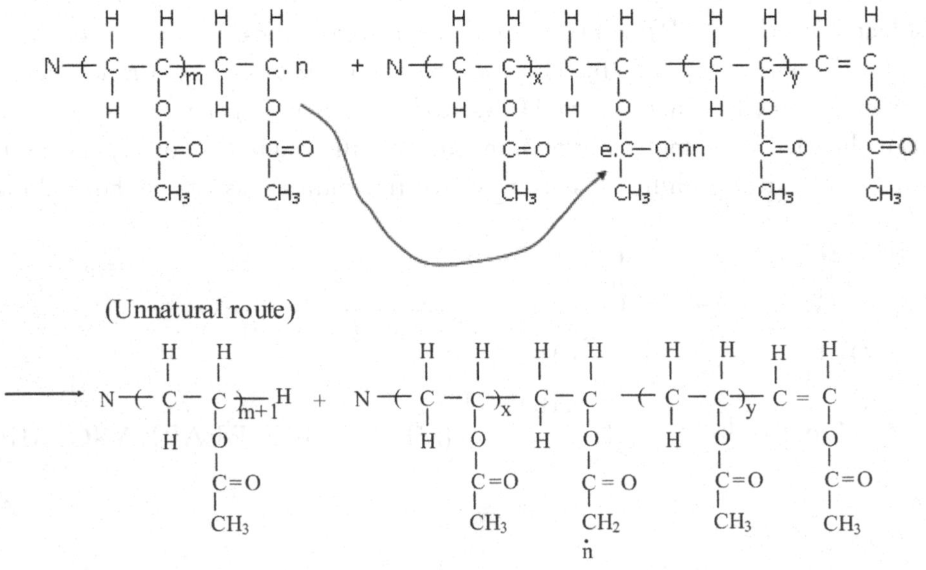

(Unnatural route)

$$ \qquad\qquad 2.54b$$

NOT FAVORED

Without the use of "the eye of the needle", the mechanism above looks favored. How can it be favored when the dead polymer is unsaturated and the C = C center is less nucleophilic than the C = O center?

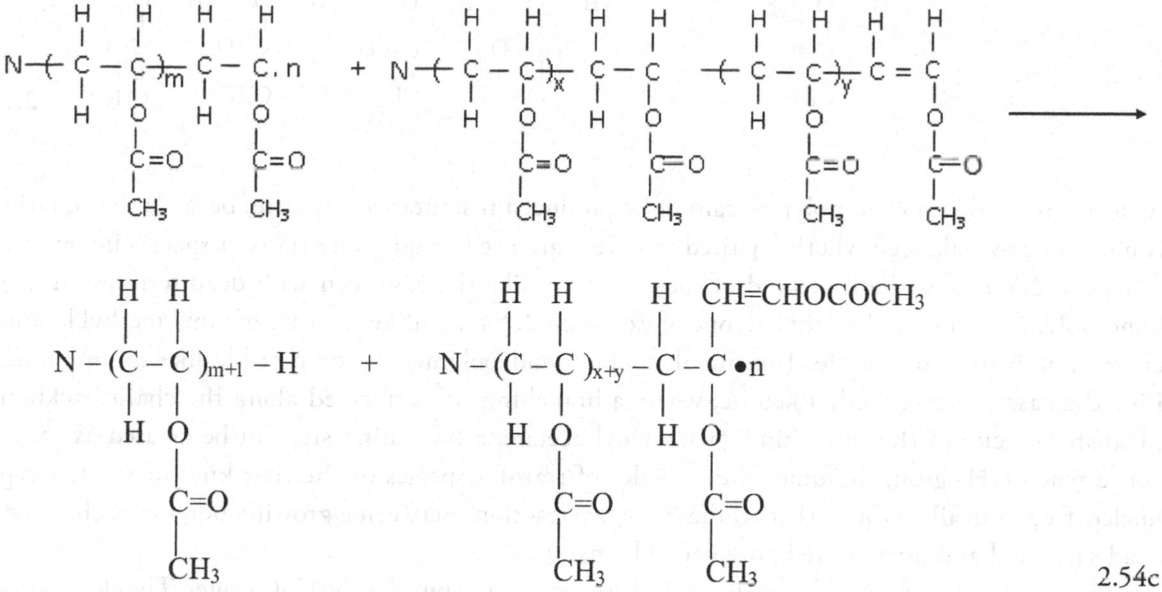

$$2.54c$$

(I)   FAVORED

However, CH_3 group in Equation 2.54b can be the one to provide transfer species despite the fact that -O-Chain group is more radical-pushing than CH_3. This can be done nucleo-free-radically if the equation is to be radically balanced. In place of the use of the dead terminal double bond polymer, dead saturated polymer is therefore used as shown below.

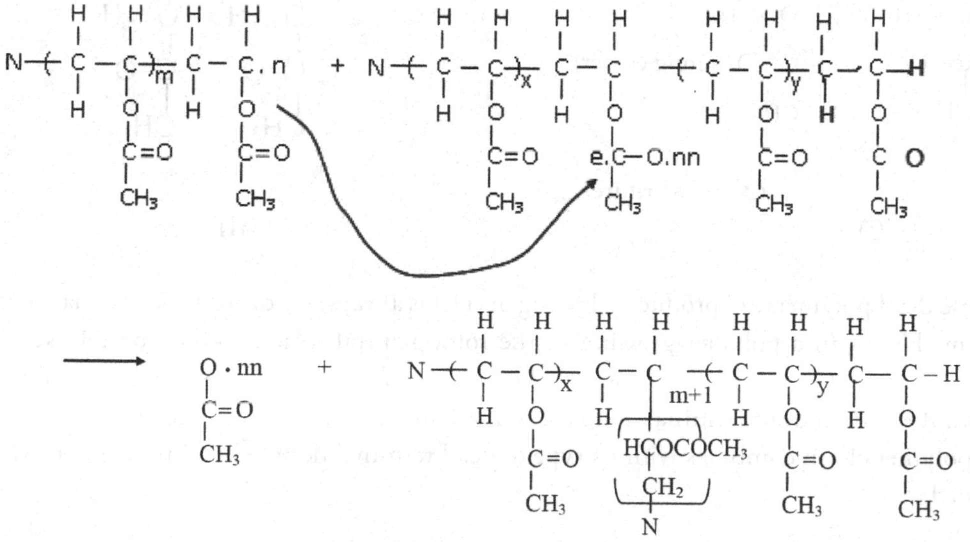

NOT RADICALLY BALANCED

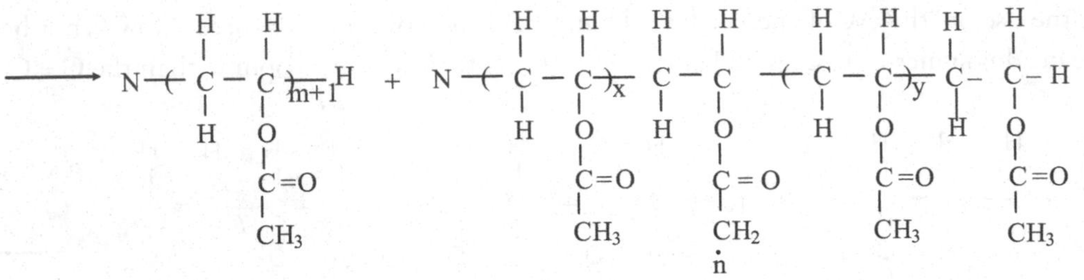

$$\longrightarrow \quad N\!\!\left(\!\!\begin{array}{cc} H & H \\ | & | \\ C & C \\ | & | \\ H & O \\ & | \\ & C\!=\!O \\ & | \\ & CH_3 \end{array}\!\!\right)_{\!m+1}\!\!H \;+\; N\!\!\left(\!\!\begin{array}{cc} H & H \\ | & | \\ C & C \\ | & | \\ H & O \\ & | \\ & C\!=\!O \\ & | \\ & CH_3 \end{array}\!\!\right)_{\!x}\!\!\begin{array}{c} H \\ | \\ C \\ | \\ H \end{array}\!\!\begin{array}{c} H \\ | \\ C \\ | \\ O \\ | \\ C\!=\!O \\ | \\ CH_2 \\ \dot n \end{array}\!\!\left(\!\!\begin{array}{cc} H & H \\ | & | \\ C & C \\ | & | \\ H & O \\ & | \\ & C\!=\!O \\ & | \\ & CH_3 \end{array}\!\!\right)_{\!y}\!\!\begin{array}{cc} H & H \\ | & | \\ C\!- & C\!-\!H \\ | & | \\ H & O \\ & | \\ & C\!=\!O \\ & | \\ & CH_3 \end{array} \qquad 2.55$$

As it seems dead branched polymer cannot be produced instantaneously here, because the equation is not radically balanced whether paired initiators are used or not. The transfer species involved in Equation 2.54c is of the first kind of the first type, like the case seen with dead terminal double bond polymer from vinyl methyl ketone of Equation 2. 54a. Unlike the case of vinyl methyl ketone, charged activation of the dead terminal double bond polymer is not possible like its monomer. Like the case of vinyl methyl ketone, where a branching site is formed along the chain backbone (Transfer species of the third kind), with vinyl acetate, a branching site can be created as shown above when CH_3 group becomes the provider of transfer species of the first kind of the fist type nucleo-free-radically. Notice that all these involve reactions between a growing polymeric chain and dead saturated and unsaturated polymeric chains.

Vinyl acetate has no transfer species of the first kind of first type for the C=C center. The electro-free-radical route, the route natural to it is not favored as already shown. However, ***for exploratory purposes,*** it will be assumed to be favored as it will reveal many things. Recalled below is the type of dead terminal double bond polymer obtained as already shown in Equation 2.20b and recalled below.

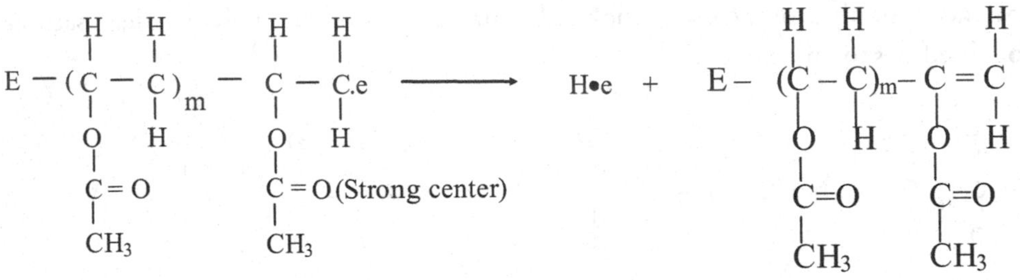

$$E\!-\!\left(\!\!\begin{array}{cc} H & H \\ | & | \\ C & C \\ | & | \\ O & H \\ | & \\ C\!=\!O & \\ | & \\ CH_3 & \end{array}\!\!\right)_{\!m}\!\!\begin{array}{cc} H & H \\ | & | \\ C & C.e \\ | & | \\ O & H \\ | & \\ C\!=\!O\,(\text{Strong center}) \\ | & \\ CH_3 & \end{array} \quad\longrightarrow\quad H\bullet e \;+\; E\!-\!\left(\!\!\begin{array}{cc} H & H \\ | & | \\ C & C \\ | & | \\ O & H \\ | & \\ C\!=\!O & \\ | & \\ CH_3 & \end{array}\!\!\right)_{\!m}\!\!\begin{array}{cc} H & H \\ | & | \\ C & =\!C \\ | & | \\ O & H \\ | & \\ C\!=\!O & \\ | & \\ CH_3 & \end{array}$$

(Natural route)

(I)   (II)   (2.20b)

When these dead polymers are produced, hydrogen (H_2) is always one of the products via Combination mechanism. How can a polymer growing in the route natural to it be killed by release of transfer species of the second kind for Nucleophiles? As seen so far, it has been release of transfer species of the first kind to give a dead terminal double bond polymer as was the case with propylene. When the growing polymer chain combines with its type of dead terminal double bond polymer, the followings are obtained.

$$(I) \quad + \quad (II) \longrightarrow$$

$$\text{NATURAL ROUTE} \qquad\qquad 2.56$$

Worthy of note is that a single chain with a branching site is obtained via the route natural to the monomer. This is almost like propylene in Equation 2.50b, but with a difference. The branching site obtained with propylene is along the main chain backbone, but here it is on the CH_3 group. These are clear indications that the monomer cannot be polymerized electro-free-radically. One can observe so far the great uniqueness of all these monomers and the marked difference between nucleo-free-radicals and electro-free-radicals. *The uniquenesses are to be expected, because Nature is too complex to comprehend using so many Universal concepts based on current developments – NEW FRONTIERS. When the Avatars said that "we have eyes but cannot see with them", it is indeed TRUE. When it is said that "Seeing is Believing", it is true in the REAL and PHYSICAL world, but not true in the COMPLEX and NATURAL world, because the world we leave in is not only real, but also imaginary- i.e. a complex world. When it is said that "WITH GOD, ALL THINGS ARE POSSIBLE", the statement is meaningless, for "WITH THE ALMIGHTY INFINITE GOD, ALL THINGS THAT WORK ACCORDING TO HIS LAWS- THE LAWS OF NATURE- ARE POSSIBLE", is the TRUTH. It is only in our world that where all things which are impossible are made possible by HUMANS and not THE CREATOR. "WITH THE ALMIGHTY INFINITE GOD, NOTHING IS IMPOSSIBLE" is the truth. Nature is incomprehensible only in the eyes of those who can see with the eye of the needle That is why the first law is "TO BE FREE AND NOT TO BE FREE". How many are born FREE? NONE! When one is born as a child, the child is in chains [Good and Bad chains]. It is only the child that can unchain him- or her-self and gradually become free under Natural operating conditions.*

Now consider the nucleo-free-radical polymerization of methyl acrylate and methyl methacrylate. If the dead polymethyl acrylate is saturated along the chain backbone, the followings are obtained.

$$\longrightarrow \quad N\!\!-\!\!(\!\!\underset{\underset{\displaystyle C=O}{|}}{\overset{\overset{\displaystyle H}{|}}{C}}\!\!-\!\!\underset{\underset{\displaystyle \underset{\displaystyle CH_3}{|}}{\overset{}{}}}{\overset{\overset{\displaystyle H}{|}}{C}})_{m+1}\!\!-\!\!CH_3 \quad + \quad N\!\!-\!\!(\underset{}{\overset{}{C}}-\underset{}{\overset{}{C}})_x \, C\!-\!C\!-\!(C\!-\!C)_y\!\!-\!\!Y \qquad 2.57a$$

NOT RADICALLY BALANCED

$$\longrightarrow \quad N\!\!-\!\!(C\!-\!C)_{m+1}\!\!-\!\!H \quad + \quad N\!\!-\!\!(C\!-\!C)_x \, C\!-\!C\!-\!(C\!- \qquad 2.57b$$

RADICALLY BALANCED

Note that the long chain substituent group on the carbon of the $C = O$ activation center is a radical-pushing group of lower capacity than OCH_3 group. When CH_3 transfer species on OCH_3 group is abstracted, the equation cannot be radically balanced. However, since NATURE abhors a vacuum and since a nuleo-free-radical growing chain is involved, a branching site is formed as shown in the last equation above, with the chain providing transfer species of the third kind, just like the case of vinyl acetate of Equation 2.55, but with a difference or the case of methyl vinyl ketone of Equation 2.51, but also with a difference. Dead terminal double bond polymers can be obtained here nucleo-free-radically like for methyl methacrylate. How obtained is important. Note that the $C = O$ center activated was not chosen randomly, but by the nature of the operating conditions as being the one of lowest capacity of all the $C=Os$. The same type as above will be obtained radically. Chargedly, to be candid, where they are all paired, it is impossible due to steric limitations. Radically, where they can both be free and paired media in character, only the free ones can be used. So far what we have seen are the free-media ones. We are yet to see the paired ones which as is obvious are difficult to use, due to steric limitations.

$$N\!\!-\!\!(C\!-\!C)_n - C\!-\!C\bullet n \quad \longrightarrow \quad N\!\!-\!\!(C\!-\!C)_n - C = C \quad + \quad H\bullet n \qquad 2.58$$

IMPOSSIBLE REJECTION

$$N\!\!-\!\!(C\!-\!C)_n - C\!-\!C\bullet n \quad \longrightarrow \quad N\!\!-\!\!(C\!-\!C)_n - C = C \quad + \quad H\bullet n \qquad 2.59$$

IMPOSSIBLE REJECTION

$$N-(\overset{\overset{\displaystyle H}{|}}{\underset{\underset{\displaystyle H}{|}}{C}}-\overset{\overset{\displaystyle H}{|}}{\underset{\underset{\displaystyle C=O}{|}}{C}})_n-\overset{\overset{\displaystyle H}{|}}{\underset{\underset{\displaystyle H}{|}}{C}}-\overset{\overset{\displaystyle H}{|}}{\underset{\underset{\displaystyle C=O}{|}}{C}}\bullet n......e\bullet X \longrightarrow N-(\overset{\overset{\displaystyle H}{|}}{\underset{\underset{\displaystyle H}{|}}{C}}-\overset{\overset{\displaystyle H}{|}}{\underset{\underset{\displaystyle C=O}{|}}{C}})_n-\overset{\overset{\displaystyle H}{|}}{\underset{\underset{\displaystyle H}{|}}{C}}-\overset{\overset{\displaystyle H}{|}}{C}=C=O \ + \ XOCH_3$$

With OCH₃ groups below the relevant carbons. (I) .... (II) .... (2.60)

FAVORED [Free or Paired]

$$N-(\overset{\overset{\displaystyle H}{|}}{\underset{\underset{\displaystyle H}{|}}{C}}-\overset{\overset{\displaystyle CH_3}{|}}{\underset{\underset{\displaystyle C=O}{|}}{C}})_n-\overset{\overset{\displaystyle H}{|}}{\underset{\underset{\displaystyle H}{|}}{C}}-\overset{\overset{\displaystyle CH_3}{|}}{\underset{\underset{\displaystyle C=O}{|}}{C}}\bullet n......e\bullet X \longrightarrow N-(\overset{\overset{\displaystyle H}{|}}{\underset{\underset{\displaystyle H}{|}}{C}}-\overset{\overset{\displaystyle CH_3}{|}}{\underset{\underset{\displaystyle C=O}{|}}{C}})_n-\overset{\overset{\displaystyle H}{|}}{\underset{\underset{\displaystyle H}{|}}{C}}-\overset{\overset{\displaystyle CH_3}{|}}{C}=C=O \ + \ XOCH_3$$

(2.61)

(SEE EQUATION 2.10) [Free or Paired]

Radically, the dead terminal double bond polymers (Ketenes) are those shown in the last two equations for these two monomers. *All the dead polymers when transfer spies of the second kinds are involved as above in the first two equations show the original characters of the originating monomers. But this cannot be the transfer species. With the transfer species of the first kind for acrylates and all males which have transfer species, the character of the dead polymer are all alike. Above Ketenes are obtained and these are also Electrophiles. They are identical to what is obtained when the monomer molecularly rearranges.*

Therefore when the living and dead polymer chains of Equations 2.60 and 2.61 are involved, the followings are obtained using (I) with free-media initiators and (II) of Equation 2.60.

$$(I) \ + \ (II) \longrightarrow N-(\overset{\overset{\displaystyle H}{|}}{\underset{\underset{\displaystyle H}{|}}{C}}-\overset{\overset{\displaystyle H}{|}}{\underset{\underset{\displaystyle C=O}{|}}{C}})_n-\overset{\overset{\displaystyle H}{|}}{\underset{\underset{\displaystyle H}{|}}{C}}-\overset{\overset{\displaystyle H}{|}}{\underset{\underset{\displaystyle C=O}{|}}{C}}-\overset{\overset{\displaystyle O}{\|}}{C}-\overset{\overset{\bullet n}{\displaystyle C}}{\underset{\underset{\displaystyle H}{|}}{}}-\overset{\overset{\displaystyle H}{|}}{\underset{\underset{\displaystyle H}{|}}{C}}-(\overset{\overset{\displaystyle H}{|}}{\underset{\underset{\displaystyle C=O}{|}}{C}}-\overset{\overset{\displaystyle H}{|}}{\underset{\underset{\displaystyle H}{|}}{C}})_n-N$$

with OCH₃ groups below relevant carbons. (2.62)

NATURAL ROUTE

The linear addition is still obvious from the equation above when the nucleo-free-radical center is linearly placed with the branch already in place. The chain as it is, has lost its electrophilic character which can be regained depending on the next mode of operation.

With the use of free-negatively charged initiator for the monomers, the followings are obtained when a strong initiator is involved using methyl methacrylate.

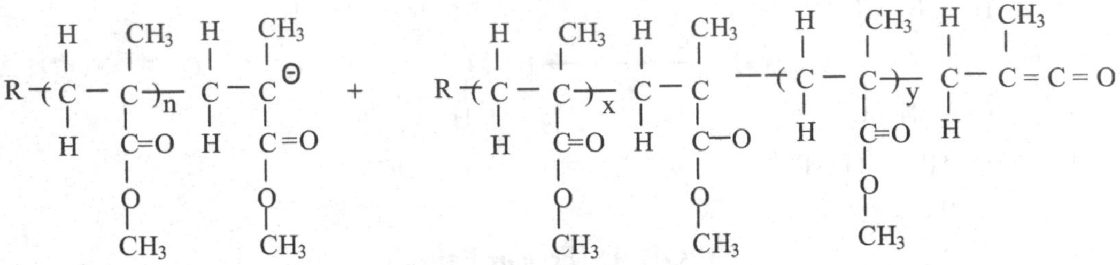

$$
\begin{array}{ccc}
 & & \\
R\!\!-\!\!\Big(\!\!C\!-\!C\!\!\Big)_{\!n}\,C\!-\!C^{\ominus} & + & R\!\!-\!\!\Big(\!\!C\!-\!C\!\!\Big)_{\!x}\,C\!-\!C\;-\!\!\Big(\!C\!-\!C\!\!\Big)_{\!y}\,C\!-\!C=C=O \\
\text{(I)} & & \text{(II)}
\end{array}
\qquad 2.63
$$

Note again that the negative charge above is of the free-paired type, i.e. e.g. t-$H_9C_4^{\ominus}$......$^{\oplus}$Li type and not of the $H_3CO^{\ominus}$......$^{\oplus}$Na type which is non-free, called ion-paired initiator where the negative charge is an anion while the positive charge is a cation. Similar living dead chain as in Equation 2.62 cannot be obtained due to steric limitations.

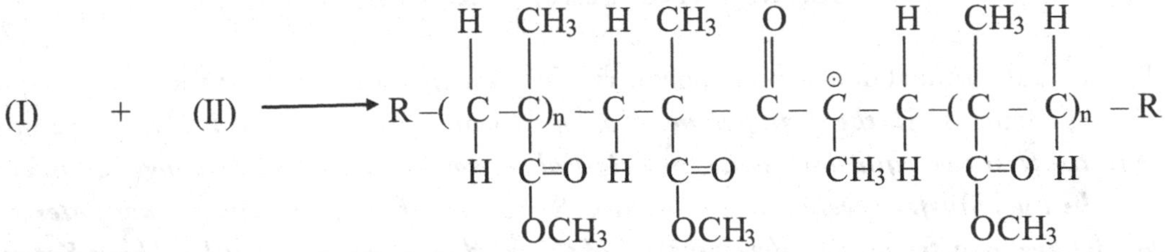

$$
\text{(I)} \quad + \quad \text{(II)} \longrightarrow R\!-\!\big(\,C\!-\!C\big)_n - C\!-\!C\;-\;C\!-\!C\!-\!C\!-\!\big(C\!-\!C\big)_n - R
$$

NATURAL ROUTE [STERIC LIMITATIONS]   2.64

When molecular rearrangement is allowed to take place in the presence of a weak initiator, the followings are to be expected.

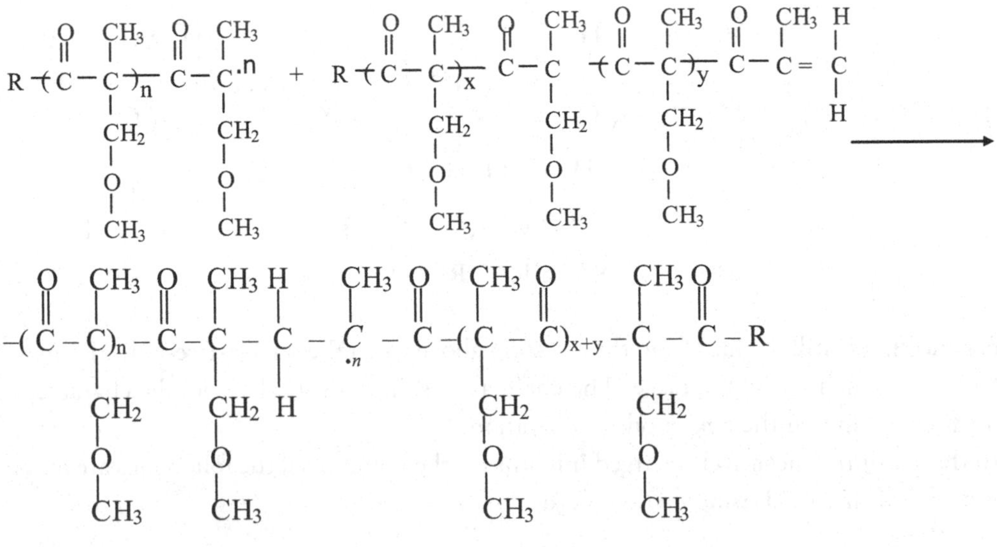

$$
\begin{array}{l}
R\!\!-\!\!\Big(\!C\!-\!C\!\!\Big)_{\!n}\,C\!-\!C^{\cdot n} \;+\; R\!\!-\!\!\Big(\!C\!-\!C\!\!\Big)_{\!x}\,C\!-\!C\;\Big(\!C\!-\!C\!\!\Big)_{\!y}\,C\!-\!C=C \longrightarrow \\[6pt]
R\!\!-\!\!\big(C\!-\!C\big)_n - C\!-\!C\;-\;C\;-\;C\;-\;C\!-\!\big(C\!-\!C\big)_{x+y} - C\;-\;C\!-\!R
\end{array}
\qquad 2.65
$$

Branching site formation is still favored after molecular rearrangement of the monomer only radically.

For monomers such as alkyl vinyl ether, the following is obtained "cationically" via Combination mechanism if the dead chain is saturated.

$$
\underset{\begin{array}{c}|\\CH_3\end{array}}{R\!\!\left[\!\!\begin{array}{c}H\\|\\C\\|\\H\end{array}\!\!-\!\!\begin{array}{c}H\\|\\C\\|\\O\end{array}\!\!\right]_{\!n}}\!\!\begin{array}{c}H\\|\\C\\|\\H\end{array}\!\!-\!\!\underset{\begin{array}{c}|\\CH_3\end{array}}{\begin{array}{c}H\\|\\C^{\oplus}\\|\\O\end{array}}
\quad + \quad
R\!\!\left[\!\!\begin{array}{c}H\\|\\C\\|\\H\end{array}\!\!-\!\!\begin{array}{c}H\\|\\C\\|\\O\\|\\CH_3\end{array}\!\!\right]_{\!x}\!\!\begin{array}{c}H\\|\\C\\|\\H\end{array}\!\!-\!\!\begin{array}{c}H\\|\\C\\|\\O\\|\\CH_3\end{array}\!\!\left[\!\!\begin{array}{c}H\\|\\C\\|\\H\end{array}\!\!-\!\!\begin{array}{c}H\\|\\C\\|\\O\\|\\CH_3\end{array}\!\!\right]_{\!y}\!\!Y
$$

$$\xrightarrow{\hspace{2cm}} \text{No transfer} \hspace{4cm} 2.66$$

In the absence of activation centers in the dead polymer, no transfer from the dead polymer can take place under Combination mechanism. The reaction above has been used to illustrate some basic fundamental principles. If the dead polymer is unsaturated like the type shown in Equation 2.42 and used along with a single growing polymer chain, then a branching site is obtained only electro-free-radically, since with positively charged centers which are always paired, steric limitations will be a problem..

2.2 Transfer species and stability in Molecular rearrangement of Activated Monomers

So far molecular rearrangement of some activated monomers where favored have been observed. There is no doubt that this phenomenon is very important since a monomer such a vinyl alcohol is unknown in the monomeric state, because acetaldehyde, the carbonyl form, is more stable,[3] When vinyl alcohol is activated the following is obtained:-

$$
\underset{\text{activated vinyl alcohol}}{H\!-\!\underset{\underset{\underset{H}{|}}{\underset{O}{|}}}{\overset{H}{\underset{|}{C}}}\!-\!\overset{H}{\underset{|}{C}}^{\!\oplus}}
\quad \xrightarrow[\text{rearrangement}]{\text{(molecular}} \quad
\underset{\text{activated acetaldehyde (more stable)}}{H\!-\!\overset{H}{\underset{\underset{H}{|}}{C}}\!-\!\overset{H}{\underset{\oplus}{C}}\!-\!O^{\ominus}}
\hspace{2cm} 2.67
$$

Thus, vinyl alcohol cannot exist in the activated state. The molecular rearrangement is favored; here both chargedly and radically, since the transfer species hydrogen is either radically or ionically bonded to the oxygen center.

One should also expect activated alkyl vinyl ethers to favor molecular rearrangement as follows:-

$$
\underset{\text{(I)}}{H\!-\!\underset{\underset{\underset{CH_3}{|}}{\underset{O}{|}}}{\overset{H}{\underset{\ominus}{C}}}\!-\!\overset{H}{\underset{|}{C}}^{\!\oplus}}
\quad \xrightarrow{\hspace{1.5cm}} \quad
\underset{\text{(II)}}{{}^{\oplus}C\!-\!O^{\ominus}\ \underset{\underset{CH_3}{|}}{\overset{\overset{H}{|}}{\underset{CH_2}{|}}}}
\hspace{3cm} 2.68
$$

(Activated methyl vinyl ether) (Activated aldehyde)

<u>NOT FAVOURED</u>

Aldehydes have been shown not to favor nucleo-free-radical and "anionic" polymerization. Alkyl vinyl ethers have also been shown not to favor nucleo-free-radical and "anionic" routes. However, when (II) of Equation 2.68 is subjected to anionic attack, the following is obtained:

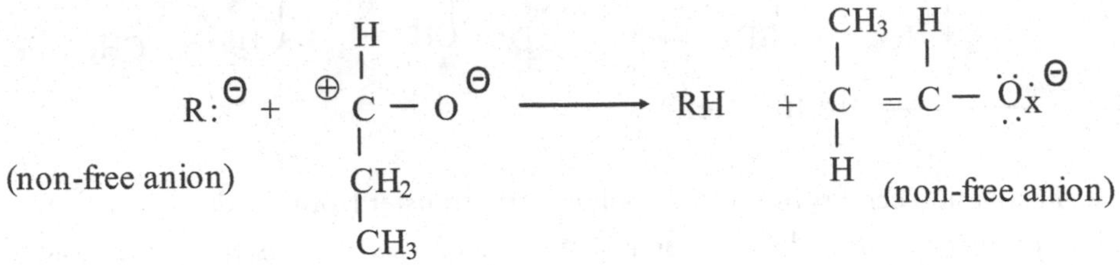

$$R{:}^{\ominus} \;+\; \overset{\oplus}{C}{-}O^{\ominus} \longrightarrow RH \;+\; C = C - \overset{..}{O}{\overset{..}{x}}{}^{\ominus}$$

(non-free anion) ⟶ (non-free anion)

(Transfer from monomer to initiator) 2.69

The transfer species abstracted $H^{\oplus}$ is different from that involved in molecular rearrangement of Equation 2.68. Hence molecular rearrangement of (I) is not favored. If all the H on the β-carbon center had been CH_3, then molecular rearrangement would have been favored. For the monomer represented below, molecular rearrangement is not also favored, since negatively charged (not anionically) initiator will not abstract what was transferred.

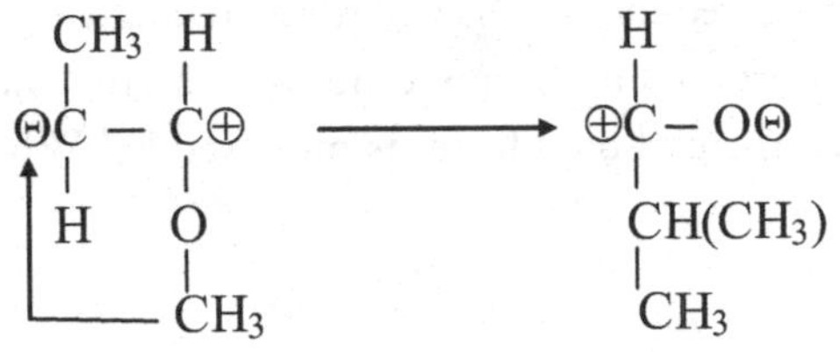

$$\overset{CH_3}{\underset{H}{{}^{\ominus}C}} - \overset{H}{\underset{O}{C^{\oplus}}} \longrightarrow \overset{H}{{}^{\oplus}C} - O^{\ominus}$$

(I) ⟶ (II) 2.70

NOT FAVORED

It is H that is abstracted as shown below for (II), instead of CH3 for (I).

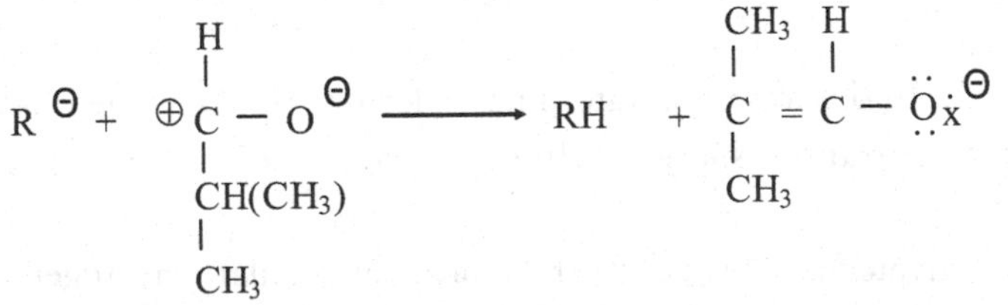

$$R^{\ominus} \;+\; \overset{\oplus}{C}{-}O^{\ominus} \longrightarrow RH \;+\; C = C - \overset{..}{O}{\overset{..}{x}}{}^{\ominus}$$

2.71

For molecular rearrangement to be favored, the same transfer species must be involved in other steps. For the monomer shown below, molecular rearrangement is favored, since when (II) is involved, the transfer species remains the same as shown below:-

83

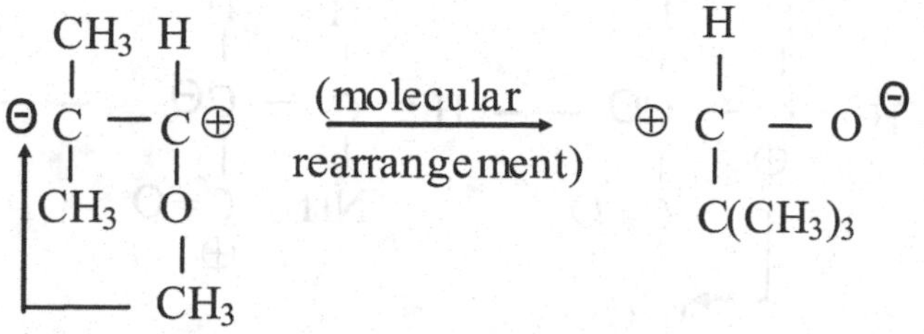

$$\underset{\text{(I)}}{\overset{\displaystyle \ominus C \overset{\displaystyle CH_3}{\underset{\displaystyle CH_3}{|}} - C \overset{\displaystyle H}{\underset{\displaystyle O}{|}} \oplus \underset{\displaystyle CH_3}{|}} \xrightarrow[\text{rearrangement)}]{\text{(molecular}} \quad \underset{\text{(II)}}{\oplus C \overset{\displaystyle H}{\underset{\displaystyle C(CH_3)_3}{|}} - O \ominus}$$

$$\tag{2.72}$$

$$\oplus C \overset{\displaystyle H}{\underset{\displaystyle C(CH_3)_3}{|}} - O \ominus \longrightarrow RCH_3 \;+\; C \overset{\displaystyle CH_3}{\underset{\displaystyle CH_3}{|}} = C \overset{\displaystyle H}{|} - \ddot{\underset{\displaystyle \cdot \cdot}{O}}_x \ominus$$

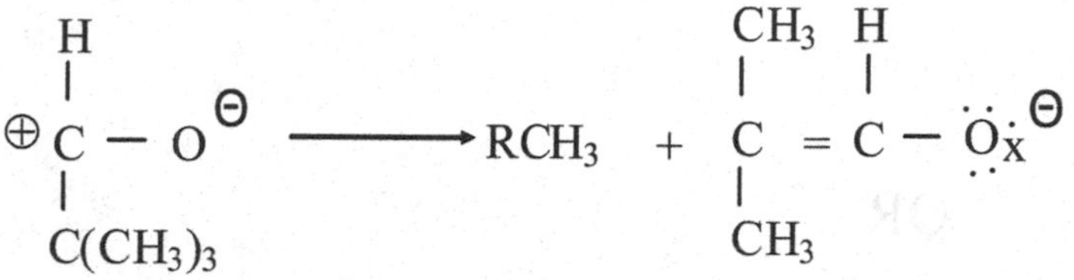

$$\tag{2.73}$$

Thus, not all vinyl ethers favor molecular rearrangement. For molecular rearrangement to be favored for alkyl vinyl ethers, *the transfer species involved <u>must be less or of equal radical-pushing capacity with the least radical-pushing substituent group carried by the active carbon center accepting the transfer species.</u>* [Don't eat more than what the body can take if you want to remain healthy]

The polymerization of acrylamide (or methacrylamide) by strong bases such as sodium, organolithium compounds, and alkoxides, in the presence of suitable solvents, have been found to yield a polymer structure (I) shown below:-

$$nCH_2 = CH - CO - NH_2 \longrightarrow \left(CH_2 - CH_2 - \overset{\displaystyle O}{\overset{\displaystyle \|}{C}} - NH \right)_n \tag{2.74}$$

(I) (Polyamide structure)

which was said to be quite unexpected.[4] The polyamide structure of the polymer was confirmed by the quantitative hydrolysis of the polymer to 3-aminopropionic acid,[4] for which a polymer-ization mechanism which has been found not to be possible was proposed.[4,5] In the first case, it is not 3-aminopropionic acid that is formed, but β-aminopropionic acid. Secondly, (I) of Equation 2.74 cannot by hydrolyzed to form an amino acid as will shortly be indicated.

It is also important to note that, when molecular rearrangement is favored, the monomer must be deactivated to ascertain that a new product has been formed. Therefore, when acryl- amide is activated the following occur:-

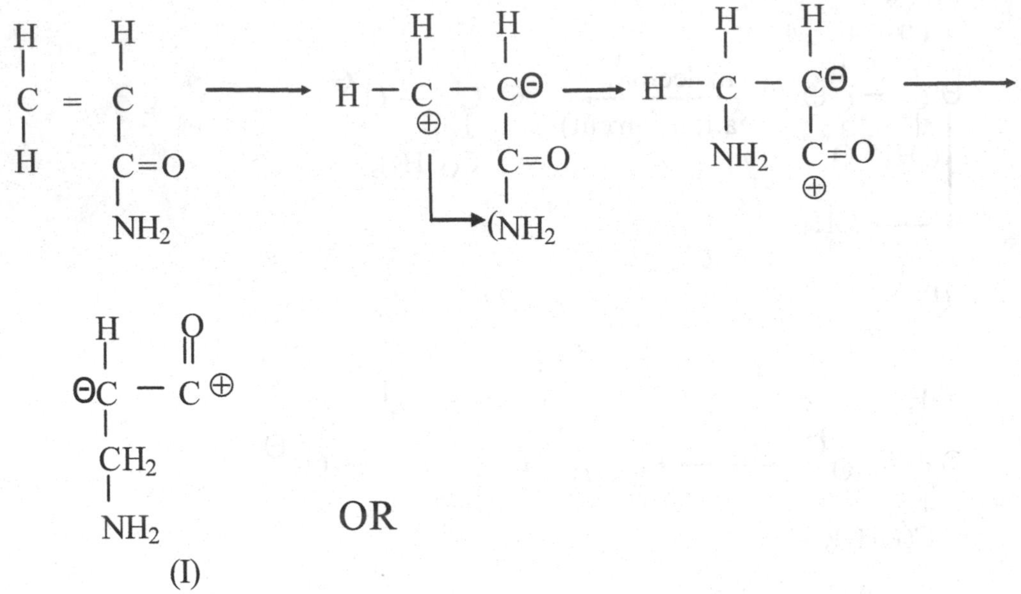

$$
\begin{array}{ccc}
\overset{\displaystyle H}{\underset{\displaystyle H}{C}} = \overset{\displaystyle H}{\underset{\displaystyle \underset{\displaystyle NH_2}{C=O}}{C}}
& \longrightarrow &
H-\overset{\displaystyle H}{\underset{\displaystyle \oplus}{C}} - \overset{\displaystyle H}{\underset{\displaystyle \underset{\displaystyle (NH_2}{C=O}}{C^{\ominus}}}
\longrightarrow
H-\overset{\displaystyle H}{\underset{\displaystyle NH_2}{C}} - \overset{\displaystyle H}{\underset{\displaystyle \underset{\oplus}{C=O}}{C^{\ominus}}}
\longrightarrow
\end{array}
$$

$$
{}^{\ominus}\overset{\displaystyle H}{C} - \overset{\displaystyle O}{\overset{\|}{C}}{}^{\oplus}
$$
$$
\underset{\displaystyle NH_2}{\overset{\displaystyle CH_2}{|}}
$$
$$
(I) \qquad\qquad OR \qquad\qquad\qquad\qquad 2.75
$$

$$
\begin{array}{ccc}
\overset{\displaystyle H}{\underset{\displaystyle H}{C}} = \overset{\displaystyle H}{\underset{\displaystyle \underset{\displaystyle NH_2}{C=O}}{C}}
& \longrightarrow &
{}^{\oplus}\overset{\displaystyle H}{\underset{\displaystyle H}{C}} - \overset{\displaystyle H}{C}{}^{\ominus}
\longrightarrow
{}^{\oplus}\overset{\displaystyle H}{\underset{\displaystyle H}{C}} - \overset{\displaystyle H}{\underset{\displaystyle H}{C}} - \overset{\displaystyle O}{\overset{\|}{C}} - N^{\ominus}
\end{array}
$$

$$
\begin{array}{ccc}
& \underset{\displaystyle H-N-H}{{}^{\oplus}C-O^{\ominus}} & \\
(II) & & (III)
\end{array}
\qquad 2.76
$$

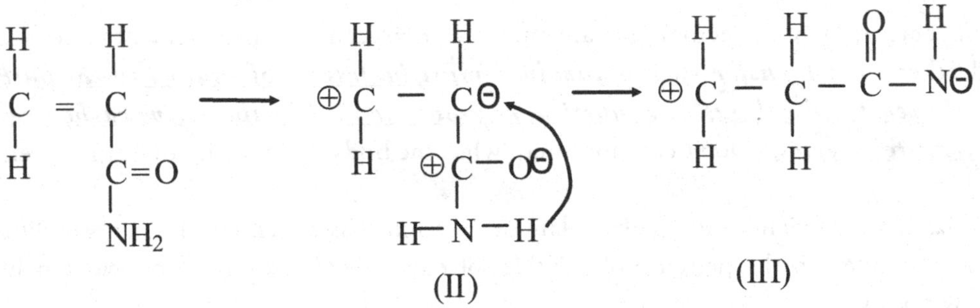

Note that first and foremost in all the considerations of molecular rearrangement, never is there a time where the transfer species is from a part of substituent group attached to an activated carbon center as shown by (II) above to some active carbon center by-passing a possible activated carbonyl center. It is impossible for (III) of Equation 2.76 to form a stable molecule by de-activation. It is (III) above that will favor the production of polyamide polymer chain structure indicated by (I) of Equation 2.74. However, since the type of transfer species indicated by (II) above is not the type concerned with the particular molecular rearrangement that has been identified, it is (I) of Equation 2.75 that is the favored activated rearrangement, which is the only rearrangement that can favor the production of an amino acid by hydrolysis as shown below, only radically, since the nitrogen center with hydrogen atoms will not allow for ionic hydrolysis.

$$R-(\underset{\underset{NH_2}{|}}{\underset{CH_2}{|}}{\overset{\overset{O}{\|}}{C}}-\overset{\overset{H}{|}}{C})_n-\underset{\underset{NH_2}{|}}{\underset{CH_2}{|}}{\overset{\overset{O}{\|}}{C}}-\overset{\overset{H}{|}}{C}-X+ \quad (n+1)\ H_2O \longrightarrow (n+1)\ [HO.nn+e.\underset{\underset{NH_2}{|}}{\underset{CH_2}{|}}{\overset{\overset{O}{\|}}{C}}-\overset{\overset{H}{|}}{C}.n] \ + \ RX \ + \ (n+1)e.H$$

$$(n+1)\ _{n.}\overset{\overset{H}{|}}{\underset{\underset{NH_2}{|}}{\underset{CH_2}{|}}{C}}-\overset{\overset{O}{\|}}{C}.e \quad + \quad (n+1)H_2O \longrightarrow [n+1](I) \ + \ (n+1)HO.nn$$

(I)

$$+(n+1)\ H.e \longrightarrow (n+1)\ HO-\overset{\overset{O}{\|}}{C}-\underset{\underset{H}{|}}{\overset{\overset{H}{|}}{C}}-\underset{\underset{H}{|}}{\overset{\overset{H}{|}}{C}}-NH_2 \qquad\qquad 2.77a$$

(β-amino propionic-acid)

In the reaction above, it is not the polymer chain that is hydrolyzed, but the monomer which has rearranged. In its activated state, it reacts with the water to give the amino acid.

Stage 1:

$$(II) \quad + \quad H_2O \quad \rightleftharpoons \quad n\bullet\overset{\overset{H}{|}}{\underset{\underset{NH_2}{|}}{\underset{CH_2}{|}}{C}}-\overset{\overset{O}{\|}}{C}-O-H \quad + \quad H\bullet e \qquad 2.77b$$

$$\longrightarrow H_2N-\underset{\underset{H}{|}}{\overset{\overset{H}{|}}{C}}-\underset{\underset{H}{|}}{\overset{\overset{H}{|}}{C}}-\overset{\overset{O}{\|}}{C}-O-H$$

β-propionic amino acid 2.77c

Overall Equation: $H_2C=CHCONH_2 \quad + \quad H_2O \longrightarrow H_2N\text{-}CH_2-CH_2-CO-OH$

The activated molecularly rearranged acrylamide could not allow water to exist in Equilibrium State of Existence, but only in Stable state of Existence. Hence the amino acid was produced via Equilibrium mechanism. The above considerations clearly send a message. Was a polymer produced after the rearrangements according to Equation 2.74? If a polymer was produced, then what type of polymer was it which when hydrolyzed broke down completely? Did the polymer break down by Depropagation or Degradation because of the environment it was existing in- the types of solvents, additives, impurities, and so on?

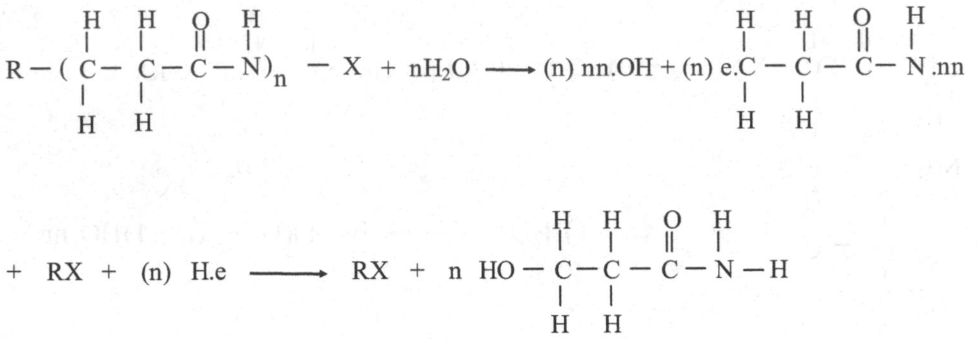

$$R-(\overset{\overset{\displaystyle H}{|}}{\underset{\underset{\displaystyle H}{|}}{C}}-\overset{\overset{\displaystyle H}{|}}{\underset{\underset{\displaystyle H}{|}}{C}}-\overset{\overset{\displaystyle O}{\|}}{C}-\overset{\overset{\displaystyle H}{|}}{N})_n - X + nH_2O \longrightarrow (n)\ nn.OH + (n)\ e.\overset{\overset{\displaystyle H}{|}}{\underset{\underset{\displaystyle H}{|}}{C}}-\overset{\overset{\displaystyle H}{|}}{\underset{\underset{\displaystyle H}{|}}{C}}-\overset{\overset{\displaystyle O}{\|}}{C}-\overset{\overset{\displaystyle H}{|}}{N}.nn$$

$$+ RX + (n)\ H.e \longrightarrow RX + n\ HO-\overset{\overset{\displaystyle H}{|}}{\underset{\underset{\displaystyle H}{|}}{C}}-\overset{\overset{\displaystyle H}{|}}{\underset{\underset{\displaystyle H}{|}}{C}}-\overset{\overset{\displaystyle O}{\|}}{C}-\overset{\overset{\displaystyle H}{|}}{N}-H \qquad 2.78a$$

(not an acid)

No dead or living polymer was ever obtained after the rearrangement, because the monomer unit above cannot be deactivated, unless done as follows.

Stage 1:

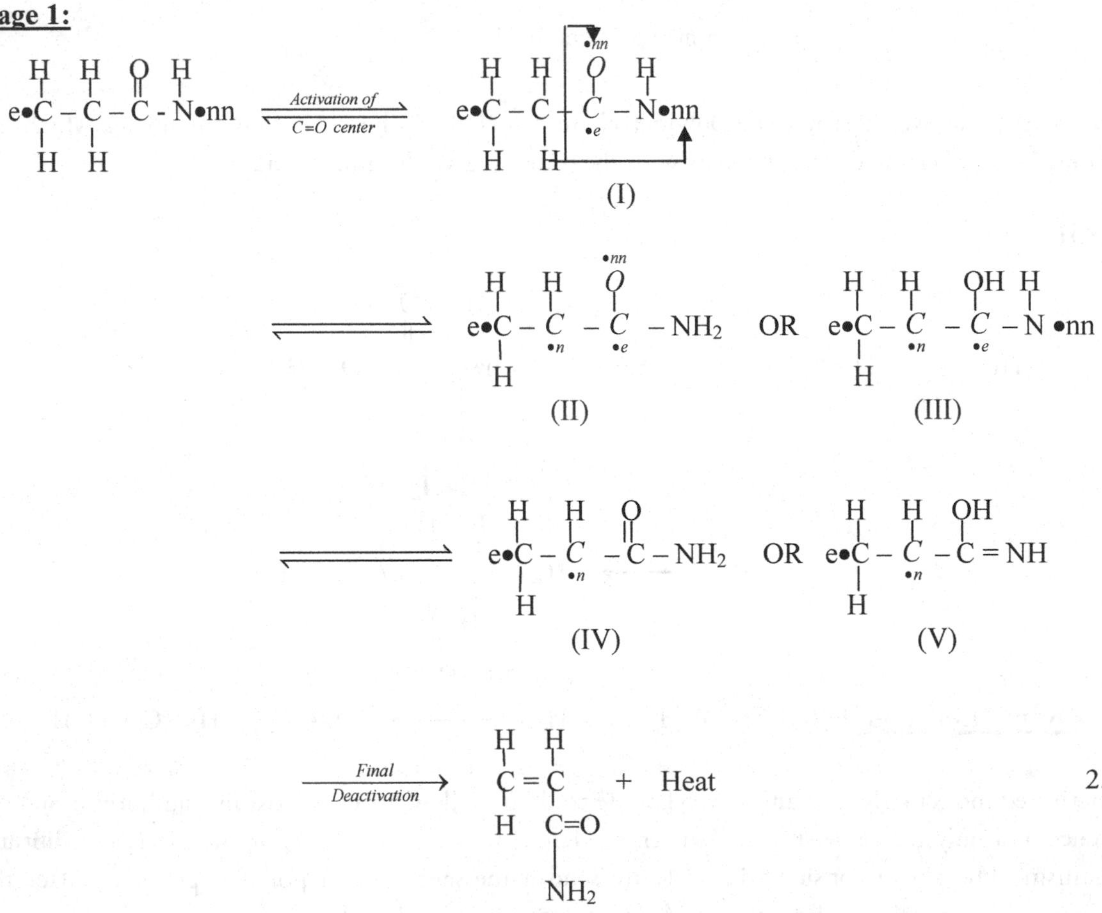

$$e\bullet\overset{\overset{\displaystyle H}{|}}{\underset{\underset{\displaystyle H}{|}}{C}}-\overset{\overset{\displaystyle H}{|}}{\underset{\underset{\displaystyle H}{|}}{C}}-\overset{\overset{\displaystyle O}{\|}}{C}-\overset{\overset{\displaystyle H}{|}}{N}\bullet nn \underset{C=O\ center}{\overset{Activation\ of}{\rightleftharpoons}} e\bullet\overset{\overset{\displaystyle H}{|}}{\underset{\underset{\displaystyle H}{|}}{C}}-\overset{\overset{\displaystyle H}{|}}{\underset{\underset{\displaystyle H}{|}}{C}}-\overset{\overset{\displaystyle \bullet nn}{\overset{\displaystyle O}{|}}}{\underset{\underset{\displaystyle \bullet e}{}}{C}}-\overset{\overset{\displaystyle H}{|}}{N}\bullet nn$$

(I)

$$\rightleftharpoons e\bullet\overset{\overset{\displaystyle H}{|}}{\underset{\underset{\displaystyle H}{|}}{C}}-\overset{\overset{\displaystyle H}{|}}{\underset{\underset{\displaystyle \bullet n}{}}{C}}-\overset{\overset{\displaystyle \bullet nn}{\overset{\displaystyle O}{}}}{\underset{\underset{\displaystyle \bullet e}{}}{C}}-NH_2 \quad OR \quad e\bullet\overset{\overset{\displaystyle H}{|}}{\underset{\underset{\displaystyle H}{|}}{C}}-\overset{\overset{\displaystyle H}{|}}{\underset{\underset{\displaystyle \bullet n}{}}{C}}-\overset{\overset{\displaystyle OH}{}}{\underset{\underset{\displaystyle \bullet e}{}}{C}}-N\bullet nn$$

(II) 　　　　　　(III)

$$\rightleftharpoons e\bullet\overset{\overset{\displaystyle H}{|}}{\underset{\underset{\displaystyle H}{|}}{C}}-\overset{\overset{\displaystyle H}{|}}{\underset{\underset{\displaystyle \bullet n}{}}{C}}-\overset{\overset{\displaystyle O}{\|}}{C}-NH_2 \quad OR \quad e\bullet\overset{\overset{\displaystyle H}{|}}{\underset{\underset{\displaystyle H}{|}}{C}}-\overset{\overset{\displaystyle H}{|}}{\underset{\underset{\displaystyle \bullet n}{}}{C}}-\overset{\overset{\displaystyle OH}{}}{C}=NH$$

(IV) 　　　　　　(V)

$$\underset{Deactivation}{\overset{Final}{\longrightarrow}} \quad \overset{\overset{\displaystyle H}{|}}{\underset{\underset{\displaystyle H}{}}{C}}=\overset{\overset{\displaystyle H}{|}}{\underset{\underset{\displaystyle C=O}{|}}{C}} \quad + \quad Heat \qquad 2.78b$$

$$\underset{NH_2}{|}$$

There is no doubt that the monomer unit above exists, via other means but not by molecular rearrangement. When it exists, it "wonderfully" rearranges to a more Stable State that which can be deactivated as shown in the Stage above back to acrylamide! How Nature operates is too much to comprehend. Note however should be taken of the origin of the monomer unit. Did it originate from the monomer shown below, i.e. the deactivated state of (V) above? YES.

Stage 1:

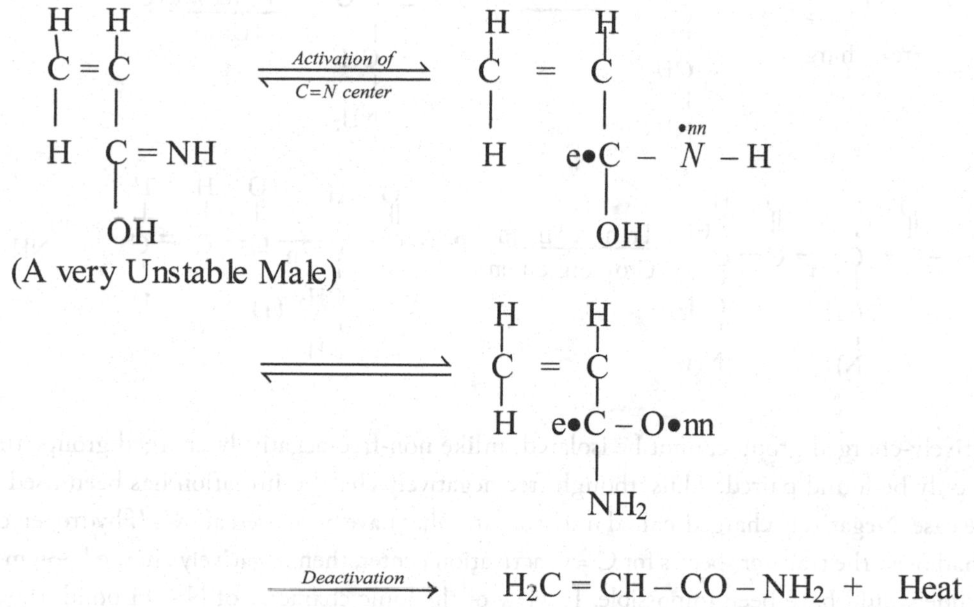

$$\underset{\text{(A very Unstable Male)}}{\begin{array}{c} \overset{H}{\underset{|}{C}} = \overset{H}{\underset{|}{C}} \\ \overset{|}{H} \quad \overset{|}{C} = NH \\ \overset{|}{OH} \end{array}} \quad \underset{C=N \text{ center}}{\overset{\text{Activation of}}{\rightleftharpoons}} \quad \begin{array}{c} \overset{H}{\underset{|}{C}} = \overset{H}{\underset{|}{C}} \\ \overset{|}{H} \quad e\bullet \overset{|}{C} - \overset{\bullet nn}{N} - H \\ \overset{|}{OH} \end{array}$$

$$\rightleftharpoons \quad \begin{array}{c} \overset{H}{\underset{|}{C}} = \overset{H}{\underset{|}{C}} \\ \overset{|}{H} \quad e\bullet \overset{|}{C} - O\bullet nn \\ \overset{|}{NH_2} \end{array}$$

$$\xrightarrow{\text{Deactivation}} \quad H_2C = CH - CO - NH_2 \; + \; \text{Heat} \qquad 2.78c$$

Worthy of note is the fact that, why should the C = N center be activated in the presence of C = C center when the monomer is a MALE? The reason is obvious-its instability with respect to C =O activation center and the fact that NH_2 is more radical or chargedly pushing than OH group. All these are clear indications of the following orders-

$$C = O \; > \; C = N \qquad ; \qquad H_2N \; > \; OH$$

Order of Nucleophilicity **Order of Radical/Charge pushing capacity** 2.79

Hence it is (II) and (IV) of Equation 2.78b that are the favored ones. Thus, (I) of Equation 2.74 cannot be hydrolyzed to an amino-acid in the same manner as described above. Hence, the molecularly rearranged monomer indicated by (I) of Equation 2.75 is the expected activated state of the monomer chargedly. The transfer species $^{\ominus}NH_2$ for the substituent group $CONH_2$ is the transfer species involved throughout the course of polymerization of the monomer chargedly as shown below:-

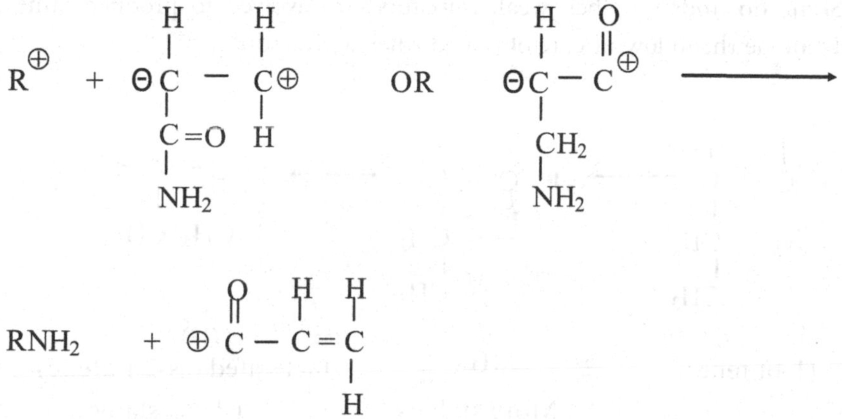

$$R^{\oplus} \; + \; \underset{\begin{array}{c}\overset{|}{C=O}\\\overset{|}{NH_2}\end{array}}{\overset{H}{\ominus}\overset{|}{C}} - \overset{H}{\underset{|}{C}}{}^{\oplus} \qquad OR \qquad \underset{\begin{array}{c}\overset{|}{CH_2}\\\overset{|}{NH_2}\end{array}}{\overset{H}{\ominus}\overset{|}{C}} - \overset{O}{\overset{\|}{C}}{}^{\oplus} \qquad \longrightarrow$$

$$RNH_2 \; + \; {}^{\oplus}\overset{O}{\overset{\|}{C}} - \overset{H}{\underset{|}{C}} = \overset{H}{\underset{|}{C}} \qquad\qquad\qquad 2.80a$$
$$\qquad\qquad\qquad\qquad \overset{|}{H}$$

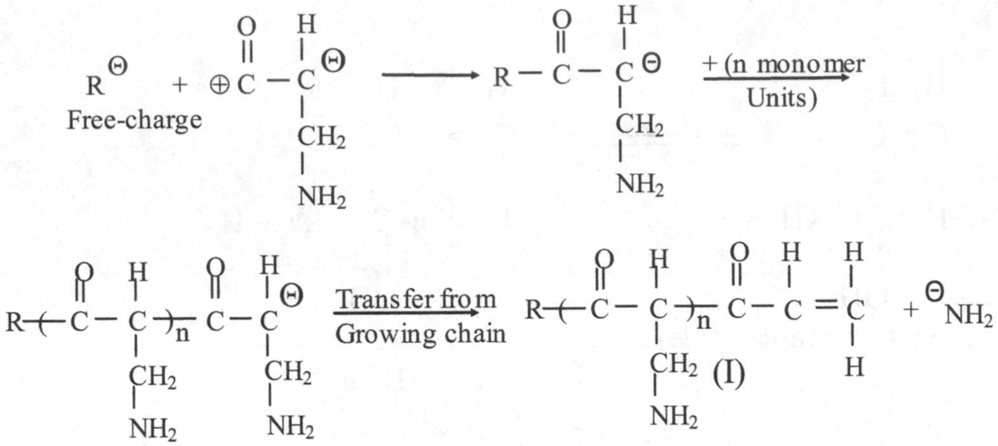

$$R^{\ominus} \ + \ {}^{\oplus}\overset{\overset{\displaystyle O}{\|}}{C} - \overset{\overset{\displaystyle H}{|}}{\underset{\underset{\displaystyle NH_2}{|}}{\underset{\overset{\displaystyle CH_2}{|}}{C}}}{}^{\ominus} \ \longrightarrow \ R - \overset{\overset{\displaystyle O}{\|}}{C} - \overset{\overset{\displaystyle H}{|}}{\underset{\underset{\displaystyle NH_2}{|}}{\underset{\overset{\displaystyle CH_2}{|}}{C}}}{}^{\ominus} \quad \xrightarrow{\text{+(n monomer}} \text{Units)}$$

Free-charge

$$R \overset{}{\underset{}{\left(\overset{\overset{\displaystyle O}{\|}}{C} - \overset{\overset{\displaystyle H}{|}}{\underset{\underset{\displaystyle NH_2}{|}}{\underset{\overset{\displaystyle CH_2}{|}}{C}}}\right)_n} \overset{\overset{\displaystyle O}{\|}}{C} - \overset{\overset{\displaystyle H}{|}}{\underset{\underset{\displaystyle NH_2}{|}}{\underset{\overset{\displaystyle CH_2}{|}}{C}}}{}^{\ominus} \quad \underset{\text{Growing chain}}{\overset{\text{Transfer from}}{\longrightarrow}} \quad R \overset{}{\left(\overset{\overset{\displaystyle O}{\|}}{C} - \overset{\overset{\displaystyle H}{|}}{\underset{\underset{\displaystyle NH_2}{|}}{\underset{\overset{\displaystyle CH_2}{|}}{C}}}\right)_n} \overset{\overset{\displaystyle O}{\|}}{C} - \overset{\overset{\displaystyle H}{|}}{C} = \overset{\overset{\displaystyle H}{|}}{\underset{\underset{\displaystyle H}{}}{C}} \ + \ {}^{\ominus}NH_2 \qquad 2.80b$$

(I)

Free-negatively-charged group cannot be isolated unlike non-free-negatively-charged group- the anion. They can only be found paired. Thus, though free negatively charged initiation has been used, it is not indeed the case. Negatively charged-paired initiators are what have been used above. If hydrogen cation on nitrogen had been the transfer species for C = C activation center, then negatively charged polymerization of acrylamide would have been impossible. In view of the ionic character of N – H bond, the nitrogen center would largely encourage branch formations as shown below.

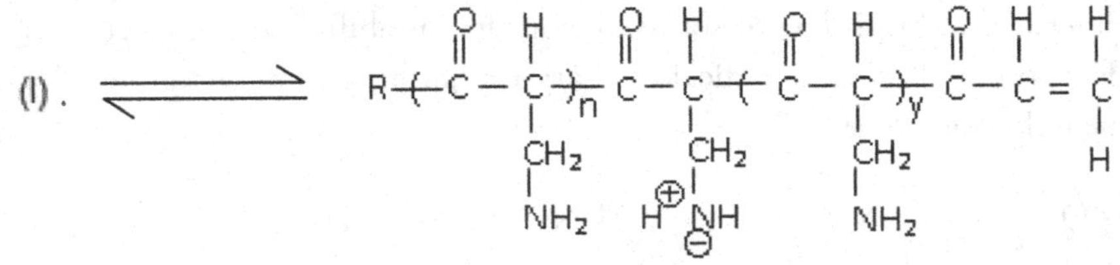

$$(I). \ \rightleftharpoons \ R \left(\overset{\overset{\displaystyle O}{\|}}{C} - \overset{\overset{\displaystyle H}{|}}{\underset{\underset{\displaystyle NH_2}{|}}{\underset{\overset{\displaystyle CH_2}{|}}{C}}}\right)_n \overset{\overset{\displaystyle O}{\|}}{C} - \overset{\overset{\displaystyle H}{|}}{\underset{\underset{\underset{\displaystyle \ominus}{\displaystyle H\overset{\oplus}{N}H}}{|}}{\underset{\overset{\displaystyle CH_2}{|}}{C}}} \left(\overset{\overset{\displaystyle O}{\|}}{C} - \overset{\overset{\displaystyle H}{|}}{\underset{\underset{\displaystyle NH_2}{|}}{\underset{\overset{\displaystyle CH_2}{|}}{C}}}\right)_y \overset{\overset{\displaystyle O}{\|}}{C} - \overset{\overset{\displaystyle H}{|}}{C} = \overset{\overset{\displaystyle H}{|}}{\underset{\underset{\displaystyle H}{}}{C}}$$

$$2.81$$

This center can be used with either STEP monomers or some specific types of ADDITION monomers to produce branched polymers. When the monomer is Addition, this can lead to Depropagation based on the operating conditions as will be shown downstream in this Volume.

There is need to throw more light into why (II) of Equation 2.76 is not favored apart from instability of the activated state. In the last chapter, 1-butene was shown not to favor molecular rearrangement to produce a different less stable monomer. So also are some other monomers where molecular rearrangement was expected. Some do undergo when weak initiators are involved to produce same monomer e.g. propylene. For 1-butene the followings are obtained when activated:-

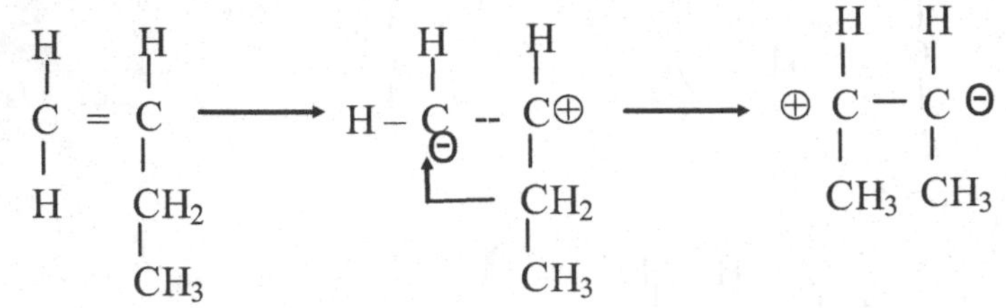

$$\overset{\overset{\displaystyle H}{|}}{\underset{\underset{\displaystyle H}{|}}{C}} = \overset{\overset{\displaystyle H}{|}}{\underset{\underset{\underset{\displaystyle CH_3}{|}}{\displaystyle CH_2}}{C}} \ \longrightarrow \ H - \overset{\overset{\displaystyle H}{|}}{\underset{\underset{\displaystyle \ominus}{}}{C}} \cdots \overset{}{\underset{\underset{\underset{\displaystyle CH_3}{|}}{\displaystyle CH_2}}{C}}{}^{\oplus} \ \longrightarrow \ {}^{\oplus}\overset{\overset{\displaystyle H}{|}}{\underset{\underset{\displaystyle CH_3}{|}}{C}} - \overset{\overset{\displaystyle H}{|}}{\underset{\underset{\displaystyle CH_3}{|}}{C}}{}^{\ominus}$$

(1-butene)	(I)	(II)
	(More stable)	(activated cis-2-butene)
		(Less stable)

2.82

89

Here, the hydrogen atom on CH_2CH_3 group is transferred, for which 2-butene is obtained, the unexpected arrangement. Therefore, the hydrogen cation will not be a transfer species. The transfer species cannot be that which is the more electropositive group on the substituent group as shown below again for 1-butene.

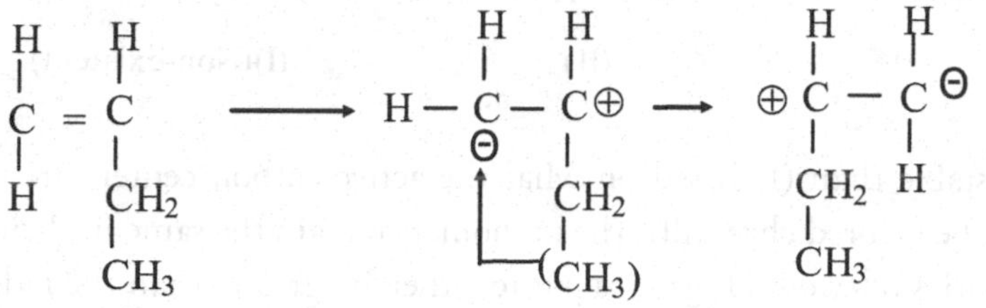

IMPOSSIBLE REARRANGEMENT   2.83

Thus, when $^{\oplus}CH_3$ group is transferred for molecular rearrangement, the same monomer is obtained. Since this cannot be abstracted during initiation, hence CH_3 cannot be the transfer species. For propylene, the following is obtained:-

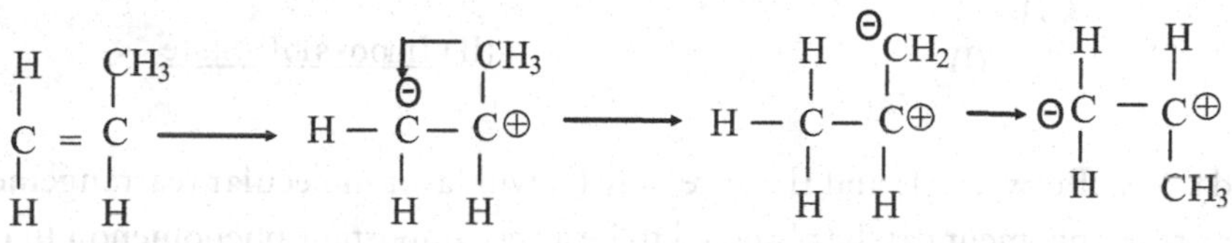

2.84

$H^{\oplus}$ from CH_3 group is transferred since CH_3 group cannot be scissioned from the active carbon centers. After the molecular rearrangement, propylene is regenerated. This can however affect polymerization times if a weak initiator is involved. Thus, while 1-butene does not undergo rearrangement, propylene does.

For isobutylene, molecular rearrangement to produce another monomer is not favored as shown below, since the same activated monomer is obtained. This does not imply that it does not

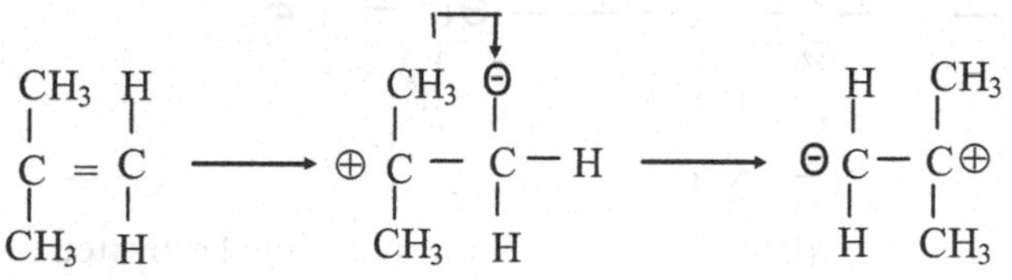

2.85

take place when the conditions exist. For the monomer of Equation 2.13 and $CF_3CH = CH_2$, molecular rearrangement is also not favored as shown below:-

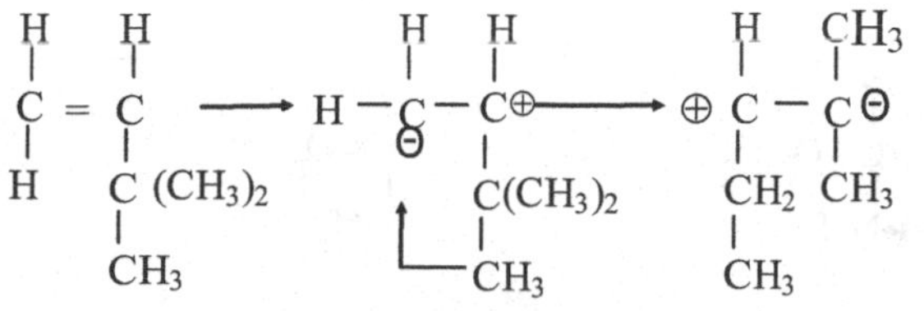

(II)   (I) (Non-existent)   2.86

(II) being far more stable than (I), does not favor the existence of (I) when transfer of transfer species is forcibly done. One cannot transfer CH_3 group to a center carrying H. Free-radically the same applies.

$$
\begin{array}{ccc}
\underset{\displaystyle\underset{H}{|}}{\overset{\displaystyle\overset{CF_3}{|}}{C}} = \underset{\displaystyle\underset{H}{|}}{\overset{\displaystyle\overset{H}{|}}{C}} & \longrightarrow & \ominus\underset{\displaystyle\underset{H}{|}}{\overset{\displaystyle\overset{CF_3}{|}}{C}} - \overset{\oplus}{\underset{\displaystyle\underset{H}{|}}{C}} - H & \longrightarrow & \oplus\underset{\displaystyle\underset{F}{|}}{\overset{\displaystyle\overset{F}{|}}{C}} - \underset{\displaystyle\underset{H}{|}}{\overset{\displaystyle\overset{CH_2F}{|}}{C}}\ominus
\end{array}
$$

$$\text{(II)} \qquad\qquad \text{(I)(Non-existent)} \qquad\qquad 2.87$$

(II) is also far more stable than (I) based on what the active carbon centers are carrying. Since (I) of both equation cannot be favored chargedly, the monomers remain the same in their activated states. The radical-pushing capacities of two CH_3 groups put together are greater than the radical-pushing capacity of CH_2 CH_3.[6] (I) of both equations should be:-

$$
\ominus\underset{\displaystyle\underset{CH_2}{\underset{|}{\underset{\displaystyle CH_3}{|}}}}{\overset{\displaystyle\overset{H}{|}}{C}} - \overset{\oplus}{\underset{\displaystyle\underset{CH_3}{|}}{\overset{\displaystyle\overset{CH_3}{|}}{C}}}
\qquad\qquad
\ominus\underset{\displaystyle\underset{F}{|}}{\overset{\displaystyle\overset{F}{|}}{C}} - \overset{\oplus}{\underset{\displaystyle\underset{H}{|}}{\overset{\displaystyle\overset{CH_2F}{|}}{C}}}
$$

$$\text{(I)} \qquad\qquad\qquad \text{(II) \underline{Impossible State}} \qquad\qquad 2.88$$

in their activated states. Between (I) and II above only (I) will favor molecular rearrangement. Thus, this type of molecular rearrangement can be observed to be a very important phenomenon in double bonded species radically or chargedly and triply bonded species only radically. Before it takes place, the monomer must be in Activated States of Existence.

Now consider the molecular rearrangement of some important monomers. Acrylamide has been considered. For methacrylamide, the following is obtained chargedly. It can also take place radically.

$$
\underset{\displaystyle\underset{H}{|}}{\overset{\displaystyle\overset{H}{|}}{C}} = \underset{\displaystyle\underset{\displaystyle\underset{NH_2}{|}}{\underset{|}{\overset{\displaystyle C=O}{|}}}}{\overset{\displaystyle\overset{CH_3}{|}}{C}}
\longrightarrow
H - \overset{\oplus}{\underset{\displaystyle\underset{\displaystyle\underset{NH_2}{}}{\underset{|}{\overset{\displaystyle C=O}{|}}}}{\overset{\displaystyle\overset{H}{|}}{C}}} - \underset{\displaystyle\underset{C=O}{}}{\overset{\displaystyle\overset{CH_3}{|}}{C}}\ominus
\longrightarrow
\ominus\underset{\displaystyle\underset{\displaystyle\underset{NH_2}{|}}{\underset{|}{\overset{\displaystyle CH_2}{|}}}}{\overset{\displaystyle\overset{CH_3}{|}}{C}} - \underset{\displaystyle}{\overset{\displaystyle\overset{O}{\|}}{C}}\oplus
$$

$$\qquad\qquad \text{(II)} \qquad\qquad \text{(a substituted activated dimethyl ketene)} \qquad 2.89$$

If the CH_3 group is replaced with CF_3, the following is obtained:-

$$
\underset{\displaystyle\underset{H}{|}}{\overset{\displaystyle\overset{H}{|}}{C}} = \underset{\displaystyle\underset{\displaystyle\underset{NH_2}{|}}{\underset{|}{\overset{\displaystyle C=O}{|}}}}{\overset{\displaystyle\overset{CF_3}{|}}{C}}
\longrightarrow
H - \overset{\oplus}{\underset{\displaystyle\underset{\displaystyle\underset{NH_2}{}}{\underset{|}{\overset{\displaystyle C=O}{|}}}}{\overset{\displaystyle\overset{H}{|}}{C}}} - \underset{\displaystyle}{\overset{\displaystyle\overset{CF_3}{|}}{C}}\ominus
\qquad \text{OR} \qquad
H - \overset{\oplus}{\underset{\displaystyle\underset{H}{|}}{\overset{\displaystyle\overset{H}{|}}{C}}} - \underset{\displaystyle\underset{\displaystyle\underset{NH_2}{|}}{\underset{|}{\overset{\displaystyle C=O}{|}}}}{\overset{\displaystyle\overset{CF_3}{|}}{C}}\ominus
\longrightarrow
$$

$$\qquad\qquad \text{(I)} \qquad\qquad\qquad\qquad \text{(II)}$$

```
   CF₃  O                  CH₂◄─               H    CF₃
   |    ||                 |    ⊕              |    |
  ⊖C —  C⊕      OR    [   ⊖C — C — F   ⟶   ⊕C — C⊖    ]      ⟶
   |                       |    |              |    |
   CH₂                     C=O  F              H    C=O
   |                       |                        |
   NH₂                     NH₂                      NH₂

  (I)′                    (III)                   (IV)
```

```
         CF₃  O
         |    ||
        ⊖C —  C⊕
         |
         CH₂
         |
         NH₂

        (V)
```
                                                                              2.90

(I) and (II) in the equation above are based on which of $CONH_2$ and CF_3 groups is more chargedly-pulling group. As it turns out however and will be explained further in the Series and Volume, the former is more chargedly-pulling than the latter (i.e., $CONH_2 > CF_3$ in charge-pulling capacity) and (V) is the molecular species favored with or without the replacement of CH_3 with CF_3 group. Also, (III) is less chargedly stable than (II) or (IV).

Molecular rearrangement for methyl methacrylate has been considered. For methyl acrylate, the following is obtained:-

```
  H    H                    H   H                     H    O
  |    |                    |   |                      |    ||
  C =  C          ⟶      H─ C — C⊖         ⟶        ⊖C — C⊕
  |    |                    |⊕  |                      |
  H    C=O                      C=O                    CH₂
       |                        |                      |
       O                        O                      O
       |                        ‖                      |
       CH₃                      CH₃                     CH₃
```
                                                                              2.91

Like the acrylamides and methyl methacrylate, substituted ketene is the monomer being polymerized for these monomers when weak initiators of charged-paired types are used. Thus, polyketones are the products of the charged-polymerization of these monomers under certain conditions, the most important of which is use of weak charged-paired initiators. Free-radically, the molecular rearrangements of these monomers are possible but not as common chargedly, though paired radical initiators are not yet commonly known to exist. Hence polyacrylamide, poly (methyl acrylate), poly (methyl methacrylate) etc. are produced from these monomers only free-radically (using the most common types of free-radical initiators generating "Catalyst") and not free-paired chargedly. What these initiators are, are unknown in Present-day Science. Chargedly, coordination initiators must be used to favor their polymerizations to produce molecularly rearranged products.

The molecular rearrangement in alkyl vinyl ethers has been considered. When favored based on the types of groups carried by the α-carbon center and the strength of the initiators active center, only polyacetals are largely produced.

With monomers such as acrylonitrile, ethylene, vinyl chloride etc., there is no transfer species to abstract for molecular rearrangement. With methyl acrylonitrile, there is no molecular rearrangement chargedly since transfer is not possible. ***One cannot transfer $H^{\ominus}$ or $H \bullet n$ as a transfer species, just like $n \bullet CH_3$ (or $n \bullet R$) or $^{\ominus}CH_3$ (or $^{\ominus}R$) types of groups. But they (but $H^{\ominus}$) can be released from a growing chain where possible.*** Hence free-radically, there is also no molecular rearrangement. With ignorance, free-radically, we may think that it is possible as shown below wherein the same activated monomer is obtained :-

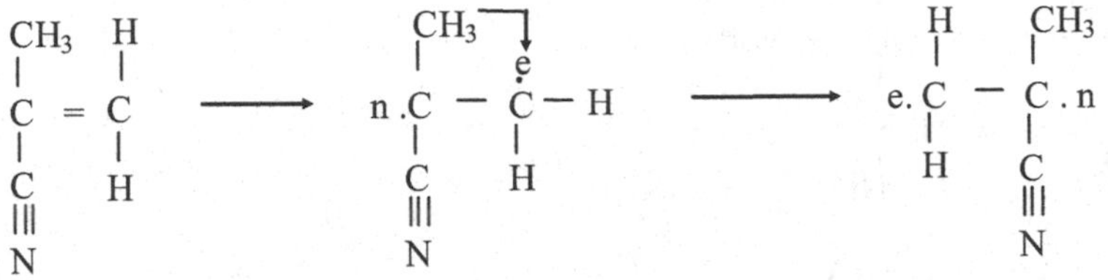

$$\text{2.92}$$

IMPOSSIBLE REARRANGEMENT

The same growing polymer chains are produced free-radically or chargedly for all acrylonitriles which cannot molecularly rearrange, whether the initiator is of the weak or strong type. Nucleo-free-radically or with the use of negatively-charged-paired initiators, it is the MALE (i.e. $Y \equiv C = C$ center) center that is involved. Electro-free-radically or with the use of positively-charged-paired initiators, it is the FEMALE (i.e. $X \equiv C \equiv N$ center) center that is involved.

Thus, when a growing polymer chain of 1-butene is to be killed, the following is obtained:-

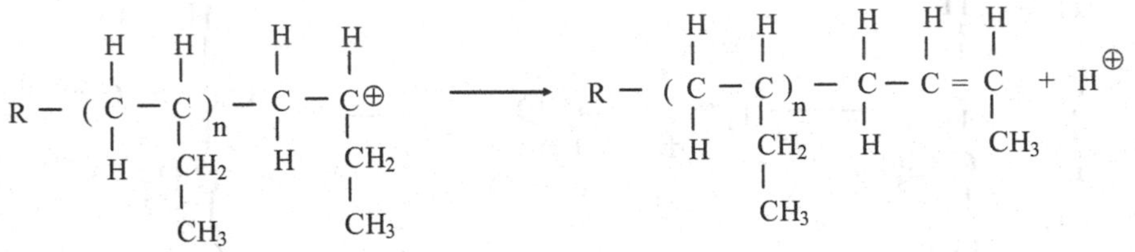

$$\text{2.93}$$

and not

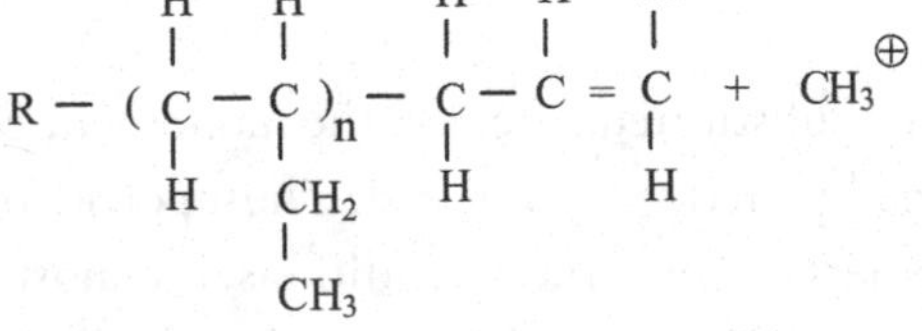

$$\text{2.94}$$

Similarly, for the growing polymer chain of (I)' of Equation 2.90, the following is obtained

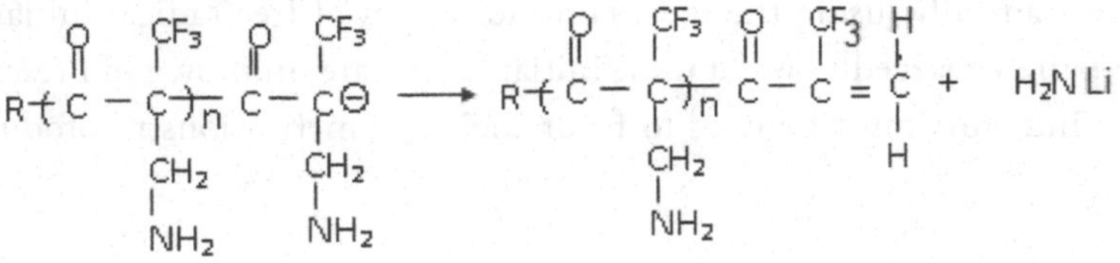

$$\text{2.95}$$

93

The initiator used above is $R^\ominus$.........$^\oplus$Li. Hence, $LiNH_2$ is one of the products, otherwise $H_2N^\ominus$ will not be released to kill the chain to make it chargedly and stoichiometrically balanced. Radically it can be released without being paired. The character of the terminal double bond section of the polymer produced is almost the same as that in the original monomer. Whereas if F is the transfer species, the following is obtained.

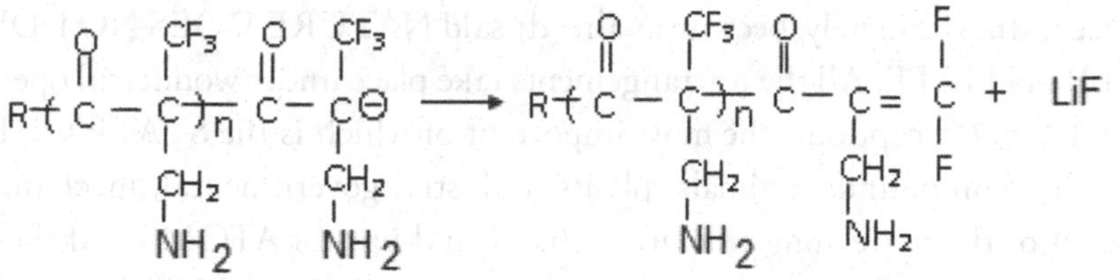

$$R \left(\overset{O}{\overset{\|}{C}} - \underset{\underset{NH_2}{\overset{|}{CH_2}}}{\overset{\overset{CF_3}{\overset{|}{}}}{\underset{|}{C}}} \right)_n \overset{O}{\overset{\|}{C}} - \underset{\underset{NH_2}{\overset{|}{CH_2}}}{\overset{\overset{CF_3}{\overset{|}{}}}{\underset{|}{C}}}{}^\ominus \longrightarrow R \left(\overset{O}{\overset{\|}{C}} - \underset{\underset{NH_2}{\overset{|}{CH_2}}}{\overset{\overset{CF_3}{\overset{|}{}}}{\underset{|}{C}}} \right)_n \overset{O}{\overset{\|}{C}} - \underset{\underset{NH_2}{\overset{|}{CH_2}}}{C} = \underset{\overset{|}{F}}{\overset{\overset{F}{\overset{|}{}}}{C}} + LiF$$

NOT FAVORED – impossible transfer   2.96

The dead terminal double bond polymer produced is not the same in character as with the original monomer. This is clear indication that the transfer species is from $CONH_2$ group and not CF_3 group. *Therefore, the charged-pulling capacity of $CONH_2$ is more than that of CF_3, because it is the richer group that gives unless when it cannot do so if the equation is to obey some of the laws of Mathematics- chargedly, radically and stoichiometrically balanced.* The monomer via $C = C$ π-bond activation is a strong electrophile as shown below.

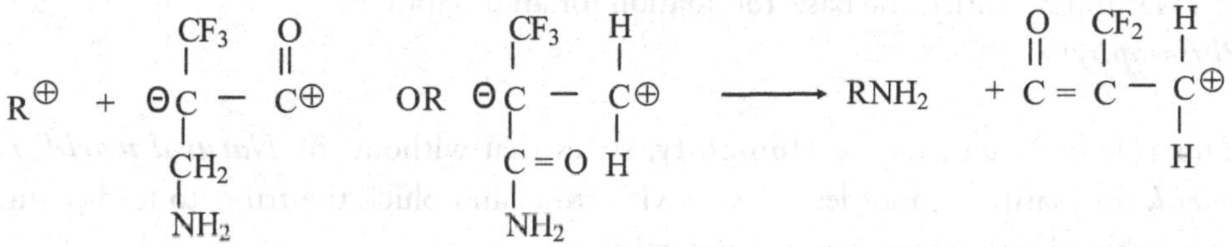

$$R^\oplus + {}^\ominus\underset{\underset{NH_2}{\overset{|}{CH_2}}}{\overset{\overset{CF_3}{\overset{|}{}}}{C}} - \overset{O}{\overset{\|}{C}}{}^\oplus \quad OR \quad {}^\ominus\underset{\underset{NH_2}{\overset{|}{C=O}}}{\overset{\overset{CF_3}{\overset{|}{}}}{C}} - \overset{\overset{H}{\overset{|}{}}}{\underset{H}{C}}{}^\oplus \longrightarrow RNH_2 + \overset{O}{\overset{\|}{C}} = \underset{\overset{|}{H}}{\overset{\overset{CF_2}{\overset{|}{}}}{C}} - \overset{\overset{H}{\overset{|}{}}}{\underset{H}{C}}{}^\oplus$$

2.97

It can clearly be observed in general for transfer species of the first kind of the first type in particular, that it is the transfer species preventing the monomer favoring opposite route, that is involved in killing a growing polymer chain in its natural route internally, and that is involved in molecular rearrangement. It is that transfer species which is the least radically-pulling (elec-tronegative) or radical-pushing (electropositive) on the substituent group. In Males, it comes from the Pulling groups, while in Females, it comes from Pushing groups.

2.3 Proposition of Rule of Chemistry

In proposing rule of Addition in Chemistry relevant to any chapter, the past from Volume (I) will always be recalled at appropriate times. In view of the importance of already known laws of conservations of Materials, Energy, Momentum, and much more particularly in Engineering, hence the main origin of all of them are identified herein. This is the Law of Conservation of transfer of transfer species which is nothing else but chemical conglomerations, made up of atoms, their elements, combinations of them and more. Without them, no Chemical (Micro-) or Polymeric (Macro-) reactions can take place. From them, we see how the Forces of Nature operate. From them, we see how Nature communicates to do what is to be expected and unexpected. From them, we begin to see not with the physical eyes, but with the "the eye of the needle".

Very great laws in continuation of what has been seen so far will emerge, because one has long ways to go, the ways which can only be seen by few. Molecular rearrangement is one of the most important rearrangements phenomena in *Tautomerism* in which living atomic or molecular species are carried along with their charges or radicals, instead of the movement of just radicals or "charges" alone, another form of rearrangements of compounds which could be of Living (Animals and Plants) or Non-Living systems. These will be shown as we move along the Series and Volumes. These rearrangements do not and never take place indiscriminately, because as already said NATURE DOES NOT DISCRIMINATE and indeed DIFFERENTIATE. All the rearrangements take place under wonderful operating conditions where there are no laws of exceptions, the most important of which is the STATE OF EXISTENCE of the SYSTEM whether compounds, animals, plants, industry, governmental, mechanical, electrical or whatever, because all of them are conglomerations (i.e. Build-ups) of ATOMS and their ELEMENTS. The types of System, their transfer species, the types of ACTIVATION CENTERS carried by them where they exist and how all these are placed are important in seeing how NATURE operates. For example, there are some monomers which exist, but cannot yet be identified, because of our limitations and great level of IGNORANCE. A very important simple one as already shown is vinyl nitrate. Then, how can one not state laws when one is dealing with brilliant Homo-sapiences (Not animals) that are so STIFF-NECKED in their ways of life and living? Regardless the types of human beings we are, the truth must be said and stated into laws without any fear of the unknown.

Rule 59: This rule of Chemistry for **Philosophy**, states that, to be a Philosopher, you must first and foremost be a Natural Scientist, the basic foundation for all disciplines.
(Laws of Philosophy)

Rule 60: This rule of Chemistry for **Humanity,** states that without the *Natural world, there is no Physical world,* for Nature cannot leave a tree with fruits and pluck the fruits to feed living and non-living systems, otherwise those systems will not exist.
(The fifth law of Nature)

Rule 61: This rule of Chemistry for **All compounds**, states that, these are combinations of atoms and their elements to build up different families of Molecules all of which have their different unique characters **(Chemical, Physical and Mathematical properties)**, with their domains and boundaries, many of which exist transiently depending on the operating conditions of where they live in; for which without their existence, nothing exists in the Physical and Natural worlds.
 (Laws of Creations for All Compounds)

Rule 62: This rule of Chemistry for **All types of reactions**, states that, since **one (integer) mole of any compound or system in a liter of a solution** contains the **same number of molecules that which is called Avogadro's number**, molecular species of the compound or system react with themselves productively or non-productively on integral (not real) molar (molecular) basis; for which the final equation for every stage is said to be **STOICHIOMETRICALLY** balanced.
(Laws of Mathematics of Equality and Inequality)

Rule 63: This rule of Chemistry for **All Chemical and Polymeric reactions**, states that, first and foremost, regardless the mechanisms (Equilibrium, Decomposition and Combination) by which they take place, they must be **STOICHIOMETRICALLY** balanced whether products are formed or not.
(Laws of Mathematics of Equality)

Rule 64: This rule of Chemistry for **All things in Life chemical and polymeric in character,** states that, **when an equation is said to be radically or chargedly balanced,** this implies that when free on the Left Hand Side (LHS) of an equation, it must be free on the Right Hand Side (RHS) of the equation and when non-free on the LHS of the equation, it must be non-free on the RHS; otherwise stable product(s) must be obtained on the RHS to balance the equation and this obtains only **via Decomposition mechanisms.**
(Laws of Math\ematics for Equality and Inequality)

Rule 65: This rule of Chemistry for **All things in Life chemical and polymeric in character**, states that, whether products are formed or not **via Equilibrium mechanisms,** the steps involved do not have to be RADICALLY OR CHARGEDLY balanced; when products are present on both sides of a chemical equation **via Decomposition mechanisms,** the steps involved have to be RADICALLY balanced; while when no dead products are obtained **via Combination mecha-nisms,** stages or steps involved have to be RADICALLY OR CHARGEDLY balanced.
(Laws of Physics and Mathematics)

Rule 66: This rule of Chemistry for **All things in Life chemical and Polymeric in character**, states that, when stable products are obtained **via Combination mechanisms**, the issue of Radical balancing does not arise, but only Stoichiometric balancing, just as when only one compound is made to exist in Decomposition state of existence to commence decomposition **via Decomposition mechanism,** the issue of Radical balancing does not arise, since a stable ptoduct is also obtained.
(Laws of Physics. Meta-Physics and Mathematics)

Rule 67: This rule of Chemistry for **All compounds and systems,** states that, all of them carry species which they use to react or interact with other compounds of different or same families and one of these species which identify with the family they belong to are called **SUBSTI-TUENT GROUPS** which are subset or same as **SUBTITUTED GROUPS** and these carry what are called **TRANSFER SPECIES** which can be part of the compound or system (e.g. $CONH_2$) or a part of a part carried by the compound (e.g. NH_2) or part of a part of a part (e.g. H).
(Laws of Creations from Nature)

Rule 68: This rule of Chemistry for **all compounds and systems,** states that, *there are many types of transfer species of many kinds and types released from compounds or systems during their reactions when attacked naturally and unnaturally;* and all these kinds and types depend on the States of Existence of the compound or system as determined by operating conditions.
(Laws of Creations from Nature)

Rule 69: This rule of Chemistry for **Addition polymerization systems,** states that, never is there a time when $H^{\ominus}$ or $\underline{H}{\bullet}n$, or $\underline{n{\bullet}CH_3}$, or $\underline{n{\bullet}R}$, or $^{\ominus}CH_3$, or $^{\ominus}R$, or $Cl{\bullet}en$, or $Cl^{\oplus}$, etc., types of groups can be used as transfer species or be abstracted as transfer species; but some (underlined) can however be released or rejected from a growing chain where possible in the route not natural to it under mild to very harsh operating conditions only radically.
(Laws of Creations for Addition polymerization systems)

Rule 70: This rule of Chemistry for **Olefinic Addition monomers,** states that, when the monomer **is activated,** the release of transfer species when it exists and when attacked unnaturally **takes place under Combination and Equilibrium mechanisms;** for which the transfer species involved is said to be of the **First kind of any type,** and when it takes place via Combination mechanism, the equation must be

radically or chargedly balanced and when it takes place via Equilibrium mechanism, the equation does not have to be radically or chargedly balanced.
(Laws of Creations from Nature)

<u>Rule 71:</u> This rule of Chemistry for **Addition polymerization systems,** states that, when a monomer or system **which has transfer species of the first kind of the first type,** grows in the route natural to it, it can be killed by release of this transfer species to give **a dead terminal double- or triple- or ketenic or cumulenic polymer of same or similar character as the originating compound or system**; the same transfer species abstracted in the route not natural to it and the same involved in Molecular rearrangement of the first kind of first type where it exists.
(Laws of Conservation of Transfer of Transfer species)

<u>Rule 72:</u> This rule of Chemistry for **Olefinic Addition monomers,** states that, when a monomer **which has no transfer species of the first kind of the first type,** grows in the route unnatural to it, it can be killed by release of transfer species to give only dead terminal double polymers or gigantic compounds or systems of the same character as the originating compound or system; for which the transfer species released from the β-carbon center is said to be of the **Second kind of first (Unnatural route) and second types (Natural route).**
(Laws of Creations for Termination-Transition)

<u>Rule 73:</u> This rule of Chemistry for **all things in Life chemical and polymeric in character,** states that, when a compound or system **with or without Activation centers is in *Stable State of Existence,*** the release of transfer species when it exists and when attacked **takes place only under Equilibrium mechanisms;** for which the transfer species is said to be transfer species of the **Second First kind of a particular type.**
(Laws of Creations for Transfer species)

<u>Rule 74:</u> This rule of Chemistry for **all things in Life chemical and polymeric in character,** states that, when a compound or system **with or without Activation center is in *Equilibrium State of Existence,*** the release of the component held in Equilibrium State of Existence to give a more stable compound or system **takes place only under Equilibrium mechanisms;** for which the component held in Equilibrium state of existence cannot be said to be transfer species, since it is the one involved in abstracting a transfer species.
(Laws of Creations for Transfer species)

<u>Rule 75:</u> This rule of Chemistry for **compounds and systems,** states that, **movements of electro-free-, electro-non-free-radicals from their carriers and with their carriers during chemical reactions either via Equilibrium or Decomposition mechanisms and some other phenonmena,** can produce different forms of energy, ranging from Electrical (Flow of current in selected fluids and solids), and Chemical (Real and Imaginary-Exothermic); for which the presence of specific different types of operating conditions based on the countless numbers of variables are involved.
(Laws of Creations in Engineering)

<u>Rule 76:</u> This rule of Chemistry for **non-ringed compounds and systems,** states that, for MALES to exist, they must carry X and Y Activation centers specially adjacently located (as already stated into law) and for FEMALES to exist, they can only have one or more X Activation centers of same or different capacities; for which for the MALES to fully exist, the X Activation center must carry one hetero-atom

(e.g. C=O, C=S, C=NH, all carrying O, S, and N respectively and etc.), otherwise no MALES can exist.
(Laws of Creations from Nature)

Rule 77: This rule of Chemistry for **ringed compounds and systems**, states that, for MALES to exist, they must carry X and Y Activation centers specially adjacently located (as already stated into law) and for FEMALES to exist, they can only have one or more X Activation centers of same or different capacities; for which for the MALES to fully exist, all of the Ys and some of the X Activation centers must carry one hetero-atom **(e.g. C=O, C=S, C=NH for Y centers
and O, S, and N, C =C for X centers);** otherwise no MALE can exist.
(Laws of Creations from Nature)

Rule 78: This rule of Chemistry for **Olefinic Addition monomers**, states that, when Molecular rearrangement of the first kind of the **first type** is allowed to take place, **the transfer species of the first type** is that coming from either alkanyl or alkylane (called alkyl groups in Present-day Science) groups **(e.g. CH_3, C_2H_5 or indeed C_nH_{2n+1} groups)** or non-alkylane groups (e.g. OH, OR, NH_2, NR_2, COOR, $CONR_2$, $CONO_3$, etc. groups) carried on the α-carbon center.**
(Laws of Molecular Rearrangements of first kind of first type)

 Rule 79: This rule of Chemistry for **Olefinic Addition monomers,** states that, when **two substituent groups of same character but different capacities {e.g. $(H_5C_2)C(CH_3)=CH_2$, $(OH)H_3CC=CH_2$} on the same carbon center of attack are present,** and Molecular rearrange-ment of the first kind and the first type is allowed to take place, the substituent group involved for releasing transfer species, **is the substituent group of greater capacity** to give a more stable Nucleophile as the driving force when favored (such as favored for the second case above, but not the first).
(Laws of Creations for Molecular rearrangement)

 Rule 80: This rule of Chemistry for **Olefinic Addition monomers,** states that, when **two substituent groups of two opposite characters {e.g. $(HOOC)C(CH_3)=CH_2$, $(H_2NOC)H_3C-C=CH_2$} on the same carbon center of attack are present (Males),** and Molecular rearrangement of the first kind of the first type is allowed to take place, the substituent group involved for releasing transfer species, is *the substituent group of radical- or -charged-pulling character* to give a more stable Electrophile as the driving force.
(Laws of Creations for Molecular rearrangement)

Rule 81: This rule of Chemistry for **Olefinic Addition monomers,** states that, when **two substituent groups of two opposite characters {e.g. $(H_3CCO)C(CH_3)=CH_2$} on the same center of attack are present,** no Molecular rearrangement can take place, if the substituent groups involved have transfer species which cannot be used; for which the compound or system is a weak Electrophile (Male). **[Acroleins]**
(Laws of Creations for Molecular rearrangement)

Rule 82: This rule of Chemistry for **Olefinic Addition monomers,** states that, when two **substi-tuent groups of different characters and different capacities {e.g. $H_2C=C(CH_3)O-COCH_3$} on the same carbon center of attack are present,** no molecular rearrangement can take place, if the substituent groups involved have transfer species which cannot be used, for which the com-pound or system is a weak Nucleophile (Female) **[Vinyl acetate].**
(Laws of Creations for Molecular rearrangement)

Rule 83: This rule of Chemistry for **Addition monomers,** states that, why monomers such as vinyl amine ($H_2C=CHNH_2$), vinyl alcohol ($H_2C=CHOH$), vinyl nitrates ($H_2C=CHONO_2$), allene ($H_2C=C=CH_2$), acetylenyl alcohol {$HC\equiv COH$}, acetylenyl amine ($HC\equiv CNH_2$),- all FEMALES, and all unique with respect to H atom, are not known to exist or when they do only favor Transient State of Existence because of the operating conditions, **is because they each molecularly rearrange to give another more stable compound or system.**
(Laws of Creations for Existence/Transformations)

Rule 84: This rule of Chemistry for **Addition monomers**, states that, where more than one nucleophilic center exists in a compound or system, *the first to be activated during reactions is the least nucleophilic that which requires the least operating conditions;* for which the following orders have begun to be identified.

$$C=O > C=N > {}^{\oplus}N=O > C=C > N=O \quad ; \quad N=O > C=C \ \textbf{(Chargedly)}$$

(Radically and some chargedly)

Order of Nucleophilicity

(Laws of Creations for Nucleophilic centers)

Rule 85: This rule of Chemistry for **Sources of Transfer species**, states that, the order of radical-pushing and radical-pulling capacities of groups as identified so far are as follows-

$$-OSO_2R > -ONO_2 > -NR_2 > -NRH > -NH_2 > -OH > -OR \quad \gg \quad -CR_3 > -CR_2H >$$

NON-FREE-GROUPS   **FREE-GROUPS**

$$-CRH_2 > -CH_3 > -CH_2X > -CX_2H > H > -CX_3 > -NO_2 > -SO_2R$$

FREE-GROUPS

Order of Radical-Pushing Capacities of Sources of Transfer species

$$-C\equiv N > -CONR_2 > -CONRH > -CONH_2 > -COOR > -COOH > -COR > -COH >$$

FREE- GROUPS

$$-N=O > -OCOR > \ .... \ \ggg F$$

NON-FREE-GROUPS

Order of Radical-Pulling Capacities of Sources of Transfer species

where the Rs are alkylane (alkyl) groups and X are non-metallic halogens (F, Cl, and Br). *(Laws of Physics)*

Rule 86: This rule of Chemistry for **Sources of Transfer species,** states that, the order of charged-pushing and charged-pulling capacities of groups as identified so far are as follows-

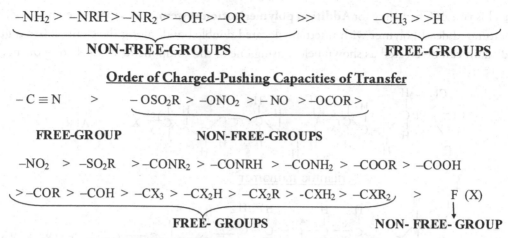

$$-NH_2 > -NRH > -NR_2 > -OH > -OR \qquad >> \qquad -CH_3 >> H$$

NON-FREE-GROUPS   **FREE-GROUPS**

Order of Charged-Pushing Capacities of Transfer

$$-C \equiv N \qquad > \qquad -OSO_2R > -ONO_2 > -NO > -OCOR >$$

FREE-GROUP   **NON-FREE-GROUPS**

$$-NO_2 > -SO_2R > -CONR_2 > -CONRH > -CONH_2 > -COOR > -COOH$$

$$> -COR > -COH > -CX_3 > -CX_2H > -CX_2R > -CXH_2 > -CXR_2 \quad > \quad F\ (X)$$

FREE- GROUPS   **NON- FREE- GROUP**

Order of Charged-Pulling Capacities of Transfer species
Order of Charged-Pulling Capacities of Sources of Transfer species

with many of the pulling groups having no ability to carry negative charges on their central atom despite their charge-pulling capabilities, where the Rs and X still remain as defined so far.
(Laws of Physics)

Rule 87: This rule of Chemistry for **Addition polymerization systems,** states that, all living polymer chains can *only grow linearly and horizontally* via Combination (Forwardly) or Equi-librium (Backwardly) mechanisms in the routes Natural or not Natural to them.
(Laws of Creations for Propagation/Growth)

Rule 88: This rule of Chemistry for **Addition polymerization systems,** states that, when living polymer chains exist in the route **not Natural to them**, they can be terminated (killed) either by presence of impurities or introduction of terminating agents or loss of transfer species of the second kind or diffusion to a dead terminal double bond or saturated polymer to give either an unsaturated dead polymer or a saturated dead polymer with or without externally located Activa-tion centers and a living polymer chain from the dead polymer with or without a branching site.
(Laws of Creations for Termination/Rebirth)

Rule 89: This rule of Chemistry for **Addition polymerization systems,** states that, when a saturated dead polymer chain is attacked by a growing polymer chain via Activation centers externally located on the dead polymer, dead saturated polymer and a living polymer chain with branching sites are obtained; for which the transfer species involved coming from the main chain backbone is said to be transfer species of the **THIRD KIND of the first type.**
(Laws of Creations for Branching site formation)

Rule 90: This rule of Chemistry for **Addition polymerization systems,** states that, when a saturated dead polymer chain is attacked by a growing polymer chain via Activation centers externally located on the dead polymer, dead saturated polymer and a living polymer chain with a branching site which is not directly placed on the main chain backbone is obtained and the transfer species involved is the first kind of the first type.
(Laws of Creations for Branching site formation)

Rule 91: This rule of Chemistry for **Addition polymerization system,** states that, when only a Z/N type of an unsaturated dead polymer with internally located double bonds along the chain exists, a growing branched chain can be created as shown below using one of the components of Z/N components-

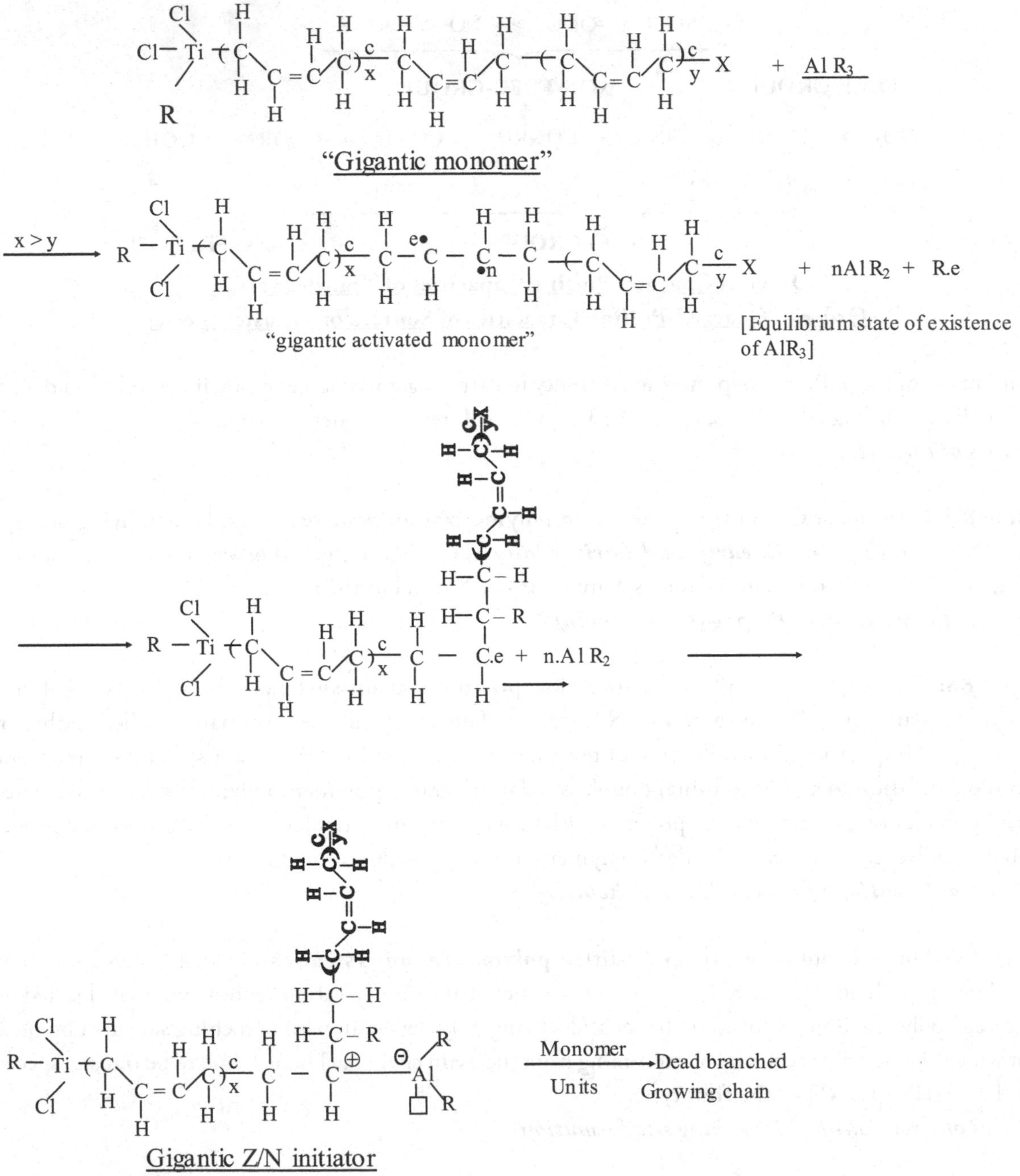

"Gigantic monomer"

$x > y$

"gigantic activated monomer"

[Equilibrium state of existence of AlR_3]

Gigantic Z/N initiator

Monomer Units → Dead branched Growing chain

for which after the electro-free-radical (R.e) has activated and added, the "gigantic monomer" being a strong nucleophile diffuses to the nucleo-free-radical counter-center to form gigantic Z/N initiator by Equilibrium mechanism *noting that it is not a branching site that is obtained here, but a dead branched growing polymer chain, in which the linear addition of monomer is still maintained.* *(Laws of Creations for Branching phenomena)*

101

Rule 92: This rule of Chemistry for **Addition polymerization systems,** states that, when a dead terminal double bond polymer chain is attacked by a growing polymer chain in the route natural to it, a saturated polymer with a branching site along the main chain backbone is obtained and this can only take place in a Free-media system.

(Laws of Creations for Branching site formation)

Rule 93: This rule of Chemistry for **Addition and Step polymerization systems,** states that, when saturated dead polymers exist for them, branching and cross-linking sites can only exist for them when they either carry externally located Activation centers which can only be used via Combination mechanisms or carry an externally located center which can readily exist in Equili-brium State of Existence and used via Equilibrium mechanisms.

(Laws of Creations for Branching site formation)

Rule 94: This rule of Chemistry for **Addition polymerization systems,** states that, the first step in the polymerization of any Addition monomer, is the *activation* of the monomer, followed by *molecular rearrangement* of the first kind of the activated monomer where possible based on the operating conditions, followed by *electrostatic rearrangement and reorientation* of the monomer in readiness for addition, followed by *abstraction of transfer species* where they exist (transfer from monomer to initiator sub-step), and finally followed by the *initiation* of the monomer, if transfer from monomer step is not favored.

(Laws of Creations for Initiation)

Rule 95: This rule of Chemistry for **Addition polymerization systems,** states that, when paired radical or charged initiators (Coordination initiators) of the *Covalent types* are involved, the two centers are real and active *(i.e. dual in character),* for which when a monomer is around, one particular active center based on the character of the monomer diffuses to the already activated monomer being initiated, while the activated monomer at the same time after *molecular rearrangement* if any, *electrostatic rearrangement and reorientation* diffuses to the counter inactive center of the coordination initiator; for which in this manner no transfer species exists other than favoring the conditions of radical or charge balancing.

(Laws of Creations for Coordination Initiators)

Rule 96: This rule of Chemistry for **Addition polymerization systems,** states that, when paired charged initiators (Coordination initiators) of the *Electrostatic types* are involved, there is only one active center-the real center *(i.e. single in character),* for which when a monomer is around, the center diffuses to the already activated monomer being initiated, while the activated monomer at the same time after *molecular rearrangement* if any, *electrostatic rearrangement and reorientation* diffuses to the imaginary counter inactive center of the coordination initiator; for which in this manner transfer species is abstracted if the route is not natural or addition follows if the route is natural provided the conditions of charge balancing is favored; noting that radically, these initiators cannot be formed.

(Laws of Creations for Coordination Initiators)

Rule 97: This rule of Chemistry for **all Addition monomers or systems,** states that, *for the occurrence of molecular rearrangements of the first kind of any type,* **the first driving force** is the existence of an Addition monomer or system with π-bond type of Activation center in which one of the active centers is carrying transfer species of the first kind of any type, followed by **the second driving force** which is the

use of weak initiators, followed by **the third driving force** which is the ability of the activated monomer after rearrangement to deactivate itself to produce a more stable monomer or molecule or system if the need arises.
(Laws of Creations for Molecular rearrangement/Transformation)

Rule 98: This rule of Chemistry for **Addition monomer or systems**, states that, during homopolymerization of any polymerizable monomer or system, diffusion of monomer when activated to initiator, or growing polymer chain is impossible, but the reverse *(Diffusion control- led mechanism),* otherwise homopolymerization will not take place; unless the initiator with one monomer unit is inside an <u>emulsified</u> mini-reactor **(Micelle)**, wherein the monomer in its own separate phase is made to diffuse to the emulsified growing polymer chain *(Mass-transfer controlled mechanism)..*
(Laws of Creations for Diffusion Controlled Mechanism)

Rule 99: This rule of Chemistry for **Addition monomers or systems,** states that, **the kinetic route favored by a monomer or system partly depends** on the character (Male or Female or "Both") of the monomer or system, which is partly determined by the type of activation center(s) carried by the monomer or system, the types and capacities of substituent groups carried by the monomer or system and the types of initiators involved.
(Laws of Creations for Addition monomers)

Rule 100: This rule of Chemistry for **Polar/Ionic compounds,** states that, **the first driving force** favoring the existence of ionic bonds is the presence of a positively charged center where the last shell is fully empty with no presence of paired unbonded or bonded radicals in the last shell (i.e. **Cation-** *Empty Hole in* its boundary), followed by **the second driving force** which is that in which the last shell of the negatively charged center a hetero-atom is full with paired unbonded or bonded radicals in the last shell that which gives it its **Polar** character, for which **the third driving force** for the ionic centers is such that the radicals involved on the Central atoms [Central hetero-atom carrying the negative non-free charge **(Anion) and Central metallic atom carrying the positive charge (Cation)**] *cannot be Hybridized; for which when they are hybridized, no ionic charges can exist.*
(Laws of Creation for Ionic Bond formations)

Rule 101: This rule of Chemistry for **Vinyl nitrate,** states that, this Addition monomer is very unstable for which the followings are obtained at normal operating conditions-

(I) Vinyl nitrate (Unstable)   (Stable)

MOLECULAR REARRANGEMENT OF THE FIRST KIND OF THE FIRST TYPE

$$
\begin{array}{c}
\text{H} \quad \text{H} \\
| \quad | \\
\text{C} = \text{C} \\
| \quad | \\
\text{H} \quad \text{O} \\
| \\
\text{NO}_2
\end{array}
\quad
\underset{\textit{Existence}}{\overset{\textit{Equilibrium State of}}{\rightleftharpoons}}
\quad
\begin{array}{c}
\text{H} \quad \text{H} \\
| \quad | \\
\text{C} = \text{C} - \text{O} \bullet nn \\
| \\
\text{H}
\end{array}
\quad + \quad e\bullet\text{NO}_2
$$

for which a ketone (II) above after molecular rearrangement of the first kind of the first type is obtained, noting that the monomer cannot be activated chargedly, due to electrostatic forces of repulsion, ONO_2 group being a charged-pulling group while radically it is a radical-pushing group of lower capacity than H; that which sends a message – that all chemical reactions take place only radically.

(Laws of Creations for Vinyl nitrate)

2.4. Conclusions

Forty three additional rules have been proposed bringing the total number to one hundred and one so far, so many of which based on the contents in Volume (I), have not been stated yet. Step by step the driving forces favoring the existence of certain phenomena such as molecular rearrangement, ionic, covalent, electrostatic, and polar bond formations, have begun to be identified. Indeed, we have only just begun.

The different types of transfer species that exist in Addition monomers are being care-fully identified in steps. Only those associated with olefinic monomers and very few hetero-atom containing monomers have begun to be put into rules. Of the three types of transfer species of the first kind, the most common seen so far is the first type in all Addition monomers considered so far. It is not however the only type of the first kind of transfer species that is associated with *the law of conservation of transfer of transfer species.* There are other types of the first kind also involved.

The non-existence of transfer from monomer Step during homopolymerization was stated into a rule where the origin of the concept of Diffusion Controlled mechanism was identified and distinguished from Mass Transfer Controlled mechanism. How to identify the characters of Addition monomers, have been clearly identified beginning from the first Volume and started to be stated into rules here in view of subsequent applications. The reasons why some monomers/ compounds are not known to exist have been identified.

References

1. http://www.freepatentsonline.com/3965081.html (1976)

2. C.R. Noller, Textbook of Organic Chemistry", W.B. Saunders Company, (1966), pg. 64.

3. C.R. Noller, "Textbook of Organic Chemistry", W.B. Saunders Company, (19660, pg. 599.

4. G. Odion, "Principles of Polymerization", McGraw-Hill Book Company, 1970), pg. 355 – 357.

5. L.W. Bush and D.S. Breslow, Macromolecules, 1: 189 (1968).

6. C.R. Noller, "Textbook of Organic Chemistry", W.B. Saunders Company, (1966), pg. 380.

Problems

2.1. (a) What is a transfer species?

(b) State the law of conservation of transfer of transfer species.

(c) Of what relevance is the law in polymerization systems?

2.2. Show the stages involved during the initiation step of the following monomers:-

$$
(i) \quad
\begin{array}{c}
H \quad CH_3 \\
| \quad\quad | \\
C = C \\
| \quad\quad | \\
Cl \; I_3 \quad H
\end{array}
\qquad
(ii) \quad
\begin{array}{c}
H \quad CH_3 \\
| \quad\quad | \\
C - C \\
| \quad\quad | \\
H \quad C = O \\
\quad\quad\quad | \\
\quad\quad\quad O \\
\quad\quad\quad | \\
\quad\quad\quad CH_3
\end{array}
\qquad
(iii) \quad
\begin{array}{c}
H \quad CH_3 \\
| \quad\quad | \\
C = C \\
| \quad\quad | \\
H \quad O \\
\quad\quad\quad | \\
\quad\quad\quad C = O \\
\quad\quad\quad | \\
\quad\quad\quad CH_3
\end{array}
\qquad
(iv) \quad
\begin{array}{c}
H \quad CH_3 \\
| \quad\quad | \\
C = C \\
| \quad\quad | \\
H \quad O \\
\quad\quad\quad | \\
\quad\quad\quad CH_3
\end{array}
$$

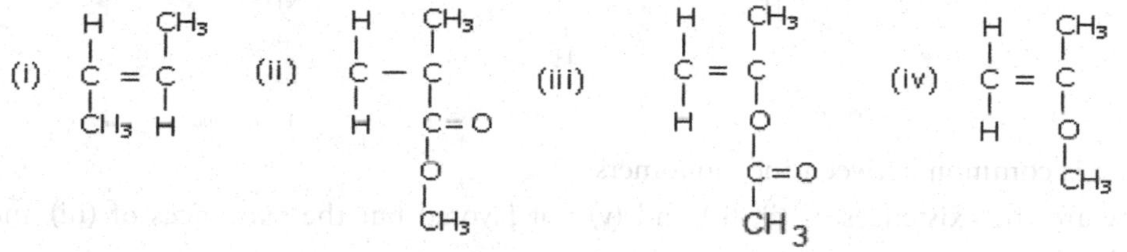

when the following weak initiators are involved:-

(a) $H^{\oplus}$ - free cationic initiator.

(b) $CH_3O^{\ominus}$- non-free anionic initiator.

(c) $H^{.n}$ – free-nucleo-free-radical.

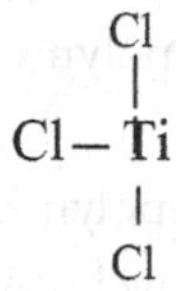

(d) $Cl - \overset{\displaystyle Cl}{\underset{\displaystyle Cl}{\overset{|}{\underset{|}{Ti}}}}$ .e - free electro-free-radical initiator or compound.

2.3. Shown below are the following monomers.

$$
(i) \quad
\begin{array}{c}
H \quad CH_3 \\
| \quad\quad | \\
C = C \\
| \quad\quad | \\
H \quad Cl
\end{array}
\qquad
(ii) \quad
\begin{array}{c}
CH_3 \quad H \\
| \quad\quad | \\
C = C \\
| \quad\quad | \\
H \quad CH \\
\quad\quad / \quad \backslash \\
\quad CH_3 \;\; CH_3
\end{array}
\qquad
(iii) \quad
\begin{array}{c}
CH_3 \quad H \\
| \quad\quad | \\
C = C \\
| \quad\quad | \\
H \quad C = O \\
\quad\quad\quad | \\
\quad\quad\quad O \\
\quad\quad\quad | \\
\quad\quad\quad CH_3
\end{array}
\qquad
(iv) \quad
\begin{array}{c}
H \quad H \\
| \quad\quad | \\
C = C \\
| \\
H
\end{array}
\;\;
\bigcirc\!\!-CH_3
$$

$$
(v) \quad
\begin{array}{c}
CF_2 \quad H \\
| \quad\quad | \\
C = C \\
| \quad\quad | \\
H \quad CH_2
\end{array}
\qquad
(vi) \quad
\begin{array}{c}
H \quad CH_3 \\
| \quad\quad | \\
C = C \\
| \quad\quad | \\
H \quad N^{\oplus} \\
\quad\quad\quad /\!\!\backslash \\
\quad\quad O \quad\;\; O^{\ominus}
\end{array}
\qquad
(vii) \quad
\begin{array}{c}
H \quad CH_3 \\
| \quad\quad | \\
C = C \\
| \quad\quad | \\
H \quad C = O \\
\quad\quad\quad | \\
\quad\quad\quad CH_3
\end{array}
\qquad
(viii) \quad
\begin{array}{c}
CH_3 \; CH_3 \\
| \quad\quad | \\
C = C \\
| \quad\quad | \\
H \quad CH_3
\end{array}
$$

106

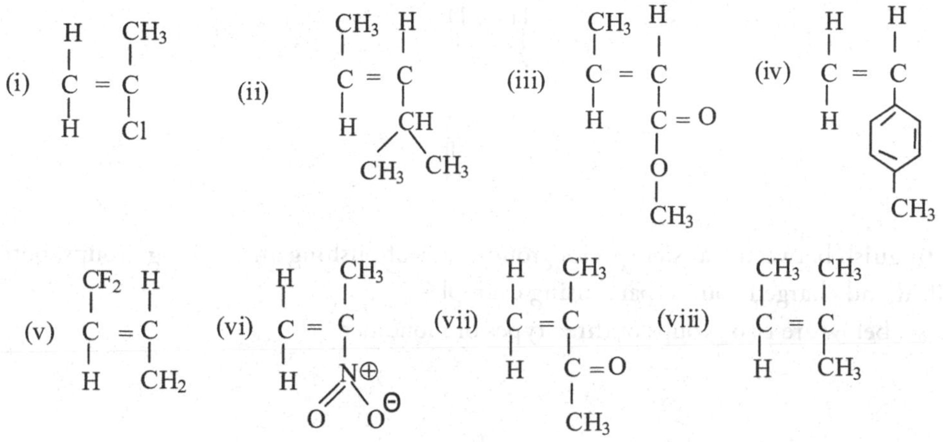

(ix) 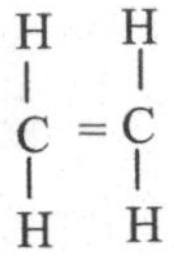
$$\begin{array}{c} H\quad H \\ |\quad\ | \\ C = C \\ |\quad\ | \\ H\quad H \end{array}$$

(x) 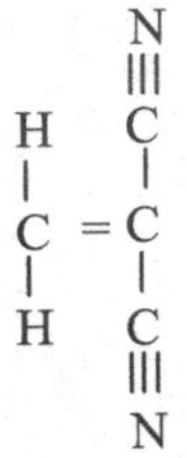
$$\begin{array}{c} N \\ ||| \\ H\quad C \\ |\quad\ | \\ C = C \\ |\quad\ | \\ H\quad C \\ ||| \\ N \end{array}$$

Identify the kinds of transfer species involved in all charged and free-radical routes for the monomers.

2.4. Shown below are the following monomers

(i) 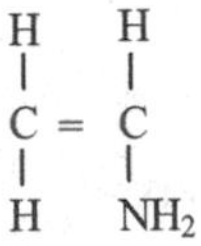
$$\begin{array}{c} H\quad H \\ |\quad\ | \\ C = C \\ |\quad\quad | \\ H\quad NH_2 \end{array}$$

(ii) 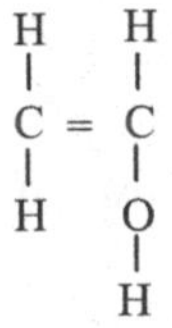
$$\begin{array}{c} H\quad H \\ |\quad\ | \\ C = C \\ |\quad\ | \\ H\quad O \\ \quad\ | \\ \quad\ H \end{array}$$

(iii) 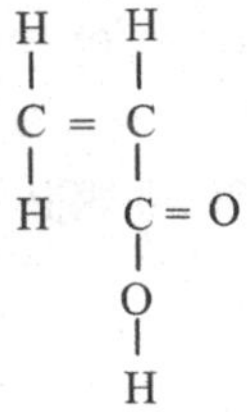
$$\begin{array}{c} H\quad H \\ |\quad\ | \\ C = C \\ |\quad\quad | \\ H\quad C = O \\ \quad\quad | \\ \quad\quad O \\ \quad\quad | \\ \quad\quad H \end{array}$$

(iv) 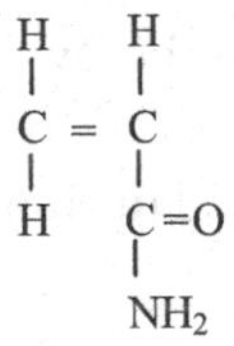
$$\begin{array}{c} H\quad H \\ |\quad\ | \\ C = C \\ |\quad\quad | \\ H\quad C=O \\ \quad\quad | \\ \quad\quad NH_2 \end{array}$$

(v) 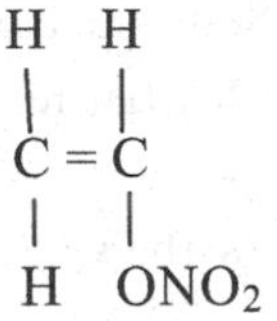
$$\begin{array}{c} H\quad H \\ |\quad\ | \\ C = C \\ |\quad\quad | \\ H\quad ONO_2 \end{array}$$

(a) What is common between the monomers?

(b) Why are the existences of (i), (ii) and (v) not favored but the existences of (iii) and (iv) are ? Explain.

(c) How can the charged polymerizations of (iii) and (iv) be made possible?

(d) Why is the free-radical polymerization of the monomers (iii) and (iv) more favored than their charged polymerization?

2.5. (a) What can be done to make (i) and (ii) of Q2.4 favor carrying a carbon-carbon backbone where possible, that is, prevent molecular rearrangement? Explain

(b) What can also be done to make (iii) and (iv) of same question favor charged polymerization without molecular rearrangement?

(c) Why does a monomer such as acrylonitrile shown below not favor positively charged polymerization via $\cdot C = C$ activation center? Under what conditions can it be polymerized positively chargedly and what type of polymeric chain will be obtained?

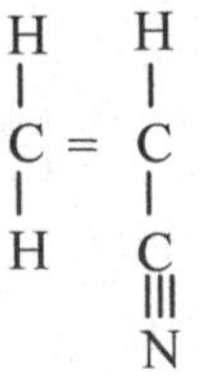

$$\begin{array}{c} H\quad H \\ |\quad\ | \\ C = C \\ |\quad\ | \\ H\quad C \\ ||| \\ N \end{array}$$

2.6. (a) Distinguish between transfer species groups, radical-pushing and pulling groups, both for their radical and charged counterparts using examples.

(b) Shown below are two groups of three types of monomers-

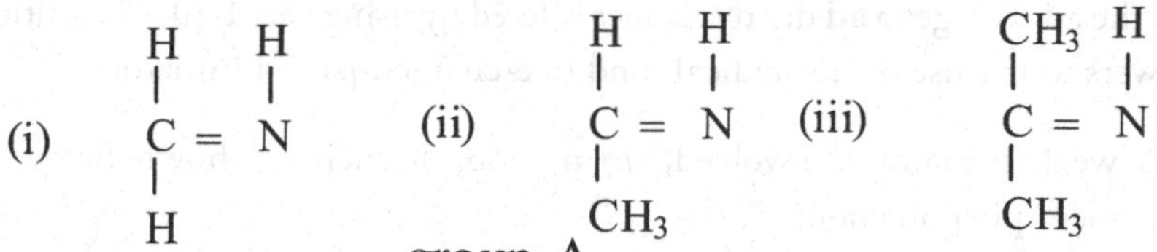

$$\text{(i)} \quad \begin{array}{c} H \\ | \\ C = N \\ | \\ H \end{array} \qquad \text{(ii)} \quad \begin{array}{c} H \quad H \\ | \quad\quad | \\ C = N \\ | \\ CH_3 \end{array} \qquad \text{(iii)} \quad \begin{array}{c} CH_3 \quad H \\ | \quad\quad | \\ C = N \\ | \\ CH_3 \end{array}$$

group A

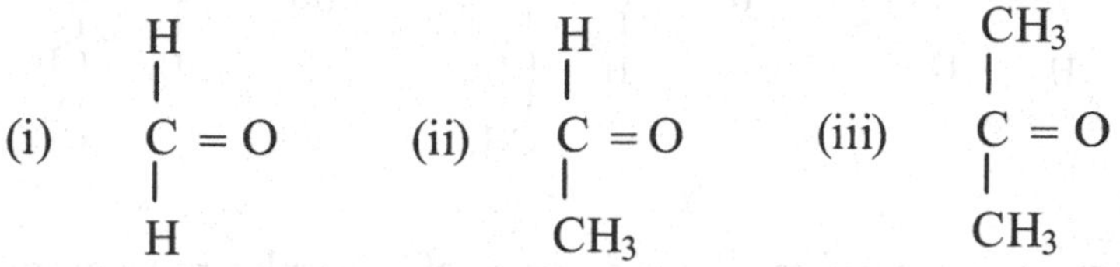

$$\text{(i)} \quad \begin{array}{c} H \\ | \\ C = O \\ | \\ H \end{array} \qquad \text{(ii)} \quad \begin{array}{c} H \\ | \\ C = O \\ | \\ CH_3 \end{array} \qquad \text{(iii)} \quad \begin{array}{c} CH_3 \\ | \\ C = O \\ | \\ CH_3 \end{array}$$

group B

Distinguish between the two groups of the three types of monomers in terms of stability and the routes favored by them when suppressed.

2.7. (a) For the monomers shown in Q2.6 (b), identify the kinds of transfer species involved chargedly.
  (b) Why can't the first group be activated chargedly or radically when not suppressed while the second group of monomers can?
  (c) Which of the monomers obey the law of conservation of transfer of transfer species? Explain. Is molecular rearrangement favored by any of the monomers? If not, explain.

2.8. (a) Referring again to the monomers listed in Q 2.6 (b), why can't free negatively charged initiators be used where possible? Why is it that non-free negatively charged (anionic) initiators cannot be used where possible for some of them?
  (b) Why can't nucleo-free-radicals be used for any of the monomers?
  (c) Identify the kinds and characters of transfer species involved radically and show how they are involved.

2.9. Shown below are two monomers

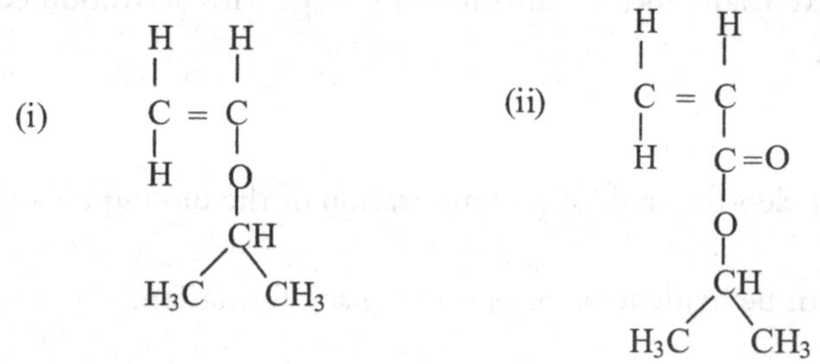

$$\text{(i)} \quad \begin{array}{c} H \quad H \\ | \quad\quad | \\ C = C \\ | \quad\quad | \\ H \quad\;\; O \\ \quad\quad | \\ \quad\quad CH \\ \quad H_3C \;\; CH_3 \end{array} \qquad\qquad \text{(ii)} \quad \begin{array}{c} H \quad H \\ | \quad\quad | \\ C = C \\ | \quad\quad | \\ H \quad\;\; C = O \\ \quad\quad\;\; | \\ \quad\quad\;\; O \\ \quad\quad\;\; | \\ \quad\quad\;\; CH \\ \quad\;\; H_3C \;\; CH_3 \end{array}$$

  (a) Distinguish between the two monomers in terms of their characters based on routes favored by them chargedly and radically.

(b) What are the advantages **and** disadvantages offered by using very bulky substituent group? Limit your answers to the use of free-radical- and free-charged-paired initiators.

2.10. (a) When a weak initiator is involved, do monomers such as shown below favor molecular rearrangement phenomenon?

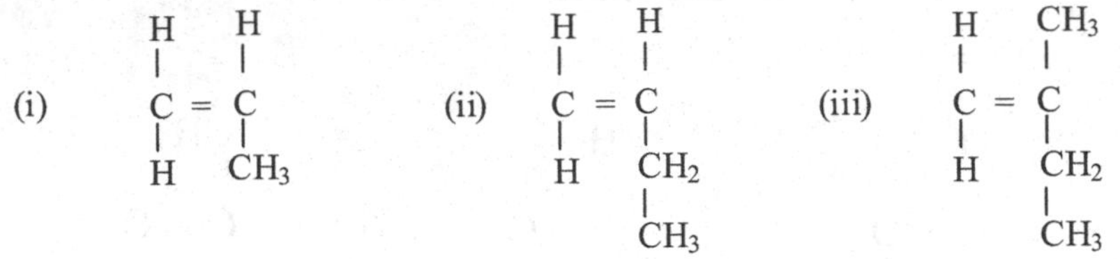

$$
(i) \quad
\begin{array}{cc}
H & H \\
| & | \\
C & = C \\
| & | \\
H & CH_3
\end{array}
\qquad
(ii) \quad
\begin{array}{cc}
H & H \\
| & | \\
C & = C \\
| & | \\
H & CH_2 \\
 & | \\
 & CH_3
\end{array}
\qquad
(iii) \quad
\begin{array}{cc}
H & CH_3 \\
| & | \\
C & = C \\
| & | \\
H & CH_2 \\
 & | \\
 & CH_3
\end{array}
$$

(b) Why does a weak initiator favor the occurrence of molecular rearrangement and a strong initiator does not?

(c) Distinguish between the use of weak and strong initiators on the polymerization of the monomers above.

2.11. (a) Why does propylene not favor existence of a branching site between a dead saturated polymer and a growing polymer chain, but does via the terminal double bond? Show how this can be done?

(b) Can species covalently bonded to the main chain backbone of olefinic monomers be readily abstracted? If it can, under what conditions?

(c) Consider the monomer shown below:-

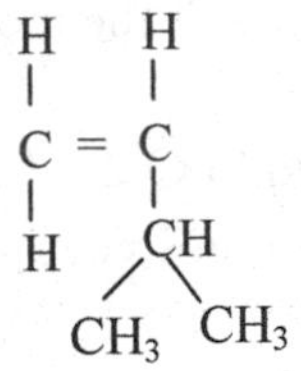

$$
\begin{array}{cc}
H & H \\
| & | \\
C & = C \\
| & | \\
H & CH \\
 & \diagup \ \diagdown \\
CH_3 & CH_3
\end{array}
$$

Define the route favored by the monomer and the transfer species involved. Does the monomer favor molecular rearrangement? If not, explain.

2.12. (a) Can you show how branching site can be formed from reaction between a dead saturated polymer with externally located substituent groups with activation center for monomers such as the acrylates?

(b) How can branching site be obtained for the acetates?

(c) Consider the nucleo-free-radical polymerization of the monomers shown below:-

Show whether they can be made to favor branching site formations.

2.13. (a) Distinguish between "Transfer from monomer sub-step" and "Transfer from dead polymer sub-step".

(b) Identify and distinguish between the two monomers shown in Q 2.12 (c) in every respect when the following initiators are involved- a) NaCN, b) ROR/BF$_3$.

2.14. Shown below is a dead polymer of vinyl acetate of chain lengths (x+y).

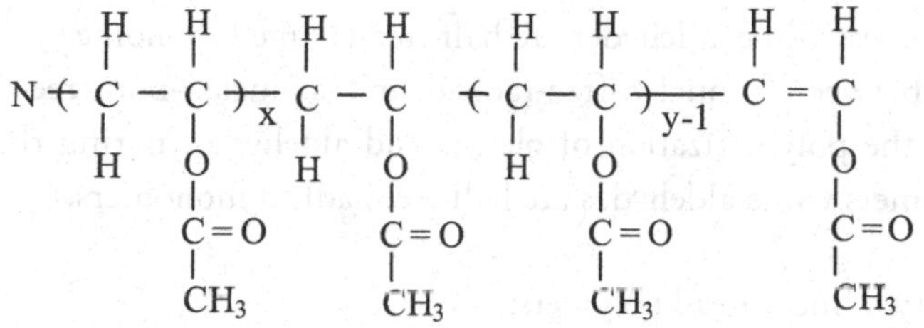

$$
N \left(\begin{array}{c} H \\ | \\ C \\ | \\ H \end{array} - \begin{array}{c} H \\ | \\ C \\ | \\ O \\ | \\ C=O \\ | \\ CH_3 \end{array}\right)_x \begin{array}{c} H \\ | \\ C \\ | \\ H \end{array} - \begin{array}{c} H \\ | \\ C \\ | \\ O \\ | \\ C=O \\ | \\ CH_3 \end{array} \left(\begin{array}{c} H \\ | \\ C \\ | \\ H \end{array} - \begin{array}{c} H \\ | \\ C \\ | \\ O \\ | \\ C=O \\ | \\ CH_3 \end{array}\right)_{y-1} \begin{array}{c} H \\ | \\ C \\ | \\ C=O \\ | \\ CH_3 \end{array} = \begin{array}{c} H \\ | \\ C \\ | \\ O \\ | \\ C=O \\ | \\ CH_3 \end{array}
$$

How was the dead polymer obtained? Which of the two Activation centers is more nucleophilic? Explain how you can add another monomer such as vinyl chloride along the chain.

2.15. (a) How does the law of conservation of transfer of transfer species affect the existence of molecular rearrangements in alkyl vinyl ethers?

(b) From the observations so far, show which of the groups is more pulling chargedly - CF_3 and $-CONH_2$ groups. What are the radical characters of the groups?

2.16. (a) Why is $-CF_3$ group a free-radical-pushing group and a free-charged-pulling group? Can $-CF_3$. $-COCOCH_3$, $-C\equiv N$, $-COOCH_3$, $-CONH_2$ groups carry a negative charge on their C centers when they can bring negative charges to the C centers carrying them? Explain.

(b) Why do -OR and $-NR_2$ groups (non-free-radical-pushing groups) behave differently from a -Cl group (a non-free-radical-pulling group)? Explain.

(c) Why are the following groups shown below

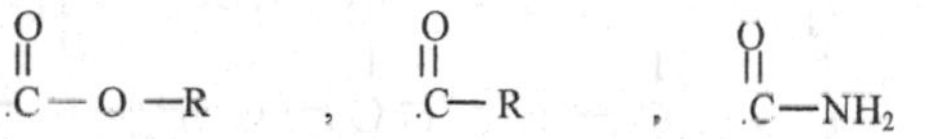

$$
\begin{array}{ccc}
O & O & O \\
|| & || & || \\
.C - O - R & , \quad .C - R & , \quad .C - NH_2
\end{array}
$$

pulling groups free-radically and chargedly?

(d) Identify the types and characters of the transfers species carried by the substituent groups above.

(e) Distinguish between the two groups shown below

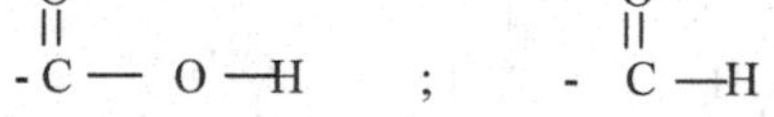

$$
\begin{array}{ccc}
O & & O \\
|| & & || \\
-C - O - H & ; & - C - H
\end{array}
$$

in terms of their charged characters.

2.17. (a) What is a transfer species of the second first kind of the first type?

(b) Below are shown two negatively-paired-charged–initiators for the polymerization of acrylamide

(i) CH_3ONa (ii) H_9C_4Li

Show which can be used to favor full polymerization of the monomer. Show how branches can be favored when used, using your own method.

2.18. (a) Distinguish between a free-negative charge and a non-free negative charge in their use as initiators for the polymerizations of olefins and aldehydes, noting that olefins are full free-charged monomers while aldehydes are half-free-charged monomers.

(b) Distinguish between a nucleo-free-radical and a nucleo-non-free-radical in their use as initiators for the polymerization of olefins and aldehydes, noting that olefins are full free-radical monomers while aldehydes are half-free-radical monomers.

2.19. Shown below are two linear dead polymers:-

(i)

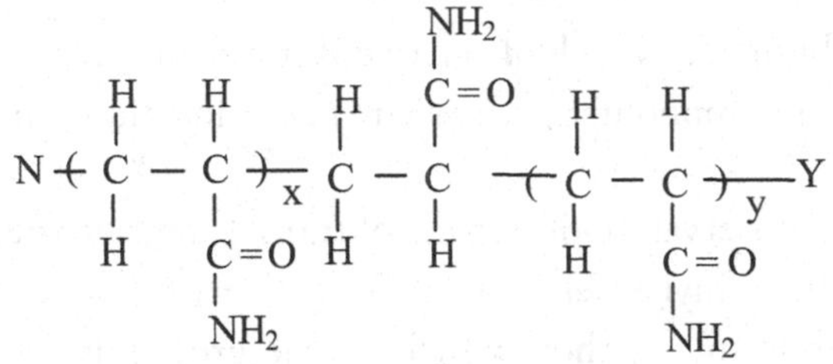

$$N\left(\begin{array}{c}H\\|\\C\\|\\H\end{array}-\begin{array}{c}H\\|\\C\\|\\C=O\\|\\NH_2\end{array}\right)_x\begin{array}{c}H\\|\\C\\|\\H\end{array}-\begin{array}{c}NH_2\\|\\C=O\\|\\C\\|\\H\end{array}\left(\begin{array}{c}H\\|\\C\\|\\H\end{array}-\begin{array}{c}H\\|\\C\\|\\C=O\\|\\NH_2\end{array}\right)_y Y$$

(ii)

$$R\left(\begin{array}{c}O\\||\\C\end{array}-\begin{array}{c}H\\|\\C\\|\\CH_2\\|\\NH_2\end{array}\right)_x\begin{array}{c}NH_2\\|\\CH_2\\|\\C\\||\\O\end{array}-\begin{array}{c}H\\|\\C\\|\\H\end{array}\left(\begin{array}{c}O\\||\\C\end{array}-\begin{array}{c}H\\|\\C\\|\\CH_2\\|\\NH_2\end{array}\right)_y\begin{array}{c}O\\||\\C\end{array}-\begin{array}{c}H\\|\\C=\end{array}\begin{array}{c}H\\|\\C\\|\\H\end{array}$$

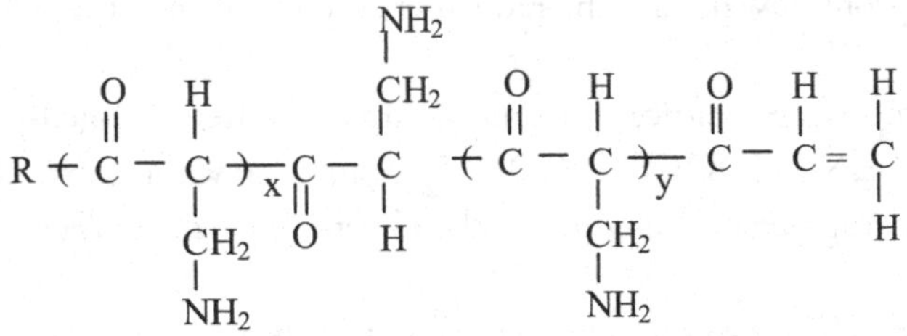

and two linear living polymers:-

(iii)

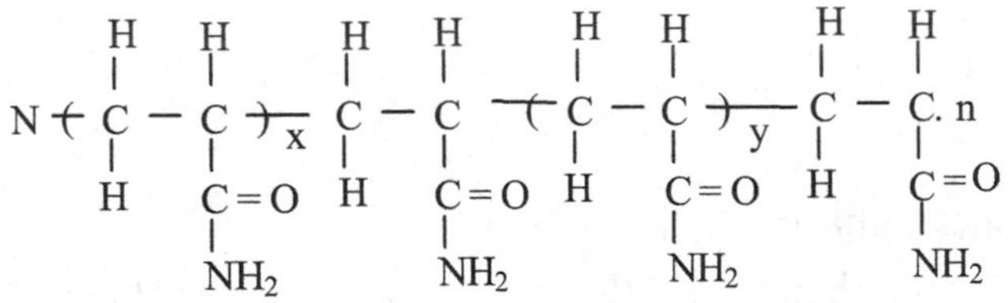

$$N\left(\begin{array}{c}H\\|\\C\\|\\H\end{array}-\begin{array}{c}H\\|\\C\\|\\C=O\\|\\NH_2\end{array}\right)_x\begin{array}{c}H\\|\\C\\|\\H\end{array}-\begin{array}{c}H\\|\\C\\|\\C=O\\|\\NH_2\end{array}\left(\begin{array}{c}H\\|\\C\\|\\H\end{array}-\begin{array}{c}H\\|\\C\\|\\C=O\\|\\NH_2\end{array}\right)_y\begin{array}{c}H\\|\\C\\|\\H\end{array}-\begin{array}{c}H\\|\\C\\|\\C=O\\|\\NH_2\end{array}.n$$

(iv)

$$R\left(\begin{array}{c}O\\||\\C\end{array}-\begin{array}{c}H\\|\\C\\|\\CH_2\\|\\NH_2\end{array}\right)_x\begin{array}{c}O\\||\\C\end{array}-\begin{array}{c}H\\|\\C\\|\\CH_2\\|\\NH_2\end{array}\left(\begin{array}{c}O\\||\\C\end{array}-\begin{array}{c}H\\|\\C\\|\\CH_2\\|\\NH_2\end{array}\right)_y\begin{array}{c}O\\||\\C\end{array}-\begin{array}{c}H\\|\\C^{\ominus}\\|\\CH_2\\|\\NH_2\end{array}\ldots\ldots\overset{\oplus}{Li}$$

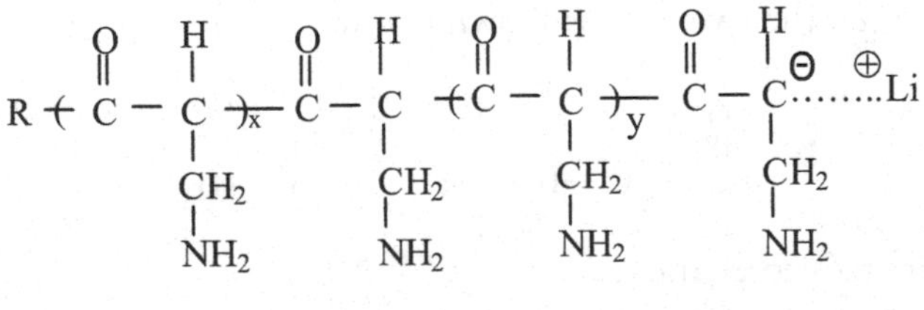

(a) Describe how the different chains were obtained.

(b) Show how branching sites can be favored for the different chains using growing polymer chains during polymerization where possible.

2.20. (a) Consider the electro-free-radical and "cationic" polymerization of acrylamide and methyl acrylate.

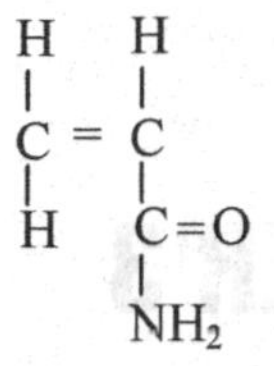

$$
\begin{array}{cc}
\begin{array}{c}
\text{H} \quad \text{H} \\
| \qquad | \\
\text{C} = \text{C} \\
| \qquad | \\
\text{H} \quad \text{C}{=}\text{O} \\
\qquad | \\
\qquad \text{NH}_2 \\
(\text{I})
\end{array}
&
\begin{array}{c}
\text{H} \quad \text{H} \\
| \qquad | \\
\text{C} = \text{C} \\
| \qquad | \\
\text{H} \quad \text{C}{=}\text{O} \\
\qquad | \\
\qquad \text{O} \\
\qquad | \\
\qquad \text{CH}_3 \\
(\text{II})
\end{array}
\end{array}
$$

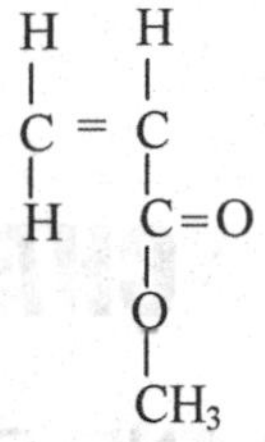

(i) Why are the routes not favored for the C = C center, but favored for the C = O center?

(ii) In (iii) of Q2.19, dead terminal double bond polymers can still be produced nucleo-free-radically when not paired. Identify the types and show which the real one is if not both.

(b) Shown below is the monomer called vinyl nitrate-

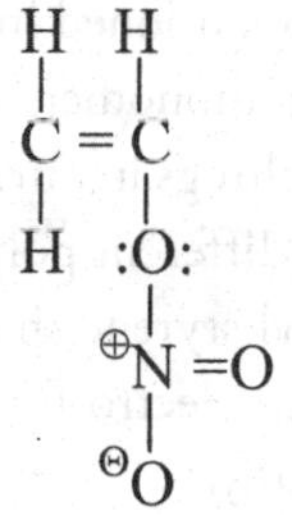

$$
\begin{array}{c}
\text{H} \quad \text{H} \\
| \qquad | \\
\text{C} = \text{C} \\
| \qquad | \\
\text{H} \quad :\text{O}: \\
\qquad | \\
\qquad \overset{\oplus}{\text{N}} {=} \text{O} \\
\qquad | \\
\qquad \overset{\ominus}{\text{O}}
\end{array}
$$

(i) What is the character of the monomer?

(i) Which of the Activation centers is more nucleophilic chargedly and radically?

(i) What is the transfer species in the monomer

(i) Why is the monomer not known to exist at STP operating conditions?

(i) What is the capacity of the NO_2 group in ONO_2 group radically and chargedly?

(i) What is the activation state of Existence chargedly and radically and therefore based on your answer why do all chemical reactions take place radically?

(ii) What does it rearrange to?

CHAPTER 3

CHARGED AND RADICAL RESONANCE STABILIZATION PHENOMENA AND EFFECTS IN OLEFINS AND POLYMERIZATION SYSTEMS

3.0 Introduction

In the first Volume, the phenomena were mentioned in passing with respect to Addition monomers. Here the phenomena as they apply to Addition monomers will begin to be consider-ed. The phenomena are so important that, they can affect so many things in our world as well as the transfer of transfer species in polymeric systems to favor the existence of different polymeric products. It is in view of the existence of such phenomena that for example α- methyl styrene shown below favors all existing free-radical and "charged"[1-8] routes, but propylene favors only the electro-free-radical and positively charged routes, despite the similarities in the two monomers structurally.

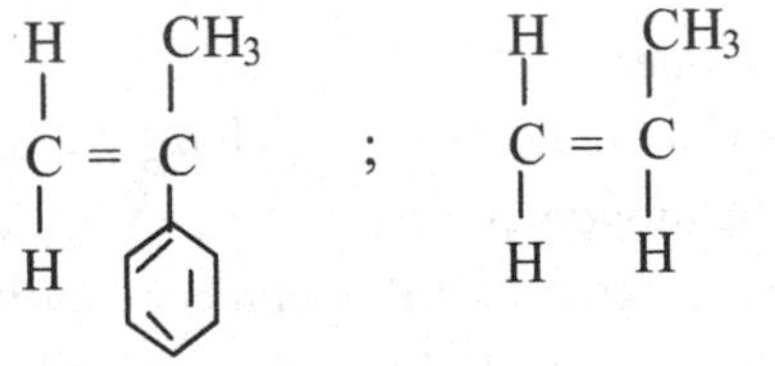

$$
\begin{array}{cccc}
\text{H} & \text{CH}_3 & \text{H} & \text{CH}_3 \\
| & | & | & | \\
\text{C} & = & \text{C} & \quad ; \quad & \text{C} & = & \text{C} \\
| & | & | & | \\
\text{H} & & \text{H} & \text{H}
\end{array}
$$

3.1

α - Methyl styrene   Propylene

[Nucleophiles] - Full Free centers

When they both undergo similar routes natural to them, different polymeric products are obtained. But when they undergo routes not natural to them polymeric products are obtained only from one of them and not from the other. Why?

For years, it has been observed, for example, that despite the similarity between acrolein and 1,3 – butadiene (that is, conjugation of the alkene and carbonyl double bond) as shown below, 1,4-type of polymerization has never been observed for acrolein.[9] Why? In order to provide resonance stabilization effects, certain conditions must be met, some of which have been slightly mentioned in passing.

$$H \quad H$$
$$| \quad |$$
$$C = C - C = O \qquad \qquad \overset{\oplus}{\underset{4}{C}} - \underset{3}{C} = \underset{}{C} - \overset{1}{O}\ominus \qquad \qquad 3.2$$
$$| \qquad |$$
$$H \qquad H$$

<u>Acrolein</u>   <u>1,4-Activated acrolein</u>
(Impossible existence via Resonance stabilization)

Without the use of universal literature data much in abundance, how can one move forward? Why the 1,4 Activated acrolein above is said to be impossible is because the 3,4-**Full free-monomer via the C = C center** has now been transformed to 1,4- 0r 1,2-**Half free-monomer**, instead of **Full free** by Activation and movement of electro-free-radicals. *It is important to recall again at this point in time that it is only the electro-radicals that diffuse all the time in the presence or absence of nucleo-radicals. Nucleo-radicals only diffuse in the absence of the electro-radicals in the system or when paired. Notice again that there is a difference between the movement of a radical and the carrier of the radical. An electro-free-radical can be removed from its carrier to form cation or positive charge, but a nucleo-non-free-radical cannot be removed from its carrier to form an anion or negative charge. This is why nucleo-radicals cannot be removed from its carrier. Only electro-radicals can. Can one do the same for charges? Can one remove the positive charge carried by $H^\oplus$ to leave what behind? What is it carrying when the hole is empty? This is what Present-day Science do, something which is impossible in the sight of the ALMIGHTY INFINITE GOD, something against HIS LAWS, the laws of Nature. However, let us for now "assume" that the existence of 1,4- charged states shown in Equations 3.2 and 3.3 below and thereafter are possible, until when we discover that it cannot be possible.*

$$H \quad H \qquad H$$
$$| \quad | \qquad |$$
$$C = C - C = C \qquad \qquad \overset{\oplus}{\underset{1}{C}} - \underset{2}{C} = \underset{3}{C} - \underset{4}{C}\ominus \qquad \qquad 3.3$$
$$| \qquad | \quad |$$
$$H \qquad H \quad H$$

<u>1,3-butadiene</u>   <u>1,4-Activated butadiene</u>
(Possible existence not via Resonance stabilization)

The 1,4 Activated butadiene above is said to be possible because the 1,2-**Full free-monomer** has been transformed to the same 1,4-**Full free-monomer** by Activation and movement of electro-free radicals.

Their effects on molecular rearrangement of olefins will also be considered. These phenomena are provided by special substituted groups with or without transfer species. While some will be identified here, more will be identified in the Series and Volumes.

3.1. Resonance Stabilization in Olefinic monomers

The phenomena of resonance stabilization have never been properly addressed for many years. Though it has been well known to exist, the frontiers for its development have never gone beyond classical organic chemistry, where much still have to be desired, in particular, radically where little or nothing is known about them (Radical) in Chemistry.

3.1.1. Charged Resonance Stabilization in Olefinic monomers

In general, a monomer with only one activation center can never be resonance stabilized. There has to be more than one activation center conjugatedly placed, whether of C = C type or C = O type or C = N type etc., before the phenomena can ever be considered.

To start with, consider the following monomers which when activated in full, one should expect the followings.

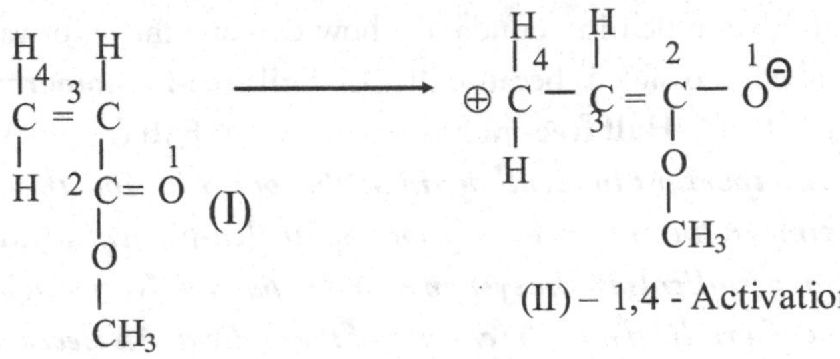

$$\text{(I)} \longrightarrow \text{(II) – 1,4 - Activation} \qquad\qquad 3.4$$

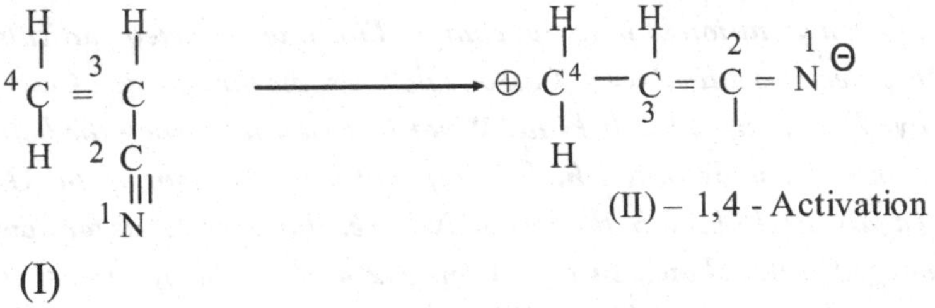

$$\text{(I)} \longrightarrow \text{(II) – 1,4 - Activation} \qquad\qquad 3.5$$

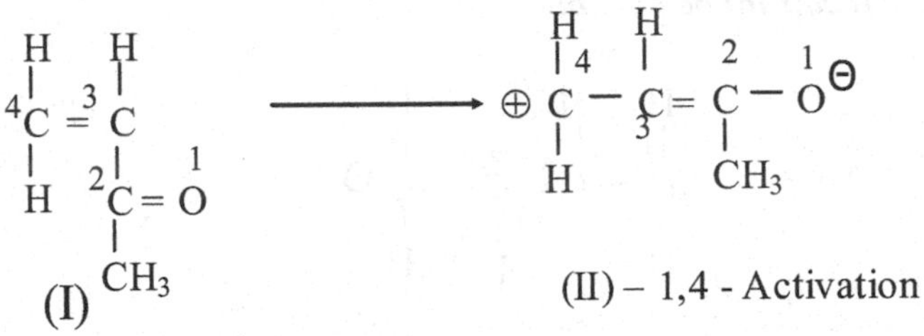

$$\text{(I)} \longrightarrow \text{(II) – 1,4 - Activation} \qquad\qquad 3.6$$

Why are full activations of the monomers above not known to exist from literature data? Is it because the C = O activation center is more difficult to activate than C = C activation center? YES. While it is partly true on the basis of type of polymerization media and initiator involved, on the other hand it is not true, since copolymers can be produced from for example *acrolein monomer* when involved via the two activation centers chargedly not in full but in part. With charged-paired initiators, copolymers are said to be produced under certain conditions positively chargedly. Also with the monomer in the last equation above (that is methyl vinyl ketone) with negatively-charged-paired initiator, copolymers can never be produced as shown below. The same too applies to acrolein for a different reason, that which is charge balancing.:

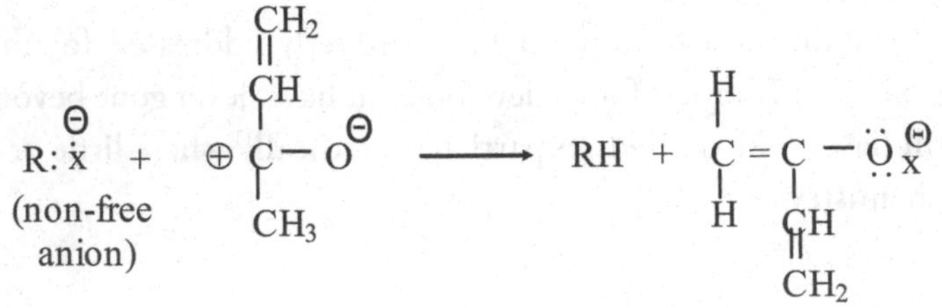

$$\text{R:}\ddot{x}\ \text{(non-free anion)} + \oplus \overset{CH_2}{\underset{CH_3}{\overset{\|}{\underset{|}{\overset{CH}{\underset{|}{C}}}}}} - O^{\ominus} \longrightarrow RH + \text{...} \qquad\qquad 3.7$$

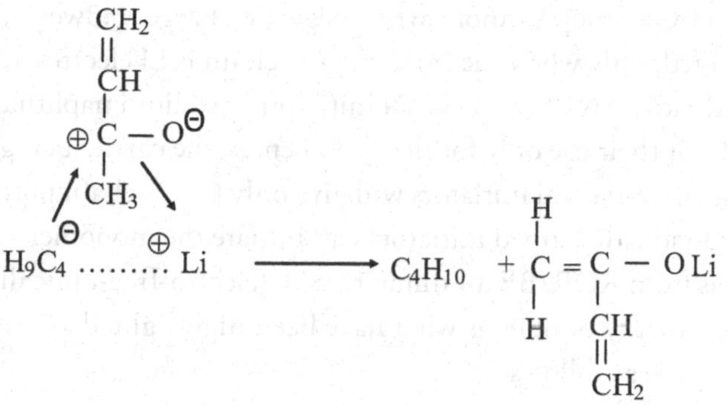

$$
\begin{array}{c}
CH_2 \\
\| \\
CH \\
| \\
\oplus\,C - O^\ominus \\
| \\
CH_3 \\
\diagup \qquad \diagdown
\end{array}
$$

$$
\underset{\ominus}{H_9C_4} \ldots\ldots \overset{\oplus}{Li} \longrightarrow C_4H_{10} +
\begin{array}{c}
\quad H \\
\quad | \\
C = C - O\,Li \\
| \quad | \\
H \quad CH \\
\quad \| \\
\quad CH_2
\end{array}
\qquad 3.8
$$

While C = O activation center cannot be polymerized, C = C activation center can. It is important to note the group from which the $H^\oplus$ has been abstracted, CH_3 group being more radical/chargedly-pushing than -CH = CH_2 group. Where two similar groups are located on the same carbon center, it is the more radical/charged-pushing or pulling group that is involved. Here in Nature, we take from the rich rather than from the poor. The transfer species involved in the reactions above are transfer species of the first kind of the first type. With the type of initiator in the last equation above, initiation will indeed be favored via the C=O center, the carrier of the chain being Li, only when paired radically, the initiator being dual in character only radically (but not chargedly).

Now with acrolein, the situation is different with negative charges as shown below.

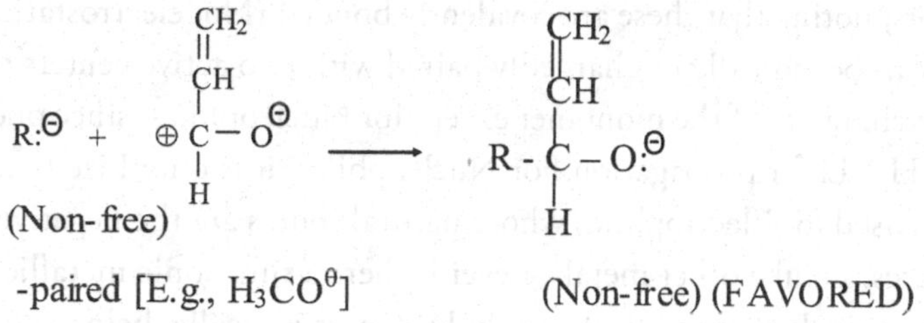

$$
R\!:^\ominus +
\begin{array}{c}
CH_2 \\
\| \\
CH \\
| \\
\oplus\,C - O^\ominus \\
| \\
H
\end{array}
\longrightarrow
\begin{array}{c}
CH_2 \\
\| \\
CH \\
| \\
R - C - O\!:^\ominus \\
| \\
H
\end{array}
\qquad 3.9a
$$

(Non-free)

-paired [E.g., H_3CO^θ] (Non-free) (FAVORED)

However, the nucleo-free and nucleo-non-free-radical routes will not be favored by it in the presence of transfer species which cannot be removed chargedly due to electrostatic forces of repulsion.

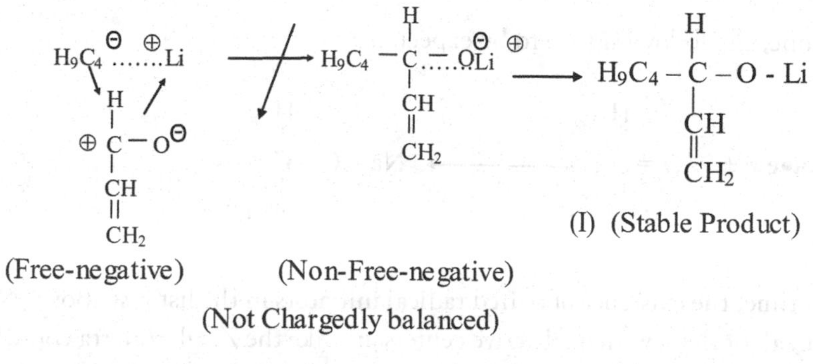

$$
\underset{\substack{\downarrow \\ \oplus\,C - O^\ominus \\ | \\ CH \\ \| \\ CH_2}}{\overset{\ominus \quad \oplus}{H_9C_4 \ldots\ldots Li}} \;\;\Big/\!\!\!\!\longrightarrow\;\;
\begin{array}{c}
\quad H \\
\quad | \\
H_9C_4 - C \ldots^\ominus\; O\,Li^\oplus \\
\quad | \\
\quad CH \\
\quad \| \\
\quad CH_2
\end{array}
\longrightarrow
\begin{array}{c}
\quad H \\
\quad | \\
H_9C_4 - C - O - Li \\
\quad | \\
\quad CH \\
\quad \| \\
\quad CH_2
\end{array}
$$

(Free-negative) (Non-Free-negative) (I) (Stable Product) 3.9b

(Not Chargedly balanced)

As is already shown and will be fully confirmed in this Volume, ***in the absence of resonance stabilization effect, -H is less or of equal radical/charge-pushing capacity than –CH = CH_2 group.*** Hence, the orientation of the monomer as shown in Equation 3.9b above is either wrong or right. However, the route is not favored, because the equation is not chargedly balanced. It is nevertheless favored anionically, because H cannot be abstracted from –HC = CH_2 group, whether the monomer is resonance stabilized or not. The carbon center carrying the $H_2C=$ group cannot carry a negative charge. However, the use of $H_9C_4{}^{\ominus}$......$^{\oplus}Li$ for polymerization is favored, only when the carrier of the chain is Li electro-free-radically as has been said. Sodium cyanide paired-radical $Na^{\bullet e}$........$^{n\bullet}C\equiv N$ initiator, or sodium naphthalene paired radical $Na^{\bullet e}$......$^{n\bullet}C_{10}H_7$ **initiator** will favor their use only for the C=O center, the carrier being Na. Thus, while free-negatively-charged and nucleo-free-radical initiators will give only C = C chain polymers alone and not copolymers, the non-free-paired-radical/charged initiators will initiate the monomer via the C=O center. Positively charged centers such as from ROR/BF$_3$ or dilute H$_2$SO$_4$ (electro-free-radically) will give copolymers via C=C and C=O centers. Worthy of note, is what have been high-lighted above. It sends a large number of messages. The messages are as follows-

(i)   While in $H_9C_4{}^{\ominus}$...$^{\oplus}Li$, negative charges can be carried by H_9C_4 group, in $Na^{\bullet e}$....$^{n\bullet}C\equiv N$ and $Na^{\bullet e}$......$^{n\bullet}C_{10}H_7$, negative charges cannot be carried by the -C≡N ($^{\ominus}C\equiv N$) and $-C_{10}H_7$ or $-C_6H_5$ ($^{\ominus}C_{10}H_7$ or $^{\ominus}C_6H_5$) centers, because of the presence of electrostatic forces of repulsion provided by π-bonds or paired unbonded radicals adjacently located.

(ii) Hence, monomers such as acetylene, nitrogen, oxygen, C ≡N centers and the likes cannot be activated chargedly, but only radically.

(iii) Hence, negative charges whose presence cannot be tolerated when adjacently located to paired unbonded radicals or π-bond cannot be resonance stabilized.

(iv) Hence, initiators such as $Na^{\bullet e}$.......$^{n\bullet}C\equiv N$ and $Na^{\bullet e}$......$^{n\bullet}C_6H_5$, can only be radically paired with two active centers; each center used based on the character of the monomer, just like Ziegler/Natta (Z/N) types of initiators; noting that these are covalently bonded (Not electrostatically bonded).

(v)   Hence, $H_9C_4{}^{\ominus}$......$^{\oplus}Li$ can be radically or chargedly paired with two active centers only radically, each used based on the character of the monomer except for Nucleophiles, since one cannot have $H_9C_4{}^{\ominus}$.......$^{\oplus}C_2H_4(C_2H_4)_nLi$ for propagation. For Nucleophiles, it is only Li$\bullet$e that will be used alone. It can readily be used for Electrophiles whose natural routes are the negatively charged or nucleo-free-radical routes. Unlike other metals, never is there a time ionic metallic elements can carry negative charges or nucleo-free-radicals. Only H does it radically, being the first member of all atoms and their elements. ***However, while Na, Li, can remove H.n as transfer species of the first kind, H cannot do it to itself.*** Hence H$_2$SO$_4$ with its H can give copolymers, but not with Li or Na.

Thus, using Na alone, the followings are to be expected.

$$Na\bullet e \;+\; \begin{array}{c} H \\ | \\ O=C \\ | \\ HC=CH_2 \end{array} \quad\longrightarrow\quad \begin{array}{c} H \\ | \\ Na-O-C\bullet e \\ | \\ HC=CH_2 \end{array} \qquad\qquad 3.10b$$

Notice for the first time, the existence of paired radical initiators in the list just above -NaOCH$_3$, NaCN, t-H$_9$C$_4$Li, NaC$_{10}$H$_7$, all of them with dual active centers just like the Ziegler/Natta coordination initiators. The Z/N case is different, because the two active centers are metallic unlike the cases above. The situation above is for vinyl formaldehyde. With vinyl ketones, the situation is different but the same with respect to resonance stabilization. With respect to routes of polymerization, while Na can readily abstract H, it like other metals cannot abstract R groups.

Now, coming back to the non-existence of (II), of Equations 3.4 – 3.6. Is it because of existence of molecular rearrangement? The answer is obviously NO, since only the first monomer (Equation

3.4) favors molecular rearrangement when weak radical/charged initiators are involved. It is the only one that carries a transfer species. Try to remove $\equiv N$ or $=O$ types of groups without removing the π-bonds. Nature will not allow it. Indeed with (II), how can molecular rearrangement provide its 4,3 –Addition monomer? One of the common factors for the three monomers is the character of their activation centers as shown below.

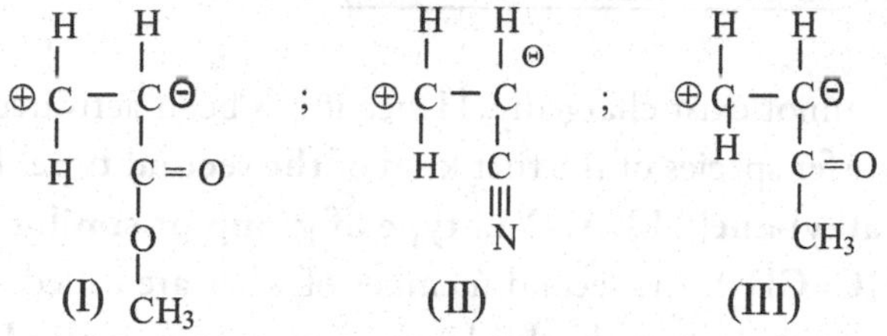

$$
\begin{array}{ccc}
\overset{H}{\underset{|}{\oplus C}} - \overset{H}{\underset{|}{C\ominus}} & \overset{H}{\underset{|}{\oplus C}} - \overset{H}{\underset{|}{\overset{\ominus}{C}}} & \overset{H}{\underset{|}{\oplus C}} - \overset{H}{\underset{|}{C\ominus}} \\
\overset{|}{H}\quad \overset{|}{C=O} & \overset{|}{H}\quad \overset{|}{C} & \overset{|}{H}\quad \overset{|}{C=O} \\
\qquad \overset{|}{O} & \qquad \overset{|||}{N} & \qquad \overset{|}{CH_3} \\
\text{(I)}\ CH_3 & \text{(II)} & \text{(III)}
\end{array}
\qquad 3.11
$$

(All electrophiles) -Full Free centers

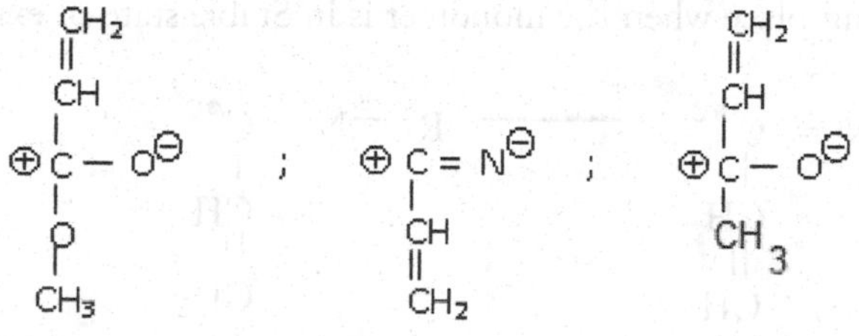

$$
\begin{array}{ccc}
\overset{CH_2}{\underset{||}{}} & & \overset{CH_2}{\underset{||}{}} \\
\overset{CH}{\underset{|}{}} & & \overset{CH}{\underset{|}{}} \\
\oplus C - O\ominus & \oplus C = N\ominus & \oplus C - O\ominus \\
\overset{|}{O} & \overset{|}{CH} & \overset{|}{CH_3} \\
\overset{|}{CH_3} & \overset{||}{CH_2} & \\
\text{(I)} & \text{(II)}^* & \text{(III)}
\end{array}
\qquad 3.12
$$

(All nucleophiles) – Half Free center

The characters of the nucleophiles are well defined when charged initiators are involved. Note the activated state of Existence of (II) above. Based on what has just been identified above, (II) in the last equation can only be activated radically. Chargedly, it cannot be activatated due to electrostatic forces of repulsion. One is going to use them radically. Where they are used chargedly, the reason is for exploratory purposes.

None will favor non-free negatively charged polymerization as shown below.

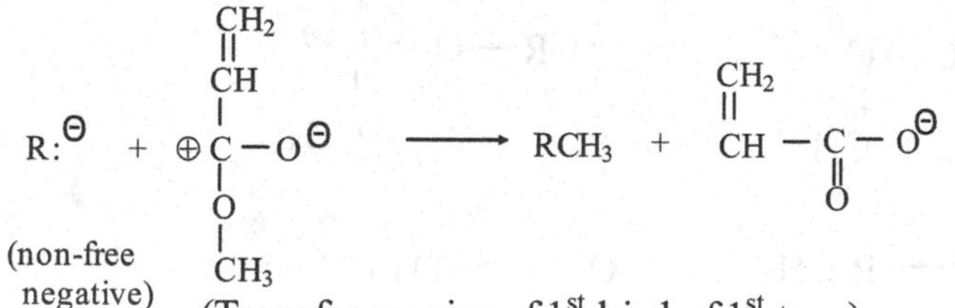

$$
R:\ominus + \begin{array}{c} \overset{CH_2}{\underset{||}{}} \\ \overset{CH}{\underset{|}{}} \\ \oplus C - O\ominus \\ \overset{|}{O} \\ \overset{|}{CH_3} \end{array} \longrightarrow RCH_3 + \begin{array}{c} \overset{CH_2}{\underset{||}{}} \\ CH - \overset{C}{\underset{||}{}} - O\ominus \\ \overset{}{O} \end{array}
\qquad 3.13
$$

(non-free negative)   (Transfer species of 1st kind of 1st type)

$$
R\oplus + \begin{array}{c} \overset{CH_2}{\underset{||}{}} \\ \overset{CH}{\underset{|}{}} \\ \oplus C - O\ominus \\ \overset{|}{O} \\ \overset{|}{CH_3} \end{array} \longrightarrow \begin{array}{c} \overset{OCH_3}{\underset{|}{}} \\ \oplus C - O - R \\ \overset{|}{CH} \\ \overset{||}{CH_2} \end{array} \text{OR ROCH}_3 + \begin{array}{c} \oplus C = O \\ \overset{|}{CH} \\ \overset{||}{CH_2} \end{array}
\qquad 3.14
$$

(free-positive)

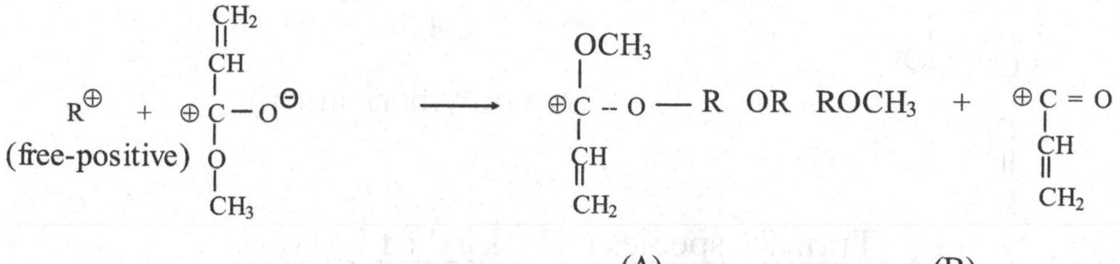

.(A)   (B)

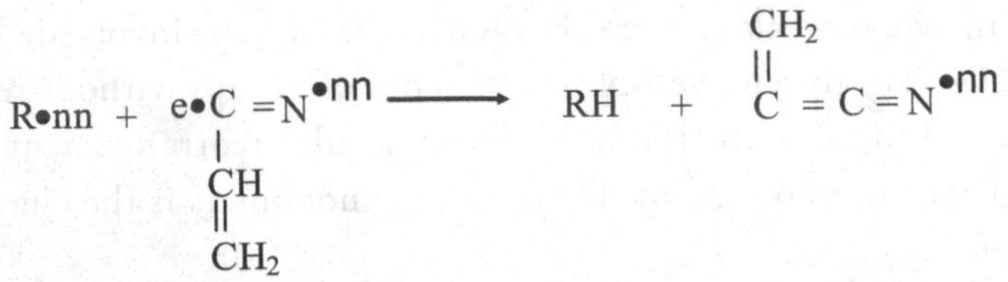

$$R\bullet nn \; + \; e\bullet C = N^{\bullet nn} \longrightarrow RH \; + \; C = C = N^{\bullet nn} \qquad 3.15$$

with the substituents: on the first C, $\overset{|}{\underset{CH_2}{CH}}$ (i.e. $-CH=CH_2$); and on the product the top CH_2 double-bonded.

(Transfer species of 1st kind of 2nd type)

Notice that (II) of Equation 3.12 cannot exist chargedly. Hence it has been activated radically. For the first time, notice what is called transfer species of the first kind of **the second type.** It is of second type because it is not coming from **an alkyl-ane[ALKANYL] type of group or similar type,** but from **an alkyl-ene[ALKENYL] group (-HC=CH$_2$),** the second member of what are called **ALKYL groups** as will be fully shown and seen downstream. In fact, the third type as an exercise is **alkyl-yne [ALKYNYL]** group (=CH$_2$). For the last but one equation above, it could be (A) or (B). With (B), it is transfer species of the second first kind taking place when the monomer is in Stable state of existence..

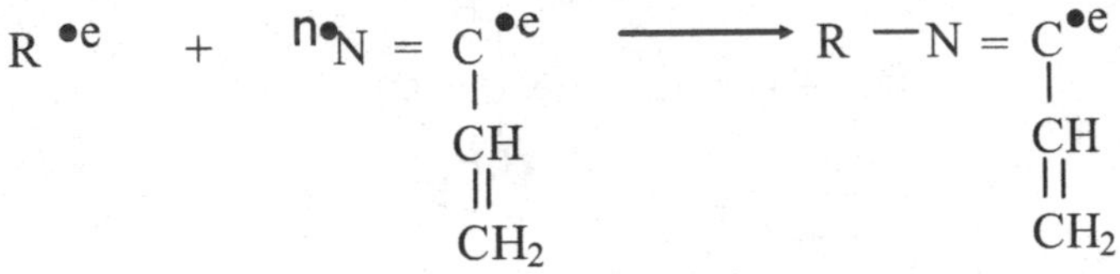

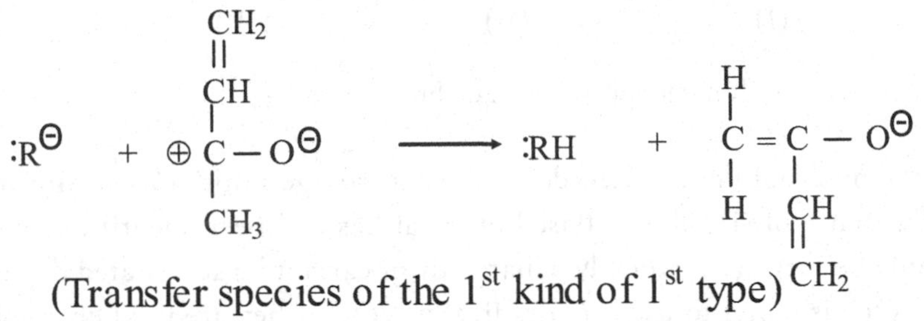

$$R^{\bullet e} \; + \; {}^{n}\bullet N = C^{\bullet e} \longrightarrow R - N = C^{\bullet e}$$

(with substituent $\overset{|}{\underset{CH_2}{CH}}$ on the carbons)

$$:R^{\ominus} \; + \; {}^{\oplus}C - O^{\ominus} \longrightarrow :RH \; + \; \overset{H}{\underset{H}{C}} = \underset{CH}{C} - O^{\ominus} \qquad 3.16$$

(Transfer species of the 1st kind of 1st type)

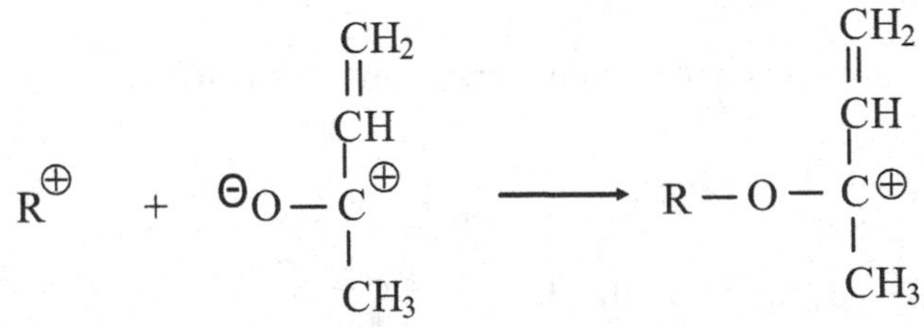

$$R^{\oplus} \; + \; {}^{\ominus}O - C^{\oplus} \longrightarrow R - O - C^{\oplus} \qquad 3.17$$

$$H_9 C_4{}^{\ominus} \ldots \ldots {}^{\oplus}Li \longrightarrow R\,CH_3 \; + \; O = C - OLi \qquad 3.18$$
$$(H_{12}C_5)$$

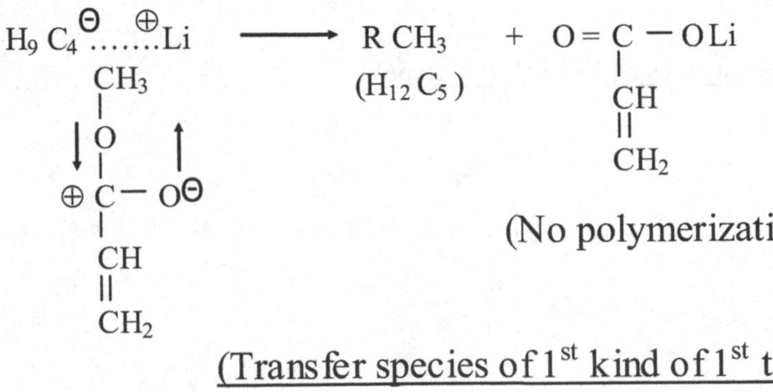

(No polymerization)

3.19

(Transfer species of 1st kind of 1st type)

119

As is obvious in Equation 3.19, *it is the charged-pushing capacity rather than electronegativity or electropositivity of a group that determines the orientation of an activated monomer around a co-ordination center.* OCH_3 group is non-free charged/radical-pushing group whether connect-ed to a carbonyl (C=O) or C=C or any center. On the other hand, R (e.g., CH_3) is a free-radical/ charged-pushing group whether connected to a carbonyl center or not. When part of a substituent group, it remains electropositive when abstracted.

It is of great importance to note the kinds of transfer species involved, since these are all being identified based on the law of conservation of transfer of transfer species. According to Equations 3.13 and 3.19, when transfer species exist on two substituent groups of same character carried by the same carbon center, the transfer species involved is that carried by the more radical- pulling or pushing group, whether it is more electronegative or electropositive than that carried by the less radical-pulling or pushing group, unless the more radical-pushing or pulling group possesses no transfer species. All these are in accordance with the natural law of co-existence; for how can Sweetness at one end of the spectrum exist without Bitterness at the other end of the spectrum? How can so-called HEAVEN exist without its co-existence with HELL? HOW? It is the richer that gives or provides the driving force as clearly indicated below, when demand by society for co-existence is required. For in Chemistry (the study of the LAWS of NATURE) just as in Life, there are RICH and POOR components.

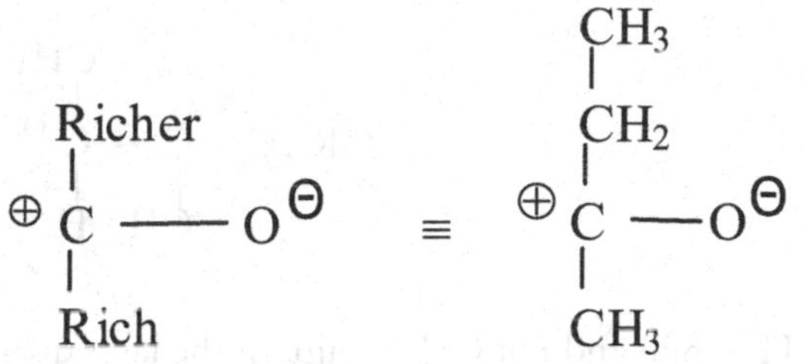

$$\underset{\text{(II)a}}{\overset{\oplus}{C}\overset{\displaystyle |}{\underset{\displaystyle |}{\overset{\text{Richer}}{\underset{\text{Rich}}{}}}}\!\!-\!\!O^{\ominus}} \quad\equiv\quad \underset{\text{(II)b}}{\overset{\oplus}{C}\overset{\displaystyle |}{\underset{\displaystyle |}{\overset{\overset{\displaystyle CH_3}{|}\ \ \ \ \ CH_2}{CH_3}}}\!\!-\!\!O^{\ominus}} \qquad\qquad 3.21$$

Richer $\equiv C_2H_5$, Rich $\equiv CH_3$, H $\equiv$ poor
$C_2H_5 > CH_3 > H$ (in radical/charged pushing capacity)

(II)b was obtained from an unstable vinyl alcohol for example, as shown below.

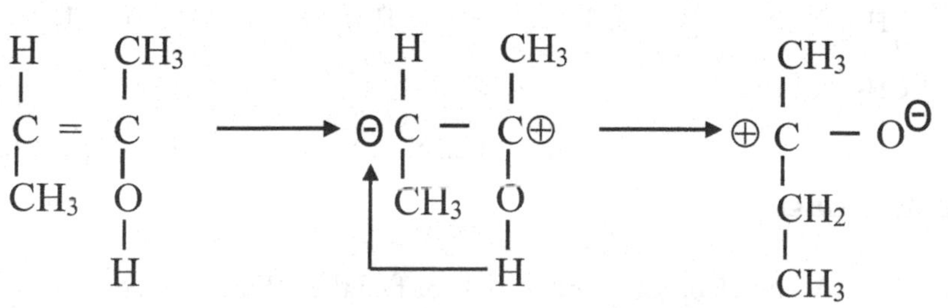

$$
\underset{\underset{CH_3}{|}}{\overset{\overset{H}{|}}{C}}=\underset{\underset{\underset{\underset{H}{|}}{O}}{|}}{\overset{\overset{CH_3}{|}}{C}}
\quad\longrightarrow\quad
\overset{\ominus}{\underset{\underset{CH_3}{|}}{\overset{\overset{H}{|}}{C}}}-\underset{\underset{\underset{\underset{H}{|}}{O}}{|}}{\overset{\overset{CH_3}{|}}{\overset{\oplus}{C}}}
\quad\longrightarrow\quad
\overset{\oplus}{\underset{\underset{\underset{CH_3}{|}}{CH_2}}{\overset{\overset{CH_3}{|}}{C}}}-O^{\ominus}
\qquad 3.22
$$

With negative charges, free and non-free, the followings are obtained.

$$\underset{\text{Non-free}}{\overset{\ominus}{R_x}} \quad + \quad \overset{CH_3}{\underset{\underset{CH_3}{\overset{|}{\underset{\oplus}{C}}}-O^{\ominus}}{\overset{|}{\underset{|}{CH_2}}}} \quad \longrightarrow \quad RH \quad + \quad \overset{H \quad CH_3}{\underset{CH_3}{\overset{|}{\underset{|}{C}}= \overset{|}{C} - O^{\ominus}}}$$

$$H_9 C_4 \overset{\ominus}{\cdots\cdots} \overset{\oplus}{Li} \quad \longrightarrow \quad C_4 H_{10} \quad + \quad \overset{H \quad CH_3}{\underset{CH_3}{\overset{|}{\underset{|}{C}} = \overset{|}{C} - O\,Li}}$$

with structure
$$\overset{CH_3}{\underset{\underset{CH_3}{\overset{|}{\underset{\oplus}{C}}-O^{\ominus}}}{\overset{|}{\underset{|}{CH_2}}}}$$

3.23

$$H_9 C_4 \overset{\ominus}{\cdots\cdots} \overset{\oplus}{Li} \quad \longrightarrow \quad C_4 H_{10} \quad + \quad \overset{H \quad CH_3}{\underset{CH_3}{\overset{|}{\underset{|}{C}} = \overset{|}{C} - O\,Li}}$$

with structure
$$\overset{CH_3}{\underset{\underset{CH_3}{\overset{|}{\underset{\oplus}{C}}-O^{\ominus}}}{\overset{|}{\underset{|}{CH_2}}}}$$

$$\text{OR} \quad O = \overset{CH_3}{\underset{CH_3}{\overset{|}{C}}} - \overset{\ominus}{\underset{H}{\overset{|}{C}}} \overset{\oplus}{\cdots} Li$$

3.24

Transfer species is from the C_2H_5 group and not CH_3 group. In the last equation above, we could have it also paired without having initiation taking place.

Now, considering conjugated 1,3-dienes, beginning with butadiene, the following is obtained:-

$$\overset{H \quad H}{\underset{\underset{\underset{1\,CH_2}{\parallel}}{\overset{2\,CH}{|}}}{\overset{\ominus}{\overset{|4}{C}} - \overset{3}{\underset{|}{C}}\oplus}} \quad \longleftrightarrow \quad \overset{H \quad H \quad\quad H}{\underset{\underset{H}{|}}{\overset{\ominus}{\overset{4|}{C}} - \overset{|}{\underset{3}{C}} = \overset{2}{\underset{H}{\overset{|}{C}}} - \overset{1}{\underset{H}{\overset{|}{C}}\oplus}}} \quad \longleftrightarrow \quad \overset{CH_2}{\underset{\underset{H}{|}}{\overset{\parallel}{\overset{CH \quad H}{\overset{\ominus}{\overset{2|}{C}} - \overset{1}{\underset{H}{\overset{|}{C}}\oplus}}}}}$$

3.25

(4, 3 – butadiene)　　　　(1, 4 –butadiene)　　　　(2, 1 – butadiene)

__(I) Less Nucleophilic center__　　　　(II)　　　　　　　(III)

(All Nucleophiles-either male or female) – Full Free centers

Because of the "symmetric" nature of 1,4-butadeiene, it is important to note that the same sym-metricity is not maintained in the 1,2- and 3,4- Addition mono-forms, since the radical/charged-pushing capacity of hydrogen which is almost nil on phenyl group,[10] is almost of the same with or lesser capacity than -CH $=CH_2$ group which in turn is of lesser capacity than the phenyl group. *Therefore, herein we move from 4,3-mono-form to 1,4-mono-form.* **As it seems, this can only take place radically, because positive**

charges cannot be moved when there is nothing to move. Negative charges cannot be moved, because though they are carrying two radicals of male and female character, two opposite radicals cannot be moved. Only movement of one radical is possible and this is only the electro-radical.

In the last equation above, movement from one mono-form to another mono-form can only take place radically. But let us continue doing it chargedly, since we have "assumed" it just as we do in Present-day Science. All the activation centers are of equal nucleophilicity. **Alkoxy, Alkyl, Alkenyl and phenyl groups** in general, are said to be "radical-pushing groups".[11] Alkyl-ene groups are radical-pushing groups of equal, greater, or lesser capacity than H, depending on what they are carrying and their use for providing resonance stabilization. They are of equal capacity with H when there is no resonance stabilization and they carry only H or specific types of groups. Alkyl group(s) is a family on its own in which alkyl-ene or alkenyl group is one of them.

For isoprene, pentadiene and methyl sorbate, the followings are obtained:-

(2,1-pentadiene)   (4,1-pentadiene)   (3,4-pentadiene)   3.26

(I)   (II)   **(III) Less nucleophilic center**

(All Weak Nucleophiles) -Full Free centers

The movement of electro-radical or "positive charge" begins not with parent 1,4-isoprene monomer, but with the less nucleophilic center. It is the less nucleophilic center that is first activated all the time. *It is from 3,4 mono-form, we move to 1,4 only radically as will become apparent as we move down the Series and Volumes.* Chargedly, it is the 3,4-mono-form that only exists since no movement seems to be possible as will become apparent and no molecular rearrangement can take place. The location of the CH_3 group in (III), accounts for why 3-carbon center is carrying a negative charge. CH_3 is a radical pushing group. That is, though the alkyl-ene group is radical-pushing, its $[H_2C=C(CH_3)-]$ capacity is less than that of H, since the group is internally located.

(2,1-pentadiene)   (4,1-pentadiene)   (3,4-pentadiene)   3.27

(I)   (II)   (III)-**Less Nucleophilic center**

(All Strong Nucleophiles) - Full Free centers

In view of the location of the CH_3 group here in (III), the 3-carbon center is carrying a positive charge. It is important to note the direction of the arrows of CH_3 group on the substituent groups $-C(CH_3)$ $= CH_2$ (Isoprene) and $CH_3CH=CH$ - in the equations above. Since all the centers are of different nucleophilicities, it is only one of 2,1- and 3,4-mono-forms that will be the first to be activated, followed by the movement of the electro-free-radical As it seems it begins here *from 3,4- to 4,1- mono-form*. So far, it can be noted that all the mono-forms of 3-dienes have the same character with their parent monomer. Note that the movement of electro-free-radicals here is different from movement of transfer species in molecular rearrangements. Transfer species are carriers of radicals and charges. It is only the 3,4- and 1,4-mono-forms, that molecularly rearrange to itself here just like in propene. This can take place chargedly and radically or indeed only radically, since the I,4-mono-form does not exist chargedly.

For a male shown below, the situation is different. The main transfer species for molecular rearrangement can only be $-OCH_3$ group and not H, being a Male. Only (II) and (III) can molecularly rearrange to give a stronger male, ketenic in character.

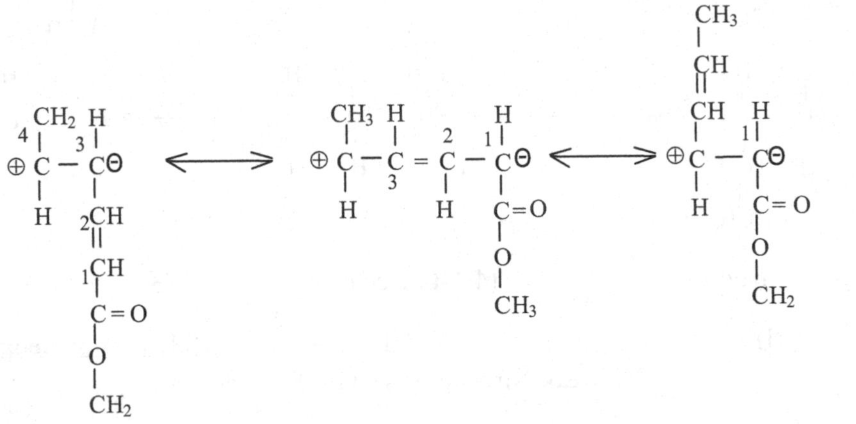

(3,4-trans methyl sorbate)   (1,4- trans methyl sorbate)   (1,2- trans-methyl sorbate)   3.28a

(I)   (II)   **(III) Less nucleophilic center**

(All strong electrophiles)

How many Ys and Xs are in this unique monomer?

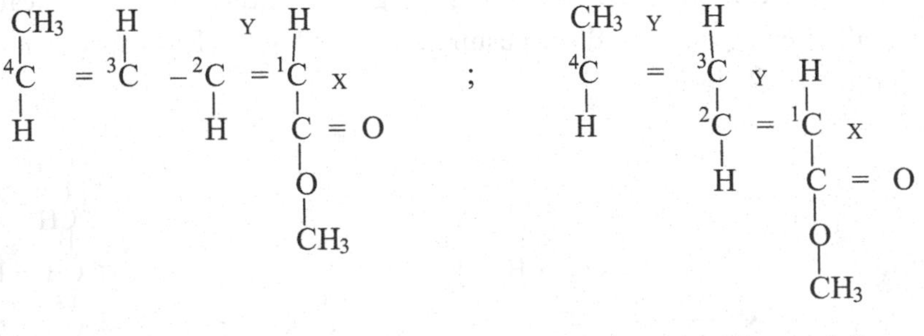

1,4-Electrophile   3,4- & 1,2-Electrophiles   3.28b

In the 1,4-mono-form, there are two C=C centers in which only one is Y; that is the one adjacent-ly located to the C=O center. ***Therefore, the least nucleophilic center is the 1,2-mono-form.*** The same applies for the cis- monomer which cannot readily be polymerized. When strong nucleo-free-radical initiators are used, it is the 1,4-mono-form that is involved. Chargedly, only the 1,2- mono-form can be obtained. When weak nucleo-free-radical initiators are involved, it is the 1,2-mono-form that is involved. Electro-free-radically, for the 1,4 –mono-form, it is the C = O center that is involved with no initiation favored, while when the initiator is weak, for the 1,2 –mono-form, it is the same C = O that is involved with no initiation possible (Transfer species of the second first kind). Notice that though these monomers have been activated chargedly so far, only one exists. They have only been used to illustrate some fundamental principles such as knowing that activation chargedly is still possible only for one mono-form. When monomers favor resonance stabilization, this can only be done RADICALLY. When activated chargedly, only one center is involved without any form of resonance stabilization. From that one center, molecular rearrangement can take place to give another mono-form.

The monomers above are trans-methyl sorbates which are reported to favor polymerization via negatively charged-paired initiators. It is only 1,2- mono-form that can do it. The cis-methyl sorbates have no character since they will not favor any charged/radical coordination routes. *However, it is from the 1,2- mono-form, we move to 1,4-mono-form radically.*

3.29

(3,4 – cis-methyl sorbate)   (1,4-cis-methyl sorbate)   (1,2 – cis- methyl sorbate)

(I)   (II)   (III) **Less nucleophilic center**

(All electrophiles – but of no character)

For chloroprene, the following should be obtained.

(1,2 – chloroprene)   (1,4 – chloroprene)   (3,4-chloroprene)

(I) Less nucleophilic center   (II)   (III)

(Non-existent)   **(Non-existent)**   **(Non-existent)**

(All nucleophiles free-radically)   3.30

Like vinyl chloride, the 1,2-mono-form of chloroprene (I) cannot undergo charged activation. For (II) and (III), this limitation is not there. The less nucleophilic center is the 1,2-mono-form which cannot be activated chargedly. Then, why can (II) and (III) exist without (I) chargedly when they are all resonance stabilized?! ***Hence, one can now begin to see why resonance stabilization cannot take place chargedly.*** Free-radically (Nucleo-free and electro-free-radical-ly), they are all polymerizable. *It is from the 1,2-mono-form, we move to 1,4-mono-form only free-radically.*

Now replacing Cl with a free-pulling group such as $COOCH_3$, the following is obtained-

$$\text{(Structure I)} \longleftrightarrow \text{(Structure II)} \longleftrightarrow \text{(Structure III)} \qquad 3.31$$

(1,2-2´- acrylated butadiene) (1,4- 2´-acrylated butadiene) (3,4-2´-acrylated butadiene)
(I) (II) (III)

The Least Nucleophilic center

(All weak electrophiles) —Full Free centers

In view of the location of $COOCH_3$ group in substituent group of (III), the 3 – carbon center is carrying a positive charge. That is, **$CH_3OOCC = CH_2$ group is a radical/charged-pushing group**. $CH_3 –C =CH_2$ group is also a radical/charged-pushing group, but of weaker capacity than H, while the above is of greater capacity than H. *It is from the 1,2-mono-form we move to 1,4-mono-form only free-radically*. 3,4-mono-form does not exist. No molecular rearrangement is possible, *since the $COOCH_3$ group is shielded* being located internally.

In identifying the characters of the resonance stabilized 1,3-dienes above, it is important to note that *weak and strong* have been used. All the two resonance stabilized mono-forms of butadiene, in which only 1,4-mono-form is symmetric, are weak nucleophiles like ethylene, since they are reported to undergo both negatively and positively charged routes. The isoprenes are even weaker nucleophiles since transfer species of the first kind exist in its 1,2-mono-form, but cannot be abstracted due to resonance stabilization effect. Resonance stabilization has provided a shield around the group being located internally. They are therefore all reported to undergo both negatively and positively charged routes which as can be seen so far is true. 1,3-pentadiene is a strong nucleophile, since it is reported to only undergo positively charged route like propylene. Trans-methyl sorbates are strong electrophiles since they are also reported to undergo negatively charged paired route. For the monomer represented in Equation 3.31, all the resonance stabilized forms are weak electrophiles since its 1,2-mono-form has transfer species which cannot be abstracted. Probably, that is the more reason why it may be said not to exist as a mono-form, that which cannot be true when a case like methyl styrene is considered. The group is shielded by resonance stabilization. The two mono-forms in Equation 3.31 are also reported to undergo both negatively and positively charged routes. What is important to note, is the fact that *the route(s) favored by the parent monomer (1,4-addition) is the same favored by their mono-forms in view of the influence of resonance stabilization and these only take place RADICALLY.*

Before identifying one of the major driving forces favoring existence of resonance stabili-zation, one will consider the phenomenon of molecular rearrangement in all the forms of 1,3-dienes. Beginning with butadiene of Equation 3.25, where $CH_2 = CH-$ group can be observed to be of greater capacity than H as a radical-pushing group, when resonance stabilization is provid-ed. In the absence of any substituent group, no molecular rearrangement is possible. Neverthe-less, based on the mono-addition product obtained when weak sterically hindered Ziegler-Natta (Z/N) initiators[12] are involved "cationically", (that is, a 4,3-Addition product, which is obtained from (I) of Equation 3.25), $CH_2 = CH-$ group being of a greater capacity than H, clear indication that (III) of Equation 3.25 is not symmetric. So also is (I). Only (II) is close to being symmetric, when indeed it is not. If it (CH_2=CH- group) was equal to H, then what is the essence of –CH=CH – group in providing resonance stabilization? Hence when resonance stabilization is not provided, they become equal. It is only then, (III) can be said to exist.

Syndiotactic 3, 4- butadiene

$$(3.32)$$

For isoprene, the following is obtained for its 1,2- monoform of Equation 3.26 (Non-existent).

(I)   (II)   (I)

$$(3.33)$$

Due to resonance stabilization not provided by the CH_2 = CH- group chargedly, the transfer species on CH_3 can be moved, for which the same monomer is obtained. For 1,4- and 3,4-mono-forms, no rearrangement is possible, because the group is internally located for the centers. Hence the transfer of transfer species is impossible for the case above in the presence or absence of resonance stabilization groups.

For 1,2-pentadiene of Equation 3.27, i.e. (I) {Non-existent}, when molecular rearrangement is considered, the followings are obtained.

(I) –1,2-mono-form   (II) - ?   (III) –1,4-diene

$$(3.34)$$

(IV)-3,4-monoform   (III) –1,4-diene.

Existence of (II) is not favored since it is less stable than the other forms and is not resonance stabilized. It is 3,4-monoform that can rearrange to (I) and not the other way round. Indeed, 1,2-mono-form can never exist since never a time will it be activated alone chargedly or radically, because that center in more nucleophilic than the 3,4- center. *One can see that ONLY TWO MONO-FORMS EXIST FOR DIENES and not three as we have been showing so far and believed to be the case in Present-day Science.* Secondly, one can see that through molecular rearrangement where possible, this can be used to identify the more nucleophilic center in a Diene. It is that center which when made to undergo molecular rearrangement, a resonance stabilized mono-form cannot be obtained. While the 1,4-mono-form rearranges back to itself, the 3,4-mono-form rearranges to the 1,4-mono-form, clear indication that the 1,4-mono-form is the more stable of the two mono-forms, since one of driving forces favoring the occurrence of molecular rearrangement is production of a more stable monomer or same monomer. Nevertheless, it is important to note that resonance stabilization groups only affect the groups carried by the carbon centers internally located, i.e., on the β- or δ -positions. It does not affect groups externally located i.e., on α-position.

Worthy of note in all the considerations so far are the followings-

i) Movement takes place from the less or least nucleophilic center to others where possible. The movement is such that, it is only the electro-radical and not the positively charged (free or non-free) one that moves no matter what the condition is. *It can move from a π- bond if there is an opposite part (a nucleo-free or non-free-radical) visibly adjacently located. It can also move when visibly present to grab a nucleo-free-radical or non-free-radical from a π-bond adjacently located.* What types of rearrangements these movements will give, (different from movement of radical/charge carrying atomic or molecular species as in Molecular rearrangement) will be shown and seen downstream in this Volume. In view of the two types of movements above, there are therefore two types of resonance stabilization phenomena of this kind. While the latter, i.e., movement of the visible electro-free-radical is said to be RESONANCE STABILIZATION OF THE FIRST KIND OF THE FIRST TYPE, the former, i.e., movement from a π-bond is said to be RESONANCE STABILIZATION OF THE FIRST KIND OF THE SECOND TYPE.

ii) In all the Dienes, the least nucleophilic center for them including 1,3-butadiene is always fixed. Based on the patterns so far, the least nucleophilic center in 1,3-Dienes Nucleophilic in character is 3,4-mono-form for those that carry radical-pushing groups. When Cl is carried, the least nucleophilic center is the 1,2-monoform. For Electrophiles it is also the 1,2-mono-form.

iii) It is only the 1,4- and 3,4-mono-forms for Nucleophiles where the group is externally located that can molecularly rearrange.

iv) When Activation centers are conjugatedly placed, they can only be activated one at a time.

v) When there is resonance stabilization between centers, their activations can only take place RADICALLY and not chargedly. If chargedly, only one center can be used with no existence of 1,4-mono-form along the chain. This does not imply that 1,4-mono-form cannot be obtained chargedly during polymerization. It can be obtained for Nucleophiles, first radically by forming a ring, followed by instantaneous opening of the ring chargedly as will be shown downstream; but not for many Electrophiles.

With methyl sorbate where the substituent groups are externally located, the followings are obtained for its mono-forms during rearrangement. Charged activation is still explored.

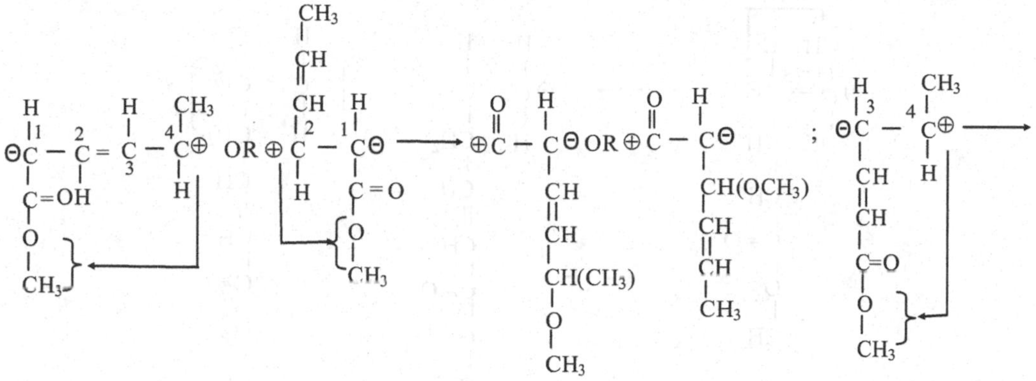

(I) 1,4-mono-form   (II) 1,2-mono-form   (III) From (I) (IV) From (II) (V) 3,4-mono-form
                                           (Stabilized)   (Non-stabilized) C

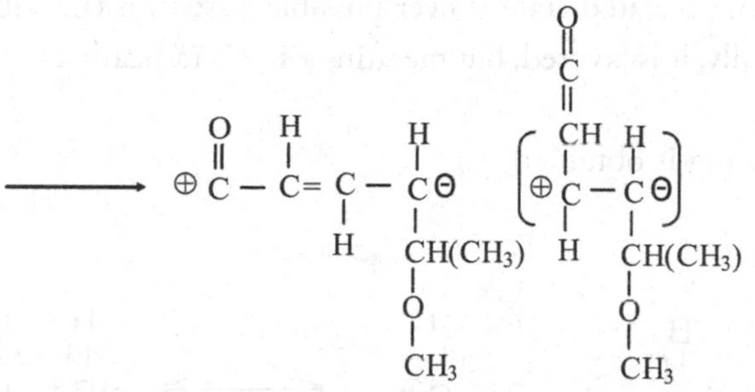

3.35

(VI) – (Stabilized)   (VII)-Radical movement rearrangement

Molecular rearrangements to produce (III) or (VI) looks favored, since (IV) is not resonance stabilized. 3,4- and 1,4-monoforms are the only two mono-forms which rearrange to give the same resonance stabilized Ketene with Electrophilic character stronger and more stable than their originating mono-forms. This is only possible in full radically from 3,4- and 1,4- mono-forms. Chargedly from only 1,2-monoform, (IV) which is non-resonance stabilized is obtained, and that is indeed the least nucleophilic center. Then, if one is to follow the same pattern as with Nucleophiles, molecular rearrangement, may not therefore be used for Electrophiles to identify the two mono-forms. Though (III) and (VI) are resonance stabilized with its second mono-form based on radical movement to give (VII) shown above, 3,4-mono-form is not one of the mono-forms. Thus, the only product obtained from (II) before the commencement of resonance stabilization, after molecular rearrangement is (IV) which is not resonance stabilized. Transfer of OCH_3 is the favored transfer species during rearrangement and not H from CH_3. When such radical/charged pulling group is present, it takes control over the whole system regardless the capacity of CH_3 types of group (radical/charged-pushing groups (Mother Nature). Note also, that since OCH_3 group is the transfer species, (IV) from (II) can rearrange to give the original 1,4-monoform. Hence, same for Nucleophiles seem to apply for Electrophiles. They are have two monoforms. On the other hand, impossible activated states are obtained when CH_3 group is involved as shown below.

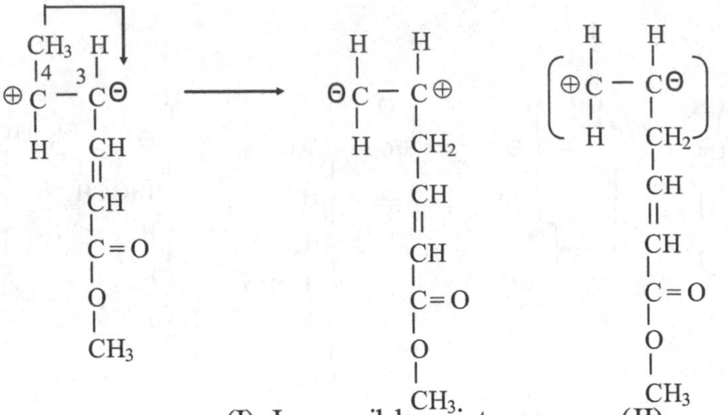

(I) Impossible existence   (II)   3.36

While (I) is obtained, (II) is the actual activated state if ever possible based on the substituted groups carried by the centers chargedly. Radically, it is favored, but meaningless. (I) radically cannot be resonance stabilized.

For acrylated butadiene, the following is obtained.

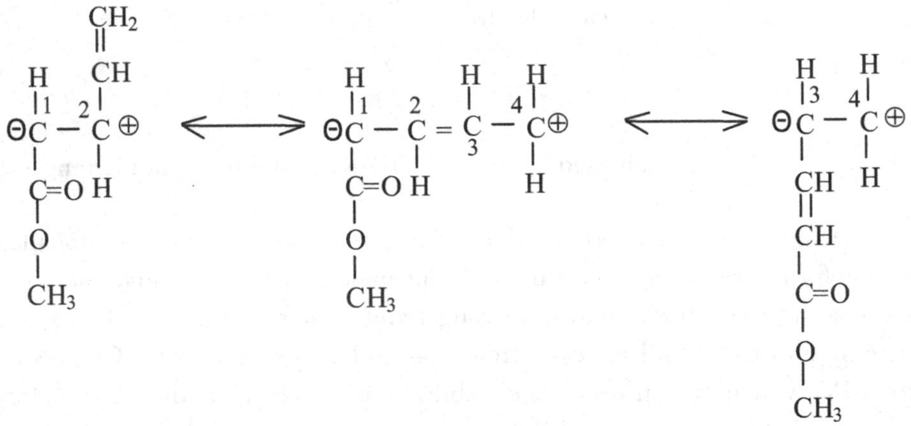

3.37

(I) (1,2-acrylated butadiene)   (II) (1,4-acrylated butadiene)   (III) (3,4-acrylated)
Less Nucleophilic center

(All strong Electrophiles)

When ideal Z/N initiators are involved negatively only1,4,-addition and 1,2-addition products are reported to be exclusively obtained. For isoprene (positively) and 2'-acrylated butadiene (negatively), 4,1-addition and 4,3-addition products are reportedly said to be obtained using present nomenclature used so far and as used in present-day Science, while for butadiene and pentadiene positively, 1,4-addition and 1,2-addition products are exclusively obtained. Based on the natural route of a monomer, nomenclature-wisely for isoprene and acrylated butadiene of Equations 3.26 and 3.31, the followings are obtained.

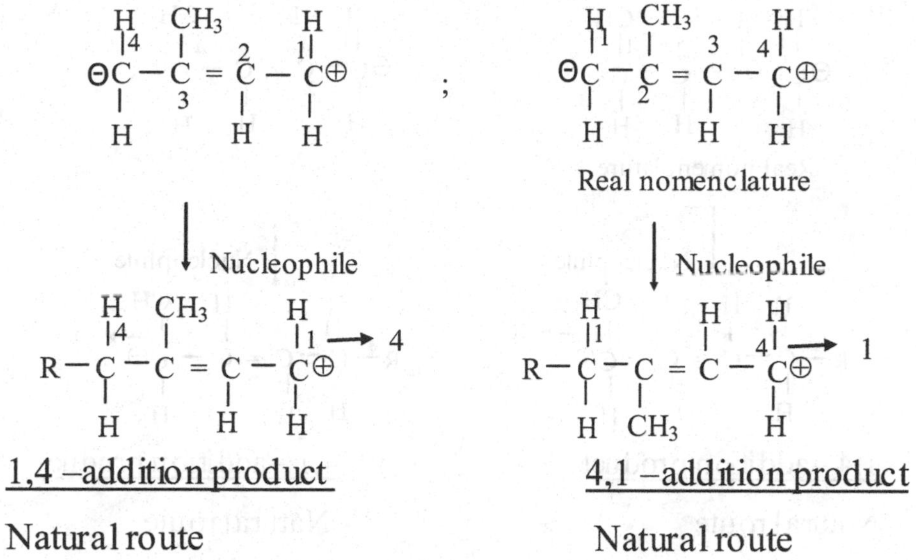

```
    H   CH3      H                              H   CH3      H
    |4   |     2  |1                            |1   |    3  4|
  ⊖C — C = C — C⊕          ;                  ⊖C — C = C — C⊕
    |    3|      |                              |    2|       |
    H     H     H                               H     H      H
```
Real nomenclature

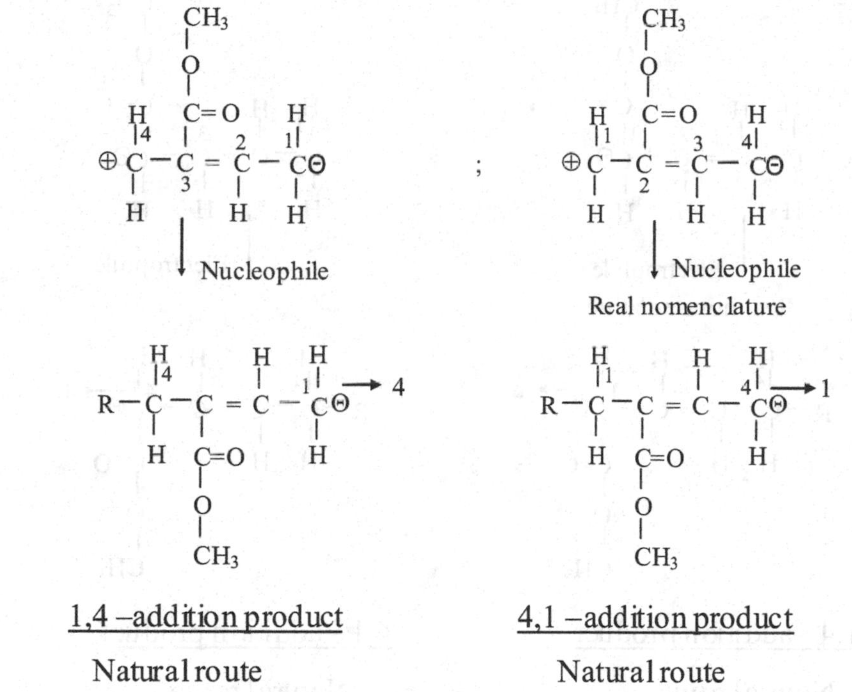

| | Nucleophile | | | Nucleophile |

```
    H   CH3        H                            H        H   H
    |4   |        1|  → 4                        |1       |   4| → 1
R — C — C = C — C⊕                          R — C — C = C — C⊕
    |            |                                |        |    |
    H   H        H                                H   CH3       H
```

__1,4 –addition product__   __4,1 –addition product__   3.38a

Natural route   Natural route

```
            CH3                                 CH3
            |                                   |
            O                                   O
            |                                   |
    H   C=O      H                         H   C=O      H
    |4   |    2  |1                        |1   |    3  4|
  ⊕C — C = C — C⊖          ;             ⊕C — C = C — C⊖
    |    3|      |                         |    2|       |
    H     H     H                          H     H      H
```

| Nucleophile | | | Nucleophile |
Real nomenclature

```
    H        H   H                            H        H   H
    |4       |   1|  → 4                       |1       |   4| → 1
R — C — C = C — C⊖                          R — C — C = C — C⊖
    |        |                                 |        |
    H   C=O      H                             H   C=O      H
            |                                           |
            O                                           O
            |                                           |
           CH3                                         CH3
```
3.38b

__1,4 –addition product__   __4,1 –addition product__

Natural route   Natural route

Since 1,3-dienes when polymerized using Z/N initiators or any initiator via their natural route, 1,4-or 3,4-or 1,2-Addition products are said to be obtained in general, this should also apply to isoprene, 2'acrylated butadiene and similar types. However for them as shown above, 4,1-, 2,1-and 4,3-Addition products are obtained using Paired-media initiators for those that favor both routes like the two cases above. For isoprene it is not 1,4- or 3,4- and 1,2- Addition products sometimes reported in the literature.[13,14] Products which should not exist are reported to be fractions of the overall products! In general, the state of confusion in the literature with respect to Dienes is incomprehensible. For 1,3-pentadiene and it's acrylated counterpart, the followings are similarly obtained.

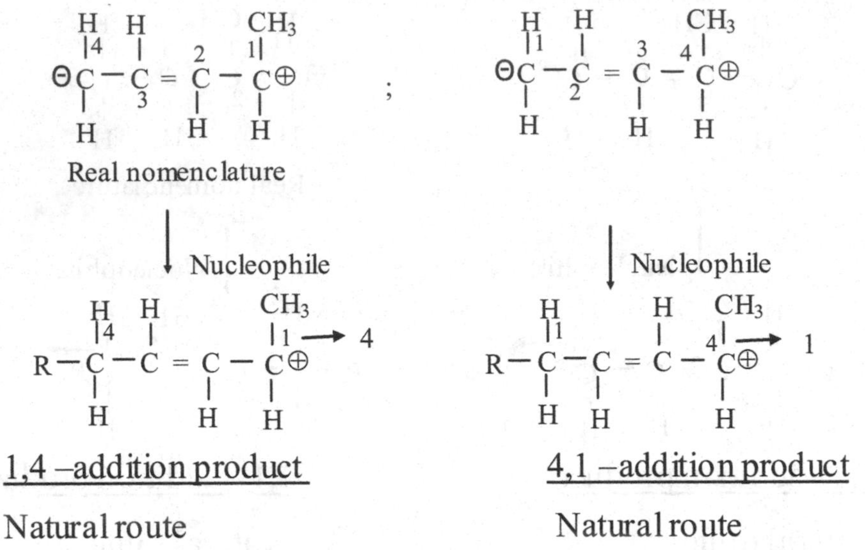

$$\overset{\ominus}{\underset{H}{\overset{H}{C}}} \overset{4}{-} \overset{H}{\underset{3}{C}} = \overset{CH_3}{\underset{H}{\overset{2}{C}}} \overset{1}{-} \overset{CH_3}{\underset{H}{\overset{\oplus}{C}}} \qquad ; \qquad \overset{\ominus}{\underset{H}{\overset{H}{C}}} \overset{1}{-} \overset{H}{\underset{2}{C}} = \overset{3}{\underset{H}{C}} \overset{4}{-} \overset{CH_3}{\underset{H}{\overset{\oplus}{C}}}$$

Real nomenclature

↓ Nucleophile   ↓ Nucleophile

1,4 –addition product   4,1 –addition product   3.39a

Natural route   Natural route

↓ Electrophile   ↓ Electrophile

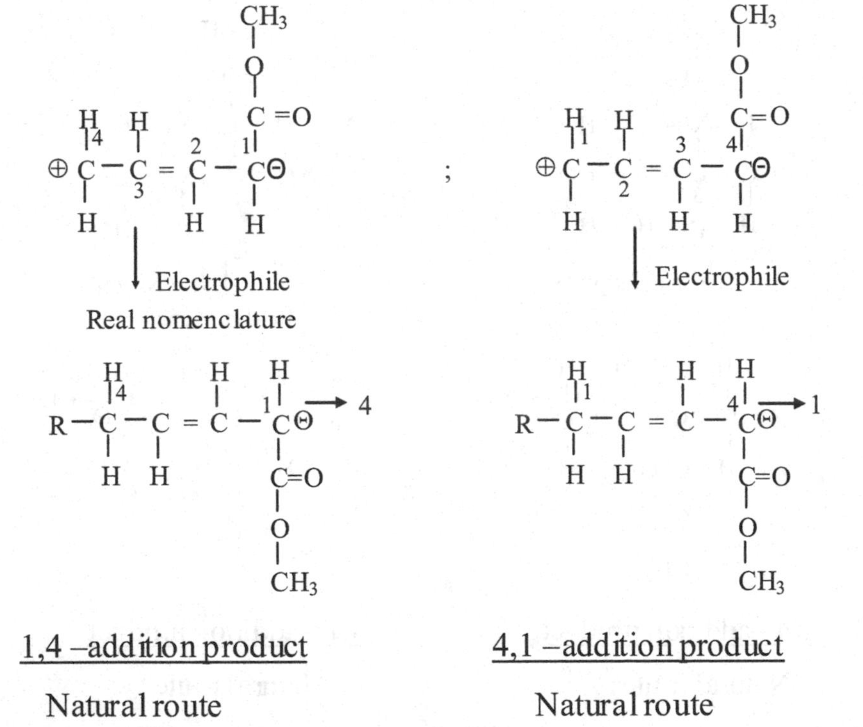

Real nomenclature

1,4 –addition product   4,1 –addition product   3.39b

Natural route   Natural route

Unlike the two cases above, the addition products are 1,4- and 3,4- Addition products for penta-diene and 1,4- and 1,2- for the acrylated diene. These are monomers which favor only the routes natural to them. When the COOCH$_3$ group is replaced with Cl (1-Chloro-1,3-butadiene), nucleo-free-radically the route not natural to it, 1,4- and 1,2- Addition products are obtained. These are just like Electrophiles, when indeed, they are Nucleophiles. Electro-free-radically, the route natural to it, 4,1- and 2,1- are the Addition products. This is unlike that of Butadiene where 1,4- and 3,4-Additions are the Addition products electro-free-radically and 4,1- and 4,3- Additions are the Addition products nucleo-free-radically. One can see, the very great wonders of Nature.

Therefore, for polymerization, via routes which are natural to the monomer, 1,4-, 3,4-, 4,1-, 4,3-, 1,4- 1,2-, 4,1-, 2,1- Addition products are involved, depending on the types of groups the monomer is carrying and the character of the monomer, noting that Z/N initiators can only be used in the route natural to the monomer.

One can thus observe some departure from the original system of nomenclature for these monomers, in view of taking into account the nucleophilic and electrophilic characters of the monomers and their natural routes. In general, common to all 1,3 –dienes are the followings:

(i)   The mono-forms's activation centers, have same character with their parent monomer and are all full-free charged/free-radical monomers as their parent monomer.

(ii)   The mono-form favor only the same route favored by their parent monomer.

(iii)   All the Addition mono-forms in a diene have the same stability in their activated states since

(iv)   Only 1,4-, 3,4-mono-forms for Nucleophiles, 1,4-, 1,2- mono-forms for Electro-philes favor molecular rearrangements where possible to produce only more stable resonance stabilized monomers.

(v)   All of them including the first member have only two mono-forms. With Trienes, there will also be three mono-forms and so on.

(vi)   The location of the substituted groups does not determine the character of the monomer, but the type.

(vii)   The location of the substituted groups does determine where the number should start, noting that never is there a time numbering begins from inside. It does not make any sense.

(viii) The location of the substituted group determines where the least number should be placed. For if it externally located, it carries the first number.

(ix)   When two substituted groups are externally located at both ends and they are of different characters, the numbering begins from the center carrying radical-pulling group, since that is the group in control of the monomer. When the groups are of the same character, the more radical-pushing or pulling group carries the first number.

All these features are made possible due to resonance stabilization effects present in the mono-forms and the manners by which chemical and polymeric reactions take place and currents are produced in fluids and solids. Of all these features, the most important is RADICAL BALANCING, i.e., existence of **full free- on both the RHS and LHS of the equation or full non-free-radical characters on both sides or Half-non-free on both sides of the equation for their resonance stabilized mono-forms.**

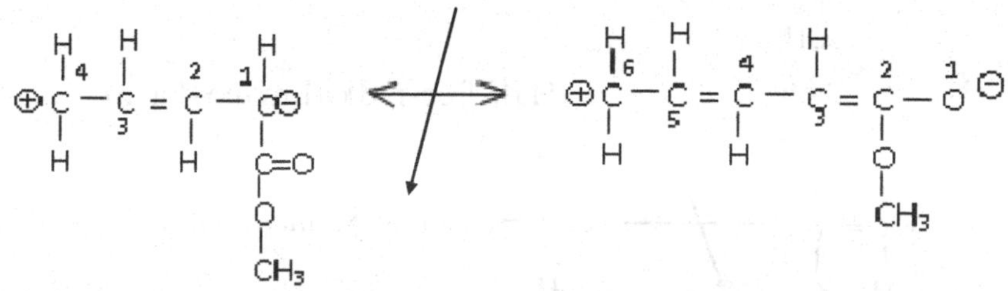

(Full Free Charge monoform)   (Half free Charged monoform)   3.40

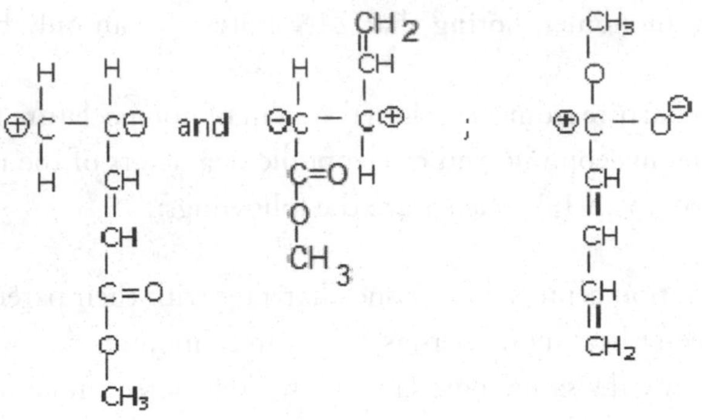

(Electrophiles and   (Nucleophile and   3.41

full free-charged mono-forms)   Half free-charged mono-form)

It is in view of the dual characters favored by the mono-forms of monomers such as acrolein, acrylonitrile, acrylates, etc. that makes the existence of 1,4- Addition polymerization, that is, existences of (II)s of Equations 3.4 - 3.6 impossible. In view of the fact that the equation is not radically balanced, hence presence of resonance stabilization is impossible. Otherwise, one would have been able to obtain 1,6- type-addition from acrylated butadiene as shown above in Equation 3.41. If it was possible, then the 1,3-diene would have become 1,3,5-triene, that which is impossible.

It can largely be observed why identification of the character of an activation center rather than the monomer is very important, since there are monomers which can favor both charged routes, there are monomers which favor one and there are those which favor none.

3.1.2. Radical Resonance Stabilization in Olefinic monomers

Considering the same following monomers of last section, one should also expect the followings.

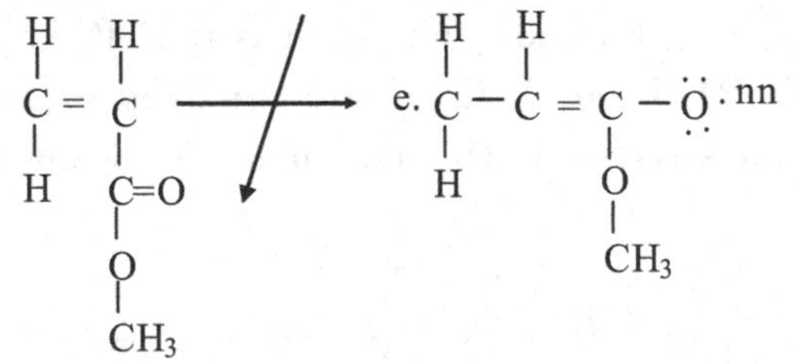

3.42

(Half free-radical mono-form)

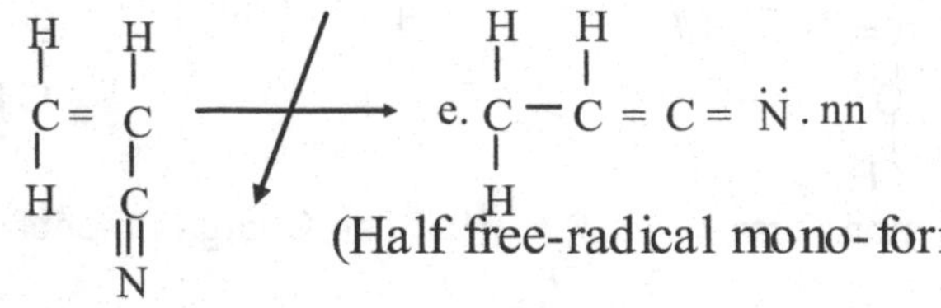

3.43

(Half free-radical mono-form)

133

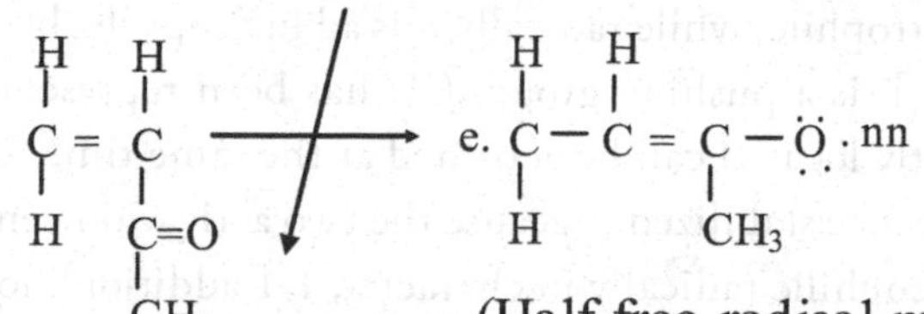

$$
\begin{array}{cc}
H & H \\
| & | \\
C = C \\
| & | \\
H & C=O \\
& | \\
& CH_3
\end{array}
\quad\xrightarrow{\quad\not\quad}\quad
\begin{array}{c}
H \quad\; H \\
| \quad\;\; | \\
e.\; C - C = C - \ddot{O}.\,nn \\
| \qquad\qquad | \\
H \qquad\quad CH_3
\end{array}
\qquad 3.44
$$

(Half free-radical mono-form)

In the first case, the existence of full free-radical activation center along with a half-free-radical activation center makes the existence of a 1,4 half – free-radical monomer impossible, even if it is possible to make C = O activation center electrophilic in character.

For the monomer shown below, the followings are obtained:-

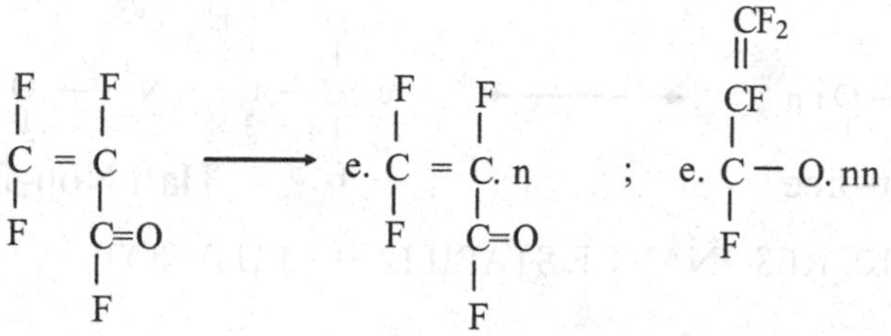

$$
\begin{array}{c}
F \quad F \\
| \quad\; | \\
C = C \\
| \quad\; | \\
F \quad C=O \\
\qquad | \\
\qquad F
\end{array}
\;\longrightarrow\;
\begin{array}{c}
F \quad F \\
| \quad\; | \\
e.\;C = C.\,n \\
| \quad\; | \\
F \quad C=O \\
\qquad | \\
\qquad F
\end{array}
\;;\;
\begin{array}{c}
CF_2 \\
\| \\
CF \\
| \\
e.\;C - O.\,nn \\
| \\
F
\end{array}
\qquad 3.45
$$

(I) (Electrophile)   (II) (Nucleophile)

(free-radically)   (non-free-radically)

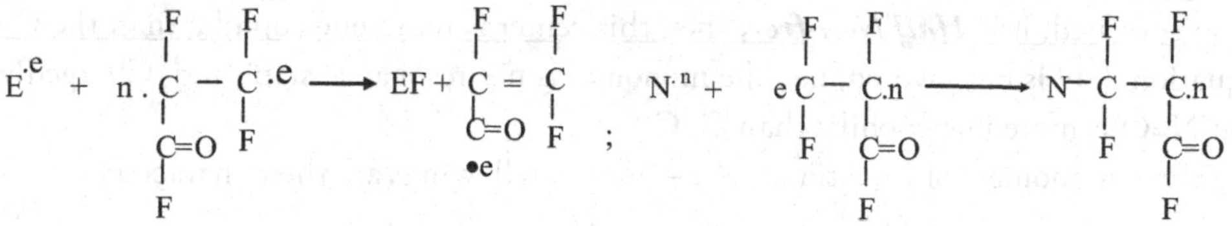

$$
E^{.e} + n.\!\!\begin{array}{c} F \;\; F \\ |\;\;| \\ C - C.e \\ | \;\; F \\ C=O \\ | \\ F \end{array}
\longrightarrow
EF + \!\!\begin{array}{c} F \;\; F \\ |\;\;| \\ C = C \\ | \;\; F \\ C=O \\ \bullet e \end{array}
\qquad
N.^{n} + \!\!\begin{array}{c} F \;\; F \\ |\;\;| \\ e\,C - C.n \\ | \;\; F \\ C=O \\ | \\ F \end{array}
\longrightarrow
\!\!\begin{array}{c} F \;\; F \\ |\;\;| \\ N-C - C.n \\ | \;\; F \\ C=O \\ | \\ F \end{array}
\qquad 3.46
$$

Favored   Favored

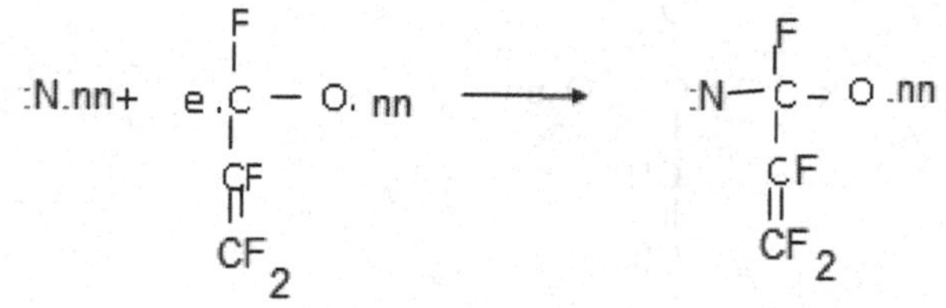

$$
:N.nn + \;\; e\,.\!\!\begin{array}{c} F \\ | \\ C - O.\,nn \\ | \\ CF \\ \| \\ CF_2 \end{array}
\longrightarrow
:N\!-\!\!\begin{array}{c} F \\ | \\ C - O.\,nn \\ | \\ CF \\ \| \\ CF_2 \end{array}
\qquad 3.47
$$

Electro-free-radically, F can be abstracted from (I), since it can be rejected nucleo-free-radically. While F cannot be abstracted from (II) as transfer species of first kind of second type electro-free-radically when activated, it is abstracted from (I) as transfer species of first kind of the first type. (I) can molecularly rearrange to give $F(CF_3)C=C=O$. Note that, the monomer above is a stronger Electrophile than $H_2C = CHCOH$. For nitroethylene, the followings are obtained:-

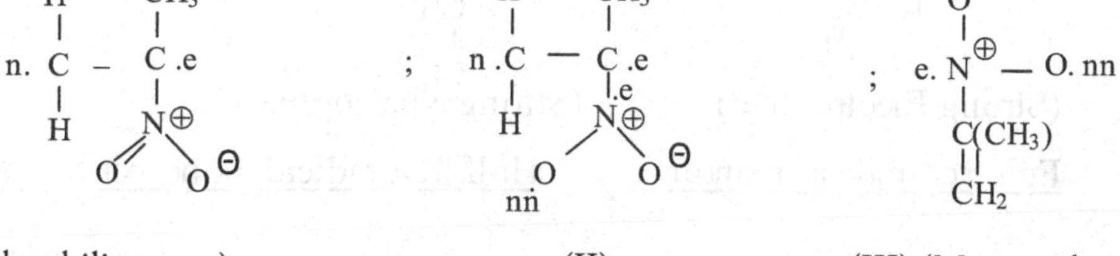

$$
\begin{array}{c}
H \quad\;\; CH_3 \\
| \qquad\; | \\
n.\; C - C\,.e \\
| \qquad\; | \\
H \qquad N^{\oplus} \\
\qquad \swarrow \;\searrow \\
\qquad O \qquad O^{\ominus}
\end{array}
\;;\;
\begin{array}{c}
H \quad\;\; CH_3 \\
| \qquad\; | \\
n.\,C - C\,.e \\
| \qquad\; |.e \\
H \qquad N^{\oplus} \\
\qquad \swarrow \;\searrow \\
\qquad O \qquad O^{\ominus} \\
\;\; nn
\end{array}
\;;\;
\begin{array}{c}
O^{\ominus} \\
| \\
e.\; N^{\oplus} - O.\,nn \\
| \\
C(CH_3) \\
\| \\
CH_2
\end{array}
\qquad 3.48
$$

(I) (Less nucleophilic center)   (II)   (III) (More nucleophilic center)

Full free-radical form   POSSIBLE   Half free-radical form

134

Chargedly, the monomer is an Electrophile, while radically it is a Nucleophile, because chargedly NO_2 is a pulling group, while radically, it is a pushing group. (II) has been represented in order to show that two activation centers adjacently located can be activated at the same time, if the initiator is very strong and the centers are not resonance stabilized. Because the two activation centers are electrophilic and nucleophilic chargedly or nucleophilic radically in character, 1,4-addition monoform cannot exist chargedly since charges cannot be moved and the radicals cannot be placed as shown in (II). In (I), the C = C center has been activated. Nucleo-free-radicals cannot initiate it, but electro-free-radicals can initiate it. The $^{\oplus}N = O$ center can be initiated electro-free-radically, but cannot be used since the center is more nucleophilic than the C=C center. Indeed, it can be resonance stabilized as shown below only if $^{\oplus}N{=}O$ is less e . C

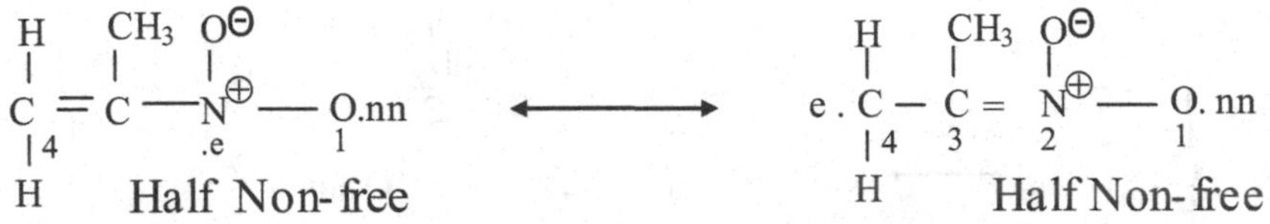

$$
\underset{\text{Half Non-free}}{\overset{H}{\underset{H}{\overset{|}{C}}}{=}\overset{CH_3}{\overset{|}{C}}{-}\overset{O^{\ominus}}{\overset{|}{\underset{.e}{N^{\oplus}}}}{-}\underset{1}{O.nn}} \quad\longleftrightarrow\quad \underset{\text{Half Non-free}}{e.\overset{H}{\underset{H}{\overset{|}{\underset{4}{C}}}}{-}\overset{CH_3}{\underset{3}{\overset{|}{C}}}{=}\overset{O^{\ominus}}{\underset{2}{\overset{|}{N^{\oplus}}}}{-}\underset{1}{O.nn}}
$$

LOOKS RESONANCE STABILIZED, BUT NOT.   3.49

It should be noted that the polar bond, an imaginary bond, does not prevent the nitrogen center from carrying a radical since the Boundary laws have not been broken. Since when the N=O center (which is less nucleophilic than the C=C center radically) is activated, it is **Full Non-free** and when the $^{\oplus}N{=}O$ center is also activated, it is **Half Non-free,** then this center is more nucleophilic than the C=C center. Hence, Equation 3.49 is not favored, i.e., the monomer is not resonance stabilized. Chargedly, it is well known that N=O is more nucleophilic than C=C.

For the three monomers of Equations 3.42 – 44, the followings are their characters.

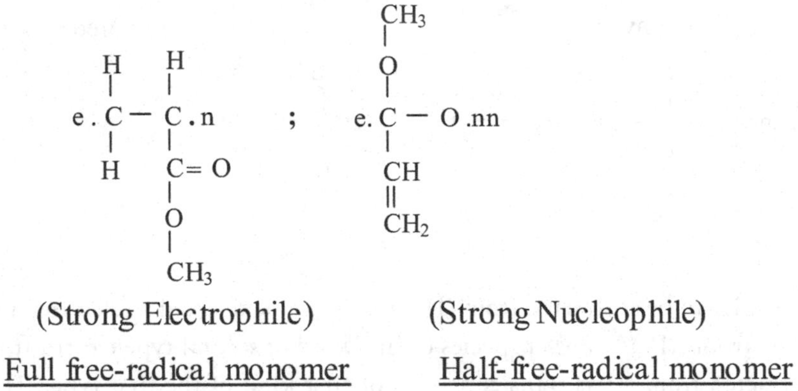

$$
e.\overset{H}{\underset{H}{\overset{|}{C}}}{-}\overset{H}{\underset{\underset{\underset{\underset{CH_3}{|}}{O}}{\overset{|}{\underset{O}{C}}}}{\overset{|}{\underset{\|}{C}}}.n \quad ; \quad e.\overset{CH_3}{\underset{\underset{\underset{CH_2}{\|}}{CH}}{\overset{|}{\underset{O}{C}}}}{-}O.nn
$$

(Strong Electrophile)   (Strong Nucleophile)

Full free-radical monomer   Half-free-radical monomer   3.50

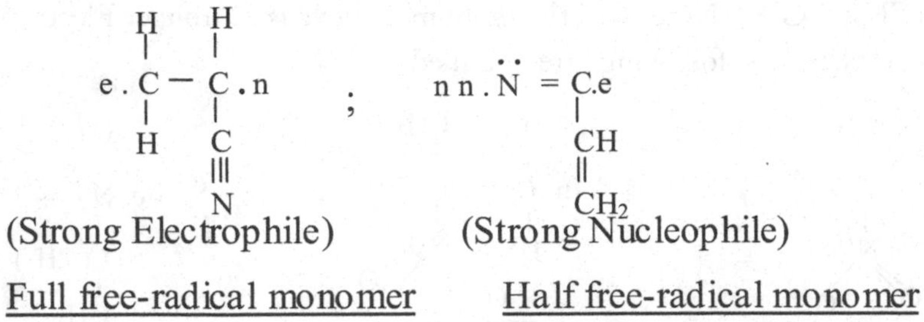

$$
e.\overset{H}{\underset{H}{\overset{|}{C}}}{-}\overset{H}{\underset{\underset{\underset{N}{\text{\scriptsize III}}}{C}}{\overset{|}{C}}}.n \quad , \quad nn.\overset{..}{N}{=}\underset{\underset{CH_2}{\|}}{\overset{}{\underset{CH}{C}}}.e
$$

(Strong Electrophile)   (Strong Nucleophile)

Full free-radical monomer   Half free-radical monomer   3.51

$$\underset{\substack{|\\H}}{\overset{\substack{H\\|}}{e.C}} - \underset{\substack{|\\C=O\\|\\CH_3}}{\overset{\substack{H\\|}}{C.n}} \quad ; \quad \underset{\substack{|\\CH\\||\\CH_2}}{\overset{\substack{CH_3\\|}}{e.C}} - O.nn$$

3.52

(Weak Electrophile)   (Strong Nucleophile)
<u>Full free radical monomer</u>   <u>Half free-radical monomer</u>

In view of the differences in character of the activation centers on the monomer and types of active centers, existences of resonance stabilization and 1,4 –addition monomers are impossible.

For 1,3-dienes, the followings are obtained:-

$$\underset{\substack{|\\H\\|\\CH\\||\\CH_2}}{\overset{\substack{H\\|4}}{n.C}} - \underset{\substack{3|\\|\\H}}{\overset{\substack{H\\|}}{C.e}} \quad\longleftrightarrow\quad \underset{\substack{|\\H}}{\overset{\substack{H\\|1}}{e.C}} - \underset{2}{\overset{}{C}} = \underset{\substack{|\\H}}{\overset{3}{C}} - \underset{\substack{|\\H}}{\overset{\substack{H\\|4}}{C.n}} \quad\cancel{\longleftrightarrow}\quad \underset{\substack{|\\CH\\||\\CH_2}}{\overset{\substack{H\\|}}{n.C^2}} - \underset{\substack{|\\H}}{\overset{\substack{H\\|}}{^1C.e}}$$

3.53

<u>(All Nucleophiles) [Either a male or female]-**BD (Butadiene) 3,4-mono-form**</u>

$$\underset{\substack{|\\H\\|\\CH\\||\\CH_2}}{\overset{\substack{H\\|1}}{n.C}} - \underset{\substack{|}}{\overset{\substack{CH_3\\|2}}{C.e}} \quad\cancel{\longleftrightarrow}\quad \underset{\substack{|\\H}}{\overset{\substack{H\\|1}}{n.C}} - \underset{2}{\overset{CH_3}{C}} = \underset{\substack{|\\H}}{\overset{3}{C}} - \underset{\substack{|\\H}}{\overset{\substack{H\\|4}}{C.e}} \quad\longleftrightarrow\quad \underset{\substack{|\\C(CH_3)\\||\\CH_2}}{\overset{}{n.C^3}} - \underset{\substack{|\\H}}{\overset{\substack{H\\|}}{^4C.e}}$$

3.54

<u>All Nucleophiles (weaker) – **IP (Isoprene) 4,3-mono-form**</u>

$$\underset{\substack{|\\H\\|\\CH\\||\\CH_2}}{\overset{\substack{CH_3\\|1}}{e.C}} - \underset{\substack{2|\\|\\}}{\overset{\substack{H\\|}}{C.n}} \quad\cancel{\longleftrightarrow}\quad \underset{\substack{|\\H}}{\overset{\substack{CH_3\\|1}}{e.C}} - \underset{2}{\overset{H}{C}} = \underset{\substack{|\\H}}{\overset{3}{C}} - \underset{\substack{|\\H}}{\overset{\substack{H\\|4}}{C.n}} \quad\longleftrightarrow\quad \underset{\substack{|\\CH\\|\\CH_3}}{\overset{}{e^3.C}} - \underset{\substack{|\\H}}{\overset{\substack{H\\|4}}{C.n}}$$

3.55

<u>All Nucleophiles (strong) - **PD (Penta-diene) 3,4-mono-form**</u>

136

```
     CH2                          H   H
      ‖                           |   |
      CH          H    H          C   C
 H    |           |  2 |  4 |     H   H
 | 1  2|          |    C = C    .e   | 3   4 |
 n . C — C . e    n . C—  ‖   C — C . e        n . C — C . e
 |    |           |    3          |   |
 C=O  H           C=O  H          CH  H
 |                |                ‖
 O                O                CH
 |                |                |
 CH3              CH3              C = O
                                   |
                                   O
                                   |
                                   CH3          3.56
```

<u>All Electrophiles (strong) – **MAD (Methyl acrylated diene) 1,2-mono-form**</u>

```
      CH3                          CH3
      |                            |
      CH            H    H   CH3    H   CH3
      ‖             | 1  2|    |    | 3   4|
      CH            C — C = C — C.e  C — C.e
 H    |            n.        3   1               n.
 | 1  2|           |    |   H                     |    |
 n . C — C . e     C=O  H                         CH   H
 |    |            |                              ‖
 C=O  H            O                              CH
 |                 |                              |
 O                 CH3                            C=O
 |                                                |
 CH3                                              O
                                                  |
                                                  CH3      3.57
```

<u>All Electrophiles (Strong) –**MS (Methyl sorbate) 1,2-mono-form**</u>

```
 H    Cl          H    Cl   H        H    H
 | 1  2|          | 1  |    | 4      | 3  | 4
 e . C — C . n    e . C — C = C — C . n   e. C — C . n
 |    |           |    2    |    |        |    |
 H    CH          H         H    H        CCl  H
      ‖                                   ‖
      CH2                                 CH2      3.58
```

<u>All Nucleophiles (weak) –**CP (Chloro-prene) 1,2-mono-form**</u>

```
      CH3               CH3
      |                 |
      O                 O
      |                 |
 H    C=O          H    C=O   H        H         H
 | 1  2|           | 1  |     | 4      | 3       | 4
 e . C — C . n     e . C — C = C — C . n  e . C ——— C . n
 |    |            |    2     |    |        |        |
 H    CH           H          H    H        C(COOCH3) H
      ‖                                     ‖
      CH2                                   CH3      3.59
```

<u>All Electrophiles (weak) – **2^L AD (2^L Acrylated diene) 2,1-mono-form**</u>

137

These are their real states, as was shown chargedly. Chargedly, it was an experiment, i.e. an exploration. While the same character is maintained for all their addition monomers, it is important to note that:-

(i) All are full-free-radical monomers.

(ii) $CH_2 = C(CH_3)$ group is a radical-pushing group of lower capacity than H.

(iii) $CH_2 = C(COOCH_3)$ group is also a radical-pushing group. So also is

(iv) $CH_2 = CCl$ group or $RHC = CH-$, all of greater capacity than H.

(v) From Equation 3.53, $CH_2 = CH-$ group is of greater capacity than H, when resonance stabilization is provided. If it was not greater, resonance stabilization cannot be provided.

(vi) $(COOCH_3) HC = CH-$, $Cl_2C = CH-$ and alike, are radical-pulling groups of far greater capacity than H and all the above, on the other side of the spectrum

For a difference, consider the monomer shown below. The followings are obtained in view of the presence of resonance stabilization when activated. This is what cannot be provided chargedly, although we have used it much more for exploratory purposes.

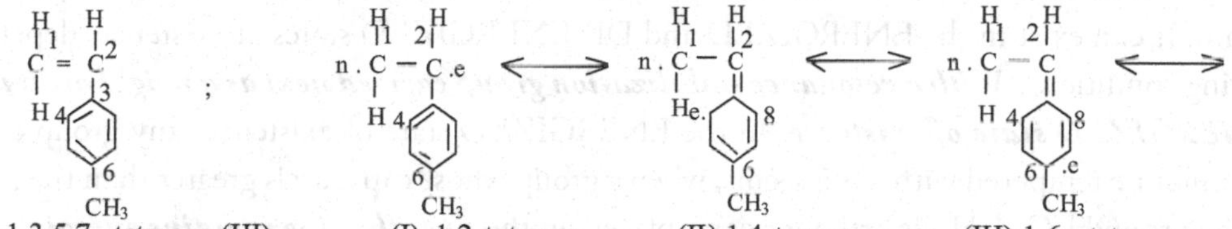

1,3,5,7 –tetraene (III)   (I) 1,2-tetraene   (II) 1,4-teraene   (III) 1,6 - tetraene

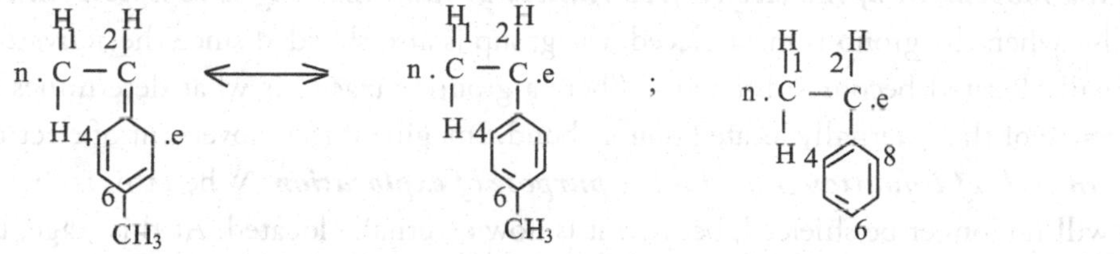

(IV) 1,8 –tetraene   (I) 1,2 –tetraene   (V) Styrene's Nomenclature   3.60

Note that the numbering shown above looks tentative. It is not, but the real one. The movement above is from (I) to (II) to (III) to (IV) and back to (I). It looks like a ***partly Closed tetraene,*** consisting of a closed Triene and an opened mono-ene. If the CH_3 group above is H, then we have styrene the first member of the family of alkene Alipharomatic compounds for which the numbering system is not the reverse of what is shown above, but the same. The numbering begins with one as shown in (V) in the last equation, goes round the ring This is clear indication that styrenes are indeed tetraenes. Note, for the first time that styrene and the members of the family are not DIENES, but a Tetraene partly ***Closed and Discrete.*** The resonance stabilization in Benzene ring is ***Closed and Continuous,*** while in Dienes, Trienes, Tetraenes they are ***Opened and Discrete.*** In Styrene, the electro-free-radical moves from a point and comes back to the same point. This is not the case with benzene. Therefore, the CH_3 group above which is internally located para-placed on the benzene ring can be observed to be resonance stabilized. The same applies to the methyl group in α-methyl styrene. With α-methyl styrene, the group is resonance stabilized radically, since it is also internally located. If the group was meta-placed, it will also be resonance stabilized since it is inside a ring carring a resonance stabilization group. Chargedly, it is not resonance stabilized as has been wrongly speculated in the past[1-8]. The origin of all the forms above is not that where all the four Activation centers are instantaneously involved in activation. It is the center which is least nucleophilic that is first activated and that center is (I) above just as for all its Nucleophilic and Electrophilic members. For all of them, it is 1,2- tetraene That is the only center that can be used chargedly without being resonance stabilized. The C=C bond is far less nucleophilic than the ones in Benzene ring. It was from (I), all the others are formed by the movement of the e• radical when visibly present or from the π-bond when not visibly present on the β-carbon center. This is not a different type of movement, but unique as will shortly become obvious. It moves to grab the nucleo-radical on an adjacently located π- bond to form another π-bond. Positive charges cannot move, because they are carrying nothing i.e. no radicals as already said. Even negative charges with two opposite radicals cannot move, as already said, because only one radical can be moved and the only one is the electro-radical (free or non-free). Thus, one can see very distinctively for the first time the great distinction between RADICALS and CHARGES. One can also see why all reactions take place far more radically (if not only radically), than chargedly. Already, we know that they can never take place ionically. ***When activated chargely, they remain in that state except where molecular rearrangement is found possible.*** The transfer species internally located cannot be resonance stabilized chargedly. Radically, they can be too resonance stabilized to prevent their removal and electro-radical can be moved to give other mono-forms.

Worthy of note at this point in time, is that Benzene is a very unique compound as will be shown downstream. It can exist in the ENERGIZED and DE-ENERGIZED states of existence, depending on the operating conditions. ***With a resonance stabilization group carried next as a neighbor, it is always in the ENERGIZED state of existence***. In the ENERGIZED state of existence, any group carried by the ring cannot be tempered with. As it seems, when a group whose capacity is greater than the resonance stabilization group $H_2C=CH-$, is ortho- or para- placed on the ring, ***the ring remains energized in such a way that the movement of the electro-free-radical around the ring is complete and the group is shielded***. Also when the group is meta-placed, the group is also shielded since the activated state of the center externally located becomes different. Where a group is placed is what determines the activated state of existence of the externally located double bond. Imagine if the movement of electro-free-radical had stopped ***in (III) of Equation 3.60, for the purpose of exploration***. When this is the case, then the H_3C group will no longer be shielded, because it is now externally located. At this stage, the monomer becomes a TRIENE. Molecular rearrangement can now commence as shown below.

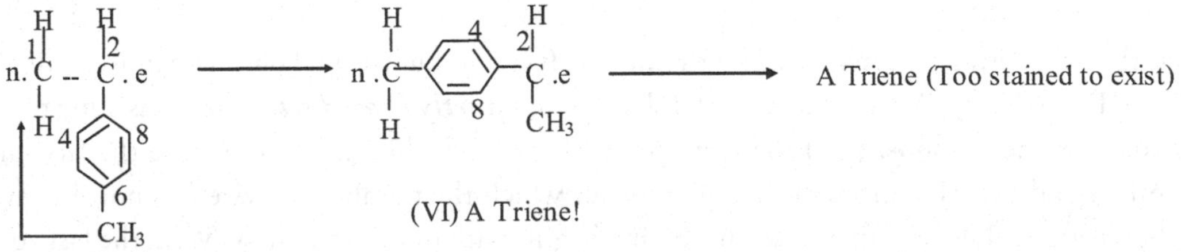

(VI) A Triene!   3.61

1,2- Tetraene

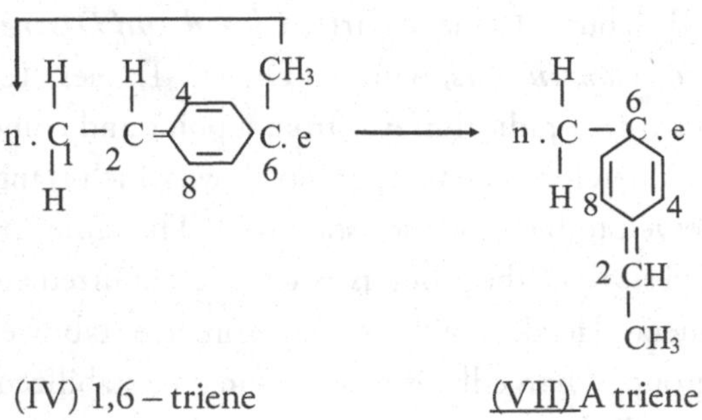

(IV) 1,6 – triene   (VII) A triene (Too strained to exist)   3.62

Though (VI) and (VII) do not seem to belong to any of the resonance stabilized members of Equation 3.61, they are all the same. Nevertheless, one can see why the continuous movement of the electro-free-radical either visibly located or from the π, must be accomplished or take place, otherwise the constant monomer units observed along the chain would not have been possible.

Unlike 1,3-dienes, all the resonance stabilized forms here seem to rearrange to produce the same mono-forms as shown below using (VII).

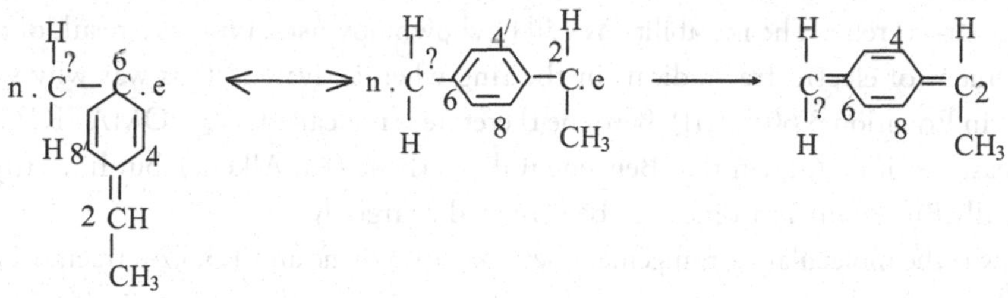

3.63

(VII) A-triene   (VI) A– triene   (VIII) A – triene

The same (VIII) is obtained for the so-called trienes when they rearrange, a clear indication of the fact that styrenes are Tetraenes, noting that (VIII) is too strained to exist.

Unlike the case above, for **α-methyl styrene shown below**, the H on CH_3 cannot be abstracted in view of the resonance stabilization provided by the phenyl group, based on its internal location. Note that unlike the case above, the phenyl group is carrying no group other than Hs in the ring. Hence, the electro-free-radical obtained when activated can readily go round and come back to where it started. Note that this is unlike the case of isoprene, because of the power of benzene ring. Note at this point in time as will be shown downstream that the phenyl group is more radical-pushing than the alkenyl group ($H_2C=CH-$) which in turn is more radical-pushing than H. Hence, when **Styrene** is activated, the carbon center carrying the benzene ring (Phenyl group) is the one carrying the electro-free-radical or positive charge.

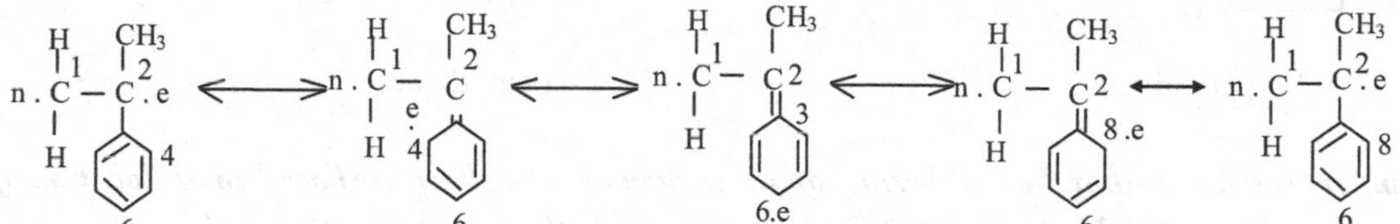

(I) 1,2-   (II) 1,4-   (III) 1,6-   (IV) 1,8-   (I) 1,2-

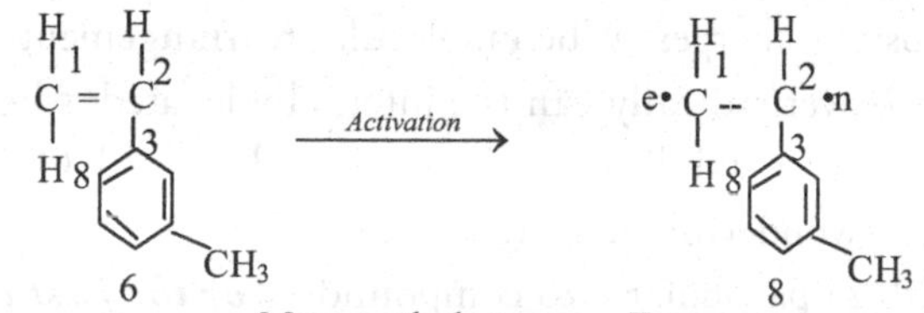

Meta-methyl styrene – Tetraene

3.64

Some of the numberings as used by some schools of thought so far are in order as will become obvious as we move along the Series and Volumes. However, one can observe that these monomers are Tetraenes never depending on where the group is placed since they are all internally located. It is only in the **α**-carbon center that groups can be externally located.

Unlike 1, 3 –dienes, the electro-free-radical is able to move and come back to where it started in the ENERGIZED state of existence of the phenyl group because though the phenyl and alkenyl groups are resonance stabilization groups, there are very big differences between them. While alkenyl groups can be used differently radically and chargedly, the phenyl group can only be used radically, because of the greatest unsaturation (Maximum) located inside the six membered ring for charges to exist as a result of great influence of electrostatic forces of repulsion. For this reason, chargedly, the CH_3 group above cannot be resonance stabilized. Eight-membered ring with more of such unsaturation (four π-bonds inside a ring) exists because of the size of the ring. However being unstable, it molecularly rearranges to give something different- styrene. The instability as will be shown downstream is as a result of the limitations placed on movement of electro-free-radicals in the ring when activated. This was why we stopped the movement in (I) in Equation 3.60 at (III). Benzene therefore is not called an AROMATIC COMPOUND for nothing. Please, let it be known that Benzene unlike ethene (An Alkene), but like ethyne cannot be activated chargedly.But its linear isomer can be activated chargedly.

Shown below is the molecular rearrangement between a 1,3 diene and 1,3,5,7-tetraene of Equation 3.61.

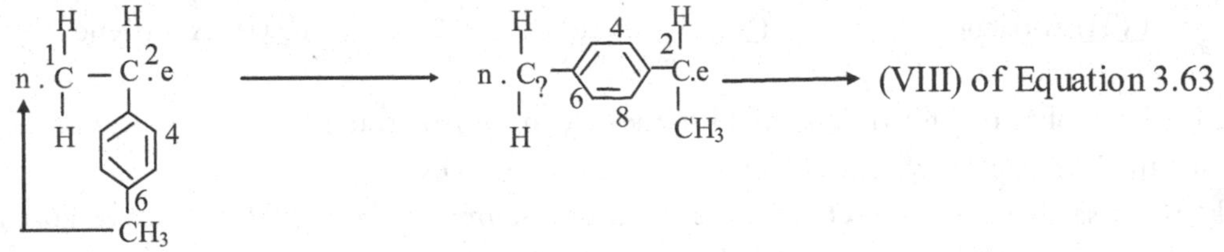

(I) Tetraene   (IV) A triene?

<u>A methyl substituted styrene (See Equation 3.61)</u>

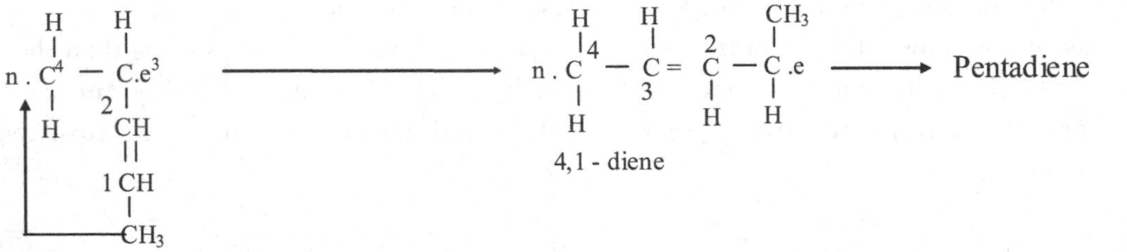

4,3 – pentadiene   1,3 – pentadiene

3.66

It is important to note that the numberings on the carbon centers above are based on the location of the substituent group and the character of the monomer. Only the in last equation is the rearrangement valid In styrene, only one para-position and two ortho-positions on the benzene ring are allowed to carry electro-free-radicals and not positive charges. When molecular rearrangement takes place, it can only do so from the α- Carbon center. Never radically can the internally located π-bonds be activated one at a time. Only radically can the resonance stabilization character be displayed. Only radically can they move internally in the ring to give the wonderful characters displayed by them to generate different colors as will be shown downstream with in particular Azo compounds. *For the first time, the styrene types of monomers are being fully boldly reclassified.* They possess only the mono- olefinic character, and not the di-, tri- and tetra-olefinic characters, for the following reasons-

(a) The Alipharomatic character of the monomer
(b) Strength of activation center externally located and conjugatedly placed on a very unsaturated ring. It is small compared to that inside the ring.
(c) The strength of the resonance stabilization group inside benzene.

From all indications, the phenyl groups have more resonance stabilization capacity than alkenyl groups, since the former contains more activation centers than the latter. Disturbing and worthy of note is that monomers such as styrene, α-methyl styrene have been reported to undergo so-called **anionic and cationic** polymerization routes[1-8]. The CH_3 group in α-methyl styrene cannot be resonance stabilized chargedly, but only radically. Radically, they say they can be polymerized[3] What this means radically cannot be explained. Nevertheless, based on the NEW SCIENCE, they can be polymerized radically (nucleo- and electro-free-radically) and only positively chargedly (not cationically). While pentadiene can be polymerized in its 3,4- mono-form chargedly when made from 1,3-Dienes (Positively) and 1,4-, and 3,4 mono-forms can be polymerized electro-free-radically.

3.2. Transfer from Monomer to Initiator

It is important to note in all the considerations so far that when a substituent group is internally located in a resonance stabilized species, it is cannot be tempered with. This is shown clearly below for styrenes and 1, 3 –dienes.

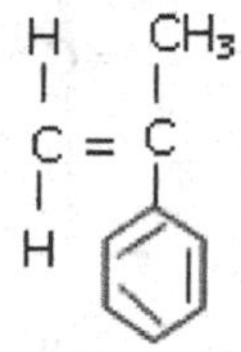

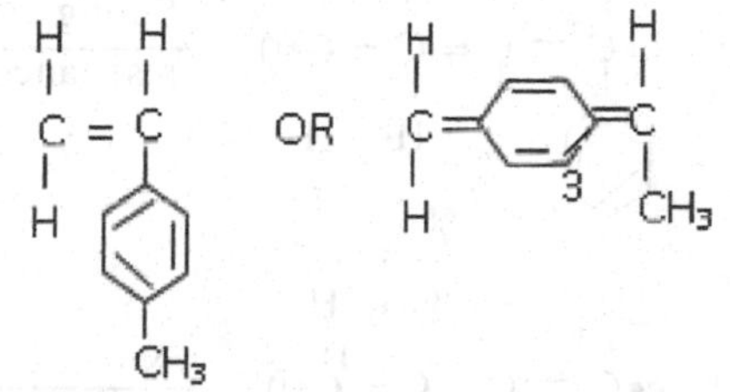

(I) <u>CH_3 group resonance stabilized</u>   (II) <u>CH_3 group also **fully resonance stabilized**</u>

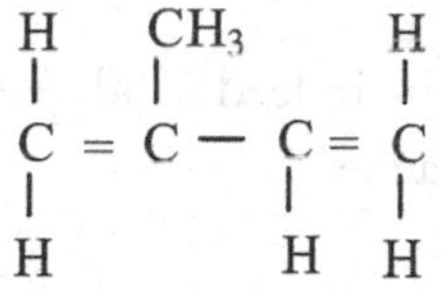

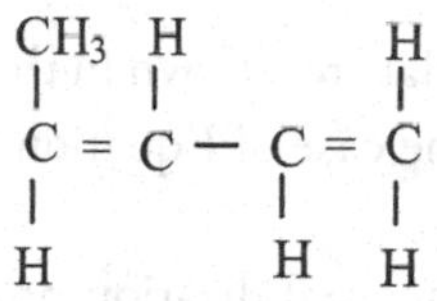

3.67

(I) <u>CH_3 group resonance stabilized</u>   (II) <u>CH_3 group not resonance stabilized</u>

3.68

The CH_3 group of (II) is resonance stabilized, because it is internally located. Shown below for pentadiene where the CH_3 is now heavily shielded. The H group located on same carbon center carrying CH_3 has

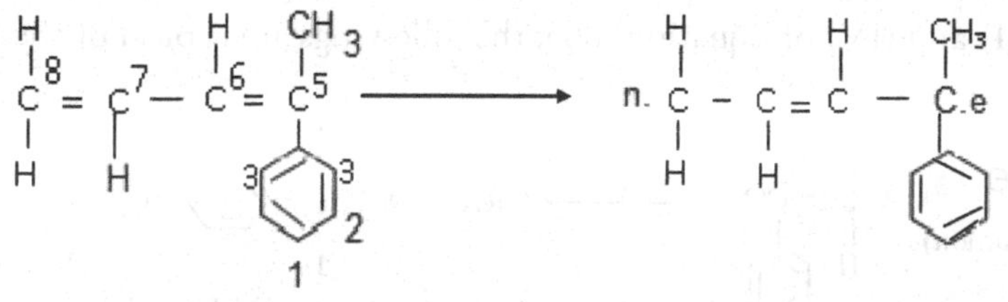

<u>A 1,3,5,7-tetraene</u>   (I)

3.69

been replaced with a resonance stabilization group in this case the phenyl group. *When an alkenyl group is involved, a 1,3, 5- triene is obtained as shown below, as opposed to 1,3, 5, 7- tetraene wrongly shown above. Based on the new style of nomenclature, it is 1,3,5,7,9-Pentaene which can be resonance stabilized state radically.*

$$
\underset{\text{A,1,3,5 - triene}}{\overset{\displaystyle \begin{array}{c} H \quad H \quad CH_3 \\ | \quad\;\; | \quad\;\; | \\ C = C - C = C\,3 \\ | \quad\;\; | \quad\;\; | \\ H \quad H \;\; 2CH \\ \qquad\quad \| \\ \qquad 1CH_2 \end{array}}{}} \quad\longrightarrow\quad n.\overset{\begin{array}{c}H \quad H \quad CH_3 \quad H\\| \quad\;\; | \quad\;\;\; | \quad\;\; |\end{array}}{C - C = C - C = C - C.e} \qquad\qquad 3.70
$$

$$
(I)
$$

The center first activated above is the 1,2- activation center, the least nucleophilic center.

When some of (I) of Equation 3.67 – 70 are involved, the followings are obtained.

$$
N\bullet^{\,n} \; + \; e\bullet C - C\bullet n \;\; \xrightarrow{\;\text{Strong}\;}_{\text{resistance}}\; N - C - C\bullet n \qquad\qquad 3.71
$$

$$
N\bullet n \; + \; e\bullet C - C = C - C\bullet n \;\; \xrightarrow{\;\text{Strong}\;}_{\text{resistance}}\; N - C - C = C - C\bullet n \qquad\qquad 3.72
$$

$$
N\bullet n \; + \; e\bullet C - C = C - C\bullet n \;\; \xrightarrow{\;\text{No}\;}_{\text{resistance}}\; N - C - C = C - C\bullet n \qquad\qquad 3.73
$$

[Note that the nomenclatures shown in the two equations above are indeed valid. As an exercise, complete the nomenclature of the case of Equation 3.72, since it is a Pentaene.]

In view of the resonance stabilization provided for the CH_3 group, transfer species of the first kind of the first type cannot be abstracted free-radically. Chargedly, this is not possible for first equation above where ***transfer from monomer step is favored.*** The same will not apply to 1,2- and 3,4-mono-forms of the second and third equations above respectively due to electrostatic forces of repulsion. Radically and in some cases such as just above chargedly, transfer from monomer step is not favored. However, in view of presence of transfer species which cannot be abstracted, the rate of initiation will be very slow (resistance). For the (I) and (IV) of Equation 3.65, the followings are typical of the reactions chargedly.

$$
R^{\ominus}_{(\text{free-anion})} \; + \; {}^{\ominus}C - C^{\oplus} \;\;\longrightarrow\;\; RH \; + \; {}^{\ominus}C - \!\!\left\langle\!\!\bigcirc\!\!\right\rangle\!\!- C = C \qquad \text{OR}
$$

$$
\text{(I) FAVORED}
$$

$$
C = \!\!\left\langle\!\!\bigcirc\!\!\right\rangle\!\!= C = C^{\ominus}
$$

$$
\text{(II) NOT FAVORED} \qquad\qquad 3.74
$$

$$R^{\ominus} \; + \; \underset{\underset{CH_3}{|}}{\overset{\overset{H}{|}}{\oplus C}} - \!\!\!\bigcirc\!\!\!- \underset{\underset{H}{|}}{\overset{\overset{H}{|}}{C\ominus}} \; \longrightarrow \; RH \; + \; \underset{\underset{H}{|}}{\overset{\overset{H\;\;\;H}{|\;\;|}}{\ominus C - C}} = \!\!\!\bigcirc\!\!\! = \underset{\underset{H}{|}}{\overset{\overset{H}{|}}{C}}$$

(free-anion)

OR

[CANNOT EXIST] (I) NOT FAVORED

$$\underset{\underset{H}{|}}{\overset{\overset{H}{|}}{-C}} - \!\!\!\bigcirc\!\!\!- \underset{\underset{H}{|}}{\overset{\overset{H}{|}}{C}} = \underset{\underset{H}{|}}{\overset{\overset{H}{|}}{C}}$$

(II) NOT FAVORED   3.75

Transfer from Monomer to Initiator

Notice what is favored above. It is such, because charges cannot be moved. Negatively, they cannot be polymerized.

Electro-free-radically or positively, the followings are obtained.

$$E^{\cdot e} \; + \; n.\underset{\underset{H}{|}}{\overset{\overset{H}{|}}{C}} - \underset{\bigcirc}{\overset{\overset{CH_3}{|}}{C}}.e \; \longrightarrow \; E - \underset{\underset{H}{|}}{\overset{\overset{H}{|}}{C}} - \underset{\bigcirc}{\overset{\overset{CH_3}{|}}{C}}.e \qquad 3.76$$

$$E^{\cdot e} \; + \; n.\underset{\underset{H}{|}}{\overset{\overset{H}{|}}{C}} - \underset{\underset{H}{|}}{\overset{\overset{H}{|}}{C}} = \underset{\underset{CH_3}{|}}{\overset{\overset{\bigcirc}{|}}{C}} - \underset{\underset{CH_3}{|}}{\overset{}{C}}.e \; \longrightarrow \; E - \underset{\underset{H}{|}}{\overset{\overset{H}{|}}{C}} - \underset{\underset{H}{|}}{\overset{\overset{H}{|}}{C}} - \underset{\underset{\bigcirc}{}}{\overset{\overset{H}{|}}{C}} - \underset{}{\overset{\overset{CH_3}{|}}{C}}.e \qquad 3.77$$

$$E^{\cdot e} \; + \; n.\underset{\underset{H}{|}}{\overset{\overset{H}{|}}{C}} - \underset{\underset{H}{|}}{\overset{\overset{CH_3}{|}}{C}} = \underset{\underset{H}{|}}{\overset{\overset{H}{|}}{C}} - \underset{\underset{H}{|}}{\overset{}{C}}.e \; \longrightarrow \; E - \underset{\underset{H}{|}}{\overset{\overset{H}{|}}{C}} - \underset{\underset{H}{|}}{\overset{\overset{CH_3}{|}}{C}} - \underset{\underset{H}{|}}{\overset{\overset{H}{|}}{C}} - \underset{\underset{H}{|}}{\overset{}{C}}.e \qquad 3.78$$

$$E^{\cdot e} \; + \; n.\underset{\underset{H}{|}}{\overset{\overset{H}{|}}{C}} - \underset{\underset{H}{|}}{\overset{\overset{H}{|}}{C}} = \underset{\underset{CH_3}{|}}{\overset{\overset{H}{|}}{C}} - \underset{}{\overset{}{C}}.e \; \longrightarrow \; E - \underset{\underset{H}{|}}{\overset{\overset{H}{|}}{C}} - \underset{\underset{H}{|}}{\overset{\overset{H}{|}}{C}} - \underset{\underset{CH_3}{|}}{\overset{\overset{H}{|}}{C}} - \underset{}{\overset{}{C}}.e \qquad 3.79$$

$$E^{\cdot e} \; + \; n.\underset{\underset{H}{|}}{\overset{\overset{H}{|}}{C}} - \underset{\underset{\bigcirc\text{-}CH_3}{}}{\overset{\overset{H}{|}}{C}}.e \; \longrightarrow \; E - \underset{\underset{H}{|}}{\overset{\overset{H}{|}}{C}} - \underset{\underset{\bigcirc\text{-}CH_3}{}}{\overset{\overset{II}{|}}{C}}.e \qquad 3.80$$

None of the above but one (Equation 3.79) have transfer species to reject to give dead terminal double bond polymers, due to resonance stabilization or absence of it. Thus, it can be observed how

144

resonance stabilization groups can affect the route favored by a monomer during initiation and the type of polymer produced.

Now consider replacing the CH_3 group with CF_3 group. Chargedly for styrene types of monomers, the followings are obtained.

$$\underset{\text{(free-anion)}}{R^{\ominus}} + \oplus \overset{H}{\underset{H}{\overset{|}{C}}} - \overset{CF_3}{\underset{\underset{\bigcirc}{|}}{C}}{\ominus} \xrightarrow[\text{Resistance}]{\text{(No}} R - \overset{H}{\underset{H}{\overset{|}{C}}} - \overset{CF_3}{\underset{\underset{\bigcirc}{|}}{C}}{\ominus} \qquad 3.81$$

$$R^{\oplus} + \ominus \overset{CF_3}{\underset{\underset{\bigcirc}{|}}{C}} - \overset{H}{\underset{H}{\overset{|}{C}}}{\oplus} \longrightarrow RF + \overset{CF_2}{\underset{\underset{\bigcirc}{||}}{C}} - \overset{H}{\underset{H}{\overset{|}{C}}}{\oplus} \qquad 3.82$$

Chargedly, it is not resonance stabilized. Here, we have transfer from monomer to initiator.

$$\underset{\text{(free-anion)}}{R^{\ominus}} + \oplus \overset{H}{\underset{H}{\overset{|}{C}}} - \overset{H}{\underset{\underset{CF_3}{\bigcirc}}{C}}{\ominus} \xrightarrow[\text{Addition}]{\text{(No Resistance to}} R - \overset{H}{\underset{H}{\overset{|}{C}}} - \overset{H}{\underset{\underset{CF_3}{\bigcirc}}{C}}{\ominus} \qquad 3.83$$

$$R^{\oplus} + \ominus \overset{H}{\underset{\underset{CF_3}{\bigcirc}}{C}} - \overset{H}{\underset{H}{\overset{|}{C}}}{\oplus} \longrightarrow RF + \oplus \overset{F}{\underset{F}{\overset{|}{C}}} - \bigcirc - \overset{H}{\underset{H}{\overset{|}{C}}} = \overset{H}{\underset{H}{\overset{|}{C}}} \qquad 3.84$$

How can the growing polymer chain of Equation 3.83 be killed when $F^{\ominus}$ is released with the paired initiator, when charges cannot move to kill the chain? A three-membered ring will be too strained to exist. Unlike the previous corresponding monomers which are nucleophiles, these look more like electrophiles chargedly. Hence the reactions above. Free-radically, the situation is different.

$$E^{\cdot e} + e \cdot \overset{CF_3}{\underset{\underset{\bigcirc}{|}}{C}} - \overset{H}{\underset{H}{\overset{|}{C}}} \cdot n \xrightarrow[\text{Resistance}]{\text{(No}} e \cdot \overset{CF_3}{\underset{\underset{\bigcirc}{|}}{C}} - \overset{H}{\underset{H}{\overset{|}{C}}} - E \qquad 3.85$$

145

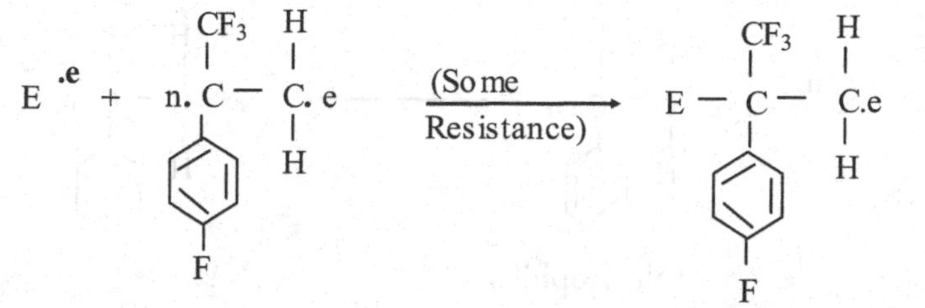

$$E \cdot^{e} + n.\overset{\displaystyle \overset{CF_3}{|}}{\underset{\displaystyle \underset{}{}}{C}} - \overset{\displaystyle \overset{H}{|}}{\underset{\displaystyle \underset{H}{|}}{C}} \cdot e \xrightarrow[\text{Resistance}]{\text{(Some)}} E - \overset{\displaystyle \overset{CF_3}{|}}{\underset{\displaystyle \underset{}{}}{C}} - \overset{\displaystyle \overset{H}{|}}{\underset{\displaystyle \underset{H}{|}}{C}} \cdot e$$

3.86

In the absence of transfer species, the two reactions above are favored with little or no resistance to addition, noting that the monomers above free-radically are all nucleophiles. They will all favor nucleo-free-radical attacks also.

For the 1,3 –dienes, the followings are obtained for pentadiene types.

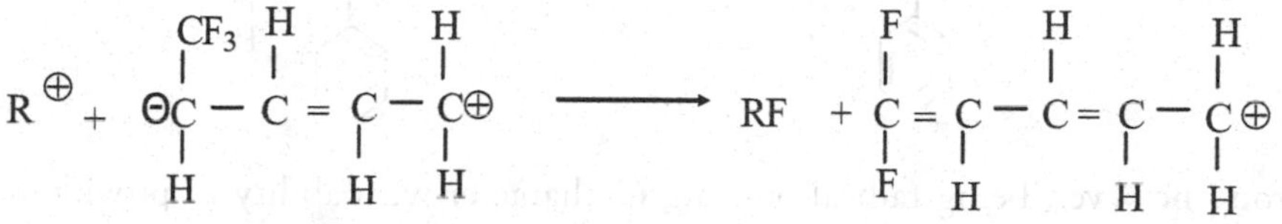

$$R^{\oplus} + \ominus C - C = C - C^{\oplus} \longrightarrow RF + C = C - C = C - C^{\oplus}$$

(Nucleophile) NON-EXISTENT

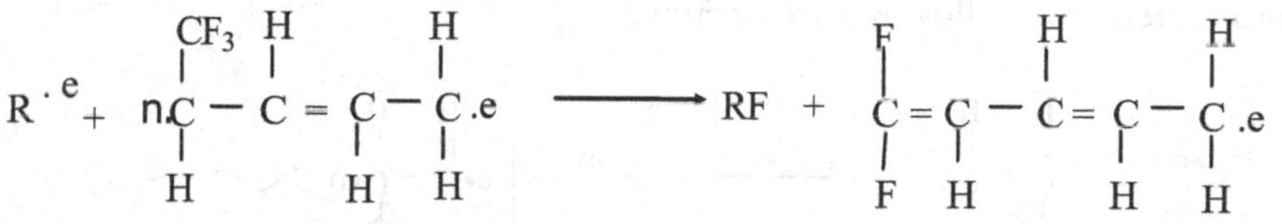

3.87

$$R \cdot^{e} + n.C - C = C - C \cdot e \longrightarrow RF + C = C - C = C - C \cdot e$$

(Nucleophile)

3.88

$$N \cdot^{n} + e.C - C = C - C \cdot n \longrightarrow N - C - C = C - C \cdot n$$

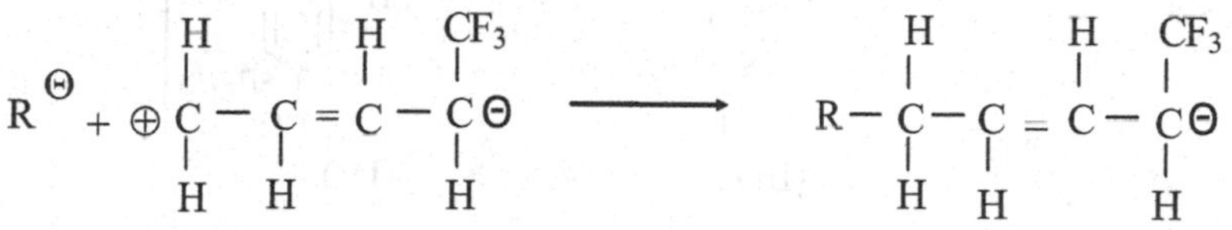

(Nucleophile)

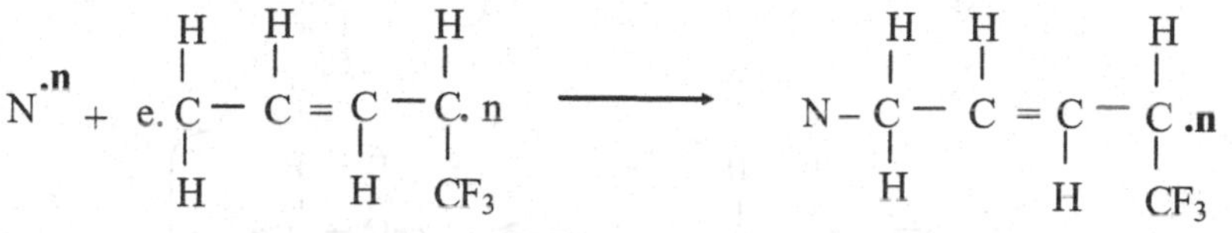

3.89

$$R^{\ominus} + \oplus C - C = C - C \ominus \longrightarrow R - C - C = C - C \ominus$$

(Nucleophile) NON-EXISTENT

3.90

The routes favored by these monomers can clearly be identified and found to be unique. Chargedly in the first and last equations above, it is 1,2-monoform that is actually involved,

For the styrene types of monomers, consider replacing CF_3 with F. Then, the followings are obtained.

$$R^{\ominus} + \overset{\displaystyle \overset{H}{|}}{\underset{\displaystyle \underset{H}{|}}{C}} = \overset{\displaystyle \overset{F}{|}}{\underset{\displaystyle \underset{}{}}{C}} \longrightarrow \text{Activation not possible}$$

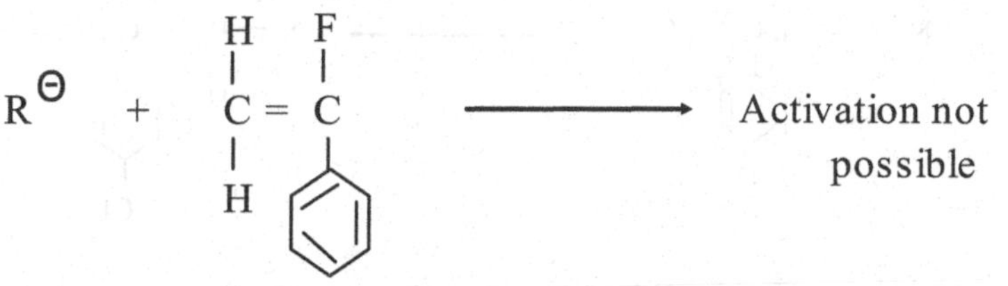

3.91

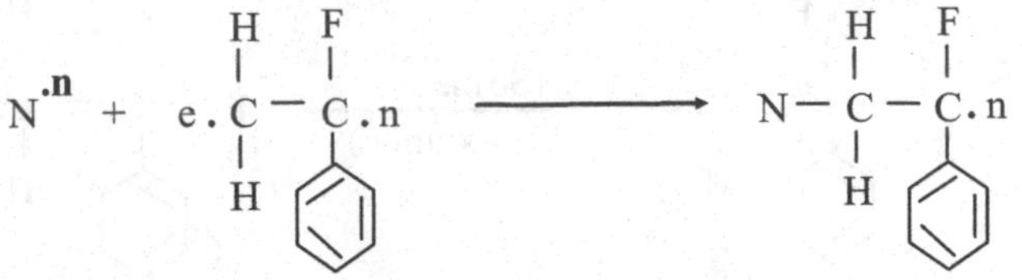

$$N^{.n} + e.\overset{\overset{H}{|}}{\underset{\underset{H}{|}}{C}} - \overset{\overset{F}{|}}{\underset{\underset{\bigcirc}{|}}{C}}.n \longrightarrow N - \overset{\overset{H}{|}}{\underset{\underset{H}{|}}{C}} - \overset{\overset{F}{|}}{\underset{\underset{\bigcirc}{|}}{C}}.n \qquad 3.92$$

(Nucleophile)

While the nucleo-free-radical is located on the center carrying the phenyl group here, in styrene the reverse is the case as shown below. The phenyl group is of far greater radical-pushing capa-city than H.

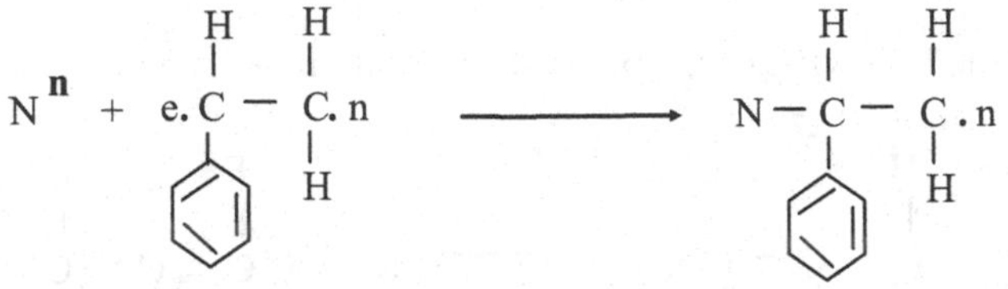

$$N^{n} + e.\overset{\overset{H}{|}}{\underset{\underset{\bigcirc}{|}}{C}} - \overset{\overset{H}{|}}{\underset{\underset{H}{|}}{C}}.n \longrightarrow N - \overset{\overset{H}{|}}{\underset{\underset{\bigcirc}{|}}{C}} - \overset{\overset{H}{|}}{\underset{\underset{H}{|}}{C}}.n \qquad 3.93$$

The phenyl group however, being radical-pushing in character with ability to provide resonance, will prefer being adjacently located to an electro-free-radical center.

For p-chloro-styrene, the followings are obtained.

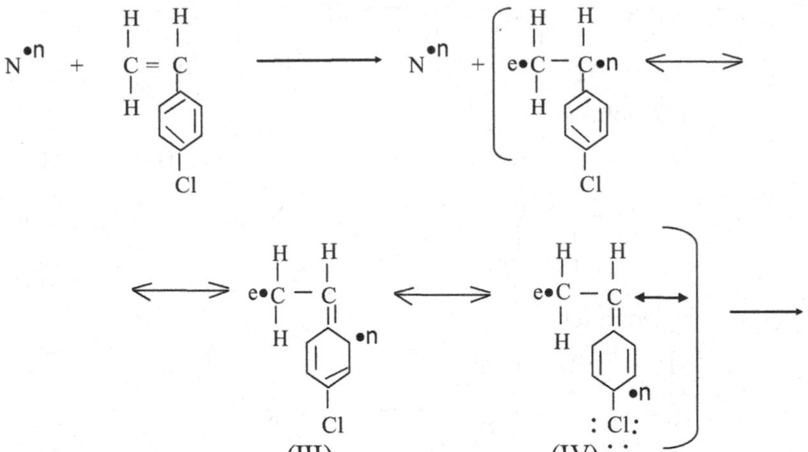

(III)   (IV)   3.94

Chargedly where resonance stabilization cannot take place, it can be activated since electrostatic forces of repulsion cannot function here, the Cl atom being distantly located. The monomer can only be resonance stabilized free-radically. Free-radically, the followings are obtained.

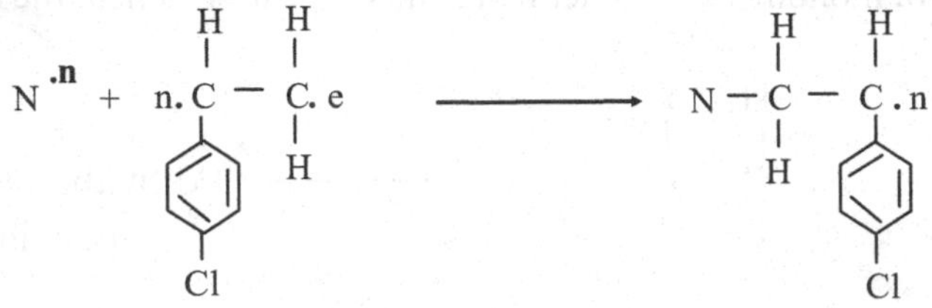

$$N^{.n} + n.\overset{\overset{H}{|}}{\underset{\underset{\bigcirc}{|}}{C}} - \overset{\overset{H}{|}}{\underset{\underset{H}{|}}{C}}.e \longrightarrow N - \overset{\overset{H}{|}}{\underset{\underset{\bigcirc}{|}}{C}} - \overset{\overset{H}{|}}{\underset{\underset{H}{|}}{C}}.n \qquad 3.95a$$

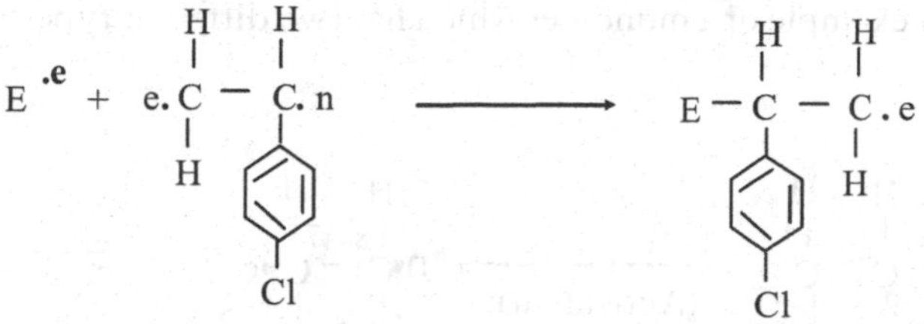

$$E \cdot^e + e.C - C.n \longrightarrow E - C - C.e$$

(with H and H on carbons, and Cl-substituted phenyl groups)

3.95b

Both routes are favored in the absence of transfer species.

It is important to note that chargedly some monomers which look as if they cannot be activated, can indeed be activated. This is more obvious by comparing the followings.

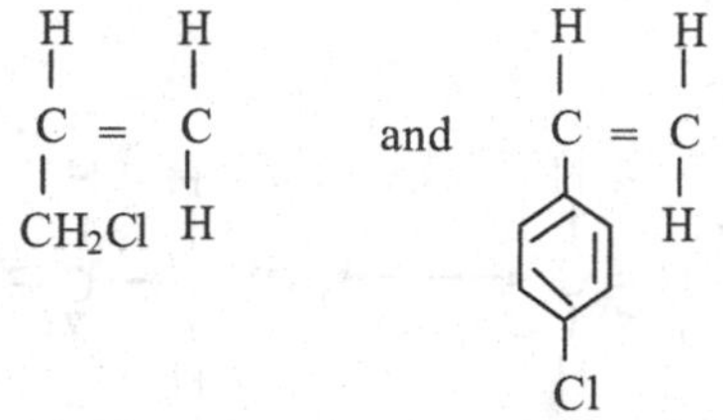

$$\underset{CH_2Cl}{\overset{H}{\underset{|}{C}}} = \underset{H}{\overset{H}{\underset{|}{C}}} \quad \text{and} \quad \underset{(Cl\text{-phenyl})}{\overset{H}{\underset{|}{C}}} = \underset{H}{\overset{H}{\underset{|}{C}}}$$

(I) (Cannot be chargedly activated)   (II) (Can be chargedly activated)   3.96

(I) cannot be chargedly activated, because it is allylic. Secondly, (II) above is almost identical to vinyl chloride when the phenyl group is removed. However it can be activated chargedly. In general, when resonance stabilization groups are present, the followings shown below in Table 3.1 are favored.

Table 3.1. Influence of Resonance stabilization groups on capacity of substituted groups externally located.

#	Group	Equivalence	Type of resonance stabilization group
1	  –⬡–Cl  –⬡–CH₃ –⬡–H	>Cl >CH$_3$ >H	Phenyl- (Only Radically)
2	–CH =CH– H –CH =CH– H$_3$ –CH =CH– Cl	>H >CH$_3$ >Cl	Mono-alkenyl- (Only Radically)
3	–CH =CH– CH=CH–CH–H –CH =CH– CH=CH–CH–Cl	>H >Cl	Di – alkenyl- (Only Radically)

Despite the equivalence or approximate equivalence, the presence of resonance stabilization groups still marks a big difference in very many respects, particularly with respect to stability of the activated monomers and polymerization times during the full course of polymerization. It is important to note the location of the substituted groups on the resonance stabilization groups. The equivalence above is only valid for substituted groups internally located, groups which are fully resonance stabilized whether ortho-, para- or meta-placed.

Divinyl benzene is an example of a monomer which has two different types of resonance stabilization groups as shown below.

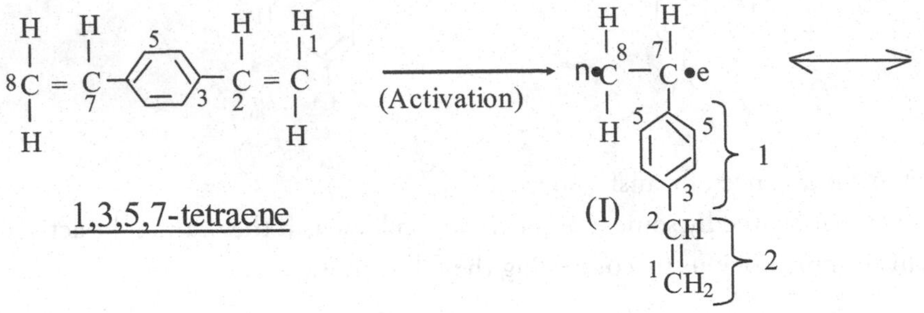

$$1,3,5,7\text{-tetraene} \qquad\qquad (I)$$

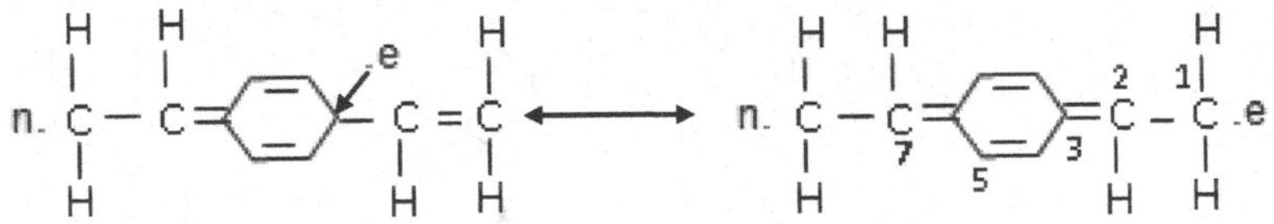

(Note that the nomenclature shown above is tentative and indeed not valid)

So also is the resonance stabilization shown above not valid, since the ring must be fully involved in providing resonance stabilization. (I) the only one that exists chargedly, carries a cross-linking site externally located. Radically and chargedly, they have no transfer species for which therefore, they will favor all routes, clear indication of the weak Nucleophilic character of the monomer. Radically it look like a tetraene, which it is, but not as shown above. The move-ment of the electro-free-radical does not get to (III) above, but comes back to where it started. Only one vinyl group is functional. It is the side that is not functional that is used for cross-linking as will be shown. Chargedly, there is only one mono-form, 1,2-. When polymerized using negative charges, there is no transfer species of the second kind, and positively there is also no transfer species. Radically, there are transfer species of second kind only nucleo-free-radically as will be shown shortly.

Universally, it is called 1,4-divinyl benzene (para-placement of the alkenyl groups). It is made to look as if the benzene ring centrally placed in ortho- or para-forms with two alkylene groups is saturated, By the name-1,4-, there is a difference between whether the two groups are ortho-, meta-, or para- placed, but this does not provide the real nomenclature of the compounds.

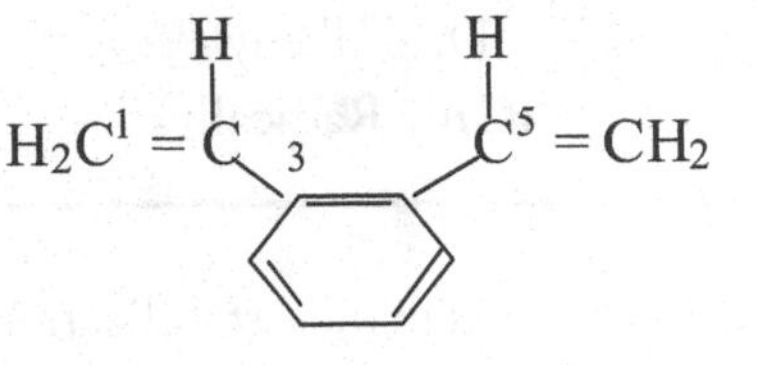

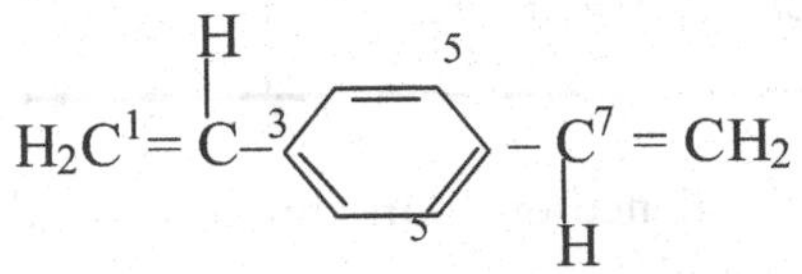

Ortho-placed 1,3,5-triene

(I)

;

Para-placed 1,3,5,7-tetraene   3.98a

(II)

149

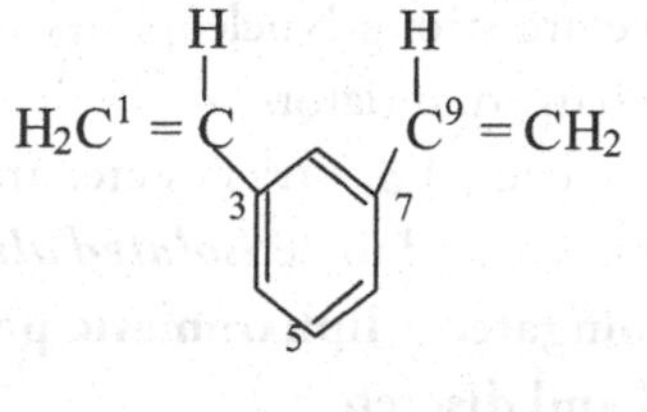

$$\text{H}_2\text{C}^1 = \overset{\overset{\displaystyle H}{|}}{C} \qquad \overset{\overset{\displaystyle H}{|}}{C^9} = \text{CH}_2$$

(III)   3.98b

(I) easily converts to naphthalene and is said not to be a component of the usual mixtures of di-vinyl benzene, though known to exist. This is unlike the case of Ortho- benzoquinone and Para-benzoquinone both MALES which exist while meta-benzoquinone does not exist as will be shown downstream. The ortho- and para- benzoquinones are resonance stabilized moving from FULL –FREE-RADICALS to FULL NON-FREE- RADICALS (unlike what we have seen so far), as will also be shown downstream. Very so-called little things which we put aside with a WAVE OF THE HAND, are where the so-called mysteries of LIFE exist and these so-called mysteries is the NEW FRONTIER- the NEW SCIENCE for humanity. The only MYSTERY in humanity in our Planet, Earth, in our Cosmos a single macro-"ATOM" where the NUCLEUS is the SUN, is THE ALMIGHTY INFINITE GOD.

With CH_3 group replacing H in number 7- or 2-carbon atom above, the CH_3 group internally located will become fully resonance stabilized, i.e. well shielded against attack.

As shown below for 1,4 –dienes in the presence of strong initiators, where resonance stabilization cannot be provided, independent activation of all the activation centers may be possible chargedly or radically. When isolatedly placed, all the centers can be activated at the same time if and only if they are of the same nucleophilicity. If they are of different nucleophii-cities only one can be activated. It is the less nucleophilic center that is first activated, whether two of them of the same capacity are present or not, like in 1,3-butadiene.

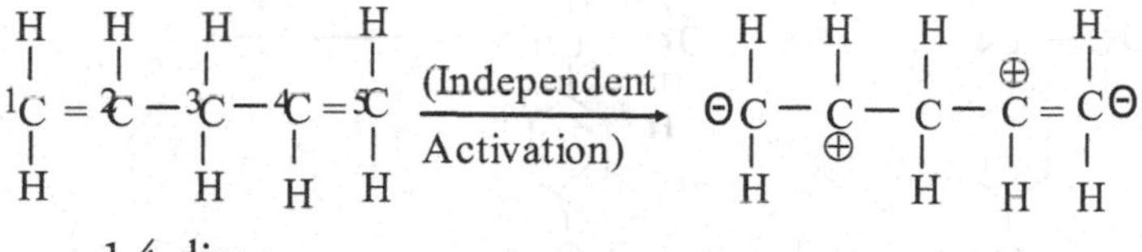

3.99

1,4-diene

POSSIBLE ACTIVATION

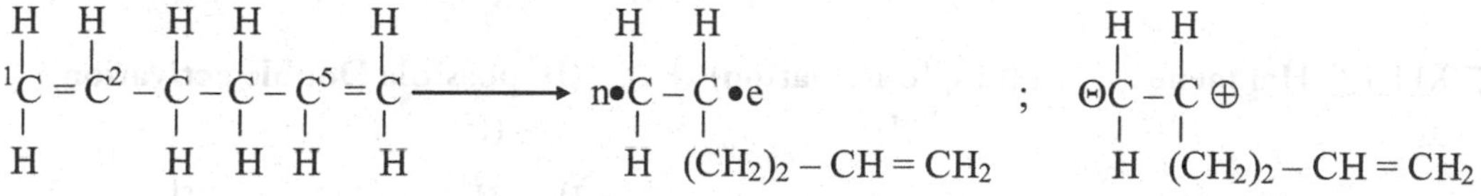

1,5-diene   (Radically)   (Chargely)

REAL ACTIVATION   3.100

The two cases above have two activation centers of the same capacity isolatedly placed. They can all be activated at the same time if a strong initiator is used. All higher dienes (1,4- upwards) can never be resonance stabilized since $-CH_2 - CH = CH_2$, $-CH_2 - CH_2 - CH = CH_2$, etc. are no resonance stabilization groups, but some can molecularly rearrange to give a resonance stabilized molecule such as the first case above, if H is the transfer species. If $-CH = CH_2$ is the transfer species, the same monomer is obtained, and this is probably the case, since both H and $-CH = CH_2$ are equal in the absence of resonance stabilization. How can an electro-radical move in them when there is no adjacently located π- bond nearby? The monomers above are strong Nucleophiles which can only be polymerized using electro-free-radicals or free-positively charged initiators.

It can thus be observed why 1,3 –dienes, 1,3,5-trienes etc. are *conjugated aliphatic poly olefins,* while 1,4 –dienes 1,5 –dienes and higher are said to be *isolated aliphatic poly olefins.* Styrene, Divinyl benzene are very few examples of **conjugated Alipharomatic poly olefins which are tetraenes and higher. As tetraene, it is fully closed and discrete.**

The corresponding equation to Equation 3.97 free-radically for divinyl benzene *in the presence of a strong initiator,* is as follows.

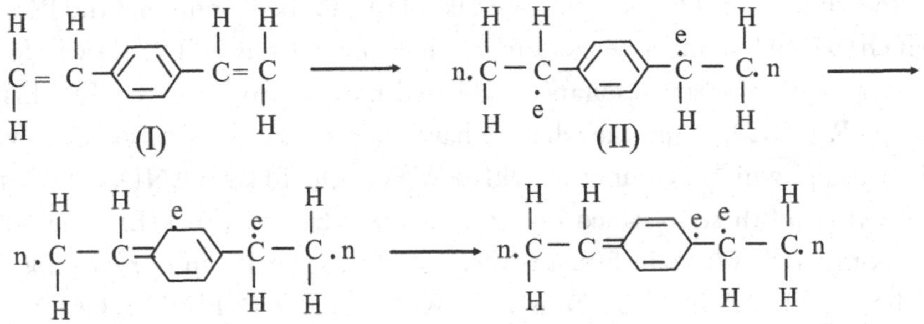

3.101

(Impossible existence)

To have two electro-free-radicals which unlike charges which repel and attract when adjacently located in a symmetric or non-symmetric manner, in a resonance stabilized monomer is still impossible. As can be observed so far, while charges can repel and attract, radicals cannot repel, and even attract. Now consider having two phenyl groups as the resonance stabilization groups adjacently located in a manner similar to alkyl-enes in 1,3,5-trienes.

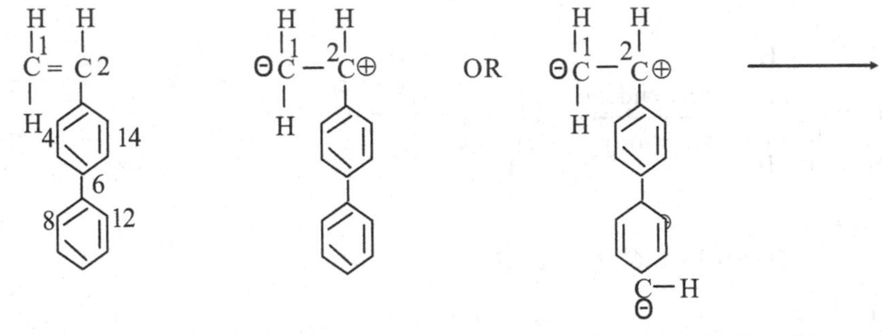

1,3,5,7,9,11,13- Heptaene   **(Single activation)**   **(Impossible Double activation)**
                                        (I)a   (II)a

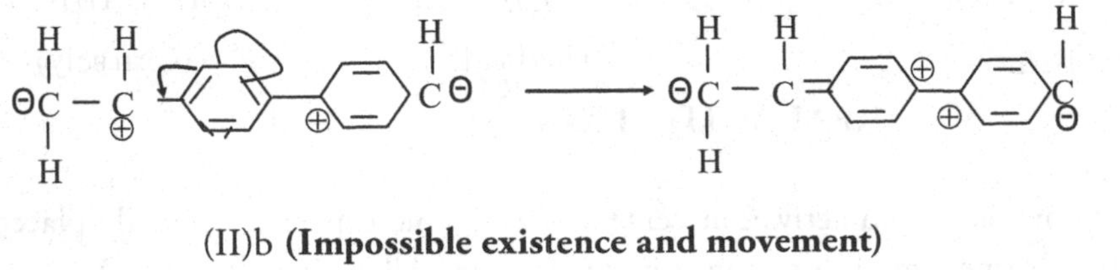

3.102

(II)b **(Impossible existence and movement)**

(Note that the nomenclature shown above is the real one if the two rings are involved.)

Like 1,3,5- trienes, the activation centers of the monomer above cannot be activated indepen-dently. They can only be activated one at a time to favor only mono-1,2-additions, (I)a, just like in styrene (1,2-additions). Unlike in divinyl benzene where the second alkenyl group is not involved if the ring has to be fully used, here the two rings are fully involved in the movement of the elctro-free-radical back to where it started. Thus, one can observe that, ***with two similar or dissimilar resonance stabilization groups adjacently located, provision of resonance stabilization is possible only radically, the manner of provision depending on the types and placements involved.*** All these new concepts are important to note for future developments and applications.

3.3 Transfer from growing polymer chain

It has been observed that when a monomer which previously did not favor a route exists, it can be made to favor the route when resonance stabilization is provided for the group which origin-ally prevented polymerization. Many examples have already been shown. For the simple exam-ples shown below, we already know what to expect for the growing polymer chains, in view of the law of conservation of transfer of transfer species and the role of resonance stabilization.

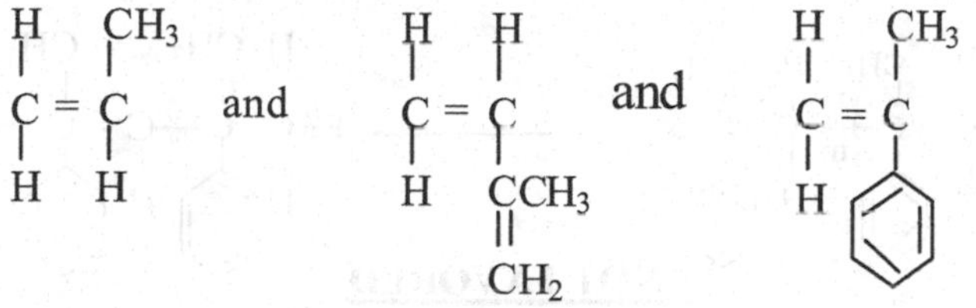

3.103

Now, consider growing polymer chains of α-methyl styrene free-radically.

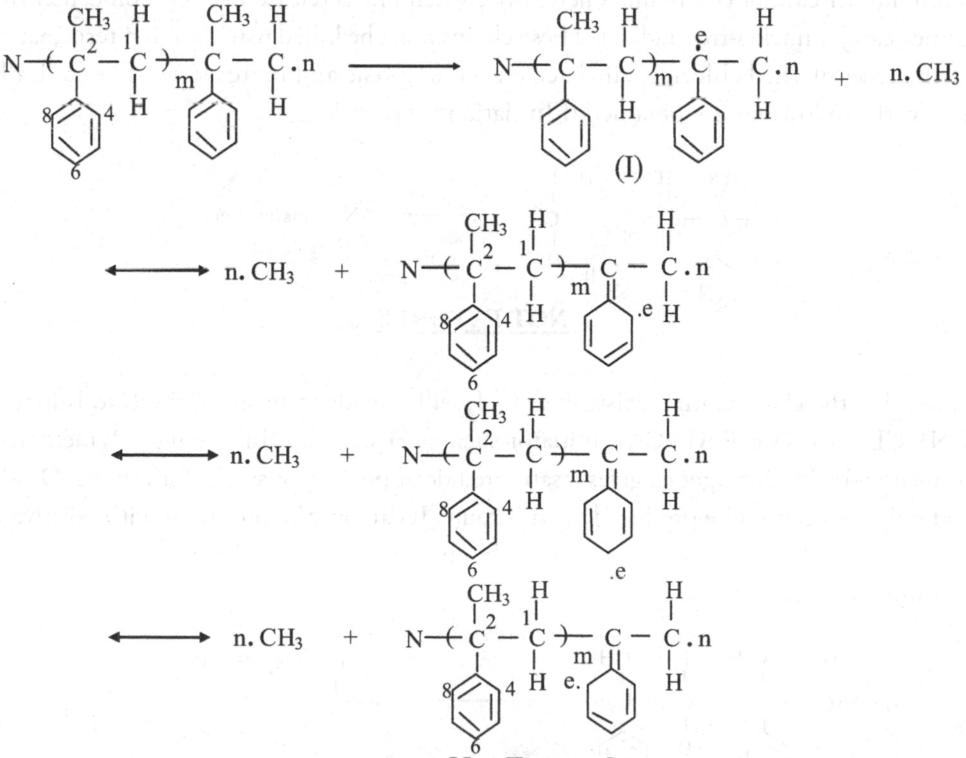

(I)

Not Favored

$$N\left(\begin{array}{c}CH_3\\|\\C\\|\\\bigcirc\end{array}-\begin{array}{c}H\\|\\C\\|\\H\end{array}\right)_m\begin{array}{c}CH_3\\|\\C\\\end{array}=\begin{array}{c}H\\|\\C\\|\\H\end{array}\quad+\quad \bigcirc \cdot n \qquad 3.104$$

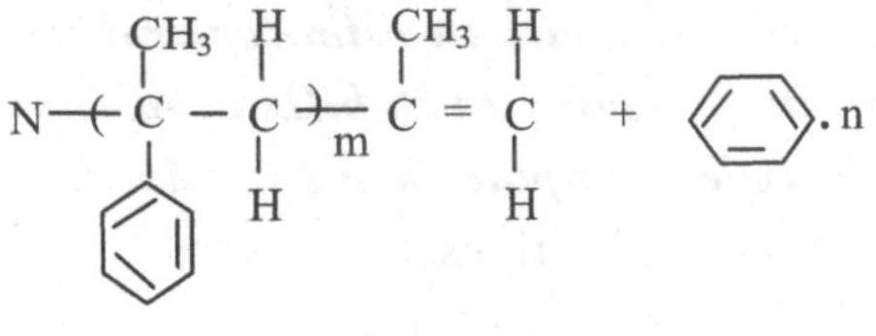

OR

(II) <u>**Not favored**</u>

Since the CH_3 group is more radical-pushing in capacity electro-free-radically than phenyl group, but nucleo-free-radically the reverse, (I) is looks favored to give a dead terminal double polymer of the same character as the original monomer produced in the route not natural to it. But since the groups carried on the phenyl group are resonance stabilized, they cannot be removed. (II) is not favored since the dead terminal double bond polymer produced is not resonance stabilized. *Rejections of transfer species of the second kind of the first type was considered here, since the route is not natural to the monomer.*

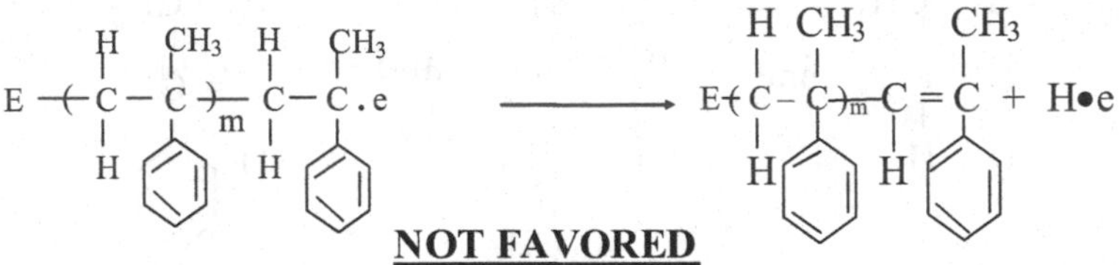

$$E\left(\begin{array}{c}H\\|\\C\\|\\H\end{array}-\begin{array}{c}CH_3\\|\\C\\|\\\bigcirc\end{array}\right)_m\begin{array}{c}H\\|\\C\\|\\H\end{array}-\begin{array}{c}CH_3\\|\\C\\|\\\bigcirc\end{array}.e \quad\longrightarrow\quad E\left(\begin{array}{c}H\\|\\C\\|\\H\end{array}-\begin{array}{c}CH_3\\|\\C\\\end{array}\right)_m\begin{array}{c}H\\|\\C\\|\\\bigcirc\end{array}=\begin{array}{c}CH_3\\|\\C\\|\\\bigcirc\end{array}\;+\; H\bullet e \qquad 3.105$$

<u>**NOT FAVORED**</u>

A dead terminal double bond resonance stabilized polymer with a stronger Nucleophilic character than the original monomer cannot be obtained here, since when H.e is released the carbon center from where released cannot carry a nucleo-free-radical. These chains must be killed using foreign terminating agents. This is like the cases of vinyl chloride, vinyl acetate, chloroprene and more electro-free-radically.

Chargedly, the followings are obtained if Initiation is favored.

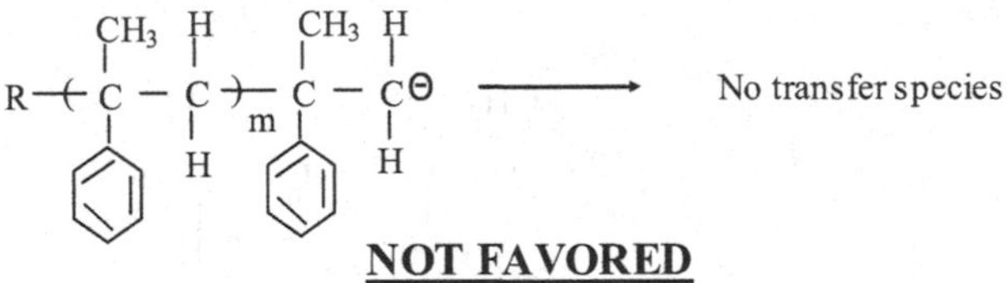

$$R\left(\begin{array}{c}CH_3\\|\\C\\|\\\bigcirc\end{array}-\begin{array}{c}H\\|\\C\\|\\H\end{array}\right)_m\begin{array}{c}CH_3\\|\\C\\|\\\bigcirc\end{array}-\begin{array}{c}H\\|\\C\\|\\H\end{array}^{\ominus} \quad\longrightarrow\quad \text{No transfer species}$$

<u>**NOT FAVORED**</u>

3.106

However, note that the chain cannot exist, since CH_3 will provide transfer species (No Initiation). With the use of NUCLEO-FREE-RADICAL initiator such as $H_9C_4{}^{.n}\ldots\ldots{}^{.e}Li^1$, living polymers which when terminated using terminating agents, gives a saturated dead polymer just like for ethene. Dead terminal double bond polymers cannot be produced via this route. It can only be produced with positively charged initiators as shown below. Hence, with negatively charged paired initiators, we have transfer from monomer to initiator, i.e. no initiation.

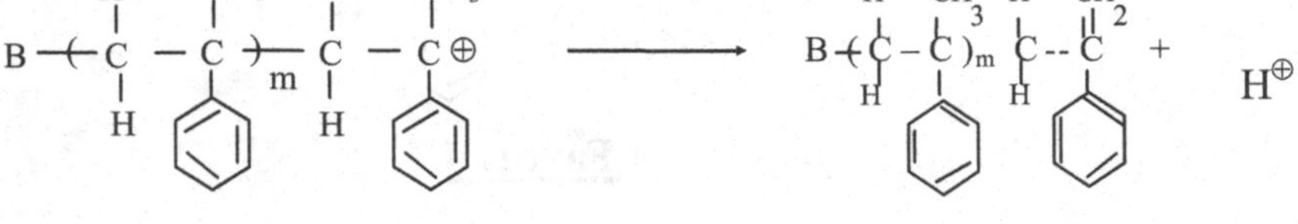

$$B\left(\begin{array}{c}H\\|\\C\\|\\H\end{array}-\begin{array}{c}CH_3\\|\\C\\|\\\bigcirc\end{array}\right)_m\begin{array}{c}H\\|\\C\\|\\H\end{array}-\begin{array}{c}CH_3\\|\\C\\|\\\bigcirc\end{array}^{\oplus} \quad\longrightarrow\quad B\left(\begin{array}{c}H\\|\\C\\|\\H\end{array}-\begin{array}{c}CH_3\\|\\C\\\end{array}\right)_m\begin{array}{c}H\\|\\C\\|\\H\end{array}--\begin{array}{c}CH_2\\||\\C\\|\\\bigcirc\end{array}\;+\; H^{\oplus}$$

3.107

This is the charged route natural to the monomer. The transfer species above is of the first kind of first type released when the chain has reached its optimum chain length based on the Glass transition temperature of the polymer in the absence of foreign agents.

For styrene nucleo-free-radically, the following is obtained.

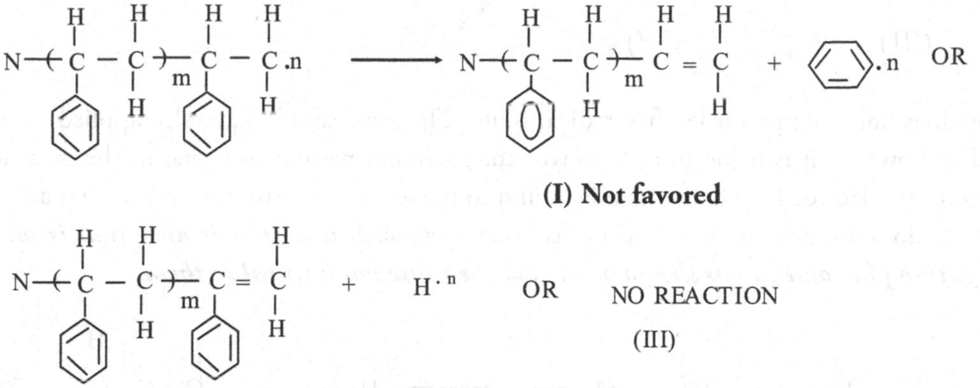

(I) Not favored

(II) Not Favored

3.108

When resonance stabilization is provided, the group providing it, has greater radical-pushing capacity than H. Apart from the fact that the capacity of $n\bullet CH_3$ is less than that of $n\bullet C_6H_5$ which in turn is less than that of $n\bullet H$ as will be shown downstream, no group on the carbon center on the phenyl center can be abstracted, It was only with α-methyl styrene, dead terminal double bond polymers was obtained above cationically. Free-radically, they cannot be obtained.

It is important at this point in time to address the issue of depropagation reactions with respect to α-methyl styrene. α-methyl styrene is thought to have a significant depropagation rate at temperatures as low as 25 °C [15]. Based on the mechanisms of depropagation associated with very unique families of monomers and a different method of Addition polymerization as will be shown in this Volume downstream, ***this is not possible nucleo-free-radically and in particular electro-free-radically***. Nucleo-free-radically, there is very strong resistance to addition as was shown in Equation 3.71. In the process, there is delayed addition and this could partly be thought to be a form of depropagation. At very high temperatures of polymerization however, the chains can be readily broken, particularly if it is a living one. When it is scissioned, this can be done homolytically and not heterolytically, only when there is no transfer species for the growing chain and temperature of polymerization is high. But at low temperatures of polymerization, this does not occur. If not for the resonance stabilization provided by the resonance stabilization groups, α-methyl styrene should be more female in character (that is nucleophilic) than styrene as shown below when compared with ethylene and propylene.

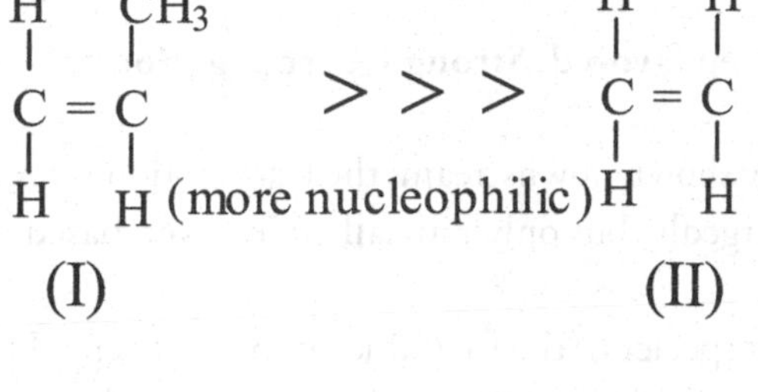

(I)   (II)

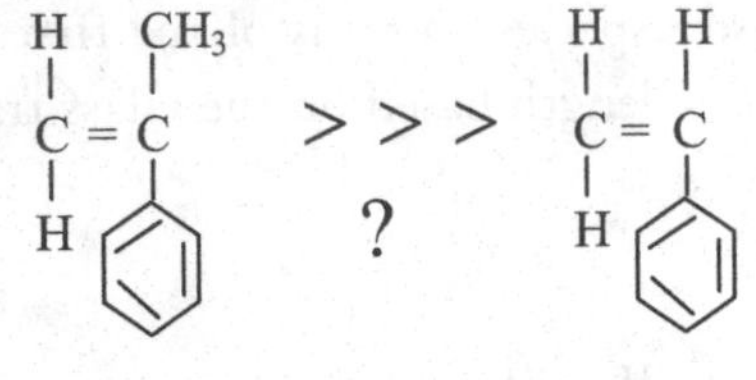

$$
\begin{array}{ccc}
\underset{\underset{\displaystyle \bigcirc}{|}{\overset{H}{\underset{|}{C}}} = \overset{CH_3}{\underset{|}{C}} & >>> & \underset{\underset{\displaystyle \bigcirc}{|}{\overset{H}{\underset{|}{C}}} = \overset{H}{\underset{|}{C}} \\
(III) & ? & (IV)
\end{array}
$$

3.109

Propylene does not undergo nucleo-free-radical route. The same should have also applied to α--methyl styrene. But however, it is being forced to favor the route not natural to it due to the presence of the phenyl group in place of H. Therefore, it is bound to possess some resistance in favoring an unnatural route, which does not exist with styrene (no transfer species). *It is that resistance that is said to be a depropagation phenomenon as shown below via the route not natural to them.*

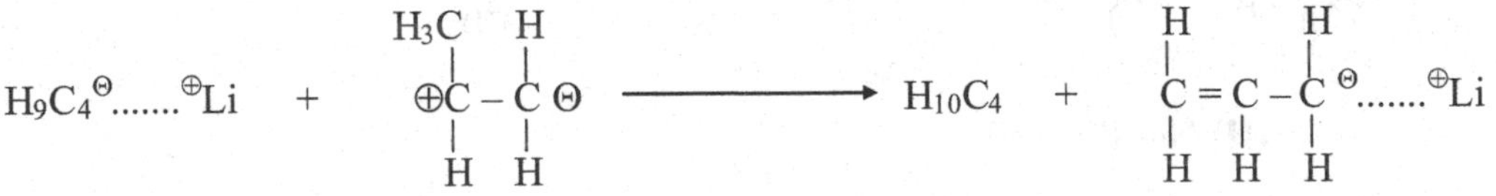

$$
H_9C_4{}^{\ominus}\text{.......}{}^{\oplus}Li \;+\; \overset{H_3C}{\underset{H}{\overset{|}{\underset{|}{\oplus C}}}} - \overset{H}{\underset{H}{\overset{|}{\underset{|}{C}}}}{}^{\ominus} \longrightarrow H_{10}C_4 \;+\; \overset{H}{\underset{H}{\overset{|}{\underset{|}{C}}}} = \overset{H}{\underset{H}{\overset{|}{\underset{|}{C}}}} - \overset{}{\underset{H}{\overset{}{\underset{|}{C}}}}{}^{\ominus}\text{.......}{}^{\oplus}Li
$$

No resistance; No Initiation Step; No Depropagation   3.110

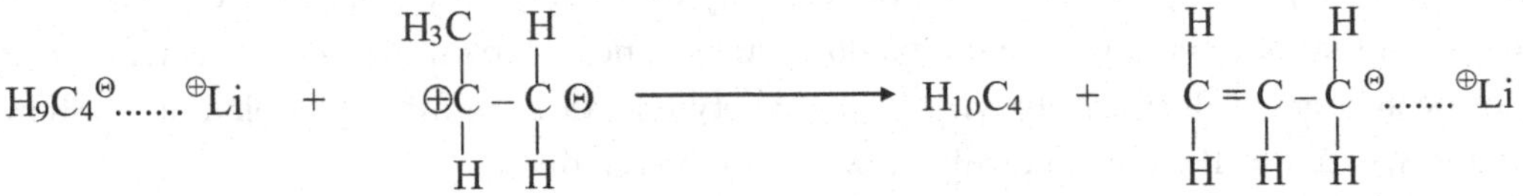

$$
H_9C_4{}^{\ominus}\text{.......}{}^{\oplus}Li \;+\; \overset{H_3C}{\underset{H}{\overset{|}{\underset{|}{\oplus C}}}} - \overset{H}{\underset{H}{\overset{|}{\underset{|}{C}}}}{}^{\ominus} \longrightarrow H_{10}C_4 \;+\; \overset{H}{\underset{H}{\overset{|}{\underset{|}{C}}}} = \overset{H}{\underset{H}{\overset{|}{\underset{|}{C}}}} - \overset{}{\underset{H}{\overset{}{\underset{|}{C}}}}{}^{\ominus}\text{.......}{}^{\oplus}Li
$$

Weak resistance; Initiation Step favored; Little or No Depropagation   3.111

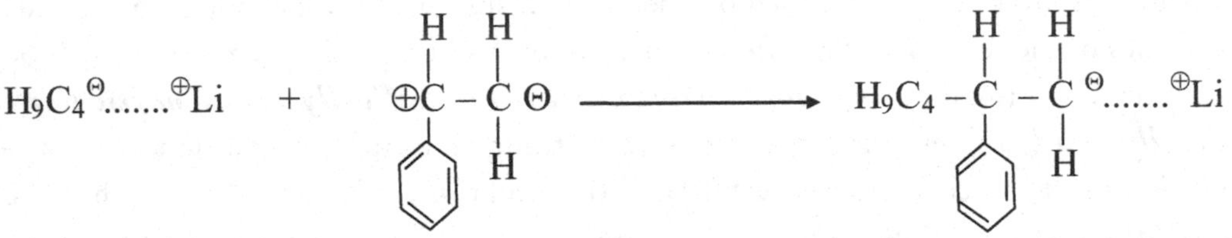

$$
H_9C_4{}^{\ominus}\text{.......}{}^{\oplus}Li \;+\; \underset{\underset{\displaystyle \bigcirc}{|}}{\oplus C} - \overset{H}{\underset{H}{\overset{|}{\underset{|}{C}}}}{}^{\ominus} \longrightarrow H_9C_4 - \underset{\underset{\displaystyle \bigcirc}{|}}{\overset{H}{\underset{|}{C}}} - \overset{H}{\underset{H}{\overset{|}{\underset{|}{C}}}}{}^{\ominus}\text{.......}{}^{\oplus}Li
$$

Weak resistance; Initiation Step favored; Little or No Depropagation   3.112

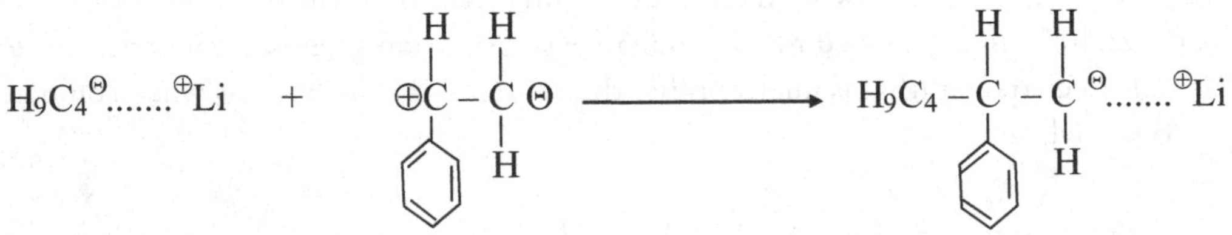

$$
H_9C_4{}^{\ominus}\text{.......}{}^{\oplus}Li \;+\; \underset{\underset{\displaystyle \bigcirc}{|}}{\overset{H}{\underset{|}{\oplus C}}} - \overset{H}{\underset{H}{\overset{|}{\underset{|}{C}}}}{}^{\ominus} \longrightarrow H_9C_4 - \underset{\underset{\displaystyle \bigcirc}{|}}{\overset{H}{\underset{|}{C}}} - \overset{H}{\underset{H}{\overset{|}{\underset{|}{C}}}}{}^{\ominus}\text{.......}{}^{\oplus}Li
$$

Strong resistance; Initiation Step favored; Strong Depropagation   3.113

As is already begun to be shown and will be fully shown downstream, the last reaction is favor-ed, because resonance stabilization cannot be provided chargedly, but only radically. However, based on the type of the chain, Li will be the carrier of the chain.

Propylene (Real name Propene) has transfer species of the first kind of the first type. Hence it cannot be polymerized with free-negatively charged paired initiator (Not anionic ion-paired initiator as used in present-day Science). Above for the first reaction, the initiator as already shown is Free-negatively charged paired initiator present due to the ionic or charged environ-ment provided for it. For this route, there is no resistance whatsoever being strongly Nucleophilic (Female). Ethylene (Real name ethene) has no transfer species of the first kind. The only one it has is of the second kind which cannot be released chargedly (i.e. as $H^{\ominus}$ which does not exist, H being an ionic metal). It has weak resistance which is large compared to monomers such as vinyl chloride or vinyl acetate which cannot be activated chargedly (Positively or negatively). Hence, though it is a weak nucleophile, it is more nucleophilic than the last two mentioned monomers. Therefore, at very high temperatures, it is bound to depropagate. Styrene is like ethene, except that it has a resonance stabilization group which has nothing to resonance stabilize apart from the H. Therefore, like ethene in the route not natural to it, it will depropagate at high temperatures as will shortly be shown, its rate equal or higher than that of ethene. α-methyl styrene unlike styrene has transfer species which cannot be released like the case of propene, because of the strong resistance or shield provided by a resonance stabilization group adjacently located. Because of the resistance in the route not natural to the monomer, hence at high temperatures less than those for styrene and ethene, it will depropagate readily far more than styrene will. The temperature of polymerization has to be greatly reduced to prevent the depropagation. That temperature which is unique to the different monomers is still a type of CEILING TEMPERA-TURE unlike the types experienced when a monomer is being polymerized in the route NATURAL to it as will be shown downstream. The monomers considered herein do not depropagate in the route NATURAL to them which is the use of Electro-free-radical initiators. Propylene (Propene) will not depropagate "cationically" (i.e., use of positively charged initiators) or electro-free-radically. So also are the others above. The temperature at which full polymeriza-tion of the monomer takes place here without Depropagation is called **"Radical Unnatural Route Polymerization (RURP) Ceiling Temperature** different from **"Radical Natural Route Polymerization" (RNRP) Ceiling Temperature.** The former deals with olefinic monomers, while the latter deals with non-olefinic monomers.

Shown below is how the depropagation takes place using α-methyl styrene at high temperatures.

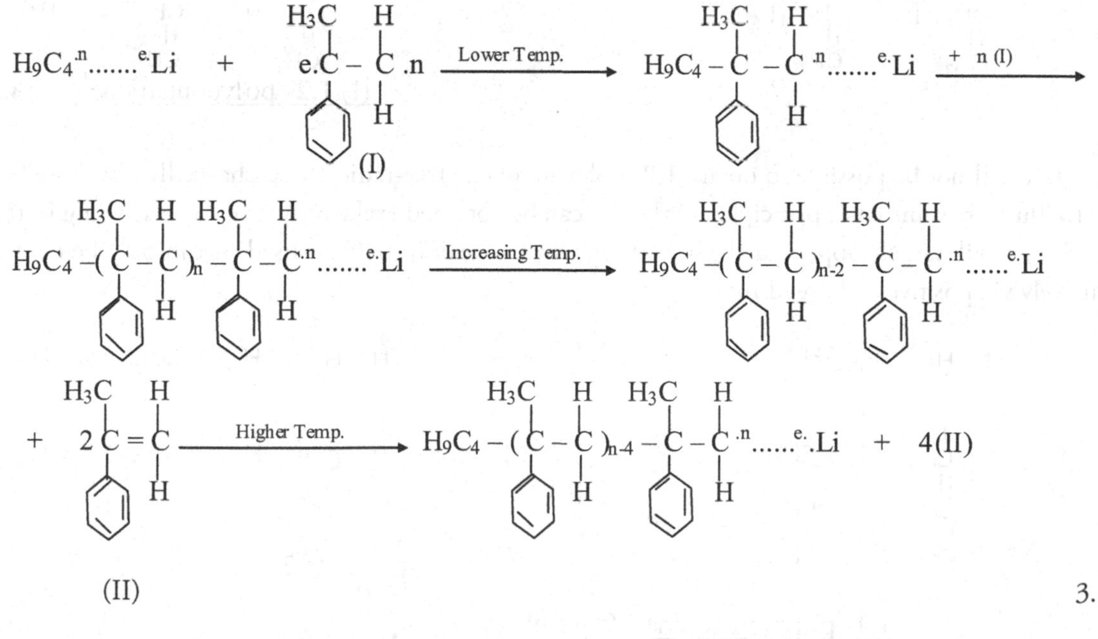

(II)   3.114

When a monomer unit is released in its activated state, it deactivates and in the process heat is released. It is the released heat during deactivation that is the source of increasing temperature as the polymerization progresses because the growing polymer has no way to terminate itself from within in the absence of transfer species of any kind. This is the major source of depro-pagation for both olefinic and non-olefinic types of monomers with this characteristic "Ceiling temperature". The release of the monomer unit is because there is no transfer species of any kind to release from the unit. With olefins, the growing polymer chains grow forwardly whereas with non-olefinic monomers where depropagation takes place, the growing polymer chain grows backwardly as we shall see downstream. With α-methyl styrene, the rate at which monomer is added to its growing polymer chain is very slow due to resistance. The same will apply to most monomers where transfer species exist, but cannot be abstracted or rejected.

For the pentadienes, the followings will prevail if 1,4 –mono-form's presence is favored.

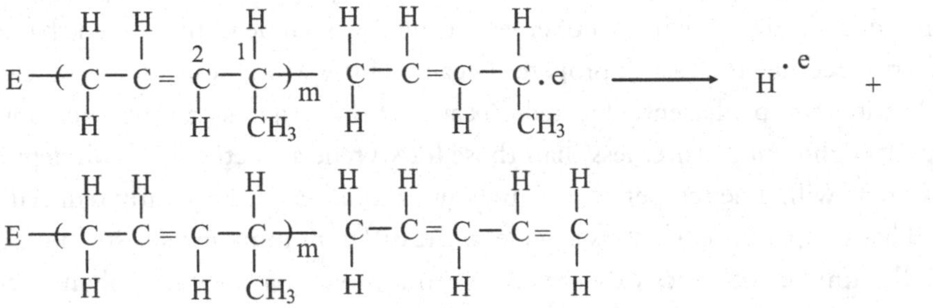

3.115

1,4-polypentadiene (Presence not favored chargedly)

It should be noted that the character of the dead terminal double bond polymer produced is the same with the original monomer. In accordance with the law of conservation of transfer of transfer species, when the driving forces exist, the growing polymer chain can be killed internally.

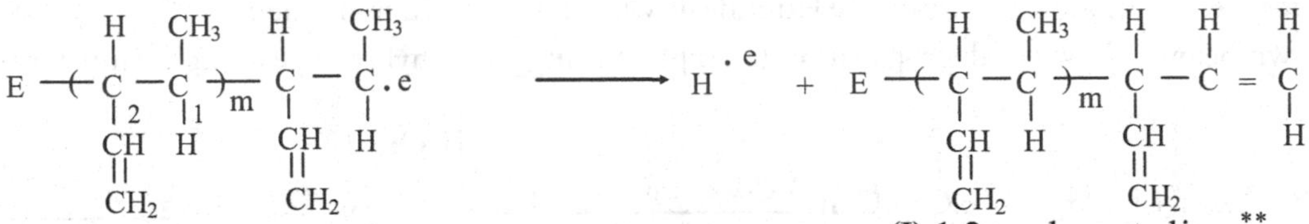

(I) 1,2- polypentadiene**   3.116

Though it will not be possible to obtain 1,2- polypentadiene free-radically or chargedly, these are being used to illustrate some basic principles. Only 1,4- can be obtained exclusively when a very strong initiator is used, 3,4- will largely appear at the beginning when a weak initiator is used. It can only be obtained exclusively via positively charged route.

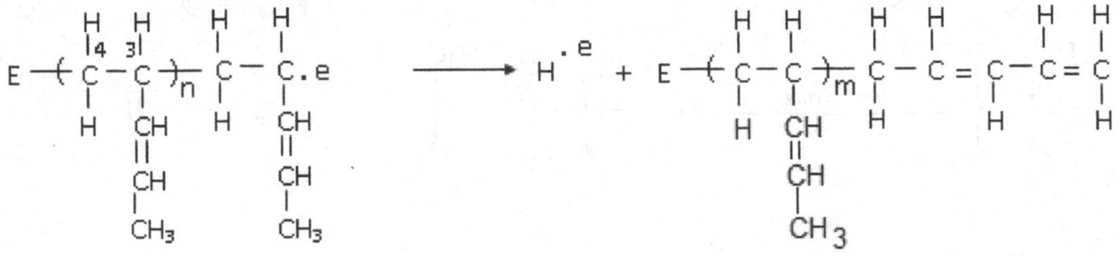

3.117

3,4- polypentadiene   (m = n)

While the 1,4- and 3,4 – dead terminal double bond polymers bear same resemblance and character with the original monomer, that for 1,2- does not seem to have a resemblance. While it seems to bear the same character as the original monomer, it is no longer resonance stabilized. It looks like an isolated diene, just like what it does when it undergoes molecular rearrangement. This is not surprising, since as we have already established, only the 1,4- and 3,4-mono-forms can undergo molecular rearrangement where possible without losing their characters and the only mono-forms. This finally sends a great message which we already differently identified. The message is that 1,2 mono-form does not exist when a monomer is resonance stabilized for this diene. In order words, radically for diene, there are only two mono-forms; and for trienes, there are three mono-forms and for tetraenes there are four mono-forms and so on. Chargedly, there is only one mono-form. In the absence or presence of molecular rearrangement, the mono-forms remain the same for Nucleophiles and Electrophiles

For growing polymer chains of the monomer of Equation 3. 92, the followings are obtained.

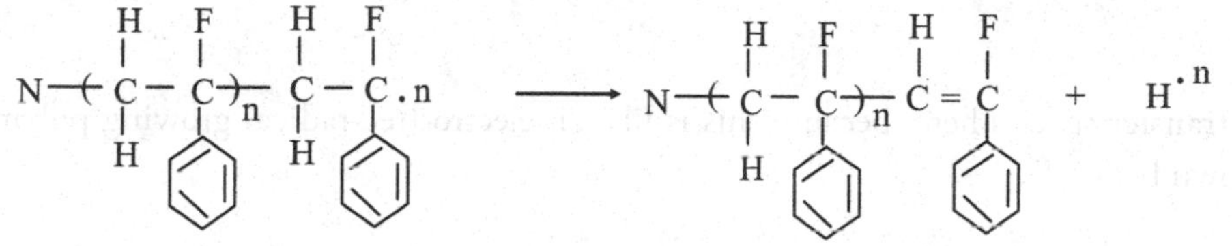

3.118

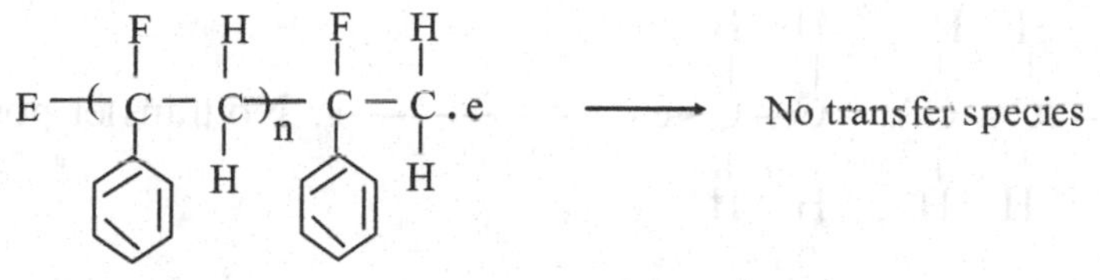

3.119

There is no transfer species, since when the phenyl group is released, the character of the monomer is lost. Free-radically the monomer is a nucleophile. Chargedly, it cannot be activated. Looking at growing polymer chains of other monomers which have been considered thus far, the followings will prevail.

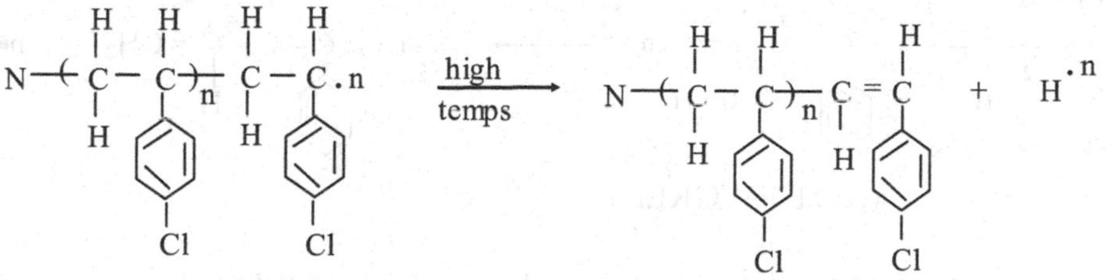

(Transfer species of second kind)

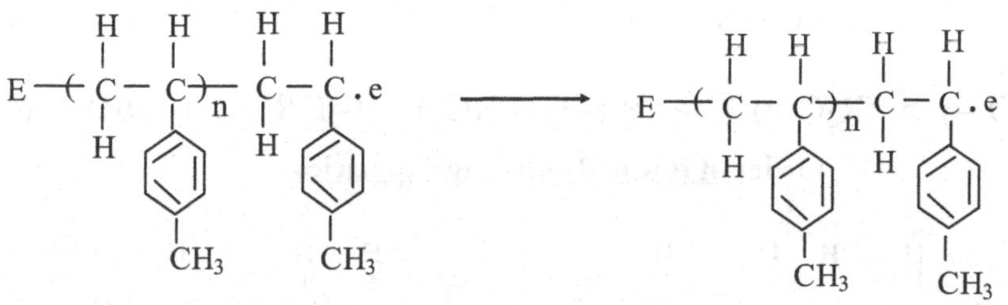

3.120

(Transfer species is resonance stabilized)

3.121

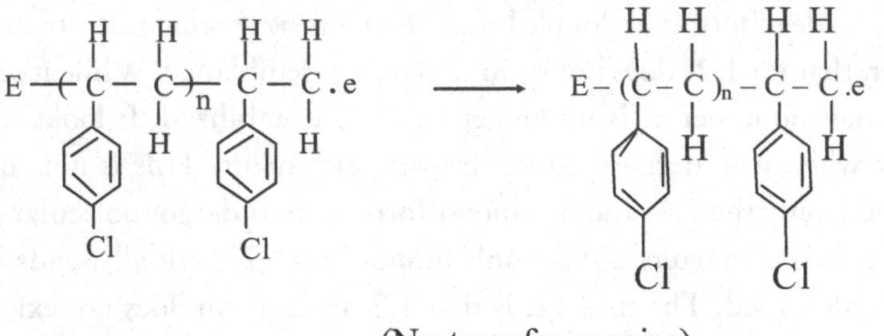

3.122

(No transfer species)

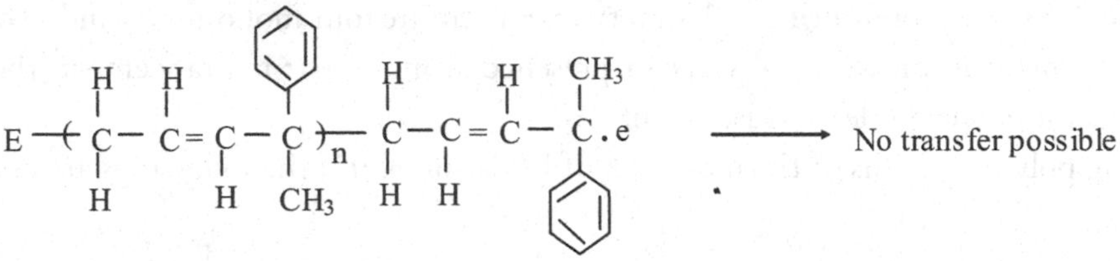

3.123

There is no transfer species here, because this is like an electro-free-radical growing polymer chain of ethene as shown below.

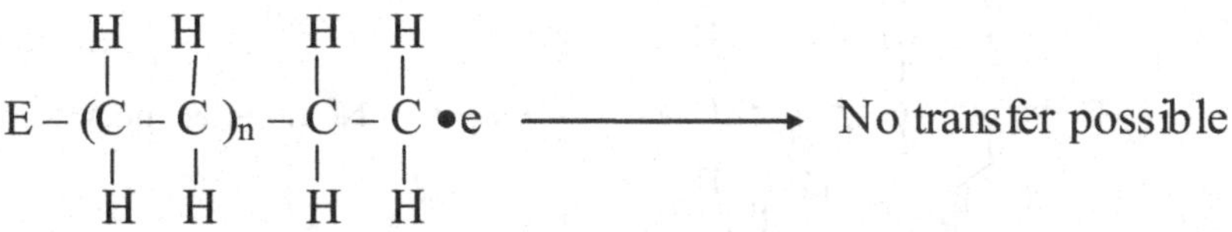

3.124

How can the C center releasing the transfer species carry a nucleo-free-radical based on the capacities of other species carried by the center? The issue is not only just the capacities of the group carried, but the type of initiator used. It is impossible. These which are growing electro-free-radically cannot depropagate, unlike the case shown below similar to the case of α-methyl styrene (see Equation 3.114).

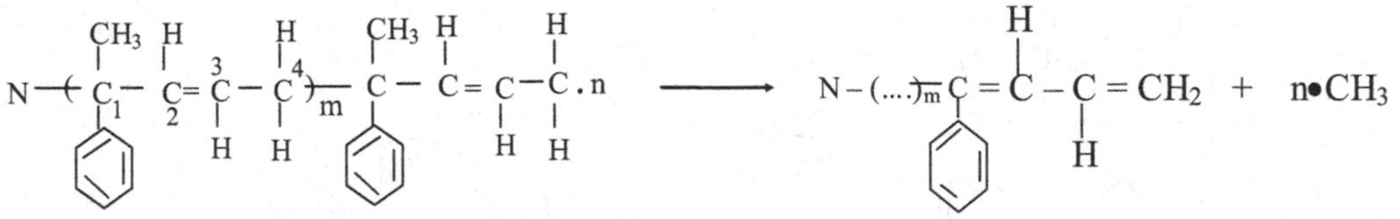

3.125

(NOT FAVORED)

A careful look at Equation 3.123 and the past, using "the eye of the needle" sends a message. Groups such as R_2C =, O =, RN= are strong radical- or "charged"- PUSHING groups of far greater capacity than the -R or -H or -OR or $-NR_2$ or $-O–NO_3$ groups.

$$(RN =) > (O =) > (R_2C =), \quad >>> (- O - NO_3) > (- NR_2) > (- OR) > (- R) > (- H)$$

<u>Order of Radical-pushing capacities</u> 3.126

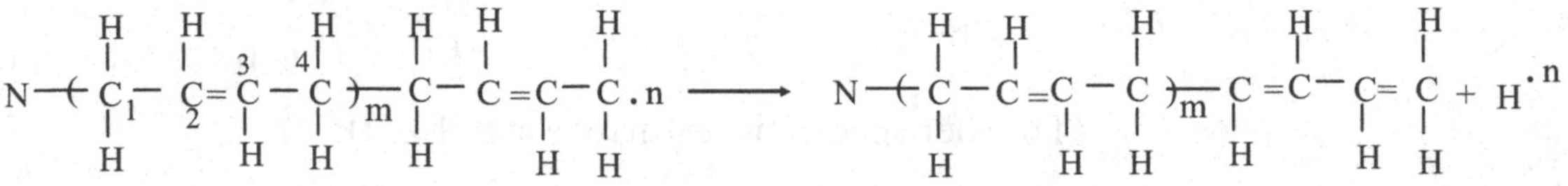

(Transfer species of the second kind) 3.127

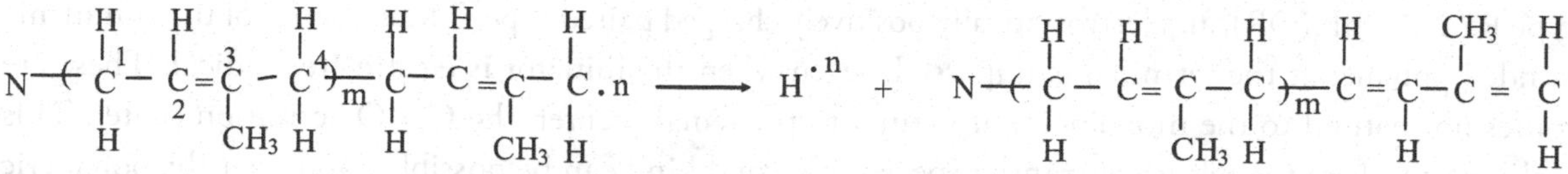

$$\text{(Transfer species of the second kind)} \qquad 3.128$$

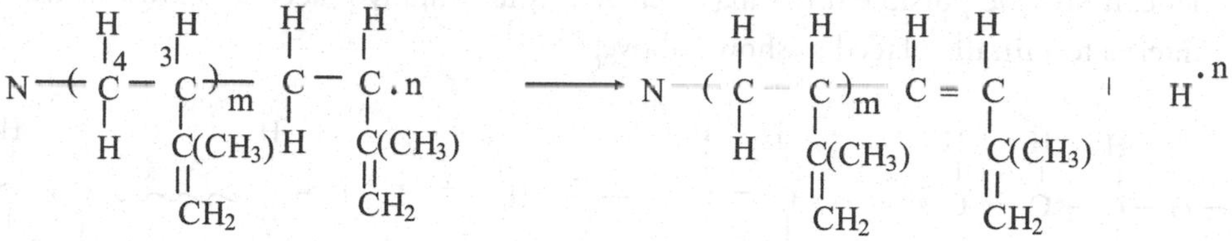

$$\text{(Transfer species of the second kind)} \qquad 3.129$$

Without any doubt, the release of H as $H^{\cdot n}$ will in most cases require higher operating conditions. The examples above show how complex NATURE operates. Indeed, we have only just begun. What one has been doing so far is highly advanced, for which without reaching the tertiary level of education in CHEMISTRY, they cannot be understood. *Yet, without this background, no discipline can be chosen. Hence, so much work will have to be done to initiate humans from primary to tertiary level of Education for which all textbooks must be thrown into the archives (and not dustbin) and rewritten, noting that without this New Chemistry at the primary and secondary levels of acquisition of knowledge in life, one cannot move forward to graduate in any discipline. otherwise our world of illusions will continue forever.*

For the growing polymer chains of the monomers shown below, the followings are obtained, in the need to search if (II) and (III) are electrophiles, that is, males.

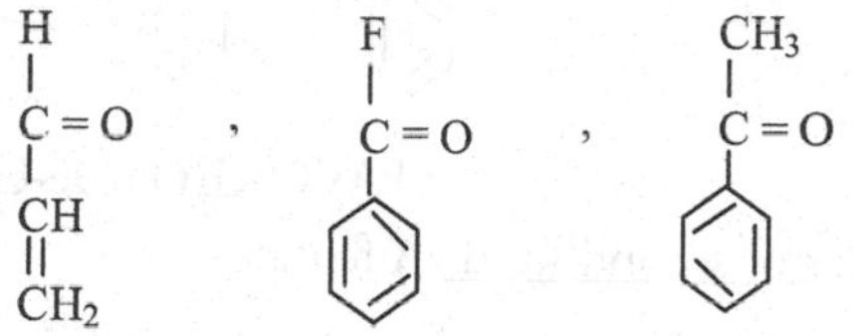

$$\qquad 3.130$$

(I) (II) (III)

Male ? ?

$$\qquad 3.131$$

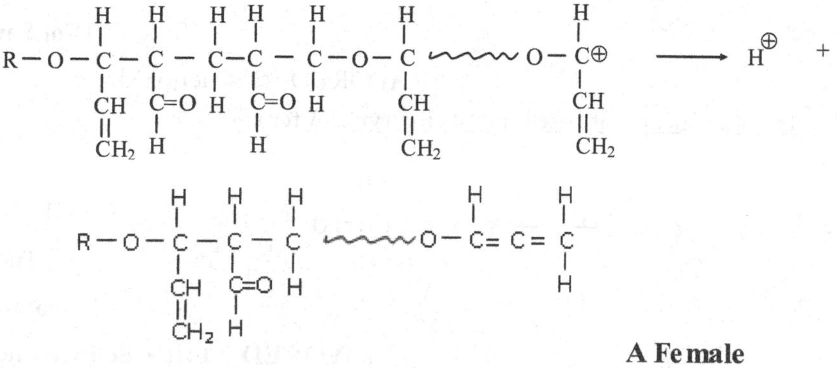

A Female

NOT FAVORED (No Transfer species)

The carbon center cannot carry a negative charge, due to electrostatic forces of repulsion between the negative charge and π-bond carried by the same C center. The initiator which has been used above is of the type R$^{\oplus}$.......$^{\ominus}$BF$_3$(OR), an electrostatically positively charged paired type. The existence of the cumulenic bonded polymer at the terminal is impossible, even when the initiator is electro-free-radical. These are routes not natural to the monomer, but natural to the female center, the C = O activation center. This is the route where the release of transfer species of second kind can be possible. Note that the polymeric chain above from acrolein is a randomly placed copolymer, from a single monomer. The release of such transfer species can only be possible if the male center was terminally placed as shown below, not when the female center is terminally placed as shown above.

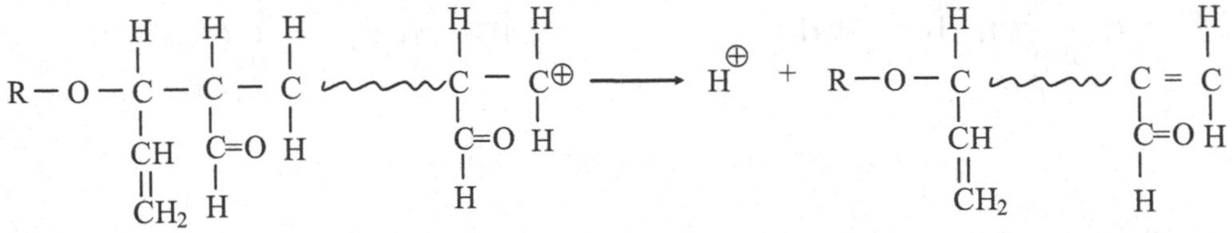

Transfer species of second kind (A male) 3.132

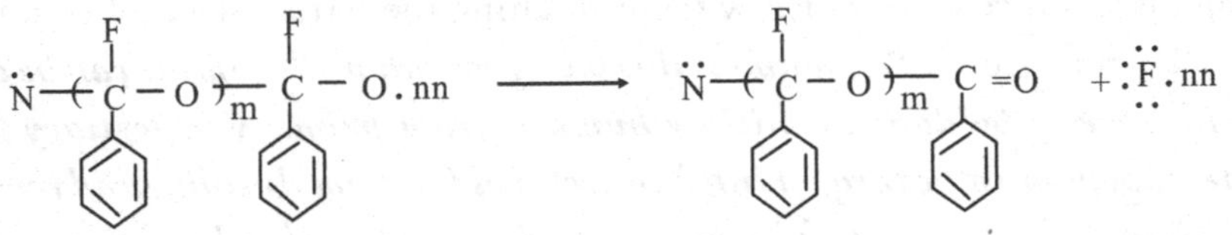

OR No Transfer

FAVORED (De-energized)

Transfer species of second kind (A female) 3.133

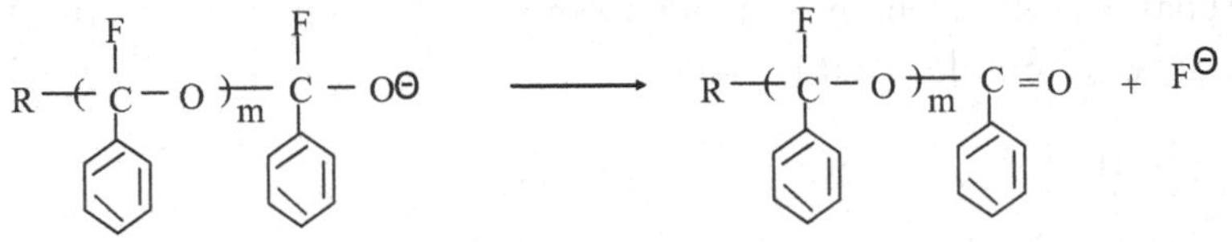

OR No Transfer

[Not Favored]

FAVORED (De-energized) 3.134

Transfer species of second kind (A female)

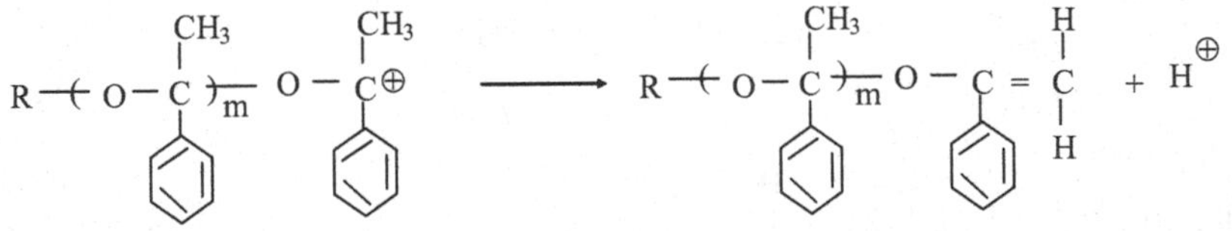

OR No Transfer

[Not Favored]

FAVORED (De-energized)

Transfer species of first kind of first type (A female) 3.135

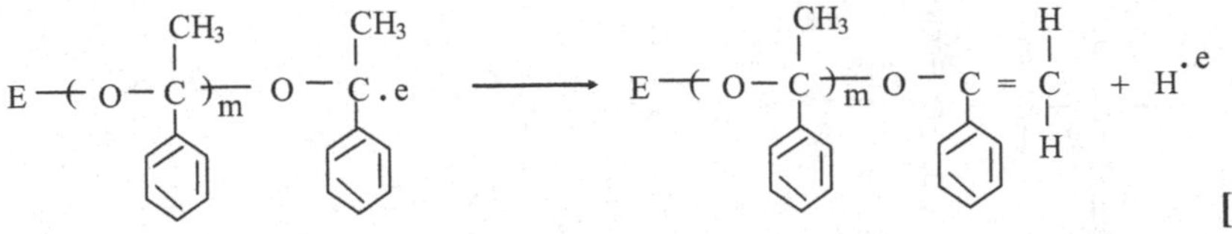

OR No Transfer

[Favored Case]

FAVORED (Only when De-energized)

Transfer species of first kind of first type (A female) 3.136

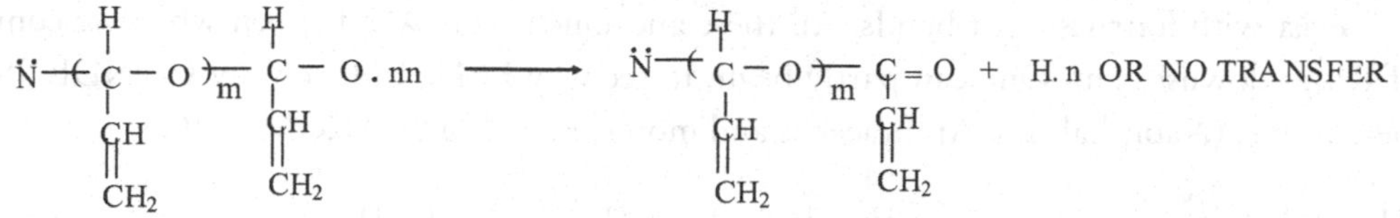

FAVORED
(Transfer species of the second kind) (A female)   3.137

Though only electro-free-radicals and positively charged centers are the only ones that can attack the C=O centers here, being the natural routes, anions and nucleo-non-free-radicals are being used here to show some basic fundamental principles. In the first reaction above (Equation 3.131), the dead cumulenic terminal bond polymer obtained bears no semblance to the character of the original monomer from which it was obtained. It cannot even undergo molecular rearrangement to give acetylenes. In fact as shown below, it cannot be activated chargedly. Only living polymers can be obtained if the terminal is carrying a nucleophilic end (i.e. a female).

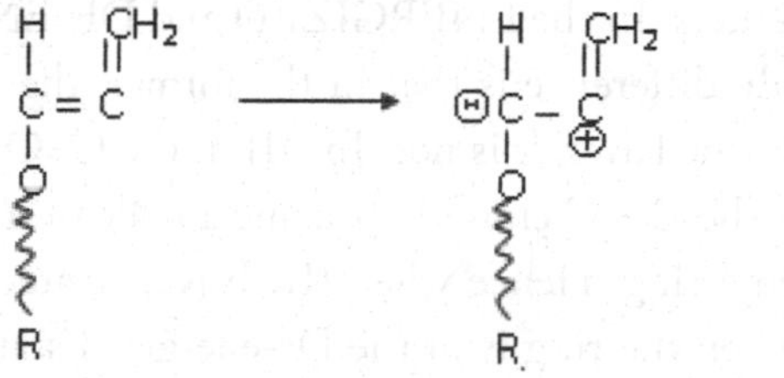

(Impossible existence)   3.138

In Equation 3.132, the dead polymer obtained bears the same character as the original monomer which is an Electrophile (Male). This is a case where the monomer unit at the terminal is electrophilic. Based on the equations above, one can observe that since the monomers are not resonance stabilized, transfer species can readily be rejected for all of them where possible. The case where the phenyl group is present, the rejection takes place when the ring is in the DE- ENERGIZED state of existence. Obviously, the C=O center being adjacently located to a C=C in the ring clearly indicates that the monomers shown in Equation 3.130 are all Electrophiles. Thus, while formaldehyde is a Nucleophile (Female), Acrolein and benzaldehyde are Electrophiles (Male)

One had expected that the alkenyl and phenyl groups are alike in all respects, since they are resonance stabilization groups. But based on the very little seen so far, BENZENE (C_6H_6) is very different from ETHENE ($H_2C=CH_2$), just as the PHENYL (-C_6H_5) group is very different from the ALKYL-ENE (ALKENYL) (-$CH=CH_2$) group. [Because there is a family in the Hydro-carbon Family Tree called ALKYLENES in which the first member is Methylene, hence it preferable to use ALKENYL in place of ALKYL-ENE.] While Benzene is very NULEOPHI-LIC, the Ethene is weakly NUCLEOPHILIC, because while the former has four Activation centers (three **complete** visible and the fourth invisible π-bonds-the Strain Energy of the ring), the latter has only one Activation center (visible). It is said to be complete because it is a six-membered ring which is so full and exists. Three-, four- or five- membered rings with three π-bonds internally located cannot exist, due to existence of what is called maximum required strain energy (Max.R.S.E). There is also what one called the Minimum required strain energy (Min.R.S.E). These are all different from Strain energy (S.E). Seven- and higher membered ones can exist with three π-bonds all more nucleophilic but less strained than the six-membered ring. If six-membered rings can carry three π-bonds, why can't seven and higher-membered ones carry them when they are conjugatedly placed? The seven-membered ring cannot exist with four visible π-bonds. All these and much more will be seen when we come to the Physics of Rings downstream. One can partly begin to see why BENZENE or any CONDENSED full six-membered rings (Naphthalenes, Anthracenes and more) are said to be AROMATICS.

(Ia)	(Ib)	(IIa)	(IIb)	(IIIa)	(IIIb)
Ethene	Benzene	Butadiene	Styrene	Vinyl ketone	Phenyl ketone
(Nucleophile)	(Nucleophile)	(Nucleophile)	(Nucleophile)	(Electrophile)	(Electrophile)

3.139

All of them are different. Only (IIIa) and (III)b are Electrophiles or Males and the reason is beca-use of the placements of C=C and C = O centers. In the ENERGIZED and DE-ENERGIZED states of existence, (III)b is an Electrophile. The only difference is that in the former, the ring (and not the monomer) is still resonance stabilized, while in the latter, it is not. In (III)a, the C=O center is far more nucleophilic than the C = C center, for which the C = C chose to become a male or Y center. In (III)b, C = O center is less nucleophilic than the Phenyl ring. Hence when (III)b is activated by any initiator, it is only the C = O center that is activated, when the ring is in the De-energized state of existence. Never is there a time when any bond in the ring can be activated. In fact the ring cannot be hydrogenated to loosen its nucleophilicity by any means, until the group carried by it is changed chemically by so many means available to us provided by NATURE, without disobeying its LAWS.

Unlike what has been seen so far, when (III)b is activated in its Energized state of existence, the followings are obtained with the resonance stabilization provided by the phenyl group.

Half-free   Half-free   Half-free   3.140

Instead of the movement from **Full**-free to **Full**-free (Nucleophiles) or **Full**-free to **Full**-non-free (Electrophiles), here we are seeing movement from **Half-free** to **Half-free** (Electrophiles). To have cases where we move from Full-free to Half-free or Half-free to Full-free is impossible if the phenomena of Resonance Stabilization are to exist. The case where we move from **Full**-non-free to **Full**-non-free exists (Nucleophiles), for example O = N – N = O (Nitric oxide). The case where we move from Full-non-free to Full-free is yet to be found.

3.4. Proposition of Rules of Chemistry

From all the considerations so far, one can observe why the need for establishment of rules of Chemistry are very important. Based on the phenomena of resonance stabilization, and application of the rules of molecular rearrangements, the existence of dead terminal double or triple or cumulenic etc. bond polymers, the concepts of transfer species of different kinds, characters of activation centers, the law of conservation of transfer of transfer species etc., one can observe how the observations which have been made over the years, but could not be adequately explained, are being clearly explained in a systematic, orderly and unquestionable manner. Why monomers favor the specific routes which have been identified experimentally and no other routes, are being clearly explained. How activation of activation centers take place under different conditions are being clearly explained. Indeed, what are contained herein above is countless to list.

Rule 102: This rule of Chemistry for **Radicals,** states that, while radicals have males and females (electro- and nucleo-), they are not influenced by electrostatic and electrodynamic forces of repulsions and attractions.
(Laws of Physics)

Rule 103: This rule of Chemistry for **Charges,** states that, while Ionic, and Covalent charges have males and females (positive and negative), they are influenced by electrostatic and electro-dynamic forces of repulsion and attraction.
(Laws of Physics)

Rule 104: This rule of Chemistry for **Compounds with Over-unsaturated Activation centers**, states that,, there are *three types of them*- the first being those that have more than one π-bond (e.g. $HC\equiv CH$, N_2, $R-C\equiv N$), the second being those that are fully unsaturated in closed rings (e.g. Benzene), the third being those that carry π-bonds with paired unbonded radicals adjacently located (e.g. O_2, N_2); for which none can be activated chargedly due to electrostatic forces of repulsion.
(Laws of Physics)

Rule 105: This rule of Chemistry for **Removal of Radicals**, states that, only electro-radicals can be removed from its carrier either to form a positive charge on its carrier if there is a receiver or to form a new π-bond if there a nucleo-radical nearby or to move from hole to hole adjacently placed to themselves and to the electro-radical for several applications too countless to list.
(Laws of Physics)

Rule 106: This rule of Chemistry for **Removal of Charges,** states that, a positive charge cannot be removed from its carrier since there is nothing to move and a negative charge cannot be re-moved from its carrier since the two radicals of opposite spins are coming from two different central atoms.
(Laws of Physics)

Rule 107: This rule of Chemistry for **Sources of electro-radicals,** states that, there are three main sources of electro-radicals- the first coming from the the last shell of electropositive atoms, the second from visible and invisible π- bonds and the third from paired unbonded radicals.
(Laws of Physics)

Rule 108: This rule of Chemistry for **all things in life,** states that, all operations positively or negatively follow the GAUSSIAN CURVE, in which at one end of the spectrum are HYDRO-CARBONS, or BITTERNESS, or SATANISM, or HELL, or BAD...... and at the other end are FLUORINATED HYDROCARBONS, or SWEETNESS, or GODLINESS, or HEAVEN, or

GOOD.......; respectively for which for any to exist, they must CO-EXIST.
(The sixth law of Nature- laws of Philosophy and Mathematics)

Rule 109: This rule of Chemistry for **an Alkene such as Propene that carry substituted or substituent groups with transfer species of the first kind of the first type,** states that, the transfer species which prevents it from undergoing polymerization in the route not natural to it, can be shielded if internally located by provision of what are called **RESONANCE STABI-LIZATION GROUPS and this can only be done radically.**
(Laws of Creations for Resonance Stabilization groups)

Rule 110: This is rule of Chemistry for **Resonance Stabilization phenomena**, states that, there are **two kinds**, in which the ***first kind*** is that which takes place when compounds with more than one Activation center either conjugatedly or cumulatively placed and one center is activated, results in movement of electro-radical to form a new one and a π-bond, continuously with limitatios, for other subsequent applications.
(Laws of Creations for Resonance Stabilization phenomena of the first kind)

Rule 111: This rule of Chemistry for **Resonance Stabilization phenomenon of the first kind**, states that, there are two types, in which the ***first type*** is that in which a visible electro-radical diffuses to grab a nucleo-radical from a π-bond adjacently located to form another π-bond, while the ***second type*** is that in which a hidden electro-radical diffuses from a π-bond to grab a visible nucleo-free-radical adjacently located to form another π-bond; while in the former a visible electro-radical is left in front, in the latter a visible nucleo-radical is left behind after movement.
(Laws of Creations for Resonance Stabilization phenomenon of the first kind)

Rule 112: This rule of Chemistry for **Resonance Stabilization phenomenon of the first kind,** states that, when Dienes are involved, only two mono-forms exist for them, while for Trienes, three mono-forms and for Tetraenes, four mono-forms exist for them, and so on.
(Laws of Creations for Dienes, Trienes, Tetraenes and Polyenes)

Rule 113: This rule of Chemistry for **Resonance Stabilization phenomenon of the first kind,** states that, due to the influence of resonance stabilization, all the mono-forms favor the same routes of polymerization, favor the same type of molecular rearrangement when allowed to take place and behave all alike when used as monomers, but not with respect to types of the products obtained from each of the mono-forms.
(Laws of Creations for Resonance Stabilization phenomenon)

Rule 114: This rule of Chemistry for **Resonance Stabilization phenomena,** states that, though resonance stabilization cannot take place chargedly, they can still be polymerized chargedly with only one mono-form, that resulting from the least nucleophilic center.
(Laws of Creations for Resonance Stabilization phenomena)

Rule 115: This rule of Chemistry for **Unsaturated Hydrocarbons with more than one Activa-tion center,** states that, there are ***three kinds*** of RESONANCE STABILIZATION GROUP -
Alkenyl (-CH=CH$_2$), Alkynyl (-C≡CH), and phenyl –C$_6$H$_5$; of which the most unique is the phenyl group.
(Laws of Creations for Resonance stabilization Groups)

Rule 116: This rule of Chemistry for **Resonance Stabilization phenomenon of the first kind,** states that, for resonance stabilization to be favored, the compound must first be activated and the center activated is the least nucleophilic herein called the **First center.**
(Laws of Creations for Resonance Stabilization First center)

Rule 117: This rule of Chemistry for **Resonance Stabilization phenomenon of the first kind,** states that, when Styrene is involved, the resonance stabilization provider is the phenyl group and the monomer is a closed tetraene that has only one mono-form.
(Laws of Creations for Styrene)

Rule 118: This rule of Chemistry for **Styrene**s, states that, when any type of group (radical-pushing or pulling grous) is placed anywhere on the ring (para-, ortho- or meta-), the resonance stabilization provider, the group is well shielded since it is internally located inside the ring; the presence of the group used mainly to determine the types of radicals or charges carried by the active centers.
(Laws of Creations for Styrenes)

Rule 119: This rule of Chemistry for **Numbering of Unsaturated hydrocarbons,** states that, the numberings on the carbon centers of an unsaturated hydrocarbon such as **Benzene**, are based on the location(s) of substituent groups, for which if not unsaturated, the numbering begins from the carbon center carrying the group if one; if two and of same character of different capacity, it begins from the carbon center carrying the group of higher radical-pushing or pulling capacity; and if two but of different character, it begins with the carbon center carrying the radical-pulling group.
(Laws of Creations for Nomenclature of Systems)

Rule 120: This rule of Chemistry for **Numbering of Unsaturated hydrocarbons**, states that, the numberings on the carbon centers of an unsaturated hydrocarbon such as **Dienes, Trienes, Tetraenes and higher,** are based on the location of the substituent group(s), for which in general the substituent group if only one is made to carry the lowest number starting from one of the terminals; and if only two of the same character, the one with the higher radical-pushing or pulling capacity carries the lowest number starting from the terminals; and if only two of different characters, the radical-pulling one carries the lowest number starting from the terminals.
(Laws of Creations for Nomenclature of Systems)

Rule 121: This rule of Chemistry for **Numbering of Unsaturated hydrocarbons,** states that, the numberings on the carbon center of an unsaturated hydrocarbon such as **Styrene** whether substituent group(s) are present on the aliphatic side of the ring or not, or on the aromatic side or not, begins from the aliphatic end externally located to give a 1,2-mono-form.
(Laws of Creations for Nomenclature of Systems)

Rule 122: This rule of Chemistry for **Dienes,** states that, while Butadiene the first member of the family has 4,3- and 4,1-mono-forms, the Nucleophilic members have 4,3- and 4,1-mono-forms and the Electrophilic members have 2,1- and 4,1-mono-forms, (both for groups externally located and based on the route natural to them).
(Laws of Creations for Dienes)

Rule 123: This rule of Chemistry for **Molecular rearrangements of the first kind of the first type in a resonance stabilized compound,** states that, molecular rearrangement takes place only when substituent group with transfer species is externally located on the monomer and this takes place only with Nucleophiles where one stage is required for the two mono-forms; for if it takes place with Electrophiles, two stages will be required for its 1,2-mono-form while one stage will be required for its 1,4-mono-form, clear indication that rearrangement may not take place with Electrophiles when resonance stabilized.
(Laws of Creations for Molecular rearrangement)

Rule 124: This rule of Chemistry for **Resonance Stabilization phenomenon of the first kind,** states that, with Chloroprene and Isoprene where the groups are internally located, the first has 1,2- and 1,4-mono-forms while the second has 3,4- and 1,4- mono-forms; for which the same pattern applies for the likes, Trienes, Tetraenes and higher.
(Laws of Creations for Dienes)

Rule 125: The rule of Chemistry for **Dienes,** states that, never can isolatedly placed Diene molecularly rearrange to give a resonance stabilized conjugatedly placed Diene; never can a resonance stabilized Diene rearrange to an isolatedly placed Diene.
(Laws of Creations for Dienes)

Rule 126: This rule of Chemistry for **Vinyl chloride and 6-Chloro-styrene** shown below, states that, while (A) shown below can be activated chargedly with no resonance stabilization taking place and can be made to favor both positively and negatively charged routes, (B) cannot be activated chargedly, due to presence of electrostatic forces of repulsion, with both (A) and (B) favoring both electro-free- and nucleo-free-radical routes.

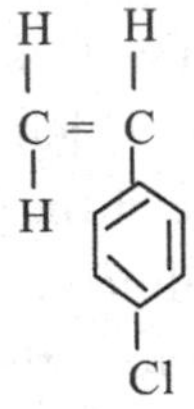

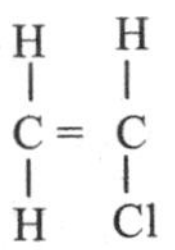

(Laws of Creations for Vinyl chloride and 6-Chloro-Styrene)

Rule 127: This rule of Chemistry for **Addition polymerization of 1,3-Dienes,** states that, poly-meric products obtained should be clearly identified by the route favored during polymerization, for which when for example chloroprene is involved, nucleo-free-radically 2,1- and 4,1- Addition products are exclusively obtained, while electro-free-radically 1,2- and 1,4- Addition products are exclusively obtained; for which the same pattern applies to others.
(Laws of Creations for Polymerization products from 1,3-Dienes)

Rule 128: This rule of Chemistry for **Aliphatic Electrophiles,** states that, based on one of the driving forces for Resonance Stabilization as applies to Mathematics and Physics, NO RESONANCE STABILIZATION can be provided for them since movement from **FULL**-FREE-RADICAL to **HALF**-FREE-RADICAL or the reverse is impossible
(Laws of Mathematics and Physics)

Rule 129: This rule of Chemistry for **Resonance Stabilization phenomenon of the first kind,** states that, when a Nucleophilic center is externally adjacently located to an Electrophilic/ Nucleophilic center (i.e. a MALE) as shown below, based on one of the driving forces for Resonance Stabilization as applies to Mathematics and Physics, RESONANCE STABILI-ZATION can be provided, since movement from **FULL**-FREE-RADICAL to **FULL**-NON-FREE-RADICAL is possible.

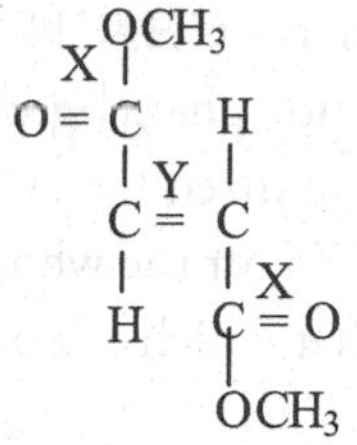

$$
\begin{array}{cc}
 & X \overset{\displaystyle OCH_3}{|} \\
O = C & H \\
| \ \ Y & | \\
C = C \\
| & | \ X \\
H & C = O \\
 & | \\
 & OCH_3
\end{array}
$$

(Laws of Mathematics and Physics)

Rule 130: This rule of Chemistry for **Aliphatic Nucleophilic Dienes, Trienes, and Tetraene,** states that, RESONANCE STABILIZATION is favored for them when the movement is from **FULL**-FREE-RADICAL to **FULL**-FREE-RADICAL or from **HALF**-FREE-RADICAL **to HALF**-FREE-RADICAL in support of the Laws of Mathematics and Physics.
(Laws of Mathematics and Physics)

Rule 131: This rule of Chemistry for **Ringed Electrophiles with functional centers**, states that, the issue of RESONANCE STABILIZATION does not arise for them if and only if only one hetero atom is present, since the functional centers do not carry π-bonds; for which when more than one is present such as O and N. then its presence may be favored only radically.
(Laws of Creations for Electrophiles)

Rule 132: This rule of Chemistry for **Ringed Electrophiles with π-bonds internally located,** states that, RESONANCE STABILIZATION is favored by them if two Males (Y) and two Females (X) centers are adjacently placed such that movement is from **FULL**-FREE-RADICAL to **FULL**-NON-FREE-RADICAL as is the case with Benzoquinone.
(Laws of Creations for Electrophiles)

Rule 133: This rule of Chemistry for **Resonance Stabilization groups**, states that, based on the types of radicals carried by their centers when activated when all the families were investigated, the following is the valid order of capacity of the major radical-pushing groups with respect to H-

$$- C_6H_5 \ > \ -CH=CH_2 \ > \ H \ > \ -C\equiv CH \ >$$
Order of Radical-pushing-Capacity in the presence of Resonance Stabilization

for which in the absence of resonance stabilization the capacities are almost equal to or less than that of H.
(Laws of Creations for Resonance Stabilization Groups)

Rule 134: This rule of Chemistry for **Benzene, Acetylene, Formaldehyde and Ethene,** states that, as first members of their different families and based on the considerations so far, the following is the order of Nucleophilic capacity-

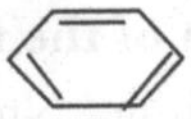

$$\langle\!\!\rangle \quad \gg \quad HC \equiv CH \quad \gg \quad H_2C = O \quad \gg \quad H_2C = CH_2$$

Order of Nucleophilic-Capacity
(Laws of Creations of First members of Some Families)

Rule 135: This rule of Chemistry for **Resonance stabilization Providers,** states that, while all the non-ringed resonance stabilization providers such as $-CH{=}CH_2$ are the last to be activated when providing resonance stabilization, the same is the case for phenyl group when providing resonance stabilization, for which while (A) below is not resonance stabilized (i.e., CH_3 group in $COCH_3$ is not shielded), (B) is resonance stabilized (i.e., CH_3 group in $COCH_3$ or the whole group can be shielded) because the phenyl group is in its Energized state of existence, that which the other groups cannot do.

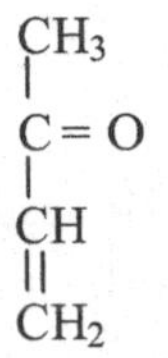

(A) 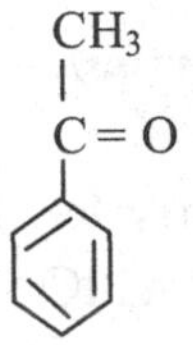   (B)

(Laws of Creations for Resonance Stabilization Providers)

Rule 136: This rule of Chemistry for the **Aromatic Ketone shown below,** states that, when activated in its De-energized state of existence, the followings are obtained with resonance stabilization provided by the phenyl group-

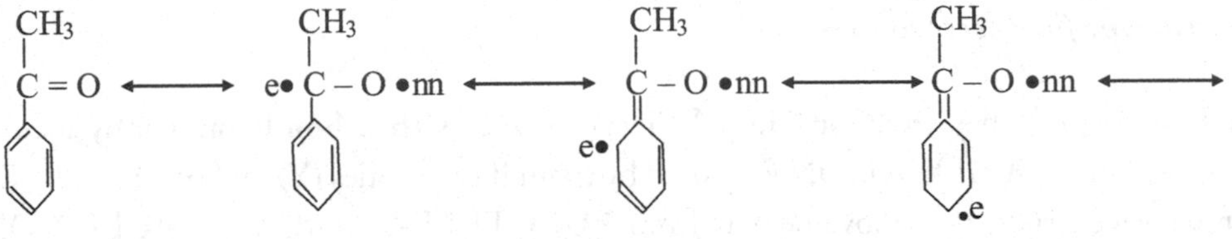

Half-free   Half-free   Half-free

for which instead of the movement from Full-free to Full-free (Nucleophiles) or Full-free to Full-non-free (Electrophiles), here the movement is from *HALF*-free to *HALF*-free (Electrophiles).
(Laws of Creations for Aromatic Ketones)

Rule 137: This rule of Chemistry for **Special radical-pushing groups**, states that, based on the Chemistry of Dienes of Nucleophilic and Electrophilic characters, the followings are valid-

$$(RN =) > (O =) > (R_2C =) \ggg (- O{-}NO_2) > (- NR_2) > (- OR) > (- R) > (- H)$$

Order of Radical-pushing capacities

(Laws of Creations for Special Radical-pushing Groups)

Rule 138: This rule of Chemistry for **Alkyl Groups,** states that, based on the types of groups seen so far, there are three types of Alkyl groups-the first coming from the Alkane family are herein called *Alkanyl groups* (e.g. $-CH_3$), while the ones coming from the Alkene family are herein called *Alkenyl groups* (e.g.

-CH=CH_2) while the ones coming from the Alkyne family are herein called ***Alkynyl groups*** (e.g. -C≡CH which cumulenically is represented by =CH_2 in order to provide a transfer species like the others). ***(Laws of Creations for Alkyl Groups)***

Rule 139: This rule of Chemistry for **Ceiling Temperatures,** states that, Ceiling temperatures the temperature at which full polymerization of a monomer takes place before Depropagation be-gins are of two kinds- the first herein called **"Radical Unnatural Route Polymerization (RURP) Ceiling Temperature** and the second herein called **"Radical Natural Route Polymerization" (RNRP) Ceiling Temperature;** while the former deals with olefinic monomers, the latter deals with non-olefinic monomers. ***(Laws of Creations for Ceiling Temperatures)***

Rule 140: This rule of Chemistry for **Resonance Stabilization Groups**, states that, when sub-stituent groups are internally and externally located on them, the followings are obvious-

H_2C=C($COOCH_3$)- > H_2C=CCl- > H_9C_4CH=CH- > H_2C=CH- > H > H_2C=C(CH_3)-
Order of Radical-pushing capacity

($COOCH_3$)CH=CH- > ClCH=CH- > FCH=CH- >
Order of Radical-pulling capacity

(Laws of Creations for Resonance Stabilization groups)

Rule 141: This rule of Chemistry for **The monomer shown below**, states that, while the monomer is an Electrophile chargedly, it is a Nucleophile radically in which unlike chargedly the N = O center is less nucleophilic than C = C center and like chargedly the ⊕N = O is more nucleophilic than C = C center and as such, it cannot be resonance stabilized as shown below-

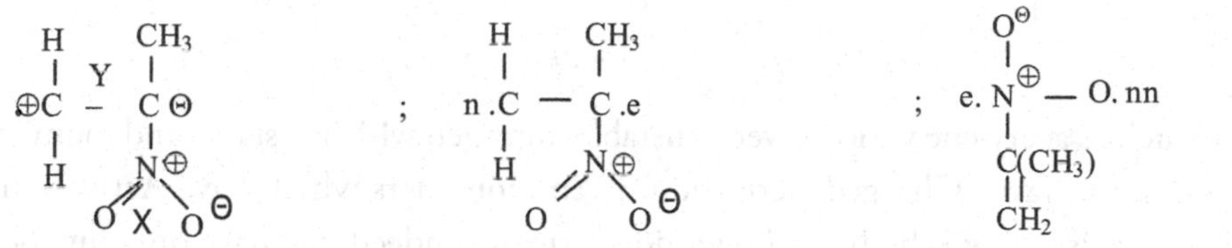

ELECTROPHILE Less nucleophilic More nucleophilic

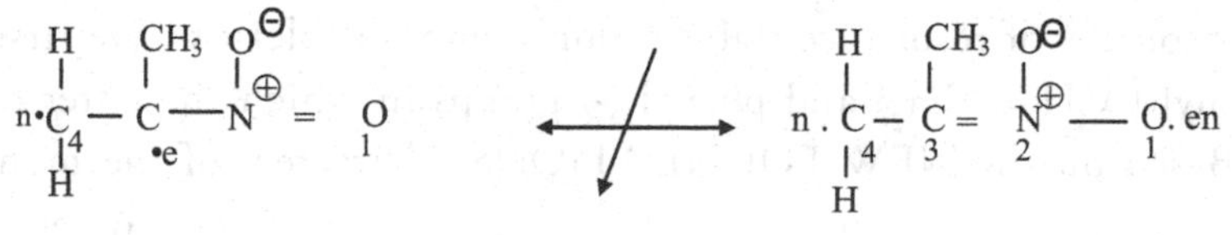

since movement is from ***FULL***-free-radical to ***HALF***-free-radical, for if it was N = O, it would have been favored moving ***HALF***-free-radical to ***HALF***-free-radical.
(Laws of Creations for α-Methyl Vinyl nitrite)

Rule 142: This rule of Chemistry for **Transfer Species**, states that, ***Transfer species of the first kind of the second type*** is that carried by Alkenyl groups (Not common) as show below for Acrylonitrile-

$$R\bullet nn \ + \ e\bullet C = N^{\bullet nn} \quad\longrightarrow\quad RH \ + \ \overset{\displaystyle CH_2}{\underset{}{\overset{\displaystyle \|}{C}}} = C = N^{\bullet nn}$$

with the left side carrying $\underset{\overset{\displaystyle \|}{CH_2}}{\overset{\displaystyle \|}{CH}}$ substituent.

<u>(Transfer species of 1st kind of 2nd type)</u>.

$$R^{\bullet e} \ + \ nn\bullet N = C^{\bullet e} \quad\longrightarrow\quad R - N = C^{\bullet e}$$

with $\underset{\overset{\displaystyle \|}{CH_2}}{\overset{\displaystyle \|}{CH}}$ on both C centers.

(Laws of Creations for Transfer Species of First Kind of Second Type)

3.5. Conclusion

Forty one rules have been proposed here to account for part of the past, present, and even the future. Rules with respect to Existence of Paired Radicals, Ceiling Temperatures and many very important concepts cannot yet be fully stated until downstream when the time is ripe to state them. The rules stated so far, are far more educative than what are contained in the text, but very simple to comprehend. In the past, even a monomer such as vinyl chloride when activated was thought to be chargedly resonance stabilized as already nullified by Equation 1.6 of chapter 1 and clearly recalled below.

$$\underset{\overset{\displaystyle |}{H}}{\overset{\displaystyle H}{C}} = \underset{\overset{\displaystyle |}{Cl}}{\overset{\displaystyle H}{C}} \quad\nrightarrow\quad \ominus \underset{\overset{\displaystyle |}{H}}{C} - \underset{\overset{\displaystyle \ddot{Cl}}{C}}{\oplus} \quad\nrightarrow\quad \ominus \underset{\overset{\displaystyle |}{H}}{C} - \underset{\overset{\displaystyle \ddot{Cl}}{\oplus}}{C} \qquad 3.141$$

Even monomers such as acetylene which is very unstable, nitrogen which is stable and many more of them have been shown not to favor Charged Activation. Even monomers which have Activation centers and cannot be activated exist, as will be fully shown downstream. Indeed, we have only just begun because NATURE is very COMPLEX. We have long begun to see Mathematics and Physics in CHEMISTRY, the greatest of the trios.

The two major sources of resonance stabilization group, for olefins have first been identified. These are the alkenyl ($-CH = CH_2$) and phenyl (-) groups in which the latter has more capacity than the former. Based on the NEW FOUNDATIONS, characters of the monomers and types of resonance stabilization groups, routes favored by them and so on, new nomenclatures for these addition monomers have emerged. Establishments of these new nomenclatures have been based on past literature data for these monomers all of which are very informative and rich, but not in place. Products should be defined by the new nomenclature, since for example polybutadienes obtained "anionically" or nucleo-free-radically are different from those obtained "cationically" or electro-free-radically and since also for example "Mr. Idiot" should not be used as a name. Nothing was created or came to existence by ACCIDENT. When a child is born, the Child is in CHAINS. It is only the Child that can unchain itself.

For the first time, the nomenclature of all phenyl containing resonance stabilized monomers is being established with little or no error based on the new concepts. The order and manner by which the activation centers are activated are being shown step by step. These cannot be conclusively established, until ringed monomers (both opening and non-opening types) have been considered. For the first time also, how transfer species are transferred in Chemical and Polymeric systems are being systematically explained. Chemical or Polymeric reactions don't take place without applying the laws of NATURE. Transfer species are never abstracted indiscriminately.

Finally, it is important to note that all the rules which have been established so far, are laws which can be found in our worldly coexistence and too broad to ignore, because ignoring them is mathematically the same as IGNORANCE. These are laws which should form the basic foundations for all disciplines including RELIGION.

References

1. D. P. Wyman, I. H. Song, "The butyllithium initiated bulk anionic polymerization of α-methylstyrene", Die Makromolekulare Chemie, Volume 115, Issue 1, pgs. 64-72. 1967, Published Online 12 Mar. 2003.

2. J. Leonard, S. L. Malhotra, "Equilibrium anionic polymerization of α-methylstyrene in p-dioxane", J. of Polymer Science Part A-1; Polymer Chemistry, Vol.2, Issue 7, pgs. 1983-1991, 1971, Published online 10 May 2003.

3. D. J. Worsfold, S. Bywater, "Anionic polymerization of α-methylstyrene", J. of Polymer Science, Volume 26, Issue 114, pgs. 299-304, 1957, Published online 19 May 2003.

4. K. M. Hui, Y. K. Org, "Kinetics of anionic polymerization of α-methylstyrene in tetrahydrofuran and toluene mixed solvent" J. of Polymer Science: Polymer Chemistry Edition, Volume 14, Issue 6, pgs. 1311-1316,, 1975, Published online 8 Apr. 2003.

5. K. M. Hui, T. L. Ng, "Kinetics of anionic polymerization of α-methylstyrene in tetra-hydrofuran", J. of Polymer Science: Polymer Chemistry Edition, 1969, published online 10 May 2003.

6. L. Toman, S. Pokomy, J. Spevacek, "Cationic polymerization of α-methylstyrene in dichloromethane initiated with system 2,5-dichloro-2,5-dimethylhexane-BCl_3. II. Continuous addition of monomer into initiator, quasiliving polymerization" http://www3.intersience.wiley.com/journal/106565455/abstract

7. J. M. Ginn, K. J. Ivin, "Kinetics of anionic polymerization of α-methylstyrene in tetrahydrofuran in the presence and absence of triethyleneglycoldimethylether", Die Macromolecukulare Chemie, Volume 139, Issue 1, pgs. 47-60, 1970, published online 12 May 2003.

8. R. A. Andrekanic, J. S. Salec, S. E. Mulhall, Y. V. Bubnov, "Polymerization of α-methylstyrene Patent 6649716" http://www freepatentsonline.com/6649716 html, 2009.

9. G. Odian, "Principle of Polymerization", McGraw-Hill Book Company, (1970), pg. 355.

10. C.R. Noller, "Textbook of Organic Chemistry," W. B Saunders Company, (1966), pg. 380.

11. G. Odian, "Principle of Polymerization", McGraw-Hill Book company, (1970), pg. 165.

12. S.N.E. Omorodion, "New Frontiers in Engineering Science and the Arts", Vol 1, Chapter 8. In Press, 2012.

13. G. Natta et al; Chim.Ind. (Milan), 40:362(1958); 41:116, 398,526,1163(1959); Makromol. Chem; 77: 114, 126 (1964).

14. G. Wilke, Angew. Chem, 68:306(1956).

15. "An Intensive short Course on Polymer Production Technology, Polymer Reaction Engineering-Part I, McMaster Univeristy, Canada, 1976.

Problems

3.1. What are charged and radical resonance stabilization phenonmena? What are the distinguishing features of the phenomena.

3.2. Consider the two monomers shown below:-

(i) 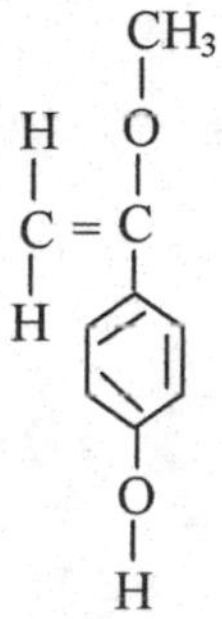
$$CH_3$$
$$\overset{\displaystyle H \quad O}{\underset{\displaystyle H}{|\quad\;|}}$$
$$C = C$$

(ii) 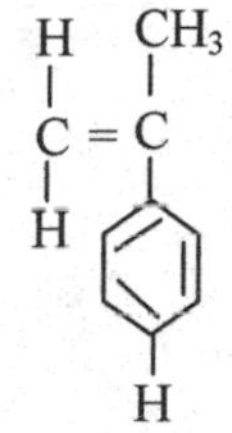
$$\overset{\displaystyle H \quad CH_3}{\underset{\displaystyle H}{|\quad\;|}}$$
$$C = C$$

(a) What are the equivalences of the two monomers?
(b) What are the transfer species in the two monomers if they exist?
(c) Show the activation of the monomers chargedly and free-radically.
(d) Show how the monomers are resonance stabilized? Explain.

3.3. "The character of an activation center is very important in many respects" Discuss.

3.4. (a) Distinguish between the two major sources of resonance stabilization groups shown below, in terms of -

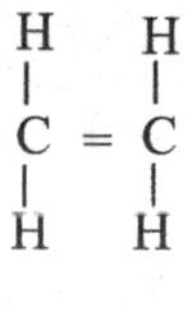

$$\overset{\displaystyle H \quad H}{\underset{\displaystyle H \quad H}{|\quad\;|}}$$
$$C = C \qquad ; \qquad \langle\!\!\!\bigcirc\!\!\!\rangle$$

(i) Character, (ii) Activations, (iii) Capacity, (iv) Polymerization characteristics, (v) Provision of resonance stabilization and (vi) Nomenclature with respect to (b) below.

(b) Compare the two above with the two below:-

174

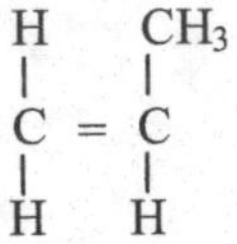

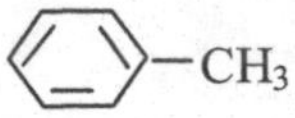

$$
\begin{array}{cc}
H & CH_3 \\
| & | \\
C & = C \\
| & | \\
H & H
\end{array}
\qquad ; \qquad
\langle\!\!\!\!\bigcirc\!\!\!\!\rangle\!-\!CH_3
$$

3.5. Consider the two compounds shown below:-

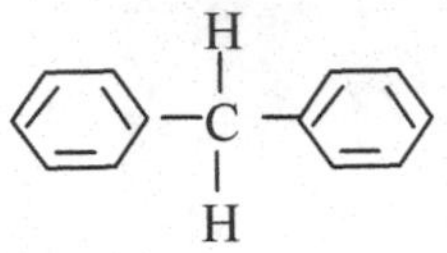

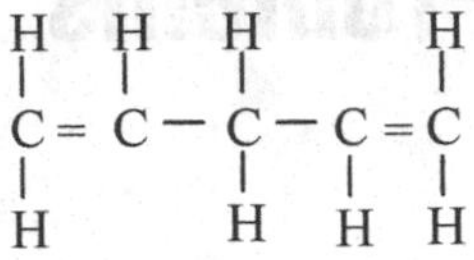

$$
\langle\!\!\!\bigcirc\!\!\!\rangle\!-\!\overset{\overset{\displaystyle H}{|}}{\underset{\underset{\displaystyle H}{|}}{C}}\!-\!\langle\!\!\!\bigcirc\!\!\!\rangle
\qquad ; \qquad
\begin{array}{ccccc}
H & H & H & & H \\
| & | & | & & | \\
C = & C - & C - & C = & C \\
| & & | & | & | \\
H & & H & H & H
\end{array}
$$

(a) Identify the characters of the compounds.
(b) Which of them is resonance stabilized?
(c) What are the transfer species in the monomers where activation is favored? Explain.
(d) Which one will favor molecular rearrangement when the conditions exist?
(e) In terms of activation, which is easier to activate, and why?

3.6. Shown below are two monomers:-

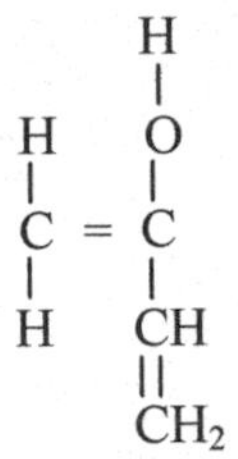

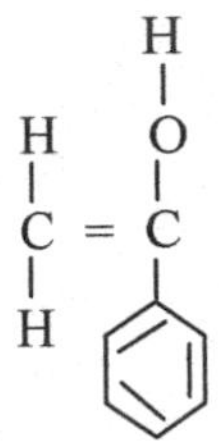

$$
\begin{array}{cc}
 & H \\
 & | \\
H & O \\
| & | \\
C = & C \\
| & | \\
H & CH \\
 & || \\
 & CH_2
\end{array}
\qquad ; \qquad
\begin{array}{cc}
 & H \\
 & | \\
H & O \\
| & | \\
C = & C \\
| & | \\
H & \langle\!\!\!\bigcirc\!\!\!\rangle
\end{array}
$$

(a) Is molecular rearrangement limited by the character of the activation centers above? Explain.
(b) Are the monomers above unstable?
(c) What are the polymeric products obtained if they are stabilized? How can they be stabilized if found to be unstable?
(d) Distinguish between the two monomers nomenclaturally.

3.7. Distinguish between the two monomers shown below in terms of –

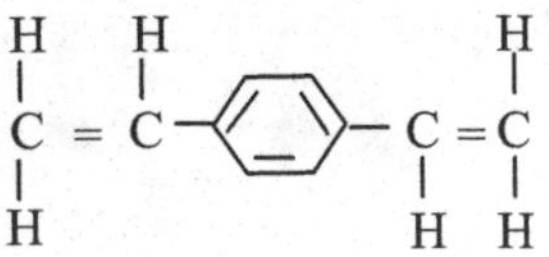

$$
\begin{array}{cccc}
H & H & & H \\
| & | & & | \\
C = & C - \langle\!\!\!\bigcirc\!\!\!\rangle - & C = & C \\
| & | & & | \\
H & & H & H
\end{array}
\quad \text{and} \quad
\begin{array}{ccccc}
H & H & & H & H \\
| & | & & | & | \\
C = & C - & C = C - & C = & C \\
| & | & & | & | \\
H & H & & H & H
\end{array}
$$

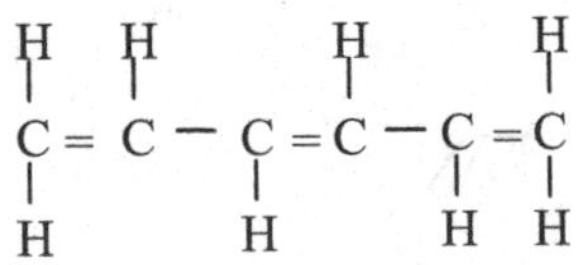

(a) Character- chargedly and free-radically.
(b) Their nomenclature.
(c) Their activations
(d) Applications.

3.8. Below are shown the following substituted groups:-

(i) $\overset{\displaystyle |}{CH_2 = CH}$ (ii) $CH_3 - CH = \overset{\displaystyle |}{CH}$ (iii) $Cl - CH = \overset{\displaystyle |}{CH}$ (iv) $CH_3OOCCH = \overset{\displaystyle |}{CH}$

(v) $CH_2 = \overset{\displaystyle |}{C}(COOCH_3)$ (vi) $CH_2 = \overset{\displaystyle |}{C}Cl$ (vii) $CH_2 = \overset{\displaystyle |}{C}(CH_3)$

(viii) $Cl-\langle\underline{\ }\rangle-$ (ix) 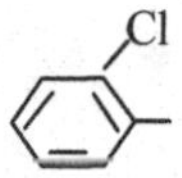 (image)   (x) 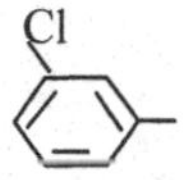 (image)

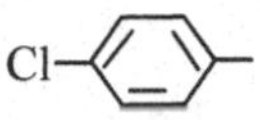

(a) Identify which are radical-pushing or pulling resonance stabilization groups chargedly and free-radically.
(b) For those which are radical-pushing or pulling, identify the order of their capacities.
(c) Identify those which have substituted groups externally or internally located.
(d) Use them as resonance stabilization groups for a C = C activation center to identify the character that will be imposed on the monomer of your choice.

3.9 (a) Distinguish between the activations of the following monomers.

(i)  (image)   (ii) 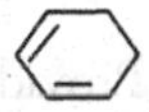 (image)   (iii)  (image)   (iv)  (image)

(v)  (image)   (vi)  (image)

(b) Identify where resonance stabilization phenomena are present and where a different type of molecular rearrangement is favored, (Use literature data)

3.10. Consider the monomer shown below.

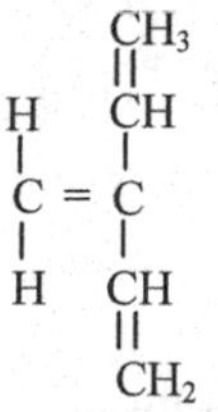

$$
\begin{array}{cc}
 & CH_3 \\
 & || \\
H & CH \\
| & | \\
C & = \ C \\
| & | \\
H & CH \\
 & || \\
 & CH_2 \\
\end{array}
$$

(a) How many activation centers exist?
(b) What are the characters of the activation centers?
(c) Establish the type of nomenclature of the mono-forms obtained from the monomer.
(d) Identify the type of transfer species present in the mono-forms. If none exists, explain why.

3.11. (a) Distinguish between the two substituent groups shown below.

(i) 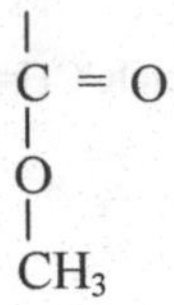
$$\begin{array}{c} | \\ C = O \\ | \\ O \\ | \\ CH_3 \end{array}$$

(ii) 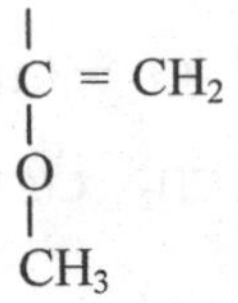
$$\begin{array}{c} | \\ C = CH_2 \\ | \\ O \\ | \\ CH_3 \end{array}$$

(b) Distinguish radically and chargedly between the rate of propagations of the two monomers shown below-

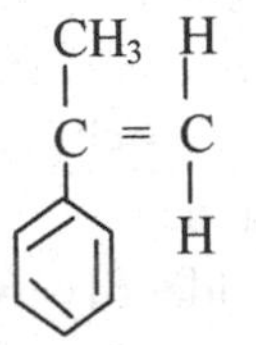

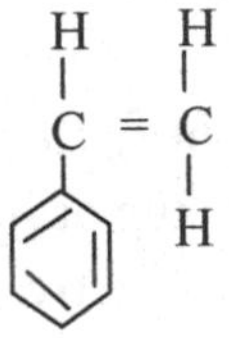

3.12 (a) How does the law of conservation of transfer of transfer species apply in the presence of resonance stabilization groups?

(b) Is the monomer (i) shown below a Tetraene or Heptaene? Provide the true nomenclatures for them.

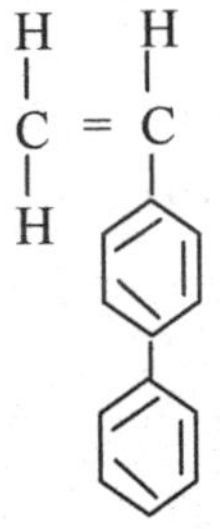

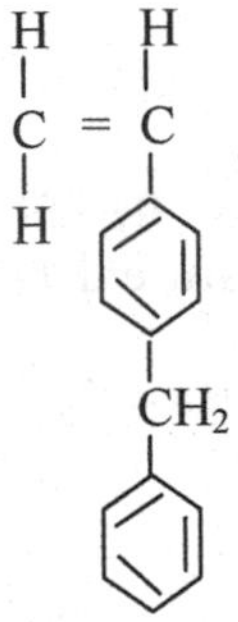

(c) Distinguish between (i) and (ii) above.

3.13. (a) How is transfer of transfer species of the second kind affected in resonance and non-resonance stabilized monomers during polymerization?

(b) What is transfer species of the first kind of the second type? Give three examples of monomers that favor its existence.

(c) Considering the monomers shown below, identify which ones have transfer species of the second kind in the routes natural and not natural to them free-radically and chargedly. Explain.

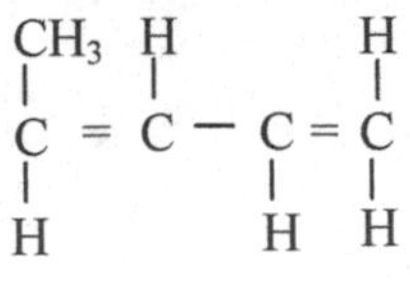

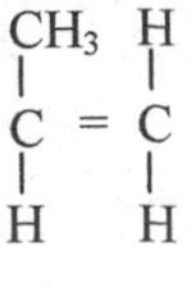

177

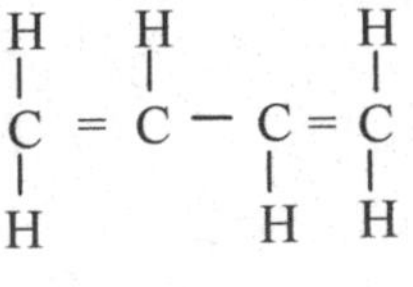

$$
\text{(i)} \quad
\begin{array}{c}
CH_3 \;\; H \\
| \quad\;\; | \\
C = C - C = C \\
| \qquad\quad | \;\; | \\
H \qquad\;\; H \;\; H
\end{array}
\qquad
\text{(ii)} \quad
\begin{array}{c}
CH_3 \;\; H \\
| \quad\;\; | \\
C = C \\
| \quad\;\; | \\
H \quad\;\; H
\end{array}
\qquad
\text{(iii)} \quad
\begin{array}{c}
H \;\; H \qquad H \\
| \;\;\; | \qquad\; | \\
C = C - C = C \\
| \qquad\quad | \;\; | \\
H \qquad\;\; H \;\; H
\end{array}
$$

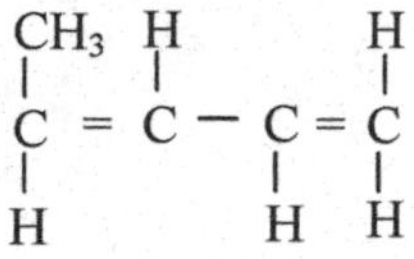

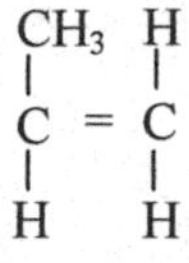

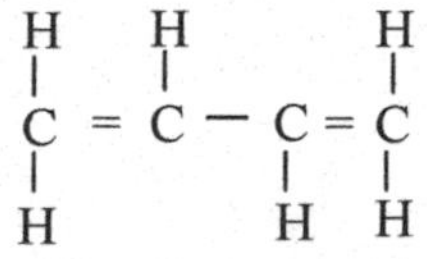

3.14 (a) Are presence of dead terminal double bond polymers favored in only one route? Explain using the monomers in Q 3.13 (c).

   (b) If the answer to (a) is No, while explaining, identify the routes where favored for the monomers above.

3.15 Shown below are some monomers.

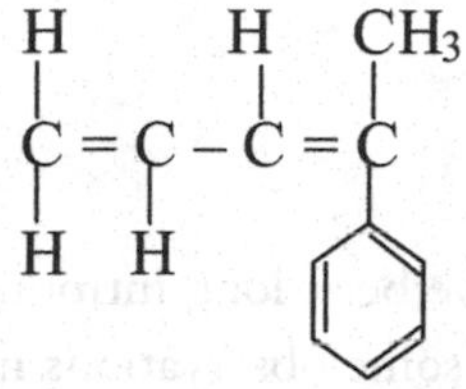

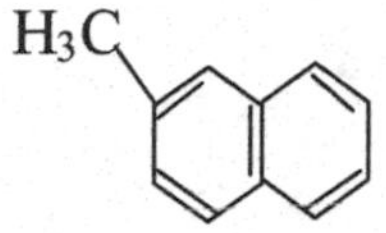

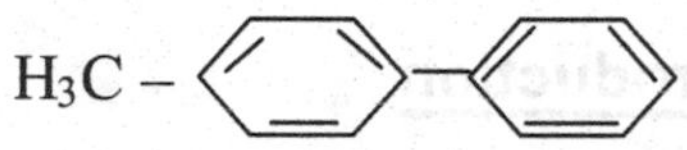

$$
\begin{array}{c}
H \qquad\; H \;\; CH_3 \\
| \qquad\;\; | \quad\;\; | \\
C = C - C = C \\
| \qquad\; | \qquad\;\; | \\
H \quad\; H \qquad\; \text{(phenyl)}
\end{array}
\quad ; \quad
\text{(methyl naphthalene)}
\quad ; \quad
H_3C - \text{(biphenyl)}
$$

Answer the following questions-

   a) What are the names of these monomers?
   b) Show how resonance stabilization is favored in them?
   c) Some humans and countries have shields and some don't. What is the essence of this?
   d) Provide the nomenclature for the different monomers

CHAPTER 4

THE CHARACTERS OF MONO-OLEFINS WITH RESPECT TO MOLECULAR REARRANGEMENT PHENOMENA.

4.0 Introduction

Though the concepts of molecular rearrangement phenomena seem to have been long introduced, and probably now thought to be known, they have only been used to explain some observations made over the years in chemical and polymeric systems. Nevertheless, there are still many observations that will demand further explanations and guiding rules.

Two extreme families of olefins have been identified in the course of the study- the alkenes and the perfluoroalkenes. These are monomers that carry largely alkanyl groups as substituent groups, e.g. CH_3, C_2H_5, CF_3, CF_2H, etc. Several families of these monomers exist. These include $CH_2 = CHR$, $CH_2 = CR_1R_2$, $CHR_1 = CHR_2$, $CH_2 = C(R)_2$, $CHR = CHR$, $CHR = C(R)_2$, $CHR_1 = CR_2R_3$, $C(R)_2 = C(R)_2$, $CH_2 = CHR_F$, $CH_2 = CR_{F1}R_{F2}$, $CHR_F = CHR_F$, $CHR_{F1} = CHR_{F2}$ ...., $CF_2 = CFR_F$, $CF_2 = C(R_F)_2$ etc. As has been observed so far, the characters of some of these classes of monomers are very unique chargedly and radically and must be known. Therefore, there is need to identify the common factors in their use as monomers.

Some useful general rules with respect to molecular rearrangement phenomena have been proposed, and these will still find useful applications here in establishing new foundations, since for example there are some monomers which cannot favor polymerization, due to steric limitations posed by the use of an unsuitable initiator, but can indeed be polymerized when the conditions are adequatedly chosen. In general, with respect to molecular rearrangement, the transfer species involved are of the first kind.

4.1 Characteristics of CH2 = CHR, CH2 = CRR, CHR = CHR, CHR = C(R)2 and C(R)2 = C(R)2 group.

All the Rs are alkanyl groups. There is need to identify which of the members undergo molecular rearrangement and which monomers within the members cannot undergo molecular rearrangement, when the conditions exist. Consider propylene and 1 - butene. While propylene does undergo to produce itself, 1-butene does not when the conditions exist.

$$
\begin{array}{ccccc}
\underset{\underset{H}{|}}{\overset{\overset{H}{|}}{C}} = \underset{\underset{CH_3}{|}}{\overset{\overset{H}{|}}{C}}
& \longrightarrow &
\ominus\underset{\underset{\mathrm{HH-C-H}}{|}}{\overset{\overset{H}{|}}{C}} - \overset{\overset{H}{|}}{C}\oplus
& \longrightarrow &
\ominus\underset{\underset{H}{|}}{\overset{\overset{H}{|}}{C}} - \underset{\underset{CH_3}{|}}{\overset{\overset{H}{|}}{C}}\oplus
\end{array}
\qquad 4.1
$$

$$
\begin{array}{ccccc}
\underset{\underset{CH_2}{|}}{\overset{\overset{H}{|}}{C}} = \underset{\underset{CH_3}{|}}{\overset{\overset{H}{|}}{C}}
& \longrightarrow &
\ominus\overset{\overset{H}{|}}{C} - \underset{\underset{CH_2}{|}}{\overset{\overset{H}{|}}{C}}\oplus
& \xcancel{\longrightarrow} &
\ominus\underset{\underset{CH_2}{|}}{\overset{\overset{H}{|}}{C}} - \overset{\overset{H}{|}}{C}\oplus
\end{array}
\qquad 4.2
$$

(CH2 = CHR)

$$
\begin{array}{ccccc}
\ominus\underset{\underset{\mathrm{HH-C-H}}{|}}{\overset{\overset{H}{|}}{C}} - \overset{\overset{H}{|}}{C}\oplus
& \xcancel{\longrightarrow} &
\ominus\underset{\underset{CH_3}{|}}{\overset{\overset{H}{|}}{C}} - \underset{\underset{CH_3}{|}}{\overset{\overset{H}{|}}{C}}\oplus
& \longrightarrow &
\ominus\underset{\underset{CH_3}{|}}{\overset{\overset{H}{|}}{C}} - \underset{\underset{CH_3}{|}}{\overset{\overset{H}{|}}{C}}\oplus
\end{array}
\qquad 4.3
$$

(I) <u>More stable</u> (II) less stable <u>(CHR = CHR)</u>

Of the three reactions above, it is only the first that is favored. In the second reaction, CH_3 is the transfer species when H is still present. This is against the law or rule of nature as already stated. Though it may seem that the same monomer is produced, there is no transfer of transfer species. In the last reaction, (II) is far less stable than (I). Hence no transfer will take place to move from more stable state (ORDERLINESS) to less stable state (DISORDERLINESS), for this is not NATURE, when the issue is not bombardment. Hence, the need to look at these families of monomers. Then, is CH_2 = CHR more stable than CHR = CHR?

Now, consider R as a secondary alkyl or alkanyl group.

$$
\begin{array}{ccccc}
\underset{\underset{\underset{CH_3 \quad CH_3}{\diagdown \diagup}}{CH}}{\overset{\overset{H}{|}}{C}} = \underset{}{\overset{\overset{H}{|}}{C}}
& \longrightarrow &
\ominus\overset{\overset{H}{|}}{C} - \underset{\underset{\underset{CH_3 \quad CH_3}{}}{CH}}{\overset{\overset{H}{|}}{C}}\oplus
& \xcancel{\longrightarrow} &
\ominus\underset{\underset{CH_3}{|}}{\overset{\overset{CH_3}{|}}{C}} - \underset{\underset{CH_3}{|}}{\overset{\overset{H}{|}}{C}}\oplus
\left(\oplus\underset{\underset{CH_3}{|}}{\overset{\overset{CH_3}{|}}{C}} - \underset{\underset{CH_3}{|}}{\overset{\overset{H}{|}}{C}}\ominus\right)
\end{array}
\qquad 4.4
$$

(III)

(I) <u>More stable</u> (II) <u>Impossible state</u>

(II) is an impossible activated state, since (III) is the true activated state. Only H can be used as a transfer species for (I). The number of substituted groups (H) on the monomer cannot be increased to produce a stronger nucleophile. It can only be reduced or left as it is for which no transfer takes place. Then, is CH_2 = CHR more stable than CHR = CHR?

With a tertiary alkyl group, the following is obtained. No molecular rearrange-

$$
\begin{array}{ccc}
\overset{\text{H}}{\underset{\text{H}}{\text{C}}} = \overset{\text{H}}{\underset{\underset{\underset{\text{CH}_3 \quad \text{CH}_3}{|}}{\text{C(CH}_3)}}{\text{C}}}
& \longrightarrow &
\overset{\text{H}}{\underset{\text{H}}{\ominus\text{C}}} - \overset{\text{H}}{\underset{\underset{\underset{\text{CH}_3 \quad \text{CH}_3}{|}}{\text{C(CH}_3)}}{\text{C}\oplus}}
\end{array}
\;\;\not\longrightarrow\;\;
\overset{\text{CH}_3}{\underset{\text{CH}_3}{\ominus\text{C}}} - \overset{\text{H}}{\underset{\underset{\text{CH}_3}{\text{CH}_2}}{\text{C}\oplus}}
\qquad 4.5
$$

(I) <u>More stable</u>   (II) less stable <u>$C(R)_2 = CHR_1$</u>

ment is favored by all $CH_2 = CHR$ group of monomers to produce another more stable monomer whether R is primary, secondary or tertiary. Then, the answer to the question is YES. Of all the members in this family, only propylene undergoes transfer of transfer species during molecular rearrangement. The same will also apply free-radically. Propylene is the least nucleophilic of all the monomers. Note that when molecular rearrangement is not favored does not imply that transfer species does do exist for the monomers above. (I) of Equations 4.3 to 4.5 will still produce dead terminal double bond polymers for their growing polymer chains. The transfer species for Equations 4.3 and 4.4 is H, while CH_3 is the transfer species for Equation 4.5, that which could not be used for rearrangement.

For $CH_2 = CRR$ members, one will begin with isobutylene.

$$
\begin{array}{ccccc}
\overset{\text{H}}{\underset{\text{H}}{\text{C}}} = \overset{\text{CH}_3}{\underset{\text{CH}_3}{\text{C}}}
& \longrightarrow &
\overset{\text{H}}{\underset{\text{H}}{\ominus\text{C}}} - \overset{\text{CH}_3}{\underset{\text{CH}_3}{\text{C}\oplus}}
& \longrightarrow &
\overset{\text{H}}{\underset{\text{H}}{\ominus\text{C}}} - \overset{\text{CH}_3}{\underset{\text{CH}_3}{\text{C}\oplus}}
\end{array}
\qquad 4.6
$$

(A)

Like propylene, this undergoes molecular rearrangement to produce the same monomer when the conditions exist. For higher alkanyl groups, the followings are obtained.

$$
\begin{array}{ccc}
\overset{\text{H}}{\underset{\text{H}}{\text{C}}} = \overset{\text{C}_2\text{H}_5}{\underset{\underset{\underset{\text{CH}_3}{|}}{\text{CH}_2}}{\text{C}}}
& \longrightarrow &
\overset{\text{H}}{\underset{\text{H}}{\ominus\text{C}}} - \overset{\text{C}_2\text{H}_5}{\underset{\underset{\underset{\text{CH}_3}{|}}{\text{CH}_2}}{\text{C}\oplus}}
\end{array}
\;\;\not\longrightarrow\;\;
\overset{\text{H}}{\underset{\text{CH}_3}{\ominus\text{C}}} - \overset{\text{C}_2\text{H}_5}{\underset{\text{CH}_3}{\text{C}\oplus}}
\qquad 4.7
$$

(B)   (I) <u>More stable</u>   <u>(II) less stable (CHR = CRR1) - R1> R</u>

<u>(CH2 = CR1R1)</u>

$$
\begin{array}{ccc}
\overset{\text{H}}{\underset{\text{H}}{\text{C}}} = \overset{\text{C}_3\text{H}_7}{\underset{\text{C}_3\text{H}_7}{\text{C}}}
& \longrightarrow &
\overset{\text{H}}{\underset{\text{H}}{\ominus\text{C}}} - \overset{\text{C}_3\text{H}_7}{\underset{\underset{\underset{\text{C}_2\text{H}_5}{}}{\text{CH}_2}}{\text{C}\oplus}}
\end{array}
\;\;\not\longrightarrow\;\;
\overset{\text{H}}{\underset{\text{C}_2\text{H}_5}{\ominus\text{C}}} - \overset{\text{C}_3\text{H}_7}{\underset{\text{CH}_3}{\text{C}\oplus}}
\qquad 4.8
$$

(C)   (I) <u>More stable</u>   <u>(II) less stable (CHR1 = CR2R) – R2 > R1 > R</u>

(CH2 = CR2R2)

No transfer of transfer species is favored. That does not mean that dead terminal double bond polymers cannot be produced. Like $CH_2 = CHR$ group, no molecular rearrangement is favored by $CH_2 = CRR$ group. Only the first member favors transfer of transfer species. Yet (C) > (B) > (A) in nucleophilic capacity, that is, (C) is more female than (B) which in turn is more female than (A).

For CHR = CHR group, consider 2 - butenes (cis- and trans- forms).

$$
\begin{array}{ccc}
\underset{H}{\overset{CH_3}{C}} = \underset{H}{\overset{CH_3}{C}} & \longrightarrow & \ominus\underset{H}{\overset{CH_3}{C}} - \underset{H}{\overset{CH_3}{C}}\oplus & \longrightarrow & \ominus\underset{H}{\overset{H}{C}} - \underset{\underset{CH_3}{|}}{\overset{H}{\underset{CH_2}{C}}}\oplus
\end{array}
\qquad 4.9
$$

$$
CHR = CHR \qquad\qquad \text{(I) \underline{Less Stable}} \qquad\qquad \text{(II) \underline{More Stable} ($CH_2 = CHR_1$) $R_1 > R$}
$$

This is the reverse of the reaction of Equation 4.3. 2–butenes (cis- or trans-) molecularly rearrange to produce 1-butene. For higher alkyls, the followings are obtained.

$$
\begin{array}{ccc}
\underset{H}{\overset{C_2H_5}{C}} = \underset{\underset{CH_3}{|}}{\overset{H}{\underset{CH_2}{C}}} & \longrightarrow & \ominus\underset{H}{\overset{\overset{CH_3}{|}}{\underset{}{\overset{CH_2}{C}}}} - \underset{\underset{CH_3}{|}}{\overset{H}{\underset{CH_2}{C}}}\oplus & \longrightarrow & \ominus\underset{CH_3}{\overset{H}{C}} - \underset{\underset{CH_3}{\underset{|}{CH_2}}}{\overset{H}{\underset{|}{CH_2}}}\oplus
\end{array}
\qquad 4.10
$$

$$
(CHR_1 = CHR_1) \qquad \begin{array}{c}\text{(I) \underline{Less stable}}\\ \underline{CHR_1 = CHR_1}\end{array} \qquad \begin{array}{c}\text{\underline{(II) More stable}}\\ \underline{(CHR = CHR_2)\ R_2 > R_1 > R}\end{array}
$$

$$
\begin{array}{ccc}
\underset{H}{\overset{C_3H_7}{C}} = \underset{\underset{C_2H_5}{|}}{\overset{H}{\underset{CH_2}{C}}} & \longrightarrow & \ominus\underset{H}{\overset{C_3H_7}{C}} - \underset{\underset{C_2H_5}{|}}{\overset{H}{\underset{CH_2}{C}}}\oplus & \longrightarrow & \ominus\underset{C_2H_5}{\overset{H}{C}} - \underset{\underset{C_3H_7}{\underset{|}{CH_2}}}{\overset{H}{\underset{|}{CH_2}}}\oplus
\end{array}
\qquad 4.11
$$

$$
(CHR_2 = CHR_2) \qquad \begin{array}{c}\text{(I) \underline{Less stable}}\\ \underline{CHR_2 = CHR_2}\end{array} \qquad \begin{array}{c}\text{\underline{(II) More stable}}\\ \underline{(CHR_1 = CHR_3)\ R_3 > R_2 > R_1}\end{array}
$$

Molecular rearrangement is favored by them to produce a more stable monomer. It is only the first member that moves from CHR = CHR to CH_2 = CHR_1, while the other members move from CHR = CHR to CHR_1 = CHR_2 ($R_2 > R_1$). Imagine the types of polymeric products obtained when a weak initiator is used to initiate polymerization.

For CHR = $C(R)_2$ group of monomers, the followings are obtained.

$$
\begin{array}{ccc}
\underset{CH_3}{\overset{H}{C}} = \underset{CH_3}{\overset{CH_3}{C}} & \longrightarrow & \ominus\underset{CH_3}{\overset{H}{C}} - \underset{CH_3}{\overset{CH_3}{C}}\oplus & \longrightarrow & \ominus\underset{H}{\overset{H}{C}} - \underset{\underset{CH_3}{\underset{|}{CH_2}}}{\overset{CH_3}{C}}\oplus
\end{array}
\qquad 4.12
$$

$$
(CHR = C(R)_2 \qquad\qquad \text{(I) \underline{Less stable}} \qquad\qquad \begin{array}{c}\text{\underline{(II) More stable}}\\ \underline{(CH_2 = CR_1 R)\ R_1 > R}\end{array}
$$

$$
\begin{array}{ccc}
\underset{C_2H_5}{\overset{H}{C}} = \underset{C_2H_5}{\overset{C_2H_5}{C}} &\longrightarrow& \ominus C - C\oplus &\longrightarrow& \ominus C - C\oplus
\end{array}
\qquad 4.13
$$

$(CHR_1 = C(R_1)_2)$ (I) Less stable   (II) More stable

$\underline{CHR_1 = C(R_1)_2}$ $\underline{(CRH = CR_1R_2)\ R_2 > R_1 > R}$

$$
\begin{array}{ccc}
\underset{C_2H_5}{\overset{H}{C}} = \underset{C_2H_5}{\overset{C_2H_5}{C}} &\longrightarrow& \ominus C - C\oplus &\longrightarrow& \ominus C - C\oplus
\end{array}
\qquad 4.14
$$

$(CHR_2 = C(R_2)_2)$ (I) Less stable   (II) More stable

$\underline{CHR_2 = C(R_2)_2}$ $\underline{CHR_1 = CR_3R_2}$

Molecular rearrangement is favored by this group, with only the first member being the one likely to favor least-sterically hindered polymerization, in view of the reduction in the number of substituent groups from three on two carbon centers to two on one carbon center.

Finally for $C(R)_2 = C(R)_2$ groups, the followings are obtained.

$$
\begin{array}{ccc}
\underset{CH_3}{\overset{CH_3}{C}} = \underset{CH_3}{\overset{CH_3}{C}} &\longrightarrow& \ominus C - C\oplus &\longrightarrow& \ominus C - C\oplus
\end{array}
\qquad 4.15
$$

(I) Less stable $\underline{CH_2 = CRR_1}$

Molecular rearrangement is favored to produce a less sterically hindered monomer.

$$
\begin{array}{ccc}
\ominus \underset{C_2H_5}{\overset{C_2H_5}{C}} = \underset{CH_2}{\overset{C_2H_5}{C}} &\longrightarrow& \ominus C - C\oplus &\longrightarrow& \ominus C - C\oplus
\end{array}
\qquad 4.16
$$

(I) Less stable   (II) More stable

$\underline{CHR_1 = CRR_2}$

Unlike the last group, the number of substituent group is reduced from four to three for primary and secondary alkanyls, but not tertiary alkanyls. All the cases considered so far, are similar in behavior free-radically. All the monomers being NUCLEOPHILES, will favor their Natural routes which are cationic and electro-free-radical routes.

4.2 Characteristics of CH2 = CR1R2, CHR1 = CHR2R3, CHR1 = CHR2, CR1R2 = CR3R4 groups

Beginning with 2 - pentene, the following is obtained.C

$$
\begin{array}{ccc}
\underset{\substack{|\\ H}}{\overset{\substack{H\\ |}}{C}} = \underset{\substack{|\\ CH_2 \\ | \\ CH_3}}{\overset{\substack{CH_3\\ |}}{C}}
& \longrightarrow &
\underset{\substack{|\\ H}}{\overset{\substack{H\\ |}}{\ominus C}} - \underset{\substack{|\\ CH_2^- \\ | \\ CH_3}}{\overset{\substack{CH_3\\ |}}{C\oplus}}
& \not\longrightarrow &
\underset{\substack{|\\ CH_3}}{\overset{\substack{H\\ |}}{\ominus C}} - \underset{\substack{|\\ CH_3}}{\overset{\substack{CH_3\\ |}}{C\oplus}}
& \quad 4.17
\end{array}
$$

(I) More Stable   (II) Less Stable

This is the reverse of the reaction of Equation 4.12. Molecular rearrangement is not favored. The same as applies to $CH_2 = C(R)_2$ applies to $CH_2 = CR_1R_2$.

With $CHR_1 = CHR_2$ members, the followings apply.

$$
\begin{array}{ccccc}
\underset{\substack{|\\ CH_3}}{\overset{\substack{H\\ |}}{C}} = \underset{\substack{|\\ H}}{\overset{\substack{C_2H_5\\ |}}{C}}
& \longrightarrow &
\underset{\substack{|\\ CH_3}}{\overset{\substack{H\\ |}}{\ominus C}} - \underset{\substack{|\\ H}}{\overset{\substack{C_2H_5\\ |}}{C\oplus}}
& \longrightarrow &
\underset{\substack{|\\ CH_3}}{\overset{\substack{H\\ |}}{\ominus C}} - \underset{\substack{|\\ H}}{\overset{\substack{C_2H_5\\ |}}{C\oplus}}
& \quad 4.18
\end{array}
$$

$$
\begin{array}{ccccc}
\underset{\substack{|\\ CH_3}}{\overset{\substack{H\\ |}}{C}} = \underset{\substack{|\\ H}}{\overset{\substack{CH_2\\ |\\ (C_2H_5)}}{C}}
& \longrightarrow &
\underset{\substack{|\\ CH_3}}{\overset{\substack{H\\ |}}{\ominus C}} - \underset{\substack{|\\ H}}{\overset{\substack{CH_2\\ |\\ (C_2H_5)}}{C\oplus}}
& \not\longrightarrow &
\underset{\substack{|\\ C_2H_5}}{\overset{\substack{H\\ |}}{\ominus C}} - \underset{\substack{|\\ H}}{\overset{\substack{C_2H_5\\ |}}{C\oplus}}
& \quad 4.19
\end{array}
$$

(I) More stable   (II) Less stable

$$
\begin{array}{cccc}
\underset{\substack{|\\ CH_3}}{\overset{\substack{H\\ |}}{C}} = \underset{\substack{|\\ CH_2\\ |\\ C_3H_7}}{\overset{\substack{H\\ |}}{C}}
& \longrightarrow &
\underset{\substack{|\\ CH_3}}{\overset{\substack{H\\ |}}{\ominus C}} - \underset{\substack{|\\ CH_2^-\\ |\\ C_3H_7}}{\overset{\substack{H\\ |}}{C\oplus}}
& \not\longrightarrow &
\underset{\substack{|\\ C_3H_7 \\ CH_2 \\ | \\ CH_3}}{\overset{\substack{H\\ |}}{\ominus C}} - \underset{}{\overset{\substack{H\\ |}}{C\oplus}}
& \left(\underset{\substack{|\\ C_3H_7}}{\overset{\substack{H\\ |}}{\oplus C}} - \underset{\substack{|\\ CH_2 \\ | \\ CH_3}}{\overset{\substack{H\\ |}}{C\ominus}} \right)
& 4.20
\end{array}
$$

(I) More stable   (II) (Non-existent) (III)

In the first reaction, molecular rearrangement is favored to produce same monomer. In the second reaction, there is no transfer of transfer species. The reaction is the reverse of the reaction of Equation 4.10. With the third reaction, existence of (II) is impossible, since the activated state of the monomer is (III).

For $CHR_1 = CR_2R_3$ group of monomers, the followings are to be expected.

$$
\begin{array}{ccccc}
\underset{\substack{|\\ CH_3}}{\overset{\substack{H\\ |}}{C}} = \underset{\substack{|\\ CH_3}}{\overset{\substack{C_2H_5\\ |}}{C}}
& \longrightarrow &
\underset{\substack{|\\ CH_3}}{\overset{\substack{H\\ |}}{\ominus C}} - \underset{\substack{|\\ CH_3}}{\overset{\substack{C_2H_5\\ |}}{C\oplus}}
& \longrightarrow &
\underset{\substack{|\\ CH_3}}{\overset{\substack{H\\ |}}{\ominus C}} - \underset{\substack{|\\ CH_3}}{\overset{\substack{C_2H_5\\ |}}{C\oplus}}
& \quad 4.21
\end{array}
$$

$$
\begin{array}{ccc}
\underset{\underset{CH_3}{|}}{\overset{\overset{H}{|}}{C}} = \underset{\underset{C_2H_5}{|}}{\overset{\overset{C_3H_7}{|}}{C}}
& \longrightarrow &
\underset{\underset{CH_3}{|}}{\overset{\overset{H}{|}}{{}^{\ominus}C}} - \underset{\underset{C_2H_5}{|}}{\overset{\overset{C_3H_7}{|}}{C^{\oplus}}}
& \not\longrightarrow &
\underset{\underset{C_2H_5}{|}}{\overset{\overset{H}{|}}{{}^{\ominus}C}} - \underset{\underset{C_2H_5}{|}}{\overset{\overset{C_2H_5}{|}}{C^{\oplus}}}
\end{array}
$$

(More Stable)   (Less Stable)   4.22

In the first reaction, transfer of transfer species is favored to produce the same monomer. The second reaction is the reverse of the reaction of Equation 4.13. Thus, in general for this group of monomers ($R_2 + R_3 > R_1$), molecular rearrangement is not favored. So far, it is favored only for their first members.

For the last group of monomers- $CR_1R_2 = CR_3R_4$, the followings are also expected.

$$\text{(structure)} \qquad 4.23$$

More stable

$$\text{(structure)} \qquad 4.24$$

More stable

Molecular rearrangement is favored by them to produce more stable monomers. Though some of the monomers will not favor polymerization after molecular rearrangement, due to steric limita-tions, the unique characteristics of the monomers are worthy of note.

4.3 Characteristics of CH2 = CHRF, CH2 = CRF RF, CHRF = CHRF, CHRF = CRFRF groups

The R_{Fs} are fluoro-alkanyls - CF_3, CH_2F, CF_2H,...., C_2F_5, etc., and others such as CH_2COOR, CH_2OR, CH_2CN etc. These are unique substituted groups, since chargedly they are charged-pulling groups and their existences on monomers are not popularly known, but free-radically they are radical-pushing groups. Beginning with the first group $CH_2 = CHR_F$, the followings are obtained chargedly, when they can be activated.

$$
\begin{array}{ccc}
\underset{\underset{H}{|}}{\overset{\overset{H}{|}}{C}} = \underset{\underset{CF_3}{|}}{\overset{\overset{H}{|}}{C}}
& \longrightarrow &
\oplus\underset{\underset{H}{|}}{\overset{\overset{H}{|}}{C}} - \underset{\underset{CF_3}{|}}{\overset{\overset{}{}}{C}}\ominus
\end{array}
\quad\cancel{\longrightarrow}\quad
\oplus\underset{\underset{F}{|}}{\overset{\overset{F}{|}}{C}} - \underset{\underset{CH_2F}{|}}{\overset{}{C}}\ominus
\qquad 4.25
$$

$$\text{(I) \underline{More stable}} \qquad\qquad \text{(II) Less \underline{stable} (Not the favored state)}$$

Since CF_3 is more chargedly-pulling than CH_2F, (I) is far more stable than (II). Molecular rearrangement is not favored.

$$
\begin{array}{ccc}
\underset{\underset{H}{|}}{\overset{\overset{H}{|}}{C}} = \underset{\underset{CF_2H}{|}}{\overset{\overset{H}{|}}{C}}
& \longrightarrow &
\oplus\underset{\underset{H}{|}}{\overset{\overset{H}{|}}{C}} - \underset{\underset{CF_2H}{|}}{\overset{}{C}}\ominus
\end{array}
\quad\cancel{\longrightarrow}\quad
\oplus\underset{\underset{F}{|}}{\overset{\overset{H}{|}}{C}} - \underset{\underset{CH_2F}{|}}{\overset{\overset{H}{|}}{C}}\ominus
\qquad 4.26
$$

$$\text{(I) \underline{More stable}} \qquad\qquad \text{(II) Less \underline{stable}}$$

Since CF_2H is more chargedly-pulling than CH_2F, (I) is also far more stable than (II). Indeed (II) of Equation 4.25 may not favor the activated state shown in view of the resultant force on the π - radicals ($CH_2F < 2F + H$). CH_2F is more chargedly-pulling than F. The difference however is small.

$$
\begin{array}{ccc}
\underset{\underset{H}{|}}{\overset{\overset{H}{|}}{C}} = \underset{\underset{CH_2F}{|}}{\overset{\overset{H}{|}}{C}}
& \longrightarrow &
\oplus\underset{\underset{H}{|}}{\overset{\overset{H}{|}}{C}} - \underset{\underset{CH_2F}{|}}{\overset{\overset{H}{|}}{C}}\ominus
\end{array}
\quad\cancel{\longrightarrow}\quad
\oplus\underset{\underset{H}{|}}{\overset{\overset{H}{|}}{C}} - \underset{\underset{CH_2F}{|}}{\overset{\overset{H}{|}}{C}}\ominus
\qquad 4.27
$$

Transfer of transfer species is not favored, with or without molecular rearrangement. Thus, chargedly none of the members here favor molecular rearrangement to produce another monomer. This clearly shows that chargedly, non-free-negative charges cannot be transferred for these Nucleophiles. *Based on the Laws of conservation of transfer of transfer species, either the monomers cannot be activated chargedly or both routes will be favored. How can one have a Nucleophile whose natural route is positively charged route, now be made to look like an Electrophile, unable to favor the route natural to it chargedly? Radically, the situation is different, such as the case of vinyl acetate which cannot be activated chargedly.*

Free-radically the followings are obtained for the members.

$$
\begin{array}{ccc}
\underset{\underset{H}{|}}{\overset{\overset{H}{|}}{C}} = \underset{\underset{CF_3}{|}}{\overset{\overset{H}{|}}{C}}
& \longrightarrow &
e.\underset{\underset{H}{|}}{\overset{\overset{H}{|}}{C}} - \underset{\underset{CF_3}{|}}{\overset{\overset{H}{|}}{C}}.n
\end{array}
\quad\longrightarrow\quad \text{Transfer species (F)}
\qquad 4.28
$$

Electro-free-radically, the monomer a Nucleophile cannot be initiated. The monomer will favor the nucleo-free-radical route with transfer species of the first kind of the first type in the route not natural to it! *For example compare –COOH with CF_3 (i.e. $e\bullet C=O$ with $e\bullet CF_2$). Both can carry positive charges, electro-free- and nucleo-free-radicals very readily; but not negative charges. As it seems the monomer above which is a Nucleophile, behaves like an Electrophile, something which is not to* *be expected. This is almost like vinyl acetate with only transfer species of the second kind and with initiation not favored elcctro-free-radically.*

$$
\begin{array}{ccc}
\underset{\underset{H}{|}}{\overset{\overset{H}{|}}{C}} = \underset{\underset{CF_2H}{|}}{\overset{\overset{H}{|}}{C}} & \longrightarrow & n.\underset{\underset{H}{|}}{\overset{\overset{H}{|}}{C}} - \underset{\underset{CF_2H}{|}}{\overset{\overset{H}{|}}{C}}.e & \longrightarrow & n.\underset{\underset{F}{|}}{\overset{\overset{F}{|}}{C}} - \underset{\underset{CH_3}{|}}{\overset{\overset{H}{|}}{C}}.e \\
 & & (I) & & (II) \text{ More stable}
\end{array}
$$

$$4.29$$

$$
\begin{array}{ccc}
\underset{\underset{H}{|}}{\overset{\overset{H}{|}}{C}} = \underset{\underset{CH_2F}{|}}{\overset{\overset{H}{|}}{C}} & \longrightarrow & n.\underset{\underset{H}{|}}{\overset{\overset{H}{|}}{C}} - \underset{\underset{CH_2F}{|}}{\overset{\overset{H}{|}}{C}}.e & \longrightarrow & n.\underset{\underset{H}{|}}{\overset{\overset{F}{|}}{C}} - \underset{\underset{CH_3}{|}}{\overset{\overset{H}{|}}{C}}.e \\
 & & (I) \text{ Less stable} & & (II) \text{ More stable}
\end{array}
$$

$$4.30$$

Since $CH_3 > CH_2F > CHF_2 > H > CF_3$ in free-radical-pushing capacity, molecular rearrange-ment of the monomers are favored when there exists transfer species of only H. Nothing could be used in CF_3. This clearly indicates that the boundary between the Nucleophilic and Electrophilic characters displayed by these monomers is the monomer carrying CF_3 group. CHF_2 and CH_2F groups make them complete Nucleophiles. One of them (CH_2F group) is allylic in character.

Now, consider replacing the CF_3 groups with higher groups. Free-radically, the follow-ings are obtained.

$$
\begin{array}{cccc}
\underset{\underset{\underset{CF_3}{|}}{\overset{}{CF_2}}}{\overset{\overset{H}{|}}{\underset{|}{C}}} = \underset{}{\overset{\overset{H}{|}}{C}} & \longrightarrow & e.\underset{\underset{\underset{CF_3}{|}}{\overset{}{CF_2}}}{\overset{\overset{H}{|}}{\underset{|}{C}}} - \underset{\underset{H}{|}}{\overset{\overset{H}{|}}{C}}.n & \not\longrightarrow & e.\underset{\underset{F}{|}}{\overset{\overset{F}{|}}{C}} - \underset{\underset{\underset{CF_3}{|}}{\overset{}{CH_2}}}{\overset{\overset{H}{|}}{\underset{|}{C}}}.n \\
A & & (I) \text{ Stable} & & (II) \text{ (Not Possible)}
\end{array}
$$

$$4.31$$

$$
\begin{array}{cccc}
\underset{\underset{\underset{CF_3}{|}}{\overset{}{CHF}}}{\overset{\overset{H}{|}}{\underset{|}{C}}} = \underset{}{\overset{\overset{H}{|}}{C}} & \longrightarrow & n.\underset{\underset{\underset{CF_3}{|}}{\overset{}{CHF}}}{\overset{\overset{H}{|}}{\underset{|}{C}}} - \underset{\underset{H}{|}}{\overset{\overset{H}{|}}{C}}.e & \overset{H}{\underset{Transfer}{\not\longrightarrow}} & n.\underset{\underset{CF_3}{|}}{\overset{\overset{F}{|}}{C}} - \underset{\underset{CH_3}{|}}{\overset{\overset{H}{|}}{C}}.e \\
(B) & & & &
\end{array}
$$

$$\downarrow \quad CF_3 \text{ transfer}$$

$$4.32$$

$$
n.\underset{\underset{\underset{CF_3}{|}}{\overset{}{CH_2}}}{\overset{\overset{H}{|}}{\underset{|}{C}}} - \underset{\underset{}{}}{\overset{\overset{H}{|}}{C}}.e
$$

$$(I) \underline{\text{(Favored)}}$$

When molecular rearrangement is favored, one of the driving forces is that the number of substituent groups is either reduced or remains the same. (A) is stable since there is no transfer species for molecular rearrangement, but favors only nucleo- free-radical route. With (B), rear-rangement is favored with the monomer being a stronger Nucleophile, since only the electro-free-radical route is favored. Since these groups can carry nucleo- and electro-free-radicals, their radical-pushing capacities when present elctro-free-radically on a monomer are as follows:-

$$CH_3 > CH_2F > CHFCF_3 > CHF_2 > H > CF(CF_3)_2 > CF_2CF_3 > CF_3 \qquad 4.33$$

<u>Order of Free-radical-pushing capacity</u>

$$4.34$$

CF_3 is the transfer species, since it is the least transfer species in radical-pushing capacity on the substituent group. However, the same monomer is obtained, so that molecular rearrangement being favored or not favored will depend on the type and strength of initiator used.

Consider replacing the CF_3 groups with larger groups. The followings will prevail for the monomers.

$$4.35$$

$$4.36$$

Based on Equation 4.33, there are transfer species above, noting that the following is also valid.

$$CH_2C_2F_5 > CH_2CF_3 > CH_2F > CH(C_2F_5)F > CH(CF_3)F > H \qquad 4.37$$

<u>Order of free-radical-pushing capacity</u>

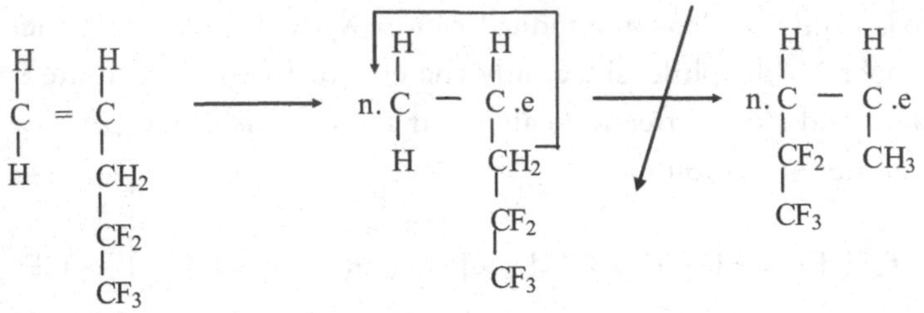

$$
\begin{array}{c}
\text{H}\quad\text{H} \\
\text{C} = \text{C} \\
\text{H}\quad\text{CH}_2 \\
\text{CF}_2 \\
\text{CF}_3
\end{array}
\longrightarrow
\left[
\begin{array}{c}
\text{H}\quad\text{H} \\
\text{n.C} - \text{C .e} \\
\text{H}\quad\text{CH}_2 \\
\text{CF}_2 \\
\text{CF}_3
\end{array}
\right]
\nrightarrow
\begin{array}{c}
\text{H}\quad\text{H} \\
\text{n.C} - \text{C .e} \\
\text{CF}_2\quad\text{CH}_3 \\
\text{CF}_3
\end{array}
\qquad 4.38
$$

<u>Not Possible</u>

If H is the transfer species, molecular rearrangement is not favored. With C_2F_5 as transfer species, the same monomer is obtained with favored molecular rearrangement based on Equation 4.37.

Chargedly, the followings are obtained for these groups.

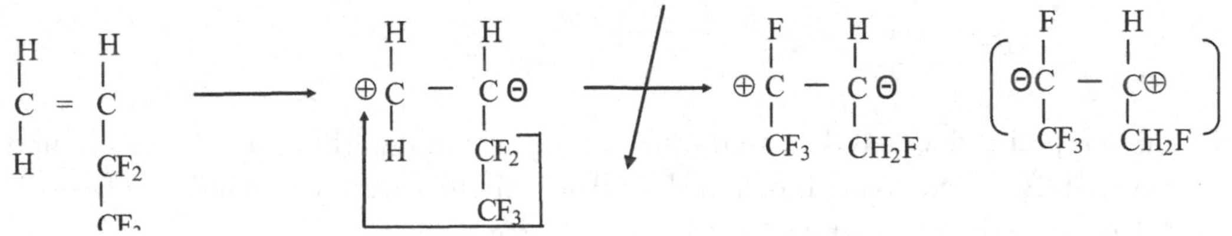

$$ 4.39 $$

(I) (II) (III)

(Impossible existence) (Impossible existence)

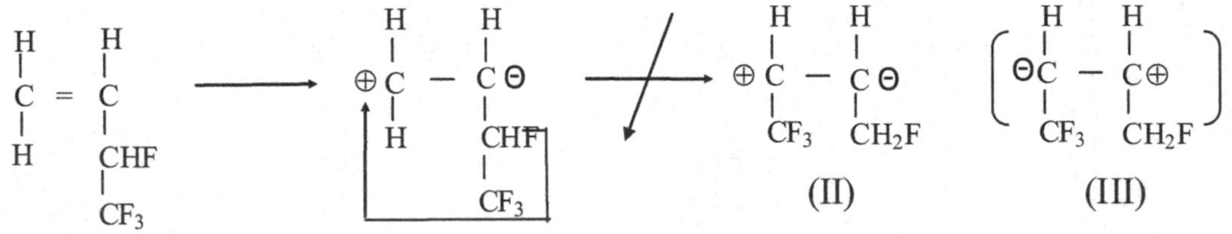

$$ 4.40 $$

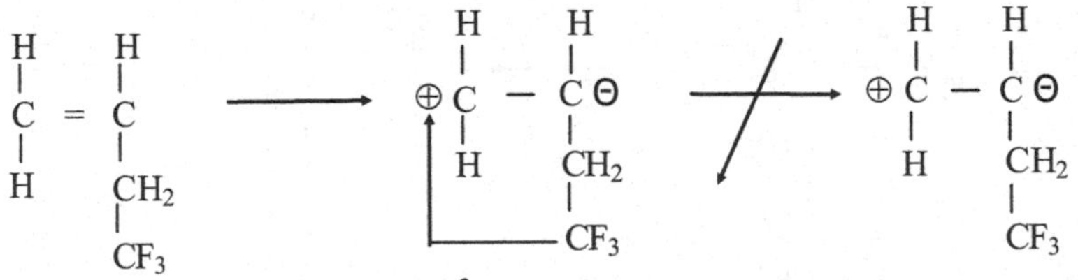

(II) (III)

$$ 4.41 $$

($^{\ominus}CF_3$ cannot exist)

In the first two equations, molecular rearrangement is not favored. The same will apply free-radically for the first case, but not for the last two cases. For the reaction of Equation 4.39, existence of (II) and (III) are impossible to the point when it is believed that such cases cannot be chargedly activated. (III) the true activated state of (II), cannot exist due to electrostatic forces of repulsion. (III) of Equation 4.40 is the true activated state of (II). For the last equation, transfer of transfer species is not favored, CF_3 cannot carry a negative charge. Whether the CF_3 group above is replaced with higher groups or not, molecular rearrangement is not favored by these monomers chargedly.

For the second group of monomers $CH_2 = C(R_F)_2$, consider the followings chargedly, with the hope that activation is possible.

(I) More stable   (II) Less stable   4.42

(I) More stable   (III) Less stable   4.43

Molecular rearrangement is not favored since F cannot be a transfer species and $CF_3 > CH_2F$ in charged-pulling capacity. For the other groups, the followings are obvious.

(I) More stable   (II) Less stable   4.44

(I) More stable   (II) Less stable   4.45

4.46

* CF_2 (and not H) transferred (Impossible)

$$
\begin{array}{ccc}
& & CF_3 \\
& & | \\
& & CF_2 \\
& & | \\
H & & CH_2 \\
| & & | \\
C & = & C \\
| & & | \\
H & & CH_2 \\
& & | \\
& & CF_2 \\
& & | \\
& & CF_3
\end{array}
\longrightarrow
\begin{array}{ccc}
& & CF_3 \\
& & | \\
& & CF_2 \\
& & | \\
H & & CH_2 \\
| & & | \\
\oplus C & - & C \ominus \\
| & & | \\
H & & CH_2 \\
& & \Big\{ \begin{array}{c} CF_3 \\ | \\ CF_3 \end{array}
\end{array}
\;\not\longrightarrow\;
\begin{array}{ccc}
& & CF_3 \\
& & | \\
& & CF_2 \\
& & | \\
H & & CH_2 \\
| & & | \\
\oplus C & - & C \ominus \\
| & & | \\
H & & CH_2 \\
& & | \\
& & CF_2 \\
& & | \\
& & CF_3
\end{array}
\qquad 4.47
$$

<u>* C_2F_5 (and not H) transferred (Impossible)</u>

In the first reaction above, molecular rearrangement is not favored, since $CF_2H > CH_2F$ in charge-pulling capacity. While transfer of transfer species is not allowed for $CH_2 = C(CHF_2)_2$ groups, it is also not favored for $CH_2 = C(CH_2R_F)_2$ groups. ***These are Nucleophiles wherein only charged- or radical-pushing groups can be transferred.***

Free-radically, the followings are to be expected.

$$
\begin{array}{ccc}
H & & CF_3 \\
| & & | \\
C & = & C \\
| & & | \\
H & & CF_3
\end{array}
\longrightarrow
\begin{array}{ccc}
H & & CF_3 \\
| & & | \\
e.\,C & - & C\,.n \\
| & & | \\
H & & CF_3
\end{array}
\longrightarrow
\text{No species to transfer}
\qquad 4.48
$$

$$
\begin{array}{ccc}
& & CF_3 \\
& & | \\
H & & CF_2 \\
| & & | \\
C & = & C \\
| & & | \\
H & & CF_2 \\
& & | \\
& & CF_3
\end{array}
\longrightarrow
\begin{array}{ccc}
& & CF_3 \\
& & | \\
H & & CF_2 \\
| & & | \\
e\,C & - & C\,.n \\
| & & | \\
H & & CF_2 \\
& & | \\
& & CF_3
\end{array}
\;\not\longrightarrow\;
\begin{array}{ccc}
F & & C_2F_5 \\
| & & | \\
e.\,C & - & C\,.n \\
| & & | \\
F & & CH_2 \\
& & | \\
& & CF_3
\end{array}
\qquad 4.49
$$

(Impossible State)

Only the nucleo-free-radical route will be favored by them, the route not natural to them.

$$
\begin{array}{ccc}
H & & CF_2H \\
| & & | \\
C & = & C \\
| & & | \\
H & & CF_2H
\end{array}
\longrightarrow
\begin{array}{ccc}
H & & CF_2H \\
| & & | \\
n.\,C & - & C\,.e \\
| & & | \\
H & & CF_2H
\end{array}
\longrightarrow
\begin{array}{ccc}
F & & CF_2H \\
| & & | \\
n.\,C & - & C\,.e \\
| & & | \\
F & & CH_3
\end{array}
\qquad 4.50
$$

<u>More stable</u>

$$
\begin{array}{ccc}
H & & CH_2F \\
| & & | \\
C & = & C \\
| & & | \\
H & & CH_2F
\end{array}
\longrightarrow
\begin{array}{ccc}
H & & CH_2F \\
| & & | \\
n.\,C & - & C\,.e \\
| & & | \\
H & & CH_2F
\end{array}
\longrightarrow
\begin{array}{ccc}
F & & CH_2F \\
| & & | \\
n.\,C & - & C\,.e \\
| & & | \\
H & & CH_3
\end{array}
\qquad 4.51
$$

<u>More stable</u>

In general, in the presence of transfer species, molecular rearrangement is favored for these families of monomers and when favored only the electro-free-radical route is favored.

For the next member of monomers ($CHR_F = CHR_F$) chargedly, the followings are obtained.

$$ \qquad 4.52 $$

(I) <u>More stable</u>   (II) <u>less stable</u> (Impossible existence)

$$ \qquad 4.53 $$

(I) <u>More stable</u>   (II) Less <u>stable</u> (Impossible existence)

Since these monomers are nucleophiles, F cannot be used as transfer species for rearrangement. Hence, molecular rearrangement is not favored for this member of monomers. ***Either both charged routes are favored or not by the monomers or they cannot be activated chargedly.***

$$ \qquad 4.54 $$

(III) <u>(Impossible transfer species)</u>

Existence of (II) is impossible, due to electrostatic forces of repulsion. F as well as H cannot be transfer species. Either both routes are favored or the monomer cannot be activated chargedly.

$$
\underset{\text{(I)}}{
\begin{array}{c}
H \\ | \\ C \\ | \\ CFH_2
\end{array}
=
\begin{array}{c}
CFH_2 \\ | \\ C \\ | \\ H
\end{array}
}
\;\longrightarrow\;
\underset{\text{(I)}}{
\Theta C
\begin{array}{c}
H \\ | \\ \\ | \\ CFH_2
\end{array}
-\;
C \oplus
\begin{array}{c}
CFH_2 \\ | \\ \\ | \\ H
\end{array}
}
\;\xrightarrow[\text{Transfer}]{H}\;
\Theta C
\begin{array}{c}
F \\ | \\ \\ | \\ H
\end{array}
-\;
C \oplus
\begin{array}{c}
H \\ | \\ \\ | \\ CH_2 \\ | \\ CFH_2
\end{array}
$$

(II) (Impossible existence)

F transfer

$$
\underset{\text{(I)}}{
\begin{array}{c}
H \\ | \\ C \\ | \\ CFH_2
\end{array}
=
\begin{array}{c}
CFH_2 \\ | \\ C \\ | \\ H
\end{array}
}
\;\longrightarrow\;
\Theta C
\begin{array}{c}
H \\ | \\ \\ | \\ CFH_2
\end{array}
-\;
C \oplus
\begin{array}{c}
CFH_2 \\ | \\ \\ | \\ H
\end{array}
\;\xrightarrow[\text{Transfer}]{H}\;
\Theta C
\begin{array}{c}
F \\ | \\ \\ | \\ H
\end{array}
-\;
C \oplus
\begin{array}{c}
H \\ | \\ \\ | \\ CH_2 \\ | \\ CFH_2
\end{array}
$$

(II) (Impossible existence)

F transfer

$$
\oplus C
\begin{array}{c}
H \\ | \\ \\ | \\ H
\end{array}
=
C \Theta
\begin{array}{c}
H \\ | \\ \\ | \\ CHF \\ | \\ CFH_2
\end{array}
$$

(III) <u>(Impossible transfer species)</u>   4.55

Like the case above, molecular rearrangement cannot take place. Either both routes are favored or not favored by these monomers or they cannot be activated chargedly.

$$
\begin{array}{c}
H \\ | \\ C \\ | \\ CFH \\ | \\ CF_3
\end{array}
=
\begin{array}{c}
CF_3 \\ | \\ CFH \\ | \\ C \\ | \\ H
\end{array}
\;\longrightarrow\;
\Theta C
\begin{array}{c}
H \\ | \\ \\ | \\ CFH \\ | \\ CF_3
\end{array}
-\;
C \oplus
\begin{array}{c}
CF_3 \\ | \\ CFH \\ | \\ \\ | \\ H
\end{array}
\;\nrightarrow\;
\oplus C
\begin{array}{c}
H \\ | \\ \\ | \\ CF_3
\end{array}
-\;
C \Theta
\begin{array}{c}
CF_3 \\ | \\ CFH \\ | \\ CHF \\ | \\ H
\end{array}
$$

(Impossible transfer species)   4.56

$$
\begin{array}{c}
H \\ | \\ C \\ | \\ CH_2 \\ | \\ CF_3
\end{array}
=
\begin{array}{c}
CF_3 \\ | \\ CH_2 \\ | \\ C \\ | \\ H
\end{array}
\;\longrightarrow\;
\Theta C
\begin{array}{c}
H \\ | \\ \\ | \\ CH_2 \\ | \\ CF_3
\end{array}
-\;
C \oplus
\begin{array}{c}
CF_3 \\ | \\ CH_2 \\ | \\ \\ | \\ H
\end{array}
\;\nrightarrow\;
\oplus C
\begin{array}{c}
H \\ | \\ \\ | \\ H
\end{array}
-\;
C \Theta
\begin{array}{c}
H \\ | \\ CH(CF_3) \\ | \\ CH_2 \\ | \\ CF_3
\end{array}
$$

4.57

(Impossible Transfer species)

Like the cases above either both routes are favored by this monomer in the absence of electro-static forces of repulsion when suitable initiators are not used or it cannot be activated chargedly.

Free-radically, the followings are to be expected.

$$
\begin{array}{ccc}
H & & H \\
| & & | \\
C & = & C \\
| & & | \\
CF_3 & & CF_3
\end{array}
\quad\longrightarrow\quad
n.\;
\begin{array}{ccc}
H & & H \\
| & & | \\
C & - & C\,.e \\
| & & | \\
CF_3 & & CF_3
\end{array}
\quad\longrightarrow\quad
\text{No species to transfer for}
\qquad 4.58
$$

molecular rearrangement.

$$
\begin{array}{ccc}
H & & H \\
| & & | \\
C & = & C \\
| & & | \\
CF_2 & & CF_2 \\
| & & | \\
CF_3 & & CF_3
\end{array}
\longrightarrow
n.\;
\begin{array}{ccc}
H & & H \\
| & & | \\
C & - & C\,.e \\
| & & | \\
CF_2 & & CF_2 \\
| & & | \\
CF_3 & & CF_3
\end{array}
\longrightarrow
n.\;
\begin{array}{ccc}
F & & H \\
| & & | \\
C & - & C\,.e \\
| & & | \\
F & & CH(CF_3) \\
& & | \\
& & CF_2 \\
& & | \\
& & CF_3
\end{array}
\qquad 4.59
$$

<u>More stable</u>

$$
\begin{array}{ccc}
H & & CF_2H \\
| & & | \\
C & = & C \\
| & & | \\
CF_2H & & H
\end{array}
\longrightarrow
n.\;
\begin{array}{ccc}
H & & CF_2H \\
| & & | \\
C & - & C\,.e \\
| & & | \\
CF_2H & & H
\end{array}
\longrightarrow
n.\;
\begin{array}{ccc}
F & & H \\
| & & | \\
C & - & C\,.e \\
| & & | \\
F & & CH_2 \\
& & | \\
& & CF_2H
\end{array}
\qquad 4.60
$$

<u>(More stable)</u>

$$
\begin{array}{ccc}
H & & CH_2F \\
| & & | \\
C & = & C \\
| & & | \\
CH_2F & & H
\end{array}
\longrightarrow
n.\;
\begin{array}{ccc}
H & & CH_2F \\
| & & | \\
C & - & C\,.e \\
| & & | \\
CH_2F & & H
\end{array}
\longrightarrow
n.\;
\begin{array}{ccc}
H & & H \\
| & & | \\
C & - & C\,.e \\
| & & | \\
F & & CH_2 \\
& & | \\
& & CH_2F
\end{array}
\qquad 4.61
$$

<u>More stable</u>

In the presence of transfer species, molecular rearrangement is favored by some of these monomers. For higher groups for the two reactions (Equations 4.56 and 57), the corresponding free-radical reactions are as follows.

$$
\begin{array}{ccc}
 & & CF_3 \\
 & & | \\
H & & CFH \\
| & & | \\
C & = & C \\
| & & | \\
CFH & & H \\
| & & \\
CF_3 & &
\end{array}
\longrightarrow
n.\;
\begin{array}{ccc}
 & & CF_3 \\
 & & | \\
H & & CFH \\
| & & | \\
C & - & C\,.e \\
| & & | \\
CFH & & H \\
| & & \\
CF_3 & &
\end{array}
\longrightarrow
n.\;
\begin{array}{ccc}
F & & H \\
| & & | \\
C & - & C\,.e \\
| & & | \\
H & & CH(CF_3) \\
& & | \\
& & CFH \\
& & | \\
& & CF_3
\end{array}
\qquad 4.62
$$

<u>More stable</u>

$$\begin{array}{ccc}
\begin{array}{c} CF_3 \\ | \\ CH_2 \\ | \\ H \quad C \\ | \quad \| \\ C = C \\ | \quad | \\ CH_2 \quad H \\ | \\ CF_3 \end{array}
& \xrightarrow{\hspace{1cm}} &
\begin{array}{c} CF_3 \\ | \\ CH_2 \\ | \\ H \quad | \\ e.\; C - C.n \\ | \quad | \\ CH_2 \quad H \;_{CF_3} \\ | \\ CF_3 \end{array}
& \xrightarrow{\hspace{1cm}} &
\begin{array}{c} H \quad H \\ | \quad | \\ n.\; C - C.e \\ | \quad | \\ H \quad CH(CF_3) \\ | \\ CH_2 \\ | \\ CF_3 \end{array}
\end{array}$$

$$\underline{\text{More stable}} \hspace{4cm} 4.63$$

Molecular rearrangement is in general favored by the monomer in this group free-radically.

For the last group of monomers in this section, the followings are obtained chargedly, if activation is possible.

$$\begin{array}{ccc}
\begin{array}{c} H \quad CF_3 \\ | \quad | \\ C = C \\ | \quad | \\ CF_3 \quad CF_3 \end{array}
& \xrightarrow{\hspace{1cm}} &
\begin{array}{c} H \quad CF_3 \\ | \quad | \\ \oplus C - C \ominus \\ | \quad | \\ CF_3 \quad CF_3 \end{array}
& \xslashedrightarrow{\hspace{1cm}} &
\begin{array}{c} F \quad CF_3 \\ | \quad | \\ \oplus C - C \ominus \\ | \quad | \\ F \quad CHF \\ | \\ CF_3 \end{array}
\end{array}$$

$$\underline{\text{More Stable}} \hspace{2cm} \underline{\text{Less stable}} \hspace{2cm} 4.64$$

$$\begin{array}{ccc}
\begin{array}{c} H \quad CF_2H \\ | \quad | \\ C = C \\ | \quad | \\ CF_2H \quad CF_2H \end{array}
& \xrightarrow{\hspace{1cm}} &
\begin{array}{c} H \quad CF_2H \\ | \quad | \\ \oplus C - C \ominus \\ | \quad | \\ CF_2H \quad CF_2H \end{array}
& \xslashedrightarrow{\hspace{1cm}} &
\begin{array}{c} H \quad CF_2H \\ | \quad | \\ \oplus C - C \ominus \\ | \quad | \\ F \quad CHF \\ | \\ CF_2H \end{array}
\end{array}$$

$$\underline{\text{More Stable}} \hspace{2cm} \underline{\text{Less stable}} \hspace{2cm} 4.65$$

$$\begin{array}{ccc}
\begin{array}{c} H \quad CH_2F \\ | \quad | \\ C = C \\ | \quad | \\ CH_2F \quad CH_2F \end{array}
& \xrightarrow{\hspace{1cm}} &
\begin{array}{c} H \quad CH_2F \\ | \quad | \\ \oplus C - C \ominus \\ | \quad | \\ CH_2F \quad CH_2F \end{array}
& \xslashedrightarrow{\hspace{1cm}} &
\begin{array}{c} H \quad CH_2F \\ | \quad | \\ \oplus C - C \ominus \\ | \quad | \\ H \quad CHF \\ | \\ CH_2F \end{array}
\end{array}$$

$$\underline{\text{More Stable}} \hspace{2cm} \underline{\text{Less stable}} \hspace{2cm} 4.66$$

Chargedly, molecular rearrangement is not favored, since F cannot be used here as transfer species. Here, either both routes are favored or not or the monomers cannot be activated charged-ly. Free-radically, the followings are obtained.

$$\begin{array}{ccc}
\begin{array}{c} H \quad CF_3 \\ | \quad | \\ C = C \\ | \quad | \\ CF_3 \quad CF_3 \end{array}
& \xrightarrow{\hspace{1cm}} &
\begin{array}{c} H \quad CF_3 \\ | \quad | \\ e.\; C - C.n \\ | \quad | \\ CF_3 \quad CF_3 \end{array}
& \xrightarrow{\hspace{1cm}} & \text{No transfer species}
\end{array} \hspace{1cm} 4.67$$

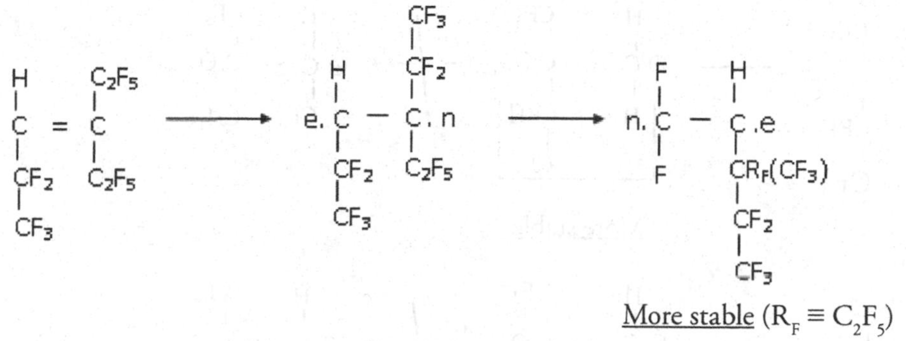

$$
\begin{array}{cc}
H & C_2F_5 \\
| & | \\
C & = C \\
| & | \\
CF_2 & C_2F_5 \\
| & \\
CF_3 &
\end{array}
\longrightarrow
\begin{array}{cc}
 & CF_3 \\
 & | \\
H & CF_2 \\
| & | \\
e.\,C & - C.\,n \\
| & | \\
CF_2 & C_2F_5 \\
| & \\
CF_3 &
\end{array}
\longrightarrow
\begin{array}{cc}
F & H \\
| & | \\
n.\,C & - C.\,e \\
| & | \\
F & CR_F(CF_3) \\
 & | \\
 & CF_2 \\
 & | \\
 & CF_3
\end{array}
\qquad 4.68
$$

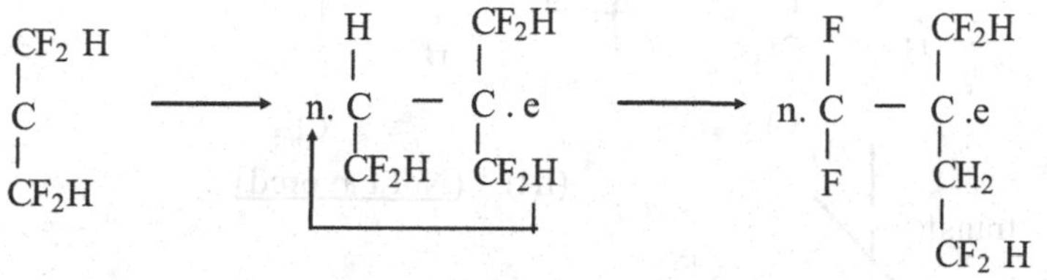

More stable ($R_F \equiv C_2F_5$)

$$
\begin{array}{c}
CF_2H \\
| \\
C \\
| \\
CF_2H
\end{array}
\longrightarrow
\begin{array}{cc}
H & CF_2H \\
| & | \\
n.\,C & - C.\,e \\
| & | \\
CF_2H & CF_2H
\end{array}
\longrightarrow
\begin{array}{cc}
F & CF_2H \\
| & | \\
n.\,C & - C.\,e \\
| & | \\
F & CH_2 \\
 & | \\
 & CF_2H
\end{array}
\qquad 4.69a
$$

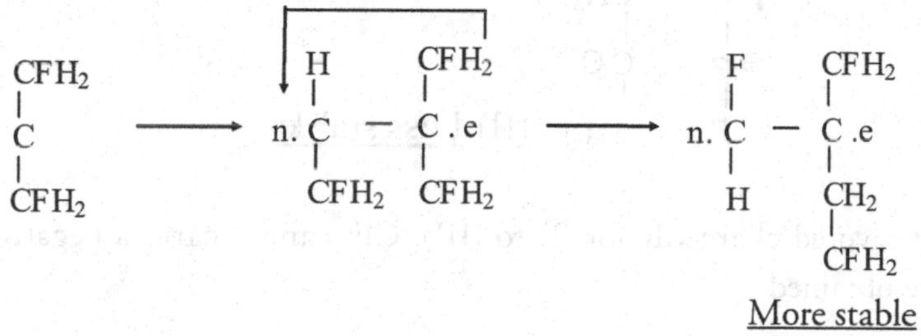

$$
\begin{array}{c}
CFH_2 \\
| \\
C \\
| \\
CFH_2
\end{array}
\longrightarrow
\begin{array}{cc}
H & CFH_2 \\
| & | \\
n.\,C & - C.\,e \\
| & | \\
CFH_2 & CFH_2
\end{array}
\longrightarrow
\begin{array}{cc}
F & CFH_2 \\
| & | \\
n.\,C & - C.\,e \\
| & | \\
H & CH_2 \\
 & | \\
 & CFH_2
\end{array}
$$

More stable

Molecular rearrangement is favored for all but the first case which is the only one that will favor the route not natural to it. So far, it can be observed how the radical/charged-pushing and pulling capacities of substituent groups and transfer species are being slowly identified using molecular rearrangement phenomena.

4.4 Characteristics of CH2 = CRF1RF 2, CHRF1 = CHRF2, CHRF1 = CRF2RF3 groups

The R_Fs are still halogenated alkanyls or similar type of groups, but of different capacities.

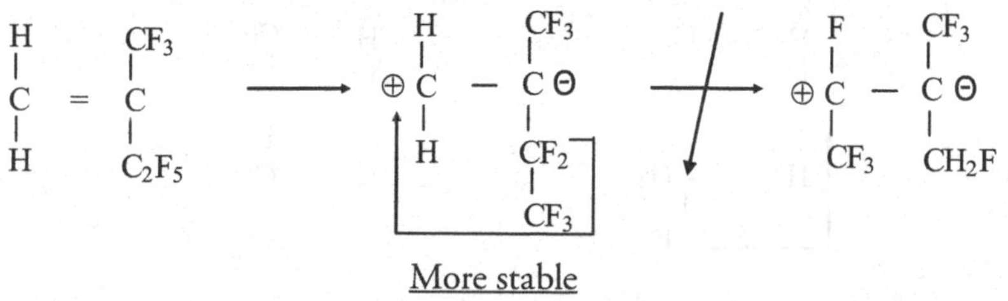

$$
\begin{array}{cc}
H & CF_3 \\
| & | \\
C & = C \\
| & | \\
H & C_2F_5
\end{array}
\longrightarrow
\begin{array}{cc}
H & CF_3 \\
| & | \\
\oplus C & - C \ominus \\
| & | \\
H & CF_2 \\
 & | \\
 & CF_3
\end{array}
\nrightarrow
\begin{array}{cc}
F & CF_3 \\
| & | \\
\oplus C & - C \ominus \\
| & | \\
CF_3 & CH_2F
\end{array}
\qquad 4.70
$$

More stable

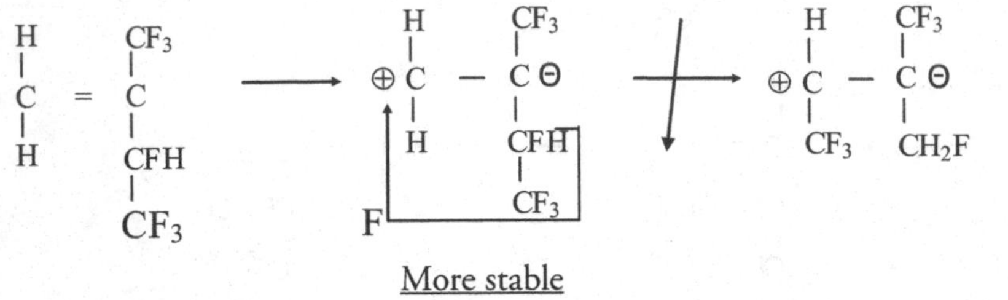

$$
\begin{array}{ccc}
\overset{\text{H}}{\underset{\text{H}}{\text{C}}} & = & \overset{\text{CF}_3}{\underset{\overset{|}{\underset{\text{CF}_3}{\text{CFH}}}}{\text{C}}}
\end{array}
\longrightarrow
\oplus\text{C} - \text{C}\ominus
\nrightarrow
\oplus\text{C} - \text{C}\ominus
$$

$$\text{(4.71)}$$

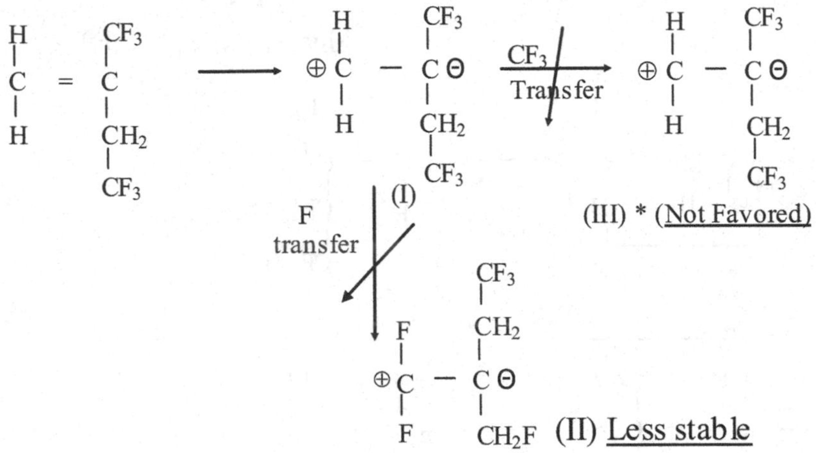

More stable

$$\text{(4.72)}$$

(I)

F
transfer

(III) * (Not Favored)

(II) Less stable

Molecular rearrangement is not favored chargedly for (I) to (III). CF_3 cannot carry a negative charge. Free-radically, the followings are obtained.

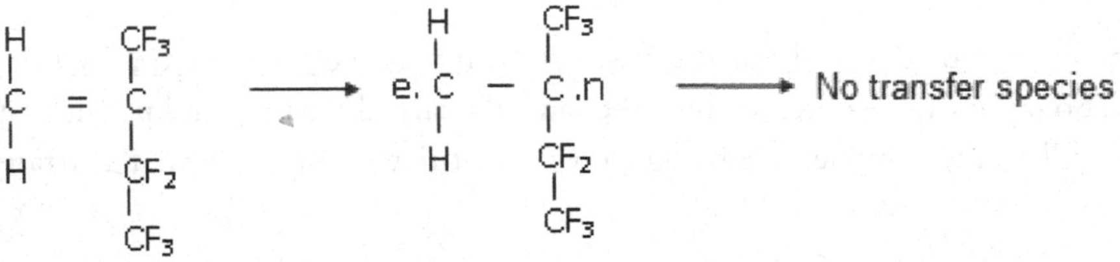

$$
\begin{array}{ccc}
\overset{\text{H}}{\underset{\text{H}}{\text{C}}} & = & \overset{\text{CF}_3}{\underset{\overset{|}{\underset{\text{CF}_3}{\text{CF}_2}}}{\text{C}}}
\end{array}
\longrightarrow
\text{e.C} - \text{C.n}
\longrightarrow
\text{No transfer species}
$$

$$\text{(4.73)}$$

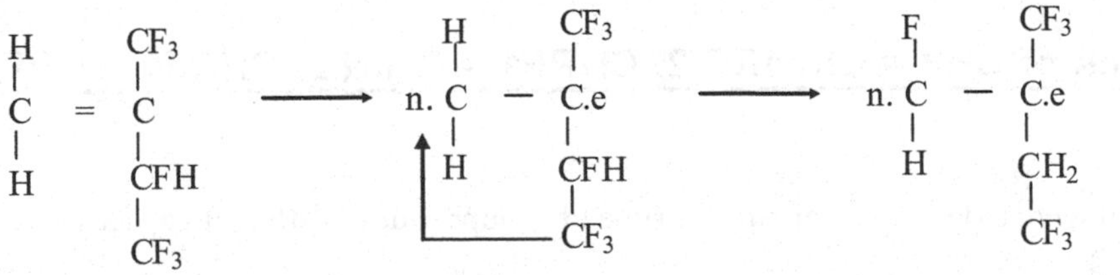

$$\text{(4.74)}$$

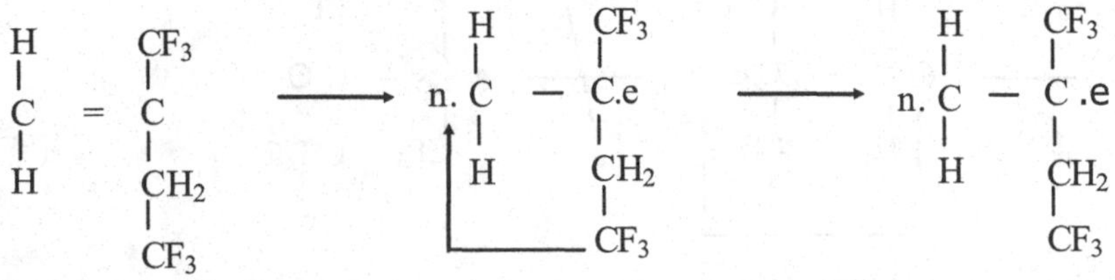

$$\text{(4.75)}$$

For the first reaction above, no transfer species exist for it. Therefore, only the nucleo-free-radical route is favored by it. For the last two reactions, there is molecular rearrangement to give a different more stable monomer for the second and the same monomer for the last. When the CF_3 is replaced with higher groups such as C_2F_5, the same applies.

For the second group of monomers $CHR_{F1} = CHR_{F2}$, the followings are to be expected.

$$
\begin{array}{c}
\overset{\displaystyle CF_3}{\underset{\displaystyle CF_3}{\underset{|}{H-C}}} = \overset{\displaystyle CF_3}{\underset{\displaystyle H}{\underset{|}{C-CF_2}}}
\longrightarrow
\underset{\text{\underline{More stable}}}{\overset{\displaystyle CF_3}{\underset{\displaystyle CF_3}{\oplus C}} - \overset{\displaystyle CF_2}{\underset{\displaystyle H}{C\ominus}}}
\quad\nearrow\!\!\!\diagdown
\quad
\overset{\displaystyle F}{\underset{\displaystyle CF_3}{\oplus C}} - \overset{\displaystyle H}{\underset{\displaystyle CHF}{\underset{\displaystyle CF_3}{C\ominus}}}
\end{array}
\qquad 4.76
$$

$$
\overset{CF_3}{\underset{CF_2H}{\underset{|}{H-C}}} = \overset{CFH}{\underset{H}{\underset{|}{C}}}
\longrightarrow
\underset{\text{\underline{More stable}}}{\overset{CFH}{\underset{CF_2H}{\oplus C}} - \overset{}{\underset{H}{C\ominus}}\ H}
\quad\nearrow\!\!\!\diagdown\quad
\overset{CHF}{\underset{CF_3}{\oplus C}} - \overset{CF_2H}{\underset{H}{C\ominus}}
\qquad 4.77
$$

$$
\overset{CF_3}{\underset{CF_3}{\underset{|}{H-C}}} = \overset{CFH}{\underset{H}{\underset{|}{C}}}
\longrightarrow
\overset{CFH}{\underset{CF_3}{\oplus C}} - \overset{}{\underset{H}{C\ominus}}\ H
\quad\nearrow\!\!\!\diagdown\quad
\overset{CHF}{\underset{CF_3}{\oplus C}} - \overset{CF_3}{\underset{H}{C\ominus}}
\qquad 4.78
$$

$$
\overset{CF_3}{\underset{CH_2F}{\underset{|}{H-C}}} = \overset{CH_2}{\underset{H}{\underset{|}{C}}}
\longrightarrow
\overset{CH_2}{\underset{CH_2F}{\oplus C}} - \overset{}{\underset{H}{C\ominus}}\ H
\quad\nearrow\!\!\!\diagdown\quad
\overset{H}{\underset{H}{\oplus C}} - \overset{H}{\underset{\underset{CH_2F}{CH(CF_3)}}{C\ominus}}
\qquad 4.79
$$

$$
\overset{CF_3}{\underset{CF_3}{\underset{|}{H-C}}} = \overset{CH_2}{\underset{H}{\underset{|}{C}}}
\longrightarrow
\overset{CH_2}{\underset{CF_3}{\ominus C}} - \overset{}{\underset{H}{C\oplus}}\ H
\longrightarrow
\overset{CH_2}{\underset{CF_3}{\ominus C}} - \overset{}{\underset{H}{C\oplus}}\ H
\qquad 4.80
$$

The CH_2CF_3 look almost allylic in character, that which chargedly and free-radically is the case. **Only the last case will favor rearrangement. Despite the rearrangement, it will not favor any of the charged routes. The others will either favor no route or cannot be activated chargedly.** The rules which have been established so far, are the rules being applied to identify the capacities of the groups and more driving forces for molecular rearrangement.

Free-radically, the followings are obtained.

$$
\begin{array}{c}
\overset{\displaystyle CF_3}{\underset{\displaystyle CF_3}{\overset{\displaystyle |}{\underset{\displaystyle |}{\overset{\displaystyle CF_2}{\underset{\displaystyle C}{\overset{\displaystyle |}{H-C=C}}}}}}}\quad\longrightarrow\quad
n.\overset{H}{\underset{CF_3}{\overset{|}{\underset{|}{C}}}}-\overset{CF_3}{\underset{H}{\overset{|}{\underset{|}{C}}_{CF_2}}}.e \quad\longrightarrow\quad
n.\overset{F}{\underset{F}{\overset{|}{\underset{|}{C}}}}-\overset{H}{\underset{CH(CF_3\,CF_3)}{\overset{|}{\underset{|}{C}}}}.e
\end{array}
\qquad 4.81
$$

<u>(More stable)</u>

Since $C_2F_5 > CF_3$ in radical-pushing capacity radically (See Equation 4.33), molecular rearrangement is favored.

$$
\begin{array}{c}
H-\overset{CFH}{\underset{CF_2H}{C}}=\overset{CF_3}{\underset{H}{C}}\quad\longrightarrow\quad
n.\,C-C.e \quad\longrightarrow\quad
n.\,C-C.e
\end{array}
\qquad 4.82
$$

<u>More stable</u>

Since $CFH(CF_3) > CF_2H$ (See Equation 4.33) in radical-pushing capacity, molecular rearrangement is favored to produce a more stable monomer.

$$
\begin{array}{c}
H-\overset{CFH}{\underset{CF_3}{C}}=\overset{CF_3}{\underset{H}{C}}\quad\longrightarrow\quad
n.\,C-C.e \quad\longrightarrow\quad
n.\,C-C.e
\end{array}
\qquad 4.83
$$

<u>More stable</u>

$$
\begin{array}{c}
H-\overset{CH_2}{\underset{CH_2F}{C}}=\overset{CF_3}{\underset{H}{C}}\quad\longrightarrow\quad
n.\,C-C.e \quad\longrightarrow\quad
n.\,C-C.e
\end{array}
\qquad 4.84
$$

<u>More stable</u>

$$
\begin{array}{c}
H-\overset{CH_2}{\underset{CF_3}{C}}=\overset{CF_3}{\underset{H}{C}}\quad\longrightarrow\quad
n.\,C-C.e \quad\longrightarrow\quad
n.\,C-C.e
\end{array}
\qquad 4.85
$$

Chargedly, molecular rearrangement is not favored for any of them except the last case. Free-radically, molecular rearrangement is favored. In general, $\underline{\textbf{CH}_2\textbf{CF}_3 > \textbf{CH(CF}_3)_2 > \textbf{C(CF}_3)_3}$ **in pushing capacity free-radically.**

For the last group of monomers $CHR_{F1} = CR_{F2}R_{F3}$, the followings are obtained.

$$
\begin{array}{ccc}
\begin{array}{cc} H & CF_3 \\ | & | \\ C & = C \\ | & | \\ CF_3 & CF_2 \\ & | \\ & CF_3 \end{array}
& \longrightarrow &
\begin{array}{cc} H & CF_3 \\ | & | \\ \oplus C & - C\ominus \\ | & | \\ CF_3 & CF_2 \\ & | \\ & CF_3 \end{array}
\end{array}
\quad\not\longrightarrow\quad
\begin{array}{cc} F & CF_3 \\ | & | \\ \oplus C & - C\ominus \\ | & | \\ CF_3 & CHF \\ & | \\ & CF_3 \end{array}
\qquad 4.86
$$

$$\underline{\text{More stable}}\ (R_{F3} > R_{F2} = R_{F1})$$

$$
\begin{array}{cc} & \underline{R_{F2}} \\ H & CF_3 \\ | & | \\ C & = C \\ | & | \\ CF_2H & CF_2 \\ \underline{R_{F1}} & | \\ & CF_3 \\ & \underline{R_{F3}} \end{array}
\quad\longrightarrow\quad
\begin{array}{cc} H & CF_3 \\ | & | \\ \oplus C & - C\ominus \\ | & | \\ CF_2H & CF_2 \\ & | \\ & CF_3 \end{array}
\quad\not\longrightarrow\quad
\begin{array}{cc} F & CF_3 \\ | & | \\ \oplus C & - C\ominus \\ | & | \\ CF_3 & CHF \\ & | \\ & CF_2H \end{array}
\qquad 4.87
$$

$$\underline{\text{More stable}}\qquad (R_{F3} > R_{F2} > R_{F1})$$

$$
\begin{array}{cc} H & CF_3 \\ | & | \\ C & = C \\ | & | \\ CF_2H & CHF \\ & | \\ & CF_3 \end{array}
\quad\longrightarrow\quad
\begin{array}{cc} H & CF_3 \\ | & | \\ \oplus C & - C\ominus \\ | & | \\ CF_2H & CHF \\ & | \\ & CF_3 \end{array}
\quad\not\longrightarrow\quad
\begin{array}{cc} H & CF_3 \\ | & | \\ \oplus C & - C\ominus \\ | & | \\ CF_3 & CHF \\ & | \\ & CF_2H \end{array}
\qquad 4.88
$$

$$\underline{\text{More stable}}\quad (R_{F3} > R_{F2} > R_{F1})$$

$$
\begin{array}{cc} & CF_3 \\ & | \\ H & CF_2 \\ | & | \\ C & = C \\ | & | \\ CH_2F & CF_2H \end{array}
\quad\longrightarrow\quad
\begin{array}{cc} & CF_3 \\ & | \\ H & CF_2 \\ | & | \\ \oplus C & - C\ominus \\ | & | \\ CH_2F & CF_2H \end{array}
\quad\not\longrightarrow\quad
\begin{array}{cc} F & CF_2H \\ | & | \\ \oplus C & - C\ominus \\ | & | \\ CF_3 & CHF \\ & | \\ & CH_2F \end{array}
\qquad 4.89
$$

$$\underline{\text{Move stable}}\qquad\qquad (R_{F3} > R_{F2} > R_{F1})$$

$$
\begin{array}{cc} H & CF_3 \\ | & | \\ C & = C \\ | & | \\ CH_2F & CF_2H \end{array}
\quad\longrightarrow\quad
\begin{array}{cc} H & CF_3 \\ | & | \\ \oplus C & - C\ominus \\ | & | \\ CH_2F & CF_2H \end{array}
\quad\not\longrightarrow\quad
\begin{array}{cc} F & CF_2H \\ | & | \\ \oplus C & - C\ominus \\ | & | \\ F & CHF \\ & | \\ & CH_2F \end{array}
\qquad 4.90
$$

$$\underline{\text{More stable}}\qquad (RF3 > RF2 > RF1)$$

$$
\begin{array}{ccc}
\text{H} & \text{CF}_3 \\
| & | \\
\text{C} & = & \text{C} \\
| & | \\
\text{CH}_2\text{F} & \text{CF}_2\text{H}
\end{array}
\longrightarrow
\begin{array}{ccc}
\text{H} & \text{CF}_2\text{H} \\
| & | \\
\oplus\,\text{C} & - & \text{C}\,\ominus \\
| & | \\
\text{CH}_2\text{F} & \text{CF}_2\text{H}
\end{array}
\;\not\longrightarrow\;
\begin{array}{ccc}
\text{F} & \text{CF}_2\text{H} \\
| & | \\
\oplus\,\text{C} & - & \text{C}\,\ominus \\
| & | \\
\text{H} & \text{CHF} \\
& | \\
& \text{CH}_2\text{F}
\end{array}
\qquad 4.91
$$

<u>More stable</u>   Less stable (RF3 = RF2 > RF1)

$$
\begin{array}{ccc}
\text{H} & \text{CF}_2\text{H} \\
| & | \\
\text{C} & = & \text{C} \\
| & | \\
\text{CF}_3 & \text{CFH}_2
\end{array}
\longrightarrow
\begin{array}{ccc}
\text{H} & \text{CF}_2\text{H} \\
| & | \\
\ominus\,\text{C} & - & \text{C}\,\oplus \\
| & | \\
\text{CF}_3 & \text{CFH}_2
\end{array}
\;\not\longrightarrow\;
\begin{array}{ccc}
\text{F} & \text{CHF}_2 \\
| & | \\
\ominus\,\text{C} & - & \text{C}\,\oplus \\
| & | \\
\text{H} & \text{CH}_2 \\
& | \\
& \text{CF}_3
\end{array}
\qquad 4.92
$$

<u>More stable</u>   (Impossible state) (RF1 > RF3 > RF2)

Molecular rearrangement is not favored by any of them, even when H is used as transfer species for them. As it seems these monomers cannot be activated chargedly. Free-radically, the follow-ings are to be expected.

$$
\begin{array}{ccc}
\text{H} & \text{CF}_3 \\
| & | \\
\text{C} & = & \text{C} \\
| & | \\
\text{CF}_3 & \text{CF}_2 \\
& | \\
& \text{CF}_3
\end{array}
\longrightarrow
\begin{array}{ccc}
\text{H} & \text{CF}_3 \\
| & | \\
\text{e.\,C} & - & \text{C .n} \\
| & | \\
\text{CF}_3 & \text{CF}_2 \\
& | \\
& \text{CF}_3
\end{array}
\longrightarrow
\text{No transfer species}
\qquad 4.93
$$

$(R_3 > R_2 = R_1)$

$R_3 \equiv CF_2 - CF_3$, and $R_2 \equiv CF_3$. Free-radically, it has been established that $R_3 > R_2$.

$$
\begin{array}{ccc}
\text{H} & \text{CF}_3 \\
| & | \\
\text{C} & = & \text{C} \\
| & | \\
\text{CF}_2\text{H} & \text{CF}_2 \\
& | \\
& \text{CF}_3
\end{array}
\longrightarrow
\begin{array}{ccc}
\text{H} & \text{CF}_3 \\
| & | \\
\text{e.\,C} & - & \text{C .n} \\
| & | \\
\text{CF}_2\text{H} & \text{CF}_2 \\
& | \\
& \text{CF}_3
\end{array}
\longrightarrow
\text{No transfer species}
\qquad 4.94
$$

$(R_1 > R_3 > R_2)$

Though there is no transfer species for rearrangement, the first will only favor the nucleo-free-radical route, while the second will favor no route.

$$
\begin{array}{ccc}
& \text{CF}_3 \\
& | \\
\text{H} & \text{CF}_2 \\
| & | \\
\text{C} & = & \text{C} \\
| & | \\
\text{CH}_2\text{F} & \text{CHF}_2
\end{array}
\longrightarrow
\begin{array}{ccc}
& \text{CF}_3 \\
& | \\
\text{H} & \text{CF}_2 \\
| & | \\
\text{e.\,C} & - & \text{C .n} \\
| & | \\
\text{CH}_2\text{F} & \text{CF}_2\text{H}
\end{array}
\longrightarrow
\text{No transfer species}
\qquad 4.95
$$

(R3 > R1 > R2)

$$
\underset{\substack{H\\|\\C\\|\\CH_2F}}{}=\underset{\substack{CF_3\\|\\CF_2\\|\\C\\|\\CHF_2}}{} \longrightarrow \; e.\underset{\substack{H\\|\\C\\|\\CH_2F}}{}-\underset{\substack{CF_3\\|\\CF_2\\|\\C.n\\|\\CF_2H}}{} \longrightarrow \; \textbf{No transfer species}
$$

$$(R1 > R3 > R2)$$

4.96

$$
\underset{\substack{H\\|\\C\\|\\CH_2F}}{}=\underset{\substack{CF_2H\\|\\C\\|\\CF_2H}}{} \longrightarrow \; e.\underset{\substack{H\\|\\C\\|\\CH_2F}}{}-\underset{\substack{CF_2H\\|\\C.n\\|\\CF_2H}}{} \longrightarrow \; n.\underset{\substack{H\\|\\C\\|\\F}}{}-\underset{\substack{H\\|\\C.e\\|\\CH\\ HF_2C\quad CF_2H}}{}
$$

$$(R1 > R3 = R2) \qquad\qquad \underline{\text{More stable}}$$

4.97

$$
\underset{\substack{H\\|\\C\\|\\CF_3}}{}=\underset{\substack{CH_2F\\|\\C\\|\\CF_2H}}{} \longrightarrow \; n.\underset{\substack{H\\|\\C\\|\\CF_3}}{}-\underset{\substack{CH_2F\\|\\C.e\\|\\CF_2H}}{} \longrightarrow \; \text{No transfer species}
$$

$$(R_3 > R_2 > R_1)$$

4.98

Though there is no rearrangement for all but one (Equation 4.97), that is the only one that will favor the electro-free-radical route, the route natural to them. For the others, if H was transferred for rearrangement, it will not be the same transfer species during initiation and termi-nation. *However, initiation is not favored nucleo-free-radically, but electro-free-radically for only the last but one equation. Only the case of Equation 4.93 will favor nucleo-free-radical route as already said.*

4.5 Characteristics of CF2 = CFRF, CF2 = CRFRF, CFRF = CFRF, CFRF = CRFRF groups.

These are fluoro-alkenes. Beginning with the first group, the followings are obtained

$$
\underset{\substack{F\\|\\C\\|\\F}}{}=\underset{\substack{F\\|\\C\\|\\CF_3}}{} \longrightarrow \; \oplus\underset{\substack{F\\|\\C\\|\\F}}{}-\underset{\substack{F\\|\\C\,\ominus\\|\\CF_3}}{}
$$

$$\text{(Existence not favored chargedly)}$$

4.99

$$
\underset{\substack{F\\|\\C\\|\\F}}{}=\underset{\substack{F\\|\\C\\|\\CF_3}}{} \longrightarrow \; n.\underset{\substack{F\\|\\C\\|\\F}}{}-\underset{\substack{F\\|\\C.e\\|\\CF_3}}{} \longrightarrow \text{No transfer species}
$$

4.100

Perfluoropropylenes are not popularly used for homopolymerization probably because the CF_3, C_2F_5, C_3F_7, etc. groups are weak free-radical-pushing groups and very unique. Chargedly, they are pulling groups of large capacities, that which makes their nucleophiles behave like electrophiles. For this reason unlike $H^{.n}$, such groups may readily be rejected from their growing polymer chains nucleo-free-radically as shown below.

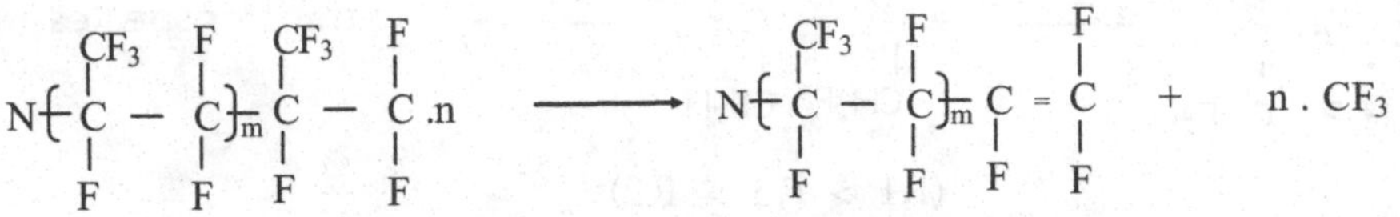

$$N \left[\begin{matrix} CF_3 \\ | \\ C \\ | \\ F \end{matrix} - \begin{matrix} F \\ | \\ C \\ | \\ F \end{matrix} \right]_{\!m} \begin{matrix} CF_3 \\ | \\ C \\ | \\ F \end{matrix} - \begin{matrix} F \\ | \\ C \\ | \\ F \end{matrix} .n \longrightarrow N \left[\begin{matrix} CF_3 \\ | \\ C \\ | \\ F \end{matrix} - \begin{matrix} F \\ | \\ C \\ | \\ F \end{matrix} \right]_{\!m} \begin{matrix} CF_3 \\ | \\ C \\ \| \end{matrix} \begin{matrix} F \\ | \\ C \\ | \\ F \end{matrix} + n \cdot CF_3$$

(Transfer species of the Second kind)　　　　　　　　　　　4.101

This is not possible electro-free-radically, wherein no transfer species exists to be rejected. How-ever, consider the following.

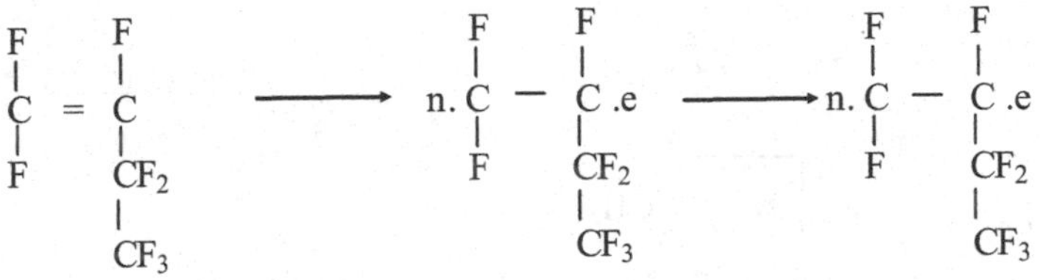

$$\begin{matrix} F \\ | \\ C \\ | \\ F \end{matrix} = \begin{matrix} F \\ | \\ C \\ | \\ CF_2 \\ | \\ CF_3 \end{matrix} \longrightarrow n \cdot \begin{matrix} F \\ | \\ C \\ | \\ F \end{matrix} - \begin{matrix} F \\ | \\ C .e \\ | \\ CF_2 \\ | \\ CF_3 \end{matrix} \longrightarrow n \cdot \begin{matrix} F \\ | \\ C \\ | \\ F \end{matrix} - \begin{matrix} F \\ | \\ C .e \\ | \\ CF_2 \\ | \\ CF_3 \end{matrix}$$

　　　　　　　　　　　4.102

Transfer of transfer species is favored here with or without any molecular rearrangement to produce dead terminal double bond polymers like propylene does electro-free-radically. The *transfer species here is e.CF$_3$*. Nucleo-free-radically, it cannot be initiated.

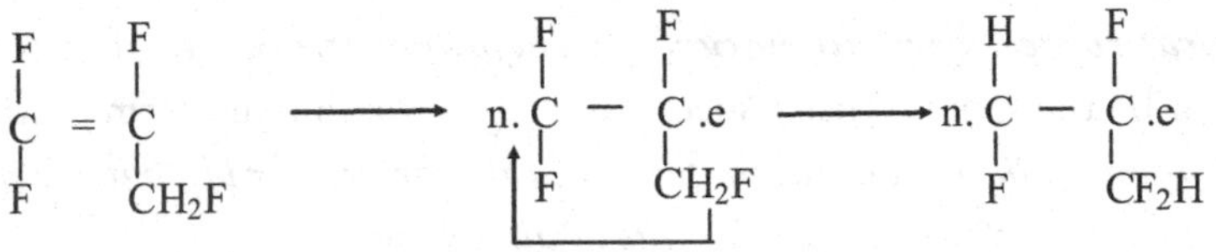

$$\begin{matrix} F \\ | \\ C \\ | \\ F \end{matrix} = \begin{matrix} F \\ | \\ C \\ | \\ CH_2F \end{matrix} \longrightarrow n \cdot \begin{matrix} F \\ | \\ C \\ | \\ F \end{matrix} - \begin{matrix} F \\ | \\ C .e \\ | \\ CH_2F \end{matrix} \longrightarrow n \cdot \begin{matrix} H \\ | \\ C \\ | \\ F \end{matrix} - \begin{matrix} F \\ | \\ C .e \\ | \\ CF_2H \end{matrix}$$

　　　　　　　　　　　More stable　　　　　　　　　　　4.103

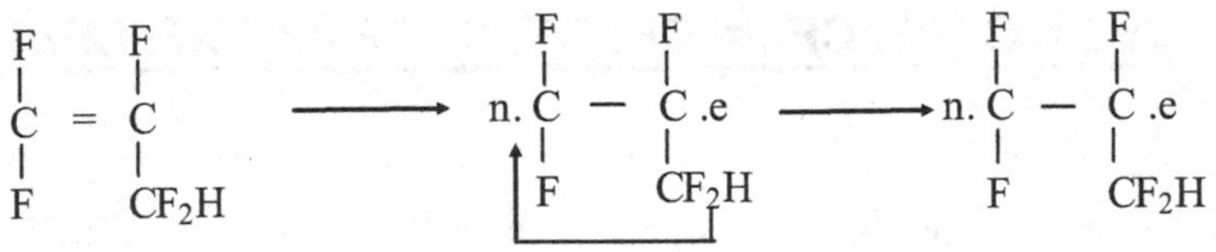

$$\begin{matrix} F \\ | \\ C \\ | \\ F \end{matrix} = \begin{matrix} F \\ | \\ C \\ | \\ CF_2H \end{matrix} \longrightarrow n \cdot \begin{matrix} F \\ | \\ C \\ | \\ F \end{matrix} - \begin{matrix} F \\ | \\ C .e \\ | \\ CF_2H \end{matrix} \longrightarrow n \cdot \begin{matrix} F \\ | \\ C \\ | \\ F \end{matrix} - \begin{matrix} F \\ | \\ C .e \\ | \\ CF_2H \end{matrix}$$

　　　　　　　　　　　4.104

$$\begin{matrix} F \\ | \\ C \\ | \\ F \end{matrix} = \begin{matrix} F \\ | \\ C \\ | \\ CH_2 \\ | \\ CF_3 \end{matrix} \longrightarrow n \cdot \begin{matrix} F \\ | \\ C \\ | \\ F \end{matrix} - \begin{matrix} F \\ | \\ C .e \\ | \\ CH_2 \\ | \\ CF_3 \end{matrix} \longrightarrow n \cdot \begin{matrix} H \\ | \\ C \\ | \\ H \end{matrix} - \begin{matrix} F \\ | \\ C .e \\ | \\ CF_2(CF_3) \end{matrix}$$

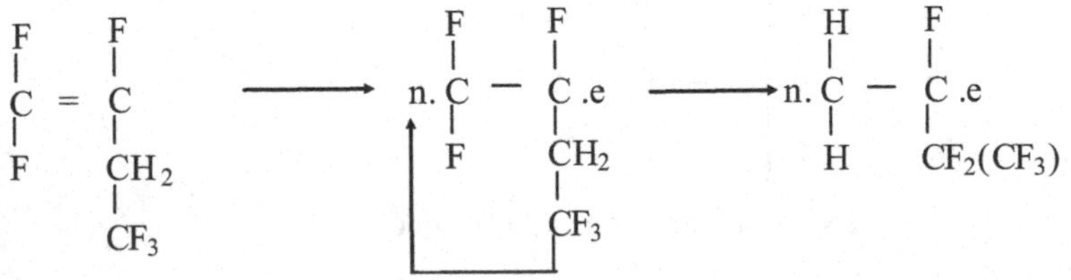

　　　　　　　　　　　4.105

More stable　　　　　(Impossible existence)

Molecular rearrangement is observed not to be favored for this group of monomers. Where favored, the same monomer is obtained as in Equation 4.104. Transfer of transfer species is favored for all of them to produce dead terminal double bond polymers electro-free-radically.

　　　This is shown below for Equations 4.102 and 4.103.

$$E-(C-C)_n-C-C \bullet e \longrightarrow E-(C-C)_n-C-C=C + e \bullet CF_3 \qquad 4.106a$$

$$E-(C-C)_n-C-C \bullet e \longrightarrow E-(C-C)_n-C-C=C + e \bullet H \qquad 4.106b$$

For the second group, the followings are to be expected.

$$C=C \longrightarrow \oplus C - C \ominus \longrightarrow \oplus C - C \ominus \qquad 4.107a$$

$$C=C \longrightarrow \oplus C - C \ominus \longrightarrow \oplus C -\cdot C \ominus \qquad 4.107b$$

<u>More stable</u>

In the first reaction, transfer of transfer species is favored with or without molecular rearrange-ment, ***noting that for the first time F is being used as transfer species.*** This is largely because of the absence of H in the system. In the second reaction, there is no need to distinguish between $F:^\ominus$ (non-free-negatively charged) and $^\ominus CF_3$ (free-negatively charged), since CF_3 cannot carry a negative charge. CF_3 is more chargedly-pulling than F, for which F is the transfer species for the C_2F_5 group "anionically", based on the laws. ***With negatively charged initiator, only F can be released for which when released, it cannot be used to initiate a monomer, but used to form a stable product. The monomers look like Electrophiles***

$$C=C \longrightarrow \oplus C - C \ominus \longrightarrow\!\!\!/ \;\; \oplus C - C \ominus \qquad 4.108$$

F $\qquad$ <u>Looks more stable</u>

$$C=C \longrightarrow \oplus C - C \ominus \longrightarrow\!\!\!/ \;\; \text{Similar to above} \qquad 4.109$$

```
    CF3                      CF3
     |                        |
F   CH2                  F   CH2
 |   |                    |   |
  C = C         ------>    ⊕C — C⊖
 |   |                    |   |
F   CH2                  F   CH2
     |                        |
    CF3                      CF3
```

4.110

No transfer species

Molecular rearrangement is not favored for the first two cases above, because of the presence of H the monomer being a Nucleophile. For the last reaction, there is no transfer species, for which either both routes are favored or they cannot be activated chargedly. They will all favor the negatively charged route in the absence of electrodynamic forces of repulsion.

Free-radically, the followings are favored.

```
F    CF3              F    CF3
|     |               |     |
 C  = C     ----->  n. C — C .e   ----->   No transfer species
|     |               |     |
F    CF3              F    CF3
```

4.111

```
     CF3                   CF3                        CF3
      |                     |                          |
F    CF2              F    CF2                 F    C2F5
 |    |                |    |                   |    |
  C = C     ----->   n. C — C.e    ----->    n. C — C .e
 |    |                |    |                   |    |
F    CF2              F    CF2                 F    C2F5
      |                     |
     CF3                   CF3
```

4.112

```
F    CF2H             F    CF2H              F    CF2H
|     |               |     |                |     |
 C  = C     ----->  n. C — C.e     ----->  n. C — C.e
|     |               |     |                |     |
F    CF2H             F    CF2H              F    CF2H
```

4.113

```
F    CH2F             F    CH2F              F    CH2F
|     |               |     |                |     |
 C  = C     ----->  n. C — C.e      --/-->  n. C — C.e
|     |               |     |                |     |
F    CH2F             F    CH2F              H    CF2H
```

More stable

4.114

```
    CF3                   CF3                      CF3
     |                     |                        |
F   CH2              F    CH2               H    CH2
 |   |                |    |                 |    |
  C = C     ----->  n. C — C.e      --/-->  n. C — C .e
 |   |                |    |                 |    |
F   CH2              F    CH2               H    CF2
     |                     |                        |
    CF3                   CF3                      CF3
```

More stable

4.115

$$
\begin{array}{c}
\text{CF}_3 \\
|\\
\text{F}\quad\text{CFH} \\
|\qquad|\\
\text{C}=\text{C} \\
|\qquad|\\
\text{F}\quad\text{CFH} \\
|\\
\text{CF}_3
\end{array}
\longrightarrow
\begin{array}{c}
\text{CF}_3 \\
|\\
\text{F}\quad\text{CFH} \\
|\qquad|\\
\text{n. C}-\text{C .e} \\
|\qquad|\\
\text{F}\quad\text{CFH} \\
\qquad\quad|\\
\qquad\quad\text{CF}_3
\end{array}
\;\not\longrightarrow\;
\begin{array}{c}
\text{CF}_3 \\
|\\
\text{F}\quad\text{CFH} \\
|\qquad|\\
\text{n. C}-\text{C .e} \\
|\qquad|\\
\text{H}\quad\text{CF}_2 \\
\qquad\quad|\\
\qquad\quad\text{CF}_3
\end{array}
\qquad 4.116
$$

<u>More stable</u>

Free-radically, molecular rearrangement is favored for the second and third cases, but no new monomer formed. For the last three cases, no rearrangement was favored, by any of the monomers. Nevertheless, all with the exception of the first will favor only the electro-free-radical route. The first will favor both routes.

For the third group of monomers- $\text{CFR}_\text{F}=\text{CFR}_\text{F}$, the followings are obtained.

$$
\begin{array}{c}
\text{F}\quad\text{F} \\
|\qquad|\\
\text{C}=\text{C} \\
|\qquad|\\
\text{CF}_3\quad\text{CF}_3
\end{array}
\longrightarrow
\begin{array}{c}
\text{F}\quad\text{F} \\
|\qquad|\\
\ominus\text{C}-\text{C}\oplus \\
|\qquad|\\
\text{CF}_3\quad\text{CF}_3
\end{array}
\qquad 4.117
$$

(Impossible existence)

Chargedly, the monomers cannot be activated. Free-radically, we have the followings.

$$
\begin{array}{c}
\text{F}\quad\text{F} \\
|\qquad|\\
\text{C}=\text{C} \\
|\qquad|\\
\text{CF}_3\quad\text{CF}_3
\end{array}
\longrightarrow
\begin{array}{c}
\text{F}\quad\text{F} \\
|\qquad|\\
\text{n. C}-\text{C .e} \\
|\qquad|\\
\text{CF}_3\quad\text{CF}_3
\end{array}
\longrightarrow \text{No transfer species}
\qquad 4.118
$$

$$
\begin{array}{c}
\text{CF}_3 \\
|\\
\text{H}\quad\text{CF}_2 \\
|\qquad|\\
\text{C}=\text{C} \\
|\qquad|\\
\text{C}_2\text{F}_5\quad\text{F}
\end{array}
\longrightarrow
\begin{array}{c}
\text{CF}_3 \\
|\\
\text{H}\quad\text{CF}_2 \\
|\qquad|\\
\text{e. C}-\text{C.n} \\
|\qquad|\\
\text{CF}_2\quad\text{F} \\
|\\
\text{CF}_3
\end{array}
\longrightarrow
\begin{array}{c}
\text{F}\quad\text{H} \\
|\qquad|\\
\text{n. C}-\text{C.e} \\
|\qquad|\\
\text{F}\quad\text{CF(CF}_3) \\
\qquad\quad|\\
\qquad\quad\text{CF}_2 \\
\qquad\quad|\\
\qquad\quad\text{CF}_3
\end{array}
\qquad 4.119
$$

<u>More stable</u>

$$
\begin{array}{c}
\text{F}\quad\text{CF}_2\text{H} \\
|\qquad|\\
\text{C}=\text{C} \\
|\qquad|\\
\text{CF}_2\text{H}\quad\text{F}
\end{array}
\longrightarrow
\begin{array}{c}
\text{F}\quad\text{CF}_2\text{H} \\
|\qquad|\\
\text{n. C}-\text{C .e} \\
|\qquad|\\
\text{CF}_2\text{H}\quad\text{F}
\end{array}
\longrightarrow
\begin{array}{c}
\text{F}\quad\text{F} \\
|\qquad|\\
\text{n. C}-\text{C .e} \\
|\qquad|\\
\text{F}\quad\text{CFH} \\
\qquad\quad|\\
\qquad\quad\text{CF}_2\text{H}
\end{array}
\qquad 4.120
$$

<u>More stable</u>

$$
\begin{array}{c}
\text{F}\quad\text{CH}_2\text{F} \\
|\qquad|\\
\text{C}=\text{C} \\
|\qquad|\\
\text{CH}_2\text{F}\quad\text{F}
\end{array}
\longrightarrow
\begin{array}{c}
\text{F}\quad\text{CH}_2\text{F} \\
|\qquad|\\
\text{n. C}-\text{C .e} \\
|\qquad|\\
\text{CH}_2\text{F}\quad\text{F}
\end{array}
\longrightarrow
\begin{array}{c}
\text{H}\quad\text{F} \\
|\qquad|\\
\text{n. C}-\text{C .e} \\
|\qquad|\\
\text{F}\quad\text{CFH} \\
\qquad\quad|\\
\qquad\quad\text{CH}_2\text{F}
\end{array}
\qquad 4.121
$$

<u>More stable</u>

$$\begin{matrix} & CF_3 \\ & | \\ F & CH_2 \\ | & | \\ C & = C \\ | & | \\ CH_2 & F \\ | \\ CF_3 \end{matrix} \longrightarrow \begin{matrix} & & CF_3 \\ & & | \\ CF_3\!\downarrow & F & CH_2 \\ & | & | \\ n.\ C & - & C\ .e \\ & | & | \\ & CH_2 & F \\ & | \\ & CF_3 \end{matrix} \longrightarrow \begin{matrix} H & F \\ | & | \\ n.\ C & - & C\ .e \\ | & | \\ H & CF(CF_3) \\ & | \\ & CH_2 \\ & | \\ & CF_3 \end{matrix} \qquad 4.122$$

<u>More stable</u>

Free-radically, molecular rearrangement is favored by the monomers, except for the first and last cases. Only the first will favor the nucleo-free-radical route, while the others will favor the use of electro-free-radicals, the route natural to them.

For the last group of monomers, the followings take place.

$$\begin{matrix} F & CF_3 \\ | & | \\ C & = C \\ | & | \\ CF_3 & CF_3 \end{matrix} \longrightarrow \begin{matrix} F\!\downarrow & F & CF_3 \\ & | & | \\ \oplus C & - & C \ominus \\ & | & | \\ & CF_3 & CF_3 \end{matrix} \ \not\to \ \begin{matrix} F & CF_3 \\ | & | \\ \oplus C & - & C \ominus \\ | & | \\ F & CF_2 \\ & | \\ & CF_3 \end{matrix} \qquad 4.123$$

<u>(More stable)</u>

$$\begin{matrix} F & C_2F_5 \\ | & | \\ C & = C \\ | & | \\ C_2F_5 & CF_2 \\ & | \\ & CF_3 \end{matrix} \longrightarrow \begin{matrix} & & CF_3 \\ & & | \\ & F & CF_2 \\ & | & | \\ \oplus C & - & C \ominus \\ & | & | \\ & CF_2 & CF_2 \\ & | & | \\ & CF_3 & CF_3 \\ F \end{matrix} \ \not\to \ \begin{matrix} F & C_2F_5 \\ | & | \\ \oplus C & - & C \ominus \\ | & | \\ CF_3 & CF_2 \\ & | \\ & CF_2 \\ & | \\ & CF_3 \end{matrix} \qquad 4.124$$

<u>(More stable)</u>

$$\begin{matrix} F & CF_2H \\ | & | \\ C & = C \\ | & | \\ CF_2H & CF_2H \end{matrix} \longrightarrow \begin{matrix} & F & CF_2H \\ & | & | \\ \oplus C & - & C \ominus \\ & | & | \\ F & CF_2H & CF_2H \end{matrix} \ \not\to \ \begin{matrix} F & CF_2H \\ | & | \\ \oplus C & - & C \ominus \\ | & | \\ H & CF_2 \\ & | \\ & CF_2H \end{matrix} \qquad 4.125$$

<u>(More stable)</u>

$$\begin{matrix} F & CH_2F \\ | & | \\ C & = C \\ | & | \\ CH_2F & CH_2F \end{matrix} \longrightarrow \begin{matrix} F\!\downarrow & F & CH_2F \\ & | & | \\ \oplus C & - & C \ominus \\ & | & | \\ & CH_2F & CH_2F \end{matrix} \ \not\to \ \begin{matrix} H & CH_2F \\ | & | \\ \oplus C & - & C \ominus \\ | & | \\ H & CF_2 \\ & | \\ & CH_2F \end{matrix} \qquad 4.126$$

<u>(More stable)</u>

$$
\begin{array}{cc}
 & CF_3 \\
 & | \\
F & CH_2 \\
| & | \\
C & = C \\
| & | \\
CH_2 & CH_2 \\
| & | \\
CF_3 & CF_3
\end{array}
\longrightarrow
\begin{array}{cc}
 & CF_3 \\
 & | \\
F & CH_2 \\
| & | \\
\ominus C & - C \oplus \\
| & | \\
CH_2 & CH_2 \\
| & | \\
CF_3 & CF_3
\end{array}
\qquad 4.127
$$

Electrostatic forces of repulsion

For the last three cases, molecular rearrangement is not favored by the monomers noting that these monomers are nucleophiles. H which should be the transfer species cannot be removed due to electrostatic forces of repulsion. Negatively or positively, they cannot be initiated. *For the first two cases which are fully fluorinated, F is the transfer species, for which the only route favored by them is the negatively charged route making them look like Electrophiles!*

Free-radically, the followings are obtained.

$$
\begin{array}{cc}
F & CF_3 \\
| & | \\
C & = C \\
| & | \\
CF_3 & CF_3
\end{array}
\longrightarrow
\begin{array}{cc}
F & CF_3 \\
| & | \\
n.C & - C.e \\
| & | \\
CF_3 & CF_3
\end{array}
\longrightarrow \text{No transfer species}
\qquad 4.128
$$

$$
\begin{array}{cc}
 & CF_3 \\
 & | \\
F & CF_2 \\
| & | \\
C & = C \\
| & | \\
C_2F_5 & CF_2 \\
 & | \\
 & CF_3
\end{array}
\longrightarrow
\begin{array}{cc}
 & CF_3 \\
 & | \\
F & CF_2 \\
| & | \\
n.C & - C.e \\
| & | \\
C_2F_5 & CF_2 \\
 & | \\
 & CF_3
\end{array}
\longrightarrow
\begin{array}{cc}
F & C_2F_5 \\
| & | \\
n.C & - C.e \\
| & | \\
F & CF(CF_3) \\
 & | \\
 & C_2F_5
\end{array}
\qquad 4.129
$$

(More stable)

$$
\begin{array}{cc}
F & CF_2H \\
| & | \\
C & = C \\
| & | \\
CF_2H & CF_2H
\end{array}
\longrightarrow
\begin{array}{cc}
F & CF_2H \\
| & | \\
n.C & - C.e \\
| & | \\
CF_2H & CF_2H
\end{array}
\longrightarrow
\begin{array}{cc}
F & CF_2H \\
| & | \\
n.C & - C.e \\
| & | \\
F & CFH \\
 & | \\
 & CF_2H
\end{array}
\qquad 4.130
$$

(More stable)

$$
\begin{array}{cc}
F & CH_2F \\
| & | \\
C & = C \\
| & | \\
CH_2F & CH_2F
\end{array}
\longrightarrow
\begin{array}{cc}
F & CH_2F \\
| & | \\
n.C & - C.e \\
| & | \\
CH_2F & CH_2F
\end{array}
\longrightarrow
\begin{array}{cc}
F & CH_2F \\
| & | \\
n.C & - C.e \\
| & | \\
H & CFH \\
 & | \\
 & CH_2F
\end{array}
\qquad 4.131
$$

(More stable)

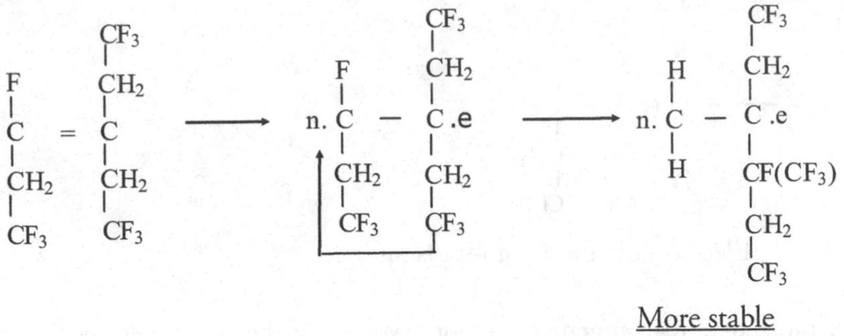

(4.132)

More stable

Molecular rearrangement is favored by all of the monomers in this group, when transfer species exist, even when the C center carrying F does not remain with it after rearrangement, like the last case. However, all except the first will favor only the electro-free-radical route.

4.6 Characteristics of CF2 = CRF1RF2, CFRF 1 = CFRF2, CFRF1 = CRF2RF3 groups.

For the cases to be considered here RF3, RF2 and RF1 are of different charged-pulling capacities and radical-pushing capacities.

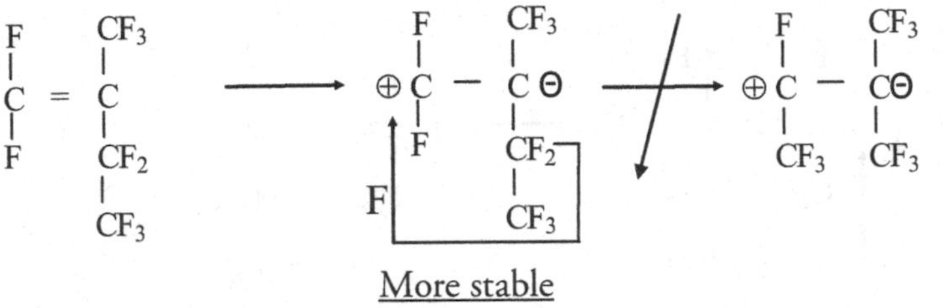

(4.133)

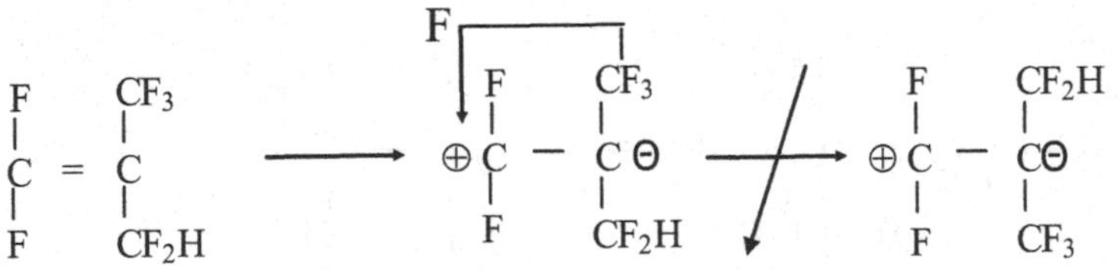

More stable

(4.134)

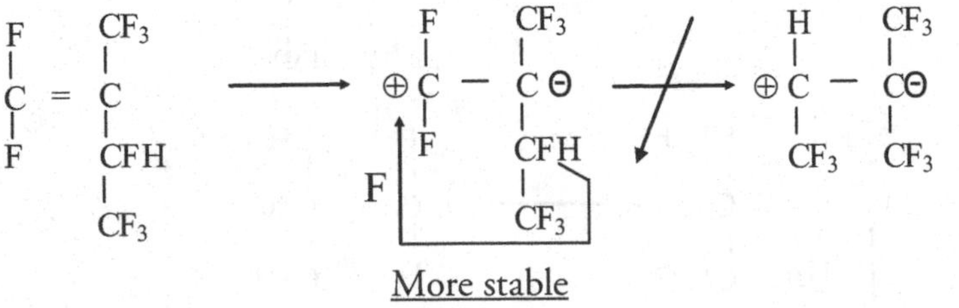

(4.135)

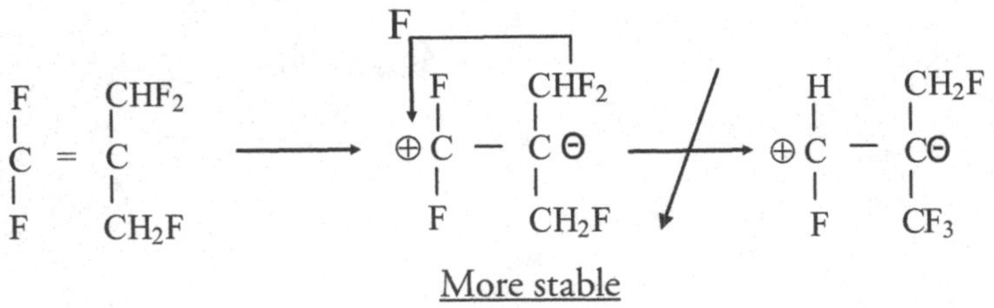

More stable

(4.136)

More stable

209

$$
\begin{array}{cc}
F & CHF_2 \\
| & | \\
C = C \\
| & | \\
F & CH_2 \\
& | \\
& CF_3
\end{array}
\longrightarrow
\begin{array}{cc}
F & CHF_2 \\
| & | \\
\oplus C - C \ominus \\
| & | \\
F & CH_2 \\
& | \\
& CF_3
\end{array}
\nrightarrow
\begin{array}{cc}
H & CF_3 \\
| & | \\
\oplus C - C \ominus \\
| & | \\
F & CH_2 \\
& | \\
& CF_3
\end{array}
\qquad 4.137
$$

$$
\begin{array}{cc}
& CF_3 \\
& | \\
F & CHF \\
| & | \\
C = C \\
| & | \\
F & CH_2 \\
& | \\
& CF_3
\end{array}
\longrightarrow
\begin{array}{cc}
& CF_3 \\
& | \\
F & CHF \\
| & | \\
\oplus C - C \ominus \\
| & | \\
F & CH_2 \\
& | \\
& CF_3
\end{array}
\nrightarrow
\begin{array}{cc}
& CF_3 \\
& | \\
H & CH_2 \\
| & | \\
\oplus C & C \ominus \\
| & | \\
CF_3 & CF_3
\end{array}
\qquad 4.138
$$

<u>More stable</u>

Chargedly, none will favor molecular rearrangement here. Nevertheless, it is important to note the sources of the transfer species and therefore the relative capacities of the groups because nothing takes place indiscriminately in Nature. All what one has been doing are numerous to list, bur most importantly is searching for the capacity orders of the groups and the characters dis-played by these monomers. **As it seems, these cannot be activated chargedly.** Their route which seems to be favored is that with the use of negative charges with F as transfer species, that which is least expected for a Nucleophile, partially hydrogenated.

Free-radically, the followings are obtained.

$$
\begin{array}{cc}
F & CF_3 \\
| & | \\
C = C \\
| & | \\
F & CF_2 \\
& | \\
& CF_3
\end{array}
\longrightarrow
\begin{array}{cc}
F & CF_3 \\
| & | \\
n.C - C\,.e \\
| & | \\
F & CF_2 \\
& | \\
& CF_3
\end{array}
\longrightarrow
\begin{array}{cc}
F & CF_3 \\
| & | \\
n.C - C\,.e \\
| & | \\
F & CF_2 \\
& | \\
& CF_3
\end{array}
\qquad 4.139
$$

$$
\begin{array}{cc}
F & CF_3 \\
| & | \\
C = C \\
| & | \\
F & CF_2H
\end{array}
\longrightarrow
\begin{array}{cc}
F & CF_3 \\
| & | \\
n.C - C\,.e \\
| & | \\
F & CF_2H
\end{array}
\longrightarrow
\begin{array}{cc}
F & CF_3 \\
| & | \\
n.C - C\,.e \\
| & | \\
F & CF_2H
\end{array}
\qquad 4.140
$$

$$
\begin{array}{cc}
F & CF_3 \\
| & | \\
C = C \\
| & | \\
F & CFH \\
& | \\
& CF_3
\end{array}
\longrightarrow
\begin{array}{cc}
F & CF_3 \\
| & | \\
n.C - C\,.e \\
| & | \\
F & CFH \\
& | \\
& CF_3
\end{array}
\nrightarrow
\begin{array}{cc}
F & CF_3 \\
| & | \\
n.C - C\,.e \\
| & | \\
H & CF_2(CF_3)
\end{array}
\qquad 4.141
$$

(More stable)

$$
\begin{array}{cc}
F & CHF_2 \\
| & | \\
C = C \\
| & | \\
F & CH_2F
\end{array}
\longrightarrow
\begin{array}{cc}
F & CHF_2 \\
| & | \\
n.C - C\,.e \\
| & | \\
F & CH_2F
\end{array}
\nrightarrow
\begin{array}{cc}
F & CHF_2 \\
| & | \\
n.C - C\,.e \\
| & | \\
H & CF_2H
\end{array}
\qquad 4.142
$$

<u>More stable</u>

$$
\begin{array}{c}
\underset{\underset{\displaystyle CH_2}{\displaystyle |}}{\overset{\displaystyle F}{\underset{\displaystyle |}{C}}} = \overset{\displaystyle CHF_2}{\underset{\displaystyle |}{\underset{\displaystyle |}{C}}} \\
\underset{\displaystyle CF_3}{}
\end{array}
\longrightarrow
\begin{array}{c}
n.\overset{\displaystyle F}{\underset{\displaystyle F}{C}} - \overset{\displaystyle CHF_2}{\underset{\displaystyle CH_2}{C}}.e \\
CF_3 \rule{0pt}{0pt} CF_3
\end{array}
\xrightarrow{} \!\!\!\!\!/
\begin{array}{c}
n.\overset{\displaystyle F}{\underset{\displaystyle H}{C}} - \overset{\displaystyle CHF_2}{\underset{\displaystyle CF_2(CF_3)}{C}}.e
\end{array}
\qquad 4.143
$$

$$\underline{\text{More stable}}$$

$$
\begin{array}{c}
\overset{\displaystyle CF_3}{\underset{\displaystyle |}{}} \\
\underset{\underset{\displaystyle CH_2}{\displaystyle |}}{\overset{\displaystyle F}{\underset{\displaystyle |}{C}}} = \overset{\displaystyle CHF}{\underset{\displaystyle |}{\underset{\displaystyle |}{C}}} \\
\underset{\displaystyle CF_3}{}
\end{array}
\longrightarrow
\begin{array}{c}
CF_3 \\
n.\overset{\displaystyle F}{\underset{\displaystyle F}{C}} - \overset{\displaystyle CHF}{\underset{\displaystyle CH_2}{C}}.e \\
CF_3
\end{array}
\xrightarrow{} \!\!\!\!\!/
\begin{array}{c}
CF_3 \\
n.\overset{\displaystyle F}{\underset{\displaystyle H}{C}} - \overset{\displaystyle CHF}{\underset{\displaystyle CF_2(CF_3)}{C}}.e
\end{array}
\qquad 4.144
$$

$$\underline{\text{More stable}}$$

Free-radically molecular rearrangement is not favored by the monomers in this group despite the fact that the number of substituent groups on the carbon center remains the same. CHEMISTRY is deeply logical in Nature.

For the second group in the list $CFR_{F1} = CFR_{F2}$, chargedly, they cannot be activated. Free-radically, the followings are obtained.

$$
\begin{array}{c}
\underset{\displaystyle CF_3}{\overset{\displaystyle F}{C}} = \underset{\underset{\displaystyle CF_3}{\displaystyle CFH}}{\overset{\displaystyle F}{C}}
\end{array}
\longrightarrow
\begin{array}{c}
n.\overset{\displaystyle F}{\underset{\displaystyle CF_3}{C}} - \overset{\displaystyle F}{\underset{\underset{\displaystyle CF_3}{\displaystyle CFH}}{C}}.e
\end{array}
\xrightarrow{} \!\!\!\!\!/
\begin{array}{c}
n.\overset{\displaystyle F}{\underset{\displaystyle H}{C}} - \overset{\displaystyle F}{\underset{\displaystyle CF_3 \, CF_3}{C}}.e
\end{array}
\qquad 4.145
$$

$$\text{(Impossible existence)}$$

$$
\begin{array}{c}
\underset{\displaystyle CF_3}{\overset{\displaystyle F}{C}} = \underset{\displaystyle CF_2H}{\overset{\displaystyle F}{C}}
\end{array}
\longrightarrow
\begin{array}{c}
n.\overset{\displaystyle F}{\underset{\displaystyle CF_3}{C}} - \overset{\displaystyle F}{\underset{\displaystyle CF_2H}{C}}.e
\end{array}
\xrightarrow{} \!\!\!\!\!/
\begin{array}{c}
n.\overset{\displaystyle F}{\underset{\displaystyle F}{C}} - \overset{\displaystyle F}{\underset{\underset{\displaystyle CF_3}{\displaystyle CFH}}{C}}.e
\end{array}
\qquad 4.146
$$

$$\underline{\text{(Impossible Transfer (H>CF}_2)}$$

$$
\begin{array}{c}
\underset{\displaystyle CF_3}{\overset{\displaystyle F}{C}} = \underset{\underset{\displaystyle CF_3}{\displaystyle CF_2}}{\overset{\displaystyle F}{C}}
\end{array}
\longrightarrow
\begin{array}{c}
n.\overset{\displaystyle F}{\underset{\displaystyle CF_3}{C}} - \overset{\displaystyle F}{\underset{\underset{\displaystyle CF_3}{\displaystyle CF_2}}{C}}.e
\end{array}
\longrightarrow
\begin{array}{c}
n.\overset{\displaystyle F}{\underset{\displaystyle F}{C}} -- \overset{\displaystyle F}{\underset{\displaystyle CF_3 \, CF_3}{C}}.e\cdot
\end{array}
\qquad 4.147
$$

$$
\begin{array}{c}
\underset{\displaystyle CF_3}{\overset{\displaystyle F}{C}} = \underset{\displaystyle CH_2F}{\overset{\displaystyle F}{C}}
\end{array}
\longrightarrow
\begin{array}{c}
n.\overset{\displaystyle F}{\underset{\displaystyle CF_3}{C}} - \overset{\displaystyle F}{\underset{\displaystyle CH_2F}{C}}.e
\end{array}
\xrightarrow{} \!\!\!\!\!/
\begin{array}{c}
n.\overset{\displaystyle H}{\underset{\displaystyle F}{C}} - \overset{\displaystyle F}{\underset{\underset{\displaystyle CF_3}{\displaystyle CFH}}{C}}.e
\end{array}
\qquad 4.148
$$

$$\text{(Impossible transfer)}$$

$$
\begin{array}{c}
\overset{F}{\underset{CF_3}{C}} = \overset{F}{\underset{CH_2}{C}} \quad \longrightarrow \quad n.\overset{F}{\underset{CF_3}{C}} - \overset{F}{\underset{CH_2}{C}}.e \quad \nrightarrow \quad n.\overset{H}{\underset{H}{C}} - \overset{F}{\underset{CF}{C}}.e \\
\underset{CF_3}{} \underset{CF_3}{} \underset{CF_3 \;\; CF_3}{}
\end{array}
\qquad 4.149
$$

(More stable)   (Impossible existence)

$$
\overset{F}{\underset{CH_2F}{C}} = \overset{F}{\underset{CF_2H}{C}} \;\longrightarrow\; e.\overset{F}{\underset{CH_2F}{C}} - \overset{F}{\underset{CF_2H}{C}}.n \;\longrightarrow\; n.\overset{F}{\underset{H}{C}} - \overset{F}{\underset{CFH}{C}}.e
$$
$$
\underset{CF_2H}{}
$$
$$
\qquad\qquad\qquad\qquad\qquad\qquad\qquad\qquad\qquad\qquad\qquad\qquad 4.150
$$

$$
\overset{F}{\underset{CH_2}{C}} = \overset{F}{\underset{CF_2H}{C}} \;\longrightarrow\; e.\overset{F}{\underset{CH_2}{C}} - \overset{F}{\underset{CF_2H}{C}}.n \;\nrightarrow\; n.\overset{H}{\underset{H}{C}} - \overset{F}{\underset{CF}{C}}.e
$$
$$
\underset{CF_3}{} \underset{CF_3}{} \underset{CF_3 \;\; CF_3H}{}
$$
$$
\qquad\qquad\qquad\qquad\qquad\qquad\qquad\qquad\qquad\qquad\qquad\qquad 4.151
$$

(Impossible existence)

Only the fully fluorinated one but one (Equation 3.150) favored molecular rearrangement. It can be noticed why the second driving force for molecular rearrangement is provision of a more stable activated monomer which can readily be deactivated when desired. ***Having an activated monomer with reduced substituent group is another driving force for molecular rearrange-ment.*** All those where molecular rearrangement was not favored, can be observe to favor no free-radical route. The fully fluorinated one of Equation 3.147 and that of Equation 3.150 were the only ones found to favor the route natural to them, the electro-free-radical route. One can observe that these are strong Nucleophiles.

The charged/radical-pushing and -pulling capacities of the groups can readily be identified. For example, worthy of note are the followings-

$$C(CF_3)_3 > CF(CF_3)_2 > CF_2CF_3 > CF_3 > F$$

Order of *charged*-pulling capacity   4.152a

$$H > C(CF_3)_3 > CF(CF_3)_2 > CF_2CF_3 > CF_3$$

Order of *radical*-pushing capacity   4.152b

$$C(CF_3)_3 > CH(CF_3)_2 > CFHCF_3 > CF_2H > F$$

Order of *charged*-pulling capacity   4.152c

$$CH_3 > CH_2F > CHF_2 > H > CF_3$$

Order of *radical*-pushing capacity   4.152d

For the last group of monomers in this section consider $CFR_{F1} = CR_{F2}R_{F3}$ group. Chargedly, the followings are expected, if charged activation is possible.

$$
\begin{array}{c}
\overset{\displaystyle F}{\underset{\displaystyle CF_2H}{C}} = \overset{\displaystyle CF_3}{\underset{\displaystyle \underset{\displaystyle CF_3}{|}CF_2}{C}}
\longrightarrow
\overset{\displaystyle F}{\underset{\displaystyle CF_2H}{\oplus C}} - \overset{\displaystyle CF_3}{\underset{\displaystyle \underset{\displaystyle CF_3}{|}CF_2}{C\ominus}}
\nrightarrow
\overset{\displaystyle F}{\underset{\displaystyle CF_3}{\oplus C}} - \overset{\displaystyle CF_3}{\underset{\displaystyle \underset{\displaystyle CF_2H}{|}CF_2}{C\ominus}}
\end{array}
\qquad 4.153
$$

$$\text{More stable } (R_{F3} > R_{F2} > R_{F1}) \; C$$

$$
\begin{array}{c}
\overset{\displaystyle F}{\underset{\displaystyle CFH_2}{C}} = \overset{\displaystyle CF_3}{\underset{\displaystyle CF_2H}{C}}
\longrightarrow
\overset{\displaystyle F}{\underset{\displaystyle CFH_2}{\oplus C}} - \overset{\displaystyle CF_3}{\underset{\displaystyle CF_2H}{C\ominus}}
\nrightarrow
\overset{\displaystyle F}{\underset{\displaystyle F}{\oplus C}} - \overset{\displaystyle CF_2H}{\underset{\displaystyle \underset{\displaystyle CFH_2}{|}CF_2}{C\ominus}}
\end{array}
\qquad 4.154
$$

$$\text{More stable}$$
$$(R_{F2} > R_{F3} > R_{F1})$$

$$
\begin{array}{c}
\overset{\displaystyle F}{\underset{\displaystyle CFH_2}{C}} = \overset{\displaystyle \overset{\displaystyle CF_3}{|}CF_3}{\underset{\displaystyle CF_2H}{C}}
\longrightarrow
\overset{\displaystyle F}{\underset{\displaystyle CFH_2}{\oplus C}} - \overset{\displaystyle \overset{\displaystyle CF_3}{|}CF_2}{\underset{\displaystyle CF_2H}{C\ominus}}
\nrightarrow
\overset{\displaystyle F}{\underset{\displaystyle CF_3}{\oplus C}} - \overset{\displaystyle CF_2H}{\underset{\displaystyle \underset{\displaystyle CFH_2}{|}CF_2}{C\ominus}}
\end{array}
\qquad 4.155
$$

$$\text{More stable} \qquad (R_{F2} > R_{F3} > R_{F1})$$

For the cases above like many encountered so far, $H^{\oplus}$ cannot be transferred due to electrostatic forces of repulsion when placed on an active center carrying negative charge.

$$
\begin{array}{c}
\overset{\displaystyle F}{\underset{\displaystyle \underset{\displaystyle CF_3}{|}CF_2}{C}} = \overset{\displaystyle CHF_2}{\underset{\displaystyle CH_2F}{C}}
\nrightarrow
\ominus\overset{\displaystyle F}{\underset{\displaystyle \underset{\displaystyle CF_3}{|}CF_2}{C}} - \overset{\displaystyle CHF_2}{\underset{\displaystyle CH_2F}{C\oplus}}
\longrightarrow
\oplus\overset{\displaystyle F}{\underset{\displaystyle CF_3}{C}} - \overset{\displaystyle F}{\underset{\displaystyle \underset{\displaystyle CH_2F}{|}CF(CHF_2)}{C\ominus}}
\end{array}
$$

$$\text{(Impossible existence)} \qquad \text{(Impossible existence)} \qquad 4.156$$
$$(R_{F1} > R_{F2} > R_{F3})$$

$$
\begin{array}{c}
\overset{\displaystyle F}{\underset{\displaystyle \underset{\displaystyle CF_3}{|}CF_2}{C}} = \overset{\displaystyle CH_2F}{\underset{\displaystyle CH_2F}{C}}
\nrightarrow
\ominus\overset{\displaystyle F}{\underset{\displaystyle \underset{\displaystyle CF_3}{|}CF_2}{C}} - \overset{\displaystyle CH_2F}{\underset{\displaystyle CH_2F}{C\oplus}}
\end{array}
\qquad 4.157
$$

$$\text{(Impossible existence) } (RF1 > RF3 = RF2) \; C$$

It can be observed that the last two monomers will not favor charged activation, in view of the relative capacities of the substituent groups and therefore presence of electrostatic forces of repulsion. Since the establishment of capacities of substituted groups is very important, this is already being systematically identified for very difficult cases. The last two cases above are almost similar to the two activated

Sunny N.E. Omorodion

monomers shown below in terms of the capacities on the two active carbon centers, but very different since they cannot be activated chargedly. Note so far that molecular rearrangement chargedly cannot take place with these monomers.

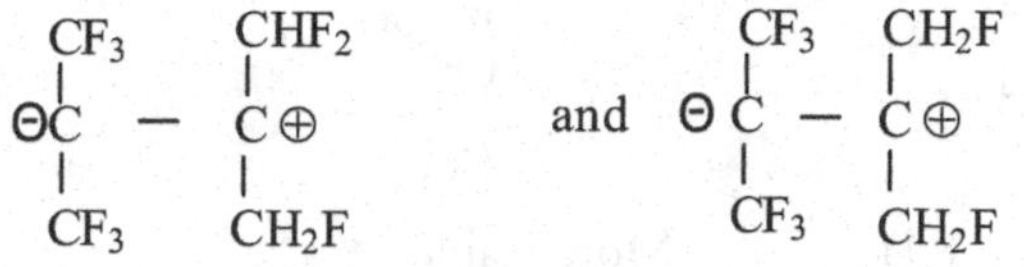

$$\overset{CF_3}{\underset{CF_3}{\ominus C}} - \overset{CHF_2}{\underset{CH_2F}{C\oplus}} \quad \text{and} \quad \overset{CF_3}{\underset{CF_3}{\ominus C}} - \overset{CH_2F}{\underset{CH_2F}{C\oplus}} \qquad 4.158$$

$$\overset{F}{\underset{CF_3}{C}} = \overset{CF_3}{\underset{\underset{CF_3}{CF_2}}{C}} \longrightarrow \overset{F}{\underset{CF_3}{\oplus C}} - \overset{CF_3}{\underset{\underset{CF_3}{CF_2}}{C\ominus}} \longrightarrow \overset{F}{\underset{CF_3}{\oplus C}} - \overset{CF_3}{\underset{\underset{CF_3}{CF_2}}{C\ominus}} \qquad 4.159$$

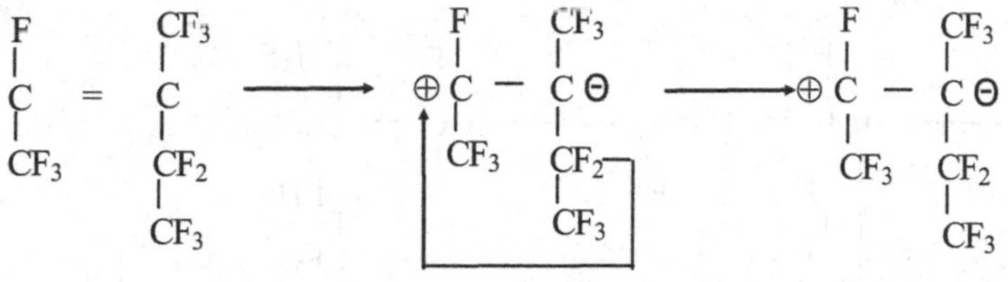

$$(R_{F3} > R_{F2} = R_{F1})$$

None of the monomers here seem to favor molecular rearrangement chargedly using F as transfer species, since the monomers are Nucleophiles, except the last case where there is no H atom.

Free-radically, the followings are obtained.

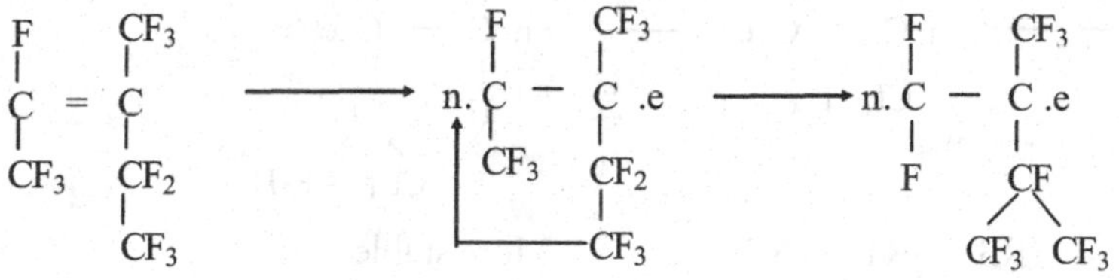

$$\overset{F}{\underset{CF_3}{C}} = \overset{CF_3}{\underset{\underset{CF_3}{CF_2}}{C}} \longrightarrow n.\overset{F}{\underset{CF_3}{C}} - \overset{CF_3}{\underset{\underset{CF_3}{CF_2}}{C}}.e \longrightarrow n.\overset{F}{\underset{F}{C}} - \overset{CF_3}{\underset{\overset{CF}{CF_3\ CF_3}}{C}}.e \qquad 4.160a$$

$$(R2 = R1 > R3) \qquad \text{More stable}$$

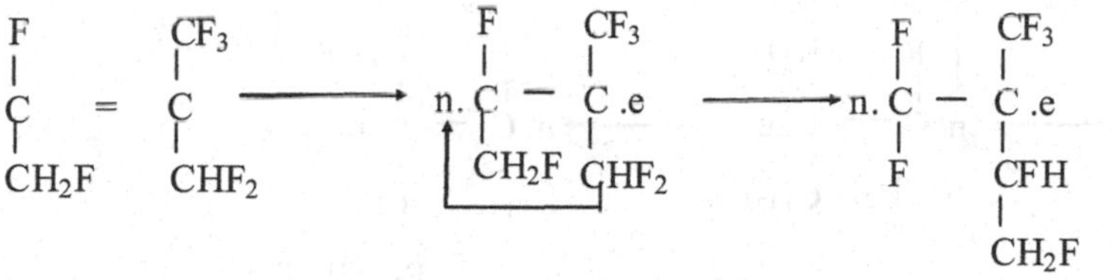

$$\overset{F}{\underset{CH_2F}{C}} = \overset{CF_3}{\underset{CHF_2}{C}} \longrightarrow n.\overset{F}{\underset{CH_2F}{C}} - \overset{CF_3}{\underset{CHF_2}{C}}.e \longrightarrow n.\overset{F}{\underset{F}{C}} - \overset{CF_3}{\underset{\overset{CFH}{CH_2F}}{C}}.e \qquad 4.160b$$

$$(R1 > R3 > R2) \qquad \underline{\text{More stable}}$$

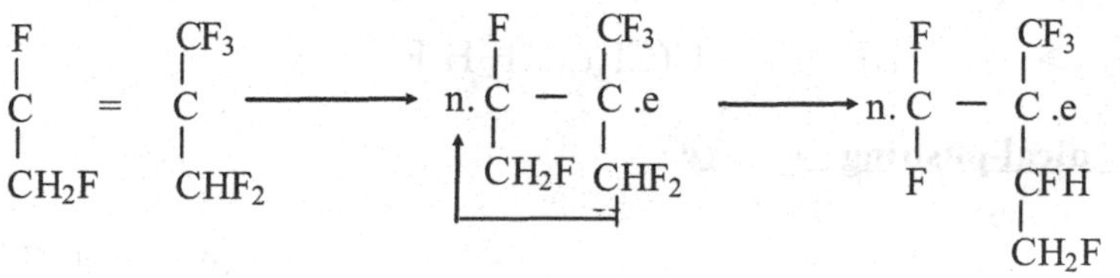

$$\overset{F}{\underset{CH_2F}{C}} = \overset{CF_3}{\underset{CHF_2}{C}} \longrightarrow n.\overset{F}{\underset{CH_2F}{C}} - \overset{CF_3}{\underset{CHF_2}{C}}.e \longrightarrow n.\overset{F}{\underset{F}{C}} - \overset{CF_3}{\underset{\overset{CFH}{CH_2F}}{C}}.e \qquad 4.161$$

$$(R1 > R3 > R2) \qquad \underline{\text{More stable}}$$

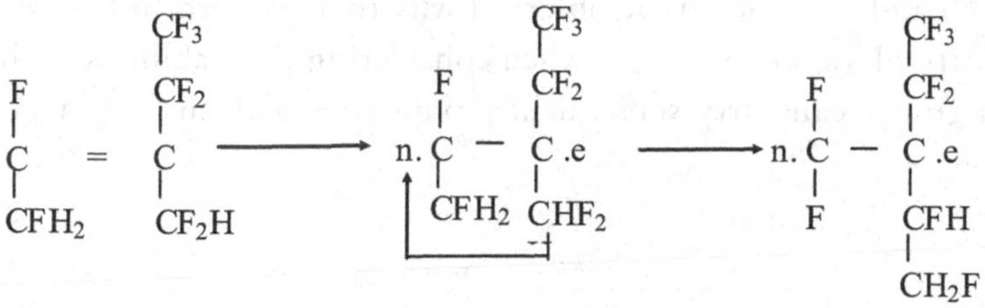

$$\overset{F}{\underset{CFH_2}{C}} = \overset{\overset{CF_3}{CF_2}}{\underset{CF_2H}{C}} \longrightarrow n.\overset{F}{\underset{CFH_2}{C}} - \overset{\overset{CF_3}{CF_2}}{\underset{CHF_2}{C}}.e \longrightarrow n.\overset{F}{\underset{F}{C}} - \overset{\overset{CF_3}{CF_2}}{\underset{\overset{CFH}{CH_2F}}{C}}.e \qquad 4.162$$

$$(R1 > R3 > R2) \qquad \underline{\text{More stable}}$$

214

$$
\begin{array}{c}
\underset{\overset{|}{CF_2}}{\overset{F}{\underset{|}{C}}} = \underset{\overset{|}{CH_2F}}{\overset{CHF_2}{\underset{|}{C}}} \longrightarrow n.\underset{\overset{|}{CF_2}}{\overset{F}{\underset{|}{C}}} - \underset{\overset{|}{CH_2F}}{\overset{CHF_2}{\underset{|}{C}}}.e \longrightarrow n.\underset{\overset{|}{H}}{\overset{F}{\underset{|}{C}}} - \underset{\overset{|}{CFH}}{\overset{CHF_2}{\underset{|}{C}}}.e \\
\underset{CF_3}{|} \qquad\qquad\qquad\qquad \underset{CF_3}{|} \qquad\qquad\qquad \underset{\underset{CF_3}{|}}{\overset{CF_2}{|}}
\end{array}
$$

$$\text{(R3} > \text{R2} > \text{R1)} \qquad \underline{\text{More stable}} \qquad\qquad 4.163$$

$$
\underset{\overset{|}{CF_2}}{\overset{F}{\underset{|}{C}}} = \underset{\overset{|}{CH_2F}}{\overset{CH_2F}{\underset{|}{C}}} \longrightarrow n.\underset{\overset{|}{CF_2}}{\overset{F}{\underset{|}{C}}} - \underset{\overset{|}{CH_2F}}{\overset{CH_2F}{\underset{|}{C}}}.e \longrightarrow n.\underset{\overset{|}{F}}{\overset{F}{\underset{|}{C}}} - \underset{\overset{|}{CFH}}{\overset{CH_2F}{\underset{|}{C}}}.e
$$

$$\text{(R3} = \text{R2} > \text{R1)} \qquad \underline{\text{More stable}} \qquad\qquad 4.164$$

$$
\underset{\overset{|}{CF_2H}}{\overset{F}{\underset{|}{C}}} = \underset{\overset{|}{CH_2F}}{\overset{\overset{CF_3}{|}}{\underset{|}{\overset{CF_2}{|}}}} \longrightarrow n.\underset{\overset{|}{CF_2H}}{\overset{F}{\underset{|}{C}}} - \underset{\overset{|}{CH_2F}}{\overset{\overset{CF_3}{|}}{\underset{|}{CH_2}}}.e \longrightarrow n.\underset{\overset{|}{H}}{\overset{H}{\underset{|}{C}}} - \underset{\underset{CF_3 \quad CF_2H}{\diagup \diagdown}}{\overset{CH_2F}{\underset{|}{C}}}.e
$$

$$\text{(R3} > \text{R1} > \text{R2)} \qquad \text{More stable} \qquad\qquad 4.165$$

$$
\underset{\overset{|}{CH_2F}}{\overset{F}{\underset{|}{C}}} = \underset{\overset{|}{CH_2F}}{\overset{\overset{CF_3}{|}}{\underset{|}{\overset{CH_2}{|}}}} \longrightarrow n.\underset{\overset{|}{CH_2F}}{\overset{F}{\underset{|}{C}}} - \underset{\overset{|}{CH_2F}}{\overset{\overset{CF_3}{|}}{\underset{|}{CH_2}}}.e \longrightarrow n.\underset{\overset{|}{H}}{\overset{H}{\underset{|}{C}}} - \underset{\underset{CF_3 \quad CH_2F}{\diagup \diagdown}}{\overset{CH_2F}{\underset{|}{C}}}.e
$$

$$\text{(R2} > \text{R3} = \text{R1)} \qquad \text{More stable} \qquad\qquad 4.166a$$

$$-C(CF_3)(CH_2F)F \quad > \quad -CH_2F \quad > \quad -C(CF_3)(CF_2H)F$$

$$\textbf{\underline{Order of radical-pushing capacity}} \qquad\qquad 4.166b$$

The conditions under which molecular rearrangement is favored can be observed. Though the monomers are nucleophiles, when the activated states in Equations 4.160a to 4.164 are compared with those of Equations 4.165 and 166a, it can be observed why there is need to look deeply at the types of substituted groups carried by active centers, when considering the stabilities of the activated monomers. The substituted groups can carry substituent groups such as F in CF_3 or H in CH_3 or none at all.

4.7 Characteristics of CF2 = CFR, CF2 = CRR, CFR = CFR, CFR = CRR groups.

Beginning with the first group, consider the followings.

$$
\begin{array}{cc}
F & F \\
| & | \\
C & = \ C \\
| & | \\
F & CH_3
\end{array}
\quad \longrightarrow \quad
\begin{array}{cc}
F & F \\
| & | \\
\ominus C & - \ C\oplus \\
| & | \\
F & CH_3
\end{array}
\qquad\qquad 4.167
$$

(Impossible existence)

Chargedly, activations of the monomers in this group are impossible, due to electrostatic forces of repulsion.

$$
\begin{array}{cc}
F & F \\
| & | \\
C & = \ C \\
| & | \\
F & CH_3
\end{array}
\longrightarrow
\begin{array}{cc}
F & F \\
| & | \\
n.\,C & - \ C\,.e \\
| & | \\
F & CH_3
\end{array}
\ \not\longrightarrow \
\begin{array}{cc}
H & F \\
| & | \\
n.\,C & - \ C\,.e \\
| & | \\
H & CF_2H
\end{array}
\qquad 4.168
$$

<u>More stable</u>

$$
\begin{array}{cc}
F & F \\
| & | \\
C & = \ C \\
| & | \\
F & CH_2 \\
 & | \\
 & CH_3
\end{array}
\longrightarrow
\begin{array}{cc}
F & F \\
| & | \\
n.\,C & - \ C\,.e \\
| & | \\
F & CH_2{}^- \\
 & | \\
 & CH_3
\end{array}
\ \not\longrightarrow \
\begin{array}{cc}
H & F \\
| & | \\
n.\,C & - \ C\,.e \\
| & | \\
CH_3 & CF_2H
\end{array}
\qquad 4.169
$$

<u>More stable</u>   <u>Impossible existence</u>

Free-radically, molecular rearrangement or even transfer of transfer species to produce the same monomer does not take place. Note that the placement of F on its active center must always show up after rearrangement.

Considering next the second group, $CF_2 = CRR$, chargedly they cannot be activated. Free-radically, the followings are obtained.

$$
\begin{array}{cc}
F & CH_3 \\
| & | \\
C & = \ C \\
| & | \\
F & CH_3
\end{array}
\longrightarrow
\begin{array}{cc}
F & CH_3 \\
| & | \\
n.\,C & - \ C\,.e \\
| & | \\
F & CH_3
\end{array}
\ \not\longrightarrow \
\begin{array}{cc}
H & CH_3 \\
| & | \\
n.\,C & - \ C\,.e \\
| & | \\
H & CF_2H
\end{array}
\qquad 4.170
$$

<u>More stable</u>

$$
\begin{array}{cc}
 & CH_3 \\
 & | \\
F & CH_2 \\
| & | \\
C & = \ C \\
| & | \\
F & CH_2 \\
 & | \\
 & CH_3
\end{array}
\longrightarrow
\begin{array}{cc}
 & CH_3 \\
 & | \\
F & CH_2 \\
| & | \\
n.\,C & - \ C\,.e \\
| & | \\
F & CH_2{}^- \\
 & | \\
 & CH_3
\end{array}
\ \not\longrightarrow \
\begin{array}{cc}
H & C_2H_5 \\
| & | \\
n.\,C & - \ C\,.e \\
| & | \\
CH_3 & CF_2H
\end{array}
\qquad 4.171
$$

Free-radically, there is no doubt that molecular rearrangement is not favored by the monomers.

For the next groups of monomers CFR = CFR, chargedly they cannot also be activated. Free-radically, the followings are to be expected.

$$\text{More stable} \qquad 4.172$$

$$\text{More stable} \qquad 4.173$$

Molecular rearrangement is not favored here with all of them favoring only the electro-free-radical route, the natural route.

With the CFR = CRR group, charged activation is still not possible. Free-radically the followings are obtained.

$$\text{More stable} \qquad 4.174$$

$$\text{More stable} \qquad 4.175$$

Like the previous group, all will only favor polymerization electro-free-radically with or without molecular rearrangement.

4.8 Characteristics of CF2 = CR1R2, CFR1 = CFR2 and CFR1 = CR2R3 groups.

Chargedly, none of the group members will favor activation. Free-radically the follow-ings are to be expected.

$$
\begin{array}{ccc}
\overset{\displaystyle F}{\underset{\displaystyle F}{|}}C = \overset{\displaystyle CH_3}{\underset{\displaystyle \underset{CH_3}{|} CH_2}{|}}C & \longrightarrow & n.\overset{\displaystyle F}{\underset{\displaystyle F}{|}}C - \overset{\displaystyle CH_3}{\underset{\displaystyle \underset{CH_3}{|} CH_2}{|}}C.e & \nrightarrow & n.\overset{\displaystyle H}{\underset{\displaystyle CH_3}{|}}C - \overset{\displaystyle CH_3}{\underset{\displaystyle CF_2H}{|}}C.e
\end{array}
$$

$$\underline{\text{More stable}}$$

4.176

$$
\begin{array}{ccc}
\overset{\displaystyle F}{\underset{\displaystyle F}{|}}C = \overset{\displaystyle CH_3}{\underset{\displaystyle CF_2H}{|}}C & \longrightarrow & n.\overset{\displaystyle F}{\underset{\displaystyle F}{|}}C - \overset{\displaystyle CH_3}{\underset{\displaystyle CF_2H}{|}}C.e & \nrightarrow & n.\overset{\displaystyle H}{\underset{\displaystyle H}{|}}C - \overset{\displaystyle CF_2H}{\underset{\displaystyle CF_2H}{|}}C.e
\end{array}
$$

$$\underline{\text{More stable }}(CF_2 = CRR_F)$$

4.177

Molecular rearrangement is not favored by the monomers in the first group.

For the second group, the followings are obtained.

$$
\begin{array}{ccc}
\overset{\displaystyle F}{\underset{\displaystyle CH_3}{|}}C = \overset{\displaystyle F}{\underset{\displaystyle \underset{CH_3}{|} CH_2}{|}}C & \longrightarrow & n.\overset{\displaystyle F}{\underset{\displaystyle CH_3}{|}}C - \overset{\displaystyle F}{\underset{\displaystyle \underset{CH_3}{|} CH_2}{|}}C.e & \nrightarrow & n.\overset{\displaystyle H}{\underset{\displaystyle CH_3}{|}}C - \overset{\displaystyle F}{\underset{\displaystyle \underset{CH_3}{|} CFH}{|}}C.e
\end{array}
$$

(Impossible existence)   4.178

$$
\begin{array}{ccc}
\overset{\displaystyle F}{\underset{\displaystyle CH_3}{|}}C = \overset{\displaystyle F}{\underset{\displaystyle CH_2F}{|}}C & \longrightarrow & e.\overset{\displaystyle F}{\underset{\displaystyle CH_3}{|}}C - \overset{\displaystyle F}{\underset{\displaystyle CH_2F}{|}}C.n & \nrightarrow & n.\overset{\displaystyle H}{\underset{\displaystyle H}{|}}C - \overset{\displaystyle F}{\underset{\displaystyle \underset{CH_2F}{|} CFH}{|}}C.e
\end{array}
$$

4.179

$$\underline{\text{More stable }}(CFR = CFR_F)\ R > R_F$$

None of the monomers will favor molecular rearrangement. All but the last one can only be polymerized electro-free-radically. The last one will favor no free-radical route.

$$
\begin{array}{ccc}
\overset{\displaystyle F}{\underset{\displaystyle CH_3}{|}}C = \overset{\displaystyle C_2H_5}{\underset{\displaystyle CH_3}{|}}C & \longrightarrow & n.\overset{\displaystyle F}{\underset{\displaystyle CH_3}{|}}C - \overset{\displaystyle CH_3}{\underset{\displaystyle \underset{CH_3}{|} CH_2}{|}}C.e & \nrightarrow & n.\overset{\displaystyle H}{\underset{\displaystyle CH_3}{|}}C - \overset{\displaystyle CH_3}{\underset{\displaystyle \underset{CH_3}{|} CFII}{|}}C.e
\end{array}
$$

$$\underline{\text{More stable }}(R_3 > R_2 = R_1)$$

4.180

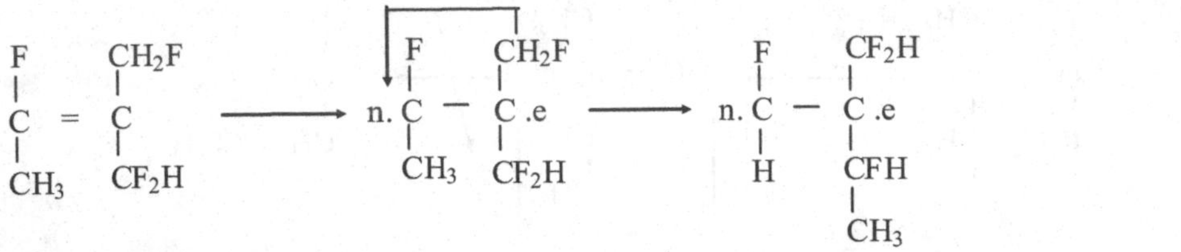

$$
\begin{array}{ccc}
\underset{|}{\overset{F}{C}} = \underset{|}{\overset{CH_2F}{C}} & \longrightarrow & \overset{F}{n.C} - \overset{CH_2F}{C.e} & \longrightarrow & \overset{F}{n.C} - \overset{CF_2H}{C.e} \\
CH_3 & CF_2H & CH_3 & CF_2H & H & CFH \\
& & & & & CH_3
\end{array}
$$

4.181

$(R > R_{F1} > R_{F2})$ \qquad \underline{More stable} $(CFR = CR_{F1} R_{F2})$

Only the last case above and the likes favor molecular rearrangement and polymerizations are favored by all of them only electro-free-radically. It is important to note that groups such as $CF_2 = CRR_F$, $CFR = CFR_F$ and $CFR = CR_{F1}R_{F2}$ have been partly considered above. However, they will still be given their due classifications.

4.9 Characteristics of CF2 = CRRF, CFR = CFRF, CFR = CRFRF, CFRF = CRRF groups

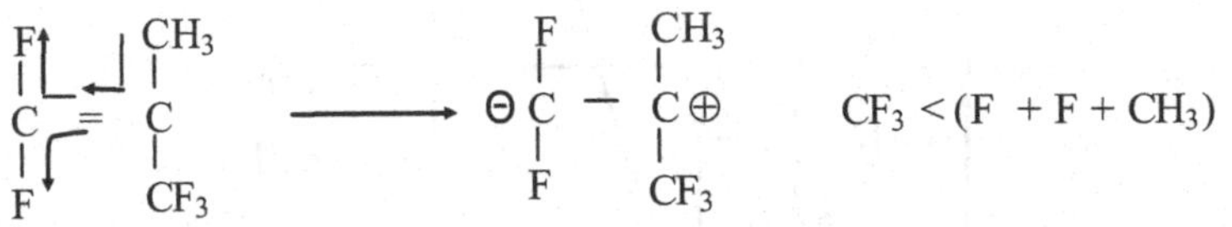

$$
\begin{array}{ccc}
\overset{F}{C} = \overset{CH_3}{C} & \longrightarrow & \overset{F}{\ominus C} - \overset{CH_3}{C \oplus} & \qquad CF_3 < (F + F + CH_3) \\
F & CF_3 & F & CF_3
\end{array}
$$

4.182

Impossible existence (if condition above is valid)

But if substituted groups such as CN, NO_2 are used in place of CF_3, the reverse will be the case, because their charged-pulling capacities are about twice as much as that of CF_3.

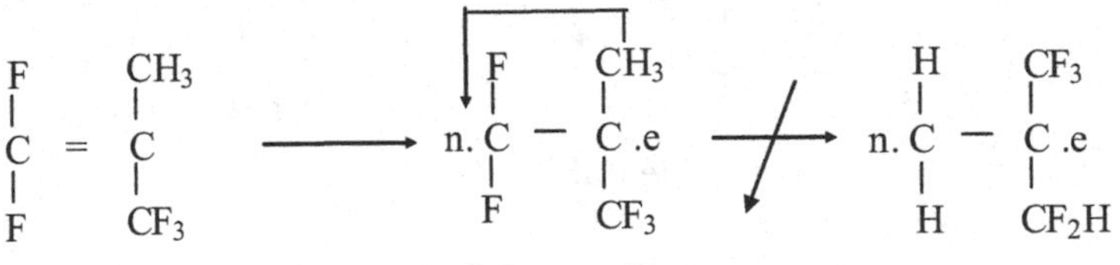

$$
\begin{array}{ccc}
\overset{F}{C} = \overset{CH_3}{C} & \longrightarrow & \overset{F}{n.C} - \overset{CH_3}{C.e} & \nrightarrow & \overset{H}{n.C} - \overset{CF_3}{C.e} \\
F & CF_3 & F & CF_3 & H & CF_2H
\end{array}
$$

4.183

\underline{More stable}

$$
\begin{array}{ccc}
\overset{F}{C} = \overset{CH_3}{C} & \longrightarrow & \overset{F}{n.C} - \overset{CH_3}{C.e} & \nrightarrow & \overset{H}{n.C} - \overset{CHF_2}{C.e} \\
F & CF_2 & F & CF_2 & H & CF_2 \\
& CH_3 & & CH_3 & & CH_3
\end{array}
$$

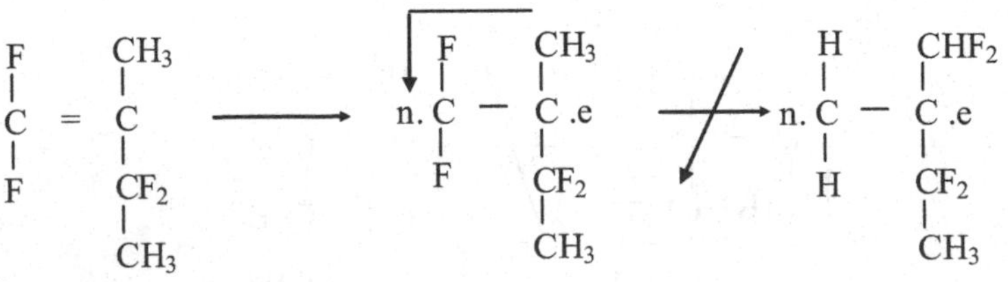

4.184

It should be noted that $-CH_3 > -CF_2(CH_3)$ in pushing-capacity, and so also $-CF_2H > -H$ in pushing-capacity free-radically.

$$
\begin{array}{cc}
F & CH_3 \\
| & | \\
C & = \ C \\
| & | \\
F & CFH \\
& | \\
& CH_3
\end{array}
\longrightarrow
\begin{array}{cc}
F & CH_3 \\
| & | \\
n.\,C & - \ C\,.e \\
| & | \\
F & CFH \\
& | \\
& CH_3
\end{array}
\xrightarrow{\quad\big/\quad}
\begin{array}{cc}
F & CH_3 \\
| & | \\
n.\,C & - \ C\,.e \\
| & | \\
CH_3 & CF_2H
\end{array}
$$

$$\text{\underline{More stable}}$$

4.185

It seems that molecular rearrangement is not favored by the monomers in this group.

Chargedly for the second group, the monomers cannot be activated. Free-radically, the followings are to be expected.

$$
\begin{array}{cc}
F & F \\
| & | \\
C & = \ C \\
| & | \\
CH_3 & CF_3
\end{array}
\longrightarrow
\begin{array}{cc}
F & F \\
| & | \\
e.\,C & - \ C\,.n \\
| & | \\
CH_3 & CF_3
\end{array}
\xrightarrow{\quad\big/\quad}
\begin{array}{cc}
H & F \\
| & | \\
n.\,C & - \ C\,.e \\
| & | \\
H & CFH \\
& | \\
& CF_3
\end{array}
$$

$$\text{(I) (More stable)} \qquad\qquad \text{(II)}$$
$$(R > R_F)$$

4.186

The transfer species transferred, is not the transfer species that will be involved during polymeri-zation of (II) above. *On the other hand, to have an electro-free-radical active C center carrying F in the absence of another radical-pulling group of equal or greater capacity than F is im-possible.* Hence molecular rearrangement is not favored.

$$
\begin{array}{cc}
F & F \\
| & | \\
C & = \ C \\
| & | \\
CH_3 & CF_2 \\
& | \\
& CH_3
\end{array}
\longrightarrow
\begin{array}{cc}
F & F \\
| & | \\
e.\,C & - \ C\,.n \\
| & | \\
CH_3 & CF_2 \\
& | \\
& CH_3
\end{array}
\xrightarrow{\quad\big/\quad}
\begin{array}{cc}
H & F \\
| & | \\
n.\,C & - \ C\,.e \\
| & | \\
H & CHF \\
& | \\
& CF_2 - CH_3
\end{array}
$$

$$\text{More stable} \quad (R > RF)$$

4.187

$$
\begin{array}{cc}
F & F \\
| & | \\
C & = \ C \\
| & | \\
CH_3 & CFH \\
& | \\
& CH_3
\end{array}
\longrightarrow
\begin{array}{cc}
F & F \\
| & | \\
e.\,C & - \ C\,.n \\
| & | \\
CH_3 & CFH \\
& | \\
& CH_3
\end{array}
\xrightarrow{\quad\big/\quad}
\begin{array}{cc}
H & F \\
| & | \\
n.\,C & - \ C\,.e \\
| & | \\
H & CFH \\
& | \\
& CFH \\
& | \\
& CH_3
\end{array}
$$

$$(R > R_F)$$

4.188a

$$-CH_3 > -CR_2F > -CHRF > -CH_2F > -CRF_2 > -CHF_2 > H > -CF_3$$
$$\text{\underline{Order of radical-pushing capacity}}$$

4.188b

As it seems from the last equation above, there will never be a time when $R_F > R$ in radical-pushing capacity. For the third group, the followings are to be expected.

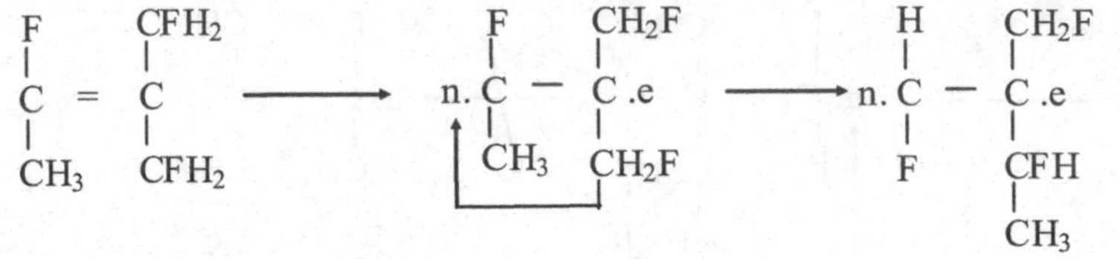

$$
\begin{array}{c}
\overset{\displaystyle F}{\underset{\displaystyle CH_3}{|}}C = \overset{\displaystyle CFH_2}{\underset{\displaystyle CFH_2}{|}}C \longrightarrow n.\overset{\displaystyle F}{\underset{\displaystyle CH_3}{|}}C - \overset{\displaystyle CH_2F}{\underset{\displaystyle CH_2F}{|}}C.e \longrightarrow n.\overset{\displaystyle H}{\underset{\displaystyle F}{|}}C - \overset{\displaystyle CH_2F}{\underset{\displaystyle CFH}{|}}C.e
\end{array} \qquad 4.189
$$

$$CH_3$$

$$R > R_{F1} = R_{F2} \qquad \underline{\text{More stable}}$$

$$
\begin{array}{c}
\overset{\displaystyle F}{\underset{\displaystyle CH_3}{|}}C = \overset{\displaystyle CH_2F}{\underset{\displaystyle CH_2}{|}}C \longrightarrow n.\overset{\displaystyle F}{\underset{\displaystyle CH_3}{|}}C - \overset{\displaystyle CH_2F}{\underset{\displaystyle CH_2}{|}}C.e \longrightarrow n.\overset{\displaystyle H}{\underset{\displaystyle H}{|}}C - \overset{\displaystyle CH_2F}{\underset{\displaystyle CF(CF_3)}{|}}C.e
\end{array}
$$

$$CF_3 \qquad\qquad CF_3 \qquad\qquad CH_3 \qquad 4.190$$

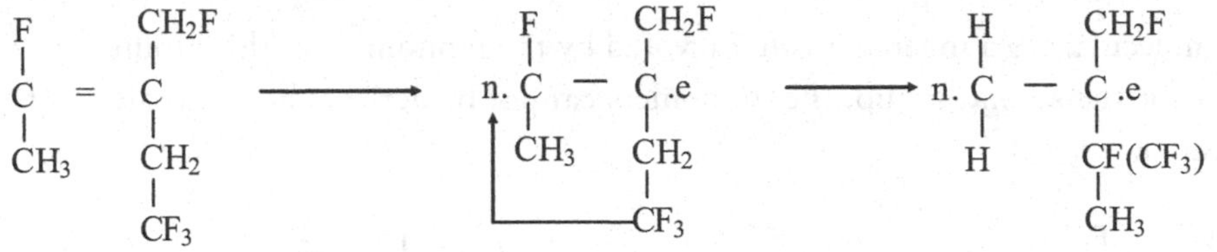

$$R > R_{F1} > R_{F2} \ (CFR = CR_{F1}R_{F2}) \qquad \underline{\text{More stable}}$$

$$
\begin{array}{c}
CH_3 \\
\overset{\displaystyle F}{\underset{\displaystyle CH_3}{|}}C = \overset{\displaystyle CF_2}{\underset{\displaystyle CF_2H}{|}}C \longrightarrow n.\overset{\displaystyle F}{\underset{\displaystyle CH_3}{|}}C - \overset{\displaystyle CF_2}{\underset{\displaystyle CF_2H}{|}}C.e \longrightarrow n.\overset{\displaystyle F}{\underset{\displaystyle F}{|}}C - \overset{\displaystyle CF_2H}{\underset{\displaystyle CF(CH_3)}{|}}C.e
\end{array}
$$

$$CH_3 \qquad 4.191$$

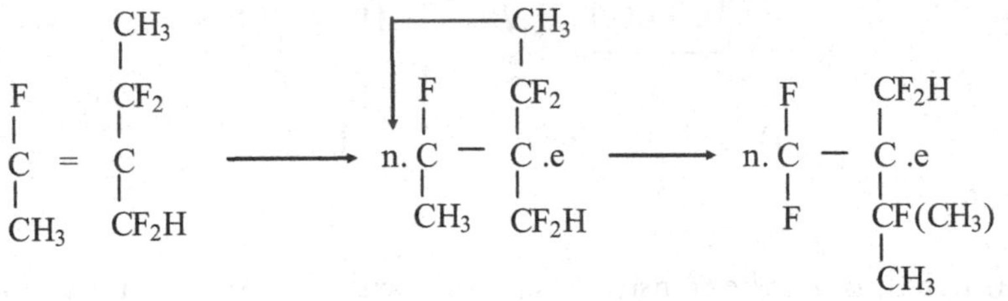

$$\text{(I) } \underline{\text{More stable}}$$

$$R > R_{F2} > R_{F1} \ (CFR = CR_{F1}R_{F2})$$

The transfer species involved above (CH_3) will be the same involved during polymerization of (I). One will wonder why H on CF_2H group cannot be the species involved, in order to have what is shown below, which is more stable than the original activated monomer. The reason is beca-

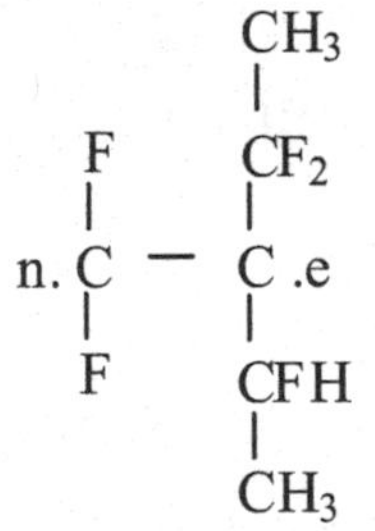

$$
\begin{array}{c}
CH_3 \\
| \\
CF_2 \\
| \\
n.\overset{\displaystyle F}{\underset{\displaystyle F}{|}}C - \overset{}{\underset{}{|}}C.e \\
| \\
CFH \\
| \\
CH_3
\end{array} \qquad 4.192
$$

use the rules have to be adhered to. ***It is the more radical/charged-pushing or -pulling group that provides the transfer species, unless the more radical/charged-pushing or -pulling group does not have transfer species.*** (See Equation 4.188b)

Now, considering the charged activations of the monomers, the followings are obtained.

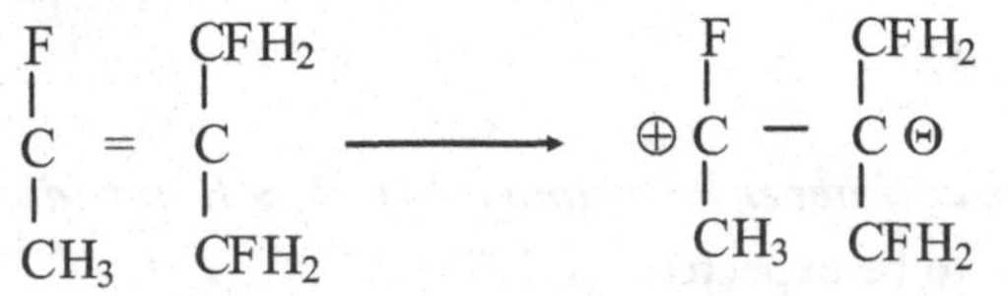

$$
\begin{array}{c}
\overset{\displaystyle F}{\underset{\displaystyle CH_3}{|}}C = \overset{\displaystyle CFH_2}{\underset{\displaystyle CFH_2}{|}}C \longrightarrow \oplus\overset{\displaystyle F}{\underset{\displaystyle CH_3}{|}}C - \overset{\displaystyle CFH_2}{\underset{\displaystyle CFH_2}{|}}C \ominus
\end{array}
$$

$$\text{No route favored} \qquad RF1 = RF2 > R \qquad\qquad 4.193$$

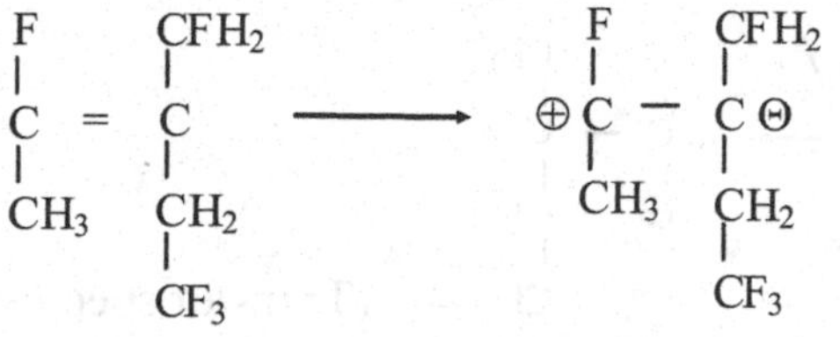

$$\begin{array}{ccc}
\overset{\displaystyle F}{\underset{\displaystyle CH_3}{|}}\!\!\!\overset{|}{\underset{|}{C}} = \overset{\displaystyle CFH_2}{\underset{\displaystyle CH_2}{\overset{|}{\underset{|}{C}}}} & \longrightarrow & \oplus\overset{\displaystyle F}{\underset{\displaystyle CH_3}{\overset{|}{\underset{|}{C}}}} - \overset{\displaystyle CFH_2}{\underset{\displaystyle CH_2}{\overset{|}{\underset{|}{C}}}}\ominus \\
& & \\
\underset{\displaystyle CF_3}{|} & & \underset{\displaystyle CF_3}{|}
\end{array} \qquad 4.194$$

No route favored   RF2 > RF1 > F (CFR=CRF1RF2)

In the absence of electrostatic forces of repulsion and molecular rearrangement, no route can be favored. As it seems, these monomers cannot be activated chargedly. ***When F is connected to an active C center alone, it seems in general that the carbon center can never carry a positive charge or an electro-free-radical, provided two groups on the active C centers are not of equal or greater charge-pulling or radical-pulling capacity than that of F.*** The monomers will therefore not favor charged polymerization.

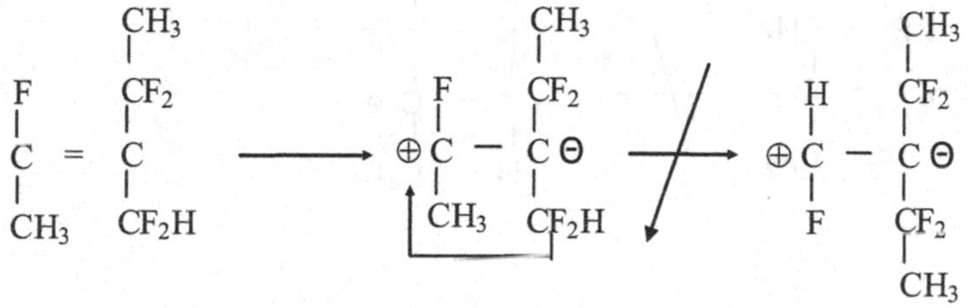

$$\begin{array}{ccccc}
\overset{\displaystyle CH_3}{|} & & \overset{\displaystyle CH_3}{|} & & \overset{\displaystyle CH_3}{|} \\
\overset{\displaystyle F}{|}\overset{\displaystyle CF_2}{|} & & \overset{\displaystyle F}{|}\overset{\displaystyle CF_2}{|} & & \overset{\displaystyle H}{|}\overset{\displaystyle CF_2}{|} \\
C = C & \longrightarrow & \oplus C - C\ominus & \not\longrightarrow & \oplus C - C\ominus \\
\overset{\displaystyle |}{CH_3}\,\overset{\displaystyle |}{CF_2H} & & \overset{\displaystyle |}{CH_3}\,\overset{\displaystyle |}{CF_2H} & & \overset{\displaystyle |}{F}\,\overset{\displaystyle |}{CF_2} \\
& & & & \overset{\displaystyle |}{CH_3}
\end{array} \qquad 4.195$$

<u>More stable</u> $R_{F2} > R_{F1} > F$ (CFR = $CR_{F1}R_{F2}$)

Like the cases above, transfer of transfer species is not favored whether F or H is involved. As a Nucleophile, what should be involved is H. The monomer here will also not favor charged poly-merization.

For the last group of monomers, the followings are obtained.

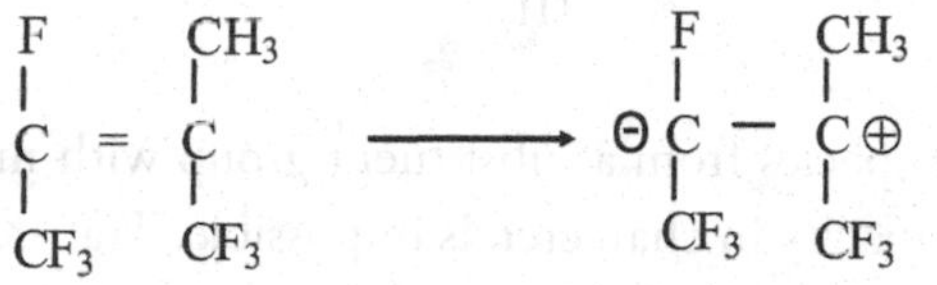

$$\begin{array}{ccc}
\overset{\displaystyle F}{|}\overset{\displaystyle CH_3}{|} & & \overset{\displaystyle F}{|}\overset{\displaystyle CH_3}{|} \\
C = C & \longrightarrow & \ominus C - C\oplus \\
\overset{\displaystyle |}{CF_3}\,\overset{\displaystyle |}{CF_3} & & \overset{\displaystyle |}{CF_3}\,\overset{\displaystyle |}{CF_3}
\end{array} \qquad 4.196$$

Impossible existence (CFRF = CRRF)

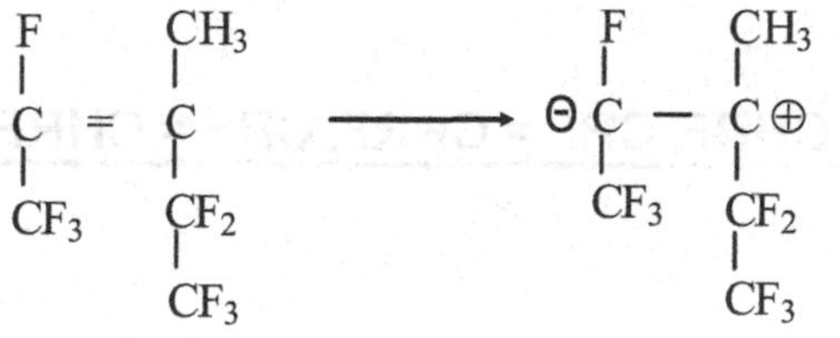

$$\begin{array}{ccc}
\overset{\displaystyle F}{|}\overset{\displaystyle CH_3}{|} & & \overset{\displaystyle F}{|}\overset{\displaystyle CH_3}{|} \\
C = C & \longrightarrow & \ominus C - C\oplus \\
\overset{\displaystyle |}{CF_3}\,\overset{\displaystyle |}{CF_2} & & \overset{\displaystyle |}{CF_3}\,\overset{\displaystyle |}{CF_2} \\
\underset{\displaystyle CF_3}{|} & & \underset{\displaystyle CF_3}{|}
\end{array} \qquad (F + CF_3 + CH_3) > C_2F_5$$

$$4.197$$

<u>Impossible existence</u> (CFR$_{F1}$ = CRR$_{F2}$)

Chargedly, the monomers cannot be activated if the inequality shown above is valid. Free-radically, the followings are obtained.

222

$$\underset{\substack{|\\ CF_3}}{\overset{\substack{F\\|}}{C}} = \underset{\substack{|\\ CF_3}}{\overset{\substack{CH_3\\|}}{C}} \quad \longrightarrow \quad n.\underset{\substack{|\\ CF_3}}{\overset{\substack{F\\|}}{C}} - \underset{\substack{|\\ CF_3}}{\overset{\substack{CH_3\\|}}{C}}.e \quad \xcancel{\longrightarrow} \quad n.\underset{\substack{|\\ H}}{\overset{\substack{H\\|}}{C}} - \underset{\substack{|\\ CFH\\|\\ CF_3}}{\overset{\substack{CF_3\\|}}{C}}.e \qquad 4.198$$

(Transfer species)

(I) More stable (II)

Though (II) looks more stable than (I), satisfying the driving forces that favor molecular rear-rangement and the law of conservation of transfer of transfer species are very important. In the absence or presence of steric limitations, the monomers above will not favor free-radical polymerization. ***It like many others will only favor the electro-free-radical route if the F atom cannot be abstracted electro-free-radically, that which is indeed not possible.***

$$\underset{\substack{|\\ CF_3}}{\overset{\substack{F\\|}}{C}} = \underset{\substack{|\\ CF_2\\|\\ CF_3}}{\overset{\substack{CH_3\\|}}{C}} \quad \longrightarrow \quad n.\underset{\substack{|\\ CF_3}}{\overset{\substack{F\\|}}{C}} - \underset{\substack{|\\ CF_2\\|\\ CF_3}}{\overset{\substack{CH_3\\|}}{C}}.e \quad \xcancel{\longrightarrow} \quad n.\underset{\substack{|\\ H}}{\overset{\substack{H\\|}}{C}} - \underset{\substack{|\\ C_2F_5}}{\overset{\substack{CF_3\\|\\ CHF\\|}}{C}}.e \qquad 4.199$$

$$\underset{\substack{|\\ CF_3}}{\overset{\substack{F\\|}}{C}} = \underset{\substack{|\\ CF_2\\|\\ CH_3}}{\overset{\substack{CH_3\\|}}{C}} \quad \longrightarrow \quad n.\underset{\substack{|\\ CF_3}}{\overset{\substack{F\\|}}{C}} - \underset{\substack{|\\ CF_2\\|\\ CH_3}}{\overset{\substack{CH_3\\|}}{C}}.e \quad \xcancel{\longrightarrow} \quad n.\underset{\substack{|\\ H}}{\overset{\substack{H\\|}}{C}} - \underset{\substack{|\\ CFH\\|\\ CF_3}}{\overset{\substack{CF_2CH_3\\|}}{C}}.e \qquad 4.200$$

(I) (II)

It is again recalled, that in general transfer of transfer species from a substituent group with more than one transfer species to centers carrying groups of less capacity in character, is impossible. Transfer species to be transferred must be of less or equal capacity than those carried by the receiving active center. No free-radical route is favored by these monomers

4.10 Characteristic of CHF = CHR, CHF = CHRF, CHF = CRRF, CRF = CHRF, CRFF = CHRF, CRF = CHR groups.

For the first group, consider the followings:-

$$\underset{\substack{|\\ F}}{\overset{\substack{H\\|}}{C}} = \underset{\substack{|\\ CH_3}}{\overset{\substack{H\\|}}{C}} \quad , \quad \underset{\substack{|\\ F}}{\overset{\substack{H\\|}}{C}} = \underset{\substack{|\\ CH_2\\|\\ CH_3}}{\overset{\substack{H\\|}}{C}} \qquad 4.201$$

Chargedly, the monomers cannot be activated. Free-radically, the followings are obtained.

$$\begin{array}{cc} H & H \\ | & | \\ C & = C \\ | & | \\ F & CH_3 \end{array} \longrightarrow \begin{array}{cc} H & H \\ | & | \\ n.C & - C.e \\ | & | \\ F & CH_3 \end{array} \quad\not\longrightarrow\quad \begin{array}{cc} H & H \\ | & | \\ n.C & - C.e \\ | & | \\ H & CH_2F \end{array} \qquad 4.202$$

<u>More stable</u>

$$\begin{array}{cc} H & H \\ | & | \\ C & = C \\ | & | \\ F & CH_2 \\ & | \\ & CH_3 \end{array} \longrightarrow \begin{array}{cc} H & H \\ | & | \\ n.C & - C.e \\ | & | \\ F & CH_2 \\ & | \\ & CH_3 \end{array} \quad\not\longrightarrow\quad \begin{array}{cc} H & H \\ | & | \\ n.C & - C.e \\ | & | \\ CH_3 & CH_2F \end{array} \quad \left(\begin{array}{cc} H & H \\ | & | \\ e.C & - C.n \\ | & | \\ CH_3 & CH_2F \end{array} \right) \qquad 4.203$$

<u>Impossible existence</u>

Free-radically molecular rearrangement is not favored by the members of the group. They will however favor only electro-free-radical route.

For the second group, the followings are obtained.

$$\begin{array}{cc} H & H \\ | & | \\ C & = C \\ | & | \\ F & CF_3 \end{array} \longrightarrow \begin{array}{cc} H & H \\ | & | \\ \oplus C & - C \ominus \\ | & | \\ F & CF_3 \end{array} \quad\not\longrightarrow\quad \begin{array}{cc} F & H \\ | & | \\ \oplus C & - C \ominus \\ | & | \\ F & CF_2H \end{array} \qquad 4.204$$

<u>Impossible</u>

$$\begin{array}{cc} H & H \\ | & | \\ C & = C \\ | & | \\ F & CF_2 \\ & | \\ & CF_3 \end{array} \longrightarrow \begin{array}{cc} H & H \\ | & | \\ \oplus C & - C \ominus \\ | & | \\ F & CF_2 \\ & | \\ & CF_3 \end{array} \quad\not\longrightarrow\quad \begin{array}{cc} F & H \\ | & | \\ \oplus C & - C \ominus \\ | & | \\ CF_3 & CHF_2 \end{array} \quad \left(\begin{array}{cc} F & H \\ | & | \\ \ominus C & - C \oplus \\ | & | \\ CF_3 & CHF_2 \end{array} \right) \qquad 4.205$$

<u>Impossible existences</u>

$$\begin{array}{cc} H & H \\ | & | \\ C & = C \\ | & | \\ F & CH_2F \end{array} \longrightarrow \begin{array}{cc} H & H \\ | & | \\ \oplus C & - C \ominus \\ | & | \\ F & CH_2F \end{array} \qquad 4.206$$

No rearrangement

Molecular rearrangement being favored, depends on the type of R_F group involved. Polymeri-zation being favored depends on absence of electrostatic forces of repulsion, the type of place-ment of the groups and the type of initiator used. These look as if they can be polymerized using negatively charged initiators in which the transfer species is F. Worthy of note so far, is that the DIRECTIONS OF ARROWS SHOWN DURING MOLECULAR REARRANGEMENT INVOLVING F TRANSFER ARE NOT TRULY AS SHOWN in all the equations. It is the carrier of the positive charge or electro-free-radical that diffuses to grab F^θ. Once only F amongst halogen atoms is grabbed, it is very difficult to release it, via Equilibrium state of existence.

Free-radically, the followings are obtained.

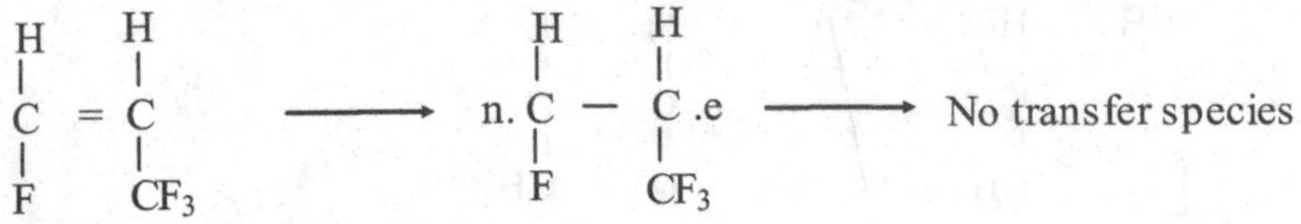

```
  H   H                    H   H
  |   |                    |   |
  C = C          ──→    n. C — C .e      ──→   No transfer species        4.207
  |   |                    |   |
  F   CF₃                  F   CF₃
```

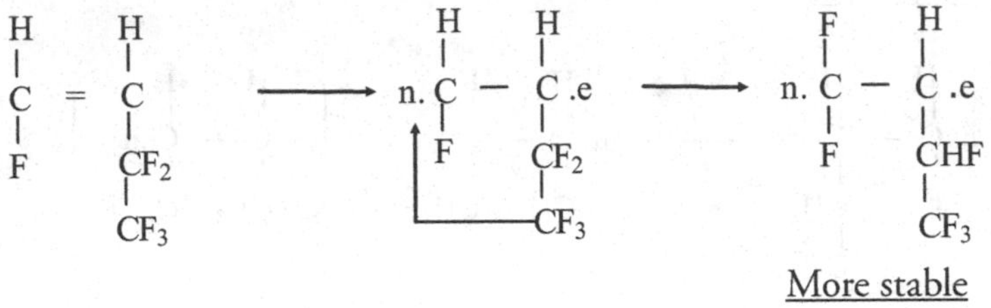

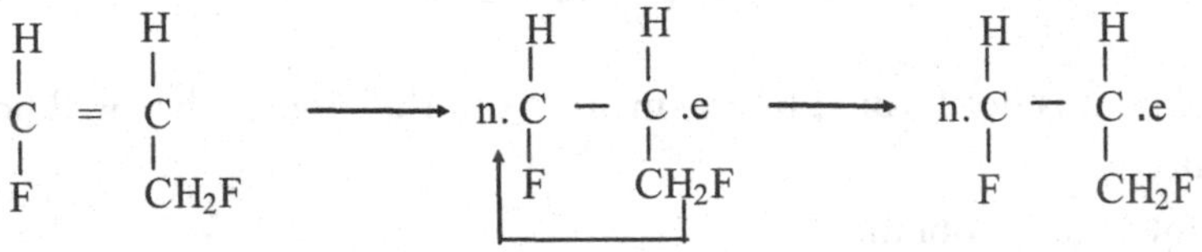

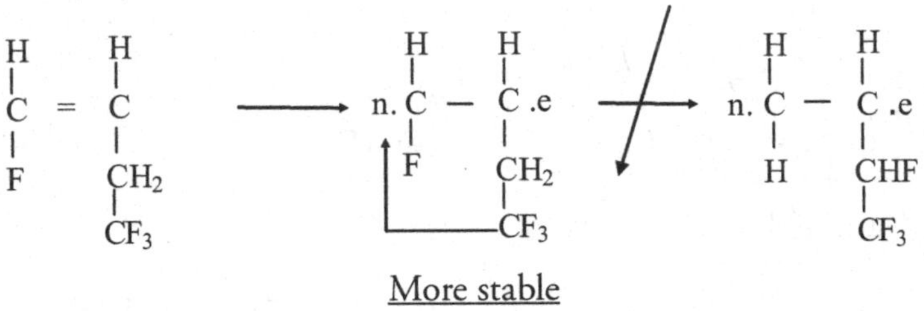

```
  H   H                    H   H                   F   H
  |   |                    |   |                   |   |
  C = C          ──→    n. C — C .e      ──→    n. C — C .e               4.208
  |   |                    |   |                   |   |
  F   CF₂                  F   CF₂                 F   CHF
      |                        └──── CF₃               |
      CF₃                                              CF₃
```
                                              <u>More stable</u>

```
  H   H                    H   H                   H   H
  |   |                    |   |                   |   |
  C = C          ──→    n. C — C .e      ──→    n. C — C .e               4.209
  |   |                    |   |                   |   |
  F   CH₂F                 F   CH₂F                F   CH₂F
```

```
  H   H                    H   H                   H   H
  |   |                    |   |                   |   |
  C = C          ──→    n. C — C .e    ─/→      n. C — C .e               4.210
  |   |                    |   |                   |   |
  F   CH₂                  F   CH₂                 H   CHF
      |                        └──── CF₃               |
      CF₃                                              CF₃
```
                    <u>More stable</u>

The behaviors of these monomers can be observed chargedly and free-radically, based on the capacities of the substituted groups.

For the third group of monomers $CHF = CRR_F$, the followings are obtained chargedly.

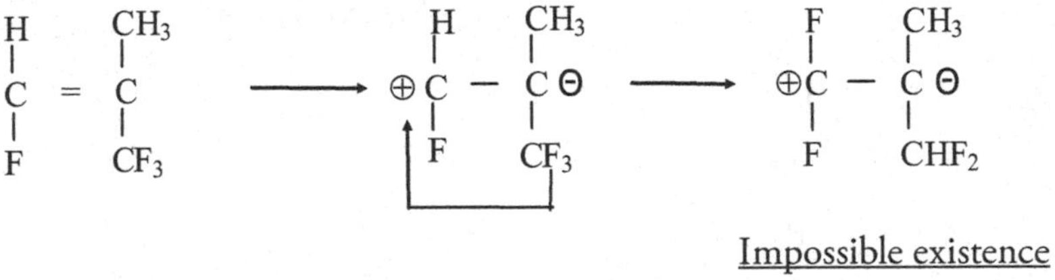

```
  H   CH₃                    H   CH₃                  F   CH₃
  |   |                      |   |                    |   |
  C = C          ──→      ⊕C — C⊖        ──→       ⊕C — C⊖                4.211
  |   |                      |   |                    |   |
  F   CF₃                    F   CF₃                  F   CHF₂
```
                                              <u>Impossible existence</u>

"Assumed": $(CF_3 + H) > (F + CH_3)$ in capacity.

If "the assumption" is not valid, charged activation will be impossible. When lower or higher R_F groups are involved, molecular rearrangement is not favored.

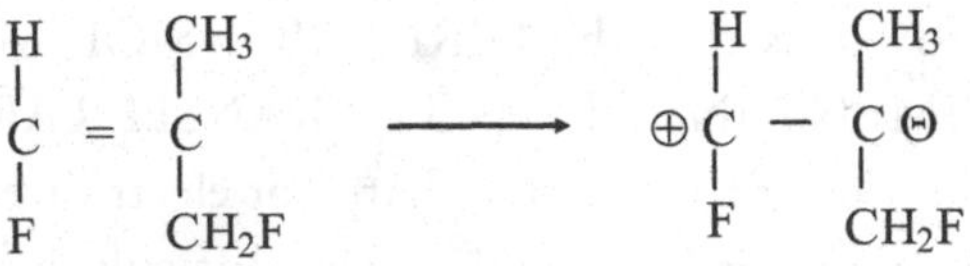

```
  H   CH₃                    H   CH₃
  |   |                      |   |
  C = C          ──→      ⊕C — C⊖                                        4.212
  |   |                      |   |
  F   CH₂F                   F   CH₂F
```

The monomer is almost analogous to the type shown below apart from the symmetry.

$$
\begin{array}{ccc}
\underset{\underset{F}{|}}{\overset{\overset{H}{|}}{C}} = \underset{\underset{CH_2F}{|}}{\overset{\overset{CH_3}{|}}{C}} & \cong & \underset{\underset{F}{|}}{\overset{\overset{H}{|}}{C}} = \underset{\underset{F}{|}}{\overset{\overset{H}{|}}{C}}
\end{array}
\qquad 4.213
$$

<u>Cannot be activated chargedly</u>

If activated chargedly in such a way that the C center carrying F is made to carry positive charge, then, they will all favor negatively charged routes, making them look like Electrophiles. Hence, it is believed that these monomers cannot be activated chargedly.

Free-radically the followings are obtained.

$$
\underset{\underset{F}{|}}{\overset{\overset{H}{|}}{C}} = \underset{\underset{CF_3}{|}}{\overset{\overset{CH_3}{|}}{C}} \longrightarrow \; n.\underset{\underset{F}{|}}{\overset{\overset{H}{|}}{C}} - \underset{\underset{CF_3}{|}}{\overset{\overset{CH_3}{|}}{C}}.e \;\; \not\longrightarrow \;\; n.\underset{\underset{H}{|}}{\overset{\overset{H}{|}}{C}} - \underset{\underset{CH_2F}{|}}{\overset{\overset{CF_3}{|}}{C}}.e
\qquad 4.214
$$

<u>More stable</u>

$$
\underset{\underset{F}{|}}{\overset{\overset{H}{|}}{C}} = \underset{\underset{CH_2F}{|}}{\overset{\overset{CH_3}{|}}{C}} \longrightarrow \; n.\underset{\underset{F}{|}}{\overset{\overset{H}{|}}{C}} - \underset{\underset{CH_2F}{|}}{\overset{\overset{CH_3}{|}}{C}}.e \;\; \not\longrightarrow \;\; n.\underset{\underset{H}{|}}{\overset{\overset{H}{|}}{C}} - \underset{\underset{CH_2F}{|}}{\overset{\overset{CH_2F}{|}}{C}}.e
\qquad 4.215
$$

<u>More stable</u>

Chargedly and free-radically, molecular rearrangement is not favored. Only electro-free-radical polymerization is favored by the monomers.

For the fourth groups of monomers $CRF = CHR_F$, the followings are obtained chargedly.

$$
\underset{\underset{CH_3}{|}}{\overset{\overset{F}{|}}{C}} = \underset{\underset{H}{|}}{\overset{\overset{CF_3}{|}}{C}} \longrightarrow \; \oplus\underset{\underset{CH_3}{|}}{\overset{\overset{F}{|}}{C}} - \underset{\underset{H}{|}}{\overset{\overset{CF_3}{|}}{C}}\ominus \;\; \not\longrightarrow \;\; \oplus\underset{\underset{F}{|}}{\overset{\overset{F}{|}}{C}} - \underset{\underset{\underset{CH_3}{|}}{CF_2}}{\overset{\overset{H}{|}}{C}}\ominus
$$

(Impossible existence) 4.216

If H is transferred from CH_3 group, no rearrangement is favored. Indeed it is only H that should be the transfer species and not F, the monomer being a nucleophile.

$$
\underset{\underset{CH_3}{|}}{\overset{\overset{F}{|}}{C}} = \underset{\underset{H}{|}}{\overset{\overset{\underset{CF_2}{|}}{CF_3}}{C}} \longrightarrow \; \oplus\underset{\underset{CH_3}{|}}{\overset{\overset{F}{|}}{C}} - \underset{\underset{H}{|}}{\overset{\overset{\underset{CF_2}{|}}{CF_3}}{C}}\ominus \;\; \not\longrightarrow \;\; \oplus\underset{\underset{CF_3}{|}}{\overset{\overset{F}{|}}{C}} - \underset{\underset{\underset{CH_3}{|}}{CF_2}}{\overset{\overset{H}{|}}{C}}\ominus
$$

4.217

<u>Impossible existence</u>

Chargedly, these monomers cannot be polymerized with negatively or positively charged-paired initiators. As nucleophiles, their natural route chargedly is the positively charged route.

Free-radically the followings are obtained.

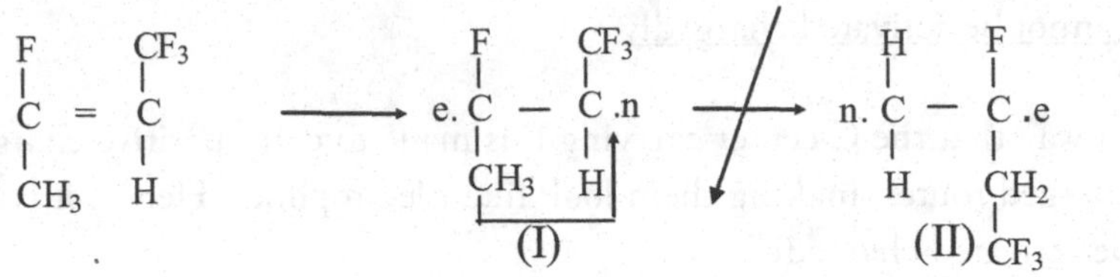

$$
\begin{array}{ccc}
\overset{\displaystyle F}{\underset{\displaystyle CH_3}{C}} = \overset{\displaystyle CF_3}{\underset{\displaystyle H}{C} } & \longrightarrow & e.\overset{\displaystyle F}{\underset{\displaystyle CH_3}{C}} - \overset{\displaystyle CF_3}{\underset{\displaystyle H}{C}}.n \\
& & (I) \\
& & \text{(More stable)}
\end{array}
\quad \nrightarrow \quad
n.\overset{\displaystyle H}{\underset{\displaystyle H}{C}} - \overset{\displaystyle F}{\underset{\displaystyle CH_2}{C}}.e \qquad 4.218
$$

(II) CF_3

One wonders why the C center carrying F is made to carry an electro-free-radical as shown above. *As it seems, the activated state of the monomer shown above is wrong, since the presence of F on the C center is paramount as a symbol that the C center must carry a nucleo-free-radical in the absence of groups such as CN, COOR.* When the monomer is properly activated, then the monomer will be found to favor both routes. Then, the monomer can no longer favor molecular rearrangement.

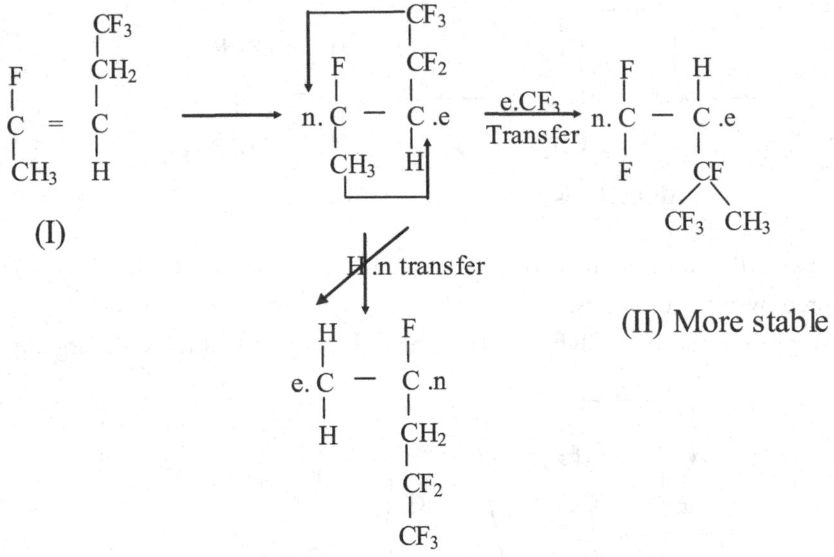

$$
\begin{array}{ccc}
\overset{\displaystyle F}{\underset{\displaystyle CH_3}{C}} = \overset{\displaystyle CF_3 \; CH_2}{\underset{\displaystyle H}{C}} & \longrightarrow & n.\overset{\displaystyle F}{\underset{\displaystyle CH_3}{C}} - \overset{\displaystyle CF_3 \; CF_2}{\underset{\displaystyle H}{C}}.e \\
(I) & &
\end{array}
\xrightarrow[\text{Transfer}]{e.CF_3}
n.\overset{\displaystyle F}{\underset{\displaystyle F}{C}} - \overset{\displaystyle H}{\underset{\displaystyle CF}{C}}.e
$$

CF$_3$ CH$_3$

(II) More stable

H.n transfer

$$
e.\overset{\displaystyle H}{\underset{\displaystyle H}{C}} - \overset{\displaystyle F}{\underset{\displaystyle CH_2}{C}}.n \\
\underset{\displaystyle CF_3}{\underset{\displaystyle |}{CF_2}}
$$

(III) <u>Impossible transfer species</u>

4.219

Existence of (III) is impossible. (II) is more stable than (I). The monomer will only favor electro-free-radical polymerization. *In general, it can observe that <u>there is need for quantitative capacities rather than qualitative capacities for substituted groups.</u>* The monomers free-radically being nucleophiles, will favor only transfer of electro-free-radicals when activated.

For the fifth group of monomers, the followings are obtained chargedly.

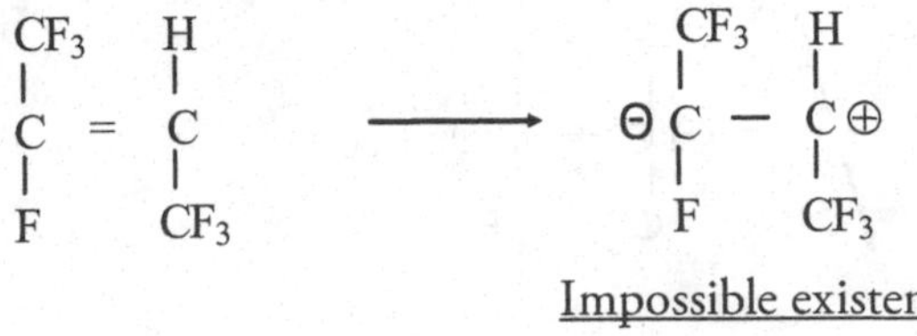

$$
\overset{\displaystyle CF_3}{\underset{\displaystyle F}{C}} = \overset{\displaystyle H}{\underset{\displaystyle CF_3}{C}} \longrightarrow \ominus\overset{\displaystyle CF_3}{\underset{\displaystyle F}{C}} - \overset{\displaystyle H}{\underset{\displaystyle CF_3}{C}}\oplus \qquad 4.220
$$

<u>Impossible existence</u> (CFR$_F$ = CHR$_F$)

Sunny N.E. Omorodion

$$
\begin{array}{ccc}
CF_3 & & H \\
| & & | \\
C & = & C \\
| & & | \\
F & & CF_2 \\
& & | \\
& & CF_3
\end{array}
\longrightarrow
\begin{array}{ccc}
CF_3 & & H \\
| & & | \\
\ominus C & - & C \oplus \\
| & & | \\
F & & CF_2 \\
& & | \\
& & CF_3
\end{array}
\qquad 4.221
$$

$$\text{\underline{Impossible existence} } (CFR_{F1} = CHR_{F2}) \; R_{F2} > R_{F1}$$

Chargedly, activation of the monomer is only possible when $R_{F2} > (CF_3 + F + H)$ in capacity.

$$
\begin{array}{ccc}
C_2F_5 & & H \\
| & & | \\
C & = & C \\
| & & | \\
F & & CF_3
\end{array}
\longrightarrow
\begin{array}{ccc}
C_2F_5 & & H \\
| & & | \\
\ominus C & - & C \oplus \\
| & & | \\
F & & CF_3
\end{array}
\qquad 4.222
$$

$$\text{\underline{Impossible existence} } (CFR_{F1} = CHR_{F2}) \; R_{F1} > R_{F2}$$

When $R_{F1} > R_{F2}$, charged activation of the monomer is impossible. Free-radically, the followings are obtained.

$$
\begin{array}{ccc}
CF_3 & & H \\
| & & | \\
C & = & C \\
| & & | \\
F & & CF_3
\end{array}
\longrightarrow
\begin{array}{ccc}
CF_3 & & H \\
| & & | \\
n.C & - & C.e \\
| & & | \\
F & & CF_3
\end{array}
\longrightarrow
\text{No transfer species nucleo-free-radically}
$$

$$(R_{F1} = R_{F2}) \qquad\qquad 4.223$$

$$
\begin{array}{ccc}
CF_3 & & H \\
| & & | \\
C & = & C \\
| & & | \\
F & & CF_2 \\
& & | \\
& & CF_3
\end{array}
\longrightarrow
\begin{array}{ccc}
CF_3 & & H \\
| & & | \\
n.C & - & C.e \\
| & & | \\
F & & CF_2 \\
& & | \\
& & CF_3
\end{array}
\longrightarrow
\begin{array}{ccc}
F & & H \\
| & & | \\
n.C & - & C.e \\
| & & | \\
F & & CF \\
& & \diagdown \\
& & CF_3 \; CH_3
\end{array}
$$

$$(RF2 > RF1) \qquad \text{more stable} \; (CFRF1 = CHRF2) \qquad 4.224$$

$$
\begin{array}{c}
CF_3 \\
| \\
CF_2 \quad H \\
| \quad\quad | \\
C = C \\
| \quad\quad | \\
F \quad\quad CF_3
\end{array}
\longrightarrow
\begin{array}{c}
CF_3 \\
| \\
CF_2 \quad H \\
| \quad\quad | \\
n.C - C.e \\
| \quad\quad | \\
F \quad\quad CF_3
\end{array}
\;\not\longrightarrow\;
\begin{array}{c}
F \quad\quad F \\
| \quad\quad | \\
e.C - C.n \\
| \quad\quad | \\
F \quad\quad CH \\
\quad\quad \diagup \diagdown \\
\quad CF_3 \; CF_3
\end{array}
$$

$$(RF1 > RF2) \qquad \text{Impossible existence} \; (CFRF1 = CHRF2) \qquad 4.225$$

The last reaction above, clearly indicates that for a nucleophile, a carrier carrying a nucleo-free-radical cannot be a transfer species for molecular rearrangement. Molecular rearrangement will never be favored. Nucleo-free-radical route will be favored by the first and last monomers, while electro-free-radical route is favored by the second monomer only after molecular rearrangement.

$$
\begin{array}{ccc}
CH_3 & & H \\
| & & | \\
C & = & C \\
| & & | \\
F & & CH_3
\end{array}
\longrightarrow
\begin{array}{ccc}
CH_3 & & H \\
| & & | \\
\ominus C & - & C \oplus \\
| & & | \\
F & & CH_3
\end{array}
\qquad 4.226
$$

$$\text{\underline{Impossible existence}}$$

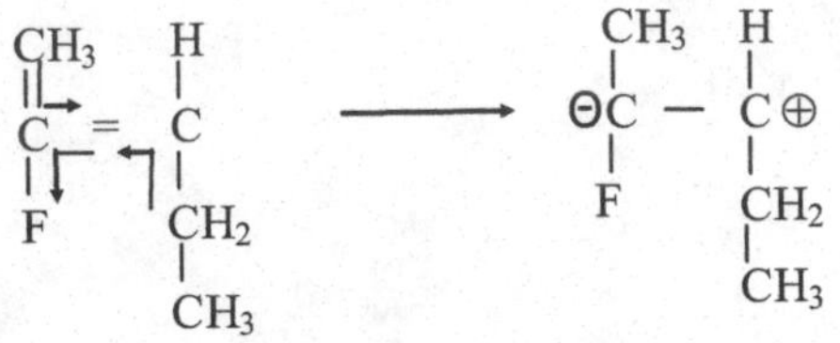

$$
\begin{array}{ccc}
\overset{CH_3}{\underset{F}{\overset{|}{\underset{|}{C}}}} = \overset{H}{\underset{\underset{|}{CH_2}}{\overset{|}{\underset{|}{C}}}} & \longrightarrow & \overset{CH_3}{\underset{F}{\overset{|}{\underset{|}{\ominus C}}}} - \overset{H}{\underset{\underset{|}{CH_2}}{\overset{|}{\underset{|}{C \oplus}}}}
\end{array}
$$

$$\underset{CH_3}{}$$

$$\underset{CH_3}{}$$

Impossible existence

4.227

$$
\begin{array}{ccc}
\overset{CH_3}{\underset{F}{\overset{|}{\underset{|}{\overset{CH_2}{\underset{|}{C}}}}}} = \overset{H}{\underset{CH_3}{\overset{|}{\underset{|}{C}}}} & \longrightarrow & \overset{C_2H_5}{\underset{F}{\overset{|}{\underset{|}{\ominus C}}}} - \overset{H}{\underset{CH_3}{\overset{|}{\underset{|}{C \oplus}}}}
\end{array}
$$

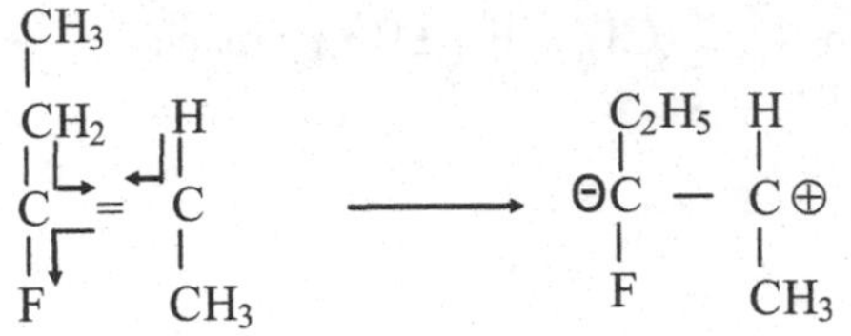

4.228

Impossible existence (CFR$_1$ = CHR$_2$) R$_1$ > R$_2$

In most cases, in view of the higher capacity of F group compared to H and alkanyl groups, it seems that charged activations of the monomers are impossible. Free-radically, the followings are obtained.

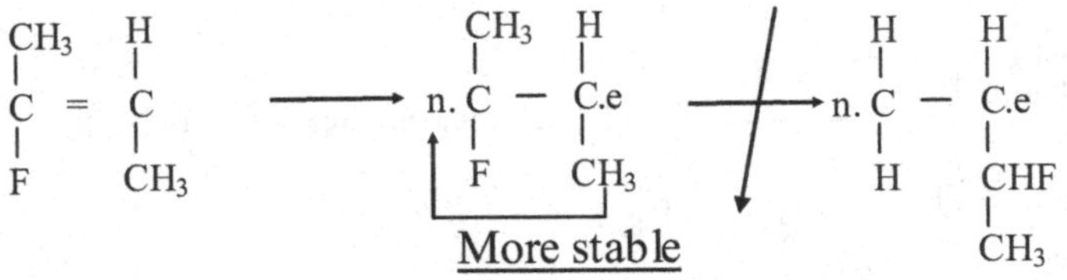

$$
\overset{CH_3}{\underset{F}{\overset{|}{\underset{|}{C}}}} = \overset{H}{\underset{CH_3}{\overset{|}{\underset{|}{C}}}} \longrightarrow n.\overset{CH_3}{\underset{F}{\overset{|}{\underset{|}{C}}}} - \overset{H}{\underset{CH_3}{\overset{|}{\underset{|}{C.e}}}} \xrightarrow{\quad\quad} n.\overset{H}{\underset{H}{\overset{|}{\underset{|}{C}}}} - \overset{H}{\underset{\underset{|}{CH_3}}{\overset{|}{\underset{|}{C.e}}}}
$$

More stable

4.229

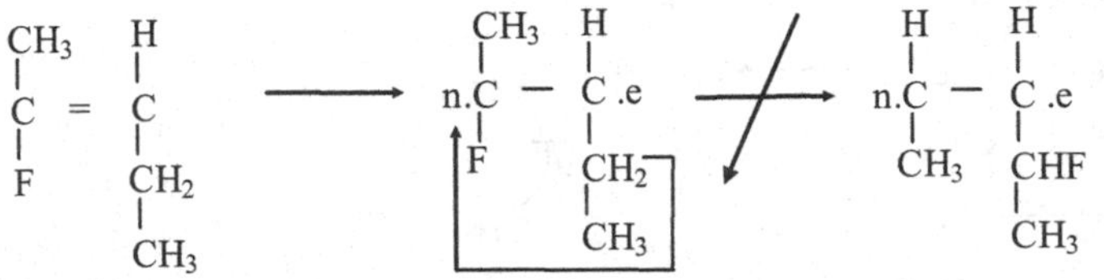

$$
\overset{CH_3}{\underset{F}{\overset{|}{\underset{|}{\overset{}{\underset{|}{C}}}}}} = \overset{H}{\underset{\underset{|}{CH_2}}{\overset{|}{\underset{|}{C}}}} \longrightarrow n.\overset{CH_3}{\underset{F}{\overset{|}{\underset{|}{C}}}} - \overset{H}{\underset{CH_2}{\overset{|}{\underset{|}{C.e}}}} \xrightarrow{\quad\quad} n.\overset{H}{\underset{CH_3}{\overset{|}{\underset{|}{C}}}} - \overset{H}{\underset{\underset{|}{CH_3}}{\overset{|}{\underset{|}{C.e}}}}
$$

$$\underset{CH_3}{}$$

More stable

4.230

$$
\overset{CH_3}{\underset{F}{\overset{|}{\underset{|}{\overset{CH_2}{\underset{|}{C}}}}}} = \overset{H}{\underset{CF_3}{\overset{|}{\underset{|}{C}}}} \longrightarrow n.\overset{CH_3}{\underset{F}{\overset{|}{\underset{|}{\overset{CH_2}{\underset{|}{C}}}}}} - \overset{H}{\underset{CH_3}{\overset{|}{\underset{|}{C.e}}}} \xrightarrow{\quad\quad} n.\overset{H}{\underset{H}{\overset{|}{\underset{|}{C}}}} - \overset{H}{\underset{\underset{|}{C_2H_5}}{\overset{|}{\underset{|}{C.e}}}}
$$

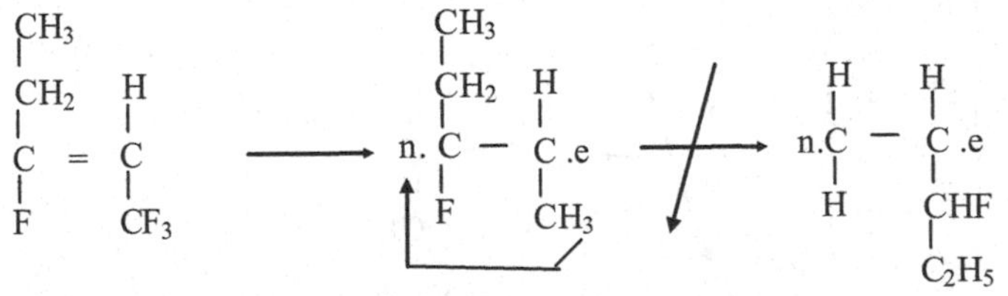

4.231

More stable

For all the monomers, only electro-free-radical polymerization is favored by them.

So far, it has been important to note amongst other concepts some very unique transfer species, where these transfer species or substituent groups such as CF_3, CF_2H, and CFH_2 cannot carry negative charges due to electrostatic forces of repulsion as shown below.

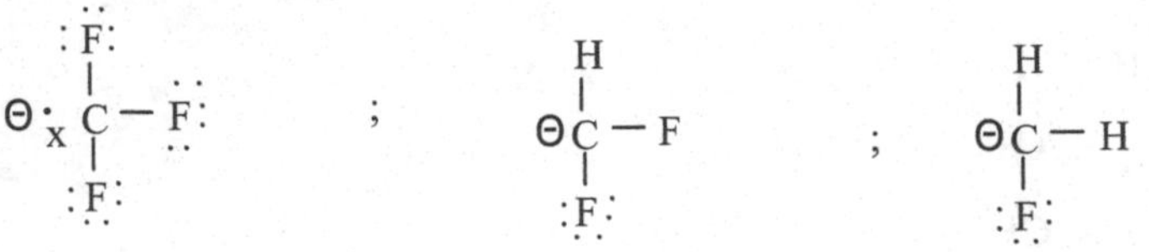

$$
\overset{..}{\underset{..}{:F:}} \qquad\qquad\qquad\qquad \overset{H}{} \qquad\qquad \overset{H}{}
$$

$$
\Theta \overset{}{\underset{x}{:}} C - \overset{..}{\underset{..}{F}:} \quad ; \quad \Theta \overset{|}{\underset{:\overset{..}{F}:}{C}} - F \quad ; \quad \Theta \overset{|}{\underset{:\overset{..}{F}:}{C}} - H
$$

Impossible states   Allylic

4.232

Hence when such groups are part of an allylic or similar type of groups, they cannot be trans-ferred or rejected chargedly.

$$
\underset{\text{(I)}}{\underset{\underset{CH_2F}{|}}{\overset{\overset{CF_3}{|}}{\overset{\overset{CH_2}{|}}{\underset{H}{\overset{H}{C}}}}} = \overset{\overset{CF_3}{|}}{\underset{H}{\overset{CH_2}{C}}} \quad \xrightarrow{\; +\; R^{\oplus}\;} \quad \overset{\oplus}{\underset{CH_2F}{\overset{H}{C}}} = \overset{\overset{CF_3}{|}}{\underset{H}{\overset{CH_2}{C}}}{}^{\ominus} \quad \longrightarrow \quad R - \underset{\underset{CF_3}{|}}{\underset{CH_2}{\overset{H}{C}}} - \overset{CH_2F}{\underset{H}{C}}{}^{\oplus} \quad \xrightarrow{\; n(I)\;}
$$

$$
R \text{WWWWW} \overset{\overset{H}{|}}{\underset{\underset{CF_3}{|}}{\underset{CH_2}{C}}} - \overset{H}{\underset{\oplus}{C}} - \overset{H}{\underset{F}{C}}{}^{\ominus} \quad + \quad H^{\oplus}
$$

$$
\underline{\text{Impossible state}} \qquad\qquad\qquad 4.233
$$

$$
R^{\ominus} + \overset{\oplus}{\underset{CH_2F}{\overset{H}{C}}} - \overset{\overset{CF_3}{|}}{\underset{H}{\overset{CH_2}{C}}}{}^{\ominus} \quad \cancel{\longrightarrow} \quad RH + \overset{\ominus}{\underset{F}{\overset{H}{C}}} - \overset{H}{\underset{\oplus}{C}} - \overset{\overset{CF_3}{|}}{\underset{H}{\overset{CH_2}{C}}}{}^{\ominus}
$$

$$
\underline{\text{Impossible state}} \qquad\qquad 4.234
$$

$$
\underset{\underset{CH_2F}{|}}{\overset{\overset{CF_3}{|}}{\overset{\overset{CFH}{|}}{\underset{H}{\overset{H}{C}}}}} = C \quad \xrightarrow{\; R^{\oplus}\;} \quad \overset{\ominus}{\underset{\underset{CF_3}{|}}{\underset{CFH}{\overset{H}{C}}}} - \overset{CH_2F}{\underset{H}{C}}{}^{\oplus} \quad \longrightarrow \quad RF + \overset{\oplus}{\underset{CF_3}{\overset{H}{C}}} - \overset{H}{\underset{\ominus}{C}} - \overset{CH_2F}{\underset{H}{C}}{}^{\oplus} \qquad 4.235
$$

$$
\Big\downarrow R^{\ominus}
$$

$$
\overset{\oplus}{\underset{H}{\overset{CH_2F}{C}}} - \overset{\overset{}{|}}{\underset{\underset{CF_3}{|}}{\overset{CFH}{C}}}{}^{\ominus} \quad \cancel{\longrightarrow} \quad RH + \overset{\ominus}{\underset{H}{\overset{F}{C}}} - \overset{\oplus}{\underset{H}{\overset{}{C}}} - \overset{\overset{H}{|}}{\underset{\underset{CF_3}{|}}{\overset{CFH}{C}}}{}^{\ominus}
$$

$$
\text{[IMPOSSIBLE STATE]}
$$

$$
\underset{\underline{\text{More stable}}}{\overset{\oplus}{\underset{H}{\overset{CH_2F}{C}}} - \underset{\underset{CF_3}{\lfloor\_}}{\overset{CFH}{C}}{}^{\ominus}} \quad \cancel{\longrightarrow} \quad \overset{\oplus}{\underset{CF_3}{\overset{H}{C}}} - \overset{\overset{H}{|}}{\underset{CH_2F}{\overset{CHF}{C}}}{}^{\ominus} \;\; ; \;\; \underset{\underline{\text{More stable}}}{\overset{\oplus}{\underset{H}{\overset{CF_3}{C}}} - \underset{\underset{CF_3}{\lfloor\_}}{\overset{CFH}{C}}{}^{\ominus}} \quad \cancel{\longrightarrow} \quad \overset{\oplus}{\underset{CF_3}{\overset{H}{C}}} - \overset{\overset{H}{|}}{\underset{CF_3}{\overset{CHF}{C}}}{}^{\ominus} \qquad 4.236
$$

$$H_2C = C(CF_2CH_3)(H) \xrightarrow{R^{\oplus}} {}^{\oplus}C(H)(CH_2F) - {}^{\ominus}C(CF_2CH_3)(H) \longrightarrow RF + {}^{\oplus}C(F)(CH_3) - {}^{\ominus}C(H) - {}^{\oplus}C(CH_2F)(H)$$

(Structures with $R^{\ominus}$ pathway)

$${}^{\ominus}C(F)(H) - {}^{\oplus}C(H) - {}^{\ominus}C(CF_2CH_3)(H) \quad + \quad RH \qquad 4.237$$

Impossible state

From all indications, **there is no doubt that when allylic groups such as CH_2F, CH_2OCOCH_3, etc., are involved as substituent groups, the monomer cannot be activated chargedly, but only radically. The same applies to groups such as CHF_2, CF_3, $CH(OCOCH_3)$ etc.** As will clearly become obvious in the Series and Volumes the followings are valid:-

$$\overset{e}{\underset{\cdot}{C_2H_5}} > \overset{e}{\underset{\cdot}{CH_3}} > \overset{e}{\underset{\cdot}{H}} > \overset{e}{\underset{\cdot}{C_2F_5}} > \overset{e}{\underset{\cdot}{CF_3}}$$

$$n.CF_3 > \overset{n}{\dot{C}_2F_5} > \overset{n}{\dot{H}} > \overset{n}{\dot{C}H_3} > \overset{n}{\dot{C}_2H_5}$$

Order of free-radical capacity   4.238

4.11 Proposition of Rules of Chemistry

Having considered molecular rearrangements in most olefinic monomers where special substituent groups of the types C_nH_{2n+1} (R), C_nF_{2n+1} (R_F) or C_nX_{2n+1} or CHX_2 or CH_2X or $C(R)_2R_F$, $C(R_F)_2R$, etc., where X is any charged/radical-pulling group, and their hybrids are involved, there is need to propose rules based on how Nature operates. In all the above consi-derations, the second driving force for molecular rearrangement as already identified, is that which involves the use of <u>weak initiators</u>. From the exercise, it was not surprising to note why these special groups of monomers were chosen for considerations. Both known and unknown monomers were considered in order to illustrate some basic fundamental principles with respect to transfer of transfer species, activation of monomers, stability concepts, substituent group types and so on.

It will be a useless exercise proposing rules for all the different groups of the olefinic monomers considered here unless where the need arises. Rather, rules will be proposed with respect to capacities of different types of substituent groups, any additional driving force(s) for molecular rearrangement of monomers and other important and common concepts cutting across Addition monomers, their polymerization systems and indeed all things in life.

Rule 143: This rule of Chemistry for **Addition monomers as applies to many things in life,** states that, *substituted groups are those* which are either saturated or unsaturated, charged/ radical-pushing or-pulling, and may carry *substituent groups* also called transfer species or may carry none. (See Rule 67). *(Laws of Creations for Substituted groups)*

Rule 144: This rule of Chemistry for **Addition monomers as applies to many things in life,** states that, when *substituted groups* carried by the active centers of some polymerizable monomers are very large, they can no longer be polymerized due to what is called **STERIC HINDRANCE OR LIMITATIONS.** *(Laws of Creations for Propagation)*

Rule 145: This rule of Chemistry for **Addition monomers as applies to many things in life,** states that, *during molecular rearrangement of the first kind,* transfer of transfer species which is greater in capacity than any of those of the same character carried by the receiving active carbon center is impossible. *(Laws of Creations for Limitations of a receiving center during Molecular rearrangement)*

Rule 146: This rule of Chemistry for **Molecular rearrangement in Addition monomers,** states that, *the fourth driving force for molecular rearrangement* is favoring the existence of a new activated monomer with *less or equal number of substituent or substituted groups*; for which the concept of symmetricity is impossible. [See Rule 97 in Chapter 2] *(Laws of Creations for Molecular rearrangement)*

Rule 147: The rule of Chemistry for **Molecular rearrangement in Addition monomers,,** states that, *where different transfer species of different characters of first kind exist on two active centers of the activated monomer, the fifth driving force* is that in which the transfer species involved is determined by the character of the monomer; for which for electrophiles (i.e. males), non-free-negatively charged or nucleo-non-free-radical species coming from the active carbon center carrying negative charge or nucleo-free-radical are involved, while for nucleo-philes (i.e., females), free-positively charged or electro-free-radical species coming from the active carbon center carrying positive charge or electro-free-radical are involved. *(Laws of Physics and Mathematics)*

Rule 148: This rule of Chemistry for **Molecular rearrangement in Addition monomers,** states that, *the sixth driving force for molecular rearrangement,* is the existence of a more stable newly activated monomer compared to the original activated monomer and the stability is marked by increase in strength of its natural active center and increase in strength in the opposite direction of the second active center. *(Laws of Creations for Propagation)*

Rule 149: This rule of Chemistry for **Addition monomers which carry allylic CH_2X and CHX_2 types of groups, where X are F, Cl, Br, OH, $OCOCH_3$, etc.** states that, these monomers cannot be activated chargedly, since they cannot undergo the route which is natural to them, but may undergo the route which is unnatural to them. *(Laws of Creations for Addition monomers)*

Rule 150: This rule of Chemistry for **Alkenes and Perfluoroalkenes,** states that, when there exist any monomer which can undergo the route which is not natural to it, but not the route which is natural to it, then such monomers cannot be activated chargedly. *(Laws of Creations for Alkenes and Perfluoroalkenes)*

Rule 151: This rule of Chemistry for **Alkenes and Perfluoroalkenes,** states that, free-radically while only the first members of their families favor both free-radical routes (i.e., routes natural and unnatural to them), all the lower members in the family favor only the route natural to them- the electro-free-radical route or none.
(Laws of Creations for Alkenes and Perfluroalkenes)

Rule 152: This rule of Chemistry for **Alkenes and Perfluoroalkenes,** states that, when they carry an F atom on an active C center and made to undergo molecular rearrangement, the new monomer obtained must still be such that carry the F atom on the same or different active C center; otherwise molecular rearrangement will not take place.
(Laws of Creations for Alkenes and Perfluoroalkenes)

Rule 153: This rule of Chemistry for **Alkenes and Perfluoroalkenes,** states that, chargedly while the fully hydrogenated ones favor only the positively charged route [NUCLEOPHILES], some special fully fluorinated ones such as shown below favor only the negatively charged route [ELECTROPHILES].

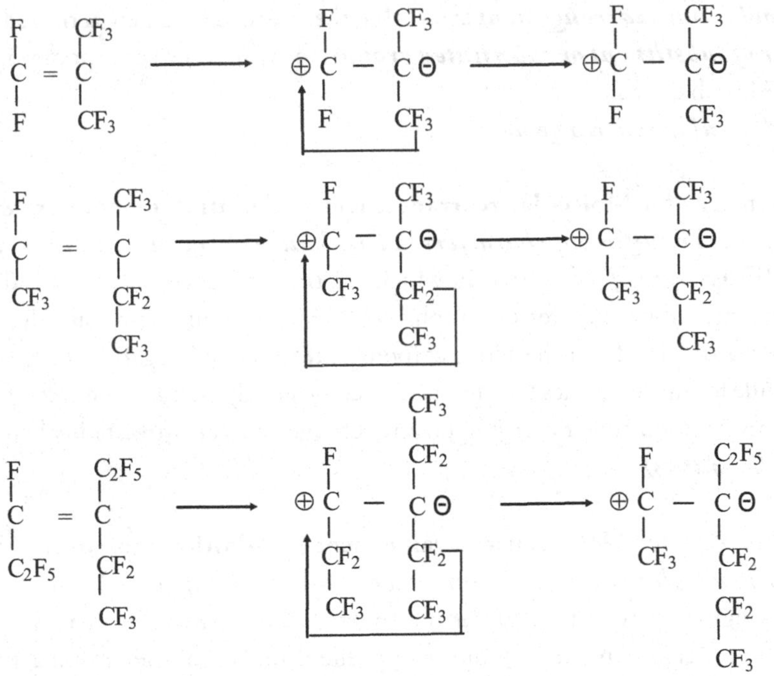

for which the activation of such monomers will be impossible chargedly since there is no visible X center carried by the monomers; while radically the situation is different.
(Laws of Creations for Alkenes and Perfluoroalkenes)

Rule 154: This rule of Chemistry for **Alkenes and Perfluoroalkenes,** states that, chargedly between the two extremes are monomers which cannot be activated due to electrostatic forces of repulsion or ability to favor route not natural to them, but that which is natural and also monomers which cannot favor any route when activated.
(Laws of Creations for Alkenes and Perfluoroalkenes)

Rule 155: This rule of Chemistry for **Addition monomers which carry alkanyl types of alkyl primary substituent groups**, states that, *electro-free-radically, their free-radical-pushing capacities can be determined from the following order:-*

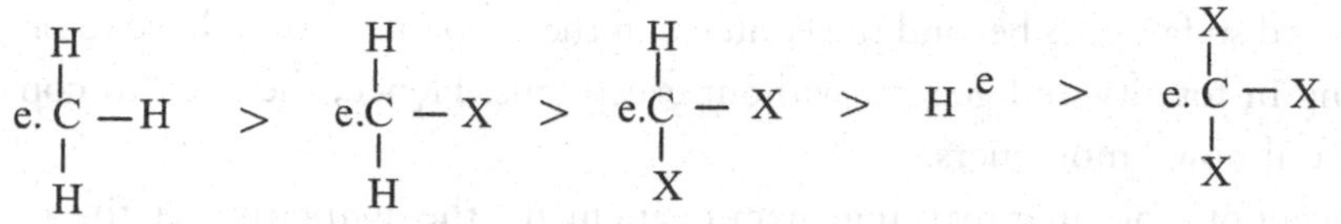

$$
\underset{\underset{H}{|}}{\overset{\overset{H}{|}}{e.C}} - H \quad > \quad \underset{\underset{H}{|}}{\overset{\overset{H}{|}}{e.C}} - X \quad > \quad \underset{\underset{X}{|}}{\overset{\overset{H}{|}}{e.C}} - X \quad > \quad H^{\cdot e} \quad > \quad \underset{\underset{X}{|}}{\overset{\overset{X}{|}}{e.C}} - X
$$

<u>Order of Free-radical-pushing capacities</u>

where Xs are halogens other than iodine (I) or fully halogenated groups. [See Rule 85]
(Laws of Creations for Addition monomers)

Rule 156: This rule of Chemistry for **Charged Initiators,** states that, all Charged Initiators are paired in character; for which existence of Free-media charged initiators in the absence of isolation is impossible, noting that only anions can be used free-medially, while cations cannot be used free-medially, due to charge balancing.
(Laws of Creations for Charged Initiators)

Rule 157: This rule of Chemistry for **Paired Initiators,** states that, Radically paired initiators carry two centers while ionically paired ones carry only one active center as shown below for some of them--

$$\underline{H_2CO}^{\ominus}......^{\oplus}Na \quad ; \quad H_3CO^{\cdot nn}.....^{e\bullet}Na \quad ; \quad ClMg^{\oplus}....^{\ominus}OC_4H_9 \quad ; \quad ClMg^{\cdot e}.........^{n\bullet}OC_4H_9$$

$$Na^{\bullet e}.......^{n\bullet}C \equiv N \quad ; \quad Na^{\bullet e}........^{n\bullet}C_{10}H_7$$

TWO ACTIVE CENTERS

for which the last two can never be chargedly paired due to electrostatic forces of repulsion and the third one can be used positively, because the positive charge on $ClMg^{\oplus}$ is not a cation.
(Laws of Creations for Initiators)

Rule 158: This rule of Chemistry for **Alkene and Perfluoroalkene families**, states that, the followings are valid for radical-/charged- pulling and pushing groups-

$$-C(CF_3)_3 \quad > \quad -CF(CF_3)_2 \quad > \quad -CF_2CF_3 \quad > \quad -CF_3 \quad > \quad -F$$

<u>Order of *charged*-pulling capacity</u>

$$C(CF_3)_3 \quad > \quad CH(CF_3)_2 \quad > \quad CFHCF_3 \quad > \quad CF_2H \quad > \quad F$$

<u>Order of *charged*-pulling capacity</u>

$$CH_3 \quad > \quad CH_2F \quad > \quad CHF_2 \quad > \quad H \quad > \quad CF_3$$

<u>Order of *radical*-pushing capacity</u>

(Laws of Creation for Substituent Groups)

4.12 Conclusions

Sixteen rules of Chemistry in Addition polymerization and chemical systems less than what to be expected have been proposed bringing the total rules so far to one hundred and fifty eight. Proposition of rules, as can be observed so far, goes beyond the contents in the chapters. If rules have to be proposed, it must be all embracing in totality and general without exceptions. Hence, the need to consider both existing or known and unknown monomers.

Using the new concept of molecular rearrangement phenomena, the qualitative relative capacities of alkanyl types of groups are being systematically identified. How some monomers which are not known to favor both free-radical and charged polymerizations, can be made to do so, are being clearly shown. On the other hand, why some cannot favor free-radical and or some or all charged polymerization routes are being explained. Free-radically, all the monomers which have been considered here are Nucleophiles. Chargedly also, the same groups of monomers are Nucleophiles, but with a great difference in terms of inability to polymerize many of them, and the favored existence of Electrophiles for some fully fluorinated ones. Charges and radicals are uniquely different particularly in terms of electrostatic forces of repulsion and limitations.

Though, it may seem that the concept of molecular rearrangement phenomena have been conclusively considered, this is just but a grain of sand, Most of the rules are based on the phenomena and general applications in science related disciplines of chemistry. For the first time, a substituent group has been clearly defined and distinguished from a substituted group for the purpose of orderliness and clear understanding of the new concepts. It should be noted also that there are some rules already identified, but not yet proposed until further justifications have been provided

References

1. C. R . Noller, "Textbook of Organic Chemistry", 3rd Edition, W.B. Saunders Company, 1966.

2. J. Hines, "Physical Organic Chemistry" 2nd Edition; McGraw- Hill Book Company, Inc. New York 1962.

Problems

4.1. Identify all the driving forces so far shown favoring the existence of molecular rearrangement phenomena of the first kind of the first type in olefinic monomers. What are the advantages offered by the molecular rearrangement phenomena?

4.2. Beginning with a weak initiator of all types for free-radical and charged polymerizations, will molecular rearrangement phenomena take place throughout the polymerization of the monomer until limiting or full (100%) conversion? Explain using examples.

4.3. State the laws involved during the phenomena of molecular rearrangement for different types of olefinic monomers where they apply.

4.4. Shown below are the following monomers.

(i) 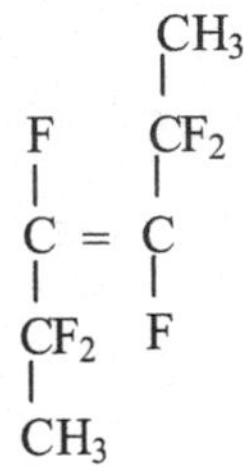
$$
\begin{array}{c}
\quad\quad CH_3 \\
\quad\quad | \\
F \quad CF_2 \\
| \quad\quad | \\
C = C \\
| \quad\quad | \\
CF_2 \quad F \\
| \\
CH_3
\end{array}
$$

(ii) 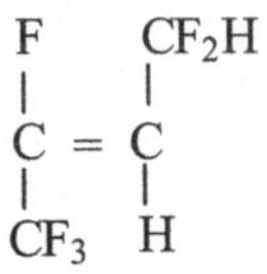
$$
\begin{array}{c}
F \quad CF_2H \\
| \quad\quad | \\
C = C \\
| \quad\quad | \\
CF_3 \quad H
\end{array}
$$

(iii) 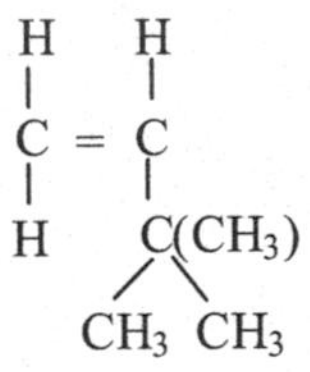
$$
\begin{array}{c}
H \quad H \\
| \quad\quad | \\
C = C \\
| \quad\quad | \\
H \quad C(CH_3) \\
\quad\quad \diagup \diagdown \\
\quad CH_3 \quad CH_3
\end{array}
$$

(iv) 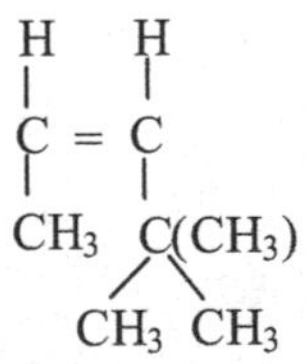
$$
\begin{array}{c}
H \quad H \\
| \quad\quad | \\
C = C \\
| \quad\quad | \\
CH_3 \quad C(CH_3) \\
\quad\quad \diagup \diagdown \\
\quad CH_3 \quad CH_3
\end{array}
$$

(v) 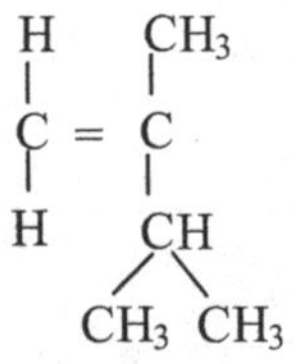
$$
\begin{array}{c}
H \quad CH_3 \\
| \quad\quad | \\
C = C \\
| \quad\quad | \\
H \quad CH \\
\quad\quad \diagup \diagdown \\
\quad CH_3 \quad CH_3
\end{array}
$$

(a) Identify the characters of the monomers free-radically and chargedly.
(b) Which of the monomers favor molecular rearrangement? Explain.
(c) Identify the transfer species of the first kind of the first type in all the monomers free-radically and chargedly.

4.5. (a) Why is perfluoro-propylene not popularly known to be used for producing homopolymers (Tg = 11°C) nucleo-free-radically?

(b) Below as some monomers.

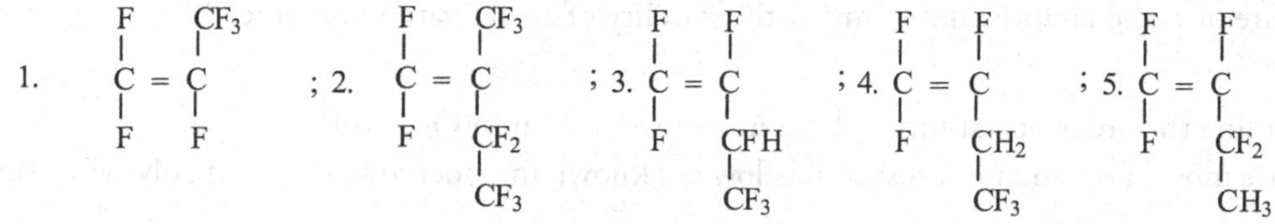

```
         F   CF3          F   CF3          F    F           F    F           F    F
         |    |           |    |           |    |           |    |           |    |
1.       C  = C     ; 2.  C  = C     ; 3.  C  = C     ; 4.  C  = C     ; 5.  C  = C
         |    |           |    |           |    |           |    |           |    |
         F    F           F   CF2          F   CFH          F   CH2          F   CF2
                               |                |                |                |
                              CF3              CF3              CF3              CH3
```

(i) Show the activations of the monomers.

(ii) Identify the order of the charged/radical-pushing or pulling-capacities of the substituent groups.

(iii) Which of the monomers will favor transfer of transfer species? Identify the transfer species in all the monomers and therefore the characters of the monomers.

4.6. Shown below are some monomers.

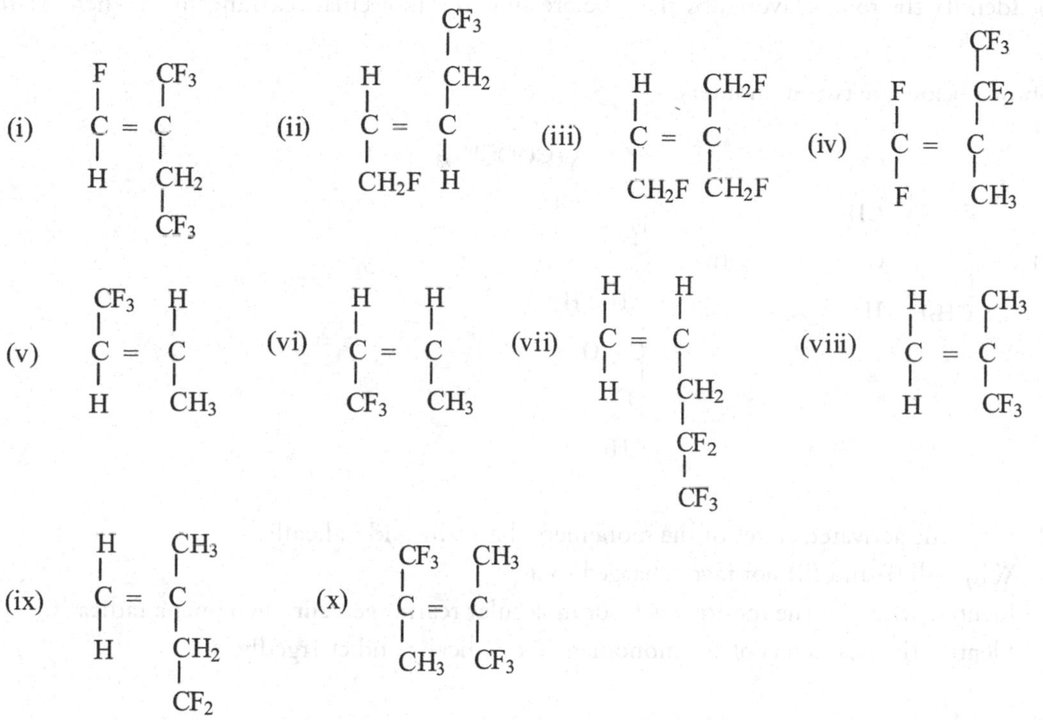

```
                                              CF3                                         CF3
                                               |                                           |
         F    CF3          H    CH2            H    CH2F           F    CF2
         |     |           |     |             |     |            |     |
(i)      C  =  C    (ii)   C  =  C      (iii)  C  =  C     (iv)   C  =  C
         |     |           |     |             |     |            |     |
         H    CH2         CH2F   H            CH2F  CH2F          F    CH3
               |
              CF3
```

```
        CF3   H            H    H                 H    H             H    CH3
         |    |            |    |                 |    |             |     |
(v)      C =  C    (vi)    C  = C       (vii)     C  = C    (viii)   C  =  C
         |    |            |    |                 |    |             |     |
         H   CH3          CF3   CH3               H   CH2            H    CF3
                                                      |
                                                     CF2
                                                      |
                                                     CF3
```

```
         H    CH3              CF3   CH3
         |     |                |     |
(ix)     C  =  C        (x)     C  =  C
         |     |                |     |
         H    CH2              CH3   CF3
               |
              CF2
```

(a) Establish the order of the pushing capacities of all the substituted groups free-radically.

(b) Establish the order of the pulling capacities of all the substituted groups chargedly.

(c) Without any activation, identify the characters of all the monomers chargedly and free-radically.

(d) Show their activated states before molecular rearrangements.

4.7. (a) When the conditions for molecular rearrangement exist, show the new activated states of the monomers in Q 4.6 free-radically and chargedly.

(b) Identify the routes favored by the monomers before and after molecular rearrangement.

4.8 (a) Distinguish between a substituted and substituent group using CF_3 and CH_3.

(b) Can a new activated state be found to be more stable than the original activated state, without undergoing molecular rearrangement?

(c) What are or is the factor(s) determining the stability of an activated monomer?

4.9. (a) Determine the order of stabilities of the monomers shown in Q4.5 (b).

(b) Why are most of the monomers shown below not known to undergo free-radical polymerizations?

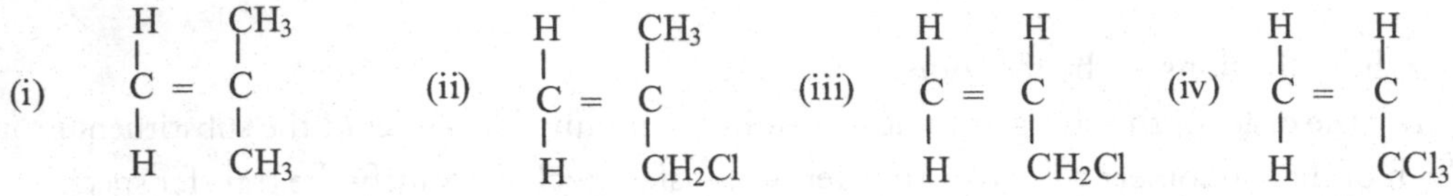

$$
\text{(i)} \quad
\begin{array}{c}
\text{H} \quad\ \text{CH}_3 \\
| \qquad\ | \\
\text{C} = \text{C} \\
| \qquad\ | \\
\text{H} \quad\ \text{CH}_3
\end{array}
\qquad
\text{(ii)} \quad
\begin{array}{c}
\text{H} \quad\ \text{CH}_3 \\
| \qquad\ | \\
\text{C} = \text{C} \\
| \qquad\ | \\
\text{H} \quad\ \text{CH}_2\text{Cl}
\end{array}
\qquad
\text{(iii)} \quad
\begin{array}{c}
\text{H} \quad\ \text{H} \\
| \qquad\ | \\
\text{C} = \text{C} \\
| \qquad\ | \\
\text{H} \quad\ \text{CH}_2\text{Cl}
\end{array}
\qquad
\text{(iv)} \quad
\begin{array}{c}
\text{H} \quad\ \text{H} \\
| \qquad\ | \\
\text{C} = \text{C} \\
| \qquad\ | \\
\text{H} \quad\ \text{CCl}_3
\end{array}
$$

(c) Which of the monomers will favor transfer of transfer species and molecular rearrangement when the conditions exist?

(d) Identify the routes favored by them before and after molecular rearrangement where possible.

4.10 Shown below are two monomers.

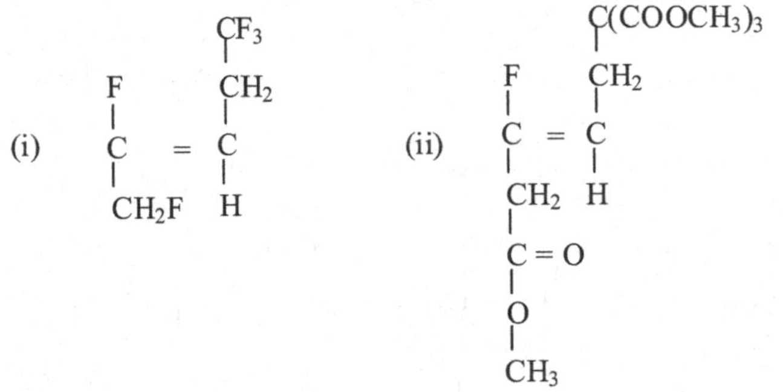

$$
\text{(i)} \quad
\begin{array}{c}
\qquad\quad \text{CF}_3 \\
\qquad\quad | \\
\text{F} \quad\ \text{CH}_2 \\
| \qquad\ | \\
\text{C} = \text{C} \\
| \qquad\ | \\
\text{CH}_2\text{F} \ \text{H}
\end{array}
\qquad
\text{(ii)} \quad
\begin{array}{c}
\qquad\quad \text{C(COOCH}_3)_3 \\
\qquad\quad | \\
\text{F} \quad\ \text{CH}_2 \\
| \qquad\ | \\
\text{C} = \text{C} \\
| \qquad\ | \\
\text{CH}_2 \ \text{H} \\
| \\
\text{C} = \text{O} \\
| \\
\text{O} \\
| \\
\text{CH}_3
\end{array}
$$

(a) Show the activated states of the monomers chargedly and radically.

(b) Why will (i) and (ii) not favor charged route?

(c) Identify which of the monomers favor molecular rearrangement phenomena radically.

(d) Identify the characters of the monomers free-radically and chargedly.

4.11. Shown below are three monomers.

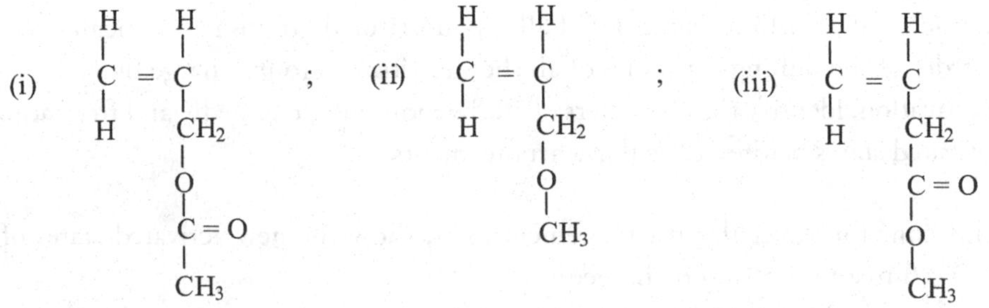

$$
\text{(i)} \quad
\begin{array}{c}
\text{H} \quad\ \text{H} \\
| \qquad\ | \\
\text{C} = \text{C} \\
| \qquad\ | \\
\text{H} \quad\ \text{CH}_2 \\
\qquad\quad | \\
\qquad\quad \text{O} \\
\qquad\quad | \\
\qquad\quad \text{C} = \text{O} \\
\qquad\quad | \\
\qquad\quad \text{CH}_3
\end{array}
\ ;
\quad
\text{(ii)} \quad
\begin{array}{c}
\text{H} \quad\ \text{H} \\
| \qquad\ | \\
\text{C} = \text{C} \\
| \qquad\ | \\
\text{H} \quad\ \text{CH}_2 \\
\qquad\quad | \\
\qquad\quad \text{O} \\
\qquad\quad | \\
\qquad\quad \text{CH}_3
\end{array}
\ ;
\quad
\text{(iii)} \quad
\begin{array}{c}
\text{H} \quad\ \text{H} \\
| \qquad\ | \\
\text{C} = \text{C} \\
| \qquad\ | \\
\text{H} \quad\ \text{CH}_2 \\
\qquad\quad | \\
\qquad\quad \text{C} = \text{O} \\
\qquad\quad | \\
\qquad\quad \text{O} \\
\qquad\quad | \\
\qquad\quad \text{CH}_3
\end{array}
$$

(a) Show the activation of the monomers free-radically and chargedly.

(b) Which of the monomers favor molecular rearrangement phenomena chargedly and radically.

(c) Identify the characters of the monomers as well as their transfer species free-radically and chargedly.

(d) Distinguish between the $OCOCH_3$, OCH_3 and $COOCH_3$ groups as part of a substituent group radically and chargedly, when they are themselves substituent groups that provide transfer species.

4.12 (a) Which of the monomers in Q 4.10. and Q 4.11 favor resonance stabilization phenomena? Explain.

(a) What group of olefinic monomers do these monomers belong to?

(b) In the absence of other factors, under what condition(s) can (ii) of Q 4.10 if charged

(c) activation is possible be made to favor charged polymerization? Will the cis-monomer favor any charged route under any conditions? Explain.

(d) Shown below are two types of monomers.

(i)

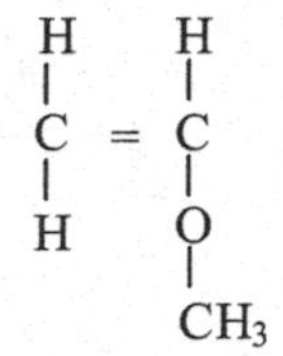

$$\begin{array}{ccc} H & & H \\ | & & | \\ C & = & C \\ | & & | \\ H & & O \\ & & | \\ & & CH_3 \end{array}$$

(ii)

$$\begin{array}{c} H \\ | \\ C = O \\ | \\ O \\ | \\ CH_3 \end{array}$$

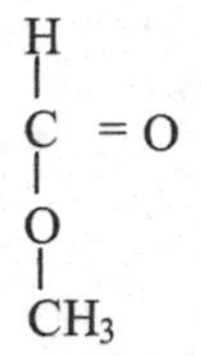

Under what conditions will OCH_3 be more readily abstracted from them free-radically and chargedly? What routes are favored by them? Why can't (i) molecularly rearrange, but (ii) can?

SECTION B
Inter-Molecular Addition Monomers
Non-Olefins

CHAPTER 5

TRANSFER OF TRANSFER SPECIES IN ALDEHYDES, KETONES AND RELATED MONOMERS

5.0 Introduction

Aldehydes and Ketones are important members of Addition monomers, since as has been observed so far, very little is known about their polymerizations. Whether they are nucleophiles or electrophiles has never been clearly defined. The Tables provided in Volume 1 for Addition monomers and their pseudo-counterparts were not just provided from the blues or literature data alone. They were provided after full realization of what a monomer, an ion, a charge, a radical, a substituted group, a substituent group, an atom, and so on are.

Unlike olefinic monomers where many of the charged routes are well established, for aldehydes and ketones, their routes chargedly have never been well defined. For example, there is no way acetaldehyde or acetone can undergo "anionic" or indeed free and non-free negatively charged polymerization reactions[1]. On the other hand, nothing is known about the radical polymerizations of these monomers. In attempting to provide true definitions of a radical, free-radical and non-free-radical, it was shown that the so-called free-radical polymerizations of some aldehydes and ketones observed over the years are no nucleo-free-radical polymerizations, but non-free-radical and electro-free-radical polymerizations.[2] Why these groups of monomers have never been known to favor free-radical polymerizations, have never been explained. From current developments, all these are clearly becoming obvious. Why some olefins favor rearrangement to produce aldehydes and ketones and sometimes the reverse has never been explained.

Like the olefinic Addition monomers, both the fluorinated and non-fluorinated extremes will be considered, in view of the diversities of routes and characters favored by the monomers and for the polymerization of these monomers. This will be done for all pseudo-Addition monomers in the Volume. In the process of trying to explain the phenomena of resonance stabilization in some olefinic monomers such as acrolein, acrylates, acrylamides, acrylonitriles, it was observed that, in the absence of resonance stabilization, an Addition monomer can possess both the Addition and pseudo-Addition features. In order words, there are some of these Addi-tion monomers, which can be considered to be aldehydes or ketones when activated via the non- C = C activation centers, that is via the C = O activation centers. These are the related monomers. Their considerations are still limited, since in most cases copolymers (random and alternating) are produced. These will be fully considered after all the present exercises are completed. Where the existence is favored will however be identified. Few thiocarbonyl monomers will also be considered. Meanwhile, ketenes, isocyanates will be excluded in the present analysis, since these monomers which are so unique deserve separate attention.

While most of the rules proposed so far, also apply to aldehydes and ketones, there are still some rules which are peculiar to aldehydes and ketones as a family of Addition monomers (Pseudo-Addition monomers). Apart from the 50 - 50 heterochain characters of the polymer backbones, all the features of Addition polymerization systems still apply. For aldehydes and ketones, like olefins there are similar kinds and types of transfer species-depending on the State of Existence and the types of substituted groups carried, noting however that unlike olefins, the number of substituted groups carried is less and the Activated States of them are FIXED. While the symmetric character of Olefins can be along the vertical and horizontal axes, the symmetric character of Formaldehyde and Ketones is along the horizontal axis, the axis of Propagation for all linear polymers. Aldehydes do not have any axis of symmetry, unlike formaldehyde.

Uniquely enough, the first members of all olefinic families of monomers, aldehydes, cumulenes, ketenes, isocyanates, acetylenes, etc. have been observed to be uniquely different from other members in their families, in that they tend to favor all the kinetic routes where possible or none at all. For example ethylene, butadiene, etc. are unique with olefins. With the family of aldehydes and ketones, formaldehyde is most unique. Like the olefins, some type of resonance stabilization groups can provide resonance stabilization for aldehydes and ketones as we have already seen (See Equation 3.140) for Males. Though, molecular rearrangement pheno-mena is not common with them, other types of rearrangement to another compound (Enolization) exist with them as will be shown downstream in this Volume.

5.1 Aldehydes

5.1.1 Charged characters of Aldehydes

Beginning with the first member, formaldehyde for which there is no substituent group, the following is obtained when activated chargedly.

$$
\begin{array}{ccc}
& H & H \\
& | & | \\
& C = O & \longrightarrow \quad \oplus C - O^{\ominus} \\
& | & | \\
& H & H \\
& & (I)
\end{array}
$$

5.1

It has no transfer species of any kind and therefore favors both charged routes (Free-positively and Non-free negatively). Unlike olefins, only non-free negative charges (called anions) can be used free-medially and pairedly. Anionically, its growing polymer chain cannot be killed from within. Since the route is not Natural to it, and there is no transfer species, like the case of α-methyl styrene, same type of Depropagation may take place here but differently. Therefore the temperature of polymerization must be reduced based on the route if polymers are to be obtained.

With positively charged paired initiators, the following is obtained for its growing polymer chain.

$$
\begin{array}{ccccc}
& H & & H & \\
& | & & | & \\
R \!-\!\!\left(\!O - C\!\right)_{\!n}\!\!- & O - & C\oplus & \longrightarrow & \text{No transfer possible} \\
& | & & | & \\
& H & & H &
\end{array}
$$

5.2

Thus, living polymers are produced both ways, in the absence of foreign transfer species or terminating agents.

Now, consider the fluorinated version of the monomer, which when activated, the following will be obtained.

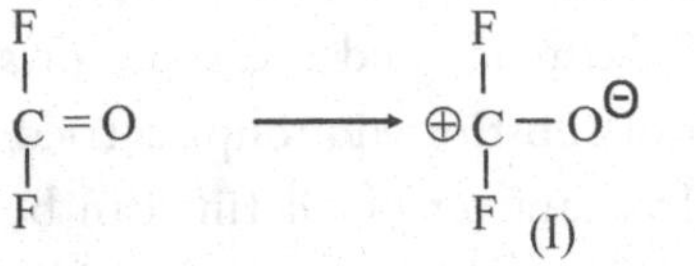

$$\underset{\underset{F}{|}}{\overset{\overset{F}{|}}{C}}=O \quad \longrightarrow \quad \oplus\underset{\underset{F}{|}}{\overset{\overset{F}{|}}{C}}-O^{\ominus} \qquad \text{(I)} \qquad\qquad 5.3$$

Anionically for its growing polymer chain, the following is obtained in the absence of electrostatic forces of repulsion between the non-free (anionic) initiator and the fluorine atoms.

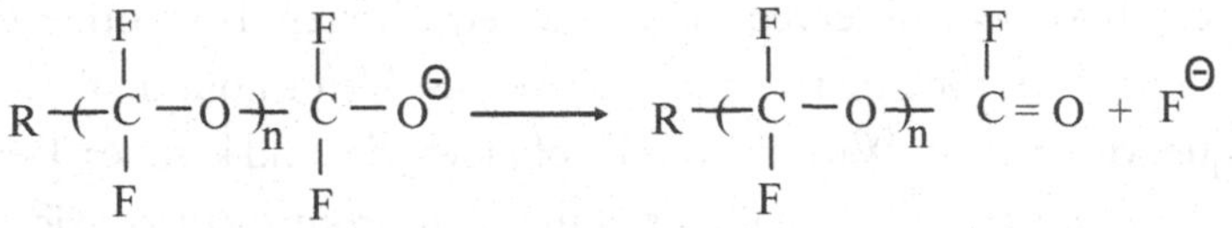

$$R\left(\underset{\underset{F}{|}}{\overset{\overset{F}{|}}{C}}-O\right)_{n}\underset{\underset{F}{|}}{\overset{\overset{F}{|}}{C}}-O^{\ominus} \quad\longrightarrow\quad R\left(\underset{\underset{F}{|}}{\overset{\overset{F}{|}}{C}}-O\right)_{n}\overset{\overset{F}{|}}{C}=O \; + \; F^{\ominus} \qquad\qquad 5.4$$

The transfer species rejected is of the second kind.

With positively charged paired initiator, the following is obtained.

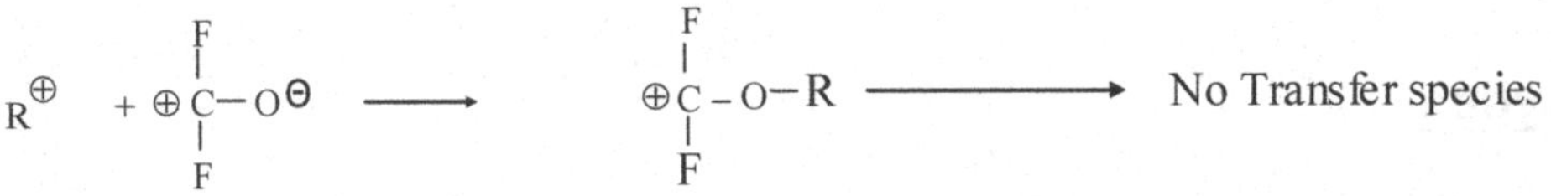

$$R^{\oplus} \; + \; \oplus\underset{\underset{F}{|}}{\overset{\overset{F}{|}}{C}}-O^{\ominus} \quad\longrightarrow\quad \oplus\underset{\underset{F}{|}}{\overset{\overset{F}{|}}{C}}-O-R \quad\longrightarrow\quad \text{No Transfer species} \qquad 5.5$$

Both routes like formaldehyde are favored. Their Natural route is the positively charged-paired initiator and not "cationic", as wrongly used in Present-day Science.

When positively-charged-paired initiators are used however, the monomer above will not favor polymerization, due to electrostatic forces of repulsion. However, for the half fluorinate monomer, polymerization will be favored, an indication of the strong nucleophilic character of the monomer.

Now, with acetaldehyde, the following is obtained.

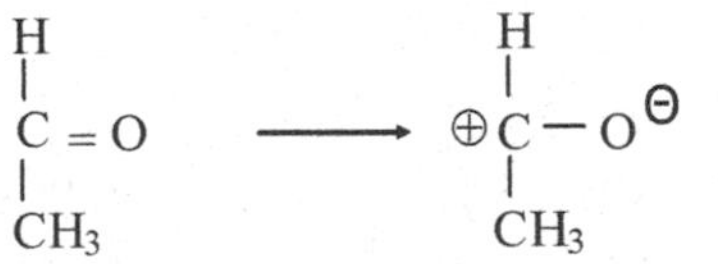

$$\underset{\underset{CH_3}{|}}{\overset{\overset{H}{|}}{C}}=O \quad\longrightarrow\quad \oplus\underset{\underset{CH_3}{|}}{\overset{\overset{H}{|}}{C}}-O^{\ominus} \qquad\qquad 5.6$$

Molecular rearrangement is not favored since vinyl alcohol is far less stable than acetaldehyde and also since the monomer is only slightly in Activated State of Existence (Not in Activated/ Equilibrium State of Existence). ***Under Activated/Equilibrium State of Existence, the monomer cannot be polymerized.*** For its growing polymer chain "cationically", when polymerization is favored, the following is obtained.

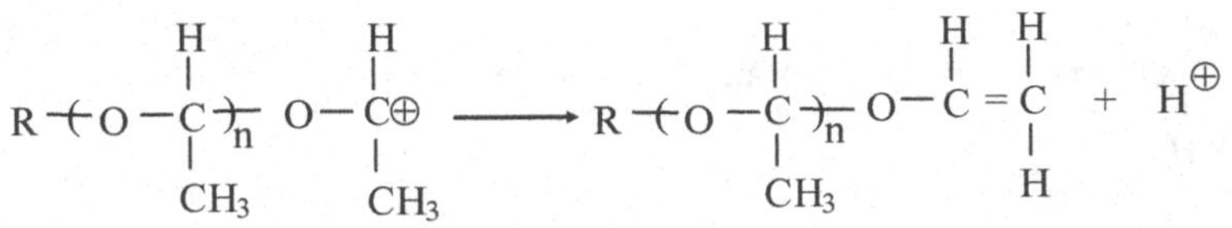

$$R\left(O-\underset{\underset{CH_3}{|}}{\overset{\overset{H}{|}}{C}}\right)_{n}O-\overset{\overset{H}{|}}{\underset{\underset{CH_3}{|}}{C}}\oplus \quad\longrightarrow\quad R\left(O-\underset{\underset{CH_3}{|}}{\overset{\overset{H}{|}}{C}}\right)_{n}O-\overset{\overset{H}{|}}{C}=\overset{\overset{H}{|}}{\underset{\underset{H}{|}}{C}} \; + \; H^{\oplus} \qquad 5.7$$

This is the same transfer species abstracted in transfer from monomer step "anionically" as shown below.

$$RO^{\ominus} \quad + \quad \overset{\oplus}{\underset{\underset{CH_3}{|}}{\overset{\overset{H}{|}}{C}}} - O^{\ominus} \quad \longrightarrow \quad ROH \quad + \quad \underset{\underset{H}{|}}{\overset{\overset{H}{|}}{C}} = C - O^{\ominus} \qquad \qquad 5.8a$$

(Non-free) -Anionically

$$R^{\ominus} \quad + \quad \overset{\oplus}{\underset{\underset{CH_3}{|}}{\overset{\overset{H}{|}}{C}}} - O^{\ominus} \quad \longrightarrow \quad RH \quad + \quad \overset{\ominus}{\underset{\underset{H}{|}}{C}} - \underset{\underset{H}{||}}{C} = O \qquad \qquad 5.8b$$

(free)

Replacing H above with F, the following is obtained.

$$RO^{\ominus} \quad + \quad \overset{\oplus}{\underset{\underset{CH_3}{|}}{\overset{\overset{F}{|}}{C}}} - O^{\ominus} \quad \longrightarrow ROH \quad + \quad \overset{5.9}{\underset{\underset{H}{|}}{\overset{\overset{H}{|}}{C}}} = \underset{\underset{F}{|}}{C} - O^{\ominus} \qquad \qquad 5.9$$

(Non-free)   (I)

$$R^{\oplus} \quad + \quad \overset{\oplus}{\underset{\underset{CH_3}{|}}{\overset{\overset{F}{|}}{C}}} - O^{\ominus} \quad \longrightarrow \quad \overset{\oplus}{\underset{\underset{CH_3}{|}}{\overset{\overset{F}{|}}{C}}} - O - R \qquad \qquad 5.10$$

"Anionically" the routes (free-media and ion-paired-media) are not favored. When positively charged-paired initiators are employed, initiation is favored in the absence of electrostatic forces of repulsion.

$$R \overset{}{(} O - \underset{\underset{CH_3}{|}}{\overset{\overset{F}{|}}{C}} \overset{}{)_n} O - \underset{\underset{CH_3}{|}}{\overset{\overset{F}{|}}{C}}^{\oplus} - - ^{\ominus}BF_3 \ OR \quad \longrightarrow \quad R \overset{}{(} O - \underset{\underset{CH_3}{|}}{\overset{\overset{F}{|}}{C}} \overset{}{)_n} O - \underset{\underset{H}{|}}{\overset{\overset{F}{|}}{C}} = \underset{\underset{H}{|}}{\overset{\overset{H}{|}}{C}} \quad + \quad BF_3 \quad + \quad ROH \qquad 5.11$$

The nucleophilic character of the monomer is still maintained despite the presence of forces of repulsion. Now, consider replacing CH_3 with CF_3; the following is obtained.

$$RO \overset{}{(} \underset{\underset{CF_3}{|}}{\overset{\overset{F}{|}}{C}} - O \overset{}{)_n} \underset{\underset{CF_3}{|}}{\overset{\overset{:\ddot{F}:}{|}}{C}} - \ddot{O}\overset{\ominus}{:} \quad \longrightarrow RO \overset{}{(} \underset{\underset{CF_3}{|}}{\overset{\overset{F}{|}}{C}} - O \overset{}{)_n} \underset{}{\overset{\overset{CF_3}{|}}{C}} = O \quad + \quad x\ddot{F}\overset{\ominus}{:} \qquad 5.12$$

The monomer can be observed to be like an electrophile. However, free-"cationically" the following is obtained.

$$R^{\oplus} \quad + \quad \overset{\oplus}{\underset{\underset{CF_3 \ (I)}{|}}{\overset{\overset{F}{|}}{C}}} - O^{\ominus} \quad \longrightarrow \quad \overset{\oplus}{\underset{\underset{F}{|}}{\overset{\overset{CF_3}{|}}{C}}} - O - R \qquad \qquad 5.13$$

It should be noted that it is the non-free (anion) $F^{\ominus}$ that is rejected in Equation 5.12 and not the free negative charge CF_3 which is not possible. When charged-paired initiators are employed both routes are not favored as shown below.

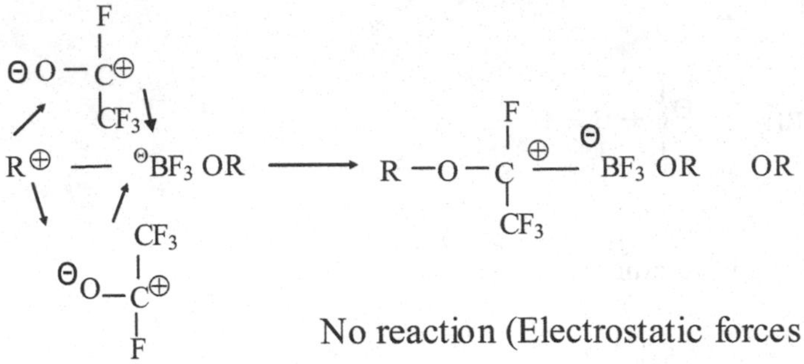

No reaction (Electrostatic forces of repulsion)   5.14

<u>Favored reaction</u>

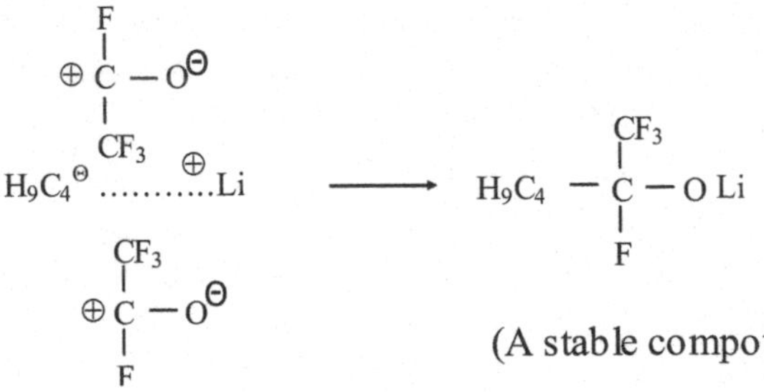

                                                                    5.15

(A stable compound, but not favored- Forces of repulsion)

Due to electrodynamic forces of repulsion between the fluorine atom on the monomer and that on the counter-ion or central center, polymerization is not favored positively. Negatively, no polymerization is favored since the negative active center is free, while the growing chain is a non-free anionic one. If $H_3CO^{\ominus}.....^{\oplus}Na$ had been used, electrostatic forces of repulsion will not allow for the initiation. Even then, the active center will be Na or Li only when paired radically, the monomer being Nucleophilic with the forces of repulsion still in place. When the F on the monomer is replaced with H, polymerization will be favored as shown below.

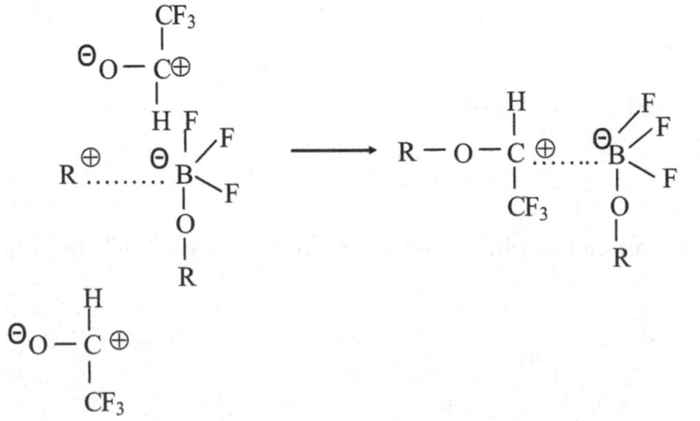

                                                                    5.16

This then implies that the monomer's nucleophilicity is still maintained, only to be partly disturbed by the presence of these non-free centered groups and limited choice of initiator. The trifluoroacetaldehyde above will also favor non-free-negatively charged routes, in the absence of transfer species of the first kind of the first type. ***This also is a clear indication of presence of Depropagation in the route not natural to it at high temperatures of polymerization.***

Fluorothiocarbonyl compounds are important monomers. Considering thiocarbonyl fluoride, the followings are obtained.

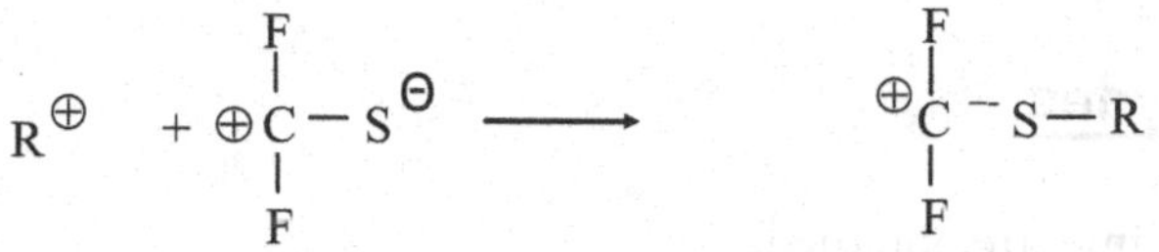

$$
R^{\oplus} \;+\; \underset{F}{\overset{F}{\oplus C}} - S^{\ominus} \;\longrightarrow\; \underset{F}{\overset{F}{\oplus C}} - S - R
$$

5.17

$$
R\!\cdots\!\overset{F}{\underset{O}{\overset{|}{\underset{|}{B}}}}\!\overset{F}{\nwarrow}\!{}^{\ominus} + \overset{\ominus}{S} - \underset{F}{\overset{F}{\underset{|}{C^{\oplus}}}} \;\longrightarrow\; \text{No reaction (Repulsion)}
$$

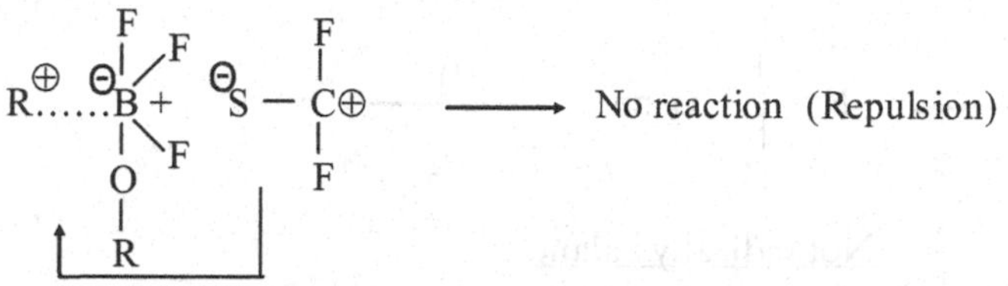

5.18

Like its oxygen counterpart, the reaction is not favored. It is only when "special" Z/N types of initiators are involved that the polymerization becomes fully favored in the absence of electrostatic forces of repulsion.

With anionic initiators, the followings are obtained.

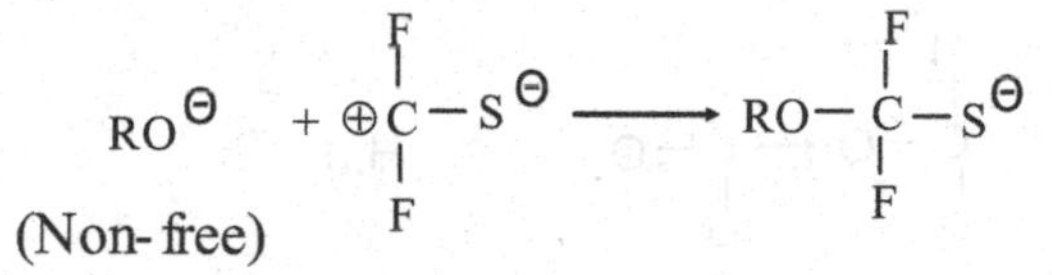

$$
RO^{\ominus} \;+\; \underset{F}{\overset{F}{\oplus C}} - S^{\ominus} \;\longrightarrow\; RO - \underset{F}{\overset{F}{C}} - S^{\ominus}
$$

(Non-free)

5.19

$$
RO \left(\underset{F}{\overset{F}{C}} - S \right)_{\!n} \underset{F}{\overset{F}{C}} - S^{\ominus} \;\longrightarrow\; RO \left(\underset{F}{\overset{F}{C}} - S \right)_{\!n} \underset{F}{\overset{F}{C}} = S \;+\; F^{\ominus}
$$

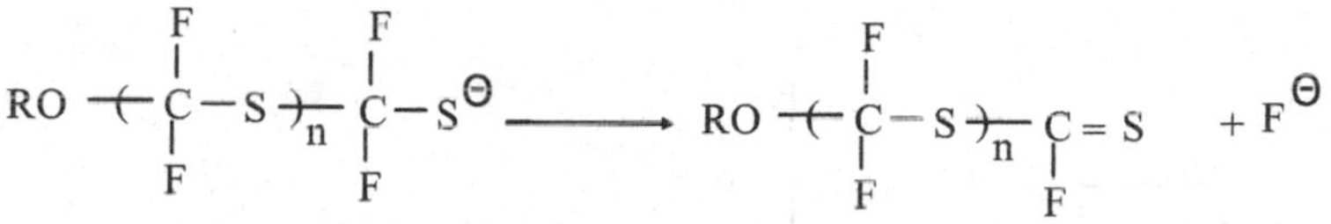

5.20

Usually the polymerizations of these monomers have to be conducted at very low polymerization temperatures, in view of the phenomena of propagation/depropagation (ceiling temperature)[1] when positively charged initiators are used, since there is no transfer species to release to kill the chain. This will not yet be considered until full consideration of another method of Addition polymerization via Equilibrium mechanism downstream. Nevertheless, it is important to note the difference between the presence of the following hetero centers along the chain shown in Figure 5.1 below.

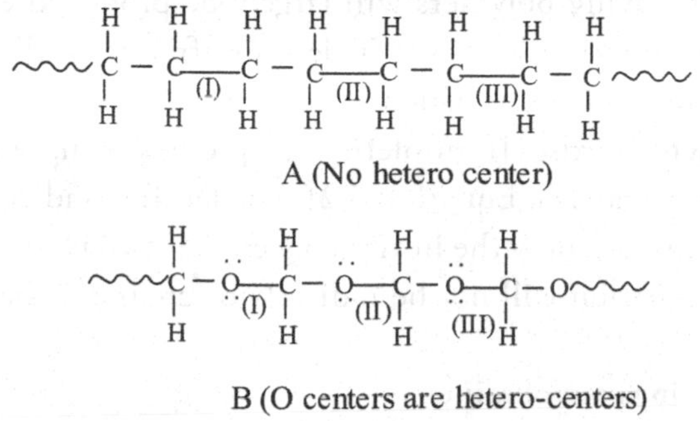

A (No hetero center)

B (O centers are hetero-centers)

Figure 5.1 Existence/non-existence of hetero-centers between repeating units in Ethylene and formaldehyde.

There are no functional centers in A and B. Functional centers only exist in RINGS. In B, the O centers are hetero-centers which if they were connected to an ionic center such as Sodium will become vulnerable to growth of the chain. Imagine if H was the one externally located, then the chain can never grow, but depropagate if temperature is high.

5.1.2. Radical characters of Aldehydes

Beginning with formaldehyde, the followings are obtained.

$$N^{\cdot n} \quad + e.\overset{\overset{\textstyle H}{|}}{\underset{\underset{\textstyle H}{|}}{C}} - O.nn \quad \longrightarrow \quad N - \overset{\overset{\textstyle H}{|}}{\underset{\underset{\textstyle H}{|}}{C}} - O.nn \quad \xrightarrow{\;+\,(I)\;}$$

$$\text{(I)} \qquad\qquad \underline{\text{Not radically balanced}} \qquad\qquad 5.21$$

$$\overset{..}{N}\cdot nn \quad + e.\overset{\overset{\textstyle H}{|}}{\underset{\underset{\textstyle H}{|}}{C}} - O.nn \quad \longrightarrow \quad \overset{..}{N} - \overset{\overset{\textstyle H}{|}}{\underset{\underset{\textstyle H}{|}}{C}} - O.nn \qquad\qquad 5.22$$

$$\overset{..}{N}\!\!-\!\!\Big(\!\overset{\overset{\textstyle H}{|}}{\underset{\underset{\textstyle H}{|}}{C}} - O\Big)_{\!n}\!\! \overset{\overset{\textstyle H}{|}}{\underset{\underset{\textstyle H}{|}}{C}} - O.nn \quad \longrightarrow \quad N\!\!-\!\!\Big(\!\overset{\overset{\textstyle H}{|}}{\underset{\underset{\textstyle H}{|}}{C}} - O\Big)_{\!n}\!\! \overset{\overset{\textstyle H}{|}}{\underset{}{C}}\!=\!O. \quad + \quad H.n$$

$$\text{Transfer species of second kind} \qquad\qquad 5.23$$

$$E^{\cdot e} \quad + nn.O - \overset{\overset{\textstyle H}{|}}{\underset{\underset{\textstyle H}{|}}{C}}.e \quad \longrightarrow \quad E - O - \overset{\overset{\textstyle H}{|}}{\underset{\underset{\textstyle H}{|}}{C}}.e \qquad\qquad 5.24$$

$$E\!\!-\!\!\Big(\!O - \overset{\overset{\textstyle H}{|}}{\underset{\underset{\textstyle H}{|}}{C}}\Big)_{\!n}\!\! O - \overset{\overset{\textstyle H}{|}}{\underset{\underset{\textstyle H}{|}}{C}}.e \quad \longrightarrow \quad \text{No transfer species} \qquad\qquad 5.25$$

Though the monomer (radically or chargedly) is a nucleophile, both electro-free-radical and nucleo-non-free-radical routes are favored by it. Living polymers will largely be produced electro-free-radically in the absence of foreign or terminating agents. Nucleo-non-free-radically, dead terminal double bond polymers along with an initiator (H.n) which cannot reinitiate are produced. These are routes which are presently not known to exist. If an olefin was present along with the formaldehyde, it will be initiated by H•n and propagate. In Equation 5.21, Nucleo-free-radically like with the use of free-negatively-charged-paired initiator, only the initiation step may be favored under Equilibrium mechanism conditions, since the equation will not be radically or chargedly balanced under Combination mechanism.

For the fluorinated counterparts, the followings are obtained.

$$\ddot{\text{N}} \cdot \text{nn} \quad + \text{e.} \underset{\underset{\text{F}}{|}}{\overset{\overset{\text{F}}{|}}{\text{C}}} - \text{O. nn} \quad \longrightarrow \quad \ddot{\text{N}} - \underset{\underset{\text{F}}{|}}{\overset{\overset{\text{F}}{|}}{\text{C}}} - \text{O. nn} \qquad\qquad 5.26$$

$$\ddot{\text{N}} \left(\underset{\underset{\text{F}}{|}}{\overset{\overset{\text{F}}{|}}{\text{C}}} - \text{O} \right) \underset{\underset{\text{F}}{|}}{\overset{\overset{\text{F}}{|}}{\text{C}}} - \text{O. nn} \quad \longrightarrow \quad \ddot{\text{N}} \left(\underset{\underset{\text{F}}{|}}{\overset{\overset{\text{F}}{|}}{\text{C}}} - \text{O} \right)_n \underset{\underset{\text{F}}{|}}{\overset{}{\text{C}}} = \text{O} \quad + \ddot{:}\ddot{\text{F}} \cdot \text{nn} \qquad 5.27$$

$$\ddot{\text{N}} \left(\underset{\underset{\text{F}}{|}}{\overset{\overset{\text{F}}{|}}{\text{C}}} - \text{O} \right) \underset{\underset{\text{F}}{|}}{\overset{\overset{\text{F}}{|}}{\text{C}}} - \text{O. nn} \quad \longrightarrow \quad \ddot{\text{N}} \left(\underset{\underset{\text{F}}{|}}{\overset{\overset{\text{F}}{|}}{\text{C}}} - \text{O} \right)_n \underset{\underset{\text{F}}{|}}{\overset{}{\text{C}}} = \text{O} \quad + \ddot{:}\ddot{\text{F}} \cdot \text{nn} \qquad 5.28$$

Like the positively charged case, the electro-free-radical route is favored with no transfer species. It may seem that, the monomer radically and chargedly has no character, when indeed it is a Nucleophile- a weak one. With formaldehyde monomer, in view of the routes favored by it, it is either a nucleophile or an electrophile. It is a Weak Nucleophile. The transfer species involved above (Equation 5.27) is of the second kind.

Now considering acetaldehyde, the followings are to be expected.

$$\text{N}^{\cdot n} \quad + \text{e.} \underset{\underset{\text{H}}{|}}{\overset{\overset{\text{CH}_3}{|}}{\text{C}}} - \text{O. nn} \quad \longrightarrow \quad \text{NH} \quad + \text{n.} \underset{\underset{\text{H}}{|}}{\overset{\overset{\text{H}}{|}}{\text{C}}} - \underset{\underset{\text{H}}{|}}{\overset{}{\text{C}}} = \text{O} \qquad\qquad 5.29$$

$$\ddot{\text{N}} \cdot \text{nn} \quad + \text{e.} \underset{\underset{\text{H}}{|}}{\overset{\overset{\text{CH}_3}{|}}{\text{C}}} - \text{O. nn} \quad \longrightarrow \quad \ddot{\text{N}}\text{H} \quad + \underset{\underset{\text{H}}{|}}{\overset{\overset{\text{H}}{|}}{\text{C}}} = \underset{\underset{\text{H}}{|}}{\overset{}{\text{C}}} - \text{O. nn} \qquad\qquad 5.30$$

$$\text{E}^{\cdot e} \quad + \text{nn.} \text{O} - \underset{\underset{\text{H}}{|}}{\overset{\overset{\text{CH}_3}{|}}{\text{C}}} \text{. e} \quad \longrightarrow \quad \text{E} - \text{O} - \underset{\underset{\text{H}}{|}}{\overset{\overset{\text{CH}_3}{|}}{\text{C}}} \text{. e} \qquad\qquad 5.31$$

$$\text{E} \left(\text{O} - \underset{\underset{\text{H}}{|}}{\overset{\overset{\text{CH}_3}{|}}{\text{C}}} \right)_n \text{O} - \underset{\underset{\text{H}}{|}}{\overset{\overset{\text{CH}_3}{|}}{\text{C}}} \text{. e} \quad \longrightarrow \quad \text{E} \left(\text{O} - \underset{\underset{\text{H}}{|}}{\overset{\overset{\text{CH}_3}{|}}{\text{C}}} \right)_n \text{O} - \underset{\underset{\text{H}}{|}}{\overset{}{\text{C}}} = \underset{\underset{\text{H}}{|}}{\overset{\overset{\text{H}}{|}}{\text{C}}} \quad + \quad \text{H}^{\cdot e} \qquad 5.32$$

The transfer species rejected from the growing polymer chain electro-free-radically to produce dead terminal double bond polymer, is the same abstracted nucleo-non-free-radically and nucleo-free-radically. Evidently, acetaldehyde is a stronger nucleophile than formaldehyde. The transfer species involved above is of the first kind of the first type.

Replacing the lone H with F, the followings are obtained.

$$\ddot{\text{N}}\cdot \text{nn} \;+\; \text{e}.\overset{\overset{\displaystyle CH_3}{|}}{\underset{\underset{\displaystyle F}{|}}{C}}-\text{O}.\text{nn} \;\longrightarrow\; \ddot{\text{N}}\text{H} \;+\; \overset{\overset{\displaystyle H}{|}}{\underset{\underset{\displaystyle H}{|}}{C}}=\overset{}{\underset{\underset{\displaystyle F}{|}}{C}}-\text{O}.\text{nn} \qquad\qquad 5.33$$

$$\text{E}^{\cdot e} \;+\; \text{e}.\overset{\overset{\displaystyle CH_3}{|}}{\underset{\underset{\displaystyle F}{|}}{C}}-\text{O}.\text{nn} \;\longrightarrow\; \text{e}.\overset{\overset{\displaystyle CH_3}{|}}{\underset{\underset{\displaystyle F}{|}}{C}}-\text{O}-\text{E} \qquad\qquad 5.34$$

The nucleo-non-free-radical route is not favored. Electro-free-radical route being favored is a clear indication of the Nucleophilic character of the monomer. Dead terminal double bond polymer is produced. The situation is however different for trifluoroactaldehyde.

$$\text{E}^{\cdot e} \;+\; \text{e}.\overset{\overset{\displaystyle CF_3}{|}}{\underset{\underset{\displaystyle H}{|}}{C}}-\text{O}.\text{nn} \;\longrightarrow\; \text{E}-\text{O}-\overset{\overset{\displaystyle CF_3}{|}}{\underset{\underset{\displaystyle H}{|}}{C}}.\text{e} \qquad\qquad 5.35$$

$$\text{E}\!\left(\!\!-\text{O}-\overset{\overset{\displaystyle CF_3}{|}}{\underset{\underset{\displaystyle H}{|}}{C}}\!\!\right)_{\!\!n}\!\!-\text{O}-\overset{\overset{\displaystyle CF_3}{|}}{\underset{\underset{\displaystyle H}{|}}{C}}.\text{e} \;\longrightarrow\; \text{No transfer species} \qquad\qquad 5.36$$

$$\ddot{\text{N}}\cdot \text{nn} \;+\; \text{e}.\overset{\overset{\displaystyle CF_3}{|}}{\underset{\underset{\underset{\displaystyle (I)}{\displaystyle H}}{|}}{C}}-\text{O}.\text{nn} \;\longrightarrow\; \ddot{\text{N}}-\overset{\overset{\displaystyle CF_3}{|}}{\underset{\underset{\displaystyle H}{|}}{C}}-\text{O}.\text{nn} \;\xrightarrow{\;+\,n\,(I)\;} \qquad 5.37$$

There is transfer species because the equation is radically balanced since a stable molecule is produced. But note that it is H that is released and not CF_3, because the capacity of CF_3 is greater than that of H which in turn is greater than that CH_3 nucleo-free-radically. Chargedly this is impossible, because $^{\ominus}CF_3$ does not exist. In the absence of foreign agents, living polymers are produced electro-free-radically, the route natural to it and anionically. If the CF_3 group is now replaced with C_2F_5 group, the followings are obtained.

$$\text{E}^{\cdot e} \;+\; \text{e}.\overset{\overset{\overset{\overset{\displaystyle CF_3}{|}}{\displaystyle CF_2}}{|}}{\underset{\underset{\displaystyle H}{|}}{C}}-\text{O}.\text{nn} \;\longrightarrow\; \text{E}-\text{O}-\overset{\overset{\displaystyle H}{|}}{\underset{\underset{\underset{\displaystyle CF_3}{|}}{\underset{\displaystyle CF_2}{|}}}{C}}.\text{e} \qquad\qquad 5.38$$

$$\text{E}\!\left(\!\!-\text{O}-\overset{\overset{\displaystyle H}{|}}{\underset{\underset{\underset{\displaystyle CF_3}{|}}{\underset{\displaystyle CF_2}{|}}}{C}}\!\!\right)_{\!\!n}\!\!-\text{O}-\overset{\overset{\displaystyle H}{|}}{\underset{\underset{\underset{\displaystyle CF_3}{|}}{\underset{\displaystyle CF_2}{|}}}{C}}.\text{e} \;\longrightarrow\; \text{E}\!\left(\!\!-\text{O}-\overset{\overset{\displaystyle H}{|}}{\underset{\underset{\underset{\displaystyle CF_3}{|}}{\underset{\displaystyle CF_2}{|}}}{C}}\!\!\right)_{\!\!n}\!\!-\text{O}-\overset{\overset{\displaystyle H}{|}}{\underset{}{C}}=\overset{\overset{\displaystyle F}{|}}{\underset{\underset{\displaystyle F}{|}}{C}} \;+\; F_3C^{\cdot e} \qquad 5.39$$

$$\overset{\cdot\cdot}{N}\cdot nn \quad + \quad e.\overset{\overset{H}{|}}{\underset{\underset{CF_3}{|}}{\underset{CF_2}{\underset{|}{C}}}}-O.nn \quad \longrightarrow \quad \overset{\cdot\cdot}{N}CF_3 \quad + \quad \overset{\overset{F}{|}}{\underset{\underset{H}{|}}{\underset{F}{C}}}=C-O.nn$$

5.40

The same transfer species abstracted (CF_3) is the same rejected from the growing polymer chain electro-free-radically. This is transfer species of the first kind of the first type. Radically, the routes have altered merely by replacing CF_3 by C_2F_5, making the monomer a stronger nucleo-phile. Thus, it can be observed how the law of conservation of transfer of transfer species strong-ly applies for these families of monomers, like for the olefinic monomers, and what it means to be first member of a "family", male or female. A "family" here is that in which they are either all males or all females, and not with both males and females, noting the uniqueness of the first member.

For the fully fluorinated counterparts, the followings should be expected.

$$E^{\cdot e} \quad + \quad e.\overset{\overset{F}{|}}{\underset{\underset{CF_3}{|}}{C}}-O.nn \quad \longrightarrow \quad e.\overset{\overset{F}{|}}{\underset{\underset{CF_3}{|}}{C}}-O^- E$$

5.41

How can F•en be released here as transfer species of the first kind when it cannot be abstracted via the nucleo-radical route?

5.42

$$\overset{\cdot\cdot}{N}.nn \quad + \quad e.\overset{\overset{F}{|}}{\underset{\underset{CF_3}{|}}{C}}-O.nn \quad \longrightarrow \quad \overset{\cdot\cdot}{N}-\overset{\overset{F}{|}}{\underset{\underset{CF_3}{|}}{C}}-O.nn$$

$$\overset{\cdot\cdot}{N}+\overset{\overset{F}{|}}{\underset{\underset{CF_3}{|}}{C}}-O+_n \overset{\overset{F}{|}}{\underset{\underset{CF_3}{|}}{C}}-O.nn \quad \longrightarrow \quad \overset{\cdot\cdot}{N}+\overset{\overset{F}{|}}{\underset{\underset{CF_3}{|}}{C}}-O+_n \overset{\overset{F}{|}}{\underset{\underset{CF_3}{|}}{C}}=O \quad + \quad \overset{\cdot\cdot}{\underset{\cdot\cdot}{:}}F.nn$$

5.43

The transfer species involved is of the second kind. While $E^{\cdot e}$ can abstract F as a substituted group only if the monomer was not activated, it cannot abstract it if the monomer is activated. It can abstract it from CF_3 group in olefins, when placed on the active C center, but cannot reject it in the route unnatural to the monomer.

$$E^{\cdot e} \quad + \quad n.\overset{\overset{CF_3}{|}}{\underset{\underset{H}{|}}{C}}-\overset{\overset{H}{|}}{\underset{\underset{H}{|}}{C}}.e \quad \longrightarrow \quad EF \quad + \quad e.\overset{\overset{F}{|}}{\underset{\underset{F}{|}}{C}}-\overset{\overset{H}{|}}{\underset{\underset{H}{|}}{C}}=\overset{\overset{H}{|}}{\underset{\underset{H}{|}}{C}} \quad OR$$

$$\overset{\overset{F}{|}}{\underset{\underset{F}{|}}{C}}=\overset{\overset{H}{|}}{\underset{\underset{H}{|}}{C}}-\overset{\overset{H}{|}}{\underset{\underset{H}{|}}{C}}.e$$

5.44

252

$$N \overset{\displaystyle H}{\underset{\displaystyle H}{-}} \overset{\displaystyle H}{\underset{\displaystyle CF_3}{C}} - \overset{\displaystyle H}{\underset{\displaystyle H}{C}}\!\!\!\!\Big)_n \overset{\displaystyle H}{\underset{\displaystyle H}{C}} - \overset{\displaystyle H}{\underset{\displaystyle CF_3}{C}} \cdot n \quad \not\longrightarrow \quad N \overset{\displaystyle H}{\underset{\displaystyle H}{C}} - \overset{\displaystyle H}{\underset{\displaystyle CF_3}{C}}\!\!\!\!\Big)_n \overset{\displaystyle H}{\underset{\displaystyle H}{C}} - \overset{\displaystyle F}{\underset{\displaystyle F}{C = C}} \quad + \quad \overset{..}{\underset{..}{:}} F \cdot nn$$

<u>Against the Conservation laws of Transfer species</u>   5.45

It is not possible for a non-free-radical to be rejected from a free-radical growing polymer chain as transfer species of first kind of the first type in the route unnatural to it. Nucleo-free-radically, the following is indeed obtained.

$$N \overset{\displaystyle H}{\underset{\displaystyle H}{-}} \overset{\displaystyle H}{\underset{\displaystyle CF_3}{C}} - \overset{\displaystyle H}{\underset{\displaystyle CF_3}{C}}\!\!\!\!\Big)_n \overset{\displaystyle H}{\underset{\displaystyle H}{C}} - \overset{\displaystyle H}{\underset{\displaystyle CF_3}{C}} \cdot n \quad \longrightarrow \quad N \overset{\displaystyle H}{\underset{\displaystyle H}{C}} - \overset{\displaystyle H}{\underset{\displaystyle CF_3}{C}}\!\!\!\!\Big)_n \overset{\displaystyle H}{C = C}\underset{\displaystyle CF_3}{} \quad + \quad H \cdot n$$

      5.46

If the F atom cannot be abstracted electro-free-radically, the following would have been obtained.

$$E \overset{\displaystyle H}{\underset{\displaystyle CF_3}{-}} \overset{\displaystyle H}{\underset{\displaystyle H}{C}} - \overset{\displaystyle H}{\underset{\displaystyle CF_3}{C}}\!\!\!\!\Big)_n \overset{\displaystyle H}{\underset{\displaystyle H}{C}} - \overset{\displaystyle H}{\underset{\displaystyle H}{C}} \cdot e \quad \longrightarrow \quad E - (\overset{\displaystyle H}{\underset{\displaystyle CF_3}{C}} - \overset{\displaystyle H}{\underset{\displaystyle H}{C}})_n - \overset{\displaystyle H}{\underset{\displaystyle H}{C}} = \overset{\displaystyle H}{\underset{\displaystyle H}{C}} \; + F_3C \bullet e \;\; \text{OR No Transfer species}$$
   [NONE FAVORED]   5.47

[Cannot be made to exist]

With the case above, that is, last equation if favored, CF_3 cannot be released as transfer species of the second kind since the route is natural to the monomer, unlike some cases such as in Equation 5.39. H cannot also be released, though the main driving force is to have a dead polymer with the same character as the originating monomer when killed from within. Hence, the last equation above is not favored. Almost like vinyl acetate, it is Equation 5.46 that is favored, that wherin the dead terminal bouble polymer obtained bears the same character as the monomer.

5.2 Ketones

5.2.1 Ionic characters of Ketones

Beginning with acetone, the followings are obtained if activation is favored.

$$RO^{\ominus} \; + \; \oplus \overset{\displaystyle CH_3}{\underset{\displaystyle CH_3}{C}} - O^{\ominus} \quad \longrightarrow \quad ROH \; + \; \overset{\displaystyle H}{\underset{\displaystyle H}{C}} = \overset{\displaystyle}{\underset{\displaystyle CH_3}{C}} - O^{\ominus}$$
(non-free)
-anion
      5.48

$$\underset{\substack{\displaystyle | \\ CH_3}}{\overset{\substack{CH_3 \\ \displaystyle |}}{\oplus C - O^{\ominus}}}$$

$$\underset{\substack{\displaystyle | \\ \oplus C - O^{\ominus} \\ \displaystyle | \\ CH_3}}{\overset{\substack{CH_3 \\ \displaystyle |}}{H_9 C_4 \overset{\ominus}{} \overset{\oplus}{\dots\dots} Li}} \longrightarrow C_4 H_{10} + \underset{\substack{\displaystyle | \quad | \\ H \quad CH_3}}{\overset{\substack{H \\ \displaystyle |}}{C = C - OLi}}$$

5.49

$$R \overset{}{\left(O - \underset{\substack{\displaystyle | \\ CH_3}}{\overset{\substack{CH_3 \\ \displaystyle |}}{C}} \right)_n} O - \underset{\substack{\displaystyle | \\ CH_3}}{\overset{\substack{CH_3 \\ \displaystyle |}}{C^{\oplus}}} \longrightarrow R \overset{}{\left(O - \underset{\substack{\displaystyle | \\ CH_3}}{\overset{\substack{CH_3 \\ \displaystyle |}}{C}} \right)_n} O - \underset{\substack{\displaystyle | \\ CH_3}}{\overset{\substack{CH_3 \\ \displaystyle |}}{C}} = \underset{\substack{\displaystyle | \\ H}}{\overset{\substack{H \\ \displaystyle |}}{C}} + H^{\oplus}$$

5.50

In Equation 4.49, the initiator cannot initiate the monomer, since in addition Li cannot be the carrier being cationic, the monomer being a Nucleophile. The negative end was used, in order to show that the monomer is a Nucleophile and not an Electrophile. Negatively, acetone cannot be polymerized. The same applies to higher members as shown.

$$\underset{\substack{\displaystyle | \\ C_2 H_5}}{\overset{\substack{CH_3 \\ \displaystyle |}}{\oplus C - O^{\ominus}}}$$

$$\underset{\substack{\displaystyle | \\ CH_2 \\ \displaystyle | \\ \oplus C - O^{\ominus} \\ \displaystyle | \\ CH_3}}{\overset{\substack{CH_3 \\ \displaystyle |}}{H_9 C_4 \overset{\ominus}{} \overset{\oplus}{\dots\dots} Li}} \longrightarrow H_{10} C_4 + \underset{\substack{\displaystyle | \quad | \\ \quad CH_3}}{\overset{\substack{H \quad CH_3 \\ \displaystyle | \quad |}}{C = C - OLi}}$$

5.51

The transfer species of the first kind of the first type involved is from the more radical-pushing group. It is taking from the rich and never the poor. With positive charges, the monomers can be polymerized, with the same transfer species involved in killing the growing chain from within.

Now considering replacing one of the CH_3 groups with CF_3, the followings are obtained.

$$\underset{\substack{\displaystyle \\ (Non\text{-}free)}}{RO^{\ominus}} + \underset{\substack{\displaystyle | \\ CH_3}}{\overset{\substack{CF_3 \\ \displaystyle |}}{\oplus C - O^{\ominus}}} \longrightarrow ROH + \underset{\substack{\displaystyle | \quad | \\ H \quad CF_3}}{\overset{\substack{H \\ \displaystyle |}}{C = C - O^{\ominus}}}$$

5.52

$$\begin{array}{c} CF_3 \\ | \\ \oplus C - O^{\ominus} \\ | \\ CH_3 \\ H_9C_4\ldots^{\ominus}\overset{\oplus}{}Li \\ | \\ CH_3 \\ | \\ \oplus C - O^{\ominus} \\ | \\ CF_3 \end{array} \longrightarrow C_4H_{10} + \begin{array}{c} H \quad CF_3 \\ | \quad | \\ C = C - OLi \\ | \\ H \end{array} \qquad 5.53$$

$$R^{\oplus} + \begin{array}{c} CF_3 \\ | \\ \oplus C - O^{\ominus} \\ | \\ CH_3 \end{array} \longrightarrow RCF_3 + \begin{array}{c} \oplus C = O \\ | \\ CH_3 \end{array} \quad OR \quad R-O-\begin{array}{c} CF_3 \\ | \\ C^{\oplus} \\ | \\ CH_3 \end{array}$$

"Covalent Type" (I) Not favored   Favored (II)   5.54

$$\begin{array}{c} CF_3 \\ | \\ ^{\ominus}O - C \\ | \\ CH_3 \quad F \\ R\ldots \overset{\ominus}{B}\diagup F \\ \diagdown F \\ O \\ | \\ CH_3 \quad R \\ | \\ ^{\ominus}O - C^{\oplus} \\ | \\ CF_3 \end{array} \quad \text{"Electrostatic type"} \longrightarrow R-O-\begin{array}{c} CF_3 \\ | \\ C^{\oplus} \\ | \\ CH_3 \end{array} \ldots \overset{\ominus}{B}\begin{array}{c} \diagup F \\ F \\ \diagdown F \\ | \\ O \\ | \\ R \end{array} \qquad 5.55$$

Negatively the monomer cannot be polymerized. With positively charged-paired covalent and electrostatic types of initiators, full polymerization is favored with the death of the growing polymer chain represented as follows

$$R\left(O-\begin{array}{c}CF_3\\|\\C\\|\\CH_3\end{array}\right)_n O-\begin{array}{c}CF_3\\|\\C^{\oplus}\\|\\CH_3\end{array} \ldots \overset{\ominus}{B}\begin{array}{c}\diagup F\\ F\\ \diagdown F\\|\\O\\|\\R\end{array} \longrightarrow R\left(O-\begin{array}{c}CF_3\\|\\C\\|\\CH_3\end{array}\right)_n O-\begin{array}{c}CF_3 \quad H\\|\quad\ |\\C = C\\|\\H\end{array} + HOR + BF_3 \qquad 5.56$$

The monomer can be observed to be strongly nucleophilic in character despite the strong presence of CF_3 group.

When the replacement of CH_3 groups is completed, the followings are obtained.

$$R^{\ominus} + \begin{array}{c}CF_3\\|\\\oplus C - O^{\ominus}\\|\\CF_3\end{array} \longrightarrow R-\begin{array}{c}CF_3\\|\\C - O^{\ominus}\\|\\CF_3\end{array} \qquad 5.57$$

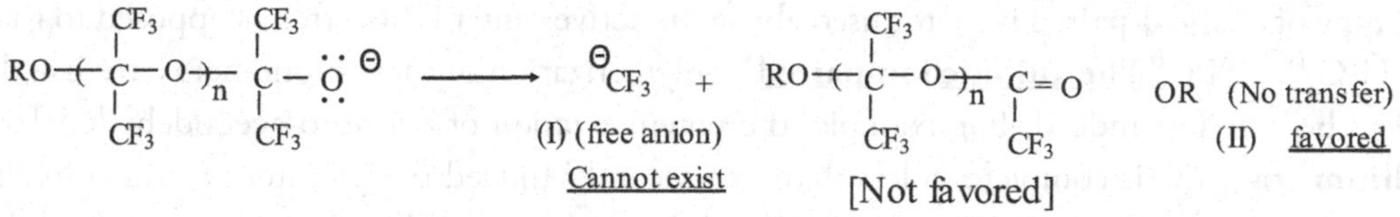

$$RO-\left(\underset{\underset{CF_3}{|}}{\overset{\overset{CF_3}{|}}{C}}-O\right)_n \overset{\overset{CF_3}{|}}{\underset{\underset{CF_3}{|}}{C}}-\overset{..}{\underset{..}{O}}{}^{\ominus} \longrightarrow \quad \overset{\ominus}{}CF_3 \quad + \quad RO-\left(\underset{\underset{CF_3}{|}}{\overset{\overset{CF_3}{|}}{C}}-O\right)_n \overset{\overset{CF_3}{|}}{\underset{\underset{CF_3}{|}}{C}}=O \qquad OR \quad (No\ transfer)$$

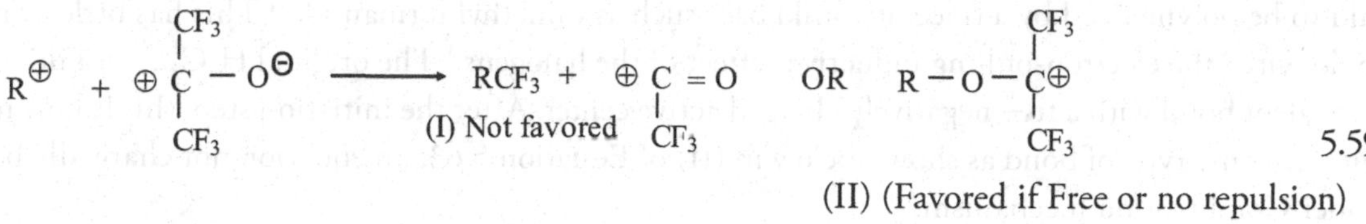

(I) (free anion)   [Not favored]   (II) <u>favored</u>

<u>Cannot exist</u>   5.58

$$R^{\oplus} \quad + \quad {}^{\oplus}\overset{\overset{CF_3}{|}}{\underset{\underset{CF_3}{|}}{C}}-O^{\ominus} \longrightarrow RCF_3 \quad + \quad {}^{\oplus}\underset{\underset{CF_3}{|}}{C}=O \qquad OR \quad R-O-\overset{\overset{CF_3}{|}}{\underset{\underset{CF_3}{|}}{C}}{}^{\oplus}$$

(I) Not favored   5.59

(II) (Favored if Free or no repulsion)

$$\overset{\ominus}{}O-\overset{\overset{CF_3}{|}}{\underset{\underset{CF_3}{|}}{C}}{}^{\oplus}$$

$$R^{\oplus} \cdots\cdots {}^{\ominus}\overset{\overset{F}{|}}{\underset{\underset{O}{|}}{B}}{\overset{\nwarrow F}{\searrow F}} \longrightarrow \quad \text{No reaction (Electrostatic forces of repulsion)}$$

$$\overset{\ominus}{}O-\overset{\overset{CF_3}{|}}{\underset{\underset{CF_3}{|}}{C}}{}^{\oplus}$$

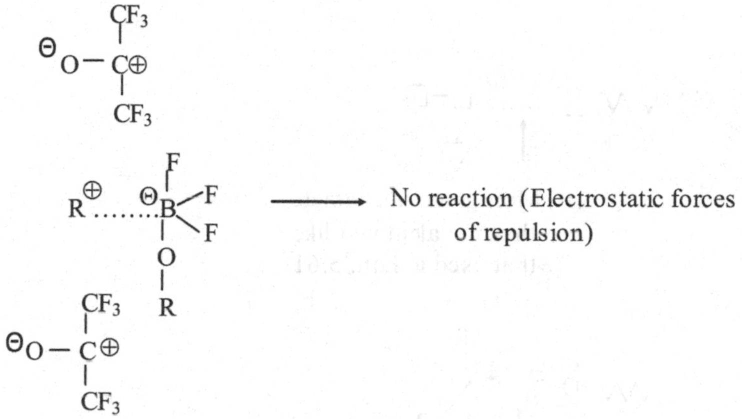

                                                     5.60

When a suitable positively charged-paired initiator is used, polymerization will be favored. It should be noted that when CF_3 is connected directly to an active carbon center, it cannot be abstracted as an anion, by a free cation since the carbon center on CF_3 cannot carry a negative charge. Hence, the reactions indicated by Equations 5.54 (I), 5.58 (I) and 5.59 (I) do not exist.

In view of the strong electrostatic forces of repulsion involved when some positively charged-paired initiators are involved, it may seem that the existence of suitable initiators for the "cationic" polymerization of these monomers is impossible. No, they exist. Anionically, they can readily be polymerized particularly at very low temperatures, since the route is unnatural to the monomer. No transfer species exist for their growing polymer chains chargedly.

Due to electrostatic forces of repulsion, the reaction below may not take place. However, it is shown herein for specific reasons. It can only take place radically.

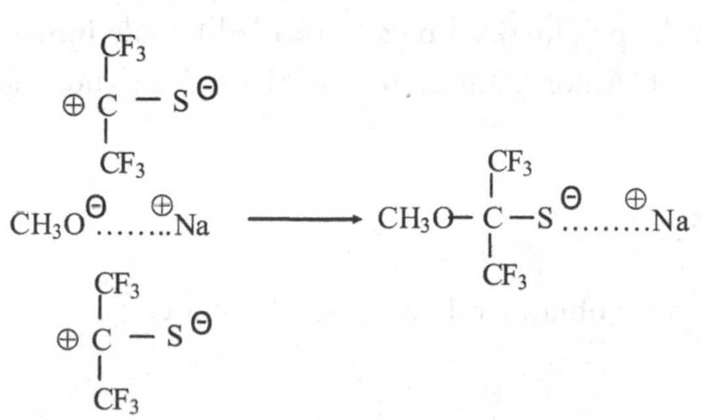

$$\begin{array}{c} {}^{\oplus}\overset{\overset{CF_3}{|}}{\underset{\underset{CF_3}{|}}{C}}-S^{\ominus} \\ CH_3O^{\ominus}\cdots\cdots{}^{\oplus}Na \end{array} \longrightarrow CH_3O-\overset{\overset{CF_3}{|}}{\underset{\underset{CF_3}{|}}{C}}-S^{\ominus}\cdots\cdots{}^{\oplus}Na \qquad 5.61$$

$$\begin{array}{c} {}^{\oplus}\overset{\overset{CF_3}{|}}{\underset{\underset{CF_3}{|}}{C}}-S^{\ominus} \end{array}$$

$$CH_3O-\left(\overset{\overset{CF_3}{|}}{\underset{\underset{CF_3}{|}}{C}}-S\right)_n \overset{\overset{CF_3}{|}}{\underset{\underset{CF_3}{|}}{C}}-S^{\ominus}\cdots\cdots{}^{\oplus}Na \longrightarrow \text{No transfer species} \qquad 5.62$$

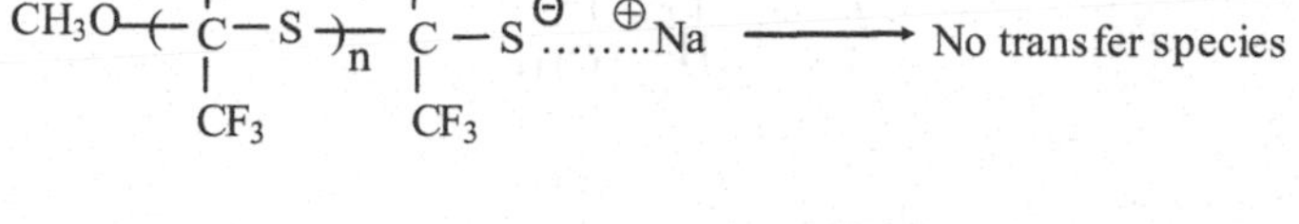

Note the type of charged-paired initiator used above. Its active center is non-free as opposed to that in for example $H_9C_4^{\ominus}$....$^{\oplus}$Li. **"The anionic ion-paired"** polymerization of these monomers have usually been observed to be very fast indeed. For example, the polymerization of trifluoro-acetaldehyde (fluoral) by **butyllithium** at - 78⁰C is complete in less than one second.[3] Indeed the initiator is radical in character wherein the active center is the lithium center. For the case above, it is Na. Thiocarbonyl fluoride is also said to be polymerized by a trace of a mild base such as dimethyl formamide.[4] This has little or nothing to do with "the electron-pulling inductive" effect of the halogens.[1] The original H_9C_4 .....Li initiator has a covalent bond with a free-negatively charged active center. After the initiation step, this is now replaced with an ionic type of bond as shown below in (II) of Equation 5.63, an equation not chargedly balanced under Combination mechanism.

$$H_9C_4 \overset{\ominus}{} \ldots\ldots\overset{\oplus}{Li}\text{—}\square \qquad\qquad \text{vvv—}\overset{..}{\underset{..}{O}}{}^{\ominus} \overset{\oplus}{Li}\text{—}\square$$

Covalent bond
(I)

(II) More ionic in character
than covalent just like
that used in Eqn. 5.61.

5.63

$$HO\overset{\ominus}{}\ldots\ldots\overset{\oplus}{N}\overset{R}{\underset{H}{\diagup}}\overset{\diagup R}{\diagdown R} \qquad\qquad \text{vvv—}\overset{..}{\underset{..}{O}}{}^{\ominus}\ldots\ldots\overset{\oplus}{N}\overset{R}{\underset{H}{\diagup}}\overset{\diagup R}{\diagdown R}$$

(I) Electrostatic bond   (II) Electrostatic bond

5.64

When the original bond is electrostatic, the same bond character is still maintained throughout the course of polymerization. When the original bond is covalent, the same bond character is maintained throughout the course of polymerization. The same applies to when the original bond is ionic. The pulling effect of the CF_3 groups assists in reducing completely to nil the nucleophi-lic character of the monomer, thus making it favor anionic route and not negatively-charged route. The rate of polymerization "anionically" cannot be fast even when the initiator shown in the last equation above was used. It is also for these reasons that the reported use of butyllithium for the polymerization of fluoral at -78⁰C in less than one second is anionically impossible. The Lithium center must have done the job electro-free-radically. The monomer being a female (Nucleophile) cannot be polymerized by a "so-called female initiator" such as butyllithium in so short a time. The initiator is not "Anionic ion-paired" initiator. It has no anionic center.

5.2.2 Radical characters of Ketones

Beginning also with acetone, the followings are obtained if activation is favored.

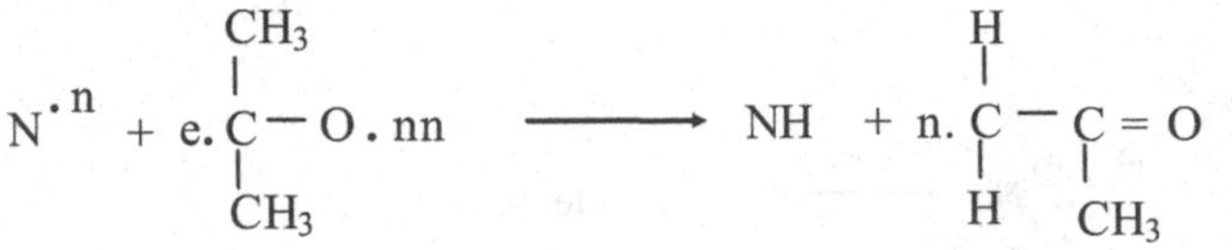

$$N\cdot^n + e.\underset{CH_3}{\overset{CH_3}{\underset{|}{\overset{|}{C}}}}\text{—}O.nn \longrightarrow NH + n.\underset{H}{\overset{H}{\underset{|}{\overset{|}{C}}}}\text{—}\underset{CH_3}{\overset{}{C}}=O$$

5.65

257

$$\overset{..}{N}\cdot nn \;+\; e.\underset{\underset{CH_3}{|}}{\overset{\overset{CH_3}{|}}{C}}-O.nn \;\longrightarrow\; \overset{..}{N}H \;+\; \underset{\underset{H}{|}}{\overset{\overset{H}{|}}{C}}=\underset{\underset{CH_3}{|}}{C}-O.nn \qquad\qquad 5.66$$

$$E^{\cdot e} \;+\; nn.O-\underset{\underset{CH_3}{|}}{\overset{\overset{CH_3}{|}}{C}}.e \;\longrightarrow\; E-O-\underset{\underset{CH_3}{|}}{\overset{\overset{CH_3}{|}}{C}}.e \qquad\qquad 5.67$$

$$E\left(O-\underset{\underset{CH_3}{|}}{\overset{\overset{CH_3}{|}}{C}}\right)_{\!n}O-\underset{\underset{CH_3}{|}}{\overset{\overset{CH_3}{|}}{C}}.e \;\longrightarrow\; E\left(O-\underset{\underset{CH_3}{|}}{\overset{\overset{CH_3}{|}}{C}}\right)_{\!n} O-\underset{\underset{CH_3}{|}}{C}=\underset{\underset{H}{|}}{\overset{\overset{H}{|}}{C}} \;+\; H\overset{e}{\cdot} \qquad 5.68$$

The equations above speak for themselves. The strong nucleophilic character of the monomer is reflected by favoring only electro-free-radical route. Since most of the free-radicals largely known to be used worldwide are of the nucleo-free-radical types, it is not surprising why this monomer has never been well known to favor free-radical polymerization. The same analysis above will apply when the size of the CH_3 groups is increased, noting that the monomer will become more nucleophilic in character as R group is increased in size.

Now replacing one of the CH_3 groups with CF_3, the followings will be expected.

$$\overset{..}{N}\cdot nn \;+\; e.\underset{\underset{CH_3}{|}}{\overset{\overset{CF_3}{|}}{C}}-O.nn \;\longrightarrow\; \overset{..}{N}H \;+\; \underset{\underset{H}{|}}{\overset{\overset{H}{|}}{C}}=\underset{\underset{CF_3}{|}}{C}-O.nn \qquad\qquad 5.69$$

$$E\left(O-\underset{\underset{CH_3}{|}}{\overset{\overset{CF_3}{|}}{C}}\right)_{\!n}O-\underset{\underset{CH_3}{|}}{\overset{\overset{CF_3}{|}}{C}}.e \;\longrightarrow\; E\left(O-\underset{\underset{CH_3}{|}}{\overset{\overset{CF_3}{|}}{C}}\right)_{\!n} O-\underset{\underset{CF_3}{|}}{C}=\underset{\underset{H}{|}}{\overset{\overset{H}{|}}{C}} \;+\; H^{\cdot e} \qquad 5.70$$

The same transfer species abstracted in transfer from monomer step is rejected from its growing polymer chain electro-free-radically, in accordance to the law of conservation of transfer of transfer species.

Finally, completing the replacement, hexafluoroacetone is obtained.

$$\overset{..}{N}\cdot nn \;+\; e.\underset{\underset{CF_3}{|}}{\overset{\overset{CF_3}{|}}{C}}-O.nn \;\longrightarrow\; \overset{..}{N}-\underset{\underset{CF_3}{|}}{\overset{\overset{CF_3}{|}}{C}}-O.nn \qquad\qquad 5.71$$

$$\overset{..}{N}\left(\underset{\underset{CF_3}{|}}{\overset{\overset{CF_3}{|}}{C}}-O\right)_{\!n}\underset{\underset{CF_3}{|}}{\overset{\overset{CF_3}{|}}{C}}-O.nn \;\longrightarrow\; N\left(\underset{\underset{CF_3}{|}}{\overset{\overset{CF_3}{|}}{C}}-O\right)_{\!n} \underset{(I)}{C}=O \;+\; n.CF_3 \;\; OR \;\; \text{(No transfer species)}$$
$$\text{(II) Not favored.}$$

[Note: $n.CF_3 > n.H > n.CH_3$ in capacity] $\qquad\qquad$ 5.72

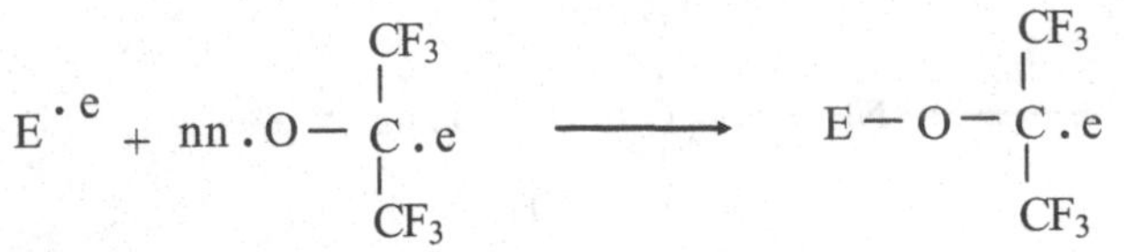

$$E \cdot e \quad + \quad nn \cdot O - \underset{\underset{CF_3}{|}}{\overset{\overset{CF_3}{|}}{C}} \cdot e \quad \longrightarrow \quad E - O - \underset{\underset{CF_3}{|}}{\overset{\overset{CF_3}{|}}{C}} \cdot e \qquad \qquad 5.73$$

$$E \left(O - \underset{\underset{CF_3}{|}}{\overset{\overset{CF_3}{|}}{C}} \right)_n O - \underset{\underset{CF_3}{|}}{\overset{\overset{CF_3}{|}}{C}} \cdot e \quad \longrightarrow \quad \text{No transfer species} \qquad 5.74$$

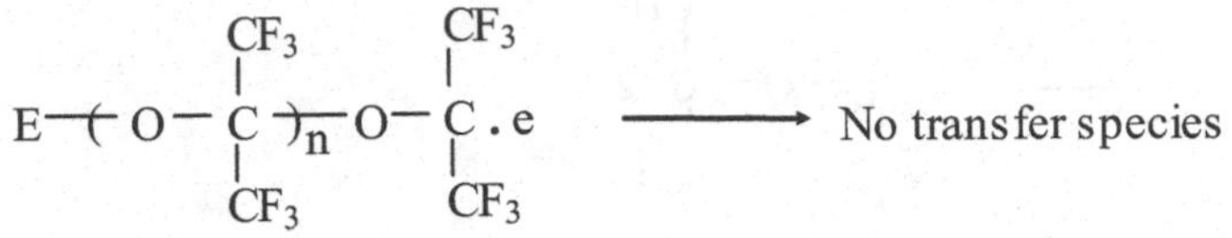

Since a nucleo-free-radical can be rejected from a nucleo-non-free-radical growing chain, *as long as a stable molecule is produced*, the CF_3 rejected is transfer species of second kind, *with much difficulty as shown by the inequality above.* For the monomer above, it can be observed that both routes are favored. However, living polymers in the absence of foreign agents are obtained only electro-free-radically, the route natural to it.

When CF_3 is replaced with larger groups, the situation is different as shown below.

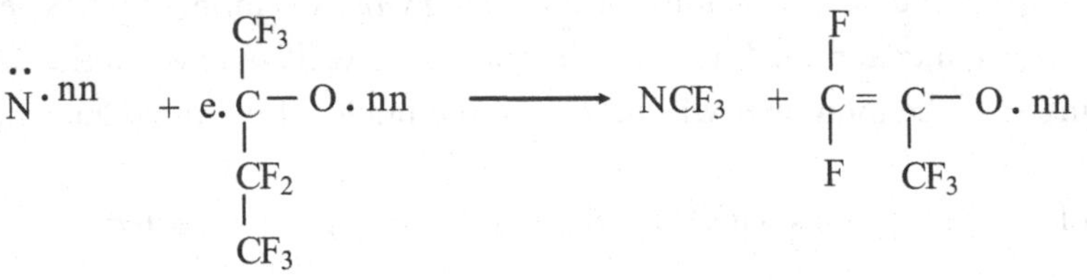

$$\overset{..}{N} \cdot nn \quad + \quad e \cdot \underset{\underset{\underset{CF_3}{|}}{CF_2}}{\overset{\overset{CF_3}{|}}{C}} - O \cdot nn \quad \longrightarrow \quad NCF_3 \quad + \quad \underset{\underset{F}{|}}{\overset{\overset{F}{|}}{C}} = \underset{\underset{CF_3}{|}}{C} - O \cdot nn \qquad 5.75$$

$$E \left(O - \underset{\underset{\underset{CF_3}{|}}{CF_2}}{\overset{\overset{CF_3}{|}}{C}} \right)_n O - \underset{\underset{\underset{CF_3}{|}}{CF_2}}{\overset{\overset{CF_3}{|}}{C}} \cdot e \quad \longrightarrow \quad E \left(O - \underset{\underset{\underset{CF_3}{|}}{CF_2}}{\overset{\overset{CF_3}{|}}{C}} \right)_n O - \underset{\underset{F}{|}}{\overset{\overset{CF_3}{|}}{C}} = C \quad + \quad {}^{e}CF_3 \qquad 5.76$$

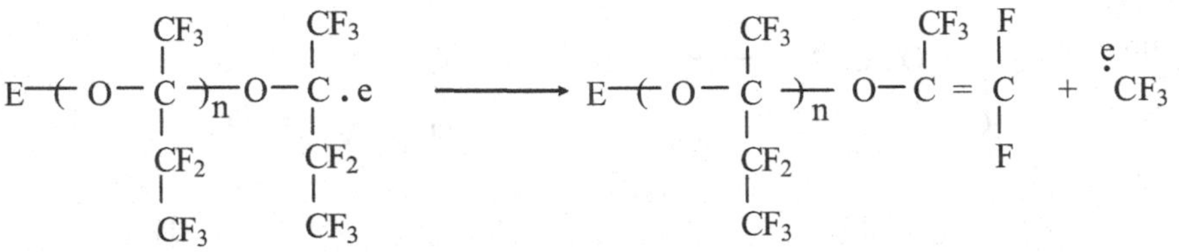

The monomers can now be observed to be indeed nucleophiles. The same transfer species abstracted is the same rejected from the growing polymer chain.

Consider replacing one of the CF_3 groups with CHF_2 or CH_2F, the followings will be expected.

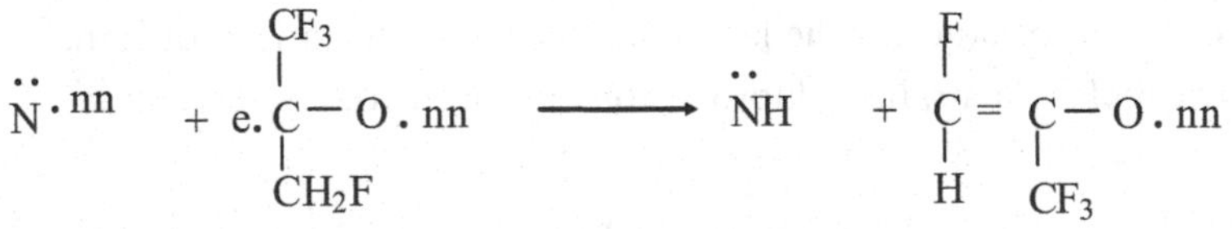

$$\overset{..}{N} \cdot nn \quad + \quad e \cdot \underset{\underset{CH_2F}{|}}{\overset{\overset{CF_3}{|}}{C}} - O \cdot nn \quad \longrightarrow \quad \overset{..}{N}H \quad + \quad \underset{\underset{H}{|}}{\overset{\overset{F}{|}}{C}} = \underset{\underset{CF_3}{|}}{C} - O \cdot nn \qquad 5.77$$

$$E \left(O - \underset{\underset{CH_2F}{|}}{\overset{\overset{CF_3}{|}}{C}} \right)_n O - \underset{\underset{CH_2F}{|}}{\overset{\overset{CF_3}{|}}{C}} \cdot e \quad \longrightarrow \quad E \left(O - \underset{\underset{CH_2F}{|}}{\overset{\overset{CF_3}{|}}{C}} \right)_n O - \underset{\underset{H}{|}}{\overset{\overset{CF_3}{|}}{C}} = C \quad + \quad H^{\cdot e} \qquad 5.78$$

$$\overset{..}{N} \cdot nn \quad + \quad e \cdot \underset{\underset{CF_2H}{|}}{\overset{\overset{CF_3}{|}}{C}} - O \cdot nn \quad \longrightarrow \quad \overset{..}{N}H \quad + \quad \underset{\underset{F}{|}}{\overset{\overset{F}{|}}{C}} = \underset{\underset{CF_3}{|}}{C} - O \cdot nn \qquad 5.79$$

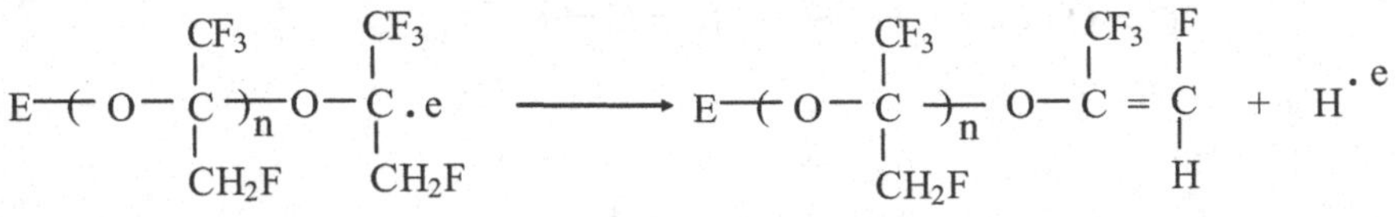

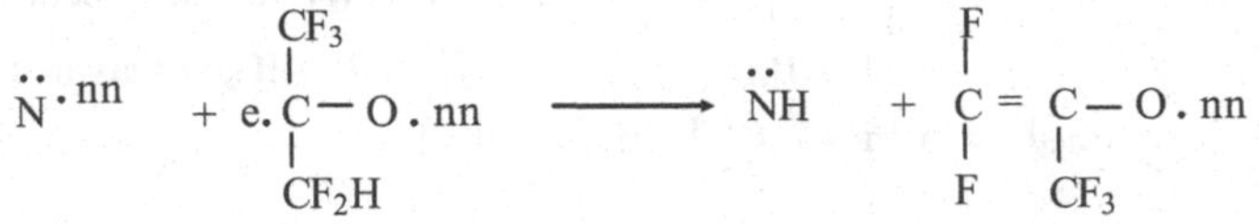

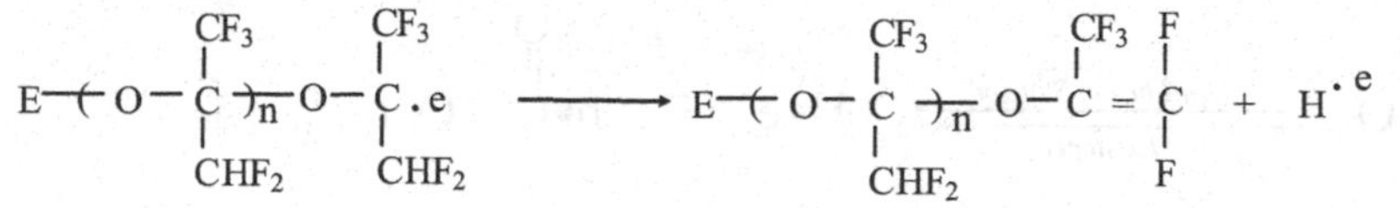

$$E + O - \underset{\underset{CHF_2}{|}}{\overset{\overset{CF_3}{|}}{C}} \rightarrow_n O - \underset{\underset{CHF_2}{|}}{\overset{\overset{CF_3}{|}}{C}} \cdot e \longrightarrow E + O - \underset{\underset{CHF_2}{|}}{\overset{\overset{CF_3}{|}}{C}} \rightarrow_n O - \underset{\underset{F}{|}}{\overset{\overset{CF_3}{|}}{C}} = \underset{\underset{F}{|}}{\overset{\overset{F}{|}}{C}} + H \cdot^e \qquad 5.80$$

The equations speak for themselves. *Hence, there have been very great objections over the years, that polymeric equations as were and are still written were completely meaningless, since they are or were neither chargedly nor radically nor stoichiometrically balanced. Though objections were raised through journals, it made no impart, for which sometimes either the objection was rejected, because it needed personal experimental support, without the realizations that none of the world's data have been provided with adequate explanation in all disciplines.* From the reactions above, it is obvious that radically, these monomers are nucleophiles.

This brings one to the beginning of the end of considerations of aldehydes and ketones and their perfluoro forms, from which it can largely be observed that only two transfer species are largely involved when activated mildly:-

(a) Of the first kind of the first type.
(b) Of the second kind.

Before however considering the ideal related monomers which are olefinic in character, there is need to use the advantages offered by the new developments to identify the capacities of substi-tuted groups.

5.3 Special substituted groups on carbonyl centers

5.3.1 Their charged characters

Beginning with pushing groups, consider using etheric groups in place of one H in form-aldehyde.

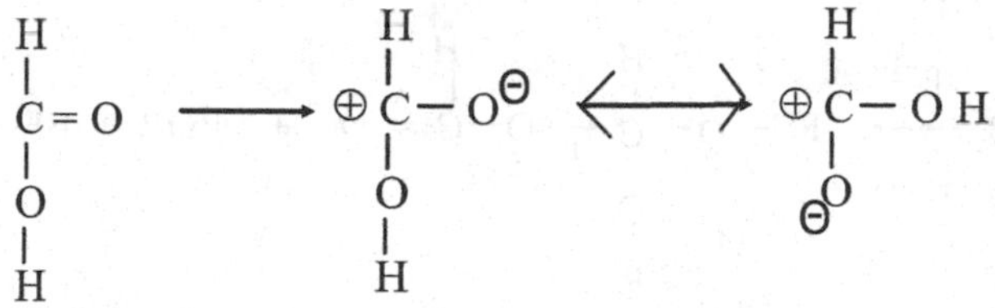

$$\underset{\underset{H}{|}}{\overset{\overset{H}{|}}{\underset{\underset{}{O}}{C}}} = O \longrightarrow \oplus \underset{\underset{H}{|}}{\overset{\overset{H}{|}}{\underset{\underset{}{O}}{C}}} - O^\ominus \longleftrightarrow \oplus \underset{\underset{\ominus O}{|}}{\overset{\overset{H}{|}}{C}} - OH \qquad 5.81$$

(Formic acid)   [Molecular rearrangement]

Transfer of transfer species seems to take place to favor however the existence of the same monomer which is still unstable as shown below.

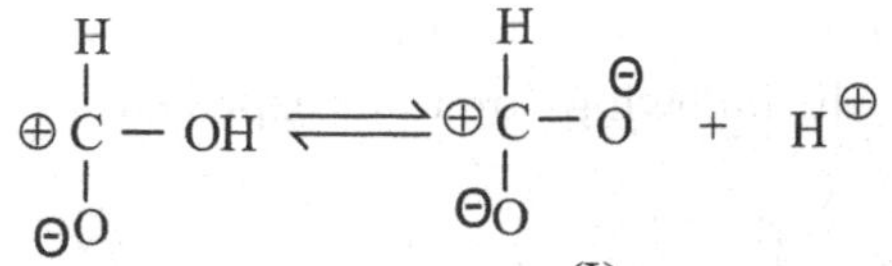

$$\oplus \underset{\underset{\ominus O}{|}}{\overset{\overset{H}{|}}{C}} - OH \rightleftharpoons \oplus \underset{\underset{\ominus O}{|}}{\overset{\overset{H}{|}}{C}} - O^\ominus + H^\oplus \qquad 5.82a$$

(I)

Very Strong ACTIVATED/EQUILIBRIUM STATE OF EXISTENCE
[Ionic character]

260

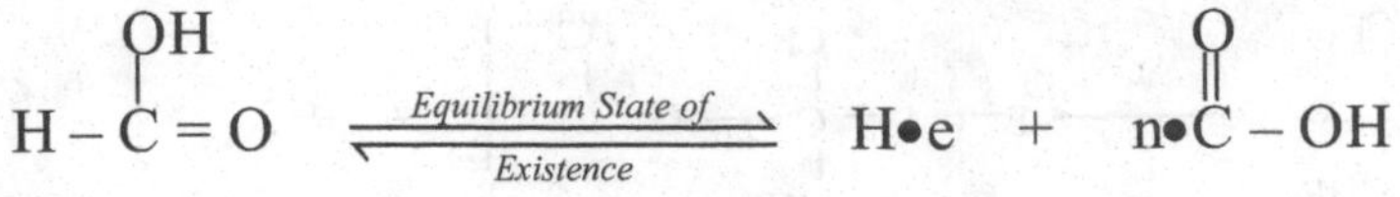

$$\underset{\substack{|\\ H-C=O}}{\overset{OH}{}} \quad \underset{\text{Existence}}{\overset{\text{Equilibrium State of}}{\rightleftharpoons}} \quad H\bullet e \; + \; n\bullet \overset{\overset{O}{\|}}{C}-OH \qquad \qquad 5.82b$$

EQUILIBRIUM STATE OF EXISTENCE

Because of what is in Equation, hence the Equilibrium state of existence of the acid is as shown above. Note that, the component (H herein) involved in Activated/Equilibrium State of Existence is the one involved as transfer species during molecular rearrangement in Activated State of Existence, different from that held in Equilibrium State of Existence in Aldehydes, Ketones and even acetic and other higher acids. COH is ionically bonded, but not $COCH_3$. With olefins, the situation is similar to above in addition to having a case where both are the same.

To stabilize (I) in Equation 5.82a, a passive catalyst can either be used or the $H^{\oplus}$ must be replaced with an alkanyl group (R) as shown below.

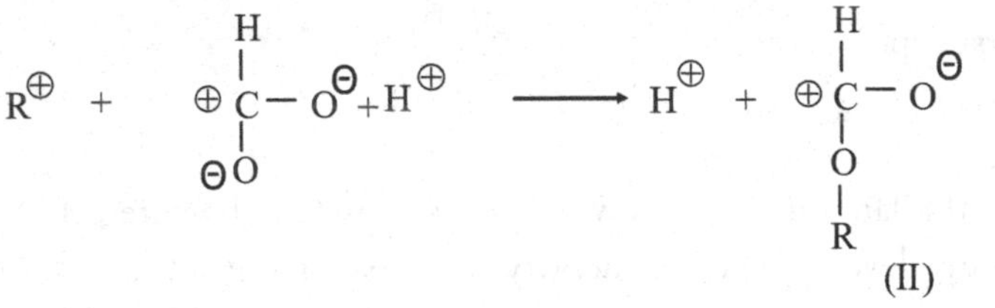

$$R^{\oplus} \; + \; \oplus\overset{\overset{H}{|}}{\underset{\underset{\ominus O}{|}}{C}}-O^{\ominus}+H^{\oplus} \longrightarrow H^{\oplus} \; + \; \oplus\overset{\overset{H}{|}}{\underset{\underset{\underset{R}{|}}{\underset{O}{|}}}{C}}-O^{\ominus}$$

$$\text{(II)} \qquad \qquad 5.83$$

$$R^{\oplus} \; + \; \oplus\overset{\overset{H}{|}}{\underset{\underset{\underset{R}{|}}{\underset{O}{|}}}{C}}-O^{\ominus} \longrightarrow R-O-\overset{\overset{H}{|}}{\underset{\underset{\underset{R}{|}}{\underset{O}{|}}}{C}}\oplus \quad \text{or} \quad ROR + \underset{\oplus}{\overset{\overset{H}{|}}{C}}=O$$

$$\textbf{(Favored)} \qquad \qquad 5.84$$

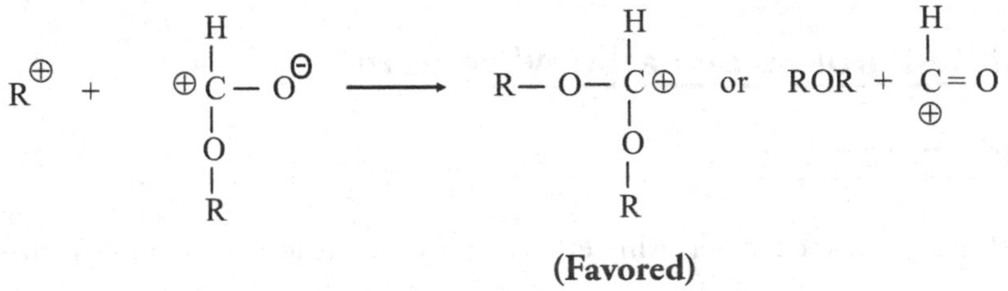

$$R\text{---}(O-\overset{\overset{H}{|}}{\underset{\underset{R}{|}}{\underset{O}{|}}}{C})_{\!n}\!\!-O-\overset{\overset{H}{|}}{\underset{\underset{R}{|}}{\underset{O}{|}}}{C}\oplus \cdots\cdots\overset{\ominus}{\underset{\underset{R}{|}}{\underset{O}{|}}}{B}\!\!\underset{F}{\overset{\overset{F}{|}}{\diagdown}}\!\! \longrightarrow R\text{---}(O-\overset{\overset{H}{|}}{\underset{\underset{R}{|}}{\underset{O}{|}}}{C})_{\!n}\!\!-O-\overset{\overset{H}{|}}{C}=O \; + \; ROR + BF_3$$

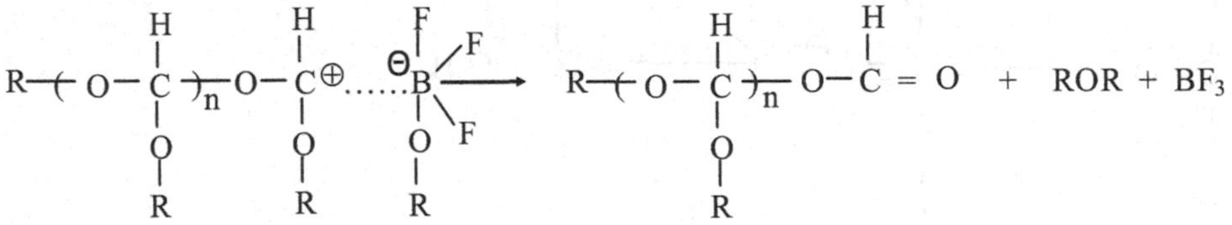

$$5.85$$

Thus, to favor the existence of the last reaction, (II) of Equation 5.83 is used as the original monomer using only positively charged-paired initiators hoping that electrostatic forces of repulsion does not prevent it. With proper choice of initiator, this limitation can be removed. With negative charges, it can be noted that the monomer cannot be polymerized, since the transfer species R will not allow it. The reaction above clearly shows that the OR group is truly a radical-pushing group, since the monomers are nucleophiles.

Considering next NH_2, NRH and NR_2 groups, the followings are to be expected.

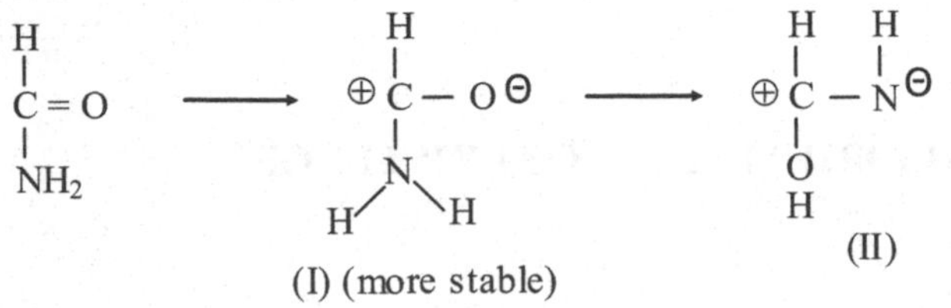

$$\underset{\underset{NH_2}{|}}{\overset{\overset{H}{|}}{C}}=O \longrightarrow \oplus\overset{\overset{H}{|}}{\underset{\underset{\underset{H}{\diagup}\underset{}{N}\diagdown H}{|}}{C}}-O^{\ominus} \longrightarrow \oplus\overset{\overset{H}{|}}{\underset{\underset{\underset{H}{|}}{\underset{O}{|}}}{C}}-\overset{\overset{H}{|}}{N}{}^{\ominus}$$

$$\text{(I) (more stable)} \qquad \qquad \text{(II)} \qquad \qquad 5.86a$$

In view of the fact that the number of substituted groups has increased from two to three, and the fact that NH_2 is more radical-pushing than OH, molecular rearrangement is not favored. Secondly, this is clear indication of the following order of capacity as Nucleophilic centers.

$$C = O > C = N$$
Order of Nucleophilic capacity 5.86b

In Nature in the absence of any external force, we generally move from a Less Nucleophilc or Electrophilic (Less Stable) to a more Nucleophilic or Electrophilic center (More Stable). This may look different from the concept of Concentration gradient in Mass Transfer or Temperature gradient in Heat transfer, when indeed they are all the same. It is (II) above that molecularly rearranges to (I) as we shall see down- stream. On the other hand, it can be reversed under different operating conditions, such as use of an external force, such as a passive catalyst.

Nevertheless, (I) is still unstable, but less so than (II) due to presence of more different ionic centers in (II) than in (I). (II) is the source of Equilibrium State of Existence with more transfer species than in (I) as shown below.

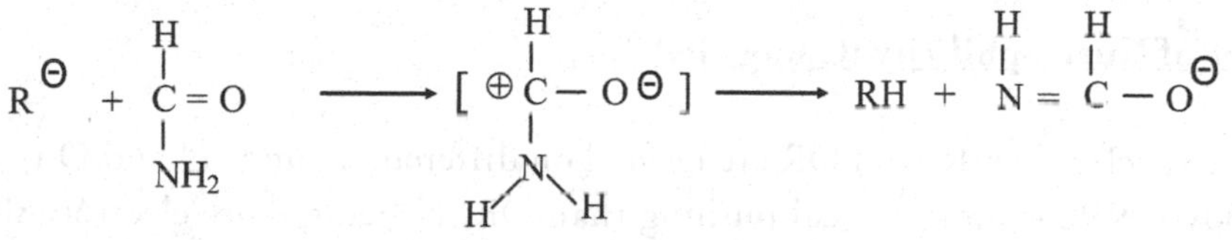

5.87

When the Hs on nitrogen center are replaced with R groups, then the following is obtained using positively charged-paired initiators.

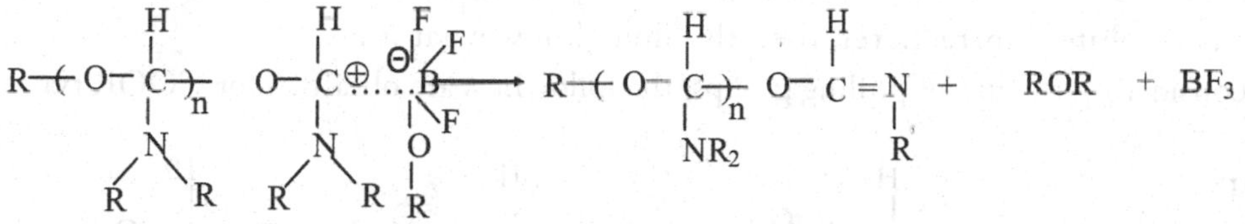

5.88a

The monomers can be observed to be also nucleophilic, since the negatively charged route is not favored. This is a stronger nucleophile than the etheric case (formic acid and etc.) in view of the presence of more transfer species in NR_2 than in OR. Notice the type of dead terminal double polymer obtained. It looks like (II) of Equation 5.86a. When a more stable (II) is polymerized positively, the followings are obtained for its growing polymer chain.

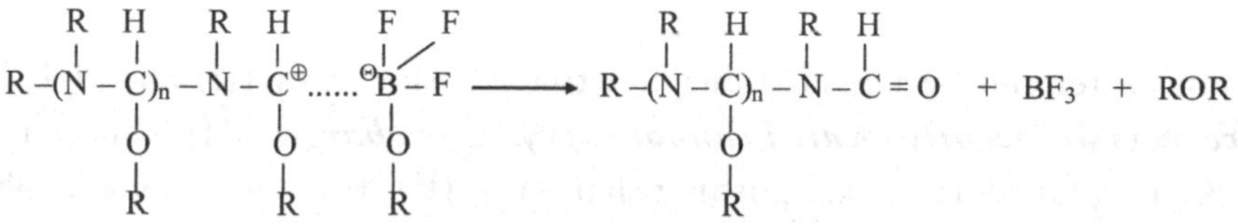

5.88b

The dead terminal double bond polymer obtained looks like (I) of Equation 5.86a. It is very stable, just like the case above. We will see more of these downstream in this Volume. The logi-cal thinking for olefins looks almost the same but different as shown below.

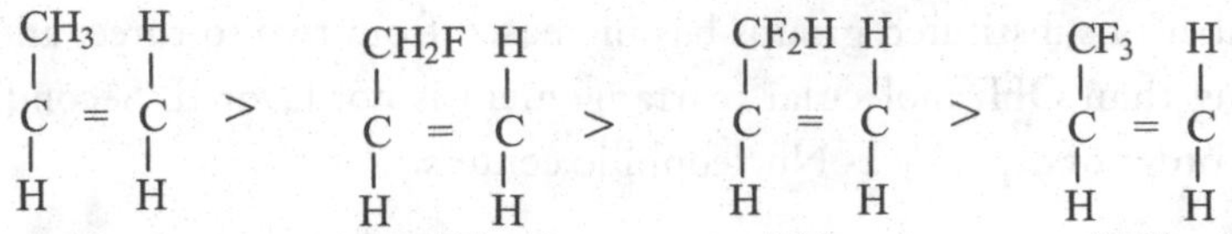

$$\underset{\text{(I) -3}}{\overset{CH_3 \; H}{\underset{H \quad H}{C=C}}} > \underset{\text{(II) -2}}{\overset{CH_2F \; H}{\underset{H \quad H}{C=C}}} > \underset{\text{(III) -1}}{\overset{CF_2H \; H}{\underset{H \quad H}{C=C}}} > \underset{\text{(IV) -0}}{\overset{CF_3 \; H}{\underset{H \quad H}{C=C}}}$$

5.89

Order of nucleophilicity (Radically)

While (I) has three transfer species, (II) has two, (III) has only one transfer species all of same type and (IV) has none. Chargedly, the following is obtained.

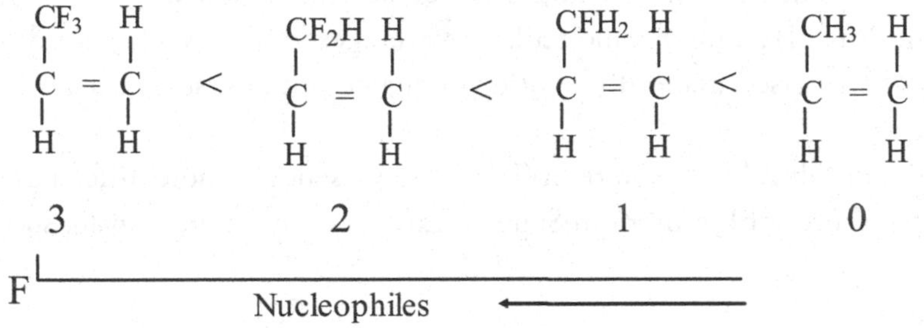

$$\underset{3}{\overset{CF_3 \; H}{\underset{H \quad H}{C=C}}} < \underset{2}{\overset{CF_2H \; H}{\underset{H \quad H}{C=C}}} < \underset{1}{\overset{CFH_2 \; H}{\underset{H \quad H}{C=C}}} < \underset{0}{\overset{CH_3 \; H}{\underset{H \quad H}{C=C}}}$$

5.90

F — Nucleophiles ←

Order of Nucleophilicity (Chargedly)

Though the transfer species on NR_2 and OR are located on different centers, N and O respect-ively unlike the cases above, NR_2 is more radical-pushing than OR, N being more electroposi-tive than O. One can see why the need for the use of "THE EYE OF THE NEEDLE" is very very important. This is not the physical eyes or the "mind", but something else located between the two physical eyes. The Medical Doctors have identified the GLAND, but don't know what it is, BECAUSE OF ABSENCE OF THE NEW FRONTIERS. Metaphysicians and Mystics call it the THIRD EYE, that which is absolutely correct. Yet, they also don't know what it is.

Now, considering the use of pulling groups, the followings are obtained for COOR types.

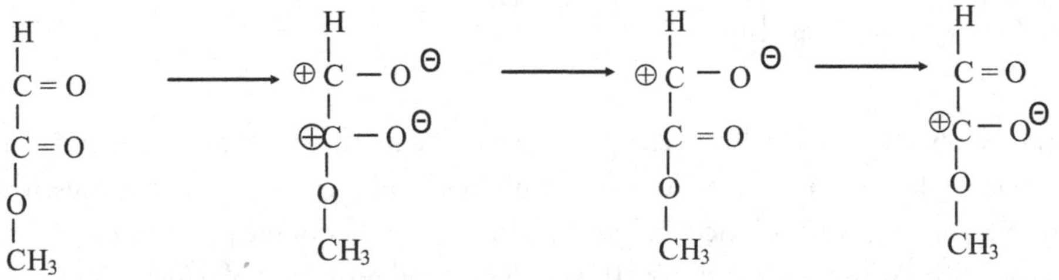

5.91

(I) Impossible existence (II) Less Nucleophilic center (III) more Nucleophilic center

Only one activation center can be activated one at a time, as shown by (II) and (III). (I) cannot exist, *since two centers adjacently located cannot carry same charge*. (III) is more nucleophilic than (II), since OCH_3 is far more radical-pushing than H of (II). It is the less nucleophilic center that will be activated all the time. However chargedly, let us see what the activation of the more nucleophilic center looks like.

$$
RO^{\ominus} + \underset{\substack{| \\ CH_3}}{\overset{\substack{H \\ | \\ C=O \\ | \\ C=O \\ | \\ O}}{}} \longrightarrow ROCH_3 + \underset{\substack{| \\ H}}{\overset{\substack{\ominus \\ O-C=O \\ | \\ C=O}}{}} \quad OR \quad RO-\underset{\substack{| \\ O \\ | \\ CH_3}}{\overset{\substack{H \\ | \\ C-O^{\ominus} \\ | \\ C=O}}{}}
$$

$$
\text{(I) Not favored} \qquad \text{(II) Favored} \qquad\qquad 5.92
$$

$$
\underset{\substack{H_9C_4 \overset{\ominus}{\dots\dots}\overset{\oplus}{Li}}}{\overset{\substack{H \\ | \\ C=O \\ | \\ \oplus C-O^{\ominus} \\ | \\ O \\ | \\ CH_3}}{}} \longrightarrow C_5H_{12} + \underset{\substack{| \\ H}}{\overset{\substack{O=C-OLi \\ | \\ C=O}}{}} \qquad \text{OR No Initiation} \qquad 5.93
$$

$$
\text{NOT FAVORED} \qquad\qquad \text{FAVORED}
$$

$$
R^{\oplus} + \underset{\substack{| \\ O \\ | \\ CH_3}}{\overset{\substack{H \\ | \\ C=O \\ | \\ \oplus C-O^{-}}}{}} \longrightarrow R-O-\underset{\substack{| \\ O \\ | \\ CH_3}}{\overset{\substack{H \\ | \\ C=O \\ | \\ C^{\oplus}}}{}}
$$

$$
\text{(Not favored)} \qquad\qquad\qquad 5.94
$$

$$
R\left(O-\underset{\substack{| \\ O \\ | \\ CH_3}}{\overset{\substack{H \\ | \\ C=O \\ | \\ C}}{}}\right)_n O-\underset{\substack{| \\ O \\ | \\ CH_3}}{\overset{\substack{H \\ | \\ C=O \\ | \\ C^{\oplus}}}{}}\underset{\substack{| \\ O \\ | \\ R}}{\overset{\substack{F \quad F \\ \ominus | \\ B \\ | \\ F}}{}} \longrightarrow R\left(O-\underset{\substack{| \\ O \\ | \\ CH_3}}{\overset{\substack{H \\ | \\ C=O \\ | \\ C}}{}}\right)_n O-\underset{}{\overset{\substack{H \\ | \\ C=O}}{}} + BF_3
$$

$$
\text{[NOT FAVORED]} \qquad\qquad\qquad 5.95
$$

Though that center will not be activated, it has been used exploratively

The monomer with $COOCH_3$ replacing one H in formaldehyde still remains nucleophilic, since there are now two different types of $C=O$ activation centers, that carrying the HCO group being more nucleophilic. Indeed, it is the less nucleophilic center that is first attacked and not the more nucleophilic center used above. The less nucleophilic center will favor both charged routes just like formaldehyde. Note the type of dead terminal double polymer produced here, the character of which is same as the original monomer. Chargedly and also radically, the case above is not resonance stabilized, since movement will be from HALF FREE to FULL NON-FREE.

Replacing COOR with $CONR_2$, the followings are obtained.

$$RO^{\ominus} + \begin{array}{c} H \\ | \\ C=O \\ | \\ {}^{\oplus}C - O^{\ominus} \\ | \\ N \\ / \ \backslash \\ R \quad R \end{array} \longrightarrow ROR + \begin{array}{c} R \\ | \\ N = C - O^{\ominus} \\ | \\ C=O \\ | \\ H \end{array} \mathbf{OR} \quad RO - \begin{array}{c} H \\ | \\ C - O^{\theta} \\ | \\ C=O \\ | \\ NR_2 \end{array} \qquad 5.96$$

Favored

$$R \left(O - \begin{array}{c} H \\ | \\ C \\ | \\ NR_2 \end{array} \right)_n O - \begin{array}{c} H \\ | \\ C=O \\ | \\ {}^{\oplus}C \\ | \\ N \\ / \ \backslash \\ R \quad R \end{array} \quad \begin{array}{c} F \quad F \\ {}^{\ominus} \backslash / \\ B \\ | \\ O \quad F \\ | \\ R \end{array} \longrightarrow R \left(O - \begin{array}{c} H \\ | \\ C \\ | \\ N \\ / \ \backslash \\ R \quad R \end{array} \right)_n O - \begin{array}{c} H \\ | \\ C=O \\ | \\ C=N \\ | \\ C=O \\ | \\ H \end{array} \begin{array}{c} R \\ | \\ \end{array} + ROR + BF_3$$

[FOR EXPLORATION] 5.97

Similar reactions as above are also obtained. In view of NR_2 being more pushing than OR, $CONR_2$ is more pulling than $COOR$ in capacity. For the less nucleophilic center, via both charged routes, there are no transfer species, a clear indication that via the anionic route, depropagation must take place if the temperature of polymerization is above the "Ceiling temperature" of the monomer. Via the real center, no transfer species exist positively or anionically.

With COR replacing $CONR_2$, the followings are obtained.

$$R^{\ominus} + \begin{array}{c} H \\ | \\ C=O \\ | \\ {}^{\oplus}C - O^{\ominus} \\ | \\ CH_3 \end{array} \quad \mathbf{OR} \quad \begin{array}{c} H \\ | \\ {}^{\oplus}C - O^{\ominus} \\ | \\ C=O \\ | \\ CH_3 \end{array} \longrightarrow$$

(I) More nucleophilic center   (II) Less nucleophilic center

$$RH + \begin{array}{c} H \\ | \\ C=C-O^{\ominus} \\ | \quad | \\ H \quad C=O \\ | \\ H \end{array} \quad \mathbf{OR} \quad RCH_3 + \begin{array}{c} O \\ \| \\ C=C-O^{\ominus} \\ | \\ H \end{array} \mathbf{OR} \quad R - \begin{array}{c} H \\ | \\ C - O^{\ominus} \\ | \\ C=O \\ | \\ CH_3 \end{array}$$

Not **Favored**   **Impossible**   **Favored** 5.98

It is impossible, because groups such as COH, $COOR$, $CONH_2$, CF_3 and more cannot carry nega-tive charges, though they are pulling groups, more so radically than chargedly.

$$
\begin{array}{c}
\underset{\overset{|}{C}=O}{\overset{H}{|}} \\
\oplus \overset{|}{C} - O^{\ominus} \\
\underset{CH_3}{|}
\end{array}
$$

$$C_4H_9^{\ominus}.......^{\oplus}Li \longrightarrow C_4H_{10} + \begin{array}{c} \overset{H}{\underset{|}{C}} = \overset{|}{\underset{|}{C}} - OLi \\ \overset{|}{H} \quad \overset{|}{C}=O \\ \overset{|}{H} \end{array} \quad OR \quad \begin{array}{c} Not \\ balanced \end{array} \qquad 5.99$$

It is the top C = O center that is actually activated, with initiations favored both ways.

$$R \left(O - \overset{\overset{\overset{H}{|}}{\overset{C=O}{|}}}{\underset{CH_3}{C}} \right)_n \quad O - \overset{\overset{\overset{H}{|}}{\overset{C=O}{|}}}{\underset{CH_3}{\overset{\oplus}{C}}} \longrightarrow R \left(O - \overset{\overset{\overset{H}{|}}{\overset{C=O}{|}}}{\underset{CH_3}{C}} \right)_n - O - \overset{\overset{|}{C=O}}{\underset{\overset{|}{H}}{C}} = \overset{\overset{H}{|}}{\underset{H}{C}} + H^{\oplus}$$

[FOR EXPLORATION] 5.100

Indeed, like all the others considered so far, the actual C = O center activated with any charged initiator is the first one. Nevertheless, note the types of explorative dead terminal double bond polymers produced here. It is a MALE, unlike the others above (See Equations 5.95, 5.97 and 5.100). A female monomer suddenly became a male, because of the unique difference between CR_3, NR_2, and OR groups (R being H, or alkanyl groups). The nucleophilic character of the monomer is weaker here than the others, a clear indication of the less pulling capacity of COR group than the others. Hence, one should expect:-

$$CONR_2 \quad > \quad COOR \quad > \quad COR \qquad 5.101$$

Order of pulling capacity radically and chargedly

These are clear indications of the fact that C=O center activated here with anionic or positively charged initiators is not the ones in the group, but the C=O center carrying the groups.

When the CH_3 above is replaced with H, the followings are obtained, when activation is made possible.

$$R^{\ominus} + \begin{array}{c} \overset{H}{\underset{|}{C}}=O \\ \oplus \overset{|}{\underset{|}{C}} - O^{\ominus} \\ \overset{|}{H} \end{array} \longrightarrow RH + \begin{array}{c} \overset{O}{\overset{||}{C}} = \overset{|}{\underset{|}{C}} - O^{\ominus} \\ \overset{|}{H} \end{array} \qquad 5.102$$

(I)   [Impossible Reaction]

$$
\begin{array}{c}
H \\
| \\
C=O \\
\oplus C - O^{\ominus} \\
| \\
H \ (I)
\end{array}
$$

$$
H_3CO^{\ominus}........\oplus Na \longrightarrow
\begin{array}{c}
H \\
| \\
C=O \\
| \\
CH_3O-C-O^{\ominus}.........\oplus Na \\
| \\
H
\end{array}
\xrightarrow{+\ n(I)}
$$

$$
CH_3O\!\!\left(\!\!\begin{array}{c} H \\ | \\ C=O \\ | \\ C - O \\ | \\ H \end{array}\!\!\right)_{\!\!n}\!\!\begin{array}{c} H \\ | \\ C=O \\ | \\ C - O^{\ominus}........\oplus Na \\ | \\ H \end{array}
\longrightarrow \text{No transfer species} \qquad 5.103
$$

$$
R\!\!\left(\!\!\begin{array}{c} H \\ | \\ O-C \\ | \\ C=O \\ | \\ H \end{array}\!\!\right)_{\!\!n}\!\!\begin{array}{c} H \\ | \\ \oplus \\ O-C \\ | \\ C=O \\ | \\ H \end{array}\!\!\begin{array}{c} F\ F \\ \ominus | / \\ B \\ / | \\ O \ F \\ | \\ R \end{array}
\longrightarrow \text{No transfer species} \qquad 5.104
$$

It is important, to note that $R^{\ominus}$ cannot abstract H from COH group, since the C in the C=O center cannot carry a negative charge due to electrostatic forces of repulsion. It can be observed that, with non-free negatively charged paired initiator, the monomer can be polymerized via any C=O center, an indication of the less nucleophilic character of the monomer than with $COCH_3$ group where both routes can still be favored using only one C=O center. This is clearly an indication that COH is less pulling than COR group. That is:-

$$
COR > COH \ ; \ COOR > COOH \ ; \ CONR_2 > CONHR > CONH_2 \qquad 5.105
$$

<u>Order of pulling capacity radically and chargedly</u>

The ability of a group to impart more electrophilic or nucleophilic character to a monomer which is largely electrophilic or nucleophilic respectively should be a measure of the pulling capacity of the group. This seems to be partly supported by the data provided for benzene quantitatively from electric moment measurements.[5]

$$
C \equiv N > COR > COH > CCl_3 > COOR > Cl > F \geq COOH \qquad 5.106
$$

<u>"Order of so-called *electron-withdrawing* capacity"</u>[5]

The word so-called has been used, because electrons only exist inside the nucleus of an atom and word withdrawing has been replaced with pulling. ***When the non-halogenated groups above are placed on benzene ring, the compounds are ELECTROPHILES, "resonance stabilized" radically and not chargedly (From Half-non-free to Half-non-free). Hence the measurements above are radically based.*** When placed on ethene (Olefins), all are ELECTROPHILES not resonance stabilized.

Now, replacing H with CN, the followings are obtained.

$$R^{\ominus} + \underset{\underset{\underset{N}{\overset{|||}{C}}}{\overset{|}{C}}}{\overset{H}{\underset{|}{C}}}=O \longrightarrow RH + \underset{\underset{N}{\overset{|||}{C}}}{C}=C=N^{\ominus} \quad OR \quad R-\underset{\underset{N}{\overset{|||}{C}}}{\overset{H}{\underset{|}{C}}}-O^{\ominus}$$

$$\text{[Impossible existence]} \qquad\qquad\qquad \text{(Favored)} \qquad\qquad 5.107$$

$$\underset{\underset{CH_3O^{\ominus}\overset{\oplus}{\dots}Na}{}}{\underset{\underset{H}{|}}{\overset{\overset{\overset{N}{|||}}{C}}{\underset{|}{C}}}-O^{\ominus}} \quad\text{(I)} \longrightarrow CH_3O-\underset{\underset{H}{|}}{\overset{\overset{\overset{N}{|||}}{\overset{|}{C}}}{C}}-O^{\ominus}\overset{\oplus}{\dots}Na$$

$$5.108$$

$$R^{\oplus} + \underset{\underset{\underset{N}{\overset{|||}{C}}}{\overset{|}{C}}}{\overset{H}{\underset{|}{C}}}=O \quad\text{(I)} \longrightarrow R-O-\underset{\underset{\underset{N}{\overset{|||}{C}}}{\overset{|}{C}}}{\overset{H}{\underset{|}{C^{\oplus}}}} \quad OR \quad R-N=\underset{\underset{\underset{H}{|}}{\overset{|}{C}}=O}{C^{\oplus}}$$

$$\text{(Not favored)}$$

$$\xrightarrow{+ n(I)} R-\left(O-\underset{\underset{\underset{N}{\overset{|||}{C}}}{\overset{|}{C}}}{\overset{H}{\underset{|}{C}}}\right)_{\overline{n}}-O-\underset{\underset{\underset{N}{\overset{|||}{C}}}{\overset{|}{C}}}{\overset{H}{\underset{|}{C^{\oplus}}}}$$

$$\textbf{\underline{FAVORED}} \qquad\qquad\qquad\qquad 5.109$$

Note that both $C \equiv N$ and $C = O$ centers are nucleophilic (X) of different capacities. If the $C \equiv N$ activation center is activated in preference to $C = O$ activation center, transfer species ($H^{\oplus}$) does not exists for its growing polymer chain with positively charged paired initiators. ***In fact as already found and stated into law,*** that center cannot be activated chargedly, but only radically. With the $C = O$ center, there is no transfer species. Anionically, the same also applies. All these are due to the strong pulling capacity of CN- indeed the strongest so far (See Equation 5.106). With NO_2 group, charged activation is favored despite the presence of polar bonds on $N = O$ center.

For a group such as $COCCl_3$, the followings should be expected.

$$\underset{\underset{\underset{CCl_3}{|}}{\overset{\overset{\overset{H}{|}}{C}=O\text{ (a)}}{\underset{|}{C}=O\text{ (b)}}}}{}\text{(I)} \quad ; \quad \underset{\underset{\underset{CH_3}{|}}{\overset{\overset{\overset{H}{|}}{C}=O}{\underset{|}{C}=O}}}{}\text{(II)} \quad ; \quad \underset{\underset{\underset{H}{|}}{\overset{\overset{\overset{H}{|}}{C}=O}{\underset{|}{C}=O}}}{}\text{(III)}$$

$$5.110$$

The first question to ask for (I) is that which of the two C = O centers is more nucleophilic. We already know for the cases of (II) and (III). Chargedly or radically, it is the first one, (a) that is more nucleophilic here. ***Thus, the nucleophilicity of a center is determined by the types of groups carried by the active center(s).***

$$R^{\oplus} \ + \ \overset{\overset{\displaystyle \overset{H}{|}}{\underset{\displaystyle \underset{CCl_3}{|}}{\overset{\oplus}{C}-O^{\ominus}}}{\underset{\displaystyle C=O}{}} \longrightarrow R-O-\overset{\overset{\displaystyle \overset{H}{|}}{\underset{\displaystyle \underset{CCl_3}{|}}{\overset{\oplus}{C}}}}{\underset{\displaystyle C=O}{}} \quad OR \quad \overset{\overset{\displaystyle \overset{H}{|}}{\underset{\displaystyle \underset{CCl_3}{|}}{\overset{\oplus}{C}-OR}}}{\underset{\displaystyle C=O}{}} \qquad 5.111$$

[Favored]

$$\begin{array}{c} CCl_3 \\ | \\ C=O \\ | \\ \overset{\oplus}{C}-O^{\ominus} \\ | \\ H \\ H_9C_4 \overset{\ominus}{\ldots\ldots} \overset{\oplus}{Li} \end{array} \longrightarrow \begin{array}{c} CCl_3 \\ | \\ C=O \\ | \\ H_9C_4 - C-O \ Li \\ | \\ H \end{array} \qquad 5.112$$

[FOR EXPLORATION]

With non-free-negatively charged paired initiator, polymerization is favored. (I) is almost like (III) for which one should expect the following order in their pulling capacities.

$$\underset{\displaystyle CH_3}{-C=O} \quad > \quad \underset{\displaystyle H}{-C=O} \quad > \quad \underset{\displaystyle CCl_3}{-C=O} \qquad 5.113$$

Order of pulling capacities radically and chargedly

5.3.2. Their radical characters

Beginning with etheric groups, the followings are to be expected

$$\overset{..}{N}\cdot nn \ + \ e.\underset{\displaystyle \underset{R}{\overset{\displaystyle |}{O}}}{\overset{\displaystyle \overset{H}{|}}{C}-O}.nn \longrightarrow \overset{..}{N}R \ + \ \underset{\displaystyle O}{\overset{\displaystyle \overset{H}{|}}{C}-O}.nn \qquad 5.114$$

$$E\overset{e}{\cdot} \ + \ nn.O-\underset{\displaystyle \underset{R}{\overset{\displaystyle |}{O}}}{\overset{\displaystyle \overset{H}{|}}{C}}.e \longrightarrow EOR \ + \ \underset{\displaystyle \dot{e}}{\overset{\displaystyle \overset{H}{|}}{C}}=O \quad OR \quad E-O-\underset{\displaystyle \underset{R}{\overset{\displaystyle |}{O}}}{\overset{\displaystyle \overset{H}{|}}{C}}.e \qquad 5.115$$

(I)   (Favored)

269

From the reactions above, since the monomer is a strong nucleophile, the OR group being a strong radical-pushing group, only the electro-free-radical route is favored. Replacing OR with NR_2 group, similar reactions are obtained for which the following is obtained for its growing polymer chain.

$$
E \left(O - \underset{\underset{\displaystyle \underset{R}{\diagup} \underset{R}{\diagdown}}{N}}{\overset{\overset{\displaystyle H}{|}}{C}} \right)_n O - \underset{NR_2}{\overset{\overset{\displaystyle H}{|}}{C}} \cdot e \longrightarrow E \left(O - \underset{\underset{\displaystyle \underset{R}{\diagup} \underset{R}{\diagdown}}{N}}{\overset{\overset{\displaystyle H}{|}}{C}} \right)_n O - \underset{\underset{\displaystyle R}{|}}{\overset{\overset{\displaystyle H}{|}}{C}} = N \quad + \quad R^{\cdot e}
$$

5.116

While it may be possible to compare OR and NR_2 groups based on the number of transfer species, it is not possible to compare OR, NR_2 and CR_3 groups based on the number of transfer species, since the first two are non-free-pushing groups while CR_3 is a free-pushing group. CH_3, CH_2Cl, $CHCl_2$ and CCl_3 can be compared together as has already been done.

Though the existence of carbonyl centers with some type of pulling groups may not be favored due to operating conditions, they will be used however to illustrate some basic principles. One will begin with $COOCH_3$.

$$
\overset{\cdot \cdot}{N} \cdot \text{nn} \quad + \quad e. \underset{\underset{\underset{\underset{CH_3}{|}}{O}}{\overset{\overset{\displaystyle C=O}{|}}{C}}}{\overset{\overset{\displaystyle H}{|}}{C}} - O \cdot \text{nn} \longrightarrow \overset{\cdot \cdot}{N} - \underset{\underset{\underset{\underset{CH_3}{|}}{O}}{\overset{\overset{\displaystyle C=O}{|}}{C}}}{\overset{\overset{\displaystyle H}{|}}{C}} - O \cdot \text{nn} \xrightarrow{\; + \; n(I) \;}
$$

(I)

$$
\overset{\cdot \cdot}{N} \left(\underset{\underset{\underset{\underset{CH_3}{|}}{O}}{\overset{\overset{\displaystyle C=O}{|}}{C}}}{\overset{\overset{\displaystyle H}{|}}{C}} - O \right)_n \underset{\underset{\underset{\underset{CH_3}{|}}{O}}{\overset{\overset{\displaystyle C=O}{|}}{C}}}{\overset{\overset{\displaystyle H}{|}}{C}} - O \cdot \text{nn} \longrightarrow \overset{\cdot \cdot}{N} \left(\underset{\underset{\underset{\underset{CH_3}{|}}{O}}{\overset{\overset{\displaystyle C=O}{|}}{C}}}{\overset{\overset{\displaystyle H}{|}}{C}} - O \right)_n \underset{\underset{OCH_3}{|}}{\overset{\overset{\displaystyle C=O}{|}}{C}} = O \quad + \quad n.H
$$

5.117

$$
E^{\cdot e} \quad + \quad e. \underset{\underset{\underset{OCH_3}{|}}{CO}}{\overset{\overset{\displaystyle H}{|}}{C}} - O \cdot \text{nn} \longrightarrow E - O - \underset{\underset{\underset{\underset{CH_3}{|}}{O}}{\overset{\overset{\displaystyle C=O}{|}}{|}}}{\overset{\overset{\displaystyle H}{|}}{C}} \cdot e \quad \text{OR} \quad E - O - \underset{\underset{\underset{CH_3}{|}}{O}}{\overset{\overset{\displaystyle CHO}{|}}{C}} \cdot e
$$

5.118

(I) (Favored)   (II)

From the reactions above, the first $C=O$ center involved is either nucleophilic or electrophilic. This is the less nucleophilic center. For the second $C=O$ activation center, the followings are obtained.

270

$$\ddot{\text{N}}\cdot \text{nn} \quad + \quad \begin{array}{c} \text{H} \\ | \\ \text{C}=\text{O} \\ | \\ \text{e}\cdot\text{C}-\text{O}.\text{nn} \\ | \\ \text{O} \\ | \\ \text{CH}_3 \end{array} \quad \longrightarrow \quad \overset{..}{\text{N}}\text{CH}_3 \quad + \quad \begin{array}{c} \overset{\text{O}}{\overset{||}{\text{C}}}-\text{O}.\text{nn} \\ | \\ \text{C}=\text{O} \\ | \\ \text{H} \end{array} \qquad 5.119$$

$$\text{E}\cdot^e \quad + \quad \begin{array}{c} \text{H} \\ | \\ \text{C}=\text{O} \\ | \\ \text{e}.\text{C}-\text{O}.\text{nn} \\ | \\ \text{O} \\ | \\ \text{CH}_3 \end{array} \quad \longrightarrow \quad \text{EOCH}_3 \quad + \quad \begin{array}{c} ^{e\cdot}\text{C}=\text{O} \\ | \\ \text{C}=\text{O} \\ | \\ \text{H} \end{array} \qquad \text{OR} \quad \text{ADDITION (None favored)}$$

[See (II) of Equation 5.118]

$$5.120$$

The C = O center here is more nucleophilic than the C = O above, even though the electro-free-radical route is also the favored route. This is the center involved electro-free-radically since the monomer is a Nucleophile. It is the first center that is involved electro-free- and nucleo-non-free-radically all the time.

With CONR$_2$ replacing COOR, the followings are obtained when the wrong center is used nucleo-non-free-radically

$$5.121$$

$$\ddot{\text{N}}\cdot \text{nn} \quad + \quad \begin{array}{c} \text{H} \\ | \\ \text{C}=\text{O} \\ | \\ \text{e}.\text{C}-\text{O}.\text{nn} \\ | \\ \text{NR}_2 \end{array} \quad \longrightarrow \quad \ddot{\text{N}}\text{R} \quad + \quad \begin{array}{c} \text{R} \\ | \\ \text{N}=\text{C}-\text{O}.\text{nn} \\ | \\ \text{C}=\text{O} \\ | \\ \text{H} \end{array}$$

$$\text{E}^{\cdot e} \quad + \quad \begin{array}{c} \text{H} \\ | \\ \text{O}=\text{C} \\ | \\ \text{nn}.\text{O}-\text{C}.\text{e} \\ | \\ \text{NR}_2 \end{array} \quad \longrightarrow \quad \begin{array}{c} \text{H} \\ | \\ \text{O}=\text{C} \\ | \\ \text{E}-\text{O}-\text{C}.\text{e} \\ | \\ \text{NR}_2 \end{array}$$

[FOR EXPLORATION]

$$5.122$$

Similar reactions as with COOR groups are obtained. Indeed, it is only the first center that is activated electro-free-radically and nucleo-non-free-radically.

With COR group, the followings are obtained.

$$\ddot{\text{N}}\cdot \text{nn} \quad + \quad \begin{array}{c} \text{H} \\ | \\ \text{C}=\text{O} \\ | \\ \text{C}=\text{O} \\ | \\ \text{CH}_3 \end{array} \quad \longrightarrow \quad \ddot{\text{N}}\text{H} \quad + \quad \begin{array}{c} \text{H} \\ | \\ \text{C}=\text{C}-\text{O}.\text{nn} \\ | \quad | \\ \text{H} \quad \text{C}=\text{O} \\ | \\ \text{H} \end{array} \quad \text{OR} \quad \begin{array}{c} \overset{..}{\text{N}}\text{CH}_3 + \overset{\text{H}}{\overset{|}{\text{O}=\text{C}=\text{C}-\text{O}.\text{nn}}} \\ \text{[Impossible Reaction]} \end{array}$$

$$5.123$$

$$E \xleftarrow{} O - \underset{\underset{H}{\overset{|}{C}=O}}{\overset{CH_3}{\overset{|}{\underset{|}{C}}}} \xrightarrow{}_n O - \underset{\underset{H}{\overset{|}{C}=O}}{\overset{CH_3}{\overset{|}{\underset{|}{C}}}} . e \longrightarrow E \xleftarrow{} O - \underset{\underset{H}{\overset{|}{C}=O}}{\overset{CH_3}{\overset{|}{\underset{|}{C}}}} \xrightarrow{}_n O - \underset{\underset{H}{\overset{|}{C}=O}}{\overset{H}{\overset{|}{C}}} = \overset{H}{C} \quad + \quad H^{\cdot e}$$

5.124

(This is a strong Male!) **NOT FAVORED**

The dead terminal double bond polymer above is a strong MALE regardless the capacity of the OR group which is far smaller than the CHO group which is a pulling group. This clearly sends the message as already shown chargedly, that this is not the center involved in activation. From a Female monomer polymerized in the route natural to it, a dead terminal double bond strong Male polymer is obtained, "because of the uniqueness of CH_3 type of group (See Equations 5.100 and 101)". Though this is an exploratory exercise, these phenomena have been observed with Molecular rearrangement, wherein some Female monomers rearrange to Males and Males rearrange to Females under unique operating conditions- type of monomer, type of initiator, temperature and pressure and more as will be shown downstream.

Now, considering the unique case where the two C = O centers are of the same capacity, that is, (H) O = C – C =O(H).

$$\ddot{N}^{\cdot nn} + e . \underset{\underset{H}{\overset{|}{C}=O}}{\overset{H}{\overset{|}{C}}} - O . nn \longrightarrow \ddot{N}H + \underset{}{\overset{O \ H}{\overset{\parallel \ |}{C}}} = C - O . nn \quad OR \quad \underset{\underset{H}{\overset{|}{C}=O}}{N - \overset{H}{\overset{|}{C}} - O} \bullet mn$$

NOT FAVORED   FAVORED   5.125

$$E \xleftarrow{} O - \underset{\underset{H}{\overset{|}{C}=O}}{\overset{H}{\overset{|}{C}}} \xrightarrow{}_n O - \underset{\underset{H}{\overset{|}{C}=O}}{\overset{H}{\overset{|}{C}}} . e \longrightarrow E \xleftarrow{} O - \underset{\underset{H}{\overset{|}{C}=O}}{\overset{H}{\overset{|}{C}}} \xrightarrow{}_n O - C = C = O \quad + \quad H^{\cdot e}$$

OR

No Transfer species

FAVORED   5.126

In the last equation above, transfer species of the first kind cannot be released, since it was not abstracted in the route not natural to it. Therefore, living polymers are produced here. Dead polymers can only be produced in the presence of terminating agents or impurities. Free-radically, in view of more pushing capacity of CH_3 than H, the HC = O group is less pulling than the $H_3CC = O$ group. The presence of the pulling group COH does alter the nucleophilic character of the monomer. However, between $COOCH_3$ and $COCH_3$ or COH groups, in view of the routes favored, one should also expect $COCH_3$ to be less pulling than $COOCH_3$.

While the monomers still maintain some nucleophilic character with COR, COOR, $CONR_2$ groups to different levels, with CN, both routes are also involved, but with that giving C = O center on the monomer a less nucleophilic character.

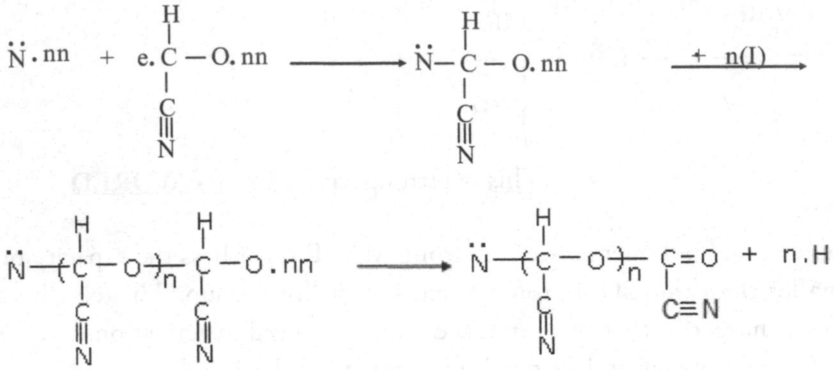

$$\ddot{N}.nn \; + \; e.\underset{\underset{N}{\overset{\overset{H}{|}}{\underset{|}{C}}}{C}}{C}=O.nn \longrightarrow \ddot{N}-\underset{\underset{N}{\overset{\overset{H}{|}}{\underset{|}{C}}}{C}}{C}-O.nn \qquad \xrightarrow{\;+\;n(I)\;}$$

$$\ddot{N}-(\underset{\underset{N}{\overset{\overset{H}{|}}{\underset{|}{C}}}{C}}{C}-O)_n \underset{\underset{N}{\overset{\overset{H}{|}}{\underset{|}{C}}}{C}}{C}-O.nn \longrightarrow \ddot{N}-(\underset{\underset{N}{\overset{\overset{H}{|}}{\underset{|}{C}}}{C}}{C}-O)_n \underset{C\equiv N}{\overset{\overset{H}{|}}{C}}=O \; + \; n.H$$

Transfer species of second kind   5.127

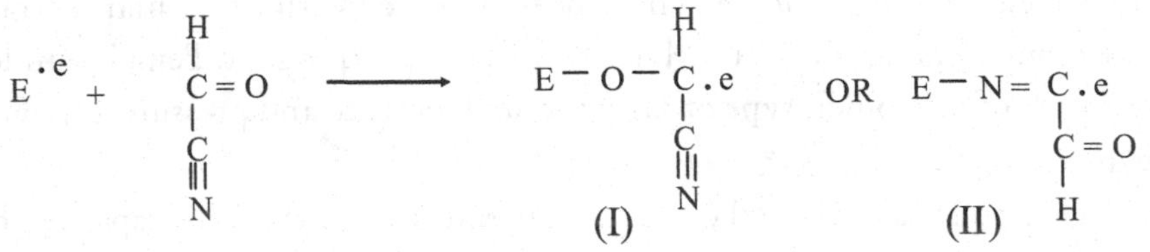

$$E^{\cdot e} \; + \; \underset{\underset{N}{\overset{\overset{H}{|}}{\underset{|}{C}}}{C}}{C}=O \longrightarrow E-O-\underset{\underset{N}{\overset{\overset{H}{|}}{\underset{|}{C}}}{C}}{C}.e \qquad OR \qquad E-N=\underset{\underset{H}{\overset{\overset{C}{|}}{=O}}}{C}.e \qquad 5.128$$

(I)   (II)

(Favored)

The C $\equiv$ N center cannot be involved, since it is far more nucleophilic than the C = O center, noting that the monomer is a Nucleophile. If it was an Electrophile where the C=O center is Y and the C$\equiv$N is the X center, then (II) of the last equation above would have been favored.

From the reactions above, it can be observed that CN group is a strong radical-pulling group for formaldehyde. The strength is further decreased in acetaldehyde as shown below, acetalde-hyde being more nucleophilic than formaldehyde. Hence formaldehyde has been used as the base for measurement.

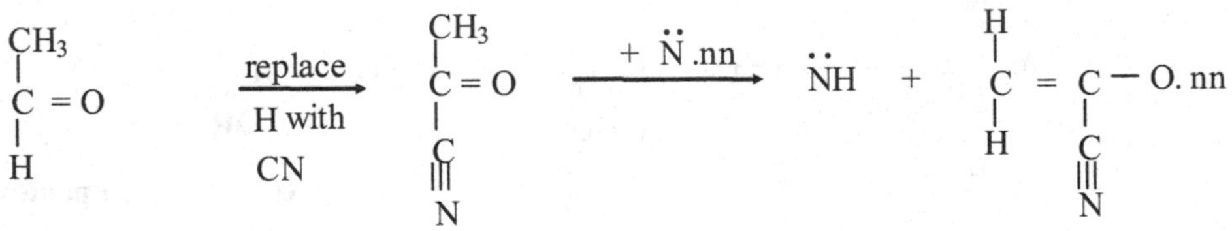

$$\underset{H}{\overset{CH_3}{\underset{|}{\overset{|}{C}=O}}} \quad \xrightarrow[\substack{H\,with \\ CN}]{replace} \quad \underset{\underset{N}{\overset{\overset{CH_3}{|}}{\underset{|}{C}}}{C}}{C}=O \quad \xrightarrow{\;+\;\ddot{N}.nn\;} \quad \ddot{N}H \; + \; \underset{\underset{N}{\overset{\overset{H}{|}}{\underset{|}{C}}}{C}}{\overset{\overset{H}{|}}{C}}=C-O.nn$$

(Favored)   5.129

This is unlike the case of Equation 5.123, where transfer species do not exist for the less nucleo-philic center. This we have seen with olefins particularly during molecular rearrangements.

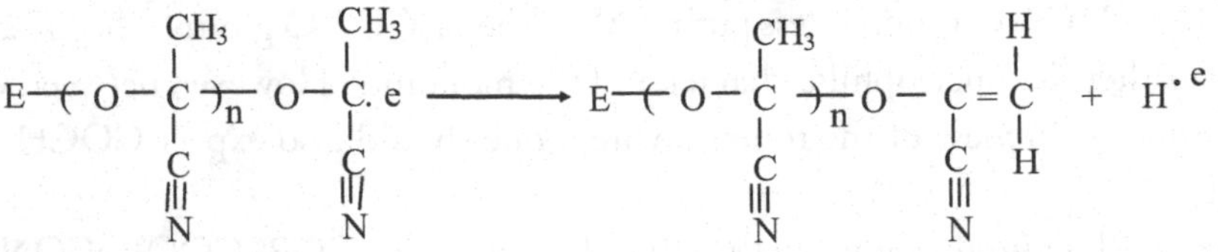

$$E-(O-\underset{\underset{N}{\overset{\overset{CH_3}{|}}{\underset{|}{C}}}{C}})_n O-\underset{\underset{N}{\overset{\overset{CH_3}{|}}{\underset{|}{C}}}{C}}{C}.e \longrightarrow E-(O-\underset{\underset{N}{\overset{\overset{CH_3}{|}}{\underset{|}{C}}}{C}})_n O-\underset{\underset{N}{\overset{\overset{H}{|}}{\underset{|}{C}}}{C}}{C}=\overset{\overset{H}{|}}{C} \; + \; H^{\cdot e}$$

[Favored]   5.130a

OR $E-(N=C)_n-N=C=C=O \; + \; e\bullet CH_3$

$$\underset{CH_3}{\overset{|}{\overset{C=O}{|}}}$$

5.130b

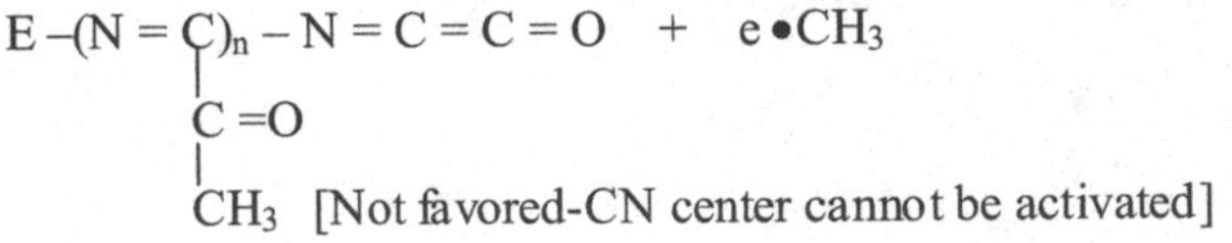

[Not favored-CN center cannot be activated]

Electro-free-radically, only one center is involved regardless the operating conditions, noting as already said that the $C \equiv N$ center which cannot be activated chargedly is more nucleophilic than the $C = O$ center. More energy would have been required to activate the $C \equiv N$ and $C = O$ centers at the same time if it was a Male, the $C \equiv N$ center requiring more energy than the $C = O$ center. Nucleo-non-free-radically for these cases, initiation is not favored for any of the centers, because of presence of transfer species of the first kind. Worthy of note is the type of dead terminal double bonded polymers obtained for both centers. The dead polymers are both MALES. Yet, the monomer is a Nucleophile (Female) which turned to an Electrophile (An acrylonitrile) after polymerization in the route natural to it.

$$
\begin{array}{ccc}
\underset{\underset{\overset{|}{C}}{\overset{RO}{\underset{\|}{|}}}{\overset{Y}{C}} = \underset{\overset{|}{C}}{\overset{H}{C}} & \longrightarrow & n\bullet \underset{\overset{|}{C}}{C} - \underset{}{C}\bullet e \quad ; \quad \underset{\overset{\|}{N}}{\overset{}{C}} = C = O \longrightarrow e\bullet C - \underset{\|}{C}\bullet n
\end{array}
\qquad 5.130c
$$

ACTIVATION OF ELECTROPHILIC CENTER

Based on all the considerations above so far, one can observe that $=CR_2$, $=O$, $=NR$, are strong radical-pushing groups while $=C=C=O$, and $=C=C=NR$ groups are stronger radical-pushing groups with the following capacities.

$$O = C = C = \quad < \quad R - N = C = C =$$

<u>Order of Radical-pushing capacities</u>

$$R - N = \quad > \quad O = \quad > \quad R_2C =$$

<u>Order of Radical-pushing capacities</u>
5.131

Though not yet stated into law, it is important to recall Equation 3. 126. These could also be charged-pushing groups,. These cannot allow negative charges to be carried by the active C centers. They look divalent in character.

It has been shown that when NO_2 group is involved free-radically, it is a free-pushing group. As shown by the reactions below, the $C = O$ center is more nucleophilic than the $N = O$ center. Hence, NO_2 group is a weak radical-pushing group in capacity.

$$
\ddot{N}\cdot nn \;+\; e.\overset{\overset{H}{|}}{\underset{\underset{\overset{\oplus}{N}}{|}}{C}} - O.nn \;\longrightarrow\; \ddot{N} - \overset{\overset{H}{|}}{\underset{\underset{\overset{\oplus}{N}}{|}}{C}} - O.nn
\qquad 5.132a
$$

$$
\ddot{N}\cdot nn \;+\; e.\overset{\overset{O^{\ominus}}{|}}{\underset{\underset{\overset{|}{C=O}}{\underset{\overset{|}{H}}{}}}{\overset{\oplus}{N}}} - O.nn \;\longrightarrow\; \ddot{N}H \;+\; O=C=\overset{\overset{O^{\ominus}}{|}}{\overset{\oplus}{N}} - O.nn \quad OR \quad :N - \overset{\overset{O^{\ominus}}{|}}{\underset{\underset{\overset{|}{C=O}}{\underset{\overset{|}{H}}{}}}{\overset{\oplus}{N}}} - O.nn
$$

(Favored)   5.132b

$$5.133$$

$$5.134$$

For both cases, both centers can be involved, depending on the type of initiator and operating conditions. But, the first to be activated all the time is the less nucleophilic center, the N=O center. For both cases, there are no transfer species to be abstracted or released, the natural route being the electro-free-radical route. H could not be abstracted, because of the strong radical-pushing capacity of O= group and the fact that C is more electropositive than N. C cannot carry a nucleo-free-radical in the presence of N. That the C=O center is more nucleophilic than the N=O center, will be shown downstream in this Volume.

One had thought that the monomer above, is the only case that is resonance stabilized, moving from *Full-non-free to Full-non-free. However above, it is movement from Half-free to Full-non-free.* Like most of the cases considered so far such as in H – CO – CO – H, the movement is from *Half-free to Full-non-free for Females*, a clear indication that these monomers cannot be resonance stabilized. The corresponding non-resonance stabilized case for Males is movement from *Full-free to Half –non-free or Half-free.* The case of O=C – C_6H_4 – C=O, a male, is movement from *Full-free to Full-non-free*, well resonance stabilized.

5.4. Related Monomers

These are largely half olefinic monomers with at least two activation centers either conjugatedly placed or isolatedly placed. Where they are conjugatedly placed, it has already been explained why the monomers do not favor 1,4 –addition polymerization or are not resonance stabilized.

5.4.1 Their Charged characters

Beginning with acrylates, the followings are to be expected for activation via C = O activation center, a situation which will rarely take place. It will be the C = C center that will be activated with free negative charges all the time.

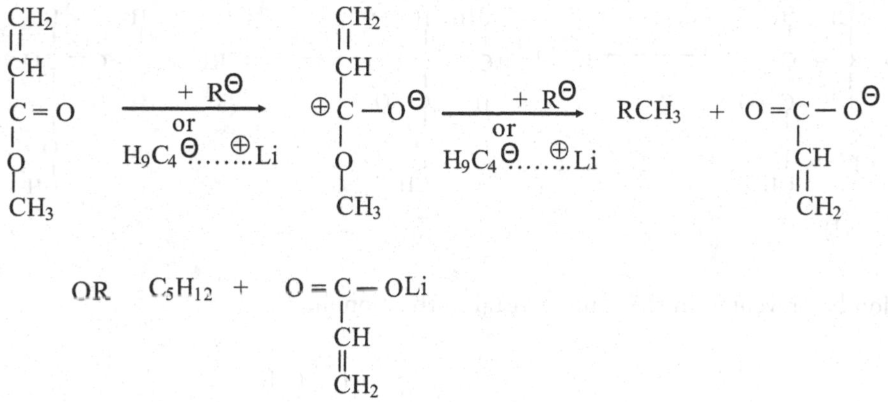

$$\underset{\begin{array}{c}CH_2\\ \parallel\\ CH\\ \mid\\ C=O\\ \mid\\ O\\ \mid\\ CH_3\end{array}}{} \xrightarrow[\substack{\text{or}\\ H_9C_4\overset{\ominus}{}\text{.....}\overset{\oplus}{}Li}]{+\ R^{\ominus}} \quad \underset{\begin{array}{c}CH_2\\ \parallel\\ CH\\ \mid\\ \oplus C-O^{\ominus}\\ \mid\\ O\\ \mid\\ CH_3\end{array}}{} \xrightarrow[\substack{\text{or}\\ H_9C_4\overset{\ominus}{}\text{.....}\overset{\oplus}{}Li}]{+\ R^{\ominus}} \quad RCH_3 \ + \ \underset{\begin{array}{c}\mid\\ CH\\ \parallel\\ CH_2\end{array}}{O=C-O^{\ominus}}$$

$$\text{OR} \quad C_5H_{12} \ + \ \underset{\begin{array}{c}\mid\\ CH\\ \parallel\\ CH_2\end{array}}{O=C-OLi} \qquad\qquad\qquad 5.135$$

$$R\left(O-\underset{\begin{array}{c}\mid\\ O\\ \mid\\ CH_3\end{array}}{\overset{\begin{array}{c}CH_2\\ \parallel\\ CH\\ \mid\end{array}}{C}}\right)_n O-\underset{\begin{array}{c}\mid\\ O\\ \mid\\ CH_3\end{array}}{\overset{\begin{array}{c}CH_2\\ \parallel\\ CH\\ \mid\end{array}}{C}}\oplus\text{.......}\overset{\ominus}{B}\underset{F}{\overset{F}{\diagup}}\diagdown F \longrightarrow R\left(O-\underset{\begin{array}{c}\mid\\ O\\ \mid\\ CH_3\end{array}}{\overset{\begin{array}{c}CH_2\\ \parallel\\ CH\\ \mid\end{array}}{C}}\right)_n O-\overset{\begin{array}{c}CH_2\\ \parallel\\ CH\\ \mid\end{array}}{C}O + ROCH_3 + BF_3$$

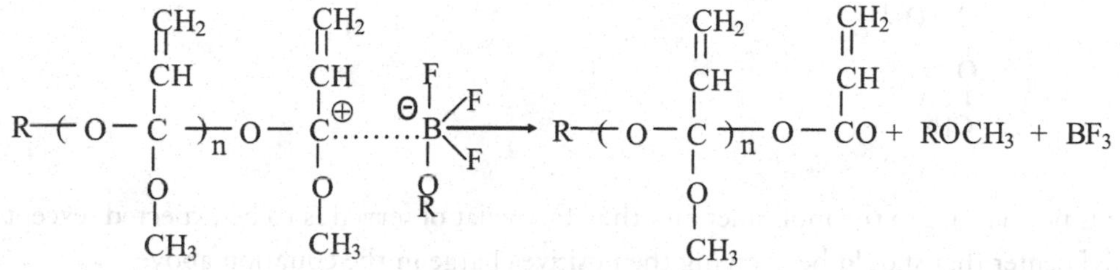

$$5.136$$

The transfer species is CH_3 since OCH_3 group is more radical-pushing than $CH = CH_2$ group. It should be noted that positively charged-paired initiators are used here. No useful molecular rearrangement can take place via $C = O$ activation center here. It is important to note that $C = O$ activation center is a strong nucleophilic center, while the $C = C$ activation center is a strong electrophilic center. Hence, copolymers cannot be produced from these monomers via any charged route when used alone.

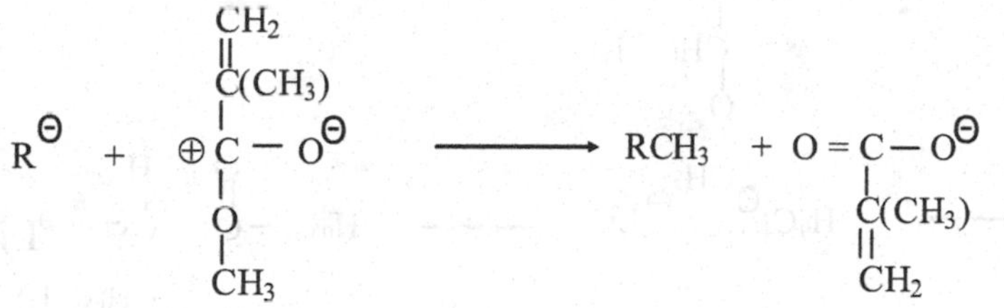

$$R^{\ominus} \ + \ \underset{\begin{array}{c}\mid\\ O\\ \mid\\ CH_3\end{array}}{\overset{\begin{array}{c}CH_2\\ \parallel\\ C(CH_3)\\ \mid\end{array}}{\oplus C}-O^{\ominus}} \longrightarrow RCH_3 \ + \ \underset{\begin{array}{c}\mid\\ C(CH_3)\\ \parallel\\ CH_2\end{array}}{O=C-O^{\ominus}}$$

$$5.137$$

Methyl methacrylate

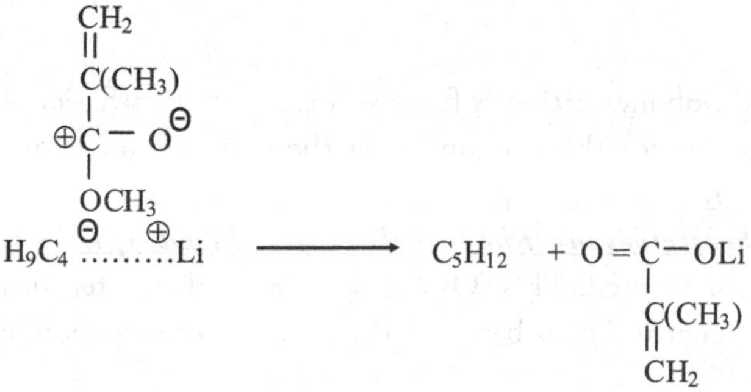

$$\underset{H_9C_4\overset{\ominus}{}\text{.........}\overset{\oplus}{}Li}{\underset{\begin{array}{c}\mid\\ OCH_3\end{array}}{\overset{\begin{array}{c}CH_2\\ \parallel\\ C(CH_3)\\ \mid\end{array}}{\oplus C}-O^{\ominus}}} \longrightarrow C_5H_{12} \ + \ \underset{\begin{array}{c}\mid\\ C(CH_3)\\ \parallel\\ CH_2\end{array}}{O=C-OLi}$$

$$5.138$$

The same transfer species as above are involved.

Now, consider the monomer shown below.

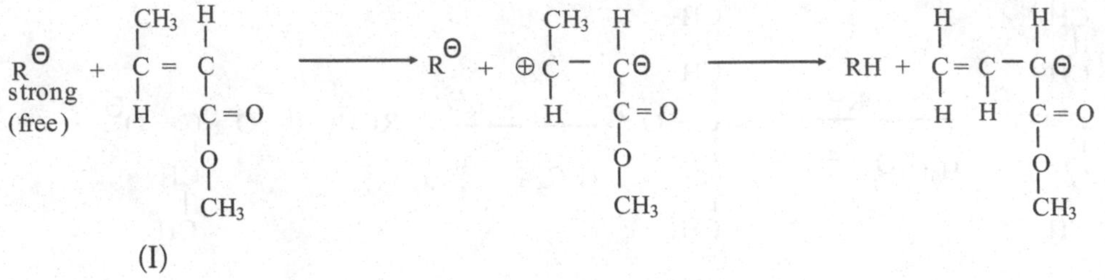

$$R^{\ominus}_{\substack{\text{strong}\\ \text{(free)}}} + \underset{\underset{\overset{|}{C}=O}{\overset{|}{CH_3}}}{\overset{\overset{CH_3}{|}}{C}} = \underset{\overset{|}{H}}{\overset{\overset{H}{|}}{C}} \longrightarrow R^{\ominus} + \underset{\underset{\overset{|}{C}=O}{\overset{|}{O}}}{\overset{\overset{CH_3}{|}}{\oplus C}} - \underset{\overset{|}{CH_3}}{\overset{\overset{H}{|}}{C}}^{\ominus} \longrightarrow RH + \underset{\underset{\overset{|}{C}=O}{\overset{|}{H}}}{\overset{\overset{H}{|}}{C}} = \underset{\overset{|}{H}}{\overset{\overset{H}{|}}{C}} - \underset{\overset{|}{CH_3}}{\overset{\overset{H}{|}}{C}}^{\ominus}$$

(I)   5.139

Why can initiation be prevented in the route natural to the monomer?

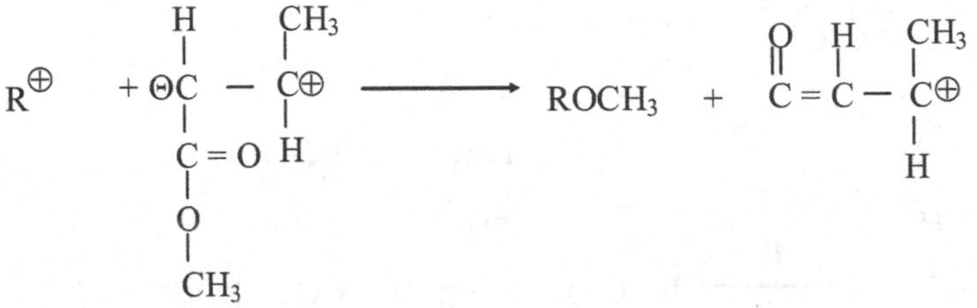

$$R^{\oplus} + \overset{\ominus}{\underset{\underset{\overset{|}{O}}{\overset{|}{C}=O}}{C}}H - \overset{CH_3}{\underset{\overset{|}{H}}{C}}\oplus \longrightarrow ROCH_3 + \overset{O}{\underset{\overset{\parallel}{C}}{}} = \overset{H}{\underset{\overset{|}{H}}{C}} - \overset{CH_3}{\underset{\overset{|}{H}}{C}}\oplus$$

5.140

This route is not natural to the monomer and therefore what observed is to be expected, except that it is the carbonyl center that should be carrying the positive charge in the equation above.

Chargedly, one may think that the monomer cannot be polymerized via C = C activation center, in the absence of molecular rearrangement. Whether molecular rearrangement is allowed or not, only free negatively charged -paired route will be favored as shown below.

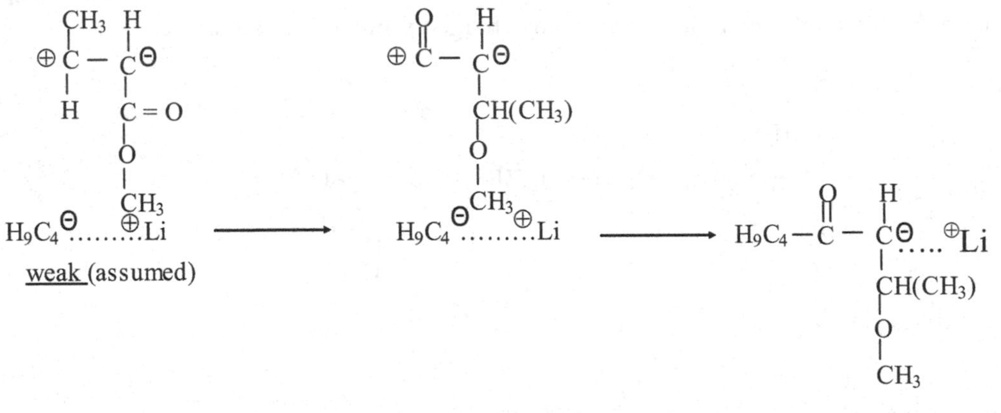

$$\underset{H_9C_4^{\ominus}........\oplus Li}{\underset{\underline{\text{weak}}\ (\text{assumed})}{\overset{\overset{CH_3}{|}}{\oplus C} - \overset{\overset{H}{|}}{C}^{\ominus}}} \longrightarrow \underset{H_9C_4^{\ominus}.........\oplus Li}{\overset{\overset{O}{\parallel}}{\oplus C} - \overset{\overset{H}{|}}{C}^{\ominus}} \longrightarrow H_9C_4 - \overset{O}{\underset{\overset{\parallel}{C}}{}} - \overset{\overset{H}{|}}{C}^{\ominus}..... {}^{\oplus}Li$$

5.141

Whether the initiator is strong or weak, polymerization is favored, except that different chains are produced. Indeed, the only transfer species for this monomer via the C=C center is only OCH$_3$ and not H as shown in Equation 5.139. ***It is probably for this reason, all charged initiators are paired in character and the groups themselves must be largely trans-placed, if initiation is to be favored.*** If cis- placed, no initiation is favored. This (OCH$_3$) is what is abstracted in the route not natural to it, and released in the route natural to it, based on the Laws of conservation of transfer of transfer species.

With activation via C = O activation center, the followings are obtained.

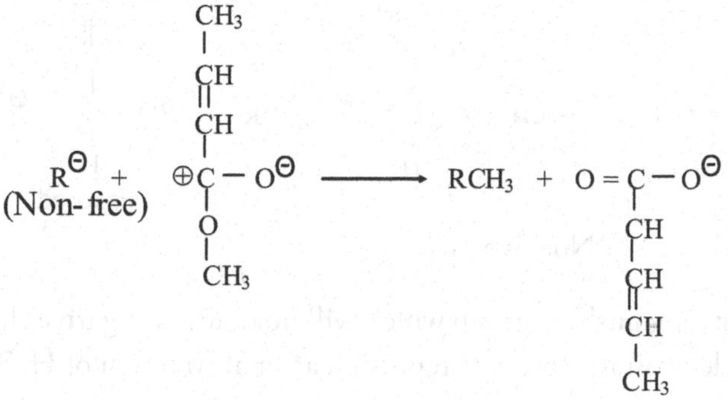

$$
\begin{array}{c}
CH_3 \\
| \\
CH \\
|| \\
CH
\end{array}
$$

$$R^{\ominus} \;+\; {}^{\oplus}C - O^{\ominus} \longrightarrow RCH_3 \;+\; O = C - O^{\ominus}$$

(Non-free), with $O\text{--}CH_3$ substituents on the carbon, giving

$$\qquad\qquad\qquad\qquad\qquad\qquad\qquad 5.142$$

$$R\!-\!(O\!-\!C)_n\,O\!-\!C^{\oplus}\cdots O^{\ominus}\!\cdots B \overset{F}{\underset{F}{\longleftarrow}} F \longrightarrow R\!-\!(O\!-\!C)_n\,O\!-\!C\!=\!O + ROCH_3 + BF_3$$

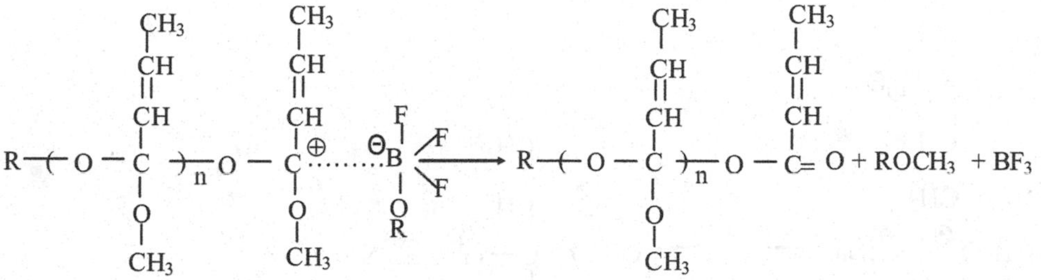

$$\qquad\qquad\qquad\qquad\qquad\qquad\qquad 5.143$$

Positively, there is polymerization via the C=O center. Thus, with the use of strong positively charged-paired initiators, only the C = O activation center will favor its use. It should be noted that $-OCH_3$ is far more radical-pushing than $-CH = CH - CH_3$, though the former is non-free while the latter is free.

With acrylamides, similar analysis as with acrylates above apply, with however strong ionic character. Copolymers cannot be obtained from a single monomer. Between the three cases shown below (I), (II), and (III), C = O of (I) is more nucleophilic than in (II) which in turn is more nucleophilic than in (III), whereas, via the C=C activation center (III) is more electrophilic

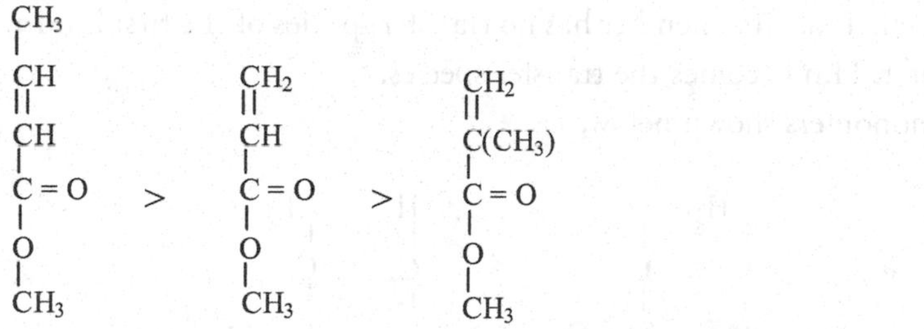

$$
\begin{array}{ccccc}
\begin{array}{c} CH_3 \\ | \\ CH \\ || \\ CH \\ | \\ C=O \\ | \\ O \\ | \\ CH_3 \end{array}
& > &
\begin{array}{c} CH_2 \\ || \\ CH \\ | \\ C=O \\ | \\ O \\ | \\ CH_3 \end{array}
& > &
\begin{array}{c} CH_2 \\ || \\ C(CH_3) \\ | \\ C=O \\ | \\ O \\ | \\ CH_3 \end{array}
\\
(I) & & (II) & & (III)
\end{array}
$$

$$\qquad\qquad\qquad\qquad\qquad\qquad\qquad 5.144$$

(from (I) of   (from methacrylate)   (from methyl methacrylate)
Eqn. 5.139)

Order of nucleophilicity

than (II) which in turn is more electrophilic than (I). This is as a result of the order of pushing capacities of the substituent groups as shown below.

$$CH_3 - CH = CH- \;>\; CH_2 = CH- \;>\; CH_2 = C(CH_3)\,-$$

Order of pushing capacity radically/chargedly

$$\qquad\qquad\qquad\qquad\qquad\qquad\qquad 5.145$$

Considering next alkanyl vinyl aldehydes and ketones, one will begin with acrolein, the first member of alkanyl vinyl aldehydes.

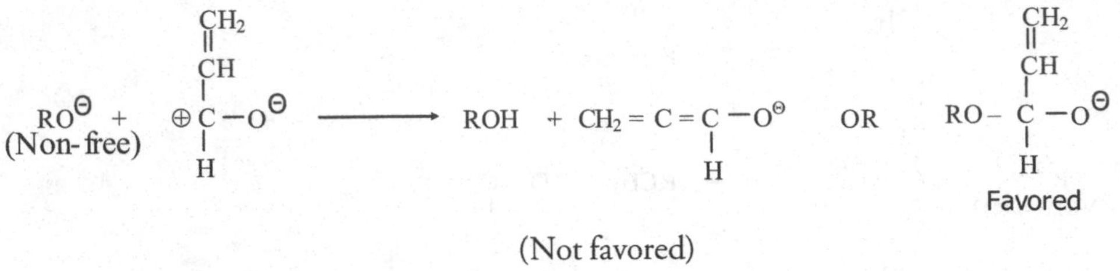

$$
\text{RO}^{\ominus} + \underset{\substack{| \\ \text{H}}}{\overset{\substack{\text{CH}_2 \\ \| \\ \text{CH} \\ |}}{\oplus \text{C} - \text{O}^{\ominus}}} \longrightarrow \text{ROH} + \text{CH}_2 = \text{C} = \underset{\substack{| \\ \text{H}}}{\text{C}} - \text{O}^{\ominus} \quad \text{OR} \quad \text{RO} - \underset{\substack{| \\ \text{H}}}{\overset{\substack{\text{CH}_2 \\ \| \\ \text{CH} \\ |}}{\text{C}}} - \text{O}^{\ominus}
$$

(Non-free)   Favored

                       (Not favored)   5.146

It is not favored, because $=\text{CH}_2$ is a very strong pushing group which will not allow a negative charge to be carried by the C center carrying it, due electrostatic forces of repulsion after abstraction of H. Therefore, initiation as shown above will be favored.

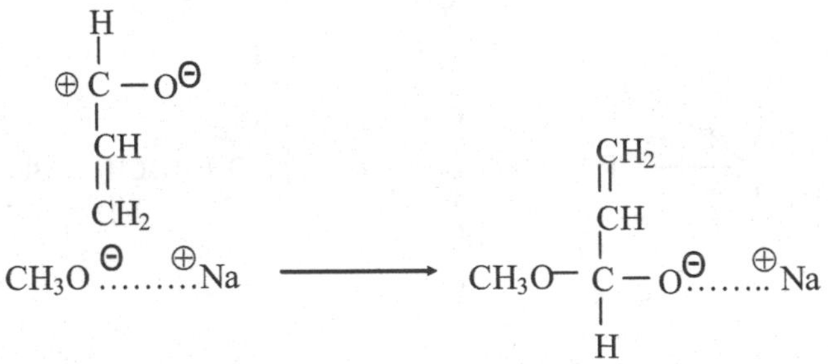

$$
\underset{\substack{| \\ \text{CH} \\ \| \\ \text{CH}_2}}{\overset{\substack{\text{H} \\ |}}{\oplus \text{C} - \text{O}^{\ominus}}} \qquad \text{CH}_3\text{O} \overset{\ominus}{\ldots\ldots} \overset{\oplus}{} \text{Na} \longrightarrow \text{CH}_3\text{O} - \underset{\substack{| \\ \text{H}}}{\overset{\substack{\text{CH}_2 \\ \| \\ \text{CH} \\ |}}{\text{C}}} - \text{O} \overset{\ominus}{\ldots\ldots} \overset{\oplus}{} \text{Na}
$$

                                                                  5.147

Since there is no resonance stabilization, H is more or equal to $-\text{CH} = \text{CH}_2$ in radical-pushing capacity. Thus, C = O activation center will favor the use of non-free-negative media and non-free-negatively charged-paired initiators. It will also favor the use of free-positively charged-paired route. The C = C activation center favors both negatively and positively charged routes, the COH group being a mild pulling group. Hence, copolymers can be obtained from a single monomer using positively-charged-paired initiators. Copolymers cannot be obtained using negatively charged initiators, because for the C= O center, non-free-negatively charged initiator is required, while for the C=C center free-negatively charged initiator is required. This first member has no transfer species of the first kind except when Na.e is used for the C = C center. H.n becomes the transfer species.

Consider the three monomers shown below,

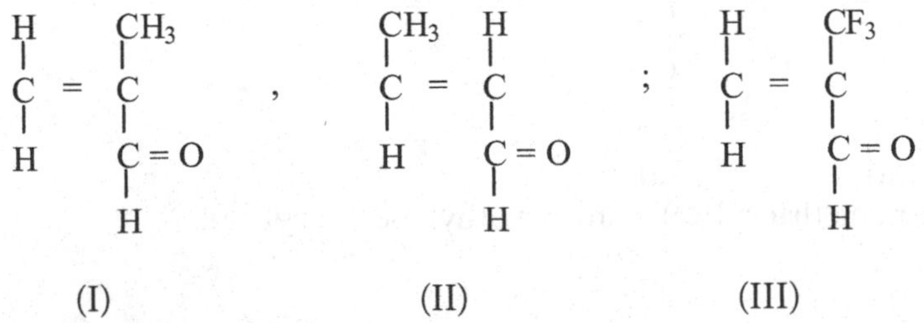

$$
\underset{\substack{| \\ \text{H}}}{\overset{\substack{\text{H} \\ |}}{\text{C}}} = \underset{\substack{| \\ \text{C} = \text{O} \\ | \\ \text{H}}}{\overset{\substack{\text{CH}_3 \\ |}}{\text{C}}} \quad , \quad \underset{\substack{| \\ \text{H}}}{\overset{\substack{\text{CH}_3 \\ |}}{\text{C}}} = \underset{\substack{| \\ \text{C} = \text{O} \\ | \\ \text{H}}}{\overset{\substack{\text{H} \\ |}}{\text{C}}} \quad ; \quad \underset{\substack{| \\ \text{H}}}{\overset{\substack{\text{H} \\ |}}{\text{C}}} = \underset{\substack{| \\ \text{C} = \text{O} \\ | \\ \text{H}}}{\overset{\substack{\text{CF}_3 \\ |}}{\text{C}}} \qquad 5.148
$$

            (I)   (II)   (III)

For (I), via C = C activation center, both routes are favored free-radically and chargedly. Like acrolein, the same analysis is obtained for (I), noting however that (I) via C = O center is less nucleophilic than in acrolein, since $\text{CH}_2 = \text{CH} -$ is more radical-pushing than $\text{CH}_2 = \text{C}(\text{CH}_3)\text{-}$. For (II), the followings are obtained. With negatively-charged-paired initiators, polymerization is favored for this trans-monomer, but not for the cis-type.

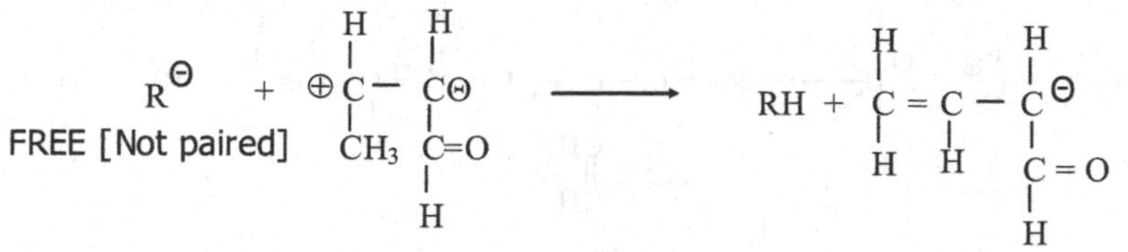

$$R^{\ominus} \quad + \quad \overset{\oplus}{\underset{\underset{H}{\overset{|}{C=O}}}{\underset{CH_3}{\overset{H}{\underset{|}{C}}}}} - \overset{H}{\underset{}{\overset{|}{C}}}{}^{\ominus} \quad \longrightarrow \quad RH \quad + \quad \overset{H}{\underset{H}{\overset{|}{C}}} = \overset{H}{\underset{}{\overset{|}{C}}} - \overset{H}{\underset{\underset{H}{\overset{|}{C=O}}}{\overset{|}{C}}}{}^{\ominus} \qquad 5.149$$

FREE [Not paired]

NOT FAVORED

With free-media-negatively charged initiator as shown above, initiation step is not favored whether cis- or trans-. However, this is not to be expected, the route being natural to it. Fortunately, free-media charged initiators do not exist. With positively charged-paired initiator, the following is supposed to be obtained in the route not natural to it.

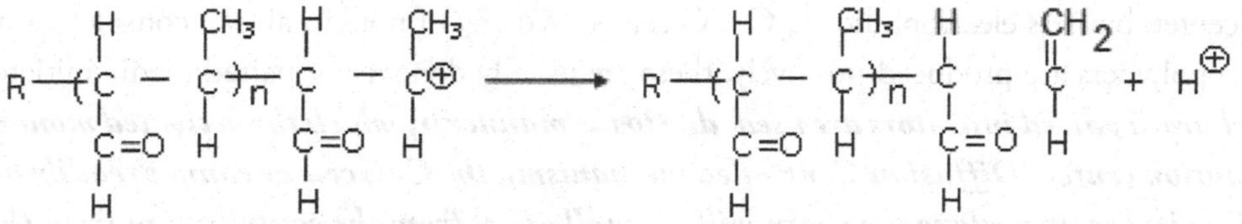

$$R \overset{}{\underset{}{\text{—}}} \!\!\left(\!\! \overset{H}{\underset{\underset{H}{\overset{|}{C=O}}}{\overset{|}{C}}} - \overset{CH_3}{\underset{H}{\overset{|}{C}}} \!\!\right)_{\!\!n}\!\! \overset{H}{\underset{\underset{H}{\overset{|}{C=O}}}{\overset{|}{C}}} - \overset{CH_3}{\underset{H}{\overset{|}{C}}}{}^{\oplus} \quad \longrightarrow \quad R \overset{}{\underset{}{\text{—}}} \!\!\left(\!\! \overset{H}{\underset{\underset{H}{\overset{|}{C=O}}}{\overset{|}{C}}} - \overset{CH_3}{\underset{H}{\overset{|}{C}}} \!\!\right)_{\!\!n}\!\! \overset{H}{\underset{\underset{H}{\overset{|}{C=O}}}{\overset{|}{C}}} - \overset{CH_2}{\underset{H}{\overset{||}{C}}} \quad + \quad H^{\oplus}$$

<u>NOT FAVORED</u>   5.150a

With polymerization being favored positively-paired via $C = C$ activation center, it may seem that the monomer is a nucleophile. How can transfer species of the first kind of the first type be released in the route not natural to the monomer? How? It is impossible. Though the route is favored, the transfer species released is of the second kind as shown below.

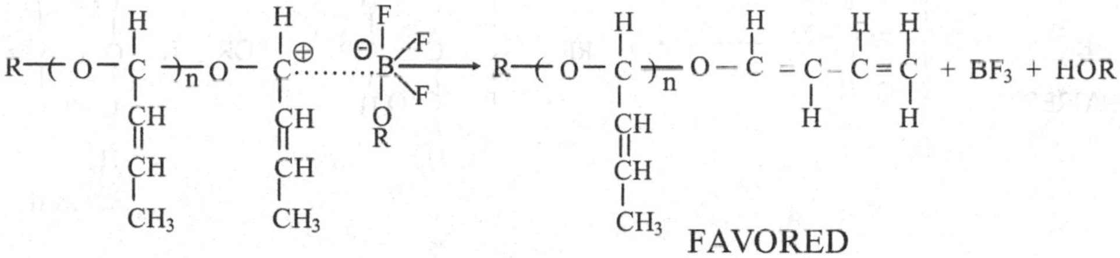

$$R \overset{}{\underset{}{\text{—}}} \!\!\left(\!\! O - \overset{H}{\underset{\underset{\underset{CH_3}{|}}{\overset{||}{CH}}}{\overset{|}{C}}} \!\!\right)_{\!\!n}\!\! O - \overset{H}{\underset{\underset{\underset{CH_3}{|}}{\overset{||}{CH}}}{\overset{\oplus}{C}}} \cdots\cdots {\overset{\ominus}{\underset{\underset{R}{\overset{|}{O}}}{B}}}\!\!\overset{\nearrow F}{\underset{\searrow F}{\text{—}F}} \longrightarrow R \overset{}{\underset{}{\text{—}}} \!\!\left(\!\! O - \overset{H}{\underset{\underset{\underset{CH_3}{|}}{\overset{||}{CH}}}{\overset{|}{C}}} \!\!\right)_{\!\!n}\!\! O - \overset{H}{\underset{}{\overset{|}{C}}} = \overset{H}{\underset{H}{\overset{|}{C}}} - \overset{H}{\underset{H}{\overset{|}{C}}} \quad + \quad BF_3 \quad + \quad HOR$$

FAVORED

                5.150b

The dead terminal double bond polymer here is the one that bears the same character as the originating monomer, not the one obtained in the last equation. This clearly sends a message, that which is the fact that the reaction of Equation 5.149 is not favored when cis-placed, but favored when trans-placed, only when the initiator is paired. It will not take place and does not exist, if Free-negatively charged initiator such as $^{\ominus}C_4H_9$ can exist in isolation. It must be paired with its positive counterpart. Hence when an initiator such as $H_9C_4{}^{\ominus}....{}^{\oplus}Li$, is used, the natural route is favored. The polymer produced is a living one with no transfer species and no possibility of producing copolymers. The negatively charged chain must be killed with terminating agents. This unfortunately is done by impurities to give polymers with broad distribution of molecular weights. Imagine the harm we do to our world when the mechanisms of how NATURE operates are unknown. Yet, we think we know, because of our desire for FAME, WEALTH and RECOGNITION! All these mean nothing in the sight of HIM. HE is an infinite bundle of KNOWLEDGE. Indeed, we have only just started.

With positively charged-paired initiator, for the monomer, in addition to the reaction of Equation 5.150b, the following is obtained.

$$R-(O-\underset{\underset{\underset{CH_3}{|}}{\overset{\overset{CH}{\|}}{CH}}}{\overset{\overset{H}{|}}{C}})_n O-\underset{\underset{\underset{CH_3}{|}}{\overset{\overset{CH}{\|}}{CH}}}{\overset{\overset{H}{|}}{C^{\oplus}}} \cdots \overset{\ominus}{B}\underset{\underset{R}{\overset{\overset{F}{|}}{O}}}{\overset{F}{\diagup}}F \longrightarrow R-(O-\underset{\underset{\underset{CH_3}{|}}{\overset{\overset{CH}{\|}}{CH}}}{\overset{\overset{H}{|}}{C}})_n O-\underset{}{\overset{\overset{H}{|}}{C}} = C - C = \underset{}{\overset{\overset{H}{|}}{C}} + BF_3 + HOR$$

FAVORED   5.151

Transfer species is released here, since it is the same abstracted from that center when negatively attacked, unlike with acrolein. Thus, ***dead terminal resonance stabilized polymer*** shown in Equation 5.151 is produced. Like (I) of Equation 5.148, copolymers cannot be obtained for this monomer negatively as already said. Largely positively charged polymerization of C = O and C = C centers (to give copolymers) and free-negatively-charged polymerization of C = C center are favored. (II) of Equation 5.148 obviously is more nucleophilic than (I) via C = O center, but less electrophilic via C = C center. Worthy of note in all the considerations so far, is that when copolymers are produced positively, these are usually done using paired-media initiators. ***When positively charged paired initiators are used, due to the manner by which the activated monomer enters the coordination center [Diffusion Controlled mechanism], the C=C center cannot readily be reached, and therefore its presence along the chain will be small apart from the contribution from the fact that the route is not natural to the monomer. Copolymers can be better obtained if free media initiators are used for (I), acrolein and not (II). Though this cannot be found chargedly, the use of dilute H_2SO_4 which is electro-free-radical when used can provide better copolymers wherein the C=C center can only be alternating placed with C=O and never alone.*** From all the considerations so far, one can observe the LOGICAL character of CHEMISTRY too much and complex to comprehend.

$$R^{\oplus} + \overset{\ominus}{\underset{\underset{\underset{H}{|}}{\overset{\overset{CF_3}{|}}{C}}}{C}} - \overset{\overset{H}{|}}{\underset{H}{C^{\oplus}}} \longrightarrow RF + \underset{\underset{\underset{H}{|}}{\overset{\overset{F}{|}}{C}}}{\overset{\overset{F}{|}}{C}} = C - \overset{\overset{H}{|}}{\underset{H}{C^{\oplus}}} \quad OR \quad R - \overset{\overset{CF_3}{|}}{\underset{\underset{\underset{H}{|}}{\overset{\overset{}{C=O}}{}}}{C}} - \overset{\overset{H}{|}}{\underset{H}{C^{\oplus}}}$$

[PAIRED]   5.152

Favored

$$R-(\overset{\overset{H}{|}}{\underset{H}{C}} - \overset{\overset{CF_3}{|}}{\underset{\underset{\underset{H}{|}}{\overset{C=O}{}}}{C}})_n \overset{\overset{H}{|}}{\underset{H}{C}} - \overset{\overset{CF_3}{|}}{\underset{\underset{\underset{H}{|}}{\overset{C=O}{}}}{C^{\ominus}}} \longrightarrow R-(\overset{\overset{H}{|}}{\underset{H}{C}} - \overset{\overset{CF_3}{|}}{\underset{\underset{\underset{H}{|}}{\overset{C=O}{}}}{C}})_n \overset{\overset{H}{|}}{\underset{H}{C}} - \overset{}{\underset{}{C}} = \overset{\overset{F}{|}}{\underset{F}{C}} + F^{\ominus}$$

(Free end)   (Not favored)   5.153

In the first reaction above, when positively charged-paired initiator is used, initiation is favored. Propagation definitely follows to give a growing polymer chain which cannot be killed from within. The reason why initiation is favored is because CHO group is more pulling than CF_3 group. CHO group with no transfer species is the component entering the coordination center. The second reaction is favored since there is no transfer species of the first kind of the first type to reject for this monomer. By this, the law of Conservation of transfer of transfer species has not been broken. Living polymer chains will be obtained positively and negatively when paired initiator are used. The law as stated is fully in place for all types of paired initiators when applied. Thus, via C = C center, the monomer is almost like (II) since copolymers can be obtained with paired-media positively charged initiator.

$$R^{\ominus} \; + \; \underset{H}{\overset{CH_2}{\underset{|}{\overset{\|}{\underset{C-O^{\ominus}}{\overset{C(CF_3)}{|}}}}}} \longrightarrow RCF_3 \; + \; \underset{H}{\overset{CH_2}{\underset{|}{\overset{\|}{\underset{C=O}{\overset{C^{\ominus}}{|}}}}}} \qquad Or \qquad ADDITION \;(Favored)$$

(I)   Impossible existence   5.154

Via the C=O center, initiation is favored, since the negative charge cannot be carried by the center. Positively, the followings are obtained.

$$R \underset{CH_2}{\underset{\|}{\underset{C(CF_3)}{|}}}\left(O - \underset{}{\overset{H}{\underset{|}{C}}} \right)_n O - \underset{CH_2}{\underset{\|}{\underset{C(CF_3)}{|}}} \overset{H}{\underset{|}{C^{\oplus}}} \longrightarrow R\left(O - \overset{H}{\underset{|}{C}}\right) O - \underset{CH_2}{\underset{\|}{\underset{C(CF_3)}{|}}} C = C = \overset{H}{\underset{|}{C}} \;\; + \;\; (+)CF_3$$

[NOT FAVORED]   5.155

It is a living polymer that is produced here. It cannot be killed from within. The dead terminal double bond polymer produced above has no semblance with the originating monomer. *Negatively, like all the others, no copolymers can be obtained, even when the centers favor the route, because the negatively charged paired route is natural to the monomer. The anionic route is not natural to the monomer. It is only via the route which is not natural to the monomer that copolymers are obtained from a single monomer.* Positively with paired media initiators, copolymers can be obtained for all of them. (I), (II), and (III) are all Electrophiles of different capacities, (III) > (I) > (II) in that order. This will be fully confirmed when radical initiators are involved.

So far, largely olefinic cases where the aldehydic features are present have been considered. Considering those in which ketonic features are present, one will begin with methyl vinyl ketone.

$$R^{\ominus} \; + \; \underset{\underset{CH_3}{\overset{|}{C=O}}}{\overset{H}{\underset{|}{C}}} = \overset{H}{\underset{}{C}} \longrightarrow R - \underset{H}{\overset{H}{\underset{|}{C}}} - \underset{\underset{CH_3}{\overset{|}{C=O}}}{\overset{H}{\underset{|}{C^{\ominus}}}} \xrightarrow{\;+\;n(I)\;}$$

(I)

$$R\left(\underset{H}{\overset{H}{\underset{|}{C}}} - \underset{\underset{CH_3}{\overset{|}{C=O}}}{\overset{H}{\underset{|}{C}}}\right)_n \underset{H}{\overset{H}{\underset{|}{C}}} - \underset{\underset{CH_3}{\overset{|}{C=O}}}{\overset{H}{\underset{|}{C^{\ominus}}}} \longrightarrow \text{No transfer species}$$

           5.156

$$R^{\oplus} \; + \; \underset{\underset{CH_3}{\overset{|}{C=O}}}{\overset{}{\underset{|}{{}^{\ominus}C}}} - \overset{H}{\underset{}{C^{\oplus}}} \longrightarrow R - \underset{\underset{CH_3}{\overset{|}{C=O}}}{\overset{H}{\underset{|}{C}}} - \overset{H}{\underset{}{C^{\oplus}}} \xrightarrow{\;+\;n(I)\;} R\left(\underset{\underset{CH_3}{\overset{|}{C=O}}}{\overset{H}{\underset{|}{C}}} - \overset{H}{\underset{}{C}}\right)_n \underset{\underset{CH_3}{\overset{|}{C=O}}}{\overset{H}{\underset{|}{C}}} - \overset{H}{\underset{}{C^{\oplus}}}$$

(I)

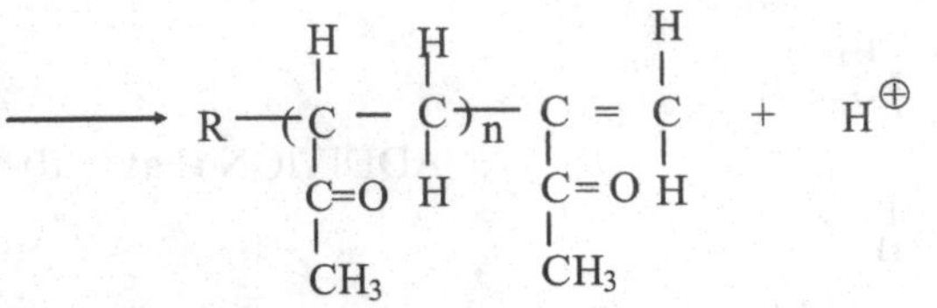

$$\longrightarrow R-(\underset{\underset{CH_3}{|}}{\overset{H}{\underset{|}{C}}} - \underset{H}{\overset{H}{\underset{|}{C}}})_n \underset{\underset{CH_3}{|}}{\overset{H}{\underset{|}{C}}} = \underset{H}{\overset{H}{\underset{|}{C}}} \quad + \quad H^{\oplus}$$

$$5.157$$

Transfer Species of the Second kind

Though both routes are favored, the monomer by virtue of the presence of a unique pulling group, is an electrophile. Hence, transfer species of the second kind is involved in the last reaction above, noting that the positive route is not natural to the C =C center, but to C = O center.

Considering activation of the C = O center, the followings are obtained.

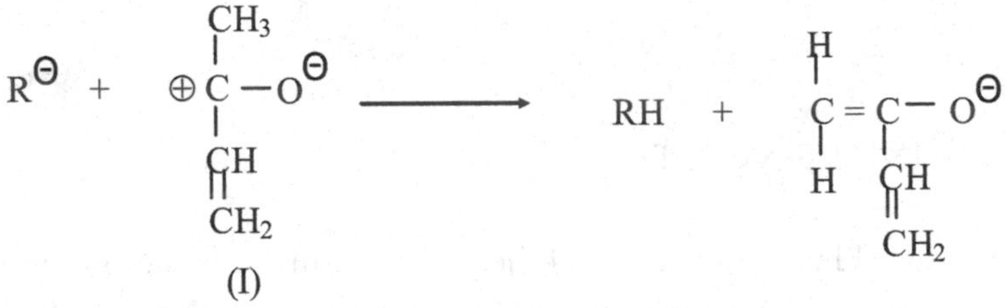

$$R^{\ominus} \quad + \quad \oplus\underset{\underset{CH_2}{\overset{\underset{||}{CH}}{|}}}{\overset{\overset{CH_3}{|}}{C}} - O^{\ominus} \quad \longrightarrow \quad RH \quad + \quad \underset{\underset{CH_2}{\overset{\underset{||}{CH}}{|}}}{\overset{H}{\underset{H}{C}}} = \underset{}{\overset{}{C}} - O^{\ominus}$$

$$(I)$$

$$5.158$$

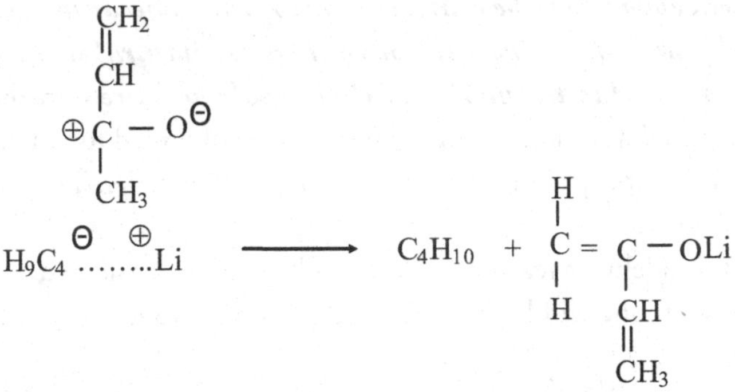

$$\underset{H_9C_4 \overset{\ominus}{}\overset{\oplus}{\dots\dots Li}}{\oplus\underset{\underset{CH_3}{|}}{\overset{\overset{\underset{|}{CH}}{\overset{||}{CH_2}}}{C}} - O^{\ominus}} \quad \longrightarrow \quad C_4H_{10} \quad + \quad \underset{\underset{CH_3}{\overset{\underset{||}{CH}}{|}}}{\overset{H}{\underset{H}{C}}} = \underset{}{\overset{}{C}} - OLi$$

$$5.159$$

$$R-(O-\underset{\underset{CH_2}{\overset{\underset{||}{CH}}{|}}}{\overset{\overset{CH_3}{|}}{C}})_n O - \underset{\underset{CH_3}{\overset{\underset{||}{CH}}{|}}}{\overset{\overset{CH_3}{|}}{C}}\oplus \quad \longrightarrow \quad R-(O-\underset{\underset{CH_2}{\overset{\underset{||}{CH}}{|}}}{\overset{\overset{CH_3}{|}}{C}})_n O - \underset{\underset{CH_3}{\overset{\underset{||}{CH}}{|}}}{\overset{\overset{CH_3}{|}}{C}} = \underset{H}{\overset{H}{\underset{|}{C}}} \quad + \quad H^{\oplus}$$

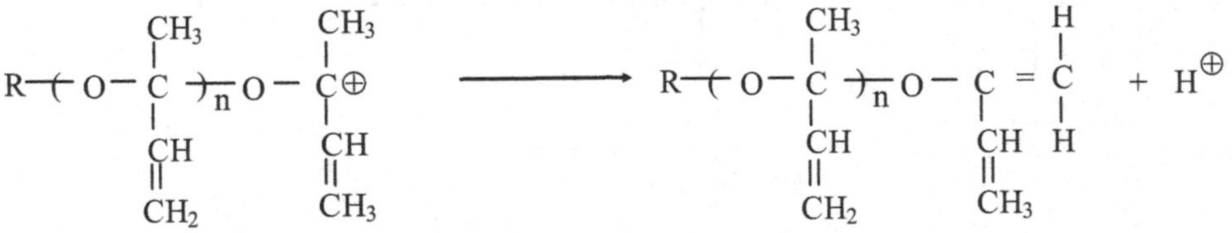

$$5.160$$

The ketonic character of the monomer (I) of Equation 5.158, is largely due to the presence of two carbon centered groups with transfer species of the first kind, while the aldehydic character of a monomer in these families, is absence of transfer species in one of the two substituted groups or indeed presence of H or F. Thus, while presence of copolymers can be favored positively with paired-media initiators such as $F_2B^{\oplus}.....^{\ominus}BF_4$, it is not favored negatively and indeed can never be favored by them. Hence, methyl vinyl ketone can be said to be more nucleophilic than acrolein as shown below via C = O center.

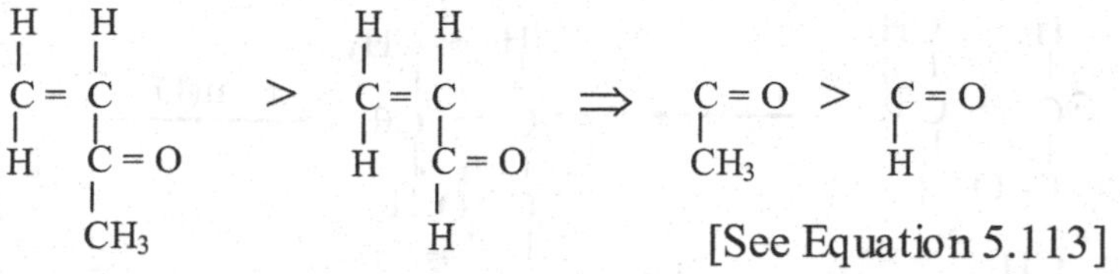

$$\begin{array}{ccc}
\underset{\underset{\displaystyle \underset{\displaystyle CH_3}{|}}{\overset{\displaystyle |}{C=O}}}{\overset{\displaystyle \overset{H}{|}}{\underset{|}{C}}=\overset{\displaystyle \overset{H}{|}}{\underset{|}{C}} & > & \underset{\underset{\displaystyle \underset{\displaystyle H}{|}}{\overset{\displaystyle |}{C=O}}}{\overset{\displaystyle \overset{H}{|}}{\underset{|}{C}}=\overset{\displaystyle \overset{H}{|}}{\underset{|}{C}}
\end{array}$$

$$\Rightarrow \quad \underset{CH_3}{\overset{|}{C}}=O \; > \; \underset{H}{\overset{|}{C}}=O \qquad \text{[See Equation 5.113]} \qquad 5.161$$

(I) (more nucleophilic CO center) (II) (more radical-pulling capacity)

In view of this realization, then $COCH_3$ is more pulling than COH as indicated by (II) above. Based on the inequality in (II) above, the first monomer is more electrophilic than the second monomer. The conclusion above is obvious, since ketones are stronger nucleophiles than aldehydes.

Considering the three monomers shown below, for (I), via $C = C$ activation center, both

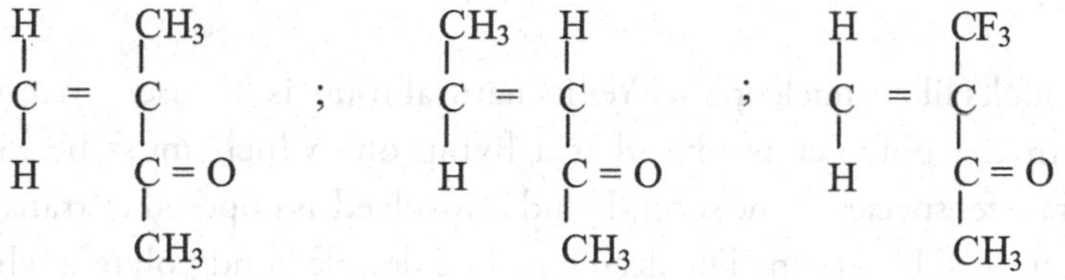

$$
\underset{\underset{\displaystyle CH_3}{|}}{\underset{C=O}{\overset{\displaystyle \overset{H}{|}}{C}}} = \underset{\displaystyle \overset{CH_3}{|}}{\overset{}{C}} \quad ; \quad
\underset{\underset{\displaystyle CH_3}{|}}{\underset{C=O}{\overset{\displaystyle \overset{CH_3}{|}}{C}}} = \underset{\displaystyle \overset{H}{|}}{\overset{}{C}} \quad ; \quad
\underset{\underset{\displaystyle CH_3}{|}}{\underset{C=O}{\overset{\displaystyle \overset{H}{|}}{C}}} = \underset{\displaystyle \overset{CF_3}{|}}{\overset{}{C}}
\qquad 5.162
$$

(I)   (II)   (III)

charged routes are favored. Like methyl vinyl ketone, the same analysis is obtained for (I), not-ing that the $C = O$ center in (I) just like in the others is more nucleophilic than that in methyl vinyl ketone. With (II), the followings are obtained, using an isolated negatively charged initiator which does not exist.

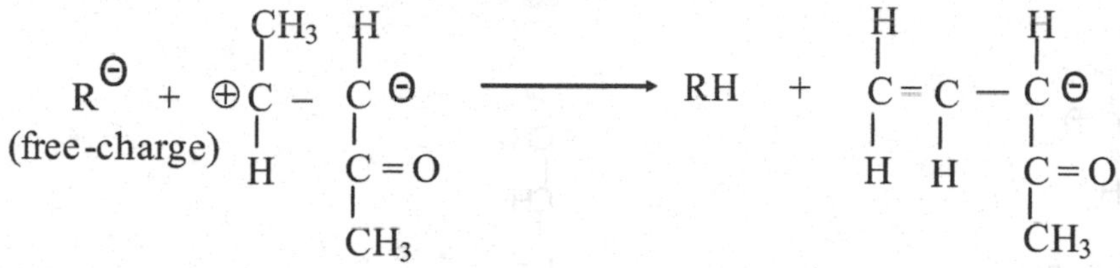

$$
R^{\ominus} + {}^{\oplus}\underset{\underset{\displaystyle CH_3}{|}}{\underset{C=O}{\overset{\displaystyle \overset{CH_3}{|}}{C}}} - \underset{\displaystyle \overset{H}{|}}{\overset{}{C}}{}^{\ominus} \longrightarrow RH + \underset{\underset{\displaystyle CH_3}{|}}{\underset{C=O}{\overset{\displaystyle \overset{H}{|}}{C}}} = \underset{\displaystyle \overset{H}{|}}{\overset{}{\underset{|}{C}}} - \underset{}{\overset{}{C}}{}^{\ominus}
$$

(free-charge)   5.163

This is not to be expected, since the monomer is an Electrophile. This is the nature of these groups of monomers where the placement of groups is important (Cis- and Trans- placements).

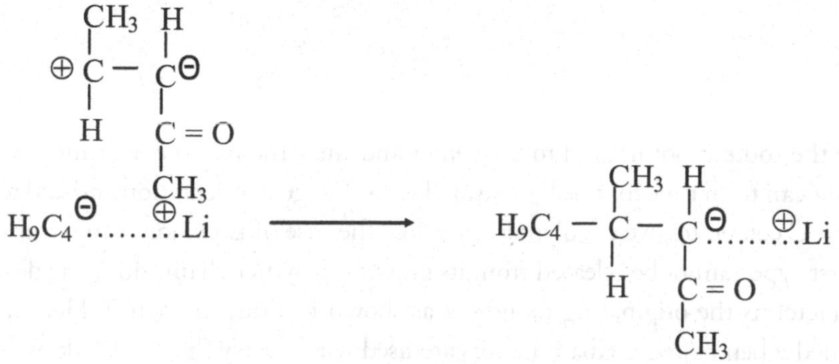

$$
\begin{array}{c}
{}^{\oplus}\underset{\underset{\displaystyle CH_3}{|}}{\underset{C=O}{\overset{\displaystyle \overset{CH_3}{|}}{C}}} - \underset{}{\overset{\displaystyle \overset{H}{|}}{C}}{}^{\ominus} \\
H_9C_4{}^{\ominus}\ldots\ldots{}^{\oplus}Li
\end{array}
\longrightarrow
\begin{array}{c}
H_9C_4 - \underset{\underset{\displaystyle CH_3}{|}}{\underset{C=O}{\overset{\displaystyle \overset{CH_3}{|}}{C}}} - \underset{}{\overset{\displaystyle \overset{H}{|}}{C}}{}^{\ominus}\ldots\ldots{}^{\oplus}Li
\end{array}
\qquad 5.164
$$

$$R^{\oplus} + {}^{\ominus}\overset{\underset{\displaystyle C=O}{|}}{\underset{\underset{\displaystyle CH_3}{|}}{C}}\overset{\displaystyle H}{-}\,C^{\oplus}\overset{\displaystyle CH_3}{\underset{\displaystyle H}{}} \longrightarrow R-\overset{\displaystyle H}{\underset{\underset{\displaystyle CH_3}{|}}{\underset{\displaystyle C=O}{|}}{C}}-C^{\oplus}\overset{\displaystyle CH_3}{\underset{\displaystyle H}{}} \quad \xrightarrow{\;+\; n(I)\;}$$

$$(I)$$

$$R-\!\left(\!\overset{\displaystyle H}{\underset{\underset{\displaystyle CH_3}{|}}{\underset{\displaystyle C=O}{|}}{C}}-\overset{\displaystyle CH_3}{\underset{\displaystyle H}{C}}\!\right)_{\!\! n}\!\overset{\displaystyle H}{\underset{\underset{\displaystyle CH_3}{|}}{\underset{\displaystyle C=O}{|}}{C}}-C^{\oplus}\overset{\displaystyle CH_3}{\underset{\displaystyle H}{}} \longrightarrow R-\!\left(\!\overset{\displaystyle H}{C}-\overset{\displaystyle CH_3}{C}\!\right)_{\!\! n}\!C=C + H^{\oplus} \qquad 5.165$$

Via C = C center, the monomer looks like a nucleophile. Yet its natural route is the use of free-negatively charged-paired initiator wherein the polymer produced is a living one which must be killed with terminating agents. Positively, transfer species of the second kind is involved as opposed to transfer species of the first kind which is there on the CH_3 group. The dead terminal double bond polymer which is of a semblance to the original monomer can be used as a branching site when specific types of initiators radical in character are used. This can only take place when paired initiators are used, but not for branching due to steric limitations. Nevertheless, (II) of Equation 5.162 can be observed to be more nucleophilic than (I). Hence, with positively charged-paired initiator, it can be seen that the C = O center in (II) will be more nucleophilic than in (I). That in (III) of Equation 5.162 will be more nucleophilic than that in (II), since $H_2C=C(CF_3)$- is more charge-pushing than $(CH_3)HC=CH$-. Radically, the reverse is the case. Positively, the C = O center will be more attacked than the C = C center.

$$\qquad\qquad\qquad 5.166$$

In view of the fact that the C = O center is the first to gain entry into the coordination center when C = C is first activated, since the route is not natural to the center and since the monomer cannot be resonance stabilized, the C = O center can from time to time be activated with change in orientation and add to the chain along with that of the C = C center, to give copolymers. Just like the case of Equation 5.165, transfer species of the first kind of the first type cannot be released from its growing polymer chain, since the dead polymer does not have same character as the originating monomer as shown in Equation 5.160. Hence, positively, copolymers can be obtained when paired media initiators are used with largely the C = O along the chain.

For the third monomer (III), the followings are to be expected.

$$R^{\oplus} \ + \ {}^{\ominus}\overset{\overset{\displaystyle CF_3}{|}}{\underset{\underset{\displaystyle CH_3}{|}}{\underset{\displaystyle C=O}{C}}} - \overset{\overset{\displaystyle H}{|}}{\underset{\displaystyle H}{C^{\oplus}}} \ \longrightarrow \ RF \ + \ \overset{\overset{\displaystyle F}{|}}{\underset{\underset{\displaystyle CH_3}{|}}{\underset{\displaystyle C=O}{C}}} = \overset{F}{C} - \overset{\overset{\displaystyle H}{|}}{\underset{\displaystyle H}{C^{\oplus}}} \qquad \text{OR} \qquad \text{ADDITION}$$

(Not favored)   (Favored)

5.167

It is said to be favored, because H_3CCO group is the first to gain entry into the coordination center, instead of CF_3, being more charged-pulling in character. With positively paired initiator, it is the $C=O$ center that is more involved with favored initiation. With free-negatively charged paired initiator, the followings are to be expected.

$$R \!-\!\!\left(\!\overset{\overset{\displaystyle H}{|}}{\underset{\displaystyle H}{C}} - \overset{\overset{\displaystyle CF_3}{|}}{\underset{\underset{\displaystyle CH_3}{|}}{\underset{\displaystyle C=O}{C}}}\!\right)_{\!\!n}\!\! \overset{\overset{\displaystyle H}{|}}{\underset{\displaystyle H}{C}} - \overset{\overset{\displaystyle CF_3}{|}}{\underset{\underset{\displaystyle CH_3}{|}}{\underset{\displaystyle C=O}{C^{\ominus}}}} \ \longrightarrow \ R \!-\!\!\left(\!\overset{\overset{\displaystyle H}{|}}{\underset{\displaystyle H}{C}} - \overset{\overset{\displaystyle CF_3}{|}}{\underset{\underset{\displaystyle CH_3}{|}}{\underset{\displaystyle C=O}{C}}}\!\right)_{\!\!n}\!\! \overset{\overset{\displaystyle H}{|}}{\underset{\displaystyle H}{C}} - \overset{F}{C} = \overset{\overset{\displaystyle F}{|}}{\underset{\underset{\displaystyle CH_3}{|}}{\underset{\displaystyle C=O}{C}}} F \ + \ F^{\ominus}$$

(Release of F is not favored)

5.168

Via $C = C$ activation center, the monomer is a strong electrophile. Via $C = O$ activation center, the followings are to be expected.

$$R^{\ominus} \ + \ {}^{\oplus}\overset{\overset{\displaystyle CH_3}{|}}{\underset{\underset{\displaystyle CH_2}{\|}}{\underset{\displaystyle C(CF_3)}{C}}} - O^{\ominus} \ \longrightarrow \ RH \ + \ \overset{\overset{\displaystyle H}{|}}{\underset{\displaystyle H}{C}} = \overset{\overset{\displaystyle}{|}}{\underset{\underset{\displaystyle CH_2}{\|}}{\underset{\displaystyle C(CF_3)}{C}}} - O^{\ominus} \qquad \text{OR ADDITION (FAVORED)}$$

5.169

$$\begin{array}{c} {}^{\oplus}\overset{\overset{\overset{\displaystyle CH_2}{\|}}{C(CF_3)}}{\underset{\underset{\displaystyle CH_3}{|}}{C}} - O^{\ominus} \\ \\ H_9C_4 {}^{\ominus} \ldots\ldots {}^{\oplus}Li \end{array} \ \longrightarrow \ H_{10}C_4 \ + \ \overset{\overset{\displaystyle H}{|}}{\underset{\displaystyle H}{C}} = \overset{}{\underset{\underset{\underset{\displaystyle CH_2}{\|}}{\underset{\displaystyle C(CF_3)}{}}}{C}} - OLi$$

[Not favored]

5.170

$$R \!-\!\!\left(\! O - \overset{\overset{\displaystyle CH_3}{|}}{\underset{\underset{\displaystyle CH_2}{\|}}{\underset{\displaystyle C(CF_3)}{C}}}\!\right)_{\!\!n}\!\! O - \overset{\overset{\displaystyle CH_3}{|}}{\underset{\underset{\displaystyle CH_2}{\|}}{\underset{\displaystyle C(CF_3)}{C^{\oplus}}}} \ \longrightarrow \ R \!-\!\!\left(\! O - \overset{\overset{\displaystyle CH_3}{|}}{\underset{\underset{\displaystyle CH_2}{\|}}{\underset{\displaystyle C(CF_3)}{C}}}\!\right)_{\!\!n}\!\! O - \overset{}{\underset{\underset{\underset{\displaystyle CH_2}{\|}}{\underset{\displaystyle C(CF_3)}{}}}{C}} = \overset{\overset{\displaystyle H}{|}}{\underset{\displaystyle H}{C}} \ + \ H^{\oplus}$$

5.171

Like (III) of Equation 5.148 (vinyl aldehydes), copolymers can be produced positively chargedly for this third case here, **which by the new definition provided for aldehydes and ketones,** is a ketone with two transfer species- H on CH_3 and F on CF_3. This third case is the most Electro-philic of the three cases of Equation 4.162, followed by (I) and followed by (II).

Finally, it will be interesting to consider few olefinic cases when the C = C and C = O activation centers are isolatedly placed. For this purpose, the following monomers will be of interest.

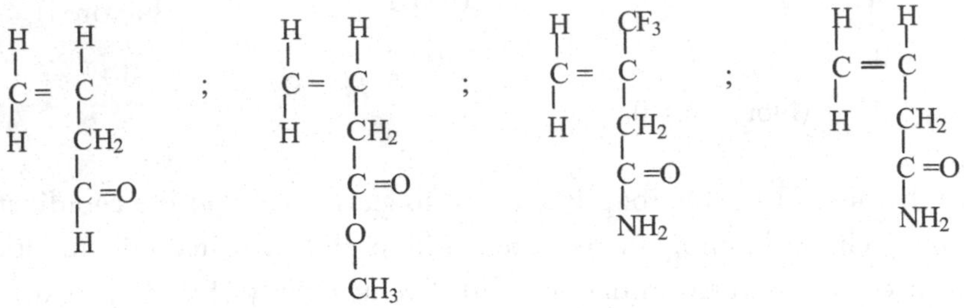

$$
\begin{array}{c}
\underset{H}{\overset{H}{|}}C = \underset{CH_2}{\overset{H}{|}}C \quad ; \quad
\underset{H}{\overset{H}{|}}C = \underset{CH_2}{\overset{H}{|}}C \quad ; \quad
\underset{H}{\overset{H}{|}}C = \underset{CH_2}{\overset{CF_3}{|}}C \quad ; \quad
\underset{H}{\overset{H}{|}}C = \underset{CH_2}{\overset{H}{|}}C
\end{array}
$$

(I) $\quad\quad$ (II) $\quad\quad$ (III) $\quad\quad$ (IV)

5.172

While (I) is an aldehyde in character, the others are "Acetic or amidic" in character via C = O activation center. Via C = C activation centers, if activation chargedly is possible, the COH, COOCH$_3$, and CONH$_2$ groups cannot be used as transfer species of the first kind positively. Their centers cannot carry negative charges. **In fact, they cannot be activated chargedly.** Via C = O activation center which is more nucleophilic, (I) is a nucleophile with CH$_2$ = CH- as transfer species. With (II), (III) & (IV), different transfer species are involved as shown below using non-paired-non-free negatively charged initiator (e.g. CH$_3$O$^{\ominus}$ i.e. an anion).

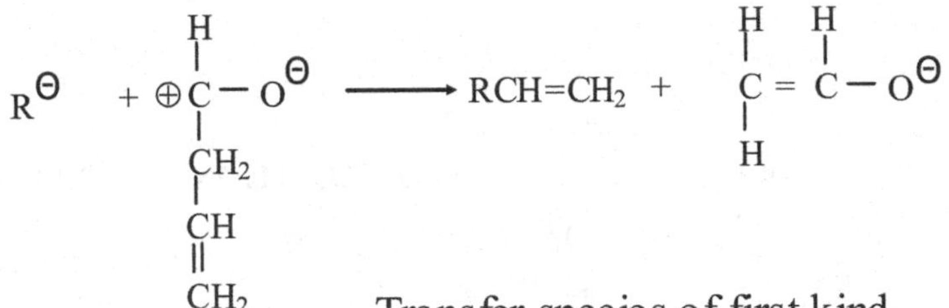

$$ R^{\ominus} \;+\; {}^{\oplus}\underset{\overset{|}{CH_2}}{\overset{\overset{|}{H}}{C}} - O^{\ominus} \longrightarrow RCH=CH_2 \;+\; \underset{\overset{|}{H}}{\overset{\overset{|}{H}\;\;\overset{|}{H}}{C = C}} - O^{\ominus} $$

with CH$_2$ and CH (double bond) and CH$_2$ chain.

Transfer species of first kind

5.173

Transfer species of first kind

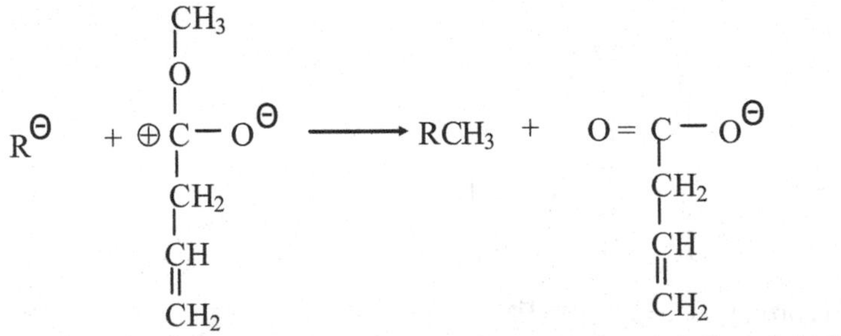

5.174

Transfer species of first kind

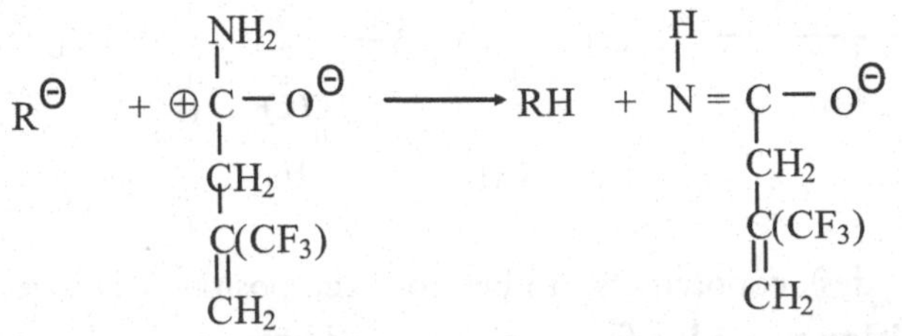

5.175

$$R^{\oplus} \quad + \quad \overset{\overset{\displaystyle NH_2}{|}}{\underset{\underset{\displaystyle CH_2}{\overset{\displaystyle |}{\underset{\displaystyle CH_2}{\|}}}}{\oplus C - O^{\ominus}}} \quad \longrightarrow \quad \overset{\overset{\displaystyle NH_2}{|}}{\underset{\underset{\displaystyle CH_2}{\overset{\displaystyle |}{\underset{\displaystyle CH_2}{\|}}}}{\oplus C - O - R}}$$

5.176

Indeed none of the reactions above take place, since the C – O center is not the center activated negatively or positively, C = O center being more nucleophilic than the C = C center. When they are isolatedly placed, only one center can be activated if they are of different nucleophilicities and that center is the C = C center, the less nucleophilic center. This is unlike when the centers are conjugatedly placed when not resonance stabilized and different from when they are reso-nance stabilized. If activation had been possible via the C = C center, the negatively charged route would have been favored. This is the route not natural to the monomer. The positively charged route could also have been favored. Hence the monomer cannot be activated chargedly. The C = O center is obviously more nucleophilic, since the natural route is favored, but however can never be used.

Thus for all the monomers above existence of copolymers via intermolecular addition is obviously impossible chargedly. *It is important to note that all the substituent groups on the monomers via C = C center are charged-pulling groups of different capacities,* noting for example that -CH$_3$ is less pushing than -CH$_2$ -CH – CH$_2$ group in the presence of resonance stabilization.

So far it can be observed that largely olefinic cases have been considered in search of their aldehydic and ketonic features. Thus, all the cases above in Equation 5.172, but (I), do not belong to the families of Aldehydes and Ketones.

5.4.2 Their radical characters

As with the charged cases, beginning with acrylates, the followings are obtained via C = O and C = C activation centers.

$$\overset{\cdot\cdot}{N}\!\cdot nn \quad + \quad e.\underset{\underset{\displaystyle CH_3}{\overset{\displaystyle |}{\underset{\displaystyle O}{|}}}}{\overset{\overset{\displaystyle CH_2}{\|}}{\underset{\overset{\displaystyle CH}{|}}{C} - O.\,nn}} \quad \longrightarrow \quad \overset{\cdot\cdot}{N}CH_3 \quad + \quad \overset{\cdot\cdot}{O}= \underset{\underset{\displaystyle CH_2}{\overset{\displaystyle \|}{\underset{\displaystyle CH}{|}}}}{C - O.\,nn}$$

5.177

$$E\cdot^e \quad + \quad e.\underset{\underset{\displaystyle CH_3}{\overset{\displaystyle |}{\underset{\displaystyle O}{|}}}}{\overset{\overset{\displaystyle CH_2}{\|}}{\underset{\overset{\displaystyle CH}{|}}{C} - O.\,nn}} \quad \longrightarrow \quad e.\underset{\underset{\displaystyle CH_3}{\overset{\displaystyle |}{\underset{\displaystyle O}{|}}}}{\overset{\overset{\displaystyle CH_2}{\|}}{\underset{\overset{\displaystyle CH}{|}}{C} - O - E}}$$

5.178

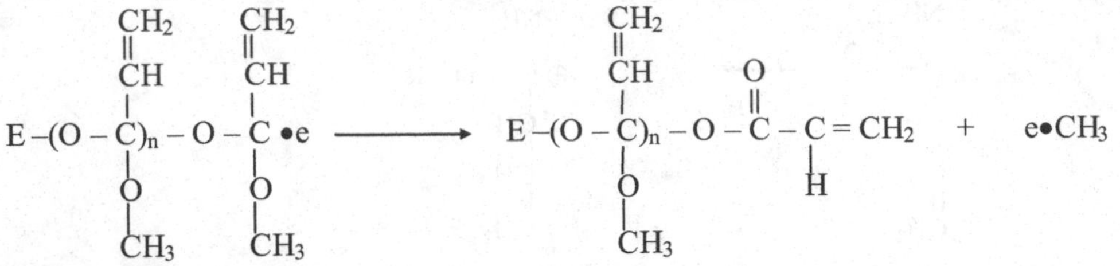

$$E-(O-C)_n-O-C \bullet e \longrightarrow E-(O-C)_n-O-C-C=CH_2 \quad + \quad e\bullet CH_3$$

5.179

Thus with methylacrylate, copolymers cannot be produced via electro-free-radical route. Only the C = C center can be used nucleo-free-radically, the natural route of the monomer. Electro-free-radically, only the C=O center can be used with adequate control.

When methyl methacrylate is involved, the situation is the same. While the C = C center favors only nucleo-free-radical route, the C = O center favors electro-free- radical route. Hence existence of copolymers cannot be favored for this type of monomer. For the special case of the monomer shown below, the followings are to be expected.

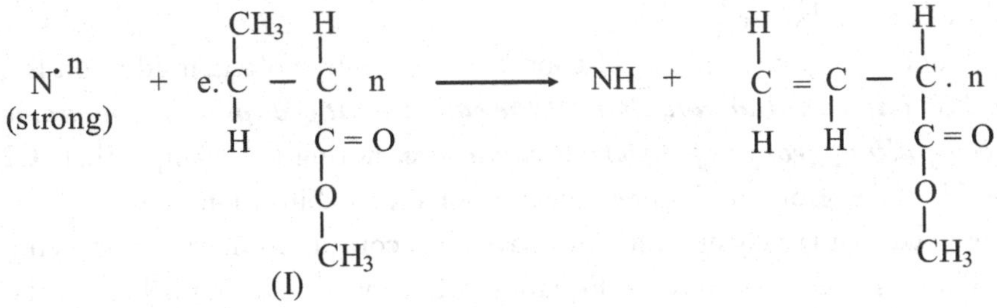

$$N^{\bullet n} \; (strong) \quad + \quad e.C-C.n \longrightarrow NH \quad + \quad C=C-C.n$$

(I)

5.180

$$E^{\bullet e} \quad + \quad n.C-C.e \longrightarrow E-OCH_3 \quad +$$

$$e\bullet C-C=C$$

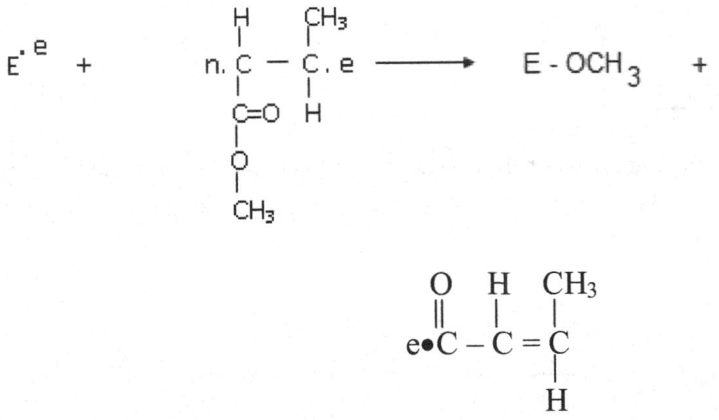

5.181

Free-radically, no route is favored in the absence of molecular rearrangement. It is no surprise why it is only known to favor polymerization using paired-negatively charged initiator. It will do the same i.e. favor initiation when paired nucleo-free-radical initiator such as $H_6C_5^{\bullet n}.....^{e\bullet}Na$ is used, giving it clear indication of its Electrophilic character. Thus, with free-media radicals no polymerization is favored.

With activation via C = O activation center the followings are to be expected.

289

$$\ddot{N}\cdot nn \quad + \quad e.\,\underset{\underset{\underset{CH_3}{|}}{\underset{||}{CH}}}{\overset{\overset{\overset{CH_3}{|}}{\overset{O}{|}}}{C}}-O.\,nn \quad \longrightarrow \quad \ddot{N}CH_3 \quad + \quad O=\underset{\underset{CH_3}{|}}{\overset{\overset{CH}{||}}{\underset{CH}{}}}C-O.\,nn$$

5.182

$$E\cdot^{e} \quad + \quad e.\,\underset{\underset{\underset{CH_3}{|}}{\underset{|}{CH}}}{\overset{\overset{\overset{CH_3}{|}}{\overset{O}{|}}}{C}}-O.\,nn \quad \longrightarrow \quad e.\,\underset{\underset{\underset{CH_3}{|}}{\underset{|}{CH}}}{\overset{\overset{\overset{CH_3}{|}}{\overset{O}{|}}}{C}}-O-E$$

5.183

Thus radically, polymerization is favored electro-free-radically. Copolymers cannot be obtained electro-free-radically as the case chargedly. The same analysis as with acrylates will also apply to acrylamides. ***Worthy of note, is that when they undergo molecular rearrangement, only the nucleo-free-radical route is favored by them.***

Considering alkyl vinyl aldehydes and ketones, the most unique of all the members, the followings are obtained for acrolein.

$$\ddot{N}\cdot nn \; + \; e.\,\underset{\underset{\underset{CH_2}{||}}{\underset{|}{CH}}}{\overset{\overset{H}{|}}{C}}-O.\,nn \; \longrightarrow \; \ddot{N}H \; + \; \underset{\underset{H}{|}}{C}=C=\underset{\underset{H}{|}}{C}-O.\,nn \quad OR \quad N-\underset{\underset{\underset{CH_2}{||}}{\underset{|}{CH}}}{\overset{\overset{H}{|}}{C}}-O.\,nn$$

[Not possible]

Favored   5.184

$$E(O-\underset{\underset{\underset{CH_2}{||}}{\underset{|}{CH}}}{\overset{\overset{H}{|}}{C}})_n O-\underset{\underset{\underset{CH_2}{||}}{\underset{|}{CH}}}{\overset{\overset{H}{|}}{C}}.e \; \longrightarrow \; E(O-\underset{\underset{\underset{CH_2}{||}}{\underset{|}{CH}}}{\overset{\overset{H}{|}}{C}})_n O-\underset{\underset{H}{|}}{\overset{\overset{H}{|}}{C}}=C=\underset{\underset{H}{|}}{\overset{\overset{H}{|}}{C}} \; + \; H.e$$

[H cannot be released]

5.185

The dead terminal double bond polymer does not need to have full semblance with the originating monomer. However, its presence is impossible. Therefore only living polymers are produced electro-free-radically via the C=O center, since that C center after release cannot carry a nucleo-free-radical in the presence of $H_2C=$ group on the C center.

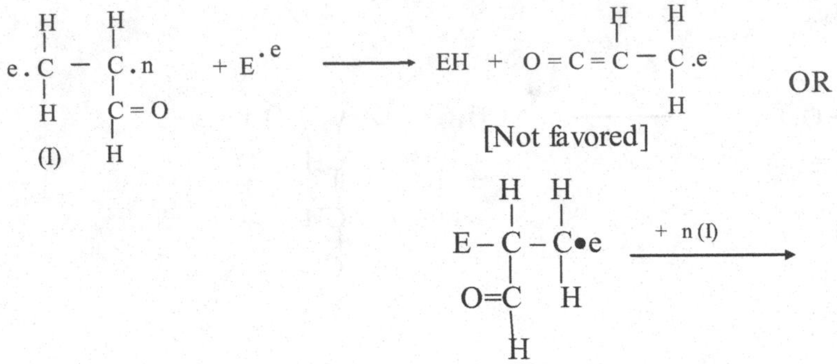

$$
\text{e} \cdot \overset{\overset{\textstyle H}{|}}{\underset{\underset{\textstyle H}{|}}{C}} - \overset{\overset{\textstyle H}{|}}{\underset{\underset{\textstyle C=O}{|}}{C}} \cdot n \quad + \text{ E}^{\cdot e} \longrightarrow \text{ EH } + \text{ O}=\text{C}=\text{C} - \overset{\overset{\textstyle H}{|}}{\underset{\underset{\textstyle H}{|}}{C}} \cdot e \qquad \mathbf{OR}
$$

(I)   H   [Not favored]

$$
\text{E} - \overset{\overset{\textstyle H}{|}}{\underset{\underset{\textstyle O=C}{|}}{C}} - \overset{\overset{\textstyle H}{|}}{\underset{\underset{\textstyle H}{|}}{C}} \bullet e \xrightarrow{\ +\ n\,(I)\ }
$$

[Favored]

$$
\text{E} - (\overset{H}{\underset{O=C}{C}} - \overset{H}{\underset{H}{C}})_n - \overset{H}{\underset{O=C}{C}} - \overset{H}{\underset{H}{C}} \bullet e \longrightarrow \text{E} - (\overset{H}{\underset{O=C}{C}} - \overset{H}{\underset{H}{C}})_n - \overset{H}{\underset{O=C}{C}} = \overset{H}{\underset{H}{C}} \ + \ e \bullet H
$$

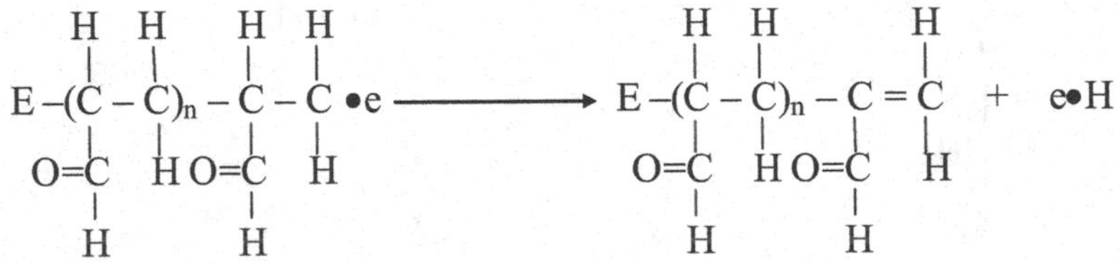

Transfer species of second kind   5.186

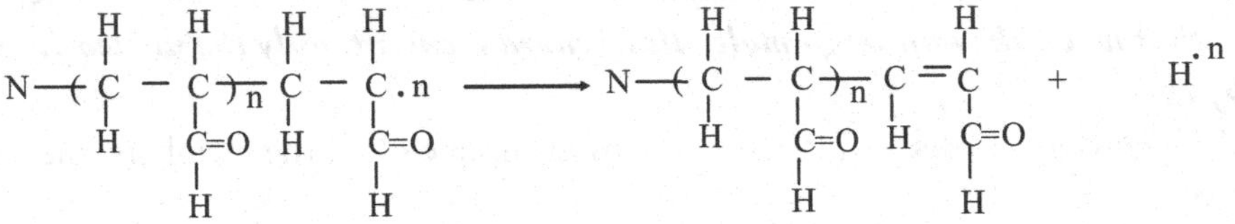

$$
\text{N} - (\overset{H}{\underset{H}{C}} - \overset{H}{\underset{C=O}{C}})_n \overset{H}{\underset{H}{C}} - \overset{H}{\underset{C=O}{C}} \cdot n \longrightarrow \text{N} - (\overset{H}{\underset{H}{C}} - \overset{H}{\underset{C=O}{C}})_n \overset{H}{\underset{H}{C}} = \overset{H}{\underset{C=O}{C}} \ + \ \text{H}^{\cdot n}
$$

Transfer species of the second kind [Not favored]   5.187

What is observed here is different from the use of charged initiators. Dead terminal double bond polymers can be produced only electro-free-radically for the C=C center. Nucleo-free-radically, the route natural to it, transfer species of the second kind cannot be released. Electro-free-radically, the release of H as an electro-free-radical carrying species takes place very readily. Just like the charged case, acrolein will also favor copolymer production electro-free-radically. Like the charged case, copolymers cannot be produced nucleo-radically, since free- and non-free are used for C=C and C=O centers respectively. The same analysis will apply to alkyl vinyl ketones only electro-free-radically, noting that the C=O center is more nucleophilic than that of acrolein. Nucleo-radically, no copolymers can be obtained and only nucleo-free-radical can be used polymerizably. Note that where copolymers are produced so far, they are random in placement.

Considering the special monomer shown below, the followings are to be expected.

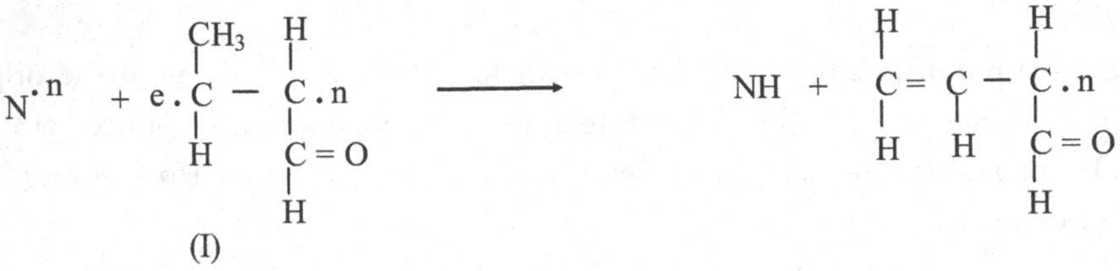

$$
\text{N}^{\cdot n} \ + \ \text{e} \cdot \overset{\overset{\textstyle CH_3}{|}}{\underset{\underset{\textstyle H}{|}}{C}} - \overset{\overset{\textstyle H}{|}}{\underset{\underset{\textstyle C=O}{|}}{C}} \cdot n \longrightarrow \text{NH } + \ \overset{\overset{\textstyle H}{|}}{\underset{\underset{\textstyle H}{|}}{C}} = \overset{\overset{\textstyle H}{|}}{\underset{\underset{\textstyle H}{|}}{C}} - \overset{\overset{\textstyle H}{|}}{\underset{\underset{\textstyle C=O}{|}}{C}} \cdot n \qquad 5.188
$$

(I)

When paired nucleo-free-radical initiator such as $H_5C_6^{\cdot n} \ldots \ldots ^{\cdot e} Na$ is used, the route is favored.

291

$$E^{\cdot e} \quad + \quad n. \underset{\underset{\underset{H}{|}}{\overset{\overset{H}{|}}{C}}=O}{\overset{\overset{H}{|}}{C}} - \overset{\overset{CH_3}{|}}{\underset{\underset{H}{|}}{C}} \cdot e \quad \longrightarrow \quad E - \underset{\underset{\underset{H}{|}}{\overset{\overset{H}{|}}{C}}=O}{\overset{\overset{H}{|}}{C}} - \overset{\overset{CH_3}{|}}{\underset{\underset{H}{}}{C}} \cdot e \quad \xrightarrow{\quad + \; n(I) \quad}$$

$$E - (\overset{\overset{H}{|}}{\underset{\underset{\underset{H}{|}}{O=C}}{C}} - \overset{\overset{CH_3}{|}}{\underset{\underset{H}{|}}{C}})_n - \overset{}{\underset{\underset{\underset{H}{|}}{O=C}}{C}} = \overset{\overset{CH_3}{|}}{\underset{H}{C}} \quad + \quad H \bullet e$$

[Transfer species of the second kind]

5.189

This is the route not natural to the monomer. Hence transfer species of second kind was released to kill the chain. ***Evidently, there is no doubt that paired free- and non-free-radical initiators exist in addition of the realization that the route natural to a monomer should in general be favored.*** Some such as vinyl acetate, Cis-methyl sorbate do not. It all depends on what type of initiator is doing the job and the type of monomer.

$$\ddot{N} \cdot nn \quad + \quad e. \underset{\underset{\underset{\underset{CH_3}{|}}{CH}}{\overset{||}{CH}}}{\overset{\overset{H}{|}}{C}} - O. nn \quad \longrightarrow \quad \overset{..}{N}H \quad + \quad \underset{H}{\overset{\overset{H}{|}}{C}} = \underset{H}{\overset{\overset{H}{|}}{C}} - \overset{\overset{H}{|}}{\underset{}{C}} = \overset{\overset{H}{|}}{\underset{}{C}} - O. nn$$

(Favored)   5.190

Initiation is not favored due to presence of transfer species of the first kind of the first type.

$$E \leftarrow O - \underset{\underset{\underset{\underset{CH_3}{|}}{CH}}{\overset{||}{CH}}}{\overset{\overset{H}{|}}{C}} \xrightarrow{}_n O - \underset{\underset{\underset{\underset{CH_3}{|}}{CH}}{\overset{||}{CH}}}{\overset{\overset{H}{|}}{C}} \cdot e \quad \longrightarrow \quad E \leftarrow O - \underset{\underset{\underset{\underset{CH_3}{|}}{CH}}{\overset{||}{CH}}}{\overset{\overset{H}{|}}{C}} \xrightarrow{}_n O - \overset{\overset{H}{|}}{\underset{}{C}} = \underset{H}{\overset{\overset{H}{|}}{C}} - \underset{H}{\overset{\overset{H}{|}}{C}} = \overset{\overset{H}{|}}{\underset{}{C}} \quad + \quad H \cdot e$$

5.191

Based on Equation 5.190, the equation above is favored. Dead terminal resonance stabilized polymer chain is obtained in the route natural to the center, but not to the monomer (see Equations 5.151 and 5.166). Nevertheless, copolymers can also be produced only electro-free-radically for this monomer.

    For the olefinic cases where the $C = C$ and $C = O$ activation centers are isolatedly placed, that is, cases shown in Equation 5.172, the situation is virtually the same radically. All the monomers are nucleophiles. Now consider the possibility of molecularly rearranging (I) of Equation 5.188.

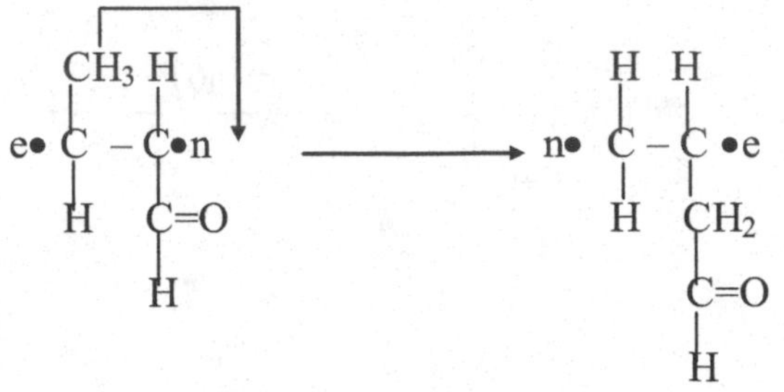

$$e\bullet \overset{\displaystyle CH_3}{\underset{\displaystyle H}{C}} - \overset{\displaystyle H}{\underset{\displaystyle \underset{\displaystyle H}{C=O}}{C}}\bullet n \quad \longrightarrow \quad n\bullet \overset{\displaystyle H}{\underset{\displaystyle H}{C}} - \overset{\displaystyle H}{\underset{\displaystyle \underset{\displaystyle \underset{\displaystyle H}{C=O}}{CH_2}}{C}}\bullet e$$

[WEAK MALE]　　　　　　　[STRONG FEMALE]

(I)　　　　　　　　　　　(II)　　　　　　　　　　　　5.192

(II) is a strong Nucleophile with two nucleophilic activation centers- C=C and C=O, in which C=O is more nucleophilic. If the C = C center is less nucleophilic, then the followings are to be expected.

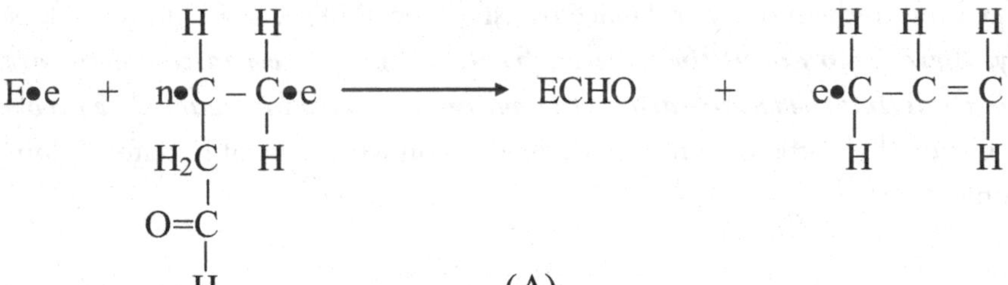

$$E\bullet e \;+\; n\bullet \overset{\displaystyle H}{\underset{\displaystyle \underset{\displaystyle \underset{\displaystyle H}{O=C}}{H_2C}}{C}} - \overset{\displaystyle H}{\underset{\displaystyle H}{C}}\bullet e \quad \longrightarrow \quad ECHO \;+\; e\bullet \overset{\displaystyle H}{\underset{\displaystyle H}{C}} - \overset{\displaystyle H}{\underset{\displaystyle H}{C}} = \overset{\displaystyle H}{\underset{\displaystyle H}{C}}$$

(A)　　　　　　　　　　　　　　　　　5.193

(III) [WRONG ACTIVATION]

(III) was used, because many will think that CH_2CHO radically is a free-radical-pulling group, when it has been shown that CH_2Cl, $CHCl_2$, CCl_3 groups are free-radical-pushing groups. Chargedly they are free-charged-pulling groups whose centers cannot carry negative charges. Hence, the (B) shown below is the use of (II) of Equation 5.192. With (III) above, notice that the route which is natural to it, is not favored, because that is not its real activated state of Existence. It is the (II) of Equation 5.192 that is the real Activated State of Existence.

$$E\bullet e \;+\; e\bullet \overset{\displaystyle H}{\underset{\displaystyle \underset{\displaystyle \underset{\displaystyle H}{O=C}}{H_2C}}{C}} - \overset{\displaystyle H}{\underset{\displaystyle H}{C}}\bullet n \quad \longrightarrow \quad E\bullet \overset{\displaystyle H}{\underset{\displaystyle H}{C}} - \overset{\displaystyle H}{\underset{\displaystyle \underset{\displaystyle \underset{\displaystyle H}{C=O}}{CH_2}}{C}}\bullet e \quad \xrightarrow{\;+\; n\,(II)\;}$$

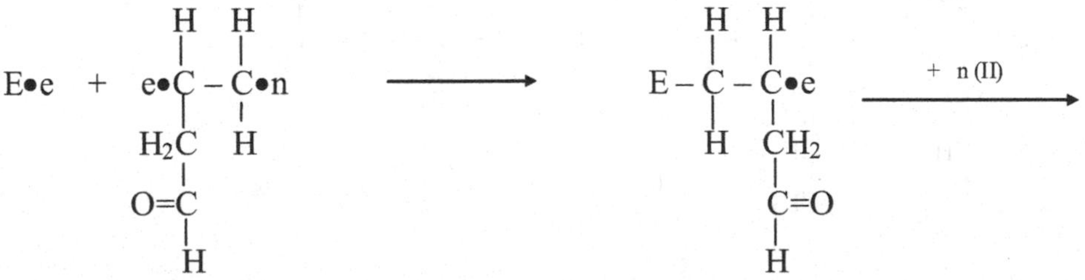

(II) –A FEMALE　　　　　　　　　　(IV)

[B]

$$E-(\underset{\underset{H}{|}}{\overset{\overset{H}{|}}{C}}-\underset{\underset{\underset{\underset{\underset{H}{|}}{C=O}}{|}}{CH_2}}{\overset{\overset{H}{|}}{C}})_n-\underset{\underset{H}{|}}{\overset{\overset{H}{|}}{C}}-\underset{\underset{\underset{\underset{H}{|}}{C=O}}{|}}{\overset{\overset{H}{|}}{C}}=\underset{\underset{\underset{H}{|}}{C=O}}{\overset{\overset{H}{|}}{C}} \quad + \quad H\bullet e$$

(V) – A MALE   5.194

(II) which is a strong Nucleophile (UNSTABLE FEMALE) favored its natural route to give a dead terminal double bond polymer which is an Electrophile (STABLE MALE) with the same character as its originating monomer. ***Indeed, the molecular rearrangement of Equation 5.192 from Male to a Female may be said not to be favored, because in (I) an Electrophile, transfer species can only come from radical-pulling groups.*** However, being a very weak Electrophile wherein such transfer species does not exist in COH, unlike in COOR, this rearrangement may be possible for Nature is too much to comprehend, taking note of the operating conditions under which all these take place. These are very useful information for the Medical Scientists, Medical Doctors and Sociologists. From the observations above, the following is obvious-

$$-CH_3 > \underline{-CH_2X} > \underline{-CHX_2} > -H > \underline{-CX_3}$$

ORDER OF RADICAL-PUSHING CAPACITY   5.195

where X is F, Cl, Br, CHO, COOH, CONH$_2$ and so on.

Chargedly, the order of the capacities of those highlighted exists but meaningless when carried by active centers, since when carried, such monomers cannot be activated chargedly, the groups being charged pulling. They can only be activated radically. Only few can be activated chargedly in the absence of electrostatic forces of repulsion. A look at Equation 5.193 when activated chargedly, shows that the transfer species cannot be removed chargedly.

5.5. Proposition of Rules and Conclusions.

Aldehydic and ketonic Addition monomers have been largely considered, in order to determine the routes favored by them, identify their unique features with respect to some of the phenomena which have so far been established. All the transfer species present with these families of monomers have been clearly identified. These include:-

(i)   Transfer species of first kind of first type.
(ii) " " " second kind.

The third kind has not yet been fully identified,

Through their identifications, how free and non-free transfer species, charged and radical transfer species are abstracted from a monomer and released or rejected from a growing polymer chain, have begun to be shown.

Rule 159: This rule of Chemistry for **Addition monomers,** states that, for a monomer to be activated and initiated, it cannot exist in Equilibrium or Activated/Equilibrium State of Exis-tence, but only in Activated State of Existence.
(Laws of Creations for Initiation)

Rule 160: This rule of Chemistry for **Addition monomers,** states that, based on the nature of compounds, the axis of symmetry for Projection is horizontally placed; for which one cannot cut across a σ-bond vertically or diagonally to create an axis of symmetry for Projection.
(Laws of Creations for Projections)

Rule 161: This rule of Chemistry for **Alkene hydrogen containing Addition monomers (i.e. those with C = C type of bond),** states that, in terms of Projections, there are three members of the family- the first member which is symmetric called **ETHENE,** the second member which is non-symmetric called the **1- and 2- ALKENES,** and the third member which can be symmetric or non-symmetric called the **ISO-ALKENES.**
(Laws of Creations of Alkene family Projections)

Rule 162: This rule of Chemistry for **Carbonyl and the Likes hydrogen containing Addition monomers [i.e. those with C = S, C = O, etc., types of bond],** states that, there are three members of the family- the first member which is symmetric called **FORMALDEHYDE,** the second member which is non-symmetric called **ALDEHYDES,** and the third member which can be symmetric or non-symmetric called the **KETONES.**
(Laws of Creations Formaldehyde/Aldehydes/Ketones families Projections)

Rule 163: This rule of Chemistry for **Olefins,** states that, all Alkenes carrying either H and C or H, C and a hetero atom such as F, Cl, O, N, S, are called OLEFINS containing ringed and non-ringed Addition monomers that form different families, in which the Hydrocarbon Family Tree is the foundation member.
(Laws of Creations for Olefins)

Rule 164: This rule of Chemistry for **Carbonyls and the Likes containing monomers,** states that, when activated unlike Olefins, the radicals or charges carried by the active centers are fixed, and independent of the substituted groups carried by the active C center.
(Laws of Creations for Carbonyls and the Likes)

Rule 165: This rule of Chemistry for **INITIATORS,** states that, there are two kinds of Initiators used for polymerization of Addition monomers- **FREE-MEDIA and PAIRED-MEDIA initiators,** in which the **FREE-MEDIA** ones are those that carry either one type of radical alone or ionic charges (Cations and Anions) alone and **are the only ones alone in the system,** and the **PAIRED-MEDIA** ones are those that carry either two types of oppositely placed paired radicals or oppositely placed paired charges (Cationic/Anionic charges, Covalent/Electrostatic positive and negative charges); for which for the free-media ones,

there is only one ACTIVE CENTER, while for the paired-media ones, there can be one or two ACTIVE CENTERS, depending on the type of pairing.
(Laws of Creations for Initiators)

Rule 166: This rule of Chemistry for **Initiation of Addition monomers,** states that, while all can be initiated by both male and female Free-media and Paired-media initiators with only one active center based on the strength of the radicals or charges carried and operating conditions such as temperature, radical or charged environment; Paired-media initiators with two active centers can only initiate the monomer according to the character of the monomer and operating conditions such as temperature, radical or charged environment.
(Laws of Creations for Initiations)

Rule 167: This rule of Chemistry for **Addition monomers,** states that, when an *Electrophilic* single Addition monomer is such that can produce copolymers (*Randomly or alternatingly placed)* from itself, this can only be done in the route which is unnatural to the monomer; while in the route natural to them and for *Nucleophiles,* this is impossible.
(Laws of Creations for Copolymerization)

Rule 168: This rule of Chemistry for **Two or more Nucleophilic Addition monomers of different capacities,** states that, when there is perfect mixing under isothermal conditions, only *Block* copolymers can be obtained, beginning with the least nucleophilic in capacity until fully consumed, followed by the next and so on, when PAIRED-MEDIA initiators are involved in the route natural to them.
(Laws of Creations for Copolymerization)

Rule 169: This rule of Chemistry for **Two or more Nucleophilic Addition monomers of different capacities,** states that, when there is perfect mixing under isothermal conditions, *both Random and Block* copolymers can be obtained, when FREE-MEDIA initiators are involved in the route natural to them.
(Laws of Creations for Copolymerization)

Rule 170: This rule of Chemistry for **Aldehydic and Ketonic Addition monomers (i.e., the C = O, C = S, etc., types of monomers),** states that, these unlike Olefins are unique families of monomers, since most of them can exist in **ACTIVATED/EQUILIBRIUM STATE** OF EXIS-TENCE, when special passive catalysts are present; for which when they do, they can molecu-larly rearrange to give another compound (E.g., acetaldehyde and vinyl alcohol).
(Laws of Creations for Aldehydes and Ketones)

Rule 171: This rule of Chemistry for **Aldehydic and Ketonic Addition monomers (i.e., the C = O, C = S, etc., types of monomers),** states that, these are NUCLEOPHILES whose natural routes are electro-free-radical and positive-free-charge routes; and unnatural routes for those that favor them are nucleo-non-free-radicals and anionic (i.e. negative non-free-charges) routes
(Laws of Creations for Aldehydes and Ketones)

Rule 172: This rule of Chemistry for **Addition monomers,** states that, *when they favor the route not natural to them and they do not have transfer species of the second kind to kill the growing chain,* depropagation

of the chain takes place increasing with increasing polymerization temperature; for which the monomer must have a ceiling temperature, below which depropaga-tion ceases.
(Laws of Propagation/Depropagation}

Rule 173: This rule of Chemistry for **Addition monomers which carry alkanyl types of alkyl groups (i.e. of the primary substituent groups),** states that, just as for halogens (See Rules 85 and 158), their free-radical-pushing capacities can be determined from the following order-

$$-CH_3 > \underline{-CH_2X} > \underline{-CHX_2} > -H > \underline{-CX_3}$$
ORDER OF RADICAL-PUSHING CAPACITY
where X is F, Cl, Br, CHO, COOH, $CONH_2$ and so on.

(Laws of Creations for Alkanyl groups)

Rule 174: This rule of Chemistry for **Transfer species,** states that, when groups such as NH_2, NR_2, OH, OR are placed on C=C centers, the monomers are *NUCLEOPHILES in* view of the radical-pushing capacities of the groups, while when they are placed on C=O centers which are conjugatedly placed on C=C centers, the monomers are *ELECTROPHILES,* in view of the radical-pulling capacities of the groups, because the Central atoms (N, O) are electronegative, which can therefore only be transferred when they carry nucleo-non-free-radicals or anions.
(Laws of Creations for Transfer Species)

Rule 175: This rule of Chemistry for **Male Addition monomers viz- RHC=CH(COR), RHC=CHCOOR, RHC=CHCONR$_2$, where R is H or an alkanyl group such as CH$_3$,** states that, of the three monomers, only the first has no transfer species of the first kind of the first type in the absence of solid ionic metallic centers, because the Central atom on the R group on CO is not a hetero atom; for which its ability to function as a Full Electrophile like the other two is impossible.
(Laws of Creations for Electrophiles)

Rule 176: This rule of Chemistry for **Male Addition monomers viz- RHC=CH(COR), RHC=CHCOOR, RHC=CHCONR$_2$, where R is H or an alkanyl group such as CH$_3$,** states that, while the last two will undergo molecular rearrangement of the first kind of the first type wherein as Electrophiles only nucleo-non-free-radicals or anions (OH, OR, NH_2, NR_2) can be transferred, the first one cannot undergo such molecular rearrangement in view of its character but can be made to undergo molecular rearrangement using the electrophilic center when the possibility exists, because Nature abhors a vacuum.
(Laws of Creations for Electrophiles)

Rule 177: This rule of Chemistry for **Male Addition monomers viz- RHC=CH(COR), RHC=CHCOOR, RHC=CHCONR$_2$, where R is H or an alkanyl group such as CH$_3$,** states that, all three types can be polymerized in their natural routes only when **PAIRED initiators** with two active centers are used and the two groups when present are trans-placed, the natural routes for the C=C center being the use of only nucleo- free-radical and negative-free-charged active centers.
(Laws of creations for Electrophiles)

Rule 178: This rule of Chemistry for **Addition monomers and all things in life,** states that, the phenomena of Molecular rearrangement are so unique i*n that when they take place in the real domain, the* *movement is from a less stable Female of a family to a more stable Female of same family or another family or a less stable Male of one family to a more stable Male of the same family or another family;* **while in the imaginary domain, movement is from a less stable male to a more stable female of same or different families or a less stable female to a more stable male of same or different families,** under very unique operating conditions.

(Laws of Physics and Metaphysics)

<u>**Rule 179:**</u> This rule of Chemistry for **Some Unique Electrophilic Addition monomers such as RHC=CH(COR),** states that, when molecular rearrangement transforms a less Male monomer to a Female monomer, it is to be expected because when the Female is polymerized in the route natural to it, the dead terminal double bond polymer produced is a Male as shown below with this family of monomers-

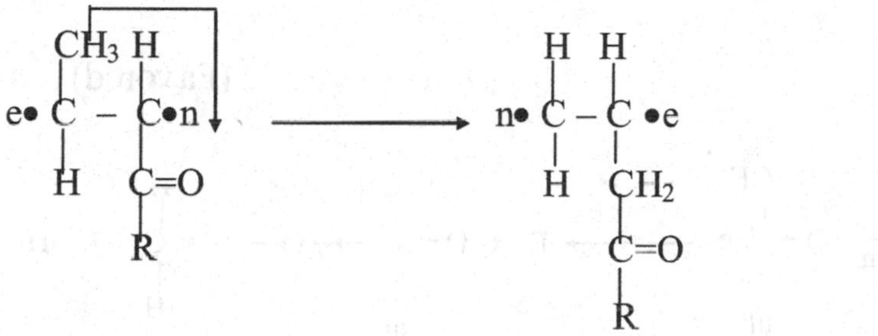

[WEAK MALE]   [STRONG FEMALE]

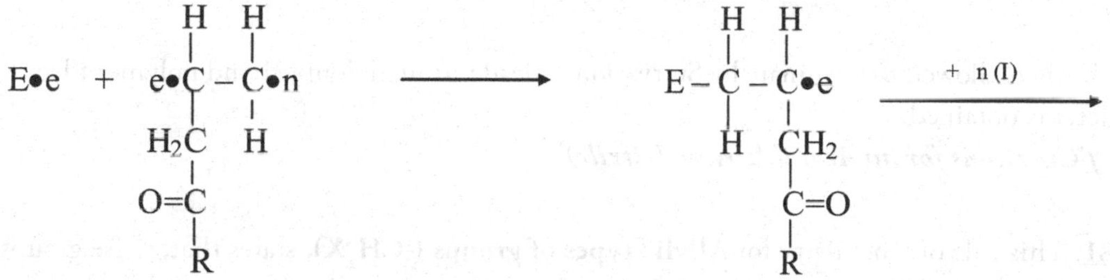

(I) [A FEMALE]

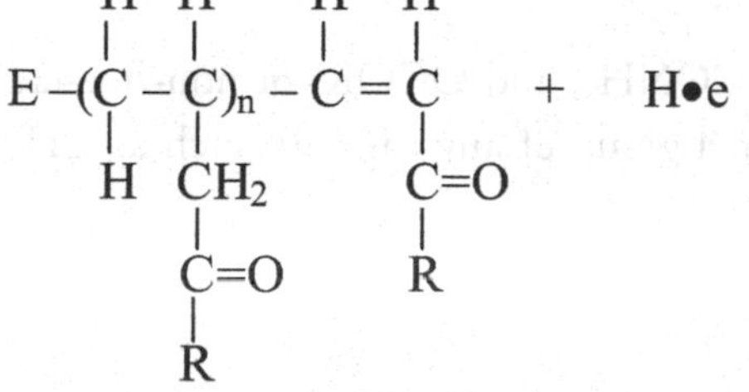

[A MALE]

R is H or an alkanyl group; noting that this can only take place radically and not chargedly because chargedly the Female cannot be activated (allylic); noting that rearrangement is imaginary
(Laws of Metaphysics)

Rule 180: This rule of Chemistry for **An Acetaldehydic Nitrilic Nucleophile shown below,** states that, when used as a monomer, the followings are obtained-

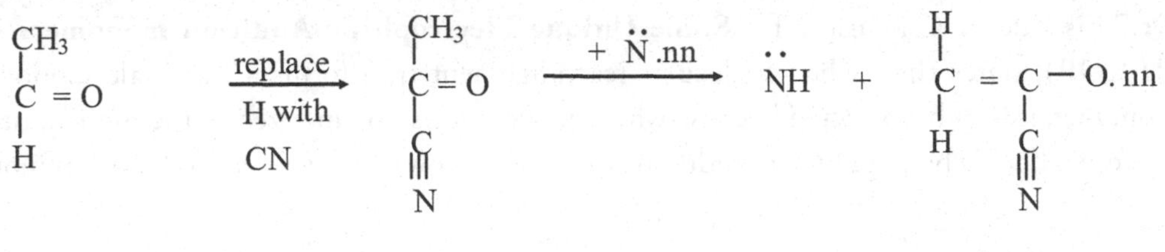

(Favored)

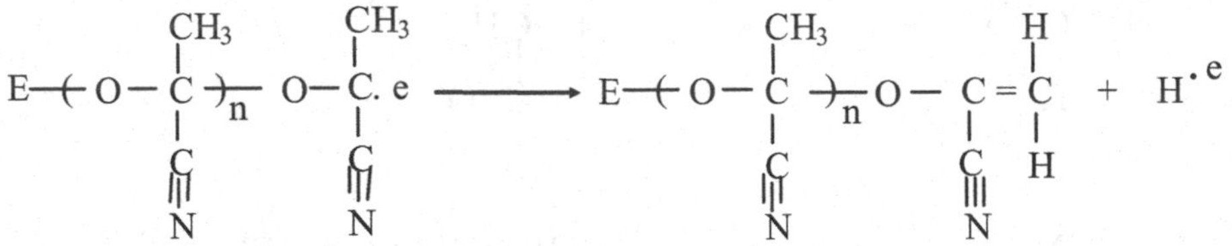

[Favored]

for which when allowed to terminate by Starvation, a dead terminal double bond polymer Elec-trophilic in character is obtained.
(Laws of Creations for an Acetaldehydic Nitrile)

Rule 181: This rule of Chemistry for **Allylic types of groups (-CH$_2$X),** states that, these groups are so unique for which the following is valid for them-

$$R \ > \ - \ CH_2X \ > \ CHX_2 \ > \ H$$
Order of radical-pushing capacity

where X is a radical pulling group (F, Cl, COOR, CONH$_2$, and OCOR) or non-free-radical-pushing group (OH, NH$_2$) and R is a radical-pushing alkanyl group of any capacity such as CH$_3$ for the case above. [See Rule 173]
(Laws of Creations for Allylic groups)

Rule 182: This rule of Chemistry for **Some Unique Nucleophilic Addition monomers such as HO – C ≡ CH,** states that, when molecular rearrangement transforms a Female monomer to a Male monomer, it is to be expected because when the Male is polymerized in the route natural to it, the dead terminal double bond polymer produced is a stronger Female as shown below with this family of monomers-

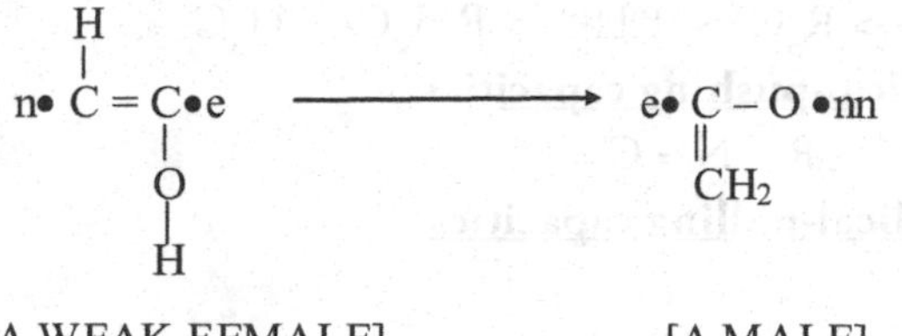

$$
\begin{array}{c}
\overset{\displaystyle H}{\underset{\displaystyle O}{\overset{\displaystyle |}{\underset{\displaystyle |}{n\bullet\ C = C \bullet e}}}} \\
\overset{\displaystyle |}{\underset{\displaystyle H}{}}
\end{array}
\quad\longrightarrow\quad
\begin{array}{c}
e\bullet\ C - O \bullet mn \\
\underset{\displaystyle CH_2}{\overset{\displaystyle \|}{}}
\end{array}
$$

[A WEAK FEMALE] [A MALE]

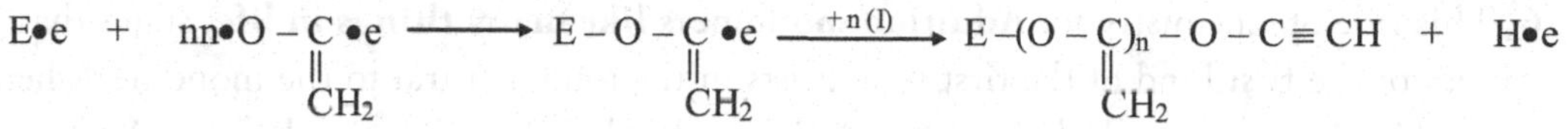

$$
E\bullet e \; + \; nn\bullet O - \underset{CH_2}{\overset{\|}{C}} \bullet e \;\longrightarrow\; E - O - \underset{CH_2}{\overset{\|}{C}} \bullet e \;\xrightarrow{+n\,(I)}\; E - (O - \underset{CH_2}{\overset{\|}{C}})_n - O - C \equiv CH \; + \; H\bullet e
$$

(I) [MALE] [A STRONGER FEMALE]

noting that this can only take place radically and not chargedly, since the female cannot be activated chargedly.
(Laws of Metaphysics)

Rule 183: This rule of Chemistry for **Addition monomers which carry alkanyl groups in which one H atom on the last or Central C atom is replaced by resonance stabilization groups,** states that, these are all free-pushing groups with the following capacities-

$$
-C_2H_5 \geq -C_2H_4-CH= CH_2 > -CH_3 \geq -CH_2 -CH = CH_2 \geq -CH_2 - CH = CH - CH = CH_2
$$
Order of free-pushing capacities of alkanyl groups (Radically and Chargedly)

Including the followings-

$$
H_3C - CH = CH - > CH_3 > H_2C = CH - > H > H_2C = C(CH_3)-
$$
Order of free-pushing capacities of alkanyl and alkenyl groups
$$
H_2C = CX- > R > H
$$

where X and R remain as already defined.
Order of free-radical-pushing capacities of alkenyl and alkanyl groups.
(Laws of Creations for Alkanyl and Alkenyl groups)

Rule 184: This rule of Chemistry for **Addition monomers which carry pulling resonance stabilization groups,** states that, the followings are the order of their pulling capacities-

$$
X_2C=CH- > XHC=C(CH_3)- > X-CH= CH - > X > X(CH_3)C=CH-
$$
Order of free-pulling capacities of alkenyl groups (Radically and Chargedly)

where X remains as already defined.
(Laws of Creations for Radical-pulling resonance stabilization group)

Rule 185: This rule of Chemistry for **Addition monomers,** states that, there are many in complex organic system which carry strong radical-pushing and pulling groups such as shown below with the following capacities-

$$R - N = > O = > R_2C = >>>> R_3C- > R_2HC- > RH_2C- > H_3C-$$

Order of Radical-pushing capacities

$$O = C = > R - N = C =$$

Order of Radical-pulling capacities

(Laws of Creations for Special Groups)

Rule 186: This rule of Chemistry for **Addition monomers like many things in life,** states that, when transfer species of the first kind of the first type exists in the route natural to the monomer when free-media or paired initiators are used, the transfer species **cannot** be rejected during Propagation as transfer species of the first kind of the first type in the route unnatural to it as clearly distinguished by two family members of Olefinic monomers below-

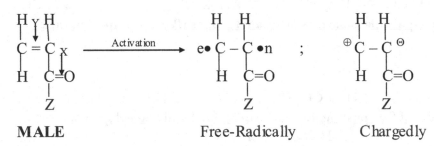

(A) NO TRANSFER SPECIES OF FIRST KIND

(where Z is OH, OR, NH₂, NHR, NR₂)

(B) TRANSFER SPECIES OF FIRST KIND (Y)

for which as it seems (A) looks like a Male with no visible female center, since there seem to exist transfer species of the first kind in the route natural to it, preventing initiation of the mono-mer, but can be rejected in the route which is not natural to the monomer as transfer species of the first kind of the first type!

(Laws of Creations for Addition monomer)

Rule 187: This rule of Chemistry for **Aldehydic and Ketonic Addition monomers,** states that, *in general there are two major types of transfer species* - the first kind of the first type and the second kind during polymerization of these monomers.

(Laws of Creations for Aldehydes and Ketones)

Rule 188: This rule of Chemistry for **Aldehydic and Ketonic Addition monomers**, states that, in general ketonic monomers are more nucleophilic than aldehydes, which in turn are more nucleophilic than formaldehyde which in turn is more nucleophilic than alkenes.

(Laws of creation for Formaldehyde, Aldehydes and Ketones)

Thirty rules have been proposed to identify with unique qualities of aldehydes and ketones, olefins, and related Addition monomers. For the first time, the distinction between free and non-free-charges and radicals are being clearly identified using the laws of conservation of transfer of transfer species. For the first time, the radical-pulling and pushing qualitative capa-cities of non-alkyl groups are being provided without the use of experimental data or measure-ments, but information gathered in books, the literature and the use of the THIRD EYE. Aldehydes and Ketones can be observed to be uniquely different from Olefinic monomers in many respects. For example, an important rule not stated is the fact that, the order of nucleophilicity of a carbonyl center (i.e., $C=O$), is measured by the types and capacity of the only two groups carried by the carbon center. It was not stated, because it is inherently present in some of the rules which have so far been stated and clearly obvious.

References

1. G. Odian, "Principles of Polymerisation," McGraw Hill Book company, (1970), pg, 345 - 351.

2. S. N. E. Omorodion, "New Frontiers in Engineering Science and the Arts", Vol 1, Chapters 6 and 7.

3. W. K. Bushfield and E. Whalley, Can. J. Chem., 43: 2289 (1965).

4. W. H. Sharkey, J. Macromol. Sci. (Chem.), A1(2): 291 (1967).

5. C. R . Noller, "Textbook of Organic Chemistry", W.B. Saunders Company, (1966), pg. 380.

Problems

5.1 Distinguish as briefly as possible between Olefins and aldehydes (formaldehyde, alkyl aldehyde, and dialkyl aldehydes.

5.2 (a) Why is molecular rearrangement phenomenon not favored by aldehydic and ketonic monomers? Explain.

(b) On the basis of transfer species, define an aldehyde and a ketone.

(c) Can some aldehydes and ketones be fully electrophilic in character? Explain.

5.3 Using C = O activation center in formaldehyde and some other rules, determine qualitatively, the radical/charge-pushing or pulling capacities of the following substituent/substituted groups-

(i) CF_3 (ii) CHF_2 (iii) CH_2F (iv) CH_3 (v) CF_2CF_3 (vi) $CHFCF_3$
(vii) CH_2CF_3 (viii) CH_2CHF_2 (ix) CH_2CH_2F (x) CH_2CH_3. Show it tabularly.

5.4 (a) Identify the additional rules which have been used to arrive at the capacity orders in Q 5.3 above.

(b) Identify the transfer species involved where present for the substituent groups above.

5.5 Distinguish between $CF_3CF = O$ and $CH_3CH = O$ as Addition monomers radically and chargedly.

5.6 (a) Will radical polymerizations of aldehydic and ketonic monomers be much slower than their charged polymerization? Explain.

(b) Shown below are some groups:-

(i) SO_2R (ii) SO_3H (iii) NO_2 (iv) F (v) HSO_4

Identify from their structures the non-free -pulling substituent group, non-free -pulling substituted group (i.e. without transfer species), free -pulling substituted group (i.e. without transfer species).

(c) Why are (ii) and (v) not popularly known to be used as substituted groups for Addition monomers?

5.7 (a) Vinyl acetate is a unique monomer. Is the acetate group a substituent or non-substituent group chargedly and radically? Explain.

(b) Using $C = O$ activation center in formaldehyde, determine qualitatively the charged / radical -pulling capacities where possible of the following groups:-

(i) $OCOCH_3$ (ii) CH_2OCOCH_2 (iii) Cl

(c) For the two groups shown below, why is $OCOCH_3$ not a non-free pushing group?

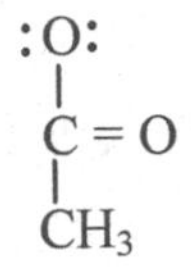

$$
\begin{array}{ccc}
:O: & & \\
| & & :O: \\
C = O & \text{and} & | \\
| & & CH_3 \\
CH_3 & &
\end{array}
$$

Non free   Non-free

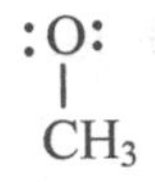

pulling group   pushing group.

5.8 Consider the two monomers shown below

$$
\text{(i)} \quad
\begin{array}{c}
H \quad H \\
| \quad | \\
N = C \\
| \\
H
\end{array}
\qquad
\text{(ii)} \quad
\begin{array}{c}
CF_2H \\
| \\
C = O \\
| \\
CF_2H
\end{array}
$$

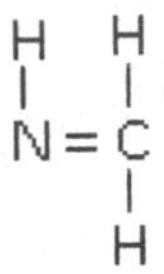

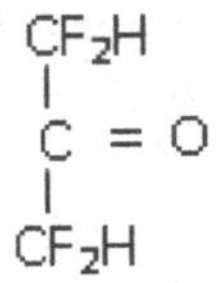

(a) Why can't (i) favor polymerization chargedly and radically?

(b) What can be done to the monomer, so that polymerization is favored?

(c) Discuss the charged and radical polymerization of (ii).

5.9 (a) Under what conditions does transfer species of the second kind exist?

(b) Are transfer species of the second kind important in terms of the law of conversation of transfer of transfer species? Explain using (i) of Q 5.8 above and compare with that of $CF_2 = O$,

5.10 (a) Between $C = S$ and $C = O$ activation centers, which is more nucleophilic? Explain

(b) Below are two activated monomers:-

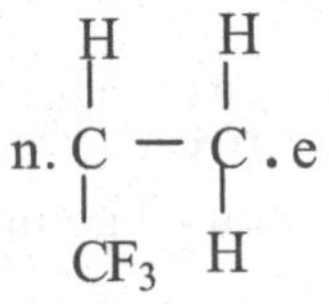

$$
\begin{array}{cc}
\begin{array}{c}
H \quad H \\
| \quad | \\
n. C - C . e \\
| \quad | \\
CF_3 \quad H
\end{array}
& ; \qquad
\begin{array}{c}
CF_3 \\
| \\
nn. N - C . e \\
\| \\
O
\end{array}
\\
\text{(I)} & \text{(II)}
\end{array}
$$

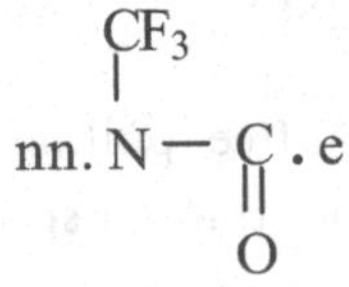

(i) Why can't F in (I) be rejected nucleo-free-radically, but can in (II) nucleo-non-free-radically? Explain using some of the laws so far estabilished.

(ii) Using the same activation center in (II), identify the charged route(s) favored by (II).

Sunny N.E. Omorodion

5.11. Consider the following monomers shown below.

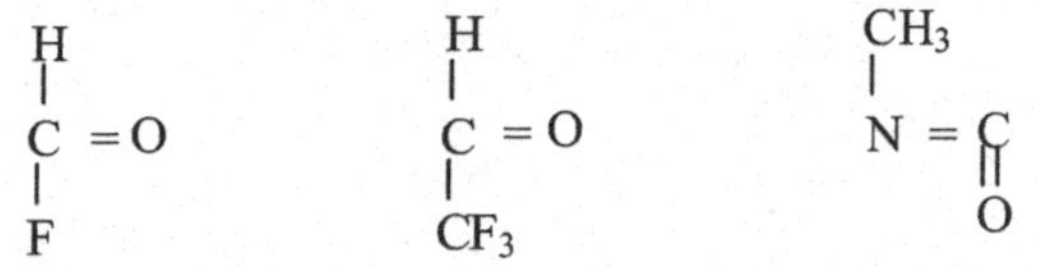

$$
\begin{array}{ccc}
\overset{\displaystyle H}{\underset{\displaystyle F}{C=O}} & \overset{\displaystyle H}{\underset{\displaystyle CF_3}{C=O}} & \overset{\displaystyle CH_3}{\underset{\displaystyle O}{N=C}} \\
(I) & (II) & (III)
\end{array}
$$

(a) Identify the cases where species of the first and second kind exist.
(b) Identify the routes favored by them radically and chargedly.
(c) Compare (III) above with (II) of Q5.10.

5.12 Distinguish between the classical organic chemistry of aldehydes and ketones as of the past or present, with current developments.

5.13 What are the main differences between charges, ions and radicals? What are also common between them?

5.14 Shown below are the following monomers:-

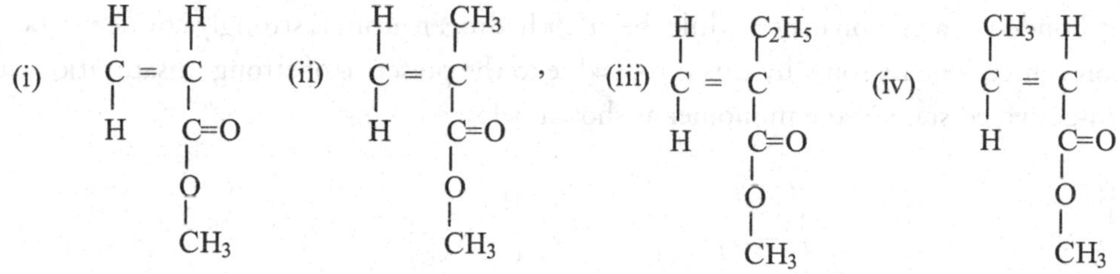

(i) $\underset{\displaystyle H}{\overset{\displaystyle H}{C}}=\underset{\displaystyle \underset{\displaystyle \underset{\displaystyle CH_3}{O}}{C=O}}{\overset{\displaystyle H}{C}}$ (ii) $\underset{\displaystyle H}{\overset{\displaystyle H}{C}}=\underset{\displaystyle \underset{\displaystyle \underset{\displaystyle CH_3}{O}}{C=O}}{\overset{\displaystyle CH_3}{C}}$, (iii) $\underset{\displaystyle H}{\overset{\displaystyle H}{C}}=\underset{\displaystyle \underset{\displaystyle \underset{\displaystyle CH_3}{O}}{C=O}}{\overset{\displaystyle C_2H_5}{C}}$ (iv) $\underset{\displaystyle H}{\overset{\displaystyle CH_3}{C}}=\underset{\displaystyle \underset{\displaystyle \underset{\displaystyle CH_3}{O}}{C=O}}{\overset{\displaystyle H}{C}}$

Identify the routes favored radically and chargedly, the order of capacities of the activation centers, and the overall order of capacities of characters of the monomers.

5.15 Shown below are the following monomers:-\

(i) $\underset{\displaystyle H}{\overset{\displaystyle H}{C}}=\underset{\displaystyle \underset{\displaystyle H}{C=O}}{\overset{\displaystyle H}{C}}$ (ii) $\underset{\displaystyle H}{\overset{\displaystyle H}{C}}=\underset{\displaystyle \underset{\displaystyle CH_3}{C=O}}{\overset{\displaystyle CH_3}{C}}$ (iii) $\underset{\displaystyle H}{\overset{\displaystyle H}{C}}=\underset{\displaystyle \underset{\displaystyle CCl_3}{C=O}}{\overset{\displaystyle C_2H_5}{C}}$

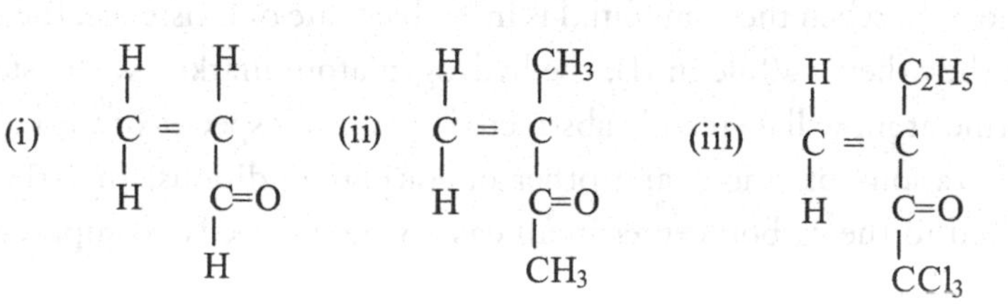

(a) Show the order of capacities of the activation centers.
(b) Under what conditions are copolymers obtained where possible from the monomers above.

CHAPTER 6

TRANSFER OF TRANSFER SPECIES IN ACETYLENES AND NITRILES

6.0 Introduction

Based on the new definition of an Addition monomer, an ion, a radical etc., the chemistry of the reactions of these monomers as provided over the years based on laboratory data, can now be readily explained without ambiguity. Acetylenes are very useful compounds, which have found very wide applications, except as monomers. Indirectly, the first member has always been the source of most olefinic Addition monomers. The reason is largely because one hydrogen atom in acetylene $HC \equiv CH$ is very loosely covalently bonded to a carbon center while the other hydrogen atom is strongly covalently bonded to a second carbon center. The reason why this is so, is due to the presence of strong unsaturation in the activated and non-activated state of the monomer as shown below.

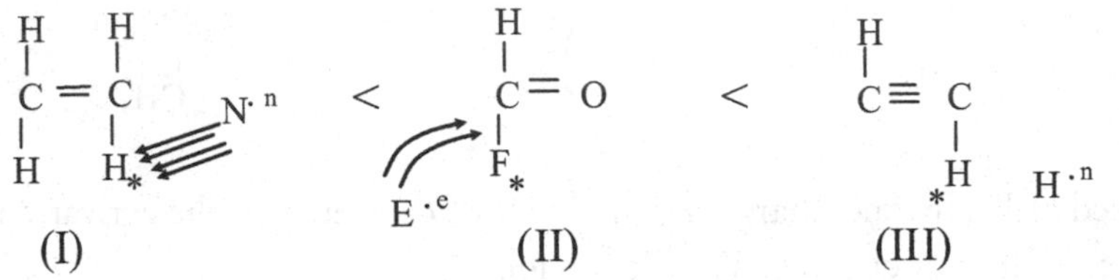

$$
\begin{array}{ccc}
\underset{(I)}{\begin{array}{c} H \quad H \\ | \quad | \\ C = C \\ | \quad | \\ H \quad H_* \end{array}} N^{\cdot n} \quad < &
\underset{(II)}{\begin{array}{c} H \\ | \\ C = O \\ | \\ F_* \end{array}} E^{\cdot e} \quad < &
\underset{(III)}{\begin{array}{c} H \\ | \\ C \equiv C \\ | \\ H_* \end{array}} H^{\cdot n}
\end{array}
\qquad 6.1
$$

<u>Increasing order of different species one of which is of the second first kind of second type</u>

Transfer species of the second first kind exist when the compound is in Stable State of Existence, the state in which acetylene cannot be, unlike the others. While in (I), the hydrogen atom marked with asterisk cannot be abstracted, in (II), the fluorine atom will be readily abstracted as transfer species of the second first kind, depending on the polymerization conditions and other operating conditions. In (III), the hydrogen atom is indeed loosely bonded to the carbon center in all cases where there is no suppressing agent (i.e., a passive catalyst) as shown below.

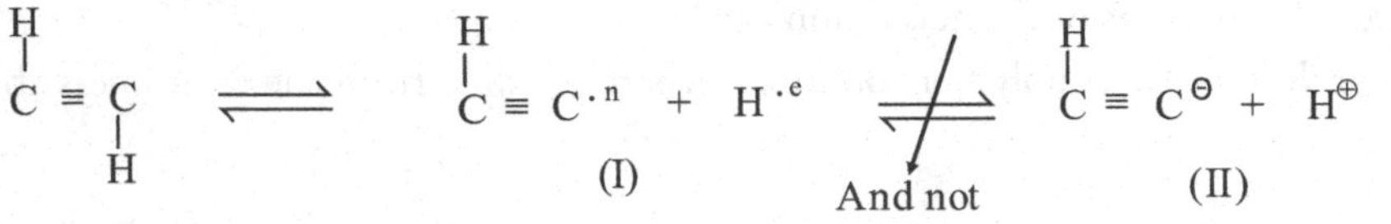

$$
\begin{array}{c} H \\ | \\ C \equiv C \\ | \\ H \end{array} \rightleftharpoons
\underset{(I)}{\begin{array}{c} H \\ | \\ C \equiv C^{\cdot n} \end{array}} + H^{\cdot e} \quad \underset{\text{And not}}{\nrightarrow} \quad
\underset{(II)}{\begin{array}{c} H \\ | \\ C \equiv C^{\ominus} \end{array}} + H^{\oplus}
\qquad 6.2
$$

The driving force for C center to carry a negative charge when the C - H bond is a product of hybridization, does not exist. It does not even exist after activation of the π-bond due to electro-dynamic forces of

307

repulsion from the second π-bond, for which charged activations of all acety-lenes is impossible. Secondly, the half-headed reversed arrows in (I) () is the symbolic representation of Equilibrium State of Existence.

Thus, when (I) is involved with any of known initiators, no polymerization is favored as shown below depending on the type. Worthy of note is that acetylene is a very strong Nucleo-phile (i.e. Female), for which the use of a female initiator is virtually impossible as shown below.

$$R^{\oplus} + \overset{\overset{\displaystyle H}{|}}{C} = C^{\cdot n} \; H^{\cdot e} \longrightarrow \text{No reaction} \qquad 6.3$$

$$R^{\ominus} + \overset{\overset{\displaystyle H}{|}}{C} \equiv C^{\cdot n} \; H^{\cdot e} \longrightarrow \text{No reaction} \qquad 6.4$$

$$N^{\cdot n} + \overset{\overset{\displaystyle H}{|}}{C} \equiv C^{\cdot n} \; H^{\cdot e} \longrightarrow NH + \overset{\overset{\displaystyle H}{|}}{C} \equiv C^{\cdot n} \xrightarrow{\;+\,(I)\;}$$

$$NH + \overset{\overset{\displaystyle H}{|}}{C} \equiv C^{\cdot n} H^{\cdot e} + \overset{\overset{\displaystyle H}{|}}{C} \equiv C^{\cdot n} \longrightarrow \text{No polymerization} \qquad 6.5$$

$$E^{\cdot e} + \underset{(I)}{\overset{\overset{\displaystyle H}{|}}{C} \equiv C^{\cdot n} \; H^{\cdot e}} \longrightarrow \underset{(II)}{\overset{\overset{\displaystyle H}{|}}{C} \equiv \underset{\underset{\displaystyle E}{|}}{C}} + \; H^{\cdot e} \xrightarrow{\;+\,(I)\;}$$

$$\underset{\underset{\displaystyle E}{|}}{\overset{\overset{\displaystyle H}{|}}{C} \equiv C} + \overset{\overset{\displaystyle H}{|}}{C} \equiv C^{\cdot n} \; H^{\cdot e} + H^{\cdot e} \longrightarrow \begin{array}{l} \text{Initiation or chemical reaction} \\ \text{May be possible} \end{array} \qquad 6.6$$

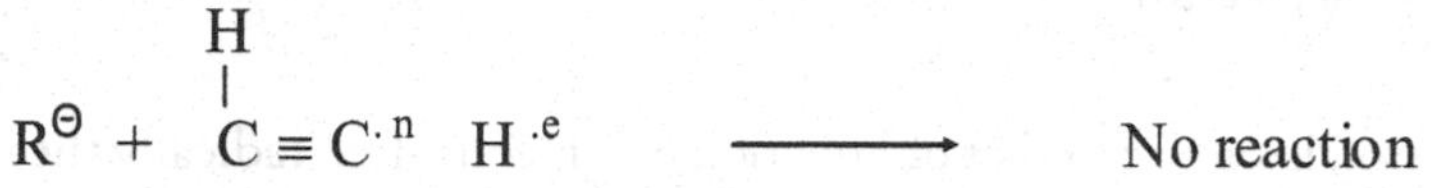

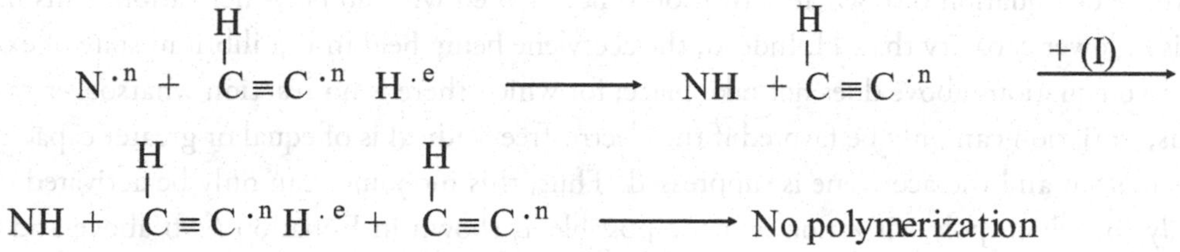

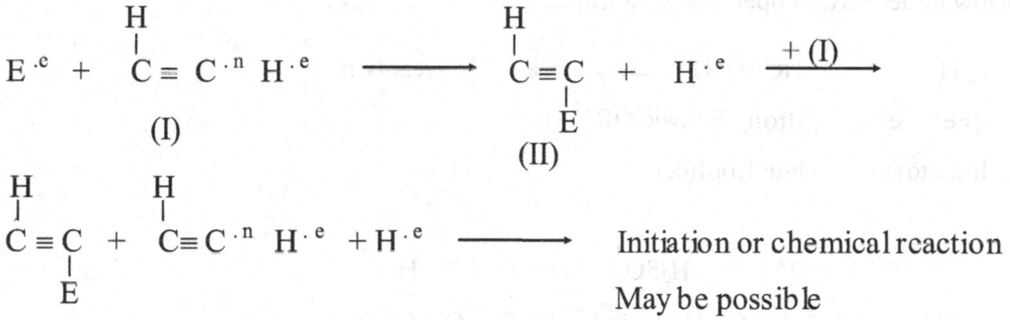

It is only in the last reaction that existence of a new product is favored. When hydrogen molecule is produced, the reaction does not go beyond the first step. This is the product (II) that can serve also as a monomer depending on what E or the initiator is. When activation of acetylene is readily favored under higher operating conditions (i.e., when suppressed by a passive catalyst) this can be done only radically as shown below.

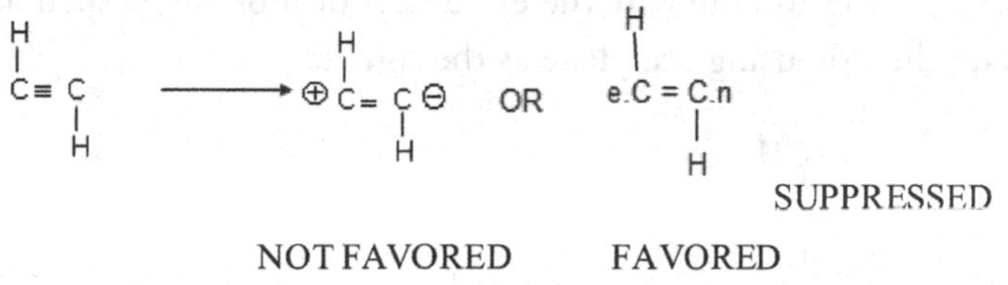

$$\overset{\overset{\displaystyle H}{|}}{\underset{\underset{\displaystyle H}{|}}{C} \equiv C} \longrightarrow \overset{\oplus}{\underset{\underset{\displaystyle H}{|}}{\overset{\overset{\displaystyle H}{|}}{C}} = C} \overset{\ominus}{} \quad OR \quad \underset{\underset{\displaystyle H}{|}}{e.\overset{\overset{\displaystyle H}{|}}{C} = C.n}$$

$$\qquad\qquad \text{NOT FAVORED} \qquad \text{FAVORED} \qquad\qquad \text{SUPPRESSED} \qquad 6.7$$

When attacked by free-radicals the followings are obtained.

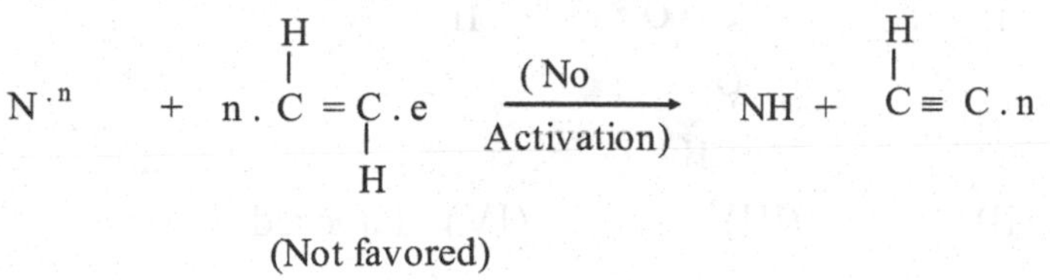

$$N^{\cdot n} + n.\overset{\overset{\displaystyle H}{|}}{\underset{\underset{\displaystyle H}{|}}{C} = C}.e \xrightarrow[\text{Activation}]{(\text{No}}\; } NH + \overset{\overset{\displaystyle H}{|}}{C} \equiv C.n \qquad 6.8$$

$$\text{(Not favored)}$$

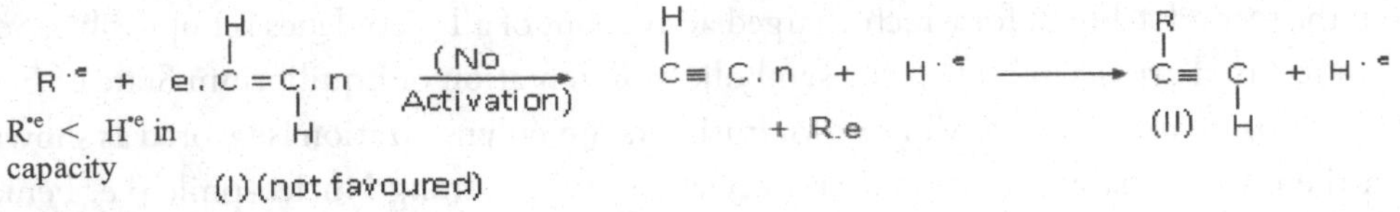

$$R^{\cdot e} + e \cdot \overset{\overset{\displaystyle H}{|}}{C} = \overset{\overset{\displaystyle }{|}}{C} \cdot n \xrightarrow[\text{Activation)}]{\text{(No}} \overset{\overset{\displaystyle H}{|}}{C} \equiv C \cdot n + H^{\cdot e} \longrightarrow \overset{\overset{\displaystyle R}{|}}{C} \equiv \overset{\overset{\displaystyle }{|}}{C} + H^{\cdot e}$$

$R^{\cdot e} < H^{\cdot e}$ in capacity $\qquad$ (I) (not favoured) $\qquad$ + R.e $\qquad$ (II) $\quad$ H $\qquad\qquad$ 6.9a

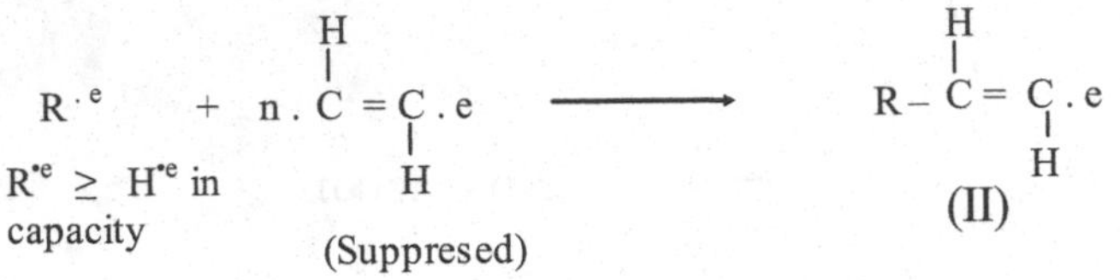

$$R^{\cdot e} + n \cdot \overset{\overset{\displaystyle H}{|}}{C} = \overset{\overset{\displaystyle }{|}}{C} \cdot e \longrightarrow R - \overset{\overset{\displaystyle H}{|}}{C} = \overset{\overset{\displaystyle }{|}}{C} \cdot e$$

$R^{\cdot e} \geq H^{\cdot e}$ in capacity $\qquad$ (Suppresed) $\qquad\qquad$ (II) $\qquad\qquad$ 6.9b

Nucleo-free-radically, the reaction is similar to that of Equation 6.5. Electro-free-radically the reactions are similar to those of Equation 6.6, where activation is not favored with no poly-merization. This happens only if $R^{\cdot e}$ is of lower capacity than H. Indeed, the acetylene being held in Equilibrium state of existence in the first two equations above does not take place; for which there is no reaction whatsoever as shown above. Thus, initiation can only be favored if the electro-free- radical is of equal or greater capacity than the hydrogen atom and the acetylene is suppressed. Thus, this monomer can only be activated electro-free-radically for which polymerization is made possible as shown in Equation 6.9b above and clearly shown below under certain operating conditions.

$$H^{\cdot n} + HC \equiv CH \longrightarrow \text{No reaction} \qquad\qquad 6.10$$

[Female $\qquad$ [Strong Female OR

Initiator] $\qquad$ Nucleophile]

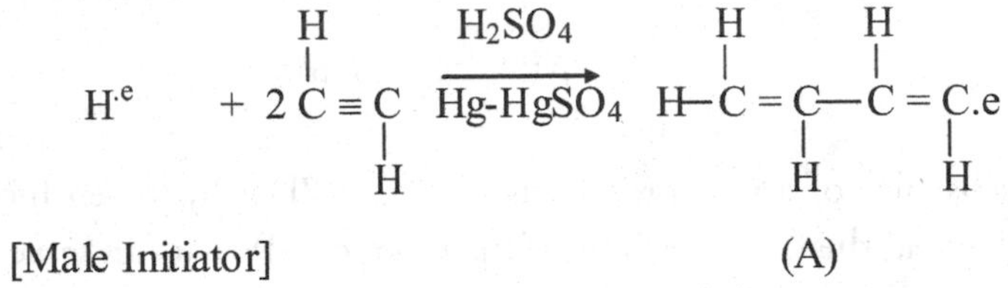

$$H^{\cdot e} + 2\overset{\overset{\displaystyle H}{|}}{C} \equiv \overset{\overset{\displaystyle }{|}}{C} \xrightarrow[\text{Hg-HgSO}_4]{\text{H}_2\text{SO}_4} \overset{\overset{\displaystyle H}{|}}{H - C} = \overset{\overset{\displaystyle H}{|}}{C} - \overset{\overset{\displaystyle H}{|}}{C} = \overset{\overset{\displaystyle }{|}}{C} \cdot e$$

[Male Initiator] $\qquad\qquad$ (A) $\qquad\qquad$ 6.11

Propagation as shown above may not be possible, if the passive catalyst used above can only suppress about fifty percent of the acetylene. A good passive catalyst is that which will fully suppress the whole acetylene.

It will therefore however not be surprising to note why the existences of monomers such as shown below are virtually impossible to obtain directly using acetylene as the source.

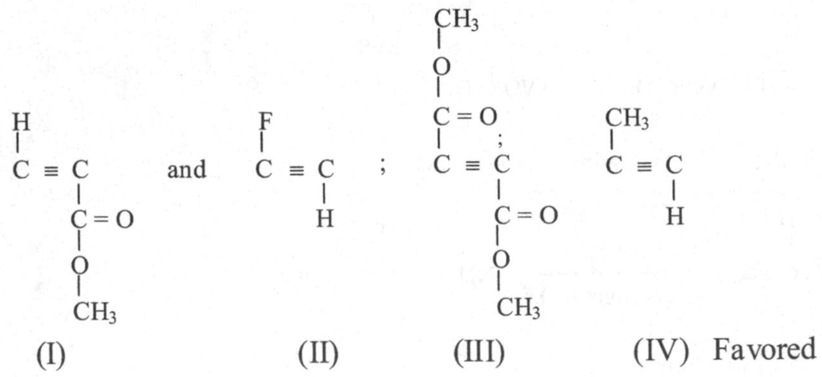

$$\underset{(I)}{\overset{\overset{\displaystyle H}{|}}{\overset{|}{C}} \equiv \overset{\overset{\displaystyle }{|}}{\underset{\underset{\underset{\underset{CH_3}{|}}{O}}{\underset{|}{C=O}}}{C}}} \quad \text{and} \quad \underset{(II)}{\overset{\overset{\displaystyle F}{|}}{\overset{|}{C}} \equiv \overset{\overset{\displaystyle }{|}}{\underset{\underset{H}{|}}{C}}} \quad ; \quad \underset{(III)}{\overset{\overset{\overset{\overset{\overset{\displaystyle CH_3}{|}}{O}}{|}}{C=O}}{|}}{\overset{|}{C}} \equiv \overset{\overset{\displaystyle }{|}}{\underset{\underset{\underset{\underset{CH_3}{|}}{O}}{\underset{|}{C=O}}}{C}}} \quad \underset{(IV) \text{ Favored}}{\overset{\overset{\displaystyle CH_3}{|}}{\overset{|}{C}} \equiv \overset{\overset{\displaystyle }{|}}{\underset{\underset{H}{|}}{C}}} \qquad 6.12$$

To obtain (I) for example, from (I) of Equation 6.2, the followings are obtained

$$(H_3COOC)_2O \;+\; \underset{\underset{H}{|}}{C} \equiv C^{\cdot n}\; H^{\cdot e} \longrightarrow CH_3OCOOH \;+\; \underset{\underset{H}{|}}{C} \equiv CCOOCH_3 \qquad 6.13$$

$$F_2 \;+\; \underset{\underset{H}{|}}{C} \equiv C.n \longrightarrow HF \;+\; \underset{\underset{H}{|}}{C} \equiv C - F \qquad 6.14$$
$$+\; H^{\cdot e}$$

The equations above are favored via Equilibrium mechanism. Also electro-free-radicals such CH_3, C_2H_5 (IV) can displace the H atom to form another product with acetylenic character. For all of them, there are several other methods which can be used to synthesize them. It is important to note from the overall analysis, that the hydrogen atom that is loosely held is that connected to the nucleo-free-radical carbon center. Hence, nucleo-free-radical initiation is not favored for acetylene a monomer which is a strong nucleophile far more than the alkenes.

Ability to use (II) of Equation 6.6 or 6.9b depends on what E or R is. If E is a metallic pushing group, that is, that without transfer species of the first kind e.g. ionic metal, then the following is obtained.

$$\underset{\underset{H}{|}}{\overset{\overset{E}{|}}{C}} \equiv C \;\rightleftharpoons\; \underset{(I)}{\overset{\overset{H}{|}}{C} \equiv C^{\cdot n}} \;+\; E^{\cdot e} \;;\; \underset{(II)}{\overset{\overset{E}{|}}{C} \equiv C.n} \;+\; H.e$$

$$E \equiv Na \qquad\qquad\qquad E \equiv e.g.\; H_3C.e \qquad 6.15$$

$$e.\underset{\underset{H}{|}}{\overset{\overset{E}{|}}{C}} = C.n \longrightarrow \underset{(I)}{\overset{\overset{E}{|}}{C} \underset{\underset{H}{|}}{\equiv} C} \qquad 6.16$$

<u>E ≡ strong Pushing group e.g. CH_3 group</u>

(II) in Equation 6.15, is still unstable, because of the presence of the second H atom. Its stability however will partly depend on the capacity of E. While (I) of Equation 6.16 if stable can favor only free-radical polymerizations in the absence of metal, (I) of Equation 6.15 cannot, until the Na metal and even the remaining H have been replaced with pushing groups of greater capacity. The strong nucleophilic character of acetylene can thus be observed since only the existence of the following types of acetylenic monomers for polymerization are favored using acetylene ($HC \equiv CH$) as the only source.

$$\underset{\underset{R}{|}}{\overset{\overset{H}{|}}{C}} \equiv C \qquad , \qquad \underset{\underset{R'}{|}}{\overset{\overset{R}{|}}{C}} \equiv C \qquad\qquad 6.17$$

(where R, R^1 are Pushing groups)

Just as very little is known about the chemistry of the reactions of acetylenes, the same applies to nitriles and indeed most or all Addition monomers as is clearly becoming obvious.

Like the first member of acetylenes ($CH \equiv CH$), the first member of nitriles cannot also in most cases favor polymerization for a different reason. The first member, a well known poison is hydrogen cyanide which cannot readily be activated because of the more favored free-radical "Equilibrium State of Existence" as shown below in its unactivated state. This is also as a result of the strong unsaturation in the monomer.

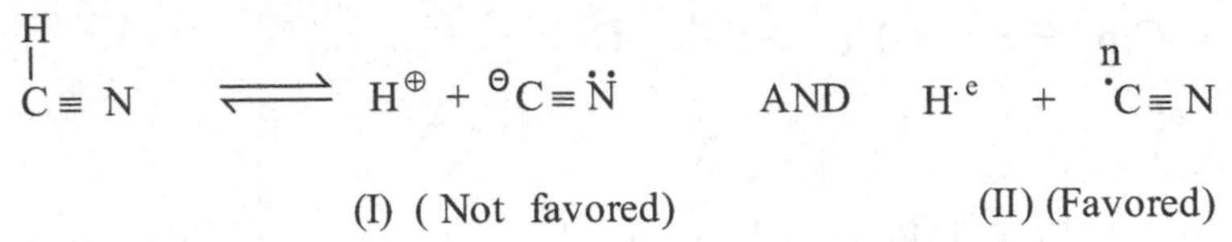

$$\overset{\overset{\text{H}}{|}}{C} \equiv N \quad \rightleftharpoons \quad H^{\oplus} + {}^{\ominus}C \equiv \overset{..}{N} \qquad AND \qquad H^{\cdot e} + \overset{n}{\overset{\cdot}{C}} \equiv N \qquad\qquad 6.18$$

$$\text{(I) (Not favored)} \qquad\qquad \text{(II) (Favored)}$$

Shown below are the strong unsaturations in these monomers, for which like Benzene, they can never be activated chargedly.

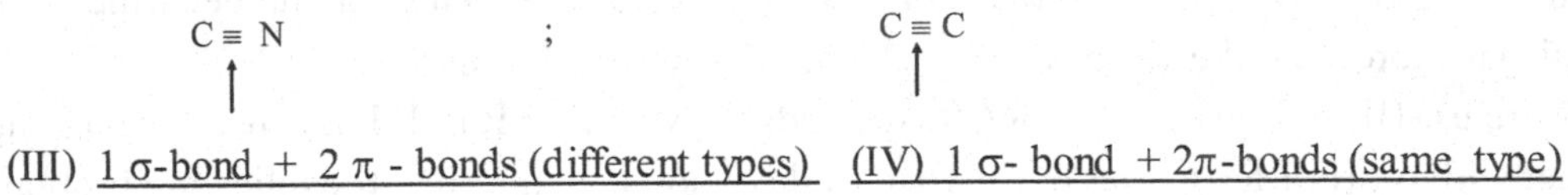

$$\overset{\uparrow}{C \equiv N} \qquad\qquad ; \qquad\qquad \overset{\uparrow}{C \equiv C} \qquad\qquad 6.19$$

$$\text{(III) } \underline{1 \, \sigma\text{-bond} + 2 \, \pi \text{ - bonds (different types)}} \quad \text{(IV) } \underline{1 \, \sigma\text{- bond} + 2\pi\text{-bonds (same type)}}$$

In (III), there is no hybridization, while in (IV) there is hybridization of the two carbon centers (sp - type). Since the carbon center of $C \equiv N$ is not hybridized in the presence of N, the C - H bond in $HC \equiv N$ cannot be ionic but covalent just as in the C - H bond in $HC \equiv CH$ where the C center is hybridized. However, the C – H bond in both are different in their covalent capacities. There is no doubt that the charged state of (I) is not favored due to electrostatic forces of repulsion. It should also be noted that $N\equiv$ and $O=$ are strong radical-pushing groups, like NR_2 and OR groups. Hence for (II) of Equation 6.18, the H has to be replaced with alkyl groups to favor full covalent existence as shown below free-radically.

$$R^{\cdot e} + H \overset{e}{\cdot} \overset{n}{\overset{..}{C}} \equiv N \quad \longrightarrow \quad \overset{\overset{R}{|}}{C} \equiv N + H^{\cdot e}$$

$$\text{(I)} \qquad\qquad 6.20$$

$$\underset{\underset{\text{Sp}^3 \text{ hybridized}}{}}{H - \overset{\overset{H}{|}}{\underset{\underset{H}{|}}{C}} - C \equiv N} \qquad \text{(for } R \equiv CH_3) \qquad\qquad 6.21$$

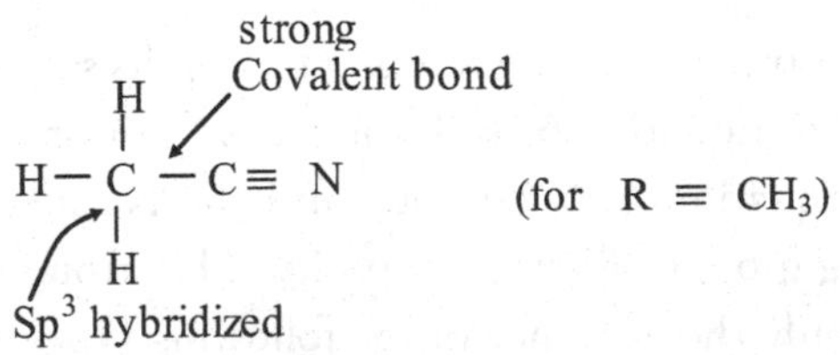

strong
Covalent bond

$$R^{\cdot n} + H \overset{e}{\overset{\ominus}{}} \overset{\ominus n}{C \equiv N} \quad \longrightarrow \quad RH + n \cdot C \equiv N \qquad\qquad 6.22$$

$$Na^{\odot} \cdot e + H \overset{e}{\overset{\ominus}{}} \overset{\odot n}{C \equiv N} \quad \longrightarrow Na^{\oplus} \cdots {}^{\ominus}CN + H^{\cdot e} \quad OR \quad Na^{\oplus \ominus}C \equiv N + H^{\oplus}$$

$$\text{(I)} \qquad\qquad\qquad \text{(II)}$$

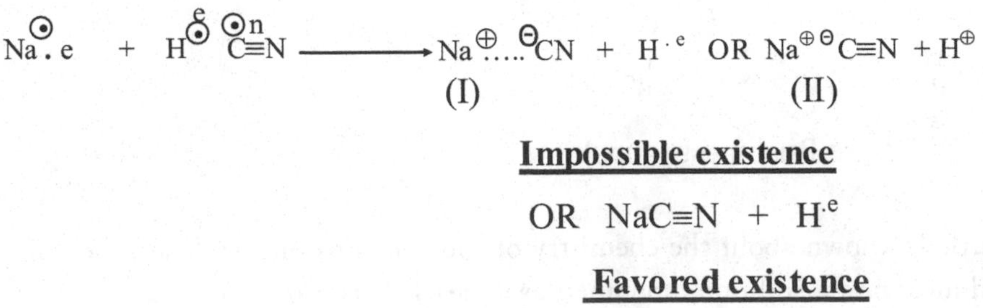

Impossible existence

$$OR \quad NaC \equiv N + H^{\cdot e} \qquad\qquad 6.23$$

Favored existence

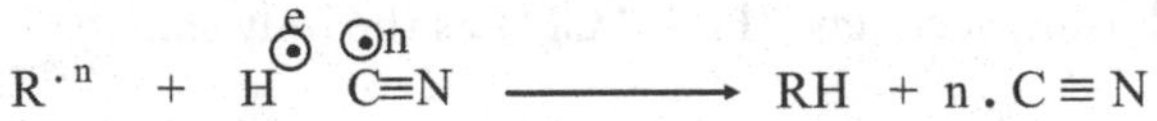

311

From the reactions above, it is obvious that

(i) (I) of Equation 6.20 is the product that can serve as a monomer for nitriles.

(ii) Monomers such as shown below can be obtained from hydrogen cyanide as a source, as well as by other means.

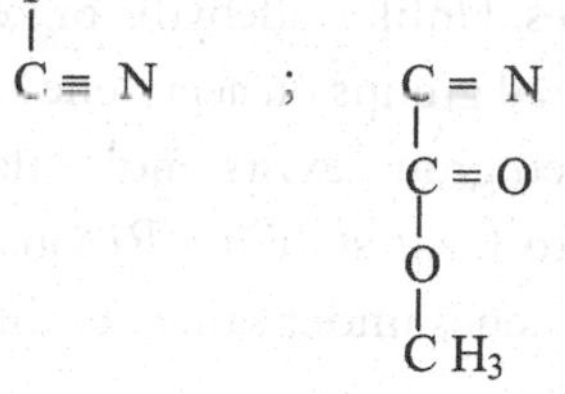

$$
\underset{\substack{|\\ C \equiv N}}{F} \quad ; \quad
\begin{array}{c}
C = N \\
| \\
C = O \\
| \\
O \\
| \\
C\,H_3
\end{array}
\qquad\qquad 6.24
$$

These are monomers with pulling groups.

(iii) (I) of Equation 6.23 has been said to be used in the past as "anionic ion-paired initiator" at very low temperatures of polymerization (-50 to - 40 °C) for acrolein[1], when indeed the initiator does not exist as already shown above. NaCN like HCN can only exist radically. Amongst "anionic ion-paired initiators" said to be used for the polymerization of acrolein viz butyllithium, sodium naphthalene, and NaCN, sodium or sodium cyanide (NaCN) in tetrahydrofuran or toluene was said to be an exception, since polymerization was observed to proceed completely through the carbonyl group which is the nucleophilic or female center in acrolein, C=C center being the electrophilic or male center. With negative free-charges as already said the C=O center in acrolein cannot be initiated. The same applies nucleo-free-radically. Electro-free-radically the C=O center is polymerized as shown below using an aldehyde.

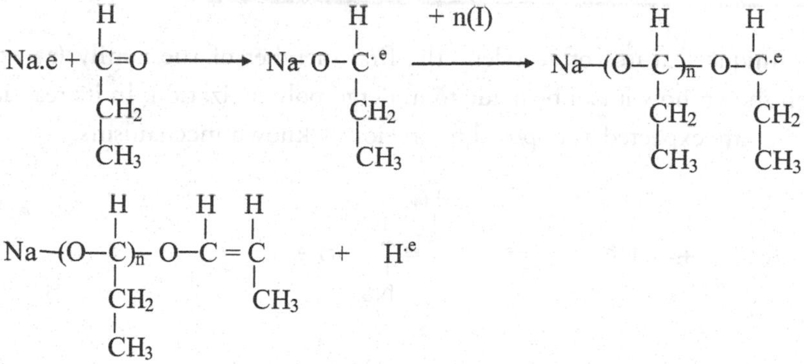

$$
\text{Na.e} + \begin{array}{c} H \\ | \\ C=O \\ | \\ CH_2 \\ | \\ CH_3 \end{array}
\longrightarrow
\text{Na}\;O-\begin{array}{c} H \\ | \\ C^{\cdot\,e} \\ | \\ CH_2 \\ | \\ CH_3 \end{array}
\xrightarrow{+\,n(I)}
\text{Na}-\left(\!O-\begin{array}{c} H \\ | \\ C \\ | \\ CH_2 \\ | \\ CH_3 \end{array}\!\right)_{\!\!n}\!\!O-\begin{array}{c} H \\ | \\ C^{\cdot\,e} \\ | \\ CH_2 \\ | \\ CH_3 \end{array}
$$

$$
\text{Na}-\left(\!O-\begin{array}{c} H \\ | \\ C \\ | \\ CH_2 \\ | \\ CH_3 \end{array}\!\right)_{\!\!n}\!\!O-\begin{array}{c} H \\ | \\ C \\ | \\ CH_2 \end{array}=\begin{array}{c} H \\ | \\ C \\ | \\ CH_3 \end{array}
\quad + \quad H^{\cdot\,e}
\qquad\qquad 6.25
$$

Unlike $t\text{-}H_9C_4^{\,\ominus}$ .....$^{\oplus}Li$ bond which is fully covalent chargedly and free-radically at two different conditions (charged and radical environments respectively), $NC^{\cdot n}$....^{e}Na bond is also fully covalent only free-radically at any condition. While $H_3C^{\cdot n}$ ....^{e}Li is covalent more radically than chargedly, $t\text{-}H_9C_4^{\,\ominus}$ ......$^{\oplus}Li$ is covalent more chargedly than radically. The reason for this is obviously the difference in the size, structure, and the electronegativity of the Central C atom in CH_3 and $t\text{-}C_4H_9$ with respect to these ionic types of metals. The covalent bond in the latter ($t\text{-}H_9C_4^{\,\ominus}$....$^{\oplus}Li$) is more elastic than in the former ($H_3C^{\ominus}$....$^{\oplus}Li$). When NaCl is frozen, the bonds between Na and Cl are largely covalent, but when in solution, they are ionic. It can thus be observed that there are indeed so many types of covalent bonds for which their existences are favored whether there is hybridization of orbitals or not, unlike ionic bond.

From the considerations above, it can be observed why nitriles when used as monomers are strong nucleophiles as will clearly and shortly become obvious. What is worthy of note about these monomers is that when polymerization is favored, the polymers obtained are as strong as steel. In addition to the transfer species already identified for aldehydes, ketones and olefinic monomers, there are two new ones associated with acetylenes and nitriles. ***These are transfer species of the first kind of the second and third types.*** Molecular rearrangement is favored by very few types of acetylene to a lesser extent than with olefins, in view of presence and location of only two substituted groups on the monomer as opposed to four on mono-olefins. It cannot however rearrange to cumulenes. Unlike aldehydic or ketonic monomers where molecular rearrangement is not favored, the two substituted groups on acetylene are located on two different active centers. With nitriles where only one substituted group exists, molecular rearrangement is also very limited, than in acetylenes, because of the drive to favor stability. Resonance stabilization phenomena do not seem to be favored by the nitriles, but acetylenes under same conditions which favor their use as monomers.

6.1. Acetylenes

There is need to recall the past with respect to the mechanisms of the reactions of acetylenes. With aldehydes and ketones, the observations of the past were clearly explained using the newly proposed concepts. **In the past up to the moment, no mechanisms existed.** Based on current developments and new definitions, only three types of mechanisms as already stated into law cut across the entire system, depending on the character of the monomers. In fact, it cuts across all things in life. One of these mechanisms-Combination has clearly been seen. Others will be slowly and skillfully disclosed as we move along the Volumes and the Series.

6.1.1. The mechanisms of the Reactions of Acetylene.

In view of the unique features of acetylene the first member of the family (as exists with other families), it has been shown how it can be made to undergo polymerization. In its reaction with molten sodium, the followings are expected as opposed to previously known mechanisms.[2]

$$
\underset{\text{(I)}}{\overset{\displaystyle H}{\underset{\displaystyle |}{C \equiv C} \cdot^{n}} \, Na^{\cdot e}} + H^{\cdot e} \longrightarrow \underset{\text{(II)}}{\overset{\displaystyle H}{\underset{\displaystyle |}{\underset{\displaystyle Na}{C \equiv C}}} + H.e}
$$

6.26

Na atom being far more electropositive or electro-free-radical than H atom, readily displaces the H from acetylene to form sodium acetylides (I), which was thought to be ionic in character, but shown not to be so, because the nucleo-free-radical on the carbon center of the acetylene is hybridized and indeed not non-free. The sodium is only loosely covalently bonded being an ionic metal.

When the acetylene reacts with a primary alkyl halide, the followings are obtained.

$$
\overset{\displaystyle H}{\underset{\displaystyle |}{C \equiv C} \cdot^{n}} H^{\cdot e} + RCl \longrightarrow \overset{\displaystyle H}{\underset{\displaystyle |}{\underset{\displaystyle R}{C \equiv C}}} + H^{\cdot e} + Cl^{\cdot nn}
$$

$$\longrightarrow \quad n . \overset{\overset{\displaystyle H}{|}}{C}=\overset{\overset{\displaystyle }{|}}{\underset{\underset{\displaystyle R}{|}}{C}}.e \ + \ H^{\cdot e} \ + \ Cl^{\cdot nn} \quad \longrightarrow \quad \overset{\overset{\displaystyle H}{|}}{C}=\overset{\overset{\displaystyle }{|}}{\underset{\underset{\displaystyle R}{|}}{\underset{\displaystyle H}{}}}.e \ + \ Cl . nn$$

$$\longrightarrow \quad \overset{\overset{\displaystyle H}{|}}{\underset{\underset{\displaystyle H}{|}}{C}}=\overset{\overset{\displaystyle Cl}{|}}{\underset{\underset{\displaystyle R}{|}}{C}} \qquad\qquad\qquad 6.27$$

(I)

If the acetylene had been fully activated, the followings would also be obtained.

$$e . \overset{\overset{\displaystyle H}{|}}{\underset{\underset{\displaystyle H}{|}}{C}}=\overset{}{\underset{}{C}}. n \ + \ R^{\cdot e} \ + \ Cl^{\cdot nn} \quad \xrightarrow{\quad \substack{\text{(Activation} \\ \text{assumed)}} \quad} \quad \overset{\overset{\displaystyle H}{|}}{\underset{\underset{\displaystyle Cl}{|}}{C}}=\overset{\overset{\displaystyle R}{|}}{\underset{\underset{\displaystyle H}{|}}{C}}$$

(II)   6.28

Note that (II) is different from (I) of Equation 6.27. (II) is the real product since the capacity of $H_3C^{\cdot e}$ is greater than that of $H^{\cdot e}$. It is also important to note that only the radical route will favor the existence of for example vinyl chloride, vinyl bromide, etc., using the activated state of the monomer (See Vol.I, Chapter 5) under some specific operating conditions. When the monomer is activated using HCl, in view of the fact that $H^{\cdot e}$ is of equal capacity with the groups carried by the monomer if the acid is concentrated, activation is favored noting that the $H^{\cdot e}$ carried by acetylene is a strong one since acetylene is strongly nucleophilic.

$$H^{\cdot e} \ + \ Cl^{nn} \ + \ \overset{\overset{\displaystyle H}{|}}{\underset{\underset{\displaystyle H}{|}}{C}}\equiv C \quad \longrightarrow \quad e. \overset{\overset{\displaystyle H}{|}}{C}=\overset{}{\underset{\underset{\displaystyle H}{|}}{C}}. n \ + \ H^{\cdot e} \ + \ Cl^{nn} \quad \longrightarrow \quad \overset{\overset{\displaystyle H}{|}}{\underset{\underset{\displaystyle H}{|}}{C}}=\overset{}{\underset{\underset{\displaystyle H}{|}}{C}}.e \ + \ Cl^{nn} \quad \longrightarrow$$

[Stabilized or Suppressed]

$$\overset{\overset{\displaystyle H}{|}}{\underset{\underset{\displaystyle H}{|}}{C}}=\overset{\overset{\displaystyle Cl}{|}}{\underset{\underset{\displaystyle H}{|}}{C}} \qquad\qquad\qquad 6.29$$

$H^{\cdot e}$ is the first to add followed by Cl .nn to form vinyl chloride. To stabilize the acetylene, a passive catalyst must be present. Dimerization could take place in the process to give chloro-prene (Neither male or female) in the presence of cuprous salt in ammonium chloride.[3]

$$H^{\cdot e} \ \ ^n \overset{\overset{\displaystyle H}{|}}{C}\equiv C \ + \ \overset{\overset{\displaystyle H}{|}}{\underset{\underset{\displaystyle H}{|}}{C}}\equiv C \quad \longrightarrow \quad \overset{\overset{\displaystyle H}{|}}{\underset{\underset{\displaystyle H}{|}}{C}}=\overset{}{\underset{}{C}}.e \ + \ ^n \overset{\overset{\displaystyle H}{|}}{C}\equiv C \quad \longrightarrow \quad ^4\overset{\overset{\displaystyle H}{|}}{\underset{\underset{\displaystyle H}{|}}{C}}=\overset{}{\underset{\underset{\displaystyle H}{|}}{C}}{}^{-2}\overset{\overset{\displaystyle H}{|}}{C}\equiv C^1$$

[Stabilized by Cu_2Cl_2/NH_4Cl]   [Vinylacetylene]

$$+ H^{\cdot e \ nn \cdot}Cl$$

$$\xrightarrow{\hspace{2cm}} \quad e.\overset{\displaystyle H}{\underset{\displaystyle H}{C}} - \overset{\displaystyle H}{\underset{\displaystyle H}{C}} - C \equiv C \quad \xrightarrow{\hspace{1cm}} \quad {}^{\cdot e}\overset{\displaystyle H}{\underset{\displaystyle H}{C}} - \overset{\displaystyle H}{\underset{\displaystyle H}{C}} = C = \overset{\displaystyle H}{C} .n \quad + \quad H^{\cdot e} \ Cl^{nn} \xrightarrow{\hspace{2cm}}$$

{Resonance Stabilization} What type?

$$Cl - \overset{\displaystyle H}{\underset{\displaystyle H}{C}} - \overset{\displaystyle H}{\underset{\displaystyle H}{C}} = C = \overset{\displaystyle H}{\underset{\displaystyle H}{C}} \xrightarrow{\hspace{1.5cm}} \quad \underset{\displaystyle e\bullet \ \overset{|}{C}H_2}{\overset{\displaystyle H \quad CH_2}{C = C}} \xrightarrow{\hspace{1.5cm}} \quad \overset{\displaystyle H \quad H}{\underset{\displaystyle H \quad \underset{\displaystyle CH_2}{C\bullet e} \ Cl\bullet nn}{C = C}}$$

[A Cumulene] Cl•nn [Electro-radicalization]

$$\overset{\displaystyle H \quad H \quad \quad H}{\underset{\displaystyle H \quad \quad Cl \quad H}{C = C - C = C}}$$

[Chloroprene] 6.30

The mechanism of the reactions as will become clear in the development, is EQUILIBRIUM mechanism with three stages. The first stage is the formation of the dimer, because of the pre-sence of the type of passive catalyst which could only stabilize about 50 percent of the acetylene. Note the manner by which the dimer was produced. The loosely held H atom initiated the stabilized acetylene to form vinylacetylene. *The dimer formed is resonance stabilized, the resonance stabilization provided by the $-C\equiv CH$ group and not by $-CH=CH_2$ group being less nucleophilic.* The second stage begins with the 3,4- center of the dimer (Not a diene, but **something else**), the less nucleophilic center, that is activated by the acid which adds to it to form a cumulene. One can observe the conditions under which cumulenes are resonance stabi-lized. It is the condition where an alkenyl and an alkynyl are conjugatedly placed. Like the Di-enes, Tri-enes, one will now call that **"something else"** highlighted above an ENEYNE (Vinylalkyne), just as we shall see the case of DI-YNE downstream. In the third stage, the unique cumulene formed wherein CH_2Cl (Not CH_3 type of group) substituent group is carried by it, is placed in Equilibrium state of existence, followed by what has been called ELECTRO-RADICALIZATION downstream and finally addition to form the isoprene. Note how the isoprene was formed. Activation took place only in the first and second stages. In the third stage, it was something else. As an exercise, the reader should compare the four types of cumulenes- $H_2C=C=CH_2$, $(H_3C)HC=C=CH_2$, $(ClH_2C)HC=C=CH_2$, and $(ROOC)HC=C=CH_2$.

It is well known that one or two moles of halogen or halogen acid add to the triple bond.

$$RC \equiv CH \begin{cases} + \ X_2 \xrightarrow{\hspace{1.5cm}} RXC = CHX \\ + \ 2X_2 \xrightarrow{\hspace{1.5cm}} RCX_2CHX_2 \\ + \ HX \xrightarrow{\hspace{1.5cm}} RXC = CH_2 \qquad \text{[X are halogens]} \\ + \ 2HX \xrightarrow{\hspace{1.5cm}} RCX_2CH_3 \end{cases}$$

6.31

The type of acetylene used above shows that it has been stabilized with the presence of R group. This is reflected in Equation 6.27 in which the product obtained {(I) of Equation 6.27} is the same as that shown above, noting that in Equation 6,27, the first stage where RCl was suppressed was the questionable stage based on Equation 6.28.

$$H^{\cdot e} \ ^{nn\cdot}X \ + \ ^{e\cdot}\underset{H}{\overset{R}{C}} = C^{\cdot n} \longrightarrow \underset{H}{\overset{R \atop |}{C}} = C^{\cdot e} \ + \ X^{\cdot nn} \longrightarrow \overset{H \quad R}{\underset{H \quad X}{C = C}} \quad \underline{+ \ HX}$$

$$(I)$$

$$H^{\cdot e} \quad n.\overset{R \ \ H}{\underset{X \ \ H}{C - C}}.e \ + X.nn \longrightarrow H - \overset{R \ \ H}{\underset{X \ \ H}{C - C}}.e \ + X.nn \longrightarrow H - \overset{R \ \ H}{\underset{X \ \ H}{C - C}} - X \qquad 6.32$$

$$(II) \qquad\qquad\qquad\qquad\qquad\qquad\qquad (III) - RCHXCH_2X$$

(II) is the real activated state of (I) if R is an alkyl group. (III) obtained above is the real product and not that shown in Equation 6.31 (i.e. $RCX_2CH_3)^2$. Indeed, (II) cannot be activated chargedly due to electrostatic forces of repulsion. It has been said that there are only three mechanisms by which Nature and indeed all things operate, be it chemical, polymeric, mechanical, behavioural, governmental, etc. All the mechanisms we have encountered so far are Combination mechanism such as during Addition polymerization (i.e. propagation) and Decomposition mechanism such as during initiator preparation from the catalyst. Indeed, we have only just seen very little of these two mechanisms so far, wherein we saw that the equations are not only stoichiometrically balanced, but also radically and chargedly balanced. These two mechanisms are common with polymeric systems. The third mechanism common with chemical reactions is Equilibrium mechanism. For example, consider the reactions of the halogen with a triple bond in acetylene as shown in Equation 6.31. Here, the halogen of choice is Cl_2.

$$Cl_2 \ + \ RC \equiv CH \longrightarrow 2Cl.nn \ + \ RC \equiv CH \longrightarrow \text{No activation possible} \quad 6.33$$

If Cl_2 is made to exist in Equilibrium state of existence, the following is obtained.

$$Cl_2 \rightleftharpoons Cl \cdot en \ + \ Cl \cdot nn$$
$$\quad\quad (Male) \quad\quad (Female)$$
$$Equilibrium \ State \ of \ existence$$
$$\qquad\qquad\qquad\qquad\qquad\qquad\qquad\qquad 6.34$$

$$Cl_2 \xrightarrow{\ Decomposition \ state \ of \ existence\ } \quad Cl \cdot nn \ + \ Cl \cdot nn$$
$$\qquad\qquad\qquad\qquad\qquad (Female) \quad (Female)$$
$$\qquad\qquad\qquad\qquad\qquad\qquad\qquad\qquad 6.35$$

$$Cl \cdot nn \ + \ Cl \cdot nn \xrightarrow{\ Combination \ State \ of \ existence\ } Cl_2 \qquad 6.36$$

Stage 1 : $\quad Cl_2 \quad \rightleftharpoons \quad\quad Cl.en \ + \ Cl.nn$

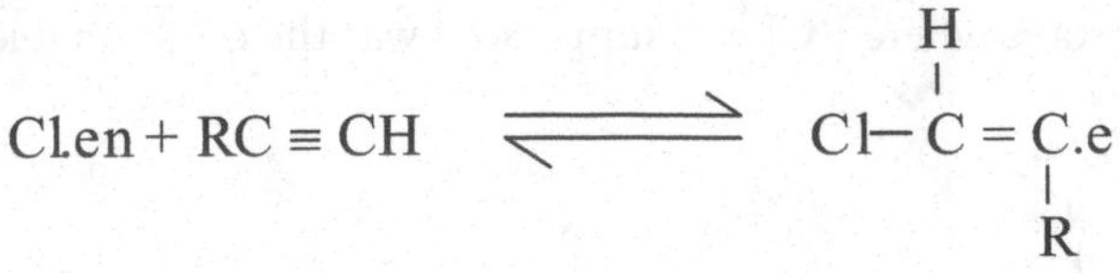

$$Cl.en + RC \equiv CH \rightleftharpoons Cl-\overset{\overset{\displaystyle H}{|}}{C}=\underset{\underset{\displaystyle R}{|}}{C}.e$$

(I)

$$(I) \quad + \quad Cl.nn \longrightarrow \overset{\overset{\displaystyle H \quad Cl}{| \quad |}}{\underset{\underset{\displaystyle Cl \quad R}{| \quad |}}{C}=C}$$

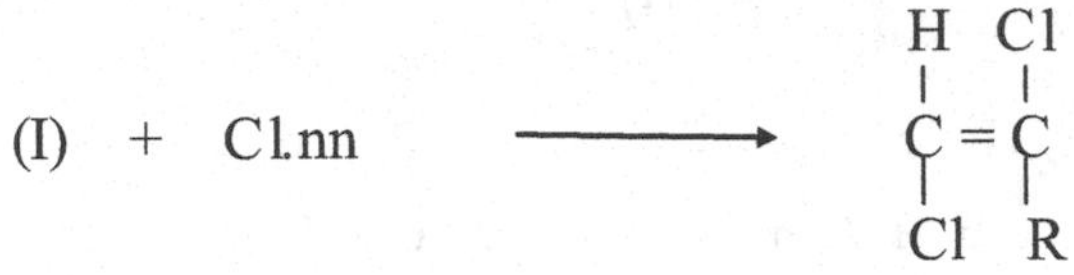

(II)   HClC=CClR   6.37

Overall equation: Cl_2 + $RC \equiv CH \longrightarrow HClC = CClR$ 6.38

This is how chemical reactions take place in Equilibrium mechanism. Just as in Combination and Decomposition mechanisms, there can be one or more stages, because everything in life or Nature take place in stages either in series or in parallel or both (STAGEWISE OPERATIONS). In the stage above, notice that there are three steps, the last step being a combination step, for without it no product can be obtained. A stage which cannot be interpreted into a mathematical language is no stage, since as already said, Mathematics is the only Natural language of commu-nication in the real and imaginary domains. Without the Equilibrium mechanism, the product above in Equations 6.31 (First reaction) and 6.38 cannot be obtained. All these are New Concepts very little of which one has begun to show since the beginning of Volume I. Indeed it is yet too early to introduce just the above New Concepts. One is forced to do it inorder to explain some of the reactions in Equation 6.31, without the need to ask questions. Imagine if the existence of Cl.en had been favored during combination mechanism, then the followings would have been obtained.

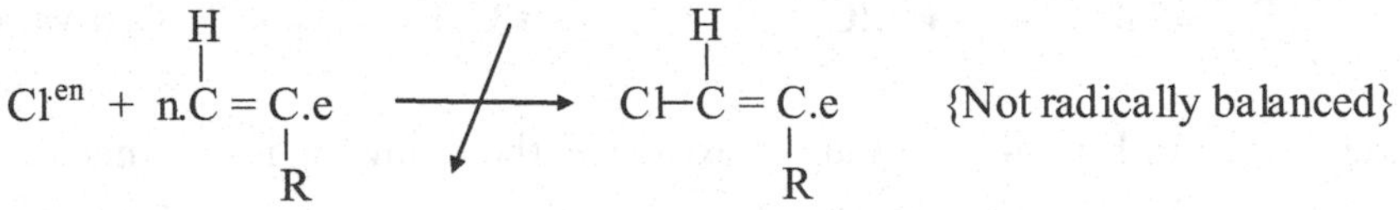

$$Cl^{en} + n.\overset{\overset{\displaystyle H}{|}}{\underset{\underset{\displaystyle R}{|}}{C}}=C.e \quad \cancel{\longrightarrow} \quad Cl-\overset{\overset{\displaystyle H}{|}}{\underset{\underset{\displaystyle R}{|}}{C}}=C.e \quad \text{\{Not radically balanced\}}$$

(Non-free-radical)   (Free-radical)   6.39

While the initiation above is not favored, because the equation is not radically balanced, in Equilibrium mechanism, where half double headed arrows are used (Equation 6.37), this is not the case for as long as the male character (.e and .en) is maintained free (.e) or non-free (.en).

With one more mole of Cl_2 in the system, the second stage follows as shown below.

Stage 2: $Cl_2 \rightleftharpoons Cl.en + nn.Cl$

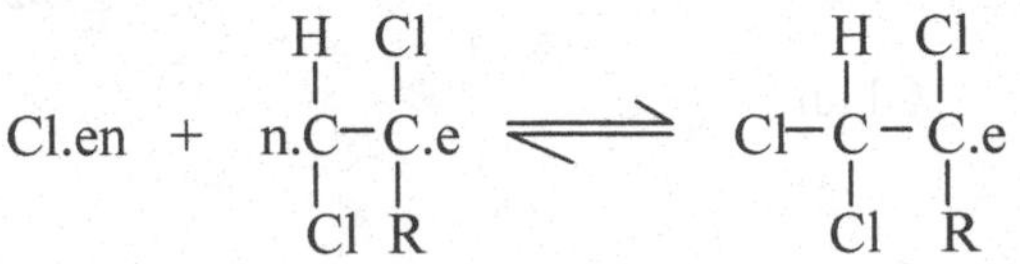

$$Cl.en \quad + \quad n.\overset{\overset{\displaystyle H \quad Cl}{| \quad |}}{\underset{\underset{\displaystyle Cl \quad R}{| \quad |}}{C}-C}.e \rightleftharpoons Cl-\overset{\overset{\displaystyle H \quad Cl}{| \quad |}}{\underset{\underset{\displaystyle Cl \quad R}{| \quad |}}{C}-C}.e$$

(III)

$$Cl.nn + (III) \longrightarrow \begin{array}{c} H \quad Cl \\ | \quad | \\ Cl{-}C{-}C{-}Cl \\ | \quad | \\ Cl \quad R \end{array} \qquad 6.40$$

$$(IV){-}Cl_2HC{-}CCl_2R$$

Overall equation: $Cl_2 + ClHC = CRCl \longrightarrow RCCl_2CHCl_2$ $\qquad 6.41$

(IV) above is the same as the product shown in Equation 6.31. It can only be obtained via Equilibrium mechanism. In all these reactions, it is the male radical or positive charge that initiates the attack all the time, except when not present in the system. (II) of Equation 6.37 cannot be activated chargedly due to electrostatic forces of repulsion. Cl_2 cannot also exist in Equilibrium state of existence in a charged state, because Cl cannot carry a positive charge in the real state, since it falls into the last group of the Periodic Table. It is members of lower groups, the more valent ones (O, N,) that can carry a positive charge of the imaginary type such as shown below for oxygen in dilute hydrochloric acid.

Stage 1: $\qquad HCl \rightleftharpoons H.e + Cl.nn$

$$H.e + H_2O \rightleftharpoons \begin{array}{c} H \\ | \\ en.O{-}H \\ | \\ H \end{array}$$

$$(I)$$

$$Cl.nn + (I) \longrightarrow \begin{array}{c} H \\ | \\ Cl^{\ominus}.......^{\oplus}O{-}H \\ | \\ H \end{array} \qquad 6.42$$

(II) Dilute hydrochloric acid

For the first time right from Volume I, we began to see what is the real Natural Sciences - Mathematics, Physics and Chemistry. Waite a miunite, we have seen nothing yet for knowledge has no end. For the first time, we have to see what is the so-called "hydroxonium ion" as in the structure for dilute hydrochloric acid, i.e. (II). It is not an ion, but an electrostatic charge, noting that the dotted bond above is an electrostatic bond, an imaginary bond. We can dilute (II) futher, since the Equilibrium state of existence, that which is fixed for all things in Nature including compounds, humans, animals, plants and more, is as follows.

$$\begin{array}{c} H \\ | \\ Cl^{\ominus}......^{\oplus}O{-}H \\ | \\ H \end{array} \rightleftharpoons H^{.e} + \begin{array}{c} H \\ | \\ Cl^{\ominus}....^{\oplus}O{-}H \\ \quad \bullet nn \end{array} \qquad 6.43a$$

In stage 1 above, notice the Combination state of existence of (II), different from the Equilibrium state of existence shown in Equation 6.43a above. The H.e can add to another water molecule in stage 2 to give what is shown below.

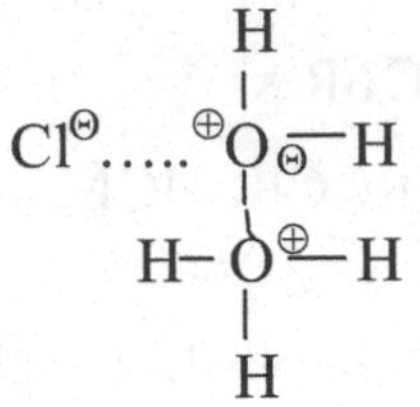

$$Cl^{\ominus}.....\overset{\overset{\displaystyle H}{|}}{\underset{}{O}}_{\ominus}^{\oplus}\!\!-\!H$$
$$H\!-\!\overset{}{\underset{\underset{\displaystyle H}{|}}{O}}{}^{\oplus}\!\!-\!H$$

(III)- More dilute hydrchloric acid.   6.43b

One can begin to see not only with the physical eyes, but also with the eye of the needle i.e. "the third eye" that which every living system is endowed with. When current is put into (III), it will be carried through by the + - + - + -.... Imaginary charges. This is one of the most important conditions an electrolyte must satisfy to be so classified. Thus, for the first time we have begun to show how current flows in some solutions.

So far, we were forced to digress based on what we saw with Equation 6.31. Now, when peroxides are used for one of the reactions of Equation 6.31, the followings are said to take place.

$$RC \equiv CH \;+\; HBr \;\xrightarrow{\;Peroxides\;}\; RCH = CHBr \qquad\qquad 6.44$$

In the absence of the peroxide, the product according to Equation 6.31 is $H_2C = CRBr$. The product in the presence of peroxide is **_possibly_** obtained as follows.

$$HOOH \longrightarrow 2HO.nn \longrightarrow 2H^{.n} + 2\, nn.O.en \xrightarrow{+\,2HBr} 2HO.nn \;+\; 2Br.en \;+\; 2H.n$$

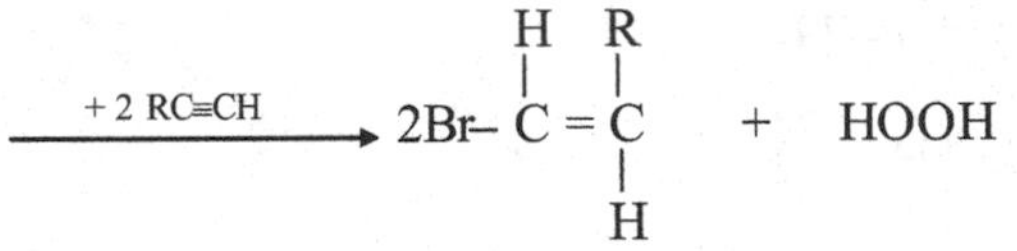

$$\xrightarrow{+\,2\,RC\equiv CH} 2Br\!-\!\overset{\overset{\displaystyle H}{|}}{C} = \overset{\overset{\displaystyle R}{|}}{\underset{\underset{\displaystyle H}{|}}{C}} \;+\; HOOH$$

$$6.45$$

Stage 1: $\qquad\qquad HOOH \longrightarrow 2HO.nn$

$$2HO.nn \longrightarrow 2H.n \;+\; 2\,nn.O.en$$

$$2\,nn.O.en \;+\; 2HBr \longrightarrow 2HO.nn \;+\; 2Br.en$$

$$2\,Br.en \;+\; 2\,n.\,\overset{\overset{\displaystyle H}{|}}{\underset{\underset{\displaystyle R}{|}}{C}} = C\,.e \;\rightleftharpoons\; 2\,Br - \overset{\overset{\displaystyle H}{|}}{\underset{\underset{\displaystyle R}{|}}{C}} = C\,.e$$

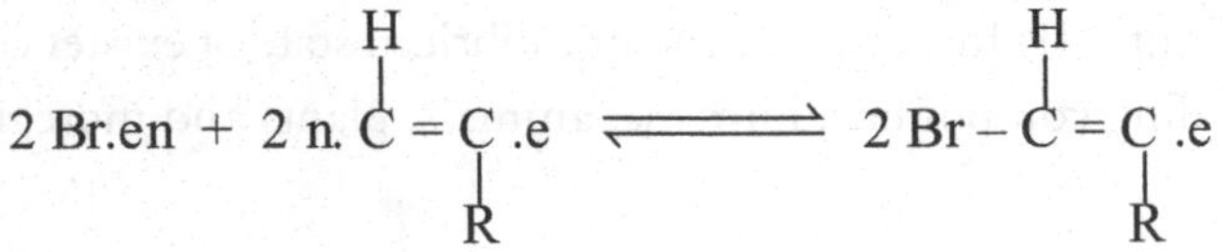

$$(A)$$

$$(A) \;+\; 2H.n \longrightarrow 2\,BrHC = CHR$$

$$2\,HO.nn \longrightarrow HOOH \qquad\qquad 6.46a$$

Overall Equation: $HOOH \;+\; 2HBr \;+\; 2RC\equiv CH \longrightarrow 2BrHC = CHR \;+\; HOOH$ 6.46b

The Decomposition mechanism above is not favored. It is that wherein a nucleo-free-radical and oxidizing oxygen are first produced during the decomposition of HOOH. In the third step of that stage the oxidizing oxygen then oxidizes the HBr which is in Stable state of existence to give Br^{en} and $HO^{\cdot nn}$. The Br^{en} then activates the acetylene (which is in a more Stable state of existence than HBr) followed by addition to $H^{\cdot n}$ to give the product in the fourth and fifth steps of the stage. At the end of that stage, HOOH is recovered thus making it behave like an active catalyst. ***Via Equilibrium mechanism, the product obtained is same as will be shown in the laws.*** The Stage above which is not favored was the so-called Anti-Markovnikov's rule in Chapter 1 (See Equation 1.65).

When ionic metallic centers are involved with the monomers, the situation is different whether the monomer has substituent groups or not, since ionic metals cannot be strongly covalently bonded to a carbon center.

$$
\begin{array}{c}
\overset{\displaystyle R'}{\underset{\displaystyle Na}{|}}\\[-2pt]
C \equiv C \; + \; R^{\cdot e} \; + \; Cl \,.nn
\end{array}
\;\longrightarrow\;
\begin{array}{c}
\overset{\displaystyle R'}{\underset{\displaystyle Na}{|}}\\[-2pt]
e\,.C = C.n \; + \; R^{\cdot e} \; + \; Cl \,.nn
\end{array}
$$

(I) $(\,R' > R \; or \; R' < R\,)$

$$
\longrightarrow\;
\begin{array}{c}
R' \quad R\\
| \qquad |\\
C = C\\
| \qquad |\\
Cl \quad Na
\end{array}
$$

(Not favored)   6.47

The activated state of (I) above is least expected, since Na is still loosely bonded to the carbon center. Therefore, the real reaction is as follows.

$$
\begin{array}{c}
\overset{\displaystyle R'}{\underset{\displaystyle Na}{|}}\\[-2pt]
C = C \; + \; R^{\cdot e} \; + \; Cl \,.nn
\end{array}
\;\longrightarrow\;
\begin{array}{c}
\overset{\displaystyle R'}{|}\\[-2pt]
C = C^{\cdot n} \; Na^{\cdot e} \; + \; R^{\cdot e} \; + \; Cl \,.nn
\end{array}
$$

(Favored)

$$
\longrightarrow\;
\begin{array}{c}
\overset{\displaystyle R'}{\underset{\displaystyle R}{|}}\\[-2pt]
C \equiv C \; + \; Na^{\cdot e} \; + \; Cl \,.nn
\end{array}
\;\longrightarrow\;
\begin{array}{c}
\overset{\displaystyle R'}{\underset{\displaystyle R}{|}}\\[-2pt]
C \equiv C \; + \; Na^{\oplus} \; + \; Cl^{\ominus}
\end{array}
$$

(I)

$$
\longrightarrow\;
\begin{array}{c}
\overset{\displaystyle R'}{\underset{\displaystyle R}{|}}\\[-2pt]
C = C \; + \; NaCl.
\end{array}
$$
                                                                        6.48

Though the last mechanism above looks like Decomposition, it is indeed an Equilibrium mechanism. From the two equations above, it is the last that is favored, since the metallic sodium with little or no pushing capacity is indeed loosely held to the carbon center. Though Na, Li etc have little or no pushing capacity, they are strongly electropositive. On the other hand, it is important to note that this applies only to ionic metals, the Groups IA, IIA and IIIA elements.

Acetylene is said to react readily with aqueous ammoniacal silver nitrate or aqueous ammoniacal cuprous chloride solution to give water insoluble carbides. They are water insoluble because Ag and Cu cannot form ionic bonds, but only other types of bonds- covalent, electro-static, and dative. The ammonias are datively bonded, while the nitrate or chloride is covalently bonded to Ag or Cu center the Central atoms. This is unlike the case with Na above. Na is covalently bonded but of a very weak type in view of the strong ionic character of Na. Thus, the mechanisms of the reactions involving the non-ionic metals above are tentatively as follows. How they take place will shortly be explained.

$$\overset{\overset{\displaystyle H}{\displaystyle |}}{C} \equiv C^{\cdot n} \quad H^{\cdot e} + Ag^{\cdot en} \quad \longrightarrow \quad \overset{\overset{\displaystyle H}{\displaystyle |}}{C} \equiv \underset{\underset{\displaystyle Ag}{\displaystyle |}}{C} + H^{\cdot e} \quad \overset{+Ag^{\cdot en}}{\longrightarrow}$$

$$\underset{\underset{\displaystyle (I)}{}}{\overset{\overset{\displaystyle Ag}{\displaystyle |}}{C} \equiv C^{\cdot n}} \quad H^{\cdot e} + Ag^{\cdot en} + H^{\cdot e} \quad \longrightarrow \quad \overset{\overset{\displaystyle Ag}{\displaystyle |}}{C} \equiv \underset{\underset{\displaystyle Ag}{\displaystyle |}}{C} + H_2 \qquad 6.49$$

Since Ag has little or no pushing capacity and most importantly non-ionic in character, the H is loosely bonded to the carbon center as shown by (I) above. The Ag-C bonds are therefore covalent in character unlike the C-H bond or C-Na bond in acetylene. It is very important to note that hydrogen molecules are produced in the process, a product which cannot be obtained if the reactions above were charged in character. Since the silver is a salt ($AgNO_3$) in ammonia, hydrogen molecule will not be formed. In the displacement reaction, this is what happens in two stages.

Stage 1:

$$HC \equiv CH \rightleftharpoons HC \equiv C^{\cdot n} + H^{\cdot e}$$

$$AgNO_3 \rightleftharpoons Ag^{en} + {}^{nn \cdot}O - NO_2$$

$$H^{\cdot e} + {}^{nn \cdot}O - NO_2 \rightleftharpoons HONO_2$$

$$HC \equiv C^{\cdot n} + {}^{en \cdot}Ag \longrightarrow HC \equiv CAg \qquad 6.50$$

Overall equation: $HC \equiv CH + AgNO_3 \longrightarrow HNO_3 + HC \equiv CAg \qquad 6.51$

Stage 2:

$$HC \equiv CAg \rightleftharpoons AgC \equiv C^{\cdot n} + H^{\cdot e}$$

$$AgNO_3 \rightleftharpoons Ag^{en} + {}^{nn \cdot}O - NO_2$$

(I)

$$** \quad H^{\cdot e} + {}^{nn \cdot}O - NO_2 \rightleftharpoons HONO_2$$

$$AgC \equiv C^{\cdot n} + {}^{en \cdot}Ag \rightleftharpoons AgC \equiv CAg \qquad 6.52$$

Overall equation: $HC \equiv CH + 2AgNO_3 \rightleftharpoons 2HNO_3 + AgC \equiv CAg \qquad 6.53$

Without the ammonia, as shown in Stage 2, no product can be obtained, since what Equation 6.53 is telling us is that $AgNO_3$ is soluble in silver acetylide and will dissolve in acetylide as they belong to almost the same family- Polar/Non-ionic. If the operating conditions of the forward and backward reactions of Equation 6.53 are made different, the reaction becomes reversible. This is similar to the case of sodium chloride (NaCl) and water (H_2O), viz. hydrochloric acid (HCl) and sodium hydroxide (NaOH). In order to make Stage 2 a productive one, the stage is completed as follows, noting that the third step of Stage 2 marked with double star will cease to exist.

$$H^{\cdot e} \;+\; \overset{\cdot\cdot}{N}H_3 \;\rightleftharpoons\; \underset{\underset{\displaystyle H \quad H}{|}}{\overset{\overset{\displaystyle H \quad H}{|}}{e\cdot N}}$$

(II)

$$(I) \;+\; (II) \longrightarrow O_2N-O^{\ominus}\ldots\ldots^{\oplus}NH_4$$

(III)- (Ammonium nitrate)   6.54

Overall equation: $2AgNO_3 + 2NH_3 + HC \equiv CH \longrightarrow AgC \equiv CAg + 2H_4NNO_3$ 6.55

If one is to access the State of the Chemistry of today, there is no doubt that all the textbooks for it and related disciplines must be thrown into the archives and not the dustbin. For the first time, we are seeing the real structure of ammonium nitrate. The bond in (III) above as in ammonium hydroxide, is an electrostatic bond (Not ionic bond). So also are the charges. The bond is an electrostatic bond, an imaginary bond. In the first stage, $AgC \equiv CH$ was first formed. In Stage 2, $AgC \equiv CAg$ was then formed. One can see from the states of existences, why non-ionic carbides are insoluble in water

   If activation of the acetylene had been favored, then the followings will occur.

$$\underset{\underset{\displaystyle H}{|}}{\overset{\overset{\displaystyle H}{|}}{e\cdot C}} = C^{\cdot n} \;+\; Ag^{\cdot en} \;\xrightarrow[+\,nn.ONO_2]{} \; \underset{\underset{\displaystyle H \quad H}{|\quad\;|}}{\overset{\overset{\displaystyle Ag \quad ONO_2}{|\quad\;\;|}}{C}} = C$$

6.56

(Not favored)

Based on Combination mechanism, the activation above is not radically balanced. Hence it is not favored. It can be observed that metals, ionic or non-ionic cannot activate acetylene. It is impor-tant to note that Ag as well as Cu is carrying electro-non-free-radicals on their centers.

   With one alkyl group present on the acetylene, only the hydrogen atom can be disp-laced, in the presence of an ionic metal or the alkyl group. For example, sodium acetylides can be formed readily by the reaction of acetylenes with highly dispersed sodium (particle size 10 - 25μ) in xylene or by reaction with sodium amide in liquid ammonia.

$$RC \equiv CH \;+\; Na^{\oplus}\,^{\ominus}NH_2 \longrightarrow RC \equiv C^{\ominus}\,^{\oplus}Na \;+\; NH_3$$

(I) Impossible existence   6.57

(I) as represented in many textbooks cannot carry ionic charges or covalent charges due to elec-trostatic forces of repulsion on the acetylene. In the sodium amide, the charges are ionic. The mechanism of the reaction is as follows.

Stage 1:

$$NaNH_2 \rightleftharpoons Na^{\cdot e} + {}^{nn\cdot}NH_2$$

$$RC \equiv CH \rightleftharpoons RC \equiv C^{\cdot n} + {}^{e\cdot}H$$

$$RC \equiv C^{\cdot n} + {}^{e\cdot}Na \rightleftharpoons RC \equiv CNa$$

$$H^{\cdot e} + nn. NH_2 \longrightarrow NH_3 \qquad\qquad 6.58$$

Overall Equation: $NaNH_2 + RC \equiv CH \longrightarrow RC \equiv CNa + NH_3 \qquad 6.59$

The mechanism can be noted to be Equilibrium mechanism of only one stage containing four steps. It is the last step that determines whether the stage is either a solubility/insolubility stage or a productive stage. Notice what has suppressed the Equilibrium State of Existence of the NH_3. It is the sodium acetylide.

Stronger alkyl groups in place of Ag, or Na, etc., can however also displace the hydrogen atom. This is one of the useful methods for introduction of the triple bond into other molecules in order to make it a monomer.

$$RC \equiv CH + R'X \longrightarrow RC \equiv CR' + HX \qquad\qquad 6.60$$

$$CH_3C \equiv C^{\ominus}{}^{\oplus}Na + C_2H_5Br \longrightarrow NaBr + CH_3C \equiv CC_2H_5$$

Sodium methylacetylide   **Methylethylacetylene *(2-pentyne)***   6.61

The mechanism of the second or last reaction has already been shown in Equation 6.48. For the first reaction, the followings are similarly obtained.

Stage 1:

$$RC \equiv CH \rightleftharpoons RC \equiv C^{\cdot n} + H^{\cdot e}$$

$$H^{\cdot e} + R'X \rightleftharpoons HX + R'^{\cdot e}$$

$$R'^{\cdot e} + {}^{n\cdot}C \equiv CR \longrightarrow RC \equiv CR' \qquad\qquad 6.62$$

Overall Equation: Same as in Equation 6.60.   (6.60)

When the two hydrogen atoms are replaced with alkyl groups, the followings is obtained.

$$\underset{\underset{R}{|}}{\overset{\overset{R}{|}}{C}} \equiv C + Ag^{\cdot ne} \longrightarrow n.\underset{\underset{R}{|}}{\overset{\overset{R}{|}}{C}} \equiv C.e + Ag^{\cdot ne} \longrightarrow \text{No reaction} \qquad 6.63$$

Since ***acetylides and carbides*** are said to be salts of the weak acid-acetylene, they are hydrolyzed by water.

$$HC \equiv C^{\cdot n}{}^{e\cdot}Na + H_2O \longrightarrow HC \equiv CH + NaOH \qquad 6.64$$

$$\underset{\underset{Ca}{\diagdown\diagup}}{C \equiv C} + 2H_2O \longrightarrow Ca(OH)_2 + HC \equiv CH \qquad\qquad 6.65$$

(I)

In present-day Science, the equations above are usually represented as charged reactions of ionic types! Indeed, these are no salts, since the charges above are not ionic but covalent and ***acetylene is not an***

acid. One can as well say that methane, ammonia are acids!, since their Equilibrium state of existence is as follows. This of course is impossible. For an acid to exist, there must be an hetero atom.

$$H_4C \rightleftharpoons H^{\cdot e} + {}^{n\cdot}CH_3 ; \qquad HC \equiv CH \rightleftharpoons HC \equiv C^{\cdot n} + H^{\cdot e}$$

$$H_3N \rightleftharpoons H^{\cdot e} + {}^{nn\cdot}NH_2 \qquad\qquad 6.66$$

The covalent character of these groups of compounds is shown by the structure of the carbide, a three membered ring, (I) of Equation 6.65. The mechanisms of the two reactions are as follows.

Stage 1: $HC \equiv CNa \rightleftharpoons Na^{\cdot e} + HC \equiv C^{\cdot n}$

$Na.e + H_2O \rightleftharpoons HO^{\cdot en} + NaH + Heat$

$HO \bullet en \rightleftharpoons nn \bullet O \bullet en + H \bullet e + Heat$

$HC \equiv C^{\cdot n} + H^{\cdot e} \rightleftharpoons HC \equiv CH$

$nn \bullet O \bullet en + NaH \rightleftharpoons HO \bullet nn + Na \bullet e + Heat$

$Na^{\cdot e} + {}^{nn\cdot}OH \longrightarrow NaOH \qquad\qquad 6.67$

Stage 1: (I) of Eqn. 6.65 $\rightleftharpoons {}^{\cdot e}Ca - C \equiv C^{\cdot n}$

${}^{n\cdot}C \equiv C - Ca^{\cdot e} + H_2O \rightleftharpoons {}^{n\cdot}C \equiv C - CaH + HO^{\cdot en} + Heat$

$HO \bullet en \rightleftharpoons nn \bullet O \bullet en + H \bullet e + Heat$

$nn \bullet O \bullet en + n \bullet C \equiv C - CaH \rightleftharpoons n \bullet C \equiv C - Ca \bullet e + nn \bullet OH + Heat$

$H \bullet e + n \bullet C \equiv C - Ca \bullet e \rightleftharpoons HC \equiv C - Ca \bullet e$

$HC \equiv C - Ca \bullet e + nn \bullet OH \longrightarrow HC \equiv C - Ca - OH \qquad\qquad 6.68a$

Stage 2: $HC \equiv C - Ca - OH \rightleftharpoons HC \equiv C \bullet n + e. Ca - OH$

${}^{e\cdot}CaOH + H_2O \rightleftharpoons HCaOH + HO \bullet en + Heat$

$HO \bullet en \rightleftharpoons nn \bullet O \bullet en + H \bullet e + Heat$

$HC \equiv C \bullet n + H \bullet e \rightleftharpoons HC \equiv CH$

$nn \bullet O \bullet en + HCaOH \rightleftharpoons HO \bullet nn + e \bullet CaOH + Heat$

$\longrightarrow Ca(OH)_2 \qquad\qquad 6.68b$

Notice that the Equilibrium state of existence of acetylide could not be suppressed in Stage 1 of Equation 6.67 and Equation 6.68b by the presence of water, because of the presence of ionic metals. Alkalis when full are very stable compounds. They cannot be used to stabilize the acetylenes. Notice that the reactions are exothermic in character, with release of heat in many steps.

Hydrogen can be added to the triple bond in the presence of hydrogen catalyst. The catalysts are members of Transition metals which cannot form hydrides with hydrogen. They are passive in the sense that they are not chemically involved in the reactions. Their pores help to break hydrogen into two parts as shown below.

Stage 1: $H_2 \xrightleftharpoons{H_2 \; Cat.} H^{\cdot e} + H^{\cdot n}$

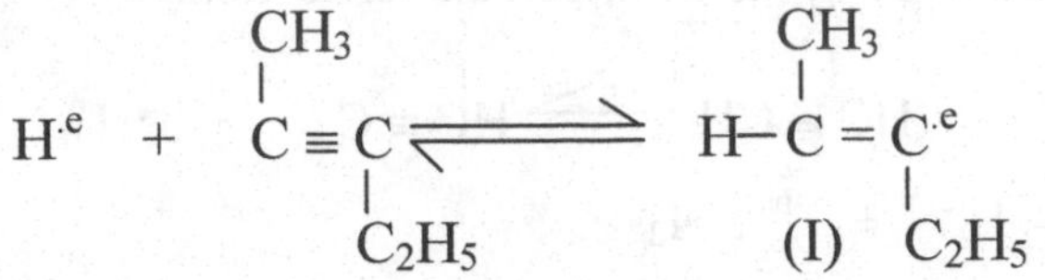

$$H^{\cdot e} + \underset{\underset{C_2H_5}{|}}{\overset{\overset{CH_3}{|}}{C}} \equiv C \rightleftharpoons \underset{\underset{(I) \; C_2H_5}{|}}{\overset{\overset{CH_3}{|}}{H - C}} = C^{\cdot e}$$

$(I) \quad + \quad H^{\cdot n} \longrightarrow HCH_3C = CC_2H_5H$ \hfill 6.69

Overall Equation: $H_2 + H_3CC \equiv CC_2H_5 \xrightarrow{H_2 \; Cat.} HCH_3C = CC_2H_5H$ \hfill 6.70

Stage 2: $H_2 \xrightleftharpoons{H_2 \; Cat.} H^{\cdot e} + H^{\cdot n}$

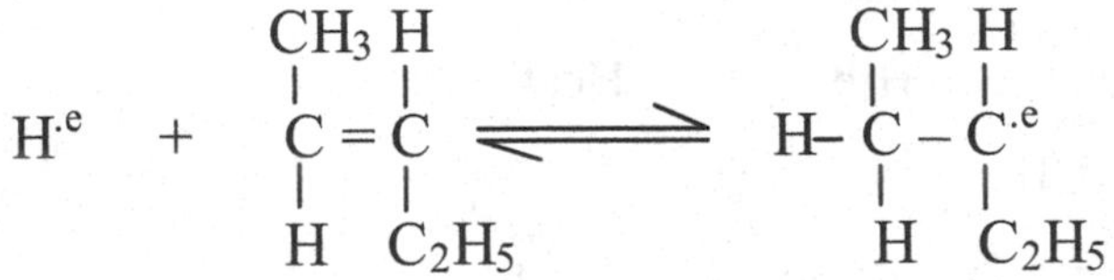

$$H^{\cdot e} + \underset{\underset{H}{|} \; \underset{C_2H_5}{|}}{\overset{\overset{CH_3}{|} \; \overset{H}{|}}{C = C}} \rightleftharpoons \underset{\underset{H}{|} \; \underset{C_2H_5}{|}}{\overset{\overset{CH_3}{|} \; \overset{H}{|}}{H - C - C^{\cdot e}}}$$

$$(II)$$

$(II) \quad + \quad H^{\cdot n} \longrightarrow CH_3CH_2 - CH_2C_2H_5$ \hfill 6.71

Overall Equation: $H_2 + HCH_3C = CHC_2H_5 \xrightarrow{H_2 \; Cat.} CH_3CH_2{}^-CH_2C_2H_5$ \hfill 6.72

The Equilibrium State of Existence of H_2 has thus been shown above probably for the first time. Reduction can also be brought about also by means of sodium in liquid ammonia.

$RC \equiv CR + 2Na + 2NH_3 \longrightarrow RHC = CHR + Na^{\oplus}{}^{\ominus}NH_2$ \hfill 6.73

Stage 1: $2NH_3 \rightleftharpoons 2H\cdot e + 2nn\cdot NH_2$

$2Na\cdot e + 2nn\cdot NH_2 \rightleftharpoons 2NaNH_2$

$2H^{\cdot e} \longrightarrow H_2$ \hfill 6.74

Overall Equation: $2Na + 2NH_3 \longrightarrow 2NaNH_2 + H_2$ \hfill 6.75

Stage 2: $NaNH_2 \rightleftharpoons Na^{\cdot e} + {}^{nn.}NH_2$

$Na^{\cdot e} + H_2 \rightleftharpoons NaH + H^{\cdot e}$

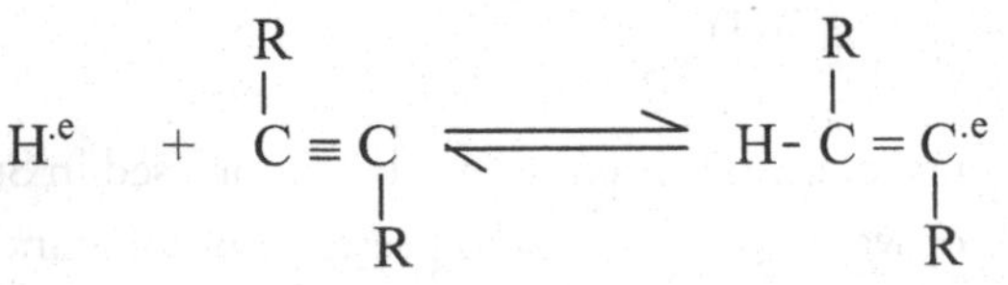

$$H^{\cdot e} + \underset{\underset{R}{|}}{\overset{\overset{R}{|}}{C}} \equiv C \rightleftharpoons \underset{\underset{R}{|}}{\overset{\overset{R}{|}}{H - C}} = C^{\cdot e}$$

$$(I)$$

$(I) \quad + \quad {}^{nn.}NH_2 \longrightarrow HRC = CR(NH_2)$ \hfill 6.76

Overall Equation: $NaNH_2 + H_2 + RC \equiv CR \longrightarrow NaH + HRC = CR(NH_2)$ \hfill 6.77

Stage 3: $NaH \rightleftharpoons Na^{\cdot e} + H^{\cdot n}$

$$Na^{\cdot e} + HRC = CR(NH_2) \rightleftharpoons NaNH_2 + HRC = \overset{\overset{\displaystyle R}{\displaystyle |}}{C}.e$$

$$HRC = RC.e + H^{\cdot n} \longrightarrow HRC = CRH \qquad\qquad 6.78$$

Overall Overall Equation: $RC \equiv CR + 2Na + 2NH_3 \longrightarrow RHC = CHR$

$$2Na^{\oplus}\ ^{\ominus}NH_2 \qquad (6.73)$$

Due to steric limitations, only the trans- monomer can be obtained here. One can observe so far that this is a NEW SCIENCE different from the one of today. Thus, one can largely observe, how these reactions take place and why the products are what they are.

Acetylene adds water in the presence of sulfuric acid and mercurous sulfate (mercury-mercuric sulfate mixture).

$$HC \equiv CH + H_2O \xrightarrow{H_2SO_4,\ Hg-HgSO_4} CH_3CHO$$
$$\text{(Acetaldehyde)} \qquad 6.79$$

$$RC \equiv CH + H_2O \xrightarrow{H_2SO_4,\ Hg-HgSO_4} RCOCH_3$$
$$\text{(A Ketone)} \qquad 6.80$$

Notice that the Equilibrium State of Existence of acetylene has been suppressed by the passive catalyst. That of water is not suppressed. The mechanisms are as follows.

Stage 1: $H_2O \rightleftharpoons H^{\cdot e} + {}^{nn\cdot}OH$

$$H^{\cdot e} + \overset{\overset{\displaystyle H}{\displaystyle |}}{\underset{\underset{\displaystyle H}{\displaystyle |}}{C}} \equiv C \rightleftharpoons \overset{\overset{\displaystyle H}{\displaystyle |}}{\underset{\underset{\displaystyle H}{\displaystyle |}}{C}} = \overset{}{\underset{\underset{\displaystyle H}{\displaystyle |}}{C}}.e$$

$$\text{(I)}$$

$$\text{(I)} + {}^{nn\cdot}OH \longrightarrow \overset{\overset{\displaystyle H \quad H}{\displaystyle |\quad|}}{\underset{\underset{\displaystyle H \quad O-H}{\displaystyle |\quad|}}{C = C}}$$

$$\text{(II) Vinyl alcohol} \qquad 6.81$$

Overall Equation: $C_2H_2 + H_2O \longrightarrow H_2C = CHOH \qquad 6.82$

Stage 1:

$$H_2O \rightleftharpoons H^{.e} + {}^{nn.}OH$$

$$H^{.e} + \begin{array}{c} H \\ | \\ C \equiv C \\ | \\ R \end{array} \rightleftharpoons \begin{array}{c} H \\ | \\ C = C^{.e} \\ | \quad | \\ H \quad R \end{array}$$

(I)

$$(I) + {}^{nn.}OH \longrightarrow \begin{array}{c} H \quad R \\ | \quad | \\ C = C \\ | \quad | \\ H \quad OH \end{array}$$

6.83

(II) A Vinyl alcohol

Overall Equation: $RC \equiv CH + H_2O \longrightarrow H_2C = CROH$

6.84

The (II) of Equations 6.81 and 6.83 are very unstable that they molecularly rearrange in the second stage as has already been shown in the Series and recalled below.

$$\begin{array}{c} H \quad H \\ | \quad | \\ C = C \\ | \quad | \\ H \quad OH \end{array} \longrightarrow \begin{array}{c} H \quad H \\ | \quad | \\ n.C - C.e \\ | \quad | \\ H \quad O \\ \quad | \\ \quad H \end{array} \longrightarrow \begin{array}{c} H \\ | \\ e.C - O.nn \\ | \\ CH_3 \end{array} \longrightarrow \begin{array}{c} H \\ | \\ C = O \\ | \\ CH_3 \end{array}$$

(Acetaldehyde)

(Unstable)   (Stable)

6.85

$$\begin{array}{c} H \quad R \\ | \quad | \\ C = C \\ | \quad | \\ H \quad OH \end{array} \longrightarrow \begin{array}{c} H \quad R \\ | \quad | \\ n.C - C.e \\ | \quad | \\ H \quad O \\ \quad | \\ \quad H \end{array} \longrightarrow \begin{array}{c} R \\ | \\ e.C - O.nn \\ | \\ CH_3 \end{array} \longrightarrow \begin{array}{c} R \\ | \\ C = O \\ | \\ CH_3 \end{array} \quad (R \le CH_3)$$

(A Ketone)

(Unstable)   (Stable)

6.86

The molecular arrangement above is of the first kind of the first type. The rearrangement can take place radically and chargedly. These form the second stage for them. One can see how other olefins can be obtained directly from acetylene. (See Chap. 5, Vol. I)

Acetylides are also known to be readily attacked by carbonyl groups. The reactions can be explained as follows.

$$\begin{array}{c} H \\ | \\ C \equiv C^{.n} Li^{.e} \end{array} + \begin{array}{c} H \\ | \\ e.C - O.nn \\ | \\ R \end{array} \longrightarrow \begin{array}{c} R \\ | \\ Li O - C - C \equiv CH \\ | \\ H \end{array}$$

(stronger   (I) (nucleophile)
nucleophile)

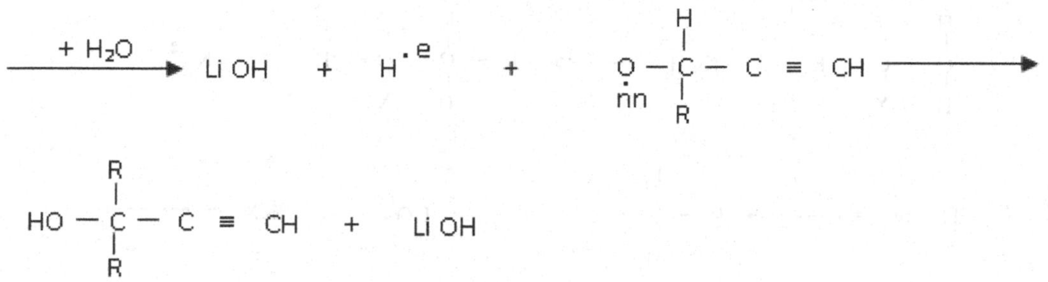

$$\xrightarrow{+ H_2O} \quad LiOH \quad + \quad H \cdot^e \quad + \quad \underset{nn}{\overset{H}{\underset{\cdot\cdot}{O}}} - \overset{H}{\underset{R}{C}} - C \equiv CH \longrightarrow$$

$$HO - \overset{R}{\underset{R}{C}} - C \equiv CH \quad + \quad LiOH$$

$$6.87$$

The mechanism of the reaction is Equilibrium in two stages. It is the aldehyde that is first activated by $Li^{\cdot e}$ followed by hydrolysis of the product from the first stage in the manners shown in Equations 6.67 and 6.68 wherein Heat is released. Alkynols are produced in the process. With a ketone, the followings are obtained.

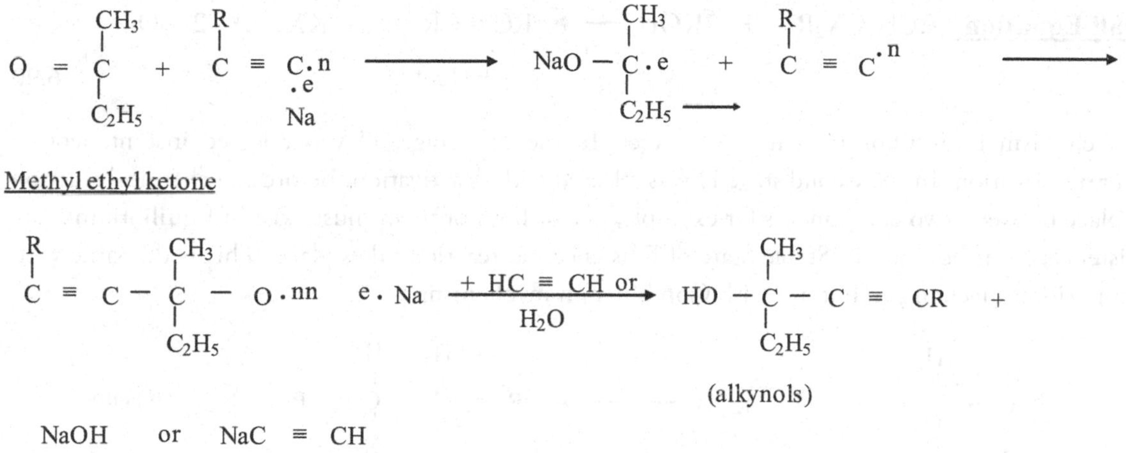

$$O = \overset{CH_3}{\underset{C_2H_5}{C}} \quad + \quad \overset{R}{\underset{\cdot e}{\underset{Na}{C}}} \equiv C \cdot n \longrightarrow NaO - \overset{CH_3}{\underset{C_2H_5}{C}} \cdot e \quad + \quad \overset{R}{C} \equiv C^{\cdot n} \longrightarrow$$

Methyl ethyl ketone

$$\overset{R}{\underset{}{C}} \equiv C - \overset{CH_3}{\underset{C_2H_5}{C}} - O \cdot nn \quad e \cdot Na \quad \xrightarrow[H_2O]{+ HC \equiv CH \; or} \quad HO - \overset{CH_3}{\underset{C_2H_5}{C}} - C \equiv CR \quad +$$

(alkynols)

$$NaOH \quad or \quad NaC \equiv CH \qquad\qquad 6.88$$

If $HC\equiv CNa$ had been used instead of $RC\equiv CNa$, similar reaction will be obtained producing 3-hydroxy - 3 -methyl-1-pentyne (methyl pentynol). The reactions above are only favored by ionic metals. It is clearly obvious that none of the reactions are ionic or charged in character as have been thought to be the case for so many years. This does imply that acetylene cannot undergo charged polymerizations.

6.1.2 "Charged characters of Acetylenes"

Already one can confidently ascertain that acetylenes with at least one radical-pushing group can be polymerized only radically, provided the capacity of the radical-pushing substituent group is strong enough to stabilize the monomer. For acetylene the first member, we have seen how some passive catalysts can stabilize it. How monomers with at least one substituent group can be obtained using acetylene as the source, have been shown. They can also be obtained from alkyl halides as shown below.

$$R - \underset{\underset{\displaystyle H}{|}}{\overset{\overset{\displaystyle H}{|}}{C}} - \underset{\underset{\displaystyle X}{|}}{\overset{\overset{\displaystyle X}{|}}{C}} - R \quad + \quad KOH \quad \longrightarrow \quad R - \underset{\underset{\displaystyle H}{|}}{\overset{\overset{\displaystyle H}{|}}{C}} - \underset{\underset{\displaystyle X}{|}}{\overset{\overset{\displaystyle X}{|}}{C}} - R \quad + \quad K \cdot ^{e}$$

$$+ \quad HO \cdot nn \quad \longrightarrow \quad R - \underset{\underset{\displaystyle H}{|}}{\overset{\overset{\displaystyle H}{|}}{C}} - \underset{\underset{\displaystyle \dot{e}}{|}}{\overset{\overset{\displaystyle X}{|}}{C}} - R \quad + \quad HO \cdot ^{nn} \quad + \quad KX \quad \longrightarrow$$

$$R - \underset{\underset{\displaystyle R}{|}}{\overset{\overset{\displaystyle H}{|}}{C}} = \underset{\underset{\displaystyle R}{|}}{\overset{\overset{\displaystyle X}{|}}{C}} - R \quad + \quad KX \quad + \quad H_2O \quad \xrightarrow{+ \ KOH} \quad \underset{\underset{\displaystyle R}{|}}{\overset{\overset{\displaystyle H}{|}}{C}} = \underset{\underset{\displaystyle R}{|}}{\overset{\overset{\displaystyle e}{|}}{C}} \quad + \quad 2KX \quad + \quad HO \cdot nn$$

$$+ \quad H_2O \quad \longrightarrow \quad \underset{\underset{\displaystyle R}{|}}{C} \equiv \underset{\underset{\displaystyle R}{|}}{C} \quad + \quad 2KX \quad + \quad 2H_2O$$

$$6.89$$

Overall Equation: $RCH_2CX_2R \ + \ 2KOH \ \longrightarrow \ RC \equiv CR \ + \ 2KX \ + \ 2H_2O$

$$+ \ Heat \qquad 6.90$$

The mechanism is Equilibrium with two stages. In the first stage, H was released instantaneously without deactivation. In the second stage H was released with deactivation. Before any chemical reaction takes place between two components for example, one or both of them must exist in Equilibrium State of Existence; for if both are in Stable State of Existence, no reaction takes place. This is the same with Decomposition mechanism, but not with Combination mechanism.

$$R - \underset{\underset{\displaystyle X}{|}}{\overset{\overset{\displaystyle H}{|}}{C}} - \underset{\underset{\displaystyle X}{|}}{\overset{\overset{\displaystyle H}{|}}{C}} - R \quad + \quad KOH \quad \longrightarrow \quad R - \underset{\underset{\displaystyle \dot{e}}{|}}{\overset{\overset{\displaystyle H}{|}}{C}} - \underset{\underset{\displaystyle X}{|}}{\overset{\overset{\displaystyle H}{|}}{C}} - R \quad + \quad HO \cdot nn$$

$$+ \quad KX \quad \longrightarrow \quad R - \underset{\underset{\displaystyle X}{|}}{\overset{\overset{\displaystyle H}{|}}{C}} = \underset{}{C} - R \quad + \quad KX \quad + \quad H_2O \quad \xrightarrow{KOH}$$

$$R - \underset{\underset{\displaystyle \dot{e}}{|}}{C} = \overset{\overset{\displaystyle \dot{n}}{|}}{C} - R \quad + \quad 2KX \quad + \quad 2H_2O \quad \longrightarrow \quad R - C \equiv C - R \quad +$$

$$2KX \quad + \quad 2H_2O \qquad\qquad\qquad 6.91$$

Overall Equation: $RHCXCXHR \ + \ 2KOH \ \longrightarrow \ RC \equiv CR \ + \ 2KX \ + \ 2H_2O$

$$+ \quad Heat \qquad 6.92$$

The mechanism like the case above is Equilibrium mechanism with two stages. Unlike the case above, H was released in the two stages with deactivation.

$$R - \underset{\underset{\displaystyle X}{|}}{\overset{\overset{\displaystyle X}{|}}{C}} - \underset{\underset{\displaystyle X}{|}}{\overset{\overset{\displaystyle X}{|}}{C}} - R \quad + \quad \underset{. \ en}{\overset{. \ nn}{Zn}} \quad \longrightarrow \quad R - \underset{\underset{\displaystyle X}{|}}{\overset{\overset{\displaystyle n.}{|}}{C}} - \underset{\underset{\displaystyle \dot{e}}{|}}{\overset{\overset{\displaystyle X}{|}}{C}} - R \quad + \quad ZnX_2$$

$$(Zinc \ dust)$$

$$\longrightarrow \quad R - \overset{\overset{e}{|}}{\underset{\underset{n}{|}}{C}} - \overset{\overset{n}{|}}{\underset{\underset{e}{|}}{C}} - R \quad + \quad 2ZnX_2 \quad \longrightarrow \quad R - C \equiv C - R \quad + \quad 2ZnX$$

$$\text{6.93}$$

Overall Equation: $RX_2CCX_2R \quad + \quad 2Zn \longrightarrow RC \equiv CR \quad + \quad 2ZnX_2 \quad + \quad Heat \qquad 6.94$

Equilibrium mechanism is operating here in four stages. For the first time from Vol (I), notice the structure of Zinc dust. Chargedly for example, the reactions represented above are impossi-ble. On the other hand as it has been said alkenes or alkyl halides can never dissociate to produce ions, since their carbon centers are sp^3 - hybridized e.g. CH_3Cl and the center carrying the positive charge is not empty.

Now, one will determine how acetylenes with pulling groups can be obtained, using some of the methods above. This is not a matter of delving into the chemistry of acetylenes, but an attempt to determine the types of acetylenes to consider and use for polymerization studies.

$$ROOC - \overset{\overset{H}{|}}{\underset{\underset{H}{|}}{C}} - \overset{\overset{Cl}{|}}{\underset{\underset{Cl}{|}}{C}} - COOR \quad + \quad KOH \longrightarrow ROOC - \overset{\overset{H}{|}}{\underset{\underset{n}{|}}{C}} - \overset{\overset{e}{|}}{\underset{\underset{Cl}{|}}{C}} - COOR \quad +$$

$$KCl \quad + \quad H_2O \xrightarrow{+ KOH} ROOC - \overset{\overset{n}{\cdot}}{\underset{}{C}} = \underset{\underset{e}{\cdot}}{C} - COOR \quad + \quad 2KCl \quad + \quad 2H_2O$$

$$\longrightarrow ROOC - C \equiv C - COOR \quad + \quad 2KCl \quad + \quad 2H_2O \quad + \quad Heat \qquad 6.95$$

Overall Equation: $ROOCCH_2CCl_2COOR \quad + \quad 2KOH \longrightarrow HOOCC \equiv CCOOH \quad + \quad 2KCl$

$$+ \quad 2H_2O \quad + \quad Heat \qquad 6.96$$

This is an Equilibrium mechanism with two stages.

$$H - \overset{\overset{H}{|}}{\underset{\underset{H}{|}}{C}} - \overset{\overset{Cl}{|}}{\underset{\underset{Cl}{|}}{C}} - COOR \quad + \quad 2KOH \longrightarrow H - \overset{\overset{n}{\cdot}}{\underset{\underset{n}{\cdot}}{C}} - \overset{\overset{e}{\cdot}}{\underset{\underset{e}{\cdot}}{C}} - COOR \quad + \quad 2KCl \quad + \quad 2H_2O$$

$$\longrightarrow H - C \equiv C - COOR \quad + \quad 2KCl \quad + \quad 2H_2O \qquad 6.97$$

Overall Equation: $H_3CCCl_2COOH \quad + \quad 2KOH \longrightarrow HC \equiv CCOOR \quad + \quad 2KCl \quad +$

$$+ \quad Heat \qquad 6.98$$

$$H - \overset{\overset{H}{|}}{\underset{\underset{H}{|}}{C}} - \overset{\overset{Cl}{|}}{\underset{\underset{Cl}{|}}{C}} - Cl \quad + \quad 2KOH \longrightarrow H - \overset{\overset{n}{\cdot}}{\underset{\underset{n}{\cdot}}{C}} - \overset{\overset{e}{\cdot}}{\underset{\underset{e}{\cdot}}{C}} - Cl \quad + \quad 2KCl \quad + \quad 2H_2O$$

$$\longrightarrow H - C \equiv C - Cl \quad + \quad 2KCl \quad + \quad 2H_2O \qquad 6.99$$

Overall Equation: CH_3CCl_3 + $2KOH$ $\longrightarrow$ $HC \equiv CCl$ + $2KCl$ + $2H_2O$

+ Heat   6.100

Here, the two bonds were formed instantaneously without deactivation.

$$R - \underset{\underset{H}{|}}{\overset{\overset{H}{|}}{C}} - \underset{\underset{Cl}{|}}{\overset{\overset{Cl}{|}}{C}} - Cl \ + \ 2KOH \ \longrightarrow \ R - \underset{\dot{n}}{\overset{\cdot}{\underset{|}{C}}} - \overset{\cdot}{\underset{\dot{e}}{C}} - Cl \ + \ 2KCl \ + \ 2H_2O$$

$$\longrightarrow \ R - C \equiv C - Cl \ + \ 2KCl \ + \ 2H_2O \qquad\qquad 6.101$$

Overall Equation: RCH_2CCl_3 + $2KOH$ $\longrightarrow$ $RC \equiv CCl$ + $2KCl$ + $2H_2O$ + Heat   6.102

Here again, the two bonds were also formed instantaneously without deactivation.

$$CHCl_2 - \underset{\underset{H}{|}}{\overset{\overset{H}{|}}{C}} - \underset{\underset{Cl}{|}}{\overset{\overset{Cl}{|}}{C}} - CH_3 \ + \ 2KOH \ \longrightarrow \ Cl H\overset{\overset{e}{\cdot}}{C} - \underset{\dot{n}}{\overset{\overset{H}{|}}{C}} - \overset{\overset{e.}{\cdot}}{\underset{\underset{Cl}{|}}{C}} - \overset{n.}{CH_2} \ + \ 2KCl$$

$$+ \ 2H_2O \ \longrightarrow \ Cl H C = \underset{\underset{Cl}{|}}{C} - \overset{}{C} = CH_2 \ + \ 2KCl \ + \ 2H_2O \qquad 6.103$$

Overall Equation: $HCl_2CH_2CCl_2CH_3$ + $2KOH$ $\longrightarrow$ $HClC = CH - CCl = CH_2$ + $2KCl$ +

$2H_2O$ + Heat   6.104

$$Cl_3C - \underset{\underset{H}{|}}{\overset{\overset{H}{|}}{C}} - \underset{\underset{Cl}{|}}{\overset{\overset{Cl}{|}}{C}} - CH_3 \ + \ 2KOH \ \longrightarrow \ Cl_2C - \overset{\overset{n}{\cdot}}{\underset{\dot{e}}{C}} - \overset{\overset{Cl}{|}}{\underset{\underset{H}{|}}{C}}_{e.} - \overset{n.}{CH_2} \ + \ H_2O \ + \ HOCCl_3$$

$$+ \ 2KCl \ \longrightarrow \ Cl_2C = CH - CCl = CH_3 \ + \ 2H_2O \ + \ 2KCl \qquad 6.105$$

Overall Equation: $Cl_3CH_2CCl_2CH_3$ + $2KOH$ $\longrightarrow$ $Cl_2C = CH - CCl = CH_2$ + $2H_2O$

+ Heat + $2KCl$ 6.106

In the last two equations, notice that 1,3-dienes were obtained as opposed to acetylenes. Thus, it can be observed that existence of acetylenes with pulling groups can be favored. So also are those with pushing groups from substituted alkanes. While the existence of (IV) of Equation 6.12 is favored directly from acetylene or indirectly from substituted alkanes, the existence of $HC \equiv CC(Cl_3)$ directly from acetylene cannot be obtained as shown from Equations 6.99 to 6.105.

$$Cl_3C - \underset{\underset{H}{|}}{\overset{\overset{H}{|}}{C}} - \underset{\underset{Cl}{|}}{\overset{\overset{Cl}{|}}{C}} - H \ + \ 2KOH \ \longrightarrow \ Cl_2C - \overset{e.}{\underset{\dot{n}}{C}} - \overset{\overset{e}{\cdot}}{\underset{\underset{Cl}{|}}{C}}^{n.} - H \ + \ 2KCl \ + \ H_2O$$

$$\longrightarrow \ Cl_2C = C = CHCl \ + \ 2KCl \ + \ 2H_2O \qquad\qquad 6.107a$$

Overall Equation: $Cl_3CCH_2CCl_2H + 2KOH \longrightarrow Cl_2C=C=CHCl + 2KCl + 2H_2O$

$+ \quad$ Heat $\qquad\qquad$ 6.107b

$$Cl_3C - \underset{\underset{Cl}{|}}{\overset{\overset{Cl}{|}}{C}} - \underset{\underset{H}{|}}{\overset{\overset{H}{|}}{C}} - H \quad + \quad KOH \quad \longrightarrow \quad Cl_2\overset{e}{C} - \underset{\underset{Cl}{|}}{\overset{\overset{Cl}{|}}{C}}.n \quad + \quad HOCH_3 \quad + \quad KCl$$

$$\longrightarrow \quad Cl- \underset{\underset{}{}}{\overset{\overset{\mathbf{Cl}}{|}}{C}} = \underset{\underset{Cl}{|}}{\overset{\overset{Cl}{|}}{C}} \quad + \quad \mathbf{CH_3OH} \quad + \quad KCl \qquad\qquad 6.108a$$

Overall Equation: $Cl_3C(Cl)_2C\text{-}CH_3 + KOH \longrightarrow Cl_2C=CCl_2 + KCl + CH_3OH$

$+ \quad$ Heat $\qquad\qquad$ 6.108b

One can observe the types of products obtained from the substituted propanes. To obtain $HC\equiv CCCl_3$ by this means, is almost impossible, because of the radical characters of CCl_3 as a radical-pushing group of less capacity than H, rich with chlorine atoms. The best means of obtaining is directly from an acetylene such as $HC\equiv CF$. The reactions of Equation 6.93 can be used to obtain $X\text{-}C \equiv C\text{-}X$ from CX_3CX_3 ($X\equiv Cl$). Having established the existence of acetylenes with pulling groups, one can now move ahead to consider the polymerizations of these monomers. But before then, it must be noted that unlike the first member of acetylenes, the first member of perfluoro-acetylenes can readily be polymerized, because the two fluorine atoms are strongly held to their C centers.

$$HC \equiv CH \longrightarrow H - C \equiv C\cdot^n H\cdot^e \quad ; \quad FC \equiv CF \longrightarrow F - C \equiv C - F$$

$$(I) \qquad\qquad\qquad\qquad\qquad (II) \qquad\qquad \text{Very stable} \qquad 6.109$$

This stability is only peculiar to F amongst all the halogen atoms.

So far in all the considerations, worthy of note is that no charged reactions have been observed. Though one has said that acetylenes and nitriles cannot be activated chargedly, for exploratory purposes consider the charged activation of $CH \equiv CR$. **EXPLORATORY PURPOSE**

$$R^{\oplus} + \overset{\ominus}{C} = \underset{\underset{C_2H_5}{|}}{\overset{\overset{H}{|}}{C}}{}^{\oplus} \longrightarrow R - \overset{\overset{H}{|}}{C} = \underset{\underset{C_2H_5}{|}}{C}{}^{\oplus} \qquad\qquad 6.110$$

$$R^{\ominus} + \overset{\ominus}{C} = \underset{\underset{\underset{CH_3}{|}}{\overset{}{CH_2}}}{\overset{\overset{H}{|}}{C}}{}^{\oplus} \longrightarrow RH + \underset{\underset{CH_3}{|}}{\overset{\overset{H}{|}}{C}} = C = C^{\ominus}$$

(Free-charge) $\qquad\qquad\qquad\qquad\qquad\qquad\qquad\qquad\qquad\qquad$ 6.111

$$R + \left(\overset{\overset{H}{|}}{C} = \underset{\underset{C_2H_5}{|}}{\overset{\overset{H}{|}}{C}} \right)_n \overset{}{C} = \underset{\underset{C_2H_5}{|}}{C}{}^{\oplus} \longrightarrow R + \left(\overset{\overset{H}{|}}{C} = \underset{\underset{C_2H_5}{|}}{\overset{\overset{H}{|}}{C}} \right)_n C = C = \underset{\underset{CH_3}{|}}{\overset{\overset{H}{|}}{C}} + H^{\oplus} \qquad 6.112$$

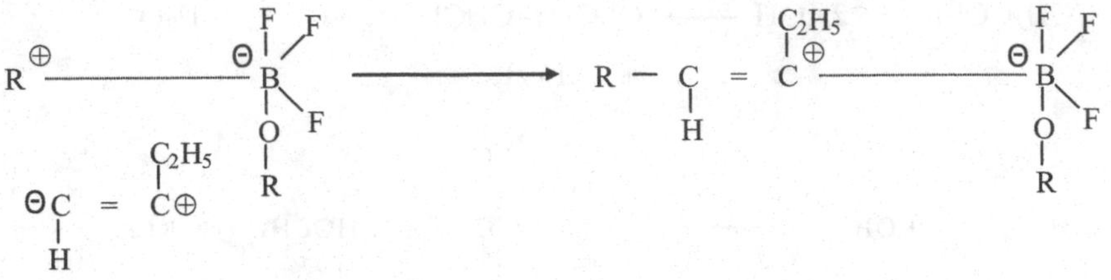

$$R^{\oplus} \underline{\hspace{3cm}} \overset{\overset{F}{\underset{|}{\overset{\ominus}{B}}}\overset{F}{\diagup}}{\underset{\underset{R}{\overset{|}{O}}}{\diagdown}F} \longrightarrow R - \overset{H}{\underset{|}{C}} = \overset{\overset{C_2H_5}{|}}{\overset{\oplus}{C}} \underline{\hspace{3cm}} \overset{\overset{F}{\underset{|}{\overset{\ominus}{B}}}\overset{F}{\diagup}}{\underset{\underset{R}{\overset{|}{O}}}{\diagdown}F}$$

$$\overset{\overset{C_2H_5}{|}}{\underset{\underset{H}{\overset{|}{C}}}{\ominus C}} = \overset{}{C\oplus}$$

$$6.113$$

$$H_9C_4 \overset{\ominus}{\underline{\hspace{3cm}}} \overset{\oplus}{\hspace{0.2cm}} Li \longrightarrow H_{10}C_4 \;+\; \overset{\overset{CH_3}{|}}{\underset{\underset{H}{\overset{|}{C}}}{C}} = C = \overset{}{\underset{\underset{H}{\overset{|}{C}}}{C}} \overset{\ominus}{\underline{\hspace{3cm}}} \overset{\oplus}{\hspace{0.2cm}} Li$$

$$\overset{\overset{CH_3}{|}}{\underset{\overset{|}{CH_2}}{}}$$
$$\overset{}{\underset{\underset{H}{\overset{|}{C}}}{\oplus C}} = C\ominus$$

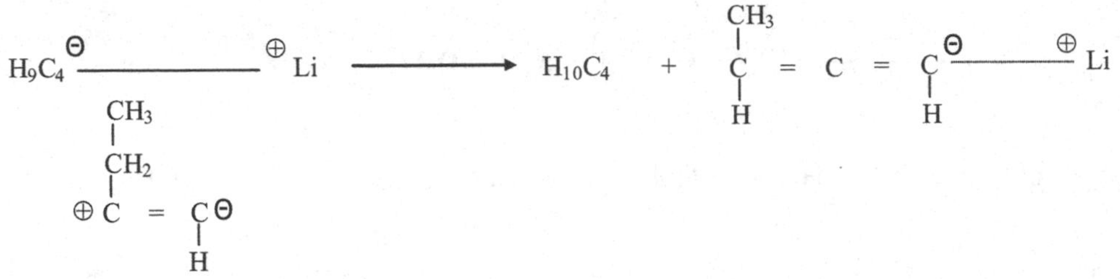

$$6.114$$

As shown by the equations above, only the positively charged routes are favored under certain conditions with involvement of transfer species of the first kind of the first type, to give a cumulenic type of terminal bond. When the second H is replaced with another alkyl group, similar reactions are obtained as shown below using only negatively charged-paired initiators. Positively for such cases, polymerization is fully favored for any strength of initiator.

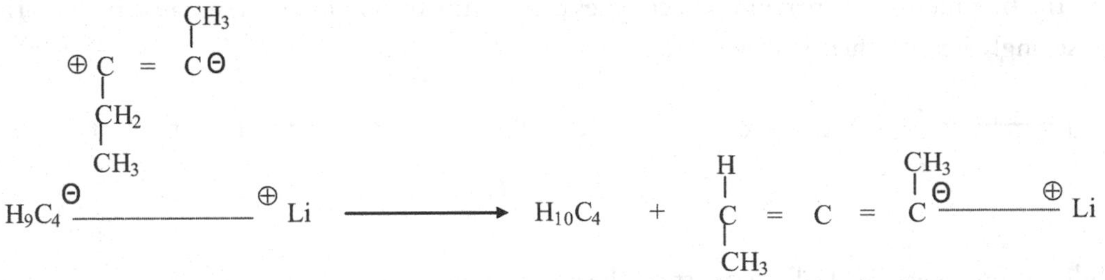

$$\overset{\overset{CH_3}{|}}{\underset{\overset{|}{CH_2}}{\underset{\overset{|}{CH_3}}{\oplus C}}} = C\ominus$$

$$H_9C_4 \overset{\ominus}{\underline{\hspace{3cm}}} \overset{\oplus}{\hspace{0.2cm}} Li \longrightarrow H_{10}C_4 \;+\; \overset{\overset{H}{|}}{\underset{\underset{CH_3}{\overset{|}{C}}}{C}} = C = \overset{\overset{CH_3}{|}}{\underset{}{C}} \overset{\ominus}{\underline{\hspace{2cm}}} \overset{\oplus}{\hspace{0.2cm}} Li \quad 6.115$$

Like olefins, with radical-pushing groups, acetylenes are nucleophiles of stronger capacity, except that acetylenes cannot be activated chargedly. With etheric groups, the followings are obtained with the first member.

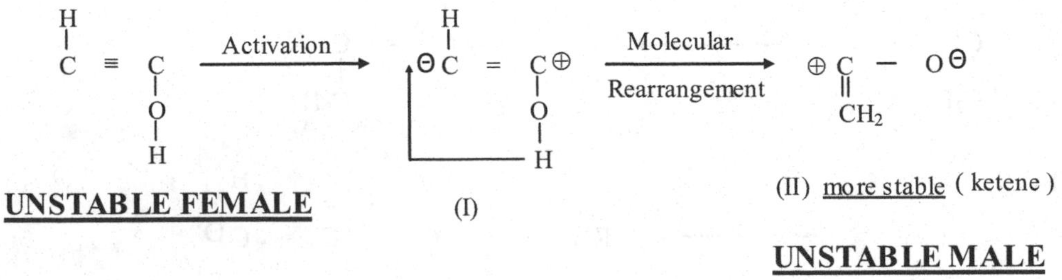

$$\overset{\overset{H}{|}}{\underset{\underset{\overset{|}{O}}{\overset{|}{C}}}{C}} \equiv \overset{}{\underset{\overset{|}{H}}{C}} \xrightarrow{\text{Activation}} \overset{\overset{H}{|}}{\underset{\underset{\overset{|}{O}}{\overset{|}{C}}}{\ominus C}} = \overset{}{\underset{\overset{|}{H}}{C\oplus}} \xrightarrow[\text{Rearrangement}]{\text{Molecular}} \overset{}{\underset{\overset{||}{CH_2}}{\oplus C}} - O\ominus$$

UNSTABLE FEMALE (I) (II) more stable (ketene)

UNSTABLE MALE 6.116

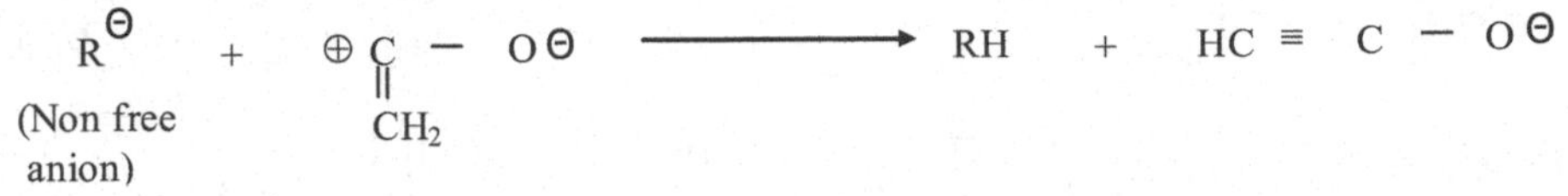

$$R\ominus \;+\; \overset{}{\underset{\overset{||}{CH_2}}{\oplus C}} - O\ominus \longrightarrow RH \;+\; HC \equiv C - O\ominus$$
(Non free anion)

$$6.117$$

Since (II) is more stable than (I), molecular rearrangement is favored for this case to produce a ketene. It will remain favored as long as OH is unchanged. The transfer species involved in the process is of *the*

first kind of the first type, but in transfer from monomer step is of ***the first kind of a third type.*** It is of the third type in view of its source – alkynyl (=CH$_2$). Negatively the monomer cannot be polymerized. Positively, for its growing polymer chain, the following is obtained.

$$R \fMinus O - \underset{\underset{CH_2}{\|}}{C} \fracslash_n O - \underset{\underset{CH_2}{\|}}{C}^{\oplus} \cdots\cdots\cdots \overset{\ominus}{} \underset{\underset{O}{\underset{|}{R}}}{\overset{F \diagdown \diagup F}{B}} \diagdown F \longrightarrow$$

$$R \fMinus O = \underset{\underset{CH_2}{\|}}{C} \fracslash_n O - C \equiv CH \quad + \quad HOR \quad + \quad BF_3 \qquad\qquad 6.118$$

It should be noted that ***dead terminal triple bond polymer*** is produced here when this type of transfer species is involved.

For higher etheric groups, the followings are obtained.

$$\underset{\underset{\underset{CH_3}{|}}{O}}{\overset{\overset{H}{|}}{C}} \equiv C \longrightarrow \overset{\ominus}{} \underset{\underset{\underset{CH_3}{|}}{O}}{\overset{\overset{H}{|}}{C}} = C^{\oplus} \longrightarrow \overset{\oplus}{} \underset{\underset{\underset{CH_3}{|}}{CH}}{\overset{\|}{C}} - O \overset{\ominus}{}$$

$$\qquad\qquad (II) \qquad\qquad\qquad\qquad\qquad (III) \;\;(Not\ favored) \qquad\qquad 6.119$$

$$R^{\ominus} \quad + \quad \overset{\oplus}{} \underset{\underset{\underset{CH_3}{|}}{CH}}{\overset{\|}{C}} - O \overset{\ominus}{} \longrightarrow RH \quad + \quad \overset{\overset{CH_3}{|}}{C} \equiv C - O \overset{\ominus}{} \qquad\qquad 6.120$$

The transfer species involved in transfer from monomer step is different from that involved in molecular rearrangement. Hence, molecular rearrangement is not favored. From the considera-tions above, it is obvious that (I) of Equation 6.116 is less nucleophilic than (II) of Equation 6.119, since (II) cannot rearrange to give ketenes.

$$\overset{\ominus}{} \underset{\underset{\underset{CH_3}{|}}{O}}{\overset{\overset{H}{|}}{C}} = C^{\oplus} \quad + \quad R \overset{\ominus}{} \longrightarrow RCH_3 \quad + \quad \overset{\overset{H}{|}}{C} \equiv C - O \overset{\ominus}{}$$

$$\qquad\qquad\qquad\qquad\qquad\qquad\qquad\qquad\qquad\qquad 6.121$$

$$R \fMinus \underset{\underset{\underset{CH_3}{|}}{O}}{\overset{\overset{H}{|}}{C}} = C \fracslash_n \underset{\underset{\underset{CH_3}{|}}{O}}{\overset{\overset{H}{|}}{C}} = C^{\oplus} \cdots\cdots\cdots \overset{\ominus}{} \underset{\underset{O}{\underset{|}{R}}}{\overset{F \diagdown \diagup F}{B}} \diagdown F \longrightarrow$$

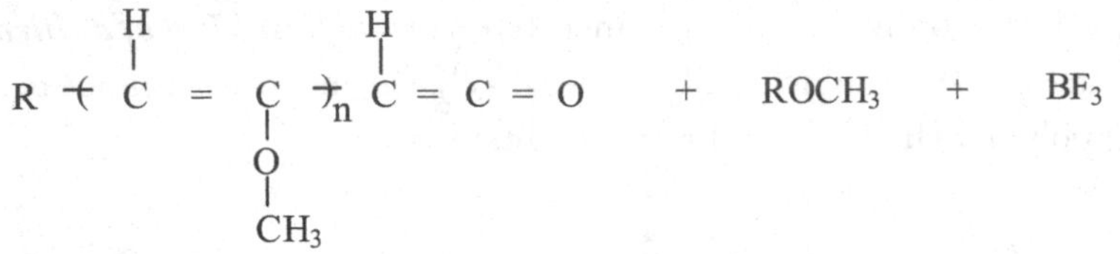

$$R \overbracket{\left(C = C \right)}_{n} C = C = O \quad + \quad ROCH_3 \quad + \quad BF_3$$

where the two carbons bear H and H, with one carbon bearing $O-CH_3$.

$$\text{6.122}$$

For higher etheric groups, molecular rearrangement is not favored. When the two Hs, in the monomer of Equation 6.116 is replaced with CH_3, molecular rearrangement is favored to produce dimethylketene. When the etheric groups are replaced with NH_2 or NHR, molecular rearrangement will be favored to produce new monomers such as shown below.

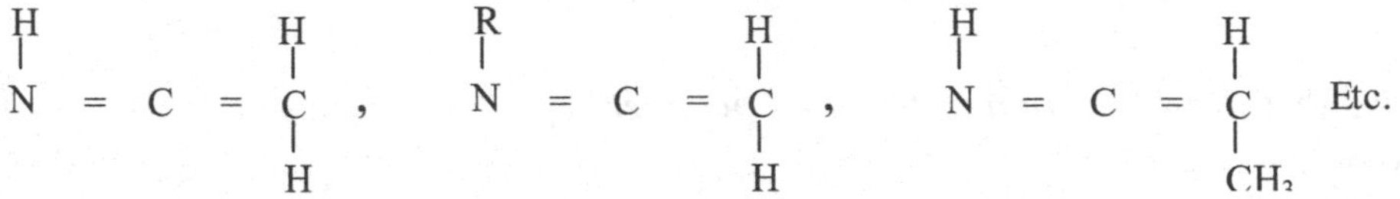

$$N = C = C \ (\text{with H, H}) \ , \quad N = C = C \ (\text{with R, H}) \ , \quad N = C = C \ (\text{with H, } CH_2) \quad \text{Etc.} \qquad 6.123$$

When groups such as NR_2 are involved, molecular rearrangement can readily be prevented depending on what the carbon center is carrying.

For monomer with at least one radical-pulling group, the followings are obtained for F.

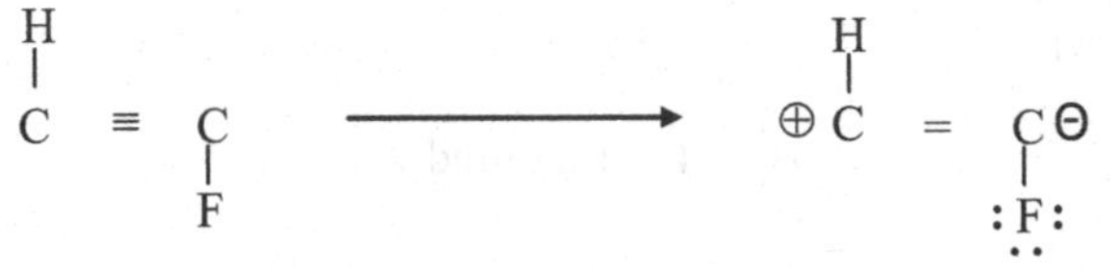

$$C \equiv C \ (\text{with H, F}) \quad \longrightarrow \quad \oplus C = C \ominus \ (\text{with H, :F:})$$

$$\text{6.124}$$

(Impossible existence)

Thus, the monomer cannot be activated chargedly, ***noting however that none of the acetylene monomers can be activated chargedly. We are just on exploration. From what we see here, Radical polymerization of these monomers will be easy to understand.*** With halogenated alkyl groups, the followings are obtained, if activation would have been possible.

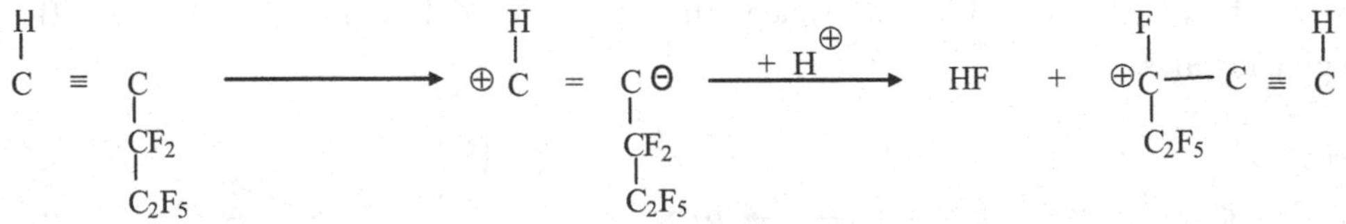

$$C \equiv C \ (H; CF_2, C_2F_5) \ \longrightarrow \ \oplus C = C \ominus \ (H; CF_2, C_2F_5) \ \xrightarrow{+ H^{\oplus}} \ HF \ + \ \oplus C - C \equiv C \ (F, C_2F_5; H) \qquad 6.125$$

$$R \overbracket{\left(C = C \right)}_{n} C = C \ominus \ \longrightarrow \ R \overbracket{\left(C = C \right)}_{n} C = C = C \ + \ F^{\ominus} \qquad 6.126$$

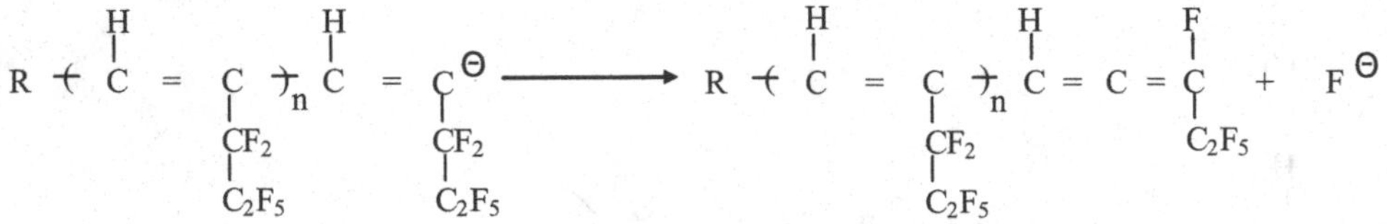

The negative route can only be favored by the monomer, making it look like an electrophile!

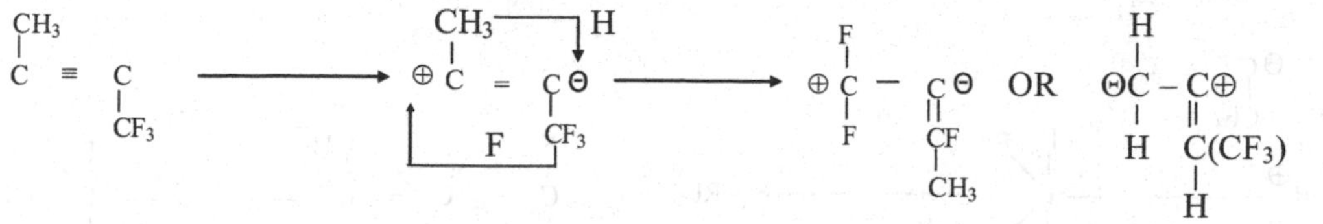

$$CH_3-C\equiv C-CF_3 \longrightarrow {}^{\oplus}C(CH_3)=C^{\ominus}(F)(CF_3)(H) \longrightarrow {}^{\oplus}C(F)(F)-C^{\ominus}=CF-CH_3 \quad OR \quad {}^{\ominus}C(H)=C^{\oplus}-C(CF_3)(H)$$

(I)	(II) Impossible existence	(III) Not favored 6.127

Molecular rearrangement is not in general favored by these monomers, except for the cases so far identified and yet to be identified. Apart from (II) being far less stable than (I), (II) cannot carry the charges indicated. On the other hand, if it has to be favored, the number of substituted groups must be less or remain the same. As a nucleophile (i.e. Female), it is $H^{\oplus}$ that should be the transfer species [i.e. (III)]. Yet, it is not favored. In general, it is allene that molecularly rearran-ges to an acetylene as shown below.

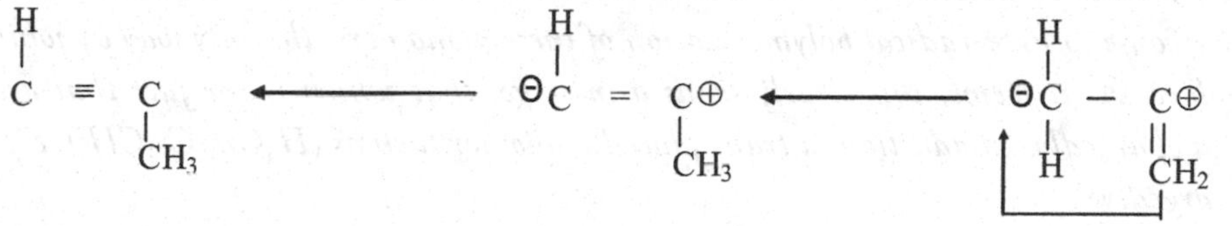

$$H-C\equiv C-CH_3 \longleftarrow {}^{\ominus}C(H)=C^{\oplus}-CH_3 \longleftarrow {}^{\ominus}C(H)-C^{\oplus}(H)=CH_2$$

	Methyl acetylene (more stable) 2-substituted groups	allen (less stable) 3-substituted groups. 6.128

(I) of Equation 6.127 will not favor the use of free-initiators, like the one of Equation 6.125. When paired initiators are involved, polymerization would have been favored, if its activation had been possible. Even then, transfer species of the first kind cannot be released in the route not natural to the monomer (See Equations 6.126) and below.

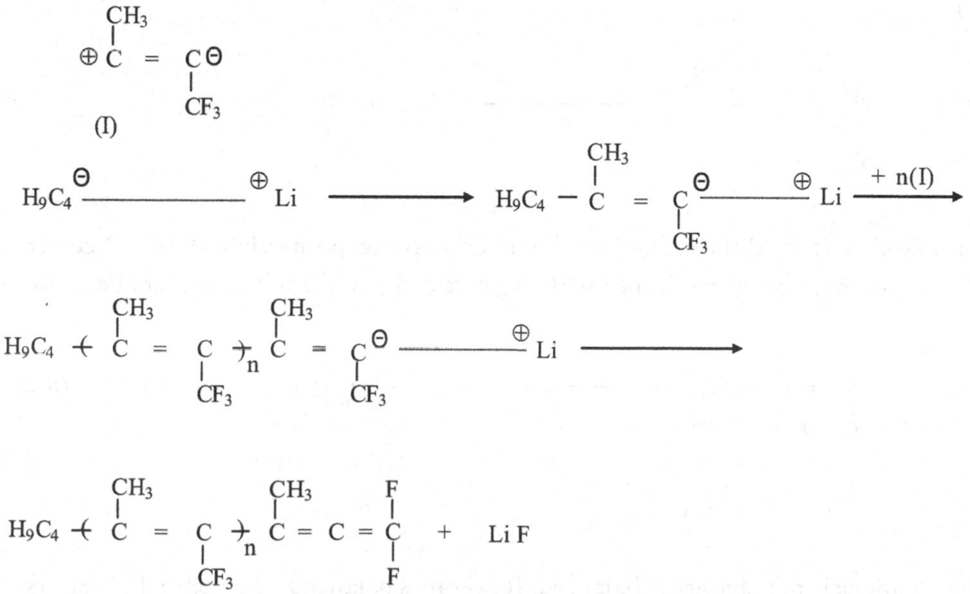

$$\overset{\text{(I)}}{{}^{\oplus}C(CH_3)=C^{\ominus}(CF_3)}$$

$$H_9C_4{}^{\ominus} \cdots {}^{\oplus}Li \longrightarrow H_9C_4-C(CH_3)=C^{\ominus}(CF_3)\cdots {}^{\oplus}Li \xrightarrow{+\,n(I)}$$

$$H_9C_4\left(C(CH_3)=C(CF_3)\right)_n C(CH_3)=C^{\ominus}(CF_3)\cdots {}^{\oplus}Li \longrightarrow$$

$$H_9C_4\left(C(CH_3)=C(CF_3)\right)_n C(CH_3)=C=C(F)(F) + LiF$$

$$6.129$$

[UNNATURAL ROUTE] – Not Favored

$$
\begin{array}{c}
\overset{\displaystyle CH_3}{\underset{\displaystyle CF_3}{\overset{|}{\underset{|}{C}}}} \\
{}^{\ominus}C = \overset{\oplus}{C}
\end{array}
$$

$$
R^{\oplus}\text{---}\overset{\ominus}{B}\begin{smallmatrix} F \\ F \\ F \end{smallmatrix} \text{ (O--R)} \longrightarrow RF \;+\; \underset{F}{\overset{F}{C}} = C = \overset{CH_3}{\overset{\oplus}{C}}\text{---}\overset{\ominus}{B}\begin{smallmatrix} F \\ F \\ F \end{smallmatrix}\text{(O--R)}
$$

6.130

[NATURAL ROUTE]-Not also favored

The trans-monomer has thus been used to show that the monomer still looks like an electrophile, despite the presence of a pushing group whose capacity is less than that of the pulling group. Under normal conditions, *via the route natural to it, there is no transfer species, while via the route not natural to it, transfer species must be abstracted if carried. Whereas, here the reverse is the case. This will be concluded when we come to Free-radical polymerization of these monomers, the only way by which their activations take place. As it seems, this clearly sends a message, that which is the fact that acetylenes cannot be activated chargedly. In addition, a trans-placed monomer such as* $(H_3C)HC=CH(CF_3)$ *cannot be polymerized chargedly.*

When the CF_3 group is replaced with $COOCH_3$ type of group, the followings are obtained, noting that, now one is dealing with an Electrophile with a great difference.

$$
\underset{\substack{| \\ CH_3}}{\overset{\substack{H \\ |}}{C}} \equiv \underset{\substack{| \\ C=O \\ | \\ O \\ | \\ CH_3}}{C} \;+\; R^{\oplus} \longrightarrow ROCH_3 \;+\; \overset{O}{\overset{||}{\oplus\, C}}\text{---}C \equiv CH \quad OR
$$

(I) Favored

$$
\underset{\substack{| \\ (A)}}{\overset{\substack{CH_3 \\ | \\ O \\ | \\ C=O \\ |}}{C}} \equiv C^{\cdot n}\, H^{\cdot e} \;+\; \underset{(II)}{R^{\oplus}} \longrightarrow \text{No reaction}
\qquad 6.131
$$

(A) does not indeed exist. And the center that should be activated positively is the C=O center. "Assuming" the monomer is activated, then with negative charges, the following is to be expected.

$$
R \!+\! \underset{\substack{| \\ C=O \\ | \\ O \\ | \\ CH_3}}{\overset{\substack{H \\ |}}{C}} = C \!+\!_n\, \underset{\substack{| \\ C=O \\ | \\ O \\ | \\ CH_3}}{\overset{\substack{H \\ |}}{C}} = C^{\ominus} \longrightarrow R \!+\! \underset{\substack{| \\ C=O \\ | \\ O \\ | \\ CH_3}}{\overset{\substack{H \\ |}}{C}} = C \!+\!_n\, C = C = C = O \;+\; {}^{\ominus}OCH_3
$$

6.132

The equation as above is not chargedly balanced. It becomes balanced when paired initiators are used. Molecular rearrangement from the C=C center has been prevented by the use of strong initiators. A ketenic type of terminal bond can be observed to be produced. In view of the strong peculiar character of $C \equiv C$ centers, it is (I) above that is favored. That is, the H atom does not need to be replaced by a substituent group before activation can be possible. As shown below, (III) and (IV) shown below can only be unactivated under the conditions indicated.

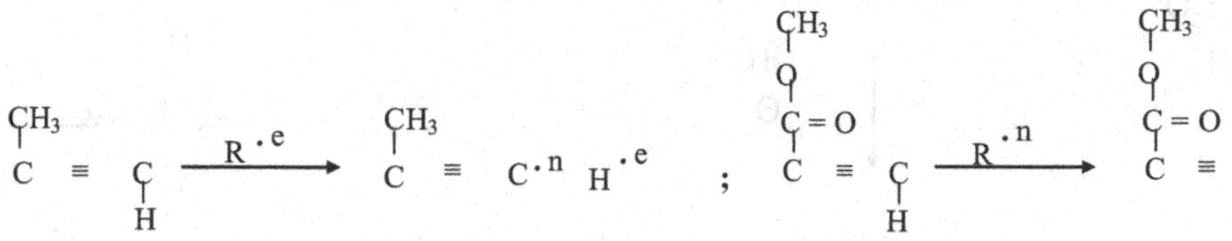

$$
\begin{array}{c}
CH_3 \\
\mid \\
C \equiv C \\
\mid \\
H
\end{array}
\xrightarrow{R \cdot e}
\begin{array}{c}
CH_3 \\
\mid \\
C \equiv C \cdot^n \; H \cdot^e
\end{array}
\quad ; \quad
\begin{array}{c}
CH_3 \\
\mid \\
O \\
\mid \\
C = O \\
\mid \\
C \equiv C \\
\mid \\
H
\end{array}
\xrightarrow{R \cdot n}
\begin{array}{c}
CH_3 \\
\mid \\
O \\
\mid \\
C = O \\
\mid \\
C \equiv C \cdot^n \; H \cdot^e
\end{array}
$$

(III) <u>Favored</u>   (IV) <u>Not Favored</u>   6.133

Nevertheless, though (IV) can readily be used, the most suited acetylenic electrophiles for polymerization are those where the hydrogen atoms are fully replaced.

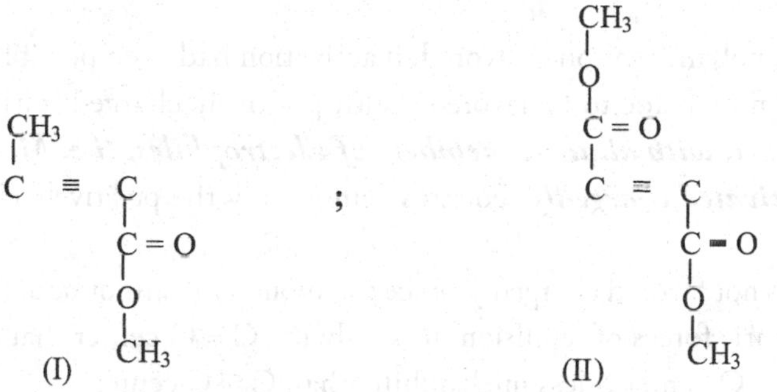

$$
\begin{array}{c}
CH_3 \\
\mid \\
C \equiv C \\
 \mid \\
 C = O \\
 \mid \\
 O \\
 \mid \\
 CH_3
\end{array}
\quad ; \quad
\begin{array}{c}
CH_3 \\
\mid \\
O \\
\mid \\
C = O \\
\mid \\
C \equiv C \\
 \mid \\
 C = O \\
 \mid \\
 O \\
 \mid \\
 CH_3
\end{array}
$$

(I)   (II)   6.134

<u>**Highly polymerizable electrophilic acetylenes**</u>

For (II) above, the followings are obtained when the wrong center is activated chargedly.

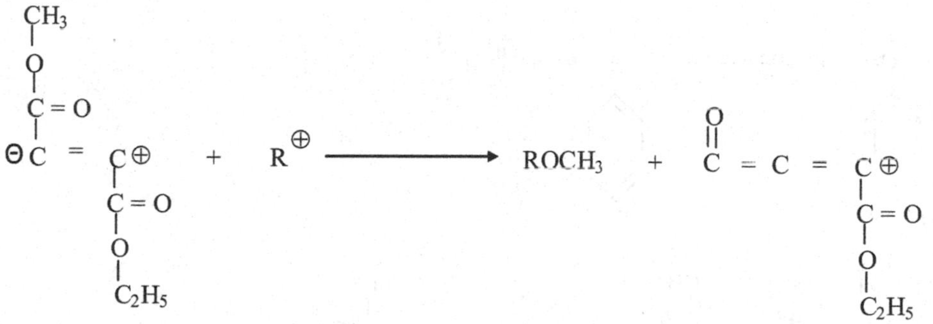

$$
\begin{array}{c}
CH_3 \\
\mid \\
O \\
\mid \\
C = O \\
\mid \\
{}^{\ominus}C \equiv C^{\oplus} \\
 \mid \\
 C = O \\
 \mid \\
 O \\
 \mid \\
 C_2H_5
\end{array}
\; + \; R^{\oplus}
\longrightarrow
ROCH_3 \; + \;
\begin{array}{c}
O \\
\parallel \\
C = C = C^{\oplus} \\
 \mid \\
 C = O \\
 \mid \\
 O \\
 \mid \\
 C_2H_5
\end{array}
$$

                          6.135

For the electrophiles above, only the negatively free-charged paired routes are favored, if activa-tion of the $C \equiv C$ center had been favored. In fact, it will rarely be favored even radically, ***because the first center that will first be activated is the C=O center being less nucleophilic than the $C \equiv C$ center,*** as will shortly be shown.

   When COR is the substituent group, the followings are obtained.

$$\text{(I)} \qquad \xrightarrow{\qquad} \qquad \xrightarrow{+ R^{\oplus}} \qquad \underline{\text{Looks favored}}$$

$$+ R^{\ominus}$$

$$RCH_3 \quad + \qquad \underline{\text{Not favored}} \qquad\qquad\qquad 6.136$$

With negatively-charged-paired initiators, polymerization is favored, if activation had been possible. Like similar olefins, only one charged route can be made to be favored. With positively charged initiator, it is the C = O center that is activated. ***In fact, with all these members of Electrophiles, (i.e. Males), it is only the C = O center that can be activated chargedly.*** For this center, only the positively charged route is favored by them.

With acetate groups, polymerization is not favored chargedly, since the monomer cannot be activated. Apart from the dual influence of electrostatic forces of repulsion, it is only the C = O center that can be activated in these monomers, since the C = O center is less nucleophilic than C ≡ C center.

Considering resonance stabilization phenomena in acetylenes, it seems that for some pushing resonance stabilization groups and all pulling stabilization groups, there should be no hydrogen atom on the monomer when the groups are involved. For example consider using phenyl groups.

$$6.137$$

IMPOSSIBLE EXISTENCE $\qquad\qquad\qquad 6.138$

339

$$6.139$$

Only the second monomer (Equation 6.138) will favor polymerization free-radically and not chargedly. The charges seen in the equation are no charges, but radicals. For the first monomer, the hydrogen atom must be replaced with radical-pulling or alkanyl groups before the monomer can favor being used for polymerization. For the first case which is very unstable, due to resonance stabilization, the phenyl group () cannot be loosely bonded being of greater capacity than $H^{\cdot e}$. The third case cannot be activated chargedly, whether the H is replaced or not. The resonance stabilization group is also not loosely bond, because of the Cl carried.

$$6.140$$

$$6.141$$

On the whole, it is important to note that whether resonance stabilization groups are present or not, the routes favored by the monomers still remain the same, unlike in olefins where there is a group to resonance stabilize.

6.1.3 Free-radical characters of Acetylenes

Like olefins, most of the polymerizable nucleophiles favor only electro-free-radical route.

$$6.142$$

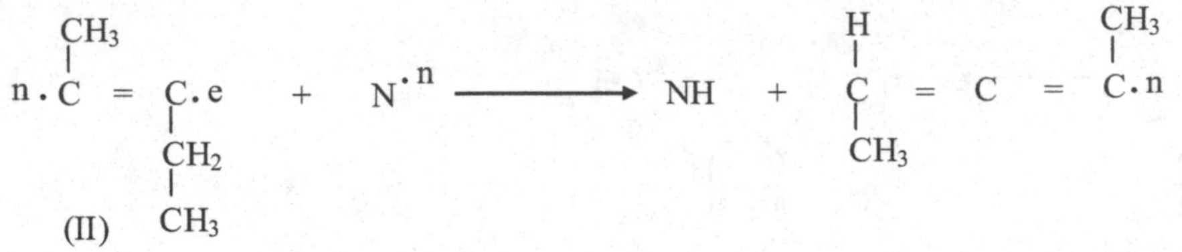

$$\underset{(II)}{n \cdot C} \underset{\substack{| \\ CH_2 \\ | \\ CH_3}}{\overset{\substack{CH_3 \\ |}}{=}} C \cdot e \ + \ N \cdot^n \longrightarrow NH \ + \ \underset{\substack{| \\ CH_3}}{\overset{\substack{H \\ |}}{C}} = C = \overset{\substack{CH_3 \\ |}}{C} \cdot n$$

$$\tag{6.143}$$

$$\underset{(III)}{n \cdot C} \underset{\substack{| \\ O \\ | \\ CH_3}}{\overset{\substack{H \\ |}}{=}} C \cdot e \ + \ N \cdot^n \longrightarrow NCH_3 \ + \ O = C = \overset{\substack{H \\ |}}{C} \cdot n$$

$$\underset{(IV)}{n \cdot C} \underset{\substack{| \\ CH_2 \\ | \\ CH_3}}{\overset{\substack{CH_3 \\ |}}{=}} C \cdot e \ + \ E \cdot^e \longrightarrow \qquad E - \overset{\substack{CH_3 \\ |}}{C} \underset{\substack{| \\ CH_2 \\ | \\ CH_3}}{=} C \cdot e$$

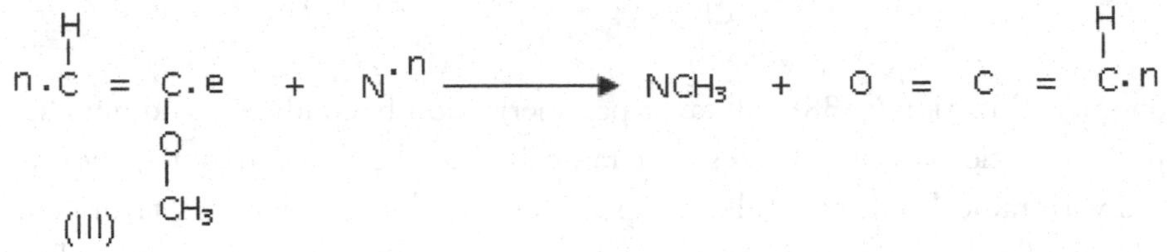

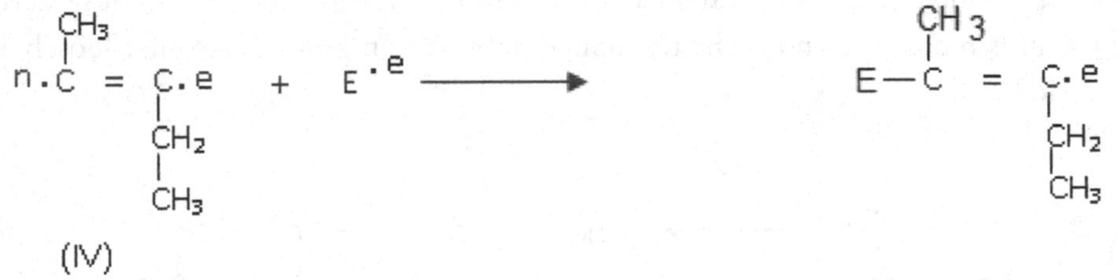

$$\tag{6.144}$$

None can be polymerized nucleo-free-radically. All can be polymerized electro-free-radically. (III) cannot favor molecular rearrangement. (II) and (I) will favor activation electro-free-radical-ly, if the electro-free-radical initiator is of greater or equal capacity than CH_3 for (I) or C_2H_5 for (II).

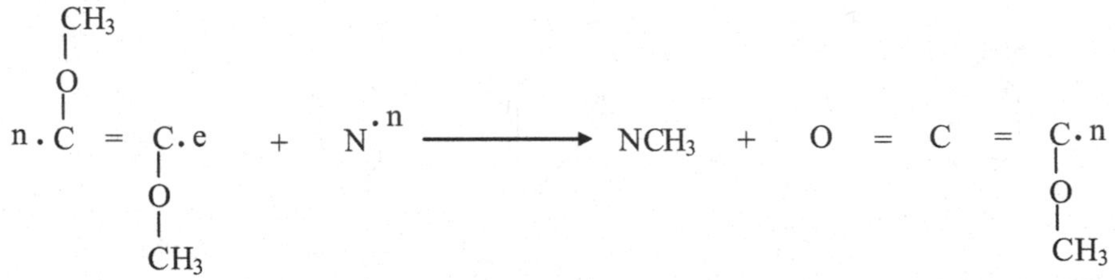

$$n \cdot C \underset{\substack{| \\ O \\ | \\ CH_3}}{\overset{\substack{CH_3 \\ | \\ O \\ |}}{=}} C \cdot e \ + \ N \cdot^n \longrightarrow NCH_3 \ + \ O = C = \underset{\substack{| \\ O \\ | \\ CH_3}}{C} \cdot n$$

$$\tag{6.145}$$

$$\underset{(weak)}{E \cdot^e} \ + \ n \cdot C \underset{\substack{| \\ O \\ | \\ CH_3}}{\overset{\substack{CH_3 \\ | \\ O \\ |}}{=}} C \cdot e \longrightarrow E - O -- \underset{\substack{\| \\ C(OCH_3) \\ | \\ CH_3}}{C} \cdot e$$

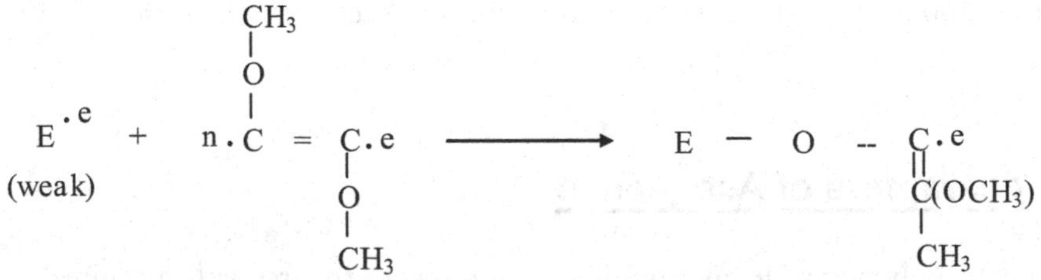

$$\tag{6.146}$$

(Molecular rearrangement)

With or without molecular rearrangement, the electro-free-radical route is favored. The OR group has transfer species which can be involved in transfer from monomer step since the group is a radical-pushing group.

For the monomer shown below, polymerization is fully favored via the route not natural to it- the nucleo-free-radical route.

341

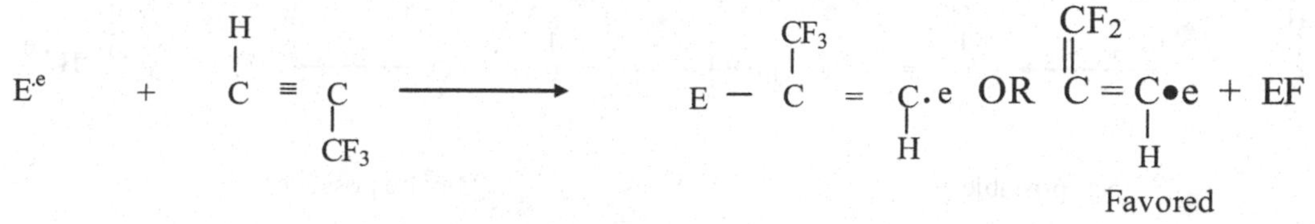

$$E^e \quad + \quad \underset{\underset{CF_3}{|}}{\overset{\overset{H}{|}}{C}} \equiv C \quad \longrightarrow \quad E - \underset{\underset{H}{|}}{\overset{\overset{CF_3}{|}}{C}} = C.e \quad OR \quad \overset{\overset{CF_2}{\|}}{C} = \underset{\underset{H}{|}}{C}\bullet e \; + \; EF$$

$$\text{Favored} \qquad 6.147$$

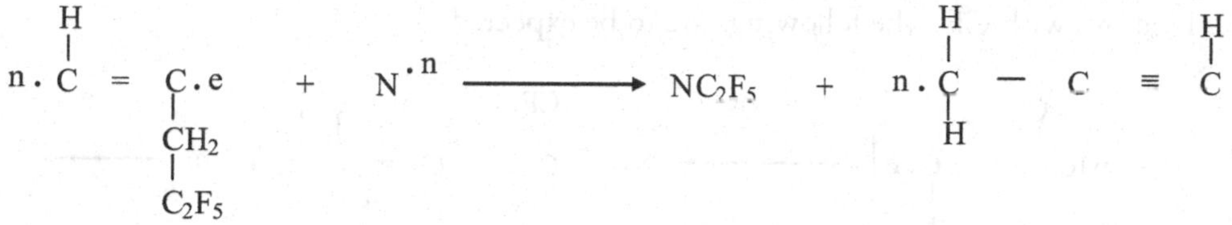

$$n. \underset{\underset{\underset{\underset{C_2F_5}{|}}{CH_2}}{|}}{\overset{\overset{H}{|}}{C}} = C.e \quad + \quad N^{.n} \quad \longrightarrow \quad NC_2F_5 \quad + \quad n.\underset{\underset{H}{|}}{\overset{\overset{H}{|}}{C}} - C \equiv \overset{\overset{H}{|}}{C}$$

$$6.148$$

When the CF_3 group in the first equation is replaced with larger groups such as shown in the last equation, only the electro-free-radical route is favored depending on the capacity of the initiator. Nevertheless, for the first case above, it can still be activated though it favors the route not natural to it, without release of F as transfer species of the first kind of the first type. For $F_3C\text{-}C\equiv C\text{-}CH_3$, this can be activated, but can only be initiated electro-free-radically when the groups are trans-placed and free-radical paired-media initiators are used.

Now, considering those with radical-pulling groups, the followings are obtained with $HC\equiv CF$, when activation of the monomer is possible

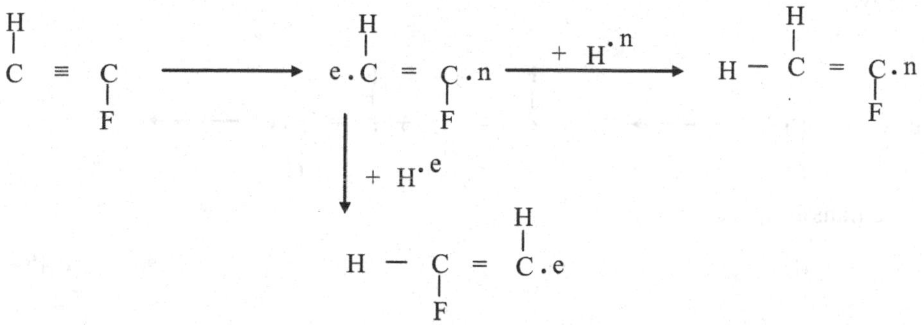

$$\underset{\underset{F}{|}}{\overset{\overset{H}{|}}{C}} \equiv C \quad \longrightarrow \quad e.\underset{\underset{F}{|}}{\overset{\overset{H}{|}}{C}} = C.n \quad \xrightarrow{+ \; H^{.n}} \quad H - \underset{\underset{F}{|}}{\overset{\overset{H}{|}}{C}} = C.n$$

$$\Big\downarrow + \; H^{.e}$$

$$H - \underset{\underset{F}{|}}{\overset{\overset{H}{|}}{C}} = C.e$$

$$6.149$$

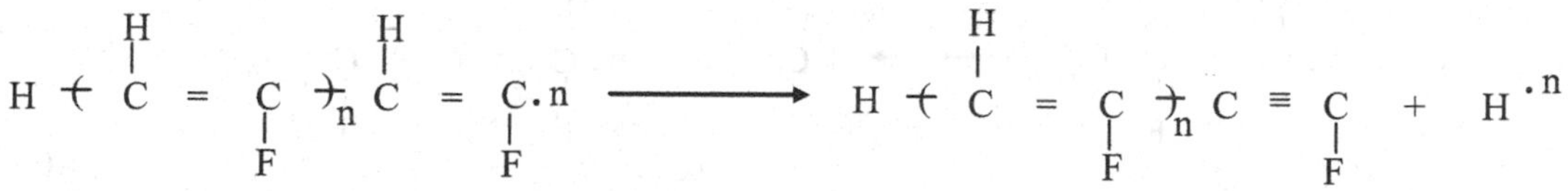

$$H \overset{}{\underset{}{\Big(}} \underset{\underset{F}{|}}{\overset{\overset{H}{|}}{C}} = C \overset{}{\underset{n}{\Big)}} \underset{\underset{F}{|}}{\overset{\overset{H}{|}}{C}} = C.n \quad \longrightarrow \quad H \overset{}{\underset{}{\Big(}} \underset{\underset{F}{|}}{\overset{\overset{H}{|}}{C}} = C \overset{}{\underset{n}{\Big)}} C \equiv \underset{\underset{F}{|}}{C} \; + \; H^{.n}$$

$$6.150$$

$$H \overset{}{\underset{}{\Big(}} \underset{\underset{H}{|}}{\overset{\overset{F}{|}}{C}} = C \overset{}{\underset{n}{\Big)}} \underset{\underset{H}{|}}{\overset{\overset{F}{|}}{C}} = C.e \quad \longrightarrow \quad \text{No transfer species}$$

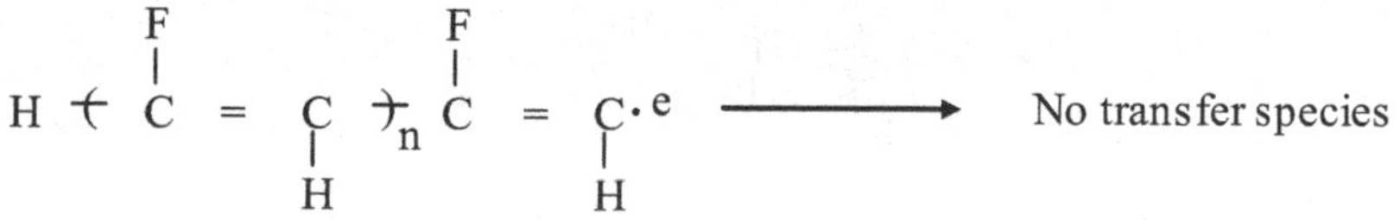

$$6.151$$

A first member which is a hybrid of the first members of the alkynes and perfluoalkynes, can be observed to undergo both free-radical routes, probably in the presence of strong stabilizing agent. The followings shown below does not exist for this first member

$$\underset{\underset{F}{|}}{\overset{\overset{H}{|}}{C}} \equiv C \longrightarrow \overset{\overset{H}{|}}{C} \equiv C.e \quad F\cdot^{nn} \quad ; \quad \underset{\underset{H}{|}}{\overset{\overset{F}{|}}{C}} \equiv C \longrightarrow \underset{\underset{F}{|}}{C} \equiv C\cdot^n H\cdot^e$$

Not possible   May be possible   6.152

Replacing the H above with CF_3, the followings are to be expected.

$$N\cdot^n \; + \; n\left[e.\overset{\overset{CF_3}{|}}{C} = \underset{\underset{F}{|}}{C}.n \right] \longrightarrow N \left(\overset{\overset{CF_3}{|}}{C} = \underset{\underset{F}{|}}{C}\right)_{n-1} \overset{\overset{CF_3}{|}}{C} = \underset{\underset{F}{|}}{C}.n \longrightarrow$$

(Nucleophile)

$$N \left(\overset{\overset{CF_3}{|}}{C} = \underset{\underset{F}{|}}{C}\right)_{n-1} \overset{}{C} \equiv \underset{\underset{F}{|}}{C} \; + \; n.CF_3$$

(I)   <u>Transfer species of second kind</u>   6.153

Since $n.CF_3$ is involved as transfer species of the second kind, homopolymerization nucleo-free-radically is favored. Its release will require very high operating conditions, since $n.CF_3 > n.H$ in radical pushing capacity nucleo-free-radically.

$$E\cdot^e \; + \; (m+1)\, n.\overset{\overset{F}{|}}{\underset{\underset{CF_3}{|}}{C}} = C.e \longrightarrow E\left(\overset{\overset{F}{|}}{\underset{\underset{CF_3}{|}}{C}} = C\right)_m \overset{\overset{F}{|}}{\underset{\underset{CF_3}{|}}{C}} = C.e \longrightarrow$$

No transfer species

(I)   6.154

$$N\cdot^n \; + \; e.\overset{\overset{CF_3}{\overset{|}{\overset{CF_2}{|}}}}{\underset{\underset{F}{|}}{C}} = C.n \longrightarrow NCF_3 \; + \; n.\overset{\overset{F}{|}}{\underset{\underset{F}{|}}{C}} - C \equiv C \underset{F}{}$$

(I)   6.155

$$E\cdot^e \; + \; (m+1) \; \overset{\overset{F}{|}}{\underset{\underset{CF_2}{\underset{|}{\underset{CF_3}{}}}}{\underset{n}{C}}} = C.e \longrightarrow E\left(\overset{\overset{F}{|}}{\underset{\underset{CF_2}{\underset{|}{CF_3}}}{C}} = C\right)_m \overset{\overset{F}{|}}{\underset{\underset{CF_2}{\underset{|}{CF_3}}}{C}} = C.e \longrightarrow$$

$$E \xleftarrow{} \left[\begin{array}{c} F \\ | \\ C \\ | \\ CF_2 \\ | \\ CF_3 \end{array} = \begin{array}{c} F \\ | \\ C \\ | \end{array} \right]_m \xrightarrow{} \begin{array}{c} F \\ | \\ C \\ | \end{array} = C = \begin{array}{c} F \\ | \\ C \\ | \\ F \end{array} + e \cdot CF_3$$

$$(I) \qquad\qquad 6.156$$

All these cases here are nucleophilic with electro-free-radical polymerization favored by them. Only the first member favors both routes

Using free-radical-pulling groups, the followings are to be expected for COOR groups.

$$e \cdot C = C \cdot n + N \cdot^n \xrightarrow{\qquad} N - C = C \cdot n$$

with H on top carbons, C=O, O, CH₃ substituents

$$(I) \qquad\qquad NOT\ FAVORED \qquad\qquad 6.157$$

$$E \cdot e + \begin{array}{c} H \\ | \\ C \\ \end{array} = \begin{array}{c} C \\ | \\ e \cdot C - O \cdot nn \\ | \\ O \\ | \\ CH_3 \end{array} \longleftrightarrow e \cdot C = C = \begin{array}{c} C - O \cdot nn \\ | \\ O \\ | \\ CH_3 \end{array} \longrightarrow e \cdot C = \begin{array}{c} C \\ \| \\ C - O - E \\ | \\ O \\ | \\ CH_3 \end{array}$$

$$(II) \qquad (III) \qquad (IV)$$

$$RESONANCE\ STABILIZED \qquad\qquad 6.158$$

$$FULLY\ FAVORED$$

Based on what has been shown in the last equation, the first equation is not favored. Since the C = O center is less nucleophilic than the C ≡ C center, it is the first to be activated all the time. That center (C = O) is a half-free monomer which is the 1,2- mono-form (II). From here reso-nance stabilization commences to give the 1,4- mono-form (III), that which is also a half-free monomer. Hence, one can observe that the 3,4- mono-form cannot exist. That is the Y center. One can observe a very great difference of this olefinic acrylates. With olefinic acrylates, the Y center is less nucleophilic than the X center, while the reverse is the case with acetylinic acrylates. Indeed, the two routes of importance, based on the resonance stabilization favored by acetylenic males are the electro-free-radical and nucleo-non-free-radical routes. Nucleo-free-radicals cannot be uses for them, because of radical balancing. Notice that the OCH₃ group is shielded.

$$\begin{array}{c} CH_3 \\ | \\ C \equiv C \\ | \\ C = O \\ | \\ O \\ | \\ CH_3 \end{array} + :N \cdot nn \longrightarrow :NH + \begin{array}{c} H \\ | \\ C = C = C \\ | \qquad\qquad \| \\ H \qquad\qquad C - O \cdot nn \\ | \\ O \\ | \\ CH_3 \end{array}$$

Electrophile

$$6.159$$

Nucleo-free- and nucleo-non-free-radically whether the initiator is weak or strong, initiation is not favored. The transfer species is provided by CH_3 group externally located on the $- C \equiv C(CH_3)$ group, noting that OCH_3 group is internally located and therefore shielded. Instead of the route natural to the monomer being favored, it is the electro-free-radical route that is favored. For exploratory purposes, imagine if molecular rearrangement is allowed for these cases.

EXPLORATORY PURPOSES

$$
\text{e. } C = C.n \quad \xrightarrow{\hspace{2cm}} \quad \text{e. } C - C.n \qquad\qquad 6.160
$$

$$
\text{e. } C = C.n \quad \xrightarrow{\hspace{2cm}} \quad \text{e. } C - C.n \qquad\qquad 6.161
$$

$$
R - C - C - C - C.n \quad \xrightarrow{\hspace{2cm}} \quad R - C - C - C - C \equiv C \quad + \quad nn.NH_2
$$

<u>Transfer species of first kind of 1st or 4th type</u>

These are other unique sets of cases where molecular rearrangements are favored when weak paired-radical initiators are involved for this trans-monomer. Now, coming back to the monomer of Equation 6.159, electro-free-radically the followings are obtained.

$$
E \cdot e + n.C = C.e \quad \xrightarrow{\hspace{2cm}} \quad E - C = C.e \quad \xrightarrow{+n(I)}
$$

$$
E - C = C - C = C.e \quad \xrightarrow{\hspace{2cm}} \quad E - C = C - C = C = C \quad + \quad H \cdot e
$$

<u>Not Favored</u>   6.163a

Since the electro-free-radical route is not natural to the monomer, transfer species of the first kind which exists (OCH_3) must be abstracted. Electro-free-radically it cannot therefore produce a dead polymer in situ, unless through the $C = O$ center with CH_3 as transfer species as shown below.

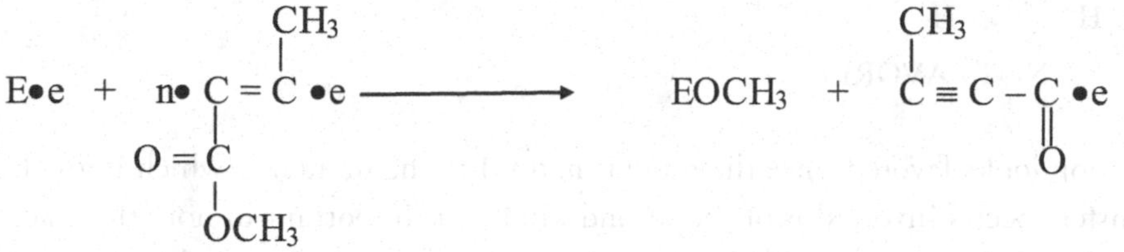

$$E\bullet e \ + \ n\bullet \overset{\displaystyle CH_3}{\underset{\displaystyle \underset{\displaystyle OCH_3}{O=C}}{C}}=C\bullet e \longrightarrow EOCH_3 \ + \ \overset{\displaystyle CH_3}{\underset{\displaystyle O}{C}} \equiv C-\overset{\displaystyle \|}{C}\bullet e$$

$$\text{6.163b}$$

$$E-(O-\overset{\displaystyle \overset{\displaystyle C(CH_3)}{\|\|}}{\underset{\displaystyle OCH_3}{C}})_n-O-\overset{\displaystyle \overset{\displaystyle C(CH_3)}{\|\|}}{\underset{\displaystyle OCH_3}{C}}\bullet e \longrightarrow E-(O-\overset{\displaystyle \overset{\displaystyle C(CH_3)}{\|\|}}{\underset{\displaystyle OCH_3}{C}})_n-O-\overset{\displaystyle \overset{\displaystyle C(CH_3)}{\|\|}}{C}=O \ + \ e\bullet CH_3$$

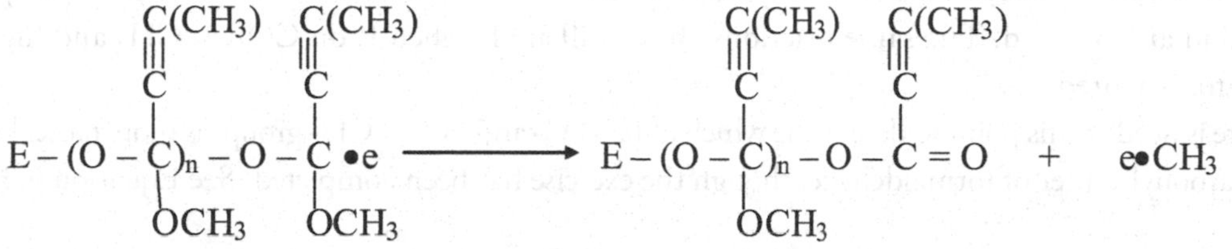

$$\text{6.163c}$$

This seems to be clear indication that $C = O$ center is of greater capacity than $C \equiv C$ center. *Indeed, one should expect the double bond between two centers with great electronegativity difference of 1 ($C = O$) should be more nucleophilic in character than a triple bond between two centers with zero electronegativity difference ($C \equiv C$). It is not, but as follows.*

$$C \equiv N \ > \ C \equiv C \ > \ C = O \ > \ C = N \ > \ C = S > C = C$$

Electronegativity $\quad$ 0.5+π $\qquad$ 0+π $\qquad$ 1.0 $\qquad$ 0.5 $\qquad$ 0 $\qquad$ 0

Difference $\qquad\qquad$ (π < 0.5 – 1.0)

ORDER OF NUCLEOPHILICITY $\qquad$ 6.163d

[Note that the π-bonds in all of them are different]

One cannot use the order of electronegativity difference or the numbers of π-bonds present as a measure of order of nucleophilicity alone, but the types of groups carried by the centers based on the routes favored and some other variables, for measurement.

With groups such as COR, the followings are obtained for R equal to H.

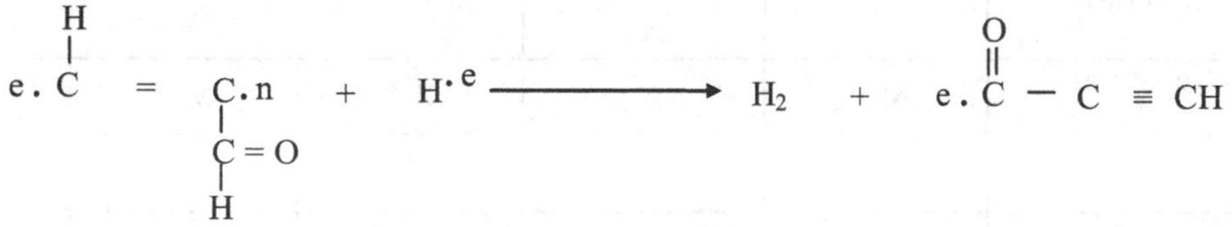

$$e.\overset{\displaystyle H}{\underset{\displaystyle \underset{\displaystyle H}{C=O}}{C}} = \overset{}{\underset{\displaystyle \underset{\displaystyle H}{C=O}}{C}}.n \ + \ H\cdot e \longrightarrow H_2 \ + \ e.\overset{\displaystyle \overset{\displaystyle O}{\|}}{C}-C\equiv CH$$

$$\text{NOT FAVORED}$$

$$+\overset{\displaystyle H}{\underset{\displaystyle \underset{\displaystyle H}{C=O}}{C}}=\overset{\displaystyle H}{\underset{\displaystyle \underset{\displaystyle H}{C=O}}{C}}+_n \overset{\displaystyle H}{C}=\overset{\displaystyle H}{C}.n \longrightarrow N+\overset{\displaystyle H}{\underset{\displaystyle \underset{\displaystyle H}{C=O}}{C}}=\overset{\displaystyle H}{\underset{\displaystyle \underset{\displaystyle H}{C=O}}{C}}+_n C=C=C=O$$

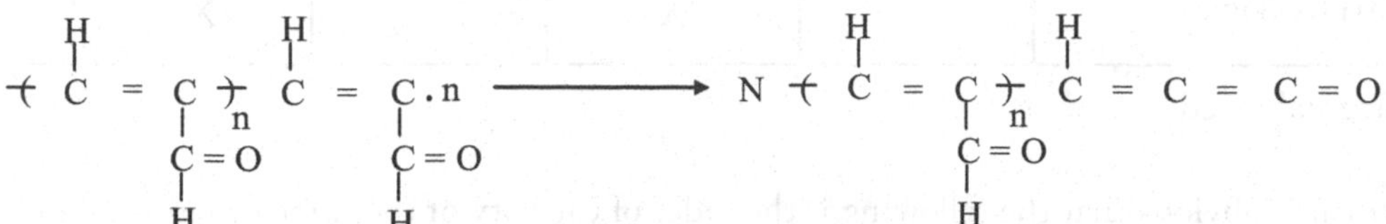

$$\text{NOT FAVORED}$$

$$\text{OR} \qquad\qquad\qquad \text{6.164}$$

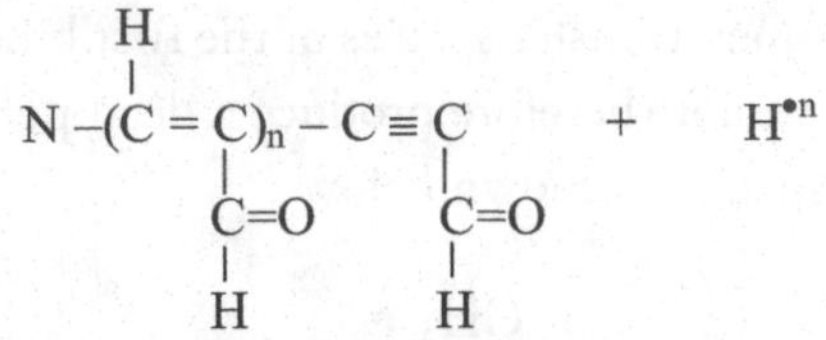

$$N-(C=C)_n - C \equiv C \quad + \quad H^{\bullet n}$$

with substituents C=O (with H) on the terminal carbons.

NOT FAVORED

$$6.165$$

This second reaction looks favored since the route is natural to the monomer which is an electro-phile, wherein the transfer species involved is of the second kind. It is important to note the reactions where nucleo-free-radical hydrogen atom or alkyl groups are involved, that is, those where transfer species of second kind are involved. The same reactions above will apply when R on COR is CH_3 and higher and the initiator is paired.

There is need at this point to determine which of $HC \equiv C-$ and $H_2C = CH-$ groups is more radical-pushing using a carbonyl center of formaldehyde, though the exercise has been completed (See Equation 6.30).

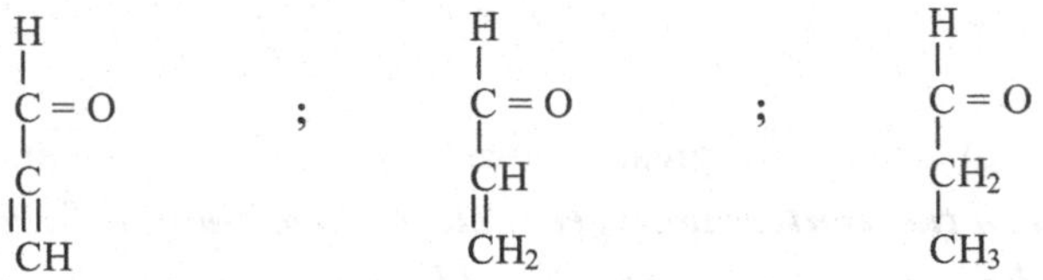

Structure (I): $C=O$ (with H) attached to $C \equiv\!\!\equiv\!\!\equiv CH$; Structure (II): $C=O$ (with H) attached to $CH = CH_2$; Structure (III): $C=O$ (with H) attached to $CH_2 - CH_3$

(I)   (II)   (III)

$$6.166$$

The routes favored by the three monomers are shown below in Table 6.1. From the data shown

Table 6.1. Routes favored by (I), (II) and (III) of Equation 6.166

Routes	(I) ($H^{\bullet e}$, $H^{\oplus}$)		(II)		(III)
	C = O	C $\equiv$ C	C = O	C = C	C = O
Nucleo-non-free-radical	√	√*	√	X	X
(Nucleo-free-radical)	X	X*	X	√	X
Electro-free-radical	√	√*	√	√	√
Paired – media (Anionic)	√	X	√	X	X
Paired – media (Negative)	X	X	X	√	X
Paired-media (Cationic)	X	X	X	X	X
Paired-media (Positive)	√	X	√	√	√
Free-media (Anionic)	√	X	√	X	X

* Resonance stabilized

in Table 6.1 it is obvious that the following is the order of capacity of the three groups:-

$$- \quad C \equiv CH \qquad < \quad -H \quad \leq \quad -CH = CH_2 \quad < \quad -CH_2-CH_3$$

Order of pushing capacity.

$$6.167$$

While the order above is valid for the groups, it is obvious from the table that acetylene is more nucleophilic than C=O activation center, which in turn is more nucleophilic than C=C activation centers. It should be noted that the groups compared are carbon-centered and belong to the following families - alkynes, alkenes and alkanes, unlike H in formaldehyde, which belongs to a separate family. The capacity above is valid in general chargedly and radically. While CN group is pulling but (C≡N is) nucleophilic, HC≡C group is pushing and also nucleophilic. While H can fully favor carrying cations under any conditions when ionic environments exist, the other groups carry only positive charges which are non-ionic and cannot be isolated as already shown in the Series.

For a hybrid of alkanyl and alkynyl groups the following is obtained.

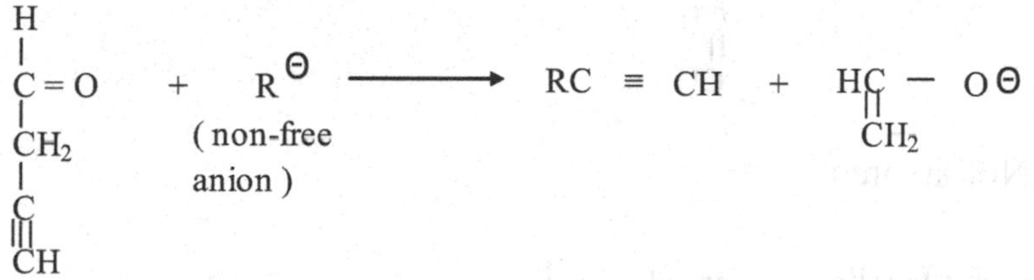

$$\begin{array}{c} H \\ | \\ C=O \\ | \\ CH_2 \\ | \\ C \\ ||| \\ CH \end{array} \quad + \quad R^\ominus \xrightarrow{} \quad RC \equiv CH \quad + \quad \begin{array}{c} HC - O^\ominus \\ || \\ CH_2 \end{array}$$

$$(\text{non-free anion})$$

$$6.168$$

The monomer in Equation 6.164 for alkynes is almost analogous to that of acrolein for alkenes. When the H in C≡CH group is replaced with CH_3 and other higher alkyl groups, the situation becomes completely different, noting that the following prevails:-

$$> \quad - C \equiv C - C_3H_7 \quad > \quad - C \equiv C - C_2H_5 \quad > \quad - C \equiv CCH_3 \quad > \quad H$$

$$> \quad - C \equiv CH \quad > \quad -C \equiv CCF_3 \quad > \quad \ldots\ldots\ldots$$

$$6.169$$

Order of radical-pushing capacity

It is the order of the H, CH_3, C_2H_5 etc. groups that determine the order shown. Consider providing resonance stabilization using alkenyl group. As shown below, it is the $-C \equiv CH$ group

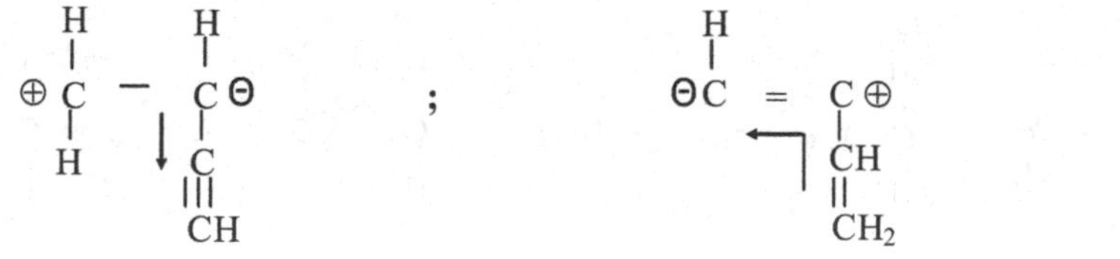

$$\begin{array}{cc} \begin{array}{c} H \quad\quad H \\ | \quad\quad\quad | \\ \oplus C - C\ominus \\ | \quad\quad\quad\downarrow | \\ H \quad\quad\quad C \\ \quad\quad\quad\quad ||| \\ \quad\quad\quad\quad CH \end{array} & ; & \begin{array}{c} H \\ | \\ \ominus C = C\oplus \\ \quad\quad | \\ \quad\quad CH \\ \quad\quad || \\ \quad\quad CH_2 \end{array} \end{array}$$

$$(I) \qquad\qquad\qquad\qquad (II) \quad \text{NOT POSSIBLE} \qquad\qquad 6.170$$

$$\begin{array}{cc} \begin{array}{c} H \quad\quad H \\ | \quad\quad\quad | \\ e.C - C.n \\ | \quad\quad\quad | \\ H \quad\quad\quad C \\ \quad\quad\quad ||| \\ \quad\quad\quad CH \end{array} & ; & \begin{array}{c} H \\ | \\ n.C = C.e \\ \quad\quad | \\ \quad\quad CH \\ \quad\quad || \\ \quad\quad CH_2 \end{array} \end{array}$$

$$(I) \qquad\qquad\qquad\qquad\qquad (II)$$

$$(\text{SEE EQN.6.30}) \qquad\qquad \text{NOT POSSIBLE} \qquad\qquad 6.171$$

that is pulling on the π-bond of the -CH=CH$_2$ group. The alkenyl (or alkyl-ene) group cannot provide resonance stabilization for the alkynyl (or alkyl-yne) group. It is the other way round. Recalling Equation 6.30, the activated states of the monomer above is favored, only for the (I)s. However, resonance stabilization can only be provided radically and never chargedly. That is, (I) of Equation 6.170 cannot be resonance stabilized. It is (I) of Equation 6.171 that can undergo resonance stabilization when stabilized. When not stabilized or suppressed, the monomer which is resonance stabilized will not exist, because H will still be loosely bonded in acetylene if the second substituent group is of less pushing capacity than H. Hence, what is shown below is not favored, because H$_2$C=CH- group is of greater capacity than H [See (II) of Equation 6.171]. If it was favored, the reactions of Equation 6.30 would not have taken place. Note that (II) of Equation 6.171 was said not to be possible because acetylene is more nucleophilic than ethene.

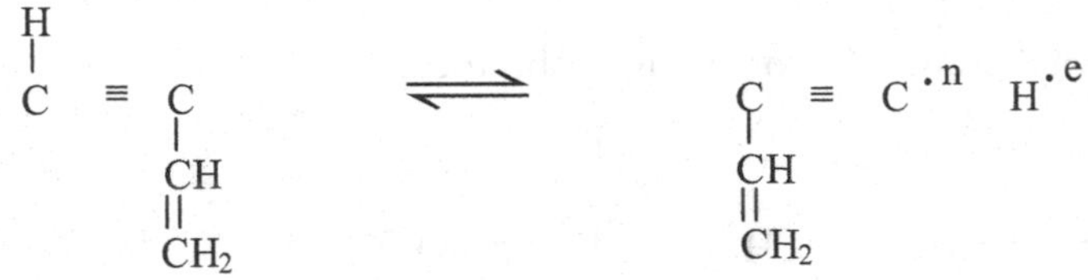

$$6.172$$

(Not favored)

When the H is however replaced, the followings are obtained.

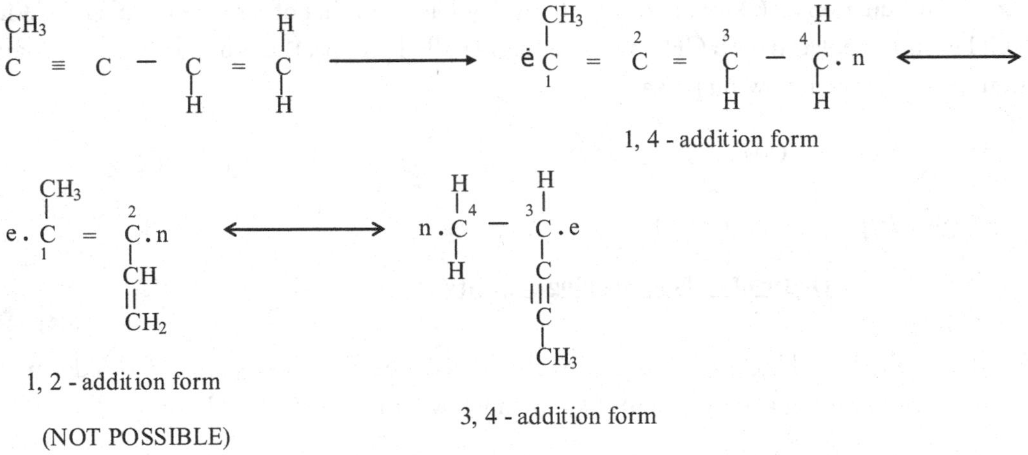

1, 4 - addition form

1, 2 - addition form

(NOT POSSIBLE)

3, 4 - addition form

$$6.173$$

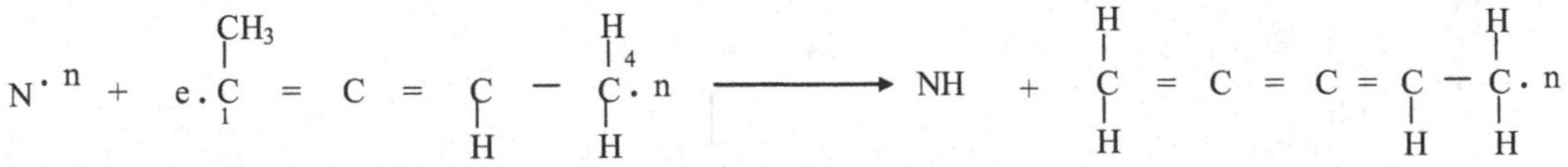

$$6.174$$

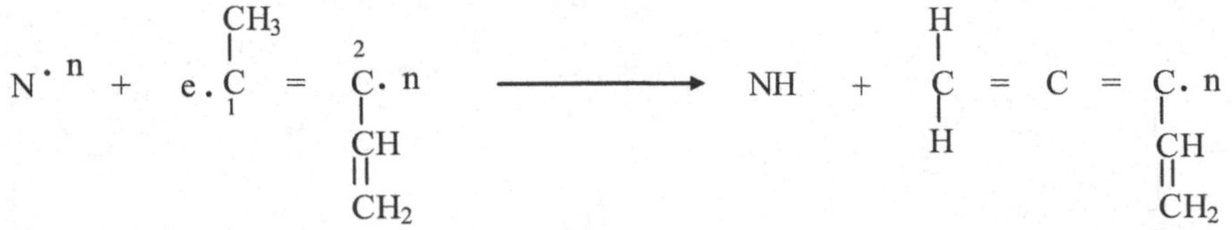

$$6.175$$

The reason why the activation of the 1,2-mono-form is said not to be possible in Equation 6.173, is because, the center first activated is the least nucleophilic center. The 1,2-center is the most nucleophilic center. The 3,4-center is the less nucleophilic center, through which the 1,4-mono-form is obtained.

349

$$N^{\bullet n} \;+\; e \cdot \overset{\overset{\displaystyle H}{|}}{\underset{\underset{\displaystyle CH_3}{\underset{\displaystyle |}{\overset{\displaystyle |}{C}}}}{\underset{|}{C}}}{}^3 - \overset{\overset{\displaystyle H}{|}}{\underset{\underset{\displaystyle H}{|}}{C}}^4 \cdot n \longrightarrow NH \;+\; \overset{\overset{\displaystyle H}{|}}{\underset{\underset{\displaystyle H}{|}}{C}} = C = C = \overset{\overset{\displaystyle H}{|}}{C} - \overset{\overset{\displaystyle H}{|}}{\underset{\underset{\displaystyle H}{|}}{C}}\cdot n$$

6.176

$$E^{\bullet e} \;+\; n \cdot \overset{2}{\underset{\underset{\underset{\displaystyle CH_2}{\parallel}}{\underset{\displaystyle CH}{|}}}{C}} = \overset{\overset{\displaystyle CH_3}{|}}{\underset{1}{C}} \cdot e \longrightarrow E - \overset{}{\underset{\underset{\underset{\displaystyle CH_2}{\parallel}}{\underset{\displaystyle CH}{|}}}{C}} = \overset{\overset{\displaystyle CH_3}{|}}{C} \cdot e$$

6.177

NOT FAVORED

Though the last equation above is not favored, because 1,2 mono-form cannot be activated, the monomer can be observed to be a strong nucleophile. The same type of reactions above cannot take place chargedly. The conditions which favor the full existence of resonance stabilization phenomena can thus be identified- the same that determine the polymerizability of the monomer.

Considering using $-C \equiv CH$ as the source of resonance stabilization for acetylene, the following will be obtained if its existence is favored- i.e., a Diyne.

$$\overset{\overset{\displaystyle H}{|}}{C} \equiv C - C \equiv \overset{\overset{}{}}{\underset{\underset{\displaystyle H}{|}}{C}} \;\rightleftharpoons\; \overset{\overset{\displaystyle H}{|}}{C} \equiv C - C \equiv C^{\bullet n} \; H^{\bullet e}$$

(I)

6.178

Note that this is not the real Equilibrium state of the existence of the Diyne above. It is that in which the $-C \equiv CH$ is the component held as opposed to H, in view of its lower capacity with respect to H. Tentatively however, the H atoms are replaced as follows-

$$R^{\bullet e} \;+\; \overset{\overset{\displaystyle H}{|}}{C} \equiv C - C \equiv C^{\bullet n} \; H^{\bullet e} \longrightarrow \overset{\overset{\displaystyle H}{|}}{C} = C - C \equiv \overset{}{\underset{\underset{\displaystyle R}{|}}{C}} \;+\; H^{\bullet e}$$

(I)

6.179

$$\overset{\overset{\displaystyle H}{|}}{C} \equiv \overset{\bullet}{C} - C \equiv \overset{}{\underset{\underset{\displaystyle R}{|}}{C}} \longrightarrow \overset{}{\underset{\underset{\displaystyle R}{|}}{C}} \equiv C - C \equiv C^{\bullet n} \; H^{\bullet e} \quad (+ \quad R^{\prime \bullet e})$$

(A) (II)

$$e \cdot \overset{}{\underset{\underset{\displaystyle R}{|}}{C}} = C = C = \overset{\overset{\displaystyle R'}{|}}{C} \cdot n \qquad + \; H.e$$

(III)

6.180

If activation of (A) is not favored, (III) is obtained when a suitable group of lesser capacity than R is used. When (A) is involved, the followings are obtained.

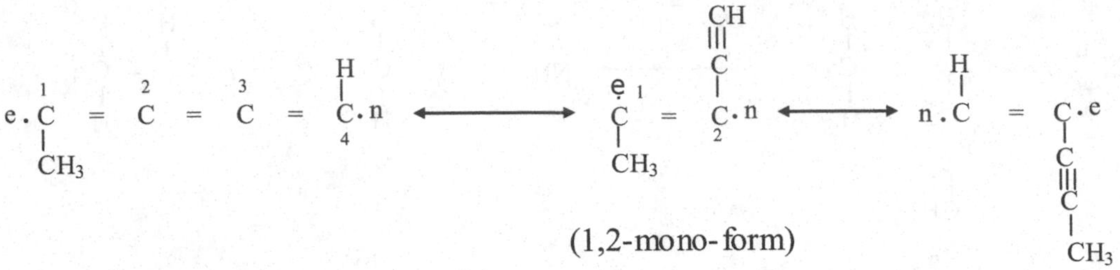

$$\underset{\underset{CH_3}{|}}{e \cdot \overset{1}{C}} = \overset{2}{C} = \overset{3}{C} = \underset{4}{\overset{H}{\overset{|}{C} \cdot n}} \longleftrightarrow \underset{\underset{CH_3}{|}}{\overset{e}{\overset{1}{C}}} = \overset{\overset{CH}{|||}}{\overset{C}{\underset{2}{C} \cdot n}} \longleftrightarrow n \cdot \overset{H}{\overset{|}{C}} = \underset{\underset{CH_3}{|}}{\overset{C}{\overset{|||}{\overset{C}{\overset{|}{C} \cdot e}}}}$$

(1,2-mono-form)

NON-EXISTENCE   6.181

It is believed that, when $R^{\cdot e}$ or $E^{\cdot e} > H$, presence of (I) is that favored, as opposed to the case shown below (B), since the $-C \equiv CH$ group is providing resonance stabilization for no group. There is no substituent groups internally located.

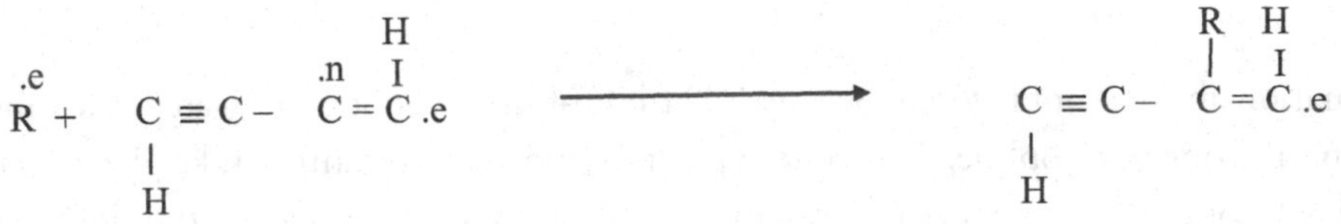

$$\overset{.e}{R} + \underset{\underset{H}{|}}{C \equiv C-} \overset{.n}{\overset{\overset{H}{|}}{C}} = \overset{I}{C} . e \longrightarrow C \equiv C- \underset{\underset{H}{|}}{\overset{\overset{R}{|}}{C}} = \overset{\overset{H}{|}}{\overset{I}{C}} . e$$

(B)   6.182

If the remaining H in (A) is replaced with F, then the followings are to be expected.

$$\underset{\underset{R^1}{|}}{C} \equiv C - C \equiv C - F \quad + \quad R^{\cdot e} \longrightarrow \underset{\underset{R^1}{|}}{C} \equiv C - C \equiv C^{\cdot e} + RF$$

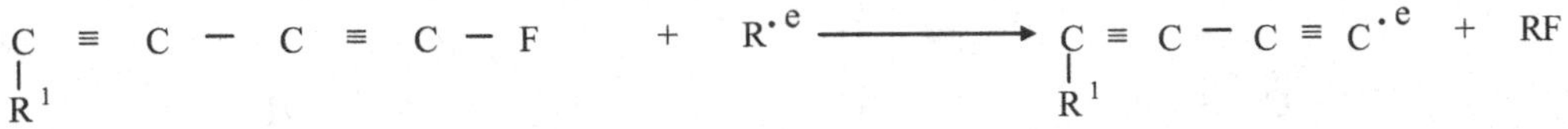

6.183

When the monomer above cannot be activated, then the fluorine atom must be replaced to favor being used for polymerization. It is however believed that the monomer can be activated when suppressed. When the replacement is complete with pushing groups, the monomer will favor only the electro-free-radical route as shown below, just as should be the case above.

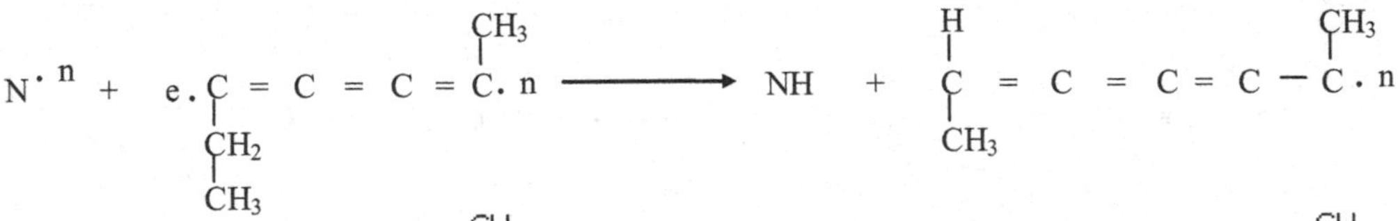

$$N^{\cdot n} + e \cdot \underset{\underset{\underset{CH_3}{|}}{CH_2}}{\overset{\overset{CH_3}{|}}{C}} = C = C = C \cdot n \longrightarrow NH + \underset{\underset{CH_3}{|}}{\overset{\overset{H}{|}}{C}} = C = C = C - \overset{\overset{CH_3}{|}}{C} \cdot n$$

6.184

$$E^{\cdot e} + n \cdot \underset{\underset{CH_3}{|}}{\overset{\overset{CH_3}{|}}{\overset{CH_2}{|}}{C}} = C = C = C \cdot e \longrightarrow E - \underset{\underset{CH_3}{|}}{\overset{\overset{CH_2}{|}}{\overset{CH_3}{|}}{C}} = C = C - C \cdot e$$

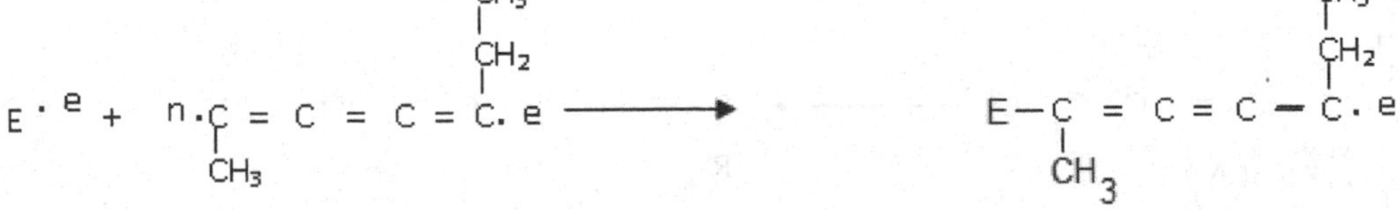

6.185

Chargedly, it cannot be activated. *It can be observed that for the first member of the family, $-C \equiv CH$ group cannot indeed be used as a resonance stabilization group alone.* It can only be used once the H has been replaced by alkylane or other groups. For those with pulling groups such as COOR and CN, the followings are to be expected exploratively.

351

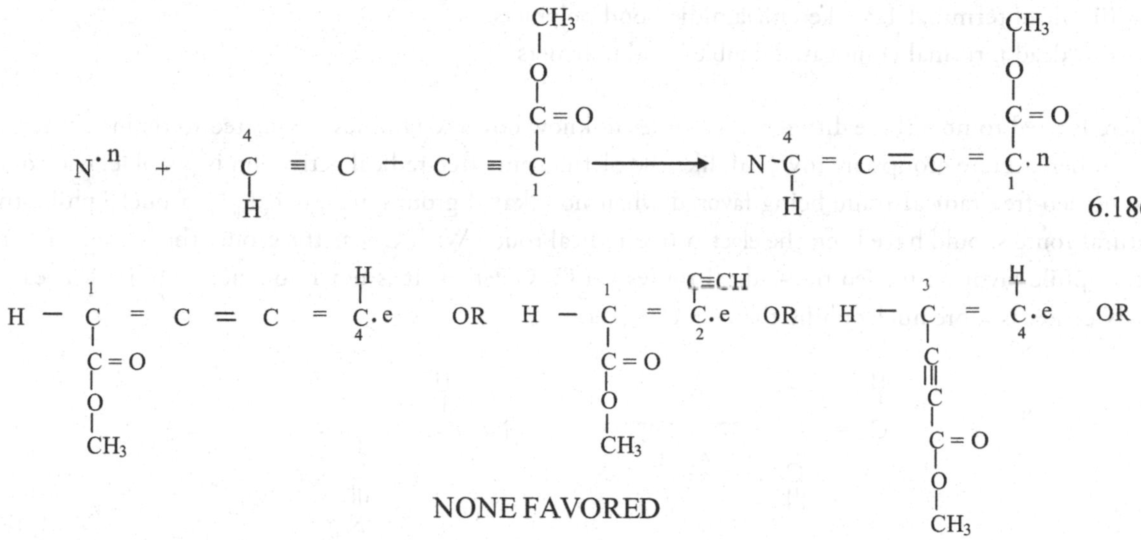

$$N^{\bullet n} \ + \ \underset{\overset{|}{H}}{\overset{4}{C}} \equiv C - C \equiv \underset{\overset{|}{\underset{O}{\overset{|}{\underset{C=O}{\overset{|}{\underset{CH_3}{}}}}}}}{\overset{1}{C}} \longrightarrow N - \underset{\overset{|}{H}}{\overset{4}{C}} = C = C = \underset{\overset{|}{\underset{O}{\overset{|}{\underset{C=O}{\overset{|}{\underset{CH_3}{}}}}}}}{\overset{1}{C}} \cdot n$$

6.186

$$H - \underset{\overset{|}{\underset{C=O}{\overset{|}{\underset{O}{\overset{|}{CH_3}}}}}}{\overset{1}{C}} = C = C = \overset{4}{C} \cdot e \quad OR \quad H - \underset{\overset{|}{\underset{C=O}{\overset{|}{\underset{O}{\overset{|}{CH_3}}}}}}{\overset{1}{C}} = \overset{C \equiv CH}{\overset{2}{C}} \cdot e \quad OR \quad H - \overset{3}{C} = \overset{4}{C} \cdot e \quad OR$$

NONE FAVORED

$$HOCH_3 \ + \ e \bullet \overset{\overset{O}{\|}}{C} - C \equiv C - C \equiv \underset{\overset{|}{H}}{C} \qquad [H\bullet e \text{ cannot carry the chain}]$$

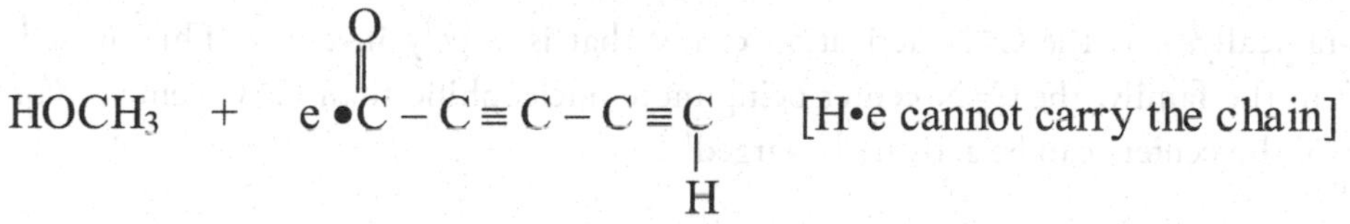

6.187

[Favored for only 3,4- and 1,4-mono-forms]

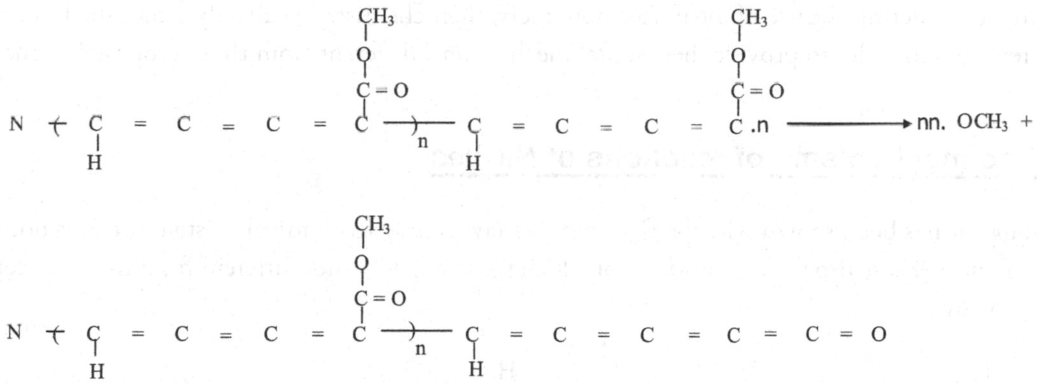

$$N \overset{}{\underset{}{\left(} \underset{\overset{|}{H}}{\overset{\overset{|}{\underset{O}{\overset{|}{\underset{C=O}{\overset{|}{CH_3}}}}}}}{C} = C = C = C \right)_n } \underset{\overset{|}{H}}{C} = C = C = C \cdot n \longrightarrow nn. \ OCH_3 +$$

$$N \overset{}{\underset{}{\left(} \underset{\overset{|}{H}}{\overset{\overset{|}{\underset{O}{\overset{|}{\underset{C=O}{\overset{|}{CH_3}}}}}}}{C} = C = C = C \right)_n } \underset{\overset{|}{H}}{C} = C = C = C = C = O$$

6.188

Indeed, it is the C=O center that is first activated, for which the numbering above is wrong. It is a 1.6- mono-form, the other mono-forms being 1,2- and 1,4- mono-forms. Nucleo-non-free-radical and electro-free-radicals routes are the routes favored. The types of dead terminal bonded polymers produced from these types of monomers, are worthy of note:-

(i)   dead terminal triple bond polymers.
(ii) dead terminal 1,3 - or 1,4-cumulenic bond polymers.
(iii) dead terminal 1,3- or 1,4- ketinic/amidic bond polymers.

This is very much unlike olefinic monomers where the followings exist.

(i)   dead terminal double bond polymers.

352

(ii) dead terminal 1,3 – ketenic/amidic bond polymers.

(iii) dead terminal conjugated double bond polymers.

There is need to note these differences in order to know how the families are related to them-selves.

When acetate groups are involved, like the olefins, only free-radical activation is possible, with only the nucleo-free-radical route being favored when no alkanyl groups are carried. As a nucleophile, the natural route should have been the electro-free-radical route. With CN as the group, the monomer is an electrophile favoring nucleo-free-radical routes via C≡C center. It is not resonance stabilized, because C≡N center is more nucleophilic than C≡C center.

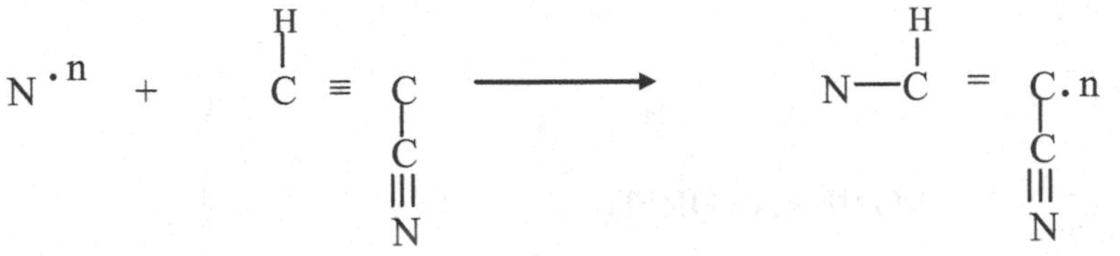

$$N^{\cdot n} \; + \; \begin{array}{c} H \\ | \\ C \equiv C \\ \quad\;\; | \\ \quad\;\; C \\ \quad\;\; ||| \\ \quad\;\; N \end{array} \longrightarrow \begin{array}{c} H \\ | \\ N—C = C\cdot n \\ \qquad\quad | \\ \qquad\quad C \\ \qquad\quad ||| \\ \qquad\quad N \end{array} \qquad 6.189$$

Electro-free-radically, it is the C≡N activation center that is largely involved. This indeed is the strongest MALE in the family, the C≡N center being more nucleophilic than C≡C center. Worthy of note is that none of the centers can be activated chargedly.

6.2 Nitriles

Before considering the use of nitriles as monomers, their chemistry as already established in the past will be revisited in order to provide their actual mechanisms, different from those proposed in the past.

6.2.1 The mechanisms of reactions of Nitriles

Though it has been shown why the first member favors only free-radical existence, it was not stated that the monomer can also favor activation, for which the character is not different from those of acetylene as shown below.

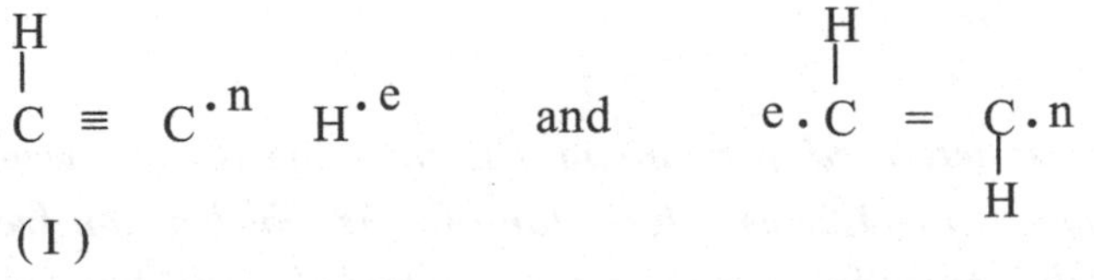

$$\begin{array}{c} H \\ | \\ C \equiv C\cdot n \quad H\cdot e \\ (I) \end{array} \qquad \text{and} \qquad \begin{array}{c} H \\ | \\ e\cdot C = C\cdot n \\ \quad\;\; | \\ \quad\;\; H \end{array}$$

$$\text{(II) Activated States (Suppressed)} \qquad 6.190$$

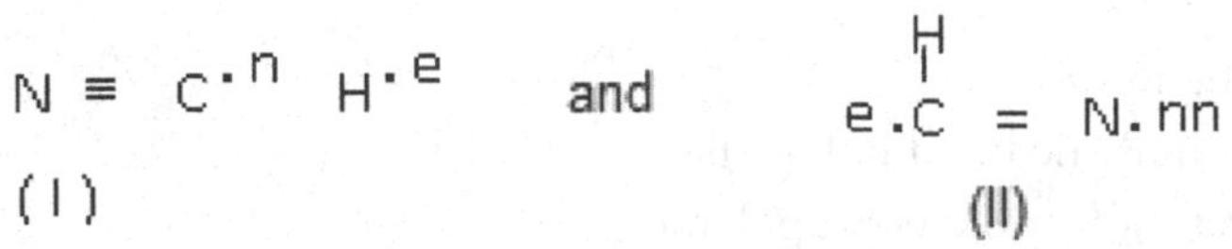

$$\begin{array}{c} N \equiv C\cdot n \quad H\cdot e \\ (I) \end{array} \qquad \text{and} \qquad \begin{array}{c} H \\ | \\ e\cdot C = N\cdot nn \\ (II) \end{array}$$

$$\text{Activated States (Suppressed)} \qquad 6.191$$

When acetylene is activated this can only be done by an activated monomer or when suppressed by a passive catalyst. When nitrile is activated, the hydrogen atom cannot be loosely held and this can be done when suppressed. While there is loose connection in its unactivated state, it does not exist when activated. Hence (I) and (II) in the last equation will behave almost like acetylene. Nevertheless, in order to provide full activation of the monomer for full polymerization, the H should preferably be replaced with other groups such as a pushing group (R), or pulling group (X).

Existence of nitriles with pulling groups can be favored when the possibility of using the type of compound shown below as source, is possible free-radically. The possibility lies only on the favored existence of (I) below.

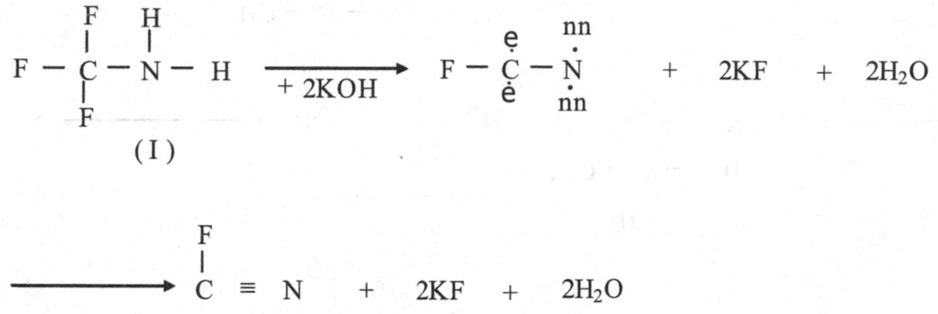

$$F - \underset{\underset{F}{|}}{\overset{\overset{F}{|}}{C}} - \underset{|}{\overset{|}{N}} - H \xrightarrow{+ 2KOH} F - \underset{\underset{e}{|}}{\overset{\overset{e}{\cdot}}{C}} - \overset{nn}{\underset{nn}{\overset{\bullet}{N}}} + 2KF + 2H_2O$$

$$\text{(I)}$$

$$\xrightarrow{} \underset{|}{\overset{\overset{F}{|}}{C}} \equiv N + 2KF + 2H_2O \qquad\qquad 6.192$$

Via Equilibrium mechanism, (I) decomposes to give KF and the product. Via Decomposition mechanism, (I) may decompose to give HF and the product.

Isobutylene is said to give N - t - butyl-formamide when it reacts with hydrogen cyanide[3]. The reaction can be explained as follows.

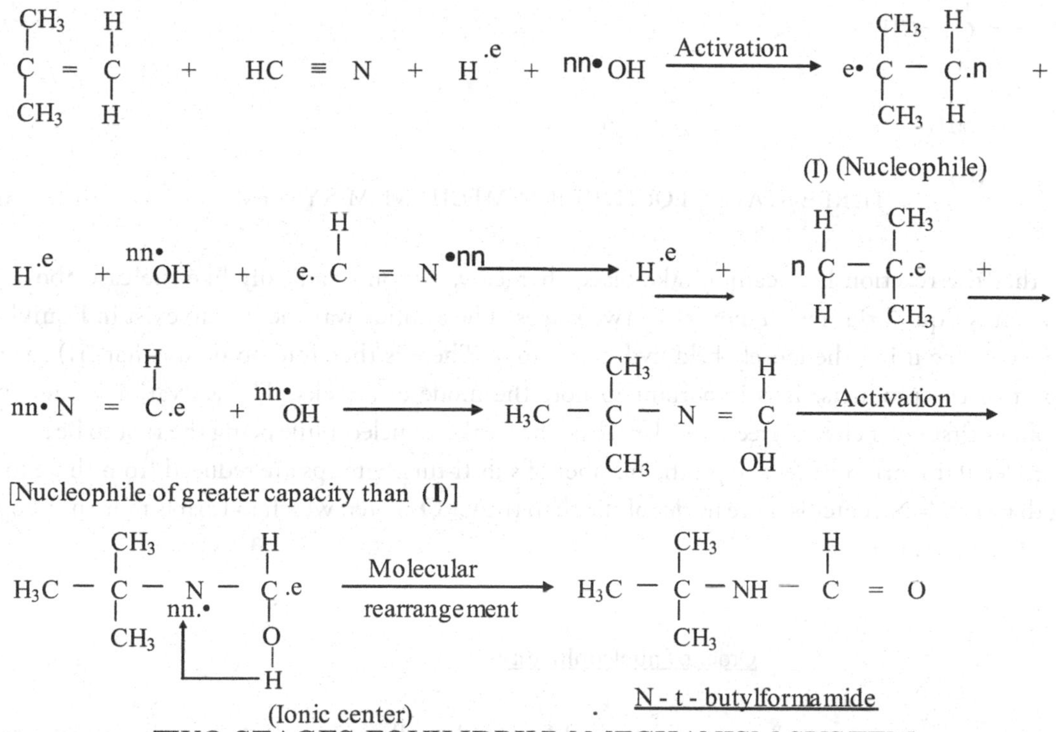

$$\underset{\underset{CH_3}{|}}{\overset{\overset{CH_3}{|}}{C}} = \underset{\underset{H}{|}}{\overset{\overset{H}{|}}{C}} + HC \equiv N + H^{.e} + nn^{\bullet}OH \xrightarrow{\text{Activation}} e\bullet \underset{\underset{CH_3}{|}}{\overset{\overset{CH_3}{|}}{C}} - \underset{\underset{H}{|}}{\overset{\overset{H}{|}}{C}}.n +$$

$$\text{(I) (Nucleophile)}$$

$$H^{.e} + nn^{\bullet}OH + e.\underset{}{\overset{\overset{H}{|}}{C}} = N^{\bullet nn} \xrightarrow{} H^{.e} + n\underset{\underset{H}{|}}{\overset{}{C}} - \underset{\underset{CH_3}{|}}{\overset{\overset{CH_3}{|}}{C}}.e +$$

$$nn\bullet N = C.e + \overset{nn\bullet}{OH} \xrightarrow{} H_3C - \underset{\underset{CH_3}{|}}{\overset{\overset{CH_3}{|}}{C}} - N = \underset{\underset{OH}{|}}{\overset{\overset{H}{|}}{C}} \xrightarrow{\text{Activation}}$$

[Nucleophile of greater capacity than **(I)**]

$$H_3C - \underset{\underset{CH_3}{|}}{\overset{\overset{CH_3}{|}}{C}} - \underset{\underset{nn.\bullet}{\overset{}{N}}}{} - \underset{\underset{O}{|}}{\overset{\overset{H}{|}}{C}}.e \xrightarrow[\text{rearrangement}]{\text{Molecular}} H_3C - \underset{\underset{CH_3}{|}}{\overset{\overset{CH_3}{|}}{C}} - NH - \underset{}{\overset{\overset{H}{|}}{C}} = O$$

$$\underset{\text{(Ionic center)}}{\overset{}{\underset{H}{|}}}$$

<u>N - t - butylformamide</u>   6.193

TWO STAGES EQUILIBRIUM MECHANISM SYSTEM

In the reaction above which can only take place radically and never chargely, HCN can be observed to favor full activation from $(CH_3)_3C^{.e}$. Notice how molecular rearrangement has taken place in the last stage. The mechanism above is indeed an Equilibrium mechanism- two stage system. Without activation of the nitrile (HC≡N), the followings are obtained.

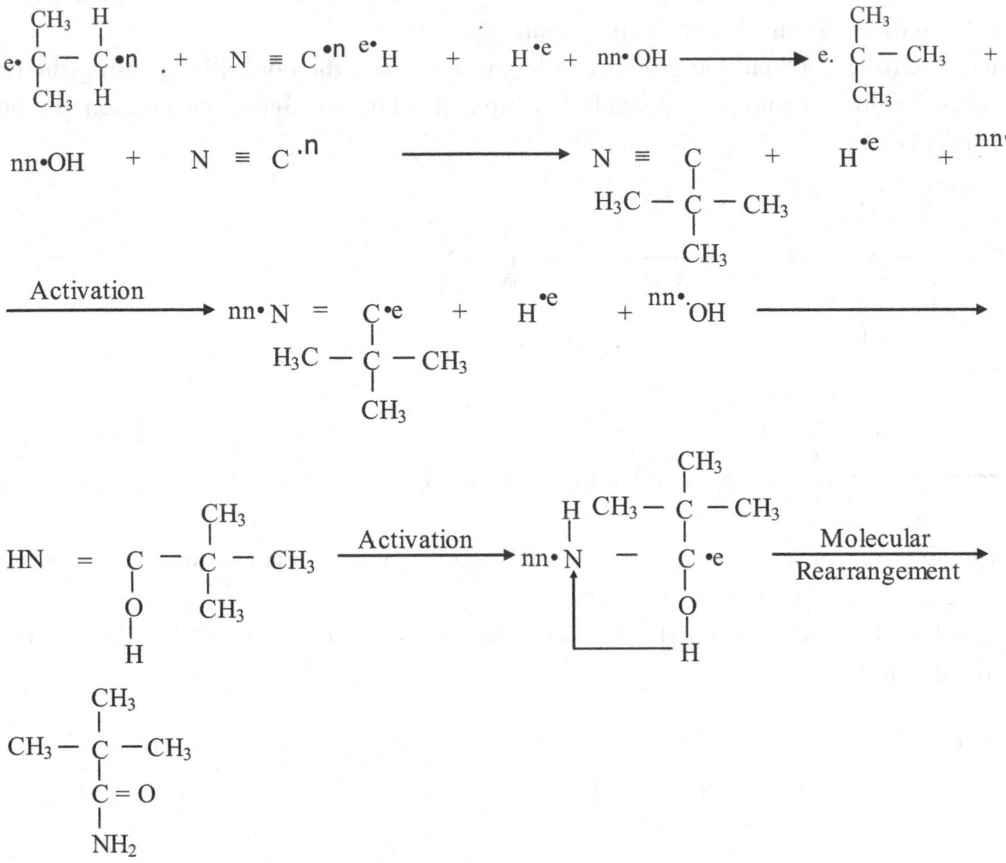

$$e\bullet \overset{\overset{\displaystyle CH_3}{|}}{\underset{\underset{\displaystyle CH_3}{|}}{C}} - \overset{\overset{\displaystyle H}{|}}{\underset{\underset{\displaystyle H}{|}}{C}}\bullet n \quad . + \quad N \equiv C^{\bullet n}\,{}^{e\bullet}H \quad + \quad H^{\bullet e} + nn\bullet OH \longrightarrow e.\ \overset{\overset{\displaystyle CH_3}{|}}{\underset{\underset{\displaystyle CH_3}{|}}{C}} - CH_3 \quad + H\bullet e$$

$$nn\bullet OH \quad + \quad N \equiv C^{\bullet n} \longrightarrow N \equiv \underset{\underset{\displaystyle CH_3}{|}}{\underset{\displaystyle H_3C - \underset{|}{C} - CH_3}{C}} \quad + \quad H^{\bullet e} \quad + \quad {}^{nn\bullet}OH$$

$$\xrightarrow{\text{Activation}} nn\bullet N = \underset{\underset{\displaystyle CH_3}{|}}{\underset{\displaystyle H_3C - \underset{|}{C} - CH_3}{C}}\bullet e \quad + \quad H^{\bullet e} \quad + \quad {}^{nn\bullet}OH \longrightarrow$$

$$HN = \overset{\overset{\displaystyle CH_3}{|}}{\underset{\underset{\displaystyle H}{|}}{\underset{\displaystyle O}{C}}} - \overset{\overset{\displaystyle CH_3}{|}}{\underset{\underset{\displaystyle CH_3}{|}}{C}} - CH_3 \xrightarrow{\text{Activation}} nn\bullet N - \overset{\overset{\displaystyle CH_3 - \overset{\overset{\displaystyle CH_3}{|}}{\underset{|}{C}} - CH_3}{\text{H}}}{\underset{\underset{\displaystyle H}{|}}{\underset{\displaystyle O}{C}}}\bullet e \xrightarrow[\text{Rearrangement}]{\text{Molecular}}$$

$$CH_3 - \overset{\overset{\displaystyle CH_3}{|}}{\underset{\underset{\displaystyle NH_2}{|}}{\underset{\displaystyle C = O}{C}}} - CH_3$$

(II)

THREE STAGES EQUILIBRIUM MECHANISM SYSTEM   6.194

Notice that the reaction here cannot take place chargedly, but only radically like the case above. The mechanism is Equilibrium with more than two stages. The cyanide was the first to exist in Equilibrium state of existence using the loosely held hydrogen atom. There is therefore no doubt that (II) may also form part of the products. It is important to note the mode of attacks. The activation centers being nucleophiles first favor electro-free-radical attacks, the weakest nucleo-phile being the first to be attacked. In the molecular rearrangement steps, the number of substi-tuted groups are reduced from three to two, noting that the C≡N center is more nucleophilic than the C=O center, which in turn is more nucleophilic than N=C activation center.

$$C \equiv N \quad > \quad C = O \quad > \quad N = C$$

<u>Order of nucleophilicity</u>   6.195

Tertiary "carbonium ions" said to be formed from either alkenes or alcohols add to nitriles to give nitrilium salts. Addition of water is said to give an N - t-alkyl amide. This can be explained as shown below for alkenes.

$$\xrightarrow{+ N \equiv CR^1} \quad R - \overset{\overset{\displaystyle R}{|}}{\underset{\underset{\displaystyle CH_3}{|}}{C}} .e \quad + \quad nn^\bullet N = \overset{\overset{\displaystyle R^1}{|}}{C} .e \quad \longrightarrow \quad R - \overset{\overset{\displaystyle R}{|}}{\underset{\underset{\displaystyle CH_3}{|}}{C}} - N = \overset{\overset{\displaystyle R^1}{|}}{C} .e$$

$$\xrightarrow[\text{or Dilute } H_2SO_4]{nn.OH \text{ from } H_2O} \quad R - \overset{\overset{\displaystyle R}{|}}{\underset{\underset{\displaystyle CH_3}{|}}{C}} - N = \overset{\overset{\displaystyle R^1}{|}}{\underset{\underset{\underset{\underset{\displaystyle H}{|}}{O}}{||}}{C}} \quad \xrightarrow{\text{Activation}} \quad R - \overset{\overset{\displaystyle R}{|}}{\underset{\underset{\displaystyle CH_3}{|}}{C}} - \overset{.nn}{N} - \overset{\overset{\displaystyle R^1}{|}}{\underset{\underset{\underset{\displaystyle O}{|}}{}}{C}} .e \quad \xrightarrow[\text{Rearrangement}]{\text{Molecular}}$$

(I)

$$R - \overset{\overset{\displaystyle R}{|}}{\underset{\underset{\displaystyle CH_3}{|}}{C}} - NH - \overset{\overset{\displaystyle R^1}{|}}{\underset{\underset{}{}}{C}} = O$$

(II)   6.196

Indeed, it is the sulfate that may first be formed. This is removed by the water to give (I) above, which molecularly rearranges to give (II). It is a three stage Equilibrium mechanism system. All the reactions described so far indeed take place radically. Unlike hydrogen cyanide which can favor only free-radical Equilibrium State of Existence, $RC \equiv N$ does not at all standard conditions.

Nitriles are also said to favor both acid and base- catalyzed hydrolysis reactions.

$$\overset{\overset{\displaystyle R}{|}}{C} \equiv N \quad \xrightarrow[\text{(dilute } H_2SO_4)]{\overset{H}{} \overset{\bullet e}{} \text{Activation}} \quad e \bullet \overset{\overset{\displaystyle R}{|}}{C} = N^{\bullet nn} \quad + \quad H^{\bullet e} \quad \longrightarrow \quad \overset{\overset{\displaystyle R}{|}}{\underset{e \bullet}{C}} = \overset{}{\underset{\underset{\displaystyle H}{|}}{N}}$$

$$\xrightarrow{nn^\bullet OH} \quad \overset{\overset{\displaystyle R}{|}}{\underset{\underset{\underset{\displaystyle H}{|}}{O}}{C}} = \overset{}{\underset{\underset{\displaystyle H}{|}}{N}} \quad \xrightarrow{\text{Activation}} \quad e. \overset{\overset{\displaystyle R}{|}}{\underset{\underset{\underset{\displaystyle H}{|}}{O}}{C}} - \overset{}{\underset{\underset{\displaystyle H}{|}}{N}}^{\bullet nn} \quad \xrightarrow[\text{Rearrangement}]{\text{Molecular}} \quad \overset{\overset{\displaystyle R}{|}}{\underset{\underset{\displaystyle NH_2}{|}}{C}} = O$$

(amide)

<u>Catalysis by acid</u> (Dilute H_2SO_4)   6.197

$$\overset{\overset{\displaystyle C}{|}}{\underset{\underset{\displaystyle R}{|}}{}} \equiv N \quad \xrightarrow{H^{\bullet e}} \quad e. \overset{\overset{\displaystyle C}{}}{\underset{\underset{\displaystyle R}{|}}{}} = N \quad \xrightarrow{\overset{nn.}{OH}} \quad \overset{\overset{\displaystyle R}{|}}{\underset{\underset{\underset{\displaystyle H}{|}}{O}}{C}} = \overset{\overset{\displaystyle H}{|}}{N} \quad \xrightarrow[\text{Molecular Rearrangement}]{\text{Activation/}}$$

$$\overset{\overset{\displaystyle R}{|}}{\underset{\underset{\displaystyle NH_2}{|}}{C}} = O \qquad \underline{\text{Catalyzed by base}} \text{ (Dilute KOH)} \qquad 6.198$$

(amide)

Indeed, as will become obvious, these reactions take place only radically. Since the monomer is a nucleophile, $^{nn \bullet}OH$ will not attack in the presence of $H^{\bullet e}$. When $^{nn \cdot}OH$ attacks in the absence of $H^{\cdot e}$, the following will be obtained for R equals CH_3.

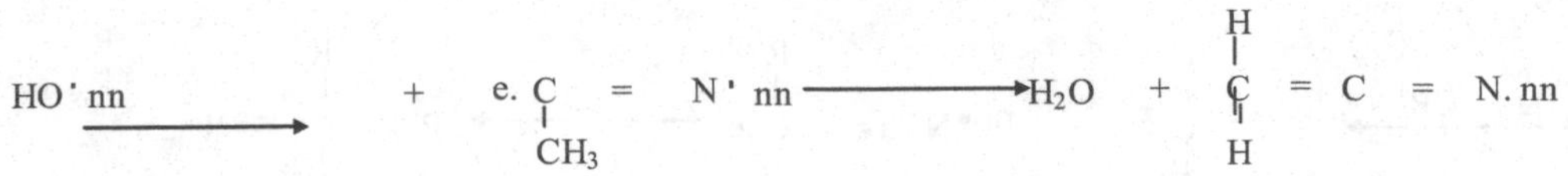

$$HO \cdot nn \xrightarrow{\hspace{2cm}} + \quad e. \underset{\underset{CH_3}{|}}{C} = N \cdot nn \xrightarrow{\hspace{2cm}} H_2O + \underset{\underset{H}{||}}{\overset{\overset{H}{|}}{C}} = C = N.nn$$

6.199

Hence, care should be taken to distinguish between an acid or base catalyzed reaction. When the amide group is further hydrolyzed in the presence of dilute hydrochloric acid, the followings are obtained. Carboxylic acid or its salt is obtained when the reaction is in the presence of a base such as aqueous sodium hydroxide.

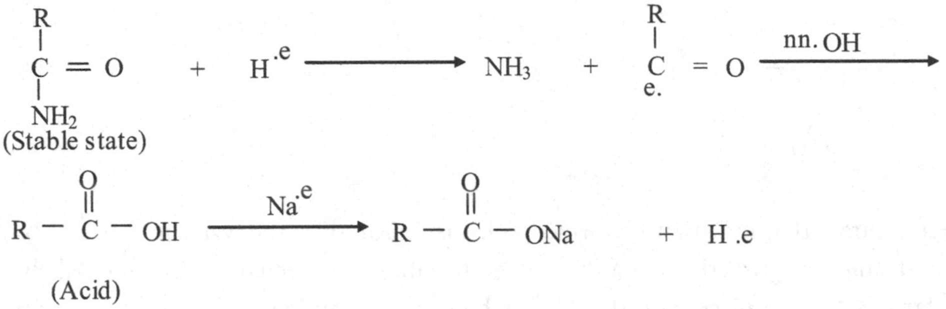

$$\underset{\underset{\underset{\text{(Stable state)}}{NH_2}}{|}}{\overset{\overset{R}{|}}{C}} = O \quad + \quad H^{.e} \xrightarrow{\hspace{2cm}} NH_3 \quad + \quad \underset{\underset{e.}{|}}{\overset{\overset{R}{|}}{C}} = O \xrightarrow{\hspace{0.5cm} nn.OH \hspace{0.5cm}}$$

$$R - \overset{\overset{O}{||}}{C} - OH \xrightarrow{\hspace{0.5cm} Na^{.e} \hspace{0.5cm}} R - \overset{\overset{O}{||}}{C} - ONa \quad + \quad H.e$$

(Acid)

6.200

It should be noted that transfer species of the second first kind of the second type is involved in the first step above. The abstraction of NH_2 is largely due to its strong ionic character and ionic environment provided by the base. If NR_2 is used in place NH_2, abstraction would similarly be favored. All these are made possible free radically, and never chargedly. These are made possi-ble since NH_2 like OH group is a pushing group of a different character. With the use of dilute HCl, a chloride was first formed, followed by hydrolysis to form the acid. Nevertheless, as we move along in the Series and Volumes, all these mechanisms will be fully understood.

Primary and Secondary alkyl cyanides are said to be readily alkylated when heated with an alkyl halide and finely powdered sodium amide. If the alkyl halide is of the form CH_3Cl, then only radical reactions are favored. If the alkyl group is that obtained from an alkene as shown below, then charged existence of the group can be made possible. The charges cannot be isolated except when ionic under Equilibrium mechanism.

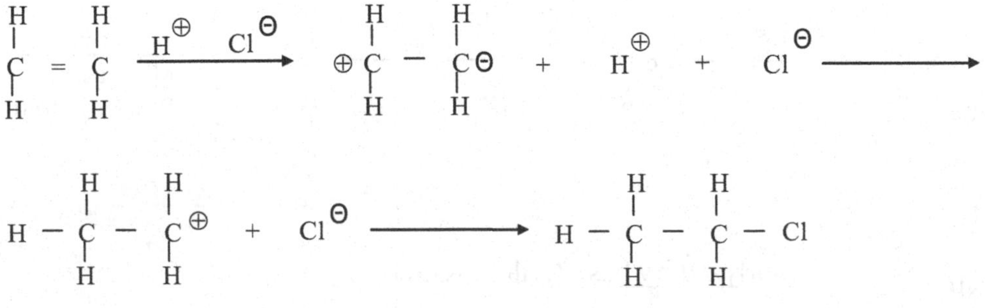

$$\underset{\underset{H}{|}}{\overset{\overset{H}{|}}{C}} = \underset{\underset{H}{|}}{\overset{\overset{H}{|}}{C}} \xrightarrow{\hspace{0.3cm} H^{\oplus} \quad Cl^{\ominus} \hspace{0.3cm}} \oplus\underset{\underset{H}{|}}{\overset{\overset{H}{|}}{C}} - \underset{\underset{H}{|}}{\overset{\overset{H}{|}}{C}}\ominus \quad + \quad H^{\oplus} \quad + \quad Cl^{\ominus} \xrightarrow{\hspace{1cm}}$$

$$H - \underset{\underset{H}{|}}{\overset{\overset{H}{|}}{C}} - \underset{\underset{H}{|}}{\overset{\overset{H}{|}}{C}}\oplus \quad + \quad Cl^{\ominus} \xrightarrow{\hspace{1cm}} H - \underset{\underset{H}{|}}{\overset{\overset{H}{|}}{C}} - \underset{\underset{H}{|}}{\overset{\overset{H}{|}}{C}} - Cl$$

(II)

6.201

Radically, the reaction is also favored. When such alkyl groups are involved, then the followings can be tried to determine the true mechanisms of the reactions above.

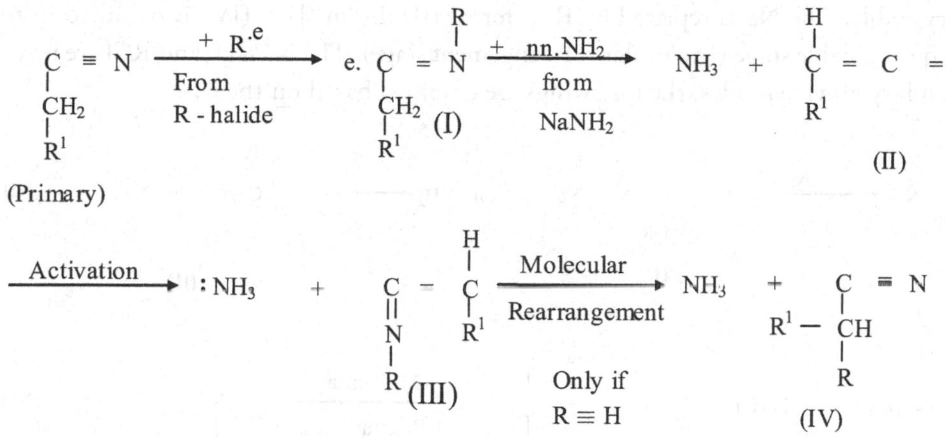

$$C \equiv N \xrightarrow[\text{From R-halide}]{+ R^{\cdot e}} \text{e.} \begin{array}{c} R \\ | \\ C = N \\ | \\ CH_2 \\ | \\ R^1 \end{array} \text{(I)} \xrightarrow[\text{from NaNH}_2]{+ \text{nn.NH}_2} NH_3 + \begin{array}{ccc} H & & R \\ | & & | \\ C & = C = N \\ | & \\ R^1 \end{array}$$

$$\begin{array}{c} | \\ CH_2 \\ | \\ R^1 \end{array}$$

(Primary)　　　　　　　　　　　　　　　　　　　　　　　　　　　(II)

$$\xrightarrow{\text{Activation}} : NH_3 + \begin{array}{c} H \\ | \\ C = C \\ || \quad | \\ N \quad R^1 \\ | \\ R \end{array} \text{(III)} \xrightarrow[\text{Rearrangement}]{\text{Molecular}} NH_3 + \begin{array}{c} C \equiv N \\ | \\ R^1 - CH \\ | \\ R \end{array}$$

Only if R ≡ H

(IV)　　　　　　6.202a

The presence of (II) above is worthy of note. Note that the nucleo-free-radical did not diffuse in the presence of an electro-free-radical to abstract H. It can only be obtained when hydrogen is first released as $H^{\cdot e}$ from (I). After the halide had added to (I), in the process of its removal by $NaNH_2$, H was released to form (II). (IV) was then obtained by molecular rearrangement via activation of the less nucleophilic center (C=C center). If it had taken place at the C=N center, then, the followings would have been obtained.

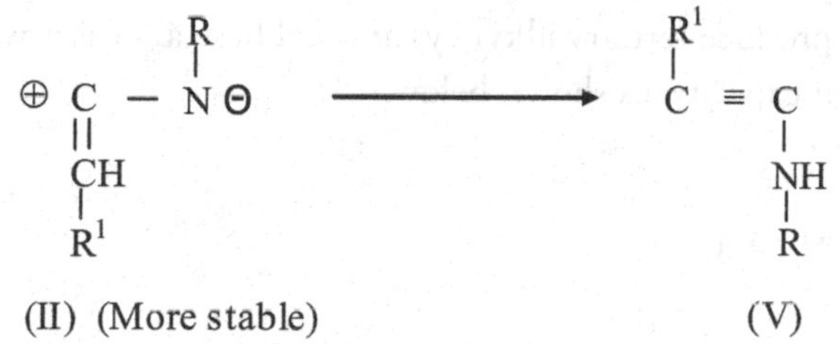

$$\begin{array}{c} R \\ | \\ \oplus C - N \ominus \\ || \\ CH \\ | \\ R^1 \end{array} \longrightarrow \begin{array}{c} R^1 \\ | \\ C \equiv C \\ | \\ NH \\ | \\ R \end{array}$$

(II) (More stable)　　　　　　　　(V)　　　　　　6.202b

It is (IV) of Equation 6.202a that is said to be indeed obtained and never (V) above which cannot take place chargedly as shown above, due to electrostatic forces of repulsion. However since RCl cannot be the one to commence the reaction in the presence of $NaNH_2$, the mechanism tentatively is therefore as follows-

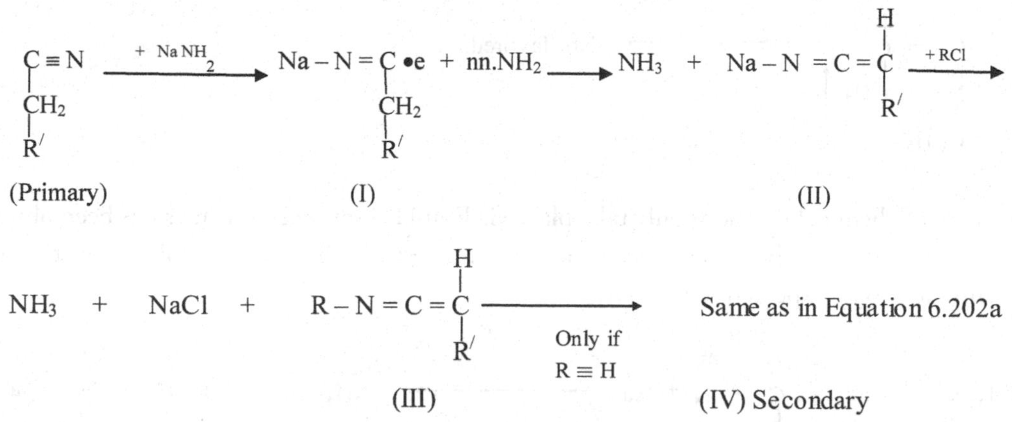

$$\begin{array}{c} C \equiv N \\ | \\ CH_2 \\ | \\ R^/ \end{array} \xrightarrow{+ \text{ Na NH}_2} Na - N = \begin{array}{c} C \bullet e \\ | \\ CH_2 \\ | \\ R^/ \end{array} + \text{nn.NH}_2 \longrightarrow NH_3 + Na - N = C = \begin{array}{c} H \\ | \\ C \\ | \\ R^/ \end{array} \xrightarrow{+ RCl}$$

(Primary)　　　　　　　(I)　　　　　　　　　　　(II)

$$NH_3 + NaCl + R - N = C = \begin{array}{c} H \\ | \\ C \\ | \\ R^/ \end{array} \xrightarrow{\text{Only if } R \equiv H} \text{Same as in Equation 6.202a}$$

(III)　　　　　　　　(IV) Secondary　　　　6.203

One cannot fully conclude that the primary has been converted to the secondary alkyl cyanides. However, it can only take place radically. $NaNH_2$ begins the reaction by activating the primary alkyl cyanide to give (II) above in one Equilibrium mechanism stage. In the second stage, the alkyl halide comes into

358

play being very stable. The Na is replaced by R to form (III). From (III), (IV) is obtained in the third stage the only questionable stage (Molecular rearrangement stage). The $NaNH_2$ and RCl are no catalysts.

 With secondary alkyl cyanides, the followings are obtained based on the past.

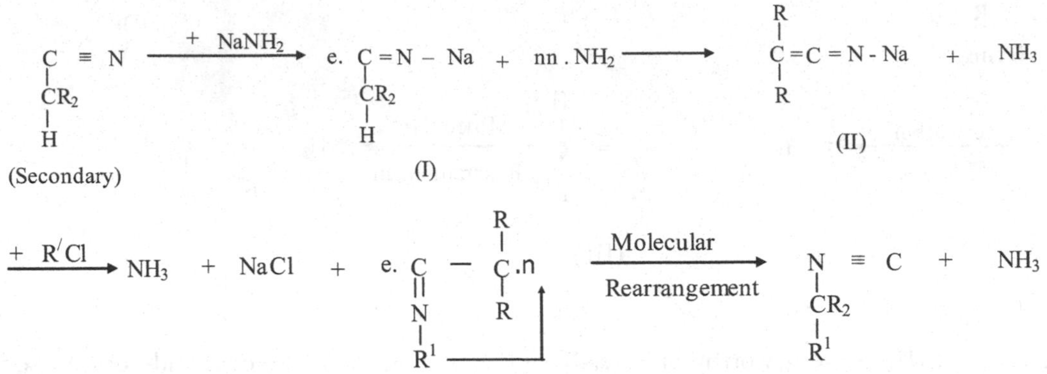

$$
\begin{array}{c}
C \equiv N \\
| \\
CR_2 \\
| \\
H
\end{array}
\xrightarrow{+ \ NaNH_2}
\text{e.}
\begin{array}{c}
C = N - Na \\
| \\
CR_2 \\
| \\
H
\end{array}
+ \ nn.NH_2
\longrightarrow
\begin{array}{c}
R \\
| \\
C = C = N - Na \\
| \\
R
\end{array}
+ \ NH_3
$$

(Secondary)   (I)   (II)

$$
\xrightarrow{+ \ R'Cl}
NH_3 \ + \ NaCl \ + \
\text{e.}
\begin{array}{c}
\quad \ R \\
\ | \\
C - C.n \\
\| \quad \ | \\
N \quad R \\
| \\
R^1
\end{array}
\xrightarrow[\text{Rearrangement}]{\text{Molecular}}
\begin{array}{c}
N \equiv C \\
\quad | \\
\quad CR_2 \\
\quad | \\
\quad R^1
\end{array}
+ \ NH_3
$$

(III) (For $R^1 \leq R$)   6.204

Looking back at (III) of Equation 6.202a, the transfer species involved in molecular rearrange-ment is of greater capacity than H and other transfer species in (III). Hence via this mechanism secondary alkyl cyanides cannot be obtained from primary alkyl cyanides. On the other hand, based on Equation 6.204, only selected secondary alkyls can be used to produce tertiary alkyl cyanides. These are cases where R` for Equation 6.204 is less than or equal to R in capacity, as shown below.

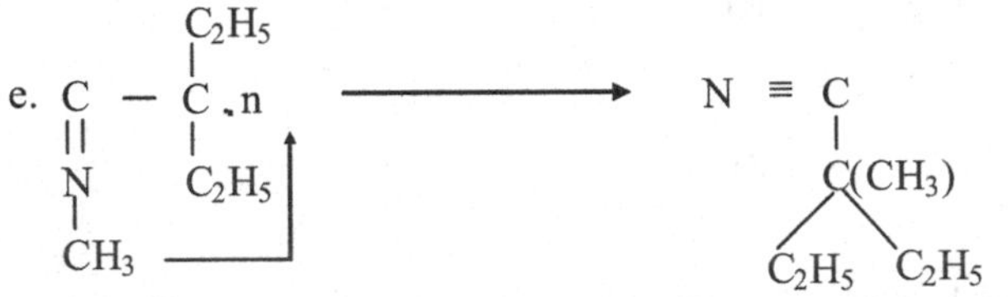

$$
\text{e.}
\begin{array}{c}
\quad \ C_2H_5 \\
\quad | \\
C - C.n \\
\| \quad | \\
N \quad C_2H_5 \\
| \\
CH_3
\end{array}
\longrightarrow
\begin{array}{c}
N \equiv C \\
\quad | \\
\quad C(CH_3) \\
\quad / \ \backslash \\
\quad C_2H_5 \ \ C_2H_5
\end{array}
$$

6.205

The capacity of CH_3 is less than those of the C_2H_5s carried by the receiving center unlike the case shown below.

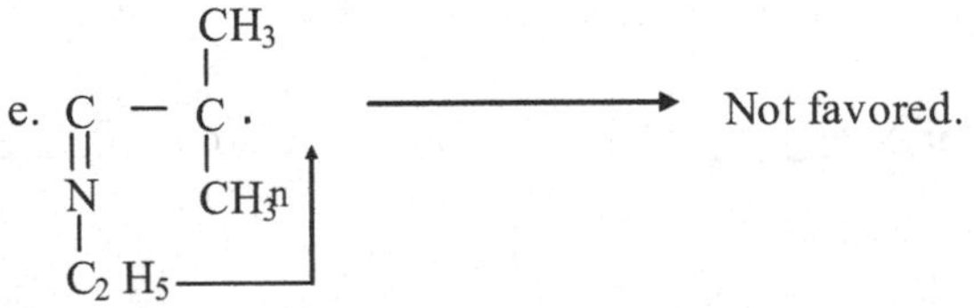

$$
\text{e.}
\begin{array}{c}
\quad \ CH_3 \\
\quad | \\
C - C. \\
\| \quad | \\
N \quad CH_3 \ n \\
| \\
C_2H_5
\end{array}
\longrightarrow
\text{Not favored.}
$$

6.206

All these types of chemical reactions only take place via Equilibrium mechanism as has been observed so far. Of the three reactants, the sodium amide has the strongest Equilibrium state of existence. Therefore the followings are to be expected tentatively.

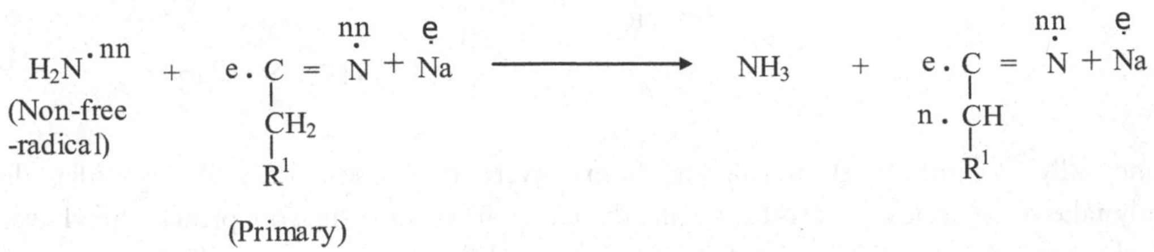

$$
H_2N^{\cdot nn} \ +
\text{e.}
\begin{array}{c}
\quad \overset{nn}{\overset{\bullet}{C}} = \overset{\underset{\ominus}{}}{N} + Na \\
| \\
CH_2 \\
| \\
R^1
\end{array}
\longrightarrow
NH_3 \ + \
\text{e.}
\begin{array}{c}
\quad \overset{nn}{\overset{\bullet}{C}} = \overset{\underset{\ominus}{}}{N} + Na \\
| \\
n.CH \\
| \\
R^1
\end{array}
$$

(Non-free
-radical)

(Primary)

<u>Transfer from monomer step</u>

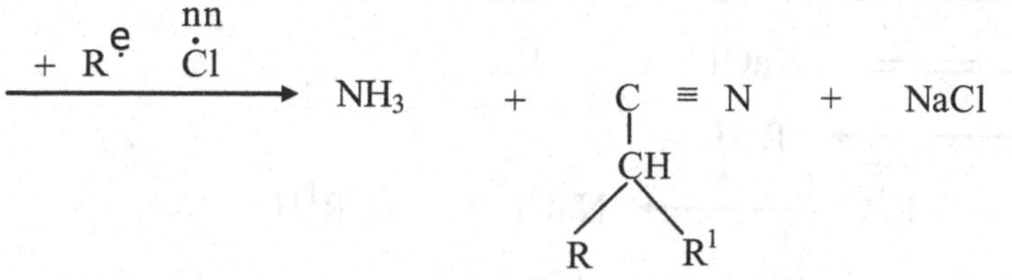

$$\xrightarrow{\quad + \; R\overset{e}{\cdot} \quad \overset{nn}{\overset{\cdot}{Cl}} \quad} \quad NH_3 \quad + \quad \underset{\underset{R \quad R^1}{\overset{|}{\underset{CH}{\diagup \diagdown}}}}{C} \equiv N \quad + \quad NaCl$$

(Secondary)   6.207

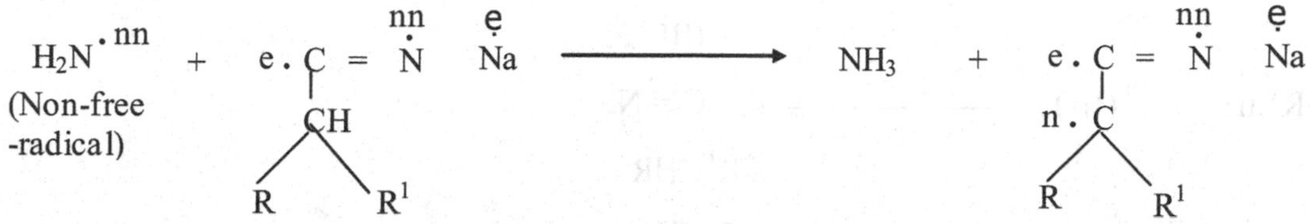

$$H_2\overset{\cdot nn}{N} \; + \; e \cdot \underset{\underset{R \quad R^1}{\overset{|}{\underset{CH}{\diagup \diagdown}}}}{\overset{nn}{\overset{\cdot}{C}}} = \overset{e}{\overset{\cdot}{N}} \; Na \quad \xrightarrow{\qquad} \quad NH_3 \; + \; e \cdot \underset{\underset{R \quad R^1}{\overset{|}{\underset{n \cdot C}{\diagup \diagdown}}}}{\overset{nn}{\overset{\cdot}{C}}} = \overset{e}{\overset{\cdot}{N}} \; Na$$

(Non-free
-radical)

<u>Transfer from monomer step</u>

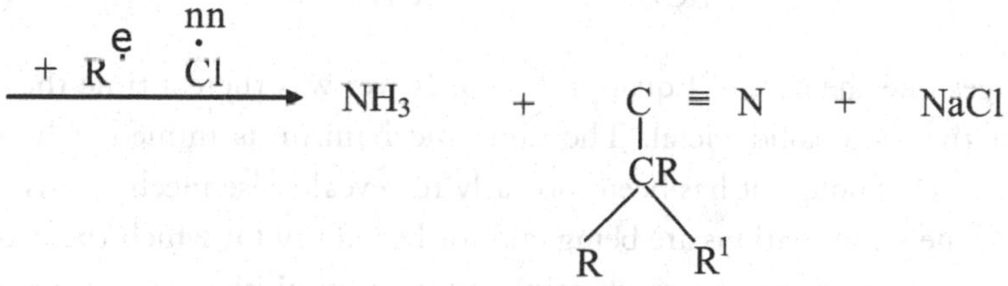

$$\xrightarrow{\quad + \; R\overset{e}{\cdot} \quad \overset{nn}{\overset{\cdot}{Cl}} \quad} \quad NH_3 \quad + \quad \underset{\underset{R \quad R^1}{\overset{|}{\underset{CR}{\diagup \diagdown}}}}{C} \equiv N \quad + \quad NaCl$$

(Tertiary)   6.208

A non-free radical cannot activate the monomer in the presence of an electro-free-radical carried by Na. The monomers being nucleophiles will prefer electro-free-radical attack before any other attack. Hence indeed, the followings are obtained in stages.

Stage 1: $NaNH_2 \; \rightleftharpoons \; Na.e \; + \; nn.NH_2$

$$Na.e \quad + \quad \underset{\underset{H-CHR}{\overset{|}{}}}{C} \equiv N \quad \rightleftharpoons \quad NaH \quad + \quad \underset{\underset{e.CHR}{\overset{|}{}}}{C} \equiv N$$

(I)

$$nn.NH_2 \; + \; (I) \quad \xrightarrow{\qquad} \quad \underset{\underset{H_2N-CHR}{\overset{|}{}}}{C} \equiv N$$

(II)   6.209

Overall Equation: $NaNH_2 \; + \; H_2RC-C \equiv N \xrightarrow{\qquad} NaH \; + \; H_2NHRC-C \equiv N$ 6.210

Stage 2: $\quad$ NaH $\rightleftharpoons$ $Na.e$ $+$ $n.H$

$\quad$ $Na.e$ $+$ R^1Cl $\rightleftharpoons$ $NaCl$ $+$ $R^1.e$

$\quad$ $R^1.e$ $+$ $n.H$ $\longrightarrow$ R^1H $\hfill$ 6.211

Overall Equation: $\quad$ NaH $+$ $R^1Cl \longrightarrow NaCl$ $+$ R^1H $\hfill$ 6.212

Stage 3: $\quad$ R^1H $\rightleftharpoons$ $H.e$ $+$ $n.R^1$

$\quad$ $H.e$ $+$ (II) $\rightleftharpoons$ $\underset{e.CHR}{C \equiv N}$ $+$ NH_3

$$\text{(III)}$$

$\quad$ $R^1.n$ $+$ (III) $\longrightarrow$ $\underset{R^1CHR}{C \equiv N}$

$$\text{(IV)} \hfill 6.213$$

Overall overall equation: $NaNH_2$ $+$ R^1Cl $+$ $\underset{RCH_2}{C \equiv N} \longrightarrow \underset{RCHR^1}{C \equiv N}$ $+$ $NaCl$ $+$ NH_3 $\hfill$ 6.214

On the whole there are three stages like the case of Equation 6.203. Never was there a time the $C\equiv N$ activated due to the presence of the ionic solid metal. The same mechanism as immediately above applies to the secondary alkyl cyanide. Though it has been too early to reveal these mechanisms, there was no choice in doing these, since new foundations are being laid for humanity for which the need for credibility is important. Chemistry is a very deep subject wherein so many possibilities may seem to show up when indeed there is one and only one where the Laws of Nature are not broken, that also wherein OPERATING CONDITIONS play the most significant role.

$\quad$ Looking at amide groups, its "tautomeric form" known as imidic acid, cannot readily be isolated because of the greater stability of the amide form as shown below.

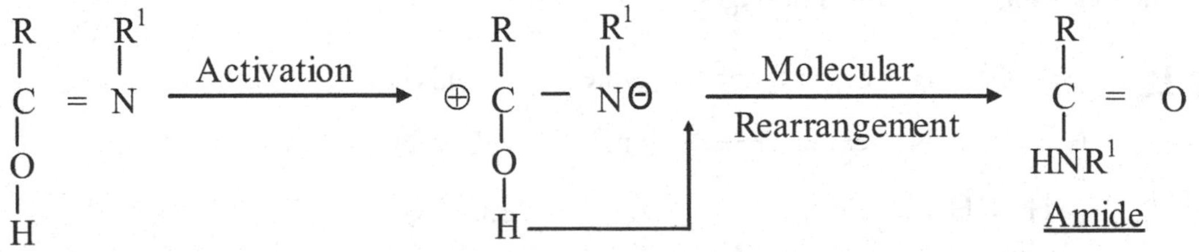

$$\underset{\text{(I) \underline{Imidic acid}}}{\overset{R \quad\quad R^1}{\underset{\underset{H}{|}}{\underset{\overset{|}{O}}{\underset{|}{C}}} = N}} \xrightarrow{\text{Activation}} \underset{\underset{H}{|}}{\overset{R \quad\quad R^1}{\oplus\, \underset{\overset{|}{O}}{\underset{|}{C}} - N\ominus}} \xrightarrow[\text{Rearrangement}]{\text{Molecular}} \underset{\underset{\underline{\text{Amide}}}{HNR^1}}{\overset{R}{\underset{|}{C}} = O}$$

$$\hfill 6.215$$

Though (I) cannot readily be isolated, their derivatives can readily be prepared, e.g., imide chlorides. Imide chloride can react with ammonia to yield amidine as follows.

$$
\begin{array}{c}
R \quad R^1 \\
| \quad\quad | \\
C = N \\
| \\
Cl
\end{array}
\longrightarrow
\begin{array}{c}
R \quad R^1 \\
| \quad\quad | \\
C = N \\
| \\
Cl
\end{array}
\xrightarrow{+\ H^{.e}}
HCl \quad + \quad
\begin{array}{c}
R \quad R^1 \\
| \quad\quad | \\
C = N \\
\quad\quad .e
\end{array}
$$

$$
\xrightarrow[nn.NH_2]{+}
\begin{array}{c}
R \quad\quad R^I \\
| \quad\quad\quad | \\
C = N \\
| \\
N \\
/ \ \backslash \\
H \quad\quad H \\
\text{Amidine}
\end{array}
\qquad\qquad 6.216
$$

The transfer species abstracted, $Cl^{\ominus}$, is of the second first kind of the first type. It is important to note why derivatives such as the chloride can be isolated, but not the acid.

6.2.2. Charged characters of Nitriles

Having considered the relevant aspect of the chemistry of these monomers, it is obvious that HCN does get activated if suppressed and the strength of initiator is adequate, only radically. On the other hand, one has said that this triple bonded compound/monomer, like acetylene, cannot be activated chargedly due to electrostatic forces of repulsion from the second π-bond adjacently located to the active center. Nevertheless, for *the purpose of exploration,* it is herein considered possible for specific reasons, part of which is also to show what has been existing in all textbooks universally, all of which are WRONG.

When fully activated, it will favor only the positively charged and electro-free-radical routes, depending on the polymerization media, HCN being *said to be strongly acidic* in character. Acid -based initiators will not be ideal. Using positively-charged-paired initiator, the followings should be expected.

$$
\begin{array}{c}
R^{\oplus} \underline{\quad\quad\quad} \overset{\ominus}{B} \overset{F}{\underset{\underset{\displaystyle O}{|}}{\underset{\displaystyle R}{\overset{\nearrow}{\underset{\searrow}{F}}}}} \\
\ \ H \\
\ominus N = \overset{\oplus}{C} \quad R \\
\\
(I)
\end{array}
\longrightarrow
R - N = \overset{\overset{\displaystyle H}{|}\ \oplus}{C} \underline{\quad\quad\quad} \overset{\ominus}{B} \overset{F}{\underset{\underset{\displaystyle O}{|}}{\underset{\displaystyle R}{\overset{\nearrow}{\underset{\searrow}{F}}}}}
\quad + \ n(I) \longrightarrow
$$

$$
R \left(N = \overset{\overset{\displaystyle H}{|}}{C} \right)_n N = \overset{\overset{\displaystyle H}{|}\ \oplus}{C} \underline{\quad\quad\quad} \overset{\ominus}{B} \overset{F}{\underset{\underset{\displaystyle O}{|}}{\underset{\displaystyle R}{\overset{\nearrow}{\underset{\searrow}{F}}}}}
\longrightarrow
\begin{array}{c}
\text{No transfer} \\
\text{species}
\end{array}
$$

$$
\qquad\qquad\qquad\qquad\qquad\qquad\qquad\qquad\qquad\qquad\qquad\qquad 6.217
$$

The $R^{\oplus}$ on the paired initiator must be greater or of equal strength with H on the monomer. With negatively charged-paired initiator, the followings are also to be expected.

$$
H_9C_4 \overset{\ominus}{\underline{\quad\quad\quad}} \overset{\oplus}{Li} \quad + \quad H \overset{\oplus}{\ } \overset{\ominus}{C} \equiv N \longrightarrow C_4H_{10} \quad + \quad Li \underline{\quad\quad} C \equiv N
$$

$$
\qquad\qquad\qquad\qquad\qquad\qquad\qquad\qquad\qquad\qquad\qquad\qquad 6.218
$$

The monomer being strongly nucleophile will not favor activations using negatively charged-paired initiators. The same will apply nucleo-free-radically.

When the H is replaced, full activation of the monomer is guaranteed independent of the strength of the initiator. Considering alkanyl groups, the followings are obtained.

$$R^{\ominus} \;+\; \overset{CH_3}{\underset{|}{\oplus C}} = N^{\ominus} \longrightarrow RH \;+\; \overset{H}{\underset{|}{\underset{H}{C}}} = C = N^{\ominus}$$

(Non-free)

$$\text{6.219}$$

$$R \overset{CH_3}{\underset{|}{+N = C}}\Big\rangle_n N = \overset{CH_3}{\underset{|}{C^{\oplus}}} \longrightarrow R \overset{CH_3}{\underset{|}{+N = C}}\Big\rangle_n N = \overset{H}{\underset{|}{\underset{H}{C}}} = C \;+\; H^{\oplus}$$

$$\text{6.220}$$

Only the positively-charged route is favored, an indication of the nucleophilic character of the monomer.

With etheric groups, the followings are to be expected. Indeed, it cannot take place chargedly.

$$\overset{\oplus}{\underset{\underset{H}{\overset{|}{O}}}{C}} = N^{\ominus} \longrightarrow \overset{\oplus}{\underset{\underset{H}{\overset{|}{N}}}{C}} - O^{\ominus} \xrightarrow{+\ R^{\ominus}} RH \;+\; {}^{\ominus}N = C = O$$

(I)   (II)   (More stable)

Isocyanic acid

$$\text{6.221}$$

$$R \overset{}{\underset{\underset{H}{\overset{|}{N}}}{+O - C}}\Big\rangle_n O - \overset{\oplus}{\underset{\underset{H}{\overset{|}{N}}}{C}} \longrightarrow R \overset{}{\underset{\underset{H}{\overset{|}{N}}}{+O - C}}\Big\rangle_n O - C \equiv N \;+\; H^{\oplus}$$

$$\text{6.222}$$

Since (II) is more stable than (I), molecular rearrangement is favored. If the C=O happens to be activated, the polymer if formed will however favor strong branch formations in view of the presence of weakly bonded hydrogen atom to a nitrogen center. Nevertheless, the hydrogen atom will have to be replaced to favor the reactions of Equation 6.222. With R group (alkanyl) replacing H in OH group, the followings are obtained, though there seems to be no restriction against molecular rearrangement being favored.

$$\overset{}{\underset{\underset{R}{\overset{|}{O}}}{C}} \equiv N \longrightarrow \overset{\oplus}{\underset{\underset{R}{\overset{|}{O}}}{C}} = N^{\ominus} \longrightarrow \overset{\oplus}{\underset{\underset{R}{\overset{|}{N}}}{C}} - O^{\ominus}$$

Cyanates

(I)   (II)   (More stable) 6.223

Since there is only one transfer species for both cases, in view of the fact that the carbonyl center is more nucleophilic than C=N center, the more pushing OR becomes, the less molecular rear-rangement is favored. That is, the less radical-pushing RO becomes or the bulkier R becomes, the more difficult it is to transfer it.

$$R \overbrace{\left(O - \underset{\underset{R}{\overset{\displaystyle N}{|}}}{\overset{\displaystyle \|}{C}} \right)_n}^{} O - \underset{\underset{R}{\overset{\displaystyle N}{|}}}{\overset{\displaystyle \|}{C}}{}^{\oplus} \longrightarrow R \overbrace{\left(O - \underset{\underset{R}{\overset{\displaystyle N}{|}}}{\overset{\displaystyle \|}{C}} \right)_n}^{} O - C \equiv N \ + \ R^{\oplus}$$

6.224

The transfer species involved are of the first kind of the third type and the molecular rearrange-ment that took place is of the first kind and the first type. It is only when molecular rearrange-ment is favored, that the polymerization of the monomer becomes possible in free-media sy-stems. ***Indeed, it is the C=N center that is activated all the time, being less nucleophilic, and not the C=O which has been used above so far. It has been used since it does appear when molecular rearrangement takes place before deactivation.***

When alkenyl groups are involved, the followings are obtained.

$$\underset{\underset{CH_2}{\overset{\displaystyle \|}{}}}{\underset{\overset{\displaystyle CH}{\overset{\displaystyle \|}{}}}{C}} \equiv N \xrightarrow{\ + \ R^{\ominus}\ } {}^{\oplus}\underset{\underset{CH_2}{\overset{\displaystyle \|}{}}}{\underset{\overset{\displaystyle CH}{\overset{\displaystyle \|}{}}}{C}} = N^{\ominus} \longrightarrow RH \ + \ {}^{\ominus}\underset{\overset{\displaystyle CH_2}{\overset{\displaystyle \|}{}}}{C} - C \equiv N$$

(I) (Acrylonitrile)   Cannot exist

6.225

$$R^{\ominus} \ + \ {}^{\oplus}\underset{\overset{\displaystyle |}{H}}{\overset{\overset{\displaystyle H}{|}}{C}} - \underset{\underset{N}{\overset{\displaystyle \|}{C}}}{C}{}^{\ominus} \longrightarrow R - \underset{\overset{\displaystyle |}{H}}{\overset{\overset{\displaystyle H}{|}}{C}} - \underset{\underset{N}{\overset{\displaystyle \|}{C}}}{C}{}^{\ominus}$$

(Free)

6.226

$$R \overbrace{\left(N = \underset{\underset{CH_2}{\overset{\displaystyle \|}{}}}{\underset{\overset{\displaystyle CH}{\overset{\displaystyle |}{}}}{C}} \right)_n}^{} N = \underset{\underset{CH_2}{\overset{\displaystyle \|}{}}}{\underset{\overset{\displaystyle CH}{\overset{\displaystyle |}{}}}{C}}{}^{\oplus} \longrightarrow R \overbrace{\left(N = \underset{\underset{CH_2}{\overset{\displaystyle \|}{}}}{\underset{\overset{\displaystyle CH}{\overset{\displaystyle |}{}}}{C}} \right)_n}^{} N = C = C = \underset{\overset{\displaystyle |}{H}}{\overset{\overset{\displaystyle H}{|}}{C}}$$

                                                                          Not possible

6.227

The alkenyl group cannot provide resonance stabilization, since both activation centers are different in character. Indeed, since the C≡N center cannot be activated chargedly, it is the C=C center that is activated with positive and negative charges. The transfer species involved above is of the first kind of the second type and it existence is impossible due to electrostatic forces of repulsion. With phenyl group, the followings are to be expected.

$$\underset{\text{(phenyl)}}{\overset{\displaystyle C \equiv N}{}} \longrightarrow {}^{\oplus}\underset{\text{(phenyl)}}{\overset{\displaystyle C = N^{\ominus}}{}} \xrightarrow{\ + \ R^{\oplus}\ } \underset{\text{(phenyl)}}{\overset{\displaystyle R - N = C^{\oplus}}{}}$$

6.228

Like HCN to some extent, it is believed that largely the positive route will be favored with no transfer species. Negative charges will not activate the C≡N center. Nevertheless, it should be noted that these cannot take place chargedly.

Considering monomers with pulling groups, the followings are obtained with F, CF_3 etc.

$$\underset{F}{C} \equiv N \xrightarrow{\ +\ R^{\oplus}\ } \underset{F}{C} \equiv N \longrightarrow RF \ +\ ^{\oplus}C \equiv N \qquad 6.229$$

$$RO\left(\underset{F}{C} = N\right)_n \underset{F}{C} = N^{\ominus} \longrightarrow RO\left(\underset{F}{C} = N\right)_n C \equiv N \ +\ F^{\ominus} \qquad 6.230$$

The transfer species involved here is of the second kind, while that involved in the first equation is of the second first kind. When activated, the route natural to it is favored.

$$R^{\oplus} \ +\ ^{\oplus}\underset{CF_3}{C} = N^{\ominus} \longrightarrow R - N = \underset{CF_3}{C}^{\oplus} \xrightarrow{\ +\ n(I)\ }$$
$$(I)$$

$$R\left(N = \underset{CF_3}{C}\right)_n N = \underset{CF_3}{C}^{\oplus} \longrightarrow \text{NO transfer species} \qquad 6.231$$

$$R\left(\underset{CF_3}{C} = N\right)_n \underset{CF_3}{C} = N\overset{\ominus}{\underset{x}{..}} \longrightarrow \text{NO transfer species} \qquad 6.232$$

Since CF_3 cannot carry negative charges, it cannot be released from a free or non-free-negative growing polymer chain. Both negative and positive routes are favored to produce only living polymers, an indication of CF_3 being more pulling than F.

With COOR, COR types of groups, the followings are obtained.

$$R^{\ominus} \ +\ \underset{\substack{C = O \\ | \\ O \\ | \\ CH_3}}{C} \equiv N \longrightarrow ^{\oplus}\underset{\substack{C = O \\ | \\ O \\ | \\ CH_3}}{C} = N^{\ominus} \ +\ R^{\ominus} \longrightarrow$$
$$\text{(Non- free)}$$

$$R - \underset{\substack{C = O \\ | \\ O \\ | \\ CH_3}}{C} = N^{\ominus} \qquad\qquad 6.233$$

$$R^{\ominus} \;+\; \overset{\oplus}{C}\!\!\!\!\begin{array}{c} N \\ \parallel\!\parallel \\ C \\ | \end{array}\!\!\!\!- O^{\ominus} \;\longrightarrow\; RCH_3 \;+\; \overset{\ominus}{O} - \overset{\displaystyle C}{\underset{\displaystyle N}{\underset{\parallel\parallel}{\overset{|}{C}}}} = O$$

(Non- free)

6.234

$$R^{\oplus} \;+\; \overset{\ominus}{N} = \overset{\oplus}{C} \;\longrightarrow\; R - N = \overset{\ominus}{C} \xrightarrow{\;+\; \text{n(I)}\;}$$

$$\begin{array}{c} | \\ C = O \\ | \\ O \\ | \\ CH_3 \end{array} \qquad \begin{array}{c} | \\ C = O \\ | \\ O \\ | \\ CH_3 \end{array}$$

(I)

$$R \;\text{-(} N = C \text{)}_n\; N = \overset{\oplus}{C} \;\longrightarrow\; \text{NO transfer species}$$

$$\begin{array}{cc} | & | \\ C = O & C = O \\ | & | \\ O & O \\ | & | \\ CH_3 & CH_3 \end{array}$$

6.235

$$R^{\oplus} \;+\; \overset{\oplus}{C}\!\!\!\!\begin{array}{c} N \\ \parallel\!\parallel \\ C \\ | \end{array}\!\!\!\!- O^{\ominus} \;\longrightarrow\; R - O - \overset{\displaystyle CH_3}{\underset{\displaystyle N}{\underset{\parallel\parallel}{\overset{|}{\underset{C}{\overset{O}{\overset{|}{C}}}}}}} \,{\oplus}$$

6.236

$$R^{\oplus} \quad\underset{\underset{\displaystyle \overset{\ominus}{O} - \overset{\oplus}{C}}{\underset{\displaystyle \underset{N}{\overset{\parallel\parallel}{C}}}{|}}{\overset{\displaystyle \overset{F}{\underset{|}{\overset{\ominus}{B}}}}{\underset{\displaystyle \underset{CH_3 \; O}{\underset{|}{\overset{|}{\;\;\;R}}}}{}}}\!\!\!\!\begin{array}{c} F \\ \diagup \\ \diagdown \\ F \end{array} \;\longrightarrow\; R - O - \overset{\displaystyle CH_3}{\underset{\displaystyle \underset{N}{\overset{\parallel\parallel}{C}}}{\underset{|}{\overset{\displaystyle \overset{O}{\overset{|}{\oplus}}}{C}}}} \quad \overset{\displaystyle \overset{F}{\underset{|}{\overset{\ominus}{B}}}}{\underset{\displaystyle \underset{R}{\underset{|}{O}}}{}}\!\!\!\!\begin{array}{c} F \\ \diagup \\ \diagdown \\ F \end{array}$$

6.237

Notice that the monomer has not been properly placed above. Nevertheless, the route is still favored, since the $C \equiv N$ center cannot be activated chargedly. On the other hand, it should be noted that, since the $C=O$ center is less nucleophilic than the $C\equiv N$ center and the monomer is a Nucleophile, never is there a time the $C\equiv N$ center is ever activated.

$$R^{\ominus} \; + \; \underset{\underset{\displaystyle CH_3}{\overset{\displaystyle |}{\underset{\displaystyle |}{C = O}}}}{\overset{\displaystyle |}{C \equiv N}} \longrightarrow \underset{\underset{\displaystyle CH_3}{\overset{\displaystyle |}{\underset{\displaystyle |}{C = O}}}}{\overset{\displaystyle \oplus}{C} = N^{\ominus}} \; + \; R^{\ominus} \longrightarrow$$

$$RCH_3 \; + \; \overset{\overset{\displaystyle C = O}{\displaystyle \|}}{C} = N^{\ominus} \hspace{4cm} 6.238$$

Though the activation of the C≡N center will not be favored, the reaction above and those of Equations 6.233 and 6.235, are being used to illustrate some basic principles. COR being a radical-pulling group can release R positively only if it is directly connected to a positive active center in free media initiator systems, that which does not exist.

$$R^{\ominus} \; + \; \overset{\overset{\displaystyle N}{\underset{\displaystyle \|\|}{C}}}{\underset{\displaystyle CH_3}{\overset{\displaystyle |}{\oplus C}}} - O^{\ominus} \longrightarrow RH \; + \; \underset{\displaystyle H}{\overset{\displaystyle H}{C}} = \underset{\overset{\displaystyle C}{\underset{\displaystyle N}{\|\|}}}{C} - O^{\ominus} \hspace{2cm} 6.239$$

$$R^{\oplus} \; + \; \ominus O - \underset{\overset{\displaystyle C}{\underset{\displaystyle N}{\|\|}}}{\overset{\overset{\displaystyle CH_3}{\displaystyle |}}{C\oplus}} \longrightarrow R - O - \underset{\overset{\displaystyle C}{\underset{\displaystyle N}{\|\|}}}{\overset{\overset{\displaystyle CH_3}{\displaystyle |}}{C\oplus}} \xrightarrow{\; + \; n(I)}$$

$$(I)$$

$$R \; \left(O - \underset{\overset{\displaystyle C}{\underset{\displaystyle N}{\|\|}}}{\overset{\overset{\displaystyle CH_3}{\displaystyle |}}{C}} \right)_n O - \underset{\overset{\displaystyle C}{\underset{\displaystyle N}{\|\|}}}{C} = \underset{\displaystyle H}{\overset{\displaystyle H}{C}} \; + \; H^{\oplus}$$

$$\hspace{11cm} 6.240$$

$$\underset{\displaystyle H_9C_4 \overset{\ominus}{\rule{2cm}{0.4pt}} \overset{\oplus}{Li}}{\overset{\overset{\displaystyle N}{\underset{\displaystyle \|\|}{C}}}{\underset{\displaystyle CH_3}{\overset{\displaystyle |}{\oplus C}} - O^{\ominus}}} \xrightarrow[\substack{\underline{Favored}\\\underline{Center}}]{} H_{10}C_4 \; + \; \underset{\displaystyle H}{\overset{\displaystyle H}{C}} = \underset{\overset{\displaystyle C}{\underset{\displaystyle N}{\|\|}}}{C} - O \rule{2cm}{0.4pt} Li$$

$$\hspace{11cm} 6.241$$

If the orientation of the monomer had been well placed above, the transfer species cannot be abstracted.

$$\begin{array}{c}
\overset{\displaystyle CH_3}{\underset{\displaystyle |}{}} \\
\overset{\displaystyle O}{\underset{\displaystyle |}{}} \\
C = O \\
| \\
\oplus C \;=\; N\ominus \\
\end{array}$$

$$H_9C_4 \overset{\ominus}{\underline{\hspace{3cm}}} \overset{\oplus}{Li} \longrightarrow H_9C_4 - \underset{\displaystyle |}{\overset{\displaystyle CH_3}{\underset{\displaystyle O}{\underset{\displaystyle |}{\underset{\displaystyle C=O}{\underset{\displaystyle |}{}}}}}\; C \;=\; N \underline{\hspace{3cm}} Li$$

(Non-Favored Center)

6.242

$$\begin{array}{c}
CH_3 \\
| \\
C = O \\
| \\
\oplus C \;=\; N\ominus \\
\end{array}$$

$$H_9C_4 \overset{\ominus}{\underline{\hspace{3cm}}} \overset{\oplus}{Li} \longrightarrow H_9C_4 - \underset{\displaystyle C=O}{\underset{\displaystyle |}{\overset{\displaystyle CH_3}{\overset{\displaystyle |}{}}}}\; C \;=\; N \underline{\hspace{3cm}} Li$$

(Non-Favored Center)

6.243

The routes favored by the monomers under normal and controlled conditions can thus be observed. The groups are observed not to reduce the nucleophilic character of C≡N activation center. Note that the initiator used above, cannot polymerize the centers.

With CN as the substituted group, the following are obtained.

$$\begin{array}{c}
C \;\equiv\; N \\
| \\
C \\
||| \\
N
\end{array}
\longrightarrow
\begin{array}{c}
\oplus C \;=\; N\ominus \\
| \\
\oplus C \\
|| \\
\ominus N
\end{array}
\quad OR \quad
\begin{array}{c}
\oplus C \;=\; N\ominus \\
| \\
C \\
||| \\
N
\end{array}$$

(I) (Not possible)   (II)   (Assumed possible

6.244

$$R^{\oplus} \;+\; \ominus N \;=\; \underset{\displaystyle N}{\underset{\displaystyle |||}{\overset{\displaystyle C^{\oplus}}{\overset{\displaystyle |}{\overset{\displaystyle C}{}}}}} \longrightarrow R \;-\; N \;=\; \underset{\displaystyle N}{\underset{\displaystyle |||}{\overset{\displaystyle C^{\oplus}}{\overset{\displaystyle |}{\overset{\displaystyle C}{}}}}}$$

6.245

$$R^{\ominus} \;+\; \underset{\displaystyle N}{\underset{\displaystyle |||}{\overset{\displaystyle \oplus C}{\overset{\displaystyle |}{\overset{\displaystyle C}{}}}}} = N\ominus \longrightarrow R \;-\; \underset{\displaystyle N}{\underset{\displaystyle |||}{\overset{\displaystyle C}{\overset{\displaystyle |}{\overset{\displaystyle C}{}}}}} = N\oplus$$

6.246

Though the two activation centers are nucleophilic, resonance stabilization cannot be favored, since movement is from Half-free to Full-Non-free. Secondly, according to (I), two centers placed side by side cannot carry two same covalent charges. Chargedly, in the absence of transfer species, both charged routes seem to be favored for this **Cyanogen.** Whether the existence of this gaseous monomer is favored or not, is not presently the issue. However, only the positively charged route will be favored in view of the strong nucleophilic character of C≡N center, if its activation had been possible.

6.2.3. Radical characters of Nitriles

This is the way the polymerizable monomers of acetylenes and nitriles can be activated. Beginning with those carrying pushing groups, consider HCN. HCN will only favor electro-free radical activation at low temperatures, using H $\cdot^e$ when fully suppressed by a passive catalyst.

$$N\cdot^n \ + \ \underset{CH_3}{e\cdot C} \ = \ N\cdot nn \ \longrightarrow \ NH \ + \ \underset{H}{\overset{H}{n\cdot C}} - C \ \equiv \ N \qquad 6.247$$

$$\ddot{N}\cdot^{nn} \ + \ \underset{CH_3}{e\cdot C} \ = \ N\cdot nn \ \longrightarrow \ \ddot{N}H \ + \ \underset{H}{\overset{H}{C}} = C \ = \ N\cdot nn \qquad 6.248$$

$$E \ \overset{}{\underset{CH_3}{+}} N = \overset{}{\underset{CH_3}{C}} \overset{}{+}_n N = \overset{}{\underset{CH_3}{C}}\cdot e \ \longrightarrow \ E \ \overset{}{\underset{CH_3}{+}} N = C \overset{}{+}_n N = C = \overset{H}{\underset{H}{C}} \ + \ H\cdot^e \qquad 6.249$$

Only the electro-free-radical route is favored by the monomers.

With etheric groups, the followings are obtained.

$$\ddot{N}\cdot^{nn} \ + \ \underset{\underset{CH_3}{|}}{\underset{O}{|}}{e\cdot C} = N\cdot nn \ \xrightarrow[\text{Rearrangement}]{\text{Molecular}} \ \ddot{N}\cdot^{nn} \ + \ \underset{\underset{CH_3}{|}}{\underset{N}{||}}{e\cdot C} - O\cdot nn \ \longrightarrow$$

$$\ddot{N}CH_3 \ + \ nn\cdot N \ = \ C \ = \ O \qquad 6.250$$

$$E \ \overset{}{\underset{\underset{CH_3}{|}}{\underset{N}{||}}{+}} O - C \overset{}{+}_n O - \underset{\underset{CH_3}{|}}{\underset{N}{||}}{C}\cdot e \ \longrightarrow \ E \ \overset{}{\underset{\underset{CH_3}{|}}{\underset{N}{||}}{+}} O - C \overset{}{+}_n O - C \ \equiv \ N \ + \ e\cdot CH_3 \qquad 6.251$$

Indeed, the center actually activated is not the C=O but the C≡N, being less nucleophilic.

With alkenyl groups, the followings are to be expected.

$$\ddot{N}\cdot^{nn} \ + \ \underset{\underset{CH_2}{\overset{||}{CH}}}{e\cdot C} = N\cdot nn \ \longrightarrow \ \ddot{N}H \ + \ CH_2 \ = \ C \ = \ C \ = \ N\cdot nn$$

(Acrylonitrile) $\qquad 6.252$

$$R \overbrace{}^{} N = \underset{\underset{CH_2}{\overset{\|}{CH}}}{C} \overbrace{}_{n} N = \underset{\underset{CH_2}{\overset{\|}{CH}}}{C}.e \longrightarrow R \overbrace{}^{} N = \underset{\underset{CH_2}{\overset{\|}{CH}}}{C} \overbrace{}_{n} N = C = C = \underset{\underset{H}{\overset{H}{|}}}{C} + H \cdot^{e}$$

Transfer species of the first kind of the second type $\qquad\qquad$ 6.253

Like the charged case, only the electro-free-radical route, the natural route is favored, except that the charged route is not possible.

With pulling groups, consider using F.

$$\underset{\underset{F}{|}}{C} \equiv N + E\cdot^{e} \longrightarrow EF + e.C \equiv N \qquad\qquad 6.254a$$

$$\underset{\underset{F}{|}}{C} \equiv N + E\cdot^{e} \longrightarrow E - N = \underset{\overset{F}{|}}{C}.e \qquad\qquad 6.254b$$

$$\overset{..}{N} \overbrace{}^{} \underset{\underset{F}{|}}{C} = N \overbrace{}_{n} \underset{\underset{F}{|}}{C} = N \cdot^{nn} \longrightarrow \overset{..}{N} \overbrace{}^{} \underset{\underset{F}{|}}{C} = N \overbrace{}_{n} C \equiv N + :\overset{..}{\underset{..}{F}}.nn$$

Transfer species of second kind $\qquad\qquad$ 6.255

Transfer species of the second kind is involved here, the route being unnatural to the monomer. When activated, the natural route is the electro-free-radical route.

$$E\cdot^{e} + e.\underset{\underset{CF_3}{|}}{C} = N.nn \longrightarrow E - N = \underset{\underset{CF_3}{|}}{C}\cdot^{e} \quad OR \quad ECF_3 + N \equiv \underset{e}{C}$$

(I) FAVORED $\qquad$ (II) $\qquad\qquad$ 6.256

$$\overset{..}{N} \overbrace{}^{} \underset{\underset{CF_3}{|}}{C} = N \overbrace{}_{n} \underset{\underset{CF_3}{|}}{C} = N.nn \longrightarrow \overset{..}{N} \overbrace{}^{} \underset{\underset{CF_3}{|}}{C} = N \overbrace{}_{n} C \equiv N + n.CF_3$$

Transfer species of second kind $\qquad\qquad$ 6.257

The transfer species involved in the last reaction is of the second kind. When CF_3 is involved, there is no release of transfer species, noting that (II) above is not favored.

$$\overset{.}{N}\cdot^{nn} + e.\underset{\underset{\underset{CF_3}{|}}{\underset{CF_2}{|}}}{C} = N.nn \longrightarrow \overset{..}{N}CF_3 + \underset{\underset{F}{|}}{\overset{\overset{F}{|}}{C}} = C = N.nn$$

6.258

$$E \overbrace{}^{} N = \underset{\underset{\underset{CF_3}{|}}{\underset{CF_2}{|}}}{C} \overbrace{}_{n} N = \underset{\underset{\underset{CF_3}{|}}{\underset{CF_2}{|}}}{C}.e \longrightarrow E \overbrace{}^{} N = \underset{\underset{\underset{CF_3}{|}}{\underset{CF_2}{|}}}{C} \overbrace{}_{n} N = C = \underset{\underset{F}{|}}{\overset{\overset{F}{|}}{C}} + e.CF_3$$

6.259

With free-radical pulling groups, the followings are obtained for COOR.

$$E \cdot^e \;+\; e \cdot \overset{\displaystyle C = N \cdot nn}{\underset{\displaystyle \underset{\displaystyle \underset{\displaystyle CH_3}{O}}{C=O}}{|}} \longrightarrow E - N = \overset{\displaystyle C \cdot e}{\underset{\displaystyle \underset{\displaystyle \underset{\displaystyle CH_3}{O}}{C=O}}{|}}$$

$$\qquad\qquad 6.260$$

$$\overset{\displaystyle \cdot\cdot}{N} + \overset{\displaystyle C = N}{\underset{\displaystyle \underset{\displaystyle \underset{\displaystyle CH_3}{O}}{C=O}}{}} +_n \overset{\displaystyle C = N \cdot nn}{\underset{\displaystyle \underset{\displaystyle \underset{\displaystyle CH_3}{O}}{C=O}}{}} \longrightarrow N + \overset{\displaystyle C = N}{\underset{\displaystyle \underset{\displaystyle \underset{\displaystyle CH_3}{O}}{C=O}}{}} +_n C \equiv N \;+\; n \cdot COOCH_3$$

<u>Transfer species of second kind</u>

$$\qquad\qquad 6.261$$

However, it is the C = O center that is activated in preference to C ≡ N center, being less nucleophilic. Hence, the reations just above are not favored.

$$E \cdot^e \;+\; e \cdot \overset{\displaystyle \overset{\displaystyle N}{\overset{\displaystyle |||}{\overset{\displaystyle C}{|}}}}{\underset{\displaystyle \underset{\displaystyle CH_3}{O}}{C}} - O \cdot nn \longrightarrow e \cdot \overset{\displaystyle \overset{\displaystyle CH_3}{\overset{\displaystyle |}{\overset{\displaystyle O}{|}}}}{\underset{\displaystyle \underset{\displaystyle N}{\overset{\displaystyle C}{|||}}}{C}} - O - E$$

$$\qquad\qquad 6.262$$

$$\overset{\displaystyle \cdot\cdot}{N} \cdot nn \;+\; e \cdot \overset{\displaystyle \overset{\displaystyle N}{\overset{\displaystyle |||}{\overset{\displaystyle C}{|}}}}{\underset{\displaystyle \underset{\displaystyle CH_3}{O}}{C}} - O \cdot nn \longrightarrow \overset{\displaystyle \cdot\cdot}{N}CH_3 \;+\; O = \overset{\displaystyle C}{\underset{\displaystyle \underset{\displaystyle N}{\overset{\displaystyle C}{|||}}}{}} = O \cdot nn$$

(Favored center)

$$\qquad\qquad 6.263$$

Like the simulated charged case (Exploration), where "cationic" initiators favor full polymeri-zation of the monomer, electro-free-radically, polymerization is favored via C = O center. For most other pulling groups COR, OCOR, CN etc., the routes favored and transfer species involved, will essentially follow the same pattern of analysis so far established, noting that the C≡N and C = O are strong nucleophilic centers. While C=O active centers carry the same type of charges or radicals irrespective of the type of substituted group involved, the C≡N and C≡C active centers carry only radicals. Only the C ≡ N center is fixed.

Alkyl cyanides can readily be reduced to primary amines by hydrogenation.

$$RC \equiv N \;+\; 2H_2 \xrightarrow{\;Rh,\,Pt,\,or\,Ni\;} RCH_2NH_2 \qquad\qquad 6.264$$

$$RC \equiv N \;+\; 4Na \;+\; 4C_2H_5OH \longrightarrow RCH_2NH_2 \;+\; 4NaOC_2H_5 \quad 6.265$$

Secondary and tertiary are formed at the same time by addition of amine to the intermediate imine, followed by loss of ammonia and further reduction or by reductive removal of the amino group.

$$RCH = NH \; + \; H_2NR \longrightarrow R-CH-NHR$$

with NH_2 below, $H_2/Cat.$ branch:

$$NH_3 \; + \; RCH = NR \xrightarrow{H_2/Cat.} RCH_2NHR \qquad 6.266$$

The secondary amine gives rise by similar series of reactions to the tertiary amine. Hydrogen can be kept in Equilibrium State of existence either by using selected Transition metals or their salts or using some chemical means.

Stage 1:
$$H_2 \xrightleftharpoons{Cat.} H.e \; + \; H.n$$

$$H.e \; + \; N \equiv \underset{R}{C} \xrightleftharpoons{Activation} H-N = \underset{R}{C}.e$$

$$H.n \; + \; (I) \longrightarrow H-N = \underset{R}{C}-H$$

$$(II) \qquad 6.267$$

Overall Equation: $\quad H_2 \; + \; N \equiv CR \longrightarrow HN = CHR \qquad 6.268$

Stage 2:
$$H_2 \xrightleftharpoons{Cat.} H.e \; + \; H.n$$

$$H.e \; + \; (II) \xrightleftharpoons{Activation} H-\underset{R}{\overset{H \quad H}{N-C}}.e$$

$$(III)$$

$$H.n \; + \; (III) \longrightarrow H-\underset{R}{\overset{H \quad H}{N-C}}-H$$

$$(IV) \qquad 6.269$$

Overall overall Equation: $\quad 2H_2 \; + \; N \equiv CR \longrightarrow H_2N - CH_2R \qquad (6.264)$

The Equilibrium mechanism is of two stages and very easy to understand. One can decide to stop in the first stage when only one mole of H_2 is used. For the second method of hydrogenation, the followings are obtained.

Stage 1: $\qquad$ $8Na \rightleftharpoons 8Na.e$

$8Na.e \quad + \quad 8C_2H_5OH \rightleftharpoons 8C_2H_5ONa \quad + \quad 8H.e$

$\qquad\qquad\qquad 8H.e \longrightarrow 4H_2$ $\qquad\qquad\qquad\qquad$ 6.270

Overall Equation: $\ 8Na + 8C_2H_5OH \longrightarrow 8C_2H_5ONa \quad + \quad 4H_2$ $\qquad$ 6.271

Stage 2: $\qquad$ $4Na \rightleftharpoons 4Na.e$

$\quad 4Na.e \quad + \quad 4H_2 \rightleftharpoons 4NaH \quad + \quad 4H.e$

$\qquad\qquad\qquad 4H.e \longrightarrow 2H_2$ $\qquad\qquad\qquad\qquad$ 6.272

Overall Equation: $\ 12Na \quad + \quad 8C_2H_5OH \longrightarrow 8C_2H_5ONa + 4NaH + 2H_2$ $\quad$ 6.273

Stage 3: $\qquad$ $NaH \rightleftharpoons Na.e \quad + \quad H.n$

$\qquad Na.e + H_2 \rightleftharpoons NaH \quad + \quad H.e$

$$H.e \ + \ N \equiv CR \rightleftharpoons \ H-\underset{\displaystyle R}{\overset{}{N}}=\overset{}{C}.e$$

$$(I)$$

$$H.n \ + \ (I) \longrightarrow \ H-N=\underset{\displaystyle R}{C}-H$$

$$(II) \qquad\qquad\qquad 6.274$$

Overall Equation: $\ 12Na \ + \ 8C_2H_5OH \ + \ N \equiv CR \longrightarrow 8C_2H_5ONa \ + \ 4NaH \ +$

$$H_2 \ + \ HN = CHR \qquad\qquad 6.275$$

Stage 4: $\qquad$ $NaH \rightleftharpoons Na.e \quad + \quad H.n$

$\qquad Na.e + H_2 \rightleftharpoons NaH \quad + \quad H.e$

$$H.e \ + \ (II) \rightleftharpoons \ H-\overset{\displaystyle H}{\underset{\displaystyle R}{N}}-\overset{\displaystyle H}{C}.e$$

$$(III)$$

$$H.n \ + \ (III) \longrightarrow \ H_2N - CH_2R \qquad\qquad 6.276$$

Overall overall Equation: $\ 12Na \ + \ 8C_2H_5OH + RC \equiv N \longrightarrow RCH_2NH_2 + 8C_2H_5ONa$

$$+ \ 4NaH \qquad\qquad 6.277$$

Compare the final overall equation above with the past as represented in Equation 6.265. Usually the use of Na/C_2H_5OH combination is one of the chemical methods used in providing H_2 in its Equilibrium State of existence, unknown to present-day Science. [See also Equations 6.73 to 6.78.] When used, the ratio of Na to the so-called absolute alcohol is always greater than one. Note how the sodium hydride was used to provide H.n and H.e. Note also how like poles have combined together to form H_2 molecule in the last steps of Stages 1 and 2. If the intermediate (II) was to be obtained, the overall equation would be as follows after three stages.

Overall Equation: $6Na \ + \ 4C_2H_5OH \ + \ RC \equiv N \longrightarrow HN = CHR \ + \ 4C_2H_5ONa$

$$+ \ 2NaH \qquad\qquad 6.278$$

For the reactions of Equation 6.266, the followings are obtained-

Stage 1: $\qquad H_2NR \ \rightleftharpoons \ H.e \ + \ nn.NHR$

$$H.e \ + \ HN = CHR \ \rightleftharpoons \ H-\overset{\overset{\displaystyle H}{|}}{N}-\underset{\underset{\displaystyle R}{|}}{\overset{\overset{\displaystyle H}{|}}{C}}.e$$

$$(I)$$

$$RHN.nn \ + \ (I) \ \longrightarrow \ H_2N-\underset{\underset{\displaystyle R}{|}}{\overset{\overset{\displaystyle H}{|}}{C}}-NHR$$

$$6.279$$

Overall reaction: $H_2NR \ + \ HN = CHR \longrightarrow R-\underset{\underset{\displaystyle NH_2}{|}}{CH}-NHR$

$$(II) \qquad\qquad 6.280$$

Stage 2: $\qquad (II) \ \rightleftharpoons \ R-\underset{\underset{\displaystyle H_2N}{|}}{\overset{\overset{\displaystyle H}{|}}{C}} - \overset{\overset{\displaystyle R}{|}}{N}.nn \ + \ H.e$

$$(III)$$

$$(III) \ \rightleftharpoons \ e.\underset{\underset{\displaystyle R}{|}}{\overset{\overset{\displaystyle H}{|}}{C}} - \overset{\overset{\displaystyle R}{|}}{N}.nn \ + \ nn.NH_2$$

$$(IV)$$

$$H.e \ + \ nn.NH_2 \ \rightleftharpoons \ NH_3$$

$$(IV) \ \xrightarrow{\text{De-activation}} \ \underset{\underset{\displaystyle R}{|}}{\overset{\overset{\displaystyle H}{|}}{C}} = \overset{\overset{\displaystyle R}{|}}{N} \ + \ \text{Heat}$$

$$6.281$$

Overall Equation: $RHCNH_2 - NHR \longrightarrow NH_3 + HRC = NR + Heat$ 6.282a

As against literature data above, one expected the following overall equation-

$$RHCNH_2 - NHR \longrightarrow NH_2R + HRC = NH + Heat \qquad 6.282b$$

Nevertheless, the stages above speak for themselves. Note the Equilibrium States of existence of the components in the first step of every stage. *It is the poorer N center with respect to H when more than one N center is present that is the source of Equilibrium state of Existence as will be encountered downstream.* They are fixed, like the finger prints in humans. Note how the Stages move in steps, just as we operate in life. One can never by-pass a stage as we tend to do in present-day Science by saying for example "that this is the rate determining step". It is all nonsense, for this is a NEW SCIENCE bound to affect all disciplines. One has begun to show in steps *"the missing links"* since the beginning of our time. So far, we have just begun.

Stage 1:
$$H_2 \underset{\longleftarrow}{\overset{H_2 \ Cat}{\longrightarrow}} H.e + H.n$$

$$H.e + \underset{Of \ Eqn \ 6.280}{(II)} \rightleftharpoons e.\overset{\displaystyle H \quad R}{\underset{\displaystyle R}{C-NH}} + NH_3$$

$$(I)$$

$$H.n + (I) \longrightarrow H-\overset{\displaystyle H \quad R}{\underset{\displaystyle R \quad H}{C-N}}$$

         6.283

Overall Equation: $H_2 + RHCNH_2 - NHR \xrightarrow{\ H_2 \ Cat.\ } NH_3 + RCH_2NHR$ 6.284

Note that ammonia is one the products. Without providing the mechanisms of these reactions, one cannot understand the Chemistry of these compounds; one cannot understand how NATURE operates. Coming back to the reactions of Equation 6.266, we have the followings.

Stage 1:
$$H_2 \underset{\longleftarrow}{\overset{H_2 \ Cat.}{\longrightarrow}} H.e + H.n$$

$$H.e + RCH = NR \xrightarrow{\ Activation\ } H-N-\overset{\displaystyle R \quad H}{\underset{\displaystyle R}{C.e}}$$

$$(I)$$

$$H.n + (I) \longrightarrow H-N-\overset{\displaystyle R \quad H}{\underset{\displaystyle R}{C-H}}$$

$$(II) \ (RCH_2NHR) \qquad 6.285$$

Overall Equation: $H_2 + RCH = NR \longrightarrow RCH_2NHR$ 6.286

For the first time, most of the things which could not be explained in the past, have begun to be explained and this will continue in the Series and Volumes. All the mechanisms which have been shown above in stages are Equilibrium mechanisms which embrace all kinds of reactions such as Combustion, Oxidation [See Appendix VI], Hydrogenation, Decomposition of some compounds [See Appendices II, III, and IV] etc.

Consider where the H group in HCN is replaced with NR_2 group. *These are dialkyl-cyanamides obtained when sodium cyanamide ($Na_2N - C \equiv N$) is said to displace "halide ion" from alkyl halides or alkyl sulfates.*

$$N \equiv C - N^{2\ominus} + RX \underset{X}{\rightleftharpoons} N \equiv C - N^{\ominus} - R \xrightarrow[X]{RX} N \equiv C - NR_2 \qquad 6.287$$

The equations above as have been written and used in present-day Science, are all meaningless. These reactions do not take place ionically or chargedly. *The nitrile group can be hydrolyzed by aqueous acids or bases to give the N, N–di-alkylcarbamic acid, which spontaneously loses carbon dioxide.*

$$R_2N - C \equiv N + 2H_2O \xrightarrow{H\oplus Or \ominus OH} NH_3 + [R_2NCOOH] \longrightarrow R_2NH + CO_2 \qquad 6.288$$

For the two reactions, the followings are obtained.

Stage 1:

$$N \equiv C - \underset{\underset{Na}{|}}{\overset{\overset{Na}{|}}{N}} \rightleftharpoons N \equiv C - N.nn \ (I) + Na.e$$

$$Na.e + RX \rightleftharpoons NaX + R.e$$

$$R.e + (I) \longrightarrow N \equiv C - \underset{}{\overset{\overset{Na}{|}}{N}} - R \ (II) \qquad 6.289$$

Overall Equation: $NCNNa_2 + RX \longrightarrow NaX + NCNRNa \qquad 6.290$

Stage 2:

$$(II) \rightleftharpoons Na.e + N \equiv C - \overset{\overset{R}{|}}{N}.nn \ (III)$$

$$Na.e + RX \rightleftharpoons NaX + R.e$$

$$R.e + (III) \longrightarrow N \equiv C - NR_2 \qquad 6.291$$

Overall Equation: $NCNRNa + RX \longrightarrow NaX + N \equiv C - NR_2 \qquad 6.292$

When ionic metals are present in compounds, they are the first to be held in Equilibrium State of existence, particularly when they are held ionically. The two stages above cannot take place chargedly, since only ionic charges can be isolated. Covalent, Electrostatic and Polar charges cannot be isolated. These are paired all the time where they exist. When the cyanamide is hydrolyzed according to Equation

6.288, the followings are obtained, noting that -NR_2 group is a strong radical-pushing group of greater capacity than –R and –NH_2 groups [NR_2 > NRH > NH_2 > R].

Stage 1:

$$H_2O \rightleftharpoons H.e + nn.OH$$

$$H.e + N \equiv C - NR_2 \rightleftharpoons \underset{\underset{NR_2}{|}}{H - N = C.e}$$

(I)

$$HO.nn + (I) \longrightarrow \underset{\underset{NR_2}{|}}{H - N = C - OH}$$

(II)   6.293

Overall Equation: $H_2O + N \equiv C - NR_2 \longrightarrow H - N = C(NR_2) - OH$ 6.294

Note the type of product obtained in this first stage. The monomer a strong female has been activated by a male initiator in the second step. It cannot undergo polymerization here, because of the presence of nn.OH. If only H.e was present in the system with the monomer, then the followings would have been obtained.

$$H.e + n(N \equiv C - NR_2) \longrightarrow \underset{\underset{NR_2}{|}}{H -(N = C)_{n-1} - N = C = NR} + R.e$$

6.295

The transfer species involved in killing the chain is of the first kind of the first type.

Stage 2:

$$H_2O \rightleftharpoons H.e + nn.OH$$

$$H.e + H - N = C(NR_2) - OH \rightleftharpoons \underset{\underset{NR_2}{|}}{H - \overset{\overset{H \quad OH}{| \quad |}}{N - C.e}}$$

(III)

$$HO.nn + (III) \longrightarrow \underset{\underset{NR_2}{|}}{H_2N - \overset{\overset{OH}{|}}{C} - OH}$$

(IV)   6.296

Overall Equation: $H_2O + H - N = C(NR_2) - OH \longrightarrow$ (IV)   6.297

Stage 3:

$$(IV) \rightleftharpoons \underset{\underset{NR_2}{|}}{H_2N - \overset{\overset{OH}{|}}{C} - O.nn} + H.e$$

(V)

$$(V) \quad \rightleftharpoons \quad e.\overset{\overset{\displaystyle NH_2}{|}}{\underset{\underset{\displaystyle NR_2}{|}}{C}} - O.nn \quad + \quad nn.OH$$

$$(VI)$$

$$H.e \quad + \quad nn.OH \quad \rightleftharpoons \quad H_2O$$

$$(VI) \quad \xrightarrow{\text{De-Activation}} \quad \overset{\overset{\displaystyle NH_2}{|}}{\underset{\underset{\displaystyle NR_2}{|}}{C}} - O \quad + \quad Heat$$

$$(VII) \hspace{8cm} 6.298$$

Stage 4: NH_3 is removed by water from (VII) to give the overall equation.

Overall Equation: $\quad (IV) \quad \longrightarrow \quad NH_3 \quad + \quad HOOCNR_2 \hspace{3cm} 6.299$

Stage 5: $\qquad (VII) \quad \rightleftharpoons \quad H.e \quad + \quad nn.\overset{}{\underset{\underset{\displaystyle NR_2}{|}}{O}} - C = O$

$$(VIII)$$

$$(VIII) \quad \rightleftharpoons \quad R_2N.nn \quad + \quad e.\overset{\overset{\displaystyle O}{\|}}{C} - O.nn$$

$$(IX)$$

$$H.e \quad + \quad nn.NR_2 \quad \rightleftharpoons \quad HNR_2$$

$$(IX) \quad \xrightarrow{\text{De-Activation}} \quad CO_2 \hspace{4cm} 6.300$$

Overall Equation: $\quad HOOCNR_2 \quad \longrightarrow \quad CO_2 \quad + \quad HNR_2 \hspace{2cm} 6.301$

Note the Equilibrium State of Existence of compounds. For example in Stage 3, it is the H of OH that is held and not H of NH_2 in that compound. These states as already said are fixed. Nature does not discriminate or work indiscriminately. Nature does not differentiate but only integrates, noting that in Mathematics the Natural language of communication, to integrate we must first differentiate just as we do to determine the rate at which a product is obtained. In our Physical part of the world, for any of them to exist, they must co-exist. Note that all the stages above are not Decomposition mechanism, but Equilibrium mechanism. The Decomposition of the compounds is not such that produced initiators as products or one of the products.

Acrylonitrile is made chiefly from either acetylene or propylene.

$$HC \equiv CH \quad + \quad H - C \equiv N \quad \xrightarrow{(CuCl, NH_4Cl, HCl) - 90^0 C} \quad H_2C = CHCN \hspace{2cm} 6.302$$

Acrylic acid usually is made commercially by the hydrocarbonylation of acetylene in the pre-sence of nickel carbonyl.

$$HC \equiv CH \;+\; CO \;+\; H_2O \xrightarrow{Ni(CO)_4, HCl} H_2C = CHCOOH \qquad 6.303$$

Stage 1:
$$H - C \equiv N \rightleftharpoons H.e \;+\; n.C \equiv N$$

$$H.e \;+\; H - C \equiv C - H \xrightarrow{\text{Activation}} \underset{(I)}{H - \overset{\displaystyle H}{\underset{\displaystyle |}{C}} = \overset{\displaystyle H}{\underset{\displaystyle |}{C}}.e}$$

$$N \equiv C.nn \;+\; (I) \longrightarrow \begin{array}{c} H \quad H \\ | \quad\; | \\ C = C \\ | \quad\; | \\ H \quad C \equiv N \end{array} \qquad 6.304$$

Overall Equation: $HCN \;+\; HC \equiv CH \xrightarrow{\text{Passive Cat.}} H_2C = CHCN$ (6.302)

While the passive catalysts- the chlorides suppressed the Equilibrium State of Existence of acetylene, they could not suppress that of hydrogen cyanide. They or HCN did also suppressed that of HCl, otherwise vinyl chloride would also have been obtained. For the last reaction, the followings are also obtained.

Stage 1:
$$H_2O \rightleftharpoons H.e \;+\; nn.OH$$

$$H.e \;+\; CO \xrightarrow{\text{Activation}} \underset{(I)}{H - \overset{\displaystyle O}{\overset{\displaystyle \|}{C}}.n}$$

$$HO.nn \;+\; (I) \longrightarrow H - COOH \qquad 6.305$$

Overall Equation: $H_2O \;+\; CO \longrightarrow HCOOH$ 6.306

Stage 2:
$$HCOOH \rightleftharpoons H.e \;+\; n.COOH$$

$$H.e \;+\; H - C \equiv C - H \xrightarrow{\text{Activation}} \underset{(II)}{H - \overset{\displaystyle H}{\underset{\displaystyle |}{C}} = \overset{\displaystyle H}{\underset{\displaystyle |}{C}}.e}$$

$$HOOC.n \;+\; (II) \longrightarrow \begin{array}{c} H \quad H \\ | \quad\; | \\ C = C \\ | \quad\; | \\ H \quad C = O \\ \quad\quad | \\ \quad\quad O \\ \quad\quad | \\ \quad\quad H \end{array} \qquad 6.307$$

Overall Equation: $HCOOH \;+\; H - C \equiv C - H \longrightarrow H_2C = CHCOOH$ 6.308

The Equilibrium State of Existence of acetylene is suppressed here by the nickel carbonyl. Worthy of note is the Equilibrium State of Existence of formic acid, the first member of the carboxylic acids. The hydrogen held is different from the other members as have already been shown and recalled below.

$$HCOOH \rightleftharpoons H.e + \underset{\underset{\|}{\overset{O}{\|}}}{n.C} - OH \qquad 6.309$$

$$HCOOCH_3 \rightleftharpoons H.e + \underset{\overset{\|}{n.C} - OCH_3}{\overset{O}{\|}} \qquad 6.310$$

$$RCOOH \rightleftharpoons H.e + R - \underset{\overset{\|}{C}}{\overset{O}{\|}} - O.nn \qquad 6.311$$

$$CH_3OH \rightleftharpoons H.e + nn.OCH_3 \qquad 6.312$$

When aqueous CH_3OH is used in place of H_2O in Equation 6.303, methyl acrylate an ester, is similarly obtained. Note the Equilibrium State of Existence of so-called organic alcohols. They look acidic in character, if not for the absence of a hetero Central atom such as S, N, Cl. In their place is the alkanyl group. Worthy of note also is that CO has an activation center shown below.

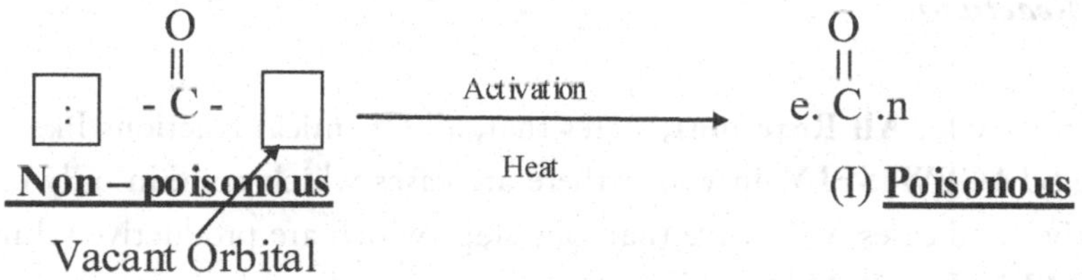

$$\underset{\substack{\textbf{Non - poisonous} \\ \text{Vacant Orbital}}}{\boxed{:} - \overset{\overset{O}{\|}}{C} - \boxed{}} \xrightarrow[\text{Heat}]{\text{Activation}} \underset{\textbf{(I) Poisonous}}{e.\overset{\overset{O}{\|}}{C}.n} \qquad 6.313$$

Carbon monoxide becomes poisonous once activated by heat. The real poison is the electro-free-radical center. All the reactions just above and states of existences take place only radically. ***All ionic compounds such as H_2O. HCl, NaOH, etc., cannot exist chargedly in Decomposition or Combination States of Existence, but only in Equilibrium State of Existence.*** It should also be noted as seen so far that, just as there are only three types of Mechanisms for all systems, so also there are three States of Existences for everything in life and indeed all systems. In addition to the three States of Existences are those in which if the compound is not in Equilibrium State of Existence, it can be in Stable State of Existence or Activated State of Existence or Activated/ Equilibrium State of Existence, if it has Activation center(s).

6.3 Proposition of Rules of Chemistry and Conclusions

Based on the new concepts of existence of free and non-free charges and radicals, free and paired media systems, their male and female counterparts, existences of activation centers and several other phenomena too countless to list, the chemistry of acetylenes and nitriles have begun to be considered to a very great extent. Why these monomers are not popularly known to favor polymerizations have become clearly obvious and how they can be made to do so have begun to be shown. Some of the most important distinguishing features of the monomers that clearly identify with them, will be stated into rules in view of the complete departure from what has been known to be the case in the past. For the first time also, one has been forced to introduce so many concepts new to humanity. One had thought in delaying the introduction of Equilibrium mechanism until later in the Series and Volumes. This could not be done since we have begun to go into the chemistry of monomeric compounds and TIME is an independent imaginary variable of the fourth dimension. ***TIME waits for NOTHING, but for SOMETHING.***

The rules which have been proposed so far, in addition to the reactions which have been known to be favored by these monomers, have been used as guides to move one step forward, making sure that the natural rules are not contravened. The first and indeed all the members of acetylenes and nitriles have been shown not to favor charged existence via their centers. When they contain other activation centers, charged activations can be favored via those centers. Unlike the alkene families where the MALES and FEMALES can be activated radically and chargedly, with alkyne families, this can only be done radically, wherein the strongest male character is provided by the NITRILE group (called herein as ***Acetylonitrile,*** the corresponding case of acrylonitrile for alkenes).

We are beginning to realize that most or all chemical reactions take place radically, despite the very important but limited existence of charges all due to electrostatic forces of repulsion and attraction, the forces holding the weightless Planets in place and together.

Rule 189: This rule of Chemistry for **All reactions,** states that, in general there are two types of Reactions-the first called **Chemical reactions** which are those that deal with **Micromolecules** and the second being **Polymeric reactions** which are those that deal with **Macromolecules.**
(Laws of Creations for All Reactions)

Rule 190: This rule of Chemistry for **All Reactions,** states that, all Chemical reactions like all other operations in life take place **STAGEWISELY,** in which there are cases which contain only one stage productive and non-productive, and cases with more than one stage which are productive taking place either in **SERIES or in PARALLEL or BOTH.**
(Laws of Creations-Stagewise operations-The Seventh law of Nature)

Rule 191: This rule of Chemistry for **WATER,** states that, this compound which is polar/ionic in character is the only compound which is both an ACID and an ALCOHOL; for without it nothing exists in humanity-

$$H_2O \underset{\text{Existence}}{\overset{\text{Equilibrium State of}}{\rightleftharpoons}} H^{\oplus} + {}^{\ominus}OH \quad OR \quad H\bullet e + nn\bullet OH$$

$$\text{ACID} \qquad\qquad \text{ALCOHOL} \qquad\qquad \text{ACID} \qquad\qquad \text{ALCOHOL}$$

noting that it was there long before our world was created, hence it can be said to be a universal solvent and neutral in character.
(Laws of Creations-Acid and Alcohol)

Rule 192: This rule of Chemistry for **Inorganic Acids,** states that, all these compounds are polar/ionic in character; for which they are composed of an ionic metal (H of Group IA) the component carrying positive free charge ***(Cations) or electro-free-radical*** and ionic non-metallic electronegative polar elements or species in which the Central atoms are hetero in character (from Groups VB, VIB and VIIB) the components carrying negative non-free charges ***(Anions) or nucleo-non-free-radicals;*** which when held in Equilibrium State of Existence, the followings are obtained for all of them-

$$HCl \xrightleftharpoons[\text{Existence}]{\text{Equilibrium State of}} H^{\oplus} + :\overset{\cdot\cdot}{\underset{\cdot\cdot}{Cl}}^{\ominus} \quad OR \quad H\bullet e + :\overset{\cdot\cdot}{\underset{\cdot\cdot}{Cl}}\bullet nn$$

$$HNO_2 \xrightleftharpoons[\text{Existence}]{\text{Equilibrium State of}} H^{\oplus} + {}^{\ominus}\overset{\cdot\cdot}{\underset{\cdot\cdot}{O}} -- N = O \quad AND\ NOT\ {}^{\ominus}N^{\oplus} - {}^{\ominus}O \quad OR$$

$$\overset{\displaystyle O}{\underset{\displaystyle \|}{}}$$

[Not possible]

$$H\bullet e + nn\bullet \overset{\cdot\cdot}{\underset{\cdot\cdot}{O}} -- N = O$$

$$H_2S \xrightleftharpoons[\text{Existence}]{\text{Equilibrium State of}} H^{\oplus} + {}^{\ominus}\overset{\cdot\cdot}{\underset{\cdot\cdot}{S}} - H \quad OR \quad H\bullet e + nn\bullet \overset{\cdot\cdot}{\underset{\cdot\cdot}{S}} - H$$

noting that none of them is carrying a carbon center
(Laws of Creations- for Inorganic Acids)

Rule 193: This rule of Chemistry for **Polar/Ionic Organic Acids**, states that, these are compounds composed of an ionic metal (H of Groups IA) the component carrying positive free charge (Cation) or electro-free-radical and an ionic non-metallic electronegative polar element O, N, S of Groups VB, VIB the component carrying negative non-free **charge (Anion) or nucleo-non-free-radical and** most importantly a CENTRAL CARBON ATOM that which gives it the Organic character (particularly when carbonyl or the like in character); which when made to exist in Equilibrium State of existence, the followings are obtained for some of them-

$$\overset{\displaystyle O}{\underset{\displaystyle \|}{H_3CCOH}} \xrightleftharpoons[\text{of Existence}]{\text{Equibrium State}} H^{\oplus} + {}^{\ominus}O\overset{\displaystyle O}{\underset{\displaystyle \|}{C}}CH_3 \quad OR \quad H\bullet e + nn\bullet O\overset{\displaystyle O}{\underset{\displaystyle \|}{C}}CH_3$$

$$H_3COH \xrightleftharpoons[\text{of Existence}]{\text{Equibrium State}} H^{\oplus} + {}^{\ominus}OCH_3 \quad OR \quad H\bullet e + nn\bullet OCH_3$$

(Laws of Creations for Polar/Ionic Organic acids)

Rule 194: This rule of Chemistry for **Polar/Non-ionic Organic Acids**, states that, these are compounds composed of an ionic metal (H of Group IA) the component carrying electro-free-radical and a non-ionic Central element C, that which gives it the Organic character, being the one carrying a nucleo-free-radical; which when made to exist in Equilibrium State of existence, the followings are obtained-

$$\overset{\displaystyle O}{\underset{\displaystyle \|}{HCOH}} \xrightleftharpoons[\text{of Existence}]{\text{Equibrium State}} H\bullet e + n\bullet \overset{\displaystyle O}{\underset{\displaystyle \|}{C}}OH$$

$$HC\equiv N \xrightleftharpoons[\text{of Existence}]{\text{Equibrium State}} H\bullet e + n\bullet C\equiv N \ ; \ HC\equiv CH \xrightleftharpoons[\text{of Existence}]{\text{Equibrium State}} H\bullet e + n\bullet C\equiv CH$$

$$H-N=C=O \xrightleftharpoons[\text{of Existence}]{\text{Equibrium State}} H\bullet e + nn\bullet N=C=O$$

(Laws of Creations for Polar/Non-Ionic Organic Acids)

Rule 195: This rule of Chemistry for *Electrostatic* **Inorganic Acids**, states that, all these compounds are polar/ionic in character; for which when held in Equilibrium State of Existence, they are composed of an ionic metal (H of Groups IA) the component carrying only electro-free-radical, and a non-ionic non-metallic polar element (O of Group VIB) the components carrying only nucleo-non-free-radicals and most importantly a CENTRAL ATOM a non-ionic non-Transition metal, examples of which are as follows-

$$\underset{\text{HCOH}}{\overset{\displaystyle O}{\|}} \quad \underset{\text{of Existence}}{\overset{\text{Equibrium State}}{\rightleftharpoons}} \quad H\bullet e \quad + \quad n\bullet \underset{\text{COH}}{\overset{\displaystyle O}{\|}}$$

$$HC\equiv N \quad \underset{\text{of Existence}}{\overset{\text{Equibrium State}}{\rightleftharpoons}} \quad H\bullet e \quad + \quad n\bullet C\equiv N \quad ; \quad HC\equiv CH \quad \underset{\text{of Existence}}{\overset{\text{Equibrium State}}{\rightleftharpoons}} \quad H\bullet e \quad + \quad n\bullet C\equiv CH$$

$$H-N=C=O \quad \underset{\text{of Existence}}{\overset{\text{Equibrium State}}{\rightleftharpoons}} \quad H\bullet e \quad + \quad nn\bullet N=C=O$$

(Laws of Creations for Polar/Non-Ionic Organic Acids)

Rule 196: This rule of Chemistry for **Ionic Inorganic Alcohol,** states that, all these compounds (also called **BASES OR ALKALIES)** are polar/ionic in character; for which when held in Equilibrium State of Existence, ionic metals (Groups IA, IIA and IIIA) are the components carrying positive free charges or electro-free-radicals and ionic non-metals of less electro-positivity than the metals (O of Group VIB) the components carrying negative non-free charges or nucleo-non-free-radicals connected to hydrogen, examples of which include-

$$NaOH \quad \underset{\text{Existence}}{\overset{\text{Equilibrium State of}}{\rightleftharpoons}} \quad Na^{\oplus} \quad + \quad {}^{\ominus}OH \quad OR \quad Na\bullet e \quad + \quad nn\bullet OH$$

$$Ca(OH)_2 \quad \underset{\text{Existence}}{\overset{\text{Equilibrium State of}}{\rightleftharpoons}} \quad Ca^{2\oplus} \quad + \quad 2{}^{\ominus}OH \quad OR \quad HO-Ca\bullet e \quad + \quad nn\bullet OH$$

$$OR \quad e\bullet Ca\bullet e \quad + \quad 2nn\bullet OH$$

(Laws of Creations for Alcohols)

Rule 197: This rule of Chemistry for **All Chemical Reactions,** states that, no reaction takes place CHARGEDLY (Ionically, Covalently, Electrostatically and Polarly) even during molecular rearrangements, but only RADICALLY.
(Laws of Creations for Chemical reactions)

Rule 198: This rule of Chemistry for **All Polymeric reactions,** states that, while few Polymeric reactions take place covalently, electrostatically and even ionically (anionic), most take place radically, while none takes place polarly.
(Laws of Creations for Polymeric reactions)

Rule 199: This rule of Chemistry for **The first member of acetylenes,** states that, *this com-pound favors only radical existence; for which one of the hydrogen atoms is loosely held radically* to its carbon center in the unactivated state all the time, that is, always in Equilibrium State of Existence in the absence of a suppressing agent or neighbor as shown below-

$$H - C \equiv C - H \quad \underset{\text{Existence}}{\overset{\text{Equilibrium State of}}{\rightleftharpoons}} \quad H\bullet e \quad + \quad n\bullet C \equiv C - H$$

(Laws of Creations for Acetylene's State of Existence)

Rule 200: This rule of Chemistry for **The first member of Acetylenes (HC≡CH),** states that, acetylene *can be activated and polymerized electro- or nucleo-free-radically when its Equili-brium State of Existence is suppressed with the use of suitable passive catalyst;* for which in the absence of these passive catalysts, the loosely held hydrogen atom must be replaced by either radical-pushing group with capacity greater than that of the hydrogen atom carried by the acetylene or a radical-pulling group.
(Laws of Creation for Acetylene)

Rule 201: This rule of Chemistry for **Acetylenes (HC ≡ CR) where R is a pushing alkanyl group such as CH$_3$,** states that, *these acetylenes are **Nucleophiles (Females)** for which if the R group is weak, it cannot be activated by R^l if less than R unless when suppressed by the use of a passive catalyst,* since they will still exist in Equilibrium State of Existence as shown below-

$$R - C \equiv C - H \quad \underset{\text{Existence}}{\overset{\text{Equilibrium State of}}{\rightleftharpoons}} \quad R - C \equiv C\bullet n \quad + \quad H\bullet e \qquad \text{[If R is weak in capacity]}$$

otherwise if R^l for the initiator is strong, the monomer can readily be activated and used as a monomer.
(Laws of Creations for Acetylene Family)

Rule 202: This rule of Chemistry for **Acetylenes (R^{l}C ≡ CR) where R^l and R are radical-pushing alkanyl groups of same or different capacities,** states that, *these acetylenes are **Nucleophiles (Females)** of stronger capacities than those from Alkenes, which can be activated only radically and can only be polymerized electro-free-radically, just like RC≡CH when activation is favored.*
(Laws of Creations Acetylene Family)

Rule 203: This rule of Chemistry for **Acetylenes of the type RC ≡ CR where R is a radical pushing group,** states that, these can be hydrogenated or reduced by means of sodium in Liquid ammonia as shown below-

$$RC \equiv CR \quad + \quad 2Na \quad + \quad 2NH_3 \quad \longrightarrow \quad RHC = CHR \quad + \quad 2Na^{\oplus}\,{}^{\ominus}NH_2 \qquad \text{(a)}$$

Stage 1: $\qquad\qquad 2NH_3 \quad \rightleftharpoons \quad 2H\bullet e \quad + \quad 2nn\bullet NH_2$

$\qquad 2Na\bullet e \quad + \quad 2nn\bullet NH_2 \quad \rightleftharpoons \quad 2NaNH_2$

$\qquad\qquad\qquad 2H^e \quad \longrightarrow \quad H_2 \qquad\qquad\qquad\qquad\qquad\qquad \text{(b)}$

Overall Equation: $\quad 2Na \quad + \quad 2NH_3 \quad \longrightarrow \quad 2NaNH_2 \quad + \quad H_2 \qquad\qquad \text{(c)}$

Stage 2:

$$NaNH_2 \rightleftharpoons Na^{.e} + {}^{nn.}NH_2$$

$$Na^{.e} + H_2 \rightleftharpoons NaH + H^{.e}$$

$$(I)$$

$$(I) + {}^{nn.}NH_2 \longrightarrow HRC = CR(NH_2) \tag{d}$$

Overall Equation: $NaNH_2 + H_2 + RC\equiv CR \longrightarrow NaH + HRC=CR(NH_2)$ $\qquad$ (e)

Stage 3: $\qquad NaH \rightleftharpoons Na^{.e} + H^{.n}$

$$Na^{.e} + HRC = CR(NH_2) \rightleftharpoons NaNH_2 + HRC = \overset{\overset{\displaystyle R}{|}}{C}.e$$

$$HRC = RC.e + H^{.n} \longrightarrow HRC = CRH \tag{f}$$

Overall Overall Equation: $RC \equiv CR + 2Na + 2NH_3 \longrightarrow RHC = CHR +$

$$2Na^{\oplus} {}^{\ominus}NH_2 \tag{a}$$

a three-stage Equilibrium mechanism system

(Laws of Creations for Acetylene Family)

Rule 204: This rule of Chemistry for **Acetylenes (HC $\equiv$ CZ) where Z is a radical-pulling group such as COOR, COR, CONR$_2$, NO$_2$, SO$_2$R (where R is a radical-pushing alkanyl group such as CH$_3$),** states that, *these acetylenes are Electrophiles (Males) of greater capacities than those from Alkenes, since the Y center (C$\equiv$C) is more nucleophilic than the X center(C=O);* **noting that** they can be activated only radically without the need to replace the H atom.

(Laws of Creations for Acetylene Family)

Rule 205: This rule of Chemistry for **Acetylenes (RC $\equiv$ CZ) where Z is a radical-pulling group such as COOR, COR, CONR$_2$, NO$_2$, SO$_2$R and R is a radical-pushing alkanyl group such as CH$_3$,** states that, *these acetylenes are strong Electrophiles (Males) which can only be activated radically;* for which when done and the initiator is strong, only the electro-free-radical route is favored by them via the C=O center for 1,2- addition mono-form or via resonance stabilization to give the 1.4- addition mono-form; with only the first member (when R$\equiv$H) favoring in addition the nucleo-non-free-radical route, the route natural to the monomer, noting that never is there a time when the C$\equiv$C center is activated, being the most nucleophilic of all the mono-forms and when R is replaced with F, CF$_3$ types of groups, the electrophilicity increases.

(Laws of Creations for Acetylene Family)

Rule 206: This rule of Chemistry for **The first member of the hybrids of acetylenes and per-fluoroacetylenes (HC≡CF),** states that, *this compound favors only radical existence; for which none of H and F atoms are loosely bonded radically* to the two carbon centers in the unactivat-ed state, and can therefore be activated to favor both free-radical routes.
(Laws of Creations for Acetylene Family)

Rule 207: This rule of Chemistry for **Acetylenes of the type RC≡CF where R is a free-radical -pushing group of the type C_nH_{2n+1},** etc., states that, *these monomers are Nucleophiles which favor electro-free-radical route;* while when R is replaced with C_nF_{2n+1}, etc. types of groups, *the monomers are weak Nucleophiles which favor both radical routes for CF_3 only, but become stronger for groups greater than CF_3, since only electro-free-radical route becomes the only route.*
(Laws of Creations for Acetylene Family)

Rule 208: This rule of Chemistry for **Acetylenes of the type FC≡CF,** states that, this monomer wherein the F atoms are strongly bonded to the active C centers is a weak Nucleophile, since it can be made to favor both free-radical routes.
(Laws of Creations for Acetylene Family)

Rule 209: This rule of Chemistry for **Acetylenes of the type $R_FC≡CH$ where R_F is a free-radical-pushing group of the type CH_2F, CHF_2, CH_2COOCH_3, $CH(CONH_2)_2$, etc.,** states that, *these monomers are Nucleophiles where the need to replace the second H atom does not arise for it to be initiated.*
(Laws of Creations for Acetylene Family)

Rule 210: This rule of Chemistry for **Acetylenes of the type RC≡CH where R is a pushing group of the type OH, NH_2,** states that, *these monomers which are females like $HOCH = CH_2$, $H_2NHC = CH_2$ are unstable;* for which when activated, can molecularly rearrange to form $CH_2 = C = O$ or $CH_2 = C = NH$ unstable Electrophiles under specific operating conditions.
(Laws of Meta-Physics) [See Rule 182]

Rule 211: This rule of Chemistry for **Acetylenes of the type $R`C ≡ CR_1$ where R` is a radical -pushing group of the type OR, NR_2, and R_1 is an alkanyl group of less or equal capacity than the R in R`,** states that, *these monomers are stable females; for which when activated can molecularly rearrange to form dialkyl male monomers, $R_2C = C= O$ or $R_2C = C = NR$ under specific operating conditions*
(Laws of Meta-Physics)

Rule 212: This rule of Chemistry for **Addition monomers of the type $H_2C=C=O$ and $H_2C= C=NH$,** states that, *these monomers are unstable Electrophiles (i.e., Males) with X and Y activation centers, in which only the latter can rearrange via Electroradicalization to give a more stable female compound —a Nitrile $H_3C - C ≡ N$*

Unstable Male Monomer Female   **Unstable Male Monomer**   Female

(Laws of Meta-Physics)

Rule 213: This rule of Chemistry for **Addition monomers of the type R$_2$C=C=O, R$_2$C=C=NR where R are Alkanyl groups,** states that, *these are stable Electrophiles (Males) in which the latter is a product of stronger females (Nitriles and acetylenes), all based on the operating conditions and the size of the groups.*

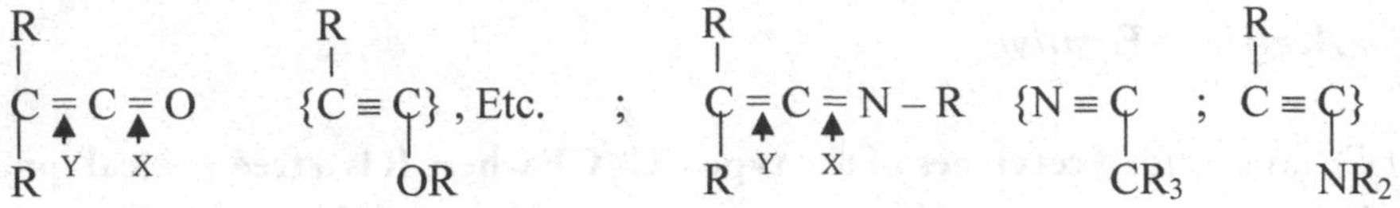

$$
\begin{array}{ccc}
\underset{\underset{R}{|}}{\overset{\overset{R}{|}}{C}} = C = O & \{C \equiv C\}, \text{Etc.} \quad ; & \underset{\underset{R}{|}}{\overset{\overset{R}{|}}{C}} = C = N-R \quad \{N \equiv C \; ; \; C \equiv C\}
\end{array}
$$

Stable Electrophile   Nucleophile   Stable Electrophile   Nucleophiles

(Laws of Meta-Physics)

Rule 214: This rule of Chemistry for **Acetylenes of the type R$_F$C $\equiv$ CF where R$_F$ is a radical-pulling group of the type COOCH$_3$, NO$_2$, COCH$_3$, etc.,** states that, *these monomers of different capacities can be activated and polymerized only free-radically via the C = O or N$^{\oplus}$-O$^{\ominus}$ center or it's resonance stabilized 1,4-monoform, since these are Electrophiles (i.e. Males) where the Y activation center is more nucleophilic than the X activation center; while with CN group, the reverse is the case, with no resonance stabilization.*

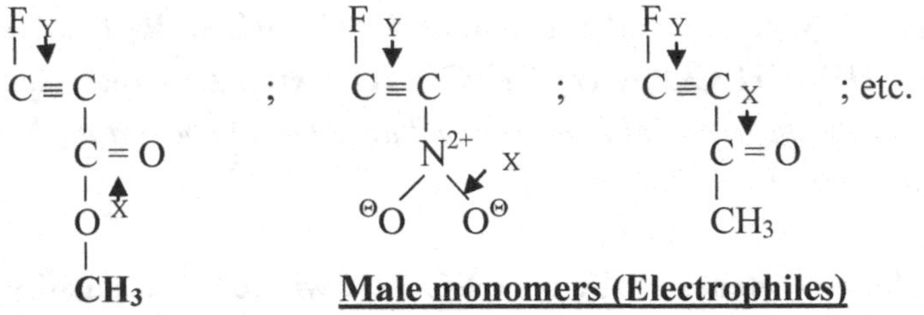

Male monomers (Electrophiles)

(Laws of Creations for Fluorinated Electrophilic Acetylenes)

Rule 215: This rule of Chemistry for **Electrophilic acetylenes (R$_F$C $\equiv$ CF, R$_F$C $\equiv$ CH) where R$_F$ is a radical-pulling group of the type COOR, COR, etc.,** states that, from the order of nucleophilicity of the Activation centers, the most Electrophilic is that carrying the CN group (Herein called **ACETYLONITRILE**), because the followings are valid and only that with CN is not resonance stabilized and will favor the nucleo-free-radical route via the C $\equiv$ C activation center, while in the others the C $\equiv$ C center cannot be used alone.

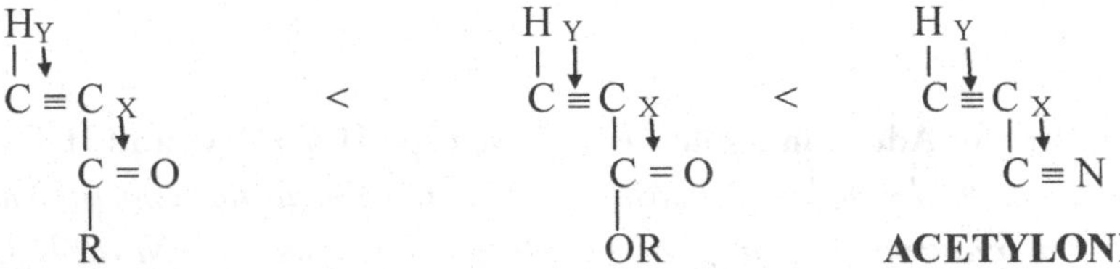

Order of Male character

$$C \equiv N > C \equiv C > C = O > C = N > C = C$$

Order of Nucleophilicity

(Laws of Creations for Electrophilic Acetylenes)

Rule 216: This rule of Chemistry for **Polymerizable Acetylenic Addition monomers,** states that, in general there are three major types of transfer species - *of the first kind of the first type, of the first kind of the third type (= CH$_2$), and of the second kind,* with each kind serving various functions for a monomer and a growing polymer chain.
(Laws of Creations for Transfer Species in Polymerization of Acetylenes)

Rule 217: This rule of Chemistry for **Acetylenic family members of Addition monomers,** states that, *transfer species of the **first kind of the third type** are those externally located on an unsaturated carbon center of a substituent group doubly bonded to the α- carbon center of the monomer; for which they are identified with* ketenes and Cumulenes such as H on =CH$_2$, =CHCH$_3$, =CHF; while transfer species of the ***first kind of the fourth and fifth types** are those externally located on an alkenyl group bonded to the α-carbon center of the monomer such as H on the CH$_3$ group of – CH=CH-CH$_3$ and on a alkynyl group singly bonded to the α-carbon center of the monomer* such as H on the CH$_3$ or C$_2$H$_5$ group of – C≡CCH$_3$, – C≡CC$_2$H$_5$, respectively, whether resonance stabilization is provided or not by them.
(Laws of Creation for Transfer Species of First kind of Third, Fourth and fifth types)

Rule 218: This rule of Chemistry for **Acetylenes which carry metals in place of two H atoms,** states that, these are called *CARBIDES* of which there are two main types of different characters- **Ionic-metallic carbides [NaC≡CNa, Ca(C≡C)] and Non-ionic-metallic carbides [AgC≡CAg, CuC≡CCu];** for which while the latter cannot be hydrolyzed by water, the former can be hydrolyzed.
(Laws of Creation for Carbides)

Rule 219: This rule of Chemistry for **Acetylenes which carry one metal in place of one H atom,** states that, these are called *ACETYLIDES* of which there are two main types of different characters- **Ionic-metallic acetylides [NaC≡CH, NaC≡CR)] and Non-ionic-metallic acety-lides [AgC≡CH, CuC≡CH, AgC≡CR].**
(Laws of Creation for Acetylides)

Rule 220: This rule of Chemistry for **Acetylenes of the types CH ≡ CH, CH ≡ CR, where Rs are pushing groups,** states that, *when combined with ionic and non-ionic metals, this can only be done radically, for which the metals cannot add to the centers by activation, but can only displace one hydrogen atom one at time to form Acetylides and Carbides.*
(Laws of Creations for Metallic Carbides and Acetylides)

Rule 221: This rule of Chemistry for **Acetylides of the types NaC ≡ CH, AgC ≡ CH,** states that, *these compounds cannot be used as monomers, since they favor strong Equilibrium State of Existence; for which for them to be used, the metal has to be replaced with pushing or pulling groups.*

$$NaC \equiv CH \rightleftharpoons Na.e \;+\; HC \equiv C.n \;;\; AgC \equiv CH \rightleftharpoons H.e \;+\; AgC \equiv C.n$$

Equilibrium State of Existence of Acetylides

(Laws of Creations for Acetylides)

<u>**Rule 222:**</u> This rule of Chemistry for **Carbides of the types Ca(C ≡ C), AgC ≡ CAg,** states that, *these compounds cannot be used as monomers, since they favor strong Equilibrium State of Existence; for which for them to be used, the metals have to be replaced with pushing or pulling groups.*

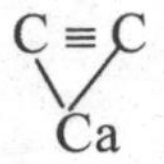

$$C \equiv C \xrightarrow{} \text{n.C} \equiv C - Ca \,.e \quad ; \quad AgC \equiv CAg \rightleftharpoons Ag.en \;+\; AgC \equiv C.n$$
$$\underset{Ca}{\diagdown\diagup}$$

(Very Stable)

<u>**Equilibrium State of Existence of some Carbides**</u>

(Laws of Creations for Carbides)

<u>**Rule 223:**</u> This rule of Chemistry for **Addition monomer with substituted groups of the types -C_2F_5, - CF = CF_2, -C ≡ CF,** states that, *based on the chemical behaviors of the groups, the followings are the order of their pushing capacities:-*

$$[\text{-CH}_2\text{- CH}_3 \; > \; \text{-CH}_3 \; > \; \text{-CH} = \text{CH}_2 \; > \; H \; > \; \text{-C} \equiv \text{CH}] \text{ See Rule 133}$$
$$\text{-CF}_2\text{- CF}_3 \; > \; \text{-CF}_3 \; > \; \text{-CF} = \text{CF}_2 \; > \; F \; > \; \text{-C} \equiv \text{CF}$$

<u>**Order of radical-pushing capacity**</u>

(Laws of Creations for Radical-pushing groups)

<u>**Rule 224:**</u> This rule of Chemistry for **Resonance stabilized compounds,** states that, when resonance stabilization is provided, the activation centers of the resonance stabilization group providing it must be of equal or greater capacity than the activation centers of the receiving center as shown; for which this can only take place radically.

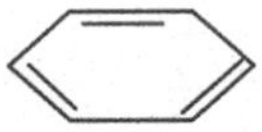

$$\langle\bigcirc\rangle \; >> \; H - C \equiv C - H \; >> \; H_2C = CH_2$$

<u>**Order of Nucleophilicity of Major Sources of Resonance Stabilization groups**</u>

(Laws of Creations for Resonance Stabilization Provider)

<u>**Rule 225:**</u> This rule of Chemistry for **Acetylenic resonance stabilization groups of the type RC ≡ C- where R is an alkyl-ane pushing group,** states that, *the order of their capacities can be obtained from the following, based on the capacities of the substituted groups.*

$$C_4H_9C \equiv C - \; > C_3H_7C \equiv C- \; > \; C_2H_5C \equiv C- \; > CH_3C \equiv C- \; > \; HC \equiv C-$$

<u>**Order of radical-pushing capacity**</u>

(Laws of Creations for Resonance Stabilization Groups)

<u>**Rule 226:**</u> This rule of Chemistry for **Dimers of acetylene of the type $HC \equiv C$ -CH=CH_2,** states that, *this compound which favors only free-radical existence, has been called **AN ENEYNE (and not Vinylacetylene),*** since it is the acetylenic group that is providing the shield or resonance stabilization for the vinyl group as shown below; for which the hydrogen atom on the acetylene is not loosely bonded to its carbon center in the unactivated state.

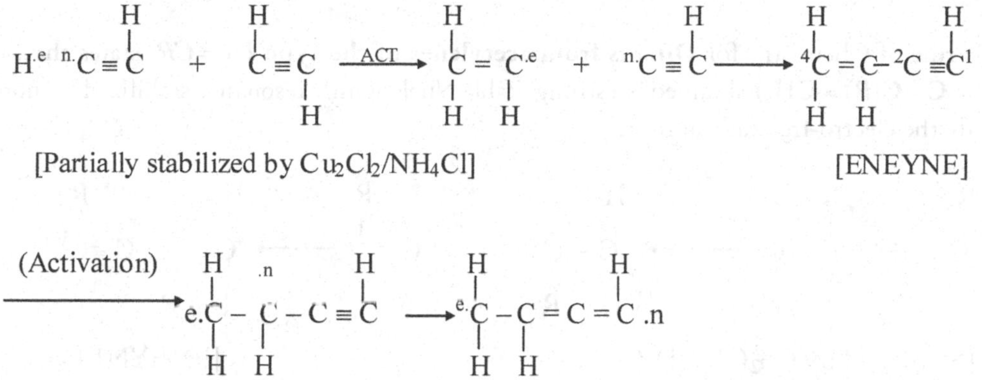

$$H^{.e \ n.}C \equiv C \quad + \quad \overset{H}{\underset{H}{C}} \equiv C \quad \xrightarrow{ACT} \quad \overset{H}{C} = \overset{.e}{\underset{H}{C}} \quad + \quad {}^{n.}C \equiv \overset{H}{\underset{H}{C}} \quad \longrightarrow \quad {}^{4}\overset{H}{C} = \overset{}{\underset{H}{C}} - {}^{2}\overset{H}{\underset{H}{C}} \equiv C^{1}$$

[Partially stabilized by Cu_2Cl_2/NH_4Cl]   [ENEYNE]

(Activation)

$$\xrightarrow{} \quad \overset{H}{\underset{H}{e.C}} - \overset{.n}{\underset{H}{C}} - C \equiv C \quad \longrightarrow \quad \overset{H}{\underset{H}{{}^{e}.C}} - \overset{H}{\underset{H}{C}} = C = \overset{H}{C}.n$$

Resonance Stabilization of the first kind of the Second type

(Laws of Creations for Eneyne)

Rule 227: This rule of Chemistry for **Chloroprene synthesis from Resonance stabilized ene by acetylene of the type HC≡C-CH=CH$_2$** states that, in the ENEYNE, the double bond which is the 3, 4-mono-form is the first to be activated being the less nucleophilic center; for which when activated the 1, 4-mono-form obtained is cumulenic in character and addition of HCl follows as shown below-

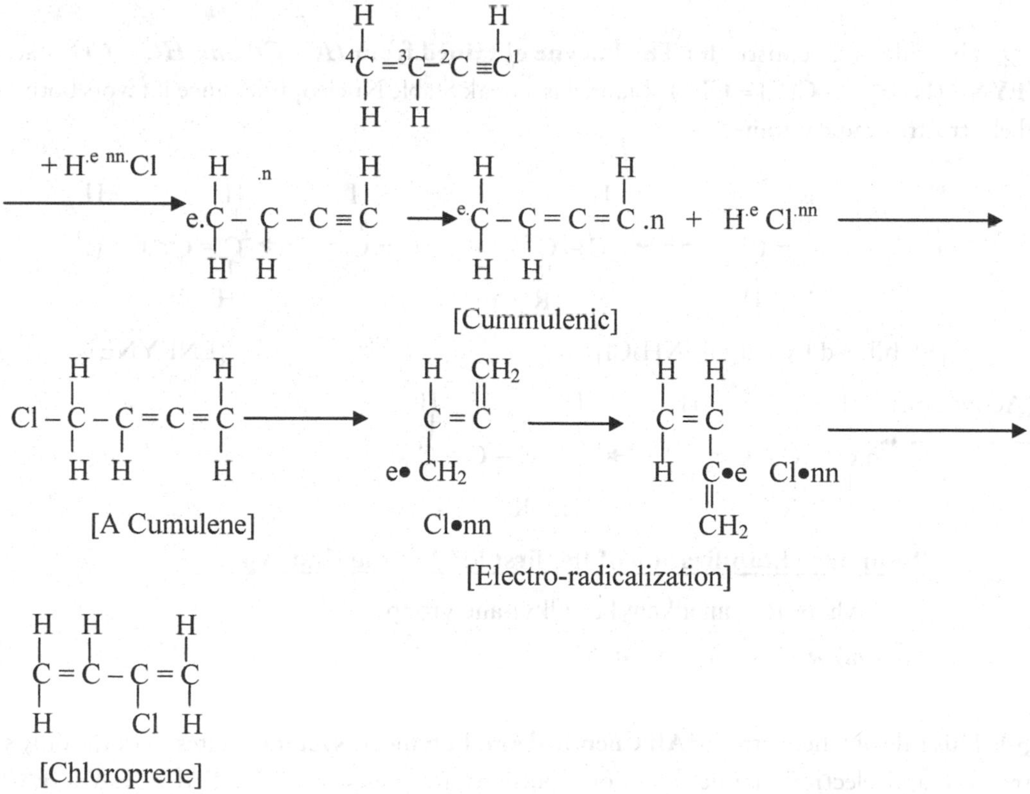

$$\overset{H}{\underset{H}{{}^{4}C}} = {}^{3}\overset{}{\underset{H}{C}} - {}^{2}\overset{H}{C} \equiv C^{1}$$

$+ H^{.e \ nn.}Cl$

$$\xrightarrow{} \quad \overset{H}{\underset{H}{e.C}} - \overset{.n}{\underset{H}{C}} - C \equiv C \quad \longrightarrow \quad \overset{H}{\underset{H}{{}^{e}.C}} - \overset{H}{\underset{H}{C}} = C = C.n \quad + \quad H^{.e}Cl^{.nn} \quad \longrightarrow$$

[Cummulenic]

$$Cl - \overset{H}{\underset{H}{C}} - \overset{H}{\underset{H}{C}} = C = \overset{H}{\underset{H}{C}} \quad \longrightarrow \quad \overset{H}{\underset{e\bullet}{C}} \overset{CH_2}{\underset{CH_2}{=}} C \quad \longrightarrow \quad \overset{H}{\underset{H}{C}} = \overset{H}{\underset{C\bullet e}{C}} \quad Cl\bullet nn \quad \longrightarrow$$
$$\qquad\qquad\qquad\qquad\qquad Cl\bullet nn \qquad\qquad\qquad\qquad CH_2$$

[A Cumulene]   [Electro-radicalization]

$$\overset{H}{\underset{H}{C}} = \overset{H}{\underset{}{C}} - \overset{}{\underset{Cl}{C}} = \overset{H}{\underset{H}{C}}$$

[Chloroprene]

a three-stage EQUILIBRIUM mechanism system, in which in the first stage the dimer was formed, followed by the second stage where 1-Chloromethyl-1,3-cumulene was formed and in the third stage, this product put in Equilibrium state of existence rearranged via a phenomenon called Electroradicalization to form the Chloroprene.

(Laws of Creations for Chloroprene synthesis from Eneyne)

Rule 228: This rule of Chemistry for **Dimers from acetylenes of the type** $HC\equiv CR$, states that, **the ENEYNE ($RC \equiv C - C(R) = CH_2$)** obtained is a strong Stable Nucleophilic resonance stabilized monomer which favors only the electro-free-radical route.

$$H^{.e} \quad ^{n.}C \equiv \overset{\overset{R}{|}}{C} \quad + \quad \overset{\overset{R}{|}}{C} \equiv \underset{\underset{H}{|}}{C} \quad \longrightarrow \quad \overset{\overset{H}{|}}{C} = \underset{\underset{R}{|}}{C}^{.e} \quad + \quad ^{n.}C \equiv \overset{\overset{R}{|}}{C} \quad \longrightarrow \quad {}^{4}\overset{\overset{H}{|}}{C} = {}^{-2}\underset{\underset{R}{|}}{C} - C \equiv C^{1}$$

[Stabilized by Cu_2Cl_2/NH_4Cl]   [ENEYNE]

(Activation)
$$\longrightarrow \quad ^{n.}\underset{\underset{H}{|}}{\overset{\overset{H}{|}}{C}} - \underset{\underset{R}{|}}{\overset{\overset{R}{|}}{C}} - C \equiv C^{.e} \quad \longrightarrow \quad ^{n.}\underset{\underset{H}{|}}{\overset{\overset{H}{|}}{C}} - \underset{\underset{R}{|}}{\overset{\overset{R}{|}}{C}} = C = C^{.e}$$

Resonance Stabilization of the first kind of the first type

where R is an alkanyl or alkyl-ane group

(Laws of Creations for the Eneyne Family)

Rule 229: This rule of Chemistry for **The Eneyne obtained from** $HC\equiv CR$ **and** $HC \equiv CH$, states that, **the ENEYNE ($HC \equiv C - C(R) = CH_2$)** obtained is a weak Stable Nucleophile, since it favors both nucleo-free and electro-free-radical routes.

$$H^{.e} \quad ^{n.}C \equiv \overset{\overset{H}{|}}{C} \quad + \quad \overset{\overset{R}{|}}{C} \equiv \underset{\underset{H}{|}}{C} \quad \longrightarrow \quad \overset{\overset{H}{|}}{C} = \underset{\underset{R}{|}}{C}^{.e} \quad + \quad ^{n.}C \equiv \overset{\overset{H}{|}}{C} \quad \longrightarrow \quad {}^{4}\overset{\overset{H}{|}}{C} = {}^{-2}\underset{\underset{R}{|}}{C} - C \equiv C^{1}$$

[Stabilized by Cu_2Cl_2/NH_4Cl]   [ENEYNE]

(Activation)
$$\longrightarrow \quad ^{n.}\underset{\underset{H}{|}}{\overset{\overset{H}{|}}{C}} - \underset{\underset{R}{|}}{\overset{\overset{H}{|}}{C}} - C \equiv C^{.e} \quad \longrightarrow \quad ^{n.}\underset{\underset{H}{|}}{\overset{\overset{H}{|}}{C}} - \underset{\underset{R}{|}}{\overset{\overset{H}{|}}{C}} = C = C^{.e}$$

Resonance Stabilization of the first kind of the first type

where R is an alkanyl or alkyl-ane group

(Laws of Creations for the Eneyne Family)

Rule 230: This rule of Chemistry for **All Chemical and Polymeric systems,** states that, the only sources of electrostatic and electrodynamic forces of repulsions are presence of i) REAL CHARGES- Ionic, Covalent and ii) Paired unbonded radicals carried by some metallic and non-metallic centers all located along the boundary.

(Laws of Creations for Electrostatic forces of Repulsion)

Rule 231: This rule of Chemistry for **Acetylene of the type HC≡C-C≡CH**, states that, this compound known as 1,3-DIYNE favors only free-radical existence; for which one acetylene group as opposed to one hydrogen atom is loosely bonded radically to its carbon center in the unactivated state, being of lower capacity than H under normal operating conditions as shown below-

$$HC≡C - C≡CH \quad \underset{\text{Existence}}{\overset{\text{Equilibrium State of}}{\rightleftharpoons}} \quad HC≡C•e \quad + \quad n•C≡CH$$

(Laws of Creations for 1,3-Diyne)

Rule 232: This rule of Chemistry for **Resonance stabilized acetylene of the type HC≡C - C≡CH,** states that, *this compound when suppressed is a **weak Nucleophile**, since it will favor both free-radical routes.*
(Laws of Creations for 1,3-Diyne)

Rule 233: This rule of Chemistry for **1,3-Diynes of the type RC≡C - C≡CH, where R is a radical -pushing group of the type C_nH_{2n+1},** states that, when suppressed, they are resonance stabilized with only two monoforms-3,4- and 1,4- mono-forms favoring only the electro-free-radical route and when not suppressed the H atom must be replaced either by radical-pushing or radical-pulling groups.
(Laws of Creations for 1,3-Diyne family)

Rule 234: This rule of Chemistry for **Resonance Stabilized Acetylenes of the type RC≡C - C≡ CH where R is a pushing group of the type OH, NH$_2$, NHR,** states that, *these monomers are unstable; for which when activated with a weak initiator, they molecularly rearrange to form Electrophiles.*
(Laws of Metaphysics)

Rule 235: This rule of Chemistry for **Acetylenes which carry substituent groups such as COOR, COR, etc. other than CN group,** states that, only their activation centers (C = O) can be activated chargedly and used positively for polymerization.
(Laws of Physics)

Rule 236: This rule of Chemistry for **Resonance stabilized acetylenes with. phenyl groups,** states that, of all the members, only the first member is unstable for which when heated, it readily decomposes to give Benzene, Carbon black as shown below-

Stage 1:

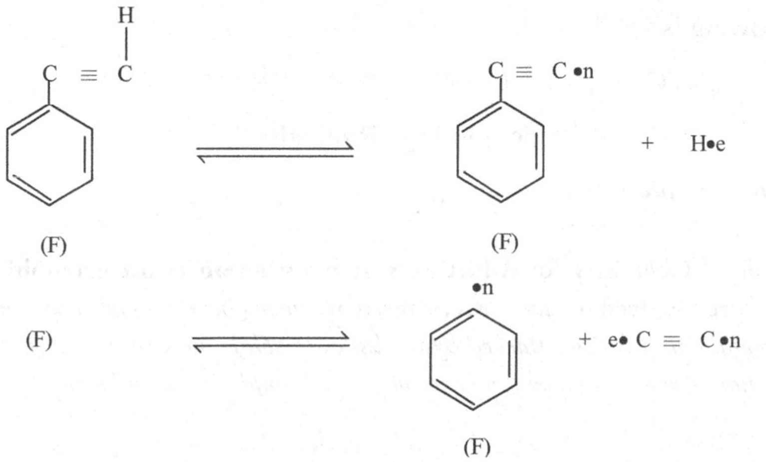

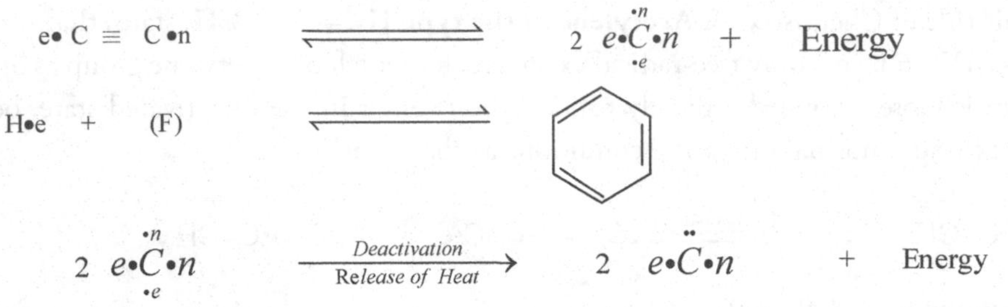

$$e \bullet C \equiv C \bullet n \quad \rightleftharpoons \quad 2 \ e \bullet \overset{\bullet n}{\underset{\bullet e}{C}} \bullet n \quad + \quad \text{Energy}$$

$$H \bullet e \quad + \quad (F) \quad \rightleftharpoons \quad \text{(benzene ring)}$$

$$2 \ e \bullet \overset{\bullet n}{\underset{\bullet e}{C}} \bullet n \quad \xrightarrow[\text{Release of Heat}]{\text{Deactivation}} \quad 2 \ e \bullet \overset{\bullet \bullet}{C} \bullet n \quad + \quad \text{Energy}$$

$$\text{Activated Carbon} \qquad\qquad\qquad \text{Carbon Black} \qquad\qquad (a)$$

Overall equation : *Phenyl Benzene* $\longrightarrow$ *Benzene* + 2*Carbon Black*

$$+ \quad Energy \,(2) \qquad\qquad (b)$$

for which the H atom on acetylene must be replaced with radical-pushing group for it to become stable. *(Laws of Creations for Acetylene Families)*

Rule 237: This rule of Chemistry for **Addition systems wherein two nucleophilic monomers of different capacities are involved in any route,** states that, *during initiation wherein none can be singly initiated <u>by a particular initiator,</u> it is the less nucleophilic monomer that is first attacked by the more nucleophilic monomer to form a couple, before the addition of the initiator or active growing center to the formed couple as shown below.*

$$N \bullet n \ + \ \overset{O^{\ominus}}{\underset{}{en \bullet S^{\oplus}}} - O \bullet nn \ + \ \overset{H_3C}{\underset{H}{e \bullet C = C \bullet n}} \longrightarrow N \bullet n \ + \ \overset{O^{\ominus} \quad H_3C}{\underset{H}{en \bullet S^{\oplus} - O - C = C \bullet n}}$$

$$\text{Less Nucleophilic} \qquad \text{More Nucleophilic} \qquad\qquad \textbf{COUPLE}$$

$$\textbf{(Unreactive)} \qquad \textbf{(Unreactive and suppressed)}$$

$$\longrightarrow \quad N - \overset{O^{\ominus} \quad H_3C}{\underset{H}{S^{\oplus} - O - C = C}} \bullet n$$

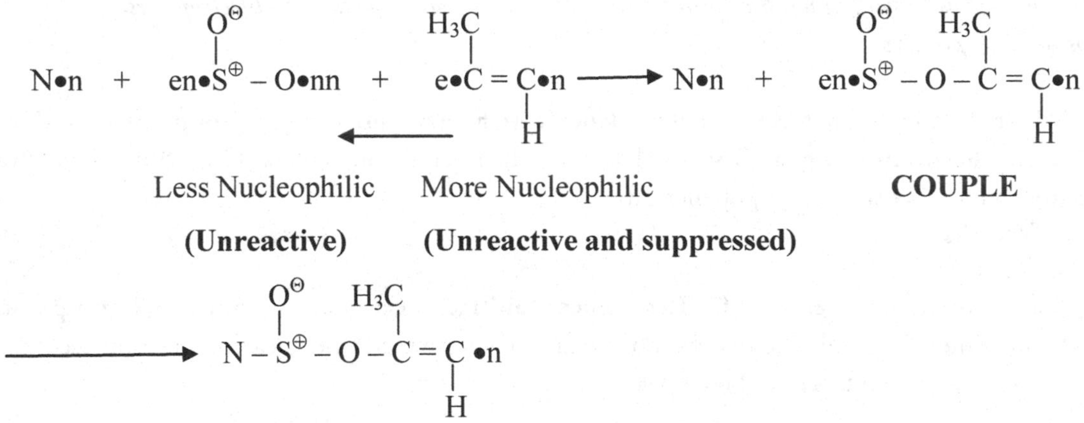

for which the following is valid-

$$C \equiv C \quad > \quad C = C \quad > \quad S = O$$

<u>Order of Nucleophilicity Radically</u>

(Laws of Creations for Addition Alternating Co-polymerization)

Rule 238: This rule of Chemistry for **Addition systems wherein two electrophilic monomers of different capacities are involved in any route,** states that, *during initiation wherein none can be initiated, it is the more electrophilic that is first attacked by the less electrophile monomer to form a couple, before the addition of the initiator or active growing center to the formed couple as shown below-*

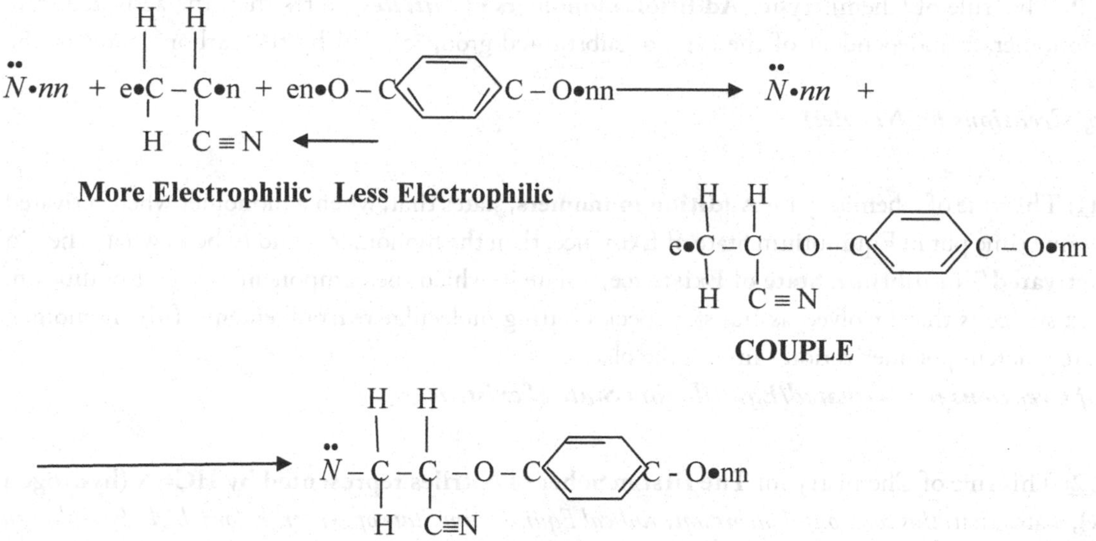

More Electrophilic   Less Electrophilic

COUPLE

noting that neither can acrylonitrile favor nucleo-non-free-radical- polymerization, nor can the benzoquinone favor its use though the initiator is natural to it, because of electrostatic forces of repulsion between the paired unbonded radicals on the initiator and on the O atom of the monomer; for which alternating placement will prevail.

(Laws of Creations for Addition Alternating Copolymerization)

Rule 239: This rule of Chemistry for **Addition systems wherein one electrophilic monomer is involved in any route,** states that, *during initiation wherein none of the activation centers can be initiated by a particular initiator, it is the MALE (i.e. Electrophilic center) that is first attacked by the FEMALE (i.e. Nucleophilic center to form a couple, before the addition of the initiator or active growing center to the formed couple as shown below.*

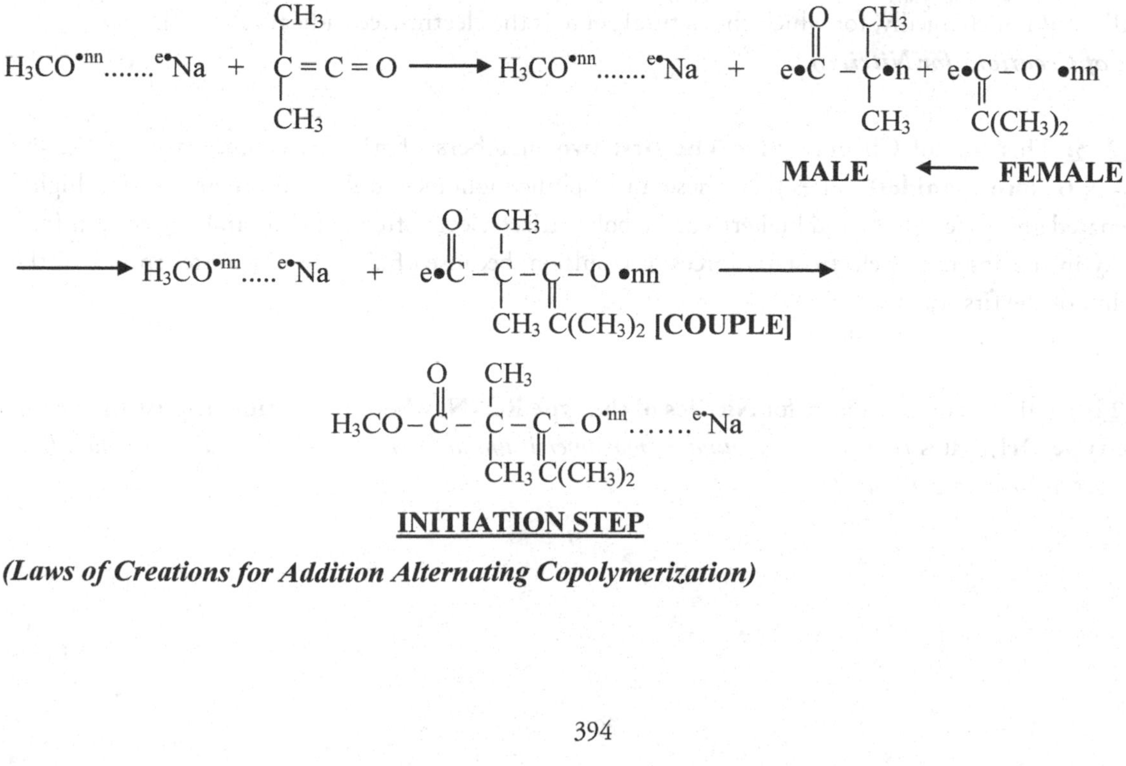

MALE ← **FEMALE**

[COUPLE]

INITIATION STEP

(Laws of Creations for Addition Alternating Copolymerization)

Rule 240: This rule of Chemistry for **Addition monomers of *Nitriles,*** states that, the activated state of the monomers is independent of the type of substituted group carried by the carbon center of the monomers.
(Laws of Creations for Nitriles)

Rule 241: This rule of Chemistry for **Addition monomers,** states that, when a monomer when activated is at the same time put in Equilibrium State of Existence, then the monomer is said to be in what is herein called **Activated/Equilibrium State of Existence,** a state in which the component held in Equilibrium state of existence is that involved as transfer species during molecular rearrangement of the monomer, and a state wherein polymerization cannot take place.
(Laws of Creations for Activated/Equilibrium State of existence)

Rule 242: This rule of Chemistry for **The first member of Nitriles represented by HC≡N (hydrogen cyanide),** states that, *this compound favors only radical Equilibrium State of existence, for which the hydrogen atom is loosely bonded to the carbon center, electro-free-radically in the unactivated state.*

$$H - C \equiv N \quad \underset{\text{\textit{Existence}}}{\overset{\text{\textit{Equilibrium State of}}}{\rightleftharpoons}} \quad H \bullet e \quad + \quad n \bullet C \equiv N$$

(Laws of Creations for Nitrile (Hydrogen Cyanide))

Rule 243: This rule of Chemistry for **HC≡N**, states that, this compound if suppressed can only be activated by initiators whose capacity is greater than or equal to $H^{\bullet e}$, otherwise the H atom must be replaced free-radically for it to become a monomer.
(Laws of Creations for Nitrile)

Rule 244: This rule of Chemistry for ***Nitriles* of the type RC≡N where R is a free-radical-pushing group of the type C_nH_{2n+1} only,** states that, these compounds are monomers which can be activated only radically and not chargedly, for which the natural route is the electro-free-radical route.
(Laws of Creations for Nitriles)

Rule 245: This rule of Chemistry for **The first two members of nitriles represented by FC≡N, $F_3CC≡N$ (fluoro cyanides),** states that, these nucleophilic monomers unlike all others carrying higher halogenated groups (e.g. C_2F_5 and higher), can be polymerized electro-free-radically and nucleo-non-free-radically in the absence of electrostatic forces of repulsion, because of absence of transfer species of the first kind of the first type.
(Laws of Creations for Nitriles)

Rule 246: This rule of Chemistry for **Nitriles of the type RC≡N, where R is a radical-pushing group of the type OH,** states that, *this compound or monomer is unstable, since when activated, it molecularly rearranges to form isocyanic acid.*

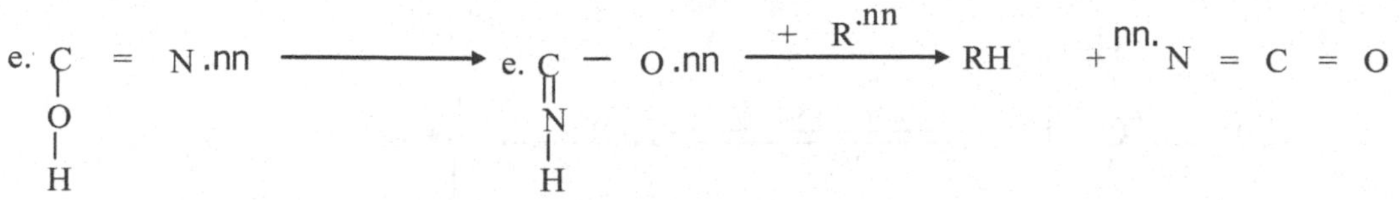

e.
$$
\begin{array}{c}
\mathrm{C} = \mathrm{N}\,.nn \\
|\ \\
\mathrm{O} \\
|\ \\
\mathrm{H}
\end{array}
\longrightarrow
\begin{array}{c}
\text{e. } \mathrm{C} - \mathrm{O}\,.nn \\
\|\ \\
\mathrm{N} \\
|\ \\
\mathrm{H}
\end{array}
\xrightarrow{\ +\ R^{.nn}\ }
\mathrm{RH} \quad +\ ^{nn.}\mathrm{N} = \mathrm{C} = \mathrm{O}
$$

(I)   (II)   (More stable)

<u>Isocyanic acid</u>

(Laws of Creations for Isocyanic acid)

Rule 247: This rule of Chemistry for **Nitriles of the type RC≡N, where R is a radical-pushing group of the NH₂ type,** states that, *this compound or monomer is slightly stable, since it is the more unstable carbodiimide (HN=C=NH) that molecularly rearranges to the nitrile.*
(Laws of Creations for Carbodiimide)

Rule 248: This rule of Chemistry for **Nitriles of the type R`C≡N where R` is a radical-push-ing group of the type OR, NR₂ (Rs are alkyl groups),** states that, *since N≡C > C = O in nucleophilic capacity, molecular rearrangement takes place when the R groups are less pushing or not bulky in size, a clear indication that the followings are valid.*

$$[-OC_3H_7 < -OC_2H_5 < -OCH_3 < -OH] < [=NC_3H_7 > = NC_2H_5 > = NCH_3 > = NH]$$
<u>Order of Radical Pushing capacity.</u>
(Laws of Creations for Radical-pushing groups)

Rule 249: This rule of Chemistry for **Nitriles of the type R_FC≡N where R_F are pulling free group of the type COOCH₃, COR, CONH₂, NO₂, SO₂R,** states that, *these compounds are monomers which favor activations chargedly and radically only via the centers of R_F; for which only the electro-free-radical and positively charged routes are favored.*
(Laws of Creations for Nitriles)

Rule 250: This rule of Chemistry for **Nitriles during Polymerization,** states that, *in general, there are four major types of transfer species- of the first kind of the first type, first kind of the second type, first kind of the third type, and second kind; with each kind serving various functions for a monomer, and a growing polymer chain.*
(Laws of Creations for Transfer Species for Polymerization of Nitriles)

Rule 251: This rule of Chemistry for **Polar/Ionic compounds such as H₂O, HCl, NaOH, NH₃, Ca(ONO₂)₂ and so on,** states that, all of them cannot be expressed ionically when they exist in DECOMPOSITION and COMBINATION States of Existence, but only in EQUILIBRIUM State of Existence; for which when they are electrostatically placed, they cannot be expressed ionically or chargedly in any State of Existence when combined with only WATER, but only radically as shown below-

$$HO^{\ominus}......\overset{\overset{\textstyle H}{|}}{\underset{\underset{\textstyle H}{|}}{O}}-H \quad \underset{Existence}{\overset{Equilibrium\ State\ of}{\rightleftharpoons}} \quad HO^{\ominus}........\overset{\overset{\textstyle H}{|}}{\underset{\underset{\textstyle H}{|}}{O}}\bullet nn \quad + \quad e\bullet H$$

Water Dimer

$$Cl^{\ominus}........\overset{\overset{\textstyle H}{|}}{\underset{\underset{\textstyle H}{|}}{O}}-H \quad \underset{Existence}{\overset{Equilibrium\ State\ of}{\rightleftharpoons}} \quad Cl^{\ominus}........\overset{\overset{\textstyle H}{|}}{\underset{\underset{\textstyle H}{|}}{O}}\bullet nn \quad + \quad e\bullet H$$

50/50 Dilute Hydrochloric Acid

$$HO^{\ominus}.....\overset{\overset{\textstyle H}{|}}{\underset{\underset{\textstyle H\ H}{|}}{N}}-H \quad \underset{Existence}{\overset{Equilibrium\ State\ of}{\rightleftharpoons}} \quad HO^{\ominus}........\overset{\overset{\textstyle H\ H}{\diagup}}{\underset{\underset{\textstyle H}{|}}{N}}\bullet nn \quad + \quad e\bullet H$$

50/50 Ammonium Hydroxide

$$HO^{\ominus}.......\overset{\overset{\textstyle H}{|}}{\underset{\underset{\textstyle H}{|}}{O}}-Na \quad \underset{Existence}{\overset{Equilibrium\ State\ of}{\rightleftharpoons}} \quad HO^{\ominus}........\overset{\overset{\textstyle H}{|}}{\underset{\underset{\textstyle H}{|}}{O}}\bullet nn \quad + \quad e\bullet Na$$

50/50 Dilute Sodium hydroxide

for which whether radically or chargedly placed, the ionic character still remains inherently hidden (i.e., imaginary).
(Laws of Creations for Polar/Ionic Compounds)

Rule 252: This rule of Chemistry for **Dead polymers from Acetylenic growing polymer chains,** states that, the followings are the types of dead terminal bonded polymers produced from them from within-

    (a) Dead terminal triple bond polymers
    (ii) Dead terminal 1,3-or 1,4- cumulenic bond polymers
    (iii) Dead terminal 1,3- or 1,4- ketinic/amidic bond polymers
(Laws of Creations for Acetylenic Dead Terminal double-bonded Polymers)

Rule 253: This rule of Chemistry for **Dead polymers from Olefinic growing polymer chains,** states that, the followings are the types of dead terminal bonded polymers produced from them from within-

    (i) Dead terminal double bond polymers
    (ii) Dead terminal 1,3 – ketinic/amidic bond polymers
    (iii) Dead terminal conjugated double bond polymers.
(Laws of Creations for Olefinic Dead Terminal double-bonded Polymers)

Rule 254: This rule of Chemistry for **Dead polymers from Nitrilic growing polymer chains,** states that, the followings are the types of dead terminal bonded polymers produced from them from within-

  (b) Dead terminal triple bond polymers
  (ii) Dead terminal 1,3 – isocyanic bonded polymers
  (iii) Dead terminal 1,3 – cumulative nitrilic bonded polymers
  (iv) Dead terminal 1,4 – cumulative nitrilic bonded polymers
(Laws of Creations for Nitrilic Dead Terminal double-bonded Polymers)

Rule 255: This rule of Chemistry for **Resonance Stabilized growing polymer chain,** states that, when there exists a substituent group that is shielded by a resonance stabilization group, transfer species of the first or second kind can never be released from it for its growing polymer chain to kill the chain; for which living chains are produced.
(Laws of Creation for Resonance Stabilized Growing polymer chain)

Rule 256: This rule of Chemistry for **Hydrobromination of $RC \equiv CH$ in the presence of hydrogen peroxide,** states that, the followings are obtained-

$$RC \equiv CH \;+\; HBr \quad \xrightarrow{\;Peroxides\;} \quad RCII = CHBr$$

whereas in the absence of hydrogen peroxide, $H_2C = CRBr$ is the product; for which the mechanism for the reaction in the presence of peroxide is as follows.
Decomposition mechanism

Stage 1: $\qquad HOOH \xrightarrow{\quad\quad} 2HO.nn$

$\qquad\qquad 2HO.nn \xrightarrow{\quad\quad} 2H.n \;+\; 2\, nn.O.en$

$\qquad 2\, nn.O.en +\; 2HBr \xrightarrow{\quad\quad} 2HO.nn \;+\; 2Br.en$

$$2\,Br.en + 2\, n.\; \overset{\displaystyle H}{\underset{\displaystyle R}{C}} = C .e \;\rightleftharpoons\; 2\,Br - \overset{\displaystyle H}{\underset{\displaystyle R}{C}} = C .e$$

$$\qquad\qquad\qquad (A) \qquad\qquad\qquad\qquad (B)$$

(B) + 2 H.n $\longrightarrow$ 2 BrHC = CHR

2 HO.nn $\longrightarrow$ HOOH

(a)

Equilibrium mechanism

Stage 1:   RC $\equiv$ CH $\rightleftharpoons$ (A)

(A) + HBr $\rightleftharpoons$ n• $\overset{\text{H}}{\underset{\text{R}}{\text{C}}}$ = C – H   +   Br •en

$\longrightarrow$ BrHC = CHR

(b)

Overall Equation: HOOH + HBr + RC$\equiv$CH $\longrightarrow$ BrHC = CHR + HOOH

for which the mechanism above is Equilibrium mechanism (b) and not (a), since oxidizing oxygen cannot exist via Decomposition mechanism; for which when HOOH is not involved as a passive catalyst, it is so involved actively in three stages with increased molar concentration (production of HOBr, HOOBr as intermediates) without the existence of itself, but only as water and oxidizing oxygen molecules (but not as HOOH).
(Laws of Creations for Decomposition mechanism)

Rule 257: This rule of Chemistry for **Synthesis of Acrylonitrile from Acetylene,** states that, the mechanism of the reaction is as follows-

$$HC \equiv CH \; + \; H - C \equiv N \; \xrightarrow{(CuCl, NH_4Cl, HCl) - 90^0 C} H_2C = CHCN$$

(d)

Stage 1:   H – C $\equiv$ N $\rightleftharpoons$ H.e   +   n. C $\equiv$ N

H.e   +   H – C $\equiv$ C – H $\xrightarrow{\text{Activation}}$ H – $\overset{\text{H}}{\underset{}{\text{C}}}$ = $\overset{\text{H}}{\underset{}{\text{C}}}$ .e

(I)

N $\equiv$ C.nn   +   (I) $\longrightarrow$ $\overset{\text{H}\;\;\text{H}}{\underset{\text{H}\;\;\text{C}\equiv\text{N}}{\text{C} = \text{C}}}$

(b)

Overall Equation: HCN   +   HC $\equiv$ CH $\xrightarrow{\text{Passive Cat.}}$ H₂C = CHCN

(c)

A single stage Equilibrium mechanism system, noting that while the passive catalysts- the chlorides suppressed the Equilibrium State of existence of acetylene, they could not suppress that of hydrogen cyanide.
(Laws of Creations for Synthesis of Acrylonitrile from Acetylene)

Rule 258: This rule of Chemistry for **Synthesis of Acrylic acid from Acetylene,** states that, *the reaction which involves the hydrocarbonylation of acetylene in the presence of nickel carbonyl is as follows-*

$$HC \equiv CH \ + \ CO \ + \ H_2O \ \xrightarrow{Ni(CO)_4, HCl} \ H_2C = CHCOOH \tag{a}$$

Stage 1:
$$H_2O \ \rightleftharpoons \ H.e \ + \ nn.OH$$

$$H.e \ + \ CO \ \xrightleftharpoons{\text{Activation}} \ H - \overset{\overset{\displaystyle O}{\|}}{C}.e$$
$$(I)$$

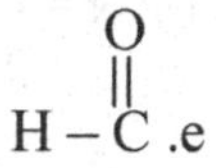

$$HO.nn \ + \ (I) \ \longrightarrow \ H - COOH \tag{b}$$

Overall Equation: $H_2O \ + \ CO \ \longrightarrow \ HCOOH$ (c)

Stage 2:
$$HCOOH \ \rightleftharpoons \ H.e \ + \ n.COOH$$

$$H.e \ + \ H - C \equiv C - H \ \xrightleftharpoons{\text{Activation}} \ H - \overset{\overset{\displaystyle H}{|}}{C} = \overset{\overset{\displaystyle H}{|}}{C}.e$$
$$(II)$$

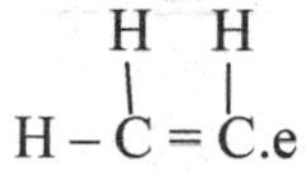

$$HOOC.n \ + \ (II) \ \longrightarrow \ \begin{array}{c} \overset{H}{|} \quad \overset{H}{|} \\ C = C \\ \overset{|}{H} \quad \overset{|}{C} = O \\ \quad \overset{|}{O} \\ \quad \overset{|}{H} \end{array} \tag{d}$$

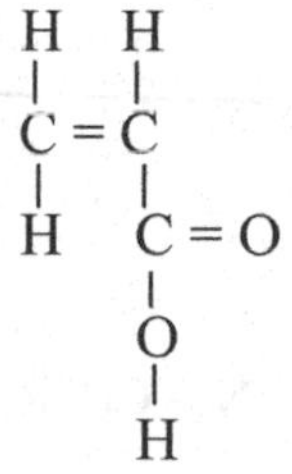

Overall Equation: $HCOOH \ + \ H - C \equiv C - H \ \longrightarrow \ H_2C = CHCOOH$ (e)

a two stage Equilibrium mechanism system, in which the Equilibrium state of existence of acetylene is suppressed by the nickel carbonyl; noting the Equilibrium State of Existence of formic acid, the first member of the carboxylic acids and the fact that when aqueous CH_3OH is used in place of H_2O, methyl acrylate an ester is similarly obtained.
(Laws of Creations for Synthesis of Acrylic Acid from Acetylene)

Rule 259: This rule of Chemistry for **Dialkycyanamides ($N \equiv C - NR_2$),** states that, when hydrolyzed by aqueous acids or bases, N, N-di-alkylcarbamic acid obtained spontaneously loses carbon dioxide as shown below-

$$R_2N - C \equiv N + 2H_2O \ \xrightarrow{H^\oplus \ Or \ \ominus OH} \ NH_3 + [R_2NCOOH] \ R_2NH + CO_2 \tag{a}$$
the mechanisms of which are as follows-

Stage 1:

$$H_2O \rightleftharpoons H.e \ + \ nn.OH$$

$$H.e \ + \ N \equiv C - NR_2 \rightleftharpoons \underset{\underset{NR_2}{|}}{H - N = C.e}$$

(I)

$$HO.nn \ + \ (I) \longrightarrow \underset{\underset{NR_2}{|}}{H - N = C - OH}$$

(II) (b)

Overall Equation: $H_2O \ + \ N \equiv C - NR_2 \longrightarrow H - N = C(NR_2) - OH$ (c)

Stage 2:

$$H_2O \rightleftharpoons H.e \ + \ nn.OH$$

$$H.e \ + \ H - N = C(NR_2) - OH \rightleftharpoons \underset{\underset{NR_2}{|}}{\overset{\overset{H \ \ OH}{| \ \ \ |}}{H - N - C.e}}$$

(III)

$$HO.nn \ + \ (III) \longrightarrow \underset{\underset{NR_2}{|}}{\overset{\overset{OH}{|}}{H_2N - C - OH}}$$

(IV)

Overall Equation: $H_2O \ + \ H - N = C(NR_2) - OH \longrightarrow (IV)$ (e)

Stage 3:

$$\textbf{(IV)} \rightleftharpoons \underset{\underset{\mathbf{NR_2}}{|}}{\overset{\overset{\mathbf{OH}}{|}}{\mathbf{H_2N - C - O.nn}}} \ + \ \mathbf{H.e}$$

(V)

$$\textbf{(V)} \rightleftharpoons \underset{\underset{NR_2}{|}}{\overset{\overset{NH_2}{|}}{e.C - O.nn}} \ + \ nn.OH$$

(VI)

$$H.e \ + \ nn.OH \rightleftharpoons H_2O$$

(a)
(b)

$$\textbf{(VI)} \xrightarrow{\text{De-Activation}} \underset{\underset{NR_2}{|}}{\overset{\overset{NH_2}{|}}{C = O}} \ + \ \text{Heat}$$

(VII) (f)
(c)

401

Stage 4:   **Water exists in Equlibrium state of existence to form NH₃ and HOOCNR₂.**

$$\text{Stage 4: } Water exists...$$

Stage 4: **Water exists in Equlibrium state of existence to form NH₃ and HOOCNR₂.**

Overall Equation: (IV) $\longrightarrow$ NH_3 + $HOOCNR_2$ (g)

Stage 5:   **(VII)** $\rightleftharpoons$ **H.e**   +   **nn.O – C = O**

$$| \\ NR_2$$

(VIII)

$$\overset{\displaystyle O}{\underset{\displaystyle \|}{}}$$

(VIII) $\rightleftharpoons$ $R_2N.nn$ + $e.C – O.nn$

(IX)

H.e   + $nn.NR_2$ $\rightleftharpoons$ HNR_2

(IX) $\xrightarrow{\text{De-Activation}}$ CO_2 +   Heat   (h)

Overall Equation: $HOOCNR_2$ $\longrightarrow$ CO_2 + HNR_2 + Heat   (i)

a five stage Equilibrium mechanism system, noting the Equilibrium State of Existence compounds and how Nature operates.

(Laws of Creations for Dialkylcyanamides)

Rule 260: This rule of Chemistry for **Imidic acid [R(OH)C=NR′],** states that, this compound cannot readily be isolated, because it readily molecularly rearranges (not via electroradicaliza-tion) to an amide as shown below-

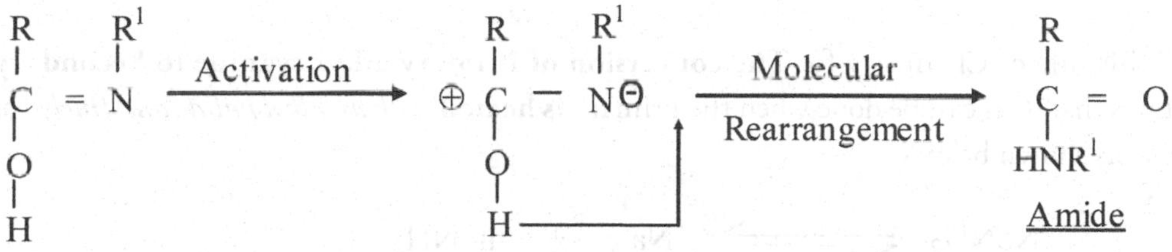

(I) Imidic acid

wherein the molecular rearrangement is of the first kind of the first type, that which takes place chargedly and radically; noting that when the amide is used as a monomer, the routes are only the electro-free-radical and positively charged routes

(Laws of Creations for Imidic acid of the Nitrile Family Tree)

Rule 261: This rule of Chemistry for **Imide chloride,** states that, when this stable derivative of Imidic acid is made to react with ammonia, Amidine is obtained as shown below-.

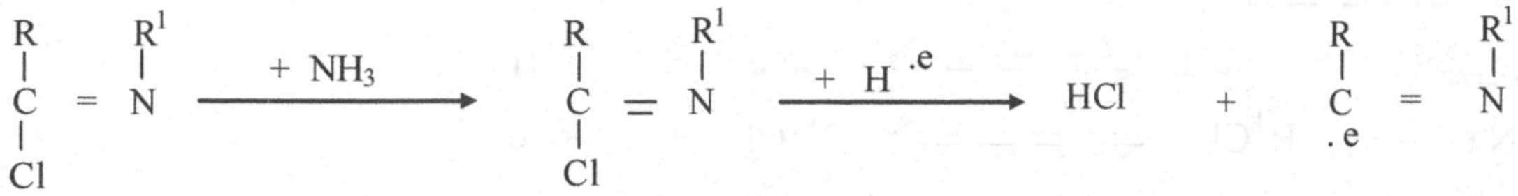

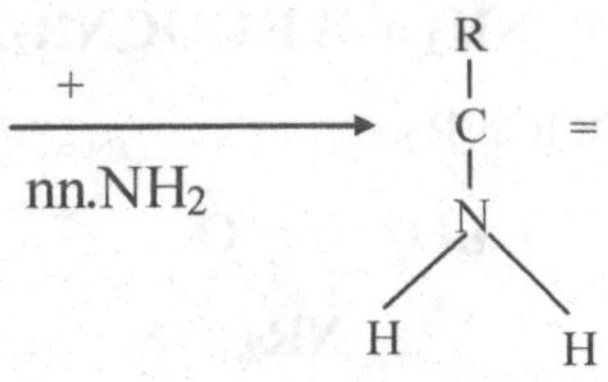

$$R \quad R^{I}$$
$$\xrightarrow{\hspace{2cm}} \quad C = N$$
$$+ \qquad \qquad \qquad |$$
$$nn.NH_2 \qquad \qquad N$$
$$\qquad \qquad \qquad H \quad H$$

Amidine

a one stage Equilibrium mechanism system in which the transfer species abstracted, $Cl^{\ominus}$ is of the second first kind of the first type.
(Laws of Creations for Imide Chloride of the Nitrile Family Tree)

Rule 262: This rule of Chemistry for **The Diaminealkane shown below,** states that, its Equili-brium state of existence is as follows-

$$R - CH - NHR \quad \rightleftharpoons \quad \begin{array}{c} H \quad R \\ | \quad | \\ R - C - N \bullet nn \\ | \\ H_2N \end{array} + \quad H.e$$
$$\qquad | \qquad \qquad$$
$$\qquad NH_2$$

EQUILIBRIUM STATE OF EXISTENCE

that in which it is the poorer N center with respect to H when more than one N center is present on a Central carbon center, that is the source of Equilibrium state of Existence.
(Laws of Creations for a Diaminealkane)

Rule 263: This rule of Chemistry for **The conversion of Primary alkyl cyanide to Secondary alkyl cyanide,** states that, this can be done when the primary **is heated** *with an alkyl halide and finely powdered sodium amide* as shown below-

Stage 1: $\qquad NaNH_2 \rightleftharpoons Na.e + nn.NH_2$

$$Na.e \quad + \quad \begin{array}{c} C \equiv N \\ | \\ H - CHR \end{array} \quad \rightleftharpoons \quad NaH \quad + \quad \begin{array}{c} C \equiv N \\ | \\ e.CHR \end{array}$$

(I)

$$nn.NH_2 \quad + \quad (I) \quad \xrightarrow{\hspace{2cm}} \quad \begin{array}{c} C \equiv N \\ | \\ H_2N - CHR \end{array}$$

(II)

Overall Equation: $NaNH_2 + H_2RC-C \equiv N \xrightarrow{\hspace{1cm}} NaH + H_2NHRC-C \equiv N$

Stage 2: $\qquad NaH \rightleftharpoons Na.e + n.H$

$$Na.e \quad + \quad R^1Cl \quad \rightleftharpoons \quad NaCl \quad + \quad R^1.e$$

$$R^1.e \quad + \quad n.H \quad \xrightarrow{\hspace{2cm}} \quad R^1H$$

Overall Equation: $NaH \ + \ R^1Cl \longrightarrow NaCl \ + \ R^1H$ (d)

Stage 3: $R^1H \rightleftharpoons H.e \ + \ n.R^1$

$$H.e \ + \ (II) \rightleftharpoons \underset{\underset{(III)}{e.CHR}}{C \equiv N} \ + \ NH_3$$

(e)

$$R^1.n \ + \ (III) \longrightarrow \underset{\underset{(IV)}{R^1CHR}}{C \equiv N}$$

Overall overall equation: $NaNH_2 \ + \ R^1Cl \ + \underset{RCH_2}{C \equiv N} \longrightarrow \underset{RCHR^1}{C \equiv N} + NaCl$

(f)

a three stage Equilibrium mechanism system, wherein the $C\equiv N$ center was never activated; noting that the same mechanism as above applies for the conversion of secondary alkyl cyanide to tertiary alkyl cyanide; noting the involvement of transfer species of the second first kind of the third type.

(Laws of Creations for Alkyl Cyanides)

Seventy five rules have been proposed to cover acetylenes, nitriles, Addition systems and much more. Systematically, the nucleophilic and electrophilic characters of activation centers are being carefully identified, through all the new concepts and phenomena. For the first time, the pushing or pulling (which indeed are rightly called radical-pushing or pulling) capacities of alkyl, olefinic, acetylenic, nitrilic and more groups have begun to be identified. All these information will be very useful when showing how different types of copolymers are produced. From all the considerations, one can begin to appreciate the importance of the phenomenon of molecular arrangement, electroradicalization yet to be fully identified and defined, and the need for proposition of rules, which as can be noticed so far are natural laws inside chemistry.

Little or nothing has in the past been known about resonance stabilization phenomena in organic compounds including monomers. While it can be observed that, nitriles like aldehydes and ketones cannot be resonance stabilized, acetylenes can be resonance stabilized with nothing and little to shield, based on the character. However, acetylenic groups can be used as resonance stabilization groups, for alkylenic groups. For the first time, it has been shown that acetylenes, nitriles and other triply bonded compounds cannot be activated chargedly. Why these and more could not be observed since the development of Science can clearly be seen, since all the things contained in the NEW FRONTIERS are new to humanity. In all the considerations so far, despite the identifications of driving forces favoring the existence of ionic bonds, situations exist where a polar/ionic compound cannot exist ionically in Equilibrium State of Existence. These are all indications of the complexity of NATURE.

References

1. G. Odian, "Principles of Polymerization", McGraw Hill Book Company, (1970), pgs. 353 - 355.

2. C. R . Noller, "Textbook of Organic Chemistry," W. B . Saunders Company, (19 a66), pgs. 146 - 151.

3. C. R . Noller, "Textbook of Organic Chemistry," W. B. Saunders Company, (1966), pgs. 236 – 240

Problems

6.1 Distinguish between

(a) acetylene and ethylene
(b) 1,3 - butadiene and 1,3 - butadyne

$$\begin{array}{ccccccc}
H & & H & & H & & H \\
| & & | & & | & & | \\
C & = & C & - & C & = & C \\
| & & | & & | & & | \\
H & & H & & H & &
\end{array}
\qquad
\begin{array}{ccccccc}
H & & & & & & H \\
| & & & & & & | \\
C & \equiv & C & - & C & \equiv & C \\
& & & & & & | \\
& & & & & & H
\end{array}$$

6.2 (a) Distinguish between transfer species of the first kind of the

(i) first type
(ii) second type
(iii) third type
Give examples.

(b) What are Free-media and Paired-media initiators?

6.3. (a) Explain why $HC \equiv CH$ and $HC \equiv N$ favor radical existence in the non-activated and activated states, but not charged existence?
   (b) Under what conditions do they favor radical existence in non-activated state? Explain
   (c) Under what conditions can they be activated radically?
   (d) Can they be activated chargedly? Explain.

6.4 Consider the following monomers shown below.

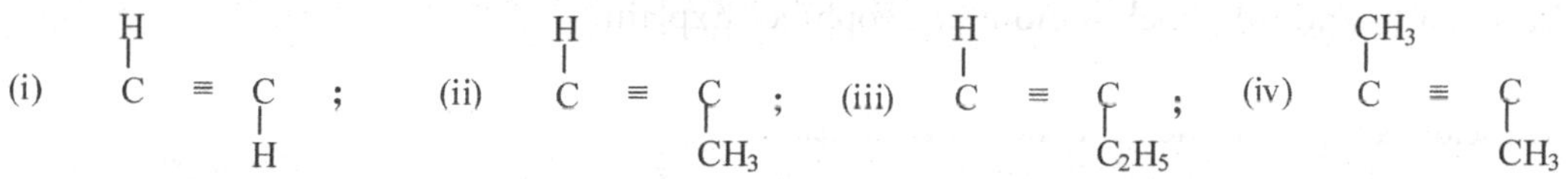

$$\text{(i)} \quad \begin{array}{c} H \\ | \\ C \\ | \\ H \end{array} = C \; ; \quad \text{(ii)} \quad \begin{array}{c} H \\ | \\ C \\ \end{array} \equiv \begin{array}{c} \\ C \\ | \\ CH_3 \end{array} \; ; \quad \text{(iii)} \quad \begin{array}{c} H \\ | \\ C \\ \end{array} \equiv \begin{array}{c} \\ C \\ | \\ C_2H_5 \end{array} \; ; \quad \text{(iv)} \quad \begin{array}{c} CH_3 \\ | \\ C \\ \end{array} \equiv \begin{array}{c} \\ C \\ | \\ CH_3 \end{array}$$

You are also provided with the three initiators shown below.

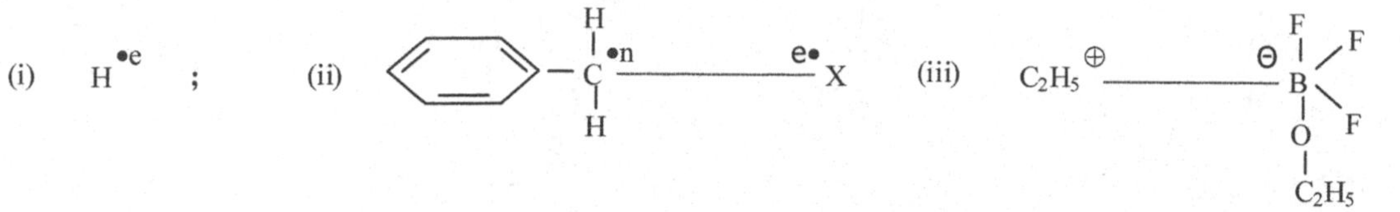

(i) $H^{\bullet e}$; (ii) [benzene ring]$- \overset{\overset{\displaystyle H}{|}}{\underset{\underset{\displaystyle H}{|}}{C}} \overset{\bullet n}{} \underline{\qquad\qquad} \overset{e\bullet}{X}$ (iii) $C_2H_5 \overset{\oplus}{} \underline{\qquad\qquad} \overset{\ominus}{B} \begin{array}{c} F \\ | \diagup F \\ \\ | \diagdown F \\ O \\ | \\ C_2H_5 \end{array}$

406

(a) Which of the monomers will favor polymerization using any of the three initiators?

(b) Show the order of the nucleophilic characters of the monomers.

(c) Compare (ii) with propylene (propene) very briefly.

6.5 Shown below are the following monomers.

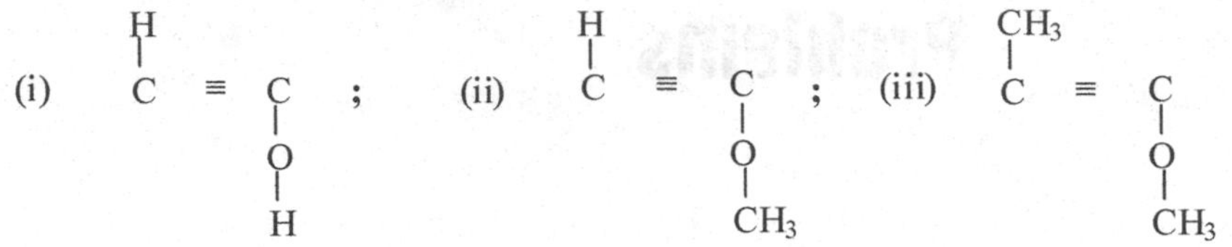

(a) Which of the monomers favor molecular rearrangement phenomenon? Explain.

(b) Using the initiators of Q 6.4, which of the monomers will favor polymerization if molecular rearrangement is allowed.

(c) What are the transfer species involved in the initiation/propagation of the monomers?

6.6 (a) With two nucleophiles of carbon-carbon types adjacently located such as shown below (i) and (ii), resonance stabilization can be provided, but with two nuclephiles of carbon-hetero atom types also adjacently located (iii) and (iv), it cannot be provided. Why?

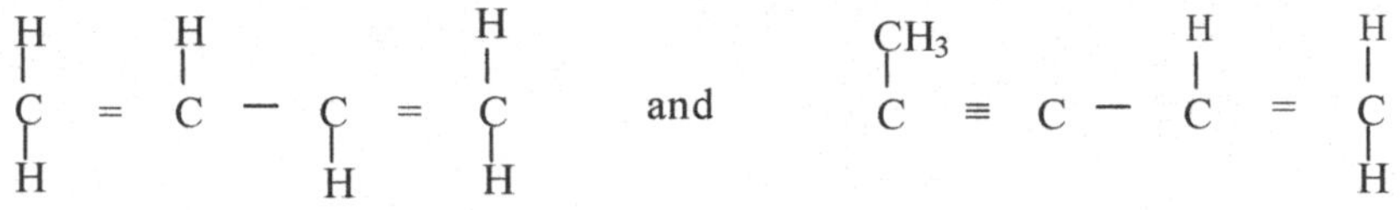

Resonance stabilization can be provided

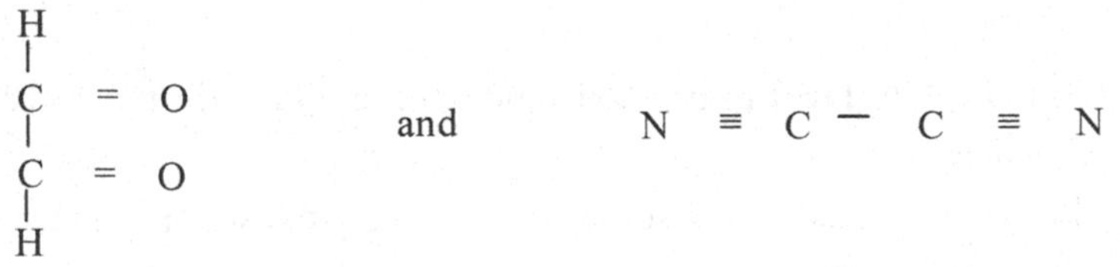

Resonance stabilization can be provided

(b) Compare the resonance stabilized forms of (ii) with those of (i).

(c) Between (i) and (ii) which is more nucleophilic? Explain.

6.7 Shown below are an amidic acid and derivatives.(i)

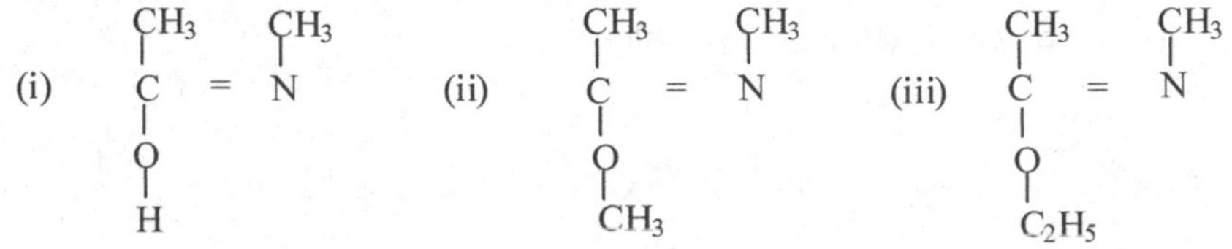

(a) Show the order of their stabilities in terms of their abilities to favor molecular rearrangement.

(b) Identify their characters and the types of transfer species involved using the various types of initiators identified so far.

6.8 Shown below are two monomers.

(i) $O = C = O$ (ii) $\underset{\underset{H}{|}}{N} = C = \overset{\overset{H}{|}}{N}$

(a) Identify the similarities between the two monomers.

(b) Why is (i) stable and (ii) is unstable? Explain

(c) Between (i) and (ii), which is more nucleophilic? Explain.

(d) What are the substituted groups in the monomers? Identify their characters and capacities.

(e) Will the stability of (ii) be affected when one H or two Hs are replaced with CH_3? Explain.

6.9 (a) Show three methods, by which acetylenes can be synthesized, based on application of the rules already proposed.

(b) Identify all the possible chemical tools involved in all the three methods.

6.10 Distinguish between the types of dead terminal bonded polymers produced from acety-lenes and those from mono-olefins.

6.11 (a) Distinguish between the types of dead terminal bonded polymers produced when transfer species of the first kinds and second kind are involved, in terms of the routes favored by them during polymerization of the originating monomers.

(b) Distinguish between transfer species of the first second kind of the
(i) first type
(ii) second type
(iii) third type
when monomers are not activated. Use examples and reactions.

6.12 (a) Why can't acetylides not favor their use as monomers?

(b) Distinguish between ionic and non-ionic metals.

(c) Why is $AgC \equiv CH$ more stable than $NaC \equiv CH$?

(d) Show if the monomer shown below favors any free-radical route.

$$\underset{\underset{CF_3}{|}}{\overset{\overset{CH_3}{|}}{C}} \equiv C$$

6.13 (a) Shown below are two classes of pulling groups:-

A   COOR ,   COR , $CONH_2$, $C \equiv N$
B NO_2 , SO_2R , C_2F_5

(i) Are they of the free types or non-free types? Identify or classify them.

(ii) What is or are the distinction(s) between the two groups?

(b) Shown below are four monomers.

$$
\begin{array}{cccc}
\underset{|}{C} \equiv N\ ; & \underset{|}{C} \equiv N\ ; & \underset{|}{C} \equiv N\ ; & \underset{|}{C} \equiv N \\
\underset{|}{C}=O & CF_3 & F & O \\
O & & & | \\
| & & & C_2H_5 \\
H & & & \\
(I) & (II) & (III) & (IV)
\end{array}
$$

Why are (II), (III) and (IV) known to exist, but not that of (I)? Explain as clearly as possible.

6.14. When propylene reacts with hydrogen bromide in the presence of peroxides, n-propyl bromide is formed instead of i-propyl bromide.

$$ CH_3CH = CH_2 + HBr \xrightarrow{\text{(Peroxides)}} CH_3\,CH_2\,CH_2\,Br $$

The same also applies to acetylenes

$$ RC \equiv CH + HBr \xrightarrow{\text{(Peroxides)}} RCH = CHBr $$

Explain tentatively the mechanism of the reactions.

6.15 (a) $ RC \equiv CH + HBr \longrightarrow RCBr = CH_2 \xrightarrow{\ +\ HBr\ } $

$ RCBr_2 - CH_3 \quad OR \quad RHCBr - CH_2Br $

(b) $ RC \equiv CH + Br_2 \longrightarrow RCBr = CHBr \xrightarrow{\ +\ Br_2\ } $

$ RCBr_2 - CHBr_2 $

(c) $ NaCN + Cl_2 \longrightarrow ClCN + NaCl $

Explain the mechanism of the three reactions above. Why can the reactions not take place chargedly? In (a), under what conditions are the two different products obtained?

6.16 (a) Why is the H atom in acetylene loosely bonded free-radically and not chargedly?

(b) Why is the H atom in nitrile (HCN) loosely bonded radically and also not chargedly? Can NaCN be used as a charged initiator in the same manner as H_9C_4Li is used?

(c) Why is the hydrogen atom in acetylene not bonded nucleo-free-radically ($H \cdot^n$), but electro-free-radically ($H \cdot^e$)?

(d) Why is H in $HC \equiv CR_F$ not loosely bonded to the carbon center radically? Explain (where R_F is a pulling group of the types COOR, CN etc.).

6.17 (a) Why is the R in $HC \equiv CR$ or $RC \equiv CR$ not loosely bonded free-radically?

(b) In the production of carbides from acetylene, hydrogen gas is produced. Is this possible chargedly? Explain the mechanism of the reaction for calcium carbide.

(c) In the production of acetylene instantaneously for welding, cutting, and cleaning operations from carbides, the carbides are hydrolyzed by water. Some radical steps are involved in the process. Identify the radical steps involved, by explaining the mechanism of the reaction.

6.18 (a) Why is it that, in acetylenes only H is the species loosely bonded to specific centers? If H is, why can't OH or OR or NH_2 etc. not favor being also loosely bonded to specific centers?

(b) What are the best acetylenic monomers to polymerize without imposing any limitation on the strength of initiator to use?

(c) Distinguish between the two acetylides and carbides shown below.

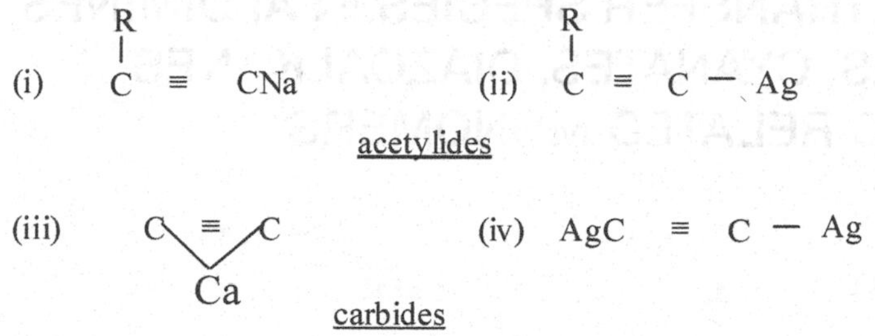

(i) $\overset{\displaystyle R}{\overset{|}{C}} \equiv CNa$ (ii) $\overset{\displaystyle R}{\overset{|}{C}} \equiv C - Ag$

acetylides

(iii) $C \overset{\equiv}{\diagdown \diagup} C$
$\qquad Ca$ (iv) $AgC \equiv C - Ag$

carbides

6.19 Show the products formed between the reactants shown below, when the conditions exist.

7.3a

(i) $CH_3C \equiv CNa$ + $C_2H_5B_r$ $\longrightarrow$

(ii) $CH_3C \equiv CNa$ + CH_3B_r $\longrightarrow$

(iii) $CH_3C \equiv CNa$ + HB_r $\longrightarrow$

(iv) $HC \equiv CNa$ + H_2O $\longrightarrow$

(v) $RC \equiv CH$ + R^1MgX $\longrightarrow$

Explain the mechanisms of the reactions.

6.20 (a) Why will $HC \equiv CH$ and $HC \equiv N$ not favor polymerization using H $\cdot^e$ alone as initiator?

(b) Why will $HC \equiv CH$ and HCN not favor polymerization using H $\cdot^n$ as initiator?

(c) Why will $FC \equiv CF$ not favor polymerization using F $\cdot^{nn}$ as initiator? Will H $\cdot^n$ favor its use for polymerization?

CHAPTER 7

TRANSFER OF TRANSFER SPECIES IN ALDIMINES, KETIMINES, CYANATES, DIAZOALKANES AND RELATED MONOMERS

7.0 Introduction

This is to complete the consideration of monomers with isolated double bonds or activation centers, since they will be very useful in application when considering cumulative double bonded monomers. Aldimines, Ketimines and their derivatives essentially belong to the same family tree, with C = N as the major activation center. Cyanates are indeed Nitriles with <u>bulky</u> etheric group as the only substituent group. Its first member, cyanic acid is so unstable that it molecularly rearranges to isocyanic acid which also is not quite stable.

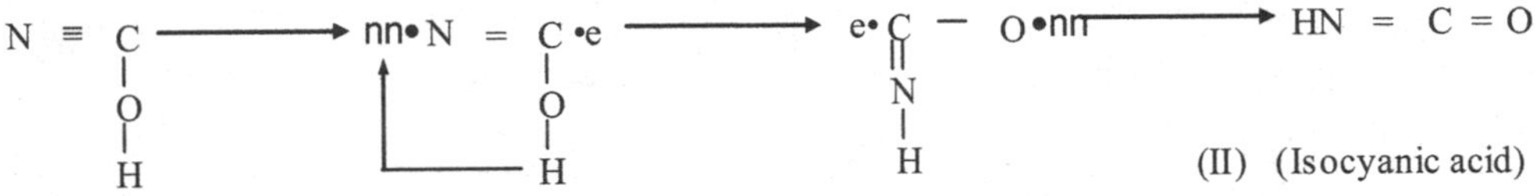

$$
N \equiv C \longrightarrow nn \bullet N = C \bullet e \longrightarrow e \bullet C - O \bullet nn \longrightarrow HN = C = O
$$

(I) (Cyanic acid)

(II) (Isocyanic acid)

7.1

It is only when H, the transfer species above is replaced with a bulky alkyl group that existence of cyanates are favored as a stable molecule or monomer. This could not be done chargedly.

All the cases to be considered herein with the exception of diazoalkanes to some extent are full pseudo-Addition monomers. Diazoalkanes are largely Addition monomers, since most of the products obtained are carbon-carbon backbones. Radically, carbon-chain products are also obtained. The simplest aliphatic diazo compound is diazomethane which is known to be a very valuable reagent. Though it will be shown downstream that the real name is Diazo-methylene, the first member of the Diazo-Alkylene (Not Alkenyl or Alkene) family, the current name will meanwhile be used. The explosive character associated with them in the gaseous state at room temperature is due to the polar character (not -ionic character) of the compound as shown below. Imagine the carbon center carrying a negative polar charge in the presence of N which is more electronegative than C. Why can't it explode when the operating conditions for handling it are not obeyed?

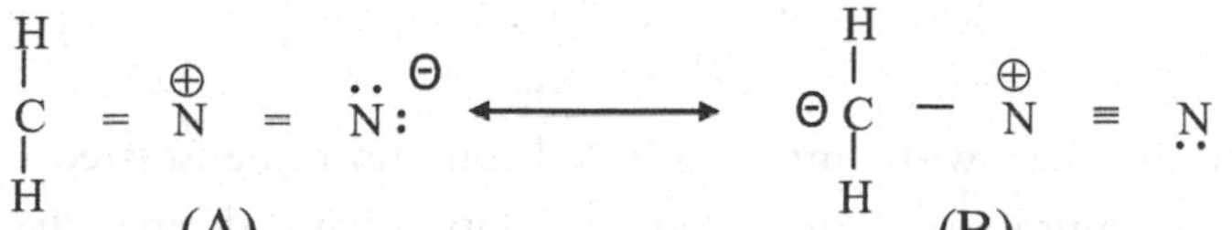

$$
\underset{\text{(A)}}{\overset{\displaystyle H}{\underset{\displaystyle H}{\overset{|}{\underset{|}{C}}}} = \overset{\oplus}{N} = \overset{..}{\underset{..}{N}}\!:^{\ominus}} \quad\longleftrightarrow\quad \underset{\text{(B)}}{^{\ominus}\overset{\displaystyle H}{\underset{\displaystyle H}{\overset{|}{\underset{|}{C}}}} - \overset{\oplus}{N} \equiv \underset{..}{N}}
$$

7.2

Polar resonance stabilization in diazomethane?

The equation as written above is not complete, because charges cannot be moved from their carriers and secondly movement begins from the Radical activated state of existence to the Polar state of existence. The absence of the radical form above makes the equation meaningless. The charges carried by the centers are not ionic but polar in character. Hence the resonance stabi-lization observed in diazoalkanes is not ionic in character, but polar as is by now obvious and will be fully seen in applications in the series. ***Polar charges (Imaginary) unlike ionic and covalent charges (Real) are not affected by forces of repulsion and attraction, because they are radical in character.*** In view of the polar character of the bond which is conjugatedly placed to an activation center (C=N) in (A), radical activation of the diazoalkanes is also possible as shown below,

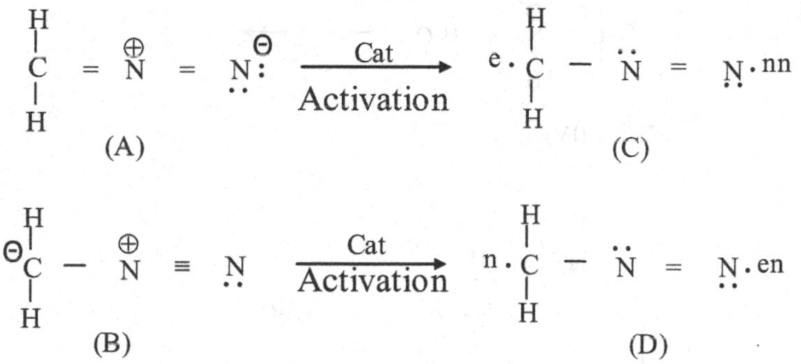

$$
\underset{\text{(A)}}{\overset{\displaystyle H}{\underset{\displaystyle H}{\overset{|}{\underset{|}{C}}}} = \overset{\oplus}{N} = \overset{..}{N}\!:^{\ominus}} \quad\xrightarrow[\text{Activation}]{\text{Cat}}\quad \underset{\text{(C)}}{e\cdot\overset{\displaystyle H}{\underset{\displaystyle H}{\overset{|}{\underset{|}{C}}}} - \overset{..}{N} = \underset{..}{N}\cdot nn}
$$

$$
\underset{\text{(B)}}{^{\ominus}\overset{\displaystyle H}{\underset{\displaystyle H}{\overset{|}{\underset{|}{C}}}} - \overset{\oplus}{N} \equiv \underset{..}{N}} \quad\xrightarrow[\text{Activation}]{\text{Cat}}\quad \underset{\text{(D)}}{n\cdot\overset{\displaystyle H}{\underset{\displaystyle H}{\overset{|}{\underset{|}{C}}}} - \overset{..}{N} = \underset{..}{N}\cdot en}
$$

7.3b

where the nitrogen atoms are not carrying more than eight radicals in their last shell. It should be noted that both polar states as used today in present-day Science are very different when activated. Only one is the real polar form. In (A), the carbon center is carrying an electro-free-radical in the presence of adjacently located nitrogen atom, that which is to be expected. In (B), the nitrogen center is carrying an electro-non-free-radical in the presence of carbon which is more electropositive. This is what makes it very explosive all the time and therefore unstable as a compound. (B) is what is used largely in the charged state with the release of N_2 all the time at all operating conditions. In the radical state, N_2 is only released when heated. As will be proved downstream using Sydnones, (B) is the only polar state of diazoalkanes. Therefore, it is not polarly resonance stabilized as shown in Equation 7.2. It is from (D) that (B) is formed in its stable state via **electro-non-free-radical movement.** It is from (C) that (A) is also formed via **electro-free-radical movement**. (A) cannot be formed from (D). (A) can only be formed if an electro-non-free- or electro-free-radical moves to grab a nucleo-radical from the paired unbonded radicals to form a negative charge instead of a bond! Usually, it is the electro-radical that moves to fill a shell and form a negative charge. Hence, (A) cannot be a polar form of diazomethane, otherwise it would have been known to favor the negatively charged route.

Aldehydes, isocyanates, aldimines, ketimines, diazoalkanes and most of these monomers do form cyclic compounds readily and most of their members have "ceiling temperatures".

7.1 Ceiling Temperatures

Reaction products of aldehydes (not ketones) with ammonia have been said to be isolated and the reaction appears to result from addition of ammonia to the carbonyl group.[1] Like hyd-rates, the initial product is unstable and loses water to give an aldimine (RCH = NH) which is said to fully polymerize to a cyclic trimer. This can be explained as follows, based on current developments.

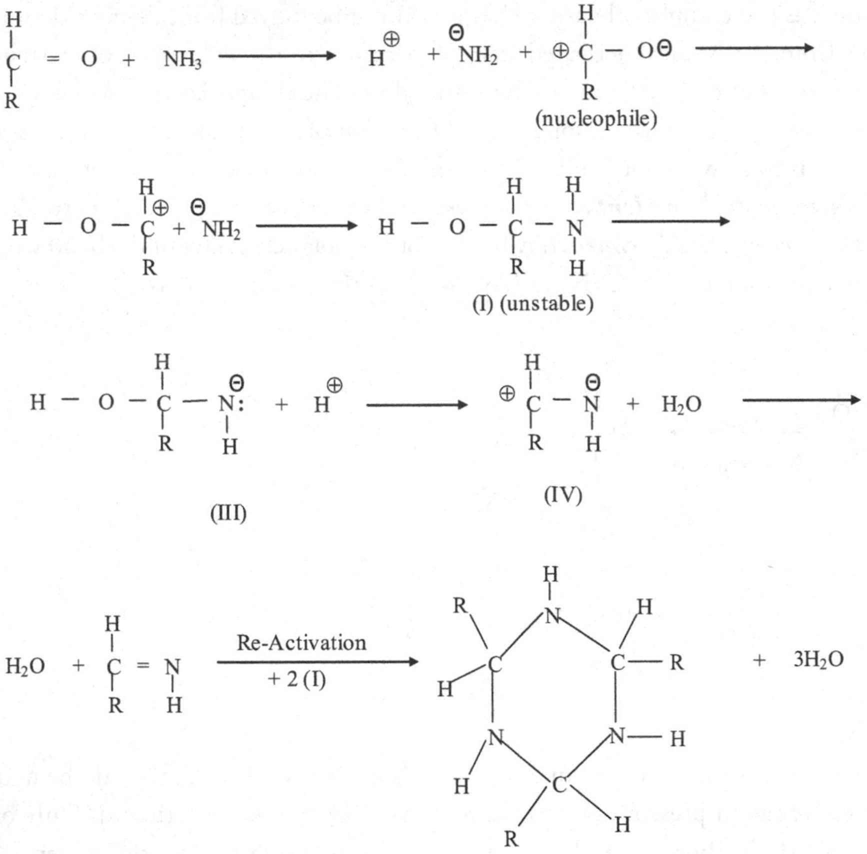

(nucleophile)

(I) (unstable)

(III)   (IV)

Re-Activation
+ 2 (I)

7.4a

Overall Equation: $3NH_3 + 3RCHO \longrightarrow 3H_2O + $ Cyclic Aldimine Trimer   7.4b

This obviously is an Equilibrium mechanism system with two stages. (I) above from the first stage is unstable. In the second stage it is held in Equilibrium State of Existence. This is then followed by loss of $^{\ominus}OH$ to form water and an aldimine. In the same stage, without deactivation, aldimines cyclizes to form a six-membered ring. It is not activated by any species, but by heat which is self-activation.

Aliphatic aldehydes, but not ketones are said to undergo acid catalyzed addition to give cyclic trimers.[1] The reaction is also said to take place slowly in the absence of added catalyst. When acid catalyzed, the followings like above are to be expected as **an Art.**

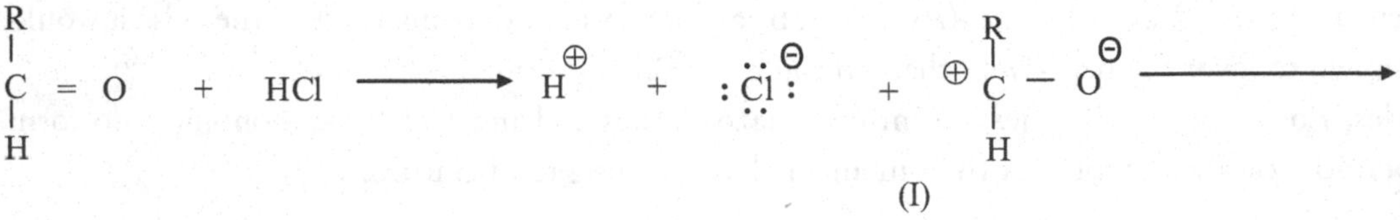

(I)

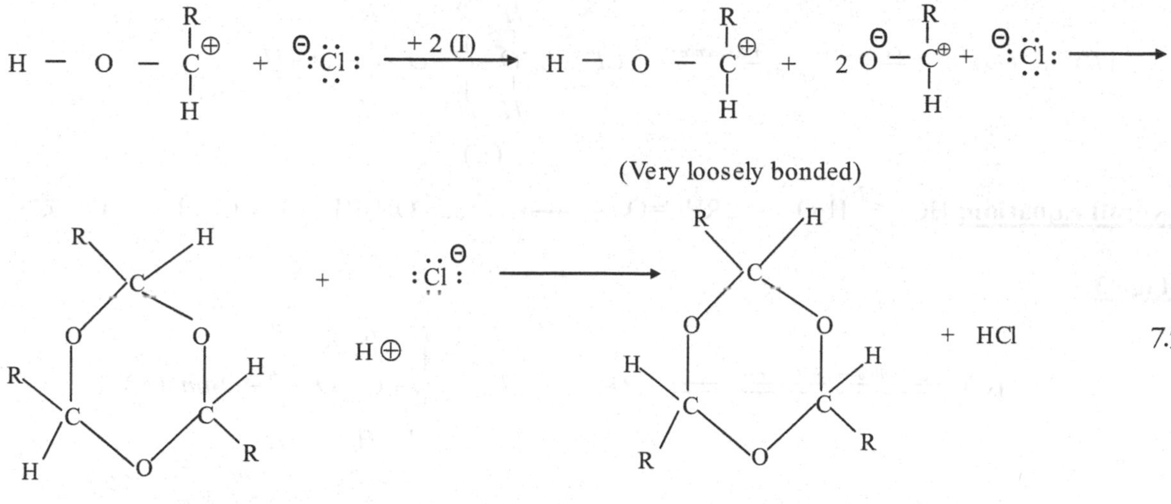

$$H - O - \overset{R}{\underset{H}{\overset{|}{C}}}{}^{\oplus} \;+\; {}^{\ominus}\!:\!\ddot{Cl}\!: \;\xrightarrow{+\,2\,(I)}\; H - O - \overset{R}{\underset{H}{\overset{|}{C}}}{}^{\oplus} \;+\; 2\, {}^{\ominus}O - \overset{R}{\underset{H}{\overset{|}{C}}}{}^{\oplus} \;+\; {}^{\ominus}\!:\!\ddot{Cl}\!: \;\longrightarrow$$

(Very loosely bonded)

$+ \quad HCl \qquad 7.5$

As a Science, that which is the New Frontiers, the followings are obtained.

Stage 1:

$$Cl^{-}........{}^{+}\!\!\underset{H}{\overset{H}{\overset{|}{O}}}\!\!-H \quad \underset{\textit{of Dilute HCl}}{\overset{\textit{Equilibrium State of Existence}}{\rightleftharpoons}} \quad H\bullet e \;+\; Cl^{-}........{}^{+}\!\!\underset{H}{\overset{H}{\overset{|}{O}}}\!\!\bullet nn \;\;(A)$$

$$H\bullet e \;+\; O=\overset{R}{\underset{H}{\overset{|}{C}}} \quad \underset{\textit{Aldehyde}}{\overset{\textit{Activation of}}{\rightleftharpoons}} \quad H-O-\overset{R}{\underset{H}{\overset{|}{C}}}\bullet e \;\;(B)$$

$\qquad\qquad\qquad\qquad\qquad\qquad\qquad\qquad\qquad\qquad\qquad\qquad 7.6\mathrm{a}$

$$(B) \;+\; (A) \quad \xrightarrow[\textit{of (C)}]{\textit{Combination State of Existence}} \quad Cl^{-}........{}^{+}\!\overset{H}{\underset{H}{\overset{|}{O}}}-\overset{R}{\underset{H}{\overset{|}{C}}}-O-H$$

$$(C)$$

Overall Equation: $HCl \;+\; H_2O \;+\; RHC{=}O \;\longrightarrow\; Cl^{\ominus}........{}^{\oplus}OH_2RHCOH \;(C) \qquad 7.6\mathrm{b}$

Stage 2:

$$(C) \quad \underset{\textit{of (C)}}{\overset{\textit{Equilibrium State of Existence}}{\rightleftharpoons}} \quad H\bullet e \;+\; Cl^{-}......{}^{+}\!\overset{H}{\underset{H}{\overset{|}{O}}}-\overset{R}{\underset{H}{\overset{|}{C}}}-O\bullet nn \;\;(D)$$

$$H\bullet e \;+\; O=\overset{R}{\underset{H}{\overset{|}{C}}} \quad \underset{\textit{Aldehyde}}{\overset{\textit{Activation of}}{\rightleftharpoons}} \quad H-O-\overset{R}{\underset{H}{\overset{|}{C}}}\bullet e \;\;(B) \qquad 7.7\mathrm{a}$$

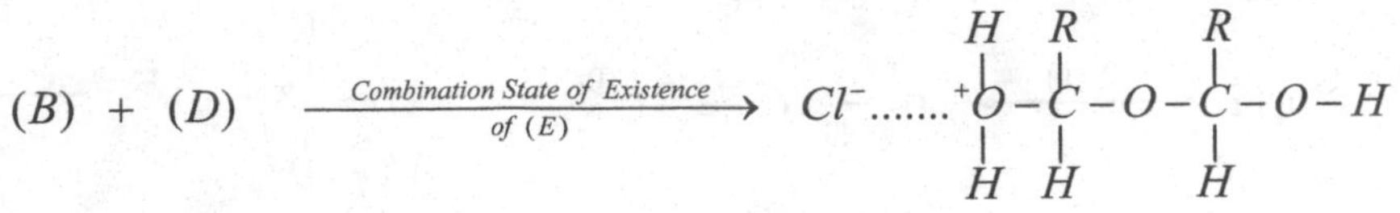

$$(B) \; + \; (D) \xrightarrow[\text{of } (E)]{\text{Combination State of Existence}} \quad Cl^- \dotsb \; {}^+\!\overset{H}{\underset{H}{O}} - \overset{R}{\underset{H}{C}} - O - \overset{R}{\underset{H}{C}} - O - H$$

$$(E)$$

Overall Equation: $HCl + H_2O + 2RHC{=}O \longrightarrow Cl^{\ominus} \dotsb {}^{\oplus}OH_2RHCORHCOH$ 7.7b

Stage 3:

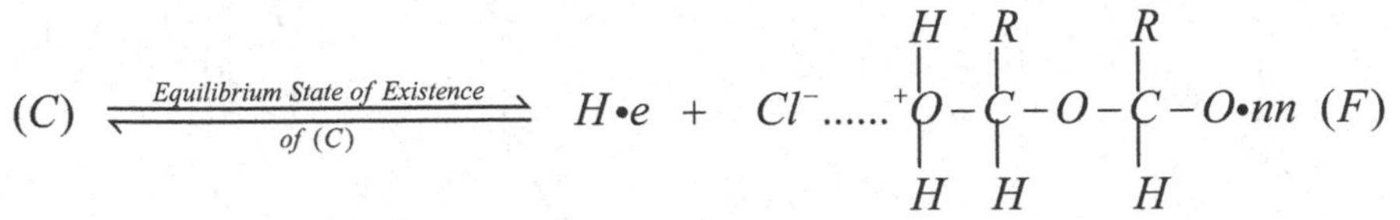

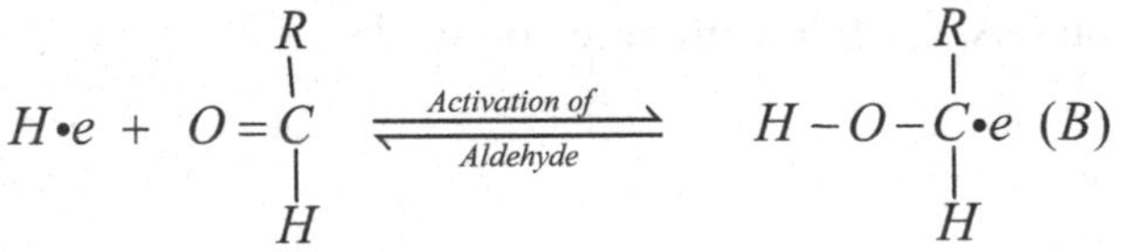

$$(C) \xrightleftharpoons[\text{of } (C)]{\text{Equilibrium State of Existence}} \quad H{\bullet}e \; + \; Cl^- \dotsb \; {}^+\!\overset{H}{\underset{H}{O}} - \overset{R}{\underset{H}{C}} - O - \overset{R}{\underset{H}{C}} - O{\bullet}nn \;\; (F)$$

$$H{\bullet}e \; + \; O{=}\overset{R}{\underset{H}{C}} \xrightleftharpoons[\text{Aldehyde}]{\text{Activation of}} \quad H - O - \overset{R}{\underset{H}{C}}{\bullet}e \;\; (B)$$ 7.8a

$$(B) \; + \; (F) \xrightarrow[\text{of } (E)]{\text{Combination State of Existence}} \quad Cl^- \dotsb \; {}^+\!\overset{H}{\underset{H}{O}} - \overset{R}{\underset{H}{C}} - O - \overset{R}{\underset{H}{C}} - O - \overset{R}{\underset{H}{C}} - O - H$$

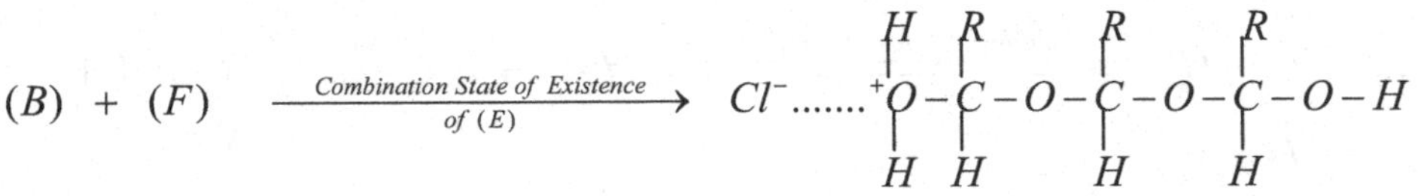

$$(G)$$

Overall Equation: $HCl + H_2O + 3RHC{=}O \longrightarrow Cl^{\ominus} \dotsb {}^{\oplus}OH_2RHCORHCORHCOH$

$$(G)$$ 7.8b

Stage 4:

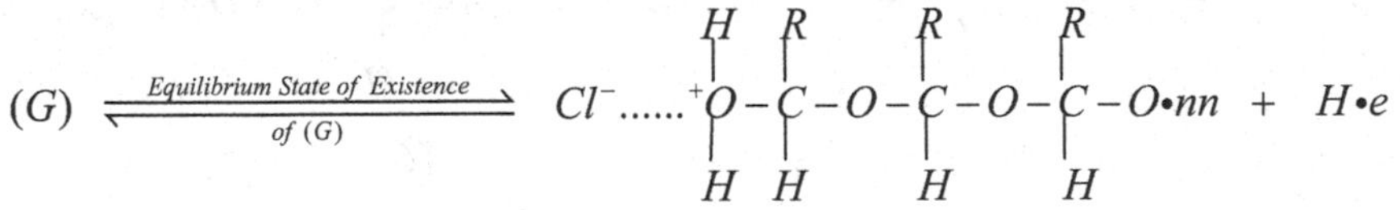

$$(G) \xrightleftharpoons[\text{of } (G)]{\text{Equilibrium State of Existence}} \quad Cl^- \dotsb \; {}^+\!\overset{H}{\underset{H}{O}} - \overset{R}{\underset{H}{C}} - O - \overset{R}{\underset{H}{C}} - O - \overset{R}{\underset{H}{C}} - O{\bullet}nn \; + \; H{\bullet}e$$

$$(H)$$

$$(H) \xrightleftharpoons{\text{Release of Dilute HCl}} \quad Cl^- \dotsb \; {}^+\!\overset{H}{\underset{H}{O}}{\bullet}nn \; + \; e{\bullet}\overset{R}{\underset{H}{C}} - O - \overset{R}{\underset{H}{C}} - O - \overset{R}{\underset{H}{C}} - O{\bullet}nn$$

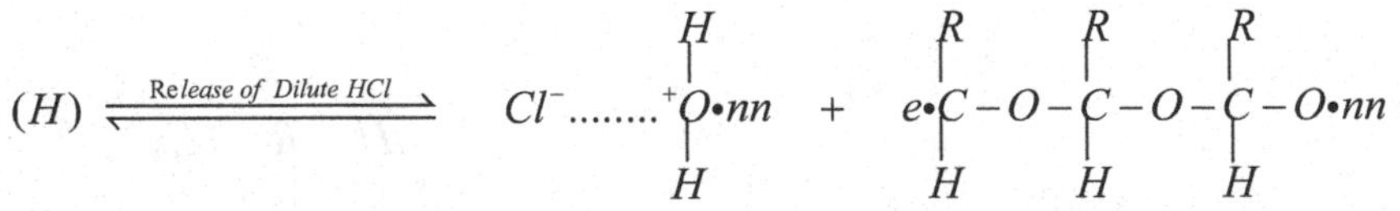

$$(I) \qquad\qquad\qquad (J)$$

$$(I) + H{\bullet}e \rightleftharpoons \qquad \text{Dilute Acid}$$

$$(J) \xrightarrow[\text{release of Heat}]{\text{Deactivation of } (J) \text{ with}} \quad RHC \overset{\displaystyle O}{\underset{\displaystyle \underset{CHR}{O \quad O}}{{} \quad}} CHR \; + \; Heat$$

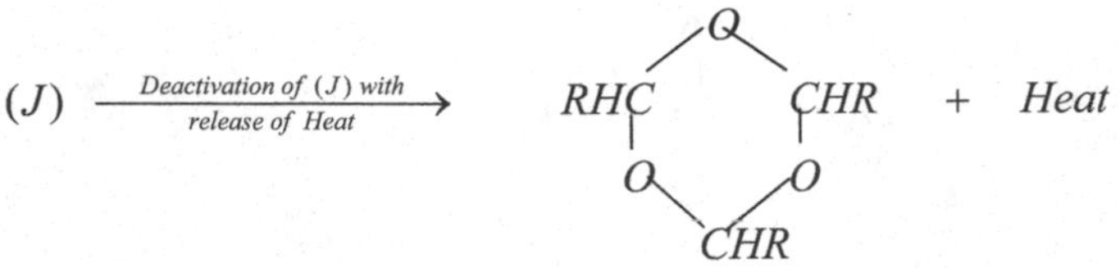

$$(K)$$ 7.9a

Overall Equation: HCl + H$_2$O + 3RHC=O ⟶ HCl + H$_2$O + (K)   7.9b

Note that concentrated HCl cannot be used, otherwise another product will be obtained. The mechanism is Equilibrium mechanism with four stages. Note that one cannot move from Stage 1 to Stage 4, as done in present-day Science. It is from the Art that the Real Science emerged for we have only just started. Currently, one has only digressed from the Art where the New Frontiers emerged in order to see where we are going. Though the reactions of Equation 7.4b was shown to take place chargedly, just like that in Equation 7.5, they all take place only radically, since charges cannot readily be isolated. In place of Equation 7.5 were Equations 7.6a to 7.9a which can only take place radically based on the structure of dilute hydrochloric acid. Most worthy of note here is a new form of Addition polymerization in Stages 1 to 3 and a form of depropagation in Stage 4 of Equation 7.9a where a suitably sized ring was formed. If there were hundreds of moles of the aldehyde in the system, polymers would have been obtained. This is how aldehydes and ketones are polymerized not "cationically" as is mistakenly believed by present-day Science and Engineering[2], but ***electro-free-radically using paired negatively charged electrostatic initiator (the dilute acid)*** and not ***anionic ion-paired initiator which does exist-e.g. RO$^{\ominus}$......$^{\oplus}$Na which really is Ionically ion-paired initiator since both centers are active chemically but not***.polymerically. The dilute HCl also has the anionic center, which cannot be used for aldehydes and Ketones. Note also herein that the mechanism is not Combination mechanism, but Equilibrium mechanism which implies that addition of monomers will be slow. In present-day Science and Engineering, it is said that the carbonyl group of formaldehyde is highly susceptible to nucleophilic attack and this monomer can be polymerized with almost any basic catalyst[2]. ***Metal alkyls, alkoxides, phenolates and carboxylates, hydrated alumina, tertiary aliphatic amines, phosphines, arsines, and nitrogen heterocyclics are among the bases which have been found to be effective catalysts.***[2] How can a nucleophilic initiator (a female) attack a female monomer? Only formaldehyde of all the aldehydes can favor both the use of male and female initiators as already shown. Others including ketones can only be polymerized electro-free-radically or the use of positively charged-paired initiators (and not cationically). Though all the initiators listed above look nucleophilic, not all are used as such. Based on what we have seen so far, let us take a look at the first six initiators above. Metallic alkyls are ***negatively charged-paired covalent initiators*** (E.g. t-H$_9$C$_4$$^{\ominus}$......$^{\oplus}$Li). Though they are dual in character (that is, has two active centers), they are so- called, because the positively charged end cannot be used for C=C monomers, since Li-(CH$_2$)$_n$H$_3$C$^{\oplus}$.......$^{\ominus}$C$_4$H$_9$ which is pairing between two C centers cannot exist, apart from the fact that the charge on the Li center is cationic. These cannot be used to polymerize the formaldehyde or other aldyhydes and ketones, because the equation during initiation will not be balanced chargedly. It is only the metallic end that can be used electro-free-radically for aldehydes and ketones for example only when free and not paired. Metallic alkoxides are ***anionic ion-paired initiators*** [Ionically ion-paired initiator] as already shown above (***RO$^{\ominus}$......$^{\oplus}$Na)***. Anionically, these can be used for only formaldehyde or halogenated ones and not for other aldehydes or ketones. Positively, the cationic end (Na$^{\oplus}$) cannot be used for any of them. The other metallic ones are like the second. Hydrated alumina and tertiary aliphatic amines are alike. These are almost like the case of dilute HCl used above (See the structure in the first step of Stage 1 of Equation 7.6a). They are ***negatively charged-paired electrostatic initiators.*** They cannot be used for these monomers chargedly, but only electro-free-radically. Shown below are the structures of the hydrated tertiary amines and alumina initiators.

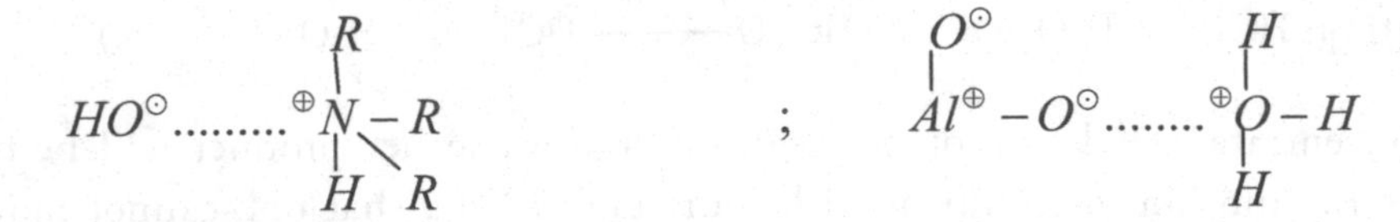

$$HO^{\ominus}........^{\oplus}\underset{\underset{R}{|}}{\overset{\overset{R}{|}}{N}}-R \qquad ; \qquad Al^{\oplus}-O^{\ominus}........^{\oplus}\underset{\underset{H}{|}}{\overset{\overset{O^{\ominus}}{|}}{O}}-H$$

$$\overset{\overset{H}{|}}{} \qquad\qquad\qquad\qquad\qquad \overset{\overset{H}{|}}{}$$

(I) Tertiary alkyl ammonium hydroxide   **(II) Hydrated alumina**   7.10

In (I), water was said to be the cocatalyst while the tertiary amine the catalyst[2]. The real case is the reverse as shown below. Indeed the structures as provided above are unknown in present-day Science. How the structure of (II) was obtained will be explained downstream in one of the Volumes. (II) is the real Aluminum hydroxide and not $Al(OH)_3$. Using (II) is like using the dilute HCl. Using (I) in the same way for polymerization, the followings are obtained.

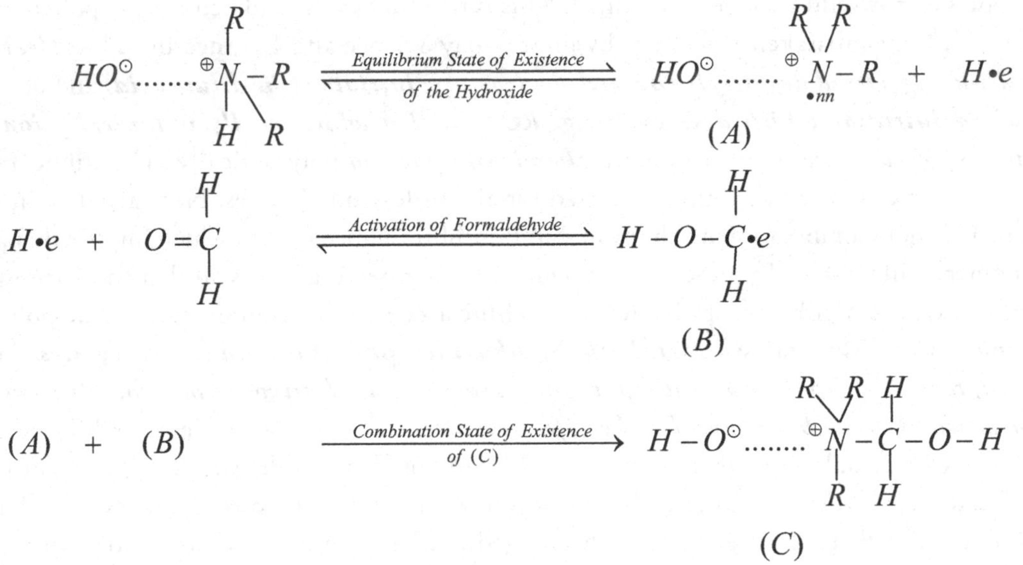

$$HO^{\ominus}........^{\oplus}\underset{\underset{H\;\;R}{|\;\;\;|}}{\overset{\overset{R}{|}}{N}}-R \quad\xrightleftharpoons[\text{of the Hydroxide}]{\text{Equilibrium State of Existence}}\quad HO^{\ominus}........^{\oplus}\underset{\bullet nn}{\overset{\overset{R\;\;R}{\backslash\;/}}{N}}-R \;\; + \;\; H\bullet e$$

$$(A)$$

$$H\bullet e \;\; + \;\; O=\underset{\underset{H}{|}}{\overset{\overset{H}{|}}{C}} \quad\xrightleftharpoons{\text{Activation of Formaldehyde}}\quad H-O-\underset{\underset{H}{|}}{\overset{\overset{H}{|}}{C}}\bullet e$$

$$(B)$$

$$(A) \;\; + \;\; (B) \quad\xrightarrow[\text{of } (C)]{\text{Combination State of Existence}}\quad H-O^{\ominus}........^{\oplus}\underset{\underset{R\;\;H}{|\;\;\;|}}{\overset{\overset{R\;\;R\;\;H}{\backslash\;/\;|}}{N}}-C-O-H$$

$$(C) \qquad\qquad 7.11a$$

Overall Equation: $HON R_3 \;\; + \;\; H_2C{=}O \xrightarrow{\hspace{3cm}} (C)$ 7.11b

Addition of formaldehyde continues in so many stages until the followings are obtained after n stages. As can be seen, water is the catalyst which provided the $H\bullet e$, the initiator.

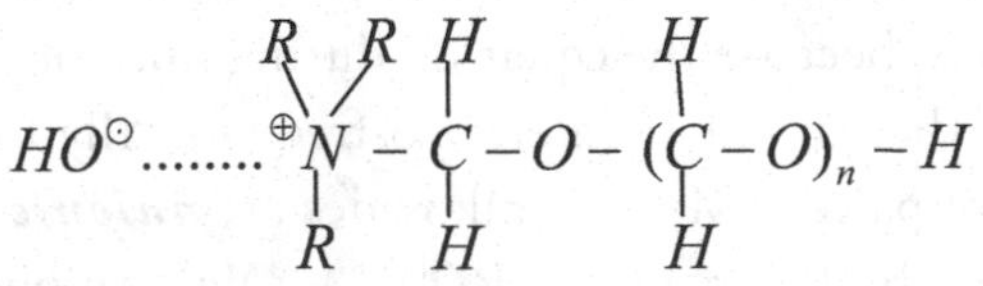

$$HO^{\ominus}........^{\oplus}\underset{\underset{R\;\;H}{|\;\;\;|}}{\overset{\overset{R\;\;R\;\;H}{\backslash\;/\;|}}{N}}-C-O-(\underset{\underset{H}{|}}{\overset{\overset{H}{|}}{C}}-O)_n-H$$

$$7.12a$$

If all the monomers are consumed and chain has not reached the optimum chain length as deter-mined by the glass-transition temperature, then depropagation commences if temperature of po-lymerization is higher than expected. This is the origin of the concept of "ceiling temperature" for monomers being polymerized in the route natural to them.

    These will readily take place as the temperature of polymerization is increased. As temperature is decreased, the rate of depropagation will decrease with decrease also in the ability of the monomer to add in stages. The monomers cannot all add in one stage, whether the mechanism is Equilibrium or Combination mechanism. On the other hand and most importantly, the nucleo-non-free-radical is present in the system, while the electro-free is adding a monomer. Thirdly, the C-N bond is strong but weaker than the C-O bond. With formaldehyde, unlike the other aliphatic aldehydes and ketones, if there are more monomers in the system, addition continues to give what is shown below via Combination mechanism if and only if the Right Hand Side (RHS) can no longer exist in Equilibrium state of existence. If at the beginning, the initiator cannot exist in Equilibrium state of existence, then only formaldehyde can be polymerized anionically via Combination mechanism.

$$HO-(\overset{\overset{\displaystyle H}{|}}{\underset{\underset{\displaystyle H}{|}}{C}}-O)_m-\overset{\overset{\displaystyle H}{|}}{\underset{\underset{\displaystyle H}{|}}{C}}-O^{\circleddash}........^{\oplus}\overset{\overset{\displaystyle R}{\diagdown}\overset{\displaystyle R}{\diagup}}{\underset{\underset{\displaystyle R}{|}}{N}}-\overset{\overset{\displaystyle H}{|}}{\underset{\underset{\displaystyle H}{|}}{C}}-O-(\overset{\overset{\displaystyle H}{|}}{\underset{\underset{\displaystyle H}{|}}{C}}-O)_n-H$$

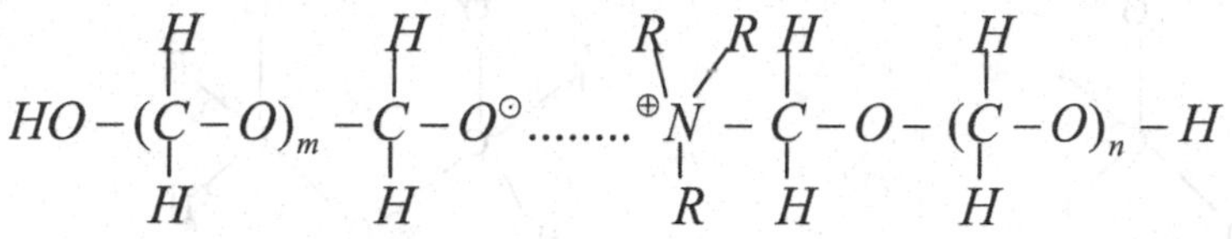

7.12b

With formaldehyde, there are two sites for polymerization. The coordination center site can readily be used without any form of degradation or depropagation of the chain. It is on the electro-free-radical side that we have depropagation taking place because of the H atom at that end. When the number of monomer units disengaging from the growing polymer chain goes beyond three or four, no rings can be formed and the reason will be shown downstream after introducing more concepts. As applies to formaldehyde, almost the same applies to fully halogenated aldehydes and ketones. With aldehydes and ketones, the coordination center cannot be used and the same applies to the alkenes. This however is what to expect when the monomer is propylene (propene). It is a one stage system

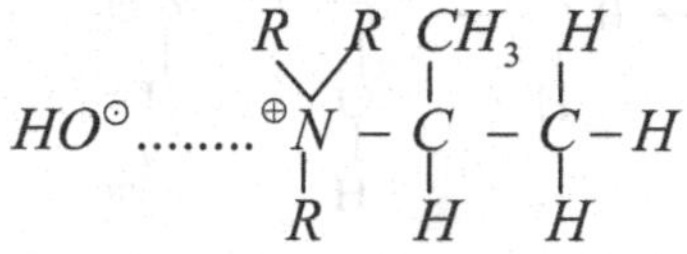

$$HO^{\circleddash}........^{\oplus}\overset{\overset{\displaystyle R}{\diagdown}\overset{\displaystyle R}{\diagup}}{\underset{\underset{\displaystyle R}{|}}{N}}-\overset{\overset{\displaystyle CH_3}{|}}{\underset{\underset{\displaystyle H}{|}}{C}}-\overset{\overset{\displaystyle H}{|}}{\underset{\underset{\displaystyle H}{|}}{C}}-H$$

(I)

7.13

with no propagation, because (I) cannot readily exist in Equilibrium State of Existence.

Coming back to the acid catalyzed reaction of Equation 7.5, in the absence of added acid

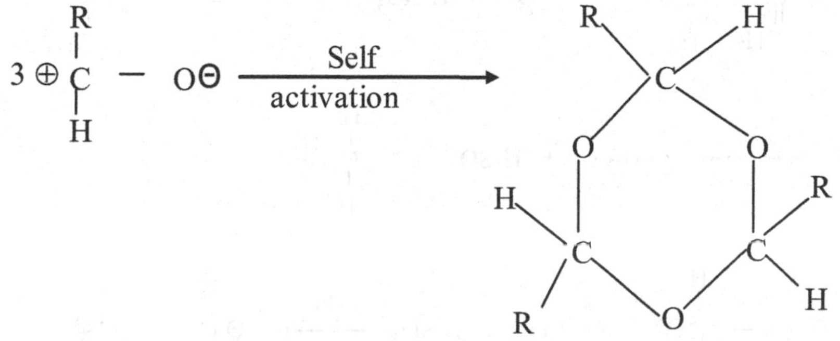

$$3\oplus \overset{\overset{\displaystyle R}{|}}{\underset{\underset{\displaystyle H}{|}}{C}}-O\circleddash \xrightarrow[\text{activation}]{\text{Self}}$$

7.14

$$3e.\underset{\underset{H}{|}}{\overset{\overset{R}{|}}{C}} - O.nn \longrightarrow e.\underset{\underset{H}{|}}{\overset{\overset{R}{|}}{C}} - O.nn \ + \ e.\underset{\underset{\leftarrow H}{|}}{\overset{\overset{R}{|}}{C}} - O.nn \ + \ e.\underset{\underset{\leftarrow H}{|}}{\overset{\overset{R}{|}}{C}} - O.nn$$

last to be activated   Second to be activated   First to be activated

7.15

It is a one stage Equilibrium mechanism process, with the monomer not in Equilibrium State of Existence or Activated/Equilibrium State of existence as will shortly be shown, but in Activated State of Existence.

With ketones, strong acids lead to addition and dehydration of them. For example, 1,3,5-trimethylbenzene (Mesitylene) is obtained from acetone as shown using present-day Science, (literature data) and current developments herein. For if the product reportedly obtained is the product at the operating conditions, then an allene must have been formed as an intermediate.

$$\left[H_2SO_4 \ + \ \underset{\underset{CH_2}{\|}}{\overset{\overset{CH_3}{|}}{C}} - O^{\ominus} \ + \ H^{\oplus} \right] \xrightarrow[\text{rearrangement}]{\text{(Molecular}} \ \overset{\ominus}{\theta}\underset{\underset{H}{|}}{\overset{\overset{H}{|}}{C}} - \underset{\underset{\underset{H}{|}}{O}}{\overset{\overset{H}{|}}{\overset{\oplus}{C}}} - \underset{\underset{H}{|}}{\overset{\overset{H}{|}}{C}}\!\!{}^{\ominus} \ +$$

$$H_2SO_4 \ + \ H^{\oplus} \longrightarrow H_2O \ + \ H_2SO_4 \ + \ \underset{\underset{H}{|}}{\overset{\overset{H}{|}}{C}} = C = \underset{\underset{H}{|}}{\overset{\overset{H}{|}}{C}}$$

<u>allene</u>

$$\xrightarrow[\text{Activation}]{\text{Self}} \ \oplus\underset{\underset{CH_2}{\|}}{\overset{}{C}} - \underset{\underset{H}{|}}{\overset{\overset{H}{|}}{C}}\!\!{}^{\ominus} \ + \ H_2O \ + \ H_2SO_4 \longrightarrow \ \ominus\underset{}{C} = \underset{\underset{CH_3}{|}}{\overset{\overset{H}{|}}{C}}\!\!{}^{\oplus} \ +$$

(V)

$$H_2SO_4 \ + \ H^{\oplus} \longrightarrow H_2O \ + \ H_2SO_4 \ + \ \underset{\underset{H}{|}}{\overset{\overset{H}{|}}{C}} = C = \underset{\underset{H}{|}}{\overset{\overset{H}{|}}{C}}$$

<u>allene</u>

$$\xrightarrow[\text{Activation}]{\text{Self}} \ \oplus\underset{\underset{CH_2}{\|}}{\overset{}{C}} - \underset{\underset{H}{|}}{\overset{\overset{H}{|}}{C}}\!\!{}^{\ominus} \ + \ H_2O \ + \ H_2SO_4 \longrightarrow \ \ominus\underset{}{C} = \underset{\underset{CH_3}{|}}{\overset{\overset{H}{|}}{C}}\!\!{}^{\oplus} \ +$$

(V)

$$H_2O \quad + \quad H_2SO_4 \quad \xrightarrow{\ +\ 2\ (I)\ }$$

[structure of a six-membered ring with CH₃, H substituents]

$$+ \quad 3H_2O \ + \ 3H_2SO_4$$

7.16

The mechanisms provided above look meaningful, but incomplete because the exercise is still an Art. As a Science the followings are obtained.

$$\begin{array}{c} CH_3 \\ | \\ C = O \\ | \\ CH_3 \end{array} \quad \underset{\textit{Existence of Acetone}}{\overset{\textit{Equilibrium State of}}{\rightleftharpoons}} \quad H\bullet e \ + \ \overset{H \quad e\bullet}{\underset{H \quad CH_3}{n\bullet C - C = O}} \ (A)$$

$$(A) \quad \underset{\textit{Acetone}}{\overset{\textit{Tautomerization of}}{\rightleftharpoons}} \quad \overset{H}{\underset{H \quad CH_3}{C = C - O\bullet nn}} \ (B) \qquad ? \quad 7.17$$

$$H\bullet e \ + \ (B) \quad \xrightarrow{\ ??\ } \quad \overset{H}{\underset{H \quad CH_3}{C = C - OH}}$$

$$(C)$$

Overall Equation: $3H_3C\text{-}CO\text{-}CH_3 \quad \xrightarrow{\textit{Tautomerization}} \quad 3H_2C=C(OH)CH_3$ *** 7.18

Worthy of note is the use of the well-known concept-Tautomerism, the mechanism of which has never been known, For example, so-called well known Azo-methane (H_3C-N=N-CH_3) is unstable and tautomerizes to Formaldehyde methyl hydrazone {H_2C=N-NH(CH_3)}[3]. How it does has never been known. As is already becoming apparent and will be fully shown downstream, when current flows in some solutions (Electrolytes) and some solids (Electrical Conductors) and when chemical reactions are taking place in the presence of nucleo-radicals, it is the electro-radical that diffuses all the time in Nature. The nucleo-radical diffuses only when there is no isolated electro-radical around (Nucleo-free-radical or nucleo-non-free-radical Initiations). Without the compound existing in **Activated/*Equilibrium State of Existence as will shortly be shown,*** this type of ***Tautomerization phenomena*** will never take place unlike what is shown in (A) of Equation 7.17. This type of Tautomerization herein called ***Enolization*** is a form of rearrangement to give a more stable molecule. This is unlike ***Molecular rearrangement*** where before it takes place, the compound must be only in the ***Activated State of Existence***. Here the monomer must be in Activated/Equilibrium state of existence. The component held in Equilibrium state of exiatence is that involved as transfer species of the first kind during molecular rearrangement.

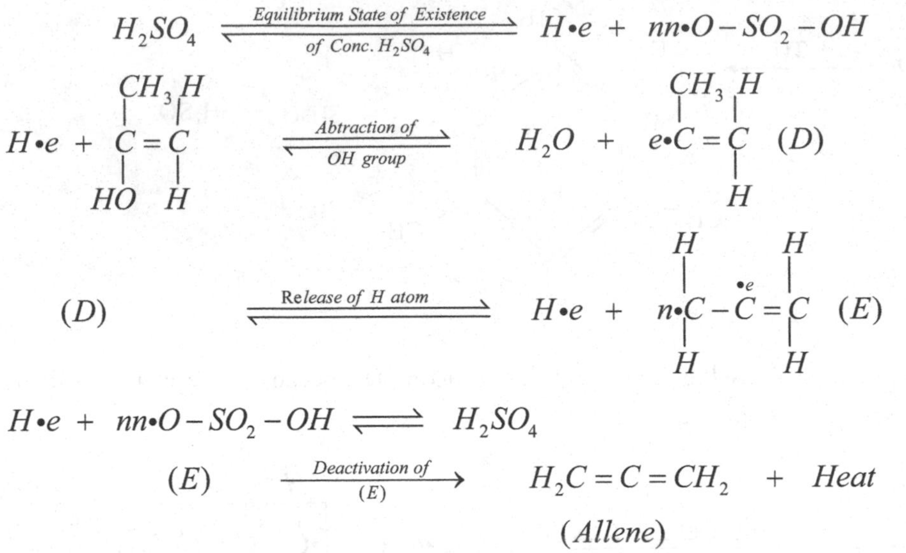

$$H_2SO_4 \; \underset{\text{of Conc. } H_2SO_4}{\overset{\text{Equilibrium State of Existence}}{\rightleftharpoons}} \; H\bullet e \;+\; nn\bullet O - SO_2 - OH$$

$$H\bullet e \;+\; \overset{\displaystyle CH_3}{\underset{\displaystyle HO}{C}} = \overset{\displaystyle H}{\underset{\displaystyle H}{C}} \;\; \underset{\text{OH group}}{\overset{\text{Abtraction of}}{\rightleftharpoons}} \;\; H_2O \;+\; e\bullet \overset{\displaystyle CH_3}{C} = \overset{\displaystyle H}{\underset{\displaystyle H}{C}} \;\; (D)$$

$$(D) \;\; \underset{}{\overset{\text{Release of H atom}}{\rightleftharpoons}} \;\; H\bullet e \;+\; n\bullet \overset{\displaystyle H}{\underset{\displaystyle H}{C}} - \overset{\displaystyle \overset{\bullet e}{}}{C} = \overset{\displaystyle H}{\underset{\displaystyle H}{C}} \;\; (E) \qquad 7.19$$

$$H\bullet e \;+\; nn\bullet O - SO_2 - OH \;\rightleftharpoons\; H_2SO_4$$

$$(E) \;\; \underset{(E)}{\overset{\text{Deactivation of}}{\longrightarrow}} \;\; H_2C = C = CH_2 \;\;+\;\; Heat$$

$$(Allene)$$

Overall Equation: $3H_2SO_4 \;+\; 3(C) \longrightarrow 3H_2O \;+\; 3H_2SO_4 \;+\; 3H_2C{=}C{=}CH_2$ 7.20

So far, note what the presence of concentrated sulfuric acid has meant. As strong as it is, the acid could not keep the acetone in Activated State of Existence, something which cannot be said to be a clear indication of the strong nucleophilic character of acetone and indeed ketones. Since dilute HCl activated aldehydes (See Equation 7.6a), concentrated H_2SO_4 or HCl should be able to activate the acetone or any ketone. In acetone with two CH_3 groups, it was readily kept in Activated/Equilibrium State of Existence by the acid to begin this Tautomerization to rearrange to give methyl vinyl alcohol which also could not be activated by the acid. In Stage 2, the acid just only removed the OH group without adding the nn•O-SO_2-OH group to (D) to replace the nn•OH group removed. This looks like transfer species of the second first kind of the second type. Instead, (D) lost an H atom to recover the acid and form (E) which deactivated to unstable Allene. Worthy of note is that the alcoholic acid could not exist in Equilibrium State of Existence in the presence of the acid but only in Stable State of Existence. Only aldehydes and ketones can undergo this type of tautomerization. Thus, when aldehydes and ketones are kept in Activated State of Existence, the transfer species of the first kind of the first type which is the H atom is held in Equilibrium State of Existence. This is the state called **Activated Equilibrium State of Existence** as shown below. When this hydrogen is held, then they all undergo what is called **Enolization** different from another type of Tautomerization phenomena called **Electroradical-ization.**

ACETALDEHYDE

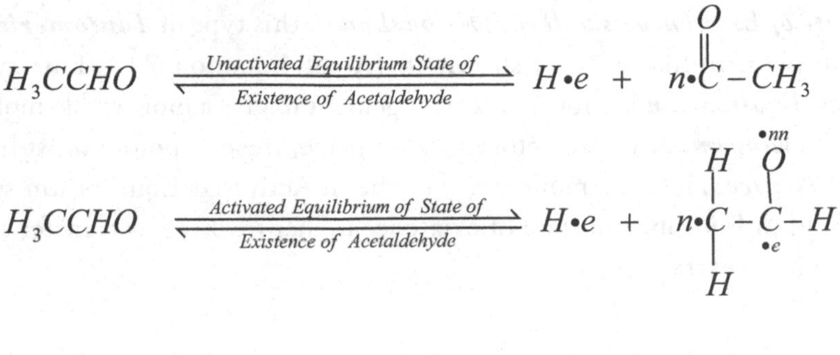

$$H_3CCHO \;\; \underset{\text{Existence of Acetaldehyde}}{\overset{\text{Unactivated Equilibrium State of}}{\rightleftharpoons}} \;\; H\bullet e \;+\; n\bullet \overset{\displaystyle O}{\overset{\|}{C}} - CH_3$$

$$H_3CCHO \;\; \underset{\text{Existence of Acetaldehyde}}{\overset{\text{Activated Equilibrium of State of}}{\rightleftharpoons}} \;\; H\bullet e \;+\; n\bullet \overset{\displaystyle H}{\underset{\displaystyle H}{C}} - \overset{\displaystyle \overset{\bullet nn}{O}}{\underset{\displaystyle \bullet e}{C}} - H \qquad 7.21$$

$$H_3CCHO \xrightleftharpoons[\text{Acetaldehyde}]{\text{Activated State of Existence of}} nn \bullet O - \underset{\underset{H}{|}}{\overset{\overset{CH_3}{|}}{C}} \bullet e$$

ALDEHYDES

$$RCH_2CHO \xrightleftharpoons[\text{Existence of Aldehydes}]{\text{Unactivated Equilibrium State of}} H \bullet e \ + \ n \bullet \overset{\overset{O}{\|}}{C} - CH_2 - R$$

$$RCH_2CHO \xrightleftharpoons[\text{Existence of Aldehydes}]{\text{Activated Equilibrium State of}} H \bullet e \ + \ n \bullet \underset{\underset{H}{|}}{\overset{\overset{R}{|}}{C}} - \underset{\underset{\bullet e}{}}{\overset{\overset{\bullet nn}{O}}{C}} - H \qquad 7.22$$

$$RCH_2CHO \xrightleftharpoons[\text{Aldehydes}]{\text{Activated State of Existence of}} nn \bullet O - \underset{\underset{H}{|}}{\overset{\overset{CH_2R}{|}}{C}} \bullet e$$

where R is an alkyl group.

KETONES

$$H_3C - CH_2\overset{\overset{O}{\|}}{C} - CH_3 \xrightleftharpoons[\text{Existence of Ketones}]{\text{Unactivated Equilibrium State of}} H_3C \bullet n \ + \ e \bullet \overset{\overset{O}{\|}}{C} - C_2H_5$$

$$R - CH_2 - \overset{\overset{O}{\|}}{C} - CH_3 \xrightleftharpoons[\text{Existence of Ketones}]{\text{Activated Equilibrium State of}} H \bullet e \ + \ n \bullet \underset{}{\overset{\overset{H}{|}}{C}} - \underset{\underset{\bullet e}{}}{\overset{\overset{\bullet nn}{O}}{C}} - CH_3$$
$$R(CH_3) \qquad 7.23$$

$$R - CH_2 - \overset{\overset{O}{\|}}{C} - CH_3 \xrightleftharpoons[\text{Ketones}]{\text{Activated State of Existence of}} nn \bullet O - \underset{}{\overset{\overset{CH_3}{|}}{C}} \bullet e$$
$$CH_2RCH_3$$

where the R here is one or more methylene groups (CH_2).

Unknown to us, is that States of Existences as already said are like finger-prints of every compound. Vinyl alcohol can *electroradicalize or undergo molecular rearrangement* depending on whether it is in Equilibrium State of Existence or Activated State of Existence respectively to give acetaldehyde which can only *enolize* in the presence of dilute aqueous alkalis or acids. Methyl vinyl alcohol ($H_2C=C(OH)CH_3$) can also *electroradicalize or molecularly rearrange* to give acetone which can only *enolize* in the presence of dilute aqueous alkalis or acids. Ethyl vinyl alcohol (**$H_2C=C(OH)$ C_2H_5**) (and other higher ones) can undergo *molecular rearrangement or electroradicalization*. Methyl ethyl ketone (and higher ones) can *enolize* to give 2-butene vinyl alcohol ($H_3CHC=C(OH)CH_3$) which can *molecularly rearrange or electroradicalize* to methyl ethyl ketone. This is reversibility under different operating condions. Recall that in Equations 7.6a to 7.9a, aldehyde did not enolize, because of the operating conditions. One can observe the great distinction between Molecular Rearrangement, Electroradicalization and Enolization phenomena. These are some parts of TAUTORIZATION phenomena. One can see how Nature operates using the "Eye of the Needle". In Stage 1 of Equation 7.17, it was thought that the Equilibrium State of Existence of the methyl vinyl alcohol in the last step of that Stage was suppressed by the presence of the concentrated acid, that which was not the case. Hence the overall Equation- Equation 7.18 was marked with triple asterisks, because that stage was indeed unproductive. On the other hand, the Equilibrium state of existence of acetone in that equation, is not the real state (See Equation 7.23). Therefore, the real Stage 1 is as follows.

Stage 1 of Equation 7.17:

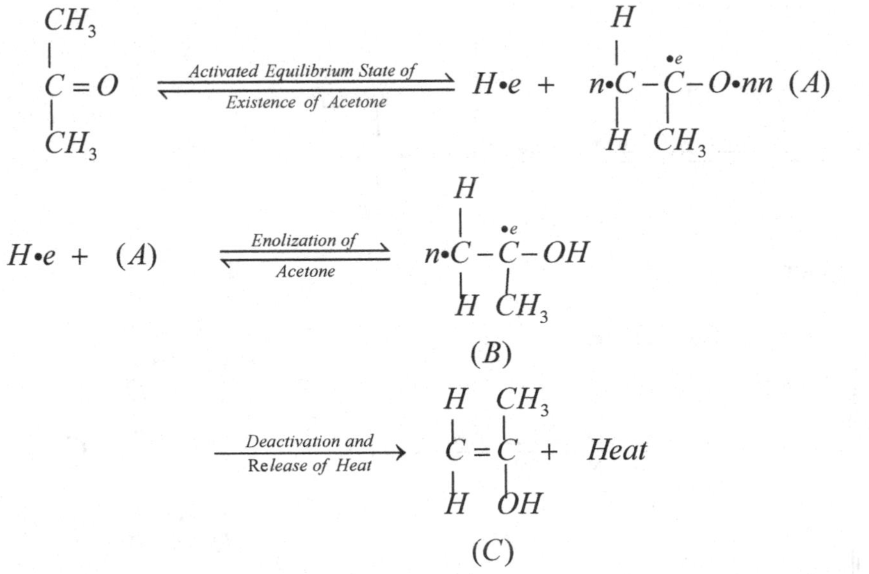

$$
\begin{array}{c}
CH_3 \\
| \\
C = O \\
| \\
CH_3
\end{array}
\xrightleftharpoons[\text{Existence of Acetone}]{\text{Activated Equilibrium State of}}
H{\bullet}e \; + \;
\begin{array}{c}
H \\
| \quad \overset{\bullet e}{} \\
n{\bullet}C - C - O{\bullet}nn \;\; (A) \\
| \quad | \\
H \;\; CH_3
\end{array}
$$

$$
H{\bullet}e \; + \; (A)
\xrightleftharpoons[\text{Acetone}]{\text{Enolization of}}
\begin{array}{c}
H \\
| \quad \overset{\bullet e}{} \\
n{\bullet}C - C - OH \\
| \quad | \\
H \;\; CH_3
\end{array}
\qquad (7.17)
$$

$$
(B)
$$

$$
\xrightarrow[\text{Release of Heat}]{\text{Deactivation and}}
\begin{array}{c}
H \;\; CH_3 \\
| \quad | \\
C = C \quad + \quad Heat \\
| \quad | \\
H \;\; OH
\end{array}
$$

$$
(C)
$$

Overall Equation: $3H_3C\text{-}CO\text{-}CH_3 \xrightarrow{\text{Tautomerization}} 3H_2C{=}C(OH)CH_3 \; + \; \text{Heat}$ (7.18)

The concentrated acid could not molecularly rearrange or keep (C) in Equilibrium State of Existence, but only in Stable State of Existence.

Stage 3:

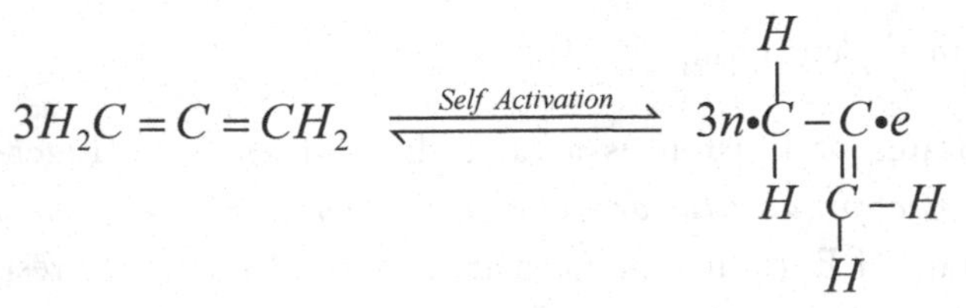

$$
3H_2C = C = CH_2
\xrightleftharpoons{\text{Self Activation}}
\begin{array}{c}
H \\
| \\
3n{\bullet}C - C{\bullet}e \\
| \quad \| \\
H \;\; C - H \\
\;\; | \\
\;\; H
\end{array}
$$

$$\xleftrightarrow[\text{i.e., H atom transfer}]{\text{Molecular Rearrangement}} \quad 3e \cdot C = C \cdot n$$

with a CH_3 group above and H below the structure.

$$\xrightarrow[\text{of Heat}]{\text{Deactivation with release}}$$

(Mesitylene structure) $\qquad$ 7.24

$$\text{(Mesitylene)}$$

<u>**Overall Equation:**</u> $3H_2SO_4 + 3H_3C\text{-}CO\text{-}CH_3 \xrightarrow[\text{Cyclization}]{\text{Molecular Rearrangement and}}$ Mesitylene +

$$3H_2SO_4 + 3H_2O \qquad 7.25$$

The acetone has been deoxygenated using a strong acid. The allene formed in Stage 2 mole-cularly rearranges to activated methyl acetylene which adds to themselves to form the ring in Stage 3. ***Thus, tricyclization taking place can be found to depend on the type of mechanism and the number of moles used. It cannot be obtained via Combination mechanism or Decom-position mechanism, but only via Equilibrium mechanism.***

It is also known that ketones are not affected appreciably by ***concentrated sodium hydrogen hydroxide*** solutions. However, the "amide ion" from for example ***sodium amide***, known to be a stronger base than "hydroxide ion", and under anhydrous conditions, converts acetone into a six membered cyclic compound known as <u>isophorone.</u>[1] As an Art using universal literature data, the mechanism can be explained as follows.

$$\overset{\oplus}{C}(CH_3)_2 - O^{\ominus} + Na^{\oplus} + NH_2^{\ominus} \longrightarrow Na^{\oplus} + \overset{\oplus}{C}(CH_3)(CH_2^{\ominus}) - O^{\ominus} + NH_3$$

(I)a $\qquad\qquad\qquad\qquad\qquad\qquad\qquad\qquad$ (I)b

$$\longrightarrow Na^{\oplus} + NH_3 + \overset{\oplus}{C}(CH_2^{\ominus})_2 - O\text{-}H \longrightarrow Na^{\oplus} + H^{\oplus} + \overset{\ominus}{N}H_2 + \overset{\oplus}{C}(CH_2^{\ominus})_2 - OH$$

(II) $\qquad\qquad\qquad\qquad\qquad\qquad\qquad\qquad\qquad\qquad$ (III)

$$\longrightarrow Na^{\oplus} + \overset{H}{\underset{H}{C}} = C = \overset{H}{\underset{H}{C}} + \overset{\ominus}{N}H_2 + H_2O \xrightarrow{+ \text{(III)}} 2NaNH_2 + 2H_2O + \text{(IV)} +$$

(IV)

$$\overset{\ominus}{\underset{H}{\overset{H}{C}}} - \overset{+2}{C} - \overset{H}{\underset{H}{\overset{}{C}}}{}^{\ominus} \xrightarrow{+ \text{(I)b}} 2NaOH + NaNH_2 + NH_3 + \overset{H}{\underset{H}{C}} = C = \overset{H}{\underset{H}{C}}$$

424

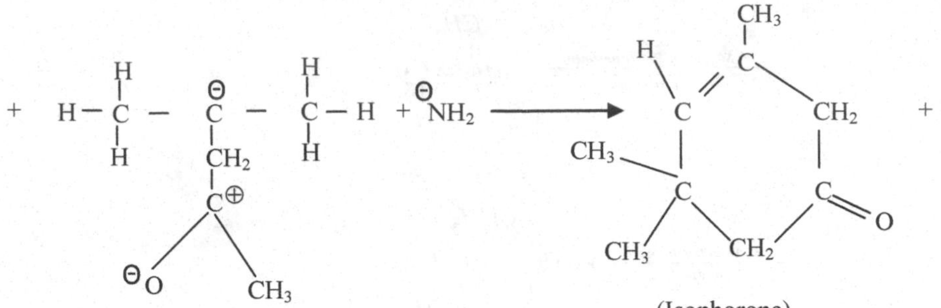

$$2NH_3 + NaNH_2 + 2NaOH \ (3NaNH_2 + 2H_2O)$$

7.26

The isophorone indicated above is different from what is probably known to be the case as shown below.

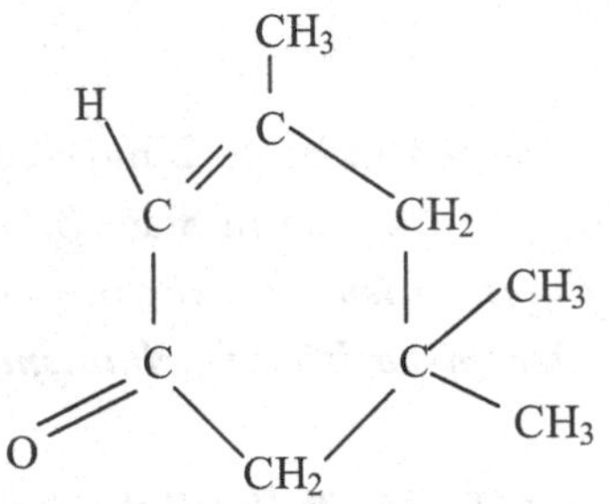

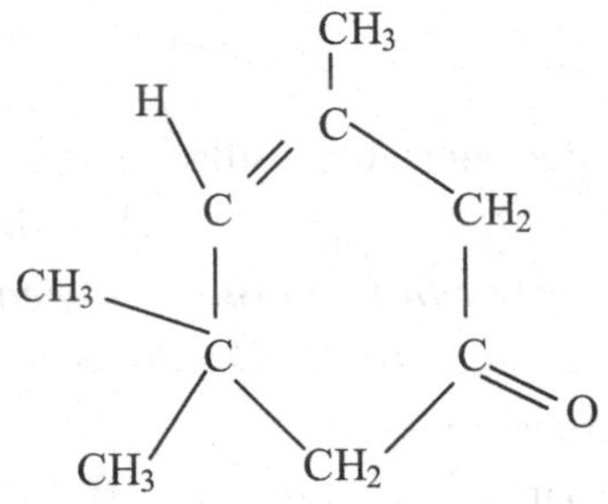

(I)   Isophorone[1]   Versus   (II)   Isophorone (Above)   7.27

Though (I) and (II) are both isometric, there must be something wrong particularly with respect to ionic expressions of chemical equations, when indeed no chemical reactions ever take place ionically. The expressions of the arrows we don't know, though we think we know. What charges are, we don't know, though we think we know? **Anything done as an Art is like living in an illusionary world and that has been our world since antiquity.** Nevertheless, without the Art, there is no beginning. An Art which is not limited by "Fear of the unknown" becomes a Natural Science (Mathematics, Physics and Chemistry), the "Alpha and Omega" of all disciplines. Now based on the New Frontiers very little of which has been seen so far, the mechanisms of the reaction above are as follows-

<u>Stage 1:</u>

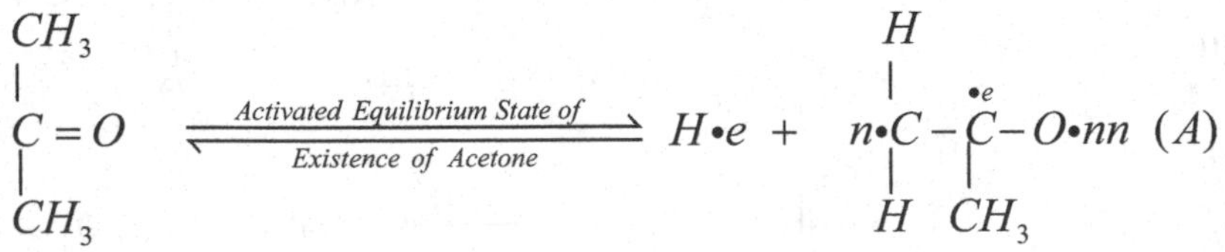

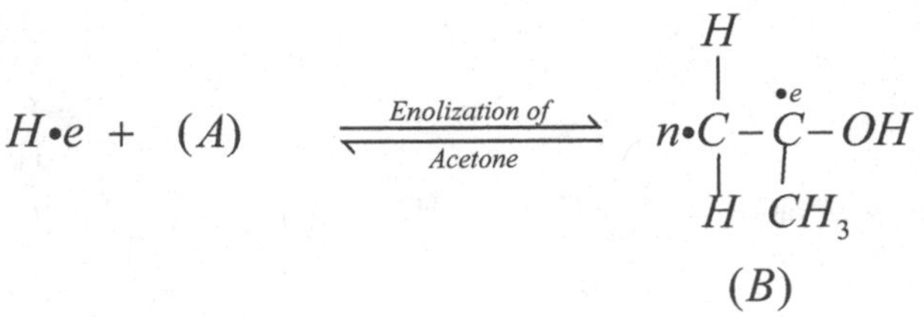

$$\xrightarrow[\text{Release of Heat}]{\text{Deactivation and}} \quad \underset{\overset{|}{H}}{\overset{\overset{|}{H}}{C}} = \underset{\overset{|}{OH}}{\overset{\overset{|}{CH_3}}{C}} \qquad \qquad 7.28$$

$$(C)$$

Overall Equation: $3H_3CCOCH_3 \xrightarrow{\text{Tautomerization}} 3H_2C=C(OH)CH_3$ 7.29

Stage 2:

$$NaNH_2 \xrightleftharpoons[\text{of Dry } NaNH_2]{\text{Equilibrium State of Existence}} Na \cdot e \; + \; nn \cdot NH_2$$

$$Na \cdot e \; + \; \underset{\overset{|}{HO}}{\overset{\overset{|}{CH_3}}{C}} = \underset{\overset{|}{H}}{\overset{\overset{|}{H}}{C}} \quad \xrightleftharpoons[\text{OH group}]{\text{Abtraction of}} \quad NaOH \; + \; e \cdot \underset{}{\overset{\overset{|}{CH_3}}{C}} = \underset{\overset{|}{H}}{\overset{\overset{|}{H}}{C}} \quad (B)$$

$$(B) \quad \xrightleftharpoons{\text{Release of H atom}} \quad H \cdot e \; + \; n \cdot \underset{\overset{|}{H}}{\overset{\overset{|}{H}}{C}} - \overset{\cdot e}{C} = \underset{\overset{|}{H}}{\overset{\overset{|}{H}}{C}} \quad (C)$$

$$H \cdot e \; + \; nn \cdot NH_2 \; \rightleftharpoons \; NH_3$$

$$(C) \quad \xrightarrow[(C)]{\text{Deactivation of}} \quad H_2C = C = CH_2$$

$$(Allene)$$

7.30

Overall Equation: $NaNH_2 \; + \; \underline{(C)} \longrightarrow NaOH \; + \; NH_3 \; + \; H_2C=C=CH_2$ 7.31

Overall Overall Equation: $NaNH_2 \; + \; 3H_3CCOCH_3 \longrightarrow NaOH \; + \; NH_3 \; + \; H_2C=C=CH_2$

$$+ \; 2H_2C=C(OH)CH_3 \quad \underline{(C)} \quad 7.32$$

Stage 3:

$$H_2C = C = CH_2 \quad \xrightleftharpoons{\text{Self Activation}} \quad n \cdot \underset{\overset{|}{H}}{\overset{\overset{|}{H}}{C}} - C \cdot e$$

$$\xrightleftharpoons[\text{i.e H atom transfer}]{\text{Molecular Rearrangement}} \quad e \cdot \underset{}{\overset{\overset{|}{CH_3}}{C}} = C \cdot n \quad (D)$$

$$(D) \; + \; 2\underset{\overset{|}{H}}{\overset{\overset{|}{H}}{C}} = \underset{\overset{|}{OH}}{\overset{\overset{|}{CH_3}}{C}} \quad \rightleftharpoons \quad n \cdot \overset{\overset{|}{H}}{C} = C - \underset{\overset{|}{CH_3}}{\overset{\overset{|}{H}}{C}} - \underset{\overset{|}{OH}}{\overset{\overset{|}{CH_3}}{C}} - \underset{\overset{|}{H}}{\overset{\overset{|}{H}}{C}} - \underset{\overset{|}{OH}}{\overset{\overset{|}{CH_3}}{C}} \cdot e$$

$$\xrightarrow[\text{Re lease of Heat}]{\text{Deactivation with}} \quad \begin{array}{c} C(CH_3) \\ HC \qquad CH_2 \\ (HO)C \qquad C(OH) \\ H_3C \qquad CH_2 \quad CH_3 \end{array} \quad (E) \qquad 7.33$$

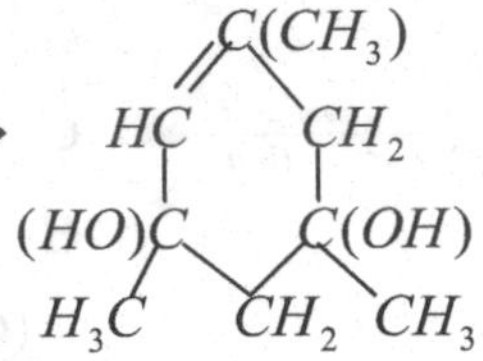

Overall Equation: $H_2C=C=CH_2 \; + \; 2H_2C=C(OH)CH_3 \; \xrightarrow[\text{Addition/Cyclization}]{\text{Molecular Rearrangement/}} \; (E)$ $\qquad$ 7.34

Stage 4:

$$(E) \updownarrow \xrightarrow[\text{of }(E)]{\text{Equilibrium State of Existence}} \begin{array}{c} C(CH_3) \\ HC \qquad CH_2 \\ gnnO-C \qquad C(OH) \\ H_3C \qquad CH_2 \quad CH_3 \end{array} \; + \; H\,ge \\ (F)$$

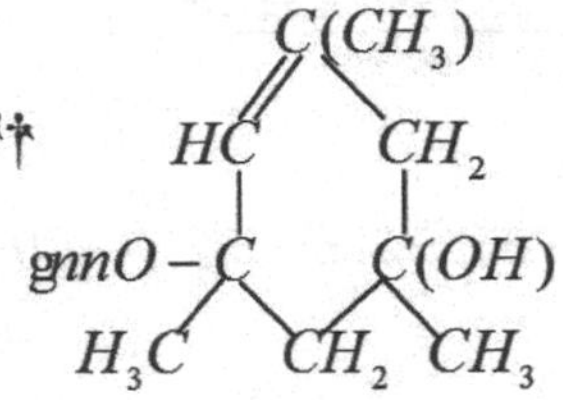

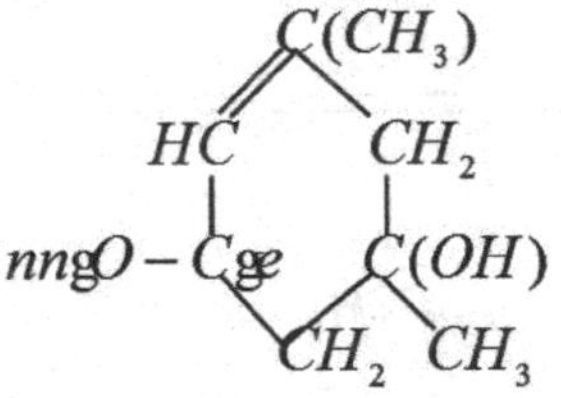

$$(F) \updownarrow \xrightarrow[\text{group}]{\text{Release of Methylide}} \begin{array}{c} C(CH_3) \\ HC \qquad CH_2 \\ nngO-C\,ge \qquad C(OH) \\ CH_2 \quad CH_3 \end{array} \; + \; ng\,CH_3 \\ (G)$$

$$H\,ge \; + \; ng\,CH_3 \updownarrow \xrightarrow[\text{Existence of } CH_4]{\text{Equilibrium State of}} \quad CH_4$$

$$(G) \xrightarrow[\text{Release of Heat}]{\text{Deactivation with}} \begin{array}{c} C(CH_3) \\ HC \qquad CH_2 \\ O=C \qquad C(OH) \\ CH_2 \quad CH_3 \end{array} \qquad 7.35 \\ (H)$$

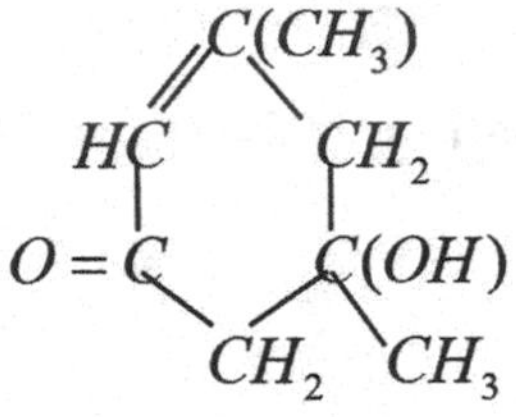

Overall Equation: $(E) \longrightarrow (H) \; + \; CH_4$ $\qquad$ 7.36

Overall Overall Equation: $NaNH_2 + 3H_3CCOCH_3 \longrightarrow NaOH \; + \; NH_3 \; + \; CH_4 \; + \; (H)$ 7.37

Stage 5:

$$CH_4 \xrightleftharpoons[\text{of Methane}]{\text{Equilibrium State of Existence}} H{\cdot}e \; + \; n{\cdot}CH_3$$

$$H{\cdot}e \; + \; NaOH \rightleftharpoons NaO{\cdot}en \; + \; H_2 \; + \; Heat$$

$$NaO{\cdot}en \rightleftharpoons Na{\cdot}e \; + \; nn{\cdot}O{\cdot}en \; + \; Heat$$

$$Na{\cdot}e \; + \; n{\cdot}CH_3 \xrightleftharpoons[\text{of } NaCH_3]{\text{Equilibrium State of Existence}} NaCH_3$$

$$nn{\cdot}O{\cdot}en \; + \; H_2 \rightleftharpoons H{\cdot}e \; + \; nn{\cdot}OH \qquad 7.38$$

$$H{\cdot}e \; + \; nn{\cdot}OH \xrightarrow[\text{by } NaCH_3]{\text{Equilibrium State suppressed}} H_2O$$

Overall Equation: $NaOH + CH_4 \longrightarrow NaCH_3 + H_2O + Heat$ 7.39

Stage 6: $NaCH_3 \underset{\text{of NaCH}}{\overset{\text{Equilibrium State of Existence}}{\rightleftharpoons}} Na{\cdot}e + n{\cdot}CH_3$

$$Na{\cdot}e + (H) \underset{\text{group}}{\overset{\text{Abstraction of OH}}{\rightleftharpoons}}$$

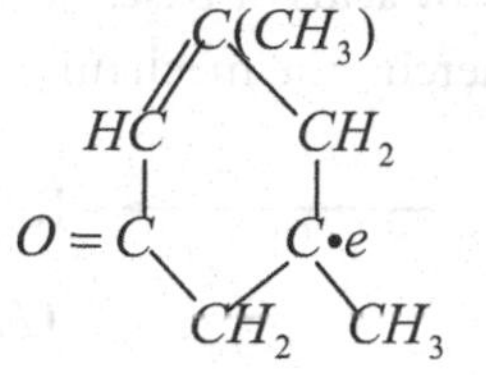

(I)

$$H_3C{\cdot}n + (I) \underset{\text{Existence of (J)}}{\overset{\text{Combination State of}}{\longrightarrow}}$$

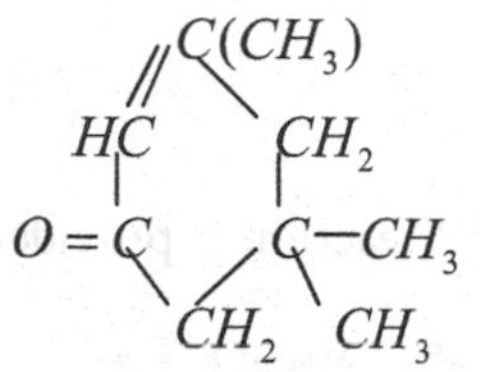

(J)   7.40

Overall Equation: $NaCH_3 + (H) \longrightarrow NaOH + $ Isophenone (J) $+ Heat$ 7.41

Overall Overall Equation: $NaNH_2 + 3H_3CCOCH_3 \longrightarrow NaOH + NH_3 + $ Isophenone $+ H_2O$ 7.42

Stage 7:

$$NH_3 \underset{\text{of Ammonia}}{\overset{\text{Equilibrium State of Existence}}{\rightleftharpoons}} H{\cdot}e + nn{\cdot}NH_2$$

$$H{\cdot}e + NaOH \rightleftharpoons NaO{\cdot}en + H_2 + Heat$$

$$NaO{\cdot}en \rightleftharpoons Na{\cdot}e + nn{\cdot}O{\cdot}en + Heat$$

$$Na{\cdot}e + nn{\cdot}NH_2 \underset{\text{of Sodium amide}}{\overset{\text{Equilibrium State of Existence}}{\rightleftharpoons}} NaNH_2$$

$$nn{\cdot}O{\cdot}en + H_2 \rightleftharpoons HO{\cdot}nn + H{\cdot}e \qquad 7.43$$

$$H{\cdot}e + nn{\cdot}OH \underset{\text{by NaNH}_2}{\overset{\text{Equilibrium State sup pressed}}{\longrightarrow}} H_2O$$

Overall Equation: $NaOH + NH_3 \longrightarrow NaNH_2 + H_2O + Heat$ 7.44

Overall Overall Equation: $NaNH_2 + 3H_3CCOCH_3 \longrightarrow NaNH_2 + $ Isophenone $2H_2O + Heat$ 7.45

As clearly shown above in stages, the true mechanism speaks for itself. "Free-ionically", NaCN or just Na have been reportedly used as initiators for polymerization of acrolein at -50 to -40°C proceeding completely through the carbonyl group, i.e. C=O center[2]. Based the mechanisms provided above, there is no doubt that whether it was conc. H_2SO_4, anhydrous $NaNH_2$, or conc. NaOH that is or was used as passive and active catalysts, the same products will be obtained *depending on the molar ratios or concentrations of the components involved, Mathematics being the natural language of communication.* In Stage 1, the catalyst is passive in control to allow for enolization. The active side of the catalyst started from Stage 2. The catalyst was finally recover-ed at the end, the time at which it was only possible. It could not be recovered after Stage 2 or Stage 4 for reasons which are obvious. ***How can a chain with an*** *initiator at the root, form a ring? How can a chain which has no root (real or imaginary), continue to grow? But some do depending on the operating conditions*

Aldehydes and ketones add to hydroxylamine, the hydroxyl derivative of ammonia, H_2NOH, to give *aldoximes and ketoximes* catalyzed by acids or bases, just in the same manner ammonia added to aldehyde (see Equations 7.4a and 7.4b). Therein, the mechanism was not fully shown in stages.

$$RCHO \ + \ H_2NOH \longrightarrow RCH=N(OH) \ + \ H_2O$$

$$\text{(An aldoxime)} \qquad\qquad 7.46a$$

$$R_2C=O \ + \ H_2NOH \longrightarrow R_2C=N(OH) \ + \ H_2O$$

$$\text{(A ketoxime)} \qquad\qquad 7.46b$$

Only the mechanism of the second reaction is provided below since the same will apply to alde-hydes.

Stage 1: $\quad H_2NOH \underset{\text{of Hydroxylamine}}{\overset{\text{Equilibrium State of Existence}}{\rightleftharpoons}} H{\bullet}e \ + \ nn{\bullet}NH(OH)$

$$H{\bullet}e \ + \ O=\overset{\displaystyle R}{\underset{\displaystyle R}{\overset{|}{\underset{|}{C}}}} \underset{\text{Addition to it}}{\overset{\text{Activation of Ketone and}}{\rightleftharpoons}} H-O-\overset{\displaystyle R}{\underset{\displaystyle R}{\overset{|}{\underset{|}{C}}}}{\bullet}e \ \ (A)$$

$$(A) \ + \ nn{\bullet}NH(OH) \underset{\text{Existence of (B)}}{\overset{\text{Combination State of}}{\longrightarrow}} H-O-\overset{\displaystyle R}{\underset{\displaystyle R}{\overset{|}{\underset{|}{C}}}}-\overset{}{\underset{\displaystyle H}{\overset{}{\underset{|}{N}}}}-OH \ \ (B)$$

$$7.47$$

Overall Equation: $H_2NOH \ + \ R_2C=O \longrightarrow HOCR_2\text{-}NH(OH)$ $\qquad 7.48$

Stage 2:

$$H-O-\overset{\displaystyle R}{\underset{\displaystyle R}{\overset{|}{\underset{|}{C}}}}-\overset{}{\underset{\displaystyle H}{\overset{}{\underset{|}{N}}}}-OH \underset{\text{of (B)}}{\overset{\text{Equilibrium State of Existence}}{\rightleftharpoons}} H{\bullet}e \ + \ H-O-\overset{\displaystyle R}{\underset{\displaystyle R}{\overset{|}{\underset{|}{C}}}}-\overset{\displaystyle OH}{\overset{|}{N}}{\bullet}nn \ \ (C)$$

$$(C) \underset{}{\overset{\text{Loss of OH group}}{\rightleftharpoons}} e{\bullet}\overset{\displaystyle R}{\underset{\displaystyle R}{\overset{|}{\underset{|}{C}}}}-\overset{\displaystyle OH}{\overset{|}{N}}{\bullet}nn \ + \ nn{\bullet}OH$$

$$H{\bullet}e \ + \ nn{\bullet}OH \rightleftharpoons H_2O$$

$$(C) \underset{\text{of Heat}}{\overset{\text{Deactivation with release}}{\longrightarrow}} \overset{\displaystyle R}{\underset{\displaystyle R}{\overset{|}{\underset{|}{C}}}}=\overset{\displaystyle OH}{\overset{|}{N}}$$

$$A \ ketoxime \qquad 7.49$$

Overall Equation: $H_2NOH \ + \ R_2C=O \longrightarrow R_2C=N(OH) \ + \ H_2O$ $\qquad 7.50$

With the use of ammonia in Equation 7.4a, it was in Stage 2, the aldimine ring was formed without allowing the aldimine to deactivate. If only one mole of ammonia was used, no ring will be formed.

The substituted hydrazines having one free amino group behave in a similar fashion as analogous to hydroxylamine. The hydrazines most used are phenylhydrazine, $C_6H_5NHNH_2$ and substituted phenylhydrazines. Why?

$$(CH_3)_2C=O \ + \ H_2NNHC_6H_5 \longrightarrow (CH_3)_2C=NNHC_6H_5 \ + \ H_2O$$

$$\textbf{Acetone phenylhydrazone} \qquad\qquad 7.51$$

Stage 1:

$$H_2N-NHC_6H_5 \underset{\textit{of the hydrazine}}{\overset{\textit{Equilibrium State of Existence}}{\rightleftharpoons}} H\bullet e \ + \ nn\bullet \overset{\overset{H}{|}}{N}-\underset{\underset{H}{|}}{N}-C_6H_5 \ (A)$$

$$H\bullet e + O=\overset{\overset{R}{|}}{\underset{\underset{R}{|}}{C}} \underset{\textit{and Addition}}{\overset{\textit{Activation of Ketone}}{\rightleftharpoons}} H-O-\overset{\overset{R}{|}}{\underset{\underset{R}{|}}{C}}\bullet e \ (B)$$

$$7.52$$

$$(B) \ + \ (A) \underset{\textit{of }(D)}{\overset{\textit{Combination State of Existence}}{\longrightarrow}} H-O-\overset{\overset{R}{|}}{\underset{\underset{R}{|}}{C}}-\overset{\overset{H}{|}}{\underset{\underset{H}{|}}{N}}-N-C_6H_5$$

$$(D)$$

Overall Equation: $R_2C=O \ + \ H_2N\text{-}NHC_6H_5 \longrightarrow (OH)CR_2\text{-}NH\text{-}NHC_6H_5$ \qquad 7.53

Stage 2:

$$(D) \underset{\textit{of }(D)}{\overset{\textit{Equilibrium State of Existence}}{\rightleftharpoons}} H\bullet e \ + \ H-O-\overset{\overset{R}{|}}{\underset{\underset{R}{|}}{C}}-\overset{\overset{\bullet nn}{}}{N}-\underset{\underset{H}{|}}{N}-C_6H_5$$

$$(E)$$

$$(E) \underset{}{\overset{\textit{Release of OH group}}{\rightleftharpoons}} HO\bullet nn \ + \ e\bullet\overset{\overset{R}{|}}{\underset{\underset{R}{|}}{C}}-\overset{\overset{NH(C_6H_5)}{|}}{N}\bullet nn$$

$$7.54$$

$$(F)$$

$$H\bullet e + nn\bullet OH \rightleftharpoons H_2O$$

$$(F) \xrightarrow[\textit{release of Heat}]{\textit{Deactivation of }(F)\textit{ and}} R_2C=N-NH(C_6H_5) \ + \ Heat$$

Overall Equation: $R_2C=O \ + \ H_2N\text{-}NHC_6H_5 \longrightarrow H_2O \ + \ R_2C=N\text{-}NH(C_6H_5) \ + \ Heat$

$$7.55$$

The need to show that ketones can be activated was necessitated by the fact that azo-alkanes (R-N=N-R) cannot be activated by any means except inside special pores which breaks down not only the π-bonds but also the σ-bonds, as will be shown downstream. For example ask yourself, how does Nature produce ammonia from N_2 and H_2 at very low operating conditions, not in the way we do industrially today in our physical world? If the phenyl group had been an alkyl group another product would have been obtained after Stage 1 above, because of the following differences in their (D)s Equilibrium State of Existence.

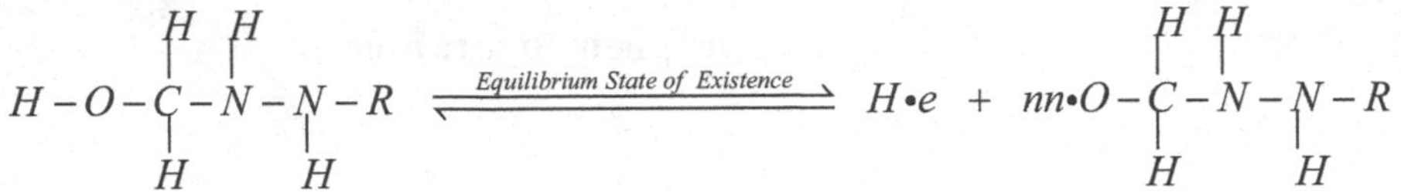

$$H-O-\underset{\underset{H}{|}}{\overset{\overset{H}{|}}{C}}-\underset{}{\overset{\overset{H}{|}}{N}}-\underset{\underset{H}{|}}{N}-R \quad \underset{\xrightarrow{\hspace{2cm}}}{\overset{Equilibrium\ State\ of\ Existence}{\rightleftharpoons}} \quad H\bullet e \ + \ nn\bullet O-\underset{\underset{H}{|}}{\overset{\overset{H}{|}}{C}}-\overset{\overset{H}{|}}{N}-\underset{\underset{H}{|}}{N}-R \qquad 7.56$$

$$H-O-\underset{\underset{H}{|}}{\overset{\overset{H}{|}}{C}}-\overset{\overset{H}{|}}{N}-\underset{\underset{H}{|}}{N}-C_6H_5 \quad \underset{\xrightarrow{\hspace{2cm}}}{\overset{Equilibrium\ State\ of\ Existence}{\rightleftharpoons}} \quad H\bullet e \ + \ H-O-\underset{\underset{H}{|}}{\overset{\overset{H}{|}}{C}}-\overset{\overset{\bullet nn}{}}{N}-\underset{\underset{H}{|}}{N}-C_6H_5 \qquad 7.57$$

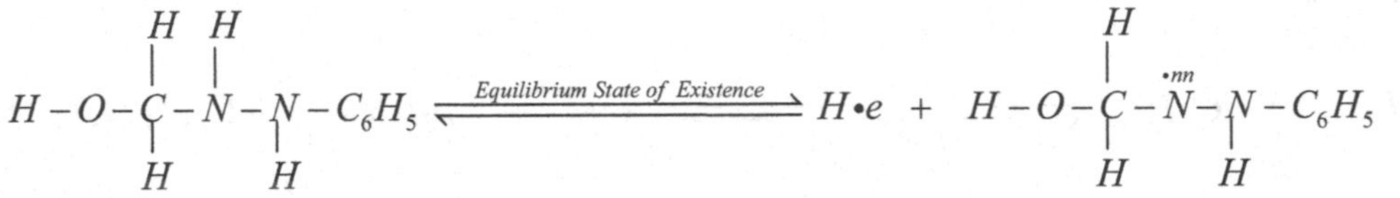

Why this is so will be explained downstream in the Volumes.

Aldimines, aldehydes, isocyanic acid, alkenes, and etc., favor existence of cyclic trimer or tetramers, not because of "self-activation", but because of how the reactions take place based on the operating conditions and molar ratios of components involved. How isocyanic acid favors cyclic trimerization will be explained in the Series and why six-membered rings are more common than four- or eight or ten-membered rings, will be explained. Ability to "self-activate" depends on how stable the monomer is and the operating conditions. From what has been seen so far above, the ability to "self-activate" has nothing to do with "low ceiling temperatures" of the polymerization of some monomers. The polymerization of isocyanates for example is said to have a low ceiling temperature and cyclic trimmer formation is the major product at temperatures above about -20 °C.[2]

Though "ceiling temperature" has not been properly defined in the literature [2,4-8], there is need to look more closely at the phenomenon with the new foundations. ***The "ceiling tempera-ture" is said to be the temperature at which the propagation and depropagation rates are equal.*** This definition is correct, but incomplete. A growing polymer chain cannot depropagate at the same time, and when there is depropagation during polymerization (a form of degradation of a polymer chain), the temperature has to be high and/or the chain must have an ionic bond at the end(s), that which is common with carbonyl, isocyanic, aldiminic, Step polymeric chains and more. Finally, the method of Addition polymerization has to be uniquely different from the well-known methods as has been shown above. On the other hand, it has been said that with the prime exception of formaldehyde the first member of the family, carbonyl monomers have low ceiling temperatures[2]. It has been observed that most of the early attempts to polymerize carbonyl compounds were carried out at temperatures that were, in retrospect, too high. The use of temperature above the "ceiling temperature" resulted in the absence of polymer formation due to unfavorable equilibrium between monomer and polymer. Carbonyl monomers were said to be successfully polymerized to high polymer when the polymerization reactions were carried out at temperatures below the ceiling temperatures of the monomers.[2] These observations for carbonyl monomers, isocyanates, etc. clearly provide another definition of "ceiling temperature" different from the one above. Hence, in the present light, *"the ceiling temperature" is that temperature where the rate of propagation of non-olefinic monomers via electro-free-radical route of polymerization using Hydrogen atom equals the rate of depropagation.* This definition is still not complete. Below that temperature, these unique family of monomers with the usual exception of their first members, can no longer depropagate when a monomer is about to be added. For such monomers, higher temperature polymerization above the ceiling temperature can effectively take place without depropagation if the chain is growing from the front, i.e., following the traditional method of Addition. Reported cases of styrenes, methyl methacrylates, etc. having high ceiling temperatures[6-8] have been addressed in Chapter 3 for olefin monomers. Since these olefinic monomers cannot favor Addition polymerization from the back, the concept of their having "ceiling temperatures" is very limited. For example, for styrenes a different phenomenon as already provided with respect to effect of resonance stabilization phenomena takes place. These are monomers which radically favor routes natural and unnatural to them. It does not take place with methyl methacrylate. When polymerized **in the route not natural to it and transfer species exist, but cannot be released, then depropagation takes place** for which the polymerization temperature must be reduced to prevent it. Most importantly, these monomers are completely different from the aldehydes and ketones where **the route is natural to the monomer with transfer species which cannot be released,** because the monomer is growing backwardly. Though, formaldehyde, trifluoroacetaldehyde, and other fully halogenated (probably except iodine) aldehydes and ketones, cannot enolize, most can undergo cyclic trimerization, such as trioxane from formaldehyde. Cyclic Trimerization and Tetramerization can also take place with even acetylene under unique operating conditions. From the considerations above, one can now start seeing the complete definition of "Ceiling Temperature". Two definitions exist, one for Non-Olefinic and the other for Olefinic monomers. They both take place only RADICALLY.

When urea is strongly heated above its melting point, it decomposes into ammonia and isocyanic acid, which is said to be tautomeric with cyanic acid[9]. In Equation 7.1, one showed how cyanic acid molecularly rearranges to isocyanic acid. Shown below are their Equilibrium States of Existence.

$$H-N=C=O \rightleftharpoons H\bullet e + nn\bullet N=\underset{O}{\overset{||}{C}} \quad ; \quad N\equiv\underset{OH}{\overset{|}{C}} \rightleftharpoons H\bullet e + N\equiv\underset{O\bullet nn}{\overset{|}{C}} \qquad 7.58$$

Isocyanic Acid   **Cyanic Acid**

The isocyanic acid is said to polymerize at once to a mixture of about 30 percent of the linear polymer, **cyamelide**, and 70 percent of the trimmer, **cyanuric acid**. Notice that both the trimmer and polymer were produced at the same time without the use of externally introduced initiators.

$$H_2NCONH_2 \xrightarrow{\text{Heat}} NH_3 + HN=C=O \rightleftharpoons N\equiv COH \; ; \; xHN=C=O \xrightarrow{\text{Heat}} \left[-NH-\overset{O}{\overset{||}{C}}- \right]_x$$

                                    Isocyanic acid   cyanic acid   30%   7.59

$$3HNCO \longrightarrow \text{(I)} \rightleftharpoons \text{(II)}$$

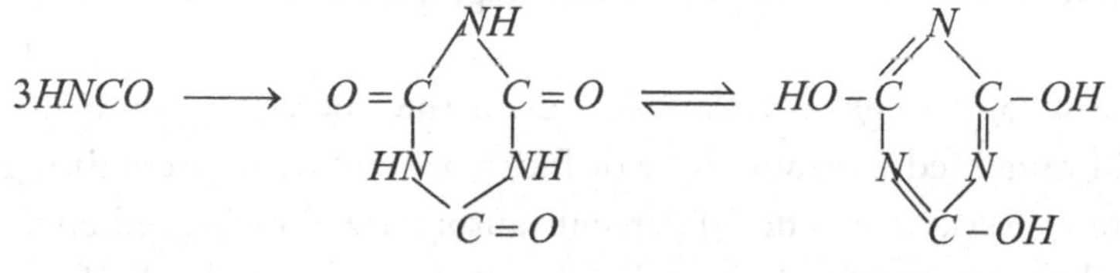

              (I)   Cyanuric acid   (II)   70%   7.60

While isocyanic acid can tautomerize to cyanic acid by Electroradicalization, it cannot enolize and molecularly rearrange to cyanic acid, because the C=O center cannot be activated

in the presence of C=N center, C=O center being more nucleophilic (see Chapter 5). Cyanic acid can molecularly rearrange and enamidize to isocyanic acid. So one can observe that isocyanic acid is more stable than cyanic acid. From the last equation above, notice that the ring either enamidized or molecularly rearranged or the two rings must have come from the two acids. How can a polymer appear without the use of an initiator? Imagine if dilute HCl is used as an initiator, the followings would be obtained both when the acid is stabilized and in Equilibrium state of existence at different times where possible. If at the beginning the acid was in stable state of existence, the Addition is by Combination mechanism. If it was in Equilibrium state of existence as is usually the case, the Addition is by Equilibrium mechanism.

$$Cl^{\ominus} - (\overset{\overset{O}{\parallel}}{C} - \overset{\overset{H}{\mid}}{N})_m - \overset{\overset{O}{\parallel}}{C} - \overset{\overset{H}{\mid}}{N}^{\ominus} \ldots\ldots \overset{\oplus}{\underset{\overset{\mid}{H}}{O}} - (\overset{\overset{H}{\mid}}{C} - \overset{\overset{O}{\parallel}}{N})_n - \overset{\overset{H}{\mid}}{C} - \overset{\overset{O}{\parallel}}{NH_2} \qquad 7.61$$

Both centers of the initiator cannot be involved at the same time. The monomer being an Electrophile, the coordination center will be the first to be used. As is clearly obvious from the end of the right hand side of the chain, there are bound to be depropagation if that center was used and the temperature is above the "ceiling temperature" of the growing polymer chain which no doubt will be higher than that of formaldehyde. When the urea was heated strongly above its melting point, the followings took place.

Stage 1:

$$H_2N\overset{\overset{O}{\parallel}}{C}NH_2 \rightleftharpoons H\bullet e + nn\bullet\overset{\overset{H}{\mid}}{N} - \overset{\overset{O}{\parallel}}{C} - NH_2$$

$$(A)$$

$$(A) \rightleftharpoons H_2N\bullet nn + nn\bullet\overset{\overset{H}{\mid}}{N} - \overset{\overset{O}{\parallel}}{C}\bullet e \quad (B)$$

$$H\bullet e + nn\bullet NH_2 \rightleftharpoons NH_3$$

$$(B) \xrightarrow[\text{Release of Heat}]{\text{Deactivation with}} H - N = C = O + Heat \qquad 7.62$$

Overall Equation: $1000H_2N\text{-}CO\text{-}NH_2 \longrightarrow 1000NH_3 + 1000H\text{-}N\text{=}C\text{=}O + Heat$ \quad 7.63

Without the activated isocyanic acid deactivating in the last step, the stage can never be complete and the stage will become unproductive. The isocyanic acid is thus in Stable state of Existence despite the presence of heat. In the second stage to follow, based on the operating conditions, about 30 percent of the isocyanic acid remained in Stable state of Existence, while the remaining 70 percent was put in Equilibrium state of Existence. The 70 percent tautomerized or indeed enolized to form activated cyanic acid. Whether tautomerized or enolized, one stage is involved. If the conversion of isocyanic acid to cyanuric acid via Enolization (Not Tautomerism (i.e., Electroradicalization)) is complete, then there will be no polymer. Hence, it is with isocyanic acid, the two products from literature data become existent.

Stage 2:

$$3H - N = C = O \rightleftharpoons 3H{\cdot}e \;+\; 3nn{\cdot}N = \overset{{\cdot}e}{\ddot{C}} - O{\cdot}nn$$

$$\rightleftharpoons 3nn{\cdot}N = \overset{{\cdot}e}{\ddot{C}} - O{\cdot}nn \;+\; 3H{\cdot}e$$

$$\rightleftharpoons 3nn{\cdot}N = \overset{{\cdot}e}{\ddot{C}} - OH$$

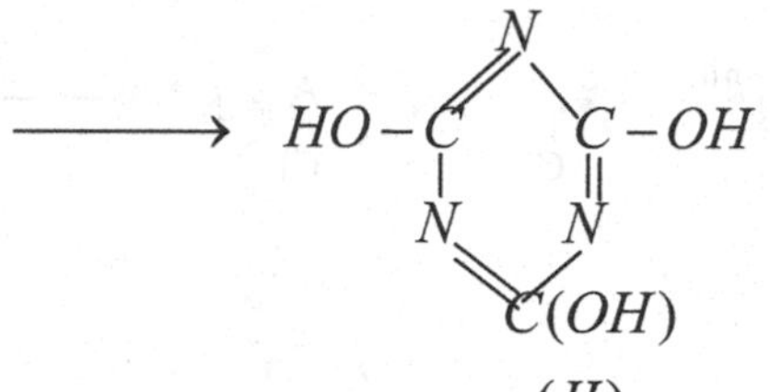

$$\longrightarrow \quad \underset{(II)}{\underset{\text{ring structure}}{HO - C \cdots C - OH}}$$

ENOLIZATION   7.64

Overall Equation: $699\,\text{H-N=C=O} \longrightarrow 233\,(II)$ 7.65

It is (II) of Equation 7.60 that is formed and this readily molecularly rearranges to give (I).

With one mole of isocyanic acid left, the initiator is prepared in the next stage as follows using either the isocyanic acid in Equilibrium state of Existence and ammonia in Stable state of existence.

Stage 3:

$$H - N = C = O \rightleftharpoons H{\cdot}e \;+\; nn{\cdot}N = C = O \;\; (A)$$

$$H{\cdot}e \;+\; NH_3 \rightleftharpoons en{\cdot}NH_4 \;\; (B)$$

$$(B) \;+\; (A) \longrightarrow O = C = N^{\ominus} \ldots\ldots^{\oplus}NH_4$$ 7.66

$$(C) \;\; Initiator$$

Overall Equation: $2\,\text{H-N=C=O} \longrightarrow (C)$ 7.67

This is the initiator used to polymerize the remaining 30 percent of the isocyanic acid in Stable state of Existence to produce a chain similar to that shown in Equation 7.61 and shown below,

$$O = C = N - (\overset{O}{\overset{\|}{C}} - \overset{H}{\overset{|}{N}})_n - \overset{O}{\overset{\|}{C}} - \overset{H}{\overset{|}{N}}{}^{\ominus} \ldots\ldots {}^{\oplus}NH_3 - (\overset{O}{\overset{\|}{C}} - \overset{H}{\overset{|}{N}})_{299-n} - \overset{O}{\overset{\|}{C}} - \overset{H}{\overset{|}{N}} - H$$ 7.68

where n can go from 0 to a small number. It is the left-hand side that is first or only used because the monomer is an Electrophile with no transfer species even when the hydrogen is replaced with alkyl groups. Addition here takes place via Combination mechanism for the left-hand side and Equilibrium mechanism for the right-hand side. With the C-N bond as opposed to C-O bond, trimerization cannot readily take place from the left-hand side. When the optimum chain length has been reached, then the right-hand side takes over with addition taking place via Equilibrium mechanism.

Like ketones, ketimines will not readily favor self-activation and trimerize. When they do, under vigorous conditions, similar products as with ketones will be obtained. Cyanogen halides are said to be stable when pure, but "polymerize" readily in the presence of free halogen to give cyanuryl halides, "the acid halides" of cyanuric acid.[9] The mechanism is obviously Equilibrium in just a single stage. The free halogen could not add to polymerize it due to electrostatic forces of repulsion coming from the adjacently located paired unbonded radicals and the π-bond. Even then, the free halogen cannot be used as an initiator alone by itself.

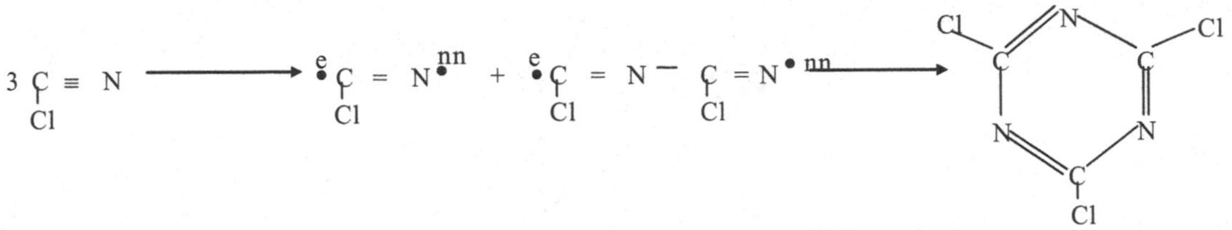

(Cyanuryl choride)   7.69

The free halogen passively keeps the cyanogen chloride in Activated state of Existence. Addition takes place from the electro-free-radical end to form a trimmer which cyclizes to form a six-membered ring by deactivation, all in one stage.

When dicyandiamide is heated in the presence of anhydrous ammonia and methyl alcohol, melamine, the cyclic trimmer of cyanamide, is formed.[9]

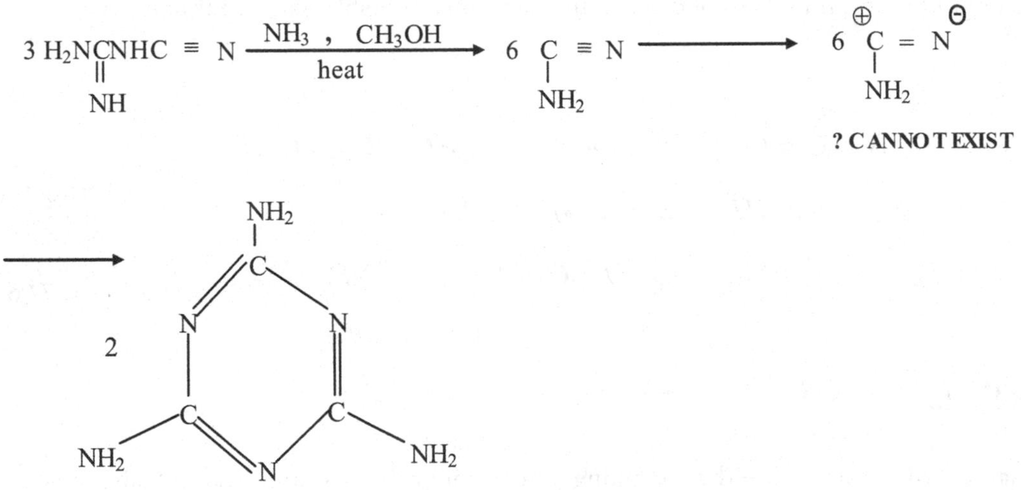

? CANNOT EXIST

Melamine   7.70

It can largely be observed how cyclization takes place in these monomers which are largely nucleophiles. One wonders why both dry ammonia and methyl alcohol are present in the system, when only one of them should do the job. We have already shown as rule in the last Chapter how carbodiimide (HN=C=NH) molecularly rearranges to give cyanamide ($H_2NC \equiv N$) the product above. We have also stated as a rule that NITRILES triply bonded cannot be activated chargedly. One can observe why question mark has been placed in the last Equation above. This clearly shows that all these reactions take place radically. At pH < 5, cyanamide is known to be stable in aqueous solution, but dimerizes to dicyandiamide at pH 7-12. What is or are the operating conditions providing the difference in pH levels? Is it WATER? Is it in addition to water the presence of acids or bases?

$$H_2NC\equiv N \;\; + \;\; H_2NC\equiv N \;\; \xrightarrow{\;pH\;7\text{--}12\;} \;\; \underset{\underset{\displaystyle NH}{\|}}{H_2NCNHC\equiv N}$$

$$\text{Dicyandiamide} \qquad\qquad 7.71$$

The dicyandiamide was obtained as follows. 50 percent of the cyanamide undergo **Enamidi-zation** another form of Tautomerization to give carbodiimide.

Stage 1:

$$H_2NC\equiv N \;\; \rightleftharpoons \;\; H\bullet e \;\; + \;\; nn\bullet N - \overset{\displaystyle \overset{H}{|}}{\overset{\cdot e}{C}} = N\bullet nn$$

$$\xrightarrow[\;\;\;\;\;\;\;\;]{Enamidization} \;\; nn\bullet N - \underset{\underset{\displaystyle NH}{\|}}{\overset{\displaystyle \overset{H}{|}}{C}}\bullet e \qquad\qquad 7.72$$

$$\xrightarrow{\;Deactivation\;} \;\; HN = C = NH \;\; + \;\; Heat$$

Overall Equation: $H_2NC\equiv N \;\; \xrightarrow{\;Enamidization\;} \;\; HN{=}C{=}NH \;\; + \;\; Heat \qquad 7.73$

One can observe that the phenomenon here is not Electroradicalization, but another form of Enolization. Like aldehydes and ketones, cyanamide can also exist in Activated/Equilibrium State of Existence, and the reason is because it has all that are needed to exist as such when the operating conditions exist. Equilibrium States of Existences are finger-prints of all compounds. Cyanamide with a triple bond is LESS nucleophilic than carbodi-imide with two double bonded C=N centers. But the full conversion is not complete. Hence the followings take place in Stage 2.

Stage 2:

$$H_2N - C \equiv N \;\; \underset{Existence}{\overset{Equilibrium\;State\;of}{\rightleftharpoons}} \;\; H\bullet e \;\; + \;\; nn\bullet N - C \equiv N^{\displaystyle \overset{H}{|}}$$

$$H\bullet e \;\; + \;\; HN = C = NH \;\; \xrightarrow{\;Activation/\,Addition\;} \;\; H - \underset{\underset{\displaystyle NH}{\|}}{\overset{\displaystyle \overset{H}{|}}{N} - C}\bullet e \;\; (B) \qquad (A)$$

$$(A) \;\; + \;\; (B) \;\; \xrightarrow{\;Combination/\,Deactivation\;} \;\; H - \overset{\displaystyle \overset{H}{|}}{N} - \underset{\underset{\displaystyle NH}{\|}}{C} - \overset{\displaystyle \overset{H}{|}}{N} - C \equiv N \qquad 7.74$$

$$\textit{Dicyan}\,\text{di}\,a\,\text{m}\,ide$$

Overall Equation: $H_2N{-}C\equiv N \;\; + \;\; HN{=}C{=}NH \;\; \xrightarrow{\;\;\;\;\;\;} \;\; Dicyandiamide \qquad 7.75$

In Equation 7.70, the methyl alcohol which is an organic acid was there to keep the pH low enough so that cyanamide when formed does not enamidize. The ammonia something which the so-called alcohol would have done, acted as an active catalyst as shown below when the dicyandiamide was heated.

Stage 1:

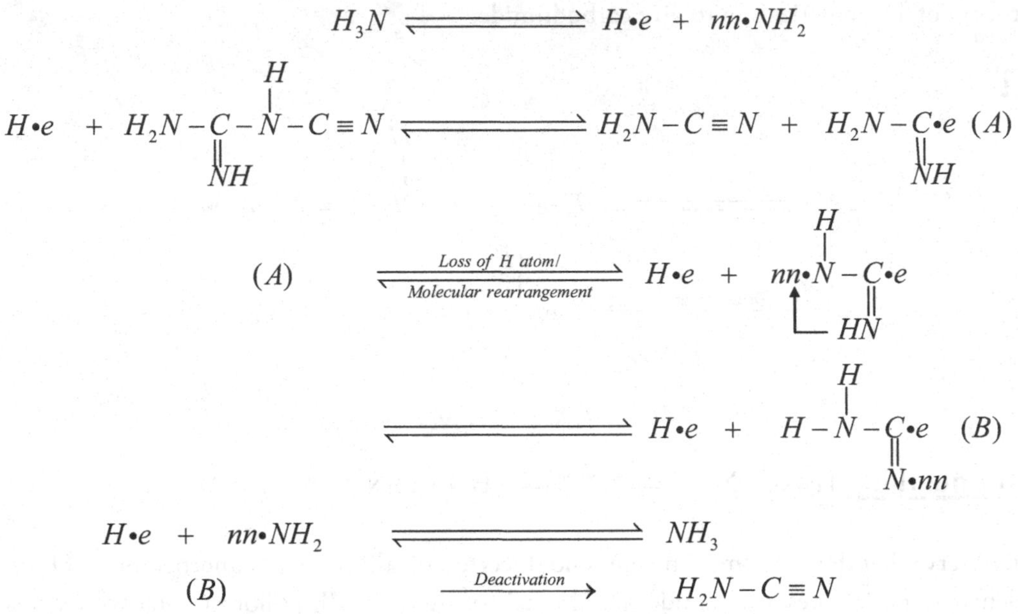

$$H_3N \; \rightleftharpoons \; H{\bullet}e \;+\; nn{\bullet}NH_2$$

$$H{\bullet}e \;+\; \underset{\underset{NH}{\|}}{H_2N-C}-\underset{\underset{}{\overset{\overset{H}{|}}{N}}}-C\equiv N \;\rightleftharpoons\; H_2N-C\equiv N \;+\; \underset{\underset{NH}{\|}}{H_2N-C}{\bullet}e \quad (A)$$

$$(A) \quad \xrightarrow[\text{Molecular rearrangement}]{\text{Loss of H atom/}} \quad H{\bullet}e \;+\; nn{\bullet}\underset{\underset{HN}{\|}}{N}-\overset{\overset{H}{|}}{C}{\bullet}e$$

$$\rightleftharpoons \quad H{\bullet}e \;+\; H-\overset{\overset{H}{|}}{N}-\underset{\underset{N{\bullet}nn}{\|}}{C}{\bullet}e \quad (B)$$

$$H{\bullet}e \;+\; nn{\bullet}NH_2 \quad \rightleftharpoons \quad NH_3$$

$$(B) \quad \xrightarrow{\text{Deactivation}} \quad H_2N-C\equiv N$$

7.76

Overall Equation: $H_3N \;+\; H_2N\text{-}(C{=}NH)\text{–}NH\text{-}C{\equiv}N \xrightarrow{\text{Heat}} H_3N \;+\; 2H_2N\text{-}C{\equiv}N$ 7.77

One wonders why the need for ammonia whose pH is high. Based on the number of moles used above, the products shown above are obtained. But with more than three moles of the dicyandi-amide and ammonia, the ring is formed in the same stage above.

Stage 1:

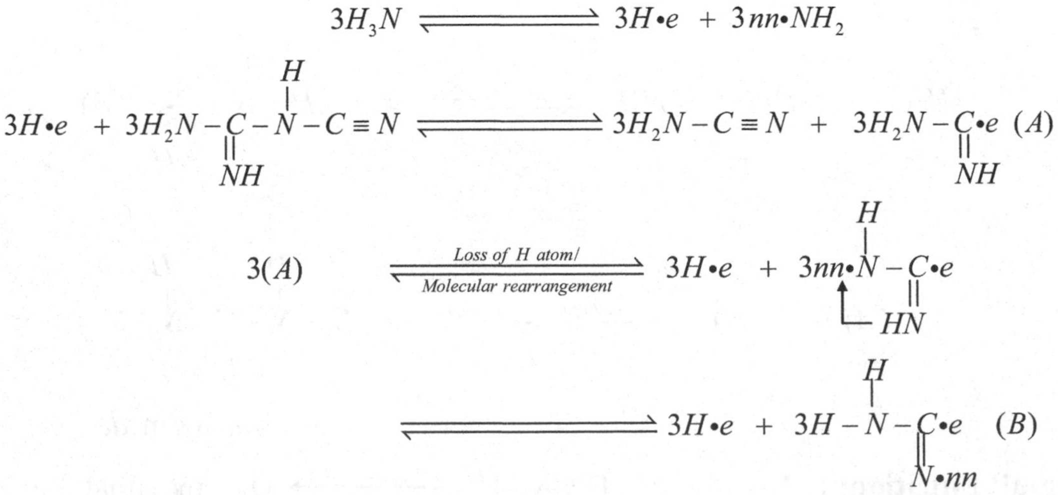

$$3H_3N \; \rightleftharpoons \; 3H{\bullet}e \;+\; 3nn{\bullet}NH_2$$

$$3H{\bullet}e \;+\; 3\underset{\underset{NH}{\|}}{H_2N-C}-\overset{\overset{H}{|}}{N}-C\equiv N \;\rightleftharpoons\; 3H_2N-C\equiv N \;+\; 3\underset{\underset{NH}{\|}}{H_2N-C}{\bullet}e \quad (A)$$

$$3(A) \quad \xrightarrow[\text{Molecular rearrangement}]{\text{Loss of H atom/}} \quad 3H{\bullet}e \;+\; 3nn{\bullet}\underset{\underset{HN}{\|}}{N}-\overset{\overset{H}{|}}{C}{\bullet}e$$

$$\rightleftharpoons \quad 3H{\bullet}e \;+\; 3H-\overset{\overset{H}{|}}{N}-\underset{\underset{N{\bullet}nn}{\|}}{C}{\bullet}e \quad (B)$$

$$3H \bullet e \ + \ 3nn \bullet NH_2 \ \rightleftharpoons \ 3NH_3$$

$$3(B) \quad \underset{\longleftarrow}{\xrightarrow{\textit{Activation/ Addition}}} \quad nn \bullet N = \underset{\underset{NH_2}{|}}{C} - N = \underset{\underset{NH_2}{|}}{C} - N = C \bullet e$$

$$\longrightarrow \quad \textit{Melamine}$$

7.78

Overall Equation: $3H_3N \ + \ 3H_2N\text{-}(C=NH)\text{-}NH\text{-}C\equiv N \ \longrightarrow \ \text{Melamine} \ + \ 3NH_3 \ +$

$$3\,H_2N - C \equiv N \qquad 7.79$$

When Trimerization takes place via Addition polymerization, it is not a polymer that is produc-ed, but a cyclic compound. When a monomer can be trimerized, it can also be polymerized. For the first time, one has seen another form of Tautomerization, called herein **Enamidization (analogous to Enolization)** limited to the families of cyanamide, aldimine and ketimines, just as Enolization is limited to the families of Aldehydes and Ketones. When an initiator such as dilute HCl is used to polymerize this monomer, the followings are obtained when the initiator is in Equilibrium state of existence.

$$Cl^{\ominus}........^{\oplus}\underset{\underset{H}{|}}{\overset{H}{O}} - \underset{\underset{H}{|}}{\overset{H}{C}} = N - (\underset{\underset{}{}}{\overset{NH_2}{C}} = N)_n - \underset{}{\overset{NH_2}{C}} = N - H \qquad 7.80$$

The monomer being strongly nucleophilic (i.e. a Female) with transfer species of the first kind of the first type, cannot be polymerized by the left-hand side, but only electro-free-radically using the right-hand side, just like the aldehydes and ketones. With this method of polymerization, this monomer has a "ceiling temperature" whose value will be lower than that of cyanogen chloride used in Equation 7.69 which can use both sides of the initiator. If the polymerization tempera-ture is greater than the "ceiling temperature", then depropagation commences.

7.2 Aldimines and Ketimines

Aldimines and ketimines are products of the following type of unstable alkene monomers shown below through molecular rearrangement or tautomerization. These are the imines of aldehydes and ketones respectively. Like formaldehyde ($O=CH_2$), the first member is presently called methanimine ($HN=CH_2$) yet to be isolated. It can as well be called formaldimine.

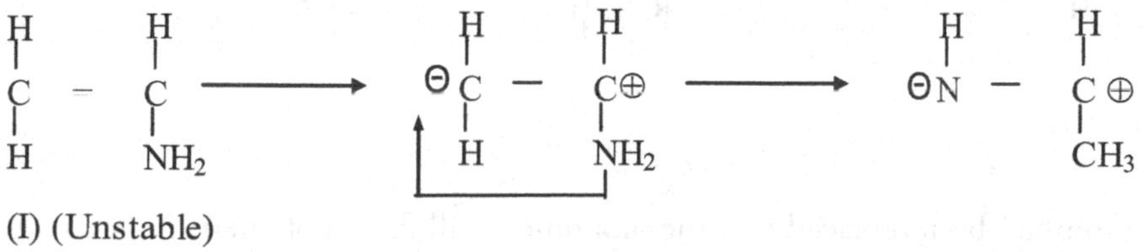

$$\underset{\underset{H}{|}}{\overset{H}{C}} - \underset{\underset{NH_2}{|}}{\overset{H}{C}} \longrightarrow \quad ^{\ominus}\underset{\underset{H}{|}}{\overset{H}{C}} - \underset{\underset{NH_2}{|}}{\overset{H}{C}}^{\oplus} \longrightarrow \quad ^{\ominus}N - \underset{\underset{CH_3}{|}}{\overset{H}{C}}^{\oplus}$$

(I) (Unstable)

<u>An Activated Aldimine</u> 7.81

$$\underset{H}{\overset{H}{C}} = \underset{NH_2}{\overset{CH_3}{C}} \longrightarrow \overset{\ominus}{C}\underset{H}{\overset{H}{-}} \overset{\oplus}{C}\underset{NH_2}{\overset{CH_3}{}} \longrightarrow \overset{\oplus}{C}\underset{CH_3}{\overset{CH_3}{-}} N\overset{H}{\ominus}$$

(II) (Unstable)

<u>An Activated Ketimine</u>   7.82

$$\underset{H}{\overset{H}{C}} = \underset{NH_2}{\overset{C_2H_5}{C}} \longrightarrow n\bullet\underset{H}{\overset{H}{C}} - \underset{NH_2}{\overset{C_2H_5}{C}}\bullet e \qquad \text{Rearrangement Impossible} \qquad 7.83$$

(III) (Stable)

$$\underset{H}{\overset{H}{C}} = \underset{\underset{CH_3 \quad CH_3}{N}}{\overset{CH_3}{C}} \longrightarrow \overset{\ominus}{C}\underset{H}{\overset{H}{-}} \overset{\oplus}{C}\underset{\underset{CH_3 \quad CH_3}{N}}{\overset{CH_3}{}} \longrightarrow \text{Rearrangement Impossible}$$

7.84

(IV) (Stable)

While (I) and (II) are very unstable (i.e., cannot favor isolated existence) (III) and (IV) are quite stable, since they will not molecularly rearrange or tautomerize. The hydrogen atoms on the nitrogen center cannot be said to be chargedly and radically responsible for the instability, since (III) cannot molecularly rearrange or tautomerize and it is therefore stable. The reason can be seen from the Law of conservation of transfer of transfer species. Only one hydrogen atom is ionically or radically bonded to the nitrogen center one at a time. Nevertheless, the aldimines and ketimines are still slightly unstable to polymerization, since the hydrogen atom on the nitrogen N

$$R^{\ominus} + \overset{\ominus}{N}\underset{CH_3}{\overset{H \quad H}{-}} \overset{\oplus}{C} \longrightarrow RH + \overset{\ominus}{C}\underset{H}{\overset{H}{-}} \underset{H}{\overset{}{C}} = N \qquad OR \qquad N = \overset{H}{\underset{CH_3}{C}}$$

$$(R^{\ominus} \equiv \text{free } \ominus) \qquad\qquad (R^{\ominus} \equiv \text{Non-free-negative-charge})$$

NOT FAVORED   7.85

$$R^{\oplus} + \underset{CH_3}{\overset{H \quad H}{N = C}} \longrightarrow H^{\oplus} + \underset{CH_3}{\overset{R \quad H}{N = C}}$$

7.86

It is when the hydrogen atom has been replaced that the monomers will favor polymerization chargedly. Radically the monomers will not favor polymerization since the hydrogen atom is also radically loosely bonded particularly when not activated.

When radical activation is favored, the followings are obtained.

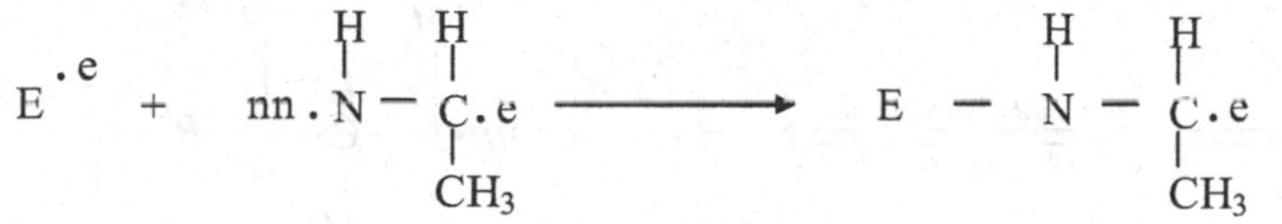

$$E \cdot^e \quad + \quad nn \cdot \overset{\displaystyle H}{\underset{\displaystyle CH_3}{N}} - \overset{\displaystyle H}{\underset{\displaystyle }{C}} \cdot e \quad \longrightarrow \quad E - \overset{\displaystyle H}{\underset{\displaystyle }{N}} - \overset{\displaystyle H}{\underset{\displaystyle CH_3}{C}} \cdot e$$

(If suppressed)

7.87

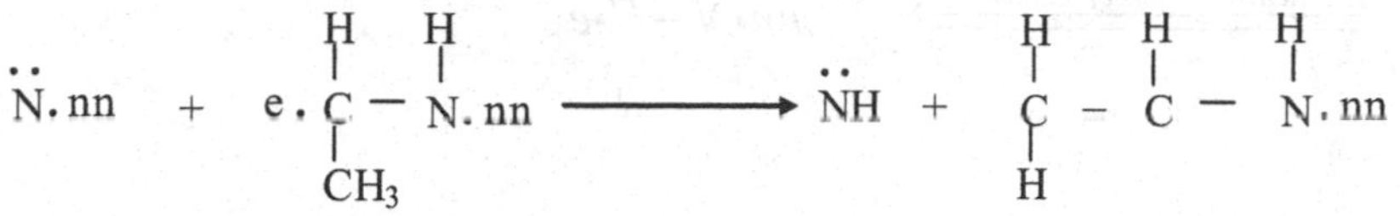

$$\overset{\displaystyle ..}{N} \cdot nn \quad + \quad e \cdot \overset{\displaystyle H}{\underset{\displaystyle CH_3}{C}} - \overset{\displaystyle H}{N} \cdot nn \quad \longrightarrow \quad \overset{\displaystyle ..}{N}H \quad + \quad \overset{\displaystyle H}{\underset{\displaystyle H}{C}} - \overset{\displaystyle H}{C} - \overset{\displaystyle H}{N} \cdot nn$$

(If suppressed)

7.88

$$E \cdot^e \quad + \quad \overset{\displaystyle H}{N} = \overset{\displaystyle H}{\underset{\displaystyle CH_3}{C}} \quad \longrightarrow \quad \overset{\displaystyle E}{N} = \overset{\displaystyle H}{\underset{\displaystyle CH_3}{C}} \quad + \quad H \cdot^e$$

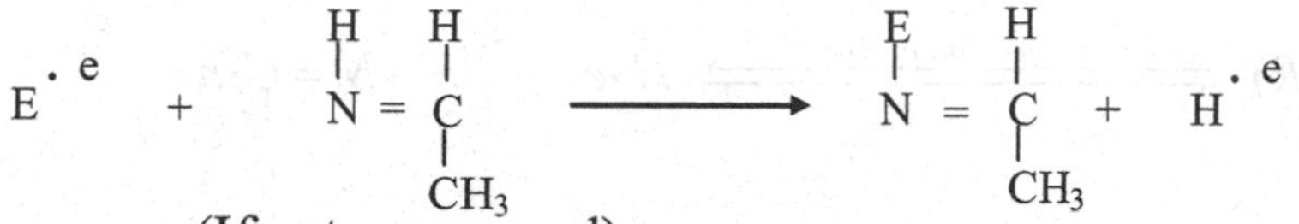

(If not suppressed)

7.89

(If not suppressed) 7.89 It is obvious that between Equations 7.87 and 7.89, the latter will prevail most of the time; that is, the monomer cannot be polymerized radically under most operating conditions. Initiation and propagation will only be favored when the monomer is suppressed by a passive catalyst.

7.2.1 Charged characters of substituted Aldimines and Ketimines.

When the H on the nitrogen center is replaced, the strong need to operate at very low polymerization temperatures does not much arise, if Electrostatic type of electro-free-radical initiators are not used (i.e. Backward addition). Beginning with alkyl substituent groups, the followings are obtained.

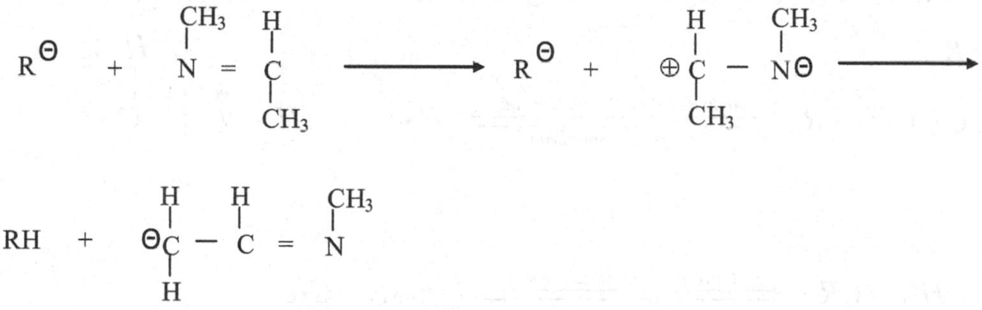

$$R^{\ominus} \quad + \quad \overset{\displaystyle CH_3}{\underset{\displaystyle CH_3}{N}} = \overset{\displaystyle H}{C} \quad \longrightarrow \quad R^{\ominus} \quad + \quad \overset{\displaystyle H}{\underset{\displaystyle CH_3}{\oplus C}} - \overset{\displaystyle CH_3}{N\ominus} \quad \longrightarrow$$

$$RH \quad + \quad \overset{\displaystyle H}{\underset{\displaystyle H}{\ominus C}} - \overset{\displaystyle H}{C} = \overset{\displaystyle CH_3}{N}$$

7.90

This unlike the aldimine above can enamidize, but cannot molecularly rearrange. Like aldehydes therefore, they also favor Activated/Equilibrium State of Existence as shown below radically. It cannot be done chargedly.

FORMALDIMINE

$$HN = CH_2 \quad \underset{\xrightarrow{\hspace{2cm}}}{\overset{Equilibrium\ State\ of\ Existence}{\rightleftharpoons}} \quad H \bullet e \quad + \quad nn \bullet N = CH_2$$

ALDIMINES

$$HN = CH(R) \quad \underset{\xrightarrow{\hspace{2cm}}}{\overset{Equilibrium\ State\ of\ Existence}{\rightleftharpoons}} \quad H \bullet e \quad + \quad nn \bullet N = CH(R)$$

$$HN = CH(CH_2)R' \xrightleftharpoons[\text{Existence when suppressed}]{\text{Activated Equilibrium State of}} H\bullet e \;+\; nn\bullet \overset{\overset{\textstyle H}{|}}{N} - \overset{\overset{\textstyle \bullet e}{\underset{\textstyle |}{C}}}{\underset{\textstyle H}{}} - \overset{\overset{\textstyle H}{|}}{\underset{\textstyle \bullet n}{C}} - R'$$

$$HN = CH(R) \xrightleftharpoons{\text{Activated State of Existence}} nn\bullet \overset{\overset{\textstyle H}{|}}{N} - \overset{\overset{\textstyle H}{|}}{\underset{\textstyle R}{C}}\bullet e$$

7.91a

ALKYL NITROGEN SUBSTITUTED ALDIMINES

$$R - N = CH(R) \xrightleftharpoons{\text{Equilibrium State of Existence}} H\bullet e \;+\; R - N = \overset{}{\underset{\textstyle R}{C}}\bullet n$$

$$R - N = \overset{}{\underset{\textstyle H}{C}} - (CH_2) - R' \xrightleftharpoons[\text{Existence}]{\text{Activated Equilibrium State of}} H\bullet e \;+\; R - \overset{\overset{\textstyle \bullet nn}{}}{\underset{\textstyle \bullet e}{N}} - \overset{\overset{\textstyle H}{|}}{\underset{\textstyle H}{C}} - \overset{\overset{\textstyle \bullet n}{}}{C} - R'$$

$$R - N = \overset{}{\underset{\textstyle H}{C}} - R \xrightleftharpoons{\text{Activated State of Existence}} nn\bullet N - \overset{\overset{\textstyle R\;\;R}{|\;\;\;|}}{\underset{\textstyle H}{C}}\bullet e$$

7.91b

KETIMINES

$$HN = CR_2 \xrightleftharpoons{\text{Equilibrium State of Existence}} H\bullet e \;+\; nn\bullet N = CR_2$$

$$HN = \overset{\overset{\textstyle R}{|}}{C} - CH_2 - R \xrightleftharpoons[\text{Existence (Suppressed)}]{\text{Activated Equilibrium State of}} H\bullet e \;+\; H - \overset{\overset{\textstyle \bullet nn}{}}{N} - \overset{\overset{\textstyle \bullet e}{}}{\underset{\textstyle R}{C}} - \overset{\overset{\textstyle H}{|}}{\underset{\textstyle \bullet n}{C}} - R$$

$$HN = CR_2 \xrightleftharpoons{\text{Activated State of Existence}} nn\bullet N - \overset{\overset{\textstyle H\;\;R}{|\;\;\;|}}{\underset{\textstyle R}{C}}\bullet e$$

7.92

ALKYL NITROGEN SUBSTITUTED KETIMINES

$$R - N = \overset{\overset{\textstyle R}{|}}{C} - R'' - CH_3 \xrightleftharpoons{\text{Equilibrium State of Existence}} R\bullet n \;+\; R - N = C - R'' - \overset{\overset{\textstyle \bullet e \quad H}{\qquad |}}{\underset{\textstyle H}{C}} - H$$

$$R - N = C - C - R \xrightleftharpoons[\text{Existence}]{\text{Activated / Equilibrium State of}} H{\cdot}e \; + \; R - \overset{{\cdot}nn}{N} - \overset{{\cdot}e}{C} - \underset{{\cdot}n}{C} - R$$

with R, H substituents on the carbons.

$$R - N = CR_2 \xrightleftharpoons{\text{Activated State of Existence}} nn{\cdot}N - C{\cdot}e$$

with R substituents.

$$7.93$$

where the R and R' are alkyl and large number of methylene groups, depending on where placed. Aldimines and ketimines will only enamidize when suppressed such that Equilibrium State of Existence is not favored from the N center. However, the alkyl nitrogen substituted aldimines... and ketimines will readily enamidize. These can be polymerized like the aldehydes and ketones using for example dilute HCl electro-free-radically. Like the aldehydes and ketones, only the right-hand side of the initiator can be used. The left-hand side cannot be used because of presence of transfer species of the first kind of the first type. This monomer will definitely have "ceiling temperatures" and undergo depropagation when the polymerization temperature is above the "ceiling temperature".

$$Cl^{\ominus} \ldots\ldots\ldots {}^{\oplus}O - (C - N)_n - C - N - H$$

with H, CH_3 substituents.

$$7.94$$

Anionically or with the use of negatively charged-paired initiator, the monomer cannot be initiated as already shown in Equation 7.90. With paired positive charges, the followings are obtained.

$$R^{\oplus} + {}^{\ominus}N - C^{\oplus} \longrightarrow R - N - C^{\oplus} \xrightarrow{+ \, n(I)}$$

with CH_3, H, CH_3 substituents.

$$R {+} N - C {\xrightarrow{}}_n N - C^{\oplus} \longrightarrow R {+} N - C {\xrightarrow{}}_n N - C = C + H^{\oplus}$$

with CH_3, H, CH_3 substituents.

$$7.95$$

The nucleophilic character of the monomer can thus be observed with transfer species of the first kind of the first type being involved. The same will apply radically and to the substituted ketimines.

Replacing the CH_3 group with OR groups, the followings are obtained.

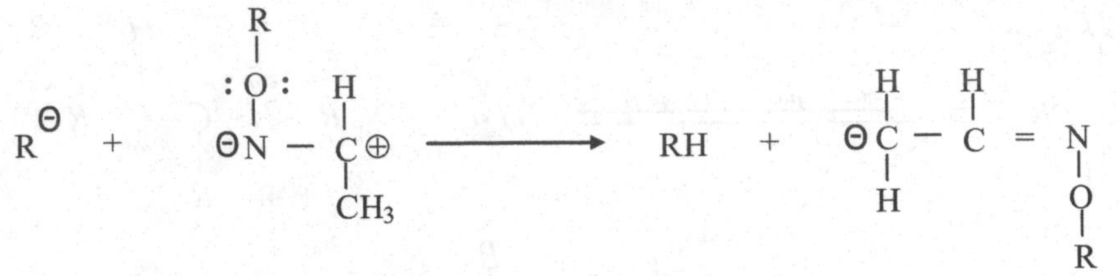

$$R^{\ominus} \;+\; {}^{\ominus}\!N - \overset{\overset{R}{|}}{\underset{\underset{CH_3}{|}}{\overset{\overset{:O:}{|}}{C}}}{}^{\oplus}\!\!\overset{H}{\underset{}{}} \longrightarrow RH \;+\; {}^{\ominus}\!\overset{H}{\underset{H}{C}} - \overset{H}{\underset{}{C}} = \overset{N}{\underset{\underset{R}{O}}{}}$$

$$\text{7.96}$$

(I)   <u>Impossible state</u>

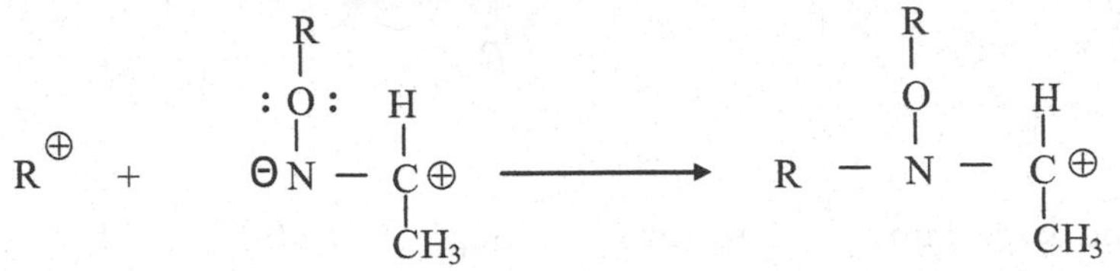

$$R^{\oplus} \;+\; {}^{\ominus}\!N - \overset{\overset{R}{|}}{\underset{\underset{CH_3}{|}}{\overset{\overset{:O:}{|}}{C}}}{}^{\oplus}\!\!\overset{H}{\underset{}{}} \longrightarrow R - \overset{\overset{R}{|}}{\underset{}{N}} - \overset{\overset{O}{|}}{\underset{\underset{CH_3}{|}}{C}}{}^{\oplus}\!\!\overset{H}{\underset{}{}}$$

(I)   <u>Impossible state</u>

$$\text{7.97}$$

Due to electrostatic forces of repulsion, the monomer cannot be activated chargedly, but only radically. If the carbon center is carrying the etheric group, the followings are to be expected.

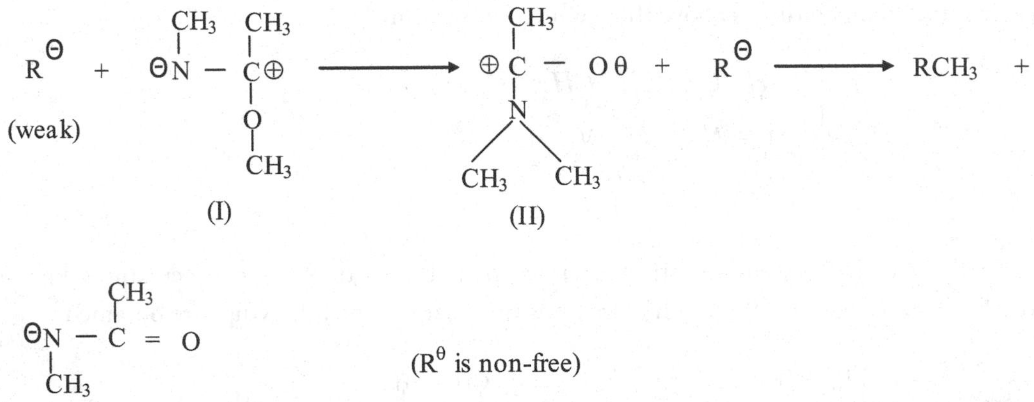

$$R^{\ominus} \;+\; {}^{\ominus}\!N - \overset{\overset{CH_3 \quad CH_3}{| \quad\quad |}}{\underset{\underset{\underset{CH_3}{|}}{O}}{C}}{}^{\oplus} \longrightarrow {}^{\oplus}\!C - O^{\ominus} \;+\; R^{\ominus} \longrightarrow RCH_3 \;+$$

(weak)   (I)   (II)

$$\overset{CH_3}{\underset{CH_3}{{}^{\ominus}\!N - C = O}} \qquad (R^{\ominus} \text{ is non-free}) \qquad \text{7.98}$$

Molecular rearrangement being favored will depend on what the nitrogen center is carrying. Nevertheless, (II) above is more stable than (I). Free positively, the monomer will not favor initiation whether molecular rearrangement is favored or not.

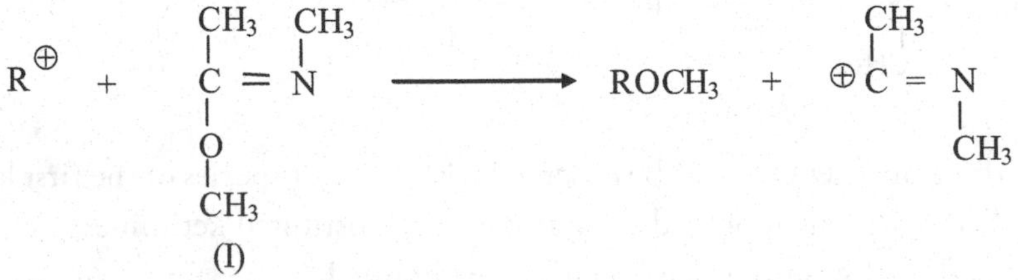

$$R^{\oplus} \;+\; \overset{\overset{CH_3 \quad CH_3}{| \quad\quad |}}{\underset{\underset{\underset{CH_3}{|}}{O}}{C = N}} \longrightarrow ROCH_3 \;+\; \overset{\overset{CH_3}{|}}{\underset{\underset{CH_3}{|}}{{}^{\oplus}\!C = N}}$$

(I)

$$\text{7.99}$$

This takes place only when the monomer is not activated, i.e., in Stable state of existence. With positively charged-paired initiators, the initiation step is favored, the monomer being a Nucleophile.

$$R^{\oplus} \;+\; \underset{\substack{| \\ N \\ \diagup \; \diagdown \\ CH_3 \quad CH_3}}{\overset{\substack{CH_3 \\ |}}{C}} = O \;\longrightarrow\; RN(CH_3)_2 \;+\; \overset{\substack{CH_3 \\ |}}{\oplus C} = O$$

Free-media

(II)   7.100

In the first reaction, transfer species of the second first kind of the first type is involved, since the route is natural to the monomer, while in the second reaction, transfer species of the second first kind of the second type is involved. The two equations above are meaningless, since positively-charged- free initiators cannot exist. Only cations are free.

Now, considering using resonance stabilization group, the followings are obtained.

$$\underset{\substack{| \\ CH \\ \| \\ CH_2}}{\overset{\substack{CH_3 \\ | \\ O \\ |}}{\oplus C}} - \overset{\substack{CH_3 \\ |}}{N^{\ominus}} \;+\; R^{\ominus} \;\longrightarrow\; RCH_3 \;+\; \overset{\ominus}{O} - \underset{\substack{| \\ CH \\ \| \\ CH_2}}{C} = \overset{\substack{CH_3 \\ |}}{N}$$

7.101

The etheric group cannot be resonance stabilized. Hence the monomer is electrophilic.

$$\underset{\substack{| \\ CH \\ \| \\ CH_2}}{\overset{\substack{CH_3 \; CH_3 \\ | \quad | \\ X}}{\oplus C}} \; N^{\ominus} \quad\quad OR \quad\quad \underset{\substack{| \\ H}}{\overset{\substack{H \quad H \\ | \quad | \\ Y}}{\oplus C}} \; \underset{\substack{| \\ C(CH_3) \\ \| \\ N \\ | \\ CH_3}}{C^{\ominus}} \;;\quad (I) \xrightarrow{+ R^{\ominus}} RH \;+\; \underset{\substack{| \\ H}}{\overset{\substack{H \\ |}}{\ominus C}} - \underset{\substack{| \\ CH \\ \| \\ CH_2}}{C} = \overset{\substack{CH_3 \\ |}}{N}$$

(I) (Nucleophile)

(II) (Electrophile)

$$(II) \xrightarrow{+ R^{\ominus}} R - \underset{\substack{| \\ H}}{\overset{\substack{H \\ |}}{C}} - \underset{\substack{| \\ C(CH_3) \\ \| \\ N \\ | \\ CH_3}}{\overset{\substack{H \\ |}}{C^{\ominus}}}$$

7.102

It can be observed that the monomer cannot be resonance stabilized, since the two activation centers above are of different characters, -RC = NR group being a pulling group. When the resonance stabilization group is placed on the nitrogen center, resonance stabilization can be provided radically, and the monomer is a complete Nucleophile. The case above is an electrophile, i.e. a male with X (Nucleophilic) and Y (Electrophilic) centers. Positively, X is attacked and the route is favored. The monomer looks like a substituted acrolein.

Now consider using pulling groups on the N center. Only those which are of the free types can be used. That is, the non-free types such as F and $OCOCH_3$ cannot be used due to electro-static forces of repulsion, when activated chargedly. They can only be used radically.

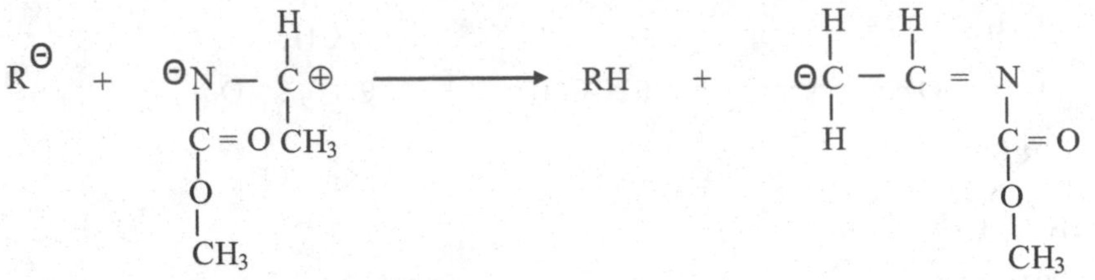

$$R^{\ominus} \quad + \quad {}^{\ominus}N - \overset{H}{\underset{\underset{\displaystyle CH_3}{\overset{\displaystyle |}{O}}}{\underset{|}{\overset{|}{C}}}}{\overset{\oplus}{C}} \quad \longrightarrow \quad RH \quad + \quad {}^{\ominus}\overset{H}{\underset{\displaystyle H}{\overset{|}{C}}} - \overset{H}{\underset{\underset{\displaystyle CH_3}{\overset{\displaystyle |}{O}}}{\underset{|}{\overset{|}{C}}}} = N \qquad \text{7.103a}$$

The monomer is a Nucleophile. The natural route is therefore supposed to be free- positively charged route. But as shown below, it is not favored.

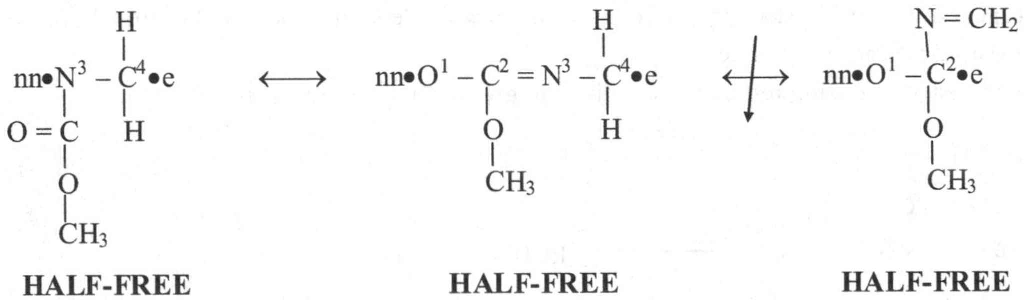

$$\underset{\displaystyle CH_3}{\underset{|}{\overset{\displaystyle O}{\underset{\|}{O = C}}}} \quad nn\bullet N^3 - \overset{H}{\underset{\displaystyle H}{\overset{|}{C^4}}}\bullet e \quad \longleftrightarrow \quad nn\bullet O^1 - \overset{}{\underset{\displaystyle CH_3}{\underset{|}{C^2}}} = N^3 - \overset{H}{\underset{\displaystyle H}{\overset{|}{C^4}}}\bullet e \quad \overset{\big\downarrow}{\longleftrightarrow}\!\!\!\!/ \quad nn\bullet O^1 - \overset{N = CH_2}{\underset{\displaystyle CH_3}{\underset{|}{C^2}}}\bullet e \qquad \text{7.103b}$$

HALF-FREE   **HALF-FREE**   **HALF-FREE**

Indeed, the OCH_3 group cannot be abstracted as will shortly be shown radically, but only charge-ly because it is radically resonance stabilized. However, when negatively-charged-paired initiators of the Electrostatic types (anionic) are involved, polymerization will be favored if the groups are trans-placed. *These clearly send a message that which is that these monomers are MALES.* If the CH_3 group is replaced with H, free-media-negatively/negatively charged-paired routes will be favored. The same as above will apply to $CONH_2$, $CONHR$, and $CONR_2$ types of groups.

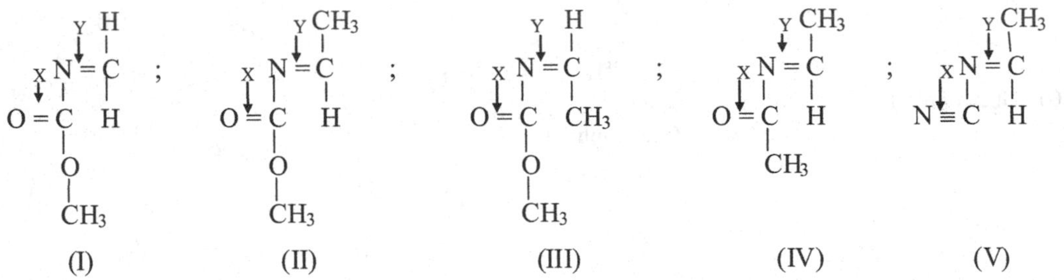

$$\underset{(I)}{\overset{\displaystyle Y \;\; H}{\underset{\displaystyle CH_3}{\underset{|}{\overset{\displaystyle O}{\underset{\|}{O = C}}}}}\;\;} \quad ; \quad \underset{(II)}{\cdots} \quad ; \quad \underset{(III)}{\cdots} \quad ; \quad \underset{(IV)}{\cdots} \quad ; \quad \underset{(V)}{\cdots}$$

ALL MALES OF DIFFERENT CAPACITIES   7.104a

(I) is a substituted formaldimine, a strong Electrophile. (II) is trans-placed more Electrophilic than (III) which is cis-placed, the only one that cannot be initiated with either free-media or paired initiators negatively. *The cis-placed monomers with one type of transfer species on the C active centers externally located can never be initiated and grow. They remain stunted, but can still be used to produce other products.* (IV) is no longer unique as with other members of unsaturated monomers where they exist wherein they are weak Males. (V) is a strong Male based on the order of nucleophilicity of the centers. <u>All activation centers are all Nucleophilic in character, that which is MOTHER NATURE. Electrophilic centers only start appearing when more than one Activation center of different nucleophilicities are conjugatedly or cumulatively placed.</u> (II) to (V) above are all Aldimines. The same apply to their Ketimines where the H on the C center is changed to an alkanyl group. Worthy of note in all the monomers above is that they are all resonance stabilized with substituent group to resonance stabilize or shield.

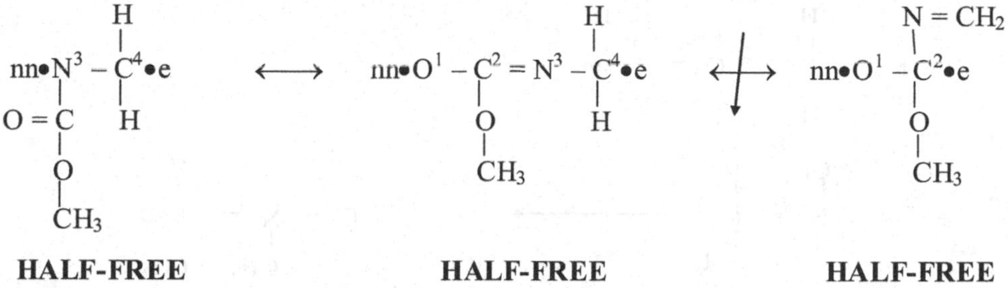

$$\underset{\substack{|\\O=\overset{|}{C}\ \ H\\|\\O\\|\\CH_3}}{nn\bullet N^3 - \overset{\overset{H}{|}}{C^4}\bullet e} \qquad \longleftrightarrow \qquad \underset{\substack{|\\O\\|\\CH_3}}{nn\bullet O^1 - C^2 = \overset{\overset{H}{|}}{N^3} - \overset{\overset{H}{|}}{C^4}\bullet e} \qquad \overset{\displaystyle\downarrow}{\longleftrightarrow\!\!\!| } \qquad \underset{\substack{|\\O\\|\\CH_3}}{nn\bullet O^1 - \overset{\overset{N=CH_2}{|}}{C^2}\bullet e}$$

HALF-FREE   **HALF-FREE**   **HALF-FREE**   7.104b

Worthy of note is that the OCH_3 group is internally located for which no transfer species exist to be abstracted, for which also the reaction of Equation 7.103b cannot take place radically, but only chargedly. Only the 3,4- and 1,4-monoforms can be observed above, based on the rules stated. The movement above is from Half-free to Half-free. The group providing resonance stabilization is the O = COR or O = CR or N ≡C groups with centers more nucleophilic than the C = N center. When the group is placed on the carbon center, no resonance stabilization can be provided.

For COR types of groups, the followings are to be expected when free-media initiators are involved. With paired radical initiators of the electrostatic types, both routes are favored. No copolymers can be produced here for this aldimine and ketimine, except for the formaldimine.

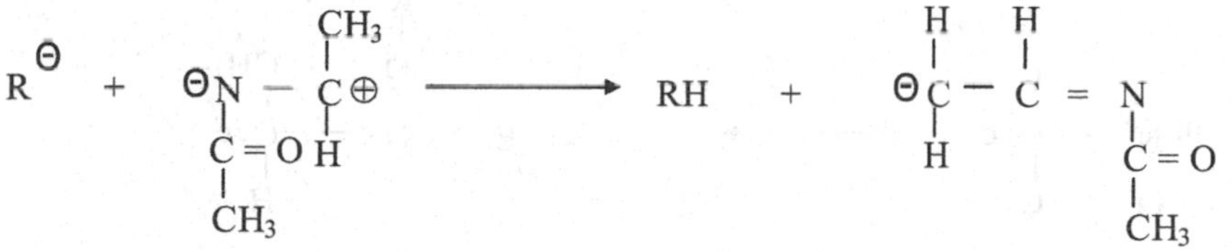

$$R^{\ominus} \ + \ \underset{\substack{|\\C=O\ H\\|\\CH_3}}{{}^{\ominus}N - \overset{\overset{CH_3}{|}}{C}^{\oplus}} \ \longrightarrow \ RH \ + \ \underset{\substack{|\\H\\\\}}{{}^{\ominus}\overset{\overset{H}{|}}{C}} - \overset{\overset{H}{|}}{C} = \underset{\substack{|\\C=O\\|\\CH_3}}{N} \qquad\qquad 7.105$$

$$R^{\oplus} \ + \ \underset{\substack{|\\C=O\ H\\|\\CH_3}}{{}^{\ominus}N - \overset{\overset{CH_3}{|}}{C}^{\oplus}} \ \longrightarrow \ R - \underset{\substack{|\\C=O\ H\\|\\CH_3}}{N} = \overset{\overset{CH_3}{|}}{C}{}^{\oplus} \qquad\qquad 7.106$$

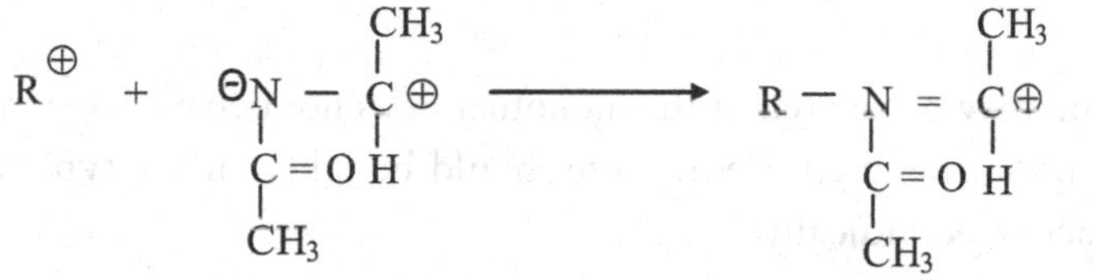

Unlike COOR type, molecular rearrangement cannot take place. The unique character of COR type of groups displayed here is due to the more electropositive tendency of the groups, despite the fact that they are stronger pulling groups; and the fact that the nitrogen center unlike oxygen center can still carry a substituted group; and finally the fact that, the negative charge carried by the nitrogen center is independent of the group carried by it.

7.2.2 Radical characters of substituted Aldimines and Ketimines

Beginning with N-methylethanimine (real name **N-methylethyleneimine)** or acetaldehyde N-methylimine, the followings are to be expected when used as a monomer.

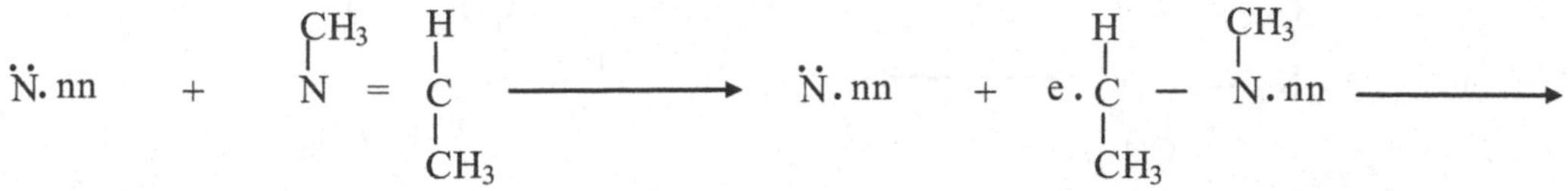

$$\ddot{N}.nn \quad + \quad \underset{\substack{|\\CH_3}}{N} = \overset{\overset{CH_3}{|}\ \overset{H}{|}}{C} \quad \longrightarrow \quad \ddot{N}.nn \quad + \quad e.\underset{\substack{|\\CH_3}}{C} - \overset{\overset{H}{|}\ \ \overset{CH_3}{|}}{N}.nn \quad \longrightarrow$$

$$NH \quad + \quad \overset{\displaystyle H}{\underset{\displaystyle H}{C}} = \overset{\displaystyle H}{C} - \overset{\displaystyle CH_3}{N.nn}$$

$$\text{7.107}$$

$$E^{\cdot e} \quad + \quad nn \cdot N = \overset{\displaystyle CH_3}{\underset{\displaystyle CH_3}{C.e}} \quad \overset{\displaystyle H}{\longrightarrow} \quad E -- N - \overset{\displaystyle H}{\underset{\displaystyle CH_3}{C.e}}$$

$$\text{7.108}$$

Radically, substituted aldimines can only be polymerized electro-free-radically being a nucleophile and these initiators are not yet as common as nucleo-free-radical initiators. It is for the same reason, propylene (propene) could not be polymerized.

Replacing the CH_3 group with OR groups, the followings are obtained, one at a time.

$$N^{\cdot n} \quad + \quad e \cdot \overset{\displaystyle H}{\underset{\displaystyle CH_3 \; O \, R}{C}} - N.nn \quad \longrightarrow \quad NH + n \cdot \overset{\displaystyle H}{\underset{\displaystyle H}{C}} - \overset{\displaystyle H}{C} = \overset{\displaystyle}{N} \; (O R)$$

$$\text{7.109}$$

$$E^{\cdot e} \quad + \quad nn. \overset{\displaystyle H}{\underset{\displaystyle O \, R}{N}} - \overset{\displaystyle H}{\underset{\displaystyle CH_3}{C.e}} \quad \longrightarrow \quad E - \overset{\displaystyle R \, O}{N} - \overset{\displaystyle CH_3}{\underset{\displaystyle H}{C.e}}$$

$$\text{7.110}$$

Electro-free-radically, polymerization may be favored, if the monomer does not behave like vinyl acetate. While the monomer above cannot be activated chargedly, note should be taken of the types of radicals carried by the active centers when activated radically.

$$\overset{..}{N}.nn \; (\text{Weak}) \; + \; e \cdot \overset{\displaystyle CH_3 \; O \; CH_3}{\underset{\displaystyle CH_3}{C}} -- N.nn \; \longrightarrow \; \overset{..}{N}.nn \; + \; e \cdot \overset{\displaystyle CH_3}{C} - O.nn \longrightarrow$$

$$\overset{\displaystyle NCH_3}{} + nn \cdot \overset{\displaystyle CH_3}{N} - \overset{\displaystyle CH_3}{C} = O$$

$$\text{7.111}$$

$$E^{\cdot e} \quad + \quad nn \cdot \overset{\displaystyle CH_3 \; O}{N} -- \overset{\displaystyle CH_3}{\underset{\displaystyle CH_3}{C.e}} \quad \longrightarrow \quad E - \overset{\displaystyle CH_3 \; O}{N} -- \overset{\displaystyle CH_3}{\underset{\displaystyle CH_3}{C.e}}$$

$$\text{7.112}$$

Electro-free-radically, initiation is favored the monomer being nucleophilic. Molecular rearran-gement can however be favored, the C=O center being more nucleophilic than the C=N center. For the more stable state, the followings are obtained electro-free-radically.

$$E^{\cdot e} + e.\underset{\underset{\underset{CH_3 \quad CH_3}{\diagdown N \diagup}}{|}}{\overset{\overset{CH_3}{|}}{C}} - O.nn \longrightarrow EN(CH_3)_2 + e.\underset{}{\overset{\overset{CH_3}{|}}{C}} = O \quad Or \quad E-O-\underset{\underset{N(CH_3)_2}{|}}{\overset{\overset{CH_3}{|}}{C}} \bullet e$$

$$\qquad\qquad\qquad\qquad\qquad\qquad\qquad (I) \qquad\qquad\qquad\qquad (II)\ Favored \qquad 7.113$$

When not activated, (I) is favored with $N(CH_3)_2$ group as transfer species of the second first kind of the second type. But when activated, (II) is obtained, the monomer being strongly nucleophilic.

In the presence of commonly known resonance stabilization groups, the followings are obtained.

$$\overset{..}{N}.nn + e.\underset{\underset{\underset{CH_2}{\|}}{CH}}{\overset{\overset{H}{|}}{C}} - \overset{CH_3}{\underset{}{N}}.nn \longrightarrow \overset{..}{N}H + \underset{\underset{H}{|}}{\overset{\overset{H}{|}}{C}} = C = \underset{}{\overset{\overset{H}{|}}{C}} - \overset{CH_3}{\underset{}{N}}.nn$$

$$\qquad\qquad\qquad\qquad\qquad\qquad\qquad\qquad\qquad\qquad\qquad\qquad\qquad\qquad\qquad 7.114$$

$$E^{\cdot e} + nn.\underset{\underset{Y}{\overset{X}{\|}}}{\overset{\overset{CH_3}{|}}{N}} \underset{}{\overset{}{-}} \underset{\underset{\underset{CH_2}{\|}}{CH}}{\overset{\overset{H}{|}}{C}}.e \longrightarrow + E - \underset{}{\overset{}{N}} -- \underset{\underset{\underset{CH_2}{\|}}{CH}}{\overset{\overset{\overset{CH_3 \quad H}{|\quad|}}{}}{C}}.e$$

$$\qquad\qquad\qquad\qquad\qquad\qquad\qquad\qquad\qquad\qquad\qquad\qquad\qquad\qquad\qquad 7.115$$

The monomer is an electrophile (Male) and the center being used above is nucleophilic. The centers however favor their natural routes. Like acrolein, it is not resonance stabilized.

Considering this unique type of monomer wherein –N = CHR group is adjacently located to C = C center. The group is a pushing group of greater capacity than -NR$_2$ group. The N = C center carried by the group is of greater nucleophilic capacity than the C = C center. Therefore one should expect the C =N center to provide resonance stabilization for the C = C center. It is possible because the movement is from Full-free to Full-free. It is not a Male, because the more nucleophilic group is a pushing group. Nevertheless, looking at the C = N center, the followings are to be expected, ***though the center that will only be activated all the time is the 3,4-mono-form C = C center and 1,4-mono-form via resonance stabilization, the monomer being a FEMALE.*** If the monomer was a strong MALE, the center activated will depend on the type of initiator activating it.

$$e.\underset{\underset{\underset{CH_2}{\|}}{CH}}{\overset{\overset{H}{|}}{C}} - \overset{}{\underset{}{N}}.nn + E^{\cdot e} \longrightarrow E - \underset{\underset{\underset{CH_2}{\|}}{CH}}{\overset{}{N}} - \underset{\underset{H}{|}}{\overset{\overset{CH_3}{|}}{C}}.e$$

$$\qquad CH_3 \quad CH$$

$$\qquad\qquad\qquad\qquad\qquad\qquad\qquad\qquad\qquad\qquad\qquad\qquad\qquad\qquad\qquad 7.116$$

$$\ddot{N}.nn \quad + \quad e.\underset{\underset{\underset{CH_2}{\|}}{\underset{CH}{|}}}{\overset{\overset{H}{|}}{C}} - N.nn \quad \longrightarrow \quad \ddot{N}H \quad + \quad \underset{\underset{H}{|}}{\overset{\overset{H}{|}}{C}} = \underset{\underset{\underset{CH_2}{\|}}{\underset{CH}{|}}}{\overset{\overset{H}{|}}{C}} - N.nn \qquad 7.117$$

$$N{\bullet}n \quad + \quad e{\bullet}\underset{\underset{\underset{CH_3}{|}}{\underset{\underset{CH}{\|}}{N}}}{\overset{\overset{H}{|}}{C}} - \underset{\underset{H}{|}}{\overset{\overset{H}{|}}{C}}{\bullet}n \quad \longrightarrow \quad NH \quad + \quad n{\bullet}\underset{\underset{H}{|}}{\overset{\overset{H}{|}}{C}} - \overset{\overset{H}{|}}{C} = N - \underset{\underset{H}{|}}{\overset{\overset{H}{|}}{C}} = C \qquad 7.118a$$

$$E{\bullet}e \quad + \quad n{\bullet}\underset{\underset{\underset{CH_3}{|}}{\underset{\underset{CH}{\|}}{N}}}{\overset{\overset{H}{|}}{C}} - \underset{\underset{H}{|}}{\overset{\overset{H}{|}}{C}}{\bullet}e \quad \longrightarrow \quad E - \underset{\underset{\underset{CH_3}{|}}{\underset{\underset{CH}{\|}}{N}}}{\overset{\overset{H}{|}}{C}} - \underset{\underset{H}{|}}{\overset{\overset{H}{|}}{C}}{\bullet}e \quad \xrightarrow{+n\,(I)} \quad E - (\underset{\underset{H}{|}}{\overset{\overset{H}{|}}{C}} - \underset{\underset{\underset{CH_3}{|}}{\underset{\underset{CH}{\|}}{N}}}{\overset{\overset{H}{|}}{C}})_n - \underset{\underset{H}{|}}{\overset{\overset{H}{|}}{C}} - \overset{\overset{H}{|}}{C} = N - \underset{\underset{H}{|}}{\overset{\overset{H}{|}}{C}} = C$$

$$(I)$$

$$+ \quad H{\bullet}e \qquad 7.118b$$

The transfer species involved above is still of the first kind and of the first type, though it is not adjacently located to the active center. It is resonance stabilized and clearly seen in the 1,4-mono-form.

For pulling groups, the followings are obtained

$$\overset{..}{N}.nn \quad + \quad e.\underset{\underset{\underset{\underset{CH_3}{|}}{\underset{O}{|}}}{\underset{C=O}{|}}}{\overset{\overset{H}{|}}{C}} - N.nn \quad \longrightarrow \quad NH \quad + \quad \underset{\underset{H}{|}}{\overset{\overset{H}{|}}{C}} = \underset{\underset{\underset{\underset{CH_3}{|}}{\underset{O}{|}}}{\underset{C=O}{|}}}{\overset{\overset{H}{|}}{C}} - N.nn$$

$$\text{FAVORED} \qquad 7.119$$

$$E^{\cdot e} \quad + \quad nn.\underset{\underset{\underset{CH_3}{|}}{\underset{O}{|}}}{N} - \overset{\overset{H}{||}}{\underset{\underset{CH_3}{|}}{C}}.e \quad \longrightarrow \quad EOCH_3 \quad + \quad \overset{\overset{O}{||}}{C} = N - \underset{\underset{CH_3}{|}}{\overset{\overset{H}{|}}{C}}.e$$

$$\text{NOT FAVORED} \qquad 7.120$$

Chargedly, it has been seen that this monomer is a MALE, i.e. an ELECTROPHILE. The monomer above is cis-placed. For the trans-placed monomer, initiation is favored nucleo-non -free-radically when paired initiators are used ($H_3CO^{\bullet nn}$.......$^{\bullet \bullet}$Na). Being resonance stabilized, the last equation above is not favored (See Equations 7.104a and 7.104b).

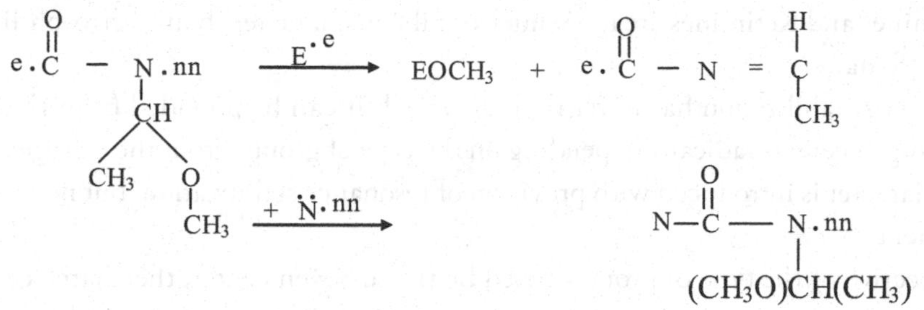

$$e \cdot \overset{\overset{O}{\|}}{C} - \underset{\underset{\underset{CH_3}{|}}{\overset{|}{CH}} \quad \underset{\underset{CH_3}{O}}{}}{\underset{}{N} \cdot nn} \quad \xrightarrow{E \cdot e} \quad EOCH_3 \quad + \quad e \cdot \overset{\overset{O}{\|}}{C} - N = \overset{\overset{H}{|}}{\underset{\underset{CH_3}{|}}{C}}$$

$$\xrightarrow{+ \ \ddot{N} \cdot nn} \quad \overset{\overset{O}{\|}}{N - C} - \underset{(CH_3O)\overset{|}{C}H(CH_3)}{N \cdot nn} \qquad\qquad 7.121a$$

Molecular rearrangement does "seem" to be favored, a clear indication that the isocyanate shown below is a MALE which cannot molecularly rearrange, noting that OH or OR group is already shielded.

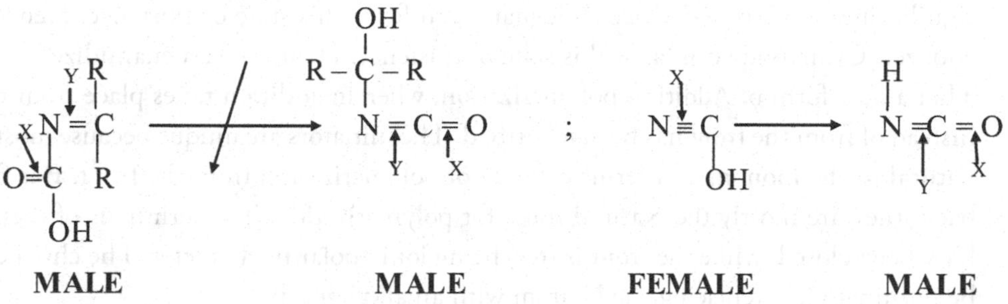

$$\underset{O=C \quad R}{\overset{x}{\underset{\underset{OH}{|}}{N}} = \overset{\overset{Y}{\downarrow} R}{C}} \quad \xrightarrow{\quad / \quad} \quad \underset{Y \quad X}{N \equiv C = O} \quad ; \quad \underset{\overset{}{OH}}{N \equiv \overset{\overset{x}{}}{C}} \quad \xrightarrow{\quad} \quad \underset{Y \quad X}{N \equiv \overset{\overset{H}{}}{C} = O}$$

MALE   **MALE**   **FEMALE**   **MALE**   7.121b

Worthy of note is that when molecular rearrangement takes place, the conjugatedly placed mono-mer can no longer be resonance stabilized. If OH group is resonance stabilized, the transfer species cannot be moved. Hence, molecular rearrangement does not take place with this type of monomer which is resonance stabilized. ***Thus, when a non-ringed Male which is resonance stabilized exist, it cannot undergo molecular rearrangement when the group is internally located.*** One can observe how complex NATURE CAN BE.

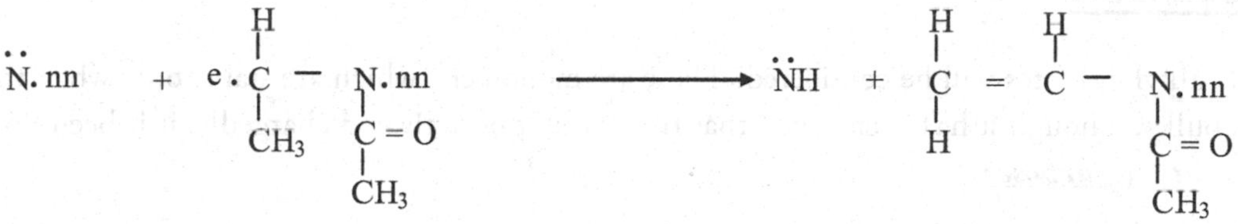

$$\ddot{N} \cdot nn \quad + \quad e \cdot \underset{\underset{\underset{CH_3}{|}}{\overset{|}{C}H_3} \quad \underset{\underset{CH_3}{|}}{\overset{|}{C}=O}}{\overset{\overset{H}{|}}{C}} - N \cdot nn \quad \xrightarrow{\hspace{2cm}} \quad \ddot{N}H \quad + \quad \underset{\underset{H}{|}}{\overset{\overset{H}{|}}{C}} = \underset{\underset{\underset{CH_3}{|}}{\overset{|}{C}=O}}{\overset{\overset{H}{|}}{C}} - N \cdot nn \qquad 7.122$$

$$E \cdot e \quad + \quad nn \cdot \underset{\underset{\underset{CH_3}{|}}{\overset{|}{C}=O}}{\overset{\overset{H}{|}}{N}} - \overset{\overset{H}{|}}{\underset{CH_3}{C} \cdot e} \quad \xrightarrow{\hspace{2cm}} \quad E - \underset{\underset{\underset{CH_3}{|}}{\overset{|}{C}=O}}{\overset{\overset{H}{|}}{N}} - \overset{\overset{H}{|}}{\underset{CH_3}{C} \cdot e} \qquad 7.123$$

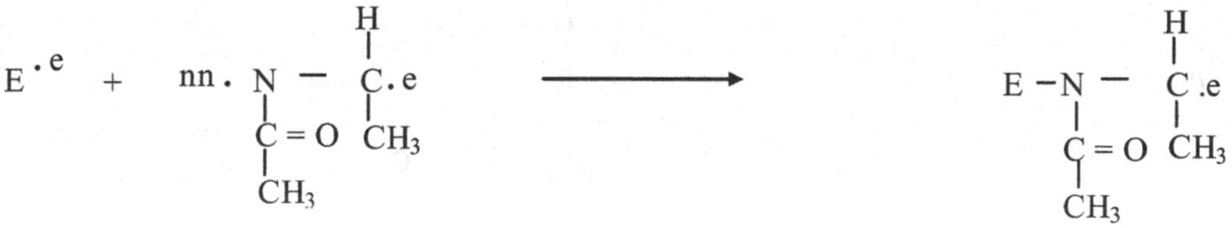

Occurrence of molecular rearrangement is not possible. Electro-free-radically, polymerization of the monomer is favored. The same applies nucleo-non-free-radically when paired provided the monomer is trans-placed in character. Copolymers can be produced here electro-free-radically.

From all the considerations, the following can be observed:-

(i)   Why Aldimines and Ketimines have never been popularly known to favor both charged and radical polymerization. The reasons are now becoming obvious.

(ii)   That Aldimines and Ketimines are more nucleophilic in character than electro-philic. That is, these are females.

(iii)   Why resonance stabilization has never been observed. It can be provided from the carbon and the nitrogen centers radically depending on the type of group. From the nitrogen center, the Male character is introduced with provision of resonance stabilization, but no molecular rearrangement.

(iv)   That irrespective of the type of group carried by the nitrogen center, the charge or radical carried by it when activated remains the same, whether the group is R, OH, OR, NH_2, NHR, etc. types of groups.

(v)   Just as Aldehydes and Ketones undergo Enolization, so also Aldimines and Ketimines undergo what has been called herein Enamidization since they can favor Activated/ Equilibrium State of Existence. Isocynates can favor this state of existence. Hence it only enolizes. Cyanamide can favor this state of existence. Hence, it can enamidize.

(vi)   That a new form of Addition polymerization, wherein addition takes place from the back instead of from the front has been identified. The initiators are unique because most are not natural to the monomer in terms of route of polymerization from the front. But from the back, they are mostly the Natural route for polymerization. The terminals of their chains have been closed, while the front is free, being ionic/polar in character. The chain can only be terminated by replacing the H atom with an alkyl group.

(vii)   Based on this new form of Addition polymerization, limited to only the Aldehydes, Ketones, Aldimines, Ketimines, and etc. types of monomers, the origin of the source of Depropagation during polymerizations of these types of monomers have been identified and "Ceiling temperature" has therefore been redefined.

(viii) Etc.

7.3.   Cyanates

Only alkyl cyanates will be considered. These are monomers, which are stable only when the alkyl group is bulky. Though it has been stated that they cannot be activated chargedly, it is been done here for ***exploratory purposes.***

$$
\begin{array}{c}
C \equiv N \\
| \\
O \\
| \\
R
\end{array}
\longrightarrow
\begin{array}{c}
{}^{\oplus}C = N^{\ominus} \\
| \\
O \\
| \\
R
\end{array}
\qquad 7.124
$$

$$
RO^{\ominus} \; + \;
\begin{array}{c}
{}^{\oplus}C = N^{\ominus} \\
| \\
O \\
| \\
R
\end{array}
\longrightarrow
ROR \; + \; {}^{\ominus}O - C \equiv N
$$

(Non-free)

$$7.125$$

$$
R^{\oplus} \; + \;
\begin{array}{c}
C \equiv N \\
| \\
O \\
| \\
R
\end{array}
\longrightarrow
ROR \; + \; {}^{\oplus}C \equiv N
$$

Free media

$$7.126$$

$$R^{\oplus} \underset{\underset{\overset{|}{N} = \overset{\oplus}{C}}{\overset{|}{O}}}{\overset{\overset{\overset{F}{|}\diagup F}{\underset{R}{\overset{\ominus}{B}}\diagdown F}}{\underset{\overset{|}{O}}{\underset{R}{\overset{|}{C}}}}}\ \ \longrightarrow\ \ R - N = \overset{\overset{R}{\overset{|}{O}}}{\overset{|}{\underset{}{C^{\oplus}}}} \underset{\overset{|}{O}}{\overset{\overset{F}{|}\diagup F}{\underset{R}{\overset{\ominus}{B}}\diagdown F}} \qquad 7.127$$

From the reactions above, only positively charged-paired initiators would have favored the charged polymerization of the monomer in the absence of electrostatic forces of repulsion and steric limitations. The transfer species involved is of the first kind of the first type.

Radically, the followings are obtained.

$$\ddot{N} \cdot nn\ +\ e \cdot \underset{\underset{R}{\overset{|}{O}}}{\overset{|}{C}} = N \cdot nn \ \longrightarrow\ \ddot{N}R\ +\ nn \cdot O - C \equiv N$$

$$E \cdot {}^e\ +\ e \cdot \underset{\underset{R}{\overset{|}{O}}}{\overset{|}{C}} = N \cdot nn \ \longrightarrow\ EOR\ +\ e \cdot C \equiv N \quad OR \quad E - N = \underset{OR}{\overset{|}{C}} \bullet e \quad 7.128$$

$$\text{(I) FAVORED WHEN STABLE} \qquad \text{(II) FAVORED}$$

$$7.129$$

(I) is favored when not activated. When activated, it is (II) that is favored. Electro-free-radically, the monomer can be polymerized in view of the fact that the route is Natural to it. For the growing polymer chain electro-free-radically and not positively, the following is obtained positively for exploratory purposes.

$$R \underset{\overset{|}{R}}{\overset{\overset{}{}}{\underbrace{\left(N = \underset{\underset{R}{\overset{|}{O}}}{\overset{|}{C}} \right)}_{n}}} N = \underset{\underset{R}{\overset{|}{O}}}{\overset{|}{C^{\oplus}}} \underset{\underset{R}{\overset{|}{O}}}{\overset{\overset{F}{|}\diagup F}{\overset{\ominus}{B}\diagdown F}} \ \longrightarrow$$

$$R \underset{}{\underbrace{\left(N = \underset{\underset{R}{\overset{|}{O}}}{\overset{|}{C}} \right)}_{n}} N = C = O\ +\ ROR\ +\ BF_3$$

$$7.130$$

The transfer species abstracted "anionically" or nucleo-non-free-radically, is the transfer species rejected from its growing polymer chain in accordance to the Conservation law of transfer of transfer species.

7.4. Diazoalkanes

Unlike monomers which have been considered so far some of which can be self-activated at temperatures above the ceiling temperatures, diazoalkanes do not have ceiling temperatures. Since these monomers are in the polar states, they cannot be activated chargedly without losing a stable compound.

There is need to visit the chemistry of these monomers, in view of their unique structures. As a matter of fact, based on what the nitrogen molecule is carrying, these are not diazo-alkanes, but diazo-alkylenes. When the Carbenes family is activated, they become Alkylenes, the first member of which is Methylene, followed by ethylene, and next propylene and so on.

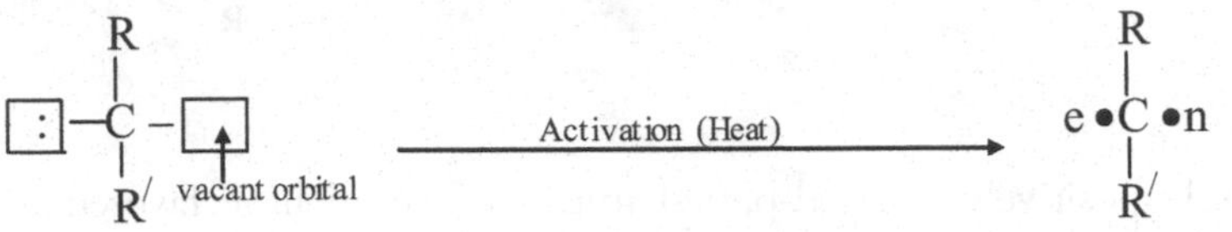

$$
\begin{array}{c}
R \\
| \\
:\!-\!C\!-\!\boxed{\uparrow} \\
| \quad \text{vacant orbital} \\
R'
\end{array}
\qquad \xrightarrow{\text{Activation (Heat)}} \qquad
\begin{array}{c}
R \\
| \\
e \bullet C \bullet n \\
| \\
R'
\end{array}
\qquad 7.131
$$

CARBENES   **ALKYLENES**

where R and R$'$ are H or alkyl groups. The Rs could be the same or different. These are the first members of "The Engineering Hydrocarbon Family Tree" as will be shown downstream [Appendix I]. The Alkylenes are quite different from the Alkenes in which the first member is ethene (Not ethylene as commonly used today despite the corrections from the International Union of Pure and Applied Chemistry), followed by propene (Not propylene), and next butene and so on. Shown below are first four members of the Alkylene family.

$$
\begin{array}{ccccccc}
\begin{array}{c} H \\ | \\ e\bullet C \bullet n \\ | \\ H \end{array}
& ; &
\begin{array}{c} H \\ | \\ e\bullet C \bullet n \\ | \\ CH_3 \end{array}
& ; &
\begin{array}{c} H \\ | \\ e\bullet C \bullet n \\ | \\ C_2H_5 \end{array}
& ; &
\begin{array}{c} CH_3 \\ | \\ e\bullet C \bullet n \\ | \\ CH_3 \end{array}
\end{array}
$$

Methylene (CH_2)   Ethylene (C_2H_4)   Propylene (C_3H_6)   Isomer of Propylene (C_3H_6) 7.132

As already said, never can an element carry two similar opposite charges on their centers. The central C atom cannot carry + and − real charges on its center at the same time, whether ionic, covalent, electrostatic or polar charges. They can only carry radicals and imaginary charges. This alone signifies to us that electrons don't exist outside the nucleus of an atom, but inside. What are called electrons today are radicals. The living alkylenes are what attach themselves to dead molecular nitrogen, N≡N, to give di-azo-alkylenes. The azo is −N=N − and "di" indicates the polar character of the compound. The Alkylenes are indeed isomers of alkenes. While propene does not have isomers notice that propylene has isomers. So also are the higher ones. We are the only ones that know where the name Diazo-alkane came from! One thing that is so clear, is that we have only just begun despite the great foundations that have been laid so far before now by very great human beings (Homo-sapience).

Diazomethane (i.e. diazomethylene –Appendix I) is said to decompose in pentane solution containing a catalytic amount of cuprous chloride to yield a less energetic form of carbene. In the presence of an alkene, it adds entirely cis to the double bond to give a cis-substituted cyclopro-pane.[3]

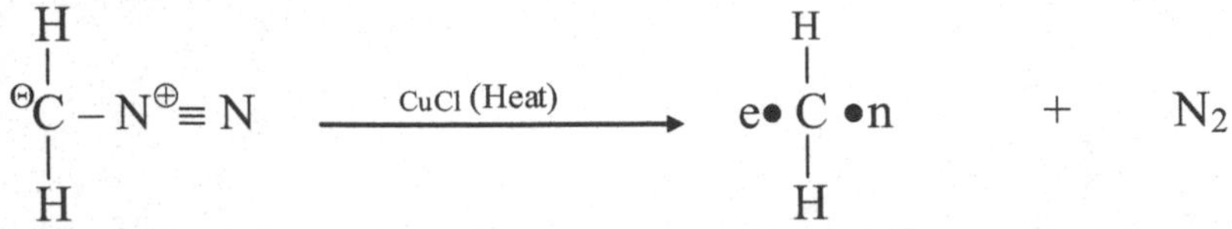

$$
\begin{array}{c}
H \\ | \\
{}^{\ominus}C - N^{\oplus} \equiv N \\ | \\ H
\end{array}
\qquad \xrightarrow{\text{CuCl (Heat)}} \qquad
\begin{array}{c}
H \\ | \\
e\bullet C \bullet n \\ | \\ H
\end{array}
\quad + \quad N_2 \qquad 7.133
$$

DECOMPOSITION MECHANISM (One Stage)

$$7.134$$

EQUILIBRIUM MECHANISM (One Stage)

The CuCl acted as a passive catalyst to increase the polar character of the diazomethane and release the N_2 via Decomposition mechanism. If it is decomposed via Equilibrium mechanism, the methylene must deactivate to give carbene. Cu has five paired unbonded radicals in the last shell and these provide the polar character. The methylene already in its activated state diffuses to activate the cis-2-Butene and add to it being more nucleophilic than the alkene or activated alkene to form (I) in one single stage which closes to form (II) via Equilibrium mechanism. If it was the nucleo-free-radical end that had diffused, transfer species of the first kind of the first type would have been abstracted and no ring will ever be formed. By the nature of Addition, trans-placement of the CH_3 groups is also possible. Imagine if there was one mole of diazo-methane and so many moles of the butene in the system, only the ring can be formed, since a living amorphous polymer cannot be obtained, due to steric limitations..

If the methylene had been weak, then there would have been enough time for the butene to undergo molecular rearrangement to obtain the following either as by-product along with cis-ring or full product.

$$7.135$$

Diazoalkanes themselves are also said to add to polarized unsaturated compounds to give heterocylic compounds. The reaction is believed to take place by way of the dipolar form of the diazo compound and is classed as a 1,3 -dipolar cycloaddition.[3] However, the reaction can be explained as follows:

$$
\begin{array}{c}
C_2H_5 \\
| \\
O \\
| \\
C=O \\
| \\
C \equiv C \\
\quad\quad | \\
\quad\quad C=O \\
\quad\quad | \\
\quad\quad O \\
\quad\quad | \\
\quad\quad C_2H_5
\end{array}
\quad + \quad
en\;\ddot{\cdot}\overset{\cdot}{N}=N-\overset{\displaystyle H}{\underset{\displaystyle \substack{| \\ C=O \\ | \\ O \\ | \\ C_2H_5}}{C}}.n
\quad\longrightarrow\quad
n.C=\overset{\displaystyle }{\underset{\displaystyle \substack{| \\ C=O \\ | \\ O \\ | \\ C_2H_5}}{C}}.e \quad +
$$

(I) (Electrophile)
Ethylacetylene
dicarboxylate

(II) (Electrophile)
Ethyl diazo
acetate

(Male)

$$
n.\overset{\displaystyle H}{\underset{\displaystyle \substack{| \\ C=O \\ | \\ O \\ | \\ C_2H_5}}{C}}-N=N.en
\quad\longrightarrow\quad
e.C=\overset{\displaystyle }{\underset{\displaystyle \substack{| \\ C=O \\ | \\ O \\ | \\ C_2H_5}}{C}}-\ddot{N}=N-\overset{\displaystyle H}{\underset{\displaystyle \substack{| \\ C=O \\ | \\ O \\ | \\ C_2H_5}}{C}}.n
\quad\longrightarrow
$$

C_2H_5 (Less Electrophilic)

Diffusing to Male →

(III)

$$
H_5C_2-O-\overset{\displaystyle O}{\overset{\|}{C}}-\underset{\displaystyle \substack{\| \\ C \\ \diagup\quad\diagdown \\ O \;\; N \\ \| \quad\quad \diagdown \\ H_5C_2-O \quad\quad\quad N}}{C}-\overset{\displaystyle H}{\underset{\displaystyle }{C}}-\overset{\displaystyle O}{\overset{\|}{C}}-O-C_2H_5
\quad\rightleftharpoons\quad
H_5C_2-O-\overset{\displaystyle O}{\overset{\|}{C}}-\underset{\displaystyle \substack{\| \\ C \\ \diagup\quad\diagdown \\ O \;\; N \\ \| \quad\quad \diagdown \\ H_5C_2-O \quad\quad\quad N}}{C}-\overset{\displaystyle .n}{C}-\overset{\displaystyle O}{\overset{\|}{C}}-O-C_2H_5
$$

(IV)

$$
+\;\;H.e \quad\rightleftharpoons\quad
H_5C_2-O-\overset{\displaystyle O}{\overset{\|}{C}}-\underset{\displaystyle \substack{\| \\ C \\ \diagup\quad\diagdown \\ O \;\; N \\ \| \quad\quad \diagdown \\ H_5C_2-O \quad\quad\quad N \\ \quad\quad\quad | \\ \quad\quad\quad H}}{C}-\overset{\displaystyle }{C}-\overset{\displaystyle O}{\overset{\|}{C}}-O-C_2H_5
$$

(V)

Ethyl pyrozole- 3, 4, 5-tricarboxylate

7.136

The ethylacetylene dicarboxylate and ethyl diazoacetate are males and as has begun to be shown and will be fully shown downstream, *it is the female that diffuses to the male or less male that diffuses to the more male or more female that diffuses to the less female.* In the absence of CuCl, N_2 could not be released from the diazo compound with many polar centers on the oxygen centers. Instead the diazo compound was placed in the radical state already shown in Equation 7.3b. (II) above which is less male diffuses with its electro-non-free-radical end to the alkene which is more male and resonance stabilized, activates it and adds to it. The electro-non-free-radical cannot abstract OCH_3, because one of the OCH_3 groups has been shielded by the resonance stabilization favored in

455

moving from Half-Full-Free to Half-Full-free, unlike in Benzoquinone also an Electrophile, since full non-free resonance stabilization could not take place in the presence of C≡C bond. The electro-free-radical end of the couple (III) formed, diffuses to the nucleo-free-radical end of the couple to form (IV), a five-membered (not seven-membered) ring unable to abstract OCH_3 group because *the group has been sheilded.* (IV) cannot molecularly rearrange because the – N = N – bond based on what the nitrogen centers are carrying cannot be activated as will be shown downstream when looking at "Nitrogen-Hydrocarbon Family Tree". How can it be activated when the two nitrogen centers are connected to carbon which is more electropositive than N? It is only a type of compound like Cl – N = N – Cl that can be activated. *Nitrogen cannot carry an electro-free-radical when a carbon center **directly connected** to it is carrying a nucleo-free-radical.* It will be **very explosive,** with release of great amount of energy. This is not like the case of diazomethane which when activated as shown in Equation 7.3b, the electro-radical is carried by N while the C is carrying the nucleo-free-radical, because the N center carrying the electro-radical which is non-free is **not directly connected** to a C center. That is why the explosive character in diazomethane is well controlled, unlike when adjacently or directly placed. (IV) rearranged to (V) *via **Electroradicalization as shown above and not by molecular rearrangement.*** Thus while methyl acrylate cannot be resonance stabilized, (I) above also an Electrophile can be resonance stabilized. One can observe the great wonders of NATURE. It is via Electroradicalization the so called Azomethane (H_3C-N=N-CH_3) whose real name is Azoethane as will be shown downstream rearranges to the more stable *formaldehyde methyl hydrazone*[3] or more correctly **methyl formaldimine** as shown below. The use of formaldehyde as part of name is wrong since there is no O in the molecule.

Stage 1:

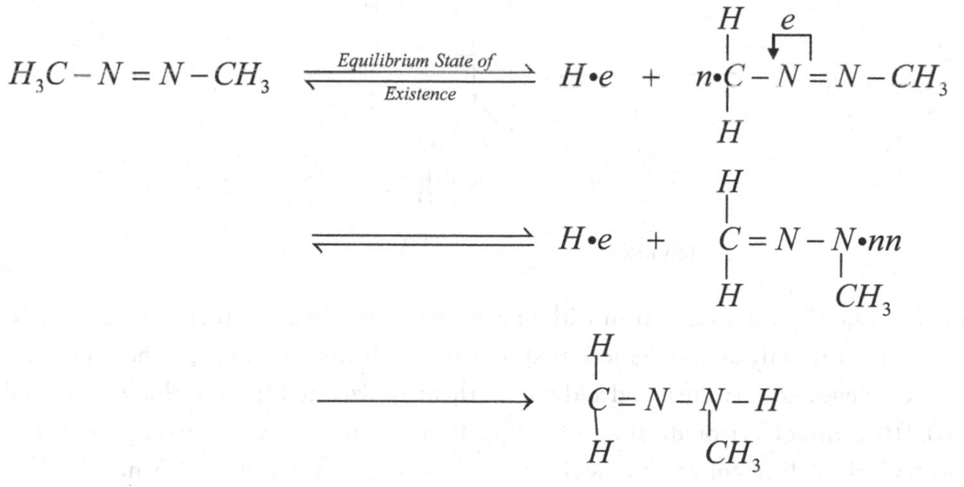

$$H_3C - N = N - CH_3 \quad \underset{\text{Existence}}{\overset{\text{Equilibrium State of}}{\rightleftharpoons}} \quad H\bullet e \quad + \quad n\bullet \overset{\overset{H}{|}}{\underset{\underset{H}{|}}{C}} - \overset{\overset{e}{\downarrow}}{N} = N - CH_3$$

$$\rightleftharpoons \quad H\bullet e \quad + \quad \overset{\overset{H}{|}}{\underset{\underset{H}{|}}{C}} = N - \underset{\underset{CH_3}{|}}{N}\bullet nn$$

$$\longrightarrow \quad \overset{\overset{H}{|}}{\underset{\underset{H}{|}}{C}} = N - \underset{\underset{CH_3}{|}}{N} - H$$

"Formaldehye methyl hydrazone"   7.137

Based on the Equilibrium State of Existence of the methyl formaldimine shown below, it cannot tautomerize and it cannot undergo molecular rearrangement when activated.

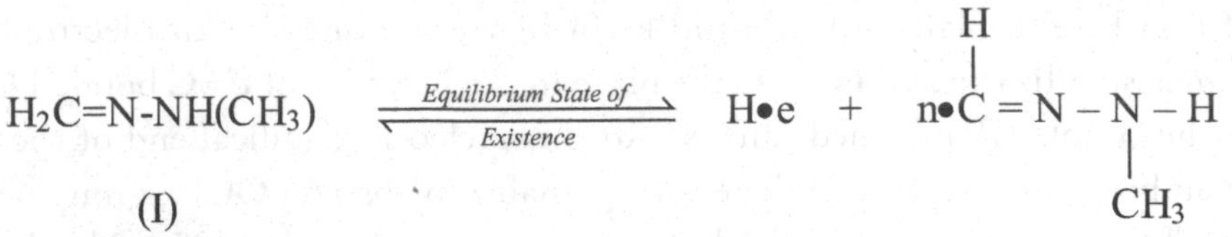

$$H_2C=N-NH(CH_3) \xrightleftharpoons[\text{Existence}]{\text{Equilibrium State of}} H\bullet e \;+\; \underset{\underset{CH_3}{|}}{n\bullet C} \overset{\overset{H}{|}}{=} N-N-H \qquad 7.138$$

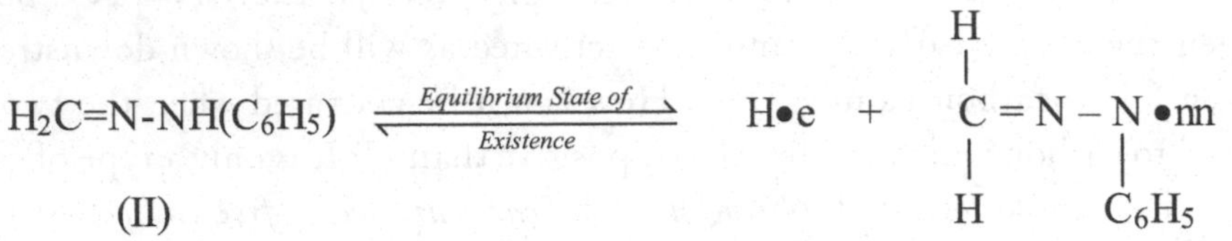

(I)

$$H_2C=N-NH(C_6H_5) \xrightleftharpoons[\text{Existence}]{\text{Equilibrium State of}} H\bullet e \;+\; \underset{\underset{H}{|}}{C} \overset{\overset{H}{|}}{=} N - \underset{\underset{C_6H_5}{|}}{N} \bullet nn \qquad 7.139$$

(II)

Based on the Equilibrium State of Existence of (II), it can electroradicalize but not molecular rearrange to give ***Azo toluene ($H_3C-N=N-C_6H_5$) which in present-day Science is called benzeneazomethane.*** The real name of (II) is ***Phenyl formaldimine.*** Compare the Equations above with Equations 7.56 and 7.57. Without searching for the finger prints of many compounds, we are just wasting time.

Diazoalkanes have also been reported to add to acetylene in ***cold ethereal solutions*** to produce heterocylic compounds.[10]

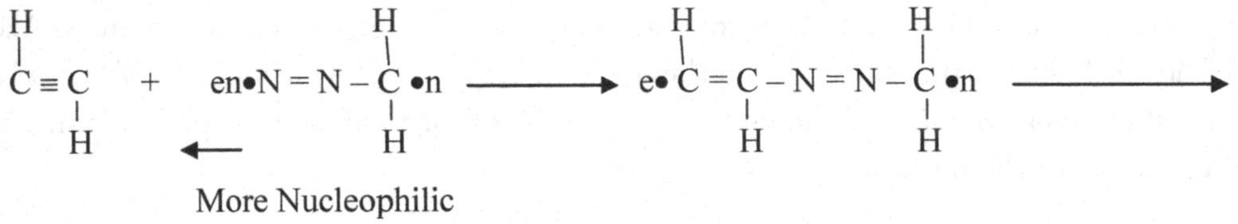

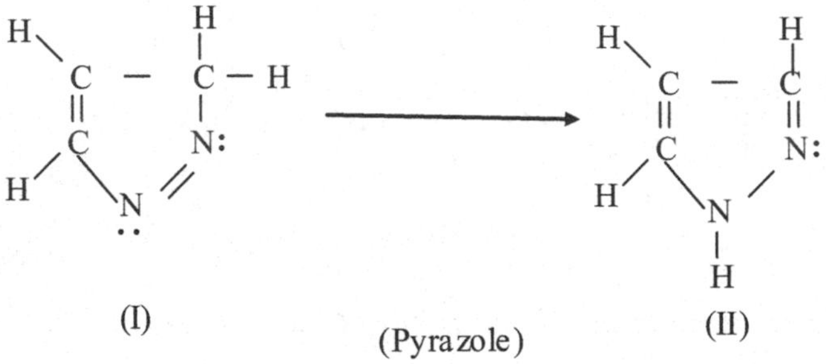

$$\underset{\underset{H}{|}}{C} \overset{\overset{H}{|}}{\equiv} C \;+\; en\bullet N = N - \underset{\underset{H}{|}}{C} \overset{\overset{H}{|}}{\bullet} n \longrightarrow e\bullet \underset{\underset{H}{|}}{C} \overset{\overset{H}{|}}{=} C - N = N - \underset{\underset{H}{|}}{C} \overset{\overset{H}{|}}{\bullet} n \longrightarrow$$

More Nucleophilic

(I)   (Pyrazole)   (II)   7.140

The mechanism is Equilibrium mechanism with one or two stages. In cold ethereal solutions, N_2 cannot be released. At best, it can only exist in its activated state which diffuses to suppress the Equilibrium State of Existence of acetylene and activate it and add to it. It them cyclized to form (I) which electroradicalized to produce (II). (II) cannot electroradicalize to give (I). It cannot molecularly rearrange to give (I). If (II) rearranges molecularly, a different product with two C=N bonds will be obtained, since the C = N bond is more nucleophilic than C = C bond.

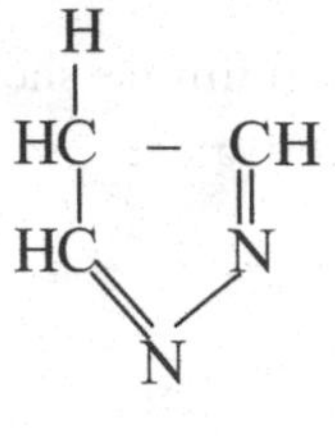

7.141

457

Diazoalkanes have also been reported to add to fully substituted acetylenes to produce derivatives of cyclopropene.[11] The reactions can be explained as follows.

$$
\begin{array}{c}
\underset{\underset{R'}{|}}{\overset{\overset{R}{|}}{C}} \equiv C \quad + \quad \underset{\underset{R'''}{|}}{\overset{\overset{R''}{|}}{\ominus C}} - N^{\oplus} \equiv N \quad \longrightarrow \quad e\bullet \underset{\underset{R}{|}}{\overset{\overset{R'}{|}}{C}} = C\bullet n \quad + \quad e\bullet \underset{\underset{R'''}{|}}{\overset{\overset{R''}{|}}{C}}\bullet n \quad + \quad N_2
\end{array}
$$

$$ (R' > R) $$

$$
\longrightarrow \quad N_2 \quad + \quad n\bullet \underset{\underset{R'''}{|}}{\overset{\overset{R''}{|}}{C}} - \underset{\underset{R'}{|}}{\overset{\overset{R}{|}}{C}} = C\bullet e \quad \longrightarrow \quad \underset{\underset{R''}{}}{\overset{\overset{R'}{|}}{C}} = \underset{\underset{R'''}{}}{\overset{\overset{R}{|}}{C}} \quad + \quad N_2
$$

7.142

This is a *one stage Equilibrium mechanism process*. It was the polar state of diazomethane that was first activated. The diazo compound being more nucleophilic than the acetylene, diffuses to activate the acetylene and add to it to form the ring. In the same manner the derivatives of cyclopropene has been obtained above, so also can derivatives of cyclopropane be obtained using alkenes.

That diazoalkanes are stronger nucleophiles than aldehydes and ketones is illustrated with the following analysis. The mechanism is Equilibrium.

$$
\begin{array}{c}
\underset{\underset{H}{|}}{\overset{\overset{R}{|}}{C}} = O \quad + \quad \underset{\underset{R}{|}}{\overset{\overset{R}{|}}{\ominus C}} - N^{\oplus} \equiv N \quad \longrightarrow \quad n\bullet \underset{\underset{R}{|}}{\overset{\overset{R}{|}}{C}}\bullet e \quad + \quad nn\bullet O - \underset{\underset{H}{|}}{\overset{\overset{R}{|}}{C}}\bullet e \quad + \quad N_2
\end{array}
$$

More Nucleophilic

$$
n\bullet \underset{\underset{R}{|}}{\overset{\overset{R}{|}}{C}} - O - \underset{\underset{H}{|}}{\overset{\overset{R}{|}}{C}}\bullet e \quad + \quad N_2 \longrightarrow \quad N_2 \quad + \quad H - \underset{\underset{O}{\diagdown}}{\overset{\overset{R}{|}}{C}} - \underset{\underset{O}{\diagup}}{\overset{\overset{R}{|}}{C}} - R
$$

7.143

Diazoalkanes are also well known to react with aldehydes to form ketones. For example, diazo-ethane (real name diazoethylene) reacts with propionaldehyde to give 3-pentanone or diethyl ketone and 1-Diazopropane reacts with propionaldehyde to 3-Hexanone in good yields.[12]

Stage 1:

$$
H - \overset{\overset{\displaystyle O}{\|}}{C} - C_2H_5 \quad \underset{Existence}{\overset{Equilibrium\ State\ of}{\rightleftharpoons}} \quad H\bullet e \quad + \quad n\bullet \overset{\overset{\displaystyle O}{\|}}{C} - C_2H_5
$$

(I)

$$
H\bullet e \quad + \quad n\bullet \underset{\underset{CH_3}{|}}{\overset{}{C}} - N = N\bullet en \quad \overset{Heat}{\rightleftharpoons} \quad H - \underset{\underset{CH_3}{|}}{\overset{\overset{H}{|}}{C}}\bullet e \quad + \quad N_2
$$

$$
H_5C_2\bullet e \quad + \quad (I) \quad \longrightarrow \quad H_5C_2 - \overset{\overset{\displaystyle O}{\|}}{C} - C_2H_5 \qquad\qquad 7.144
$$

(3-Pentanone)

458

Overall Equation: $HCOC_2H_5 \ + \ H(CH_3)C\text{-}N = N \longrightarrow H_5C_2 - CO - C_2H_5 \ + \ N_2$ 7.145

Stage 1:

$$H - \overset{\overset{\displaystyle O}{\|}}{C} - C_2H_5 \ \underset{Existence}{\overset{Equilibrium\ State\ of}{\rightleftharpoons}} \ H\bullet e \ + \ n\bullet \overset{\overset{\displaystyle O}{\|}}{C} - C_2H_5$$

$$(I)$$

$$H\bullet e \ + \ n\bullet \overset{\overset{\displaystyle H}{|}}{\underset{\underset{\displaystyle C_2H_5}{|}}{C}} - N = N\bullet en \ \overset{Heat}{\rightleftharpoons} \ H - \overset{\overset{\displaystyle H}{|}}{\underset{\underset{\displaystyle C_2H_5}{|}}{C}} \bullet e \ + \ N_2$$

$$H_7C_3 \bullet e \ + \ (I) \ \longrightarrow \ H_7C_3 - \overset{\overset{\displaystyle O}{\|}}{C} - C_2H_5 \qquad\qquad 7.146$$

$$\textbf{(3-Hexanone)}$$

Overall Equation: $HCOC_2H_5 \ + \ H(C_2H_5)C\text{-}N = N \longrightarrow H_7C_3 - CO - C_2H_5 \ + \ N_2$ 7.147

Stage 1:

$$H - \overset{\overset{\displaystyle O}{\|}}{C} - C_2H_5 \ \underset{Existence}{\overset{Equilibrium\ State\ of}{\rightleftharpoons}} \ H\bullet e \ + \ n\bullet \overset{\overset{\displaystyle O}{\|}}{C} - C_2H_5$$

$$(I)$$

$$H\bullet e \ + \ n\bullet \overset{\overset{\displaystyle CH_3}{|}}{\underset{\underset{\displaystyle CH_3}{|}}{C}} - N = N\bullet en \ \overset{Heat}{\rightleftharpoons} \ H - \overset{\overset{\displaystyle CH_3}{|}}{\underset{\underset{\displaystyle CH_3}{|}}{C}} \bullet e \ + \ N_2$$

$$H_5C_2 \bullet e \ + \ (I) \ \longrightarrow \ H - \overset{\overset{\displaystyle CH_3}{|}}{\underset{\underset{\displaystyle CH_3}{|}}{C}} - \overset{\overset{\displaystyle O}{\|}}{C} - C_2H_5$$

$$7.148$$

$$\textbf{(Isopropylethyl ketone)}$$

Overall Equation: $HCOC_2H_5 \ + \ (CH_3)_2C\text{-}N = N \longrightarrow (H_3C)_2CH - CO - C_2H_5 \ + \ N_2$ 7.149

Stage 1:

$$H - \overset{\overset{\displaystyle O}{\|}}{C} - C_6H_5 \ \underset{Existence}{\overset{Equilibrium\ State\ of}{\rightleftharpoons}} \ H\bullet e \ + \ n\bullet \overset{\overset{\displaystyle O}{\|}}{C} - C_6H_5$$

$$(I)$$

$$H\bullet e \ + \ n\bullet \overset{\overset{\displaystyle H}{|}}{\underset{\underset{\displaystyle CH_3}{|}}{C}} - N = N\bullet en \ \overset{Heat}{\rightleftharpoons} \ H - \overset{\overset{\displaystyle H}{|}}{\underset{\underset{\displaystyle CH_3}{|}}{C}} \bullet e \ + \ N_2$$

$$H_5C_2 \bullet e \ + \ (I) \ \longrightarrow \ H_5C_2 - \overset{\overset{\displaystyle O}{\|}}{C} - C_6H_5$$

$$\textbf{(Phenylethyl ketone)}$$

Overall Equation: $HCOC_6H_5 \ + \ H(CH_3)C\text{-}N = N \longrightarrow H_5C_2 - CO - C_6H_5 \ + \ N_2$ 7.151

Either rings or ketones or mixtures of them are formed, depending on the operating conditions and the types of components involved. When benzene ring is present, the ketone cannot be strongly activated and therefore mostly ketones are the product. For ketones to be formed, observe that the aldehyde exists in Equilibrium State of Existence and not Activated/Equilibrium State of Existence. As is clearly obvious in Equation 7.143, in a single stage Equilibrium mecha-nism process, before a ring can be formed, the aldehyde must be kept in Activated State of Existence which depends on the use of heat or a hot radical.

So far, worthy of note is that all the reactions of diazoalkylenes (so-called diazoalkanes) are radical in character. Also never was there a time the aldehydes were kept in **Activated/Equil-ibrium State of Existence.** Based on the mechanisms provided above, when diazoalkylenes react with ketones, both rings and higher ketone will be obtained as products. How are these higher ketones obtained? The reactions of 5-nonanone with diazoethane in the presence of various activation catalysts which gave **homologation products** shown below and epoxide, were examined.[13]

$$H_9C_4 - CO - C_4H_9 \underset{\text{Catalyst}}{\overset{CH_3CHN_2}{\rightleftharpoons}} H_9C_4 - CO - CH(CH_3) - C_4H_9 \text{ (I)}$$

$$+ \quad H_9C_4 - CH(CH_3) - CO - CH(CH_3) - C_4H_9 \text{ (II)}$$

$$+ \quad H_9C_4 - CO - CH(CH_3) - CH(CH_3) - C_4H_9 \text{ (III)}$$

Homolo-gation Products

$$+ \quad$$

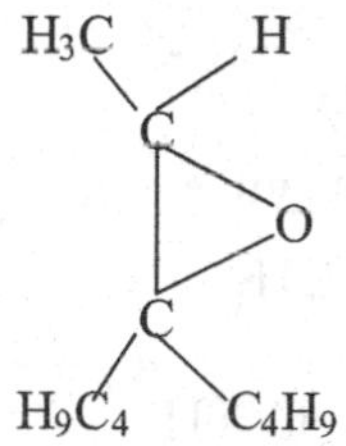

Epoxide (IV)

7.152

When catalysts such as alcohols and lithium chloride were used, no products were obtained because 5-nonanone is very stable. When $BF_3.O(C_2H_5)_2$ in ether was used at -20-0°C, less than 2 percent of the homologation products were obtained. When $(CH_3)_3Al$ in CH_2Cl_2 was used at -78°C, 52 percent of 5-nonanone reacted to give 92 percent of (I) and 8 percent of (IV). At 0°C, 50 percent reacted to give 91 percent (I), 1 percent (II) and 8 percent (IV). When $(i-C_4H_9)_3Al$ in CH_2Cl_2 was used at -78°C, 65 percent reacted to give 92 percent (I), 1 percent (II) and 7 percent (IV). When $(C_2H_5)_2AlCl$ in CH_2Cl_2 was used at -78°C, 74 percent reacted to give 63 percent (I), 30 percent (II), 7 percent (III) and no (IV). This is summarized in Table 7.1 below.

Table 7.1 Reactions of 5-Nonanone with Diazoethane in the presence of some catalysts.

Types of Catalyst	Temp. °C	% of Nonanone Consumed	% of Homologation Products			% Epoxides
			I	II	III	IV
Alcohol & LiCl		0	0	0	0	0
$BF_3O(C_2H_5)$ in Ether	-20-0		Less than 2%			
$(CH_3)_3Al$ in CH2Cl2	-78 0	52 50	92 91	0 1	0 0	8 8

(i-C4H9)3Al in CH2Cl2	-78	65	92	1	0	7
(C2H5)2AlCl in CH2Cl2	-78	74	63	30	7	0

One can observe that these catalysts look as if they are selective. At these low temperatures, N_2 was released. So far, we have identified the mechanisms for formation of (IV) but not for the homologation products. The yield of (IV) is very small because at such low temperatures only a very small fraction of the ketone was activated. Notice that in order to obtain (I), only one side of the ketone was involved. For (II), two sides of the ketone were involved. For (III), only one side was involved with presence of more than one ethylene group. For the formation of (I), the following is the mechanism.

Stage 1:

$$H_9C_4 - \overset{\overset{\displaystyle O}{\|}}{C} - C_4H_9 \underset{\text{\tiny Existence}}{\overset{\text{\tiny Equilibrium State of}}{\rightleftharpoons}} H_9C_4 \bullet n \quad + \quad e\bullet \overset{\overset{\displaystyle O}{\|}}{C} - C_4H_9$$

$$\text{(I)} \qquad\qquad\qquad \text{(II)}$$

$$e\bullet \overset{\overset{\displaystyle H}{|}}{\underset{\underset{\displaystyle CH_3}{|}}{C}} - N = N\bullet nn \overset{\text{\tiny Heat}}{\rightleftharpoons} e\bullet \overset{\overset{\displaystyle H}{|}}{\underset{\underset{\displaystyle CH_3}{|}}{C}} \bullet n \quad + \quad N_2$$

$$\text{(II)} \quad + \quad n\bullet \overset{\overset{\displaystyle H}{|}}{\underset{\underset{\displaystyle CH_3}{|}}{C}} \bullet e \rightleftharpoons H_9C_4 - \overset{\overset{\displaystyle O}{\|}}{C} - \overset{\overset{\displaystyle H}{|}}{\underset{\underset{\displaystyle CH_3}{|}}{C}} \bullet e$$

$$\text{(III)}$$

$$\text{(III)} \quad + \quad \text{(I)} \longrightarrow H_9C_4 - \overset{\overset{\displaystyle H}{|}}{\underset{\underset{\displaystyle CH_3}{|}}{C}} - \overset{\overset{\displaystyle O}{\|}}{C} - C_4H_9$$

$$\underline{\textbf{(I)}}\textbf{-(n-butyl-iso-hexyl ketone)} \qquad 7.153$$

Overall Equation: $H_9C_4COC_4H_9 \; + \; H(CH_3)C\text{-}N = N \longrightarrow H_9C_4 - CO - CH(CH_3)\,C_4H_9$

$$+ \quad N_2 \qquad 7.154$$

For (II), the followings are to be expected beginning with the 5-nanonone, with the next stage as Stage 2.

Stage 2:

$$H_9C_4 - \overset{\overset{\displaystyle H}{|}}{\underset{\underset{\displaystyle CH_3}{|}}{C}} \overset{\overset{\displaystyle O}{\|}}{C} - C_4H_9 \underset{\text{\tiny Existence}}{\overset{\text{\tiny Equilibrium State of}}{\rightleftharpoons}} H_9C_4\bullet n \quad + \quad e\bullet \overset{\overset{\displaystyle O}{\|}}{C} - \overset{\overset{\displaystyle H}{|}}{\underset{\underset{\displaystyle CH_3}{|}}{C}} - C_4H_9$$

$$\text{(I)} \qquad\qquad\qquad \text{(II)}$$

$$e\bullet \overset{\overset{\displaystyle H}{|}}{\underset{\underset{\displaystyle CH_3}{|}}{C}} - N = N\bullet nn \quad \underset{}{\overset{Heat}{\rightleftharpoons}} \quad e\bullet \overset{\overset{\displaystyle H}{|}}{\underset{\underset{\displaystyle CH_3}{|}}{C}} \bullet n \quad + \quad N_2$$

$$(II) \quad + \quad n\bullet \overset{\overset{\displaystyle H}{|}}{\underset{\underset{\displaystyle CH_3}{|}}{C}} \bullet e \quad \rightleftharpoons \quad H_9C_4 - \overset{\overset{\displaystyle H}{|}}{\underset{\underset{\displaystyle CH_3}{|}}{C}} - \overset{\overset{\displaystyle O}{\|}}{C} - \overset{\overset{\displaystyle H}{|}}{\underset{\underset{\displaystyle CH_3}{|}}{C}} \bullet e$$

$$(III)$$

$$(III) \quad + \quad (I) \quad \longrightarrow \quad H_9C_4 - \overset{\overset{\displaystyle H}{|}}{\underset{\underset{\displaystyle CH_3}{|}}{C}} - \overset{\overset{\displaystyle O}{\|}}{C} - \overset{\overset{\displaystyle H}{|}}{\underset{\underset{\displaystyle CH_3}{|}}{C}} - C_4H_9$$

(II)-(di-iso-hexyl ketone)   7.155

Overall Equation: $H_9C_4CO\ CH(CH_3)\ C_4H_9\ +\ H(CH_3)C\text{-}N = N \longrightarrow N_2\ +$

$$H_9C_4\ CH(CH_3) - CO - CH(CH_3)\ C_4H_9 \qquad 7.156$$

For (III), the followings are similarly obtained.

Stage 1:

$$H_9C_4 - \overset{\overset{\displaystyle O}{\|}}{C} - C_4H_9 \quad \underset{Existence}{\overset{Equilibrium\ State\ of}{\rightleftharpoons}} \quad H_9C_4\bullet n \quad + \quad e\bullet \overset{\overset{\displaystyle O}{\|}}{C} - C_4H_9$$

$$(I)$$

$$2e\bullet \overset{\overset{\displaystyle H}{|}}{\underset{\underset{\displaystyle CH_3}{|}}{C}} - N = N\bullet nn \quad \overset{Heat}{\rightleftharpoons} \quad 2\ e\bullet \overset{\overset{\displaystyle H}{|}}{\underset{\underset{\displaystyle CH_3}{|}}{C}} \bullet n \quad + \quad 2N_2$$

$$H_9C_4 - \overset{\overset{\displaystyle O}{\|}}{C}\bullet c \quad + \quad 2n\bullet \overset{\overset{\displaystyle H}{|}}{\underset{\underset{\displaystyle CH_3}{|}}{C}} \bullet e \quad \rightleftharpoons \quad H_9C_4 - \overset{\overset{\displaystyle O}{\|}}{C} - \overset{\overset{\displaystyle H}{\downarrow}}{\underset{\underset{\displaystyle CH_3}{|}}{C}} - \overset{\overset{\displaystyle H}{|}}{\underset{\underset{\displaystyle CH_3}{|}}{C}} \bullet e$$

$$(II)$$

$$(II) \quad + \quad (I) \quad \longrightarrow \quad H_9C_4 - \overset{\overset{\displaystyle H}{|}}{\underset{\underset{\displaystyle CH_3}{|}}{C}} - \overset{\overset{\displaystyle H}{|}}{\underset{\underset{\displaystyle CH_3}{|}}{C}} - \overset{\overset{\displaystyle O}{\|}}{C} - C_4H_9$$

(III)-(n-butyl-iso-octyl ketone)   7.157

Overall Equation: $H_9C_4COC_4H_9\ +\ 2H(CH_3)C\text{-}N = N \longrightarrow H_9C_4 - CO - [CH(CH_3)]_2C_4H_9$

$$+\ \ 2N_2 \qquad 7.158$$

All what the passive catalysts were able to accomplish was to keep the aldehydes and ketones in Equilibrium State of Existence. Their capacity to do it was limited. Only $ClAl(C_2H_5)_2$ of all the aluminum compounds was able to keep about 75 percent of the 5-nonanone in Equilibrium State of Existence. Hence, there was enough to produce 7 percent of (III) which the others could not do. Even then, most of the reactors commonly used today both in the industries and academia are not the real ones, wherein for example, one cannot have perfect contact or mixing between components as exist in Nature. The reasons why these catalysts are different can be seen from their Equilibrium States of Existence which may be their existing state at such low temperatures or may be in Stable states of existence.

$$AlR_3 \quad \xrightleftharpoons[\text{Existence}]{\text{Equilibrium State of}} \quad R{\bullet}e \;+\; n{\bullet}AlR_2 \quad (I)$$

$$ClAlR_2 \quad \xrightleftharpoons[\text{Existence}]{\text{Equilibrium State of}} \quad R_2Al{\bullet}e \;+\; Cl{\bullet}nn \quad (II)$$

$$R^{\oplus}{\ldots\ldots\ldots}{}^{\circleft}\overset{\overset{F}{\diagup}\,\overset{F}{\diagdown}}{\underset{OR}{B}}{-}F \quad \xrightleftharpoons[\text{Existence}]{\text{Equilibrium State of}} \quad RO{\bullet}nn \;+\; R^{\oplus}{\ldots\ldots}{}^{\circleft}\underset{\bullet e}{B}{-}F \quad (III)$$

7.159

Only Ionic metals (Groups IA and IIA) and Ionic Transition metals (Group IIIA), with the exception of H cannot carry female i.e. nucleo-free or non-free radicals on their centers. It is forbidden. Other metals can. Hence it not surprising to see Al in (I) above carrying a nucleo-free-radical, free because there are no paired unbonded radicals on the aluminum center. One cannot use the (III) for backward polymerization, for so many reasons which are obvious or can be seen from all the considerations so far. Indeed, one has already started preparing *an Encyclopedia for States of Existences of compounds-Equilibrium, Combination, Decomposition, Activated and Activated/Equilibrium.* For example in Glycerol with three hydroxyl groups, there is only one particular hydrogen that is held in Equilibrium State of Existence. What is that hydrogen which is the finger print of the glycerol, for Nature is first and foremost ORDERLINESS? (III) is one of the initiators used for polymerizing diazoalkanes. Indeed, notice that in (III), negative charge on the Boron center is adjacently located to O and Fs. No repulsion takes place, because the negative charge is imaginary, radical in character. All initiators from the Al components are Electrostatic of the positive type also, as will be shown downstream.

The homologation of cyclopentanone with diazoalkanes was also said to result in formation of undesired cycloheptanones at the expense of the intermediary cyclohexanone, because it is believed that the reactivity of cyclic ketones is in the order of cyclohexanone > cyclopentanone > cycloheptanone > cyclooctanone.[13,14] The organoaluminum-promoted single homologation of cyclopentanone was effected with (trimethylsilyl)diazomethane, where the single-homologanated cyclohexanone was said to be successfully trapped as its trimethylsilyl enol ether.[13]

(I)   (II)   (III)   (IV)

7.160

When $(CH_3)_3Al$ was used at -20-0°C, 68 percent of the cyclic ketone was consumed, to give 96 percent of (I), 2 percent each of (II) and (IV). When $BF_3O(C_2H_5)_2$ was used at -20°C, 35 percent of the ketone was consumed, to give 64 percent of (I), 23 percent of (II), 10 percent of (III), and 3 percent of (IV).

Though downstream in the Series and Volumes, the Chemistry and Physics of ringed compounds will be provided, one cannot waite to provide the mechanisms of the reactions above for the purpose of credibility.

Stage 1:

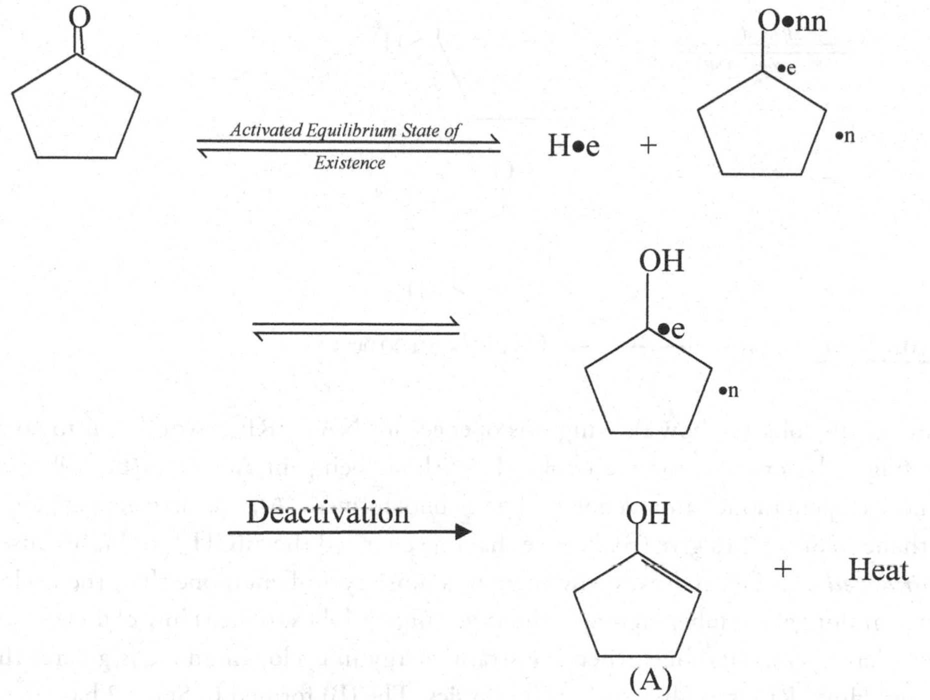

$$\text{(cyclopentanone)} \underset{\substack{\text{Activated Equilibrium State of}\\\text{Existence}}}{\rightleftharpoons} H \bullet e \quad + \quad \text{(cyclopentanone radical)}$$

$$\rightleftharpoons \quad \text{(cyclopentanol radical)}$$

$$\xrightarrow{\text{Deactivation}} \quad \text{(A, cyclopentenol)} \quad + \quad \text{Heat} \qquad 7.161$$

(A)

(ENOLIZATION)

Overall Equation: $66\ H_8C_5O \longrightarrow 66(A) + \text{Heat} \qquad 7.162$

Stage 2:

$$\underset{H}{\overset{Z}{n\bullet C - N = N \bullet cn}} \rightleftharpoons \underset{H}{\overset{Z}{e\bullet C \bullet n}} + N_2 \qquad \{Z \text{ is } H \text{ or } Si(CH_3)_3 \text{ group}\} \qquad 7.163$$

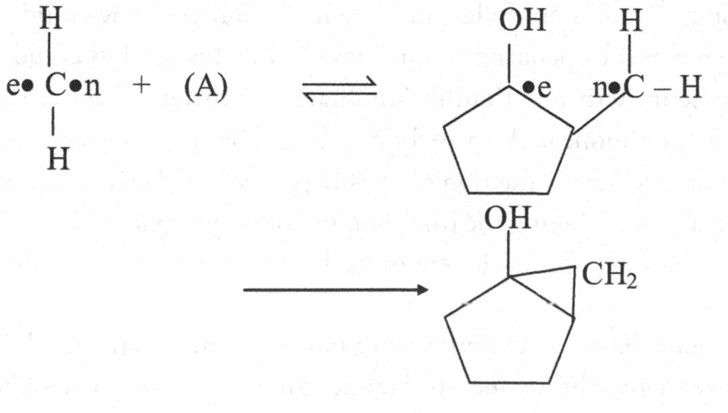

$$\underset{H}{e\bullet C\bullet n} + (A) \rightleftharpoons \text{(intermediate)} $$

$$\xrightarrow{\qquad} \text{(B)} \qquad 7.164$$

(B)

Overall Equation: $66\,\text{Diazomethane} + 66(A) \longrightarrow 66(B) + 66\,N_2 \qquad 7.165$

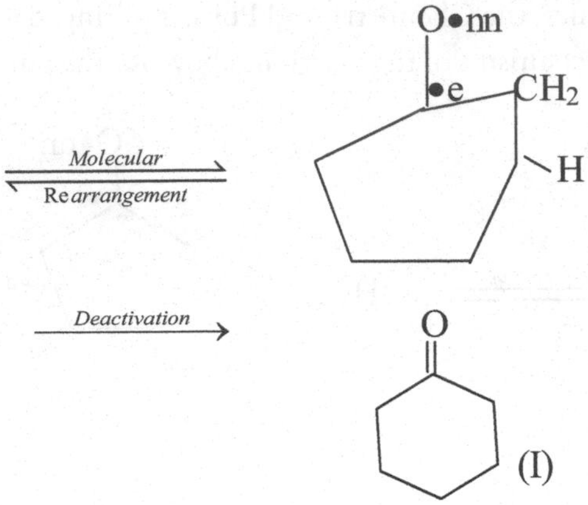

$$
\begin{array}{c}
O\bullet nn \\
\bullet e \quad CH_2 \\
\diagdown H
\end{array}
$$

$$\underset{Re\mathit{arrangement}}{\overset{Molecular}{\rightleftarrows}}$$

$$\overset{Deactivation}{\longrightarrow}$$

$$
O
$$

(I)   7.166

<u>Overall Equation:</u>   (B) $\longrightarrow$ Cyclohexanone   7.167

One can wonderfully observe how the ring was opened, for NATURE is wonderful to compre-hend. In the first stage, the cyclopentanone enolized. Without being in Activated/Equilibrium State of Existence, the cyclopentanone cannot enolize. The alcohol formed, (A), was then activated by the diazo-siliconicmethane in Stage 2 to give (B). Notice that one changed the $Si(CH_3)_3$ to H, because *"**one was** sterically hindered"*. In fact that was why there was more cyclo-heptanone than the cyclohexanone, although one cannot get cycloheptanone without getting cyclohexanone. How can cyclohexanone be more reactive than cyclopentanone, when the strain energy in cyclopentanone is greater than that in cyclohexanone? How? Rings are like springs in Physics. The (B) formed in Stage 2 has two rings. The three-membered ring in addition to having a common boundary with the five-membered ring and what it is carrying, is so heavily strained to exist as a ring. With only one point of scission located inside, the ring opens up and molecularly rearranges to form the cyclohexanone in Stage 3. It is a three stage process unlike one stage for non-ringed ketones, **i.e. homologation of cyclic ketones is far slower than homologation of non-ringed ketones.**

After Stage 3, Stage 4 follows to produce cycloheptanone. Only a small fraction of the cyclo-hexanone was kept in Activated Equilibrium State of Existence when the alkylated aluminum ca-talyst was used. The cycloheptanone is obtained in the same manner as above in three stages bringing the total number of stages to six. With this catalyst, Stage 7 could not follow because no cycloheptanone could be kept in Activated Equilibrium State of Existence at the operating condi-tions. While this catalyst could only keep about 66 percent of the cyclopentanone in Activated Equilibrium State of Existence, the second catalyst only kept about 33 percent of the cyclopen-tanone in Activated Equilibrium State of Existence. Unlike the first catalyst, the second catalyst was able to keep a fraction of cycloheptanone in Activated/Equilibrium State of Existence to produce cyclooctanone, bringing the total number of stages to nine. Note that when the first ring is formed, the conversion based on the cyclopentanone kept in Activated/Equilibrium State of Existence would have been 100 percent if there was no diazoalkane left in the system. This means that cyclohexanone can be obtained alone, based on the number of moles of diazoalkane used. While the alcohol can only molecularly rearrange to give the ketone, the ketone can only enolize to give the alcohol. Notice that the molecular rearrangement in Stage 3 is different from all the ones we have seen so far in ringed and non-ringed compounds, for without it the six-membered ring cannot be formed. The transfer species involved looks like the first kind of the first type. It would have been if a ring was not involved. This is also different from moving from (A) of Equation 7.161 to cyclopentanone. For the case of Stage 3 and that shown below, it is transfer species of **the first kind of the sixth type** identical to that of moving from vinyl alcohol to acetaldehyde..

465

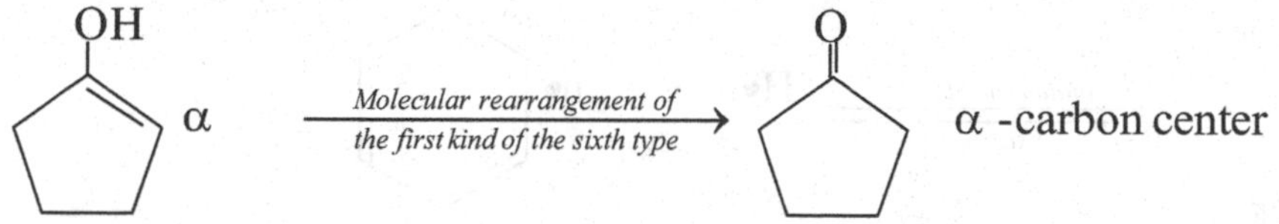

7.168

The other molecular rearrangement which is important during insertion of alkylene groups inside a ring as shown in Stage 3 of Equation 7.166 will be called **Molecular Rearrangement of the first kind of the sixth type,** because of the uniqueness of ringed compounds particularly with living systems.

At the same time when the cyclohexanone was being formed, (IV) of Equation 7.160 was also being formed, both taking place in parallel. (IV) is obtained by only those which favored being activated without having the ability to exist in Activated/Equilibrium State of Existence. Of the remaining Stable fractions, only a small one could be activated not by the catalyst, but the diazoalkane. The very small fraction in the use of alkylated aluminum is the remaining 2 percent consumed as follows.

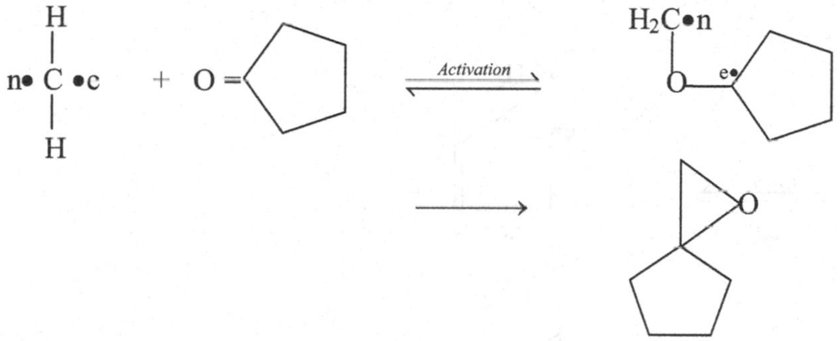

Overall Equation: H_2CN_2 + H_8C_5O ⟶ N_2 + (IV)   7.169

By the time cyclohexanone was formed, (IV) had already been formed in parallel. The diazo-alkane did have to waite for the cyclopentanone to enolize since the solvents assisted by the catalyst was essentially doing the job with little assistance from the diazoalkane. *As has already been said, there is no doubt these compounds do communicate with themselves, organize themselves before they begin their jobs, just like how the Ziegler/Natta initiators operate when polymerizable male and female monomers come around.*

Cycloheptatriene is said to be obtained in 87 percent yield by the cuprous chloride-catalyzed reaction of benzene with diazomethane.[15]

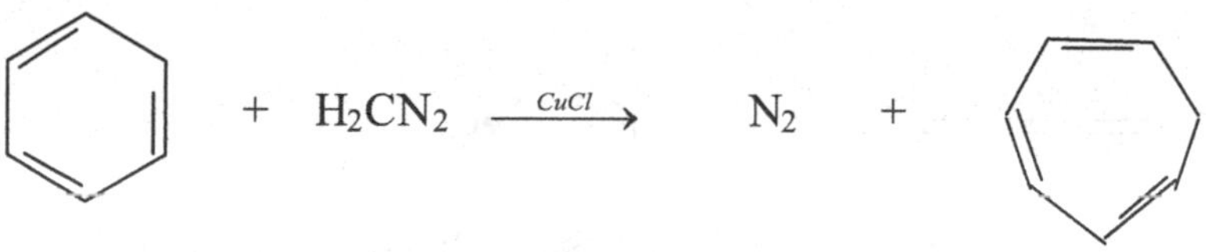

7.170

Benzene is highly nucleophilic. It is therefore a unique female. It has no point of scission. Apart from the three double bonds of equal capacity carried by it, it has invisible π-bonds inside the ring called strain energy which is still less than what is called the Maximum required Strain Energy (MaxRSE), otherwise it cannot exist as a ring. How can you open such a ring and expand it? Only one π-bond can be activated one at a time. It is resonance stabilized only radically and never chargedly. Like acetylene, it is quite unstable under most operating conditions, that is, it is always in Equilibrium State of Existence as shown below.

466

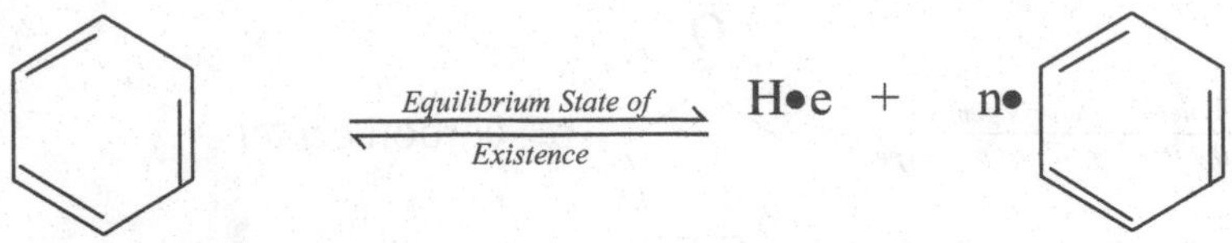

7.171

In this state of existence, if it was more nucleophilic than the diazomethane, the product cannot be obtained. But, since the diazomethane is more nucleophilic than the strong benzene, it will stabilize it. The benzene can no longer exist in Equilibrium State of Existence. In fact diazoalk-anes are more nucleophilic than any monomer considered so far.

Stage 1:

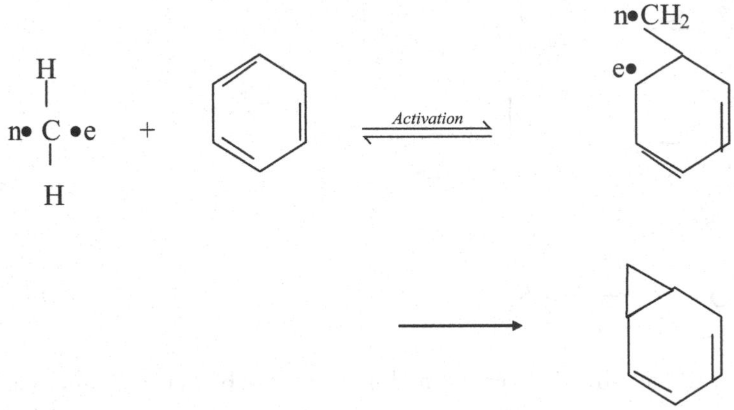

(A)

Overall Equation: H_2CN_2 + H_6C_6 ⟶ N_2 +   (A)   7.172

(A) is very unstable to be isolated, because the three-membered ring is fused to an unsaturated six-membered ring. The three membered ring has only one point of scission and the ring opens as shown below in Stage 2.

Stage 2:

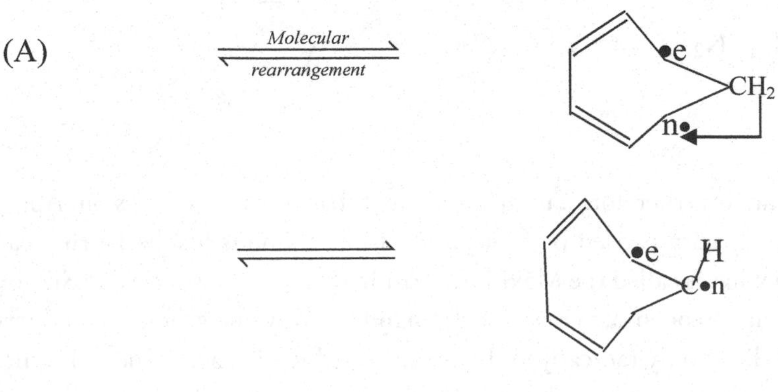

467

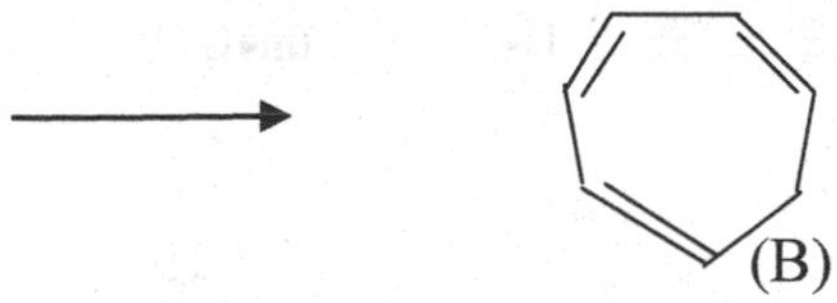

$$\longrightarrow \quad \text{(B)} \qquad\qquad 7.173$$

Overall Equation: $\quad H_2CN_2 \quad + \quad H_6C_6 \quad \longrightarrow \quad N_2 \quad + \quad H_8C_7 \qquad\qquad 7.174$

Looking at Stage 2, one can again observe another form of molecular rearrangement. It also looks like transfer species of the first kind of the first type. In this unique case, notice that the transfer species is coming from the methylene. If the diazoalkane had been $(CH_3)_2CN_2$, the molecular rearrangement will never take place, unless the hydrogen atoms on the benzene ring are changed to CH_3 or some of them are changed to radical-pulling groups. The laws of Nature are wonderful to comprehend. In our physical world, we do something else such as taking from the poor to give to the rich or choosing a "rich" fool to be a leader. This type of molecular rearrangement similar to the type shown in Stage 3 of Equation 7.166 is also called ***Molcular Rearrangement of the first kind of the sixth type***, peculiar only to ringed compounds. ***There is no OH group here; hence CH_2 group became the source of transfer species.***

The synthesis of diazoalkanes is of great interest in terms of the mechanisms which have been proposed in the past.[3] N - Methyl - N - nitroso amides are said to react readily with aqueous potassium hydroxide to give diazomethane.

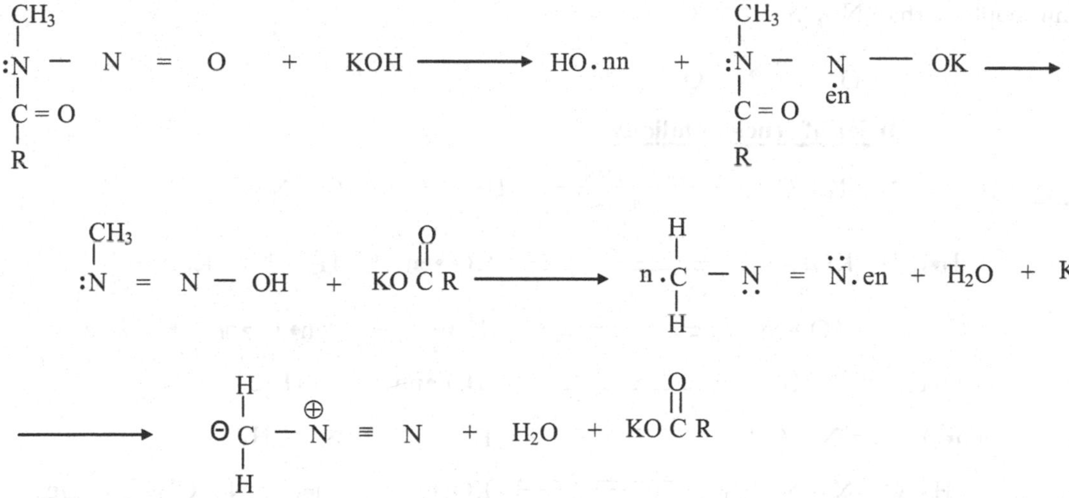

$$
\begin{array}{c}
CH_3 \\
| \\
:N \;-\; N = O \quad + \quad KOH \;\longrightarrow\; HO\cdot nn \quad + \quad :N \;-\; N \;-\; OK \longrightarrow \\
| \qquad\qquad\qquad\qquad\qquad\qquad\qquad\qquad | \qquad\quad \dot{e}n \\
C = O \qquad\qquad\qquad\qquad\qquad\qquad\qquad\qquad C = O \\
| \qquad\qquad\qquad\qquad\qquad\qquad\qquad\qquad\qquad | \\
R \qquad\qquad\qquad\qquad\qquad\qquad\qquad\qquad\qquad R
\end{array}
$$

$$
\begin{array}{c}
CH_3 \qquad\qquad\qquad\qquad O \qquad\qquad\qquad\qquad\qquad H \\
| \qquad\qquad\qquad\qquad\quad || \qquad\qquad\qquad\qquad\qquad | \\
:N \;=\; N \;-\; OH \quad + \quad KO\,C\,R \;\longrightarrow\; n\cdot C \;-\; N \;=\; \ddot{N}\cdot en \;+\; H_2O \;+\; K(\\
\qquad\qquad\qquad\qquad\qquad\qquad\qquad\qquad\qquad\qquad | \\
\qquad\qquad\qquad\qquad\qquad\qquad\qquad\qquad\qquad\qquad H
\end{array}
$$

$$
\begin{array}{c}
\qquad\qquad\qquad H \qquad\qquad\qquad\qquad\qquad\qquad\qquad\qquad O \\
\qquad\qquad\qquad | \quad \oplus \qquad\qquad\qquad\qquad\qquad\qquad\qquad\; || \\
\longrightarrow \quad {}^{\ominus}C \;-\; N \;\equiv\; N \quad + \quad H_2O \quad + \quad KO\,C\,R \qquad\qquad 7.175 \\
\qquad\qquad\qquad | \\
\qquad\qquad\qquad H
\end{array}
$$

The provision of the mechanism above was an Art, because things which cannot take place naturally were made to take place. Unlike some cases which have been shown here, we cannot know the number of Stages, we cannot see what is actually going on even with the use of the "eye of the needle", because that is the world we live in. With the use of the current development that which is the real SCIENCE, the followings are the mechanisms.

Stage 1: $H_2O \xrightleftharpoons[\text{Existence}]{\text{Equilibrium State of}} H\bullet e + nn\bullet OH$

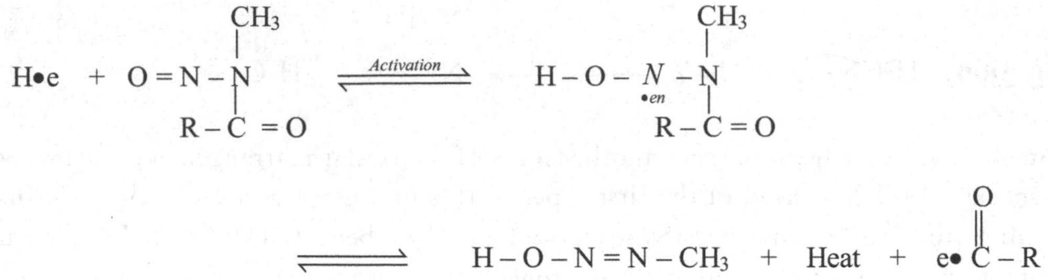

$$H\bullet e + \underset{\underset{R-C=O}{|}}{\overset{\overset{CH_3}{|}}{O=N-N}} \xrightleftharpoons{\text{Activation}} \underset{\underset{R-C=O}{|}}{\overset{\overset{CH_3}{|}}{H-O-\underset{\bullet en}{N}-N}}$$

$$\rightleftharpoons H-O-N=N-CH_3 + \text{Heat} + e\bullet \overset{\overset{O}{\|}}{C}-R$$

$$R-\overset{\overset{O}{\|}}{C}\bullet e + nn\bullet OH \longrightarrow RCOOH \qquad\qquad 7.176$$

Overall Equation: $H_2O + O=N\text{-}N(CH_3)RCO \longrightarrow H\text{-}O\text{-}N=N\text{-}CH_3 + RCOOH +$

Large amount of HEAT   7.177

How can nitrogen carry an electro-radical whether free or not in the presence of CH_3, and shockingly H? That is why very large amount of heat must be released and a stable molecule is obtained without deactivation in the third step of the stage. Notice the center activated, a clear indication of $C = O$ being more nucleophilic than $N=O$.

$$C = O \quad > \quad N = O \qquad\qquad 7.178$$

Order of Nucleophilicity

Stage 2: $H-O-N=N-CH_3 \xrightleftharpoons[\text{Existence}]{\text{Equilibrium State of}} H\bullet e + nn\bullet O-N=N-CH_3$

$$H\bullet e + KOH \rightleftharpoons KO\bullet en + H_2 + \text{Heat}$$

$$KO\bullet en \rightleftharpoons K\bullet e + en\bullet O\bullet nn + \text{Heat}$$

$$nn\bullet O\bullet en + H_2 \rightleftharpoons HO\bullet nn + H\bullet e$$

$$H\bullet e + nn\bullet O-N=N-CH_3 \rightleftharpoons H-O-N=N-CH_3$$

$$K\bullet e + H-O-N=N-CH_3 \rightleftharpoons KOH + en\bullet N=N-CH_3 + \text{Heat}$$

$$H_3C-N=N\bullet en \rightleftharpoons \textbf{Heat} + H_3C\bullet e + N_2$$

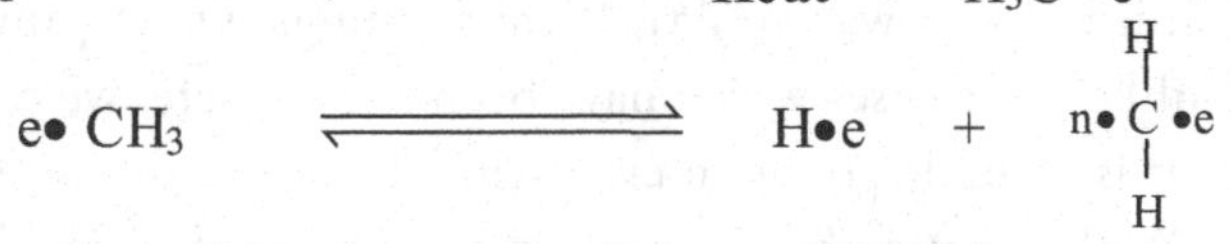

$$e\bullet CH_3 \rightleftharpoons H\bullet e + \underset{\underset{H}{|}}{\overset{\overset{H}{|}}{n\bullet C\bullet e}}$$

$$H\bullet e + nn\bullet OH \rightleftharpoons H_2O$$

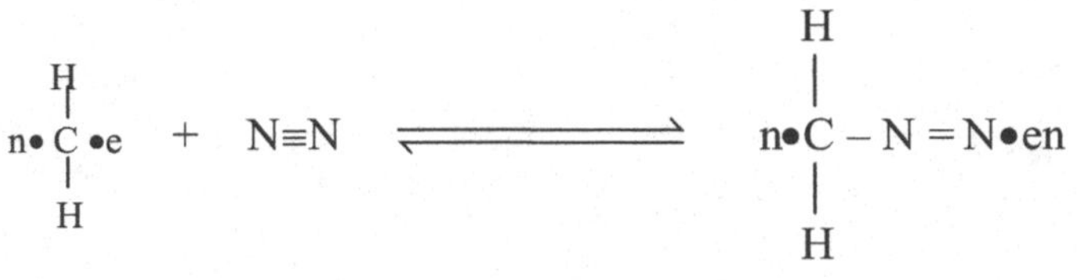

$$\underset{\underset{H}{|}}{\overset{\overset{H}{|}}{n\bullet C\bullet e}} + N\equiv N \rightleftharpoons \underset{\underset{H}{|}}{\overset{\overset{H}{|}}{n\bullet C-N=N\bullet en}}$$

(A)

469

$$(A) \longrightarrow \overset{H}{\underset{H}{\overset{|}{C}}} {}^{\ominus} - \overset{\oplus}{N} \equiv N \ + \ \textbf{Heat}$$

Overall Equation: $KOH + H_2O + O=N-NCH_3RCO \longrightarrow$ Large amount   7.179

$$HEAT + KOH + H_2O + H_2CN_2 + RCOOH \qquad 7.180$$

Heat was again released in many steps in Stage 2 for which the diazomethane was instantaneous-ly produced in a particular state, because it is forbidden for nitrogen to be carrying an electro-radical in the presence of C and H. Heat released here is far more than that released in Stage 1. Notice that dilute KOH indeed was an active catalyst. On the whole there are two stages.

Of the various N - methyl - N - nitroso amides that have been used, N, N`- dinitroso- N, N"- dimethylterephthalamide is said to be the most convenient because it is available commercially in a stabilized form.

$$(II) + H_2O \longrightarrow KO\overset{O}{\overset{||}{C}} - \bigcirc - \overset{O}{\overset{||}{C}}OK + H_3C - \dot{N} = N - OH + \qquad$$

(III)

$$\qquad\qquad\qquad\qquad\qquad\qquad\qquad\qquad 7.181$$

$$(I) + H_2O \longrightarrow (III) + 2 (II) + 2H_2O$$

There are sides of N –methyl – N – nitroso amides here separated by benzene. One side is first worked on in two stages as was done above to give the overall equation.

Overall Equation: $KOH + H_2O + O=N-NCH_3CO-H_4C_6-COCH_3N-N=O \longrightarrow$

$$KOH + H_2O + H_2CN_2 + HOOC-H_4C_6-COCH_3N-N=O \qquad 7.182$$

Many authors believe that 2 moles of KOH is required for the reaction, just because of lack of understanding of mechanisms of reactions. Indeed no Chemist including the author anywhere has any understanding of mechanisms of reactions. We think we know without realizing that we know nothing including the author. Even if there were more than one mole of dilute KOH in the system, only one side can be used based on how the products can be obtained. The dilute KOH recovered above begins to work on the second side bringing the total number of stages to four with the final overall equation.

Final Overall Equation: $KOH + H_2O + O=N-NCH_3CO-H_4C_6-COCH_3N-N=O \longrightarrow$

$$KOH + H_2O + 2H_2CN_2 + HOOC-H_4C_6-COOH \qquad 7.183$$

7.4.1. Charged features of Diazoalkanes

Beginning with diazomethane, the followings are to be expected.

$$R^{\ominus} + \overset{H}{\underset{H}{\overset{|}{\ominus C}}} - \overset{\oplus}{\underset{\underset{\cdot\cdot}{N}}{\overset{|||}{N}}} \quad OR \quad \overset{\oplus}{N} = N\overset{\ominus}{\underset{CH_2}{\overset{||}{:}}} \longrightarrow RH + \overset{H}{\underset{}{\overset{|}{\ominus C}}} = \overset{\oplus}{N} = \overset{\ominus}{N}$$

$$\text{(I)} \qquad\qquad \text{(II)} \qquad\qquad \underline{\text{Polar "anionic" species}}$$

NOT FAVORED

<u>Transfer species of the first kind of the third type</u>   7.184

(II) which was thought to be one of the polar forms of diazomethane in present-day Science was put into the equation above for exploratory purpose. Transfer species for (I) and (II) which are different are not particularly important since the nitrogen molecule must be released in the process, because the positive charge is polar and imaginary. It is the positive side that gives it its imaginary character and that side can therefore not be used. For negatively charged electrostatic initiator, it is the positive side that gives it its imaginary character and that side can therefore not be used. For positively charged electrostatic initiator, it is the positive side that gives it its real character and therefore that side can be used. However, with negatively charged free-paired initiator, the followings may take place.

$$R^{\ominus} + \overset{H}{\underset{CH_3}{\overset{|}{\ominus C}}} - \overset{\oplus}{N} \equiv N \longrightarrow RH + \overset{H}{\underset{H \ H}{\overset{|}{C}}} = C^{\ominus} + N_2$$

$$\text{(I)} \qquad\qquad\qquad\qquad\qquad \text{(II)} \qquad\qquad\qquad 7.185$$

IMPOSSIBLE EXISTENCE

$$\text{(II)} + \text{(I)} \longrightarrow \overset{H \ H}{\underset{H \ H}{\overset{|\ \ |}{C=C}}} + \overset{H}{\underset{H \ H}{\overset{|}{C}}} = C^{\ominus} + N_2$$

$$7.186$$

Note that the double bond is formed instantaneously, because a center cannot carry two opposite charges. The transfer species involved is of the first kind of the first type. But notice that in the continuous abstraction of transfer species, instead of ethylene being formed, it is ethene its isomer that is continuously formed. The $R^{\ominus}$ used above is the type carried by $H_4C_9{}^{\ominus}.......{}^{\oplus}Li$, that is, free negatively charged center.

Thus, this can possibly be used for the synthesis of alkenes only radically. Chargedly, the reaction cannot take place, since (II) above cannot exist chargedly but only radically. The mechanism above radically is Combination mechanism which has its own rules different from those of Decomposition and Equilibrium mechanisms.

In view of the strong nucleophilic character of the monomers, free positively charged-paired initiators can readily be employed, if it is strong enough to activate it.

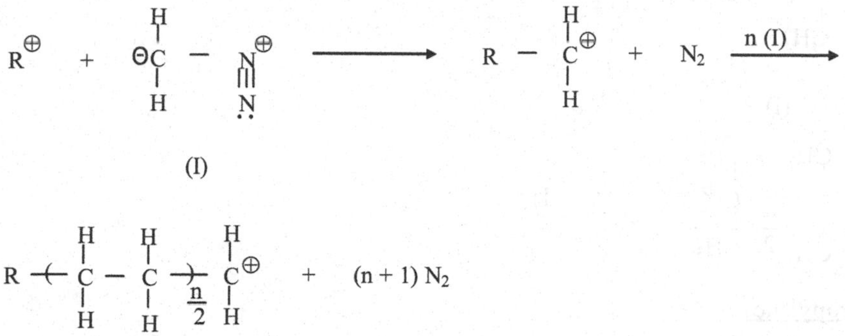

(I)

$$7.187$$

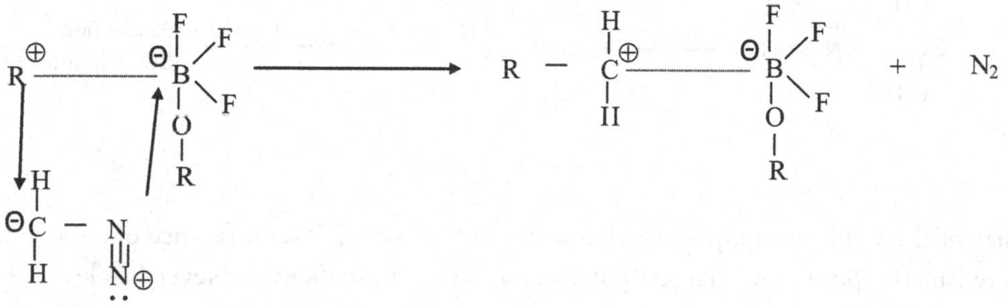

$$7.188$$

Positively, the polymerization of the monomer is readily favored, to produce linear poly-methylene (and not polyethylene) and nitrogen. Why this is so, will become obvious.

Replacing one of the H atoms in diazomethane, with an alkyl group, the followings are obtained. The $R^{\oplus}$ used below is BF_3/ROR or other suitable paired media initiator.

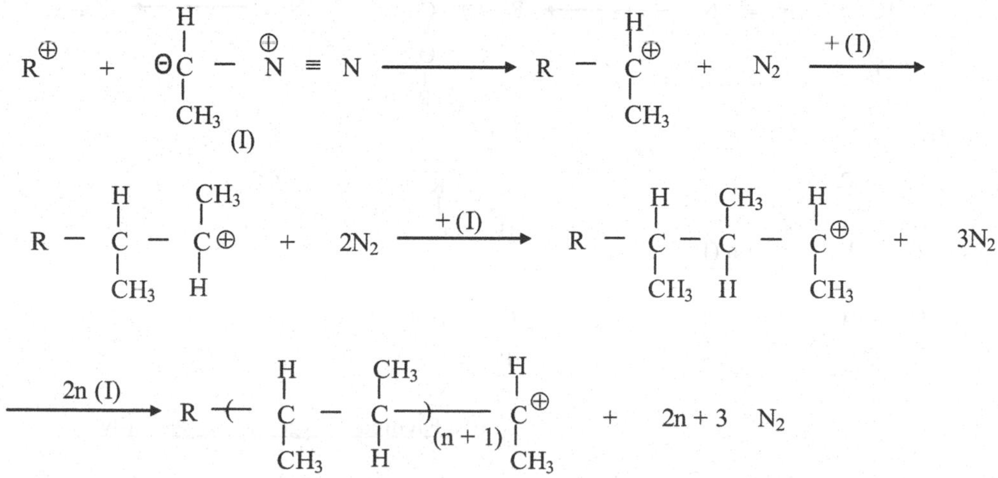

(I)

Polyethylene

$$7.189$$

Due to steric limitations more of poly (Trans 2-butene) than poly (Cis-2-butene) will be produced, depending on the strength of initiator. With the replacement of the two hydrogen atoms, with alkyl groups, the use of very strong initiators or special coordination initiators will be desirable, depending on the size of the alkyl groups.

$$R^{\oplus} \ + \ {}^{\ominus}\!\!\underset{CH_3}{\overset{CH_3}{C}}\!\!-\ \overset{\oplus}{N} \equiv N \ \longrightarrow \ R-\underset{CH_3}{\overset{CH_3}{\overset{|}{C}}}{}^{\oplus} \ + \ N_2 \ \xrightarrow{\ +(I)\ }$$

(Strong)

(I)

$$R \underset{CH_3 \quad CH_3}{\overset{CH_3 \quad CH_3}{\left(\!\! \overset{|}{\underset{|}{C}} - \overset{|}{\underset{|}{C}} \!\!\right)}}_{\tfrac{n}{2}} \underset{CH_3}{\overset{CH_3}{\overset{|}{C}}}{}^{\oplus} \ + \ (n+1)\,N_2$$

<u>(Poly propylene)</u>   7.190

$$R^{\oplus} \ + \ {}^{\ominus}\!\!\underset{C_3H_7}{\overset{CH_3}{C}}\!\!-\ \overset{\oplus}{N} \equiv N \ \longrightarrow \ R-\underset{C_3H_7}{\overset{CH_3}{\overset{|}{C}}}{}^{\oplus} \ + \ N_2 \ \xrightarrow{\ +(I)\ } \ \begin{array}{l}\text{No reaction}\\ \text{(Steric hindrance)}\end{array}$$

(I)   7.191

Due to the size of the C_3H_7 group, propagation may not be favored. The influence of steric limitations can be observed in the positively charged polymerization of diazoalkanes. Neverthe- less, what is far better is the use of paired initiators with reservoirs. One can thus observe what are the real polyethylene, polypropylene and so on. Though the structures of polyethene, polymethylene, polycyclopropane look the same, they are still different.

When one H in diazomethane is replaced with OR groups, the followings will be obtained exploratively.

$$R^{\oplus} \ + \ {}^{\ominus}\!\!\underset{\underset{R}{\overset{|}{\underset{..}{O}}}}{\overset{H}{C}}\!\!-\ \overset{\oplus}{N} \equiv \overset{..}{N} \ \longrightarrow \ R-\underset{\underset{R}{\overset{|}{O}}}{\overset{H}{\overset{|}{C}}}{}^{\oplus} \ + \ N_2 \ \xrightarrow{\ +(I)\ } \ 2N_2 \ +$$

(I)

$$R-\underset{\underset{R}{\overset{|}{O}}}{\overset{H}{\overset{|}{C}}} - \underset{H}{\overset{O-R}{\overset{|}{C}}}{}^{\oplus} \ \xrightarrow{\ +n\,(I)\ } \ R\underset{\underset{R}{\overset{|}{O}}}{\overset{H}{\left(\!\!\overset{|}{C}\right.}} - \underset{H}{\overset{O-R}{\overset{|}{C}}}\underset{\tfrac{n}{2}}{\left.\!\!\right)} \underset{\underset{R}{\overset{|}{O}}}{\overset{H}{\overset{|}{C}}} - \underset{H}{\overset{O-R}{\overset{|}{C}}}{}^{\oplus} \ + \ (n+2)\,N_2$$

(II) <u>Favored chargedly and radically</u>   7.192

Due to electrodynamic forces of repulsion and steric limitations, largely syndiotactic placement will be favored, if the monomer had existed polarly. ***Electrostatic forces of repulsion between the real side of this type of polar bond (i.e., the negatively charged side) and the paired unbonded radicals on the***

O center adjacently located, must exist since that center is still active being real. Only the positive side cannot repel or attract being imaginary. Though, the monomer above can exist in the radical state, it cannot exist in the polar state, the state being the stable state. Therefore, cases where we have groups such as Cl, NH_2, OCOR in place of OR, cannot exist polarly. Their diazoalkanes can only exist radically, if during their synthesis, the radical state is the last step of a stage, wherein no deactivation is possible.

When one H is replaced with radical-pulling groups such as COOR, the followings are to be expected.

$$R^{\ominus} \ + \ \underset{\substack{C=O \\ | \\ O \\ | \\ (I) \ CH_3}}{{}^{\ominus}C} \overset{H}{\underset{|}{-}} \overset{\oplus}{N} \equiv \ddot{N} \ \longrightarrow \ RCH_3 \ + \ \underset{\substack{C=O \\ | \\ O^{\ominus}}}{{}^{\ominus}C} \overset{H}{\underset{|}{-}} \overset{\oplus}{N} \equiv N \ \ OR \ NO \ REACTION$$

$$(II) \ NOT \ FAVORED \qquad\qquad 7.193a$$

$$R^{\oplus} \ + \ \underset{\substack{C=O \\ | \\ O \\ | \\ CH_3}}{{}^{\ominus}C} \overset{H}{\underset{|}{-}} N^{\oplus} \equiv N \ \longrightarrow \ \underset{\substack{{}^{\oplus}C-O-R \\ | \\ O \\ | \\ CH_3}}{{}^{\ominus}C} \overset{H}{\underset{|}{-}} N^{\oplus} \equiv N \ \ OR \ ROCH_3 \ + \ \underset{H}{\overset{O}{C}} = C^{\oplus} + N_2$$

(II) NOT FAVORED $\qquad$ **(III) FAVORED**

RADICALLY

$$7.193b$$

Negatively, addition is impossible being strongly nucleophilic. Via the C = O center, no initiation is possible. *Though the C = O is less nucleophilic than the diazo center, positively as will be shown downstream, initiation is favored due to the orientation of the monomer in the coordination center. N_2 is released in the process.* The C = O center is not attacked, because the diazo compound is already in an activated state and the C = O center is the last to gain entry into the coordination center.

$$C^{\ominus} - {}^{\oplus}N \equiv N \ > \ N \equiv N \ > \ C \equiv N \ > \ C \equiv C \ > \ C = O$$

ORDER OF NUCLEOPHILICITY $\qquad\qquad$ 7.194

The same analysis as above will apply when COOR is replaced with $CONH_2$, and COR types of groups, noting that these are Electrophiles.

Now using a radical-pulling group such as CF_3, we have the followings.

$$R - (\underset{\substack{| \\ CF_3}}{\overset{\substack{H \\ |}}{C}} - \underset{\substack{| \\ H}}{\overset{\substack{CF_3 \\ |}}{C}})_{n-1} - \underset{\substack{| \\ CF_3}}{\overset{\substack{H \\ |}}{C}}{}^{\oplus} \quad OR \quad RF \ + \ \underset{\substack{| \\ F}}{\overset{\substack{F \\ |}}{{}^{\oplus}C}} - \underset{\substack{| \\ H}}{\overset{\substack{| \\ }}{C}}{}^{\ominus} - N^{\oplus} \equiv N$$

FAVORED $\qquad\qquad$ 7.195

Due to electrostatic/electrodynamic forces of repulsion and steric limitations, largely syndiotactic placement will be favored via the positively charged-paired route. ***It takes place, because $\equiv$N is a strong radical-/charged-pulling group***. CF_2 group is the last group to gain entry to the coordi-nation center. Electro-free-radically, there is no initiation.

Of what relevance will provision of resonance stabilization groups be to the diazo com-pound? It is not possible. It is the diazo-center that should provide resonance stabilization. How-ever, using alkyl groups, the followings are obtained.

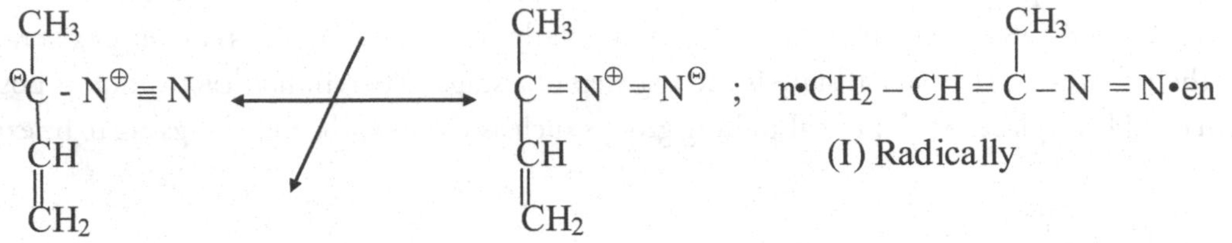

$$
\begin{array}{c}
CH_3 \\
{}^{\ominus}\overset{|}{C} - N^{\oplus} \equiv N \\
\overset{|}{CH} \\
\overset{\parallel}{CH_2}
\end{array}
\quad \longleftrightarrow \quad\quad\quad
\begin{array}{c}
CH_3 \\
\overset{|}{C} = N^{\oplus} = N^{\ominus} \\
\overset{|}{CH} \\
\overset{\parallel}{CH_2}
\end{array}
\quad ; \quad
\begin{array}{c}
CH_3 \\
n \bullet CH_2 - CH = \overset{|}{C} - N = N \bullet en \\
\text{(I) Radically}
\end{array}
$$

7.196

Resonance stabilization cannot be provided for CH_3 group or any other group, due to the presence of the polar bond or charges. In addition, as already shown in Equation 7.3a, it is not polarly resonance stabilized. It is from the polar state, the radical state is obtained via activation. Therefore, the existence of (I) above radically is not possible. It is the diazo compound that should provide resonance stabilization for the alkene, the monomer being a complete Nucleophile. With Electrophiles from dienes, the provider of resonance stabilization is the alkene center more nucleophilic than the Y center. **Thus, this polar monomer can never be provided with resonance stabilization by a resonance stabilization group. It is important to note also that molecular rearrangement cannot take place in a polar monomer.**

7.4.2 Radical features of Diazoalkanes

It has been observed that all the chemical reactions considered with respect to diazo-alkanes so far take place only radically, yet their polymerizations considered so far can take place chargedly. Based on the radical activated state of diazoalkanes, only electro-free-radicals, nucleo-free-radicals and electro-non-free-radicals can be involved in their homopolymerizations.

Diazomethane (Diazomethylene) is the first member of diazoalkanes (diazoalkylenes) family and like other first member of other families, it is very unique. At very low temperatures of the order of 10 K, it does form diazirines and alkenes such as shown below for diazoethylene and diazopropylene. However, it should be noted that diazirine the first member exists but very unstable.

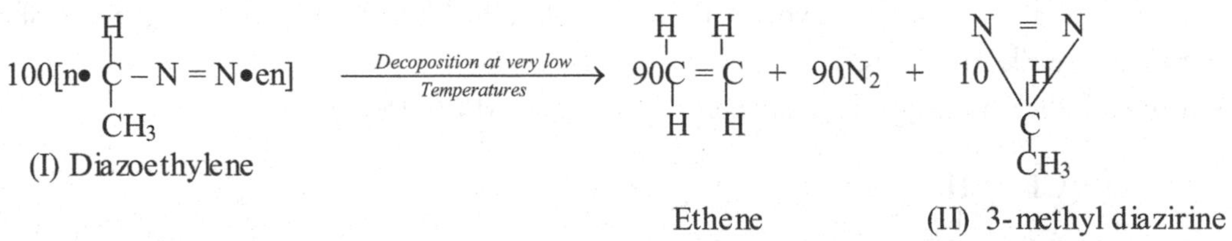

$$
100[n \bullet \overset{\overset{\displaystyle H}{|}}{\underset{\underset{\displaystyle CH_3}{|}}{C}} - N = N \bullet en]
\xrightarrow[\text{Temperatures}]{\text{Decoposition at very low}}
90 \overset{\overset{\displaystyle H}{|}}{C} = \overset{\overset{\displaystyle H}{|}}{\underset{\underset{\displaystyle H}{|}}{C}} \underset{\underset{\displaystyle H}{}}{} + 90N_2 + 10
\begin{array}{c}
N = N \\
\diagdown \;\; \diagup H \\
\overset{\parallel}{C} \\
\overset{|}{CH_3}
\end{array}
$$

| (I) Diazoethylene | | Ethene | (II) 3-methyl diazirine |

7.197

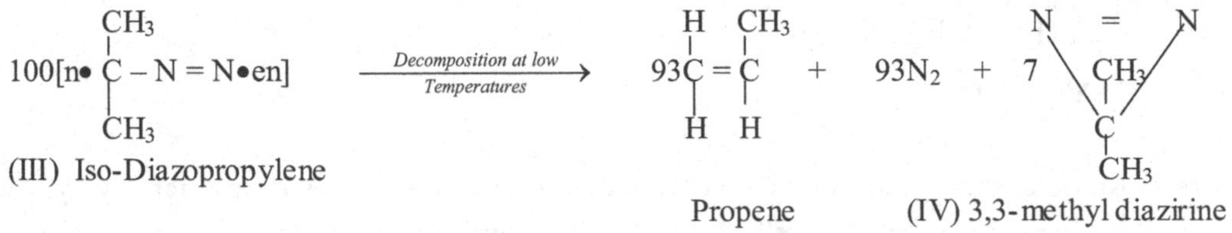

$$
100[n \bullet \overset{\overset{\displaystyle CH_3}{|}}{\underset{\underset{\displaystyle CH_3}{|}}{C}} - N = N \bullet en]
\xrightarrow[\text{Temperatures}]{\text{Decomposition at low}}
93 \overset{\overset{\displaystyle H}{|}}{C} = \overset{\overset{\displaystyle CH_3}{|}}{\underset{\underset{\displaystyle H}{|}}{C}} + 93N_2 + 7
\begin{array}{c}
N = N \\
\diagdown \;\; CH_3 \\
\overset{\parallel}{C} \\
\overset{|}{CH_3}
\end{array}
$$

| (III) Iso-Diazopropylene | | Propene | (IV) 3,3-methyl diazirine |

7.198

$$100[n\bullet \overset{\displaystyle H}{\underset{\displaystyle C_2H_5}{C}} - N = N\bullet en] \xrightarrow[\text{Temperature}]{\textit{Decomposition at low}} 90\overset{\displaystyle H}{\underset{\displaystyle H}{C}} = \overset{\displaystyle H}{\underset{\displaystyle CH_3}{C}} + 95N_2 + 5$$

(V) n-Diazopropylene

+ **5(Z)**

$$\begin{array}{c} N = N \\ \diagdown \ \diagup \ H \\ \underset{\displaystyle C_2H_5}{C} \end{array}$$

(VI) 3-ethyl diazirine

[where Z is a cycloalkane]

7.199

The numbers used are meaningful, but arbitrarily chosen. The mechanisms of decomposition are usually by either Equilibrium or Decomposition or both. *If a compound to be decomposed cannot exist in Equilibrium State of Existence, then the mechanism is exclusively Decom-position mechanism. If the compound cannot exist in Stable State of Existence, the mechanism is exclusively Equilibrium mechanism.* Whether the compound has activation centers or not, the rules remain the same. Where a compound has a fraction in Equilibrium State of Existence and remaining fraction in Stable State of Existence based on the operating conditions, *then decomposition takes place in parallel via Equilibrium and Decomposition mechanisms.* For the case of diazoalkanes, since the diazoalkanes are in activated state of existence, decomposition can take place either exclusively by Decomposition mechanism or Equilibrium mechanism or both, like the case for decomposition of Azoalkanes which take place by both mechanisms. For the cases above, the followings are obtained.

Stage 1:

$$n\bullet \overset{\displaystyle H}{\underset{\displaystyle CH_3}{C}} - N = N\bullet en \longrightarrow N_2 + e\bullet \overset{\displaystyle H}{\underset{\displaystyle CH_3}{C}} \bullet n\,(A) + \text{Heat}$$

$$(A) \longrightarrow n\bullet \overset{\displaystyle H}{\underset{\displaystyle H}{C}} = \overset{}{\underset{\displaystyle H}{C}} + H\bullet e$$

(B)

$$\rightleftharpoons \qquad H_2C = CH_2 \qquad\qquad 7.200$$

Via Equilibrium mechanism (FAVORED)

Stage 1:

$$n\bullet \overset{\displaystyle H}{\underset{\displaystyle CH_3}{C}} - N = N\bullet en \rightleftharpoons N_2 + e\bullet \overset{\displaystyle H}{\underset{\displaystyle CH_3}{C}} \bullet n\ (A) + \text{Heat}$$

$$(A) \rightleftharpoons n\bullet \overset{\displaystyle H}{\underset{\displaystyle H}{C}} - \overset{\bullet n}{\underset{\displaystyle H}{C}}\bullet e + H\bullet e$$

$$\rightleftharpoons \quad n\bullet \overset{\overset{\displaystyle H}{|}}{C} - \overset{\overset{\displaystyle H}{|}}{C}\bullet e$$

$$\xrightarrow[\text{Heat}]{\text{Deactivation/}} \quad H_2C = CH_2 \qquad\qquad 7.201$$

At low temperatures, diazoalkanes can be observed to largely exist in the radical state, after activation of the polar form. For the fractions (small) which did not decompose, they cyclize to form the (II), (IV) and (VI) above. If the operating conditions are higher than the low temperature range, either the (B) above further decomposes to release H$\bullet$n to produce H$_2$ and acetylene or part of the ethene formed decomposes to give H$_2$ and acetylene. The decomposition of the acetylene may continue depending on the use of higher operation typical of the type used in the Petroleum industry. It should at this point in time be pointed out that **decompositions on ordinary surfaces** is different from **decompositions inside pores** of for example Transition metals. Different products are obtained in both cases with very different operating conditions. The same mechanism as above also apply to iso- propylene. For n-propylene, the (Z) is supposed to be obtained as follows via Equilibrium mechanism.

Stage 1:

$$n\bullet \overset{\overset{\displaystyle H}{|}}{\underset{\underset{\displaystyle C_2H_5}{|}}{C}} - N = N \bullet en \rightleftharpoons e\bullet \overset{\overset{\displaystyle H}{|}}{\underset{\underset{\displaystyle C_2H_5}{|}}{C}} \bullet n \quad + \quad N_2 \quad + \quad \text{Heat}$$

$$\text{(B)}$$

$$\xrightleftharpoons[\text{Re}\,arrangement]{\text{Molecular}} \quad e\bullet \overset{\overset{\displaystyle H}{|}}{C} \bullet n$$

Impossible movement

$$H - \overset{}{\underset{}{C}} - H$$
$$H - \overset{}{\underset{}{C}} - H$$
$$H$$

$$\rightleftharpoons \quad e\bullet \overset{\overset{\displaystyle H}{|}}{\underset{\underset{\displaystyle H}{|}}{C}} - \overset{\overset{\displaystyle H}{|}}{\underset{\underset{\displaystyle H}{|}}{C}} - \overset{\overset{\displaystyle H}{|}}{\underset{\underset{\displaystyle H}{|}}{C}} \bullet n$$

$$\longrightarrow \quad H_2C - CH_2 \qquad + \quad \text{Heat}$$
$$\diagdown\diagup$$
$$CH_2 \qquad\qquad 7.202$$

$$\text{(Z)} \quad \text{NOT FAVORED}$$

Unlike what was shown in Equation 7.199, cyclopropane cannot be obtained. The main product will always remain to be the alkene. Now consider the case of diazo n-butylene and diazo iso-butylene.

Stage 1:

$$n\bullet \overset{\overset{\displaystyle H}{|}}{\underset{\underset{\displaystyle C_3H_7}{|}}{C}} - N = N\bullet en \quad\underline{\hspace{3cm}}\quad e\bullet \overset{\overset{\displaystyle H}{\downarrow}}{\underset{\underset{\underset{\displaystyle C_2H_5}{|}}{CH_2}}{C}}\bullet n \quad + \quad N_2 \quad + \quad Heat$$

(Diazo n-butylene)

(C)

$$(C) \quad\underline{\hspace{3cm}}\quad \overset{\overset{\displaystyle H}{|}}{\underset{\underset{\displaystyle H}{|}}{C}} = \overset{}{\underset{\underset{\displaystyle H}{|}}{C}}\bullet e \quad + \quad H_5C_2\bullet n$$

$$\rightleftharpoons \quad H_2C = CH(C_2H_5) \qquad\qquad 7.203$$

Stage 1:

$$e\bullet \overset{\overset{\displaystyle H}{|}}{\underset{\underset{\displaystyle C_3H_7}{|}}{C}} - N = N\bullet nn \quad\rightleftharpoons\quad e\bullet \overset{\overset{\displaystyle H}{\downarrow}}{\underset{\underset{\displaystyle C_3H_7}{|}}{C}}\bullet n \quad + \quad N_2 \quad + \quad Heat$$

(A)

$$(A) \quad\rightleftharpoons\quad H - \overset{\overset{\displaystyle \bullet n}{}}{\underset{\underset{\displaystyle \bullet e}{}}{C}} -- \overset{\overset{\displaystyle H}{|}}{\underset{\underset{\displaystyle H}{|}}{C}} - \overset{\overset{\displaystyle H}{|}}{\underset{\underset{\displaystyle H}{|}}{C}} - \overset{\overset{\displaystyle H}{|}}{\underset{\underset{\displaystyle H}{|}}{C}} - H \quad \text{Impossible movement}$$

$$\rightleftharpoons\quad e\bullet \overset{\overset{\displaystyle H}{|}}{\underset{\underset{\displaystyle H}{|}}{C}} - \overset{\overset{\displaystyle H}{|}}{\underset{\underset{\displaystyle H}{|}}{C}} - \overset{\overset{\displaystyle H}{|}}{\underset{\underset{\displaystyle H}{|}}{C}} - \overset{\overset{\displaystyle H}{|}}{\underset{\underset{\displaystyle H}{|}}{C}}\bullet n \qquad\qquad 7.204$$

$$\underline{\hspace{3cm}}\quad \overset{\displaystyle H_2C - CH_2}{\underset{\displaystyle H_2C - CH_2}{\big\downarrow\quad |}}$$

NOT FAVORED

Overall Equation: 100(Diazo n-butylene) $\longrightarrow$ 97N$_2$ + 97(1-Butene)

+ 3(3-propyl diazirine) 7.205

Just like n-diazo propylene, cyclobutane cannot be obtained. Hence the order of conversion level is as indicated in the overall equations. When it was diazo n-propylene, cyclopropane was thought to be one of the products. With diazo n-butylene, cyclobutane was also thought to be one of the products. These cannot be obtained for n-diazo-alkylenes, unless if H is made to move as H•n instead of as an atom H•e.

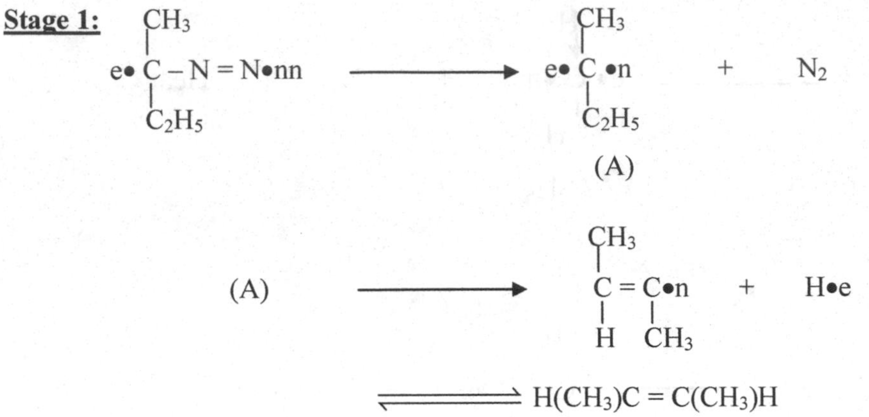

Stage 1:

$$e\bullet \overset{\displaystyle CH_3}{\underset{\displaystyle C_2H_5}{C}} - N = N\bullet nn \longrightarrow e\bullet \overset{\displaystyle CH_3}{\underset{\displaystyle C_2H_5}{C}}\bullet n \quad + \quad N_2$$

(A)

$$(A) \longrightarrow \overset{\displaystyle CH_3}{\underset{\displaystyle H}{C}} = \overset{}{\underset{\displaystyle CH_3}{C}}\bullet n \quad + \quad H\bullet e$$

$$\rightleftharpoons H(CH_3)C = C(CH_3)H$$

Cis-/Trans- 2-Butene   7.206

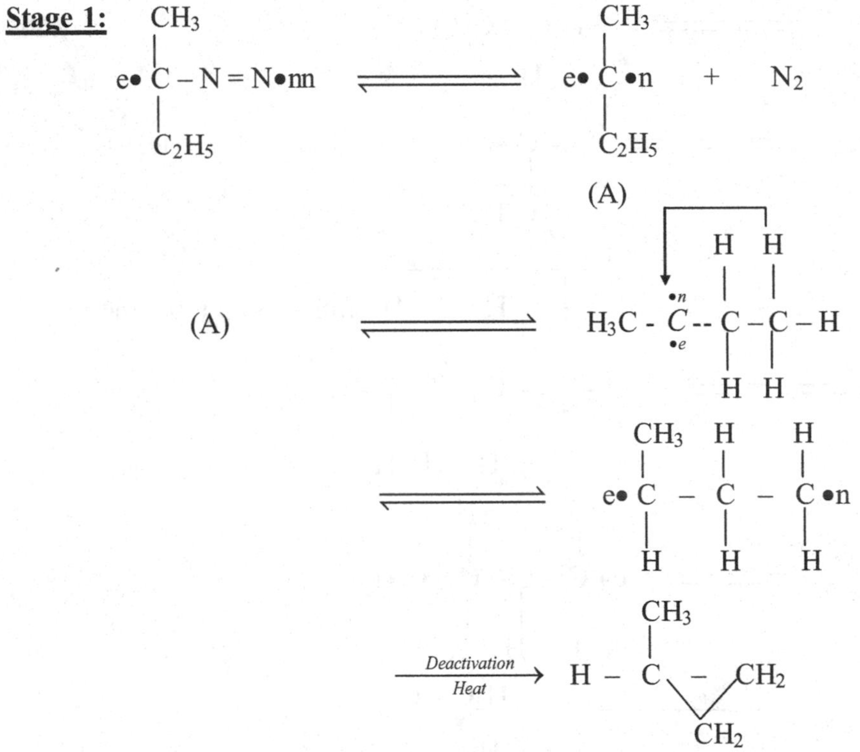

Stage 1:

$$e\bullet \overset{\displaystyle CH_3}{\underset{\displaystyle C_2H_5}{C}} - N = N\bullet nn \rightleftharpoons e\bullet \overset{\displaystyle CH_3}{\underset{\displaystyle C_2H_5}{C}}\bullet n \quad + \quad N_2$$

(A)

$$(A) \rightleftharpoons H_3C - \overset{}{\underset{\bullet e}{\overset{\bullet n}{C}}} -- \overset{\displaystyle H}{\underset{\displaystyle H}{C}} - \overset{\displaystyle H}{\underset{\displaystyle H}{C}} - H$$

$$\rightleftharpoons e\bullet \overset{\displaystyle CH_3}{\underset{\displaystyle H}{C}} - \overset{\displaystyle H}{\underset{\displaystyle H}{C}} - \overset{\displaystyle H}{\underset{\displaystyle H}{C}}\bullet n$$

$$\xrightarrow[Heat]{Deactivation} H - \overset{\displaystyle CH_3}{\underset{\displaystyle CH_2}{C}} - CH_2$$

Methylcyclopropane

FAVORED   7.207

Overall Equation: 100(Diazo iso-butylene) $\longrightarrow$ 97N$_2$ + 94(2-Butene) + 3(Methyl-Cyclo-propane) + 3(Methyl ethyl diazirine)   7.208

Notice that one step has been taken above during the molecular rearrangement to give the ring which has been reported in the past to be one of the products during the decomposition of Diazo n-butylene.[16] Indeed this product cannot be obtained from diazo-n-butylene even if more than one step had been taken in Step 2 of Equation 7.204 as shown below.

Sunny N.E. Omorodion

Stage 1:

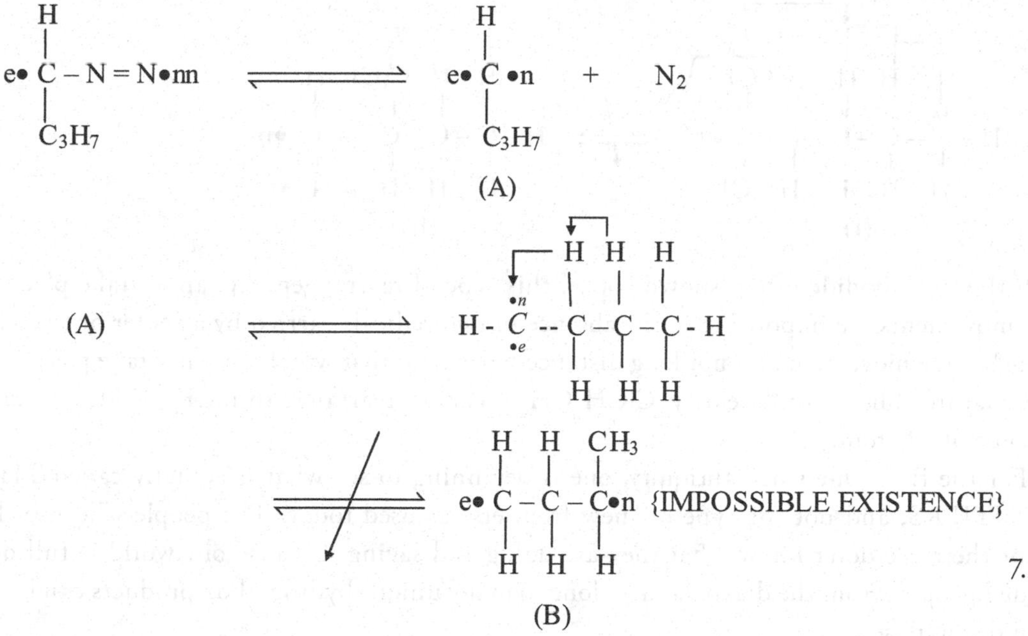

$$e\bullet \overset{\displaystyle H}{\underset{\displaystyle C_3H_7}{C}} - N = N\bullet nn \quad \rightleftharpoons \quad e\bullet \overset{\displaystyle H}{\underset{\displaystyle C_3H_7}{C}}\bullet n \;+\; N_2$$

(A)

$$(A) \quad \rightleftharpoons \quad H - \overset{\bullet n}{\underset{\bullet e}{C}} -- \overset{H}{\underset{H}{C}} - \overset{H}{\underset{H}{C}} - \overset{H}{\underset{H}{C}} - H$$

$$\cancel{\rightleftharpoons} \quad e\bullet \overset{H}{\underset{H}{C}} - \overset{H}{\underset{H}{C}} - \overset{CH_3}{\underset{H}{C}}\bullet n \quad \{\text{IMPOSSIBLE EXISTENCE}\}$$

7.209

(B)

The existence of (B) above is impossible because the nucleo-free-radical end is carrying the more radical-pushing group, that which should be carried by the electro-free-radical end being the largest group. The movement of H.e is as shown above, but not from the extreme end to the other end, a movement which is not possible. Unlike with diazo-n-butylene, the step by step movement of H•e is favored for diazo-iso-butylene. **Hence, the Molecular rearrangement is said to be of the _First kind of the seventh type,_ wherein the movement of H or the like is very slow, and they are transfer species.** The same side by side movement as above cannot apply to the cases shown below in the absence of ACTIVATION center.

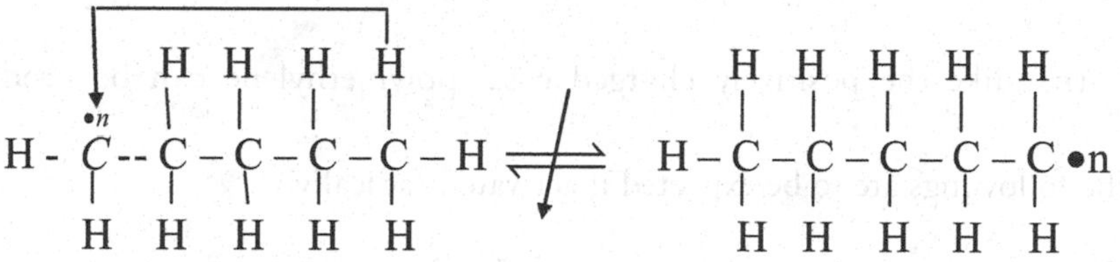

$$H - \overset{\bullet n}{C} -- \overset{H}{\underset{H}{C}} - \overset{H}{\underset{H}{C}} - \overset{H}{\underset{H}{C}} - \overset{H}{\underset{H}{C}} - H \quad \cancel{\rightleftharpoons} \quad H - \overset{H}{\underset{H}{C}} - \overset{H}{\underset{H}{C}} - \overset{H}{\underset{H}{C}} - \overset{H}{\underset{H}{C}} - \overset{H}{\underset{H}{C}}\bullet n$$

(I) $H_{11}C_5\bullet n$ (I) $H_{11}C_5\bullet n$ 7.210

The same alkylide is produced, though such movement is impossible.

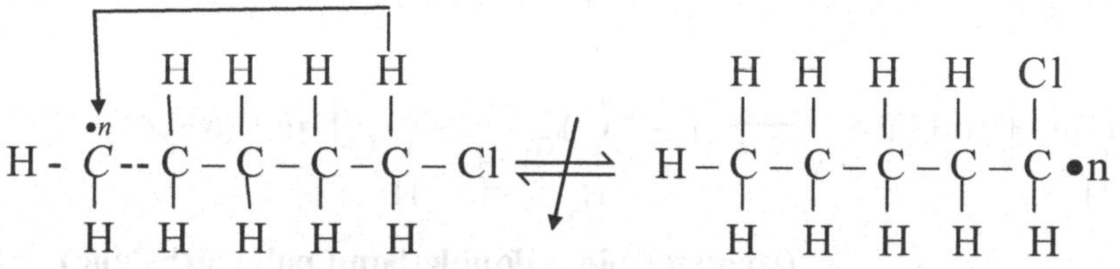

$$H - \overset{\bullet n}{C} -- \overset{H}{\underset{H}{C}} - \overset{H}{\underset{H}{C}} - \overset{H}{\underset{H}{C}} - \overset{H}{\underset{H}{C}} - Cl \quad \cancel{\rightleftharpoons} \quad H - \overset{H}{\underset{H}{C}} - \overset{H}{\underset{H}{C}} - \overset{H}{\underset{H}{C}} - \overset{H}{\underset{H}{C}} - \overset{Cl}{\underset{H}{C}}\bullet n$$

(I) $n\bullet CH_2C_4H_8Cl$ (II) $n\bullet CHClC_4H_9$ 7.211

480

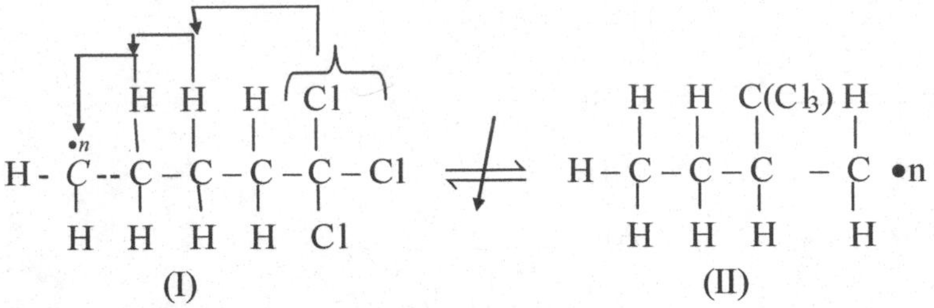

$$H - \overset{\bullet n}{\underset{\underset{H}{|}}{\overset{\overset{H}{|}}{C}}} - \overset{\overset{H}{|}}{\underset{\underset{H}{|}}{C}} - \overset{\overset{H}{|}}{\underset{\underset{H}{|}}{C}} - \overset{\overset{H}{|}}{\underset{\underset{H}{|}}{C}} - \overset{\overset{Cl}{|}}{\underset{\underset{Cl}{|}}{C}} - Cl \quad \ne \quad H - \overset{\overset{H}{|}}{\underset{\underset{H}{|}}{C}} - \overset{\overset{H}{|}}{\underset{\underset{H}{|}}{C}} - \overset{\overset{C(Cl_3)}{|}}{\underset{\underset{H}{|}}{C}} - \overset{\overset{H}{|}}{\underset{\underset{H}{|}}{C}} \bullet n \qquad 7.212$$

$$\text{(I)} \qquad\qquad\qquad\qquad\qquad\qquad \text{(II)}$$

With the two alkylide group shown above, this type of rearrangement cannot take place, because such movements are impossible in the absence of two radicals carried by a center or two centers. It is side by side movement and not long distance movement that which does not take place in Nature, since Nature abhors non-linearity. $ClCH_2CH_2.n$ cannot rearrange to $n.CHClCH_2$, because of the presence of Cl atom.

For the first time since antiquity, one is beginning to see what is actually called THE NEW FRONTIERS, and not the type of "new frontiers" as used today. The people who use them and accept their use don't know what they are doing and saying, because our world is full of narrow minded people. From the diazoalkanes alone so many unique hydrocarbon products can be obtained when decomposed.

When used as a monomer, the followings are to be expected free-radically (noting that diazoalkanes are strong nucleophiles) beginning with diazomethylene.

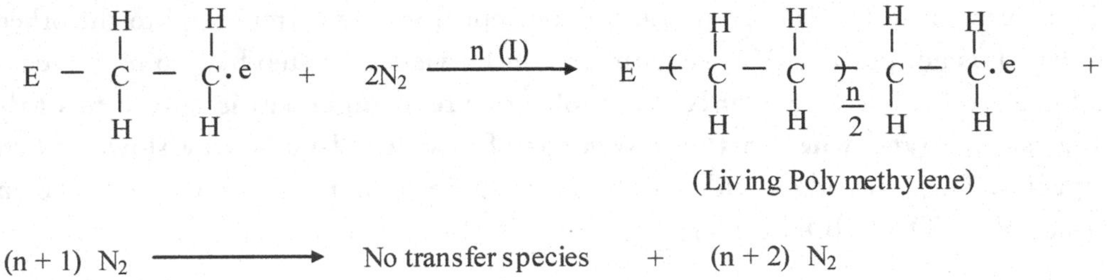

$$E - \overset{\overset{H}{|}}{\underset{\underset{H}{|}}{C}} - \overset{\overset{H}{|}}{\underset{\underset{H}{|}}{C}} \cdot e \quad + \quad 2N_2 \quad \xrightarrow{n\,(I)} \quad E \Big(\overset{\overset{H}{|}}{\underset{\underset{H}{|}}{C}} - \overset{\overset{H}{|}}{\underset{\underset{H}{|}}{C}} \Big)_{\frac{n}{2}} \overset{\overset{H}{|}}{\underset{\underset{H}{|}}{C}} - \overset{\overset{H}{|}}{\underset{\underset{H}{|}}{C}} \cdot e \quad +$$

$$\text{(Living Polymethylene)}$$

$$(n+1)\ N_2 \quad \longrightarrow \quad \text{No transfer species} \quad + \quad (n+2)\ N_2 \qquad 7.213$$

Thus, it can be observed, that like the positively charged case, polymethylene can be produced electro-free-radically.

Nucleo free-radically, the followings are to be expected if activated radically.

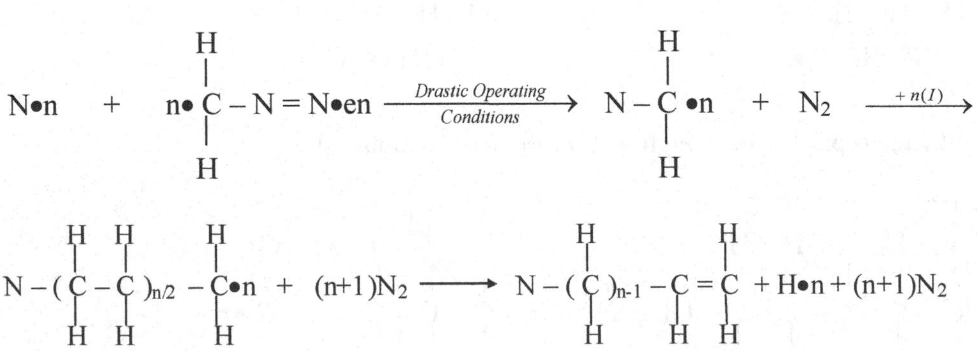

$$N\bullet n \quad + \quad n\bullet \overset{\overset{H}{|}}{\underset{\underset{H}{|}}{C}} - N = N\bullet en \quad \xrightarrow[\text{Conditions}]{\textit{Drastic Operating}} \quad N - \overset{\overset{H}{|}}{\underset{\underset{H}{|}}{C}} \bullet n \quad + \quad N_2 \quad \xrightarrow{+\,n(I)}$$

$$N - \Big(\overset{\overset{H}{|}}{\underset{\underset{H}{|}}{C}} - \overset{\overset{H}{|}}{\underset{\underset{H}{|}}{C}} \Big)_{n/2} - \overset{\overset{H}{|}}{\underset{\underset{H}{|}}{C}}\bullet n \quad + \quad (n+1)N_2 \quad \longrightarrow \quad N - \Big(\overset{H}{\underset{H}{C}} \Big)_{n-1} - \overset{\overset{H}{|}}{C} = \overset{\overset{H}{|}}{\underset{\underset{H}{|}}{C}} \quad + H\bullet n + (n+1)N_2$$

(Dead terminal double bond polymethylene) 7.214

Nucleo-free-radically like ethene, polymerization is favored with release of N_2 molecules. How-ever, being strongly nucleophilic, far more so than ethene and using an initiator which is not natural to it, it will need **far harsher operating conditions** *than that used for ethene- very high pressures in thousands and very high*

481

temperatures, to make nucleo-free-radical polymerization possible. Hence, diazomethane (diazomethylene) has never been known to favor the use of the commonly known initiators, such as initiators from benzoyl peroxide. Like a nucleo-free-radical growing polymer chain, this can release transfer species of the second kind to kill the chain to give dead terminal double bond polymers.

However, if the wrong polar state can exist not as diazomethane but something else, then when activated radically, the followings are obtained.

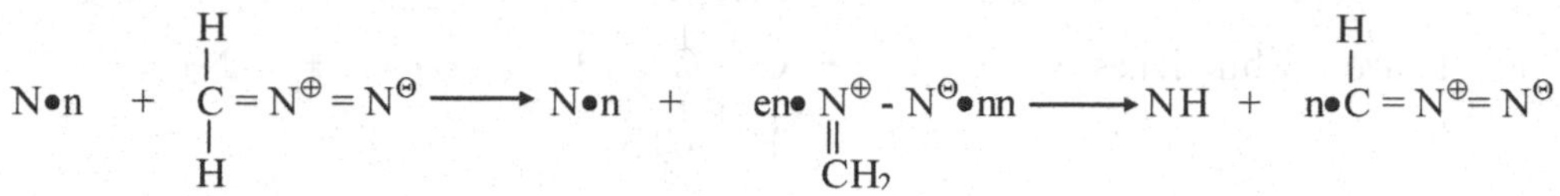

$$N{\bullet}n \;+\; \underset{\underset{H}{|}}{\overset{\overset{H}{|}}{C}}=N^{\oplus}=N^{\ominus} \longrightarrow N{\bullet}n \;+\; en{\bullet}\,\underset{\underset{CH_2}{\|}}{N^{\oplus}}\text{-}N^{\ominus}{\bullet}nn \longrightarrow NH \;+\; n{\bullet}\overset{\overset{H}{|}}{C}=N^{\oplus}=N^{\ominus}$$

$$7.217$$

Though, it shows clearly that the route is not natural to the monomer, the question is can a single polar center be activated whether it has a π-bond or not? It seems that the answer is yes via the π-bond, except that by the time the initiator is about to add if activated by heat, N_2 has been releas-ed. Nevertheless, worthy of note from the equation above is that the center activated is not the C =N center, but the N = N center, a clear indication of the followings-

$$N^{\oplus}=N^{\theta} \;>\; C=N \;>\; N=N \qquad\qquad 7.218$$

Order of Nucleophilicity

When one H atom is replaced with an R group, only the electro-free-radical route will be favored by them. Nucleo-free-radicals and nucleo-non-free-radicals if N_2 is not released will abstract transfer species of the first kind of the first type as shown below.

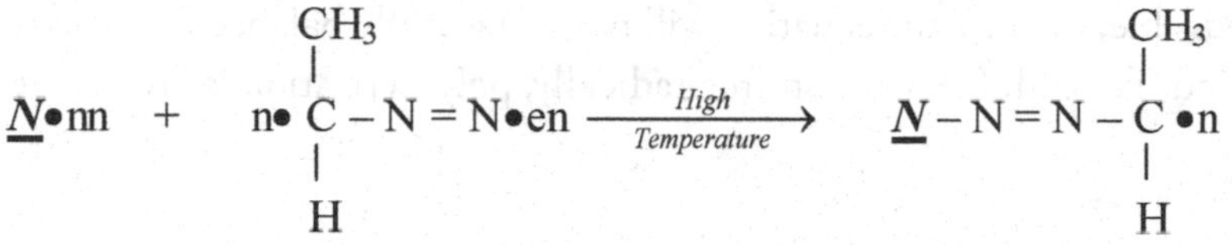

$$\underline{N}{\bullet}nn \;+\; n{\bullet}\,\underset{\underset{H}{|}}{\overset{\overset{CH_3}{|}}{C}}\text{-}N=N{\bullet}en \;\xrightarrow[\text{Temperature}]{\text{High}}\; \underline{N}\text{-}N=N\text{-}\underset{\underset{H}{|}}{\overset{\overset{CH_3}{|}}{C}}{\bullet}n$$

NOT RADICALLY BALANCED $\qquad\qquad 7.219$

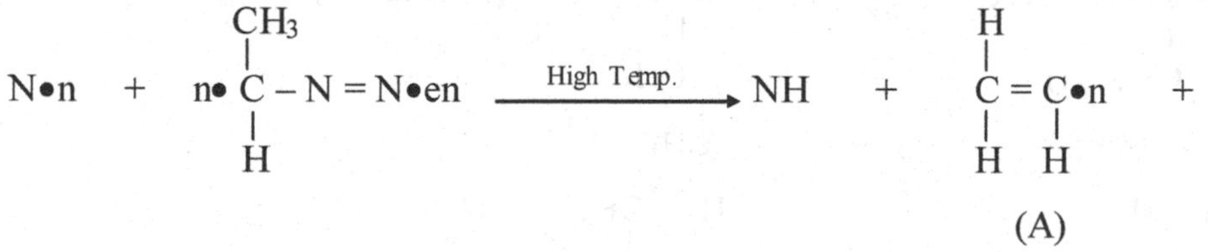

$$N{\bullet}n \;+\; n{\bullet}\,\underset{\underset{H}{|}}{\overset{\overset{CH_3}{|}}{C}}\text{-}N=N{\bullet}en \;\xrightarrow{\text{High Temp.}}\; NH \;+\; \underset{\underset{H\;\;H}{|\;\;|}}{\overset{\overset{H}{|}}{C}}=C{\bullet}n \;+\;$$

$$\qquad\qquad (A) \qquad\qquad 7.220a$$

$$(A) \;+\; n{\bullet}\underset{\underset{H}{|}}{\overset{\overset{CH_3}{|}}{C}}\text{-}N=N{\bullet}en \;\longrightarrow\; \underset{\underset{H\;\;H}{|\;\;|}}{\overset{\overset{H\;\;H}{|\;\;|}}{C}}=C \;+\; (A) \;+\; N_2$$

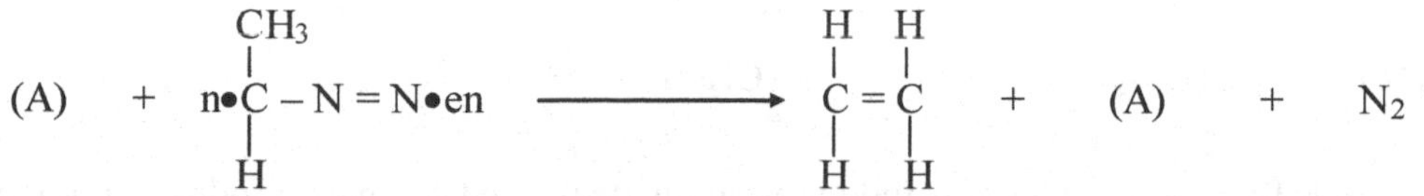

$$\qquad\qquad 7.220b$$

Overall Equation: $N{\bullet}n \;+\; 100(CH_3)HCN_2 \longrightarrow NH \;+\; 99H_2C=CH_2 \;+\; 100N_2$

$$+\;\;(A) \qquad\qquad 7.221$$

$$\text{N}\bullet\text{n} \quad + \quad \text{n}\bullet\overset{\overset{\displaystyle CH_3}{|}}{\underset{\underset{\displaystyle C_2H_5}{|}}{C}} - N = N\bullet\text{en} \quad \xrightarrow{High\ Temp.} \quad NH \quad + \quad \overset{\overset{\displaystyle CH_3}{|}}{\underset{\underset{\displaystyle H\quad CH_3}{|\quad|}}{C}} = C\bullet\text{n} \quad + \quad N_2$$

$$\text{(B)}$$

7.222a

$$\text{(B)} \quad + \quad \text{Diazo iso-butylenes} \quad \longrightarrow \quad \overset{\overset{\displaystyle CH_3\ H}{|\quad|}}{\underset{\underset{\displaystyle H\quad CH_3}{|\quad|}}{C}} = C \quad + \quad \text{(B)} \quad + \quad N_2$$

7.222b

Overall Equation: $\text{N}\bullet\text{n} \quad + \quad 100\,(CH_3)_2CN_2 \longrightarrow \quad NH \quad + \quad 99(CH_3)HC = CH(CH_3)$

$$+ \quad 100N_2 \quad + \quad \text{(B)}$$

7.223

If activation to the radical state is favored by the initiator which is not natural to it, then alkene monomers can readily be obtained from diazoalkylenes. If diazo n-butylene had been used, then 2-butene would be produced. Electro-free-radically we have the followings.

$$\text{E}\bullet\text{e} \quad + \quad \text{n}\bullet\overset{\overset{\displaystyle CH_3}{|}}{\underset{\underset{\displaystyle H\ \ (I)}{|}}{C}} - N = N\bullet\text{en} \quad \xrightarrow[Temperature]{Fairly\ Low} \quad E - \overset{\overset{\displaystyle CH_3}{|}}{\underset{\underset{\displaystyle H}{|}}{C}} - N = N \bullet\text{en}$$

NOT RADICALLY BALANCED

7.224

In the presence of heat and a passive catalyst, polymerization is favored electro-free-radically. Without releasing N_2, polymerization is not possible, because the equation will not be radically balanced. Nucleo-non-free-radically, polymerization is not favored. Electro-non-free-radically, polymerization is favored at very low temperatures.

$$\text{E}\bullet\text{e} \quad + \quad \text{n}\bullet\overset{\overset{\displaystyle H}{|}}{\underset{\underset{\displaystyle CH_3\ \ (I)}{|}}{C}} - N = N\bullet\text{en} \quad \xrightarrow[Heat]{CuCl\ or} \quad E - \overset{\overset{\displaystyle H}{|}}{\underset{\underset{\displaystyle CH_3}{|}}{C}} \bullet\text{e} \quad + \quad N_2 \xrightarrow{+\,n(I)} \quad (n+1)N_2$$

$$\overset{\displaystyle CH_3\ \ (I)}{}$$

$$E-(\overset{\overset{\displaystyle H}{|}}{\underset{\underset{\displaystyle CH_3}{|}}{C}} - \overset{\overset{\displaystyle CH_3}{|}}{\underset{\underset{\displaystyle H}{|}}{C}})_{n/2} - \overset{\overset{\displaystyle H}{|}}{\underset{\underset{\displaystyle CH_3}{|}}{C}} \bullet\text{e} \quad \longrightarrow \quad E-(\overset{\overset{\displaystyle H}{|}}{\underset{\underset{\displaystyle CH_3}{|}}{C}} - \overset{\overset{\displaystyle CH_3}{|}}{\underset{\underset{\displaystyle H}{|}}{C}})_{n/2} - \overset{\overset{\displaystyle H}{|}}{\underset{\underset{\displaystyle H}{|}}{C}} = \overset{\overset{\displaystyle H}{|}}{\underset{\underset{\displaystyle H}{|}}{C}} \quad + \quad H\bullet\text{e} \quad + \quad (n+1)N_2$$

7.225

In the absence of steric limitations, initiation and propagation are favored to finally produce a randomly placed dead terminal double bond polymer. Otherwise, syndiotactic placement will largely be favored.

Replacing the two H atoms with R groups will not alter the analysis obtained so far. Using OR groups, the followings are obtained exploratively.

$$E - \underset{\underset{CH_3}{\overset{|}{O}}}{\overset{\overset{H}{|}}{\underset{|}{C}}}.e \quad + \quad N_2 \quad \longrightarrow \quad E - \underset{\underset{CH_3}{\overset{|}{O}}}{\overset{\overset{H}{|}}{\underset{|}{C}}}.e \quad + \quad N_2 \quad \xrightarrow{+ n\,(I)}$$

$$E \underset{\underset{CH_3}{\overset{|}{O}}}{\overset{\overset{H}{|}}{\leftarrow}} \underset{\underset{H}{\overset{|}{O}}}{\overset{\overset{CH_3}{|}}{C}} \underset{\frac{n}{2}}{\overset{}{\rightarrow}} \underset{\underset{CH_3}{\overset{|}{O}}}{\overset{\overset{H}{|}}{C}}.e \quad + \quad (n+1)\,N_2 \quad \longrightarrow \qquad\qquad 7.226a$$

$$E\bullet e \quad + \quad n\bullet \underset{\underset{OR}{|}}{\overset{\overset{CH_3}{|}}{C}} - N = N\bullet en \quad \longrightarrow \quad E - \underset{\underset{OR}{|}}{\overset{\overset{CH_3}{|}}{C}}\bullet e \quad + \quad N_2 \quad \xrightarrow{+n(I)}$$

$$E - (\underset{\underset{OR}{|}}{\overset{\overset{CH_3}{|}}{C}} - \underset{\underset{CH_3}{|}}{\overset{\overset{OR}{|}}{C}})_{n/2} - \underset{}{\overset{\overset{CH_3}{|}}{C}} = O \quad + \quad R\bullet e \qquad\qquad 7.226b$$

Dead terminal aldehydic or ketonic polymers are produced with release of transfer species of the first kind of the first type, ***noting that one has already shown that the monomer cannot exist in a polar state, but only in the radical state.***

Different situations apply to CF_3, COOR and COR types of groups. With CF_3 type of group, the followings are to be expected.

$$N\bullet n + n\bullet \underset{\underset{CF_3}{|}}{\overset{\overset{H}{|}}{C}} - N = N\bullet en \longrightarrow N - \underset{\underset{CF_3}{|}}{\overset{\overset{H}{|}}{C}}\bullet n + N_2 \qquad\qquad 7.227a$$

$$N\bullet n + n\bullet \underset{\underset{CF_3}{|}}{\overset{\overset{CH_3}{|}}{C}} - N = N\bullet en \longrightarrow NH + \underset{\underset{H}{|}}{\overset{\overset{H}{|}}{C}} = \underset{\underset{CF_3}{|}}{C}\bullet n + N_2 \quad . \qquad 7.227b$$

$$F_2Fe\bullet en + n\bullet \underset{\underset{CF_3}{|}}{\overset{\overset{H}{|}}{C}} - N = N\bullet en \longrightarrow FeF_3 \quad + \quad \underset{\underset{F}{|}}{\overset{\overset{F}{|}}{C}} = \underset{}{\overset{\overset{H}{|}}{C}} - N = N \bullet en \qquad 7.227c$$

$$N\bullet nn + n\bullet \underset{\underset{CF_3}{|}}{\overset{\overset{CH_3}{|}}{C}} - N = N\bullet en \longrightarrow N - N = N - \underset{\underset{CF_3}{|}}{\overset{\overset{CH_3}{|}}{C}}\bullet n \qquad\qquad 7.227d$$

NOT RADICALLY BALANCED

The unique character of CF_3 types of groups can be observed from the equations above. In Equations 7.227a, initiation is favored, noting that the route is not Natural to the monomer. In the Natural route, the followings are obtained.

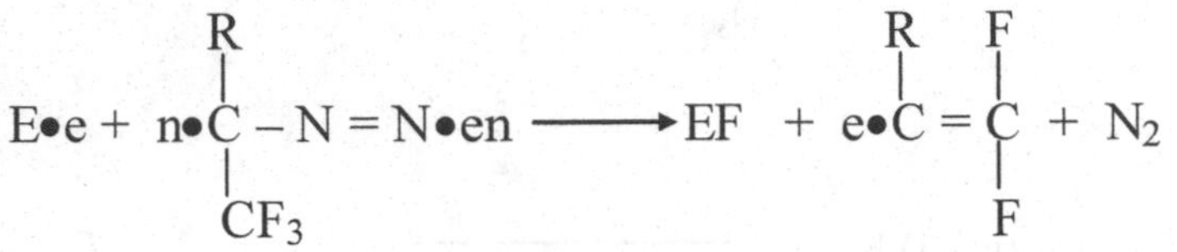

$$E\bullet e + \underset{\underset{CF_3}{|}}{\overset{\overset{R}{|}}{n\bullet C}} - N = N\bullet en \longrightarrow EF + e\bullet \underset{}{\overset{\overset{R}{|}}{C}} = \underset{\underset{F}{|}}{\overset{\overset{F}{|}}{C}} + N_2 \qquad 7.228a$$

(where R is H or CH₃ or higher ones)

The analyses as above will only slightly differ when the CF_3 group is replaced with COOR group. It is only the positively charged- and electro-free-radical-paired route that will be favored, noting that the C=O center cannot be activated and used when N_2 is released when coordination initiators are involved. The situation is different when CF_3 is changed to higher groups such as C_2F_5 and R. The electro- or nucleo-free-radical free-media route can never be favored. One can see the great uniqueness of CH_3 and CF_3 groups in different ways. With COR types of groups, almost the same as with COOR type of group will apply in particular chargedly. Electro-free-radically, initiation is favored when paired, and nucleo-free-radically, when N_2 is released.

When resonance stabilization groups are introduced, no resonance can be provided. Therefore a look at the monomer clearly indicates that it is an Electrophile, i.e. a male. Hence, it cannot be resonance stabilized.

$$\begin{matrix} \overset{H}{|} & \overset{Y}{\downarrow} & \overset{H}{|} \\ C & = & C & X \\ \underset{H}{|} & & \underset{n\bullet C - N = N\bullet en}{|} & \downarrow \\ & & \underset{H}{|} \end{matrix} \qquad \equiv \qquad \begin{matrix} H \\ | \\ n\bullet C - N = N\bullet en \\ | \\ CH \\ || \\ CH_2 \end{matrix} \qquad 7.228b$$

Electro-free-radically and nucleo-free-radically, the monomer can be polymerized. When used nucleo-free-radically, the polar section is not activated as shown above, It is only activated electro-free-radically.

7.5 Related Monomers

Related monomers are those which have polar or electrostatic bonds and/or those which are strongly nucleophilic. Monomers with polar or electrostatic bonds include sulfur dioxide, phosphines which themselves are strong nucleophiles. Others to be considered include nitroso compounds, quinones, carbon monoxide, etc. These are monomers which will be very useful during copolymerization studies. There is therefore the need to determine their characters and the routes favored by them. Unlike most of the monomers which have been considered thus far, these are cases which cannot be activated chargedly.

7.5.1 Sulfur dioxides (SO2)

As has been shown already in the series SO_2 has the following structural configuration.

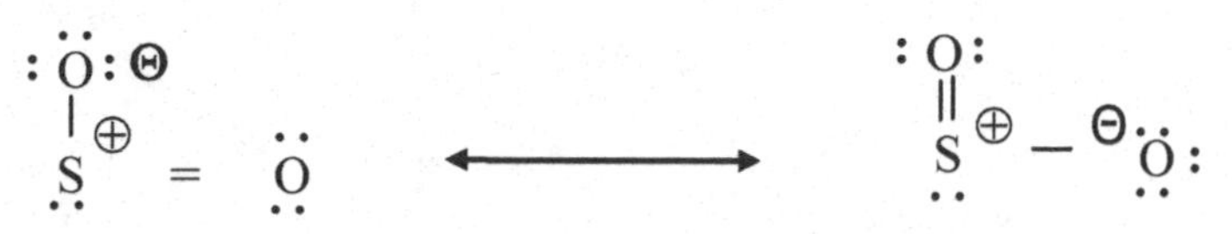

$$
\begin{array}{ccc}
\text{(I)a} & \longleftrightarrow & \text{(I)b}
\end{array}
\qquad 7.229
$$

(polar resonance stabilization)

Unlike diazoalkanes, the polar sites cannot be used for charged polymerizations, because the S center cannot carry more than eight radicals in its last shell and most importantly cannot release a transfer species to form a stable molecule. Chargedly, the S = O activation center cannot be activated due to electrostatic forces of repulsion. Even when the sulfur dioxide is of the substituted type shown below, the positively charged polar sites cannot be used for polymeri-zation. Same applies to the negatively charged polar sites, because the S center cannot carry more than eight radicals in its last shell. In fact, what is shown below cannot be activated chargedly.

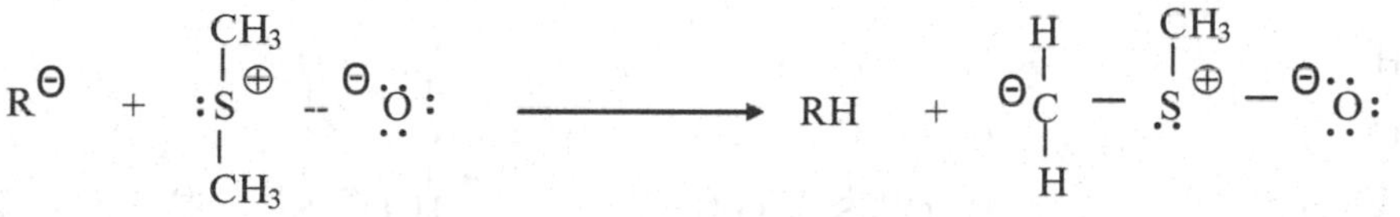

$$
7.230
$$

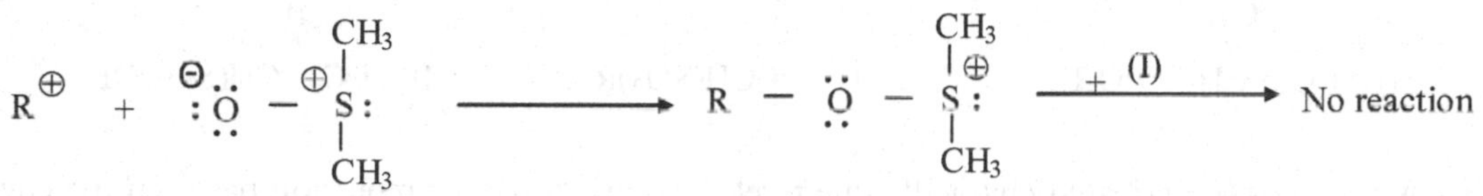

$$
7.231
$$

There is no propagation, since the S center cannot accept more than eight radicals in the last shell, based on the Period to which it belongs to in the Periodic Table. On the other hand, there is no stable molecule to reject in the process. Notice clearly that in all the configurations shown, no more than 2 radicals can be carried by the last shell of hydrogen, and no more than eight radicals can be carried by the last shells of C, S, and O. In the first equation, there is no abstraction, since the monomer is not activated.

The SO_2 will only favor radical activation as shown below, since there is an activation center.

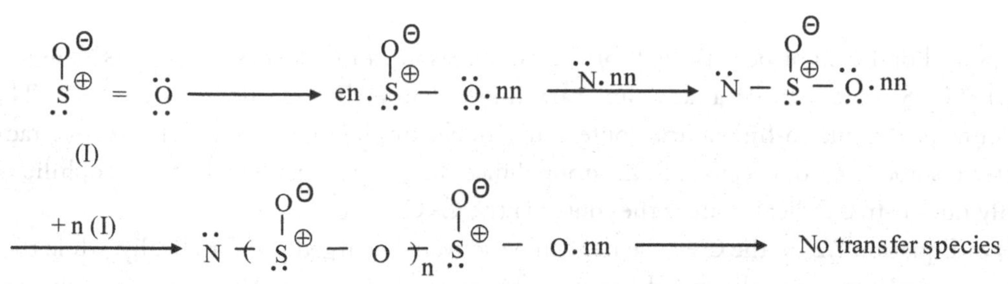

$$
7.232
$$

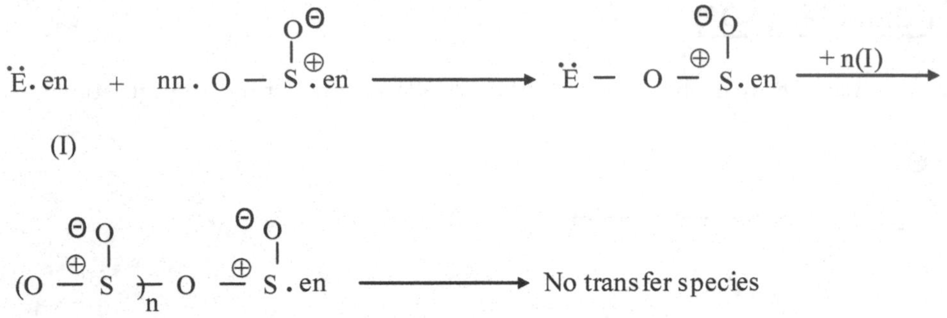

$$\ddot{E}.en \;+\; nn.O - \overset{\overset{\displaystyle O^{\ominus}}{|}}{\underset{}{S}}^{\oplus}.en \;\longrightarrow\; \dot{E} - O - \overset{\overset{\displaystyle ^{\ominus}O}{|}}{\underset{}{S}}^{\oplus}.en \;\xrightarrow{\;+\,n(I)\;}$$

(I)

$$(O - \overset{\overset{\displaystyle ^{\ominus}O}{|}}{\underset{}{S}}^{\oplus})_{\overline{n}}\, O - \overset{\overset{\displaystyle ^{\ominus}O}{|}}{\underset{}{S}}^{\oplus}.en \;\longrightarrow\; \text{No transfer species} \qquad 7.233$$

Though the monomer is a nucleophile, it will readily favor both the non-free-radical routes, with no involvement of transfer species of any kind. It is so (fifty/fifty) because the radical types carried by the active centers are fixed. It cannot favor nucleo- or electro-free-radical routes, unless under certain conditions such as during copolymerizations. Notice that the polymeric products are still carrying the polar bonds outside the main chain backbone. We have seen that the polar bonds here cannot be used as activation centers chargedly or radically.

So far, it has not been possible to introduce a group on this monomer without adding something or replacing one of the oxygen atoms, in order to see their effects. However, consider introducing a resonance stabilization group such as $-CH = CH_2$.

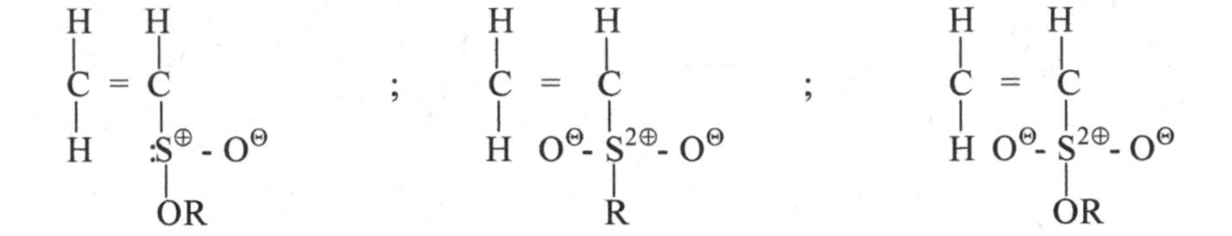

$$\begin{array}{ccc}
\overset{\displaystyle H \quad H}{\underset{}{}} & \overset{\displaystyle H \quad H}{\underset{}{}} & \overset{\displaystyle H \quad H}{\underset{}{}} \\
C = C & C = C & C = C \\
H \quad :S^{\oplus}\!-O^{\ominus} & \;;\; \quad H \;\; O^{\ominus}\!-S^{2\oplus}\!-O^{\ominus} & \;;\; \quad H \;\; O^{\ominus}\!-S^{2\oplus}\!-O^{\ominus} \\
OR & R & OR \qquad\qquad 7.234
\end{array}$$

(I) $H_2C = CH(SO)OR$ (II) $H_2C = CH(SO_2)R$ (III) $H_2C = CH(SO_2)OR$

(I) as shown cannot be activated chargedly due to electrostatic forces of repulsion between the covalent negative charge when activated and paired unbonded radicals on the sulfur center. Only the C = C bond can be activated radically as shown below.

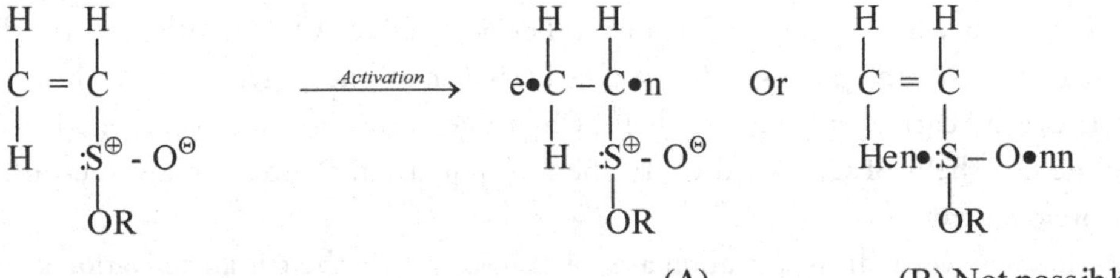

$$\begin{array}{ccc}
\overset{\displaystyle H \quad H}{\underset{}{}} & \overset{\displaystyle H \quad H}{\underset{}{}} & \overset{\displaystyle H \quad H}{\underset{}{}} \\
C = C & e\bullet C - C\bullet n & C = C \\
H \quad :S^{\oplus}\!-O^{\ominus} & \xrightarrow{\text{Activation}} \quad H \quad :S^{\oplus}\!-O^{\ominus} \quad \text{Or} & Hen\bullet :S - O\bullet nn \\
OR & OR & OR \\
& (A) & (B)\ \text{Not possible} \qquad 7.235
\end{array}$$

The single polar bond cannot be activated for the same reason as for the case of SO_2 as shown clearly above in (B). The S center cannot at any time carry more than eight radicals in its last shell. The C = C center will favor the nucleo-free-radical route being an electrophile i.e. a male. Electro-free-radically, polymerization is not favored, because of OR group. Since the polar center which is nucleophilic cannot be used, only nucleo-free-radical route is the route via the C=C center.

For (II) of Equation 7.234, the C = C center can be activated chargedly and radically, while the polar center cannot be activated radically and the polar charges cannot be used. When activated chargedly or free-radically, the followings are to be expected.

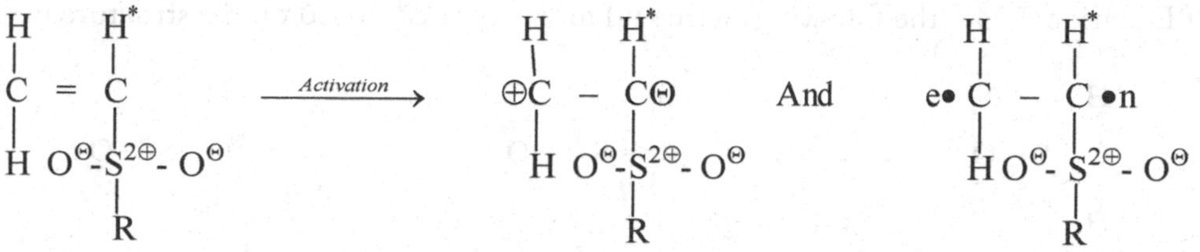

$$\underset{\underset{R}{\overset{\displaystyle H}{|}}}{\overset{\displaystyle H}{\underset{|}{C}}} \overset{H^*}{\underset{|}{=}} C \quad \xrightarrow{\text{Activation}} \quad \oplus C - C^\ominus \quad \text{And} \quad e\bullet C - C \bullet n$$

Below the arrows, the structures:

$$\underset{\underset{R}{\overset{\displaystyle H}{|}} O^\ominus - S^{2\oplus} - O^\ominus}{\overset{\displaystyle H}{\underset{|}{C}} = \underset{|}{C} \overset{H^*}{}}$$

$$\underset{\underset{R}{\overset{\displaystyle H}{|}} O^\ominus - S^{2\oplus} - O^\ominus}{\oplus C - C^\ominus \overset{H^*}{}}$$

$$\underset{\underset{R}{\overset{\displaystyle H}{|}} O^\ominus - S^{2\oplus} - O^\ominus}{e\bullet C - C\bullet n \overset{H^*}{}}$$

| (C) | (D) | 7.236 |

With the use of positively charged initiators, the polar center which is nucleophilic cannot be used. Therefore, one should expect the C = C center to favor attack since there is no transfer species. The same applies electro-free-radically noting that this monomer resembles acrolein and its alkyl form (H_2C = CHCRO). Like acrolein, it should favor the negatively charged and nucleo-free-radical routes the natural routes. Unlike acrolein, no copolymers can be obtained. But when the H^* is replaced with C $\equiv$ N group, the positively charged route will no longer be favored via the C = C center. The polymerizable X-center will then become the C $\equiv$ N center only radically. Thus, this monomer with or without the CN group will favor not only the negatively charged route[17], but also the nucleo-free-radical route. Nothing prevents the monomer from favoring the nucleo-free-radical route unless it cannot be activated radically. The analysis of (III) is almost identical to that of (II). Most important is the fact that they are all electrophiles, i.e. males. The analysis so far has not fully informed us about the order of nucleophilicity of S = O activation center with respect to another center. In fact, the center has not appeared, noting that unlike propene it is less nucleophilic since propene cannot favor any of the routes favored by SO_2. What has appeared so far is the polar center, the type which cannot be activated. There is need to know these, because as already maintained, Nature is Orderliness, having rules which have no except-ion, unlike the rules of the physical world where the EYE OF THE NEEDLE is never applied.

Now, consider the following three unique compounds, first members of their different families shown below.

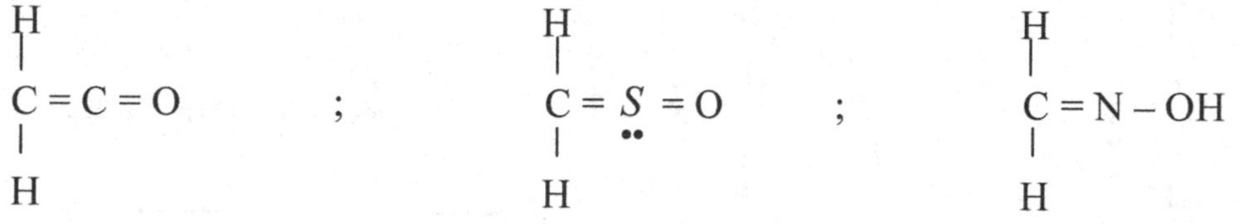

$$\underset{\underset{|}{\overset{|}{H}}}{\overset{\displaystyle H}{\underset{|}{C}} = C = O} \qquad ; \qquad \underset{\underset{|}{\overset{|}{H}}}{\overset{\displaystyle H}{\underset{|}{C}} = \underset{\bullet\bullet}{S} = O} \qquad ; \qquad \underset{\underset{|}{\overset{|}{H}}}{\overset{\displaystyle H}{\underset{|}{C}} = N - OH}$$

| (I) | (II) ?? | (III) | 7.238 |

The N center cannot carry another double bond or even a polar bond. Hence (III) is as shown. We have already seen much of (I). (III)- formaldoxime a product from hydroxylamine and formaldehyde, cannot molecularly rearrange or electroradicalize as will be shown shortly. There is however a clear indication of the stronger nucleophilic character of C = N center with respect to the N = O center. It is the $H_3C - N$ = O that can electroradicalize to (III). As will shortly be shown, the N = O center based on what it is carrying cannot be activated (See Equation 7.176). Since C is more electropositive than N, one should therefore expect the followings.

$$C = N \quad > \quad N = O \qquad\qquad 7.239$$

Order of Nucleophilicity

For (II) of Equation 7.238, the followings were said to be expected, based on the structure in (II)[18].

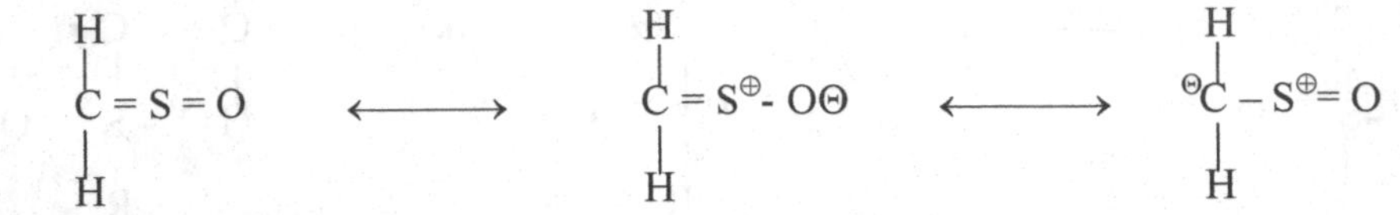

$$
\overset{\displaystyle H}{\underset{\displaystyle H}{\overset{|}{\underset{|}{C}}}} = S = O
\qquad\longleftrightarrow\qquad
\overset{\displaystyle H}{\underset{\displaystyle H}{\overset{|}{\underset{|}{C}}}} = S^{\oplus}\text{-}O\ominus
\qquad\longleftrightarrow\qquad
\overset{\displaystyle H}{\underset{\displaystyle H}{\overset{|}{\underset{|}{{}^{\ominus}C}}}} - S^{\oplus}= O
$$

| (II) Neutral sulfine | (A) Sulfylene | (B) Sulfylide | 7.240 |

NON-EXISTENT

Though sulfine a S-oxide of thioformaldehyde exists, (II) cannot be the real structure, because the S center cannot carry more than eight radicals in its last shell. In (II) of Equation 7.238, the S center is carrying ten radicals in its last shell. Secondly, polar bonds cannot appear from a compound which is in a neutral state as in (II). Though S and C are equi-electropositive, they are still uniquely different based on the Group to which they belong to in the Periodic Table. The real structure of (II) is (A) which is clearly shown below.

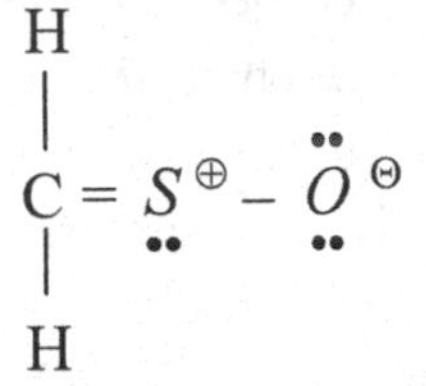

$$
\overset{\displaystyle H}{\underset{\displaystyle H}{\overset{|}{\underset{|}{C}}}} = \overset{..}{\underset{..}{S}}{}^{\oplus} - \overset{..}{\underset{..}{O}}{}^{\ominus}
$$

(I)   7.241

In the last shell of S of (B), there are eight radicals. In the last shell of C in carbon monoxide, there are six radicals with one vacant orbital. In the last shell of S of (II), there are ten radicals that which necessitated the need for polar bond formation. CO which was used to compare with it, is not polar. Thus, **sulfines** are polar compounds which cannot molecularly rearrange or electroradicalize. One can say that it is polar/radical/polar resonance stabilized almost like diazoalkylenes which is radical/polar.

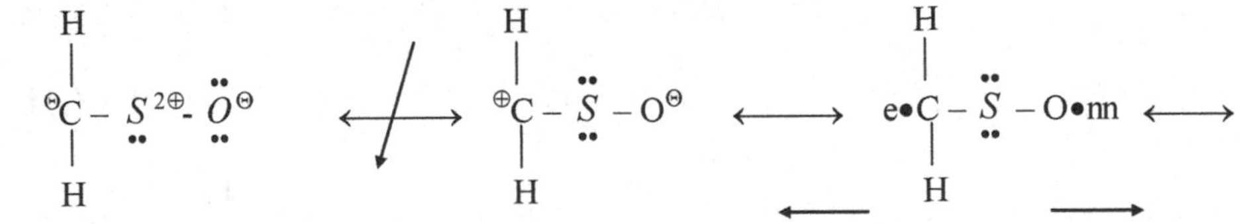

$$
\overset{\displaystyle H}{\underset{\displaystyle H}{\overset{|}{\underset{|}{{}^{\ominus}C}}}} - S^{2\oplus}\text{-}\overset{..}{\underset{..}{O}}{}^{\ominus}
\qquad\longleftrightarrow\!\!\!/\qquad
\overset{\displaystyle H}{\underset{\displaystyle H}{\overset{|}{\underset{|}{{}^{\oplus}C}}}} - \overset{..}{\underset{..}{S}} - O^{\ominus}
\qquad\longleftrightarrow\qquad
\overset{\displaystyle H}{\underset{\displaystyle H}{\overset{|}{\underset{|}{e\bullet C}}}} - \overset{..}{\underset{..}{S}} - O\bullet nn
\qquad\longleftrightarrow
$$

| (I) Not favored | (II) | (III) |

$$
\overset{\displaystyle H}{\underset{\displaystyle H}{\overset{|}{\underset{|}{C}}}} = \underset{..}{S}{}^{\oplus} - O^{\ominus}
\qquad\longleftrightarrow\!\!\!/\qquad
\overset{\displaystyle H}{\underset{\displaystyle H}{\overset{|}{\underset{|}{{}^{\ominus}C}}}} - \underset{..}{S}{}^{\oplus} = O
$$

7.242

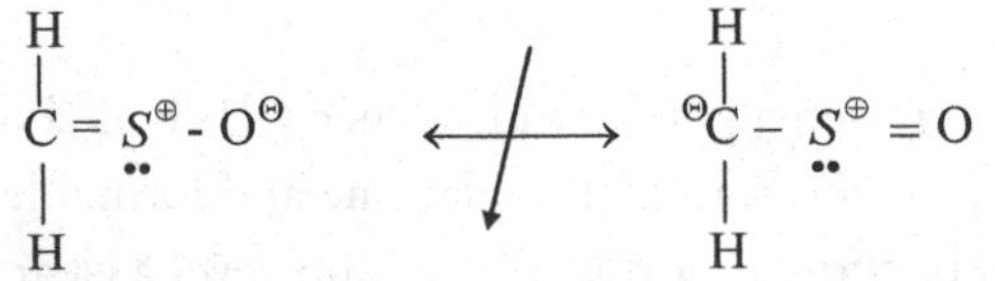

| (IV) Sulfine (Sulfylene) | (V) Not favored (Sulfylide) |

(IV) and (V) were (A) and (B) of Equation 7.240 respectively. Thus, this sulfine, (IV) or (A), has three resonance stabilized forms, (II), (IV) and a radical form (III). (II) is its form in a neutral environment, while (IV) is its form in an acidic environment. When both are activated, (III) is obtained. In fact (III) can cyclize to give a three membered ring as shown below.

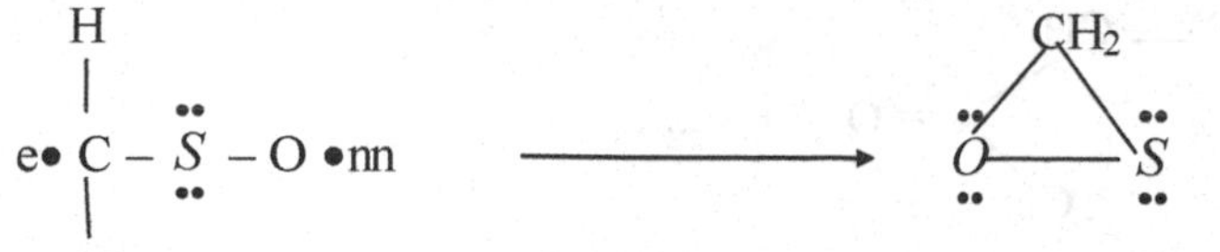

$$\begin{array}{c} H \\ | \\ e\bullet\; C - \overset{\bullet\bullet}{\underset{\bullet\bullet}{S}} - O \;\bullet nn \\ | \\ H \end{array} \quad\longrightarrow\quad (A) \qquad\qquad 7.243$$

As will be shown downstream, the three-membered ring has two functional centers identified by the presence of paired unbonded radicals, O and S of which none can be used and it is a ring which can only instantaneously open itself when heated. It has an invisible π-bond which is called the **Strain Energy**. It has only one point of scission and this is the S – O bond, and not the C – S or C – O bond. It therefore opens instantaneously by itself with a small amount of heat since the Minimum Required Strain Energy (MRSE) is small.

Stage 1:

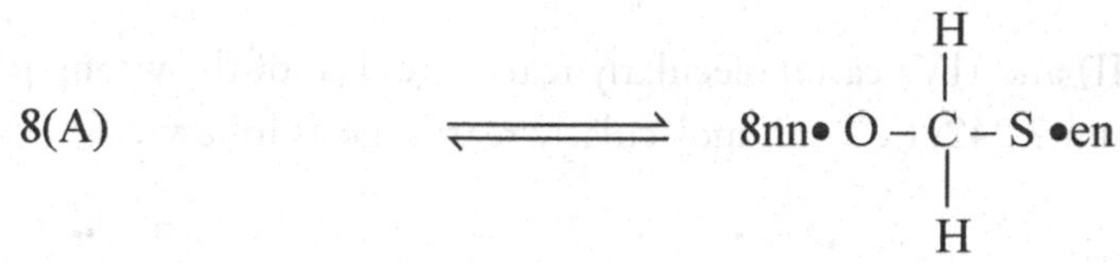

$$8(A) \quad\rightleftharpoons\quad \begin{array}{c} H \\ | \\ 8nn\bullet\; O - C - S \;\bullet en \\ | \\ H \end{array}$$

$$(B)$$

$$8(B) \quad\rightleftharpoons\quad 8H_2C = O \;+\; 8\,en\bullet\, S\, \bullet nn \;+\; \text{HEAT}$$

$$8\, en\bullet\, S\, \bullet nn \xrightarrow{\text{Deactivation}} \quad \text{(S ring)} \;+\; \text{Heat}$$

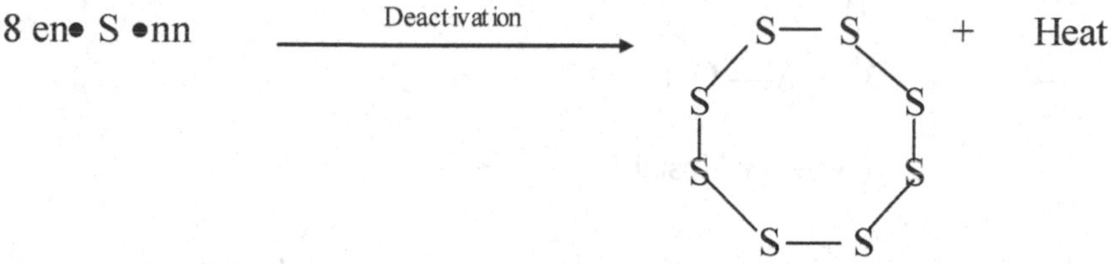

$$7.244$$

Rhombic sulfur

Overall Equation: $8(A) \longrightarrow 8H_2C = O \;+\; \text{Rhombic sulfur} \;+\; \text{Large Heat} \qquad 7.245$

Just as we have oxidizing *oxygen elements* (en• O •nn) to be shown downstream when provid-ing the mechanisms of Combustion and Oxidation, so also we have *sulfurizing sulfur elements* (en• S •nn) which can readily be obtained from rhombic sulfur and other smaller sized rings of sulfur or from other chemical compounds which one will call *sulfurizing agents just like oxidizing agents.* Indeed the sulfine itself is a sulfurizing agent. The mechanism above was how Tropothione (cycloheptatrienethione) S-oxide shown below was converted to tropone in part by oxidation using m-chloroperbenzoic acid (m-CPBA).[18] The (cycloheptatrienethione) S-oxide was only obtained from tropothione by oxidation.

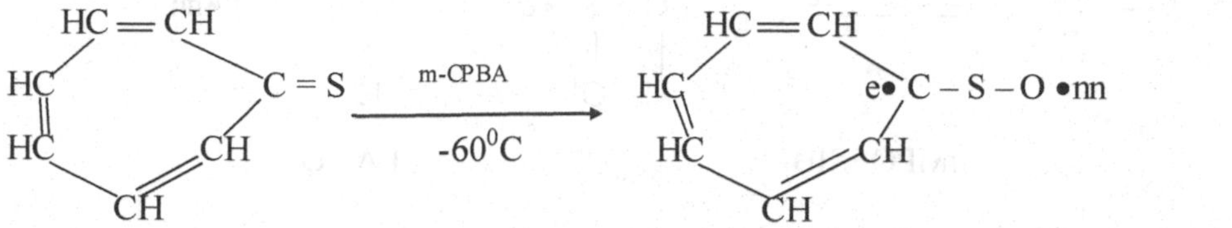

$$\text{(tropothione ring, } C=S) \xrightarrow[-60^0C]{\text{m-CPBA}} \text{(ring, } e\bullet\, C - S - O\, \bullet nn)$$

Tropothione   Tropothione S-oxide (Dark red needles)

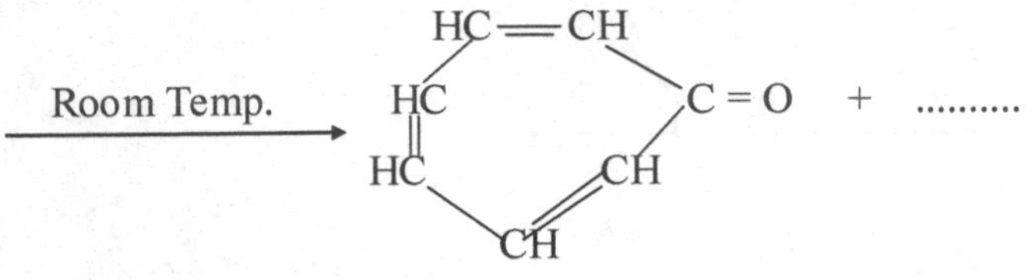

$$\xrightarrow{\text{Room Temp.}}$$

Tropone

+ ....

7.246

The fact that S – O was the point of scission of (A) of Equation 7.243 in Stage 1 of Equation 7.244, clearly indicates that-

$$C = O \;>\; C = S \;\gg\; S = O \qquad\qquad 7.247$$

Order of Nucleophilicity

None of the real polar forms (II) and (IV) can molecularly rearrange. But of the wrong polar forms (I) and (V) of the sulfine in Equation 7.242, (V), can molecularly rearrange as follows,

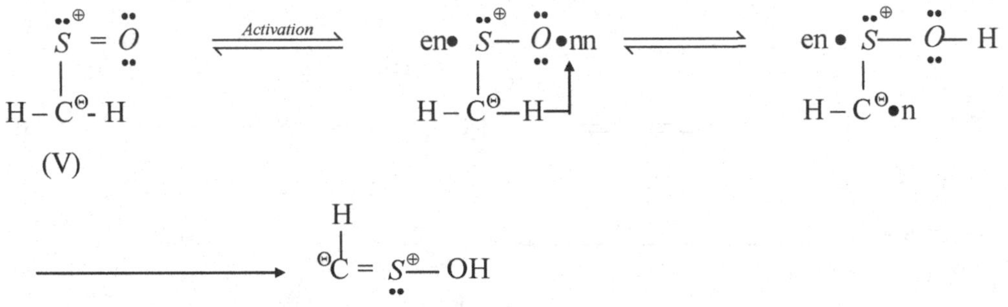

(V)

(A) Very Unstable

7.248

(A) is very unstable and indeed not known to exist. ***Exploration is a wonderful experience.*** It readily tautamerizes back to (IV). The activated center can be used in different forms. The presence of the polar charge did not prevent the transfer of transfer species of the ***first kind of the third type***. The polar bond is like a π-bond, and also like a π-bond, it can only be used under certain conditions. In the Equation above, the center will not favor the nucleo-non-free-radical route, but only the electro-non-free-radical route a clear indication of its nucleophilic character. Exploratively, this was one of the compounds which was thought to exist as one of the sulfines, and changed to a different compound which on its own came back to the real sulfine. When the real polar case, (IV) of Equation 7.242 is activated via the π-bond alone, the following is obtained.

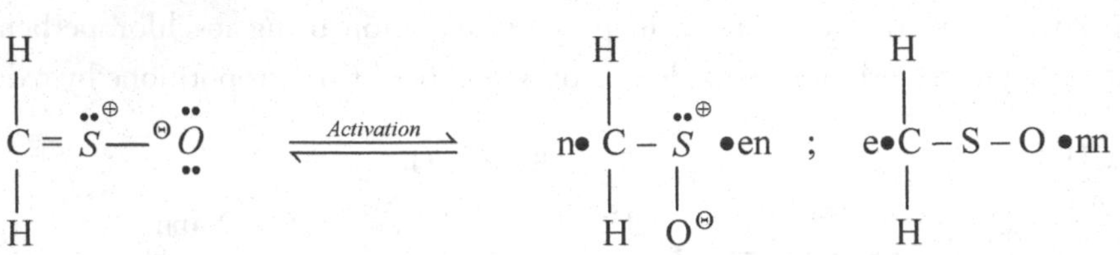

IMPOSSIBLE FAVORED 7.249

This was not the state in which it was used when cycloheptatrienethione S-oxide was added to 1-morpholinocyclopentane in chloroform solvent at room temperature.[18] It is said to be impossible above, because based on Equation 7.247, the C = S cannot be activated in the presence of a polar S=O being more nucleophilic The C center being in the fully unsaturated ring, makes the ring to tautomerize to give the product obtained in which the polar bond was not shown in the structure of the product obtained. One can observe how Nature operates. It is too much to comprehend. Clearly, one can see why we need the use of the **"eye of the needle"**

7.5.2 Phosphines and carbon monoxide

To be very candid, most of the structures given for phosphorus compounds are not correct because the phosphorus atom cannot carry more than eight radicals in its last shell. It cannot use the 3d orbitals for bonding, based on the Period it belongs to in the Periodic Table. Every atom has its domain and boundary just like everything in life including living systems. For example, the real structure of phosphorus penta-chloride PCl_5 is as shown below.

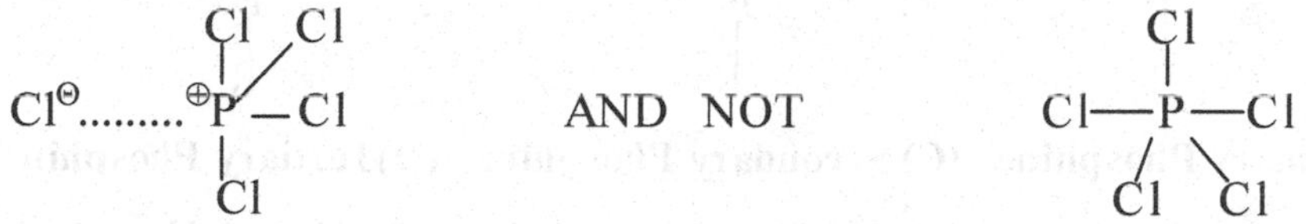

(A) –PCl₅ from Cl₂/PCl₃   AND NOT   **(B) Non Existent**

7.250

The reasons why Nature put polar and electrostatic bonds in place, are because the laws of a boundary must not be contravened or broken. These bonds are complex as already said, while covalent and ionic bonds are real those which we can see. In (A), there are eight radicals in the last shell of P. The dotted line you see is an electrostatic bond, an imaginary bond. It is neither here nor there. In (B), there are ten radicals, noting that in both, all the five radicals have been used for their creation. Having ten radicals is against the boundary laws of Nature for the Period P belongs to in the Periodic Table. Even the structure of the so-called hexafluorophosphate "ion" $PF_6^{\ominus}$ is unknown. In fact the negative charge carried by it is not an ion. The structures of the two types of P_2Cl_{10} a dimer are unknown. Shown below are the real structures of some of them.

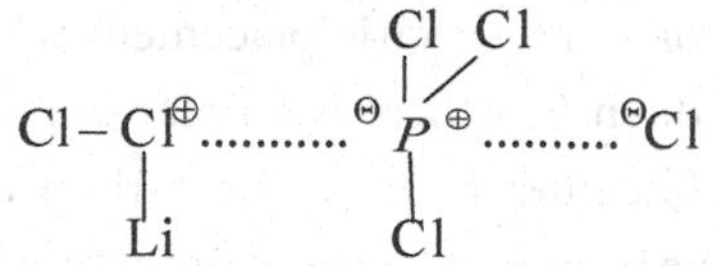

(A)-From LiCl/PCl₅   **(B)-From LiCl/PCl₃**

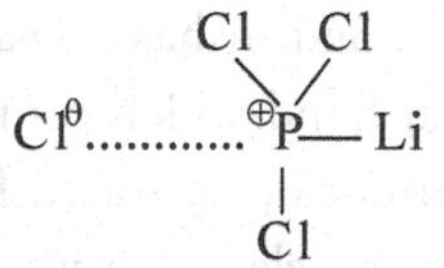

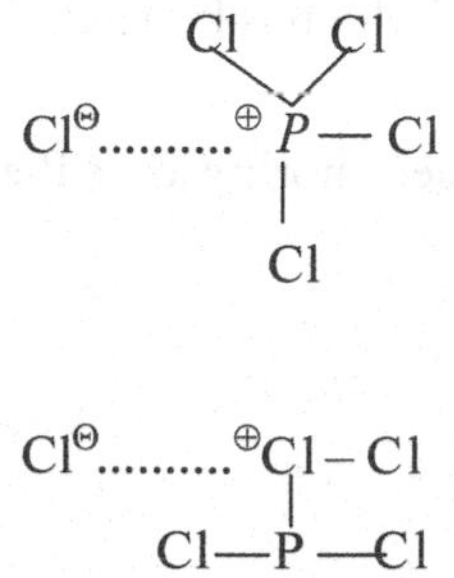

(C) From PCl₃/Cl₂   **(D) P₂Cl₁₀ from 2PCl₃/2Cl₂**

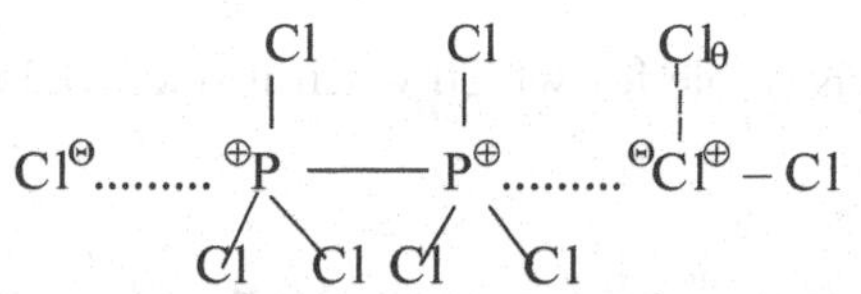

7.251

492

All the dotted bonds carried by them are electrostatic bonds, while the others are covalent bonds. The charges carried by them are electrostatic charges very different from ionic charges. Notice that the last shell of P cannot carry more than eight radicals. When one started these writings, one was using "electron", because the elites have been so conditioned to believe that electrons exist outside the nucleus. With time during the re-conditioning process, one changed to use of the real thing-radicals, because electrons only exist differently inside the nucleus made of two mirror images-Matter and Anti-matter.

The structures of the acids of phosphorus are in order, because the polar bonds are shown and well recognized. So also are the structures of phosphine (PH_3), its alkylated derivatives- primary phosphines (PH_2R), secondary phosphines (PHR_2) and tertiary phosphines (PR_3), its aryl derivatives- $P(C_6H_5)_3$, and so on. But there are some phosphines whose structures are not known. Shown below are the structures of some of these phosphines.

(A) Phosphine (B) Primary Phosphine (C) Secondary Phosphine (D)Tertiary Phosphine

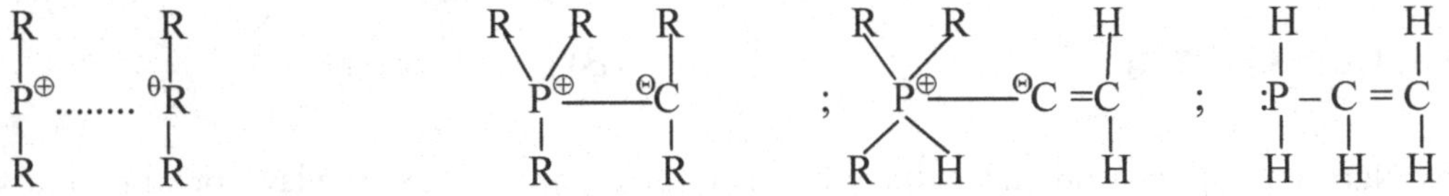

(E) PR$_5$ (R$_2$/PR$_3$)   (F) PR$_3$CR$_2$ (G) PR$_3$C$_2$H$_3$ (H) PH$_2$C$_2$H$_3$

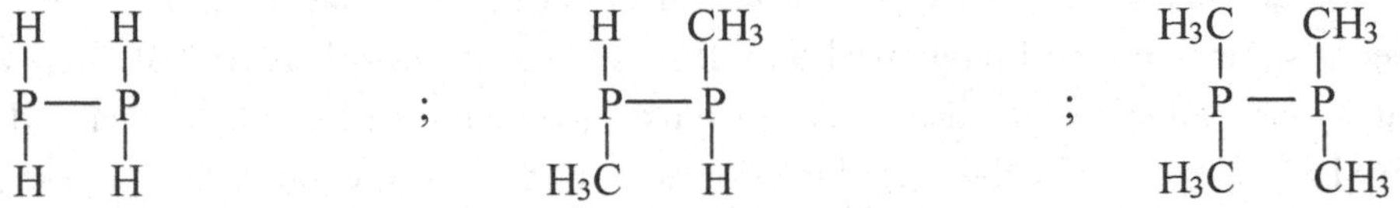

7.252

(I) Phosphorus tetrahydride   (J) 1,2- dimethylide phosphorus hydride   (K) Phosphorus tetramethylide

Only (E) carries an electrostatic bond. It has no vacant orbitals for stereospecific placement when used as an initiator. (F) carries a polar bond which cannot be activated. In (G) there is a polar bond and an activation center that which cannot be activated from the C = C center because the carbon center is connected to a P center which is more electropositive than C. For the same reason (H) cannot be activated. *Hence PH$_2$, unlike NH$_2$ is not known as a substituent group.* (I) which can also be called diphosphorus tetrahydride or diphosphine is indeed phosphorus tetrahydride and not PH_4 which is believed to have sp^3 hybride orbitals[19]. While the Equilibrium State of Existence of NH_3 is that where H•e is the component held, that for PH_3 is stable for which when it does, the following is obtained, noting that P and H are equi-electropositive.

$$PH_3 \quad \xrightleftharpoons[\text{Existence}]{\text{Equilibrium State of}} \quad H\bullet n \quad + \quad en\bullet PH_2 \qquad\qquad 7.253$$

$$NH_3 \quad \xrightleftharpoons[\text{Existence}]{\text{Equilibrium State of}} \quad H\bullet e \quad + \quad nn\bullet NH_2 \qquad\qquad 7.254$$

(J) and (K) are diphosphines. Thus, unlike ammonia which is polar/ionic, phosphine is polar/non-ionic. It is less polar than water. Therefore, it cannot solubilize in water like ammonia does. One can see why alkyl phosphines cannot be made from phosphine and alkyl halides as is done for ammonia.

So far we are yet to see whether a polar bond can be activated. One source of phosphine is said to be by "hydrolyzing" aluminum phosphide with dilute acids.

$$AlP \quad + \quad 3HCl \quad \longrightarrow \quad AlCl_3 \quad + \quad PH_3 \qquad\qquad 7.255$$

Without knowing the structure of any compound, the mechanism of a reaction in which the compound is involved cannot be explained.

Stage 1: $\qquad\qquad HCl \quad \rightleftharpoons \quad H\bullet e \quad + \quad Cl\bullet nn$

$$H\bullet e \quad + \quad Al^{2\oplus}\!\!-\!\!^{2\ominus}P \quad \underset{Heat}{\overset{Activation}{\rightleftharpoons}} \quad H\bullet e \quad + \quad e\bullet Al^{\oplus}\!\!-\!\!^{\ominus}\overset{\bullet\bullet}{P}\bullet nn$$

$$\rightleftharpoons \quad H\text{-}P^{\ominus}\!\!-\!\!^{\oplus}Al\bullet e \quad (A) \qquad\qquad 7.256$$

$$(A) \quad + \quad Cl\bullet nn \quad \longrightarrow \quad H\text{-}P^{\ominus}\!\!-\!\!^{\oplus}Al\text{-}Cl$$

Overall Equation: $HCl \quad + \quad AlP \quad \longrightarrow \quad H\text{-}P^{\ominus}\text{-}^{\oplus}Al\text{-}Cl \qquad 7.257$

Stage 2: $\qquad\qquad HCl \quad \rightleftharpoons \quad H\bullet e \quad + \quad Cl\bullet nn$

$$H\bullet e \quad + \quad H\text{-}P^{\ominus}\text{-}^{\oplus}Al\text{-}Cl \quad \overset{Activation}{\longrightarrow} \quad H\bullet e \quad + \quad nn\bullet \overset{\displaystyle H}{\underset{\displaystyle Cl}{\overset{|}{\underset{|}{P\text{-}Al}}}}\bullet e$$

$$\rightleftharpoons \quad \overset{\displaystyle Cl}{\overset{|}{H_2P\text{-}Al}}\bullet e \quad (B)$$

$$(B) \quad + \quad Cl\bullet nn \quad \longrightarrow \quad H_2P\text{-}Al\text{-}Cl_2 \qquad\qquad 7.258$$

Overall Equation: $HCl \quad + \quad HPAlCl \quad \longrightarrow \quad H_2P\text{-}Al\text{-}Cl_2 \qquad 7.259$

Stage 3: $\qquad\qquad HCl \quad \rightleftharpoons \quad H\bullet e \quad + \quad Cl\bullet nn$

$$H_2P\text{-}Al\text{-}Cl_2 \quad \rightleftharpoons \quad nn\bullet PH_2 \quad + \quad e\bullet AlCl_2$$

$$Cl_2Al\bullet e \quad + \quad nn\bullet Cl \quad \rightleftharpoons \quad AlCl_3$$

$$H\bullet e \quad + \quad nn\bullet PH_2 \quad \longrightarrow \quad PH_3 \qquad\qquad 7.260$$

Final Overall Equation: $3HCl \quad + \quad AlP \quad \longrightarrow \quad PH_3 \quad + \quad AlCl_3 \qquad 7.261$

The structure of AlP is not $Al \equiv P$ or $Al^{\oplus} = {}^{\ominus}P$, because as will be shown downstream, ***metals cannot carry double or triple bonds.*** Its polar bonds here were activated, because the boundary laws were not contravened. P and Al cannot carry more than eight radicals in the last shell as dictated by Argon of the third Period where they belong. In the reactions above, there are three stages and the mechanism is Equilibrium without any form of hydrolysis. Though dilute acid was said to be used, based on the mechanism above, water was never involved. *It should be noted that under standard temperature and pressure, compounds which have activation centers in form of a gas or liquid or solid state do not exist in Activated State of Existence, not even diazo-methylene. They exist in either Electrostatic or Polar or Ionic or Stable States of Existence for them to remain harmless to their environments in particular to Living systems.* Hence, AlP is seen to exist in the polar state.

Some other important sources of phosphine, are from the reaction of white phosphorus with concentrated aqueous sodium hydroxide and from the direct combination of hydrogen with white phosphorus.

$$4P \ + \ 3NaOH \ + \ 3H_2O \longrightarrow 3NaH_2PO_2 \ + \ PH_3 \qquad 7.262$$

All the atoms in the Periodic Table have their elements, some we can see, others we can't see. As we move along the Series and Volumes one will start showing the structures of those we can see, for Carbon Black, Activated carbon, Coke, Coal, the different types of sulfur, phosphorus and so on. White phosphorus[19] is one of the elements of phosphorus whose structure is shown below.

$$\text{en} \bullet \overset{\bullet\bullet}{\underset{\bullet en}{P}} \bullet \text{nn} \qquad 7.263$$

(A) White phosphorus

Stage 1:

$$H_2O \rightleftharpoons H\bullet e \ + \ nn\bullet OH$$

$$H\bullet e \ + \ (A) \rightleftharpoons H - \overset{\bullet\bullet}{\underset{\bullet en}{P}} \bullet en \ (B) \qquad 7.264$$

$$(B) \ + \ nn\bullet OH \longrightarrow H - \overset{\bullet\bullet}{\underset{\bullet en}{P}} - OH \ (C)$$

Overall Equation: $\quad H_2O \ + \ P \longrightarrow HPOH \qquad 7.265$

Stage 2:

$$H - \overset{\bullet\bullet}{\underset{\bullet en}{P}} - OH \rightleftharpoons H - \underset{\bullet\bullet}{P} - O\bullet nn \ + \ H\bullet e$$

$$H\bullet e \ + \ NaOH \rightleftharpoons H_2 \ + \ NaO\bullet en \ + \ Heat$$

$$NaO\bullet en \rightleftharpoons Na\bullet e \ + \ nn\bullet O\bullet en \ + \ Heat$$

$$nn\bullet O\bullet en \ + \ H_2 \rightleftharpoons HO\bullet nn \ + \ H\bullet e$$

$$HO\bullet nn \ + \ H\bullet e \rightleftharpoons H_2O$$

$$H - P - O\bullet nn \ + \ Na\bullet e \longrightarrow H - P - ONa \qquad 7.266a$$

Overall Equation: $8Na + 8H_2O + 8P \longrightarrow 8H - P - ONa \ + \ 8H_2O \ + \ Heat \qquad 7.266b$

Stage 3:

$$2H - \overset{\cdot\cdot}{\underset{\bullet en}{P}} - ONa \; + \; 2H_2O \; \rightleftharpoons \; 2H - \overset{\displaystyle OH}{\underset{\displaystyle |}{P}} - ONa \; + \; 2H\bullet e$$

$$2H\bullet e \longrightarrow H_2 \qquad\qquad\qquad 7.267a$$

Overall Equation: $\quad 8HPONa + 8H_2O \longrightarrow 8NaH_2PO_2 + 4H_2 \qquad 7.267b$

Stage 4: $\quad 2(A) \; + \; 2H_2 \; \rightleftharpoons \; 2H - \underset{\bullet en}{P} \bullet nn \; + \; 2H\bullet e$

$$\longrightarrow \; 2H - \overset{\cdot\cdot}{\underset{\bullet en}{P}} - H \; (D) \qquad\qquad 7.268$$

Overall Equation: $\quad 2P \; + \; 2H_2 \longrightarrow 2PH_2 \qquad\qquad 7.269$

Stage 5:

$$2H - \overset{\cdot\cdot}{\underset{\bullet en}{P}} - H \; + \; 2H_2 \; \rightleftharpoons \; 2H - \overset{\displaystyle H}{\underset{\displaystyle \cdot\cdot}{\underset{|}{P}}} - H \; + \; 2H\bullet e$$

$$2H\bullet e \longrightarrow H_2 \qquad\qquad\qquad 7.270$$

Overall Equation: $\quad 2PH_2 \; + \; 2H_2 \longrightarrow 2PH_3 \; + \; H_2 \qquad 7.271$

Overall Equation (Stgs 4 & 5): $\quad 2P \; + \; 4H_2 \longrightarrow 2PH_3 \; + \; H_2 \qquad 7.272$

Final Overall Equation: $\quad 8NaOH \; + \; 8H_2O \; + \; 10P \longrightarrow$

$$8NaH_2PO_2 \; + \; 2PH_3 \; + \; H_2 \; + \; Heat \qquad 7.273$$

One can imagine how an equation like that shown in Equation 7.262 as commonly represented in many text books can be very deceptive. There are so many cases of such equations scattered here and there in all Chemistry and Chemistry related textbooks. Hydrogen one of the products of the reaction is never shown. ***Unknown is that without H the Head of the family of all elements, no other elements can exist.*** Not even the Transition metals which cannot form hydrides with H can exist. H opened the door for the other elements to exist. That is the way it is for all first members of different families of compounds, and that is Nature.

Stages 4 and 5 have shown to us how direct combination of hydrogen with the white phos-phorus gives the phosphine. If there had not been enough P in the system, then phosphine would not have been obtained, i.e., the reactions would have stopped in Stage 3 to give sodium phos-phonous acid and hydrogen. This would have required 2 moles of NaOH instead of the 8 moles which must be used if phosphine was to be one of the products in the presence of more P. Imagine if H_2 catalysts were present, the situation will be different. Indeed, from what we have seen so far our laboratory equipments have to be rebuilt and same applies to the entire labora-tories and the same applies to most of the Chemical and Chemical related Industries and so on. Apart from what one can see above, imagine if (C) and (D) of Equations 7.264 and 2.768, both carrying electro-non-free-radical were gaseous inside an opened laboratory, it will be deadly, but not as deadly as if it was an electro-free-radical. So far, one can observe that phosphine and some of their family members cannot be used as monomers.

Carbon monoxide has a π- bond that can be used, for example in the presence of Mg at very high temperatures despite the presence of paired unbonded radicals still present on the C center in the presence of a vacant 2p orbital on the carbon center. The vacant orbital just as exists in Zinc (Zn) metal to give Zinc dust when activated, is another form of Activation center; the others being some π-bonds, some polar bonds and some functional centers on rings. Recalled below as has already been shown, is the unique case of carbon monoxide which is gaseous under standard temperature and pressure (STP).

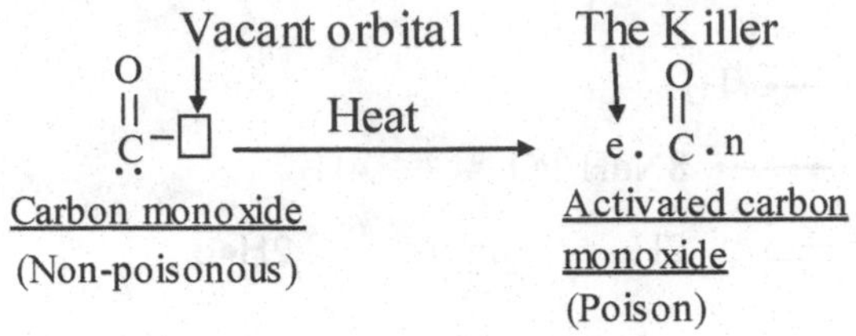

$$
\underset{\substack{\text{Carbon monoxide}\\ \text{(Non-poisonous)}}}{\overset{\overset{\displaystyle O}{\underset{\displaystyle \|}{}}}{\underset{\displaystyle ..}{C}}-\overset{\overset{\text{Vacant orbital}}{\downarrow}}{\Box}} \xrightarrow{\text{Heat}} \underset{\substack{\text{Activated carbon}\\ \text{monoxide}\\ \text{(Poison)}}}{\overset{\overset{\text{The Killer}}{\downarrow \overset{\displaystyle O}{\underset{\displaystyle \|}{}}}}{e.\, C.\, n}}
$$

7.274

At STP and low temperatures, it is non-poisonous. But when heated such as in running a hydro-carbon driven car in a closed environment such as a closed garage, it becomes a poison and the real poison is the electro-free-radical the side always diffusing all the time looking for something to attack. The heat activates it. If there was no vacant orbital in the last shell, it cannot be activated. The radicals carried by the carbon centers are free-radicals in the absence of any more paired unbonded radicals on the carbon center. If the carbon monoxide was in the liquid state, it will be less or non-poisonous depending on how used. Don't let it come in contact with a reactive compound, because it will flow in and do something positive or negative or flow out if no reaction. If it was a solid, it will not be poisonous unless when soluble.

In the absence of any additional activation center specially placed, it is a nucleophile of weak capacity, because it can undergo initiation step nucleo-free- and electro-free-radically, since it has no transfer species. However, being a nucleophile, it will be far easier to activate electro-free-radically. While it has never been reported to homopolymerize, homopolymerization of Carbon mono-sulfide has in recent years been reported[20]. Without the use of initiators, they cannot add continuously to themselves. At best, cyclic rings may be obtained such as shown below, the existence of the ring depending on its Maximum Required Strain Energy (Max. R. S. E.). For if exceeded, the ring can never exist. It has to be extremely large for it to exist.

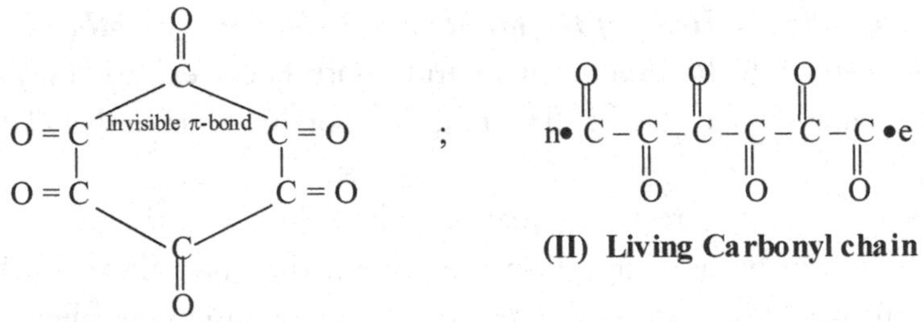

(I) represents a hexacarbonyl ringed compound with O=C groups, invisible π-bond; (II) represents a living carbonyl chain.

(II) Living Carbonyl chain

(I) A Hexacarbonyl ringed compound [Too strained to exist]

7.275

If CO bond in other non-ringed compounds is more nucleophilic than CS bond in other non-ringed compounds, then the activation center in CO should be more nucleophilic than the activation center in CS, since CS has shown signs of polymer formation than CO.

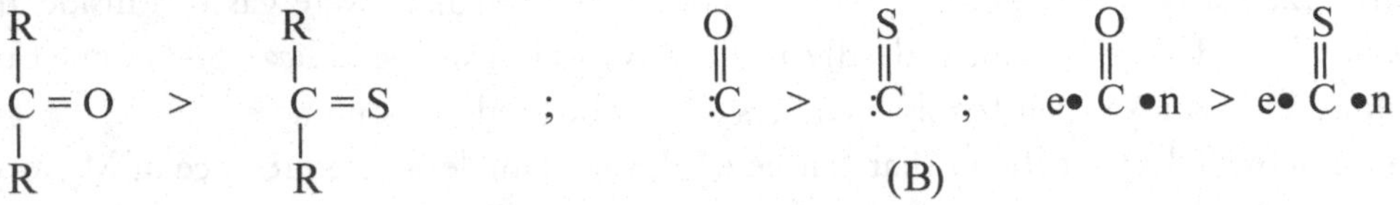

$$
\underset{\text{(A)}}{\overset{R}{\underset{R}{|}}{C}=O \;>\; \overset{R}{\underset{R}{|}}{C}=S} \quad ; \quad \overset{\overset{O}{\|}}{:C} \;>\; \overset{\overset{S}{\|}}{:C} \quad ; \quad \underset{\text{(B)}}{e\bullet \overset{\overset{O}{\|}}{C}\bullet n \;>\; e\bullet \overset{\overset{S}{\|}}{C}\bullet n}
$$

Order of Nucleophilicity

7.276

The fact that the activation center in CO is more nucleophilic than the activation center in CS, hence one will require more harser operating conditions to initiate the activation center in CO. Note that of all the atoms and elements of Group VIb, C cannot form double bonds with the others except with C, O and S because all the others are metallic and more electropositive than C. At best, the metals can only form polar bonds with C. With ringed compounds where the (A) types exist as electrophiles, the order is not reversed as shown below.

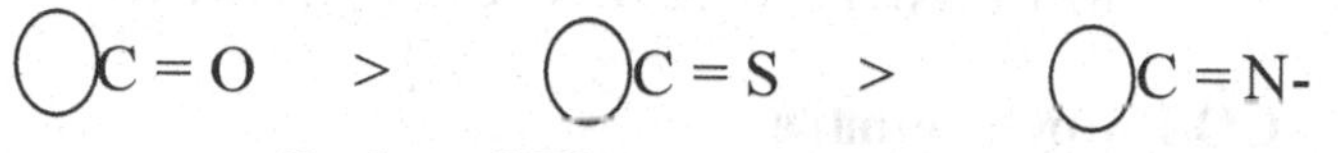

$$ \bigcirc\!\!\!\!\text{C} = \text{O} \quad > \quad \bigcirc\!\!\!\!\text{C} = \text{S} \quad > \quad \bigcirc\!\!\!\!\text{C} = \text{N-} $$

Order of Electrophilicity

7.277

Note that the activation center in CO and CS in (B) of Equation 7.276 cannot be activated chargedly, because both charges cannot be on the same C center and the negative charge cannot stay on the carbon center due to electrostatic forces of repulsion coming from the π-bond. Thus CO can be polymerized with the oxygens syndiotactically placed as shown in (II) of Equation 7.275. The (II) when formed is one of the many cancer causing agents in addition to what it does on O_2 taken in. It is almost impossible to close it to give for example (I). However if very large in size, it may become stable and be the seat for formation of lumps in in particular living systems.

CO has been known to favor some copolymerization reactions with vinyl monomers such as vinyl acetate[21], some cyclic monomers such as epoxide[22], some azidiridines such as ethylene-imine, propylenimine[23] and some allene monomers such as arylallenes[24]. By the nature of CO, when first activated, it will be the one to diffuse to activate and add to the other monomer to give alternating placement most of the time, unless the comonomer favors homo-polymerization with the initiator. This will be fully explained downstream when the driving forces for alternating placement in Addition polymerization systems are provided. Vinyl acetate as has been shown, cannot be activated chargedly and CO cannot also be activated chargedly. Therefore, its copoly-merization with CO is not a charged one[21].

CO/Vinyl acetate

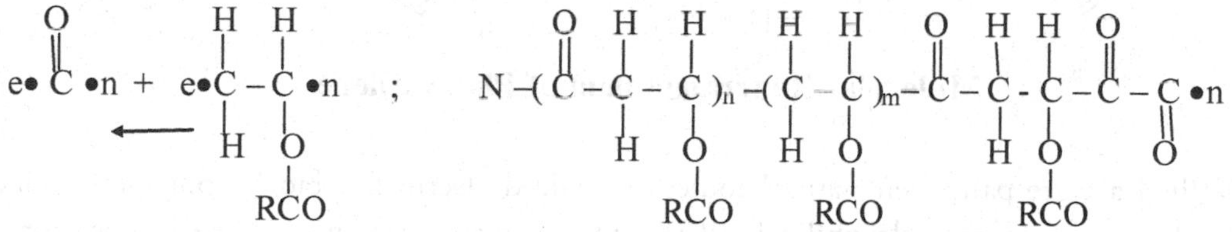

NON-ALTERNATING COPOLYMER

7.278

CO/Propylene oxide

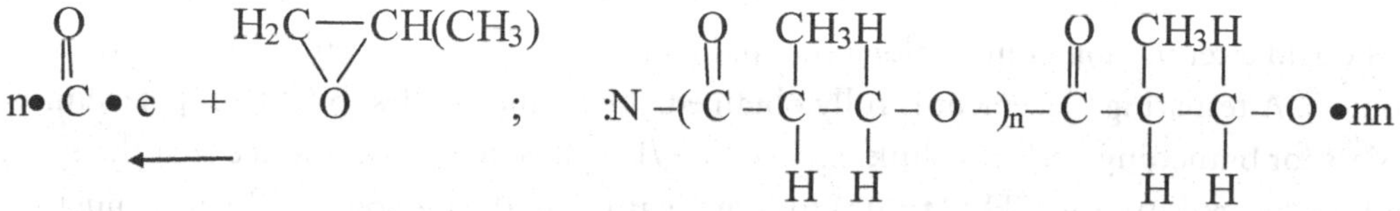

ALTERNATING COPOLYMER

7.279

CO$_2$/Ethylene oxide

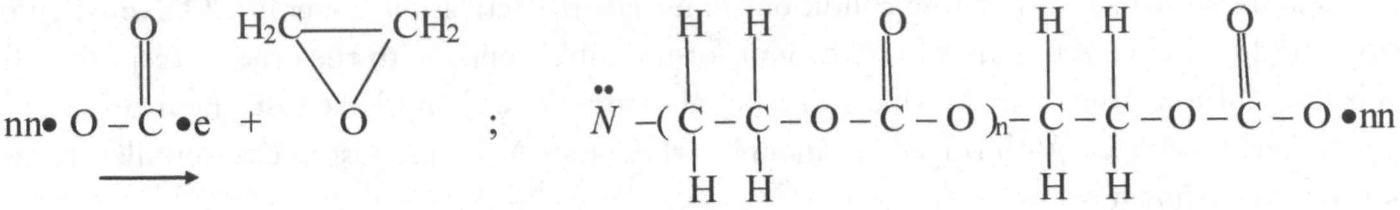

$$nn\bullet O - \overset{\overset{\textstyle O}{\|}}{C} \bullet e \;+\; \overset{\displaystyle H_2C \!-\!\!\!-\!\!\!- CH_2}{\underset{O}{\diagdown \, \diagup}} \;;\; \overset{..}{N} - (\overset{H}{\underset{H}{C}} - \overset{H}{\underset{H}{C}} - O - \overset{\overset{\textstyle O}{\|}}{C} - O\,)_n - \overset{H}{\underset{H}{C}} - \overset{H}{\underset{H}{C}} - O - \overset{\overset{\textstyle O}{\|}}{C} - O \bullet nn$$

MILD ALTERNATING COPOLYMER   7.280

CO/Propyleneimine

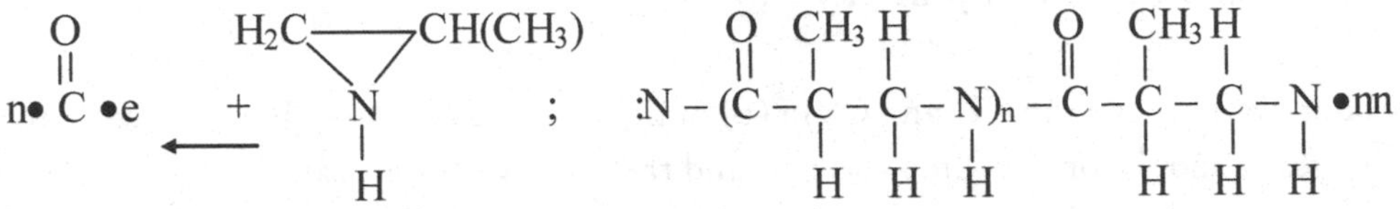

$$n\bullet \overset{\overset{\textstyle O}{\|}}{C} \bullet e \;+\; \overset{\displaystyle H_2C \!-\!\!\!-\!\!\!- CH(CH_3)}{\underset{\underset{H}{N}}{\diagdown \, \diagup}} \;;\; :N - (\overset{\overset{\textstyle O}{\|}}{C} - \overset{CH_3}{\underset{H}{C}} - \overset{H}{\underset{H}{C}} - N)_n - \overset{\overset{\textstyle O}{\|}}{C} - \overset{CH_3}{\underset{H}{C}} - \overset{H}{\underset{H}{C}} - N \bullet nn$$

ALTERNATING COPOLYMER   7.281

CO/Phenylallene

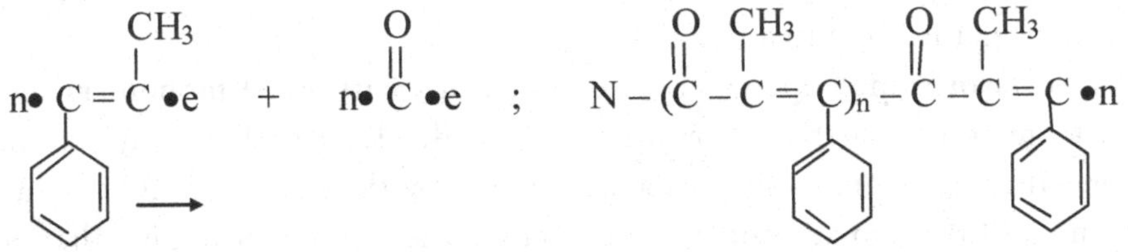

$$n\bullet \overset{CH_3}{\underset{\displaystyle \bigcirc}{C}} = C \bullet e \;+\; n\bullet \overset{\overset{\textstyle O}{\|}}{C} \bullet e \;;\; N - (\overset{\overset{\textstyle O}{\|}}{C} - \overset{CH_3}{\underset{\displaystyle \bigcirc}{C}} = C)_n - \overset{\overset{\textstyle O}{\|}}{C} - \overset{CH_3}{\underset{\displaystyle \bigcirc}{C}} = C \bullet n$$

ALTERNATING COPOLYMER   7.282

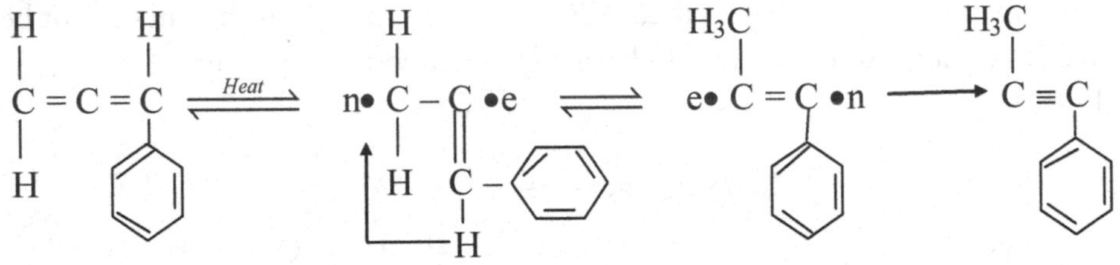

$$\overset{H}{\underset{H}{C}} = C = \overset{H}{\underset{\displaystyle \bigcirc}{C}} \;\underset{\xleftarrow{\hspace{1em}}}{\xrightarrow{\;Heat\;}}\; n\bullet \overset{H}{\underset{\underset{\displaystyle -H}{H-C-\bigcirc}}{C}} - \overset{H}{\underset{\displaystyle \bigcirc}{C}} \bullet e \;\rightleftharpoons\; e\bullet \overset{H_3C}{\underset{\displaystyle \bigcirc}{C}} = C \bullet n \;\rightarrow\; \overset{H_3C}{\underset{\displaystyle \bigcirc}{C}} \equiv C$$

Molecular Rearrangement of Phenylallene   7.283

With CO/Vinyl acetate pair, their natural routes are indeed electro-free-radical polymerization. But because Vinyl acetate is more nucleophilic, it diffuses to CO to add to it and produce minor alternating placement. From time to time, more of vinyl acetate than CO will appear on their own along the chain. For CO/Propylene oxide pair, both monomers are each unreactive to the initiator which is not their natural route. The instantaneously opened ring diffuses to CO and adds it to form a couple which forms the carboxylic acid ester monomer units along the chain nucleo-non-free-radically, the route unnatural to both monomers. Alternating placement is fully obtained. The same applies to CO/Propyleneimine pair which has sites for branching and cross-linking. For CO$_2$/Ethylene oxide pair, notice that the initiator is also nucleo-non-free-radical (e.g. CH$_3$O$\bullet$nn). This is not natural to the monomers which are nucleophiles. It can polymerize the ring when opened instantaneously as well as the CO$_2$ under harsh operating conditions. Hence, alternating placement is said to be mild. It can open the ring via the functional center, the center for electro-radicals. Nevertheless, notice that CO$_2$ can also be used as a comonomer as shown above to give polycarbonate ester and indeed as a monomer.

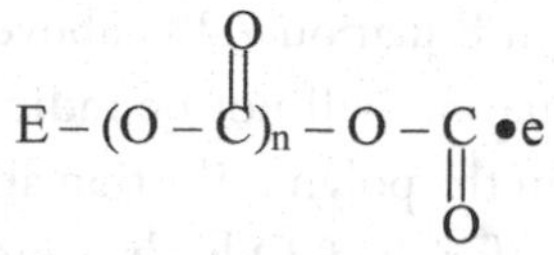

$$E - (O - \overset{\displaystyle O}{\underset{\displaystyle \parallel}{C}})_n - O - \overset{\displaystyle \bullet e}{\underset{\displaystyle \parallel}{\underset{\displaystyle O}{C}}}$$

Poly(Carbon dioxide)   7.284

Like nucleo-non-free-radical initiators, electro-free-radical initiators are not commonly used or even known to exist. But they are there in abundance. Indeed over 95 percent of all the radical initiators used in the Industries and Academia are nucleo-free-radical initiators. Note that radical catalysts are those which when they decompose, produce only one type of radical, male or female alone and not both. Some exist alone by themselves (e.g. e•TiCl$_3$, Na•e).

For CO/Phenylallene pair, the allene which cannot be resonance stabilized by the presence of benzene ring first molecularly rearranges as shown in Equation 7. 283 to give methylphenyl acetylene which is far more stable and more nucleophilic than CO. The methylphenyl acetylene can also electroradicalize under different operating conditions (e.g. presence of some Transition transition metal complexes). Being more nucleophilic, hence it diffused to the CO to add to it. The methylphenyl acetylene has transfer species of the first kind and first type carried by the CH$_3$ group. Left alone it cannot be polymerized nucleo-free-radically. *It should be recalled that the C = C double bonds always alluded to in specifying order of nucleophilicity are indeed not for C = C centers carried by monomers such as ethene, vinyl chloride, vinyl acetate, etc. female monomers which can in addition favor routes not natural to them and methyl acrylate and so on- male monomers which can favor both electro-free-radical (unnatural) and nucleo-free-radical (natural) routes,* **but for the character of the family.** These (the males) are less nucleophilic than the activation center in CO. The activation center in CO is however far less nucleophilic than C = O center, since for example acetone can form alternating placement with CO nucleo-non-free-radically. When ketones or water or both are used as solvents in alternating copolymerization of CO with propylene as shown below[25], it does not mean that the Ketones are unreactive with the CO. All depends on the operating conditions, since as shown below, nucleo-free-radicals (and not nucleo-non-free-radicals) are involved. Even the water itself is reactive with ketone. ***All operations in life depend on the OPERATING CONDITIONS.***

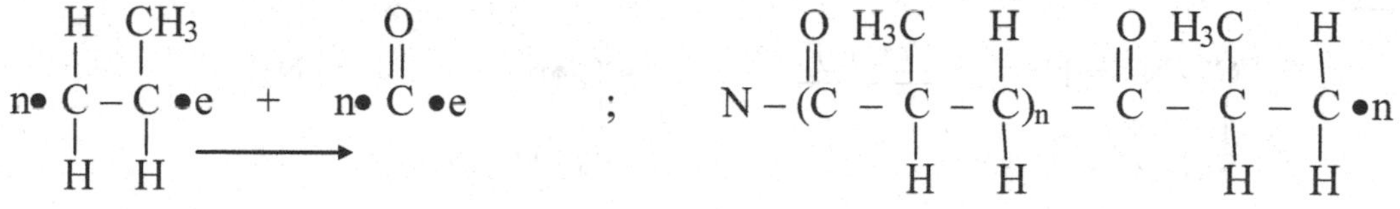

$$n\bullet \underset{\displaystyle \overset{\displaystyle |}{H}}{\overset{\displaystyle \overset{\displaystyle H}{|}}{C}} - \underset{\displaystyle \overset{\displaystyle |}{H}}{\overset{\displaystyle \overset{\displaystyle CH_3}{|}}{C}} \bullet e \ + \ n\bullet \overset{\displaystyle \overset{\displaystyle O}{\parallel}}{C} \bullet e \quad \xrightarrow{\hspace{1cm}} ; \quad N - (\overset{\displaystyle \overset{\displaystyle O}{\parallel}}{C} - \underset{\displaystyle \overset{\displaystyle |}{H}}{\overset{\displaystyle \overset{\displaystyle H_3C}{|}}{C}} - \underset{\displaystyle \overset{\displaystyle |}{H}}{\overset{\displaystyle \overset{\displaystyle H}{|}}{C}})_n - \overset{\displaystyle \overset{\displaystyle O}{\parallel}}{C} - \underset{\displaystyle \overset{\displaystyle |}{H}}{\overset{\displaystyle \overset{\displaystyle H_3C}{|}}{C}} - \underset{\displaystyle \overset{\displaystyle |}{H}}{\overset{\displaystyle \overset{\displaystyle H}{\setminus}}{C}} \bullet n$$

ALTERNATING PLACEMENT   7.285

If the CO had diffused to the propene, the couple formed would not have favored alternating placement or be initiated. Notice that it is polymeric ketone that is produced. It is therefore conclusively believed that ketone a strong nucleophile is far more nucleophilic than CO. Shown below are the two different types of possible couples from CO/Acetone pair.

$$e\bullet \underset{\displaystyle \overset{\displaystyle |}{CH_3}}{\overset{\displaystyle \overset{\displaystyle CH_3}{|}}{C}} - O - \overset{\displaystyle \overset{\displaystyle O}{\parallel}}{C} \bullet n \quad ; \quad e\bullet \overset{\displaystyle \overset{\displaystyle O}{\parallel}}{C} - \underset{\displaystyle \overset{\displaystyle |}{CH_3}}{\overset{\displaystyle \overset{\displaystyle CH_3}{|}}{C}} - O \bullet nn$$

                                                                 7.286

(I) CO diffusing to Acetone   **(II) Acetone diffusing to CO   [FAVORED]**

(I) will not favor nucleo-free-radical route, the initiator used in Equation 7.285 above. (II) will also not favor the use of nucleo-free-radical initiators, because the equation will not be radically balanced. It is (II) that is favored. One can see why ketone was not involved in the polymeriza-tion above in the absence of nucleo-non-free-radicals. The acetone was used as a solvent for the CO both being Polar/Non-ionic. Water is Polar/Ionic, while propene is Non-polar/Non-ionic. It is good to know all these when choosing solvents, unless the need arises to form precipitates.

From all the considerations so far, the followings are obvious.

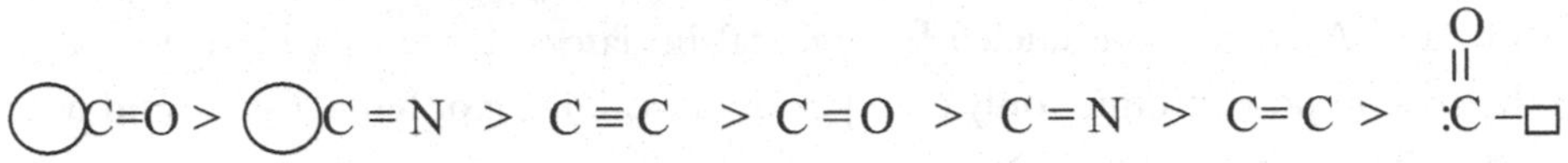

$$\bigcirc\!C{=}O > \bigcirc\!C = N > C \equiv C > C = O > C = N > C = C > \overset{\overset{O}{\|}}{:C - \square}$$

Order of Nucleophilicity   7.287

In recent years, (trimethylsilyl) diazomethane was reported to be selectively carbonylated to (trimethylsilyl) ketene at 10^0C under atmospheric pressure of carbon monoxide in the presence of octacarbonyl dicolbalt as the catalyst[26]. The catalyst here is passive in character and not active, because one wonders how the rate of (trimethylsilyl) ketene can be first order with respect to both (trimethylsilyl) diazomethane and octacarbonyl dicobalt and negative first order with respect to CO. If first order with respect to the catalyst, then it has been active in the process. Shown below is the real structure of the catalyst.

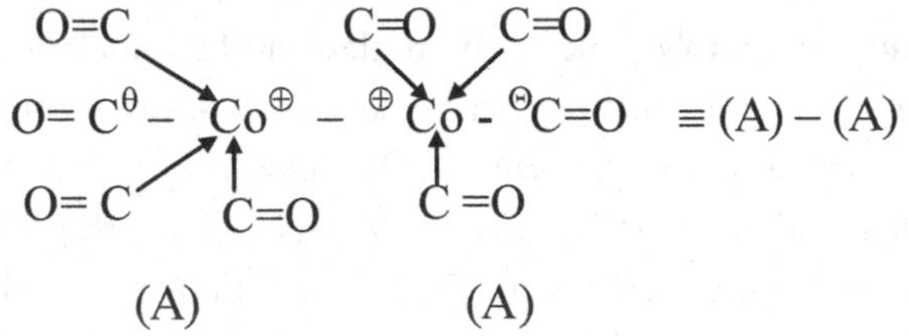

(A)   (A)

Octacarbonyl dicobalt   7.288

Stage 1:

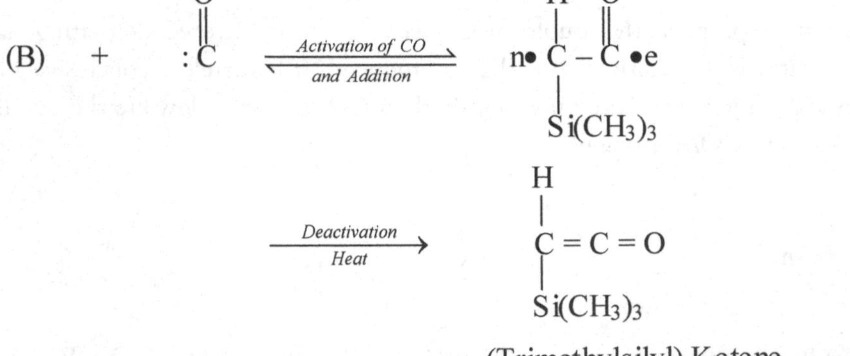

(B)

7.289

(Trimethylsilyl) Ketene

501

Overall Equation: $CO + HC(Si(CH_3)_3)N_2 \xrightarrow{(A)\text{-}(A)} N_2 + HC(Si(CH_3)_3) = C = O$

7.290

Based on the mechanism which is Equilibrium, and the overall equation, the catalyst is not active and as will be shown downstream when these stages whether for Equilibrium, Combination, or Decomposition mechanisms are being interpreted into Mathematical language, the reaction is first order each to both CO and (trimethylsilyl) diazomethane. How can the ketene be produced if the catalyst was active? In the stage above, it was the diazomethane that diffused to the CO, a clear indication that the diazo compound is more nucleophilic than the activation center in CO.

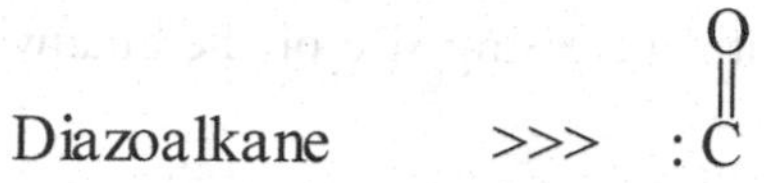

$$Diazoalkane \quad >>> \quad :\overset{\overset{\textstyle O}{\|}}{C}$$

Order of Nucleophilicity

7.291

7.5.3. Nitroso compounds

The structural configuration for these compounds is as follows, for which only non-free-radical activation would have been favored.

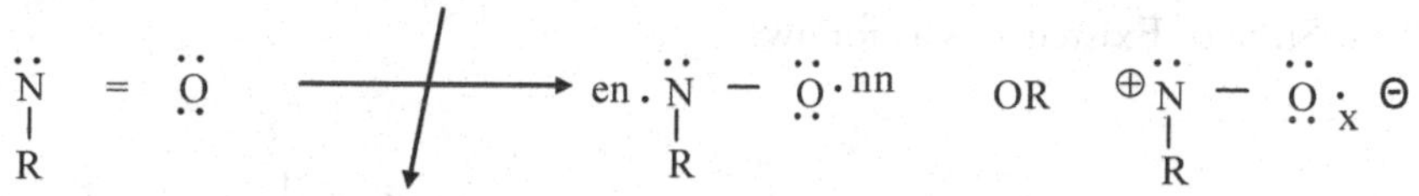

$$\underset{R}{\overset{\cdot\cdot}{N}} = \overset{\cdot\cdot}{\underset{\cdot\cdot}{O}} \quad \nrightarrow \quad en\cdot\underset{R}{\overset{\cdot\cdot}{N}} - \overset{\cdot\cdot}{\underset{\cdot\cdot}{O}}\cdot nn \quad OR \quad \overset{\oplus}{\underset{R}{\overset{\cdot\cdot}{N}}} - \overset{\cdot\cdot}{\underset{\cdot\cdot}{O}}_x \ominus$$

7.292

Chargedly, it cannot be activated in the presence electrostatic forces of repulsion. Activation is not favored because the N center is connected next to a C center which is more electropositive than N. It can only be activated if groups such as halogens, H, Phenyl, OH, OR, NH$_2$, NHR, NR$_2$ and etc. are carried by it. However, some of them can undergo Electroradicalization as shown below for formamide (CH_3-N=O). [See before Equation 7.238 and thereafter]

Stage 1:

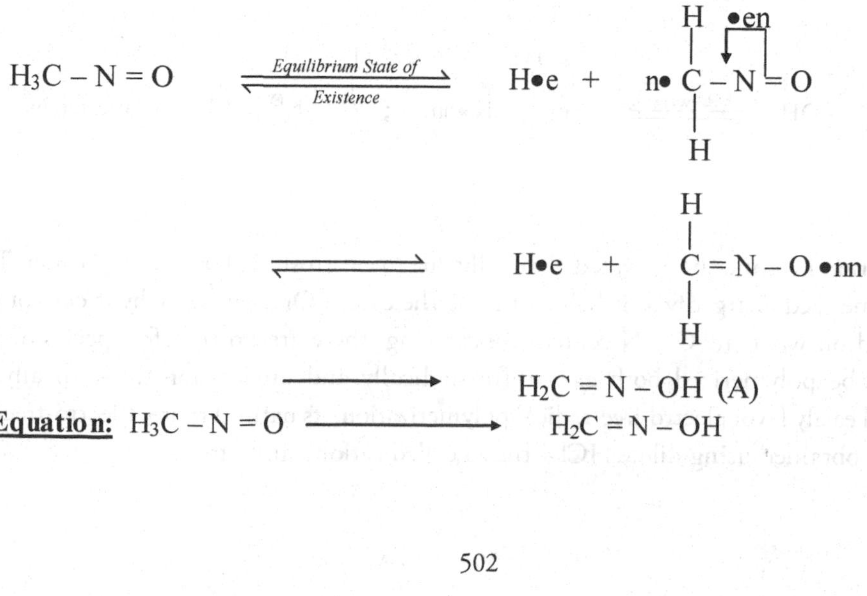

$$H_3C - N = O \quad \underset{Existence}{\overset{Equilibrium\ State\ of}{\rightleftharpoons}} \quad H\bullet e + n\bullet \overset{\overset{\textstyle H \quad \bullet en}{|}}{\underset{\underset{\textstyle H}{|}}{C}} - N = O$$

$$\rightleftharpoons \quad H\bullet e + \overset{\overset{\textstyle H}{|}}{\underset{\underset{\textstyle H}{|}}{C}} = N - O \bullet nn$$

$$\longrightarrow \quad H_2C = N - OH \quad (A) \qquad\qquad 7.293$$

Overall Equation: $H_3C - N = O \longrightarrow H_2C = N - OH$

7.294

This is not enolization, but Electroradicalization, because the N = O center was not activated and cannot be activated with what the N center is carrying. Some other higher members when made to exist in Equilibrium State of Existence, may also electroradicalize.

$$H_5C_2 - N = O \quad \underset{\text{Existence}}{\overset{\text{Equilibrium State of}}{\rightleftharpoons}} \quad H\bullet e \quad + \quad n\bullet \overset{\overset{\displaystyle H}{|}}{\underset{\underset{\displaystyle CH_3}{|}}{C}} - N = O$$

$$(I) \hspace{8cm} (B) \hspace{2cm} 7.295$$

The same indeed applies to so-called Azo-alkanes. $H_3C - N = N - CH_3$, $H_3C - N = N - C_2H_5$ can all electroradicalize, their ability decreasing drastically with increasing size of the alkanyl group [See Appendices II and III].

$$H_5C_5 - N = N - C_2H_5 \quad \underset{\text{Existence}}{\overset{\text{Equilibrium State of}}{\rightleftharpoons}} \quad H\bullet e \quad + \quad n\bullet \overset{\overset{\displaystyle CH_3}{|}}{\underset{\underset{\displaystyle H}{|}}{C}} - N = N - C_2H_5$$

$$\hspace{11cm} (C) \hspace{2cm} 7.296$$

The (A) of Equation 7.293 which is formaldehyde oxime the first member of the family of Ketoximes, cannot molecularly rearrange to formamide regardless the operating conditions. It cannot electroradicalize because its Equilibrium State of Existence is as follows.

$$H_2C = N - OH \quad \underset{\text{Existence}}{\overset{\text{Equilibrium State of}}{\rightleftharpoons}} \quad H\bullet e \quad + \quad n\bullet \overset{\overset{\displaystyle H}{|}}{C} = N - OH \hspace{1.5cm} 7.297$$

The product of the rearrangement from (C) of Equation 7.296 can rearrange by Enamidization to give alkenes carrying NH(NHR) groups. Hence, it is believed that it is not only when CH_3 group is carried in R – N = N – R, that Electroradicalization is made to take place. Hence, acetaldehyde oxime can be obtained from (B) of Equation 7.295, if and only if (I) can be made to exist in Equilibrium state of existence. Formaldehyde oxime unlike formamide is a potential monomer. When activated, the followings are obtained.

$$H_2C = N - OH \quad \overset{\text{Activation}}{\rightleftharpoons} \quad e\bullet \overset{\overset{\displaystyle H}{|}}{\underset{\underset{\displaystyle H \;\; OH}{|\;\;|}}{C}} - N \bullet nn \quad ; \quad \overset{\oplus}{\underset{\underset{\displaystyle H \;\; H}{|\;\;|}}{C}} - \overset{\ominus}{\underset{\underset{\displaystyle \underset{\bullet\bullet}{O}:}{|}}{N}} \left.\begin{array}{c} \\ \\ \end{array}\right\} \begin{array}{l} \text{Electrostatic forces} \\ \\ \text{of Repulsion} \end{array} \hspace{1cm} 7.298$$

Worthy of note is that it cannot be activated chargedly due to electrostatic forces of repulsion. Therefore, it cannot be polymerized chargedly as is believed to be the case.[27] One can see why it cannot rearrange molecularly. Based on what the C = N centers are carrying, there are no transfer species of any kind. Therefore, it can be polymerized both electro-free-radically and nucleo-non-free-radically. Being a nucleophile, it will easily favor electro-free-radical polymerization, its natural route. Electro-free-radically, the followings are obtained using dilute HCl – the so called cationic initiator.

Stage 1:

$$Cl^{\ominus}..........^{\oplus}\overset{\overset{\displaystyle H}{|}}{\underset{\underset{\displaystyle H}{|}}{O}}-H \quad \underset{\text{Existence}}{\overset{\text{Equilibrium State of}}{\rightleftharpoons}} \quad H\bullet e \quad + \quad Cl^{\ominus}..........^{\oplus}\overset{\overset{\displaystyle H}{|}}{\underset{\underset{\displaystyle H}{|}}{O}}\bullet nn$$

$$(A)$$

$$H\bullet c \quad + \quad nn\bullet \overset{\overset{\displaystyle H}{|}}{\underset{\underset{\displaystyle HO}{|}}{N}}-\overset{\overset{\displaystyle}{}}{\underset{\underset{\displaystyle H}{|}}{C}}\bullet e \quad \rightleftharpoons \quad H-\overset{\overset{\displaystyle H}{|}}{\underset{\underset{\displaystyle HO}{|}}{N}}-\overset{\overset{\displaystyle}{}}{\underset{\underset{\displaystyle H}{|}}{C}}\bullet e$$

$$(B) \qquad\qquad\qquad (C)$$

$$(C) \quad + \quad (A) \quad \longrightarrow \quad Cl^{\ominus}..........^{\oplus}\overset{\overset{\displaystyle H}{|}}{\underset{\underset{\displaystyle H}{|}}{O}}-\overset{\overset{\displaystyle H}{|}}{\underset{\underset{\displaystyle H}{|}}{C}}-\overset{\overset{\displaystyle}{}}{\underset{\underset{\displaystyle OH}{|}}{N}}-H$$

$$(D) \qquad\qquad\qquad 7.299$$

Stages 2 to n:

$$(D) \quad \underset{\text{Existence}}{\overset{\text{Equilibrium State of}}{\rightleftharpoons}} \quad Cl^{\ominus}.........^{\oplus}\overset{\overset{\displaystyle H}{|}}{\underset{\underset{\displaystyle H}{|}}{O}}-\overset{\overset{\displaystyle H}{|}}{\underset{\underset{\displaystyle H}{|}}{C}}-\overset{\overset{\displaystyle}{}}{\underset{\underset{\displaystyle OH}{|}}{N}}\bullet nn \quad + \quad H\bullet e$$

$$(E)$$

$$(E) \quad + \quad n(B) \quad \rightleftharpoons \quad H\bullet e \quad nn\bullet \overset{\overset{\displaystyle H}{|}}{\underset{\underset{\displaystyle HO}{|}}{N}}-\overset{\overset{\displaystyle H}{|}}{\underset{\underset{\displaystyle H}{|}}{C}}-(\overset{\overset{\displaystyle}{}}{\underset{\underset{\displaystyle HO}{|}}{N}}-\overset{\overset{\displaystyle H}{|}}{\underset{\underset{\displaystyle H}{|}}{C}})_{n-1}-\overset{\overset{\displaystyle H}{|}}{\underset{\underset{\displaystyle H}{|}}{O}}^{\oplus}....^{\ominus}Cl$$

$$(F)$$

$$\longrightarrow \quad Cl^{\ominus}.........^{\oplus}\overset{\overset{\displaystyle H}{|}}{\underset{\underset{\displaystyle H}{|}}{O}}-(\overset{\overset{\displaystyle H}{|}}{\underset{\underset{\displaystyle H}{|}}{C}}-\overset{\overset{\displaystyle}{}}{\underset{\underset{\displaystyle OH}{|}}{N}})_{n}-\overset{\overset{\displaystyle H}{|}}{\underset{\underset{\displaystyle H}{|}}{C}}-\overset{\overset{\displaystyle}{}}{\underset{\underset{\displaystyle OH}{|}}{N}}-H$$

Poly(formaldehyde oxime)

$$7.300$$

Overall Equation: Dilute HCl + n(Formaldehyde oxime) $\longrightarrow$ Poly (formaldehyde oxime)

$$7.301$$

The first stage is the Initiation step while the Equilibrium stage-wise addition step is the Propa-gation step. Based on the RHS terminal end of the chain, one can observe why polymerization cannot take place above 60°C, the Ceiling Temperature of the monomer[27]. Below 60°C, the chain will continue to grow until the optimum chain length dictated by the Glass Transition tempe-rature of the polymer is reached. Whether reached or not the chain must be terminated by replacing the H on the N center with something like CH_3 group. The LHS of the initiator cannot be used, because the monomer cannot be activated chargedly. Unlike aldoximes and ketoximes, formaldehyde oxime can favor the nucleo-non-free-radical route because of absence of transfer species of the first kind of the first type carried by alkyl groups. All these monomers have Ceiling temperatures and unlike the first member can therefore undergo Enamidization as follows using acetaldoxime as an example.

Stage 1:

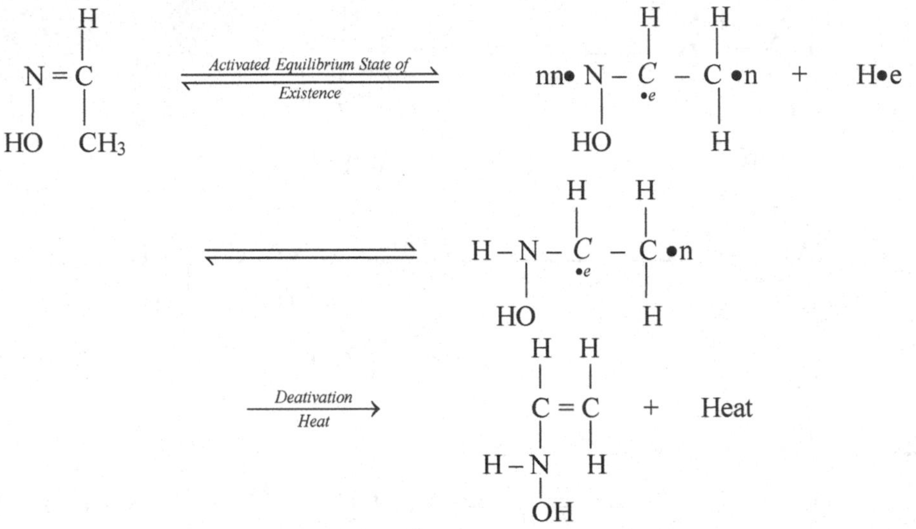

(A) Very unstable

7.302

(A) seems to be so unstable at normal conditions. Hence, it has never been known to exist. It molecularly rearranges back to the acetaldoxime as follows.

Stage 1:

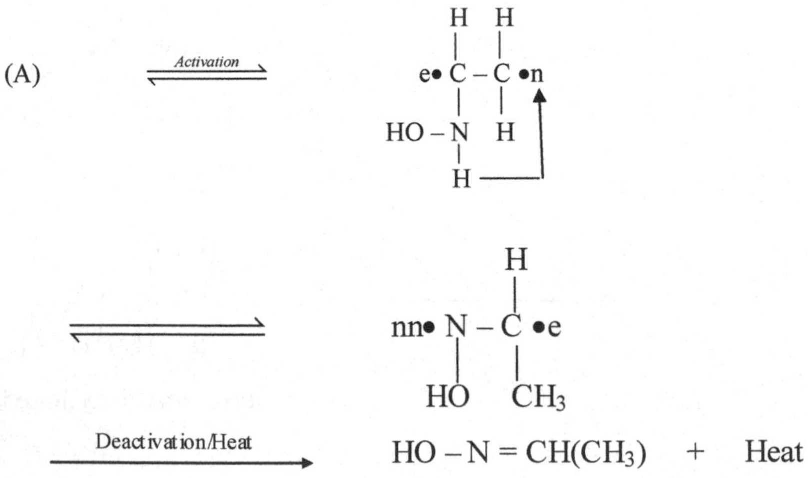

$$HO - N = CH(CH_3) \quad + \quad Heat$$

7.303

Thus, one can observe to some extent what goes on when the nitrogen center in N = O is carrying an alkanyl group.

In place of R group, consider H. The compound obtained is called ***Nitroxyl (HN=O).*** It is well known in the gas phase.[28] Like the other members, it is Polar/Non-ionic as shown below.

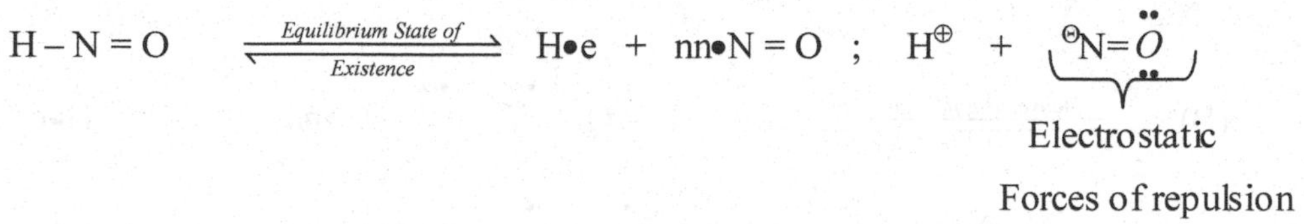

$$H-N=O \quad \underset{\text{Existence}}{\overset{\text{Equilibrium State of}}{\rightleftharpoons}} \quad H\bullet e \ + \ nn\bullet N=O \ ; \quad H^{\oplus} \ + \ {}^{\ominus}\ddot{N}=\ddot{O}$$

Electrostatic

Forces of repulsion

7.304

Since the real part of electrostatic charges do repel and attract, the followings shown below shows that it cannot still behave like an acid radically.

Stage 1:

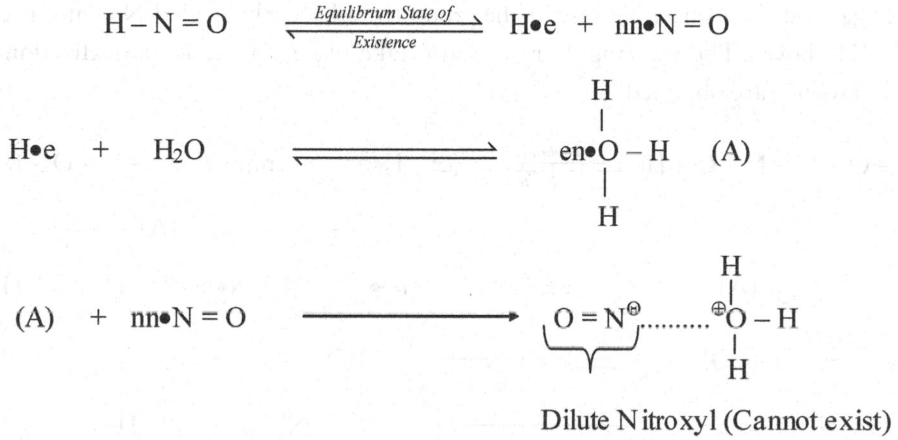

$$H-N=O \quad \underset{\text{Existence}}{\overset{\text{Equilibrium State of}}{\rightleftharpoons}} \quad H\bullet e \ + \ nn\bullet N=O$$

$$H\bullet e \ + \ H_2O \quad \rightleftharpoons \quad en\bullet \overset{\displaystyle H}{\underset{\displaystyle H}{\mid}}O-H \quad (A)$$

$$(A) \ + \ nn\bullet N=O \quad \longrightarrow \quad O=N^{\oplus}\cdots\cdots{}^{\oplus}\overset{\displaystyle H}{\underset{\displaystyle H}{\mid}}O-H$$

Dilute Nitroxyl (Cannot exist) 7.305a

Nitrosyl is very reactive. It readily dimerizes as shown below.

Stage 1: $\quad H-N=O \quad \rightleftharpoons \quad H\bullet e \ + \ nn\bullet N=O$

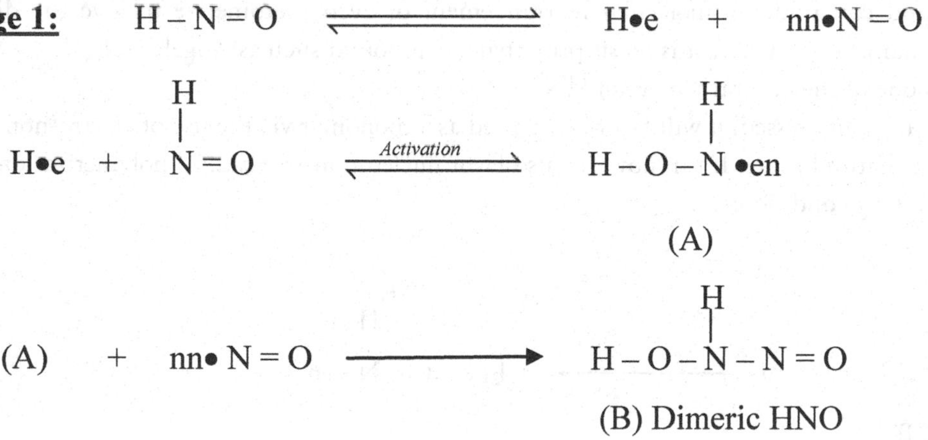

$$H\bullet e \ + \ \overset{\displaystyle H}{\underset{}{\mid}}N=O \quad \overset{\text{Activation}}{\rightleftharpoons} \quad H-O-\overset{\displaystyle H}{\underset{}{\mid}}N \bullet en$$

(A)

$$(A) \ + \ nn\bullet N=O \quad \longrightarrow \quad H-O-\overset{\displaystyle H}{\underset{}{\mid}}N-N=O$$

(B) Dimeric HNO 7.305b

This dimer is known to loose water to give N_2O. This it does as follows using the New Found-ations which is "The New Frontiers".

Stage 2:

$$(B) \quad \xrightleftharpoons[\text{of Existence}]{\text{Activated Equilibrium State}} \quad H-O-\overset{\bullet nn}{N}-\overset{\bullet en}{N}-O\bullet nn \quad + \quad H\bullet e$$

$$\xrightleftharpoons{\text{Enolization}} \quad H-O-\overset{\bullet nn}{N}-\overset{\bullet en}{N}-O-H$$

$$\xrightarrow{\text{Deactivation/Heat}} \quad H-O-N=N-O-H \quad + \quad \text{Heat}$$

$$\text{An Azo-compound} \qquad 7.305c$$

The dimer has undergone Enolization phenomenon to give an Azo hydroxyl compound. Note that the N = O center was again activated here because of the group the adjacently located N center is carrying (H in HNO and OH above). The rearrangement can also take place via Electroradicalization. When decomposed, the followings are obtained.

Stage 1:

$$H-O-N=N-O-H \quad \xrightleftharpoons[\text{Existence}]{\text{Equilibrium State of}} \quad H\bullet e \quad + \quad nn\bullet\, O-N=N-O-H$$

$$(A)$$

$$(A) \qquad \xrightleftharpoons{\text{Heat}} \quad nn\bullet\, O-N=N\bullet en \quad + \quad nn\bullet OH$$

$$H\bullet\, e \quad + \quad nn\bullet\, OH \quad \rightleftharpoons \quad H_2O$$

$$nn\bullet\, O-N=N\bullet en \quad \longrightarrow \quad O^{\ominus}-N^{\oplus}\equiv N \quad + \quad \text{Heat} \qquad 7.306c$$

Overall Equation: $HO\text{-}N=N\text{-}OH \quad \longrightarrow \quad H_2O \quad + \quad N_2O \quad + \quad \text{Heat} \qquad 7.306d$

The azo-compound can undergo molecular rearrangement or even tautomerize to give the dimer. Based on the mechanisms provided, it is no surprise that a compound such as Angeli's salt(Na – O – N = N – O – Na) is one of the general sources of HNO.

If nitroxyl is fully suppressed, it will favor being used as a monomer via the use of electro-non-free-radical initiator, its natural route. It can however also favor nucleo-non-free-radical polymerization, but under higher operating conditions.

Initiation Step:

$$Cl_2Fe\bullet en \quad + \quad H-N=O \quad \longrightarrow \quad Cl_2Fe-O-\overset{\overset{\textstyle H}{|}}{N}\bullet en \qquad 7.307a$$

Propagation Step:

$$Cl_2Fe-O-\overset{\overset{\textstyle H}{|}}{N}\bullet en \quad + \quad n(H-N=O) \quad \longrightarrow \quad Cl_2Fe-(O-\overset{\overset{\textstyle H}{|}}{N})_n-O-\overset{\overset{\textstyle H}{|}}{N}\bullet en \qquad 7.307b$$

In the absence of transfer species for the growing chain, it cannot be killed from within. Termi-nating agents must be used to kill the chain. Notice what has been used as initiator above. This is what iron does to O_2 of the air under heat to give a polymeric soft looking product which we seem to ignore (Corrosion). Nucleo-non-free-radically, there is transfer species of the second kind, since a nucleo-free radical can be released from a nucleo-non-free-radical growing chain as long as a stable molecule is produced. The release is however faster when the H is changed to a halogen atom such as F. F and H will be released as transfer species of the second kind.

$$\overset{\displaystyle F}{\overset{\displaystyle |}{\ddot{N}}} - (N-O)_n - \overset{\displaystyle |}{\underset{\displaystyle F}{N}} - N - O \bullet nn \longrightarrow \overset{\displaystyle F}{\overset{\displaystyle |}{\ddot{N}}} - (N-O)_n - N = O \quad + \quad F \bullet nn$$

$$\text{7.308a}$$

With OR types of groups, whether molecular rearrangement takes place or not, the same compound remains, noting that based on the group carried, it can be activated. It is Polar/Non-ionic even when R an alkyl group is H (Polar/Ionic).

$$H - O - N = O \quad \underset{\text{Existence}}{\overset{\text{Equilibrium State of}}{\rightleftharpoons}} \quad H \bullet e \quad + \quad nn \bullet O - N = O \quad ; \quad H^{\oplus} + \quad {}^{\ominus}\ddot{O} - \ddot{N} = N$$

Electrostatic forces
of repulsion

$$\text{7.308b}$$

It dimerizes as shown below.

Stage 1: $H - O - N = O \quad \rightleftharpoons \quad H \bullet e \quad + \quad nn \bullet O - N = O$

$H \bullet e \quad + \quad H - O - N = O \quad \rightleftharpoons \quad H - O - N^{\bullet en} - O - H$

(A)

$O = N - O \bullet nn \quad + \quad (A) \quad \longrightarrow \quad \begin{array}{l} H - O - N - O - H \\ \qquad\quad | \\ \qquad\quad O - N = O \quad (B) \end{array}$

$$\text{7.309}$$

which at 0^0C, breaks down to blue Dinitrogen trioxide as shown below.

Stage 1:   (B) $\rightleftharpoons$ $H \bullet e \quad + \quad \begin{array}{l} nn \bullet O - N - O - H \\ \qquad\qquad | \\ \qquad\qquad O - N = O \end{array}$

(C)

(C) $\rightleftharpoons$ $HO \bullet nn \quad + \quad \begin{array}{l} en \bullet N - O \bullet nn \\ \qquad\quad | \\ \qquad\quad O - N = O \quad (D) \end{array}$

$H \bullet e \quad + \quad nn \bullet OH \quad \rightleftharpoons \quad H_2O$

(D) $\longrightarrow$ $O = N - O - N = O \quad + \quad$ Heat

Dintrogen trioxide (Blue)

$$\text{7.310a}$$

Overall Equation: $2HON = O \longrightarrow H_2O \quad + \quad O=NON=O \quad + \quad$ Heat   7.310b

Note from above that the following is valid

$$-OH \ < \ -O-N=O \quad ; \quad -H \ > \ -N=O$$

At room temperature, it exists in Equilibrium state of existence being unstable and the color becomes brown as shown below.

Stage 1: $O=N-O-N=O \rightleftharpoons O=N\bullet nn + en\bullet O-N=O$

                                                   (Brown)

$$O=N-O\bullet en \rightleftharpoons O=N\bullet en + nn\bullet O\bullet en + \text{Heat}$$

$$O=N\bullet en + nn\bullet N=O \rightleftharpoons O=N-N=O$$

$$nn\bullet O\bullet en + O=N-N=O \longrightarrow O=\overset{\oplus}{\underset{|}{N}}-O^{\theta} + \text{Heat}$$

                                                    $N=O$

                                                  (A)   7.311a

Note that (A) was formed instantaneously, since the N center cannot carry more than eigth radicals in the last shell.

Stage 2: $O=N-O-N=O \rightleftharpoons O=N\bullet nn + en\bullet O-N=O$

                                                   (Brown)

$$O=N-O\bullet en \rightleftharpoons O=N\bullet en + nn\bullet O\bullet en + \text{Heat}$$

$$O=N\bullet en + nn\bullet N=O \rightleftharpoons O=N-N=O$$

$$nn\bullet O\bullet en + (A) \longrightarrow O=\overset{\oplus}{N}-{}^{\theta}O + \text{Heat}$$

                                         $O=\overset{|}{\underset{}{N}}{}^{\oplus}-{}^{\theta}O$ (B)

                               Dinitrogen tetroxide (Colourless)   7.311b

Overall Equation: $2O=N-O-N=O \longrightarrow O=N-N=O + O_2N-NO_2 + \text{Heat}$

                                                               7.311c

The chemistry on nitrous acid in textbooks are worthy of note filled with great information all with blind interpretations. That is how all data from all disciplines have blind interpretations. One can observe the unique families of N/O compounds and indeed how Nature operates. It is no surprise why most or all chemical reactions take place radically, because Nature amongst others abhors a vacuum. In the second stage, instead of the oxidizing oxygen oxidizing O=N-N=O as it did in the first stage, it completed the oxidation of (A) being more stable. As has been shown and will be fully shown downstream, alkyl groups are a collection of alkyl-ane [alkanyl] groups (e.g. $e\bullet CH_3$), alkyl-ene [alkenyl] groups (e.g. $e\bullet CH = CH_2$), and alkyl-yne [alkynyl] groups (e.g. $e\bullet C \equiv CH$). Whether $e\bullet CH_3$ or $n\bullet CH_3$, because of our level of ignorance which is too great to comprehend, we today call them alkyl groups. Because of the STIFF NECKEDNESS of humanity, one has been using alkyl groups, electrons, ethylene, and so many too countless to mention, to carry humanity along and begin the CHANGES in steps. One hopes the words like Nitrosonium ion, Nitrous acidinium ion, Nitrite ion should never be used, because N=O can never carry ionic charges and can never be activated chargedly.

Now, with this type of groups (OR), there is transfer species of the first kind of the first type, carried by the OR group. When nucleo-non-free-radical initiators not natural to the route of the monomer are used, the followings are to be expected.

$$\ddot{N}\cdot nn \quad + \quad en\cdot \overset{\displaystyle |}{\underset{\displaystyle \overset{|}{O} \;\; CH_3}{N}} - \ddot{O}\cdot nn \quad \longrightarrow \quad NCH_3 \quad + \quad O = N - O\cdot nn$$

$$\text{7.312}$$

Nucleo-non-free-radically, the monomers cannot be polymerized. Electro-non-free-radically, the route natural to the monomer, the followings are obtained.

$$Fe - (O - \underset{\displaystyle OCH_3}{N})_n - O - \underset{\displaystyle OCH_3}{N} \bullet en \quad \longrightarrow \quad Fe - (O - N)_n - O - \underset{\displaystyle OCH_3}{N} = O \quad + \quad e\bullet CH_3$$

$$\text{7.313}$$

Thus, one can observe how an electro-free-radical can been obtained from an electro-non-free-radical, just as a nucleo-free-radical was obtained from a nucleo-non-free-radical (e.g. Benzyol peroxide or peroxides decompositions).

Using C_2F_5 and higher types of groups which are less radical-pushing than H, their monomers cannot be activated, because of the C center adjacently located. Only when the group is CF_3 will activation take place, because of absence of transfer species of the first kind. Indeed, whether the group is more radical-pushing or less radical-pushing than H, activation can never be favored. The same as for CF_3 applies to radical-pulling groups such as COOR, $CONR_2$ and so on, since no transfer species exists. On the other hand, the C=O center being more nucleophilic than the N= O center, can never be used for activation, the monomer being Nucleophilic. For COR groups, no activation may be possible.

7.5.4. Quinones

There are only two isomers of cyclohexadienedione. These are 1,4-benzoquinone or para-benzoquinone or 1,4-dione or Quinone and 1,2-benzoquinone or ortho-benzoquinone or 1,2-dione

The meta- form cannot exist, because a structure with two double bonds in the ring cannot be obtained for it as shown below.

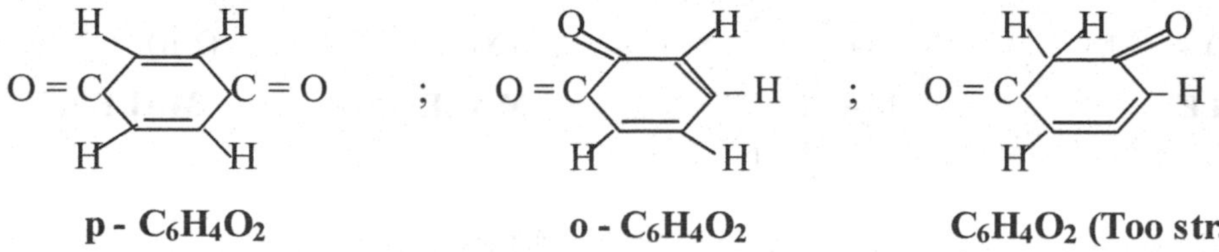

p - $C_6H_4O_2$;   **o - $C_6H_4O_2$**   ;   **$C_6H_4O_2$ (Too strained)**   7.314

In the meta- position, $C_6H_4O_2$ cannot be obtained, because it is too strained to exist, since two double bonds are cumulatively placed inside the ring carrying two C = Os. Hence only the para- and ortho- are the only ones that can be obtained from the oxidation of para-dihydroxybenzene and ortho-dihydroxybenzene respectively. The meta-dihydroxybenzene cannot be oxidized to give meta-benzo-quinone, because it does not exist. For the first time, one can observe that some compounds exist where only para- and their ortho-positions exist, but not the meta-position. This is clear indication of the similarity of both para- and ortho-positions and their difference from meta-position. All these will be shown downstream as we move along when the mechanisms of Oxidation and Combustion are provided *(Appendix VI)*. For the first time, one will not only see how oxidizing agents operate, but also the structure of oxidizing oxygen molecule just as we have for molecular oxygen.

When the para- and ortho- isomers are activated, C = C being less nucleophilic than C = O will be the one to be first activated, noting that these quinones are MALES unlike what we have in some cyclic and non-cyclic compounds as shown below.

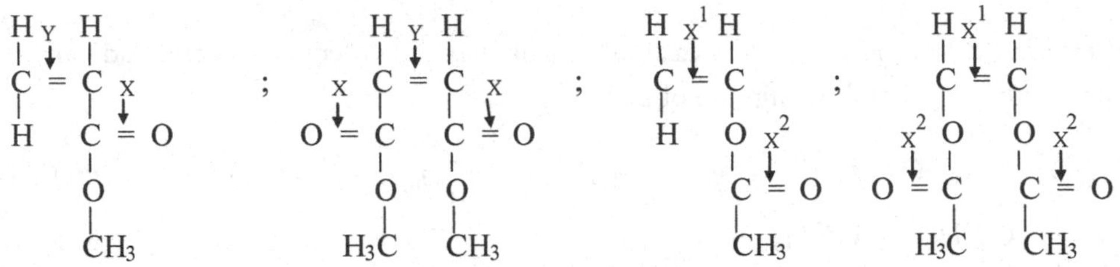

(I)   (II)   (III)   (IV)

MALE   **MALE**   **FEMALE**   **FEMALE**

$$(x^2 > x^1) \qquad\qquad 7.315$$

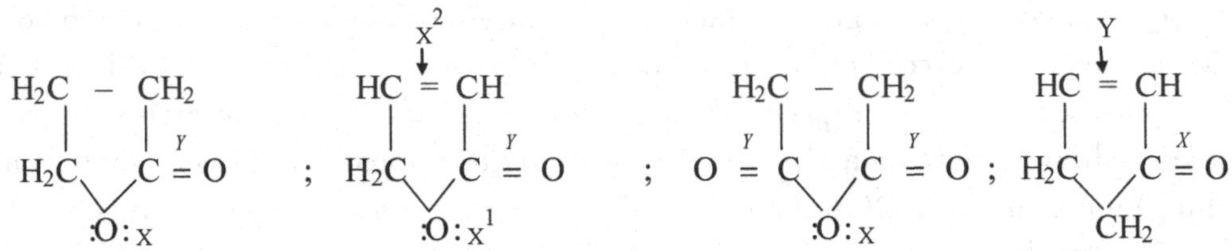

(V)   (VI)   (VII)   (VIII)

MALE   **MALE**   **MALE**   **MALE**

$$(x^2 > x^1)$$

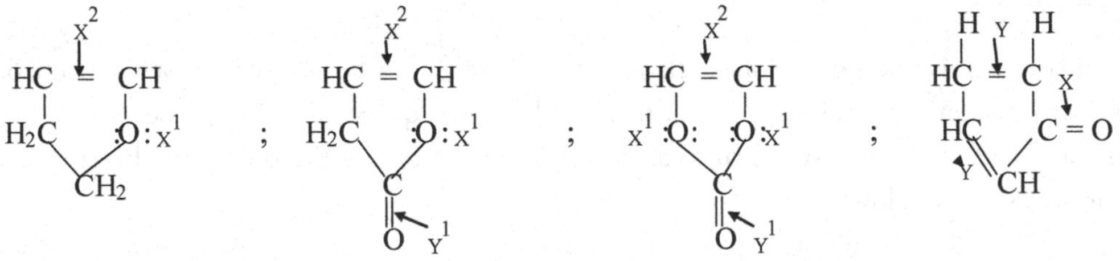

(IX)   (X)   (XI)   (XII)

FEMALE   **MALE**   **MALE**   **MALE**

$$(x^2 > x^1) \qquad\qquad 7.316$$

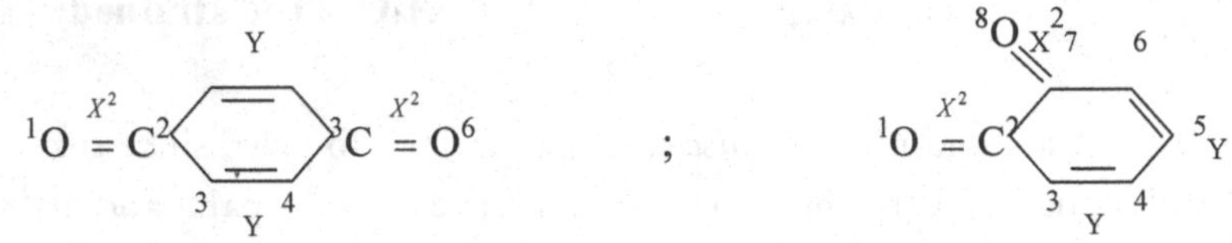

(A)   (B)

MALE   **MALE**   7.317

Just as we have here for Oxygen, so also we have when the O is replaced with S and NH or NR. None of them but (II) in Equations 7.315 is resonance stabilized. Another one that is also resonance stabilized in the groups of Equation 7.316, not shown is maleic anhydride. With the same types of groups carried by them and manners of arrangements, there seems to be more males in ringed compounds than in non-ringed compounds, because of the additional presence of functional centers in hetero-ringed compounds. What indeed makes the C = O, C = S, C = NH centers become electrophilic, that is, Male or Y, is the presence of functional centers adjacently located to them. Via the functional centers, they are electrophilic. Unlike in (VIII) and (XII) of Equation 7.316, (A) and (B) above are resonance stabilized as shown below using (A), the quinone our main focus.

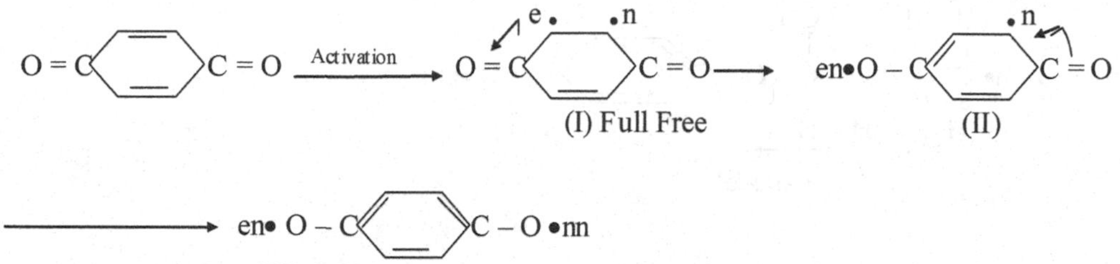

(I) Full Free   (II)

(III) **COMPLETE ACTIVATION –Full Non-Free**   7.318

The resonance stabilization here is of the first kind and of the first and second types. It begins with first type in (I) and ends with the second type in (II) to give the complete 1,6- mono-form. There are two mono-forms here– 3,4- and 1,6- mono-forms for the p-benzoquinone. For the ortho-benzoquinone, there are also two mono-forms- 3,4- and 1,8-mono-forms. One can begin to see the marked difference between the para- and ortho- benzoquinones. Chargedly, only the Y centers can be activated one at a time- the 3,4- center, being less nucleophilic. The two Y centers here are of the same nucleophilicity. If a radical pushing group is placed on one of them, then only one center becomes the least nucleophilic.

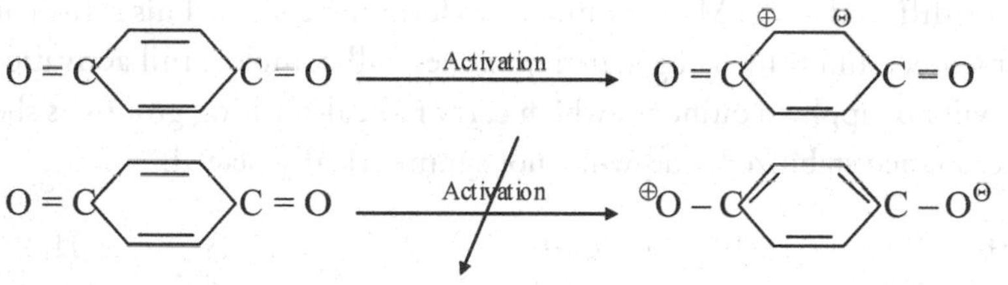

                                                                            7.319

Worthy of note is that O is carrying a positive charge in the presence of C and H which are more electropositive elements. The reasons are obvious. Radically, the electro-free-radical did not move from full- free monomer to half-free monomer, or from half-free to full-non-free monomer, but *from full-free to full-non-free monomer*. Based on the nature of (II) of Equation 7.318, it cannot be used as an oxidizing agent unless the Hs are changed to radical-pulling groups such as halogens. Unlike CO_2 where one single C atom is carrying two O atoms, here two carbon centers are involved. In view of the fact that the two carbon centers are inside an aromatic ring, this compound can be called *Aromatic Carbon dioxide.* In place of CO_2 released during combustion of aliphatic hydrocarbons, it is the quinones and carbon dioxide that are released during combustion of *"Alipharomatic"* hydrocarbons. Take note of the use of alipharomatic instead of aromatic as will become apparent. For example, while benzene is aromatic, toluene is alipharomatic (Methyl benzene or Phenyl methane). The monomer in its activated state is like O_2, but with a great difference. As an electrophile or male, the natural route is nucleo-non-free- radical route. It is however a weak electrophile, because it can also be polymerized electro-non-free-radically. It has no transfer species. All the groups are internally located and are shielded.

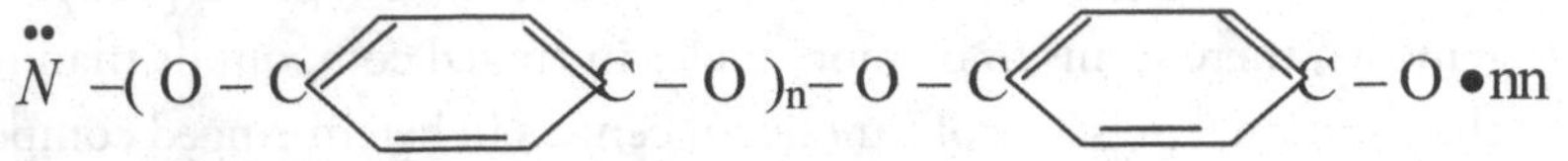

$$\ddot{N} - (O - C \langle \underline{\hspace{1cm}} \rangle C - O)_n - O - C \langle \underline{\hspace{1cm}} \rangle C - O \bullet nn \qquad 7.320$$

Because of the presence of $- O - O -$ bond between benzene rings along the living chain, it is better to use the monomer as a comonomer. This will be fully explained downstream in the Series and Volumes. However, consider very briefly its copolymerization with 1,3-pentadiene.

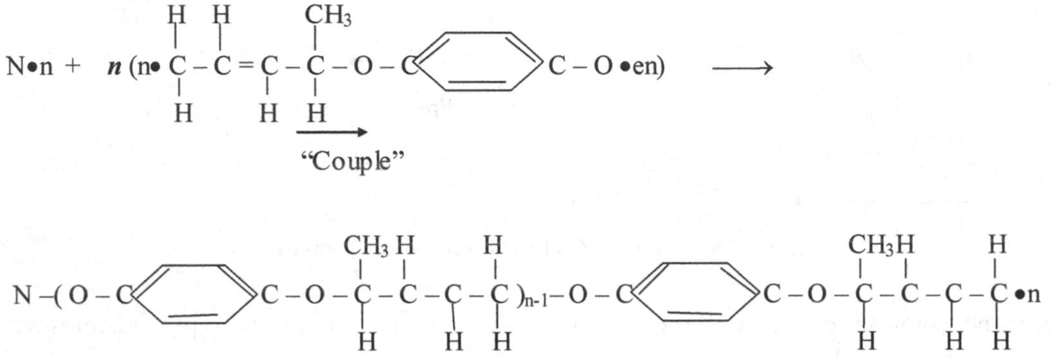

$$N\bullet n \; + \; n \; (n\bullet \underset{H}{\overset{H}{C}} - \underset{}{\overset{H}{C}} = C - \underset{H}{\overset{CH_3}{C}} - O - C \langle \underline{\hspace{1cm}} \rangle C - O \bullet en) \;\; \longrightarrow$$

$$\xrightarrow{\text{``Couple''}}$$

$$N - (O - C \langle \underline{\hspace{1cm}} \rangle C - O - \underset{H}{\overset{CH_3}{C}} - \underset{H}{\overset{H}{C}} - \underset{H}{\overset{}{C}} - \underset{}{\overset{H}{C}})_{n-1} - O - C \langle \underline{\hspace{1cm}} \rangle C - O - \underset{H}{\overset{CH_3}{C}} - \underset{H}{\overset{H}{C}} - \underset{H}{\overset{}{C}} - \underset{}{\overset{H}{C}} \bullet n$$

$$\qquad\qquad 7.321$$

Nucleo-free-radically, quinone cannot be polymerized. Unlike 1,3-butadiene where harsh operating conditions are required, the natural route being electro-free-radical route, pentadiene cannot be polymerized nucleo-free-radically. The natural route of Quinone is nucleo-non-free-radical route. Hence full alternating placement will be obtained, both monomers being unreactive to the route. Note that it was the Female (pentadiene) that diffused to the Male (Quinone) to form the couple. This is the natural order. While Quinone the first member and fully halogenated quinones will complete full activation shown in Equation 7.318, the same will not apply to quinones which carry radical-pushing groups as shown below, since the groups are not resonance stabilized, and when not symmetrically placed.

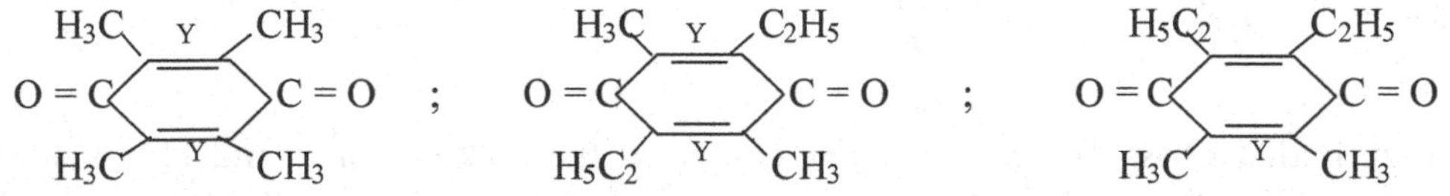

$$\begin{array}{ccccc}
H_3C \;\; Y \;\; CH_3 & & H_3C \;\; Y \;\; C_2H_5 & & H_5C_2 \;\; C_2H_5 \\
O = C \langle \;\; \rangle C = O & ; & O = C \langle \;\; \rangle C = O & ; & O = C \langle \;\; \rangle C = O \\
H_3C \;\; Y \;\; CH_3 & & H_5C_2 \;\; Y \;\; CH_3 & & H_3C \;\; Y \;\; CH_3 \\
(I)^* & & (II)^* & & (III)
\end{array} \qquad 7.323$$

Tetramethyl-quinone   2,5-diethyl-3,6-dimethyl-quinone   2,3-dimethyl-5,6-diethyl-quinone

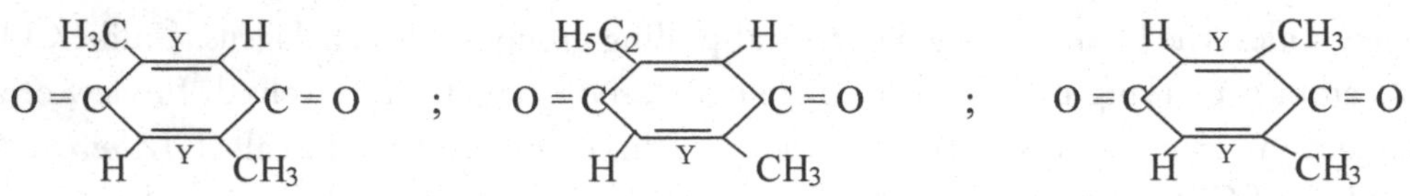

$$\begin{array}{ccccc}
H_3C \;\; Y \;\; H & & H_5C_2 \;\; H & & H \;\; Y \;\; CH_3 \\
O = C \langle \;\; \rangle C = O & ; & O = C \langle \;\; \rangle C = O & ; & O = C \langle \;\; \rangle C = O \\
H \;\; Y \;\; CH_3 & & H \;\; Y \;\; CH_3 & & H \;\; Y \;\; CH_3 \\
(IV)^* & & (V) & & (VI)^*
\end{array} \qquad 7.324$$

2,5-dimethyl-quinone   2-ethyl-5-methyl-quinone   2,6-dimethyl-quinone

(VII)
2-methyl-3-ethyl-quinone

(VIII)
2-methyl-6-ethyl-quinone

(IX)   7.325
2-methyl-quinone

(X)*
Tetrachloro-quinone

(XI)*
2,5-dichloro-quinone

(XII)   7.326
2,3,5-trichloro-quinone

Note that the word Quinone is indeed p-benzoquinone as already indicated, because of the symmetric form of the structure more so than o-benzoquinone. As an exercise, what has been done above for p-benzoquinone should be done for o-benzoquinone, because now we are using "the eye of the needle" and not just the physical eyes. All along, notice that it has not just been THE REAL/IMAGINARY SCIENCE, but also THE PHILOSOPHY OF HUMANITY. That is why the trios- Mathematics, Physics and Chemistry are called THE NATURAL SCIENCES and no other discipline. They are not Physical Sciences which also exist. They are not Applied Sciences which also exist. They are not Life Sciences which also exist. They are not Social Sciences which also exist. They are not ARTS OR AN ART which also exists. They are everything, for without them nothing exists including RELIGION.

In the last four equations above, the center which can be activated have been shown applying the foundations which have been laid so far. For we already know that ethene is far more nucleophilic than tetrachloroethene and 2-butene is more nucleophilic than propene which is more nucleophilic than ethene. We already know that the less nucleophilic center is first act-ivated before the more nucleophilic one. The same applies if they were electrophiles. Where two centers exist whether of equal capacity or not, conjugatedly, cumulatively or isolatedly placed or not, only one center can be activated one at a time. Never can two centers be activated at the same time when conjugatedly or cumulatively placed whether symmetrically placed or not. Only when they are isolatedly placed can two centers be activated at the same time if and only if they are of the same capacity. For those marked with single asterisk any of the Y center can be activated, but not all can be resonance stabilized. Those that are not marked have only one center which can be activated and not all can be resonance stabilized. The issue of whether the compound exists or not is not the focus, but something else, that which Nature is. It is no surprise therefore why a Quinone compound such as 2, 5-dimethyl-p-benzo-quinone, (IV) of Equation 7.324, cannot copolymerize with vinyl monomers[29]. It cannot form the couple shown in Equation 7.321; **because it cannot be fully activated since the groups carried are not symmetrically placed.** Thus, when activated using same initiator as used for copolymerization, the followings are obtained.

514

$$\text{NH} \quad + \quad \text{O}=\text{C} \overset{\overset{\text{H}-\text{C}-\text{H}}{\|}}{\underset{\underset{\text{CH}_3}{}}{\diamond}} \text{C}=\text{O}$$

7.327

Thus, it cannot favor nucleo-free-radical polymerization the route natural to the center, because of transfer species of the first kind of the first type. It will not favor the electro-free-radical route, because of the presence of C=O center. All the above took place because it is not resonance stabilized, since the two activation centers are not symmetrically placed like in (VI). Quinones can be observed to be very unique group of monomers.

7.6 Proposition of Rules of Chemistry and Concluding Remarks

From all the considerations so far, one has begun to reach the BREAKING POINT in the introduction of the NEW CHEMISTRY. Without any doubt, this chapter is the most important chapter in this VOLUME for which so many rules will unquestionably be stated and proposed. These are no rules, but indeed the Laws of CREATION, in which the laws of NATURE have explicitly been shown so far. One can observe the importance of activation centers in molecular compounds. New types which will be fully applied to identify others later in the Series were introduced here with ringed compounds. These are called functional centers and invisible π-bonds, vacant orbitals, paired unbonded radicals. The types which we have been very familiar with, are of the non-ringed types such as $C \equiv C$, $N \equiv N$, $C \equiv N$, $C=C$, $C = O$, $C = N$, $N = O$, $N = N$, $S = O$, $O = O$, etc. types and the ringed ones. How all of them are used to provide MALE and FEMALE characters of monomer have begun to be fully shown, for which most MALE CHARACTERS are from RINGED compounds.

For the first time, new types of molecular rearrangement phenomena unique to diazo-alkanes and ringed compounds have begun to be identified. How some rings are opened has begun to be shown. For the first time, the mechanisms of Electroradicalization, Enolization and newly identified one called Enamidization have been provided.

For the first time also, new and real definition for "ceiling temperature" of some families of monomers has been provided. In the process, for the first time, a new type of Addition polymeri-zation system has been identified. How depropagation takes place during propagation with these families of monomers have begun to be explained. It is because of their occurrences that "ceiling temperatures" arose, a temperature below which depropagation stops. The new Addition poly-merization system unlike the others so far encountered takes place via Equilibrium mechanisms with no release of small by-molecular products as we have in Step polymerization systems.

It has never been known that diazoalkanes could favor radical existence, because what radicals are were unknown. For the first time, how some diazoalkanes can be made to favor electro-free-radical, nucleo-free-radical, and electro-non-free-radical polymerizations have begun to be provided. The polymeric products largely obtained are poly (diazoalkanes) and mostly polyalkylenes. Chargedly, diazoalkylenes have been confirmed and identified to favor the positive routes. In view of the living character of the growing polymer chains with transfer species of the first kind, new names have been identified for the polymeric products. For example polyethylene is not obtained from ethene, but from diazomethylene, since it is the methylene group that is indeed involved. Polyethene is that obtained from ethene. Poly-propylene is that obtained from diazopropylene, while polypropene is that obtained from propene and so on. Why this is so will become clear in the next chapter.

Rule 264: This rule of Chemistry for **All things,** states that, when chemical, polymeric or other types of reactions take place outside their Nuclei, they do so *only along the boundaries of atoms/elements or of compounds or systems all of which are conglomeration and compositions of different atoms/elements;* for which the domains only serve to determine the size, the capacity of the compound or system and the physical state of the compound or system.
(Laws of Creations for Boundaries-the eighth law of Nature)

Rule 265: This rule of Chemistry for **All things**, states that, though polymeric reactions can take place chargedly and radically [See Rule 198] and chemical reactions take place only radically [See Rule 197], since charges of any type have their origins from radicals, all reactions in all Systems along the boundary whether Chemical, Polymeric, Electrical, Communications, Industrial, Governmental, etc., or not, take place only **RADICALLY.**
(Laws of Creations for Radicality- The ninth Law of Nature)

Rule 266: This rule of Chemistry for **Formation of Couples**, states that, before a couple can be formed between two or more monomers, they must first and foremost be in Activated state of existence.
(Laws of Creations for formation of Couples)

Rule 267: This rule of Chemistry for **Charges**, states that, while *Electrostatic charges (Complex) have males and females, the imaginary sides are not affected or influenced by electrostatic forces of repulsion and attraction and their bonds can never be activated.*
(Laws of Metaphysics)

Rule 268: This rule of Chemistry for **Addition monomers,** states that, monomers called alky-lenes in present-day Science are not alkylenes, but ALKENES; of which the first member of the family is ETHENE ($H_2C = CH_2$).
(Laws of Creations for ALKENE family renaming)

Rule 269: This rule of Chemistry for **Addition monomers,** states that, there are two kinds of ALKYLENES, the first being ALKYLENES (Living radical carrying species) and the second being ALKYL-ENES, alkyl groups from ALKENES; for which the latter are called *ALKENYL since Alkylene family exist as a member of the Hydrocarbon Family Tree.*
f Creations for Alkenyl/Alkylene family renaming)

Rule 270: This rule of Chemistry for **Addition monomers,** states that, for all ALKYL groups, there are indeed two types-*Alkyl and Alkylide.*

$$-CH = CH_2 \qquad ; \qquad \begin{matrix} H & H \\ | & | \\ C & = C \\ | & | \\ H & H \end{matrix} \qquad ; \qquad \begin{matrix} H \\ | \\ e\bullet\, C\, \bullet n \\ | \\ CH_3 \end{matrix}$$

e•CH = CH$_2$, n•CH = CH$_2$

Ethenyl Ethenylide

ALKYL-ENE (or Alkenyl) **ALKENE** **ALKYLENE**

(Ethenyl group) (Ethene) (Ethylene)

(Laws of Creations for Alkenyl/Alkylene family renaming)

Rule 271: This rule of Chemistry for **Addition monomers,** states that, though the present-day Science call the structure shown below as Diazo-Alkanes, the real name based on the structures which are truly in place, is DIAZO-ALKYLENE,

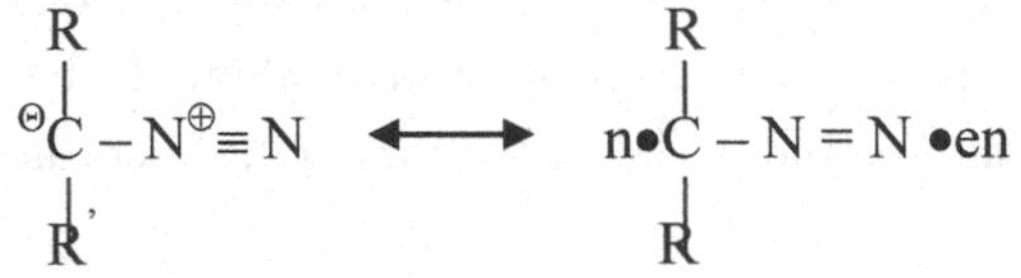

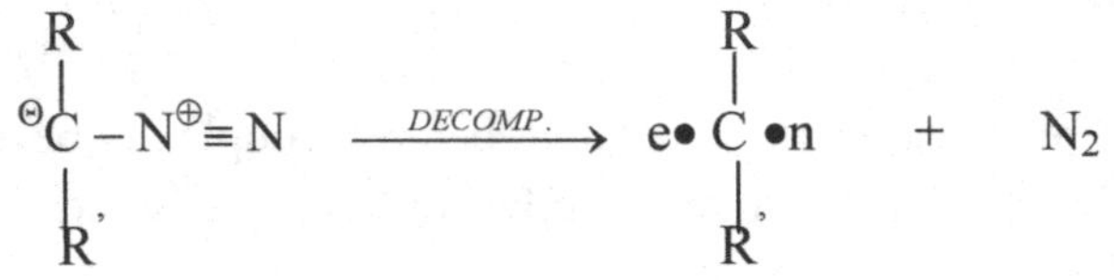

$$\overset{R}{\underset{R'}{\overset{|}{\underset{|}{C}}}} {}^{\ominus} - N^{\oplus} \equiv N \quad \longleftrightarrow \quad n\bullet\overset{R}{\underset{R}{\overset{|}{\underset{|}{C}}}} - N = N \bullet en$$

$$\overset{R}{\underset{R'}{\overset{|}{\underset{|}{C}}}} {}^{\ominus} - N^{\oplus} \equiv N \quad \xrightarrow{DECOMP.} \quad e\bullet \overset{R}{\underset{R'}{\overset{|}{\underset{|}{C}}}} \bullet n \quad + \quad N_2$$

DIAZO-ALKYLENE ALKYLENE

where R and R´ are some specific radical pulling or pushing groups if the polar state must exist.
(Laws of Creations for renaming of Diazo-alkanes)

Rule 272: This rule of Chemistry for **All Systems,** states that, without knowing the Structure of any Compound or System, the mechanisms of the behavior of the System cannot be explained, for in it lies **the identity, the real and imaginary characters** of the compound or system.
(Laws of Creations for Structures of Compounds or Systems).

Rule 273: This rule of Chemistry for **Resonance Stabilization phenomena,** states that, the *second kind* of Resonance Stabilization is *Radical/Polar Resonance stabilizations;* for which the first two types are *Radical/Polar and Polar/Radical/Polar;* noting that *Polar resonance stabilization does exist* since the origin of all the Polar forms is the Radical state as clearly indicated with examples from the following compounds-

i)   **Diazo-Alkylene compounds**

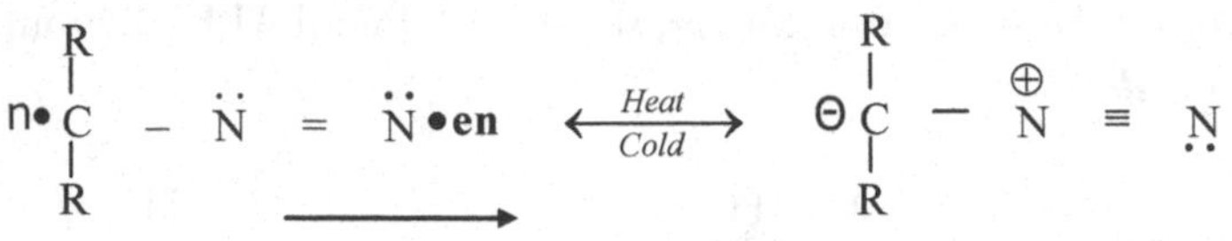

$$n\bullet \overset{R}{\underset{R}{\overset{|}{\underset{|}{C}}}} - \overset{..}{N} = \overset{..}{N}\bullet en \quad \xleftrightarrow{\substack{Heat \\ Cold}} \quad {}^{\ominus}\overset{R}{\underset{R}{\overset{|}{\underset{|}{C}}}} - \overset{\oplus}{N} \equiv \underset{..}{N}$$

Radical/Polar resonance stabilization of 1st type

$$\underset{R}{\overset{R}{|}}C = \overset{\oplus}{N} = \overset{\ominus}{\underset{..}{N}:} \longrightarrow e \cdot \underset{R}{\overset{R}{|}}C - \overset{..}{N} = \underset{..}{N} \cdot nn$$

NON – EXISTENT   **NON-EXPLOSIVE STATE**

ii) Monosulfuroxy-Alkylene compounds

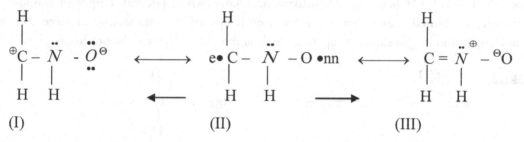

(I) Sulfylone   (II)   (III) Sulfylene

Neutral environment   Acidic environment

Polar/Radical/Polar resonance stabilization of type (II)

iii) Nitroxyl methylene

$$\overset{H}{\underset{H}{|}}\overset{\oplus}{C} - \overset{..}{N} - \overset{..}{\underset{..}{O}}{}^{\ominus} \longleftrightarrow e\bullet \overset{H}{\underset{H}{|}}C - \overset{..}{N} - O \bullet nn \longleftrightarrow \overset{H}{\underset{H}{|}}C = \overset{\oplus}{N} - {}^{\ominus}O$$

(I)   (II)   (III)

Neutral environment   Acidic environment

Polar/Radical/Polar resonance stabilization of type (II)

(Laws of Creations for Polar/Radical Resonance Stabilization Phenomena)

<u>**Rule 274:**</u> This rule of Chemistry for **Radical/Polar Resonance Stabilized Addition mono-mers,** states that, their stable state of existence is the Polar form which when heated or in the presence of a passive catalyst gets activated to the radical form for which when they do, they can add to form cyclic rings as part of products in the absence of other components in their vicinity, as shown below-

For Monosulfuroxy-methylene

$$e\bullet \overset{H}{\underset{H}{|}}C - \overset{..}{\underset{..}{S}} - O \bullet nn \quad \longleftrightarrow \quad$$

(A) Sulfuroxyrine

For Diazo-methylene

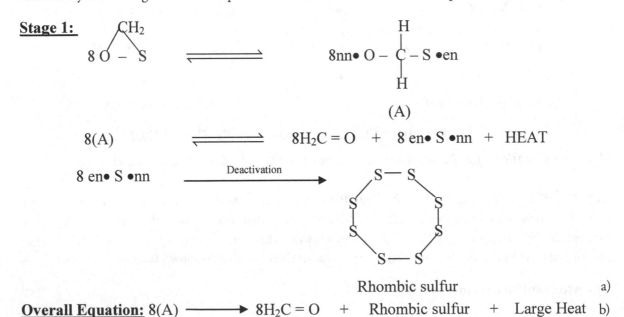

(B) Diazirine

For Nitroxyl methylene

(C) Oxaazacyclopropane

for which the formation of such rings is stronger when the H atoms are replaced with radical-pushing groups for only these members of families of Nucleophiles which can only be opened instantaneously without the use of functional centers or π-bond type of activation center.
(Laws of Creations for Radical/Polar Resonance Stabilization Compounds)

Rule 275: This rule of Chemistry for **Addition monomers where tricyclic ring from mono-sulfuroxy-methylene is formed,** states that, when the ring is opened instantaneously in large concentrations, formaldehyde and ringed sulfur compounds such as rhombic sulfur are the products-

Stage 1:

(A)

$8(A) \rightleftharpoons 8H_2C = O + 8\text{ en}\bullet S \bullet nn + HEAT$

$8 \text{ en}\bullet S \bullet nn \xrightarrow{\text{Deactivation}}$

Rhombic sulfur   a)

Overall Equation: $8(A) \longrightarrow 8H_2C = O + $ Rhombic sulfur $ + $ Large Heat   b)

for which just as exists *oxidizing oxygen elements* (en$\bullet$ O $\bullet$nn), so also there exists *sulfurizing sulfur elements* (en$\bullet$ S $\bullet$nn) which can readily be obtained from rhombic sulfur and other smaller sized rings of sulfur or from other compounds such as above herein called *sulfurizing agents (just like oxidizing agents).*
(Laws of Creations for Mono-Sulfuroxy methylene)

Rule 276: This rule of Chemistry for **Addition monomers where tricyclic ring from diazo-alkylenes is formed,** states that, when the ring is opened instantaneously, N_2 is released with formation of an alkylene.

Stage1:

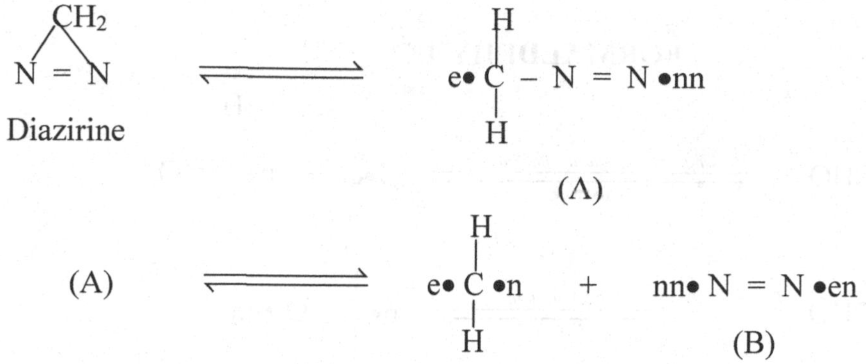

$$\underset{\text{Diazirine}}{\overset{\displaystyle CH_2}{\underset{\displaystyle N = N}{\diagdown\diagup}}} \quad\rightleftharpoons\quad e\bullet \underset{\displaystyle H}{\overset{\displaystyle H}{C}} - N = N \bullet nn$$

$$(A)$$

$$(A) \quad\rightleftharpoons\quad e\bullet \underset{\displaystyle H}{\overset{\displaystyle H}{C}} \bullet n \quad + \quad nn\bullet N = N \bullet en$$

$$(B)$$

$$(B) \quad\longrightarrow\quad N_2 \quad + \quad \text{Heat} \qquad\qquad\qquad \text{i)}$$

Overall equation: N_2CH_2 (Diazirine) $\longrightarrow$ Methylene $+$ N_2 $+$ Heat ii)

(Laws of Creations for Diazo-alkylenes)

Rule 277: This rule of Chemistry for **Addition monomers where tricyclic ring from Nitroxyl methylene is formed,** states that, when the ring is opened instantaneously, formaldehyde and the first member of Azo compound are formed as shown below.

Stage1:

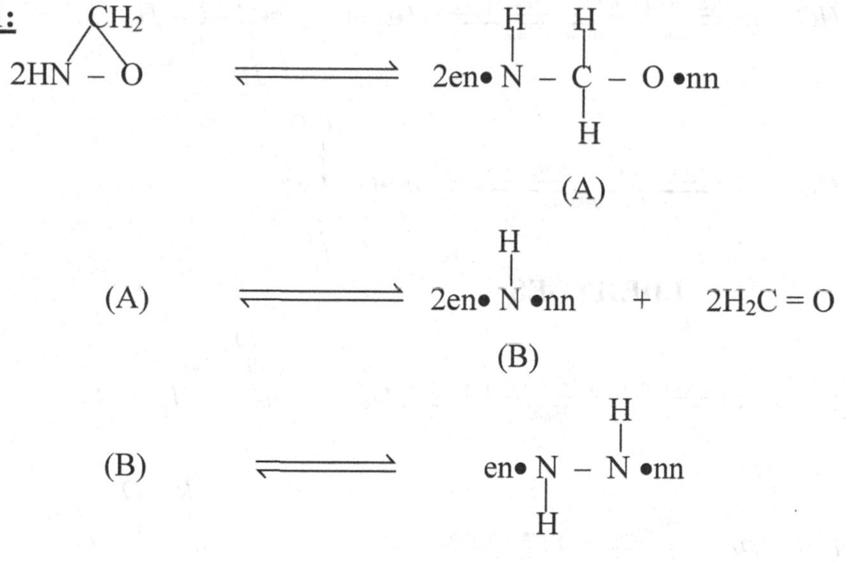

$$\underset{\displaystyle 2HN - O}{\overset{\displaystyle CH_2}{\diagdown\diagup}} \quad\rightleftharpoons\quad 2en\bullet \underset{\displaystyle H}{\overset{\displaystyle H}{N}} - \underset{\displaystyle H}{\overset{\displaystyle H}{C}} - O \bullet nn$$

$$(A)$$

$$(A) \quad\rightleftharpoons\quad 2en\bullet \overset{\displaystyle H}{N} \bullet nn \quad + \quad 2H_2C = O$$

$$(B)$$

$$(B) \quad\rightleftharpoons\quad en\bullet \underset{\displaystyle H}{\overset{\displaystyle H}{N}} - N \bullet nn$$

$$(C)$$

$$(C) \quad\longrightarrow\quad HN = NH \quad + \quad \text{Heat} \qquad\qquad \text{i)}$$

Overall equation: $2NOCH_2$ $\longrightarrow$ $HN = NH$ $+$ $2H_2C = O$ $+$ Heat ii)

for if formaldehyde was not released instantaneously (or N_2H_2 is not formed instaneously), then the stage will become unproductive if HN=NH is to be one of the products, noting the presence of a Nitrogenizing nitrogen [en•(NH)•nn] almost analogous to oxidizing oxygen or sulfurizing sulfur

(Laws of Creations for Nitroxyl alkylenes)

Rule 278: This rule of Chemistry for **Formaldehyde, Aldehydes and Ketones,** states that, these unique monomers like many others of its kind, other than being in Stable State of Existence, can favor the following State of Existences-

FORMALDEHYDE

$$HCHO \quad \underset{}{\overset{\text{Unactivated Equilibrium State of}}{\underset{\text{Existence}}{\rightleftarrows}}} \quad H\bullet e \; + \; n\bullet \overset{\displaystyle H}{\underset{}{|}}C = O$$

$$HCHO \quad \underset{}{\overset{\text{Activated State of}}{\underset{\text{Existence}}{\rightleftarrows}}} \quad e\bullet \overset{\displaystyle H}{\underset{\displaystyle H}{C}} - O \bullet nn$$

ACETALDEHYDE

$$H_3CCHO \quad \underset{}{\overset{\text{Unactivated Equilibrium State of}}{\underset{\text{Existence of Acetaldehyde}}{\rightleftarrows}}} \quad H\bullet e \; + \; n\bullet \overset{\displaystyle O}{\overset{\|}{C}} - CH_3$$

$$H_3CCHO \quad \underset{}{\overset{\text{Activated Equilibrium   State of}}{\underset{\text{Existence of Acetaldehyde}}{\rightleftarrows}}} \quad H\bullet e \; + \; n\bullet \overset{\displaystyle H}{\underset{\displaystyle H}{C}} - \overset{\overset{\bullet nn}{\displaystyle O}}{\underset{\bullet e}{C}} - H$$

$$H_3CCHO \quad \underset{}{\overset{\text{Activated State of Existence of}}{\underset{\text{Acetaldehyde}}{\rightleftarrows}}} \quad nn\bullet O - \overset{\displaystyle CH_3}{\underset{\displaystyle H}{C}}\bullet e$$

ALDEHYDES

$$RCH_2CHO \quad \underset{}{\overset{\text{Unactivated Equilibrium State of}}{\underset{\text{Existence of Aldehydes}}{\rightleftarrows}}} \quad H\bullet e \; + \; n\bullet \overset{\displaystyle O}{\overset{\|}{C}} - CH_2 - R$$

$$RCH_2CHO \quad \underset{}{\overset{\text{Activated Equilibrium State of}}{\underset{\text{Existence of Aldehydes}}{\rightleftarrows}}} \quad H\bullet e \; + \; n\bullet \overset{\displaystyle R}{\underset{\displaystyle H}{C}} - \overset{\overset{\bullet nn}{\displaystyle O}}{\underset{\bullet e}{C}} - H$$

$$RCH_2CHO \quad \underset{}{\overset{\text{Activated State of Existence of}}{\underset{\text{Aldehydes}}{\rightleftarrows}}} \quad nn\bullet O - \overset{\displaystyle CH_2R}{\underset{\displaystyle H}{C}}\bullet e$$

where R is an alkyl group.

KETONES

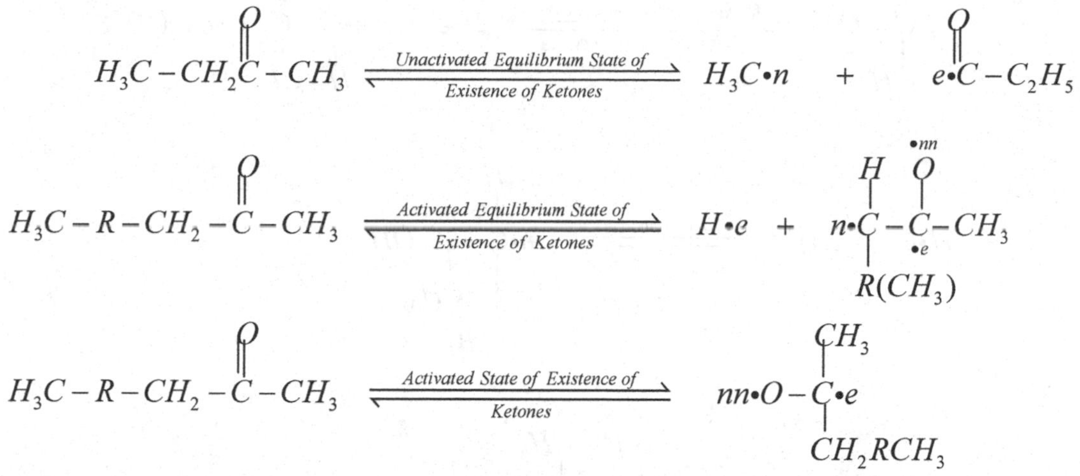

$$H_3C - CH_2\overset{\overset{O}{\parallel}}{C} - CH_3 \underset{\text{Existence of Ketones}}{\overset{\text{Unactivated Equilibrium State of}}{\rightleftharpoons}} H_3C\bullet n \quad + \quad e\bullet\overset{\overset{O}{\parallel}}{C} - C_2H_5$$

$$H_3C - R - CH_2 - \overset{\overset{O}{\parallel}}{C} - CH_3 \underset{\text{Existence of Ketones}}{\overset{\text{Activated Equilibrium State of}}{\rightleftharpoons}} H\bullet e \quad + \quad n\bullet\overset{\overset{H}{\mid}}{\underset{\underset{R(CH_3)}{\mid}}{C}} - \overset{\overset{\bullet nn}{O}}{\underset{\bullet e}{C}} - CH_3$$

$$H_3C - R - CH_2 - \overset{\overset{O}{\parallel}}{C} - CH_3 \underset{\text{Ketones}}{\overset{\text{Activated State of Existence of}}{\rightleftharpoons}} nn\bullet O - \overset{\overset{CH_3}{\mid}}{\underset{\underset{CH_2RCH_3}{\mid}}{C}}\bullet e$$

where the R here is one or more methylene groups (CH_2).

(Laws of Creation for States of Existence for part of C/H/O family tree)

Rule 279: This rule Chemistry for **Hydrazones,** states that, the Equilibrium State of Existence of two carrying either alkylane or phenyl group are as follows-

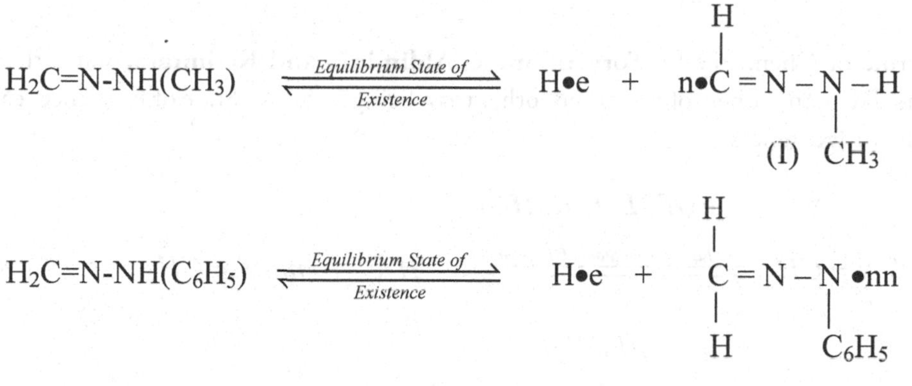

$$H_2C=N-NH(CH_3) \underset{\text{Existence}}{\overset{\text{Equilibrium State of}}{\rightleftharpoons}} H\bullet e \quad + \quad n\bullet\overset{\overset{H}{\mid}}{C}=N-\overset{\overset{\mid}{\mid}}{\underset{\underset{CH_3}{\mid}}{N}}-H$$
$$\text{(I)}$$

$$H_2C=N-NH(C_6H_5) \underset{\text{Existence}}{\overset{\text{Equilibrium State of}}{\rightleftharpoons}} H\bullet e \quad + \quad \overset{\overset{H}{\mid}}{\underset{\underset{H}{\mid}}{C}}=N-\overset{\overset{\mid}{\mid}}{\underset{\underset{C_6H_5}{\mid}}{N}}\bullet nn$$
$$\text{(II)}$$

for which based on the Equilibrium State of Existence of (II), it can "tautomerize" i.e., electroradicalize but not molecularly rearrange to give Azo toluene ($H_3C - N = N - C_6H_5$) which present-day Science calls benzeneazomethane, while (I) cannot.
(Laws of Creations for Hydrazones)

Rule 280: This rule of Chemistry for **Aldehydic and Ketonic types of Addition monomers which can be made to exist in ACTIVATED/EQUILIBRIUM STATE OF EXISTENCE at higher operating conditions,** states that, when they do in the presence of an acid or a base as a passive catalyst, they rearrange to give another organic compound acidic in character which bears the same character with the original monomer; for which the rearrangement is called **ENOLIZATION** as shown below, for without the additional Activated State of Existence, the rearrangement cannot take place.

$$
\begin{array}{c}
CH_3 \\
| \\
C = O \\
| \\
CH_3
\end{array}
\xrightleftharpoons[\text{Existence of Acetone}]{\text{Activated Equilibrium State of}}
H\bullet e \;+\;
\begin{array}{c}
H \\
| \quad \overset{\bullet e}{} \\
n\bullet C - C - O\bullet nn \\
| \quad | \\
H \quad CH_3
\end{array}
\;(A)
$$

$$
H\bullet e \;+\; (A) \quad
\xrightleftharpoons[\text{Acetone}]{\text{Enolization of}}
\begin{array}{c}
H \\
| \quad \overset{\bullet e}{} \\
n\bullet C - C - OH \\
| \quad | \\
H \quad CH_3
\end{array}
$$

$$(B)$$

$$
\xrightarrow[\text{Release of Heat}]{\text{Deactivation and}}
\begin{array}{c}
H \quad CH_3 \\
| \quad | \\
C = C \\
| \quad | \\
H \quad OH
\end{array}
\;+\; \text{Heat}
$$

$$(C)$$

DEFINITION OF ENOLIZATION   [A single stage]

(Laws of Creations for Enolization)

<u>**Rule 281:**</u> This rule of Chemistry for **Formaldimine, Aldimines and Ketimines,** states that, these unique monomers like many others of their kind, other than being in Stable State of Existence, can favor the following State of Existences-

FORMALDIMINE

$$
HN = CH_2 \xrightleftharpoons[]{\text{Equilibrium State of Existence}} H\bullet e \;+\; nn\bullet N = CH_2
$$

ALDIMINES

$$
HN = CH(R) \xrightleftharpoons[]{\text{Equilibrium State of Existence}} H\bullet e \;+\; nn\bullet N = CH(R)
$$

$$
HN = CH(CH_2)R' \xrightleftharpoons[\text{Existence when sup pressed}]{\text{Activated Equilibrium State of}} H\bullet e \;+\;
\begin{array}{c}
H \quad\;\; H \\
| \quad\;\; | \\
nn\bullet N - \overset{\bullet e}{C} - \underset{\bullet n}{C} - R' \\
| \\
H
\end{array}
$$

$$
HN = CH(R) \xrightleftharpoons[]{\text{Activated State of Existence}}
\begin{array}{c}
H \quad H \\
| \quad | \\
nn\bullet N - C\bullet e \\
| \\
R
\end{array}
$$

$$HN = CH(R) \quad \xrightleftharpoons{\text{Activated State of Existence}} \quad nn\bullet N - \underset{\underset{R}{|}}{\overset{\overset{H}{|}}{C}}\bullet e \;\; (\text{with second } H)$$

ALKYL NITROGEN SUBSTITUTED ALDIMINES

$$R - N = CH(R) \quad \xrightleftharpoons{\text{Equilibrium State of Existence}} \quad H\bullet e \;+\; R - N = \underset{R}{\overset{|}{C}}\bullet n$$

$$R - N = \underset{\underset{H}{|}}{C} - (CH_2) - R' \quad \xrightleftharpoons[\text{Existence}]{\text{Activated Equilibrium State of}} \quad H\bullet e \;+\; R - \overset{\bullet nn}{N} - \underset{\bullet e}{C} - \overset{\bullet n}{\underset{\underset{H}{|}}{C}} - R'$$

$$R - N = \underset{\underset{H}{|}}{C} - R \quad \xrightleftharpoons{\text{Activated State of Existence}} \quad nn\bullet N - \overset{\overset{R}{|}}{\underset{\underset{H}{|}}{C}}\bullet e \;(R)$$

KETIMINES

$$HN = CR_2 \quad \xrightleftharpoons{\text{Equilibrium State of Existence}} \quad H\bullet e \;+\; nn\bullet N = CR_2$$

$$HN = \overset{\overset{R}{|}}{C} - CH_2 - R \quad \xrightleftharpoons[\text{Existence (Suppressed)}]{\text{Activated Equilibrium State of}} \quad H\bullet e \;+\; H - \overset{\bullet nn}{N} - \underset{\underset{R}{|}}{\overset{\bullet e}{C}} - \overset{\overset{H}{|}}{\underset{\bullet n}{C}} - R$$

$$HN = CR_2 \quad \xrightleftharpoons{\text{Activated State of Existence}} \quad nn\bullet N - \overset{\overset{H}{|}}{\underset{\underset{R}{|}}{\overset{\overset{R}{|}}{C}}}\bullet e$$

ALKYL NITROGEN SUBSTITUTED KETIMINES

$$R - N = \overset{\overset{R}{|}}{C} - R'' - CH_3 \quad \xrightleftharpoons{\text{Equilibrium State of Existence}} \quad R\bullet n \;+\; R - N = \overset{\bullet e}{C} - R'' - CH_3$$

$$R - N = \overset{\overset{R}{|}}{C} - \overset{\overset{H}{|}}{\underset{\underset{H}{|}}{C}} - R \quad \xrightleftharpoons[\text{Existence}]{\text{Activated Equilibrium State of}} \quad H\bullet e \;+\; R - \overset{\bullet nn}{N} - \underset{\underset{R}{|}}{\overset{\bullet e}{C}} - \overset{\overset{H}{|}}{\underset{\bullet n}{C}} - R$$

where the R and R' are alkyl and large number of methylene groups, depending on where placed.
(Laws of Creations for States of Existence of part of C/H/N Family tree)

<u>**Rule 282:**</u> This rule of Chemistry for **Cyanamides, Aldimines and Ketimines types of Addition monomers which can be made to exist in ACTIVATED EQUILIBRIUM STATE OF EXISTENCE at higher operating conditions,** states that, when they do, they rearrange to give another organic compound acidic/aminic in character which bears the same character with the original monomer; for which the rearrangement is called <u>**ENAMIDIZATION**</u> as shown below for cyanamide; for without the additional Activated State of Existence, the rearrangement cannot take place.

<u>**Stage 1:**</u>

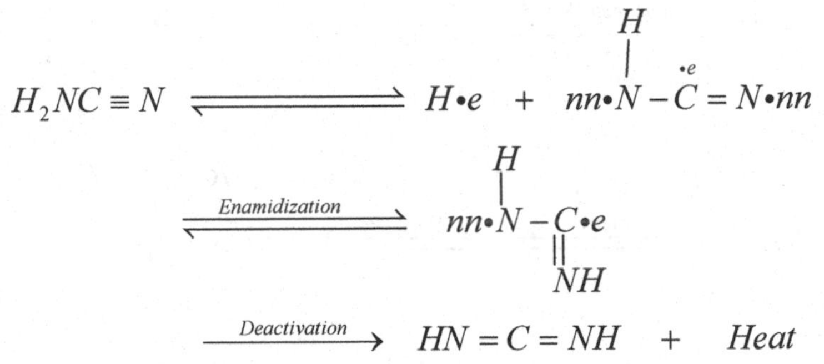

$$H_2NC \equiv N \;\rightleftharpoons\; H\bullet e \;+\; nn\bullet N - \overset{\overset{\displaystyle H}{|}}{C} \overset{\bullet e}{=} N\bullet nn$$

$$\xrightarrow{\text{\textit{Enamidization}}} \quad nn\bullet N - \overset{\overset{\displaystyle H}{|}}{\underset{\underset{\displaystyle NH}{\parallel}}{C}}\bullet e$$

$$\xrightarrow{\text{\textit{Deactivation}}} \quad HN = C = NH \;+\; Heat$$

DEFINITION OF ENAMIDIZATION (or Imidization)

(Laws of Creations for Enamidization)

<u>**Rule 283:**</u> This rule of Chemistry for **Substituted Hydrazines,** states that, their Equilibrium State of Existence are as follows-

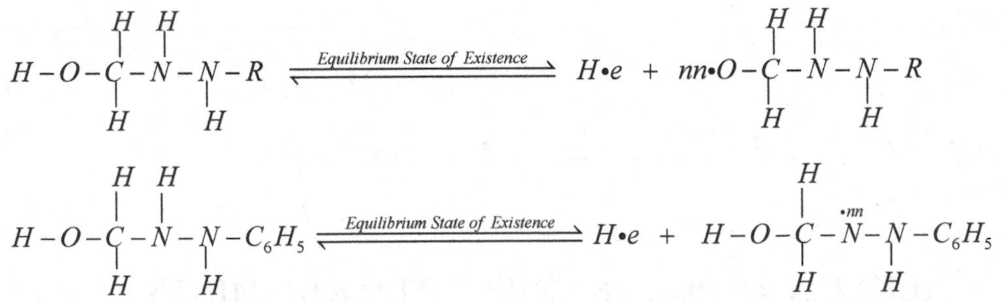

$$H-O-\overset{\overset{\displaystyle H}{|}}{\underset{\underset{\displaystyle H}{|}}{C}} - \overset{\overset{\displaystyle H}{|}}{N} - \overset{\overset{\displaystyle H}{|}}{N} - R \xrightarrow{\text{\textit{Equilibrium State of Existence}}} H\bullet e \;+\; nn\bullet O - \overset{\overset{\displaystyle H}{|}}{\underset{\underset{\displaystyle H}{|}}{C}} - \overset{\overset{\displaystyle H}{|}}{N} - \overset{\overset{\displaystyle H}{|}}{N} - R$$

$$H-O-\overset{\overset{\displaystyle H}{|}}{\underset{\underset{\displaystyle H}{|}}{C}} - \overset{\overset{\displaystyle H}{|}}{N} - N - C_6H_5 \xrightarrow{\text{\textit{Equilibrium State of Existence}}} H\bullet e \;+\; H - O - \overset{\overset{\displaystyle H}{|}}{\underset{\underset{\displaystyle H}{|}}{C}} - \overset{\overset{\displaystyle \bullet nn}{N}}{} - \overset{\overset{\displaystyle H}{|}}{N} - C_6H_5$$

where R is an alkanyl group. [See rule 279]

(Laws of Creations for Substituted Hydrazines)

<u>**Rule 284:**</u> This rule of Chemistry for **Cyanamide ($H_2NC \equiv N$),** states that, it readily dimerizes as shown below to give dicyandiamide, in which as shown below 50 percent of the cyanamide undergoes Enamidization to give carbodiimide which combines with the remaining 50 percent kept in Equilibrium state of existence in two Stages via Equilibrium mechanism.

$$H_2NC \equiv N \quad + \quad H_2NC \equiv N \quad \xrightarrow{pH\ 7-12} \quad \underset{\underset{NH}{\|}}{H_2NCNHC \equiv N}$$

Dicyandiamide   a)

Stage 1:

$$H_2NC \equiv N \quad \rightleftharpoons \quad H \bullet e \quad + \quad nn \bullet \overset{H}{\underset{|}{N}} - \overset{\bullet e}{C} = N \bullet nn$$

$$\rightleftharpoons \xrightarrow{Enamidization} \quad nn \bullet \overset{H}{\underset{|}{N}} - \underset{\underset{NH}{\|}}{C} \bullet e \qquad\qquad\qquad i)$$

$$\xrightarrow{Deactivation} \quad HN = C = NH$$

Overall Equation: $H_2NC \equiv N \xrightarrow{Enamidization} HN = C = NH$ ii)

and since Cyanamide with a triple bond is more nucleophilic than carbodiimide, the next stage follows-

Stage 2:

$$H_2N - C \equiv N \quad \xrightarrow[Existence]{Equilibrium\ State\ of} \quad H \bullet e \quad + \quad nn \bullet \overset{H}{\underset{|}{N}} - C \equiv N$$

$$(A)$$

$$H \bullet e \quad + \quad HN = C = NH \quad \xrightarrow{Activation/Addition} \quad H - \overset{H}{\underset{|}{N}} - \underset{\underset{NH}{\|}}{C} \bullet e \quad (B) \qquad\qquad iii)$$

$$(A) \quad + \quad (B) \quad \xrightarrow{Combination/Deactivation} \quad H - \overset{H}{\underset{|}{N}} - \underset{\underset{NH}{\|}}{C} - \overset{H}{\underset{|}{N}} - C \equiv N$$

Dicyandiamide

Overall Equation: $H_2N\text{-}C \equiv N \quad + \quad HN = C = NH \longrightarrow$ Dicyandiamide   iv)

(Laws of Creations for Dimerization of Cyanamide)

Rule 285: This rule of Chemistry for **Electrostatic Initiators and compounds,** states that, these are paired centered compounds carrying imaginary bonds, for which there are two types-

$$\underset{R}{\overset{Cl}{\underset{|}{Al^{\oplus}}}}\cdots\cdots\overset{\ominus}{\underset{Cl}{\overset{Cl}{\underset{|}{Ti}}}}-Cl \quad , \quad \underset{F}{\overset{F}{\underset{|}{B^{\oplus}}}}\cdots\cdots\overset{\ominus}{\underset{F}{\overset{F}{\underset{|}{B}}}}-F \quad ; \quad RO^{\ominus}\cdots\cdots\overset{H\;R}{\underset{H}{\overset{|\;/}{N}}}-R \quad , \quad Cl^{\ominus}\cdots\cdots\overset{H}{\underset{H}{\overset{|}{O}}}-H$$

Positively Charged Electrostatic Initiators   Negatively Charged Electrostatic Initiators

(Laws of Creations for Electrostatic Initiators/Compounds)

Rule 286: This rule of Chemistry for **Non-ionic metallic hydroxides,** states that, the structure of hydrated alumina is the structure of aluminum hydroxide and not $Al(OH)_3$ as used currently and the same applies to other members in the same Group family. [See Rule 195]

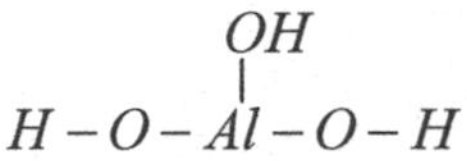

$$H-O-\overset{\overset{OH}{|}}{Al}-O-H \qquad ; \qquad \overset{\overset{O^{\ominus}}{|}}{Al^{\oplus}}-O^{\ominus}\cdots\cdots\overset{\overset{H}{|}}{\underset{H}{\overset{\oplus}{O}}}-H$$

Not the structure of Alu min um
hydroxide

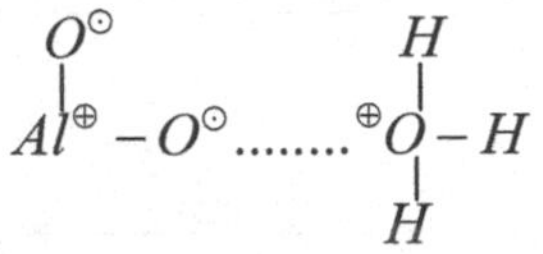

Hydrated alu min a

(Laws of Creations for Hydrated alumina)

Rule 287: This rule of Chemistry for **Electrostatically Negatively Charged-paired Initiators,** states that, these initiators when they exist in Equilibrium state of existence can be used to polymerize Half-free monomers such as C = O, C = N, C = S types in the route natural to them by Equilibrium mechanism from the imaginary center electro-free-radically in such a manner where the growing polymer chains grow backwardly as shown below, not from the Electrostatic centers where the growing polymer chains can only grow forwardly.

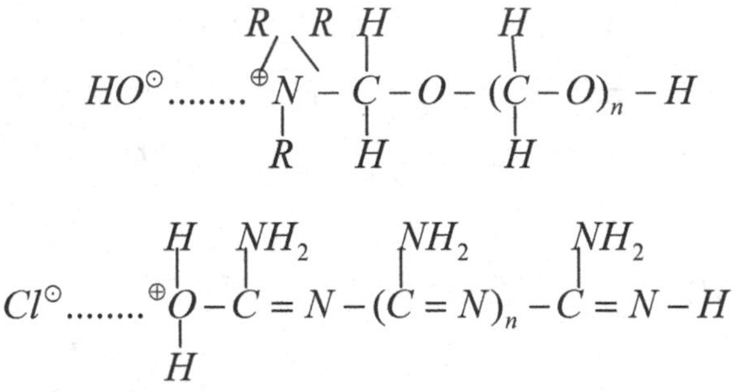

$$HO^{\ominus}\cdots\cdots\overset{R\;R\;H}{\underset{R\;\;\;H}{\overset{\backslash/\;|\;|}{\underset{|\;\;\;|}{N}}}}-\overset{|}{C}-O-(\overset{\overset{H}{|}}{\underset{H}{C}}-O)_n-H$$

$$Cl^{\ominus}\cdots\cdots\overset{H\;\;NH_2}{\underset{H}{\overset{|\;\;\;|}{\underset{|}{O}}}}-\overset{}{C}=N-(\overset{\overset{NH_2}{|}}{C}=N)_n-\overset{\overset{NH_2}{|}}{C}=N-H$$

<u>**Addition Polymerization via Backward movement of growing chain.**</u>

(Laws of Creations for Addition Polymerization-Backward Addition)

Rule 288{This rule of Chemistry for **C = O, C = N, C = S types of Addition monomers**, states that, only their first members will favor the use of the forward and backward initiating centers of Electrostatic negatively charged initiators, one at a time depending on the operating conditions as follows-

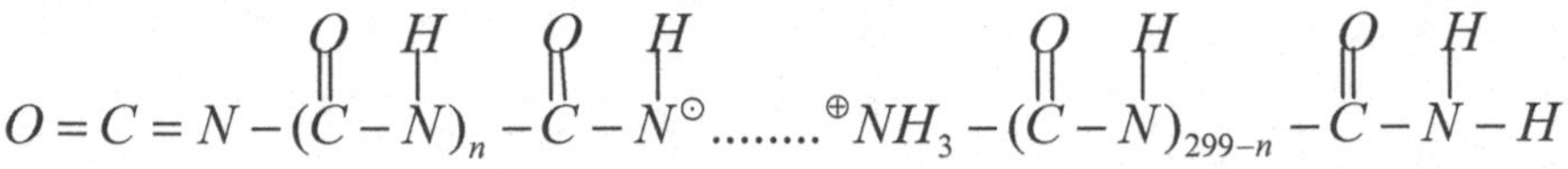

$$O=C=N-(\overset{\overset{O}{\|}}{C}-\overset{\overset{H}{|}}{N})_n-\overset{\overset{O}{\|}}{C}-N^{\ominus}\cdots\cdots\overset{\oplus}{N}H_3-(\overset{\overset{O}{\|}}{C}-\overset{\overset{H}{|}}{N})_{299-n}-\overset{\overset{O}{\|}}{C}-\overset{\overset{H}{|}}{N}-H$$

Initiator: From Isocyanic acid and NH₃.

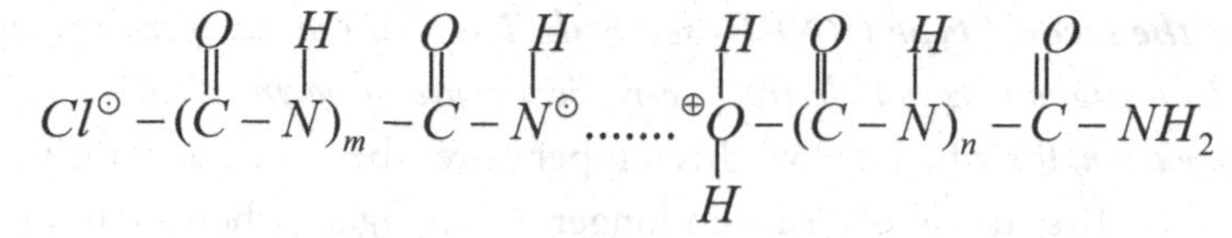

$$Cl^{\ominus} - (C-N)_m - C - N^{\ominus} \ldots\ldots {}^{\oplus}O - (C-N)_n - C - NH_2$$

with the structure showing O and H groups above each C-N unit, and an H below the O.

Initiator: Dilute hydrochloric acid.

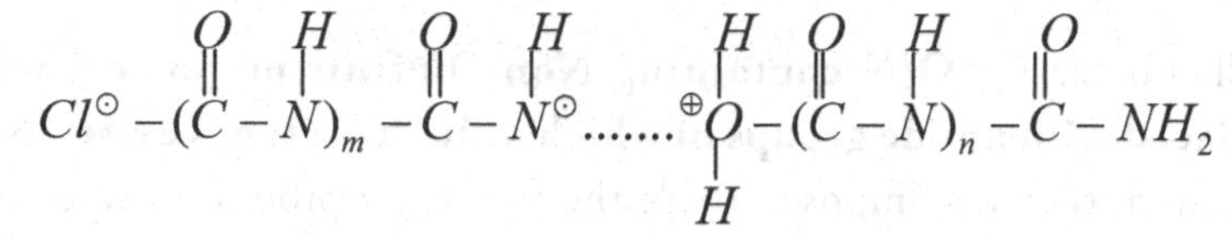

$$Cl^{\ominus} - (C-N)_m - C - N^{\ominus} \ldots\ldots {}^{\oplus}O - (C-N)_n - C - NH_2$$

Initiator: Dilute hydrochloric acid.

noting that forward addition (Combination mechanism) *takes place when the initiator is in Stable state of existence, while backward addition* (Equilibrium mechanism) *takes place when the initiator is in Equilibrium state of existence.*

(Laws of Creations for Addition Polymerization-Forward and Backward Additions)

Rule 289: This rule of Chemistry for **O or N containing monomers of the type C = O, C = N, i.e. Aldehydes and ketones, aldimines and ketimines and the likes,** states that, while the C = O types do not have Oxygen center carrying substituted types of groups, the C = N types have Nitrogen center carrying substituted types of groups to give them very unique characters, some examples of which are shown below-

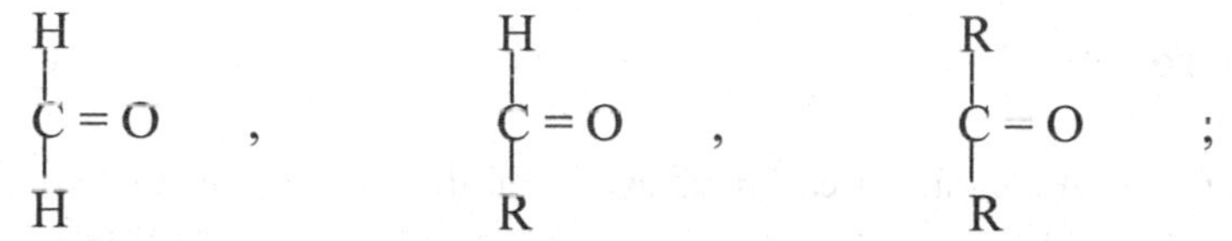

H C = O H	H C = O R	R C – O R ;
Formaldehyde	**Aldehydes**	**Ketones**
H H C = N H	H H C = N R	R H C = N R ,
Formaldimine	**Aldimines**	**Ketimines**
H R C = N H	H R C = N R	R R C = N R
Formaldimines	**Aldimines**	**Ketimines**

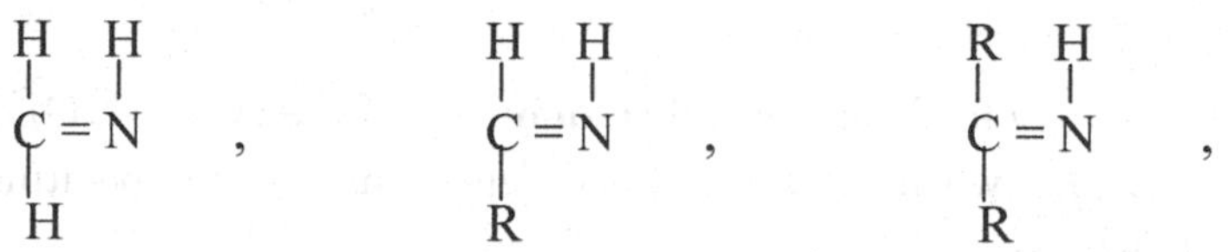

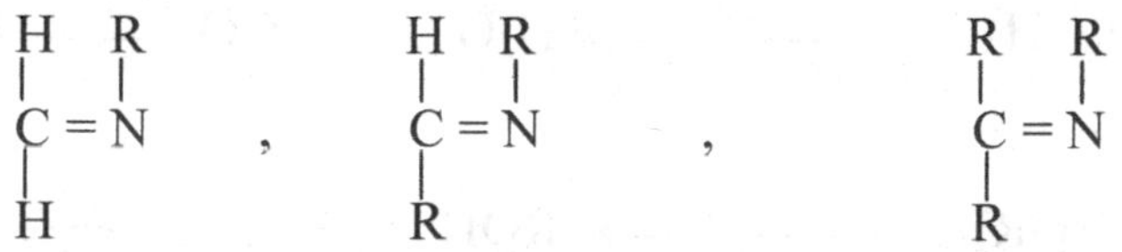

Nitrogen pushing-carrying substituted Formaldimines, Aldimines, and Ketimines

(Laws of Creations for Half-Free Non-Olefinic Addition monomers)

Rule 290: This rule of Chemistry for **O or N or S containing Non-Olefinic monomers which undergo polymerization backwardly using Electrostatic Negatively Charged-paired Initiators,** states that,

"the ceiling temperature of the second type (RNRP- see Rule 139)" is that temperature where the rate of propagation of the non-olefinic monomers via electro-free-radical route of polymerization using Hydrogen atom equals the rate of depropagation; for which below this temperature, these unique family of monomers with the usual exception of their first members, can no longer depropagate when monomers are about to be added.
(Laws of Creations for Ceiling Temperature)

<u>Rule 291:</u> This rule of Chemistry for **O/N containing Non-Olefinic monomers with alkanyl groups in KETONES replaced with amide groups of which only three types exist (-NH$_2$, -NHR, -NR$_2$)**, states that, when heated, they decompose to give the following products important in living systems-

$$H_2NCONH_2 \xrightarrow{\text{Heat}} NH_3 + HN=C=O \longrightarrow \text{Polymer} + \text{Ring}$$

Primary Urea   **Isocyanic acid**

$$HRNCONH_2 \xrightarrow{\text{Heat}} NH_2R + HN=C=O \longrightarrow \text{Polymer} + \text{Ring}$$

Primary/Secondary Urea   **Isocyanic acid**

$$HRNCONHR \xrightarrow{\text{Heat}} NH_2R + RN=C=O$$

Secondary Urea   **Alkyl isocyanate**

$$HRNCONR_2 \xrightarrow{\text{Heat}} NHR_2 + RN=C=O$$

Secondary/tertiary Urea   **Alkyl isocyanate**

$$R_2NCONR_2 \xrightarrow{\text{Heat}} \text{Nothing, just Activation}$$

Tertiary Urea

for which only the primary and secondary ureas may have the ability to rearrange to a different compound via Molecular rearrangement and Enolization depending on the operating conditions.
(Laws of Creations in Living and Non-living Systems)

<u>Rule 292:</u> This rule of Chemistry for **Non-Olefinic monomers of the types (HO)$_2$C = O, (HO)(OR) C = O, (RO)$_2$C = O,** states that, when these are heated depending on the operating conditions, they decompose to give the followings-

$$(HO)CO(OH) \xrightarrow{\text{Heat}} H_2O + CO_2 \longrightarrow \text{Ring}$$

Carbonic acid

$$(HO)CO(OR) \xrightarrow{\text{Heat}} ROH + CO_2 \longrightarrow \text{Ring}$$

Mono-alkyl substituted Carbonic acid

$$(RO)CO(OR) \xrightarrow{\text{Heat}} \text{Nothing, just Activation}$$

Di-alkyl substituted Corbonate

for which whether rearrangement takes place or not the same mono-form is obtained.
(Laws of Creations in living and Non-living Systems)

Sunny N.E. Omorodion

<u>**Rule 293:**</u> This rule of Chemistry for **Ringed compounds,** states that, when they (E.g. Tricyclic, Tetracylic, Pentacyclic, Hexacylic types) are formed, this can only be done via EQUILIBRIUM mechanism and no other mechanisms.
(Laws of Creations for Ringed compounds)

<u>**Rule 294:**</u> This rule of Chemistry for Acid **catalyzed reaction of acetaldehyde,** states that, the mechanisms of the reaction are as shown below in the presence of dilute hydrochloric acid-

<u>**Stage 1:**</u> **Initiation Step**

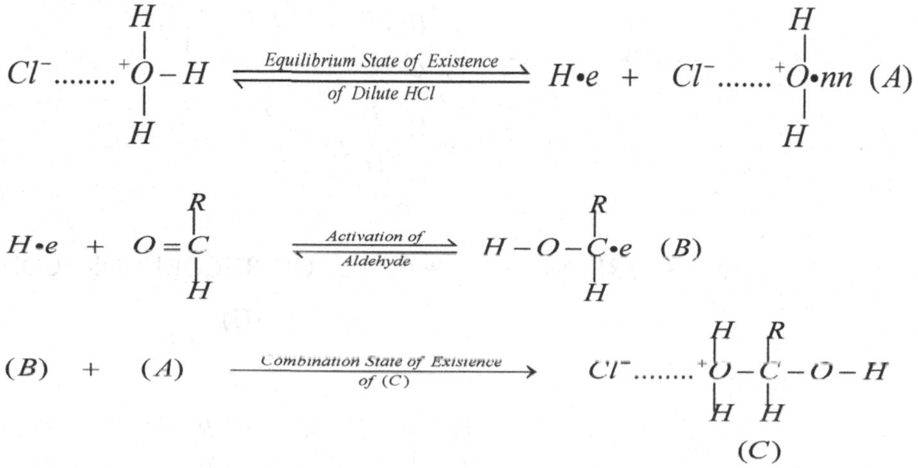

i)

ii)

$$Cl^-........^+O-H \xrightleftharpoons[\text{of Dilute HCl}]{\text{Equilibrium State of Existence}} H\bullet e + Cl^-........^+O\bullet nn \quad (A)$$

$$H\bullet e + O=C \xrightleftharpoons[\text{Aldehyde}]{\text{Activation of}} H-O-C\bullet e \quad (B)$$

$$(B) + (A) \xrightarrow[\text{of (C)}]{\text{Combination State of Existence}} Cl^-........^+O-C-O-H \quad (C)$$

<u>**Overall Equation:**</u> HCl + H₂O + RHC=O ⟶ Cl⁻........⁺OH₂RHCOH (C)

<u>**Stage 2:**</u> **Propagation begins**

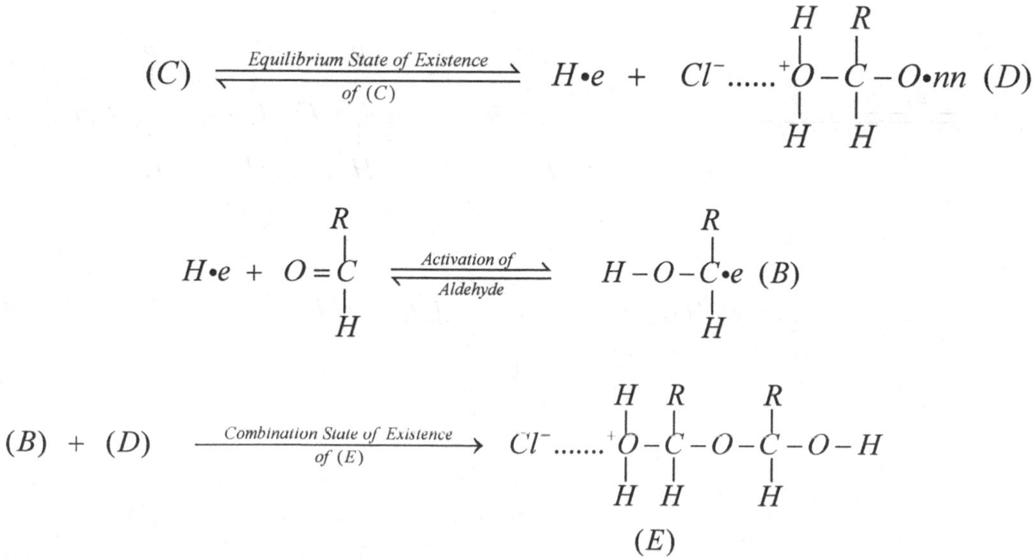

iii)

iv)

$$(C) \xrightleftharpoons[\text{of (C)}]{\text{Equilibrium State of Existence}} H\bullet e + Cl^-......^+O-C-O\bullet nn \quad (D)$$

$$H\bullet e + O=C \xrightleftharpoons[\text{Aldehyde}]{\text{Activation of}} H-O-C\bullet e \quad (B)$$

$$(B) + (D) \xrightarrow[\text{of (E)}]{\text{Combination State of Existence}} Cl^-........^+O-C-O-C-O-H \quad (E)$$

<u>**Overall Equation:**</u> HCl + H₂O + 2RHC=O ⟶ Cl⁻......⁺OH₂RHCORHCOH (E)

Stage 3:

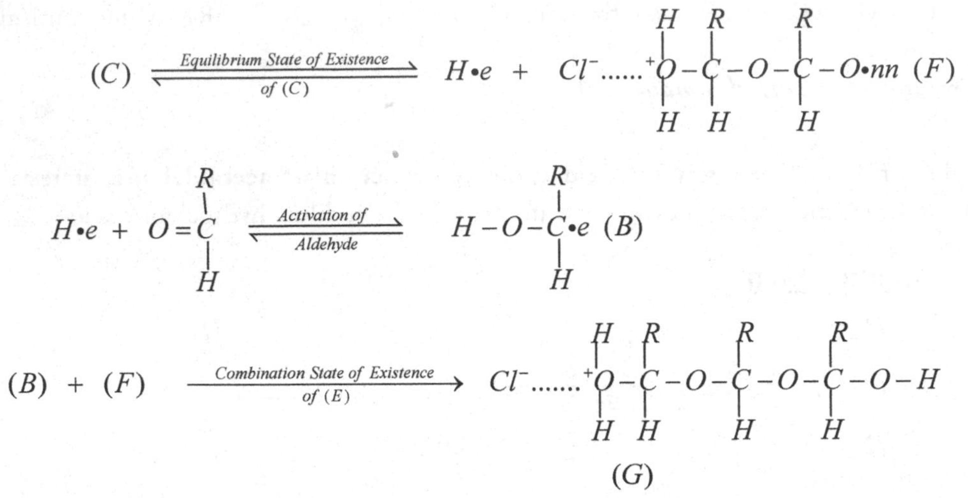

$$(C) \underset{\text{of } (C)}{\overset{\text{Equilibrium State of Existence}}{\rightleftharpoons}} H \bullet e \ + \ Cl^- ......^+ \overset{H}{\underset{H}{\overset{|}{O}}} - \overset{R}{\underset{H}{\overset{|}{C}}} - O - \overset{R}{\underset{H}{\overset{|}{C}}} - O \bullet nn \quad (F)$$

$$H \bullet e \ + \ O = \overset{R}{\underset{H}{\overset{|}{C}}} \underset{\text{Aldehyde}}{\overset{\text{Activation of}}{\rightleftharpoons}} H - O - \overset{R}{\underset{H}{\overset{|}{C}}} \bullet e \quad (B) \qquad \text{v)}$$

$$(B) \ + \ (F) \xrightarrow{\underset{\text{of } (E)}{\text{Combination State of Existence}}} Cl^- ....^+ \overset{H}{\underset{H}{\overset{|}{O}}} - \overset{R}{\underset{H}{\overset{|}{C}}} - O - \overset{R}{\underset{H}{\overset{|}{C}}} - O - \overset{R}{\underset{H}{\overset{|}{C}}} - O - H$$
$$(G)$$

Overall Equation: $HCl \ + \ H_2O \ + \ 3RHC=O \longrightarrow Cl^{\ominus}......^{\oplus}OH_2RHCORHCORHCOH$
$$(G) \qquad \qquad \text{vi)}$$

Stage 4:

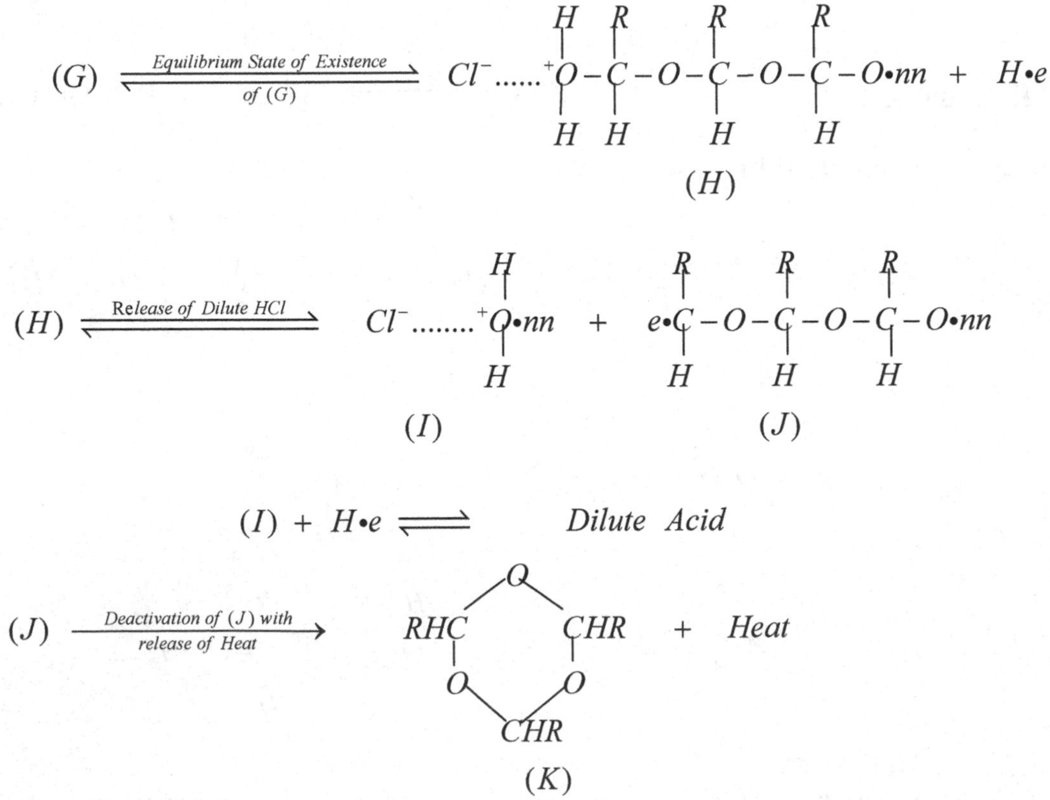

$$(G) \underset{\text{of } (G)}{\overset{\text{Equilibrium State of Existence}}{\rightleftharpoons}} Cl^- ......^+ \overset{H}{\underset{H}{\overset{|}{O}}} - \overset{R}{\underset{H}{\overset{|}{C}}} - O - \overset{R}{\underset{H}{\overset{|}{C}}} - O - \overset{R}{\underset{H}{\overset{|}{C}}} - O \bullet nn \ + \ H \bullet e$$
$$(H)$$

$$(H) \underset{}{\overset{\text{Release of Dilute HCl}}{\rightleftharpoons}} Cl^- ....^+ \overset{H}{\underset{H}{\overset{|}{O}}} \bullet nn \ + \ e \bullet \overset{R}{\underset{H}{\overset{|}{C}}} - O - \overset{R}{\underset{H}{\overset{|}{C}}} - O - \overset{R}{\underset{H}{\overset{|}{C}}} - O \bullet nn$$
$$(I) \qquad \qquad \qquad (J)$$

$$(I) \ + \ H \bullet e \rightleftharpoons \qquad \text{Dilute Acid}$$

$$(J) \xrightarrow{\underset{\text{release of Heat}}{\text{Deactivation of } (J) \text{ with}}} \quad RHC \overset{O}{\underset{O}{<}} CHR \ + \ \text{Heat}$$

$$(K) \qquad \qquad \text{vii)}$$

Overall Equation: $HCl \ + \ H_2O \ + \ 3RHC=O \longrightarrow HCl \ + \ H_2O \ + \ (K)$

a four stage Equilibrium mechanism system in which Addition polymerization in Stages 1 to 3 took place and in Stage 4 a suitably sized trioxane ring was forced to be formed in the absence of more monomers in the vicinity or the system
(Laws of Creations for Cyclization of Aldehydes)

Rule 295: This rule of Chemistry for **Acid catalyzed reaction of Ketones,** states that, in the presence of concentrated sulfuric acid, the followings are obtained-

Stage 1:

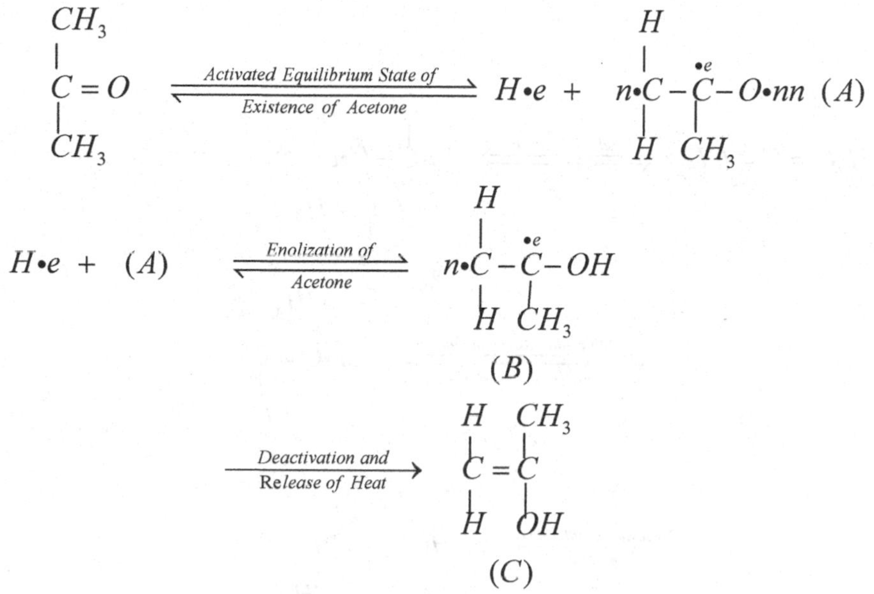

$$CH_3-C(CH_3)=O \xrightleftharpoons[\text{Existence of Acetone}]{\text{Activated Equilibrium State of}} H\bullet e \ + \ n\bullet C(H)(H)-C(CH_3)-O\bullet nn \quad (A)$$

$$H\bullet e \ + \ (A) \xrightleftharpoons[\text{Acetone}]{\text{Enolization of}} n\bullet C(H)(H)-C(CH_3)-OH \quad \text{(B)} \qquad \text{i)}$$

$$\xrightarrow[\text{Release of Heat}]{\text{Deactivation and}} \ \begin{matrix} H & CH_3 \\ C & = & C \\ H & OH \end{matrix} \quad \text{(C)}$$

Overall Equation: $3H_3C\text{-}CO\text{-}CH_3 \xrightarrow{\text{Enolization}} 3H_2C\text{=}C(OH)CH_3$ ii)

in which in the first stage, the ketone was enolized to give (C), the operating conditions for enolization provided by the presence of concentrated H_2SO_4, followed by attack of (C) by the

Stage 2:

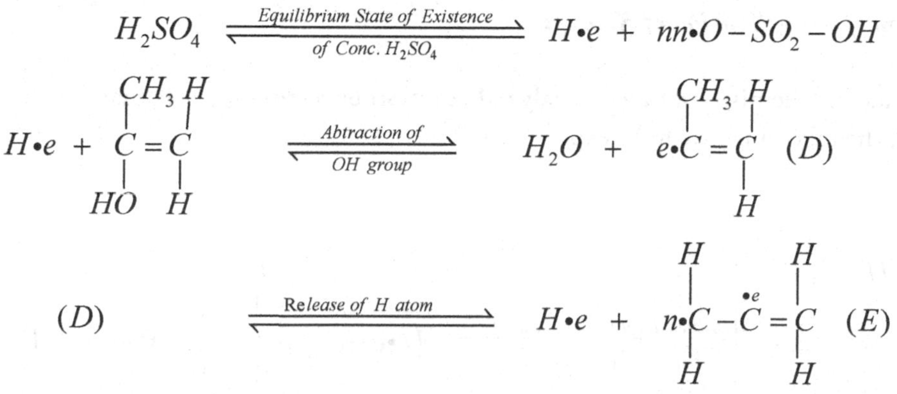

$$H_2SO_4 \xrightleftharpoons[\text{of Conc. } H_2SO_4]{\text{Equilibrium State of Existence}} H\bullet e \ + \ nn\bullet O-SO_2-OH$$

$$H\bullet e \ + \ \begin{matrix} CH_3 & H \\ C & = & C \\ HO & H \end{matrix} \xrightleftharpoons[\text{OH group}]{\text{Abtraction of}} H_2O \ + \ \begin{matrix} CH_3 & H \\ e\bullet C & = & C \\ & H \end{matrix} \ (D)$$

$$(D) \xrightleftharpoons{\text{Release of H atom}} H\bullet e \ + \ \begin{matrix} H & & H \\ n\bullet C & - & C & = & C \\ H & & H \end{matrix} \ (E) \qquad \text{iii)}$$

$$H \bullet e \ + \ nn \bullet O - SO_2 - OH \ \rightleftharpoons \ H_2SO_4$$

$$(E) \quad \xrightarrow[\;(E)\;]{Deactivation \ of} \quad H_2C = C = CH_2 \ + \ Heat$$

$$(Allene)$$

Overall Equation: $3H_2SO_4 \ + \ 3(C) \longrightarrow 3H_2O \ + \ 3H_2SO_4 \ + \ 3H_2C{=}C{=}CH_2$ iv)

same concentrated sulfuric acid which was in large concentration to give (D) which lost an atom of H to give the acid and an allene which is unstable and rearranges in Stage 3 to form methyl acetylene which in view of the numbers of moles present added sequentially in three hidden steps-

Stage 3:

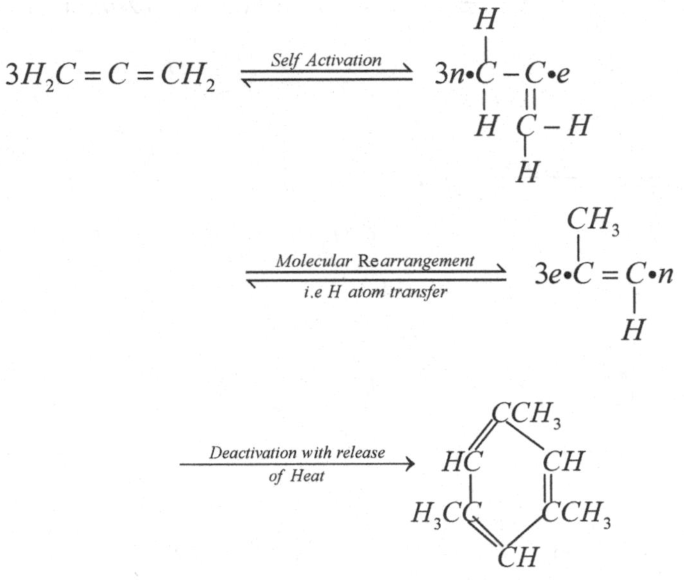

$$3H_2C = C = CH_2 \quad \xrightarrow{Self \ Activation} \quad 3n\bullet\overset{\displaystyle H}{\underset{\displaystyle H}{\overset{\displaystyle |}{\underset{\displaystyle |}{C}}}} - \overset{\displaystyle }{\underset{\displaystyle C-H}{\overset{\displaystyle \|}{C}}}\bullet e$$ xi)

$$\xrightarrow[i.e \ H \ atom \ transfer]{Molecular \ Rearrangement} \quad 3e\bullet\overset{\displaystyle CH_3}{\underset{\displaystyle H}{\overset{\displaystyle |}{\underset{\displaystyle |}{C}}}} = C\bullet n$$

$$\xrightarrow[of \ Heat]{Deactivation \ with \ release} \quad \text{(Mesitylene ring structure)}$$ v)

$$(Mesitylene)$$

Overall Equation: $3H_2SO_4 \ + \ 3H_3C{-}CO{-}CH_3 \quad \xrightarrow[Cyclization]{Molecular \ Rearrangement \ and} \quad \text{Mesitylene} \ +$ vi)

to form mesitylene.
(Laws of Creations for Cyclization of Ketones)

Rule 296: This rule of Chemistry for **base catalyzed reaction of Ketones,** states that, in the presence of sodium amine, the followings are to be expected-

Stage 1:

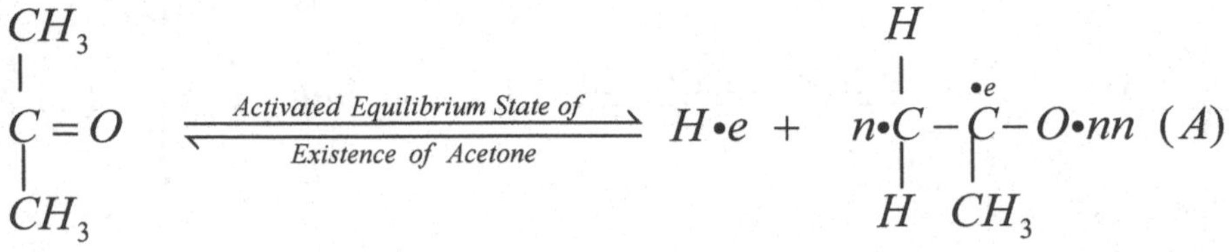

$$\overset{\displaystyle CH_3}{\underset{\displaystyle CH_3}{\overset{\displaystyle |}{\underset{\displaystyle |}{C}}}} = O \quad \xrightarrow[Existence \ of \ Acetone]{Activated \ Equilibrium \ State \ of} \quad H\bullet e \ + \ n\bullet\overset{\displaystyle H}{\underset{\displaystyle H}{\overset{\displaystyle |}{\underset{\displaystyle |}{C}}}} - \overset{\displaystyle \bullet e}{\underset{\displaystyle CH_3}{\overset{\displaystyle }{\underset{\displaystyle |}{C}}}} - O\bullet nn \ (A)$$

$$H\bullet e \;+\; (A) \;\xrightleftharpoons[\text{Acetone}]{\text{Enolization of}}\; n\bullet C\overset{\displaystyle \overset{H}{|}}{\underset{\displaystyle \underset{H}{|}}{-}}\overset{\displaystyle \bullet e}{C}\overset{}{\underset{\displaystyle \underset{CH_3}{|}}{-}}OH$$

$$(B)$$

$$\xrightarrow[\text{Release of Heat}]{\text{Deactivation and}}\; \overset{\displaystyle \overset{H}{|}}{\underset{\displaystyle \underset{H}{|}}{C}}=\overset{\displaystyle \overset{CH_3}{|}}{\underset{\displaystyle \underset{OH}{|}}{C}} \qquad\qquad \text{i)}$$

$$(C)$$

Overall Equation: $3H_3CCOCH_3 \xrightarrow{\text{Enolization}} 3H_2C=C(OH)CH_3$ ii)

Stage 2:

$$NaNH_2 \;\xrightleftharpoons[\text{of Dry NaNH}_2]{\text{Equilibrium State of Existence}}\; Na\bullet e \;+\; nn\bullet NH_2$$

$$Na\bullet e \;+\; \overset{\displaystyle \overset{CH_3}{|}}{\underset{\displaystyle \underset{HO}{|}}{C}}=\overset{\displaystyle \overset{H}{|}}{\underset{\displaystyle \underset{H}{|}}{C}} \;\xrightleftharpoons[\text{OH group}]{\text{Abtraction of}}\; NaOH \;+\; e\bullet\overset{\displaystyle \overset{CH_3}{|}}{C}=\overset{\displaystyle \overset{H}{|}}{\underset{\displaystyle \underset{H}{|}}{C}} \quad (B)$$

$$(B) \;\xrightleftharpoons{\text{Release of H atom}}\; H\bullet e \;+\; n\bullet\overset{\displaystyle \overset{H}{|}}{\underset{\displaystyle \underset{H}{|}}{C}}-\overset{\displaystyle \bullet e}{C}=\overset{\displaystyle \overset{H}{|}}{\underset{\displaystyle \underset{H}{|}}{C}} \quad (C)$$

$$H\bullet e \;+\; nn\bullet NH_2 \;\rightleftharpoons\; NH_3$$

$$(C) \;\xrightarrow[(C)]{\text{Deactivation of}}\; H_2C=C=CH_2$$

$$(Allene) \qquad\qquad \text{iii)}$$

Overall Equation: $NaNH_2 \;+\; \underline{(C)} \longrightarrow NaOH \;+\; NH_3 \;+\; H_2C=C=CH_2$ iv)

Overall Overall Equation: $NaNH_2 \;+\; 3H_3CCOCH_3 \longrightarrow NaOH \;+\; NH_3 \;+\; H_2C=C=CH_2$

$$+\; 2H_2C=C(OH)CH_3 \quad \underline{(C)} \qquad \text{v)}$$

Stage 3:

$$H_2C=C=CH_2 \;\xrightleftharpoons{\text{Self Activation}}\; n\bullet\overset{\displaystyle \overset{H}{|}}{\underset{\displaystyle \underset{H}{|}}{C}}-\overset{\displaystyle \overset{}{}}{\underset{\displaystyle \underset{\overset{\displaystyle \underset{H}{|}}{C}-H}{\|}}{C}}\bullet e$$

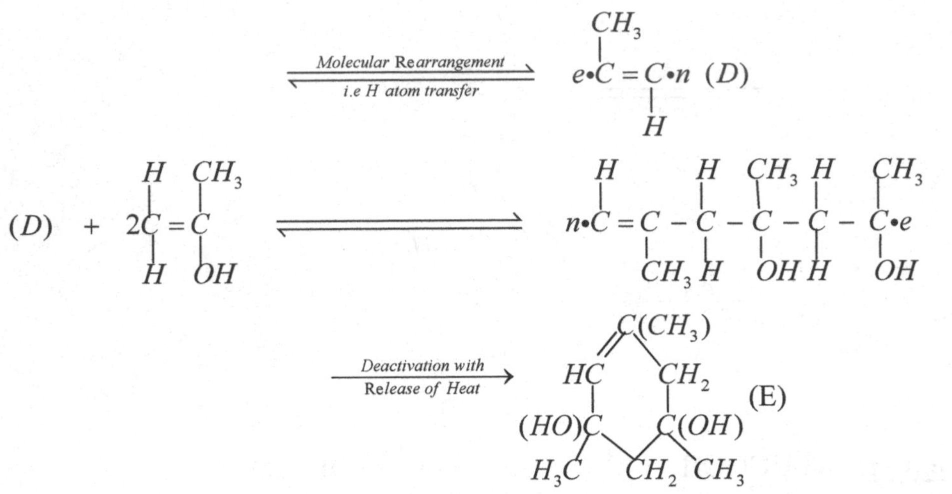

$$\xrightarrow[\text{i.e H atom transfer}]{\text{Molecular Rearrangement}} \quad e \cdot C = C \cdot n \quad (D)$$

with CH_3 above and H below the central carbons.

$$(D) \quad + \quad 2\,{\overset{H}{\underset{H}{C}}} = {\overset{CH_3}{\underset{OH}{C}}} \quad \rightleftharpoons \quad n \cdot {\overset{H}{C}} = {\overset{H}{\underset{CH_3}{C}}} - {\overset{H}{\underset{H}{C}}} - {\overset{CH_3}{\underset{OH}{C}}} - {\overset{H}{\underset{H}{C}}} - {\overset{CH_3}{\underset{OH}{C}}} \cdot e$$

$$\xrightarrow[\text{Release of Heat}]{\text{Deactivation with}} \quad (E)$$

ring structure: $C(CH_3)$ (double bond) HC — CH_2, $(HO)C$ — $C(OH)$, H_3C CH_2 CH_3 (E)

vi)

Overall Equation: $H_2C=C=CH_2 \ + \ 2H_2C=C(OH)CH_3 \xrightarrow[\text{Addition/Cyclization}]{\text{Molecular Rearrangement/}} \ (E)$

vii)

Stage 4:

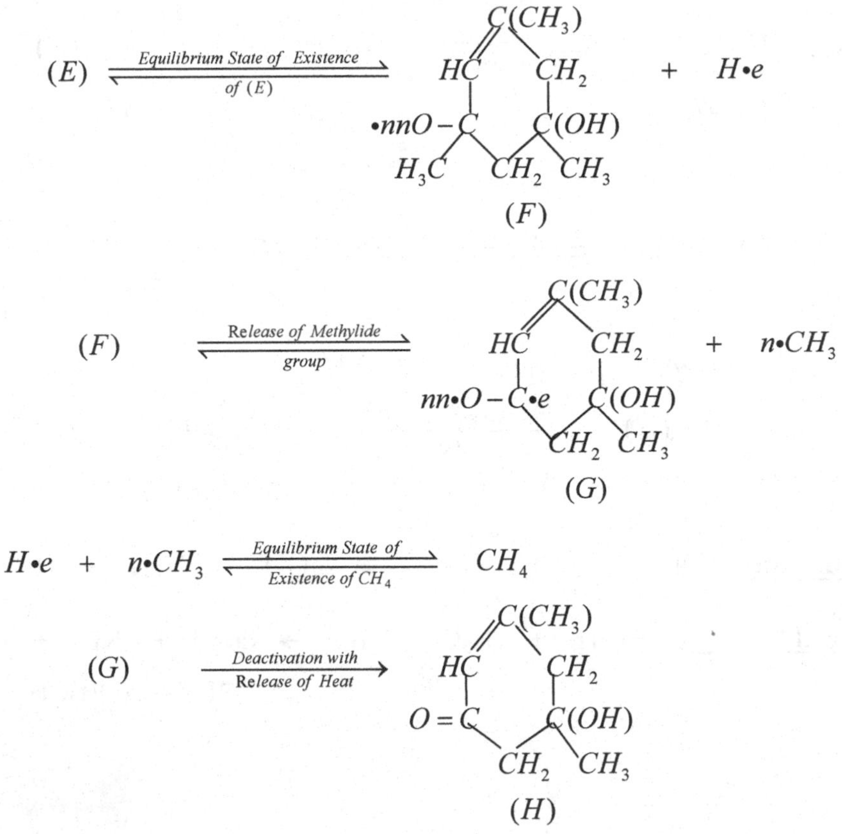

$$(E) \xrightarrow[\text{of } (E)]{\text{Equilibrium State of Existence}} \quad \text{[ring } F\text{]} \quad + \quad H \cdot e$$

ring structure (F): $C(CH_3)$ (double bond) HC — CH_2, $\cdot nnO - C$ — $C(OH)$, H_3C CH_2 CH_3

viii)

$$(F) \xrightarrow[\text{group}]{\text{Release of Methylide}} \quad \text{[ring } G\text{]} \quad + \quad n \cdot CH_3$$

ring structure (G): $C(CH_3)$ (double bond) HC — CH_2, $nn \cdot O - C \cdot e$ — $C(OH)$, CH_2 CH_3

$$H \cdot e \quad + \quad n \cdot CH_3 \xrightarrow[\text{Existence of } CH_4]{\text{Equilibrium State of}} \quad CH_4$$

$$(G) \xrightarrow[\text{Release of Heat}]{\text{Deactivation with}} \quad \text{[ring } H\text{]}$$

ring structure (H): $C(CH_3)$ (double bond) HC — CH_2, $O = C$ — $C(OH)$, CH_2 CH_3

Overall Equation: $(E) \longrightarrow (H) \ + \ CH_4$

ix)

Overall Overall Equation: $NaNH_2 + 3H_3CCOCH_3 \longrightarrow NaOH \ + \ NH_3 \ + \ CH_4 \ + \ (H)$

x)

Stage 5:

$$CH_4 \;\underset{\text{of Methane}}{\overset{\text{Equilibrium State of Existence}}{\rightleftharpoons}}\; H\bullet e \;+\; n\bullet CH_3$$

$$H\bullet e \;+\; NaOH \;\rightleftharpoons\; NaO\bullet en \;+\; H_2 \;+\; Heat$$

$$NaO\bullet en \;\rightleftharpoons\; Na\bullet e \;+\; nn\bullet O\bullet en \;+\; Heat$$

$$Na\bullet e \;+\; n\bullet CH_3 \;\underset{\text{of } NaCH_3}{\overset{\text{Equilibrium State of Existence}}{\rightleftharpoons}}\; NaCH_3$$

$$nn\bullet O\bullet en \;+\; H_2 \;\rightleftharpoons\; H\bullet e \;+\; nn\bullet OH$$

$$H\bullet e \;+\; nn\bullet OH \;\underset{\text{by } NaCH_3}{\overset{\text{Equilibrium State suppressed}}{\longrightarrow}}\; H_2O$$

Overall Equation: $NaOH \;+\; CH_4 \longrightarrow NaCH_3 \;+\; H_2O \;+\; Heat$ xii)

Stage 6: $\quad NaCH_3 \;\underset{\text{of NaCH}}{\overset{\text{Equilibrium State of Existence}}{\rightleftharpoons}}\; Na\bullet e \;+\; n\bullet CH_3$

$$Na\bullet e \;+\; (H) \;\underset{\text{group}}{\overset{\text{Abstraction of OH}}{\rightleftharpoons}}\; (I) \;+\; NaOH$$

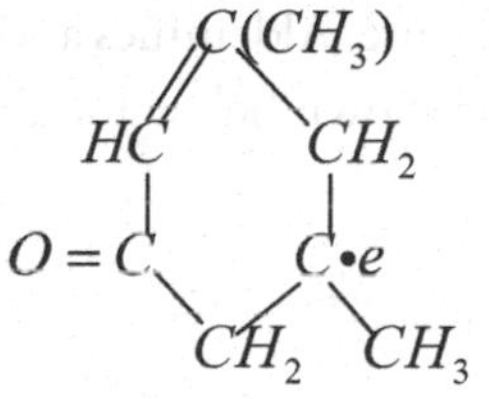

(I)

$$H_3C\bullet n \;+\; (I) \;\underset{\text{Existence of } (J)}{\overset{\text{Combination State of}}{\longrightarrow}}\; (J)$$ xiii)

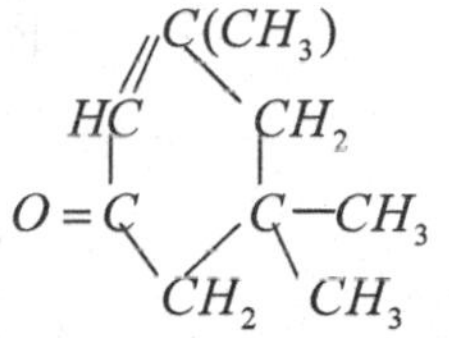

Overall Equation: $NaCH_3 \;+\; (H) \longrightarrow NaOH \;+\; Isophenone$ xvi)

Overall Overall Equation: $NaNH_2 \;+\; 3H_3CCOCH_3 \longrightarrow NaOH \;+\; NH_3 \;+\; Isophenone \;+\; H_2O$ xv)

Stage 7:

$$NH_3 \;\underset{\text{of Ammonia}}{\overset{\text{Equilibrium State of Existence}}{\rightleftharpoons}}\; H\bullet e \;+\; nn\bullet NH_2$$

$$H\bullet e \;+\; NaOH \;\rightleftharpoons\; NaO\bullet en \;+\; H_2 \;+\; Heat$$

$$NaO\bullet en \;\rightleftharpoons\; Na\bullet e \;+\; nn\bullet O\bullet en \;+\; Heat$$

$$Na\bullet e \;+\; nn\bullet NH_2 \;\underset{\text{of Sodium amide}}{\overset{\text{Equilibrium State of Existence}}{\rightleftharpoons}}\; NaNH_2$$

$$nn\bullet O\bullet en \;+\; H_2 \;\rightleftharpoons\; HO\bullet nn \;+\; H\bullet e$$ xvi)

$$H\bullet e \ + \ nn\bullet OH \ \xrightarrow[\text{by } NaNH_2]{\text{\textit{Equilibrium State} sup \textit{pressed}}} \ H_2O$$

Overall Equation: $NaOH \ + \ NH_3 \longrightarrow NaNH_2 \ + \ H_2O \ + \ Heat$ xvii)

Overall Overall Equation: $NaNH_2 \ + 3H_3CCOCH_3 \longrightarrow NaNH_2 \ + \ Isophenone$

$$2H_2O \ + \ Heat \qquad \text{xviii)}$$

seven stage Equilibrium mechanism system, taking note of the reaction between NaOH and methane in Stage 5, reactions between NaOH and ammonia in Stage 7 where the only one mole of $NaNH_2$ used was recovered, for which reason only one mole of the enol formed was converted to allene which rearranged in Stage 3 to start adding to remaining enols to form an unstable ring (E), which was finally converted to (J) the isophenone.

(Laws of Creations for Cyclization of Ketones)

<u>**Rule 297:**</u> This rule of Chemistry for **Addition of Ammonia to formaldehyde, Aldehydes, and Ketones,** states that, formaldimine, Aldemines and Ketimines are produced as well as cyclic trimers based on the operating conditions as shown below for aldehydes-

<u>**Stage 1:**</u> NH_3 $H\bullet e \ + \ nn\bullet NH_2$

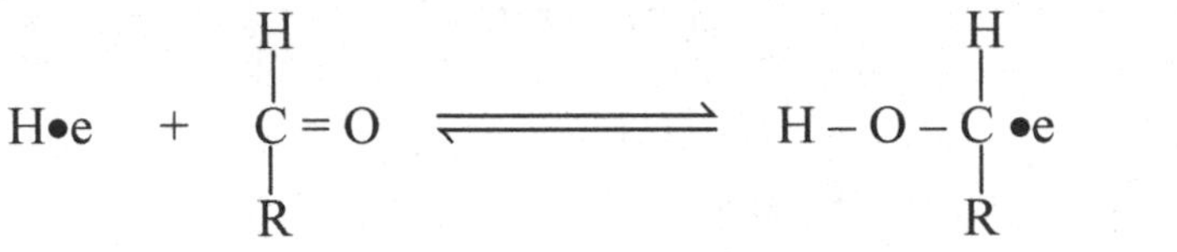

$$H\bullet e \ + \ \overset{\displaystyle H}{\underset{\displaystyle R}{C}}=O \ \rightleftharpoons \ H-O-\overset{\displaystyle H}{\underset{\displaystyle R}{C}}\bullet e$$

$$(A)$$

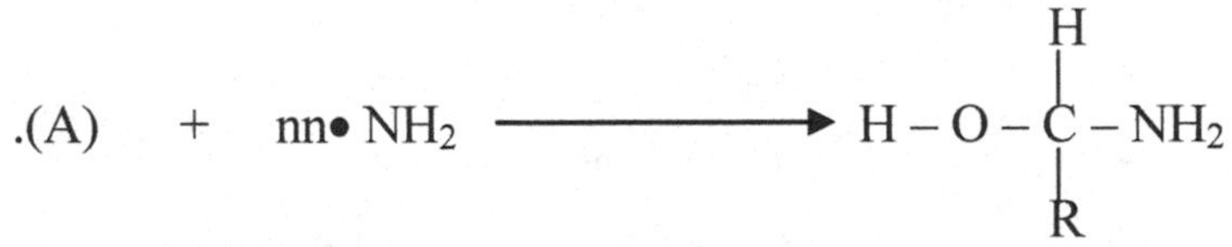

$$.(A) \ + \ nn\bullet NH_2 \longrightarrow H-O-\overset{\displaystyle H}{\underset{\displaystyle R}{C}}-NH_2$$

$$(B) \qquad\qquad\qquad \text{i)}$$

<u>**Stage 2:**</u>

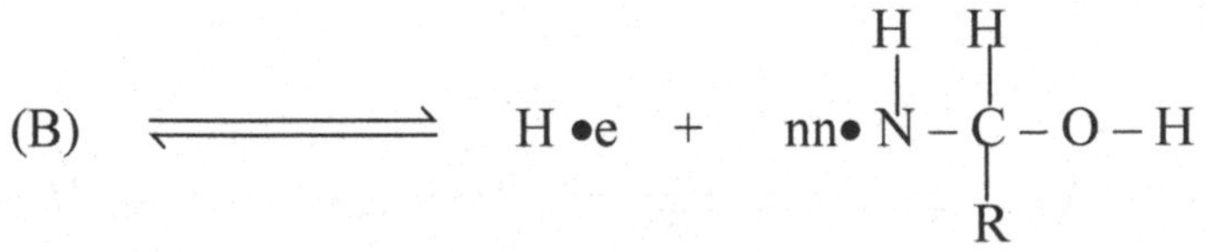

$$(B) \ \rightleftharpoons \ H\bullet e \ + \ nn\bullet \overset{\displaystyle H \ \ H}{\underset{\displaystyle R}{N-C}}-O-H$$

$$(C)$$

$$(C) \ \rightleftharpoons \ HO\bullet nn \ + \ nn\bullet \overset{\displaystyle H \ \ H}{\underset{\displaystyle R}{N-C}}\bullet e$$

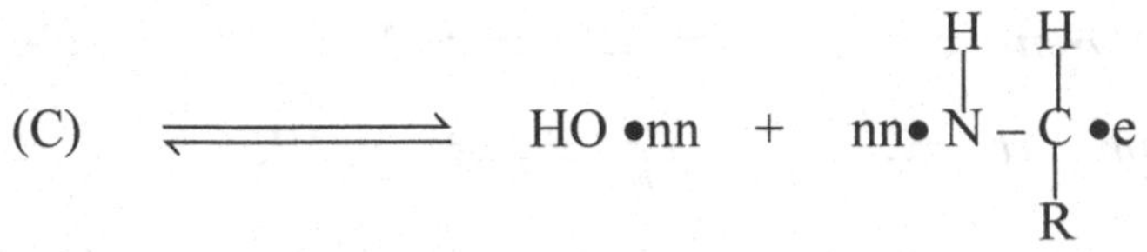

$$(D)$$

$$H\bullet e \ + \ nn\bullet OH \ \rightleftharpoons \ H_2O$$

$$(D) \longrightarrow HN=CHR \ + \ Heat \qquad\qquad \text{ii)}$$

Stage 3:

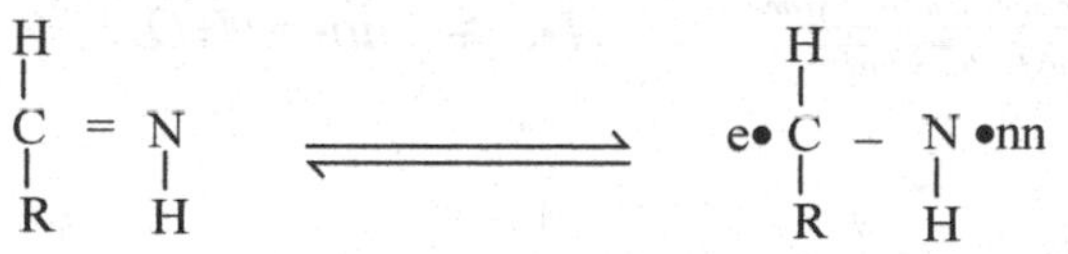

$$\underset{R}{\overset{H}{\underset{|}{C}}} = \underset{H}{N} \;\; \rightleftharpoons \;\; e\bullet \underset{R}{\overset{H}{\underset{|}{C}}} - \underset{H}{\overset{|}{N}} \bullet nn$$

(E)

$$(E) \;\; + \;\; 2HN = CHR \;\; \longrightarrow$$

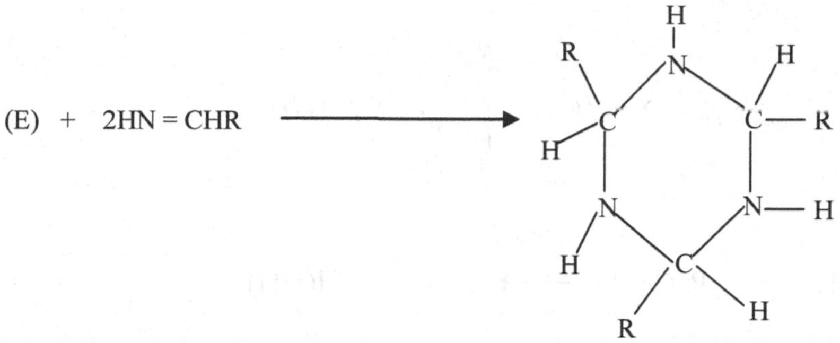

iii)

Overall Equation: $3NH_3 \;\; + \;\; 3RCHO \;\; \longrightarrow \;\; 3H_2O \;\; + \;\;$ Cyclic Aldimine Trimer

$+ \;\;$ Heat

iv)

a three stage Equilibrium mechanism system, in which formaldimines were formed in Stage 2 which heat cyclized to give a tricyclic amine.
(Laws of Creations for cyclization of Aldehydes/NH₃)

Rule 298: This rule of Chemistry for **Addition of hydroxylamine to formaldehyde, aldehydes and Ketones,** states that, Formaldoxime, Aldoxime and Ketoxime are produeed as shown below depending on the operating conditions-

$$H_2C{=}O \;\; + \;\; H_2NOH \;\; \longrightarrow \;\; H_2C{=} N(OH) \;\; + \;\; H_2O$$
(A Formaldoxime)

$$RCHO \;\; + \;\; H_2NOH \;\; \longrightarrow \;\; RCH{=}N(OH) \;\; + \;\; H_2O$$
(An aldoxime)

$$R_2C{=}O \;\; + \;\; H_2NOH \;\; \longrightarrow \;\; R_2C{=}N(OH) \;\; + \;\; H_2O$$
(A ketoxime)

with only the mechanism of the third reaction provided below since the same will apply to aldehydes.

538

Stage 1: $H_2NOH \xrightleftharpoons[\text{of Hydroxylamine}]{\text{Equilibrium State of Existence}} H\bullet e + nn\bullet NH(OH)$

$$H\bullet e + O=\overset{\overset{R}{|}}{\underset{\underset{R}{|}}{C}} \xrightleftharpoons[\text{Addition to it}]{\text{Activation of Ketone and}} H-O-\overset{\overset{R}{|}}{\underset{\underset{R}{|}}{C}}\bullet e \quad (A)$$

$$(A) + nn\bullet NH(OH) \xrightarrow[\text{Existence of }(B)]{\text{Combination State of}} H-O-\overset{\overset{R}{|}}{\underset{\underset{R}{|}}{C}}-\overset{}{\underset{\underset{H}{|}}{N}}-OH \quad (B)$$

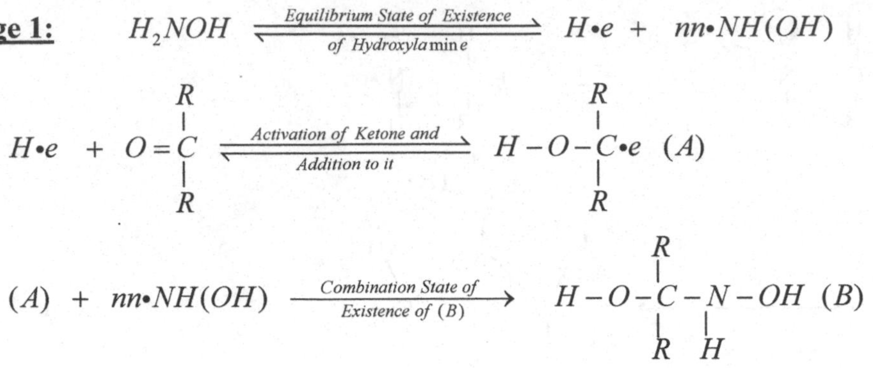

i)

Overall Equation: $H_2NOH + R_2C=O \longrightarrow HOCR_2\text{-}NH(OH)$ ii)

Stage 2:

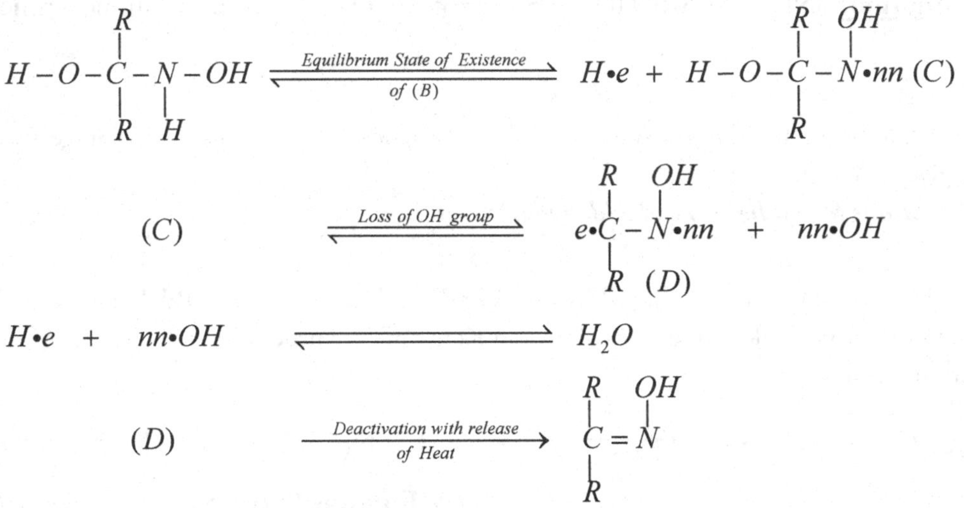

$$H-O-\overset{\overset{R}{|}}{\underset{\underset{H}{|}}{C}}-\overset{}{\underset{\underset{}{}}{N}}-OH \xrightleftharpoons[\text{of }(B)]{\text{Equilibrium State of Existence}} H\bullet e + H-O-\overset{\overset{R\;\;OH}{|\;\;\;|}}{\underset{\underset{R}{|}}{C}}-N\bullet nn \quad (C)$$

$$(C) \xrightleftharpoons[]{\text{Loss of OH group}} e\bullet\overset{\overset{R\;\;OH}{|\;\;\;|}}{\underset{\underset{R}{|}}{C}}-N\bullet nn + nn\bullet OH \quad (D)$$

iii)

$$H\bullet e + nn\bullet OH \xrightleftharpoons[]{} H_2O$$

$$(D) \xrightarrow[\text{of Heat}]{\text{Deactivation with release}} \overset{\overset{R\;\;OH}{|\;\;\;|}}{\underset{\underset{R}{|}}{C}}=N$$

A ketoxime

Overall Equation: $H_2NOH + R_2C=O \longrightarrow R_2C=N(OH) + H_2O$ iv)

for which this is a clear indication of the strong influence and relationship between C = O and C = N families of compounds via ammonia and hydroxylamine.
(Laws of Creations between Families of Compounds)

Rule 299: This rule of Chemistry for **Formaldimines, Aldimines and Ketimines,** states that, when the H on the Nitrogen center is changed to NH_2 group, their original families change as shown below for formaldimine and aldimines-

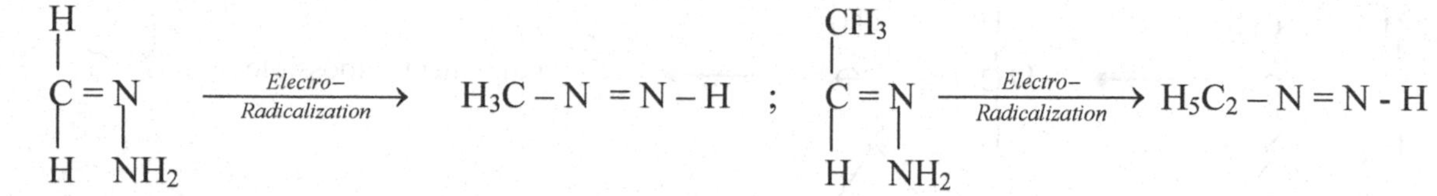

$$\underset{\underset{\displaystyle H}{|}}{\overset{\overset{\displaystyle H}{|}}{C}} = \underset{\underset{\displaystyle NH_2}{|}}{N} \xrightarrow[\textit{Radicalization}]{\textit{Electro-}} H_3C - N = N - H \quad ; \quad \underset{\underset{\displaystyle H}{|}}{\overset{\overset{\displaystyle CH_3}{|}}{C}} = \underset{\underset{\displaystyle NH_2}{|}}{N} \xrightarrow[\textit{Radicalization}]{\textit{Electro-}} H_5C_2 - N = N - H$$

Hydrazone   **Azo-methane**   **Acetaldehyde hydrazone**   **Azo-ethane**

for which the same for aldimines apply to ketimines.

(Laws of Creations- Between Families of Compounds)

Rule 300: This rule of Chemistry for **Some compounds during rearrangements,** states that, while Molecular rearrangement phenomenon involves movement from Less Stable State of Existence to a More Stable State of Existence, Enolization/Enamidization phenomena involve the reverse movement for the purpose of orderliness based on different operating conditions, *a clear indication of **the great importance of the influence of Operating Conditions.***

(Laws of Creations for influence of Operating Conditions)

Rule 301: This rule of Chemistry for **Aldimines and Ketimines,** states that, the origin of these compounds/monomers are vinyl amines of the primary and secondary types those which are very unstable; for which they molecularly rearrange as shown below-

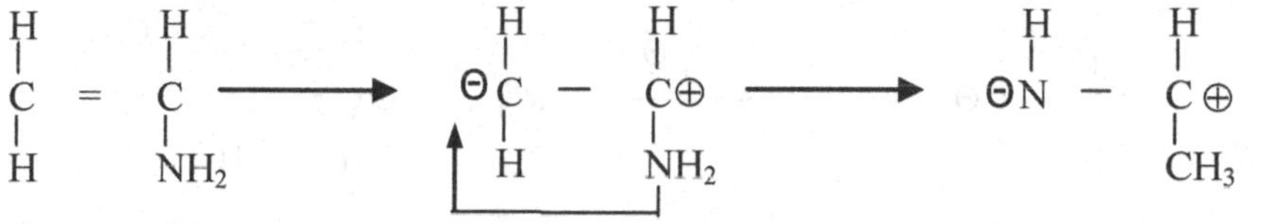

$$\underset{\underset{\displaystyle H}{|}}{\overset{\overset{\displaystyle H}{|}}{C}} = \underset{\underset{\displaystyle NH_2}{|}}{\overset{\overset{\displaystyle H}{|}}{C}} \longrightarrow \overset{\ominus}{\underset{\underset{\displaystyle H}{|}}{\overset{\overset{\displaystyle H}{|}}{C}}} - \overset{\oplus}{\underset{\underset{\displaystyle NH_2}{|}}{\overset{\overset{\displaystyle H}{|}}{C}}} \longrightarrow \overset{\ominus}{\underset{}{N}} - \overset{\oplus}{\underset{\underset{\displaystyle CH_3}{|}}{\overset{\overset{\displaystyle H}{|}}{C}}}$$

(I) (Unstable)

<u>An Activated Aldimine</u>   i)

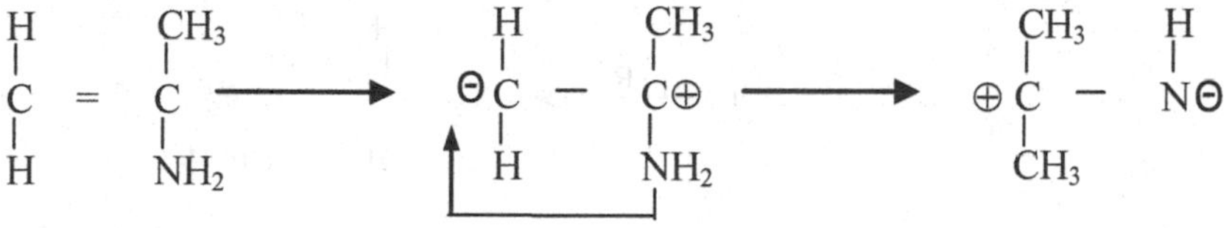

$$\underset{\underset{\displaystyle H}{|}}{\overset{\overset{\displaystyle H}{|}}{C}} = \underset{\underset{\displaystyle NH_2}{|}}{\overset{\overset{\displaystyle CH_3}{|}}{C}} \longrightarrow \overset{\ominus}{\underset{\underset{\displaystyle H}{|}}{\overset{\overset{\displaystyle H}{|}}{C}}} - \overset{\oplus}{\underset{\underset{\displaystyle NH_2}{|}}{\overset{\overset{\displaystyle CH_3}{|}}{C}}} \longrightarrow \overset{\oplus}{\underset{\underset{\displaystyle CH_3}{|}}{\overset{\overset{\displaystyle CH_3}{|}}{C}}} - \overset{\ominus}{\underset{}{N}}\overset{\overset{\displaystyle H}{|}}{}$$

(II) (Unstable)

<u>An Activated Ketimine</u>   ii)

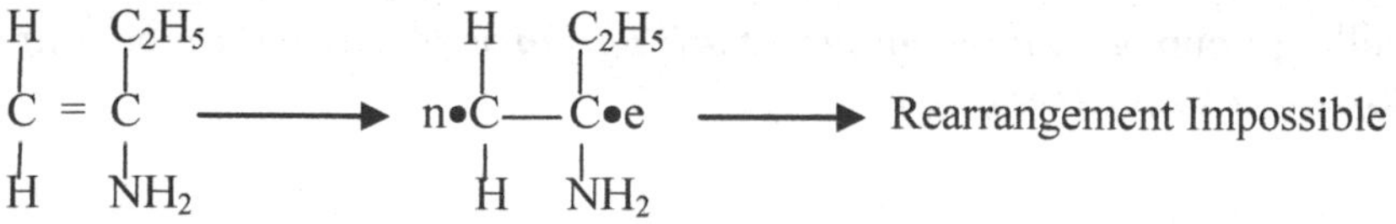

$$\underset{\underset{\displaystyle NH_2}{|}}{\overset{\overset{\displaystyle H}{|}}{C}} = \underset{\underset{\displaystyle NH_2}{|}}{\overset{\overset{\displaystyle C_2H_5}{|}}{C}} \longrightarrow \underset{\underset{\displaystyle NH_2}{|}}{n\bullet\overset{\overset{\displaystyle H}{|}}{C}} - \underset{\underset{\displaystyle NH_2}{|}}{\overset{\overset{\displaystyle C_2H_5}{|}}{C}}\bullet e \longrightarrow \text{Rearrangement Impossible}$$

(III) (Stable)   iii)

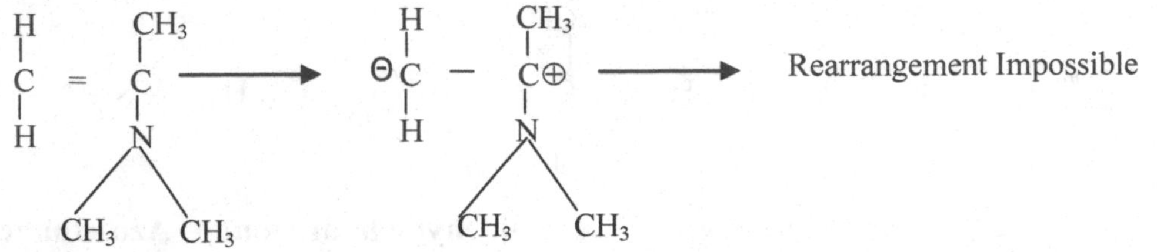

iv)

(IV) (Stable)

for which while (I) and (II) are very unstable (i.e. cannot favor isolated existence), (III) and (IV) are quite stable, since they will not molecularly rearrange or electroradicalize or enalmidize; for which the following is valid-

$$C = N \quad > \quad C = C$$

Order of Nucleophilicty

(Laws of Creations for Vinyl amines)

Rule 302: This rule of Chemistry for **Alkyl nitrogen substituted Aldimine where one H atom on the C center is replaced by a vinyl or alkenyl group,** states that, this monomer is an electrophile as shown below chargedly-

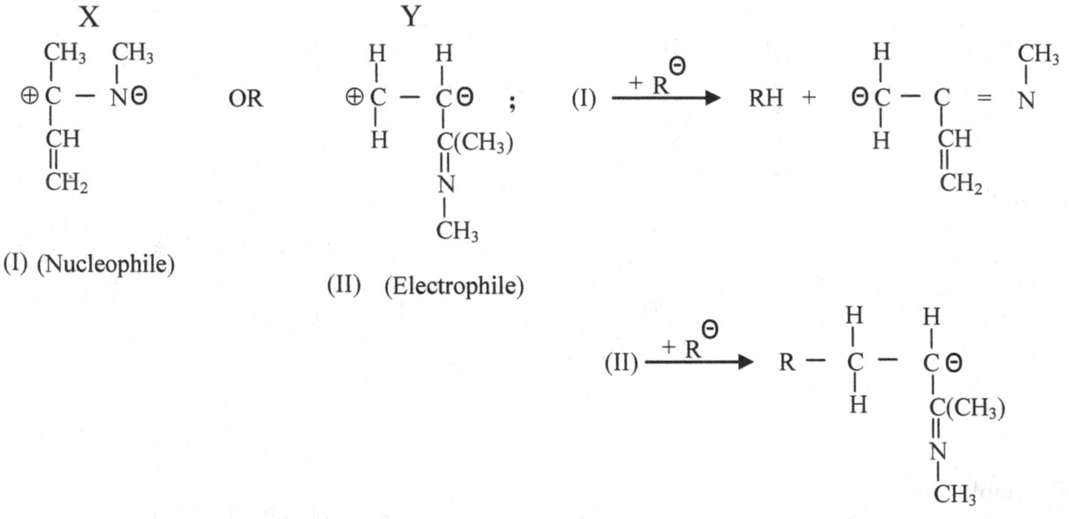

(I) (Nucleophile)

(II)   (Electrophile)

for which -RC = NR group is a pulling group, *with no resonance stabilization* provided for the monomer. *(Laws of Creations for N-Substituted Vinyl aldimine)*

Rule 303: This rule of Chemistry for **Formaldimine, Aldimine and Ketimines,** states that, when the H on the Nitrogen center is replaced with pulling groups such as $-COOR$, $-CONH_2$, $-COR$, $-CN$, and the like, the monomers are all Electrophiles (MALES) of different capacities as shown below-

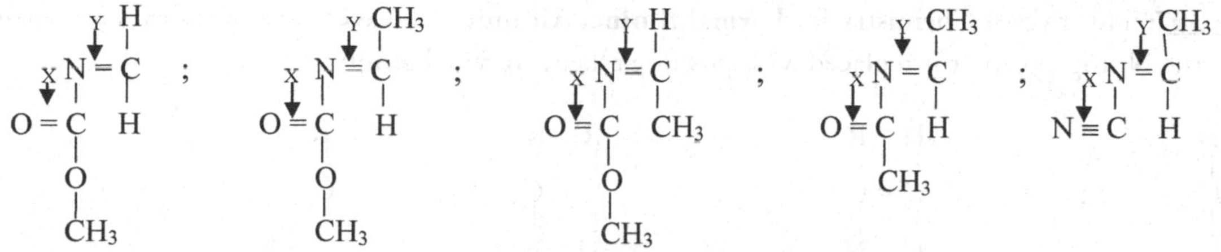

ALL MALES OF DIFFERENT CAPACITIES

for which (I) is a substituted formaldimine, a strong Electrophile. (II) is trans-placed more Elec-trophilic than (III) which is cis-placed, (IV) like in most other families is a Weak Electrophile and (V) is the strongest electrophile, all substituted aldimines as apply to substituted Ketimines. *(Laws of Creations for Electrophilic Formaldi- Aldi- and Keti- Mines)*

Rule 304: This rule of Chemistry for **All cis-placed monomers with both types of transfer species on both sides of the active centers,** states that, these monomers can never be initiated and grow with either free media or paired media initiators; *for which they cannot be used as monomers, but as compounds to produce other products.*
(Laws of Creations for Cis-placed Monomers)

Rule 305: This rule of Chemistry for **Monomers which have cis- and trans- placed substi-tuent groups with both types of transfer species on both sides of the active centers,** states that, when Molecular rearrangement takes place, free-media initiators can initiate and polymerize them only in the route Natural to them.
(Laws of Creations for Trans- or Cis-placed Electrophiles)

Rule 306: This rule of Chemistry for **Formaldimine, Aldimine and Ketimines,** states that, when the H on the Nitrogen center is replaced with pulling groups such as –COOR, -CONH$_2$, -COR, -CN, and the like, the monomers are not only Males of different capacities, but are found to be all resonance stabilized as shown below-

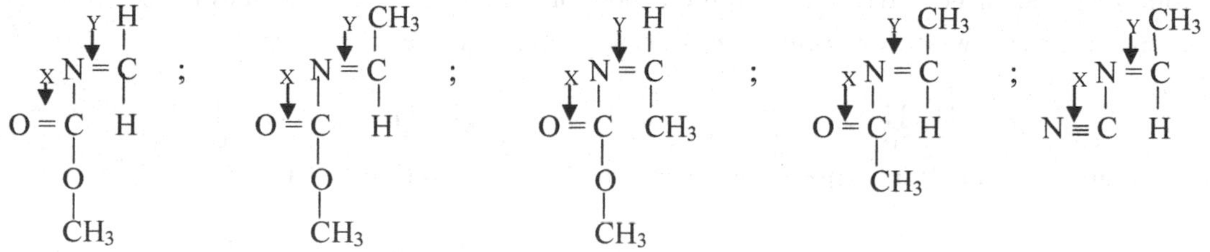

ALL MALES OF DIFFERENT CAPACITIES

with two mono-forms-3,4- and 1,4- mono-forms; for which it can be observed that the OCH$_3$ group is shielded.
(Laws of Creation for Resonance Stabilization of a Non-Olefinic Electrophiles)

Rule 307: This rule of Chemistry for **Formaldimine, Aldimine and Ketimines,** states that, when the H on the Nitrogen center is replaced with pushing alkenyl or vinyl group such as

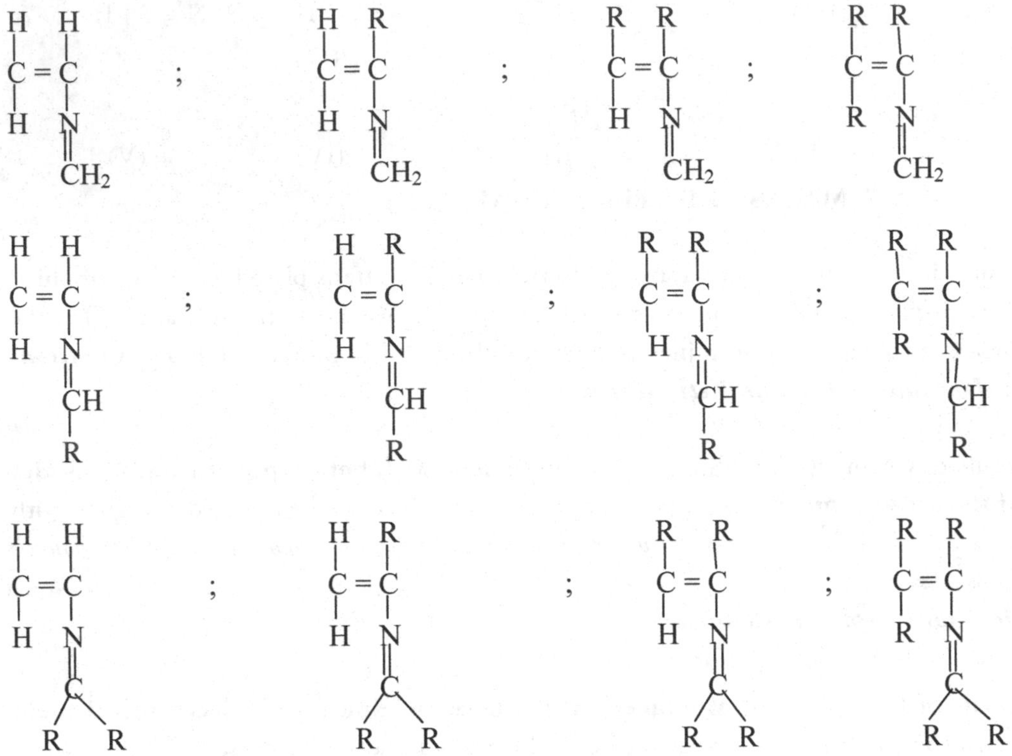

$$
\begin{array}{cccc}
\begin{array}{c} H\ \ H \\ |\ \ \ | \\ C=C \\ |\ \ \ | \\ H\ \ N \\ \ \ \ \ \| \\ \ \ \ \ CH_2 \end{array} &
\begin{array}{c} H\ \ R \\ |\ \ \ | \\ C=C \\ |\ \ \ | \\ H\ \ N \\ \ \ \ \ \| \\ \ \ \ \ CH_2 \end{array} &
\begin{array}{c} R\ \ R \\ |\ \ \ | \\ C=C \\ |\ \ \ | \\ H\ \ N \\ \ \ \ \ \| \\ \ \ \ \ CH_2 \end{array} &
\begin{array}{c} R\ \ R \\ |\ \ \ | \\ C=C \\ |\ \ \ | \\ R\ \ N \\ \ \ \ \ \| \\ \ \ \ \ CH_2 \end{array}
\end{array}
$$

$$
\begin{array}{cccc}
\begin{array}{c} H\ \ H \\ |\ \ \ | \\ C=C \\ |\ \ \ | \\ H\ \ N \\ \ \ \ \ \| \\ \ \ \ \ CH \\ \ \ \ \ | \\ \ \ \ \ R \end{array} &
\begin{array}{c} H\ \ R \\ |\ \ \ | \\ C=C \\ |\ \ \ | \\ H\ \ N \\ \ \ \ \ \| \\ \ \ \ \ CH \\ \ \ \ \ | \\ \ \ \ \ R \end{array} &
\begin{array}{c} R\ \ R \\ |\ \ \ | \\ C=C \\ |\ \ \ | \\ H\ \ N \\ \ \ \ \ \| \\ \ \ \ \ CH \\ \ \ \ \ | \\ \ \ \ \ R \end{array} &
\begin{array}{c} R\ \ R \\ |\ \ \ | \\ C=C \\ |\ \ \ | \\ R\ \ N \\ \ \ \ \ \| \\ \ \ \ \ CH \\ \ \ \ \ | \\ \ \ \ \ R \end{array}
\end{array}
$$

$$
\begin{array}{cccc}
\begin{array}{c} H\ \ H \\ |\ \ \ | \\ C=C \\ |\ \ \ | \\ H\ \ N \\ \ \ \ \ \| \\ \ \ \ \ C \\ \ \ \ /\ \backslash \\ R\ \ \ \ R \end{array} &
\begin{array}{c} H\ \ R \\ |\ \ \ | \\ C=C \\ |\ \ \ | \\ H\ \ N \\ \ \ \ \ \| \\ \ \ \ \ C \\ \ \ \ /\ \backslash \\ R\ \ \ \ R \end{array} &
\begin{array}{c} R\ \ R \\ |\ \ \ | \\ C=C \\ |\ \ \ | \\ H\ \ N \\ \ \ \ \ \| \\ \ \ \ \ C \\ \ \ \ /\ \backslash \\ R\ \ \ \ R \end{array} &
\begin{array}{c} R\ \ R \\ |\ \ \ | \\ C=C \\ |\ \ \ | \\ R\ \ N \\ \ \ \ \ \| \\ \ \ \ \ C \\ \ \ \ /\ \backslash \\ R\ \ \ \ R \end{array}
\end{array}
$$

Vinyl Imines

for which, they are all FEMALES, i.e. NUCLEOPHILES, and are *Resonance stabilized,* the group providing resonance stabilization being the imines $-N=CH_2$, $-N=CHR$, $-N=CR_2$; noting that group shielded is the R group internally located.
(Laws of Creations for Vinyl Imines)

Rule 308: This rule of Chemistry for **Vinyl Imines,** states that, while their formaldimines 3,4- and 1,4- mono-forms favor both free-radical routes, those of Aldimines and Ketimines favor only the route Natural to them as shown below using a simple case of an Aldimine.

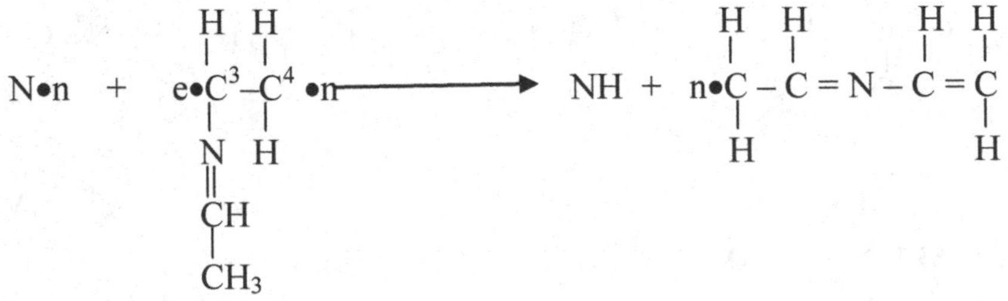

$$
N\bullet n\ +\ e\bullet\overset{}{C^3}-\overset{}{C^4}\bullet n \longrightarrow NH\ +\ n\bullet C-C=N-C=C
$$

(structure: $N\bullet n + e\bullet C^3(H)(N=CH-CH_3)-C^4(H)(H)\bullet n \rightarrow NH + n\bullet CH_2-CH=N-CH=CH_2$)

543

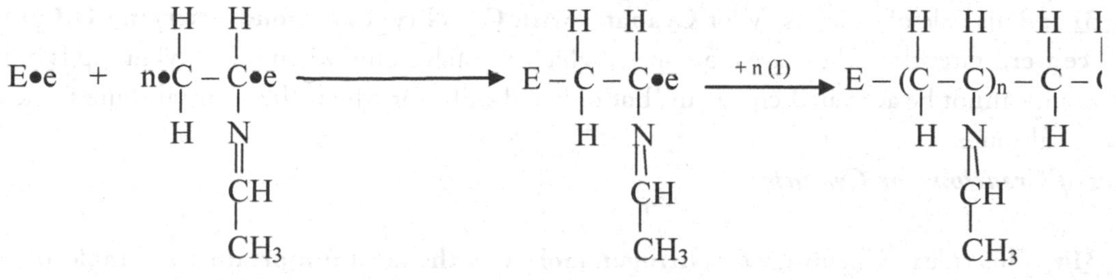

$$
E \bullet e \;+\; n \bullet \underset{\underset{\underset{\underset{CH_3}{|}}{CH}}{\underset{|}{N}}}{\overset{\overset{H}{|}}{\underset{|}{C}}} - \underset{\underset{}{}}{\overset{\overset{H}{|}}{\underset{H}{C}}} \bullet e \longrightarrow E - \underset{\underset{}{}}{\overset{\overset{H}{|}}{\underset{H}{C}}} - \underset{\underset{\underset{\underset{CH_3}{|}}{CH}}{\underset{|}{N}}}{\overset{\overset{H}{|}}{\underset{|}{C}}} \bullet e \xrightarrow{+ n\,(I)} E - (\underset{\underset{\underset{\underset{CH_3}{|}}{CH}}{\underset{|}{N}}}{\overset{\overset{H}{|}}{\underset{|}{C}}} - \overset{\overset{H}{|}}{\underset{\underset{H}{|}}{C}})_n - \overset{\overset{H}{|}}{\underset{|}{C}} - \overset{\overset{H}{|}}{C}
$$

(I)

for which the transfer species involved above is still of the first kind and of the first type, as clearly seen in the 1,4-mono-form, the more stable of the two mono-forms where steric hindrance is no limitation.
(Laws of Creations for Vinyl Imines)

Rule 309: This rule of Chemistry for **Vinyl Imines,** states that, only those vinyl groups which carry transfer species internally located such as R' in $-CR' = CH_2$, $- CR' = CHR$, $- CR' = CR_2$, can be shielded by the resonance stabilization group whose capacity is as shown below-

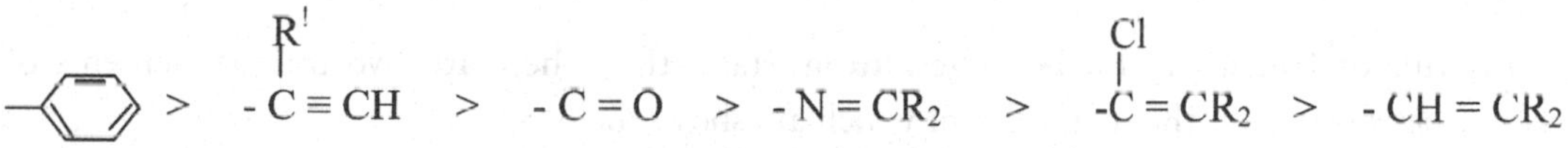

$$
-\langle \text{C}_6\text{H}_x \rangle \;>\; -\overset{\overset{R'}{|}}{C} \equiv CH \;>\; - C = O \;>\; - N = CR_2 \;>\; -\overset{\overset{Cl}{|}}{C} = CR_2 \;>\; - CH = CR_2
$$

Order of Resonance Stabilization capacities   [R' is OR, R, NH_2, NR_2,....]

(Laws of Creations for Resonance Stabilization Groups)

Rule 310: This rule of Chemistry for **Formaldimine, Aldimine and Ketimines,** states that, where when the H on the Nitrogen center is replaced with pulling groups such as $-COOR$, $-CONH_2$, $-COR$, $-CN$, and the like, the monomers were found not only to be Males of different capacities, but also found to be all resonance stabilized (See Rule 306); they can never molecularly rearrange.
(Laws of Creations for Non-Olefinic Electrophiles)

Rule 311: This rule of Chemistry for **Olefinic and Non-Olefinic monomers that undergo de-propagation during polymerization forwardly and backwardly,** states that, their Living growing polymer chains when depropagation is eliminated cannot be killed from within, but by external means using different methods for both of the types of monomers. [See Rule 290]
(Laws of Creations for Termination of RURP and RNRP Monomers)

Rule 312: This rule of Chemistry for **Paired Radical initiators,** states that, these can only exist between centers with covalent and ionic types of bonds and not from centers which are electrostatically bonded and the two centers must be such with **large electropositive/electro- negative potential difference, such as that between Na (0.9) and C (2.5) if not of the same or different metals or O (3.5) and Na (0.9); for if of the same or different metals, the condition above does not apply.**
(Laws of Creations for Paired-Radical Initiators)

Rule 313: This rule of Chemistry for **Cyanates with C ≡ N types of bonds carrying OR groups on the C center,** states that, these are monomers, which are stable only when the alkyl group is bulky, and that which cannot be activated chargedly, but only radically, for which the Natural route is the electro-free-radical route.
(Laws of Creations for Cyanates)

Rule 314: This rule of Chemistry for **Nitrogen molecule, the most important basic molecule in Life**, states that, this molecule with triple bond (N ≡ N) cannot be activated chargedly, but only radically, for which to break the bond, one has to leave our environment to another environment which is SURFACE CHEMISTRY that which takes place in capillary pores carried by some Transition metallic compounds or their atoms/elements.
(Laws of Creations for Nitrogen)

Rule 315: This rule of Chemistry for **All ATOMS,** states that, the number of elements present for them, depends on the valency state of the Atom and the type of Atom in terms of the Group to which it belongs to in the Periodic Table, noting that ionic metallic atoms (but H) do not have elements.
(Laws of Creations- for Atoms and their Elements)

Rule 316: This rule of Chemistry for **Nitrogen atom,** states that, there are two unique elements of Nitrogen being electronegative, the structures of which are shown below-

$$
\overset{\bullet nn}{\underset{\bullet nn}{:N\bullet nn}} \quad ; \quad \overset{\bullet en}{\underset{\bullet en}{:N\bullet nn}} \quad ; \quad \overset{\bullet nn}{\underset{\bullet nn}{:N\bullet en}}
$$

The Atom   *Elements of N – Agents*

for which, for the high electronegativity (like the case of O) and high valence state of N, the elements can easily be identified in SURFACE CHEMISTRY.
(Laws of Creations for Elements of N Atom)

Rule 317: This rule of Chemistry for **Compounds such as R – N = N – R', where R and R' are alkyl types of groups of different or same capacities greater than H (some less than H),** states that, the activation center <u>cannot be activated radically or chargedly</u>, because the Carbon center(s) when adjacently located to the Nitrogen centers will not allow it, since C is more electropositive than N; noting however that the π-bond can only be broken not by activation but by other means including the pores of Transition metals (SURFACE CHEMISTRY) beginning from the R group(s).
(Laws of Creations for Orderliness in Electronegativity)

Rule 318: This rule of Chemistry for **Compounds such as R – N = N – R' where R and R' are radical-pushing such as NH₂, OH, types and radical-pulling such as Cl, F, and the likes,** states that, the activation center <u>can be activated only radically,</u> because the centers adjacently located to the Nitrogen center is less electropositive than N.
(Laws of Creations for Orderliness in Electronegativity)

Rule 319: This rule of Chemistry for **Family members of compounds called DIAZENES OR DIIMINES of which AZO-compounds are members,** states that, only the first member of those carrying special radical-pushing groups and special radical-pulling groups can be activated radically, based on the operating conditions, as shown below-

$$H_3C - N = N - CH_3 \quad ; \quad H - N = N - H \quad ; \quad Cl - N = N - Cl \quad ; \quad HO - N = N - OH$$

    Polar/Non-ionic   Polar/Non-ionic   Very Polar/Non-ionic   Polar/Ionic

    Cannot be Activated   **Can be Activated only Radically**

(Laws of Creations for N containing compounds)

Rule 320: This rule of Chemistry for **Diazo-Alkylenes and the likes which are radically /polarly resonance stabilized,** states that, these are LIVING ACTIVATED monomers only RADICALLY in the presence of highly polar passive catalysts when very low operating conditions are involved.
(Laws of Creations for Diazo-alkylenes)

Rule 321: This rule of Chemistry for **Diazo-Alkylenes which are radically/polarly resonance stabilized,** states that, chemically never is there a time when they react chargedly but only radically; for which when heat is applied in the absence of a catalyst, they explode to release alkylenes; while polymerically, they can react chargedly only positively but not ionically.
(Laws of Creations for Diazo-alkylenes)

Rule 322: This rule of Chemistry for **Diazo-Alkylenes which are radically/polarly resonance stabilized,** states that, these are unique monomers which in the presence of unique passive catalysts such as cuprous chloride, AlR_3, $ClAlR_2$, ROR/BF_3 shown below, decompose at very low temperatures to release alkylenes.

$$CuCl \quad \underset{Existence}{\overset{Equilibrium\ State\ of}{\rightleftharpoons}} \quad Cl\bullet nn + Cu\bullet en$$

$$AlR_3 \quad \underset{Existence}{\overset{Equilibrium\ State\ of}{\rightleftharpoons}} \quad R\bullet e + n\bullet AlR_2 \quad (I)$$

$$ClAlR_2 \quad \underset{Existence}{\overset{Equilibrium\ State\ of}{\rightleftharpoons}} \quad R_2Al\bullet e + Cl\bullet nn \quad (II)$$

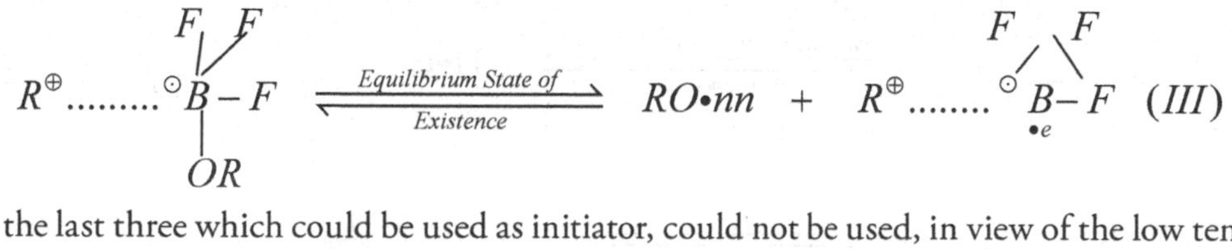

$$R^{\oplus}.........^{\circleddash}\!\!\overset{\overset{F\diagdown \diagup F}{|}}{\underset{|}{B}}\!\!\!-F \quad \underset{Existence}{\overset{Equilibrium\ State\ of}{\rightleftharpoons}} \quad RO\bullet nn + R^{\oplus}.........^{\circleddash}\!\!\overset{\overset{F\diagup \diagdown F}{}}{\underset{\bullet e}{B}}\!\!\!-F \quad (III)$$

$$OR$$

in which the last three which could be used as initiator, could not be used, in view of the low temperature (-20^0C).
(Laws of Creations for Diazo-alkylenes)

Rule 323: This rule of Chemistry for **Homologenation phenomena,** states that, this is addition of CH_2 groups to unsaturated monomers or compounds either to produce higher homologs of the compound or to expand the size of a ring.
(Laws of Creations for Homologenation phenomena)

Rule 324: This rule of Chemistry for **Addition of Alkylenes to unsaturated monomers – Alkenes, Alkynes, Olefines, Non-olefines, ringed or non-ringed,** states that, the concept of this type of Addition is called **HOMOLOGENATION of the first kind and the first type,** the source of the Alkylenes from Diazo- Alkylenes *in the presence of specific passive catalysts* wherein the compound to be homologenated must be in **EQUILIBRIUM STATE OF EXISTENCE.**
(Laws of Creations for Homologenation phenomena)

Rule 325: This rule of Chemistry for **Addition of Alkylenes to unsaturated monomers – Alkenes, Alkynes, Olefins, Non-olefines, ringed or non-ringed,** states that, the concept of this type of Addition is called **HOMOLOGENATION of the first kind and second type,** the source of the Alkylenes from Diazo-Alkylenes *in the presence of passive catalysts* wherein the compound to be homologenated must be in **ACTIVATED/EQUIL. STATE OF EXISTENCE.**
(Laws of Creations for Homologenation phenomena)

Rule 326: This rule of Chemistry for **HOMOLOGENATION of the first kind of the first type,** states that, when Formaldehyde or the like is involved, Methyl aldehyde or the like is obtained and when Aldehydes or the likes are involved, Methyl ketones or the likes are obtained and when ketones or the likes are involved, higher homologs are obtained.
(Laws of Creations for Homologenation of the first kind of first type))

Rule 327: This rule of Chemistry for **HOMOLOGENATION of the first kind of the second type,** states that, when a cyclic Ketone such as cyclopentanone is involved, the ring is increased stepwisely in size from for example five-membered to six-membered and then to seven-mem-bered rings and so on in that natural order, depending on the number of moles of the Diazo-Alkylenes present in the system, presence of perfect mixing and the type of passive catalyst used. *(Laws of Creations for enlargement of rings by Homologenation)*

Rule 328: This rule of Chemistry for **Cyclic ketones such as cyclopentanone,** states that, when it is involved during ring expansion, the first step is as shown below, wherein the ring first undergoes ENOLIZATION-

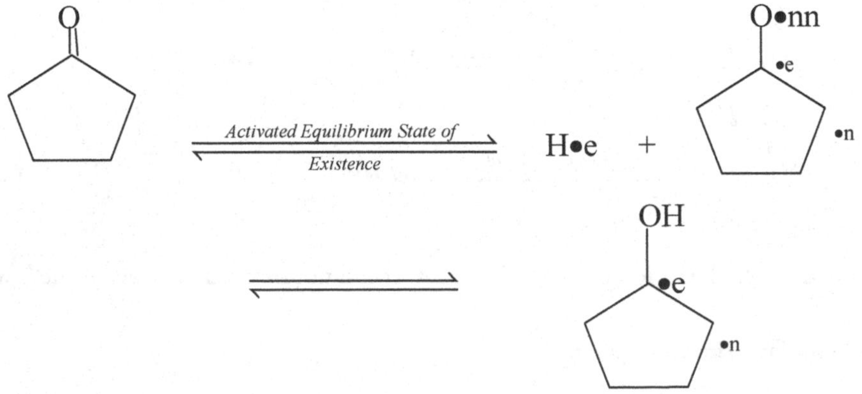

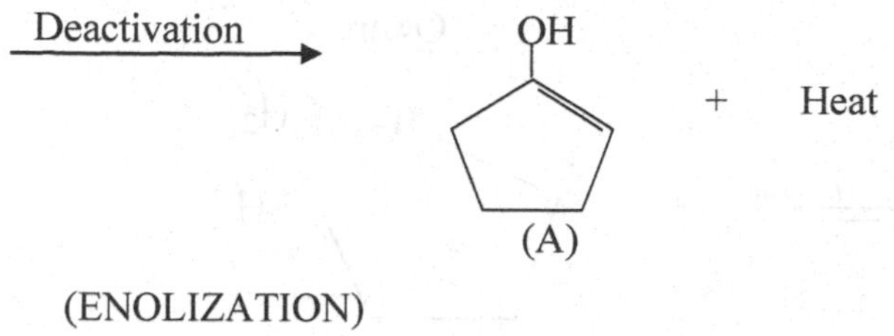

(ENOLIZATION)

for which the concept of Enolization is not limited to linear ketones, but also to cyclic ketones. *(Laws of Creations for Enolization phenomena)*

Rule 329: This rule of Chemistry for **HOMOLOGENATION of the first kind of the second type,** states that, when formaldehyde, aldehydes, ketones, and the likes are involved, cyclic ethers of different sizes can be obtained, prominent in the list being tricyclic in size.
(Laws of Creations for Cyclic Ethers)

Rule 330: This rule of Chemistry for **HOMOLOGENATION of the first kind of the second type,** states that, when ringed ketones and the likes are involved at low operating conditions, the rings formed are placed with a common boundary as shown below for which the smaller mem-bered ring is too strained to exist.

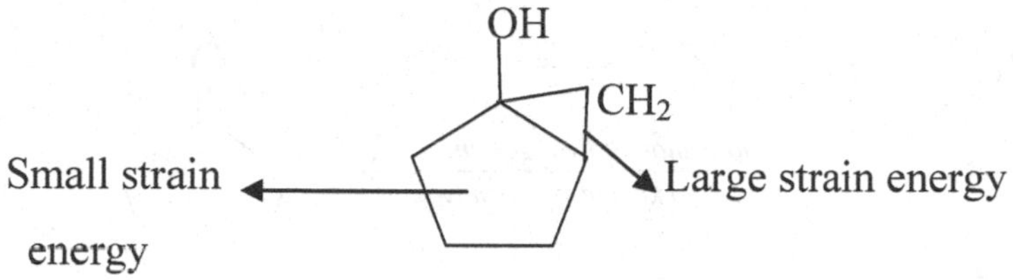

(B)

(Laws of Creations for Homologenation of first kind of second type)

Rule 331: This rule of Chemistry for **HOMOLOGENATION of the first kind of the second type,** states that, when two condensed rings with transfer species of the first kind is involved, the smaller ring opens instantaneously, followed by **Molecular rearrangement of the first kind of the SIXTH type** as shown below for enlargement, since the transfer species involved is of the first kind of the sixth type.

Stage 1:

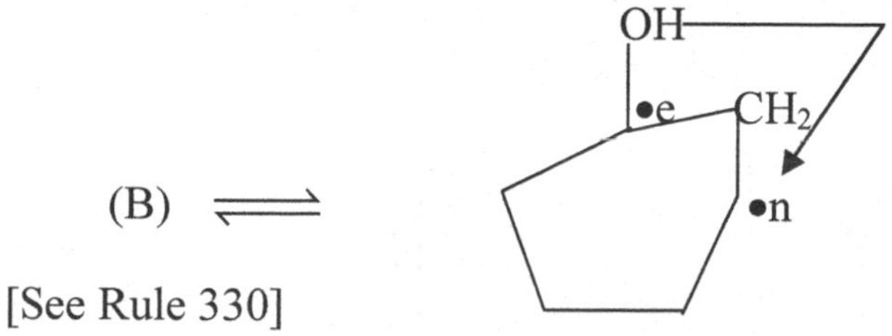

(B) $\rightleftharpoons$

[See Rule 330]

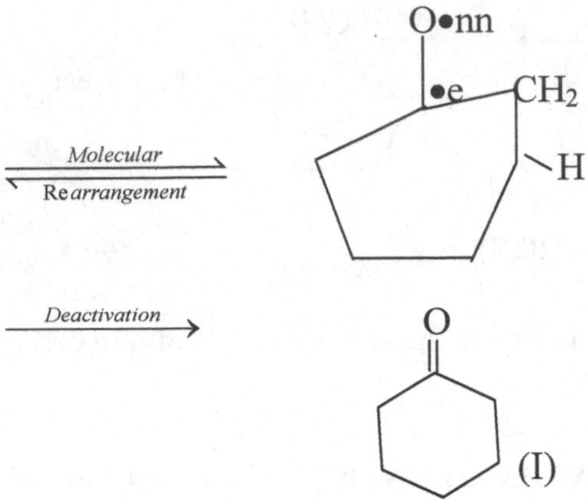

$$\xrightleftharpoons[\text{Re\,arrangement}]{\text{Molecular}}$$

$$\xrightarrow{\text{Deactivation}}$$

(I)

(Laws of Molecular rearrangement of first kind of the SIXTH type)

<u>Rule 332:</u> This rule Chemistry for **Ringed alcohol,** states that, during enlargement of the ring using Diazo-alkylenes, just as how non-ringed ketones, are obtained via Molecular rearrange-ment of the vinyl alcohols because the alcohols are unstable, so also it takes place with ringed alcohols for which the Molecular rearrangement is herein called **Molecular rearrangement of the first kind of the SIXTH type** as shown below-

$$\xrightarrow[\text{of the first kind of the sixth type}]{\text{Molecular rearrangement}}$$

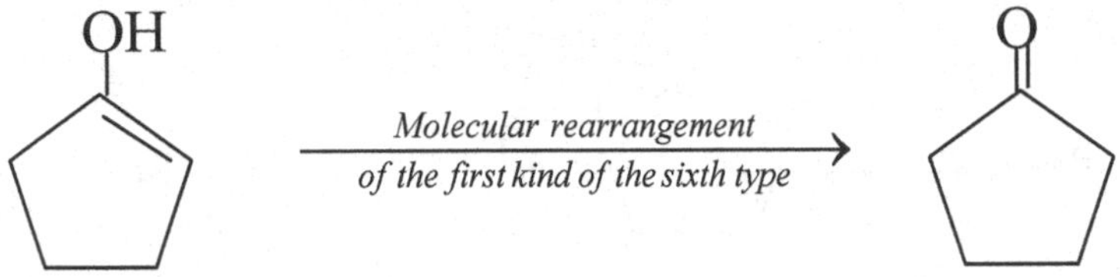

since the transfer species is carried by a group on a ring transferred without by-passing a CH_2 type of group as done during ring expansion.
(Laws of Molecular rearrangement of first kind of the sixth type)

<u>Rule 333:</u> This rule of Chemistry for **HOMOLOGENATION of the first kind of the second type for enlargement of unsaturated Hydrocarbon (H/C) family rings such as Benzene ring to cyloheptatriene,** states that, the transfer species involved in the second stage is of *the first kind of the SIXTH type* as shown below-

Stage 1:

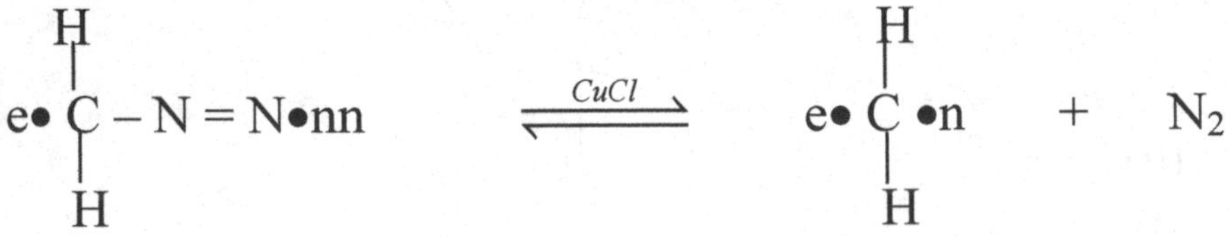

$$e\bullet \; \overset{\overset{\textstyle H}{|}}{\underset{\underset{\textstyle H}{|}}{C}} - N = N\bullet nn \quad \xrightleftharpoons{CuCl} \quad e\bullet \overset{\overset{\textstyle H}{|}}{\underset{\underset{\textstyle H}{|}}{C}} \bullet n \quad + \quad N_2$$

549

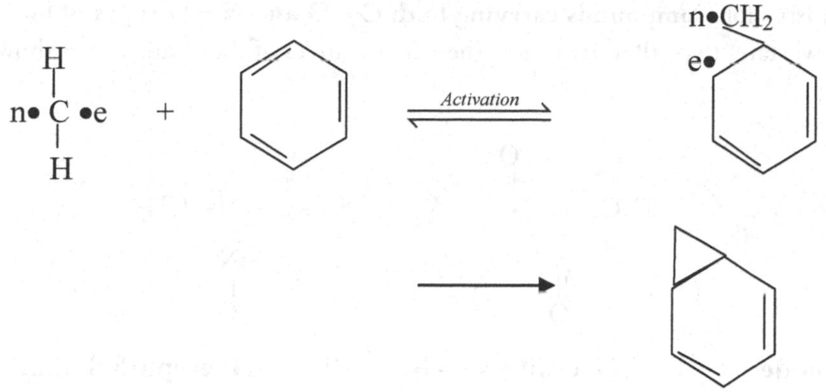

(A)

Stage 2:

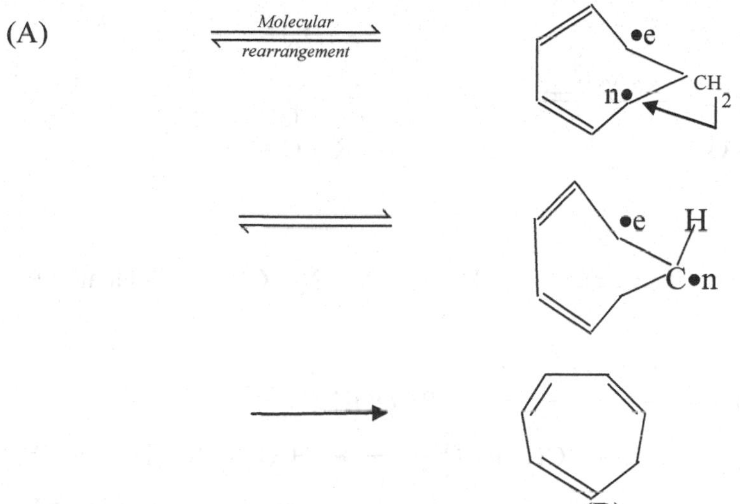

(B)

Overall Equation: H_2CN_2 + H_6C_6 $\longrightarrow$ N_2 + H_8C_7

for which the transfer species involved is that coming from the methylene and if the "diazoalkane" had been $(CH_3)_2CN_2$, the molecular rearrangement will never take place, unless the hydrogen atoms on the benzene ring are changed to CH_3 or some of them are changed to radical-pulling groups and this new type of molecular rearrangement is herein called **Molecular Rearrangement of the first kind of the SIXTH type**, peculiar only to ringed hydrocarbon compounds.
(Laws of Creations for Molecular Rearrangement of the first kind of the SIXTH type)

Rule 334: This rule of Chemistry for **Addition monomers,** states that, *during Combination mechanism, when two or more activated monomers of same nucleophilic capacity are present, they can diffuse and add to themselves when external forces such as mixing or heat or unreactive neighbors are present (Forced Diffusion),* with real Diffusion made possible only when the activated monomers are of different capacities *in the presence of a non-reactive initiator* wherein the possibility of formation of a Couple exist and when made possible, it allows the initiator to become either reactive or remain non-reactive to the couple; for which when reactive, alternating placements can be made possible.
(Laws of Creations of "Couples")

<u>**Rule 335:**</u> This rule of Chemistry for **Compounds carrying both C = O and N = O types of bonds in them such as shown below,** states that, they are one of the major sources of diazo-alkenes as shown below for one of them-

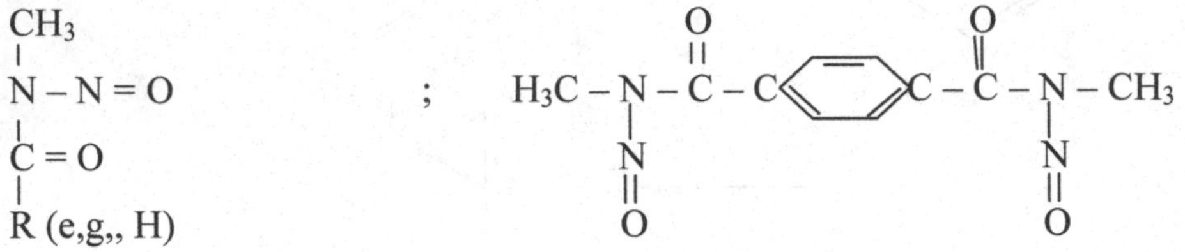

$$
\begin{array}{c}
CH_3 \\
| \\
N - N = O \\
| \\
C = O \\
| \\
R \ (e.g., H)
\end{array}
\qquad ; \qquad
\begin{array}{c}
O \qquad\qquad O \\
\| \qquad\qquad \| \\
H_3C - N - C - C\!\!\!\diagup\!\!\!\diagdown\!\!\!\diagup C - C - N - CH_3 \\
| \qquad\qquad\qquad | \\
N \qquad\qquad\qquad N \\
\| \qquad\qquad\qquad \| \\
O \qquad\qquad\qquad O
\end{array}
$$

N-methyl – N – Nitroso amide　　　　**N, N$'$ dinitroso – N, N$''$ dimethyl terephthalatamide**

i)

<u>**Stage 1:**</u>　　　　$H_2O \underset{Existence}{\overset{Equilibrium\ State\ of}{\rightleftharpoons}} H\bullet e \ + \ nn\bullet OH$

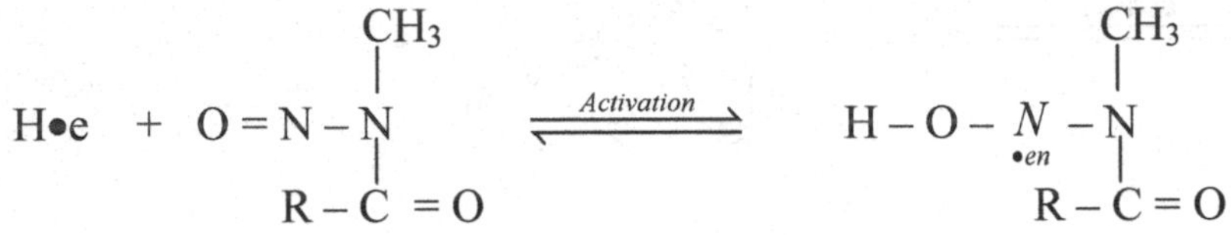

$$
H\bullet e \ + \ O = N - \overset{\displaystyle CH_3}{\underset{\displaystyle R - C = O}{\overset{|}{\underset{|}{N}}}} \underset{}{\overset{Activation}{\rightleftharpoons}} \ H - O - \underset{\bullet en}{N} - \overset{\displaystyle CH_3}{\underset{\displaystyle R - C = O}{\overset{|}{\underset{|}{N}}}}
$$

$$
\rightleftharpoons \ H - O - N = N - CH_3 \ + \ Heat \ + \ e\bullet \overset{\displaystyle O}{\overset{\|}{C}} - R
$$

$$
R - \overset{\displaystyle O}{\overset{\|}{C}}\bullet e \ + \ nn\bullet OH \longrightarrow RCOOH \qquad\qquad \textbf{ii)}
$$

<u>**Overall Equation:**</u> $H_2O \ + \ O=N-N(CH_3)RCO \longrightarrow H-O-N=N-CH_3 \ + \ RCOOH \ +$

Large amount of HEAT 　　　**iii)**

<u>**Stage 2:**</u>　$H - O - N = N - CH_3 \underset{Existence}{\overset{Equilibrium\ State\ of}{\rightleftharpoons}} H\bullet e \ + \ nn\bullet O - N = N - CH_3$

$$H\bullet e \ + \ KOH \rightleftharpoons KO\bullet en \ + \ H_2 \ + \ Heat$$

$$KO\bullet en \rightleftharpoons K\bullet e \ + \ en\bullet O\bullet nn \ + \ Heat$$

$$nn\bullet O\bullet en \ + \ H_2 \rightleftharpoons HO\bullet nn \ + \ H\bullet e$$

$$H\bullet e \ + \ nn\bullet O - N = N - CH_3 \rightleftharpoons H - O - N = N - CH_3$$

$$K\bullet e \ + \ HO - N = N - CH_3 \rightleftharpoons KOH \ + \ en\bullet N = N - CH_3 \ + \ Heat$$

$$H_3C - N = N\bullet en \rightleftharpoons \textbf{Heat} \ + \ H_3C\bullet e \ + \ N_2$$

$$e\bullet CH_3 \rightleftharpoons H\bullet e \ + \ \underset{\displaystyle H}{\overset{\displaystyle H}{\overset{|}{\underset{|}{n\bullet C\bullet e}}}}$$

$$H\bullet e \ + \ nn\bullet OH \rightleftharpoons H_2O$$

$$n\bullet \overset{\overset{\displaystyle H}{|}}{\underset{\underset{\displaystyle H}{|}}{C}} \bullet e \;+\; N\equiv N \;\rightleftharpoons\; n\bullet \overset{\overset{\displaystyle H}{|}}{\underset{\underset{\displaystyle H}{|}}{C}} - N = N\bullet en$$

(A)

$$(A) \longrightarrow \;\;^{\ominus}\overset{\overset{\displaystyle H}{|}}{\underset{\underset{\displaystyle H}{|}}{C}} - \overset{\oplus}{N}\equiv N \;+\; \textbf{Heat}$$

iv)

Overall Equation: $KOH + H_2O + O=N\text{-}NCH_3RCO \longrightarrow$ Large amount of

HEAT $+$ KOH $+$ H_2O $+$ H_2CN_2 $+$ RCOOH v)

a two stage Equilibrium mechanism system for which the dilute KOH was an active catalyst with large amount of Heat released in four steps of the second stage, noting that based on Stage 1 wherein N = O center was activated in preference to C= O center clearly indicates that the following is valid-

$$C = O \;>\; N = O$$
Order of Nucleophilicity

(Laws of Creations for Synthesis of Diazoalkenes)

Rule 336: This rule of Chemistry for **Compounds of Diazos, nitrogen molecule, Cyanates, alkynes and cabonyls,** states that, the following is the order of Nucleophilicity of the centers-

$$C^{\ominus} - {}^{\oplus}N\equiv N \;>\; N\equiv N \;>\; C\equiv N \;>\; C\equiv C \;>\; C = O$$

ORDER OF NUCLEOPHILICITY

(Laws of Creations for Order of Nucleophilic centers)

Rule 337: This rule of Chemistry for **Compounds,** states that, *if a compound which can decompose cannot exist in Equilibrium State of Existence, then the mechanism of Decomposition is exclusively Decomposition mechanism and if the compound in Stable State of Existence can be made to exist in Equilibrium state of existence, then the mechanism of Decomposition is exclusively Equilibrium mechanism.*
(Laws of Creations for Decomposition of Compounds)

Rule 338: This rule of Chemistry for **Decompositions of compounds,** states that, **decomposi-tions in our environment (INDUSTRIAL CHEMISTRY)** is different from **decompositions under the surface or inside pores of Transition metals (SURFACE CHEMISTRY)**, for which different products are obtained in both cases with very different operating conditions-high in our environment and very low in the surface.
(Laws of Creations for Decomposition of Compounds)

Rule 339: This rule of Chemistry for **Decomposition of Diazo-alkylenes,** states that, different products are obtained from them depending on the type of Alkylene carried and the operating conditions; for which for the second member of the family the followings are to be expected-

552

Via Decomposition mechanism (FAVORED)

Diazo n-ethylene

Stage 1:

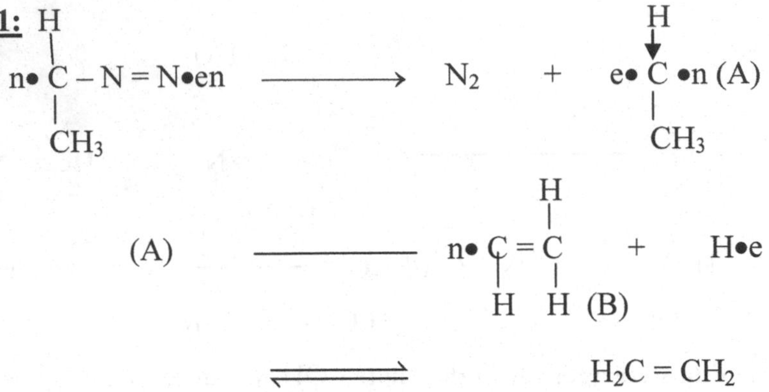

$$n\bullet \overset{\displaystyle H}{\underset{\displaystyle CH_3}{C}} - N = N\bullet en \longrightarrow N_2 \ + \ e\bullet \overset{\displaystyle \overset{H}{\downarrow}}{\underset{\displaystyle CH_3}{C}} \bullet n \ (A)$$

$$(A) \longrightarrow n\bullet \overset{\displaystyle H}{\underset{\displaystyle H}{C}} = \overset{\displaystyle H}{\underset{\displaystyle H}{C}} \ + \ H\bullet e$$

$$(B)$$

$$\rightleftharpoons \qquad H_2C = CH_2 \qquad\qquad i)$$

Via Equilibrium mechanism (FAVORED)

Stage 1:

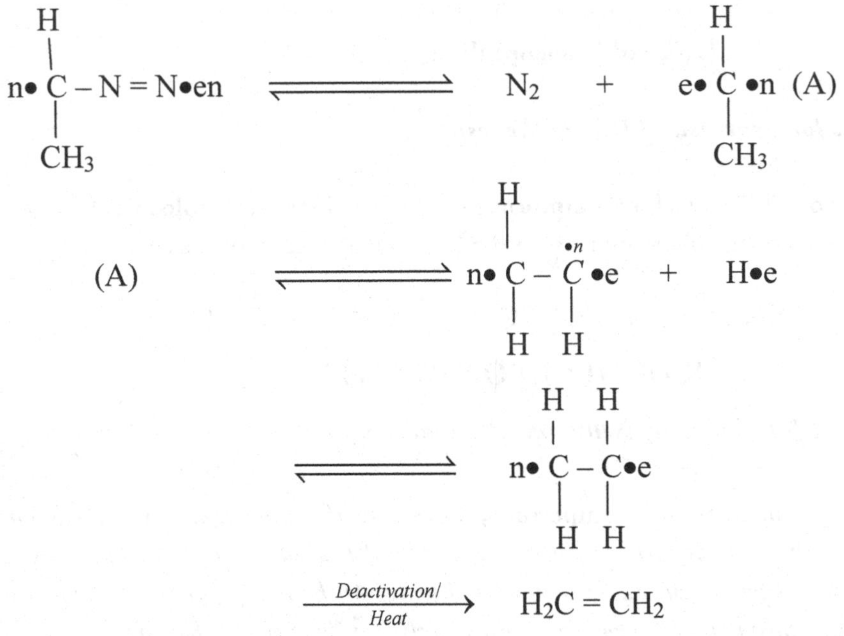

$$n\bullet \overset{\displaystyle H}{\underset{\displaystyle CH_3}{C}} - N = N\bullet en \ \rightleftharpoons \ N_2 \ + \ e\bullet \overset{\displaystyle H}{\underset{\displaystyle CH_3}{C}} \bullet n \ (A)$$

$$(A) \ \rightleftharpoons \ n\bullet \overset{\displaystyle H}{\underset{\displaystyle H}{C}} - \overset{\displaystyle \overset{\bullet n}{}}{\underset{\displaystyle H}{C}} \bullet e \ + \ H\bullet e$$

$$\rightleftharpoons \ n\bullet \overset{\displaystyle H}{\underset{\displaystyle H}{C}} - \overset{\displaystyle H}{\underset{\displaystyle H}{C}} \bullet e$$

$$\xrightarrow[\textit{Heat}]{\textit{Deactivation/}} \quad H_2C = CH_2$$

noting that this can take place by molecular rearrangement herein called <u>molecular rearran-gement of the first kind of the seventh type</u>; noting in addition the possible existence of diazirines if N_2 is not released for a very small fraction.
(Laws of Creations for Decomposition of Diazo-alkylenes)

<u>**Rule 340:**</u> This rule of Chemistry for **Decomposition of Diazo-alkylenes,** states that, different products are obtained from them depending on the type of Alkylene carried and the operating conditions; for which for the n- and iso- forms of the third member of the family, the followings are to be expected-

Diazo n- propylene

Stage 1:

$$n\bullet \overset{\displaystyle H}{\underset{\displaystyle C_2H_5}{C}} - N = N \bullet en \quad \rightleftharpoons \quad e\bullet \overset{\displaystyle H}{\underset{\displaystyle C_2H_5}{C}} \bullet n \quad + \quad N_2$$

$$(B)$$

$$(B) \quad \rightleftharpoons \quad e\bullet \overset{H}{C} = \overset{H}{\underset{H}{C}} \quad + \quad n\bullet CH_3$$

$$(A)$$

$$\longrightarrow \quad \overset{H}{\underset{H_3C}{C}} = \overset{H}{\underset{H}{C}} \qquad\qquad\qquad i)$$

noting that this cannot take place by molecular rearrangement, but by release of n.CH₃ group as transfer species from the C₂H₅ group of the second kind of the second type-

Diazo iso-propylene

Stage 1:

$$n\bullet \overset{\displaystyle CH_3}{\underset{\displaystyle CH_3}{C}} - N = N \bullet en \quad \rightleftharpoons \quad N_2 \quad + \quad e\bullet \overset{\displaystyle CH_3}{\underset{\displaystyle CH_3}{C}} \bullet n \quad (A)$$

$$(A) \quad \rightleftharpoons \quad n\bullet \overset{\displaystyle H}{\underset{\displaystyle H}{C}} - \overset{\displaystyle \bullet n}{\underset{\displaystyle CH_3}{C}} \bullet e \quad + \quad H\bullet e$$

$$\rightleftharpoons \quad n\bullet \overset{\displaystyle H}{\underset{\displaystyle H}{C}} - \overset{\displaystyle H}{\underset{\displaystyle CH_3}{C}} \bullet e$$

$$\xrightarrow[\text{Heat}]{\text{Deactivation/}} \quad H_2C = CHCH_3$$

Propene $\qquad\qquad$ ii)

noting that this can take place either by molecular rearrangement or release of H.e as transfer species of the first kind of the seventh type-

Stage 1:

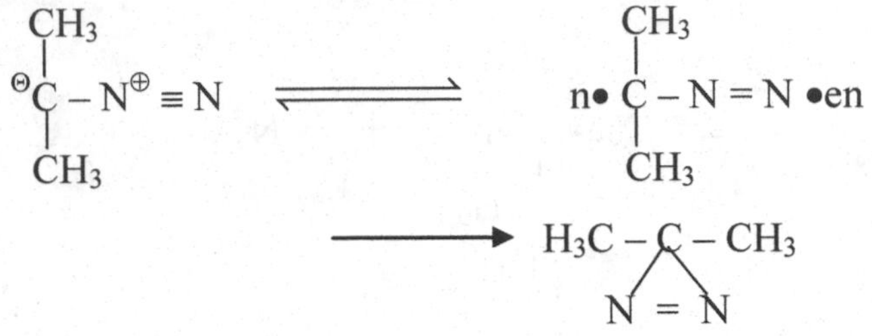

$$^{\ominus}\underset{\underset{CH_3}{|}}{\overset{\overset{CH_3}{|}}{C}} - N^{\oplus} \equiv N \quad \rightleftharpoons \quad n\bullet \underset{\underset{CH_3}{|}}{\overset{\overset{CH_3}{|}}{C}} - N = N \bullet en$$

$$\longrightarrow \quad H_3C - \underset{\underset{N \;=\; N}{\diagup \diagdown}}{C} - CH_3$$

Di-Methyl diazirine   iii)

for which no form of molecular rearrangement takes place here; noting also the possible exist-ence of diazirine for the diazo-n-propylene.

(Laws of Creations for Diazo propylene/Transfer species of the first kind of the seventh type)

Rule 341: This rule of Chemistry for **Decomposition of Diazo-alkylenes,** states that, different products are obtained from them depending on the type of Alkylene carried and the operating conditions; for which for the iso-form of the fourth member of the family, the followings are to be expected-

Diazo-iso-butylene

Via Decomposition mechanism

Stage 1:

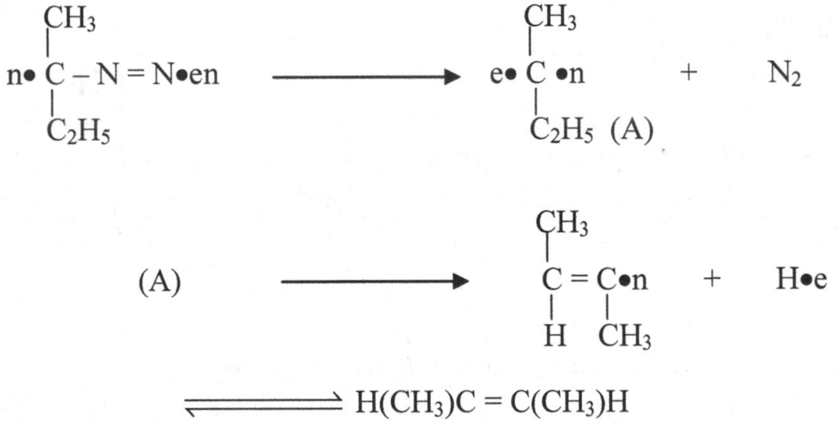

$$n\bullet \underset{\underset{C_2H_5}{|}}{\overset{\overset{CH_3}{|}}{C}} - N = N\bullet en \quad \longrightarrow \quad e\bullet \underset{\underset{C_2H_5 \;\; (A)}{|}}{\overset{\overset{CH_3}{|}}{C}} \bullet n \quad + \quad N_2$$

$$(A) \quad \longrightarrow \quad \underset{\underset{H \quad CH_3}{|\qquad|}}{\overset{\overset{CH_3}{|}}{C}} = C\bullet n \quad + \quad H\bullet e$$

$$\rightleftharpoons \quad H(CH_3)C = C(CH_3)H$$

Cis-/Trans- 2-Butene   i)

noting that this can take place either by molecular rearrangement or by release of H.e as transfer species of first kind of the seventh type when the mechanism is via Equilibrium mechanism; and if the operating condition is such that allows for continued movement of H.e, this is followed by-

Via Equilibrium mechanism

Stage 1:

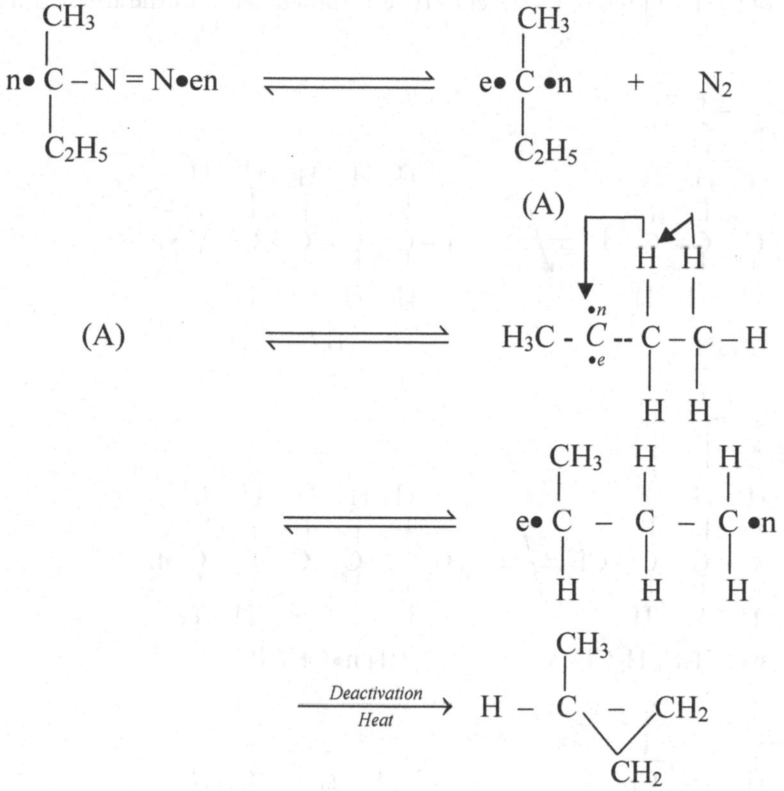

$$n\bullet \underset{\underset{C_2H_5}{|}}{\overset{\overset{CH_3}{|}}{C}} - N = N\bullet en \quad \rightleftharpoons \quad e\bullet \underset{\underset{C_2H_5}{|}}{\overset{\overset{CH_3}{|}}{C}} \bullet n \quad + \quad N_2$$

$$(A)$$

$$(A) \quad \rightleftharpoons \quad H_3C - \overset{\bullet n}{\underset{\bullet e}{C}} \text{--} \overset{\overset{H}{|}}{\underset{\underset{H}{|}}{C}} - \overset{\overset{H}{|}}{\underset{\underset{H}{|}}{C}} - H$$

$$\rightleftharpoons \quad e\bullet \underset{\underset{H}{|}}{\overset{\overset{CH_3}{|}}{C}} - \underset{\underset{H}{|}}{\overset{\overset{H}{|}}{C}} - \underset{\underset{H}{|}}{\overset{\overset{H}{|}}{C}} \bullet n$$

$$\xrightarrow[\text{Heat}]{\textit{Deactivation}} \quad H - \underset{\underset{CH_2}{\diagdown \diagup}}{\overset{\overset{CH_3}{|}}{C}} - CH_2$$

Methylcyclopropane ii)

Stage 1:

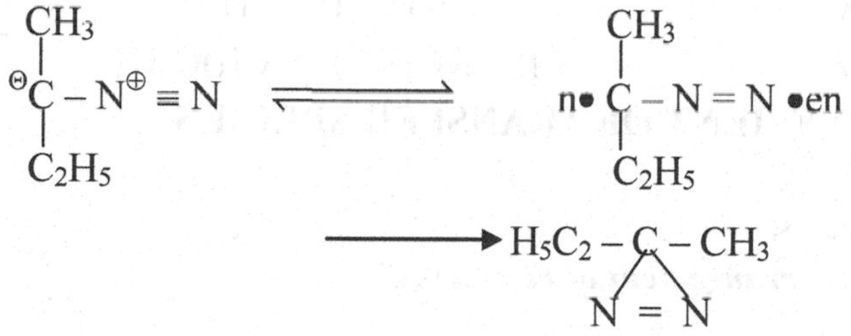

$$\overset{\ominus}{\underset{\underset{C_2H_5}{|}}{\overset{\overset{CH_3}{|}}{C}}} - N^{\oplus} \equiv N \quad \rightleftharpoons \quad n\bullet \underset{\underset{C_2H_5}{|}}{\overset{\overset{CH_3}{|}}{C}} - N = N \bullet en$$

$$\longrightarrow \quad H_5C_2 - \underset{\underset{N = N}{\diagdown \diagup}}{C} - CH_3$$

Methyl-Ethyl diazirine iii)

Overall Equation: 100(Diazo iso-butylene) $\longrightarrow$ 97N$_2$ + 94(2-Butene) + 3(Methyl-Cyclopropane) + 3(Methyl Ethyl diazirine) iv)

for which only some iso-alkylenes can undergo the rearrangement shown above in ii) that in which H electro-free-radically is transfered from one C center to the next C center in such a manner that the Laws of rearrangement are not broken, herein called **Molecular rearrange-ment of the First kind of the seventh type;** noting the types of products obtained from Diazo iso-butylene under special operating conditions, the fractions and stages involved.

(Laws of Creations for Molecular rearrangement of First kind of the seventh type)

Rule 342: This rule of Chemistry for **Alkylide groups,** states that, these groups cannot undergo Molecular rearrangement of the first kind of the seventh type as shown below in the absence of activation centers-

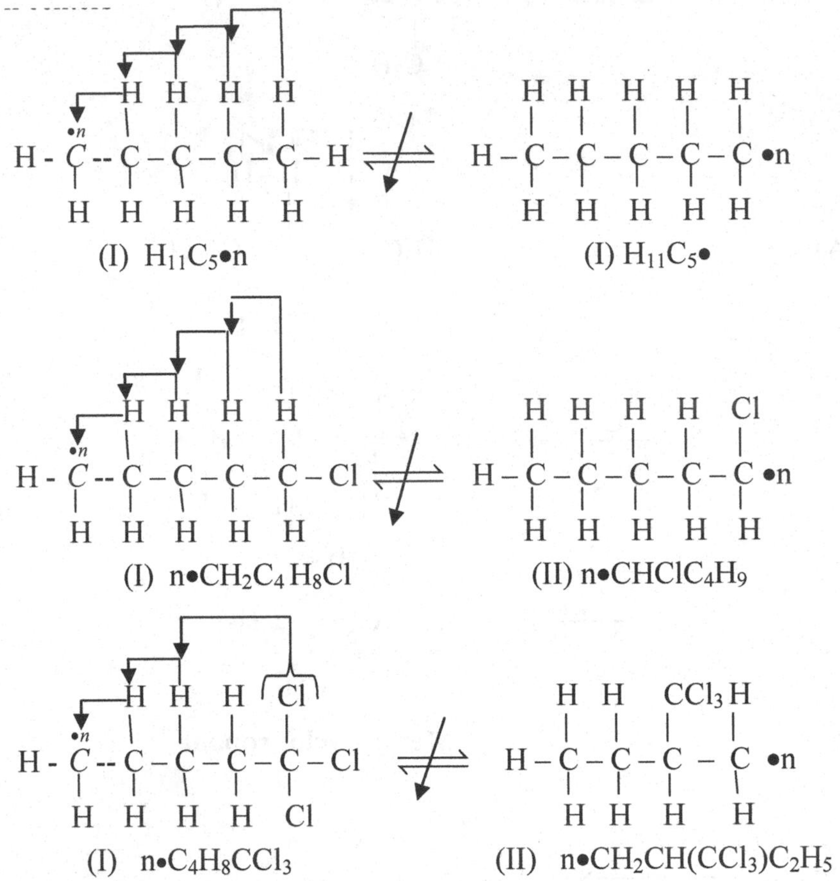

(I) $H_{11}C_5 \bullet n$ (I) $H_{11}C_5 \bullet$

(I) $n \bullet CH_2C_4H_8Cl$ (II) $n \bullet CHClC_4H_9$

(I) $n \bullet C_4H_8CCl_3$ (II) $n \bullet CH_2CH(CCl_3)C_2H_5$

IMPOSSIBLE MOVEMENT OF TRANSFER SPECIES

since NATURE abhors such movements.
(Laws of Creations for Molecular rearrangement of First Kind)

Rule 343: This rule of Chemistry for **Diazo-alkylenes,** states that, these are the strongest or one of the strongest NUCLEOPHILES of all monomers for which the routes favored by them are essentially the natural ones chargedly and radically when used as monomers, independent of what the carbon center is carrying, and for which the following order is inherently obvious-

$$C = N \quad > \quad N = N$$

Order of Nucleophilicity
(Laws of Creations for Order of Nucleophilicities of Centers)

Rule 344: This rule of Chemistry for **Diazo-alkylenes which carry resonance stabilization groups such as $-CH = CH_2$, $-C_6H_5$,** states that, these cannot resonance stabilize the Diazo-alkylene group, but make the compound an Electrophile (MALE) as shown below independent of the operating conditions whether N_2 is released or not-

557

$$
\begin{array}{c}
\overset{\displaystyle H}{\underset{\displaystyle H}{\overset{|}{\underset{|}{C}}}} = \overset{\displaystyle Y}{\underset{\displaystyle n\bullet C - N = N\bullet en}{\overset{\downarrow}{\underset{|}{\overset{|}{C}}}}} \quad \overset{X}{\underset{\downarrow}{}} \qquad \equiv \qquad n\bullet \overset{\displaystyle H}{\underset{\displaystyle CH_2}{\overset{|}{\underset{\parallel}{\overset{C - N = N\bullet en}{\underset{CH}{|}}}}}}
\end{array}
$$

$$
\begin{array}{c}
\overset{\displaystyle H}{\underset{\displaystyle H}{\overset{|}{\underset{|}{C}}}} = \overset{\displaystyle Y}{\underset{\displaystyle {}^{\ominus}C - N^{\oplus} \equiv N}{\overset{\downarrow}{\underset{|}{\overset{|}{C}}}}} \quad \overset{X}{\underset{\downarrow}{}} \qquad \equiv \qquad {}^{\ominus}\overset{\displaystyle H}{\underset{\displaystyle CH_2}{\overset{|}{\underset{\parallel}{\overset{C \qquad N^{\oplus} \equiv N}{\underset{CH}{|}}}}}}
\end{array}
$$

for which due to the presence of a living Activation center adjacently located to the less Nucleophilic center, only that center (the alkylene center) is involved all the time.
(Laws of Creations for Diazo-alkylenes)

Rule 345: This rule of Chemistry for **Perfluoro Diazoalkylene,** states that, the monomer's existence is not favored polarly, since there will be electrostatic forces of repulsion between the paired unbonded radicals of the F center and the real part of the polar charges.

$$
{}^{\ominus}\overset{\displaystyle :F}{\underset{\displaystyle :F}{\overset{|}{\underset{|}{C}}}} - N^{\oplus} \equiv N \qquad \rightleftharpoons \qquad n\bullet \overset{\displaystyle :F}{\underset{\displaystyle :F}{\overset{|}{\underset{|}{C}}}} - N = N \bullet en
$$

Cannot exist

and the same applies to the OCOR, and OR, NH_2 types of groups; for if they exist radically based on how synthesized, the radical state will be such that can never be deactivated to the polar state as the second to the last step in the stage (See iv) in Rule 335).
(Laws of Creations for Perfluoro-diazoalkylene)

Rule 346: This rule of Chemistry for **All alkenes which carry C_nF_{2n+1} group as a substituent group in the manner wherein the active carbon center carrying it is carrying a nucleo-free-radical,** states that, such a monomer when activated has t*ransfer species in the route natural to it and since based on the Laws of Conservation of transfer of transfer species such transfer species cannot be released in the route not natural to it, then this can be clear indication that such a*

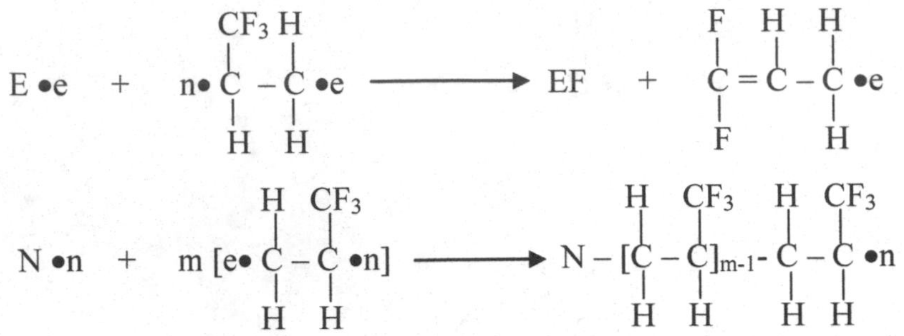

$$E \bullet e \ + \ n \bullet \overset{\overset{\displaystyle CF_3}{|}}{\underset{\underset{\displaystyle H}{|}}{C}} - \overset{\overset{\displaystyle H}{|}}{\underset{\underset{\displaystyle H}{|}}{C}} \bullet e \longrightarrow EF \ + \ \overset{\overset{\displaystyle F}{|}}{\underset{\underset{\displaystyle F}{|}}{C}} = \overset{\overset{\displaystyle H}{|}}{C} - \overset{\overset{\displaystyle H}{|}}{\underset{\underset{\displaystyle H}{|}}{C}} \bullet e$$

$$N \bullet n \ + \ m \left[e \bullet \overset{\overset{\displaystyle H}{|}}{\underset{\underset{\displaystyle H}{|}}{C}} - \overset{\overset{\displaystyle CF_3}{|}}{\underset{\underset{\displaystyle H}{|}}{C}} \bullet n \right] \longrightarrow N - [\overset{\overset{\displaystyle H}{|}}{\underset{\underset{\displaystyle H}{|}}{C}} - \overset{\overset{\displaystyle CF_3}{|}}{\underset{\underset{\displaystyle H}{|}}{C}}]_{m-1} - \overset{\overset{\displaystyle H}{|}}{\underset{\underset{\displaystyle H}{|}}{C}} - \overset{\overset{\displaystyle CF_3}{|}}{\underset{\underset{\displaystyle H}{|}}{C}} \bullet n$$

Transfer species of first kind cannot be released

monomer may still be activated radically (and even chargedly), since almost similar types of monomers such as vinyl acetate exist or ***cannot be activated at all, since the transfer species is of the first kind, unlike the case of vinyl acetate.***
(Laws of Creations for Activation of Monomers)

Rule 347: This rule of Chemistry for **Diazo-alkylenes,** states that, when one of the two H atoms is replaced with C_nF_{2n+1}, since the route natural to it is not favored radically, the diazo-alkylene can only be polymerized chargedly via positively charged route, the only route visibly present.
(Laws of Creations for Diazo-alkylenes)

Rule 348: This rule of Chemistry for **Diazoalkylenes,** states that, when one H atom is replaced with **COOR, CONH$_2$** types of groups, their presence make them Electrophiles, though the character is not shown, noting that whether N$_2$ is released or not, diazo-alkylene is more nucleophilic than the C = O center as shown below-

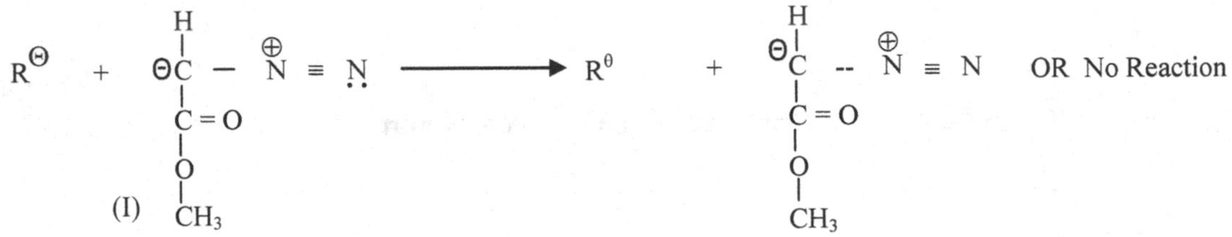

$$R^{\ominus} \ + \ \overset{\ominus}{\underset{\underset{\underset{\underset{CH_3}{|}}{O}}{\underset{|}{C=O}}}{C}} \overset{\overset{\displaystyle H}{|}}{-} \overset{\oplus}{\underset{\underset{\bullet\bullet}{N}}{N}} \equiv N \longrightarrow R^{\theta} \ + \ \overset{\ominus}{\underset{\underset{\underset{\underset{CH_3}{|}}{O}}{\underset{|}{C=O}}}{C}} \overset{\overset{\displaystyle H}{|}}{--} \overset{\oplus}{N} \equiv N \quad \text{OR No Reaction}$$

(I) \hspace{8cm} i)

$$R^{\oplus} \ + \ \overset{\ominus}{\underset{\underset{\underset{\underset{CH_3}{|}}{O}}{\underset{|}{C=O}}}{C}} \overset{\overset{\displaystyle H}{|}}{-} N^{\oplus} \equiv N \longrightarrow \overset{\ominus}{\underset{\underset{\underset{\underset{CH_3}{|}}{O}}{\underset{|}{C-O-R}}}{C}} \overset{}{-} N^{\oplus} \equiv N \ \text{OR} \ ROCH_3 + \overset{\overset{\displaystyle O}{\|}}{C} = \overset{\overset{\displaystyle H}{|}}{C^{\oplus}} + N_2$$

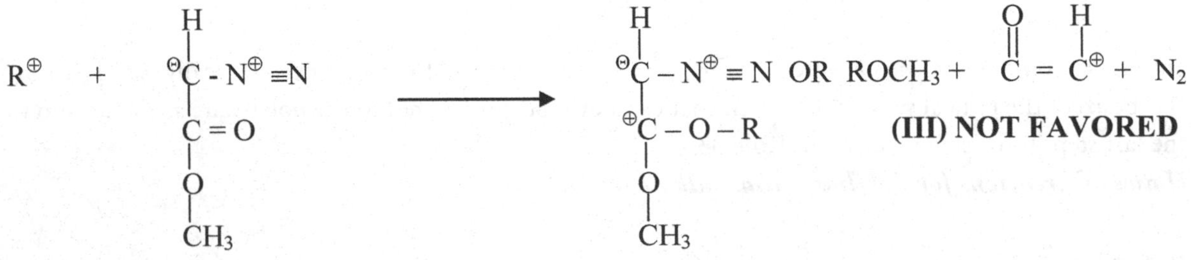

(III) NOT FAVORED

(II) NOT FAVORED \hspace{6cm} ii)

for which negatively, addition or abstraction is impossible being strongly nucleophilic with no visible positive center, and the same applies nucleo-free-radically, while elctro-free-radically no initiation is possible, but with positively charged-paired initiator, initiation is favored since the $\equiv$N *group is a very strong radical-pulling group*; for which the followings are valid-

$$\equiv N \;\gg\; -C \equiv N \;>\; -COOR \;>\; -CONH_2 \;>\; -COR \;>\; Cl$$

<u>ORDER OF CHARGE-PULLING CAPACITY</u>   iii)

(Laws of Creations for Diazo-alkylenes)

<u>Rule 349:</u> This rule of Chemistry for **Diazoalkylenes,** states that, when one H atom is replaced with **COR** types of groups, these are also Electrophiles as shown below-

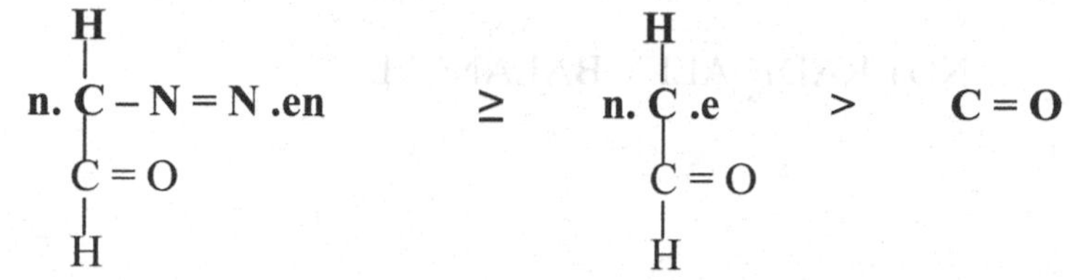

$$
\begin{array}{ccc}
& \text{H} & & \text{H} \\
& | & & | \\
\text{n. } & \text{C} - \text{N} = \text{N .en} & & \text{n. } \text{C .e} \\
& | & \geq & | \\
& \text{C} = \text{O} & > & \text{C} = \text{O} \qquad\qquad \text{C} = \text{O} \\
& | & & | \\
& \text{H} & & \text{H}
\end{array}
$$

    (I) Electrophile   (II) Electrophile

<u>ORDER OF NUCLEOPHILICITY OF THE CENTERS</u>

for which nucleo-free-radically and with negatively charged-paired initiators, no initiation is favored, while electro-free-radically, copolymers via the C = O and e. C .n centers cannot exist even under mild to higher operating conditions if N_2 is to be released and with positively charged-paired initiators only homopolymers can be obtained.

(Laws of Creations for Diazo-alkylenes)

<u>Rule 350:</u> This rule of Chemistry for **Diazoalkylenes,** states that, when one H atom is replaced with **OR type of groups,** if its existence is favored radically, then the followings are to be expected-

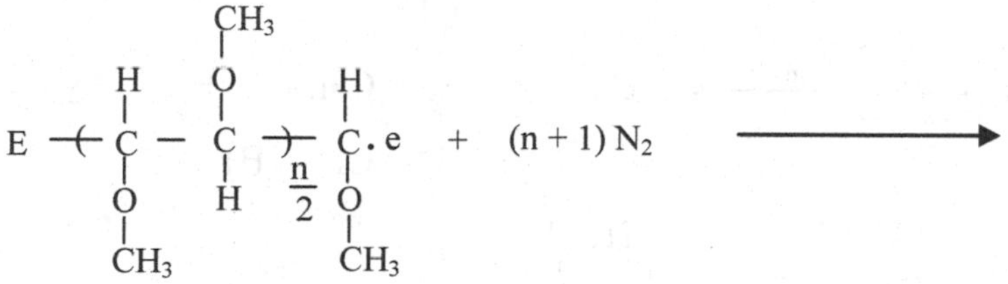

$$
E \left(\begin{array}{c} \text{H} \\ | \\ \text{C} \\ | \\ \text{O} \\ | \\ \text{CH}_3 \end{array} - \begin{array}{c} \text{CH}_3 \\ | \\ \text{O} \\ | \\ \text{C} \\ | \\ \text{H} \end{array} \right)_{\frac{n}{2}} \begin{array}{c} \text{H} \\ | \\ \text{C .e} \\ | \\ \text{O} \\ | \\ \text{CH}_3 \end{array} \; + \; (n+1)\, N_2 \quad\longrightarrow
$$

                                                                     i)

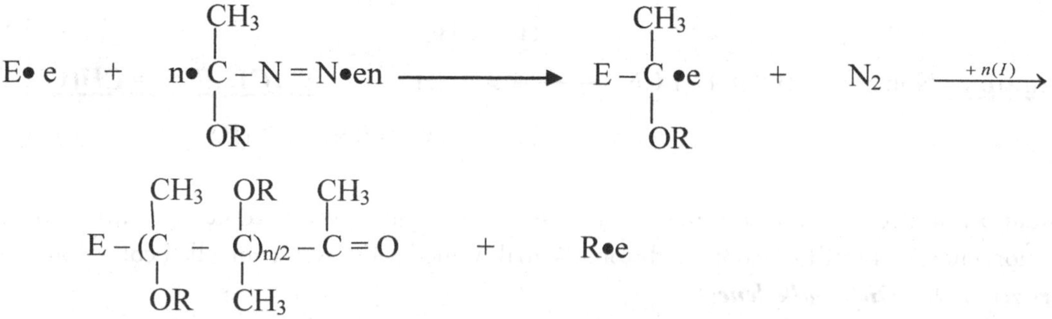

$$
E\bullet e \;+\; n\bullet \begin{array}{c} \text{CH}_3 \\ | \\ \text{C} - \text{N} = \text{N} \bullet \text{en} \\ | \\ \text{OR} \end{array} \longrightarrow \; E - \begin{array}{c} \text{CH}_3 \\ | \\ \text{C} \bullet e \\ | \\ \text{OR} \end{array} \;+\; N_2 \xrightarrow{\;+\,n(I)\;}
$$

$$
E - \left(\begin{array}{c} \text{CH}_3 \\ | \\ \text{C} \\ | \\ \text{OR} \end{array} - \begin{array}{c} \text{OR} \\ | \\ \text{C} \\ | \\ \text{CH}_3 \end{array} \right)_{n/2} - \begin{array}{c} \text{CH}_3 \\ | \\ \text{C} = \text{O} \end{array} \;+\; R\bullet e \qquad\qquad\qquad\qquad \text{ii)}
$$

for which dead terminal aldehydic or ketonic polymers are produced with release of transfer species of the first kind of the first type; noting that polarly they do not exist. [See Rule 345]

(Laws of Creations for Diazo-alkylenes)

<u>**Rule 351:**</u> This rule of Chemistry for **Diazo-alkylenes,** states that, this family can be used as a very good source of members of the Alkene family as shown below-.

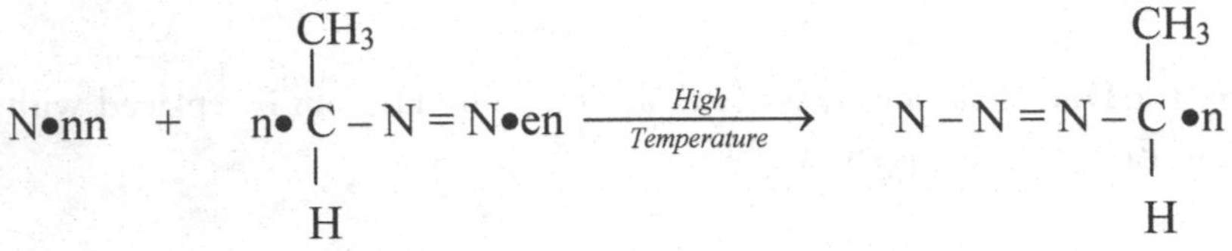

$$
N\bullet nn \;+\; n\bullet \overset{\overset{\displaystyle CH_3}{|}}{\underset{\underset{\displaystyle H}{|}}{C}} - N = N \bullet en \;\xrightarrow[\textit{Temperature}]{\textit{High}}\; N - N = N - \overset{\overset{\displaystyle CH_3}{|}}{\underset{\underset{\displaystyle H}{|}}{C}} \bullet n
$$

NOT RADICALLY BALANCED i)

Synthesis of Ethylene

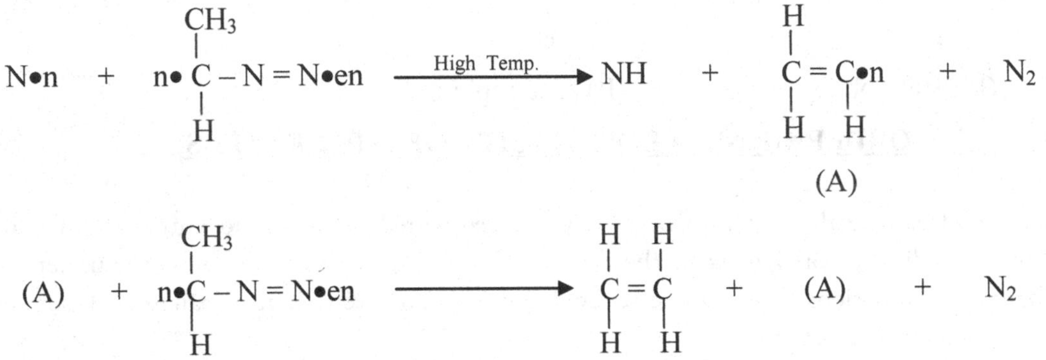

$$
N\bullet n \;+\; n\bullet \overset{\overset{\displaystyle CH_3}{|}}{\underset{\underset{\displaystyle H}{|}}{C}} - N = N\bullet en \;\xrightarrow{\text{High Temp.}}\; NH \;+\; \overset{\overset{\displaystyle H}{|}}{\underset{\underset{\displaystyle H}{|}}{C}} = \overset{}{\underset{\underset{\displaystyle H}{|}}{C}}\bullet n \;+\; N_2
$$

(A) ii)

$$
(A) \;+\; n\bullet \overset{\overset{\displaystyle CH_3}{|}}{\underset{\underset{\displaystyle H}{|}}{C}} - N = N\bullet en \;\longrightarrow\; \overset{\overset{\displaystyle H}{|}}{\underset{\underset{\displaystyle H}{|}}{C}} = \overset{\overset{\displaystyle H}{|}}{\underset{\underset{\displaystyle H}{|}}{C}} \;+\; (A) \;+\; N_2
$$

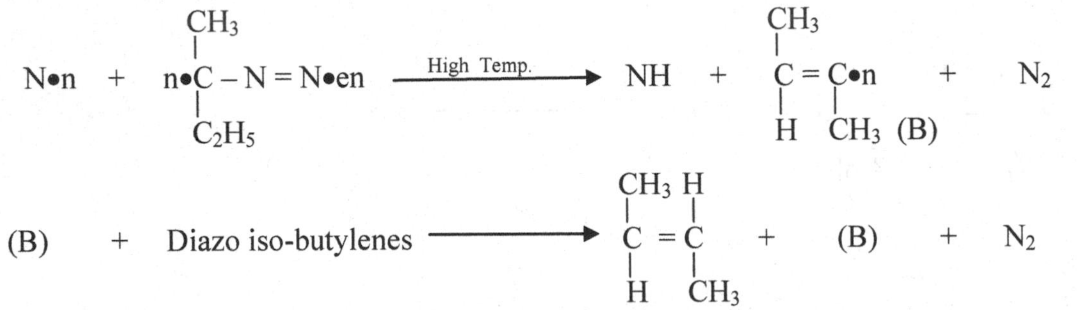

iii)

Overall Equation: $N\bullet n \;+\; 100(CH_3)HCN_2 \longrightarrow NH \;+\; \underline{\mathbf{99H_2C = CH_2}} \;+\; 100N_2$
$+\; (A)$ iv)

Synthesis of 2-Butene

$$
N\bullet n \;+\; n\bullet \overset{\overset{\displaystyle CH_3}{|}}{\underset{\underset{\displaystyle C_2H_5}{|}}{C}} - N = N\bullet en \;\xrightarrow{\text{High Temp.}}\; NH \;+\; \overset{\overset{\displaystyle CH_3}{|}}{\underset{\underset{\displaystyle H}{|}}{C}} = \overset{}{\underset{\underset{\displaystyle CH_3}{|}}{C}}\bullet n \;+\; N_2
$$

(B) i)

$$
(B) \;+\; \text{Diazo iso-butylenes} \;\longrightarrow\; \overset{\overset{\displaystyle CH_3}{|}}{\underset{\underset{\displaystyle H}{|}}{C}} = \overset{\overset{\displaystyle H}{|}}{\underset{\underset{\displaystyle CH_3}{|}}{C}} \;+\; (B) \;+\; N_2
$$

ii)

Overall Equation: $N\bullet n \;+\; 100\,(CH_3)_2CN_2 \longrightarrow NH \;+\; \underline{\mathbf{99(CH_3)HC = CH(CH_3)}}$
$+\; 100N_2 \;+\; (B)$ iii)

for if the initiator is added after the decomposition of the diazoalkylenes in the route not natural to it, then alkene monomers can readily be obtained from diazoalkylenes under well controlled conditions.
(Laws of Creations for Diazo-alkylenes)

<u>**Rule 352:**</u> This rule of Chemistry for **Sulfur dioxide (SO_2),** states that, this is a FULL-NON-FREE monomer which can only be activated radically, weakly NUCLEOPHILIC in character since it favors both Nucleo-non-free-radical and Electro-non-free-radical (Natural route) routes.
(Laws of Creation for SO2)

Rule 353: This rule of Chemistry for **Nitrogen containing monomers such as N = O wherein the Nitrogen center is carrying groups in which the Central atom is C- radical-pushing in character of greater capacity than H**, states that, just like in Azo-alkanes, these monomers cannot be activated chargedly and also radically.
(Laws of Creations for N=O Activation Centers)

Rule 354: This rule of Chemistry for **Nucleophilic Centers,** states that, based on the analysis of the following three unique compounds, first members of their different families shown below,

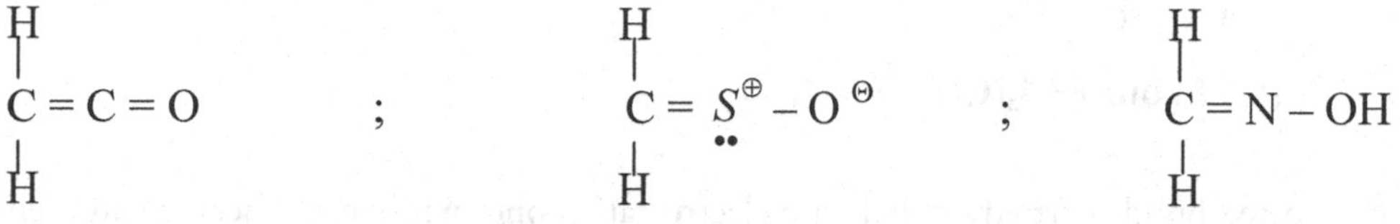

$$
\underset{\underset{H}{|}}{\overset{\overset{H}{|}}{C}} = C = O \qquad ; \qquad \underset{\underset{H}{|}}{\overset{\overset{H}{|}}{C}} = \overset{\bullet\bullet}{\underset{}{S}}{}^{\oplus} - O^{\ominus} \qquad ; \qquad \underset{\underset{H}{|}}{\overset{\overset{H}{|}}{C}} = N - OH
$$

$$
\text{(I)} \qquad\qquad\qquad \text{(II)} \qquad\qquad\qquad \text{(III)}
$$

the following order of Nucleophilicity is valid-

$$
C = O \;>\; C = N \;>\; C = S \;>\; C = C \;>\; S = O \;>\; N = (
$$

Order of Nucleophilicity
(Laws of Creations for Order of Nucleophilic Centers)

Rule 355: This rule of Chemistry for **Halogenated phosphorus compounds,** states that, since every atom has its domain and boundary just like everything in life including living systems, the real structure of some phosphorus halides are as shown below.

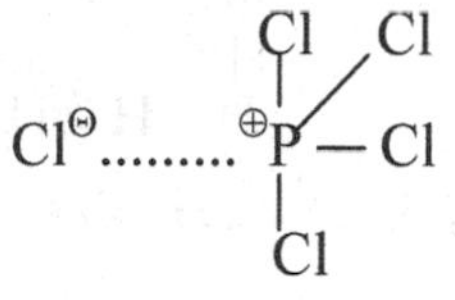

(A) –PCl₅ from Cl₂/PCl₃   AND NOT   **(B) Non Existent**

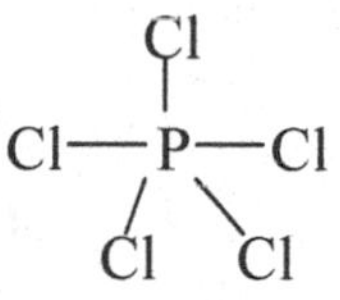

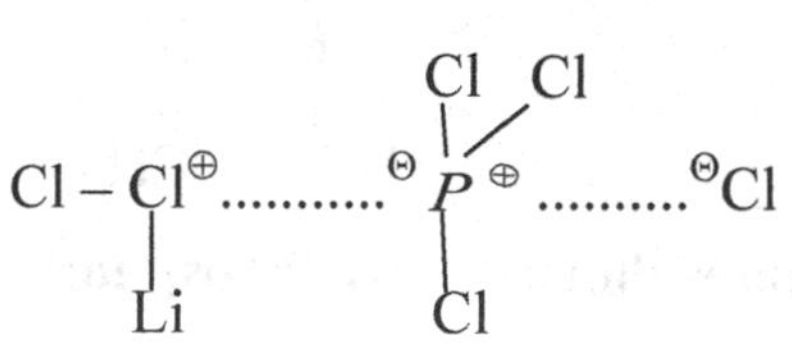

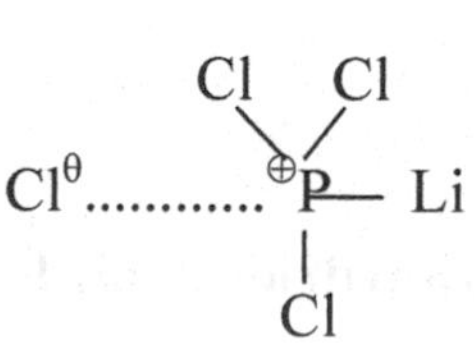

(A)-From LiCl/PCl₅   **(B)-From LiCl/PCl₃**

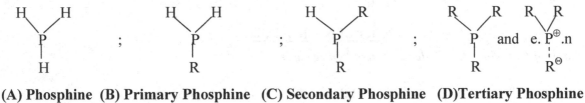

(D) P_2Cl_{10} from $2PCl_3/2Cl_2$

$Cl^{\ominus}$..........$^{\oplus}Cl - Cl$
$Cl - P - Cl$

(C) From PCl_3/Cl_2

where all the dotted bonds carried by them are electrostatic bond, while the others are covalent bonds.
(Laws of Creations for Structures of Halogenated P Compounds)

<u>**Rule 356:**</u> This rule of Chemistry for **Phosphorus compounds carrying free-pushing groups,** states that, the structures of some of these compounds called PHOSPHINES are as shown below-

(A) Phosphine **(B) Primary Phosphine** **(C) Secondary Phosphine** **(D)Tertiary Phosphine**

(E) PR_5 (R_2/PR_3) **(F) PR_3CR_2** **(G) $PR_3C_2H_3$** **(H) $PH_2C_2H_3$**

(I) Phosphorus tetrahydride **(J) 1,2- dimethylide phosphorus hydride** **(K) Phosphorus tetramethylide**

(Laws of Creations for Structures of P/C/H Compounds)

<u>**Rule 357:**</u> This rule of Chemistry for **Carbon carrying Phosphines particularly with respect to those carrying alkenyl Activation centers,** states that, none of them can be activated, since P a semi metal is more electroposive than C and *PH_2 group cannot be used as a substituent group of the non-free- radical-pushing type as used with NH_2 group.*
(Laws of Creations for Phosphine group as Substituent group)

Rule 358: This rule of Chemistry for **Phosphine and Ammonia,** states that, based on their Equilibrium State of Existences shown below, while phosphine is Polar/Non-ionic, less polar than water, ammonia is Polar/Ionic also less polar than water-

$$PH_3 \quad \underset{\text{Existence}}{\overset{\text{Equilibrium State of}}{\rightleftharpoons}} \quad H{\bullet}n \quad + \quad en{\bullet}PH_2$$

$$NH_3 \quad \underset{\text{Existence}}{\overset{\text{Equilibrium State of}}{\rightleftharpoons}} \quad H{\bullet}e \quad + \quad nn{\bullet}NH_2$$

phosphine cannot miscibilize in water like ammonia does and alkyl phosphines cannot be made from phosphine and alkyl halides as is done for ammonia; for which H cannot be abstracted as an atom from alkanyl group placed on P.
(Laws of Creations for PH$_3$ and NH$_3$ Families)

Rule 359: This rule of Chemistry for **metallic compounds,** states that, these compounds cannot carry π-bonds but only polar bonds in its place, for which the structure of Aluminum phosphide for example is as shown below and its polar bond can be activated like π-bonds also as shown below in its reaction with HCl-

$$Al^{\oplus} = {}^{\ominus}P \qquad ; \qquad Al \equiv P$$

NON EXISTENT

Stage 1: $\qquad HCl \quad \rightleftharpoons \quad H{\bullet}e \quad + \quad Cl{\bullet}nn$

$$H{\bullet}e \quad + \quad Al^{2\oplus}{-\!\!-}^{2\ominus}P \quad \underset{\text{Heat}}{\overset{\text{Activation}}{\rightleftharpoons}} \quad H{\bullet}e \quad + \quad e{\bullet}Al^{\oplus}{-\!\!\!-}^{\ominus}\overset{\bullet\bullet}{P}{\bullet}nn$$

$$\rightleftharpoons \quad H{-}P^{\ominus}{-\!\!\!-}^{\oplus}Al{\bullet}e \quad (A)$$

$$(A) \quad + \quad Cl{\bullet}nn \quad {-\!\!\!-\!\!\!-\!\!\!-} \quad H{-}P^{\ominus}{\rightarrow}{}^{\oplus}Al{-}Cl$$

Overall Equation: $HCl \quad + \quad AlP \quad {-\!\!\!-\!\!\!-\!\!\!\longrightarrow} \quad H{-}P^{\ominus}{-}{}^{\oplus}Al{-}Cl$ i)

Stage 2: $\qquad HCl \quad \rightleftharpoons \quad H{\bullet}e \quad + \quad Cl{\bullet}nn$

$$H{\bullet}e \quad + \quad H{-}P^{\ominus}{-}{}^{\oplus}Al{-}Cl \quad \overset{\text{Activation}}{\rightleftharpoons} \quad H{\bullet}e \quad + \quad nn{\bullet}\overset{\displaystyle H}{\underset{\displaystyle Cl}{\overset{\downarrow}{P{-}Al}}}{\bullet}e$$

$$\overset{\displaystyle Cl}{\underset{\displaystyle |}{}}$$

$$\rightleftharpoons \quad H_2P - Al \bullet e \quad (B)$$

$$(B) \quad + \quad Cl\bullet nn \quad \longrightarrow \quad H_2P - Al - Cl_2$$

<u>**Overall Equation:**</u> $HCl \quad + \quad HPAlCl \longrightarrow H_2P - Al - Cl_2$ \hfill ii)

<u>**Stage 3:**</u> $\qquad HCl \rightleftharpoons H\bullet e \quad + \quad Cl\bullet nn$

$$H\bullet e \quad + \quad H_2P - Al - Cl_2 \rightleftharpoons PH_3 \quad + \quad e\bullet AlCl_2$$

$$Cl_2Al\bullet e \quad + \quad nn\bullet Cl \longrightarrow AlCl_3$$

<u>**Final Overall Equation:**</u> $3HCl \quad + \quad AlP \longrightarrow PH_3 + \quad AlCl_3$ \hfill iii)

nothing first and foremost the structure of AlP, secondly that polar bonds can be activated as long as the boundary laws were not contravened since P and Al cannot carry more than eight radicals in the last shell as dictated by Argon of the third Period where they belong, and thirdly in the reactions above there are three stages and the mechanism is Equilibrium with no hydrolysis.
(Laws of Creations for Polar Character of Metals)

Rule 360: This rule of Chemistry for **Compounds that carry activation centers whether they are in the gaseous or liquid or solid states,** states that, under standard temperature and pressure, they do not exist in Activated State of Existence, not even diazo-Alkylenes, but only in either Electrostatic or Polar or Ionic or Stable States of Existence for them to remain harmless to their environments in particular to Living systems.
(Laws of Creations for Compounds)

Rule 361: This rule of Chemistry for **Numbers of elements of an atom,** states that, the number of elements of an atom, depends on the valence state of the atom, for which for example, for carbon atom there are at least four of them-carbon black, activated carbon, coal, and coke and for phosphorus, there are at least three of them- the white, red, and the black; for which many of them exist as liquid and solid forms depending on the operating conditions, with only the structure of White phosphorus the most important of the members of phosphorus shown below.

$$en \bullet \overset{\displaystyle \cdot\cdot}{\underset{\displaystyle \bullet en}{P}} \bullet nn$$

(A) **White phosphorus**

(Laws of creations for Elements of Atoms)

Rule 362: This rule of Chemistry for **Carbon monoxide,** states that, this compound has two activation centers - a π- bond that can be used at very **high operating conditions** being strongly nucleophilic and a 2p vacant orbital on the C center with paired unbonded radicals in the last orbital, far less nucleophilic than the first (the π-bond); for which at Standard Temperature and Pressure and low temperatures it is non-poisnous, but when heated, the less nucleophilic center gets activated as shown below-.

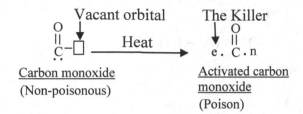

Carbon monoxide
(Non-poisonous)

Activated carbon
monoxide
(Poison)

for which as simple as it is, CO has two different types of activation centers.
(Laws of Creations for CO)

Rule 363: This rule of Chemistry for **CO and CS centers,** states that, when they are involved in chemical reactions, the following is the order of their nucleophilicity-

$$
\underset{R}{\overset{R}{C}}=O \quad > \quad \underset{R}{\overset{R}{C}}=S \quad ; \quad \overset{O}{:C} \quad > \quad \overset{S}{:C} \quad ; \quad e\bullet \overset{O}{C}\bullet n \quad > \quad e\bullet \overset{S}{C}\bullet n
$$

        (A)   (B)

Order of Nucleophilicity

(Laws of Creations for Order of Nucleophilicity of CO and CS centers)

Rule 364: This rule of Chemistry for **CO and CS types of centers when placed in a ring,** states that, the centers become Electrophiles when adjacently located to a functional center and their orders are as follows-

$$
\bigcirc C=S \quad > \quad \bigcirc C=N \quad > \quad \bigcirc C=O
$$

Order of Electrophilicity

(Laws of Creations for CO, CS, CN centers on Rings)

Rule 365: This rule of Chemistry for **Activation centers of different types,** states that, the following order of nucleophilic capacity is apparent-

$$
\bigcirc C=O > \bigcirc C=N \quad > \quad C\equiv C \quad > C=O > C=N \quad > \quad C=C \quad > \quad :\overset{O}{C}-\square
$$

Order of Nucleophilicity

(Laws of Creations for Activation centers)

Rule 366: This rule of Chemistry for **Passive catalysts whose structures are unknown,** state that, using the case (countless numbers of cases) of Octacarbonyl dicobalt, the real structure is as follows-

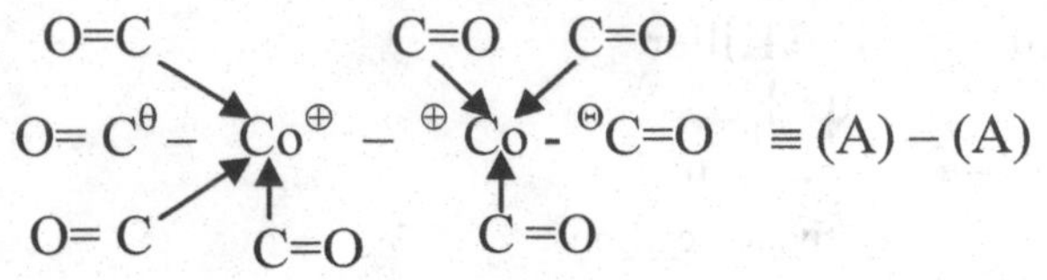

$$O= C^{\ominus} - \underset{\underset{C=O}{|}}{\overset{\overset{O=C}{\diagdown}}{Co}}^{\oplus} - {}^{\oplus}\underset{\underset{C=O}{|}}{\overset{\overset{C=O \quad C=O}{\diagup}}{Co}} - {}^{\ominus}C=O \quad \equiv (A) - (A)$$

(A) (A)

Octacarbonyl dicobalt

AND NOT

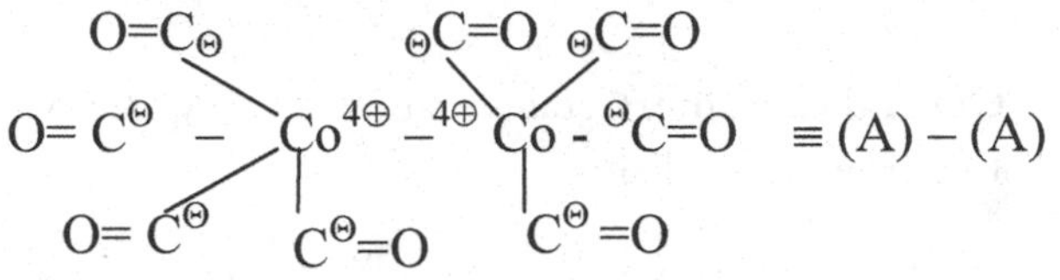

$$O= C^{\ominus} - \underset{\underset{C^{\ominus}=O}{|}}{\overset{\overset{O=C_{\ominus}}{\diagdown}}{Co}}^{4\oplus} - {}^{4\oplus}\underset{\underset{C^{\ominus}=O}{|}}{\overset{\overset{{}_{\ominus}C=O \quad {}_{\ominus}C=O}{\diagup}}{Co}} - {}^{\ominus}C=O \quad \equiv (A) - (A)$$

(A) (A)

(Octacarbonyl dicobalt)

__(Laws of Creations for Structures of a Cobalt catalyst)__

__Rule 367:__ This rule of Chemistry for **(Trimethylsilyl) diazomethane**, states that, this silicon containing monomer can selectively be carbonylated to (trimethylsilyl) ketene at 10^0C under atmospheric pressure of carbon monoxide in the presence of octacarbonyl dicolbalt as the catalyst as shown below-

__Stage 1:__

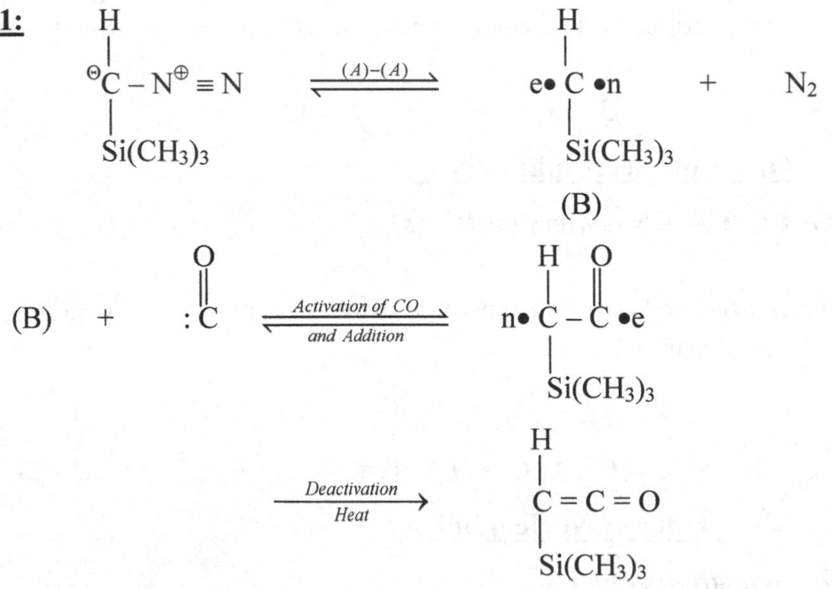

$$\underset{Si(CH_3)_3}{\overset{H}{\underset{|}{\overset{|}{{}^{\ominus}C}}}} - N^{\oplus} \equiv N \quad \underset{}{\overset{(A)-(A)}{\rightleftharpoons}} \quad e\bullet \underset{Si(CH_3)_3}{\overset{H}{\underset{|}{\overset{|}{C}}}} \bullet n \quad + \quad N_2$$

(B)

$$(B) \quad + \quad :\overset{O}{\underset{}{\overset{\|}{C}}} \quad \underset{and\ Addition}{\overset{Activation\ of\ CO}{\rightleftharpoons}} \quad n\bullet \underset{Si(CH_3)_3}{\overset{H}{\underset{|}{\overset{|}{C}}}} - \overset{O}{\overset{\|}{C}} \bullet e$$

$$\underset{Heat}{\overset{Deactivation}{\longrightarrow}} \quad \underset{Si(CH_3)_3}{\overset{H}{\underset{|}{\overset{|}{C}}}} = C = O$$

(Trimethylsilyl) Ketene   i)

__Overall Equation:__ $CO \ + \ HC(Si(CH_3)_3)N_2 \xrightarrow{(A)-(A)} N_2 \ + \ HC(Si(CH_3)_3) = C = O$

ii)

a one stage Equilibrium mechanism system wherein the catalyst is a passive one probably to suppress the polar character of the silicon containing diazomethane; noting that since the diazomethane diffused to the CO, this clearly indicates that the following is valid-

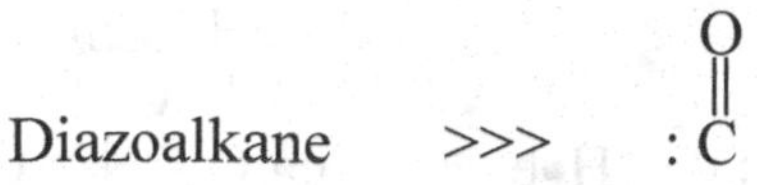

$$\text{Diazoalkane} \quad >>> \quad : \overset{\overset{\displaystyle O}{\|}}{C}$$

Order of Nucleophilicity

(Laws of Creation for Trimethylsilyl diazomethylene)

Rule 368: This rule of Chemistry for **Carbon monoxide (CO) and carbon monosulfide (CS),** states that, these are very unique monomers which favor their use less for homopolymerization than for copolymerization with selected ringed and non-ringed monomers to favor mostly ALTERNATING placement as shown below for some using CO-

CO/Ethylene oxide

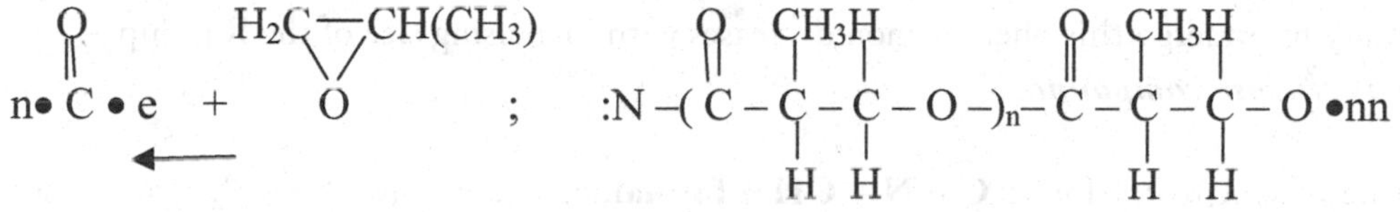

ALTERNATING COPOLYMER   i)

CO/Phenylallene

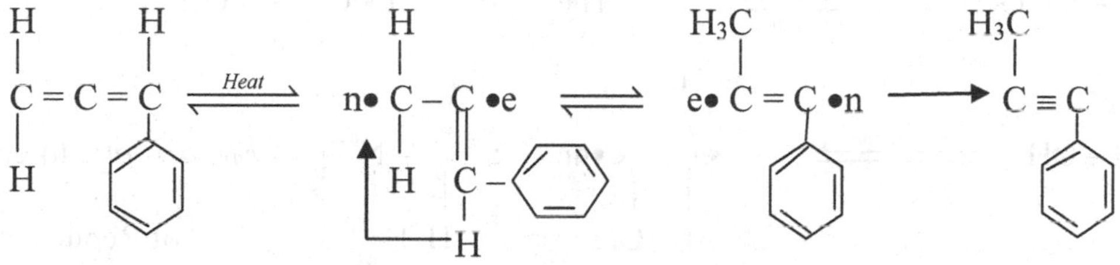

Molecular Rearrangement of Phenylallene

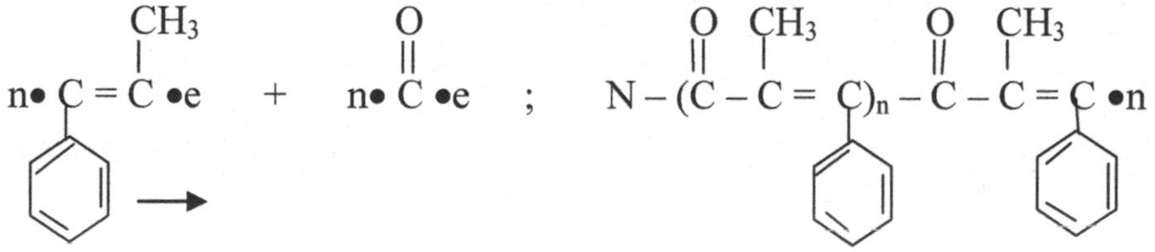

ALTERNATING COPOLYMER   ii

(Laws of Creations for CO and CS monomers)

Rule 369: This rule of Chemistry for **Carbon monoxide (CO) and carbon monosulfide (CS),** states that, when their homopolymerization is favored, this can readily be done electro-free-radically and never chargedly, since all of them are NUCLEOPHILES electrodynamically limited during propagation.
(Laws of Creations for CO and CS monomers)

<u>**Rule 370:**</u> This rule of Chemistry for **Nitroso-compounds of the type R – N = O where R is a free-pushing group such as CH$_3$,** states that, these compounds which cannot be activated radically or chargedly, can be made to do so when they can rearrange via Electroradicalization as shown below-

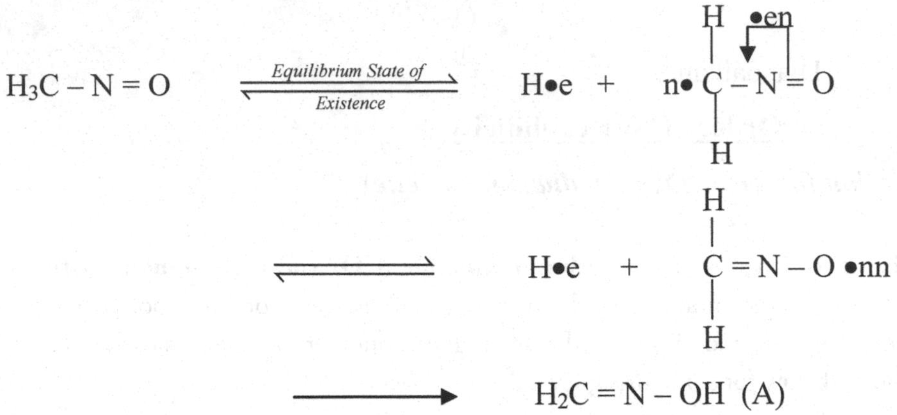

$$H_3C - N = O \quad \underset{\text{Existence}}{\overset{\text{Equilibrium State of}}{\rightleftarrows}} \quad H\bullet e \; + \; n\bullet \underset{|}{\overset{|}{C}} - N = O$$

$$\rightleftarrows \quad H\bullet e \; + \; \underset{|}{\overset{|}{C}} = N - O \bullet nn$$

$$\longrightarrow \quad H_2C = N - OH \;\; (A)$$

<u>**Overall Equation:**</u> $H_3C - N = O \longrightarrow H_2C = N - OH$

for which their ability to undergo this phenomenon decreases with increasing size of the R group.
(Laws of Creations Nitroso compounds)

<u>**Rule 371:**</u> This rule of Chemistry for **H$_2$C = N – OH a formaldehyde oxime,** states that, it cannot electroradicalize because its Equilibrium and Activated States of Existence are as follows-

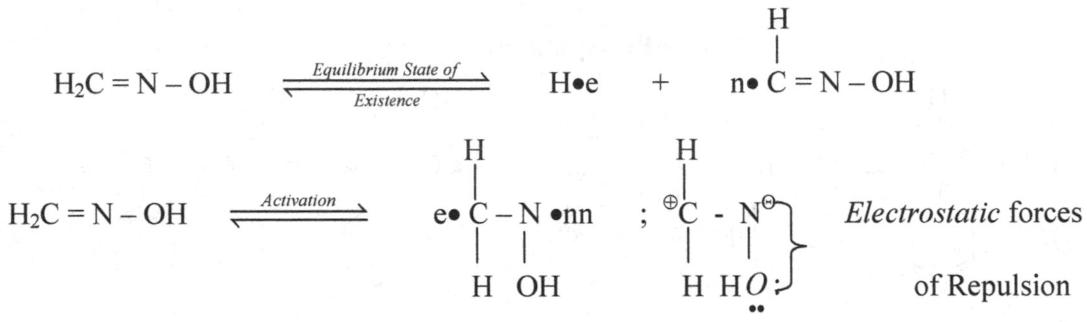

$$H_2C = N - OH \quad \underset{\text{Existence}}{\overset{\text{Equilibrium State of}}{\rightleftarrows}} \quad H\bullet e \; + \; n\bullet \overset{|}{C} = N - OH$$

$$H_2C = N - OH \quad \overset{\text{Activation}}{\rightleftarrows} \quad e\bullet \underset{|}{\overset{|}{C}} - N \bullet nn \; ; \; \overset{\oplus}{\underset{|}{C}} - N^{\ominus} \Big\} \; \textit{Electrostatic} \text{ forces}$$

of Repulsion

noting that it can be used as a monomer nucleo-non-free-radically and electro-free-radically the route natural to it as shown below-

<u>**Stage 1:**</u>

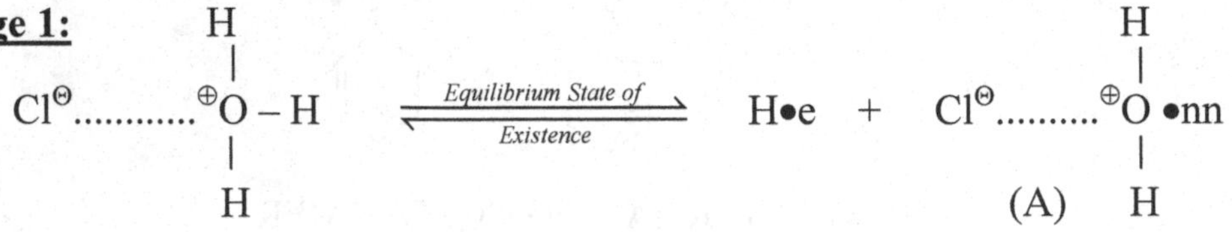

$$Cl^{\ominus} \ldots\ldots\ldots \overset{\oplus}{O} - H \quad \underset{\text{Existence}}{\overset{\text{Equilibrium State of}}{\rightleftarrows}} \quad H\bullet e \; + \; Cl^{\ominus} \ldots\ldots\ldots \overset{\oplus}{O} \bullet nn$$

(A)

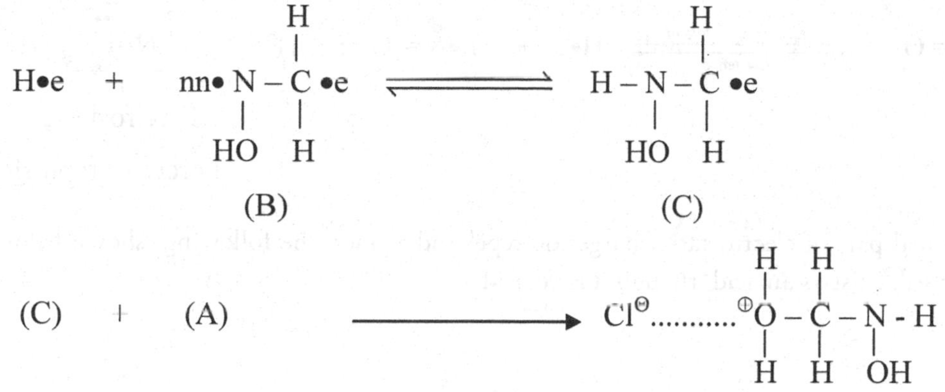

$$H \bullet e \quad + \quad nn\bullet \, N - C \bullet e \quad \rightleftharpoons \quad H - N - C \bullet e$$

with substituents (H above N, HO and H below) for (B) and (C)

$$
\begin{array}{ccc}
& H & H \\
& | & | \\
nn\bullet & N - C & \bullet e \\
& | & | \\
& HO & H
\end{array}
\qquad
\begin{array}{ccc}
& H & H \\
& | & | \\
H - & N - C & \bullet e \\
& | & | \\
& HO & H
\end{array}
$$

(B) (C)

$$(C) \quad + \quad (A) \quad \longrightarrow \quad Cl^{\ominus}\ldots\ldots\ldots \overset{\oplus}{O} - C - N - H$$

with H H above, H H OH below

(D) Initiation Step i)

Stages 2- n:

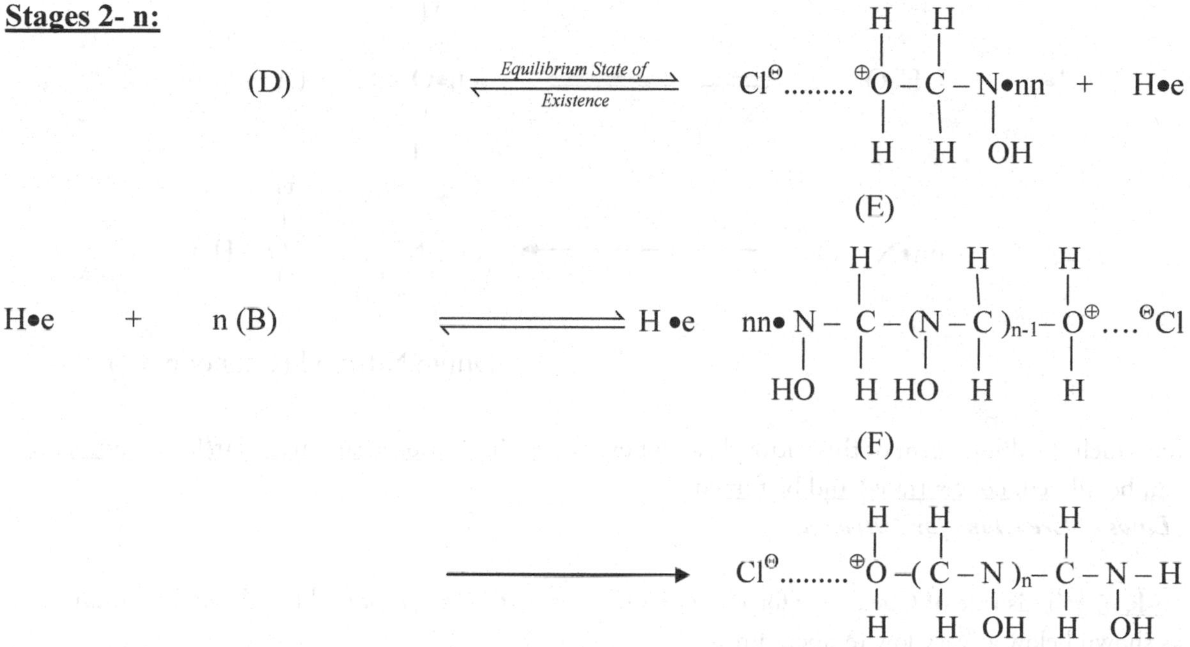

$$(D) \quad \underset{Existence}{\overset{Equilibrium\ State\ of}{\rightleftharpoons}} \quad Cl^{\ominus}\ldots\ldots\ldots \overset{\oplus}{O} - C - N\bullet nn \quad + \quad H\bullet e$$

with H H above, H H OH below

(E)

$$H\bullet e \quad + \quad n\,(B) \quad \rightleftharpoons \quad H\bullet e \quad nn\bullet \, N - C - (N - C)_{n-1} - O^{\oplus}\ldots.^{\ominus}Cl$$

with H H H above, HO H HO H H below

(F)

$$\longrightarrow \quad Cl^{\ominus}\ldots\ldots\ldots \overset{\oplus}{O} - (C - N)_n - C - N - H$$

with H H H above, H H OH H OH below

Poly(formaldehyde oxime)

Propagation Step ii)

Overall Equation: Dilute HCl + n(Formaldehyde oxime) $\longrightarrow$ Poly (formaldehyde oxime)

.iii)

(Laws of Creations for Formaldehyde oxime)

Rule 372: This rule of Chemistry for **Formaldehyde, Aldehyde and Ketone oximes,** states that, while forlmadehyde oxime cannot Enamidize, aldehyde and ketone oximes can enamidize, based on the operating conditions, one of which is ability to put them in Activated/Equilibrium state of existence and stabilized the Imines when formed.
(Laws of Creations for Formaldehyde/Aldehyde/Ketone oximes)

Rule 373: This rule of Chemistry for **R– N = O,** states that, when the R group is replaced with H, the compound obtained called **Nitroxyl (HN=O)** well known in the gas phase is polar/non-ionic as shown below.

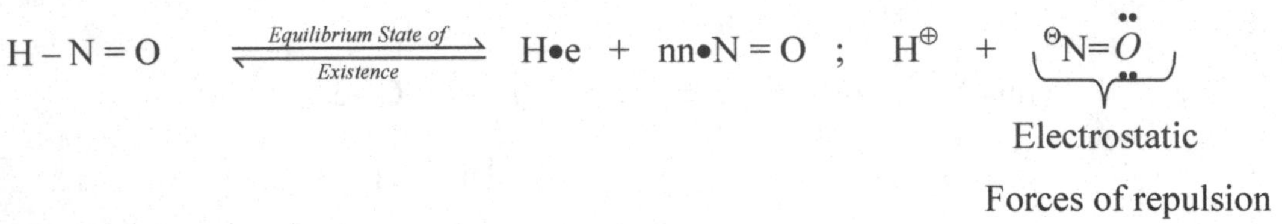

$$H - N = O \quad \underset{\text{Existence}}{\overset{\text{Equilibrium State of}}{\rightleftharpoons}} \quad H\bullet e \; + \; nn\bullet N = O \; ; \quad H^{\oplus} \; + \; \underbrace{^{\ominus}N = \overset{\bullet\bullet}{\underset{\bullet\bullet}{O}}}$$

Electrostatic

Forces of repulsion

and since the real part of electrostatic charges do repel and attract, the followings shown below shows that it cannot still exist as an acid, though it is an acid-

Stage 1:

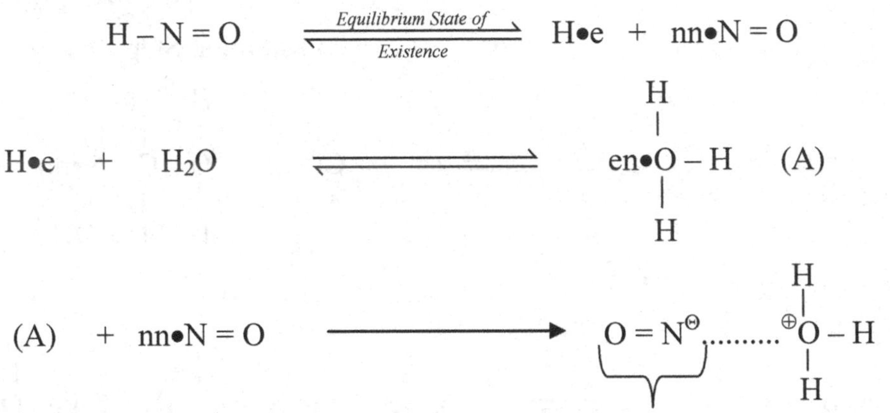

$$H - N = O \quad \underset{\text{Existence}}{\overset{\text{Equilibrium State of}}{\rightleftharpoons}} \quad H\bullet e \; + \; nn\bullet N = O$$

$$H\bullet e \; + \; H_2O \quad \rightleftharpoons \quad \overset{\displaystyle H}{\underset{\displaystyle H}{\overset{|}{\underset{|}{en\bullet O - H}}}} \quad (A)$$

$$(A) \; + \; nn\bullet N = O \quad \longrightarrow \quad \underbrace{O = N^{\ominus}}\cdots\cdots\cdots\overset{\displaystyle H}{\underset{\displaystyle H}{\overset{|}{\underset{|}{^{\oplus}O - H}}}}$$

Dilute Nitroxyl (Cannot exist)

for which the dilute form of this nitroxyl cannot exist; <u>clear indication that only Polar/Ionic compounds can be diluted, concentrated and be fumed</u>.
(Laws of Creations for Nitroxyl)

Rule 374: This rule of Chemistry for **H – N = O,** states that, this compound has the ability to dimerize as shown below at very low temperatures-

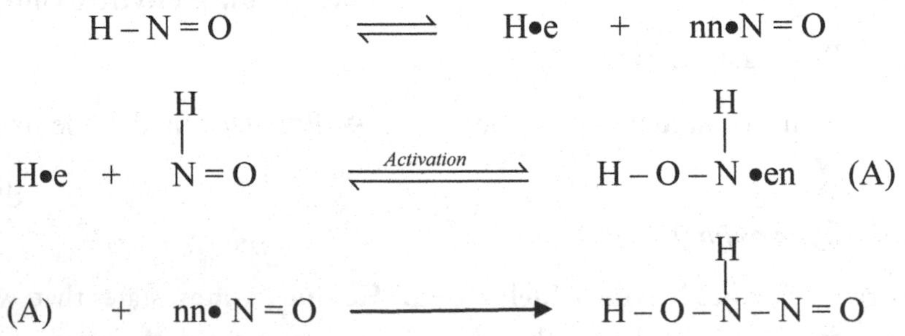

$$H - N = O \quad \rightleftharpoons \quad H\bullet e \; + \; nn\bullet N = O$$

$$H\bullet e \; + \; \overset{\displaystyle H}{\overset{|}{N = O}} \quad \overset{\text{Activation}}{\rightleftharpoons} \quad \overset{\displaystyle H}{\overset{|}{H - O - N}} \bullet en \quad (A)$$

$$(A) \; + \; nn\bullet N = O \quad \longrightarrow \quad \overset{\displaystyle H}{\overset{|}{H - O - N - N = O}}$$

(B) Dimeric HNO

(Laws of Creations for Dimerization of Nitroxyl)

Rule 375: This rule of Chemistry for **The Dimer of H – N = O,** states that, when the dimer is heated, the following takes place- [Dimer is HONHN=O (B)]

(B) $\xrightleftharpoons[\text{of Existence}]{\text{Activated Equilibrium State}}$ $\quad H-O-\overset{\bullet nn}{N}-\overset{\bullet en}{N}-O\bullet nn \quad + \quad H\bullet e$

$\xrightleftharpoons{\text{Enolization}}$ $\quad H-O-\overset{\bullet nn}{N}-\overset{\bullet en}{N}-O-H$

$\xrightarrow{\text{Deactivation/Heat}}$ $\quad H-O-N=N-O-H \quad + \quad Heat$

An Azo-compound

for which it undergoes Enolization to give a unique Azo-compound which can be activated only radically.
(Laws of Creations for Dimers of Nitroxyl)

Rule 376: This rule of Chemistry for **The Dimer of H – N = O,** states that, when the dimer is heated, after the formation of the Azo-compound, the followings further take place-

Stage 1: $\quad H-O-N=N-O-H \xrightleftharpoons[\text{Existence}]{\text{Equilibrium State of}} H\bullet e \quad + \quad nn\bullet O-N=N-O-H$

(A)

(A) $\xrightleftharpoons{\text{Heat}}$ $\quad nn\bullet O -N=N\bullet en \quad + \quad nn\bullet OH$

$H\bullet e \quad + \quad nn\bullet OH \xrightleftharpoons{\quad\quad} H_2O$

$nn\bullet O-N=N\bullet en \xrightarrow{\quad\quad} O^{\ominus}-N^{\oplus}\equiv N \quad + \quad Heat \quad\quad$ i)

Overall Equation: $HO\text{-}N=N\text{-}OH \xrightarrow{\quad\quad} H_2O \quad + \quad N_2O \quad + \quad Heat \quad\quad$ ii)

taking note very importantly of the second step in the stage where OH was released as a transfer species from a very unique center, because a polar bond is about to be formed, for which water is released along with the formation of dinitrogen oxide.
(Laws of Creations for Azo hydrogen peroxide)

Rule 377: This rule of Chemistry for **H – N = O,** states that, if the nitroxyl is fully suppressed, it will favor being used as a monomer via the use of electro-non-free-radical initiator, its natural route and can also favor nucleo-non-free-radical polymerization, but under higher operating conditions as shown below-

Initiation Step:

$Cl_2Fe \bullet en \quad + \quad H-N=O \xrightarrow{\quad\quad} Cl_2Fe-O-\overset{\displaystyle H}{\overset{\displaystyle |}{N}} \bullet en \quad\quad$ i)

Propagation Step:

$Fe-O-\overset{\displaystyle H}{\overset{\displaystyle |}{N}} \bullet en \quad + \quad n(H-N=O) \xrightarrow{\quad\quad} Fe-(O-\overset{\displaystyle H}{\overset{\displaystyle |}{N}})_n-O-\overset{\displaystyle H}{\overset{\displaystyle |}{N}} \bullet en \quad\quad$ ii)

Initiation Step:

$$H_3CO \bullet nn \ + \ H-N=O \longrightarrow H_3CO - \overset{\overset{\displaystyle H}{|}}{N} - O \bullet nn \qquad\qquad \text{iii)}$$

Propagation Step:

$$H_3CO - \overset{\overset{\displaystyle H}{|}}{N} - O \bullet nn \ + \ n(H-N=O) \longrightarrow H_3CO - (\overset{\overset{\displaystyle H}{|}}{N} - O)_n - N = O \ + \ H \bullet n \quad \text{iv)}$$

for which the release of transfer species of second kind is faster when the H is changed to a halogen atom such as F.

$$\overset{\displaystyle ..}{N} - (N-O)_n - \overset{\overset{\displaystyle F}{|}}{\underset{\underset{\displaystyle F}{|}}{N}} - O \bullet nn \longrightarrow \overset{\displaystyle ..}{N} - (N-O)_n - N = O \ + \ F \bullet nn$$

$$\qquad\qquad\qquad\qquad\qquad\qquad\qquad\qquad\qquad\qquad\qquad\qquad\qquad\qquad\qquad \text{v)}$$

(Laws of Creations for Nitroxyl)

Rule 378: This rule of Chemistry for **Nitrous acid (H – O – N = O),** states that, this compound is an acid (inorganic) only radically since it is polar/non-ionic, as shown below-

$$H - O - N = O \ \underset{\textit{Existence}}{\overset{\textit{Equilibrium State of}}{\rightleftharpoons}} \ H \bullet e \ + \ nn \bullet O - N = O \ ; \ H^{\oplus} \ + \ {}^{\ominus}O - \overset{\displaystyle ..}{N} = N$$

[See Rule 192]   Electrostatic forces of repulsion

(Laws of Creations for Nitrous acid)

Rule 379: This rule of Chemistry for **Nitrous acid,** states that, it like Nitroxyl has the ability to dimerize as shown below at very low temperatures-

Stage 1: $H - O - N = O \ \rightleftharpoons \ H \bullet e \ + \ nn \bullet O - N = O$

$H \bullet e \ + \ H - O - N = O \ \rightleftharpoons \ H - O - N^{\bullet en} - O - H$

$$(A)$$

$$O = N - O \bullet nn \ + \ (A) \longrightarrow H - O - \overset{\overset{\displaystyle }{}}{\underset{\underset{\displaystyle O - N = O}{|}}{N}} - O - H \quad (B) \qquad \text{i)}$$

which at 0^0C, breaks down to blue Dinitrogen trioxide as shown below-

Stage 1:   (B) $H \bullet e$ + $nn \bullet O - N - O - H$
                                                                      $|$
                                                                      $O - N = O$

                                                                      (C)

(C) $\rightleftharpoons$ $HO \bullet nn$ + $en \bullet N - O \bullet nn$
                                                                $|$
                                                                $O - N = O$

                                                                (D)

$H \bullet e$ + $nn \bullet OH$ $\rightleftharpoons$ H_2O

(D) $\longrightarrow$ $O = N - O - N = O$ +   Heat
                                      Dintrogen trioxide (Blue)   ii)

Overall Equation: $2HON = O$ $\longrightarrow$ H_2O + $O = NON = O$ +   Heat   iii)

from which the following can be seen to be valid-

$$-OH \; < \; -O - N = O \quad ; \quad -H \; > \; -N = O$$

Order of radical-pushing capacity   iv)

(Laws of Creations for Dimer of Nitrous acid)

Rule 380: This rule of Chemistry for **DInitrogen trioxide,** states that, at room temperature, it exists in Equilibrium state of existence being unstable and the color becomes brown and finally break down to colorless dinitrogen tetroxide and nitric oxide as shown below.

Stage 1: $O = N - O - N = O$ $\rightleftharpoons$ $O = N \bullet nn$ + $en \bullet O - N = O$
                                                                            (Brown)

          $O = N - O \bullet en$ $\rightleftharpoons$ $O = N \bullet en$ + $nn \bullet O \bullet en$ +   Heat

   $O = N \bullet en$ + $nn \bullet N = O$ $\rightleftharpoons$ $O = N - N = O$

$nn \bullet O \bullet en$ + $O = N - N = O$ $\longrightarrow$ $O = N^{\oplus} - {}^{\theta}O$ +   He at
                                                                    $|$
                                                                    $N = O$

                                                                    (A)   i)

Stage 2: $O = N - O - N = O$ $\rightleftharpoons$ $O = N \bullet nn$ + $en \bullet O - N = O$
                                                                            (Brown)

          $O = N - O \bullet en$ $\rightleftharpoons$ $O = N \bullet en$ + $nn \bullet O \bullet en$ +   Heat

   $O = N \bullet en$ + $nn \bullet N = O$ $\rightleftharpoons$ $O = N - N = O$ [Dinitrogen dioxide]

$$nn\bullet O \bullet en \quad + \quad (A) \quad \longrightarrow \quad O = \overset{|}{N}^{\oplus} - \,^{\ominus}O \quad + \quad Heat$$

$$O = N^{\oplus} - \,^{\ominus}O$$

(B)

Dinitrogen tetraoxide (Colourless)   ii)

Overall Equation: $2O = N - O - N = O \longrightarrow O = N - N = O + O_2N - NO_2 +$ Heat   iii)

noting that in the second stage, instead of the oxidizing oxygen oxidizing O=N-N=O as it did in the first stage, it completed the oxidation of (A) being more stable to give (B).
(Laws of Creations for Nitrous acid)

Rule 381: This rule of Chemistry for **R – O – N = O types of compound,** states that, these compounds are strong Nucleophiles which can only be activated radically, wherein the natural route is the use of electro-non-free-radical.
(Laws of Creations Alkyl nitrite)

Rule 382: This rule of Chemistry for $C_nF_{n+1} - N = O$, ROOC – N =O, ROC – N = O types of monomers, states that, these compounds can be activated radically and not chargedly via the N = O activation center, except when for the first, the group is higher than CF_3; while for the last two cases, they are resonance stabilized, favoring both non-free-radical routes with great limitations- presence of -O-O- bonds along the chain, just like in O_2 and benzoquinone at high temperatures.
(Laws of Creations for $R_FN=O$ types of Monomers)

Rule 383: This rule of Chemistry for **Quinones,** states that, there are only two isomers of cyclohexadienedione and these are 1,4-benzoquinone or para-benzoquinone or 1,4-dione or Quinone and 1,2-benzoquinone or ortho-benzoquinone or 1,2-dione; for which the meta- form cannot exist, because a structure with two double bonds in the ring cannot be obtained for it as shown below.

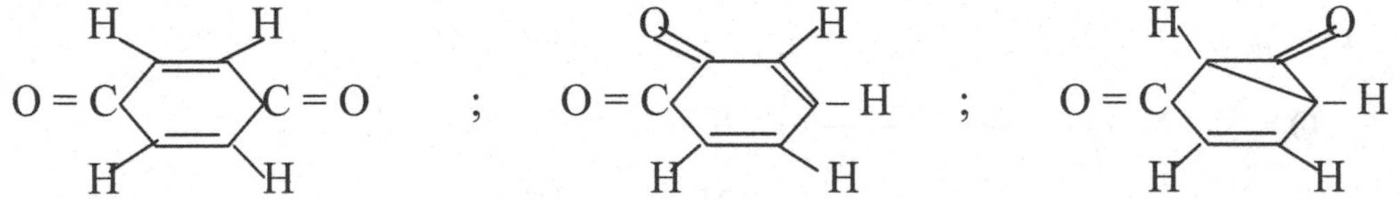

p - $C_6H_4O_2$ o - $C_6H_4O_2$ $C_6H_4O_2$

m-$C_6H_4O_2$ Cannot exist

for which ever way the meta is placed, it cannot exist, due to excessive SE in the ring(s).

(Laws of Creations for Quinones)

Rule 384: This rule of Chemistry for **All Systems with Activation centers,** states that, all Activation centers wherever they exist are FEMALES, i.e., NUCLEOPHILIC in character whether they are Nucleophiles (Female) or Electrophiles (Males); for which it can be seen that our world created by the ALMIGHTY INFINITE GOD is controlled or owned by MOTHER NATURE (As a "gift").
(Laws of Creations for Our World-The tenth law of NATURE)

575

Rule 385: This rule of Chemistry for **All Systems with Activation centers,** states that, though all Activation centers are NUCLEOPHILIC in character, shown below are examples of NUCLEOPHILIC centers forced to be ELECTOPHILIC in character by a stronger or weaker NUCLEOPHILIC center adjacently located to it; for which the main existence of MALES can be found with RINGED compounds-

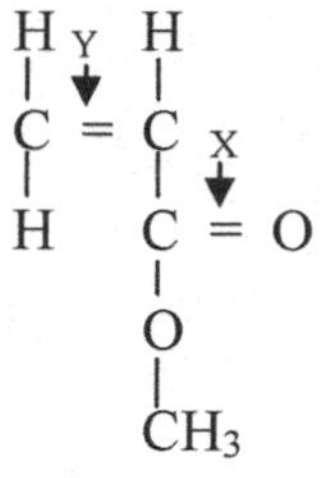

(I)   (II)   (III)   (IV)

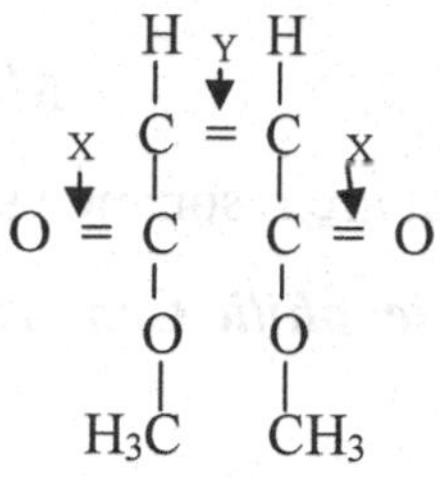

MALE   **MALE**   **FEMALE**   **FEMALE**

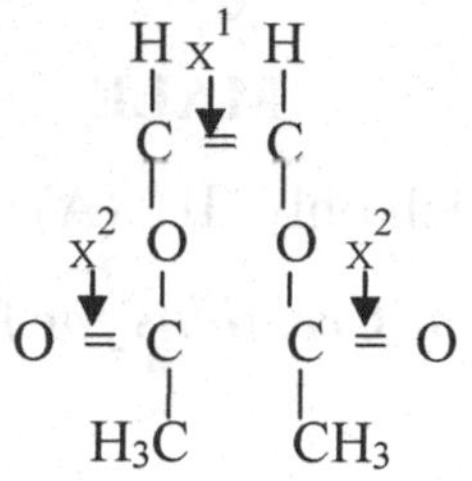

$$(x^2 > x^1)$$

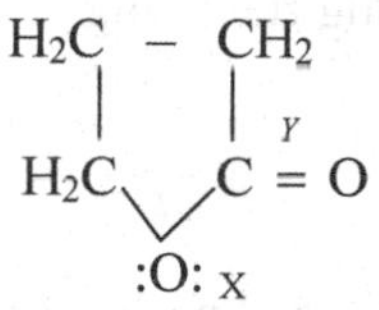

(V)   (VI)   (VII)   (VIII)

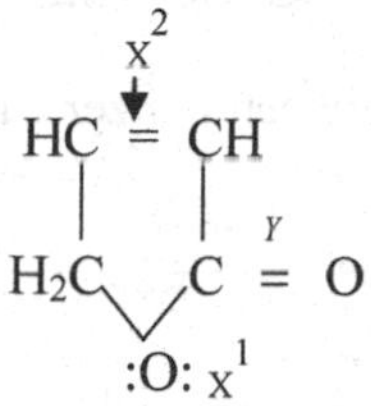

MALE   **MALE**   **MALE**   **MALE**

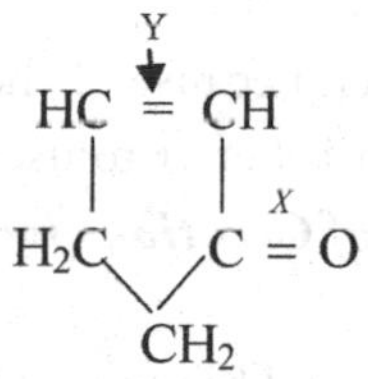

$$(x^2 > x^1)$$

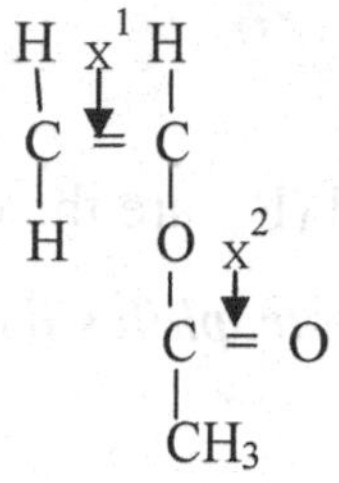

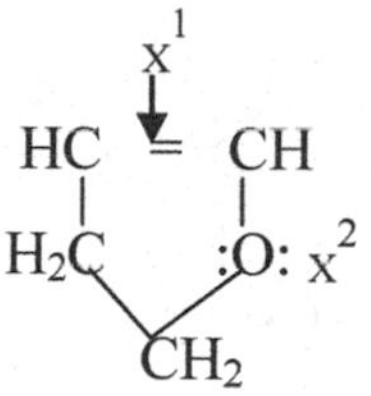

(IX)   (X)   (XI)   (XII)

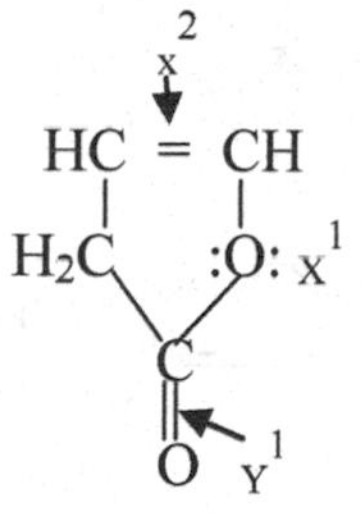

FEMALE   **MALE**   **MALE**   **MALE**

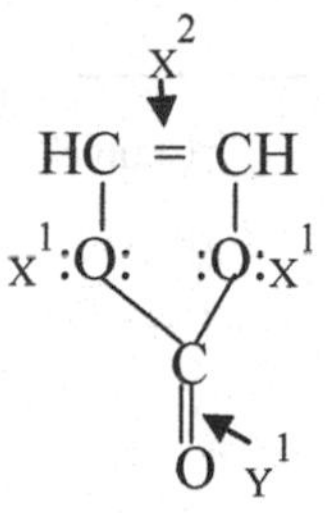

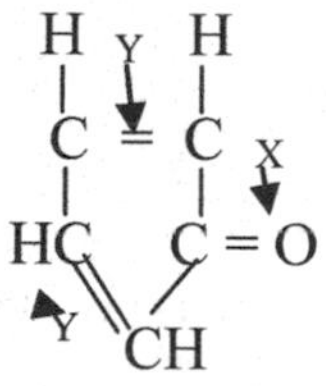

$$(x^2 > x^1)$$

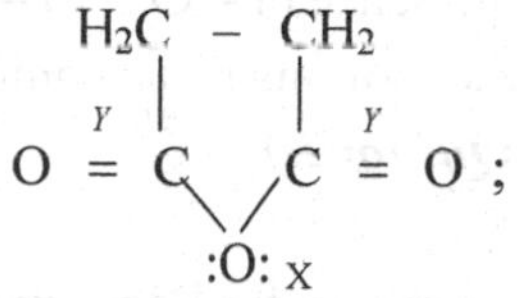

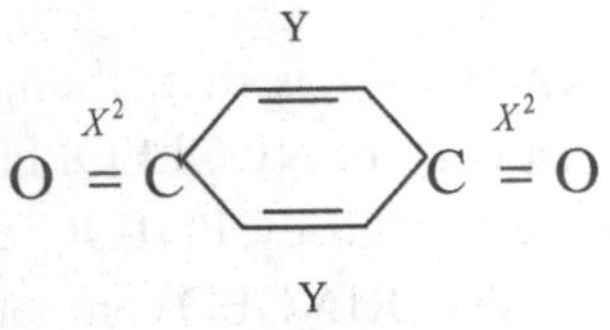

$$O = C \overset{X^2}{\diagdown} \underset{Y}{\overset{Y}{\bigcirc}} \overset{X^2}{\diagup} C = O$$

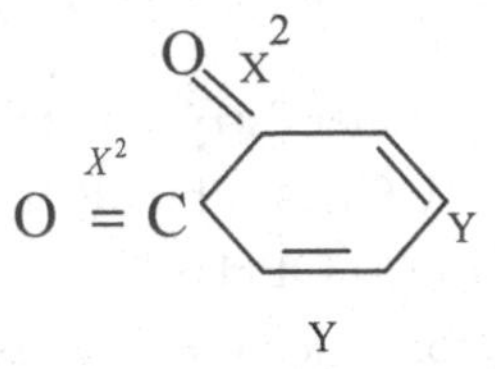

$$\underset{Y}{\overset{O}{\diagdown} X^2} \quad O = C \overset{X^2}{\diagdown} \underset{Y}{\bigcirc} Y$$

(A)   (B)

MALE   **MALE**

for which only (II), (A) and (B), are the ones that are resonance stabilized.

(Laws of Creations for Electrophiles from Nucleophilic Centers)

<u>**Rule 386:**</u> This rule of Chemistry for **Quinones,** states that, like all resonance stabilized mono-mers, there are only two mono-forms- the 3,4- and 1,6-mono-forms; for which when polymerization is favored, it is the 1,6-mono-form that is involved as shown below nucleo-non-free-radically;

$$\overset{\bullet\bullet}{N} -(O - C \qquad\qquad C - O)_n - O - C \qquad\qquad C - O \bullet nn$$

for which because of the presence of $- O - O -$ bond between benzene rings along the living chain, it is better to use the monomer as a comonomer.

(Laws of Creations for Quinones)

<u>**Rule 387:**</u> This rule of Chemistry for **Para-quinone,** states that, when copolymerized with for example 1,3-Pentadiene, full alternating copolymers are obtained nucleo-free-radically as shown below-

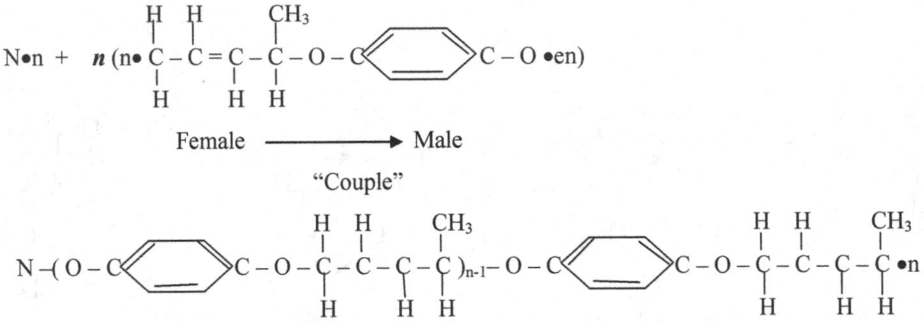

$$N\bullet n \; + \; \boldsymbol{n}\,(n\bullet \overset{H}{\underset{H}{\overset{|}{C}}} - \overset{H}{\underset{|}{\overset{|}{C}}} = \overset{}{C} - \overset{CH_3}{\underset{H}{\overset{|}{C}}} - O - C\!\!\bigcirc\!\!C - O \bullet en)$$

Female $\longrightarrow$ Male

"Couple"

$$N-(O - C\!\!\bigcirc\!\!C - O - \overset{H}{\underset{H}{\overset{|}{C}}} - \overset{H}{\underset{H}{\overset{|}{C}}} - \overset{}{\underset{}{C}} - \overset{CH_3}{\underset{H}{\overset{|}{C}}})_{n-1} - O - C\!\!\bigcirc\!\!C - O - \overset{H}{\underset{H}{\overset{|}{C}}} - \overset{H}{\underset{H}{\overset{|}{C}}} - \overset{}{\underset{}{C}} - \overset{CH_3}{\underset{H}{\overset{|}{C}}} \bullet n$$

for which the formation of the Couple was made possible, because the two monomers are unreactive to the initiator.

(Laws of Creations for Copolymerization using p-Quinone)

<u>**Rule 388:**</u> This rule of Chemistry for **CO$_2$ and p-Quinone,** states that, while the former is Alphatic and Nucleophilic in character, the latter is Aromatic and Electrophilic in character; for which both can be used as monomers or comonomers depending on the operating conditions.

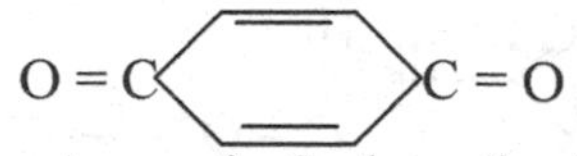

$$O = C \diagup\!\!\!\!\overline{}\!\!\!\!\diagdown C = O \qquad\qquad O = C = O$$

Aromatic Carbon dioxide $\qquad\qquad$ Aliphatic Carbon dioxide

(Laws of Creations for Aliphatic and Aromatic CO_2)

Rule 389: This rule of Chemistry for **Quinones carrying radical-pushing or pulling groups,** states that, only quinones which carry groups that are symmetrically placed, can be resonance stabilized as shown below for those marked with asterisk, while those where the groups are not symmetrically placed cannot be resonance stabilized and used as a monomer; noting that this applies not only to p-quinone, but also to o-quinone where the two carbonyl centers are adjacently placed separated by the two internally located π-bonds symmetrically placed-

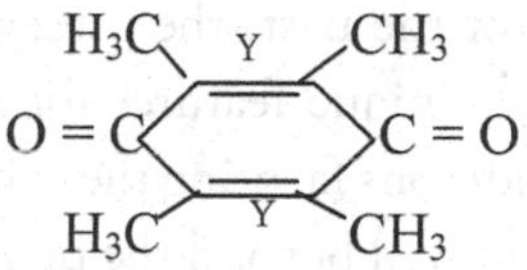

(I)*	(II)	(III)
Tetramethyl-quinone	**2,5-diethyl-3,6-dimethyl-quinone**	**2,3-dimethyl-5,6-diethyl-quinone**

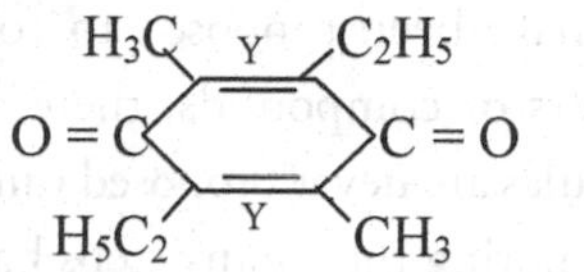

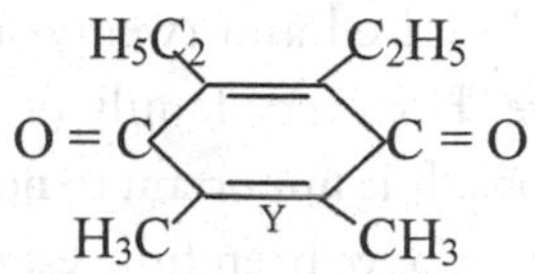

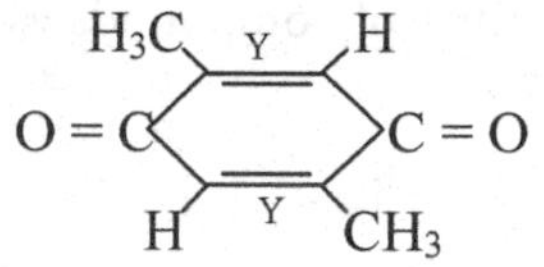

(IV)	(V)	(VI)*
2,5-dimethyl-quinone	**2-ethyl-5-methyl-quinone**	**2,6-dimethyl-quinone**

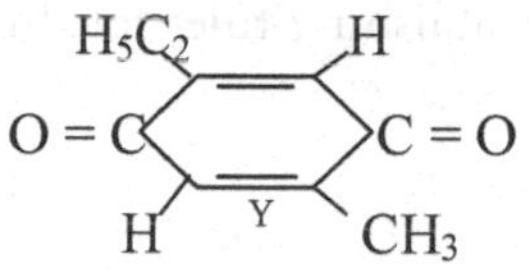

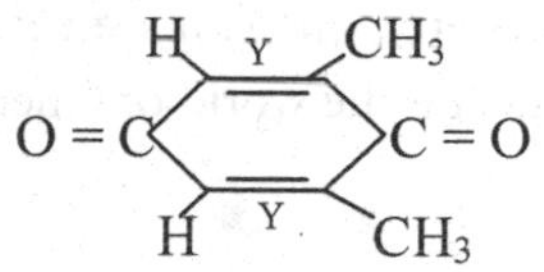

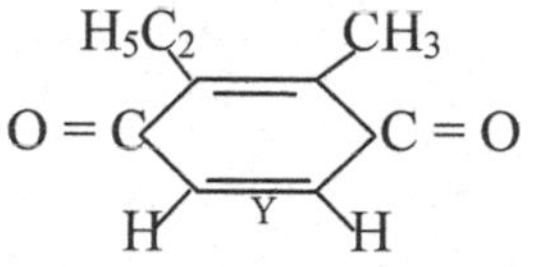

(VII)	(VIII)	(IX)
2-methyl-3-ethyl-quinone	**2-methyl-6-ethyl-quinone**	**2-methyl-quinone**

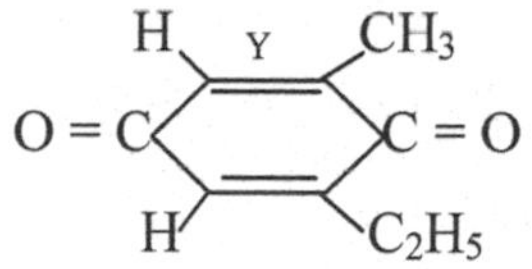

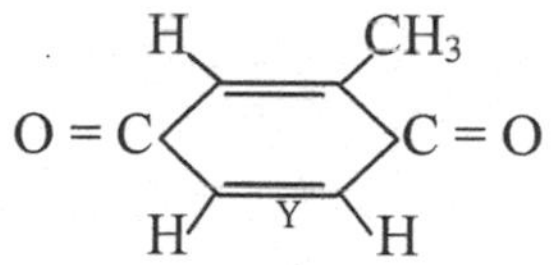

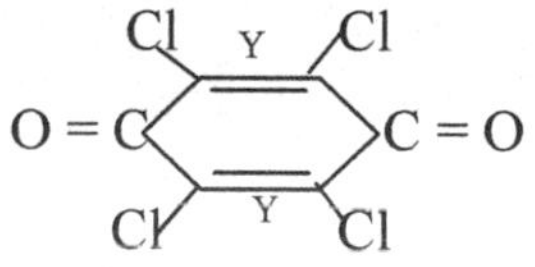

(X)*	(XI)	(XII)
Tetrachloro-quinone	**2,5-dichloro-quinone**	**2,3,5-trichloro-quinone**

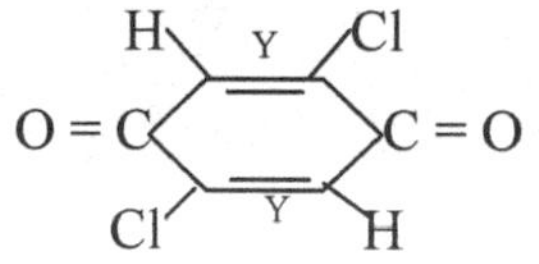

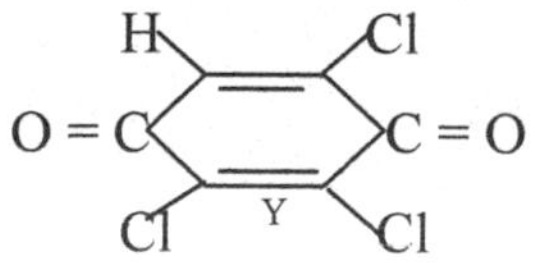

for which for those which cannot be resonance stabilized, the followings are obtained using (IV) above. when activated using same initiator as used for copolymerization-

$$N\bullet n \;+\; O=C\underset{H}{\overset{H_3C}{\diagdown}}\!\!\diagup\!\!\diagdown C=O \longrightarrow N\bullet n \;+\; O=C\underset{H}{\overset{H_3C\;\bullet e\quad\bullet n\;H}{\diagdown}}\!\!\diagup\!\!\diagdown C=O \longrightarrow$$

$$NH \;+\; O=C\diagup\!\!\diagdown C=O$$

noting that the o- quinone unlike the p- quinone cannot be used as a monomer.

(Laws of Creations for Quinones)

One hundred and twenty-six rules have been proposed to cover part of the past, the present and the future. For every family of monomers or compounds, there are always unique features for future applications. It is important to note that rules are never proposed until all conditions favoring the existence of such rules have been fully established, noting that many steps have been moved beyond the present to justify such propositions. Of most of the Addition monomers considered so far, diazoalkanes have been the most unique. Based on some of the misunderstand-ings which have been made in the past with respect to definitions such as ceiling temperatures, polymeric products of diazoalkanes, the characters of these monomers etc., one can see the need for establishing rules for different families of compounds as they operate inside the world of Chemistry.

References

1. C. R. Noller, "Textbook of organic chemistry", W. B. Saunders Company, (1966), pgs. 201 -205.

2. G. Odian, "Principles of Polymerisation", McGraw-Hill Book Company, (1970), pgs. 346-351.

3. C. R. Noller, "Textbook of organic chemistry", W. B. Saunders Company, (1966), pgs. 249-250.

4. F. S. Dainton and K. J. Ivin, Quart. Rev. (London), 12 : 61 (1958).

5. F. S. Dainton and K. J. Ivin, Trans. Faraday Soc. 46 : 331 (1950).

6. L. A. Wall, Soc Plastic Eng. J., 16 : 1 (1960).

7. H. McCormick, J. Polymer Sci., 25 : 488 (1957).

8. D. J. Worsfold and S. Bywater, J. Polymer Sci., 26 : 299 (1957)

9. C. R. Noller, "Textbook of organic chemistry", W. B. Saunders Company, (1966), pgs. 284 - 289.

10. I. L. Finar, "Organic Chemistry", Vol. 2, Third Edition For E. L. B. S (1964), pg. 421.

11. C. R. Noller, "Textbook of organic chemistry", W. B. Saunders Company, (1966), pg. 641.

12. R. Smith, "Notes-Reaction of Diazoethane and 1-Diazopropanne with Aliphatic Aldehydes", J. Org. Chem., 1960, 25 (3), pp. 453-454.

13. K. Maruoka, A. B. Concepcion and H. Yamamoto, "Organoaluminum-Promoted Homologation of Ketones with Diazoalkanes" J. Org. Chem., 1994, 59 (17), pp4725-4726.

14. C. D. Gutsche, Org. React., 1954, 8, 364.

15. C. R. Noller, "Textbook of organic chemistry", W. B. Saunders Company, (1966), pg. 652.

16. J. M. Figuera, J. M. Perez and A. P. Wolf, "Photolysis of diazo-n-butane. Study of the unimolecular decomposition of activated intermediates and energy partitioning", J. Chem. Soc., Faraday Trans. 1, 1975, 71. 1905-1917.

17. F. Rodriquez, "Principles of Polymer Systems", McGraw-Hill Book Company, (1970), pg. 72.

18. T. Machiguechi, T. Hasegawa and H. Otani, "Tropothione S-Oxide: The First Example of a Sulfin Charge Reversion", J. Am. Chem. Soc., 1994,116, 407-408.

19. C. R. Noller, "Textbook of organic chemistry", W. B. Saunders Company, (1966), pgs. 270-274.

20. J. H. Chou and T. B. Rauchfuss, "Solvatothermal Routes to Poly(Carbon Monosulfide)s Using Kinetically Stabilized Precursors", J. Am. Chem. Soc., 1997, Vol.119, 4537-4538.

21. T. Kochi, A. Nakamura, H. Ida and K. Nozaki, "Alternating Copolymerization of Vinyl Acetate with Carbon Monoxide", J. Am. Chem. Soc.,2007, 129 (25), pp. 7770-7771.

22. N. Kyoko and N. Koji, "Precision polymer synthesis: Alternating copolymerization of epoxide with carbon dioxide and carbon monoxide", Expected Materials for the Future (L4328A), Vol. 6; NO.8; 2006, pp. 50-58.

23. T. Kagiya, S. Narisawa, T. Ichida, K. Fukui, H. Yokota and M. Kondo, "Alternative copolymerization of aziridines and carbon monoxide by Gamma-ray irradiation", J. Polym. Sci. Part A-1: Polymer Chemistry, Vol. 4, Issue 2, 1966, pgs. 293-299.

24. J. C. Choi, I. Yamaguchi, K. Osakada and T. Yamamoto, "Alternating copolymerization of arylallenes with carbon monoxide catalyzed by a π-alkylrhodium complex. Synthesis of new polyketones with regulated structure and molecular weight", Macromolecules, 1998, Vol. 31, n°25, pp. 8731-8736.

25. P. K. Hanna, A. M. Piotrowski, M Andrzej, "Copolymer of carbon monoxide and propylene produced in water/ketone solvent", United States Patent 5270441, 1993.

26. N. Ungvari, T. Kegl and F. Ungvary, "Octacarbonyl dicobalt-catalyzed selective carbonylation of (trimethylsilyl) diazomethane to obtain (trimethylsilyl) ketene", J. Molecular Catalysis A: Chemical, Vol. 219, Issue 1, 2004, pgs. 7-11.

27. D. D. Parker, A. B. Padias and H. K. Hall, "Synthesis and polymerization studies of Formaldehyde oxime and its derivatives", J. Polym. Sci. Part A, Polymer Chemistry, 2000, Vol. 38, n°10, pp. 1866-1872.

28. http://www.absoluteastronomy.com/topics/Nitroxyl

29. C. F. Hauser and N. L. Zutty, "Quinone copolymerization. 1. Reactions of p-chloranil, p-benzoquinone, and 2,5-dimethyl-p-benzoquinone with vinyl monomers under free-radical initiation", J. Polym. Sci. Part A-1: Polymer Chemistry, Online-2003, Vol.8, Issue 6, pgs. 1385-1401

Problems

7.1. Define the following terms-

(a) "Self-activation of a monomer".
(b) Activated/Equilibrium State of Existence.
(c) Ceiling temperature of a monomer.
(d) Molecular rearrangement of the fourth, fifty, and sixth types.

7.2. (a) Determine the increasing order of the ceiling temperatures of the following monomers or compounds-
Aldimine, formaldehyde, isocyanates, acetaldehyde, acetone, cyanogen chloride and cyanamide.
(b) Why do full free radical monomers not have ceiling temperatures?
(c) Under what conditions is existence of molecular rearrangement made possible?

7.3. (a) Why are diazoalkylenes not known to undergo radical polymerization reactions?
(b) Distinguish between the charged and radical polymerizations of diazoalkylene.
(c) What makes diazoalkylene so unique?

7.4. Azo-"methane" $CH_3N = NCH_3$ is said to be unstable at 200°C to yield free-radicals and N_2. The methyl radical may recombine to form ethane. But the methylide radicals obtained can be used as initiators.

(i) Explain why it is unstable at that temperature?
(ii) What is the type of radical most readily produced? Explain tentatively the mechanism of the reaction.
(iii) Under what conditions will the free-radicals abstract hydrogen and initiate free-radical polymerizations.
(iv) When a stream of inert gas saturated with tetramethyllead vapor passed through a silica tube is heated at one spot, what happens?
(v) From the names provided for the polymeric products of charged polymerization of diazomethylene (CH_2N_2), what name changes should be made for Azo-Alkanes? What should $CH_3N = NH$ be called?

7.5. (a) Distinguish between the reactions of aldimines and ketimines with a strong acid.
(b) Compare with that of acetaldehyde and acetone.

7.6. Shown below are some derivatives of aldimines and ketimines -(i)

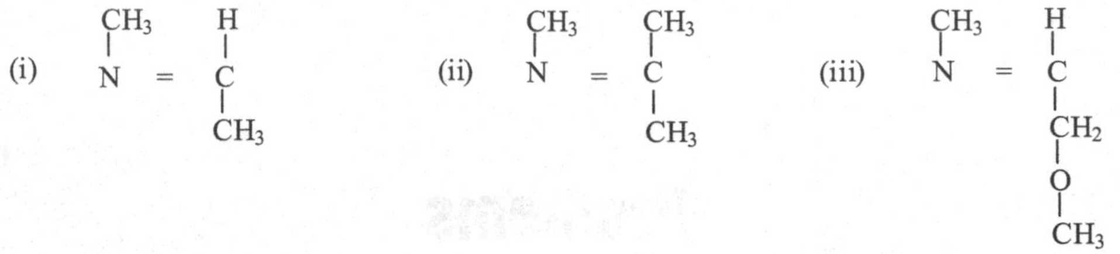

(i)
$$CH_3 \quad H$$
$$N = C$$
$$CH_3$$

(ii)
$$CH_3 \quad CH_3$$
$$N = C$$
$$CH_3$$

(iii)
$$CH_3 \quad H$$
$$N = C$$
$$CH_2$$
$$O$$
$$CH_3$$

(a) Show how some or all of these are obtained from alkyl vinyl amines.

(b) Identify the route(s) favored by the monomers.

(c) Can they provide resonance stabilization to another group or can resonance stabilization be provided for the monomers? Explain.

7.7. (a) Under what conditions can alkyl cyanates exist as stable monomers? Explain using equations.

(b) Why are etheric nitriles (alkyl cyanates) important amongst the nitriles?
Identify the route favored by alkyl cyanates and the kinds and types of transfer species involved with them.

(c) Compare cyanamide with the corresponding cyanate, in terms of molecular rearrangement, cyclization and character.

7.8. (a) Compare and contrast the radical polymerization of ethylene with that of diazomethylene.

(b) What is the driving force favoring the formation of rings between the two states shown below?

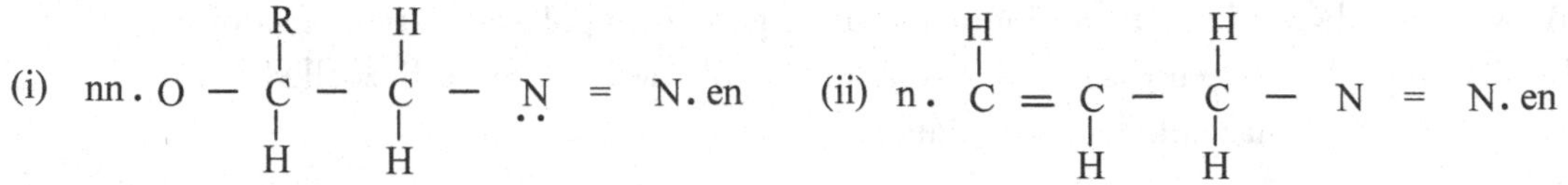

(i)
$$R \quad H$$
$$nn.O - C - C - N = N.en$$
$$H \quad H$$

(ii)
$$H \quad H$$
$$n.C = C - C - N = N.en$$
$$H \quad H$$

Can these states actually exist?

(c) Distinguish between the types of molecular rearrangements where possible for the activated states shown below –

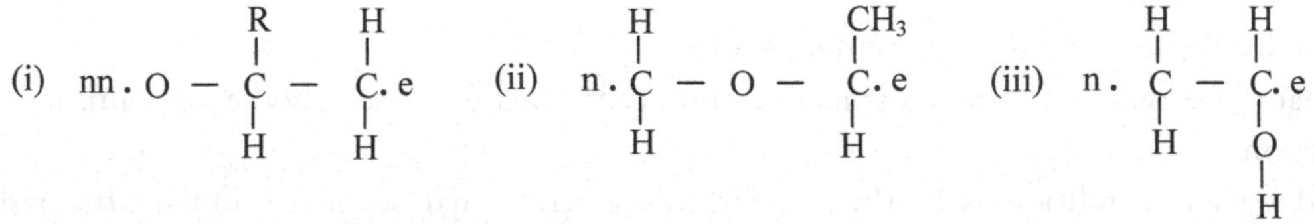

(i)
$$R \quad H$$
$$nn.O - C - C.e$$
$$H \quad H$$

(ii)
$$H \quad CH_3$$
$$n.C - O - C.e$$
$$H \quad H$$

(iii)
$$H \quad H$$
$$n.C - C.e$$
$$H \quad O$$
$$H$$

7.9. (a) Show that diazomethane is more nucleophilic than acetaldehyde using the law of diffusion controlled mechanism and known products obtained from their reactions.

(b) Why can diazoalkylenes not favor being resonance stabilized using external resonance stabilization groups?

(c) What type of resonance stabilization exists in diazoalkylenes? Explain.

7.10. (a) Show whether aldehydes can be obtained from the following two activated states-?

(i) 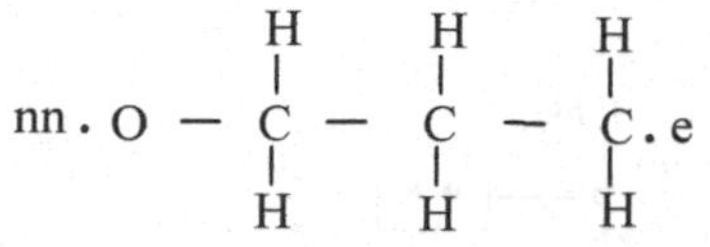 nn . $O - \overset{\displaystyle H}{\underset{\displaystyle H}{C}} - \overset{\displaystyle H}{\underset{\displaystyle H}{C}} - \overset{\displaystyle H}{\underset{\displaystyle H}{C}}$. e

(ii) 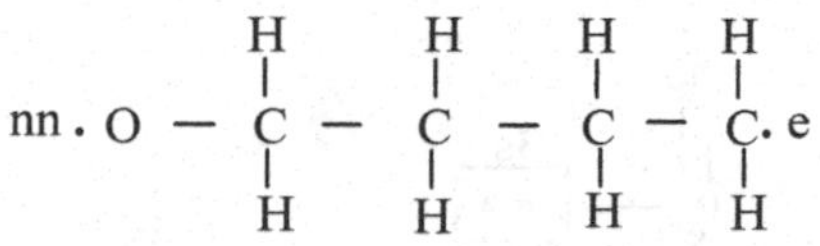 nn . $O - \overset{\displaystyle H}{\underset{\displaystyle H}{C}} - \overset{\displaystyle H}{\underset{\displaystyle H}{C}} - \overset{\displaystyle H}{\underset{\displaystyle H}{C}} - \overset{\displaystyle H}{\underset{\displaystyle H}{C}}$. e

(b) Distinguish between the charged and radical activations of the following activation centers –

$$C = C \ , \quad C \equiv C \ , \quad C = N \ , \quad C - N \ , \quad C = O \ , \quad C - S \ , \quad S = O \text{,}$$

$$N = O \ , \quad N \equiv N \quad \text{and} \quad O = O$$

7.11. (a) Shown below is the charged activated state of a quinone in Present-day Science -

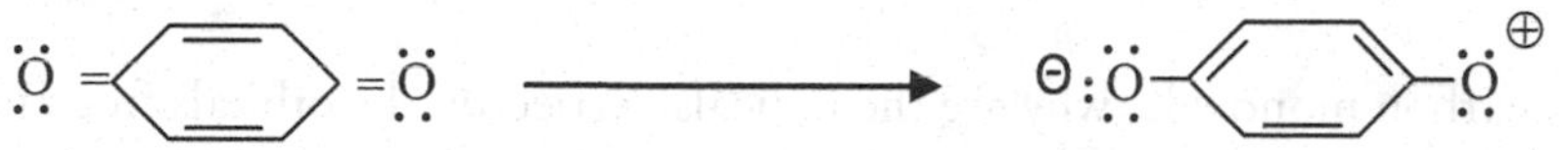

(i)   What are the types of charges carried by the oxygen centers?
(ii) Can the monomer be resonance stabilized?
(iii) Why can the charged activated state not exist?
(b)   Shown below are two monomers - a full free-radical monomer and full non-free-radical monomer.

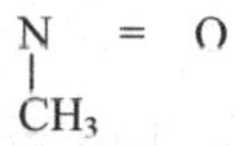

$$N = O \\ \underset{\displaystyle CH_3}{|}$$

Full non-free-radical monomer

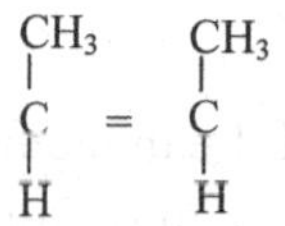

$$\overset{\displaystyle CH_3}{\underset{\displaystyle H}{C}} = \overset{\displaystyle CH_3}{\underset{\displaystyle H}{C}}$$

Full free-radical monomer

Distinguish between the radical initiation steps of the monomers where possible using strong electro- and nucleo-free and non-free radicals for both.

7.12. (a) Distinguish between the followings when used as monomers under different operating conditions where possible.

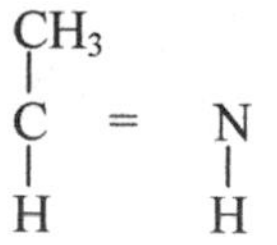

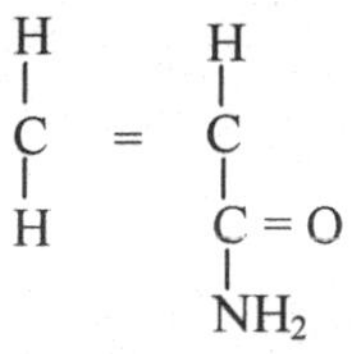

$$\overset{\displaystyle CH_3}{\underset{\displaystyle H}{C}} = \overset{\displaystyle N}{\underset{\displaystyle H}{N}}$$

$$\overset{\displaystyle H}{\underset{\displaystyle H}{C}} = \overset{\displaystyle H}{\underset{\displaystyle \underset{\displaystyle NH_2}{|}}{\underset{\displaystyle C = O}{C}}}$$

(b) Distinguish between the two classes of activated centers -

$$C = N \ , \quad C \equiv N \ , \quad C = S \quad \text{and} \quad N = O \ , \quad S = O \ ,$$

$$N \equiv N \ , \quad O = O \ \cdot$$

(c) Why is the C = O center in CO not commonly used as an activation center?

7.13. Shown below are two different types of phosphines.

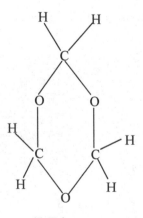

(a) How many radicals are in the last shell of phosphorus? What is the valence state of phosphorus in the phosphines?
(b) Can any of them or indeed all types of phosphines be activated?
(c) Can quaternary phosphonium salt be obtained as we have for quaternary ammon-ium salt? Explain.

7.14. In the activated state of carbon monoxide, why are the radicals carried by the orbitals free radicals and not non-free radicals? In what orbitals are the free radicals located? Under what conditions is carbon monoxide a poison? If the CO was in the liquid or solid state, will it be poisonous? Explain.

7.15. Shown below is an important and one of the commonest monomers.

(a) Show the activated states of the monomer radically and chargedly.
(b) While fluorinated aldehydes can be polymerized, why is CO_2 not popularly known to be used as a monomer?
(c) How nucleophilic is CO_2 compared to aldehydes and ketones?
(d) Is CO_2 activated when it is inside the human system? Explain.
(e) Why is the existence of the compound (II) shown below not popularly known?

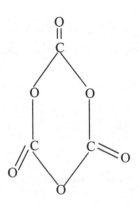

(I) Trioxane

(II) Unknown compound

7.16. The reactions of Grignard reagents (RMgX) with O_2 or CO_2, followed by addition of mineral acid (HX) produces alcohol or carboxylic acid respectively as shown below.

(i) $\quad O_2 \quad + \quad RMgX \xrightarrow{\hspace{1cm}} XMg\,O-O-R \xrightarrow{+\,RMgX} 2R-O\,MgX \xrightarrow{+\,2HX}$

$\qquad\qquad\qquad 2ROH \quad + \quad 2MgX_2$

(ii) $\quad CO_2 \quad + \quad RMgX \xrightarrow{\hspace{1cm}} RCOOMgX \xrightarrow{+\,HX} RCOOH \quad + \quad MgX_2$

where R is an alkyl group and X is a halogen atom. Though it too early to answer the question, since what Grignard reagent is has not been fully shown, explain tentatively the mechanism of the reactions above based on very little current developments so far. (Clue-Grignard reagents are dimeric in character).

7.17. Shown below are four different types of diazoalkylenes.

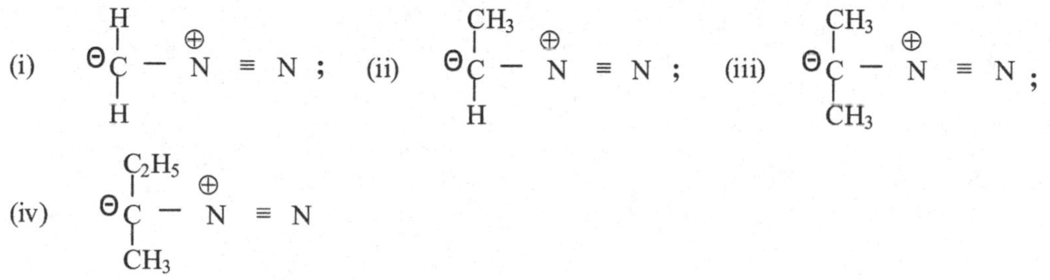

(i) $\quad {}^{\ominus}\!\!\underset{H}{\overset{H}{C}} - \overset{\oplus}{N} \equiv N \;;\quad$ (ii) $\quad {}^{\ominus}\!\!\underset{H}{\overset{CH_3}{C}} - \overset{\oplus}{N} \equiv N \;;\quad$ (iii) $\quad {}^{\ominus}\!\!\underset{CH_3}{\overset{CH_3}{C}} - \overset{\oplus}{N} \equiv N \;;$

(iv) $\quad {}^{\ominus}\!\!\underset{CH_3}{\overset{C_2H_5}{C}} - \overset{\oplus}{N} \equiv N$

(a) Show using equations how the use of free-radical and charged paired- initiators give different types of polymeric products under different operating conditions.
(b) Compared the products with those obtained from corresponding types of olefines.
(c) In what order will the polymerizations of the monomers be favored chargedly?

7.18. Why the use of the words "MOTHER NATURE" associated with chemical compounds with ACTIVATION centers? Discuss.

7.19 Discuss extensively the distinctions between Forward and Backward Addition polymerization systems.

7.20 (a) Based on the NEW FRONTIERS, begin to list some of the driving forces encountered so far favoring the existence of Alternating placement in ADDITION polymerization systems.
(b) Lists all the different types of Initiators used in ADDITION polymerization systems.

7.21. (a)   Radically, show the types of products obtained from the monomers of Q 7.17.

(b)   Identify the characters of the monomers by identifying the route(s) favored by them.

(c)   Why can't diazoalkylenes be deactivated to produce stable diazoalkylenes without polar bonds at high operating conditions?

(d)   Shown below is an imaginary or unknown monomer.

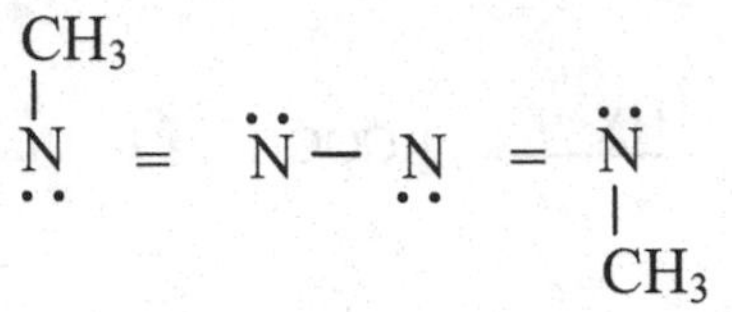

$$\overset{\displaystyle CH_3}{\underset{\displaystyle \cdot\cdot}{N}} \; = \; \overset{\cdot\cdot}{\underset{\cdot\cdot}{N}} - \overset{}{\underset{\cdot\cdot}{N}} \; = \; \overset{\cdot\cdot}{\underset{\displaystyle CH_3}{N}}$$

What can you say about the monomer based on the New Frontiers?

CHAPTER 8

TRANSFER OF TRANSFER SPECIES IN CUMULENCES

8.0 Introduction

If three or more consecutive carbon atoms are joined by double bonds, then the compound is said to be a cumulative double bonded compound. Compounds having two or more double bonds joining adjacent carbon are called cumulenes as shown below.

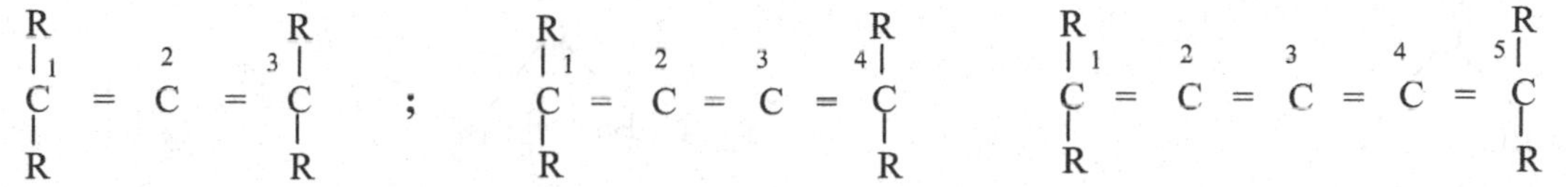

$$
\underset{R}{\overset{R}{\underset{|}{\overset{|}{C}}}} _1 = \overset{2}{C} = \underset{R}{\overset{R}{\underset{|}{\overset{3|}{C}}}} \quad ; \quad \underset{R}{\overset{R}{\underset{|}{\overset{|}{C}}}} _1 = \overset{2}{C} = \overset{3}{C} = \underset{R}{\overset{R}{\underset{|}{\overset{4|}{C}}}} \quad \underset{R}{\overset{R}{\underset{|}{\overset{|}{C}}}} _1 = \overset{2}{C} = \overset{3}{C} = \overset{4}{C} = \underset{R}{\overset{R}{\underset{|}{\overset{5|}{C}}}}
$$

8.1

     1,3 - Cumulenes   1,4 - Cumulenes   1,5 - Cumulenes

(where Rs which can be different are substituted groups)
When the R groups are not the same, then the normenclature begins to change.

Very little is known about these compounds when used as monomers. The first member of 1,3 - cumulenes is the allene (propadiene)[1], which can readily be synthesized by series of reactions from for example glycerol via largely radical reactions as shown below.

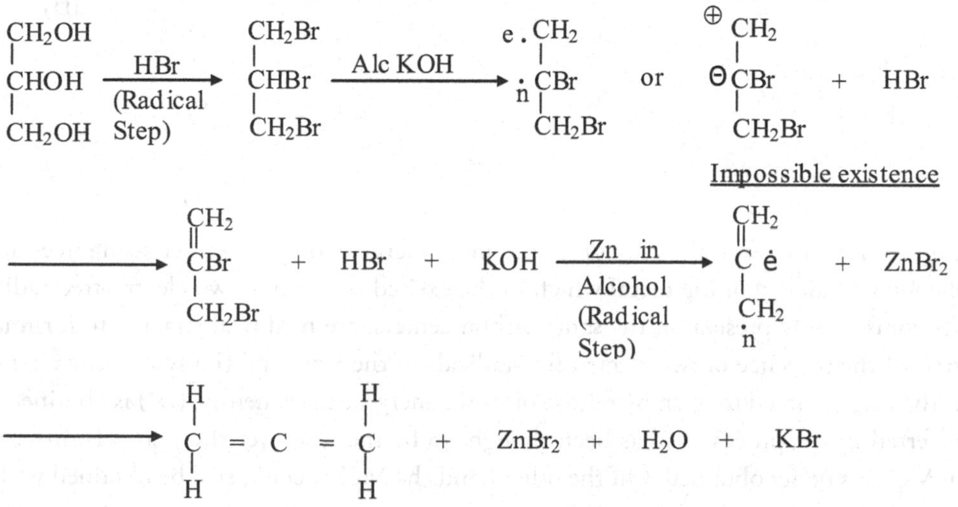

$$
\begin{array}{c}
CH_2OH \\
| \\
CHOH \\
| \\
CH_2OH
\end{array}
\xrightarrow[\text{(Radical Step)}]{HBr}
\begin{array}{c}
CH_2Br \\
| \\
CHBr \\
| \\
CH_2Br
\end{array}
\xrightarrow{\text{Alc KOH}}
\begin{array}{c}
e \cdot CH_2 \\
| \\
\overset{.}{\underset{n}{C}}Br \\
| \\
CH_2Br
\end{array}
\quad \text{or} \quad
\begin{array}{c}
\oplus CH_2 \\
| \\
\ominus CBr \\
| \\
CH_2Br
\end{array}
+ \; HBr \; + \; KOH
$$

$$\underline{\text{Impossible existence}}$$

$$
\longrightarrow
\begin{array}{c}
CH_2 \\
\| \\
CBr \\
| \\
CH_2Br
\end{array}
+ \; HBr \; + \; KOH
\xrightarrow[\text{Alcohol (Radical Step)}]{\text{Zn in}}
\begin{array}{c}
CH_2 \\
\| \\
C \; \dot{e} \\
| \\
\underset{n}{\overset{.}{C}H_2}
\end{array}
+ \; ZnBr_2
$$

$$
\longrightarrow
\underset{H}{\overset{H}{\underset{|}{\overset{|}{C}}}} = C = \underset{H}{\overset{H}{\underset{|}{\overset{|}{C}}}}
\; + \; ZnBr_2 \; + \; H_2O \; + \; KBr
$$

8.2

In the absence of nucleo-free-radical in the system, the last step above is favored. That last step can never be ionic, since Zn itself is a non-ionic metal. Indeed none of the steps can be ionic. Indeed the mechanism for the reactions above is Equilibrium with many stages.

A general synthesis of allenes is said to involve addition of dibromocarbene, generated from bromoform and potassium t-butoxide, to an olefin followed by reaction of the dibromocyclopropane with magnesium in ether or various organometallic reagents.[1]

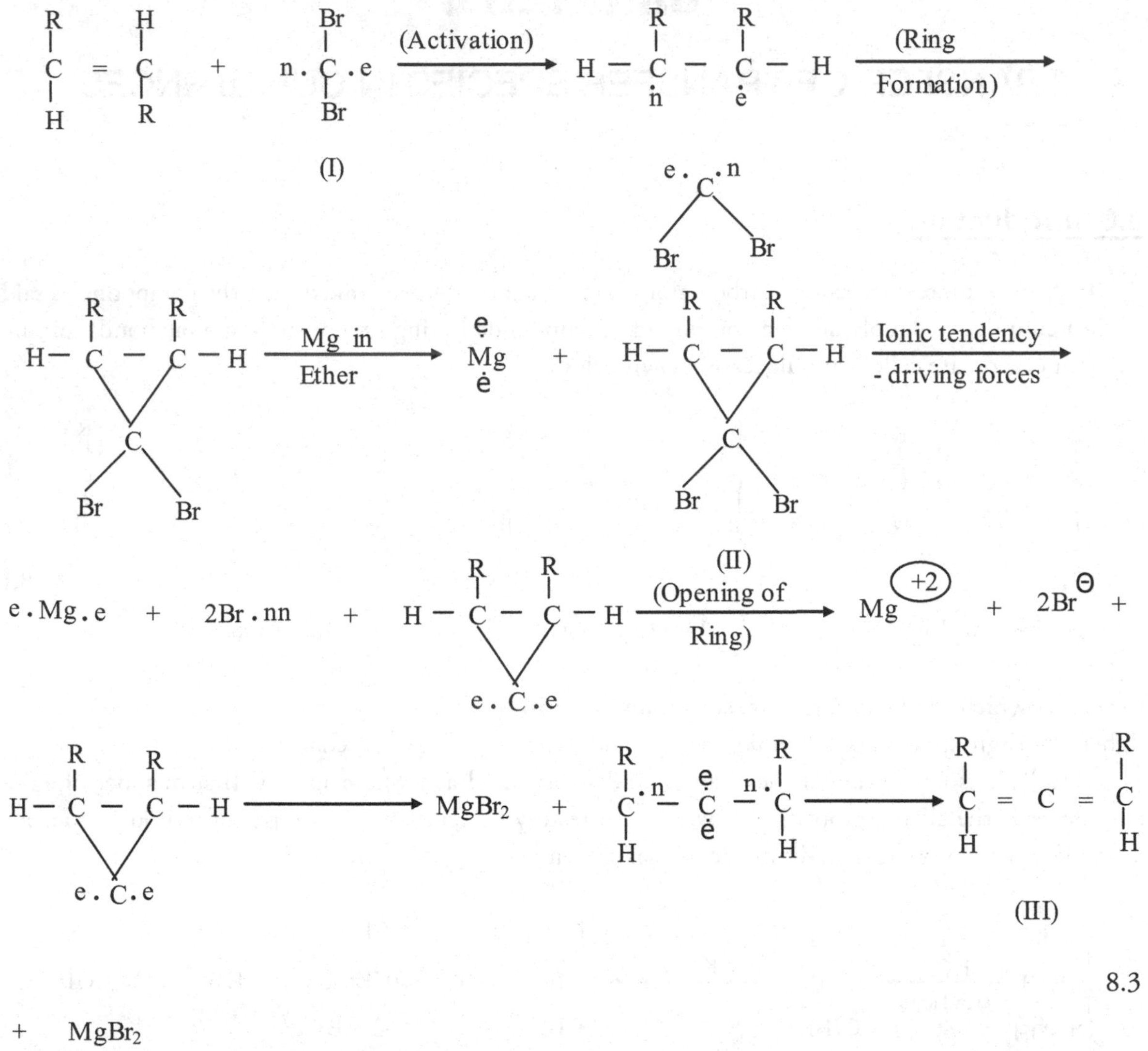

8.3

+ $MgBr_2$

It is important to note that all the major reactions are radical in character. The carbene (I) for example cannot carry charges or ions, but radicals. Though ringed monomers can only be opened using free-radical means, due to the ionic tendency of Mg metal which in the excited state carry two electro-free-radicals, the two bromine atoms readily present on the same carbon centers, are readily abstracted to form ionic $Mg^{2\oplus} 2Br^{\ominus}$. In view of the presence of two electro-free-radicals in the ring, and the symmetric character of the monomer, the ring is forced to open by release of strain energy. Subsequently, (III) is obtained. No radicals are transferred in the process, as has been thought to be the case over the years. Transfer only takes place when $MgBr_2$ is being obtained. On the other hand the $MgBr_2$ could still be obtained without any transfer of radicals, depending on the operating conditions energywise. When ringed compounds are considered downstream, this will become obvious.

Though in general, little is known about the chemistry of these monomers, allene has been reported to readily undergo rearrangement.[1] The reaction of allene with sodium gives sodium methylacetylide as follows.

$$
2\,\overset{\displaystyle H}{\underset{\displaystyle H}{C}} = C = \overset{\displaystyle H}{\underset{\displaystyle H}{C}} \xrightarrow[\text{In ether}]{2Na \cdot e} 2\,n \cdot \overset{\displaystyle H}{\underset{\displaystyle H}{C}} - \overset{\displaystyle }{\underset{\displaystyle CH_2}{C}} \cdot e + 2Na \xrightarrow[\text{rearrangement)}]{\text{(Molecular}}
$$

$$
2Na\,^{\cdot e} + 2\,n \cdot C \equiv \underset{\displaystyle CH_3}{C} + 2\,H \cdot e \longrightarrow 2NaC \equiv C(CH_3) + H_2 \qquad 8.4
$$

The monomer is first activated through any of the π-bonds, since it is symmetric. This is then followed by molecular rearrangement, involving transfer species of the first kind of the third type to form the acetylene. Then, this is followed by Na attack, in the process of which hydrogen molecule is released. All these take place by Equilibrium mechanism only radically in many stages.

Allenes can be observed to be closely related to the acetylenes and olefins. In cumulenes, in view of the cumulative character of the double bonds, only one double bond can be activated one at a time. With symmetrically placed 1,3-cumulence, only one mono-form exists as already shown for allene and below for (III) of Equation 8.3. For non-symmetrically placed 1,3 - cumulenes, just as for the symmetric case, only one mono-form exists. Unlike acetylene, cumu lenes undergo some form of molecular rearrangements

$$
\overset{\displaystyle CH_3}{\underset{\displaystyle H}{\overset{1}{C}}} = \overset{2}{C} = \overset{\displaystyle CH_3}{\underset{\displaystyle H}{\overset{3}{C}}} \longrightarrow \ominus\overset{\displaystyle CH_3}{\underset{\displaystyle H}{\overset{1}{C}}} - \overset{2}{\underset{\displaystyle \overset{3}{CH}}{C}}\oplus \quad OR \quad \oplus\overset{2}{\underset{\displaystyle \overset{1}{CH}}{C}} - \overset{\displaystyle CH_3}{\underset{\displaystyle H}{\overset{3}{C}}}\ominus
$$
$$
\underset{\displaystyle CH_3}{} \qquad\qquad \underset{\displaystyle CH_3}{}
$$

CANNOT BE USED-NOT POSSIBLE   8.5

to produce more stable monomers in terms of character and reduced number of substituted groups. Also, unlike acetylenes which cannot be activated chargedly, cumulenes can be activated chargedly with more numbers of transfer species associated with them. Like acetylenes, very little is known about the use of cumulenes as monomers. In view of the cumulative character of the double bonds, a different type of resonance stabilization phenomenon in cumulenes exists internally for some of the higher cumulenes.

Unlike acetylene, none of the hydrogen atoms in allene is loosely bonded to the carbon centers, so that both free-radical and charged activations of allene and other 1,3 -cumulenes are readily favored. While the external carbon centers in allene are sp^2- hybridized, the internal carbon center is sp-hybridized; while the active carbon centers in acetylene are sp- hybridized.

8.1. 1,3 - Cumulenes (Allenes)

1,3 - cumulenes are allenes of which the first member is allene. Having been introduced to some important aspects of these monomers, there is now the need to consider the transfer of transfer species.

8.1.1. Charged Character of Allenes

When allene is activated using a strong free negatively charged initiator, the followings are obtained, taking note that there are indeed none as we explore herein.

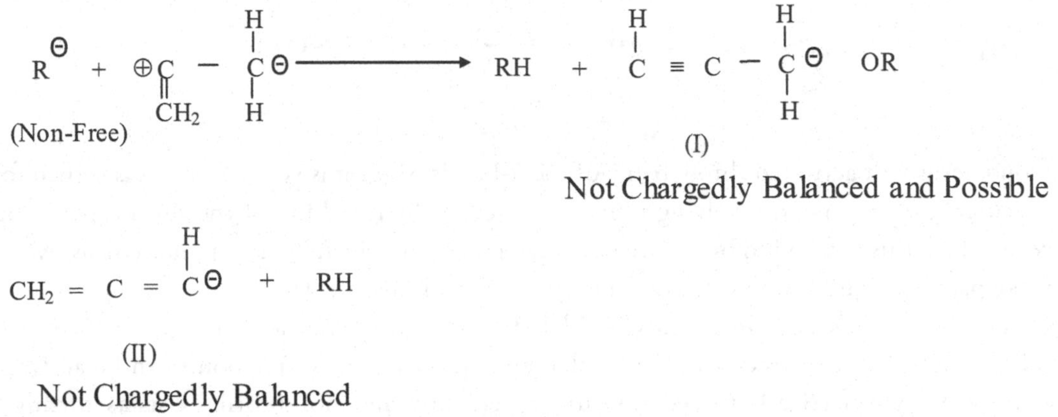

(Non-Free)

(I)

Not Chargedly Balanced and Possible

(II)

Not Chargedly Balanced

8.6

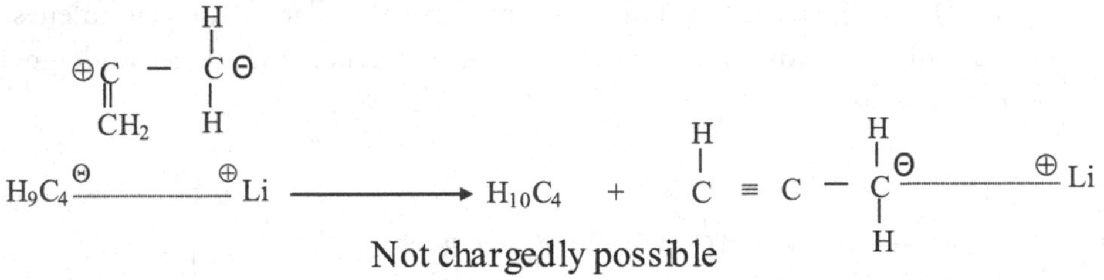

Not chargedly possible

8.7

Negatively, Initiation is impossible. Even then $H^{\oplus}$ cannot be abstracted in all the equations above, since a covalent negative charge cannot sit on the C center after abstraction due to electrostatic forces of repulsion. The reactions above can only take place radically. Hence with the type of group carried, though activation is favored chargedly, it will not be activated chargedly due to presence of transfer species which cannot be removed. A NUCLEOPHILIC MONOMER CANNOT BE MADE ELECTROPHILIC, when the monomer has transfer species. The Nucleophilic monomer cannot therefore favor the use of Negatively charged initiator. **Hence, one can begin to observe that cumulenes cannot be activated chargedly like acetylenes when radical-pushing groups with transfer species are carried. When groups where no transfer species exist in the route not natural to it, then it can be activated and used chargedly.** Radically therefore, the transfer species involved is of the first kind of the third type. When molecular rearrangement is favored which should be the case, whether the initiator is strong or weak, Initiation is also impossible chargedly. In fact, when molecular rearrangement is favored, this can only be done radically.

With strong positively charged and electro-free-radical initiators, the followings are obtained.

$$R \left(\underset{\underset{H}{|}}{\overset{\overset{H}{|}}{C}} - \underset{\underset{CH_2}{||}}{\overset{}{C}} \right)_{\!n} \underset{\underset{H}{|}}{\overset{\overset{H}{|}}{C}} - \underset{\underset{CH_2}{||}}{\overset{}{C}}{}^{\oplus} \longrightarrow R \left(\underset{\underset{H}{|}}{\overset{\overset{H}{|}}{C}} - \underset{\underset{CH_2}{||}}{\overset{}{C}} \right)_{\!n} \underset{\underset{H}{|}}{\overset{\overset{H}{|}}{C}} - C \equiv CH \qquad OR$$

(I) FAVOURED RADICALLY

$$R \left(\underset{\underset{H}{|}}{\overset{\overset{H}{|}}{C}} - \underset{\underset{CH_2}{||}}{\overset{}{C}} \right)_{\!n} \underset{\underset{H}{|}}{\overset{}{C}} = C = \underset{\underset{H}{|}}{\overset{\overset{H}{|}}{C}} + H^{\oplus}$$

(II) FAVORED CHARGEDLY

8.8

In (I), transfer species of the first kind of the third type is involved that which is impossible chargedly but only radically, while in (II) transfer species of second kind is involved that which can only take place chargedly when transfer species of the first kind and third type is present, is also impossible. Since the route is natural to the monomer, transfer species of the second kind cannot be involved. Hence, (II) cannot exist. Hence, the monomer cannot be activated chargedly.

When molecular rearrangement is favored, then the followings are obtained.

$$H_5C_2{}^{\oplus} \quad\text{------}\quad {}^{\ominus}\underset{\underset{R}{|}}{\overset{}{\underset{|}{\overset{}{\underset{O}{B}}}}}\overset{\overset{F}{|}}{{\diagdown}}{\diagup}^{F}_{F} \qquad | \qquad {}^{e}\cdot H \quad {}^{n}\cdot C \equiv \underset{\underset{CH_3}{|}}{C} \longrightarrow \text{No polymerization.}$$

8.9

As is already obvious, the acetylene cannot be activated chargedly, but only radically. Radically, using free or paired-electro-free-radical initiators, the followings are to be expected.

$$E^{\bullet e} + n\bullet \underset{\underset{CH_3}{|}}{\overset{\overset{H}{|}}{C}} = C \bullet e \longrightarrow E - \underset{\underset{CH_3}{|}}{\overset{\overset{H}{|}}{C}} = C \bullet e \xrightarrow{\;+\,(I)\;} E \!\left(\underset{\underset{CH_3}{|}}{\overset{\overset{H}{|}}{C}} = C \right)_{\!n} - \underset{\underset{H}{|}}{\overset{\overset{H}{|}}{C}} = C = \underset{\underset{H}{|}}{\overset{\overset{H}{|}}{C}} + H\bullet e$$

8.10

With free or paired initiators with no reservoir, the polymers obtained will have irregular placements and therefore amorphous in character. In general, to have stereo-regulating electro-free-radical paired initiator may not be difficult to find, if the nucleo-radical centers do have or carry vacant orbitals or reservoirs. Nevertheless, comparing the last equation with Equation 8.8, one can observe how Nature operates, too much to comprehend. It can clearly be observed that transfer species of the first kind of the first type is involved to produce a cumulenic terminal bonded polymer. There is no doubt that when an allene is activated, it readily undergoes molecular rearrangement most of the time where possible. For example, (III) of Equation 8.3 can molecularly rearrange, but not when the second H atom is changed to R greater than CH_3.

Now, replacing one of the hydrogen atoms with a pushing alkyl group, the followings are obtained.

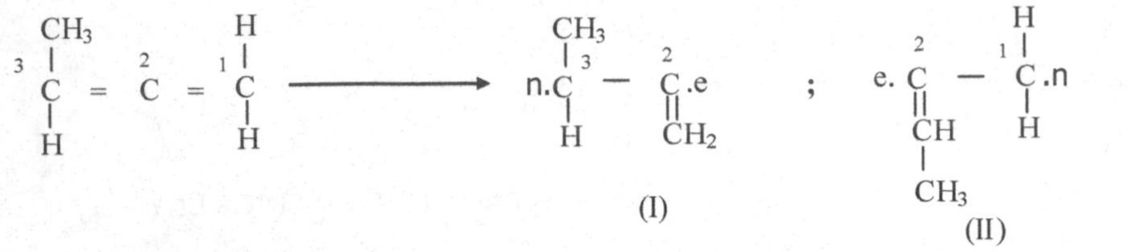

$$
\begin{array}{c}
CH_3 \qquad\quad H \qquad\quad H \\
3\ | \qquad 2\ | \qquad 1\ | \\
C\ =\ C\ =\ C \qquad\longrightarrow\qquad n.C\ -\ C.e \qquad ; \qquad e.\ C\ -\ C.n \\
| \qquad\qquad | \qquad\qquad | \qquad\qquad\qquad\quad || \qquad\qquad | \\
H \qquad\qquad H \qquad\qquad H\ \ CH_2 \qquad\qquad CH \quad H \\
\qquad\qquad\qquad\qquad\qquad\qquad\qquad\qquad\qquad\qquad\qquad | \\
\qquad\qquad\qquad\qquad\qquad\qquad (I) \qquad\qquad\qquad\qquad CH_3 \\
\end{array}
$$

(with labels 3 C, 2 C, 1 C on left; n.C(3) — C(2).e on (I); e.C(2) — C(1).n on (II))

(I)

(II)   8.11

(Nucleophiles)

Only one center can be activated above. One exists while the other does not exist. The radicals carried by the active carbon centers are based on the following order of capacities –

$$(CH_3)HC\ =\ \quad>\quad H_2C\ =\ \quad>>\quad H_3C\ -\quad>\quad H\ -$$

Order of radical-pushing capacities   8.12

It is (II) that will be activated all the time being less Nucleophilic than (I). One would have thought that (I) of Equation 8.11 is the real one since it favors molecular rearrangement as shown below to give (III) which is more stable than (I) in Nucleophilicity.

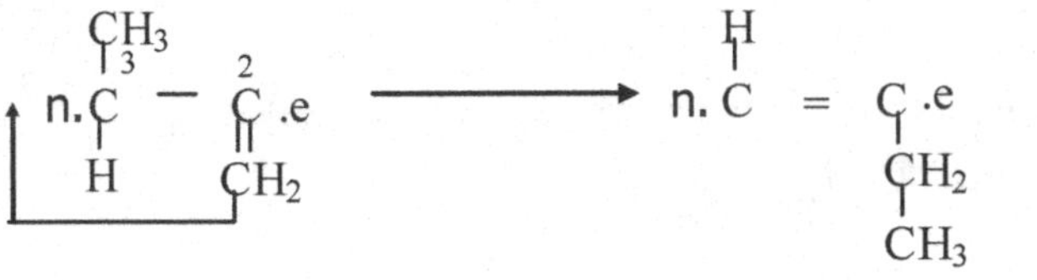

$$
\begin{array}{c}
CH_3 \qquad\qquad\qquad\qquad\qquad H \\
3\ | \qquad 2 \qquad\qquad\qquad\qquad\quad | \\
n.C\ -\ C.e \qquad\longrightarrow\qquad n.\ C\ =\ C.e \\
| \qquad\ \ || \qquad\qquad\qquad\qquad\qquad\qquad | \\
H \qquad CH_2 \qquad\qquad\qquad\qquad\quad CH_2 \\
\qquad\qquad\qquad\qquad\qquad\qquad\qquad\qquad\qquad | \\
\qquad\qquad\qquad\qquad\qquad\qquad\qquad\qquad\ CH_3 \\
\end{array}
$$

(I)   (III) More stable   8.13

Nevertheless, this center is never used. It is only being used for exploratory purposes and as an eye opener. (II) the real center involved, favor molecular rearrangement as shown below in the presence or absence of a weak initiator.

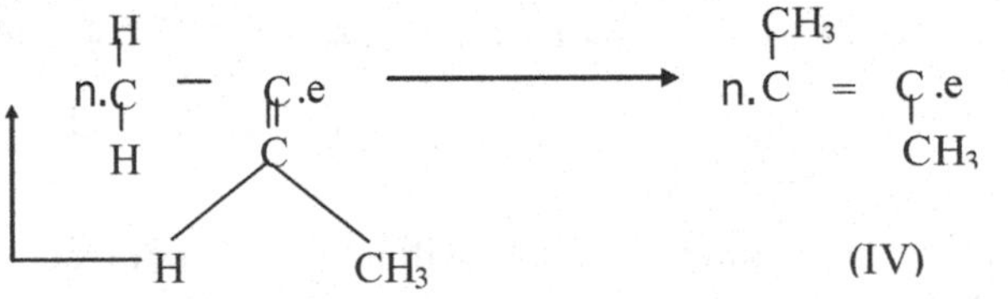

$$
\begin{array}{c}
H \qquad\qquad\qquad\qquad\qquad\qquad CH_3 \\
| \qquad\qquad\qquad\qquad\qquad\qquad\qquad | \\
n.C\ -\ C.e \qquad\longrightarrow\qquad n.C\ =\ C.e \\
| \qquad\ \ || \qquad\qquad\qquad\qquad\qquad\qquad | \\
H \qquad\ C \qquad\qquad\qquad\qquad\qquad\ CH_3 \\
\qquad\quad / \ \backslash \\
\qquad H \quad CH_3 \qquad\qquad\qquad\qquad (IV) \\
\end{array}
$$

(II)   8.14

(IV) is more nucleophilic than (II), since C ≡ C activation center is more nucleophilic than C = C activation center, wherein the following relationship has been found to be valid in Equation 8.12.

$$CH_3\ -\ CH\ =\ \quad>>\quad CH_3 \quad \text{(Radical-pushing capacity)}$$ 8.15

Thus, one mono-form can strongly be observed to favor molecular rearrangement to acetylenic monomers when the conditions exist.

Considering the two mono-forms for polymerization, the followings are obtained in the absence of molecular rearrangement.

$$R^{\cdot n} + n.\underset{\underset{H}{|}}{\overset{\overset{CH_3}{|}}{C}} - \overset{\overset{}{\|}}{\underset{\underset{CH_2}{}}{C}}.e \longrightarrow RH + \underset{}{\overset{\overset{H}{|}}{C}} \equiv C - \underset{\underset{CH_3}{|}}{\overset{\overset{H}{|}}{C}}.n \qquad 8.16$$

$$R^{\cdot n} + n.\underset{\underset{H}{|}}{\overset{\overset{H}{|}}{C}} - \underset{\underset{\underset{CH_3}{|}}{CH}}{\overset{\overset{}{\|}}{C}}.e \longrightarrow RH + \underset{\underset{CH_3}{|}}{\overset{}{C}} \equiv C - \underset{\underset{H}{|}}{\overset{\overset{H}{|}}{C}}.n \qquad 8.17$$

$$R - \underset{\underset{CH_3}{|}}{\overset{\overset{H}{|}}{C}} - \underset{\underset{CH_2}{}}{\overset{\overset{}{\|}}{C}} - \underset{\underset{H}{|}}{\overset{\overset{H}{|}}{C}} - \underset{\underset{\underset{CH_3}{|}}{CH}}{\overset{\overset{}{\|}}{C}} \xrightarrow{\hspace{2cm}} \underset{\underset{CH_3}{|}}{\overset{\overset{H}{|}}{C}} - \underset{\underset{CH_2}{}}{\overset{\overset{}{\|}}{C}}.e \longrightarrow H^e +$$

$$R - \underset{\underset{CH_3}{|}}{\overset{\overset{H}{|}}{C}} - \underset{\underset{CH_2}{}}{\overset{\overset{}{\|}}{C}} - \underset{\underset{H}{|}}{\overset{\overset{H}{|}}{C}} - \underset{\underset{\underset{CH_3}{|}}{CH}}{\overset{\overset{}{\|}}{C}} \xrightarrow{\hspace{2cm}} \underset{\underset{CH_3}{|}}{\overset{\overset{H}{|}}{C}} - C \equiv \overset{\overset{H}{|}}{C} \qquad 8.18$$

$$R - \underset{\underset{CH_3}{|}}{\overset{\overset{H}{|}}{C}} - \underset{\underset{CH_2}{}}{\overset{\overset{}{\|}}{C}} - \underset{\underset{H}{|}}{\overset{\overset{H}{|}}{C}} - \underset{\underset{\underset{CH_3}{|}}{CH}}{\overset{\overset{}{\|}}{C}} \xrightarrow{\hspace{2cm}} \underset{\underset{H}{|}}{\overset{\overset{H}{|}}{C}} - \underset{\underset{\underset{CH_3}{|}}{CH}}{\overset{}{C}}.e \longrightarrow$$

$$R \xrightarrow{\hspace{3cm}} \underset{\underset{H}{|}}{\overset{\overset{H}{|}}{C}} - C \equiv \overset{\overset{CH_3}{|}}{C} + H^e \qquad 8.19$$

Nevertheless, in the two last equations, only one mono-form is the real monoform that should be involved and that mono-form is (II) of Equation 8.11 with the dead terminal double bond looking like the one in Equation 8.19. Though, the transfer species involved are essentially the same for the two mono-forms - of the first kind of the third type, acetylenic terminal bonded polymers can be observed to be produced. It is when molecular rearrangement is favored that the cumulenic terminal bonded polymers are produced.

When the two hydrogen atoms located on same carbon center are replaced, the followings are to be expected. The exploration still continues.

$$\underset{\underset{CH_3}{|}}{\overset{\overset{CH_3}{|}}{\underset{3}{C}}} = \underset{2}{C} = \underset{1}{\underset{\underset{H}{|}}{\overset{\overset{H}{|}}{C}}} \longrightarrow \underset{\underset{CH_3}{|}}{\overset{\overset{CH_3}{|}}{\underset{3}{\Theta C}}} - \underset{2}{\underset{\underset{CH_2}{}}{\overset{\overset{}{\|}}{C^\oplus}}} \quad ; \quad \underset{2}{\overset{}{\underset{\overset{}{\|}}{\oplus C}}} - \underset{1}{\underset{\underset{H}{|}}{\overset{\overset{H}{|}}{C^\Theta}}}$$

$$\underset{(I)}{} \qquad \underset{CH_3 \quad CH_3}{\overset{\overset{}{\diagup\diagdown}}{C}}$$

$$\text{(Nucleophiles)} \qquad \text{(II)} \qquad 8.20$$

Only (I) of the two mono-forms favors molecular rearrangement as shown below, and that same (I) is not the center that will be first activated.

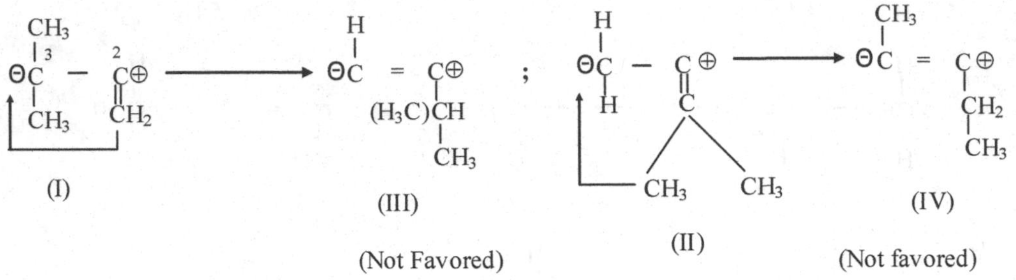

(I)

(III)

(Not Favored)

(II)

(IV)

(Not favored)

8.21

When (III) will involve same transfer species, (IV) will involve different type of transfer species in transfer from monomer step. As strong nucleophiles like the last two cases just considered, only positively charged routes would have been favored if charged activation would have been possible. For its growing polymer chain, the followings are to be expected in the absence of molecular rearrangement which cannot take place.

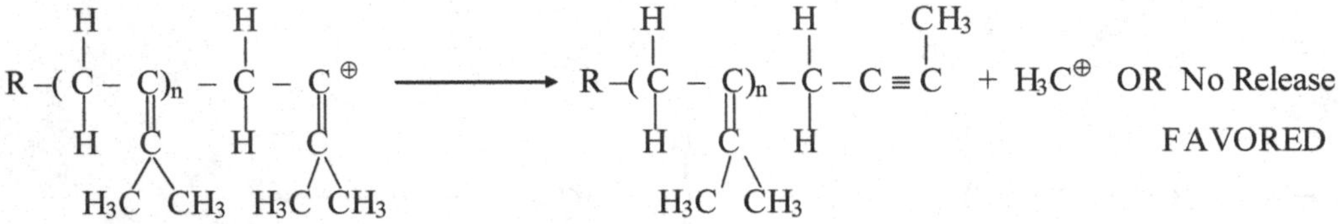

FAVORED

8.22

One has used the center supposed to be activated. Since the transfer species cannot be abstracted in the route not natural to it, it could not be rejected from the growing polymer chain in the route natural to it, all due to electrostatic forces of repulsion and more. Hence, the monomer cannot be activated chargedly.

Now considering the symmetrically placed case of Equation 8.5, the followings are obtained, though one has already said that it cannot be activated chargedly.

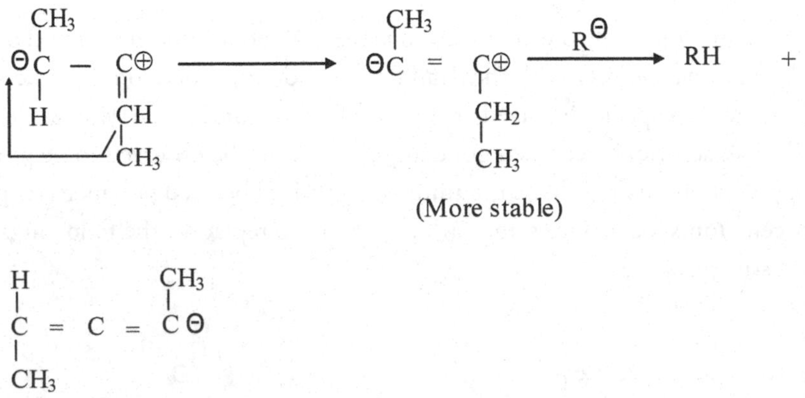

(More stable)

8.23

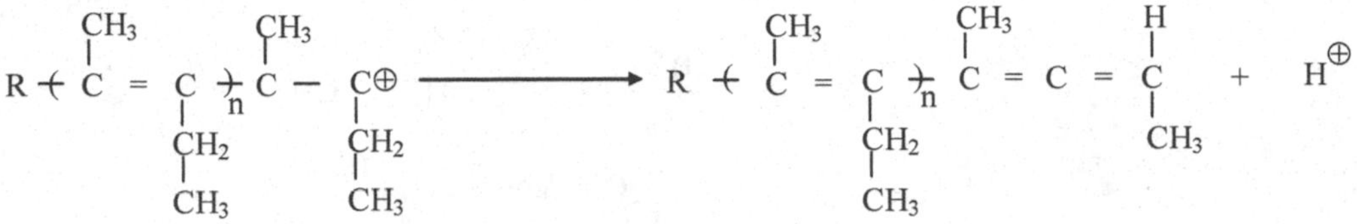

8.24

595

Molecular rearrangement which is said to be favored here cannot take place chargedly, but only radically. As an Adventurer in the Natural world, one is using it for exploratory purposes. The transfer species involved in molecular rearrangement is of the first kind of the third type, but that involved during polymerizations is of the first kind of the first type. This is peculiar to cumulenes, noting that the transfer species are of the same kind and same species.

With the last symmetrically placed type, similar behaviors are obtained, taking note that indeed all these take place only radically.

$$
\begin{array}{l}
\overset{\displaystyle CH_3}{\underset{\displaystyle CH_3}{C}} = C = \overset{\displaystyle CH_3}{\underset{\displaystyle CH_3}{C} } \longrightarrow \ominus\overset{\displaystyle CH_3}{C} - \overset{\oplus}{\underset{\displaystyle C}{C}} \longrightarrow \ominus\overset{\displaystyle CH_3}{C} = \overset{\oplus}{\underset{\displaystyle C(CH_3)}{C}}
\end{array}
$$

(with CH$_3$, CH$_3$ branches and CH$_3$ CH$_3$)

(More stable)   8.25

Molecular rearrangement is favored for these cases for which the following orders of capacities are valid.

$$
CH_3 - CH = \quad >> \quad CH_3 - CH_2 -- \quad ; \quad \overset{\displaystyle CH_3}{\underset{\displaystyle CH_3}{C}} = \quad >> \quad (CH_3) \overset{\displaystyle CH_3}{\underset{\displaystyle CH_3}{-C} } -
$$

Order of Radical pushing capacity   8.26

Thus, for symmetrically placed allenes, molecular rearrangement is highly favored when the conditions exist. For non-symmetrically placed allenes, it is also favored for some of them as shown with another allene below.

$$
\overset{\displaystyle CH_3}{\underset{\displaystyle H}{\overset{3}{C}}} = \overset{2}{C} = \overset{1}{\underset{\displaystyle \underset{\displaystyle CH_3}{CH_2}}{\overset{\displaystyle H}{C}}} \longrightarrow \ominus\overset{\displaystyle \overset{1}{CH_3}}{\underset{\displaystyle H}{C}} - \overset{\oplus}{\underset{\displaystyle \underset{\displaystyle C_2H_5}{CH}}{\overset{2}{C}}} \quad ; \quad \overset{\oplus}{\underset{\displaystyle \underset{\displaystyle CH_3}{CH}}{\overset{2}{C}}} - \overset{\ominus}{\underset{\displaystyle \underset{\displaystyle CH_3}{CH_2}}{\overset{\displaystyle \overset{3}{H}}{C}}}
$$

(I) (Nucleophile)   (II) (Nucleophile)

(The only one that can be activated)   8.27

$$
\ominus\overset{\displaystyle \overset{1}{CH_3}}{\underset{\displaystyle H}{C}} - \overset{\oplus}{\underset{\displaystyle \underset{\displaystyle C_2H_5}{CH}}{\overset{2}{C}}} \longrightarrow \ominus\overset{\displaystyle C_2H_5}{C} = \overset{\oplus}{\underset{\displaystyle \underset{\displaystyle CH_3}{CH_2}}{C}} \quad ; \quad \overset{\oplus}{\underset{\displaystyle \underset{\displaystyle CH_3}{CH}}{C}} - \overset{\ominus}{\underset{\displaystyle H}{\overset{\displaystyle \overset{\displaystyle CH_3}{CH_2}}{C}}} \longrightarrow \ominus\overset{\displaystyle CH_3}{C} = \overset{\oplus}{\underset{\displaystyle \underset{\displaystyle \underset{\displaystyle CH_3}{CH_2}}{CH_2}}{C}}
$$

(I) THE FAVORED CASE   (II)   (More stable)

8.28

In the absence of steric hindrance, only the positively charged route would have favored the polymerization of these nucleophiles if activation had been favored chargedly. For the non-symmetric cases, of the two mono-forms, only one can be activated and that is the less nucleophilic center.

Now, considering etheric groups, the followings are obtained for OH, noting that we are still exploring in a charged environment where there are countless numbers of repulsive forces.

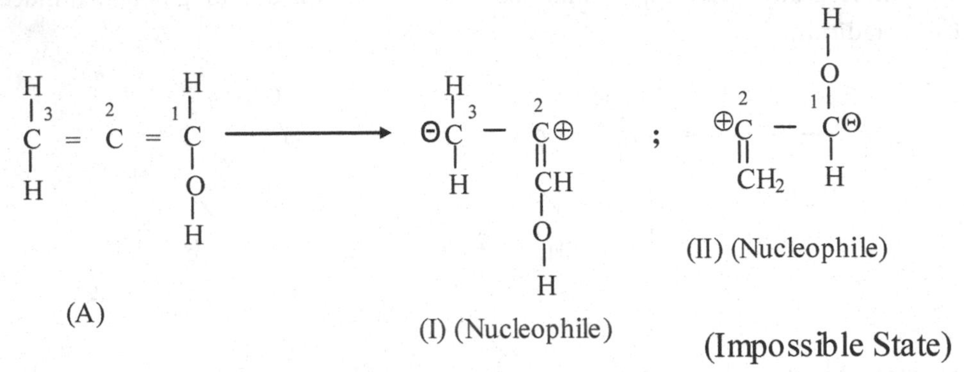

$$8.29$$

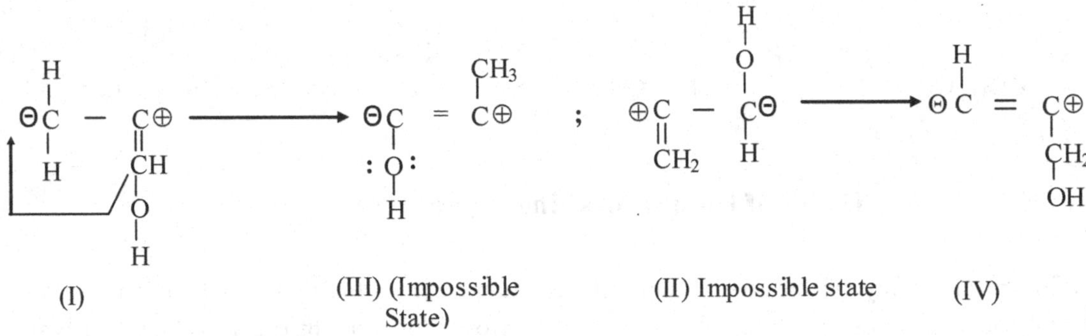

(A)

(I) (Nucleophile)   (II) (Nucleophile)

(Impossible State)

$$8.30$$

(I)   (III) (Impossible State)   (II) Impossible state   (IV)

If molecular rearrangement had been allowed chargedly, only (I) would have been used, for which it cannot take place and not even by radical means. For the same reasons, it cannot release transfer species as shown below.

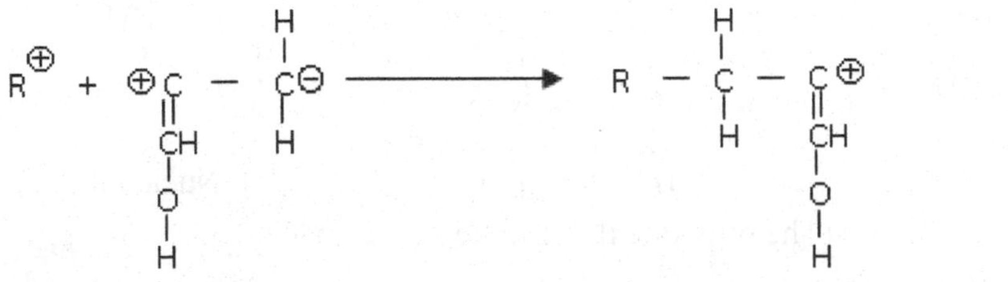

$$8.31$$

Cannot release transfer species

Thus, (I) cannot be polymerized chargedly under any conditions, but only radically without molecular rearrangement.

When the OH is replaced with OR groups, the followings are obtained.(B)

$$
\begin{array}{c}
\overset{H}{\underset{|}{\overset{3}{C}}} = \overset{2}{C} = \overset{1}{\overset{H}{\underset{|}{C}}} \\
\overset{|}{H} \qquad\qquad \overset{|}{O} \\
\qquad\qquad\qquad \overset{|}{CH_3}
\end{array}
\longrightarrow
\begin{array}{c}
{}^{\ominus}\overset{3}{\overset{H}{\underset{|}{C}}} - \overset{2}{\overset{\oplus}{C}} \\
\overset{|}{H} \qquad \overset{\|}{CH} \\
\qquad\qquad \overset{|}{O} \\
\qquad\qquad \overset{|}{CH_3}
\end{array}
\quad ; \quad
\begin{array}{c}
{}^{\oplus}\overset{2}{C} - \overset{1}{\overset{H}{\underset{|}{C}}}{}^{\ominus} \\
\overset{|}{CH_2} \qquad \overset{|}{O} \\
\qquad\qquad \overset{|}{CH_3}
\end{array}
$$

(B)   (I)   (II) Impossible state   8.32

$$
\begin{array}{c}
{}^{\ominus}\overset{H}{\underset{|}{C}} - \overset{\oplus}{C} \\
\overset{|}{H} \qquad \overset{\|}{CH} \\
\qquad\quad \overset{|}{O} \\
\qquad\quad \overset{|}{CH_3}
\end{array}
\longrightarrow
\begin{array}{c}
{}^{\ominus}C = \overset{\oplus}{C} \\
\overset{|}{O} \quad \overset{|}{CH_3} \\
\overset{|}{CH_3}
\end{array}
\quad ; \quad
\begin{array}{c}
{}^{\ominus}\overset{H}{\underset{\|}{C}} - \overset{\oplus}{C} \\
\overset{|}{CH_2} \quad \overset{|}{O} \\
\qquad \overset{|}{CH_3}
\end{array}
\xrightarrow{\;\not\;}
\begin{array}{c}
{}^{\oplus}\overset{H}{\underset{|}{C}} - O{}^{\ominus} \\
\overset{|}{C}(CH_3) \\
\overset{\|}{CH_2}
\end{array}
$$

(I)   (III) (Impossible State)   (II)   (IV)   8.33

IMPOSSIBLE REARRANGEMENT

Like the first member of the family, the same applies here. From all the observations so far the followings are valid.

$$
= CH(OH) \quad > \quad = CH(OCH_3) \quad > \quad = CH(OC_2H_5) \quad > \quad Etc
$$

$$\underline{\text{Order of pushing capacity}} \qquad\qquad 8.34$$

When the etheric groups are symmetrically placed, the followings are to be expected.

$$
\begin{array}{c}
\overset{H}{\underset{|}{\overset{3}{C}}} = \overset{2}{C} = \overset{1}{\overset{H}{\underset{|}{C}}} \\
\overset{|}{O} \qquad\qquad \overset{|}{O} \\
\overset{|}{CH_3} \qquad\qquad \overset{|}{CH_3}
\end{array}
\longrightarrow
\begin{array}{c}
{}^{\ominus}\overset{3}{C} - \overset{2}{\overset{\oplus}{C}} \\
\overset{|}{O} \qquad \overset{\|}{CH} \\
\overset{|}{CH_3} \qquad \overset{|}{O} \\
\qquad\qquad \overset{|}{CH_3}
\end{array}
$$

8.35

Impossible state

Chargedly, the monomer cannot be activated, due to electrostatic forces of repulsion apart from the fact that it cannot take place at all.

Finally with other non-free-radical-pushing groups, such as NH_2, NHR, NR_2, etc. groups, the followings are obtained.

$$
\begin{array}{c}
\overset{H}{\underset{|}{\overset{3}{C}}} = \overset{2}{C} = \overset{1}{\overset{H}{\underset{|}{C}}} \\
\overset{|}{NH_2} \qquad\qquad \overset{|}{H}
\end{array}
\longrightarrow
\begin{array}{c}
{}^{\ominus}\overset{1}{\overset{H}{\underset{|}{C}}} - \overset{2}{\overset{\oplus}{C}} \\
\overset{|}{H} \qquad \overset{\|}{CH} \\
\qquad\quad \overset{|}{NH_2}
\end{array}
\quad ; \quad
\begin{array}{c}
{}^{\oplus}\overset{2}{C} - \overset{3}{\overset{NH_2}{\underset{|}{C}}}{}^{\ominus} \\
\overset{|}{CH_2} \qquad \overset{|}{H}
\end{array}
$$

        (I)   Impossible existence

            (II)

8.36

The situation is not different from the case of OH, OR types of groups. Since NH_2 is more radical-pushing than OH group as already shown, the followings are valid -

$$= CH(NR_2) \; > \; = CH(NRH) \; > \; = CH(NH_2) \; > = CH(OH) \; > \; = CH(OR) \; > \; = CH_2 \; ...etc.$$

ORDER OF RADICAL-PUSHING CAPACITY 8.37

$$- CH_2(NR_2) \qquad > \qquad - CH_2(NHR) \qquad > \qquad - CH_2NH_2$$

<u>Order of radical-pushing capacity</u> 8.38

Having considered important pushing groups in order to establish an order, pulling groups will now be considered, beginning with replacement of one H atom in allene with a fluorine atom.

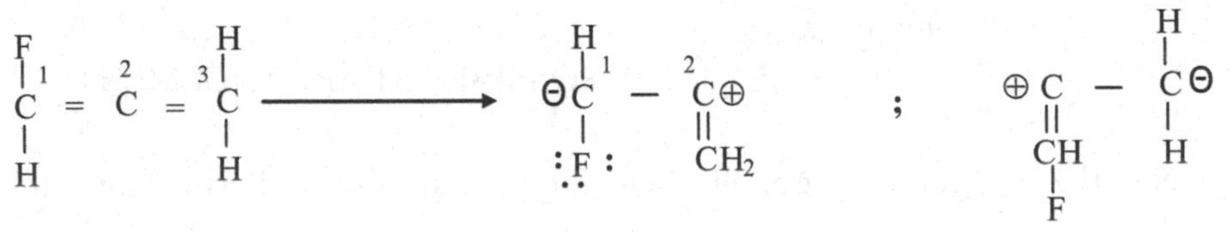

(I) Impossible
Existence

(II) 8.39

(I) is the less nucleophilic center, the center that will be the first to be activated. Chargedly, the monomer cannot be activated. The same as above will apply to $OCOCH_3$ groups. When the fluorine atoms are symmetrically placed, no activation can take place chargedly.

Now replacing F with CF_3, the followings are to be expected.

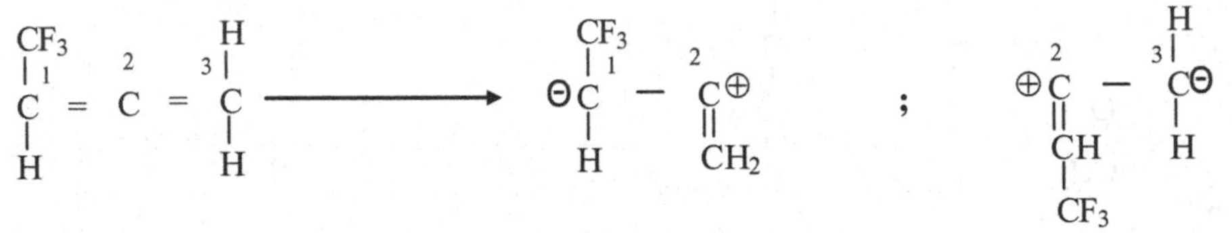

(I)   (II)   8.40

The less Nucleophilic center is (I), for which even the trans-form would not have been polymerized with positively charged paired initiators. As it seems, such monomers can never be initiated free-radically, unless when the paired types are used.

When the CF_3 groups are symmetrically placed, the situation remains the same.

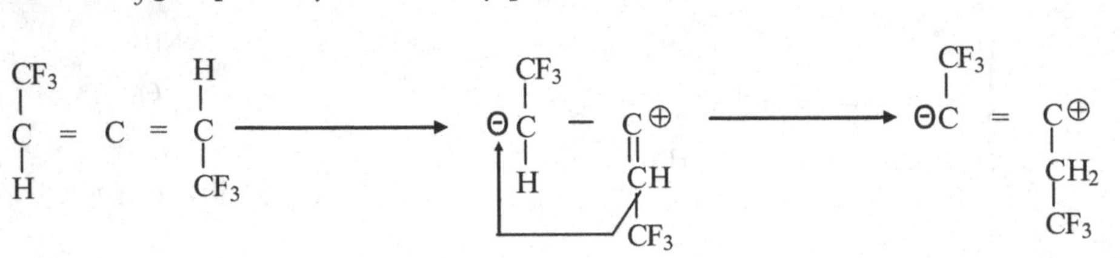

(Not favored) 8.41

$$R^{\oplus} \quad + \quad \overset{\overset{CF_3}{|}}{\underset{\underset{CF_3}{\overset{|}{CH}}}{\overset{\ominus}{C}}} = \overset{\overset{\oplus}{C}}{\underset{CH}{\overset{||}{C}}} \quad \longrightarrow \quad RF \quad + \quad \overset{\overset{F}{|}}{\underset{F}{C}} = \overset{C}{\underset{H}{\overset{|}{C}}} - \overset{\overset{\oplus}{C}}{\underset{\underset{CF_3}{\overset{|}{CH}}}{\overset{||}{C}}} \qquad 8.42$$

Since $-CH_2CF_3$ is taken to be less pulling than $-CF_3$ group chargedly, molecular rearrangement is not favored. No charged route is therefore possible. In fact, chargedly like all the others, it cannot be activated.

Using COOR as pulling groups in place of CF_3, the followings are obtained.

$$\overset{\overset{CH_3}{|}}{\underset{\underset{H}{\overset{|}{C}}}{\overset{O}{\underset{\underset{1}{C=O}}{|}}}} \underset{1}{=} \overset{}{\underset{2}{C}} = \overset{}{\underset{\underset{H}{\overset{|}{C}}}{\overset{H}{\underset{3}{|}}}} \longrightarrow \overset{\overset{CH_3}{|}}{\underset{\underset{H}{\overset{|}{C}}}{\overset{O}{\underset{\underset{1}{C=O}}{|}}}} \overset{\ominus}{\underset{H}{\overset{|}{C}}} - \overset{\overset{\oplus}{C}}{\underset{CH_2}{\overset{||}{C}}} \quad ; \quad \overset{\oplus}{\underset{\underset{C=O}{\overset{|}{CH}}}{\overset{}{C}}} - \overset{\overset{H}{\underset{3}{|}}}{\underset{H}{\overset{|}{C}}}{\overset{\ominus}{C}}$$

[Center first activated]

(I) (Electrophile)

(II) (Electrophile)   8.43

Unlike the cases considered so far, the monomer looks Electrophilic in character based on the placement of the groups. Unlike olefins, no conclusions can be made until their free-radical polymerizations are considered, since these monomers cannot be activated chargedly.

With COR, whether the situation is different or not, is immaterial. However, when simulatively activated chargedly, the followings would have been expected.

$$\overset{\overset{CH_3}{|}}{\underset{\underset{H}{\overset{|}{C}}}{\overset{C=O}{\underset{1}{|}}}} \underset{1}{=} \overset{}{\underset{2}{C}} = \overset{}{\underset{\underset{H}{\overset{|}{C}}}{\overset{H}{\underset{3}{|}}}} \longrightarrow \overset{\ominus}{\underset{\underset{CH_3}{\overset{|}{C=O}}}{\overset{|}{C}}} - \overset{\oplus}{\underset{CH_2}{\overset{||}{C}}} \quad ; \quad \overset{\oplus}{\underset{\underset{CH_3}{\overset{|}{C=O}}}{\overset{||}{C}}} - \overset{\overset{H}{\underset{3}{|}}}{\underset{H}{\overset{|}{C}}}{\overset{\ominus}{C}}$$

(I) (Electrophile)

(II)   8.44

Unlike the cases above, (I) and (II) favor the cationic route in the absence of rearrangement. It is only (I) that is indeed activated.

Having considered the unique charged qualities of allenes in terms of their use as monomers, there is need to consider the effects of resonance stabilization group on these monomers, although we already know that it cannot be provided chargedly.

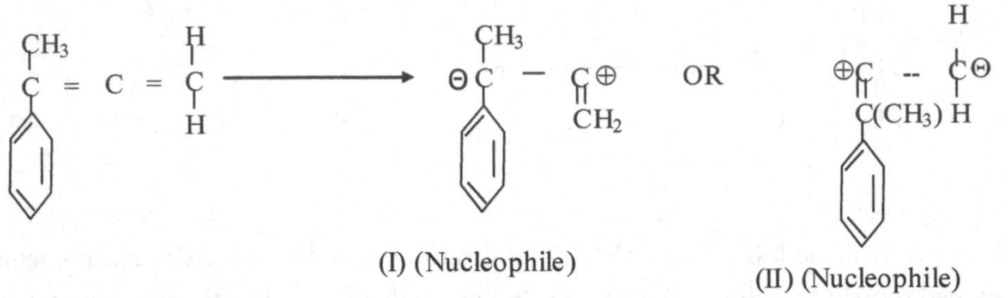

(I) (Nucleophile)   (II) (Nucleophile)   8.45

Of the two mono-forms, none will favor molecular rearrangement. Since, one has already established that resonance stabilization cannot take place chargedly, one will consider their radical characters to a large extent herein.

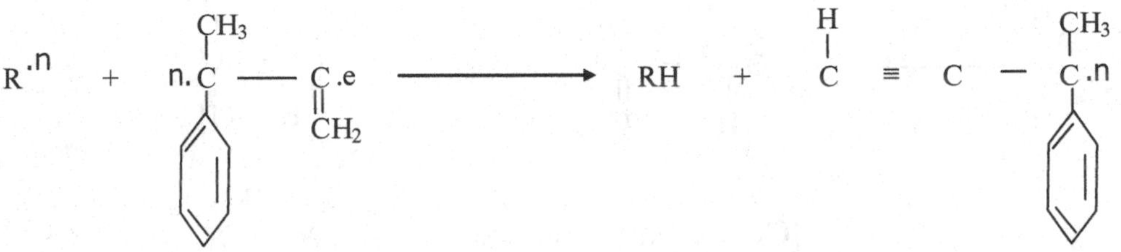

8.46

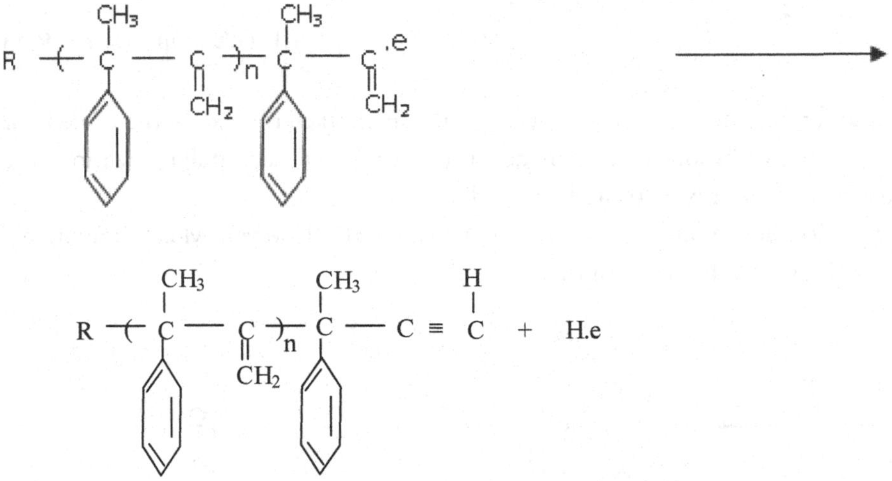

8.47

Though resonance stabilization seems to be provided, the need for it does not arise, except when CH_3 group is changed to CF_3 group. Unlike the case of α-methyl styrene where both routes are favored, here the route favored is only the natural route. *Indeed, the activation center used above is not the one supposed to be used, since it is more Nucleophilic than the second center.* That second center (II) of Equation 8.45 is used as follows.

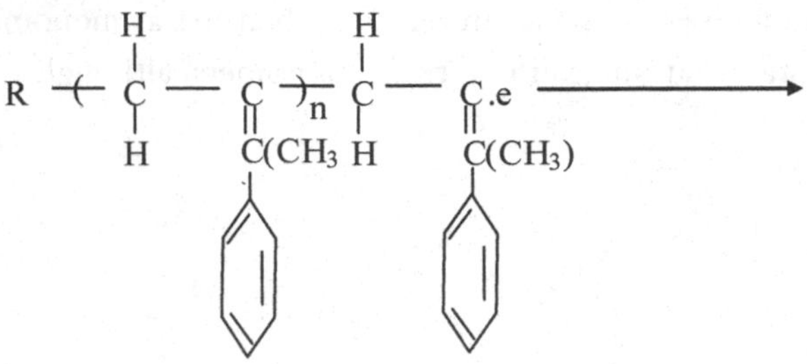

$$R \left(\begin{matrix} H \\ | \\ C \\ | \\ H \end{matrix} - \begin{matrix} H \\ | \\ C \\ \| \\ C(CH_3) \end{matrix} \right)_n \begin{matrix} H \\ | \\ C \\ | \\ H \end{matrix} - C \equiv C - CH_3 \quad + \quad e$$

$$8.48$$

With the monomer above, homopolymerization is not favored nucleo-free-radically, since the effect of resonance stabilization is not felt. The same applies to (III) and (IV) below.

$$\begin{matrix} H \\ | \\ C \end{matrix} = C = \begin{matrix} H \\ | \\ C \\ | \\ H \end{matrix} \quad ; \quad \begin{matrix} H \\ | \\ C \end{matrix} = C = \begin{matrix} H \\ | \\ C \end{matrix}$$

(III)   (IV)   8.49

$$\begin{matrix} H \\ | \\ C \end{matrix} - C - \begin{matrix} H \\ | \\ C \end{matrix} \longrightarrow \quad n.\begin{matrix} H \\ | \\ C \end{matrix} - \begin{matrix} C.e \\ \| \\ CH \end{matrix} \longrightarrow n.\begin{matrix} C \end{matrix} = \begin{matrix} CH_2 \\ | \\ C.e \end{matrix}$$

$$8.50$$

Indeed, in the reaction above, it is H that is transferred as opposed to the phenyl group only free- radically, since H is less radical-pushing than the phenyl group. Invariably, resonance stabiliza-tion can be seen to be provided for cumulenes, just in the same way as in styrene, except that there is no group to be shielded. Only the electro-free-radical route is favored by it.

With alkenyl groups, the following are to be expected.

$$\begin{matrix} CH_3 \\ | \\ C \\ | \\ CH \\ \| \\ CH_2 \end{matrix} = C = \begin{matrix} H \\ | \\ C \\ | \\ H \end{matrix} \longrightarrow n.\begin{matrix} CH_3 \\ | \\ C \\ | \\ CH \\ \| \\ CH_2 \end{matrix} - \begin{matrix} C.e \\ \| \\ CH_2 \end{matrix} \longleftrightarrow n.\begin{matrix} H \\ | \\ C \\ | \\ H \end{matrix} - \begin{matrix} H \\ | \\ C.e \\ | \\ CCH_3 \\ \| \\ CH_2 \end{matrix} \longleftrightarrow$$

(Not the center activated)

$$(A)$$

$$n.\overset{\overset{\displaystyle H}{|}}{\underset{\underset{\displaystyle H}{|}}{C}}{}^{1} - \overset{\overset{\displaystyle H}{|}}{\underset{}{C}}{}^{2} = \overset{3}{\underset{\underset{\displaystyle CH_3}{|}}{C}} - \overset{4}{\underset{\underset{\displaystyle CH_2}{||}}{C}}.e$$

(Nucleophile)   8.51a

As can be observed, resonance stabilization can be provided for alkenyl group, ***because H is less radical pushing than –CR=C=CH₂ group***. (A) seems to be the least nucleophilic center. If the less nucleophilic center in the cumulene was used, the followings are to be expected.

$$\begin{array}{c}
\underset{\underset{\underset{\displaystyle CH_2}{||}}{CH}}{\overset{\overset{\displaystyle CH_3\quad H}{|\qquad|}}{C=C=C}}\underset{\displaystyle |}{\overset{}{}} \\ \end{array}
\longrightarrow
\begin{array}{c}
e\bullet C - C \bullet n \\ \end{array}
\quad ; \quad
\begin{array}{c}
e\bullet C - C \bullet n \\ \end{array}
\overset{|}{\longleftrightarrow}\!\!\!/
\begin{array}{c}
e\bullet C - C = C - C \bullet n \\ \end{array}
$$

(B) (Nucleophile)   (C)   8.51b

(C) above is (A) wrongly activated. Thus with (B), the alkenyl group can be observed not to have the ability to provide resonance stabilization capabilities when the CH_3 group is placed on the same C center carrying the alkenyl group. Based on the last but one equation, the cumulenic group can provide it for the alkenyl group wherein the CH_3 group placed on cumulene is shielded. When the position of the alkenyl group is changed, then the followings are obtained.

$$\begin{array}{c}
\overset{\overset{\displaystyle CH_3\quad H}{|\qquad|}}{C=C=C}\\ \end{array}
\longrightarrow
\begin{array}{c}
n\bullet C - C \bullet e \\ \end{array}
\longleftrightarrow
\begin{array}{c}
n\bullet C - C = C - C \bullet e \\ \end{array}
$$

(A) (Nucleophile)   8.51c

Though the CH_3 group is not shielded, it does not provide transfer species. Resonance stabilization is still provided by the cumulene. The two mono-forms have transfer species. Both monomers are Nucleophiles. From the analysis so far, the following is valid-

$$(CH_3)HC=C=CH- \quad > \quad H_2C=C=CH- \quad > \quad H_2C=C=(CH_3)C- \quad > \quad H$$

<u>Order of radical-pushing capacity</u>   8.51d

From all the considerations so far, resonance stabilization cannot be provided for 1,3-cumulenes when radical-pushing groups are placed on the same C center carrying a phenyl group. When radical-pulling groups are placed on the same C center, resonance stabilization can then be provided. The phenyl group is the provider of resonance stabilization for the cumulene. For the alkenyl group the provider of resonance stabilization is the cumulene. From all the above, the followings are the resonance stabilization capacities of the following groups-

$$
\bigcirc\!\!-\!\!\!\!- \quad > \quad -C \equiv CH \quad > \quad \mathbf{H_2C = C = \overset{\overset{\textstyle H}{|}}{C}-} \quad > \quad H_2C = \overset{\overset{\textstyle H}{|}}{C}- \qquad 8.52a
$$

wherein only a part of the highlighted one can be used, since it can only be used when specific groups are carried by the 1,3-cumulene.

$$
-CR = C = CH_2 \quad > \quad -H \quad > \quad HC \equiv C -
$$

$$
\textbf{Order of radical-pushing capacity} \qquad\qquad 8.52b
$$

8.1.2. Free - radical Characters of Allenes

When allene is activated free –radically, the followings are obtained.

$$
\overset{\overset{\textstyle H}{|}}{\underset{\underset{\textstyle H}{|}}{C}} = C = \overset{\overset{\textstyle H}{|}}{\underset{\underset{\textstyle H}{|}}{C}} \longrightarrow n \cdot \overset{\overset{\textstyle H}{|}}{\underset{\underset{\textstyle H}{|}}{C}} - \overset{\overset{}{}}{\underset{\underset{\textstyle CH_2}{||}}{C}} \cdot e \longrightarrow n \cdot \overset{\overset{\textstyle H}{|}}{C} = \overset{}{\underset{\underset{\textstyle CH_3}{|}}{C}} \cdot e
$$

$$
\text{(Nucleophile)}
$$

$$
\xrightarrow{\; + \; H \cdot^n \;} n \cdot \overset{\overset{\textstyle H}{|}}{\underset{\underset{\textstyle H}{|}}{C}} - C \equiv \overset{\overset{\textstyle H}{|}}{C} \quad + \quad H_2
$$

$$
\qquad\qquad\qquad\qquad\qquad\qquad 8.53
$$

$$
n \cdot \overset{\overset{\textstyle H}{|}}{C} = \underset{\underset{\textstyle CH_3}{|}}{C} \cdot e \quad + \quad e \cdot CH_3 \longrightarrow H_3C - \overset{\overset{\textstyle H}{|}}{C} = \underset{\underset{\textstyle CH_3}{|}}{C} \cdot e \xrightarrow{\; + \; n \,(I) \;}
$$
$$
\quad (I)
$$

$$
H_3C + \overset{\overset{\textstyle H}{|}}{C} = \underset{\underset{\textstyle CH_3}{|}}{C} \overset{}{\underset{\underset{\textstyle n}{}}{+}} \overset{\overset{\textstyle H}{|}}{C} = \overset{\overset{\textstyle CH_3}{|}}{C} \cdot e \longrightarrow
$$

$$
H_3C + \overset{\overset{\textstyle H}{|}}{C} = \underset{\underset{\textstyle CH_3}{|}}{C} \overset{}{\underset{\underset{\textstyle n}{}}{+}} \overset{\overset{\textstyle H}{|}}{C} = C = \overset{\overset{\textstyle H}{|}}{\underset{\underset{\textstyle H}{|}}{C}} \quad + \quad H \cdot^e \qquad 8.54
$$

In the first case, only electro-free-radical will polymerize these monomers. Secondly, the electro-free-radical must be of equal or greater radical-pushing capacity than the H group carried by the monomer for polymerization to be favored. Thirdly, metallic electro-free-radical cannot be used, since it will displace the H atom and form an acetylide. Like the charged case which was used for exploratory purposes, allene is a strong nucleophile.

Now, replacing one of the hydrogen atoms with a radical-pushing alkyl group, the followings are obtained.

$$
\underset{\text{(I) (Nucleophile)}}{\qquad} \qquad \text{or} \qquad \underset{\text{(II) (Nucleophile)}}{\qquad} \qquad 8.55
$$

It is only (II) that will be first activated being less nucleophilic. However, (I) and (II) will favor molecular rearrangement. With or without molecular rearrangement, both can be polymerized only electro-free-radically.

$$
N^{\cdot n} + \; e. \; \ldots \longrightarrow \; NH \; + \; HC \equiv C - \underset{H}{\overset{CH_3}{C}} . n \qquad 8.56
$$

$$
E^{\cdot e} + \; n. \; \ldots \longrightarrow \; E - \underset{CH_3}{\overset{H}{C}} = \underset{CH_2}{C} . e \qquad 8.57
$$

Using the real center, the followings are obtained when activated.

$$
8.58
$$

With molecular rearrangement, both favor the electro-free-radical route. The radicals carried by the active centers are based on the orders of capacities of group as already shown. For the first time in the NEW FRONTIERS, new substituted groups have been systematically identified. Such groups which are radical-pushing and have been identified so far include –

$$
O = , \quad N \equiv , \quad HN = , \quad CH_2 = , \quad CH \equiv , \quad CH_3CH = , \quad \underline{CH_3C} \equiv \quad \text{etc} \qquad 8.59
$$

At this point in time therefore, there is need to give names to these new groups. While those of the first three above may not be clear-cut, those with carbon centers can adequately be given names. **For the first and third, the following names can be proposed as follows – oxidyl and aminyl groups.** CH_3, C_2H_5, C_3H_7 etc. are **alkylane or alkanyl** groups from alkanes. For alkenes and alkynes, what then should be the names of the groups derived from them? If the alkyl groups from alkanes have been called alkanyl groups and those from alkene [not alkylenes (e.g. methylene)] [2,3] and alkynes have been called, **alkenyl and alkynyl groups** respectively, then what should the remaining groups above be called? While the fourth, fifth and sixth cases can be called **cumunyl groups**, the last highlighted does not exist. The second which exists with diazoalkylene polarly connected can be called **nitryl group.** Care has been taken not to interfere with already existing names and nomenclature except where the strong need arises. *With due respect to the International Organizations, though so much have been done by the bodies, one has limited respect for them, because they, bodies of the greatest Scientists universally, **know nothing** with respect to what **an Atom, compounds, radicals, charges-ionic and non-ionic, Activation centers, Mechanisms of reactions, States of Existence and too many to list, are** [See Appendices I to VI]. This is partly one of the origins of the NEW FRONTIERS.*

When the two H atoms located on one carbon center are replaced with CH_3 groups, no molecular rearrangement can take place. It can only be polymerized electro-free-radically. One can observe the unique characters of cumulenes-very strongly nucleophilic, far more so than the first member of acetylenes-acetylene, since the first member of allenes-allene rearranges to methyl acetylene.

With the CH_3 groups now symmetrically placed, the followings are obtained.

$$\text{(equation 8.60)}$$

$$\text{(equation 8.61)}$$

$$\text{(equation 8.62)}$$

Nucleo-free-radically, this monomer like the others cannot be polymerized. Thus, it can be observed that none of these monomers with the radical-pushing groups considered so far can favor nucleo-free-radical polymerizations. They only favor electro-free-radical polymerizations.
(I)(Less nucleophilic)

Replacing one H in allene with F, the followings are to be expected.

$$\text{(equation 8.63)}$$

(I)(Less nucleophilic)   (II) (More nucleophilic)

Molecular rearrangement is favored by both mono-forms. In the absence of molecular rearrangement, the followings are obtained.

$$N^{\bullet\, n} \;+\; e \cdot \underset{\underset{CH_2}{\|}}{C} - \underset{\underset{H}{|}}{\overset{\overset{F}{|}}{C}} \cdot n \longrightarrow NH \;+\; \overset{\overset{H}{|}}{C} \equiv C - \underset{\underset{H}{|}}{\overset{\overset{F}{|}}{C}} \cdot n$$

FAVORED CENTER

8.64

$$N^{\bullet\, n} \;+\; n \cdot \underset{\underset{H}{|}}{\overset{\overset{H}{|}}{C}} - \underset{\underset{CFH}{\|}}{C} \cdot e \longrightarrow NH \;+\; \overset{\overset{F}{|}}{C} \equiv C - \underset{\underset{H}{|}}{\overset{\overset{H}{|}}{C}} \cdot n$$

8.65

When molecular rearrangement takes place, which may definitely be required in order to minimize or eliminate branch formations, only electro-free-radical route is still favored. Like other cases so far considered whether molecular rearrangement takes place or not, ***random copolymers*** cannot be produced even if the initiator is free media in character.

With fluorine atoms symmetrically placed, the followings are obtained.

$$E + \left(\overset{\overset{F}{|}}{C} = \underset{\underset{CH_2F}{|}}{C} \right)_n C = \underset{\underset{CH_2F}{|}}{\overset{\overset{F}{|}}{C}} \cdot e \longrightarrow$$

$$E + \left(\overset{\overset{F}{|}}{C} = \underset{\underset{CH_2F}{|}}{C} \right)_n C = C = \underset{\underset{F}{|}}{\overset{\overset{H}{|}}{C}} \;+\; H^{\bullet\, e}$$

8.66

The same will apply almost identically when F is replaced with CF_3 when molecular rearrangement takes place.

$$\underset{\underset{H}{|}}{\overset{\overset{CF_3}{|}}{C}} = C = \underset{\underset{H}{|}}{\overset{\overset{CF_3}{|}}{C}} \longrightarrow n \cdot \underset{\underset{H}{|}}{\overset{\overset{CF_3}{|}}{C}} - \underset{\underset{CH}{\|}}{\underset{\underset{CF_3}{|}}{C}} \cdot e \longrightarrow n \cdot \overset{\overset{H}{|}}{C} = \underset{\underset{CF_3}{|}}{\underset{\underset{CH(CF_3)}{|}}{C}} \cdot e$$

(Nucleophile)   8.67

$$E + \left(\overset{\overset{H}{|}}{C} = \underset{\underset{CH}{|}}{C} \right)_n C = \underset{\underset{CH}{|}}{\overset{\overset{H}{|}}{C}} \cdot e \longrightarrow$$
$$\qquad\qquad\quad CF_3 \quad CF_3 \qquad CF_3 \quad CF_3$$

$$E + \left(\overset{\overset{H}{|}}{C} = \underset{\underset{CH}{|}}{C} \right)_n C = C = \underset{\underset{CF_3}{|}}{\overset{\overset{H}{|}}{C}} \;+\; e \cdot CF_3$$
$$\qquad\qquad\qquad CF_3 \quad CF_3$$

8.68

It should be noted that CF_3 is the transfer species being less radical-pushing in capacity than H.

(I)(Less nucleophilic) (II) (Nucleophile)   8.69

(III) (L nucleophilic) (IV) (Nucleophile)   8.70

(I) and (III) will not molecularly rearrange. (II) and (IV) above can rearrange and these are the ones that will favor electro-free-radical polymerization in the absence of molecular rearrangc-mcnt. Yet, these are the ones that cannot be initiated. Polymerization of (I) and (III) is possible only when free-radical paired initiators are used and only when the groups are trans-placed and this will take place only nucleo-free-radically. When cis-placed, there is no polymerization.

Now, considering etheric substituent groups, the followings are to be expected for OH group.

(I) (Less nucleophile)   (II) (More nucleophilc)   8.71

(I) (Nucleophile)   (III) (Impossible)   (II)   (IV)   8.72

(I) is less Nucleophilic than (II) and that is the center that is first activated and will continue to be activated all the time whether there is perfect mixing or not and whether the initiator is free media type or not. (I) cannot molecularly rearrange. Only (II) can. Nevertheless, for both centers, whether molecular rearrangement is favored or not, only the natural route-electro-free-radical route is favored. Since the monomer itself can be activated radically, it cannot undergo Tautomerization, except when it exists in Equilibrium State of Existence without Activation as will be shown in the Appendices and when stated into law.

For OR groups, the same as above seem to apply. The same will also apply to NH_2, NHR,

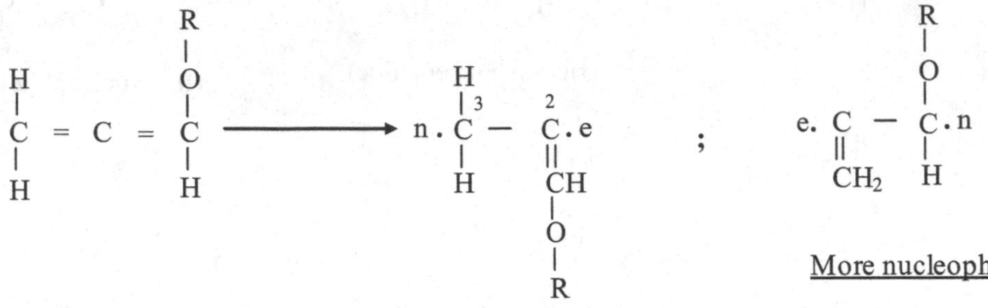

$$
\begin{array}{ccccc}
 & & R & & \\
 & & | & & \\
 H & & O & & \\
 | & & | & & \\
 C & = & C & = & C \\
 | & & & & | \\
 H & & H & & \\
\end{array}
\longrightarrow
\begin{array}{ccc}
 & H & \\
 & |\ 3 & 2 \\
 n\,.\,C & - & C\,.\,e \\
 | & & | \\
 H & & CH \\
 & & | \\
 & & O \\
 & & | \\
 & & R \\
\end{array}
\qquad ; \qquad
\begin{array}{ccc}
 & R & \\
 & | & \\
 & O & \\
 & | & \\
 e.\ C & - & C\,.\,n \\
 \| & & | \\
 CH_2 & & H \\
\end{array}
$$

<u>**More nucleophile**</u>

Less nucleophile   8.73

and NR_2 types of groups.

So far only nucleophiles have been considered. From the route favored by them, it is therefore no surprise to note why cumulenes have never been known to favor free-radical polymerizations, just like the case of propene. Considering allenes with radical-pulling groups, the followings are obtained for COOR, COR and CN types of groups.

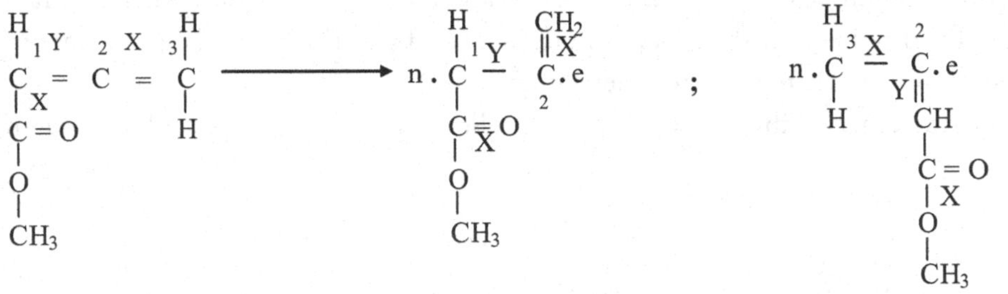

$$
\begin{array}{ccccc}
 H & & & & H \\
 |\ 1\ Y & & 2 & X & 3| \\
 C & = & C & = & C \\
 |\ X & & & & | \\
 C = O & & & & H \\
 | & & & & \\
 O & & & & \\
 | & & & & \\
 CH_3 & & & & \\
\end{array}
\longrightarrow
\begin{array}{ccc}
 H & & CH_2 \\
 |\ 1\ Y & & \|\ X \\
 n\,.\,C & - & C\,.\,e \\
 | & & 2 \\
 C \!\equiv\! O & & \\
 |\ \overline{X} & & \\
 O & & \\
 | & & \\
 CH_3 & & \\
\end{array}
\qquad ; \qquad
\begin{array}{ccc}
 H & & \\
 |\ 3\ X & 2 & \\
 n\,.\,C & - & C\,.\,e \\
 | & Y\| & \\
 H & & CH \\
 & & | \\
 & & C = O \\
 & & |\ X \\
 & & O \\
 & & | \\
 & & CH_3 \\
\end{array}
$$

(I) (Electrophile)

(II) (Electrophile)   8.74

With the exception of $-C \equiv N$ group, these are weak Electrophiles. Originally the Activation centers in cumulene are all Nucleophilic. Since cumulene is more Nucleophilic than acetylene, hence the two Activation centers in cumulene broke down to Y and X centers with the presence of COOR group adjacently located to it. One can observe that only (I) can be the center first activated and indeed always the case when nucleo-free-radical initiators are present. When present, only the paired ones will initiate the monomer. When electro-free-radicals are present, it is the $C = O$ center, the X center to Y, the most nucleophilic center that will be activated. Both (I) and (II) can molecularly rearrange as shown below.

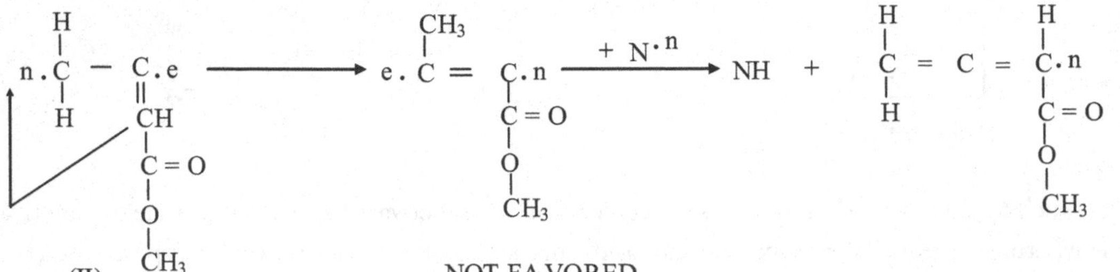

$$
\begin{array}{ccc}
 & H & \\
 & | & \\
 n\,.\,C & - & C\,.\,e \\
 | & & \| \\
 H & & CH \\
 & & | \\
 & & C = O \\
 & & | \\
 & & O \\
 & & | \\
 (II) & CH_3 & \\
\end{array}
\longrightarrow
\begin{array}{ccc}
 CH_3 & & \\
 | & & \\
 e.\ C & = & C\,.\,n \\
 & & | \\
 & & C = O \\
 & & | \\
 & & O \\
 & & | \\
 & & CH_3 \\
\end{array}
\xrightarrow{+\ N\,\cdot\,n}
NH\ +\!
\begin{array}{ccccc}
 H & & & & H \\
 | & & & & | \\
 C & = & C & = & C\,.\,n \\
 | & & & & | \\
 H & & & & C = O \\
 & & & & | \\
 & & & & O \\
 & & & & | \\
 & & & & CH_3 \\
\end{array}
$$

NOT FAVORED   8.75

$$\begin{array}{c}
\text{H} \quad \text{CH}_2 \\
\text{n}\bullet\ \overset{||}{\underset{|}{\text{C}}} - \text{C}\ \bullet\text{e} \\
\text{C}=\text{O} \\
| \\
\text{O} \\
| \\
\text{CH}_3
\end{array}
\quad\longrightarrow\quad
\begin{array}{c}
\text{O} \quad \text{H} \\
\text{e}\bullet\ \overset{||}{\text{C}} - \underset{|}{\text{C}}\ \bullet\text{n} \\
\text{COCH}_3 \\
|| \\
\text{CH}_2
\end{array}
\quad\xrightarrow{\ +\ \text{N}^{\bullet\text{n}}\ }\quad
\begin{array}{c}
\text{O} \quad \text{H} \\
\text{N} - \overset{||}{\text{C}} - \underset{|}{\text{C}}\ \bullet\text{n} \\
\text{COCH}_3 \\
|| \\
\text{CH}_2
\end{array}$$

(I) 8.76

Without any doubt, (I) is the real Electrophile, in particular after it has undergone molecular rearrangement. Electro-free-radical route is not favored here. Nevertheless, paired nucleo-free-radical initiator must be used for polymerization to be favored.

From all the considerations so far in the volume, though it has been stated that polar and electrostatic charges do not repel, since they are imaginary and products of radicals, this does not mean that these forces do not exist with them. The real parts do attract and repel like the cases of ionic and covalent charge; but not the imaginary part. The imaginary part can only repel and attract in the imaginary domain. In order words, there are two types of Electro-dynamic/static forces of repulsion and attraction – THE REAL and THE IMAGINARY ONES. Imagine if radicals of same or different character do not repel and attract, then they will be adding to themselves indiscriminately, something which is against the Laws of Physics in Nature. One is not dealing with the real world alone which universally all the journals know about, but in addition with the imaginary or more correctly the complex world as first and fully established in the Laws of Mathematics as seen inside Chemistry. All what is needed is the use of the eyes of the NEEDLE which is complex.

When COR groups are involved, the followings are to be expected.

$$\begin{array}{c}
\text{H} \quad \text{Y} \qquad \text{X} \ \text{H} \\
\overset{|}{\underset{1}{\text{C}}}\ \overline{\text{X}}\ \overset{}{\underset{2}{\text{C}}} = \overset{}{\underset{3}{\text{C}}} \\
\text{C}=\text{O} \qquad \text{H} \\
| \\
\text{CH}_3
\end{array}
\quad\longrightarrow\quad
\begin{array}{c}
\text{H} \quad \text{Y} \\
\text{n}\cdot\overset{|}{\underset{1}{\text{C}}}\ \overline{\text{X}}\ \underset{2}{\text{C}}\cdot\text{e} \\
\text{C}=\text{O} \ \ \text{CH}_2 \\
| \\
\text{CH}_3
\end{array}
\quad ; \quad
\begin{array}{c}
\text{H} \\
\text{e}_{\text{Y}}\overset{||}{\text{C}}\ \overset{\text{X}}{\overline{}}\ \overset{|}{\text{C}}\cdot\text{n} \\
\text{CH} \quad \text{H} \\
| \ \text{X} \\
\text{C}=\text{O} \\
| \\
\text{CH}_3
\end{array}$$

(I) (Electrophile) (II) (Electrophile) 8.77

This group so far is not new to us. It is (I) that is largely used, preferably the trans-placed one if it is to be polymerized in the route natural to it using paired radical nucleo-free-radical initiators. Electro-free-radically, the C=O center is the center natural to it for polymerization. Since the Y center has no transfer species electro-free-radically, the possibility of having random copolymers for this monomer, like the case of acroleins cannot be ruled out, but only electro-free-radically.

When the group is $-\text{C} \equiv \text{N}$, the monomer becomes more Electrophilic only when the groups are trans-placed and paired radical initiators are used.

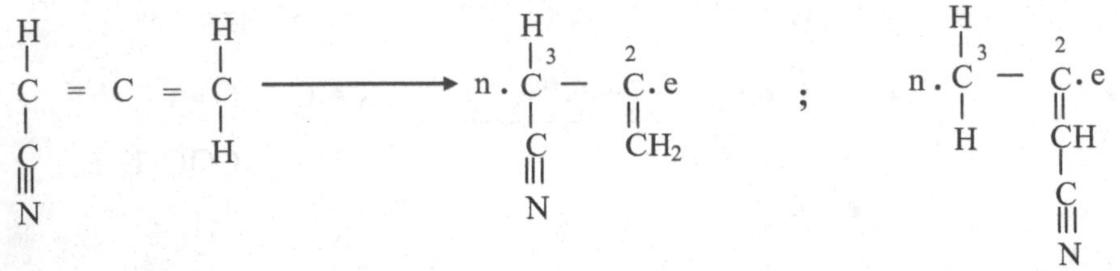

$$
\begin{array}{ccc}
H & H & H \\
| & | & | \\
C = C = C & \longrightarrow & n.\overset{3}{C} - \overset{2}{C}.e \\
| & | & \| \quad \| \\
C & H & C \quad CH_2 \\
\| & & \| \\
N & & N
\end{array}
\qquad ; \qquad
\begin{array}{c}
H \\
| \\
n.\overset{3}{C} - \overset{2}{C}.e \\
| \quad \| \\
H \quad CH \\
\quad | \\
\quad C \\
\quad \| \\
\quad N
\end{array}
$$

(I) (Electrophile)   (II) (Electrophile)   8.78

Both favor electro-free-radical route via the C≡N center and no molecular rearrangement can take place. When free media initiators are used, the nucleo-free-radical route cannot be favored. It is only (I) that exists.

Consider cases where the substituted radical-pulling groups are symmetrically placed. The same analysis as above will also apply. When molecular rearrangement is allowed to take place, the ketene produced can readily be polymerized with free media nucleo-free-radical initiator. Without rearrangement, only paired nucleo-free-radical initiator will favor polymeriza-tion when the groups are trans-placed. The transfer species involved for the molecular rearrangment and for the growing

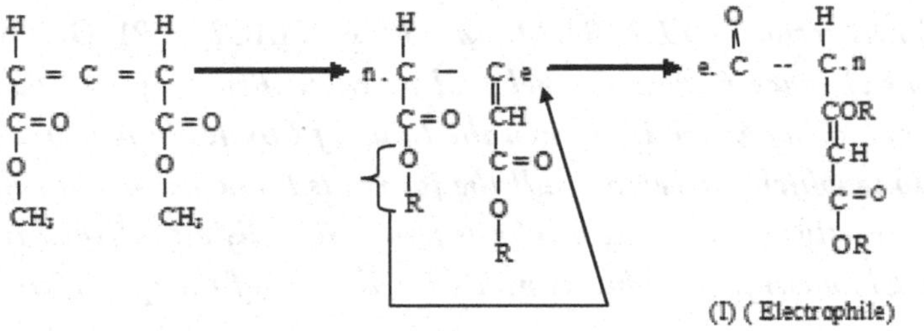

$$
\begin{array}{ccc}
H & H & H \\
| & | & | \\
C = C = C & \longrightarrow & n.C - C.e \longrightarrow e.C -- C.n \\
| & | & | \quad | \\
C=O & C=O & C=O \; CH \\
| & | & | \quad | \\
O & O & O \; C=O \\
| & | & | \quad | \\
CH_3 & CH_3 & R \quad O \\
 & & \qquad R
\end{array}
$$

(I) (Electrophile)   8.79

polymer chain is of the first kind of the first type. From all the considerations so far, one can observe how unique Nature operates. It is too complex to comprehend and so logical. Imagine what to expect with COR and CN groups where molecular rearrangement cannot take place for a family of monomers which cannot be activated chargedly. For their growing polymers, living polymers are produced as shown below for the two of them.

[With Molecular Rearrangement]

$$
\begin{array}{c}
O \; H \quad\; O \; H \\
\| \; | \quad\;\; \| \; | \\
N-(C-C)_n^s - C - C \bullet n \longrightarrow N -(C-C)_n^s - C - C = C = C - C = O \;+\; NaOR \\
\quad | \qquad\quad | \qquad\qqu\qquad | \qquad\qquad | \\
COR \quad\; COR \qquad\qquad COR \qquad\quad OR \\
\| \qquad\quad \| \qquad\qquad\qquad \| \\
CH \qquad\; CH \qquad\qquad\qquad CH \\
| \qquad\qquad | \qquad\qquad\qquad | \\
C=O \quad\; C=O \qquad\qquad\quad C=O \\
| \qquad\qquad | \qquad\qquad\qquad | \\
OR \qquad\; OR \qquad\qquad\qquad OR
\end{array}
$$

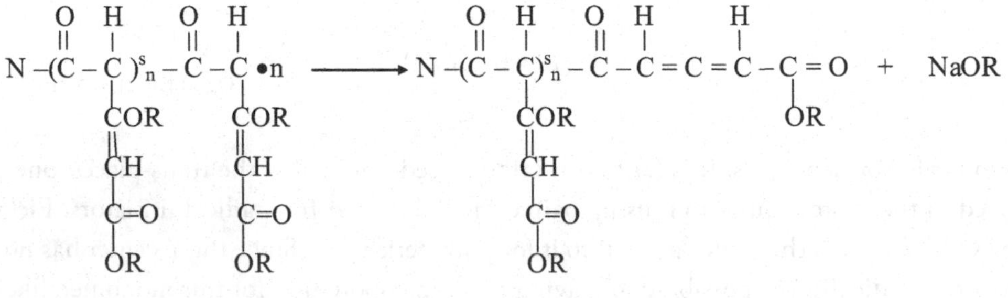

(where N is H_5C_6 in $H_5C_6^{\bullet n}$.........$^{e\bullet}Na$)   8.80

[Without Molecular rearrangement]

$$N-(C-C)_n^s-C-C\bullet n \longrightarrow N-(C-C)_n^s-C-C=C=O \quad + \quad NaOR$$

(where N is H_5C_6 in $H_5C_6^{\bullet n}$.........$^{e\bullet}Na$) 　　　　8.81

$$N-(C-C)_n-C-C\bullet n \longrightarrow \text{No Transfer species}$$

(where N is H_5C_6 in $H_5C_6^{\bullet n}$.........$^{e\bullet}Na$) 　　　　8.82

$$N-(C-C)_n-C-C\bullet n \longrightarrow \text{No Transfer species}$$

(where N is H_5C_6 in $H_5C_6^{\bullet n}$.........$^{e\bullet}Na$) 　　　　8.83

Largely syndiotactic placements will be obtained due to the influence of steric limitations and electrodynamic forces of repulsion and indeed presence of two vacant orbitals on the Na center. *One can observe that whichever way, because one is Human, we have a long way to go. All what one is doing is laying NEW FOUNDATIONS for humanity, that which makes one to bleed with tears every day in pains, because the more we know, the more we come to realize that we know NOTHING.*

Finally considering prefluorocumulences, the followings are obtained free-radically.

$$C = C = C \longrightarrow n \cdot C - C \cdot e \longrightarrow \text{No transfer species} \quad 8.84$$

$$E(\overset{F}{\underset{F}{C}} - \underset{CF_2}{C})_n \overset{F}{\underset{F}{C}} - \underset{CF_2}{C} \cdot e \longrightarrow \text{No transfer species} \quad 8.85$$

$$N \left(C - \underset{\underset{F}{|}}{\overset{\overset{F}{|}}{C}} \right)_n \overset{\overset{F}{|}}{\underset{\underset{F}{|}}{C}} - \overset{\overset{F}{|}}{\underset{\underset{F}{|}}{C}} \cdot n \longrightarrow \quad \text{No transfer species} \qquad 8.86$$

*For the first time, a "symmetric cumulene" which does not favor molecular rearrangement is being encountered only radically. For this peculiar case, as can be observed from the reactions above, the monomer like the olefinic case has no character, but unlike the olefinic case has fixed radical centers when activated since **no symmetric cumulene exists when activated.** Both free-radical routes are favored with no transfer species. It is indeed the only case, since when the Fs are replaced with CF_3 or $COOCH_3$, the followings are obtained.*

$$\underset{\underset{CF_3}{|}}{\overset{\overset{CF_3}{|}}{C}} = C = \underset{\underset{CF_3}{|}}{\overset{\overset{CF_3}{|}}{C}} \longrightarrow n \cdot \underset{\underset{CF_3}{|}}{\overset{\overset{CF_3}{|}}{C}} - \underset{\underset{CF_3 \quad CF_3}{\diagup \diagdown}}{\overset{\overset{}{\|}}{C}} \cdot e \longrightarrow n \cdot \overset{\overset{CF_3}{|}}{C} = \underset{\underset{CF_3 \quad CF_3}{\diagup \diagdown}}{\overset{\overset{}{|}}{C}} \cdot e$$

$$\text{(Nucleophile)} \qquad 8.87$$

This is a Nucleophile. With free media initiators, no free-radical route is favored whether molecular rearrangement takes place or not. Even when molecular rearrangement takes place, the monomer cannot be polymerized free medially, but only when special free-radical paired initiators are used nucleo-free-radiccally the route not natural to the monomer. (See Equation 8.70)

$$\underset{\underset{CH_3}{|}}{\overset{\overset{CH_3}{|}}{\underset{|}{\overset{|}{\underset{|}{\overset{|}{C}}}}}} \quad \cdots \longrightarrow n \cdot C - C \cdot e$$

Very weak Electrophile $\qquad$ 8.88

Existence of this type of monomer is greatly limited by electrodynamic forces of repulsion and steric limitations. With free media free-radical initiators, no route is favored because OCH_3 and $COOR$ groups are transfer species electro-free-radically and nucleo-free-radically respectively. With the use of paired free-radical initiators, polymerization may be favored if the less sterically hindered side is the first to gain entry into the free-radical coordination center. When it is allowed to undergo molecular rearrangement, it will then favor nucleo-free-radical polymerization in the presence of minor steric limitations.

From all the considerations so far, there are so many things which cannot be ignored. Free-radically, the order of the radical-pushing capacities of the halogenated and other types of alkylene groups is obviously as follows –

$$CH_2 = \quad > \quad CFH = \quad > \quad CF_2 =$$

Order of free-radical-pushing capacity

8.89

$$H_2C = \quad > \quad ROOCCH = \quad > \quad (ROOC)_2C =$$

Order of Free-radical-pushing capacity

8.90

$$F_2C = \quad > \quad (ROOC)_2C = \quad > \quad (ROC)_2C = \quad > \quad (NC)_2C =$$

Order of Free-radical-pushing capacity

8.91

$$FHC = \quad > \quad ROOCCH = \quad > \quad ROCCH = \quad > \quad NCCH =$$

Order of Free-radical-pushing capacity

8.92

All these groups are very strong free-radical-pushing groups despite the presence of radical-pulling groups carried by them. This is almost like the allylic groups.

With C_2F_5 replacing CF_3 above, the same analysis almost applies. The role of resonance stabilization groups free-radically has already been fully considered when charged characters of this monomer were considered in the last sub-section.

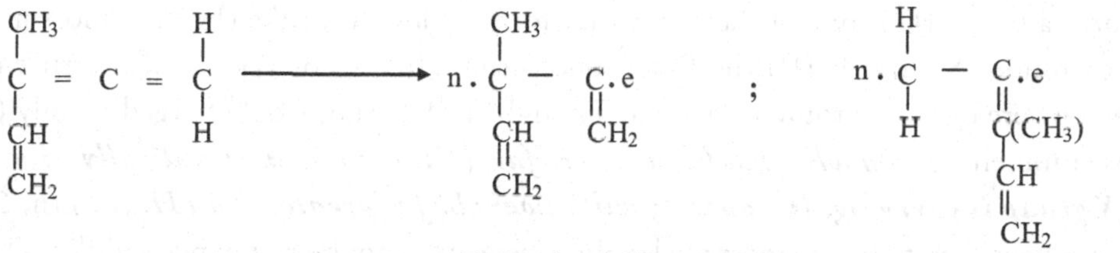

(I) (Nucleophile)   (II) (Nucleophile)   8.93

As already shown, (I) and (II) above are not the centers involved when the monomer is activated. It is the alkenyl center that is first activated and for this case resonance stabilization takes place with cumulene being the provider.

8.2. 1,4 - Cumulenes

Though these monomers are yet unknown, dead terminal 1,4-cumulenic bonded polymers have been observed when acetylene was the focus of attention (See section 6.1.3 of Chapter 6). There is therefore the need to acknowledge the existence of these monomers which without doubt will pose serious steric limitations during polymerizations. There is also the need to consider these monomers, in order to extend the New Frontiers of natural Laws being proposed and to ascertain that no rules so far proposed have been broken.

8.2.1. Charged Characters of 1,4 - Cumulenes

By the present nomenclature, the first member of 1,4-cumulences is as shown below. If Acetylenes and 1,3-cumulenes cannot be activated chargedly, the same should also apply to 1,4-cumulenes. Nevertheless the charged characters are being considered for exploratory purposes and as a guide to free-radical characters.

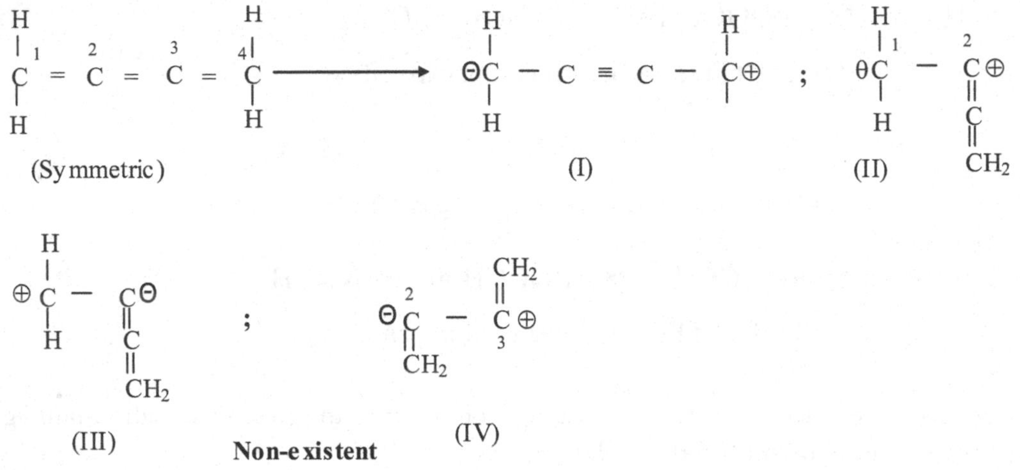

Worthy of note, is that all the mono-forms with the exception of (IV) seem as if resonance stabilized radically. (II) is one of them, **because H₂C=C= group is a free-radical-pushing group of greater capacity than H.** We know that an electro-free-radical can move from the π-bond to grab a nucleo-free-radical to form a bond. (I) is one of the two mono-forms, though unlike (IV) the most nucleophilic, it favors both routes. So also is (II) the first mono-form. This is not typical of the first member of 1,3-cumulene where only one route is favored. Though (IV) cannot exist chargedly, only (I) and (II) exist for this cumulene. ***From what has been seen so far, (III) cannot exist radically, since whatever X the =CXX group is carrying, its capacity will always be far greater than H.*** (IV) the 2,3- mono-form which has been said to be the most nucleophilic center cannot be resonance stabilize. Therefore, it is (II) that is the first to appear. (I) cannot exist instantaneously, because only one activation center can be activated one at a time. Chargedly, (II) would have been the only mono-form, if activation chargedly would have been possible. Therefore, 1,4-cumulene cannot be resonance stabilized chargedly. If (II) which can exist and (IV) which cannot exist are made to undergo an artificial molecular rearrangement, the followings would have been obtained.

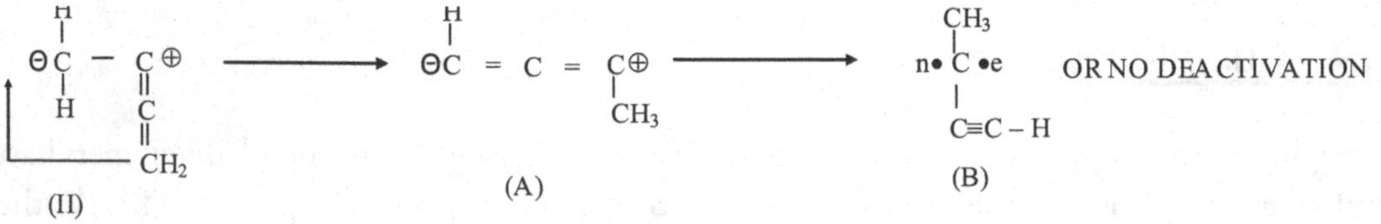

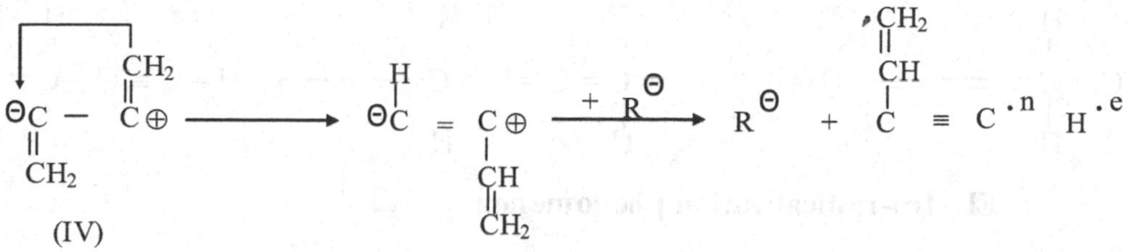

$$\overset{\ominus}{C} - \overset{\overset{\displaystyle CH_2}{\|}}{C}\oplus \longrightarrow \overset{\ominus}{C} \underset{\underset{\displaystyle CH_2}{\overset{\displaystyle \|}{CH}}}{=} C\oplus \xrightarrow{+ \ R^{\ominus}} R^{\ominus} + \overset{\overset{\displaystyle \cdot CH_2}{\|}}{\underset{}{C}} \equiv C \cdot^n \ H\cdot^e$$

(IV)   (Possible Activated State)   8.96

By artificial, is mearnt that (II) cannot indeed undergo the rearrangement in the absence of transfer species. Transfer species will exist when one H atom is changed to for example CH_3. For the last reaction, molecular rearrangement may take place, based on the laws already stated. It is the $HC\equiv C-$ group that provides resonance stabilization for Alkenyl group. Thus, it may seem that the two mono-forms (1,2- and 1,4-) can favor being used as Addition monomers, without the need of replacing the H atom(s) with other group(s), unlike the case with acetylene.

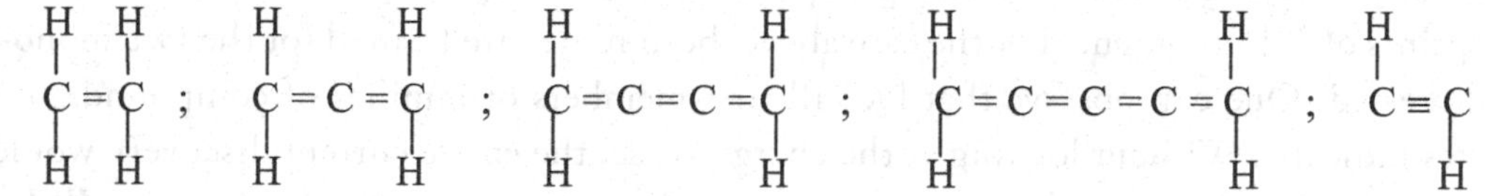

2C/4H	3C/4H	**4C/4H**	5C/4H	**2C/2H**
(I)	(II)	(III)	(IV)	(V)
H/C: 2	1.33	1.0	0.8	1.0
C/H: 0.5	0.75	1.0	1.25	1.0

8.97

Based on the ratio of C to H in the compounds, there is no doubt that (III) and (V) are close in character, except that (III) is far more nucleophilic than (V). These are supposed to be first members of their families. (II) is not, since it molecularly rearranges to become a second member of another family (V). While (V) is very unstable, since it exists all the time when not suppressed in Equilibrium State of Existence, (III) seems to be also unstable most likely existing in Activated State of Existence or Equilibrium state of existence all the time, except when suppressed as shown below.

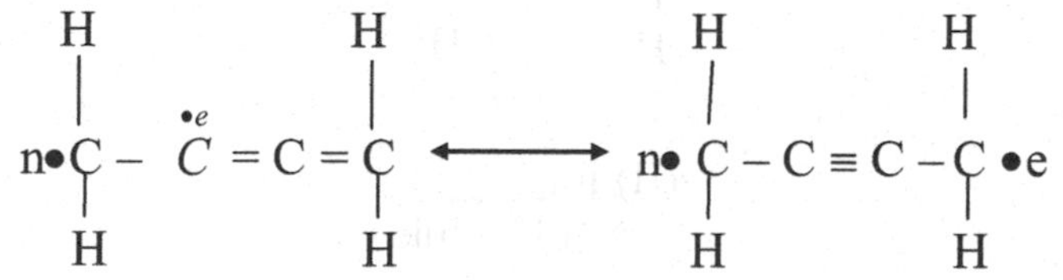

From (II) of Eqn 8.94   From (I) of Eqn 8.94

Resonance stabilization phenomenon   8.98a

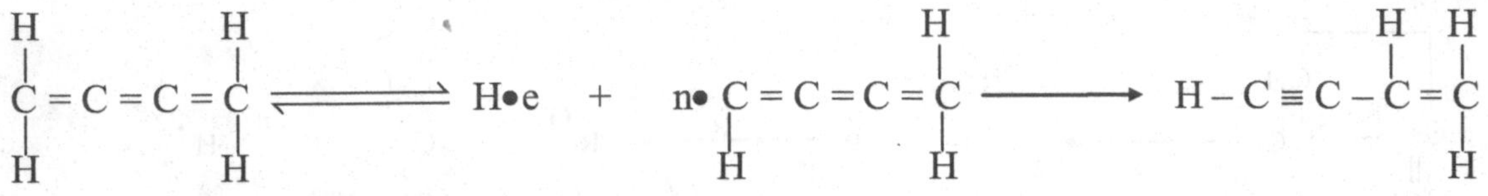

$$H_2C=C=C=CH_2 \rightleftharpoons H\bullet e \;+\; n\bullet C=C=C=CH_2 \longrightarrow H-C\equiv C-CH=CH_2$$

Electro-radicalization phenomenon 8.98b

This is 1.4-cumulene. The phenomenon above is not Molecular rearrangement, or Enolization or Enamidization, but Electroradicalization, wherein a new compound is formed. It is one of the Tautomeric methods. It is not a situation of resonance stabilization taking place but something else. As is already obvious, Electroradicalization is that which involves movement of electro-radicals when present, to grab a nucleo-radical from a π-bond adjacently located or movement of electro-radical from a π-bond to grab a nucleo-radical visibly present and adjacently located. One has already stated this into one of the rules, that which is Physics.

While (I) of Equation 8.94 as represented in Equation 8.98a is one of the mono-forms of all of 1,4-cumulenes regardless the substituent groups carried by them, the last equation above is one of the blue-prints of 1,4-cumulene. For the case above, both routes are favored for the two mono-forms when suppressed. One can observe that like all first members of families of compounds, it favors both routes radically. Without looking at the charged case, the entire current discovery would have been impossible. ***The 1,4-cumulenes seem to exist in Equilibrium state of existence all the time, except when suppressed.*** The (A) obtained in Equation 8.95 for the 1,2- mono-form was indeed obtained in its Activated/Equilibrium state of existence. It cannot be deactivated. Radically, it is an activated Carbene (B) of Equation 8.95 that is obtained. Hence, it does not exist. None of the mono-forms of the 1,4-cumulene can be kept in Activated/ Equilibrium state of existence when there is no transfer species..

There is need however at this point in time to digress and clearly distinguish between transfer species of the **second first kind** of different types, by using examples.

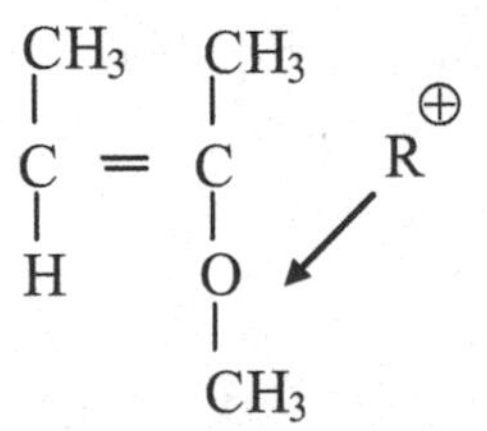

(I) <u>First type</u>
Nucleophile
Natural route

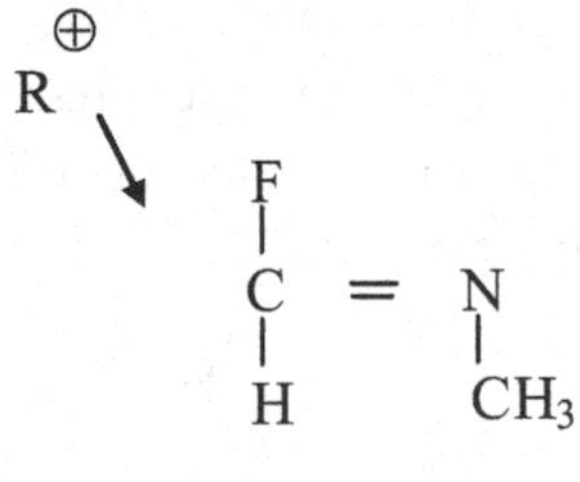

(II) <u>First type</u>
Nucleophile
Natural route

8.99

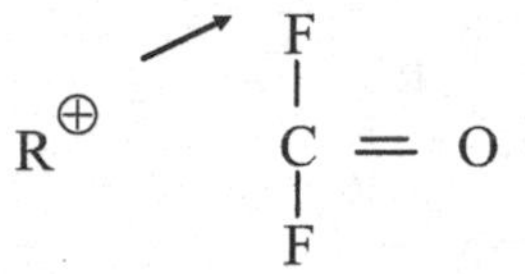

$$R^{\oplus} \nearrow \quad \underset{\underset{F}{|}}{\overset{\overset{F}{|}}{C}} = O \qquad ; \qquad R^{\oplus} \searrow \quad \underset{\left\{\begin{matrix} O \\ | \\ R \end{matrix}\right.}{\overset{\overset{H}{|}}{C}} = O$$

(III) <u>Second type</u>   (IV) <u>Second type</u>
Nucleophile   Nucleophile
Natural route   Natural route   8.100

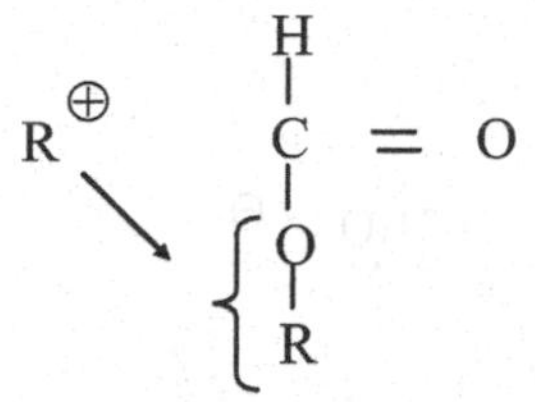

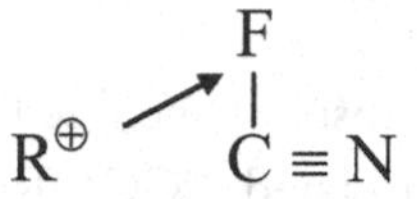

$$R^{\oplus} \nearrow \quad \underset{\overset{F}{|}}{C} \equiv N \qquad ; \qquad E^{\bullet e} \quad \underset{\left\{\begin{matrix} O \\ | \\ C_4H_9 \end{matrix}\right.}{C} \equiv N$$

(V) **<u>Third type</u>**   (VI) **<u>Third type</u>**

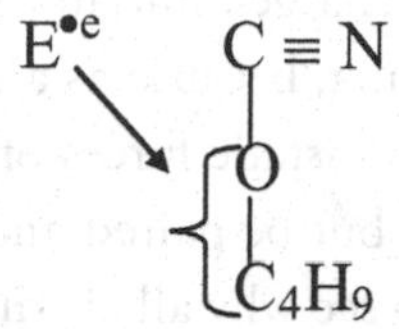

Nucleophile   Nucleophile

Natural route   Natural route   8.101

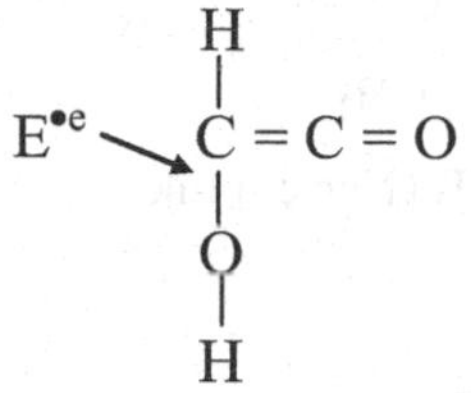

$$E^{\bullet e} \quad \underset{\underset{\overset{|}{H}}{O}}{\overset{\overset{H}{|}}{C}} = C = O \qquad ; \qquad E^{\bullet e} \searrow \quad \underset{\underset{\overset{|}{CH_3}}{\overset{|}{C}=O}}{\overset{\overset{H}{|}}{C}} = \underset{\overset{|}{H}}{\overset{\overset{H}{|}}{C}} \qquad ; \qquad \underset{\underset{\overset{|}{CH_3}}{\overset{|}{C}=O}}{\overset{\overset{H}{|}}{C}} = \underset{\overset{|}{O} \searrow E^{\bullet e}}{\overset{\overset{H}{|}}{C}}$$

(VII) **<u>Fourth type</u>**   (VIII) **<u>First type</u>**   (IX) **<u>Second type</u>**

Electrophile   Nucleophile   Nucleophile

Unnatural route   Unnatural route   Natural route   8.102

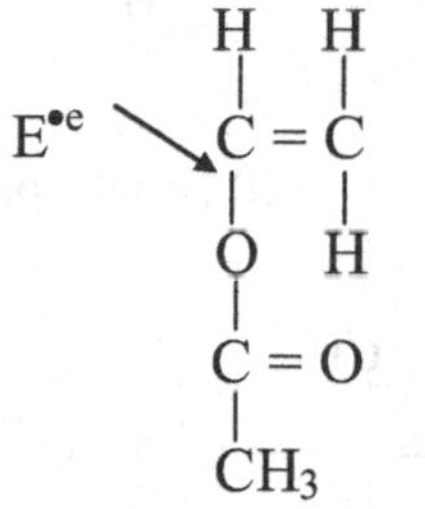

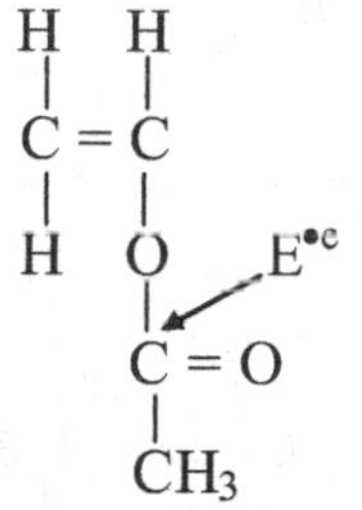

These transfer species can only be abstracted when the compound or monomer is in STABLE state of Existence. It should be noted that the $R^{\oplus}$ involved above are cations which can be isolated. The positively charged ones are paired and this cannot be used for the reactions above. Hence in place charges are the radicals which can be isolatedly placed. No R^{θ} or N.n is involved in all the abstractions above. These only abstract only when no males are around or when paired where the need arises. Only anions can be isolatedly placed as shown below.

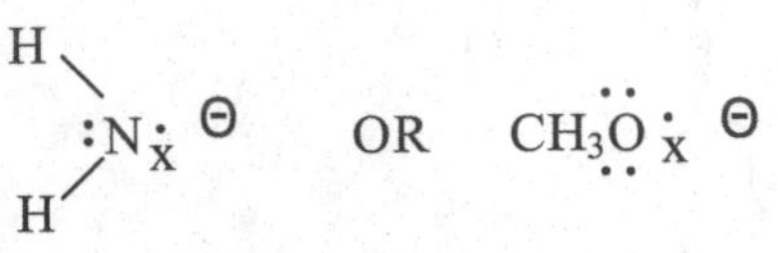

New Frontiers in Sciences, Engineering and the Arts

$$\underset{H}{\overset{H}{\diagdown}} \ddot{N} \overset{\ominus}{\underset{x}{:}} \quad OR \quad CH_3\ddot{O}\underset{x}{\overset{\ominus}{:}}$$

Free - initiator
Non - free center

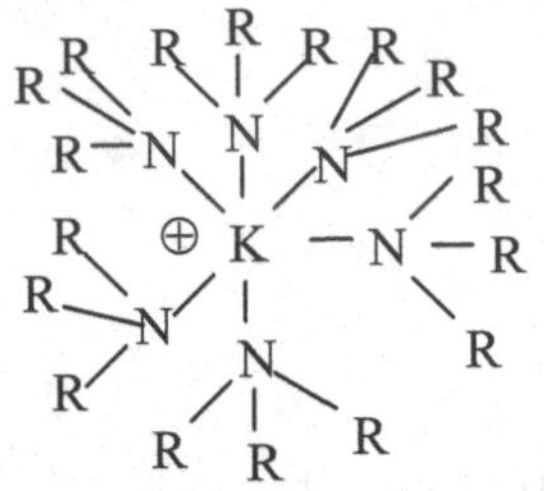

(Datively shielded from
polymerization zone)   8.103

A free centered free-negatively charged initiator such as $^\ominus R$ cannot be used to abstract the free hydrogen atom located on a non-free center, because as already shown they cannot be isolated. $^\ominus COOR$, $^\ominus COR$, $^\ominus C\equiv N$ cannot exist due to electrostatic forces of repulsion. They can only carry positive charges which cannot themselves be isolated, but be paired and these do not exist also. However, they can carry only free-radicals. One can begin to see why all chemical reactions take place only radically.

Now replacing one of the hydrogen atoms with an alkylane (alkanyl) group, the followings are obtained.

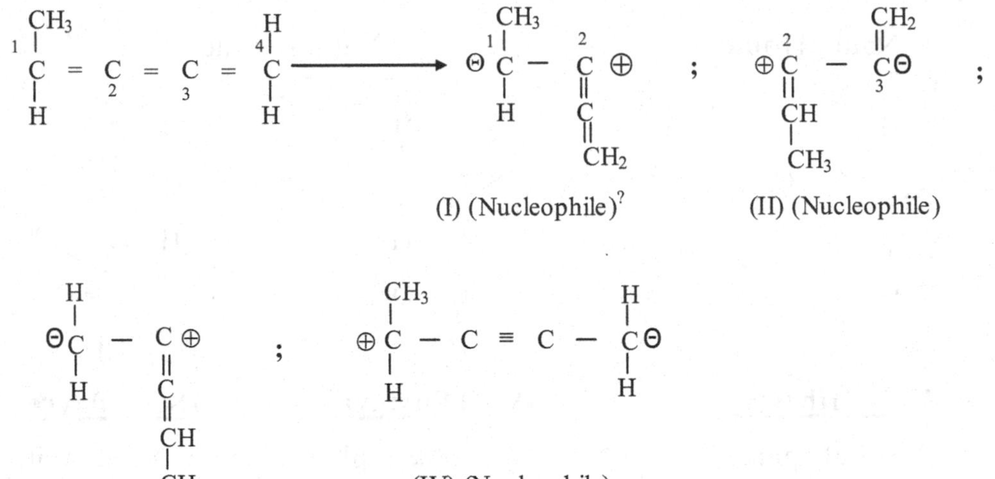

(I) (Nucleophile)?   (II) (Nucleophile)

(III) (Nucleophilic)   (IV) (Nucleophile)   8.104

It may seem that four resonance stabilized mono-forms exist as opposed to just two. Like the symmetric case, (I) and (II) which do not exist do not favor molecular rearrangement. (III) and (IV) seem to favor it as shown below for them using (IV).

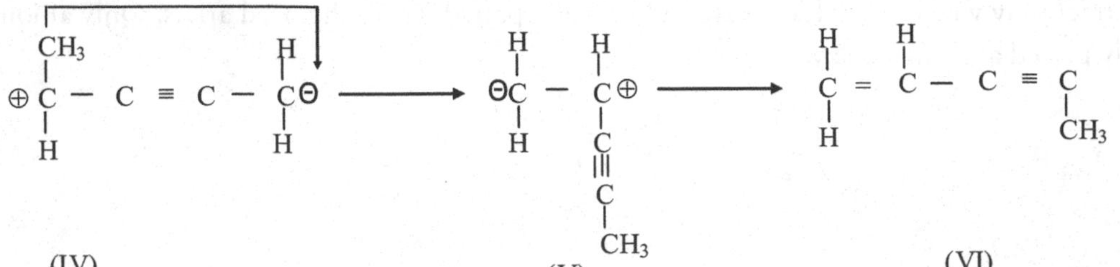

(IV)   (V)   (VI)   8.105

619

Indeed, (I) cannot be a mono-form, since it is the only one that will favor both routes instead of the natural one. While (III) can resonance stabilize free-radically to (IV), (I) cannot as shown below.

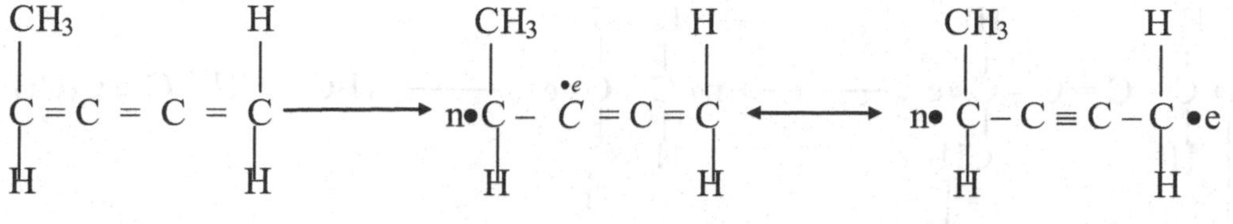

$$CH_3 \quad H \qquad CH_3 \quad H \qquad CH_3 \quad H$$
$$C=C=C=C \longrightarrow n\bullet C - \overset{\bullet e}{C}=C=C \longleftrightarrow n\bullet C-C\equiv C-C\bullet e$$
$$H \quad\quad H \qquad\quad H \quad\quad H \qquad\quad H \quad\quad H$$

[Impossible Existence]   8.106a

Unlike the symmetric case, the phenomenon cannot take place here. This clearly indicates that (I) cannot exist, because the followings are valid-

$$(H_3C)HC=C= \quad > \quad H_2C=C= \quad > \quad (H_3C)HC= \quad > \quad H_2C= \quad >> \quad H$$

Order of radical pushing capacity 8.106b

Hence, for the case in question, the least nucleophilic center is (III) of Equation 8.104 and not (I) with question mark shown in that equation. Note that all these cannot take place chargedly, but only radically. (III) which is the 3,4-mono-form undergoes resonance stabilization phenomenon as shown below to give the 1,4-mono-form (IV).

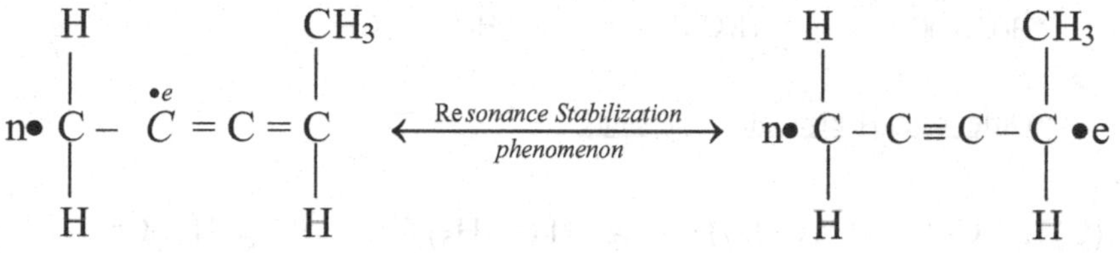

$$H \quad\quad CH_3 \qquad\qquad H \quad\quad CH_3$$
$$n\bullet C - \overset{\bullet e}{C}=C=C \underset{\text{phenomenon}}{\overset{\text{Resonance Stabilization}}{\longleftrightarrow}} n\bullet C-C\equiv C-C\bullet e$$
$$H \quad\quad H \qquad\qquad\quad H \quad\quad H$$

(III) of Eqn. 8.104   (IV) of Eqn. 8.104

RESONANCE STABILIZATION PHENOMENON   8.107

When (III) is made to undergo molecular rearrangement, the following is obtained.

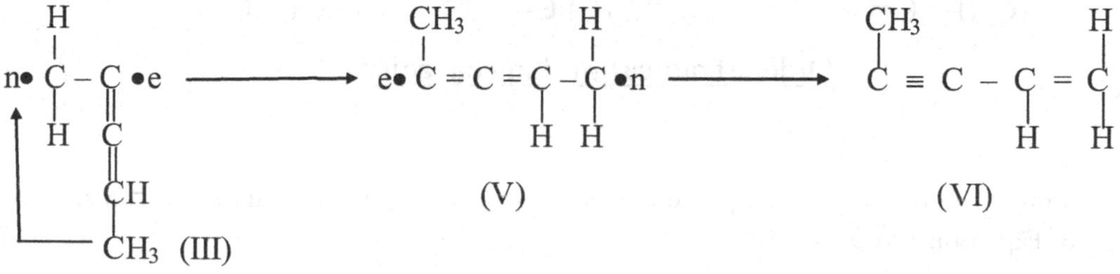

$$H \qquad\qquad\qquad CH_3 \quad H \qquad\qquad CH_3 \quad H$$
$$n\bullet C-C\bullet e \longrightarrow e\bullet C=C=C-C\bullet n \longrightarrow C\equiv C-C=C$$
$$H \quad C \qquad\qquad\qquad H \quad H \qquad\qquad\qquad H \quad H$$
$$CH \qquad\qquad\qquad (V) \qquad\qquad\qquad (VI)$$
$$CH_3 \;\;(III)$$

8.108

Notice where the transfer species is coming from. It is coming from the CH_3 group. It is only when one H atom is replaced in $H_2C=C=$ group that transfer species begin to appear. Hence, Equation 8.106b is valid. Compare the (VI) above with (VI) of Equation 8.105 obtained chargedly, noting that chargedly, that (IV) cannot exist.

620

When (IV) is made to undergo molecular rearrangement radically, the followings are also to be expected.

$$n\bullet \overset{\displaystyle H}{\underset{\displaystyle H}{C}} - C \equiv C - \overset{\displaystyle H}{\underset{\displaystyle CH_3}{C}} \bullet e \longrightarrow n\bullet \overset{\displaystyle H}{\underset{\displaystyle H}{C}} - \overset{\displaystyle H}{\underset{\displaystyle C}{C}} \bullet e \longrightarrow H_2C = CH - C \equiv C(CH_3)$$

(IV)   (VII)   8.109

It molecularly rearranges to give a product similar to that obtained when acetylene dimerizes in the presence passive catalysts as already shown. All these depend on the operating conditions. If the operating condition is such that molecular rearrangement is not allowed to take place, the polymer products obtained for them are uniquely different, for which for all of them the natural route is the only favored route-electro-free-radical route. This is unlike the first member which favors both routes.

Based on the analysis so far, it is therefore of interest to note that the larger the R group in place of CH_3 on the monomer, the followings are valid.

Etc. $<$ $H_2C = $ $<$ $H(CH_3)C = $ $<$ $H(C_2H_5)C = $ $<$ $H(C_3H_7)C = $ $<$

<u>Order of radical-pushing capacity</u>

8.110

$(C_3H_7)_2C= \; > \; (C_2H_5)_2C= \; > \; H\,(C_3H_7)\,C= \; > \; H\,(C_2H_5)\,C= \; > \; (CH_3)_2C= \; >$ etc.

<u>Order of radical-pushing capacity</u>   8.111

$(C_3H_7)_2C= C= \; > \; (C_2H_5)_2C=C= \; > \; H(C_3H_7)C= C= \; > \; H(C_2H_5)C= C= \; >$ etc.

<u>Order of radical-pushing capacity</u>   8.112

$(C_3H_7)_2C=C= $ $>$ $(C_3H_7)_2C= $ $>>$ $(C_3H_7)_3C-$

<u>Order of radical-pushing capacity</u>   8.113

Before moving ahead from this unsymmetrical cumulene, there is need to look at the type of dead polymers obtained from the growing polymer chains of (III) and (V) of Equation 8.108 and from (IV) and (VII) of Equation 8.109.

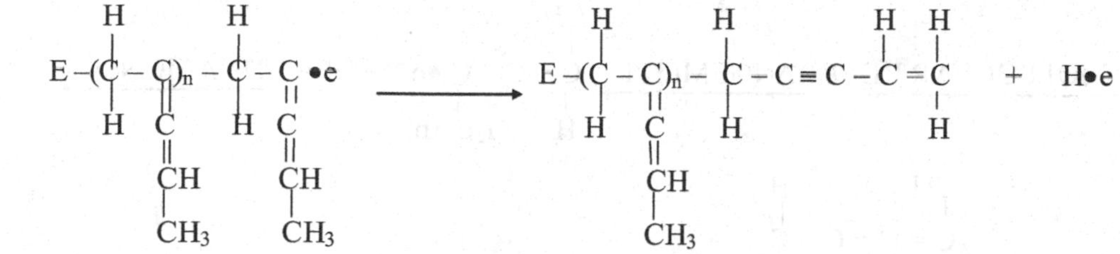

$$E-(\underset{\underset{\overset{|}{CH}}{\overset{|}{\underset{CH_3}{C}}}}{\overset{H}{\underset{|}{\overset{|}{C}}}} - \underset{\underset{CH}{\overset{|}{\underset{CH_3}{C}}}}{\overset{H}{\underset{|}{\overset{|}{C}}}})_n - \overset{H}{\underset{H}{\overset{|}{C}}} - \underset{}{C} \bullet e \longrightarrow E-(\underset{\underset{\overset{|}{CH}}{\overset{|}{\underset{CH_3}{C}}}}{\overset{H}{\underset{|}{\overset{|}{C}}}})_n - \overset{H}{\underset{H}{\overset{|}{C}}} - C \equiv C - \overset{H}{\underset{H}{\overset{|}{C}}} = \overset{H}{\underset{H}{\overset{|}{C}}} \quad + \quad H \bullet e$$

$$\textbf{(III)} \qquad\qquad\qquad\qquad \textbf{(VIII)} \qquad\qquad\qquad 8.114$$

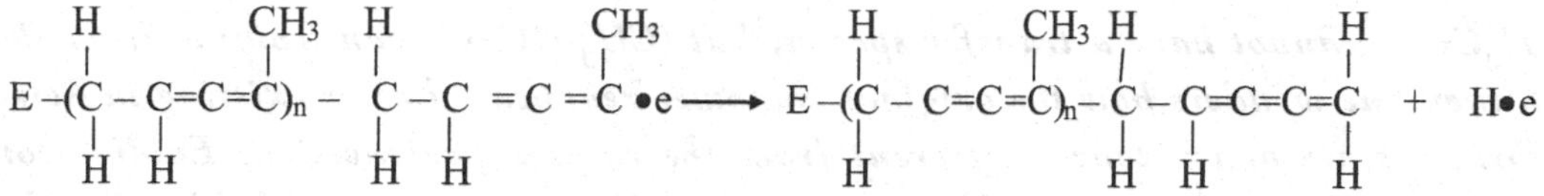

$$E-(\underset{\underset{H}{|}}{\overset{H}{\overset{|}{C}}} - \underset{H}{\overset{CH_3}{C=C=C}})_n - \underset{\underset{H}{|}}{\overset{H}{\overset{|}{C}}} - \underset{H}{\overset{CH_3}{C}} = C = \overset{}{C} \bullet e \longrightarrow E-(\underset{\underset{H}{|}}{\overset{H}{\overset{|}{C}}} - C=C=C)_n - \underset{\underset{H}{|}}{\overset{H}{\overset{|}{C}}} - \underset{H}{\overset{CH_3}{C}} - C=C=C=\overset{H}{\underset{H}{C}} \quad + \quad H \bullet e$$

$$\textbf{(V)} \qquad\qquad\qquad\qquad \textbf{(IX)} \qquad\qquad\qquad 8.115$$

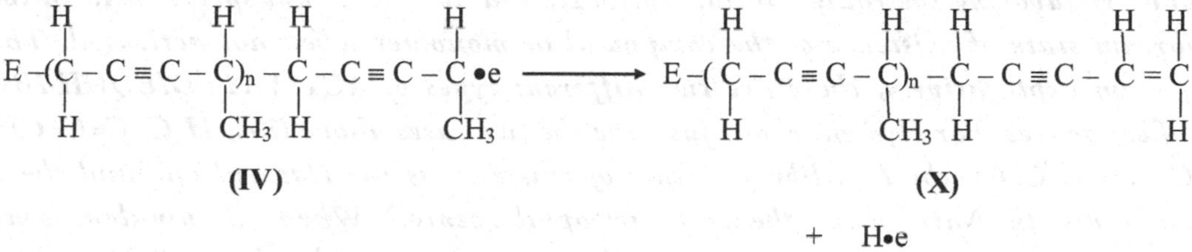

$$E-(\underset{\underset{H}{|}}{\overset{H}{\overset{|}{C}}} - C \equiv C - \underset{CH_3}{\overset{H}{\overset{|}{C}}})_n - \underset{\underset{H}{|}}{\overset{H}{\overset{|}{C}}} - C \equiv C - \overset{}{C} \bullet e \longrightarrow E-(\underset{\underset{H}{|}}{\overset{H}{\overset{|}{C}}} - C \equiv C - \underset{CH_3}{\overset{H}{\overset{|}{C}}})_n - \underset{\underset{H}{|}}{\overset{H}{\overset{|}{C}}} - C \equiv C - \underset{H}{\overset{H}{\overset{|}{C}}} = \overset{H}{\underset{H}{C}}$$

$$\textbf{(IV)} \qquad\qquad\qquad\qquad\qquad \textbf{(X)}$$

$$+ \quad H \bullet e \qquad\qquad 8.116$$

$$E-(\underset{\underset{\overset{|}{C}}{\overset{|}{\underset{CH_3}{C}}}}{\overset{H}{\underset{|}{\overset{|}{C}}}} - \underset{\underset{C}{\overset{|}{\underset{CH_3}{C}}}}{\overset{H}{\underset{|}{\overset{|}{C}}}})_n - \underset{\underset{\overset{|}{C}}{\overset{|}{\underset{CH_3}{C}}}}{\overset{H}{\underset{|}{\overset{|}{C}}}} - \overset{H}{\underset{H}{\overset{|}{C}}} \bullet e \longrightarrow E-(\underset{\underset{\overset{|}{C}}{\overset{|}{\underset{CH_3}{C}}}}{\overset{H}{\underset{|}{\overset{|}{C}}}} - \underset{H}{\overset{H}{\overset{|}{C}}})_n - \underset{\underset{\overset{|}{C}}{\overset{|}{\underset{CH_3}{C}}}}{\overset{H}{\underset{|}{\overset{|}{C}}}} - \overset{H}{\underset{H}{\overset{|}{C}}} = C = C = \overset{H}{\underset{H}{C}} \quad + \quad H \bullet e$$

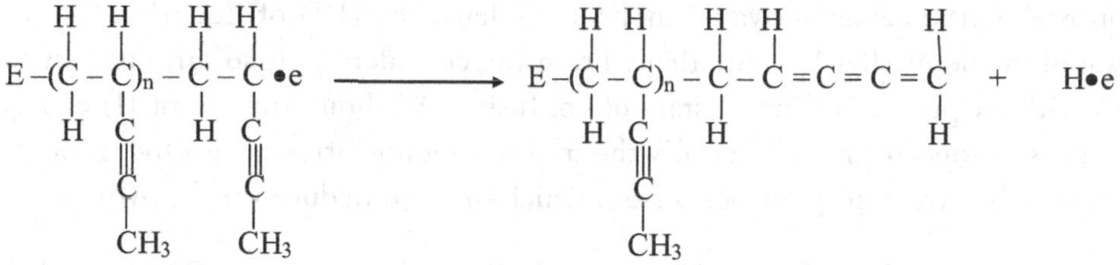

$$\textbf{(VII)} \qquad\qquad\qquad\qquad \textbf{(XI)} \qquad\qquad\qquad 8.117$$

From all the types of dead terminal *cumulenic, dead terminal eneyne polymers produced,* one can see that this 1,4-alkyl substituted cumulene is a product of eneyne family when no molecular rearrangement takes place. When (III) or (IV) of Equation 8.104 had been forced to rearrange as was done for the first member in Equation 8.95, then what is shown below would have been obtained.

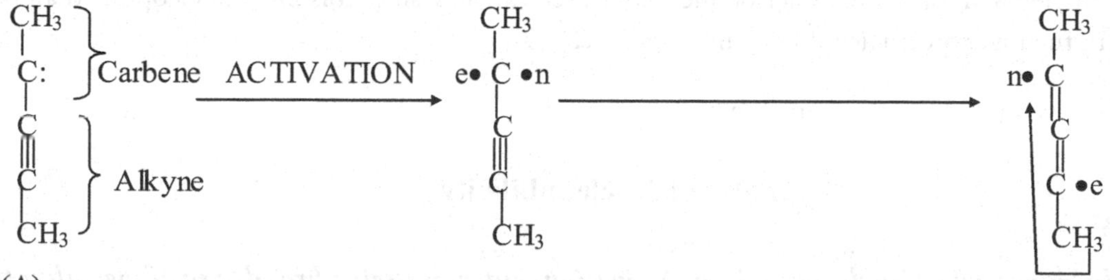

$$\left.\begin{array}{c} CH_3 \\ | \\ C: \end{array}\right\}\text{Carbene} \quad \underset{}{\text{ACTIVATION}} \quad e\bullet \overset{CH_3}{\underset{}{\overset{|}{C}}} \bullet n \quad \longrightarrow \quad n\bullet \overset{CH_3}{\underset{}{\overset{|}{C}}}$$

$$\left.\begin{array}{c} C \\ \parallel \\ C \\ | \\ CH_3 \end{array}\right\}\text{Alkyne} \qquad\qquad \begin{array}{c} C \\ \parallel \\ C \\ | \\ CH_3 \end{array} \qquad\qquad \begin{array}{c} \parallel \\ C \bullet e \\ | \\ CH_3 \end{array}$$

(A)

$$\underrightarrow{\text{MOLECULAR REARRANGEMENT}} \quad \overset{\displaystyle CH_3}{\underset{\displaystyle H}{\overset{|}{\underset{|}{C}}}} = C = \overset{\displaystyle}{\underset{\displaystyle H_2C \bullet n}{\overset{|}{C \bullet e}}} \quad \underrightarrow{\text{DEACTIVATION}}$$

$$\overset{\displaystyle CH_3}{\underset{\displaystyle H}{\overset{|}{\underset{|}{C}}}} = C = C = \overset{\displaystyle H}{\underset{\displaystyle H}{\overset{|}{\underset{|}{C}}}}$$

8.118

Since $H_2C=C=$ cannot have a transfer species, but $(CH_3)HC=C=$ can (coming from the CH_3 group), then one wonders how the original monomer reappeared after existing in Activated/ Equilibrium state of existence, different from the cases experienced in Enolization and Enamidization. In Enolization and Enamidization, the species (H atom) held in Equilibrium state of existence are transfer species used for molecular rearrangement, that which prevents the monomer favoring the route not natural to it, and not the actual species held in the real Equilibrium state of existence of the compound or monomer when not activated. Thus, as it may seem exploratively, there are two different types of ACTIVATED/EQUILIBRIUM states of existences. For Hydrocarbons just like the two cases above (i.e., $H_2C=C=C=CH_2$ and $H_2C=C=C=CH(CH_3)$, the Equilibrium state of existence is the Natural one and the center activated is also the Natural one (the least nucleophilic center). When 1,3-cumulene is made to exist in this state of existence, the same methyl acetylene can also be obtained. Note that while (II) of Equation 8.95 is the center activated in 1,4-cumulene, the (III) of Equation 8.104 is also the center activated in the methyl 1,4-cumulene. From the considerations so far, one can see how Nature operates without putting us into a state of confusion. Without any doubt (I) of Equation 8.104 cannot exist as a mono-form. In fact, only the true two mono-forms exist. One is using all the possible mono-forms for exploratory purposes from which one can deduce the followings-

$$RHC^{\upsilon} - N^{\oplus} \equiv N \;\; > \;\; RHC=C=C=CH_2 \;\; > \;\; H_2C=C=C=CH_2 \;\; > \; RC \equiv CH > etc.$$

Order of Nucleophilicity

8.119

$$R_2C=C=C=CH_2 \;\; > \; RHC=C=C=CH_2 > R_2C=C=CH_2 > RHC=C=CH_2 > etc.$$

Order of Nucleophilicity

8.120

Diazo-alkylenes are still more nucleophlic than the cumulenes, since it is more nucleophilc than benzene which in turn is more nucleophlic than cumulene.

$$Diazo\text{-}alkylenes > Benzene > 1,4\text{-}Cumulenes \; > 1,3\text{-}Cumulenes \; > \;\; Acetylenes$$

Order of Nucleophilicity

8.121

Establishing order of Nucleophilicity of Activation centers is itself a broad area of specialization for which experimental data under specific operating conditions is of great necessity. It is an enormous task, based on the foundations being laid.

Now replacing another H atom symmetrically on the external C center with CH_3 whether cis- or trans-placed, we have the followings.

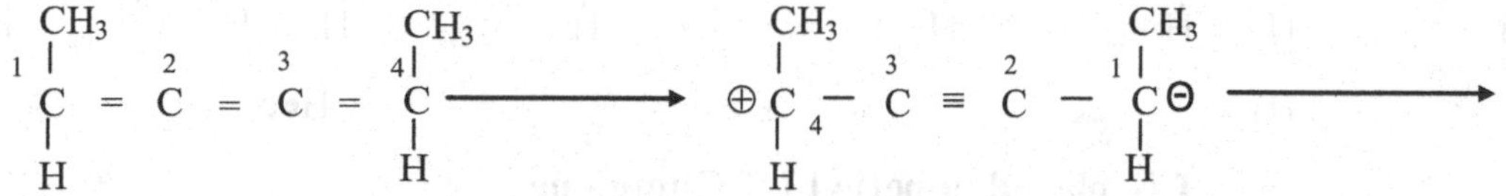

$$
\underset{\text{(Symmetric)}}{\overset{\overset{\displaystyle CH_3}{|}}{\underset{\underset{\displaystyle H}{|}}{\overset{1}{C}}} = \overset{2}{C} = \overset{3}{C} = \underset{\underset{\displaystyle H}{|}}{\overset{\overset{\displaystyle CH_3}{4|}}{C}}} \longrightarrow \underset{\text{(I) (Nucleophile)}}{\oplus\underset{\underset{\displaystyle H}{|}}{\overset{4}{C}} - \overset{3}{C} \equiv \overset{2}{C} - \underset{\underset{\displaystyle H}{|}}{\overset{\overset{\displaystyle CH_3}{1|}}{C}}\ominus} \longrightarrow
$$

$$
\underset{\text{(II) (Nucleophile)}}{\ominus\underset{\underset{\displaystyle H}{|}}{\overset{\overset{\displaystyle CH_3}{|4}}{C}} - \underset{\underset{\underset{\displaystyle CH_3}{|}}{\overset{\displaystyle CH}{|}}}{\overset{\overset{\displaystyle C}{||}}{\overset{3}{C}\oplus}}} \quad ; \quad \underset{\underset{\text{(Non-Existent)}}{\text{(III) (Nucleophile)}}}{\oplus\underset{\underset{\displaystyle H}{|}}{\overset{\overset{\displaystyle CH_3}{1|}}{C}} - \underset{\underset{\underset{\displaystyle CH_3}{|}}{\overset{\displaystyle CH}{|}}}{\overset{\overset{\displaystyle C}{||}}{\overset{2}{C}\ominus}}} \quad ; \quad \underset{\text{(IV) (Nucleophile)}}{\ominus\underset{\underset{\displaystyle CH}{||}}{\overset{2}{C}} - \underset{\underset{\underset{\displaystyle CH_3}{|}}{\overset{\displaystyle CH}{|}}}{\overset{\overset{\overset{\displaystyle CH_3}{|}}{CH}}{\overset{||}{\overset{3}{C}\oplus}}}}
$$

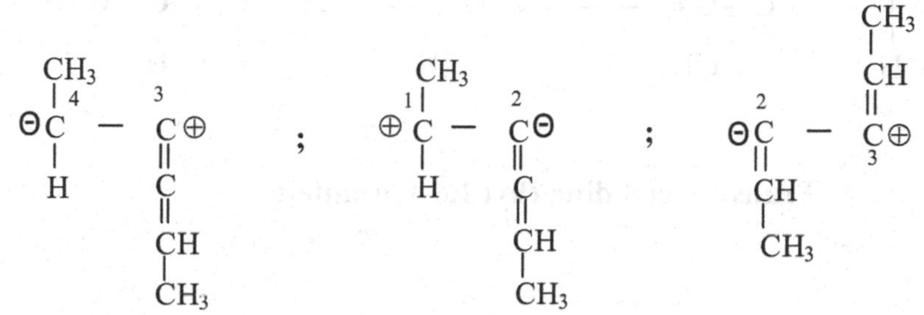

$$8.122$$

Like the first member, it is (II) and (I) that matter as shown below only free-radically.

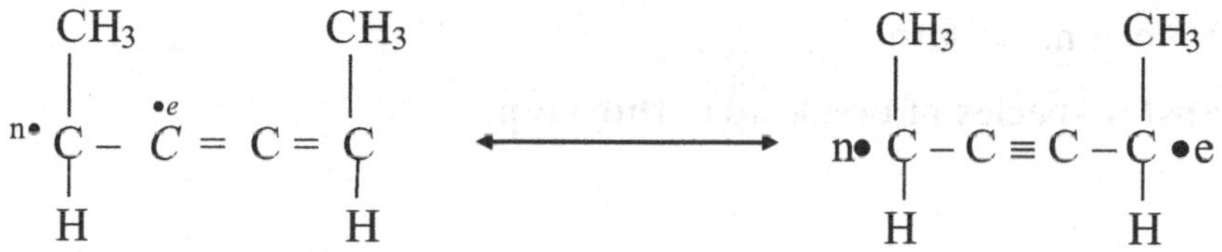

$$
\underset{\text{(II)}}{n\bullet\underset{\underset{\displaystyle H}{|}}{\overset{\overset{\displaystyle CH_3}{|}}{C}} - \overset{\bullet e}{C} = C = \underset{\underset{\displaystyle H}{|}}{\overset{\overset{\displaystyle CH_3}{|}}{C}}} \quad\longleftrightarrow\quad \underset{\text{(I)}}{n\bullet\underset{\underset{\displaystyle H}{|}}{\overset{\overset{\displaystyle CH_3}{|}}{C}} - C \equiv C - \underset{\underset{\displaystyle H}{|}}{\overset{\overset{\displaystyle CH_3}{|}}{C}}\bullet e}
$$

Resonance stabilization phenomenon $\qquad$ 8.123

Nucleo-free-radically, it cannot be polymerized due to presence of transfer species of the first kind of the first type as shown below.

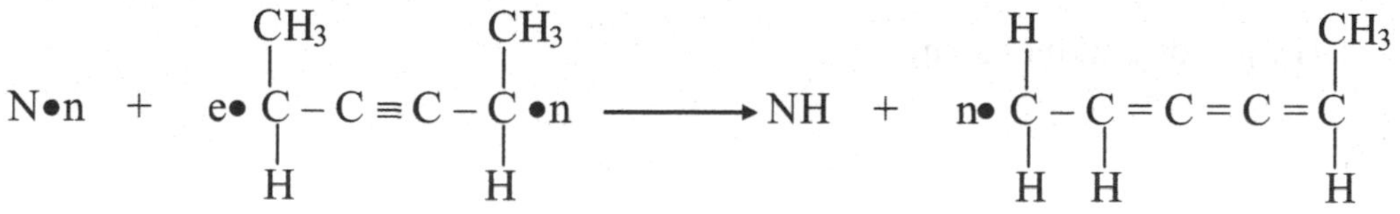

$$
N\bullet n \;+\; e\bullet\underset{\underset{\displaystyle H}{|}}{\overset{\overset{\displaystyle CH_3}{|}}{C}} - C \equiv C - \underset{\underset{\displaystyle H}{|}}{\overset{\overset{\displaystyle CH_3}{|}}{C}}\bullet n \;\longrightarrow\; NH \;+\; n\bullet\underset{\underset{\displaystyle H}{|}}{\overset{\overset{\displaystyle H}{|}}{C}} - \underset{\underset{\displaystyle H}{|}}{C} = C = C = \underset{\underset{\displaystyle H}{|}}{\overset{\overset{\displaystyle CH_3}{|}}{C}}
$$

Transfer species of first kind and first type $\qquad$ 8.124

Electro-free-radically, the followings are to be expected.

$$\underset{H}{\overset{CH_3}{E-(\overset{|}{\underset{|}{C}}-C\equiv C-\overset{CH_3}{\underset{|}{C}})_n}}-\overset{CH_3}{\underset{H}{\overset{|}{C}}}-C\equiv C-\overset{CH_3}{\underset{H}{\overset{|}{C}}}\bullet e \longrightarrow \underset{H}{\overset{CH_3}{E-(\overset{|}{\underset{|}{C}}-C\equiv C-\overset{CH_3}{\underset{H}{\overset{|}{C}}})_n}}-\overset{CH_3}{\underset{H}{\overset{|}{C}}}-C\equiv C-\overset{H}{\underset{H}{\overset{|}{C}}}=\overset{H}{C}$$

(I) $+\ \ H\bullet e$

Cis-placed dimethyl 1,4-Cumulene   8.125

$$E-(\overset{CH_3}{\underset{H}{C}}-C\equiv C-\overset{H}{\underset{CH_3}{C}})_n-\overset{CH_3}{\underset{H}{C}}-C\equiv C-\overset{H}{\underset{CH_3}{C}}\bullet e \longrightarrow E-(\overset{CH_3}{\underset{H}{C}}-C\equiv C-\overset{H}{\underset{CH_3}{C}})_n-\overset{CH_3}{\underset{H}{C}}-C\equiv C-\overset{H}{\underset{H}{C}}=\overset{H}{\underset{H}{C}}$$

(I) $+\ \ H\bullet e$

Trans-placed dimethyl 1,4-Cumulene   8.126

$$N\bullet n \ +\ e\bullet \overset{\overset{CH_3}{\overset{|}{CH}}}{\underset{\underset{CH_3}{\underset{|}{CH}}}{C}}-C\bullet n \longrightarrow NH \ +\ n\bullet \overset{CH_3}{C}=C=C=\overset{H}{\underset{CH_3}{C}}$$

(IV)-Not a mono-form

Transfer species of first kind of third type   8.127

$$E-(\overset{\overset{CH_3}{\overset{|}{CH}}}{\underset{\underset{CH_3}{\underset{|}{CH}}}{C}}-C)_n-\overset{\overset{CH_3}{\overset{|}{CH}}}{\underset{\underset{CH_3}{\underset{|}{CH}}}{C}}-C\bullet e \longrightarrow E-(\overset{\overset{CH_3}{\overset{|}{CH}}}{\underset{\underset{CH_3}{\underset{|}{CH}}}{C}}-C)_n-\overset{\overset{CH_3}{\overset{|}{CH}}}{\underset{\underset{CH_3}{\underset{|}{CH}}}{C}}-C\equiv C \ +\ H\bullet e$$

(IV) Not a mono-form   8.128

When (IV) undergoes molecular rearrangement, the followings are obtained.

$$
\begin{array}{ccccc}
\underset{\underset{\underset{|}{CH}}{\overset{CH_3}{|}}}{n\bullet\ C} - \underset{\underset{\underset{CH_3}{|}}{CH}}{C\ \bullet e} & \longrightarrow & \underset{\underset{\underset{\underset{CH_3}{|}}{CH}}{\overset{CH_3}{|}}}{n\bullet\ C} = \underset{}{C\ \bullet e} & \longrightarrow & \underset{}{C} \equiv \underset{\underset{CH_3}{|}}{C} - \overset{\overset{H}{|}}{C} = \overset{\overset{H}{|}}{C}
\end{array}
$$

(IV)   (V)   8.129

Whether rearrangement is favored or not, the two mono-forms already identified are the only existing mono-forms. Invariably, what is involved in polymerization of these monomers is the 1,4-activated mono-form.

With etheric groups, the followings are obtained, starting with the most unstable first member.

$$
\begin{array}{cccc}
\overset{\overset{H}{|}}{\underset{\underset{H}{|}}{C}} \overset{1}{=} C \overset{2}{=} C \overset{3}{=} \overset{\overset{H}{|}}{\underset{\underset{H}{|}}{C}}{}^{4} & \longrightarrow & \ominus\overset{\overset{H}{|}}{\underset{\underset{H}{|}}{C}}{}^{1} - \underset{\underset{CH_2}{\|}}{C}\oplus{}^{2} & ; & \ominus\underset{\underset{\underset{H}{|}}{CH}}{C}{}^{2} - \underset{}{C}\oplus{}^{3}
\end{array}
$$

(I) (Nucleophile)   (II) (Nucleophile)

[Impossible existence]

$$
\begin{array}{ccc}
\ominus\overset{\overset{H}{|}}{\underset{\underset{H}{|}}{C}}{}^{4} - \underset{\underset{\underset{\underset{H}{|}}{O}}{\underset{CH}{|}}}{C}\oplus{}^{3} & ; & \oplus\overset{\overset{H}{|}}{\underset{\underset{\underset{H}{|}}{O}}{C}} - C \equiv C - \overset{\overset{H}{|}}{\underset{\underset{H}{|}}{C}}\ominus
\end{array}
$$

(III) (Nucleophile)   (IV) (Nucleophile)   8.130

(I) cannot exist due to electrostatic forces of repulsion. (III) is the least nucleophilic of all the centers. Hence, it undergoes resonance stabilization phenomenon as follows.

$$
\underset{\underset{H}{|}}{\overset{\overset{H}{|}}{n\bullet\ C}} - \overset{\bullet e}{C} = C = \underset{\underset{H}{|}}{\overset{\overset{OH}{|}}{C}} \underset{\underset{phenomenon}{\xrightarrow{\hspace{1cm}}}}{\overset{Resonance\ Stabilization}{\xleftarrow{\hspace{1cm}}}} \underset{\underset{H}{|}}{\overset{\overset{H}{|}}{n\bullet\ C}} - C \equiv C - \underset{\underset{H}{|}}{\overset{\overset{OH}{|}}{C}}\bullet e
$$

RESONANCE STABILIZATION PHENOMENON   8.131

626

It is (IV) that is used for polymerization as shown below radically.

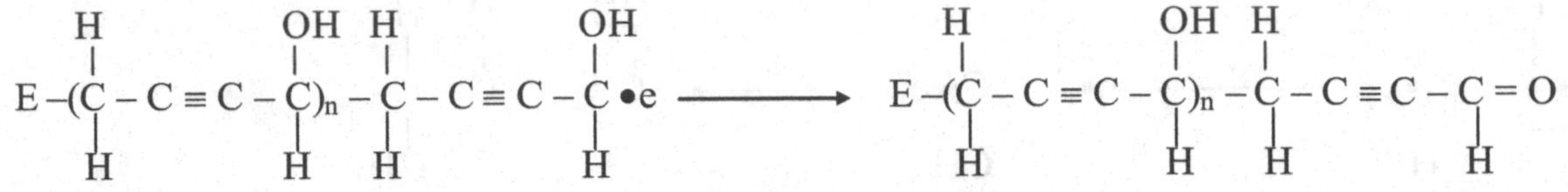

$$E-(\overset{\overset{\displaystyle H}{|}}{\underset{\underset{\displaystyle H}{|}}{C}}-C\equiv C-\overset{\overset{\displaystyle OH}{|}}{\underset{\underset{\displaystyle H}{|}}{C}})_n-\overset{\overset{\displaystyle H}{|}}{\underset{\underset{\displaystyle H}{|}}{C}}-C\equiv C-\overset{\overset{\displaystyle OH}{|}}{\underset{\underset{\displaystyle H}{|}}{C}}\bullet e \longrightarrow E-(\overset{\overset{\displaystyle H}{|}}{\underset{\underset{\displaystyle H}{|}}{C}}-C\equiv C-\overset{\overset{\displaystyle OH}{|}}{\underset{\underset{\displaystyle H}{|}}{C}})_n-\overset{\overset{\displaystyle H}{|}}{\underset{\underset{\displaystyle H}{|}}{C}}-C\equiv C-\overset{\overset{\displaystyle OH}{|}}{\underset{\underset{\displaystyle H}{|}}{C}}=O$$

$$+ \; H\bullet e \qquad\qquad 8.132$$

The transfer species involved is the same that is used for molecular rearrangement as shown below. This is the state it may largely exist in at STP or higher operating conditions.

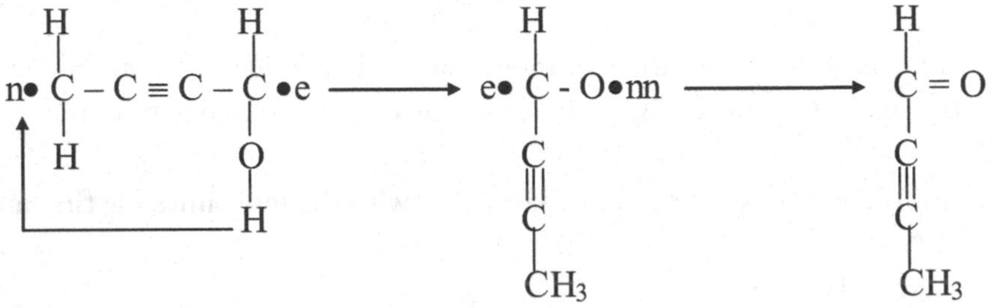

$$n\bullet \overset{\overset{\displaystyle H}{|}}{\underset{\underset{\displaystyle H}{|}}{C}}-C\equiv C-\overset{\overset{\displaystyle H}{|}}{\underset{\underset{\displaystyle O}{|}}{C}}\bullet e \longrightarrow e\bullet \overset{\overset{\displaystyle H}{|}}{\underset{\underset{\displaystyle CH_3}{||}}{C}}-O\bullet nn \longrightarrow \overset{\overset{\displaystyle H}{|}}{\underset{\underset{\displaystyle CH_3}{||}}{C}}=O$$

$$\text{An Aldehyde} \qquad 8.133$$

An aldehyde with ***transfer species of the first kind of the fifth type*** is obtained. This can enolize back to alcohol depending on the operating conditions. It may be easier to do this with a ketone, because of the following order.

$$H_5C_2-CH=CH-\;>\;H_5C_2-\;>\;H_3C-CH=CH-\;>\;H_3C-\;\geq\;H_3C-C\equiv C-\;>$$

Order of Radical-pushing capacity $\qquad\qquad$ 8.134

One knows what to expect, when OH group is changed to OR. The molecular rearrangement above will not take place. One also knows what to expect when the OH group is changed to NH_2. Aldimines begin to appear.

When only one H is replaced by a radical pulling group, the character of the monomer changes. Only special cases will be considered here for 1,4 cumulenes.

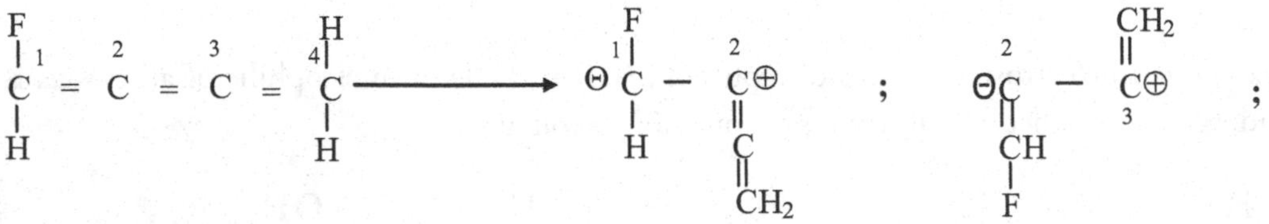

$$\overset{\overset{\displaystyle F}{|}}{\underset{\underset{\displaystyle H}{|}}{\underset{1}{C}}}=\underset{2}{C}=\underset{3}{C}=\overset{\overset{\displaystyle H}{|}}{\underset{\underset{\displaystyle H}{|}}{\underset{4}{C}}}\longrightarrow \ominus\overset{\overset{\displaystyle F}{|}}{\underset{\underset{\displaystyle H}{|}}{\underset{1}{C}}}-\underset{\underset{\underset{\displaystyle CH_2}{||}}{C}}{\overset{2}{C}}\oplus \;\; ; \;\; \ominus\underset{\underset{\underset{\displaystyle F}{|}}{CH}}{\overset{2}{C}}-\underset{3}{\overset{\overset{\displaystyle CH_2}{||}}{C}}\oplus \;\; ;$$

(I) (Impossible
Existence)

(II) (Impossible existence)

$$\underset{H}{\overset{H}{\underset{|}{\Theta C}}} \;-\; \underset{\underset{F}{\overset{|}{CH}}}{\overset{\overset{3}{\oplus}}{\underset{\parallel}{\underset{\parallel}{C}}}} \qquad ; \qquad \underset{H}{\overset{F}{\underset{|}{\Theta C}}} \;-\; C \;\equiv\; C \;-\; \underset{H}{\overset{H}{\underset{|}{C\oplus}}}$$

(III)   (IV) (Impossible
                              Existence)   8.135

Chargedly, the monomer cannot be activated.

$$\underset{H}{\overset{F}{\underset{|}{C}}} = C = C = \underset{H}{\overset{F}{\underset{|}{C}}} \longrightarrow \underset{H}{\overset{F}{\underset{|}{\Theta C}}} - \underset{\underset{F}{\overset{|}{CH}}}{\overset{\oplus}{\underset{\parallel}{C}}} \quad ; \quad \overset{\oplus}{\underset{\parallel}{C}} - \underset{F}{\overset{\overset{F}{|}}{\underset{\parallel}{\underset{|}{CH}}}} \overset{CH_2}{\underset{C\Theta}{}} \quad ;$$

   (Symmetric)   (II) Impossible existence

                                   (I) (Impossible existence)

$$\underset{H}{\overset{F}{\underset{|}{\Theta C}}} - \underset{\underset{F}{\overset{|}{CH}}}{\overset{\oplus}{\underset{\parallel}{C}}} \quad ; \quad \underset{H}{\overset{F}{\underset{|}{\Theta C}}} - C \equiv C - \underset{H}{\overset{F}{\underset{|}{C\oplus}}}$$

(III) Impossible   (IV) (Impossible
       Existence   Existence)   8.136

None will favor activation chargedly. All these can only be activated radically.

$$\underset{CF_3}{\overset{H}{\underset{|}{\overset{1}{C}}}} = \overset{2}{C} = \overset{3}{C} = \underset{H}{\overset{H}{\underset{|}{\overset{4}{C}}}} \longrightarrow \underset{CF_3}{\overset{H}{\underset{|}{\overset{1}{\Theta C}}}} - \underset{\underset{CH_2}{\overset{\parallel}{C}}}{\overset{2}{\underset{\parallel}{C\oplus}}} \quad ; \quad \overset{2}{\underset{\underset{CF_3}{\overset{|}{CH}}}{\underset{\parallel}{\oplus C}}} - \underset{3}{\overset{CH_2}{\underset{C\Theta}{\parallel}}} \quad ;$$

                                   (I) (Nucleophile)   (II) (Nucleophile)

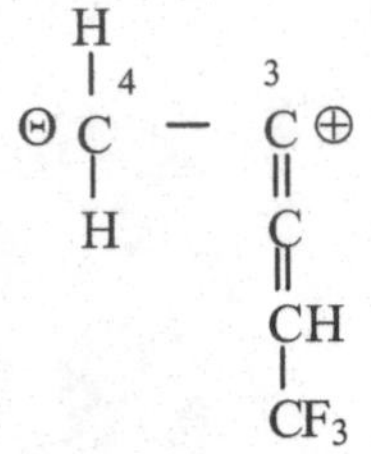

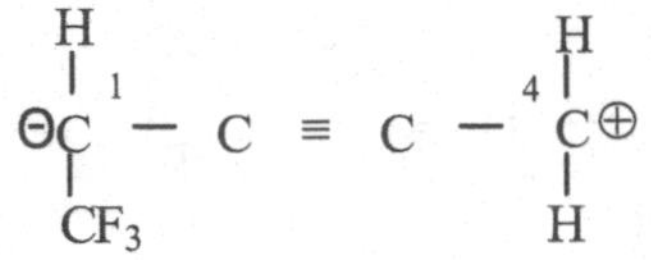

$$\Theta\overset{H}{\underset{H}{\overset{|}{\underset{|}{C}}}}\!\!{}^{4}\!-\overset{3}{\underset{\underset{CF_3}{\overset{|}{CH}}}{\overset{||}{\overset{|}{C}}}}\!\oplus \qquad ; \qquad \Theta\overset{H}{\underset{CF_3}{\overset{|}{C}}}\!\!{}^{1}\!-C\equiv C-\overset{H}{\underset{H}{\overset{|}{\underset{|}{C}}}}\!\!{}^{4}\!\oplus$$

$$\qquad\qquad\qquad\qquad\qquad\qquad\qquad\qquad\text{(IV) (Nucleophile)}$$

$$\text{(III) (Nucleophile)} \qquad\qquad\qquad\qquad\qquad\qquad\qquad 8.137$$

The least nucleophilic center is (I). This is resonance stabilized to (IV) only radically.

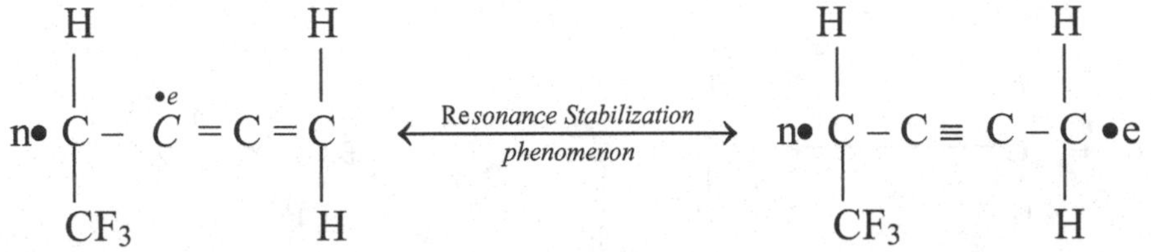

$$n\bullet\overset{H}{\underset{CF_3}{\overset{|}{C}}}-\overset{\bullet e}{\overset{H}{\underset{H}{\overset{|}{\underset{|}{C}}}}}=C=\overset{H}{\underset{H}{\overset{|}{\underset{|}{C}}}} \quad \overset{Resonance\ Stabilization}{\underset{phenomenon}{\longleftrightarrow}} \quad n\bullet\overset{H}{\underset{CF_3}{\overset{|}{C}}}-C\equiv C-\overset{H}{\underset{H}{\overset{|}{\underset{|}{C}}}}\bullet e$$

RESONANCE STABILIZATION PHENOMENON 8.138

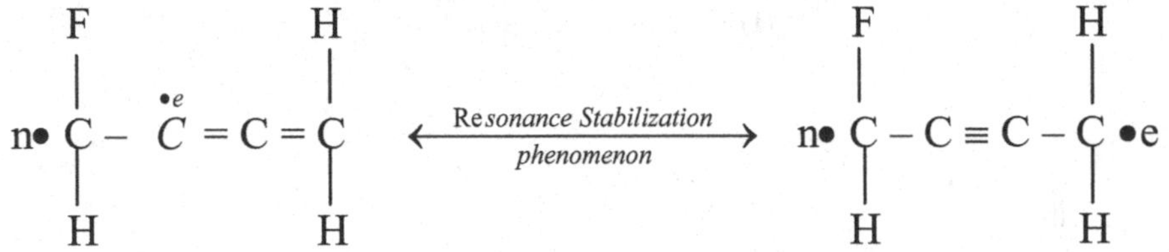

$$n\bullet\overset{F}{\underset{H}{\overset{|}{\underset{|}{C}}}}-\overset{\bullet e}{\overset{H}{\underset{H}{\overset{|}{\underset{|}{C}}}}}=C=\overset{H}{\underset{H}{\overset{|}{\underset{|}{C}}}} \quad \overset{Resonance\ Stabilization}{\underset{phenomenon}{\longleftrightarrow}} \quad n\bullet\overset{F}{\underset{H}{\overset{|}{\underset{|}{C}}}}-C\equiv C-\overset{H}{\underset{H}{\overset{|}{\underset{|}{C}}}}\bullet e$$

RESONANCE STABILIZATION PHENOMENON 8.139

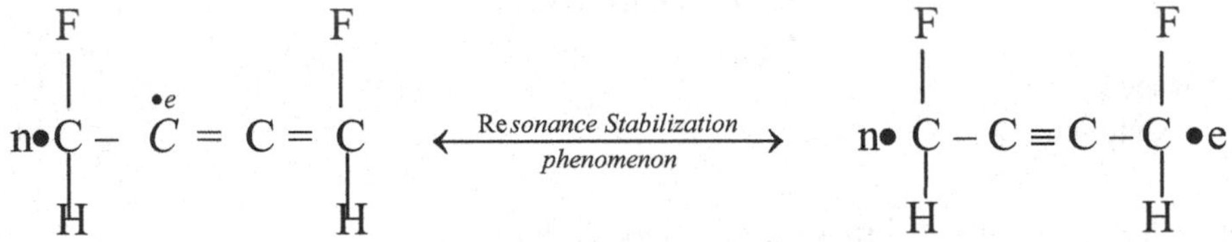

$$n\bullet\overset{F}{\underset{H}{\overset{|}{\underset{|}{C}}}}-\overset{\bullet e}{\overset{F}{\underset{H}{\overset{|}{\underset{|}{C}}}}}=C=\overset{F}{\underset{H}{\overset{|}{\underset{|}{C}}}} \quad \overset{Resonance\ Stabilization}{\underset{phenomenon}{\longleftrightarrow}} \quad n\bullet\overset{F}{\underset{H}{\overset{|}{\underset{|}{C}}}}-C\equiv C-\overset{F}{\underset{H}{\overset{|}{\underset{|}{C}}}}\bullet e$$

Resonance stabilization phenomenon 8.140

On the whole, based on the experience of the past, all the monomers above are beginning to loose their strong nucleophilic characters. For the monomer of Equation 8.138, the route not natural to it is favored, while the route natural to it cannot be favored. This is almost like the case of vinyl acetate, but with a difference. Such monomers unlike vinyl acetate type should not be;

Now, consider the case shown below.

$$\underset{H}{\overset{CF_3}{\underset{|}{\overset{|}{C}}}} \overset{2}{=} C \overset{3}{=} C \overset{4}{=} \underset{H}{\overset{CH_3}{\underset{|}{\overset{|}{C}}}} \longrightarrow \underset{H}{\overset{CF_3}{\underset{|}{\overset{|}{\Theta C}}}} \overset{1}{-} \underset{\underset{CH}{\overset{\|}{C}}}{\overset{2}{C}\oplus} \quad ; \quad \underset{CH}{\overset{2}{\Theta C}} \overset{}{-} \underset{CF_3}{\overset{\overset{CH_3}{\overset{|}{CH}}}{\overset{\|}{C}}\oplus_3} \quad ;$$

$$(I) \quad CH_3 \qquad\qquad (II)$$

$$\underset{H}{\overset{CH_3}{\underset{|}{\overset{|}{\Theta C}}}} \overset{3}{-} \underset{\underset{CF_3}{\overset{\|}{\overset{CH}{\overset{\|}{C}}}}}{\overset{}{C}\oplus} \quad ; \quad \underset{H}{\overset{CF_3}{\underset{|}{\overset{|}{\Theta C}}}} \overset{1}{-} C \equiv C \overset{4}{-} \underset{H}{\overset{CH_3}{\underset{|}{\overset{|}{C}}}\oplus}$$

$$(III) \quad CF_3 \qquad\qquad\qquad (IV) \qquad\qquad\qquad\qquad 8.141$$

Only (I) can be used, since it is the least nucleophilic center.

$$\underset{H}{\overset{CF_3}{\underset{|}{\overset{|}{n\bullet C}}}} - \underset{H}{\overset{\bullet e}{\overset{}{C}}} = C = \underset{H}{\overset{CH_3}{\underset{|}{\overset{|}{C}}}} \quad \underset{phenomenon}{\overset{Resonance\ Stabilization}{\longleftrightarrow}} \quad \underset{H}{\overset{CF_3}{\underset{|}{\overset{|}{n\bullet C}}}} - C \equiv C - \underset{H}{\overset{CH_3}{\underset{|}{\overset{|}{C}}}}\bullet e$$

RESONANCE STABILIZATION PHENOMENON 8.142

No route can be favored by it, unless when the groups are trans-placed. For this monomer, like electrophiles, only the use of special initiators for the trans-form, will polymerize them in the route not natural to them i.e., nucleo-free-radically. These are nucleophiles but behave like electrophiles.

With the use of COOR, $CONH_2$, COR, and CN types of groups, we are going to see if Electrophiles exist with 1,4-Cumulenes recalling Equations 8.119 to 8.121.

$$\underset{\underset{\underset{CH_3}{\overset{|}{O}}}{\overset{|}{\underset{}{C=O}}}}{\overset{H}{\underset{|}{\overset{|}{C}}}} \overset{2}{=} C \overset{3}{=} C \overset{4}{=} \underset{H}{\overset{H}{\underset{|}{\overset{|}{C}}}} \longrightarrow \underset{\underset{CH_3}{\overset{|}{O}}}{\overset{H}{\underset{\overset{|}{C=O}}{\overset{|}{\Theta C}}}} \overset{1}{-} \underset{CH_2}{\overset{\overset{2}{C}\oplus}{\overset{\|}{C}}} \quad ; \quad \underset{\underset{CH_3}{\overset{|}{O}}}{\overset{2}{\underset{\overset{|}{C=O}}{\underset{\overset{|}{CH}}{\overset{\|}{\Theta C}}}}} \overset{}{-} \overset{\overset{CH_2}{\overset{\|}{}}}{\overset{}{C}\oplus_3} \quad ;$$

$$(I)\ (Electrophile) \qquad\qquad CH_3$$

$$(II)\ (Nucleophile)$$

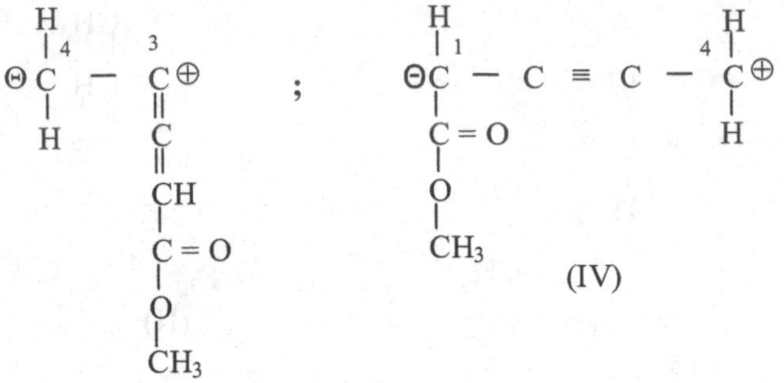

$$\overset{H}{\underset{H}{\overset{|}{\underset{|}{\Theta C}}}} \overset{4}{} - \overset{3}{\underset{\underset{\underset{\underset{\underset{CH_3}{|}}{O}}{\underset{|}{C=O}}}{\underset{\|}{CH}}}{\overset{\oplus}{\underset{\|}{C}}}} \quad ; \quad \overset{H}{\underset{\underset{\underset{CH_3}{|}}{\overset{|}{\underset{\|}{C=O}}}}{\overset{|}{\Theta C}}} \overset{1}{} - C \equiv C - \overset{H}{\underset{H}{\overset{4}{\underset{|}{\overset{|}{C}}}}}{\overset{\oplus}{}}$$

(III) (Nucleophile)   (IV)   8.143

The least nucleophilic center is (I). It is from (I), (IV) is obtained via resonance stabilization phenomenon only radically.

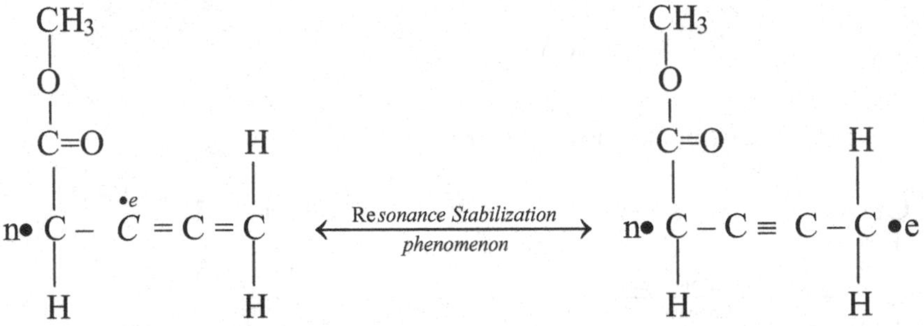

RESONANCE STABILIZATION PHENOMENON   8.144

The 1,2-mono-form is indeed an electrophile. Recent world data have made so many of these unique observations[4-8], which could not be explained, because nothing is known about all the new concepts in the NEW FRONTIERS. Indeed, all the current-day knowledge about cumulenes are sufficient enough to make any human in the field go crazy. That is the way it has been since antiquity. That the 1,2-mono-form is electrophilic clearly indicates that its cumulatively placed bonds can be changed by resonance stabilization phenomenon and transformed to the 1,4-mono-form. All along so far, we can see how NATURE operates based on very new concepts too countless to list; new concepts which are bound to affect the foundations on which humanity has been existing with since antiquity. Cations, Anions, and other real charges do not exist with the families of Cumulenes, Acetylenes and more in a sane world. They only exist in the insane world. Even compounds are not properly named. For example how can the compound shown below be wrongly named?[4]

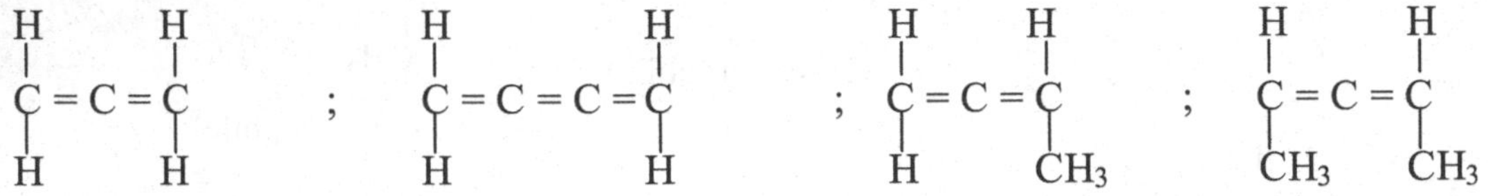

Propa-**diene**   Buta-**triene**   Buta-**1,2-diene**   Penta-**2,3-diene**

(I)   (II)   (III)   (IV)   8.145

None of them is a DIENE OR TRIENE for which the names above are wrong and absolute nonsense and the reason is obvious. What CHEMISTRY is, has never been known since antiquity. Where the numbers in (III) and (IV) are coming from cannot be explained. Based on the current developments in the NEW FRONTIERS, we already know what are DIENES, TRIENES, TETRATRIENES etc. Then, what are the real names of the above? Why the use of propa, buta, penta? Is it because of the number of carbon atoms present, when we know when and where to apply those types of nomenclature? This is absolute madness and a state of CONFUSION! Even then, from that state emerges ORDERLINESS.

The 1,4-mono-form will readily favor its natural route as a male when nucleo-free-radical initiators in particular of the paired types are involved. When the initiator is weak with respect to this type of monomer which is unique in character, molecular rearrangement will definitely take place to produce a stronger Electrophile as shown below.

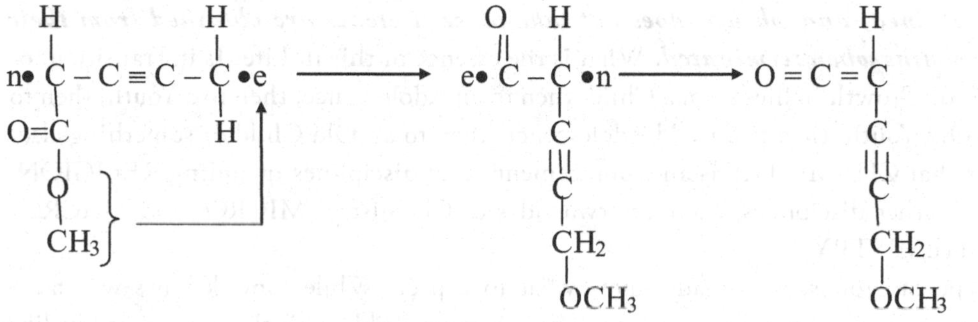

(IV)　　　　　　　(V)　　　　　(VI) KETENE　　　　8.146

8.147

With electro-free-radical initiators, the center of attack will not be the second C = C center but the C =O center. Nucleo-free-radically after the molecular rearrangement, the followings are obtained.

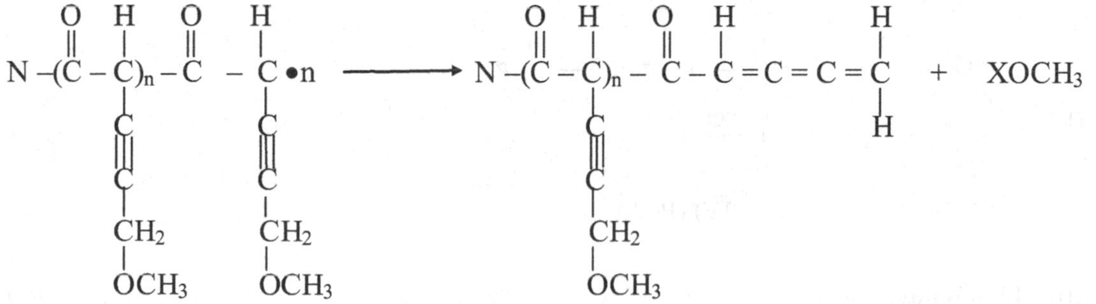

TRANSFER SPECIES OF THE FIRST KIND OF THE *FIFTH TYPE*　　8.148

The polymerization will be highly favored particularly when paired nucleo-free-radical initiator is used. As shown above there must be a counter center to receive OCH_3 group to give an additional stable molecule. Notice that the dead terminal polymer produced is the same as the original monomer. In the absence of molecular rearrangement, the followings are obtained for its growing polymer chain. Recall, that for such resonance stabilized males, one has said that molecular rearrangement cannot take place, in particular with respect to its 1,2-mono-form.

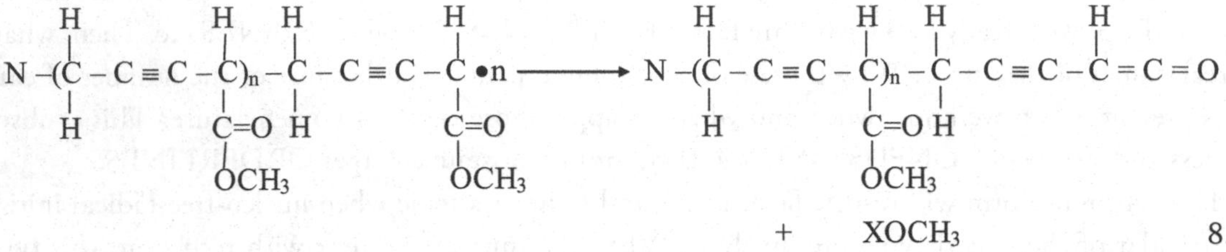

$$N-(C-C\equiv C-C)_n-C-C\equiv C-C\bullet n \longrightarrow N-(C-C\equiv C-C)_n-C-C\equiv C-C=C=O$$

$$+ \quad XOCH_3 \qquad\qquad 8.149$$

Without any doubt, one can observe that these are MALES or ELECTROPHILES. *When rearrangement takes place with these families of monomers carrying COOR and CONH₂, Cumulenes are obtained, and when it does not take place, Ketenes are obtained from their growing polymer chains when terminated.* What is the essence of this in Life? Is it Transition, or "To be born again" or Growth in life- from a Child, then to an Adolescence, then to a Youth, then to an Adult, then to Old Youth, then to an Old Adolescence, then to an Old Child or something else? One can see so far that CHEMISTRY is an embracement of all disciplines including RELIGION. By the way, like all other disciplines, there are two sides of Chemistry- MICRO- and MACRO-MOLECULAR CHEMISTRY.

With COR type of groups, we already know what to expect. While some ketenes will favor ALTERNATING placement from a single monomer, this is not possible with this monomer just like with acrolein. At best, only random placement can be obtained.

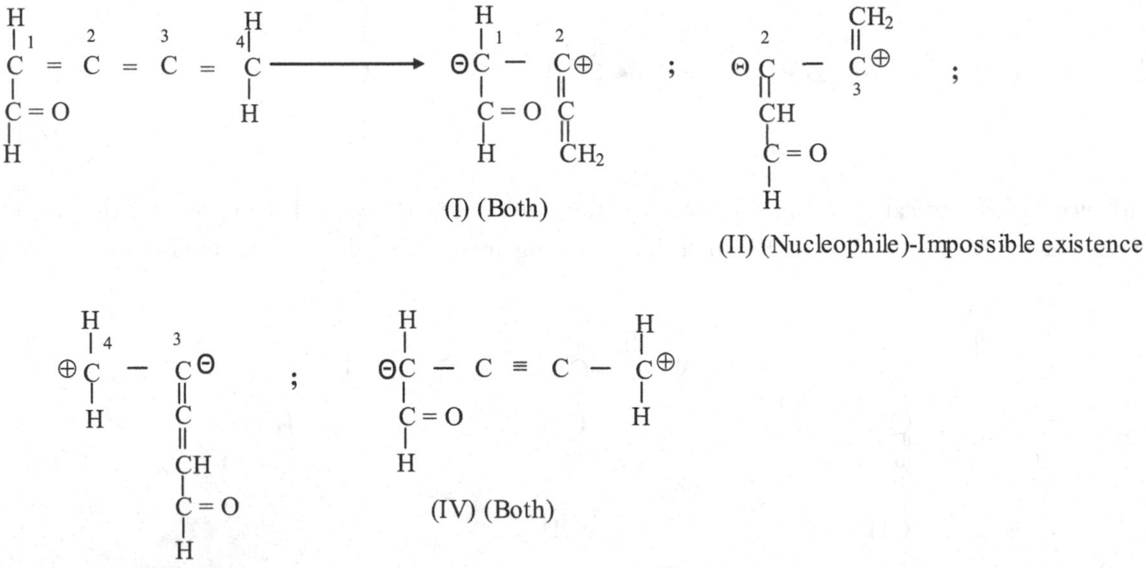

(I) (Both)

(II) (Nucleophile)-Impossible existence

(III) (Electrophile)

(IV) (Both)

$$8.150$$

(I) as in other similar cases is the least nucleophilic center. From (I), (IV) is obtained as follows.

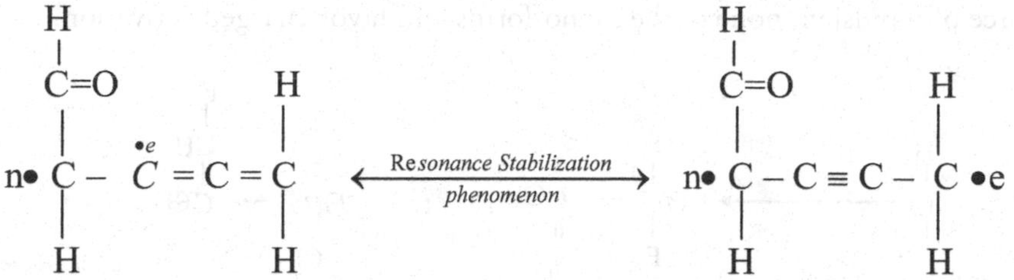

$$
\underset{\text{WEAK ELECTROPHILE}}{
n\bullet \overset{\displaystyle \overset{H}{|} \atop \overset{C=O}{|}}{C} - \overset{\bullet e}{C} = C = \overset{\displaystyle \underset{|}{H}}{C} }
\quad
\underset{\text{RESONANCE Stabilization phenomenon}}{\longleftrightarrow}
\quad
\underset{\text{WEAK ELECTROPHILE}}{
n\bullet \overset{\displaystyle \overset{H}{|} \atop \overset{C=O}{|}}{C} - C \equiv C - \overset{\displaystyle \underset{|}{H}}{C} \bullet e }
$$

RESONANCE STABILIZATION PHENOMENON 8.151

This is similar to the case of acrolein, where at very low temperatures, the C= O center is exclusively involved electro-free-radically, and at higher temperatures, random copolymers are produced positively and not cationically. Electro-free-radically this may be possible, but not when the carrier of the electro-free-radical is an ionic metal. Nucleo-free-radically only the homopolymer is obtained as shown below.

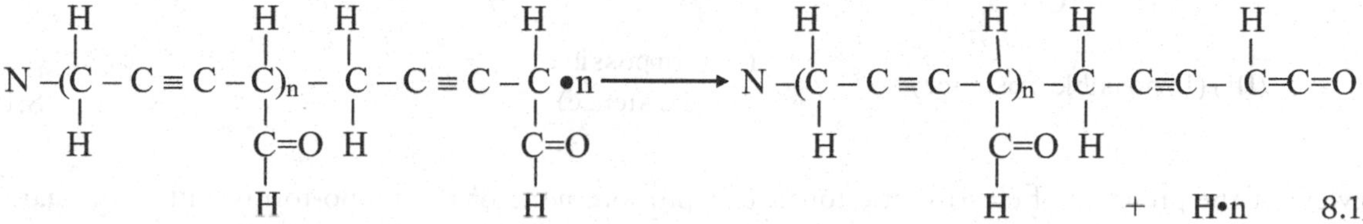

$$
N -(C - C \equiv C - C)_n - C - C \equiv C - C \bullet n \longrightarrow N -(C - C \equiv C - C)_n - C - C \equiv C - C = C = O \quad + \quad H\bullet n \quad 8.152
$$

Transfer species of the first kind of the first type as opposed to the second kind was released to kill the growing polymer chain, since the route is natural to the monomer, to give a ketene. If H.n is difficult to be released being very great in capacity compared to R.n, then living polymer is produced. ***Indeed, H.n cannot be released as transfer species of the first kind shown above.***

For the extreme cumulene, that is, perfluorocumulene, the followings are obtained.

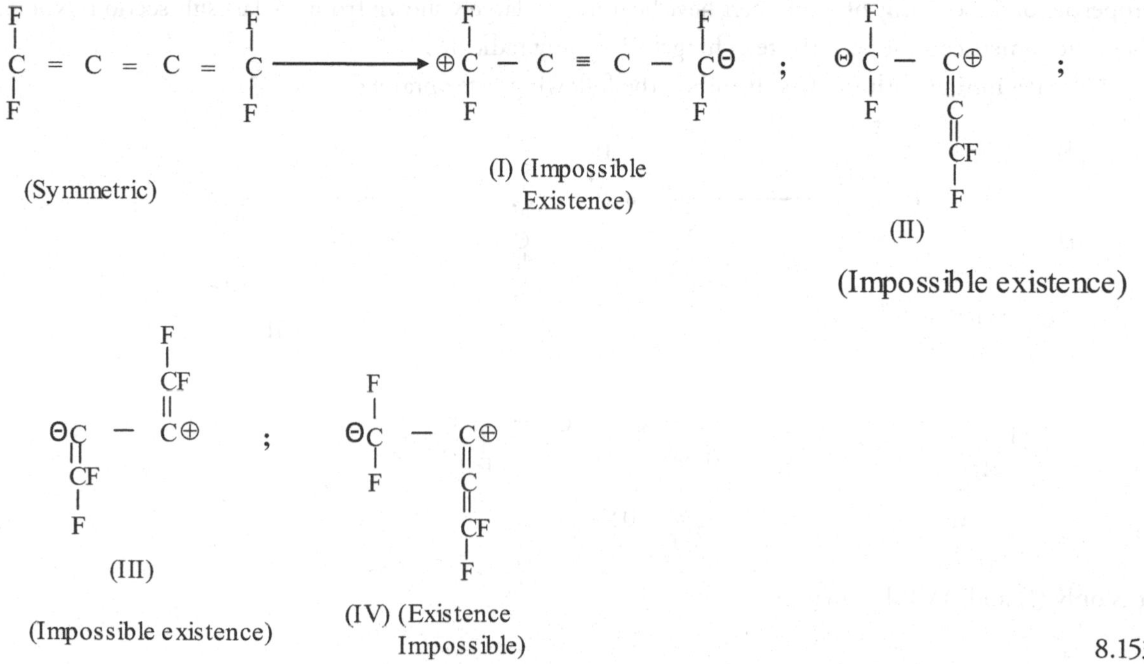

$$
\underset{\text{(Symmetric)}}{
\overset{\displaystyle \overset{F}{|}}{\underset{\underset{F}{|}}{C}} = C = C = \overset{\displaystyle \overset{F}{|}}{\underset{\underset{F}{|}}{C} }
}
\longrightarrow
\underset{\substack{\text{(I) (Impossible}\\\text{Existence)}}}{
\oplus\overset{\displaystyle \overset{F}{|}}{\underset{\underset{F}{|}}{C}} - C \equiv C - \overset{\displaystyle \overset{F}{|}}{\underset{\underset{F}{|}}{C}}\ominus
}
\quad ; \quad
\underset{\substack{\text{(II)}\\\text{(Impossible existence)}}}{
\ominus\overset{\displaystyle \overset{F}{|}}{\underset{\underset{F}{|}}{C}} - \overset{C\oplus}{\underset{\displaystyle \underset{\underset{F}{|}}{CF}}{\|\atop C\|}}
}
\quad ;
$$

$$
\underset{\substack{\text{(III)}\\\text{(Impossible existence)}}}{
\ominus\overset{\displaystyle \overset{CF}{\|} \atop \underset{\underset{F}{|}}{CF}}{C} - C\oplus
}
\quad ; \quad
\underset{\substack{\text{(IV) (Existence}\\\text{Impossible)}}}{
\ominus\overset{\displaystyle \overset{F}{|}}{\underset{\underset{F}{|}}{C}} - \overset{C\oplus}{\underset{\displaystyle \underset{\underset{F}{|}}{CF}}{\|\atop C\|}}
}
\quad\quad 8.153
$$

Due to electrostatic force of repulsion, none of the mono-forms will favor charged activation.

$$
\underset{\underset{F}{|}}{\overset{\overset{F}{|}}{\underset{1}{C}}} = \overset{2}{C} = \overset{3}{C} = \underset{\underset{F}{|}}{\overset{\overset{H}{|}}{\underset{4}{C}}} \longrightarrow \;\; {}^{\ominus}\underset{\underset{F}{|}}{\overset{1}{C}} - \overset{2}{\underset{\underset{\underset{\underset{F}{|}}{CH}}{\parallel}}{C^{\oplus}}} \;\; ; \;\; {}^{\ominus}\overset{2}{\underset{\underset{CF_2}{\parallel}}{C}} - \overset{3}{\underset{\underset{F}{|}}{\overset{\overset{\overset{F}{|}}{CH}}{C^{\oplus}}}} \;\; ;
$$

(I) (Impossible Existence)   (II) (Impossible existence)

$$
{}^{\ominus}\overset{4}{\underset{\underset{F}{|}}{C}} - \overset{3}{\underset{\underset{\underset{CF_2}{\parallel}}{\overset{\overset{C}{\parallel}}{}}}{C^{\oplus}}} \;\; ; \;\; {}^{\ominus}\underset{\underset{F}{|}}{C} - C \equiv C - \underset{\underset{F}{|}}{\overset{\overset{H}{|}}{C^{\oplus}}}
$$

(III) (Impossible existence)   (IV) (Impossible Existence)   8.154

In view of the presence of electrostatic forces of repulsion, none of the mono-forms will favor charged activation and this we shall see very shortly in the next sub-section, noting that we have covered a large section of radical characters of this family of monomers which cannot be activated chargedly.

8.167

8.2.2. Free-radical Characters of 1,4 - cumulenes

Only few examples will be considered here since the methods of analysis of the characters and properties of these family of monomers have begun to be largely shown from the last sub-section. Notably, these monomers cannot be activated chargely, but only radically.

Now beginning with the first member, the followings are obtained.

$$
\underset{\underset{H}{|}}{\overset{\overset{H}{|}}{\underset{1}{C}}} = \overset{2}{C} = \overset{3}{C} = \underset{\underset{H}{|}}{\overset{\overset{H}{|}}{\underset{4}{C}}} \longrightarrow n\cdot\overset{4}{\underset{\underset{H}{|}}{C}} - \overset{3}{\underset{\underset{\underset{CH_2}{\parallel}}{\overset{\overset{C}{\parallel}}{}}}{C\cdot e}} \;\; ; \;\; e\cdot\underset{\underset{H}{|}}{\overset{\overset{H}{|}}{\overset{1}{C}}} - \overset{2}{\underset{\underset{\underset{CH_2}{\parallel}}{\overset{\overset{C}{\parallel}}{}}}{C\cdot n}} \;\; ;
$$

(Symmetric)   (I)   (II)

$$
n\cdot\overset{2}{\underset{\underset{CH_2}{\parallel}}{C}} - \overset{3}{\underset{}{\overset{\overset{CH_2}{\parallel}}{C}\cdot e}} \;\; ; \;\; e\cdot\underset{\underset{H}{|}}{\overset{\overset{H}{|}}{\overset{1}{C}}} - C \equiv C - \overset{4}{\underset{\underset{H}{|}}{\overset{\overset{H}{|}}{C}}}\cdot n
$$

(III)   (IV)   8.155

It is only (I) and (IV) that matter.

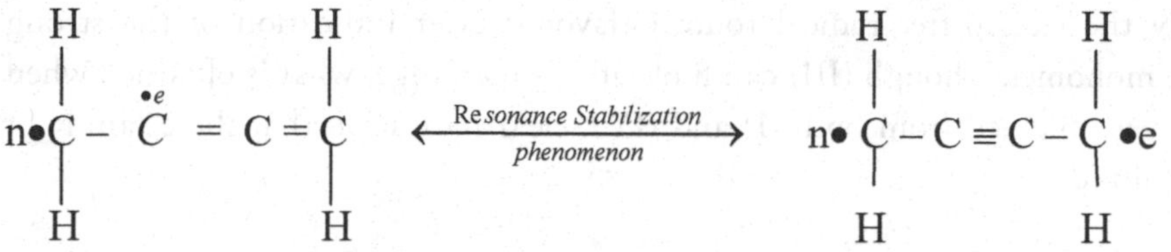

Resonance stabilization phenomenon (8.98)

It is not the 1,2 –mono-form that is used, but the 3,4 –mono-form that is used. None of them has transfer species of the first kind of the first type. Only the 2,3-mono-form has transfer species that which is not to be expected. Electro-free-radically, for the 1,4-mono-form, there is no transfer species. But nucleo-free-radically, there is transfer species of the second kind as shown below.

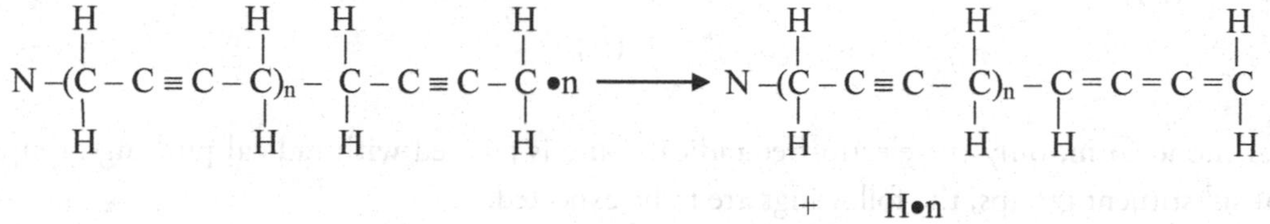

8.156

This is like ethene, for which far more drastic operating conditions may be required to release such transfer species. This will not be the case for another symmetric case with visible transfer species as shown below.

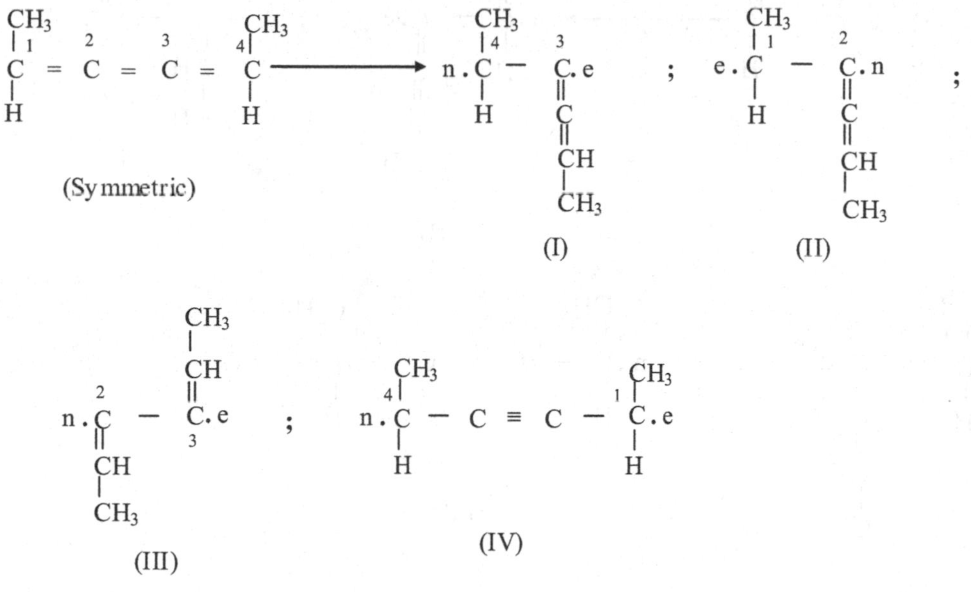

8.157

Again, like the case above, only (I) and (IV) are relevant as recalled below.

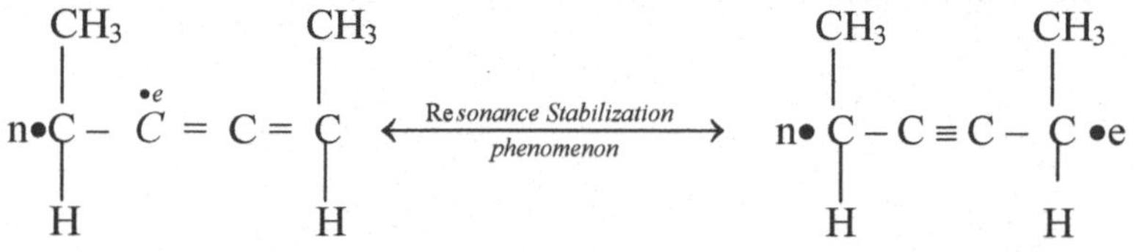

(I) (IV)

Resonance Stabilization phenomenon (8.123)

Unlike the case above, only the electro-free-radical route is favored clear indication of the strong nucleophilic character of the monomer. Though (III) can molecularly rearrange, what is obtained when it does is different from what is obtained from both (I) and (IV), clear indication that these family of monomers are resonance stabilized.

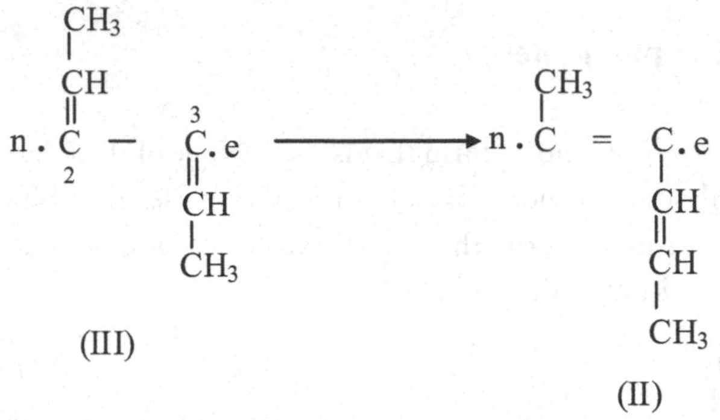

```
        CH3                              CH3
        |                                |
        CH                               CH
        ||           3                   |
   n . C   —    C . e    ———————→   n . C  =  C . e
      2         ||                        |
                CH                        CH
                |                         ||
                CH3                       CH
                                          |
       (III)                              CH3

                                         (II)                    8.158
```

With 1,4-mono-form, only the electro-free-radical route is favored with radical pushing groups. With different substituent groups, the followings are to be expected.

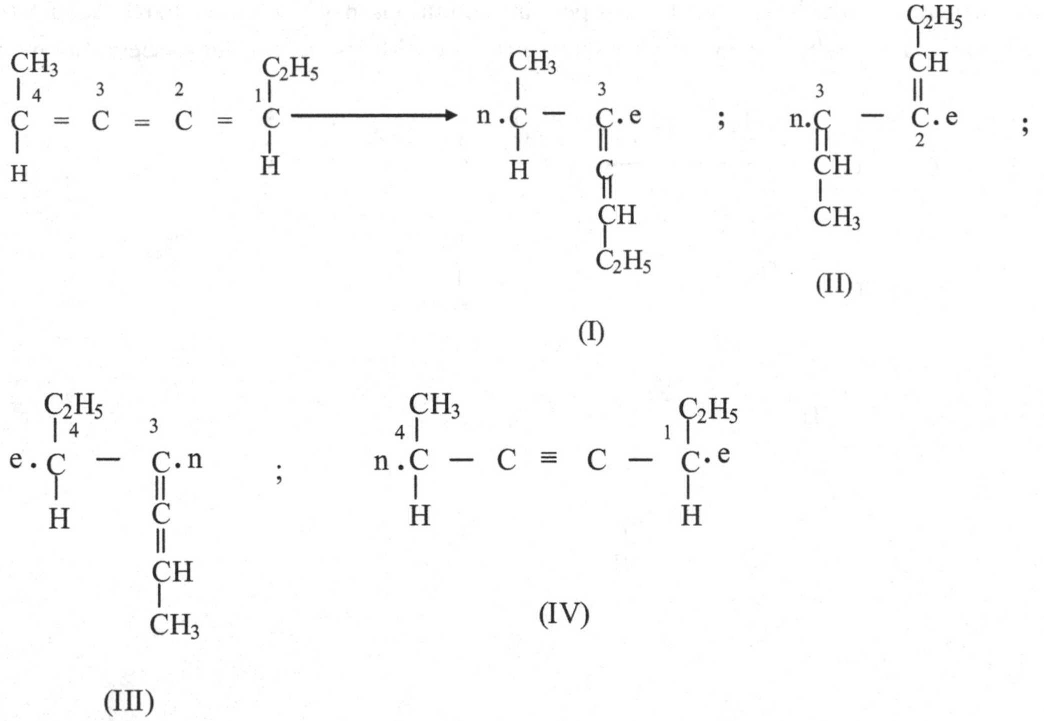

```
                                                                        C2H5
   CH3                    C2H5         CH3                                |
    | 4    3     2      1 |             |        3                        CH
    C  =  C  =  C    =    C  ——→   n . C  —  C . e    ;    n . C   —   C . e    ;
    |                     |             |     ||            ||          2
    H                     H             H     C             CH
                                              ||            |
                                              CH            CH3
                                              |
                                              C2H5                (II)

                                              (I)
```

```
     C2H5                        CH3                      C2H5
      | 4      3                 4 |                      1 |
  e . C   —   C . n      ;   n . C  —  C  ≡  C   —   C . e
      |       ||                 |                       |
      H       C                  H                       H
              ||
              CH
              |                        (IV)
              CH3

      (III)                                                       8.159
```

It is only (I) that is of relevance from which (IV) is obtained as shown below.

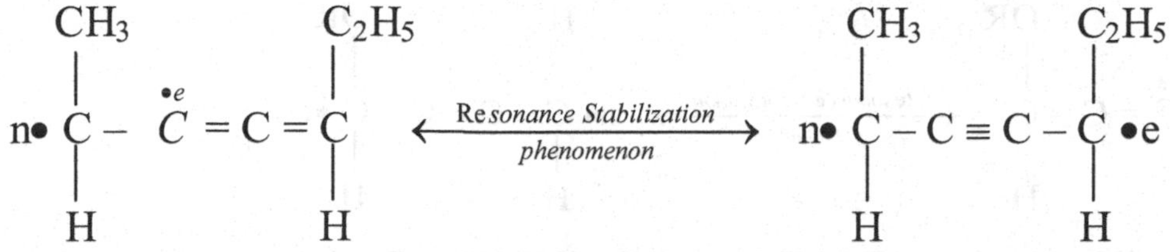

$$
n\bullet \underset{\underset{H}{|}}{\overset{\overset{CH_3}{|}}{C}} - \overset{\bullet e}{\underset{\underset{H}{|}}{\overset{\overset{C_2H_5}{|}}{C}}} = C = \underset{\underset{H}{|}}{C} \quad \xleftarrow[\text{phenomenon}]{\text{Resonance Stabilization}} \quad n\bullet \underset{\underset{H}{|}}{\overset{\overset{CH_3}{|}}{C}} - C \equiv C - \underset{\underset{H}{|}}{\overset{\overset{C_2H_5}{|}}{C}} \bullet e
$$

$$(I) \qquad\qquad\qquad\qquad\qquad\qquad\qquad\qquad (IV)$$

RESONANCE STABILIZATION PHENOMENON   8.160

Only the electro-free-radical route is favored by the monomer as shown below.

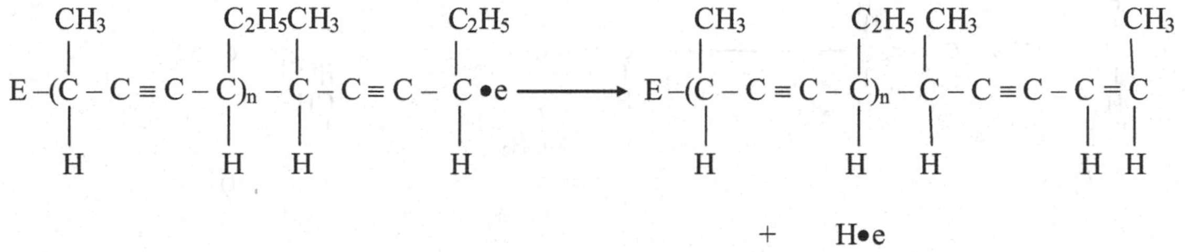

$$
E-(\overset{\overset{CH_3}{|}}{\underset{\underset{H}{|}}{C}} - C \equiv C - \overset{\overset{C_2H_5}{|}}{\underset{\underset{H}{|}}{C}})_n - \overset{\overset{CH_3}{|}}{\underset{\underset{H}{|}}{C}} - C \equiv C - \overset{\overset{C_2H_5}{|}}{\underset{\underset{H}{|}}{C}} \bullet e \longrightarrow E-(\overset{\overset{CH_3}{|}}{\underset{\underset{H}{|}}{C}} - C \equiv C - \overset{\overset{C_2H_5}{|}}{\underset{\underset{H}{|}}{C}})_n - \overset{\overset{CH_3}{|}}{\underset{\underset{H}{|}}{C}} - C \equiv C - \overset{\overset{C_2H_5}{|}}{\underset{\underset{H}{|}}{C}} = \overset{\overset{CH_3}{|}}{\underset{\underset{H}{|}}{C}}
$$

$$+ \qquad H\bullet e \qquad\qquad 8.161$$

The dead terminal polymer produced is eneyne in character.

With etheric groups, the followings are to be expected.

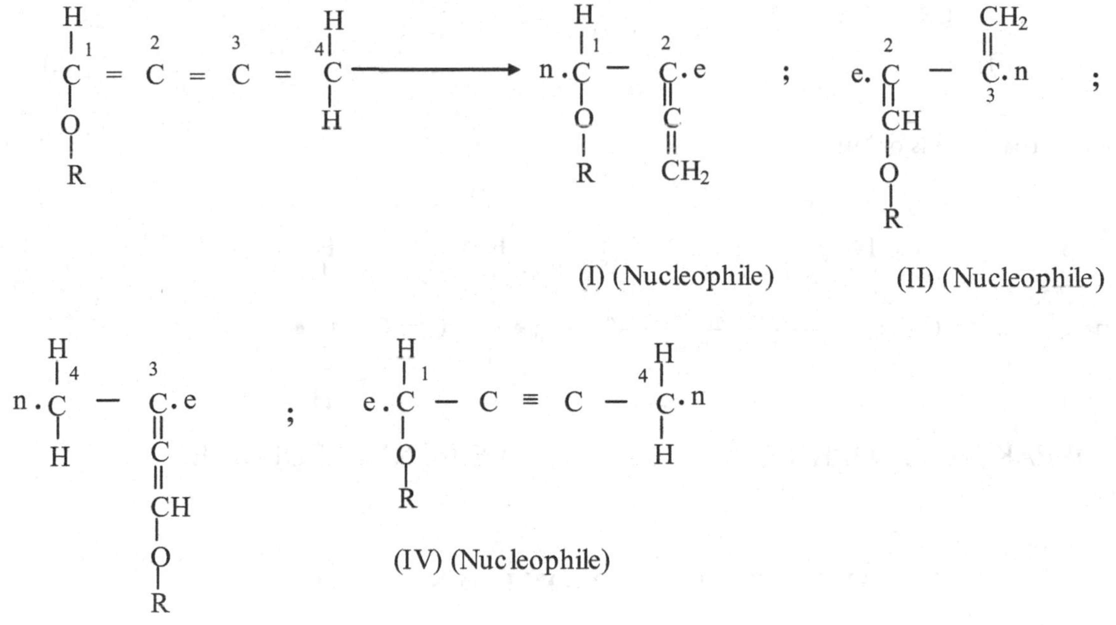

$$
\underset{\underset{R}{\overset{|}{O}}}{\overset{\overset{H}{\overset{|}{1}}}{C}} \overset{2}{=} C \overset{3}{=} C \overset{4}{=} \underset{\underset{H}{|}}{\overset{\overset{H}{|}}{C}} \longrightarrow n\cdot\underset{\underset{R}{\overset{|}{O}}}{\overset{\overset{H}{\overset{|}{1}}}{C}} \overset{2}{-} \underset{\underset{CH_2}{\overset{\|}{C}}}{C}\cdot e \quad ; \quad e\cdot\underset{\underset{R}{\overset{|}{O}}}{\overset{2}{\underset{\underset{CH}{\|}}{C}}} \overset{}{-} \underset{\underset{3}{}}{\overset{\overset{CH_2}{\|}}{C}}\cdot n \quad ;
$$

$$(I)\ (Nucleophile) \qquad\qquad (II)\ (Nucleophile)$$

$$
n\cdot\underset{\underset{H}{|}}{\overset{\overset{H}{\overset{|}{4}}}{C}} \overset{3}{-} \underset{\underset{\underset{R}{\overset{|}{O}}}{\overset{\|}{CH}}}{\overset{\overset{}{\underset{\|}{C}}}{C}}\cdot e \quad ; \quad e\cdot\underset{\underset{R}{\overset{|}{O}}}{\overset{\overset{H}{\overset{|}{1}}}{C}} - C \equiv C - \underset{\underset{H}{|}}{\overset{\overset{H}{\overset{|}{4}}}{C}}\cdot n
$$

$$\qquad\qquad\qquad (IV)\ (Nucleophile)$$

$$(III)\ (Nucleophile) \qquad\qquad\qquad\qquad\qquad 8.162$$

It is only (III) that is relevant, from which (IV) is obtained.

$$n\bullet \overset{\overset{\textstyle H}{|}}{\underset{\underset{\textstyle H}{|}}{C}} - \overset{\overset{\textstyle \bullet e}{}}{C} = C = \overset{\overset{\textstyle OR}{|}}{\underset{\underset{\textstyle H}{|}}{C}} \quad \xleftarrow{\;\;\underset{\textit{phenomenon}}{\textit{Resonance Stabilization}}\;\;} \quad n\bullet \overset{\overset{\textstyle H}{|}}{\underset{\underset{\textstyle H}{|}}{C}} - C \equiv C - \overset{\overset{\textstyle OR}{|}}{\underset{\underset{\textstyle H}{|}}{C}} \bullet e$$

RESONANCE STABILIZATION PHENOMENON 8.163

As an exercise, replace R with H and find out what will be obtained. The only route of polymerization is the use of electro-free-radicals. For symmetrically placed OR groups, the route still remains electro-free-radical route. Using F in place of one H atom, the followings are obtained.

$$\overset{\overset{\textstyle F}{|}}{\underset{\underset{\textstyle H}{|}}{\underset{1}{C}}} = \overset{2}{C} = \overset{3}{C} = \overset{\overset{\textstyle H}{|}}{\underset{\underset{\textstyle H}{|}}{\underset{4}{C}}} \longrightarrow n.\overset{\overset{\textstyle F}{|}}{\underset{\underset{\textstyle H}{|}}{\underset{1}{C}}} - \overset{2}{\underset{\underset{\underset{\textstyle CH_2}{\parallel}}{\underset{\textstyle C}{|}}}{C}}.e \quad ; \quad n.\overset{2}{\underset{\underset{\underset{\textstyle F}{|}}{\underset{\textstyle CH}{|}}}{C}} - \overset{\overset{\textstyle CH_2}{\parallel}}{\underset{3}{C}}.e \quad ;$$

$$\qquad\qquad\qquad\qquad\qquad (I) \qquad\qquad\qquad\qquad (II)$$

$$e.\overset{\overset{\textstyle H}{|}}{\underset{\underset{\textstyle H}{|}}{\underset{4}{C}}} - \overset{3}{\underset{\underset{\underset{\underset{\textstyle F}{|}}{\textstyle CH}}{\parallel}}{C}}.n \quad ; \quad n.\overset{\overset{\textstyle F}{|}}{\underset{\underset{\textstyle H}{|}}{\underset{4}{C}}} - C \equiv C - \overset{\overset{\textstyle H}{|}}{\underset{\underset{\textstyle H}{|}}{\underset{1}{C}}}.e$$

$$\qquad (III) \qquad\qquad\qquad\qquad (IV) \qquad\qquad\qquad\qquad\qquad\qquad 8.164$$

It is from (I) that (IV) is obtained.

$$n\bullet \overset{\overset{\textstyle F}{|}}{\underset{\underset{\textstyle H}{|}}{C}} - \overset{\overset{\textstyle \bullet e}{}}{C} = C = \overset{\overset{\textstyle H}{|}}{\underset{\underset{\textstyle H}{|}}{C}} \quad \xleftarrow{\;\;\underset{\textit{phenomenon}}{\textit{Resonance Stabilization}}\;\;} \quad n\bullet \overset{\overset{\textstyle F}{|}}{\underset{\underset{\textstyle H}{|}}{C}} - C \equiv C - \overset{\overset{\textstyle H}{|}}{\underset{\underset{\textstyle H}{|}}{C}} \bullet e$$

WEAK NUCLEOPHILE **WEAK NUCLEOPHILE**

RESONANCE STABILIZATION PHENOMENON 8.165

$$\overset{\overset{\textstyle F}{|}}{\underset{\underset{\textstyle H}{|}}{C}} = C = C = \overset{\overset{\textstyle F}{|}}{\underset{\underset{\textstyle H}{|}}{C}} \longrightarrow n.\overset{\overset{\textstyle F}{|}}{\underset{\underset{\textstyle H}{|}}{C}} - \overset{\underset{\underset{\underset{\underset{\textstyle F}{|}}{\textstyle CH}}{\parallel}}{\underset{\textstyle C}{}}}{C}.e \quad ; \quad n.\overset{\overset{\textstyle F}{|}}{\underset{\underset{\textstyle H}{|}}{C}} - \overset{\underset{\underset{\underset{\underset{\textstyle F}{|}}{\textstyle CH}}{|}}{\underset{\textstyle C}{\parallel}}}{C}.e \quad ;$$

$$\qquad (Symmetric) \qquad\qquad\qquad\qquad\qquad (I) \qquad\qquad\qquad\qquad (II)$$

$$
\begin{array}{ccc}
& F & \\
& | & \\
& CH & \\
& \| & \\
n\,.\,C & - & C\,.\,e \\
\| & & \\
CH & & \\
| & & \\
F & &
\end{array}
\qquad ; \qquad
\begin{array}{ccccc}
& F & & & F \\
& | & & & | \\
n\,.\,C & - & C & \equiv & C & - & C\,.\,e \\
& | & & & | \\
& H & & & H
\end{array}
$$

(III)   (IV)   8.166

$$
\begin{array}{ccc}
F & & F \\
| & \overset{\bullet e}{} & | \\
n\bullet C - & C = C = C & \\
| & & | \\
H & & H
\end{array}
\quad \xleftrightarrow[\text{phenomenon}]{\text{Resonance Stabilization}} \quad
\begin{array}{ccc}
F & & F \\
| & & | \\
n\bullet C - C \equiv C - C \bullet e \\
| & & | \\
H & & H
\end{array}
$$

(I)   (IV)

Resonance stabilization phenomenon   8.167

For the last two cases above, both nucleo- and electro-free-radical routes are favored. If other mono-forms had been used for initiation, the situation would have been completely different. Hence, in general for 1,4-cumulenes to be obtained, resonance stabilization phenomenon must take place when they are activated based on the applications of the Laws of Nature.

$$
\begin{array}{ccccc}
CF_3 & & & & H \\
| & 2 & 3 & & | \\
{}^1C & = C & = C & = {}^4C & \\
| & & & & | \\
H & & & & H
\end{array}
\longrightarrow
\begin{array}{cc}
CF_3 & \\
|\,{}^1 & {}^2 \\
n\,.\,C & - C\,.\,e \\
| & \| \\
H & C \\
& \| \\
& CH_2
\end{array}
\quad ; \quad
\begin{array}{cc}
& CH_2 \\
{}^2 & \| \\
n\,.\,C & - C\,.\,e \\
\| & {}^3 \\
CH & \\
| & \\
CF_3 &
\end{array}
\quad ;
$$

(Unsymmetric)   (I) (Electrophile)   (III) (Nucleophile)

$$
\begin{array}{cc}
H & \\
|\,{}^4 & {}^3 \\
n\,.\,C & - C\,.\,e \\
| & \| \\
H & C \\
& \| \\
& CH \\
& | \\
& CF_3
\end{array}
\quad ; \quad
\begin{array}{ccccc}
CF_3 & & & & H \\
|\,{}^1 & & & & |\,{}^4 \\
n\,.\,C & - C & \equiv & C & - C\,.\,e \\
| & & & & | \\
H & & & & H
\end{array}
$$

(II) (Electrophile)   (IV) (Electrophile)

8.168

$$
\begin{array}{ccc}
CF_3 & & H \\
| & \overset{\bullet e}{} & | \\
n\bullet C - & C = C = C & \\
| & & | \\
H & & H
\end{array}
\quad \xleftrightarrow[\text{phenomenon}]{\text{Resonance Stabilization}} \quad
\begin{array}{ccc}
CF_3 & & H \\
| & & | \\
n\bullet C - C \equiv C - C \bullet e \\
| & & | \\
H & & H
\end{array}
$$

(I) WEAK NUCLEOPHILE   (IV) WEAK NUCLEOPHILE

RESONANCE STABILIZATION PHENOMENON   8.169

Even the use of paired radical initiators will not make it favor its Natural route, being a Nucleophile. However, it like vinyl acetate will favor the nucleo-free-radical route if activated.

$$\overset{4}{C}\overset{\displaystyle CF_3}{\underset{\displaystyle H}{|}} = \overset{3}{C} = \overset{2}{C} = \overset{1}{C}\overset{\displaystyle CH_3}{\underset{\displaystyle H}{|}} \longrightarrow n.\overset{4}{C}\overset{\displaystyle CF_3}{\underset{\displaystyle H}{|}} - \overset{3}{C}.e \quad ; \quad n.\overset{3}{C} - \overset{\displaystyle CH_3}{\underset{}{|}}\overset{}{C}.e_2 \quad ;$$

(Unsymmetric)

with the C_3 carbon bearing
$$\underset{\displaystyle CH_3}{\overset{\displaystyle C}{\underset{\displaystyle CH}{\underset{\displaystyle CH_3}{|}}}}$$

(I) (None)

(III) (None)

$$e.\overset{1}{C}\overset{\displaystyle CH_3}{\underset{\displaystyle H}{|}} - \overset{2}{C}.n \quad ; \quad n.\overset{4}{C}\overset{\displaystyle CF_3}{\underset{\displaystyle H}{|}} - C \equiv C - \overset{1}{C}\overset{\displaystyle CH_3}{\underset{\displaystyle H}{|}}.e$$

(IV) (None)

(II) (None)   8.170

$$n\bullet \overset{\displaystyle CF_3}{\underset{\displaystyle H}{C}} - \overset{\displaystyle \bullet e}{C} = C = \overset{\displaystyle CH_3}{\underset{\displaystyle H}{C}} \quad \underset{\text{\textit{phenomenon}}}{\overset{\textit{Resonance Stabilization}}{\longleftrightarrow}} \quad n\bullet \overset{\displaystyle CF_3}{\underset{\displaystyle H}{C}} - C \equiv C - \overset{\displaystyle CH_3}{\underset{\displaystyle H}{C}} \bullet e$$

(I) WEAK NUCLEOPHILE   (IV) WEAK NUCLEOPHILE

RESONACE STABILIZATION PHENOMENON   8.171

$$\overset{1}{C}\overset{\overset{\displaystyle CF_3}{\displaystyle |}}{\overset{\displaystyle CF_2}{\underset{\displaystyle H}{|}}} = \overset{2}{C} = \overset{3}{C} = \overset{4}{C}\overset{\displaystyle H}{\underset{\displaystyle H}{|}} \longrightarrow n.\overset{1}{C}\overset{\displaystyle C_2F_5}{\underset{\displaystyle H}{|}} - \overset{2}{C}.e \quad ; \quad n.\overset{2}{C} - \overset{\displaystyle CH_2}{\underset{}{|}}\overset{}{C}.e_3 \quad ;$$

with
$$\underset{\displaystyle CH_2}{\overset{}{C}}$$
(I) (Electrophile)

and
$$\underset{\displaystyle CF_3}{\overset{\displaystyle CH}{\underset{\displaystyle CF_2}{|}}}$$

(III) (Nucleophile)

$$n.\overset{4}{C}\overset{\displaystyle H}{\underset{\displaystyle H}{|}} - \overset{3}{C}.e \quad ; \quad n.\overset{1}{C}\overset{\displaystyle H}{\underset{\displaystyle C_2F_5}{|}} - C \equiv C - \overset{4}{C}\overset{\displaystyle H}{\underset{\displaystyle H}{|}}.e$$

with the C_3 bearing
$$\underset{\displaystyle C_2F_5}{\overset{\displaystyle C}{\underset{\displaystyle CH}{|}}}$$

(IV) (Electrophile)

(II) (Nucleophile)   8.172

641

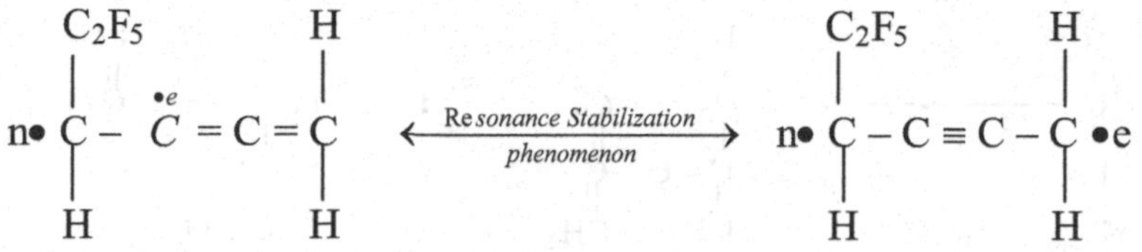

$$
\begin{array}{ccc}
\text{C}_2\text{F}_5 & & \text{H} \\
| & \overset{\bullet e}{} & | \\
\text{n}\bullet\text{C} - \text{C} = \text{C} = \text{C} & & \\
| & & | \\
\text{H} & & \text{H}
\end{array}
\quad
\xleftrightarrow[\text{phenomenon}]{\text{Resonance Stabilization}}
\quad
\begin{array}{cc}
\text{C}_2\text{F}_5 & \text{H} \\
| & | \\
\text{n}\bullet\text{C} - \text{C} \equiv \text{C} - \text{C} \bullet e \\
| & | \\
\text{H} & \text{H}
\end{array}
$$

(I) WEAK NUCLEOPHILE   (IV) WEAK NUCLEOPHILE

RESONANCE STABILIZATION PHENOMENON   8.173

For these families of monomers, the use of paired initiators is not possible in the route natural to the monomer. Since it will favor the route not natural to it, the monomer therefore is unique.

For radical-pulling groups, the followings are obtained. The need to show them as has been done so far is to distinguish it from the charged case which does not exist. Radically, we see the reality.

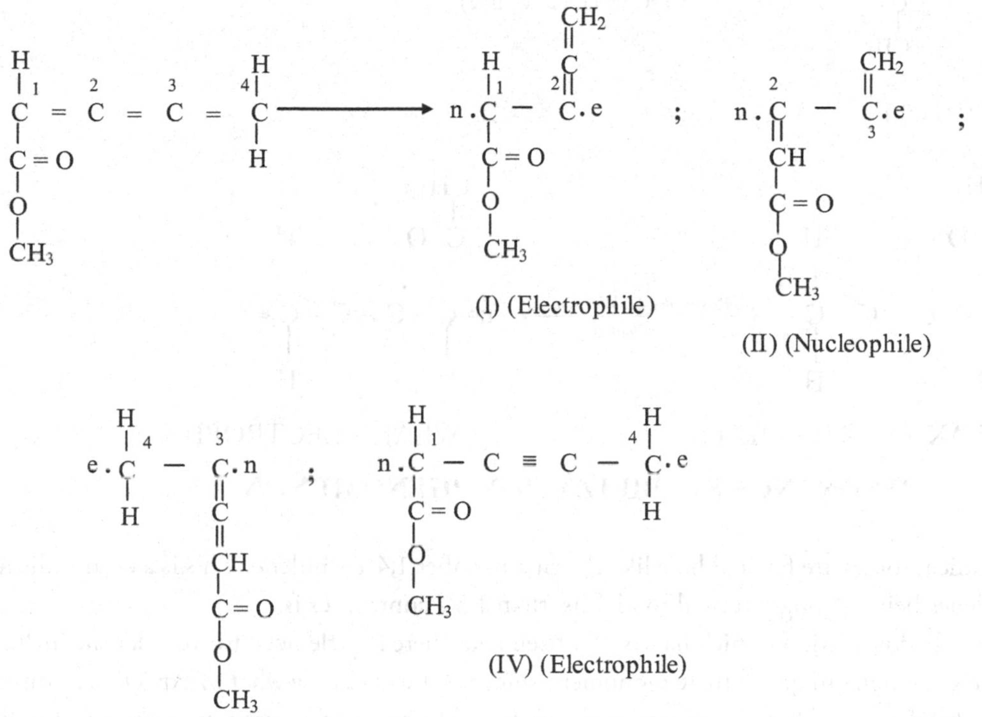

(I) (Electrophile) ; (II) (Nucleophile)

(III) (Electrophile) ; (IV) (Electrophile)

8.174

Recalled below is what has already been shown, noting that these are Electrophiles.

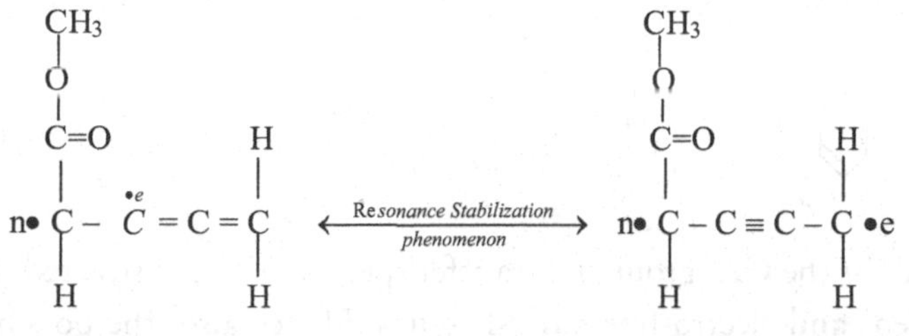

ELECTROPHILE   ELECTROPHILE

RESONANCE STABILIZATION PHENOMENON   (8.144)

642

$$
\begin{array}{c}
\overset{H}{\underset{|}{}}\\
\overset{1}{C} = \overset{2}{C} = \overset{3}{C} = \overset{4}{\underset{|}{C}} \\
\underset{|}{C}=O \quad\quad\quad H \\
CH_3
\end{array}
\longrightarrow
\quad
n \cdot \overset{H}{\underset{|}{\overset{1}{C}}} - \overset{2}{C} \cdot e
\quad ; \quad
n \cdot \overset{2}{C} - \overset{CH_2}{\underset{3}{C}} \cdot e \quad ;
$$

Structure (I): $n \cdot C(H)(C=O\,CH_3) - C(CH_2) \cdot e$ — **(I) (Electrophile)**

Structure (III): $n \cdot C - C(CH_2)(=) \cdot e$ with $CH, C=O, CH_3$ — **(III) (Nucleophile)**

$$
e \cdot \overset{H}{\underset{H}{\overset{4}{C}}} - \overset{3}{C} \cdot n
\quad ; \quad
n \cdot \overset{H}{\underset{C=O}{\overset{1}{C}}} - C \equiv C - \overset{H}{\underset{H}{\overset{4}{C}}} \cdot e
$$

with branches $C, CH, C=O, CH_3$ (left) and CH_3 (center)

(IV) (Electrophile)

(III) (Both)   8.175

$$
\begin{array}{cc}
CH_3 & \\
\underset{|}{C}=O \quad\quad H & \\
n \bullet \overset{\bullet e}{C} - C = C = C & \\
\underset{|}{H} \quad\quad H &
\end{array}
\quad
\xleftrightarrow[\text{phenomenon}]{\textit{Resonance Stabilization}}
\quad
\begin{array}{c}
CH_3 \\
\underset{|}{C}=O \quad\quad H \\
n \bullet C - C \equiv C - C \bullet e \\
H \quad\quad\quad H
\end{array}
$$

WEAK ELECTROPHILE   **WEAK ELECTROPHILE**

RESONANCE STABILIZATION PHENOMENON   8.176

Both free-radical routes are favored here like the first member 1,4-cumulene. This is a clear indication of 1,4 - cumulenes being strongly related to olefins, than 1,3 - cumulenes is.

Based on the foundations which have so far been laid, there is little need to consider the influence of resonance stabilization groups on these monomers, since we already know what to expect. 1,4-cumulene is a strong Nucleophile, for which a stronger nucleophile will be required to provide resonance stabilization for the monomer. One of the best candidates is the phenyl group from benzene a compound which is more nucleophilic than the cumulene.

$$
\begin{array}{c}
CH_3 \quad\quad H \\
C = C = C = C \\
\text{(phenyl)} \quad H
\end{array}
\longrightarrow
\quad
e \bullet \overset{CH_3}{C} - C \equiv C - \overset{H}{C} \bullet n
$$

with phenyl group and H 8.177

Due to resonance stabilization provided for the CH_3 group, no transfer species can be abstracted. Hence, the monomer will favor both the nucleo- and electro-free-radical routes. Hence also, the possibility of the monomer having a Ceiling temperature in the route not natural to it is bound to be present. When an ethene is used in place of the phenyl group, the provider of resonance stabilization will become the 1,4-cumulene $-CHC=C=CH_2$, with possibility of shielding two groups internally located.

8.3. 1,5 - Cumulenes

Whether the possible existences of these monomers are favored or not is not presently the issue, in an attempt to establish orderliness. While they may not be known to exist now, this may not be the case in the future, particularly after laying all these foundations.

8.3.1. Charged Characters of 1,5 - Cumulenes

Beginning with the first member, the followings are to be expected.

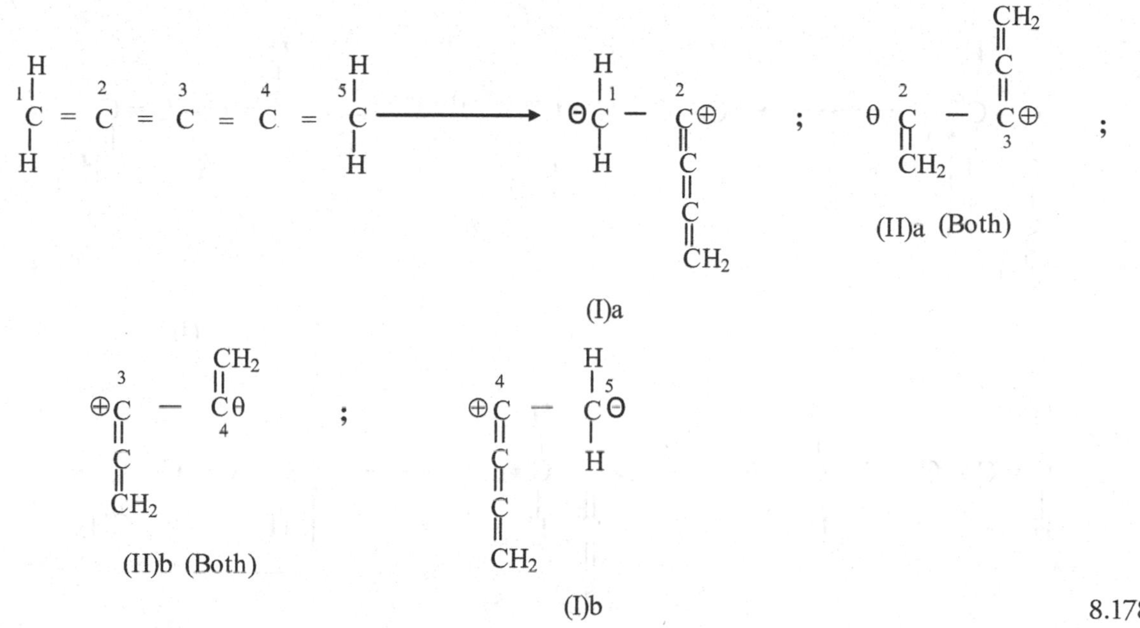

(I)a

(II)a (Both)

(II)b (Both)

(I)b

8.178

Unlike the 1,4 - cumulenes, only two types of existing and non-existing mono-forms are possible here - 1,2 - or 5,4 - and 2,3 - or 4,3 -. From the charges carried by the active centers, it is obvious that $CH_2 = C$ = is far greater in capacity than CH_2 = as radical-pushing groups. Note the charges carried by the active centers of (I)a or (II)a. Since acetylenes, 1,3-cumulenes and 1,4-cumulenes have been observed not to favor charged activations, the same also apply to 1,5-cumulenes. However, as one did for the others one is using it here for exploratory purposes.

Left alone, 1,5-cumulene can be activated across the bonds as one saw with 1,4-cumulene, only radically. Of the two mono-forms above, only one is relevant as shown below.

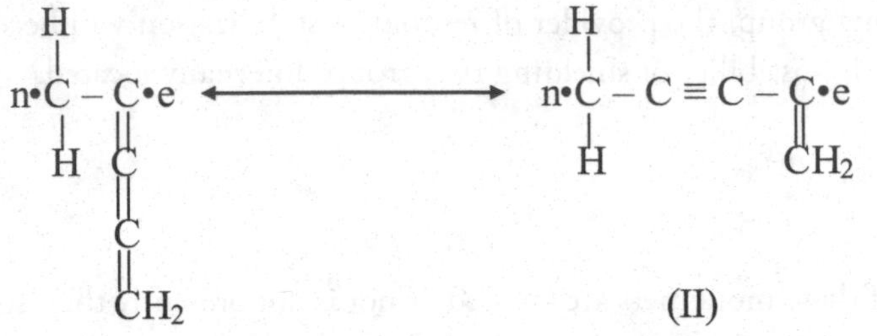

(I) RESONANCE STABILIZATION 8.179

While in 1,3-cumulene, there is one mono-form, in 1,5-cumulene, there are two mono-forms. Despite the fact that 1,3-cumulene and 1,5-cumulene favor only route natural to them, while 1,4-cumulene favors both routes, 1,5-cumulene is more nucleophilic than 14-cumulene which in turn is more nucleophilic than 1,3-cumulene. The nucleophilic characters between different families are determined by the number of activation centers carried by a monomer and the type of activation center.

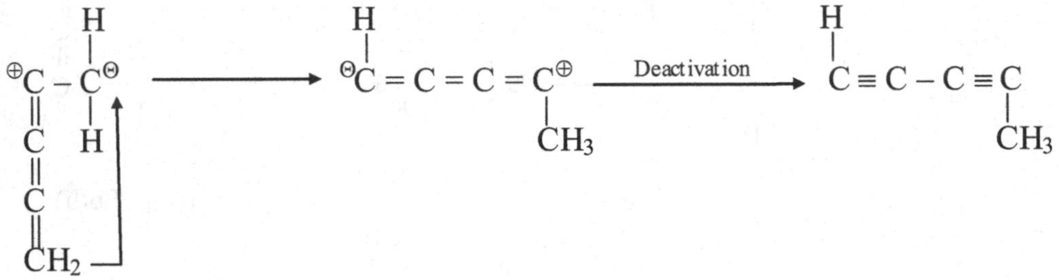

(I) (III) 8.180a

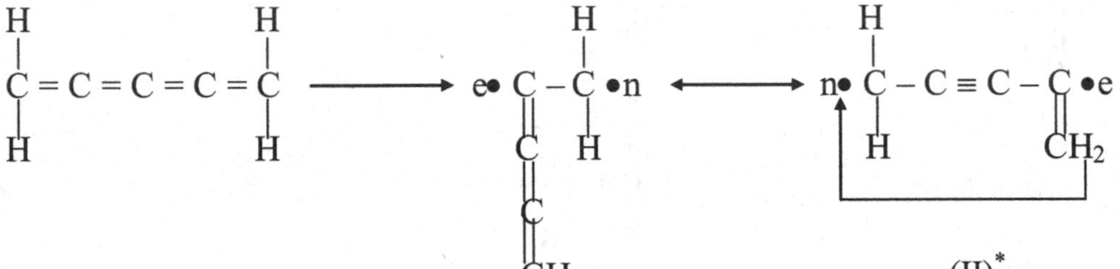

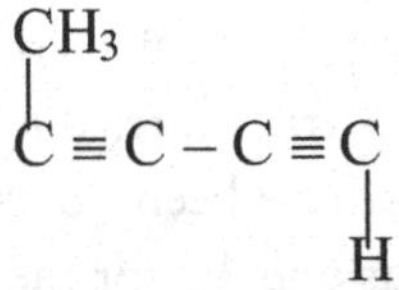

(III) 8.180b

Note that the molecular rearrangement in the first equation cannot take place chargedly as shown. They can only take place radically as shown in the last equation where the activated monomer resonance stabilized to give (II)* which molecularly rearranged to give the same (III) above. Thus, one can observe two monoforms radically-(I) and (II)* only for this monomer radically. The (III) obtained is a Diyne resonance stabilized but with nothing to shield. That is the monomer from the first member of 1,5-cumulene.

Now, when one of the H atoms is replaced with CH_3 group, the followings are obtained.

645

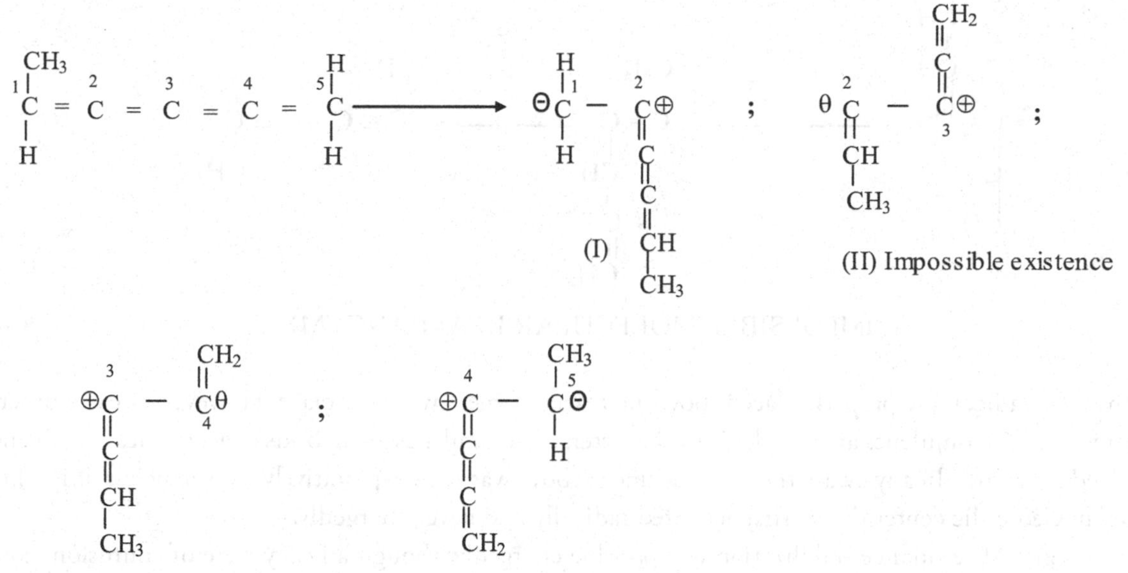

$$
\underset{\substack{1}}{\overset{CH_3}{\underset{H}{C}}} = \underset{2}{C} = \underset{3}{C} = \underset{4}{C} = \underset{5}{\overset{H}{\underset{H}{C}}} \longrightarrow \ominus\underset{\substack{1\\H}}{\overset{H}{C}} - \underset{2}{\overset{\oplus}{\underset{\substack{C\\ \parallel \\ C \\ \parallel \\ CH \\ | \\ CH_3}}{C}}} \quad ; \quad \theta\underset{2}{\underset{\substack{\parallel\\CH\\|\\CH_3}}{C}} - \underset{3}{\overset{\oplus}{\underset{}{C}}} \quad ;
$$

(I) (II) Impossible existence

$$
\oplus\underset{\substack{\parallel\\C\\\parallel\\CH\\|\\CH_3}}{\overset{3}{C}} - \underset{4}{\overset{CH_2}{\underset{}{\overset{\parallel}{C}\theta}}} \quad ; \quad \oplus\underset{\substack{\parallel\\C\\\parallel\\CH_2}}{\overset{4}{C}} - \underset{\substack{|\\H}}{\overset{CH_3}{\underset{5}{C}\ominus}}
$$

(III) Impossible existence   (IV) Non-existent   8.181

(I) molecular rearranges and deactivates as follows.

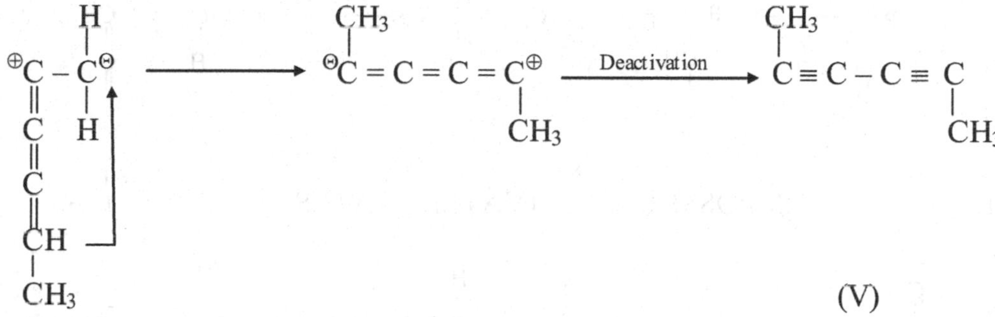

$$
\oplus\underset{\substack{\parallel\\C\\\parallel\\C\\\parallel\\CH\\|\\CH_3}}{C} - \underset{\substack{|\\H}}{\overset{H}{C}\ominus} \longrightarrow \ominus\underset{}{\overset{CH_3}{\underset{}{C}}} = C = C = \underset{\substack{|\\CH_3}}{\overset{\oplus}{C}} \xrightarrow{\text{Deactivation}} \underset{}{\overset{CH_3}{\underset{}{C}}} \equiv C - C \equiv \underset{\substack{|\\CH_3}}{C}
$$

(V)

(I) MOLECULAR REARRANGEMENT 8.182

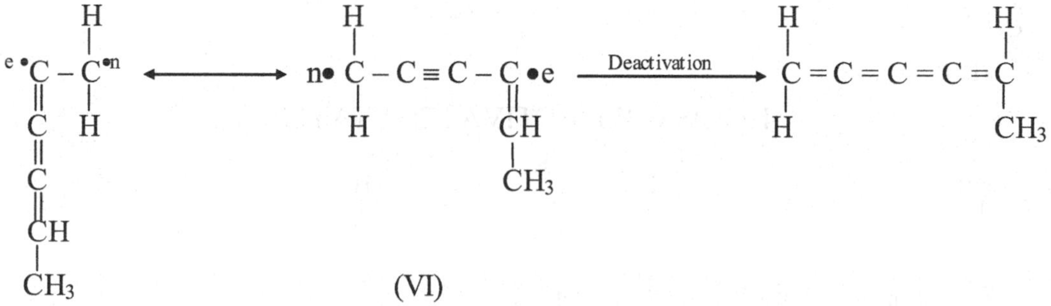

$$
e\bullet\underset{\substack{\parallel\\C\\\parallel\\C\\\parallel\\CH\\|\\CH_3}}{\overset{H}{C}} - \underset{\substack{|\\H}}{\overset{\bullet n}{\underset{}{C}}} \longleftrightarrow n\bullet\underset{\substack{|\\H}}{\overset{H}{C}} - C \equiv C - \underset{\substack{\parallel\\CH\\|\\CH_3}}{C}\bullet e \xrightarrow{\text{Deactivation}} \underset{\substack{|\\H}}{\overset{H}{C}} = C = C = C = \underset{\substack{|\\CH_3}}{\overset{H}{C}}
$$

(VI)

(I) RESONANCE STABILIZATION PHENOMENON 8.183

(V) is the real monomer and so also is (VI) all obtained from (I) which is the only one favored. Based on the routes favored by them, these are Nucleophiles for all R groups.

(II) of Equation 8.181 when wrongly activated can undergo molecularly rearrangement, but not resonance stabilized as shown below, only radically.

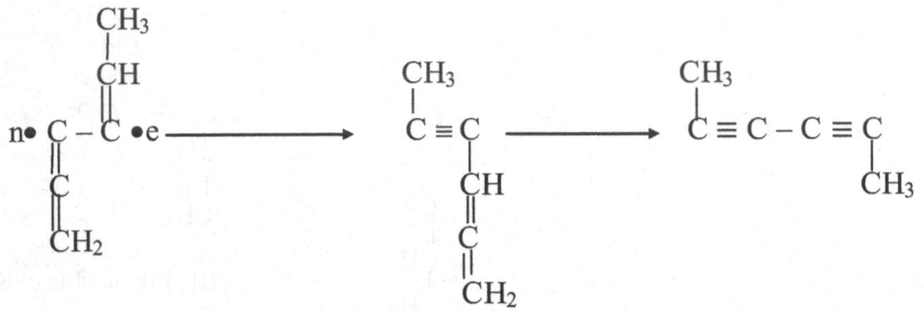

IMPOSSIBLE MOLECULAR REARRANGEMENT   8.184

When the radicals are properly placed above, no rearrangement will take place. However, one can observe that while 1,5-cumulenes are acetylenic in character, 1,4-cumulenes are hybrids of acetylenes and alkenes. Indeed as we are already aware the rearrangement above was used exploratively. Nevertheless, it is (I) the least nucleophilic center that is first activated radically and never chargedly.

Imagine if resonance stabilization was possible chargedly, though a heavy state of confusion would arise, at the end, one will gain a lot from the exercise.

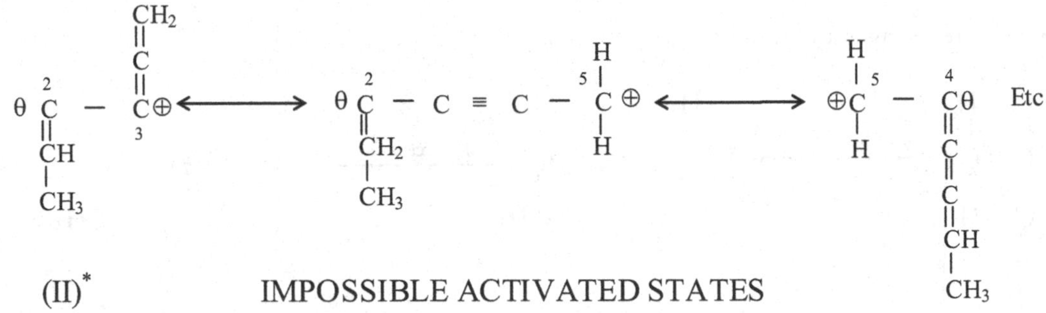

(II)*   IMPOSSIBLE ACTIVATED STATES   8.185

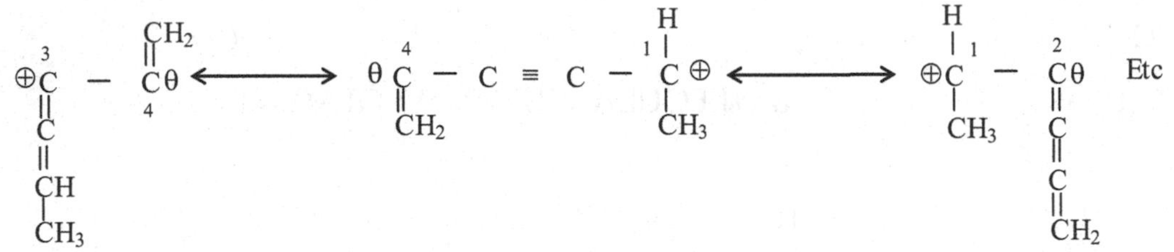

(III)*

IMPOSSIBLE ACTIVATED STATES   8.186

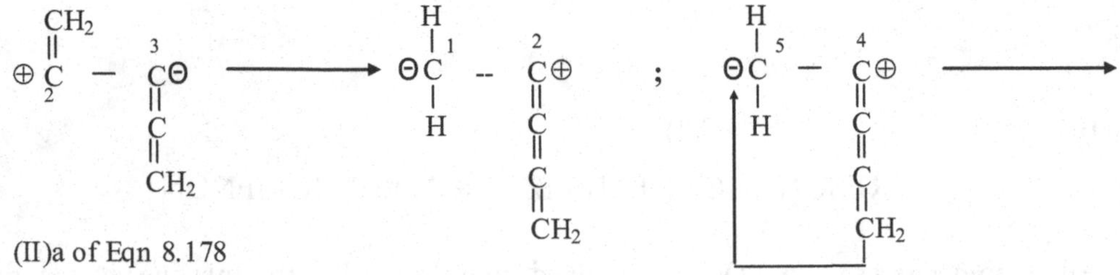

(II)a of Eqn 8.178

(I)b of Eqn. 8.178

647

$$\overset{\ominus}{\underset{1}{C}} = C \equiv C = \overset{\overset{H}{|}}{\underset{\underset{CH_3}{|}}{\overset{4}{C}}}{}^{\oplus}$$

IMPOSSIBLE ACTIVATED STATES   8.187

The first member can be found to be strongly nucleophilic. So also are the lower members for which the following order is to be expected.

$$\underset{\underset{H}{|}}{\overset{\overset{H}{|}}{C}} = C = C = C = \overset{\overset{C_2H_5}{|}}{\underset{\underset{H}{|}}{C}} > \underset{\underset{H}{|}}{\overset{\overset{H}{|}}{C}} = C = C = C = \overset{\overset{CH_3}{|}}{\underset{\underset{H}{|}}{C}} > \underset{\underset{H}{|}}{\overset{\overset{H}{|}}{C}} = C = C = C = \overset{\overset{H}{|}}{\underset{\underset{H}{|}}{C}}$$

<u>Order of Nucleophilicity</u>   8.188

For the first member, as we have seen from Equations 8.97 and 8.180a, there will be need to replace a loosely held hydrogen atom. This was seen with the second member in the family in (V) of Equation 8.182 where the Diyne obtained is carrying no H atom [$H_3C-C\equiv C-C\equiv C-CH_3$].

When two radical-pushing groups symmetrically placed are used in place of two hydrogen atoms, the followings are obtained.

(I) molecularly rearranges and resonance stabilizes as follows.

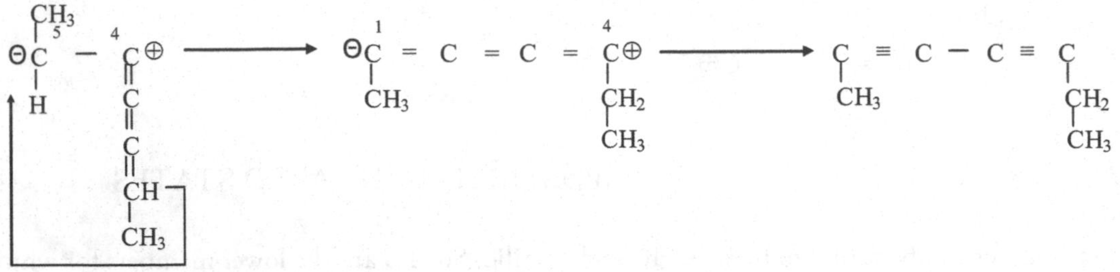

$$\overset{CH_3}{\underset{H}{\Theta C}} - C^{\oplus} \longrightarrow \Theta C = C = C = C^{\oplus} \longrightarrow C \equiv C - C \equiv C$$

(I)   MOLECULAR REARRANGEMENT   8.190

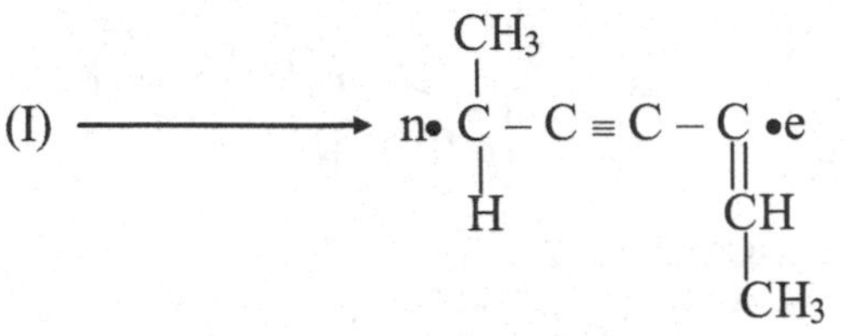

$$(I) \longrightarrow n\bullet \overset{CH_3}{\underset{H}{C}} - C \equiv C - \overset{}{C} \bullet e$$

RESONANCE STABILIZATION PHENOMENON   8.191

When the CH_3 group in Equation 8.181 is replaced with OR group, then the followings are obtained.

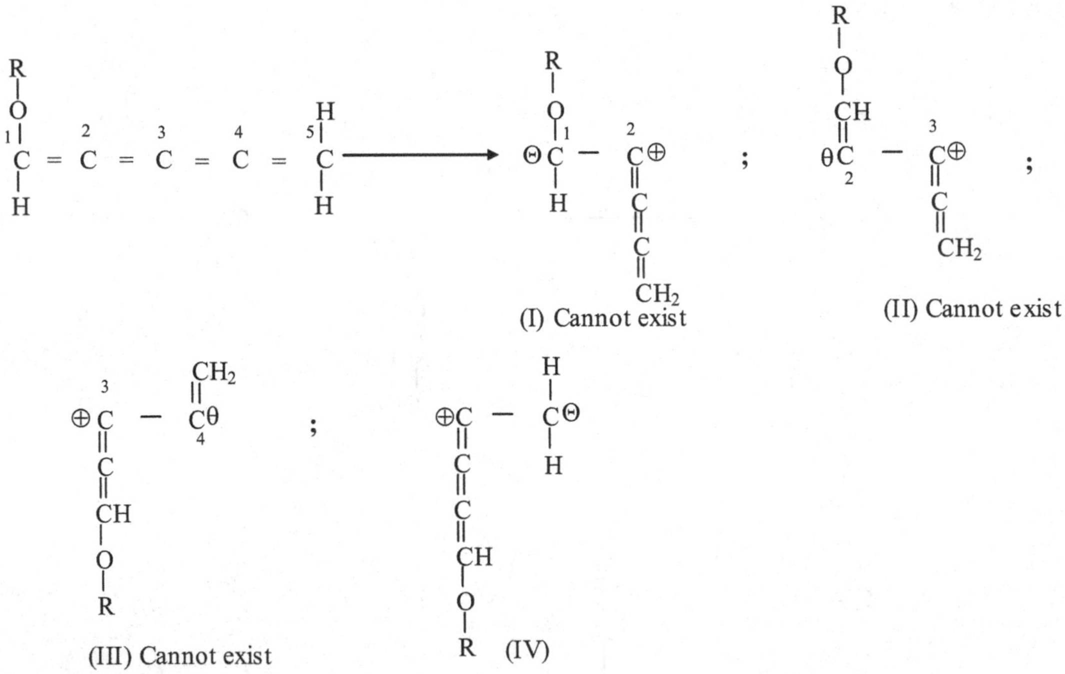

$$\overset{R}{\underset{H}{\overset{|}{\underset{|}{O}}}} \quad \overset{1}{C} = \overset{2}{C} = \overset{3}{C} = \overset{4}{C} = \overset{5}{\underset{H}{\overset{H}{C}}} \longrightarrow$$

(I) Cannot exist   (II) Cannot exist

(III) Cannot exist   (IV)   8.192

The center that is first and only activated is (IV), being the least nucleophilic. (I), (II) and (III) do not exist due to electrostatic forces of repulsion. Radically, they do not also exist. Radically, (II) cannot molecularly rearrange or be resonance stabilized. (III) can molecularly rearrange, and can be resonance stabilized. (IV) which cannot molecularly rearrange because the pushing capacity of OR group is greater than CH_3 group, can however be resonance stabilized as shown below.

$$(IV) \longrightarrow n\bullet \overset{\overset{\displaystyle H}{|}}{\underset{\underset{\displaystyle H}{|}}{C}} - C \equiv C - \overset{\overset{\displaystyle }{}}{\underset{\underset{\displaystyle CH}{\parallel}}{C}} \bullet e$$

$$(V) \quad \underset{\displaystyle OR}{|}$$

$$8.193$$

(V) is the only mono-form that can be used for homopolymerization electro-free-radically as shown below.

$$E - \overset{\overset{\displaystyle H}{|}}{\underset{\underset{\displaystyle H}{|}}{C}} - C \equiv C - \overset{\overset{\displaystyle }{}}{\underset{\underset{\underset{\displaystyle OR}{|}}{CH}}{C}} \bullet e \quad \xrightarrow{+\ n(V)} \quad E - (\overset{\overset{\displaystyle H}{|}}{\underset{\underset{\displaystyle H}{|}}{C}} - C \equiv C - \overset{\overset{\displaystyle }{}}{\underset{\underset{\underset{\displaystyle OR}{|}}{CH}}{C}})_n - \overset{\overset{\displaystyle H}{|}}{\underset{\underset{\displaystyle H}{|}}{C}} - C \equiv C - \overset{}{\underset{\underset{\displaystyle OR}{|}}{C}} \equiv C$$

$$+ \quad H \bullet e \qquad\qquad 8.194$$

Worthy of note, is the type of dead terminal doubly triple bond polymer produced. One can see that Chemistry is a wonderful NETWORK system. When the OR group is replaced with OH, the analysis remains the same. The monomer will be stable from that end, unlike the case of vinyl alcohol.

It is largely for these important observations that one hydrogen atom in the first member of 1,5 cumulene is bonded radically, while where one substituent is present, no H atom is loosely bonded.

$$\overset{\overset{\displaystyle H}{|}}{\underset{\underset{\displaystyle H}{|}}{C}} = C = C = C = \overset{\overset{\displaystyle H}{|}}{\underset{\underset{\displaystyle H}{|}}{C}} \longrightarrow \overset{}{\underset{}{C}} \equiv C - C \equiv \overset{}{\underset{\underset{\displaystyle CH_3}{|}}{C}} \ \rightleftharpoons \ H\bullet e \ + \ n\bullet C \equiv C - C \equiv \overset{}{\underset{\underset{\displaystyle CH_3}{|}}{C}}$$

$$8.195$$

These observations are supported by the increasing levels of unsaturation present as one moves from C = C to C ≡ C and then to C = C = C and to C = C = C = C in that order, based on H to C ratio. This has already been partially shown in Equation 8.97 and fully recalled below.

$$\overset{\overset{\displaystyle H}{|}}{\underset{\underset{\displaystyle H}{|}}{C}} = \overset{\overset{\displaystyle H}{|}}{\underset{\underset{\displaystyle H}{|}}{C}} \quad ; \quad \overset{\overset{\displaystyle H}{|}}{C} \equiv \overset{}{\underset{\underset{\displaystyle H}{|}}{C}} \quad ; \quad \overset{\overset{\displaystyle H}{|}}{\underset{\underset{\displaystyle H}{|}}{C}} = C = \overset{\overset{\displaystyle H}{|}}{\underset{\underset{\displaystyle H}{|}}{C}} \quad ; \quad \overset{\overset{\displaystyle H}{|}}{\underset{\underset{\displaystyle H}{|}}{C}} = C = C = \overset{\overset{\displaystyle H}{|}}{\underset{\underset{\displaystyle H}{|}}{C}}$$

$$\text{(I)} \qquad\qquad \text{(II)} \qquad\qquad \text{(III)} \qquad\qquad \text{(IV)}$$

Ratio of :	2 : 1	;	1 : 1	;	4/3 : 1	;	1 : 1
H to C	Replace none		Replace one or two H atoms		Replace one		Replace one

$$\overset{\overset{\displaystyle H}{|}}{\underset{\underset{\displaystyle H}{|}}{C}} = C = C = C = \overset{\overset{\displaystyle H}{|}}{\underset{\underset{\displaystyle H}{|}}{C}} \qquad ; \qquad \overset{\overset{\displaystyle H}{|}}{\underset{\underset{\displaystyle H}{|}}{C}} = C = C = C = C = \overset{\overset{\displaystyle H}{|}}{\underset{\underset{\displaystyle H}{|}}{C}}$$

$$\text{(V)} \qquad\qquad\qquad\qquad \text{(VI)}$$

Ratio of :   4/5 : 1   ;   2/3 : 1

H to C   <u>Replace one</u>   <u>Replace one</u>   8.196

From the analysis so far, the order of nucleophilicity of the centers can therefore be represented as follows.

$$(VI) \quad > \quad (V) \quad > \quad (IV) \quad > \quad (III) \quad > \quad (II) \quad > \quad (I)$$

<u>Order of Nucleophilicity</u>   8.197a

$$H\text{-}C \equiv C\text{-}C \equiv C\text{-}CH=CH_2 > H\text{-}C \equiv C\text{-}C \equiv C\text{-}CH_3 > H\text{-}C \equiv C\text{-}CH=CH_2 > H\text{-}C \equiv C\text{-}CH_3 > H\text{-}C \equiv C\text{-}H >$$

(VI)   (V)   (IV)   (III)   (II)

<u>ORDER OF NUCLEOPHILICITY</u>   8.197b

The replacement of H atoms is not based on the ratios of H to C atoms in the compound, but something else. 1,6-cumulene above has one H atom to replace, because when activated and made to exist in Equilibrium state of existence, the followings are obtained.

ACTIVATION OF 1,6-CUMULENE   8.198a

ELECTRORADICALIZATION   8.198b

Now, considering 1,5-cumulenes with one or more radical-pulling groups, the followings are obtained for F.

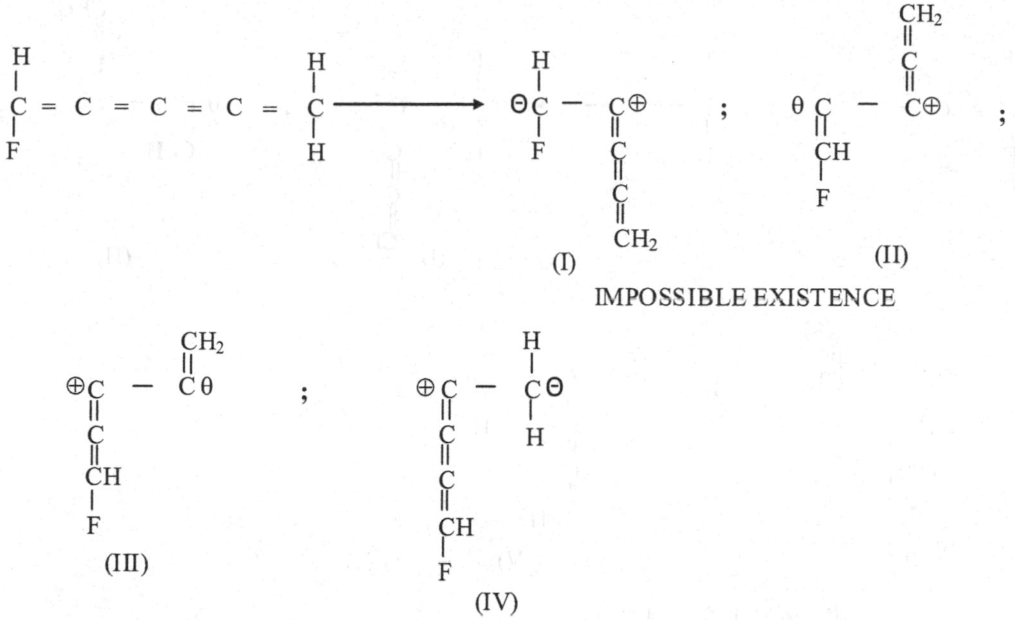

$$
\begin{array}{c}
\overset{\displaystyle H}{\underset{\displaystyle F}{|}} \\
C = C = C = C = \overset{\displaystyle H}{\underset{\displaystyle H}{C}}
\end{array}
\longrightarrow
\begin{array}{c}
\overset{\displaystyle H}{\underset{\displaystyle F}{|}} \\
\ominus C - C^{\oplus} \\
\text{(chain)}
\end{array}
$$

(I)   (II)

IMPOSSIBLE EXISTENCE

(III)   (IV)   8.199

(I) is the least nucleophilic center, that which can only be activated radically. When resonance stabilized, the followings are obtained.

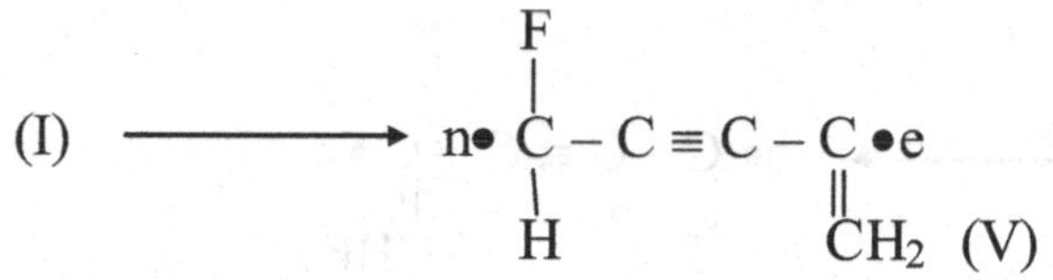

$$
(I) \longrightarrow n \bullet \overset{\displaystyle F}{\underset{\displaystyle H}{|}} C - C \equiv C - \overset{\displaystyle }{\underset{\displaystyle CH_2}{C}} \bullet e
$$

(V)   8.200

It is (V) that is used for polymerization of this monomer. One knows what to expect when F is symmetrically placed.

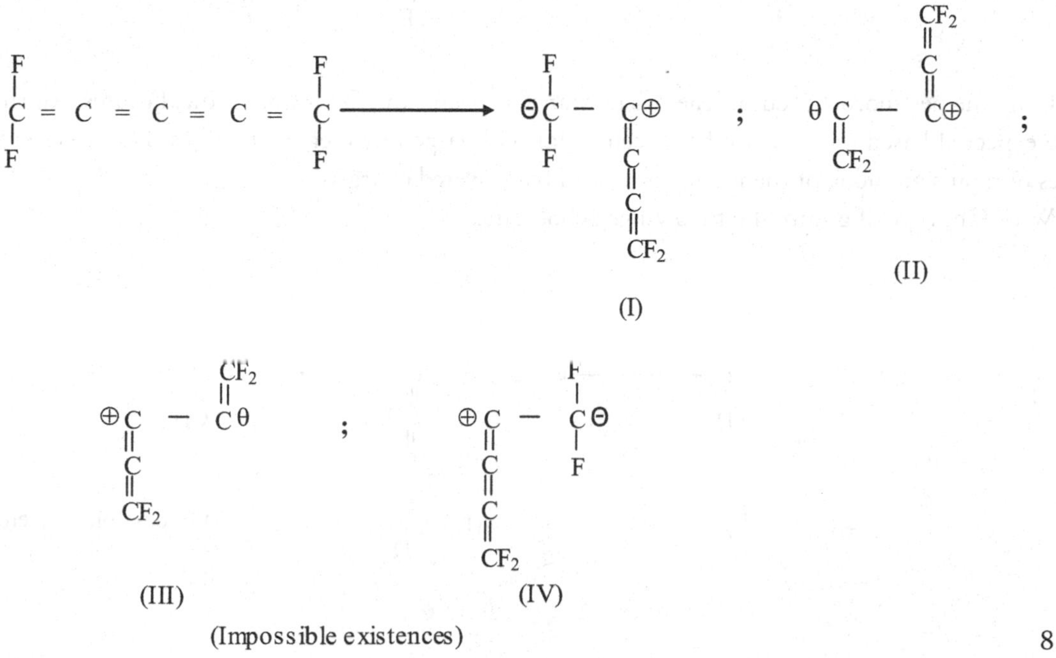

$$
\begin{array}{c}
\overset{\displaystyle F}{\underset{\displaystyle F}{|}} \\
C = C = C = C = \overset{\displaystyle F}{\underset{\displaystyle F}{C}}
\end{array}
\longrightarrow
$$

(I)   (II)

(III)   (IV)

(Impossible existences)   8.201

652

$$
\begin{matrix} F \\ | \\ C \\ | \\ H \end{matrix} = C = C = C = \begin{matrix} F \\ | \\ C \\ | \\ H \end{matrix} \longrightarrow \ominus\begin{matrix} F \\ | \\ C \\ | \\ H \end{matrix} - \begin{matrix} C \oplus \\ || \\ C \\ || \\ C \\ || \\ CFH \end{matrix} \quad ; \quad \theta\begin{matrix} CFH \\ || \\ C \\ || \\ C \\ | \\ CFH \end{matrix} - C \oplus \quad ;
$$

(I) (II)

$$
\oplus\begin{matrix} C \\ || \\ C \\ || \\ CFH \end{matrix} - \begin{matrix} CFH \\ || \\ C\theta \end{matrix} \quad ; \quad \oplus\begin{matrix} C \\ || \\ C \\ || \\ C \\ || \\ CFH \end{matrix} - \begin{matrix} F \\ | \\ C\theta \\ | \\ H \end{matrix}
$$

(III) (IV)

(Impossible existences) 8.202

For them, the followings are obtained radically.

$$
\begin{matrix} F \\ | \\ C \\ | \\ H \end{matrix} = C = C = C = \begin{matrix} F \\ | \\ C \\ | \\ H \end{matrix} \xrightarrow{\text{ACTIVATION}} n\bullet \begin{matrix} F \\ | \\ C \\ | \\ H \end{matrix} - C \equiv C - \begin{matrix} C \\ || \\ CHF \end{matrix}\bullet e
$$

8.203

$$
\begin{matrix} F \\ | \\ C \\ | \\ F \end{matrix} = C = C = C = \begin{matrix} F \\ | \\ C \\ | \\ F \end{matrix} \xrightarrow{\text{ACTIVATION}} n\bullet \begin{matrix} F \\ | \\ C \\ | \\ F \end{matrix} - C \equiv C - \begin{matrix} C \\ || \\ CF_2 \end{matrix}\bullet e
$$

8.204

Only the perfluoro 1,5-cumulene has no transfer species and favor both radical routes. All these are to be expected based on what we have seen so far with large members of families. Due to electrostatic forces of repulsion, none of the mono-forms can be activated chargedly.

With CF_3 type of group, the followings are obtained.

$$
\begin{matrix} CF_3 \\ _1| \\ C \\ | \\ H \end{matrix} \overset{2}{=} C \overset{3}{=} C \overset{4}{=} \overset{5}{\underset{}{C}} = \begin{matrix} H \\ | \\ C \\ | \\ H \end{matrix} \longrightarrow \ominus\begin{matrix} CF_3 \\ _1| \\ C \\ | \\ H \end{matrix} - \begin{matrix} C \oplus \\ || \\ C \\ || \\ C \\ | \\ CH \\ | \\ H \end{matrix} \quad ; \quad \theta\overset{2}{\underset{}{\begin{matrix} C \\ || \\ CH \\ | \\ CF_3 \end{matrix}}} - \begin{matrix} CH_2 \\ || \\ C \\ || \\ C\oplus \\ _3 \end{matrix} \quad ;
$$

(I) (II) Impossible existence

653

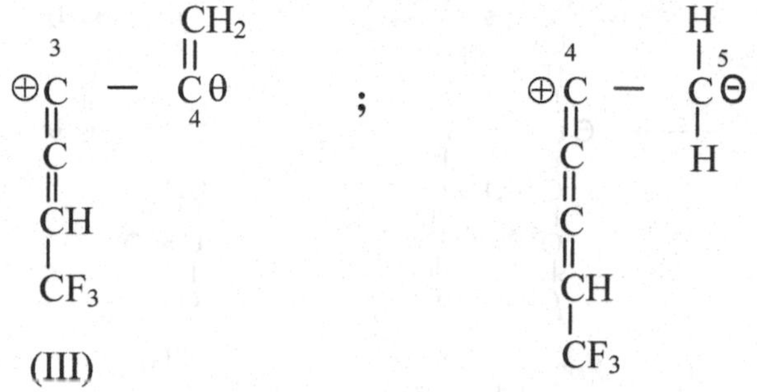

$$\overset{3}{\oplus C} - \overset{CH_2}{\underset{4}{\overset{\|}{C}\ominus}} \quad ; \quad \overset{4}{\oplus C} - \overset{H}{\underset{5}{\overset{|}{C}\ominus}}$$

(where the first structure has C=C, CH, CF₃ chain below C3, and second structure has C=C, C=C, CH, CF₃ chain below)

$$\overset{3}{\oplus C} - \overset{CH_2}{\underset{4}{\overset{\|}{C}\ominus}} \quad ; \quad \overset{4}{\oplus C} - \overset{H}{\underset{5}{\overset{|}{C}\ominus H}}$$

(III)   (IV)

Impossible existence   8.205

The least nucleophilic center is (I). Radically, it resonance stabilizes when activated to the 1,4 mono-form as shown below.

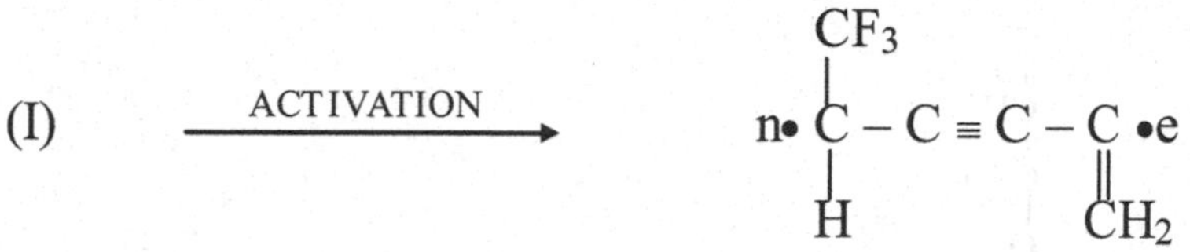

$$(I) \xrightarrow{\text{ACTIVATION}} n \bullet \overset{CF_3}{\underset{H}{\overset{|}{C}}} - C \equiv C - \overset{}{\underset{CH_2}{\overset{\|}{C}}} \bullet e$$

(V)   8.206

To polymerize this, free electro-free-radical initiators cannot be used. Only paired initiators can be used nucleo-free-radically. The same will apply when the second H is changed to CH_3 as shown below.

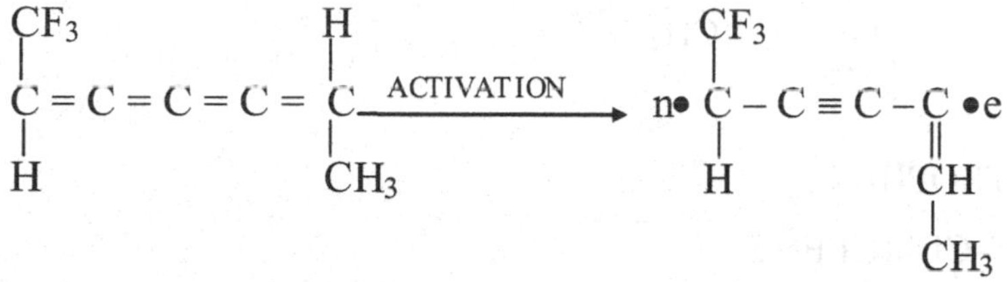

$$\overset{CF_3}{\underset{H}{\overset{|}{C}}} = C = C = C = \overset{H}{\underset{CH_3}{\overset{|}{C}}} \xrightarrow{\text{ACTIVATION}} n \bullet \overset{CF_3}{\underset{H}{\overset{|}{C}}} - C \equiv C - \overset{}{\underset{\underset{CH_3}{|}}{\overset{\|}{C}{CH}}} \bullet e$$

8.207

$$\overset{R_F}{\underset{H}{\overset{|}{C}}} = C = C = C = \overset{H}{\underset{R}{\overset{|}{C}}} \xrightarrow{\text{ACTIVATION}} n \bullet \overset{R_F}{\underset{H}{\overset{|}{C}}} - C \equiv C - \overset{}{\underset{\underset{R}{|}}{\overset{\|}{C}{CH}}} \bullet e$$

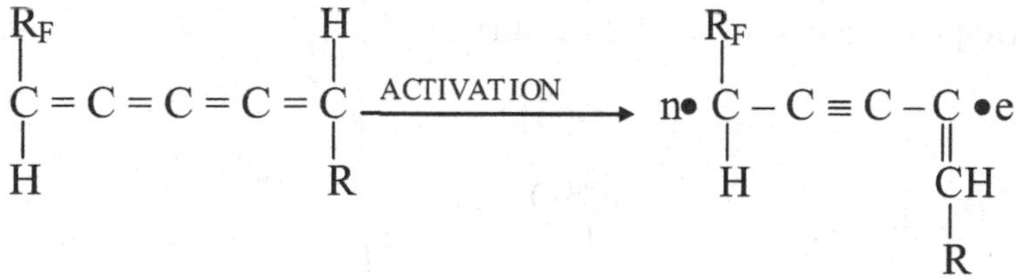

(where $R_F \equiv C_nF_{2n+1}$, $R \equiv C_nH_{2n+1}$)   8.208

With the introduction of radical-pulling groups in place of H above, we have almost the same scenario as shown below.

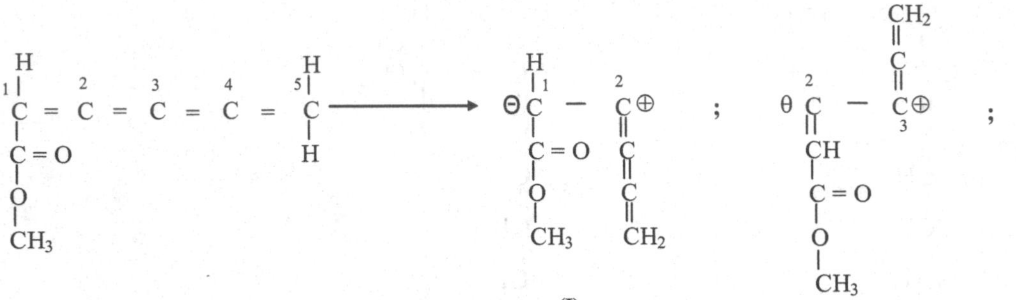

$$
\begin{array}{c}
\overset{H}{\underset{1}{\overset{|}{C}}} = \overset{2}{C} = \overset{3}{C} = \overset{4}{C} = \overset{H}{\underset{5}{\overset{|}{C}}} \\
| \qquad\qquad\qquad\qquad\quad | \\
C = O \qquad\qquad\qquad\quad H \\
| \\
O \\
| \\
CH_3
\end{array}
\longrightarrow
\begin{array}{c}
\Theta \overset{H}{\underset{1}{\overset{|}{C}}} - \overset{2}{C}\oplus \\
| \qquad\quad || \\
C = O \quad C \\
| \qquad\quad || \\
O \quad\quad C \\
| \qquad\quad || \\
CH_3 \quad CH_2
\end{array}
\quad ; \quad
\begin{array}{c}
\qquad\quad CH_2 \\
\qquad\quad || \\
\qquad\quad C \\
\theta \overset{2}{C} - \overset{3}{C}\oplus \\
|| \\
CH \\
| \\
C = O \\
| \\
O \\
| \\
CH_3
\end{array}
\quad ;
$$

(I)

(II) Impossible existence

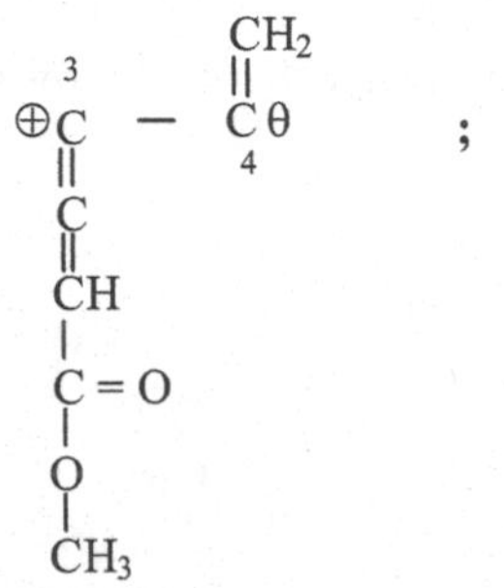

$$
\begin{array}{c}
\qquad\quad CH_2 \\
\qquad\quad || \\
\oplus \overset{3}{C} - \overset{}{C}\theta \\
|| \qquad\quad 4 \\
C \\
|| \\
CH \\
| \\
C = O \\
| \\
O \\
| \\
CH_3
\end{array}
\quad ; \quad
\begin{array}{c}
\qquad\quad\quad \overset{H}{\underset{5}{\overset{|}{}}} \\
\oplus \overset{4}{C} - \overset{}{C}\theta \\
|| \qquad\quad | \\
C \qquad\quad H \\
|| \\
C \\
|| \\
CH \\
| \\
C = O \\
| \\
O \\
| \\
CH_3
\end{array}
$$

(III) Impossible existence

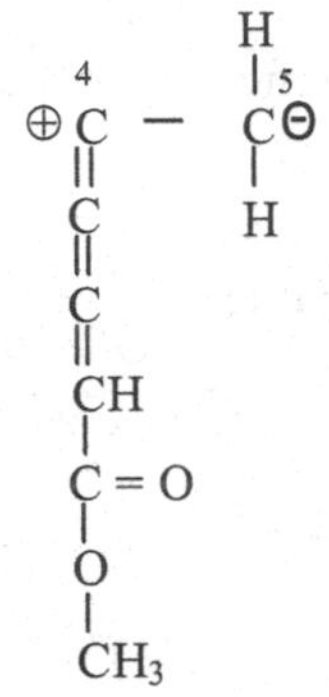

(IV)

8.209

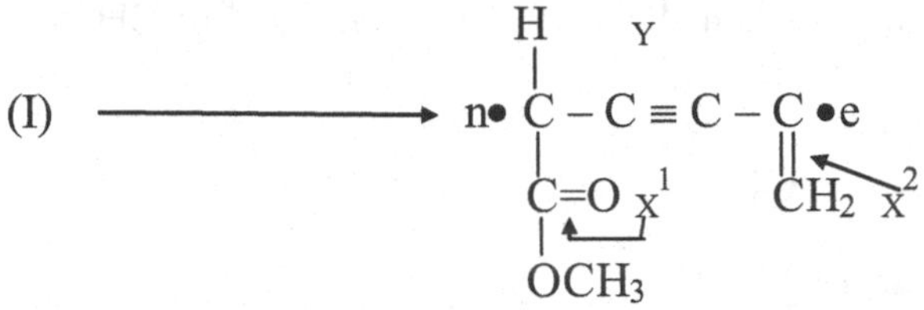

$$(I) \longrightarrow
\begin{array}{c}
H \qquad Y \\
| \qquad\quad | \\
n\bullet \overset{}{C} - C \equiv C - \overset{}{C} \bullet e \\
| \qquad\qquad\qquad || \\
C = O \;\; x^1 \qquad CH_2 \;\; x^2 \\
| \\
OCH_3
\end{array}
$$

ELECTROPHILE

FROM 1,5-CUMULENE

8.210

Recalled below again is the corresponding case of the 1,4-cumulene.

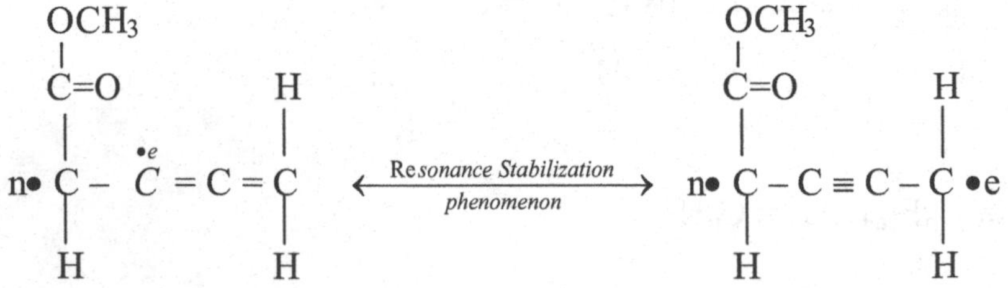

$$
\begin{array}{c}
OCH_3 \\
| \\
C = O \qquad\qquad H \\
| \qquad\qquad\qquad | \\
n\bullet \overset{}{C} - \overset{\bullet e}{C} = C = \overset{}{C} \\
| \qquad\qquad\qquad\quad | \\
H \qquad\qquad\qquad H
\end{array}
\xleftrightarrow[\text{phenomenon}]{\textit{Resonance Stabilization}}
\begin{array}{c}
OCH_3 \\
| \\
C = O \qquad\qquad H \\
| \qquad\qquad\qquad | \\
n\bullet \overset{}{C} - C \equiv C - \overset{}{C} \bullet e \\
| \qquad\qquad\qquad\quad | \\
H \qquad\qquad\qquad H
\end{array}
$$

ELECTROPHILE　　　　　　　　　ELECTROPHILE

RESONANCE STABILIZATION PHENOMENON

FROM 1,4-CUMULENE

(8.144)

The male from 1,5-cumulene is far less electrophilic than the male from 1,4-cumulene particularly when the groups are cis-placed. To polymerize nucleo-free-radically the male center in 1,5-cumulene, firstly, the groups must be trans-placed. Secondly, when trans-placed, paired nucleo-free-radical initiators must be used. Electro-free-radically, it is X^1 center that is involved instead of the X^2 center.

$$
\begin{array}{c}
\underset{\underset{N}{\overset{\overset{H}{|}}{\underset{\|\|}{C}}}{C}} = C = C = C = \underset{\overset{|}{H}}{\overset{\overset{H}{|}}{C}} \longrightarrow \ominus \underset{\underset{N}{\overset{\|\|}{C}}}{C} - \underset{\underset{CH_2}{\overset{\|}{C}}}{\overset{\oplus}{C}} \quad ; \quad \theta\ \underset{\underset{CN}{\overset{\|}{CH}}}{C} - \underset{\overset{CH_2}{\overset{\|}{C}}}{\overset{\oplus}{C}} \quad ;
\end{array}
$$

$$
\begin{array}{c}
\oplus\underset{\underset{CN}{\overset{\|}{CH}}}{\overset{\|}{C}} - \underset{\overset{CH_2}{\overset{\|}{}}}{C\theta} \quad ; \quad \oplus\underset{\underset{CN}{\overset{\|}{CH}}}{\overset{\|}{C}} - \underset{\overset{H}{|}}{C\ominus}
\end{array}
$$

(I)   (II)   (III)   (IV)   8.211

$$
(I) \longrightarrow n\bullet \underset{\underset{C\equiv N}{|}}{\overset{\overset{H}{|}}{C}} - C \equiv C - \underset{\overset{CH_2}{\|}}{C} \bullet e
$$

WEAK ELECTROPHILE   8.212

This is by far a weaker electrophile than the case above, despite the limitations already indicated, because what is obvious is the following order in capacities of the following Activation centers.

$$C \equiv N \quad > \quad 1,5 - \quad > \quad 1,4 - \quad > \quad 1,3 - \quad \text{Cumulenes}$$

<u>Order of Nucleophilicity</u>   8.213

What makes the C = O center unique is when it is present in COR group as opposed to COOR or $CONH_2$ group. With COR group, there is no transfer species, while with the others, there are transfer species.

$$
\begin{array}{c}
\underset{\underset{H}{\overset{\overset{H}{|}}{\underset{|}{C}}}}{\overset{1}{C}} = \overset{2}{C} = \overset{3}{C} = \overset{4}{C} = \underset{\overset{|}{H}}{\overset{\overset{H}{5|}}{C}} \longrightarrow \ominus\underset{\underset{H}{\overset{\|}{C=O}}}{\overset{1}{C}} - \underset{\underset{CH_2}{\overset{\|}{C}}}{\overset{2}{\overset{\oplus}{C}}} \quad ; \quad \oplus\underset{\underset{H}{\overset{\|}{C=O}}}{\overset{2}{\overset{\|}{C}}} - \underset{\overset{CH_2}{\overset{\|}{C}}}{\overset{3}{C\ominus}} \quad ;
\end{array}
$$

(I)   (II)

656

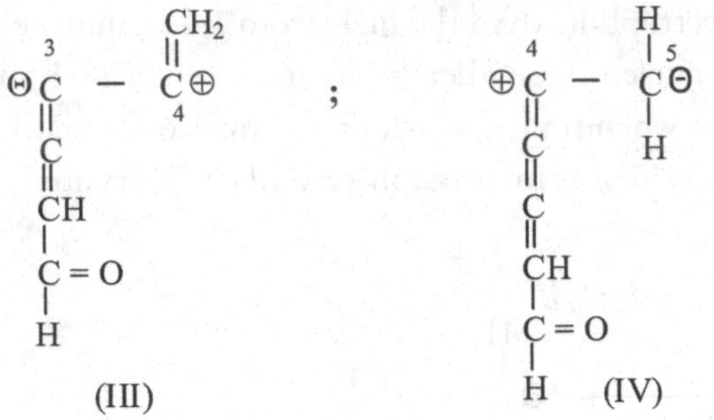

$$\underset{\text{(III)}}{\overset{\overset{\overset{3}{\text{CH}_2}}{\|}}{\ominus\text{C}-\overset{4}{\text{C}}\oplus}} \quad ; \quad \underset{\text{(IV)}}{\overset{\overset{4}{\oplus}\text{C}-\overset{\overset{5}{H}}{\text{C}}\ominus}{}}$$

8.214

$$\text{(I)} \longrightarrow n\bullet \overset{H}{\underset{\underset{H}{C=O}}{C}}-C\equiv C-\overset{\underset{CH_2}{\|}}{C}\bullet e$$

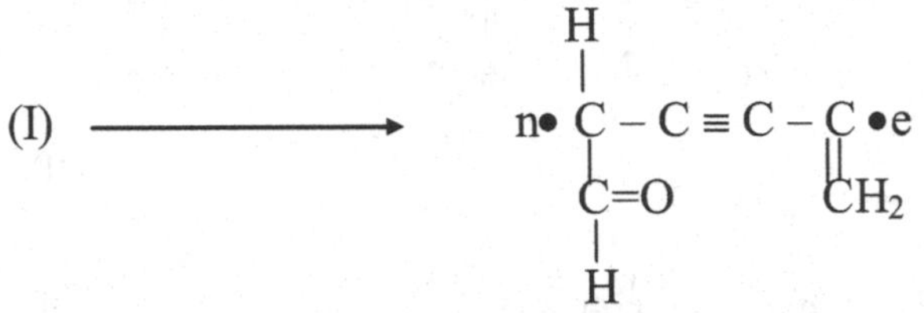

WEAK ELECTROPHILE 8.215

No copolymers can be produced here. Its behavior still remains almost like others wherein the use of paired initiators is essential if their natural route must be favored.

8.3.2 Free-radical Characters of 1,5 - cumulenes

With the first member, the followings are obtained when activated.

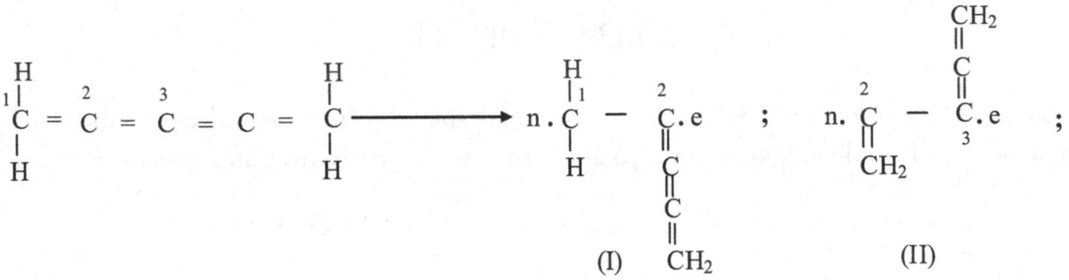

$$\overset{H}{\underset{H}{\overset{1}{C}}}=\overset{2}{C}=\overset{3}{C}=C=\overset{H}{\underset{H}{C}} \longrightarrow n.\overset{H}{\underset{H}{\overset{1}{C}}}-\overset{2}{\underset{\underset{CH_2}{C}}{C}}.e \quad ; \quad n.\overset{2}{\underset{\underset{CH_2}{\|}}{C}}-\overset{\overset{\overset{CH_2}{\|}}{C}}{\underset{3}{C}}.e \quad ;$$

(I) (II) 8.216

We have seen a great deal of these characters in the last sub-section for exploratory purposes, since it has already been established that these monomers cannot be activated charged. For this reason, only very important states will be shown herein. (II) Above does not exist. For this family, there are only two mono-forms.

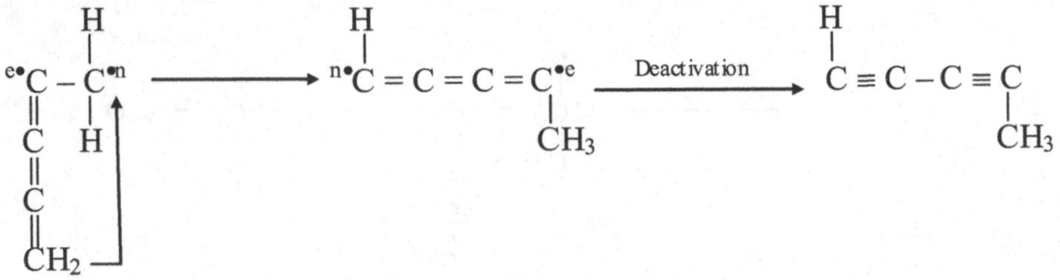

$$e\bullet\overset{H}{\underset{\underset{CH_2}{C}}{C}}-\overset{H}{\underset{}{C}}\bullet n \longrightarrow n\bullet C=C=C=C\bullet e \xrightarrow{\text{Deactivation}} \overset{H}{C}\equiv C-C\equiv C$$

(I) (III) 8.217

657

$$\overset{CH_3}{\underset{H}{\overset{1}{C}}} = \overset{2}{C} = \overset{3}{C} = \overset{4}{C} = \overset{5}{\underset{H}{\overset{H}{C}}} \longrightarrow n \cdot \overset{CH_3}{\underset{H}{\overset{1}{C}}} - \underset{\underset{CH_2}{\overset{\|}{C}}}{\overset{2}{C}} \cdot e \quad ; \quad n \cdot \overset{2}{\underset{\underset{CH_3}{CH}}{C}} - \underset{CH_2}{\overset{\overset{CH_2}{\overset{\|}{C}}}{\overset{3}{C}}} \cdot e \quad ;$$

$$\text{(I)} \qquad\qquad\qquad\qquad \text{(II)}$$

$$e \cdot \overset{3}{\underset{\underset{CH_3}{\overset{CH}{\overset{\|}{C}}}}{C}} - \overset{\overset{CH_2}{\overset{\|}{C}}}{\overset{4}{C}} \cdot n \quad ; \quad e \cdot \overset{4}{\underset{\underset{CH_3}{\overset{CH}{\overset{\|}{C}}}}{C}} - \overset{5}{\underset{H}{\overset{H}{C}}} \cdot n$$

$$\text{(III)} \qquad\qquad\qquad \text{(IV)} \qquad\qquad\qquad\qquad 8.218$$

(IV) is one of the two mono-forms.

$$e\bullet \overset{H}{\underset{\underset{\underset{CH_3}{\overset{\|}{CH}}}{\overset{\|}{C}}}{C}} - \overset{\bullet n}{\underset{H}{\overset{H}{C}}} \longrightarrow e\bullet \overset{CH_3}{\overset{|}{C}} = C = C = \overset{\bullet n}{\underset{CH_3}{\overset{|}{C}}} \xrightarrow{\text{Deactivation}} \overset{CH_3}{\overset{|}{C}} \equiv C - C \equiv \underset{CH_3}{\overset{|}{C}}$$

$$\qquad\qquad\qquad\qquad\qquad\qquad \text{(V)}$$

$$\text{(I)} \qquad\qquad \text{MOLECULAR REARRANGEMENT} \qquad\qquad 8.219$$

$$e\bullet \overset{H}{\underset{\underset{\underset{CH_3}{\overset{\|}{CH}}}{\overset{\|}{C}}}{C}} - \overset{\bullet n}{\underset{H}{\overset{H}{C}}} \longrightarrow n\bullet \overset{H}{\underset{H}{\overset{|}{C}}} - C \equiv C - \underset{\underset{CH_3}{\overset{|}{CH}}}{C} \bullet e \xrightarrow{\text{Deactivation}} \overset{H}{\underset{H}{\overset{|}{C}}} = C = C = C = \overset{H}{\underset{CH_3}{\overset{|}{C}}}$$

$$\qquad\qquad\qquad\qquad \text{(VI)}$$

$$\text{(I)} \qquad\qquad \text{RESONANCE STABILIZATION PHENOMENON} \qquad\qquad (8.183)$$

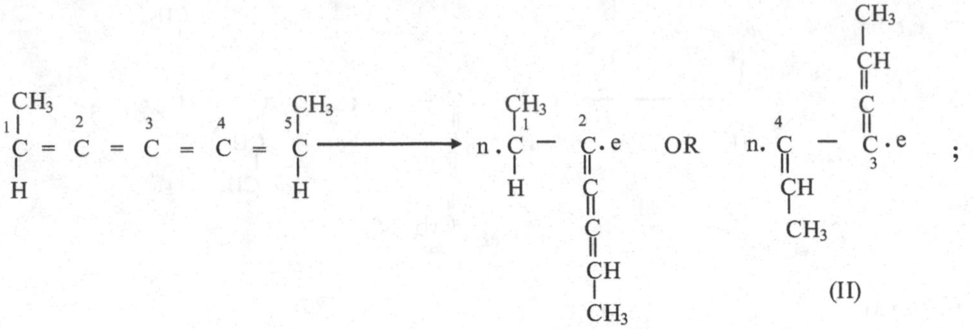

$$\underset{\underset{H}{\overset{CH_3}{\underset{|}{\overset{|}{{}_1C}}}}}{} = \underset{}{\overset{2}{C}} = \underset{}{\overset{3}{C}} = \underset{}{\overset{4}{C}} = \underset{\underset{H}{\overset{CH_3}{\underset{|}{\overset{|}{{}_5C}}}}}{} \longrightarrow n . \underset{\underset{CH_3}{\overset{CH_3}{\underset{|}{\overset{|}{{}_1C}}}}}{} - \overset{2}{C} . e \quad OR \quad n . \underset{\underset{CH_3}{\overset{CH_3}{\underset{|}{\overset{|}{C}}}}}{} - \overset{4}{C} . e$$

(I) (II)

8.220

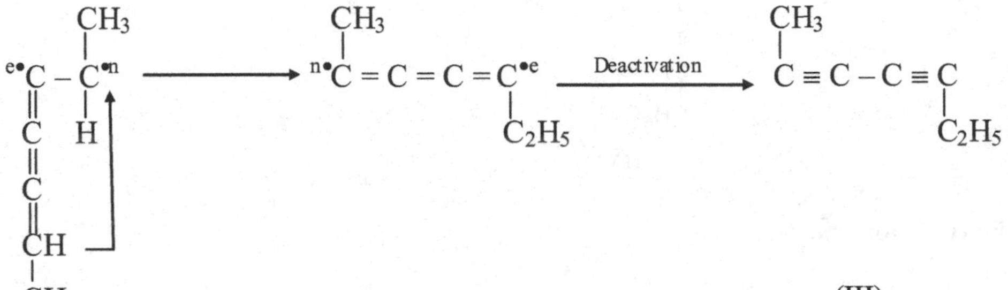

MOLECULAR REARRANGEMENT

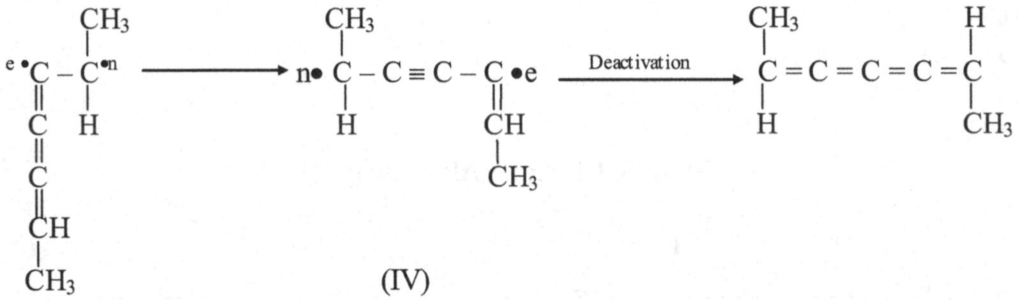

(I) → (III)

8.221

$$e\bullet \overset{CH_3}{\underset{\underset{CH_3}{\overset{|}{\underset{|}{C}}}}{\overset{|}{C}}} - \overset{CH_3}{\underset{H}{\overset{|}{C}}}\bullet n \longrightarrow n\bullet \overset{CH_3}{\underset{\underset{CH_3}{\overset{|}{\underset{|}{C}}}}{\overset{|}{C}}} = C = C = \overset{CH_3}{\underset{C_2H_5}{\overset{|}{C}}}\bullet e \xrightarrow{\text{Deactivation}} \overset{CH_3}{\underset{C_2H_5}{\overset{|}{C}}} \equiv C - C \equiv \overset{}{C}$$

(III)

RESONANCE STABILIZATION PHENOMENON

(I) → (IV)

8.222

Beginning from the first member, all these can be seen to be strong Nucleophiles favoring only the electro-free-radical route.

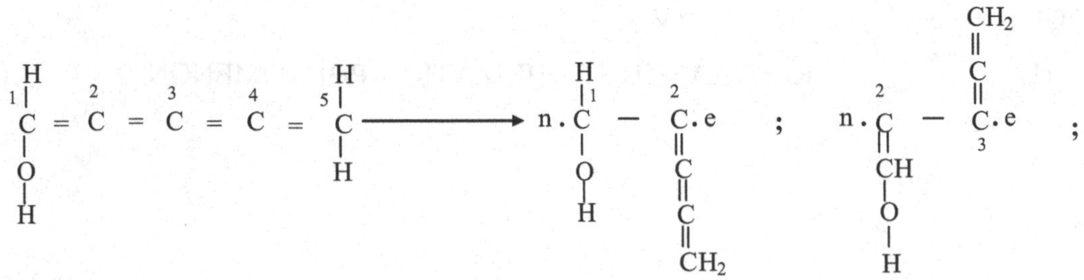

$$\underset{\underset{H}{\overset{H}{\underset{|}{\overset{|}{O}}}}}{{}_1C} = \overset{2}{C} = \overset{3}{C} = \overset{4}{C} = \underset{\underset{H}{\overset{H}{\underset{|}{\overset{|}{}}}}}{{}_5C} \longrightarrow n . C - C . e \quad ; \quad n . C - C . e \quad ;$$

(I) (Nucleophiles) (II)

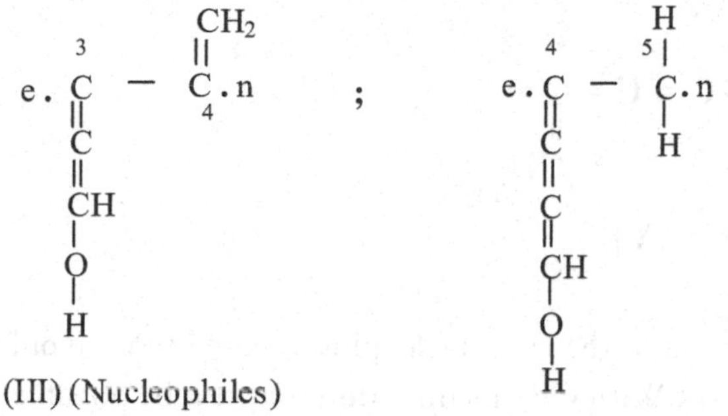

$$
e \cdot \underset{\underset{\underset{\underset{H}{|}}{O}}{\overset{\overset{3}{C}}{\underset{\|}{CH}}}}{\overset{\overset{CH_2}{\|}}{C}} - \underset{4}{C} \cdot n \quad ; \quad e \cdot \underset{\underset{\underset{\underset{H}{|}}{O}}{\overset{4}{C}}}{\overset{C}{\underset{\|}{CH}}} - \underset{\overset{5}{\overset{|}{H}}}{\overset{H}{\underset{|}{C}}} \cdot n
$$

(III) (Nucleophiles)   (IV)   8.223

Though four mono-forms have been shown above, only one exists. The others cannot exist.
(IV) which is the least nucleophilic center, cannot molecularly rearrange because the pushing capacity of OH or OR group is greater than CH_3 group. It can however be resonance stabilized as shown below.

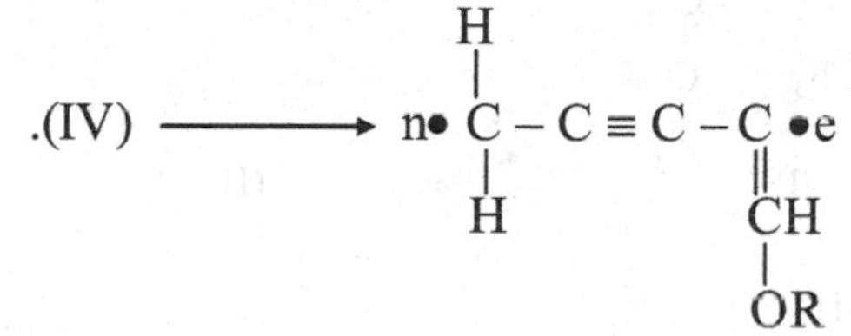

$$
.(IV) \longrightarrow n\bullet \underset{\underset{H}{|}}{\overset{\overset{H}{|}}{C}} - C \equiv C - \underset{\underset{\underset{OR}{|}}{CH}}{\overset{\|}{C}} \bullet e
$$

(V)   (8.193)

This is the mono-form that is involved for polymerization and as can be seen, it is Nucleophilic in character.

When one H in 1,5 - cumulene is replaced with F or CF_3 types of groups, one knows what to expect.

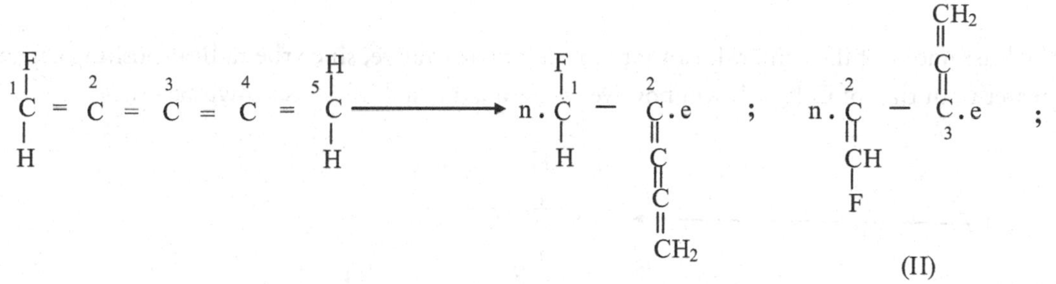

$$
\underset{\underset{H}{|}}{\overset{\overset{F}{|}}{\underset{1}{C}}} = \underset{2}{C} = \underset{3}{C} = \underset{4}{C} = \underset{\underset{H}{|}}{\overset{\overset{H}{\|}}{\underset{5}{C}}} \longrightarrow n . \underset{\underset{H}{|}}{\overset{\overset{F}{|}}{\underset{1}{C}}} - \underset{\underset{\underset{CH_2}{\|}}{\overset{C}{\underset{\|}{C}}}}{\overset{}{\underset{2}{C}}} \cdot e \quad ; \quad n . \underset{\underset{\underset{F}{|}}{\overset{C}{\underset{\|}{CH}}}}{\overset{}{\underset{2}{C}}} - \underset{3}{\overset{\overset{CH_2}{\|}}{C}} \cdot e \quad ;
$$

(I)   (II)

$$
e . \underset{\underset{\underset{F}{|}}{\overset{C}{\underset{\|}{CH}}}}{\overset{\overset{3}{C}}{}} - \underset{4}{\overset{\overset{CH_2}{\|}}{C}} \cdot n \quad ; \quad e . \underset{\underset{\underset{H}{|}}{\overset{C}{\underset{\|}{CF}}}}{\overset{\overset{4}{C}}{}} - \underset{\overset{5}{\overset{|}{H}}}{\overset{H}{\underset{|}{C}}} \cdot n
$$

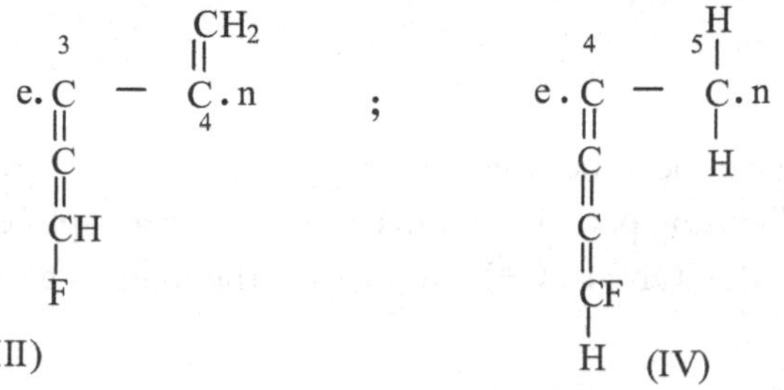

(III)   (IV)   8.224

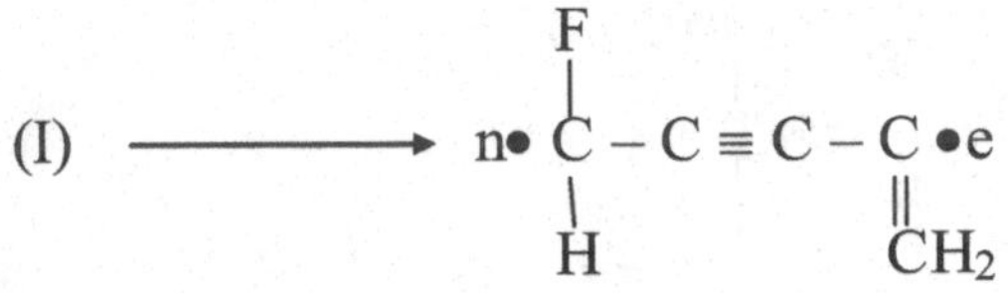

$$(I) \longrightarrow n\bullet \overset{\displaystyle F}{\underset{\displaystyle H}{C}} - C \equiv C - \underset{\displaystyle \overset{\|}{CH_2}}{C} \bullet e$$

$$(V) \hspace{5cm} (8.200)$$

The monomer still is Nucleophilic. (I) is of course the least nucleophilic center and the only center that will be involved all the time for this monomer. With CF_3, its initiation may not be possible.

With C_nF_{n+1} types of groups, the followings are obtained.

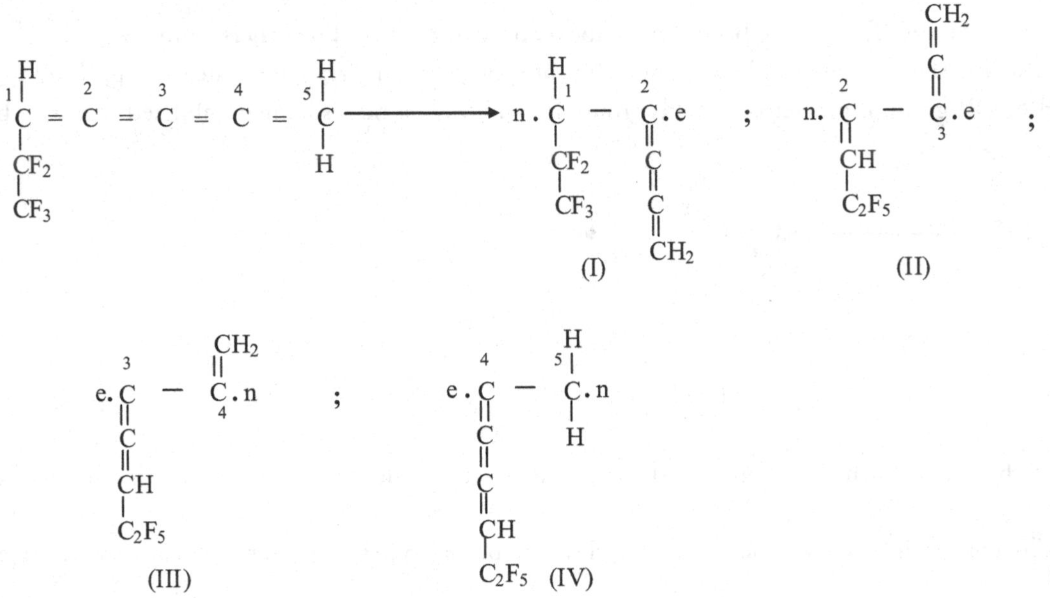

$$\underset{1}{\overset{\displaystyle H}{C}} = \overset{2}{C} = \overset{3}{C} = \overset{4}{C} = \underset{5}{\overset{\displaystyle H}{C}} \longrightarrow n.\overset{\displaystyle H}{\underset{\displaystyle \underset{CF_3}{CF_2}}{\overset{1}{C}}} - \overset{2}{\underset{\underset{CH_2}{C}}{C}}.e \quad ; \quad n.\overset{2}{\underset{\underset{C_2F_5}{CH}}{C}} - \overset{\overset{CH_2}{\|}}{\underset{3}{C}}.e \quad ;$$

with substituents CF_2, CF_3 on C_1 and H on C_5.

$$(I) \hspace{4cm} (II)$$

$$e.\overset{3}{\underset{\underset{C_2F_5}{CH}}{C}} - \overset{\overset{CH_2}{\|}}{\underset{4}{C}}.n \quad ; \quad e.\overset{4}{\underset{\underset{C_2F_5}{CH}}{C}} - \overset{5}{\underset{\displaystyle H}{\overset{\displaystyle H}{C}}}.n$$

$$(III) \hspace{4cm} (IV) \hspace{3cm} 8.225$$

(I) is the least nucleophilic center. It cannot molecularly rearrange, since the radical-pushing capacity of H is greater than that of C_nF_{n+1}. It will however be resonance stabilized as shown below.

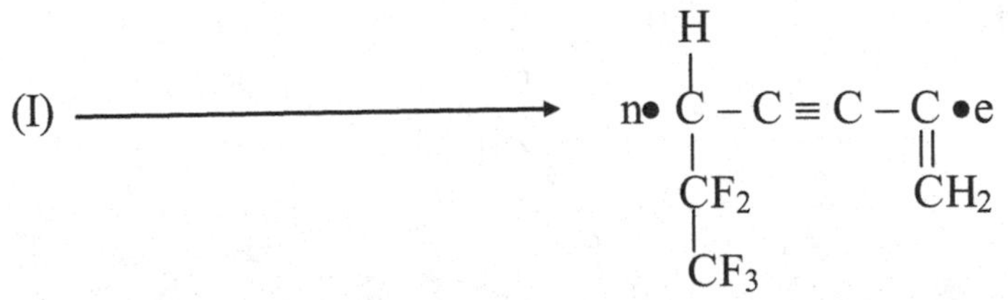

$$(I) \longrightarrow n\bullet \overset{\displaystyle H}{\underset{\displaystyle \underset{CF_3}{CF_2}}{C}} - C \equiv C - \underset{\overset{\|}{CH_2}}{C} \bullet e$$

$$8.226$$

Electro-free-radically, F is the transfer species. Nucleo-free-radically H is the transfer species. To polymerize the monomer electro-free-radically, nothing can be done whether the groups are cis- or trans-placed. It can only be polymerized nucleo-free-radically using paired initiators only when trans-placed.

For radical-pulling groups such as COOR, $CONH_2$, $C\equiv N$ and COR, the followings are obtained.

$$
\begin{array}{c}
\overset{H}{\underset{1}{|}} \quad\quad\quad\quad\quad\quad \overset{H}{\underset{5}{|}} \\
C = \overset{2}{C} = \overset{3}{C} = \overset{4}{C} = C \\
| \qquad\qquad\qquad\qquad\qquad | \\
C = O \qquad\qquad\qquad\qquad H \\
| \\
O \\
| \\
CH_3
\end{array}
\longrightarrow
\quad n.\overset{H}{\underset{|}{\underset{1}{C}}} - \overset{2}{C}.e
\quad ; \quad
n.\overset{2}{C} - \overset{CH_2}{\underset{||}{C}}.e \quad ;
$$

where (I) bears $C=O$ / O / CH_3 and $C\!=\!\!=\!C\!=\!\!=\!CH_2$ group; (II) bears CH / $C=O$ / O / CH_3 group and $CH_2\!=\!C$ group.

$$
e.\overset{3}{C} - \overset{CH_2}{\underset{||}{\underset{4}{C}}}.n \quad ; \qquad
e.\overset{4}{C} - \overset{H}{\underset{|}{\overset{5}{C}}}.n
$$

(III) bears C / CH / $C=O$ / O / CH_3 ; (IV) bears H / C / C / CH / $C=O$ / O / CH_3

(Electrophiles)   8.227

$$
\begin{array}{c}
\overset{H}{\underset{1}{|}} \quad\quad\quad\quad\quad\quad \overset{H}{\underset{5}{|}} \\
C = \overset{2}{C} = \overset{3}{C} = \overset{4}{C} = C \\
| \qquad\qquad\qquad\qquad\qquad | \\
C = O \qquad\qquad\qquad\qquad H \\
| \\
H
\end{array}
\longrightarrow
\quad n.\overset{H}{\underset{|}{\underset{1}{C}}} - \overset{2}{C}.e
\quad ; \quad
n.\overset{2}{C} - \overset{CH_2}{\underset{||}{C}}.e \quad ;
$$

(I) bears $C=O$ / H and $C\!=\!\!=\!C\!=\!\!=\!CH_2$; (II) bears CH / $C=O$ / H and $CH_2\!=\!C$

$$
e.\overset{3}{C} - \overset{CH_2}{\underset{||}{\underset{4}{C}}}.n \quad ; \qquad
e.\overset{4}{C} - \overset{H}{\underset{|}{\overset{5}{C}}}.n
$$

(III) bears C / CH / $C=O$ / H ; (IV) bears H / C / C / CH / $C=O$ / H

(Electrophiles)   8.228

Of all the mono-forms above, only the first one exists for each of them and indeed all of them. All these have been largely covered in the last sub-section. The methods of analysis are now clearly obvious. The strong influence of molecular rearrangement and resonance stabilization phenomena can be observed with these cumulenes. There are some of the existing mono-forms which in fact cannot be isolated, due to the fact that they are very unstable. With the strong role of molecular rearrangements, they rearrange to another compound. The New Frontiers being laid have largely been made possible, due to new definitions provided for a monomer, step monomer, Addition monomer, after realization of what radicals, charges, ions, Activation centers are not only in terms of female and male characters but also in terms of free and non-free characters, and so much more countless to comprehend and list.

Right from Chapter 1 to the present Chapter, it has not been an easy task, because NATURE is very complex. For one to understand how NATURE operates, a large spectrum of compounds must be investigated at both ends of the spectrum of the Gaussian curve. Indeed, to be very candid, we have only just begun. Imagine what we are going to see when one comes to Living systems-Plants, Animals and Homo Sapience, and what created them and what they use to live –Carbohydrates, Oils, Fat, Proteins, Cellulose, Amino acids, the Genes, and too much countless to list, all with their goods and bads depending on operating conditions. We have only just begun. Nevertheless, it should be noted that without these first two Volumes, all what follows can never be understood. These are the foundations for HUMANITY which since antiquity have never existed. One uses combustion every day and do not know how such energy is generated. One uses other forms of Energy every day and do not know how it is generated. One sees every day and does not know what one is seeing and what weeing is. One smells shit every day and does not know what one is smelling and what smelling is. One hears every day and does not know what one is hearing and what hearing is. One is in pains every day without knowing what pain is and how it feels. It is too countless to least for which one cannot afford to live humanly in this type of world. That is the origin of THE NEW FRONTIERS and not "the new frontiers" as stupidly used everyday for everything for the purpose of money.

With respect to monomers or compounds which cannot be activated chargedly such as Benzene, acetylene, cumulene derivatives, (Not those which cannot be activated chargedly and radically based on the type of groups carried such as in N = N, N = O), Table 8.1 below shows the so-called Magnitude and direction of Electric Dipole Moments of their Derivatives.

Table 8.1 Decreasing Order of Magnitude and Directions of Electric Dipole Moment of Benzene and the like compounds Derivatives 9

#	Group A in them E.g. (Benzene-A)	Electric Moment of them E.g. (Benzene-A)	Direction of Moment E.g. $C_6H_5 - A$ Or $C_6H_5 - A$
1	$R_2C = C=$	No Value	Radical pushing
3	$H_2C = C =$	No Value	Radical pushing
4	$HFC = C =$	No Value	Radical pushing
5	$F_2C = C =$	No Value	Radical pushing
6	$RNH_2C =$	No Value	Radical pushing
7	$HORC =$	No Value	Radical pushing
8	$R_2C =$	No Value	Radical pushing
9	$HRC =$	No Value	Radical pushing
10	$H_2C =$	No Value	Radical pushing
11	$N(CH_3)_2$	1.68	Radical pushing
12	NH_2	1.53	Radical pushing

13	OH ?	1.45	Radical pushing
14	OCH_3 ?	1.38	Radical pushing
15	$C(CH_3)_3$	0.83	Radical pushing
16	$CH(CH_3)_2$	0.79	Radical pushing
17	$CH_2(CH_3)$	0.59	Radical pushing
18	CH_3	0.36	Radical pushing
19	**H ?**	**0.00**	Radical pushing
20	COOH	− 1.6	**Radical pulling BOUNDARY**
21	F	− 1.6	Radical pulling
22	Cl	− 1.7	Radical pulling
23	Br	− 1.8	Radical pulling
24	CH_2Cl	− 1.8	Radical pulling
25	$COOC_2H_5$	− 1.9	Radical pulling
26	$CHCl_3$	− 2.0	Radical pulling
27	CCl_3	− 2.1	Radical pulling
28	CHO	− 2.8	Radical pulling
29	$COCH_3$	− 3.0	Radical pulling
30	NO	− 3.1	Radical pulling
31	SO_3H	− 3.8	Radical pulling
32	NO_2	− 4.3	Radical pulling
33	CN	− 4.4	Radical pulling

$\xrightarrow{\quad\quad} \quad \equiv \quad$ Radical pushing $\quad ; \quad \xleftarrow{\quad\quad} \quad \equiv \quad$ Radical pulling

In the Table, based on the data provided from the past[9], notice that only the two cases marked with question mark are questionable, because like the other cases, one expects, OR to be more radical pushing than OH group. Secondly, H having a value of zero is questionable, unless it is neither radical-pushing nor radical-pulling. As a unique ionic electropositive element it is radical-pushing, of very low capacity, since its members in the Group (Li, Na, K, ...) do not push or pull. Thirdly, the first ten cases and more listed in the Table have never been known in the past and up to the present moment. They are being shown for the first time, and this is indeed Nobel, Novel, Scholastic, Magnus Opium, etc. in character, just as countless numbers of cases seen so far. ***For if one is to deny himself or herself, it will stand as one of the greatest sins in HUMANITY.*** The minus sign in the Table denotes pulling capacity and not a mathematical negative symbol. Hence CN (4.4) is said to be more radical pulling than F (1.6).

8.4 Proposition of Rules of Chemistry and Conclusions

Based on the New Frontiers in chemistry, the cumulenes have been considered beyond normal expectations. Some of the little known about 1,3-cumulenes in the past with respect to their reactions have been explained using different mechanisms. The characters of the mono-forms from the cumulenes have been clearly defined. From the analysis, why mono-forms have never been popularly known to undergo charged polymerizations have been fully explained. To some extent there is no doubt that the cumulenes are not common, since many of them can be observed to readily favor molecular rearrangements. Secondly, they can also be observed to be somehow sterically hindered.

Whether some of these forces exist or not is not the issue, since there is need to justify the rules which so far have been proposed and establish new foundation in terms of the properties and characters of these monomers or compounds.

Rule 390: This rule of Chemistry for **All Odd and Even membered Cumulenes,** states that, while some of these unlike Dienes, Trienes, but like acetylenes, isocyanates and the like with triple bonds cannot be activated CHARGEDLY due to presence of electrostatic forces of repulsion, some can however be so activated to produce polymers such as with 1,3-cumulenic electrophiles (e.g., $H_2C=C=CHCOCH_3$), using $H_9C_4^{\theta}$........$^{\oplus}$Na as initiator.
(Laws of Creations from Physics and Chemistry in Chemistry)

Rule 391: This rule of Chemistry for **1,3-Cumulenes**, states that, unlike first members of many families of monomers, the first member here is strongly Nucleophilic favoring the route only natural to it; to the point of making it look like the second member of the Female **Alkyne family**.
(Laws of Creations for 1,3-Cumulene)

Rule 392: This rule of Chemistry for **1,3-Cumulene or allene,** states that, this can readily be synthesized by series of reactions from for example *Glycerol* via only radical reactions as shown below-
ust (Zn) + (G) ZnBr$_2$ + 1,3-Cumulene + Heat x)
a seven-stage Equilibrium mechanism system in which the first part was three stages, the second part with two stages and the third part with two stages, noting how and where heats are released in these system as well as finger-prints of the metallic compounds.
(Laws of Creations for 1,3-Cumulene)

Stage 1:   HBr ⇌ H •e + Br •nn

H •e + CH$_2$OH ⇌ H$_2$O + CH$_2$OH
              | |
            CHOH   e• CH
              | |
            CH$_2$OH   CH$_2$OH (A)

(A) + Br •nn ⟶ CH$_2$OH
                          |
                        CHBr
                          |
                        CH$_2$OH   i)

Stage 2:   HBr ⇌ H •e + Br •nn

H •e + CH$_2$OH ⇌ H$_2$O + e• CH$_2$
              | |
            CHBr   CHBr
              | |
            CH$_2$OH   CH$_2$OH (B)   ii)

665

(B) + Br •nn $\longrightarrow$ CH$_2$Br
| CHBr
| CH$_2$OH

Stage 3: HBr $\rightleftharpoons$ H •e + Br •nn

H •e + CH$_2$Br $\rightleftharpoons$ H$_2$O + CH$_2$Br
| CHBr | CHBr
| CH$_2$OH   e• CH$_2$ (C)

(C) + Br •nn $\longrightarrow$ CH$_2$Br
| CHBr
| CH$_2$Br (D)   iii)

Overall Equation: 3HBr + Glycerol $\longrightarrow$ 3H$_2$O + (D)   iv)

Stage 4: CH$_3$OH $\rightleftharpoons$ H• e + nn• OCH$_3$

H •e = KOH $\rightleftharpoons$ H$_2$ + KO •en + Heat

KO •en $\rightleftharpoons$ K •e + nn• O •en + Heat

nn• O •en + H$_2$ $\rightleftharpoons$ HO •nn + H •e

H •e + (D) $\rightleftharpoons$ HBr + e• CH$_2$
| CHBr
| CH$_2$Br (E)

(E) $\rightleftharpoons$ H •e + e• CH$_2$
| n• CBr
| CH$_2$Br (F)

K •e + nn• OCH$_3$ $\rightleftharpoons$ KOCH$_3$

H •e + nn• OH $\rightleftharpoons$ H$_2$O

(F) $\longrightarrow$ CH$_2$
‖ CBr + Heat
| CH$_2$Br (G)   v)

Stage 5: $KOCH_3 \rightleftharpoons K \bullet e + nn \bullet OCH_3$

$\qquad\qquad H_2O \rightleftharpoons H \bullet e + nn \bullet OH$

$H \bullet e + nn \bullet OCH_3 \rightleftharpoons H_3COH$

$K \bullet e + nn \bullet OH \longrightarrow KOH$ $\qquad\qquad\qquad\qquad$ vi)

Overall Equation: $(D) + KOH + CH_3OH \longrightarrow (G) + HBr + KOH + H_3COH$

$\qquad\qquad\qquad\qquad\qquad\qquad\qquad\qquad + \text{Heat}$ $\qquad\qquad$ vii)

Stage 6: $nn \bullet Zn \bullet en + (G) \rightleftharpoons BrZn \bullet nn + \underset{\underset{e\bullet\ CH_2\ \ (H)}{\overset{\displaystyle\|}{CBr}}}{CH_2}$

$\longrightarrow \underset{\underset{CH_2ZnBr\ \ (I)}{\overset{\displaystyle\|}{CBr}}}{CH_2}$ $\qquad\qquad\qquad\qquad$ viii)

Stage 7: $(I) \rightleftharpoons \underset{\underset{n\bullet\ CH_2}{\overset{\displaystyle\|}{CBr}}}{CH_2} + en \bullet ZnBr$

$\qquad\qquad\qquad\qquad (J)$

$(J) \rightleftharpoons \underset{\underset{n\bullet\ CH_2\ \ (K)}{\overset{\displaystyle\|}{C \bullet e}}}{CH_2} + Br \bullet nn$

$BrZn \bullet en + Br \bullet nn \rightleftharpoons ZnBr_2$

$(E) \longrightarrow \underset{\underset{CH_2}{\overset{\displaystyle\|}{\underset{\displaystyle\|}{C}}}}{CH_2} + \text{Heat}$ $\qquad\qquad\qquad\qquad\qquad$ ix)

Overall Equation: Zinc Dust (Zn) + (G) $\longrightarrow$ $ZnBr_2$ + 1,3-Cumulene + Heat $\qquad$ x)

a seven-stage Equilibrium mechanism system in which the first part was three stages, the second part with two stages and the third part with two stages, noting how and where heats are released in these system as well as finger-prints of the metallic compounds.
(Laws of Creations for 1,3-Cumulene)

Rule 393: This rule of Chemistry for **1,3-Cumulenes,** states that, unique substituent groups of the type $R_2C=$, carried by them is herein called **CUMUNYL** group-

$$R_2C = \qquad ; \qquad (R_F)_2C = \qquad ; \qquad R_FRC=$$

<u>**CUMUNYL GROUPS**</u>

(Laws of Creations for Cumunylic groups)

Where R is any alkanyl group and R_F is a radical-pulling group

Rule 394: This rule of Chemistry for **All Cumulenes,** states that, when activated the radicals carried by the active centers are fixed.

(Laws of Creations for Cumulenes)

Rule 395: This rule of Chemistry for **1,3-Cumulenes,** states that, for any type of substituent groups carried by their carbon centers, it is only in the symmetrically placed ones that any of the activation centers can be activated, while for the non-symmetrically placed ones, only one activation center can be activated as shown below-

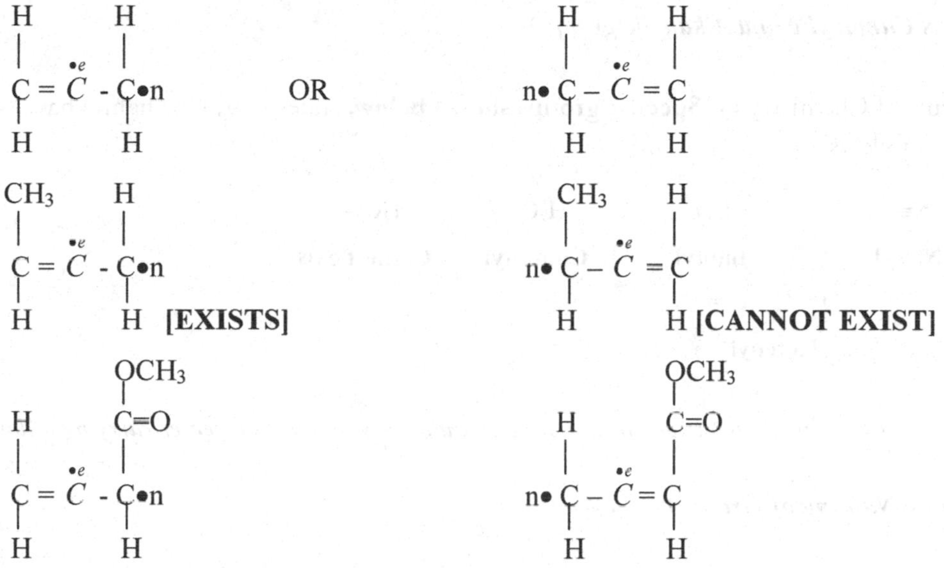

[first row diagrams]

                                    OR

[EXISTS]   [CANNOT EXIST]

EXISTS   CANNOT EXIST

(Laws of Creations for 1,3-Cumulenes)

Rule 396: This rule of Chemistry for **Cumunyl Groups**, states that, the following order is valid for them-

$$= CH(OH) \quad > \quad = CH(OCH_3) \quad > \quad = CH(OC_2H_5) \quad > \quad Etc$$

<u>Order of Radical-pushing capacity</u>

(Laws of Creations for Cumunyl groups)

Rule 397: This rule of Chemistry for **Cumunyl Groups,** states that, the following order is valid for them-

$$= CH(NR_2) \quad > \quad = CH(NRH) \quad > \quad = CH(NH_2) \quad > \quad = C(OH)_2 \quad > \quad = CH(OR) \quad > \quad = C(OR)_2 \ ...etc.$$

<u>**ORDER OF RADICAL-PUSHING CAPACITY**</u>

(Laws of Creations for Cumunyl groups)

Rule 398: This rule of Chemistry for **Allylic Groups,** states that, the following order is valid for them-

$$- CH_2(NR_2) \quad > \quad - CH_2(NHR) \quad > \quad - CH_2NH_2$$

<u>Order of radical-pushing capacity</u>

(Laws of Creations for Allylic Groups)

Rule 399: This rule of Chemistry for **Cumunyl and Alkanyl groups,** states that, the following orders are valid for them-

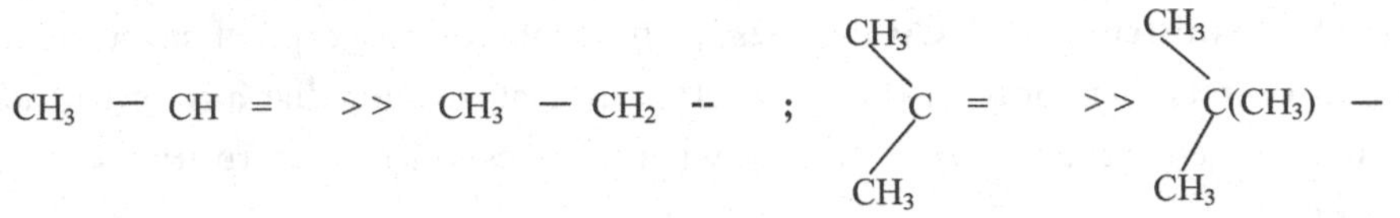

$$CH_3 - CH = \quad >> \quad CH_3 - CH_2 \cdots \quad ; \quad \overset{CH_3}{\underset{CH_3}{C}} = \quad >> \quad \overset{CH_3}{\underset{CH_3}{C(CH_3)}} -$$

<u>Order of radical- pushing capacity</u>

(Laws of Creations Cumunylic and Alkanylic groups)

Rule 400: This rule of Chemistry for **Specific groups shown below,** states that, new names have been assigned to them as follows-

$O =$; $N \equiv$ $HN =$; $H_2C =$; $HC \equiv$

Oxidyl   **Nitryl**   **Iminyl**   **Cumunyl**   **Cannot exist**

$R - N = C =$; $O = C = C =$

Cyanyl   **Ketenyl**

for which those said not to exist are those which cannot be cumulatively placed to a center carrying a double bond.

(Laws of Creations-Naming of Groups)

Rule 401: This rule of Chemistry for **Electrodynamic/Static forces of Repulsion and Attrac-tion,** states that, there are **two types of** Electro-dynamic/static forces of repulsion and attraction– **THE REAL and THE IMAGINARY ONES;** *for which those existing with Covalent charges, π-bonds, ions and paired unbonded radicals are REAL, while those existing with radicals, imaginary part of electrostatic and polar charges are IMAGINARY the world being a complex one.*

(Laws of Creations – the eleventh law of Nature)

Rule 402: This rule of Chemistry for **1,3-Cumulenes with at least one or two non-free-radical-pushing groups (e.g. OH, OR, NH$_2$, NHR, NR$_2$),** states that, these monomers which can readily be activated radically and cannot molecularly rearrange, are strong NUCLEO-PHILES.

(Laws of Creations for 1,3-Cumulenes)

Rule 403: This rule of Chemistry for **1,3-Cumulenes with four non-free-radical-pushing groups (e.g. OH, OR, NH$_2$, NHR, NR$_2$),** states that, since these monomers when activated favor both free-radical routes, this clearly indicates that these monomers are weak Nucleophiles for which their propagations will be sterically hindered with strong influence of electrodynamic forces of repulsion.

(Laws of Creations for 1,3-Cumulenes)

Rule 404: This rule of Chemistry for **Allene,** states that, when activated or when made to exist in Equilibrium state of existence, the followings are obtained-

$$
\begin{array}{c}
H\quad H\\
|\quad\quad|\\
C=C=C\\
|\quad\quad|\\
H\quad H
\end{array}
\rightleftharpoons
H\bullet e \;+\;
\begin{array}{c}
\quad\quad H\\
\quad\quad|\\
n\bullet C=C=C\\
\quad\quad|\\
\quad\quad H
\end{array}
\longleftrightarrow
H\bullet e \;+\;
\begin{array}{c}
\quad\quad H\\
\quad\quad|\\
C\equiv C-C\bullet n\\
\quad\quad|\\
\quad\quad H
\end{array}
\longrightarrow
$$

$$H - C \equiv C - CH_3$$

ELECTRORADICALIZATION

$$
\begin{array}{c}
H\quad H\\
|\quad\quad|\\
C=C=C\\
|\quad\quad|\\
H\quad H
\end{array}
\longrightarrow
\begin{array}{c}
\quad H\\
\quad |\\
n\bullet C - C\bullet e\\
\quad |\quad\;\|\\
\quad H\;\;CH_2
\end{array}
\longrightarrow
\begin{array}{c}
\quad H\\
\quad |\\
n\bullet C = C\bullet e\\
\quad\quad\;\;|\\
\quad\quad\mathbf{CH_3}
\end{array}
\longrightarrow
H - C \equiv C - CH_3
$$

MOLECULAR REARRANGEMENT

for which the same methyl acetylene is obtained.
(Laws of Creations for 1,3-Cumulene)

Rule 405: This rule of Chemistry for **1,3-Cumulenes with one non-free-radical-pushing group of the type OH,** states that, when not activated, but made to exist in Equilibrium State of Existence, the followings are obtained-

$$
\begin{array}{c}
OH\quad H\\
|\quad\quad|\\
C=C=C\\
|\quad\quad|\\
H\quad H
\end{array}
\rightleftharpoons
H\bullet e \;+\;
\begin{array}{c}
O\bullet nn\quad H\\
|\quad\quad|\\
C=C=C\\
|\quad\quad|\\
H\quad H
\end{array}
\xrightarrow{\text{ELECTRORADICALIZATION}}
\begin{array}{c}
H\quad H\\
|\quad\quad|\\
C = C\\
|\quad\quad|\\
H\quad C=O\\
\quad\quad|\\
\quad\quad H
\end{array}
$$

for which ELECTRORADICALIZATION takes place to produce a weak Electrophile-Acrolein.
(Laws of Creations for 1,3-Cumulene-ol)

Rule 406: This rule of Chemistry for **1,3-Cumulenes with two non-free-radical-pushing group of the type OH,** states that, when not activated, but made to exist in Equilibrium State of Existence, the followings are obtained-

$$
\begin{array}{c}
OH\quad OH\\
|\quad\quad|\\
C - C = C\\
|\quad\quad|\\
H\quad H
\end{array}
\rightleftharpoons
H\bullet e \;+\;
\begin{array}{c}
O\bullet nn\quad OH\\
|\quad\quad|\\
C = C - C\\
|\quad\quad|\\
H\quad H
\end{array}
\xrightarrow{\text{ELECTRORADICALIZATION}}
\begin{array}{c}
OH\quad H\\
|\quad\quad|\\
C - C\\
|\quad\quad|\\
H\quad C=O
\end{array}
$$

$$
\begin{array}{c}
OH\quad H\\
|\quad\quad|\\
C = C = C\\
|\quad\quad|\\
OH\quad H
\end{array}
\rightleftharpoons
H\bullet e \;+\;
\begin{array}{c}
O\bullet nn\quad H\\
|\quad\quad|\\
C = C = C\\
|\quad\quad|\\
OH\quad H
\end{array}
\xrightarrow{\text{ELECTRORADICALIZATION}}
\begin{array}{c}
H\quad HH\\
|\quad\quad|\\
C = C\\
|\quad\quad|\\
H\quad C=O\\
\quad\quad|\\
\quad\quad OH
\end{array}
$$

for which ELECTRORADICALIZATION takes place to produce strong Electrophiles-an Acrolein and Acrylic acid.
(Laws of Creations for 1,3-Cumulene-diols)

Rule 407: This rule of Chemistry for **1,3-Cumulenes with three non-free-radical-pushing groups of the type OH,** states that, when not activated, but made to exist in Equilibrium State of Existence, the followings are obtained-
OH OH O•nn OH

$$
\begin{array}{ccc}
\text{OH} & \text{OH} \\
| & | \\
\text{C} = \text{C} = \text{C} \\
| & | \\
\text{H} & \text{OH}
\end{array}
\rightleftharpoons
\text{H•e} \; + \;
\begin{array}{ccc}
\text{O•nn} & \text{OH} \\
| & | \\
\text{C} = \text{C} = \text{C} \\
| & | \\
\text{OH} & \text{H}
\end{array}
\xrightarrow{\text{ELECTRORADICALIZATION}}
\begin{array}{ccc}
\text{OH} & \text{H} \\
| & | \\
\text{C} = \text{C} \\
| & | \\
\text{H} & \text{C=O} \\
 & | \\
 & \text{OH}
\end{array}
$$

for which ELECTRORADICALIZATION takes place to produce a weak Electrophile-an Acrylic acid.
(Laws of Creations for 1,3-Cumulene-triol)

Rule 408: This rule of Chemistry for **1,3-Cumulenes with four non-free-radical-pushing group of the type OH**, states that, when not activated, but made to exist in Equilibrium State of Existence, the followings are obtained-

$$
\begin{array}{ccc}
\text{OH} & \text{OH} \\
| & | \\
\text{C} = \text{C} = \text{C} \\
| & | \\
\text{OH} & \text{OH}
\end{array}
\rightleftharpoons
\text{H•e} \; + \;
\begin{array}{ccc}
\text{O•nn} & \text{OH} \\
| & | \\
\text{C} = \text{C} = \text{C} \\
| & | \\
\text{OH} & \text{OH}
\end{array}
\xrightarrow{\text{ELECTRORADICALIZATION}}
\begin{array}{ccc}
\text{OH} & \text{H} \\
| & | \\
\text{C} = \text{C} \\
| & | \\
\text{OH} & \text{C=O} \\
 & | \\
 & \text{OH}
\end{array}
$$

for which ELECTRORADICALIZATION takes place to produce a strong Electrophile-an Acrylic acid.
(Laws of Creations for 1,3-Cumulene-tetraol)

Rule 409: This rule of Chemistry for **1,3-Cumulenes with one non-free-radical-pushing group of the types NH$_2$, NHR (Not NR$_2$),** states that, when not activated, but made to exist in Equilibrium State of Existence, the followings are obtained-

$$
\begin{array}{ccc}
\text{NH}_2 & \text{H} \\
| & | \\
\text{C} = \text{C} = \text{C} \\
| & | \\
\text{H} & \text{H}
\end{array}
\rightleftharpoons
\text{H•e} \; + \;
\begin{array}{ccc}
\text{HN•nn} & \text{H} \\
| & | \\
\text{C} = \text{C} = \text{C} \\
| & | \\
\text{H} & \text{H}
\end{array}
\xrightarrow{\text{ELECTRORADICALIZATION}}
\begin{array}{ccc}
\text{H} & \text{H} \\
| & | \\
\text{C} = \text{C} \\
| & | \\
\text{H} & \text{C=NH} \\
 & | \\
 & \text{H}
\end{array}
$$

or which ELECTRORADICALIZATION takes place to produce weak Electrophile-vinyl formaldimine.
(Laws of Creations for 1,3-Cumulene imine)

Rule 410: This rule of Chemistry for **1,3-Cumulenes with two non-free-radical-pushing groups of the types NH₂, NHR (Not NR₂),** states that, when not activated, but made to exist in Equilibrium State of Existence, the followings are obtained-

$$
\begin{array}{c}
\underset{H}{\overset{NH_2}{C}} = C = \underset{H}{\overset{NH_2}{C}} \quad \rightleftharpoons \quad H\bullet e \;+\; \underset{H}{\overset{HN\bullet nn}{C}} = C = \underset{H}{\overset{NH_2}{C}} \quad \xrightarrow{\text{ELECTRORADICALIZATION}} \quad \underset{H}{\overset{H_2N}{C}} = \overset{H}{\underset{C=NH}{C}} \\
\qquad\qquad\qquad\qquad\qquad\qquad\qquad\qquad\qquad\qquad\qquad\qquad\qquad H
\end{array}
$$

$$
\begin{array}{c}
\underset{NH_2}{\overset{NH_2}{C}} = C = \underset{NH_2}{\overset{H}{C}} \quad \rightleftharpoons \quad H\bullet e \;+\; \underset{NH_2}{\overset{HN\bullet nn}{C}} = C = \underset{H}{\overset{H}{C}} \quad \xrightarrow{\text{ELECTRORADICALIZATION}} \quad \underset{H}{\overset{H}{C}} = \overset{H}{\underset{C=NH}{C}} \\
\qquad\qquad\qquad\qquad\qquad\qquad\qquad\qquad\qquad\qquad\qquad\qquad\qquad\qquad NH_2
\end{array}
$$

for which ELECTRORADICALIZATION takes place to produce Electrophiles-a vinyl formaldimine and vinyl cyanimic acid.
(Laws of Creations for 1,3-Cumulene-di-imine)

Rule 411: This rule of Chemistry for **1,3-Cumulenes with three non-free-radical-pushing groups of the types NH₂, NHR (Not NR₂),** states that, when not activated, but made to exist in Equilibrium State of Existence, the followings are obtained-

$$
\begin{array}{c}
\underset{H}{\overset{NH_2}{C}} = C = \underset{NH_2}{\overset{NH_2}{C}} \quad \rightleftharpoons \quad H\bullet e \;+\; \underset{H}{\overset{H_2N}{C}} = C = \underset{NH_2}{\overset{HN\bullet nn}{C}} \quad \xrightarrow{\text{ELECTRORADICALIZATION}} \quad \underset{H}{\overset{H_2N}{C}} = \overset{H}{\underset{C=NH}{C}} \\
\qquad\qquad\qquad\qquad\qquad\qquad\qquad\qquad\qquad\qquad\qquad\qquad\qquad NH_2
\end{array}
$$

for which ELECTRORADICALIZATION takes place to produce weak Electrophile- a vinyl cyanimic acid.
(Laws of Creations for 1,3-Cumulene-tri-imine)

Rule 412: This rule of Chemistry for **1,3-Cumulenes with four non-free-radical-pushing groups of the types NH₂, NHR (Not NR₂),** states that, when not activated, but made to exist in Equilibrium State of Existence, the followings are obtained-

$$
\begin{array}{c}
\underset{NH_2}{\overset{NH_2}{C}} = C = \underset{NH_2}{\overset{NH_2}{C}} \quad \rightleftharpoons \quad H\bullet e \;+\; \underset{NH_2}{\overset{HN\bullet nn}{C}} = C = \underset{NH_2}{\overset{NH_2}{C}} \quad \xrightarrow{\text{ELECTRORADICALIZATION}} \quad \underset{H_2N}{\overset{H_2N}{C}} = \overset{H}{\underset{C=NH}{C}} \\
\qquad\qquad\qquad\qquad\qquad\qquad\qquad\qquad\qquad\qquad\qquad\qquad\qquad NH_2
\end{array}
$$

for which ELECTRORADICALIZATION takes place to produce an Electrophile-a vinyl cyanamic acid.
(Laws of Creations for 1.3-Cumulene-tetra-imine)

672

Rule 413: This rule of Chemistr y for **1,3-cumulenes with one non-free-radical-pulling group (e.g. F, Cl, OCOCH$_3$, etc.),** states that, these monomers which favor only the electro-free-radical route after molecular rearrangement, are strong NUCLEOPHILES.
(Laws of Creations for 1.3-Cumulenes)

Rule 414: This rule of Chemistry for **1,3-cumulenes with at least two non-free-radical-pulling groups (e.g. F, Cl, OCOCH$_3$, etc.) symmetrically or non-symmetrically placed,** states that, these monomers which favor only electro-free-radical route after molecular rearrangement, are strong NUCLEOPHILES.
(Laws of Creations for 1,3-Cumulenes)

Rule 415: This rule of Chemistry for **1,3-cumulenes with four non-free-radical-pulling groups (e.g. F, Cl, OCOCH$_3$, etc.),** states that, since these monomers, but one (OCOCH$_3$) when activated favor both free-radical routes, this clearly indicates that these monomers are weak Nucleophiles for which their propagations will be sterically hindered with strong influence of electrodynamic forces of repulsion.
(Laws of Creations for 1,3-Cumulenes)

Rule 416: This rule of Chemistry for **1,3-cumulenes with at least one or two free-radical-pushing groups of lower capacity than H (e.g. C$_n$F$_{2n+1}$),** states that, since these monomers which are Nucleophiles cannot favor any free-radical route but only when trans-placed and with paired initiators nucleo-free-radically, shows that these are weak Electrophiles.
(Laws of Creations for 1,3-Cumulenes)

Rule 417: This rule of Chemistry for **1,3-cumulenes with four free-radical-pushing groups (e.g. C$_n$F$_{2n+1}$),** states that, these monomers wherein the transfer species are F electro-free-radically and C$_n$F$_{2n+1}$ nucleo-free-radically, can be used as a monomer only when paired initiators are used nucleo-freee-radically and when the groups are trans-placed, clear indication of an Electrophilic character.
(Laws of Creations for Perfluorinated 1,3-Cumulenes)

Rule 418: This rule of Chemistry for **1,3-cumulenes with at least one or two free-radical-pulling groups (e.g. COOR, CONH$_2$),** states that, these monomers show their STRONG ELECTROPHILIC characters either when they molecularly rearrange or when paired initiators are involved only when the groups are trans-placed.
(Laws of Creations for 1,3-Cumulenes)

Rule 419: This rule of Chemistry for **1,3-cumulenes with at least one or two free-radical-pulling groups (e.g. COR),** states that, these monomer, show their WEAK ELECTROPHILIC characters only when paired initiators are involved when the groups are trans-placed; for which the possibility of producing random copolymers electro-free-radically may exist.
(Laws of Creations for 1,3-Cumulenes)

Rule 420: This rule of Chemistry for **1,3-cumulenes with at least one or two free-radical-pulling group (e.g. CN),** states that, these monomers show their STRONG ELECTROPHILIC characters only when paired initiators are involved and the groups are trans-placed.
(Laws of Creations for 1,3-Cumulenes)

Rule 421: This rule of Chemistry for **1,3-Cumulenes,** states that, these monomers or compounds can be the provider of resonance stabilization and can also be resonance stabilized, for which the following relationships are valid-

$$\text{(benzene ring)} \quad > \quad H-C \equiv C- \quad > \quad \underline{H_2C=C=C-} \overset{\overset{\textstyle H}{|}}{} \quad > \quad H_2C=C- \overset{\overset{\textstyle H}{|}}{}$$

ORDER OF RESONANCE STABILZATION CAPACITY

wherein the highlighted one is used in part when specific groups are carried by the 1,3-cumulene and these groups are herein called *Cumulenyl groups-*

$$(CH_3)HC=C=CH- \quad > \quad H_2C=C=CH- \quad > \quad H_2C=C=(CH_3)C- \quad > \quad H-$$

Order of radical-pushing capacity

(Laws of Creations for Cumulenyl groups)

Rule 422: This rule of Chemistry for **Cumunyl groups,** states that, groups such as $H_2C=$, $RHC=$, $R_2C=$, $R_FHC=$, $(R_F)_2C=$, where Rs are radical-pushing groups and R_Fs are radical-pulling groups, are strong cumunylic radical-pushing groups with the following capacities –

$$CH_2= \quad > \quad CFH= \quad > \quad CF_2=$$

Order of Free-radical-pushing capacity

$$H_2C= \quad > \quad ROOCCH= \quad > \quad (ROOC)_2C=$$

Order of Free-radical-pushing capacity

$$F_2C= \quad > \quad (ROOC)_2C= \quad > \quad (ROC)_2C= \quad > \quad (NC)_2C=$$

Order of Free-radical-pushing capacity

$$FHC= \quad > \quad ROOCCH= \quad > \quad ROCCH= \quad > \quad NCCH=$$

Order of Free-radical-pushing capacity

$$R_2C= \quad > \quad RHC= \quad > \quad H_2C= \quad > \quad R_FHC= \quad > \quad (R_F)_2C= \quad >> \quad H-$$

Order of radical-pushing capacities of cumunylic groups

(Laws of Creations for Cumunylic groups) [See Rules 396 & 397]

Rule 423: This rule of Chemistry for **Isomerism,** states that, ISOMERISM is a unique concept for which when ACTIVATION centers are present and involved, they take place via various mechanisms all classified as **TAUTOMERISM** which include Molecular Rearrangement, Enoli-zation, Enamidization, Electroradicalization and more; and when no ACTIVATION centers are present, they take place by placement of the elements in accordance to their valences to form different structural compounds herein called **VALENCE ISOMERISM.**

(Laws of Creations for ISOMERISM)

<u>**Rule 424:**</u> This rule of Chemistry for **Alkenyls, Alkynyls, Cumulenyls and Cumunyls,** states that, the orders of their first members in radical-pushing capacities are as follows-

$$H_2C= \quad >> \quad H_2C=C=CH- \quad > \quad H_2C=CH- \quad > \quad H- \quad > \quad HC\equiv C-$$

<u>**Cumunyl**</u>   <u>**Cumulenyl**</u>   <u>**Alkenyl**</u>   <u>**Alkynyl**</u>

<u>**Order of radical-pushing capacities**</u>

(Laws of Creations for Alkyl groups)

<u>**Rule 425:**</u> This rule of Chemistry for **1,3-Cumulenes,** states that, these have their origins from the **Alkynes family** which in turn have their origins from the **Alkenes family** which in turn have their origins from the **Alkanes family** which in turn have their origins from the **Alkylenes family.**
(Laws of Creations for 1,3-Cumulenes)

<u>**Rule 426:**</u> This rule of Chemistry for **1,3-cumulenes,** states that, based on the analysis so far it is therefore of interest to note that the larger the R group is in place of CH_3 on the monomer, the followings are valid.

$$(C_3H_7)_2C= \quad > \quad (C_2H_5)_2C= \quad > \quad H(C_2H_7)C= \quad > \quad H(C_2H_5)C= \quad > \quad (CH_3)_2C=$$

<u>Order of radical-pushing capacity</u>

$$H(C_3H_7)C= \quad > \quad H(C_2H_5)C= \quad > \quad H(CH_3)C= \quad > \quad H_2C=$$

<u>Order of radical-pushing capacity</u>

Laws of Creations for Cumunyl groups)

<u>**Rule 427:**</u> This rule of Chemistry for **1,4-Cumulenes,** states that, unlike 1,3-Cumulenes, the first member of the family like most families, favor both radical routes under different operating conditions as shown below-

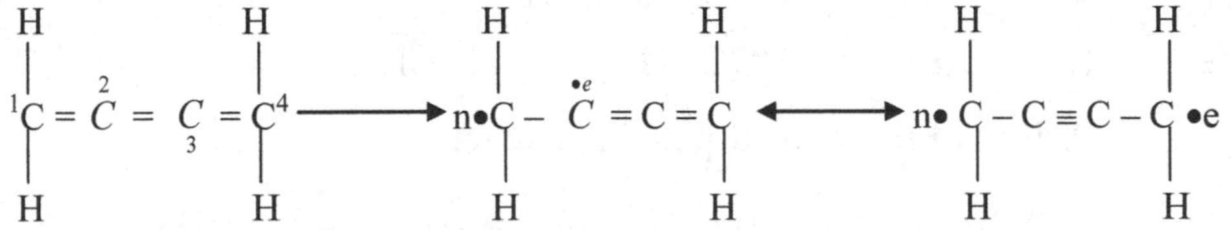

Resonance Stabilization phenomenon

for which there are two mono-forms-3,4—mono-form and 1,4—mono-form that is used, both of which can have transfer species of the first kind of the first type; electro-free-radically and nucleo-free-radically depending on the types of groups carried by the active centers, with transfer species of the second kind for the case above as shown below, just like ethene.

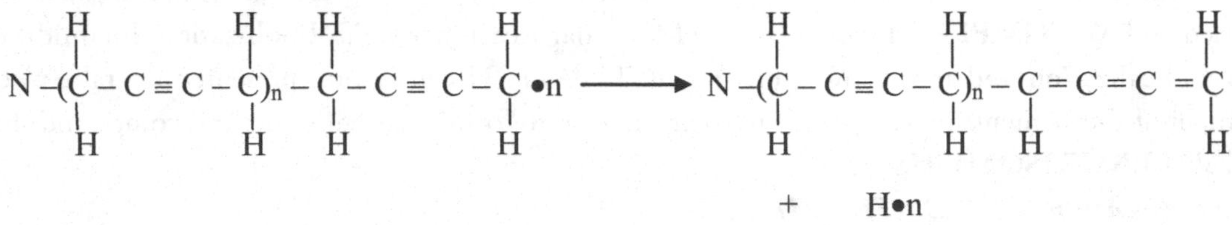

$$+ \quad H\bullet n$$

(Laws of Creations for 1,4-Cumulene)

Rule 428: This rule of Chemistry for **Cumulenes,** states that, based on abilities for provision of resonance stabilization, the following groups are resonance stabilization groups with their corresponding capacities-

$H_2C=C=C=C=C=C= C= > H_2C=C=C=C=C= > H_2C=C=C= > H_2C= \gg H_{2n+1}C_n-$

ODD

RESONANCE STABILIZATION GROUPS {Cumunyls}

$H_2C=C=C=C=C=C= > H_2C=C=C=C= > H_2C=C= \ggg H-$

EVEN

RESONANCE STABILIZATION GROUPS {Cumunyls}

$H_2C=C=C=C= > H_2C=C=C= > H_2C=C= > H_2C= \ggg H-$

EVEN   ODD   EVEN   ODD

RESONANCE STABILIZATION GROUPS   {Cumunyls}

$H_2C=C=C=CH- > H_2C=C=CH- > H_2C=CH- \gg H-$

RESONANCE STABILIZATION GROUPS {Cumulenyls)

(Laws of Creations for Resonance stabilization groups)

Rule 429: This rule of Chemistry for **1,4-Cumulenes,** states that, when the monomer or compound is made to exist in Equilibrium State of existence, the followings are to be expected-

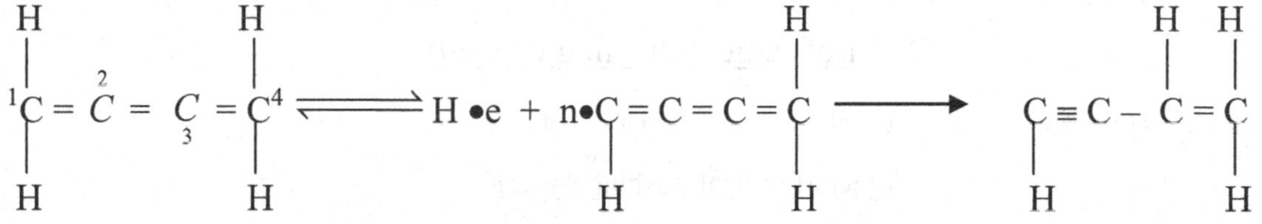

Electroradicalization phenomenon

for which, it can be observe that an Eneyne is obtained, clear indication that this is a dimer of the Alkyne family.

(Laws of Creations of 1.4-Cumulene)

Rule 430: This rule of Chemistry for **1.4-Cumulenes** of the type $HRC=C=C=CH_2$ where R is an alkanyl group, states that, these monomers are strong Nucleophiles; for which the followings take place when activated-

$$n\bullet \overset{\text{H}}{\underset{\text{H}}{C}} - \overset{\bullet e}{C} = C = \overset{\text{CH}_3}{\underset{\text{H}}{C}} \quad\longleftrightarrow\quad n\bullet \overset{\text{H}}{\underset{\text{H}}{C}} - C \equiv C - \overset{\text{CH}_3}{\underset{\text{H}}{C}} \bullet e$$

RESONANCE STABILIZATION PHENOMENON

$$n\bullet \overset{\text{H}}{\underset{\text{H}}{C}} - C \equiv C - \overset{\text{H}}{\underset{\text{CH}_3}{C}} \bullet e \quad\longrightarrow\quad n\bullet \overset{\text{H}}{\underset{\text{H}}{C}} - \overset{\text{H}}{\underset{\text{C}}{C}} \bullet e \quad\longrightarrow\quad H_2C = CH - C \equiv C(CH_3)$$

(A)

for which when molecular rearrangement takes place, dimers of acetylenes {e.g. (A)] are produced.
(Laws of Creations for 1,4-Cumulenes)

<u>**Rule 431:**</u> This rule of Chemistry for **1.4-Cumulenes** of the type $HRC=C=C=CH_2$ where R is an alkanyl group, states that, based on their analysis it is therefore of interest to note that the larger the R group in place of CH_3 on the monomer, the followings are valid-

$$(C_3H_7)_2C= \quad > \quad (C_2H_5)_2C= \quad > \quad H(C_3H_7)C= \quad > H(C_2H_5)C= \quad > \quad (CH_3)_2C= \quad > etc.$$

<u>Order of radical-pushing capacity</u>

$$H(C_3H_7)C = \quad > \quad H(C_2H_5)C = \quad > \quad H(CH_3)C = \quad > \quad H_2C =$$

<u>Order of radical-pushing capacity</u>

$$(C_3H_7)_2C= C= \quad > \quad (C_2H_5)_2C= C= \quad > \quad H(C_3H_7)C=C= \quad > \quad H(C_2H_5)C= C= \quad > \quad etc.$$

<u>Order of radical-pushing capacity</u>

$$(C_3H_7)_2C=C= \quad > \quad (C_3H_7)_2C= \quad >> \quad (C_3H_7)_3C-$$

<u>Order of radical-pushing capacity</u>

wherein the $H_2C=C=$ types of groups are herein called ***Di-cumunyl groups***. [See Rule 426]
(Laws of Creations for 1,4-Cumulenes)

<u>**Rule 432:**</u> This rule of Chemistry for some monomers, states that, the followings are their orders of Nucleophilicity-

$$RHC^{\theta} - N^{\oplus} \equiv N \;>\; RHC = C = C = CH_2 \;>\; H_2C = C = C = CH_2 \;>\; RHC = C{=}CH_2$$

Order of Nucleophilicity

$$R_2C{=} C{=} C{=}CH_2 \;>\; RHC{=}C{=}C{=}CH_2 \;>\; R_2C{=}C{=}CH_2 \;>\; RHC{=}C{=}CH_2 \;>\; etc.$$

Order of Nucleophilicity

Diazo-alkylene   >   Benzene   >   1,4-Cumulene   >   1,3-Cumulene   >   Acetylene

Order of Nucleophilicity

(Laws of Creations for Addition monomers)

Rule 433: This rule of Chemistry for **1,4-Cumulenes** with an alkanyl group, states that, these monomers can be said to have their origins from Carbenes and Alkynes, all of the Hydrocarbon family Tree, as shown below-

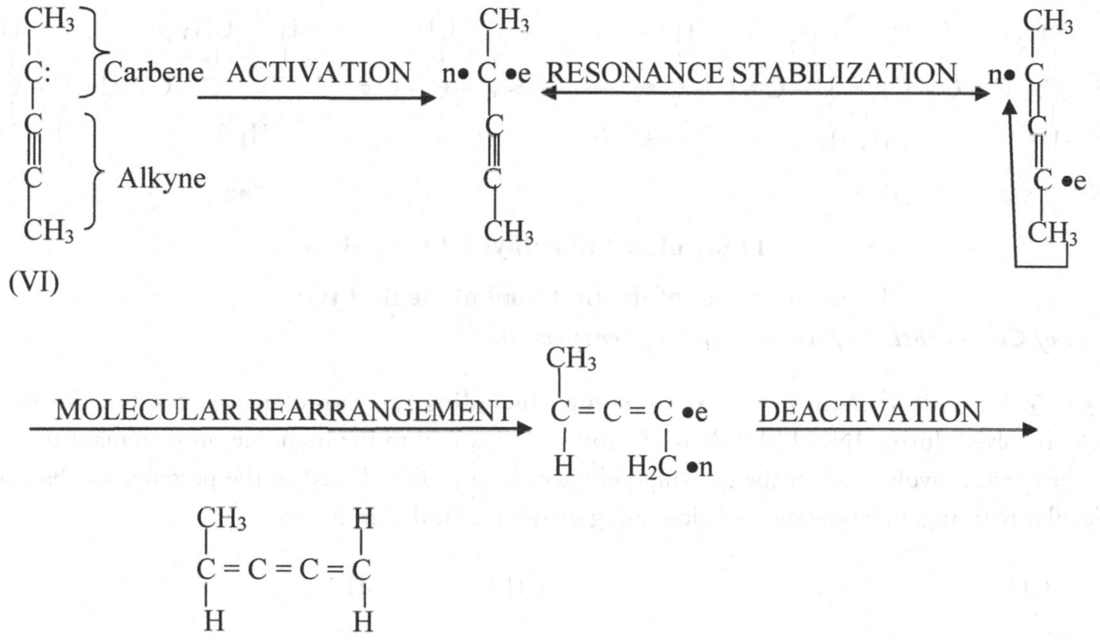

(VI)

(Laws of Creations for 1,4-Cumulenes)

Rule 434: This rule of Chemistry for **All monomers including 1,4-Cumulenes,** states that, based on the Laws of Conservation of transfer of transfer species, the transfer species involved during INITIATION in the route not Natural to the monomer must remain the same transfer species involved when the growing polymer chain is to be killed- TERMINATION in the presence or absence of molecular rearrangements as shown below using a 1,4-cumulenes-

WITH RESONANCE STABILIZATION PHENOMENON

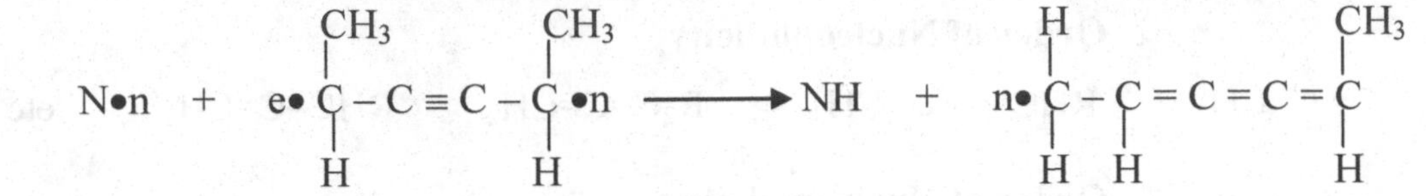

$$
N\bullet n \;+\; e\bullet \overset{\displaystyle CH_3}{\underset{\displaystyle H}{C}} - C \equiv C - \overset{\displaystyle CH_3}{\underset{\displaystyle H}{C}} \bullet n \longrightarrow NH \;+\; n\bullet \overset{\displaystyle H}{\underset{\displaystyle H}{C}} - \overset{\displaystyle }{\underset{\displaystyle H}{C}} = C = C = \overset{\displaystyle CH_3}{\underset{\displaystyle H}{C}}
$$

Transfer species of first kind of the first type

$$
E\!-\!\Big(\overset{\displaystyle CH_3}{\underset{\displaystyle H}{C}} - C \equiv C - \overset{\displaystyle CH_3}{\underset{\displaystyle H}{C}}\Big)_n \!-\! \overset{\displaystyle CH_3}{\underset{\displaystyle H}{C}} - C \equiv C - \overset{\displaystyle CH_3}{\underset{\displaystyle H}{C}} \bullet e \longrightarrow E\!-\!\Big(\overset{\displaystyle CH_3}{\underset{\displaystyle H}{C}} - C \equiv C - \overset{\displaystyle CH_3}{\underset{\displaystyle H}{C}}\Big)_n \!-\! \overset{\displaystyle CH_3}{\underset{\displaystyle H}{C}} - C \equiv C - \overset{\displaystyle CH_3}{\underset{\displaystyle H}{C}} = \overset{\displaystyle H}{\underset{\displaystyle H}{C}}
$$

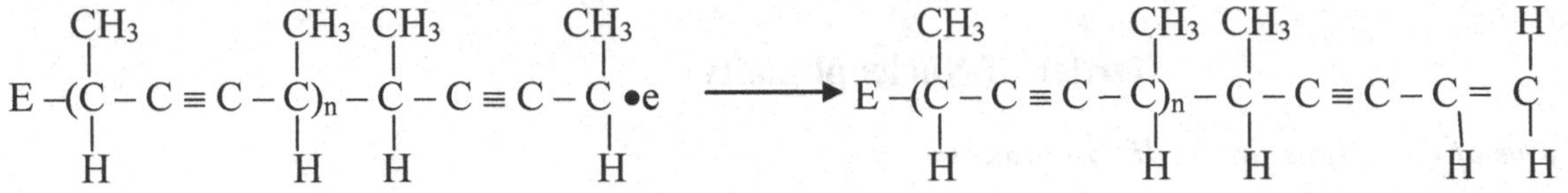

(I)   + H•e

Cis-placed dimethyl 1,4-Cumulene

Transfer species of the first kind of the first type

$$
E\!-\!\Big(\overset{\displaystyle CH_3}{\underset{\displaystyle H}{C}} - C \equiv C - \overset{\displaystyle H}{\underset{\displaystyle CH_3}{C}}\Big)_n \!-\! \overset{\displaystyle CH_3}{\underset{\displaystyle H}{C}} - C \equiv C - \overset{\displaystyle H}{\underset{\displaystyle CH_3}{C}} \bullet e \longrightarrow E\!-\!\Big(\overset{\displaystyle CH_3}{\underset{\displaystyle H}{C}} - C \equiv C - \overset{\displaystyle H}{\underset{\displaystyle CH_3}{C}}\Big)_n \!-\! \overset{\displaystyle CH_3}{\underset{\displaystyle H}{C}} - C \equiv C - \overset{\displaystyle H}{\underset{\displaystyle H}{C}} = \overset{\displaystyle H}{\underset{\displaystyle H}{C}}
$$

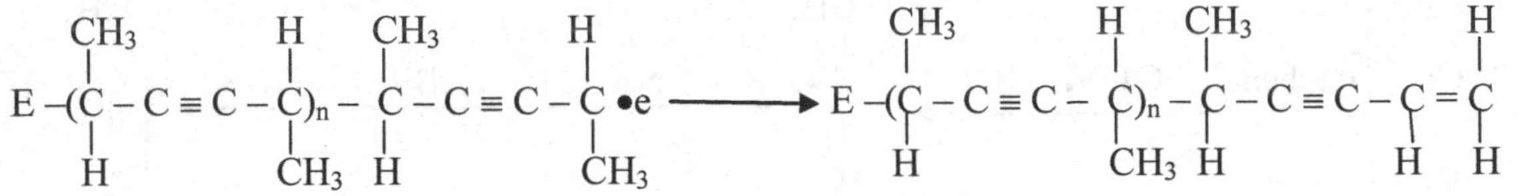

(I)   + H•e

Trans-placed dimethyl 1,4-Cumulene

Transfer species of the first kind of the first type

(Laws of Conservation of transfer of transfer species)

Rule 435: This rule of Chemistry for **All monomers including 1,4-Cumulenes,** states that, the transfer species involved during INITIATION in the route not Natural to the monomer must remain the same transfer species involved when the growing polymer chain is to be killed in the presence or absence of molecular rearrangements as shown below using unsymmetrical cumulenes-

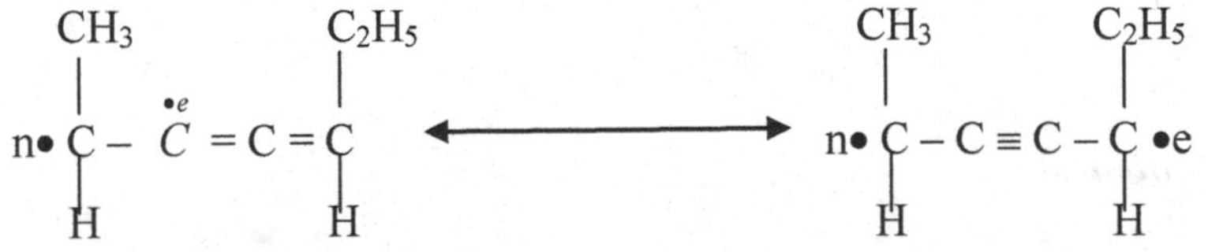

$$
n\bullet \overset{\displaystyle CH_3}{\underset{\displaystyle H}{C}} - \overset{\displaystyle \overset{\bullet e}{}}{\underset{\displaystyle }{C}} = C = \overset{\displaystyle C_2H_5}{\underset{\displaystyle H}{C}} \quad\longleftrightarrow\quad n\bullet \overset{\displaystyle CH_3}{\underset{\displaystyle H}{C}} - C \equiv C - \overset{\displaystyle C_2H_5}{\underset{\displaystyle H}{C}} \bullet e
$$

RESONANCE STABILIZATION PHENOMENON

WITH RESONANCE STABILIZATION PHENOMENON-1,4-MONO-FORM

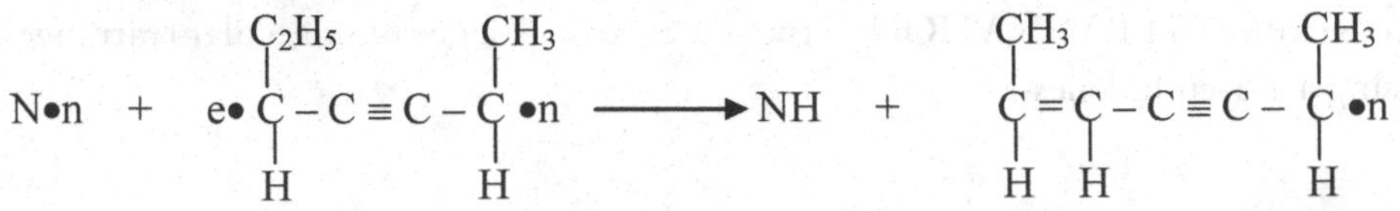

$$
N\bullet n \;+\; e\bullet \overset{\displaystyle C_2H_5}{\underset{\displaystyle H}{C}} - C \equiv C - \overset{\displaystyle CH_3}{\underset{\displaystyle H}{C}} \bullet n \longrightarrow NH \;+\; \overset{\displaystyle CH_3}{\underset{\displaystyle H}{C}} = \overset{\displaystyle }{\underset{\displaystyle H}{C}} - C \equiv C - \overset{\displaystyle CH_3}{\underset{\displaystyle H}{C}} \bullet n
$$

Transfer species of first kind of the first type

$$\text{E} -(\underset{\underset{\text{H}}{|}}{\overset{\overset{\text{CH}_3}{|}}{\text{C}}} - \text{C} \equiv \text{C} - \underset{\underset{\text{H}}{|}}{\overset{\overset{\text{C}_2\text{H}_5}{|}}{\text{C}}})_n - \underset{\underset{\text{H}}{|}}{\overset{\overset{\text{CH}_3}{|}}{\text{C}}} - \text{C} \equiv \text{C} - \underset{\underset{\text{H}}{|}}{\overset{\overset{\text{C}_2\text{H}_5}{|}}{\text{C}}} \bullet e \longrightarrow \text{E} -(\underset{\underset{\text{H}}{|}}{\overset{\overset{\text{CH}_3}{|}}{\text{C}}} - \text{C} \equiv \text{C} - \underset{\underset{\text{H}}{|}}{\overset{\overset{\text{C}_2\text{H}_5}{|}}{\text{C}}})_n - \underset{\underset{\text{H}}{|}}{\overset{\overset{\text{CH}_3}{|}}{\text{C}}} - \text{C} \equiv \text{C} - \underset{\underset{\text{H}}{|}}{\overset{\overset{\text{CH}_3}{|}}{\text{C}}} = \underset{\underset{\text{H}}{|}}{\text{C}}$$

$$+ \quad \text{H} \bullet e$$

Transfer species of the first kind of the first type

WITHOUT RESONANCE STABILIZATION PHENONMENON-1,2-MONO-FORM

$$\text{N} \bullet n + n \bullet \underset{\underset{\underset{\underset{\text{C}_2\text{H}_5}{|}}{\text{CH}}}{\overset{\overset{\text{CH}_3}{|}}{\underset{\underset{\text{C}}{\|}}{\text{C}}}}}{\overset{}{\underset{\underset{\text{H}}{|}}{\text{C}}}} - \underset{}{\text{C}} \bullet e \longrightarrow \text{NH} + n \bullet \underset{\underset{\text{CH}_3}{|}}{\overset{\overset{\text{H}}{|}}{\text{C}}} - \text{C} \equiv \text{C} - \underset{\underset{\text{CH}_3}{|}}{\overset{\overset{\text{H}}{|}}{\text{C}}} = \underset{\overset{\text{H}}{|}}{\text{C}}$$

Transfer species of first kind of first type

taking note of the same dead terminal end and identical types.
(Laws of Conservation of transfer of transfer species)

Rule 436: This rule of Chemistry for **1,4-Cumulene of the type H$_2$C=C=C=CHOH**, states that, when activated, the followings are obtained-

$$n \bullet \underset{\underset{\text{H}}{|}}{\overset{\overset{\text{H}}{|}}{\text{C}}} - \overset{\overset{\bullet e}{}}{\underset{\underset{\text{H}}{|}}{\text{C}}} = \text{C} = \underset{\overset{\text{OH}}{|}}{\text{C}} \quad \longleftrightarrow \quad n \bullet \underset{\underset{\text{H}}{|}}{\overset{\overset{\text{H}}{|}}{\text{C}}} - \text{C} \equiv \text{C} - \underset{\underset{\text{H}}{|}}{\overset{\overset{\text{OH}}{|}}{\text{C}}} \bullet e$$

RESONANCE STABILIZATION PHENOMENON

$$\text{E} -(\underset{\underset{\text{H}}{|}}{\overset{\overset{\text{H}}{|}}{\text{C}}} - \text{C} \equiv \text{C} - \underset{\underset{\text{H}}{|}}{\overset{\overset{\text{OH}}{|}}{\text{C}}})_n - \underset{\underset{\text{H}}{|}}{\overset{\overset{\text{H}}{|}}{\text{C}}} - \text{C} \equiv \text{C} - \underset{\overset{\text{OH}}{|}}{\text{C}} \bullet e \longrightarrow \text{E} -(\underset{\underset{\text{H}}{|}}{\overset{\overset{\text{H}}{|}}{\text{C}}} - \text{C} \equiv \text{C} - \underset{\underset{\text{H}}{|}}{\overset{\overset{\text{OH}}{|}}{\text{C}}})_n - \underset{\underset{\text{H}}{|}}{\overset{\overset{\text{H}}{|}}{\text{C}}} - \text{C} \equiv \text{C} - \underset{\overset{}{}}{\text{C}} = \text{O}$$

$$+ \quad \text{H} \bullet e$$

for which the transfer species involved is of the first kind of the first type.
(Laws of Creations for 1,4-Cumulenes)

Rule 437: This rule of Chemistry for **1,4-Cumulene of the type H$_2$C=C=C=CHOH,** states that, when it is unstable it molecularly rearranges as follows-

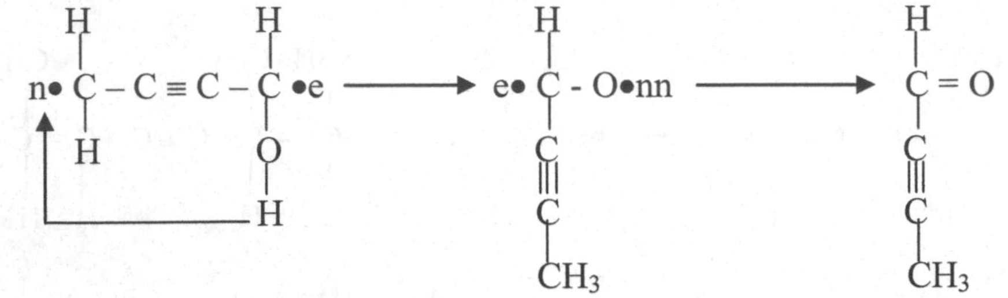

$$n\bullet \overset{\overset{\textstyle H}{|}}{\underset{\underset{\textstyle H}{|}}{C}} - C \equiv C - \overset{\overset{\textstyle H}{|}}{\underset{\underset{\textstyle O}{|}}{C}} \bullet e \longrightarrow e\bullet \overset{\overset{\textstyle H}{|}}{C} - O\bullet nn \longrightarrow \overset{\overset{\textstyle H}{|}}{C} = O$$

An Aldehyde

for which an aldehyde which is an ELECTROPHILE and resonance stabilized is obtained.
(Laws of Creations for 1,4-Cumulenes)

Rule 438: This rule of Chemistry for **1,4-cumulenes of the types H$_2$C=C=C=CHNH$_2$, H$_2$C=C=C=CNHR,** states that, these monomers are unstable for which they molecularly rearrange as shown below using one of them-

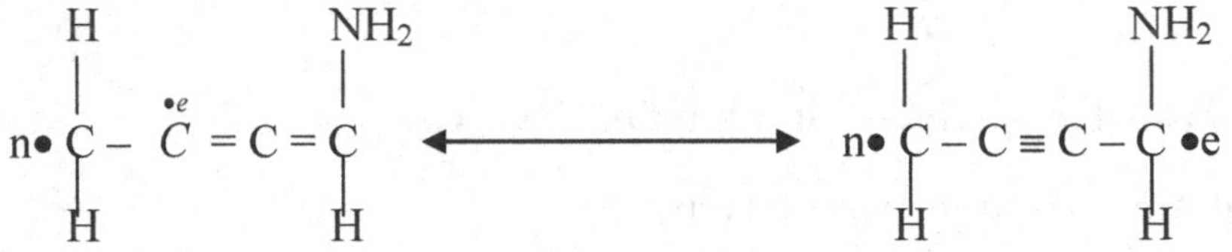

$$n\bullet \overset{\overset{\textstyle H}{|}}{\underset{\underset{\textstyle H}{|}}{C}} - \overset{\bullet e}{C} = C = \overset{\overset{\textstyle NH_2}{|}}{\underset{\underset{\textstyle H}{|}}{C}} \longleftrightarrow n\bullet \overset{\overset{\textstyle H}{|}}{\underset{\underset{\textstyle H}{|}}{C}} - C \equiv C - \overset{\overset{\textstyle NH_2}{|}}{\underset{\underset{\textstyle H}{|}}{C}} \bullet e$$

RESONANCE STABILIZATION PHENOMENON

$$n\bullet \overset{\overset{\textstyle H}{|}}{\underset{\underset{\textstyle H}{|}}{C}} - C \equiv C - \overset{\overset{\textstyle H}{|}}{\underset{\underset{\textstyle NH_2}{|}}{C}} \bullet e \longrightarrow e\bullet \overset{\overset{\textstyle H}{|}}{C} - N\bullet nn \longrightarrow \overset{\overset{\textstyle H}{|}}{C} = N$$

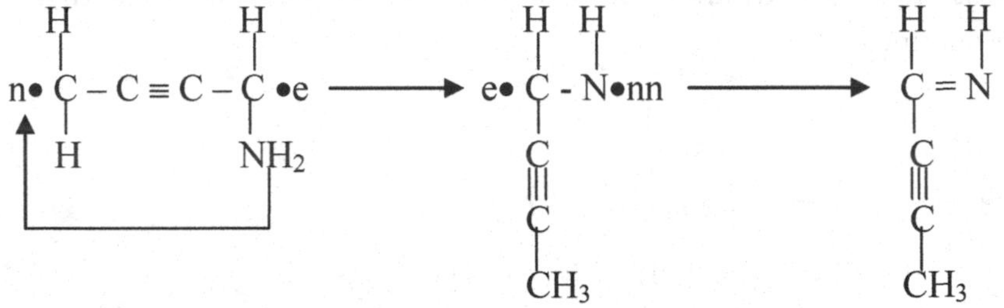

Transfer species of 1st type   **Transfer species of 5th type**   **An Aldimine**

for which an aldimine which is an ELECTROPHILE and resonance stabilized is obtained.
(Laws of Creations for 1,4-Cumulenes)

Rule 439: This rule of Chemistry for **1,4-Cumulenes of the types H$_2$C=C=C=CHOH, H$_2$C=C=C=CHNH$_2$, H$_2$C=C=C=CHNHR,** states that, when they molecularly rearrange, both the Aldehyde and Aldimine can undergo Enolization and Enamidization respectively at specific operating conditions as shown below using the last case-

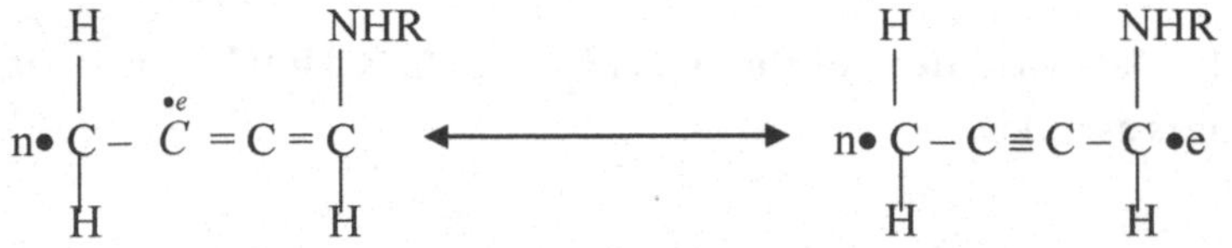

$$n\bullet \overset{\overset{\textstyle H}{|}}{\underset{\underset{\textstyle H}{|}}{C}} - \overset{\bullet e}{C} = C = \overset{\overset{\textstyle NHR}{|}}{\underset{\underset{\textstyle H}{|}}{C}} \longleftrightarrow n\bullet \overset{\overset{\textstyle H}{|}}{\underset{\underset{\textstyle H}{|}}{C}} - C \equiv C - \overset{\overset{\textstyle NHR}{|}}{\underset{\underset{\textstyle H}{|}}{C}} \bullet e$$

RESONANCE STABILIZATION PHENOMENON

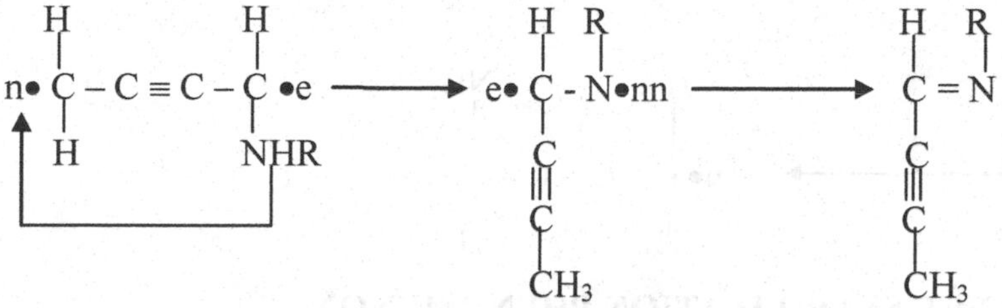

$$n\bullet \overset{\overset{\displaystyle H}{|}}{\underset{\underset{\displaystyle H}{|}}{C}} - C \equiv C - \overset{\overset{\displaystyle H}{|}}{\underset{\underset{\displaystyle NHR}{|}}{C}} \bullet e \longrightarrow e\bullet \overset{\overset{\displaystyle H}{|}}{\underset{\underset{\displaystyle CH_3}{|}}{C}} - N\bullet nn \longrightarrow \overset{\overset{\displaystyle H}{|}}{\underset{\underset{\displaystyle CH_3}{|}}{C}} = N$$

MOLECULAR REARRANGEMENT PHENOMENON

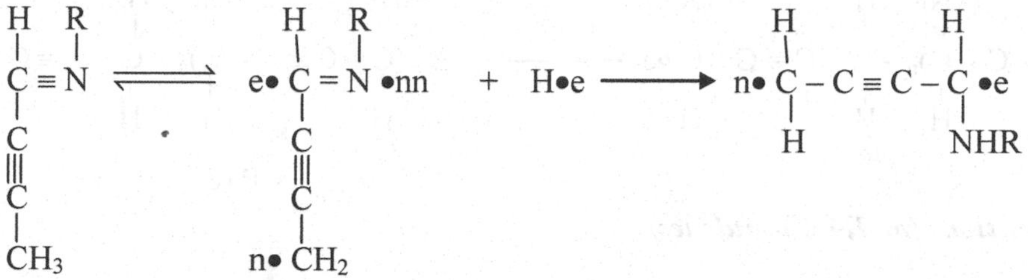

$$\overset{\overset{\displaystyle H \quad R}{| \quad |}}{\underset{\underset{\displaystyle CH_3}{|}}{C}} \equiv N \rightleftharpoons e\bullet \overset{\overset{\displaystyle H \quad R}{| \quad |}}{\underset{\underset{\displaystyle n\bullet CH_2}{|}}{C}} = N \bullet nn \quad + \quad H\bullet e \longrightarrow n\bullet \overset{\overset{\displaystyle H}{|}}{\underset{\underset{\displaystyle H}{|}}{C}} - C \equiv C - \overset{\overset{\displaystyle H}{|}}{\underset{\underset{\displaystyle NHR}{|}}{C}} \bullet e$$

ENAMIDIZATION PHENOMENON

all taking place under different operating conditions.
(Laws of Creations for 1,4-Cumulenes)

Rule 440: This rule of Chemistry for **Activated/Equilibrium state of existence phenomena**, states that, there are two types of these phenomena, both types being that in which the Equilibrium state of existence is that in which the component held is that which is involved in the Law of Conservation of transfer of transfer species when the monomer is activated, such as in Enolization (such as in aldehydes and ketones) and Enamidization (such as in aldemines and ketimines) for ringed and non-ringed cases.
(Laws of Creations for Activated/Equilibrium states of existences phenomena)

Rule 441: This rule of Chemistry for **1,4-Cumulenes of the types $H_2C=C=C=CHOR$, $H_2C=C=C=CHNR_2$,** states that, these monomers are very stable to the point where they cannot molecularly rearrange to produce Aldehydes and Aldimines and can suitably be polymerized as shown below-

FOR $H_2C=C=C=CHOR$

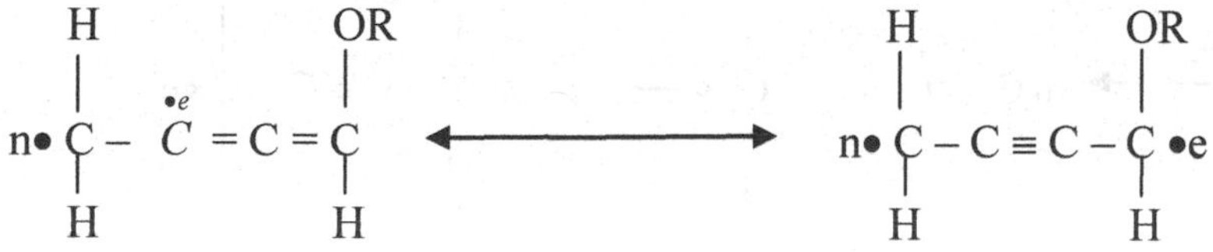

$$n\bullet \overset{\overset{\displaystyle H}{|}}{\underset{\underset{\displaystyle H}{|}}{C}} - \overset{\overset{\displaystyle}{\bullet e}}{C} = C = \overset{\overset{\displaystyle OR}{|}}{\underset{\underset{\displaystyle H}{|}}{C}} \quad \longleftrightarrow \quad n\bullet \overset{\overset{\displaystyle H}{|}}{\underset{\underset{\displaystyle H}{|}}{C}} - C \equiv C - \overset{\overset{\displaystyle OR}{|}}{\underset{\underset{\displaystyle H}{|}}{C}} \bullet e$$

RESONANCE STABILIZATION PHENOMENON

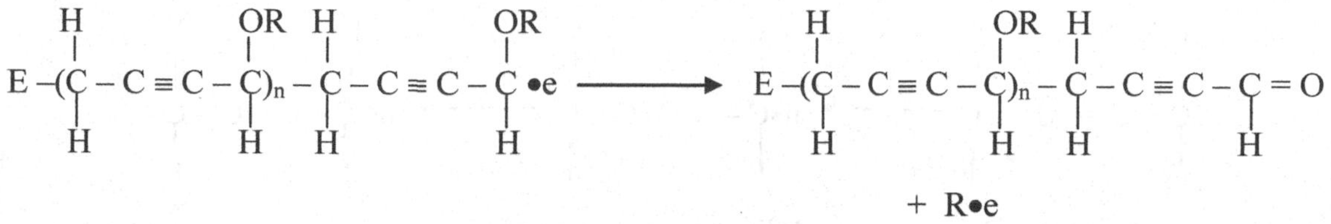

$$E - (\overset{\overset{\displaystyle H}{|}}{\underset{\underset{\displaystyle H}{|}}{C}} - C \equiv C - \overset{\overset{\displaystyle OR}{|}}{\underset{\underset{\displaystyle H}{|}}{C}})_n - \overset{\overset{\displaystyle H}{|}}{\underset{\underset{\displaystyle H}{|}}{C}} - C \equiv C - \overset{\overset{\displaystyle OR}{|}}{\underset{\underset{\displaystyle H}{|}}{C}} \bullet e \longrightarrow E - (\overset{\overset{\displaystyle H}{|}}{\underset{\underset{\displaystyle H}{|}}{C}} - C \equiv C - \overset{\overset{\displaystyle OR}{|}}{\underset{\underset{\displaystyle H}{|}}{C}})_n - \overset{\overset{\displaystyle H}{|}}{\underset{\underset{\displaystyle H}{|}}{C}} - C \equiv C - \overset{\overset{\displaystyle OR}{|}}{\underset{\underset{\displaystyle H}{|}}{C}} = O$$

$$+ \ R\bullet e$$

682

FOR H₂C=C=C=CHNR₂

$$
\overset{\displaystyle H}{\underset{\displaystyle H}{n\bullet C}} - \overset{\displaystyle}{\overset{\bullet e}{C}} = C = \overset{\displaystyle NR_2}{\underset{\displaystyle H}{C}} \quad\longleftrightarrow\quad \overset{\displaystyle H}{\underset{\displaystyle H}{n\bullet C}} - C \equiv C - \overset{\displaystyle NR_2}{\underset{\displaystyle H}{C}} \bullet e
$$

RESONANCE STABILIZATION PHENOMENON

$$
E-\overset{\displaystyle H}{\underset{\displaystyle H}{(C}}-C\equiv C-\overset{\displaystyle NR_2}{\underset{\displaystyle H}{C)_n}}-\overset{\displaystyle H}{\underset{\displaystyle H}{C}}-C\equiv C-\overset{\displaystyle NR_2}{\underset{\displaystyle H}{C}}\bullet e \longrightarrow E-\overset{\displaystyle H}{\underset{\displaystyle H}{(C}}-C\equiv C-\overset{\displaystyle NR_2}{\underset{\displaystyle H}{C)_n}}-\overset{\displaystyle H}{\underset{\displaystyle H}{C}}-C\equiv C-\overset{\displaystyle R}{\underset{\displaystyle H}{C}}=N
$$

+ R•e

(Laws of Creations for 1,4-Cumulenes)

Rule 442: This rule of Chemistry for **Alkyl groups,** states that, based on the observations so far, the followings order are to be expected-

$$H_3C - CH = C = CH- \; > \; H_3C - CH = CH - \; > \; H_3C- \; \geq \; H_3C - C \equiv C- \; >$$

Order of Radical-pushing capacity

(Laws of Creations for Alkyl-pushing groups)

Rule 443: This rule of Chemistry for 1,4-Cumulenes of the types $HFC=C=C=CH_2$, $FHC=C=C$ $=CFH$, $F_2C=C=C=CF_2$, states that, when their activations are made possible based on the operating conditions, the followings are obtained-

$$
\overset{\displaystyle F}{\underset{\displaystyle H}{n\bullet C}} - \overset{\bullet e}{C} = C = \overset{\displaystyle H}{\underset{\displaystyle H}{C}} \quad\longleftrightarrow\quad \overset{\displaystyle F}{\underset{\displaystyle H}{n\bullet C}} - C \equiv C - \overset{\displaystyle H}{\underset{\displaystyle H}{C}} \bullet e
$$

Resonance stabilization phenomenon

$$
\overset{\displaystyle F}{\underset{\displaystyle H}{C}} = \overset{\displaystyle F}{\overset{2}{C}} = \overset{\displaystyle F}{\underset{\displaystyle H}{\underset{3}{C}}} = \overset{\displaystyle F}{\underset{\displaystyle H}{C}} \longrightarrow \overset{\displaystyle F}{\underset{\displaystyle H}{n\bullet C}} - \overset{\bullet e}{C} = C = \overset{\displaystyle F}{\underset{\displaystyle H}{C}} \quad\longleftrightarrow\quad \overset{\displaystyle F}{\underset{\displaystyle H}{n\bullet C}} - C \equiv C - \overset{\displaystyle F}{\underset{\displaystyle H}{C}} \bullet e
$$

Resonance stabilization phenomenon

$$
\overset{\displaystyle F}{\underset{\displaystyle F}{C}} = C = C = \overset{\displaystyle F}{\underset{\displaystyle F}{C}} \longrightarrow \overset{\displaystyle F}{\underset{\displaystyle F}{n\bullet C}} - \overset{\bullet e}{C} = C = \overset{\displaystyle F}{\underset{\displaystyle F}{C}} \longrightarrow \overset{\displaystyle F}{\underset{\displaystyle F}{n\bullet C}} - C \equiv C - \overset{\displaystyle F}{\underset{\displaystyle F}{C}} \bullet e
$$

Resonance stabilization phenomenon

for which they favor both free-radical routes, the monomers being weak Nucleophiles.
(Laws of Creations for 1.4-Cumulenes)

Rule 444: This rule of Chemistry for **1,4-Cumulenes of the types $R_F HC=C=C=CHR$, where R_F are radical-pushing groups of less capacity than H such as $C_n F_{2n+1}$ and R is $C_n H_{2n+1}$**, states that, these are weak Nucleophiles-

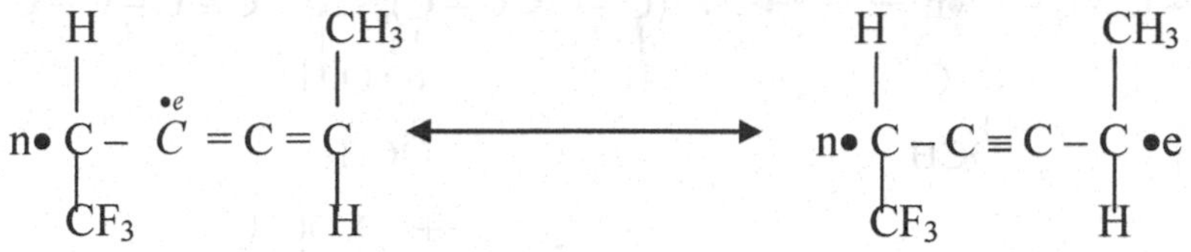

RESONANCE STABILIZATION PHENOMENON

which for any $C_n F_{2n+1}$, no route can be favored free-radically but only with paired media initiators nucleo-free-radically when the groups are trans-placed.
(Laws of Creations for 1,4-Cumulenes)

Rule 445: This rule of Chemistry for **1,4-Cumulenes of the type $R_F HC=C=C=CH_2$ where R_F is COOR or $CONH_2$,** states that, these are Electrophiles which become stronger when they molecularly rearrange as shown below-

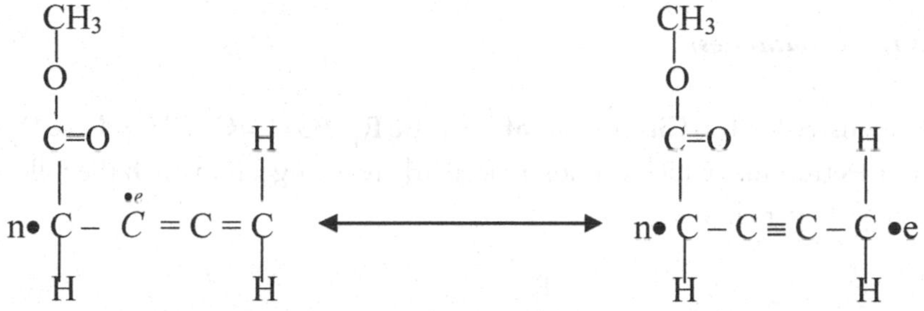

ELECTROPHILE   ELECTROPHILE
RESONANCE STABILIZATION PHENOMENON

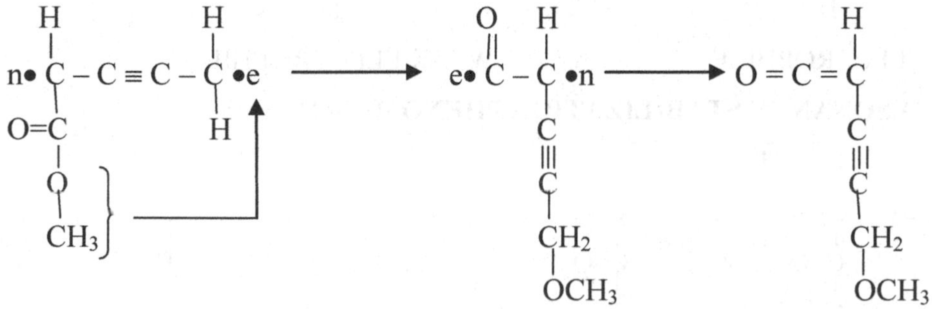

.(A)   KETENE
MOLECULAR RERRANGEMENT PHENOMENON

(Laws of Creations for 1,4-Cumulenes)

684

Rule 446: This rule of Chemistry for **1,4-Cumulenes of the type $R_FHC=C=C=CH_2$** where R_F is COOR or $CONH_2$, states that, these are strong Electrophiles for which the followings are obtained for their growing polymer chains when paired initiators $N^{\bullet n}..........^{e\bullet}X$ are used-

WITHOUT MOLECULAR REARRANGEMENT

$$N-(\overset{\overset{\textstyle H}{|}}{\underset{\underset{\textstyle H}{|}}{C}}-C\equiv C-\overset{\overset{\textstyle H}{|}}{\underset{\underset{\textstyle C=O}{|}}{C}})_n-\overset{\overset{\textstyle H}{|}}{\underset{\underset{\textstyle OCH_3}{|}}{C}}-C\equiv C-\overset{\overset{\textstyle H}{|}}{\underset{\underset{\textstyle OCH_3}{C=O}}{C}}\bullet n \longrightarrow N-(C-C\equiv C-C)_n-C-C\equiv C-C=C=O$$

$$+ \quad XOCH_3$$

WITH MOLECULAR REARRANGEMENT

$$N-(\overset{O}{C}-C)_n-\overset{O}{C}-C\bullet n \longrightarrow N-(\overset{O}{C}-C)_n-\overset{O}{C}-C=C=C=\overset{H}{C} \quad + \quad XOCH_3$$

[Transfer species of FIFTH type]

(Laws of Creations for 1,4-Cumulenes)

Rule 447: This rule of Chemistry for **1,4-Cumulenes of the type $R_FHC=C=C=CH_2$** where R_F is COR, states that, these are weak Electrophiles which cannot molecularly rearrange, for which the followings are obtained for their growing polymer chains-

$$n\bullet \overset{\overset{R}{|}}{\underset{\underset{H}{|}}{C}}-\overset{\overset{\bullet e}{C}}{\underset{\underset{H}{|}}{C}}=C=C \quad \longleftrightarrow \quad n\bullet \overset{\overset{R}{|}}{\underset{\underset{H}{|}}{C}}-C\equiv C-\overset{\overset{H}{|}}{\underset{\underset{H}{|}}{C}}\bullet e$$

WEAK ELECTROPHILE   WEAK ELECTROPHILE

RESONANCE STABILIZATION PHENOMENON

$$N-(C-C\equiv C-C)_n-C-C\equiv C-C\bullet n \longrightarrow N(C-C\equiv C-C)_n-C-C\equiv C-C=C$$

$$+ \quad R\bullet n$$

for which transfer species of the first kind of the first type is forced to be released to kill the growing polymer chain to give it a ketenic character; ***indeed it cannot be released,*** since it is not largely a transfer species of the first kind.
(Laws of Creations for 1,4-Cumulenes)

Rule 448: This rule of Chemistry for 1,4-Cumulenes of the type $R_F HC=C=C=CH_2$ where R_F is CN, states that, this is a strong Electrophile which cannot molecularly rearrange, for which the followings are obtained for its growing polymer chain-

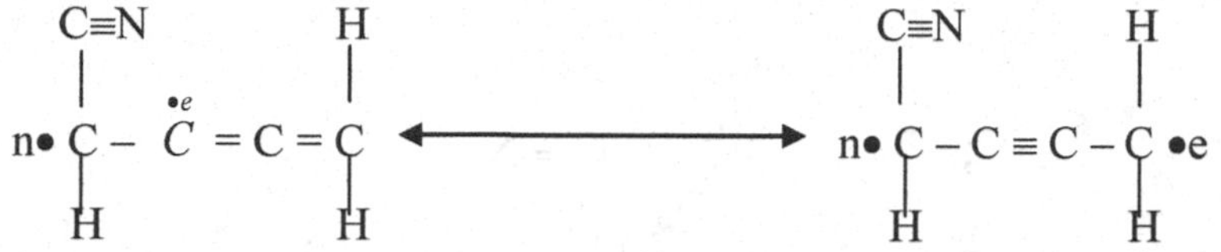

STRONG ELECTROPHILE   STRONG ELECTROPHILE
RESONANCE STABILIZATION PHENOMENON

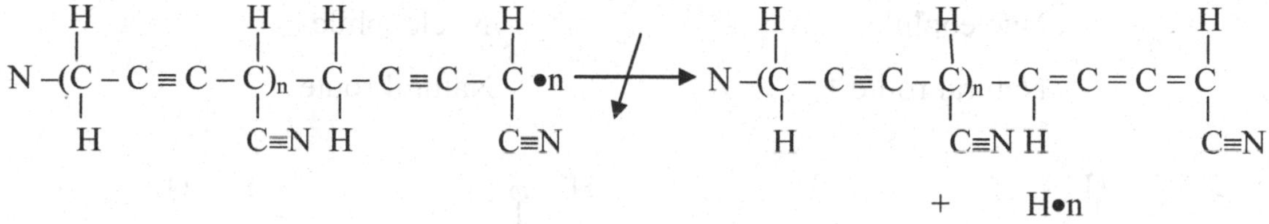

$$+ \quad H\bullet n$$

for which transfer species of the second kind cannot be released to kill the growing polymer chain, since the route is natural to the monomer.
(Laws of Creations for 1,4-Cumulenes)

Rule 449: This rule of Chemistry for **Transfer species of the second first kind,** states that, there are indeed different types as distinguished, listed and shown below-

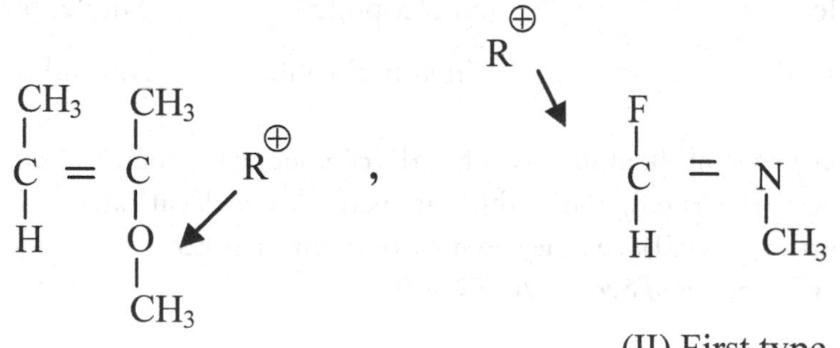

(I) <u>First type</u>
    Nucleophile
    Natural route

(II) <u>First type</u>
     Nucleophile
     Natural route

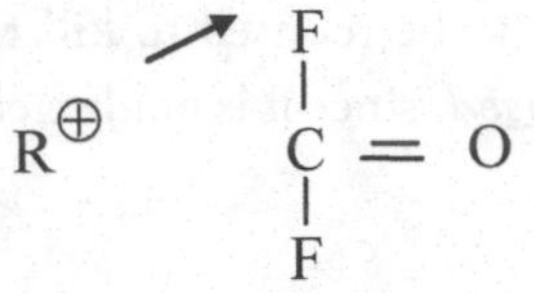

$$R^{\oplus} \quad \overset{F}{\underset{F}{\overset{|}{\underset{|}{C}}}} = O$$

(III) <u>Second type</u>
Nucleophile
Natural route

$$R^{\oplus} \quad \overset{H}{\underset{R}{\overset{|}{\underset{O}{\overset{C}{\Big\{}}}}}} = O$$

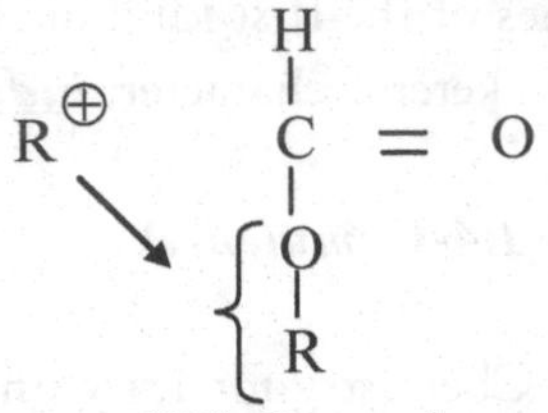

(IV) <u>Second type</u>
Nucleophile
Natural route

$$R^{\oplus} \quad \overset{F}{\underset{}{\overset{|}{C}}} \equiv N$$

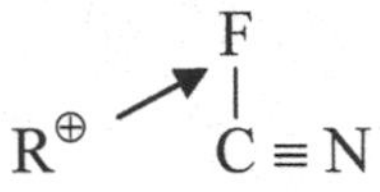

(V) **<u>Third type</u>**

Nucleophile

Natural route

$$E^{\bullet e} \quad \overset{C \equiv N}{\underset{C_4H_9}{\Big\{ O}}$$

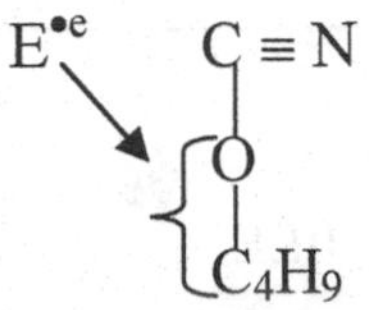

(VI) **<u>Third type</u>**

Nucleophile

Natural route

$$E^{\bullet e} \quad \overset{H}{\underset{\underset{H}{O}}{\overset{|}{C}}} = C = O$$

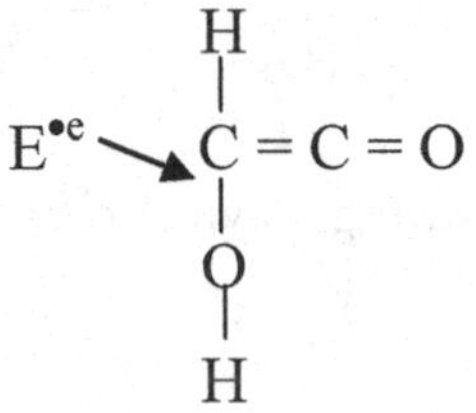

(VII) **<u>Fourth type</u>**

Electrophile

Unnatural route

$$E^{\bullet e} \quad \overset{H}{\underset{\underset{CH_3}{\overset{O}{\underset{\|}{C}}}}{\overset{|}{C}}} = \overset{H}{\underset{H}{\overset{|}{C}}}$$

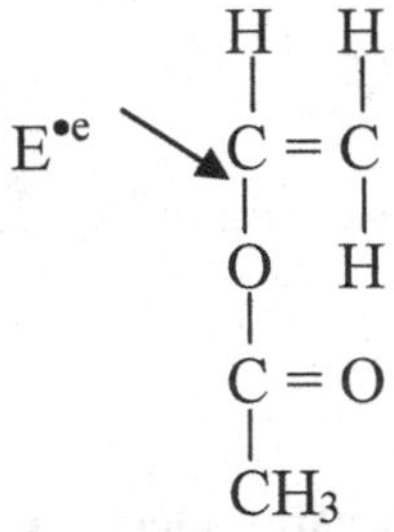

(VIII) **<u>Fisrt type</u>**

Nucleophile

Unnatural route

$$\overset{H}{\underset{H}{\overset{|}{C}}} = \overset{H}{\underset{\underset{CH_3}{\overset{O}{\underset{\|}{C}}}}{\overset{|}{C}}} \quad E^{\bullet e}$$

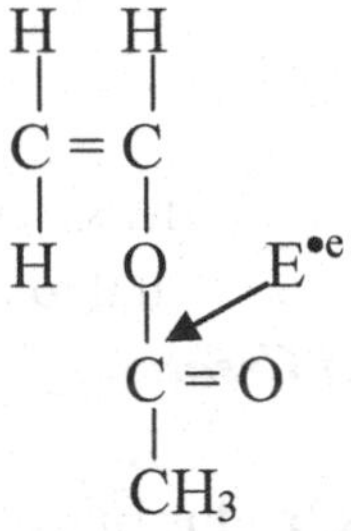

(IX) **<u>Second type</u>**

Nucleophile

Natural route

noting that these transfer species can only be abstracted when the compound or monomer is in STABLE state of Existence, taking note that it is only those which are natural, the classification applies, clear indication that there is no fourth type (VII), i.e., they do not exist with MALES.
(Laws of Creations for Transfer species of Second first kind)

<u>Rule 450:</u> This rule of Chemistry for **Anions i.e. non-free negatively charged initiators in a free-media,** states that, they can only be isolated as shown below-

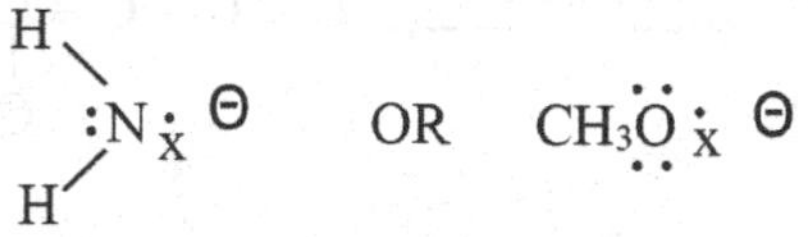

Free – media
initiator anion

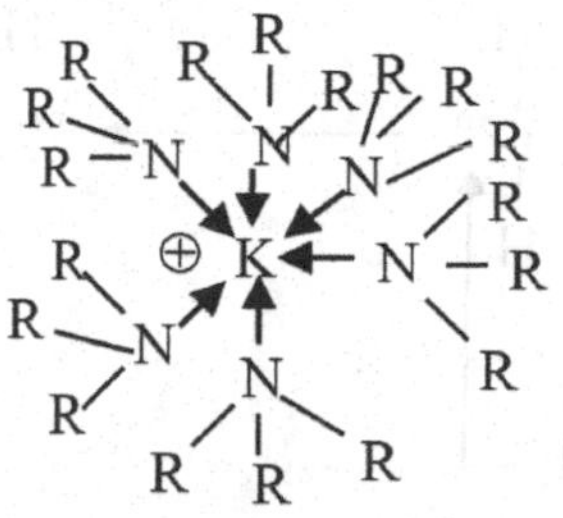

(Datively shielded from
polymerization zone)

(Laws of Creations for Free-media Anions)

Rule 451: This rule of Chemistry for **A free centered free-negatively charged initiator such as $^{\ominus}R$**, states that, this cannot be used to abstract the free hydrogen atom located on a non-free center, because they cannot be isolated, noting that $^{\ominus}COOR$, $^{\ominus}COR$, $^{\ominus}C\equiv N$ cannot exist due to electrostatic forces of repulsion and they can only carry positive charges which cannot themselves be isolated, but be paired and these do not also exist; for which only free-radicals can be carried by them.
(Laws of Creations for Negatively Charged Centers)

Rule 452: This rule of Chemistry for **1,4-Cumulenes,** states that, one of the best candidates for provision of resonance stabilization is the phenyl group from benzene a compound which is more nucleophilic than the cumulene -

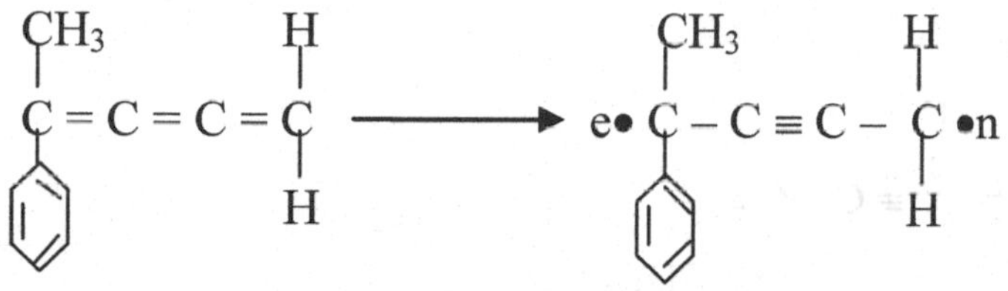

for which due to resonance stabilization provided for the CH_3 group, no transfer species can be abstracted and the monomer will favor both the nucleo- and electro-free-radical routes with the possibility of the monomer having a Ceiling temperature in the route not natural to it; for which also when the phenyl group is replaced with alkenyl group, the cumulene becomes the reosonace stabilization provider with the CH_3 group still shielded.
(Laws of Creations for 1,4-Cumulenes)

Rule 453: This rule of Chemistry for **1,5-Cumulene**, states that, this monomer which cannot be activated chargedly has a first member which unlike first members of most families, but like 1,3-Cumulene is strongly Nucleophilic as shown below-

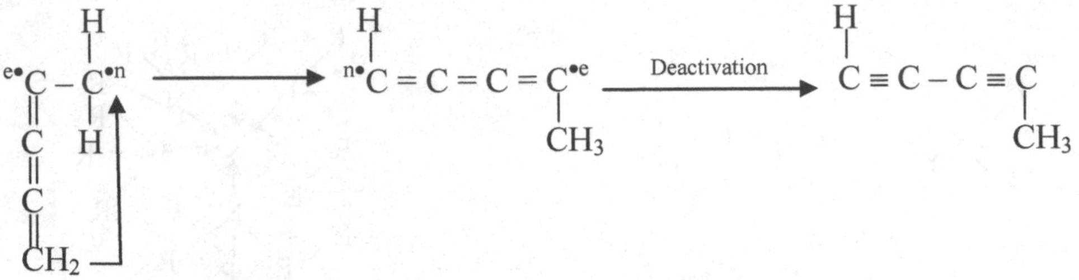

$$e\bullet C - C\bullet n \;(\text{with } H, C, C, CH_2 \text{ branches}) \longrightarrow \; n\bullet C = C = C = C\bullet e \;(\text{with } H, CH_3) \xrightarrow{\text{Deactivation}} \; C \equiv C - C \equiv C \;(\text{with } H, CH_3)$$

(Laws of Creations for 1,5-Cumulene)

Rule 454: This rule of Chemistry for **1,5-Cumulene,** states that, when made to exist in Equili-brium state of existence, the followings are to be expected-.

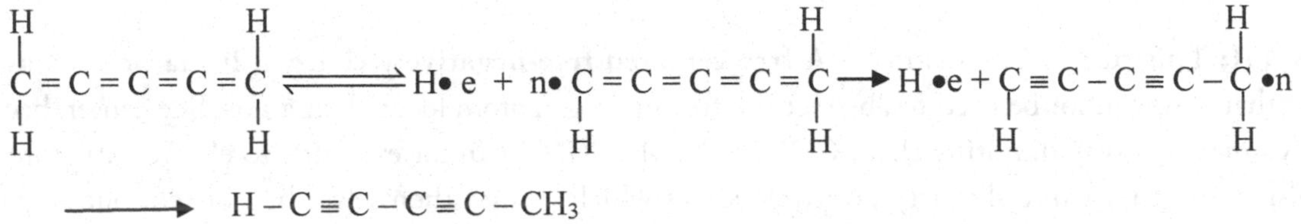

$$C = C = C = C = C \;(\text{with } H,H,H,H) \rightleftharpoons H\bullet e + n\bullet C = C = C = C = C \;(\text{with } H,H,H) \longrightarrow H\bullet e + C \equiv C - C \equiv C - C\bullet n \;(\text{with } H,H,H)$$

$$\longrightarrow H - C \equiv C - C \equiv C - CH_3$$

for which, after Electroradicalization one can observe that it belongs to the DI-YNE family.
(Laws of Creations for 1,5-Cumulene)

Rule 455: This rule of Chemistry for **1,5-Cumulene**, states that, this monomer is unstable, since when mildly heated, it decomposes to carbon black and methyl acetylene as shown below-.

Stage 1:

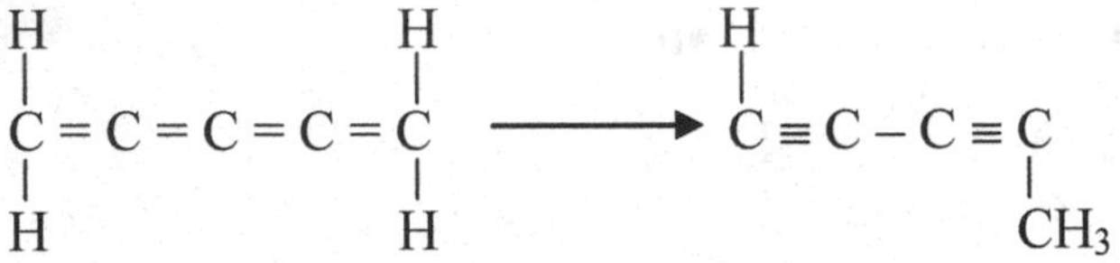

$$C = C = C = C = C \;(\text{with } H,H,H,H) \longrightarrow C \equiv C - C \equiv C \;(\text{with } H, CH_3)$$

(A)

This takes place via Electroradicalization or molecular rearrangement in two or more steps.

Stage 2:

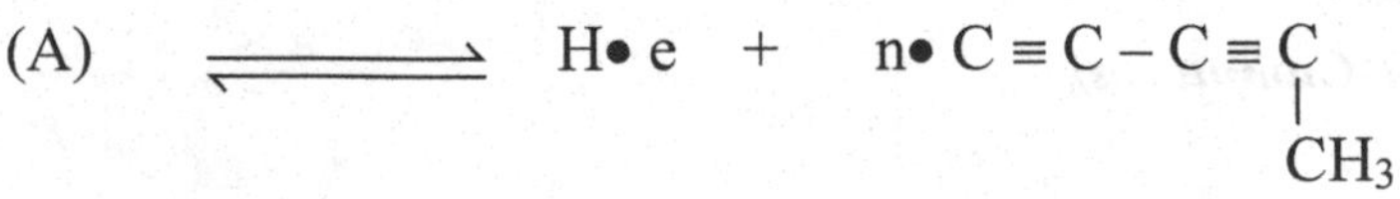

$$(A) \;\rightleftharpoons\; H\bullet e \;+\; n\bullet C \equiv C - C \equiv C \;(\text{with } CH_3)$$

(B)

$$(B) \;\rightleftharpoons\; 2\, e\bullet C\bullet n \;+\; n\bullet C \equiv C \;(\text{with } CH_3)$$

(C)   (D)

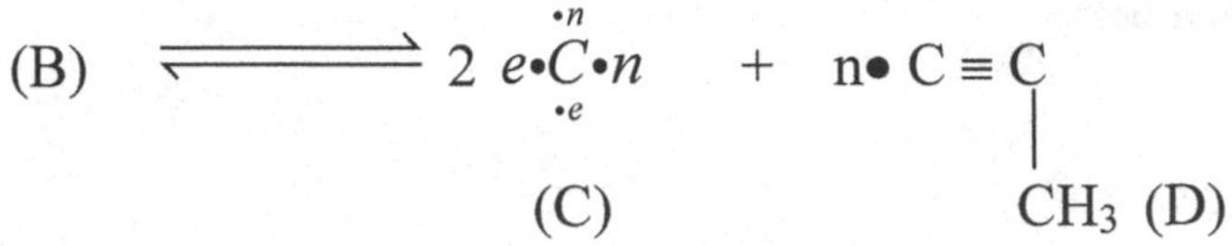

$$H \bullet e \quad + \quad (D) \quad \rightleftharpoons \quad H - C \equiv C - CH_3$$

$$(C) \quad \longrightarrow \quad 2: \overset{\bullet n}{C} \bullet e \quad \text{(Carbon Black)}$$

Overall Equation: 1,5-Cumulene $\longrightarrow$ 2C $\quad + \quad H - C \equiv C - CH_3$

(Laws of Creations for 1,5-Cumulene)

Rule 456: This rule of Chemistry for **1,5-Cumulenes,** states that, while 1,3-Cumulene has one mono-form, 1,4- Cumulenes have two mono-forms, 1,5-Cumulenes have two mono-forms as shown below-

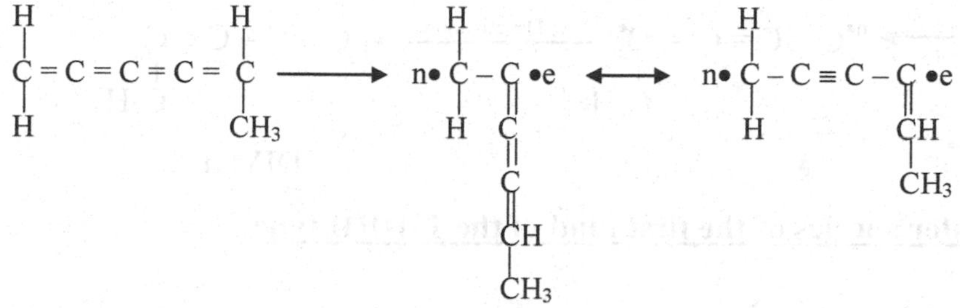

4,5-mono-form   2, 5-mono-form

RESONANCE STABILIZATION

for which one can observe that in the absence of a resonace stabilization provider, while 1,3-Cumulene is not resonance stabilized, 1,4-, 1,5- and higher Cumulenes are resonance stabilized.

(Laws of Creations for 1,5-Cumulenes)

Rule 457: This rule of Chemistry for 1,5-Cumulenes, states that, when made to carry one radical-pushing group in place of H, the followings are to be expected-

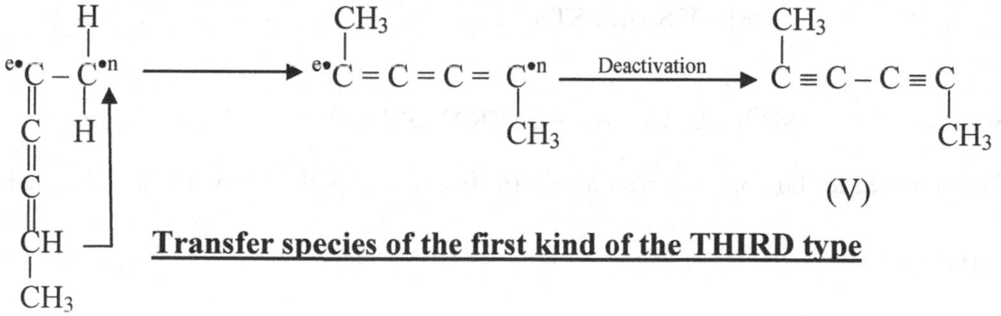

(V)

Transfer species of the first kind of the THIRD type

MOLECULAR REARRANGEMENT

for which they molecularly rearrange to produce a more stable Nucleophilic Diyne.

(Laws of Creations for 1,5-Cumulenes)

Rule 458: This rule of Chemistry for **1,5-Cumulenes of the types RHC=C=C=C=CH$_2$,** states that, the followings is the order of their Nucleophilicities-

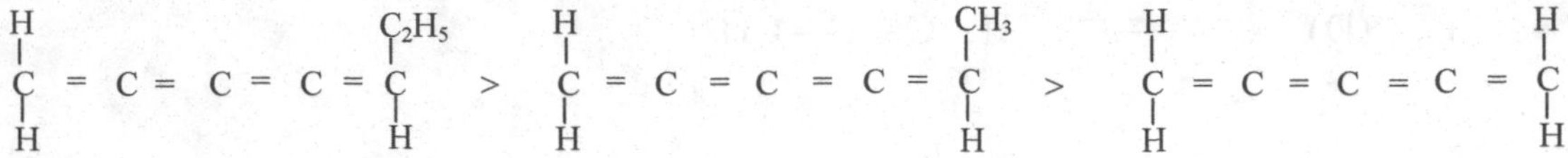

$$
\underset{H}{\overset{H}{C}} = C = C = C = \underset{H}{\overset{C_2H_5}{C}} \quad > \quad \underset{H}{\overset{H}{C}} = C = C = C = \underset{H}{\overset{CH_3}{C}} \quad > \quad \underset{H}{\overset{H}{C}} = C = C = C = \underset{H}{\overset{H}{C}}
$$

<u>Order of nucleophilicity</u>

(Laws of Creations for 1,5-Cumulenes)

Rule 459: This rule of Chemistry for **1,5-Cumulenes of the types RHC=C=C=CHR,** states that, these like the others with radical-pushing groups are strong Resonance stabilized Nucleophiles- Diynes as shown below for one of them-

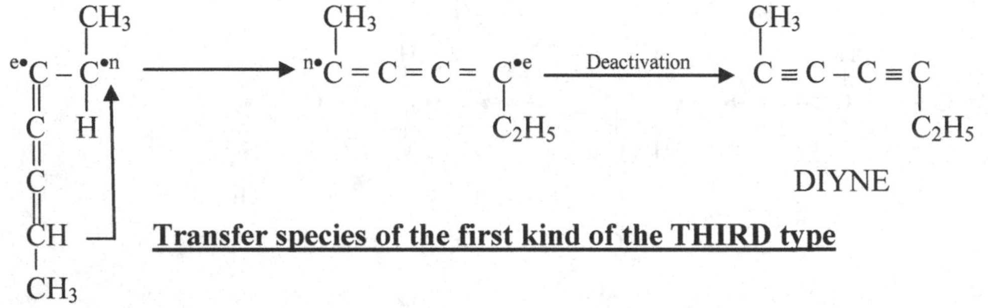

Transfer species of the first kind of the THIRD type

(I)   MOLECULAR REARRANGEMENT

(Laws of Creations for 1,5-Cumulenes)

Rule 460: This rule of Chemistry for **1,5-Cumulenes of the types (RO)HC=C=C=C=CH$_2$,** states that, the followings are obtained for their state of Existence and use as a Monomer-

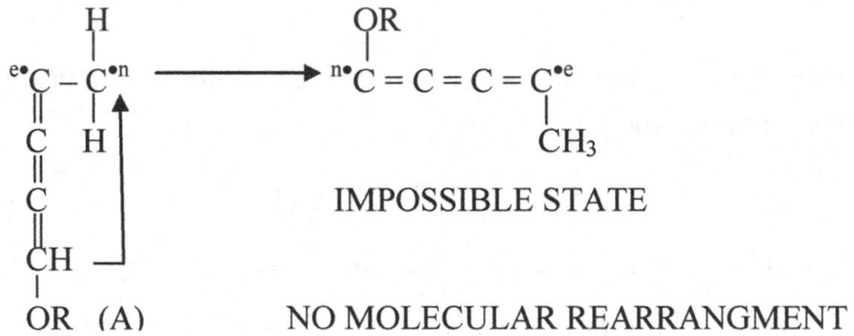

IMPOSSIBLE STATE

OR (A)   NO MOLECULAR REARRANGMENT

for which (B) when used for homopolymerization electro-free-radically, the followings are obtained-

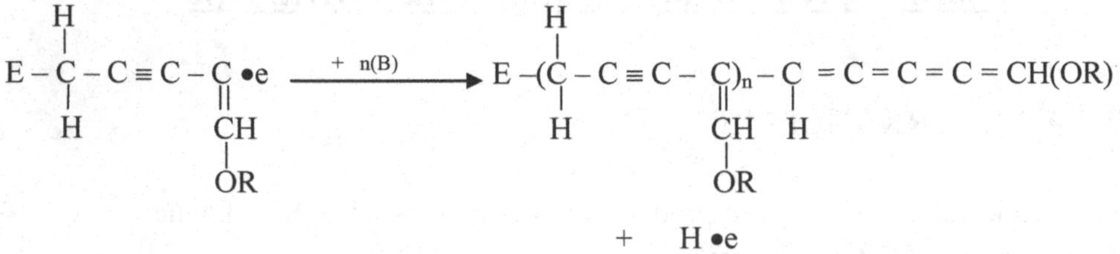

+   H •e

(Laws of Creations for 1,5-Cumulenes)

Rule 461: This rule of Chemistry for **ALIPHATIC HYDROCARBON FAMILY MEMBERS with no activation centers,** states that, the decreasing levels of saturation present as one moves along from the first member of the ALKANES to very large members, based on H to C ratio is as follows-

	CH_4 ;	C_2H_6 ;	C_3H_8 ;	C_4H_{10} ;	C_5H_{12} ;	C_6H_{14} ;
RATIO OF H:C	4	3	2(2/3)	2(1/2)	2(2/5)	2(1/3)
Diff:	1	(1/3)	(1/6)	(1/10)	(1/15)	(1/21)

	C_7H_{16} ;	C_8H_{18} ;	C_9H_{20} ;	$C_{10}H_{22}$;	$C_{11}H_{24}$;	$C_{12}H_{26}$;
RATIO OF H:C	2(2/7)	2(1/4)	2(2/9)	2(1/5)	2(2/11)	2(1/6)
Diff:	(1/28)	(1/36)	(1/45)	(1/55)	(1/66)	
RATIO OF H:C			2(2/89)	2(1/45)		2
Diff:			(1/4005)			0

for which it can be seen that the Alkane family has its limitations that which seems to be close to 2 as marked by the Carbene family.
(Laws of Creations for Saturated Hydrocarbons)

Rule 462: This rule of Chemistry for ALIPHATIC HYDROCARBON FAMILY MEMBERS with activation centers, states that, the increasing levels of unsaturation present as one moves from C = C to C ≡ C and then to C = C = C and to C = C = C = C in that order, based on H to C ratio is as follows-

	(I)		(II)		(III)		(IV)
Ratio of :	2 : 1	;	1 : 1	;	4/3 : 1	;	1 : 1
H to C	Replace none		Replace one or two H atoms		Replace one		Replace one

	(V)		(VI)	i)
Ratio of :	4/5 : 1	;	2/3 : 1	
H to C	Replace one		Replace one	

692

for which unlike the order of unsaturation, the order of nucleophilicity of the centers based on current development can be represented as follows.

$$(VI) \quad > \quad (V) \quad > \quad (IV) \quad > \quad (III) \quad > \quad (II) \quad > \quad (I)$$

Order of nucleophilicity   ii)

$$H\text{-}C{\equiv}C\text{-}C{\equiv}C\text{-}CH{=}CH_2 > H\text{-}C{\equiv}C\text{-}C{\equiv}C\text{-}CH_3 > H\text{-}C{\equiv}C\text{-}CH{=}CH_2 > H\text{-}C{\equiv}C\text{-}CH_3 > H\text{-}C{\equiv}C\text{-}H >$$

(VI)   (V)   (IV)   (III)   (II)

<u>ORDER OF NUCLEOPHILICITY</u>   iii)

noting that the replacement of H atoms is partly based on the ratios of H to C atoms in the compound as apparent with (I) and (III) where the ratio is above 1 and the order of nucleophili-city.
(Laws of Creations for Unstable Unsaturated Hydrocarbons)

Rule 463: This rule of Chemistry for **Cumunyl Groups**, states that, the followings are valid for them in radical-pushing capacity-

$H_2C{=}C{=}C{=}$ > $H_2C{=}C{=}$ > $H_2C{=}$

Tri-cumunyl   Di-cumunyl   Mono-Cumunyl

<u>Order of radical-pushing capacity</u>

(Laws of Creations for Cumunyl groups)

Rule 464: This rule of Chemistry for **1,5-Cumulenes of the types HFC=C=C=C=CH$_2$, FHC=C=C=C=CHF,** states that, these are NUCLEOPHILES for which the followings are obtained when activated-

$$
\begin{array}{c}
\underset{\overset{|}{H}}{\overset{\overset{F}{|}}{C}} = C = C = C = \underset{\overset{|}{H}}{\overset{\overset{H}{|}}{C}} \xrightarrow{\text{ACTIVATION}} n\bullet \underset{\overset{|}{H}}{\overset{\overset{F}{|}}{C}} - C \equiv C - \underset{\overset{||}{CH_2}}{C} \bullet e
\end{array}
$$

$$
\begin{array}{c}
\underset{\overset{|}{H}}{\overset{\overset{F}{|}}{C}} = C = C = C = \underset{\overset{|}{H}}{\overset{\overset{F}{|}}{C}} \xrightarrow{\text{ACTIVATION}} n\bullet \underset{\overset{|}{H}}{\overset{\overset{F}{|}}{C}} - C \equiv C - \underset{\overset{||}{CFH}}{C} \bullet e
\end{array}
$$

(Laws of Creations for Fluorinated 1,5-Cumulenes)

Rule 465: This rule of Chemistry for **1,5-Cumulenes of the types F$_2$C=C=C=C=CF$_2$,** states that, these are neither NUCLEOPHILES or ELECTROPHILES for which the followings are obtained when activated-

$$\underset{F}{\overset{F}{\underset{|}{\overset{|}{C}}}} = C = C = C = \underset{F}{\overset{F}{\underset{|}{\overset{|}{C}}}} \xrightarrow{\text{ACTIVATION}} n\bullet \underset{F}{\overset{F}{\underset{|}{\overset{|}{C}}}} - C \equiv C - \underset{CF_2}{\overset{|}{C}} \bullet e$$

(Laws of Creations for Perfluoro-1,5-Cumulene)

Rule 466: This rule of Chemistry for **1,5-Cumulenes of the types** $HF_3CC=C=C=C=CH_2$ **(I)**, $HCF_3C=C=C=C=CHCH_3$, states that, these are NUCLEOPHILES which cannot be initiated when activated in free-media, but with paired- media initiators nucleo-free-radically only when trans- placed-

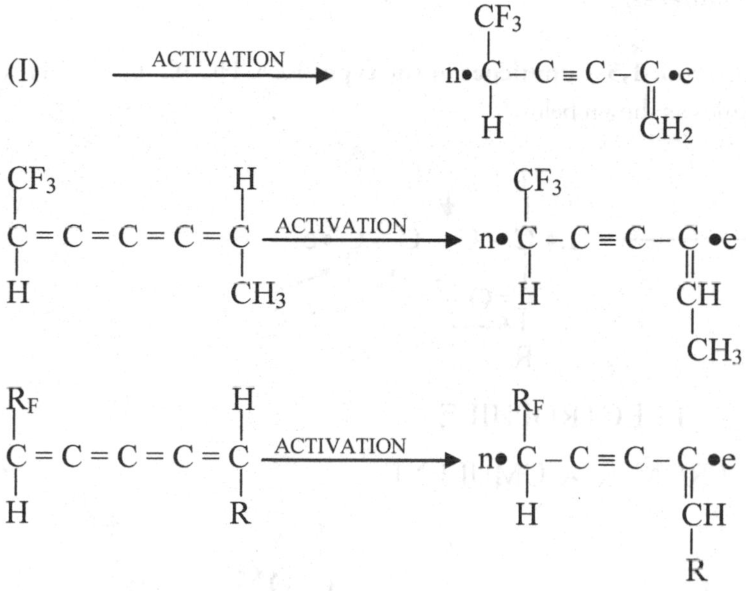

$$(I) \xrightarrow{\text{ACTIVATION}} n\bullet \underset{H}{\overset{CF_3}{\underset{|}{\overset{|}{C}}}} - C \equiv C - \underset{CH_2}{\overset{|}{C}} \bullet e$$

$$\underset{H}{\overset{CF_3}{\underset{|}{\overset{|}{C}}}} = C = C = C = \underset{CH_3}{\overset{H}{\underset{|}{\overset{|}{C}}}} \xrightarrow{\text{ACTIVATION}} n\bullet \underset{H}{\overset{CF_3}{\underset{|}{\overset{|}{C}}}} - C \equiv C - \underset{\underset{CH_3}{\overset{|}{CH}}}{\overset{|}{C}} \bullet e$$

$$\underset{H}{\overset{R_F}{\underset{|}{\overset{|}{C}}}} = C = C = C = \underset{R}{\overset{H}{\underset{|}{\overset{|}{C}}}} \xrightarrow{\text{ACTIVATION}} n\bullet \underset{H}{\overset{R_F}{\underset{|}{\overset{|}{C}}}} - C \equiv C - \underset{\underset{R}{\overset{|}{CH}}}{\overset{|}{C}} \bullet e$$

(where $R_F \equiv C_nF_{2n+1}$, $R \equiv C_nH_{2n+1}$)

(Laws of Creations for Fluorinated 1,5-Cumulenes)

Rule 467: This rule of Chemistry for **1,5-Cumulenes of the types** $ROOCHC=C=C=C=CH_2$ **(I)**, states that, when compared with the corresponding case of 1,4-Cumulenes, the followings are obtained when activated-

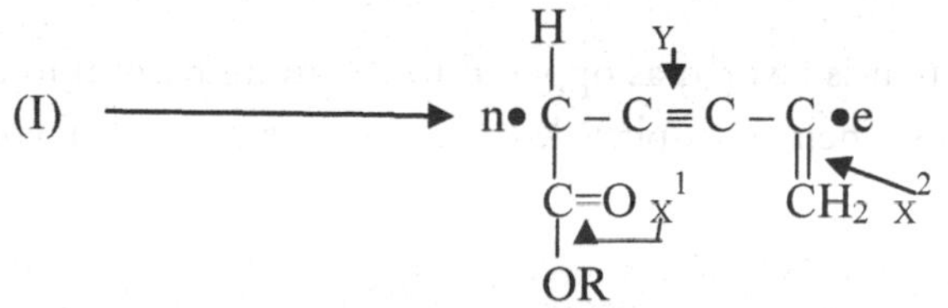

$$(I) \longrightarrow n\bullet \underset{\underset{OR}{\overset{|}{C=O}}{}^1}{\overset{H}{\underset{|}{\overset{|}{C}}}} - C \equiv C - \underset{CH_2 \; x^2}{\overset{|}{C}} \bullet e \qquad \overset{Y}{\downarrow}$$

ELECTROPHILE

FROM 1,5-CUMULENE

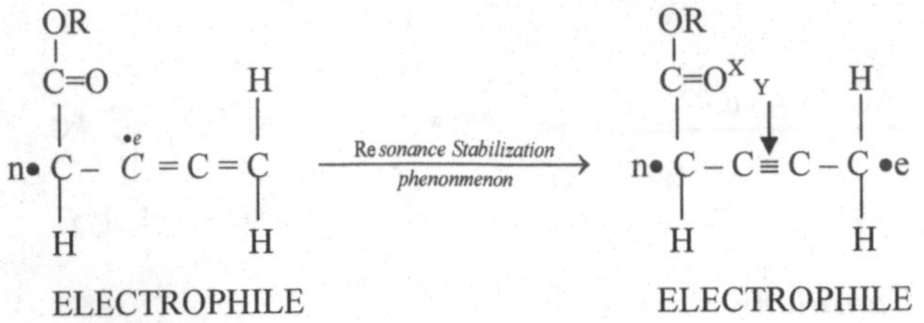

$$
\begin{array}{ccc}
\text{OR} & & \text{H} \\
| & & | \\
\text{C=O} & & \\
| & \overset{\bullet e}{} & \\
n\bullet \text{C} - \overset{\bullet e}{C} = \text{C} = \text{C} & \xrightarrow{\ \textit{Resonance Stabilization}\ \textit{phenonmenon}\ } & \\
| & & | \\
\text{H} & & \text{H}
\end{array}
$$

FROM 1.4-CUMULENE (left): ELECTROPHILE

Right structure: OR / C=O^X Y / n• C – C ≡ C – C •e / H ... H — ELECTROPHILE

for which, it can be observed that 1,4-Cumulene equivalences are more Electrophilic than 1,5-Cumulenes above and the need to use paired nucleo-free-radical initiators for their polymeri-zation is inevitable.
(Law of Creations for 1,5-Cumulenes)

Rule 468: This rule of Chemistry for **1,5-Cumulenes of the types** $ROCHC=C=C=CH_2$ (I), states that, these are Weak Electrophiles as shown below-

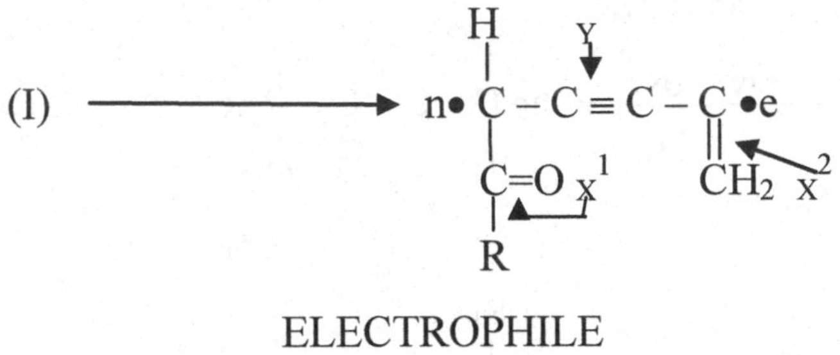

$$
(I) \longrightarrow
\begin{array}{c}
\text{H} \quad\quad \text{Y} \\
| \quad\quad \downarrow \\
n\bullet \text{C} - \text{C} \equiv \text{C} - \text{C} \bullet e \\
| \quad\quad\quad\quad \| \\
\text{C=O}\ X^1 \quad \text{CH}_2\ X^2 \\
| \\
\text{R}
\end{array}
$$

ELECTROPHILE

FROM 1,5-CUMULENE

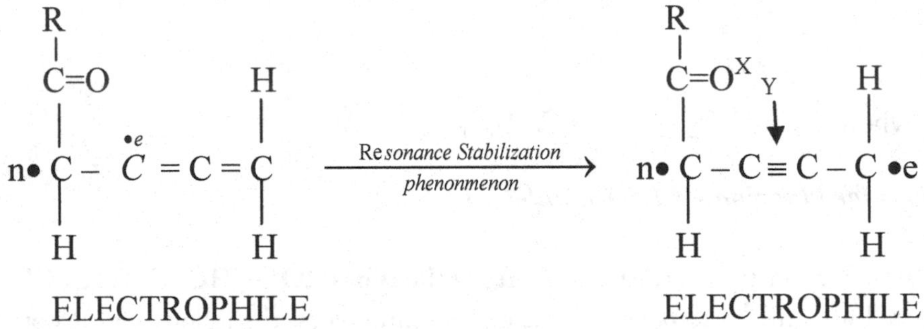

$$
\begin{array}{ccc}
\text{R} & & \text{H} \\
| & & | \\
\text{C=O} & & \\
| & \overset{\bullet e}{} & \\
n\bullet \text{C} - \overset{\bullet e}{C} = \text{C} = \text{C} & \xrightarrow{\ \textit{Resonance Stabilization}\ \textit{phenonmenon}\ } & \\
| & & | \\
\text{H} & & \text{H}
\end{array}
$$

ELECTROPHILE (left)

Right structure: R / C=O^X Y / n• C – C ≡ C – C •e / H ... H — ELECTROPHILE

FROM 1,4-CUMULENE

for which since the route not natural to it that is favored as opposed to the route natural to it with free-media, but not with paired-media initiators when trans-placed, the monomer can still favor its use for producing copolymers electro-free-radically.
(Laws of Creations for 1,5-Cumulenes)

Rule 469: This rule of Chemistry for **1,5-Cumulenes of the type** $NCHC=C=C=CH_2$ (I), states that, these are weak Electrophiles for which the followings are obtained when activated.

$$\text{(I)} \longrightarrow \quad n\bullet \overset{\overset{\displaystyle C\equiv N}{|}}{\underset{\underset{\displaystyle H}{|}}{C}} - C \equiv C - \overset{}{\underset{\underset{\displaystyle CH_2}{\|}}{C}} \bullet e$$

WEAK ELECTROPHILE

(Laws of Creations for 1,5-Cumulenes)

Rule 470: This rule of Chemistry for **Nitriles and Cumulenes,** states that, the following order in capacities of their Activation centers prevail-

$$C \equiv N \quad > \quad 1,5- \quad > \quad 1,4- \quad > \quad 1,3-$$

Cumulenes

Order of Nucleophilicity

for which unless special groups are placed on some cumulenes, the nitrile group cannot provide extended resonance stabilization for the monomers.

(Laws of Creations for Nitriles and Cumulenes)

Rule 471: This rule of Chemistry for **the first member of 1,6-Cumulene,** states that, the same orderliness as seen with 1,4-Cumulene almost apply to it; for which when activated, the followings are obtained.

$$\overset{\overset{\displaystyle H}{|}}{\underset{\underset{\displaystyle H}{|}}{C}} = C = C = C = C = \overset{\overset{\displaystyle H}{|}}{\underset{\underset{\displaystyle H}{|}}{C} } \quad \xrightarrow{\text{ACTIVATION}} \quad n\bullet \overset{\overset{\displaystyle H}{|}}{\underset{\underset{\displaystyle H}{|}}{C}} - C \equiv C - C \equiv C - \overset{\overset{\displaystyle H}{|}}{\underset{\underset{\displaystyle H}{|}}{C}} \bullet e$$

ACTIVATION OF 1,6-CUMULENE

(Laws of Creations for 1,6-Cumulene)

Eighty two rules have been proposed with respect to the past, present and future considerations and applications. New substituted groups never known to exist in the past have been identified with cumulenes, not only in terms of their characters, but also in terms of their capacities qualitatively. Alkyl groups have been additionally identified as follows – alkanyl (e.g. -CH_3), alkenyl (e.g. -$CH=CH_2$) and alkynyl (e.g. -$C\equiv CH$) for those obtained from alkanes, alkenes and alkynes respectively, cumulenyl (e.g. –$CH=C=CH_2$, -$CH=C=C=CH_2$), cumunyl (e.g. $H_2C=$, $H_2C=C=$, $H_2C=C=C=$,) from Cumulenes. New resonance stabilization groups have been identified with them which are inside the cumulenes (e.g $H_2C=C=C=$, $H_2C=C=$,) herein called the cumunyls and outside the cumulenes (e.g. $H_2C=C=C=CH-$, $H_2C=C=CH-$) herein called the cumulenyl. Rules are not stated based on numbers, but based on their contents, orderliness, further applications and non-replications.

References

1. C. R. Noller, "Textbook of Organic Chemistry," W. B. Saunders Company, (1966), pg. 581.

2. C. R. Noller,: ibid, pg. 250.

3. J. Hines, "Physical Organic Chemistry," 2nd Edition, McGraw Hill Book Company, Inc. New York (1962), pgs. 492 - 494.

4. **http://en.wikipedia.org/wiki/Cumulene,** 2009

5. N. Islam, T. Col, T. Iwasawa, M. Nishiuchi, Y. Kawamura, "Thermal cyclotrimerization of tetraphenyl [5] cumulene (tetraphenylhexapentaene) to a tricylodecadiene derivative", Chem. Commun. (Camb), 2009 Feb 7; (5): 574-6.

6. M. C. McCarthy, M. J. Travers, A Kovacs, W. Chen, S. E, Novick, C. A. Cottleb, and P. Thaddeus, "Detection and Characterization of the Cumulene Carbenes H_2C_5 and H_2C_6", Science 24 January 1997, Vol 275, no 5299, pp. 518-520.

7. B. Bildsein, M. Schweiger, H. Kopacka, Karl-Hans Qngania, and K. Wurst, "Cationic and Neutral [4]-Cumulenes C=C=C=C=C with Five Cumulated Carbons and Three to Four Ferrocenyl Termini", Organometaitics, 1998, 17, (12), pp. 2414-2424.

8. E. M. Shustorovich and N. A. Popov, "The electronic structure and properties of cumulene systems" Zhurnal Strukturnoi Khimil, Vol. 5, No. 5, pp. 770-776, 1964.

9. C. R. Noller, "Textbook of Organic Chemistry," W. B. Saunders Company, (1966), pg. 380.

Problems

8.1. How are the cumulene related to alkylenes, alkenes and acetylenes? Discuss.

8.2. Distinguish between the types of dead terminal bonded polymers obtained for 1,3 – and 1,4-cumulenes, where transfer species of the first kind of the first type is involved during polymerization.

8.3. (a) Is there any symmetric cumulene? Explain.
   (b) Identify the routes favored by the three cumulenes shown below free-radically.

(i) 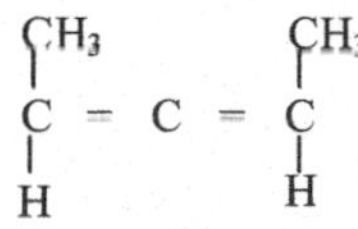
$$\begin{array}{ccccc} CH_3 & & CH_3 & & CH_3 \\ | & & | & & | \\ C & = & C & = & C \\ | & & & & | \\ H & & & & H \end{array}$$

(ii) 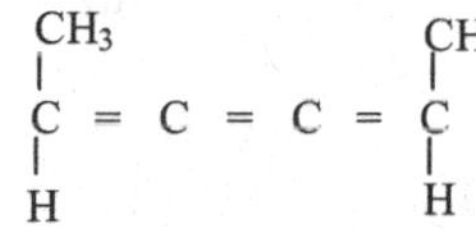
$$\begin{array}{ccccccc} CH_3 & & & & & & CH_3 \\ | & & & & & & | \\ C & = & C & = & C & = & C \\ | & & & & & & | \\ H & & & & & & H \end{array}$$

(iii) 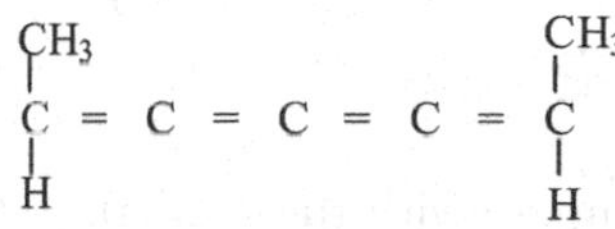
$$\begin{array}{ccccccccc} CH_3 & & & & & & & & CH_3 \\ | & & & & & & & & | \\ C & = & C & = & C & = & C & = & C \\ | & & & & & & & & | \\ H & & & & & & & & H \end{array}$$

   (c) Distinguish between (ii) and cis-2-butene.
   (d) Number the active centers for their different mono-forms.

8.4. (a) What are common between the two monomers shown below -

(i) 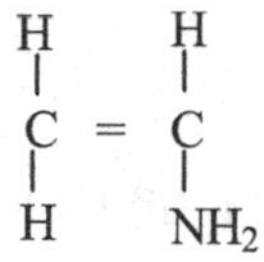
$$\begin{array}{cc} H & H \\ | & | \\ C = C \\ | & | \\ H & NH_2 \end{array} \quad \text{and} \quad \begin{array}{ccc} NH_2 & & H \\ | & & | \\ C = C = C \\ | & & | \\ H & & H \end{array}$$

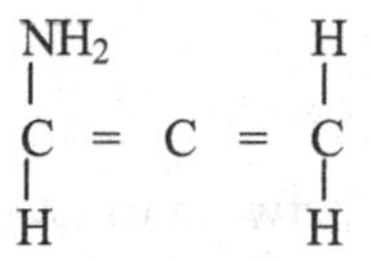

(ii) 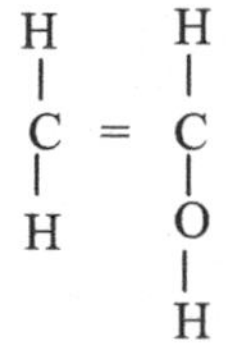
$$\begin{array}{cc} H & H \\ | & | \\ C = C \\ | & | \\ H & O \\ & | \\ & H \end{array} \quad \text{and} \quad \begin{array}{ccc} H & & H \\ | & & | \\ C = C = C \\ | & & | \\ H & & O \\ & & | \\ & & H \end{array}$$

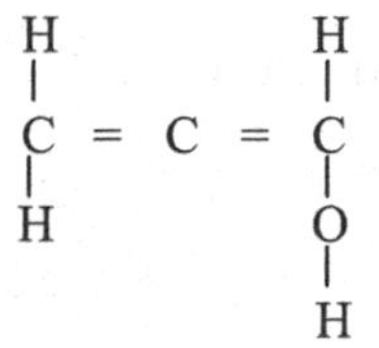

   (b) Distinguish between the initiations of olefins, 1,3 - cumulene and 1,4 – cumulene when only one hydrogen atom is replaced with $COCH_3$ group radically.

8.5. (a) Explain why transfer species $H^{\oplus}$, cannot be abstracted from the substituted group in (i) and (ii), but only $H^{\cdot e}$?

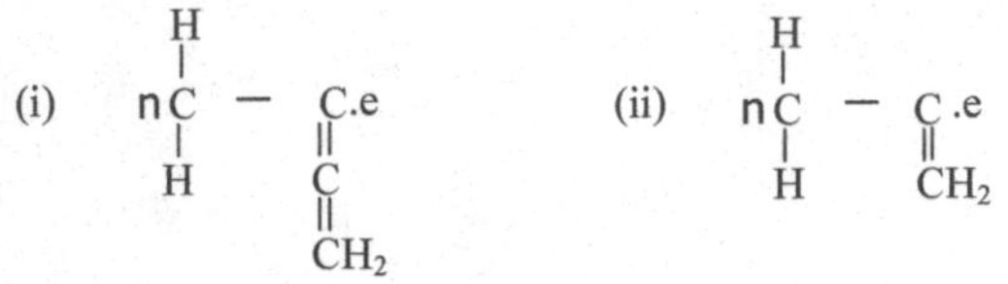

(i) $n\overset{\underset{\displaystyle |}{H}}{\underset{\underset{\displaystyle CH_2}{\overset{\displaystyle \|}{C}}}{\overset{\displaystyle |}{C}}} - \overset{\displaystyle}{\underset{\displaystyle}{C}}.e$ (ii) $n\overset{\underset{\displaystyle |}{H}}{\underset{\underset{\displaystyle CH_2}{}}{\overset{\displaystyle |}{C}}} - \overset{\displaystyle}{\underset{\displaystyle \|}{C}}.e$

(b) Does (i) have transfer species? What can be done to make (i) favor its use as a strong nucleophilic monomer?

8.6. (a) Distinguish between the three cumulenes via the routes favored by their mono-forms radically.

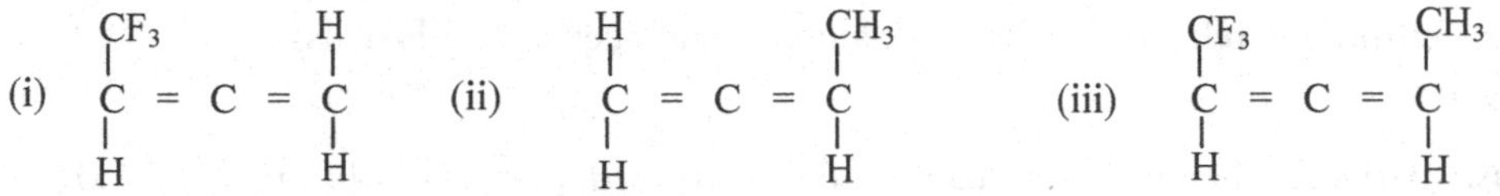

(i) $\overset{\underset{\displaystyle |}{\overset{\displaystyle |}{CF_3}}}{\underset{\displaystyle H}{C}} = C = \overset{\underset{\displaystyle |}{\overset{\displaystyle |}{H}}}{\underset{\displaystyle H}{C}}$ (ii) $\overset{\underset{\displaystyle |}{\overset{\displaystyle |}{H}}}{\underset{\displaystyle H}{C}} = C = \overset{\underset{\displaystyle |}{\overset{\displaystyle |}{CH_3}}}{\underset{\displaystyle H}{C}}$ (iii) $\overset{\underset{\displaystyle |}{\overset{\displaystyle |}{CF_3}}}{\underset{\displaystyle H}{C}} = C = \overset{\underset{\displaystyle |}{\overset{\displaystyle |}{CH_3}}}{\underset{\displaystyle H}{C}}$

(b) Do the same replacing CF_3 with C_2F_5 for (i) and (iii).

(c) Do the same for the corresponding olefins shown below

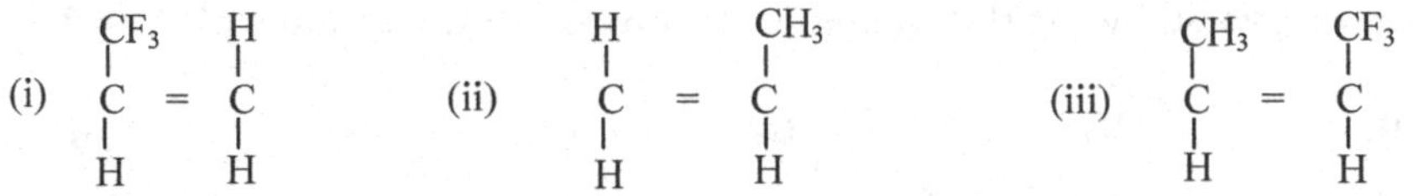

(i) $\overset{\underset{\displaystyle |}{\overset{\displaystyle |}{CF_3}}}{\underset{\displaystyle H}{C}} = \overset{\underset{\displaystyle |}{\overset{\displaystyle |}{H}}}{\underset{\displaystyle H}{C}}$ (ii) $\overset{\underset{\displaystyle |}{\overset{\displaystyle |}{H}}}{\underset{\displaystyle H}{C}} = \overset{\underset{\displaystyle |}{\overset{\displaystyle |}{CH_3}}}{\underset{\displaystyle H}{C}}$ (iii) $\overset{\underset{\displaystyle |}{\overset{\displaystyle |}{CH_3}}}{\underset{\displaystyle H}{C}} = \overset{\underset{\displaystyle |}{\overset{\displaystyle |}{CF_3}}}{\underset{\displaystyle H}{C}}$

and compare with those of (a).

8.7. Show how the following compounds can be synthesized from cumulenes
(a) Sodium methylacetylene
(b) Ethyl methylacetylene
(c) $HC \equiv CCH_2F$
(d) $HC \equiv CCH_2OH$

8.8. (a) Show the similarities or differences between (i) COH and CHNH, (ii) $COCH_3$ and $CHNCH_3$ as substituent groups. Why are the $CHNCH_3$ and CHNH types not popularly known to exist as part of a monomer?
(b) Below are shown the following compounds(i)

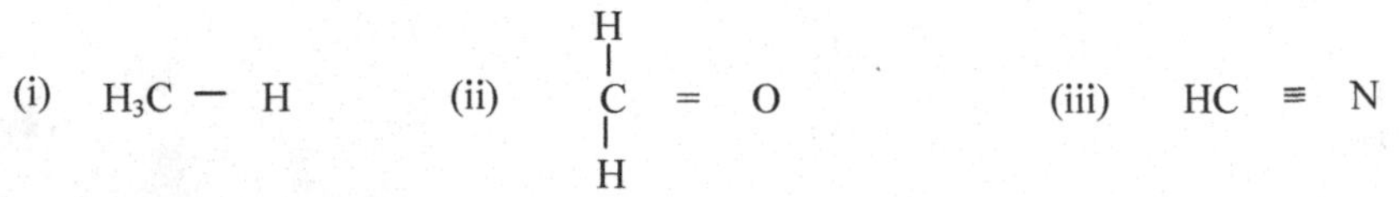

(i) $H_3C - H$ (ii) $\overset{\underset{\displaystyle |}{\overset{\displaystyle |}{H}}}{\underset{\displaystyle H}{C}} = O$ (iii) $HC \equiv N$

Identify the types of alkyl group if present, present on them. Name the compounds and identify their families. Replace one H atom in each of them with three substituent groups of your choice and identify what the compounds are.

8.9. (a) Below are shown the followings -

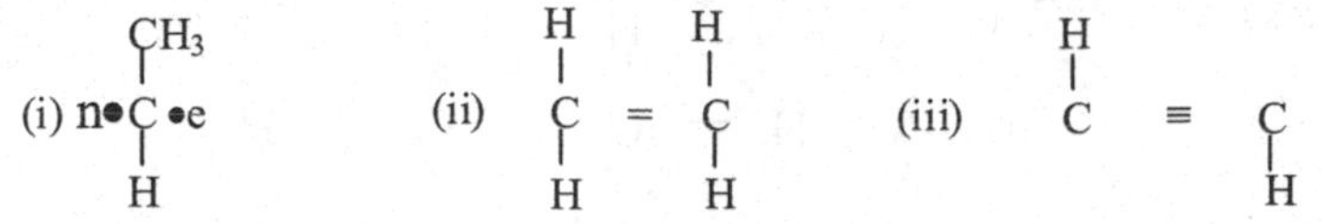

(i) n•C•e
CH₃
H

(ii)
H   H
C = C
H   H

(iii)
H
C ≡ C
H

"Ethylene"   Ethylene   Acetylene

If (i) above is to be identified as an ethylene, then based on methane, ethane, propane etc. for alkane series, what should be the ideal names for (ii) and (iii) which belong to alkene and alkyne series respectively? Explain.

(b) Based on the new nomenclature, should propylene, 2-butene, and butyne etc., exist as names for olefinic and acetylenic monomers respectively? What names should be given to the types of monomers shown below.

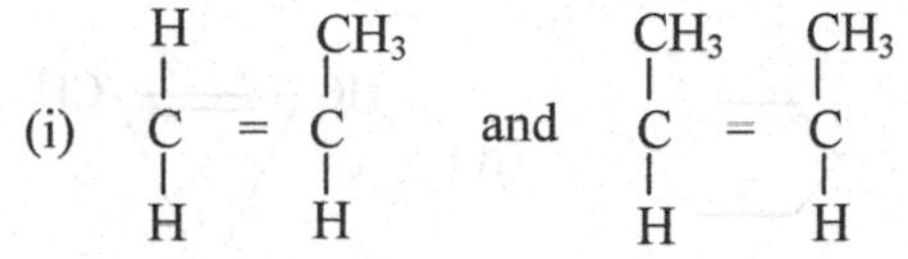

(i)
H   CH₃
C = C
H   H
Propylene
(propene)

and

CH₃   CH₃
C = C
H   H
β - Butylene
(2 - Butene)

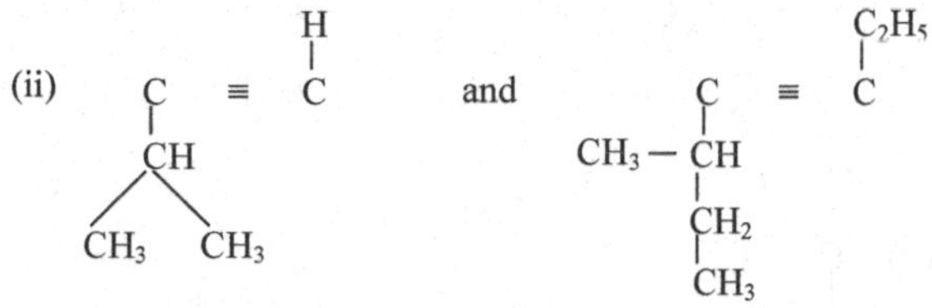

(ii)
C ≡ C
        H
CH
CH₃ CH₃
i- Propylacetylene
(Methylbutyne)

and

C ≡ C
        C₂H₅
CH₃ — CH
CH₂
CH₃
Ethyl - s - butylacetylene
(5 - methyl - 3 - heptyne)

The names is parenthesis are their nomenclatures by the international system, while the first names are said to be derivatives of alkenes and acetylene.

(c) Identify the names of the following compounds.

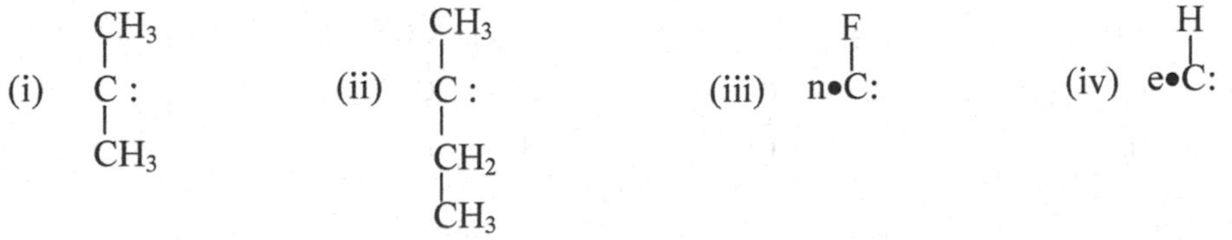

(i)
CH₃
C :
CH₃

(ii)
CH₃
C :
CH₂
CH₃

(iii)
F
n•C:

(iv)
H
e•C:

8.10. (a) Based on the new nomenclature, do butylenes used for some butenes exist?
Explain. If the answer is no after providing explanations what then should they be called?
(b) Shown below are monomers -

(i)
$$CH_3$$
$$|$$
$$C = O$$
$$|$$
$$H$$

(Acetaldehyde)

(ii)
$$\begin{array}{ccc} H & & H \\ | & & | \\ H - C & - & C - H \\ & \diagdown & \diagup \\ & O & \end{array}$$

(Ethylene oxide)

If the ideal name of (i) had been ethylene oxide, then what should be the name for (ii)?

(c) butadiene is a product of cyclo-butene, propadiene is not a product of cyclo-propene. Explain.

8.11. (a) Do propylene and pentadiene exist as names for non-ringed (linear) alkene monomers?

(b) Based on the new nomenclature, do the names for the following monomers have to change? If the answer is No, explain. Cyclobutadiene

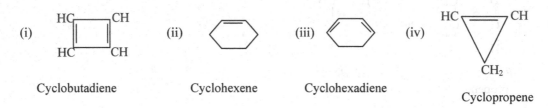

(i) Cyclobutadiene   (ii) Cyclohexene   (iii) Cyclohexadiene   (iv) Cyclopropene

8.12. Shown below are three monomers.

(i)
$$CF_3 \quad CF_3$$
$$| \qquad |$$
$$C = C = C$$
$$| \qquad |$$
$$H \qquad H$$

(ii)
$$CF_3$$
$$|$$
$$C \equiv C$$
$$|$$
$$CF_3$$

(iii)
$$CF_3 \quad CF_3$$
$$| \qquad |$$
$$C = C$$
$$| \qquad |$$
$$H \qquad H$$

(a) Show the routes favored by them, chargedly where possible and free-radically.

(b) Which of the three monomers is most nucleophilic?

(c) Replace one of the CF_3s, with C_2F_5 and repeat (a) and (b).

8.13. Shown below are three monomers.

(i)
$$CH_3 \quad CH_3$$
$$| \qquad |$$
$$C = C = C$$
$$| \qquad |$$
$$H \qquad H$$
;

(ii)
$$CH_3$$
$$|$$
$$C \equiv C$$
$$|$$
$$CH_3$$

(iii)
$$CH_3 \quad CH_3$$
$$| \qquad |$$
$$C = C$$
$$| \qquad |$$
$$H \qquad H$$

(a) Show the routes favored by them, chargedly where possible and free-radically.

(b) Which of the three monomers is most nucleophilic?

(c) Replace one of the CH_3s, with H and repeat (a) and (b).

8.14. Shown below are three monomers

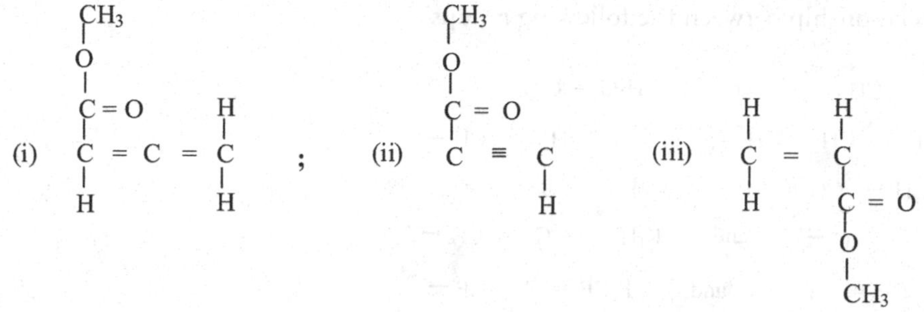

(i)
$$
\begin{array}{ccc}
 & CH_3 & \\
 & | & \\
 & O & \\
 & | & \\
 & C=O & H \\
 & | & | \\
C = C = C \\
| & & | \\
H & & H
\end{array}
$$

;   (ii)
$$
\begin{array}{ccc}
 & CH_3 & \\
 & | & \\
 & O & \\
 & | & \\
 & C=O & \\
 & | & \\
C \equiv C \\
 & | \\
 & H
\end{array}
$$

(iii)
$$
\begin{array}{cc}
H & H \\
| & | \\
C = C \\
| & | \\
H & C=O \\
 & | \\
 & O \\
 & | \\
 & CH_3
\end{array}
$$

(a) Which of the monomers can be activated very readily?

(b) Show the routes favored by them, chargedly where possible and free-radically.

(c) Which of the three monomers is most Electrophilic?

(d) Replace one H on other external carbon center with CH_3 and repeat (b) and (c).

8.15. Shown below are three monomers

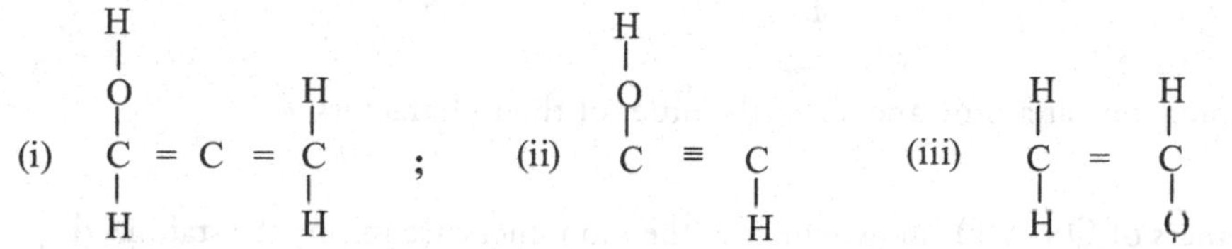

(i)
$$
\begin{array}{cc}
H & \\
| & \\
O & H \\
| & | \\
C = C = C \\
| & | \\
H & H
\end{array}
$$

;   (ii)
$$
\begin{array}{c}
H \\
| \\
O \\
| \\
C \equiv C \\
 | \\
 H
\end{array}
$$

(iii)
$$
\begin{array}{cc}
H & H \\
| & | \\
C = C \\
| & | \\
H & O
\end{array}
$$

(a) What are common between the three monomers?

(b) Replacing H on OH with CH_3, show the routes favored by them, chargedly where possible and free-radically.

(c) Which of the three monomers, after rearrangement where possible is not nucleophilic?

8.16. For the three monomers shown in Q8.15, replace the OH group with NH_2.

(a) What are common between the three monomers?

(b) Should any H atom on N center be replaced to favor the existence of a monomer for the three cases?

(c) Show the routes favored by them, chargedly where possible and radically.

(d) Which of the three monomers is most nucleophilic?

8.17. (a) Shown below are some Cumunyl groups.

(i) $CF_2 =$,   (ii) $RCH =$,   (iii) $CF_3HC =$,   (iv) $(CF_3)_2C =$,

(v) $C_2F_5HC =$,   (vi) $HCF_3 =$,   (vii) $F_2C = C = C =$,   (viii) $FHC = C = C =$,

(ix) $H_2C =$,   (ix) $H_2C = C = C =$

What is the order of their radical-pushing capacities?

(b) Which of $H_2C=$, $H_2C=C=$, $H_2C=C=C=$ groups are resonance stabilization groups? Explain.

(c) Can the active centers of these substituent groups carry charges? Explain by providing what they can indeed carry.

8.18 (a) What is the relationship between the following groups?

(i) $H_2C = C = CH -$ and $H_2C = C = C =$

(ii) $H_2C = CH - CH = CH -$ and $H_2C = CH -$

(iii) $H_2C = CH - CH = CH -$ and $H_2C = C =$

(iv) $R - C \equiv C -$ and $R_1HC = C = CH -$

(v) $F_5C_2 C \equiv C -$ and $F_2CF = C = CF -$

(b) Show the activated states of the followings monomers.CH_3

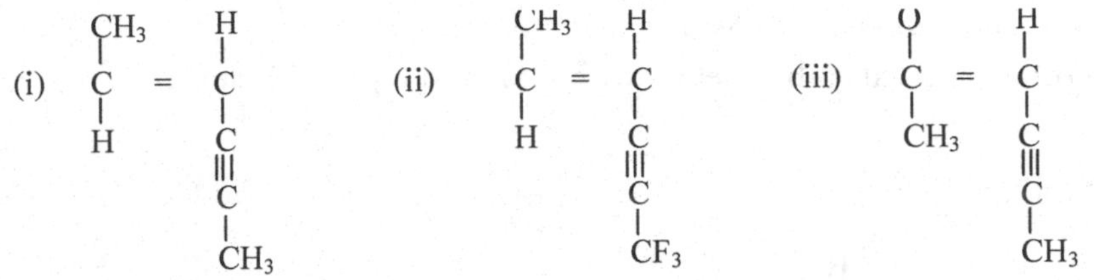

(i)
$$
\begin{array}{c}
CH_3 \quad H \\
| \qquad | \\
C = C \\
| \qquad | \\
H \qquad C \\
\quad\;\; ||| \\
\quad\;\; C \\
\quad\;\; | \\
\quad\;\; CH_3
\end{array}
$$

(ii)
$$
\begin{array}{c}
CH_3 \quad H \\
| \qquad | \\
C = C \\
| \qquad | \\
H \qquad C \\
\quad\;\; ||| \\
\quad\;\; C \\
\quad\;\; | \\
\quad\;\; CF_3
\end{array}
$$

(iii)
$$
\begin{array}{c}
O \qquad H \\
|| \qquad | \\
C = C \\
| \qquad | \\
CH_3 \quad C \\
\quad\;\; ||| \\
\quad\;\; C \\
\quad\;\; | \\
\quad\;\; CH_3
\end{array}
$$

Use both activation centers one at a time and show the order of their characters.

8.19 (a) For the monomers of Q8.18(b), show which of the monomers are resonance stabilized.

(b) Explain why they are no MALES? Name the compounds.

(c) What is common or different between $- C \equiv N$ and $- C \equiv C - C_2F_5$ as substituted groups and as activation centers, that is,

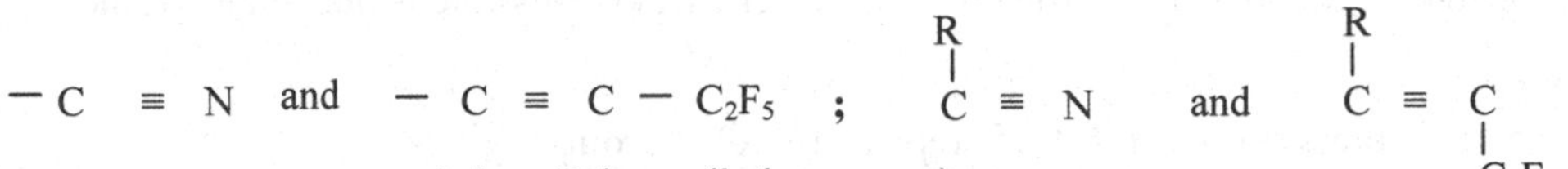

$$- C \equiv N \quad \text{and} \quad - C \equiv C - C_2F_5 \quad ; \quad \begin{array}{c} R \\ | \\ C \equiv N \end{array} \quad \text{and} \quad \begin{array}{c} R \\ | \\ C \equiv C \\ | \\ C_2F_5 \end{array}$$

(where R is an alkylane group)

(d) Can $HC \equiv$ and $N \equiv$ exist as substituted group for an activation center? Explain.

8.20 (a) Why can first members of 1,6 - cumulene and that of 1,5 - cumulene not form stable molecules as shown below?

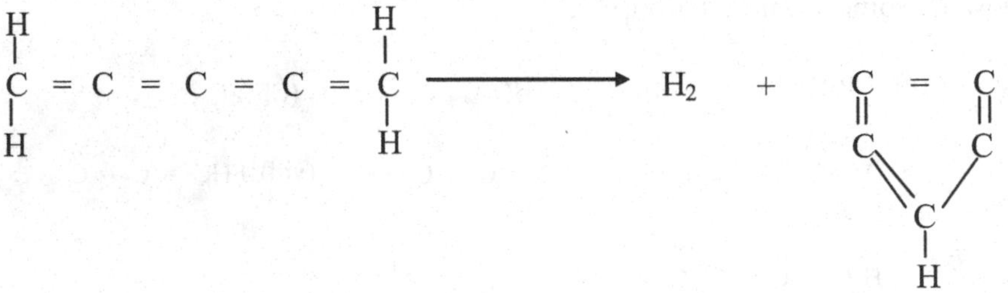

$$
\begin{array}{c}
H \qquad\qquad\qquad\qquad H \\
| \qquad\qquad\qquad\qquad\quad | \\
C = C = C = C = C \\
| \qquad\qquad\qquad\qquad\quad | \\
H \qquad\qquad\qquad\qquad H
\end{array}
\longrightarrow
H_2 \;+\;
\begin{array}{c}
C = C \\
|| \quad\; || \\
C \qquad C \\
\;\backslash \quad / \\
\;\; C \\
\;\; | \\
\;\; H
\end{array}
$$

(Cannot exist)

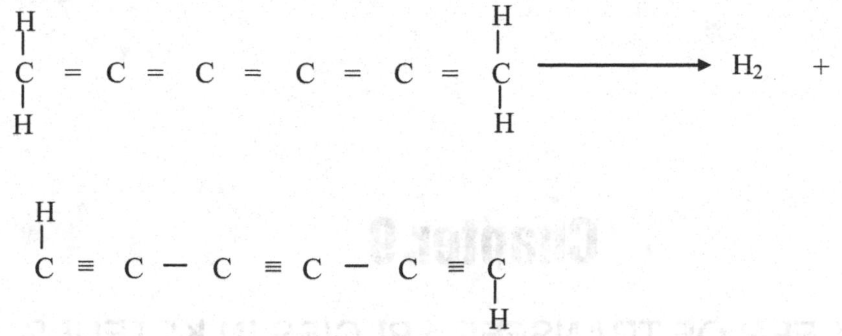

$$
\begin{array}{c}
\overset{\displaystyle H}{\underset{\displaystyle H}{\overset{|}{\underset{|}{C}}}} = C = C = C = C = \overset{\displaystyle H}{\underset{\displaystyle H}{\overset{|}{\underset{|}{C}}}} \longrightarrow H_2 \quad +
\end{array}
$$

$$
\overset{\displaystyle H}{\overset{|}{C}} \equiv C - C \equiv C - C \equiv \overset{}{\underset{\displaystyle H}{\underset{|}{C}}}
$$

(b) From all the considerations so far, why is it that 1,3 -cumulenes have never been known to be useful as monomers?

(c) In general, what is unique about the mono-forms of 1,4 - cumulenes?

8.21. (a) Shown below are the following pairs of substituted groups

(i) $H_2C =$ and $H_3C -$ (ii) $(F_2C)CH =$ and $CF_3H_2C -$

(iii) $(CH_3OOC)_2C =$ and $(CH_3OOC)_3C -$

Distinguish between their characters radically and chargedly when used as substituted groups on a monomer.

(b) What is unique about OH, OR, NH_2, and NR_2 groups when used as substituent groups on cumulenes?

(c) What is or are common between olefins and 1,4 - cumulenes?

Chapter 9

TRANSFER OF TRANSFER SPECIES IN KETENES, ISOCYANATES, IMINES, DIIMINES AND DIKETENES

9.1 Introduction

Just as very little is known about cumulenes, the same applies to Ketenes, Cumulenic Aldimines, Ketimines, Imines, Dimines and Isocyanates. Olefins, Acetylenes, "Diazoalkanes" and Cumulenes are indeed *ideal or traditional* Addition monomers while aldehydes, ketones, nitriles, isocyanates, aldimines and ketimines etc. are *pseudo*-Addition monomers, 1,3-ketenes are part ideal addition and part pseudo-Addition monomers.

During the considerations so far, with completely new concepts and dimensions introduced with respect to understanding of chemistry of compounds, a climax was reached in the last chapter, where new names were proposed for mono-olefins, mono and di- acetylenes. From the problems section of that chapter, new names were proposed for other members of di-alkenes/olefins.

Olefins are alkenes which include ethenes (mono-olefins), butadienes (di-olefins), hexatrienes (tri-olefins) etc. and perfluoroalkenes which include perfluoroethenes (mono-olefins), perfluorobutadienes (di-olefins), perfluorohaxatrienes (tri-olefins etc.). The first member of ethenes is ethene (not ethylene), while other members are derivatives of ethene. The first member of butadienes is butadiene while other members are derivatives of butadiene. Hence, names such as ethylene, propylene, pentadiene etc. do not exist with olefins as monomeric names.

"Acetylenes" are alkynes which also include ethynes (mono-alkynes), butadi-ynes (di - alkynes), hexatri-ynes (tri-alkynes) etc. and their perfluorinated versions. The first member of ethynes is ethyne (acetylene), while other members are derivatives of ethyne. The first member of butadi-ynes is butadi-yne while other members are derivatives of butadi-yne. Thus, while ethene for example is limited to alkenes part of olefins family; propenes, butenes, pentenes, hexenes etc. apply to alkenes, olefins and cyclic monomers. The same also apply to propynes, butynes, pentynes, hexynes etc. as shown below.

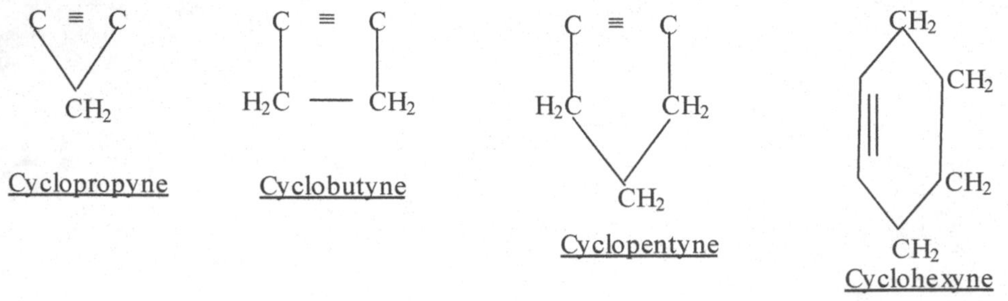

Cyclopropyne   Cyclobutyne   Cyclopentyne   Cyclohexyne   9.1

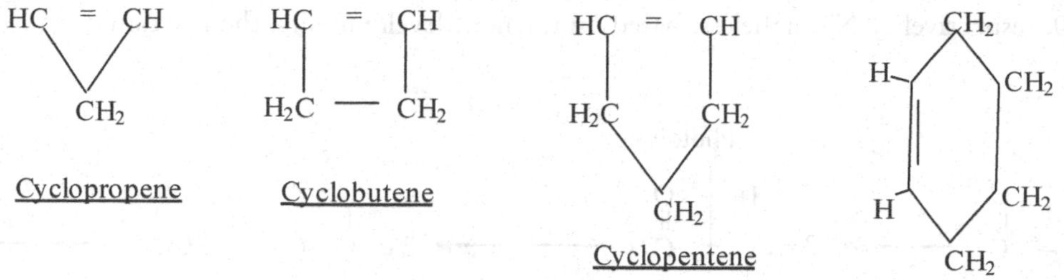

HC $=$ CH
\
/
CH$_2$

Cyclopropene

HC $=$ CH
| |
H$_2$C $—$ CH$_2$

Cyclobutene

HC $=$ CH
\ |
H$_2$C   CH$_2$
\   /
CH$_2$

Cyclopentene

CH$_2$
H— \   CH$_2$
| |
H / CH$_2$
CH$_2$

Cyclohexene   9.2

Whether the first group of monomers exists or not is not presently the issue.

Formaldehyde was also identified as methylene oxide, while acetaldehyde was said to be ethylene oxide. Hydrogen cyanide was also identified as methylyne nitride. Whether these new dimensions will apply to ketenes, isocyanates and so on will remain to be seen. Like cumulenes, for specific reasons, 1,3- 1,4- 1,5- ketenes and isocyanates will be considered as shown below.

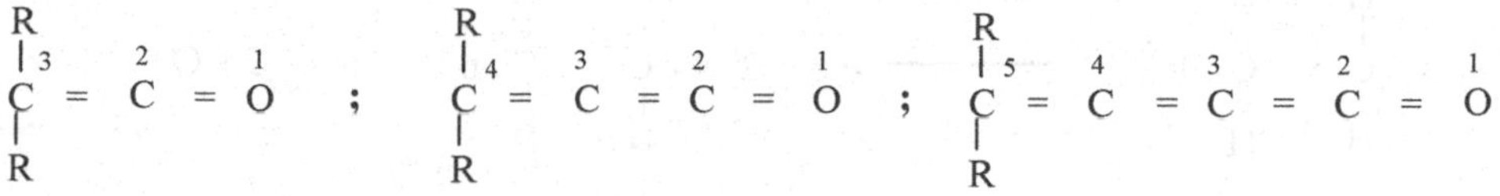

R
|3   2   1
C $=$ C $=$ O   ;
|
R

R
|4   3   2   1
C $=$ C $=$ C $=$ O   ;
|
R

R
|5   4   3   2   1
C $=$ C $=$ C $=$ C $=$ O
|
R

1, 3 - Ketene   1, 4 - cumulenic ketene   1,5 - cumulenic ketene   9.3

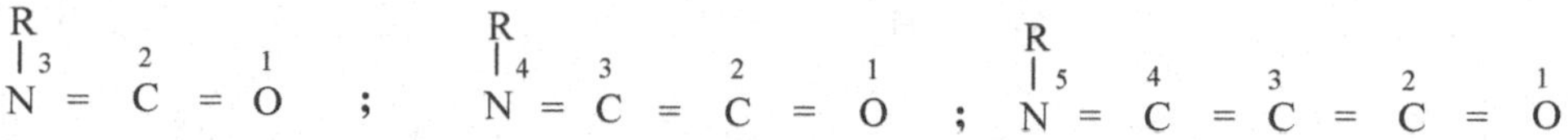

R
|3   2   1
N $=$ C $=$ O   ;

R
|4   3   2   1
N $=$ C $=$ C $=$ O   ;

R
|5   4   3   2   1
N $=$ C $=$ C $=$ C $=$ O

1, 3 - Isocyanate   1, 4 - ketenic isocyanate   1,5 - ketenic isocyanate   9.4

Ketene has been reported to be produced from the pyrolysis or photolysis of diazomethane in the presence of carbon monoxide.[1] The reaction can be explained as follows.

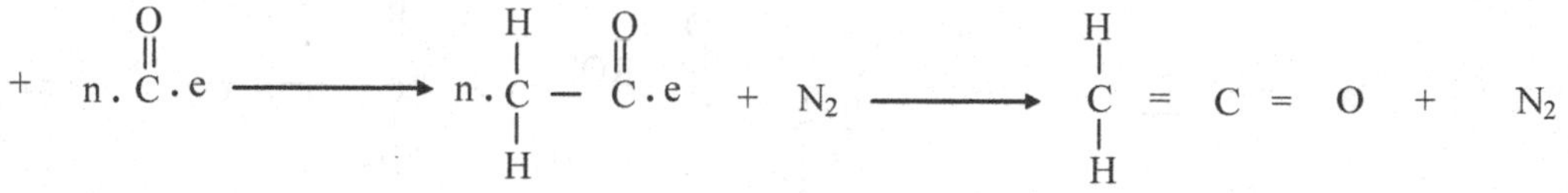

O
||
+ n.C.e ⟶ n.C$—$C.e + N$_2$ ⟶ C $=$ C $=$ O + N$_2$

(Less nucleophilic)   Ketene

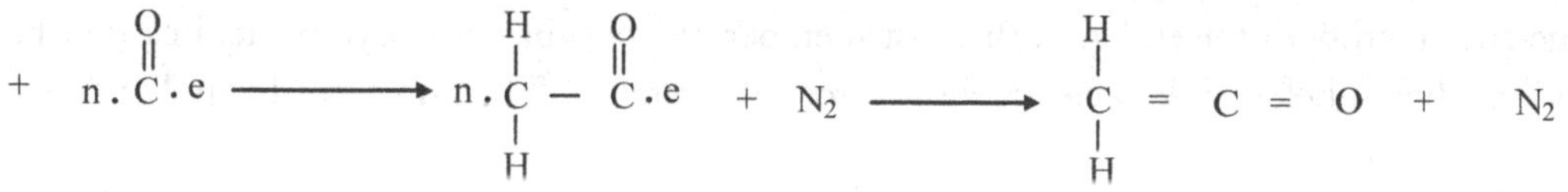

O
||
+ n.C.e ⟶ n.C$—$C.e + N$_2$ ⟶ C $=$ C $=$ O + N$_2$

(Less nucleophilic)   Ketene   9.5

The methylene can be observed to be the species diffusing to the activated carbon monoxide. This is not different from what we have seen in Chapter 7, for which the reaction can be observed to be free-radical in character. Photolysis of ketenes is also said to yield twice as much carbon monoxide as ethylene and a mechanism was proposed.[2] This was however disproved on the basis of the fact that the rate of carbon monoxide formation can be almost halved by the addition of more ethylene [3] and on the basis of the fact that with pure ketene, the quantum yields of carbon monoxide and ethylene are approximately 2.0 and 1.0, respectively. [4] Nevertheless, based on the new developments, the reaction can be explained as follows

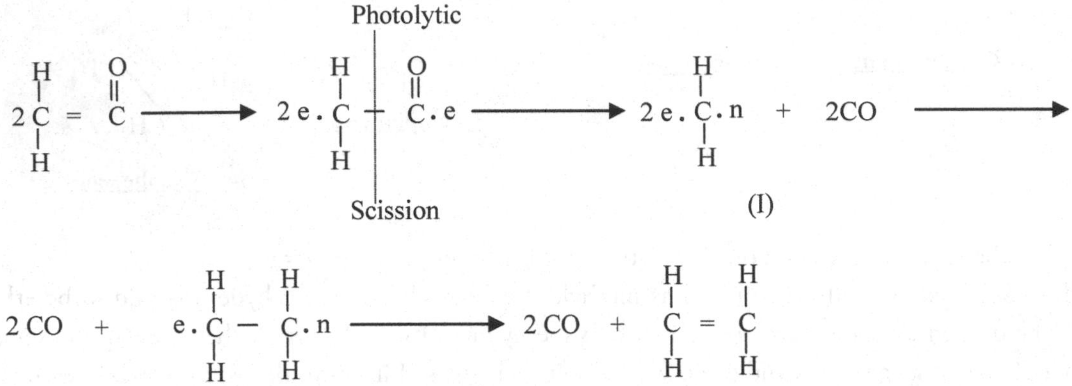

$$2\ \underset{\overset{|}{H}}{\overset{\overset{H}{|}}{C}} = \underset{}{C} \overset{O}{\overset{||}{}} \longrightarrow 2\ e\cdot\underset{\overset{|}{H}}{\overset{\overset{H}{|}}{C}} \underset{\underset{\text{Scission}}{}}{\overset{\overset{\text{Photolytic}}{|}}{|}} \overset{O}{\overset{||}{C}}\cdot e \longrightarrow 2\ e\cdot\underset{\overset{|}{H}}{\overset{\overset{H}{|}}{C}}\cdot n\ +\ 2CO \longrightarrow$$

$$(I)$$

$$2\ CO\ +\ e\cdot\underset{\overset{|}{H}}{\overset{\overset{H}{|}}{C}} - \underset{\overset{|}{H}}{\overset{\overset{H}{|}}{C}}\cdot n \longrightarrow 2\ CO\ +\ \underset{\overset{|}{H}}{\overset{\overset{H}{|}}{C}} = \underset{\overset{|}{H}}{\overset{\overset{H}{|}}{C}} \qquad\qquad 9.6$$

$$(I)\ +\ e\cdot\underset{\overset{|}{H}}{\overset{\overset{H}{|}}{C}} - \underset{\overset{|}{H}}{\overset{\overset{H}{|}}{C}}\cdot n \longrightarrow 2\ e\cdot\underset{\overset{|}{H}}{\overset{\overset{H}{|}}{C}} - \underset{\overset{|}{H}}{\overset{\overset{H}{|}}{C}}\cdot n\ +\ 2\ CO$$

$$\longrightarrow 2\ CO\ +\ 2\ \underset{\overset{|}{H}}{\overset{\overset{H}{|}}{C}} = \underset{\overset{|}{H}}{\overset{\overset{H}{|}}{C}} \qquad\qquad 9.7$$

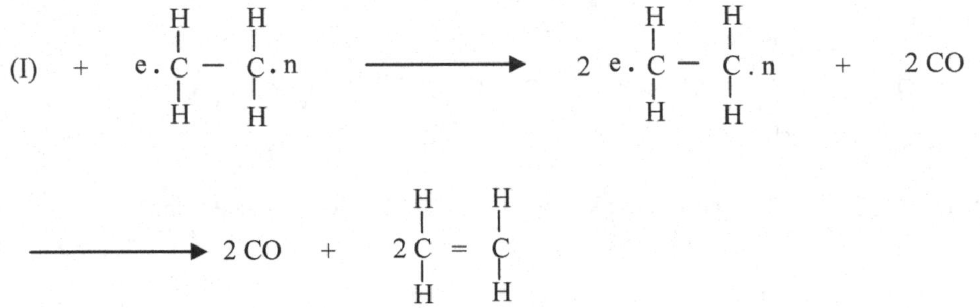

Very little is known about the routes favored when ketenes are involved. The favored existence of copolymers of the alternating types has provided better understanding of the monomers. The routes where existences of copolymers are favored have never been clearly defined. However, the first member of ketenes, like acetylene, favors being in Equilibrium state of existence. This is due to strong unsaturation of the C = C = O bonds wherein the H : C ratio is 1 : 1.

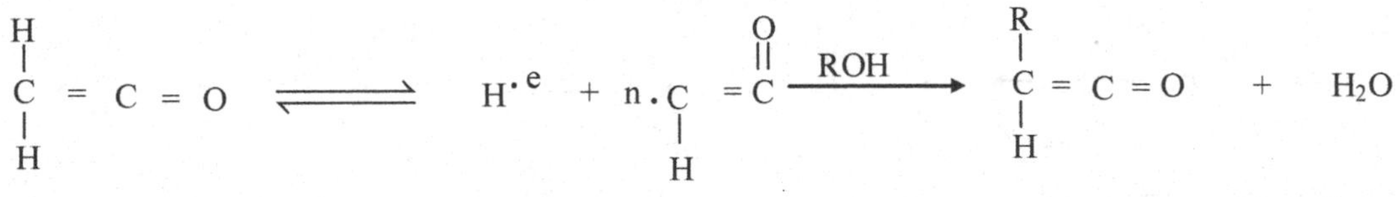

$$\underset{\overset{|}{H}}{\overset{\overset{H}{|}}{C}} = C = O \rightleftharpoons H\cdot e\ +\ n\cdot\underset{\overset{|}{H}}{\overset{}{C}} = \underset{}{C}\overset{O}{\overset{||}{}} \xrightarrow{ROH} \underset{\overset{|}{H}}{\overset{\overset{R}{|}}{C}} = C = O\ +\ H_2O \qquad 9.8$$

Like the first member of acetylenes, the first member of isocyanates, isocyanic acid cannot favor ionic existence as shown below due to electrostatic forces of repulsion. The same should apply to ketene. They can only exist radically.

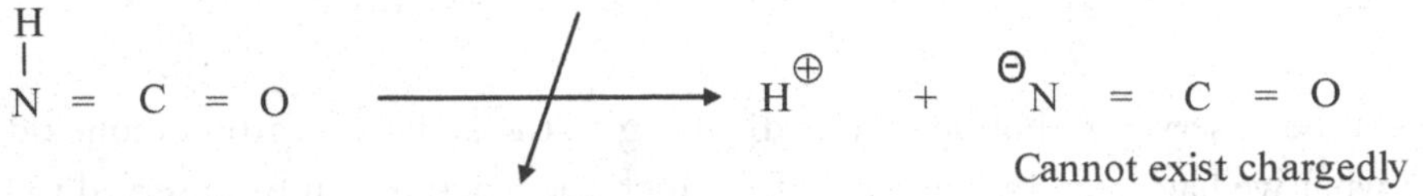

$$\underset{\overset{|}{H}}{\overset{\overset{H}{|}}{N}} = C = O \xrightarrow{\hspace{1cm}/\hspace{1cm}} H^{\oplus}\ +\ {}^{\ominus}N = C = O$$

Cannot exist chargedly $\qquad\qquad$ 9.9

Hence, the hydrogen on the nitrogen center must be replaced by for example an alkyl group before polymerization can fully take place if the monomer is not suppressed by other means.

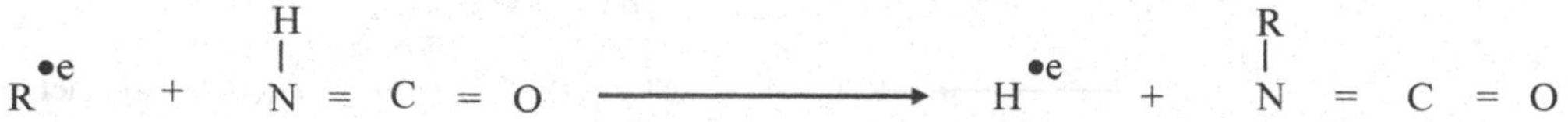

$$R \overset{\bullet e}{} \quad + \quad \overset{\overset{H}{|}}{N} = C = O \quad \longrightarrow \quad H \overset{\bullet e}{} \quad + \quad \overset{\overset{R}{|}}{N} = C = O \qquad 9.10$$

The isocyanates, like the ketenes are Males (Electrophiles) in character, for which the C=N center is Y and the C=O center is X. Y is less Nucleophilic than the C = O center.

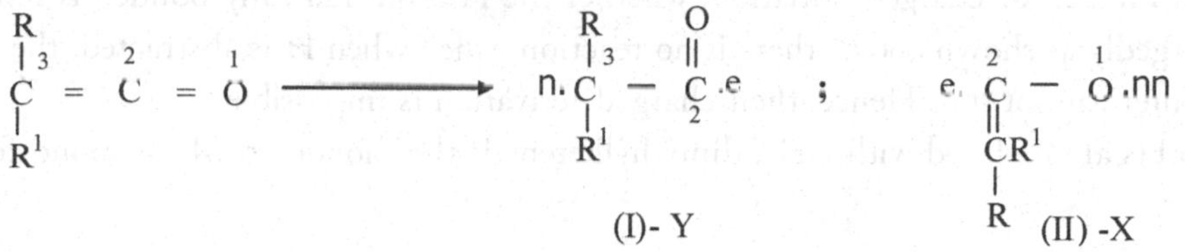

$$\overset{\overset{R}{|}}{\underset{\overset{|}{R^1}}{C}} \overset{3}{=} \overset{2}{C} = \overset{1}{O} \quad \longrightarrow \quad n.\overset{\overset{R}{|}}{\underset{\overset{|}{R^1}}{C}} \overset{3}{-} \overset{\overset{O}{||}}{\underset{2}{C}} .e \quad ; \quad e. \overset{2}{C} \overset{}{-} \overset{1}{O} .nn$$

$$\text{(I)- Y} \qquad\qquad \overset{}{\underset{R}{\overset{|}{\underset{}{CR^1}}}} \text{ (II) -X} \qquad 9.11$$

Notice why the above is an Electrophile compared to what is shown below. Though the presence of the C = C center is important, its presence is not imminent for a Male to exist. It is however absent in the case shown below.

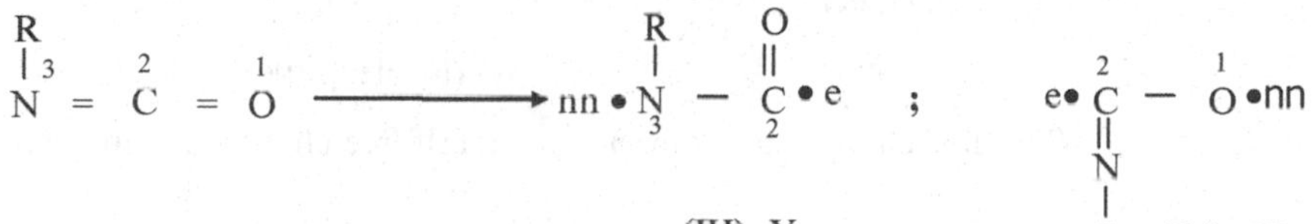

$$\overset{\overset{R}{|}}{N} \overset{3}{=} \overset{2}{C} = \overset{1}{O} \quad \longrightarrow \quad nn \bullet \overset{R}{\underset{3}{N}} \overset{}{-} \overset{\overset{O}{||}}{\underset{2}{C}} \bullet e \quad ; \quad e\bullet \overset{2}{C} \overset{}{-} \overset{1}{O} \bullet nn$$

$$\text{(III) Y} \qquad\qquad \overset{}{\underset{R}{\overset{||}{\underset{}{N}}}} \text{ (IV) X} \qquad 9.12$$

Notice that since Y < X in Nucleophilicity, Y is the center first activated whether the initiator is male or female in character. Note that N and O are electronegative. C, like S and Iodine (I) are along the boundary in the Periodic Table, that is, neither electronegative nor electropositive. Yet, the three are different, when for example we compare R-N=C=O with R-N=C-S. Since the C=N center can be made to favor both routes, ketenes are more electrophilic than isocynates.

9.1. 1,3 - Ketenes

Though the radical polymerizations of these monomers are unknown, these will be considered.

9.1.1. Charged characters of 1,3-Ketenes

Beginning with the first member, the followings are obtained when activated by another activated monomer or H·e

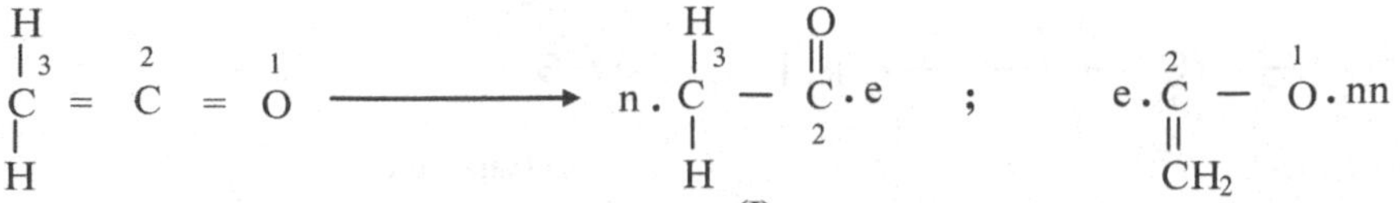

$$\overset{\overset{H}{|}}{\underset{\overset{|}{H}}{C}} \overset{3}{=} \overset{2}{C} = \overset{1}{O} \quad \longrightarrow \quad n.\overset{\overset{H}{|}}{\underset{\overset{|}{H}}{C}} \overset{3}{-} \overset{\overset{O}{||}}{\underset{2}{C}} .e \quad ; \quad e.\overset{2}{C} \overset{}{-} \overset{1}{O} .nn$$

$$\text{(I)} \qquad\qquad \overset{}{\underset{}{\overset{||}{CH_2}}} \text{(II)} \qquad 9.13$$

None of the mono-forms can favor molecular rearrangement. In fact, (II) is the source of $HC \equiv COH$ by enolization. Though we have seen that the monomer can be activated chargedly, but difficult to use, like what one has been doing, we are going to use it for exploration.

$$R^{\ominus} \; + \; \overset{\oplus}{C}\overset{\displaystyle |}{\underset{CH_2}{||}} - O^{\ominus} \longrightarrow RH \; + \; \overset{\ominus}{C}\overset{\displaystyle \overset{H}{|}}{=} C = O \quad OR \quad \text{No reaction}$$

Cannot exist

Transfer species of first kind of the third type   9.14

The ketene does not favor charged initiation, whether the H atom radically bonded is replaced or not. Therefore chargedly as shown above, there is no reaction, since when H is abstracted, the negative charge on the C center cannot stay. Hence, their charged activation is impossible.

When the two Hs are replaced with CH_3, dimethylketene is the monomer, whose mono-forms are as shown.

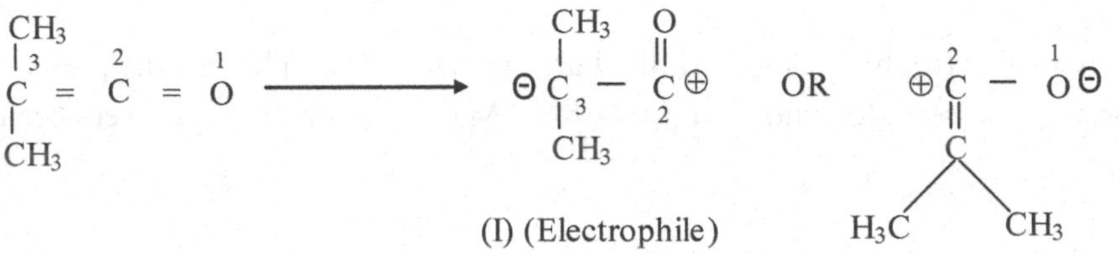

$$\overset{CH_3}{\underset{CH_3}{\overset{|}{\underset{|}{C}}}} = \overset{2}{C} = \overset{1}{O} \longrightarrow \overset{CH_3 \;\; O}{\underset{CH_3}{\overset{| \quad\; ||}{\underset{|}{\ominus C_3 - C_2^{\oplus}}}}} \quad OR \quad \overset{2}{\overset{\oplus}{C}} - \overset{1}{O}^{\ominus}$$

(I) (Electrophile)   (II) (Nucleophile)

(Full free charged monomer)   (Half free charged monomer)   9.15

It is important to note that the two mono-forms belong to two different families. (I) is a full free-charged or radical monomer. From all indications as has already been shown, it is obvious that the following prevails.

$$= O \quad >> \quad H_2C =$$

<u>Order of radical-pushing capacity</u>   9.16

This has been confirmed in the last chapter. It is for this reason that (I) is carrying the charges indicated. It is also for this reason that (I) is far less nucleophilic in character than (II) or that it is an electrophile. Since charged activation can be favored, with free-negative-charges, only (I) can be polymerized. With Non-free negative charges, none can be polymerized.

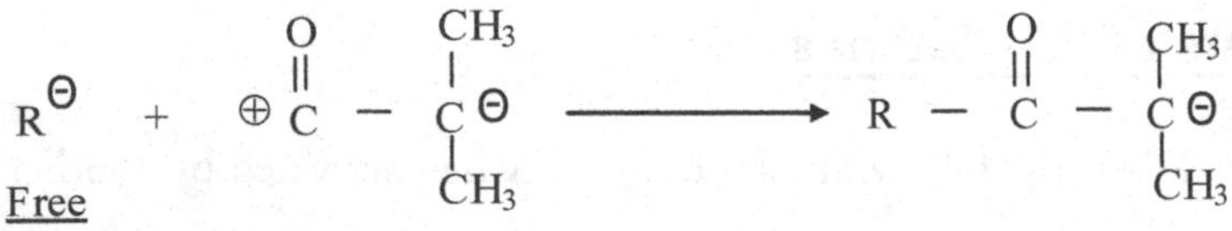

$$R^{\ominus} \; + \; \overset{O \quad\;\; CH_3}{\underset{CH_3}{\overset{||\quad\; |}{\underset{|}{\oplus C - C^{\ominus}}}}} \longrightarrow \overset{O \quad\;\; CH_3}{\underset{CH_3}{\overset{||\quad\; |}{\underset{|}{R - C - C^{\ominus}}}}}$$

Free   9.17

$$R^{\ominus} \; + \; \overset{\oplus}{C}\overset{\displaystyle ||}{\underset{C}{}} - O^{\ominus} \longrightarrow RCH_3 \; + \; \overset{CH_3}{\overset{|}{\ominus C}} = C = O$$

Free   Cannot exist   9.18

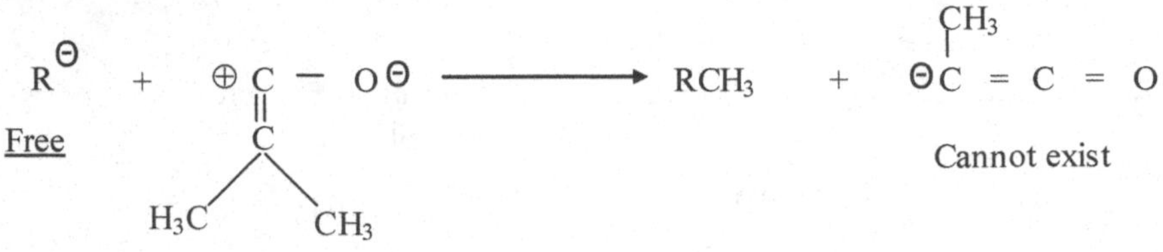

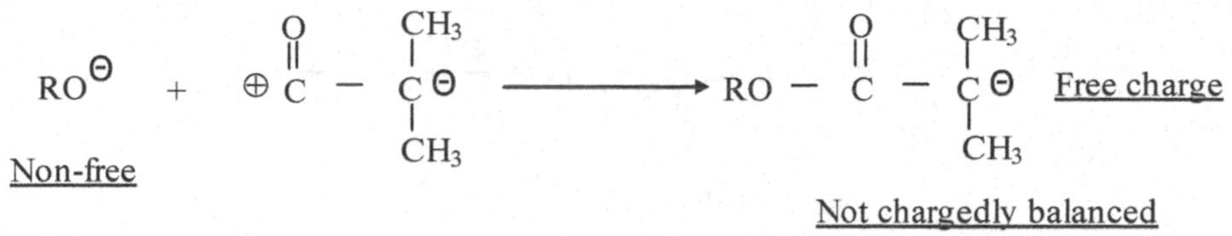

$$RO^{\ominus} \quad + \quad \overset{\oplus}{C} - \underset{\underset{CH_3}{|}}{\overset{\overset{O}{\|}}{C}} \overset{CH_3}{\underset{}{|}} {C}^{\ominus} \quad \longrightarrow \quad RO - \overset{\overset{O}{\|}}{C} - \underset{CH_3}{\overset{CH_3}{|}}{C}^{\ominus} \quad \underline{Free\ charge}$$

Non-free

$$\underline{Not\ chargedly\ balanced} \qquad\qquad 9.19$$

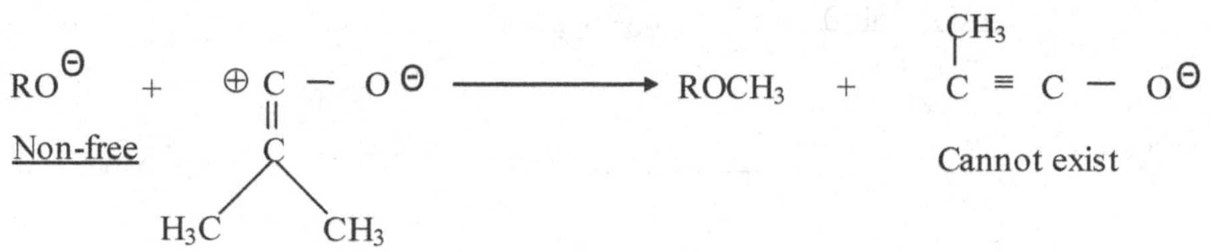

$$RO^{\ominus} \quad + \quad \overset{\oplus}{C} - O^{\ominus} \quad \longrightarrow \quad ROCH_3 \quad + \quad \overset{CH_3}{\underset{}{|}}{C} \equiv C - O^{\ominus}$$

Non-free $\qquad\qquad\qquad$ Cannot exist

(with $\underset{H_3C \qquad CH_3}{\overset{\|}{C}}$ on the ketene)

$$\qquad\qquad\qquad\qquad\qquad\qquad\qquad\qquad\qquad 9.20$$

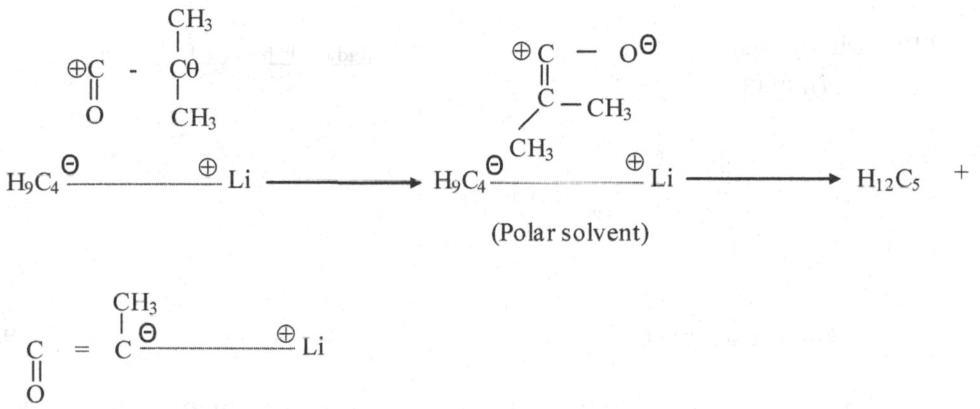

$$\underset{\underset{O}{\|}}{\overset{\oplus}{C}} - \underset{CH_3}{\overset{CH_3}{|}}{C}{\ominus}$$

$$H_9C_4 {\ominus} \underline{\qquad\qquad} {\oplus} Li \quad \longrightarrow \quad H_9C_4 {\ominus} \underline{\qquad\qquad} {\oplus} Li \quad \longrightarrow \quad H_{12}C_5 \quad +$$

with $\underset{\underset{CH_3}{\overset{|}{C}-CH_3}}{\overset{\oplus C - O^{\ominus}}{\|}}$

(Polar solvent)

$$\underset{\overset{O}{\|}}{C} = \overset{CH_3}{\underset{}{|}}{C} {\ominus} \underline{\qquad\qquad} {\oplus} Li \qquad\qquad\qquad 9.21$$

$$\underset{\underset{CH_3}{|}}{\overset{\overset{O}{\|}}{\oplus C} - {C}{\ominus}}$$

$$H_9C_4 {\ominus} \underline{\qquad\qquad} {\oplus} Li \quad \underset{\underset{solvent}{Non\text{-}polar}}{\longrightarrow} \quad H_9C_4 - \overset{\overset{O}{\|}}{C} - \underset{CH_3}{\overset{CH_3}{|}}{C}{\ominus} \underline{\qquad\qquad} {\oplus} Li$$

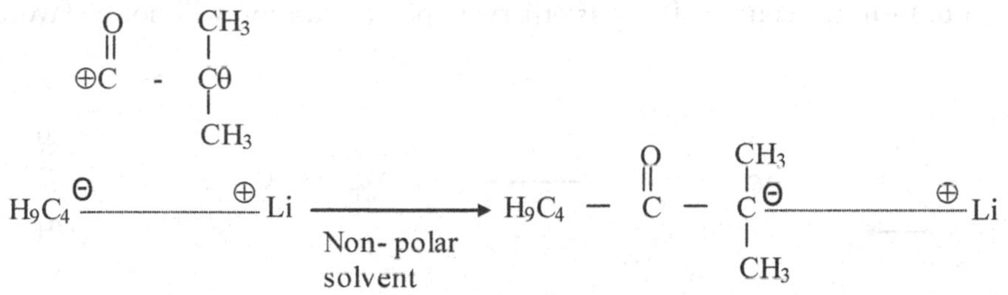

$$\textbf{FULLY FAVORED CHARGEDLY} \qquad\qquad 9.22$$

From all the explorations above, one has shown that Ketenes can be activated chargedly, and can be polymerized chargedly only for the Y center and this can also be done RADICALLY. In the last equations above, the initiator above is dual in character only radically and not chargedly. The Li center is for the C=O center (X) radically, since cations cannot be used, while the H_9C_4 center is for the C = C (Y).

Existence of alternating copolymers is favored only when nucleo-non-free-radical initiators are involved as shown below. This is the initiator which cannot polymerize any of the activation centers as has already been shown in previous chapters.

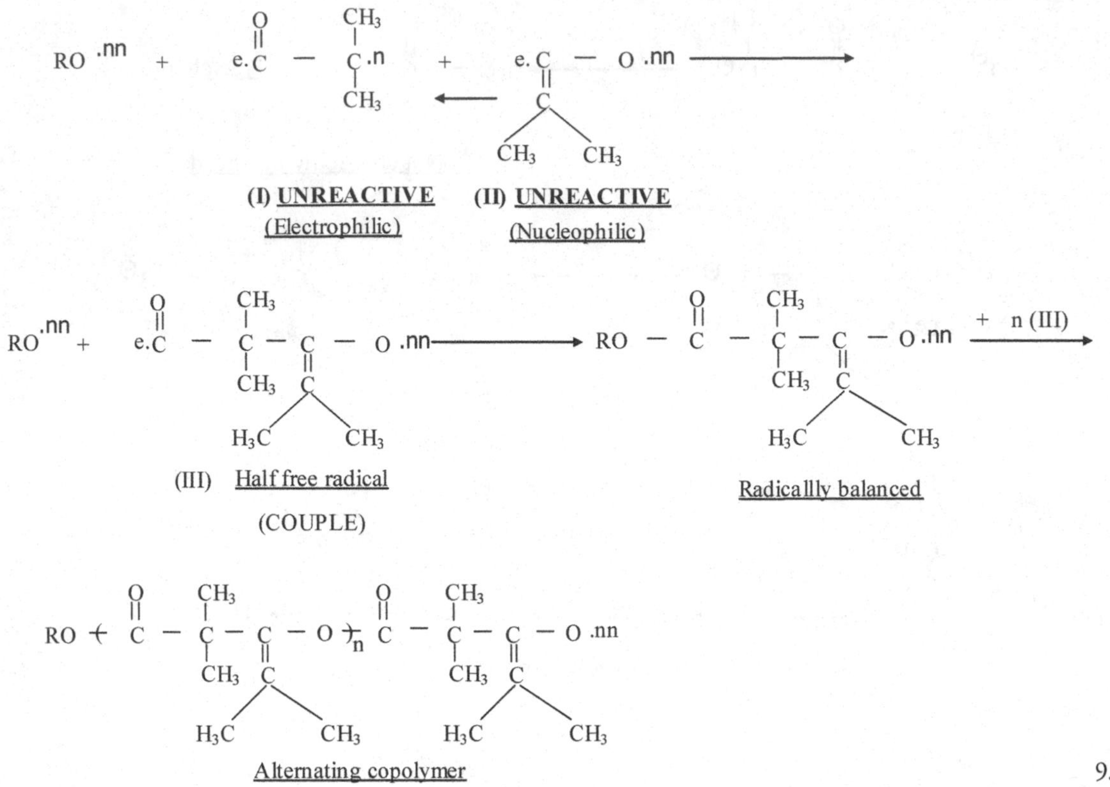

$$RO^{.nn} + e.\overset{\overset{O}{\|}}{C} - \overset{\overset{CH_3}{|}}{\underset{\underset{CH_3}{|}}{C}}.n + e.\overset{\overset{O}{\|}}{\underset{\underset{CH_3}{|}\ \underset{CH_3}{}}{C}} - O.nn \longrightarrow$$

(I) <u>UNREACTIVE</u>   **(II) <u>UNREACTIVE</u>**
(Electrophilic)   (Nucleophilic)

$$RO^{.nn} + e.\overset{\overset{O}{\|}}{C} - \overset{\overset{CH_3}{|}}{\underset{\underset{CH_3}{|}}{C}} - \overset{\overset{C}{\|}}{\underset{\underset{H_3C\quad CH_3}{}}{C}} - O.nn \longrightarrow RO - \overset{\overset{O}{\|}}{C} - \overset{\overset{CH_3}{|}}{\underset{\underset{CH_3}{|}}{C}} - \overset{\overset{C}{\|}}{\underset{\underset{H_3C\quad CH_3}{}}{C}} - O.nn \xrightarrow{+\ n\,(III)}$$

(III)   <u>Half free radical</u>   <u>Radicallly balanced</u>
(COUPLE)

$$RO\left(\overset{\overset{O}{\|}}{C} - \overset{\overset{CH_3}{|}}{\underset{\underset{CH_3}{|}}{C}} - \overset{\overset{C}{\|}}{\underset{\underset{H_3C\ CH_3}{}}{C}} - O\right)_n \overset{\overset{O}{\|}}{C} - \overset{\overset{CH_3}{|}}{\underset{\underset{CH_3}{|}}{C}} - \overset{\overset{C}{\|}}{\underset{\underset{H_3C\ CH_3}{}}{C}} - O.nn$$

<u>Alternating copolymer</u>   9.23

(II) being a nucleophile diffuses to (I) which is an electrophile to form a "COUPLE". Two moles of the monomer are required to form the couples. If it was otherwise polymerization will not be favored as shown below.

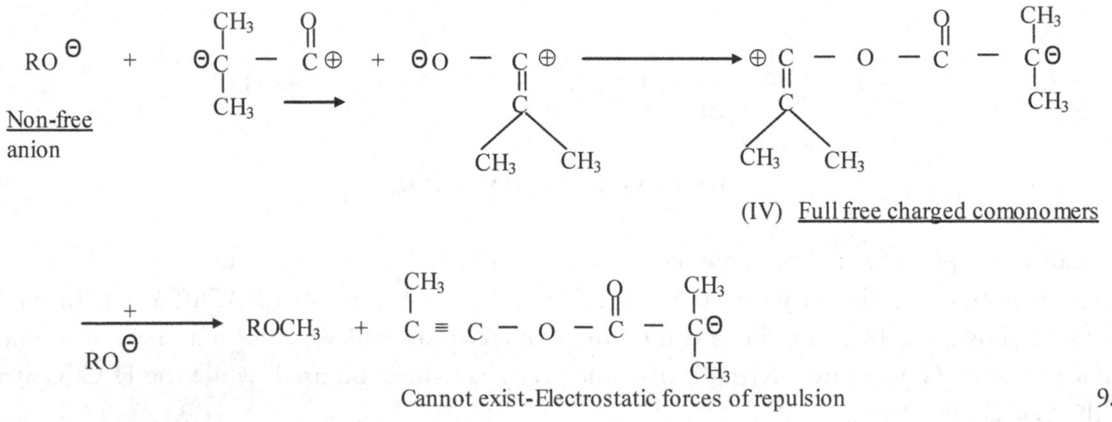

$$RO^{\ominus} + \overset{\overset{CH_3}{|}}{\underset{\underset{CH_3}{|}}{^{\ominus}C}} - \overset{\overset{O}{\|}}{C^{\oplus}} + {^{\ominus}O} - \overset{\overset{C^{\oplus}}{}}{\underset{\underset{CH_3\ CH_3}{}}{C}} \longrightarrow {^{\oplus}C} - O - \overset{\overset{O}{\|}}{\underset{\underset{CH_3\ CH_3}{}}{C}} - \overset{\overset{CH_3}{|}}{\underset{\underset{CH_3}{|}}{C^{\ominus}}}$$

<u>Non-free</u>
anion

(IV)   <u>Full free charged comonomers</u>

$$\xrightarrow[RO^{\ominus}]{+} ROCH_3 + \overset{\overset{CH_3}{|}}{C} \equiv C - O - \overset{\overset{O}{\|}}{C} - \overset{\overset{CH_3}{|}}{\underset{\underset{CH_3}{|}}{C^{\ominus}}}$$

Cannot exist-Electrostatic forces of repulsion   9.24

With nucleo-non-free-radical initiator, no reaction is possible with (IV); the same transfer species will be abstracted. On the other hand, if there had been none, the equation will not be radically balanced with continuous addition of (IV). For alternating copolymerization, it is important to note as will become clearer in volume (IV), that there are more than two driving forces favoring their existences, three most important of which are-

711

(i) Two co-monomers must be unreactive to the initiator, that is going to produce the copolymer.

(ii) Two co-monomers, if of the same character must be such that when the more nucleophilic diffuses to the less nucleophilic monomer, there is no transfer species to abstract.

(iii) In every propagation step, a "COUPLE", must first be formed before addition and this can only take place via Combination mechanism.

With OH or OR replacing H, the followings are to be expected, noting that these monomers cannot be activated chargedly as also shown below.

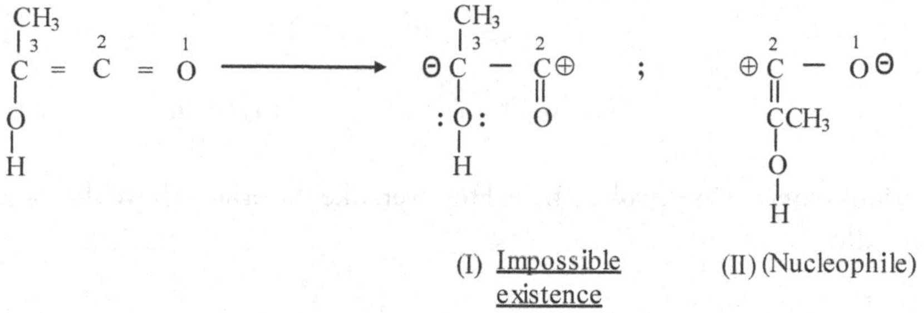

$$
\begin{array}{c}
CH_3 \\
\overset{|3}{C} = \overset{2}{C} = \overset{1}{O} \\
| \\
O \\
| \\
H
\end{array}
\longrightarrow
\begin{array}{c}
\ominus \overset{3}{C} - \overset{2}{C}\oplus \\
| \quad\quad \| \\
:O: \quad O \\
| \\
H
\end{array}
\quad ; \quad
\begin{array}{c}
\oplus \overset{2}{C} - \overset{1}{O}\ominus \\
\| \\
CCH_3 \\
| \\
O \\
| \\
H
\end{array}
$$

(I) <u>Impossible existence</u>   (II) (Nucleophile)

9.25

Like the case above, no molecular rearrangement is favored. Like cumulenes, it can be observed here that O = is more radical-pushing than OH or NH_2 etc. groups, all being non-free in character. Since (I) cannot be activated chargedly, due to electrostatic forces of repulsion, the monomer above cannot undergo charged reactions.

With OR groups, the same analyses apply. With NR_2 or NH_2 groups, similar analyses are obtained. Thus, it can be observed that unlike other cases considered so far, the hydrogen atom on the oxygen or nitrogen center is not a disturbing factor. The mono-forms cannot be obtained chargedly.

With radical-pulling groups, consider the halogens.

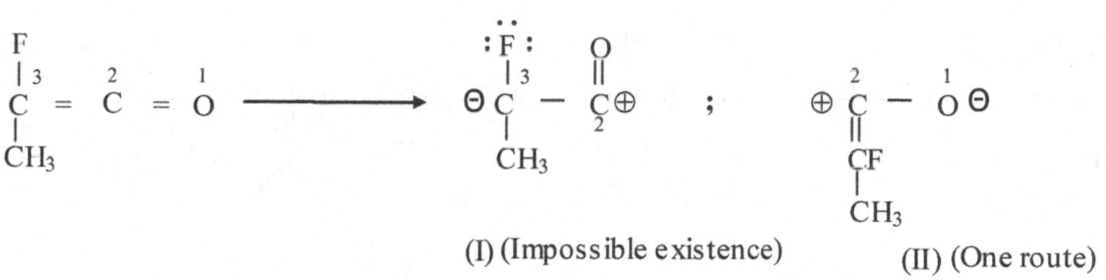

$$
\begin{array}{c}
F \\
\overset{|3}{C} = \overset{2}{C} = \overset{1}{O} \\
| \\
CH_3
\end{array}
\longrightarrow
\begin{array}{c}
:\overset{..}{F}: \quad O \\
\overset{|3}{} \quad\quad \| \\
\ominus C - C\oplus \\
| \quad\quad 2 \\
CH_3
\end{array}
\quad ; \quad
\begin{array}{c}
\oplus \overset{2}{C} - \overset{1}{O}\ominus \\
\| \\
CF \\
| \\
CH_3
\end{array}
$$

(I) (Impossible existence)   (II) (One route)

9.26

(II) seems to favor the negatively-charged route because CH_3 cannot be abstracted, due to electrostatic forces of repulsion. Hence, both mono-forms cannot be activated chargedly.

$$
RO^{\ominus} \quad + \quad
\begin{array}{c}
\oplus C - O\ominus \\
\| \\
CF \\
| \\
CH_3
\end{array}
\longrightarrow
ROCH_3 \quad + \quad
\begin{array}{c}
\oplus C - O\ominus \\
\| \\
\ominus C \\
| \\
:\overset{..}{F}:
\end{array}
$$

(Not favored)   (Existence not favored)

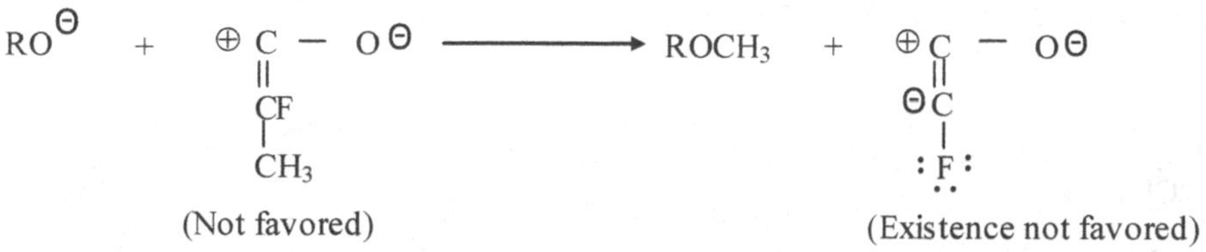

9.27

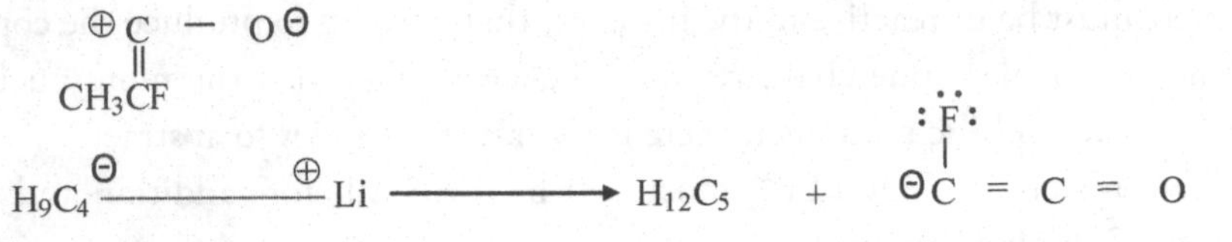

$$\underset{CH_3CF}{\overset{\oplus C \ - \ O^{\ominus}}{\|}}$$

$$H_9C_4 \underline{\qquad\qquad}\overset{\ominus}{} \overset{\oplus}{}Li \longrightarrow H_{12}C_5 \ + \ \overset{:\ddot{F}:}{\underset{\ominus C}{|}} = C = O$$

$$\underline{\text{Impossible existence}} \qquad\qquad 9.28$$

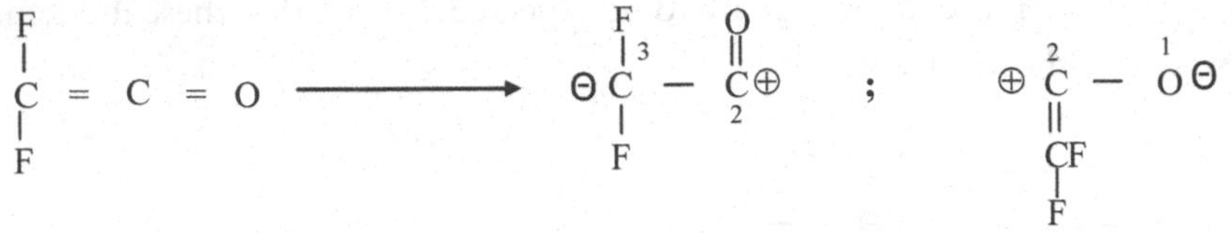

$$\underset{F}{\overset{F}{\underset{|}{\overset{|}{C}}}} = C = O \longrightarrow \overset{F}{\underset{F}{\overset{|}{\underset{|}{\ominus C}}}} {}^3 - \overset{O}{\overset{\|}{C}}^{\oplus}_2 \quad ; \quad \overset{\oplus}{} {}^2C - \overset{1}{O}^{\ominus} \underset{\underset{F}{\overset{|}{CF}}}{\overset{\|}{}}$$

$$\text{(I) (Impossible existence)} \qquad \text{(II) (Both)} \qquad 9.29$$

Only the carbonyl activation center can be involved here. However, like the others above the mono-forms cannot be activated chargedly.

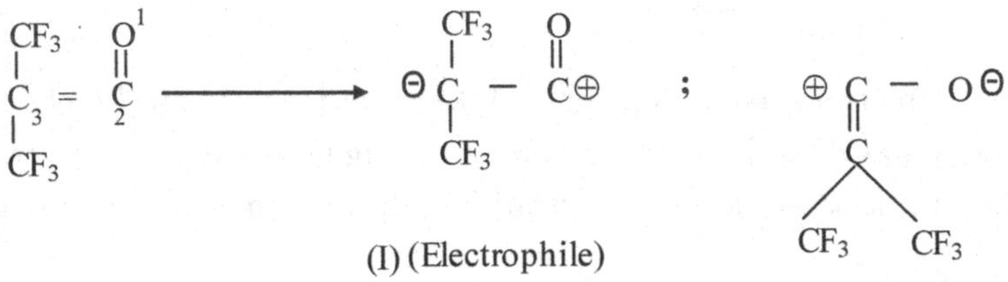

$$\underset{CF_3}{\overset{CF_3}{\underset{|}{\overset{|}{C}}}}{}_3 = \overset{O}{\overset{\|}{C}}{}^1_2 \longrightarrow \underset{CF_3}{\overset{CF_3}{\underset{|}{\overset{|}{\ominus C}}}} - \overset{O}{\overset{\|}{C}}{}^{\oplus} \quad ; \quad \overset{\oplus}{} \underset{\underset{CF_3 \quad CF_3}{\overset{|}{C}}}{\overset{\|}{C}} - O^{\ominus}$$

$$\text{(I) (Electrophile)} \qquad\qquad \text{(II)(Nucleophile)} \qquad 9.30$$

None favors molecular rearrangement. Unlike dimethylketene, only (I) can be involved to produce a ketone only with negatively charged-paired initiators when activated. Now consider the case below.

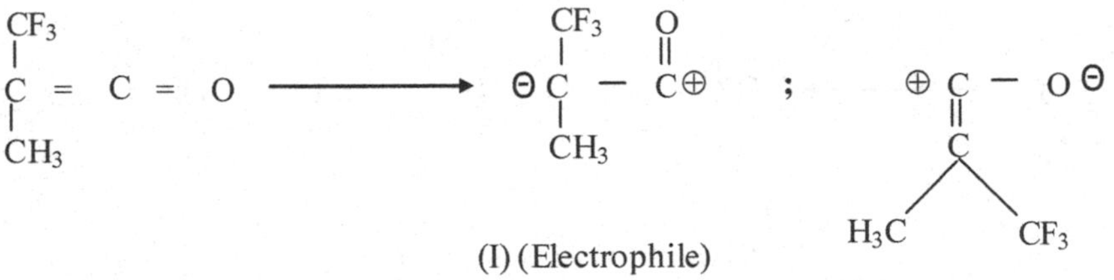

$$\underset{CH_3}{\overset{CF_3}{\underset{|}{\overset{|}{C}}}} = C = O \longrightarrow \underset{CH_3}{\overset{CF_3}{\underset{|}{\overset{|}{\ominus C}}}} - \overset{O}{\overset{\|}{C}}{}^{\oplus} \quad ; \quad \overset{\oplus}{} \underset{\underset{H_3C \qquad CF_3}{\overset{|}{C}}}{\overset{\|}{C}} - O^{\ominus}$$

$$\text{(I) (Electrophile)} \qquad\qquad \text{(II) (Nucleophile) – In part} \qquad 9.31$$

(I) is an electrophile because free negatively charged-paired initiators will favor its full polymerization. (II) above will not be as shown below.

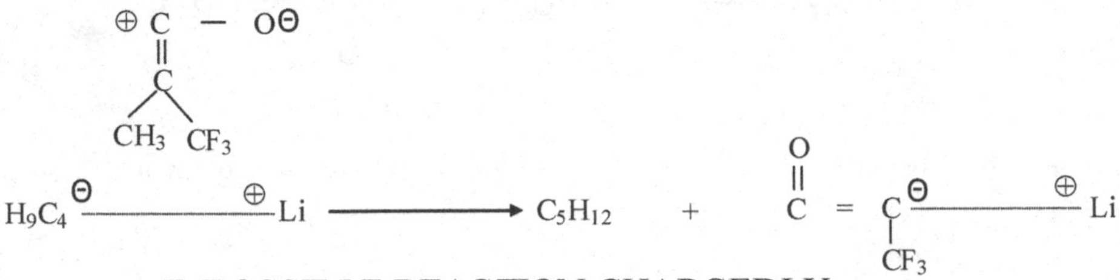

$$\underset{CH_3 \quad CF_3}{\overset{\oplus C \ - \ O^{\ominus}}{\overset{\|}{\underset{\overset{|}{C}}{}}}}$$

$$H_9C_4 \underline{\qquad\qquad}\overset{\ominus}{} \overset{\oplus}{}Li \longrightarrow C_5H_{12} \ + \ \underset{CF_3}{\overset{O}{\underset{|}{\overset{\|}{C}}}} = C \overset{\ominus}{} \underline{\qquad\qquad}\overset{\oplus}{}Li$$

$$\textbf{IMPOSSIBLE REACTION CHARGEDLY} \qquad 9.32$$

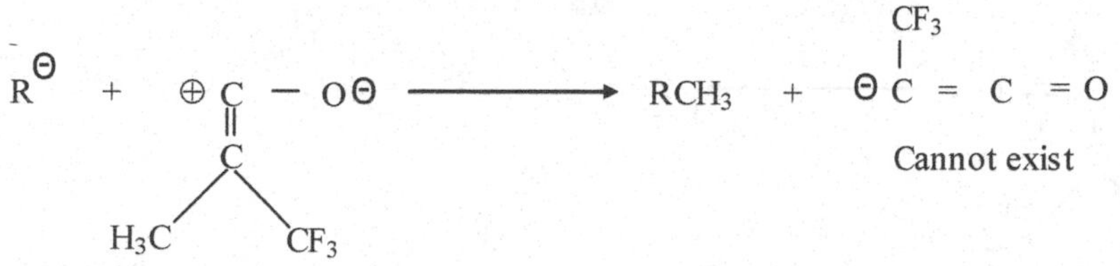

$$R^{\ominus} \ + \ \overset{\oplus}{\underset{\underset{H_3C}{\overset{\parallel}{\underset{}{C}}}\diagdown CF_3}{C}} - O^{\ominus} \longrightarrow RCH_3 \ + \ \overset{\ominus}{C} = C = O \quad (CF_3)$$

Cannot exist

$$9.33$$

The reason why initiation is not favored with negatively-charged-paired initiator is due to the fact that $(CH_3)(CF_3)C =$ group is a radical-pushing group like all others. These are important exercises being done chargedly, so that we know what to expect when rightly activated radically. One can imagine the enormous damage which has been done to CHEMISTRY since antiquity.

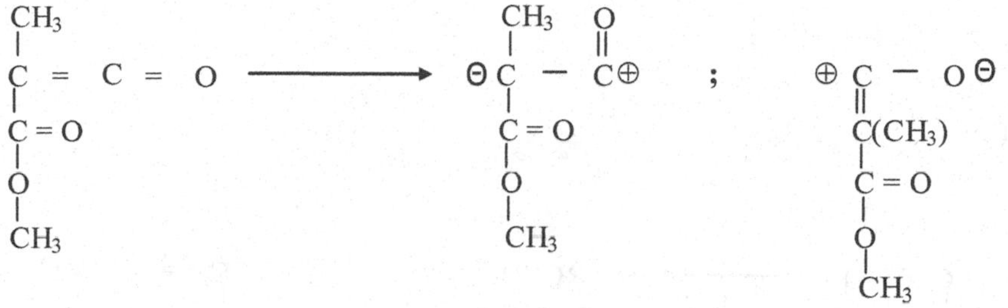

$$\begin{array}{ccc}
\underset{\underset{\underset{CH_3}{|}}{\overset{\overset{O}{|}}{C=O}}}{\overset{\overset{CH_3}{|}}{C}}=C=O & \longrightarrow & \overset{\ominus}{\underset{\underset{\underset{CH_3}{|}}{\overset{\overset{O}{|}}{C=O}}}{C}} - \overset{O}{\overset{\parallel}{C}}{}^{\oplus} \quad ; \quad \overset{\oplus}{\underset{\underset{\underset{CH_3}{|}}{\overset{\overset{C=O}{|}}{C(CH_3)}}}{C}} - O^{\ominus}
\end{array}$$

(I)(Electrophile)   (II)(Nucleophile)   9.34

(I) favors transfer of transfer species $\ominus OCH_3$ to produce same activated monomer via molecular rearrangement.

$$\begin{array}{ccc}
\underset{\underset{\underset{H}{|}}{\overset{\overset{C=O}{|}}{C=O}}}{\overset{\overset{CH_3}{|}}{C}}=C=O & \longrightarrow & \overset{\ominus}{\underset{\underset{\underset{H}{|}}{\overset{\overset{C=O}{|}}{C=O}}}{C}} - \overset{O}{\overset{\parallel}{C}}{}^{\oplus} \quad ; \quad \overset{\oplus}{\underset{\underset{\underset{H}{|}}{\overset{\overset{C=O}{|}}{C(CH_3)}}}{C}} - O^{\ominus}
\end{array}$$

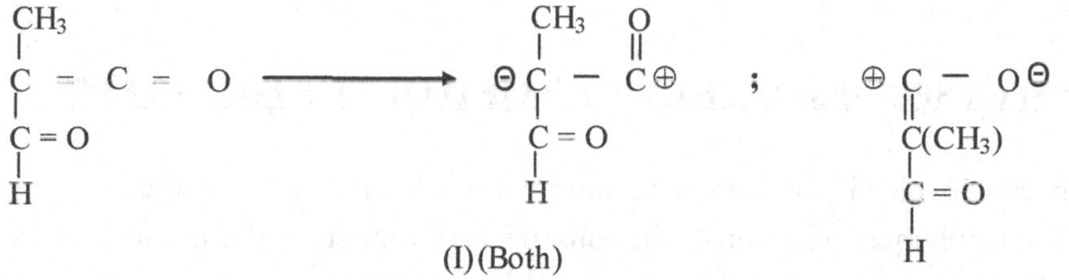

(I)(Both)

(II)(Nucleophile)   9.35

None favors molecular rearrangement. It is important to note that in all polymerizable cases, it is the less nucleophilic mono-form that shows stronger electrophilic tendency. The reason is because O = has no transfer species. For the growing polymer chain of (I) above, the following is obtained.

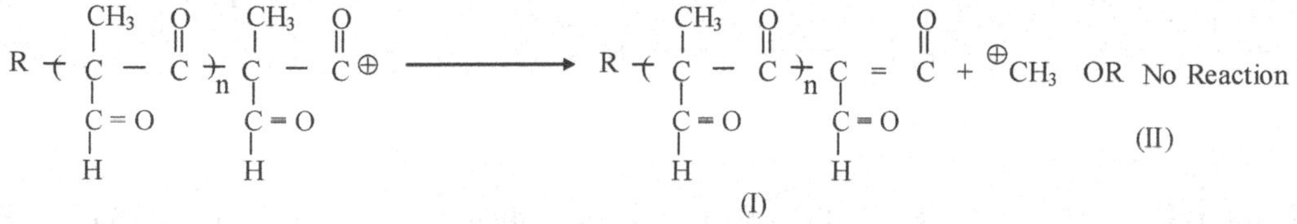

$$R \left(\!\!\begin{array}{c} \underset{\underset{H}{|}}{\overset{\overset{CH_3}{|}}{C}} - \overset{O}{\overset{\parallel}{C}} \\ {}_{\underset{C=O}{}} \end{array}\!\!\right)_n \!\!\begin{array}{c} \underset{\underset{H}{|}}{\overset{\overset{CH_3}{|}}{C}} - \overset{O}{\overset{\parallel}{C}}{}^{\oplus} \\ {}_{\underset{C=O}{}} \end{array} \longrightarrow R \left(\!\!\begin{array}{c} \underset{\underset{H}{|}}{\overset{\overset{CH_3}{|}}{C}} - \overset{O}{\overset{\parallel}{C}} \\ {}_{\underset{C=O}{}} \end{array}\!\!\right)_n \!\!\begin{array}{c} \underset{\underset{H}{|}}{\overset{}{C}} = \overset{O}{\overset{\parallel}{C}} \\ {}_{\underset{C=O}{}} \end{array} + {}^{\oplus}CH_3 \quad \text{OR No Reaction}$$

(I)   (II)

9.36

Since the route is not natural to the monomer, transfer species of the second kind is involved. The reason why mono-form of (II) of Equation 9.30 is said to favor part polymerization positively, is because of the presence of a new type of transfer species as shown below.

$$R^{\oplus} \;+\; \underset{\underset{F_3C \quad CF_3}{\underset{(I)}{\diagdown C \diagup}}}{\overset{C}{\underset{\|}{C}} = O} \;\longrightarrow\; RF \;+\; \overset{\overset{F \qquad CF_3}{|\qquad |}}{\underset{\underset{F}{|}}{\oplus C - C}} = C = O$$

9.37

$$R^{\oplus} \;+\; \underset{\underset{CF_2 \quad CF_3}{\underset{|}{\underset{CF_3}{\diagdown C \diagup}}}}{\overset{C}{\underset{\|}{C}} = O} \;\longrightarrow\; RF \;+\; \overset{\overset{CF_3 \quad CF_3}{|\qquad |}}{\underset{\underset{F}{|}}{\oplus C - C}} = C = O$$

(II)

9.38

$$R^{\oplus} \;+\; \underset{\underset{\underset{\underset{CH_3}{|}}{O}}{\underset{\underset{|}{C = O}}{\underset{C(COOCH_3)}{\overset{\|}{C}}}}}{\overset{C}{\underset{\|}{C}} = O} \;\longrightarrow\; ROCH_3 \;+\; \underset{\underset{\underset{\underset{CH_3}{|}}{O}}{\underset{\underset{|}{C = O}}{\underset{C = O}{C}}}}{\overset{O}{\underset{\|}{\oplus C}} - } \; C = C = O$$

(III)

9.39

TRANSFER SPECIES OF THE 2<u>ND</u> 1<u>ST</u> KIND OF THE FIFTH TYPE

These are transfer species involved when the monomer cannot be activated chargedly. They are only used when chemical (not polymeric reactions) reactions are involved, since the monomers are in Stable State of Existence. It is important to note that the state of the positive charges produced in Equations 9.37 - 9.39 is only possible when free cationic initiators or species are involved. They cannot be produced with charged-paired initiators. For the growing polymer chain of (I) of Equation 9.31 when the CH_3 is replaced with CF_3, the following is obtained with negatively charged initiator. (I) unlike (II) seems to be an electrophile, since there is no nucleophilic tendency.

$$R \overset{\overset{O \quad CF_3 \quad O \quad CF_3}{\| \quad | \quad \| \quad |}}{\left(C - \underset{\underset{CF_3}{|}}{C} \right)_n C - \underset{\underset{CF_3}{|}}{C}^{\ominus}} \;\longrightarrow\; R \overset{\overset{O \quad CF_3 \quad O \quad CF_3 \quad F}{\| \quad | \quad \| \quad | \quad |}}{\left(C - \underset{\underset{CF_3}{|}}{C} \right)_n C - C = \underset{\underset{F}{|}}{C}} \;+\; F^{\ominus}$$

9.40

The transfer species of the first kind of the first type is involved here. It is important to note the type of terminal ends of the dead polymer above. It should be noted that all the transfer species which have so far been identified are largely those involved under normal polymerization conditions which in general is mild, unlike those which are involved under abnormal operating conditions such as used in for example the synthesis of ketenes from decomposition of acetic acid vapor at 700 - 720°C or from acetone at 700°C.[4] These are in many cases those that involved some use of only homolytic scission or cutting as already indicated during photosynthesis or pyrolysis or irradiations radically. It can be observed in general that for every family of monomers, there must be something unique that identify with it.

9.1.2. Radical characters of 1,3-Ketenes

Resonance stabilization of these monomers has not yet been addressed. This will be done after considering the radical characters of these monomers. Beginning with the first member, the followings are to be expected.

$$\begin{array}{c}
H \\
| \\
C = C = O \\
| \\
H
\end{array}
\longrightarrow
\begin{array}{c}
H \quad O \\
| \quad \| \\
n.C - C.e \\
| \\
H
\end{array}
\quad ; \quad
\begin{array}{c}
O \\
\| \\
e.C - O.nn \\
\| \\
CH_2
\end{array}
\qquad 9.41$$

(I)   (II)

$$N^{\cdot n} + \begin{array}{c} H \quad O \\ | \quad \| \\ n\,C = C \\ | \\ H^{\cdot e} \end{array} \longrightarrow NH + \begin{array}{c} H \\ | \\ n\,C = C = O \end{array} \qquad 9.42$$

$$N^{\cdot n} + \begin{array}{c} e.C - O.nn \\ \| \\ CH_2 \end{array} \longrightarrow NH + \begin{array}{c} H \\ | \\ n.C = C = O \end{array} \qquad 9.43$$

$$\ddot{N}^{\cdot nn} + \begin{array}{c} e.C - O.nn \\ \| \\ CH_2 \end{array} \longrightarrow \ddot{N}H + \begin{array}{c} H \\ | \\ C \equiv C - O.nn \end{array} \qquad 9.44$$

$$2\,E^{\cdot e} + 2\,O = C = \begin{array}{c} H \\ | \\ C^{\cdot n} \end{array} H^{e} \longrightarrow H_2 + 2\,O = C = \begin{array}{c} H \\ | \\ C \end{array} -- E \quad 9.45$$

Due to its strong Equilibrium state of Existence, noting that, the monomer is an electrophile, electro-free-radical polymerizations are not favored. After replacement of one H atom, the monomer can still not be polymerized, since the last hydrogen atom is still loosely bonded radically, except when suppressed.

$$\begin{array}{c}
CH_3 \\
| \\
C = C = O \\
| \\
H
\end{array}
\xrightarrow{\text{HOR}}
\begin{array}{c}
CH_3 \\
| \\
n.\,C = C = O \\
| \\
H.e
\end{array}
\longrightarrow
H_2O +
\begin{array}{c}
CH_3 \\
| \\
C = C = O \\
| \\
R
\end{array}
\qquad 9.46$$

With dimethyl ketene, the followings are to be expected.

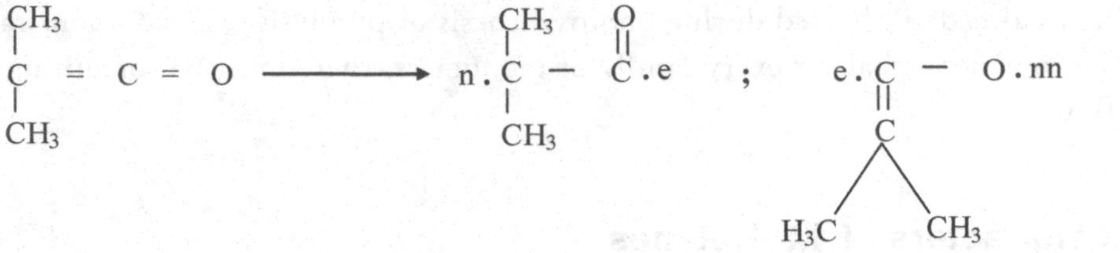

$$\underset{\underset{CH_3}{|}}{\overset{\overset{CH_3}{|}}{C}} = C = O \longrightarrow n.\underset{\underset{CH_3}{|}}{\overset{\overset{CH_3}{|}}{C}} - \overset{\overset{O}{\|}}{C}.e \quad ; \quad e.\overset{\overset{}{}}{C} - O.nn$$

(I) (Electrophile) (II) (Nucleophile) 9.47

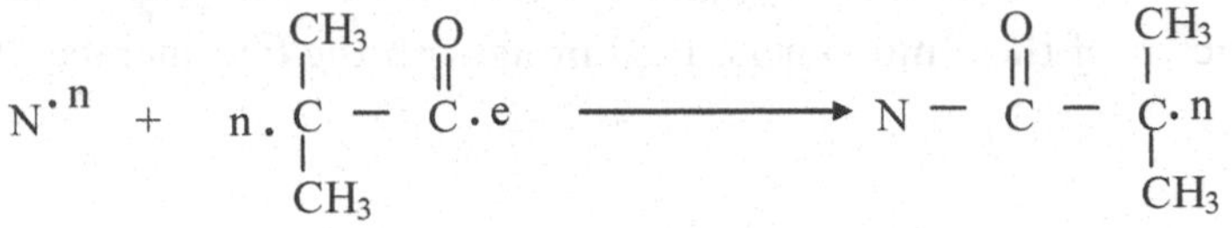

$$N^{\cdot n} + n.\underset{\underset{CH_3}{|}}{\overset{\overset{CH_3}{|}}{C}} - \overset{\overset{O}{\|}}{C}.e \longrightarrow N - \overset{\overset{O}{\|}}{C} - \underset{\underset{CH_3}{|}}{\overset{\overset{CH_3}{|}}{C}} .n$$

9.48

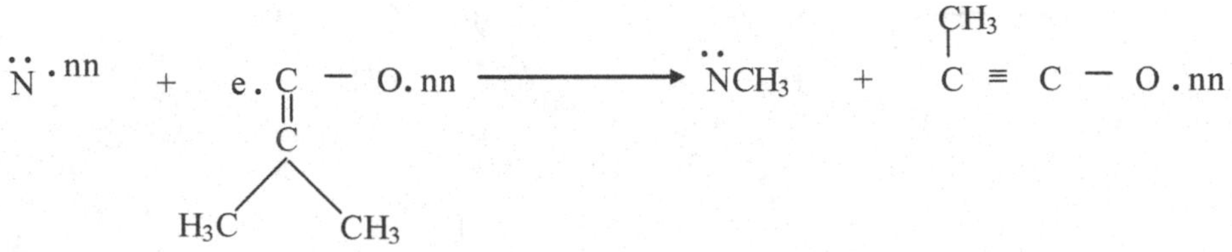

$$\overset{..}{N}{}^{\cdot nn} + e.\overset{\overset{}{}}{C} - O.nn \longrightarrow \overset{..}{N}CH_3 + \overset{\overset{CH_3}{|}}{C} \equiv C - O.nn$$

9.49

$$E{\bullet}e + nn{\bullet}O - C{\bullet}e \longrightarrow E - O - C{\bullet}e$$

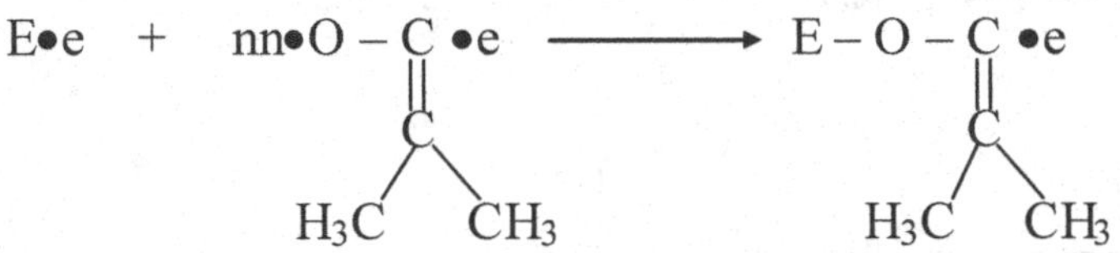

9.50

Thus, two different homopolymers are produced here using nucleo-free-radicals and electro-free-radicals coming from ionic metal such as Na- polyketenes and polyacetals respectively. With other types of electro-free-radicals, random copolymers are obtained. Thus (I) is truly an electrophile. Like what has been shown, only nucleo-non-free-radicals-free or paired can produce alternating copolymers.

With etheric groups, the followings are to be expected.

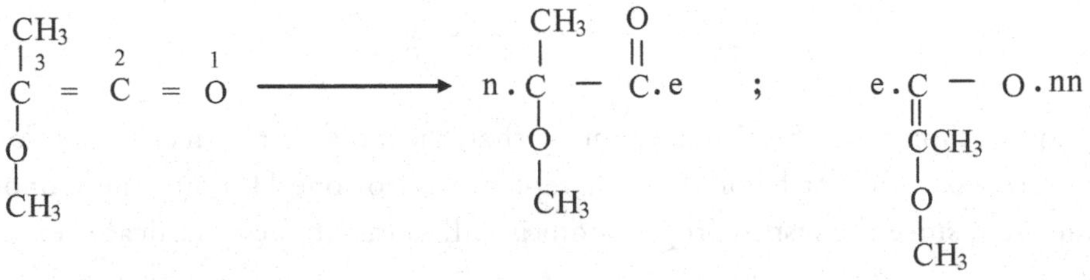

$$\underset{\underset{\underset{CH_3}{|}}{O}}{\overset{\overset{CH_3}{|}}{\underset{|}{C}}} \overset{2}{=} C \overset{1}{=} O \longrightarrow n.\underset{\underset{\underset{CH_3}{|}}{O}}{\overset{\overset{CH_3}{|}}{C}} - \overset{\overset{O}{\|}}{C}.e \quad ; \quad e.\overset{\overset{}{}}{C} - O.nn$$

(I) (Electrophile) (II) (Nucleophile) 9.51

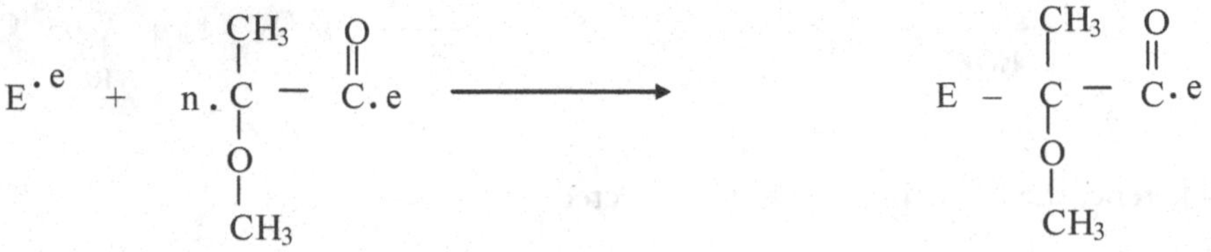

$$E^{\cdot e} + n.\underset{\underset{\underset{CH_3}{|}}{O}}{\overset{\overset{CH_3}{|}}{C}} - \overset{\overset{O}{\|}}{C}.e \longrightarrow E - \underset{\underset{\underset{CH_3}{|}}{O}}{\overset{\overset{CH_3}{|}}{C}} - \overset{\overset{O}{\|}}{C}.e$$

9.52

$$N \overset{n}{\cdot} \quad + \quad n.\underset{\underset{CH_3}{|}}{\overset{\overset{CH_3}{|}}{C}} - \underset{\underset{CH_3}{|}}{\overset{\overset{O}{||}}{C}}.e \quad \longrightarrow \quad N - \underset{\underset{CH_3}{|}}{\overset{\overset{O}{||}}{C}} - \underset{\underset{CH_3}{|}}{\overset{\overset{CH_3}{|}}{C}}.n$$

9.53

$$N \overset{n}{\cdot} \quad or \quad \overset{..}{N}\cdot^{nn} \quad + \quad e.\underset{\underset{\underset{CH_3}{|}}{O}}{\underset{|}{\overset{\overset{O}{||}}{C}} - \underset{\underset{CH_3}{|}}{\overset{\overset{}{CCH_3}}}}\; O.nn \quad \longrightarrow \quad NCH_3 \quad + \quad \overset{\overset{CH_3}{|}}{\underset{n}{C}} = C = O$$

or

$$\overset{..}{N}CH_3$$

OR

$$\underset{\underset{CH_3}{|}}{\overset{}{C}} \equiv C - O.nn$$

<u>Transfer species of 1st kind of 3rd type</u>

9.54

It can be observed why the mono-forms bear the characters shown in Equation 9.51. So far, worthy of note is that only one mono-form can favor molecular rearrangement.

$$\underset{\underset{NR_2}{|}}{\overset{\overset{CH_3}{|}}{C}} = C = O \quad \longrightarrow \quad n.\underset{\underset{NR_2}{|}}{\overset{\overset{CH_3}{|}}{C}} - \overset{\overset{O}{||}}{C}.e \quad ; \quad e.\underset{\underset{\underset{NR_2}{|}}{CCH_3}}{\overset{\overset{O}{||}}{C}} - O.nn$$

(I) (Electrophile)   (II) (Nucleophile)

9.55

The analysis is similar to the etheric case above, with however no possibility of molecular rearrangement for any of the mono-forms.

With fluorinated ketenes, the followings are obtained.

$$\underset{\underset{F}{|}}{\overset{\overset{F}{|}}{C}} = C = O \quad \longrightarrow \quad n.\underset{\underset{F}{|}}{\overset{\overset{F}{|}}{C}} - \overset{\overset{O}{||}}{C}.e \quad ; \quad e.\underset{\underset{CF_2}{|}}{\overset{\overset{O}{||}}{C}} - O.nn$$

(I) (Both routes)   (II) (Both routes)

9.56

In the absence of any transfer species of any kind, living polymers are largely produced here. But with CF_3 replacing F, the followings are obtained.

$$\underset{\underset{CF_3}{|}}{\overset{\overset{CF_3}{|}}{C}} = C = O \quad \longrightarrow \quad n.\underset{\underset{CF_3}{|}}{\overset{\overset{CF_3}{|}}{C}} - \overset{\overset{O}{||}}{C}.e \quad ; \quad e.\overset{\overset{O}{||}}{C} - O.nn$$

$$\underset{F_3C \qquad CF_3}{\overset{\overset{}{C}}{\diagup\diagdown}}$$

(I) (Electrophile)   (II) (Nucleophile)

9.57

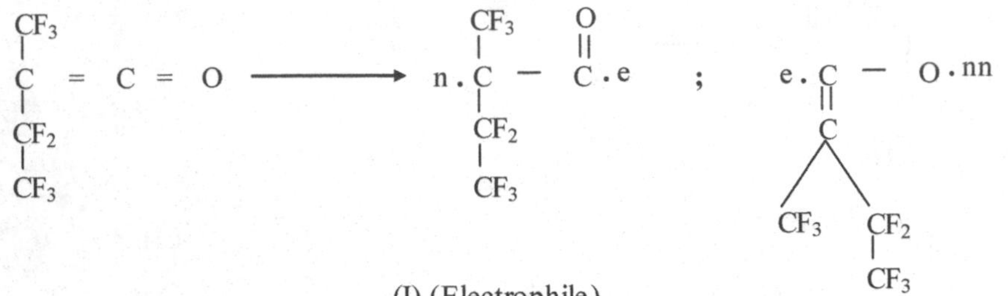

$$
\begin{array}{c}
CF_3 \\
| \\
C = C = O \\
| \\
CF_2 \\
| \\
CF_3
\end{array}
\longrightarrow
\quad
\begin{array}{c}
CF_3 \quad O \\
| \quad\quad || \\
n\,.\,C\; -\; C\,.\,e \\
| \\
CF_2 \\
| \\
CF_3
\end{array}
\quad ; \quad
\begin{array}{c}
O \\
|| \\
e\,.\,C\; -\; O\,.\,nn \\
| \\
C \\
/ \; \backslash \\
CF_3 \quad CF_2 \\
\quad\quad | \\
\quad\quad CF_3
\end{array}
$$

$$\text{(I) (Electrophile)} \qquad\qquad \text{(II) (Nucleophile)} \qquad 9.58$$

It can be observed that no two different monomers are the same. The (I)s, favor only nucleo-free-radical polymerization. Both (I)s of Equations 9.57 and 9.58 do not favor the electro-free-radical route as shown below.

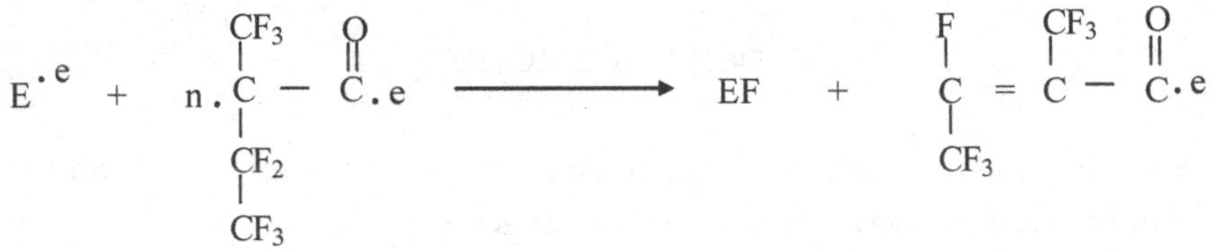

$$
E\,^{\cdot\,e} \; + \;
\begin{array}{c}
CF_3 \quad O \\
| \quad\quad || \\
n\,.\,C\; -\; C\,.\,e \\
| \\
CF_2 \\
| \\
CF_3
\end{array}
\longrightarrow
\quad EF \; + \;
\begin{array}{c}
F \quad CF_3 \quad O \\
| \quad\; | \quad\quad || \\
C = C \; - \; C\,.\,e \\
| \\
CF_3
\end{array}
$$

$$9.59$$

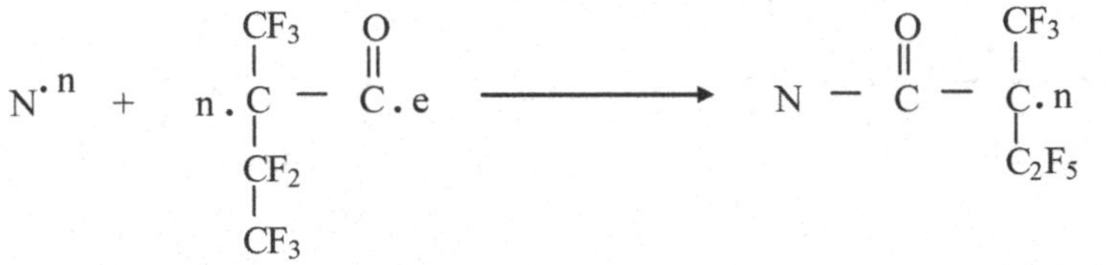

$$
N\,^{\cdot\,n} \; + \;
\begin{array}{c}
CF_3 \quad O \\
| \quad\quad || \\
n\,.\,C\; -\; C\,.\,e \\
| \\
CF_2 \\
| \\
CF_3
\end{array}
\longrightarrow
\quad
\begin{array}{c}
O \quad CF_3 \\
|| \quad\; | \\
N\; -\; C\; -\; C\,.\,n \\
| \\
C_2F_5
\end{array}
$$

$$9.60$$

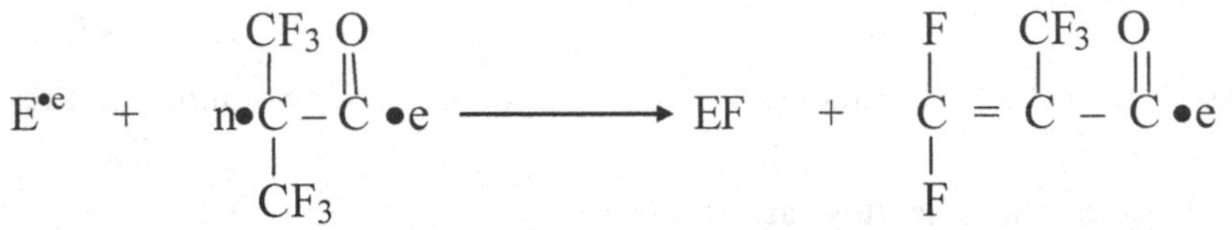

$$
E\,^{\bullet e} \; + \;
\begin{array}{c}
CF_3 \; O \\
| \quad || \\
n\bullet C - C \bullet e \\
| \\
CF_3
\end{array}
\longrightarrow
\quad EF \; + \;
\begin{array}{c}
F \quad CF_3 \; O \\
| \quad\; | \quad || \\
C = C \; - \; C \bullet e \\
| \\
F
\end{array}
$$

$$9.61$$

$$
N\,^{\bullet n} \; + \;
\begin{array}{c}
O \; CF_3 \\
|| \quad | \\
e\bullet C - C \bullet n \\
| \\
CF_3
\end{array}
\longrightarrow
\quad
\begin{array}{c}
O \; CF_3 \\
|| \quad | \\
N - C - C \bullet n \\
| \\
CF_3
\end{array}
$$

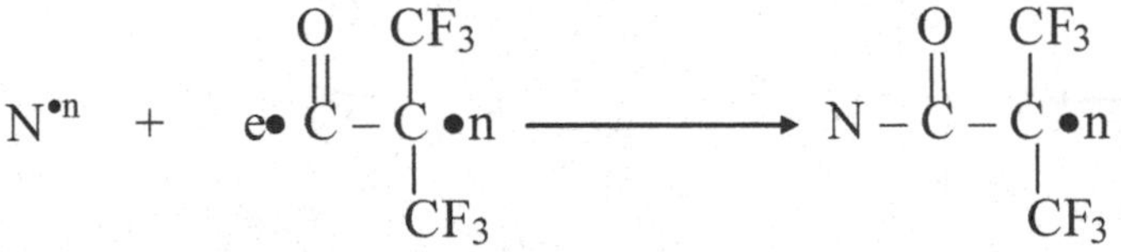

$$9.62$$

The transfer species abstracted in Equations 9.59 and 9.61, are the same. Based on the laws already stated, F is less radical-pushing than CF_3, which in turn is less radical-pushing than C_2F_5 and so on (See Table 8.1). Chargedly, the reverse is the case and meaningless due to electrostatic forces of repulsion, because for example, CF_3 cannot carry a negative charge, undermining the fact that the monomers herein cannot be activated chargedly. The same applies to other groups such as COOR, $CONH_2$, and CN. Sometimes, one is forced to finally conclude that all chemical reactions can only take place RADICALLY and never chargedly.

For the case of COOR types of groups, the followings are to be expected.

$$
\begin{array}{c}
\text{CH}_3 \\
| \\
\text{C} = \text{C} = \text{O} \\
| \\
\text{C} = \text{O} \\
| \\
\text{O} \\
| \\
\text{CH}_3
\end{array}
\longrightarrow
\begin{array}{c}
\text{CH}_3 \quad \text{O} \\
| \qquad || \\
\text{n . C} - \text{C . e} \\
| \\
\text{C} = \text{O} \\
| \\
\text{O} \\
| \\
\text{CH}_3
\end{array}
\quad ; \quad
\begin{array}{c}
\text{e . C} - \text{O . nn} \\
|| \\
\text{CCH}_3 \\
| \\
\text{C} = \text{O} \\
| \\
\text{O} \\
| \\
\text{CH}_3
\end{array}
$$

(I) (Electrophile) (II) (Nucleophile) 9.63

For (II), the nucleo-radical routes are not favored due to presence of transfer species of first kind of third type. The same applies to (I) electro-free-radically with different transfer species.

$$
\begin{array}{c}
\text{CH}_3 \\
| \\
\text{C} = \text{C} = \text{O} \\
| \\
\text{C} = \text{O} \\
| \\
\text{CH}_3
\end{array}
\longrightarrow
\begin{array}{c}
\text{CH}_3 \quad \text{O} \\
| \qquad || \\
\text{n . C} - \text{C . e} \\
| \\
\text{C} = \text{O} \\
| \\
\text{CH}_3
\end{array}
\quad ; \quad
\begin{array}{c}
\text{e . C} - \text{O . nn} \\
|| \\
\text{CCH}_3 \\
| \\
\text{C} = \text{O} \\
| \\
\text{CH}_3
\end{array}
$$

(I) (Both) (II) (Nucleophile) 9.64

$$
\text{E} \cdot \text{e} \; + \;
\begin{array}{c}
\text{CH}_3 \quad \text{O} \\
| \qquad || \\
\text{n . C} - \text{C . e} \\
| \\
\text{C} = \text{O} \\
| \\
\text{CH}_3
\end{array}
\longrightarrow
\begin{array}{c}
\text{CH3} \quad \text{O} \\
| \qquad || \\
\text{E} - \text{C} - \text{C . e} \\
| \\
\text{C} = \text{O} \\
| \\
\text{CH3}
\end{array}
$$

9.65

$$
\text{N} \cdot \text{n} \; + \;
\begin{array}{c}
\text{e C} - \text{O . nn} \\
|| \\
\text{CCH}_3 \\
| \\
\text{C} = \text{O} \\
| \\
\text{CH}_3
\end{array}
\longrightarrow
\text{NCH3} \; + \;
\begin{array}{c}
\text{n . C} = \text{C} = \text{O} \\
| \\
\text{COCH3}
\end{array}
$$

9.66

$$
\text{E} \leftarrow \text{O} - \begin{array}{c} \text{C} \\ || \\ \text{CCH}_3 \\ | \\ \text{C} = \text{O} \\ | \\ \text{CH}_3 \end{array} \rightarrow_n \text{O} - \begin{array}{c} \text{C . e} \\ || \\ \text{CCH}_3 \\ | \\ \text{C} = \text{O} \\ | \\ \text{CH}_3 \end{array}
\longrightarrow
\text{E} \leftarrow \text{O} - \begin{array}{c} \text{C} \\ || \\ \text{CCH}_3 \\ | \\ \text{C} = \text{O} \\ | \\ \text{CH}_3 \end{array} \rightarrow_n \text{O} - \text{C} \equiv \begin{array}{c} \text{C} \\ | \\ \text{COCH3} \end{array}
$$

$+ \quad \text{e . CH3}$

Transfer species of the 1st kind of third type 9.67

Ketenes are strongly electrophilic in character. It has the nucleophilic mono-form and the electrophilic mono-form.

9.1.3 Resonance stabilization in 1,3 ketenes

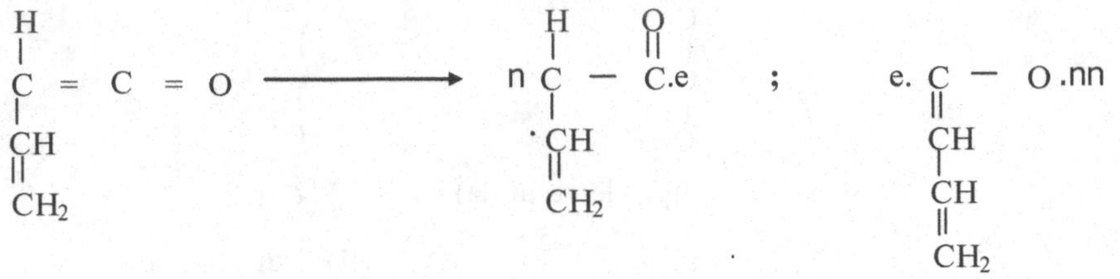

$$
\begin{array}{c}
\text{H}\\
|\\
\text{C} = \text{C} = \text{O}\\
|\\
\text{CH}\\
||\\
\text{CH}_2
\end{array}
\quad\longrightarrow\quad
n\;\begin{array}{c}
\text{H}\quad\text{O}\\
|\qquad||\\
\text{C} - \text{C.e}\\
|\\
\cdot\text{CH}\\
||\\
\text{CH}_2
\end{array}
\quad;\quad
\text{e.}\;\begin{array}{c}
\text{C} - \text{O.nn}\\
||\\
\text{CH}\\
|\\
\text{CH}\\
||\\
\text{CH}_2
\end{array}
$$

(I) (Both)   (II) (Nucleophile)   9.68

Considering (I) for resonance stabilization, the followings are obtained.

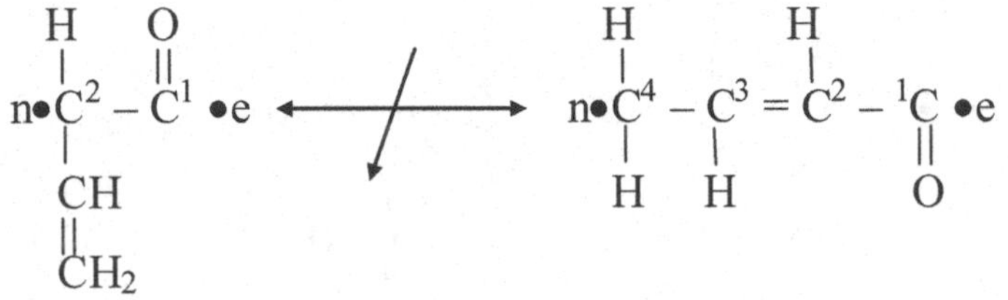

$$
n\bullet\begin{array}{c}
\text{H}\quad\text{O}\\
|\qquad||\\
\text{C}^2 - \text{C}^1\;\bullet e\\
|\\
\text{CH}\\
||\\
\text{CH}_2
\end{array}
\quad\longleftarrow\!\!\!\big/\!\!\!\longrightarrow\quad
n\bullet\begin{array}{c}
\text{H}\qquad\quad\text{H}\\
|\qquad\qquad|\\
\text{C}^4 - \text{C}^3 = \text{C}^2 - {}^1\text{C}\;\bullet e\\
|\qquad|\qquad\qquad\;||\\
\text{H}\quad\text{H}\qquad\quad\text{O}
\end{array}
$$

FULL-FREE MONOMERS   9.69a

Since the alkenyl group is the least nucleophilic center, all the above are not favored as shown below for the favored one.

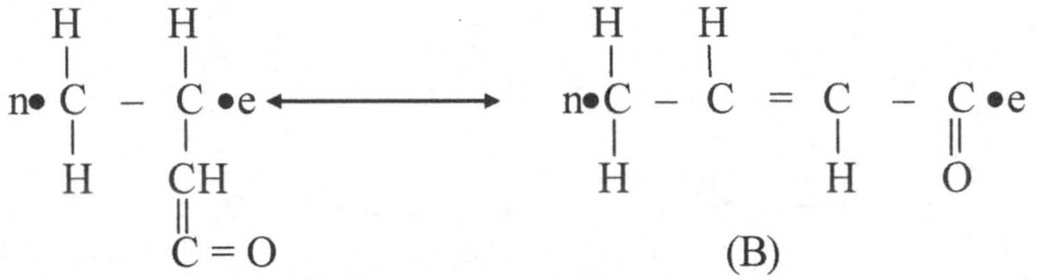

$$
n\bullet\begin{array}{c}
\text{H}\quad\text{H}\\
|\qquad|\\
\text{C} - \text{C}\;\bullet e\\
|\qquad|\\
\text{H}\quad\text{CH}\\
\qquad||\\
\qquad\text{C} = \text{O}
\end{array}
\quad\longleftrightarrow\quad
n\bullet\begin{array}{c}
\text{H}\quad\text{H}\\
|\qquad|\\
\text{C} - \text{C} = \text{C} - \text{C}\;\bullet e\\
|\qquad|\qquad\quad|\qquad||\\
\text{H}\qquad\text{H}\qquad\text{O}
\end{array}
$$

(A)   RESONANCE STABILIZATION   (B)   9.69b

Because of the strong radical-pushing capacity of O= group, (B) above is favored from that center and not from the center (I) of Equation 9.68. Hence – CH=C=O group can provide resonance stabilization being a radical-pushing group. The Male character of the monomer is fully present, since the alkenyl center has added to the Y center to become bigger, while the C=O center which is now cumulatively placed to the Central C atom is the X. As indicated above, it seems that only the MALE center can be activated in view of how the FEMALE center is cumulatively placed to the MALE center, unlike what exists with Olefins where they are conjugatedly placed. Unlike 1,3-Ketene where two mono-forms exist, here, there are also three mono-forms- (A) and (B) above and (II) of Equation 9.68. (A) and (B) mono-forms- the Y center favor both free-radical routes, while the third one favors only the route natural to it. **Notice how one has numbered the centers in Equation 9.69a** Notice also that because of the unique presence of the alkenyl group whose capacity is less than the other centers, the H atom is no longer loosely bonded.

For (II) of Equation 9.68, the following is obtained only radically.

721

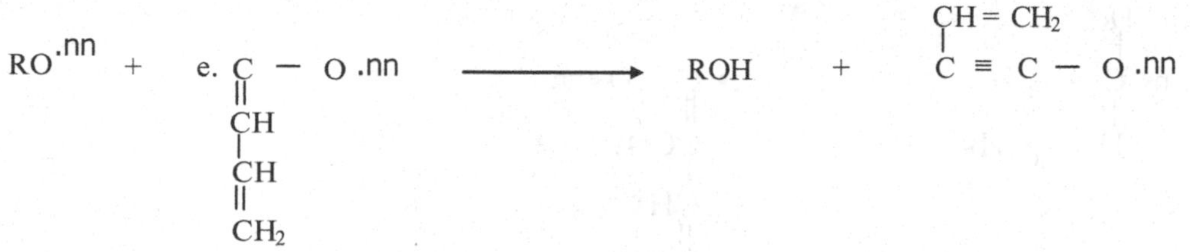

$$RO^{.nn} \quad + \quad \text{e. } \underset{\underset{\underset{CH_2}{||}}{\underset{CH}{|}}}{\overset{\overset{||}{O}}{C}} - O \text{ .nn} \quad \longrightarrow \quad ROH \quad + \quad \underset{\underset{C}{|}}{\overset{\overset{CH=CH_2}{|}}{C}} \equiv C - O \text{ .nn}$$

9.70

The C=O center is more nucleophilic than all the other two centers. Since the C=O is not a part of the resonance stabilization phenomenon, transfer species exists for that center as shown above. But with the Y center as shown in Equation 9.69b, the H abstracted above cannot be abstracted in (A), being shield.

With the presence of R group in place of H, the followings are obtained for R ≡ CH_3 or CF_3.=

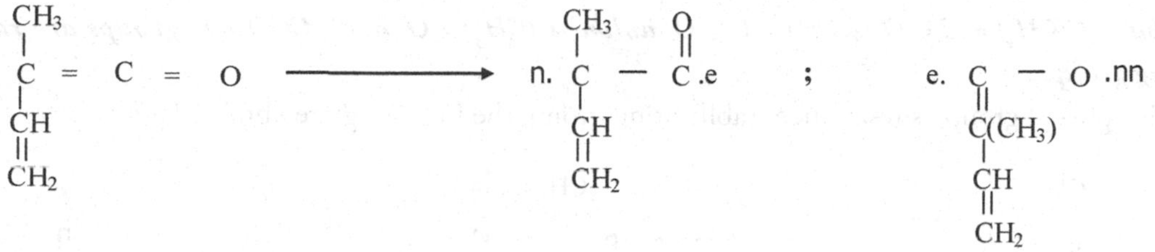

$$\underset{\underset{CH_2}{||}}{\underset{CH}{|}}\overset{\overset{CH_3}{|}}{C} = C = O \quad \longrightarrow \quad \text{n. } \underset{\underset{CH_2}{||}}{\underset{CH}{|}}\overset{\overset{CH_3}{|}}{C} - \overset{\overset{O}{||}}{C} \text{ .e} \quad ; \quad \text{e. } \underset{\underset{\underset{CH_2}{||}}{\underset{CH}{|}}}{\underset{C(CH_3)}{|}}\overset{\overset{||}{O}}{C} - O \text{ .nn}$$

(I) (Both)   (II) (Nucleophile)   9.71

Like the case above, this is resonance stabilized. Shield can therefore be provided for CH_3 group as shown below.

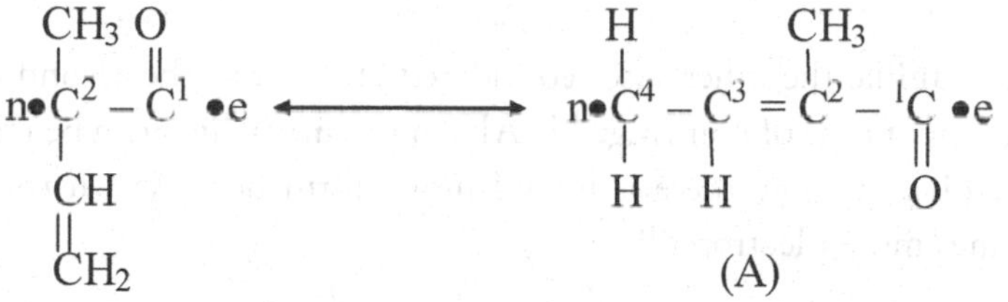

$$n\bullet\underset{\underset{CH_2}{||}}{\underset{CH}{|}}\overset{\overset{CH_3}{|}}{C^2} - \overset{\overset{O}{||}}{C^1} \bullet e \quad \longleftrightarrow \quad n\bullet\underset{H}{\overset{H}{C^4}} - \underset{H}{\overset{H}{C^3}} = \overset{\overset{CH_3}{|}}{C^2} - {}^1\underset{O}{\overset{||}{C}} \bullet e$$

(A)

CANNOT BE RESONANCE STABILIZED   9.72

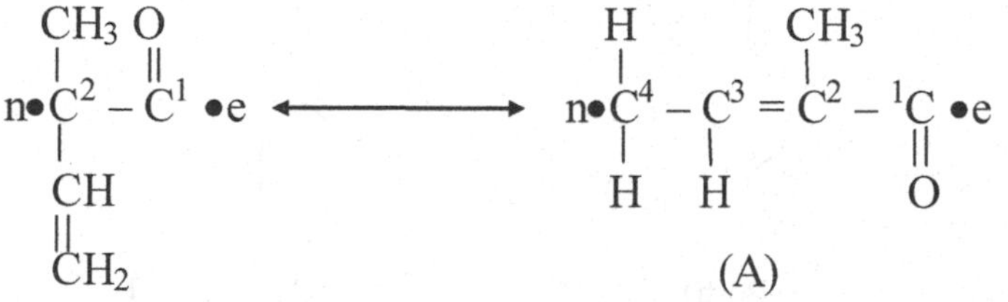

$$n\bullet\underset{\underset{CH_2}{||}}{\underset{CH}{|}}\overset{\overset{O}{||}}{C^2} - \overset{\overset{CH_3}{|}}{C^1} \bullet e \quad \longleftrightarrow \quad n\bullet\underset{H}{\overset{H}{C^4}} - \underset{H}{\overset{H}{C^3}} = \overset{\overset{CH_3}{|}}{C^2} - {}^1\underset{O}{\overset{||}{C}} \bullet e$$

(A)

CANNOT BE RESONANCE STABILIZED   9.73

The two mono-forms for the monomer above are as shown below using the first case.

$$
\begin{array}{ccc}
& \overset{H}{\underset{|}{\underset{\overset{|}{H}}{}}} \; \overset{H}{\underset{\overset{|}{CCH_3}}{}} & \\
n\bullet C - C \bullet e & ; & e\bullet C - O\bullet nn
\end{array}
$$

(I)   (II)   9.74

It is only (I) that can be resonance stabilized to give the third mono-form.

Both free-radical routes are favored by (I) since both groups are resonance stabilized. (II) is Nucleophilic with H_2C=CH- as transfer species of the first kind of the third type, while (I) is Electrophilic. ***Nevertheless, note that*** $-C(CH_3) = C = O$, $-C(H) = C = O$, ***unlike*** $-C(CH_3) = O$, ***and*** $-C(H) = O$ ***groups are radical-pushing groups.***

With phenyl group as resonance stabilization group, the following are obtained.

(I) (Both)   (II) (Nucleophile)   9.75

Only (I) can be resonance stabilized, unlike the other cases considered above. For the second or third time, we are again encountering a different type of non-ringed MALE monomer which can be resonance stabilized via only the male center. Like Styrene, there is no 1,4-mono-form here. When the CH_3 is changed to CF_3, the monomer becomes more Electrophilic.

(I) (Both)   (II) (Nucleophile)   9.76

Chargedly, both centers can exist, but cannot be used due to electrostatic forces of repulsion. Radically, (I) like other cases above favors both the nucleo-free-radical and electro-free-radical routes being resonance stabilized. Looking at the (I) of the two last equations above, the last case will have lower Ceiling temperature, since it has transfer species which cannot be abstracted.

9.2. 1,4- Cumulenic ketenes

The first member of these unknown cases is (I) below.

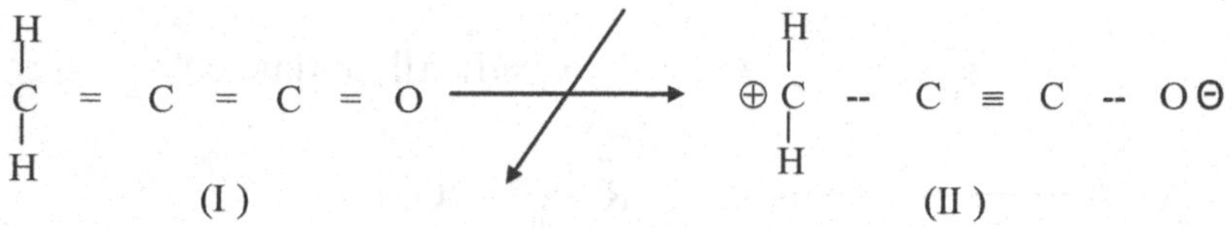

$$\text{(I)} \longrightarrow \text{(II)} \qquad 9.77$$

It has the 1,4- cumulenic character when the oxygen groups (O =) is replaced with HC = and has 1,3- ketenic character when the methylene groups (CH_2 =) is replaced with two hydrogen atoms.

9.2.1. Charged characters 1,4- cumulenic ketenes

It has been established that most of these members of compounds or monomers can be activated chargedly but cannot be used. These can only be done radically. However, for exploratory purposes, one is considering it as if possible.

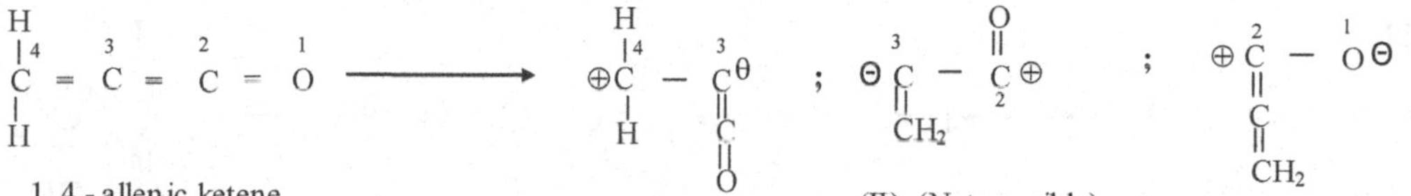

1, 4 - allenic ketene

(I) (Not possible)   (II) (Not possible)   (III) (Both)
(Both)   [Both]   9.78

(II) is the least nucleophilic center. (I) cannot be resonance stabilized. Chargedly, (I) and (II) cannot be activated due to electrostatic forces of repulsion. (I) indeed cannot exist, since it is not the least nucleophilic center. (III) is the X center which cannot be resonance stabilized favoring both routes. It cannot be resonance stabilized, because it is not the least nucleophilic center. Unlike 1,3-ketenes, it can be observed that alternating copolymer cannot be produced here, because (II) and (III) favor both radical routes. (II) favors both free-radical routes, while (III) favors electro-free and nucleo-non-free-radical routes.

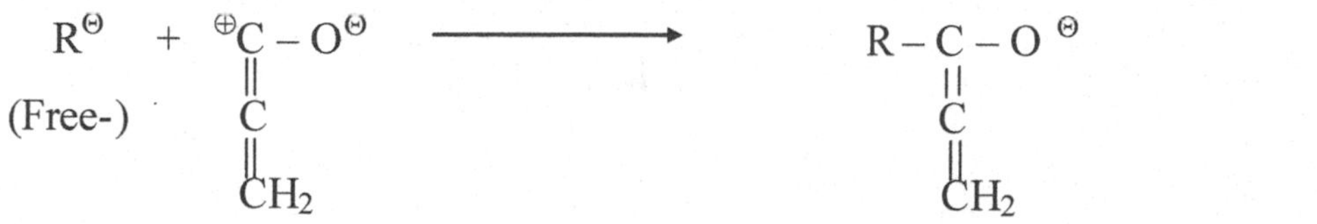

(Free-)

Not chargedly balanced   9.79

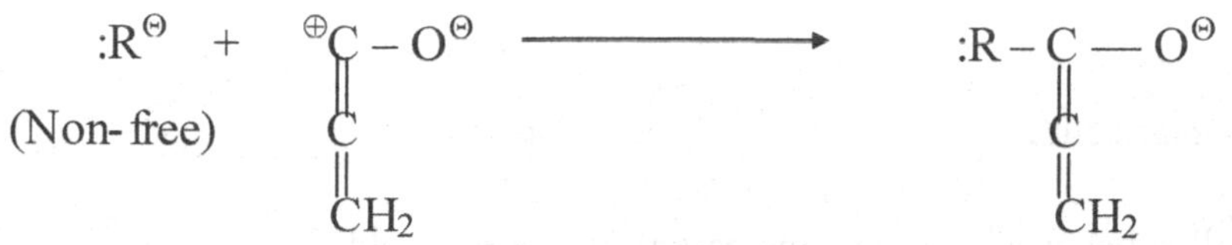

(Non- free)

9.80

Radically, the followings are obtained

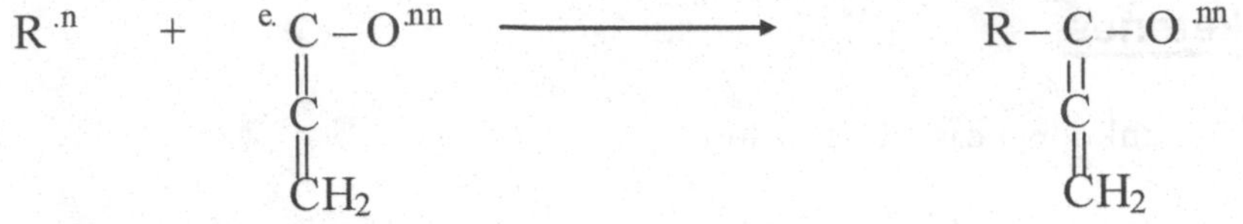

$$R^{\cdot n} \quad + \quad \overset{e.}{C} - O^{.nn} \quad \longrightarrow \quad R - C - O^{.nn}$$

$$\underset{\substack{\| \\ C \\ \| \\ CH_2}}{} \qquad\qquad\qquad \underset{\substack{\| \\ C \\ \| \\ CH_2}}{}$$

Not radically balanced   9.81

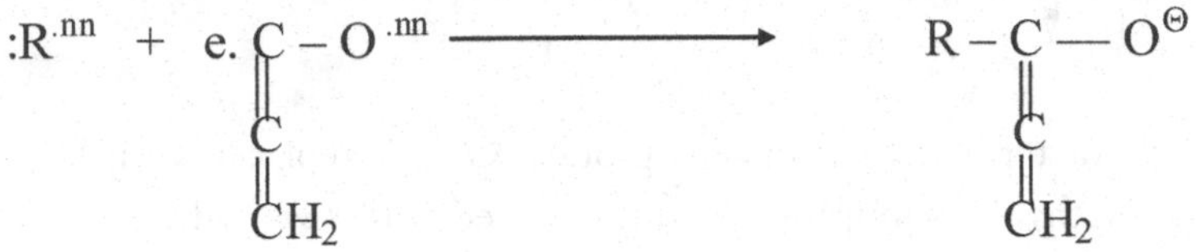

$$:R^{nn} \quad + \quad e.\,C - O^{.nn} \quad \longrightarrow \quad R - C - O^{\ominus}$$

$$\underset{\substack{\| \\ C \\ \| \\ CH_2}}{} \qquad\qquad\qquad \underset{\substack{\| \\ C \\ \| \\ CH_2}}{}$$

     9.82

Based on the H to C ratios at least one to two of the H atoms is loosely bonded radically to the carbon center one at a time. Therefore, the two hydrogen atoms must be replaced before any full polymerization of the monomer can be possible. The smaller the H:C ratio, the greater the presence of telomers. While 1,3-Ketene can be made to undergo electroradicalization to give acetylenyl alcohol ($HC \equiv COH$) only when suppressed, 1.4-Ketene can readily do it to give an equivalent of acrolein as shown below. (A) formed below is stable, since the H atom on acetylene cannot be held in Equilibrium state of existence.

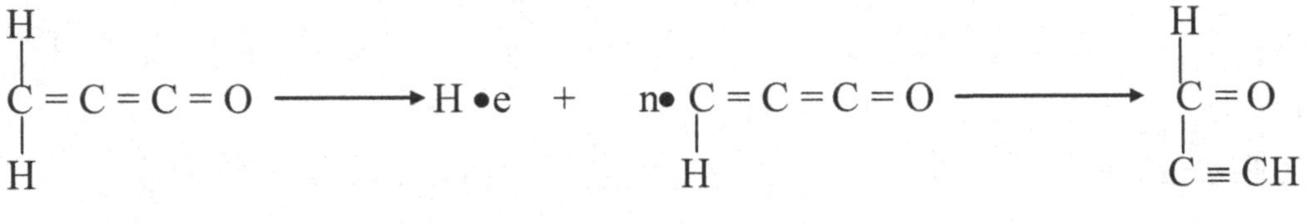

$$\underset{\substack{| \\ H}}{\overset{\substack{H \\ |}}{C}} = C = C = O \quad \longrightarrow \quad H \bullet e \quad + \quad n \bullet \underset{\substack{| \\ H}}{C} = C = C = O \quad \longrightarrow \quad \underset{\substack{| \\ C \equiv CH}}{\overset{\substack{H \\ | \\ C = O}}{}}$$

(A)   9.83

The second H atom may also be replaced as shown below.

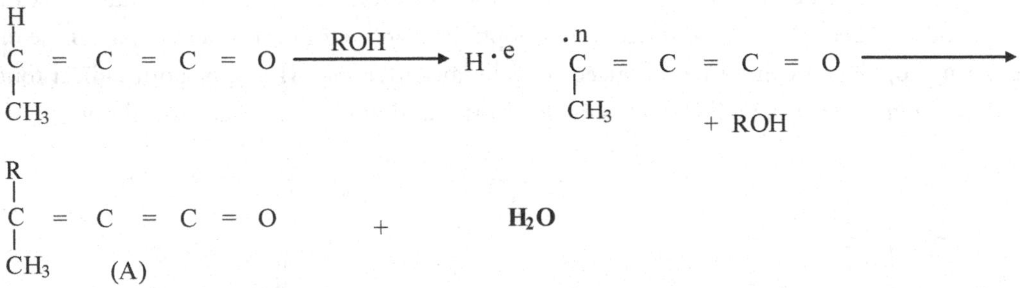

$$\underset{\substack{| \\ CH_3}}{\overset{\substack{H \\ |}}{C}} = C = C = O \quad \xrightarrow{ROH} \quad H \cdot e \quad \overset{.n}{\underset{\substack{| \\ CH_3}}{C}} = C = C = O \quad \longrightarrow$$

$$+ \; ROH$$

$$\underset{\substack{| \\ CH_3}}{\overset{\substack{R \\ |}}{C}} = C = C = O \quad + \quad \mathbf{H_2O}$$

(A)   9.84

(A) above which is very stable has two mono-forms like that of Equation 9.78. From one of the so-called mono-forms such as (I) of Equation 9.78, the following relationship is valid.

$$O = C = \quad \gg \quad O = CH - \quad ; \quad\quad O = \quad \gg \quad HO -$$

<u>Radical-puling group</u>   <u>Radical-pushing group</u>   9.85

See Equation 9.16 as recalled below where the same pattern has been established.
Notice the unique qualities of these monomers.

When the R group in (A) of Equation 9.84 is CH_3, the followings are to be expected.

$$= O \quad >> \quad H_2C =$$
<u>Order of radical-pushing capacity</u> (9.16)

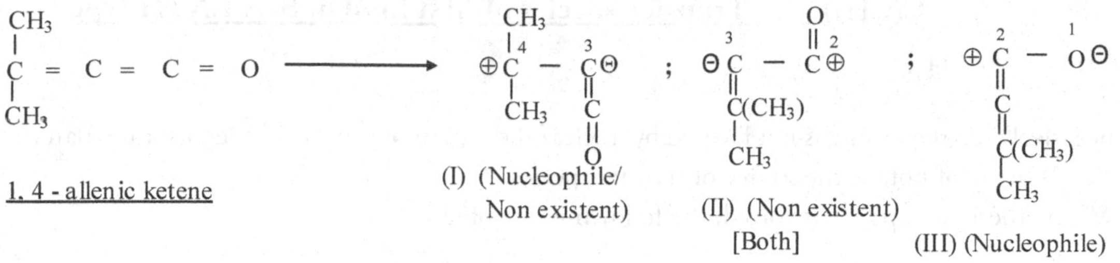

1, 4 - allenic ketene → (I) (Nucleophile/ Non existent) ; (II) (Non existent) [Both] ; (III) (Nucleophile)

9.86

(I) does not exist. (I) would have favored the positively charged route if not for electrostatic forces of repulsion. (II) is the first center, the Y center which favors both free-radical routes. This however is not a clear indication of the weak electrophilic character of the monomers though (III) the second mono-form is a full Nucleophile, since transfer species of the seventh type has been created with the presence of CH_3 group in place of H. Using (I) exploratively, free positively charged routes will be involved as follows. Non-free positive charges cannot be used, since the equation will not be chargedly balanced. Same will apply radically.

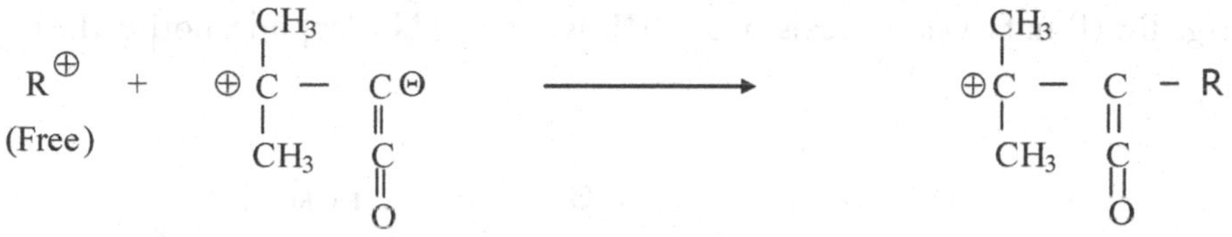

9.87

For (II) which indeed is the first mono-form, free-negative charges can be used as shown below.

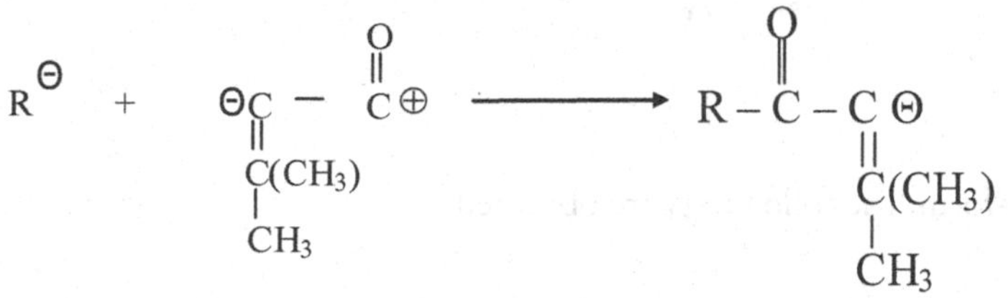

9.88

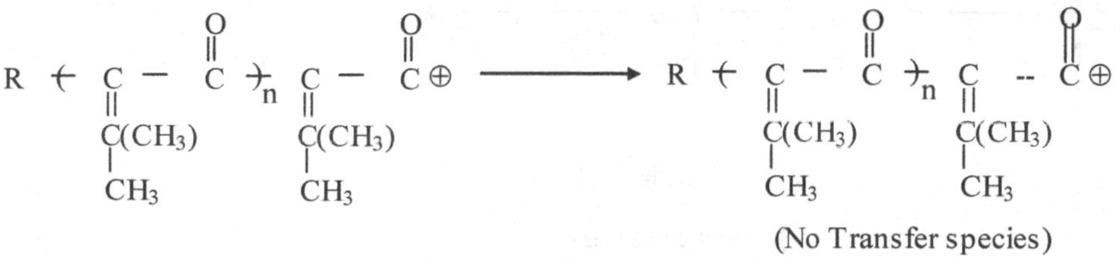

(No Transfer species) 9.89

For (III) of Equation 9.86, the second mono-form which is X, the following is obtained.

$$R^{\ominus} \ + \ \oplus\overset{\displaystyle |}{\underset{\displaystyle |}{C}}-O\ominus \longrightarrow RH \ + \ \ominus\overset{\displaystyle H}{\underset{\displaystyle H}{C}}-\overset{\displaystyle CH_3}{C}=C=C=O$$

with the left carbon bearing:
$$\begin{array}{c} C \\ \| \\ C(CH_3) \\ | \\ H_3C \end{array}$$

Transfer species of first kind of SEVENTH type   9.90

The negatively charged route is not favored by it, clear indication of a stronger Electrophile than the first member. Worthy of note is the source of transfer species.

When etheric groups are involved, the followings are obtained.

$$\overset{\displaystyle CH_3}{\underset{\displaystyle |}{C}}=C=C=O \longrightarrow \oplus C - C^{\ominus} \ ; \ \ominus C - C^{\oplus} \ ; \ \oplus C - O\ominus$$

(with substituents O–CH$_3$, C=O, C(CH$_3$), O, CH$_3$ groups as drawn)

(I) (Non-existent)   (II) (Non-existent-Both) (III) (Nucleophile)   9.91

The "charges" carried by the active centers are similar to the cases considered so far. (I) and (II) will exist radically, but not chargedly. (I) in fact cannot exist at all. (III) is a strong Nucleophile, noting the transfer species involved.

$$R^{\ominus} \ + \ \oplus\overset{\displaystyle |}{\underset{\displaystyle |}{C}}-O\ominus \ \rightleftharpoons \ \overset{\displaystyle |}{\underset{\displaystyle |}{C}}-O\ominus \ + \ RCH_3$$

Free   (with C(CH₃), O, CH₃ substituents)   (with C(CH₃), O substituents)

(III)   9.92

With (II) which is the first mono-form, the followings are obtained.

$$\ominus C - C^{\oplus} \longrightarrow \oplus\overset{\displaystyle CH_3}{C}=C^{\ominus} \xrightarrow{\ + \ R^{\oplus}}$$

(II)   (V) (more stable)

$$ROCH_3 \ + \ O=C=C=\overset{\displaystyle CH3}{C^{\oplus}}$$ 9.93

727

Because of the nature of (II) above wherein the OCH_3 is a transfer species, one will almost be mistakenly led to conclude that (II) is not one of the mono-forms. However, it is the first center to be activated, which means that no route will be favored by this monomer when allowed to undergo molecular rearrangement only radically and not chargedly as shown above. In fact, when (V) above is deactivated, the $C\equiv C$ cannot be easily activated, being more nucleophilic than the $C=O$ center. ***From all the considerations so far, it is obvious that $O = C = C = C =$ group (ODD) is a radical-pulling group more so than $O = C =$ group (ODD), while $O =$ group is a radical-pushing group.*** None but the first can provide resonance stabilization as will shortly be shown.

(III) is said to be a nucleophile, since it will not favor the free and non-free-negatively-charged routes. This was shown in Equation 9.92 for the free negatively charged one. Therein the transfer species involved was clearly identified done via Equilibrium mechanism, in order for the equation to be chargedly balanced and not as shown below because OCH_3 is far more radical-pushing than CH_3 group.

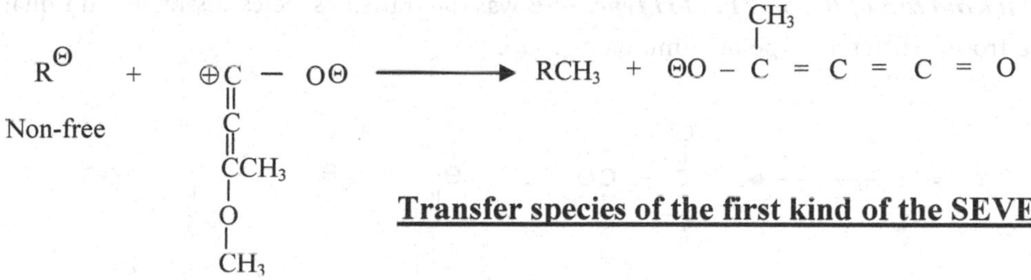

$$R^{\ominus} \quad + \quad \underset{\substack{\| \\ C \\ \| \\ CCH_3 \\ | \\ O \\ | \\ CH_3}}{\oplus C} - O^{\ominus} \longrightarrow RCH_3 \;+\; \ominus O - \overset{\overset{\displaystyle CH_3}{|}}{C} = C = C = O$$

Non-free

<u>Transfer species of the first kind of the SEVENTH type</u>

$$9.94$$

For the non-free-negatively charged one, the mechanism is Combination since the equation becomes chargedly balanced as shown below. The same transfer species is still involved. All the above can only take place radically in the same pattern, but not chargedly due to electrostatic forces of repulsion.

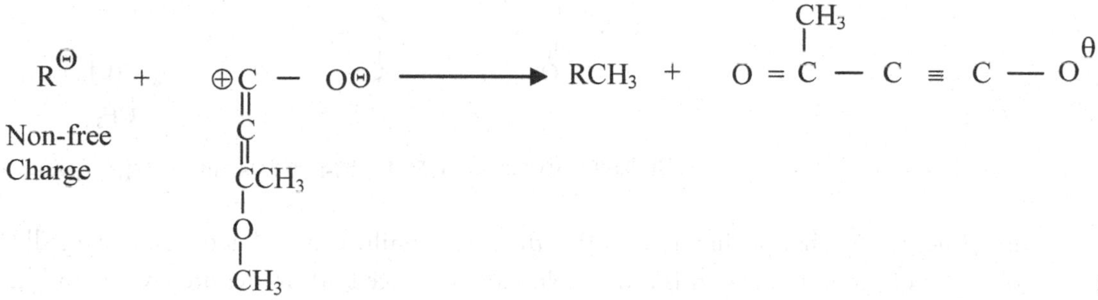

$$R^{\ominus} \quad + \quad \underset{\substack{\| \\ C \\ \| \\ CCH_3 \\ | \\ O \\ | \\ CH_3}}{\oplus C} - O^{\ominus} \longrightarrow RCH_3 \;+\; O = \overset{\overset{\displaystyle CH_3}{|}}{C} - C \equiv C - O^{\theta}$$

**Non-free
Charge**

$$9.95$$

In view of the laws which have been stated, it is CH_3 from OCH_3 that is the transfer species and not from the CH_3 group or the group itself. If it was the group itself, i.e., the other CH_3 group, then the growing chain cannot be killed from within as shown below.

$$R \left(O - \underset{\substack{\| \\ C \\ \| \\ \underset{|}{\underset{O}{CCH_3}} \\ | \\ CH_3}}{C} \right)_n O - \underset{\substack{\| \\ C \\ \| \\ \underset{|}{\underset{O}{CCH_3}} \\ | \\ CH_3}}{C} \oplus \longrightarrow R(-O-\underset{\substack{\| \\ C \\ \| \\ \underset{|}{\underset{O}{CCH_3}} \\ | \\ CH_3}}{C})_n - O - C \equiv C - \underset{\substack{CH_3 \\ |}}{C} = O$$

$$+ \quad H_3C^{\oplus} \qquad\qquad 9.96$$

The only favored route by the $C = O$ center is only the positively charged one and the transfer species involved is only from the rich $-OCH_3$ group and not CH_3 group. The transfer species involved still remains of the *first kind but of the SEVENTH type*. This was the transfer species abstracted in Equation 9.95. It is coming from a different type of cumunylic group.

$$\underset{\substack{CH_3 \\ | }}{C} = C = C = O \longrightarrow \oplus \overset{CH_3}{\underset{CF_3}{C}} - \underset{\substack{\| \\ C \\ \| \\ O}}{C} \ominus \quad ; \quad \ominus C - \underset{\substack{\| \\ C(CF_3) \\ | \\ CH_3}}{\overset{O}{C}} \oplus \quad ; \quad \oplus C - \underset{\substack{\| \\ C(CH_3) \\ | \\ CF_3}}{C} O \ominus$$

$$\text{(I) (Nucleophile)} \qquad \text{(II) (Electrophile)} \qquad \text{(III) (Nucleophile)} \qquad 9.97$$

$$\underset{\substack{CF_3 \\ | }}{C} = C = C = O \longrightarrow \oplus \overset{CF_3}{\underset{CF_3}{C}} - \underset{\substack{\| \\ C \\ \| \\ O}}{C} \ominus \quad ; \quad \ominus C - \underset{\substack{\| \\ C(CF_3) \\ | \\ CF_3}}{\overset{O}{C}} \oplus \quad ; \quad \oplus C - \underset{\substack{\| \\ C(CF_3) \\ | \\ CF_3}}{C} O \ominus$$

$$\text{(I) (Both)-non existent} \qquad \text{(II) (Both)-non esistent} \qquad \text{(III) (Both)} \quad 9.98$$

With the use of negatively charged initiators, (II) is the Electrophilic center if activation chargedly had been possible. (III) of the last equation is a weak Nucleophile since both routes are favored unlike the first case above.

$$R \left(\underset{\substack{\| \\ C \\ \| \\ \underset{|}{\underset{CF_3}{C(CF_3)}}}}{C} - O \right)_n \underset{\substack{\| \\ C \\ \| \\ \underset{|}{\underset{CF_3}{C(CF_3)}}}}{C} - O \ominus \longrightarrow \text{No Transfer species}$$

(non-free growing chain) $\qquad\qquad 9.99$

While the case above has no transfer species, (III) of Equation 9.97 is a stronger Nucleophile, because it has *transfer species of the first kind of the seventh type.*

$$
\begin{array}{c}
CH_3 \\
| \\
C = C = C = O \\
| \\
C = O \\
| \\
O \\
| \\
CH_3
\end{array}
\longrightarrow
\overset{\oplus}{C} \overset{\begin{array}{c}O \\ \| \\ C\end{array}}{\underset{\begin{array}{c}CH_3 \\ | \\ C=O \\ | \\ O \\ | \\ CH_3\end{array}}{-}} C^{\ominus}
\quad ; \quad
\overset{\ominus}{C} - \overset{\begin{array}{c}O \\ \| \\ \end{array}}{C}{}^{\oplus}
\quad ; \quad
\overset{\oplus}{C} - O^{\ominus}
$$

(I) (Nucleophile)   (II) (Electrophile)   (III) (Nucleophile) 9.100

$$
\begin{array}{c}
CH_3 \\
| \\
C = C = C = O \\
| \\
C = O \\
| \\
H
\end{array}
\longrightarrow
\overset{\oplus}{C} - C^{\ominus}
\quad ; \quad
\overset{\ominus}{C} - C^{\oplus}
\quad ; \quad
\overset{\oplus}{C} - O^{\ominus}
$$

(I) (Nucleophile)   (II) (Electrophile)   (III) (Nucleophile) 9.101

The routes favored by them have been indicated by their natural characters for which the last case is less Electrophilic. ***Worthy of note here is that the Nucleophilic center is (III) and not the C = O center externally located being more nucleophilic than the cumulatively placed one.*** With respect to resonance stabilization, these will be visited radically very shortly.

Thus, in general, there are ***two mono-forms for 1,4- cumulenic ketenes***. The resonance stabilization believed to take place from the nucleophilic center never indeed takes place. In fact, none as can be seen so far can be activated chargedly. With them, there are two types of activation centers full free-charged (II) and half free-charged (III) types. With alkyl vinyl ketones or methyl acrylates etc., there are also the two types of activation centers. These are also not resonance stabilized, since they are of two different characters as recalled below.

$$
\begin{array}{c}
H \quad\quad H \\
| \quad\quad\quad | \\
C = C \\
| \quad\quad\quad | \\
CH_3 \quad C=O \\
\quad\quad\quad | \\
\quad\quad\quad C_2H_5
\end{array}
\quad ; \quad
\overset{\oplus}{C_4} - \underset{3}{C} = \underset{2}{C} - O^{\ominus}
\quad\longleftrightarrow\!\!\!\!/\!\!\!\!\longrightarrow\quad
\overset{\oplus}{C_2} - O^{\ominus}_{1}
$$

A vinyl ketone   (I) 1,4 - Impossible existence   (II) 1,2 -Nucleophile

Non - resonance stabilized   9.102

730

9.2.2. Radical characters of 1,4-cumulenic ketenes

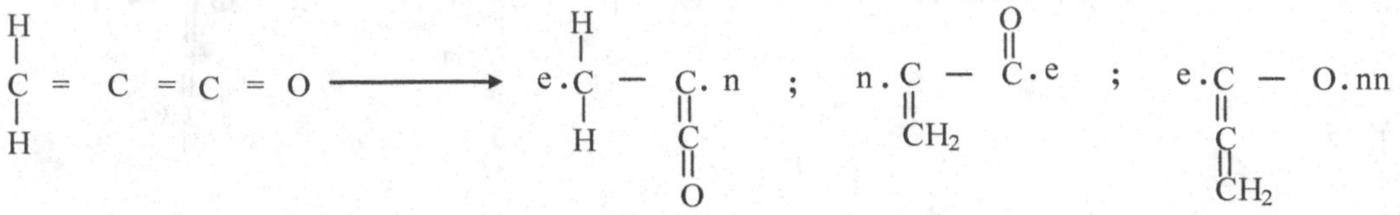

$$\begin{matrix} H \\ | \\ C \\ | \\ H \end{matrix} = C = C = O \longrightarrow \quad e \cdot \overset{H}{\underset{\underset{O}{\overset{||}{C}}}{\underset{|}{C}}} - \overset{}{\underset{}{C}} \cdot n \quad ; \quad n \cdot \overset{O}{\underset{\underset{CH_2}{||}}{C}} - \overset{}{\underset{}{C}} \cdot e \quad ; \quad e \cdot \overset{}{\underset{\underset{CH_2}{\overset{C}{||}}}{\underset{}{C}}} - O \cdot nn$$

$$(I)\ Both \qquad\qquad (II)\ Both\ \ Y \qquad (III)\ Both\ \ X \qquad 9.103$$

Like the same charged case, the O = and O = C = groups can still be observed to be two different groups, the former radical-pushing while the latter is radical-pulling. However, since both routes are favored by the two mono-forms, it cannot be said to be a strong Electrophile. Even when one H atom is replaced with radical-pushing group, its Electrophilic character is further reduced; unless when the groups are trans-placed. Hence, these monomers have never been popularly known. However, based on the C/H ratio, the need to change one or two H atoms is inevitable being highly unstable.

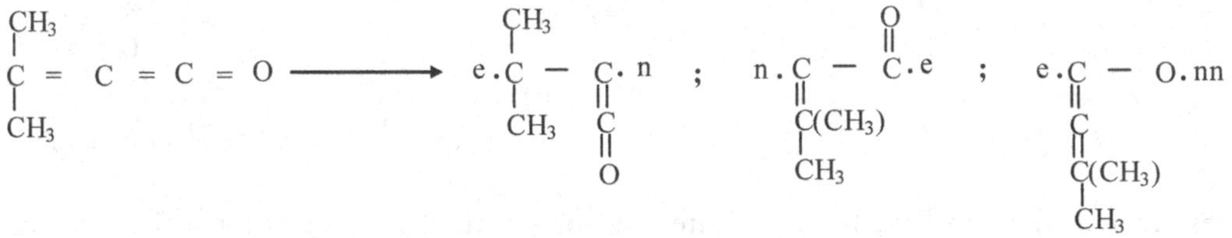

$$\begin{matrix} CH_3 \\ | \\ C \\ | \\ CH_3 \end{matrix} = C = C = O \longrightarrow \quad e \cdot \overset{CH_3}{\underset{\underset{O}{\overset{C}{||}}}{\underset{|}{C}}} - \overset{}{\underset{}{C}} \cdot n \quad ; \quad n \cdot \overset{O}{\underset{\underset{CH_3}{|}}{\underset{\underset{}{C(CH_3)}}{\overset{||}{C}}}} - \overset{}{\underset{}{C}} \cdot e \quad ; \quad e \cdot \overset{}{\underset{\underset{\underset{CH_3}{|}}{\underset{C(CH_3)}{\overset{C}{||}}}}{\underset{}{C}}} - O \cdot nn$$

$$(I)\ (Nucleophile) \qquad (II)\ (Both) \qquad (III)\ (Nucleophile) \qquad 9.104$$

With the introduction of CH_3 group, the routes favored can clearly be identified. (I) is nucleophilic and not one of the mono-forms. (II) is the Y center being the least nucleophilic. (III) unlike (I) has transfer species of the *first kind of the seventh type* as shown below.

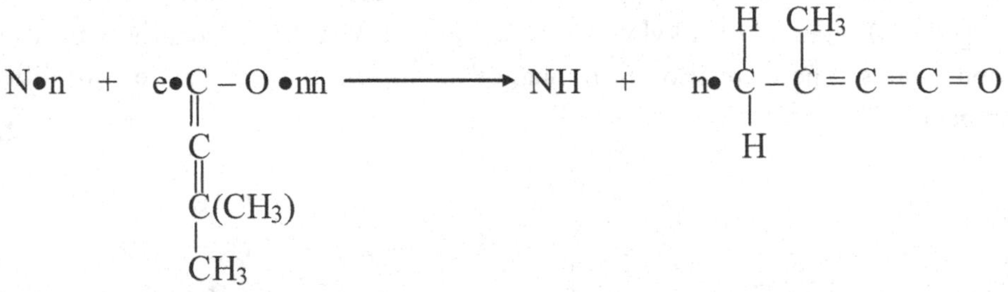

$$N \bullet n \ + \ e \bullet \overset{}{\underset{\underset{\underset{CH_3}{|}}{\underset{C(CH_3)}{\overset{C}{||}}}}{\underset{}{C}}} - O \bullet nn \longrightarrow NH \ + \ n \bullet \overset{H \quad CH_3}{\underset{\underset{H}{|}}{\underset{|\quad\ |}{C}}} - C = C = C = O$$

TRANSFER SPECIES OF THE 1st KIND OF 7th TYPE 9.105

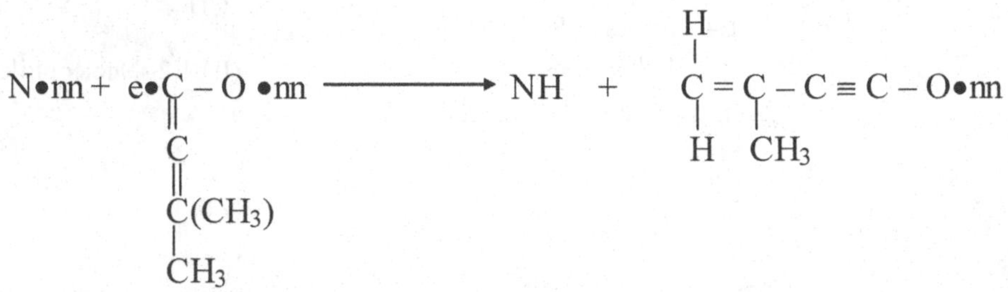

$$N \bullet nn + \ e \bullet \overset{}{\underset{\underset{\underset{CH_3}{|}}{\underset{C(CH_3)}{\overset{C}{||}}}}{\underset{}{C}}} - O \bullet nn \longrightarrow NH \ + \ \overset{H}{\underset{\underset{H\quad CH_3}{|\quad\ |}}{\underset{|}{C}}} = C - C \equiv C - O \bullet nn$$

TRANSFER SPECIES OF IST KIND OF THE SEVENTH TYPE 9.106

$$E\bullet e + (n+1)\ e\bullet \underset{\underset{\underset{CH_3}{|}}{\overset{\overset{C(CH_3)}{||}}{\overset{\overset{C}{||}}{C}}}{C} - O\ \bullet nn \longrightarrow E\text{-}(\curvearrowright)_n - O - C \equiv C - \underset{\underset{CH_3}{|}}{C} = \underset{\underset{H}{|}}{\overset{\overset{H}{|}}{C}} + H\bullet e$$

9.107

It is the last equation above that is possible and favored. For the first time, one can observe the existence of a Strong Electrophile (1,4-Cumulenic Ketenes) in view of the strong character- (II) of Equation 9.104. The Nucleophilic center carried by this electrophile is a strong female.

With etheric groups, the followings are to be expected.

$$\underset{\underset{CH_3}{|}}{\overset{\overset{CH_3}{|}}{C}} = C = C = O \longrightarrow e.\underset{\underset{CH_3}{|}}{\overset{\overset{CH_3}{|}}{C}} - \overset{\overset{O}{||}}{C}. n \quad ; \quad n.\underset{\underset{\underset{CH_3}{|}}{\overset{\overset{O}{|}}{C(CH_3)}}}{\overset{\overset{O}{||}}{C}} - C.e \quad ; \quad e.\underset{\underset{\underset{CH_3}{|}}{\overset{\overset{O}{|}}{C(CH_3)}}}{\overset{\overset{O}{||}}{C}} - O.nn$$

(I) (Nucleophile) (II) (Electrophile) (III) (Nucleophile) 9.108

(I) will favor polymerization as a Nucleophile. (II) is the Y center. (III) favors only the electro-free-radical route. Hence, the Electrophile is a weak Electrophile. Activation will always start from (II).

$$\underset{\underset{F}{|}}{\overset{\overset{F}{|}}{C}} = C = C = O \longrightarrow e.\underset{\underset{F}{|}}{\overset{\overset{F}{|}}{C}} - \overset{\overset{O}{||}}{C}. n \quad ; \quad n.\overset{\overset{O}{||}}{\underset{\underset{CF_2}{}}{C}} - C.e \quad ; \quad e.\overset{\overset{O}{||}}{\underset{\underset{CF_2}{}}{C}} - O.nn$$
(A)

(I) (Both) (II) (Electrophile) (III) (Both) 9.109

One can observe the unique character of the fluorinated monomer here. F is transfer species for only (II). This is the nature of the logical network of chemical compounds.

$$\underset{\underset{\underset{CH_3}{|}}{\overset{\overset{C-O}{|}}{O}}}{\overset{\overset{CH_3}{|}}{\underset{4}{C}}} = \underset{3}{C} = \underset{2}{C} = \underset{1}{O} \longrightarrow e.\underset{\underset{\underset{CH_3}{|}}{\overset{\overset{C=O}{|}}{O}}}{\overset{\overset{CH_3}{|}}{C}} - \overset{\overset{O}{||}}{C}. n \quad ; \quad n.\underset{\underset{\underset{CH_3}{|}}{\overset{\overset{CCH_3}{|}}{O}}}{\overset{\overset{O}{||}}{C}} - C.e \quad ; \quad e.\underset{\underset{\underset{CH_3}{|}}{\overset{\overset{CCH_3}{|}}{O}}}{\overset{\overset{O}{||}}{C}} - O.nn$$

(I) (Nucleophile) (II) (Electrophile) (III) (Nucleophile) 9.110a

Imagine if (II) which is the first mono-form, is made to undergo "molecular rearrangement", then the followings would have been obtained.

$$n\bullet \overset{\overset{\displaystyle O}{\|}}{C} - \overset{\overset{\displaystyle CCH_3}{|}}{\underset{\underset{\underset{\underset{CH_3}{|}}{O}}{C=O}}{C}} \bullet e \longrightarrow e\bullet \overset{\overset{\displaystyle O}{\|}}{C} - \overset{\overset{\displaystyle CH_3}{|}}{\underset{\underset{\underset{\underset{CH_3}{|}}{O}}{C=O}}{C}} = C\bullet n \longrightarrow \overset{\overset{\displaystyle C=O}{}}{\underset{\underset{\underset{\underset{CH_3}{|}}{O}}{C=O}}{\overset{|}{\underset{|}{C}} - CH_3}}$$

$$e\bullet \overset{}{\underset{\underset{\underset{\underset{CH_3}{|}}{O}}{C=O}}{C}} \bullet n$$

9.110b

In the equation above, OCH_3 is a transfer species, noting that the molecular rearrangement above is indeed possible. The group was carried from $COOCH_3$ group which is the real transfer species provider here. However, worthy of note at this point in time is that $-CH=C=O$ group unlike $-COOCH_3$ group is a free-radical-pushing group, while $-CH=C=C=O$ is a radical-pulling group with the following capacity-

ODD: $O = C = C = C = $ $> $ $O = C = $ $>>$

$$-\underset{R}{\overset{|}{C}} = C = C = O \quad > \quad -\underset{H}{\overset{|}{C}} = C = C = O \quad > \quad -\underset{R}{\overset{|}{C}} = O \quad > \quad -\underset{H}{\overset{|}{C}} = O$$

<u>**Order of radical-pulling capacity**</u>   9.111a

EVEN: $O = C = C = C = C =$ $>$ $O = C = C =$ $>>$

$$-\underset{R}{\overset{|}{C}} = C = C = C = O \quad > \quad -\underset{R}{\overset{|}{C}} = C = O$$

<u>**Order of Radical-pushing capacity**</u>   9.111b

$$O = \quad > \quad H_2C = C = \quad > \quad H_2C =$$

<u>**Order of Radical-pushing capacity**</u>   9.111c

The Odd and Even has to do with the numbers of C atoms carried by the O center. One can observe that 1,4-cumulenic ketene has one Electrophilic center (Y) - 2, 3- mono-form regardless of the types of groups carried by the 4-carbon center. But for the case of Equation 9.110a, the situation may be different. (I) can be further resonance stabilized in the same manner as in benzoquinone as follows-

$$e.\overset{\overset{\displaystyle CH_3}{|}}{\underset{\underset{\underset{\underset{CH_3}{|}}{O}}{C=O}}{C}} - \overset{\overset{}{}}{\underset{\underset{\underset{O}{\|}}{C}}{C}} .n \longleftrightarrow en. O - \overset{\overset{\displaystyle CH_3}{|}}{\underset{\underset{\underset{CH_3}{|}}{O}}{C}} = \overset{}{C} - \overset{\overset{}{}}{\underset{\underset{\underset{O}{\|}}{C}}{C}} .n \longleftrightarrow en. O - \overset{\overset{\displaystyle CH_3}{|}}{\underset{\underset{\underset{CH_3}{|}}{O}}{C}} = \overset{}{C} - C \equiv C - O .nn$$

         (A)   (B)

Full free-radical   **Full Non-free-radical**   9.112

Unfortunately however, (I) or (A) above does not exist as one of the mono-forms. With movement from full Free-radical state to Full Non-free-radical state, resonance stabilization like the case of Benzoquinone would have been favored here, noting the presence of one Y and two Xs of different capacities. It is (B) that is the most stable and that will largely be used only non-free-radically preferably for copolymerization, in view of vulnerable -O-O- bond along the main chain backbone. The same as above would have applied to COR, $CONH_2$ types of groups, but not $C \equiv N$ if they had existed. All these are being done for exploratory purposes

For (III) of Equation 9.109, one knows what to expect when the Fs are replaced with $OCOCH_3$ groups (R_F). It is a little different particularly for the Nucleophilic center as shown below.

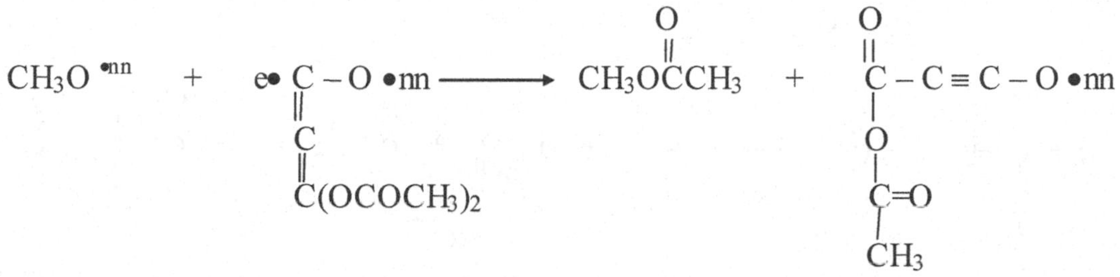

$$CH_3O \ ^{\bullet nn} \ + \ e\bullet \underset{\underset{C(OCOCH_3)_2}{\overset{\|}{C}}}{\overset{\|}{\underset{\|}{C}}}-O \bullet nn \longrightarrow CH_3O\overset{O}{\overset{\|}{C}}CH_3 \ + \ \underset{\underset{CH_3}{\overset{\|}{C=O}}}{\overset{\overset{O}{\overset{\|}{C}}}{\underset{|}{C}}}-C \equiv C-O \bullet nn$$

$$9.113$$

This indeed is Nucleophilic. The transfer species involved is of the first kind of the seventh type already identified in Equation 9.106.

With CF_3 group replacing H or F, the followings are to be expected.

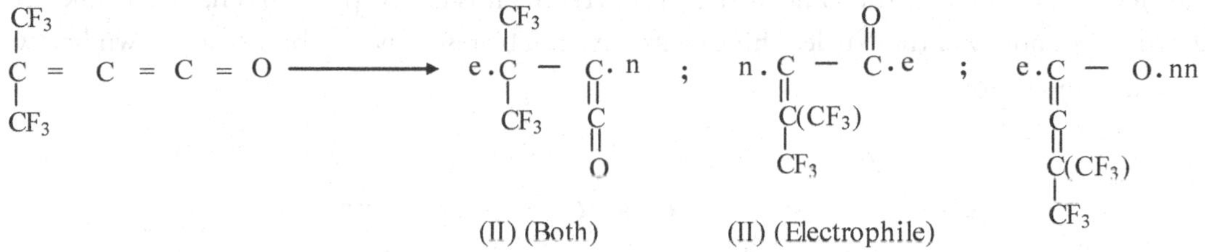

$$\underset{CF_3}{\overset{CF_3}{\overset{|}{C}}} = C = C = O \longrightarrow e\cdot\underset{CF_3}{\overset{CF_3}{\overset{|}{\underset{|}{C}}}}-\underset{\overset{\|}{O}}{\overset{\|}{C}}\cdot n \ ; \ n\cdot\underset{\overset{C(CF_3)}{\underset{CF_3}{|}}}{\overset{\|}{C}}-\overset{O}{\overset{\|}{C}}\cdot e \ ; \ e\cdot\underset{\overset{C(CF_3)}{\underset{CF_3}{|}}}{\overset{\|}{C}}-O\cdot nn$$

$$\text{(II) (Both)} \qquad \text{(II) (Electrophile)} \qquad \text{(III) (Both)} \qquad 9.114$$

Based on the experiences gathered so far, one can see the order of capacity of the Electrophilic centers, the type of Nucleophilic center, the absence of Resonance stabilization from the Electrophilic center and more. Once the characters of the groups carried are known, one begins to observe ***complex logical network*** of chemical systems. Mathematics is not the one but CHEMISTRY. CHEMISTRY IS LOGIC. Imagine what to expect when CF_3 is replaced with a higher group such as C_2F_5 group.

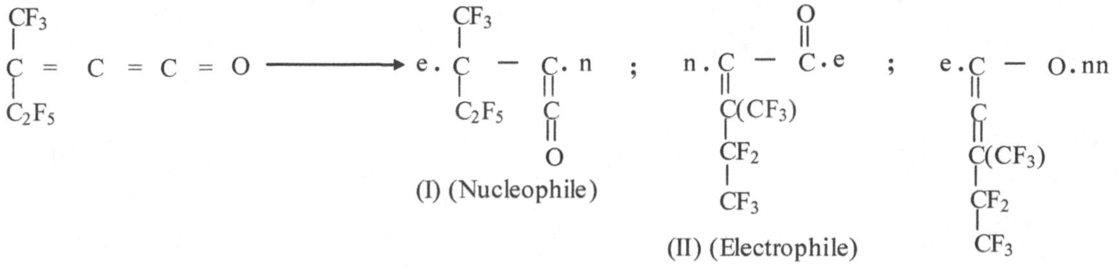

$$\underset{C_2F_5}{\overset{CF_3}{\overset{|}{C}}} = C = C = O \longrightarrow e\cdot\underset{C_2F_5}{\overset{CF_3}{\overset{|}{\underset{|}{C}}}}-\underset{\overset{\|}{O}}{\overset{\|}{C}}\cdot n \ ; \ n\cdot\underset{\overset{C(CF_3)}{\underset{\overset{CF_2}{\underset{CF_3}{|}}}{|}}}{\overset{\|}{C}}-\overset{O}{\overset{\|}{C}}\cdot e \ ; \ e\cdot\underset{\overset{C(CF_3)}{\underset{\overset{CF_2}{\underset{CF_3}{|}}}{|}}}{\overset{\|}{C}}-O\cdot nn$$

$$\text{(I) (Nucleophile)} \qquad \text{(II) (Electrophile)} \qquad \text{(III) (Nucleophile)} \qquad 9.115$$

(I) is nucleophilic. So also is (III) nucleophilic and that is (X), since nucleo-non-free-radically, it cannot be polymerized as shown below.

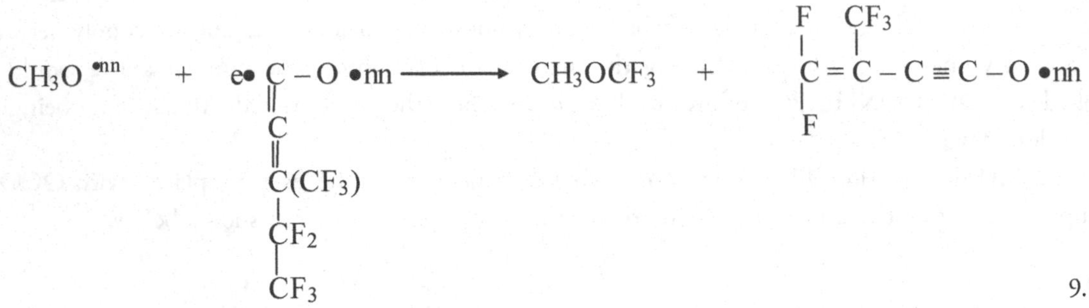

9.116

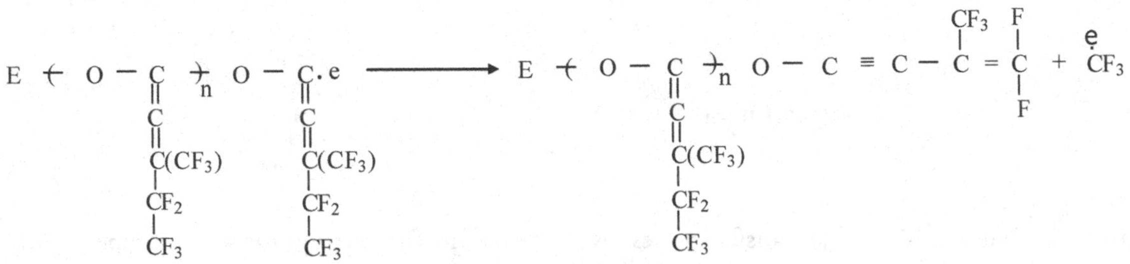

<u>Transfer species of first kind of the seventh type type</u>

9.117

The transfer species above is the same as that from (I) which is electrophilic in character. Like all cases considered so far, not even the Nucleophilic center (X) can be resonance stabilized as shown below along with the 4,3 – mono-form.

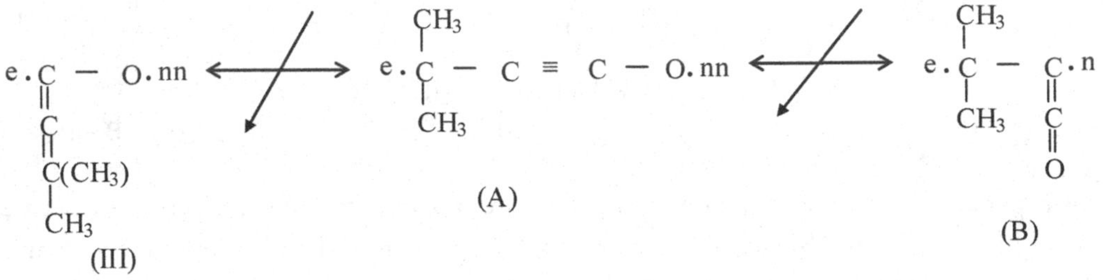

9.118

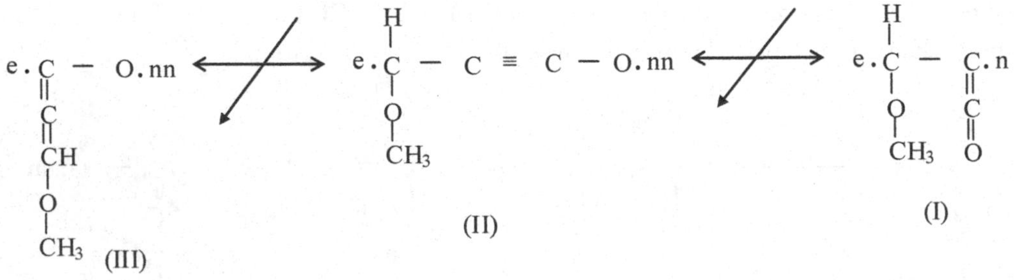

9.119

$$e \cdot \underset{\underset{C(R_F)_2}{\overset{\|}{C}}}{\overset{\|}{C}} - O.nn \quad \xleftrightarrow{\quad\not\quad} \quad e \cdot \underset{R_F}{\overset{R_F}{\underset{|}{\overset{|}{C}}}} - C \equiv C - O.nn \quad \xleftrightarrow{\quad\not\quad} \quad e \cdot \underset{R_F}{\overset{R_F}{\underset{|}{\overset{|}{C}}}} - \underset{\underset{O}{\overset{\|}{C}}}{\overset{\|}{C}}.n$$

$$(III) \qquad\qquad (II) \qquad\qquad (I) \qquad 9.120$$

$$e \cdot \underset{\underset{\underset{\underset{CH_3}{|}}{O}}{\underset{\underset{\|}{C=O}}{\overset{\|}{CH}}}}{\overset{\|}{C}} - O.nn \quad \xleftrightarrow{\quad\not\quad} \quad e \cdot \underset{\underset{\underset{CH_3}{|}}{\underset{O}{C=O}}}{\overset{H}{\underset{|}{C}}} - C \equiv C - O.nn \quad \xleftrightarrow{\quad\not\quad} \quad e \cdot \underset{\underset{\underset{CH_3}{|}}{\underset{O}{C=O}}}{\overset{H}{\underset{|}{C}}} - \overset{\overset{O}{\|}}{\underset{\|}{C}}.n$$

$$(III) \qquad\qquad (II) \qquad\qquad (I) \qquad 9.121$$

$$e \cdot \underset{\underset{\underset{CF_3}{|}}{\overset{\|}{C(CF_3)}}}{\overset{\|}{C}} - O.nn \quad \xleftrightarrow{\quad\not\quad} \quad e \cdot \underset{CF_3}{\overset{CF_3}{\underset{|}{\overset{|}{C}}}} - C \equiv C - O.nn \quad \xleftrightarrow{\quad\not\quad} \quad e \cdot \underset{CF_3}{\overset{CF_3}{\underset{|}{\overset{|}{C}}}} - \underset{\underset{O}{\overset{\|}{C}}}{\overset{}{C}}.n$$

$$(III) \qquad\qquad (II) \qquad\qquad (I) \qquad 9.122$$

$$e \cdot \underset{\underset{\underset{\underset{CF_3}{|}}{\underset{CF_2}{|}}}{\overset{\|}{C(CF_3)}}}{\overset{\|}{C}} - O.nn \quad \xleftrightarrow{\quad\not\quad} \quad e \cdot \underset{C_2F_5}{\overset{CF_3}{\underset{|}{\overset{|}{C}}}} - C = C - O.nn \quad \xleftrightarrow{\quad\not\quad} \quad e \cdot \underset{C_2F_5}{\overset{CF_3}{\underset{|}{\overset{|}{C}}}} - \underset{\underset{O}{\overset{\|}{C}}}{\overset{}{C}}.n$$

$$(III) \qquad\qquad (II) \qquad\qquad (I) \qquad 9.123$$

It may seem as if for the first time one is seeing an Electrophile being resonance stabilized from the X center. This is not possible. At this point in time, it is important to note that when resonance stabilization is to be provided for any compound, one begins all the time from the least nucleophilic center and never from any center. Movement of electro-radicals begins all the time from the least nucleophilic center.

9.2.3. External resonance stabilization for 1,4-cumulenic ketenes

Note that external resonance stabilizations can be provided for a system, which originally is not resonance stabilized. This is only possible radically via the Electrophilic center if the monomer is an Electrophile. Handling the alkenyl group as an alkanyl group, the followings are obtained.

$$CH_3\\ |\\ C = C = C = O\\ |\\ CH\\ ||\\ CH_2$$
$$\longrightarrow$$
$$e.\ \underset{\underset{CH_2}{\overset{\overset{CH_3}{|}}{|}}}{C} - \underset{\overset{||}{O}}{\overset{||}{C}}\ _{.n}$$
;
$$n.\underset{\underset{CH_2}{\overset{\overset{||}{C(CH_3)}}{||}}}{C} - \overset{\overset{O}{||}}{C}._{e}$$
;
$$e.\ \underset{\underset{CH_2}{\overset{\overset{CCH_3}{||}}{|}}}{C} - O._{nn}$$

(I) (Nucleophile)   (II) (Not existent)   (III) (Nucleophile)   9.124

Since indeed, the alkenyl group is carrying an activation center which is the least nucleophilic, the followings are obtained.

$$e{\bullet}\underset{\underset{\overset{\overset{C}{||}}{\underset{O}{||}}}{\overset{\overset{H}{|}}{C(CH_3)}}}{\overset{\overset{H}{|}}{C}} - \underset{\overset{|}{}}{C}{\bullet}n \longleftrightarrow e{\bullet}\underset{\overset{|}{H}}{\overset{\overset{H}{|}}{C}} - \overset{\overset{H}{|}}{C} = \overset{\overset{CH_3}{|}}{C} - \underset{\underset{O}{||}}{\overset{\overset{C}{||}}{C}}{\bullet}n$$

(B) Full free-radical

(A) Full free-radical   **RESONANCE STABILIZED**   9.125

Since the movement is from Full free-radical to Full-free-radical, (A) is resonance stabilized. Like 1,3-Ketene, there are three mono-forms. These are (A) above which is the Y center, its resonance stabilized mono-form (B) and (III) of Equation 9.124 the X center. *Notice that –C(R) = C = C = O is a radical-pulling group of greater capacity than –C(R) = O, both ODD; –C(R) = C = O and –C(R) = C = C = C = O both EVEN are radical-pushing groups.*

When the single H in the first member is replaced with groups, the same analysis applies. With OCH_3 type of group, the situation is the same with resonance stabilization provided only from the Electrophilic center where possible.

$$\underset{\underset{CH_2}{\overset{\overset{CH}{||}}{|}}}{\overset{\overset{O}{|}}{\underset{}{\overset{\overset{CH_3}{|}}{C}}}} = C = C = O$$
$$\longrightarrow$$
$$e.\ \underset{\overset{|}{H}}{\overset{\overset{H}{|}}{C}} - \underset{\underset{\overset{\overset{C}{||}}{O}}{\overset{\overset{C}{||}}{}}}{\overset{\overset{H}{|}}{C}}._{n}$$
;
$$e.\ \underset{\underset{CH_2}{\overset{\overset{CH}{||}}{|}}}{\overset{\overset{C(OCH_3)}{||}}{C}} - O._{nn}$$

(I) (Electrophile)   (III) (Nucleophile) 9.126

That (III) is a strong nucleophile is shown by the reaction below for the X center

737

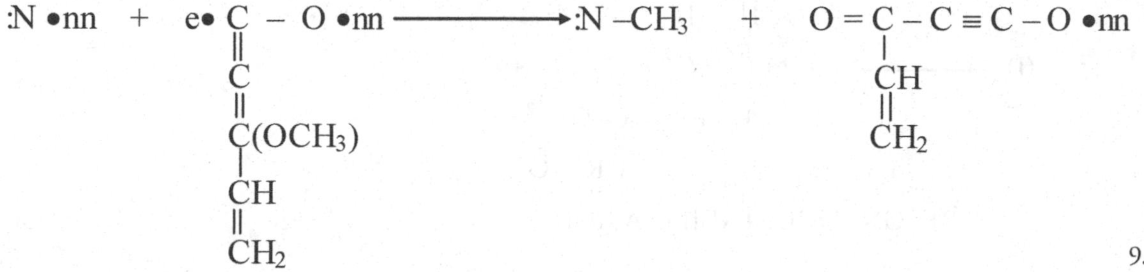

$$:N \bullet nn \quad + \quad e\bullet C - O \bullet nn \longrightarrow :N - CH_3 \quad + \quad O = C - C \equiv C - O \bullet nn$$

$$\begin{array}{c} \| \\ C \\ \| \\ C(OCH_3) \\ | \\ CH \\ \| \\ CH_2 \end{array} \qquad\qquad\qquad \begin{array}{c} | \\ CH \\ \| \\ CH_2 \end{array}$$

$$\text{9.127}$$

With (I), OCH_3 group is not a transfer species electro-free-radically, because as shown below, $- CX = C = C = O$ is a resonance stabilization group.

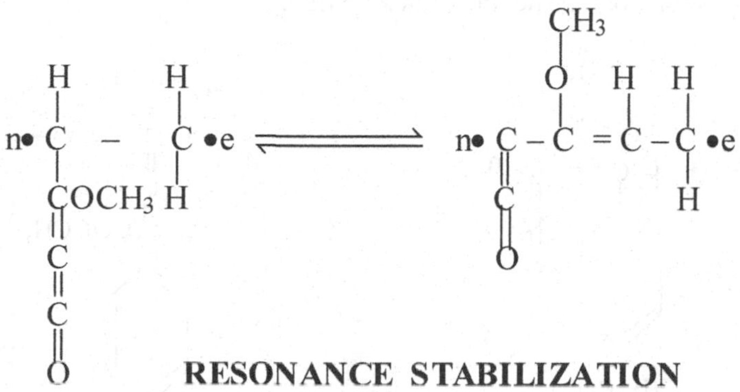

RESONANCE STABILIZATION 9.128

The transfer species of the first kind of the seventh type (CH_3) is involved in the first case, while in the second case transfer species of the first kind of the <u>first type</u> could not be abstracted, because they are resonance stabilized. It is of the <u>first type</u>, because it is coming from a group similar to COOR type of group, except that this can provide resonance stabilization here $-C(R) = C = C = O$ a diketenyl group. Evidently, the group is a resonance stabilization group of the radical-pulling type. The transfer species looks like the first type.

When groups such as COOR, COR, and CN are put in place of H, the followings are to be expected using COOR.

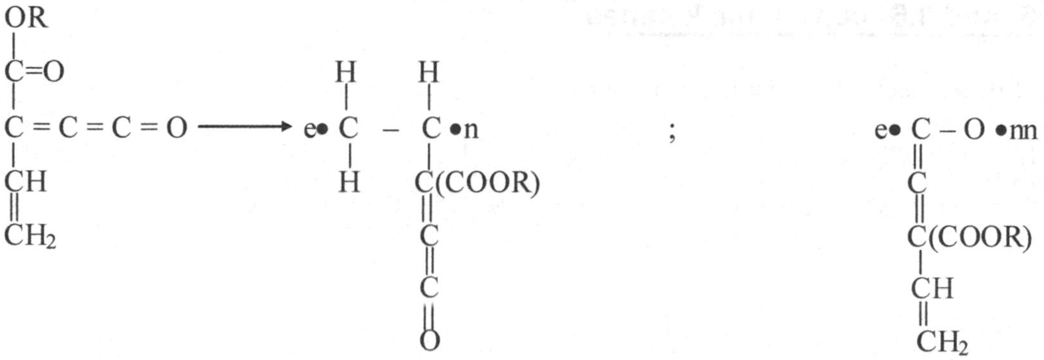

(I) Both   **(II) Nucleophile**   9.129

(I) can be resonance stabilized as shown below.

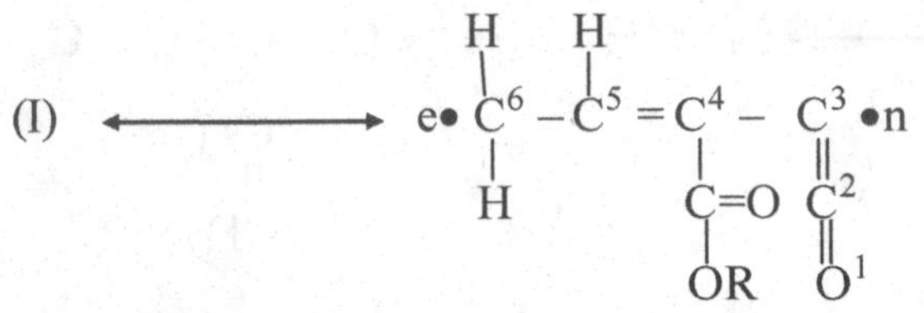

$$(I) \longleftrightarrow e\bullet \overset{\overset{H}{|}}{\underset{\underset{H}{|}}{C^6}} - \overset{\overset{H}{|}}{C^5} = C^4 - \overset{\overset{C=O}{||}}{\underset{OR}{}}\ \overset{\overset{C^2}{||}}{\underset{O^1}{}} \bullet n$$

RESONANCE STABILIZATION　　　　　　　　　　9.130

Notice the numbering of the centers. Originally, it was 1,2,3,4. Now it is to 6 in view of the presence of the alkenyl group. When a phenyl group is used, the numbering is further increased and the situation is completely different as has already been shown with 1,3-Ketene. It is only the phenyl group that can provide resonance stabilization if possible, being the most nucleophilic.

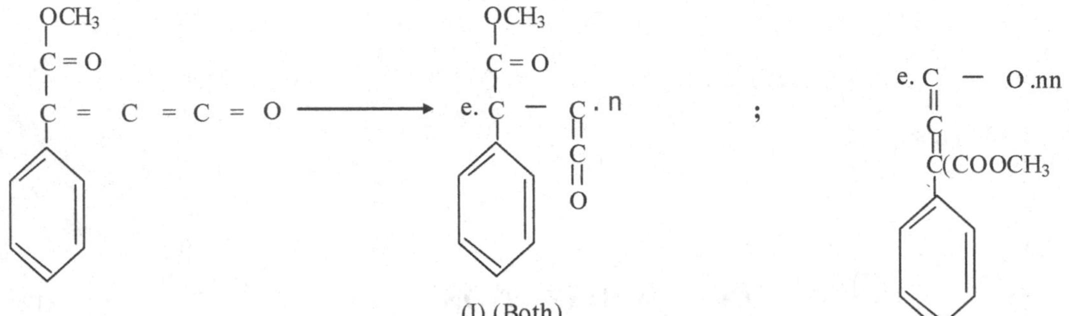

(I) (Both)　　　　　　　(III) Both　　9.131

One will think that if one had used CH_3 group in place of COOR above, because of the presence of transfer species H on CH_3 group, it cannot be abstracted, because it is shielded. As a matter of fact, (I) above is not the least nucleophilic center, and therefore does not exist. Since the least nucleophilic center is the Y center- the 2,3- mono-form, no resonance stabilization can be provided by the phenyl group. When electro-free-radicals are involved, the center most involved is the cumulatively placed C = O center (III) and not the one externally located.

9.3. 1,5- and 1,6- cumulenic ketenes

The first members of these unknown cases are-

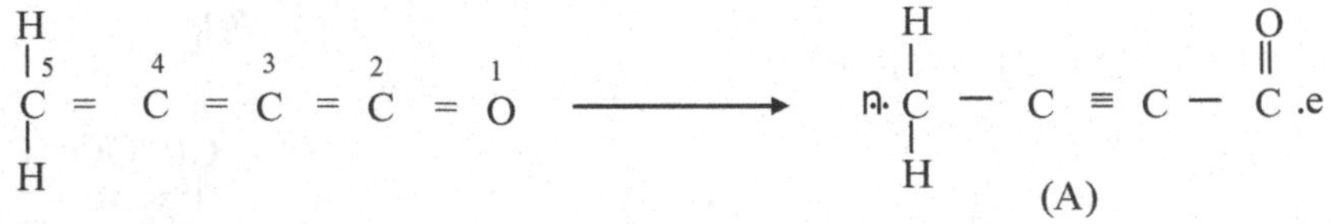

(A)　　　　　9.132

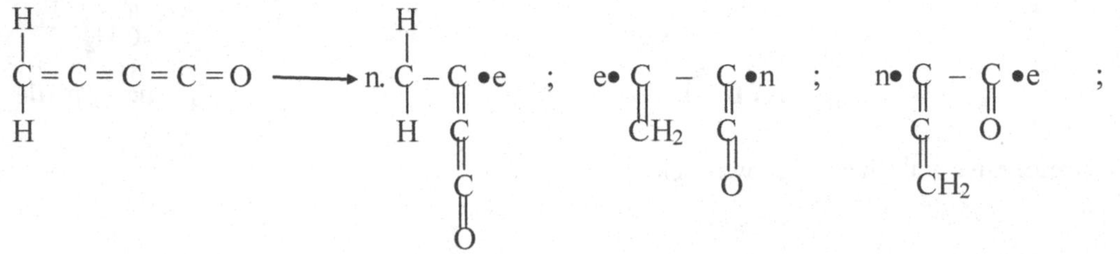

(I) Both　　　(II) Nucleophile　　　(III) Both, Y

739

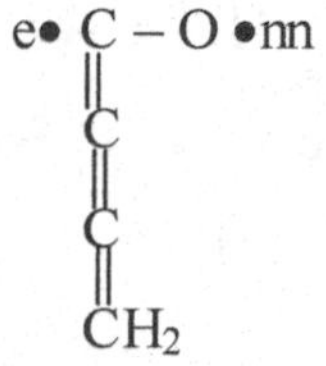

$$e\bullet \overset{\displaystyle C - O \bullet nn}{\underset{\underset{CH_2}{\overset{\|}{C}}}{\overset{\|}{C}}}$$

$$\text{(IV) Nucleophile, X} \qquad\qquad 9.133$$

(III) is the least Nucleophilic center and that center is the first Y though it favors both routes. As it seems, (III) like (I) is resonance stabilized as shown by the first equation above. (IV) which cannot be resonance stabilized is a complete Nucleophile. There are therefore three mono-forms here-(III), (A) from (III) and (IV), unlike the case with 1,3- and 1,4-Ketenes that have two mono-forms. The need to replace one of the two H atoms is however inevitable, based on the H/C ratio. All along it has been taken for granted that a passive catalyst to suppress its Equilibrium state of existence is present in the system.

$$\underset{\underset{H}{\overset{\|}{|}}}{\overset{\overset{H}{\overset{|}{}}}{C}} = \overset{5}{C} = \overset{4}{C} = \overset{3}{C} = \overset{2}{C} = \overset{1}{O} \qquad \longleftrightarrow \qquad n.\underset{\overset{\|}{CH_2}}{\overset{\|}{C}} - C \equiv C -- \overset{\overset{\displaystyle O}{\|}}{C}.e$$

$$\text{1,6-Ketene} \qquad\qquad\qquad\qquad\qquad \text{(B)} \qquad\qquad\qquad 9.134$$

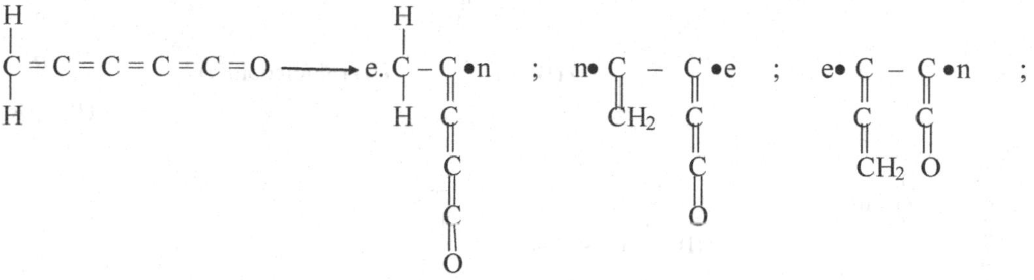

$$\overset{\overset{H}{|}}{\underset{\underset{H}{|}}{C}} = C = C = C = C = O \longrightarrow e.\underset{\underset{\underset{\underset{\underset{O}{\|}}{C}}{\overset{\|}{C}}}{\underset{H}{|}}}{\overset{\overset{H}{|}}{C}} - \overset{}{C}\bullet n \quad ; \quad n\bullet \underset{\underset{\underset{\underset{O}{\|}}{C}}{\overset{\|}{CH_2}}}{\overset{\|}{C}} - \overset{}{C}\bullet e \quad ; \quad e\bullet \underset{\underset{CH_2}{\overset{\|}{C}}}{\overset{\|}{C}} - \underset{\underset{O}{\overset{\|}{C}}}{\overset{}{C}}\bullet n \quad ;$$

$$\text{(I) Both} \qquad\qquad \text{(II) Both} \qquad\qquad \text{(III) Both}$$

$$n\bullet \underset{\underset{\underset{CH_2}{\overset{\|}{C}}}{\overset{\|}{C}}}{\overset{}{C}} - \overset{\overset{\displaystyle O}{\|}}{C}\bullet e \quad ; \quad e\bullet \underset{\underset{\underset{CH_2}{\overset{\|}{C}}}{\overset{\|}{C}}}{\overset{}{C}} - O \bullet nn$$

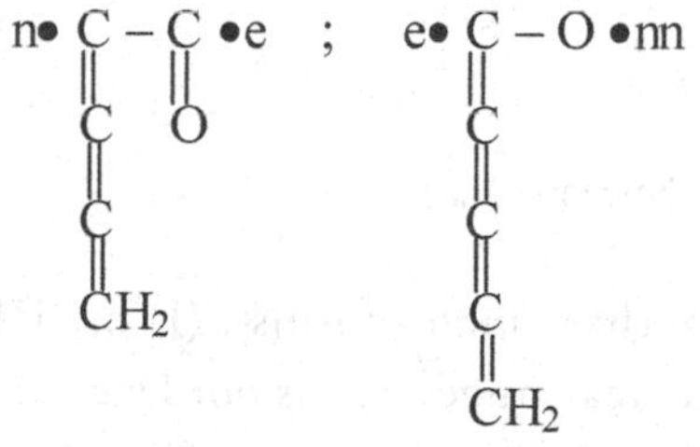

$$\text{(IV) [Both], Y} \qquad\qquad \text{(V) [Both], X} \qquad\qquad 9.135$$

(B) of Equation 9.134 exists when (IV) is resonance stabilized. (V) cannot be resonance stabilized. (I) which is not the least nucleophilic center is resonance stabilized as shown below. However, it is not one of the mono-forms. The main mono-forms are (IV), (B) and (V). (II), (III) and (B) could also be the secondary Male center. (I) cannot be a mono-form, because what is shown below cannot take place.

(I) $\longleftarrow\not\longrightarrow$ $e\bullet \overset{\overset{\displaystyle H}{|}}{\underset{\underset{\displaystyle H}{|}}{C}} - C \equiv C - \overset{\overset{\displaystyle C}{\|}}{\underset{\underset{\displaystyle O}{\|}}{C}}\bullet n$

(A)

NOT RESONANCE STABILIZED   9.136

In view of the reduced H to C ratios, two hydrogen atoms are loosely bonded radically to the carbon center. Hence they must be replaced one at a time, before the compounds can favor being used as monomers.

Replacing the two H atoms with radical-pushing groups, the following are to be expected.

$\overset{\overset{\displaystyle CH_3}{|}}{\underset{5}{C}} = \overset{4}{C} = \overset{3}{C} = \overset{2}{C} = \overset{1}{O}$ where the C also bears O–R

$\longrightarrow$

$n.\overset{5}{\underset{\underset{R}{:O:}}{\overset{\overset{CH_3}{|}}{C}}} - \overset{4}{\underset{\underset{O}{\overset{\|}{C}}}{\overset{\overset{C}{\|}}{C}}}.e$;

(I) (Both)

$e.\overset{4}{\underset{\underset{R}{\overset{O}{|}}}{\underset{C(CH_3)}{C}}} - \overset{\overset{\overset{O}{\|}}{C}}{\underset{3}{C}}.n$;

(II) (Nucleophile)

$n\overset{3}{\underset{\underset{R}{\overset{O}{|}}}{\underset{C(CH_3)}{C}}} - \overset{\overset{\overset{O}{\|}}{C}}{\underset{2}{C}}.e$

(III) (Both), Y

$e\overset{2}{\underset{\underset{\underset{\underset{R}{\overset{O}{|}}}{C(CH_3)}}{\overset{\overset{C}{\|}}{C}}}{C}} - \overset{1}{O}.nn$;

(IV) (Nucleophile), X

(I) $\longleftrightarrow$ $n\bullet \overset{\overset{CH_3}{|}}{\underset{\underset{R}{\overset{O}{|}}}{C}} -- C \equiv C -- \overset{\overset{C}{\|}}{\underset{O}{}}\bullet e$

(B)

9.137

Thus, there are three mono-forms-, (I) or (III), its resonance stabilized mono-form (B) and (II) or (IV). Molecular rearrangement is not favored by (II) or (IV) if R is greater than CH_3 and cannot be resonance stabilized from the most nucleophilic center(s). (IV) has transfer species of the first kind of the third type, *since the transfer species from H_2C=C=C= and H_2C= (third type) is different from the transfer species from $(H_3C)HC$=C= and $(CH_3)HC$=C=C= (seventh type) types of groups.*

By now, it should be obvious that groups such as O=C=, O=C=C=C= (ODD), and groups such as O=C=C=, O=C=C=C=C= (EVEN) and higher can provide resonance stabilization, both differently.

Fom the analysis so far, the following conclusions which have partly been made can be confidently made-

(i)
$$O =$$

<u>Central radical-pushing group</u> 9.138

$$O = C = C = C = C = \quad > \quad O = C = C = \quad > \quad O =$$

<u>Order of Radical-pushing capacity (EVEN)</u> 9.139a

(ii) $\quad O = C = C = C = C = C = \quad > \quad O = C = C = C = \quad > \quad O = C =$

<u>Order of Radical-pulling capacity (ODD)</u> 9.139b

(iii) $\quad -CR=C = C = C = C = O \quad > \quad -CR = C = C = O \quad > \quad -CR = O$

Order of Radical-pulling capacity (ODD) 9.140a

$$= O \quad \gg \quad -CR = C = C = C = O \quad > \quad -CR = C = O$$

Order of Radical-pushing capacity (EVEN) 9.140b

The ODD and EVEN refer to the number of C atoms in each group. It is important to note why it is radical-pushing when EVEN and why it is radical-pulling when ODD with these types of groups.

Now, consider a 1,4 ketenic equivalent of butadiene.

(structure I) $\quad$ versus $\quad$ (structure II)

(I) (symmetric) $\qquad\qquad$ (II) (symmetric) 9.141

While (I) can be resonance stabilized, so also is (II) though it is an electrophile. The movement is from Full free-radical to Full free-radical from the least nucleophilic center.

(structures III, IV, V)

(III) <u>Electrophile</u> $\qquad$ (IV) $\qquad$ (V) 9.142

In the resonance stabilized forms above, (IV) is a mono-form the X center which cannot be resonance stabilized, being the most nucleophilic and the monomer being an Electrophile.

The corresponding case of Equation 9.137 of 1,5- ketenes for 1,6 - ketenes are as follows, if its existence is favored.

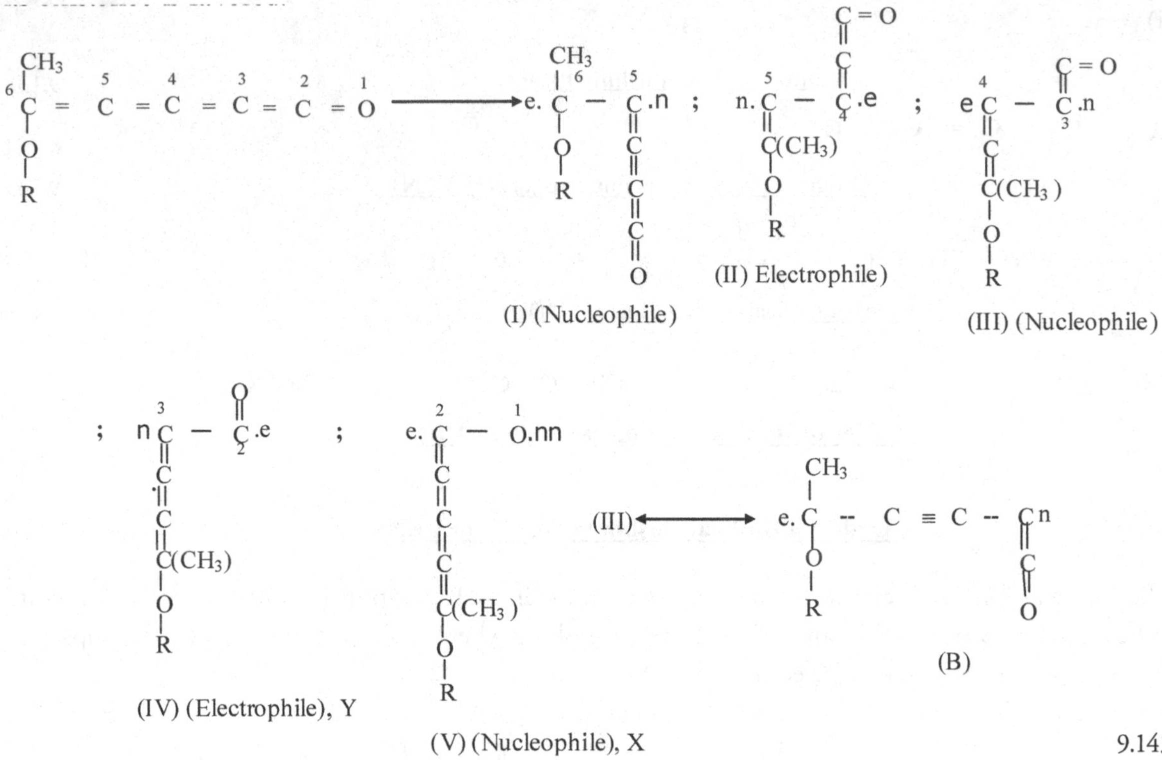

$$\underset{\underset{R}{|}}{\overset{\overset{CH_3}{|}}{\overset{6}{C}}} = \overset{5}{C} = \overset{4}{C} = \overset{3}{C} = \overset{2}{C} = \overset{1}{C} = O \longrightarrow$$

(I) (Nucleophile)

e.$\overset{CH_3}{\overset{6}{C}} - \overset{5}{C}$.n ; (II) Electrophile)

n.$\overset{5}{C} - \overset{}{C}$.e ;

e$\overset{4}{C} - \overset{}{C}$.n

(III) (Nucleophile)

; n$\overset{3}{C} - \overset{2}{C}$.e ; e.$\overset{2}{C} - \overset{1}{O}$.nn

(IV) (Electrophile), Y

(V) (Nucleophile), X

(III) $\longleftrightarrow$ e.$\overset{CH_3}{C} -- C \equiv C -- \overset{}{C}$n

(B)

9.143

Note the similarities and differences in character between the mono-forms. Like the 1,5-case, there are three mono-forms- (II) or (IV), its resonance stabilized mono-form and (III) or (V) which is strongly nucleophilic with transfer species of the first kind of the seventh type.

Finally, consider only the 1,5- ketenes for radical-pulling groups. The followings are obtained for $COOCH_3$ and CH_3 groups in place of the two H atoms.

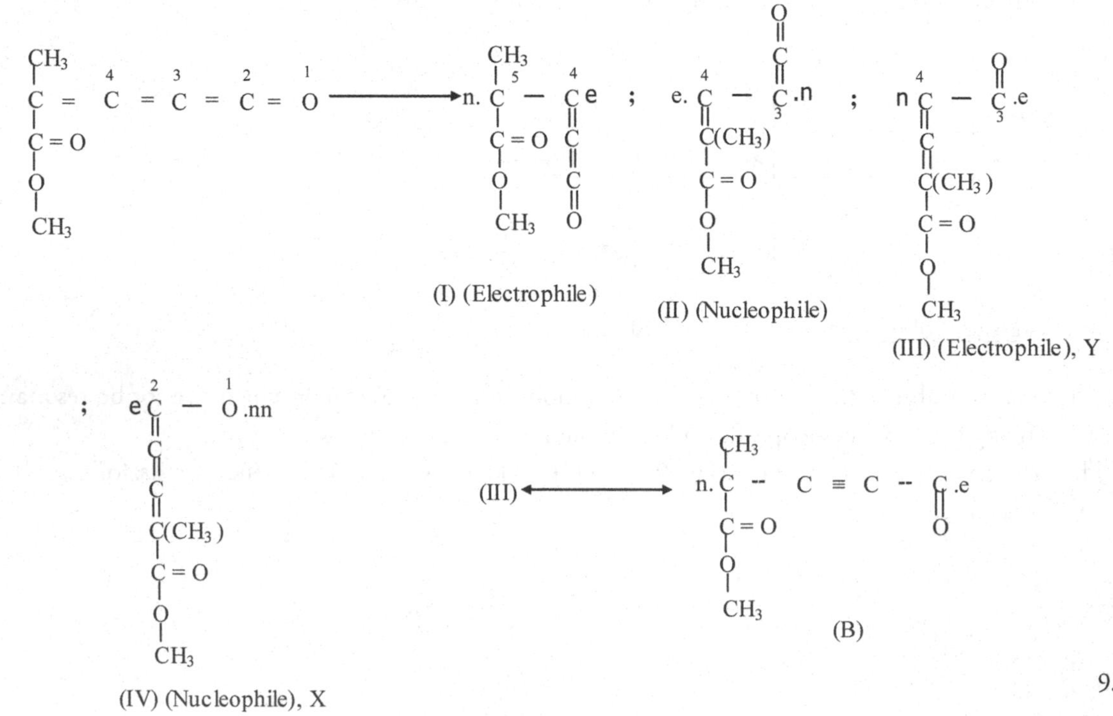

$$\underset{\underset{CH_3}{\overset{|}{O}}}{\overset{\overset{CH_3}{|}}{\overset{4}{C}}} = \overset{3}{C} = \overset{2}{C} = \overset{1}{C} = O \longrightarrow$$

n.$\overset{CH_3}{\overset{5}{C}} - \overset{4}{C}$e ;

(I) (Electrophile)

e.$\overset{4}{C} - \overset{}{C}$.n ;

(II) (Nucleophile)

n$\overset{4}{C} - \overset{}{C}$.e

(III) (Electrophile), Y

; e$\overset{2}{C} - \overset{1}{O}$.nn

; (III) $\longleftrightarrow$ n.$\overset{CH_3}{C} -- C \equiv C -- \overset{}{C}$.e

(IV) (Nucleophile), X

(B)

9.144

743

With the three mono-forms- (III), its resonance stabilized one (B) and (IV), one can observe the strong Electrophilic character of this monomer. It has an X center and a Y center. Both (I) and (III) can be resonance stabilized. The nucleophilic center to be used electro-free-radically is that shown above (IV) and not the C = O center conjugatedly placed to the ketene. Just as (IV) is to (III) so also (II) is to (I), but the former being only the first to be used via the routes natural to them. With corresponding 1,6-Ketene, the situation is almost similar, but with different transfer species. This will be identical to the case of 1,4-Ketene shown in Equation 9.110a. Thus, based on the foundations which have been laid for 1,3- and 1,4-Ketenes, one knows what to expect when resonance stabilization groups are externally located and when other groups are put in place of the H atoms.

9.4. 1,3- Isocyanates

There is still the need to revisit the chemistry of these monomers, in order to explain some past observations, using current developments. In Chapter 7, for example one showed that they cannot be activated chargedly as shown in Equation 7.1. They can only be activated radically. It is also said that, though isocyanic acid is tautomeric, two series of stable alkyl derivatives (esters) should be possible, the alkyl cyanates $ROC \equiv N$, and the alkyl isocyanates $O = C = NR$.[6] The formers have been isolated only when the alkyl group is sufficiently bulky to prevent trimerization. Obviously by now from Chapter 7, it is clear that, it is not Electroradicalization that is present, but molecular rearrangement. The alkyl isocyanates are said to be stable compounds. Hence, cyanic acids molecularly rearranged as shown only radically and not chargedly.

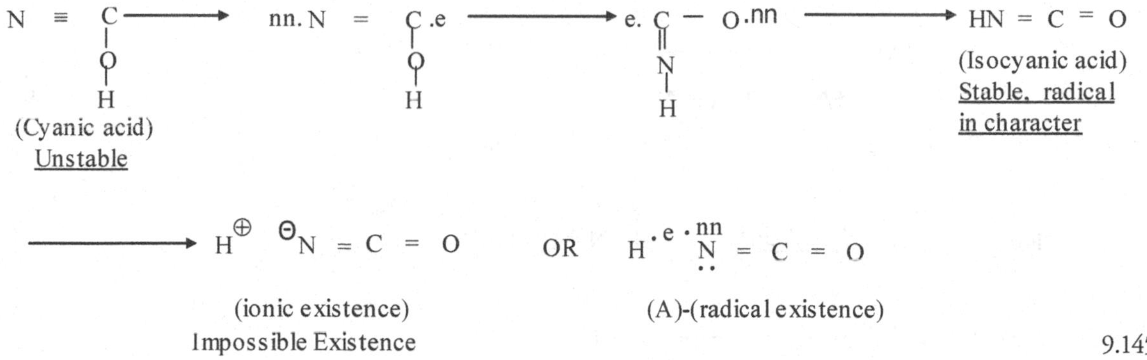

$$N \equiv C \longrightarrow nn. N = C .e \longrightarrow e. C - O .nn \longrightarrow HN = C = O$$

with structures:
- (Cyanic acid) Unstable — C bonded to O, O bonded to H
- middle form — C bonded to O, O bonded to H
- third form — C double bonded to N, N bonded to H
- (Isocyanic acid) Stable, radical in character

$$\longrightarrow H^{\oplus} {}^{\ominus}N = C = O \quad OR \quad H^{\cdot e} \cdot^{nn} N = C = O$$

(ionic existence)
Impossible Existence

(A)-(radical existence)

$$9.145$$

Thus, the hydrogen atom is loosely bonded radically to the nitrogen center, depending on the operating conditions. The activation center, which is over unsaturated (H:C ratio = 1:1) can only be activated when suppressed. With different operating conditions, (A) can be enolized to Cyanic acid.

Thus, when urea is strongly heated above its melting temperature, it decomposes to ammonia and cyanic acid. That isocyanic acid itself is not stable, is indicated by the fact that it is said to favor polymerization at once to a mixture of about 30 percent of linear polymer (cyamelide) and 70 percent of the trimmer (cyanuric acid).[6] The mechanism of these reactions have already been provided in Chapter 7 (See Equations 7.58 to 7.68). Therein, it was shown that isocynanic acid tautomerized.

Stage 1:

$$3H - N = C = O \rightleftharpoons 3H \bullet e \ + \ 3nn \bullet N = \overset{\bullet e}{C} - O \bullet nn$$

$$\rightleftharpoons 3\,nn \bullet N = \overset{\bullet e}{C} - O \bullet nn \ + \ 3H \bullet e$$

$$\rightleftharpoons 3nn \bullet N = \overset{\bullet e}{C} - OH$$

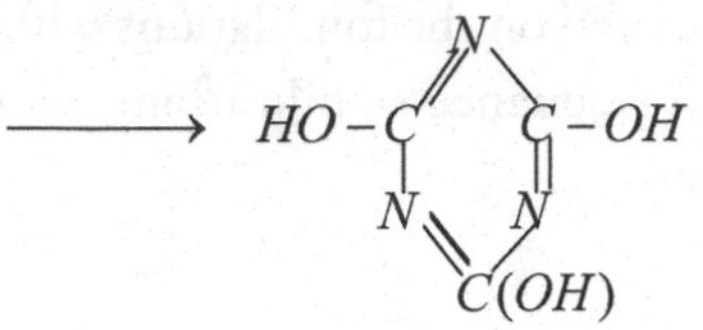

$$\longrightarrow HO - C \overset{\displaystyle \diagup\!\!\!\diagup}{} \cdots C - OH$$

(with ring)

$$N \qquad N$$
$$C(OH)$$

$$(II) \hspace{6cm} 9.146a$$

Alcohols or amines readily add to isocyanic acid and isocyanates to yield urethans as follows.

$$HN = C = O \ + \ HOR \longrightarrow H_2NCOOR$$

Alkyl Carbamates OR Urethans $\hspace{3cm}$ 9.146b

Stage 1: $\quad H - N = C = O \rightleftharpoons H \bullet e \ + \ nn \bullet N = \overset{\bullet e}{C} - O \bullet nn$

$$\rightleftharpoons nn \bullet N = \overset{\bullet e}{C} - O \bullet nn \ + \ H \bullet e$$

$$\rightleftharpoons nn \bullet N = \overset{\bullet e}{C} - OH$$

$$(A)$$

$$(A) \longrightarrow N \equiv C - OH \ + \ Heat$$

ENOLIZATION $\hspace{5cm}$ 9.147

Stage 2: $\quad ROH \rightleftharpoons H \bullet e \ + \ nn \bullet OR$

$$H \bullet e \ + \ (A) \xrightarrow{\text{ACTIVATION}} \begin{array}{c} OH \\ | \\ H - N = C \bullet e \end{array}$$

$$(B)$$

$$RO \bullet nn \ + \ (B) \longrightarrow \begin{array}{c} H - N = C - OR \\ | \\ OH \end{array}$$

$$(C) \hspace{5cm} 9.148$$

Stage 3:

$$\begin{array}{c} H - N = C - OR \\ | \\ OH \end{array} \rightleftharpoons \begin{array}{c} H \quad OR \\ | \quad \ | \\ nn \bullet N - C \bullet e \\ | \\ OH \end{array}$$

$$\rightleftharpoons \begin{array}{c} OR \\ | \\ H_2N - C \bullet e \\ | \\ O \bullet nn \end{array}$$

$$\longrightarrow H_2N - COOR$$

MOLECULAR REARRANGEMENT $\hspace{3cm}$ 9.149

The final overall equation still remains as Equation 9.146. In the first stage, due to the presence of the alcohol, the isocyanide enolized. In the second stage, the cyanate was suppressed and activated by the alcohol to give (C). The cyanate could not undergo molecular rearrangement back to the isocyanic acid. In the third stage, (C) molecularly rearranged to give the product. ***It was the H in OH group that was moved and not the R in OR group; a clear indication that OH group is richer or of greater radical-pushing capacity than OR group.*** At the beginning, one thought otherwise. It is a three stage Equilibrium mechanism system specifically used to show that OH is more radically-pushing than OR group.

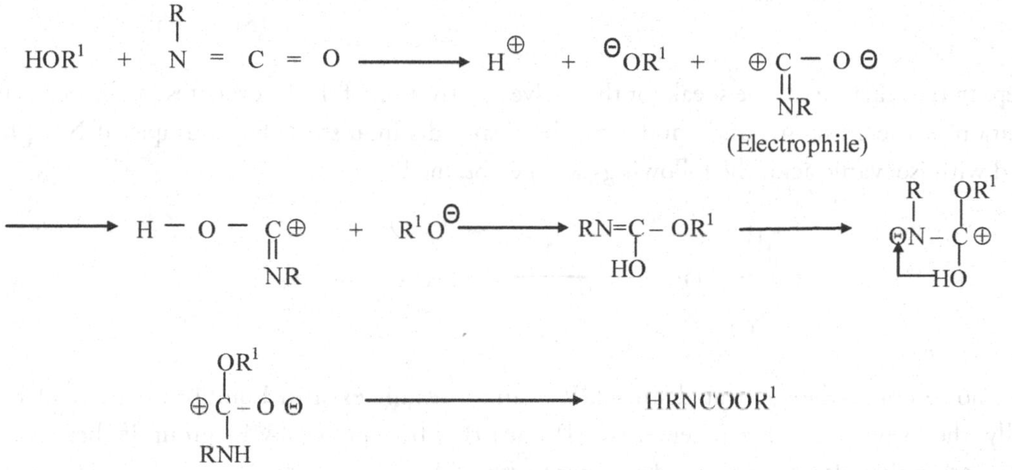

$$HOR^1 \quad + \quad \overset{\overset{\textstyle R}{|}}{N} = C = O \longrightarrow H^\oplus \quad + \quad {}^\ominus OR^1 \quad + \quad \underset{\underset{\textstyle NR}{||}}{\oplus C - O \ominus}$$

$$(\text{Electrophile})$$

$$\longrightarrow H - O - \underset{\underset{\textstyle NR}{||}}{C \oplus} \quad + \quad R^1 O^\ominus \longrightarrow \underset{\underset{\textstyle HO}{|}}{RN = C - OR^1} \longrightarrow \overset{\overset{\textstyle R \quad OR^1}{| \quad |}}{\ominus N - C \oplus} \overset{}{\underset{\textstyle HO}{\rule{0pt}{0pt}}}$$

$$\overset{\overset{\textstyle OR^1}{|}}{\underset{\underset{\textstyle RNH}{|}}{\oplus C - O \ominus}} \longrightarrow HRNCOOR^1 \qquad\qquad 9.150$$

This is a two stage Equilibrium mechanism process, noting that C=O is more nucleophilic than C=N and with an electro-free-radical, the C = O was first activated; for which the reactions can take place either chargedly or radically, though in general they take place radically all the time. Indeed, if the C=N center was the first to be activated, the same products will still be obtained, taking place in only one stage.

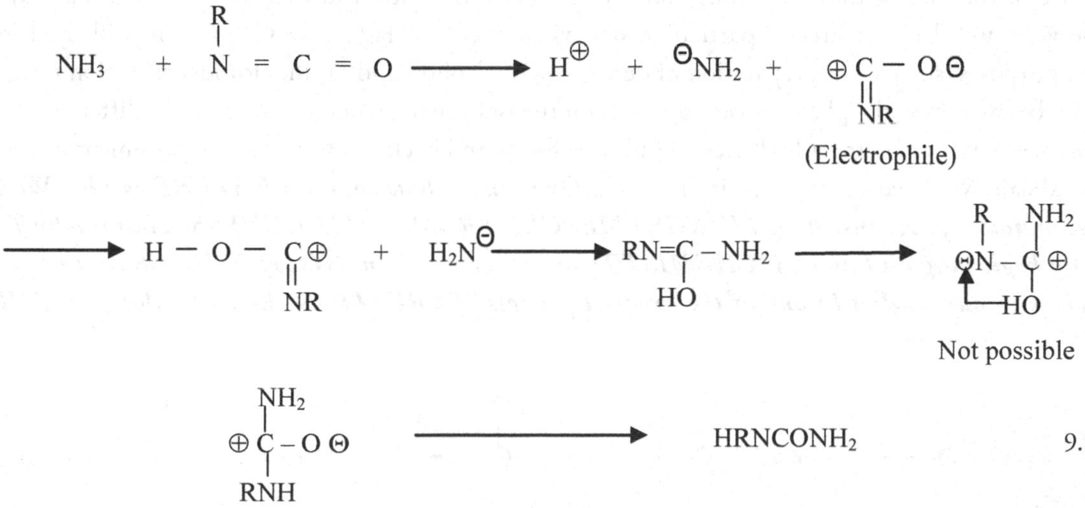

$$NH_3 \quad + \quad \overset{\overset{\textstyle R}{|}}{N} = C = O \longrightarrow H^\oplus \quad + \quad {}^\ominus NH_2 \quad + \quad \underset{\underset{\textstyle NR}{||}}{\oplus C - O \ominus}$$

$$(\text{Electrophile})$$

$$\longrightarrow H - O - \underset{\underset{\textstyle NR}{||}}{C \oplus} \quad + \quad H_2 N^\ominus \longrightarrow \underset{\underset{\textstyle HO}{|}}{RN = C - NH_2} \longrightarrow \overset{\overset{\textstyle R \quad NH_2}{| \quad |}}{\ominus N - C \oplus} \overset{}{\underset{\textstyle HO}{\rule{0pt}{0pt}}}$$

$$\text{Not possible}$$

$$\overset{\overset{\textstyle NH_2}{|}}{\underset{\underset{\textstyle RNH}{|}}{\oplus C - O \ominus}} \longrightarrow HRNCONH_2 \qquad\qquad 9.151a$$

The mechanism shown below is the favored mechanism, because in the rearrangement above NH_2 should be the source of transfer species and not OH group.

$$NH_3 \quad + \quad \overset{\overset{R}{|}}{N} = C = O \quad \longrightarrow \quad H^{\oplus} \quad + \quad {}^{\ominus}NH_2 \quad + \quad \overset{\overset{R}{|}}{{}^{\ominus}N} - \underset{\underset{O}{\|}}{C}^{\oplus}$$

(Nucleophile)

$$NH_3 \quad + \quad \overset{\overset{R}{|}}{N} = C = O \quad \longrightarrow \quad H^{\oplus} \quad + \quad {}^{\ominus}NH_2 \quad + \quad \overset{\overset{R}{|}}{{}^{\ominus}N} - \underset{\underset{O}{\|}}{C}^{\oplus}$$

(Nucleophile)   9.151b

The steps in the reactions above speak for themselves. Activation of the isocyanates can be observed. Ionic dissociation of the ammonia and amides can be clearly distinguished. For example, if NR^1_3 had been involved with isocyanic acid, the followings will be obtained.

$$NR^1_3 \quad + \quad {}^{\oplus}\overset{\overset{NH}{\|}}{C} - O^{\ominus} \quad \longrightarrow \quad \text{No reaction} \qquad 9.152$$

(I)

There is no reaction when activated since NR^1_3 cannot readily exist in Equilibrium State of Existence. Secondly, the H on the nitrogen centers of (I) must first be replaced by R^1 group if there is to be any reaction noting that electrostatic bonds cannot be formed.

$$RN = C = O \quad + \quad H_2NR^1 \quad \longrightarrow \quad RNHCONHR^1 \qquad 9.153$$

$$RN = C = O \quad + \quad HNR^1_2 \quad \longrightarrow \quad RNHCONR^1_2 \qquad 9.154$$

Because of the ease of these reactions and the formation of solid products, the isocyanates are known to be very useful in Nature. In particular, phenyl isocyanates ($H_5C_6N = C = O$) are still used for so many purposes, such as for preparation of derivatives of alcohols and amines for identification purposes. Shown below is how the phenyl group operates on the isocyanate, wherein we can see a different form of resonance stabilization from Half-free to Full-free forms and back to Half free, via movement of electro-free-radicals. We have already seen it in Vol (I). *One can see how complex NATURE can be. We have yet seen nothing, because ALL THINGS EMERGED FROM NOTHINGNESS. That is why TIME has no beginning and no end. CHEMISTRY has its origin from NOTHINGNESS as we begin to progress in our studies in one of the greatest planets of CREATION- EARTH along the MILKY WAY in our Universe.*

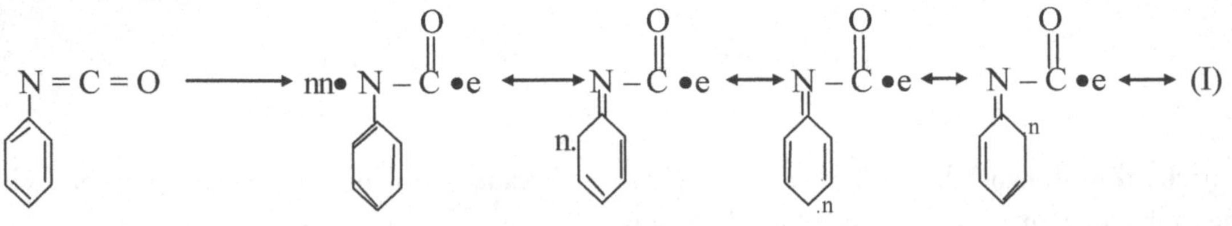

RESONANCE STABILIZATION   9.155

(I) is Half-free, while the others are Full-free only radically, and finally back to Half–free where it started. They can never take place chargedly. (II) and (IV) can form an additional ring either from the right or the left of the spectrum as shown below under equilibrium.

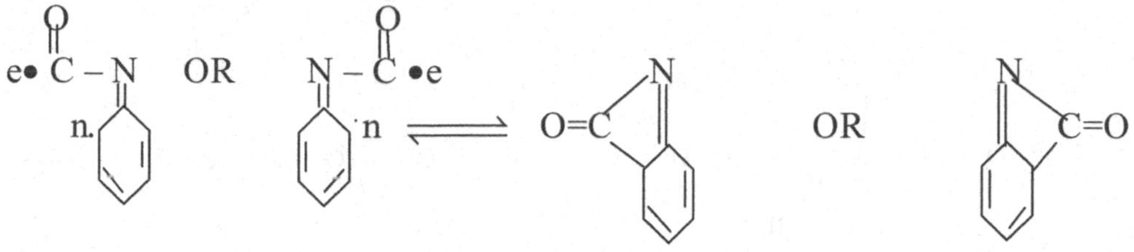

(II)   (IV)   (II)A   (IV)A   9.156

(II)A and (IV)A are fused six- and four-membered rings in which the four-membered ring is highly strained and notably MALE in character. Depending on the operating conditions, the ring can open very explosively destroying all things around its environment. The rings should not be allowed to exist when used analytically. It can be preserved based on the operating conditions and used as so desired if ever it exists.

Hydrolysis of isocyanates yields the carbonic acid which is unstable and decomposes spontaneously to amine and carbon dioxide.

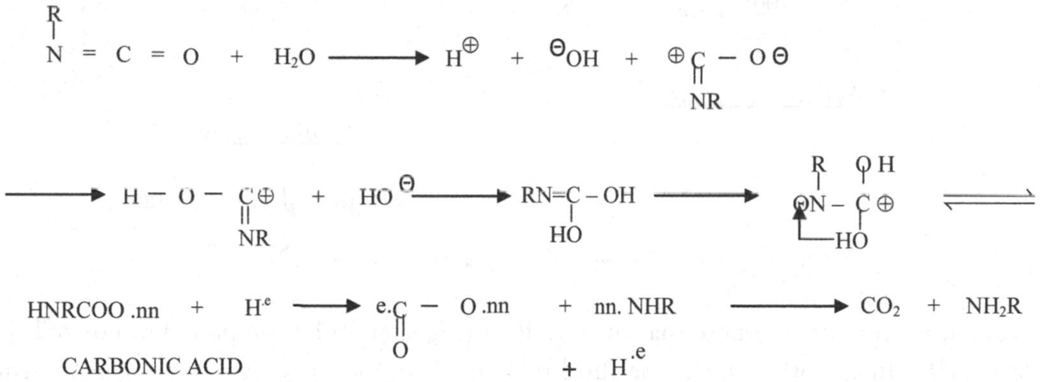

9.157

This is a three stage Equilibrium mechanism process, noting that the isocyanates are Electro-philes. The first stage ends with formation of the carbonic acid which in the second stage decomposes to amine and carbon dioxide. Though the reactions above have been shown halfway chargedly, they never do really take place chargedly, but only radically. It must be observed that in all or most of the reactions which have been considered so far in visiting the past, the N = C activation center should be the first to be largely involved, being less nucleophilic. As an Electrophile, the C =N center can be involved nucleo-non-free-radically and electro-free-radically, while the C = O center can only be involved electro-free-radically.

9.4.1. Charged characters of 1,3-Isocyanates

For its first member, isocyanic acid, the use of $H^{\oplus}$ as initiator is impossible, since cations can never be used as initiators due to radical balancing. On the other hand charges cannot be used, due to electrostatic forces of repulsion. Though these reactions take place only radically, one is looking at these chargedly for exploratory purposes. Through exploration, so many things are revealed

Nevertheless, for the favored existence of high molecular weight polymers, the H on nitrogen center must be replaced with wide range of groups including non-free radical-pulling groups such as F, and $OCOCH_3$ (acetate groups). Non-free radical-pushing groups such as OH, OR, NH_2, NHR etc. can be used in place of R groups.

$$
\overset{..}{N} = C = O \;,\quad \overset{:\overset{..}{F}:}{\underset{..}{N}} = C = O \;;\quad \overset{\overset{R}{|}}{\underset{..}{N}} = C = O\;,\quad :N = C = O
$$

(I) with $:\overset{}{O}:$, $C = O$, CH_3 ; (II) ; (III) ; (IV) with $\overset{:N}{\diagup\diagdown}\; H\; H$

<u>Cannot Favor Charged activations</u>   9.158

The (I), (II), (III) and (IV) N = C activation centers can only be activated radically. It will be useful to compare (I), (II), (III) and (IV) with nitrogen molecule and sodium azide, since sodium azide and N_2 cannot favor charged existence.

$$
\underset{..}{N} \equiv \overset{..}{N} \xrightarrow{\text{Activation}} ne.\overset{..}{N} = \underset{..}{N}.nn \;;\quad Na^{\oplus} \quad ^{x}_{\bullet}N^{\bullet}_{\bullet} \; ^{+}_{+} \; N \; ^{+}_{+} \; ^{\bullet}_{\bullet}\overset{\oplus}{N}^{\ominus}_{\bullet}
$$

Polar bond

(I) <u>Activated nitrogen</u>

(II)<u>Sodium azide</u>   9.159

Impossible existence

$$
NaN_3 \;\rightleftharpoons\; Na \bullet e \;\; nn\bullet N = N^{\oplus} = {}^{\ominus}N
$$
                  9.160

In the azide, it is important to note that the middle nitrogen atom has no paired unbonded radicals in the last shell. On the other hand, the third non-covalent bond between two nitrogen centers is not ionic, but polar bonds in character (imaginary bond). Therefore, sodium azide is Polar/Non-ionic, since the negative charge on N cannot stay due to electrostatic forces of repulsion between the charge and the adjacently located covalent bond. Many years ago, one worked with the use of NaN_3 as an additive to develop a solvent for prevention of absorption of polyacrylamide on silica surfaces for many applications (PhD thesis). It worked without realizing the mechanism of the process after more than two years of working with all kinds of chemicals which could not stop the absorption. This is the way it is with all laboratory experiments and researches universally. ***Hence the required need for one to go to the laboratory to check if the new foundations being laid for humanity as demanded by all journals universally is meaningless, for they like the author do not know what they have been doing and still doing.*** One saw it many years ago and therefore decided to change the gear with great patience following the advices given by some of the GREAT MASTERS from some EDITORS of very great journals.

Considering isocyanates with radical-pushing alkylane groups, the followings are obtained.

$$\underset{\text{(I)}}{\overset{\displaystyle CH_3}{\underset{\displaystyle O}{N = C}}} \quad + \quad R^{\ominus} \xrightarrow[\text{Non-free \quad temp}]{\text{(very Low)}} \quad \overset{\displaystyle CH_3}{\underset{\displaystyle O}{\ominus N - C^{\oplus}}} \quad + \quad R^{\ominus} \longrightarrow R - \overset{\displaystyle O}{\overset{\|}{C}} - \underset{\displaystyle CH_3}{N^{\circleddash}}$$

9.161

$$\overset{\overset{\displaystyle CH_3}{|}}{\underset{\nearrow \quad \| \quad \searrow}{\oplus C - N\ominus}} \\ \underset{H_9C_4{}^{\ominus} \quad\quad\quad Li^{\oplus}}{} \longrightarrow \text{No Reaction}$$

9.162

$$R^{\oplus} + \overset{\displaystyle CH_3}{\underset{\displaystyle O}{\ominus N - C^{\oplus}}} \longrightarrow R - \overset{\displaystyle CH_3}{\underset{\displaystyle O}{N - C^{\oplus}}} \xrightarrow{+ \; n \, (I)}$$

$$(I)$$

$$R + \overset{\displaystyle CH_3}{\underset{\displaystyle O}{N - C}}\overset{}{\underset{n}{\big)}} \overset{\displaystyle CH_3}{\underset{\displaystyle O}{N - C^{\oplus}}} \longrightarrow H_3C^{\oplus} \;\mid\; \text{Dead Terminal double bond polymer}$$

9.163

The monomer is observed to favor the use of non-free negatively-charged initiators. Polymeri-zation is not favored with negatively-charged-paired initiator, but only with positively charged paired-media initiators and anionically charged paired initiator. With positively charged initiators, initiation is only favored for the conditions indicated where dead terminal double bond polymers are produced since the route is not natural to that center in the monomers. For the case above, CH_3 was therefore released as transfer species of the second kind. With free or paired media anionic initiators, living polymers are produced. Due to ready cyclization of the monomer when activated, polymerization temperatures should be as low as possible when polymerized anoinically the route natural to the N = C center if the media is free. With positively charged-paired initiator, the center actually involved is the C=O center, with transfer species of the first kind of the third type which cannot be released, due to electrostatic forces of repulsion. Hence, that center cannot be activated chargedly

With etheric groups, the followings are to be expected.

$$\underset{\substack{\text{Non-free}\\\text{charge}}}{R^{\ominus}} \quad + \quad \overset{\displaystyle CH_3}{\underset{\displaystyle O}{\overset{\displaystyle |}{\underset{\displaystyle \|}{\overset{nn}{\bullet N} - \overset{}{C} \bullet e}}}} \longrightarrow \text{No reaction}$$

9.164

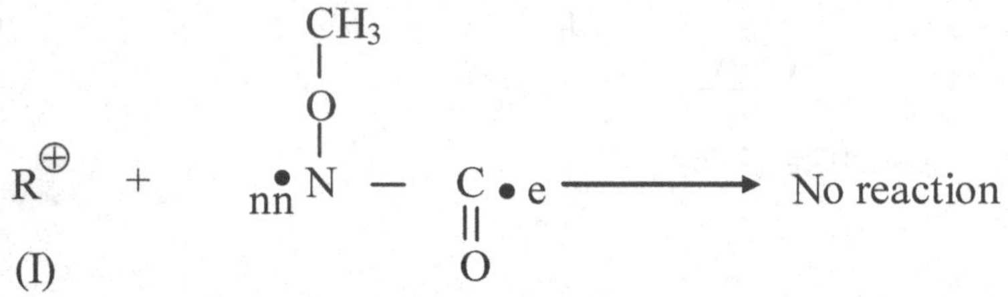

$$R^{\oplus} + \overset{\bullet}{nn}N - \overset{\overset{\displaystyle CH_3}{|}\,\overset{\displaystyle O}{|}}{\underset{\underset{\displaystyle O}{||}}{C}}\bullet e \longrightarrow \text{No reaction} \qquad 9.165$$

(I)

Both routes are readily favored only with the use of radical initiators, since charged activation cannot take place with the etheric group carried due to electrostatic forces of repulsion.

When radical-pulling groups are involved, the followings are obtained for COOCH$_3$ group.

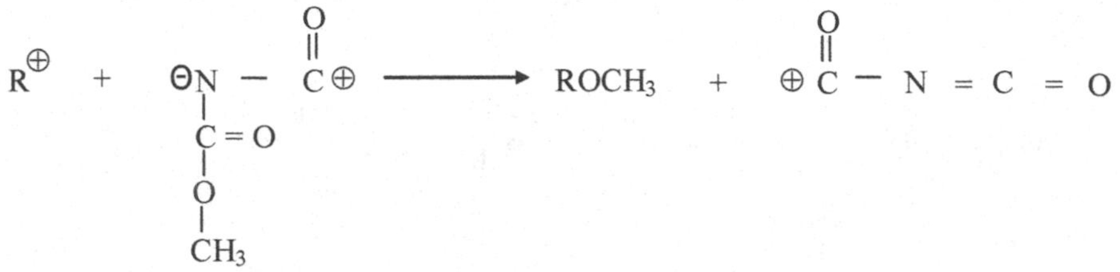

$$R^{\oplus} + \overset{\ominus}{N} - \overset{O}{\underset{||}{C}}^{\oplus} \longrightarrow ROCH_3 + \overset{O}{\underset{||}{\oplus C}} - N = C = O \qquad$$

(I) Electrophile $\qquad$ 9.166

$$R + \left(\overset{O}{\underset{||}{C}} - N \right)_n \overset{O}{\underset{||}{C}} - N^{\ominus} \longrightarrow R + \left(\overset{O}{\underset{||}{C}} - N \right)_n \overset{O}{\underset{||}{C}} - N = C = O + {}^{\ominus}OCH_3$$

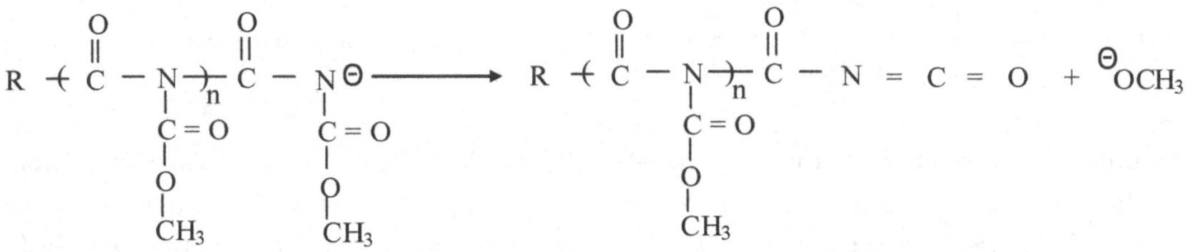

<u>Transfer species of 1st kind of 1st type</u> $\qquad$ 9.167

It can be observed that the anionic routes are favored. The same will apply to CONH$_2$ types of groups. With these groups, notice that the C = N center is truly the Y, while the C = O is the X, which cannot be used chargedly. Therefore, electro-free-radically, the C = O center will be the center activated. Evidently, these are Electrophiles. With COR types of groups, the followings are obtained.

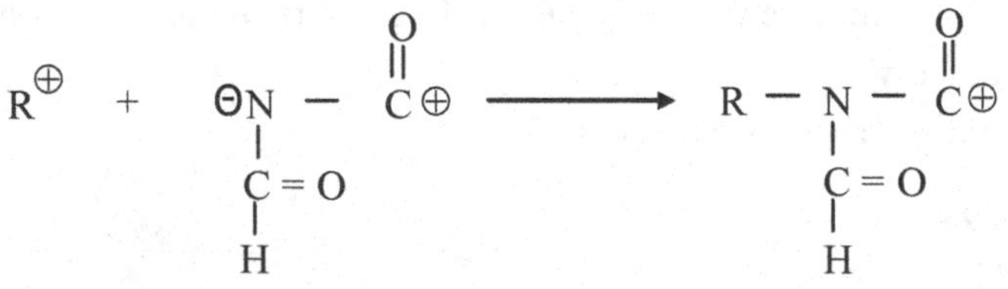

$$R^{\oplus} + \overset{\ominus}{N} - \overset{O}{\underset{||}{C}}^{\oplus} \longrightarrow R - N - \overset{O}{\underset{||}{C}}^{\oplus} \qquad$$

(I) (Both) $\qquad$ 9.168

$$R^{\ominus} + \overset{O}{\underset{||}{\oplus C}} - N^{\ominus} \longrightarrow R - \overset{O}{\underset{||}{C}} - N^{\ominus} \qquad$$

<u>Non-free anion</u> $\qquad$ 9.169

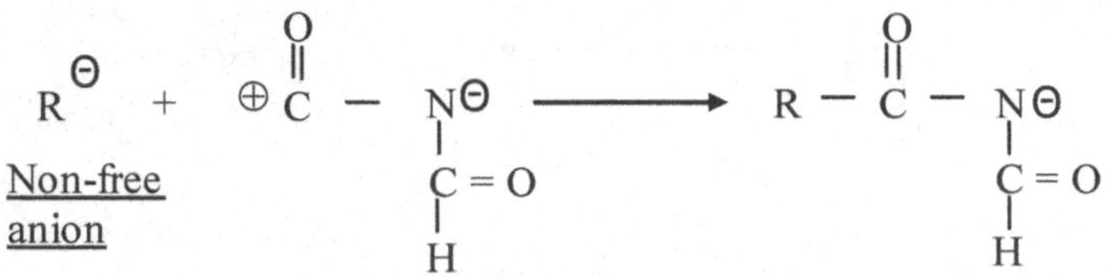

Both routes are observed to be favored by the monomer with COR groups, with absence of transfer species for their chains. With radical-pulling groups, the Electrophilic character remains.

751

If a solvent readily polar to favor activation of the C=O center, is used anionically the monomer of Equation 9.161 will favor only existence of homopolymers via N = C activation center.

$$R^{\ominus} \quad + \quad \overset{2}{\underset{\underset{\underset{CH_3}{|}}{\overset{|}{N}}}{\overset{\oplus}{C}}} - \overset{1}{O^{\ominus}} \longrightarrow RCH_3 \quad + \quad \overset{\ominus}{\underset{\bullet\bullet}{N}} = C = O$$

$$\underline{\text{Non-free}}$$

(Non-existent- Electrostatic forces of repulsion)

(II) (Nucleophile)   9.170

With paired-media positively charged initiators, random copolymers cannot be obtained, because the route is only favored by the C = N center. In general, these can only be obtained from a single Electrophile where possible.

$$R - O - \underset{\underset{\underset{CH_3}{|}}{\overset{|}{N}}}{\overset{\overset{CH_3}{|}}{\underset{\|}{C}}} - N - \underset{\overset{\|}{O}}{\overset{\|}{C}} \text{ WWW } O - \overset{\oplus}{\underset{\underset{CH_3}{|}}{\underset{\|}{C}}} \longrightarrow R - O - \underset{\underset{CH_3}{|}}{\overset{\|}{\underset{N}{C}}} \text{ WWW } O - C \equiv N$$

(Non-existent)

$$+ \quad \overset{\oplus}{CH_3}$$

RANDOM COPOLYMERIZATION (Not favored)   9.171

Unlike dimethylketene, alternating copolymers cannot be obtained since both mono-forms are half-free-monomers and more nucleophilic in character. The growing chain above cannot be terminated from the C = O center due to electrostatic forces of repulsion. It can however be terminated from the C = N center. Hence, it was said not to be favored. For the cases with radical-pulling groups such as COOR, $CONH_2$, existence of alternating copolymers is not favored with negatively charged-free-media and paired-media initiators, since the carbonyl center will not be activated chargedly.

$$\overset{\oplus}{\underset{\underset{\underset{\underset{CH_3}{|}}{\overset{|}{O}}}{\underset{\|}{C=O}}}{\overset{\overset{O}{\|}}{C}}} - \overset{\ominus}{N} + \overset{\oplus}{\underset{\underset{\underset{\underset{\underset{CH_3}{|}}{\overset{|}{O}}}{\overset{|}{C=O}}}{\overset{|}{N}}}{\underset{\|}{C}}} - O^{\ominus} \longrightarrow CH_3O - \underset{\underset{\underset{\underset{OCH_3}{|}}{\overset{|}{C=O}}}{\overset{|}{N}}}{\overset{\|}{C}} - O^{\ominus} + \overset{O}{\underset{\underset{\underset{O}{\|}}{\overset{\oplus}{C}}}{\overset{\|}{C}}} = N - \overset{\oplus}{C}$$

9.172

With positively charged-paired initiator, none of the centers will favor polymerization, since Isocyanates are Electrophiles wherein the N = C center is Y and C = O center which cannot be activated chargedly is X. This is made stronger when CF_3 or a radical-pulling group is put in place of CH_3, i.e., $F_3C - N = C = O$.

With resonance stabilization groups, the followings are to be expected only radically.

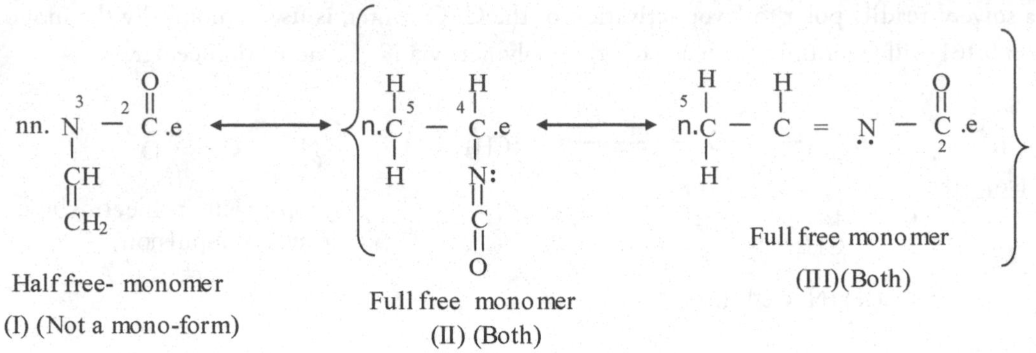

$$nn.\underset{\substack{\text{\small CH}\\\text{\small $\parallel$}\\\text{\small CH}_2}}{\overset{3}{N}}-\overset{\overset{\text{\small O}}{\parallel}}{\underset{2}{C}}.e \quad\longleftrightarrow\quad \left\{ n.\underset{\substack{|\\\text{\small H}}}{\overset{\overset{\text{\small H}}{|}}{\underset{5}{C}}}-\underset{\substack{|\\\text{\small N:}\\\text{\small $\parallel$}\\\text{\small C}\\\text{\small $\parallel$}\\\text{\small O}}}{\overset{\overset{\text{\small H}}{|}}{\underset{4}{C}}}.e \quad\longleftrightarrow\quad n.\underset{\substack{|\\\text{\small H}}}{\overset{\overset{\text{\small H}}{|}}{\underset{5}{C}}}-\underset{\substack{|\\\text{\small H}}}{\overset{\overset{\text{\small H}}{|}}{C}}=\underset{\cdot\cdot}{N}-\underset{2}{\overset{\overset{\text{\small O}}{\parallel}}{C}}.e \right\}$$

Half free- monomer   Full free monomer

(I) (Not a mono-form)   Full free monomer   (III)(Both)

            (II) (Both)

          RESONANCE STABILIZATION   9.173

One can observe that even the − N = C = O group itself is a radical-pushing resonance stabilization group, an indication that the center first activated is the C = C center in (II). Movement from (II) to (III) is by resonance stabilization wherein if H on Carbon center 4 was CH_3, it will be shielded as shown below.

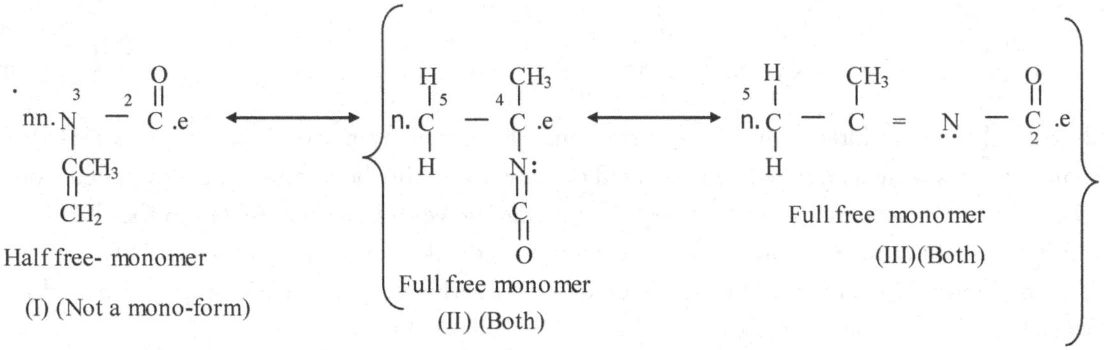

$$\overset{\cdot}{}\quad nn.\underset{\substack{\text{\small CCH}_3\\\text{\small $\parallel$}\\\text{\small CH}_2}}{\overset{3}{N}}-\underset{}{\overset{\overset{\text{\small O}}{\parallel}}{\underset{2}{C}}}.e \quad\longleftrightarrow\quad \left\{ n.\underset{\substack{|\\\text{\small H}}}{\overset{\overset{\text{\small H}}{|}}{\underset{5}{C}}}-\underset{\substack{|\\\text{\small N:}\\\text{\small $\parallel$}\\\text{\small C}\\\text{\small $\parallel$}\\\text{\small O}}}{\overset{\overset{\text{\small CH}_3}{|}}{\underset{4}{C}}}.e \quad\longleftrightarrow\quad n.\underset{\substack{|\\\text{\small H}}}{\overset{\overset{\text{\small H}}{|}}{\underset{5}{C}}}-\underset{}{\overset{\overset{\text{\small CH}_3}{|}}{C}}=\underset{\cdot\cdot}{N}-\underset{2}{\overset{\overset{\text{\small O}}{\parallel}}{C}}.e \right\}$$

Half free- monomer   Full free monomer

(I) (Not a mono-form)   (III)(Both)

        Full free monomer

        (II) (Both)

      RESONANCE STABILIZATION   9.174

Thus, for the mono-forms chargedly, they cannot be resonance stabilized even when they favor presence of same character. With activation via C = O activation centers, the following is obtained chargedly and radically. In fact, it takes place only radically as displayed in the equation below.

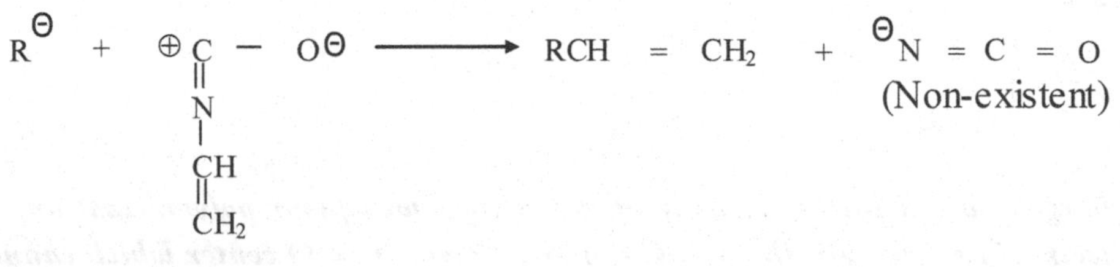

$$R^{\ominus} \quad + \quad \overset{\oplus}{\underset{\substack{\text{\small N}\\\text{\small |}\\\text{\small CH}\\\text{\small $\parallel$}\\\text{\small CH}_2}}{C}}-O^{\ominus} \quad\longrightarrow\quad RCH = CH_2 \quad + \quad \overset{\ominus}{N}=C=O$$

                  (Non-existent)

                          9.175

As will clearly become obvious, such monomers above cannot be activated chargedly.

9.4.2 Radical characters of 1,3- Isocyanates

Beginning with those with radical-pushing groups, the followings are obtained.

$$E^{\bullet e} \quad + \quad nn\bullet O - \underset{\underset{CH_3}{|}}{\overset{\parallel}{\underset{N}{C}}} \bullet e \quad \longrightarrow \quad E - O - \underset{\underset{CH_3}{|}}{\overset{\parallel}{\underset{N}{C}}} \bullet e \qquad 9.176$$

$$\overset{\cdot\cdot}{N}\cdot^{nn} \quad + \quad e\cdot \underset{\underset{CH_3}{|}}{\overset{\parallel}{\underset{N}{C}}} - O\cdot^{nn} \quad \longrightarrow \quad NCH_3 \quad + \quad \underset{nn}{\overset{\cdot\cdot}{N}} = C = O \qquad 9.177$$

<u>Transfer species of 1st kind of third type</u>

$$E \overset{}{\underset{}{(}} O - \underset{\underset{CH_3}{|}}{\overset{\parallel}{\underset{N}{C}}} \overset{}{\underset{n}{)}} O - \underset{\underset{CH_3}{|}}{\overset{\parallel}{\underset{N}{C}}}\cdot e \quad \longrightarrow \quad E \overset{}{\underset{}{(}} O - \underset{\underset{CH_3}{|}}{\overset{\parallel}{\underset{N}{C}}} \overset{}{\underset{n}{)}} O - C \equiv N \quad + \quad e\cdot CH_3$$

<u>Transfer species of 1st kind of third type</u> $\qquad$ 9.178

Only the C = O activation center favors electro-free- radical polymerization. Indeed, the C=N can also favor the route, for which when possible, random copolymers are obtained

$$e\cdot \underset{\underset{CH_3}{|}}{\overset{\overset{O}{\parallel}}{C}} - \underset{\underset{CH_3}{|}}{\overset{}{\underset{O}{\underset{|}{N}}}}\cdot nn \quad + \quad E\cdot^{e} \quad \longrightarrow \quad E - \underset{\underset{CH_3}{|}}{\overset{}{\underset{O}{\underset{|}{N}}}} - \overset{\overset{O}{\parallel}}{C}\cdot e$$

Unnatural route $\qquad$ 9.179

$$N\cdot^{nn} \quad + \quad e\cdot \underset{\underset{CH_3}{|}}{\overset{\overset{O}{\parallel}}{C}} - \underset{\underset{CH_3}{|}}{\overset{}{\underset{O}{\underset{|}{N}}}}\cdot nn \quad \longrightarrow \quad N - \overset{\overset{O}{\parallel}}{C} - \underset{\underset{CH_3}{|}}{\overset{}{\underset{O}{\underset{|}{N}}}}\cdot nn \qquad 9.180$$

With OCH_3 replacing CH_3, both routes are favored with living polymers produced in the absence of foreign agents. The same applies to the C = O activation center. The last equation above is the route natural to the monomer, for which electro-free-radically, the first equation-Equation 9.179 will not take place. If it takes place, random copolymers are produced.

$$E\cdot^{e} \quad + \quad nn.\underset{\underset{\underset{CH_3}{|}}{\overset{|}{O}}}{\underset{\overset{|}{C=O}}{N}} - \overset{\overset{O}{\parallel}}{C}\cdot e \quad \longrightarrow \quad EOCH_3 \quad + \quad e\cdot\overset{\overset{O}{\parallel}}{C} - N = \overset{\overset{O}{\parallel}}{C} \qquad 9.181$$

$$\ddot{N} \xleftarrow{} (C - N \xrightarrow{}_n C - N. \, nn \longrightarrow \ddot{N} \xleftarrow{} (C - N \xrightarrow{}_n C - N = C = O \; + \; nn \cdot OCH_3$$

with pendant groups:
- first C: O (double bond), below N: C=O, O, CH₃
- second C: O (double bond), below N: C=O, O, CH₃
- product C: O (double bond), below N: C=O, O, CH₃

<u>Transfer species of 1ˢᵗ kind of 1ˢᵗ type</u>   9.182

$$nn \cdot \ddot{N} - \underset{\underset{O}{\|}}{C} \cdot e \; + \; E \cdot e \longrightarrow EF \; + \; e \cdot \underset{\underset{F}{|}}{\overset{\overset{F}{|}}{C}} - \ddot{N} = C = O$$

(first N bears CF₃ group; product C bears F above and F below)

<u>Transfer species of 1ˢᵗ kind of first type</u>   9.183

Like the non-free and free charged cases, only the nucleo-non-free-radical route is favored for COOCH₃, CONH₂ types of groups.

$$E \cdot e \; + \; nn \cdot \underset{\underset{H}{|}}{\overset{}{N}} - \underset{\overset{O}{\|}}{C} \cdot e \longrightarrow E -- \underset{\underset{H}{|}}{\overset{}{N}} - \underset{\overset{O}{\|}}{C} \bullet e$$

(N bears C=O, H below; product same)   9.184

$$\ddot{N} \xleftarrow{} (C - N \xrightarrow{}_n C - N. \, nn \longrightarrow \ddot{N} \xleftarrow{} (C - N \xrightarrow{}_n C - N . nn$$

(each C bears O double bond; each N bears C=O and H below)   9.185

Both electro-free and nucleo-non-free-radical routes are favored for COR type of groups.

$$\ddot{N} \xleftarrow{} (C - O \xrightarrow{}_n C - O. \, nn \longrightarrow \ddot{N} \xleftarrow{} (C - O \xrightarrow{}_n C = O \; + \; nn \cdot OCH_3$$

OR No Transfer Species
NONE FAVORED

(each C bears O double bond; each C bears N, C=O, O, CH₃ pendant)   9.186

$$\ddot{N} \xleftarrow{} (C - O \xrightarrow{}_n C - O. \, nn \longrightarrow \ddot{N} \xleftarrow{} (C - O \xrightarrow{}_n C = O \; + \; n \cdot H$$

NOT FAVORED
OR No Transfer Species
NOT FAVORED

(each C bears O double bond; each C bears N, C=O, H pendant)   9.187

Indeed, nucleo-non- free-radically, the C=O center is rarely involved. It is the C=N center that is used, because COOR and COH groups are transfer species of the first kind of the third type nucleo- free- and non-free-radically.

755

When resonance stabilization groups are provided, the situation has already been considered wherein it was shown that the C=O center is the Nucleophilic center. Hence the type of reaction shown below can never take place. [See Equations 9.173 to 9.175]

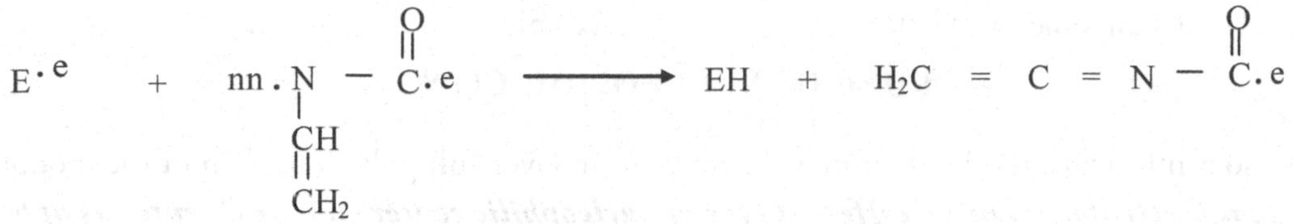

$$E \cdot^e \quad + \quad nn \cdot \underset{\underset{\overset{\|}{CH_2}}{\overset{\|}{CH}}}{N} - \overset{\overset{O}{\|}}{C} \cdot e \quad \longrightarrow \quad EH \quad + \quad H_2C = C = N - \overset{\overset{O}{\|}}{C} \cdot e$$

<u>Transfer species of 1st kind of 2nd type</u>   9.188

As already shown, $-N = C = O$ is a radical-pushing resonance stabilization group, for which existence of the reaction above is not favored.

With these groups of monomers, new activation centers have so far been identified. These include

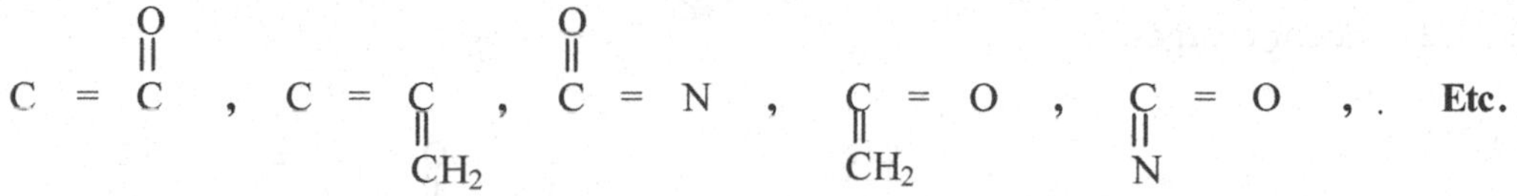

$$C = \overset{\overset{O}{\|}}{C} \quad , \quad C = \underset{CH_2}{\overset{\overset{O}{\|}}{C}} \quad , \quad \underset{CH_2}{\overset{\overset{O}{\|}}{C}} = N \quad , \quad \underset{N}{\overset{\overset{\|}{C}}{C}} = O \quad , \quad \overset{\overset{\|}{C}}{\underset{N}{C}} = O \quad , \quad \textbf{Etc.}$$

9.189

which are uniquely different from C = C, C = N and C = O activation centers associated with olefins, aldimines and ketimines, aldehydes and ketones. The new or former activation centers are far more nucleophilic than the latter and their order of capacities can be represented as follows.

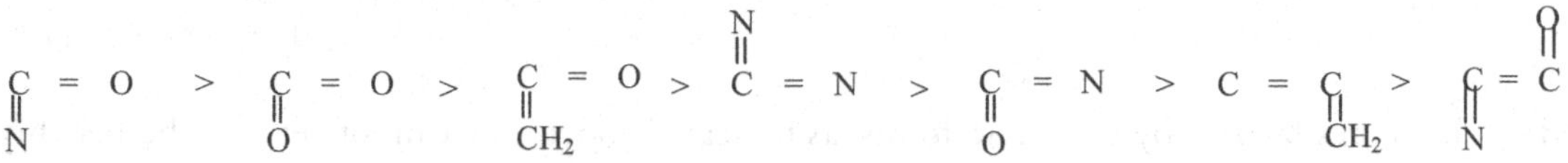

$$\underset{N}{\overset{\overset{\|}{C}}{C}} = O \quad > \quad \underset{O}{\overset{\overset{\|}{C}}{C}} = O \quad > \quad \underset{CH_2}{\overset{\overset{\|}{C}}{C}} = O \quad > \quad \overset{\overset{N}{\|}}{C} = N \quad > \quad \underset{O}{\overset{\overset{\|}{C}}{C}} = N \quad > \quad C = \underset{CH_2}{\overset{\overset{\|}{C}}{C}} \quad > \quad \underset{N}{\overset{\overset{\|}{C}}{C}} = C$$

<u>Order of Nucleophilicity</u>   9.190

Unlike olefins, the characters of the monomers carrying these activation centers are not determined by the routes favored. *Just as –RC=C=C=O group is more radical-pulling than -COR group, so also –N=C=O group is more radical-pushing than –N=CH$_2$ group.*

9.5. 1,4- Cumulenic isocyanates

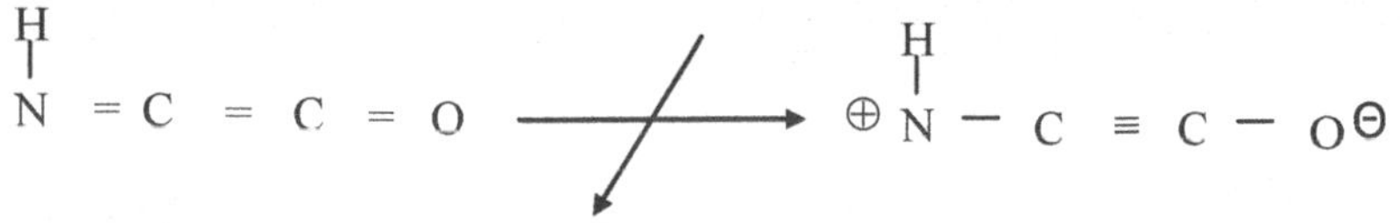

$$\overset{\overset{H}{|}}{N} = C = C = O \quad \longrightarrow\!\!\!/\!\!\!\longrightarrow \quad \oplus \overset{\overset{H}{|}}{N} - C \equiv C - O^{\ominus}$$

9.191

The first member of this unknown case is

It has the 1,3- ketenic character when HN = group is replaced with two hydrogen atoms (more nucleophilic than ketene) and isocyanic acid character when the O = C = group is replaced with = O group (more nucleophilic than isocyanic acid).

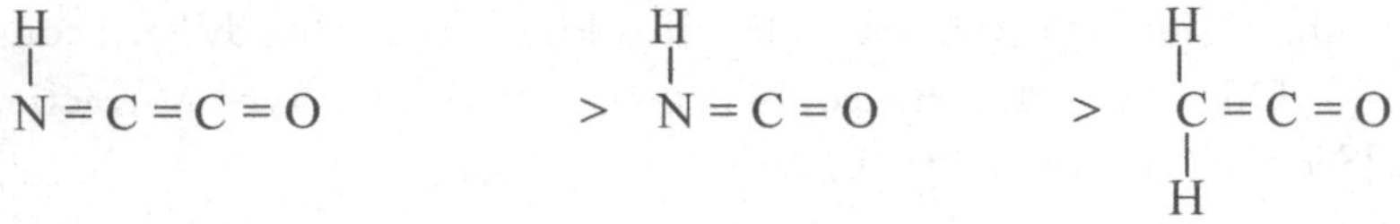

$$\underset{\text{1,4- Cumulenic isocyanic acid}}{\overset{\overset{\displaystyle H}{|}}{N}=C=C=O} \quad > \quad \underset{\text{Isocyanic acid}}{\overset{\overset{\displaystyle H}{|}}{N}=C=O} \quad > \quad \underset{\text{Ketene}}{\overset{\overset{\displaystyle H}{|}}{\underset{\underset{\displaystyle H}{|}}{C}}=C=O}$$

ORDER OF NUCLEOPHILICITY 9.192

Like 1,3- isocyanic acids, the H atom must be replaced to favor full polymerization of the monomer. *It is indeed an Electrophile with a different type of nucleophilic center- the C=C center as opposed to the C=O center, which is the most nucleophilic.*

9.5.1. Charged characters of 1,4- Cumulenic isocyanates

Beginning with alkylane groups, the followings are obtained without allowing the nitrogen center carry positive covalent charges.

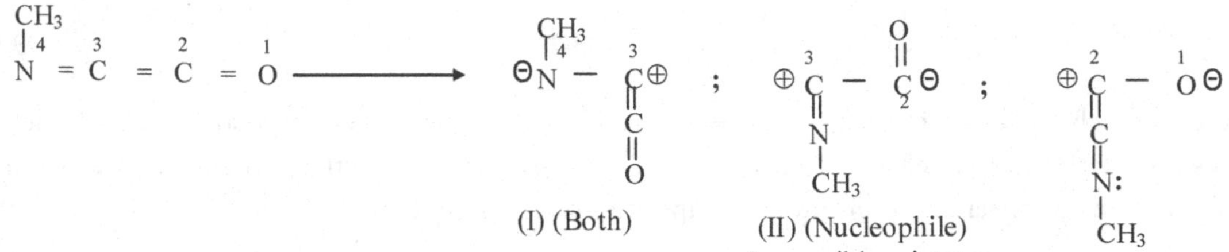

(I) (Both)   (II) (Nucleophile) Impossible existence   (III) (Nucleophile)   9.193

Based on the routes favored by the mono-forms, as indicated above, one cannot yet say whether this is a Nucleophile or an Electrophile. Whether the groups are radical-pushing or radical pulling, the charges or indeed the radicals carried by the active centers are fixed. For example, N which is electronegative cannot be made to carry a positive non-free charge or an electro non-free-radical in the presence of C atom directly bonded to it, otherwise, the result will be very explosive to comprehend. Hence, the activation centers are carrying what are shown above. Only (II) and (III) have the potentials of favoring molecular rearrangement.

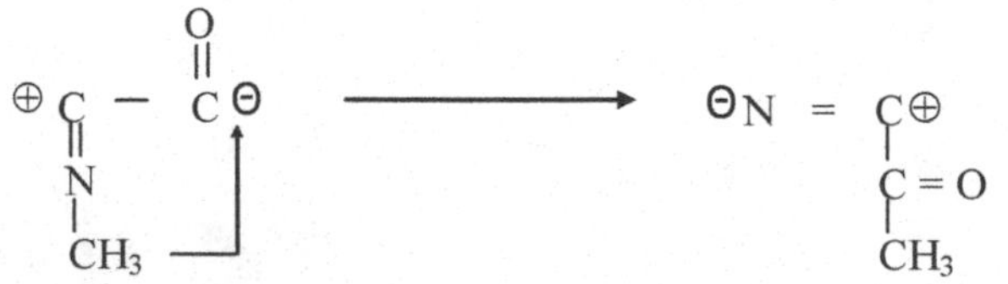

(II)   (IV) (Nucleophile) Cannot exist chargedly   9.194

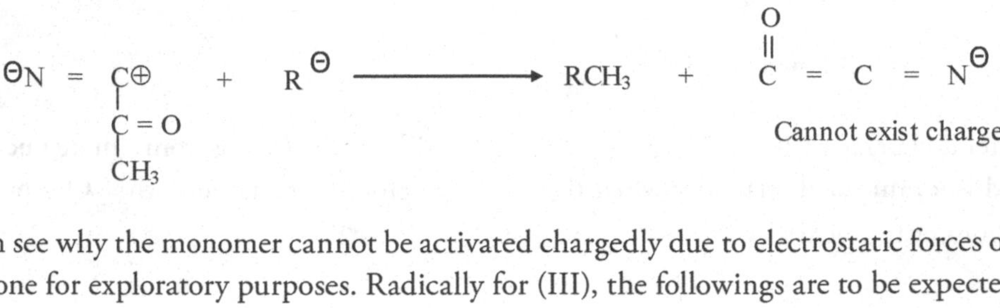

Cannot exist chargedly   9.195

One can see why the monomer cannot be activated chargedly due to electrostatic forces of repulsion. It is being done for exploratory purposes. Radically for (III), the followings are to be expected.

$$e \bullet \overset{\displaystyle C}{\underset{\displaystyle \underset{\displaystyle \underset{\displaystyle CH_3 \textemdash}{N}}{\overset{\Vert}{C}}}{C}} - O \bullet nn \longrightarrow n \bullet \overset{\displaystyle H}{\underset{\displaystyle H}{C}} - N = C = C^{\bullet e} - OH \longrightarrow \overset{\displaystyle H}{\underset{\displaystyle H}{C}} = N - C \equiv \underset{\displaystyle OH}{C}$$

$$H \bullet e$$

9.196

(III) cannot be resonance stabilized based on the types of radicals carried by the mono-form. The only center first activated is (II) the least nucleophilic center. That center is found to be strongly nucleophilic, in particular after molecular rearrangement to give (IV) whose presence is favored radically in Equation 9.194. From the foregoing, the followings are obvious-

$$CH_3 \textemdash N = \qquad >> \qquad = O$$

Order of radical-pushing capacity 9.197

$$O = C = \qquad > \qquad CH_3 \textemdash N = C = C = C = \qquad > \qquad CH_3 \textemdash N = C =$$

Order of radical-pulling capacity (ODD) 9.198

$$\underset{\underset{\underset{N}{\overset{\vert}{}}= \overset{3}{C} = \overset{2}{C} = \overset{1}{O}}{\overset{\vert}{\underset{}{O}}}}{\overset{CH_3}{\overset{\vert}{}}} \longrightarrow$$

(I) Both ; (II) Both ; (III) Nucleophile 9.199

Molecular rearrangement is not favored. It is important to note the characters obtained for the different families of monomers by substituted groups on the three mono-forms, (III) is most nucleophilic, followed by (I) and lastly (II). Since none can favor activated existence chargedly, this isocyanate cannot be activated chargedly, but only radically. It should also be noted that (II) the first mono-form (Y) favors both free-radical routes.

With radical-pulling groups, consider COOR groups.

$$\underset{\underset{\underset{CH_3}{\overset{\vert}{O}}}{\overset{\vert}{C = O}}}{\overset{4}{N} = \overset{3}{C} = \overset{2}{C} = \overset{1}{O}} \longrightarrow$$

(I) (Electrophile) ; (II) (Nucleophile) ; (III) (Both) 9.200

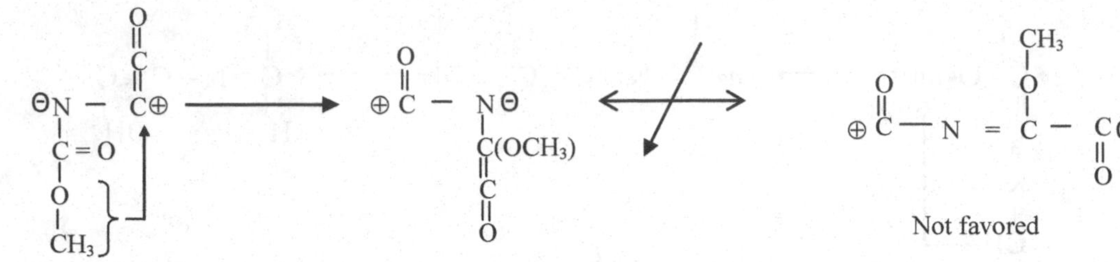

(I) (Electrophiles)   (IV)   Not favored   9.201

The negative charge in (IV) being non-free, it cannot be resonance stabilized. The main mono- form is (II) which seems to favor both routes and cannot be activated chargedly.

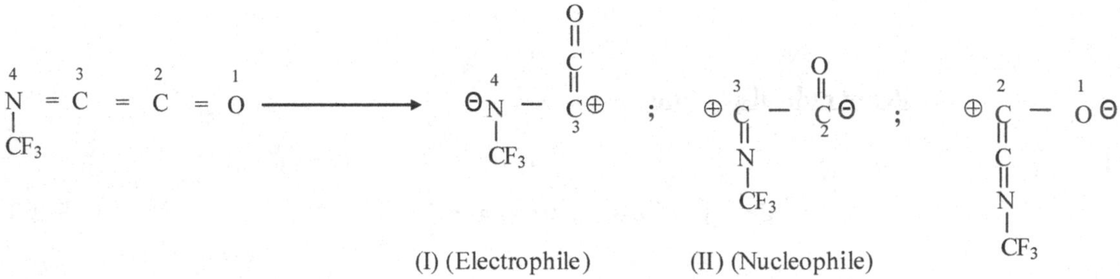

(I) (Electrophile)   (II) (Nucleophile)

(III) (Both)   9.202

Like the case above, only (I) can be considered for molecular rearrangement only radically. The main mono-form (II) is Nucleophilic while (I) is Electrophilic just like the case above.

In all the analysis, if the nitrogen center had been allowed to carry positive covalent charges, then charged activation of the monomers would be impossible.

9.5.2. Free-radical characters of 1,4-Cumulenic isocyanates

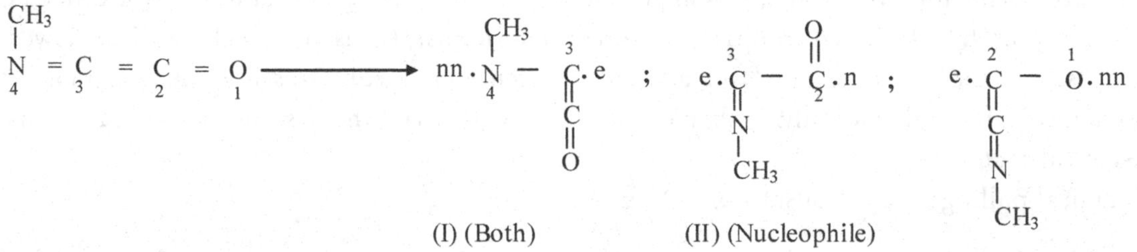

(I) (Both)   (II) (Nucleophile)

(III) (Nucleophile)   9.203

Like the charged case, most of the mono-forms have transfer species. Also like the charged case, molecular rearrangement is favored by (II) and (III), and the order of capacities deduced for the substituted groups remain the same. (II) is the main center first activated. ***Alternating copolymers cannot be obtained here nucleo-non-free-radically.*** Replacing CH_3 group with OR types, the following are to be expected.

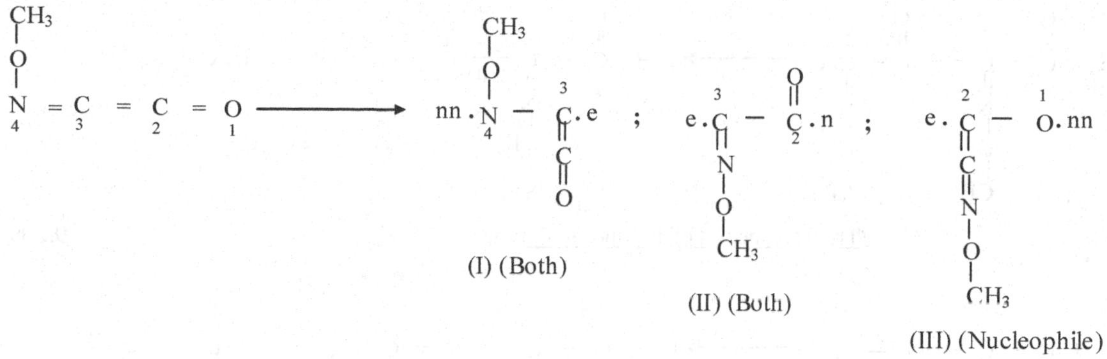

$$\underset{4}{N} = \underset{3}{C} = \underset{2}{C} = \underset{1}{O} \quad \overset{CH_3}{\overset{|}{\overset{O}{|}}} \longrightarrow \quad nn \cdot \underset{4}{N} - \overset{CH_3}{\overset{|}{\overset{O}{|}}} \underset{\substack{| \\ C \\ \| \\ O}}{C} \cdot e \quad ; \quad e \cdot \overset{3}{C} - \overset{O}{\overset{\|}{\underset{2}{C}}} \cdot n \quad ; \quad e \cdot \overset{2}{C} - \overset{1}{O} \cdot nn$$

(I) (Both)

(II) (Both)

(III) (Nucleophile)   9.204

Nevertheless, (III) still remains the most nucleophilic of the three mono-forms. However, the main mono-form is (II) that which is the Y center.

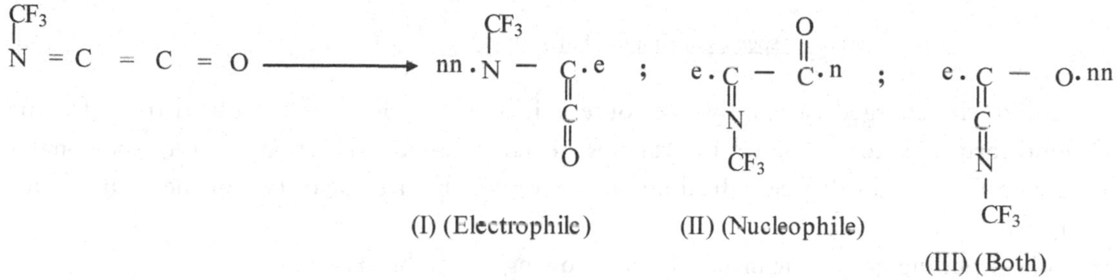

$$\overset{CF_3}{\overset{|}{N}} = C = C = O \quad \longrightarrow \quad nn \cdot \overset{CF_3}{\overset{|}{N}} - \underset{\substack{| \\ C \\ \| \\ O}}{C} \cdot e \quad ; \quad e \cdot C - \overset{O}{\overset{\|}{C}} \cdot n \quad ; \quad e \cdot C - O \cdot nn$$

(I) (Electrophile)

(II) (Nucleophile)

(III) (Both)   9.205

$$N = C = C = O \quad \longrightarrow \quad nn \cdot N - \overset{O}{\overset{\|}{\underset{\substack{CF_2 \\ CF_3}}{C}}} \cdot e \quad ; \quad e \cdot C - \overset{O}{\overset{\|}{C}} \cdot n \quad ; \quad e \cdot C - O \cdot nn$$

(I) (Electrophile)

(II) (Nucleophile)

(III) (Nucleophile) 9.206

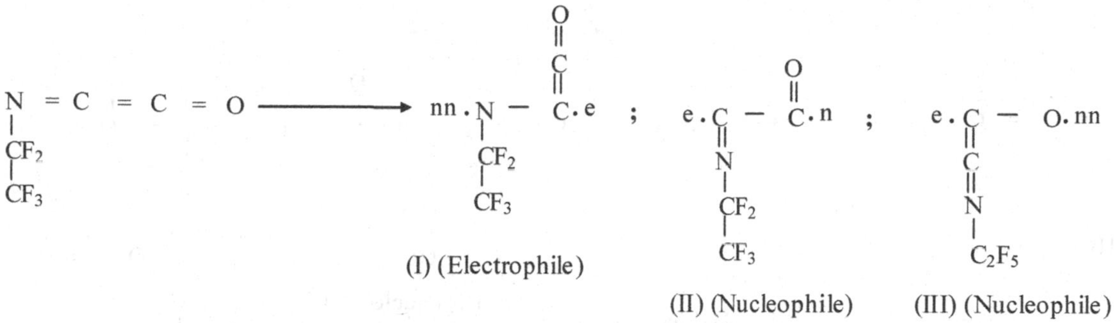

The (I)s, favor only the nucleo-non-free route, based on the followings.

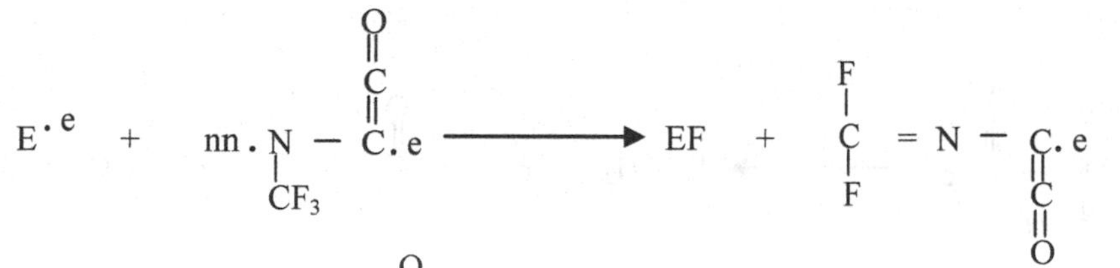

$$E \cdot e \ + \ nn \cdot \underset{CF_3}{\overset{O}{\overset{\|}{N}}} - \underset{CF_3}{C} \cdot e \longrightarrow EF \ + \ \overset{F}{\underset{F}{\overset{|}{C}}} = N - \underset{\substack{C \\ \| \\ O}}{C} \cdot e$$

9.207

$$E \cdot e \ + \ nn \cdot N - \underset{\substack{CF_2 \\ CF_3}}{C} \cdot e \longrightarrow EF \ + \ \overset{F}{\underset{CF_3}{\overset{|}{C}}} = N - \underset{\substack{C \\ \| \\ O}}{C} \cdot e$$

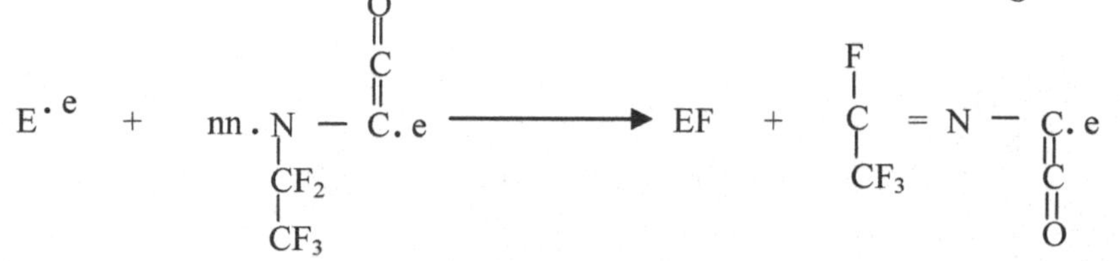

9.208

It is surprising to note that the same transfer species is involved for CF_3 and C_2F_5 groups. (II) and (III) have different transfer species as indicated below.

$$E \xleftarrow{} \left(\overset{\overset{\displaystyle O}{\|}}{C} - \underset{\underset{\underset{CF_3}{|}}{\underset{CF_2}{|}}}{\overset{\overset{\displaystyle N}{\|}}{C}} \right)_n \overset{\overset{\displaystyle O}{\|}}{C} - \overset{}{C}.e \longrightarrow E \xleftarrow{} \left(\overset{\overset{\displaystyle O}{\|}}{C} - \underset{\underset{\underset{CF_3}{|}}{\underset{CF_2}{|}}}{\overset{\overset{\displaystyle N}{\|}}{C}} \right)_n \overset{\overset{\displaystyle O}{\|}}{C} - C \equiv N + e.C_2F_5$$

<u>Transfer species of 1st kind of 3rd type</u>   9.209

$$E \xleftarrow{} \left(O - \underset{\underset{\underset{C_2F_5}{|}}{\underset{N}{\|}}}{\overset{\overset{\displaystyle O}{\|}}{C}} \right)_n O - C.e \longrightarrow E \xleftarrow{} \left(O - \underset{\underset{\underset{C_2F_5}{|}}{\underset{N}{\|}}}{\overset{\overset{\displaystyle O}{\|}}{C}} \right)_n O - C \equiv C - N = \overset{\overset{\displaystyle F}{|}}{\underset{\underset{F}{|}}{C}} + e.CF_3$$

<u>Transfer species of first kind of 7th type</u>   9.210

Radically, unlike the charged case, only one route is involved – electro-free-radical route for the identical mono-forms. Hence, while 1,3-isocyanates are half-free-radical monomers, 1,4- isocya-nates and above are half-free and full-free-radical monomers of which only one or two of the mono-forms are relevant.

When radical-pulling groups are involved, the followings are to be expected.

$$\overset{4}{N} = \overset{3}{C} = \overset{2}{C} = \overset{1}{O} \longrightarrow$$

with substituents $\underset{\underset{\underset{CH_3}{|}}{\underset{O}{|}}}{\underset{C=O}{|}}$ on N

$$nn.\overset{4}{N} - \underset{\underset{\underset{\underset{CH_3}{|}}{\underset{O}{|}}}{\underset{C=O}{|}}}{\overset{\overset{\overset{\displaystyle O}{\|}}{C}}{\underset{3}{C}}}.e \quad ; \quad e.\overset{3}{\underset{\underset{\underset{OCH_3}{|}}{\underset{C=O}{|}}}{C}} - \overset{1}{\underset{2}{C}}.n \quad ; \quad e.\overset{2}{\underset{\underset{\underset{\underset{CH_3}{|}}{\underset{O}{|}}}{\underset{C=O}{|}}}{C}} - \overset{1}{O}.nn$$

(I) (Electrophile)   (II) (Nucleophile)   (III) (Both)   9.211

$$E^{.e} + nn.\overset{}{\underset{\underset{\underset{\underset{CH_3}{|}}{\underset{O}{|}}}{\underset{C=O}{|}}}{N}} - \overset{\overset{\overset{\displaystyle O}{\|}}{C}}{C}.e \longrightarrow EOCH_3 + e.\overset{\overset{\displaystyle O}{\|}}{C} - N = C = C = O$$

(I)   9.212

761

$$\ddot{N} \quad + \quad e.\underset{\substack{| \\ C \\ \| \\ N \\ | \\ C=O \\ | \\ O \\ | \\ CH_3}}{C} - O.nn \quad \longrightarrow \quad \dot{N} - \underset{\substack{| \\ C \\ \| \\ N \\ | \\ C=O \\ | \\ O \\ | \\ CH_3}}{C} - O.nn \quad \xrightarrow{\;+\; n\,(III)\;}$$

$$(III)$$

$$\ddot{N} \overset{\substack{| \\ C \\ \| \\ N \\ | \\ C=O \\ | \\ O \\ | \\ CH_3}}{+} \underset{}{\Big(\,C} - O\,\Big)_n \underset{\substack{| \\ C \\ \| \\ N \\ | \\ C=O \\ | \\ O \\ | \\ CH_3}}{C} - O.nn \quad \longrightarrow \quad \ddot{N} \overset{\substack{| \\ C \\ \| \\ N \\ | \\ C=O \\ | \\ O \\ | \\ CH_3}}{+} \Big(\,C - O\,\Big)_n \underset{\substack{| \\ C \\ \| \\ N}}{C} = O \;+\; n\,.\,COOCH_3$$

<u>Not Favoured</u>   9.213

Unlike 1,4 - ketenes, in view of the fact the nitrogen center cannot carry electro-non-free-radicals under the present conditions, the reaction immediately above is not favored. The same order of capacities of substituted groups involved chargely also apply here.

9.5.3 Resonance stabilization in 1,4- Cumulenic isocyanates

$$\underset{\substack{| \\ CH \\ \| \\ CH_2}}{N} = C = C = O \quad \longrightarrow \quad \underset{\substack{| \\ CH \\ \| \\ CH_2}}{\overset{\ominus}{N}} - \overset{\substack{O \\ \| \\ C}}{C}{}^{\oplus} \;;\; \overset{\oplus}{\underset{\substack{\| \\ N \\ | \\ CH \\ \| \\ CH_2}}{C}} - \overset{\substack{O \\ \| \\ C}}{C}{}^{\ominus} \;;\; \overset{\oplus}{\underset{\substack{\| \\ C \\ \| \\ N \\ | \\ CH \\ \| \\ CH_2}}{C}} - O^{\ominus}$$

(I) (Both)   (II) (Nucleophile)   (III) (Nucleophile)   9.214

With resonance stabilization groups, the followings are to be expected only radically and only for (I). In fact none of the mono-forms above exist, because the least nucleophilic center is the alkenyl group. So far, one is beginning to identify new types of groups such as RN=C= which like O= C= is radical-pulling, and RN=C=C= which like O=C=C= is radical-pushing.

762

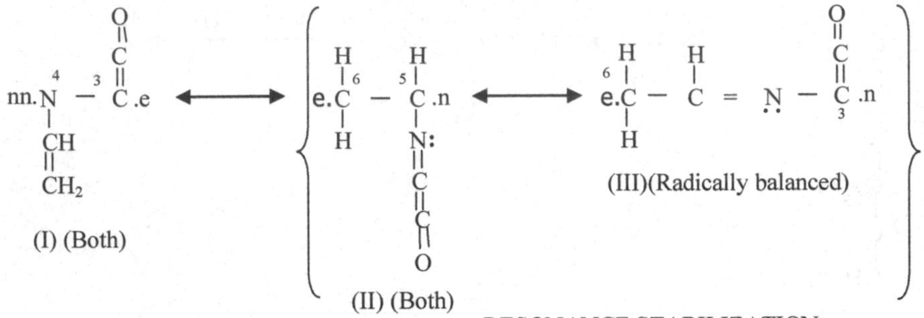

$$
\begin{array}{l}
\underset{\substack{|\\ \mathrm{CH}\\ \|\\ \mathrm{CH_2}}}{nn.\overset{4}{N}} - \overset{3}{\underset{\|}{\overset{O}{C}}}.e \quad\longleftrightarrow\quad
\left\{
\begin{array}{l}
e.\overset{6}{\underset{|}{\overset{H}{C}}} - \overset{5}{\underset{\substack{|\\ N:\\ \|\\ C\\ \|\\ O}}{\overset{H}{C}}}.n \quad\longleftrightarrow\quad
e.\overset{6}{\underset{|}{\overset{H}{C}}} - \overset{H}{\underset{H}{C}} = \underset{..}{N} - \overset{O}{\underset{\|}{C}_3}.n \\
\qquad\qquad\qquad\qquad\qquad\qquad \text{(III)(Radically balanced)}\\
\text{(II) (Both)}
\end{array}
\right.
\end{array}
$$

(I) (Both)

RESONANCE STABILIZATION   9.215

One can observe that the $-N=C=C=O$ group itself is a radical-pulling resonance stabilization group. In Equation 9.173, $-N=C=O$ was shown to be a radical-pushing resonance stabilization group in which movement just like above from (II) to (III) is by resonance stabilization wherein if H on Carbon center 4 of 1,3- was CH_3, or CF_3 or on Carbon center 5 above (1,4-), their transfer species will be shielded as shown below for 1,3-Isocyanate.

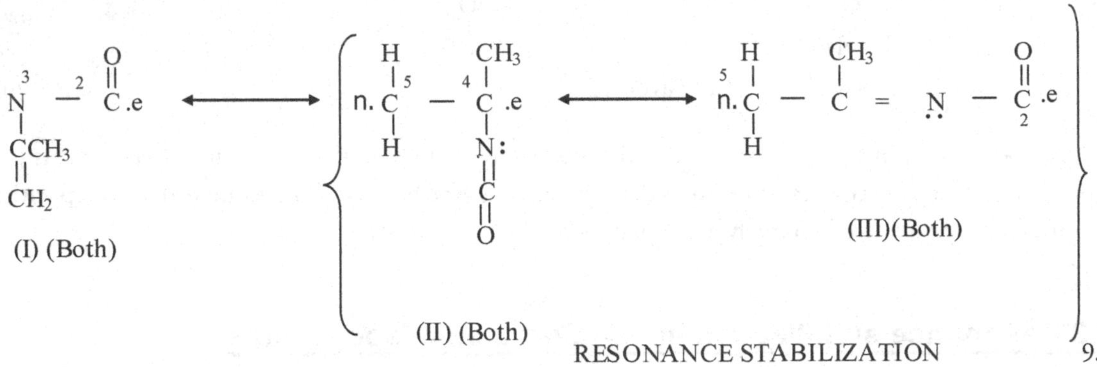

$$
\begin{array}{l}
\underset{\substack{|\\ \mathrm{CCH_3}\\ \|\\ \mathrm{CH_2}}}{\overset{3}{N}} - \overset{2}{\underset{\|}{\overset{O}{C}}}.e \quad\longleftrightarrow\quad
\left\{
\begin{array}{l}
n.\overset{5}{\underset{|}{\overset{H}{C}}} - \overset{4}{\underset{\substack{|\\ N:\\ \|\\ C\\ \|\\ O}}{\overset{CH_3}{C}}}.e \quad\longleftrightarrow\quad
n.\overset{5}{\underset{|}{\overset{H}{C}}} - \overset{CH_3}{C} = \underset{..}{N} - \overset{O}{\underset{\|}{C}_2}.e \\
\qquad\qquad\qquad\qquad\qquad\qquad \text{(III)(Both)}\\
\text{(II) (Both)}
\end{array}
\right.
\end{array}
$$

(I) (Both)

RESONANCE STABILIZATION   9.216

Like $-N=C=O$ group, $-N=C=C=O$ group is also a resonance stabilization group, but of radical-pushing and pulling types respectively. Thus, one should expect the following order in capacity for these types of groups.

$$-\underset{.}{N}=C=C=C=O \;>\; -N=C=O \;>\; -N=CH_2 \;>\; -N=CF_2$$

<u>Order of Radical-pushing capacity [ODD]</u>   9.217a

$$O=C=C=C=C=N- \;>\; O=C=C=N- \;>\; O=N-$$

<u>Order of Radical-pulling capacity [EVEN]</u>   9.217b

$$RN{=}C{=}C{=}C{=}C{=}C{=} \;>\; RN{=}C{=}C{=}C{=} \;>\; RN{=}C{=} \;>>\; RN{=}CR-$$

<u>Order of Radical-pulling capacity [ODD]</u>   9.217c

$$RN{=} \;>\; RN{=}C{=}C{=}C{=}C{=} \;>\; RN{=}C{=}C{=}$$

<u>Order of Radical-pushing capacity [EVEN]</u>   9.217d

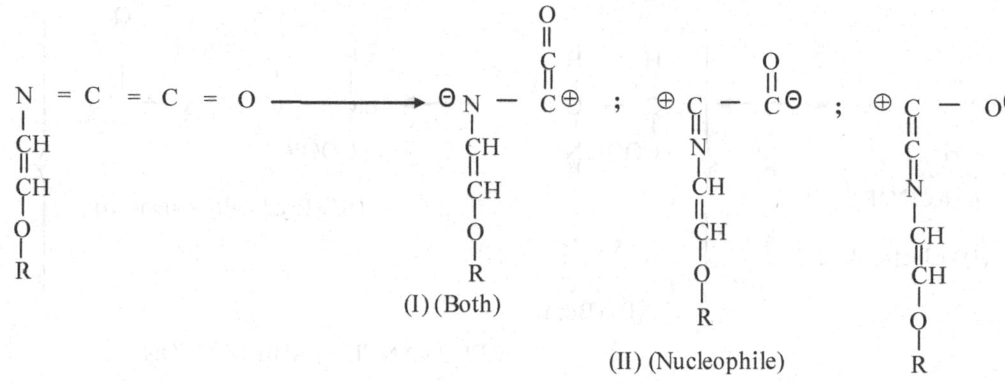

$$
\begin{array}{c}
N = C = C = O \\
|\\
CH \\
\|\\
CH \\
|\\
O \\
|\\
R
\end{array}
\quad\longrightarrow\quad
\underset{\substack{|\\CH\\\|\\CH\\|\\O\\|\\R}}{\overset{\overset{O}{\underset{\|}{}}\overset{C}{\underset{\|}{}}}{\ominus N - C\oplus}}
\;;\;
\underset{\substack{|\\N\\|\\CH\\\|\\CH\\|\\O\\|\\R}}{\overset{\overset{O}{\underset{\|}{}}}{\oplus C - C\ominus}}
\;;\;
\underset{\substack{\|\\C\\\|\\N\\|\\CH\\\|\\CH\\|\\O\\|\\R}}{\oplus C - O\ominus}
$$

(I) (Both)   (II) (Nucleophile)   (III) (Nucleophile)   9.218

None of the above is resonance stabilized. As a matter of fact, none of the mono-forms above exist, because like the case above, the least nucleophilic center still remains the alkenyl group.

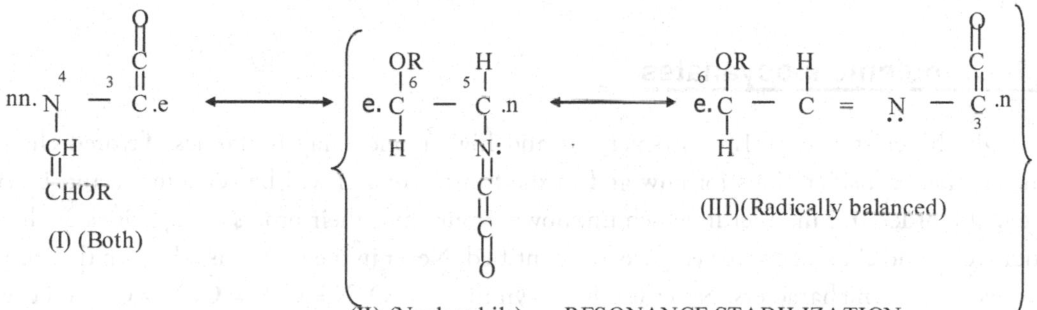

$$
\begin{array}{c}
\overset{\overset{O}{\|}}{\underset{\substack{|\\CH\\\|\\CHOR}}{nn.\overset{4}{N} - \overset{3}{C}.e}}
\end{array}
\quad\longleftrightarrow\quad
\left\{
\begin{array}{ccc}
\overset{6}{OR}\;\; \overset{5}{H} & & \overset{\overset{O}{\|}}{C} \\
e.\underset{H}{C} - \underset{\substack{N:\\\|\\C\\\|\\C\\\|\\O}}{C}.n
& \longleftrightarrow &
e.\underset{H}{\overset{OR}{C}} - C = \overset{..}{N} - \overset{}{C}.n \\
& & \text{(III)(Radically balanced)}
\end{array}
\right\}
$$

(I) (Both)   (II) (Nucleophile)   RESONANCE STABILIZATION   9.219

The OR group cannot be shielded, since it is externally located. However, the group is a radical-pulling resonance stabilization group.

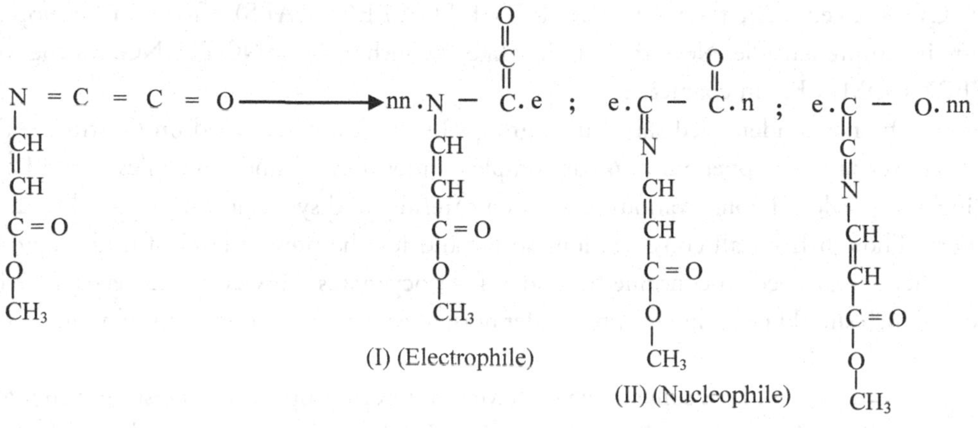

$$
\begin{array}{c}
N = C = C = O \\
|\\
CH \\
\|\\
CH \\
|\\
C = O \\
|\\
O \\
|\\
CH_3
\end{array}
\quad\longrightarrow\quad
\underset{\substack{|\\CH\\\|\\CH\\|\\C=O\\|\\O\\|\\CH_3}}{\overset{\overset{O}{\|}}{nn.N - C.e}}
\;;\;
\underset{\substack{|\\N\\|\\CH\\\|\\CH\\|\\C=O\\|\\O\\|\\CH_3}}{\overset{\overset{O}{\|}}{e.C - C.n}}
\;;\;
\underset{\substack{\|\\C\\\|\\N\\|\\CH\\\|\\CH\\|\\C=O\\|\\O\\|\\CH_3}}{e.C - O.nn}
$$

(I) (Electrophile)   (II) (Nucleophile)   (III) Nucleophile   9.220

None of the mono-forms above exist. As shown above, the monomer is an Electrophile which can be resonance stabilized.

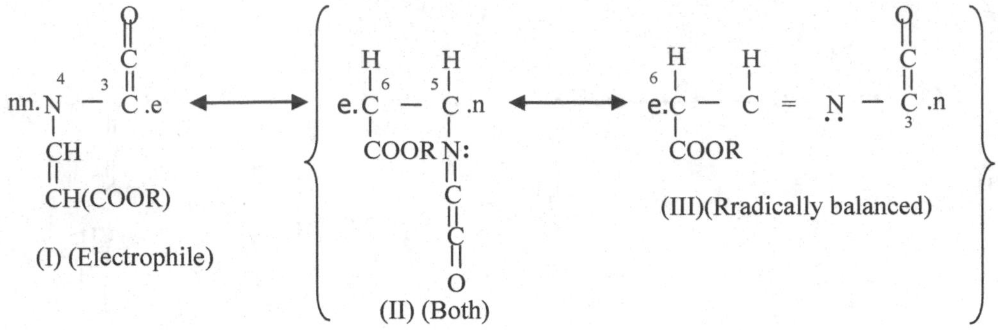

$$
\text{nn.}\underset{3}{\overset{4}{N}} - \overset{\overset{O}{\underset{\|}{C}}}{C}.e \quad\longleftrightarrow\quad \left\{ e.\overset{\overset{6}{H}}{\underset{\underset{COOR}{|}}{C}} - \overset{\overset{5}{H}}{\underset{\underset{\underset{\underset{O}{\|}}{C}}{N:}}}{C}.n \quad\longleftrightarrow\quad e.\overset{\overset{6}{H}}{\underset{\underset{COOR}{|}}{C}} - \overset{\overset{H}{|}}{C} = \overset{..}{N} - \overset{\overset{O}{\underset{\|}{C}}}{\underset{3}{C}}.n \right\}
$$

CH
‖
CH(COOR)

(I) (Electrophile)

(II) (Both)

(III)(Rradically balanced)

RESONANCE STABILIZATION   9.221

The group in the monomer has resonance stabilized the alkenyl group thereby providing a wide range for the Y center. *As it seems so far, in the absence of resonance stabilization groups, the C=C center is Y while the C=N conjugatedly placed to it is the X center. The C=O center is never activated or involved in such reactions being the most nucleophilic center.*

9.6 1,5- Cumulenic isocyanates

Though the existence of 1,5 - isocyanates and higher ones may be far less favored than 1,4 - isocyanates, their considerations for now and in the future are and will be very important, in view of the systematic order and manner in which unknown groups and their orders of capacities qualitatively, "quantitatively" and characters are being clearly identified. Never in the past, was it known that activation centers have their own characters. Never was it known that C = O, N = C, C = C, N ≡ C, centers etc. are nucleophilic (female) in characters, neither were the order of nucleophilicities known. Why most of the activation centers are more female in character than male, is a different subject matter. The unknown questions are too numerous to list to the point where one wonders how to communicate with the so-called "intellectuals" including the author, all which are a grain of sand in the sight of THE ALMIGHTY INFINITE GOD. Even then their (SO-CALLED INTELLECTUALS) efforts in development of humanity are incomprehensible. Nevertheless, they like the author know NOTHING, for the world we live in is VERY COMPLEX in character.

Never were the newly identified substituted groups known to exist. Based on the firm realization of the fact that every natural phenomenon has complete orderliness, foundation rules which have been proposed (including rules of conservation), have been carefully and systematically applied to delve into New Frontiers. Though from all considerations so far and for the present level of understanding and presentation, there is no need to continue beyond 1,4 – isocyanates. However, the needs arise because standing rules which should never fail to apply under normal or abnormal conditions of natural existence, have to be proposed, without exceptions or failures.

Though there are general laws some of which have so far been proposed, interestingly enough every family of monomers have laws specifically unique to them. It is from these laws that all the different types of chemical reactions which have been observed over the years, but could not adequately be explained, can now be explained in a systematic order without bringing in elements of doubts and exceptions. The first member of 1,5 - isocyanate (isocyanic acid) shown below, can be observed to illusionarily belong to many families such as-

$$\overset{H}{\underset{1}{N}} = \overset{2}{C} = \overset{3}{C} = \overset{4}{C} = \overset{5}{O} \longrightarrow \overset{H}{\ominus N} - C \equiv C - \overset{O}{C}\oplus$$

"RESONANCE STABILIZATION"

9.222

1,3-cumulenes when = O and HN = are replaced with two hydrogen atoms (more nucleophilic); 1,3 - isocyanates when = C = C = O group is replaced with O = group (more nucleophilic); 1,3- ketenes when HN = C = is replaced with two hydrogen atoms (more nucleophilic) and etc., than the lower members. Like the lower members, the hydrogen atom on the nitrogen center must be replaced with the same similar special groups, before polymerizations can be favored. See Equations 9.190 and 9.192 and below is an extension of it.

$$\overset{H}{\underset{}{N}} = C = C = C = O \quad > \quad \overset{H}{\underset{}{N}} = C = C = O \quad > \quad \overset{H}{\underset{}{N}} = C = O \quad > \quad \overset{H}{\underset{H}{C}} = C = O$$

| 1,5- Cumulenic isocyanic acid | 1,4-Cumulenic isocyanic acid | Isocyanic acid | Ketene |

ORDER OF NUCLEOPHILICITY

9.223

9.6.1 Charged and Radical characters of 1, 5 – isocyanates

9.224

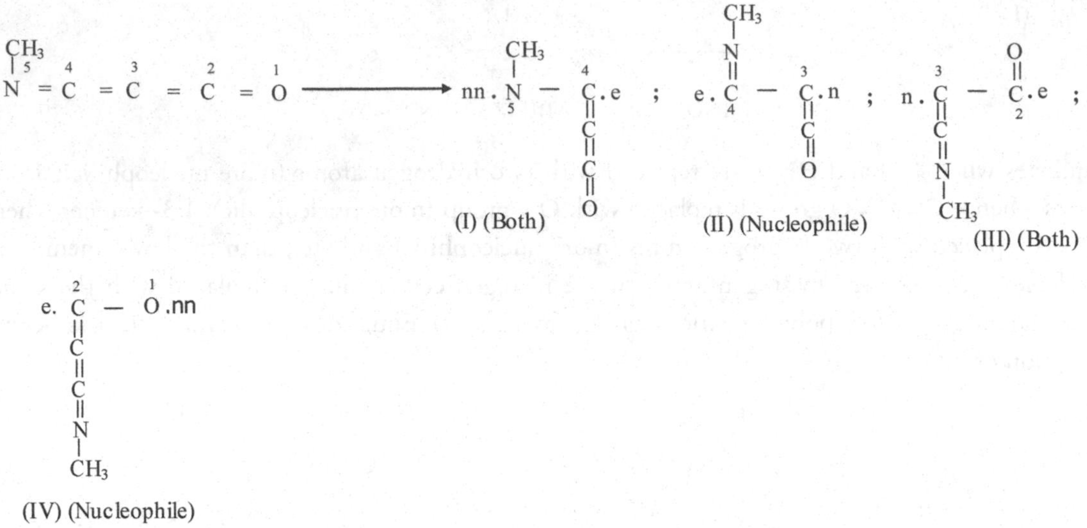

$$\text{CH}_3\text{-}\overset{5}{N}=\overset{4}{C}=\overset{3}{C}=\overset{2}{C}=\overset{1}{O} \longrightarrow$$

(I) (Both)   (II) (Nucleophile)   (III) (Both)

(IV) (Nucleophile)

9.225

The more nucleophilic character of (II) and (IV) activation centers than the other activation centers, can generally be observed for all the isocyanates. The least nucleophilic center is (II) which happens to favor only the nucleo- free-radical route.

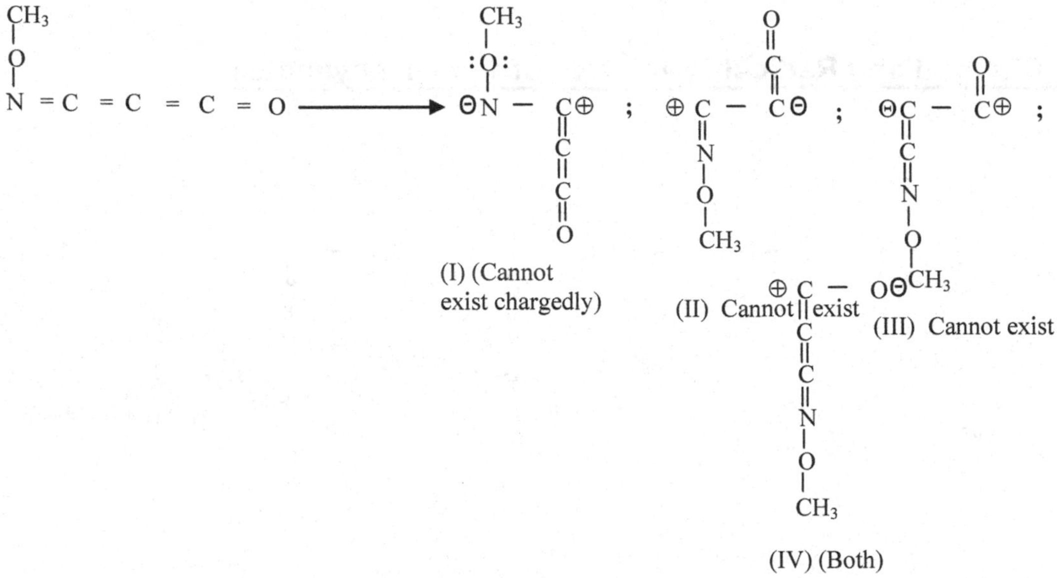

(I) (Cannot exist chargedly)   (II) Cannot exist   (III) Cannot exist

(IV) (Both)   9.226

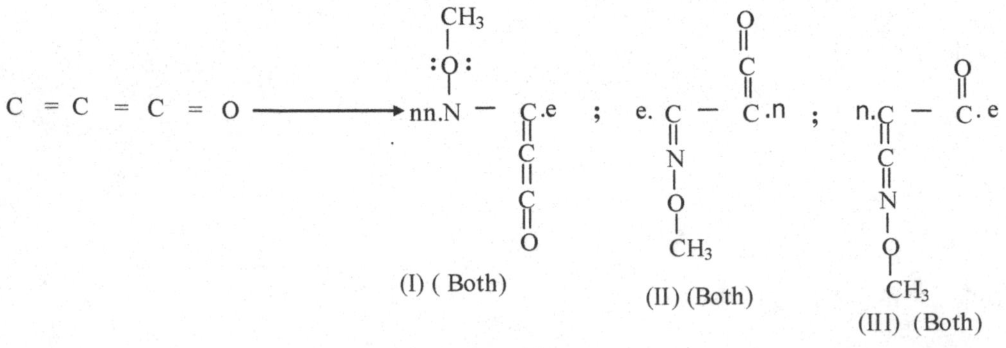

(I) (Both)   (II) (Both)   (III) (Both)

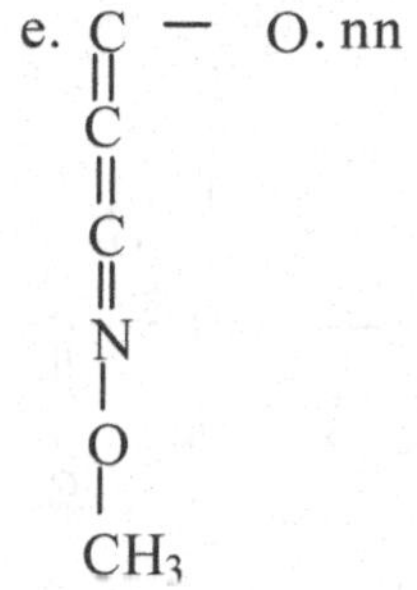

$$
\begin{array}{c}
\text{e. } C \; - \; O.\,nn \\
\parallel \\
C \\
\parallel \\
C \\
\parallel \\
N \\
\mid \\
O \\
\mid \\
CH_3
\end{array}
$$

(IV) (Both)   9.227

For all these, (I) and (II) are the first mono-forms if Electrophilic in character. As Nucleophiles, only (II) will be the mono-form. (I) as can be observed so far is resonance stabilized as already shown in Equation 9,222 only radically and shown below for the case above. (IV) cannot be resonance stabilized. However, (III) the second Y center can be resonance stabilized, just like (I). Its corresponding X center is (IV). This second Electrophilic center can never be used.

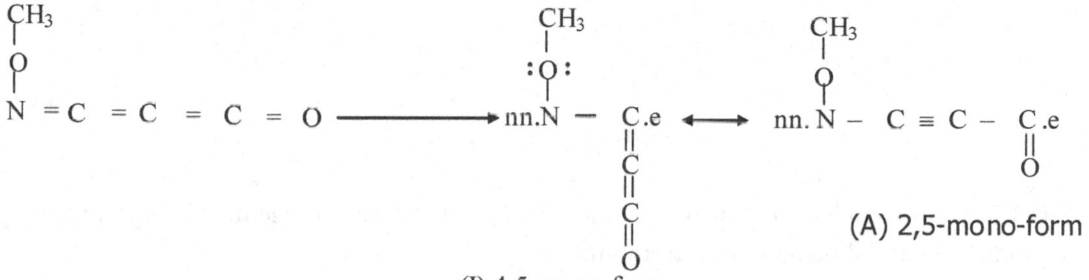

(I) 4,5-mono-form

(A) 2,5-mono-form

RESONANCE STABILIZATION   9.228

(A) above is the mono-form for (I) and as it seems, this is a "Y" center for all Isocyanates, just like in some ringed electrophiles. The (IV)s of some of those Equations can undergo molecular rearrangement where possible as Nucleophilic centers since they have the potentials of favoring it as shown below for the 1,4-Isocynates. It cannot take place with the 5,4- or 2,3- types with OR O

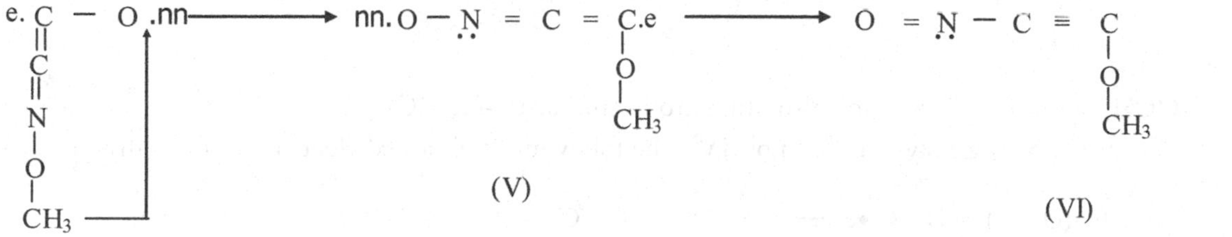

(IV) (9.199)   9.229

Note that (VI) cannot be activated chargedly or radically after the molecular rearrangement. Hence, the 1,2-center cannot indeed be activated or used an an X center.

Now, considering radical-pulling groups in place of H on nitrogen centers, the followings are obtained for COOCH$_3$ group.

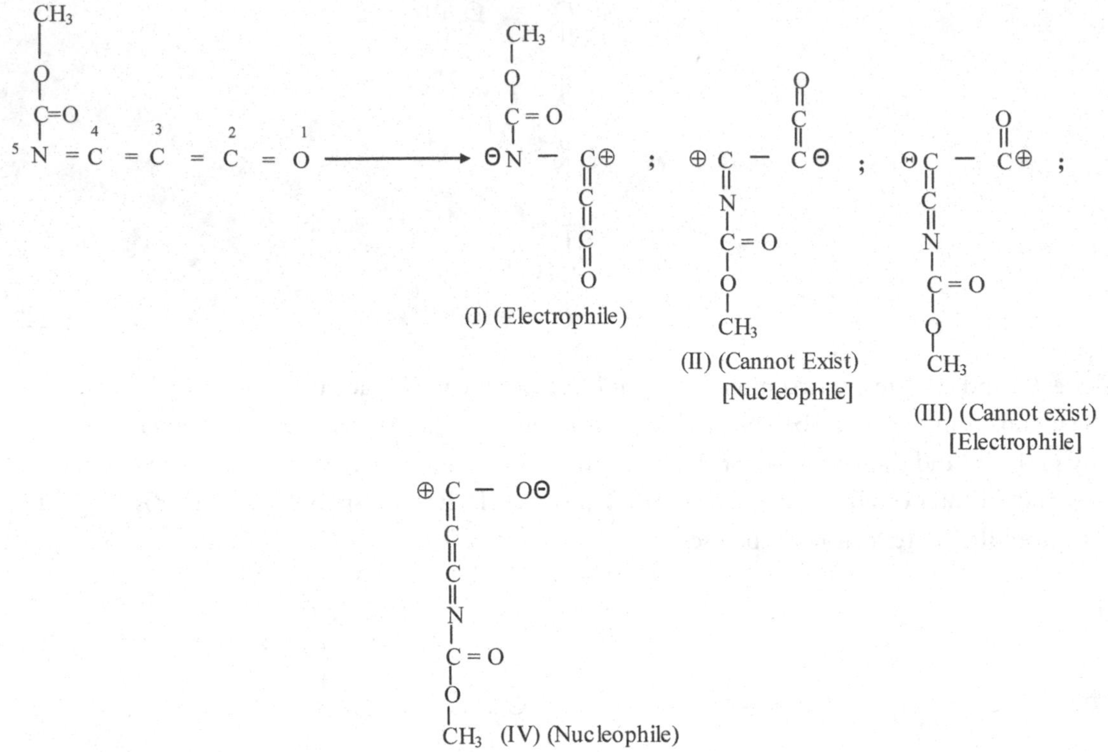

$$
\begin{array}{c}
CH_3 \\
| \\
O \\
| \\
\overset{4}{C}=O \quad \overset{3}{\;} \quad \overset{2}{\;} \quad \overset{1}{\;} \\
\overset{5}{N} = C = C = C = O
\end{array}
\longrightarrow
$$

(I) (Electrophile) ; (II) (Cannot Exist) [Nucleophile] ; (III) (Cannot exist) [Electrophile] ;

(IV) (Nucleophile)

9.230

Indeed its 4,5-mono-form has the potential of undergoing molecular rearrangement being an Electrophile and the product obtained being more electrophilic.

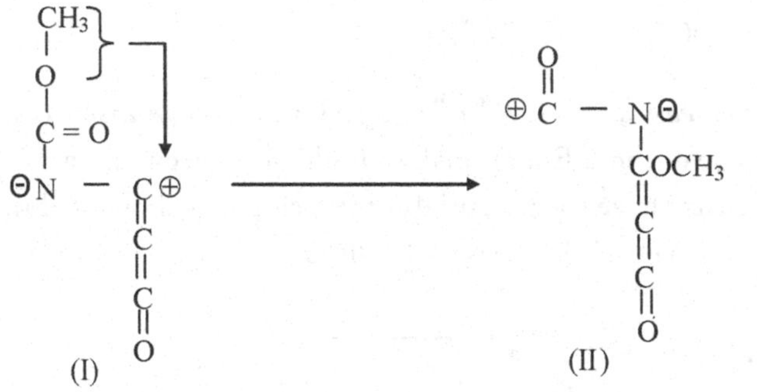

(I) $\longrightarrow$ (II)

9.231

$-C(OCH_3) = C = C = O$ is a radical-pulling group similar to $-COOCH_3$.

For the growing polymer chain of (IV), the following is obtained electro-free radically.

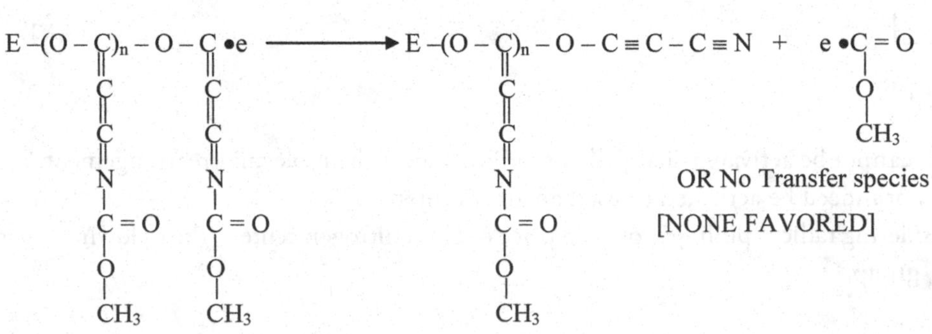

$$
E-(O-C)_n-O-C \cdot e \longrightarrow E-(O-C)_n-O-C \equiv C-C \equiv N \; + \; e \cdot C = O
$$

OR No Transfer species
[NONE FAVORED]

[Not favored]

9.232

769

One can see that transfer species involved is of the first kind of the third type, that which is not possible, because the C=O center can never be activated. If activated however, note that the –COOCH$_3$ can still be released as an electro-free-radical despite being a free-radical-pulling group. Though -OCH$_3$ or –NH$_2$ are free-radical-pushing groups (i.e., en.OCH$_3$), they can only be released as a nucleo-non-free-radical, because of what the central atom is carrying. One can also see many sources of branching sites electro-free-radically (C=O center) along the sides of the chain if such centers exist for other systems, but not for the case above since the center will never be used, whether under acidic operating conditions or not.

Having considered important aspects of 1,5-isocyanates, the pattern of analysis are obvious, with greater opportunity being offered in distinguishing between free and non-free charges, radical-donating and -withdrawing groups, radical-pushing and -pulling groups, electro-negativity and electropositivity and so on.

Based on the observations so far, having considered cumulenic ketenes, ketenic isocyanates, there is need to consider cumulenic aldimines or ketimines. Both ends of isocyanates are ***iminic and carbonylic*** in character. Both ends of ketenes are ***corbonylic and cumulenic*** in character. One will now consider aldimines, where both ends are ***iminic and cumulenic*** in character, as distinguished below.

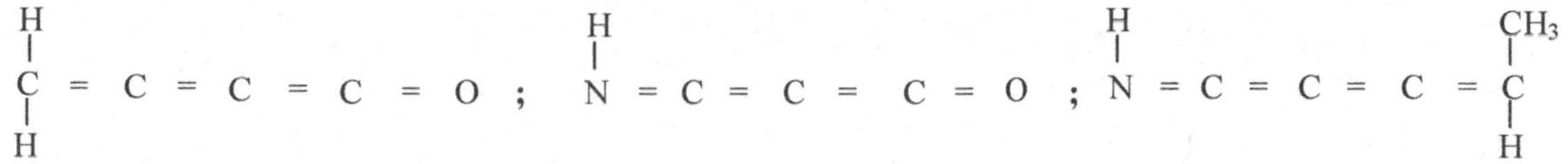

$$
\begin{array}{ccccccc}
\overset{\displaystyle H}{\underset{\displaystyle H}{C}} = C = C = C = O & ; & \overset{\displaystyle H}{N} = C = C = C = O & ; & \overset{\displaystyle H}{N} = C = C = C = \overset{\displaystyle CH_3}{\underset{\displaystyle H}{C}}
\end{array}
$$

1, 5 - ketenes	1, 5 - Isocyanate	1, 5 - Aldimine
(i) One carbonyl end	(i) One carbonyl end	(i) One cumulenic end
(ii) One cumulenic end	(ii) One iminic end	(ii) One iminic end 9.233

At the beginning, it was so to speak assumed or thought that 1,3-, 1,4-, 1,5- and so on Isocyanates were NUCLEOPHILES. But based on the considerations so far, like the Cumulenic Ketenes and Aldimines, Isocyanates are ELECTROPHILES becoming more so when the H on the N central atom is changed to radical-pulling groups with transfer species.

9.7. Cumulenic ketemines

The first members of these unknown families of monomers are as shown below.

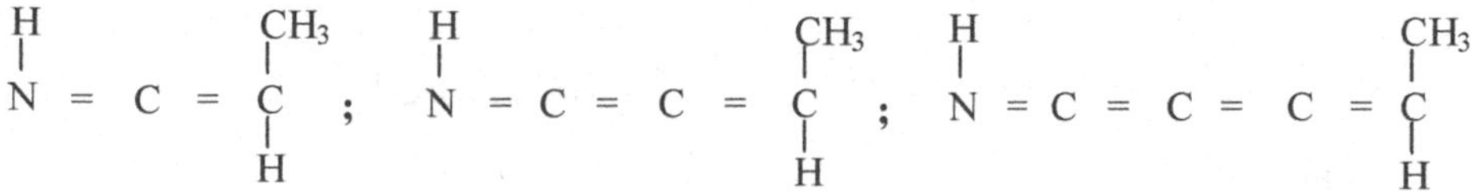

$$
\overset{\displaystyle H}{N} = C = \overset{\displaystyle CH_3}{\underset{\displaystyle H}{C}} \; ; \; \overset{\displaystyle H}{N} = C = C = \overset{\displaystyle CH_3}{\underset{\displaystyle H}{C}} \; ; \; \overset{\displaystyle H}{N} = C = C = C = \overset{\displaystyle CH_3}{\underset{\displaystyle H}{C}}
$$

1, 3 - type	1, 4 - type	1, 5 - type

Cumulenic Aldimines   9.234

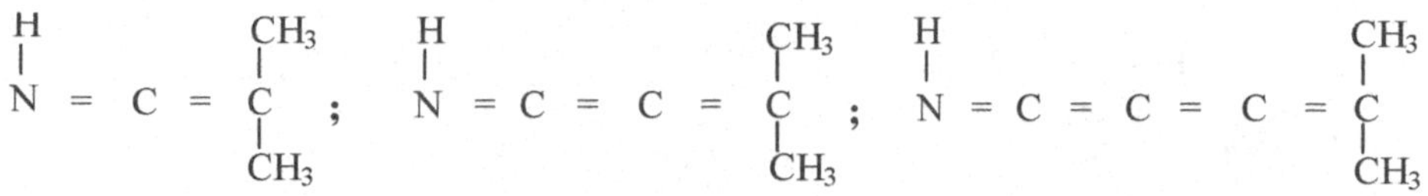

$$
\overset{\displaystyle H}{N} = C = \overset{\displaystyle CH_3}{\underset{\displaystyle CH_3}{C}} \; ; \; \overset{\displaystyle H}{N} = C = C = \overset{\displaystyle CH_3}{\underset{\displaystyle CH_3}{C}} \; ; \; \overset{\displaystyle H}{N} = C = C = C = \overset{\displaystyle CH_3}{\underset{\displaystyle CH_3}{C}}
$$

1, 3 - type	1, 4 - type	1, 5 - type

Cumulenic ketemines   9.235

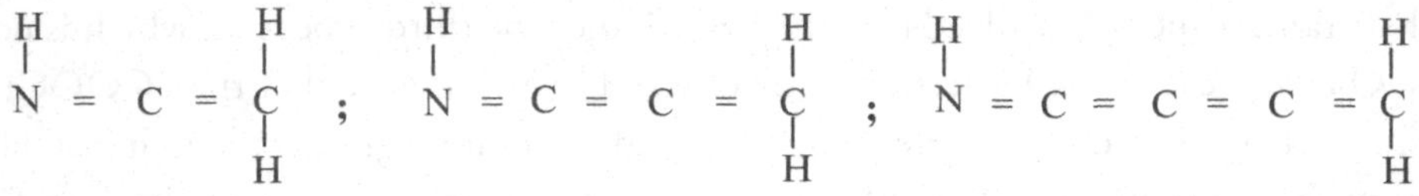

$$
\begin{array}{ccccc}
\underset{\underset{H}{|}}{\overset{\overset{H}{|}}{N}} = C = \underset{\underset{H}{|}}{\overset{\overset{H}{|}}{C} & ; & \underset{\underset{}{}}{\overset{\overset{H}{|}}{N}} = C = C = \underset{\underset{H}{|}}{\overset{\overset{H}{|}}{C} & ; & \underset{}{\overset{\overset{H}{|}}{N}} = C = C = C = \underset{\underset{H}{|}}{\overset{\overset{H}{|}}{C}
\end{array}
$$

1, 3 - type	1, 4 - type	1, 5 - type

Cumulenic formaldimines   9.236

The names above are indeed based on what the external carbon center is carrying, like in Olefins. Some may call all of them ketamines wherein the C center is said to be carrying =CR$_2$ groups.

Ketemines will be considered. Aldimines will behave almost alike. The last group will not be considered, since their mono-forms are very unstable as shown below. ***Whether one is correct or not, it should be noted that KNOWLEDGE IS NOT EXCLUSIVE TO ANY HUMAN EXCEPT THE ALMIGHTY INFINITE GOD.*** It has no Beginning and no End.

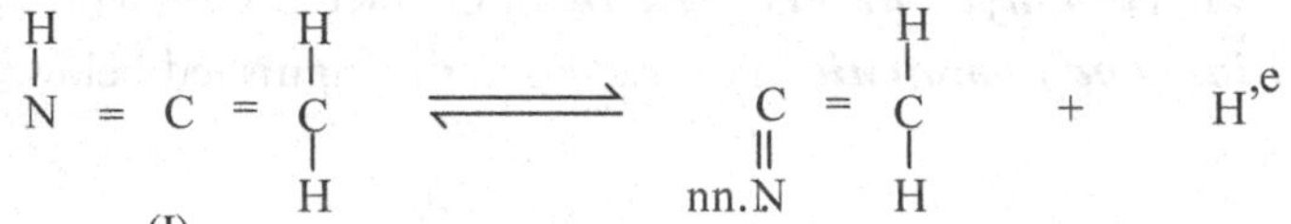

(I)   9.237

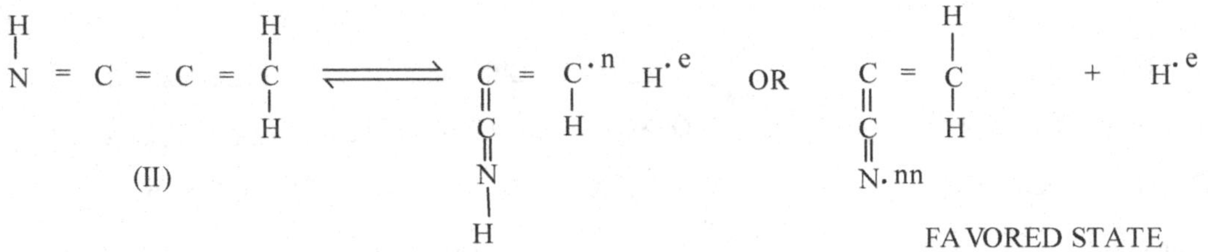

(II)   FAVORED STATE   9.238

When suppressed by a passive catalyst, they cannot molecularly rearrange as shown below. The center involved for rearrangement could be any of the centers. For the less or least nucleophilic center, the followings are to be expected.

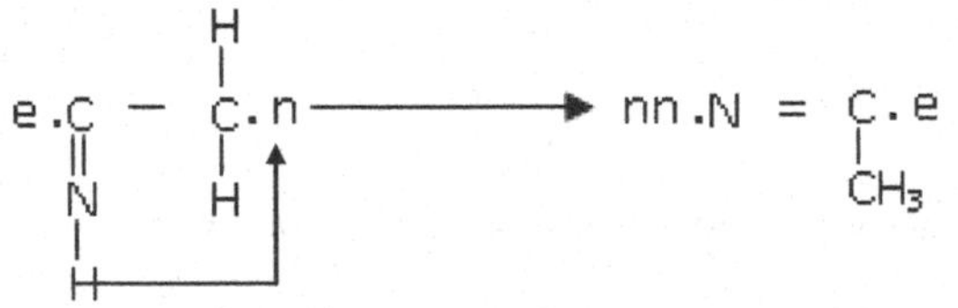

(I)   [Not favored]   9.239a

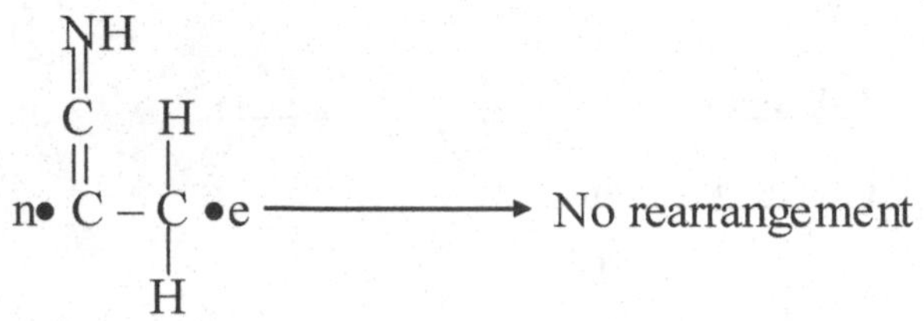

No rearrangement

1,4-type   9.239b

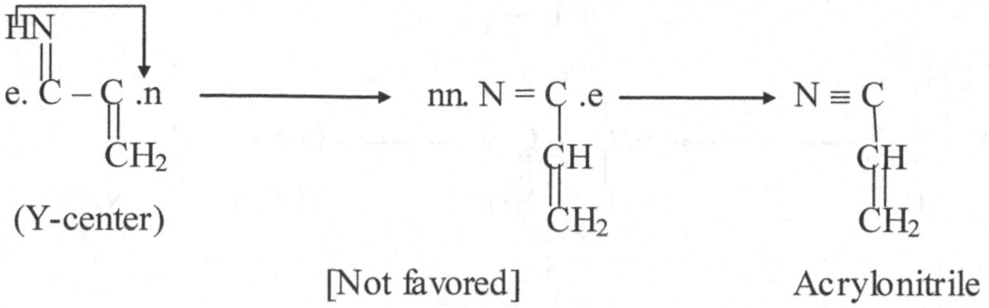

$$e. \quad \underset{CH_2}{\overset{\overset{HN}{\parallel}}{C}} - C .n \quad \longrightarrow \quad nn. \underset{\underset{CH_2}{\overset{|}{C}H}}{N} = C .e \quad \longrightarrow \quad \underset{\underset{CH_2}{\overset{|}{C}H}}{N} \equiv C$$

(Y-center)   [Not favored]   Acrylonitrile   9.239c

The 1,3- type will not rearrange, because that is the Y center. It can however electroradicalize. The 1,4-type cannot, because HN=C= is a radical-pulling group when the externally located C = C is activated. When the internally located C= C center is activated, it cannot molecularly rearrange to Acrylonitrile, because this is the least nucleophilic center and is the Y center.

For 1,5 type, the followings are to be expected.

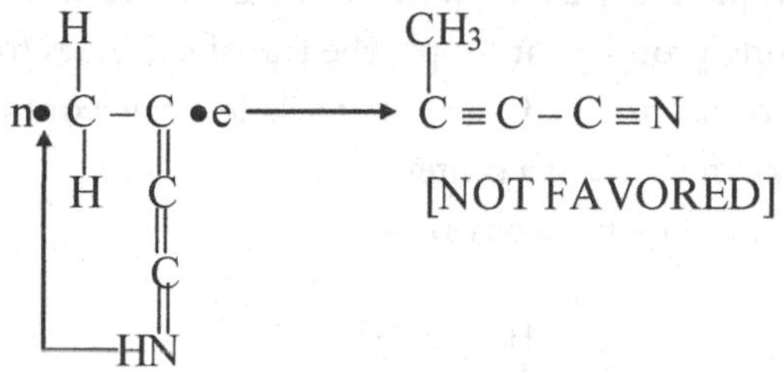

$$n\bullet \underset{\underset{\underset{HN}{|}}{\overset{|}{C}}}{\overset{\overset{H}{|}}{C}} - \underset{\overset{|}{C}}{C} \bullet e \quad \longrightarrow \quad \overset{\overset{CH_3}{|}}{C} \equiv C - C \equiv N$$

[NOT FAVORED]

9.240a

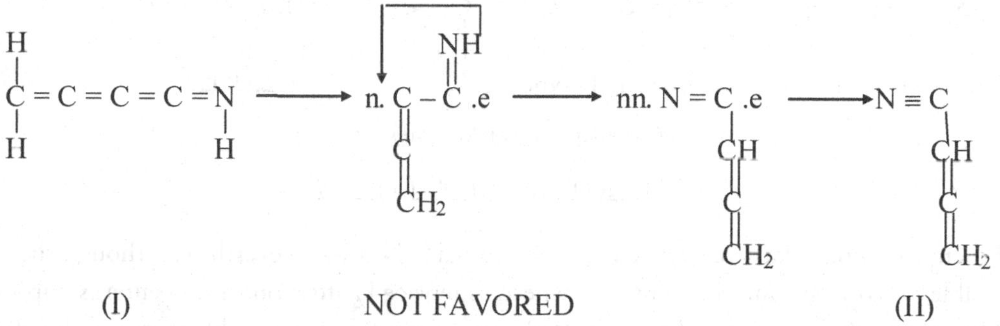

$$\underset{\overset{|}{H}}{\overset{\overset{H}{|}}{C}} = C = C = C = \underset{\overset{|}{H}}{N} \quad \longrightarrow \quad n. \underset{\underset{CH_2}{\overset{|}{C}}}{\overset{\overset{NH}{\parallel}}{C}} - C .e \quad \longrightarrow \quad nn. \underset{\underset{CH_2}{\overset{|}{C}H}}{N} = C .e \quad \longrightarrow \quad \underset{\underset{CH_2}{\overset{|}{C}H}}{N} \equiv C$$

(I)   NOT FAVORED   (II)   9.240b

This is not the least nucleophile center, the center cumulatively placed to the N = C center. This is the Y center which cannot be used, because the 4,5- mono-form is the least nucleophilic center- the first Y center. The fact that it is the Y center is shown below using the 1,3- type.

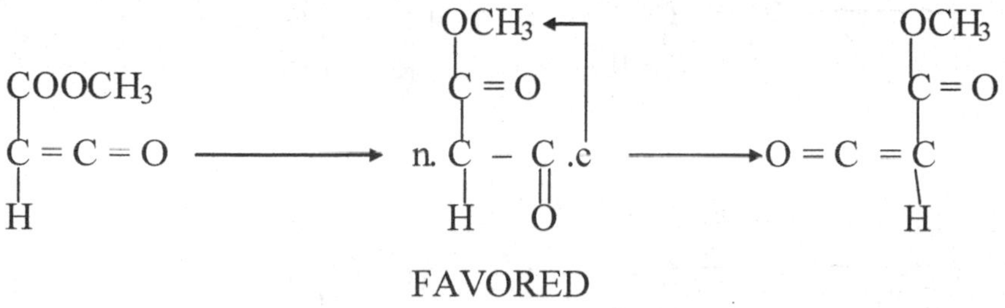

$$\underset{\overset{|}{H}}{\overset{\overset{COOCH_3}{|}}{C}} = C = O \quad \longrightarrow \quad n. \underset{\overset{|}{H}}{\overset{\overset{OCH_3}{|}}{\underset{\overset{|}{}}{C}}} - \underset{\overset{\parallel}{O}}{C} .c \quad \longrightarrow \quad O = C = \underset{\overset{|}{H}}{\overset{\overset{OCH_3}{|}}{\underset{\overset{|}{}}{C}}}$$

9.241a

FAVORED

$$
\begin{array}{c}
\text{COOCH}_3 \\
| \\
\text{C} = \text{C} = \text{N} \\
| \quad\quad | \\
\text{H} \quad\quad \text{H}
\end{array}
\longrightarrow
\begin{array}{c}
\text{OCH}_3 \leftarrow\!\rceil \\
| \\
\text{C} = \text{O} \\
| \\
\text{n. C} - \text{C .e} \\
| \quad\quad \| \\
\text{H} \quad\quad \text{NH}
\end{array}
\longrightarrow
\begin{array}{c}
\text{H} \\
| \\
\text{O} = \text{C} = \text{C} \\
| \\
\text{H}_3\text{CO} - \text{C} = \text{NH}
\end{array}
\quad\quad 9.241b
$$

$$
\begin{array}{c}
\text{H} \\
| \\
\text{C} = \text{C} = \text{N} \\
| \quad\quad | \\
\text{H} \quad\quad \text{H}
\end{array}
\longrightarrow
\begin{array}{c}
\text{H} \\
| \\
\text{mn. N} - \text{C .e} \\
\quad\quad \| \\
\text{CH}_2 \\
\uparrow\!\!\lfloor\underline{\quad\quad}\rfloor
\end{array}
\longrightarrow
\begin{array}{c}
\text{H} \\
| \\
\text{C} \equiv \text{C} \\
\quad\quad | \\
\text{NH}_2
\end{array}
$$

$$
\text{(A)} \quad\quad\quad\quad \text{FAVORED} \quad\quad\quad\quad \text{(B)}
$$

For the 1,3 type, notice the center that rearranged to give (B) of the last equation above as opposed to the nitrile of Equation 9.239a or that of Equation 9.240b. When the Y center is involved as shown in the first equation above, it is the radical-pulling groups that supply the transfer species for rearrangement or for the Conservation law during polymerization. The C = C center is less nucleophilic than the C = N center which in turn is less nucleophilic than the C =O center.

From all the considerations so far, the following is obvious-

$$
\begin{array}{c}
\text{H} \quad\quad\quad \text{H} \quad\quad\quad \text{H} \\
| \quad\quad\quad | \quad\quad\quad | \\
\text{N} = \text{C} = \text{C} \\
| \\
\text{H}
\end{array}
<
\begin{array}{c}
\text{H} \quad\quad\quad\quad\quad\quad \text{H} \\
| \quad\quad\quad\quad\quad\quad | \\
\text{N} = \text{C} = \text{C} = \text{C} \\
| \\
\text{H}
\end{array}
<
\begin{array}{c}
\text{H} \quad\quad\quad\quad\quad\quad\quad\quad\quad \text{H} \\
| \quad\quad\quad\quad\quad\quad\quad\quad\quad | \\
\text{N} = \text{C} = \text{C} = \text{C} = \text{C} \\
| \\
\text{H}
\end{array}
$$

$$
\underline{1, 3 - \text{type}} \quad\quad\quad \underline{1, 4 - \text{type}} \quad\quad\quad \underline{1, 5 - \text{type}}
$$

$$
\underline{\text{Cumulenic formaldimines}}
$$

$$
\underline{\text{ORDER OF NUCLEOPHILICITY}} \quad\quad\quad\quad 9.242
$$

None of them can molecularly rearrange to give more stable Nitriles. Nevertheless, though the existence of more stable mono-forms may be favored, the need to replace H atom on the N center is important due to Equilibrium state of existence. Hence, for both cumulenic aldimines and ketemines, the H atom on the nitrogen center must be replaced, before they can be used as iminic monomers.

Considering alkylane groups, the followings are obtained.

$$
\begin{array}{c}
\text{CH}_3 \quad\quad\, _3\text{CH}_3 \\
| \quad\quad\quad | \\
\text{N} = \text{C} = \text{C} \\
_1 \quad\,_2 \quad\quad | \\
\quad\quad\quad \text{CH}_3
\end{array}
\longrightarrow
\begin{array}{c}
\text{CH}_3 \\
| \\
\ominus\text{N} - \, _2\text{C}\oplus \\
_1 \quad\quad \| \\
\text{C(CH}_3) \\
| \\
\text{CH}_3
\end{array}
\quad ; \quad
\begin{array}{c}
\text{CH}_3 \\
_2 \qu\quad\, _3| \\
\oplus\text{C} - \text{C}\ominus \\
\| \quad\quad | \\
\text{N} \quad\quad \text{CH}_3 \\
| \\
\text{CH}_3
\end{array}
$$

$$
\text{(I) (Nucleophile)} \quad\quad\quad \text{(II) (Nucleophile)} \quad\quad 9.243
$$

$$
\begin{array}{c}
\text{CH}_3 \quad\quad \text{CH}_3 \\
| \quad\quad\quad | \\
\text{N} = \text{C} = \text{C} \\
| \\
\text{CH}_3
\end{array}
\longrightarrow
\begin{array}{c}
\text{CH}_3 \\
| \\
\text{n . N} - \text{C .e} \\
\| \\
\text{C(CH}_3) \\
| \\
\text{CH}_3
\end{array}
\quad ; \quad
\begin{array}{c}
\text{CH}_3 \\
| \\
\text{e . C} - \text{C . n} \\
\| \quad\quad | \\
\text{N} \quad\quad \text{CH}_3 \\
| \\
\text{CH}_3
\end{array}
$$

$$
\text{(III) Nucleophile} \quad\quad\quad \text{(IV) Nucleophile} \quad\quad 9.244
$$

In the absence of radical-pulling group, only the Nucleophilic center (III) can favor molecular rearrangement as shown below only radically. It cannot take place chargedly.

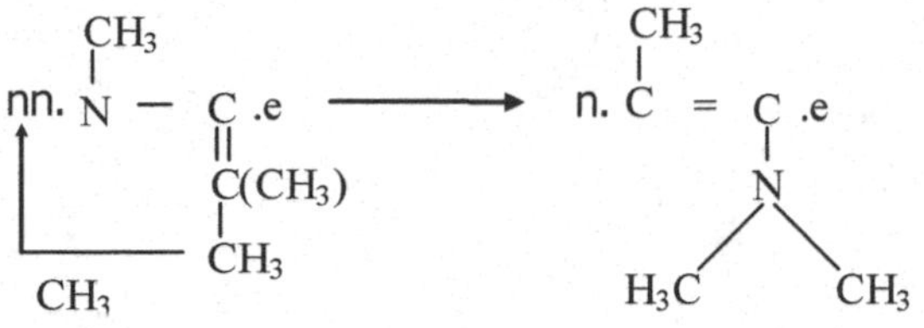

$$
\begin{array}{ccc}
\text{CH}_3 & & \text{CH}_3 \\
| & & | \\
\text{nn. N} - \text{C .e} & \longrightarrow & \text{n. C} = \text{C .e} \\
\quad\quad || & & | \\
\quad\quad \text{C(CH}_3) & & \text{N} \\
\quad\quad | & & \diagup\;\diagdown \\
\quad\quad \text{CH}_3 & & \text{H}_3\text{C}\quad\text{CH}_3 \\
\text{CH}_3 & & \\
\text{(I)} & & \text{(II) (Nucleophile)}
\end{array}
$$

(I)   (II) (Nucleophile)   FAVORED   9.245

When the Electrophilic center which seems to favor electro-free-radical routes, are involved, the followings are obtained.

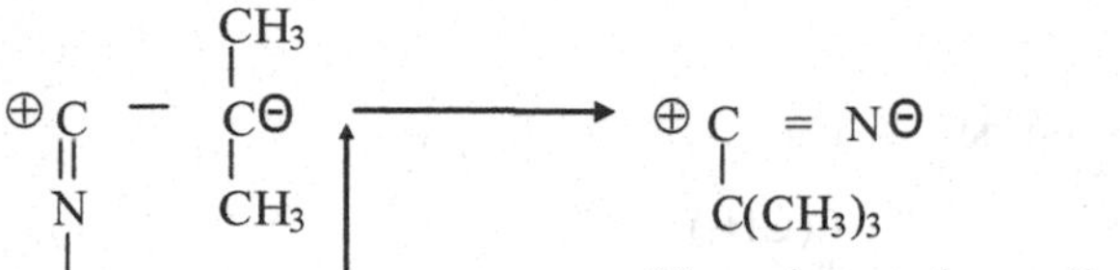

$$
\begin{array}{ccc}
& \text{CH}_3 & \\
& | & \\
\overset{\oplus}{\text{C}} - \overset{\ominus}{\text{C}} & & \overset{\oplus}{\text{C}} = \text{N}^{\ominus} \\
|| \quad | & \longrightarrow & | \\
\text{N} \quad \text{CH}_3 & & \text{C(CH}_3)_3 \\
| & & \\
\text{CH}_3 \quad\quad \text{CH}_3 & & \text{Nonexistent chargedly}
\end{array}
$$

(II)   IMPOSSIBLE REARRANGEMENT   9.246a

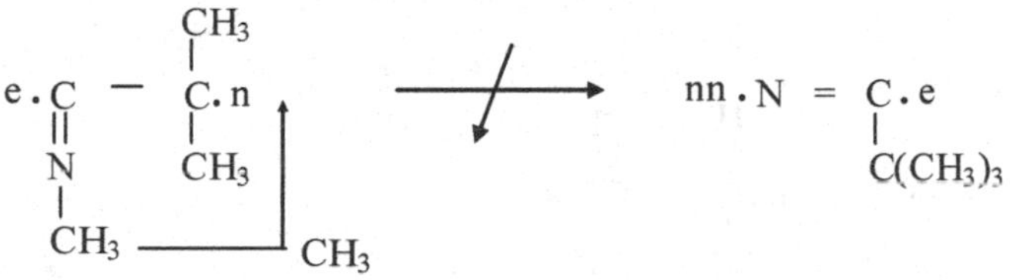

$$
\begin{array}{ccc}
& \text{CH}_3 & \\
& | & \\
\text{e . C} - \text{C . n} & & \text{nn . N} = \text{C . e} \\
|| \quad | & \cancel{\longrightarrow} & | \\
\text{N} \quad \text{CH}_3 & & \text{C(CH}_3)_3 \\
| & & \\
\text{CH}_3 \quad\quad \text{CH}_3 & &
\end{array}
$$

(II) The X center   NO REARRANGEMENT   9.246b

The rearrangement above can only be possible if the center is NUCLEOPHILIC, that which seems to be not the case from the investigations so far.

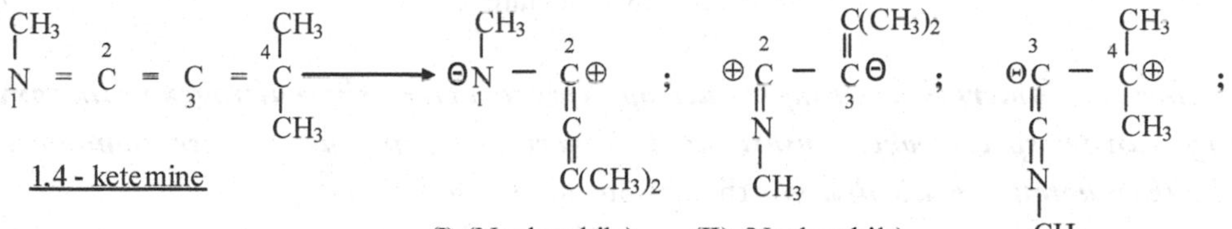

$$
\begin{array}{c}
\text{CH}_3 \quad\quad\quad \text{CH}_3 \quad\quad \text{CH}_3 \quad\quad\quad\quad\quad \text{C(CH}_3)_2 \quad\quad \text{CH}_3 \\
|\quad\;\; 2 \quad\;\; 4| \quad\quad\; | \quad\; 2 \quad\quad\quad\; 2 \quad\quad || \quad\; 3 \quad\; 4| \\
\text{N} = \text{C} = \text{C} = \text{C} \longrightarrow {}^{\ominus}\text{N} - \overset{\oplus}{\text{C}} \; ; \; \overset{\oplus}{\text{C}} - \overset{\ominus}{\text{C}} \; ; \; {}^{\ominus}\text{C} - \overset{\oplus}{\text{C}} \; ; \\
1 \quad\;\; 3 \quad\quad | \quad\quad 1 \quad\quad || \quad\quad\quad || \quad\; 3 \quad\quad || \quad\quad | \\
\quad\quad\quad\quad\quad \text{CH}_3 \quad\quad\quad\quad\quad \text{C} \quad\quad\;\; \text{N} \quad\quad\;\; \text{C} \quad\quad \text{CH}_3 \\
\quad\quad\quad\quad\quad\quad\quad\quad\quad\quad\quad || \quad\quad\quad | \quad\quad\quad || \\
\quad\quad\quad\quad\quad\quad\quad\quad\quad\quad \text{C(CH}_3)_2 \quad \text{CH}_3 \quad\; \text{N} \\
\quad\quad\quad\quad\quad\quad\quad\quad\quad\quad\quad\quad\quad\quad\quad\quad\quad\quad\; | \\
\quad\quad\quad\quad\quad\quad\quad\quad\quad\quad\quad\quad\quad\quad\quad\quad\quad\quad \text{CH}_3
\end{array}
$$

1,4 - ketemine   (I) (Nucleophile)   (II) (Nucleophile)   (III) Cannot exist

                               Cannot exist chargedly   Chargely(NUCLEOPHILE)

$$
\begin{array}{c}
\text{CH}_3 \quad\quad\quad\quad \text{CH}_3 \\
| \quad\quad\quad\quad\quad 4| \\
{}^{\ominus}\text{N} - \text{C} \equiv \text{C} - \overset{\oplus}{\text{C}} \\
1 \quad\quad\quad\quad\quad\quad | \\
\quad\quad\quad\quad\quad\quad\quad \text{CH}_3
\end{array}
$$

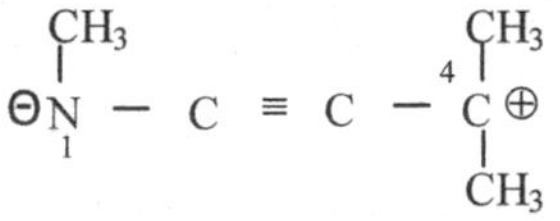

(IV) (Nucleophile) Obtained from (I)   9.247

On the basis of the developments, (II) is the Y center. (I) is the X center which cannot be resonance stabilized to (IV) as shown above. (III) does not exist. Here, only (I) can undergo molecular rearrangement as shown below radically. This is possible if it is the X center. Yet, it did not resonance stabilize to give (IV), because it is not the least nucleophilic center!

(I)   9.248

FAVORED

For (II) and (III) centers, centers wherein resonance stabilization cannot take place, the followings are obtained when made to rearrange exploratively.

(II)   9.249

Real activated state   9.250

Unlike ketenes, those monomers show stronger nucleophilic characters, since nitrogen unlike oxygen center can carry transfer species when unsaturated. Nevertheless, in general these monomers are less electrophilic than ketenes, radically. Worthy of note, are the followings-

i)   The different transfer species involved in all the mono-forms.
ii) The full and half-free characters carried by them.
iii) The fact that some of them favor RESONANCE STABILIZATION phenomenon to give more stable mono-forms, wherein electronegative elements cannot carry electro-non free-radicals for which movement is always from Half Free to Half free,

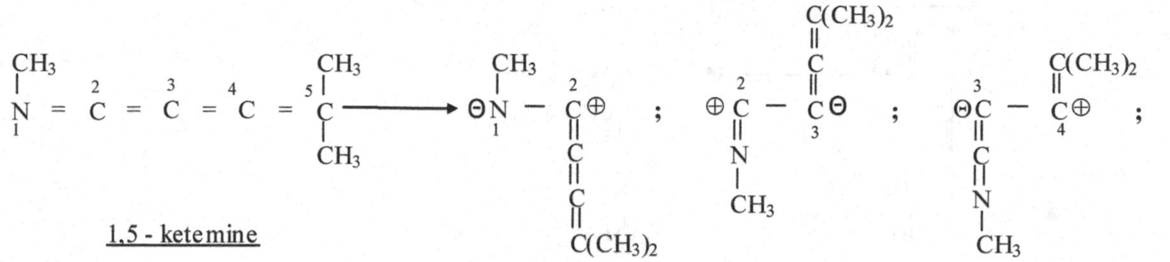

$$CH_3 - N_1 = C_2 = C_3 = C_4 = C_5(CH_3)(CH_3) \longrightarrow$$

structures:

$$(I)\ Nucleophile \qquad ^{\ominus}N_1 - C_2^{\oplus} \ (C=C=C(CH_3)_2,\ CH_3)$$

$$(II)\ [Nucleophile] \qquad ^{\oplus}C_2 - C_3^{\ominus} \ (C(CH_3)_2,\ N-CH_3)$$

$$(III)\ First\ X\ center \qquad ^{\ominus}C_3 - C_4^{\oplus} \ (C(CH_3)_2,\ N-CH_3)$$

1,5 - ketemine

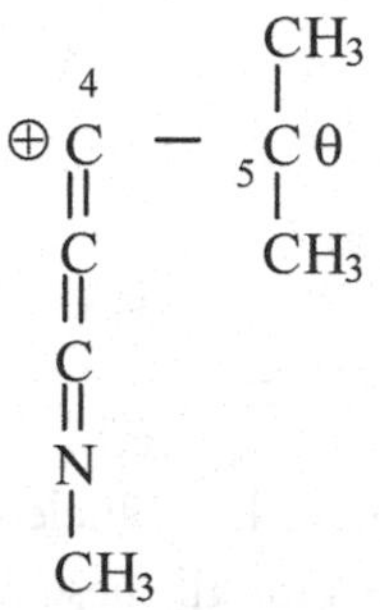

$$(IV)\ First\ Y\ center\ [Nucleophile] \qquad ^{\oplus}C_4 - {}_5C^{\ominus}(CH_3)(CH_3) \ \cdots$$

9.251

All seem to favor molecular rearrangement as shown below, only radically, since chargedly they cannot be activated. (II) and (III) cannot exist chargedly.

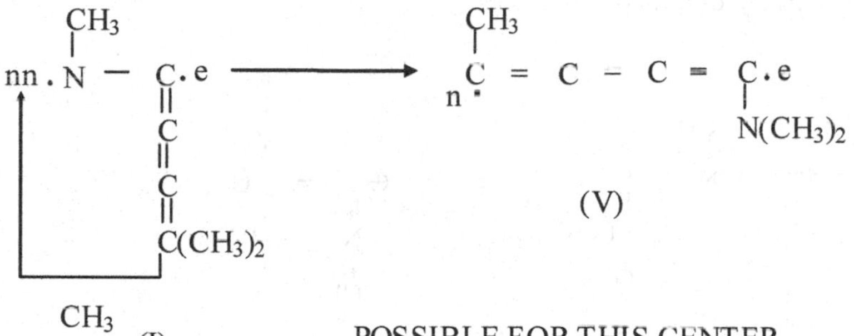

$$nn.N - C.e \ (CH_3,\ C=C=C(CH_3)_2) \quad (I) \longrightarrow {}_nC \cdot = C - C = C.e \ (CH_3,\ N(CH_3)_2) \quad (V)$$

(I)   POSSIBLE FOR THIS CENTER

9.252

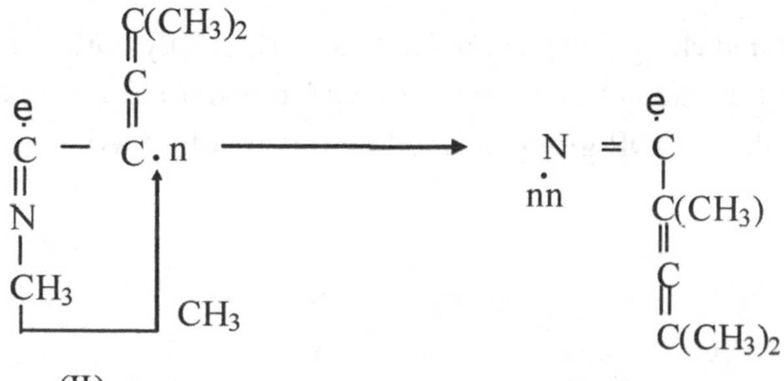

$$\overset{e}{C} - C.n \ (C(CH_3)_2,\ N-CH_3,\ CH_3) \quad (II) \longrightarrow \underset{nn}{N} = \overset{e}{C} \ (C(CH_3),\ C=C(CH_3)_2)$$

(II)   **(VI) Not possible for this center, Y**

9.253

$$
\begin{array}{c}
\text{H}_3\text{C} \qquad \text{C(CH}_3)_2 \\
\text{n} \bullet \text{C} - \overset{||}{\text{C}} \bullet \text{e} \\
\text{C} \\
|| \\
\text{N} \\
| \\
\text{CH}_3 \\
\text{(III)}
\end{array}
\qquad \longrightarrow \qquad
\begin{array}{c}
\text{CH}_3 \\
| \\
\text{n} \bullet \text{C} = \text{C} \bullet \text{e} \\
| \\
\text{CCH}_3 \\
|| \\
\text{N} \\
| \\
\text{CH}_3
\end{array}
$$

POSSIBLE FOR THIS CENTER

9.254

$$
\begin{array}{c}
\text{CH}_3 \\
| \\
\text{e} \bullet \text{C} - \text{C} \bullet \text{n} \\
|| \qquad | \\
\text{C} \quad \text{CH}_3 \\
| \\
\text{C} \\
|| \\
\text{N} \\
| \\
\text{CH}_3 \\
\text{(IV)}
\end{array}
\qquad \longrightarrow \qquad
\begin{array}{c}
\underset{nn}{\bullet\bullet}\text{N} = \text{C} = \text{C} = \text{C} \bullet \text{e} \\
| \\
\text{C(CH}_3)_3
\end{array}
\quad ;
$$

(VII) Not favored for this center.

9.255

For the mono-forms of Equation 9.251 like all the others, observe that they behave like Nucleophiles, making it look as if it is a clear indication that they are Nucleophilic in characters for all these families increasing as we move higher along the chain of each family. If this was the case, then only one center can be activated- the least nucleophilic center, that is (II) cumulatively placed down to the C = N center. To make the members of the family ELECTROPHILIC, one of the H atoms on the C atom must be changed to a radical-pulling group with transfer species.

With OR group replacing CH_3 group on the carbon center, the situation remains the same. Using only the 1,3 - types, the followings are obtained.

$$
\begin{array}{c}
\text{CH}_3 \qquad\quad \text{R} \\
| \qquad\qquad | \\
\text{N} = \text{C} = \underset{3}{\overset{}{\text{C}}} \\
{\scriptstyle 1} \quad {\scriptstyle 2} \quad | \\
\qquad\qquad \text{O} \\
\qquad\qquad | \\
\qquad\qquad \text{R}
\end{array}
\longrightarrow
\begin{array}{c}
\text{CH}_3 \\
| \\
{}^{\ominus}\text{N} - \overset{2}{\underset{}{\text{C}}}{}^{\oplus} \\
{\scriptstyle 1} \quad || \\
\qquad \text{C(OR)} \\
\qquad | \\
\qquad \text{O} \\
\qquad | \\
\qquad \text{R}
\end{array}
\quad ; \quad
\begin{array}{c}
\qquad\qquad \text{R} \\
\qquad\qquad | \\
\qquad\quad :\!\text{O}\!: \\
{}^{\oplus}\overset{2}{\text{C}} - \overset{3}{\underset{}{\text{C}}}{}^{\ominus} \\
|| \qquad | \\
\text{N} \quad :\!\text{O}\!: \\
| \qquad | \\
\text{CH}_3 \quad \text{R}
\end{array}
$$

(I) (Both)   (II) (Impossible existence)

9.256

Because of (II), the monomer cannot be activated chargedly, but only radically. Radically both of them are Nucleophiles. (I) also favors the nucleo-non-free-radical route the route which is not natural to it. With radical-pulling group such as CF_3 group in place of OR group, the followings are obtained.

$$
\begin{array}{c}
\text{CH}_3 \qquad\quad \text{CF}_3 \\
| \qquad\qquad | \\
\text{N} = \text{C} = \underset{3}{\overset{}{\text{C}}} \\
{\scriptstyle 1} \quad {\scriptstyle 2} \quad | \\
\qquad\qquad \text{CF}_3
\end{array}
\longrightarrow
\begin{array}{c}
\text{CH}_3 \\
| \\
{}^{\ominus}\text{N} - \overset{2}{\underset{}{\text{C}}}{}^{\oplus} \\
{\scriptstyle 1} \quad || \\
\qquad \text{C(CF}_3) \\
\qquad | \\
\qquad \text{CF}_3
\end{array}
\quad ; \quad
\begin{array}{c}
\qquad\qquad \text{CF}_3 \\
\qquad\qquad | \\
{}^{\oplus}\overset{2}{\text{C}} - \overset{3}{\underset{}{\text{C}}}{}^{\ominus} \\
| \qquad | \\
\text{N} \quad \text{CF}_3 \\
| \\
\text{CH}_3
\end{array}
$$

(I) (Nucleophile)   (II) (None)

9.257

(II), which cannot be polymerized chargedly or radically, cannot also undergo molecular rearrangement radically.

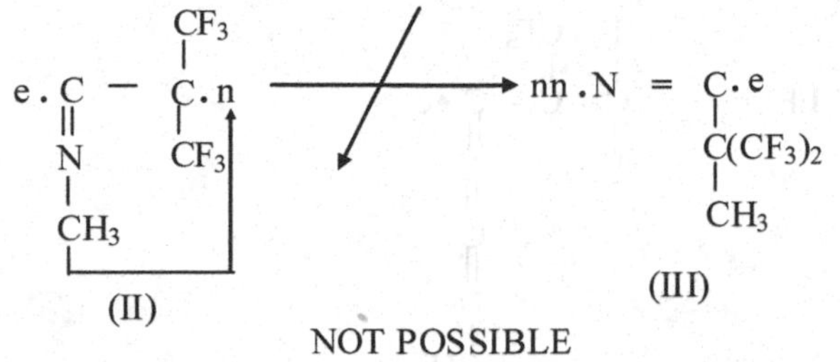

$$e \cdot C - C \cdot n \xrightarrow{\quad\not\quad} nn \cdot N = C \cdot e$$

with structure (II) having CF_3 group and N-CH_3 branch, and (III) having $C(CF_3)_2$ and CH_3 groups.

(II)   (III)

NOT POSSIBLE   9.258

In (III), CF_3 is the transfer species and not CH_3. As the Y center, it is F that should be the transfer species. With 1,5 - ketimine, the followings are obtained.

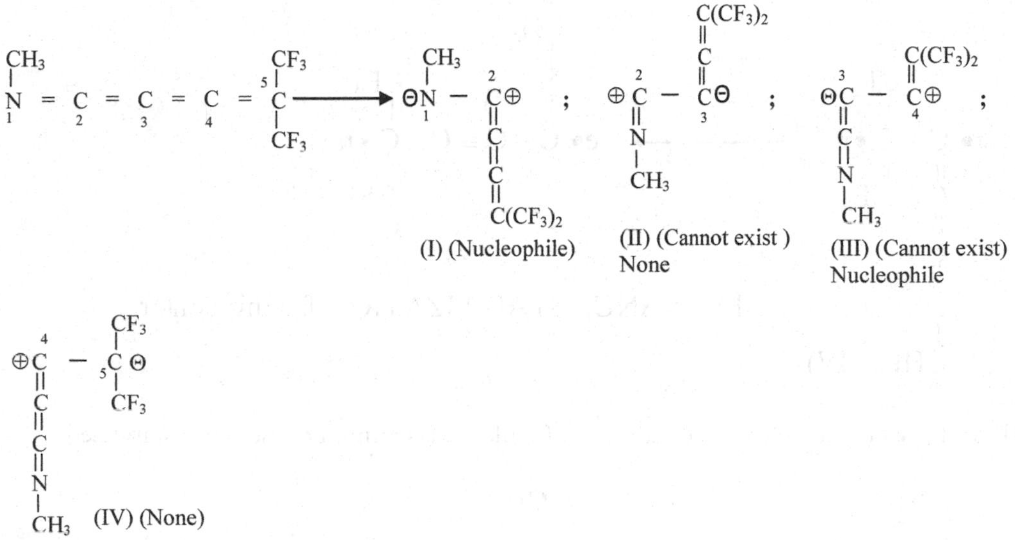

$$CH_3$$
$$N = C = C = C = C$$
$$1 \quad 2 \quad 3 \quad 4 \quad 5$$

giving (I) (Nucleophile), (II) (Cannot exist) None, (III) (Cannot exist) Nucleophile, and

(IV) (None)   9.259

With CF_3 types of groups, the strong Nucleophilic character is reduced. With (IV), the first Y center, the followings are obtained radically.

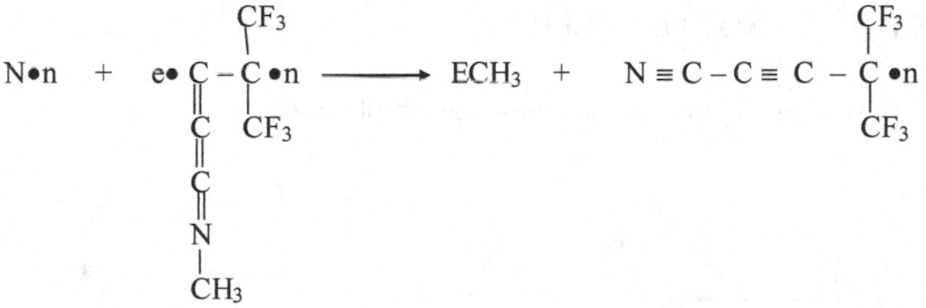

$$N \bullet n \; + \; e \bullet C - C \bullet n \longrightarrow ECH_3 \; + \; N \equiv C - C \equiv C - C \bullet n$$

9.260

$$\colon N \bullet nn \; + \; e \bullet C - C \bullet n \longrightarrow \colon NCH_3 \; + \; nn \bullet N = C = C = C = C$$

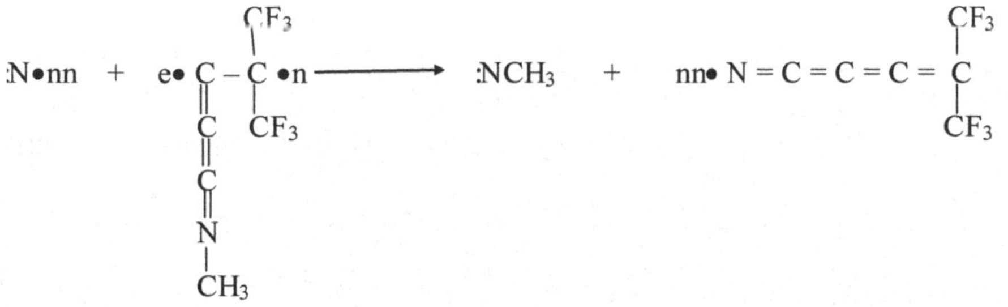

9.261

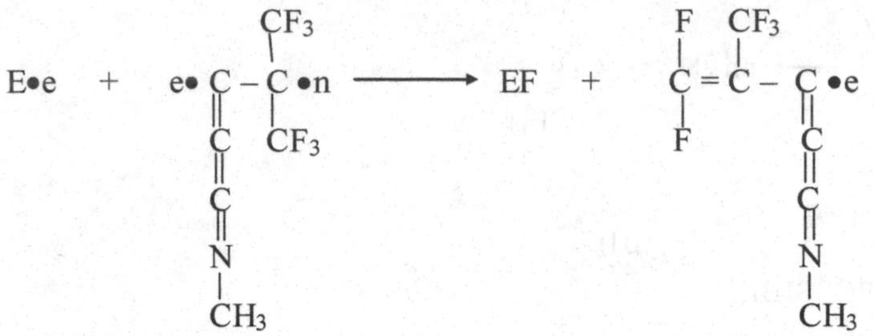

$$E\bullet e \ + \ e\bullet \underset{\underset{\underset{\underset{CH_3}{|}}{N}}{\overset{\overset{\overset{CF_3}{|}}{C}}{\underset{||}{C}}}{\overset{}{C}} - \underset{\overset{CF_3}{|}}{C}\bullet n \ \longrightarrow \ EF \ + \ \underset{\underset{F}{|}}{\overset{\overset{F}{|}}{C}} = \underset{\overset{CF_3}{|}}{C} - \underset{\underset{\underset{\underset{CH_3}{|}}{N}}{\overset{\overset{C}{||}}{C}}}{C}\bullet e \qquad 9.262$$

The only mono-forms not favoring both routes are (II) and (IV), clear indication of the weak Nucleophilic character of the monomer in view of the presence of halogenated groups.

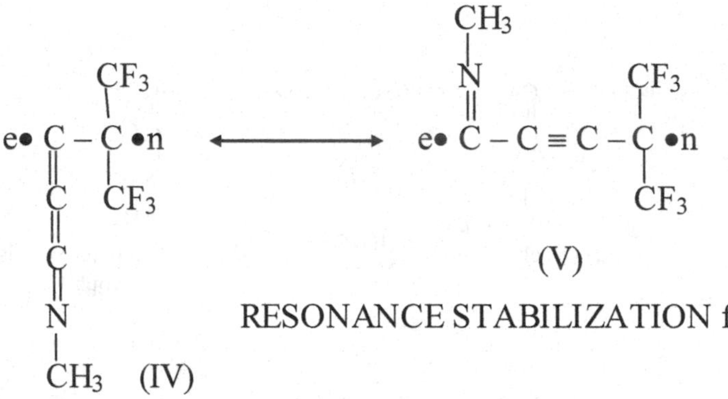

$$e\bullet \underset{\underset{\underset{\underset{CH_3}{|}}{N}}{\overset{\overset{C}{||}}{C}}}{\overset{\overset{CF_3}{|}}{C}} - \underset{\overset{CF_3}{|}}{C}\bullet n \quad \longleftrightarrow \quad e\bullet \overset{\overset{\overset{\overset{CH_3}{|}}{N}}{||}}{C} - C \equiv C - \underset{\overset{CF_3}{|}}{\overset{\overset{CF_3}{|}}{C}}\bullet n$$

(V)

RESONANCE STABILIZATION for this center

(IV)   9.263

Only (II) and (IV) can be resonance stabilized. (I) like (III) cannot be resonance stabilized.

$$nn\bullet \underset{\underset{\underset{C(CF_3)_2}{}}{\overset{\overset{C}{||}}{C}}}{\overset{\overset{CH_3}{|}}{N}} - C\bullet e \quad \longleftrightarrow \quad nn\bullet \overset{\overset{CH_3}{|}}{N} - C \equiv C - \underset{\underset{CF_3}{|}}{\overset{\overset{C(CF_3)}{|}}{C}}\bullet e$$

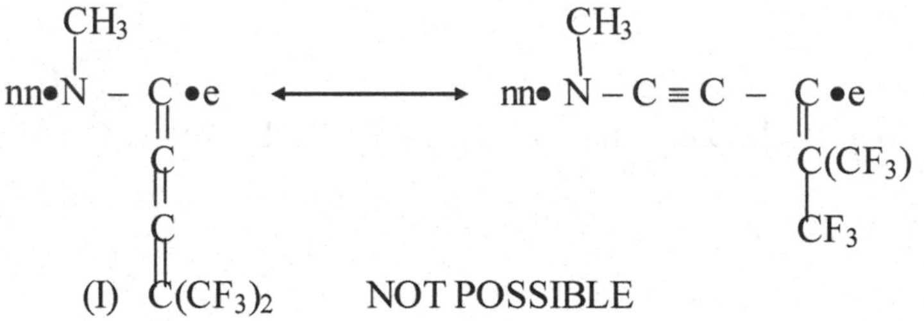

(I)   NOT POSSIBLE   9.264

With the nitrogen center carrying CF_3 group, the followings are obtained.

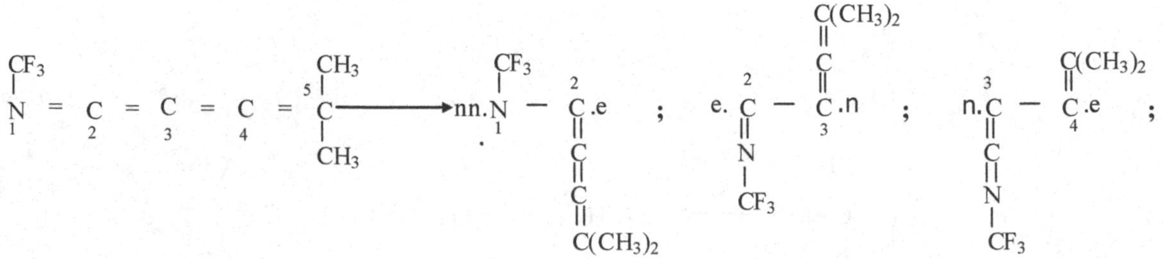

$$\underset{1}{\overset{\overset{CF_3}{|}}{N}} = \underset{2}{C} = \underset{3}{C} = \underset{4}{C} = \underset{\underset{CH_3}{|}}{\overset{\overset{CH_3}{|}}{C}}5 \longrightarrow nn.\underset{1}{\overset{\overset{CH_3}{|}}{N}} - \underset{\underset{C(CH_3)_2}{}}{\overset{\overset{C}{||}}{C}}.e \ ; \ e.\underset{\underset{\underset{CF_3}{|}}{N}}{\overset{\overset{C(CH_3)_2}{||}}{C}}2 - C.n3 \ ; \ n.C2 - \underset{\underset{CF_3}{|}}{\overset{\overset{C(CH_3)_2}{||}}{C}}.e4 \ ;$$

(I) (None)   (II) (Nucleophile)   (III) (None)

779

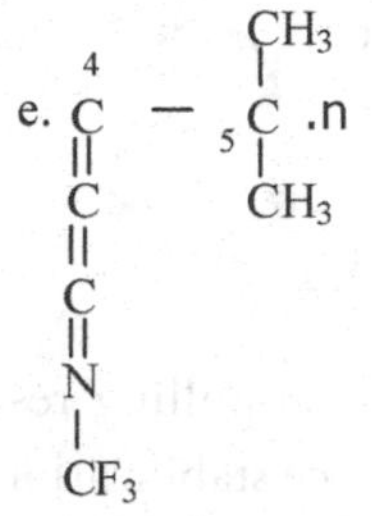

$$\text{e.} \quad \overset{4}{\underset{\displaystyle\underset{\displaystyle\underset{\displaystyle\underset{\displaystyle CF_3}{\overset{|}{N}}}{\overset{||}{C}}}{\overset{||}{\underset{\displaystyle C}{\overset{||}{C}}}}}{C} - \overset{\displaystyle CH_3}{\underset{5}{\underset{\displaystyle CH_3}{\overset{|}{\underset{\displaystyle}{C}}}}} .n$$

(IV) (Nucleophile)   9.265

(I) and (III) have the potentials of favoring molecular rearrangement. (II) and (IV) can be resonance stabilized. In the absence of molecular rearrangement, (I) and (III) favor no route. (I) cannot be resonance stabilized. When, it rearranges, the transfer species involved is not the same abstracted. The Y center is (IV) while the X center is (III).. The cumulenic ketemines are strongly related to acetylenes far more than to nitriles.

Nevertheless, based on all the considerations above, the following orders are to be expected for the groups shown below.

$$- C(CH_3) = C = C = N(CH_3) \quad > \quad .- C(CH_3) = N(CH_3) \quad > \quad .- CH = NH$$

Order of Radical-pulling capacity [ODD]   9.266a

$$- N(CH_3)_2 \quad > \quad - C(CH_3) = C = C = C = N(CH_3) \quad > \quad .- C(CH_3) = C = N(CH_3)$$

Order of Radical-pushing capacity [EVEN]   9.266b

$$RN^- C = C = C = C = C = \quad > \quad RN = C = C = C = \quad > \quad RN = C = \quad >> \quad RN = CR -$$

Order of Radical-pulling capacity [ODD]

$$RN = \quad > \quad RN = C = C = C = C = \quad > \quad RN = C = C =$$

Order of Radical-pushing capacity [EVEN]

$$- C \equiv C - N(CH_3)_2 \quad > \quad - C(CH_3) = C = C(CH_3)_2 \quad > \quad - C \equiv C - C(CH_3)_3$$

Order of Radical-pushing capacity   9.267

In order words, the CH_3 group internally located in $-C(CH_3) = C = C(CH_3)_2$ or $- C(CH_3) = C = NCH_3$ groups is shielded when used as radical-pushing group. This is very much like in alkenyl groups as distinguished below.

$$- C(\underline{CH_3}) = C(CH_3)_2 \quad \text{or} \quad - \underline{CH} = CH_2 \quad \textbf{versus} \quad - C(\underline{CH_3}) = C = C(CH_3)_2 \quad \text{or} \quad -- \underline{CH} = C = CH_2$$

Resonance stabilization groups   9.268

In all of them, the groups underlined are resonance stabilized when resonance stabilization is in place. Thus, there is no doubt that the groups shown below are radical-pushing resonance stabilization groups for these families of monomers.

$$= C = C = N - CH_3 \;, \quad = C = C(CH_3)_2 \;, \quad = C = C = C(CH_3)_3 \;,$$

$$- C \equiv C - C(CH_3)_3 \;, \quad - C \equiv C - N(CH_3)_2$$

RADICAL-PUSHING RESONANCE STABILIZATION GROUPS   9.269

Just as we have radical-pushing resonance stabilization groups so also exist radical-pulling resonance stabilization groups. Thus one can observe that $- C \equiv CH$ group which is a resonance stabilization group when suppressed is a radical-pulling type, in view of its capacity with respect to H. (Recall that this was called Resonance stabilization of the first kind of the second type).

It must also be noted that -

(i) $HC \equiv C\,(OH)$ molecularly rearranges to the first member of ketenes - ketene, $H_2C = C = O$ from female to a weak male. The weak male can enolize to the female under different operating conditions. The same takes place with vinyl alcohol which molecularly rearranges to acetaldehyde, that which also enolizes back to vinyl alcohol under different operating conditions.

(ii) $N \equiv C(OH)$ molecularly rearrangements to isocyanic acid- $HN = C = O$.

(iii) $HC \equiv C(NH_2)$ molecularly rearranges to the first member of aldimines - formaldimine, $H_2C = C = NH$ which is very unstable and rearranges to methyl nitride $N \equiv C(CH_3)$ via electroradicalization.

(iv)   The first member of the families of $\overset{\displaystyle R}{\overset{\displaystyle |}{N}} = C = \overset{\displaystyle R}{\overset{\displaystyle |}{N}}$, that is, $NH = C = NH$, can enamidize to a nitrile, $N \equiv C(NH_2)$.

(v)   Allene $H_2C = C = CH_2$ molecularly rearranges to acetylenes etc.

The expected observations above are supported by the followings -

$$H_2N - \quad > \quad - OH \quad > \quad - CH_3 \quad ; \quad H - N = \quad > \quad O = \quad > \quad H_2C =$$

Order of radical-pushing capacities   9.270

Before concluding this section, there is need to consider other few examples and know the effect of providing external resonance stabilization groups on 1, 3 - types.

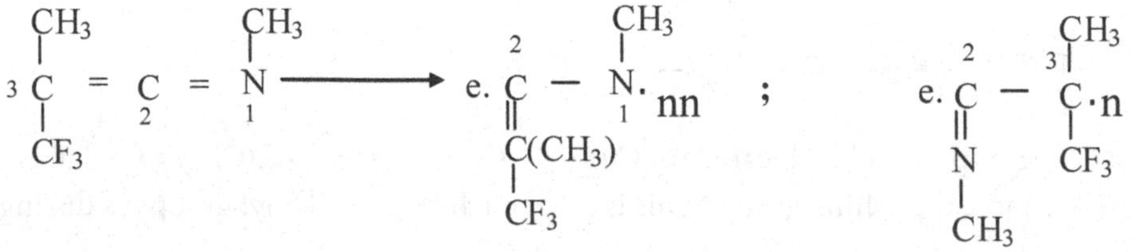

(I) (Nucleophile)   (II) (None)   9.271

(I) favors molecular rearrangement as shown below, being the X center. (II) does not, being the Y center, unless when CF_3 is the provider of transfer species (F). It may have no radical-pulling group to use, but it is there inherently.

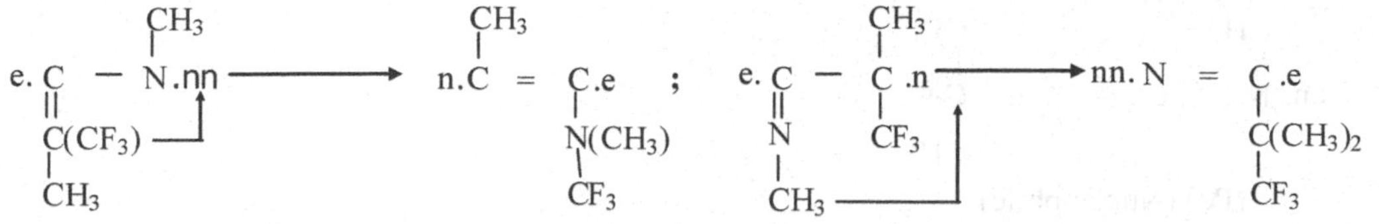

$$\underset{\substack{| \\ CH_3}}{\overset{\substack{CH_3 \\ |}}{e.\ C\ -\ N.nn}} \longrightarrow n.C\ =\ \underset{\substack{| \\ N(CH_3) \\ | \\ CF_3}}{C.e} \quad ; \quad \underset{\substack{|| \\ N \\ | \\ CH_3}}{\overset{\substack{CH_3 \\ |}}{e.\ C\ -\ }}\underset{\substack{| \\ CF_3}}{C.n} \longrightarrow nn.\ N\ =\ \underset{\substack{| \\ C(CH_3)_2 \\ | \\ CF_3}}{C.e}$$

(I)   (II) (Nucleophile)   (II)   (IV) (Nucleophile)

FAVORED   NOT FAVORED   9.272

When (II) is made to undergo rearrangement as shown above, the transfer species involved will not be the same abstracted. However, for these families which look strongly Nucleophilic, are indeed Electrophiles. (II) will be the center first activated and that center can only be used with paired free-radical initiators when the groups are trans-placed (i.e., pushing groups on one side and pulling or less pushing group on the other side).

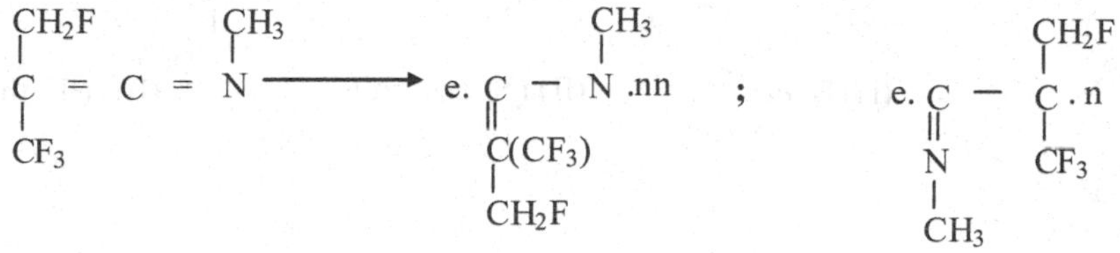

$$\underset{\substack{| \\ CF_3}}{\overset{\substack{CH_2F \\ |}}{C}}\ =\ \underset{}{C}\ =\ \overset{\substack{CH_3 \\ |}}{N} \longrightarrow e.\ \underset{\substack{|| \\ C(CF_3) \\ | \\ CH_2F}}{C}\ -\ N\ .nn \quad ; \quad e.\ \underset{\substack{|| \\ N \\ | \\ CH_3}}{C}\ -\ \underset{\substack{| \\ CF_3}}{\overset{\substack{CH_2F \\ |}}{C}}.n$$

(I) (Nucleophile)   (II) (None)   9.273

Only (I) favors molecular rearrangement. (II) can rearrange, if CF_3 is the provider of transfer species when possible. (I) is the (X) center, while (II) is the (Y) center.

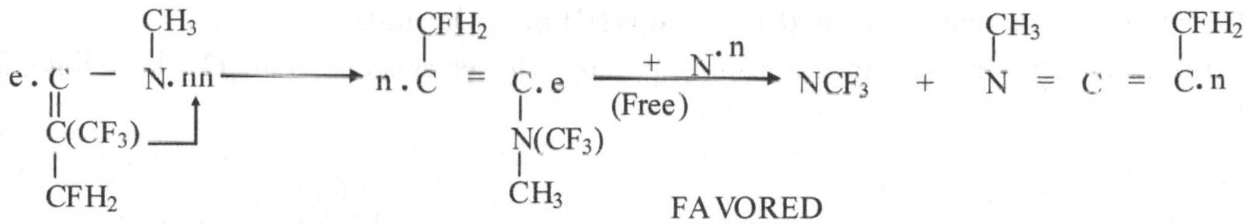

$$e.\underset{\substack{| \\ C(CF_3) \\ | \\ CFH_2}}{\overset{\substack{CH_3 \\ |}}{C}}-N.nn \longrightarrow n.C\ =\ \underset{\substack{| \\ N(CF_3) \\ | \\ CH_3}}{\overset{\substack{CFH_2 \\ |}}{C.e}} \xrightarrow[\text{(Free)}]{+\ N^{.\,n}} NCF_3\ +\ \overset{\substack{CH_3 \\ |}}{N}\ =\ C\ =\ \overset{\substack{CFH_2 \\ |}}{C.n}$$

FAVORED   9.274

$$e.\underset{\substack{| \\ C(CF_3) \\ | \\ CFH_2}}{\overset{\substack{CH_3 \\ |}}{C}}-N.nn \longrightarrow n.C\ =\ \underset{\substack{| \\ N(CF_3) \\ | \\ CH_3}}{\overset{\substack{CFH_2 \\ |}}{C.e}} \xrightarrow[\text{(Non-free)}]{+\ N^{.\,nn}} NCF_3\ +\ nn, N\ --\ \overset{\substack{CH_3 \\ |}}{C}\ \equiv\ \overset{\substack{CFH_2 \\ |}}{C}$$

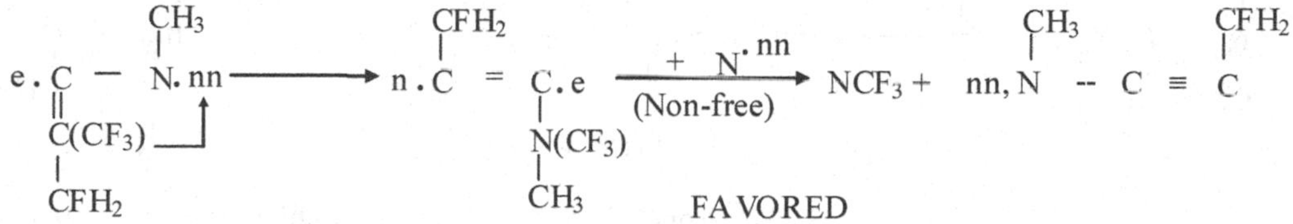

FAVORED   9.275

All the rearrangements said to be favored by the N = C centers do indeed take place, because the C = N centers are less nucleophilic than Acetylenes (C≡C). The real and first center activated all the time is the C = C center.

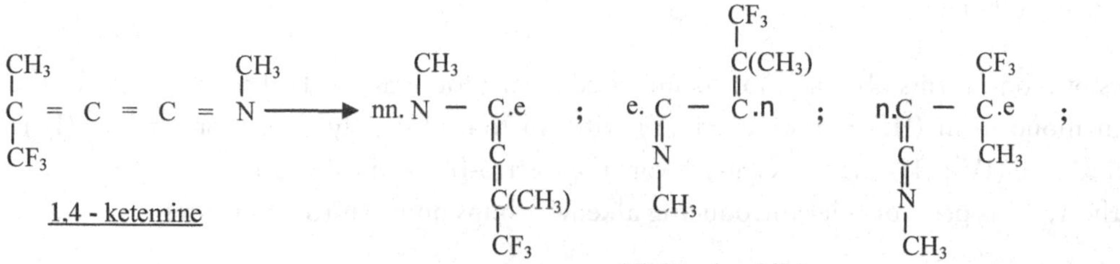

$$\underset{\substack{| \\ CF_3}}{\overset{\substack{CH_3 \\ |}}{C}}=C=C=\overset{\substack{CH_3 \\ |}}{N} \longrightarrow nn.\ N\ -\ \underset{\substack{|| \\ C \\ || \\ C(CH_3) \\ | \\ CF_3}}{C.e} \quad ; \quad e.\ \underset{\substack{|| \\ N \\ | \\ CH_3}}{\overset{\substack{C(CH_3) \\ ||}}{C}}-\overset{\substack{CF_3 \\ |}}{C}.n \quad ; \quad n.\underset{\substack{|| \\ C \\ | \\ N \\ | \\ CH_3}}{C}-\underset{\substack{| \\ CH_3}}{\overset{\substack{CF_3 \\ |}}{C.e}} \quad ;$$

<u>1,4 - ketemine</u>

(I) (Nucleophile)   (II) (Nucleophile)   (III) (Nucleophile)

$$
\begin{array}{c}
CH_3 \\
| \\
nn.\ N - C \equiv C - \overset{CF_3}{\underset{CH_3}{\overset{|}{C}.e}}
\end{array}
$$

(IV) (Nucleophile)   9.276

(IV) was obtained via resonance stabilization from (I), that which is impossible since it is not the least nucleophilic center. It can only be obtained from (III) which is the Y center. (II) is the X center which can molecularly rearrange to give a Nitrile. (I) the most nucleophilic center cannot be used.

$$
\underset{CF_3}{\overset{CF_3}{C}} = C = C = N \longrightarrow nn.\ N - \underset{C(CF_3)_2}{\overset{CH_3}{C}.e} \quad ; \quad e.\ C - \underset{CH_3}{\overset{C(CF_3)_2}{C}.n} \quad ; \quad n.C - \underset{CH_3}{\overset{CF_3}{C}.e} \quad ;
$$

.(I) (Both)   (II) (Nucleophile)   (III) (Both)

$$
nn.\ N - C \equiv C - \underset{CF_3}{\overset{CF_3}{C}.e}
$$

(IV) (Both)   9.277

(IV) was obtained via resonance stabilization from (III) and not from (I), since (III) is the (Y) center. As is beginning to become clear, the natural route to the monomer is the nucleo-non-free-radical route. Electro-free-radically, the center activated and used is (II) the (X) center.

With the N center wearing radical-pushing group of lower capacity than H, the followings are obtained.

$$
\underset{CH_3}{\overset{CH_3}{C}} = C = C = N \longrightarrow nn.\ N - \underset{C(CH_3)_2}{\overset{CF_3}{C}.e} \quad ; \quad e.\ C - \underset{CF_3}{\overset{C(CH_3)}{C}.n} \quad ; \quad n.C - \underset{CH_3}{\overset{CH_3}{C}.e} \quad ;
$$

(I) (None)   (II) (Nucleophile)   (III) (None)

$$
nn.\ N - C \equiv C - \underset{CH_3}{\overset{CH_3}{C}.e}
$$

(IV) (None)   9.278

The types of mono-forms obtained for them, based on the foundations laid so far speak for themselves. The main mono-form (III) is the center (Y) with no route for polymerization. It like (I) resonance stabilized to give (IV). (II) the (X) center favors the electro-free-radical route.

For the 1, 3 - types, consider introducing alkenyl groups on the nitrogen center.

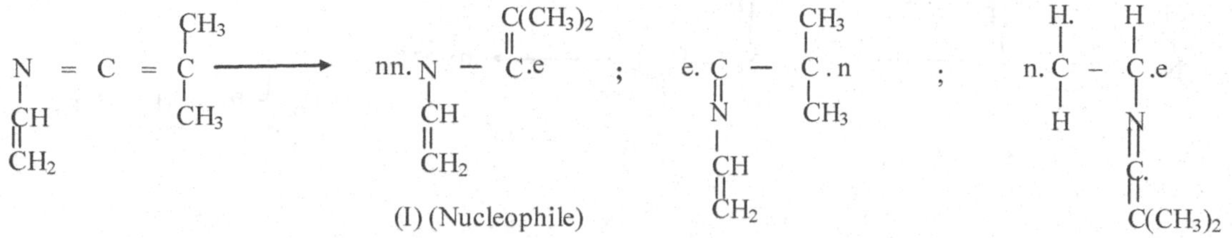

$$
\begin{array}{c}
CH_3 \\
| \\
N = C = C \\
| \qquad\quad | \\
CH \qquad CH_3 \\
\| \\
CH_2
\end{array}
\longrightarrow
\begin{array}{c}
C(CH_3)_2 \\
\| \\
nn. N - C.e \\
| \\
CH \\
\| \\
CH_2
\end{array}
\quad ; \quad
\begin{array}{c}
CH_3 \\
| \\
e. C - C.n \\
\| \qquad | \\
N \quad CH_3 \\
| \\
CH \\
\| \\
CH_2
\end{array}
\quad ; \quad
\begin{array}{c}
H. \quad H \\
| \qquad | \\
n. C - C.e \\
| \qquad | \\
H \qquad N \\
\qquad\quad \| \\
\qquad\quad C. \\
\qquad\quad | \\
\qquad C(CH_3)_2
\end{array}
$$

(I) (Nucleophile)   (II) (Nucleophile)   (III) Nucleophile.

9.279

(III) is the only mono-form above, since the alkenyl group is carrying the least nucleophilic center. It is also resonance stabilized with CH_3 group as transfer species nucleo-free-radically. Therefore, it is a Nucleophile. (I) cannot be resonance stabilized, in view of the half free and full free characters of the mono-forms. Though (I) and (II) do not exist as mono-forms, based on the manner of operations in our world today, they are still useful in many ways in particular with respect to how NATURE operates. If the alkenyl group is placed on the C atom to replace the CH_3 group moved to the N center, the monomer will still remain Nucleophilic and resonance stabilized with $- C(CH_3) = C = N(CH_3)$ as the group providing it.

It is important to note that, while $- N = CH_2$ group is radical-pushing, so also is $- N = C = CH_2$ group radical-pushing of different capacities.

$$- N = C = C = C = C = CH_2 \; > \; - N = C = C = C = CH_2 \; > \; - N = C = C = CH_2$$

$$> \quad - N = C = CH_2 \quad > \quad - N = CH_2$$

<u>Radical-pushing capacity of groups</u>   9.280a

$$- CR = C = C = C = C = NR \; > \; - CR = C = C = NR \quad > \quad - CR = NR$$

<u>Radical-pulling capacity of groups [ODD]</u>   9.280a

$$- NR_2 \quad > \quad - CR = C = C = C = NR \quad > \quad - CR = C = NR$$

Radical-pushing capacity of groups [EVEN]   9.280d

9.8. Cumulenic di-imines

In continuation of the present studies on these families of monomers, there is need to consider the di-imines, which have the following first members.

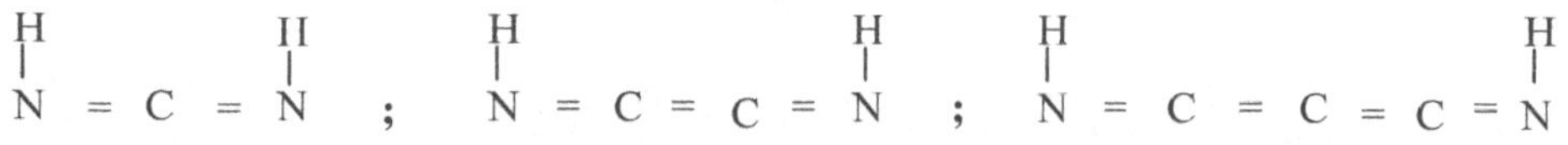

$$
\begin{array}{c}
H \qquad\qquad H \\
| \qquad\qquad | \\
N = C = N
\end{array}
\quad ; \quad
\begin{array}{c}
H \qquad\qquad H \\
| \qquad\qquad | \\
N = C = C = N
\end{array}
\quad ; \quad
\begin{array}{c}
H \qquad\qquad\qquad H \\
| \qquad\qquad\qquad | \\
N = C = C = C = N
\end{array}
$$

<u>1,3 - di-imine</u>   <u>1,4 - di-imine</u>   <u>1,5 - di-imine</u>   9.281

None of these seems to favor any stable state of existence, in view of the loosely bonded hydrogen atom one at a time, radically. Instead, nitriles may be thought to be largely obtained as shown below via molecular rearrangements only radically when suppressed with a passive catalyst.

$$\overset{H}{\underset{}{N}} = C = \overset{H}{\underset{}{N}} \longrightarrow nn. \ N - \underset{\underset{\underset{H}{|}}{\overset{||}{N}}}{\overset{H}{\underset{}{C}}.e \longrightarrow nn. \ N = \underset{NH_2}{\overset{}{C}}.e \left.\begin{array}{l} \\ \\ \end{array}\right\} \text{Radical-pushing}$$

9.282

$$\overset{H}{\underset{}{N}} = C = C = \overset{H}{\underset{}{N}} \longrightarrow e. \ C - \underset{\underset{\underset{H}{|}}{\overset{||}{N}}}{\overset{\overset{\overset{H}{|}}{\overset{||}{N}}}{\underset{}{C}}}.n \longrightarrow nn. \ N = \underset{\underset{NH}{\overset{||}{CH}}}{\overset{}{C}}.e \left.\begin{array}{l} \\ \\ \end{array}\right\} \text{Radical-pulling}$$

"FAVORED ONE"

9.283a

$$\overset{H}{\underset{}{N}} = C = C = \overset{H}{\underset{}{N}} \longrightarrow nn. N - \underset{\underset{\underset{H}{|}}{\overset{||}{N}}}{\overset{H}{\underset{}{C}}.e \nrightarrow nn. N = C = \underset{NH_2}{\overset{}{C}}.e$$

CANNOT BE DEACTIVATED

AN EXPLORATORY CENTER

9.283b

$$\overset{H}{\underset{}{N}} = C = C = C = \overset{H}{\underset{}{N}} \longrightarrow e. C - \underset{\underset{\underset{H}{|}}{\overset{||}{N}}}{\overset{\overset{\overset{NH}{||}}{\overset{||}{C}}}{\underset{}{C}}}.n \longrightarrow nn. N = \underset{\underset{NH}{\overset{||}{\underset{CH}{\overset{}{C}}}}}{\overset{}{C}}.e \left.\begin{array}{l} \\ \\ \end{array}\right\} \text{Radical-pushing}$$

"THE FAVORED ONE"

9.284

$$\overset{H}{\underset{}{N}} = C = C = C = \overset{H}{\underset{}{N}} \longrightarrow e. C - \overset{.H}{\underset{}{N}}.nn \longrightarrow$$

(with vertical chain: $\overset{||}{C}$, $\overset{||}{C}$, $\overset{||}{N}$, $\overset{|}{H}$)

$$nn. N = C = C = \underset{NH_2}{\overset{}{C}}.e \longrightarrow N \equiv C - C \equiv \underset{NH_2}{\overset{}{C}}$$

(unstable)

AN EXPLORATORY CENTER

9.285

In fact for all cases above, the existences of their mono-forms seems to be impossible if the H on the nitrogen center is not replaced. When suppressed, all the Di-imines may rearrange to give Nitriles. Notice that, while the nitriles of the first and fourth ones above, i.e., 1,3- and 1,5- (Equations 9.282 and 9.284) and 1,7- Di-imines are carrying radical-pushing groups, 1,4-, 1,8- Di-imines are carrying radical-pulling groups. 1,6- Di-imine cannot rearrange. This clearly indicates that, 6 is a unique number-odd and even in character and that all of them are Nucleophiles, if and only if the C = C center is not a Y center. As will be seen downstream, the monomers are indeed ELECTROPHILES,

with the exception of only the first member. _But, meanwhile it will be "assumed" that they are Nucleophiles for exploratory purposes._

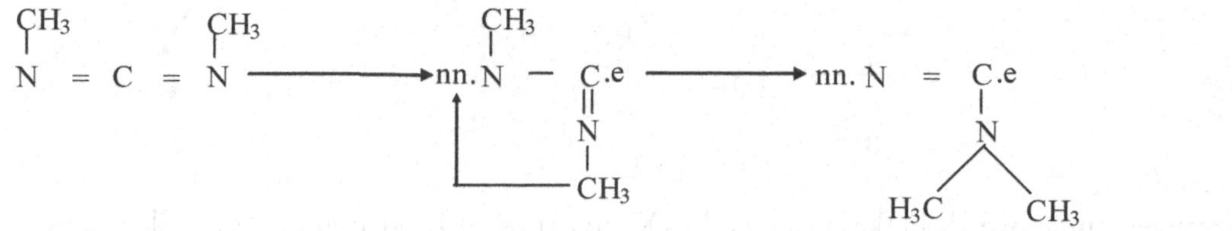

9.286

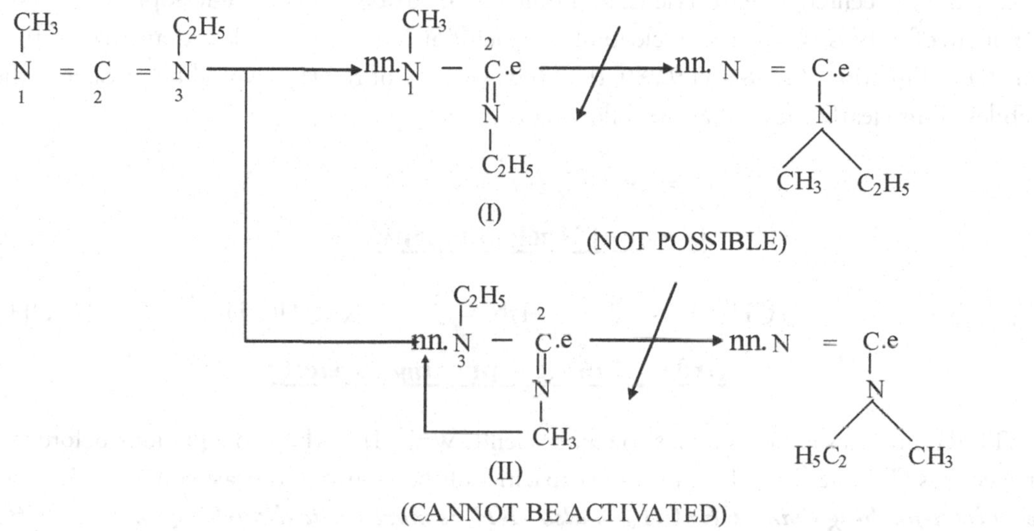

(I)

(NOT POSSIBLE)

(II)

(CANNOT BE ACTIVATED)

9.287

Only the symmetric Di-imines can undergo molecular rearrangement when the R group is not bulky, to give a stable Nitrile. The unsymmetric Di-imines cannot rearrange as shown above for only 1,3-Di-imines. In (I) above, the less nucleophilic center was activated, but the transfer species is greater than what the N center is carrying. Hence, it cannot be transferred. In (II), where rearrangement seems to be favored, the wrong center has been activated. Hence, molecular rearrangement can never take place. (I) will be used just as it is, favoring only the electro-free-radical route, for which there is no doubt that it is a Nucleophile. Chargedly, these monomers cannot be activated.

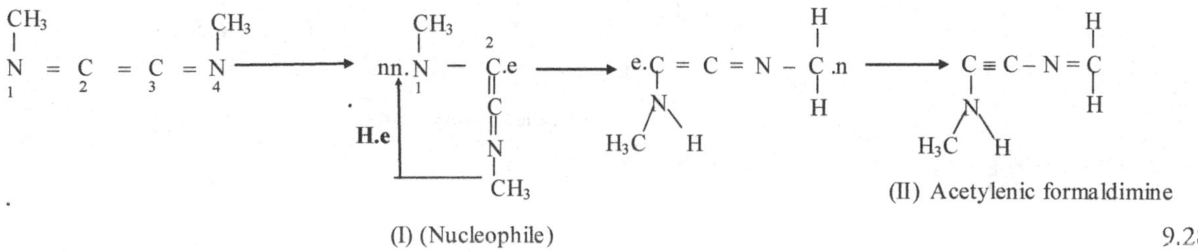

(I) (Nucleophile)

(II) Acetylenic formaldimine

9.288

As a Nucleophile, the rearrangement above does not take place. As an Electrophile, it does if the center is (X). They are only shown for exploratory purpose, from which for example capacities of these groups can be seen with the use of the EYE OF THE NEEDLE.

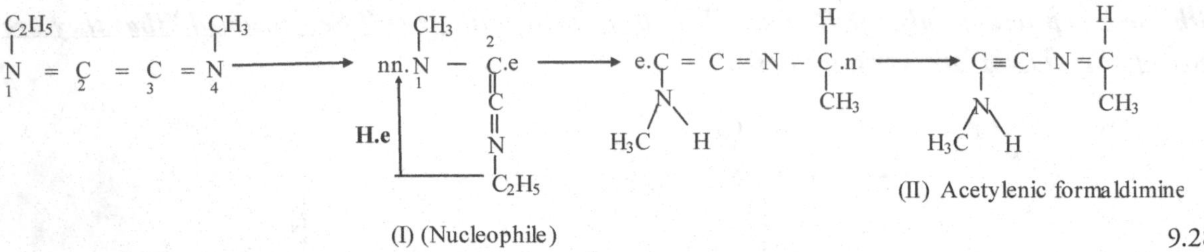

$$
\begin{array}{c}
\underset{1}{\overset{C_2H_5}{N}} = \underset{2}{\overset{CH_3}{C}} = \underset{3}{C} = \underset{4}{\overset{CH_3}{N}} \longrightarrow \quad \underset{1}{nn.N} - \overset{2}{C}.e \longrightarrow e.C = C = N - \overset{H}{C}.n \longrightarrow C \equiv C - N = \overset{H}{C}
\end{array}
$$

(I) (Nucleophile) (II) Acetylenic formaldimine

9.289

The rearrangement above also takes place as the (X) center of an Electrophile. As an Electrophile, the C = C center is the (Y) center. The Acetylenic aldimine obtained above is more nucleophilic than the case above. Worthy of note is that if the nucleophilic capacity of N=C center is less than that of the C≡C center in (II) of Equations 9.288 and 9.289, then the existence of the (II)s would be possible if they are Electrophiles. This clearly shows that the following is valid.

$$C \equiv C > C = O > C = N$$

Order of Nucleophilicity

9.290a

$$(CH_3)_2N - \quad > \quad (CH_3)HN - \quad > \quad H_2N - \quad > \quad -N=CH(CH_3) \quad > \quad -N=CH_2$$

Order of radical-pushing capacity

9.290b

How NATURE operates is too complex to comprehend. With (I) in the two equations before this, the transfer species is H. Note the radical-pushing capacities of the groups in the last equation. *Just as -CH₃ is more radical-pushing than –CH=CH₂, so also –NH₂ is more radical-pushing than –N=CH₂.* The C = N centers are all Nucleophilic in character.

This family of monomers can be observed to be more nucleophilic than cumulenic imines, though their existences are also unknown.

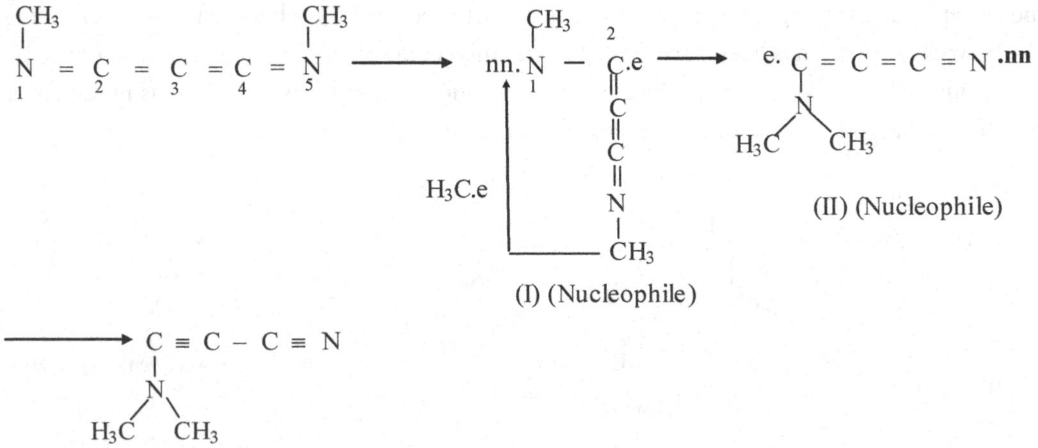

$$
\underset{1}{\overset{CH_3}{N}} = \underset{2}{C} = \underset{3}{C} = \underset{4}{C} = \underset{5}{\overset{CH_3}{N}} \longrightarrow \quad \underset{1}{nn.N} - \overset{2}{C}.e \longrightarrow e.C = C = C = N^{.nn}
$$

(I) (Nucleophile) (II) (Nucleophile)

$$\longrightarrow C \equiv C - C \equiv N$$

9.291a

Note again that the rearrangement shown above, like all the cases of Equations 9.288 and 9.289 take place for this mono-form, because that center (N = C) is the X center. However as shown above, more stable acetylenic nitrile more Nucleophilic than that of Equation 2.285 are obtained.

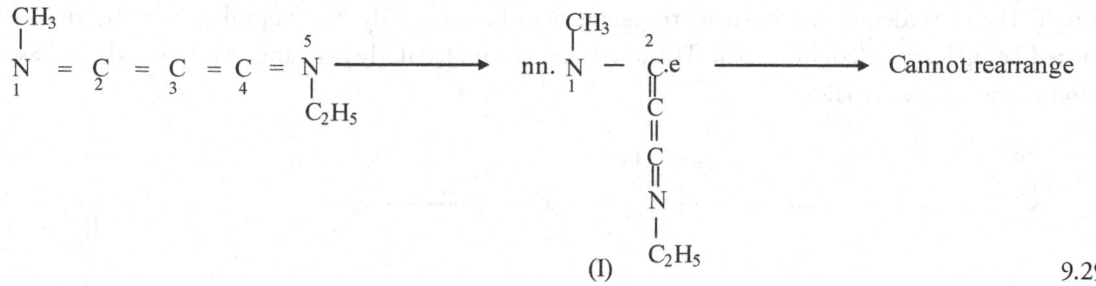

9.291b

While the 1,4-Di-imine cannot be resonance stabilized when activated via the N=C activation center, the same does not apply to the 1,5- Di-imine. While 1,4- can molecularly rearrange, 1,5- cannot when the groups are not symmetrically placed.

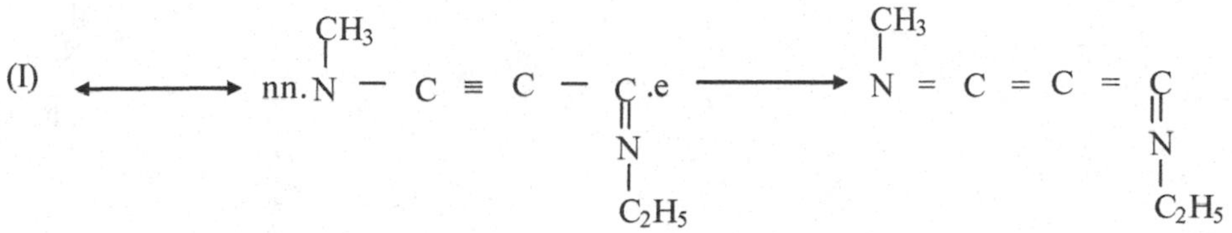

9.292

However, when activated via the C = C center cumulatively placed to the C =N center, no resonance stabilization is possible, but molecularly rearrangement is possible if the monomer is a Nucleophile as shown below. Indeed, it cannot take place since the center is the Y center, i.e., an Electrophile.

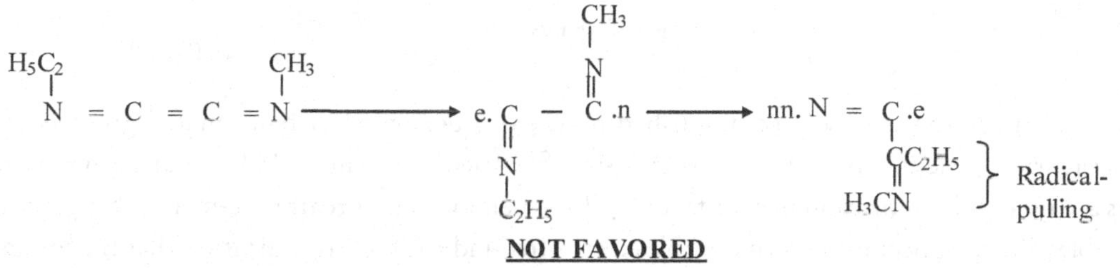

9.293a

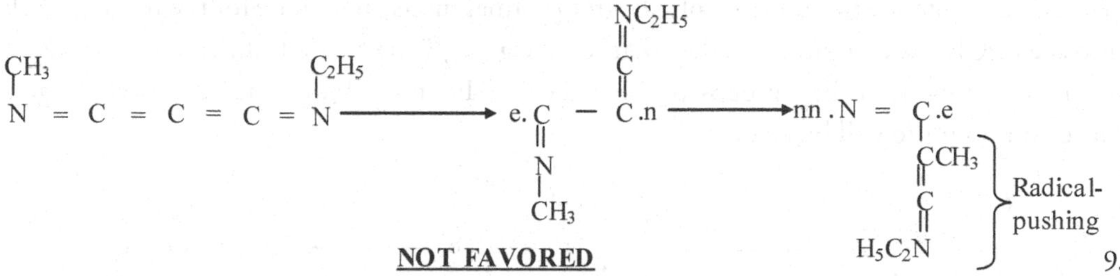

9.293b

Now introducing radical-pushing groups of lower capacity than H, the followings are obtained for CF_3 type.

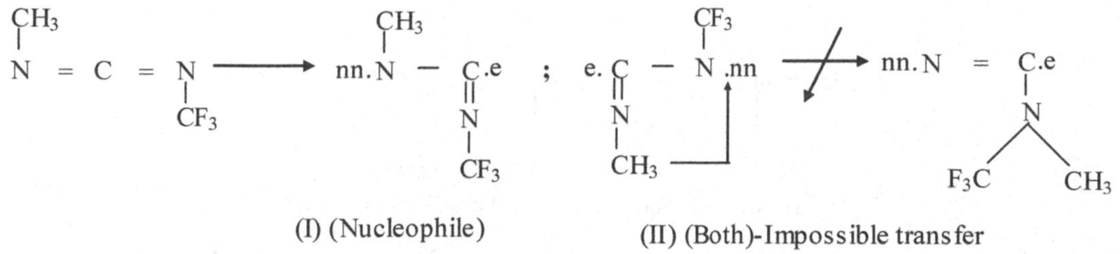

9.294

Though (I) can undergo molecular rearrangement to become fully Nucleophilic, it is (II) that is first activated being less nucleophilic than (I) and (II) cannot molecularly rearrange.as shown above based on the foundations already laid.

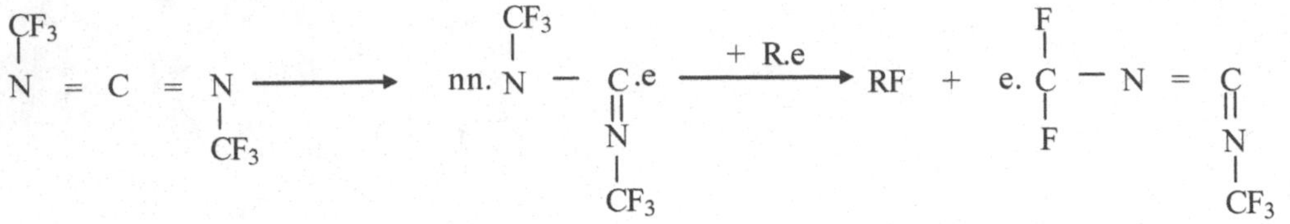

(None)   9.295a

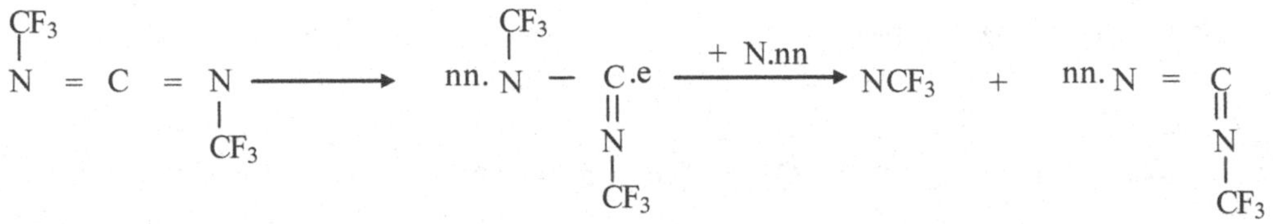

(None)   9.295b

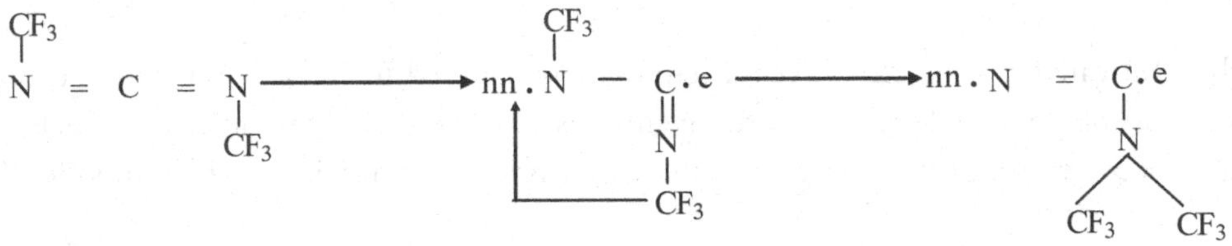

FAVORED   (Nucleophile)   9.296

It is important to note the highly distinguished nature of fluorinated or halogenated groups. Without molecular rearrangement, no route is favored using free-media initiators. When paired media radical initiators are used polymerization may be favored. But when molecular rearrange-ment takes place to give a more stable Nitrile, it becomes polymerizable electro-free-radically, clear indication that the monomer is a Nucleophile, despite the full presence of CF_3 groups. Imagine as an exercise for the reader the placement of F in place of CF_3. When groups higher than CF_3, e.g., C_2F_5 are carried, the same as above will apply. But when two different $C_{2n}F_{2n-1}$ groups are carried e.g., CF_3 and C_2F_5, no molecular rearrangement can take place and no route will be favored.

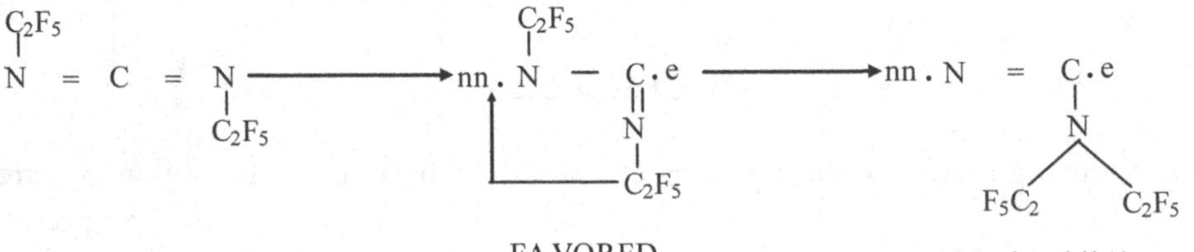

FAVORED   (Nucleophile))   9.297

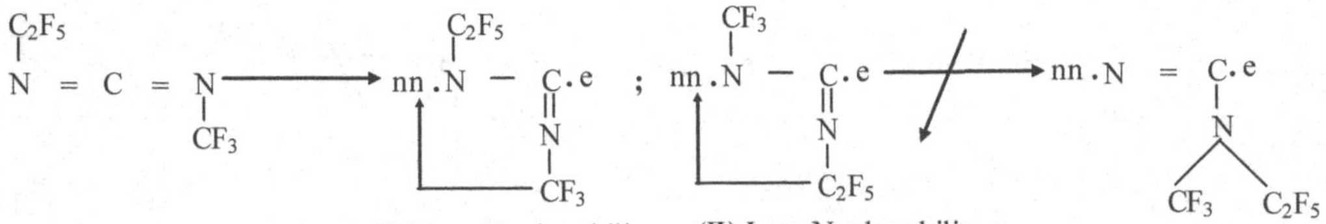

.(I) More Nucleophilic   (II) Less Nucleophilic

IMPOSSIBLE TRANSFER   9.298

789

It is important to note the highly distinguished nature of fluorinated or halogenated groups, noting that so far one has been using free-media initiators. With the use of paired initiators, the characters become so well defined.

Imagine if the monomers were taken to be Electrophiles wherein the C = N center is Y and the C = C or second N = Y center is X as if the monomer is isocyanic in character with these types of groups, then the followings would have been expected during molecular rearrangement.

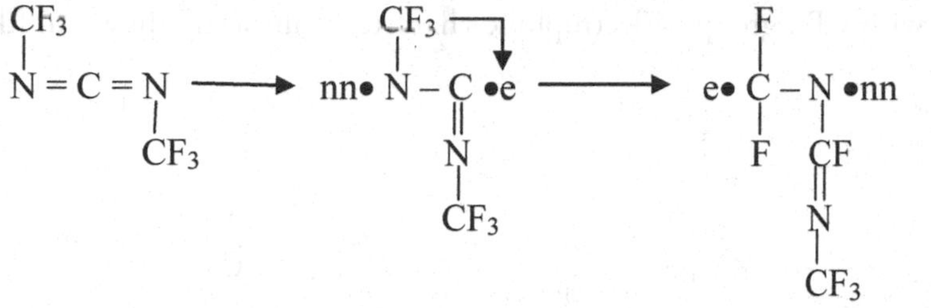

$$
\begin{array}{ccc}
\underset{\underset{CF_3}{|}}{\overset{\overset{CF_3}{|}}{N}}=C=N & \longrightarrow & nn\bullet\ N-\underset{\underset{\underset{CF_3}{|}}{\overset{\|}{N}}}{\overset{\overset{CF_3}{|}}{C}}\bullet e & \longrightarrow & e\bullet\ \underset{\underset{\underset{\underset{CF_3}{|}}{\overset{\|}{N}}}{\overset{\overset{\overset{F}{|}}{}}{C}}-N\bullet nn
\end{array}
$$

(Electrophile ?)   9.299

$$
\underset{\underset{C_2F_5}{|}}{\overset{\overset{CF_3}{|}}{N}}=C=N \longrightarrow nn\bullet\ N-\underset{\underset{\underset{C_2F_5}{|}}{\overset{\|}{N}}}{\overset{\overset{CF_3}{|}}{C}}\bullet e \longrightarrow e\bullet\ \underset{\underset{\underset{\underset{C_2F_5}{|}}{\overset{\|}{N}}}{\overset{\overset{\overset{F}{|}}{CF}}{}}}{\overset{\overset{F}{|}}{C}}-N\bullet nn
$$

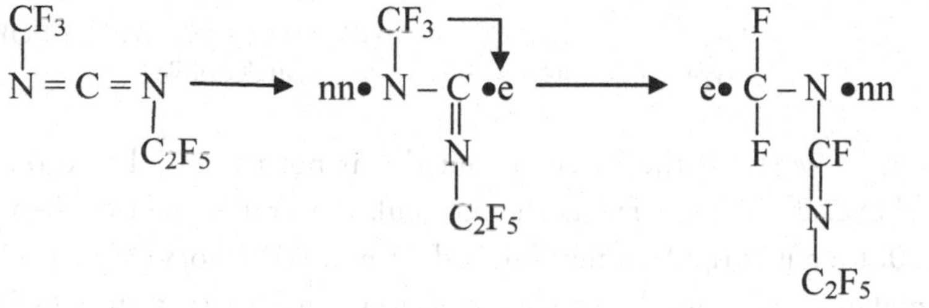

(Electrophile ?)   9.300

As it seems, the reactions above cannot be favored, because these are Nucleophiles. Only the nucleo-non-free-radical route will be favored by both centers formed after rearrangement. Note that the use of 1,3- as an example above is wrong, because there is no C = C center.

With radical-pulling groups, the followings are to be expected.

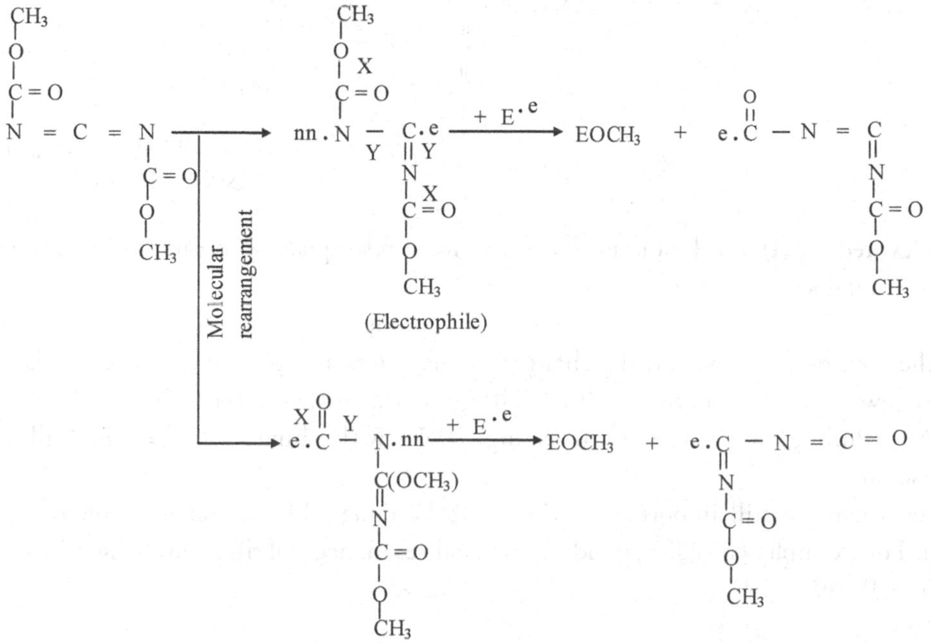

9.301

9.302

When molecular rearrangement is allowed to take place, while the Y center above favors the route natural to it (Nucleo-non-free-radical), the X center which is C = O present after the rearrangement above favors the electro-free-radical route, that which is a __Complete Electro-phile__; complete in the sense that both the Y and X centers can only be used in the route natural to them. Note however that the above will not take place, since __resonance stabilization must take place after activation, for which no molecular rearrangement can take place, since OCH__$_3$ __becomes shielded and the 1,4-mono-form is now made to favor both routes.

For the 1,4- Di-imine, with CF$_3$, groups, Electrophilic character is instantly shown unlike the case of 1,3- Di-imine.

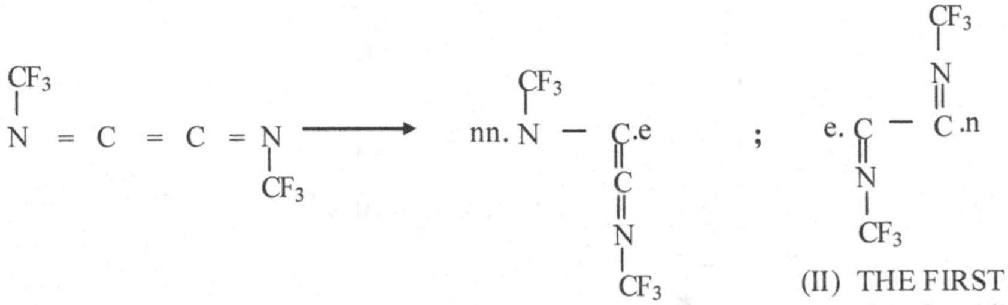

(I) Electrophile ; (II) THE FIRST MONO-FORM [Nucleophile] 9.303

Naturally, (I) is the X center, while (II) is the Y center. But this is not the case based on the routes favored by them as shown above. If (I) above favors the molecular rearrangement of the type shown in Equation 9.299 and 9.300, then it is the Y center; but it does not. (II) favors only the electro-free-radical route, that which makes it nucleophilic in character and can also rearrange to Nitriles. (I) cannot be resonance stabilized. Despite all these observations, if it is an Electrophile, (I) should be the X center, while (II) should be the Y center. When C$_2$F$_5$ and higher are put in place of CF$_3$, the situation is slightly different.

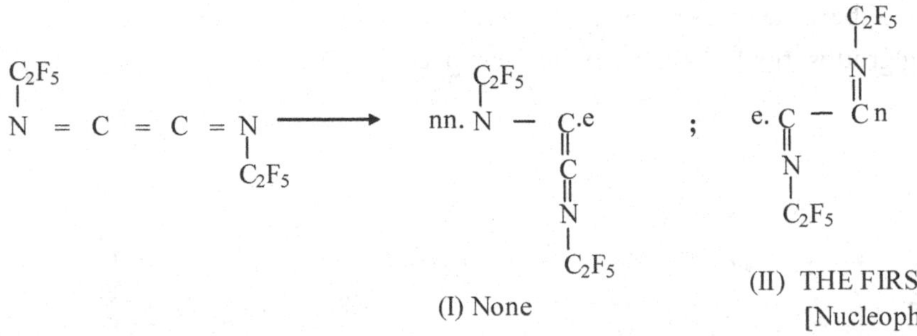

(I) None ; (II) THE FIRST MONO-FORM [Nucleophile] 9.304

No route is favored for (I), until molecular rearrangement takes place according to Equations 9.299 and 9.300, and when it does, only the nucleo-non-free-radical route is favored, making the center Electrophilic in character, when indeed it is not. It rearranges as a Nucleophile.

From the considerations so far, the changes in characters and phenomena taking place should be obvious. No new groups have been identified. The groups identified so far include

= C = NR (Pulling), = C = C = NR (Pushing), = N – R (Pushing), – CR = NR (Pulling), types of groups and so on.

However, there are still important works yet to be done with respect to resonance stabilization phenomena. For example, consider providing external resonance stabilization using alkenyl group and no other, for 1,3 - type.

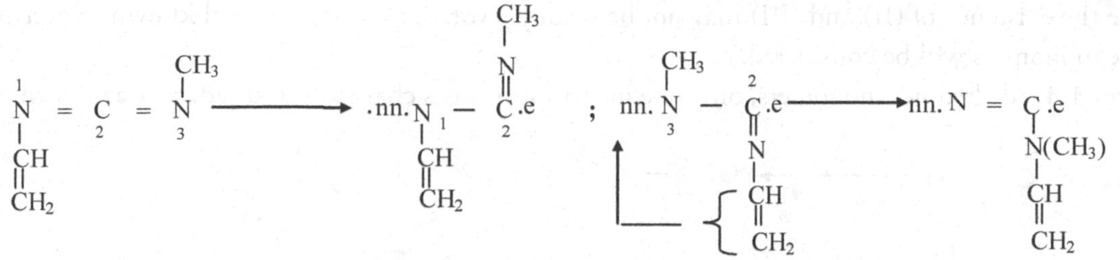

$$
\begin{array}{c}
\overset{1}{N} = \overset{CH_3}{\underset{2}{C}} = \overset{CH_3}{\underset{3}{N}} \\
| \\
CH \\
\| \\
CH_2
\end{array}
\longrightarrow
\cdot nn.\overset{CH_3}{\underset{\substack{| \\ CH \\ \| \\ CH_2}}{\underset{1}{N}}} - \overset{CH_3}{\underset{2}{C}}.e
\quad ; \quad
nn.\overset{CH_3}{\underset{\substack{| \\ N \\ | \\ CH \\ \| \\ CH_2}}{\underset{3}{N}}} - \overset{}{\underset{}{C}}.e
\longrightarrow
nn.\overset{}{N} = \overset{}{\underset{\substack{N(CH_3) \\ | \\ CH \\ \| \\ CH_2}}{C}}.e
$$

(II) (Nucleophile)   (III) (Nucleophile)

THE SECOND CENTER   9.305a

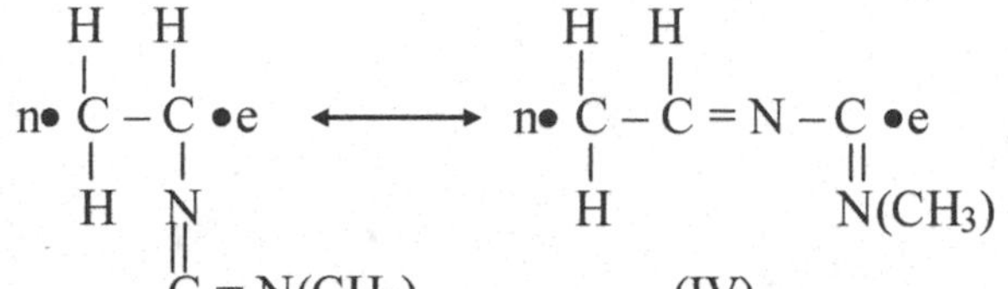

$$
n\bullet \overset{H}{\underset{\substack{| \\ H \\ | \\ N \\ \| \\ C = N(CH_3)}}{C}} - \overset{H}{\underset{}{C}} \bullet e
\quad \longleftrightarrow \quad
n\bullet \overset{H}{\underset{\substack{| \\ H}}{C}} - \overset{H}{\underset{\substack{\| \\ N(CH_3)}}{C}} = N - C \bullet e
$$

(IV)

(I) Nucleophile and Resonance Stabilized   9.305b

In view of the presence of full-free and half-free characters, resonance stabilization is not favored when the presence of the alkenyl group is not taken into consideration. It is only favored for one of the mono-forms, (I) the center which is the least nucleophilic. Nevertheless, it is important to note that (II) which is not one of the mono-forms cannot be resonance stabilized as shown below to give the same (IV) above.

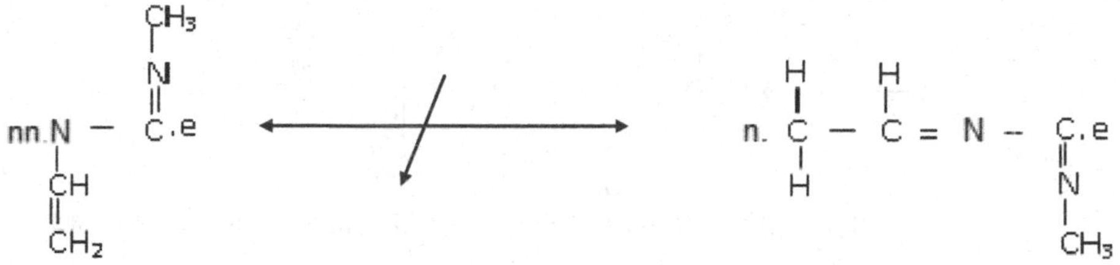

$$
nn.N - \overset{CH_3}{\underset{\substack{| \\ N \\ \| \\ CH \\ \| \\ CH_2}}{C}}.e
\quad \longleftarrow\!\!\!/\!\!\!\longrightarrow \quad
n.\overset{H}{\underset{\substack{| \\ H}}{C}} - \overset{H}{\underset{}{C}} = N - \overset{}{\underset{\substack{\| \\ N \\ | \\ CH_3}}{C}}.e
$$

NO RESONANCE STABILIZATION   9.306

It is not resonance stabilized, because, the movement is from Half-Non-free to Full- Free, though the same transfer species is involved.

9.9. Cumulenic di- carbonyls

To finally complete this chapter, one of the most unique members of the families and the last member is to be considered. These are di - carbonyl monomers which have no transfer species. The first member of the family is CO_2 with the C center divided into two parts. The members of this family of monomers include the following –

$$
\overset{1}{O} = \overset{2}{C} = \overset{3}{O} \quad ; \quad O = C = C = O \quad ; \quad O = C = C = C = O
$$

(I) (1,3 - di-carbonyl)
Carbon dioxide   (II) (1,4- di-carbonyl)   (III) (1,5 - di -carbonyl)   9.307

While the existence of (II) and (III) may not be readily favored, that of (I) is well known. Nevertheless, all the monomers will be considered.

For 1,3 – di-ketones monomers, only one mono-form exists chargedly and radically as shown below.

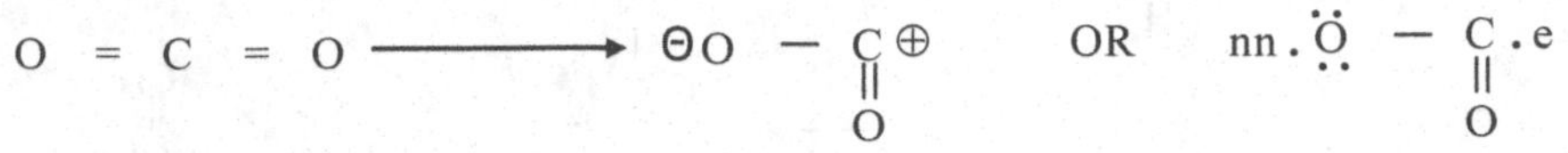

$$O = C = O \longrightarrow \overset{\ominus}{O} - \underset{\underset{O}{\|}}{\overset{\oplus}{C}} \qquad OR \qquad nn.\overset{..}{\underset{..}{O}} - \underset{\underset{O}{\|}}{C}.e$$

(Half-free charged or radical monomer)   9.308

The activation center is more nucleophilic than that of the first member of ketenes as shown below.

$$\overset{\ominus}{O} - \underset{\underset{O}{\|}}{\overset{\oplus}{C}} \qquad > \qquad \overset{\ominus}{O} - \underset{\underset{CH_2}{\|}}{\overset{\oplus}{C}}$$

(I)   (II)

Order of nucleophilicity   9.309

(I) is more nucleophilic than (II), since O = group is more radical-pushing than H_2C = group. Nevertheless, unlike (II) which enolizes, (I) cannot and tends to favor both charged and radical routes, in view of the absence of transfer species in = O group. As can be observed, the monomer can favor the use of non-free negatively charged initiators, free-media non-free negatively charged, paired-media positively charged initiators, electro-free-radical (the natural route) and nucleo-non-free-radical initiators.

With the 1,4 - di - monomers, there are only two mono-forms as shown below only radically.

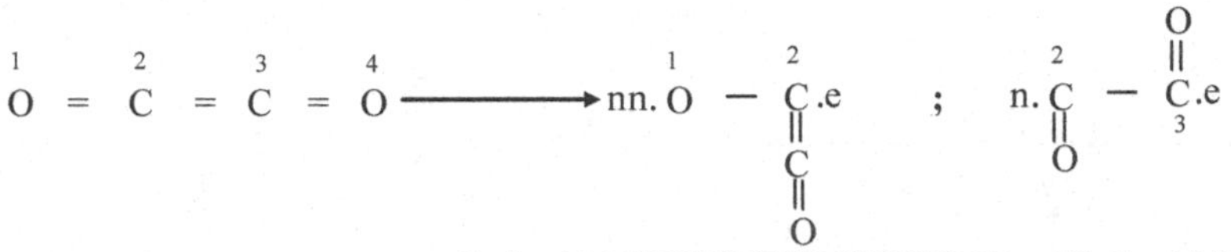

$$\overset{1}{O} = \overset{2}{C} = \overset{3}{C} = \overset{4}{O} \longrightarrow nn.\overset{1}{O} - \underset{\underset{\underset{O}{\|}}{\underset{C}{\|}}}{\overset{2}{C}}.e \quad ; \quad n.\underset{\underset{O}{\|}}{\overset{2}{C}} - \underset{\overset{\|}{O}}{\overset{\overset{O}{\|}}{C}}.e_{3}$$

(I) (both) MORE NUCLEOPHILIC   (II) (both) LESS NUCLEOPHILIC   9.310

None of the mono-forms can be internally resonance stabilized since the O center cannot carry a radical like the case of Benzoquinone, since to move here from Full-free (II) above to Full-non-free is impossible. ***It should be noted that O = C = group is a radical-pulling group.*** It is (II) that is largely involved all the time when activated, because it is the less Nucleophilic center. Why the existence of this compound or monomer, unlike the existence of benzoquinone is not known is clearly becoming obvious. (II) above is trans in character wherein the influence of electrostatic forces of repulsion is weaker than if it was cis placed. The existence of this compound can be seen during the oxidation of vinyl alcohol as shown below.

Oxidation of vinyl alcohol

Stage 1: $O_2 \rightleftharpoons 2nn\cdot O\cdot en$

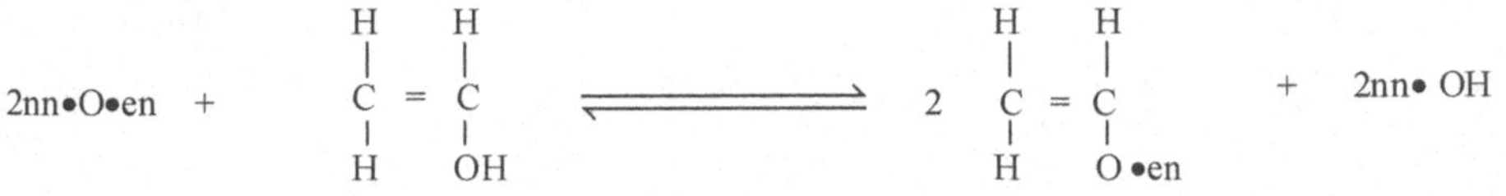

$$2nn\cdot O\cdot en \; + \; \underset{H}{\overset{H}{\underset{|}{\overset{|}{C}}}} = \underset{OH}{\overset{H}{\underset{|}{\overset{|}{C}}}} \rightleftharpoons 2 \; \underset{H}{\overset{H}{\underset{|}{\overset{|}{C}}}} = \underset{O\cdot en}{\overset{H}{\underset{|}{\overset{|}{C}}}} \; + \; 2nn\cdot OH$$

(A)

793

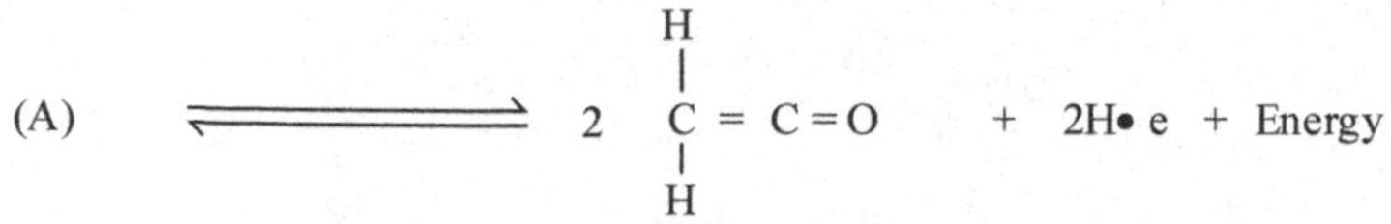

$$
\text{(A)} \quad \rightleftharpoons \quad 2 \; \underset{H}{\overset{H}{C}} = C = O \quad + \quad 2H\bullet e \;+\; Energy
$$

(B)

$$
2H\bullet e \;+\; 2HO\bullet nn \longrightarrow 2H_2O \qquad\qquad 9.311
$$

$\underline{Overall\ equation}: 2H_2C = CH(OH) + O_2 \longrightarrow 2H_2C = C = O + 2H_2O + Energy$

Stage 2:

$$O_2 \rightleftharpoons 2nn\bullet O\bullet en$$

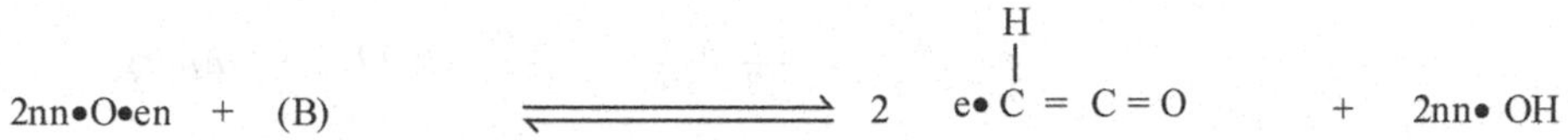

$$
2nn\bullet O\bullet en \;+\; (B) \quad \rightleftharpoons \quad 2 \; e\bullet \underset{H}{\overset{H}{C}} = C = O \quad + \quad 2nn\bullet OH
$$

(C)

$$
\longrightarrow \quad 2 \; \underset{OH}{\overset{H}{C}} = C = O
$$

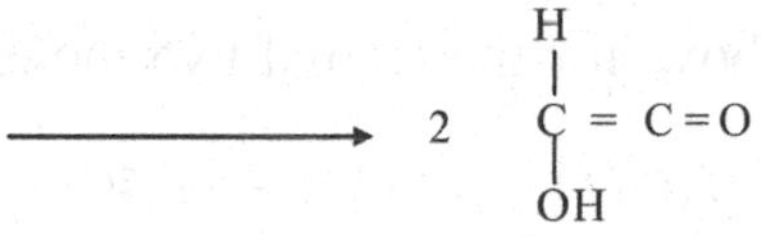

(D) \hfill 9.313

$\underline{Overall\ equation}: 2H_2C = CH(OH) + 2O_2 \longrightarrow H(OH)C = C = O + 2H_2O + Energy$ \hfill 9.314

Stage 3:

$$O_2 \rightleftharpoons 2nn\bullet O\bullet en$$

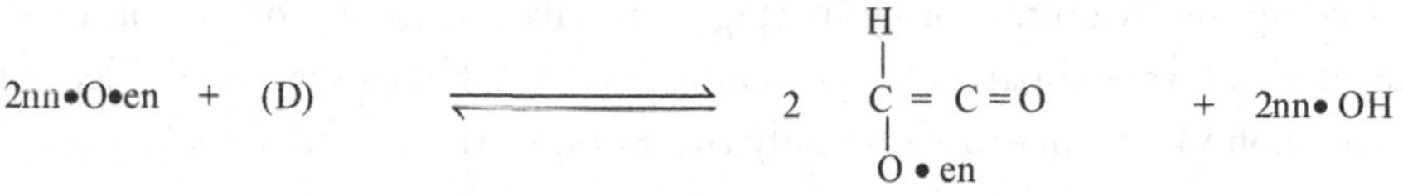

$$
2nn\bullet O\bullet en \;+\; (D) \quad \rightleftharpoons \quad 2 \; \underset{O\bullet en}{\overset{H}{C}} = C = O \quad + \quad 2nn\bullet OH
$$

(E)

$$
\text{(E)} \quad \rightleftharpoons \quad 2 \; O = C = C = O \quad + \; 2H\bullet e + Energy
$$

(F)

$$2H\bullet e \;+\; 2nn\bullet OH \longrightarrow 2H_2O$$

$\underline{Overall\ equation}: 2H_2C = CH(OH) + 3O_2 \longrightarrow 2O = C = C = O + 4H_2O + Energy$ \hfill 9.316

Stage 4:

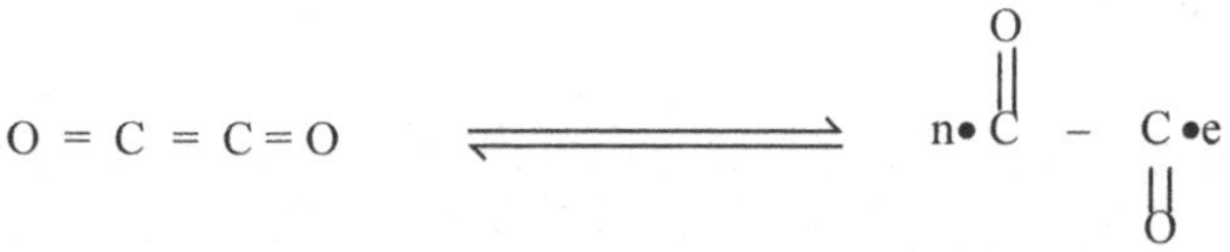

$$
O = C = C = O \quad \rightleftharpoons \quad n\bullet \overset{O}{\underset{\;}{\overset{\|}{C}}} - C\bullet e \;\; \overset{\|}{\underset{O}{}}
$$

(G)

$$
\rightleftharpoons \quad e\bullet \overset{O}{\overset{\|}{C}}\bullet n \quad + \quad \overset{O}{\overset{\|}{C}} :
$$

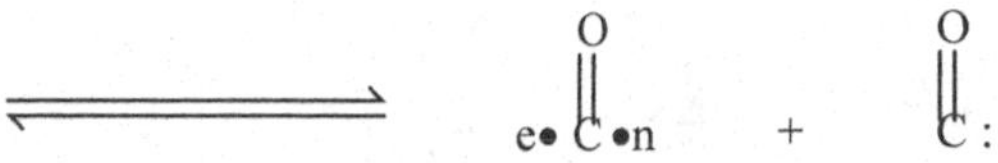

(H)

$$(H) \xrightarrow[\text{Release of Energy}]{\text{Deactivation}} \quad :\overset{O}{\underset{||}{C}} \quad + \quad Energy \qquad 9.317$$

$$\underline{Overall\ equation}: 2H_2C = CH(OH) \ + \ 3\underset{--}{O_2} \longrightarrow 4CO \ + \ 4H_2O \ + \ Energy \qquad 9.318$$

Stage 5:

$$\underset{--}{O_2} \xrightleftharpoons[\text{Oxygen molecule}]{\text{Equilibruim State Existence of Oxidising}} 2en \cdot \ddot{O} \cdot nn$$

$$2en \cdot O \cdot nn + \ 2C = O \xrightleftharpoons{\text{Oxidation}} \quad 2e \bullet \overset{O}{\underset{||}{C}} - O \bullet nn$$
$$\qquad\qquad (Stabilized) \qquad\qquad\qquad\qquad (B)$$

$$(B) \xrightarrow[\text{Release of Energy}]{\text{Deactivation}} \quad 2CO_2 \ + \ Energy \qquad 9.319$$

$$\underline{Overall\ Equation}: \underset{--}{O_2} + 2CO \longrightarrow 2CO_2 \ + \ Energy \qquad 9.320$$

Stage 6: Same as Stage 5 above for the second two moles of CO.

$$\underline{Overall\ equation}: 2H_2C = CH(OH) \ + \ 5\underset{--}{O_2} \longrightarrow 4CO_2 \ + \ 4H_2O \ + \ Energy\,(5) \qquad 9.321$$

The oxidizing oxygen was obtained not in situ, but from a stage in the use of a particular oxidizing agent. As shown above, there are six stages on the whole, noting the presence of the 1,4- di-carbonyl in the third stage. The (B) formed in Stage 1, a Ketene is an Electrophile. The (F) in Stage 3, a diketone is also another type of ELECTROPHILE, which is very unstable and cannot be oxidized. Hence, based on the operating conditions, it decomposes instantaneously in Stage 4 to give two moles of carbon monoxide with release of energy. Stages 5 and 6 are in parallel. As can be observed so far, this is a weak Electrophile.

With 1, 5 - di -carbonyl monomers, there are also only two mono-forms as shown below.

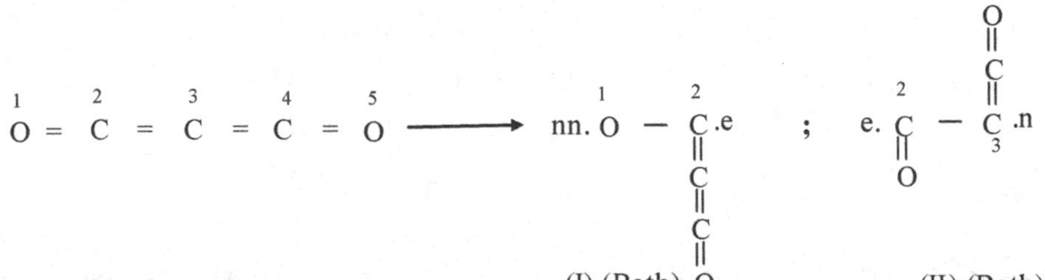

$$\underset{1}{O} = \underset{2}{C} = \underset{3}{C} = \underset{4}{C} = \underset{5}{O} \longrightarrow \quad nn.\underset{1}{O} - \underset{2}{\overset{}{C}}.e \quad ; \quad e.\underset{}{\overset{}{C}} - \underset{3}{\overset{}{C}}.n$$

$$(I)\ (Both) \qquad\qquad\qquad (II)\ (Both) \qquad 9.322$$

Only (I) could have been resonance stabilized, while (II) which looks like the Y center cannot be resonance stabilized as shown below.

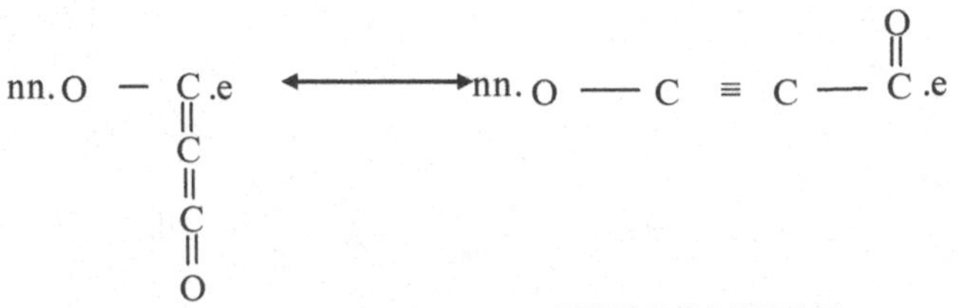

$$nn.O - \overset{}{\underset{}{C}}.e \longleftarrow \longrightarrow nn.O - C \equiv C - \overset{O}{\underset{||}{C}}.e$$

$$.(I)\ [More\ Nucleophilic] \qquad \underline{\textbf{NOT POSSIBLE}} \qquad\qquad 9.323$$

$$\text{e} \bullet \overset{\overset{\displaystyle O}{\parallel}}{\underset{\underset{\displaystyle O}{\parallel}}{C}} - C \bullet n \quad \longleftrightarrow \not\longrightarrow \quad nn \bullet O - C \equiv C - \overset{\overset{\displaystyle O}{\parallel}}{C} \bullet e$$

(II) Full- free   Half free   9.324

Though the same mono-form is obtained above, this is only possible through (I). Evidently, = C= C = O is a resonance stabilization group which is a radical-pushing group. Nevertheless, in general, it can be observed that the same routes are favored by all the cases. (I) is more nucleophilic than (II). Therefore, it is (II) that is first activated all the time. Unlike quinones as recalled below, the equations below are not favored since the movement is from Half-free to Full-non-free, and since <u>one cannot move from full-free to Full-non-free here.</u> Try it as an exercise.

$$O = C = C = O \longrightarrow en \cdot O - C \equiv C - O \cdot nn$$
$$\text{IMPOSSIBLE ACTIVATION}$$

                                                     9.325

$$\underset{\underset{\displaystyle CH_3}{|}}{N} = C - \overset{\overset{\displaystyle CH_3}{|}}{C} = N \xrightarrow{\text{(Not possible)}} en \cdot \overset{\overset{\displaystyle CH_3}{|}}{N} - C \equiv C - \underset{\underset{\displaystyle CH_3}{|}}{N} \cdot nn$$

                                                     9.326

                              (I)   (II)

COMPLETE ACTIVATION   9.327

Like para- and ortho-quinones which are resonance stabilized, existence of full non-free-radical mono-forms can be observed not to be possible for 1,4- dicarbonyl and 1,4 - diimines. Thus, for 1,4 - dicarbonyl, there are indeed two mono-forms as already shown with all the members having no transfer species.

9.10. Proposition of Rules of Chemistry and Concluding remarks

Having considered these unique and very important families of monomers - ketenes, isocyanates, imines, diimines and carbonyls, one will now proceed to propose rules which indeed are natural laws. With these families of monomers, a large number of substituted groups all radical in character, have been identified.

<u>Rule 472:</u> This rule of Chemistry for **Addition monomers,** states that, Olefins, Acetylenes, "Diazoalkanes" and Cumulenes are indeed *ideal or traditional* Addition monomers while aldehydes, ketones, nitriles, isocyanates, imines. are *pseudo*-Addition monomers and 1,3-ketenes, 1,3-imines, 1,3-isocyanates,

1,3-di-imines, 1,3-di-carbonyl and their cumulenes are either pseudo- or ideal, or part ideal Addition and part pseudo- Addition monomers.
(Laws of Creations for Addition monomers)

Rule 473: This rule of chemistry for **Addition monomers which are non-olefinic in character such as Most Cumulenes, Nitriles, Isocynates, imines, di-imines and their cumulenic members,** states that, none of them can fully be activated chargedly due to electrostatic forces of repulsion, but only radically; for which chargedly their considerations is important for purposes of finding those which can be activated chargedly and used, using the eye of the needle.
(Laws of Creations for Addition monomers)

Rule 474: This rule of Chemistry for **Ideal or traditional Cyclic Addition Hydrocarbon monomers,** states that, shown below are the real names of Cyclic Alkenes and Alkynes-

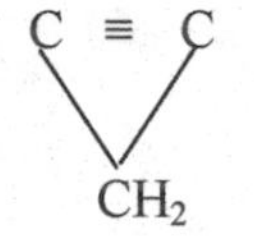

Cyclopropyne 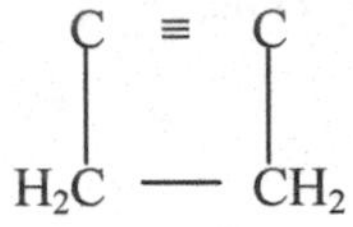 Cyclobutyne 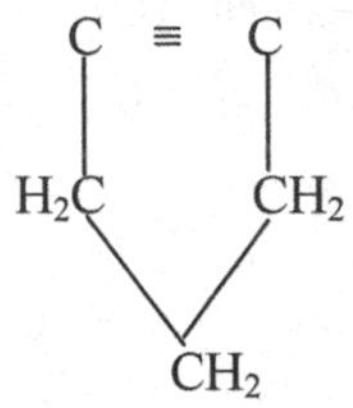 Cyclopentyne 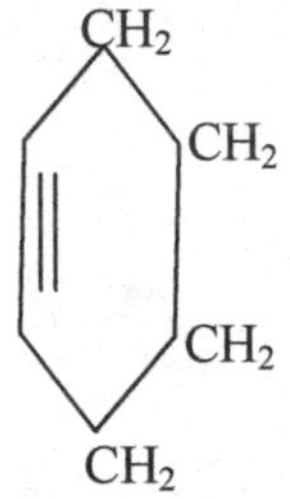 Cyclohexyne

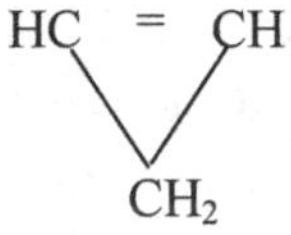

Cyclopropene 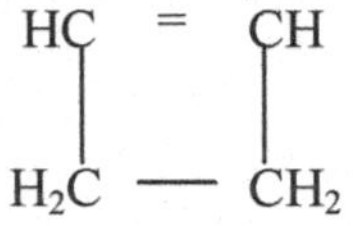 Cyclobutene 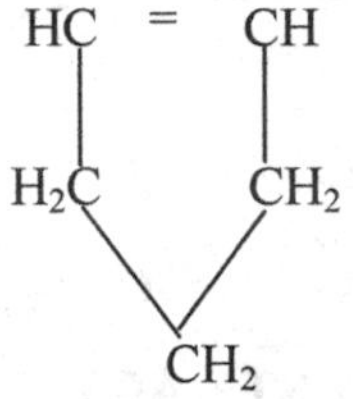 Cyclopentene 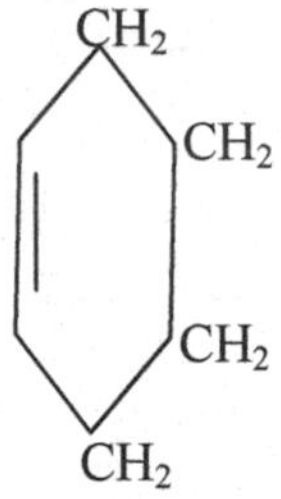 Cyclohexene

wherein some of the first group of monomers may not exist due to the presence of Max. Required Strain Energy in their rings.
(Laws of Creations for Cyclic Hydrocarbons)

Rule 475: This rule of Chemistry for **1,3-Ketene,** states that, this family of compounds/ monomers of C/O as Central atoms, contain two mono-forms with different characters based on the types of substituent groups carried by the C center for which the first member is shown below-

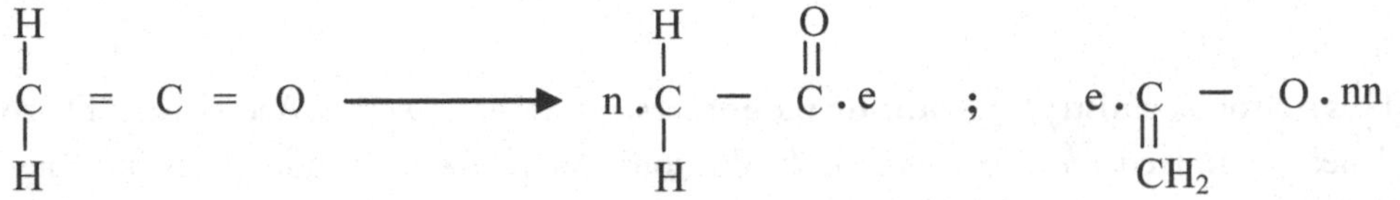

(I) (Electrophile)   (II) (Nucleophile)   i)

797

from which it can be observed that the O = and H$_2$C = groups are radical-pushing groups; for which this first member cannot be used as a monomer, until either when suppressed or when one or two of the H atoms have been replaced based on the followings-.

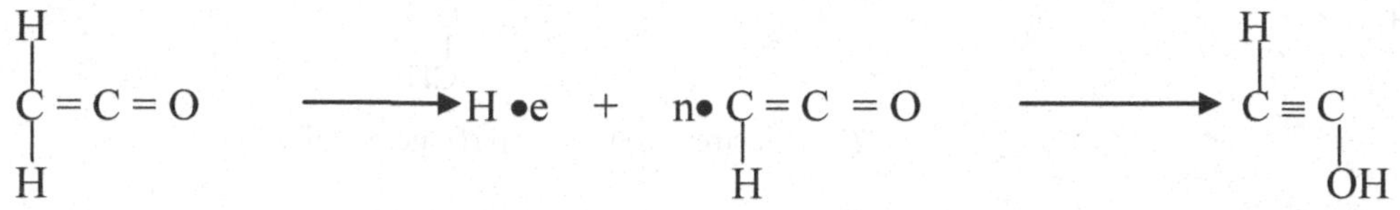

$$
\begin{array}{c}
H \\
| \\
C = C = O \\
| \\
H
\end{array}
\longrightarrow H \bullet e \; + \; n\bullet
\begin{array}{c}
C = C = O \\
| \\
H
\end{array}
\longrightarrow
\begin{array}{c}
H \\
| \\
C \equiv C \\
| \\
OH
\end{array}
$$

$$(II) \qquad\qquad ii)$$

wherein when in Equilibrium state of existence, it undergoes Electro-radicalization phenomenon to give an acetylenyl alcohol (II) above.
(Laws of Creations for 1,3-Ketene)

Rule 476: This rule of Chemistry for **1,3-Ketenes,** states that, these are ELECTROPHILES which when polymerized nucleo-free-radically *polyketones* are obtained as shown below for the initiation step, for which living polymers are obtained with no transfer species; noting that this can take place radically and chargedly-

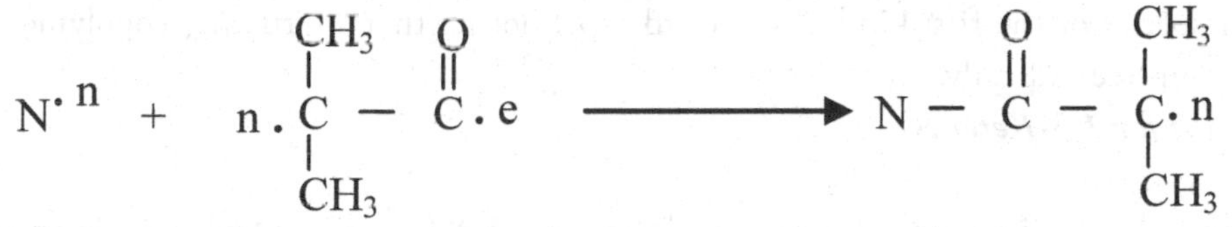

$$
N \cdot^n \; + \; n \cdot
\begin{array}{c}
CH_3 \\
| \\
C \\
| \\
CH_3
\end{array}
-
\begin{array}{c}
O \\
\| \\
C \cdot e
\end{array}
\longrightarrow
N -
\begin{array}{c}
O \\
\| \\
C
\end{array}
-
\begin{array}{c}
CH_3 \\
| \\
C \cdot n \\
| \\
CH_3
\end{array}
$$

noting that alternating copolymers can be obtained nucleo-non-free-radically.
(Laws of Creations for 1,3-Ketenes)

Rule 477: This rule of Chemistry for **1,3-Ketenes,** states that, these are ELECTROPHILES which when polymerized electro-free-radically, *polyacetals* are obtained as shown below for the initiation step, for which dead ter**minal** double bond polymers are obtained after release of transfer species of the first kind of the third type, noting that this can only be done radically and never chargedly, when ionic metals are involved

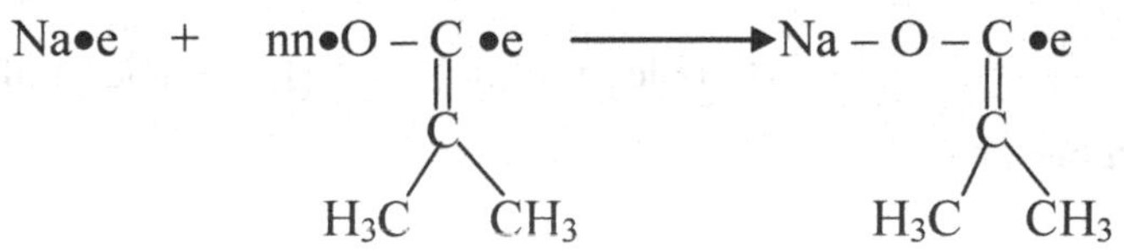

$$
Na\bullet e \; + \; nn\bullet O - \underset{\underset{\displaystyle H_3C \quad CH_3}{\underset{\displaystyle \diagdown \; C \; \diagup}{\|}}}{C} \bullet e \longrightarrow Na - O - \underset{\underset{\displaystyle H_3C \quad CH_3}{\underset{\displaystyle \diagdown \; C \; \diagup}{\|}}}{C} \bullet e
$$

(Laws of Creations for 1,3-Ketenes)

Rule 478: This rule of Chemistry for **1,3-Ketenes of the type R(OR)C = C = O where Rs are alkylane groups,** states that, with these etheric groups, the followings are to be expected-

$$
\underset{\underset{CH_3}{|}}{\overset{\overset{CH_3}{|}}{C}} = \overset{2}{C} = \overset{1}{O} \longrightarrow n.\underset{\underset{CH_3}{|}}{\overset{\overset{CH_3}{|}}{C}} - \overset{\overset{O}{||}}{C}.e \quad ; \quad e.\overset{\overset{O}{||}}{C} - O.nn
$$

(I) (Electrophile)   (II) (Nucleophile)   i)

$$
E\cdot^e + nn.O - \underset{\underset{HO}{\diagup}\underset{CH_3}{\diagdown}}{\overset{\overset{}{||}}{C}}.e \longrightarrow E -- O - \underset{\underset{HO}{\diagup}\underset{CH_3}{\diagdown}}{\overset{\overset{}{||}}{C}}.e
$$

Transfer species e.CH$_3$

ii)

$$
N\cdot^n + n.\underset{\underset{CH_3}{|}}{\overset{\overset{CH_3}{|}}{C}} - \overset{\overset{O}{||}}{C}.e \longrightarrow N - \overset{\overset{O}{||}}{C} - \underset{\underset{CH_3}{|}}{\overset{\overset{CH_3}{|}}{C}}.n
$$

No Transfer species

iii)

for which electro-free-radically, dead random copolymers *(polyacetals/ketones)* are produced with CH$_3$ group as transfer species of the first kind of the third type; noting that alternating copolymers can be obtained nucleo-non-free-radically.
(Laws of Creations for 1,3-Ketenes)

Rule 479: This rule of Chemistry for **1,3-Ketenes of the type H$_2$NRC = C = O where R is an alkylane group,** states that, the mono-forms bear the characters as those with OR group for which *polyketones and polyacetals* are the main products and for these, alternating copolymers can also be obtained nucleo-non-free-radically.

$$
\underset{\underset{NR_2}{|}}{\overset{\overset{CH_3}{|}}{C}} = C = O \longrightarrow n.\underset{\underset{NR_2}{|}}{\overset{\overset{CH_3}{|}}{C}} - \overset{\overset{O}{||}}{C}.e \quad ; \quad e.\underset{\underset{NR_2}{|}}{\overset{\overset{O}{||}}{C}} - O.nn
$$

(I) (Electrophile)   (II) (Nucleophile)

(Laws or Creations for 1,3-Ketenes)

Rule 480: This rule of Chemistry for **1,3-Ketenes of the type F$_2$C = C = O,** states that, the mono-forms have the following characters; for which in the absence of transfer species of any

$$
\underset{\underset{F}{|}}{\overset{\overset{F}{|}}{C}} = C = O \longrightarrow n.\underset{\underset{F}{|}}{\overset{\overset{F}{|}}{C}} - \overset{\overset{O}{||}}{C}.e \quad ; \quad e.\underset{\underset{CF_2}{}}{\overset{\overset{}{||}}{C}} - O.nn
$$

(I) (Both)   (II) (Both)

799

kind, living ***polyacetals and polyketones*** are the products via routes natural to the centers and no alternating copolymers can be obtained.
(Laws of Creations for 1.3-Ketenes)

Rule 481: This rule of Chemistry for **1,3-Ketenes of the type $R_F R_F^I C = C = O$ where R_F and R_F^I are** $C_n F_{2n+1}$, states that, the characters of the mono-forms are as follows-

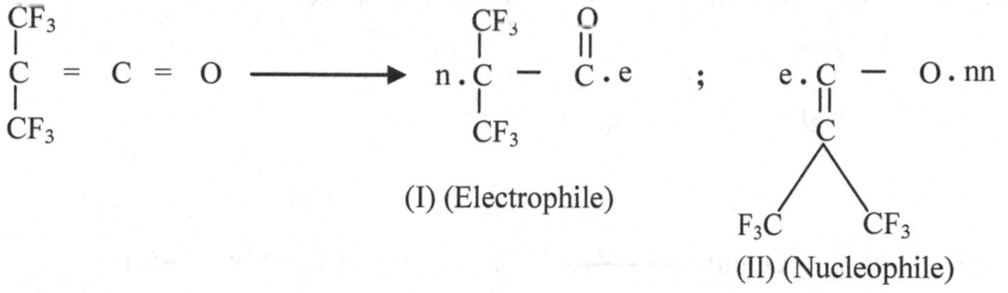

(I) (Electrophile)   (II) (Nucleophile)   i)

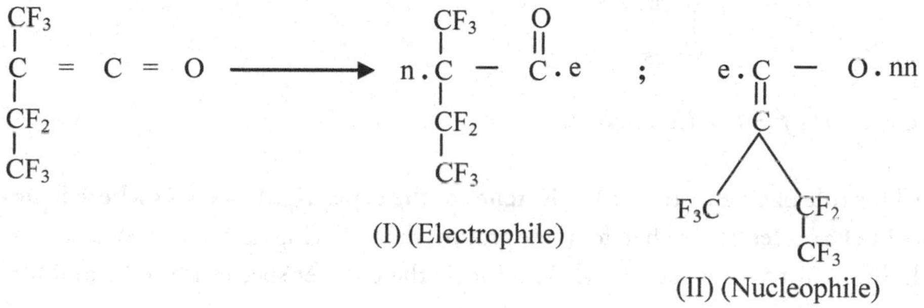

(I) (Electrophile)   (II) (Nucleophile)   ii)

for which for the fact that the (I)s do not favor the electro-free-radical route as shown bel

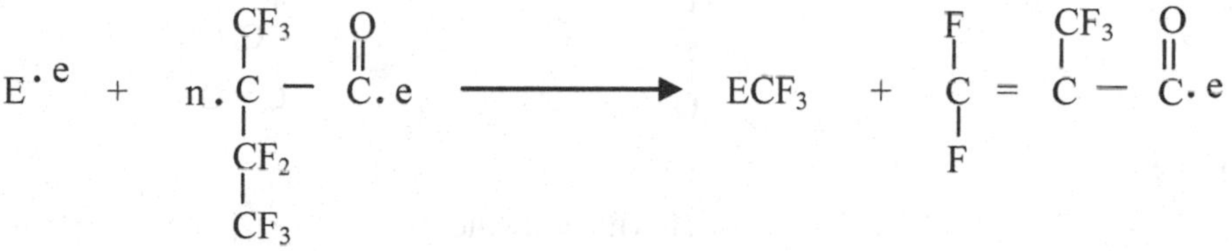

iii)

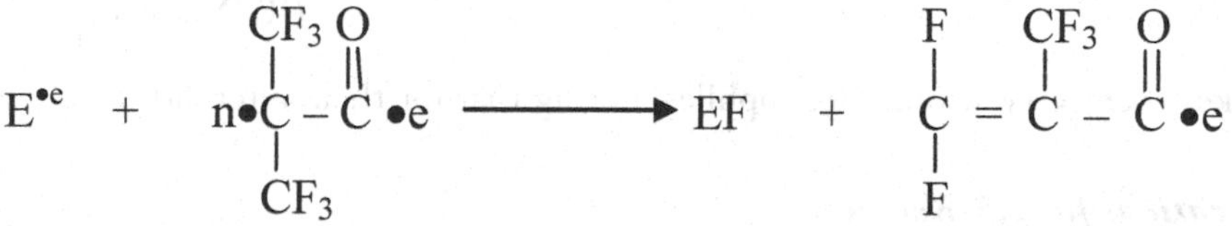

iv)

and the (II) do not favor the nucleo-radical routes as also shown below, makes these members very Electrophilic in character, noting that for these alternating copolymers cannot be obtained.

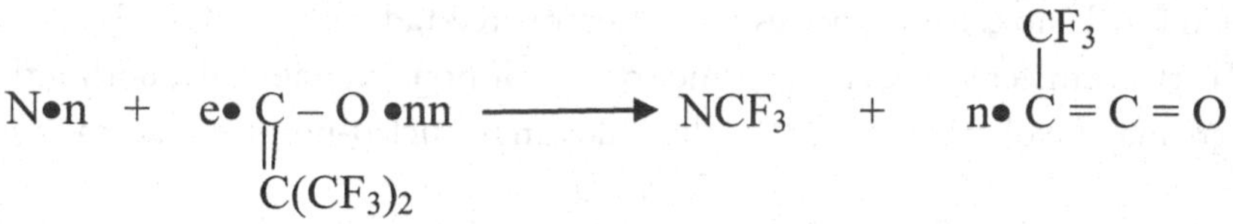

i)

$$\text{N} \bullet \text{nn} + \text{e} \bullet \underset{\underset{\text{C}(CF_3)_2}{\|}}{\text{C}} - \text{O} \bullet \text{nn} \longrightarrow NCF_3 + \underset{\overset{CF_3}{|}}{\text{C}} \equiv \text{C} - \text{O} \bullet \text{nn}$$

ii)

$$\text{N} \bullet \text{n} + \text{e} \bullet \underset{\underset{\underset{C_2F_3}{|}}{\underset{CCF_3}{\|}}}{\text{C}} - \text{O} \bullet \text{nn} \longrightarrow NCF_3 + \text{n} \bullet \underset{\overset{C_2F_5}{|}}{\text{C}} = \text{C} = \text{O}$$

iii)

$$\text{N} \bullet \text{nn} + \text{e} \bullet \underset{\underset{\underset{C_2F_5}{|}}{\underset{CCF_3}{\|}}}{\text{C}} - \text{O} \bullet \text{nn} \longrightarrow NCF_3 + \underset{\overset{C_2F_5}{|}}{\text{C}} \equiv \text{C} - \text{O} \bullet \text{nn}$$

iv)

(Laws of Creations for 1,3-Ketenes)

Rule 482: This rule of Chemistry for **1,3-Ketenes of the types $R_FRC = C = O$ where R_F is COOR, and R is alkylane in character,** states that, for these groups, the followings are to be expected; for the two mono-forms with different transfer species, wherein for (I) the transfer species is OCH_3 and for (II) it is CH_3

$$\underset{\underset{\underset{\underset{CH_3}{|}}{O}}{\underset{\underset{C=O}{|}}{|}}}{\overset{\overset{CH_3}{|}}{\text{C}}} = \text{C} = \text{O} \longrightarrow \underset{\underset{\underset{\underset{CH_3}{|}}{O}}{\underset{\underset{C=O}{|}}{|}}}{\overset{\overset{CH_3}{|}}{\text{n} \cdot \text{C}}} - \text{C.e} \quad ; \quad \underset{\underset{\underset{\underset{CH_3}{|}}{O}}{\underset{\underset{C=O}{|}}{|}}}{\underset{CCH_3}{\overset{\overset{O}{\|}}{\text{e} \cdot \text{C}}}} - \text{O. nn}$$

(I) (Electrophile)   (II) (Nucleophile)

and this makes them very strong Electrophiles, noting that for them alternating copolymers cannot be obtained.
(Laws of Creations for 1,3-Ketenes)

Rule 483: This rule of Chemistry for **1,3-Ketenes of the types $R_FRC = C = O$ where R_F is COR, and R is alkylane in character**, states that, for these groups, the followings are to be expected; for the two mono-forms, for which for (I) no transfer species exists electro-free-radically if $n.CH_3$ is not large enough to be weak and for (II) the transfer species is CH_3 nucleo free- or non-free radically, noting that for these, alternating copolymers may be obtained based on the followings nucleo-non-free-radically-

$$H.n \; > \; n.CH_3 \; > \; n.C_2H_5 \; > \; n.C_3H_7 \; .... \gg \; n.\,C_{10}H_{21}$$

Order of radical-pushing capacity nucleo-free-radically *i)*

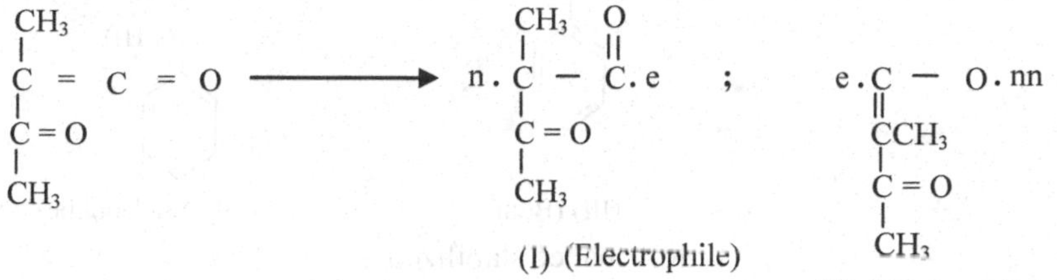

(I) (Electrophile)

(II) (Nucleophile) *ii)*

(Laws of Creation- for 1,3-ketenes)

Rule 484: This rule of Chemistry for **1,3-Ketenes of the types $R_FRC = C = O$ where R_F is $C\equiv N$, and R is alkylane in character,** states that, for these groups, the followings are to be expected; for the two mono-forms as shown below, for which while no transfer species exists electro-free-radically for (I), for (II) CH_3 is the transfer species of the first kind of the third type nucleo-radically, noting that alternating copolymers may be obtained here in view of the very strong radical-pulling and highest nucleophilic capacity of $C\equiv N$ group and center in the monomer.

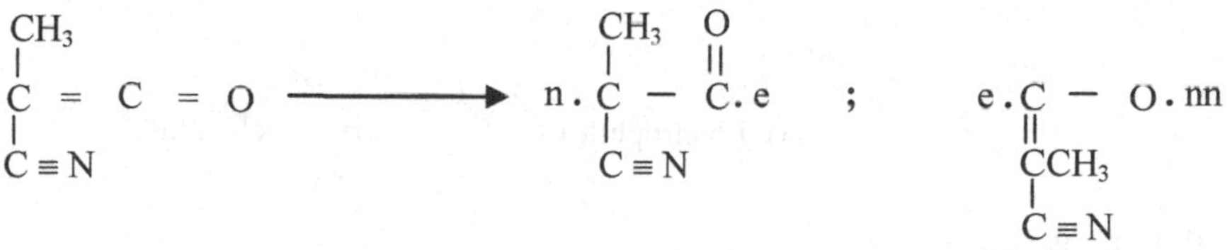

(I) (Both)

(II) (Nucleophile)

(Laws of Creations for 1,3-Ketenes)

Rule 485: This rule of Chemistry for **1,3-Ketenes with resonance stabilization groups such as shown below,** states that,

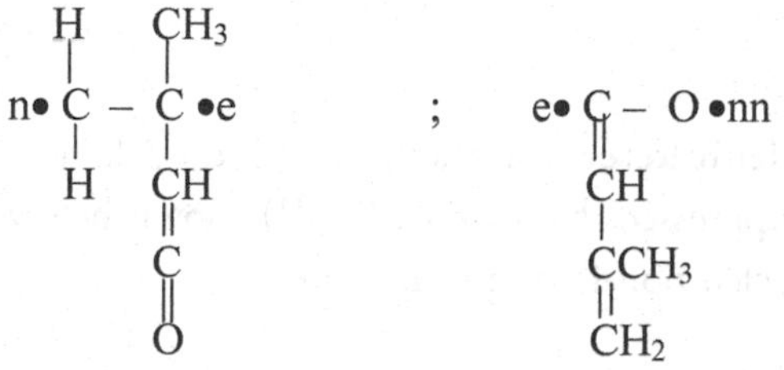

(I)

(II)

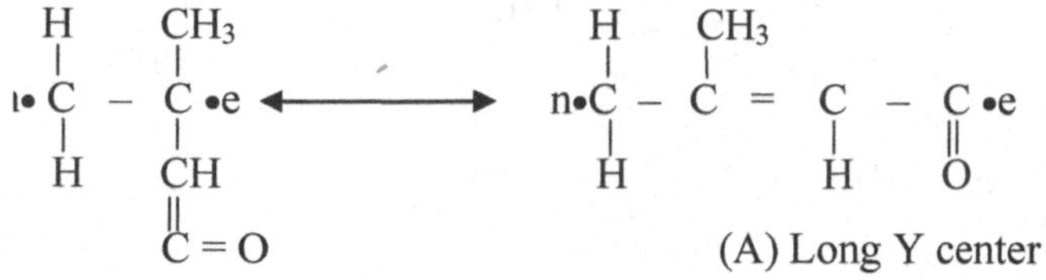

(A) Long Y center

RESONANCE STABILIZATION

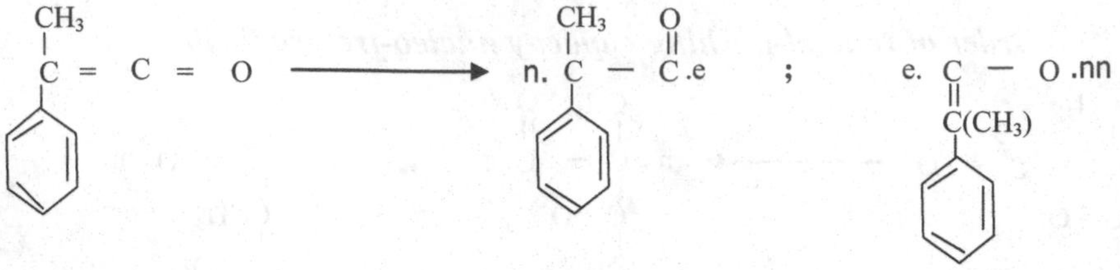

$$CH_3\text{-C}=C=O \longrightarrow n.\overset{CH_3}{\underset{}{C}}-\overset{O}{\underset{}{C}}.e \quad ; \quad e.\,C-O\,.nn$$

(III) (Both)   (IV)(Nucleophile)

Resonance stabilized

only (I11) a non-ringed Electrophile is resonance stabilized via only the male center the provider being the phenyl group, unlike (I) which is also an Electrophile resonance stabilized as shown in the second equation, the provider being the monomer group; noting that when the CH_3 is changed to CF_3, the monomer becomes more Electrophilic, i.e.,

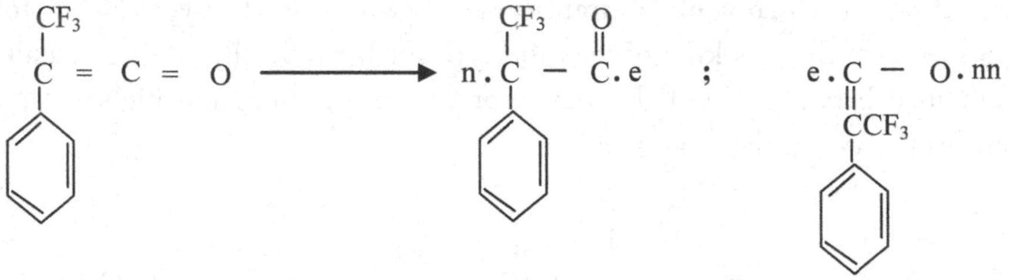

$$CF_3\text{-C}=C=O \longrightarrow n.\overset{CF_3}{\underset{}{C}}-\overset{O}{\underset{}{C}}.e \quad ; \quad e.\,C-O.\,nn$$

(I) (Electrophile)   (II)(Nucleophile)   i)

for which, the following is valid-

$$O = C = \quad >> \quad O = C(OH)\text{-} \quad > \quad O = C(CH_3)\text{-} \quad > \quad O = C(H)-$$

Order of radical-pulling capacity   ii)

noting that alternating copolymers may be difficult be obtained for the case immediately above nucleo-non-free-radically, due to steric limitations.

(Laws of Creation for 1,3-Ketenes)

Rule 486: This rule of Chemistry for **1,4- Cumulenic ketene,** states that, unlike 1,3-ketene, alternating co-polymers cannot be produced from it when suppressed, because while (II) shown below favors both free-radical routes, (III) favors electro-free and nucleo-non-free-radical routes.

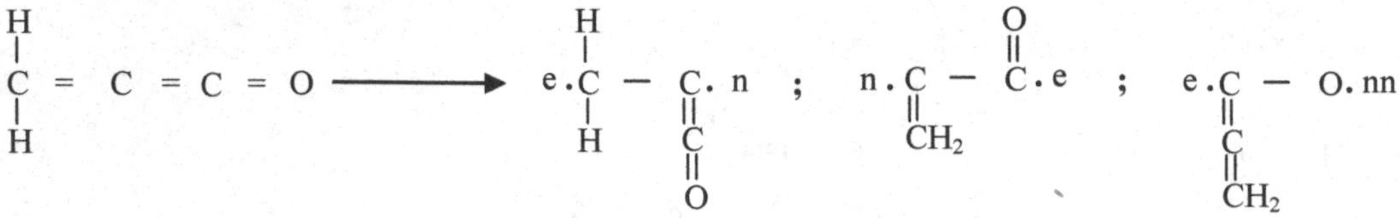

$$\text{H}_2\text{C}=C=C=O \longrightarrow e.\overset{H}{\underset{H}{C}}-\overset{C}{\underset{}{C}}.n \quad ; \quad n.\overset{}{\underset{CH_2}{C}}-\overset{O}{\underset{}{C}}.e \quad ; \quad e.\overset{}{\underset{CH_2}{C}}-O.\,nn$$

(I)   (II)   (III)   i)

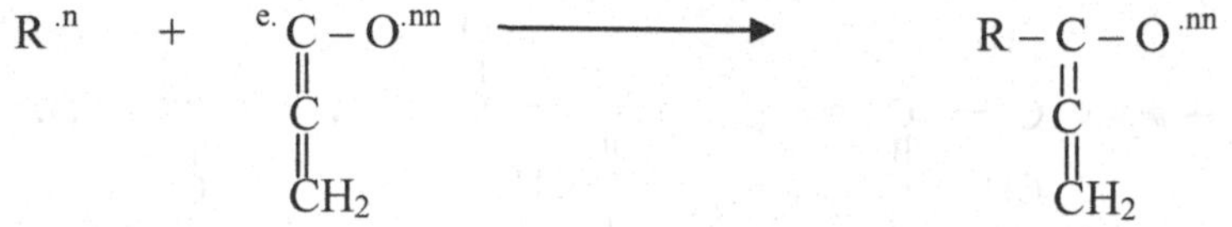

$$R^{.n} \quad + \quad {}^{e.}C-O^{.nn} \longrightarrow R-C-O^{.nn}$$

with C and CH₂ below each central C, and:

$$R^{.n} + {}^{e.}\underset{\underset{CH_2}{\overset{\|}{C}}}{\overset{\|}{C}}{-}O^{.nn} \longrightarrow R-\underset{\underset{CH_2}{\overset{\|}{C}}}{\overset{\|}{C}}{-}O^{.nn}$$

Not radically balanced   ii)

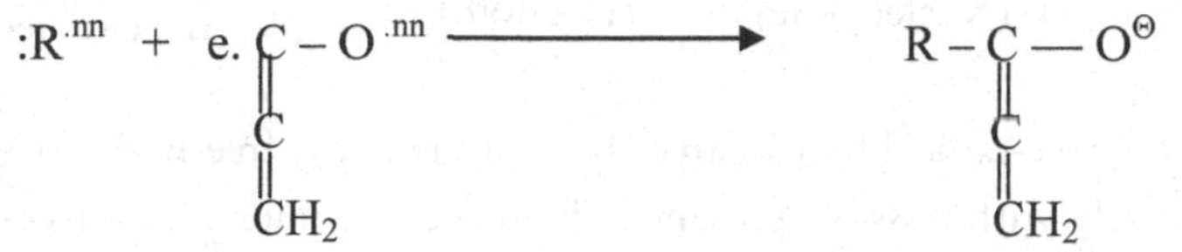

$$:R^{nn} + e.\underset{\underset{CH_2}{\overset{\|}{C}}}{\overset{\|}{C}}{-}O^{.nn} \longrightarrow R-\underset{\underset{CH_2}{\overset{\|}{C}}}{\overset{\|}{C}}{\!\!-\!\!}O^{\ominus}$$

iii)

(Laws of Creations for 1,4-Cumulenic Ketene)

Rule 487: This rule of Chemistry for **1,4-Cumulenic Ketenes,** states that, this family of compounds/ monomers of C/O as Central atoms, when suppressed, contain two mono-forms (II) the (Y) center and (III) the (X) center with different characters based on the types of substituent groups carried by the C center for which the first member is as shown below-

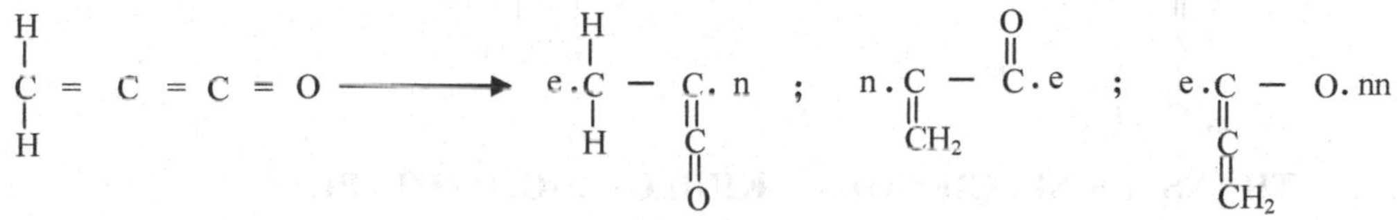

$$\underset{\underset{H}{|}}{\overset{\overset{H}{|}}{C}} = C = C = O \longrightarrow e.\underset{\underset{H}{|}}{\overset{\overset{H}{|}}{C}} - \underset{\underset{O}{\|}}{\overset{\|}{C}}.n \quad ; \quad n.\underset{\underset{CH_2}{\|}}{\overset{\|}{C}} - \overset{\overset{O}{\|}}{C}.e \quad ; \quad e.\underset{\underset{CH_2}{\|}}{\overset{\|}{C}} - O.nn$$

(I)   (II)   (III)   i)

from which it can be observed that the O = and O = C = groups are radical pushing and pulling groups respectively; for which this first member cannot be used as a monomer, until one or two of the H atoms have been replaced based on the followings-.

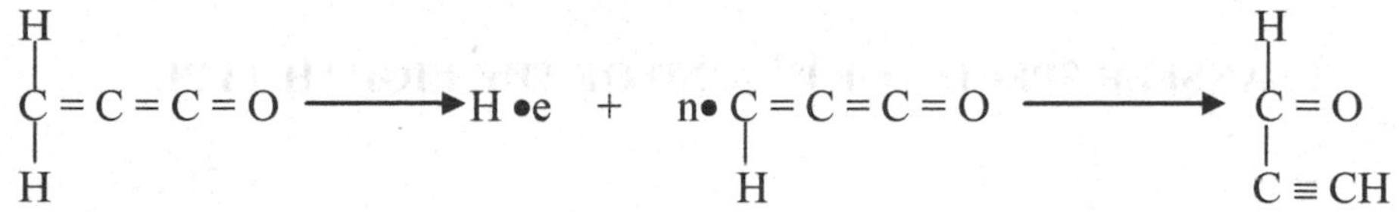

$$\underset{\underset{H}{|}}{\overset{\overset{H}{|}}{C}} = C = C = O \longrightarrow H \bullet e \quad + \quad n\bullet \underset{\underset{H}{|}}{C} = C = C = O \longrightarrow \underset{\underset{C \equiv CH}{|}}{\overset{\overset{H}{|}}{C}} = O$$

(II)   ii)

wherein when in Equilibrium state of existence, it undergoes Electro-radicalization phenomenon to give an equivalent of an unstable acrolein (II) above.
(Laws of Creations for 1,4-Cumulenic Ketenes)

Rule 488: This rule of Chemistry for **1,4-Cumulenic Ketenes,** states that, these like 1,3-ketenes are Electrophiles whose two mono-forms cannot be resonance stabilized.
(Laws of Creations for 1,4-Cumulenic Ketenes)

Rule 489: This rule of Chemistry for **1,4-Cumulenic Ketene with radical-pushing group of the type R,** states that, the followings are obtained for its mono-forms, for which the least Nucleophilic center is (II), the first center to be activated herein identified as the Y center followed by (III) the corresponding X center-

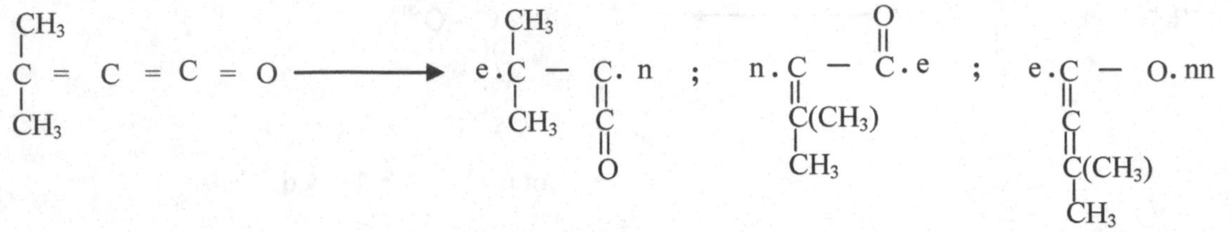

$$\begin{array}{c} CH_3 \\ | \\ C = C = C = O \\ | \\ CH_3 \end{array} \longrightarrow \begin{array}{c} CH_3 \\ | \\ e \cdot C - C \cdot n \\ | \quad | \\ CH_3 \quad C \\ \quad \| \\ \quad O \end{array} \quad ; \quad \begin{array}{c} O \\ \| \\ n \cdot C - C \cdot e \\ \| \\ C(CH_3) \\ | \\ CH_3 \end{array} \quad ; \quad \begin{array}{c} e \cdot C - O \cdot nn \\ \| \\ C \\ \| \\ C(CH_3) \\ | \\ CH_3 \end{array}$$

(I) (Nucleophile)   (II) (Both)   (III) (Nucleophile)

for which (II) which though favors both free-radical routes can only be used nucleo-free-radically while (III) can mostly be used electro-free-radically with possible presence of alternation nucleo-non-free-radically.
(Laws of Creations for 1,4-Cumulenic Ketenes)

Rule 490: This rule of Chemistry for **1,4-Cumulenic Ketene with radical-pushing group of the type R,** states that, the Nucleophilic character of the X center is ascertained as shown below,

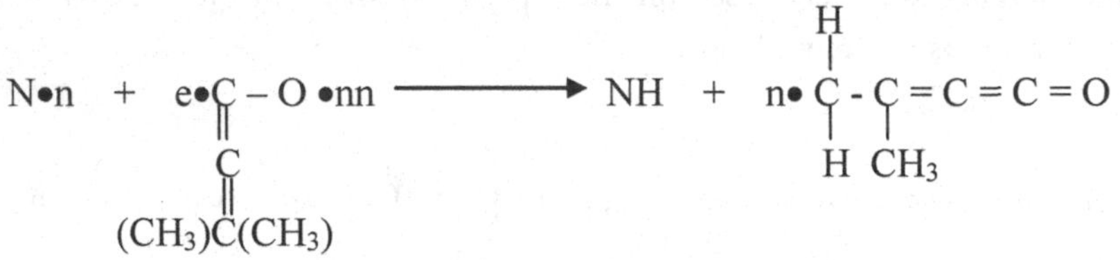

$$N\bullet n \; + \; \begin{array}{c} e \bullet C - O \bullet nn \\ | \\ C \\ \| \\ (CH_3)C(CH_3) \end{array} \longrightarrow NH \; + \; \begin{array}{c} H \\ | \\ n \bullet C - C = C = C = O \\ | \quad | \\ H \quad CH_3 \end{array}$$

TRANSFER SPECIES OF 1ST KIND OF EIGHTH TYPE

i)

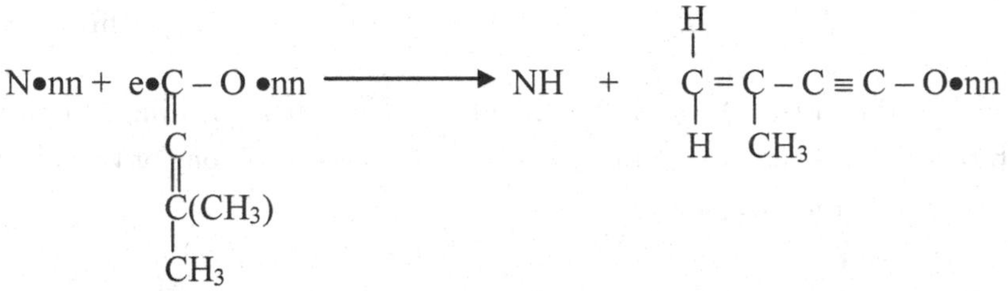

$$N\bullet nn + \begin{array}{c} e \bullet C - O \bullet nn \\ | \\ C \\ \| \\ C(CH_3) \\ | \\ CH_3 \end{array} \longrightarrow NH \; + \; \begin{array}{c} H \\ | \\ C = C - C \equiv C - O \bullet nn \\ | \quad | \\ H \quad CH_3 \end{array}$$

TRANSFER SPECIES OF IST KIND OF THE EIGHTH TYPE

ii)

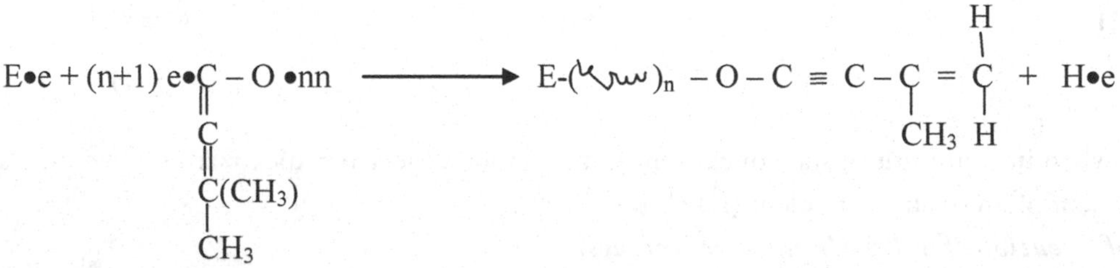

$$E\bullet e + (n+1) \begin{array}{c} e \bullet C - O \bullet nn \\ | \\ C \\ \| \\ C(CH_3) \\ | \\ CH_3 \end{array} \longrightarrow \begin{array}{c} H \\ | \\ E\text{-}(\sim\sim)_n - O - C \equiv C - C = C + H\bullet e \\ | \quad | \\ CH_3 \quad H \end{array}$$

iii)

noting the type of transfer species involved- herein called *Transfer species of the first kind of the eighth type.*
(Laws of Creations for 1,4-Cumulenic Ketenes)

Rule 491: This rule of Chemistry for **1,4-Cumulenic Ketenes with etheric groups shown below,** states that, the followings are to be expected-

iii)

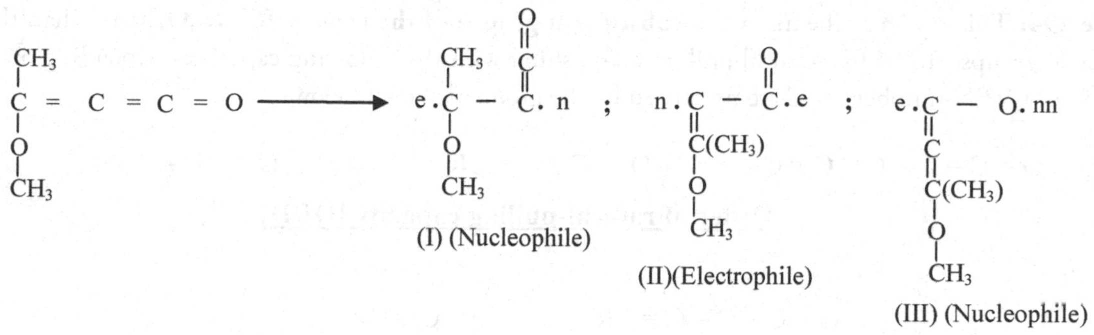

$$CH_3-\underset{\underset{CH_3}{|}}{\underset{|}{O}}{C}=C=C=O \longrightarrow$$

(I) (Nucleophile)

$e.\overset{\underset{CH_3}{|}}{\underset{|}{O}}{C}-\overset{O}{\overset{||}{C}}.n$;

$n.\overset{||}{C}-\overset{\underset{\underset{CH_3}{|}}{\underset{O}{|}}}{\underset{C(CH_3)}{|}}{C}.e$

(II)(Electrophile)

$e.\overset{O}{\overset{||}{C}}-O.nn$
$\underset{\underset{CH_3}{|}}{\underset{O}{|}}{\underset{C(CH_3)}{||}}$

(III) (Nucleophile)

for which (I) will favor polymerization electro-free radically only, (II) will favor only the use of nucleo-free-radicals, clear indication of a real Electrophile and (III) favors only the electro-free-radical route; for which if molecular rearrangement is favored, it is only by (II) making the use of paired radical initiator a necessity; otherwise only the electro-free-radical route will be favored.
(Laws of Creations for 1,4-Cumulenic Ketenes)

Rule 492: This rule of Chemistry for **1,4-Cumulenic Ketenes shown below,** states that, the followings are to be expected-

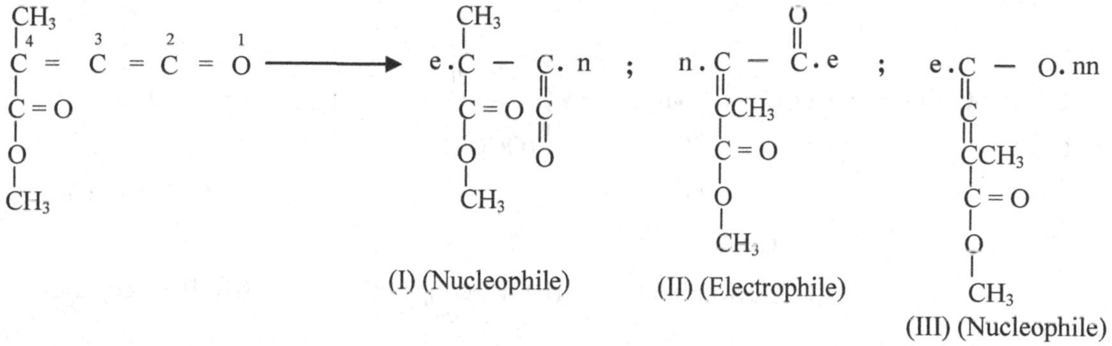

$$\underset{\underset{CH_3}{|}}{\overset{\overset{CH_3}{|}}{\underset{4}{C}}}=\underset{3}{C}=\underset{2}{C}=\underset{1}{O} \longrightarrow$$

(I) (Nucleophile)

$e.\overset{\overset{CH_3}{|}}{\underset{\underset{CH_3}{|}}{\underset{O}{|}}{\underset{C=O}{||}}}{C}-\overset{O}{\overset{||}{C}}.n$

(II) (Electrophile)

$n.\overset{||}{C}-\overset{\underset{\underset{CH_3}{|}}{\underset{O}{|}}}{\underset{C=O}{|}}{\underset{CCH_3}{||}}{C}.e$

(III) (Nucleophile)

$e.\overset{O}{\overset{||}{C}}-O.nn$

for which (II) is an Electrophilic center with $COOCH_3$ as transfer species of the third type and (III) the corresponding Nucleophilic center with H as transfer species of the first kind of the eighth type, clearly indicating the strong Electrophilic character of the monomer.
(Laws of Creations for 1,4-Cumulenic Ketenes)

Rule 493: This rule for Chemistry for **Substituent groups of the types = C.... = O,** states that, these types of groups can be free-radical-pulling or -pushing with the following capacities depending on the ODD or EVEN numbers of C atoms carried by the group as shown below-

$$O=C=C=C=C=C= \quad > \quad O=C=C=C= \quad > \quad O=C=$$

Order of radical-pulling capacity [ODD]   i)

$$O=C=C=C=C= \quad > \quad O=C=C= \quad > \quad O=$$

Order of radical-pushing capacity [EVEN]   ii)

(Laws of Creations for Substituent Groups)

Rule 494: This rule for Chemistry for **Substituent groups of the type -CR = ...O,** states that, these types of groups can be free-radical-pulling or -pushing with the following capacities depending on the ODD or EVEN numbers of C atoms carried by the group as shown below-

$$O = C = C = C = C = CR - \quad > \quad O = C = C = CR - \quad > \quad O = CR -$$

<u>Order of radical-pulling capacity [ODD]</u>   i)

$$O = C = C = C = CR - \quad > \quad O = C = CR - \quad >$$

<u>Order of radical-pushing capacity [EVEN]</u>   ii)

where R is a radical-pushing group.
(Laws of Creations for Substituent Groups)

Rule 495: This rule of Chemistry for **1,4-Cumulative Ketenes of the type ($R_{F)2}C = C = C = O$ where R_F is $OCOCH_3$,** states that, this is a strong Electrophile as shown below-

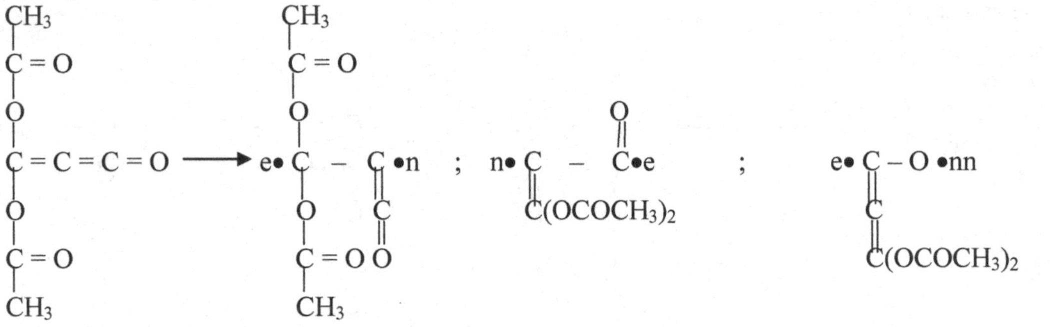

       (I) [Nucleophile]   (II) [Electrophile]   (III) [Nucleophile]

for which there are two mono-forms (II) and (III) and the transfer species involved is of the first kind of the eighth type for the Nucleophilic center [$COCH_3$] and of the first kind of the third type for the Electrophilic center [$OCOCH_3$].
(Laws of Creations for 1,4-Cumulative Ketenes)

Rule 496: This rule of Chemistry for **Perfluoro 1,4-Cumulenic Ketenes,** states that, this is a strong Electrophile as shown below-

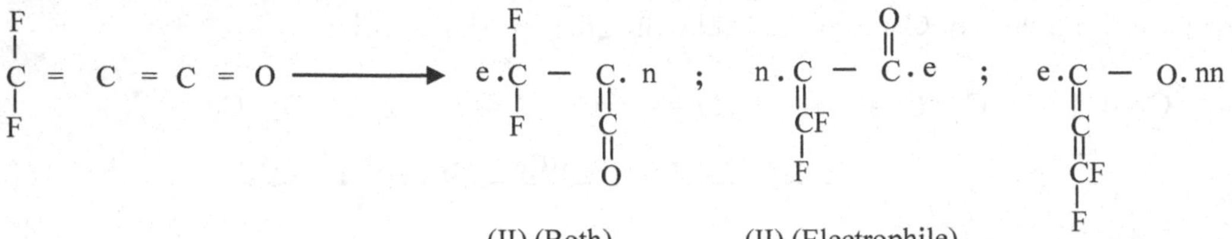

     (II) (Both)   (II) (Electrophile)   (III) (Both)

for which there are two mono-forms (II) and (III), with transfer species of the first kind of the third type only for the Electrophilic center [F] and none for the Nucleophilic center.
(Laws of Creations for 1,4-Cumulenic Ketenes)

Rule 497: This rule of Chemistry for **1,4-Cumulenic Ketenes of the type $(CF_3)_2C=C=C=O$,** states that, this is a strong Electrophile as shown below-

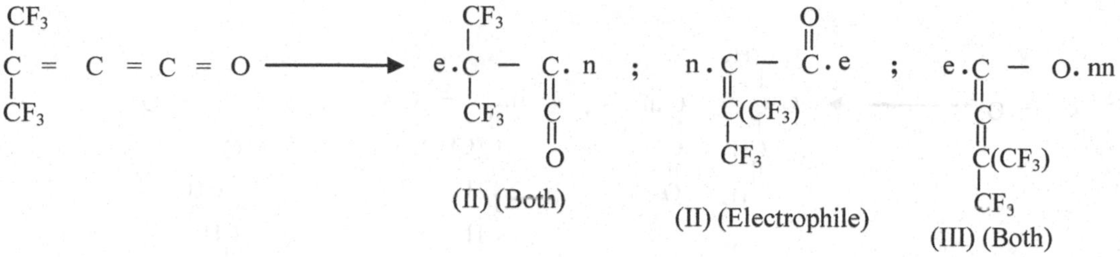

$$
\begin{array}{c}
CF_3 \\
| \\
C = C = C = O \\
| \\
CF_3
\end{array}
\longrightarrow
\begin{array}{c}
CF_3 \\
| \\
e \cdot C - C \cdot n \\
| \quad \| \\
CF_3 \quad C \\
\| \\
O
\end{array}
\; ; \;
\begin{array}{c}
O \\
\| \\
n \cdot C - C \cdot e \\
\| \\
C(CF_3) \\
| \\
CF_3
\end{array}
\; ; \;
\begin{array}{c}
O \\
\| \\
e \cdot C - O \cdot nn \\
\| \\
C \\
| \\
C(CF_3) \\
| \\
CF_3
\end{array}
$$

(II) (Both)   (II) (Electrophile)   (III) (Both)

for which two mono-forms exist- (II) and (III) with transfer species of the first kind of the third type only for the Electrophilic center $[CF_3]$ and none for the Nucleophilic center.
(Laws of Creations for 1,4- Cumulenic Ketenes)

Rule 498: This rule of Chemistry for **All things in life,** states that, the most Complex Logical Network System exists only in CHEMISTRY, since CHEMISTRY is the MOTHER of all disciplines in Humanity, for without it no other discipline exists.
(Laws of Creations in Philosophy)

Rule 499: This rule of Chemistry for **1,4-Cumulenic Ketenes of the types $(R_F)R'_F C=C=C=O$ where R_F is CnF_{2n+1} types of groups,** states that, these are strong Electrophiles as shown below-

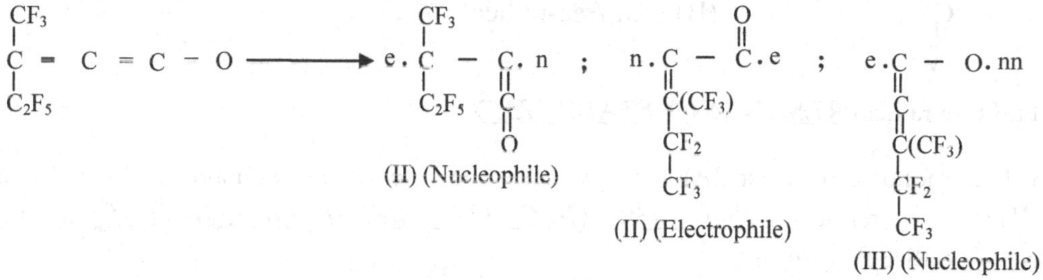

$$
\begin{array}{c}
CF_3 \\
| \\
C = C = C - O \\
| \\
C_2F_5
\end{array}
\longrightarrow
\begin{array}{c}
CF_3 \\
| \\
e \cdot C - C \cdot n \\
| \quad \| \\
C_2F_5 \quad C \\
\| \\
O
\end{array}
\; ; \;
\begin{array}{c}
O \\
\| \\
n \cdot C - C \cdot e \\
\| \\
C(CF_3) \\
| \\
CF_2 \\
| \\
CF_3
\end{array}
\; ; \;
\begin{array}{c}
O \\
\| \\
e \cdot C - O \cdot nn \\
\| \\
C \\
| \\
C(CF_3) \\
| \\
CF_2 \\
| \\
CF_3
\end{array}
$$

(II) (Nucleophile)   (II) (Electrophile)   (III) (Nucleophile)

for which (III) is nucleophilic, since nucleo-non-free-radically, it cannot be polymerized as shown below-

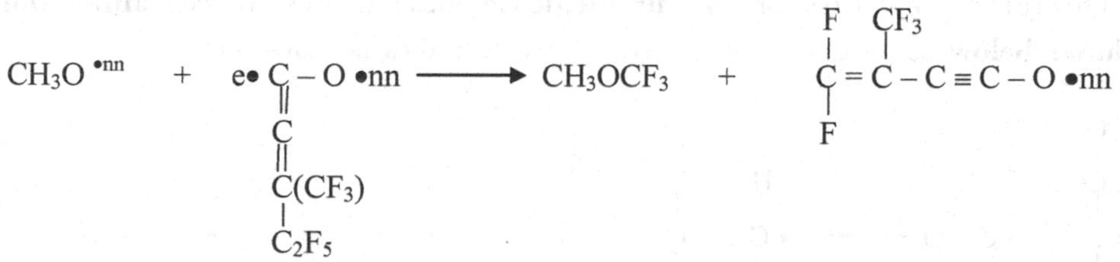

$$
CH_3O^{\bullet nn} +
\begin{array}{c}
e \bullet C - O \bullet nn \\
| \\
C \\
\| \\
C(CF_3) \\
| \\
C_2F_5
\end{array}
\longrightarrow
CH_3OCF_3 +
\begin{array}{c}
F \quad CF_3 \\
| \quad | \\
C = C - C \equiv C - O \bullet nn \\
| \\
F
\end{array}
$$

i)

$$
\begin{array}{c}
E + O - C \\
\quad\quad \| \\
\quad\quad C \\
\quad\quad \| \\
\quad\quad C(CF_3) \\
\quad\quad | \\
\quad\quad CF_2 \\
\quad\quad | \\
\quad\quad CF_3
\end{array}_n
\begin{array}{c}
O - C \cdot e \\
\| \\
C \\
\| \\
C(CF_3) \\
| \\
CF_2 \\
| \\
CF_3
\end{array}
\longrightarrow
\begin{array}{c}
E + O - C \\
\quad\quad \| \\
\quad\quad C \\
\quad\quad \| \\
\quad\quad C(CF_3) \\
\quad\quad | \\
\quad\quad CF_2 \\
\quad\quad | \\
\quad\quad CF_3
\end{array}_n
\begin{array}{c}
CF_3 \quad F \\
| \quad | \\
O - C \equiv C - C = C \\
\quad\quad\quad | \\
\quad\quad\quad F
\end{array}
+ \overset{e}{CF_3}
$$

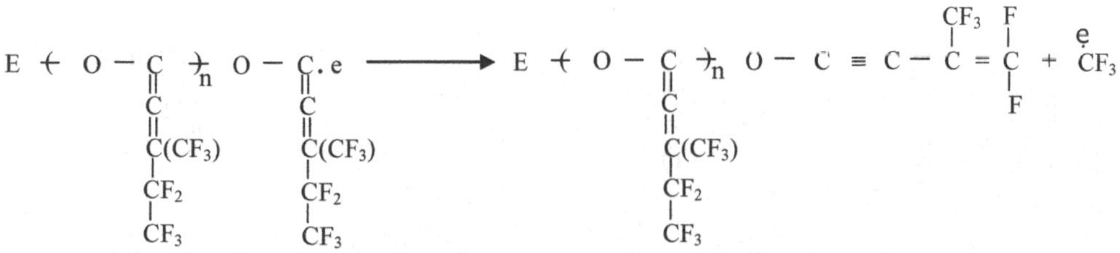

Transfer species of first kind of the eighth type

ii)

(Laws of Creations for 1,4-Cumulenic Ketenes)

Rule 500: This rule of Chemistry for **1,4-Cumulenic Ketenes with resonance stabilization group such as shown below,** states that, when the alkenyl group is used in place of an alkanyl group, the followings are obtained.

$$CH_3 \quad C = C = C = O \longrightarrow$$

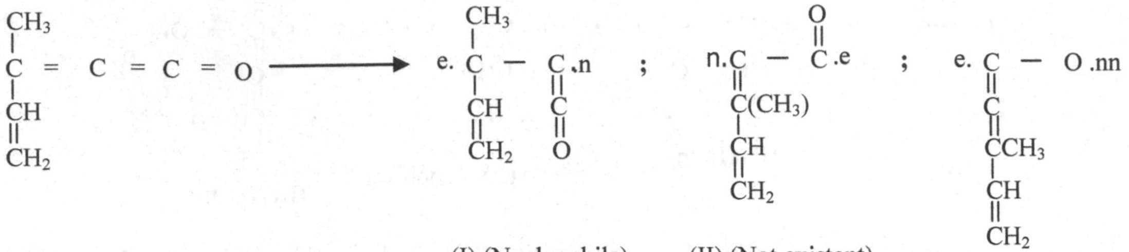

(I) (Nucleophile)   (II) (Not existent)   (III) (Nucleophile)   i)

without the realization that the alkenyl group is carrying an activation center which is the least nucleophilic, and therefore the followings are obtained-

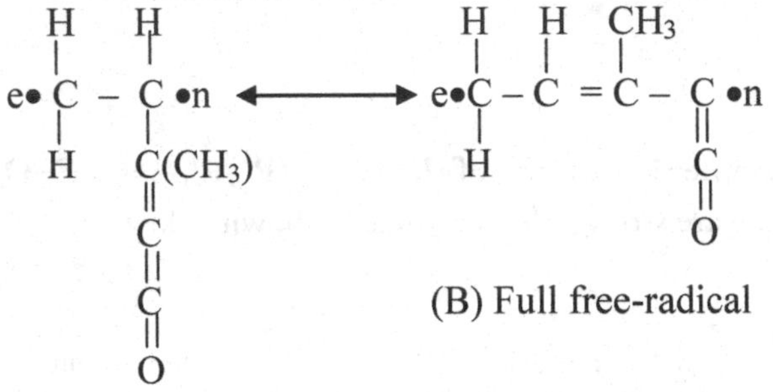

(B) Full free-radical

(A) Full free-radical **RESONANCE STABILIZED**   ii)

for which three mono-forms exist- (A) above which is the Y center, its resonance stabilized mono-form (B) and (III) the X center noting *that $-C(R) = C = C = O$ is a radical-pulling group of greater capacity than $-C(R) = O$.* (See Rule 494)
(Laws of Creations for 1,4-Cumulenic Ketenes)

Rule 501: This rule of Chemistry for **1,4-Cumulenic Ketenes with resonance stabilization group of the type shown below,** states that, these are strong Electrophiles as shown below-

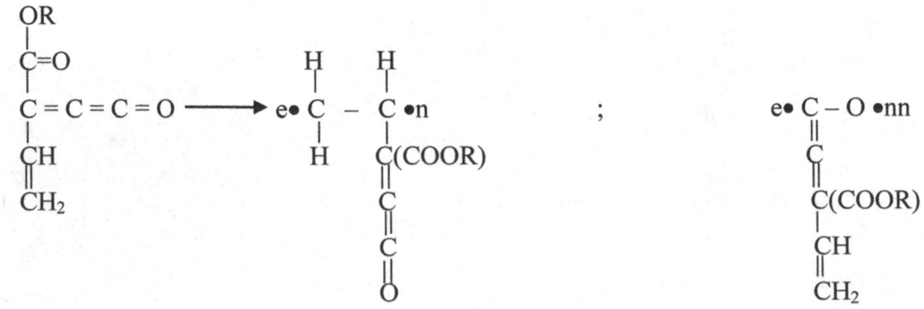

(I) Both   **(II) Nucleophile**   i)

for which (I) is resonance stabilized as shown below and found to favor the nucleo-free-radical route and (II) also found to favor the electro-free-radical route.

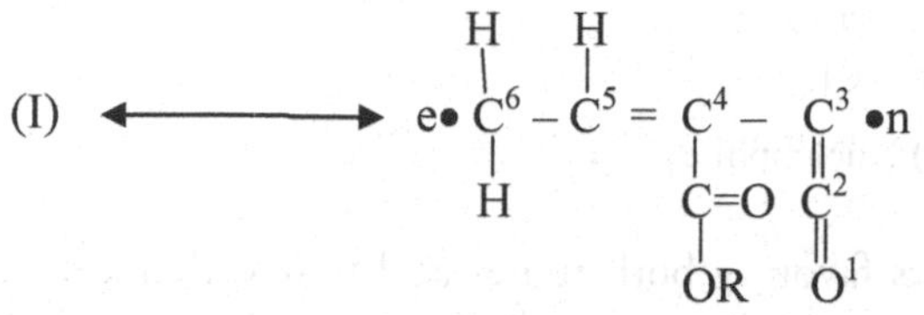

$$\text{(I)} \longleftrightarrow \quad e\bullet \overset{\overset{H}{|}}{C^6} - \overset{\overset{H}{|}}{C^5} = \overset{\overset{}{|}}{\underset{\underset{OR}{\overset{|}{C=O}}}{C^4}} - \overset{}{\underset{\underset{O^1}{\overset{||}{C^2}}}{C^3}} \bullet n$$

RESONANCE STABILIZATION

ii)

noting that the COOR group, like any other group is sheilded and that, it is the monomer group that is involved in providing resonance stabilization for the alkenyl group.
(Laws of Creations for 1,4-Cumulenic Ketenes)

Rule 502: This rule of Chemistry for **1,4-Cumulenic Ketenes with resonance stabilization group such as shown below,** states that, when the group is phenyl group, the followings are obtained.-

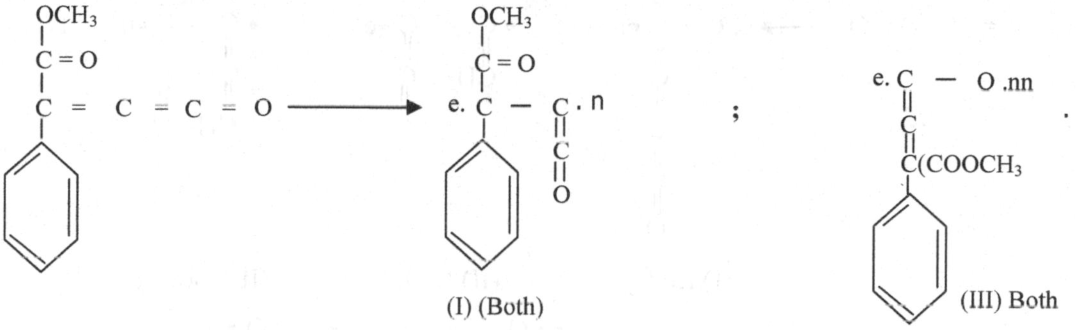

(I) (Both)   ;   (III) Both

for which it can be observed that though the phenyl types of groups looks as if it can provide resonance stabilization for the monomer through (I), the phenyl group can still provide resonance stabilization, since via the least nucleophilic center, the 2,3-center, the Y center not shown above, the $COCH_3$ group is still shielded.
(Laws of Creations for 1,4-Cumulenic Ketenes)

Rule 503: This rule of Chemistry for **the first member of 1,5-Cumulenic Ketenes,** states that, for it, the followings are obvious when suppressed-

(A)

i)

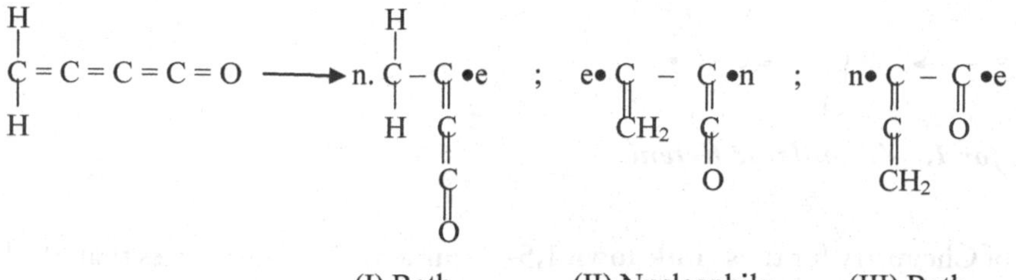

(I) Both   (II) Nucleophile   (III) Both

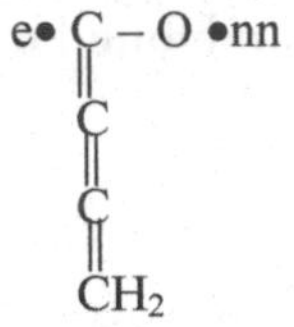

$$e \bullet \underset{\underset{\underset{\overset{\|}{CH_2}}{\overset{\|}{C}}}{\overset{\|}{C}}}{C} - O \bullet nn$$

(IV) Nucleophile   ii)

for which (I) and (III) are the Y centers favoring both routes and that which are resonance stabilized as shown by (A) above, (II) and (IV) are the corresponding X centers for each of them favoring only the electro-free-radical route, noting that for this case there is need to replace one or two H atoms with substituent groups.

(Laws of Creations for 1,5-Cumulenic Ketene)

Rule 504: This rule of Chemistry for **the first member of 1,6-Cumulenic Ketenes,** states that, for it, the followings are obvious when suppressed-

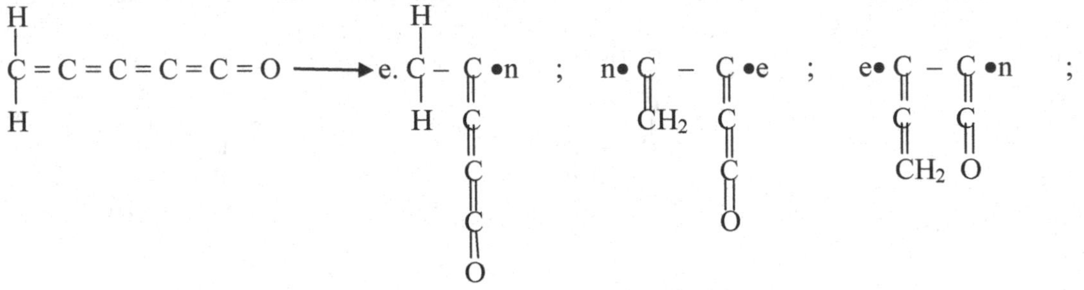

$$\underset{H}{\overset{H}{C}} = C = C = C = C = O \longrightarrow e. \underset{H}{\overset{H}{C}} - \underset{\underset{\underset{\overset{\|}{O}}{\overset{\|}{C}}}{\overset{\|}{C}}}{C} \bullet n \quad ; \quad n \bullet \underset{\overset{\|}{CH_2}}{C} - \underset{\underset{\overset{\|}{O}}{\overset{\|}{C}}}{C} \bullet e \quad ; \quad e \bullet \underset{\overset{\|}{CH_2}}{C} - \underset{\overset{\|}{O}}{C} \bullet n \quad ;$$

(I) Both   (II) Both   (III) Both

$$n \bullet \underset{\underset{\overset{\|}{CH_2}}{\overset{\|}{C}}}{C} - \underset{\overset{\|}{O}}{C} \bullet e \quad ; \quad e \bullet \underset{\underset{\underset{\overset{\|}{CH_2}}{\overset{\|}{C}}}{\overset{\|}{C}}}{C} - O \bullet nn$$

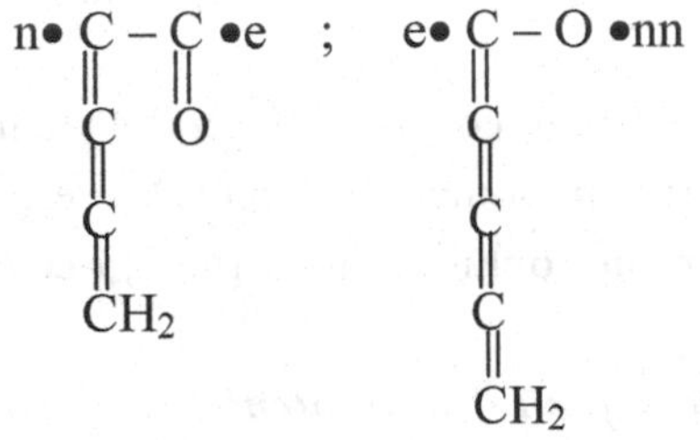

(IV) Both   (V) Both   i)

for which it is only (II) and (IV) [and not along with (III) and (V)] that can undergo resonance stabilization, with (II) being the first Y center and (III) being the corresponding X center; noting that for it to be used as a monomer, one or two H atoms must be replaced with substituent groups; noting that (II) and (IV) are the only ones that can be resonance stabilized as shown below using (IV).

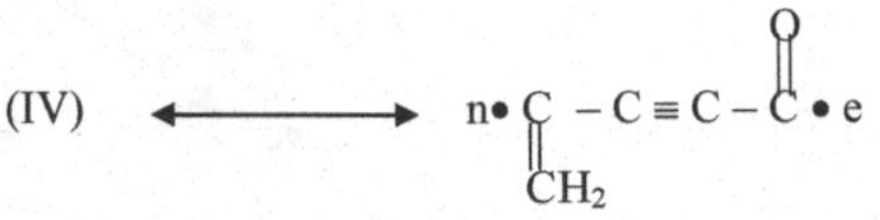

$$(IV) \quad \longleftrightarrow \quad n \bullet \underset{\overset{\|}{CH_2}}{C} - C \equiv C - \overset{\overset{O}{\|}}{C} \bullet e \qquad \qquad ii)$$

(Laws of Creations for 1,6-Cumulenic Ketene)

Rule 505: This rule of Chemistry for **these unknown 1,5-Cumulenic Ketenes,** states that, the followings are obvious with replacement of the two H atoms with radical-pushing groups,

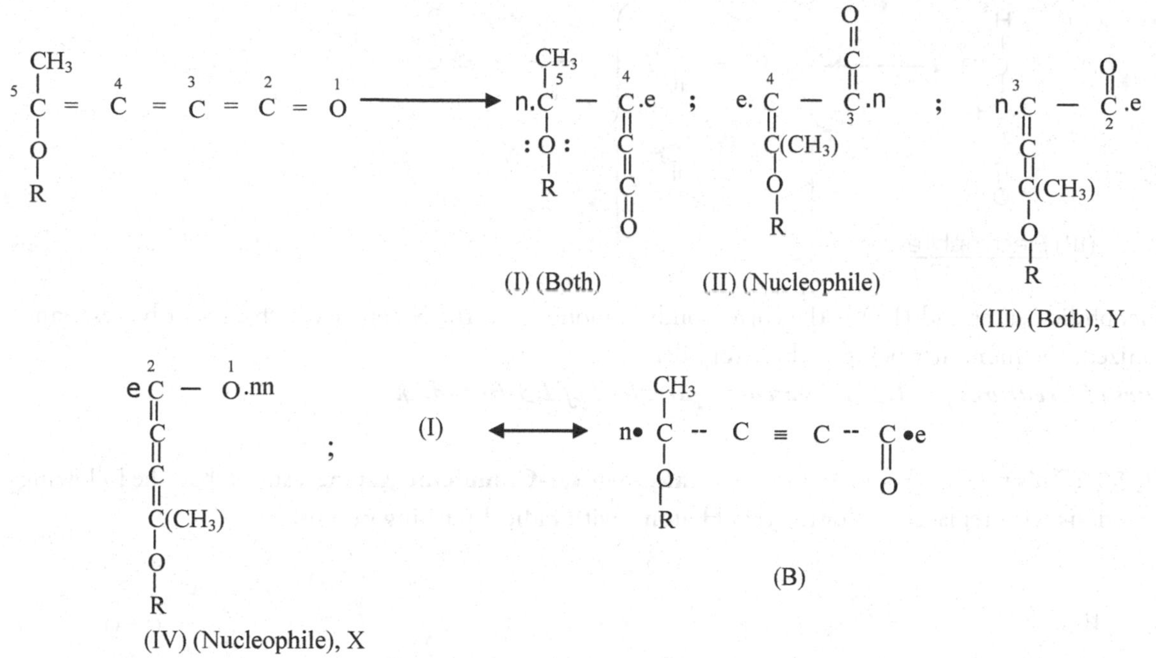

$$\begin{array}{c} CH_3 \\ | \\ {}_5C = {}^4C = {}^3C = {}^2C = {}^1O \\ | \\ O \\ | \\ R \end{array} \longrightarrow \begin{array}{c} CH_3 \\ |\,5 \\ n.C - C.e \\ \quad | \quad || \\ :O: \quad C \\ | \quad || \\ R \quad C \\ \quad || \\ \quad O \end{array} \quad ; \quad \begin{array}{c} O \\ || \\ C \\ 4 \quad || \\ e.C - C.n \\ | \quad 3 \\ C(CH_3) \\ | \\ O \\ | \\ R \end{array} \quad ; \quad \begin{array}{c} O \\ || \\ 3 \quad C \\ n.C - C.e \\ || \quad 2 \\ C \\ || \\ C(CH_3) \\ | \\ O \\ | \\ R \end{array}$$

$$\text{(I) (Both)} \qquad\qquad \text{(II) (Nucleophile)} \qquad\qquad \text{(III) (Both), Y}$$

$$\begin{array}{c} 2 \qquad 1 \\ e\,C - O.nn \\ || \\ C \\ || \\ C \\ || \\ C(CH_3) \\ | \\ O \\ | \\ R \end{array} \quad ; \qquad\qquad \text{(I)} \longleftrightarrow \begin{array}{c} CH_3 \\ | \\ n\bullet C -- C \equiv C -- C \bullet e \\ | \qquad\qquad\qquad || \\ O \qquad\qquad\qquad O \\ | \\ R \end{array}$$

$$\text{(B)}$$

$$\text{(IV) (Nucleophile), X} \hspace{6cm} \text{i)}$$

for which only (I) and (III) can be resonance stabilized and (IV) can molecularly rearrange as shown below (though not required the monomer being an Electrophile)-

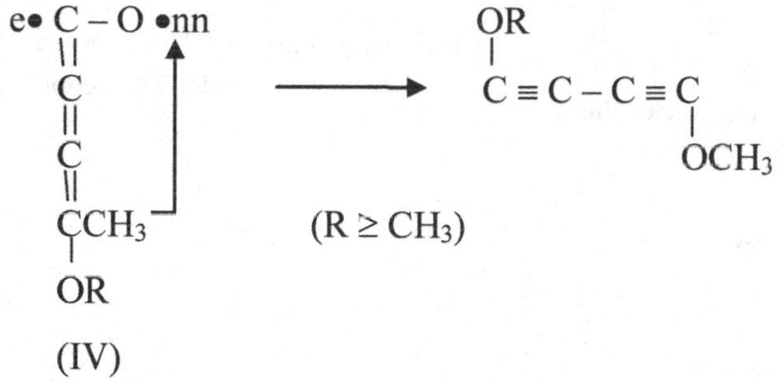

$$\begin{array}{c} e\bullet C - O \bullet nn \\ || \\ C \\ || \\ C \\ || \\ C CH_3 \\ | \\ OR \end{array} \longrightarrow \begin{array}{c} OR \\ | \\ C \equiv C - C \equiv C \\ | \\ OCH_3 \end{array}$$

$$\qquad\qquad (R \geq CH_3)$$

$$\text{(IV)} \hspace{10cm} \text{ii)}$$

with (I) and (II) as the first Y and X centers respectively, centers that will be the first to be used nucleo-free-radically and electro-free-radically.
(Laws of Creations for 1,5-Cumulenic Ketenes)

Rule 506: This rule of Chemistry for **a 1,3- di-ketenic equivalent of butadiene,** states that, for the two symmetric cases, the followings are to be expected

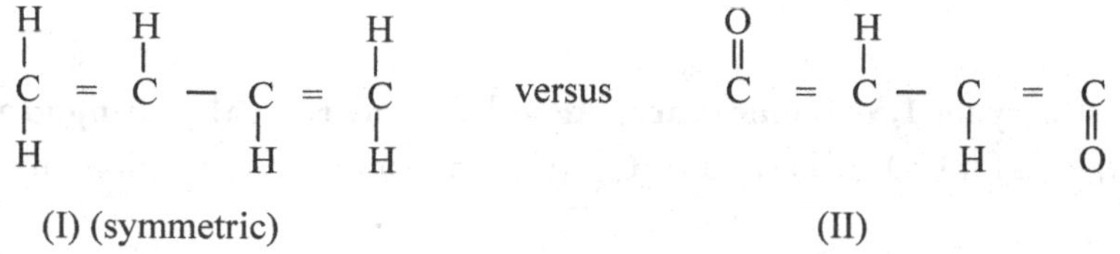

$$\begin{array}{c} H \quad\quad H \qquad\quad H \\ | \qquad | \qquad\qquad | \\ C = C - C = C \\ | \qquad\qquad | \quad | \\ H \qquad\qquad H \quad H \end{array} \qquad \text{versus} \qquad \begin{array}{c} O \quad\quad H \\ || \qquad | \\ C = C - C = C \\ | \qquad\qquad | \quad || \\ H \qquad\qquad H \quad O \end{array}$$

$$\text{(I) (symmetric)} \hspace{4cm} \text{(II)} \hspace{4cm} \text{i)}$$

for which while (I) a Nucleophile can be resonance stabilized, so also is (II) which is an Electrophile, noting that the movement is from Full free-radical to Full free-radical from the least

(III) Electrophile   (IV)   (V)   ii)

nucleophilic center and (IV) is the corresponding mono-form the X center which cannot be resonance stabilized, the monomer being an Electrophile.

(Laws of Creations for 1,3- Di-ketenic equivalent of 1,3-Butadiene)

Rule 507: This rule of Chemistry for **these unknown 1,6-Cumulenic Ketenes,** states that, the followings are obvious with replacement of the two H atoms with radical-pushing groups,

(I) (Nucleophile)   (II) (Electrophile)   (III) (Nucleophile)

(IV) (Electrophile)   (V) (Nucleophile)

for which two Electrophiles can be observed to be carried by members of this family, (II) and (IV), with (II) being the first to be used in the routes natural to them.

(Laws of Creation for 1,6-Cumulenic Ketenes)

Rule 508: This rule of Chemistry for **1,5-Ketenes carrying at least one radical-pulling group**, states that, the followings are obtained for $COOCH_3$ and CH_3 groups in place of the two H atoms-

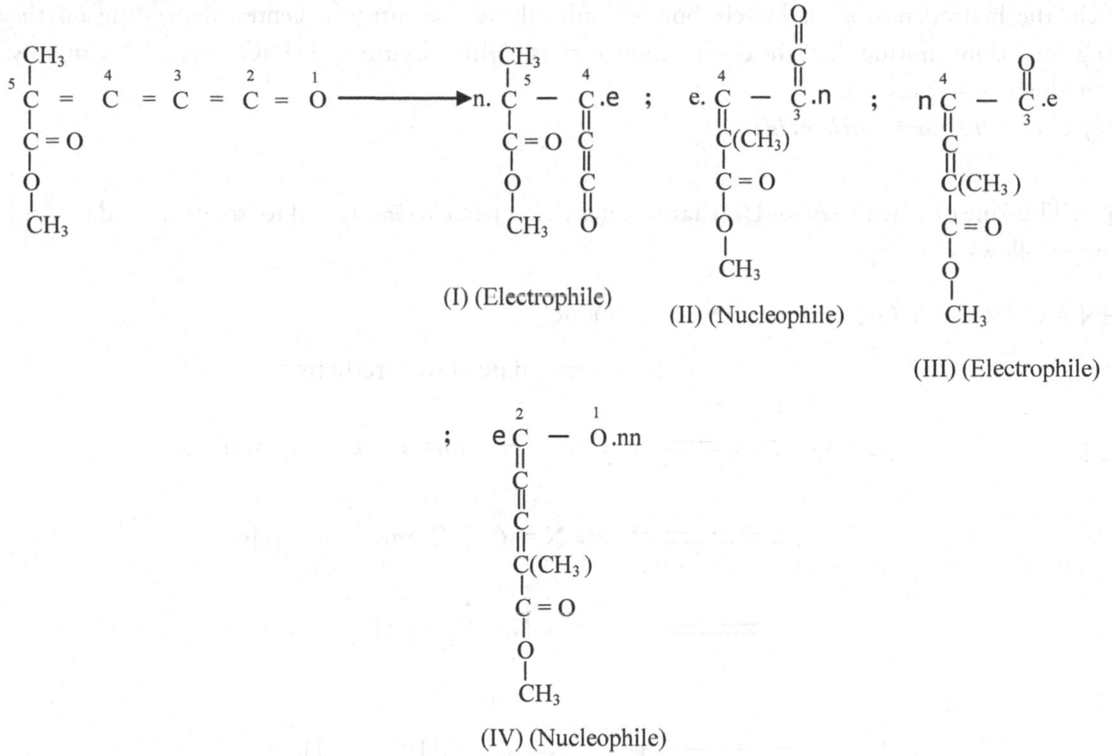

$$\underset{5}{\overset{CH_3}{\underset{|}{C}}} = \underset{4}{C} = \underset{3}{C} = \underset{2}{C} = \underset{1}{O}$$
with substituents $C=O$, O, CH_3 on carbon 5

$$\longrightarrow n.\ \underset{5}{\overset{CH_3}{\underset{|}{C}}} - \underset{4}{C}.e \quad ;$$

(I) (Electrophile)

$$e.\ \underset{4}{C} - \underset{3}{C}.n \quad ;$$

(II) (Nucleophile)

$$n\ \underset{4}{C} - \underset{3}{\overset{O}{\overset{||}{C}}}.e$$

(III) (Electrophile)

$$;\quad e\ \underset{2}{C} - \underset{1}{O}.nn$$

(IV) (Nucleophile)

for which two Electrophiles can be observed to be carried by members of this family, (I) and (II), being the first to be used in the routes natural to them.
(Laws of Creations for 1,5-Cumulenic Ketenes)

Rule 509: This rule of Chemistry for **Alkyl Cyanates,** states that, though they can only be activated radically, two series of stable alkyl derivatives (esters) exist, the *alkyl cyanates* $ROC \equiv N$, and the *alkyl isocyanates* $O = C = NR,$ in which the former can been isolated only when the alkyl group is sufficiently bulky to prevent molecular rearrangement and cyclo-trimerization.
(Laws of Creations for Alkyl Cyanates)

Rule 510: This rule of Chemistry for **Cyanic acid,** states that, since alkyl isocyanates are said to be stable compounds, cyanic acid molecularly rearranges or electroradicalizes as shown only radically and not charged based on the operating conditions.

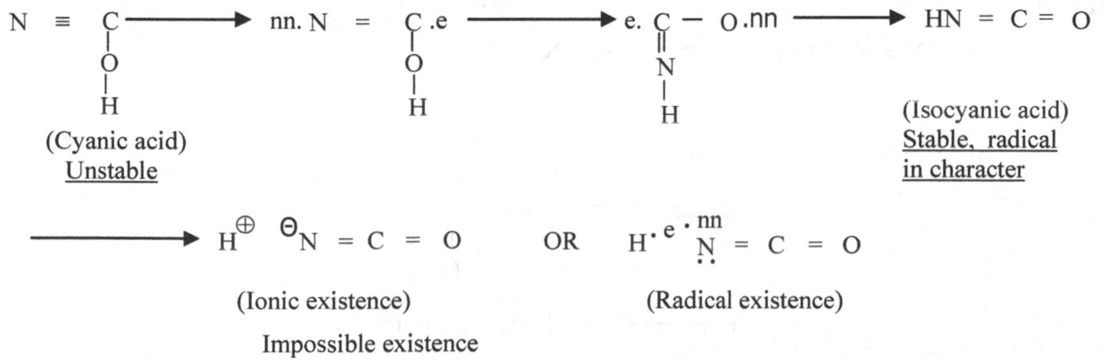

$$N \equiv C \longrightarrow nn.\ N = C.e \longrightarrow e.\ C - O.nn \longrightarrow HN = C = O$$

(Cyanic acid)
Unstable

(Isocyanic acid)
Stable, radical
in character

$$\longrightarrow H^{\oplus}\ {}^{\ominus}N = C = O \qquad OR \qquad H^{\bullet}\ e\ \overset{\bullet\ nn}{\underset{\bullet\bullet}{N}} = C = O$$

(Ionic existence)

Impossible existence

(Radical existence)

for which, the hydrogen atom is loosely bonded radically to the nitrogen center, depending on the operating conditions, noting that, the compound which is highly unsaturated (H:C ratio = 1:1) can only be activated when suppressed.

(Laws of Creations for Cyanic acid)

Rule 511: This rule of Chemistry for **Urethans,** states that, alcohols readily add to isocyanic acid to yield urethans as follows.

$$HN = C = O \; + HOR \longrightarrow H_2NCOOR$$

$$\text{Alkyl Carbamates OR Urethans} \qquad\qquad \text{i)}$$

Stage 1:
$$H - N = C = O \rightleftharpoons H\bullet e \;+\; nn\bullet N = \overset{\bullet e}{C} - O \bullet nn$$

$$\rightleftharpoons nn\bullet N = \overset{\bullet e}{C} - O \bullet nn \;+\; H\bullet e$$

$$\rightleftharpoons nn\bullet N = \overset{\bullet e}{C} - OH$$

$$(A)$$

$$(A) \longrightarrow N \equiv C - OH \;+\; \text{Heat} \qquad\qquad \text{ii)}$$

$$\textbf{ENOLIZATION} \qquad (A')$$

Stage 2:
$$ROH \rightleftharpoons H\bullet e \;+\; nn\bullet OR$$

$$H\bullet e \;+\; (A') \xrightleftharpoons{\text{ACTIVATION}} H - N = \overset{OH}{\underset{|}{C}} \bullet e$$

$$(B)$$

$$RO\bullet nn \;+\; (B) \longrightarrow H - \underset{\underset{OH}{|}}{N} = C - OR$$

$$(C) \qquad\qquad \text{iii)}$$

Stage 3:

$$H - N = \underset{\underset{OH}{|}}{C} - OR \rightleftharpoons nn\bullet \underset{\underset{OH}{|}}{\overset{\overset{H \quad OR}{|\quad\;|}}{N} - C} \bullet e$$

$$\rightleftharpoons H_2N - \overset{\overset{OR}{|}}{\underset{\underset{O \bullet nn}{|}}{C}} \bullet e$$

$$\longrightarrow H_2N - COOR$$

$$\textbf{MOLECULAR REARRANGEMENT} \qquad\qquad \text{iv)}$$

in which in the first stage, due to the presence of the alcohol, the isocyanide enolized and in the second stage, the cyanate was suppressed and activated by the alcohol to give (C) which in the third stage, molecularly rearranged to give the product; for which it was the H in OH group that was moved and not the R in OR group, ***clear indication that OH group is richer or of greater radical-pushing capacity than OR group.***
(Laws of Creations for Urethans)

Rule 512: This rule of Chemistry for **1,3-Isocyanantes,** states that, in general, these are ***Electro-philes*** for which for them represented as $R - N = C = O$ (where R is any substituent group), the $N = C$ center is Y (Electrophilic) and the $C = O$ center is X (Nucleophilic).
(Laws of Creations for 1,3-Isocyanates)

Rule 513: This rule of Chemistry for **Urethans,** states that, alcohols readily add to isocyanates to yield urethans as follows-

a supposedly wrong two stage Equilibrium mechanism process, since C=O is more nucleophilic than C=N; for it is the C = N that is first activated and used as such to give the product; for which the reactions can take place both chargedly or radically, but indeed takes place only radically.
(Laws of Creations for Urethans)

Rule 514: This rule of Chemistry for **N-alkylureas,** states that, amines readily add to isocyanates to yield Urea as follows-

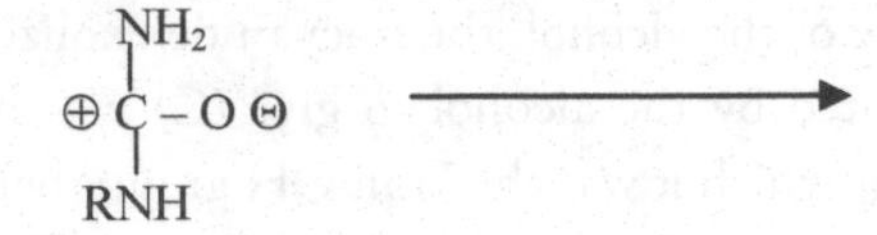

$$\underset{RNH}{\overset{NH_2}{\underset{|}{\overset{|}{\oplus C - O \ominus}}}} \longrightarrow HRNCONH_2 \qquad i)$$

for which the mechanism shown below is the real mechanism, since the molecular rearrangement above is wrong, NH_2 group being more radical-pushing than OH group-

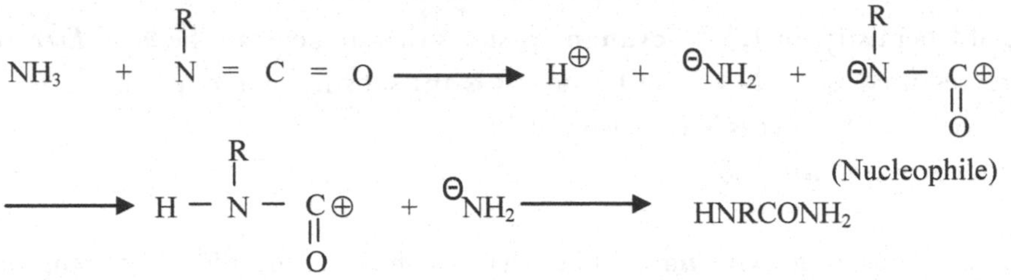

$$NH_3 \quad + \quad \overset{R}{\underset{|}{N}} = C = O \longrightarrow H^{\oplus} \quad + \quad {}^{\ominus}NH_2 \quad + \quad \overset{R}{\underset{|}{\ominus N}} - \underset{\underset{O}{\|}}{C\oplus}$$

$$\text{(Nucleophile)} \qquad ii)$$

$$\longrightarrow H - \overset{R}{\underset{|}{N}} - \underset{\underset{O}{\|}}{C\oplus} \quad + \quad {}^{\ominus}NH_2 \longrightarrow HNRCONH_2$$

a one stage wequilibrium process (like the case shown for urethans), for which how the isocyanates are activated can be observed and the same will also apply only to primary and secondary amines, but not tertiary amines.

$$RN = C = O \quad + \quad H_2NR^1 \longrightarrow RNHCONHR^1$$

$$RN = C = O \quad + \quad HNR^1_2 \longrightarrow RNHCONR^1_2 \qquad iii)$$

(Laws or Creations for N-alkylureas)

<u>**Rule 515:**</u> This rule of Chemistry for **Phenyl isocyanates ($H_5C_6N = C = O$),** states that, shown below is how the phenyl group operates on the isocyanate, wherein we can see very rapid movement of electro-free-radical from Half-free (I) to Full-free forms and back to Half-free (I) as shown below-

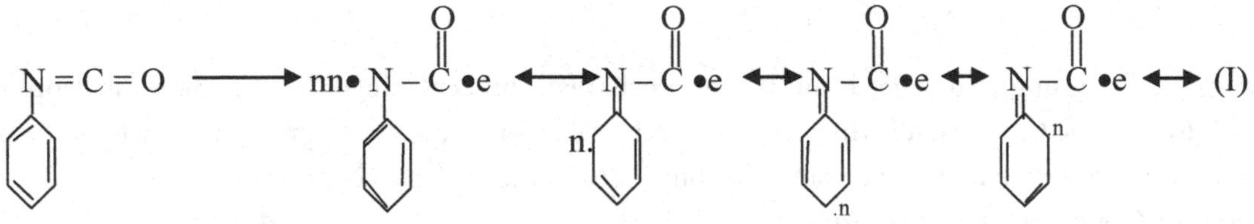

RESONANCE STABILIZATION $\qquad$ i)

noting that the movement is of no consequence since no group is being shielded (the phenyl group being the resonance stabilization provider), and (II), and (IV) may form fused rings either from the right or left of the spectrum as shown below held in Equilibrium-

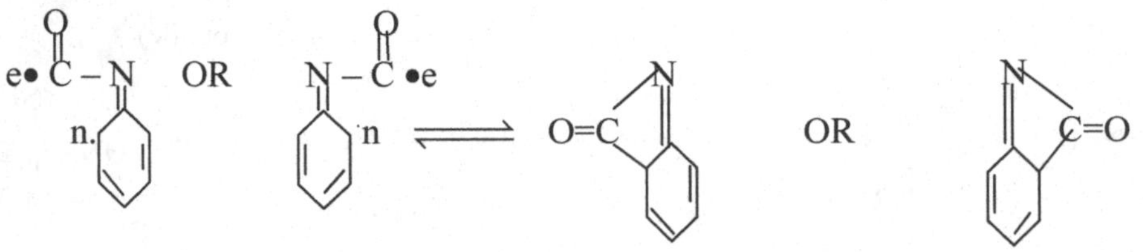

817

for which (II)A and (IV)A are fused six- and four-membered rings in which the four-membered ring is highly strained to exist and the ring is MALE in character.
(Laws of Creations for 1,3-Phenyl isocyanates)

Rule 516: This rule of Chemistry for **Carbamic acid,** states that, Hydrolysis of isocyanates yields the carbamic acid which is unstable and decomposes spontaneously to amine and carbon dioxide in two stages via Equilibrium mechanism as shown below-

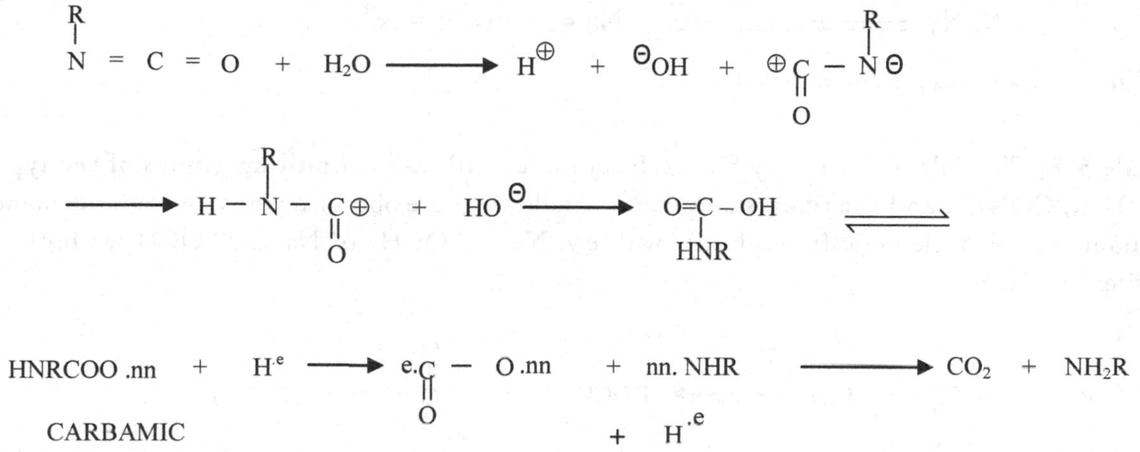

$$
\overset{\overset{\displaystyle R}{\displaystyle |}}{N} = C = O \; + \; H_2O \; \longrightarrow \; H^{\oplus} \; + \; {}^{\ominus}OH \; + \; {}^{\oplus}\underset{\underset{\displaystyle O}{\displaystyle \|}}{C} - \overset{\overset{\displaystyle R}{\displaystyle |}}{N}{}^{\ominus}
$$

$$
\longrightarrow \; H - \overset{\overset{\displaystyle R}{\displaystyle |}}{N} - \underset{\underset{\displaystyle O}{\displaystyle \|}}{C}{}^{\oplus} \; + \; HO^{\ominus} \; \longrightarrow \; \underset{\underset{\displaystyle HNR}{\displaystyle |}}{O = C - OH} \; \rightleftharpoons
$$

$$
HNRCOO \,.nn \; + \; H^{.e} \; \longrightarrow \; e.\underset{\underset{\displaystyle O}{\displaystyle \|}}{C} - O \,.nn \; + \; nn. \, NHR \; \longrightarrow \; CO_2 \; + \; NH_2R
$$

CARBAMIC $\qquad\qquad\qquad\qquad\qquad + \; H^{.e}$

noting that though the reactions have been shown chargedly, they can only take place radically with release of energy.
(Laws of Creations for Carbamic acid)

Rule 517: This rule of Chemistry for **Isocyanates carrying special substituent groups such as shown below,** states that, these cannot be activated chargedly due to electrodynamic/static forces of repulsion, for which it will be useful to compare (I), (II), (III), and (IV) with sodium azide and

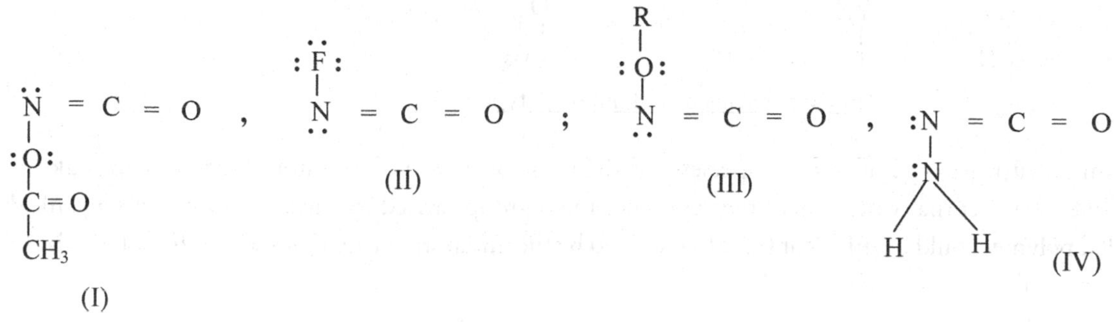

$$
\begin{array}{cccc}
\overset{\overset{\displaystyle \cdot\cdot}{\displaystyle \ddot{N}}}{\underset{\underset{\displaystyle CH_3}{\displaystyle |}}{\underset{\underset{\displaystyle C=O}{\displaystyle |}}{\underset{\underset{\displaystyle :\ddot{O}:}{\displaystyle |}}{}}}} = C = O & \overset{\overset{\displaystyle :\ddot{F}:}{\displaystyle |}}{\underset{\underset{\displaystyle \cdot\cdot}{\displaystyle \ddot{N}}}{}} = C = O & \overset{\overset{\displaystyle \overset{\overset{\displaystyle R}{\displaystyle |}}{:\ddot{O}:}}{\displaystyle |}}{\underset{\underset{\displaystyle \cdot\cdot}{\displaystyle \ddot{N}}}{}} = C = O & :N = C = O
\end{array}
$$

$\qquad$ (I) $\qquad\qquad\qquad$ (II) $\qquad\qquad\qquad$ (III) $\qquad\qquad$ (IV)

<u>Cannot Favor Charged activations</u> $\qquad\qquad\qquad\qquad\qquad\qquad\qquad$ i)

nitrogen molecule since sodium azide and N_2 cannot favor charged existence.

Polar bond

$$\overset{..}{N} \equiv \overset{..}{N} \xrightarrow{\text{Activation}} en \cdot \overset{..}{N} = \overset{..}{N} \cdot nn \quad ; \quad Na^{\oplus} \quad \overset{\ominus}{\underset{x}{\cdot N}} \overset{+}{\underset{+}{\bullet}} \overset{\oplus}{N} \overset{\ominus}{\underset{+}{\bullet}} \overset{\oplus}{\underset{+}{N}}$$

(I) <u>Activated nitrogen</u>   (II)<u>Sodium azide</u>   ii)
                                          Impossible existence

$$NaN_3 \rightleftharpoons Na \bullet e \quad nn \bullet N = N^{\oplus} = {}^{\ominus}N \qquad \text{iii)}$$

(Laws of Creations for Isocyanates)

<u>**Rule 518:**</u> This rule of Chemistry for **1,3-Isocyanates with radical pulling groups of the types COOR, CONH$_2$, and the likes,** states that, the followings are obtained when used as a monomer anionically, or nucleo-non-free-radically with e.g. $Na^{\oplus}....^{\ominus}OCH_3$ or $Na^{\bullet e}....^{nn\bullet}OCH_3$ as initiator using COOR-

(I) Electrophile   i)

<u>Transfer species of 1st kind of 1st type</u>   ii)

for which with the type of transfer species and the initiator used, the initiator is regenerated making it look like a catalyst that which is not the case, since the R group carried by the dead terminal isocyanically bonded polymer could be different from that carried by the initiator; noting that indeed the polymerization above can take place chargedly (anionically) and take place more radically (Using $H_3CO^{\bullet nn}.....^{e\bullet}Na$).
(Laws of Creations for 1,3-Isocyanates)

<u>**Rule 519:**</u> This rule of Chemistry for **1,3-Isocyanates and all other Cumulenic Isocyanates,** states that, it is only when the Y center is carrying radical-pushing groups of lower capacity than H with transfer species of the first kind such as C_nF_{2n+1} or radical-pulling groups with transfer species of the first kind, that molecular rearrangement can be made to take place, while with the X center, they cannot be made to take place even when radical-pushing groups with transfer species of the first kind is carried.
(Laws of Creations for Isocyanates)

Rule 520: This rule of Chemistry for **1,3- Isocyanates 1,4-, 1,5- and higher Cumulenic Isocyanates,** states that, while no center can undergo resonance stabilization phenomenon with 1,3-Isocyanates, with Cumulenic Isocyanates, only the least nucleophilic center, the Y center can be made to undergo this phenomenon and *this is only possible from 1,6- Cumulenic Isocyanates and upwards.*
(Laws of Creations for Isocyanates)

Rule 521: This rule of Chemistry for **alkenyl resonance stabilization group in place of H in isocyanic acid,** states that, the followings are to be expected only radically, from which one can observe that the $-N = C = O$ group is a free-radical-pushing resonance stabilization group in

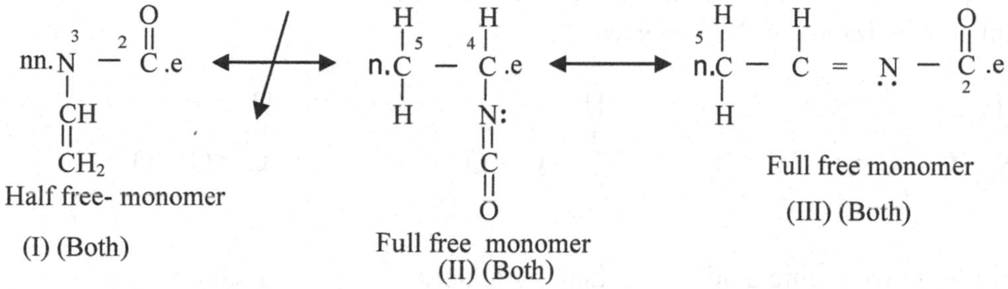

which movement from (I) to (II) is impossible and movement from (II) to (III) is by resonance stabilization, wherein if H on Carbon center 4 was CH_3, it will be shielded as shown below-

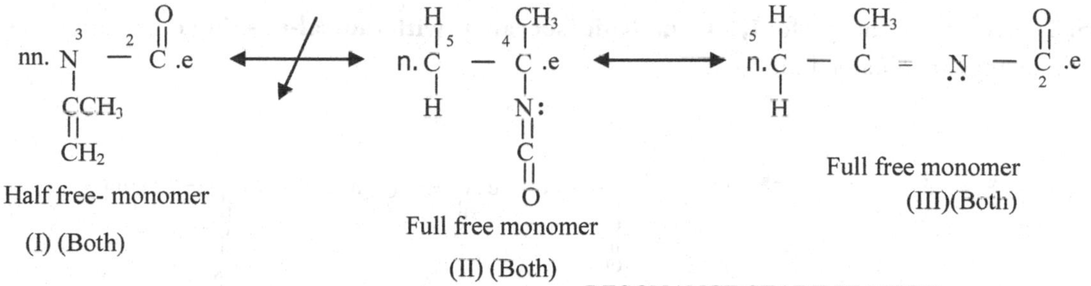

noting that while (III) is the Y, the $C = O$ center is the X center and the isocyanate group is the resonance stabilization provider.
(Laws of Creations for 1,3-Isocyanates)

Rule 522: This rule of Chemistry for **Unique activation centers shown below,** states that, the following is the order of their nucleophilicity, based on the type of groups carried by them-

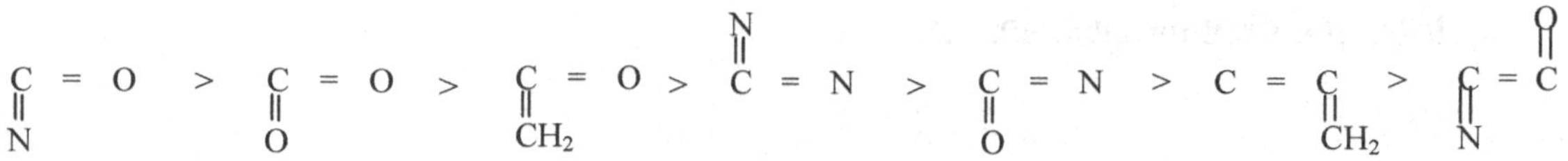

Order of Nucleophilicity

(Laws of Creations for Cumulatively placed Activation Centers)

Rule 523: This rule of Chemistry for **1,4-Cumulenic Isocyanates,** states that, in general, these are *Electrophiles* for which for them represented as $R - N = C = C = O$ (where R is any substituent group), the $N = C$ center is Y (Electrophilic), the $C = C$ center is X (Nucleophilic) and the $C = O$ is a second most nucleophilic center never used; noting that here the $C = C$ center is the least nucleophilic center and that the 1,4-Isocyanates are more nucleophilic or less electrophilic than 1,3-Isocyanates.
(Laws of Creations for 1,4-Cumulenic Isocyanates)

Rule 524: This rule of Chemistry for **the first member of 1,4-Cumulenic Isocyanates (1,4-Isocyanic acid),** states that, it has the 1,3- ketenic character when $HN =$ group is replaced with two hydrogen atoms and isocyanic acid character when the $O = C =$ group is replaced with $O =$ group (more nucleophilic than isocyanic acid), for which the following order is valid and like 1,3-isocyanic acids, the H atom must be replaced to favor full polymerization of the monomer-

$$
\begin{array}{ccccc}
\underset{|}{\overset{H}{\underset{N}{|}}} = C = C = O & > & \underset{|}{\overset{H}{\underset{N}{|}}} = C = O & > & \underset{\overset{|}{H}}{\overset{H}{\underset{|}{C}}} = C = O
\end{array}
$$

1,4- Cumulenic isocyanic acid   Isocyanic acid   Ketene

Order of Nucleophilicity

(Laws of Creations for 1,4-Isocyanic acid)

Rule 525: This rule of Chemistry for **1,4-Cumulenic Isocyanate with radical-pushing R groups,** states that, the followings are obtained for them-

$$
\underset{4}{\overset{CH_3}{\underset{|}{N}}} = \underset{3}{C} = \underset{2}{C} = \underset{1}{O} \longrightarrow nn \cdot \underset{4}{\overset{CH_3}{\underset{|}{N}}} - \underset{\underset{O}{\overset{||}{C}}}{\overset{|}{C}} \cdot e \quad ; \quad e \cdot \overset{3}{C} - \underset{2}{\overset{\overset{O}{||}}{C}} \cdot n \quad ; \quad e \cdot \overset{2}{C} - \overset{1}{O} \cdot nn
$$

(I) (Both)   (II) (Nucleophile)   (III) (Nucleophile)   i)

for which the mono-form (I) the (Y) center has no transfer species and (II) the (X) center may not favor molecular rearrangement, and from which the order of capacities deduced for the substituted groups are as follows-

$$CH_3 - N = \quad >> \quad = O$$

ii)

Order of radical-pushing capacity

$$O = C = \quad > \quad CH_3 - N = C =$$

Order of radical-pulling capacity   iii)

noting that alternating copolymers can be obtained nucleo-free-radically.
(Laws of Creations for 1,4 –Cumulenic Isocyanates)

Rule 526: This rule of Chemistry for **1,4-Cumulenic Isocyanates with radical-pushing OR groups**, states that, the followings are obtained for them-

$$
\begin{array}{c}
CH_3 \\
| \\
O \\
| \\
N = C = C = O \\
4 \quad 3 \quad 2 \quad 1
\end{array}
\longrightarrow
\begin{array}{c}
CH_3 \\
| \\
O \\
| \\
nn \cdot N - \underset{4}{C} \cdot e \\
\quad \| \\
\quad C \\
\quad \| \\
\quad O
\end{array}
\;;\;
\begin{array}{c}
O \\
\| \\
e \cdot \underset{3}{C} - \underset{2}{C} \cdot n \\
\quad \| \\
\quad N \\
\quad | \\
\quad O \\
\quad | \\
\quad CH_3
\end{array}
\;;\;
\begin{array}{c}
O \\
\| \\
e \cdot \underset{2}{C} - \underset{1}{O} \cdot nn \\
\quad \| \\
\quad C \\
\quad \| \\
\quad N \\
\quad | \\
\quad O \\
\quad | \\
\quad CH_3
\end{array}
$$

(I) (Both)   (II) (Both)   (III) (Nucleophile)

for which (I) is the (Y) center and (II) the least nucleophilic center is the (X) center and (III) the most nucleophilic center is never used.
(Laws of Creations for 1,4 –Cumulenic Isocyanates)

Rule 527: This rule of Chemistry for **1,4-Cumulenic Isocyanates with radical-pushing C_nF_{2n+1} groups**, states that, the followings are obtained for them-

$$
\begin{array}{c}
CF_3 \\
| \\
N = C = C = O
\end{array}
\longrightarrow
\begin{array}{c}
CF_3 \\
| \\
nn \cdot N - C \cdot e \\
\quad \| \\
\quad C \\
\quad \| \\
\quad O
\end{array}
\;;\;
\begin{array}{c}
O \\
\| \\
e \cdot C - C \cdot n \\
\quad \| \\
\quad N \\
\quad | \\
\quad CF_3
\end{array}
\;;\;
\begin{array}{c}
O \\
\| \\
e \cdot C - O \cdot nn \\
\quad \| \\
\quad C \\
\quad \| \\
\quad N \\
\quad | \\
\quad CF_3
\end{array}
$$

(I) (Electrophile)   (II) (Nucleophile)   (III) (Both)   i)

$$
\begin{array}{c}
N = C = C = O \\
| \\
CF_2 \\
| \\
CF_3
\end{array}
\longrightarrow
\begin{array}{c}
O \\
\| \\
C \\
\| \\
nn \cdot N - C \cdot e \\
| \\
CF_2 \\
| \\
CF_3
\end{array}
\;;\;
\begin{array}{c}
O \\
\| \\
e \cdot C - C \cdot n \\
\quad \| \\
\quad N \\
\quad | \\
\quad CF_2 \\
\quad | \\
\quad CF_3
\end{array}
\;;\;
\begin{array}{c}
O \\
\| \\
e \cdot C - O \cdot nn \\
\quad \| \\
\quad C \\
\quad \| \\
\quad N \\
\quad | \\
\quad C_2F_5
\end{array}
$$

(I) (Electrophile)   (II) (Nucleophile)   (III) (Nucleophile)   ii)

for which the (I)s, favor only the nucleo-non-free-radical route, based on the followings.

$$
E \cdot e \;+\;
\begin{array}{c}
O \\
\| \\
C \\
\| \\
nn \cdot N - C \cdot e \\
| \\
CF_3
\end{array}
\longrightarrow
EF \;+\;
\begin{array}{c}
F \\
| \\
C = N - C \cdot e \\
| \quad\quad \| \\
F \quad\quad C \\
\quad\quad \| \\
\quad\quad O
\end{array}
$$

iii)

$$E \cdot {}^e \; + \; nn \cdot \overset{\displaystyle \overset{O}{\underset{\|}{C}}}{\underset{\underset{CF_3}{|}}{\underset{CF_2}{|}}} \overset{\|}{N} - C \cdot e \; \longrightarrow \; EF \; + \; \overset{\displaystyle \overset{F}{|}}{\underset{CF_3}{|}} C = N - \overset{C}{\underset{\underset{O}{\|}}{C}} \cdot e \qquad \text{iv)}$$

from which one can say that (I) is the Y center while (II) is the X center, both favoring routes natural to them.
(Laws of Creations for 1,4-Cumulenic Isocyanates)

<u>**Rule 528:**</u> This rule of Chemistry for **1,4-Cumulenic Isocyanates with radical-pushing C_nF_{2n+1} groups,** states that, the followings are obtained for them-

$$\underset{\underset{N}{|}}{CF_3}\; N = C = C = O \; \longrightarrow \; nn \cdot \overset{\displaystyle CF_3}{\underset{\underset{O}{\|}}{\underset{C}{\|}}} N - C \cdot e \; ; \; e \cdot C - C \cdot n \; ; \; e \cdot C - O \cdot nn$$

(I) (Electrophile) (II) (Nucleophile) (III) (Both)   i)

$$\underset{\underset{CF_3}{|}}{\underset{CF_2}{|}} N = C = C = O \; \longrightarrow \; nn \cdot \overset{\displaystyle \overset{O}{\underset{\|}{C}}}{\underset{\underset{CF_3}{|}}{\underset{CF_2}{|}}} N - C \cdot e \; ; \; e \cdot C - C \cdot n \; ; \; e \cdot C - O \cdot nn$$

(I) (Electrophilic) (II) (Nucleophile) (III) (Nucleophile)   ii)

for which the (II)s and (III)s have different transfer species as indicated below-

$$E \overset{\displaystyle O}{\underset{\displaystyle}{(} C -} \overset{\displaystyle O}{\underset{\displaystyle N}{\underset{CF_2}{|}}} C \overset{\displaystyle}{)_n} \overset{\displaystyle O}{C} - C \cdot e \; \longrightarrow \; E \overset{\displaystyle O}{(} C - C)_n C - C \equiv N \; + \; e \cdot C_2F_5$$

<u>Transfer species of 1st kind of 3rd type</u>   iii)

$$E \leftarrow C - \underset{\underset{CF_3}{\overset{\|}{N}}}{\overset{\overset{O}{\|}}{C}} \overset{}{\longrightarrow_n} \underset{\underset{CF_3}{\overset{\|}{N}}}{\overset{\overset{O}{\|}}{C}} - C.e \longrightarrow E \leftarrow \underset{\underset{CF_3}{\overset{\|}{N}}}{\overset{\overset{O}{\|}}{C}} - C \overset{}{\longrightarrow_n} \overset{\overset{O}{\|}}{C} - C \equiv N + e.CF_3$$

<u>Transfer species of 1st kind of 3rd type</u>

iv)

$$E + O - \underset{\underset{\underset{C_2F_5}{|}}{\overset{\|}{N}}}{\overset{\overset{\|}{C}}{\underset{\|}{C}}} \overset{}{\longrightarrow_n} O - C.e \longrightarrow E + O - \underset{\underset{\underset{C_2F_5}{|}}{\overset{\|}{N}}}{\overset{\overset{\|}{C}}{\underset{\|}{C}}} \overset{}{\longrightarrow_n} O - C \equiv C - N = \underset{F}{\overset{F}{C}} + e.CF_3$$

<u>Transfer species of first kind of 8th type</u>

v)

for which absence of transfer species for the CF_3 case for C=O center and the fact that the entire group is transfer species for the (II) centers, indicates that the C=O center is never involved.
(Laws of Creations for 1,4-Cumulenic Isocyanates)

Rule 529: This rule of Chemistry for **1,4-Cumulenic Isocyanates with radical-pulling COOR groups,** states that, the followings are obtained for them-

$$\underset{\underset{\underset{CH_3}{|}}{\overset{|}{O}}}{\overset{\overset{4}{N}}{\underset{\underset{O}{\|}}{\overset{|}{C}=O}}} \overset{3}{=} C \overset{2}{=} C \overset{1}{=} O \longrightarrow nn.\overset{4}{N} - \underset{\underset{\underset{CH_3}{|}}{\overset{|}{O}}}{\overset{\overset{O}{\|}}{\overset{|}{C}=O}} \overset{\overset{O}{\|}}{C}.e \quad ; \quad e.\overset{3}{\underset{\underset{\underset{OCH_3}{|}}{\overset{\|}{N}}}{\overset{\overset{O}{\|}}{C}=O}} - \overset{O}{\overset{\|}{C}}.n \quad ; \quad e.\overset{2}{\underset{\underset{\underset{CH_3}{|}}{\overset{|}{O}}}{\overset{\|}{C}=O}} - \overset{1}{\overset{}{O}}.nn$$

(I) (Electrophilic)   (II) (Nucleophilic)   (III) (Both)

i)

$$E^{.e} + nn.\overset{4}{N} - \underset{\underset{\underset{CH_3 \; (I)}{|}}{\overset{|}{O}}}{\overset{\overset{O}{\|}}{\overset{|}{C}=O}} \overset{\overset{O}{\|}}{C}.e \quad \xrightarrow{\;\;/\;\;} \quad EOCH_3 + e.\overset{O}{\overset{\|}{C}} - N = C = C = O$$

ii)

for which OCH_3 is not a transfer species, because (I) is resonance stabilized.
(Laws of Creations for 1,4-Cumulenic Isocyanates)

Rule 530: This rule of Chemistry for **1,4-Cumulative Isocyanates with alkenyl resonance stabilization groups,** states that, the followings are obtained for them-

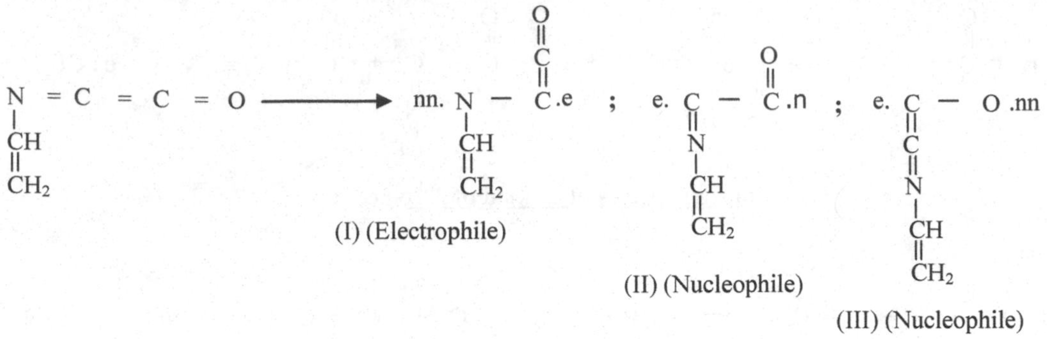

$$N = C = C = O \longrightarrow$$

with CH and CH₂ groups attached to N.

Structures labeled:

(I) (Electrophile) nn. N — C.e (with CH, CH₂ and O double bonded to C)

(II) (Nucleophile) e. C — C.n (with O, N, CH, CH₂)

(III) (Nucleophile) e. C — O .nn (with C, N, CH, CH₂)

i)

for which the followings are to be expected only radically and only for (A) shown below the center which is least nucleophilic to give (B) the Y center with (III) as the corresponding X center-

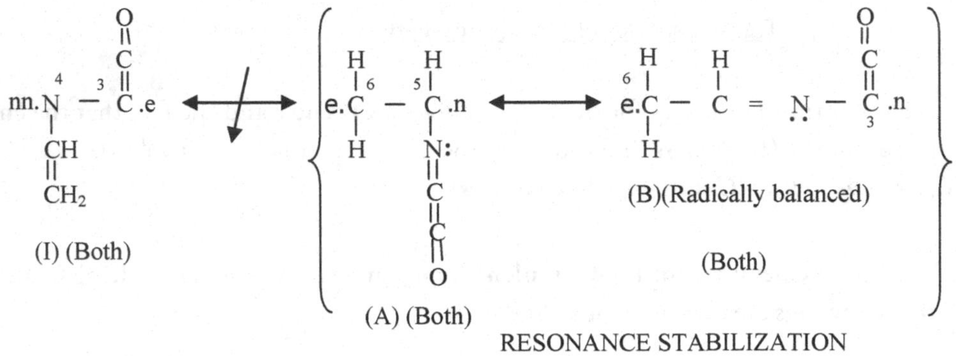

$$\text{nn.} \overset{4}{N} - \overset{3}{C}.e \quad \longleftrightarrow \!\!\! / \!\!\! \longrightarrow \quad \left\{ e.\overset{6}{C} - \overset{5}{C}.n \quad \longleftrightarrow \quad e.\overset{6}{C} - C = \ddot{N} - \overset{3}{C}.n \right\}$$

(I) (Both)

(A) (Both)

(B)(Radically balanced)

(Both)

RESONANCE STABILIZATION

ii)

from which, one can observe that even the $-N = C = C = O$ group itself is ***a non-free-radical-pushing resonance stabilization group of lower capacity than H, just like -C≡CH group*** with movement from (I) to (A) being impossible and movement from (A) to (B) is by resonance stabilization.
(Laws of Creations for 1,4-Cumulenic Isocyanates)

Rule 531: **This rule of Chemistry for 1,4-Cumulenic Isocyanates with alkenyl resonance stabilization carrying groups to be shielded,** states that, the followings are obtained for them-

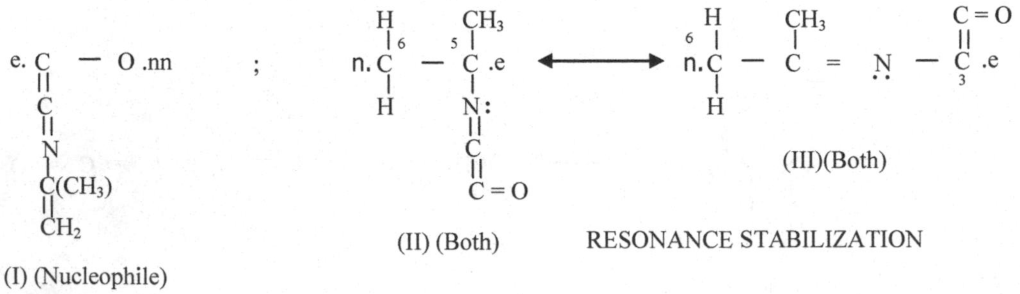

$$e. \; C \; — \; O \; .nn \quad ; \quad n.\overset{6}{C} - \overset{5}{C}.e \quad \longleftrightarrow \quad n.\overset{6}{C} - C = \ddot{N} - \overset{3}{C}.e$$

(I) (Nucleophile) with C, N, C(CH₃), CH₂

(II) (Both) with H, CH₃, N:, C, C=O

(III)(Both) with H, CH₃, C=O

RESONANCE STABILIZATION

for which (I) is the X center and (III) is the Y center, noting that the CH_3 carried by the alkenyl group has been shielded, the resonance stabilization provider being the isocyanate group.
(Laws of Creation for 1,4-Cumulenic Isocyanates)

Rule 532: This rule of Chemistry for **Isocyanates,** states that, based on the chemistry of these monomers, the followings in general are valid for them with respect to the groups carried -

$$-N = C = C = C = O \quad > \quad -N = C = O \quad >$$

Order of Radical-pulling capacity [ODD]   i)

$$O = C = C = C = C = N - \quad > \quad O = C = C = N - \quad > \quad O = N-$$

Order of Radical-pushing capacity [EVEN]   ii)

$$RN{=}C{=}C{=}C{=}C{=}C{=} \quad > \quad RN{=}C{=}C{=}C{=} \quad > \quad RN{=}C{=} \quad >> \quad RN{=}CR -$$

Order of Radical-pulling capacity [ODD]   iii)

$$RN{=} \quad > \quad RN{=}C{=}C{=}C{=}C{=} \quad > \quad RN{=}C{=}C{=}$$

Order of Radical-pushing capacity [EVEN]   iv)

noting the groups which can provide resonance stabilization.
(Laws of Creations for Substituent groups)

Rule 533: This rule of Chemistry for **1,5-Cumulenic Isocyanates,** states that, in general, these are *Electrophiles* for which for them represented as $R - N = C = C = C = O$ (where R is any substituent group), the $N = C$ center is Y^1 (Electrophilic), the $C = C$ center next to it is X^1 (Nucleophilic) and the next $C = C$ is a second Y^2 and the $C = O$ is a second most nucleophilic center never used is X^2; noting that here the X^1 center is the least nucleophilic center followed by Y^2 and then Y^1 and lastly X^2; noting the presence of two very different types of Electrophiles carried by this monomer and the fact that 1,5-Isocyanates are more nucleophilic than 1,4-Isocyanates.
(Laws of Creations for 1,5-Cumulenic Isocyanates)

Rule 534: This rule of Chemistry for **1,5-Cumulenic Isocyanic acid,** states that, it has the 1,4- ketenic character when HN = group is replaced with two hydrogen atoms (more nucleophilic than the ketene) and isocyanic acid character when the $O = C = C =$ group is replaced with $O =$ group (more nucleophilic than isocyanic acid), for which the following order is valid and like 1,3- and 1,4-isocyanic acids, the H atom must be replaced to favor full polymerization of the monomer-

$$\overset{\displaystyle H}{\underset{\displaystyle |}{N}} = C = C = C = O \quad > \quad \overset{\displaystyle H}{\underset{\displaystyle |}{N}} = C = C = O \quad > \quad \overset{\displaystyle H}{\underset{\displaystyle |}{\underset{\displaystyle |}{\underset{\displaystyle H}{C}}}} = C = C = O$$

1,5- Cumulenic isocyanic acid   1,4-Cumulenic isocyanic acid   1,4-Cumulenic ketene

Order of Nucleophilicity

(Laws of Creations for 1,5-Isocyanic acid)

Rule 535: This rule of Chemistry for **1,5-Cumulenic Isocyanate with radical-pushing R groups,** states that, the followings are obtained for them-

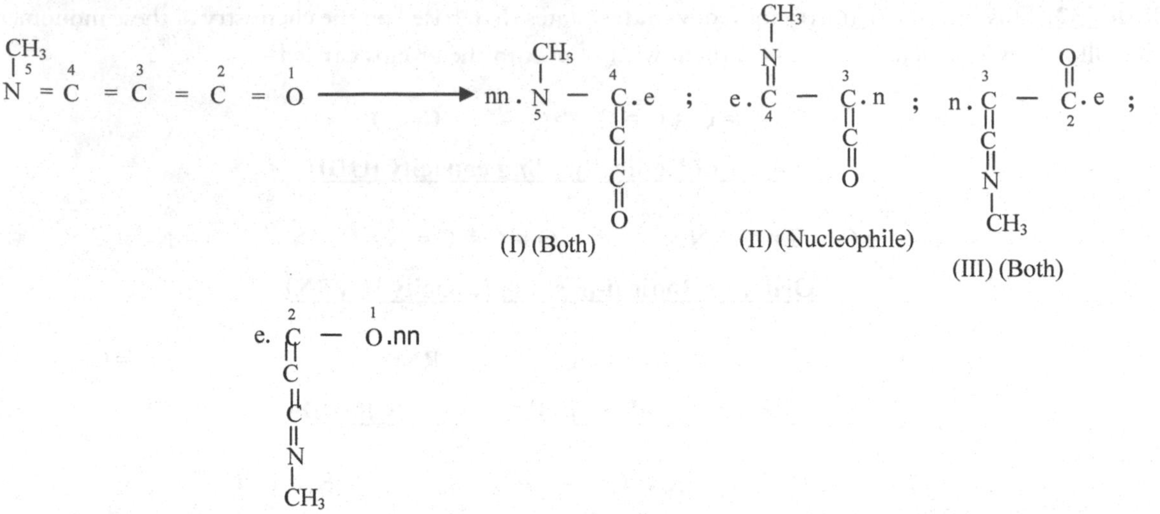

$$\underset{5}{\overset{CH_3}{\underset{|}{N}}} = \underset{4}{C} = \underset{3}{C} = \underset{2}{C} = \underset{1}{O} \longrightarrow$$

nn . $\underset{5}{\overset{CH_3}{\underset{|}{N}}}$ — $\underset{4}{C}$.e (with C‖C‖C‖O chain below) ;

e . $\underset{4}{C}$ — $\underset{3}{C}$.n (with CH₃–N= above, C‖O below) ;

n . $\underset{3}{C}$ — $\underset{2}{C}$.e (with C‖N–CH₃ below, O above) ;

(I) (Both)   (II) (Nucleophile)   (III) (Both)

e . $\underset{2}{C}$ — $\underset{1}{O}$.nn (with C‖C‖N–CH₃ chain below)

for which only (II) and (IV) have transfer species and can be made to undergo molecular rearrangement; noting the presence of two types of Electrophiles on the monomer, the first being the only ones that can be used.

(Laws of Creations for 1,5-Cumulenic Isocyanates) [See Rule 533]

Rule 536: This rule of Chemistry for **1,5-Cumulenic isocyanates**, states that, for radical-pulling groups in place of H on nitrogen centers, the followings are obtained for COOCH₃ group-.

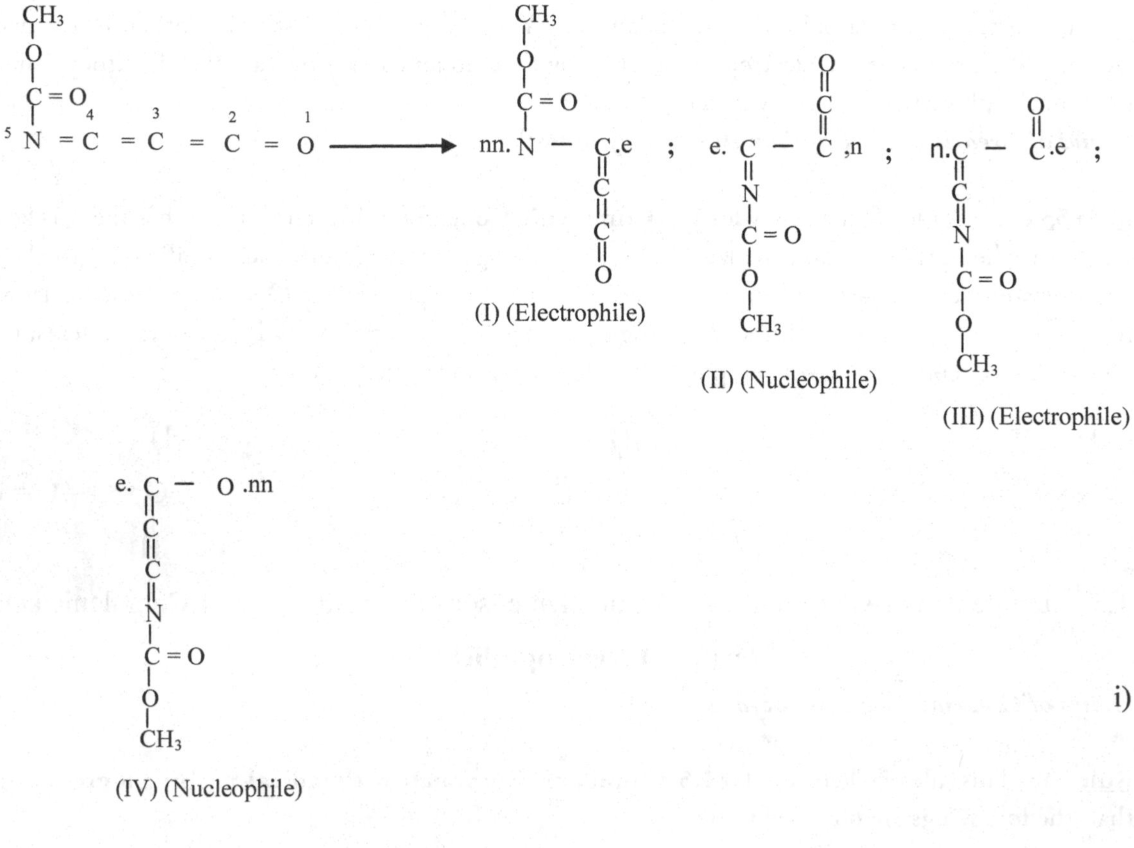

$$\underset{5}{\overset{\displaystyle \overset{CH_3}{\underset{|}{\overset{O}{\underset{|}{\overset{\|}{C=O}}}}}}{N}} = \underset{4}{C} = \underset{3}{C} = \underset{2}{C} = \underset{1}{O} \longrightarrow$$

nn . N — C.e (with C‖C‖C‖O chain below; CH₃–O–C=O above) ;

(I) (Electrophile)

e . C — C ,n (with C=O, N, C=O, O, CH₃ substituents) ;

(II) (Nucleophile)

n . C — C.e (with O above, C‖C‖N–C=O–O–CH₃ chain) ;

(III) (Electrophile)

e . C — O .nn (with C‖C‖N–C=O–O–CH₃ chain below)

(IV) (Nucleophile)

i)

n which only (I) and (III) have the potential of undergoing molecular rearrangements as shown for (I) below, noting that (I) is the only one that can be made to undergo resonance stabilization, radically-.

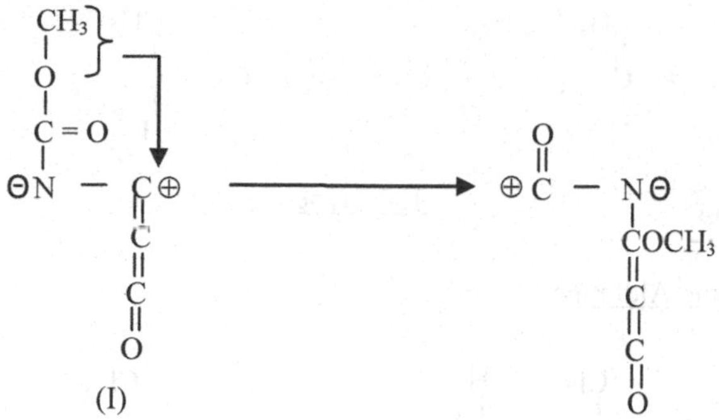

(I)

(A)

ii)

for which the following order is obtained-

$$-\overset{\displaystyle OCH_3}{\underset{\displaystyle |}{C}} = O \quad < \quad -\overset{\displaystyle OCH_3}{\underset{\displaystyle |}{C}} = C = C = O \quad < \quad -\overset{\displaystyle OCH_3}{\underset{\displaystyle |}{C}} = C = C = C = C = O$$

Order of Radical-pulling capacity (ODD)

iii)

$$-\overset{\displaystyle OCH_3}{\underset{\displaystyle |}{C}} = C = O \quad < \quad -\overset{\displaystyle OCH_3}{\underset{\displaystyle |}{C}} = C = C = C = O \quad < \quad -\overset{\displaystyle OCH_3}{\underset{\displaystyle |}{C}} = C = C = C = C = C = O$$

Order of Radical-pushing capacity (EVEN)

iv)

(Laws of Creations for 1,5-Cumulenic Isocyanates)

Rule 537: This rule of Chemistry for **Cumulenic Ketenes, Isocyanates and Aldimines,** states that, the characters of these compounds can be identified by what they carry at both ends as

1, 5 - ketenes	1, 5 - Isocyanate	1, 5 - Formaldimine
(i) One carbonyl end	(i) One carbonyl end	(i) One cumulenic end
(ii) One cumulenic end	(ii) One iminic end	(ii) One iminic end

i)

$$HN = \quad > \quad O = \quad > \quad H_2C =$$

Order of radical-pushing capacity

ii)

Isocyanates   >   Imines   >   Ketenes

Order of Nucleophilicity

iii)

(Laws of Creations for Cumulenic Ketenes, Isocyanates, and Imines)

Rule 538: This rule of Chemistry for **Cumulenic Formaldimines, Aldimines, and Ketimines,** states that, the first members of these unknown families of compounds are as shown below-

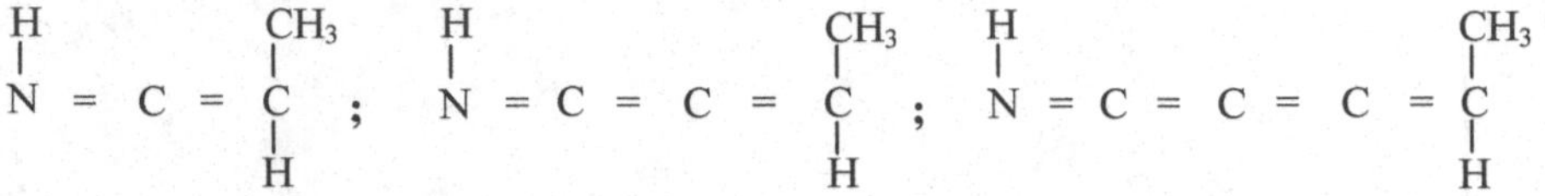

$$
\underset{\text{1, 3 - type}}{\overset{\displaystyle H}{\underset{\displaystyle H}{N = C = C}} \overset{\displaystyle CH_3}{}} \quad ; \quad \underset{\text{1, 4 - type}}{\overset{\displaystyle H}{\underset{\displaystyle H}{N = C = C = C}} \overset{\displaystyle CH_3}{}} \quad ; \quad \underset{\text{1, 5 - type}}{\overset{\displaystyle H}{\underset{\displaystyle H}{N = C = C = C = C}} \overset{\displaystyle CH_3}{}}
$$

Cumulenic Aldimines   i)

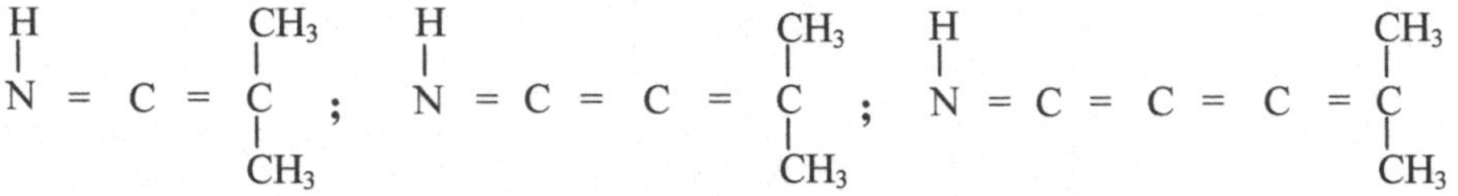

$$
\underset{\text{1, 3 - type}}{\overset{\displaystyle H}{\underset{\displaystyle CH_3}{N = C = C}} \overset{\displaystyle CH_3}{}} \quad ; \quad \underset{\text{1, 4 - type}}{\overset{\displaystyle H}{\underset{\displaystyle CH_3}{N = C = C = C}} \overset{\displaystyle CH_3}{}} \quad ; \quad \underset{\text{1, 5 - type}}{\overset{\displaystyle H}{\underset{\displaystyle CH_3}{N = C = C = C}} \overset{\displaystyle CH_3}{}}
$$

Cumulenic ketimines   ii)

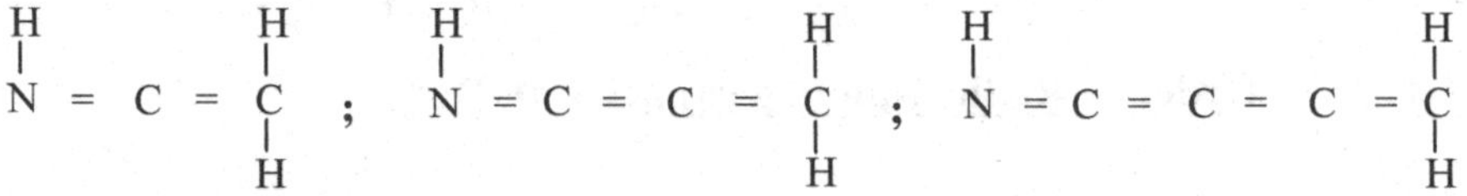

$$
\underset{\text{1, 3 - type}}{\overset{\displaystyle H}{\underset{\displaystyle H}{N = C = C}} \overset{\displaystyle H}{}} \quad ; \quad \underset{\text{1, 4 - type}}{\overset{\displaystyle H}{\underset{\displaystyle H}{N = C = C = C}} \overset{\displaystyle H}{}} \quad ; \quad \underset{\text{1, 5 - type}}{\overset{\displaystyle H}{\underset{\displaystyle H}{N = C = C = C = C}} \overset{\displaystyle H}{}}
$$

Cumulenic formaldimines   iii)

for which for all of them, the H on the Nitrogen center must be replaced for them to be used as monomers and for formaldimines, one or two H atoms on the C center may in addition be replaced, depending on the operating conditions; and in addition the order of Nucleophilicity for all cases above is as shown below-

$$
\text{1,5-type} \quad > \quad \text{1,4-type} \quad > \quad \text{1,3- type} \quad\quad\quad \text{iv)}
$$

$$
\text{Ketimines} \quad > \quad \text{Aldimines} > \quad \text{Formaldimines} \quad\quad\quad \text{v)}
$$

Order of Nucleophilicity

(Laws of Creations in Cumulenic Imines)

Rule 539: This rule of Chemistry for **One of the five forces in Nature in the real and imaginary domains as reflected in Physics that which is SMELL,** states that, the first driving force for identification of Smell is *the composition and types of the components involved [Real]*, and the second driving force is *how the components are placed or bonded [Real and Imaginary]* followed by a third driving force which is *the physical state of the components [Real]*; for which there are specific numbers of dependent variables involved for the existences of SWEET, TOXIC, BITTER, KILLER, FRIENDLY, and etc. types of Smells.
(Laws of Creations for the Physics of Smell)

Sunny N.E. Omorodion

Rule 540: This rule of Chemistry for **One of the five forces in Nature in the real and imaginary domains as reflected in Physics that which is TASTE,** states that, the first driving force for identification of Taste is *the composition and type of the components present [Real],* and a second driving force is *how the components are placed or bonded [Real and Imaginary]* followed by a third driving force which is *the type of solvent present with the components [Real and Imaginary];* for which there are specific numbers of dependent variables involved for the existences of SWEETNESS, BITTERNESS, AND FRIENDLINESS.
(Laws of Creations for the Physics of Taste)

Rule 541: This rule of Chemistry for **Two of the forces of Nature that which are SMELL and TASTE,** states that, for a compound to be TOXIC, HAVE SWEET AND BITTER SMELLS, and FRIENDLY, it must be such that must have the strong ability to exists in EQUILIBRIUM STATE OF EXISTENCE and probably DECOMPOSITION state of existence.
(Laws of Creations for the Physics of Smell)

Rule 542: This rule of Chemistry for **Nature,** states that, KNOWLEDGE IS NOT EXCLUSIVE TO ANY HUMAN EXCEPT THE ALMIGHTY INFINITE GOD since it has no Beginning and no End and the more you know, the more you come to realize that you know NOTHING, since ALL THINGS BEGAN FROM NOTHIGNESS.
(Laws of Creations for Humanity in Philosophy)

Rule 543: This rule of Chemistry **for the first members of 1,3- and 1,4-Cumulenic Formal-dimines,** states that, the followings are obtained for them as examples under different operating conditions-
Normal operating conditions:

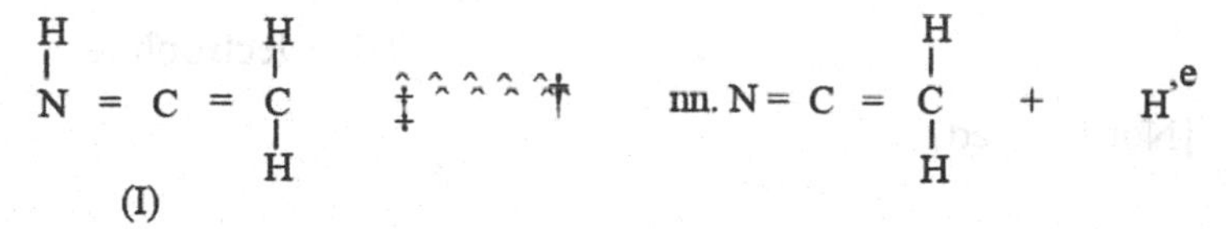

i)

ii)

for which this could rearrange by Electroradicalization to give nitrile and acrylonitrile;
Presence of a Special Passive Catalyst:
for which when they are suppressed, they may rearrange and when they do, the center involved for rearrangement is the least nucleophilic center as shown for (I) and (II) below for 1,3-, 1,4- and 1,5- respectively,

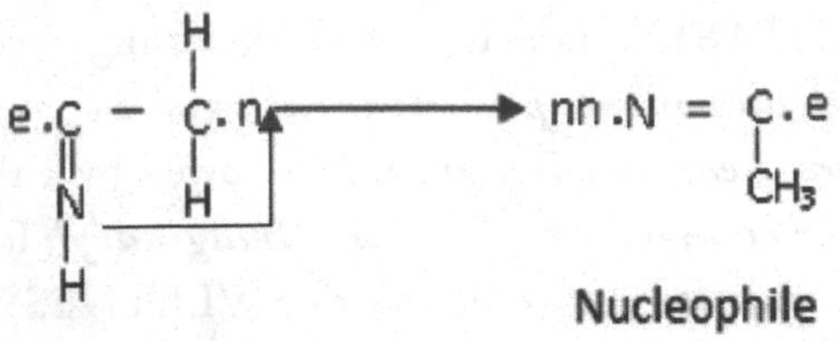

$$\underset{H}{\underset{\|}{\underset{N}{\underset{|}{H}}}}\text{e.}\underset{\|}{C} - \overset{H}{\underset{H}{\underset{|}{C}}}.\text{n} \longrightarrow \text{nn.N} = \underset{CH_3}{\underset{|}{C}}.\text{e}$$

Nucleophile

(I)

[Not favored]   iii)

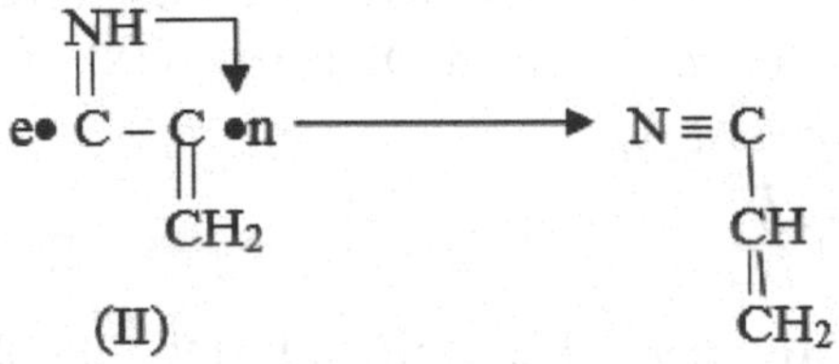

$$\overset{NH}{\underset{\|}{}}\text{e}\bullet C - \underset{CH_2}{\underset{|}{C}}\bullet\text{n} \longrightarrow N \equiv \underset{\underset{CH_2}{\|}}{\underset{|}{C}}\atop{CH}$$

(II) CH_2 **An Electrophile**

[Not favored]   iv)

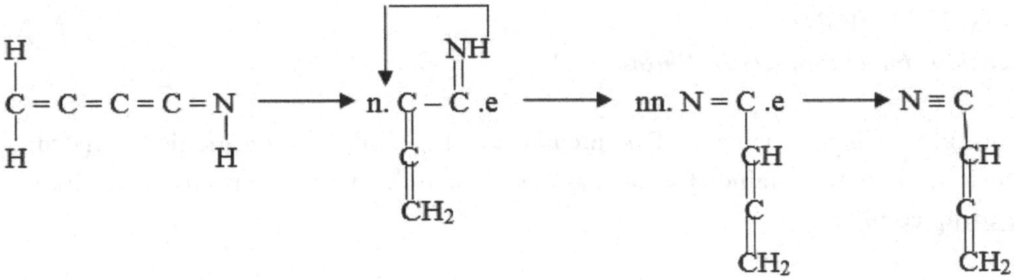

$$\overset{H}{\underset{H}{\underset{|}{C}}} = C = C = C = \overset{}{\underset{H}{\underset{|}{N}}} \longrightarrow \text{n.}\,C - \underset{\underset{CH_2}{\|}}{\underset{|}{C}}.\text{e} \longrightarrow \text{nn.}\,N = \underset{\underset{CH_2}{\|}}{\underset{|}{C}}.\text{e} \longrightarrow N \equiv \underset{\underset{CH_2}{\|}}{\underset{|}{C}}\atop{CH}$$

(I)   (II) Electrophile

[Not favored]   v)

where the center activated herein called the Y center is the least nucleophilic center, the center cumulatively placed to the N = C, and for this reason, only radical-pulling group can be transferred; noting that the 1,5- is dual in character, i.e., has two Ys and two Xs.
(Laws of Creations for Cumulenic Imines)

Rule 544: This rule of Chemistry for **1.3-Cumulenic Ketimines shown below**, states that, when same alkanyl groups are carried, the followings are obtained chargedly and radically-

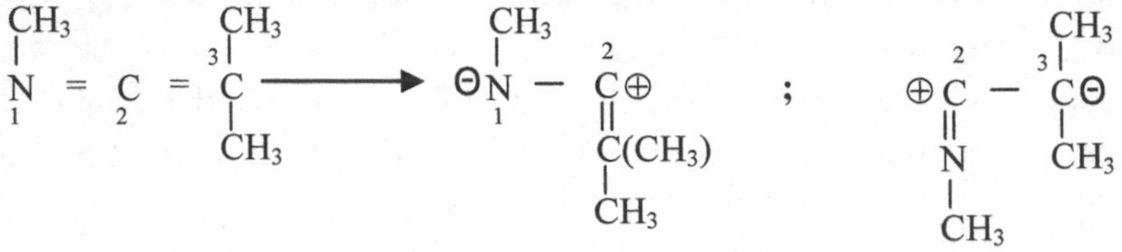

$$\underset{1}{N} = \underset{2}{C} = \overset{CH_3}{\underset{CH_3}{\underset{|}{C}}} \longrightarrow \overset{CH_3}{\underset{1}{\ominus N}} - \underset{\underset{CH_3}{\underset{|}{C(CH_3)}}}{\overset{}{C\oplus}} \quad ; \quad \oplus \underset{\underset{CH_3}{\underset{|}{N}}}{\overset{}{C}} - \overset{CH_3}{\underset{CH_3}{\underset{|}{C\ominus}}}$$

(I) (Nucleophile)   (II) (Nucleophile)

i)

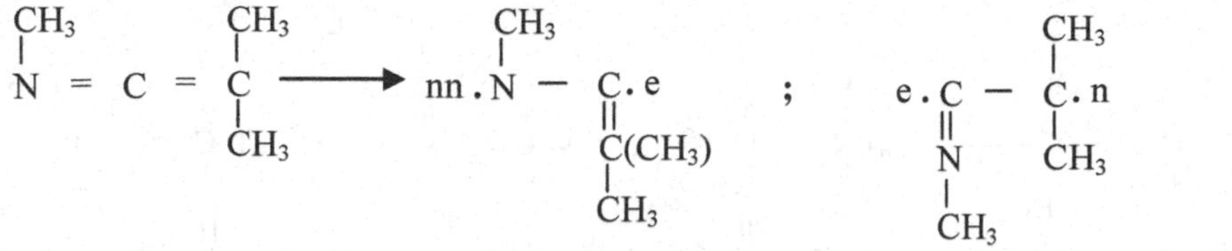

$$\underset{CH_3}{\overset{CH_3}{N}} = C = \underset{CH_3}{\overset{CH_3}{C}} \longrightarrow nn.N - \underset{\underset{CH_3}{\overset{|}{C(CH_3)}}}{\overset{||}{C}}.e \quad ; \quad e.\underset{\overset{|}{N}}{\overset{||}{C}} - \underset{CH_3}{\overset{CH_3}{C}}.n$$

(III) (Nucleophile)   (IV) (Nucleophile)

ii)

for which only (III) can molecularly rearrange as follows only radically-

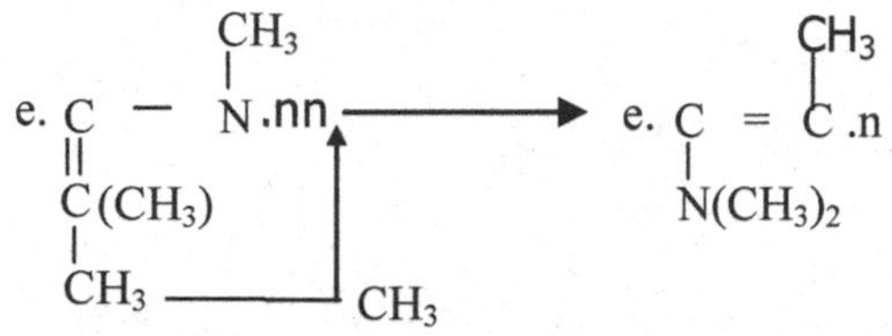

$$e.\underset{\overset{||}{C(CH_3)}}{\overset{}{C}} - N.nn \longrightarrow e.\underset{N(CH_3)_2}{\overset{CH_3}{C}} = \overset{CH_3}{C}.n$$

(III)

iii)

and charged activation can be seen to be impossible.
(Laws of Creations for 1,3-Cumulenic Imines)

Rule 545: This rule of Chemistry for **1.4-Cumulenic Ketimines shown below**, states that, when same alkanyl groups are carried, the followings are obtained radically-

$$\underset{1}{\overset{CH_3}{N}} = \underset{3}{\overset{}{C}} = \underset{CH_3}{\overset{CH_3}{C}} \longrightarrow nn.N - \underset{C(CH_3)_2}{\overset{||}{C}}.e \quad ; \quad e.\underset{\overset{|}{CH_3}}{\overset{||}{C}} - \overset{}{C}.n \quad ; \quad n.\underset{\overset{||}{N}}{\overset{C(CH_3)_2}{C}} - \overset{CH_3}{C}.e \quad ;$$

1,4 - ketimine

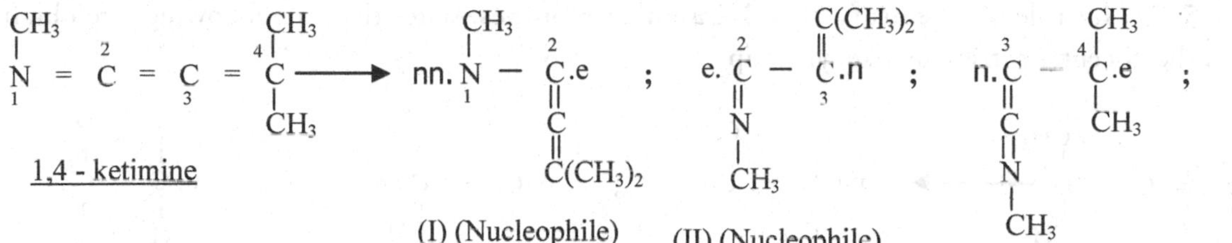

(I) (Nucleophile)   (II) (Nucleophile)   (III) (Nucleophile)

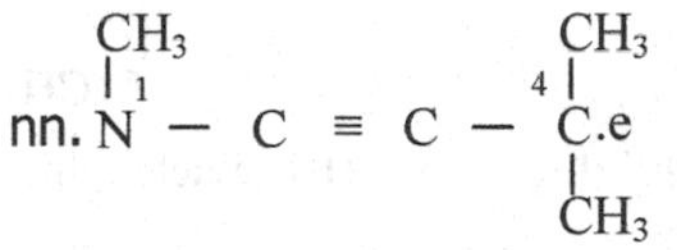

$$nn.\underset{1}{\overset{CH_3}{N}} - C \equiv C - \underset{CH_3}{\overset{CH_3}{C}}.e$$

(IV) (Nucleophile) Obtained from (I)

[NOT POSSIBLE]

in which (III) the least nucleophilic center is the Y center [which cannot resonance stabilize to give (IV)], though it is nucleophilic above; (II) is the X center and note that (IV) cannot be obtained from (I), i.e., be resonance stabilized, being the most nucleophilic center.
(Laws of Creations for 1,4-Cumulenic Ketimines)

Rule 546: This rule of Chemistry for **1.5-Cumulenic Ketimines shown below**, states that, when different radical-pushing groups are carried, the followings are obtained radically-

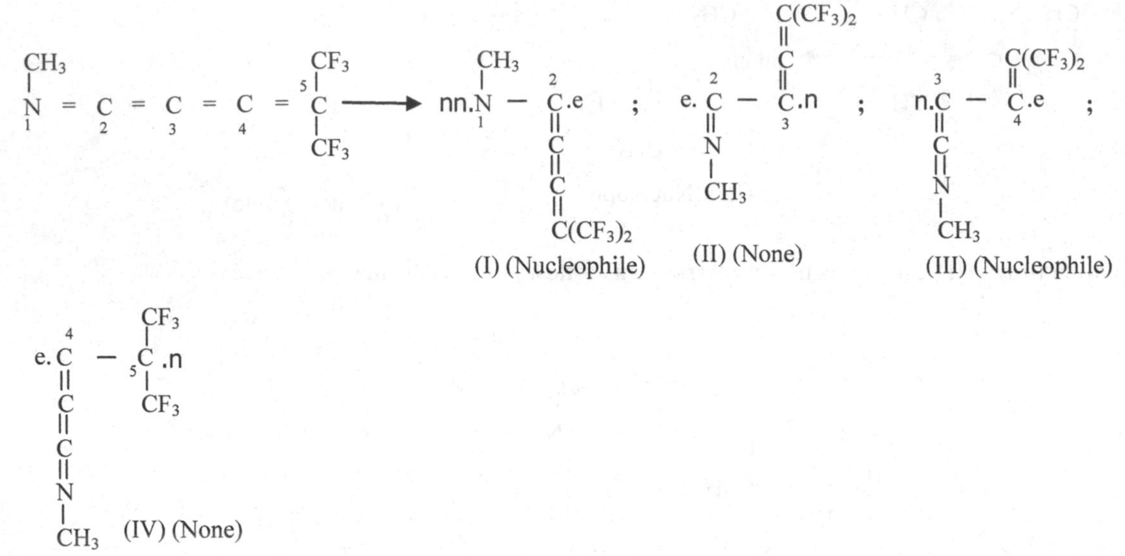

$$
\overset{CH_3}{\underset{1}{N}} = \overset{}{\underset{2}{C}} = \overset{}{\underset{3}{C}} = \overset{}{\underset{4}{C}} = \overset{5}{\underset{}{\overset{CF_3}{\underset{CF_3}{C}}}} \longrightarrow
$$

$$
nn.\overset{CH_3}{\underset{1}{N}} - \overset{2}{\underset{}{C}}.e \quad ; \quad e.\overset{2}{\underset{}{C}} - \overset{C(CF_3)_2}{\underset{3}{C}}.n \quad ; \quad n.\overset{3}{\underset{}{C}} - \overset{C(CF_3)_2}{\underset{4}{C}}.e \quad ;
$$

(I) (Nucleophile) (II) (None) (III) (Nucleophile)

$$
e.\overset{4}{\underset{}{C}} - \overset{CF_3}{\underset{5}{\overset{}{\underset{CF_3}{C}}}}.n
$$

(IV) (None)

in which (IV) is the Y center which can undergo resonance stabilization, (III) is the correspond-ing X center; (II) is a second Y center while (I) is the corresponding nucleophilic center, two centers which are never used.
(Laws of Creations for 1,5-Cumulenic Imines)

Rule 547: This rule of Chemistry for **1,3-Cumulenic Imines,** states that, the followings are obtained when the N center carries an alkenyl group-

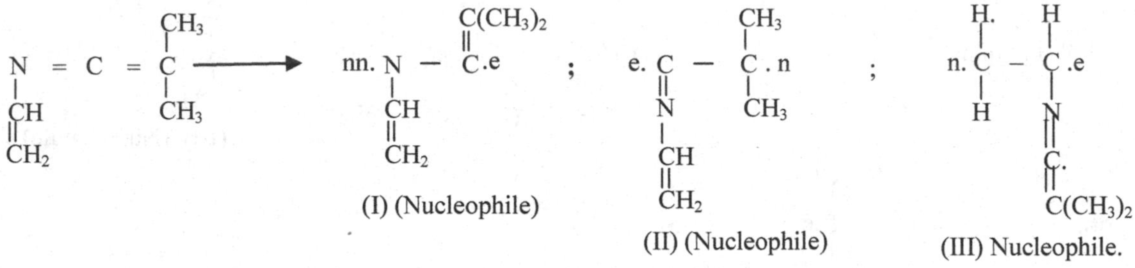

$$
\overset{}{\underset{\overset{\displaystyle CH}{\overset{\|}{CH_2}}}{N}} = C = \overset{CH_3}{\underset{CH_3}{C}} \longrightarrow
$$

$$
nn. \overset{}{\underset{\overset{\displaystyle CH}{\overset{\|}{CH_2}}}{N}} - \overset{C(CH_3)_2}{\underset{}{C}}.e \quad ; \quad e. \overset{}{\underset{N}{C}} - \overset{CH_3}{\underset{CH_3}{C}}.n \quad ;
$$

(I) (Nucleophile) (II) (Nucleophile) (III) Nucleophile.

(III) being the only mono-form above, since the alkenyl group is carrying the least nucleophilic center, resonance stabilized with CH_3 group as transfer species nucleo-free-radically; for which therefore this is a Nucleophile and if the alkenyl group is placed on the Central C atom to replace one of the CH_3 groups now moved to the N center, the monomer will still remain a Nucleophile and resonance stabilized with $- C(CH_3) = C = N(CH_3)$ as the group providing it.
(Laws of Creations for 1,3-Cumulenic Imines)

Rule 548: This rule of Chemistry for **The compounds shown below**, states that, based on the following observations-

(i) $HC \equiv C(OH)$ molecularly rearranges to the first member of ketenes - ketene, $H_2C = C = O$ from female to a weak male. The weak male can enolize to the female under different operating conditions. The same takes place with vinyl alcohol which molecularly rearranges to acetaldehyde, that which also enolizes back to vinyl alcohol under different operating conditions.

(ii) $N \equiv C(OH)$ molecularly rearranges to isocyanic acid- $HN = C = O$.

(iii) $HC \equiv C(NH_2)$ molecularly rearranges to the first member of aldimines - formaldimine, $H_2C = C = NH$ which is very unstable and rearranges to methyl nitride $N \equiv C(CH_3)$.

(iv) The first member of the families of $RN = C = NR$, that is, $HN = C = NH$, can enamidize to a nitrile, $N \equiv C(NH_2)$.

(v) Allene $H_2C = C = CH_2$ molecularly rearranges to acetylenes etc.

the followings are valid -

$$H_2N - \quad > \quad - OH \quad > \quad -CH_3 \qquad ; \qquad H - N = \quad > \quad O = \quad > \quad H_2C =$$

Order of radical-pushing capacities

(Laws of Creations for substituent groups)

Rule 549: This rule of Chemistry for **Some groups carried by Aldimines, Ketimines and the likes,** states that, the followings are the order of their capacities- . $- C(CH_3) = N(CH_3)$

$$- C(CH_3) = C = C = N(CH_3) \qquad > \qquad . - C(CH_3) = N(CH_3) \qquad > \qquad . - CH = NH$$

Order of Radical-pulling capacity [ODD]   i)

$$NH_2 \quad > \quad - C(CH_3) = C = C = C = N(CH_3) \quad > \quad . - C(CH_3) = C = N(CH_3)$$

Order of Radical-pushing capacity [EVEN]   ii)

$$RN=C=C=C=C=C= \quad > \quad RN=C=C=C= \quad > \quad RN=C= \quad >> \quad RN=CR -$$

Order of Radical-pulling capacity [ODD]   iii)

$$RN= \quad > \quad RN=C=C=C=C= \quad > \quad RN=C=C=$$

Order of Radical-pushing capacity [EVEN]   iv)

$$- C \equiv C - N(CH_3)_2 \quad > \quad - C(CH_3) = C = C(CH_3)_2 \quad > \quad - C \equiv C - C(CH_3)_3$$

Order of Radical-pushing capacity   v)

$$- C \equiv C - N(CH_3)_2 \quad > \quad - C(CH_3) = C = N(CH_3)$$

Order of Radical-pushing capacity   vi)

$$- N = C = C = C = C = CH_2 \quad > - N = C = C = C = CH_2 \quad >$$

$$- N = C = C = CH_2 \quad > \quad - N = C = CH_2 \quad > \quad - N = CH_2$$

Radical-pushing capacity of groups   vii)

$$- CR = C = C = C = C = NR \quad > \quad - CR = C = C = NR \quad > \quad - CR = NR$$

Radical-pulling capacity of groups [ODD]   viii)

$$- NR_2 \quad > \quad - CR = C = C = C = NR \quad > \quad - CR = C = NR$$

Radical-pushing capacity of groups [EVEN]   ix)

most of which are resonance stabilization groups.
(Laws of Creations for Substituents groups)

<u>**Rule 550:**</u> This rule of Chemistry for **Resonance stabilization groups from Cumulenic Imines,** states that, the following groups are resonance stabilization groups-

$$= C = C = N - CH_3 \;, \quad = C = C(CH_3)_2 \;, \quad = C = C = C(CH_3)_3 \;,$$

$$- C \equiv C - C(CH_3)_3 \;, \quad - C \equiv C - N(CH_3)_2$$

$$- C(\underline{CH_3}) = C(CH_3)_2 \;\; \text{or} \;\; - C\underline{H} = CH_2 \quad \textbf{versus} \quad - C(\underline{CH_3}) = C = C(CH_3)_2 \;\; \text{or} \;\; -- C\underline{H} = C = CH_2$$

<u>**RADICAL-PUSHING RESONANCE STABILIZATION GROUPS**</u>

where the groups underlined are shielded.
(Laws of Creations for Resonance stabilization groups)

<u>**Rule 551:**</u> This rule of Chemistry for Di-**imines which have the following first members**, states that, none of them seem to favor any stable state of existence in view of the loosely bonded

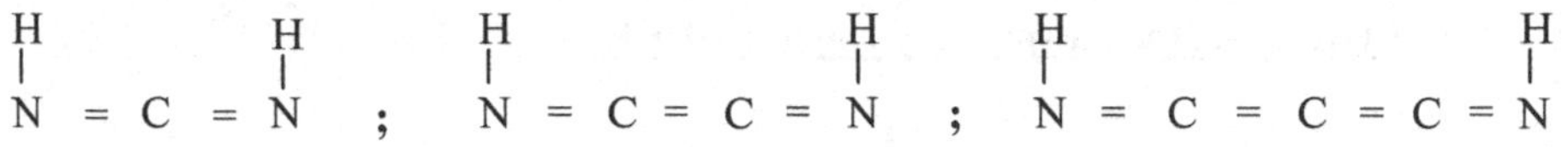

1,3 - di-imine   1,4 - di-imine   1,5 - di-imine   i)

hydrogen atom one at a time radically, for which nitriles are largely obtained only from the first member (1,3-di-imine) as shown below via molecular rearrangements only radically-

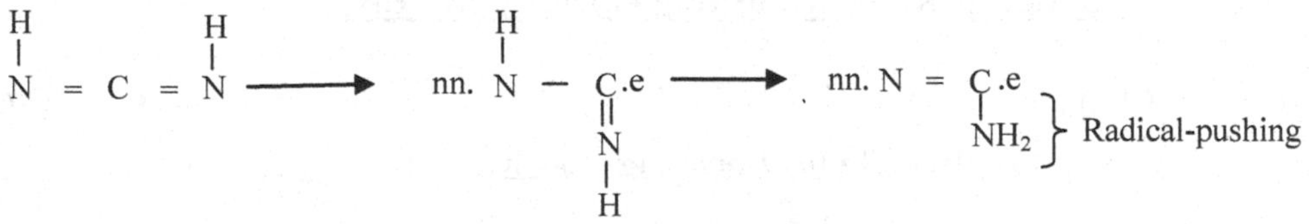

ii)

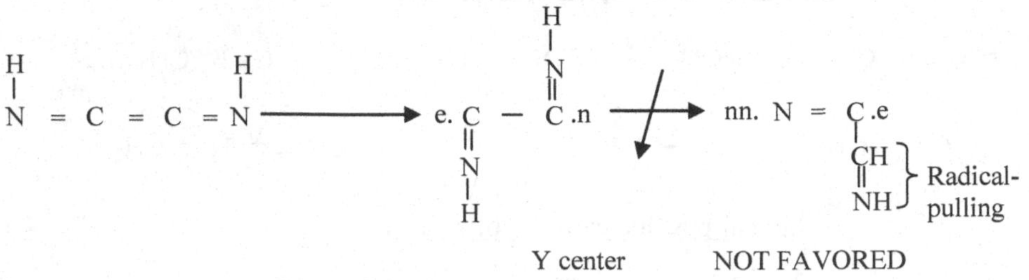

Y center   NOT FAVORED

iii)

$$ \overset{H}{\underset{|}{N}} = C = C = C = \overset{H}{\underset{|}{N}} \longrightarrow e.\ \underset{\substack{|| \\ N \\ | \\ H}}{C} - C.n \quad \xrightarrow{\hspace{2cm}} \quad nn.\ N = C.e $$

with:
- NH (double bond above C above C, e. C above N)
- Y center
- NOT FAVORED
- $\left.\begin{array}{c} CH \\ \| \\ C \\ \| \\ NH \end{array}\right\}$ Radical-pushing

iv)

for which if the H atom is replaced with F atom, no rearrangement will take place, since the N center cannot carry an electro-non-free-radical in the presence of C adjacently placed, noting that while the first member is a Nucleophile, the others are Electrophiles with X and Y centers,.
(Laws of Creations for Di-imines)

Rule 552: This rule Chemistry for **1,3 di-imines carrying alkyl radical-pushing groups,** states that, the followings are obtained for them when activated-

$$ \overset{CH_3}{\underset{|}{N}} = C = \overset{CH_3}{\underset{|}{N}} \longrightarrow nn.\ \overset{CH_3}{\underset{|}{N}} - \underset{\substack{|| \\ N \\ | \\ CH_3}}{C}.e \longrightarrow nn.\ N = \underset{\substack{| \\ N \\ H_3C \quad CH_3}}{C}.e $$

i)

$$ \overset{CH_3}{\underset{\underset{1}{|}}{N}} = \underset{2}{C} = \overset{C_2H_5}{\underset{\underset{3}{|}}{N}} $$

$$ nn.\overset{CH_3}{\underset{\underset{1}{|}}{N}} - \underset{\substack{|| \\ N \\ | \\ C_2H_5}}{\overset{2}{C}}.e \quad \xcancel{\longrightarrow} \quad nn.\ N = \underset{\substack{| \\ N \\ CH_3 \quad C_2H_5}}{C}.e $$

(I) (NOT POSSIBLE)

$$ nn.\overset{C_2H_5}{\underset{\underset{3}{|}}{N}} - \underset{\substack{|| \\ N \\ | \\ CH_3}}{\overset{2}{C}}.e \quad \xcancel{\longrightarrow} \quad nn.\ N = \underset{\substack{| \\ N \\ H_5C_2 \quad CH_3}}{C}.e $$

(II) (CANNOT BE ACTIVATED)

ii)

for which it can be observed that *only the symmetric Di-imines can undergo molecular rearrangement when the R group is not bulky, to give a stable Nitrile,* while the unsymmetric ones cannot and when activated, it is (I) above which is nucleophilic that is favored and not (II).
(Laws of Creations for 1,3-Di-imines)

Rule 553: This rule of Chemistry for **Cumulenic Di-imines,** states that, based on their Chemis-try, the followings are valid-.

$$C \equiv C > C = O > C = N$$

Order of Nucleophilicity i)

$$-N=CH(CH_3) \quad > \quad -N=CH_2 \quad > \quad (CH_3)_2N- \quad > \quad (CH_3)HN- \quad > \quad H_2N-$$

Order of radical-pushing capacity ii)

(Laws of Creations for Substituent groups)

<u>**Rule 554:**</u> This rule of Chemistry for **1,4-di-imines carrying alkyl radical-pushing groups,** states that, the followings are obtained for them when activated via the first center, the center cumulatively placed to the C =N center-

(I) (Nucleophile) (II) (Nucleophile) i)

(I) (Nucleophile) (II) (Nucleophile) ii)

for which only the (I)s will molecularly rearrange with H as transfer species, noting that (I) is the X center, and (II) is the Y center as in imines; clear indication that these are Electrophiles, despite the fact that they are Nucleophilic in character.

(Laws of Creations for 1,4-Cumulenic Di-imines)

<u>**Rule 555:**</u> This rule of Chemistry for **1,5-Di-imines carrying alkyl radical-pushing groups,** states that, the followings are obtained for them when activated via the first center, the center cumulatively placed to the C =N center-

$$\underset{1}{N} = \underset{2}{C} = \underset{3}{C} = \underset{4}{C} = \underset{5}{N} \longrightarrow$$

(with CH_3 on N_1 and C_2H_5 on N_5)

nn. $\underset{1}{N} - \underset{}{C}.e$ (I) with CH_3 on N, chain $C=C$, $C=N$, C_2H_5

; e. $\underset{}{C} - \underset{3}{C}.n$ (II) with $C_2H_5-N=C$ top, CH_3 on N

(I) Nucleophile   **(II) Nucleophile**

n. $\underset{3}{C} - \underset{4}{C}.e$ (III) with $C_2H_5-N=C$, and $C=N-CH_3$

; e. $\underset{}{C} - \underset{5}{N}.nn$ (IV) with C_2H_5 on N, chain $C=C$, $C=N-CH_3$

    (III)   **(IV)**

<u>Nucleophiles</u>   i)

$$\underset{}{N} = \overset{X.}{C} = \overset{Y}{C} = \underset{}{C} = \underset{}{N}$$ (with CH_3 on first N, C_2H_5 on last N)   ii)

for which the X and Y centers have been shown in the second equation, though nucleophilic characters are displayed by all the mono-forms (I) to (IV); for which (II) is the first center to be activated with no rearrangement possible for (I).

(Laws of Creations for 1,5-Cumulenic Di-imines)

<u>Rule 556:</u> This rule of Chemistry for **1,3-di-imines with two different radical pushing groups on the nitrogen centers,** states that, for them the followings are to be expected-

$$\underset{}{N} = \underset{}{C} = \underset{}{N} \longrightarrow$$ (with CH_3 on first N, CF_3 on last N)

nn.$\underset{}{N} - \underset{}{C}.e$ **(I) (Does not exist)** (with CH_3 on N, $C=N-CF_3$)

; e. $\underset{}{C} - \underset{}{N}nn$ **(II) (None)** (with $C=N-CH_3$, CF_3 on N) ⟶ nnN $= \underset{}{C}.e$ **Impossible movement** (with N bonded to CF_3 and CH_3)

for which (I) does not exist since the least nucleophilic center is the center first activated and (II) which exists cannot rearrange to give Nitriles.

(Laws of Creations for 1,3- Di-imines)

838

Rule 557: This rule of Chemistry for **1,3-di-imines with radical pushing groups (C_nF_{2n+1}) of lower capacity than H on the nitrogen centers**, states that, for them the followings are to be expected for n equals 1 -

i)

NOT FAVORED   ("Electrophile")   ii)

FAVORED   (Nucleophile)   iii)

for which it is important to note in the second equation above, that two centers looking alike of equal or different nucleophilicity cumulatively or conjugatedly placed cannot have X and Y characters, and that a route is favored after rearrangement in iii).
(Laws of Creations for 1,3- Di-imines)

Rule 558: This rule of Chemistry for **1,3-di-imines with radical pushing groups (C_nF_{2n+1}) of lower capacity than H on the nitrogen centers,** states that, for them the followings are to be expected for n equals 2-

FAVORED   (Nucleophile)

for which no route can be observed to be favored by them in the absence of molecular rearrange-ment for these types of monomers.
(Laws of Creations for 1,3- Di-imines)

Rule 559: This rule of Chemistry for **1,3-di-imines with radical pushing groups (C_nF_{2n+1}) of lower capacity than H on the nitrogen centers,** states that, for them the followings are to be expected for n equals 3 and above-

$$
\begin{array}{c}
C_3F_7 \\
| \\
N = C = N \\
| \\
C_3F_7
\end{array}
\longrightarrow
\begin{array}{c}
C_2F_5 \\
| \\
CF_2 \longrightarrow \\
nn\bullet N - C \bullet e \\
\quad \| \\
\quad N \\
\quad | \\
\quad C_3F_7
\end{array}
\longrightarrow
\begin{array}{c}
\quad F \quad Y \\
\quad | \quad \downarrow \\
e\bullet C - N \bullet nn \\
\quad | \quad | \\
C_2F_5 \quad CF \\
\quad \quad | \\
\quad \quad N \longleftarrow X \\
\quad \quad | \\
\quad \quad C_3F_7
\end{array}
$$

NOT POSSIBLE   (None)

for which when they rearrange as Electrophiles, no polymerization can be favored, since the center and the whole monomer are nucleophilic and as such they can only rearrange as a Nucleophile, with C_3F_7 as transfer species.
(Laws of Creations for 1,3- Di-imines).

Rule 560: This rule Chemistry for **1,3-di-imines with radical pulling groups (COOR) on the nitrogen centers,** states that, for them the followings are to be expected-

$$
\begin{array}{c}
CH_3 \\
| \\
O \\
| \\
C = O \\
| \\
N = C = N \\
\quad | \\
\quad C = O \\
\quad | \\
\quad O \\
\quad | \\
\quad CH_3
\end{array}
\xrightarrow[\text{rearrangement}]{\text{Molecular}}
\begin{array}{c}
CH_3 \\
| \\
O \quad X \\
| \\
C = O \\
| \\
nn.N - C.e \\
\quad Y \| Y \\
\quad N \\
\quad | \ X \\
\quad C = O \\
\quad | \\
\quad O \\
\quad | \\
\quad CH_3 \\
\text{(Electrophile)}
\end{array}
\xrightarrow{+ E\cdot e}
EOCH_3 \ +
\begin{array}{c}
\quad O \\
\quad \| \\
e.C - N - C \\
\quad \quad | \quad \| \\
\quad \quad N \\
\quad \quad | \\
\quad \quad C = O \\
\quad \quad | \\
\quad \quad O \\
\quad \quad | \\
\quad \quad CH_3
\end{array}
$$

$$
\begin{array}{c}
\quad X \ O \ Y \\
\quad \| \ \| \\
e.C - N. nn \\
\quad | \\
\quad C(OCH_3) \\
\quad \| \\
\quad N \\
\quad | \\
\quad C = O \\
\quad | \\
\quad O \\
\quad | \\
\quad CH_3
\end{array}
\xrightarrow{+ E\cdot e}
EOCH_3 \ +
\begin{array}{c}
e.C - N = C = O \\
\| \\
N \\
| \\
C = O \\
| \\
O \\
| \\
CH_3
\end{array}
$$

NONE FAVORED

which when molecular rearrangement is allowed to take place, while the Y center above favors the route natural to it (Nucleo-non-free-radical), the X center which is C = O present after the rearrangement above favors the electro-free-radical route, that which is a <u>Complete Electrophile</u>, complete in the sense that both the Y and X centers can only be used in the route natural to them; noting however that the above will not take place, since <u>resonance stabilization must take place after activation, for which no molecular rearrangement can take place, since OCH₃ becomes shielded and the 1,4-mono-form is now made to favor both routes.</u>
(Laws of Creations for Unique Electrophiles in 1,3- Di-imines)

<u>Rule 561:</u> This rule of Chemistry for **1,4-di-imines with radical pushing groups on the nitrogen centers,** states that, for them the followings are to be expected-

(I) (Both)   (II) (Nucleophile)   (III) (None)

wherein (II) the least nucleophilic center (Y) cannot be made to molecularly rearrange and (I) the most nucleophilic center cannot exist, and (III) the X center cannot also rearrange; unless the reverse is the case, that is, that wherein the (Y) center is more nucleophilic than the (X) center
(Laws of Creations for 1,4-Cumulenic Di-imines)

<u>Rule 562:</u> This rule of Chemistry for **1,4-di-imines with radical pushing groups (C_nF_{n+1}) of lower capacity than H on the nitrogen centers,** states that, for them the followings are to be expected for n = 1-

(I) Electrophiles   (II) (Nucleophile)

wherein (II) the least nucleophilic center (Y) cannot be made to molecularly rearrange and (I) the X center cannot also rearrange, with both favoring the route not natural to them; unless the reverse is the case, that is that where (I) is the (Y) and (II) is the (X) center, that which indeed is not the case.
(Laws of Creations for 1,4-Cumulenic Di-imines)

<u>Rule 563:</u> This rule of Chemistry for **1,4-di-imines with radical pushing groups (C_nF_{n+1}) of lower capacity than H on the nitrogen centers,** states that, for them the followings are to be expected for n = 2-

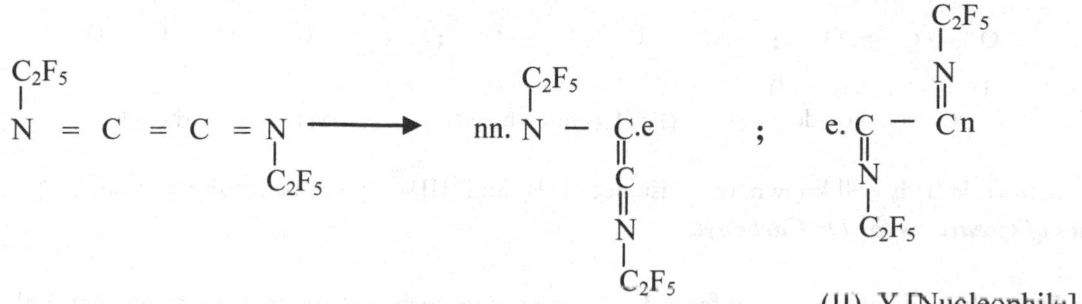

(I) None

(II) Y [Nucleophile]

for which for (I) the X center, polymerization is favored only after undergoing molecular rearrangement and for (II) the less nucleophilic center (Y), molecular rearrangement cannot take place, and found to favor the electro-free-radical route the route not natural to it, unless the reverse is the case, that is that where (I) is the (Y) and (II) is the (X) center, that which indeed is not the case.
(Laws of Creations for 1,4-Cumulenic Di-imines)

Rule 564: This rule of Chemistry for **1,3-di-imines with an external resonance stabilization alkenyl group,** states that, the followings are to be expected-

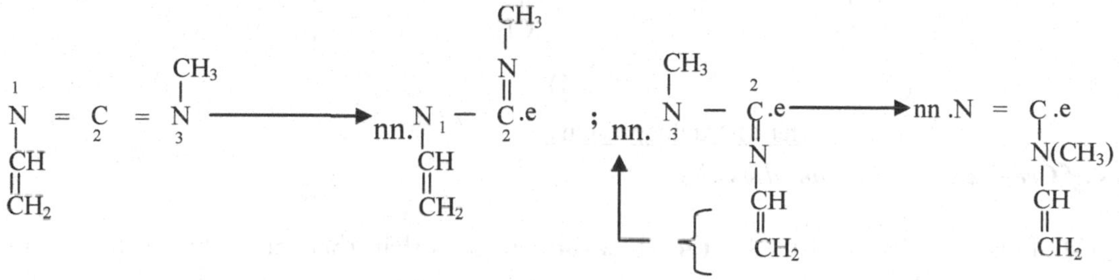

(II) (Nucleophile)

(III) (Nucleophile)

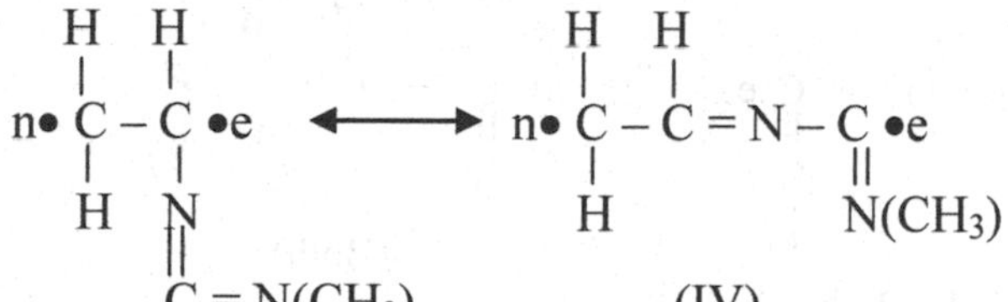

(IV)

(I) Nucleophile and Resonance Stabilized

for which in view of the fact that the alkenyl group is carrying the least nucleophilic center, (I) above is the first center activated and resonance stabilized to (IV) which is a complete Nucleophile.
(Laws of Creations for 1,3-Cumulenic Di-imine)

Rule 565: This rule Chemistry for **Unknown families of di-carbonyls,** states that, these are compounds or monomers which have no transfer species, with the first member of the family being Carbon dioxide, –

$$\overset{1}{O} = \overset{2}{C} = \overset{3}{O} \quad ; \quad O = C = C = O \quad ; \quad O = C = C = C = O$$

(I) (1,3 - di-carbonyl)
Carbon dioxide   (II) (1,4- di-carbonyl)   (III) (1,5 - di -carbonyl)

for which while (I) is well known, the existence of (II) and (III) may not be readily favored.
(Laws of Creations for Di-Carbonyls)

Rule 566: This rule of Chemistry for **1,3 - di ketenes or carbonyl monomers,** states that, only one mono-form exists chargedly and radically as shown below-

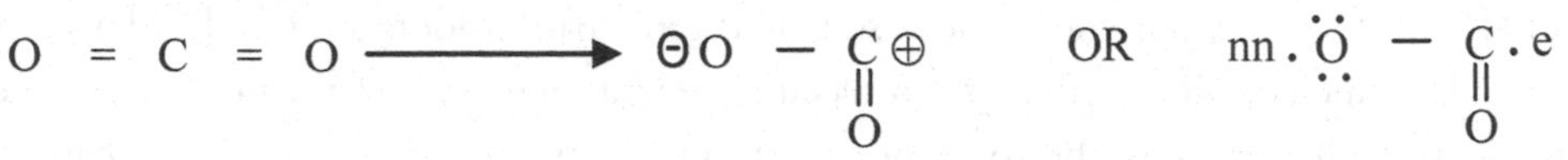

$$O = C = O \longrightarrow \ominus O - \underset{\underset{O}{\|}}{C} \oplus \qquad OR \qquad nn.\overset{..}{\underset{..}{O}} - \underset{\underset{O}{\|}}{C}.e$$

(Half-free charged or radical monomer)

for which, the activation center is more nucleophilic than that of the first member of ketenes as shown below, $= O$ group being more radical-pushing than $= CH_2$ group.

$$\ominus O - \underset{\underset{O}{\|}}{C} \oplus \qquad > \qquad \ominus O - \underset{\underset{CH_2}{\|}}{C} \oplus$$

(I)   (II)
Order of Nucleophilicity
(Laws of Creations for Carbon dioxide)

Rule 567: This rule of Chemistry for **1,4-di-Carbonyl,** states that, there are only two mono-forms as shown below only radically-

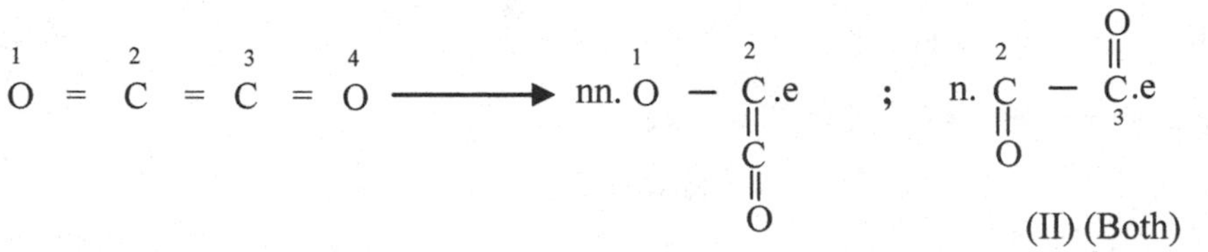

$$\overset{1}{O} = \overset{2}{C} = \overset{3}{C} = \overset{4}{O} \longrightarrow nn.\overset{1}{O} - \underset{\underset{\underset{O}{\|}}{\underset{C}{\|}}}{\overset{2}{C}}.e \quad ; \quad n.\overset{2}{\underset{\underset{O}{\|}}{C}} - \underset{3}{\overset{\overset{O}{\|}}{C}}.e$$

(I) (Both)   (II) (Both)
Less Nucleophilic center

for which, none of the mono-forms can be internally resonance stabilized since one cannot move from Half-free to Full non-free, with (I) looking like an X center while (II) looks like a Y center; noting that $O = C =$ group is a radical-pulling group of greater capacity than $O = CR -$.
(Laws of Creations for 1,4-Di-Carbonyl)

Rule 568: This rule of Chemistry for **1,4-Di-carbonyl,** states that, the existence of this compound can be seen during the oxidation of vinyl alcohol as shown below

Oxidation of vinyl alcohol

Stage 1:
$$O_2 \rightleftharpoons 2nn \cdot O \cdot en$$

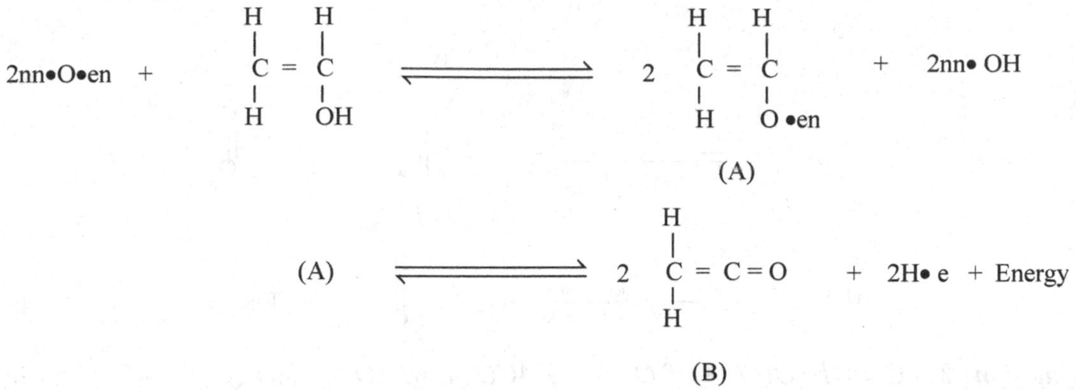

$2nn \cdot O \cdot en \;+\;$ (H₂C=CH(OH) with structure) $\rightleftharpoons$ 2 (H₂C=CH(O·en)) $+\;2nn \cdot OH$

(A)

(A) $\rightleftharpoons$ 2 (H₂C=C=O) $+\;2H \cdot e \;+\; Energy$

(B)

$2H \cdot e \;+\; 2HO \cdot nn \longrightarrow 2H_2O$ i)

Overall equation $: 2H_2C = CH(OH) \;+\; O_2 \longrightarrow 2H_2C = C = O \;+\; 2H_2O \;+\; Energy$ ii)

Stage 2:
$$O_2 \rightleftharpoons 2nn \cdot O \cdot en$$

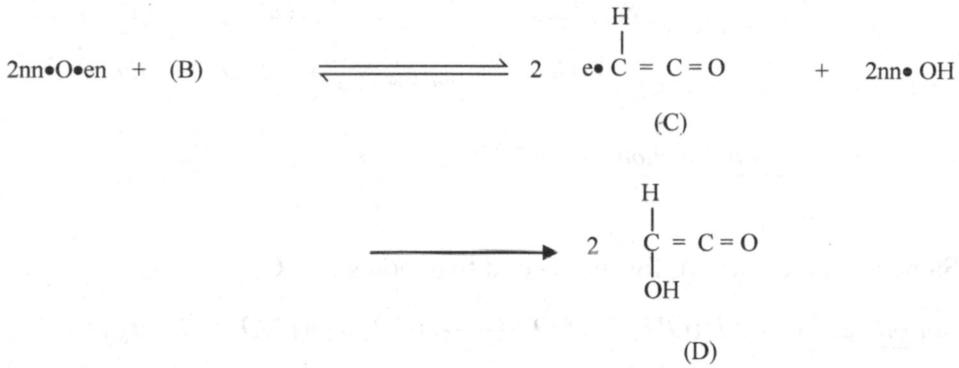

$2nn \cdot O \cdot en \;+\;$ (B) $\rightleftharpoons$ 2 (e· C=C=O with H) $+\;2nn \cdot OH$

(C)

$\longrightarrow$ 2 (C=C=O with H and OH)

(D) iii)

Overall equation $: 2H_2C = CH(OH) + 2O_2 \longrightarrow H(OH)C = C = O \;+\; 2H_2O \;+\; Energy$ iv)

Stage 3:
$$O_2 \rightleftharpoons 2nn \cdot O \cdot en$$

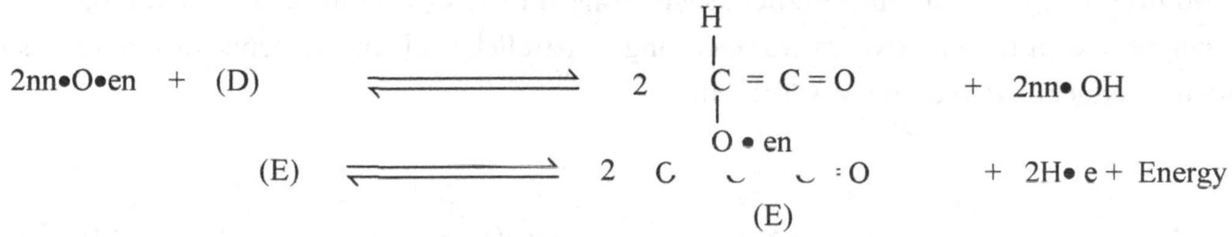

$2nn \cdot O \cdot en \;+\;$ (D) $\rightleftharpoons$ 2 (C=C=O with H and O·en) $+\;2nn \cdot OH$

(E) $\rightleftharpoons$ 2 C ⌣ ⌣ ∶O $+\;2H \cdot e + Energy$

(E)

$2H \cdot e \;+\; 2nn \cdot OH \longrightarrow 2H_2O$ v)

Overall equation $: 2H_2C = CH(OH) \;+\; 3O_2 \longrightarrow 2O = C = C = O \;+\; 4H_2O \;+\; Energy$ vi)

Stage 4:

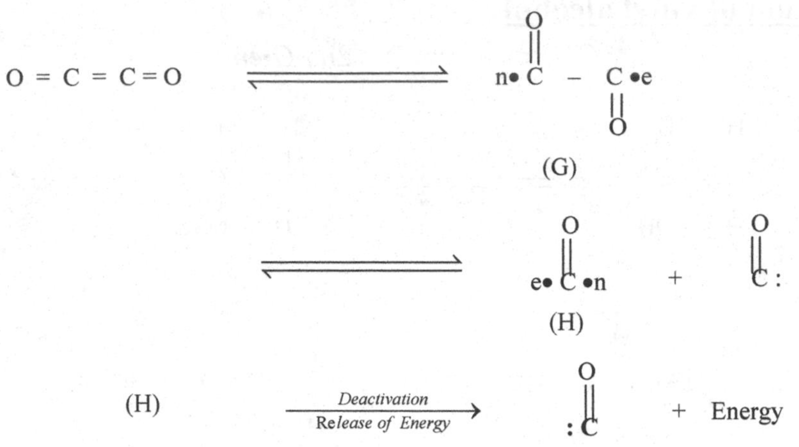

$$O = C = C = O \quad \rightleftharpoons \quad n\bullet \overset{\overset{\displaystyle O}{\|}}{C} - \overset{\underset{\displaystyle O}{\|}}{C} \bullet e$$

(G)

$$\rightleftharpoons \quad e\bullet \overset{\overset{\displaystyle O}{\|}}{C} \bullet n \quad + \quad \overset{\overset{\displaystyle O}{\|}}{C} :$$

(H)

$$(H) \quad \xrightarrow[\text{Release of Energy}]{\text{Deactivation}} \quad : \overset{\overset{\displaystyle O}{\|}}{C} \quad + \quad Energy \qquad \text{vii)}$$

$$\underline{Overall\ equation}: 2H_2C = CH(OH) \ + \ 3O_2 \longrightarrow 4CO + 4H_2O + Energy \qquad \text{viii)}$$

Stage 5:

$$O_2 \xrightarrow[\text{Oxygen molecule}]{\text{Equilibruim State Existence of Oxidising}} 2en\cdot\ddot{O}\cdot nn$$

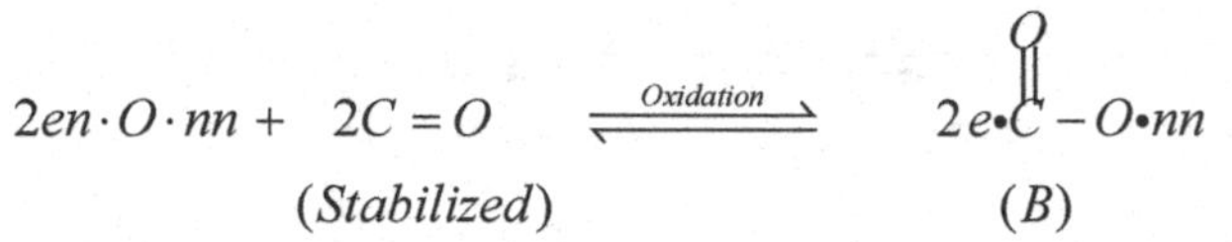

$$2en\cdot O \cdot nn + \ 2C = O \quad \xrightarrow{\text{Oxidation}} \quad 2e\bullet\overset{\overset{\displaystyle O}{\|}}{C} - O\bullet nn$$

$$\text{(Stabilized)} \qquad\qquad\qquad \text{(B)}$$

$$(B) \quad \xrightarrow[\text{Release of Energy}]{\text{Deactivation}} \quad 2CO_2 \ + \ Energy \qquad \text{ix)}$$

$$\underline{Overall\ Equation}: O_2 + 2CO \longrightarrow 2CO_2 \ + \ Energy \qquad \text{x)}$$

Stage 6: Same as Stage 5 above for the second two moles of CO.

$$\underline{Overall\ equation}: 2H_2C = CH(OH) \ + \ 5O_2 \longrightarrow 4CO_2 + 4H_2O + Energy(5) \qquad \text{xi)}$$

noting the presence of the 1,4- di-carbonyl in the third stage of a six-stage Equilibrium mechanism system wherein the (B) formed in Stage 1, a Ketene is an Electrophile and the (F) in Stage 3, **a diketone is also another type of ELECTROPHILE,** which is very unstable and cannot be oxidized and based on the operating conditions decomposes instantaneously in Stage 4 to give two moles of carbon monoxide with release of energy; and in the last two stages operating in parallel, with the presence of two excess moles of oxidizing oxygen, carbon dioxide was formed.
(Laws of Creations for 1,4-Di-Carbonyl)

Rule 569: This rule of Chemistry for **Quinones and 1,4-Carbonyls,** states that, unlike quinones aromatic carbon dioxide which can be resonance stabilized, this cannot be resonance stabilized based on the equation shown below, since when the C = C center the first to be activated is activated, no movement of radicals can take place-

845

$$O = C = C = O \quad \longleftrightarrow\!\!\!\!\!/\!\!\!\!\!\longrightarrow \quad en \cdot O - C \equiv C - O \cdot nn$$

Cannot exist

Laws of Creations for 1,4-Di-Carbonyl)

Rule 570: This rule of Chemistry for **1,5-di-Carbonyl monomer,** states that, there are also only two mono-forms as shown below-

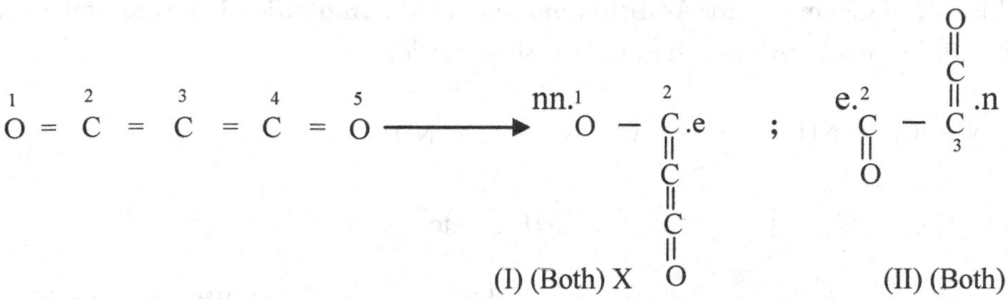

$$\underset{1}{O} = \underset{2}{C} = \underset{3}{C} = \underset{4}{C} = \underset{5}{O} \longrightarrow \underset{nn.^1}{O} - \underset{2}{C}.e \quad ; \quad \overset{e.^2}{C} - \overset{O}{\underset{3}{C}}.n$$

(I) (Both) X   (II) (Both) Y

for which (I) is the X center while (II) is the Y center, with none favoring resonance stabilization phenomenon, since the only one that can be made to be resonance stabilized is (II), the less nucleophilic center, that which is not possible since movement is from Full free- to Half free-, as shown below-

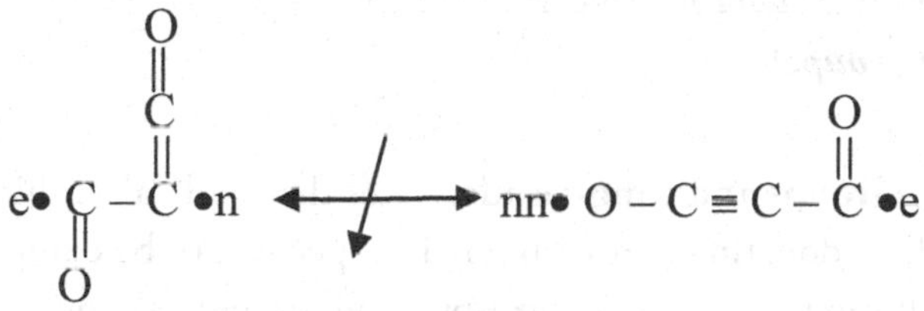

$$e\bullet \, C - C \bullet n \quad \longleftrightarrow\!\!\!\!\!/\!\!\!\!\!\longrightarrow \quad nn\bullet \, O - C \equiv C - C \bullet e$$

(II)   Full Free   (III) Half Free

(Laws of Creations for 1,5-Di – Carbonyl)

Rule 571: This rule of Chemistry for **"Electron-donating and withdrawing groups",** states that, molecules or groups which in present-day Science are called Electron-donating compounds or groups are indeed called ***RADICAL DONATING*** compounds or groups; for which they are so-called when two paired unbonded radicals are accepted by a vacant orbital of another Central atom to form ***DATIVE BONDS***, and for which the acceptor of the paired unbonded radicals are called ***RADICAL WITHDRAWING*** compounds or groups, both of which are different from ***RADICAL PUSHING and RADICAL-PULLING*** groups.
(Laws of Creations for Radical-donating and Withdrawing compounds/groups)

Rule 572: This rule of Chemistry for **Addition monomers of cumulenic di-imines,** states that, groups which contain **EVEN**-membered di-imines such as shown below-

$$- \ddot{N} = \ddot{N}H \, , \quad - \ddot{N} = C = C = \ddot{N}H \, ,$$

$$- \ddot{N} = C = C = C = C = \ddot{N}H \, , \quad etc.$$

are radical-pushing and donating groups of different capacities as shown below-

$$(-\ddot{N} = C = C = C = C = NH\ ) \quad > \quad (-\ddot{N} = C = C = NH\ ) \quad > \quad (-\underset{\cdot\cdot}{N} = NH\ )$$

<u>Radical -pushing and Donating di-iminic group</u>

(Laws of Creations for Substituent groups)

Rule 573: This rule of Chemistry for **Addition monomers of cumulenic di-imines**, states that, groups which contain ODD-membered di-imines such as shown below-

$$-\ddot{N} = C\ = \ddot{N}H\ , \quad -\ddot{N} = C = C = C = \ddot{N}H\ ,$$

$$-\ddot{N} = C = C = C\ = C = C = \ddot{N}H\ , \quad etc.$$

are also radical-pushing, donating, and resonance stabilization groups of different capacities as shown below-

$$(-\ddot{N} = C = C = C = \ddot{N}H\ ) \quad > \quad (-\ddot{N} = C\ = \ddot{N}H\ ) \quad .> \quad -\ddot{N}H_2$$

<u>Radical-pushing and Donating groups</u>

(Laws of Creations for Substituent groups)

Rule 574: This rule of Chemistry for **Groups and compounds,** states that, radical-pushing and pulling groups are no compounds, while radical donating and withdrawing species can be compounds or even atoms and groups, noting that not all groups that are radical-pushing are radical-donating and not all groups that are radical-pulling are radical-withdrawing.
(Laws of Creations in Physics from Chemistry)

Rule 575: This rule of Chemistry for **Groups, atoms and compounds,** states that, when a single radical is withdrawn or accepted, ionic, electrostatic or polar bonds are formed with formation of charges, while when two radicals are withdrawn or accepted, dative bonds are formed with no formation of charges.
(Laws of Creations in Physics from Chemistry)

Rule 576: This rule of Chemistry for **Groups, atoms, and compounds,** states that, when a single radical is donated, ionic, electrostatic or polar bonds are formed with formation of charges, while when two radicals are donated, dative bonds are formed with no formation of charges.
(Laws of Creations in Physics from Chemistry)

Rule 577: This rule of Chemistry for **Addition monomers**, states that, when a single radical is pushed, it is a nucleo-radical that is pushed, while when two radicals are pushed, it is a covalent negative charge that is pushed.
(Laws of Creations in Physics from Chemistry)

Rule 578: This rule of Chemistry for **Addition monomers,** states that, when a single radical is pulled, it is the nucleo-radical that is pulled, while when two radicals are pulled, it is a covalent negative charged that is pulled.
(Laws of Creations in Physics from Chemistry)

Rule 579: This rule of Chemistry for **Addition monomers,** states that, since a group that is radical-pushing cannot be radical-pulling, therefore no group can both be radical-pushing and

radical-pulling at the same time.
(Laws of Creations in Physics from Chemistry)

Rule 580: This rule of Chemistry for **Metallic groups,** states that, a metallic group which can donate radical(s) cannot be a radical-pushing group; while a metallic group which can withdraw radical(s) cannot be a radical-pulling group.
(Laws of Creations in Physics from Chemistry)

Rule 581: This rule of Chemistry for **Addition monomers of ketenes, isocyanates, imines, di-imines and di-carbonyls,** states that, the order of nucleophilicity of the families can be obtained from the following -

$$C \equiv N \quad > \quad \text{Isocyanates} \quad > \quad \text{Di - carbonyls} \quad > \quad \text{Di - imines} \quad > \quad \text{Ketenes}$$

$$> \quad \text{Mono - imines} \quad > \quad C \equiv C \quad > \quad \text{Cumulenes} \quad > \quad C = O$$

<u>Order of Nucleophicity</u>

Laws of Creations for Addition monomers)

Rule 582: This rule of Chemistry for **Ketenes, Isocyanates and Imines when used as monomers,** states that, in general there are five major transfer species- of the first kind of the first type, of the first kind of the second type, of the first kind of the third type, and of the first kind of the eighth type, with each kind serving different functions for a monomer, a growing polymer chain and a dead polymer.
(Laws of Creations for Transfer species)

One hundred and eleven additional rules have been proposed to mark the beginning of a new beginning of the third millennium in humanity. The need for the drastic change has been inevitable, because for too long we have been in complete darkness universally. Though we think we are advancing technologically, the situation has been such that our future generations will have no place to live in. This cannot continue to exist anymore. In recent years, humanity has been addressing these issues in a wonderful way, without realizing that the keys for these solutions are in "ones" hands. It has not been an easy task to release them, because of the type of world we live in – a COMPLEX world. Though knowledge is not exclusive to any being, we have only just begun, because no one is perfect, all having eaten from that forbidden tree.

We will find very useful applications of what have been shown with the first two Volumes not only in present day chemistry but related and all other disciplines including Religion from in particular the rules which have been proposed. It is important to note that, it is only when rules are proposed that a complete understanding is provided, in view of the complex but simple nature of the developments, which is boundless. Why some compounds exist and some do not, have been explained. Molecular rearrangement phenomena can be observed to be very important. So also are resonance stabilization and activation phenomena and so on.

For the first time, a clear distinction between radical-pushing, radical-pulling, radical-withdrawing and donating groups have been made. Thus, some radical-pushing and pulling groups can be radical-donating, and when they donate radicals in a single manner, charged bonds are formed. The groups that donate radicals doubly are limited to non-free groups. Radical-withdrawing groups do not seem to be radical-pushing groups, due to electrostatic forces of repulsion (e.g. F, Cl, $OCOCH_3$). $COOCH_3$, $CONH_2$, CN groups are radical-pulling groups, but cannot carry negative charges, and are not radical-withdrawing. Radical-withdrawing groups can also be formed with non-free radical-pushing groups such as OH, OR, NH_2 etc. When a single radical is withdrawn, charged bonds are formed. When two radicals are withdrawn, no charged bonds are formed. Hence, donating characteristics are limited to electrostatic, polar and ionic bond formations when a single radical is involved or limited to dative bonds when two radicals are involved, while pushing and pulling characteristics are limited to covalent charge and radical formations.

For the first time in Humanity, a NEW SCIENCE has emerged- a SCIENCE which is no longer AN ART. For the first time, one can see the great importance of THE NATURAL SCIENCES-Mathematics, Physics and Chemistry in which the greatest of the trios that gave birth to all disciplines is CHEMSTRY. It is in Chemistry, we see Mathematics, Physics and other disciplines as has been clearly shown so far. Even then, we have seen nothing yet. These we will see as we move to other remaining Volumes-ten in number. Nevertheless, the foundations have been laid for which all the disciplines have new definitions different from current-day definitions. Without correct definitions, no discipline can move forward. This has been the case since antiquity of existence of Homo sapience.

References

1. H. Staudinger and O. Kupfer; Ber., 45, 501 (1912); J. Hines, Physical Organis Chemistry", 2nd Edition, McGraw-Hill Book Company, Inc. New York (1962). Pgs. 492 - 494.

2. R. G. W. Norrish, H. G. Crone, and O. Salmarsh, J. Chem. Soc., 1533 (1933).

3. G. B. Kistiakowsky and N. W. Roseberg, J. Am. Chem. Soc., 72, 321 (1950).

4. A. N. Strachan and W.A. Noyes, Jr., J. Am. Chem. Soc., 76, 3258 (1954).

5. C. R. Noller, "Textbook of Organic Chemistry", W.B. Saunders Company, (1966), pg. 172.

6. C. R. Noller, ibid, pgs. 285 - 286.

Problems

9.1. Compare and contrast very briefly between isocyanates and ketenes.

9.2. What have been the advantages associated with or offered by the considerations of 1, 4 - ketenes, 1, 4 - isocyanates and above? Discuss elaborately.

9.3. (a) Why are transfer species of the second kind not known to exist with Isocyanates, Imines, Di-imines and Ketenes.
   (b) Distinguish between all the different types of transfer species of the second first kind. Under what conditions are they involved?

9.4. (a) Distinguish between the different types of resonance stabilization phenomena so far identified.
   (b) Identify the different types of resonance stabilization groups, giving examples.

9.5. (a) Which of the monomers shown below cannot be resonance stabilized?

(i) 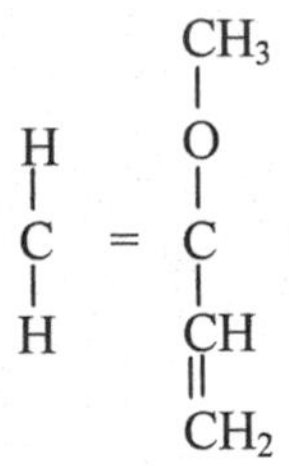
$$
\begin{array}{ccc}
 & CH_3 & \\
 & | & \\
H & O & \\
| & | & \\
C & = & C \\
| & & | \\
H & & CH \\
 & & || \\
 & & CH_2
\end{array}
$$

(ii) 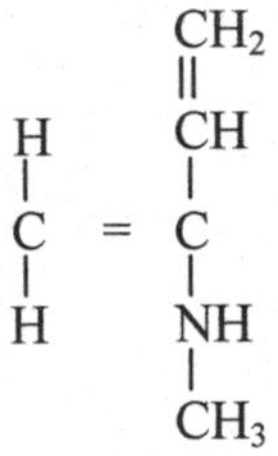
$$
\begin{array}{ccc}
 & CH_2 & \\
 & || & \\
H & CH & \\
| & | & \\
C & = & C \\
| & & | \\
H & & NH \\
 & & | \\
 & & CH_3
\end{array}
$$

(iii) 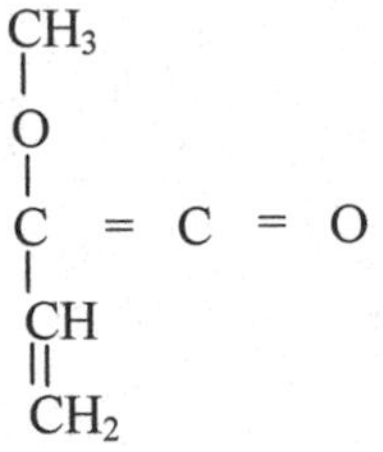
$$
\begin{array}{ccc}
CH_3 & & \\
| & & \\
O & & \\
| & & \\
C & = C = O \\
| & & \\
CH & & \\
|| & & \\
CH_2 & &
\end{array}
$$

(b) Distinguish between the use of the following substituent groups as resonance stabilization groups and for radical-pushing or pulling groups.

(i) 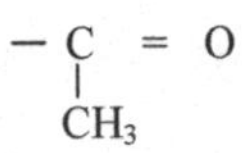 $- \underset{\underset{CH_3}{|}}{C} = O$ (ii) 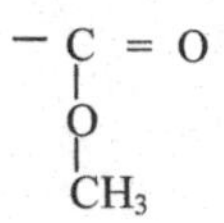 $- \underset{\underset{\underset{CH_3}{|}}{O}}{C} = O$ (iii) 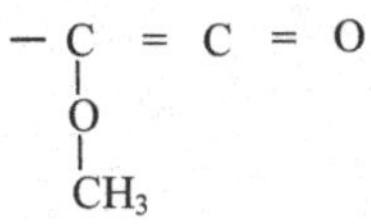 $- \underset{\underset{\underset{CH_3}{|}}{O}}{C} = C = O$ (iv) 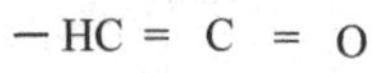 $- HC = C = O$

(v) 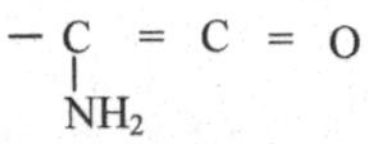 $- \underset{\underset{NH_2}{|}}{C} = C = O$ (vi) 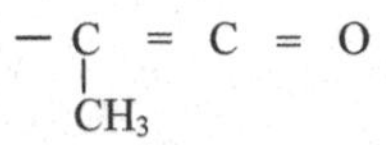 $- \underset{\underset{CH_3}{|}}{C} = C = O$ (vii) 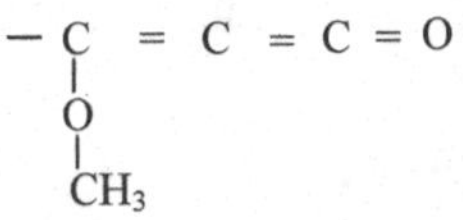 $- \underset{\underset{\underset{CH_3}{|}}{O}}{C} = C = C = O$

(viii) $- \underset{\underset{CH_3}{|}}{C} = C = C = O$ (ix) 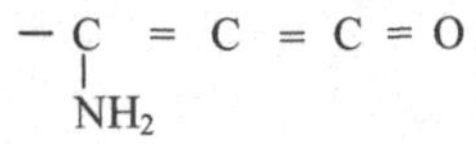 $- \underset{\underset{NH_2}{|}}{C} = C = C = O$

851

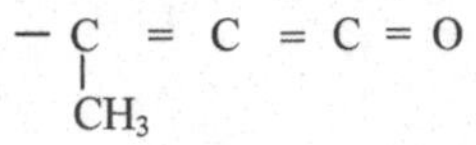

(c) Show the order of their capacities as radical-pulling or pushing groups.

9.6 Shown below is a 1,5 - dimethyl ketene.

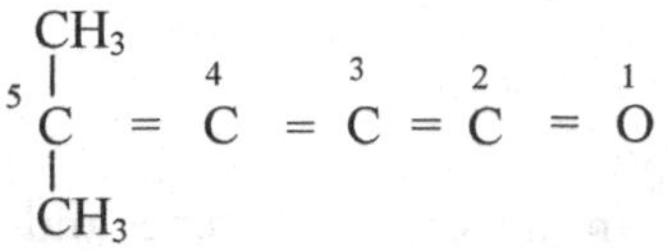

$$\overset{CH_3}{\underset{CH_3}{\overset{|}{\underset{|}{\overset{5}{C}}}}} = \overset{4}{C} = \overset{3}{C} = \overset{2}{C} = \overset{1}{O}$$

(a) Show the activated states of the mono-forms radically and chargedly where possible.

(b) Identify the routes favored by them. In the process, identify the transfer species.

(c) Distinguish the routes favored from those of 1,3 - dimethyl ketene.

9.7. (a) Identify the unique qualities of nitric oxide as a substituted group radically and chargedly.

(b) Distinguish between the use of $- HC = C = O$ and $- N = C = O$ as radical-pushing groups.

(c) While $- NR_2$, and $- NO$ groups satisfy being used as radical-donating groups, $- NO_2$, does not. Explain why this is so. How do they satisfy this condition?

9.8. (a) $- COR$, groups are radical-pulling groups, but seem to show some electropositive tendencies. Explain what this means and why this is so?

(b) Distinguish between the three families of substituted groups shown below in terms of their characters, names, capacities, resonance stabilization capabilities, and give examples of families of monomer to which they belong.

(i) $O =$, $O = C =$, $O = C = C =$, $O = C = C = C =$, Etc.

(ii) $CH_3 - N =$, $CH_3 - N = C =$, $CH_3 - N = C = C =$, Etc.

(iii) $CH_3OOC - N =$, $CH_3OOC - N = C =$, $CH_3OOC - N = C = C =$, Etc.

(c) Can the active centers carried by these groups carry charges? Explain

9.9. (a) Compare and contrast between the two sets of groups-

(i) $CH_3 - N = C =$ and $CH_3 - N = CH -$

(ii) $CH_3OOC - N = C =$ and $CH_3OOC - N = CH -$

b) Shown below are the following monomers (A) and (B).

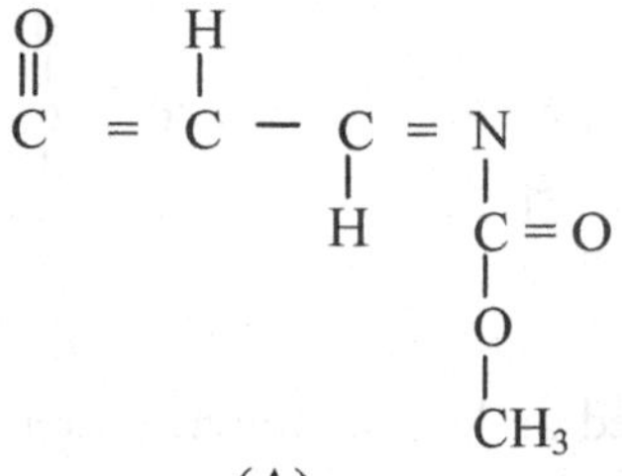

$$\underset{\underset{\overset{|}{H}}{}}{\overset{\overset{\displaystyle O}{\|}}{C}} = \overset{\overset{\displaystyle H}{|}}{C} - \underset{\underset{\underset{\underset{CH_3}{|}}{\overset{|}{O}}}{\overset{\displaystyle C = O}{|}}}{C} = N \qquad \text{and} \qquad \overset{\overset{\displaystyle O}{\|}}{C} = \overset{\overset{\displaystyle H}{|}}{C} - \overset{\overset{\displaystyle H}{|}}{C} = \underset{\underset{\underset{\underset{CH_3}{|}}{\overset{|}{O}}}{\overset{\displaystyle C = O}{|}}}{\overset{\overset{\displaystyle H}{|}}{C}}$$

(A)   (B)

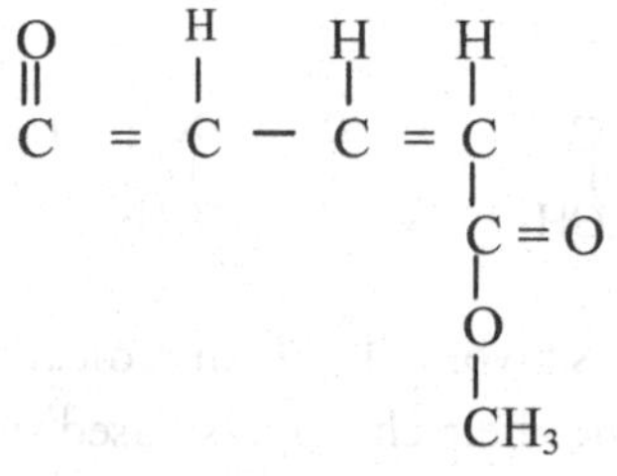

(i) Are the two monomers related?
    (ii) Identify the resonance stabilization groups and their types in the monomers if any.
    (iv) Identify the characters of all the mono-forms, based on those which are polymerizable, radically and chargedly.

9.10. (a) Shown below are the 1,4 - activated states of 1,4 - cumulenes, 1,4 - ketenes and 1,4 - isocyanates.

(i) 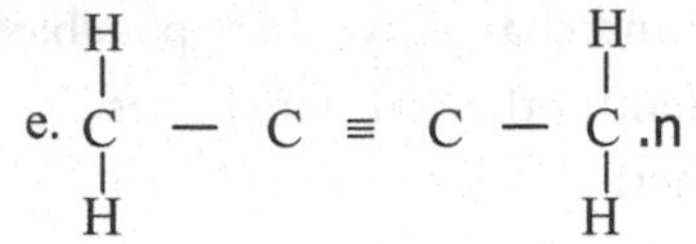   e. $\overset{\displaystyle H}{\underset{\displaystyle H}{C}} - C \equiv C - \overset{\displaystyle H}{\underset{\displaystyle H}{C}}$ .n   (ii)    e. $\overset{\displaystyle H}{\underset{\displaystyle H}{C}} - C \equiv C - O$ .nn

(iii) 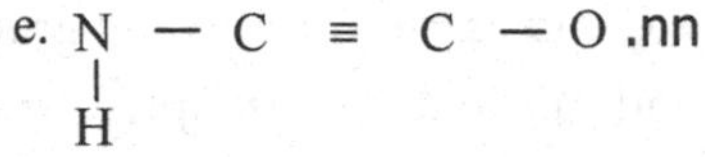   e. $\underset{\displaystyle H}{N} - C \equiv C - O$ .nn

State why the existence of (ii) and (iii) are not favored; while that of (i) is favored? Why is it that (ii) and (iii) are not resonance stabilized with their mono-forms. Compared (i), (ii) and (iii) with ethene (ethylene), formaldehyde or methylene oxide and a nitric oxide molecule.

(b) Compare the monomers (A) and (B) in Q.9.9(b) with monomers (C) and (D) shown below.

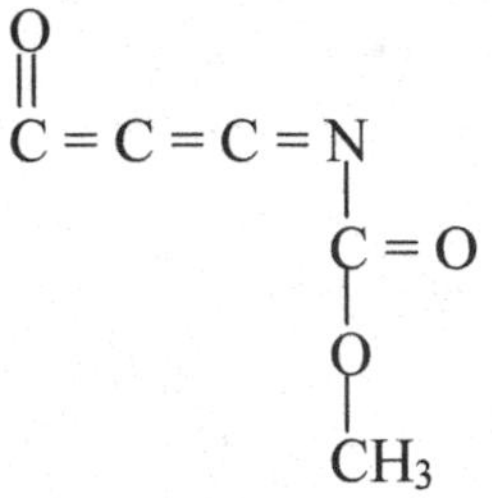

$$\overset{\displaystyle O}{\overset{\displaystyle \|}{C}} = C = C = \underset{\displaystyle \underset{\displaystyle \underset{\displaystyle CH_3}{O}}{\underset{\displaystyle |}{C=O}}}{N} \quad \text{and} \quad \overset{\displaystyle O}{\overset{\displaystyle \|}{C}} = N - C \equiv \underset{\displaystyle \underset{\displaystyle \underset{\displaystyle CH_3}{O}}{\underset{\displaystyle |}{C=O}}}{C}$$

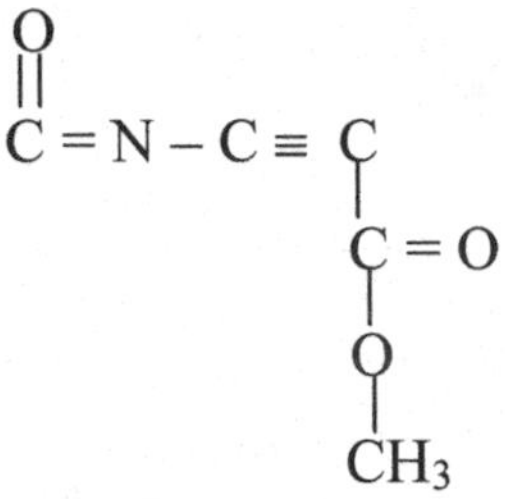

       (C)   (D)

9.11. Compare and contrast between
    (a) Cumulenic ketenes and cumulenic aldimines or ketimines.
    (b) $N = C$ and $N \equiv C$ activation centers.
    (c) OH and NH_2 substituent groups.

9.12. For the three polymerizable cumulenic aldimines shown below-

(i) 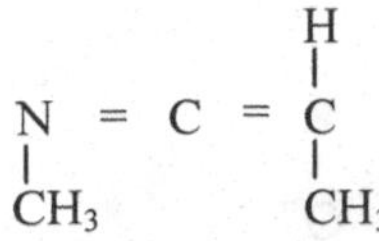 $N = C = \overset{\displaystyle H}{\underset{\displaystyle CH_3}{C}}$ $\underset{\displaystyle CH_3}{}$

(ii) 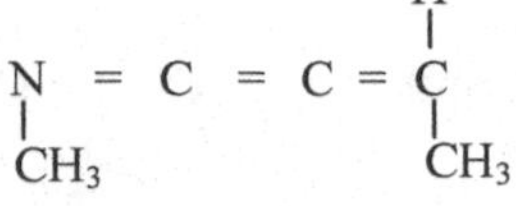 $\underset{\displaystyle CH_3}{N} = C = C = \overset{\displaystyle H}{\underset{\displaystyle CH_3}{C}}$

(iii) 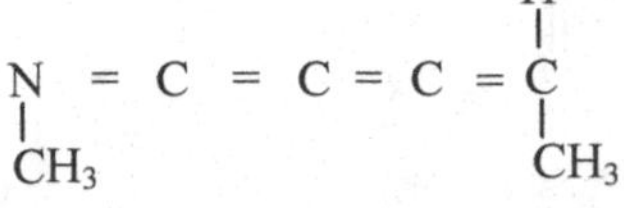 $\underset{\displaystyle CH_3}{N} = C = C = C = \overset{\displaystyle H}{\underset{\displaystyle CH_3}{C}}$

(a) Determine the routes favored by them radically.
(b) Therefore, determine their characters based on the routes favored and compare with their natural characters.

(c) Which one of them can be internally resonance stabilized? If none, explain.

9.13. (a) What is a radical-donating group? Can radical-pushing and radical- pulling groups be radical-donating groups? Explain.

(b) What is a radical-withdrawing group? Can radical-pushing and radical-pulling groups be radical-withdrawing groups? Explain.

(c) Shown below are some groups -

$$F, \quad H, \quad CCl_3, \quad CH_2Cl, \quad NHR, \quad OH, \quad NO, \quad NO_2, \quad COOCH_3 \quad \text{and} \quad SO_2H$$

Tabularly show which are radical-donating, pushing, pulling and withdrawing groups.

9.14. (a) Shown below are two monomers -

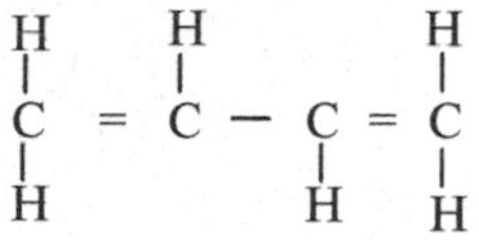

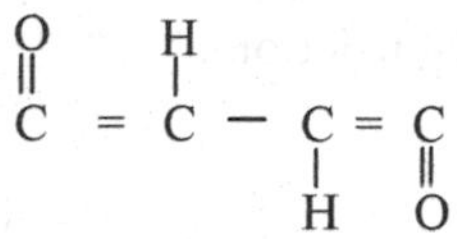

(i)   (ii)

(i) Can (i) and (ii) be resonance stabilized?

(ii) What are the transfer species present in (ii) if any and establish the character of (ii)?

(b) Convert the following groups to acetylenic resonance stabilization groups where possible.

(i) $O = C = CH-$ (ii) $H_2C = C = CH-$ (iii) $(CH_3)HC = C = CH-$

(c) Show the order of the radical-pushing capacities of the acetylenic groups where they exist.

9.15. (a) Shown below are two activated mono-forms of a 1, 5 - Cumulene -

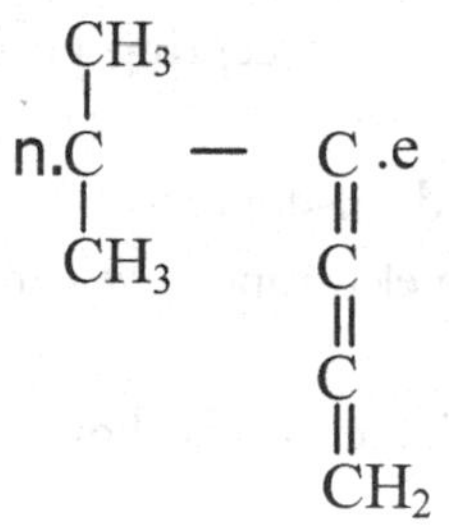

(i)   (ii)

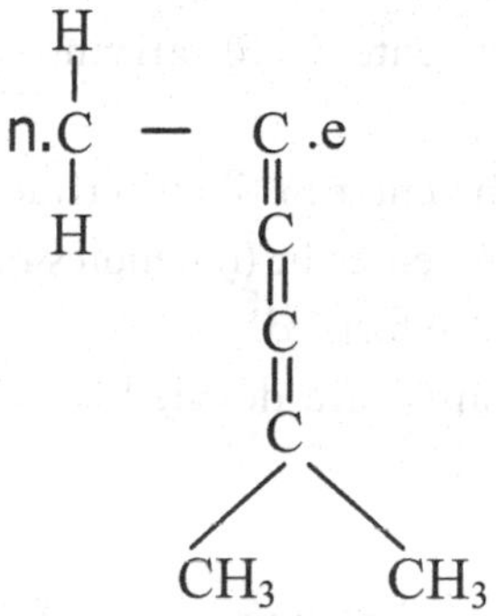

Which one molecularly rearranges? Explain.

(b) Shown below are two activated mono-forms of a 1,5 – Cumulene whether they exist or not –

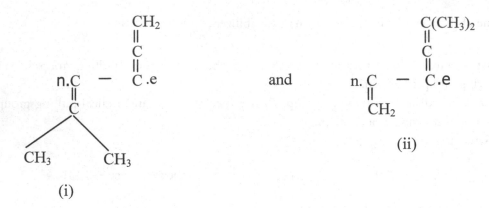

Which one molecularly rearranges? Explain. Explain why none of the above can exist.

(c) Show how the growing polymer chains of the two mono-forms shown below can be killed internally.

9.16. Shown below is the following reaction-.

$$
\begin{array}{ccc}
& CH_3 & \\
& | & \\
n.C & = & C.e \\
& & | \\
& & C(CH_3) \\
& & || \\
& & C \\
& & || \\
& & CH_2 \\
& (i) &
\end{array}
\qquad \text{and} \qquad
\begin{array}{ccc}
& CH_3 & \\
& | & \\
n.C & = & C.e \\
& & | \\
& & C \\
& & ||| \\
& & C \\
& & | \\
& & CH_2 \\
& & | \\
& & CH_3 \\
& & (ii)
\end{array}
$$

(i) What type of bond is carried by the oxygen center in (ii)?

(ii) Why is the $^{\ominus}OHSO_4$ counter-ion not isolatedly placed?

(iii) Is the dotted bond in (ii), polar, ionic or electrostatic in character? Explain.

(iv) Is the oxygen center in (i) carrying cationic donating radicals (two) or accepting a cation $H^{\oplus}$? Explain.

(v) Is the oxygen center in (i) an activation or functional center or radical-donor?

(vi) If the oxygen center in (i) denotes radicals to vacant orbital of an electropositive center, then what bonds are formed?

(vii) When radicals(s) are donated or withdrawn to form polar or ionic bonds, how many are involved-one or two?

9.17 (a) Shown below are three monomers of di-imines -

$$
\begin{array}{ccc}
CH_3 & & \\
| & & \\
N & = C = N \\
& & | \\
& & CH_3
\end{array}
\quad ; \quad
\begin{array}{ccc}
C_2H_5 & & \\
| & & \\
N & = C = N \\
& & | \\
& & CH_3
\end{array}
\quad ; \quad
\begin{array}{ccc}
H & & \\
| & & \\
N & = C = N \\
& & | \\
& & CH_3
\end{array}
$$

\qquad (A) \qquad\qquad\qquad (B) \qquad\qquad\qquad (C)

(i) Which is most unstable?

(ii) Does anyone exist as shown? If not, what products are favored by them?

(iii) Identify the order of their nucleophilicity.

(iv) Why does the route favored by them identify with their character?

(b) Compare (A) above in terms of nucleophilicity and routes favored, with the monomer shown below.

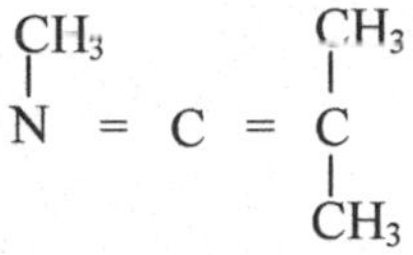

$$
\begin{array}{ccccc}
CH_3 & & & CH_3 & \\
| & & & | & \\
N & = & C & = & C \\
& & & | & \\
& & & CH_3 &
\end{array}
$$

9.18. (a) Shown below is a compound and a species -

$$
R - \overset{..}{\underset{..}{O}} - R \qquad\qquad - O - R
$$

$$
\text{(I)} \qquad\qquad\qquad \text{(II)}
$$

(i) What are the groups carried by them?

(ii) Which of them can donate radicals to form electrostatic bonds?

(iii) Which of them can donate radicals to form dative bonds?

(iv) Which of them is radical-donating and radical-withdrawing?

(b) Can you identify one of the major driving forces favoring the existence of

(i) A dative bond.

(ii) A polar or electrostatic bond.

Use examples to illustrate your understanding.

9.19. (a) Shown below are three monomers.

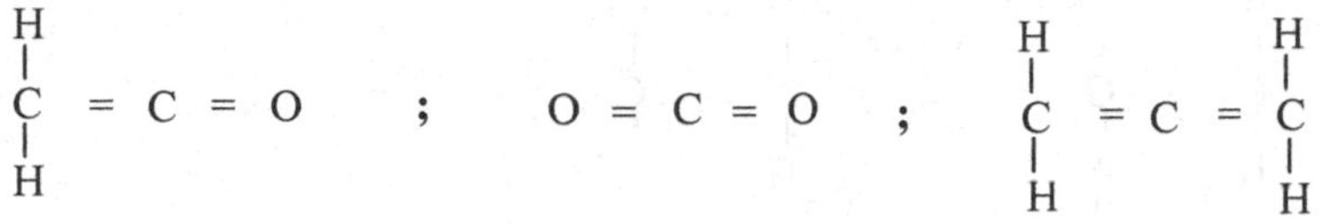

$$
\begin{array}{ccc}
H & & \\
| & & \\
C & = C = O & \quad ; \\
| & & \\
H & &
\end{array}
\qquad
O = C = O \quad ; \qquad
\begin{array}{ccc}
H & & H \\
| & & | \\
C & = C = & C \\
| & & | \\
H & & H
\end{array}
$$

$$
\text{(I)} \qquad\qquad\qquad\qquad \text{(II)} \qquad\qquad\qquad \text{(III)}
$$

(i) How many mono-forms exist for the monomers?

(ii) State the order of nucleophilicity of their activation centers in the absence of molecular rearrangement and indicate which can favor rearrangement.

(iii) Does molecular rearrangement increase the order the nucleophilicity of a monomer? Explain.

(iv) What are the routes favored by the monomers above when suppressed?

(b) Shown below are two monomers

$$
\begin{array}{ccc}
CH_3 & & \\
| & & \\
C & = C = O & \quad ; \\
| & & \\
CH_3 & &
\end{array}
\qquad
\begin{array}{ccc}
CF_3 & & \\
| & & \\
C & = C = O \\
| & & \\
CF_3 & &
\end{array}
$$

$$
\text{(I)} \qquad\qquad\qquad\qquad \text{(II)}
$$

856

(i) Identify the routes favored by their mono-forms.

(ii) Which one will favor the existence of alternating copolymers? Explain how the copolymers are obtained.

9.20. (a) Shown below are two activated monomers.

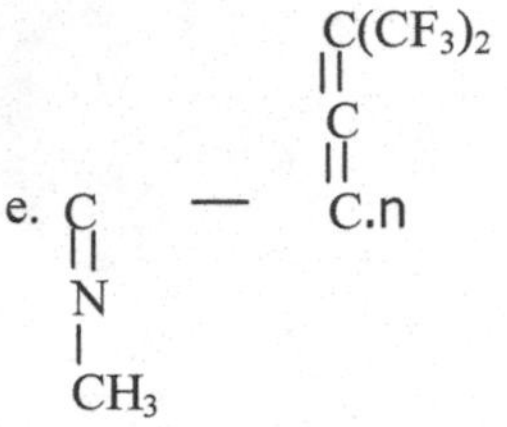

$$
\begin{array}{ll}
\text{e.} &
\begin{array}{c}
\quad\quad C(CF_3)_2 \\
\quad\quad \| \\
\quad\quad C \\
\quad\quad \| \\
C \;-\; C.n \\
\| \\
N \\
| \\
CH_3
\end{array}
\qquad\qquad
\text{e.} \;
\begin{array}{c}
\quad\quad CF_3 \\
\quad\quad | \\
C \;-\; N.nn \\
\| \\
N \\
| \\
CH_3
\end{array}
\end{array}
$$

(i) 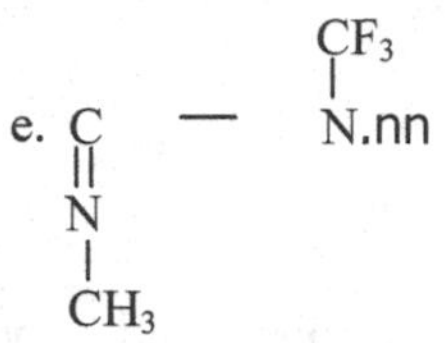   (ii)

(i) Show what routes are favored by them using free-radical initiators, in the absence and presence where possible of molecular rearrangements.

(ii) Show the routes favored using radical-paired initiators, in the absence and presence where possible of molecular rearrangements.

(iii) Of the substituent groups carried by (i) or (ii), which one has a greater capacity chargedly and radically?

(iv) Which of the mono-forms where possible can be internally resonance stabilized? Show what is obtained when the new mono-form obtained is molecularly rearranged?

(b) Distinguish between a polar charged resonance stabilization group and a non-charged resonance stabilization group.

9.21. (a) Shown below are three monomers.

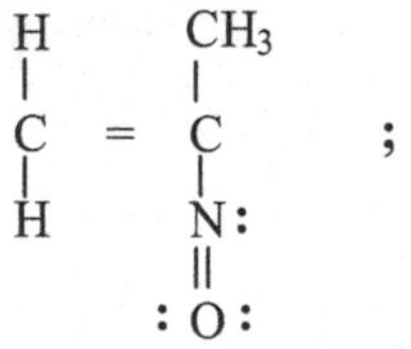

$$
\begin{array}{ccccc}
\begin{array}{c}
H \quad\; CH_3 \\
| \quad\;\; | \\
C = C \\
| \quad\;\; | \\
H \quad\; N: \\
\quad\quad \| \\
\quad\; : O:
\end{array}
& ; &
\begin{array}{c}
H \quad\; CH_3 \\
| \quad\;\; | \\
C = C \\
| \quad\;\; | \\
H \quad\; N: \\
\quad\quad \| \\
\quad\quad C \\
\quad\quad \| \\
\quad\quad O
\end{array}
& ; &
\begin{array}{c}
H \quad\; CH_3 \\
| \quad\;\; | \\
C = C \\
| \quad\;\; | \\
H \quad\; N^{\oplus} \\
\quad\; \diagup \;\; \diagdown \\
\quad O \quad\quad O^{\ominus}
\end{array}
\\
(I) & & (II) & & (III)
\end{array}
$$

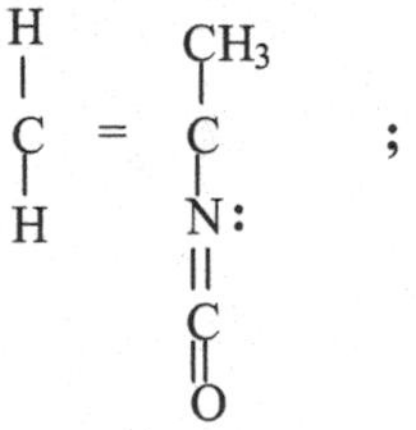

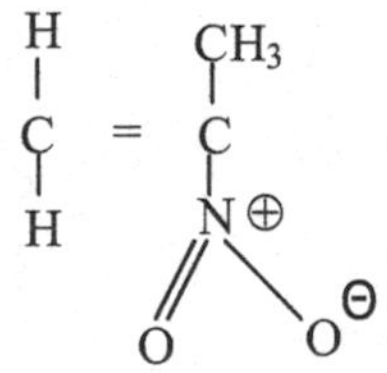

(i) Identify the types of substituted group carried by the monomers.

(ii) Which of them can favor charged polymerizations?

(iii) What are the characters of the monomers chargedly or radically? How does the radical character affect (i) and (iii)?

(b) Distinguish between full free-charged, and half free-charged monomers?

9.22. Show tabularly the characteristics of cumulenic ketenes, cumulenic isocyanates, cumulenic imines, cumulenic di-imines and cumulenic di-carbonyls.

APPENDICES

APPENDICES

Appendix I

THE HYDROCARBON FAMILY TREE

Abstract

This is not about the chemistry of hydrocarbons that which is so well known universally (as an ART and not a SCIENCE), but about something else based on current developments[1-7], that which is a New Science. The roots of this Engineering tree are composed of Carbon and Hydro-gen atoms and their elements. The inside of the stem of the Engineering tree is composed of *the first members of all the hydrocarbon families and the elements of H and C*. From them grew so many main branches which include only the first members of **Aliphatic compounds** comprising *the Alkylenes, Carbenes, Alkanes, Alkenes, Alkynes, Cumulenes, Conjugated compounds, "Car-bocyclic" or Alicyclic compounds,* and the **Alipharomatic compounds** comprising Benzene, Biphenyl, Condensed Benzene compounds, *Arylalkanes, Arylalkenes, Arylalkynes, and the Arylalicyclic compounds and corresponding cases for condensed Benzene compounds.* These main branches also carry with them the elements of H and C. These main branches gave birth to so many sub-branches which are the lower members of each family and some of the sub-branches gave birth to other sub-branches. Based on the foundation, all family members are interrelated. The first member of all the different families is uniquely different from all the other lower members in the same family that belongs to her. In the process of identifying this Engineering tree, the great significance of the Alkylenes and Carbenes families and the true structures of diamond, graphite, coke and much more were provided.

Keywords.

1. Hydrocarbons.
2. Activation Centers of the X types.
3. Molecular rearrangement.
4. Addition Polymerization.
5. Hydrogenation.

1.0 INTRODUCTION

The Chemistry of Hydrocarbons is not new to us, because universally the data which have been gathered on them are countless to comprehend. We have not only been able to identify most of their sources, but been able to synthetically produce many of them and use them far beyond comprehension. Even then, we have only just started. It is known that all the families of the hydrocarbons can be found in Nature, underground and in plants and animals. Known and unknown to us is that even the backbones of the body system of plants and animals are built with five- and six-membered ringed and cross-linked carbon/hydrogen compounds, with so many of them carrying hetero atoms either internally or externally located. Simple examples are the membranes, the chromosomes, the genes, the lungs, the heart, and much more.[8]

The hydrocarbon family tree in question here however are only carbon and hydrogen containing compounds with no hetero atoms such as O, N, and S. The carbon centers are fully hybridized when used. Therefore, in the absence of hetero atoms they are Non-polar/Non-ionic compounds.[5] When unsaturated and stable, they can only undergo chemical and Addition polymerization reactions either free-radically or chargedly (Only covalently). When saturated, they only undergo chemical reactions free-radically. They cannot carry ionic, polar or electro-static charges within themselves. They cannot also carry non-free-radicals. They can only generate Addition monomers which are all female in character, that is, carry only X-center(s).[1-4] They are one of the major sources of heat forms of energy in our world today.

The mechanisms by which we produce them synthetically are unknown, though we think we know them. Why and how for example methane is cracked to give carbon black and hydrogen while some of the other members of the same family are cracked to give alkenes, hydrogen, coke and much more are unknown, though we think we know. What hydrogen catalysts do, we don't know, but we know that there exists such catalysts. Why acetylene the first member of the alkynes is unstable up to the point where it cannot be used as a monomer in the absence of Stabilizers, but most of the other members in the family are stable and can be used as monomers are unknown, though we think we know based on experimental data. Why hydro-carbons cannot mix with water is unknown, but we think we know. How, many of them combust to produce different kinds of heat are unknown, yet some of us think we know. How, many of them are oxidized to produce different kinds of products are unknown, yet we know that oxidizing agents such as fuming nitric acid, vanadium pentoxide, potassium permanganate, hydrogen peroxide exist. Why allene the first member of the family of cumulenes is unstable and rearranges to methyl acetylene or some monomer rearrange to give another more stable mono-mer are unknown. Instead, we come with misguided or unimaginable rules (such as Markov-nikov's rule)[9] to explain them! As simple as it may seem, we do not yet know what an atom is, but we think we know so much about it. On the surface i.e. outside the nucleus, what are hydrogen and carbon atoms and their elements? These we must know before we can begin to see what hydrocarbons are based on the new foundations.[1-7]

2.0 The Nature of Hydrogen and Carbon atoms, their elements and molecules.

Hydrogen atom is a gaseous metal. As an electropositive atom, like its other members in the family, it carries one electro-free- radical[1] (Not an electron) as shown below in the last shell, its boundary.

$$H \cdot e \; ; \; Li \cdot e \; ; \; Na \cdot e \; ; \; K \cdot e \; ; \; Rb \cdot e \; ; \; Cs \cdot e \; ; \; Fr \cdot e \quad [e \equiv \text{electro-free-radical}]$$

The atoms of Group IA

$$H \cdot n \qquad [n \equiv nucleo - free - radical]$$

The elements of Group IA

Ap1.1

It is ionic in character just like its other lower members in Group IA of the Periodic Table. As the first member, it has very peculiar characters. It is the only one that can carry a nucleo-free-radical[1] (Not an anion) as shown above. It is the only one in the Group that can form a molecule as shown below, because the last shell of the H atoms are full in H_2, unlike the others such as in Li_2.

$$H{\bullet}e \quad + \quad H{\bullet}e \xrightarrow{\hspace{3cm}} H_2 \hspace{3cm} Ap1.2a$$

$$Li{\bullet}e \quad + \quad Li{\bullet}e \xrightarrow{\hspace{3cm}} Li_2$$

$$\text{Incomplete last shell} \hspace{3cm} Ap1.2b$$

$$H_2 \underset{\textit{Existence of } H_2}{\overset{\textit{Equilibrium State of}}{\rightleftharpoons}} \quad H.e \quad + \quad H.n$$

$$\text{[The Atom]} \hspace{2cm} \text{[The Hydride]}$$

$$Ap1.2c$$

H.n is what is called the Hydride. None of them is this Group can carry negative charges whether ionic, covalent, or polar but electrostatic such as in Grignard's reagent[3]. They can only carry positive charges of the ionic, covalent, electrostatic or polar types. All these are new to present day Science. If these are not known, then how can we move forward? Without the electro-free-radical on the hydrogen center, no combustion and oxidation can take place. Without the electro-free-radical on the hydrogen center, there are no acids and organic alcohols. Without hydrogen, enzymes cannot exist. Without hydrogen, there will be no hydrocarbons. Without hydrogen, there will be no water and life will not be complete or indeed exist. Without hydrogen, other atoms cannot exist. When the only radical in H, unlike other members in the same Group, is removed, what is left behind is the Nucleus.

Carbon, Sulfur and iodine form the boundary between metals and non-metals, because their electronegativity is 2.5 (if the spectrum of electronegativity of 0 to 5 is used). While iodine is a metal, carbon and sulfur have metals amongst their elements. In Group IVb, Carbon the first member, is quite unique in character from the other members- Silicon (Si), Germanium (Ge), Tin (Sn), and lead (Pb) which are metallic to different levels. Carbon is both metallic and non-metallic in character. The elements of carbon atom are Activated carbon, Carbon Black, Coke, activated Carbon Black, Charcoal and Coal[6]. These are shown below.

$$\textit{Ground State Carbon} - \overset{nn}{\underset{}{:}}\overset{\cdot}{C}\cdot nn \quad ; \quad \textit{Excited Carbon} - \overset{n}{\underset{n}{n\cdot\overset{\cdot}{C}\cdot n}} \; (\textit{Charcaol})$$

$$(A) - \textit{The Atom} \hspace{4cm} (B)$$

$$\textit{Ground State Carbon} - \overset{nn}{\underset{}{:}}\overset{\cdot}{C}\cdot nn \xrightarrow{\textit{Activation}} \overset{n}{\underset{n}{e\cdot\overset{\cdot}{C}\cdot n}}$$

$$(C) - (\textit{Activated Carbon})$$

$$Ap1.3$$

$$Carbon\ Black - \overset{\overset{nn}{\cdot}}{:}C \cdot en \underset{}{\overset{Activation}{\rightleftharpoons}} \qquad e \cdot \overset{\overset{n}{\cdot}}{C} \cdot e$$

$$(D) \qquad\qquad\qquad\qquad\qquad\qquad \underset{n}{}$$

$$(E) - (Activated\ Carbon\ Black)$$

$$\overset{e}{Coal - e \cdot \overset{\cdot}{C} \cdot e}(Semi-metallic\ Carbon)\ ;\quad Coke - \quad \overset{en}{:\overset{\cdot}{C} \cdot en}\ (Metallic\ Carbon)$$

$$\underset{n}{}\quad (Activated\ Coke)$$

$$(F) \qquad\qquad\qquad\qquad\qquad (G) \qquad\qquad\qquad Ap1.4$$

[nn ≡ nucleo-non-free-radical ; en ≡ electro-non-free-radical]

The carbon center alone cannot carry Real charges; hence these structures carry radicals only. Ground-state carbon (A), Excited carbon or Charcoal (B), Activated carbon (C) and Carbon black (D) are some of the ones that can be used to produce aliphatic hydrocarbons. These are the different elements of carbon.

(D), (E) and (F) are the core units for molecular carbon compounds which include Diamonds, Graphite, Coals, and Cokes. All of them have limited amounts of impurities such as O, S, and N inside and outside, covalently bonded in their structures. Diamond contains only very small amount of N.[10] The same applies to Graphite. Coal contains all the three including H.[11, 12] Indeed Coal unlike crude oil is a compact stratified mass derived from plants that have suffered partial decay and have been subjected to various degrees of heat and pressure. Most normal bonded coals are believed to have originated in peat swamps. The substances *peat, lignite, soft and bituminous coal, and anthracite or hard coal*s are progressive stages of metamorphosis in which the ratio of the amount of carbon to the amount of other elements increases. Coke may contain similar components as in coal but to a far lesser extent or may not contain them, because soft or bituminous coal when heated to a sufficiently high temperature (350 -1000° C) in the absence of air (Carbonization or Destructive Distillation of Coal) volatile products are formed, and a residue of *impure carbon* (called Coke) remains. When the volatile products cool at ordinary temperature, a portion condenses to a black viscous liquid known as Coal tar. One ton of coal yields about 1500 pounds of *Coke*, 8 gallons of *Coal Tar*, and 10,000 cubic feet of *Coal gas.* The Coal gas consists chiefly of hydrogen and methane in about equal volumes, along with some carbon monoxide, ethane, ethylene, benzene, carbon dioxide, oxygen, and nitrogen, and smaller amounts of cyclo-pentadiene, toluene, naphthalene, water vapor, ammonia, hydrogen sulfide, hydrogen cyanide, cyanogen, and nitric oxide. Over 215 individual compounds have been isolated from Coal tar, the first and largest fraction being naphthalene.[13] Nevertheless, it is important to note that coals of different ages have different chemical compo-sitions, and therefore different structures. Even within a certain age group or rank of coal, such as lignite or bituminous coals, the structure may vary depending on the environment in which a particular coal was formed.[11] Coal and Coke are porous while diamond and graphite are non-porous. *Coke unlike coal is a smokeless industrial and household fuel.* Diamond can burn if subjected to a high temperature in the presence of oxygen.[10] Then, the question is how do these molecules of carbon burn completely in the absence of hydrogen covalently bonded to a carbon center? Otherwise they must carry electro-free-radical(s) in some of their carbon centers at the corners. Does coke actually burn?

Diamond has an extremely low thermal expansion, and chemically inert with respect to most acids and alkalis.[10] Based on the structures provided for coal[11,12], it is resistant to acids, but not alkalis.[14] Diamond is an exceptional thermal conductor- 4 times better than copper.[10] *While the carbon centers in diamond is sp³ hybridized, those in graphite are sp² hybridized.* The diamond structure is cubic and the crystals are highly symmetric with cubic space groupings. The crystal structures of graphite are described by hexagonal lattices.[15] Just as coke is related to coal, so also is diamond to graphite. Synthetic industrial diamonds are produced using High Pressure High Temperature (HPHT) Synthesis method. In the HPHT synthesis, **graphite and a metallic catalyst** are placed in a hydraulic press under high temperatures and pressures. Over a period of a few hours the graphite converts to diamond too flawed to be used as gemstones.[10] They are however useful as edges on cutting tools and drill-bits. While Diamonds do not conduct electricity well (some are however semi-conductors),[10] graphite has the ability to conduct electri-city.[15] **Catalysts made Coke** are said to be known to have some aromatic carbons and known to *dehydrogenate* to form **graphite type aromatic** condensate. About 60% of **coke** was **graphite** type.[16] In the investigation of Micro Carbon Structure of Coke and Semi-coke using Hand-picking Method, it was revealed that high quality coke had uniform carbon structure, but the structures of inferior coke were heterogeneous. It was recognized that crystallinity of coke is influenced by the kind of mineral matter and its distribution in addition to type of organic matter such as **soft carbon and hard carbon.**[17] Without any doubt, an example of the soft carbon is (D) of Equation Ap1.4. All the ones shown in Equation Ap1.3 are also soft. The hard carbons are (E), (F) and (G) of Equation Ap1.4.

The structures of diamond, graphite and coal seem to be well known.[15,18] Since diamond with a very regular structure can combust, cannot conduct electricity and the carbon centers are sp³, clearly indicates that molecular carbon diamond is saturated with tetravalent carbon centers internally located and some trivalent carbon centers externally located at corners as shown below in Figure Ap1.1

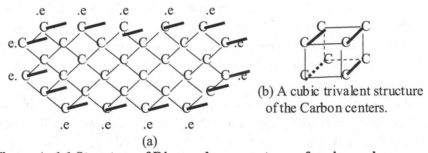

(a)

(b) A cubic trivalent structure of the Carbon centers.

Figure Ap1.1 <u>Structure of Diamond on an outer surface in one layer.</u>

The element (E) of Equation Ap1.4 or a combination of (C) and (F) or even best suited (E) and (F) can be used to obtain this molecule. These are the coke or activated Black carbon element, metallic coal and Activated Carbon. Worthy of note are the presence of electro-free radicals along the boundaries of the diamond. These can do no harm because it is not mobile, since diamond is a solid. It cannot leave the diamond whose molecular weight is very large.

The fact that graphite is sp² hybridized, can conduct electricity, hexagonal in structure, clearly indicates the presence of six-membered condensed ringed compounds arranged in different layers not connected to themselves as shown below in Figure Ap1.2.

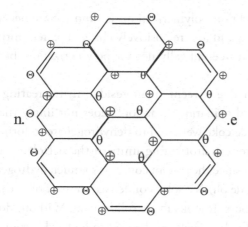

Figure Ap1.2. <u>Structure of Graphite for a single layer with polar bonds on the carbon Centers.</u>

Here only (D) or combination of (C) and (F) can be used to obtain the molecule above. Recall from above that close to amorphous diamond can be obtained from graphite in the presence of a metallic catalyst, high temperature and high pressure (HPHT)[10]. Worthy of note in the structure above are the presence of polar bonds between two carbon centers both inside and along the boundaries just like the bonds between some metals, since metals do not carry double bonds. For without them, the rings cannot exist. Also present are just two free-radicals along the boundaries. It is in view of their presence and the polar bonds in particular, that graphite is able to conduct electricity. One has already started to show how current flows in fluids using the negative type of electrostatic bonds[1-6]. Polar bonds like the electrostatic bonds are imaginary bonds[3]. From the structure, one can start to see how it transforms to diamond, through the middle of the hexagonal structure. Graphite is indeed hexagonal in structure. Electro-radicals are what jumps from hole to hole when current flows. It goes into the system through the negative end. Hence, one cannot use electrostatic bond of the positive type to conduct electricity in fluids.

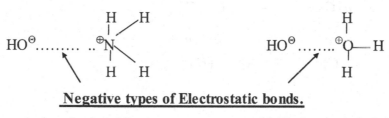

<u>Negative types of Electrostatic bonds.</u>

(a) **<u>Ammonium hydroxide</u>**   (b) **<u>Hydroxonium hydroxide</u>**   Ap1.5

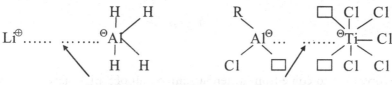

<u>Positive type of Electrostatic bonds.</u>

(c) **<u>Lithium Aluminum hydride (LiAlH$_4$)</u>**   (d) **<u>TiCl$_4$/AlR$_3$ Combination at ratio 2 : 1</u>**.

Ap1.6

Some of these can be used as initiators for Addition polymerization. (a) and (b) are negatively charged paired initiators since that is the active center. (c) and (d) are positively charged paired initiators in which (c) cannot be used cationically, but only nucleo-free-radically not for polymerization, but for chemical reactions..

The fact that the structure of coke from five percent spent residue hydrotreating catalysts was investigated based on "ruthenium ion catalyzed oxidation" reaction[15] does not imply that coke carries hydrogen atoms on them. In fact, catalysts made coke were said to dehydrogenate to form graphite type aromatic condensate.[15] Without doubt coke from hydrotreating units in the petroleum industry must contain many impurities in particular hydrogen since the system contains so much hydrogen as one of the major products. Coke from coal (Not from crude oil) is a solid condensed mass of only carbon. It burns red hot without smoke and release of water vapors. It looks like a solid metal. Without doubt therefore, (F) or (G) must be one of the major sources of coke which can be said to be hard carbon[17]. When used alone the followings are obtained.

$$x \quad \{e \bullet \overset{\overset{\textstyle e}{\bullet}}{\underset{\underset{\textstyle n}{\bullet}}{C}} \bullet e\} \quad \longrightarrow \quad \overset{\overset{\textstyle e}{\bullet}}{C} = \overset{\overset{\textstyle e}{\bullet}}{\underset{\underset{\textstyle e}{\bullet}}{C}} - (\overset{\overset{\textstyle e}{\bullet}}{\underset{\underset{\textstyle e}{\bullet}}{C}})_{x-4} - \overset{\overset{\textstyle e}{\bullet}}{\underset{\underset{\textstyle e}{\bullet}}{C}} - \overset{\overset{\textstyle e}{\bullet}}{\underset{\underset{\textstyle e}{\bullet}}{C}} \bullet e \qquad \text{Ap1.7}$$

(A) Transparent thin hard carbon rod.

To add to (A), one has to bring in nucleo-free-radical carrying carbon centers. Linear branches standing alone may also be formed making it look almost identical to diamond without cubes. This is one of the origins of Nano-technology. When used with activated carbon made from ground state carbon, the followings are obtained.

$$e \bullet \overset{\overset{\textstyle e}{\bullet}}{\underset{\underset{\textstyle n}{\bullet}}{C}} \bullet e \quad + \quad n \bullet \overset{\overset{\textstyle n}{\bullet}}{\underset{\underset{\textstyle e}{\bullet}}{C}} \bullet n \quad \longrightarrow \quad \overset{\overset{\textstyle e}{\bullet} \quad \overset{\textstyle e}{\bullet}}{C = C}_{\underset{\underset{\textstyle n \quad n}{\bullet \quad \bullet}}{}} \qquad \text{Ap1.8}$$

$$\qquad (F) \qquad\qquad (C) \qquad\qquad\qquad (H)$$

$$e \bullet \overset{\overset{\textstyle n}{\bullet}}{\underset{\underset{\textstyle n}{\bullet}}{C}} \bullet e \quad + \quad n \bullet \overset{\overset{\textstyle e}{\bullet}}{\underset{\underset{\textstyle e}{\bullet}}{C}} \bullet n \quad \longrightarrow \quad \overset{\overset{\textstyle e}{\bullet} \quad \overset{\textstyle e}{\bullet}}{C = C}_{\underset{\underset{\textstyle n \quad n}{\bullet \quad \bullet}}{}} \qquad \text{Ap1.9}$$

$$\qquad (D) \qquad\qquad (D) \qquad\qquad\qquad (H)$$

Indeed, (H) can be observed to come from different sources, all of which have been used for diamond and graphite. Thus, it can be observed that (F), (D) and (C) are all interrelated. It is the same (H) that is the source of different types of coals which are bed rock of coke. The structures of the different types of coal are shown below in Figure Ap1.3.

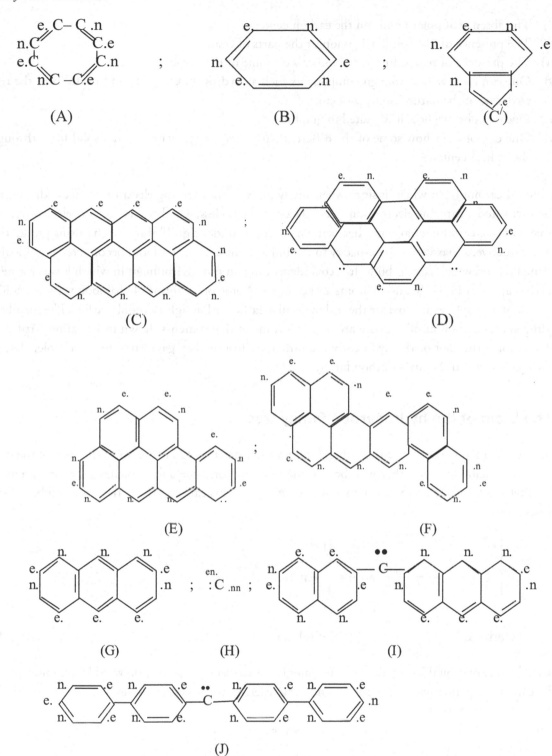

Figure Ap1.3. Structures of some of the different forms of Coal.

Worthy of note are the followings-

i)   The presence of electro- and nucleo-free-radicals along the boundaries of the different parts of coke. They cannot form a bond, because the Maximum Required Strain Energy (Max. R S E)[1-6] of the ring will be exceeded making it impossible for the ring to exist.

ii) The absence of polar bonds on the carbon centers.

iii) The presence of Carbon Black amongst the parts of coal.

iv) The presence of paired unbonded radicals on some of the carbon centers.

v) One can observe how foreign components such as hydrogen, O, S, N add to it to form the real coal, one of the natural sources of coke.

vi) One can observe how it is related to graphite.

vii) One can observe how some of the different parts merge together to form a solid mass through the radical centers.

Coke indeed cannot burn when lit since when singly placed, it is carrying electro-non-free-radicals and O_2 when activated is carrying electro-non-free-radicals. Nevertheless, whether it is carrying electro-free- or non-free-radicals, is immaterial, since heat can be generated under Equilibrium mechanisms productive or non-productive. It has very large capacity to retain large amounts of heat and become red hot. Coal is the foundation on which coke is built. It is coal (depending on the environment in which it was formed) that is the carrier of H, O, S, and N atoms and elements. Some other forms of coal look very much like coke. Diamond, graphite and coal are the real molecular carbons. Though all the above look like members of hydrogen and carbon families, they are not. It is some of the elements shown in Equations Ap1.2 to Ap1.4 that form the root of the hydrocarbon family tree. Just as they gave birth to their molecules, so also they gave birth to the hydrocarbon family tree.

2.1 The Stem of the hydrocarbon family tree.

All the first members of the hydrocarbon families and elements of H and C form the stem of the tree, because from them emerged other members of the tree that form the sub-branches. The first member of the Alkylenes family is methylene which when not activated is called carbene the first member of the Carbenes family.

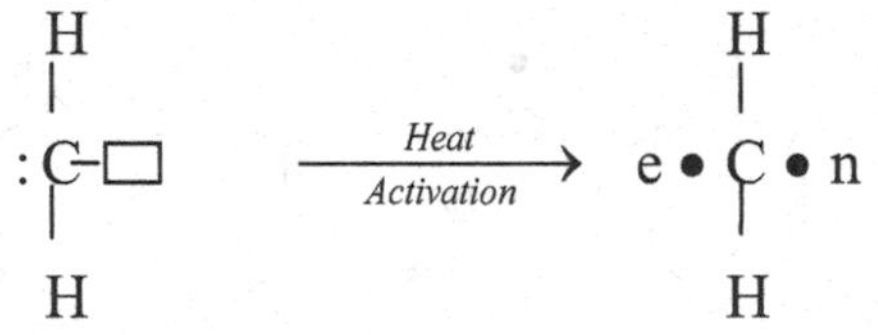

$$\underset{\text{Carbene}}{\overset{\overset{\displaystyle H}{|}}{\underset{\overset{\displaystyle |}{H}}{:C-\square}}} \qquad \xrightarrow[\text{Activation}]{\text{Heat}} \qquad \underset{\text{Methylene}}{\overset{\overset{\displaystyle H}{|}}{\underset{\overset{\displaystyle |}{H}}{e\bullet C\bullet n}}} \qquad\qquad\qquad\qquad \text{Ap1.10}$$

Without the vacant orbital left on the carbon center in the carbene, no activation would have taken place. Unlike carbene, the methylene is a poison and very reactive in the gaseous state. These can readily be obtained using for example (D) and (E) of Equation 1.4 in the presence of the element of hydrogen and its atom (both to be called elements of H) as shown below.

$$\underset{\text{(Carbon Black)}}{\overset{\overset{\displaystyle en}{\overset{\displaystyle \bullet}{}}}{:C\bullet nn}} \;+\; \underset{\text{(Atom)}}{H\bullet e} \;+\; \underset{\text{(Hydride)}}{H\bullet n} \;\longrightarrow\; \underset{\text{(Carbene)}}{\overset{\overset{\displaystyle H}{|}}{:C-H}} \qquad\qquad \text{Ap1.11}$$

$$e \cdot C \cdot n \ (\overset{e}{\underset{n}{\cdot}}) \quad + \quad H \cdot e \quad + \quad H \cdot n \quad \longrightarrow \quad e \cdot \overset{\displaystyle H}{\underset{\displaystyle H}{C}} \cdot n$$

(Act. Carbon Black)   (Atom)   (Hydride)   (Methylene)   Ap1.12

The mechanism of Addition is Equilibrium mechanism.[1] Hydrogen catalysts or surface chemistry must be required for the reactions above in order to make it a productive stage.

From the carbene or methylene, methane the first member of the alkane family is obtained as follows using the element and atom of hydrogen.

$$\overset{\displaystyle H}{:C-H} \quad + \quad H \cdot e \quad + \quad H \cdot n \quad \xrightarrow{\text{Activation}} \quad H - \overset{\displaystyle H}{\underset{\displaystyle H}{C}} - H$$

(Carbene)   (Atom)   (Hydride)   (Methane)   Ap1.13

$$e \cdot \overset{\displaystyle H}{\underset{\displaystyle H}{C}} \cdot n \quad + \quad H \cdot e \quad + \quad H \cdot n \quad \longrightarrow \quad H - \overset{\displaystyle H}{\underset{\displaystyle H}{C}} - H$$

(Methylene)   (Atom)   (Hydride)   (Methane)   Ap1.14

From the methylene and the carbene, the first member of the alkene family is obtained as follows.

$$n \cdot \overset{\displaystyle H}{\underset{\displaystyle H}{C}} \cdot e \quad + \quad \overset{\displaystyle H}{:C-H} \quad \xrightarrow{\text{Activation}} \quad n \cdot \overset{\displaystyle H}{\underset{\displaystyle H}{C}} - \overset{\displaystyle H}{\underset{\displaystyle H}{C}} \cdot e$$

$$\xrightarrow{\text{Deactivation}} \quad \overset{\displaystyle H}{\underset{\displaystyle H}{C}} = \overset{\displaystyle H}{\underset{\displaystyle H}{C}}$$

(A) (Ethene)   Ap1.15

Note that the carbene would have been the first member to be called Methene. But since the Carbene which has an activation center different from the others in the family, hence it is not the first member. The activation center of all the others is the π-bond, while the activation center of the carbene is the vacant orbital. *The carbene is the first member of a different family of hydrocarbons.* Note that (A) above has been truly identified as ethene and not ethylene as still used in many schools of taught including the Industries, because ethylene is next to methylene in the Alkylene family as will shortly be shown and indeed becoming obvious.

The first member of the alkyne family can be obtained as follows using two moles of activated carbon black and hydrogen atom and its element.

$$n \bullet \overset{\overset{e}{\bullet}}{C} \bullet e \;+\; n \bullet \overset{\overset{n}{\bullet}}{C} \bullet e \;\longrightarrow\; n \bullet C \equiv C \bullet e \;\xrightarrow[+H \bullet n]{+H \bullet e}\; H - C \equiv C - H$$

(*Acetylene*)   Ap1.16

In a similar fashion the first member of the cumulene family is obtained as follows using two moles of methylene and carbon black. The same applies for conjugated dienes in which the first member is butadiene. This can be obtained using acetylene (Ethyne) in place of carbon black.

$$e \bullet \overset{\overset{H}{|}}{\underset{\underset{H}{|}}{C}} \bullet n \;+\; :\overset{\bullet}{C} \bullet nn \;\xrightarrow{Activation}\; \overset{\overset{H}{|}}{C} = \overset{\overset{e}{\bullet}}{\underset{\underset{n}{\bullet}}{\underset{|}{C}}} \;\xrightarrow{+ \, Methylene}\; \overset{\overset{H}{|}}{\underset{\underset{H}{|}}{C}} = C = \overset{\overset{H}{|}}{\underset{\underset{H}{|}}{C}}$$

(*Allene*)   Ap1.17

$$n \bullet \overset{\overset{H}{|}}{\underset{\underset{H}{|}}{C}} \bullet e \;+\; n \bullet \overset{\overset{H}{|}}{C} = C \bullet e \;\xrightarrow[and \; activated]{C_2H_2 \; sup \; pressed}\; \overset{\overset{H}{|}}{C} = \overset{\overset{H}{|}}{C} - \overset{\overset{n}{\bullet}}{C} \bullet e \;\xrightarrow{+ \, Methylene}\; \overset{\overset{H}{|}}{C} = \overset{\overset{H}{|}}{C} - \overset{\overset{H}{|}}{C} = \overset{\overset{H}{|}}{C}$$

(*Butadiene*)   Ap1.18

To obtain 1, 3, 5-triene, two moles of acetylene and two moles of methylene will be required. Indeed acetylene tetramerizes to cyclooctatetraene and butadiene dimerizes to give 1,5-Cyclo-octadiene.

H₂C — CH₂ HC CH HC CH H₂C —CH₂	HC═ CH HC CH HC CH HC═ CH
2 Butadienes	4 Acetylenes (Ethyne)
1,5-Cyclooctadiene	Cyclooctatetraene Ap1.19

Cyclooctatetraene (tub-shaped) is not like benzene (planar) which is aromatic. Indeed cyclo-octatetraene is unstable, since it readily rearranges to give styrene.

In the same manner, the first members of the cycloalkane family are obtained as follows using either ethene and methylene or three moles of methylene.

$$\overset{\overset{H \; H}{| \; |}}{\underset{\underset{H \; H}{| \; |}}{C = C}} \;+\; e \bullet \overset{\overset{H}{|}}{\underset{\underset{H}{|}}{C}} \bullet n \;\xrightarrow{Activation}\; e \bullet \overset{\overset{H \; H \; H}{| \; | \; |}}{\underset{\underset{H \; H \; H}{| \; | \; |}}{C - C - C}} \bullet n \;\longrightarrow\; H - \overset{\overset{H \; H}{| \; |}}{C - C} - H$$

(*Cyclopropane*)   *Ap*1.20

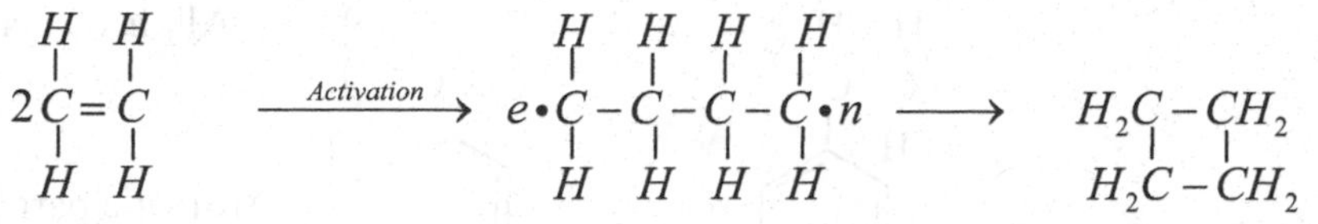

$$2 \overset{\overset{H}{|}}{\underset{\underset{H}{|}}{C}} = \overset{\overset{H}{|}}{\underset{\underset{H}{|}}{C}} \quad \xrightarrow{\text{Activation}} \quad e \bullet \overset{\overset{H}{|}}{\underset{\underset{H}{|}}{C}} - \overset{\overset{H}{|}}{\underset{\underset{H}{|}}{C}} - \overset{\overset{H}{|}}{\underset{\underset{H}{|}}{C}} - \overset{\overset{H}{|}}{\underset{\underset{H}{|}}{C}} \bullet n \quad \longrightarrow \quad \begin{matrix} H_2C - CH_2 \\ | \qquad | \\ H_2C - CH_2 \end{matrix}$$

$$(Cyclobu\tan e) \qquad\qquad Ap1.21$$

Cyclopentanes, cyclohexane up to eight-membered rings are obtained in a similar manner. The first members of Cycloalkene family are obtained as follows using acetylene and methylene.

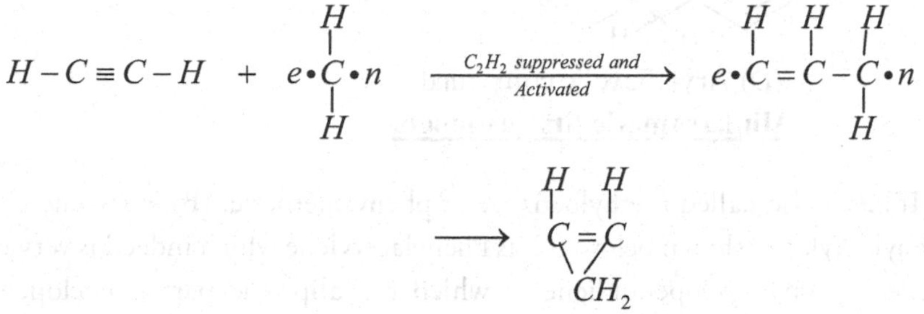

$$H - C \equiv C - H \quad + \quad e \bullet \overset{\overset{H}{|}}{\underset{\underset{H}{|}}{C}} \bullet n \quad \xrightarrow[\text{Activated}]{C_2H_2 \text{ suppressed and}} \quad e \bullet \overset{\overset{H}{|}}{C} = \overset{\overset{H}{|}}{C} - \overset{\overset{H}{|}}{\underset{\underset{H}{|}}{C}} \bullet n$$

$$\longrightarrow \quad \begin{matrix} \overset{H}{|} \quad \overset{H}{|} \\ C = C \\ \diagdown \quad \diagup \\ CH_2 \end{matrix}$$

$$(Cyclopropene)$$

$$Ap1.22$$

For cyclobutene, the followings are similarly obtained.

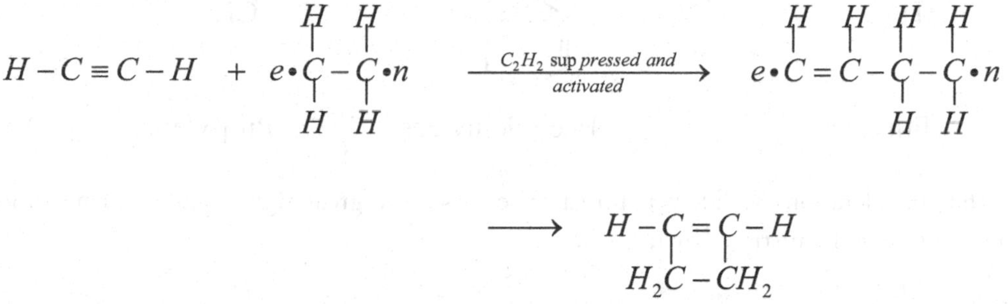

$$H - C \equiv C - H \quad + \quad e \bullet \overset{\overset{H}{|}}{\underset{\underset{H}{|}}{C}} - \overset{\overset{H}{|}}{\underset{\underset{H}{|}}{C}} \bullet n \quad \xrightarrow[\text{activated}]{C_2H_2 \text{ sup pressed and}} \quad e \bullet \overset{\overset{H}{|}}{C} = \overset{\overset{H}{|}}{C} - \overset{\overset{H}{|}}{\underset{\underset{H}{|}}{C}} - \overset{\overset{H}{|}}{\underset{\underset{H}{|}}{C}} \bullet n$$

$$\longrightarrow \quad \begin{matrix} H - C = C - H \\ | \qquad | \\ H_2C - CH_2 \end{matrix}$$

$$(Cyclobutene)$$

$$Ap1.23$$

Notice that so far the reactions have revolved around the first members of each family and the elements of carbon and hydrogen and only the aliphatic part of hydrocarbons have been considered. Notice so far that we have not cracked or done anything to any type of hydrocarbon. It is like a nano building network system. We have only been activating.

For the first member of the aromatic hydrocarbons, one should know what to expect. Three activated molecules of acetylene, when suppressed give benzene. After benzene, the other main family which also forms a main branch is naphthalene followed by anthracene which also is another main branch. The last two are condensed aromatic compounds which can similarly be built based on the foundations laid so far. As a matter of fact there are indeed three parts-Aliphatic, Aromatic and Alipharomatic, since Benzene, Naphthalene, Anthracene and other condensed Benzene rings must have sub-branches i.e. families of their own. Alipharomatic is a hybrid of the Aliphatic and Aromatic, examples of which are shown below.

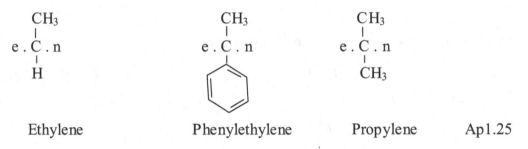

(A) Arylalkane   (B) Arylalkene   (C) Arylalkyne

(D) Arylalicyclic compound
Alipharomatic first members   Ap1.24

(A), the first member is Toluene also called methylbenzene or phenylmethane. (B) is styrene also called phenylethene and not phenylethylene[19] shown below. (C) is Phenylacetylene which indeed is very unstable. (D) is **Fluorene** also called dibenzocyclopentadiene in which the aliphatic part is cyclopentadiene. Ethylene is of the Alkylene family. The next member after ethylene is propylene.

$$CH_3$$
$$|$$
$$e \cdot C \cdot n$$
$$|$$
$$H$$

$$CH_3$$
$$|$$
$$e \cdot C \cdot n$$

$$CH_3$$
$$|$$
$$e \cdot C \cdot n$$
$$|$$
$$CH_3$$

Ethylene   Phenylethylene   Propylene   Ap1.25

From all the considerations so far, it is no surprise to see the great significance and mean-ing of methylene in our present chemistry of hydrocarbons.

Alkylenes' Family: C_nH_{2n} ; CH_2-Methylene, C_2H_4-Ethylene, C_3H_6-Propylene, ...
Alkanes' Family: C_nH_{2n+2}; CH_4,-Methane, C_2H_6,-Ethane, C_3H_8-Propane, C_4H_{10}-Butanes, ...
Alkenes' Family: $C_{n+1}H_{2(n+1)}$; C_2H_4-Ethene, C_3H_6-Propene, C_4H_8-Butenes,..
Alkynes' Family: $C_{n+1}H_{2n}$; C_2H_2 -Acetylene (Ethyne), C_3H_4-Methylacetylene (Propyne), C_4H_6-Dimethylacetylene (Butynes), C_5H_8-Ethylmethylacetylene (Pentynes),...........
Carbene's Family: C_nH_{2n}; CH_2- Carbene, C_2H_4- Methylcarbene, C_3H_6-Dimethylcarbene,.....
Cycloalkanes' Family: $C_{n+2}H_{2(n+2)}$; C_3H_6-Cyclopropane, C_4H_8-Cyclobutane, C_5H_{10}-Cyclopentane,...
Cyclopropanes' Family: $C_{n+2}H_{2(n+2)}$; C_3H_6-Cyclopropane, C_4H_8-Cyclomethylpropane, C_5H_{10}-Two isomers-1,1-Dimethylpropane and 1,2-Dimethylpropane,......
Cycloalkenes' Family: $C_{n+2}H_{2n+2}$; C_3H_4-Cyclopropene, C_4H_6-Cyclobutene, C_5H_8-Cyclopentene,....
Etc.

The significance of the methylene group has long been recognized.[20] Pyrolysis or photolysis of diazomethane is said to give nitrogen and two reactive species of composition CH_2, one of which, designated as methylene, has two unpaired *electrons* (triplet state), whereas the other, designated as carbene, has a pair of *electrons* (singlet state) and an empty orbital. Carbene in an excited state appears to be the initial product, which decays to the more stable methylene.[21]

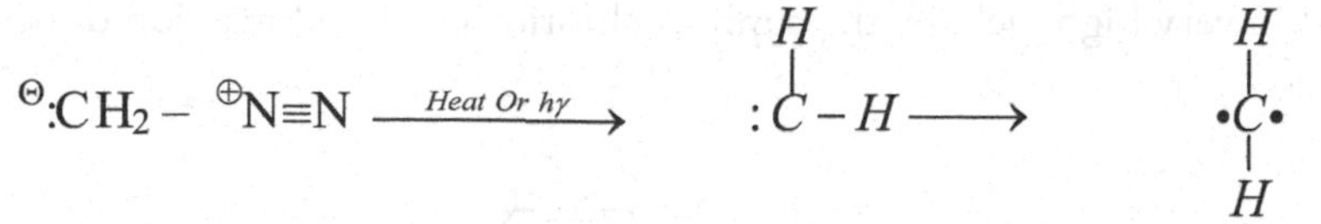

$$^{\ominus}:CH_2 - ^{\oplus}N \equiv N \xrightarrow{\text{Heat Or } h\gamma} \quad \overset{H}{\underset{}{:C-H}} \longrightarrow \quad \overset{H}{\underset{H}{\cdot C \cdot}} \qquad \text{Ap1.26}$$

First of all like in all other textbooks, there are so many things wrong with the last statement above as well as the equation. Though it looks meaningful, it is just an ART and not a SCIENCE. Though the real name is diazomethylene, one will continue to use diazomethane [H_3C-N=N-H] for now. The radicals are not identified. The resonance stabilized character of diazomethane is not identified. Carbene cannot be excited but activated. There is nothing like a singlet or triplet state. The vacant or empty orbital is not shown. Carbene is more stable than methylene and not the opposite. It is also not true that they cannot be readily isolate. However, they can serve as intermediates in reactions of diazomethane induced by heat or light such as in the following reactions.

1.

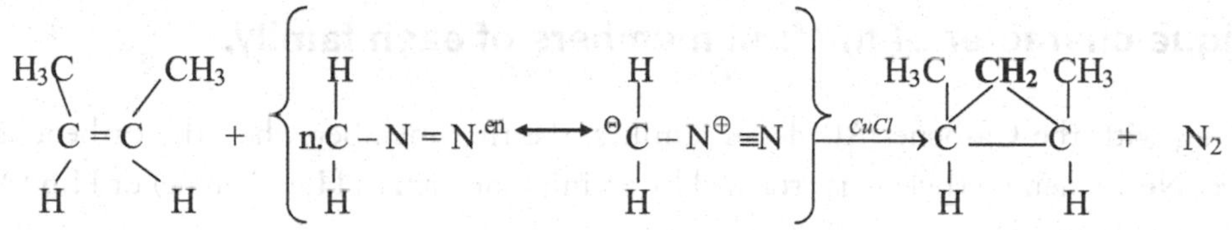

$$\underset{H}{\overset{H_3C}{>}}C=C\underset{H}{\overset{CH_3}{<}} + \left\{ n.\overset{H}{\underset{H}{C}}-N=N^{\cdot en} \longleftrightarrow \overset{\ominus}{\overset{H}{\underset{H}{C}}}-N^{\oplus} \equiv N \right\} \xrightarrow{CuCl} \overset{H_3C}{\underset{H}{\overset{}{C}}}\underline{\quad}\overset{CH_2\ CH_3}{\underset{H}{\overset{}{C}}} + N_2$$

$$\qquad\qquad\qquad\qquad\qquad (A) \qquad\qquad\qquad (B) \qquad\qquad\qquad\qquad\qquad\qquad\qquad \text{Ap1.27}$$

The charges above are polar charges and not ionic charges. Chargedly, it is (B) that is attacked with a positive charge releasing N_2 instantaneously after attachment. Radically diazomethane is decomposed as follows in the presence of the passive catalyst, CuCl, polar in character.

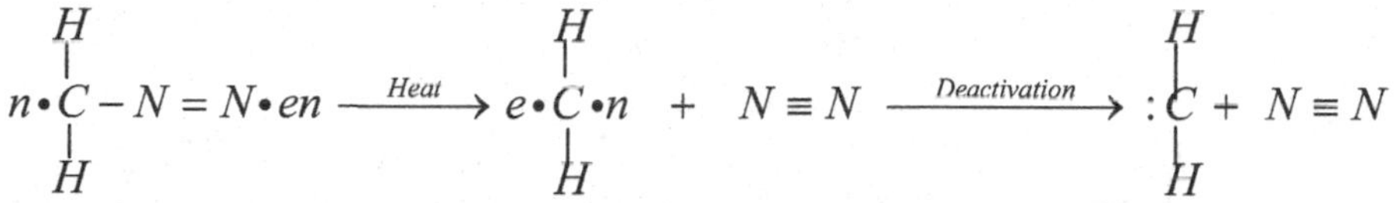

$$n\cdot\overset{H}{\underset{H}{C}}-N=N\cdot en \xrightarrow{\text{Heat}} e\cdot\overset{H}{\underset{H}{C}}\cdot n + N \equiv N \xrightarrow{\text{Deactivation}} \overset{H}{\underset{H}{:C}} + N \equiv N$$

$$\qquad\qquad (A) \qquad\qquad\qquad\qquad\qquad\qquad\qquad\qquad Carbene \qquad \text{Ap1.28}$$

Diazomethane cannot react ionically, but only chargedly (Covalent) and radically and indeed no productive chemical reactions take place ionically as wrongly shown in many Chemistry textbooks. (A) indeed is what has been used to activate the alkene using the electro-free-radical end and the mechanism is Equilibrium mechanism. The polar ones cannot be used because N_2 cannot be released instantaneously leaving the C center carrying two opposite charges.[1-6]

2. Cycloheptanone is made by ring enlargement of cyclohexanone by means of the diazomethane procedure for inserting a methylene group adjacent to the carbonyl group

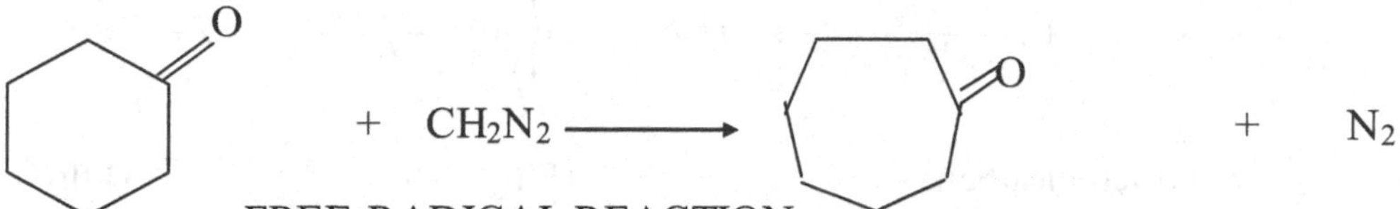

$$+ \quad CH_2N_2 \longrightarrow \qquad + \quad N_2$$

FREE-RADICAL REACTION

Cyclohexanone   Cycloheptanone   Ap1.29

3. Cycloheptatriene is obtained in very high yield by the cuprous chloride catalyzed reaction of benzene with diazomethane.

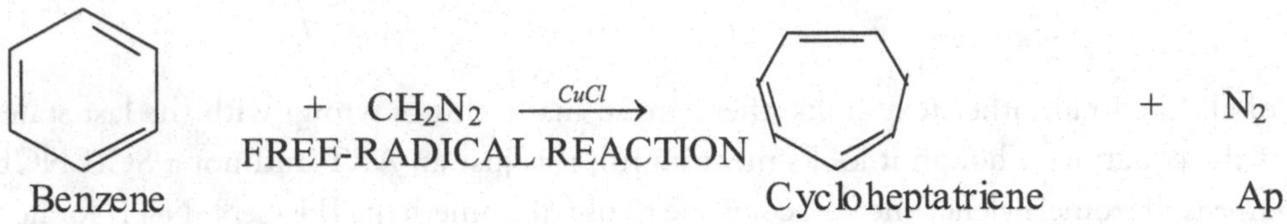

Benzene   + CH_2N_2 $\xrightarrow{CuCl}$ FREE-RADICAL REACTION   Cycloheptatriene   + N_2 Ap1.30

Diazomethane a yellow highly toxic gas is just the first member of diazoalkanes just as methane is the first member of alkanes. Being the first member, hence it is very unstable. It can explode.

All the members of the Stem are very important, in particular the first member of all the main branches that form the different families of the tree. Diazomethane is not one of them.

2.2 Unique character of the first members of each family.

Starting with the Carbenes/Alkylenes families, the first member when the carbene is activated is methylene. Now when methylene is attached by an initiator such as H.e (A male) or H.n (A female), the followings are obtained.

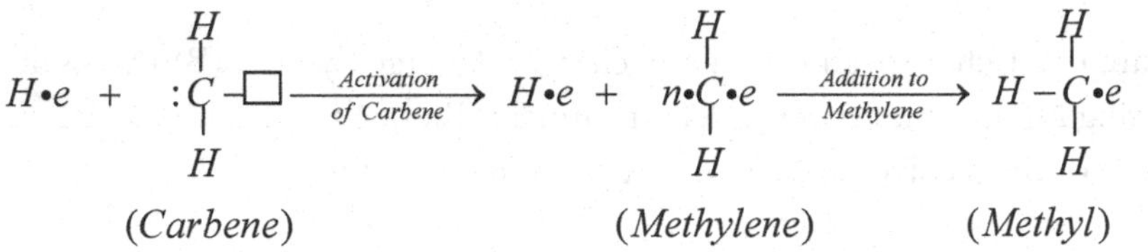

(Carbene)   (Methylene)   (Methyl)   Ap1.31

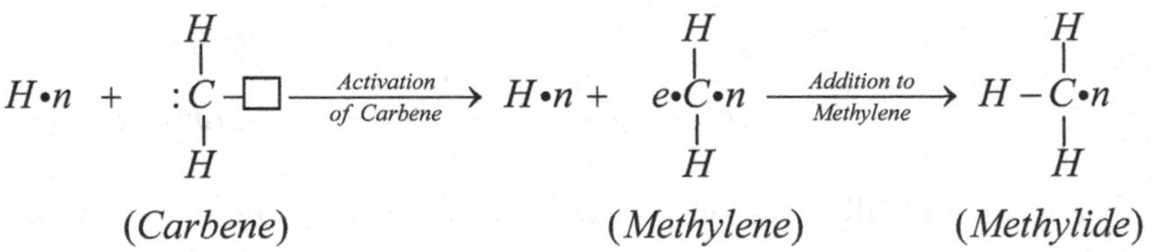

(Carbene)   (Methylene)   (Methylide)

Ap1.32

Both routes are favored by it to produce methyl and methylide groups. Just as we have hydride for hydrogen atom, so also we have methylide for methyl group. Now consider the next member of this same family-Ethylene. The same imaginary initiators used above are to be used here.

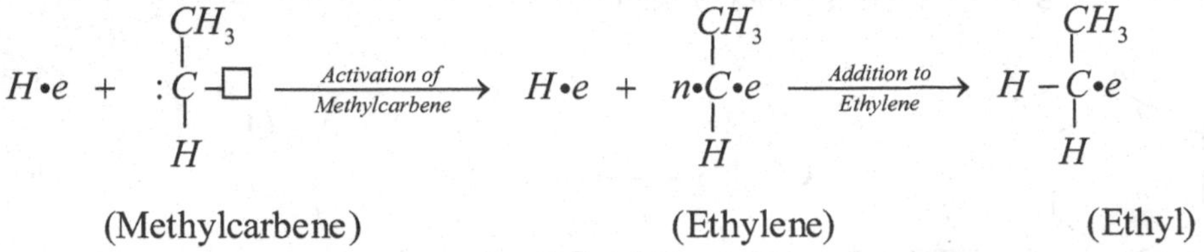

(Methylcarbene)   (Ethylene)   (Ethyl)

Ap1.33

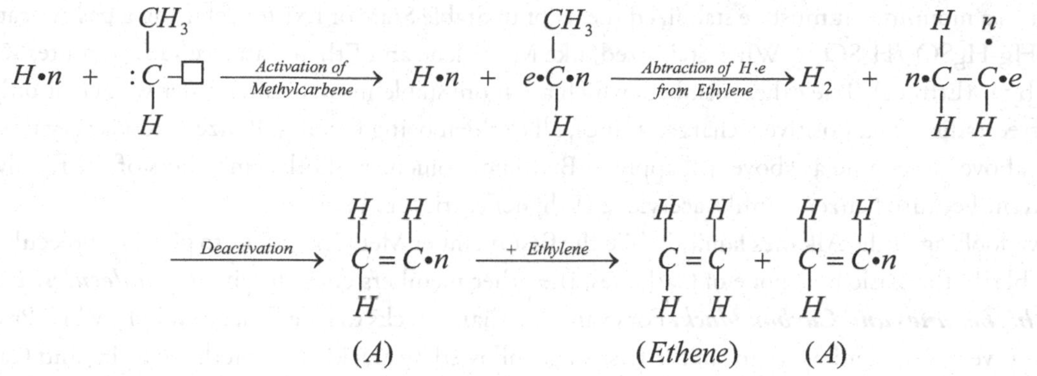

$$H \bullet n \ + \ \overset{\overset{\displaystyle CH_3}{|}}{\underset{\underset{\displaystyle H}{|}}{:C}} - \square \quad \xrightarrow[\text{Methylcarbene}]{\text{Activation of}} \quad H \bullet n \ + \ \overset{\overset{\displaystyle CH_3}{|}}{\underset{\underset{\displaystyle H}{|}}{e \bullet C \bullet n}} \quad \xrightarrow[\text{from Ethylene}]{\text{Abtraction of } H \cdot e} \quad H_2 \ + \ \overset{\overset{\displaystyle H \quad n}{| \quad \bullet}}{\underset{\underset{\displaystyle H \quad H}{| \quad |}}{n \bullet C - C \bullet e}}$$

$$\xrightarrow{\text{Deactivation}} \quad \overset{\overset{\displaystyle H \quad H}{| \quad |}}{\underset{\underset{\displaystyle H}{|}}{C = C \bullet n}} \quad \xrightarrow{+ \text{ Ethylene}} \quad \overset{\overset{\displaystyle H \quad H}{| \quad |}}{\underset{\underset{\displaystyle H \quad H}{| \quad |}}{C = C}} \ + \ \overset{\overset{\displaystyle H \quad H}{| \quad |}}{\underset{\underset{\displaystyle H}{|}}{C = C \bullet n}}$$

$$(A) \qquad\qquad\qquad (Ethene) \qquad (A)$$

Ap1.34

Overall Equation: $H.n \ + \ 2\text{Ethylene (C}_2\text{H}_4) \longrightarrow H_2 \ + \ \text{Ethene (C}_2\text{H}_4) \ + \ \text{(A) } H_3C_2.n$

Ap1.35

Worthy of note are the followings-

i) The ethylene favored the male route being female to give ethyl group.

ii) The ethylene did not favor the female route because of the presence of transfer species of the first kind of the first type.[1-6] However, in the process of the reaction, hydrogen molecule and ethene were produced along with (A) carrying a radical similar to the initiator.

iii) That Carbenes or Alkylenes are female monomers as shown when diazomethane is used as an Addition monomer, since the positively charged route is favored by it.[22]

To use H.e as initiator, it must be the only one present in the system i.e. in the absence of H.n type. The same applies to H.n if it is to be used alone. Luckily, H.n types of initiators are commonly available, for example the ones from benzoyl peroxide. When used, benzene carrying a nucleo-free-radical along with CO_2 is produced in place of H•n. The male initiator (H.e) which is unknown in present day Science is also commonly available, for example sodium (Na.e). It cannot form a molecule with itself. Thus, nucleo-free-radically when isomers of Propylene are used, Propene is produced. This clearly sends messages of relationships between the Alkylenes and Alkenes families.

Next as already shown with the Alkenes family[1-6], while ethene favors all routes including electro-free-radical and nucleo-free-radical routes, propene and all the other members favor only electro-free-radical and positively charged routes (i.e. male routes). It does not favor the cationic route. Nucleo-free-radically and negatively charged routes, unlike the case above, the same monomer is produced with Alkenes. Alkylenes cannot carry two opposite charges on the same carbon center. In general it is not possible. Hence Carbenes cannot be activated chargedly. These monomers are all females of different capacities. The first member (Methylene) like the case above (Ethene) has distinguished itself from the other members. The same as for Alkenes also apply to butadiene and the other members of the family of 1,3-Dienes and other families of conjugated dienes.

Next, considering the Alkynes family, the first member acetylene (Real name Ethyne) compared to the other lower members in the same family is very unstable. It exists in Equilibrium State of Existence all the time as shown below. For this reason, it cannot be used as a monomer.

$$H - C \equiv C - H \quad \xrightleftharpoons[\text{of Existence}]{\text{Equilibrium State}} \quad H \bullet e \ + \ n \bullet C \equiv C - H \qquad\qquad \text{Ap1.36}$$

To use it as a monomer, it must be stabilized (i.e. kept in Stable State of Existence) using a passive catalyst such as Hg-HgSO$_4$/H$_2$SO$_4$[23]. When stabilized, like Methylene and Ethene, all free-radical routes will be favored by it (Ethyne). The other members which are more stable in increasing order will favor only the electro-free-radical and positively charged routes, all of them being females. Benzene too is very unstable just like above. The same as above will apply to Benzene, Toluene and other members of the family to a lesser extent because benzene unlike acetylene (Ethyne) carries six H atoms.

Now, looking at the Alkanes family, while the first member Methane cracks to give H$_2$ molecules and Carbon black, the basic backbone of Carbenes, the other members crack to give ***H$_2$ molecules, Ethene mostly, higher ones and Carbon black.*** For example Ethane cracks to give Ethene and H$_2$, while Propane cracks to give Ethene and Methane in the first stage followed by cracking of Methane to H$_2$ and Carbon black in the second stage at higher operating conditions. n-Butane will crack in the first stage to give Ethene and Ethane, followed by cracking of Ethane to Ethene and H$_2$ at higher operating conditions in the second stage. Isomers of Butane can also be cracked. Notice the presence of the first member of Alkenes mostly as one of the products of the cracking of all crackable higher Alkanes. With Cycloalkanes families, while their first members will favor all routes when their rings are opened instantaneously, the other members will favor only the electro-free-radical and positively charged routes. The larger the size of the ring, the greater the temperature required to open the rings. One can thus see the uniqueness of first members of these families.

2.3 The main branches and sub-branches of the Hydrocarbon family tree.

The first members of all hydrocarbons distribute themselves orderly to form the main branches, carrying with them all the elements of H and C. It is the name of the first member in general that identifies with the name of the main branch. For example, in the past the methylene family was called Methylene and this was latter called the Carbenes. Herein, one has called it the **Alkylene family**. Acetylenes family was later changed to the alkyne family as we continued to advance herein. The same applies to other members in which for some, their real names have started to be properly identified for them such as the case with Olefins. In the main branches of the aromatics, apart from the first members, they also like the others carry along with them the aliphatic components through the use of the elements of H and C, so that there are indeed no aromatic parts of the main branches exclusive to each one of them. In order words, there are two main types of branches- Aliphatic and Alpharomatic parts. The same applies to the other condensed aromatic rings which have their own main branches. Some of the main branches include-

Alkylenes containing only methylene and the elements of H and C,
Carbenes containing only carbene and the elements of H and C,
Alkanes containing only methane and the elements of H and C,
Alkenes containing only ethene and the elements of H and C,
Conjugated dienes containing only butadiene and the elements of H and C.
Conjugated trienes containing only hexatriene and the elements of H and C,
Cumulative dienes containing only Allene and elements of H and C,
Cumulative polydienes containing only 1,4- 1,5- and so on and elements of H and C,
Alkynes containing only acetylene (Real name Ethyne) and the elements of H and C,
Tricycloalkanes containing only cyclopropane and the elements of H and C,
Tetracycloalkanes containing only cyclobutane and the elements of H and C,
:

:

Duodecacycloalkanes containing only the first member and elements of H and C.
Tricycloalkenes containing only cyclopropene and elements of H and C,........ETC.

This continues down to the alipharomatic main branches starting with Benzenes, Biphenyls, Naphthalenes, Anthracenes, Phenanthrenes, Naphthacenes, Dibenzanthrecenes, Ben-zo-pyrenes and more yet unknown. Since these main branches depend on the stem, they are all interrelated.

From the main branches emerge the other members of the same family as sub-branches. These are the fruits. For example, the alkane's main branch has as sub-branches, the ethane, propane, butanes, pentanes, and so on. These may have sub-sub-branches growing from them. Those that have isomers emerge together from the same main-branch as sub-branches. All these sub-branches carry with them the elements of H and C and use them for creating branches from sub-branches where possible and necessary.

The first two main branches- the Carbenes and Alkylenes, are very active in particular the Alkylenes. These also have sub-branches all of which are used along with the elements of H and C to form the barks of the stem, barks of the main branches and sub-branches. This is a network of cross-linked non-porous elastomeric membrane looking may be like that shown below in Figure Ap1.4.

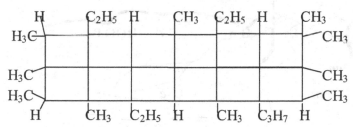

Figure Ap1.4. <u>Typical type of Bark for the Stem, Main branches and Sub-branches.</u>

It could be a single layer or multiple layers of membranes stacked together. Apart from the use of these two main branches to form the barks of the tree, they can also be used to flower and give colors to the tree based on the type of radicals carried by them. The barks are what make the stem to stand firm and look like a tree. They help to hold the branches and sub-branches together.

From all the considerations so far ***concisely presented***, Figure Ap1.5 below shows the hydrocarbon family tree, noting that this is not a classification of hydrocarbons. The tree begins with the roots, followed by the Stem. The stem begins to emerge carrying at the top the families of Carbene and Alkylenes, since these will be continually used to cover the tree and its body as the tree grows. On one side of the tree are the Aliphatic hydrocarbons, while on the other side are the Alipharomatic hydrocarbons. Once the stem is formed, the branches begin to appear from the top in a downward fashion. On the Aliphatic side, the first branch after Alkylenes is the Alkanes family while on the Alipharomatic side, the first branch after Carbenes is the Benzenes family. Carbenes should in fact be on the Aliphatic side replacing Alkylene. Alkylene was indeed used and well identified in view of the letters **ALK** for all of them. From the branches other members of the family starting from the second begin to appear as sub-branches, noting that the first member in the family (Methane or Benzene is the main branch. This is how Nature operates with respect to building of Families for Living and so-called Non-Living systems..

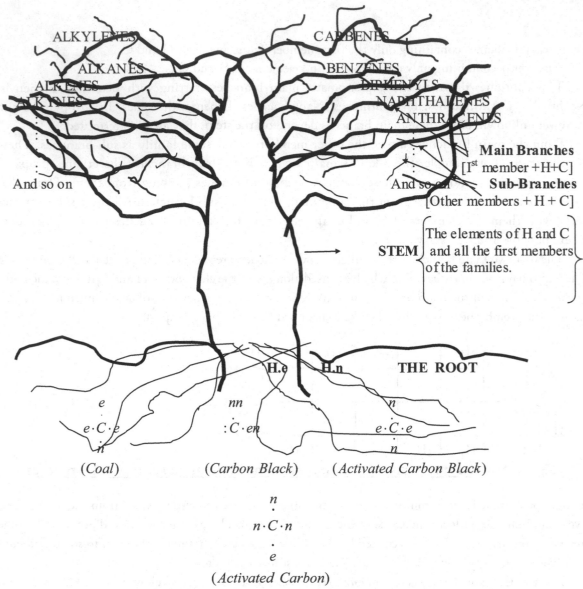

ALKYLENES
ALKANES
ALKENES
ALKYNES

And so on

CARBENES
BENZENES
DIPHENYLS
NAPHTHALENES
ANTHRACENES

And so on

Main Branches
[1st member +H+C]
Sub-Branches
[Other members + H + C]

The elements of H and C
STEM and all the first members
of the families.

H.e   H.n   **THE ROOT**

e
$e \cdot C \cdot e$
n
(*Coal*)

nn
$: C \cdot en$
(*Carbon Black*)

n
$e \cdot C \cdot e$
n
(*Activated Carbon Black*)

n
$n \cdot C \cdot n$
e

(*Activated Carbon*)

Figure Ap1.5 The Hydrocarbon Family Tree.

Note the presence of additional elements of Carbon in the root- Coal and Activated Carbon. From them, Activated Carbon Black and some alkyl groups of choice can be obtained.

3.0   Conclusions

This is not a Natural tree, but an Engineering tree, since Engineering is one's ability to copy Nature for the comfort of humanity applying four basic fundamental principles- the Natural Sciences, the Social Sciences, the Arts, and Imaginative capabilities. In the process, since the focus is only for Hydrogen and Carbon containing compounds, one started by looking at the elements of H and C atoms, the vehicles required to lay the foundations for the tree. One went further to look at the molecules of H and C atoms. For the first time, the true structures of Diamond, Graphite, Coke, and Coal were provided based on universal data, "personal observations" and the foundations being laid.

For the first time, we began to see the true names of the members of the hydrocarbon families most of which we already seem to know and the most important of which are Carbenes (Unactivated CH_2,...... CR_2) and Alkylenes (Activated CH_2......CR_2). For the first time, we saw how these members can be used to produce the Alkenes. For the first time also, we have shown that within the unsaturated members (The Alkenes, Alkynes, Cumulenes, Cyclo- alkenes, Conjugated dienes Alipharomatics and so on), there are no males, but all females. Males are only present when hetero-atoms are present at specific locations. Even coal (though shown in the figure above) which is not a strong member of the hydrocarbon family does not have males. Coal is like crude oil and gases, both of which are Natural sources of hydrocarbons. One can see coal at the bottom above- the Root. Diamond, Graphite, and other minerals like Coal are no members of the Tree.

The compositions of the stem were shown to contain all the first members of the families and the elements of H and C. How the first members are obtained using the elements of H and C, and Carbenes' first member were shown in systematic order. These first members as the stem grows start to distribute themselves to form different branches for the tree. They carry with them all the tools required to build the sub-branches and these are the elements of H and C. They also carry Carbene and Methylene along with them into their main branches. The first members of the different families which are unique with respect to the other members in the same family do not have sub-branches. They remain in their main branches which in few cases bear their names or a symbolic name. The sources for the barks and flowering of the tree were identified. From the tree, we can pluck whatever hydrocarbons desired, for we know that it will grow back again and again.

References

1. S. N. E. Omorodion, *"Chemistry and Humanity-Challenges to be faced as we advance in the Third Millennium: New Classifications for Radicals."*, Nigerian Journal of Applied Science, Vol 26, (2008), Pgs. 146-158.

2. S. N. E. Omorodion, *"Chemistry and Humanity-Challenges to be faced as we advance in the Third Millennium: The impacts of new classifications for Radicals."* Nigerian Journal of Applied Science, Vol.27, (2009), Pgs. 33-43.

3. S. N. E. Omorodion, *"Chemistry and Humanity- Challenges to be faced as we advance in the Third Millennium: New Classifications for Bonds and Charges."* Nigerian Journal of Applied Science, Vol 27, (2009), Pgs. 94-104.

4. S. N. E. Omorodion, *"Chemistry and Humanity-Challenges to be faced as we advance in the Third Millennium: The impacts of the new classifications for Bonds and Charges."* Nigerian Journal of Applied Science, Vol 28, (2010). Pgs. 37-45.

5. S. N. E. Omorodion, *"Chemistry and Humanity-Challenges to be faced as we advance in the Third Millennium: New classifications for Compounds."* Nigerian Journal of Applied Science, Vol 28, (2010), Pgs. 80-87.

6. S. N. E. Omorodion, *"Chemistry and Humanity-Challenges to be faced as we advance in the Third Millennium: The impacts of the new classifications for compounds."* Nigerian Journal of Applied Science, Vol 28, (2010), Pgs. 128-134.

7. S. N. E. Omorodion, *"Mechanism of the production of Phthalic anhydride from ortho-xylene using Combustion/Oxidation."*, Appendix VI, Vol II.

8. S. N. E. Omorodion, *"New Frontiers in Engineering, Sciences and the Arts"* Vol. VI, under review before going to press.

9. C. R. Noller, *"Textbook of Organic Chemistry"*, W. B. Saunders Company, 1966, pgs. 83-85.

10. *"Chemistry of Diamond-Part 2: Properties and Types of Diamond"*, http://chemistry.about.com/cs/geochemistry/a/aa071801a.htm, 2009.

11. MatSc 101: *Coal Structure*, http://www.ems.psu.edu/~radovic/coal structure.html, 2009.

12. D. G. Levine, R. H. Schlosberg, and B. G. Silbernagel, *"Understanding the Chemistry and Physics of Coal structure (A Review)"*, Proc. Natl. Acad. Sci. USA, Vol.79. pp.3365-3370, 1982.

13. C. R. Noller, *"Textbook of Organic Chemistry"*, W. B. Saunders Company, 1966, pgs. 389-391.

14. H. W. Sternberg, R. Raymond, and F. K. Schweighardt, *"Acid-Base Structure of Coal-Derived Asphaltenes"*, Science 4, Vol. 188. No. 4183, pp. 49-51, 1975.

15. *"The structure of Diamond and Graphite,"* (2003) http://phycomp.technion.ac.il/~anastasy/teza/teza/node3.html, 2009.

16. H. Zhang, Y. Yan, Z. Cheng, W. Sun, *"Structural Analysis of Coke on Used Catalysts during Residue Hydrotreating by Ruthenium Ion Catalyzed Oxidation Reaction"*, Petroleum Science and Technology, Vol. 27, Issue 1, pgs. 33-45 2009.

17. M. Kayoko, A. Daisuke, T. Tatsuya, M. Hiroshi, Y. Tetsuya, T. Takayuki, *"Investigation of Micro Carbon Structure of Coke and Semi-Coke Using Hand-picking Method"*, Tetsu to Hagane, F0332A, Vol. 92; No.3 pgs152-156, 2006.

18. *"The Diamond Structure"*, http://www.uncp.edu/home/mcclurem/lattice/diamond.htm, 2009.

19. C. R. Noller, *"Textbook of Organic Chemistry"*, W. B. Saunders Company, 1966, pg. 483.

20. C. R. Noller, *"Textbook of Organic Chemistry"*, W. B. Saunders Company, 1966, pg. 75.

21. C. R. Noller, *"Textbook of Organic Chemistry"*, W. B. Saunders Company, 1966, pgs. 178, 250.

22. G. Odian, *"Principles of Polymerization"*, McGraw-Hill Book Company, 1970, pg. 357.

23. C. R. Noller, *"Textbook of Organic Chemistry"*, W. B. Saunders Company, 1966, pgs. 149.

Appendix II

THE MECHANISMS OF NON-CATALYZED AND CATALYZED DECOMPOSITIONS OF AZO-HYDROCARBONS. PART I: AZO-METHANE

Abstract

Based on the new foundations[1-7], particularly with respect to the hydrocarbon family tree[8], contained herein are new classifications for Diazenes (or Diimines or "Diimides"). The first member called Diazene (H – N = N – H)[9] has no carbon element and therefore not a hydro-carbon, but yet has an Azo character (– N = N –). When decomposed non-catalytically, N_2 and H_2 are the products[6] and the mechanism can either be Equilibrium or Decomposition or both. The decompositions of family members when possible involve both Equilibrium and Decom-position mechanisms with varying types of products. When decomposed using passive catalysts, the products obtained are completely different from non-catalyzed. H, alkyl, and aryl groups carried by the nitrogen centers are radical-pushing groups. The types of decomposition depend on the types of groups carried with respect to their radical-pushing or radical-pulling characters, resonance stabilization characters, and whether the groups are the same or not.

Keywords:

1. Radical-pushing groups.
2. Decomposition mechanism.
3. Equilibrium mechanism.
4. Molecular rearrangement by Electro-radical movement.
5. Azo-Alkanes, Azo-Alkenes and Azo-Alkynes.

1.0 INTRODUCTION

The dawn has arrived for one to turn the wheel of knowledge backwards in order to move forward, based on the new foundations, wherein in particular a Radical has been well defined.[1] The need arises because since antiquity, we have not yet begun the study of the Real Science. It has only been an Art still filled with so many fears of the unknown. That big and enormous gap must be bridged.

What are Azo-compounds? What are Diazenes? What are Azo-hydrocarbons? What are Azo- hetero-hydrocarbons? If $H_3C - N = N - CH_3$ (I) is called Azo-methane[10], and $R_2CH - N = N - C_6H_5$ (II) is called benzeneazoalkanes[11], then what should $H_3C - N = N - C_2H_5$ (III) be called? Should it be called methaneazoethane like (II)? Indeed, none of the names above for (I), (II) and (III) is correct. There are so many schools of thought for the definition of an Azo-compound, just as for everything. One school says "Azo-compounds are compounds bearing *the functional group* $R - N = N - R'$, in which R and R' can be either aryl or alkyl. ***The N = N group is called an azo group,*** although the parent compound, HNNH, is called diimide. The name azo comes from *azote*, the French name of nitrogen that is derived from the Greek *a* (not) + *zoe* (to live)".[12] A compound can never be a functional group based on the definition above. What the azo carries are functional groups (i.e. difunctional) all of different kinds and types. It is what N_2 carries in its activated state that determines the character of the compound. Another school of thought says "In azo compounds nitrogen is bound to nitrogen and there is one carbon-nitrogen bond for each nitrogen"[10]. Therefore, does this mean that when H or other non-carbon elements or groups are connected to the nitrogen centers, they cannot be classified as azo-compounds? This definition is then only limited to Azo-Hydrocarbons or Hetero/Metal-containing hydro-carbons. For the case of H, the compound is called diazene and the family called Diazenes, for example $t-C_4H_9 - N = N - t-C_4H_9$ (IV) can also be called di-tert-butyldiazene[13] and not azo-t-butane, just as the so-called azomethane can be rightly called di-methyldiazene or indeed azo-ethane. The Medical school of thought defines azo compounds as "one of many organic aromatic compounds containing the divalent chromophore, $- N = N -$"[14]. Another school of thought define azo compound as "any organic chemical compound in which the azo group ($- N = N -$) is part of the molecular structure. The atomic groups attached to the nitrogen atoms may be of any organic class; but the commercially important azo compounds, those that make up more than half the commercial dyes, have the benzene group or its derivatives as the attached groups (aromatic azo compounds)"[15] This is closest to the real definition.

What makes N_2 so unique in our world? When activated alone it becomes an azo, $- N = N -$ i.e. nn •N=N• en. Can the azo be further activated when it is carrying other components other than the unique metallic H which is gaseous under most conditions above -252.87^0 C (B pt. of N is -196^0 C)? Yes, it can be activated either in the presence of Transition metal catalysts such as Pt (111) by desorption[16] or as long as it is carrying N components or other elements of less electro-negativity than N on both sides, such as some halogen elements, $Cl - N = N - Cl$ (V). But when one of the central elements of a group next to the N centers is more electropositive than nitrogen, the π-bond on N = N cannot be activated. $CH_3C_6H_4 - N = N - Cl$ (VI) mistakenly called a diazonium salt[17] is an azo salt which cannot be activated because of the presence of the carbon center on the aryl group. In the presence of cuprous chloride in HCl and heat, it decomposes with the evolution of N_2 and formation of p-chlorotoluene, that is, the components the azo compound is carrying[17].

The two groups carried by the azo could be the same or different. They could be radical-pushing or radical-pulling groups[1-6]. It is the group that largely determines if the azo compound is water soluble or organic soluble such as shown below.

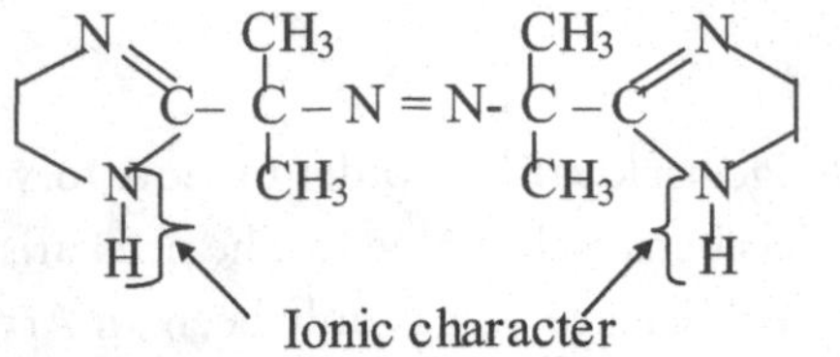

POLAR/IONIC

(VII) 2 ,2´-Azobis 2-(2-imidazolin-2-yl) propane

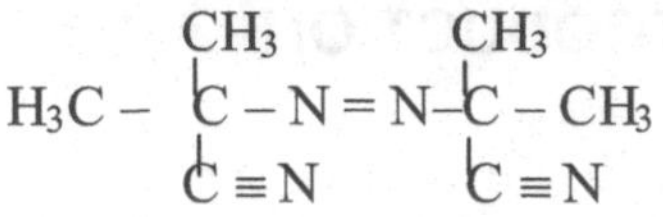

POLAR/NON-IONIC

(VIII) 2-2´ Dicyano-2, 2´
-azopropane
(Azobisisobutyronitrile)

ApII.1

Just like water, ammonia, methanol and much more which are polar/ionic, so also is (VII) above. Hence, (VII) a white powder[18] will dissolve and insolubilize in water or dissolve and solubilize in methanol or any polar/ionic solvent. (VIII) which is very well known as one of the sources of free-radical initiators is polar/non-ionic and therefore dissolves in organic solvents which are polar/non-ionic. The groups carried by the nitrogen centers in (VII) and (VIII) are radical-push-ing groups. Like Benzoyl peroxide, the free-radical initiators generated from them are nucleo-free-radicals and nucleo-non-free-radicals[1-6]. They are free because the centers carrying them have no paired unbonded radicals on them and non-free when carried by the O or N centers. They are nucleo- (i.e. female) because the groups are radical-pushing groups whether they are carrying CN group or not[19]. Indeed, the reason is because of the mechanisms involved. When (VIII) is decomposed, N_2 is released and either initiators are produced or a stable tetramethyl-succinonitrile (IX) shown below is formed or both are produced. *Initiators are largely produced when monomers are present.* When they are not present, then (IX) is largely formed, noting *what it takes for two like poles to attract and form a compound (Metaphysics).* In Equilibrium mechanism, this is usually the last step, if the stage is to be productive.[1-6]

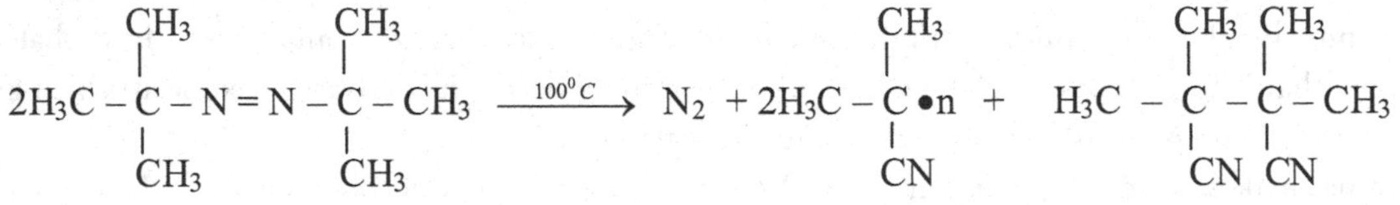

$$2H_3C - \underset{\underset{CH_3}{|}}{\overset{\overset{CH_3}{|}}{C}} - N = N - \underset{\underset{CH_3}{|}}{\overset{\overset{CH_3}{|}}{C}} - CH_3 \xrightarrow{100^0 C} N_2 + 2H_3C - \underset{\underset{CN}{|}}{\overset{\overset{CH_3}{|}}{C}} \bullet n + H_3C - \underset{\underset{CN}{|}}{\overset{\overset{CH_3}{|}}{C}} - \underset{\underset{CN}{|}}{\overset{\overset{CH_3}{|}}{C}} - CH_3$$

(VIII) Initiator (IX) ApII.2

It is important to note that the equation above is different from the present-day ones[10]. Just like the last but two statements above, if the male radical had been present as one of the products, then no polymers can be produced.

Though the main focus here is on azo-hydrocarbons, there was need to turn the wheel backwards, so that we can begin to see with the eye of the needle. Azo-hydrocarbons are those which carry only hydrogen and carbon elements on both sides of the azo via the carbon element. The hydrocarbon family tree[8] contains all the different families of the two main sections- The Aliphatic and Alipharomatic sections. In general, it seems the major families which have been of great interest universally have been only the Alkanes, and Arylalkanes. Of the Azo-Alkane family, the first member is $H - N = N - CH_3$ which is indeed the azo-methane (i.e. *methyl-diazene*).

2.0 The Azo-Alkanes.

They all gained their roots from the Diazene (H – N = N – H) just by adding CH_2. The CH_2 in question is the activated carbene or methylene. Based on the new classifications, Table ApII.1 below shows the first five members of Diazenes and Azo-alkanes. All diazenes are azo- compounds. Only the first member is not an azo-alkane, but azo-hydrogen, just as Cl-N=N-Cl is azo-chlorine. How can one call H_3C-N=N-CH_3 azo-methane when a nitrogen is not carrying H?

Table ApII.1 New classifications of Azo-Alkanes.

ALKANES	DIAZENES		AZO
#	STRUCTURE	NAME	NAME
1*	H-N=N-H	Diazene	Azo-Hydrogen
1	H-N=N-CH$_3$	Methyldiazene	Azo-Methane
2	H-N=N-C$_2$H$_5$	1-Ethyldiazene 2-Ethyldiazene (See Table ApII.2)	Azo-Ethane
3	H-N=N-C$_3$H$_7$ CH$_3$ H-N=N-C-H CH$_3$	n-Propyldiazene iso-Propyldiazene	Azo-n-Propane Azo-iso-Propane
4	H-N=N-C$_4$H$_9$ CH$_3$ H-N=N-C-C$_2$H$_5$ H CH$_3$ H-N=N-CH$_2$-C-H CH$_3$ CH$_3$ H-N=N-C-CH$_3$ CH$_3$	n-Butyldiazene s-Butyldiazene iso-Butyldiazene ? t-Butyldiazene	Azo-n-Butane Azo-s-Butane Azo-iso-Butane Azo-t-Butane

5	H-N=N-C$_5$H$_{11}$	n-Pentyldiazene	Azo-n-Pentane
	C$_2$H$_5$ H-N=N-C-C$_2$H$_5$ H	s-Pentyldiazene	Azo-s-Pentane
	CH$_3$ CH$_3$ H-N=N-C - C-H H CH$_3$	s-iso-Pentyldiazene	Azo-s-iso-Pentane
	C$_2$H$_5$ H-N=N-CH$_2$-C-H CH$_3$	iso-Pentyldiazene	Azo-iso-Pentane
	C$_2$H$_5$ H-N=N-C-CH$_3$ CH$_3$	t-Pentyldiazene	Azo-t-Pentane

- Note that this is not an Azo-alkane.

Table ApII.2 below shows the isomers of Azo-alkanes shown in the Table above. Azo-alkanes are a single family member of Azo-hydrocarbons.

Table ApII.2 Isomers of Azo-Alkanes.

#	Azo-Alkanes	Isomers	Name
1	Azo-Methane (H-N=N-CH$_3$)	None	Methyldiazene
2	Azo-Ethane (H-N=N-C$_2$H$_5$)	H$_3$C-N=N-CH$_3$	Dimethyldiazene
3	Azo-Propane (H-N=N-C$_3$H$_7$)	H H i) H$_3$C-N=N-C-C-H H H H ii) H$_3$C-N=N-C-H CH$_3$	Methyl-1-ethyldiazene Methyl-2-ethyldiazene

4	Azo-Butane[*] (H-N=N-C$_4$H$_9$)	i) H$_3$C-N=N-C$_3$H$_7$ CH$_3$ ii) H$_3$C-N=N-C-H CH$_3$ iii) 1-H$_5$C$_2$-N=N-1-C$_2$H$_5$ iv) 2-H$_5$C$_2$-N=N-2-C$_2$H$_5$	Methylpropyldiazene Methyl-iso-propyldiazene Di-1-ethyldiazene Di-2-ethyldiazene
5	Azo-Pentane[*] (H-N=N-C$_5$H$_{11}$)	i) H$_3$C-N=N-C$_4$H$_9$ ii) H$_3$C-N=N-s-C$_4$H$_9$ iii) H$_3$C-N=N-iso-C$_4$H$_9$ iv) H$_3$C-N=N-t-C$_4$H$_9$ v) 1-H$_5$C$_2$-N=N-C$_3$H$_7$ vi) 2-H$_5$C$_2$-N=N-C$_3$H$_7$ vii) 1-H$_5$C$_2$-N=N-iso-C$_3$H$_7$ viii) 2-H$_5$C$_2$-N=N-iso-C$_3$H$_7$	Methyl-n-butyldiazene Methyl-s-butyldiazene Methyl-iso-butyl-diazene Methyl-t-butyldiazene 1-Ethyl-n-propyldiazene 2-Ethyl-n-propyldiazene 1-Ethyl-iso-propyl-diazene 2-Ethyl-iso-propyl-diazene

[*]Their isomers in Table ApII.1 are isomers in this Table.

Note the unique presence of 1- and 2- ethyl based on the horizontal placement of CH$_3$ group. All other higher members follow the same order as above. Imagine what is called Azo-propane (H$_7$C$_3$-N=N-C$_3$H$_7$) is indeed one of the isomers of Azo-Hexane (H-N=N-C$_6$H$_{13}$). So also is so-called Azo-butane an isomer of Azo-Octane. Table ApII.3 below shows the first members of some of the important members of the family members of the Azo-Hydrocarbons.

Table ApII.3 First members of the families of Azo-Hydrocarbons

#	Family member	First member of the Family	Name
1	Azo-Alkylenes	$H - \overset{\bullet e}{\underset{\bullet n}{C}} - N = N - H$	Azo-Methylene
2	Azo-Alkanes	H-N=N-CH$_3$	Azo-Methane
3	Azo-Alkenes	$H - N = N - C = \overset{H}{\underset{H}{C}}$ (with H below left C)	Azo-Ethene

4	Azo-Alkynes	H-N=N-C≡C-H	Azo-Ethyne or Azo-Acetylene
5	Azo-Conjugated Dienes	H H H H-N=N-C=C-C=C H H	Azo-Butadiene
6	Azo-Cumulative Dienes	H H H-N=N-C-C=C=C H	Azo-Allene
7	Azo-Isolated Dienes	H-N=N-CH=CH-CH$_2$-CH=CH$_2$	Azo- 1,4-pentadiene
8	Azo- Tricycloalkanes	H CH$_2$ H-N=N-C CH$_2$	Azo-Cyclopropane
9	Azo-Tricycloalkenes	H-N=N-C=C-H CH$_2$	Azo-Cyclo Propene
10	Azo-Benzene	H-N=N-C$_6$H$_5$	Azo-Benzene
11	Azo-Biphenyl	H-N=N-C$_6$H$_4$-C$_6$H$_5$	Azo-Biphenyl
12	Azo-Naphthalene	H-N=N-C$_{10}$H$_7$	Azo-Naphthalene

There are many azo-hydrocarbons above not yet known or where no attentions have yet been placed on. Shown below are some other important azo-hydrocarbons also known and unknown.

$$H - \overset{\bullet e}{\underset{\bullet n}{C}} - N = N - \overset{\bullet n}{\underset{\bullet e}{C}} - H$$

(A) Dimethylidynediazene

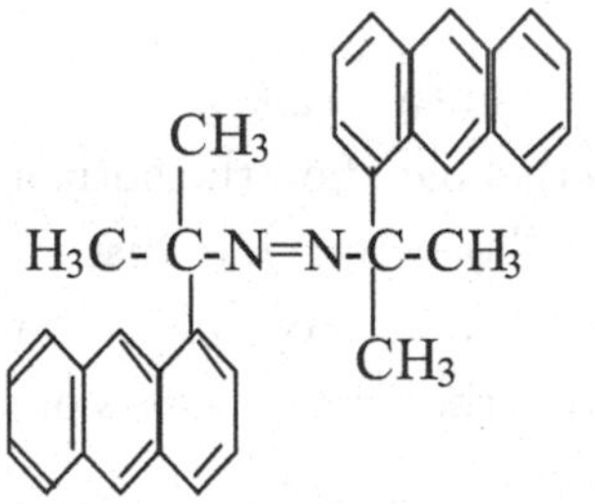

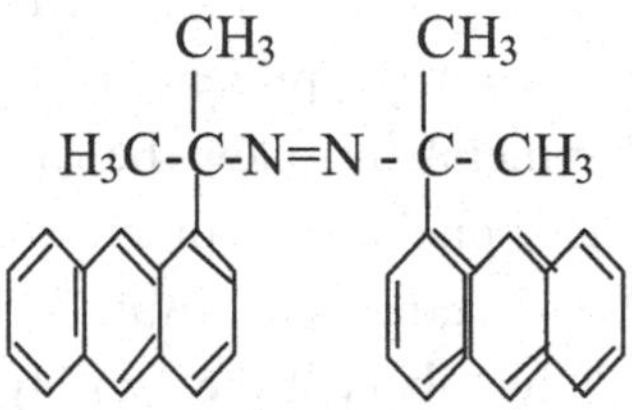

B) **Trans-dimethylanthracyldimethyl-Diazene**

(C) **Cis-dimethylanthracyldimethyl-Diazene**

(D) Divinyldiazene

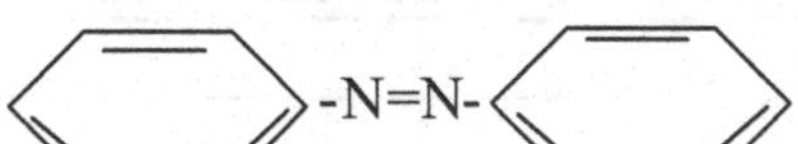

(E) Biphenyldiazene

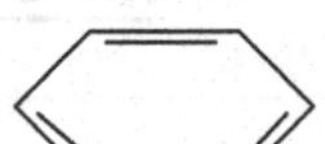

(F) Azo-Toluene

ApII.3

(A), (D), and (E) above are isomers of Azo-Acetylene, Azo-Butadiene, and Azo-Biphenyl respectively. (B) and (C) are two new anthracene-containing azo-alkanes reported to absorb UV light 600 times more strongly than simple azo-alkanes. The thermolysis of these compounds in the presence of monomers affords florescent labeled polymer. They were also reported as the first azo-alkanes to undergo *induced decomposition* in solution.[20] One has called the compounds as Trans- and

Cis-dimethylanthracyldimethyldiazene. The presence of two dimethyls in the name is disturbing. (E) is not Azo-Benzene, (D) is not Diethenediazene and (A) is not Dimethylene-diazene as used today. (F) is Azo-Toluene or Methylphenyldiazene and not benzene-azo-methane[10], said to be obtained from formaldehyde phenylhydrazone by tautomerization in solutions of hexane or carbon tetrachloride. Indeed, it is not even by molecular rearrangement as shown below, but by electro-radicalization involving movement of electro-radical from a π-bond.

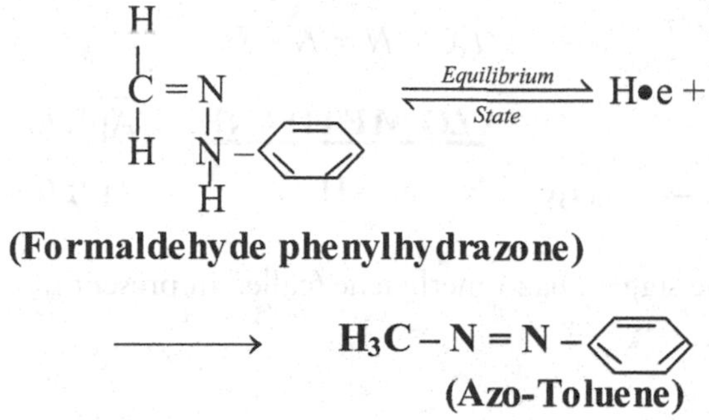

(Formaldehyde phenylhydrazone)

$$H_3C - N = N - \langle \rangle \qquad \text{ApII.4}$$
(Azo-Toluene)

This is different from molecular rearrangement phenomena, since no activation has taken place here. It happens when the compound is placed in Equilibrium State of existence, noting that it is always the electro-free or non-free radical that moves all the time, just as takes place in flow of current in solutions and solids.[1-7] It has long been identified as Electro-radicalization in the Volumes.

There is need to ascertain the existence of the groups carried by the Nitrogen center in Table ApII.3 for #1-Azo-Methylene and #3-Azo-Ethene for example.

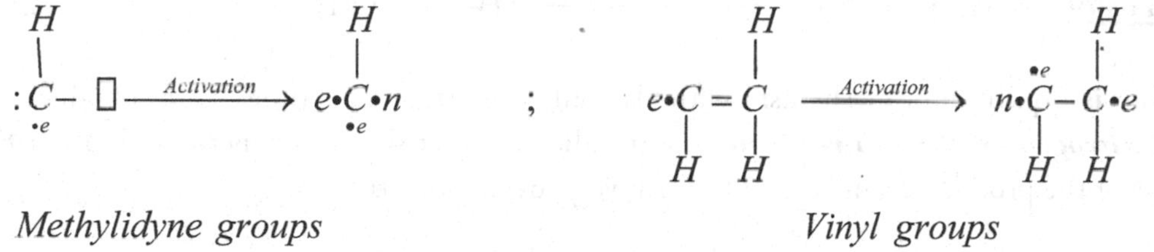

Methylidyne groups ; *Vinyl groups* ApII.5

The existence of methylidyne and vinyl species have long be known. These were identified during the adsorption and reaction of ethylene on clean Pd (100) at 300 K[21], in the temperature programmed study of acetylene on a clean, H-covered, and O-covered Pt (III) surface[22], and in the reactions of diazomethane on a clean, H-covered and O-covered Pt (III) surface[23]. So also is the existence of "ethylidyne" species during the reaction of ethylene with clean and carbide-modified Mo (110) in the temperature range 260-350 K.[24]

2.1 Azo-Methane

Azo-methane can be synthesized by any of the two methods shown below.

i) **Stage 1:**

$$H - N = N - H \xrightleftharpoons[\text{of Diazine}]{\text{Equilibrium State of Existence}} H{\bullet}e + nn.N = N - H$$

$$H{\bullet}e + \overset{\overset{H}{|}}{\underset{\underset{H}{|}}{:C}} - \square \xrightleftharpoons[\text{Carbene}]{\text{Activation of}} H{\bullet}e + n{\bullet}\overset{\overset{H}{|}}{\underset{\underset{H}{|}}{C}}{\bullet}e \rightleftharpoons H_3C{\bullet}e$$

$$H_3C{\bullet}e + nn{\bullet}N = N - H \xrightarrow[\text{of Azo-methane}]{\text{Combination State of Existence}} H_3C - N = N - H$$

AZO-METHANE   ApII.6a

Overall Equation: H-N=N-H   + H_2C ⟶ $H_3C - N = N - H$ ApII.6b

The mechanism is Equilibrium mechanism with only one stage. Diazo-methylene (called in present-day Science diazo-methane) can be a source of the methylene.

ii) **Stage 1:**

$$CH_4 \xrightleftharpoons[\text{of Methane}]{\text{Equilibrium State of Existence}} H{\bullet}e + n.CH_3$$

$$H{\bullet}e + N \equiv N \xrightleftharpoons[\text{Heat}]{\text{Activation of Nitrogen}} H - N = N{\bullet}en$$

$$H - N = N{\bullet}en + n{\bullet}CH_3 \xrightarrow[\text{of Azo-Methane}]{\text{Combination State of Existence}} H - N = N - CH_3$$

ApII.7a

$$(AZO - METHANE)$$

Overall Equation: CH_4 + N_2 ⟶   H-N=N-CH_3 ApII.7b

The mechanism is Equilibrium mechanism with also only one stage. This time, it was obtained from *methane* and *nitrogen* to give *Azo-methane* a clear indication that this is azo-methane and not H_3C-N=N-CH_3 one of the products from N_2 and C_2H_6 at very low temperatures.

2.2 Non-Catalyzed Decomposition of Azo-Methane

Ninety percent of the azo-methane has been put in Equilibrium State of Existence at particular operating conditions, with the remaining ten percent in Stable State of Existence.

Stage 1:

$$H_3C - N = N - H \xrightleftharpoons[\text{of Azo-methane}]{\text{Equilibrium State of Existence}} H{\bullet}e + nn{\bullet}N = N - CH_3$$

$$nn.N = N - CH_3 \xrightleftharpoons[\text{Higher Temperature}]{\text{More Heat i.e.}} nn{\bullet}N = N{\bullet}en + n{\bullet}CH_3$$

$$H{\bullet}e + n{\bullet}CH_3 \xrightleftharpoons[\text{of Methane}]{\text{Equilibium State of Existence}} CH_4$$

ApII.8

$$nn{\bullet}N = N{\bullet}en \xrightarrow[\text{of Energy}]{\text{De-activation with release}} N_2$$

Overall Equation: 900H_3C-N=N-H ⟶   900CH_4 +   900N_2 ApII.9

If the azo double bond can be activated, then the azo-methane should molecularly rearrange to give formaldehyde hydrazone (H_2C=N-NH_2) as shown below.

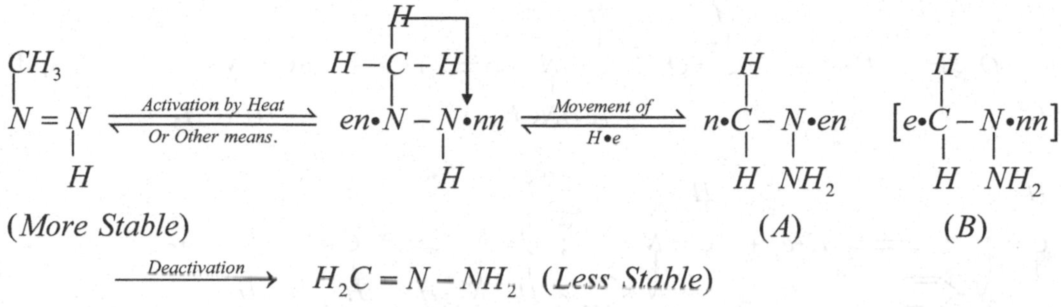

$$CH_3 \atop N = N \xrightleftharpoons[\text{Or Other means.}]{\text{Activation by Heat}} \quad en\cdot N - N\cdot nn \atop H \xrightleftharpoons[\text{H}\cdot e]{\text{Movement of}} \quad n\cdot C - N\cdot en \atop H \ \ NH_2 \quad [e\cdot C - N\cdot nn] \atop H \ \ NH_2$$

(*More Stable*) $\qquad\qquad$ (*A*) $\qquad$ (*B*)

$$\xrightarrow{\text{Deactivation}} \quad H_2C = N - NH_2 \quad (\textit{Less Stable})$$

(C) (Formaldehyde hydrazone) $\qquad\qquad\qquad$ ApII.10

IMPOSSIBLE MOLECULAR REARRANGEMENT

Though the two isomers are said to be ***very close in energy***, methyldiazene is 0.4Kcal/mol more stable than formaldehyde hydrazone.[25] "Tautomerization" of *methyldiazene* to formaldehyde-hydrazone in Ruthenium and Osmium complexes, has in the past been reported[26]. The pores of the Transition metals broke the π-bond. Nevertheless, (A) above is not the real activated state of formaldehyde hydrazone and therefore cannot exist. Its activated state is (B) as will shortly be shown. Therefore, (C) must have been formed instantaneously without the appearance of (A), that which is impossible. The movement of H and activation of the π-bond taking place inside the pores or cages, is possible. In order to show further that in the absence of a passive catalyst, -N=N- bond can never be activated, consider the conversion of dimethyldiazene to formaldehyde methylhydrazone and then to methylhydrazine hydrochloride and formaldehyde when boiled with aqueous hydrochloric acid.[10]

Stage 1:

$$H_3C - N = N - CH_3 \xrightleftharpoons[\text{of Dimethyldiazene}]{\text{Equilibrium State of Existence}} \quad H\cdot e \ + \ n\cdot CH_2 - N = N - CH_3$$

ApII.11

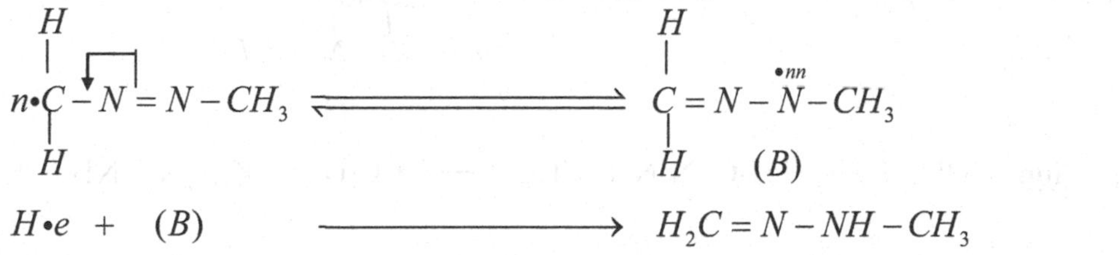

$$\begin{array}{c} H \\ | \\ n\cdot C - N = N - CH_3 \\ | \\ H \end{array} \xrightleftharpoons \begin{array}{c} H \\ | \\ C = N - \overset{\bullet nn}{N} - CH_3 \\ | \\ H \quad (B) \end{array}$$

$$H\cdot e \ + \ (B) \xrightarrow{\hspace{3cm}} H_2C = N - NH - CH_3$$

(*C*) *Formaldehyde methylhydrazone*

Overall Equation: $\quad H_3C\text{-}N\text{=}N\text{-}CH_3 \xrightarrow{\hspace{2cm}} H_2C\text{=}N\text{-}NH\text{-}CH_3 \qquad$ ApII.12

Without the rearrangement above, dimethyldiazene cannot be hydrolyzed, because the –N=N- bond cannot ordinarily be activated in the absence or presence of passive catalysts. This re-arrangement takes place when dimethyldiazene is being decomposed, but not for methyldiazene. For azo-aryls compounds, it is the reverse as already shown in Equation ApII.4 and therefore more stable than azo-alkanes. Shown below are the Equilibrium states of existences of formaldehyde, formaldehyde methylhydrazone, and formaldehyde phenylhydrazone.

$$\underset{H}{\overset{H}{C}}=O \; \rightleftharpoons \; H{\bullet}e \; + \; n{\bullet}C=O \;\; ; \;\; \underset{H\,HNCH_3}{\overset{H}{C}}=N \; \rightleftharpoons \; H{\bullet}e \; + \; n{\bullet}\overset{H}{C}=N \atop HNCH_3 \qquad\qquad \text{ApII.13}$$

$$\underset{\underset{H}{N}\langle\bigcirc\rangle}{\overset{H}{C}}=N \;\rightleftharpoons\; H{\bullet}e \;+\; \overset{H}{C}=N \atop \underset{H\;\;{\bullet nn}}{N}\langle\bigcirc\rangle \;\;;\;\; \underset{H\;\underset{H}{N}-H}{\overset{H}{C}}=N \;\rightleftharpoons\; \overset{{\bullet}n}{C}=N \atop \underset{H\;\underset{H}{N}-H}{} \;+\; H{\bullet}e \qquad \text{ApII.14}$$

Note the distinct difference between alkyl and phenyl groups. The phenyl group is almost like H in radical-pushing capacity, except when resonance stabilization is being provided by the phenyl groups.

Stage 2:

$$Cl^{-}......^{+}OH_3 \; \underset{Dilute\;HCl}{\overset{Equilibrium\;State\;of\;Existence}{\rightleftharpoons}} \; H{\bullet}e \;+\; Cl^{-}........\overset{+}{\underset{{\bullet nn}}{O}}H_2$$

$$H{\bullet}e \;+\; \underset{H\;HNCH_3}{\overset{H}{C}}=N \; \underset{methylhydrazone}{\overset{Activation\;of\;Formaldehyde}{\rightleftharpoons}} \; H{\bullet}e \;+\; e{\bullet}\underset{H\;HNCH_3}{\overset{H}{C}}-N{\bullet}nn$$

$$\rightleftharpoons \; H-\underset{H_3CNH\;H}{\overset{H}{N}-C}{\bullet}e \;\;(A)$$

$$Cl^{-}........\overset{+}{\underset{{\bullet nn}}{O}}H_2 \;\rightleftharpoons\; Cl{\bullet}nn \;+\; H_2O$$

$$Cl{\bullet}nn \;+\; (A) \;\longrightarrow\; Cl-\underset{H\;\underset{H}{N}-CH_3}{\overset{H}{C}}-\overset{H}{N}-H \qquad \text{ApII.15}$$

<u>Overall Equation:</u> $HCl + H_2O + H_2C{=}N{-}NH{-}(CH_3) \longrightarrow H_2O + ClCH_2NH{-}NH\,(CH_3)$

$$\text{ApII.16}$$

Stage 3:

$$H_2O \;\rightleftharpoons\; H{\bullet}e \;+\; nn{\bullet}OH$$

$$H{\bullet}e \;+\; ClCH_2NH-NH(CH_3) \;\rightleftharpoons\; HCl \;+\; e{\bullet}CH_2NH-NH(CH_3)$$

$$HO{\bullet}nn \;+\; e{\bullet}CH_2NH-NH(CH_3) \;\longrightarrow\; HO-CH_2NH-NH(CH_3) \qquad \text{ApII.17}$$

<u>Overall Equation:</u> $H_2O + ClCH_2NH{-}NH(CH_3) \longrightarrow HCl + HO{-}CH_2NH{-}NH(CH_3)\backslash$

$$\text{ApII.18}$$

Water alone cannot activate the hydrazine, since the hydrazine will remain in Equilibrium state of existence. Hence, it is acid hydrolysis.

Stage 4:

$$HO-CH_2NH-NH(CH_3) \rightleftharpoons H \cdot e + nn \cdot O-CH_2NH-NH(CH_3)$$

$$nn \cdot O - \overset{\overset{H}{|}}{\underset{\underset{H}{|}}{C}} - NH - NH(CH_3) \rightleftharpoons nn \cdot O \overset{\overset{H}{|}}{\underset{\underset{(C)}{}}{C}} \cdot e + nn \cdot NH - NH(CH_3)$$

$$H \cdot e + nn \cdot NH - NH(CH_3) \rightleftharpoons H_2N - NH(CH_3)$$

$$(C) \xrightarrow{Deactivation} O = CH_2 + Heat$$

ApII.19

Overall Equation: $HOCH_2NH\text{-}NH(CH_3) \longrightarrow O{=}CH_2 + H_2N\text{-}NH(CH_3) +$

Heat   ApII.20

Stage 5:

$$HCl \rightleftharpoons H \cdot e + Cl \cdot nn$$

$$H \cdot e + H_2N - NH(CH_3) \rightleftharpoons en \cdot NH_3 - NH(CH_3)$$

$$Cl \cdot nn + en \cdot NH_3 - NH(CH_3) \longrightarrow Cl^- \ldots\ldots^+ NH_3 - NH(CH_3)$$

Methylhydrazine hydrochloride   ApII.21

Overall overall equation: $H_3C\text{-}N{=}N\text{-}CH_3 + H_2O + HCl \longrightarrow O{=}CH_2 + Cl^{\ominus}\ldots\ldots^{\oplus}NH_3\text{-}NH(CH_3)$

\+ Heat   ApII.22

ii) DECOMPOSITION MECHANISM

Very large fractions in Equilibrium state of Existence (Ninety percent) have been consumed, leaving the remaining fraction in Stable state of Existence. This is consumed in parallel with Equilibrium mechanism. However, regardless the fraction in Stable state of Existence, the same products are obtained at the specific operating conditions as shown below.

Stage 1a:

$$H - N = N - CH_3 \longrightarrow N_2 + H \cdot e + n \cdot CH_3$$

$$H \cdot e + n \cdot CH_3 \rightleftharpoons CH_4$$

ApII.23a

Overall Equation: $5H\text{-}N{=}N\text{-}CH_3 \longrightarrow 5N_2 + 5CH_4$ ApII.23b

Overall overall equation: $70H\text{-}N{=}N\text{-}CH_3 \longrightarrow 70N_2 + 70CH_4$ ApII.24

Stage 1b:

$$2H-N=N-CH_3 \longrightarrow 2N_2 + 2H{\bullet}e + 2n{\bullet}CH_3$$

$$2n \cdot CH_3 \longrightarrow 2H \cdot n + 2e \cdot \underset{\underset{H}{|}}{\overset{\overset{H}{|}}{C}} \cdot n$$

$$(A)$$

ApII.25a

$$2(A) \longrightarrow H_4C_2$$

$$2H \cdot e + 2H \cdot n \rightleftharpoons 2H_2$$

Overall Equation: $30H\text{-}N{=}N\text{-}CH_3 \longrightarrow 30N_2 + 30H_2 + 15H_4C_2$ ApII.25b

Overall overall equation: $100H\text{-}N{=}N\text{-}CH_3 \longrightarrow 100N_2 + 70CH_4 + 30H_2 + 15H_4C_2$

ApII.26

Stage 1b requires higher operating condition than Stage 1a. Hence more moles were involved in Stage 1a than Stage 1b. Thus, methyldiazene which is now to be called Azo-methane has been decomposed via Decomposition mechanism to produce not only methane and nitrogen as main product, but also ethene and hydrogen. In the first stage (Stage 1a). seventy percent of the met-hyldiazene in Stable State of Existence was consumed to produce mainly its identity-CH_4 and N_2. In the second stage (Stage 1b), which required higher operating conditions to decompose n.CH_3, thirty percent was used to produce ethene, hydrogen and nitrogen. Two moles as opposed to one mole in the first stage, were required here in the second stage. Worthy of note is that no initiators could be obtained. The same would have applied if the reaction was Bimolecular as opposed to Unimolecular reactions of Stages 1a and 1b, but differently as shown below.

Stage 1:

$$2H-N=N-CH_3 \longrightarrow 2N_2 + 2H{\bullet}e + 2n{\bullet}CH_3$$

$$2H{\bullet}e + 2H-N=N-CH_3 \longrightarrow 2H_2 + 2N_2 + 2e{\bullet}CH_3$$

$$2e{\bullet}CH_3 \longrightarrow 2e \cdot \underset{\underset{H}{|}}{\overset{\overset{H}{|}}{C}} {\bullet} n + 2H{\bullet}e$$

$$(A)$$

$$2(A) \longrightarrow H_2C=CH_2$$

$$2H{\bullet}e + 2n{\bullet}CH_3 \rightleftharpoons 2CH_4$$

ApII.27a

Overall equation: $100H\text{-}N{=}N\text{-}CH_3 \longrightarrow 100N_2 + 50CH_4 + 25H_2C{=}CH_2 + 50H_2$

ApII.27b

Notice the difference between Equation ApII.26 (Unimolecular) and Equation ApII.27a (Bi-molecular). Since the decomposition of $n\bullet CH_3$ in Stage 1b of Equation ApII.25 is disturbing at such low operating conditions, it is the above that is favored.

It should be noted that both Equilibrium and Decomposition mechanisms are taking place at the same time, side by side, that is, in parallel. So also are Stages 1a and 1b taking place in parallel. Addition of Equations ApII.9 and ApII.26 gives the final overall equation.

Final Overall Equation: $1000H_3C\text{-}N\text{=}N\text{-}H \longrightarrow 1000N_2 + 970CH_4 + 15C_2H_4 +$

$$30H_2 \qquad\qquad \text{ApII.28}$$

Addition of Equation ApII.9 and ApII.27b also gives the following overall equation.

Final Overall Equation: $1000H_3C\text{-}N\text{=}N\text{-}H \longrightarrow 1000N_2 + 950CH_4 + 25C_2H_4 +$

$$50H_2 \qquad\qquad \text{ApII.29}$$

No initiator could be obtained both ways. It is the last case that is favored, based on what has been said analytically and the significance of bimolecular operating conditions.

2.3 Catalyzed Decomposition of Azo-Methane

When Azo- methane is adsorbed on clean Pt (III) surface or nitrogenase active site FeMo cofactor[27], the followings are to be expected.

Stage 1:

$$H-N=N-CH_3 \xrightarrow[\text{pores}]{\text{Activation inside the}} \quad nn\bullet \overset{\overset{\textstyle H}{|}}{N}-\underset{\underset{\textstyle CH_3}{|}}{N}\bullet en \quad (A)$$

$$\text{ApII.30}$$

$$(A) \xrightarrow{\text{Breakage of }\sigma-bond} nn\bullet\overset{\overset{\textstyle H}{|}}{\underset{\bullet\bullet}{N}}\bullet en \quad + \quad nn\bullet\overset{\overset{\textstyle CH_3}{|}}{\underset{\bullet\bullet}{N}}\bullet en$$

Overall Equation: $10H\text{-}N\text{=}N\text{-}CH_3 \longrightarrow 10N\text{-}H + 10N\text{-}CH_3$ ApII.31

Stage 2:

$$nn\bullet\overset{\overset{\textstyle CH_3}{|}}{N}\bullet en \xrightarrow{\text{Release of H atom}} H\bullet e + nn\bullet N \overset{\overset{\textstyle H-C-H}{||}}{} + Heat$$

$$(B)$$

$$(B) \xrightarrow{\text{Release of H hydride}} H\bullet n + H-C\equiv N + Heat$$

$$H\bullet e + H\bullet n \underset{\text{of }H_2}{\overset{\text{Equilibrium State of Existence}}{\rightleftharpoons}} H_2 \qquad\qquad \text{ApII.32}$$

Overall Equation: $10N\text{-}CH_3 \longrightarrow 10H_2 + 10H\text{-}C\equiv N$ ApII.33

Overall overall equation: $10H\text{-}N\text{=}N\text{-}CH_3 \longrightarrow 10H_2 + 10H\text{-}C\equiv N + 10N\text{-}H$

$$\text{ApII.34}$$

Stage 3:

$$2H-C \equiv N \xrightarrow{\text{\textit{Inside the Pores of Pt(III)}}} 2e \bullet \overset{\overset{H}{|}}{C} = N \bullet nn$$

$$2e \bullet \overset{\overset{H}{|}}{C} = N \bullet nn \longrightarrow nn \bullet N = \overset{\overset{H}{|}}{C} - \overset{\overset{H}{|}}{C} = N \bullet nn \quad (E)$$

$$(E) \longrightarrow N \equiv C - C \equiv N \quad + \quad 2H \bullet e$$

$$(\textit{Cyanogen})$$

$$2H \bullet e \longrightarrow H_2$$

ApII.35

Overall Equation: $4H\text{-}C \equiv N \longrightarrow 2N \equiv C\text{-}C \equiv N \quad + \quad 2H_2$ ApII.36

Overall overall equation: $10H\text{-}N=N\text{-}CH_3 \longrightarrow 12H_2 \ + \ 6H\text{-}C \equiv N \ + \ 10N\text{-}H \ +$
$$2N \equiv C\text{-}C \equiv N \qquad \text{ApII.37}$$

Stage 4:

$$H \bullet e \ + \ nn \bullet \overset{\overset{H}{|}}{\underset{\bullet\bullet}{N}} \bullet en \longrightarrow H - \overset{\overset{H}{|}}{\underset{\bullet\bullet}{N}} \bullet en \ (D) \qquad \text{ApII.38}$$

$$(D) \ + \ n \bullet H \xrightarrow{\text{\textit{Combination State of Existence}} \atop \text{\textit{of Ammonia}}} NH_3$$

Overall Equation: $10H_2 \ + \ 10H\text{-}N \longrightarrow 10NH_3$ ApII.39

Overall overall equation: $10H\text{-}N=N\text{-}CH_3 \longrightarrow 2H_2 \ + \ 2N \equiv C\text{-}C \equiv N \ + \ 10NH_3 \ + \ 6H\text{-}C \equiv N$

ApII.40

Very different types of products are observed to be obtained in the presence of these passive catalysts. In the first stage, the azo-methane is broken down to give two unique products, N-H and N-CH$_3$ both carrying non-free-radicals on the N center. In the second stage, the entire second product was consumed to give H$_2$ and HCN. It is important to note how components are released. The hydrogen molecules produced here are not in Stable state of Existence, because of the presence of these unique catalysts. Hence the reactions are as indicated in Stage 4 where ammonia is finally produced. In the third stage of four steps, cyanogen was obtained to produce more hydrogen. The ease of any stage is however partly determined by the number of steps involved and any stage which cannot be interpreted into Mathematical language without the need for necessary or unnecessary assumptions is no stage.

3.0 Conclusions.

For the first time, Azo-compounds have been renamed to match not only the types of products obtained when decomposed where possible or how they are obtained, but also for the purpose of orderliness. This is like the renaming of Alkenes family members (e.g. ethylene to ethene, propylene to propene) when the Alkylene family exists, of which the first member is methylene, followed by ethylene, and by propylene and so on.

For the first time, Azo-hydrocarbons have been re-classified, in which the real Azo-hydrocarbons were shown. All what have been universally called Azo-hydrocarbons are just isomers of their real or primary Azo-hydrocarbons. Alkyls-, vinyls-, aryls-, etc., diazenes are the primary azo-hydrocarbons. Any compound containing the – N=N – group is an AZO-compound regardless what the two ends are carrying. It is what they are carrying that determines the family to which they belong.

In this first part of their decompositions, one started with the first member of Azo-Alkanes, i.e., azo-methane. There are countless numbers of operating conditions involved in their decompositions-presence or absence of passive catalysts, temperature, pressure, methods of heating, the type of reactors used for their decompositions and more. In the absence of catalyst, the operating conditions are mild as was applied herein, noting that one is not cracking the azo-compound (harsh conditions). In the presence of passive catalyst, the operating conditions are far milder, yet the products look like cracking products and their mechanisms are different. The products obtained obviously depend in part on the operating temperatures and pressures. However, the products obtained catalytically and non-catalytically are uniquely different. The passive catalysts are Transition metal types. For the first time, the mechanisms of these complex unimolecular decomposition reactions have begun to be provided as a Science and not as an Art as has been the case since antiquity.

References

1. S. N. E. Omorodion, *"Chemistry and Humanity-Challenges to be faced as we advance in the Third Millennium: New Classifications for Radicals."*, Nigerian Journal of Applied Science, Vol 26, (2008), Pgs. 146-158.

2. S. N. E. Omorodion, *"Chemistry and Humanity-Challenges to be faced as we advance in the Third Millennium: The impacts of new classifications for Radicals."* Nigerian Journal of Applied Science, Vol.27, (2009), Pgs. 33-43..

3. S. N. E. Omorodion, *"Chemistry and Humanity- Challenges to be faced as we advance in the Third Millennium: New Classifications for Bonds and Charges."* Nigerian Journal of Applied Science, Vol 27, (2009), Pgs. 94-104.

4. S. N. E. Omorodion, *"Chemistry and Humanity-Challenges to be faced as we advance in the Third Millennium: The impacts of the new classifications for Bonds and Charges."* Nigerian Journal of Applied Science, Vol 28, (2010). Pgs. 37-45.

5. S. N. E. Omorodion, *"Chemistry and Humanity-Challenges to be faced as we advance in the Third Millennium: New classifications for Compounds."* Nigerian Journal of Applied Science, Vol 28, (2010), Pgs. 80-87.

6. S. N. E. Omorodion, *"Chemistry and Humanity-Challenges to be faced as we advance in the Third Millennium: The impacts of the new classifications for compounds."* Nigerian Journal of Applied Science, Vol 28, (2010), Pgs. 128-134.

7. S. N. E. Omorodion, *"New Frontiers in Engineering, Sciences and the Arts"* Vol. VI, under review before going to press.

8. S. N. E. Omorodion, "The Hydrocarbon Family Tree." Appendix I, Vol.(II).

9. "Definition of Diimide", **http://www.google.com.ng/search?hl=en&defl=en&q=define**: Diimide&ei=YSoUStWDR8... (2009).

10. C. R. Noller, *"Textbook of Organic Chemistry"*, W. B. Saunders Company, 1966, pgs. 247-250.

11. C. R. Noller, *"Textbook of Organic Chemistry"*, W. B. Saunders Company, 1966, pg. 205.

12. "Azo compounds" **http://en.wikipedia.org/wiki/Azo_compounds** (2009).

13. "Process for preparing azoalkane mixture containing at least one unsymmetrical azoalkane", United States Patent 4604455, (1986).

14. "Azo compounds-definition of Azo compounds in the Medical dictionary", Mosby's Medical Dictionary, 8th edition. (2009).

15. "Azo compounds",**http://www.britannica.com/EBchecked/topic/46915/azo-compound** (2009).

16. P. Berlowitz, B. L. Yang, J. B. Butt, and H. H. Kung, "Reactions of azomethane on a clean, H-covered, and O-covered Pt (111) surface", Surface Science, Volume 17i, issue i (1986) pgs. 69-82.

17. A. I. Vogel, "A Text-Book of Practical Organic Chemistry", English Language Book Society and Longmans, Green & CO LTD, 3rd Ed., 1955, pg.592.

18. "2, 2'-Azobis[2-(2-imidazolin-2-yl)propane]disulfate dehydrate", **http://www.wako-chem.co.jp/specialty/waterazo/VA-046B.htm** (2009).

19. C. R. Noller, *"Textbook of Organic Chemistry"*, W. B. Saunders Company, 1966, pg. 380.

20. P. S. Engel, H. Wu, and W. B. Smith, "Strongly UV Absorbing Bifunctional Azoalkanes", Org. Lett., 3(20), 2001, pgs. 3145-3148.

21. E. M. Stuve, R. J. Madix, and C. R. Brundle, "The adsorption and reaction of ethylene on clean and oxygen covered Pd (100)", Surface Science, Volumes 152-153. Part 1, (1985) pgs. 532-542.

22. C. E. Megiris, P. Berlowitz, J.B. Butt, and H. H. Kung, "Temperature programmed desorption study of acetylene on a clean, H-covered, and O-covered Pt (111) surface", Surface Science, Volume159, Issue 1, (1985) pgs. 184-198.

23. P. Berlowitz, B. L. Yang, J. B. Butt, and H. H. Kung, "Reactions of diazomethane on a clean, H-covered, and O-covered Pt (111) surface", Surface Science, Volume 159, issues 2-3 (1985) pgs. 540-554.

24. B. Friuhberger and J. G. Chen, "Reaction of Ethylene with clean and Carbide-Modified Mo (110): Converting Surface Reactivities of Molybdenum to Pt –Group Metals" J. Am. Chem. Soc., 118 (46), (1996) pgs. 11599-11609.

25. M. L. McKee, "Theoretical study of several diazaalkenes and diazaalkyl radicals", J. Am. Chem. 112 (22), (1990) pgs. 7957-7961.

26. G. Albertin, S. Antoniutti, A. Bacchi, F. De Marchi, and G. Pellzzi, "Tautomerization of methyldiazene to formaldehyde-hydrazone in ruthenium and osmium complexes", Inorg. Chem, 44 (24), (2005) pgs. 8947-8954.

27. B. M. Barney, D. Lukoyanov, T. Yang, D. R. Dean, B. M. Hoffman, and L. C. Seefeldt, **http://www.pnas.org/content/103/46/17113.short** (2006).

Appendix III

THE MECHANISMS OF NON-CATALYZED AND CATALYZED DECOMPOSITIONS OF AZO-HYDROCARBONS. PART II: AZ-ETHANE. (NON-CATALYZED)

Abstract

Herein contains the non-catalyzed decomposition of Azo-ethane ($H-N=N-C_2H_5$) and its isomer ($H_3C-N=N-CH_3$). Based on the new Classifications for Azo-hydrocarbons[1], the isomer which today is called azo-methane has been renamed azo-ethane, a product of N_2 and ethane. When decomposed, different products are obtained depending on the method of decomposition and operating conditions. Of all the operating conditions, temperatures and pressures are the most important. In all the methods, heats in different forms are involved. Herein, the mechanisms for the thermal decomposition as used in the past[2] are provided in order to explain how products reportedly obtained are obtained. The provisions will also apply to others where other forms of heat are involved non-catalytically based on their operating conditions.

Keywords:

1. Radical-pushing groups.
2. Equilibrium mechanism.
3. Decomposition mechanism.
4. Molecular rearrangement by Electro-radical movement.
5. Dimethyldiazene.

1.0 INTRODUCTION

The dawn has arrived for change. In the last part[1], Azo-hydrocarbons were re-classified. So also were the Azo-alkanes re-named for the purpose of orderliness, that which is Nature[3]. Based on the new classification and re-naming, the real azo-ethane in question is ethyldiazene ($H-N=N-C_2H_5$). This unlike azo-methane has two isomers as shown below.

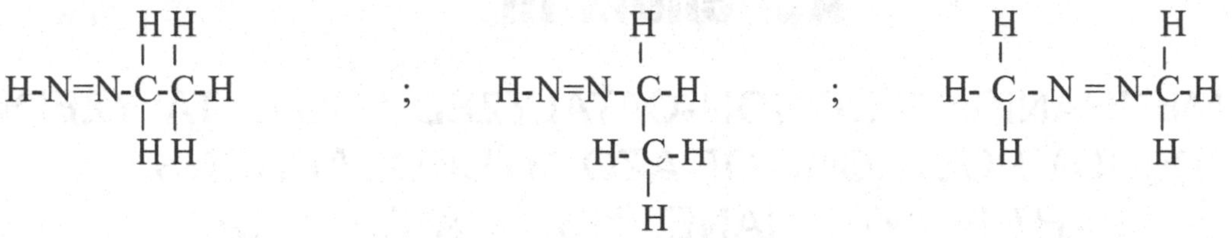

$$
\begin{array}{ccc}
\underset{\substack{|\;\;| \\ H\;H}}{\overset{\substack{H\;H \\ |\;\;|}}{H-N=N-C-C-H}} & ; & \underset{\substack{| \\ H-C-H \\ | \\ H}}{\overset{\substack{H \\ |}}{H-N=N-C-H}} & ; & \underset{\substack{| \\ H}}{\overset{\substack{H \\ |}}{H-C}}-N=N-\underset{\substack{| \\ H}}{\overset{\substack{H \\ |}}{C-H}} \quad \text{ApIII.1}
\end{array}
$$

(A) 1-Ethyldiazene   **(B) 2-Ethyldiazene**   **(C) Dimethyldiazene**

(A) and (B) are the same currently, because Equilibrium State of Existence for both of them is from the nitrogen center carrying H. But when the H is replaced with C_2H_5, then (B) becomes a true isomer as will be shown downstream. Therefore, the only isomer of azo-ethane is (C). What today is called azo-methane ($H_3C-N=N-CH_3$) is an isomer of azo-ethane ($H-N=N-C_2H_5$) and what today is called azo-ethane ($H_5C_2-N=N-C_2H_5$) is also an isomer of azo-butane ($H-N=N-C_4H_9$).

In 1973 published online 2004, thermal decomposition of "azomethane" ($H_3C-N=N-CH_3$) studied in **a static system** at temperatures between **250⁰ and 320⁰C** and at pressures between **5 and 402 torr. (0.0066 and 0.53 Atm.)**, with particular attention to identification of products was studied.[2] Major products reportedly obtained in decreasing order of importance, were *nitrogen, methane, ethane, methylethyldiimide, dimethylhydrazone, propane, tetramethyl-hydrazine, ethylene, methylpropyldiimide, and methylethylhydrazone.* Formations of *short chain polymers* were also reported. Mechanisms which looked meaningful as an Art were provided.

In 1971, **a shock tube** coupled to a time-of-flight mass spectrometer was used to study the reactions of methyl radicals at temperatures much higher than before. The radicals were formed by heating mixtures containing "azomethane" or dimethylmercury highly diluted in neon to between **1120 and 1400 K** in **reflected shock waves**. The products were *methane, ethane and ethylene (Ethene).* The effect of adding ethane and ethylene to the reaction mixture were also investigated. The proposed possible mechanism for the hydrocarbon product formation was Chain mechanism, of which the main reactions were given as follows[4]-

1. $CH_3 + CH_3 \longrightarrow C_2H_6$; **2.** $CH_3 + C_2H_6 \longrightarrow C_2H_5 + CH_4$,

3. $C_2H_5 \longrightarrow C_2H_4 + H$; **4.** $H + C_2H_6 \longrightarrow C_2H_5 + H_2$

ApIII.2

As can be observed with this type of heat at very high temperatures, with the exception of ethylene no cracking of the products took place. However, this provision is still an Art and not a Science.

In 1962, in determination of rate of decomposition of "azomethane", pyrolysis of "azomethane" was carried out in **a shock tube** over the temperature range **800-1300 K.** Concentrations of "azomethane" of 1-3% in argon were used. The products obtained were only *N_2 and C_2H_6* and the mechanism was said to be *Unimolecular* as opposed to *Chain mechanism.*[5] Even at **500 K and 60mm** pressure, the same products were obtained and the rate of decom-position due to the

radical chain was said to be approximately equal to the ***unimolecular*** rate[6]. In another study in 1937, the thermal decomposition of aliphatic azo compounds, particularly that of "azo-methane" was considered as examples of simple ***unimolecular*** reactions. The decomposition was carried out in **a special reaction bulb** placed in an **electrically heated air bath** provided with a temperature controller. The products obtained were _unreacted "azo-methane", N_2, methane, ethane, and traces of ethene._[2] These were carried out in order to find out if the theories of unimolecular reaction as proposed by some authors[8-10] who believed that the decomposition of "azomethane" is almost entirely represented by the equation $CH_3NNCH_3 = C_2H_6 + N_2$, was correct.

In recalling some of the past above, the operating conditions, type of reactors used, type of heat and the products obtained were differently highlighted. It is well known that frequently "azomethane" cannot easily be isolated, because it isomerizes spontaneously to the hydrazone.[11] "Azomethane" is also well known to be unstable above 200^0C, and yields methyl radicals and nitrogen. According to the equation below, radicals were unknown.

$$CH_3-N=N-CH_3 \longrightarrow N_2 + 2CH_3\bullet \qquad\qquad ApIII.3$$

The "methyl radicals" will combine to give ethane, but in the presence of other molecules, *they may abstract hydrogen or initiate free-radical chain reaction if present alone in the system.*[11]

The issue of whether the mechanism is a chain reaction or unimolecular reaction should not arise if the mechanisms of how Nature operates were known. Only three mechanisms exist in Nature-Equilibrium, Decomposition and Combination. ***However, the issue is still important,*** since decomposition of ammonia[12] is Bimolecular-

$$2NH_3 \longrightarrow N_2 + 3H_2. \qquad\qquad .ApIII.4$$

In Bimolecular reactions, a fraction of the component is kept in Equilibrium State of Existence, while the remaining fraction is in Stable State of Existence, with decomposition involving both of them dependently.[12] In Unimolecular reactions, the component is fully kept either in Stable or Equilibrium States of Existence or part of the component is kept in Equilibrium State of Existence decomposing all alone by itself while the remaining fraction is in Stable State of Existence decomposing also all alone by itself, i.e., independently via Decomposition state of existence. For example, consider the decomposition of the first two members of azo-alkanes family bimolecularly via Equilibrium mechanism only.

2.0 Unimolecular and Bimolecular mechanisms in Equilibrium mechanism systems for Methyldiazene:

Stage 1:

$$H-N=N-CH_3 \rightleftharpoons H\bullet e + nn\bullet N=N-CH_3$$

$$H\bullet e + H_3C-N=N-H \rightleftharpoons CH_4 + en\bullet N=N-H$$

$$en\bullet N=N-H \rightleftharpoons N_2 + H\bullet e + Heat$$

$$nn\bullet N=N-CH_3 \xrightarrow{Heat} nn\bullet N=N\bullet en + n\bullet CH_3$$

$$H \bullet e \ + \ n \bullet CH_3 \rightleftharpoons CH_4$$

$$nn \bullet N = N \bullet en \xrightarrow{\text{Deactivation}} N_2 \ + \ Heat$$

Overall Equation: $2H_3C\text{-}N\text{=}N\text{-}H \longrightarrow 2CH_4 \ + \ 2N_2$ ⟶ ApIII.5

It is important to compare the above with Equation ApII.8 which is unimolecular.
For Dimethyldiazene:

Stage 1:

$$H_3C - N = N - CH_3 \rightleftharpoons H \bullet e \ + \ n \bullet CH_2 - N = N - CH_3$$

$$H \bullet e \ + \ H_3C - N = N - CH_3 \rightleftharpoons CH_4 \ + \ en \bullet N = N - CH_3$$

$$en \bullet N = N - CH_3 \rightleftharpoons N_2 \ + \ e \bullet CH_3 \ + \ Heat$$

$$n \bullet CH_2 - N = N - CH_3 \xrightarrow{Heat} n \bullet \overset{H}{\underset{H}{C}} \bullet e \ + \ nn \bullet N = N \bullet en \ + \ n \bullet CH_3$$

$$H_3C \bullet e \ + \ n \bullet CH_3 \xrightarrow[\text{at above B pt of } C_2H_6]{\text{Combination State of Existence}} C_2H_6$$

$$nn \bullet N = N \bullet en \xrightarrow{\text{Deactivation}} N_2 \qquad \text{ApIII.6}$$

$$UNREACTIVE$$

Presence of more than one single double-headed arrow at the end of a stage clearly indicates that the stage is unreactive and therefore unproductive, since it cannot be interpreted mathematically. For a stage to be productive in Equilibrium mechanism, only one single double headed arrow should appear in the last step. As shown above, only the first member seems to favor bimolecular decomposition not in the manner to be expected, because when the overall equation is divided by 2, the reaction looks unimolecular. This is unlike the case of Equation ApIII.4. Now, if there was no sufficient heat coming from the third step to decompose $nn \bullet N\text{=}N\text{-}CH_3$ in the fourth step of Equation ApIII.5, then the stage will become reactive, stable and unproductive, that which is called solubilization[13] as shown below.

Stage 1:

$$H - N = N - CH_3 \rightleftharpoons H \bullet e \ + \ nn \bullet N = N - CH_3$$

$$H \bullet e \ + \ H_3C - N = N - H \rightleftharpoons CH_4 \ + \ en \bullet N = N - H$$

$$en \bullet N = N - H \rightleftharpoons N_2 \ + \ H \bullet e \ + \ Heat \qquad \text{ApIII.7}$$

$$H \bullet e \ + \ nn \bullet N = N - CH_3 \rightleftharpoons H - N = N - CH_3$$

$$(\textit{Stable, reactive and unproductive})$$

Overall Equation: $2H\text{-}N\text{=}N\text{-}CH_3 \rightleftharpoons N_2 \ + \ CH_4 \ + \ H\text{-}N\text{=}N\text{-}CH_3 \qquad \text{ApIII.8}$

In the same manner, if there was no sufficient heat coming from the third step to decompose $n\bullet CH_2\text{-}N\text{=}N\text{-}CH_3$ in the fourth step of Equation ApIII.6, then the stage becomes productive. $H_5C_2\text{-}N\text{=}N\text{-}CH_3$ along with N_2 and CH_4 become the products as shown below.

Stage 1:

$$H_3C-N=N-CH_3 \rightleftharpoons H\bullet e \;+\; n\bullet CH_2-N=N-CH_3$$

$$II\bullet e \;+\; H_3C-N=N-CH_3 \rightleftharpoons CH_4 \;+\; en\bullet N-N-CH_3$$

$$en\bullet N=N-CH_3 \rightleftharpoons N_2 \;+\; e\bullet CH_3 \;+\; Heat$$

$$H_3C\bullet e \;+\; n\bullet CH_2-N=N-CH_3 \longrightarrow H_5C_2-N=N-CH_3$$

ApIII.9

Overall Equation: $2H_3C\text{-}N\text{=}N\text{-}CH_3 \longrightarrow CH_4 \;+\; N_2 \;+\; H_5C_2\text{-}N\text{=}N\text{-}CH_3$ ApIII.10

Unimolecularly or bimolecularly, all azo-alkanes where they exist decompose when heated above specific temperatures, based on the number of moles present. Hence the decomposition of azo-alkanes cannot yet be said to be unimolecular in character or not. It is not a Chain reaction, but complex with many stages.

One will now begin with the decomposition of Ethyldiazene herein called azo-ethane and its isomer dimethyldiazene presently mistakenly called azomethane. Monoalkyldiazenes are known to be exceedingly good hydrogen-atom donors toward alkyl radicals.[14]

2.1 Azo-Ethanes

There are two isomers of azo-ethane. These are –

$$H-N=N-C_2H_5 \;\xrightarrow[\text{of ethyldiazene}]{\text{Equilibrium State of Existence}}\; II\bullet e \;+\; nn\bullet N=N-C_2H_5$$

ApIII.11

(A) *Ethyldiazene*

$$H_3C-N=N-CH_3 \;\xrightarrow[\text{of Dimethyldiazene}]{\text{Equilibrium State of Existence}}\; H\bullet e \;+\; n\bullet CH_2-N=N-CH_3$$

ApIII.12

(B) *Dimethyldiazene*

2.1.1 Syntheses of Azo-ethanes

i) Using one mole of ethene along with diazene, the followings are to be expected.

Stage 1:

$$H-N=N-H \;\rightleftharpoons\; H\bullet e \;+\; nn\bullet N=N-H$$

(A)

$$H\bullet e \;+\; n\bullet \underset{\overset{|}{H}}{\overset{\overset{H}{|}}{C}}-\underset{\overset{|}{H}}{\overset{\overset{H}{|}}{C}}\bullet e \;\rightleftharpoons\; H_5C_2\bullet e$$

$$H_5C_2 \bullet e \ + \ (A) \quad \longrightarrow \quad H - N = N - C_2H_5$$

AZO-ETHANE   ApIII.13

Overall Equation: H-N=N-H + C_2H_4 (Ethene) $\longrightarrow$ H-N=N-C_2H_5 ApIII.14

ii) Using ethane and N_2 at above boiling point of ethane, the followings are to be expected.

$$H_6C_2 \rightleftharpoons H \bullet e \ + \ n \bullet C_2H_5$$

Stage 1: $\quad H \bullet e \ + \ N \equiv N \underset{\text{Heat}}{\overset{\text{Activation}}{\rightleftharpoons}} H - N = N \bullet en$ ApIII.15

$$H - N = N \bullet en \ + \ n \bullet C_2H_5 \longrightarrow H - N = N - C_2H_5$$

Overall Equation: C_2H_6 + $N \equiv N$ $\longrightarrow$ H-N=N-C_2H_5 ApIII.16

i) For its isomer, the followings are to be expected using one mole of methylene.

$$H - N = N - CH_3 \rightleftharpoons H \bullet e \ + \ nn \bullet N = N - CH_3$$

$$(B)$$

Stage 1:

$$H \bullet e \ + \ n \bullet \overset{\displaystyle H}{\underset{\displaystyle H}{\overset{|}{\underset{|}{C}}}} \bullet e \ \rightleftharpoons \ e \bullet CH_3$$

$$H_3C \bullet e \ + \ (B) \quad \longrightarrow \quad H_3C - N = N - CH_3$$
ApIII.17

Overall Equation: H-N=N-CH_3 + CH_2 (Methylene) $\longrightarrow$ H_3C-N=N-CH_3 ApIII.18

ii) Using C_2H_6 at below boiling point, the followings are to be expected.

Stage 1:

$$H_6C_2 \underset{\text{between Mpt. and Bpt. of Ethane}}{\overset{\text{One of the Equilibrium States at}}{\rightleftharpoons}} H_3C \bullet e \ + \ n \bullet CH_3$$

$$H - N = N - CH_3 \rightleftharpoons H \bullet e \ + \ nn \bullet N = N - CH_3$$

$$H \bullet e \ + \ n \bullet CH_3 \rightleftharpoons CH_4$$

$$H_3C - N = N \bullet nn \ + \ e \bullet CH_3 \longrightarrow H_3C - N = N - CH_3$$

ApIII.19

Overall Equation: H_6C_2 + H-N=N-CH_3 $\longrightarrow$ CH_4 + H_3C-N=N-CH_3 ApIII.20

The reaction is favored if the Equilibrium States of Existence of methyldiazene and ethane are not suppressed. One of the greatest sources of methylene is diazomethylene which in present-day Science is mistakenly called diazo-methane.[15] Based on the synthesis above, one can see why they are called azo-ethanes. In place of methyldiazene above, one can also use N_2.

2.1.2 Decomposition of Azo-ethanes

(a) **Ethyldiazene**

i) **Equilibrium mechanism.**

Stage 1:

$$H_5C_2 - N = N - H \underset{of\ Azo-ethane}{\overset{Equilibrium\ State\ of\ Existence}{\rightleftharpoons}} H\bullet e + nn\bullet N = N - C_2H_5$$

$$nn\bullet N = N - C_2H_5 \underset{Higher\ Temperature}{\overset{More\ Heat\ i.e.}{\rightleftharpoons}} nn\bullet N = N\bullet en + n\bullet C_2H_5 \qquad ApIII.21a$$

$$H\bullet e + n\bullet C_2H_5 \underset{of\ Ethane}{\overset{Equilibium\ State\ of\ Existence}{\rightleftharpoons}} C_2H_6$$

$$nn\bullet N = N\bullet en \underset{of\ Energy}{\overset{De-activation\ with\ release}{\longrightarrow}} N_2$$

Overall Equation: $16C_2H_5\text{-}N\text{=}N\text{-}H \longrightarrow 16N_2 + 16C_2H_6 \qquad ApIII.21b$

The presence of ethane as one of the products signals to us that what has been decomposed is Azo-ethane.

Very large fraction (Eighty percent) has been chosen to exist in Equilibrium state of Existence than Stable state of Existence, over a range of operating conditions. The stable fraction is decomposed as follows.

ii) **Decomposition mechanism.**

Stage 1a:

$$H - N = N - C_2H_5 \longrightarrow N_2 + H\bullet e + n\bullet C_2H_5$$

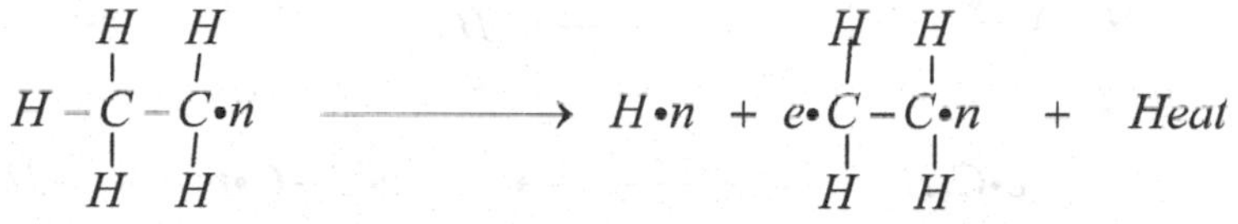

$$\begin{array}{c} H\ \ H \\ |\ \ \ | \\ H - C - C\bullet n \\ |\ \ \ | \\ H\ \ H \end{array} \longrightarrow H\bullet n + \begin{array}{c} H\ \ H \\ |\ \ \ | \\ e\bullet C - C\bullet n \\ |\ \ \ | \\ H\ \ H \end{array} + Heat$$

$$(A)$$

$$(A) \longrightarrow H_4C_2 + Heat$$

$$H\bullet e + H\bullet n \rightleftharpoons H_2$$

$$ApIII.22a$$

Overall Equation: $H\text{-}N\text{=}N\text{-}C_2H_5 \longrightarrow N_2 + H_2 + H_2C\text{=}CH_2 \qquad ApIII.22b$

Stage 1b:

$$H - N = N - C_2H_5 \longrightarrow N_2 + H\bullet e + n\bullet C_2H_5$$

$$H\bullet e + n\bullet C_2H_5 \rightleftharpoons C_2H_6$$

$$ApIII.23a$$

Overall Equation: $3H\text{-}N=N\text{-}C_2H_5 \longrightarrow 3N_2 + 3C_2H_6$ ApIII.23b

Overall overall equation: $4H\text{-}N=N\text{-}C_2H_5 \longrightarrow 4N_2 + H_2 + H_2C=CH_2 + 3C_2H_6$

ApIII.23c

The two stages above are in parallel, i.e., they are taking place at the same time. Seventy-five percent were decomposed mildly, while twenty-five percent were operated under milder operating conditions demanded by the presence of $n.C_2H_5$. Adding Equations ApIII.21b to ApIII.23c, the following overall equation is obtained.

Final overall equation: $20\,H\text{-}N=N\text{-}C_2H_5 \longrightarrow 20N_2 + H_2 + 19C_2H_6 + H_2C=CH_2$

ApIII.24

Only ethane is the hydrocarbon amongst the products, a clear indication that the compound decomposed is indeed Azo-ethane. It is only when the isomer is being decomposed that other hydrocarbons begin to appear as will become apparent. Ethene was only produced during the Decomposition mechanism at higher operating conditions. No initiator could be obtained. Notice that H_2, C_2H_6 both produced under equilibrium and N_2 are parts of the products.

The case above is Unimolecular since only one mole was used throughout in the decom-position mechanisms. Now considering the case where the reaction is Bimolecular, the followings are to be expected.

Stage 1:

$$H-N=N-C_2H_5 \longrightarrow N_2 + H{\bullet}e + n{\bullet}C_2H_5$$

$$H{\bullet}e + H-N=N-C_2H_5 \longrightarrow H_2 + N_2 + e{\bullet}C_2H_5$$

$$e{\bullet}C_2H_5 \longrightarrow e{\bullet}\overset{\displaystyle H}{\underset{\displaystyle H}{C}}-\overset{\displaystyle H}{\underset{\displaystyle H}{C}}{\bullet}n + H{\bullet}e$$

$$(A)$$

$$(A) \longrightarrow H_2C=CH_2 + Heat$$

$$H{\bullet}e + n{\bullet}C_2H_5 \rightleftharpoons C_2H_6$$

ApIII.25a

Overall equation: $4H\text{-}N=N\text{-}C_2H_5 \longrightarrow 4N_2 + 2C_2H_6 + 2H_2C=CH_2 + 2H_2 + Heat$

ApII.25b

Notice the difference between this equation and that of Equation ApIII.23c. Adding Equations ApIII.21b to ApIII.25b, the following overall equation is obtained.

Final overall equation: $20\,H\text{-}N=N\text{-}C_2H_5 \longrightarrow 20N_2 + 2H_2 + 18C_2H_6 + 2H_2C=CH_2$

$$+ \; Heat \qquad ApIII.25c$$

Stage 1 above is the actual mechanism, since the presence of the second step of Stage 1a of Equation ApIII.22a is disturbing. No initiator could be obtained. The mechanism is Bimolecular and not unimolecular.

(b) Dimethyldiazene

In the absence of H directly connected to N, the situation changes. The nitrogen centers carry the same group-CH_3 which has only one carbon element. Two mechanisms just as the cases considered so far come into play both taking place in parallel- Decomposition and Equilibrium mechanisms. Starting with Equilibrium mechanism, the followings are obtained.

i) Equilibrium mechanism

Stage 1:

$$H_3C - N = N - CH_3 \underset{\text{of Diethyldiazene}}{\overset{\text{Equilibrium State of Existence}}{\rightleftharpoons}} H \bullet e \; + \; n \bullet CH_2 - N = N - CH_3$$

$$n \bullet CH_2 - N = N - CH_3 \underset{\text{Higher Temperature}}{\overset{\text{More Heat i.e.}}{\rightleftharpoons}} n \bullet \overset{\displaystyle H}{\underset{\displaystyle H}{C}} \bullet e \; + \; nn \bullet N = N \bullet en \; + \; n \bullet CH_3$$

(Methylene)

$$H \bullet e \; + \; n \bullet CH_3 \underset{\text{of Methane}}{\overset{\text{Equilibrium State of Existence}}{\rightleftharpoons}} CH_4$$

$$nn \bullet N = N \bullet en \xrightarrow[\text{Release of Heat}]{\text{De-activation}} N \equiv N \qquad\qquad \text{ApIII.26}$$

Overall Equation: $24 H_3C\text{-}N\text{=}N\text{-}CH_3 \longrightarrow 24CH_4 \; + \; 24N_2 \; + \; 24\text{Methylene}$

$$\text{ApIII.27}$$

From the products, ***the methylene (or carbine) formed keeps part of the dimethyldiazene that were in stable state of existence back in equilibrium state of existence and used as follows.*** It is carbene that is actually formed along with the formation of H-N=N-CH_3. This decomposed in the next stage to give CH_4 and N_2. Then, Stage 2a and others shown below follow.

Stage 2:

$$H_3C - N = N - CH_3 \;\; \rightleftharpoons \;\; H \bullet e \; + \; n \bullet CH_2 - N = N - CH_3$$

$$H \bullet e \; + \; n \bullet \overset{\displaystyle H}{\underset{\displaystyle H}{C}} \bullet e \;\; \rightleftharpoons \;\; H_3C \bullet e$$

$$H_3C \bullet e \; + \; n \bullet CH_2 - N = N - CH_3 \longrightarrow H_5C_2 - N = N - CH_3 \;\; (A) \; Methylethyldiazene$$

$$\text{ApIII.28}$$

Overall Equation: $14H_3C-N=N-CH_3$ + 14Methylene $\longrightarrow$ $14H_5C_2-N=N-CH_3$

ApIII.29

Overall overall equation: $38H_3C-N=N-CH_3 \longrightarrow 24CH_4$ + $24N_2$ + $14H_5C_2-N=N-CH_3$

+ 10Methylene ApIII.30

With the formation of (A), this is followed by the formation of the next diazene as follows.

Stage 2b:

$$H_3C-N=N-CH_3 \rightleftharpoons H{\cdot}e + n{\cdot}CH_2-N=N-CH_3$$

$$H{\cdot}e + n{\cdot}\overset{H}{\underset{H}{C}}{\cdot}e \rightleftharpoons H_3C{\cdot}e$$

$$H_3C{\cdot}e + n{\cdot}\overset{H}{\underset{H}{C}}{\cdot}e \rightleftharpoons H_5C_2{\cdot}e$$

$$H_5C_2{\cdot}e + n{\cdot}CH_2-N=N-CH_3 \longrightarrow C_3H_7-N=N-CH_3$$

(B) *Methylpropyldiazene* ApIII.31

Overall Equation: $2H_3C-N=N-CH_3$ + 4Methylene $\longrightarrow$ $2H_7C_3-N=N-CH_3$

ApIII.32

In the process of dimethyldiazene rearranging, it consumes the remaining methylene as follows.

Stage 2c:

$$H_3C-N=N-CH_3 \rightleftharpoons H{\cdot}e + n{\cdot}CH_2-N=N-CH_3$$

$$H{\cdot}e + n{\cdot}\overset{H}{\underset{H}{C}}{\cdot}e \rightleftharpoons H_3C{\cdot}e$$

$$n{\cdot}\overset{H}{\underset{H}{C}}-N=N-CH_3 \xrightarrow[\text{free-radical from the } \pi-\text{bond}]{\text{Movement of electro-non}} \overset{H}{\underset{H}{C}}=N-\overset{\cdot nn}{N}-CH_3 \quad (D)$$

$$(D) + e{\cdot}CH_3 \longrightarrow \overset{H}{\underset{H}{C}}=N-\overset{CH_3}{N}-CH_3$$

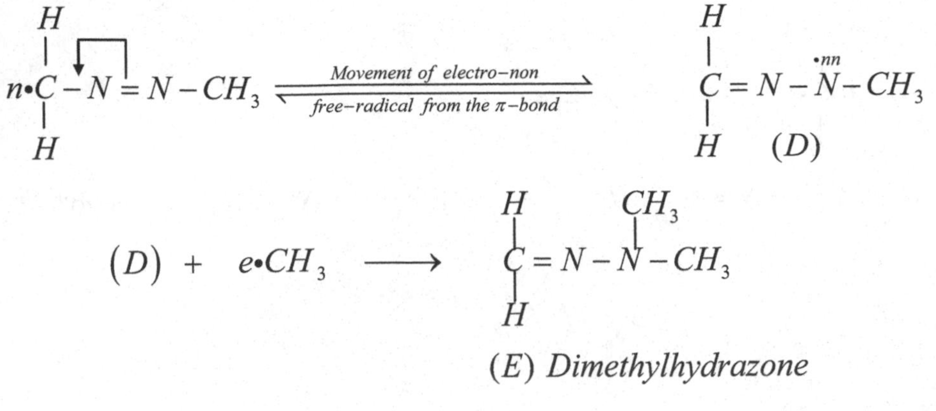

ApIII.33

(E) *Dimethylhydrazone*

Overall Equation: $6 H_3C-N=N-CH_3$ $+$ 6 Methylene $\longrightarrow$ 6 Dimethylhydrazone

$\qquad\qquad\qquad\qquad\qquad\qquad\qquad\qquad\qquad\qquad\qquad\qquad\qquad\qquad$ ApIII.34

Overall overall equation: $46 H_3C-N=N-CH_3 \longrightarrow 24CH_4 + 24N_2 + 14CH_3-N=N-C_2H_5$

$\qquad\qquad\qquad\qquad\qquad + 2H_7C_3-N=N-CH_3 + 6H_2C=N-N(CH_3)_2$ $\qquad$ ApIII.35

All the methylenes produced in the first stage, were fully consumed in the last three stages operating at the same time in parallel. Hence they were numbered as 2a, 2b, and 2c. It is shocking to observe the presence of two azo-alkanes called methylethyldiazene[2], or indeed one of the isomers of Azo-propane in Stage 2a and methylpropyldiazene[2] or indeed one of the isomers of Azo-butane in Stage 2b as two of the products of the decomposition of dimethyl-diazene. The first is an isomer of formaldehyde dimethylhydrazone. The second, in view of its low concentration and the presence of stronger components, cannot exist in Equilibrium state of Existence. It is suppressed and therefore placed in Stable state of Existence all the time. Amongst the products in Stage 1 were 24 moles of methylene which cannot readily combine with them-selves to form ethene or polyethene (Polyethylene), otherwise they cannot be used one after the other as has been done so far. Of the fourteen moles of methylethyldiazene formed, only a small fraction will exist in Equilibrium state of Existence. So far all the remaining dimethyldiazene left after consuming fourty-six moles are in Stable state of Existence. So also are the N_2, and CH_4 in Stable state of Existence.

Stage 3:

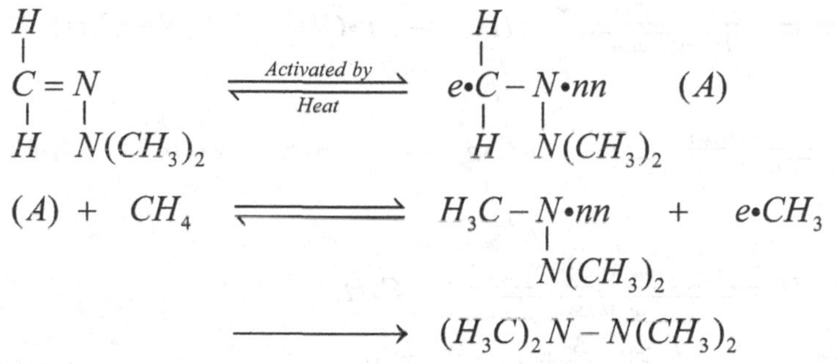

$$\begin{array}{c} H \\ | \\ C = N \\ | \quad | \\ H \quad N(CH_3)_2 \end{array} \underset{Heat}{\overset{Activated\ by}{\rightleftharpoons}} \begin{array}{c} H \\ | \\ e{\cdot}C - N{\cdot}nn \\ | \quad | \\ H \quad N(CH_3)_2 \end{array} \quad (A)$$

$$(A) + CH_4 \rightleftharpoons \begin{array}{c} H_3C - N{\cdot}nn \\ | \\ N(CH_3)_2 \end{array} + e{\cdot}CH_3$$

$$\longrightarrow (H_3C)_2 N - N(CH_3)_2 \qquad\qquad\qquad \text{ApIII.36}$$

Overall Equation: $2CH_4 + 2H_2C=N-N(CH_3)_2 \longrightarrow 2(CH_3)_2N-N(CH_3)_2$ $\qquad$ ApIII.37

Overall overall equation: $46H_3C-N=N-CH_3 \longrightarrow 22CH_4 + 24N_2 + 14CH_3-N=N-C_2H_5$

$\qquad\qquad + 2(CH_3)_2N-N(CH_3)_2 + 2H_7C_3-N=N-CH_3 + 4H_2C=N-N(CH_3)_2$ $\qquad$ ApIII.38

The presence of tetramethylhydrazine was also identified as one of the products of decomposi-tion of dimethyldiazene in the temperature range 290-340^{0}C in 1939.[16] Dimethyldiazene could not be used along with methane and methylene as shown below, because the dimethyldiazene like other alkyldiazenes cannot be activated.

Impossible Stage:

$$CH_4 \rightleftharpoons H{\cdot}e + n{\cdot}CH_3$$

$$H{\cdot}e + n{\cdot}\underset{\underset{H}{|}}{\overset{\overset{H}{|}}{C}}{\cdot}e \rightleftharpoons e{\cdot}CH_3$$

$$H_3C{\cdot}e + \underset{\underset{CH_3}{|}}{\overset{\overset{CH_3}{|}}{N}}=N \xrightleftharpoons[\text{Dimethyldiazene}]{\textit{Forced Activation of}} H_3C-\underset{\underset{CH_3}{|}}{\overset{\overset{CH_3}{|}}{N}}-N{\cdot}en \quad (B)$$

$$\text{ApIII.39}$$

$$H_3C{\cdot}n + (B) \longrightarrow H_3C-\underset{\underset{CH_3}{|}}{N}-\overset{\overset{CH_3}{|}}{N}-CH_3$$

Now is the time and turn of methylethyldiazene to operate, with the small fraction in Equili-brium state of existence. Based on the operating conditions, there is no doubt in one's mind that all these compounds communicate with themselves before their reactions begin.

Stage 4:

$$H_3C-N=N-C_2H_5 \xrightleftharpoons[\text{of Diethyldiazene}]{\textit{Equilibrium State of Existence}} H{\cdot}e + n{\cdot}CH_2-N=N-C_2H_5$$

$$n{\cdot}CH_2-N=N-C_2H_5 \xrightleftharpoons[\textit{Higher Temperature}]{\textit{More Heat i.e.}} n{\cdot}\underset{\underset{H}{|}}{\overset{\overset{H}{|}}{C}}{\cdot}e + nn{\cdot}N=N{\cdot}en + n{\cdot}C_2H_5$$

$$H{\cdot}e + n{\cdot}C_2H_5 \xrightleftharpoons[\text{of Methane}]{\textit{Equilibrium State of Existence}} C_2H_6$$

$$nn{\cdot}N=N{\cdot}en \xrightarrow[\textit{Release of Heat}]{\textit{De-activation}} N \equiv N$$

$$\text{ApIII.40}$$

Overall Equation: $2\,H_3C\text{-}N=N\text{-}C_2H_5 \longrightarrow 2\,N_2 + 2\,C_2H_6 + 2\,\text{Methylene}$

$$\text{ApIII.41}$$

The methylene formed is consumed immediately by the remaining fraction still in equilibrium state of existence.

Stage 5:

$$H_3C - N = N - C_2H_5 \quad \rightleftharpoons \quad H{\cdot}e \quad + \quad n{\cdot}CH_2 - N = N - C_2H_5$$

$$H{\cdot}e \quad + \quad n{\cdot}\overset{\overset{H}{|}}{\underset{\underset{H}{|}}{C}}{\cdot}e \quad \rightleftharpoons \quad H_3C{\cdot}e$$

$$n{\cdot}\overset{\overset{H}{|}}{\underset{\underset{H}{|}}{C}} - N = N - C_2H_5 \quad \xrightleftharpoons[\text{free-radical from the } \pi\text{-bond}]{\text{Movement of electro-non}} \quad \overset{\overset{H}{|}}{\underset{\underset{H}{|}}{C}} = N - \overset{\cdot nn}{N} - C_2H_5$$

$$(C)$$

$$(C) \quad + \quad e{\cdot}CH_3 \quad \longrightarrow \quad \overset{\overset{H}{|}}{\underset{\underset{H}{|}}{C}} = N - \overset{\overset{CH_3}{|}}{N} - C_2H_5$$

$$(D) \; \textit{Methylethylhydrazone}$$

ApIII.42

Overall Equation: $2\, H_3C\text{-}N\text{=}N\text{-}C_2H_5 \; + \; 2\,\text{Methylene} \; \longrightarrow \; 2\, H_2C\text{=}N\text{-}N\text{-}(CH_3)\, C_2H_5$

ApIII.43

Overall overall equation: $46 H_3C\text{-}N\text{=}N\text{-}CH_3 \; \longrightarrow \; 22 CH_4 \; + \; 26 N_2 \; + \; 10 CH_3\text{-}N\text{=}N\text{-}C_2H_5 \; + \; 2 C_2H_6 \; + \; 2 H_2C\text{=}N\text{-}N\text{-}(CH_3)\, C_2H_5 \; + \; 2(CH_3)_2 N\text{-}N\,(CH_3)_2 \; + \; 2 H_7C_3\text{-}N\text{=}N\text{-}CH_3 \; + \; 4 H_2C\text{=}N\text{-}N\,(CH_3)_2$

ApIII.44

In general, in movement of radicals either for chemical reactions or for rearrangements or for flow of current, it is the electro-radicals that diffuse all the time when present. The nucleo-radicals diffuse only when the electro-radical is absent. The so-called Tautomerization which indeed is Electroradicalization can now be seen to be movement of an electro-radical from a π-bond to form another bond with a nucleo-radical when present and form a more stable compound. ***Methyldiazene cannot tautomerize.***[1] ***So also are diethyldiazene (*** $H_5C_2\text{-}N\text{=}N\text{-}C_2H_5$ ***), dipropyldiazene (*** $H_7C_3\text{-}N\text{=}N\text{-}C_3H_7$ ***) and so on. These are cases which do not carry*** CH_3 ***group connected to the nitrogen centers. Not all that carry*** CH_3 ***tautomerize or isomerize, e.g.*** $H_3C\text{-}N\text{=}N\text{-}H$ ***and*** $HO\text{-}N\text{=}N\text{-}CH_3$. Indeed it is nitroso amines (R-NH-N=O) that tautomerizes to the azo-compound R-N=N-OH (known to decompose to ROH and N_2)[17] mistakenly called alkyl diazo-hydroxide[18]. If it was not an azo-compound, the products in parenthesis obtained cannot be obtained.

Thus, forty-six moles of dimethyldiazene have so far been consumed. These are the fractions in Equilibrium State of Existence. Fractions of the fractions of dimethyldiazene, methylethyl-diazene and methylpropyldiazene left in Stable state of existence decompose at the same time as follows.

ii) **Decomposition mechanism**

a) **Dimethyldiazene**

Stage 1a:

$$H_3C-N=N-CH_3 \xrightarrow[\text{of Diethyldiazene}]{\text{Decomposition State of Existence}} N_2 + e{\bullet}CH_3 + n{\bullet}CH_3$$

$$H_3C{\bullet}e + n{\bullet}CH_3 \underset{\text{at temp. between Bpt. and Mpt}}{\overset{\text{Equilibrium State of Existence}}{\rightleftharpoons}} C_2H_6$$

$$OR$$

$$H_3C{\bullet}e + n{\bullet}CH_3 \xrightarrow{\text{Above Bpt of Ethane}} C_2H_6$$

ApIII.45

Overall Equation: $14H_3C\text{-}N=N\text{-}CH_3 \xrightarrow{\hspace{3cm}} 14C_2H_6 + 14N_2$ ApIII.46

Stage 1b:

$$2H_3C-N=N-CH_3 \xrightarrow[\text{of Diethyldiazene}]{\text{Decomposition State of Existence}} 2N_2 + 2e{\bullet}CH_3 + 2n{\bullet}CH_3$$

$$2e{\bullet}CH_3 \xrightarrow{\hspace{3cm}} 2H{\bullet}e + 2e{\bullet}\overset{\displaystyle H}{\underset{\displaystyle H}{\overset{|}{\underset{|}{C}}}}{\bullet}n$$

$$(A)$$

$$2(A) \xrightarrow{\hspace{3cm}} H_2C=CH_2$$

$$2H{\bullet}e + 2n{\bullet}CH_3 \rightleftharpoons 2CH_4$$

ApIII.47a

Overall Equation: $6H_3C\text{-}N=N\text{-}CH_3 \xrightarrow{\hspace{3cm}} 6N_2 + 6CH_4 + 3H_2C=CH_2$

ApIII.47b

Stages 1a and 1b are taking place at the same time in parallel. Based on the operating conditions it is possible that ethane in the last step of Stage 1a, was produced under combination mecha-nism. The operating conditions have been such that seventy percent of the dimethyldiazene were kept in Equilibrium state of Existence, leaving remaining thirty percent in Stable state of Existence. Of the thirty percent, seventy percent were decomposed mildly while the remaining thirty percent were decomposed at higher operating conditions, based on the types of reactors used (absence of perfect mixing); for which the overall equation is as follows-

Overall overall equation: $20\ H_3C\text{-}N=N\text{-}CH_3 \xrightarrow{\hspace{2cm}} 20N_2 + 14C_2H_6 + 6CH_4 + 3C_2H_4$

ApIII.48

So far notice that the hydrocarbons have largely been produced via Equilibrium in the last step of a Decomposition stage. They have never been produced via the combination of two electro-free-radical carrying species such as $2H_3C \bullet e$ combining together to give C_2H_6 as a last step in either Equilibrium or Decomposition mechanism systems. No nucleo-free-radicals could be produced as done when Azobisisobutyronitile shown as (VIII) in Equation ApII.1 is used. The reason could partly be the difference in the groups symmetrically or non-symmetrically carried by the N centers.

Though one had expected to see where most or all the methyl electro-free-radicals combine together to form ethane, this has not been the case here so far. One of the Laws of Physics- "Like poles repel while unlike poles attract" does not fully apply to radicals. In Metaphysics, the situation is different. With Mtaphysics inside Chemistry, one can observe that like poles can also attract to form a stable compound and this takes place only under specific conditions as we have observed so far in particular with respect to the manners by which Equilibrium mechanism operates.[3,19-21] The like unlike poles combination reaction above in Stage 1a of Equation ApIII.45 is one of the Combination States of Existence of Ethane. Other combination States of existence of ethane are as follows

$$H_3C \bullet n + e \bullet CH_3 \xrightleftharpoons[\text{existence of Ethane}]{\text{Equilibrium state of}} C_2H_6 \ [At\ temperature\ of\ below\ Boilung\ po\text{int}\ an\ d$$

$$in\ particular\ M.pt]$$

$$H_3C \bullet e + n \bullet CH_3 \xrightarrow[\text{of Ethane}]{\text{Combination State of Existence}} C_2H_6 [At\ above\ Boiling\ po\text{int}\ of\ Ethane]$$

$$H \bullet e + n \bullet C_2H_5 \xrightleftharpoons[\text{of Ethane}]{\text{Equilibrium State of Existence}} C_2H_6 \ [At\ above\ melting\ po\text{int} - THE\ FINGER\ PRINT]$$

$$ApIII.49$$

These information are based in part on the data provided for boiling-points and melting-points versus number of carbon atoms for alkane family members.[22]

In place of Stages 1a and 1b of Equations ApIII.45 and ApIII.47 where ethane, methane, N_2 and ethene were produced, the followings may take place.

Stage 1:

$$2H_3C - N = N - CH_3 \longrightarrow 2N_2 + 2n \bullet CH_3 + 2e \bullet CH_3$$

$$2H_3C \bullet e + 2H_3C - N = N - CH_3 \longrightarrow 2H_4C + 2e \bullet \overset{\overset{\displaystyle H}{|}}{\underset{\underset{\displaystyle H}{|}}{C}} - N = N - CH_3$$

$$(A)$$

$$2(A) \longrightarrow 2e \bullet \overset{\overset{\displaystyle H}{|}}{\underset{\underset{\displaystyle H}{|}}{C}} \bullet n + 2N_2 + 2e \bullet CH_3$$

$$(B)$$

$$2(B) \longrightarrow H_2C = CH_2$$

$$H_2C \bullet e + e \bullet CH_3 \longrightarrow C_2H_6$$

Overall Equation: $4H_3C-N=N-CH_3 \longrightarrow 2CH_4 + CH_2=CH_2 + 4N_2 + C_2H_6$

$$+ 2n \bullet CH_3 \qquad\qquad ApIII.51$$

(A) above was reported to be the initiator which produced polymers at high pressure and low temperature.[2]

Overall overall equation: $20\ H_3C-N=N-CH_3 \longrightarrow 20N_2 + 5C_2H_6 + 10CH_4 + 5C_2H_4$

$$+ 10n \bullet CH_3 \qquad\qquad ApIII.52$$

Worthy of note is the presence of a nucleo-free-radical initiator which must have been the so-called initiator which produced polymers identified above. Identified above is an electro-free-radical initiator (A) which cannot polymerize in the presence of $n \bullet CH_3$ already present in the first step of the stage, followed by release of N_2 istantaneously and $e \bullet CH_3$ shown in that first step of the stage. It is important to compare the overall overall equations - ApIII.48 and ApIII.52. ***While the former is Unimolecular in character with ranges of operating conditions involved due to mixing, the latter is Bimolecular with one singular operating condition-molar ratio.***

All the dimethyldiazene have been consumed to give the following products all of which are in Stable states of Existence. Based on the overall equation of Equation ApIII.48 for Decom-position, the followings are obtained when Equations ApIII.44 is added to ApIII.48.

OVERALL EQUATION: $66H_3C-N=N-CH_3 \longrightarrow 26CH_4 + 46N_2 + 10CH_3-N=N-C_2H_5$

$+ 2C_2H_4 + 18C_2H_6 + 2H_2C=N-N(CH_3)C_2H_5 + 2(CH_3)_2N-N(CH_3)_2 + 2H_7C_3-N=N-CH_3$

$+ 4H_2C=N-N(CH_3)_2 \qquad\qquad ApIII.53$

When the overall equation of Equation ApIII.52 is added to Equation ApIII.44, the followings are obtained.

OVERALL EQUATION: $66H_3C-N=N-CH_3 \longrightarrow 32CH_4 + 46N_2 + 10CH_3-N=N-C_2H_5$

$+ 5C_2H_4 + 7C_2H_6 + 2H_2C=N-N-(CH_3)C_2H_5 + 2(CH_3)_2N-N(CH_3)_2 + 2H_7C_3-N=N-CH_3$

$+ 4H_2C=N-N\ (CH_3)_2 + $ **$10n \bullet CH_3$** $\qquad\qquad ApIII.54$

Note that ratios of components in Stable and Equilibrium states of existences, have been arbitrarily chosen, based on the operating conditions and the types of products reportedly obtained. The influence of OPERATING CONDITIONS for all things in our world is in-comprensible. When polymerizable radicals are reported to be obtained, they are usually nucleo-radical in character with respect to the use of azo-hydrocarbons.

c) Methylethyldiazene (An isomer of Azo-propane)

About thirty percent of this compound was placed in Equilibrium State of Existence and con-sumed in Stages 4 and 5 of Equations AppIII.40 and AppIII.42 respectively. Part of the remainder in Stable State of Existence is decomposed as follows unimolecularly.

Stage 1a:

$$H_3C - N = N - C_2H_5 \longrightarrow N_2 + n{\cdot}CH_3 + e{\cdot}C_2H_5$$

$$n{\cdot}CH_3 + e{\cdot}C_2H_5 \longrightarrow C_3H_8$$

ApIII.57a

Overall Equation: $2H_3C\text{-}N\text{=}N\text{-}C_2H_5 \longrightarrow 2N_2 + 2C_3H_8$ ApIII.57b

Propane can indeed be observed to be one of the products of the thermal decomposition of dimethyldiazene.[2]

Stage 1b:

$$H_3C - N = N - C_2H_5 \longrightarrow N_2 + n{\cdot}CH_3 + e{\cdot}C_2H_5$$

$$e{\cdot}\overset{\displaystyle H}{\underset{\displaystyle H}{C}}-\overset{\displaystyle H}{\underset{\displaystyle H}{C}}-H \longrightarrow \overset{\displaystyle H}{\underset{\displaystyle H}{C}}=\overset{\displaystyle H}{\underset{\displaystyle H}{C}} + H{\cdot}e + Heat$$

ApIII.58a

$$H{\cdot}e + n{\cdot}CH_3 \rightleftharpoons CH_4$$

Overall Equation: $2H_3C\text{-}N\text{=}N\text{-}C_2H_5 \longrightarrow 2N_2 + 2CH_4 + 2C_2H_4$ ApIII.58b

Overall overall equation: $4H_3C\text{-}N\text{=}N\text{-}C_2H_5 \longrightarrow 4N_2 + 2CH_4 + 2C_3H_8 + 2C_2H_4$

ApIII.59

The two stages above are taking place in parallel. [See Equation ApIII.25a] Now, is the time to see how far so far for the decomposition of dimethyldiazene an isomer of Azo-ethane. Adding Equation ApIII.44 to ApIII.59. the followings are obtained.

Overall overall equation: $66H_3C\text{-}N\text{=}N\text{-}CH_3 \longrightarrow 34CH_4 + 50N_2 + 6CH_3\text{-}N\text{=}N\text{-}C_2H_5$
$+ 7C_2H_4 + 7C_2H_6 + 2H_2C\text{=}N\text{-}N\text{-}(CH_3) C_2H_5 + 2(CH_3)_2N\text{-}N (CH_3)_2 + 2H_7C_3\text{-}N\text{=}N\text{-}$
$CH_3 + 4H_2C\text{=}N\text{-}N (CH_3)_2 + 2C_3H_8 + 10n{\cdot}CH_3$ ApIII.60

Bimolecularly, the followings are obtained.

Stage 1:

$$H_3C - N = N - C_2H_5 \longrightarrow N_2 + n{\cdot}CH_3 + e{\cdot}C_2H_5$$

$$H_5C_2{\cdot}e + H_3C - N = N - C_2H_5 \longrightarrow H_8C_3 + N_2 + e{\cdot}C_2H_5$$

$$H_5C_2{\cdot}e \longrightarrow e{\cdot}\overset{\displaystyle H}{\underset{\displaystyle H}{C}}-\overset{\displaystyle H}{\underset{\displaystyle H}{C}}{\cdot}n + H{\cdot}e$$

$$(A)$$

ApIII.61a

$$(A) \longrightarrow H_2C = CH_2 + Heat$$

$$H{\cdot}e + n{\cdot}CH_3 \rightleftharpoons CH_4$$

Overall Equation: $4H_3C\text{-}N\text{=}N\text{-}C_2H_5 \longrightarrow 4N_2 + 2C_3H_8 + 2H_2C\text{=}CH_2 + 2CH_4$

$$\text{ApIII.61b}$$

This indeed is the favored mechanism, that which is Bimolecular. Compare this last equation with Equation ApIII.59. Worthy of note is that no initiator is still present. When this equation is added to Equation ApIII.44, the followings are obtained.

Overall overall equation: $66H_3C\text{-}N\text{=}N\text{-}CH_3 \longrightarrow 34CH_4 + 50N_2 + 6CH_3\text{-}N\text{=}N\text{-}C_2H_5$

$+ \; 7C_2H_4 + 7C_2H_6 + 2H_2C\text{=}N\text{-}N\text{-}(CH_3)\,C_2H_5 + 2(CH_3)_2N\text{-}N\,(CH_3)_2 + 2H_7C_3\text{-}N\text{=}N\text{-}$

$CH_3 + 4H_2C\text{=}N\text{-}N\,(CH_3)_2 + 2C_3H_8 + 10n\cdot CH_3$

$$\text{ApIII.62}$$

With presence of two moles of methylpropyldiazene, there is no doubt that one mole will be decomposed, since none has stayed in Equilibrium state of existence.

d) Methylpropyldiazene (An isomer of Azo-butane)

Stage 1:

$$H_3C - N = N - C_3H_7 \longrightarrow N_2 + H_3C\cdot n + e\cdot C_3H_7$$

$$e\cdot\overset{\displaystyle H}{\underset{\displaystyle H}{C}} - \overset{\displaystyle H}{\underset{\displaystyle H}{C}} - \overset{\displaystyle H}{\underset{\displaystyle H}{C}} - H \longrightarrow e\cdot\overset{\displaystyle H}{\underset{\displaystyle H}{C}} - \overset{\displaystyle H}{\underset{\displaystyle H}{C}}\cdot n + e\cdot CH_3$$

$$(A)$$

$$(A) \longrightarrow H_2C = CH_2 + Heat$$

$$H_3C\cdot e + n\cdot CH_3 \longrightarrow C_2H_6$$

$$\text{ApIII.63}$$

Overall Equation: $H_3C\text{-}N\text{=}N\text{-}C_3H_7 \longrightarrow N_2 + C_2H_4 + C_2H_6 + Heat$

$$\text{ApIII.64}$$

Note that in the second step above, only CH_3 group could be released and not hydrogen, because when hydrogen is released, this is what happens in its unseen activated state.

$$e\cdot\overset{\displaystyle H}{\underset{\displaystyle H}{C}} - \overset{\displaystyle H}{\underset{\displaystyle CH_3}{C}}\cdot n \; \textit{would inherently have been obtained instead of} \; n\cdot\overset{\displaystyle H}{\underset{\displaystyle H}{C}} - \overset{\displaystyle H}{\underset{\displaystyle CH_3}{C}}\cdot e \; \textit{that which is impossible.}$$

Unlike all the cases encountered so far, the mechanism here is Unimolecular. With one mole of methylpropyldiazene left. The following is the overall equation.

Final overall equation: $66H_3C-N=N-CH_3 \longrightarrow 34CH_4 + 51N_2 + 6H_5C_2-N=N-CH_3$

$+ \ 8C_2H_4 + H_7C_3-N=N-CH_3 + 8C_2H_6 + 4H_2C=N-N(CH_3)_2 + 2(H_3C)_2 N-N(CH_3)_2 +$

$2H_2C=N-N(CH_3)C_2H_5 + 2C_3H_8 + 10n \bullet CH_3$ ApIII.65

Based on the mechanisms provided, one can observe how Nature operates. How all the products reportedly obtained are obtained have been clearly shown using the new foundations which indeed is the Real Science and not an Art. The fact that no additional products were obtained, does not mean that this is the end. All depends on the operating conditions. If any polymer is to be produced, it has to come from no other source other than ethene (ethylene) in the electro-free-radical route, the natural route. It is not easy to polymerize it nucleo-free-radically under the operating conditions. No ethene was produced in Equilibrium mechanism. The ethane produced could not be cracked, because the operating conditions are mild, that wherein the temperature and pressure were relatively low. The source of ethene was during the decomposition mechanism of dimethyldiazene, methylethyldiazene and methylpropyldiazene. The products obtained in de-creasing order of number of moles are nitrogen, methane, ethane, ethene, methylethyldiazene, dimethylhydrazone, propane, tetramethylhydrazine, methylpropyl-hydrazone and methylpropyl-diazene.

Unlike azo-ethane (i.e. ethyldiazene) which produced only N_2, ethane, ethene, and hydrogen, its isomer has no hydrogen, but has far more products all of great importance. The types of products obtained all depend on the operating conditions, and the method used in decomposition. The fact that there was more methane than ethane in the product does not mean that the compound is Azo-methane. Without any doubt "the operating conditions" is the most dominant and important of all the new concepts, for without it nothing takes place in life.[3]

3.0 Conclusions.

The Scientific mechanism of the non-catalytic decomposition of what today is called azo-methane has been provided herein. Indeed, based on the new foundations, the real name as used herein is azo-ethane. What today are called azo-alkanes are not the real azo-alkanes, but the isomers of the real azo-alkanes which must carry H on one of the nitrogen centers of the azo.[1] These are well known as Diazenes (H-N=N-R, where R is an alkyl or aryl group). The same applies to some of the groups which can carry hetero atoms-COOR or groups such as CN. The azo ethane in question herein is $H-N=N-C_2H_5$ and its isomer dimethyldiazene ($H_3C-N=N-CH_3$).

Not only were the mechanisms of their decompositions considered, so also were their syntheses for the dual purpose of identifying with their real names. In the process of providing the mechanisms of their syntheses (using what the azo-ethanes carry) and their decompositions, many new concepts were introduced using examples the mechanism of Tautomerization (Electroradicalization), Bimolecular versus Unimolecular reactions, Non-existence of impossible activated states, the fact that not all their N=N centers can be activated, and so on. It was shown that these reactions are Bimolecular with few being Unimolecular, complex, and stage-wise.

During the decompositions of the real azo-ethane, i.e. ethyldiazene, the products obtained were largely N_2 and ethane. When its isomer, dimethyldiazene was decomposed so many pro-ducts were obtained, the three main ones being N_2, CH_4, ethane and nucleo-free-radicals. It was the only one where nucleo-free-radicals were obtained, in view of the symmetric character of placement of the groups. Like the real azo-ethane, ethene was also produced. Unlike the real azo-ethane, H_2 and nucleo-free-radicals were not produced and higher azo-compounds (H$_3$C-N=N-C$_2$H$_5$ and H$_3$C-N=N-C$_3$H$_7$) which can readily tautomerize were produced. Only dimethyldiazene and methylethyldiazene had the strength based on the operating conditions to exist in Equilibrium state of existence and tautomerize with methylene to produce formaldehyde dimethylhydrazone and formaldehyde methylethylhydrazone. Propane was also one of the products. If there had been enough methylpropyldiazene, butane would have been produced. Without the fraction of dimethylhydrazone in Activated state of Existence and of methane in Stable State of Existence, tetramethylhydrazine would not have been produced.

References

1. S. N. E. Omorodion, *"The Mechanisms of Non-catalyzed and Catalyzed Decompositions of Azo-Hydrocarbons- Part I: Azo-methane"*, Appendix (II), Vol II.

2. Y. Parquin, W. Forst, *"The thermal decomposition of azomethane. III. Kinetics of non-inhibited reaction."*, International Journal of Chemical Kinetics, Vol. 5, Issue 4, (1973), pgs. 691-714.

3. S. N. E. Omorodion, *"Chemistry and Humanity-Challenges to be faced as we advance in the Third Millennium: New Classifications for Radicals"*, Nigerian Journal of Applied Science, Vol 26, (2008), Pgs. 146-158.

4. T. C. Clark, T. P. J. Izod, G. B. Kistiakowsky, *"Reactions of Methyl Radicals Produced by the Pyrolysis of Azomethane or Ethane in Reflected Shock Waves."* J. Chem. Phys., Vol. 54, Issue 3, (1971), pg.1295.

5. G. Chiltz, C. F. Aten Jr., and S. H. Bauer, *"Rate of Decomposition of Azomethane in a Shock tube."*, J. Phys. Chem., 66 (8), (1962), pgs. 1426-1431.

6. C. Steel, and A. F. Trotman-Dickenson, J. Chem. Soc. 975 (1959).

7. E. W. Riblett, and L. C. Rubin, *"Thermal Decomposition of Azomethane."*, J. Am. Chem. Soc., 59 (8), (1937), pgs. 1537-1540.

8. Rice and Ramsperger, J. Am. Chem. Soc., 49, 1617, (1927).

9. Rice and Ramsperger, ibid., 50, 617, (1928).

10. Kassel, J. Phys. Chem., 22, 225, (1928).

11. C. R. Noller, *"Textbook of Organic Chemistry"*, W. B. Saunders Company, 1966, pg. 248.

12. S. N. E. Omorodion, *"Chemistry and Humanity-Challenges to be faced as we advance in the Third Millennium: The impacts of the new classifications for compounds"*, Nigerian Journal of Applied Science, Vol 28, (2010), Pgs. 128-134.

13. S. N. E. Omorodion, *"Chemistry and Humanity-Challenges to be faced as we advance in the Third Millennium: New classifications for Compounds"*, Nigerian Journal of Applied Science, Vol 28, (2010), Pgs. 80-87.

14. A. G. Myers, M. Movassaghi, and B. Zheng, *"Mechanistic studies of free-radical fragmentation of monoalkyl diazenes."*, Tetrahedron Letters, Vol. 38, Issue 37, (1997), pgs.6569-6572.

15. S. N. E. Omorodion, *"The Hydrocarbon Family Tree"*, Appendix (I), Vol II.

16. H. A. Taylor, and F. P. Jahn, *"The Thermal Decomposition of Azomethane."*, J. Chem. Phys. 7, 470 (1939).

17. E. Wilberg, N. Wilberg, and A. F. Holleman, *"Inorganic Chemistry"*, (2001), pg. 666.

18. C. R. Noller, *"Textbook of Organic Chemistry"*, W. B. Saunders Company, 1966, pg. 228.

19. S. N. E. Omorodion, *"Chemistry and Humanity-Challenges to be faced as we advance in the Third Millennium: The impacts of new classifications for Radicals."* Nigerian Journal of Applied Science, Vol 27, (2009), Pgs. 34-43.

20. S. N. E. Omorodion, *"Chemistry and Humanity- Challenges to be faced as we advance in the Third Millennium: New Classifications for Bonds and Charges."* Nigerian Journal of Applied Science, Vol 27, (2009), Pgs. 94-104.

21. S. N. E. Omorodion, *"Chemistry and Humanity-Challenges to be faced as we advance in the Third Millennium: The impacts of the new classifications for Bonds and Charges."* Nigerian Journal of Applied Science, Vol 28, (2010), Pgs. 37-45.

22. C. R. Noller, *"Textbook of Organic Chemistry"*, W. B. Saunders Company, 1966, pg. 62-64.

Appendix IV

THE MECHANISMS OF NON-CATALYZED AND CATALYZED DECOMPOSITIONS OF AZO-HYDROCARBONS. PART III: AZO-ETHANE. (CATALYZED)

Abstract

In Part II[1], the mechanisms of non-catalyzed decomposition of azo-ethane were provided. Herein is the case for its catalyzed decomposition. The catalysts in question are passive and they come from the family of Transition metals, in particular Group VIII A elements – The *Transition transition Metals Triads*[2]. These are metals which cannot form hydrides with hydrogen. The most commonly used ones are Platinum (Pt), Palladium (Pd), Rhodium (Rh), Ruthenium (Ru), Iron (Fe), and Osmium (Os). The products obtained during decompositions of azo-ethane in the presence of these catalysts are completely different from those obtained in their absence. Contained also herein is the unique case of azo-methane said to undergo Tautomerization to formaldehyde hydrazone in the presence of Ruthenium and Osmium complexes.[3] In further support of the mechanisms provided in the presence of these catalysts, decompositions of N_2 and diazo-methane were also considered. The reactions are Unimolecular.

Keywords:

1. Porous Passive Catalysts.
2. Unimolecular Decomposition mechanism.
3. Bimolecular Decomposition mechanism.
4. Tautomerization.
5. Diazo-methane/Nitrogen decompositions.

1.0 INTRODUCTION

So far, it can be observed that since the new classifications for Radicals[2] and the new concepts[2,4-8] were provided, things that could not be explained before since antiquity have begun to be explained not as an Art, but as a Science. Knowing what radicals are is the Key to greater understanding of how Nature operates. As has been said in the past[2], no chemical reaction takes place ionically. They all take place radically. Only few take place chargedly (covalently, electrostatically, and polarly and not ionically) polymerically. Covalent and ionic charges are real, while electrostatic and polar charges are real and imaginary. Ionic charges can be isolated, while the others cannot be isolated.[5] How and why these charges exist have been explained. All radicals can be isolated.

So far, how Equilibrium and Decomposition reactions take place side by side or not have begun to be explained. When and why they take place has also begun to be explained. Without one or more compounds existing in Equilibrium State of Existence, Equilibrium mechanism can never take place. One or two components can be in Equilibrium State of Existence and yet be unproductive. With Decomposition mechanism, the situation is different. For any compound to be decomposed via this mechanism, it must first exist in Stable and Decomposition States of Existences. We have seen the case of a compound decomposing via Equilibrium mechanism unimolecularly with the use of methyldiazene (Azo-methane), ethyldiazene (Azo-ethane), and dimethyldiazene (An isomer of Azo-ethane). We have seen the case of a compound decomposing via Decomposition mechanism bimolecularly and unimolecularly with the use of the azo-alkanes above. We have seen the case of a compound decomposing via Equilibrium mechanism bimolecularly with the use of ammonia.[8] We have also shown the case where two compounds decompose via Decomposition mechanism bimolecularly using Aluminum trialkyl (AlR_3) and Titanium tetrachloride ($TiCl_4$). These are compounds which favor existing in Stable State than Equilibrium State of Existence. When in stable state of existence, they can only be made to exist in Decompositon state of existence, based on the operating conditions.

$$2NH_3 \xrightarrow[\text{Bimolecular reactions}]{\text{Equilibrium Mechanism}} 3H_2 \;+\; N_2 \qquad \text{ApIV.1}$$

$$AlR_3 \;+\; TiCl_4 \xrightarrow[\text{Bimolecular reactions}]{\text{Decomposition Mechanism}} Cl_2TiR \;+\; R_2TiCl_2 \qquad \text{ApIV.2}$$

We have also noted that products obtained when azo-hydrocarbons decompose via Decom-position mechanism are different from those obtained via Equilibrium mechanism. All these are Non-Porous On the Surface reactions or Non-Porous Laboratory Chemistry, like most other chemical or polymeric reactions. However, there are other reactions which take place in a different environment such as inside very tiny pores carried by some metals. Their pores can break down many compounds almost to their elements depending on the operating conditions. For example cracking of hydrocarbons carried out in the Petroleum industries at very harsh operating conditions can be easily carried out inside these special pores at normal to mild operating conditions. The products obtained therefore inside the pores are uniquely different from those obtained on non-porous surfaces. When these pores are covered by compounds such as H_2, O_2 or N_2, the products obtained are also uniquely different depending on the fraction of coverage of the pores and the type of compound used in covering the pores.

In a more recent study (2000)[9], the thermal and UV photo-induced decomposition of dimethyldiazene on *Rh (III)* was investigated by means of reflection adsorption infrared spectroscopy (RAIRS) and temperature-programmed desorption spectroscopy (TPD). The thermal decomposition data from RAIRS revealed that the dimethyldiazene adsorbs in trans-configuration mode on ***Rh (III)*** at **90 K**, decomposing exclusively by N-N bond scission, yielding $\underline{H_2, N_2, C_2N_2}$ *and traces of* $\underline{HCN}$. Upon UV irradiation at **90 K**, adsorbed dimethyl-diazene was said to undergo tautomerization, forming *formaldehyde methylhydrazone,* $CH_3NHN=CH_2$. From the post-irradiation TPD data, there was significant suppression of C_2N_2 and N_2 formation, and the appearance of new products-*methylamine (CH_3NH_2) and methane (CH_4)*. C-N bond scission was said to also occur in the illuminated chemisorbed layer at **90 K.** Worthy of note here, is that if during thermal decomposition, the scission was mainly via the N-N bond, then N_2 should not appear as one of the products, unless some decompositions have taken place on the surface and not inside the pore.

In another important recent study (1986) made available online in 2002[10], the reactions of dimethyldiazene were studied on a Clean, H-covered, and O-covered ***Pt (III)*** surface at differ-rent coverages of azo-methane and ratios of hydrogen or oxygen to dimethyldiazene, by tem-perature programmed desorption (TPD). On a Clean Pt (III) surface, dehydrogenation, together with breaking of the N=N bond to produce *HCN and* H_2 were said to be the major reactions. Small amounts of *methylamine and cyanogen* were also observed. On the H-covered surface at high hydrogen coverage, *methane* was the dominant product, in addition to H_2. The absorbed hydrogen was said to promote breaking of the C-N bond. In addition, the dehydrogenation products *HCN as well as methylamine* were also observed. On the O-covered surface, *CO, CO_2, H_2O,* and *small amounts of NO* were observed together with *the products typical for a Clean surface.*

In the last part of this series, the mechanism of Tautomerization was provided and azo-alkanes which tautomerize were identified. Methyldiazene was found not to be one of them. However, in a more recent publication (2005)[2], tautomerization of the methyldiazene to formaldehyde hydrazone in Ruthenium and Osmium complexes was reported. The two com-plexes are too complex to describe herein. However the centers of action were largely the Transition metals. The methyldiazene did not tautomerize, but underwent the decomposition reactions inside the pores. It was from the products that the formaldehyde hydrazone was formed via Decomposition mechanisms as will be shown here.

These Transition metals have very much been used in the past without caring to know what they do. Finally (1985), the reactions of diazomethane on a Clean Pt (III) produced methane, ethylene and hydrogen[11]. The same products were observed together with nitrogen for the H-covered surface. On the O-covered surface, products characteristic to Clean surface were observed together with oxidation products- water, CO, and CO_2.

Pt (III), Pd (100) and Rh (III) are some of the Transition metals used as Hydrogen catalyst. Since they do not form hydrides with hydrogen, the pores in them break the hydrogen molecule into the male and female radicals as shown below. It keeps the hydrogen in Equilibrium state of Existence.

$$H_2 \xrightleftharpoons[\textit{Equilibrium State of Existence of } H_2]{\textit{Pt (III) } H_2 \textit{ Catalyst}} \quad H{\bullet}e \quad + \quad H{\bullet}n$$

$$\qquad\qquad\qquad\qquad\qquad\qquad\qquad (\textit{Atom}) \qquad (\textit{Hydride})$$

$$H{\bullet}e \quad + \quad H{\bullet}n \xrightleftharpoons[H_2]{\textit{Equilibrium State of Existence of}} \quad H_2$$

$$H_2 \xrightarrow{\textit{Decomposition State of Existence of } H_2} \quad H{\bullet}e \quad + \quad H{\bullet}e$$

$$\qquad\qquad\qquad\qquad\qquad\qquad\qquad (\textit{Atom}) \qquad\quad (\textit{Atom})$$

$$H{\bullet}e \quad + \quad H{\bullet}e \xrightarrow{\textit{Combination State of Existence of } H_2} \quad H_2$$

$$H{\bullet}n \quad + \quad H{\bullet}n \xrightarrow{\textit{Combination State of Existence of } H_2} \quad H_2$$

The last reaction takes place only in the absence of H.e in the system, while the first reaction before the last, takes place whether H.n is present or not, because hydrogen is electropositive. The Pt (III) is not there for nothing. It is a passive catalyst. It does even more during the decom-position of these compounds.

2.0 CATALYZED DECOMPOSITION MECHANISM

One will now begin with the decomposition of azo-ethane and its isomer, starting with the isomer as a guide. There are no on-the-surface reactions at the beginning, because all the azo-ethane has been fully absorbed into the pores. If not all are absorbed into the pores i.e., inside the surface, then on-the-surface reactions between H_2 if present as one of the products from inside and unabsorbed azo-ethane takes place to produce CH_4 and N_2 as will be shown here.

2.1 Azo-Ethanes

i) $H_3C-N=N-CH_3$
a) **Clean Pt (III) surface**

The azo ethane goes into the pores wherein the π-bond is first broken. This is then followed by the breaking of the σ-bond as follows.

Stage 1:

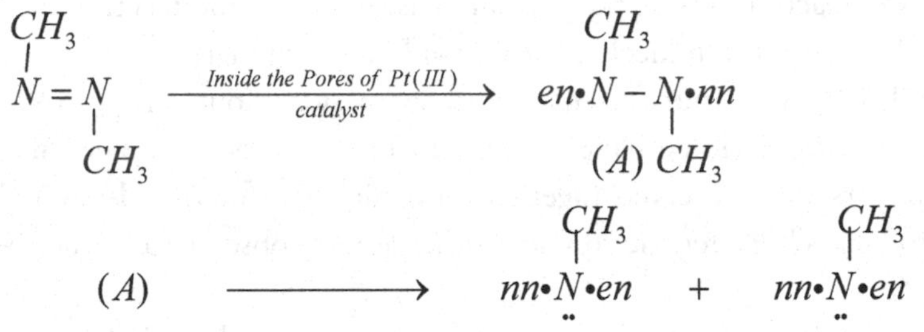

$$\begin{array}{c} CH_3 \\ | \\ N = N \\ | \\ CH_3 \end{array} \xrightarrow[\text{catalyst}]{\text{Inside the Pores of Pt(III)}} \begin{array}{c} CH_3 \\ | \\ en{\cdot}N - N{\cdot}nn \\ | \\ (A)\ CH_3 \end{array} \qquad \text{ApIV.6}$$

$$(A) \longrightarrow \overset{CH_3}{\underset{\cdot\cdot}{nn{\cdot}N{\cdot}en}} + \overset{CH_3}{\underset{\cdot\cdot}{nn{\cdot}N{\cdot}en}}$$

Overall Equation: $10H_3C\text{-}N=N\text{-}CH_3 \longrightarrow 20N\text{-}CH_3$ $\qquad$ ApIV.7

Stage 2:

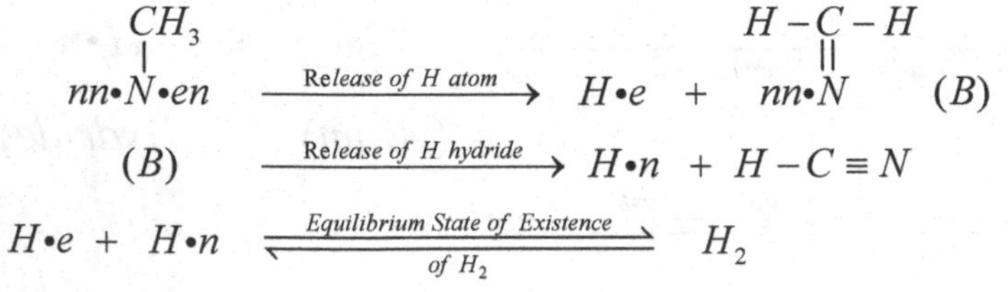

$$\begin{array}{c} CH_3 \\ | \\ nn{\cdot}N{\cdot}en \end{array} \xrightarrow{\text{Release of H atom}} H{\cdot}e + \begin{array}{c} H - C - H \\ \| \\ nn{\cdot}N \end{array} \quad (B)$$

$$(B) \xrightarrow{\text{Release of H hydride}} H{\cdot}n + H - C \equiv N$$

$$H{\cdot}e + H{\cdot}n \underset{\text{of } H_2}{\overset{\text{Equilibrium State of Existence}}{\rightleftharpoons}} H_2 \qquad \text{ApIV.8}$$

Overall Equation: $18N\text{-}CH_3 \longrightarrow 18H_2 + 18H\text{-}C\equiv N$ $\qquad$ ApIV.9

Overall overall equation: $10H_3C\text{-}N=N\text{-}CH_3 \longrightarrow 2N\text{-}CH_3 + 18H_2 + 18H\text{-}C\equiv N$
$$\text{ApIV.10}$$

Stage 3:

$$H\bullet e \ + \ nn\bullet \overset{\overset{\displaystyle CH_3}{|}}{\underset{\displaystyle ..}{N}}\bullet en \ \longrightarrow \ H - \overset{\overset{\displaystyle CH_3}{|}}{N}\bullet en \quad (C)$$

$$H - \overset{\overset{\displaystyle CH_3}{|}}{N}\bullet en \ + \ n\bullet H \ \longrightarrow \ H - \overset{\overset{\displaystyle CH_3}{|}}{N} - H$$

$$\text{(Methylamine)} \hspace{4cm} \text{ApIV.11}$$

Overall Equation: $2N\text{-}CH_3 \ + \ 2H_2 \ \longrightarrow \ 2H_2NCH_3$ \hspace{2cm} ApIV.12

Overall overall equation: $10H_3C\text{-}N\text{=}N\text{-}CH_3 \ \longrightarrow \ 16H_2 \ + \ 18H\text{-}C\text{≡}N \ + \ 2H_2N\text{-}CH_3$

$$\text{ApIV.13}$$

Stage 4:

$$2H - C \equiv N \xrightarrow{\textit{Inside the Pores of Pt (III)}} 2e\bullet \overset{\overset{\displaystyle H}{|}}{C} = N\bullet nn$$

$$2e\bullet \overset{\overset{\displaystyle H}{|}}{C} = N\bullet nn \ \longrightarrow \ nn\bullet N = \overset{\overset{\displaystyle H}{|}}{C} - \overset{\overset{\displaystyle H}{|}}{C} = N\bullet nn \quad (E)$$

$$(E) \ \longrightarrow \ N \equiv C - C \equiv N \ + \ 2H\bullet e$$

$$\text{(Cyanogen)} \hspace{3cm} \text{ApIV.14}$$

$$2H\bullet e \ \longrightarrow \ H_2$$

Overall Equation: $2H\text{-}C\text{≡}N \ \longrightarrow \ N\text{≡}C\text{-}C\text{≡}N \ + \ H_2$ \hspace{1.5cm} ApIV.15

Overall overall equation: $10H_3C\text{-}N\text{=}N\text{-}CH_3 \ \longrightarrow \ 17H_2 \ + \ 16H\text{-}C\text{≡}N \ + \ N\text{≡}C\text{-}C\text{≡}N$

$$+ \ 2H_2N\text{-}CH_3 \hspace{2cm} \text{ApIV.16}$$

Very different types of products are observed to be indeed obtained in the presence of this passive catalyst [Pt (III)]. In the first stage, the azo-ethane is broken down to give a unique product, $N\text{-}CH_3$ carrying non-free-radicals on the N center. In the second stage, ninety percent of this product was consumed to give products small part of which are consumed downstream. If all had been consumed, then there will be no methylamine. It is important to note how components are released. The hydrogen molecules produced here are not in Stable state of Existence, because of the presence of Pt (III) catalyst. Hence the reactions are as indicated in Stage 3 where the methylamine is produced. ***In the third step of the fourth stage, cyanogen was finally obtained and the reason is because components with double or triple bonds are always kept in Activated States of Existence inside the pores.*** The ease of a stage is partly determined by the number of steps involved and any stage which cannot be interpreted into Mathematical language without making necessary or unnecessary assumptions is no stage.

b) H-covered Pt (III) surface

On the H-covered surface at high hydrogen coverage, the system is already saturated with so much hydrogen molecules existing in Equilibrium state of Existence, i.e. as $H \bullet e$ and $H \bullet n$. On the other hand, only little or no pores will readily be available for the adsorption of the Azo-ethane. Therefore most of the decomposition will take place on the surface as follows in the presence of the hydrogen elements.

A) On the surface.

<u>**Stage 1:**</u>

$$H \bullet e \ + \ H_3C - N = N - CH_3 \ \longrightarrow \ CH_4 \ + \ N_2 \ + \ e \bullet CH_3$$

$$H_3C \bullet e \ + \ H \bullet n \ \longrightarrow \ CH_4 \qquad\qquad , \qquad ApIV.17$$

<u>**Overall Equation:**</u> $90H_2 \ + \ 90H_3C\text{-}N\text{=}N\text{-}CH_3 \ \longrightarrow \ 180CH_4 \ + \ 90N_2$ ApIV.18

B) Inside the pores.

The small fractions that have access into the pores decompose as already shown in four stages of Equations ApIV.6 to ApIV.16 and the overall overall equation (Equation ApIV.16) is recalled below.

<u>**Overall overall equation:**</u> $10H_3C\text{-}N\text{=}N\text{-}CH_3 \ \longrightarrow \ 17H_2 \ + \ 16H\text{-}C\equiv N \ + \ N\equiv C\text{-}C\equiv N$

$$+ \ \ 2H_2N\text{-}CH_3 \qquad (ApIV.16)$$

<u>**Final overall equation:**</u> $100H_3C\text{-}N\text{=}N\text{-}CH_3 \ + \ 90H_2 \longrightarrow 180CH_4 \ + \ 90N_2 \ + 16H\text{-}C\equiv N$

$$17H_2 \ + \ 2H_2N\text{-}CH_3 \ + \ \ N\equiv C\text{-}C\equiv N \quad ApIV.19$$

Ninety percent of the azo-compound was consumed on the surface while ten percent went into the pores. One can observe why and how the hydrogen is involved. Indeed methane is the major product, noting that little hydrogen is produced. With the presence of N_2 and H_2, one wonders why ammonia is not one of the products. The reason will become obvious very shortly.

c) O-covered Pt (III) surface.

When molecular oxygen goes into the pores, like H_2, it is broken down to give oxidizing oxygen as shown below.

$$O_2 \quad \underset{}{\overset{Pt\,(III)\ catalyst}{\rightleftharpoons}} \quad 2nn \bullet \overset{\bullet\bullet}{\underset{\bullet\bullet}{O}} \bullet en \qquad\qquad ApIV.20$$

(Molecular oxygen)   *(Oxidizing oxygen element)*

Note that the molecular oxygen above is not the oxidizing oxygen molecule which is a ring that is always in Equilibrium State of Existence.[1]

A) On the surface.

Stage 1:

$$3nn \bullet O \bullet en \quad + \quad H_3C - N = N - CH_3 \longrightarrow \underset{\underset{OH \quad (A)}{|}}{\overset{\overset{OH}{|}}{HO - C - N = N - CH_3}}$$

$$(A) \longrightarrow \underset{\underset{OH}{|}}{\overset{\overset{OH}{|}}{HO - C \bullet e}} \quad + \quad nn \bullet N = N - CH_3$$

$$(B)$$

$$(B) \longrightarrow H_2O + CO + HO \bullet en$$

$$HO \bullet en + nn \bullet N = N - CH_3 \longrightarrow HO - N = N - CH_3$$

ApIV.21

Overall Equation: $30nn \bullet O \bullet en + 10H_3C\text{-}N{=}N\text{-}CH_3 \longrightarrow 10H_2O + 10CO$

$$+ \quad 10HO\text{-}N{=}N\text{-}CH_3 \quad \text{ApIV.22}$$

Stage 2:

$$3nn \bullet O \bullet en \quad + \quad H_3C - N = N - OH \longrightarrow \underset{\underset{OH \quad (C)}{|}}{\overset{\overset{OH}{|}}{HO - C - N = N - OH}}$$

$$(C) \longrightarrow \underset{\underset{OH}{|}}{\overset{\overset{OH}{|}}{HO - C \bullet e}} \quad + \quad nn \bullet N = N - OH$$

$$(D)$$

$$(D) \longrightarrow H_2O + CO + HO \bullet en$$

$$HO \bullet en + nn \bullet N = N - OH \longrightarrow HO - N = N - OH \qquad \text{ApIV.23}$$

Overall Equation: $30nn \bullet O \bullet en + 10H_3C\text{ -}N{=}N\text{-}OH \longrightarrow 10H_2O + 10CO + 10HO\text{-}N{=}N\text{-}OH$

ApIV.24

Overall overall equation: $60nn \bullet O \bullet en + 10H_3C\text{-}N{=}N\text{-}CH_3 \longrightarrow 20\,H_2O + 20\,CO$

$$+ \quad 10\,HO\text{-}N{=}N\text{-}OH \qquad \text{ApIV.25}$$

It is worthy of note that HO-N=N-OH is an azo-compound[12], which when decomposed inside the pores in the presence of Oxygen may give nitric acid, nitrous acid, nitric oxide and water.

Stage 3:

$$nn{\cdot}O{\cdot}en \; + \; HO-N=N-OH \; \longrightarrow \; HO{\cdot}nn \; + \; O=N-N=O \; + \; H{\cdot}e$$

$$HO{\cdot}nn \; + \; 2H{\cdot}e \; \rightleftharpoons \; H_2O$$

ApIV.26

Overall Equation: $10HO\text{-}N{=}N\text{-}OH \; + \; 10nn{\bullet}O{\bullet}en \longrightarrow 10(O{=}N\text{-}N{=}O) \; + \; 10H_2O$

ApIV.27

Only a fraction of the CO can be activated, based on the operating conditions. Fifty percent is now used in the next stage, leaving the non-poisonous CO with its vacant orbital.

Stage 4:

$$nn{\cdot}O{\cdot}en \; + \; n{\cdot}\overset{\overset{O}{\|}}{C}{\cdot}e \; \longrightarrow \; CO_2$$

ApIV.28

Overall Equation: $10CO \; + \; 10nn{\bullet}O{\bullet}en \longrightarrow 10CO_2$ ApIV.29

Overall overall equation: $80nn{\bullet}O{\bullet}en \; + \; 10H_3C\text{-}N{=}N\text{-}CH_3 \longrightarrow 30H_2O \; + \; 10CO \; +$

$$10CO_2 \; + \; 10(O{=}N\text{-}N{=}O) \quad \text{ApIV.30}$$

B) Inside the pores.

The small fractions that have access into the pores decompose as already shown in four stages of Equations ApIV.6 to ApIV.16 and the overall overall equation (Equation ApIV.16) is recalled below.

Overall overall equation: $10H_3C\text{-}N\text{-}N\text{-}CH_3 \longrightarrow 17H_2 \; + \; 16H\text{-}C{\equiv}N \; + \; N{\equiv}C\text{-}C{\equiv}N$

$$+ \; 2H_2N\text{-}CH_3 \quad (\text{ApIV.16})$$

Here again, ninety percent at the operating conditions of fraction of surface covered by O_2 of the azo-ethane decomposed leaving ten percent inside the pores. Worthy of note inside the pores is that the H_2, HCN, and $H_2N\text{-}CH_3$ formed cannot be oxidized, because they are in Equilibrium State of Existence. Oxidation only takes place when the components to be oxidized are in Stable State of Existence[13]. The final overall equation therefore is as follows.

Final Overall Equation: $360(O_2) \; + \; 100H_3C\text{-}N{=}N\text{-}CH_3 \longrightarrow 270H_2O \; + \; 90CO \; +$

$90CO_2 \; + \; 90(O{=}N\text{-}N{=}O) \; + \; 17H_2 \; + \; 16H\text{-}C{\equiv}N \; + \; N{\equiv}C\text{-}C{\equiv}N \; + \; 2H_2N\text{-}CH_3$

ApIV.31

One can thus observe the large number of product obtained. These are completely different from the conventional non-catalytic processes. The number of moles of the products obtained may vary based on the fractions chosen for study. Indeed, based on Universal data and ones sense of judgment and experiences, one knows the fractions to be expected. Above so far, one has only considered the isomer of azo-ethane. The next to be considered is the real azo-ethane.

b) Ethyldiazene (Azo-ethane)
i) Clean Pt (III) surface.

When Azo- ethane is adsorbed on clean Pt (III) surface, the followings are to be expected.

Stage 1:

$$H - N = N - C_2H_5 \xrightarrow{\substack{\text{Activation inside the} \\ \text{pores}}} nn \bullet \overset{\overset{\displaystyle H}{|}}{N} - \underset{\underset{\displaystyle C_2H_5}{|}}{N} \bullet en \quad (A)$$

ApIV.32

$$(A) \xrightarrow{\text{Breakage of } \sigma - bond} nn \bullet \overset{\overset{\displaystyle H}{|}}{\underset{\bullet\bullet}{N}} \bullet en \quad + \quad nn \bullet \overset{\overset{\displaystyle C_2H_5}{|}}{\underset{\bullet\bullet}{N}} \bullet en$$

$$(A) \xrightarrow{\text{Breakage of } \sigma - bond} nn \bullet \overset{\overset{\displaystyle |}{N}}{\underset{\bullet\bullet}{N}} \bullet en \quad + \quad nn \bullet \overset{\overset{\displaystyle |}{N}}{\underset{\bullet\bullet}{N}} \bullet en$$

Overall Equation: $10H\text{-}N\text{=}N\text{-}C_2H_5 \longrightarrow 10N\text{-}H + 10N\text{-}C_2H_5$ ApIV.33

Stage 2:

$$\overset{\overset{\displaystyle C_2H_5}{|}}{nn \bullet N \bullet en} \xrightarrow{\text{Release of H atom}} H \bullet e + nn \bullet N \quad \overset{\displaystyle H - \overset{\overset{\displaystyle ||}{}}{C} - CH_3}{} \quad (B)$$

$$(B) \xrightarrow{\text{Release of } n.CH_3} H_3C \bullet n + H - C \equiv N + Heat$$

$$H \bullet e + H_3C \bullet n \xrightleftharpoons[\substack{\text{of } H_2}]{\text{Equilibrium State of Existence}} H_4C$$

ApIV.34

Overall Equation: $10N\text{-}C_2H_5 \longrightarrow 10CH_4 + 10H\text{-}C\equiv N$ ApIV.35

Overall overall equation: $10H\text{-}N\text{=}N\text{-}C_2H_5 \longrightarrow 10CH_4 + 10H\text{-}C\equiv N + 10N\text{-}H$

ApIV.36

Stage 3:

$$2H - C \equiv N \xrightarrow{\text{Inside the Pores of Pt(III)}} 2e \bullet \overset{\overset{\displaystyle H}{|}}{C} = N \bullet nn$$

$$2e \bullet \overset{\overset{\displaystyle H}{|}}{C} = N \bullet nn \longrightarrow nn \bullet N = \overset{\overset{\displaystyle H}{|}}{C} - \overset{\overset{\displaystyle H}{|}}{C} = N \bullet nn \quad (C)$$

$$(C) \longrightarrow N \equiv C - C \equiv N \, (Cyanogen) + 2H \bullet e$$

$$2He \longrightarrow H_2$$

ApIV.37

Overall Equation: $10H\text{ - }C\equiv N \longrightarrow 5N\equiv C\text{-}C\equiv N + 5H_2$ ApIV.38

Overall overall equation: $10H\text{-}N\text{=}N\text{-}C_2H_5 \longrightarrow 5H_2 + 10CH_4 +$

$$5N\equiv C\text{-}C\equiv N$$

ApIV.39

Stage 4:

$$H \cdot e + nn \cdot \overset{\overset{H}{|}}{\underset{\cdot\cdot}{N}} \cdot en \longrightarrow H - \overset{\overset{H}{|}}{\underset{\cdot\cdot}{N}} \cdot en \quad (D)$$

$$(D) + n \cdot H \xrightarrow[\text{of Ammonia}]{\text{Combination State of Existence}} NH_3 \qquad \text{ApIV.40a}$$

Overall Equation: $5H_2 + 5H\text{-}N \longrightarrow 5NH_3$ ApIV.40b

Overall overall equation: $10H\text{-}N{=}N\text{-}C_2H_5 \longrightarrow 5N{\equiv}C\text{-}C{\equiv}N + 5NH_3 + 10CH_4$

$+ \ 5N\text{-}H$ ApIV.42

Stage 5:

$$H \cdot e + nn \cdot \overset{\overset{H}{|}}{\underset{\cdot\cdot}{N}} \cdot en \longrightarrow H - \overset{\overset{H}{|}}{\underset{\cdot\cdot}{N}} \cdot en \quad (D)$$

$$(D) + n \cdot CH_3 \xrightarrow[\text{of the Amine}]{\text{Combination State of Existence}} NH_2CH_3 \qquad \text{ApIV.41a}$$

Oveall Equation: $5CH_4 + 5H\text{-}N \longrightarrow 5H_2NCH_3$ ApIV.41b

Overall overall equation: $10H\text{-}N{=}N\text{-}C_2H_5 \longrightarrow 5N{\equiv}C\text{-}C{\equiv}N + 5NH_3 + 5CH_4$

$+ \ 5NH_2CH_3$ ApIV.42

Very different types of products are observed to be obtained in the presence of this passive catalyst [Pt (III)]. In the first stage, the azo-methane is broken down to give two unique products, N-H and N-C_2H_5 both carrying non-free-radicals on the N center. In the second stage, hundred percent of the second product was consumed to provide CH_4 to be partly used down-stream. It is important to note how components are released. The H_2 and CH_4 molecules produced here are not in Stable state of Existence, because of the presence of Pt (III) catalyst. Hence, the reactions are as indicated in Stages 4 and 5 where ammonia and methyl amine are produced. In the third step of the third stage, cyanogen was obtained to produce the H_2 consumed in Stage 4. No hydrogen cyanide could be produced here. Note that not all the azo-ethane will be decomposed via this mechanism, because some will still exist in Equilibrium state of Existence.

ii) H-covered Pt (III) surface
As already said, most of the decompositions will take place on the surface as follows in the presence of the hydrogen elements.
  A) On the surface.

Stage 1:

$$H \cdot e + H_5C_2 - N = N - H \longrightarrow e \cdot C_2H_5 + N_2 + H_2$$

$$H \cdot n + H_5C_2 \cdot e \longrightarrow C_2H_6$$

ApIV.43

Overall Equation: $90H_2 + 90H\text{-}N{=}N\text{-}C_2H_5 \longrightarrow 90C_2H_6 + 90N_2 + 90H_2$

ApIV.44

B) Inside the pores.

The small fractions that have access into the pores decompose as already shown in four stages of Equations ApIV.32 to ApIV.42 and the overall overall equation (Equation ApIV.42) is recalled below.

Overall overall equation: $10H\text{-}N=N\text{-}C_2H_5 \longrightarrow 5CH_4 + 5N\equiv C\text{-}C\equiv N +$

$$5NH_3 + 5H_3CNH_2 \quad (ApIV.40c)$$

Final overall equation: $100H\text{-}N=N\text{-}C_2H_5 \longrightarrow 90C_2H_6 + 90N_2 + 5NH_2CH_3$

$$5H_3N + 5N\equiv C\text{-}C\equiv N \quad ApIV.45$$

Ninety percent of the azo-compound was consumed on the surface while ten percent went into the pores. Indeed ethane is the major product, noting that no hydrogen is produced at the end.

iii) O-covered Pt (III) surface.

When molecular oxygen goes into the pores, like H_2, it is broken down to give oxidizing oxygen as already shown.

B) On the surface.

Stage 1:

$$5nn\bullet O\bullet en + H_5C_2-N=N-H \longrightarrow \begin{matrix} OH\ OH \\ | \quad | \\ HO-C-C-N=N-H \\ | \quad | \\ OH\ OH \end{matrix} \quad (A)$$

$$(A) \longrightarrow \begin{matrix} OH\ OH \\ | \quad | \\ HO-C-C\bullet e \\ | \quad | \\ OH\ OH \end{matrix} + nn\bullet N=N-H$$

$$(B)$$

$$(B) \longrightarrow 2H_2O + 2CO + HO\bullet en$$

$$HO\bullet en + nn\bullet N=N-H \longrightarrow HO-N=N-H$$

$$ApIV.46$$

Overall Equation: $5nn\bullet O\bullet en + H\text{-}N=N\text{-}C_2H_5 \longrightarrow 2H_2O + 2CO +$

$$HO\text{-}N=N\text{-}H \quad ApIV.47$$

Stage 2:

$$nn{\cdot}O{\cdot}en \ + \ H-N=N-OH \ \longrightarrow \ HO{\cdot}nn \ + \ N_2 \ + \ en{\cdot}OH$$

$$HO{\cdot}en \ \longrightarrow \ H{\cdot}e \ + \ nn{\cdot}O{\cdot}en \qquad\qquad \text{ApIV.48}$$

$$H{\cdot}e \ + \ nn{\cdot}OH \ \rightleftharpoons \ H_2O$$

Overall Equation: $\quad H-N=N-OH \ \longrightarrow \ N_2 \ + \ H_2O \qquad\qquad$ ApIV.49

Overall overall equation: $10nn{\cdot}O{\cdot}en \ + \ 2H\text{-}N=N\text{-}C_2H_5 \longrightarrow 6H_2O \ + \ 4CO \ + \ 2N_2$

ApIV.50

Only a fraction of the CO can be activated, based on the operating conditions. Fifty percent is now used in the next stage, leaving the non-poisonous CO, each with its vacant orbital.

Stage 3:

$$nn{\cdot}O{\cdot}en \ + \ n{\cdot}\overset{\overset{\displaystyle O}{\|}}{C}{\cdot}e \ \longrightarrow \ CO_2 \qquad\qquad \text{ApIV.51}$$

Overall Equation: $2CO \ + \ 2nn{\cdot}O{\cdot}en \ \longrightarrow \ 2CO_2 \qquad\qquad$ ApIV.52

Overall overall equation: $540nn{\cdot}O{\cdot}en \ + \ 90H\text{-}N=N\text{-}C_2H_5 \longrightarrow 270H_2O \ + \ 90CO \ +$

$$90CO_2 \ + \ 90N_2 \qquad \text{ApIV.53}$$

B) Inside the pores.

The small fractions that have access into the pores decompose as already shown in four stages of Equations ApIV.32 to ApIV.42 and the overall overall equation (Equation ApIV.42) is recalled below.

Overall overall equation: $10H\text{-}N=N\text{-}C_2H_5 \ \longrightarrow \ 5CH_4 \ + \ 5N{\equiv}C\text{-}C{\equiv}N \ +$

$$5NH_3 \ + \ 5H_3CNH_2 \quad (\text{ApIV.42})$$

Here again, ninety percent at the operating conditions of fraction of surface covered by O_2 of the azo-ethane decomposed leaving ten percent inside the pores. Worthy of note inside the pores is that of all the products formed, none can be oxidized, when they are in Equilibrium state of existence. Oxidation only takes place when the components to be oxidized are in Stable State of Existence[17]. The final overall equation therefore is as follows.

Final Overall Equation: $270(O_2) \ + \ 100H\text{-}N=N\text{-}C_2H_5 \ \longrightarrow \ 270H_2O \ + \ 90CO \ +$

$90CO_2 \ + \ 90N_2 \ + \ 5H_3CNH_2 \ + \ 5N{\equiv}C\text{-}C{\equiv}N \ + \ 5NH_3 \ + \ 5CH_4 \qquad$ ApIV.54

One can thus observe the large number of products obtained. These are completely different from the conventional non-catalytic processes. Imagine if the operating conditions had been a bit higher and there was H_2 still present as a product in the system, the N_2 would have broken down as follows.

Stage 1:

$$H \bullet e \; + \; nn \bullet N = N \bullet en \xrightarrow[\text{the pores}]{N_2 \text{ activated inside}} H - N = N \bullet en$$

$$H \quad N = N \bullet en \; + \; n \bullet H \xrightarrow[\text{Existence of Diazene}]{\text{Combination State of}} H - N = N - H$$

ApIV.55

Stage 2:

$$H \bullet e \; + \; nn \bullet \overset{\overset{H}{|}}{N} - \overset{\overset{H}{|}}{N} \bullet en \xrightarrow[\text{by the pores}]{\text{Diazene activated}} H - \overset{\overset{H}{|}}{N} - \overset{\overset{H}{|}}{N} \bullet en \; (A)$$

$$(A) \; + \; n \bullet H \xrightarrow[\text{Existence of Hydrazine}]{\text{Combination State of}} H - \overset{\overset{H}{|}}{N} - \overset{\overset{H}{|}}{N} - H$$

ApIV.56

Stage 3:

$$H - \overset{\overset{H}{|}}{N} - \overset{\overset{H}{|}}{N} - H \xrightarrow[\text{the pores}]{\text{Broken down inside}} H - \overset{\overset{H}{|}}{N} \bullet en \; + \; nn \bullet \overset{\overset{H}{|}}{N} - H$$

$$(B) \qquad\qquad (C)$$

$$H - \overset{\overset{H}{|}}{N} \bullet en \; + \; n \bullet H \xrightarrow[\text{Existence of Ammonia}]{\text{Combination State of}} NH_3 \; [\textit{Stable fraction}]$$

$$H \bullet e \; + \; nn \overset{\overset{H}{|}}{N} - H \xrightleftharpoons[\text{Existence of Ammonia}]{\text{Equilibrium State of}} NH_3 \; [\textit{Equilibrium fraction}]$$

ApIV.57

Overall Equation: $\quad 3H_2 \; + \; N_2 \longrightarrow 2NH_3$ $\qquad$ ApIV.58

One can observe how Nature operates in particular underground just as in living systems. Above ground, these take place mostly via Equilibrium mechanism. We have only just begun. The reduction of N≡N (Dinitrogen) at very low operating conditions {very far from that used in Haber-Bosch process (350-550°C and 150-350 atmosphere)} has been a great subject area for many years[14,15]. Provision of the mechanism has always been an art. From the art, the real Science has emerged.

2.2 So-called Tautomerization of methyldiazene.

As has been maintained and shown, methyldiazene or azo-methane cannot molecularly re-arrange even in its activated state inside the pores or tautomerize[1]. This is contrary to recent publication[3]. Recalled below is the final equation obtained when methyldiazene was broken down inside the pores of the Ruthenium and Osmium complexes [See Equation ApII.40][1].

Overall overall equation: $10H\text{-}N\text{=}N\text{-}CH_3 \longrightarrow 2H_2 \; + \; 2N\text{≡}C\text{-}C\text{≡}N \; + \; 10NH_3 \; +$

$$6H\text{-}C\text{≡}N \qquad ApIV.59$$

From the NH_3 and H-C≡N in the products formed, the formaldehyde hydrazone was obtained as follows.

Stage 1:

$$NH_3 \xrightarrow[\text{Existence of Ammonia}]{\text{Decomposition State of}} H\bullet n \;+\; en\bullet NH_2$$

$$H_2N\bullet en \;+\; nn\bullet N = \overset{\overset{H}{|}}{C}\bullet e \xrightarrow[\text{Activated HCN}]{\text{Addition of (A) to already}} H_2N - N = \overset{\overset{H}{|}}{C}\bullet e \quad (A) \qquad \text{ApIV.60}$$

$$(A) \;+\; n\bullet H \xrightarrow[\text{Existence of formaldehyde hydrazone}]{\text{Combination State of}} H_2N - N = CH_2$$

Overall Equation: $\quad NH_3 \;+\; \text{H-C≡N} \xrightarrow{\hspace{3cm}} \text{H}_2\text{N-N=CH}_2 \qquad \text{ApIV.61}$

This is a product that can never be obtained via Equilibrium mechanism using the same react-ants. The products in Equation ApIV.59 were obtained from four stages wherein ammonia was the last to be formed. There is no doubt that as the ammonia was being formed, the formation of the formaldehyde hydrazone commenced just in one stage. There is also no doubt from the experience gathered so far that apart from the known nitrogen atom, there are other elements of nitrogen as already shown[2] and recalled below.

$$nn\bullet \overset{\bullet nn}{\underset{\bullet\bullet}{N}} \bullet nn \quad ; \quad nn\bullet \overset{\bullet nn}{\underset{\bullet\bullet}{N}} \bullet en \quad ; \quad en\bullet \overset{\bullet nn}{\underset{\bullet\bullet}{N}} \bullet en \quad ; \quad en\bullet \overset{\bullet en}{\underset{\bullet\bullet}{N}} \bullet en$$

$$\text{The Atom} \qquad \text{Less stable N} \qquad \text{Least Stable N} \quad \text{May not exist}$$

$$\text{ApIV.62}$$

THE ELEMENTS OF NITROGEN

2.3 Decomposition of Diazo-methane on Pt (III)

There is need to support the mechanisms being provided for these systems wherein the only mechanisms are all Decomposition mechanisms. With these entire hydrocarbon-containing compounds like others, the products obtained will depend in particular on the temperatures and pressures, that is, on whether it is Liquid-phase or Gas-phase decomposition. More products are obtained in the gas-phase than in the liquid-phase. One will consider decompositions on a Clean and H-covered surfaces at **290-400°C**[16] (Gas-phase).

   i) Clean Pt (III) surface.

Stage 1:

$$n\bullet \overset{\overset{H}{|}}{\underset{\underset{H}{|}}{C}} - N = N\bullet en \xrightarrow{\hspace{3cm}} e\bullet \overset{\overset{H}{|}}{\underset{\underset{H}{|}}{C}}\bullet n \;+\; N \equiv N \qquad \text{ApIV.63}$$

Overall Equation: $200H_2C\text{-}N_2 \longrightarrow 200H_2C + 200N_2$ ApIV.64

Stage 2:

$$en\bullet N = N\bullet nn \xrightarrow[\text{Existence of } N_2]{\text{Decomposition State of}} en\bullet \overset{\bullet en}{\underset{\bullet\bullet}{N}}\bullet nn + nn\bullet \overset{\bullet en}{\underset{\bullet\bullet}{N}}\bullet nn$$
  ApIV.65

$\qquad\qquad\qquad\qquad\qquad\qquad\qquad\qquad\quad$ (A) $\qquad\qquad$ (B)

Overall Equation: $50N_2 \longrightarrow 50(A) + 50(B)$ ApIV.66

Overall overall equation: $200H_2C\text{-}N_2 \longrightarrow 200H_2C + 50(A) + 50(B) + 150N_2$

  ApIV.67

Just like H_2, the N_2 has been split into the male and female parts, except that more heat will be required for N_2 than for H_2. Fifty percent of these are used in the next two stages as follows.

Stage 3a:

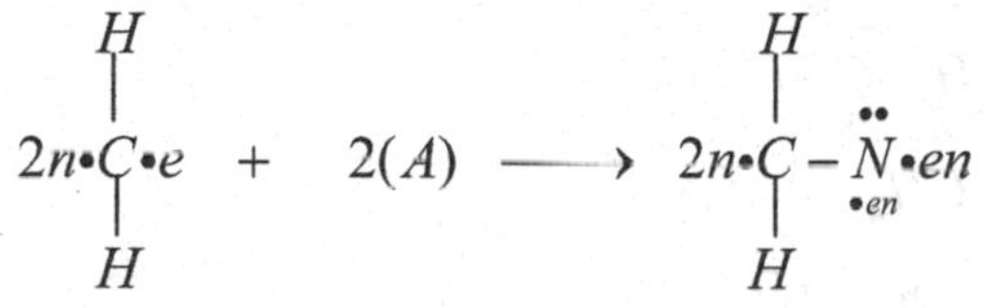

$$2n\bullet\overset{H}{\underset{H}{C}}\bullet e + 2(A) \longrightarrow 2n\bullet\overset{H}{\underset{H}{C}} - \overset{\bullet\bullet}{\underset{\bullet en}{N}}\bullet en$$

$$\longrightarrow 2H\bullet e + 2H - C \equiv N$$

$$2H\bullet e \longrightarrow H_2$$
  ApIV.68

Overall Equation: $50H_2C + 50(A) \longrightarrow 25H_2 + 50H\text{-}C\equiv N$ ApIV.69

Overall overall equation: $200H_2C\text{-}N_2 \longrightarrow 150H_2C + 150N_2 + 50(B) + 25H_2$

$\qquad\qquad\qquad\qquad\qquad\qquad\qquad\qquad\qquad + 50H\text{-}C\equiv N$ ApIV.70

Stage 3b:

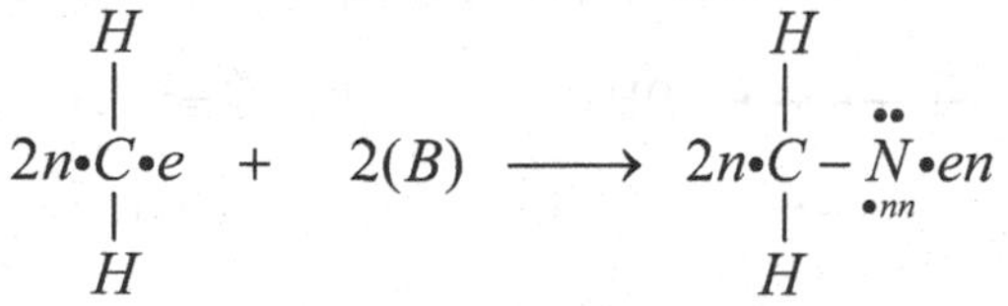

$$2n\bullet\overset{H}{\underset{H}{C}}\bullet e + 2(B) \longrightarrow 2n\bullet\overset{H}{\underset{H}{C}} - \overset{\bullet\bullet}{\underset{\bullet nn}{N}}\bullet en$$
  ApIV.71

Overall Equation: $50H_2C + 50(B) \longrightarrow 25H_2 + 50H\text{-}C\equiv N$ ApIV.72

The 3a and 3b is to indicate the parallel character of the stages, taking place at the same time.

Overall overall equation: $200H_2C\text{-}N_2 \longrightarrow 50H_2 + 100H\text{-}C\equiv N + 150N_2 + 100H_2C$

  ApIV.73

Stage 4:

$$\underset{H}{\overset{H}{e{\bullet}\overset{|}{\underset{|}{C}}{\bullet}n}} \longrightarrow H{\bullet}e \;+\; \underset{{\bullet}n}{\overset{H}{e{\bullet}C{\bullet}n}} \;\;(C)$$

ApIV.74

$$(C) \longrightarrow H{\bullet}n \;+\; \underset{{\bullet}n}{\overset{{\bullet}e}{e{\bullet}C{\bullet}n}} \;(Activated\ Carbon\ Black)$$

$$H{\bullet}e \;+\; n{\bullet}H \;\rightleftharpoons\; H_2$$

Overall Equation: $100H_2C \longrightarrow 100H_2 \;+\; 100C\ (\text{Activated Carbon Black})$

ApIV.75

Overall overall equation: $200H_2C\text{-}N_2 \longrightarrow 150H_2 \;+\; 100C \;+\; 100H\text{-}C{\equiv}N \;+\; 150N_2$

ApIV.76

Stage 5a:

$$2H{\bullet}e \;+\; \underset{{\bullet}n}{\overset{{\bullet}e}{e{\bullet}C{\bullet}n}} \longrightarrow \underset{{\bullet}e}{\overset{{\bullet}e}{H-C-H}} \;(D)$$

ApIV.77

$$2H{\bullet}n \;+\; (D) \longrightarrow CH_4$$

Overall Equation: $40H_2 \;+\; 20C \longrightarrow 20CH_4$

ApIV.78

Overall overall equation: $200H_2C\text{-}N_2 \longrightarrow 110H_2 \;+\; 80C \;+\; 100H\text{-}C{\equiv}N \;+\; 150N_2$

$+\;20CH_4$

ApIV.79

Stage 5b:

$$H{\bullet}e \;+\; 2n{\bullet}\underset{{\bullet}e}{\overset{{\bullet}n}{C}}{\bullet}e \longrightarrow \underset{{\bullet}e}{\overset{{\bullet}n}{H-C}}-\underset{{\bullet}n}{\overset{{\bullet}e}{C}}{\bullet}e \;(E)$$

ApIV.80

$$(E) \;+\; n{\bullet}H \longrightarrow \underset{{\bullet}e}{\overset{{\bullet}n}{H-C}}-\underset{{\bullet}n}{\overset{{\bullet}e}{C}}-H \;(F)$$

Overall Equation: $40H_2 \;+\; 80C \longrightarrow 40H\text{-}C{\equiv}C\text{-}H\ (F)$

ApIV.81

Overall overall equation: $200H_2C\text{-}N_2 \longrightarrow 70H_2 \;+\; 40H\text{-}C{\equiv}C\text{-}H \;+\; 100H\text{-}C{\equiv}N$

$+\;150N_2 \;+\; 20CH_4$

ApIV.82

After formation of acetylene in its double activated state, this followed by the addition of hydrogen in its equilibrium State of Existence to give ethane and ethene in parallel.

Stage 6a:

$$H{\bullet}e \;+\; \underset{{\bullet}e}{\overset{{\bullet}n}{H-C}}-\underset{{\bullet}n}{\overset{{\bullet}e}{C}}-H \longrightarrow \underset{{\bullet}e}{\overset{{\bullet}n}{H-C}}-\overset{{\bullet}e}{\underset{\underset{H}{|}}{C}}-H \;(G)$$

ApIV.83

$$(G) \;+\; n{\bullet}H \longrightarrow H_2C = CH_2$$

Overall Equation: $20H_2 + 20H\text{-}C\equiv C\text{-}H \longrightarrow 20H_2C=CH_2$ ApIV.84

Overall overall Equation: $200H_2C\text{-}N_2 \longrightarrow 50H_2 + 100H\text{-}C\equiv N + 150N_2$

$+ 20H\text{-}C\equiv C\text{-}H + 20H_2C=CH_2 + 20CH_4$ ApIV.85

Stage 6b:

$$2H\bullet e + H\overset{\bullet n}{\underset{\bullet e}{-C-}}\overset{\bullet e}{\underset{\bullet n}{C-}}H \longrightarrow H-\overset{H}{\underset{\bullet e}{\underset{|}{C}}}-\overset{\bullet e}{\underset{|}{C}}-H \quad (G)$$

$$(G) + 2n\bullet H \longrightarrow C_2H_6$$

ApIV.86

Overall Equation: $30H_2 + 15H\text{-}C\equiv C\text{-}H \longrightarrow 15C_2H_6$ ApIV.87

Overall overall equation: $200H_2C\text{-}N_2 \longrightarrow 20H_2 + 5H\text{-}C\equiv C\text{-}H + 100H\text{-}C\equiv N$

$+ 20H_2C=CH_2 + 150N_2 + 20CH_4 + 15C_2H_6$ ApIV.88

All the stages that take place under the indicated operating conditions have been clearly displayed. Note that cyanogen is not one of the products here. The system is not a Series system, but a Series/Parallel system. *The percentages which may be exact or not exact, have been chosen based on the operating conditions and the desire to show how all the products reportedly obtained[16,17] have been obtained at the operating conditions.* When diazirine (i.e. cyclic diazomethane) is decomposed, the same products as above are obtained. At low operating conditions however, with respect to temperatures, the products obtained are less. At **180 K**, for diazirine decomposition on Pt (III), *methane, ethylene and some so-called partially hydro-genated carbonaceous residue* were the products detected[18]. In order words, Stages 2, 3a and 3b did not take place, i.e. N_2 did not break down. Presence of N_2 was not indicated as a product. In Liquid-phase X-ray decomposition at **103-120 K,** *N_2 residual C and N incorporated as carbide and nitride together with thermally stable CH_2 (ads)* were reported as products[19]. Some believe that methylene species combine together to form ethylene[11] or even a polymer. If they do, then methylene will never exist. It is only the diazomethane that can be polymerized positively and electro-free-radically when nucleo-free-radicals are not present. Any polymers found as product in the system are coming from undecomposed diazomethane. The presence of H•n from hydrogen in such systems, will however not allow the chain to grow.

ii) H-covered Pt (III) surface.

The same reaction as above will take place inside the pore faster than above. Since azo-methane is always either in a polar state or radical state as already shown in Stage 1 of Equation ApIV.63, nothing can be abstracted from it. When activated in the polar state, the radical state is obtained. Therefore no reaction takes place outside the pore, except with the formation of azo-methane as shown below.

Stage 1:

$$H\bullet n + en\bullet N = N - \overset{H}{\underset{H}{\underset{|}{\overset{|}{C}}}}\bullet n \longrightarrow H - N = N - \overset{H}{\underset{H}{\underset{|}{\overset{|}{C}}}}\bullet e$$
$$(A)$$

ApIV.89

$$(A) + H\bullet e \longrightarrow H - N = N - CH_3$$

Stage 2:

$$H \bullet e \; + \; H_3C-N=N-H \longrightarrow H_2 \; + \; N_2 \; + \; e \bullet CH_3$$

$$H_3C \bullet e \; + \; H \bullet n \longrightarrow CH_4$$

ApIV.90

Overall Equation: $2H_2 \; + \; H_2C\text{-}N_2 \longrightarrow CH_4 \; + \; N_2 \; + \; H_2$ ApIV.91

Thus, more N_2 are indeed produced[11]. More methane is also produced. O-cover Pt (III) surface is not considered herein because diazo-compounds in their radical activated states cannot be oxidized. It is the products formed when decomposed that are oxidized.

3.0 Conclusions.

In provision of the mechanisms of catalytic decomposition of azo-ethane and its isomer, more supports for the new foundations being laid were necessary. For the first time, it has been shown that azo-methane (H-N=N-CH$_3$), unlike the isomer of azo-ethane (H$_3$C-N=N-CH$_3$), cannot tautomerize. None of them can molecularly rearrange. Like azo-methane, azo-ethane (H-N=N-C$_2$H$_5$) cannot also tautomerize. For the first time, the mechanisms of how N_2 is broken down to combine with hydrogen to form ammonia at very low operating conditions have been provided. This is just one of the countless ways Nature operates.

The unique and remarkable characters of Transition and Transition-transition metals have begun to be shown. Just as what some of them do to H$_2$, so also applies to N$_2$, O$_2$, Cl$_2$. and more. They break them into two parts- the male and the female, that is, they keep them in Equilibrium States of Existence. They also keep any compound with activation centers in Activated States of Existence all the time inside their pores. So far, we have been using all the pores together as one site and not as two or more different sites for hydrocarbons and non-hydrocarbons[20].

For the first time, the mechanisms of decomposition of diazomethane were provided, because of the varied types of products obtained for it in these pores at varied operating conditions. *From all considerations so far, rather than have these catalysts put inside very big reactors in very small pieces, new reactors very small in size can now begin to be designed to copy Nature.*

References

1. S. N. E. Omorodion, *"The Mechanisms of Non-catalyzed and Catalyzed Decompositions of Azo-Hydrocarbons- Part I: Azo-methane"*, Appendix II, Vol (II).

2. S. N. E. Omorodion, *"Chemistry and Humanity-Challenges to be faced as we advance in the Third Millennium: New Classifications for Radicals"*, Nigerian Journal of Applied Science, Vol 26, (2008), Pgs. 146-158.

3. G. Albertin, S. Antonlutti, A. Bacchi, F. De Marchi, G. Pelizzi, *"Tautomerization of Methyl-diazene to Formaldehyde-hydrazone in Ruthenium and Osmium Complexes"*, Inorg, Chem., (2005), 44 (24), pgs. 8947-8954.

4. S. N. E. Omorodion, *"Chemistry and Humanity-Challenges to be faced as we advance in the Third Millennium: The impacts of new classifications for Radicals"*, Nigerian Journal of Applied Science, Vol 27, (2009), Pgs. 34-43.

5. S. N. E. Omorodion, *"Chemistry and Humanity- Challenges to be faced as we advance in the Third Millennium: New Classifications for Bonds and Charges"*, Nigerian Journal of Applied Science, Vol 27, (2009), Pgs. 94-104.

6. S. N. E. Omorodion, *"Chemistry and Humanity-Challenges to be faced as we advance in the Third Millennium: The impacts of the new classifications for Bonds and Charges"*, Nigerian Journal of Applied Science, Vol 28, (2010), Pgs. 37-45

7. S. N. E. Omorodion, *"Chemistry and Humanity-Challenges to be faced as we advance in the Third Millennium: New classifications for Compounds"*, Nigerian Journal of Applied Science, Vol 28, (2010), Pgs. 80-87.

8. S. N. E. Omorodion, *"Chemistry and Humanity-Challenges to be faced as we advance in the Third Millennium: The impacts of the new classifications for compounds"*, Nigerian Journal of Applied Science, Vol 28, (2010), Pgs. 128-134.

9. A. Kiss, R. Barthos, and J. Kiss, *"Thermal and UV photo-induced decomposition of azo-methane on Rh (III)"*, Phys. Chem. Chem. Phys., (2002), 2, pg. 4237-4241.

10. P. Berlowitz, B. L. Yang, L. B. Butt, and H. H. Kung, *"Reactions of azomethane on a clean, H-covered, and O-covered Pt (III) surface"*, Surface Science, Vol. 171, Issue 1, (1986), pgs. 69-82.

11. P. Berlowitz, B. L. Yang, L. B. Butt, and H. H. Kung, *"Reactions of diazomethane on a clean, H-covered, and O-covered Pt (III) surface"*, Surface Science, Vol. 159, Issue 2-3, (1985), pgs. 540-554.

12. E. Wilberg, N. Wilberg, and A. F. Holleman, *"Inorganic Chemistry"*, (2001), pg. 666.

13. S. N. E. Omorodion, *"Mechanism of the production of Phthalic anhydride from Ortho-Xylene Using Combustion/Oxidation."*, Appendix VI, Vol II.

14. R. R. Schrock, *"Reduction of Dinitrogen"*, **http://www.pnas.org/content/103/46/17087. extract?ck=nck&cited** by=yes&legid=pnas; 103... (2009).

15. P. Avenier, M. Taoufik, A. Lesage, X. Solans-Monfort, A. Baudouin, A. de Mallmann, I. Veyre, J. M. Basset, O. Eisenstein, L. Emsley, and E. A. Quadrelli, "Dinitrogen Dissociation on an Isolated Surface Transition Atom", Science (2007), Vol. 317, no. 5841, pgs. 1056-1060.

16. W. J. Dunning, C. C. McCain, J. Chem. Soc., B, (1966), pgs. 68-72.

17. R. C. Brady III, R. Pettit, J. Am. Chem. Soc., (1980), 102 (19), pgs. 6181-6182.

18. S. S. Monim, and P. H. McBreen, Surface Science, Vol. 264, Issue 3, (1992), pgs. 341-353.

19. P. M. Loggenberg, L. Carlton, R. G. Copperthwaite, G. J. Hutchings, Surface Science, Vol. 184, Issues 1-2, (1987), pgs. 1339-1344.

20. G. S. Blackman, C. T. Kao, B. E. Bent., M. A. Van Hove, G. A. Sormorjai, Surface Science, Vol. 207, Issue 1, (1988), pgs. 66-88.

Appendix V

THE MECHANISMS OF PREPARATION OF CHITOSAN MEMBRANES FROM CHITIN

Abstract

The missing link between the Arts and Science, that which is essentially provision of mechanisms of systems including chemical, biochemical and polymeric reactions[1-6] are used herein to explain the synthesis of chitosan from chitin. The chitosan is then cross-linked to form a membrane using a cross-linking agent such as glutaraldehyde and an aromatic di-aldehyde[7]. How these are done are shown herein. These have been carried out not only to identify some errors of the past, but also to show how Nature operates. Once the mechanisms of reactions are known, it opens new doors.

Keywords

1. Chitosan polymer.
2. Chitin polymer.
3. Cross-linking agents.
4. Step and Pseudo-Step polymerization systems
5. Membranes.

1.0 INTRODUCTION

Our world in the last fifty years before 2010, the period of so-called industrial revolution has been the dumping ground for all kinds of pollutants including highly toxic heavy metals such as mercury (Hg), lead (Pb), Cadmium (Cd), Nickel (Ni), Astinum (As) and tin (Sn).[8-11] These come from our industries (industrial effluents), vehicles and many products we throw away after use and can be found in surface and underground waters, for which we cannot drink water again as we used to do so many years ago. The level of pollutants in our environment has become so alarming, to the point where Nature has temporarily refused to handle it anymore and very soon Nature will force us to start wearing gas masks made of membranes anywhere we go in-order to breath-in clean air. All these are because we do not know what we are doing. But, we think we know.

Unknown to present-day Science is that only Group IA, IIA and IIIA metals can carry ionic charges, such as $H^\oplus$ in water, $Na^\oplus$ in sodium chloride. These are cations which cannot be used for polymerization. In fact, none of the heavy metals listed above in the last paragraph can carry ionic charges when and when not compounded. The metals when alone carry only electro-radicals which can be free or non-free. Some can be excited and some can be activated. Other metals carry covalent, electrostatic and polar charges when compounded. While covalent and ionic charges are real, electrostatic and polar charges are both real and imaginary (i.e., complex). All these are required to be known when removing these heavy metals from their salts when not alone.[1-4] Removal of heavy metals from water is very important today in order to protect the health of plants and animals and our environment. In search of an efficient and cheap method, usually waste waters containing toxic metals are treated today either by addition of anions to precipitate the metals as insoluble salts[8] or by membrane filtration or by activated carbon adsorption and co-precipitation/adsorption, or by ion-exchange resin. Of all these methods, membrane filtration seems to be the most attractive, because these exist in Nature and as already said, Engineering is one's ability to copy Nature applying four basic fundamental principles.[1] In general, membranes can be found most in living systems. From their studies, one can use them to synthesize identical membranes for use in most industrial reactors (either attached as a sub-unit or internally located) and in Unit separation systems in the pharmaceutical industries and obviously in the health care industry.

Chitosan [Poly-(1→4)-β-D-glucosamine, CS] is a derivative of chitin. Chitin the major component of the arthropod shells, such as shrimp, crawfish, prawn, krill, lobster, snail shell, crab, and the structural substance of insects and fungi are polysaccharides which contain nitrogen. Other products of the shell are proteins and sodium carbonate. Chitin is one of the most abundant natural regeneratable resources. The structure of chitin appears to be identical with that of cellulose except that the acetylamino group, CH_3CONH –, replaces the hydroxyl at C-2.[12]

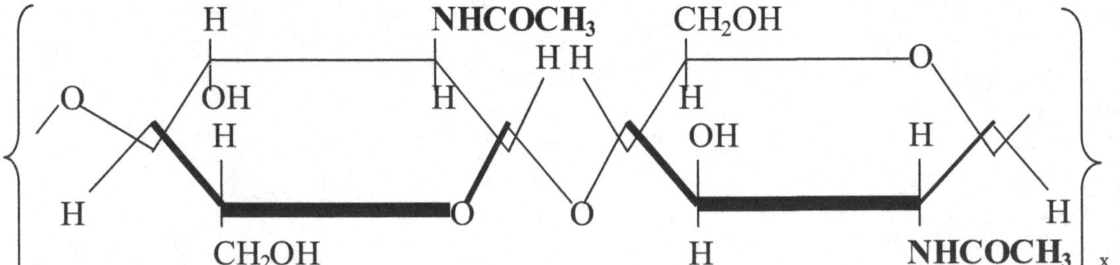

Chitin

ApV.1

The structural formula of chitosan is also shown below. These are their monomer units. They are

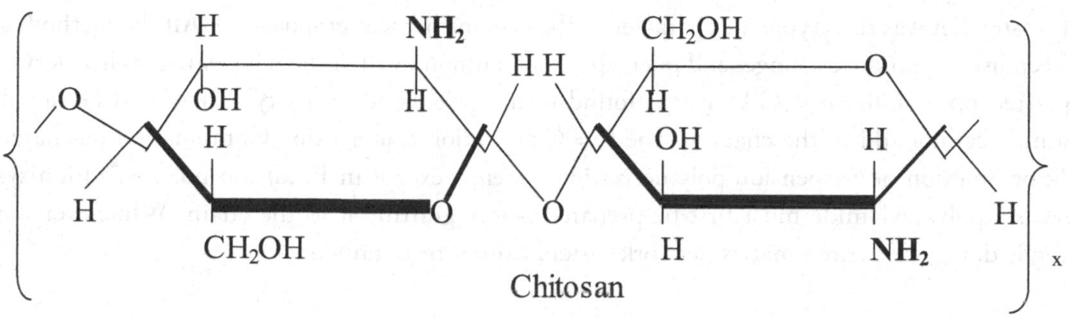

Chitosan

ApV.2

both polar/ionic polymers and therefore can fully dissolve and solubilize in polar/ionic solvents such as water, acetic acid and NaOH.[1-6] Chitosan used as a biomedical material has begun to gain more and more attention,[13,14] because of its non-toxic, biocompatibility, biodegradability, as well as good membrane-formation-spanning performances[15], the functions of anticoagulability of the blood, accelerator for the wound healing, cell delivery vehicle, and in spinal fusion. For the various applications of chitosan, different material configurations, such as membranes, microspheres, sponges, and grafting (for ion-exchange capabilities, elasticity, sorbancy, thermal resistance, and resistance to microbial attacks onto a polymer)[16] are required. All these have different microporous structures and matrices, the most outstanding of which are the membranes. Microspheres and sponges are porous, but with no matrix structure. Grafting produces non-porous structures with no matrix.

The difference between chitin and chitosan has been highlighted in Equations ApV.1 and ApV.2. From the structures, it is obvious that the *Equilibrium state of existence* of chitosan with two hydrogen atoms on the nitrogen center is far stronger than that of chitin with only one hydrogen atom. That is, it is easier to use chitosan for cross-linking than to use chitin which all the time is in *Stable state of Existence*.[1-6] Therefore, chitosan is formed through N-deacetylation of the chain molecule by hydrolysis. It is through the nitrogen center that cross-linking can take place. Presently, porous membranes are made by

i)   Phase inversion technique.
ii) Thermally induced phase separation (TIPS) technique.[17-20]
iii) Cryogenic induced phase separation (CIPS).[21]

Of the three methods, only the last used cross-linking agent-glutaraldehyde shown below along with its aromatic counterpart[7]. In the use of the agent[21], its importance was not highlighted in the

Glutaraldehyde   Phthalaldehyde

Cross-linking agents   ApV.3

paper. Instead, it was the cryogenic character of the system that was emphasized. All the methods above have been used to produce sponges and microsphere. Common to all of them is casting. When acrylamide was grafted onto chitosan (CGA), it was found to increase its adsorbtivity[22]. It should be noted that acrylamide cannot add to the chain via the C = C activation center using Combination mechanism in a Bulk or Solution or Suspension polymerization systems, except in Emulsion polymerization system. Otherwise, polyacrylamide must first be prepared before grafting it to the chain. Whichever way the grafting is done, the desired matrix network system cannot be obtained.

3.0 Synthesis of Chitosan and the membrane.

One will begin with provision of the real mechanisms for chitosan production. The chitin will be represented as shown below highlighting the point of action.

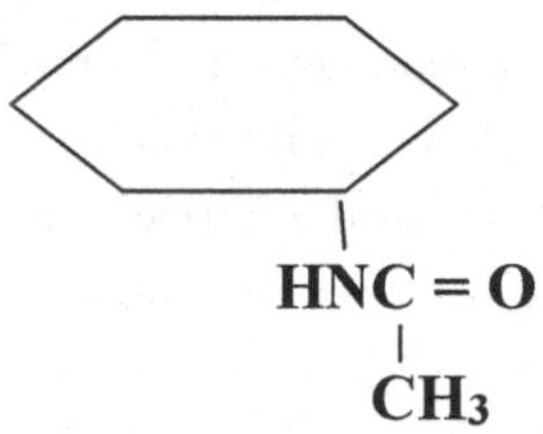

$$O$$
$$\|$$
OR CH_3CNHX

HNC = O   (where X is the ring)
|
CH_3 ApV.4

3.1 Mechanisms of Chitosan preparation.

In most cases basic hydrolysis is used, that is, heating of the chitin with 40% aqueous sodium hydroxide solution for a specified long period of time. The mixture obtained is cooled, filtered and washed with distilled water. Note that water like acetic acid is a solvent for chitosan. It was air dried and weighed. Without any doubt some of the products have been washed away! The said impure chitosan was purified by dispersing it in 10% aqueous acetic acid, centrifuged, and treated with drop wise addition of 40% aqueous sodium hydroxide, until a white flocculant precipitate was obtained at pH of 6.8. This was followed by centrifugation, and the product washed repeatedly with water, ether, ethanol respectively and allowed to dry. The product is said to be pure chitosan[22].

Stage 1:   NaOH $\rightleftharpoons$ Na.e +   nn.OH

ONa
|
Na.e + CH_3CONHX $\xrightarrow[\text{(Heat)}]{\text{Activation}}$ $CH_3C.e$
|
NHX

(I)

$$HO.nn \quad + \quad (I) \quad \longrightarrow \quad CH_3\overset{\overset{\displaystyle ONa}{|}}{\underset{\underset{\displaystyle NHX}{|}}{C}} - OH$$

$$(II) \hspace{5cm} ApV.5$$

Overall equation: $NaOH \quad + \quad CH_3CONHX \longrightarrow (II)$ $\hspace{2cm}$ ApV.6

Stage 2:

$$ApVI.8$$

$$(II) \quad \rightleftharpoons \quad CH_3\overset{\overset{\displaystyle O.nn}{|}}{\underset{\underset{\displaystyle NHX}{|}}{C}} - OH \quad + \quad Na.e$$

$$(III)$$

$$(III) \quad \rightleftharpoons \quad CH_3\overset{\overset{\displaystyle O.nn}{|}}{\underset{\underset{\displaystyle OH}{|}}{C}}.e \quad + \quad nn.NHX$$

$$(IV)$$

$$Na.e \quad + \quad nn.NHX \quad \rightleftharpoons \quad NaNHX$$

$$(IV) \quad \xrightarrow{\text{Deactivation}} \quad CH_3COOH \text{ (Acetic acid)} \quad + \quad Heat \hspace{1cm} ApV.7$$

Overall equation: $(II) \quad \longrightarrow \quad NaNHX \quad + \quad CH_3COOH \quad + \quad Heat \hspace{1cm} ApV.8$

Overall overall equation: $NaOH \quad + \quad CH_3CONHX \longrightarrow NaNHX \quad + \quad CH_3COOH$

$$+ \quad Heat \hspace{2cm} ApV.9$$

Notice the presence of acetic acid in the product in equimolar ratio with the other compounds. We should recall that we started with 40% aqueous NaOH. Since there is more water than NaOH, the first stage above indeed has more steps than shown for the stage, wherein Heat is generated, because water will be involved. However, this was not shown, because all the activations and products will still remain the same. With the presence of water in the system, the next stage follows.

Stage 3:   Na –NHX $\underset{\substack{\textit{of Stage 2}}}{\overset{\substack{\textit{See Step 3}}}{\rightleftharpoons}}$ Na.e   +   nn.NHX

Na.e   + H_2O $\rightleftharpoons$ NaH   +   H.e   +   nn•O•en   +   Heat

XHN.nn   +   e.H $\rightleftharpoons$ XNH_2

Na•e   +   nn•OH $\longrightarrow$ NaOH   ApV.10

Overall equation: NaNHX   + H_2O $\longrightarrow$ NaOH   + XNH_2 ApV.11

Overall overall equation: NaOH   + CH_3CONHX + H_2O $\overset{Heat}{\longrightarrow}$ NaOH + Heat

+ CH_3COOH + XNH_2 (Chitosan)   ApV.12

So far, three stages are involved. At this point, the acetic acid is neutralized by the sodium hydroxide present as follows.

Stage 4: CH_3COOH $\rightleftharpoons$ H.e   +   nn.OOCCH$_3$

H.e   +   NaOH $\rightleftharpoons$ H_2 +   NaO.en   +   Heat

NaO.en $\rightleftharpoons$ Na .e   +   en. O .nn   +   Heat

nn. O .en   + H_2 $\rightleftharpoons$ HO .nn   +   H .e

Na.e   + nn.OOCCH$_3$ $\rightleftharpoons$ $NaOOCCH_3$ (Organic salt)

H .e   +   nn. OH $\longrightarrow$ H_2O ApV.13a

Overall equation: HOOCCH$_3$ +   NaOH $\longrightarrow$ $NaOOCCH_3$ + H_2O +   Heat

ApV.13b

Overall overall equation: NaOH   + CH_3CONHX + H_2O $\overset{Heat}{\longrightarrow}$ CH_3COONa +

H_2O + XNH_2 (Chitosan)   +   Heat   ApV.14

Sodium acetate and water are produced along with the chitosan. NaOH has been used as a catalyst for the hydrolysis. Water alone could not be used, because H.e unlike Na.e is not strong enough to activate the C = O center. A suitable organic solvent can be used to remove the $NaOOCCH_3$ from the mixture by centrifugation. Based on the mechanisms of the reaction above, one can see how the chitosan can be removed, noting that the water and acetic acid are solvents to it. The water and the organic solvent can be removed by decompression.

3.2 Mechanism of the preparation of a suitable membrane.

Herein, one is going to use a cross-linking agent to bridge many chitosan linear polymer chains to form a porous three dimensional network system, which on further heating gives an infusible amorphous looking material insoluble in organic solutions. Instead of a solid, our desire is make it in the amorphous form, flexible enough for varied applications. To do this, depends on how the chitin was removed from the shell[23]. This is done by maintaining the shell at a high temperature for a sufficient time. The shells are then rapidly cooled, for example by plunging into liquid nitrogen, so that most of the chitin in the shells remain in the amorphous form. The "quenched" shells are then *deproteinized* by the use of dilute NaOH and heating and *deminera-lized* (to remove $CaCO_3$ using dilute HCl at room temperature) to produce chitosan.[23] The two monomer units syndiotactically placed in Equation ApV.2 will now be shortened as follows while the center of action is highlighted.

$$(I) \hspace{8cm} ApV.15$$

Notice that the NH_2 groups are syndiotactically placed. The chitosan as obtained in the last section after removing the sodium acetate is used directly with water as solvent for the membrane preparation.

Stage 1:

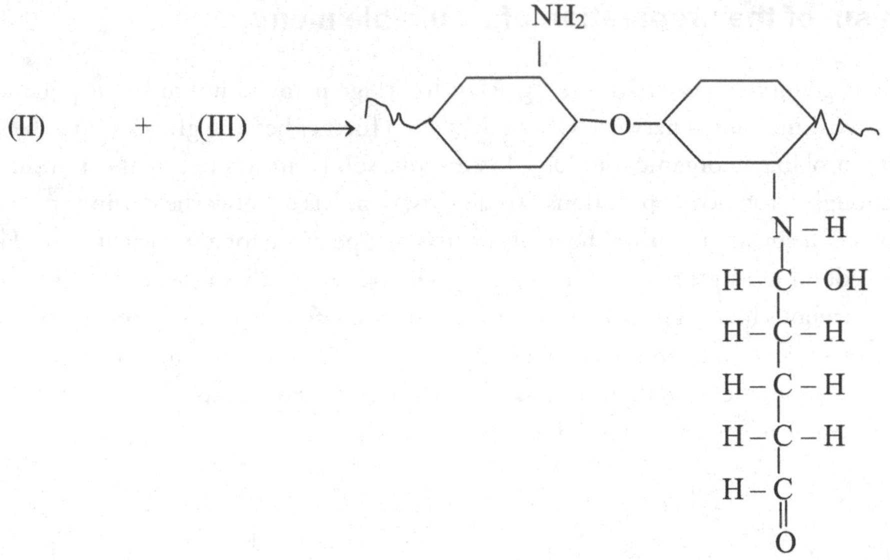

$$
\text{(II)} \quad + \quad \text{(III)} \longrightarrow
$$

(Structure showing two hexagonal rings connected by $-O-$, with NH_2 attached to the left ring, and a side chain attached to the right ring:)

$$
\begin{array}{c}
N-H \\
H-C-OH \\
H-C-H \\
H-C-H \\
H-C-H \\
H-C \\
\parallel \\
O
\end{array}
$$

(IV)   ApV.16

Overall equation: (I) + $OCHCH_2CH_2CH_2CHO$ $\longrightarrow$ (IV)   ApV.17

One began with the bottom part of the axis of the linear chain of chitosan. Addition of glutar-aldehyde continues in each stage on that side of the chain until completed. Let the number of stages be *m* a very large number. The *m* could be hundreds or thousands of monomer units of glutaraldehyde. This is then followed by addition of this cross-linking agent to the other side of the same chain. Unknown in general to Polymer scientists and engineers, is that Addition polymerization takes place by Combination mechanisms while Step polymerization takes by Equilibrium mechanisms, as shown above. Indeed, the case above is Pseudo-Step polymerization, because no small by-molecular products are released for every addition of monomer[24], whereas in Step polymerization, small molecular by-product such as water is released for every addition of monomer. The monomers in Addition polymerization carry what are called activation centers visible (π-bonds, vacant orbitals) and invisible (Strain energy inside a ring) as well as functional centers (in some ringed monomers), while the monomers in Step polymerization carry what are called functional groups (at least two of them). When there are more than two functional groups, then branching and cross-linking begin to take place when the right components are involved. Such is the case here.

(V)   Ap V.18

Addition of glutaraldehyde monomers continue on the other side of the chain in m + 1 Stages. These are done for all the chains simultaneously in more or less than 2n(m +1) stages to give the network shown in Figure ApV.1, where n is the number of chitosan chains. The chains cannot be of equal length. A similar type of network is shown for phthalaldehyde in Figure ApV.2. The former is aliphatic while the latter is aromatic.

When phthalic acid is used as cross-linking agent, the polymerization becomes Step polymerization, since water is released for every phthalic acid consumed. One can observe from the mechanisms provided, what a membrane looks like. It must be a cross-linked network chemically obtained and not that which is chemically/physically or physically obtained by for example casting. Chitosan itself is not a membrane whether grafted or not, casted or not, whether it involves the replacement of the water in the chitosan solution by organic solvents or not, but a linear water-soluble polymer with branching and cross-linking centers along the chain. When grafted, it is not a membrane. Microspheres and sponges are no membranes whether porous or not. Their ability to adsorb metals lies on their very large polar/ionic character. The greater the existence of a polar/ionic environment, the greater the adsorption capacity. Porous membranes both filter and adsorb.

Figure ApV.1 **A cross-linked matrix network of chitosan using glutaraldehyde as cross-linking agent.**

Figure ApV.2 **A cross-linked matrix network of chitosan using phthalaldehyde as cross-linking agent.**

3.3 Comparisons of types of porous chitosan membranes.

Step polymerization reactions are very slow compared to Addition polymerization reactions, because of the difference in their mechanisms of polymerization. Everything in life take place in stages.[1] For both Step and Addition polymerization systems, there are countless numbers of stages. Hence sufficient time is required to complete the cross-linking between chains in Step polymerization systems.

The shell from which the chitin was obtained is a solid non-porous type of many layers of membranes glued together by calcium carbonate to form a solid mass. One can use the membranes in Figures ApV.1 and ApV.2 in a similar manner while still retaining its porous character, just like a cloth. Of the two types of membranes shown in the figures, the aromatic type is more closed than the aliphatic type. In place of the glutaraldehyde of glutaric acid, one can also use dialdehydes from malonic and succinic acids in order to reduce the size of the pores.

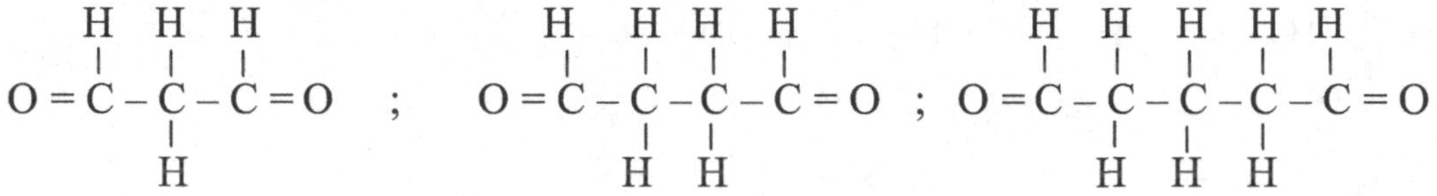

$$
\underset{\text{Malonaldehyde}}{O=\overset{\overset{\displaystyle H}{|}}{C}-\overset{\overset{\displaystyle H}{|}}{\underset{\underset{\displaystyle H}{|}}{C}}-\overset{\overset{\displaystyle H}{|}}{C}=O}
\quad ; \quad
\underset{\text{Succinaldehyde}}{O=\overset{\overset{\displaystyle H}{|}}{C}-\overset{\overset{\displaystyle H}{|}}{\underset{\underset{\displaystyle H}{|}}{C}}-\overset{\overset{\displaystyle H}{|}}{\underset{\underset{\displaystyle H}{|}}{C}}-\overset{\overset{\displaystyle H}{|}}{C}=O}
\quad ; \quad
\underset{\text{Glutaraldehyde}}{O=\overset{\overset{\displaystyle H}{|}}{C}-\overset{\overset{\displaystyle H}{|}}{\underset{\underset{\displaystyle H}{|}}{C}}-\overset{\overset{\displaystyle H}{|}}{\underset{\underset{\displaystyle H}{|}}{C}}-\overset{\overset{\displaystyle H}{|}}{\underset{\underset{\displaystyle H}{|}}{C}}-\overset{\overset{\displaystyle H}{|}}{C}=O}
\qquad \text{ApV.19}
$$

4.0 Conclusions

Because of the great importance of chitosan and its membranes in the bio- medical field and their use for waste water treatment plants and based on existing universal data, there was need to provide the mechanisms of the reactions of their synthesis. Most of the time, many of us are only interested in getting products by doing this or that. But how the products are obtained is of no interest to us. This is the missing link. Chitin is polar/ionic. So also are the products obtained from it.

In the production of the membrane, the use and the choice of cross-linking agent is very important. The methods for production and the type of cross-linking agent used will determine the porosity of the membranes obtained. For both of them, the mechanism of their synthesis is Equilibrium mechanism. For chitosan four stages are involved while for the membranes, the stages are countless.

References

1. S. N. E. Omorodion, *"Chemistry and Humanity-Challenges to be faced as we advance in the Third Millennium: New Classifications for Radicals."*, Nigerian Journal of Applied Science, Vol 26, (2008), Pgs. 146-158.

2. S. N. E. Omorodion, *"Chemistry and Humanity-Challenges to be faced as we advance in the Third Millennium: The impacts of new classifications for Radicals."* Nigerian Journal of Applied Science, Vol.27, (2009), Pgs. 33-43.

3. S. N. E. Omorodion, *"Chemistry and Humanity- Challenges to be faced as we advance in the Third Millennium: New Classifications for Bonds and Charges."* Nigerian Journal of Applied Science, Vol 27, (2009), Pgs. 94-104.

4. S. N. E. Omorodion, *"Chemistry and Humanity-Challenges to be faced as we advance in the Third Millennium: The impacts of the new classifications for Bonds and Charges."* Nigerian Journal of Applied Science, Vol 28, (2010). Pgs. 37-45.

5. S. N. E. Omorodion, *"Chemistry and Humanity-Challenges to be faced as we advance in the Third Millennium: New classifications for Compounds."* Nigerian Journal of Applied Science, Vol 28, (2010), Pgs. 80-87.

6. S. N. E. Omorodion, *"Chemistry and Humanity-Challenges to be faced as we advance in the Third Millennium: The impacts of the new classifications for compounds."* Nigerian Journal of Applied Science, Vol 28, (2010), Pgs. 128-134.

7. S. N. E. Omorodion, *"Mechanism of the production of Phthalic anhydride from ortho-xylene using Combustion/Oxidation."*, Appendix VI, Vol II.

8. A. R. W. Jackson and J. M. Jackson, (1996), *"Environmental Science"*, Longman, London. pp.: 289-336.

9. S. E. Manahan, (1994), *"Environmental Chemistry"*, 6th Ed., Lewis Publishing, London, pp.: 676-677.

10. R. Miroslav and N. B. Vladimir, (1999), *"Practical Environmental Analysis."* The Royal Society of Chemistry, UK, 6; 7-267.

11. S. Ricordel, S. Taha, I. Cisse and G.Dorange, (2001), *"Heavy metals removal by adsorption onto peanut husk carbon: characterization, kinetic study and modeling."* Separation and Purification Technol.; 24: 389-401.

12. C. R. Noller, *"Textbook of Organic Chemistry"*, W. B. Saunders Company, 1966, pg. 346.

13. R. A. A. Muzzarelli, (1977), Chitin, Pregamon Press: New York.

14. J. Yan, (1984), Chinese Sci. Bull.; 11:26.

15. T. D. Rathke, S. M. Hudson, (1994), Macromol. Chem. Phys., C34(3): 375.

16. E. Matjevic, and I. Stryker., 1966, *"Physical and Chemical properties of graft copolymerization."* Discussions Faraday Soc.; 42: 187-192.

17. A. J. Castro, O. P. III, United States Patent 4247498.

18. D. R. Lloyd, K. E. Kinzer, H. S. Tseng, (1990), J. Membrane Sci.; 52: 239.

19. P. Pan, W. J. Li, (1995), J. Membrane Sci. Technol.; 15(1): 1.

20. W. J. Li, Y. Yuan, I. Cabasso, (1995), Chinese J. Polym. Sci.; 13(1): 7.

21. Z. Y. Gu, P. H. Xue, and W. J. Li, (2001), *"Preparation of the Porous Chitosan Membrane by Cryogenic Induced Phase Separation."* Polym. Adv. Technol. 12: 665-669.

22. A. Jideowno, J. M. Okuo, and P. O. Okolo, (2007), *"Sorption of Some Heavy Metal Ions by Chitosan and Chemically Modified Chitosan",* Trends in Applied Sciences Research 2 (3): 211-217.

23. P. D. Mukherjee, (2009), *"Methods for preparing Chitin or Chitosan"*, United States Patent 6310188.

24. S. N. E. Omorodion, *"New Frontiers in Sciences, Engineering and The Arts-Vol.1- Introduction to New Classifications of Polymeric Systems and New Concepts in Chemistry"* In press, (2012).

Appendix VI

MECHANISM OF THE PRODUCTION OF PHTHALIC ANHYDRIDE FROM ORTHO-XYLENE USING COMBUSTION/OXIDATION

Abstract

Herein contains the mechanism of the combustion and oxidation of o-xylene, to produce phthalic anhydride using the new foundations which have been laid.[1-6] Apart from the great importance of phthalic anhydride in the Polymer industry, the need arises not only to distinguish between combustion and oxidation, but also to show how Nature operates and remove many misconceptions as exist in present-day Science. The mechanisms are Equilibrium multi-stage systems in series and sometimes in parallel. In some of them catalysts such as vanadium pentoxide used herein, hydrogen peroxide, are involved.

Keywords

1. Combustion.
2. Oxidation.
3. Stabilizing or de-energizing catalysts.
4. Equilibrium mechanism.
5. Vanadium pentoxide.

1.0 INTRODUCTION

Over 50 per cent of phthalic anhydride is used for the manufacture of synthetic resins of which the glyptal (glycerol and phthalic anhydride) type is the simplest. Since both phthalic acid and glycerol are polyfunctional, heating a mixture gives polymeric esters.[7] Over 30 per cent of phthalic anhydride is converted to the methyl, ethyl, n-butyl, 2-ethylhexyl, and higher alkyl esters of phthalic acid, which are used as plasticizers of synthetic polymers, especially polyvinyl chloride. Methyl phthalate (dimethyl phthalate, *DMP*) is an effective insect repellent.

Chemistry as already said deals with the study of the Laws of Nature in the real and imaginary domains. This is all based on the ways by which chemical reactions take place.[1] One has begun to introduce their real and imaginary domains.[2] Never at any point in time since antiquity has the true mechanisms of chemical reactions, polymeric reactions, bio-chemical reactions and indeed for all systems ever been properly provided. This has been the missing link, a big gap which one has begun to bridge. All we do is only to go to the laboratory or the field and collect data to interpret using the Natural language of communication-Mathematics. These however, have been very useful to us. Of all data, the most important are the Analytical data. During the synthesis of phthalic anhydride from o-xylene, it not known whether what is taking place, is oxidation or combustion or both as shown below for o-xylene or napthalene.[7]

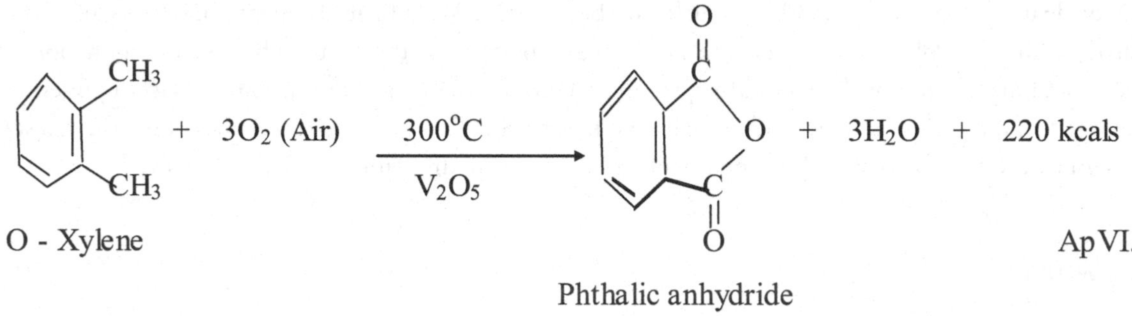

O - Xylene   Phthalic anhydride   Ap VI.1

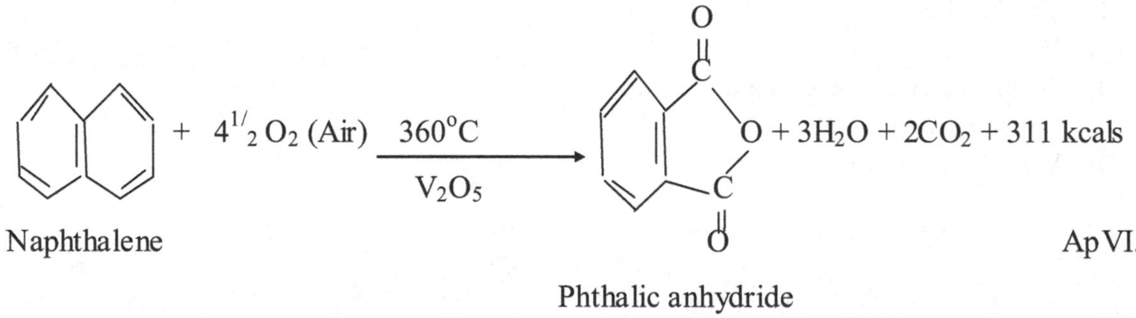

Naphthalene   Phthalic anhydride   Ap VI.2

Though the case of naphthalene has been shown above, our main focus herein is the first reaction involving o-xylene. It is said that the oxidation of o-xylene is more readily controlled than the oxidation of naphthalene, because less than half as much energy is evolved per mole of anhydride formed. Prior to the development of this process by Gibbs in 1917, sulfuric acid in the presence of mercuric sulfate had been used as oxidizing agent.[7]

Without analytical methods of analysis, the "exactness" of the equations above would not have been possible. Even then, the presence of "half" in the second equation above sends a message that oxidizing oxygen is involved using exactly nine moles of molecular oxygen. On the other hand, one wonders why vanadium pentoxide is present in the equations above. Is it used as an oxidizing agent or a stabilizing agent to suppress the Equilibrium state of existence of the o-xylene or naphthalene or as both? Its presence is

959

not required for their combustion, unless the pentoxide is present to enhance their Equilibrium states of existence. To combust them, they must exist in Equilibrium state of existence and to oxidize them, they must exist in Stable state of existence.[2]

2.0 Mechanism of combustion of o-xylene.
One wonders what the vanadium pentoxide is doing. In its absence, the followings are obtained.

Stage 1:

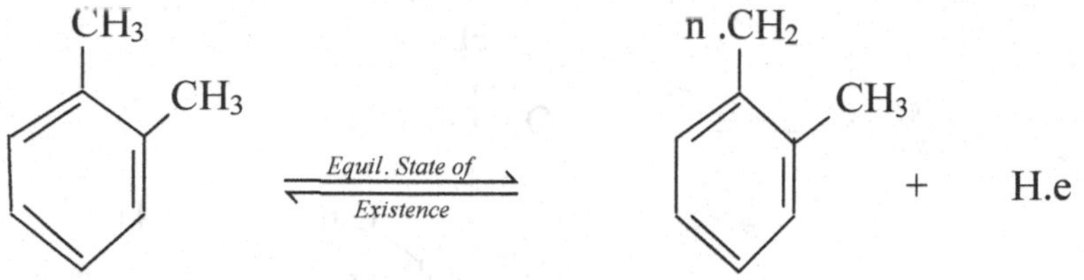

$$H.e \; + \; O = O \quad \underset{}{\overset{\textit{Activation}}{\rightleftharpoons}} \quad H - O - O.en$$

$$H - O - O.en \; + \; (I) \quad \longrightarrow \quad H - O - O - CH_2$$

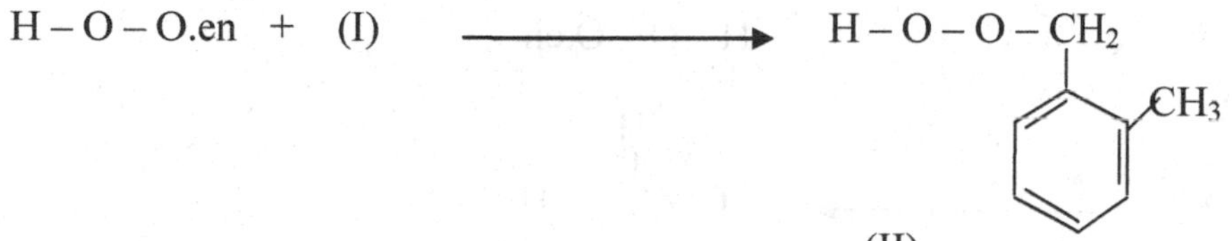

(II)   ApVI.3

Overall equation: o-xylene $+$ O_2 $\longrightarrow$ (II)   ApVI.4

Without o-xylene existing in equilibrium state, it cannot be combusted. The peroxide (II) formed decomposes in the next stage as follows.

Stage 2:

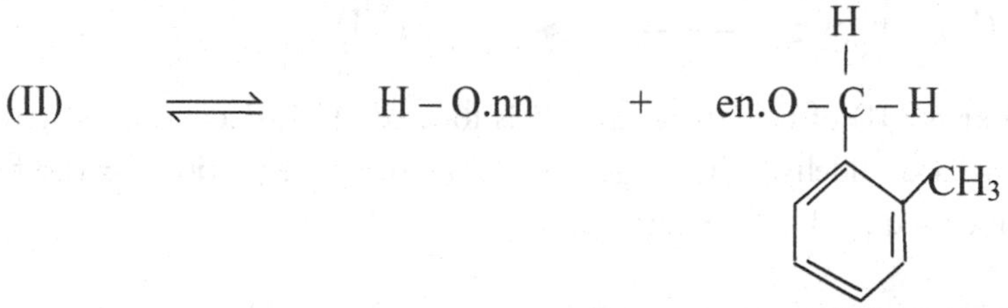

$$(II) \quad \rightleftharpoons \quad H - O.nn \quad + \quad en.O - \overset{\overset{\displaystyle H}{|}}{C} - H$$

(III)

$$(III) \quad \rightleftharpoons \quad H.e \quad + \quad O = \overset{\overset{\displaystyle H}{|}}{C} \quad + \quad Energy$$

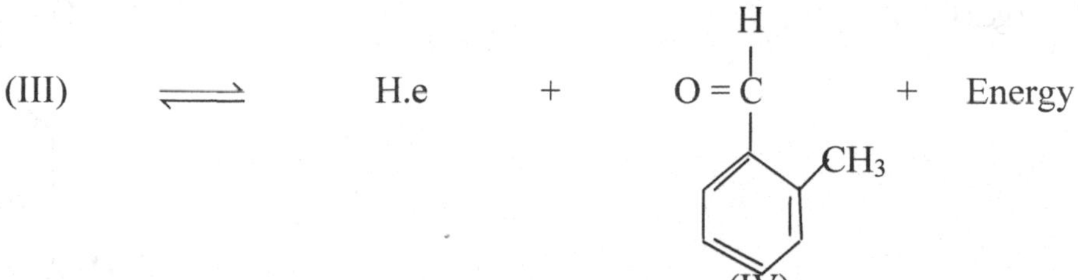

(IV)

$$H.e \; + \; nn.OH \quad \longrightarrow \quad H_2O \qquad\qquad ApVI.5$$

Overall equation: (II) $\longrightarrow$ H_2O $+$ (IV) $+$ Energy   ApVI.6

Energy is released in the second step of this stage, because oxygen is carrying an electro-non-free-radical in the presence of so many carbon and hydrogen atoms which are far more electropositive than the oxygen.[2] After two stages, the overall equation is as follows.

Overall overall equation: o-xylene + O_2 ⟶ H_2O + (IV) + Energy   ApVI.7

(IV) is an aldehyde. In the presence of more molecular oxygen from air, the next stage follows.

Stage 3:

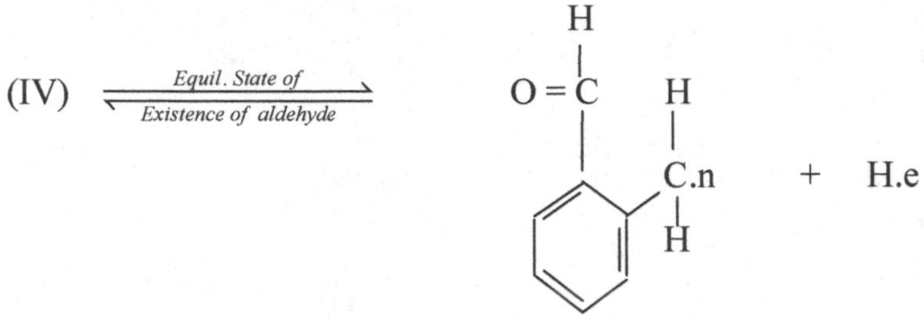

$$(IV) \xrightleftharpoons[\text{Existence of aldehyde}]{\text{Equil. State of}}$$

(V) WRONG STATE

H.e + O = O ⇌ (Activation) H – O – O.en

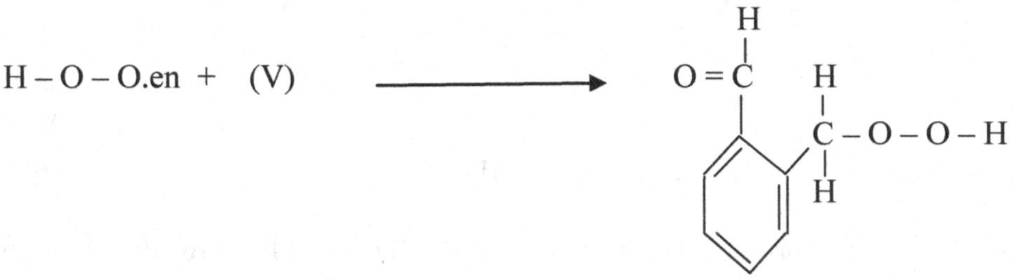

H – O – O.en + (V) ⟶

(VI)

Overall equation: (V) + O_2 ⟶ (VI)   ApVI.9

Another peroxide similar to (II) of Equation ApVI.3 is formed. This decomposes again in the manner shown in Stage 2 to give a di-aldehyde in **Stage 4** with the overall equation for the four stages shown below. Energy is also liberated in this fourth stage.

Overall overall equation: o-xylene + $2O_2$ ⟶ $2H_2O$ + (structure) + 2Energy

[Four Stages]

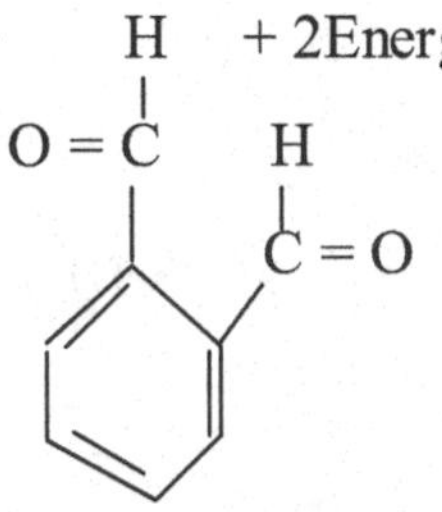

(VII) A di-aldehyde

ApVI.10

It is important to note at this point in time, that it is only we that make things that which are impossible possible, because the Equilibrium state of existence of (IV) above in Stage 3 is not the

961

real Equilibrium state of existence, that is, its natural finger-print. Using the state as shown above, the product of combustion will eventually be phthalic anhydride, that which is impossible. The H held in its Natural Equilibrium state of existence is from aldehydic side (CHO) if de-energized and not from CH$_3$ group.

With the presence of more oxygen molecules in the system, combustion continues as follows exploratively.

Stage 5:   (VII) $\rightleftharpoons$ H.e   + 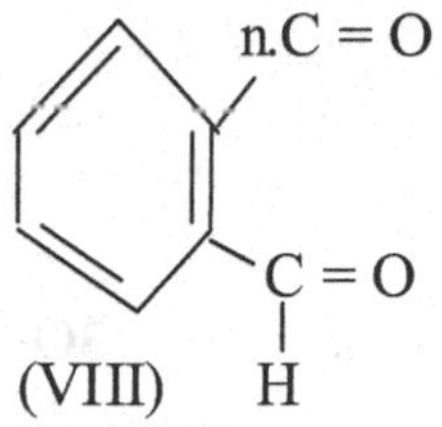
                                                    (VIII)

        H.e   +   O = O $\xrightarrow{\text{Activation}}$ H – O – O.en

    H – O – O.en   +   (VIII) $\longrightarrow$ 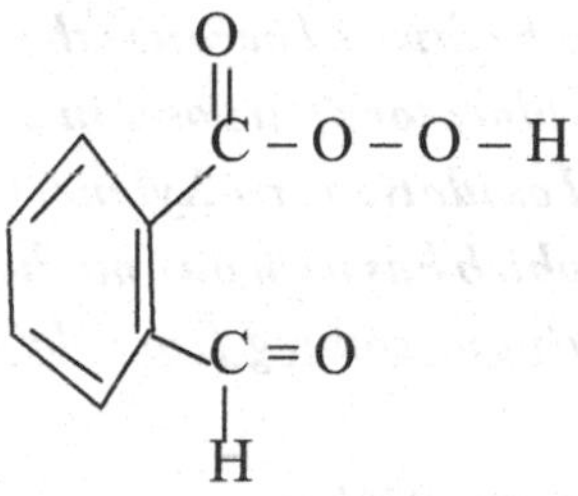
                                                (IX) (Third peroxide)   ApVI.11

Overall equation: (VII)   +   O$_2$ $\longrightarrow$ (IX)   ApVI.12

The third peroxide formed in the combustion process, decomposes as follows in Stage 6.

Stage 6:   (IX) $\rightleftharpoons$ H – O.en   + 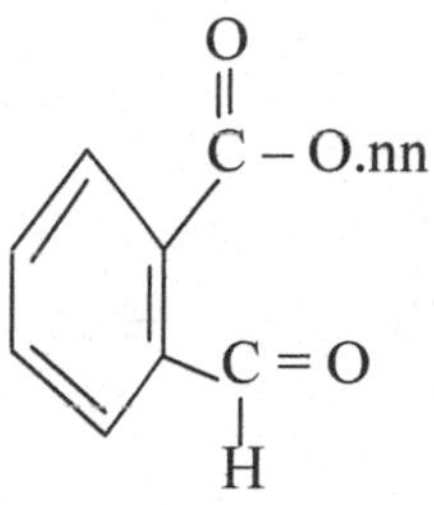
                                                        (X)

        H – O.en $\rightleftharpoons$ H.e   +   en.O.nn   +   Energy
                                    [Oxidizing oxygen atom]

    nn.O.en   +   (X) $\rightleftharpoons$ H – O.nn   + 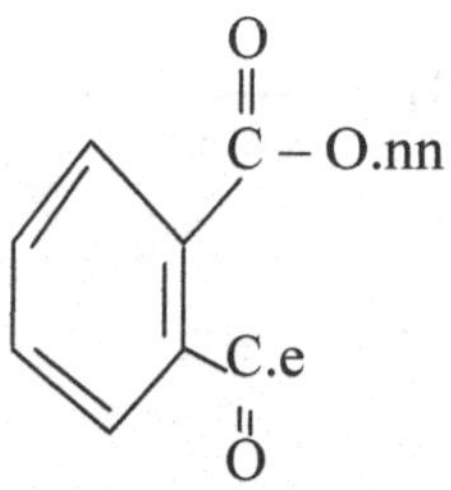
                                                    (XI)

$$H.e \;+\; nnOH \;\rightleftharpoons\; H_2O$$

(XI) $\longrightarrow$

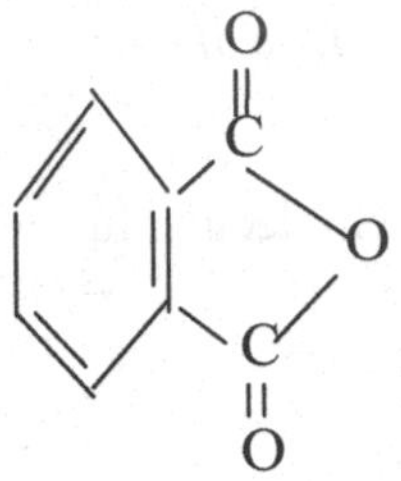

(XII) Phthalic anhydride   ApVI.13a

Overall equation: o-xylene $+$ $3O_2$ $\longrightarrow$ (XII) $+$ $3H_2O$ $+$ Energy   ApVI.13b

NOT FAVORED AND IMPOSSIBLE

It has been said not to be favored because the wrong Equilibrium state of existence for (IV) has been used in Stage 3 for exploratory purpose, in order to show that the same product cannot be obtained from combustion and oxidation of o-Xylene. One has made something which is not naturally possible possible above, that which has been our modus operadi universally. The real products of combustion are indeed, Benzoquinone coming from the aromatic side, water, CO_2 coming from the aliphatic side and energy.[8]

Worthy of note are the followings-

i   Indeed exactly three moles of molecular oxygen from the air was involved. It is the limit of the combustion, since phthalic anhydride cannot be combusted. It can only be activated as shown below.

Stage 1: H_2O $\rightleftharpoons$ H.e $+$ nn.OH

H.e $+$ (XII) $\xrightarrow{\;\text{Activation}\;}$

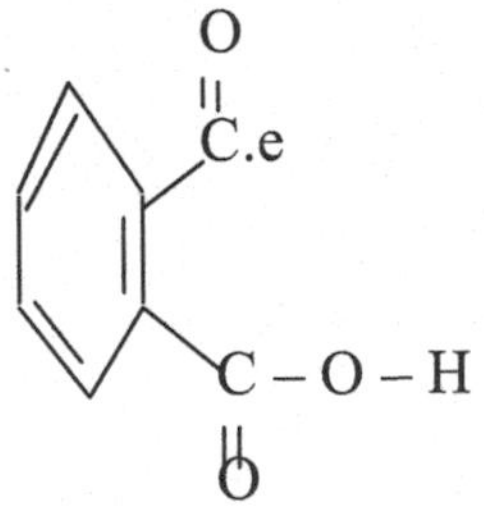

(XIII)

HO.nn $+$ (XIII) $\longrightarrow$

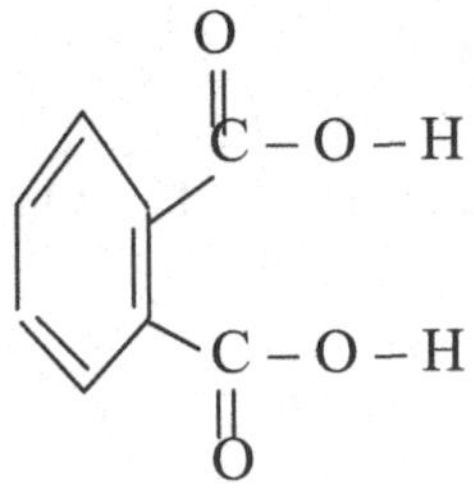

(XIV) Phthalic acid   ApVI.14a

Overall equation: H_2O $+$ (XII) $\xrightarrow[>250^\circ C]{\text{Heat}}$ (XIV)   ApVI.14b

The temperature of activation must be higher than that required for its reversibility.[6,7] In the presence of some catalyst such as Chromium-Sodium salt at 220°C, the product for the reaction above would have been benzoic acid.[7]

$$\text{(phthalic anhydride)} + H_2O \xrightarrow[220^\circ]{Cr-Na\ salt} \text{(Benzoic acid)} + CO_2 \qquad ApVI.15$$

The catalyst here is active, for it is NaOH that actually opens the ring to release CO_2, after which the catalyst is recovered as will shortly be shown for the process in question. When ammonia is used in place of water above in the absence of any catalyst but heating, Phthalimide is obtained.

$$\text{(phthalic anhydride)} + NH_3 \xrightarrow{Heat} \text{(Phthalimide)} NH + H_2O \qquad ApVI.16$$

One can observe the great importance of phthalic anhydride apart from its cross-linking capabi-lity. When the phthalic acid is heated, the followings are obtained.

Stage 1:

$$(XV) \longrightarrow (XII) - Phthalic\ anhydride \qquad ApVI.17$$

Overall equation:[7] $(XIV) \xrightarrow{Heat,\ 180^\circ C} H_2O + (XII) \qquad ApVI.18$

Different operating conditions must be involved for the forward and backward reactions.

ii   On the whole, six stages were involved.

iii   Notice that, though molecular oxygen was used, in the last stage (Stage 6) of Equation ApVI.13a, oxidizing oxygen which was to be produced was used in that same stage. Hence, both combustion and oxidation were involved for this case, like the cases of aliphatic hydrocarbons.

iv   Note the Equilibrium States of existence of the compounds here in which one of them has been wrongly used [(IV) in Stage 3. Others are correct and are fixed.

v   All along the vanadium pentoxide was not directly involved. Indirectly, it is acting as a passive catalyst with the ability to suppress the Equilibrium state of existence of what is to be combusted, that which is not desired. It also has the ability of providing oxidizing oxygen, that is, it is an oxidizing agent.

Like vanadium pentoxide, sulfuric acid in the presence of mercuric sulfate is a stabilizing agent. Unlike vanadium pentoxide, the H_2SO_4/Hg – $HgSO_4$ combination is not an oxidizing agent. For example, acetylene which is very unstable adds water in the presence of sulfuric acid and mercurous sulfate (mercury – mercuric sulfate mixture) to give acetaldehyde.[9]

$$HC \equiv CH \quad + \quad H_2O \quad \xrightarrow{H_2SO_4, \; Hg-HgSO_4} \quad CH_3CHO$$

(Acetaldehyde)

ApVI.19

$$RC \equiv CH \quad + \quad H_2O \quad \xrightarrow{H_2SO_4, \; Hg-HgSO_4} \quad RCOCH_3$$

(A Ketone)

ApVI.20

$$HC \equiv CH \; \underset{Existence}{\overset{Equilibrium\;State\;of}{\rightleftharpoons}} \; H \bullet e \; + \; n \bullet C \equiv CH$$

$$RC \equiv CH \; \underset{Existence}{\overset{Equilibrium\;State\;of}{\rightleftharpoons}} \; H \bullet e \; + \; n \bullet C \equiv CR$$

ApVI.21

For acetylene to be used as a monomer either one or the two of the hydrogen atoms must be replaced by suitable groups to make it remain in Stable state of Existence. Notice that the Equilibrium State of Existence of acetylene has been suppressed by the passive catalyst. That of water is not suppressed. The mechanisms are as follows.

Stage 1: $H_2O \; \underset{Existence}{\overset{Equilibrium\;State\;of}{\rightleftharpoons}} \;$ H.e   +   nn.OH

$$H.e \; + \; HC \equiv CH \quad \underset{sup\,pressed}{\overset{Acetylene}{\rightleftharpoons}} \quad \begin{array}{c} H \\ | \\ H-C=C.e \\ | \\ H \end{array}$$

(I)

$$HO.nn \; + \; (I) \quad \longrightarrow \quad \begin{array}{c} H \quad O-H \\ | \qquad | \\ C = C \\ | \qquad | \\ H \quad H \end{array}$$

(II) Vinyl alcohol- Very unstable

ApVI.22

Overall equation: H_2O + $HC \equiv CH$ $\xrightarrow[\text{Passive cat.}]{\text{Stabilizing agent}}$ (II) ApVI.23

The vinyl alcohol being unstable, undergoes molecular rearrangement of the first kind of the first type as follows in the presence of heat of the reaction.[5]

Stage 2:

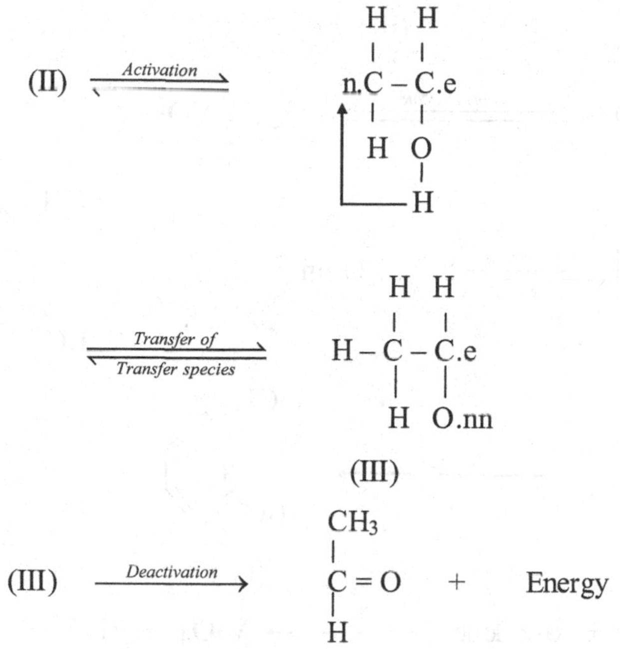

(II) $\xrightleftharpoons{\text{Activation}}$

$$\begin{array}{c} H \quad H \\ | \quad\ | \\ n.C - C.e \\ | \quad\ | \\ H \quad O \\ \qquad | \\ \qquad H \end{array}$$

$\xrightleftharpoons[\text{Transfer species}]{\text{Transfer of}}$

$$\begin{array}{c} H \quad H \\ | \quad\ | \\ H - C - C.e \\ | \quad\ | \\ H \quad O.nn \end{array}$$

(III)

(III) $\xrightarrow{\text{Deactivation}}$

$$\begin{array}{c} CH_3 \\ | \\ C = O \quad + \quad Energy \\ | \\ H \end{array}$$

ApVI.24

Overall Equation: H_2O + $HC \equiv CH$ $\xrightarrow[\text{a PassiveCat.}]{\text{Stabilizing agent}}$ H_3CCHO ApVI.25

With the use of this passive stabilizing agent, acetylene can then be readily oxidized.

3.0 Mechanisms of Oxidation of o-xylene.

It is important to know the role of the vanadium pentoxide, since its presence is not there for nothing. For one, it has the ability of suppressing the Equilibrium state of existence of the o-xylene. Then how will the o-xylene function in a stabilized state of existence in the presence of molecular oxygen of the air? This implies that the vanadium pentoxide is providing something in the form of a particular oxygen at 360°C.

Stage 1:

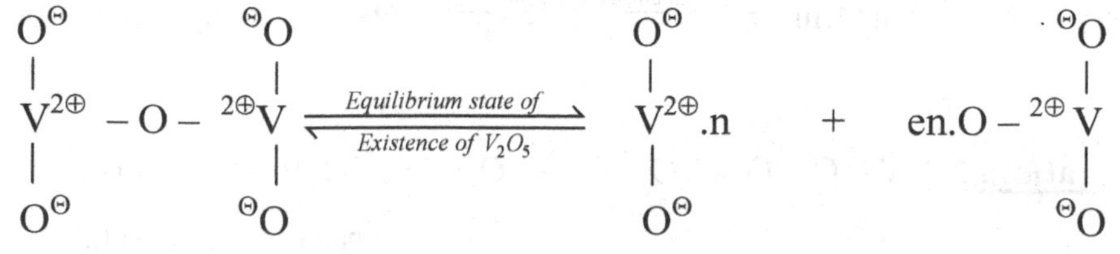

$$\begin{array}{ccc} O^{\ominus} \qquad\ ^{\ominus}O & & O^{\ominus} \qquad\ ^{\cdot\,\ominus}O \\ | \qquad\quad | & & | \qquad\quad | \\ V^{2\oplus} - O - {}^{2\oplus}V & \xrightleftharpoons[\text{Existence of } V_2O_5]{\text{Equilibrium state of}} & V^{2\oplus}.n \quad + \quad en.O - {}^{2\oplus}V \\ | \qquad\quad | & & | \qquad\qquad\qquad | \\ O^{\ominus} \qquad\ ^{\ominus}O & & O^{\ominus} \qquad\qquad\ ^{\ominus}O \end{array}$$

Vanadium pentoxide   (I)   (II)

$$(II) \quad \underset{\underset{Energy}{\longleftarrow}}{\overset{Release\ of}{\longrightarrow}} \quad \overset{\overset{O^{\ominus}}{|}}{\underset{\underset{O^{\ominus}}{|}}{V^{2\oplus}.e}} \quad + \quad nn.O.en \quad + \quad Energy$$

$$(III)$$

$$(I) \quad + \quad (III) \quad \underset{\underset{Existence\ of\ V_2O_4}{\longleftarrow}}{\overset{Equilibrium\ State\ of}{\longrightarrow}} \quad O_2V - VO_2$$

$$nn.O.en \quad + \quad \text{[o-xylene structure]} \quad \underset{\underset{o-xylene}{\longleftarrow}}{\overset{Oxidation\ of}{\longrightarrow}} \quad HO.nn \quad + \quad \text{[e.CH}_2 \text{ structure]}$$

$$\longrightarrow \quad \text{[HOCH}_2 \text{ structure]}$$

$$(IV) \qquad\qquad ApVI.26$$

Overall equation: $\quad V_2O_5 + \text{o-xylene} \quad \longrightarrow \quad V_2O_4 + (IV) + \text{Energy} \quad$ ApVI.27

Stage 2: $\qquad O_2V - VO_2 \quad \rightleftharpoons \quad O_2V.e \quad + \quad n.VO_2$

$$O_2V.e \quad + \quad O = O \quad \underset{\underset{O_2}{\longleftarrow}}{\overset{Activation\ of}{\longrightarrow}} \quad O_2V - O - O.en$$

$$O_2V - O - O.en \quad + \quad n.VO_2 \quad \longrightarrow \quad O_2V - O - O - VO_2 \qquad ApVI.28$$

Overall equation: $\quad V_2O_4 \quad + \quad O_2 \quad \longrightarrow \quad O_2V - O - O - VO_2 \qquad$ ApVI.29

Stage 3: $\quad 2O_2V - O - O - VO_2 \quad \underset{\underset{of\ peroxide}{\longleftarrow}}{\overset{Equil.\ State\ of\ Exist.}{\longrightarrow}} \quad 2O_2V - O.en \quad + \quad 2nn.O - VO_2$

$$2O_2V - O.en \quad \rightleftharpoons \quad 2O_2V.e \quad + \quad 2en.O.nn \quad + \quad 2Energy$$

$$2en.O.nn \quad \underset{\underset{Existence\ of\ Oxidizing\ O_2}{\longleftarrow}}{\overset{Equilibrium\ State\ of}{\longrightarrow}} \quad \boldsymbol{O_2} \text{ (Oxidizing oxygen molecule)}$$

$$2O_2V.e \quad + \quad 2nn.O - VO_2 \quad \longrightarrow \quad 2O_2V - O - VO_2 \qquad ApVI.30$$

Overall Equation: $2O_2V - O - O - VO_2 \longrightarrow \boldsymbol{O_2} \quad + \quad 2O_2V - O - VO_2 \quad + \quad 2Energy$

$$\text{(Vanadium pentoxide)} \qquad ApVI.31$$

Overall overall equation: $2V_2O_5 \quad + \quad \text{2o-xylene} \quad + \quad 2O_2 \quad \longrightarrow$

$$2V_2O_5 \quad + \quad 2(IV) \quad + \quad \boldsymbol{O_2} \quad + \quad 4Energy \qquad ApVI.32$$

So far, we have gone through three stages. Worthy of note are the followings-

1. The real structures of V_2O_5 and vanadium tetra-oxide (V_2O_4). They carry polar bonds as opposed to double bonds, being metallic.
2. Observe the Equilibrium States of Existence of the compound in particular that of V_2O_5.
3. One has distinguished the oxidizing oxygen molecule from the molecular oxygen molecule from air by high-lighting it. Indeed molecular oxygen has been used here not for combustion, but for recovering vanadium pentoxide just as the components of air do to recover enzymes. Indeed the vanadium pentoxide is an active catalyst.
4. Observe the large amounts of energy released so far and how they are released.
5. Stage 3 dictated to us the exact molar ratios to use to get the desired product at high conversion levels, not the way we operate in present-day Science.
6. Now, we have two sources of oxidizing oxygen molecule – $2V_2O_5$ and O_2. It is O_2 that is first used since it is already there. This is going to be used in another stage, whereas the one produced in Stage 1 was used in-situ in the same stage. Note why it could not be produced as a product in that stage.

Stage 4: O_2 $\underset{\text{Existence of } O_2}{\overset{\text{Equil. State of}}{\rightleftharpoons}}$ 2nn.O.en

2nn.O.en + 2(IV) $\overset{\text{Oxidation}}{\rightleftharpoons}$ 2 [HOCH$_2$ / e.CH$_2$ ring structure] + 2nn.OH

$\longrightarrow$ 2 [HOCH$_2$ / HOCH$_2$ ring structure]

(V)   ApVI.33

Overall equation: O_2 + 2(IV) $\longrightarrow$ 2(V)   ApVI.34

Stage 5: $2O_2V - O - VO_2$ $\rightleftharpoons$ $2O_2V.n$ + $2en.O - VO_2$

$2O_2V - O.en$ $\rightleftharpoons$ $2en.O.nn + 2e.VO_2 + 2Energy$

2nn.O.en + 2(V) $\rightleftharpoons$ 2 [OH / e.CH / HOCH$_2$ ring structure] + 2nn.OH

(VI)

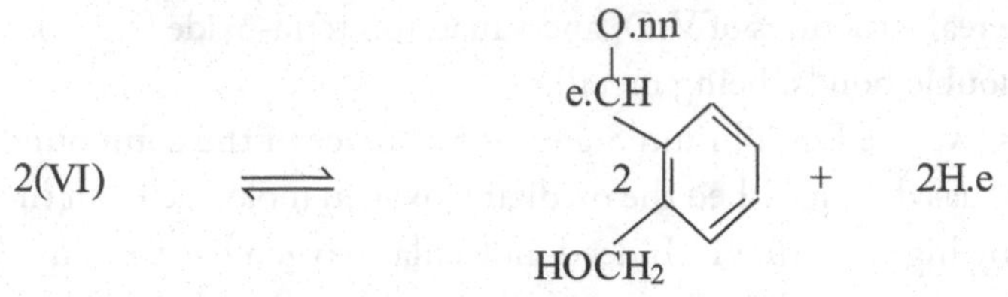

$$2(VI) \rightleftharpoons 2 \underset{HOCH_2}{\overset{\overset{\textstyle O.nn}{|}}{\underset{}{e.CH}}} \text{(benzene ring)} + 2H.e$$

(VII)

$$2H.e + 2nn.OH \rightleftharpoons 2H_2O$$

$$2O_2V.n + 2e.VO_2 \rightleftharpoons 2O_2V - VO_2$$

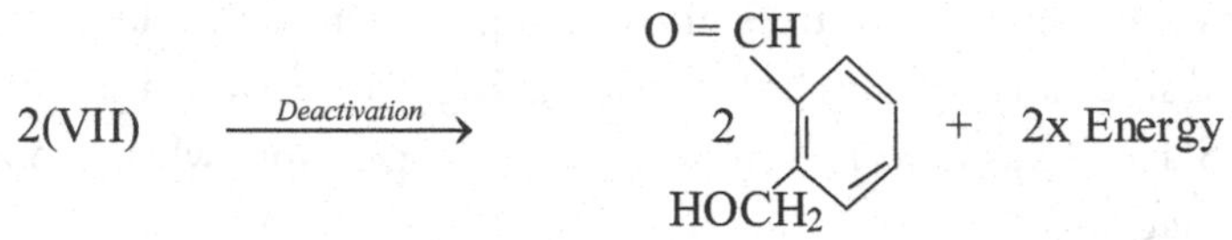

$$2(VII) \xrightarrow{\textit{Deactivation}} 2 \underset{HOCH_2}{\overset{\overset{\textstyle O=CH}{}}{\underset{}{}}} \text{(benzene ring)} + 2x \text{ Energy}$$

(VIII)   ApVI.35

Overall equation: $2V_2O_5 + 2(V) \longrightarrow 2H_2O + 2V_2O_4 + 2(VIII)$

$$+ 2(1 + x) \text{ Energy} \qquad\qquad ApVI.36$$

Using another two moles of molecular oxygen from air, V_2O_5 is recovered again as already shown in Stages 2 and 3, i.e.

Stage 6: The same as Stage 2.

Stage 7: The same as Stage 3.

For these two stages where V_2O_5 is recovered, the overall equation is as follows-

Overall equation: $2V_2O_4 + 2O_2 \longrightarrow 2V_2O_5 + O_2 + 2\text{Energy} \qquad ApVI.37$

Now the oxidation of (VIII) continues as follows.

Stage 8: $\qquad\qquad O_2 \; \underset{\textit{Existence of } O_2}{\overset{\textit{Equil. State of}}{\rightleftharpoons}} \; 2nn.O.en$

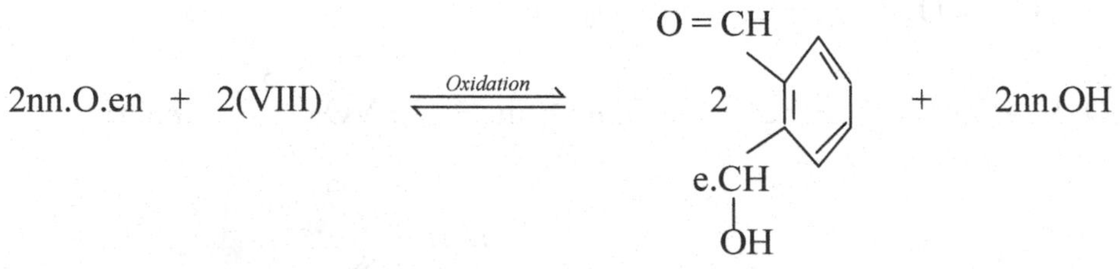

$$2nn.O.en + 2(VIII) \xrightarrow{\textit{Oxidation}} 2 \underset{\underset{OH}{|}}{\underset{e.CH}{\overset{O=CH}{}}} \text{(benzene ring)} + 2nn.OH$$

(IX)

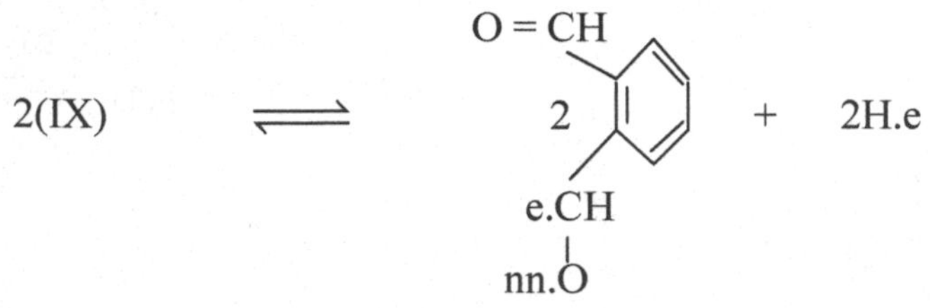

$$2(IX) \rightleftharpoons 2 \underset{\underset{nn.O}{|}}{\underset{e.CH}{\overset{O=CH}{}}} \text{(benzene ring)} + 2H.e$$

(X)

969

$$2(X) \xrightarrow{\quad Deactivation \quad} 2 \underset{O=CH}{\overset{O=CH}{\bigodot}} + 2x \text{ Energy}$$

$$(XI) \qquad\qquad\qquad \text{ApVI.38}$$

Overall equation: $O_2 + 2(VIII) \longrightarrow 2H_2O + 2(XI) + 2x\text{Energy}$ ApVI.39

Overall overall equation: $2V_2O_5 + 2o\text{-xylene} + 4O_2 \text{ (AIR)} \longrightarrow$

$$2V_2O_5 + 2(XI) + 4(2+x)\text{ Energy} + 4H_2O \qquad \text{ApVI.40}$$

Imagine the amount of energy generated so far, from two main sources- during release of oxidizing oxygen and the deactivation steps (Xs), noting that energy is required to activate the oxygen molecules. The (XI) formed here is a di-aromatic aldehyde, a very useful product. One can decide to stop the oxidation at this point, by quenching the reaction. This is a point where the vanadium pentoxide has been used as a catalyst active in character, i.e.

$$2o\text{-xylene} + 4O_2 \xrightarrow{\quad V_2O_5 \text{ } Cat. \text{ } 360^o C \quad} 2(XI) \text{ [a di-aldehyde]} + 4(2+x)\text{ Energy} + 4H_2O \qquad \text{ApVI.41}$$

However, with the presence of V_2O_5 and at the operating conditions, the oxidation of the di-aldehyde continues in the next stage.

Stage 9: $2O_2V - O - VO_2 \rightleftharpoons 2O_2V.n + 2en.O - VO_2$

$$2O_2V - O.cn \rightleftharpoons 2en.O.nn + 2e.VO_2 + 2\text{Energy}$$

$$2nn.O.en + 2(XI) \rightleftharpoons 2\underset{O=CH}{\overset{\overset{O}{\parallel}}{\underset{e.C}{\bigodot}}} + 2nn.OH$$

$$(XII)$$

$$2O_2V.e + 2n.VO_2 \rightleftharpoons 2V_2O_4$$

$$(XII) + nn\bullet OH \longrightarrow 2\underset{O=CH}{\overset{\overset{O}{\parallel}}{\underset{HO-C}{\bigodot}}}$$

$$(XIII) \qquad\qquad \text{ApVI.42}$$

Overall equation: $2V_2O_5 + 2(XI) \longrightarrow 2V_2O_4 + 2(XIII)$

$$+ 2 \text{ Energy} \qquad \text{ApVI.43}$$

The vanadium pentoxide is regenerated once again as already shown by Stages 2 and 3 and Equation ApVI.37. When this is done in **Stages 10 and 11,** the overall equation is as follows.
Stage 10: Same as Stage 2.
Stage 11: Same as Stage 3.

Overall overall equation: $2V_2O_5$ + 2o-xylene + $6O_2$ (AIR) $\longrightarrow$

$\qquad\qquad\qquad 2V_2O_5$ + 2(XIII) + 2(5 + 2x) Energy + $4H_2O$ + O_2 $\qquad$ ApVI.44

The oxidizing oxygen molecule is now used in the next stage as follows.

Stage 12: $\qquad O_2$ $\underset{\text{Existence of } O_2}{\overset{\text{Equil. State of}}{\rightleftharpoons}}$ 2nn.O.en

2nn.O.en + 2(XIII) $\overset{\text{Oxidation}}{\rightleftharpoons}$ 2 [structure] + 2nn.OH

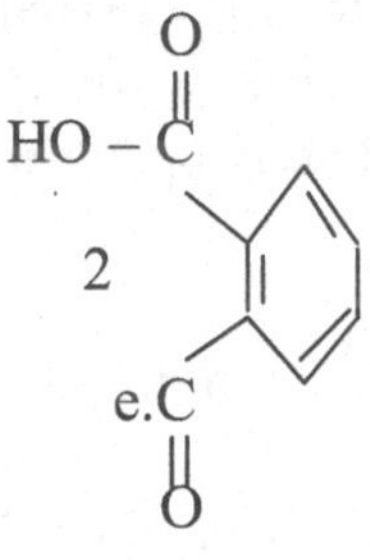

(XIV

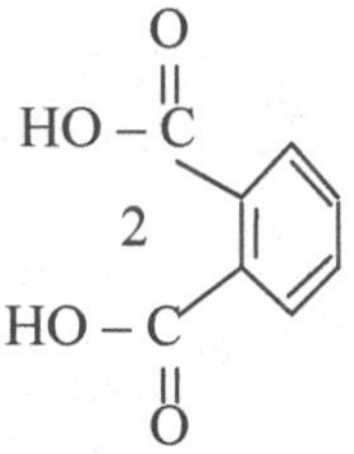

(XV) Phthalic acid $\qquad\qquad$ ApVI.45

Overall overall equation: $2V_2O_5$ + 2o-xylene + $6O_2$ (AIR) $\longrightarrow$

$\qquad\qquad 2V_2O_5$ + 2(XV) + 2(5 + 2x) Energy + $4H_2O$ $\qquad$ ApVI.46

Stage 13:

2(XV) $\rightleftharpoons$ 2 [structure] + 2H.e

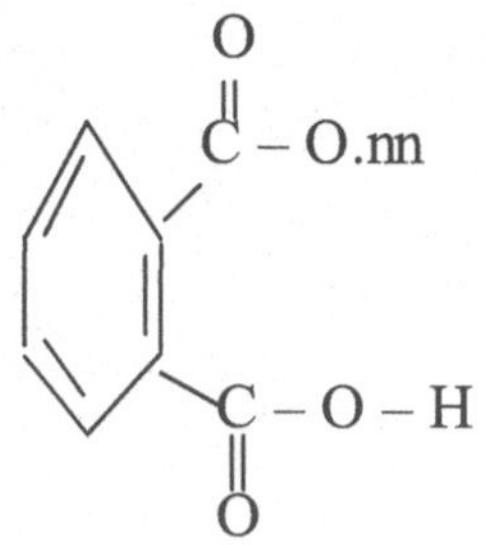

971

(XVI)

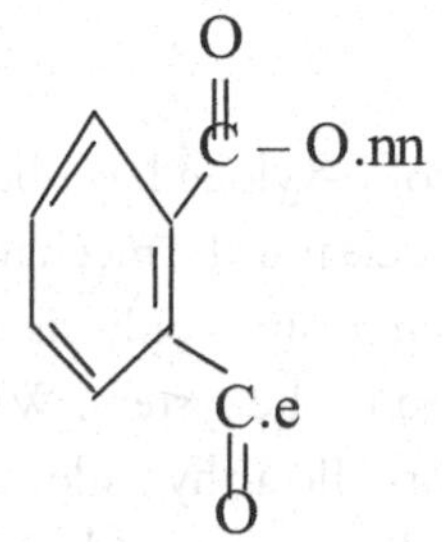

2(XVI) ⟶

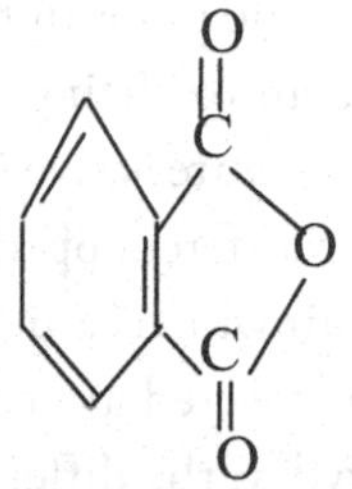

(XVII) **Phthalic anhydride**   ApVI.47

Overall overall equation: $2V_2O_5$ + 2o-xylene + $6O_2$ (AIR) $\xrightarrow{360^\circ C}$

$$2V_2O_5 + 2(5 + 2x)\ Energy + 6H_2O + 2(XVII) \qquad ApVI.48$$

The number of moles as indicated from the mechanism must be used in a well designed reactor where there is perfect mixing in-order to obtain full conversion. *[Don't divide both sides of the equation by two]* Indeed, worthy of note are the followings-

a) The large amounts of energy released.

b) That V_2O_5 is indeed a passive and an active catalyst playing *the roles of a stabilizing agent passively, an oxidizing agent actively and an active catalyst.* Many oxidizing agents do not operate this way.

c) The stage-wise operations of some chemical reactions (such as combustion, oxidation, hydrogenation and more or indeed Nature), for which one cannot bypass a particular stage and move to the end, as done in present-day Science with the use of rate determining steps and so on. Every stage is very important and every stage is step-wisely operated in a fixed order. For every reaction, there is one and only one mechanism, just as for every problem, there is one and only one solution, and for every solution there are countless numbers of "unnecessary problems".

d) Indeed the mechanism of the reaction based on the operating conditions is complete OXIDATION with no combustion taking place, unless part of the o-xylene cannot be stabilized by V_2O_5. If part cannot be stabilized, then both combustion and oxidation will take place side by side in parallel, but not the type of combustion shown above, since when the real combustion takes place benzoquinone must be one of the major products.

4.0 Conclusions

The mechanisms of oxidation of o-xylene have been provided for the first time. At the beginning, based on Equation ApVI.1, it was not clear if the mechanism is either combustion or oxidation. Shockingly, based on the analysis above, the mechanism is only via oxidation if phthalic acid is to be the product for as long as vanadium pentoxide is present in the system. When absent, the mechanism is only via combustion, but not as shown above, because phthalic anhydride cannot be obtained via combustion.

The passive character of vanadium pentoxide is its ability to suppress the Equilibrium state of existence of components such as the o-xylene, and the active part is its ability to partake in the chemical reaction process and be released at specific point in time as desired. Therefore, the V_2O_5 is an active and passive catalyst and at the same time an oxidizing agent of a different type. Some are only active and some are only passive. Some like enzymes are active, passive, and optically active and carriers of different phenomena. All of them operate under different operating conditions. It is interesting to note how Nature operates. The molecular oxygen from air which cannot oxidize, but only combust was still very important. For without it, the V_2O_5 cannot be recovered after doing its job. In the same manner too, without the components of air we breathe-in everyday, the different types of countless enzymes cannot be recovered.

If we do not understand the mechanisms of operations of systems, then nothing can be done orderly. The same applies to all disciplines. We have only just begun to turn the wheel backwards in-order to move forward. Without the analytical method in all disciplines, acquisition of knowledge would have been impossible.

References

1. S. N. E. Omorodion, *"Chemistry and Humanity-Challenges to be faced as we advance in the Third Millennium: New Classifications for Radicals."*, Nigerian Journal of Applied Science, Vol 26, (2008), Pgs. 146-158.

2. S. N. E. Omorodion, *"Chemistry and Humanity-Challenges to be faced as we advance in the Third Millennium: The impacts of new classifications for Radicals."* Nigerian Journal of Applied Science, Vol.27, (2009), Pgs. 33-43.

3. S. N. E. Omorodion, *"Chemistry and Humanity- Challenges to be faced as we advance in the Third Millennium: New Classifications for Bonds and Charges."* Nigerian Journal of Applied Science, Vol 27, (2009), Pgs. 94-104.

4. S. N. E. Omorodion, *"Chemistry and Humanity-Challenges to be faced as we advance in the Third Millennium: The impacts of the new classifications for Bonds and Charges."* Nigerian Journal of Applied Science, Vol 28, (2010). Pgs. 37-45.

5. S. N. E. Omorodion, *"Chemistry and Humanity-Challenges to be faced as we advance in the Third Millennium: New classifications for Compounds."* Nigerian Journal of Applied Science, Vol 28, (2010), Pgs. 80-87.

6. S. N. E. Omorodion, *"Chemistry and Humanity-Challenges to be faced as we advance in the Third Millennium: The impacts of the new classifications for compounds."* Nigerian Journal of Applied Science, Vol 28, (2010), Pgs. 128-134.

7. C. R. Noller, *"Textbook of Organic Chemistry"*, W. B. Saunders Company, 1966, pgs. 473-475.

8. S. N. E. Omorodion, "The Beginning of a New Dawn for Humanity. (Introduction to the world of Micro- and Macro- molecular Chemistry)", in Press, 2014.

9. C. R. Noller, *"Textbook of Organic Chemistry"*, W. B. Saunders Company, 1966, pg. 149.

Index

U

Urea 529
Urethans 815

V

Vanadium pentoxide 966
Vinyl nitrate 103
Vinyl imines 543-544

W

White phosphorus 565

X

Y

Z

Zero law of Nature 30

rinted in the United States
y Bookmasters